	To convert from	to	Multiply by
Moment	dyne-centimeter	newton-meter (N-m)	$1.000\,000\ (10^{-7})$*
	pound-foot	newton-meter (N-m)	$1.355\,818$
Power	Btu/hour	watt (W)	$2.930\,711\ (10^{-1})$
	erg/second	watt (W)	$1.000\,000\ (10^{-7})$*
	foot-pound/second	watt (W)	$1.355\,818$
	horsepower (550 ft–lbs)	watt (W)	$7.456\,999\ (10^{2})$
Pressure and Stress	atmosphere	pascal (Pa)	$1.013\,25\ (10^{5})$
	bar	pascal (Pa)	$1.000\,000\ (10^{5})$*
	centimeter of mercury (0°C)	pascal (Pa)	$1.333\,22\ (10^{3})$
	dyne/centimeter^2	pascal (Pa)	$1.000\,000\ (10^{-1})$*
	pound/foot^2	pascal (Pa)	$4.788\,026\ (10^{1})$
	pound/inch^2 (psi)	pascal (Pa)	$6.894\,757\ (10^{3})$
Speed	foot/second	meter/second (m/s)	$3.048\,000\ (10^{-1})$*
	inch/second	meter/second (m/s)	$2.540\,000\ (10^{-2})$*
	kilometer/hour	meter/second (m/s)	$2.777\,778\ (10^{-1})$
	knot	meter/second (m/s)	$5.144\,444\ (10^{-1})$
	mile/hour	meter/second (m/s)	$4.470\,400\ (10^{-1})$*
Volume	foot^3	meter^3 (m^3)	$2.831\,685\ (10^{-2})$
	gallon (U.S. liquid)	meter^3 (m^3)	$3.785\,412\ (10^{-3})$
	inch^3	meter^3 (m^3)	$1.638\,706\ (10^{-5})$
	liter	meter^3 (m^3)	$1.000\,000\ (10^{-3})$*
	quart (U.S. liquid)	meter^3 (m^3)	$9.463\,529\ (10^{-4})$

*Denotes an exact quantity.

STATICS AND DYNAMICS

Combined Version

SECOND EDITION

STATICS AND DYNAMICS

Combined Version

JERRY H. GINSBERG
School of Mechanical Engineering
Georgia Institute of Technology

JOSEPH GENIN
Department of Mechanical Engineering
New Mexico State University

JOHN WILEY & SONS

New York
Chichester
Brisbane
Toronto
Singapore

Library of Congress Cataloging in Publication Data

Ginsberg, Jerry H., 1944–
Statics; and, Dynamics.

Includes indexes.
1. Statics. 2. Dynamics. I. Genin, Joseph, 1930–
II. Ginsberg, Jerry H., 1944– . Dynamics. 1984.
III. Title. IV. Title: Dynamics.
TA351.G56 1984b 620.1 83-14564

ISBN 0-471-09745-4

Printed in the United States of America

10 9 8 7 6 5 4 3 2 1

PREFACE

Statics and Dynamics are the first courses in the engineering mechanics sequences. Their primary function is to help the student develop an understanding of the physical laws governing the response of engineering systems to forces and motion. Moreover, these subjects enhance the student's reasoning power as applied to the field of engineering, in that the student learns to solve problems logically, using the concept of mathematical models for physical systems. A successful mechanics experience is a strong motivation for further study of engineering.

Because of the strategic location of mechanics in the engineering curriculum, it is imperative that students be provided with maximum reinforcement in their instructional experience. Several of the features of this book are specifically intended to address this matter. The chapters treat the conventional topics for courses on statics and dynamics, but they have been written to be largely self-contained in order to minimize the amount of cross-referencing necessary to develop the material. This approach was also chosen because it is our desire to let the book communicate directly to the student, with a minimum of amplification and clarification required for the instructor.

Our treatment of the subject matter is founded on the recognition that the learning experience is a gradual process, as emphasized by the dual nature of the topics forming a first course in engineering mechanics. For example, in statics one set of topics—such as the treatments of force resultants, the laws of equilibrium, and the virtual work and energy principles—deals primarily with the development of the methodology of solving general problems. The other set of topics—such as the treatments of structural analysis, distributed forces, and the laws of friction—deals with the applications of the fundamental laws to specific systems and situations. Topics that may be omitted without a loss of continuity have been indicated by an asterisk preceding their title.

In both *Statics* and *Dynamics* the chapters that focus on the methodology associated with general principles are presented in a novel way. To increase comprehension of the basic principles and techniques, we precede each derivation with a discussion of why the principle is needed. The derivations are then keynoted by remarks regarding common systems that illustrate the implications of the derived concepts, as well as critical comments regarding their usefulness. This is then followed by one or more solved examples that directly illustrate how the derived principle or technique is employed.

After a broad body of basic concepts has been developed in a group of sections dealing with methodology, the question of their synthesis into a consistent procedure for solving general problems is addressed. This begins with the presentation of a set of sequential steps detailing the multiple operations necessary for the solution of general problems. These steps are merely a logical sequence to follow. (They are certainly not the only possible sequence.) With these steps, where appropriate, we indicate places where errors are commonly made. By presenting a systematic approach to problem solving we hope to enhance the student's senses of logic and deductive reasoning. Thus, the steps are not intended for memorization. Instead, as a student gains proficiency in problem solving, the procedures will be performed intuitively.

The steps are then employed in a series of three or more solved problems—called illustrative problems—whose solution requires the synthesis of the concepts previously developed. The illustrative problems are cross-referenced directly to the steps for problem solving. This allows the student to isolate a particular aspect of the solution that may prove troublesome. Numerous homework problems are presented after both examples and illustrative problems; in general, those following the latter are broader in scope.

The sole difference between the approach outlined above and the approach for the applications sections is that steps for solving problems are not presented in the latter. Emphasis in the discussion and solutions of examples in the applications sections is placed on the development of a logical approach using the basic methodology, and each group of solved examples is followed by a broad range of homework problems.

The problem-solving techniques primarily employ vector components with respect to orthogonal directions. We have chosen this approach for a variety of reasons. Most students are more comfortable treating the trigonometry of right triangles, and components are far more amenable to applications involving three-dimensional systems. Also, the process by which problems are formulated and solved is well attuned to modern trends toward computerization in areas such as CAD/CAM, structural and dynamic design, and vibrational analysis.

In order to emphasize the significance of vectors, we have used overbars to denote vector quantities. We have found that the only virtue of the current convention of using boldface letters to denote vectors is that it is easier for the printer to set into type. Boldface offers no convenient companion notation for the student to use for handwritten work. This inconsistency can cause students to become careless in notation, with the eventual result of major conceptual errors. Overbars in a text, by contrast, tend to reinforce the proper approach, because quantities are written by the student in the same way that they are read.

The development of physical understanding in mechanics is addressed in the solved problems, as well as in the formal text material. Where appropriate, in these problems care is taken to discuss the qualitative aspects of the solutions. It is in this capacity, as well as in the task of deriving basic principles, that we feel that diagrams depicting vector operations are most useful. Also, we have avoided letting mathematical sophistication and rigor overwhelm the physical aspects of the problem; each solution is implemented with the aid of only that level of mathematics appropriate to solve the problem at hand. Hopefully, this overall philosophy will give the student better insight into how an engineer thinks.

Note from the table of contents that the conventional ordering of the subject matter has been retained. Nevertheless, the viewpoint and treatment of many topics are not contained in other texts. In particular, the systematization of the methodology for treating virtual work and energy methods in statics and the treatment of the dynamics of rigid bodies in three-dimensional motion are entirely new and accessible to students with a broad range of backgrounds. Also, the application of static equilibrium principles to internal force analysis in beams and cables (Chapter 7) should provide valuable reference material for further study. The same is true for the treatments of orbital motion and vibrations contained in the dynamics portion of this volume.

Two considerations entered into the sequencing of the homework problems in each section. We tried to group problems according to the similarity of their fundamental characteristics and the resulting similarity in the techniques for solving the problem. Within each group the problems are ordered from those involving a direct application of the basic principles, to those of a more subtle nature. We have marked with an asterisk the most challenging problems in order to alert the instructor. For this reason, homework problems in a section do not necessarily appear in order of ascending difficulty.

A final technical matter for consideration here is the question of physical units. We strongly believe that the SI system of metric units is the best for engineering. Nevertheless, it is clear that a total conversion to SI is not likely to occur in the United States in the near future because a large number of engineering systems (and the tools by which they are constructed and maintained) have already been designed and built according to the U.S. Customary system of units. We have, therefore, apportioned the numerical problems approximately equally between the two systems of units. This should provide a sufficient number of homework problems for courses employing either system solely.

We gratefully acknowledge the encouragement of our colleagues, whose helpful comments proved invaluable. In particular, we wish to thank our former colleague at Purdue University, Dr. Robert W. Fox, with whom we had many delightful discussions of teaching philosophies. We also thank him as well as Hsu-Chiang Miao, Mosaad A. Foda, Karan B. Sangha, John C. Gagliardi, Joseph F. Gaziano, Michael D. Greenberg, Krishnan Pagalthivarthi, and Hugh J. Wellington at the Georgia Institute of Technology for their assistance in solving problems. We are indebted to Rona A. Ginsberg for her participation in all phases of the manuscript preparation, as well as for her excellent editorial comments. Finally, we note the knowledgeable assistance of the Wiley staff. Most especially, we are grateful to William Stenquist, for his support and enthusiasm for our work.

We would like to thank the following people for their valuable comments and suggestions provided while reviewing the text: Donald E. Carlson, University of Illinois; the late Thomas Carmondy, University of Arizona; Matthew Ciesla, New Jersey Institute of Technology; Charles M. Krousgrill, Purdue University; William M. Lee, U.S. Naval Academy; D.C. Leigh, University of Kentucky, Charles E. Long; Robert G. Oakberg, Montana State University; Arthur Petrou, University of Lowell; John T. Tielking, Texas A & M University; and Thomas Zilka, San Francisco State University.

Jerry H. Ginsberg
Joseph Genin

CONTENTS

CHAPTER 1

FUNDAMENTAL CONCEPTS

OBJECTIVES

1. To construct and understand diagrams that describe the operations of addition and subtraction of vectors, as well as multiplication of a vector by a scalar.

2. To use the laws of trigonometry to evaluate the result of a series of operations involving the addition and subtraction of vectors, as well as the multiplication of a vector by a scalar.

3. To understand the relationship between units of force and mass in the SI system of units, and in the U.S. Customary system of units.

4. To convert the magnitudes of quantities between U.S. Customary units and SI units.

A. MECHANICS

A quantity of primary importance for the design of many engineering systems is *force*. We can understand the concept of a force qualitatively by considering our sense of touch. For instance, the sensation we experience when our fingertips contact an object is a measure of how hard and in which direction we push or pull that object. The study of the way in which bodies are influenced by forces is the essence of the subject called *mechanics*.

Mechanics may be divided into two broad categories, *statics* and *dynamics*. *Statics* is the study of physical systems that remain at rest under action of a set of forces. *Dynamics* is the study of physical systems in motion. This text is devoted to the study of *statics*.

A key phrase used above is "physical system." By it we mean a collection of basic atomic elements that combine to form bodies. In the field of mechanics, for the convenience of problem solving, we tend to categorize these collections of elements into three broad groupings: particles, rigid bodies, and deformable bodies.

PARTICLE A body whose dimensions are negligible is said to be a particle. Conceptually, a particle occupies only a single point in space.

RIGID BODY A body occupying more than one point in space is said to be rigid if all of the constituent elements of matter within the body are always at fixed distances from each other.

DEFORMABLE BODY A body is said to be deformable if its constituent elements of matter experience changes in their distances from each other that are significant to the problem being investigated.

Clearly there is a certain amount of ambiguity in these definitions. For instance, is the group of molecules that compose a gas a system of particles or is it a deformable body? Another basic question we could ask is whether any body can correctly be considered to be rigid, for all real materials deform when forces are exerted upon them.

There are no absolute answers to these equations, for the *model* of the system we form depends on what knowledge we wish to gain about the response of the system. This leads to the next topic, which is the modeling process.

B. THE MODELING PROCESS

The general approach in the static analysis of a physical system is to consider first the nature of the forces acting on the system and the type of information we seek from the subsequent analysis. On the basis of those considerations we construct a conceptual model of the system. That is, we consciously consider its components to be either particles, rigid bodies, or deformable bodies. Then, by applying the laws of mechanics we determine the interrelationships among the forces acting on the system. An analysis of these forces should provide us with the information we desire. This general

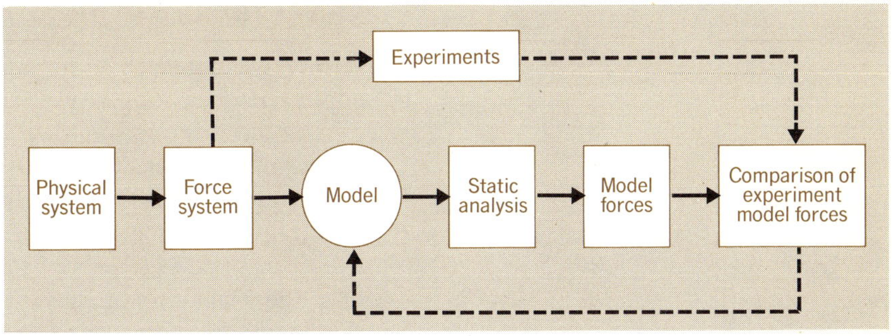

FIGURE 1

approach to the subject of statics can be summarized in the conceptual diagram shown in Figure 1.

The dashed paths in Figure 1 illustrate a key feature of the modeling process. To be sure of the validity of the model initially chosen to represent the physical system, it is necessary that the model display any relevant phenomena that are experimentally observed. If not, the model must be improved. Hence, modeling is an educated trial-and-error procedure. It is, in part, an art based on prior experience and physical intuition. On the other hand, the modeling process is also a science based on a knowledge of the function that the system is designed to perform, and of the analytical methods available for studying the phenomena exhibited by the system.

As an example, consider an object in the shape of a sphere. If this sphere is a small ball bearing resting on a flat level surface, as a first attempt one would tend to model it as a particle occupying a single point in space. We shall see in later chapters that this model would result in a successful analysis.

Suppose that we now step on this ball bearing, placing the ball bearing at the edge of the sole of our shoe. Our later studies will show that considering the ball bearing as a particle leaves us unable to explain why it tends to slide out from underfoot when we press on it harder and harder. In this case the nature of the forces acting on the ball bearing changes drastically from the case where it was merely resting on the ground. Modeling the ball bearing as a rigid body, as opposed to a single particle, would allow us to resolve this problem.

Continuing with this example, suppose that instead of stepping on a steel ball bearing, we were stepping on a tennis ball. Needless to say, the shape of the ball would be altered considerably as we stepped on it. In this case, the model of a rigid body should then be discarded in favor of the model of a deformable body.

From the foregoing examples we see that the model chosen to represent the physical system is a crucial element in the solution process. In this text, a first course on the statics of physical systems, we focus on the particle and rigid body models. The relationship between the forces acting on a body and its deformation are considered in the subject of mechanics of materials, which builds on the techniques we will develop here.

C. VECTOR QUANTITIES—FORCES

Earlier we noted that our sense of touch provides us with information about the direction, as well as the intensity, with which we push or pull on an object. In mechanics we mathematically characterize a force by both its *magnitude* (intensity) and its *direction.*

A quantity that can be characterized completely by a numerical value is said to be a *scalar*. Length, area, volume, mass, temperature, time, voltage, and energy are examples of scalars. Scalar quantities are numbers, so they obey the usual laws of algebra.

Many quantities occur in mechanics that require a description of their direction, as well as the numerical value of their magnitude. In addition to force, other quantities of this type are position, velocity, and acceleration, all of which are *vectors*. Because vectors have both *magnitude* and *direction*, they have a unique algebra. We shall develop that algebra later.

The reader may have previously learned to characterize a vector as having the properties of magnitude, direction, and sense. In this context direction is a line along which the action is taking place and sense is the heading. For example, a car moves along a north–south road (direction) headed south (sense). We will consistently consider *direction* or *sense* to imply both words.

FIGURE 2

Pictorially, a vector may be represented by a directed line segment, that is, by a straight arrow. The length of the arrow, to any convenient scale, represents the magnitude of the vector, and the direction of the arrow is the direction of the vector. Thus, in Figure 2 we show a vector quantity $\bar{A}$.

In this book we will denote vector quantities by placing a bar over the symbol adopted for the vector. Thus, the vector in Figure 2 is $\bar{A}$. In contrast, when we wish to refer to the magnitude of a vector, such as the magnitude of $\bar{A}$, we will do so by writing either $|\bar{A}|$ or simply A. Furthermore, because of the significant role of forces in mechanics, an arrow representing a force will be represented more boldly in diagrams than arrows representing other types of vector quantities.

For some vector quantities, a specification of their magnitude and direction describes them completely; that is, their action is not confined to, or associated with, a unique line in space. Such quantities are said to be *free vectors*. An example of a free vector is the statement that a person walked two kilometers (2 km) in a southeasterly direction, which is a description of the displacement resulting from a motion. Velocity and acceleration are other free vectors.

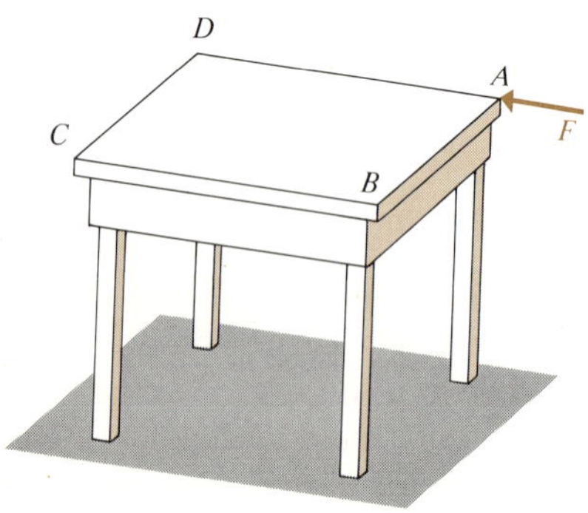

FIGURE 3

In contrast, when describing a force we must provide some specification of where the force is applied. To see this, consider the horizontal force $\bar{F}$ applied to corner A of the table shown in Figure 3. If the force $\bar{F}$ is sufficient to move the table, it will push the table forward and rotate the table counterclockwise as viewed looking downward. (Try it.) On the other hand, if a force having the same magnitude and direction as $\bar{F}$ is applied at corner B (as opposed to applying $\bar{F}$ at corner A), the result is a clockwise rotation. Hence, we can conclude that in this case we also need to specify the point of application of the force. The vectors described in the foregoing example are *fixed vectors*, for each had a unique point of application.

At times we may regard a force as a *sliding vector*, which is a vector that acts anywhere along a unique line in space. In such cases we describe the force by saying that it is collinear with a specific straight line, called its *line of action*, and then give its magnitude and sense. The actual point of application of the force along its line of action is then considered to be immaterial to the problem. For example, in Figure 3 if we choose to regard $\bar{F}$ as a sliding vector, we can imagine it being applied at point *A or D*; we would get the same resulting motion for the table.

In order to gain some insight into the seemingly ambiguous cases of fixed vectors and sliding vectors, consider a stalled car that is to be moved with a tow truck. As shown in Figure 4a, the truck may pull on the front bumper with a horizontal force $\bar{F}_A$. Alternatively, as shown in Figure 4b, it may push on the rear bumper with a horizontal force $\bar{F}_B$.

FIGURE 4a

FIGURE 4b

In this case the forces $\bar{F}_A$ and $\bar{F}_B$ are fixed vectors, having specific points of application, although they share the same line of action and have the same sense. If these forces also have the same magnitude, they will produce the same motion of the automobile. Hence, if the resulting motion was what we wanted to know, we could also treat the force applied to either bumper as a sliding vector.

Suppose now that we consider what happens if the tow truck that is attempting to move the stalled vehicle exerts too large a force. If $\bar{F}_A$ is too large, the front bumper will be torn off, whereas if $\bar{F}_B$ is too large, the rear bumper will be crushed. To account for what happened, we would have to consider the specific point of application of the force. In other words, the force could then be regarded only as a fixed vector.

These observations are a result of the *principle of transmissibility*, which states that

> The effect on a rigid body, in its entirety, of a force external to the body depends only on the magnitude, direction, and line of action of the force.

When we consider questions such as the relationship between the external forces acting on a rigid body, or the motion of a rigid body resulting from a set of external forces, we may apply the principle of transmissibility. However, if we are concerned with the effect of forces on a specific region within a rigid body, we may *not* invoke this principle. The forces acting on the body must then be treated as fixed vectors. Also, it should be noted that the principle of transmissibility pertains only to the body to which the force is actually applied. We will regard the principle of transmissibility as an axiom, even though a proof is contained in the *Dynamics* text.

Thus far we have discussed forces without really considering how they are generated. There are two broad categories of forces: those resulting from the *contact* of two bodies, and those associated with a *force field* created by the remote interaction of two bodies. Typical of the latter are force fields associated with gravity and electromagnetism. We will have more to say about gravity later in this chapter.

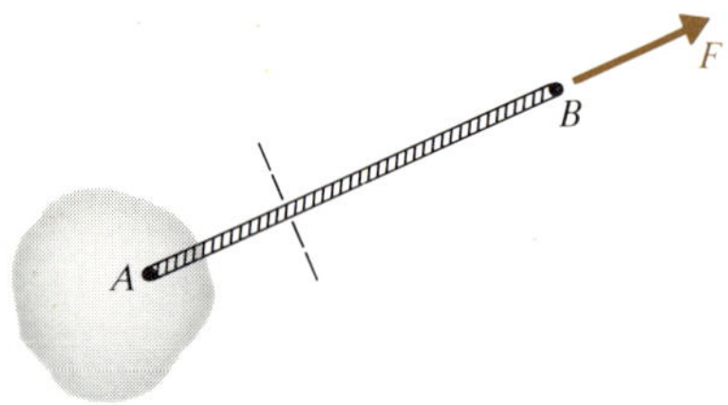

FIGURE 5

An important type of contact force is one exerted by a lightweight *cable* when it is fastened to a body. This situation is depicted in Figure 5, where a cable (or string) is fastened to an arbitrary body at point A, with a force $\bar{F}$ applied at its opposite end B. The line of action of $\bar{F}$ is line AB connecting the ends of the cable, with $\bar{F}$ pulling on the end of the cable. The magnitude F of this force is called the *tension* in the cable because it is transmitted uniformly along the length of the cable. That is, if we ask what force is applied on the region to the left of the dashed line in Figure 5 by the portion of the cable to its right, we would find that it is the same as force $\bar{F}$.

A more subtle feature of $\bar{F}$ is that it can pull on end B with any magnitude up to the breaking force of the cable, but it cannot push on it. When we try to push on end B, the cable becomes slack. The foregoing discussion contains a *modeling assumption*. The model we considered was for a cable that is inextensible.

Cables made of any material will elongate due to a tension force. A common example is a rubber band. When the stretching of the cable is significant, we will say that the cable is *elastic*. An important parameter for an elastic cable is its elongation Δ, which is the difference between its length L in the position of interest and the length it has when it is unloaded; the latter is the unstretched length L_0. Hence,

$$\boxed{\Delta = L - L_0} \qquad (1)$$

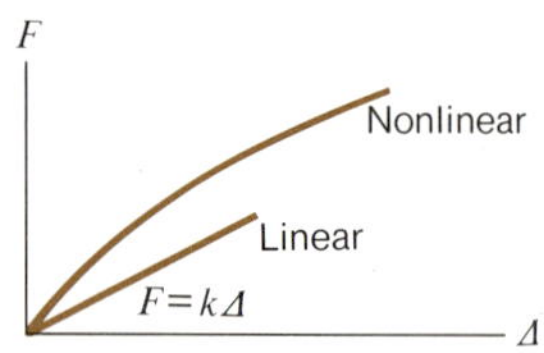

FIGURE 6

The tension in an elastic cable is directly related to its elongation, as depicted by the two examples plotted in Figure 6. If the tension is not proportional to the elongation, we obtain the curved line corresponding to a *nonlinear* elastic cable. The straight-line plot represents a *linear* elastic cable, for which

$$\boxed{F = k\Delta} \qquad (2)$$

The proportionality constant k is called the *stiffness* of the cable, because it is a measure of the tension that the cable must carry in order to obtain a given elongation.

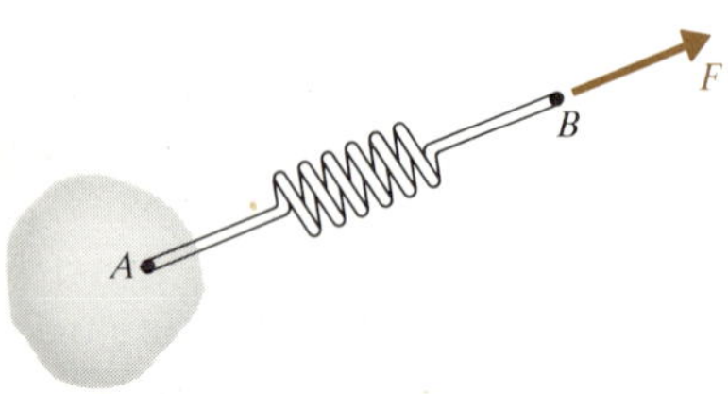

FIGURE 7

The case of a coil spring, such as the one in Figure 7, is identical to that of an elastic cable, with an important exception. Many springs will not become slack when a force $\bar{F}$ in the opposite sense from the one shown is applied at end B. Such a force is said to be *compressive*, because it shortens the spring. The compressive force is transmitted to end A, resulting in the spring pushing on the body with a force directed along the line of action BA. It follows then that eq. (1) is also valid for a linear spring, with negative Δ denoting a shortening ($L < L_0$) and negative F denoting compression.

D. VECTOR ALGEBRA

1. Addition of Vectors

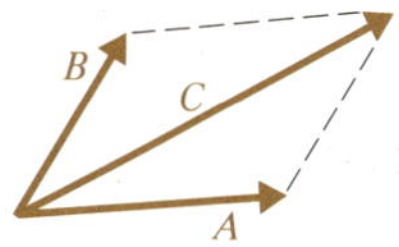

FIGURE 8

Not all quantities that have magnitude and direction are vectors. The final criterion for identifying that a quantity is a vector is that it must add by the *parallelogram law*. The vector sum $\overline{C} \equiv \overline{A} + \overline{B}$, which is called the *resultant*, is obtained by drawing the vectors $\overline{A}$ and $\overline{B}$ with their tails coincident. This forms two sides of a parallelogram. The resultant $\overline{C}$ is then the diagonal vector of this parallelgram whose tail is coincident with those of $\overline{A}$ and $\overline{B}$, as depicted in Figure 8.

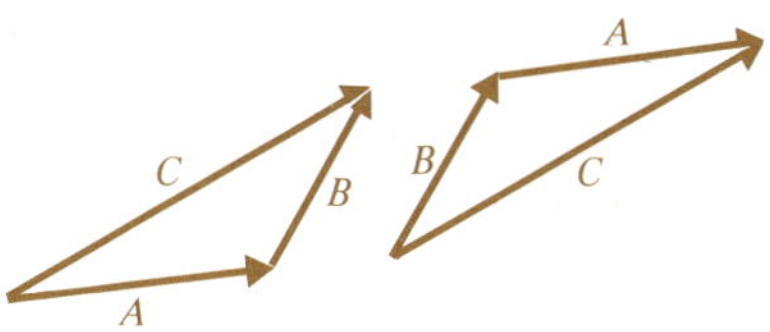

FIGURE 9

Notice that the order of addition is not important, so $\overline{A} + \overline{B} \equiv \overline{B} + \overline{A}$. This equality suggests an alternative rule for adding vectors. To do so we form a triangle by placing the tail of one vector at the head of the other. The resultant $\overline{C} = \overline{A} + \overline{B}$ is then the third side of the triangle, extending from the tail of the first vector to the head of the second. This is depicted in Figure 9 for the same vectors as were used in Figure 8.

2. Multiplication by a Scalar

The expression "double the force" means that the magnitude of the force is increased by a factor of 2, but the direction of the force is unchanged. Mathematically, we state this by letting q be a scalar number and $\overline{F}$ be an arbitrary force. The resulting vector $q\overline{F}$ is defined to be parallel to $\overline{F}$. If q is positive, then $q\overline{F}$ is in the same sense as $\overline{F}$, whereas when q is negative $q\overline{F}$ is in the opposite sense from $\overline{F}$. The (positive) magnitude of the resulting vector is defined to be

$$|q\overline{F}| \equiv |q|\,|\overline{F}|$$

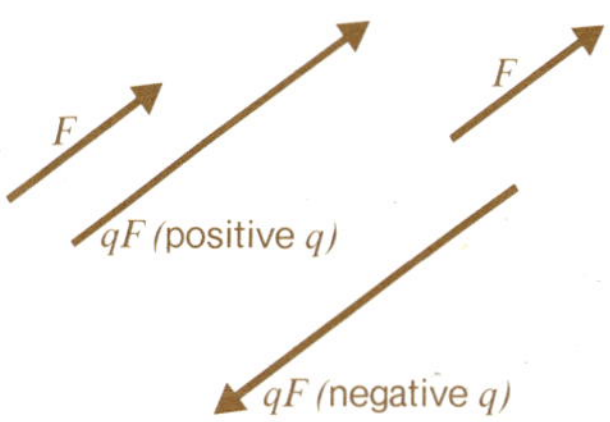

FIGURE 10

This is illustrated in Figure 10.

The *distributive law* follows from the definition of multiplication by a scalar. If $\overline{A}$ and $\overline{B}$ are vectors and q is a scalar, then

$$q(\overline{A} + \overline{B}) = q\overline{A} + q\overline{B}$$

3. Subtraction of Vectors

One rule for subtraction of vectors is a corollary of the addition rule, based on the fact that $\overline{D} = \overline{A} - \overline{B}$ is the same as $\overline{D} = \overline{A} + (-\overline{B})$. According to the rule for multiplication by a scalar, $-\overline{B}$ is a vector having the same magnitude as $\overline{B}$, but opposite direction. In Figure 11, a parallelogram whose sides are $\overline{A}$ and $-\overline{B}$ is constructed for the vectors that first appeared in Figure 8. The difference $\overline{D} = \overline{A} - \overline{B}$ is then the diagonal vector whose tail is coincident with those of $\overline{A}$ and $-\overline{B}$.

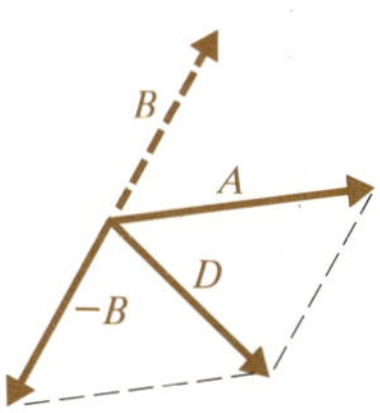

FIGURE 11

Note that the difference $\overline{D}$ forms the minor diagonal in the parallelogram of Figure 8. This suggests that an alternative rule for constructing $\overline{D}$ is to

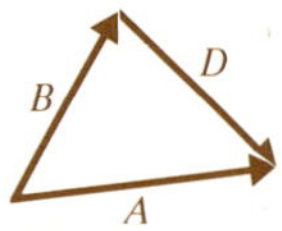

FIGURE 12

place the two vectors $\bar{A}$ and $\bar{B}$ so that their tails coincide. The difference then goes from the head of the negative vector to the head of the positive one, as shown in Figure 12. This figure also demonstrates that $\bar{B} + \bar{D} \equiv \bar{B} + (\bar{A} - \bar{B}) = \bar{A}$, as it should.

The foregoing vector operations are illustrated in the following example. It will be noted that here, as in the rest of the text, we do not solve problems by making measurements off a scale drawing.

EXAMPLE 1

Given two vectors $\bar{A}$ and $\bar{B}$, where $|\bar{A}| = 27$ directed east and $|\bar{B}| = 15$ directed 40° north of east, determine: (a) $\bar{A} + \bar{B}$, (b) $\bar{B} - \bar{A}$, (c) $\bar{A} - 5\bar{B}$.

Solution

For the purpose of drawing the vectors let us consider a horizontal line to the right to define the easterly direction and an upward line to be north. The given vectors are then as shown in the sketch.

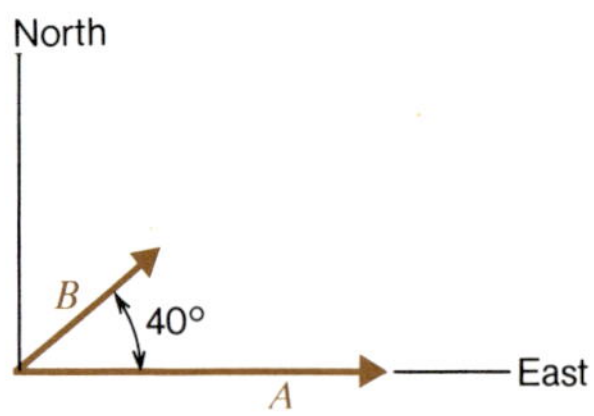

Part a

We will sum using the triangle construction. Moving the tail of $\bar{B}$ to the head of $\bar{A}$ gives the resultant $\bar{C}$ shown in the second sketch. The magnitude of $\bar{C}$ is found from the law of cosines:

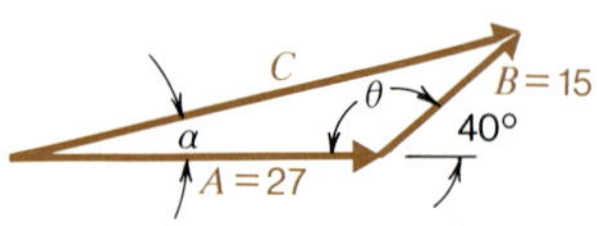

$$|\bar{C}|^2 = |\bar{A}|^2 + |\bar{B}|^2 - 2|\bar{A}|\,|\bar{B}| \cos \theta$$
$$= (27)^2 + (15)^2 - 2(27)(15) \cos 140^\circ$$
$$|\bar{C}| = 39.68$$

From the law of sines we obtain the direction of $\bar{C}$.

$$\frac{\sin \alpha}{|\bar{B}|} = \frac{\sin \theta}{|\bar{C}|}$$

$$\sin \alpha = \frac{15}{39.68} \sin 140^\circ$$

$$\alpha = 14.06^\circ$$

Hence

$$\bar{C} = \bar{A} + \bar{B} = 39.68 \text{ directed } 14.06^\circ \text{ north of east}$$

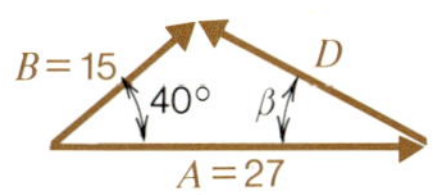

Part b

The triangle for evaluating $\bar{D} = \bar{B} - \bar{A}$ is formed by placing the vectors tail to tail. Then $\bar{D}$ goes from the head of $\bar{A}$ (the negative vector) to the head of $\bar{B}$, as shown in the sketch.

The law of cosines now gives

$$|\bar{D}|^2 = |\bar{A}|^2 + |\bar{B}|^2 - 2|\bar{A}|\,|\bar{B}| \cos 40^\circ$$
$$= (27)^2 + (15)^2 - 2(27)(15) \cos 40^\circ$$
$$|\bar{D}| = 18.26$$

From the law of sines we then find

$$\frac{\sin \beta}{|\bar{B}|} = \frac{\sin 40^\circ}{|\bar{D}|}$$

$$\sin \beta = \frac{15}{18.26} \sin 40^\circ$$

$$\beta = 31.87^\circ$$

Hence

$$\bar{D} = \bar{B} - \bar{A} = 18.26 \text{ directed } 31.87^\circ \text{ north of west}$$

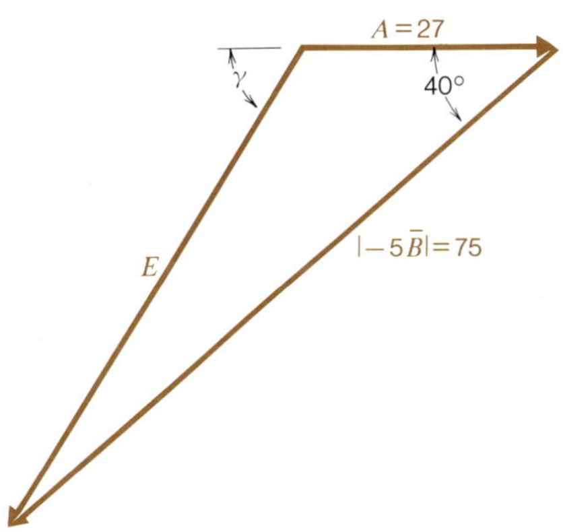

Part c

We may regard $\bar{A} - 5\bar{B}$ to be either $\bar{A} + (-5\bar{B})$ or $\bar{A} - (5\bar{B})$. Let us consider it as the former. Multiplying $\bar{B}$ by -5 gives a vector of magnitude 75 directed opposite of $\bar{B}$, that is, directed 40° south of west. Thus, the triangle construction for $\bar{E} = \bar{A} + (-5\bar{B})$ is as shown in the sketch.

The law of cosines gives

$$|\bar{E}|^2 = |\bar{A}|^2 + |-5\bar{B}|^2 - 2|\bar{A}|\,|-5\bar{B}| \cos 40^\circ$$
$$= (27)^2 + (75)^2 - 2(27)(75) \cos 40^\circ$$
$$|\bar{E}| = 57.02$$

whereas the law of sines gives

$$\frac{\sin(180^\circ - \gamma)}{|-5\bar{B}|} = \frac{\sin 40^\circ}{|\bar{E}|}$$

$$\sin(180^\circ - \gamma) = \frac{75}{57.02} \sin 40^\circ$$

$$180^\circ - \gamma = 57.72, 122.28^\circ$$

From the computation of $|\bar{E}|$ we know that $|-5\bar{B}|$ is the longest side in the triangle. It then follows that the corresponding vertex angle, $180^\circ - \gamma$, must be the largest interior angle. Hence, choosing $180^\circ - \gamma = 122.28$, we find

$$\gamma = 57.72^\circ$$

so

$$\bar{E} = \bar{A} - 5\bar{B} = 57.02 \text{ directed } 57.72^\circ \text{ south of west}$$

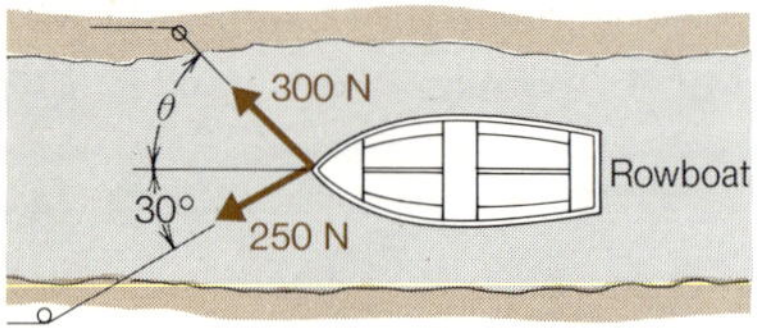

PROBS. 1.1 AND 1.3

HOMEWORK PROBLEMS

1.1 and 1.2 The angle $\theta = 30°$. Determine the magnitude and direction of the resultant of the two forces shown (N ≡ newton).

1.3 and 1.4 The angle $\theta = 55°$. Determine the magnitude and direction of the resultant of the forces shown (N ≡ newton).

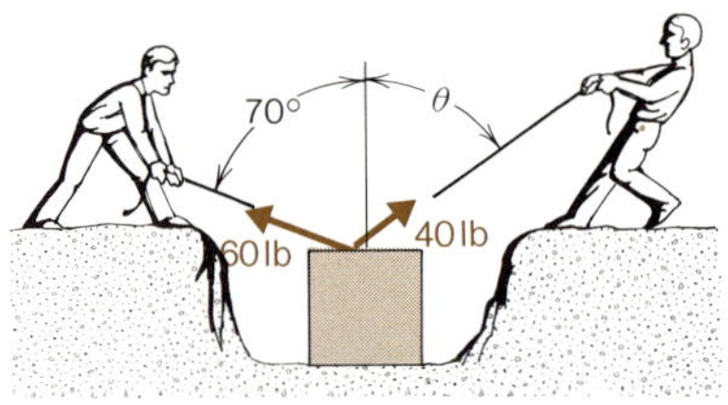

PROBS. 1.2 AND 1.4

In Problems 1.5 through 1.12; $\bar{A} = 8$ due north, $\bar{B} = 12$ at 45° north of east, $\bar{C} = 15$ at 30° south of east, $\bar{D} = 18$ at 15° west of north.

1.5 Find (a) $\bar{A} + \bar{B}$, (b) $\bar{A} - \bar{B}$

1.6 Find (a) $\bar{B} + \bar{C}$, (b) $\bar{B} - \bar{C}$

1.7 Find (a) $3\bar{A} + 2\bar{B}$, (b) $\bar{A}/3 - 2\bar{B}$

1.8 Find (a) $\bar{A} + \bar{B} + \bar{C}$, (b) $\bar{A} - \bar{B} + \bar{C}$

1.9 Find $\bar{C} + \bar{D}$ and $2\bar{C} - \bar{D}$, then show that $2(\bar{C} + \bar{D}) - (2\bar{C} - \bar{D}) = 3\bar{D}$

1.10 Find $\bar{A} + \bar{B} + \bar{C} + \bar{D}$

1.11 Find $\bar{A} + \bar{B} - \bar{C} + \bar{D}$

1.12 Find the vector $\bar{E}$ for which $\bar{A} + \bar{B} - \bar{C} + \bar{E} = \bar{0}$

E. NEWTON'S LAWS

The laws of motion stated by Sir Isaac Newton (1642–1727) in his historic work *Principia* (1687) are the foundation for the study of engineering mechanics. Here we state his laws in modern language.

FIRST LAW A particle will remain at rest or move with constant speed along a straight line, unless it is acted upon by a resultant force.

SECOND LAW When a resultant force is exerted upon a particle, the acceleration of that particle is in the direction of the force, and the magnitude of the acceleration is proportional to the magnitude of the force.

THIRD LAW Each force exerted upon a particle is the result of an interaction with another particle, the forces of action and reaction being equal in magnitude, opposite in direction, and collinear.

A system that is at rest or moving with a constant velocity is a special case of an accelerating system, so the first law is included in the second. One reason for stating the first law separately is that it is the foundation for the study of Statics. We will also make extensive use in Statics of the third law, because it will assist us in characterizing the forces acting on a system.

The second law is employed in Statics solely to discuss the force of gravity and to establish a consistent system of units for describing physical quantities. The second law may be reworded to state that the resultant force $\bar{F}$ acting on a particle is proportional to the acceleration $\bar{a}$ of the particle; the *mass* m of the particle is the factor of proportionality. The mathematical statement of the second law is therefore the following *vector equation*:

$$\boxed{\bar{F} = m\bar{a}} \tag{3}$$

A remarkable aspect of Newton's laws is that, as stated, they apply only to a particle. Part of the study of mechanics is devoted to extending these laws to deal with rigid and deformable bodies.

F. SYSTEMS OF UNITS

The quantities appearing in Newton's second law, $\bar{F} = m\bar{a}$, involve measurements of the dimensions of length, time, force, and mass, which we denote for brevity as L, T, F, and M, respectively. The dimensions chosen for the measurement of these four quantities cannot be defined independently. They must obey the *law of dimensional homogeneity*, which requires that the physical units of all terms in an equation be identical. Stated in colloquial language "apples cannot be equated to oranges."

In terms of the four basic measurements, the dimensions of acceleration are L/T^2. Dimensional homogeneity of eq. (3) requires that

$$F = M\frac{L}{T^2}$$

It follows that the units for any three of the four quantities, L, T, M, and F, may be selected freely, with the units of the fourth quantity then derived by satisfying Newton's second law. Historically, the units by which length and time are measured are well defined, and, through the efforts of international standards committees, universally accepted.

This leaves us with the choice of defining either a standard unit of mass or of force, and then deriving the units for the undefined quantity. A simple relationship between force and mass units is obtained by applying Newton's second law to a particle that is falling freely in a vacuum at the surface of the earth. The magnitude of the free-fall acceleration is denoted as g.

Until we can look at the matter in more detail in the *Dynamics* text, let us define the *weight* W of an object to be the magnitude of the gravitational attractive force exerted between the object and the earth. Newton's second law then gives

$$\boxed{W = mg} \tag{4}$$

The mass of a particle m is independent of the location where the experiment is performed. Thus a system of units where the unit of mass is defined and the force unit is derived is said to be *absolute*. In contrast, the free fall acceleration of a particle depends on the location of the particle on the earth's surface; therefore, its weight also depends on the location. A system of units in which the force unit is defined is said to be *gravitational*. Presently, one set of units that is in wide usage is the SI (Standard International) system, which is a metric absolute system. Another system in common use is the U.S. Customary system, which is a gravitational system.

With the exception of specialized topics in dynamics (e.g., orbital motion), it will suffice to solve problems using $g = 9.807$ meters/second^2 (m/s^2) (metric) and $g = 32.17$ feet/second^2 (ft/s^2) (U.S. Customary) as average values.

In Table 1 we summarize the commonly used systems of units and indicate how the derived units are obtained. The basic units in each system are defined with respect to standard bodies or with respect to physical phenomena. For example, a kilogram is the mass of a particular bar of platinum alloy. On the other hand, a meter is defined as 1,650,763.73 wavelengths in a vacuum of the orange-red line of krypton 86.

TABLE 1
SYSTEM OF UNITS

SYSTEM	LENGTH	TIME	MASS	FORCE
Metric absolute (SI)	Meter (m)	Second (s)	Kilogram (kg)	Newton (N)[a] = kg-m/s^2
Metric gravitational	Meter (m)	Second (s)	Metric slug[a] = kg-s^2/m	Kilogram (kg)
British absolute	Foot (ft)	Second (s)	Pound (lb)	Poundal (pdl)[a] = lb-ft/s^2
British gravitational (U.S. Customary)	Foot (ft)	Second (s)	Slug[a] = lb-s^2/ft	Pound (lb)

[a] Derived quantity

The choice for systems of units presented in Table 1 is further complicated by the fact that certain multiples and submultiples of the basic units in each system are given their own names, such as the mile, which is 5280 feet, or the dyne, which is one hundred-thousandth of a newton. The SI (Standard International) system has been introduced in order to simplify this situation.

Essentially, the SI system is the metric absolute system with only one unit being used to describe each type of quantity. The SI units for the dimensions of length, time, mass, and force are meters, seconds, kilograms, and newtons, respectively. Then, if one desires, decimal multiples and submultiples of the basic units may be indicated by the appropriate prefixes. Thus, in the SI system lengths 0.2 meter (m) and 200 millimeters (mm) are identical and correct. On the other hand, $2(10^{-3})$ newtons (N) or equivalently 2 millinewtons (mN) is the correct description for a force whose magnitude is 200 dynes. Lists of the SI units and of the preferred prefixes may be found in Appendix A.

Many of the examples and problems in this text will be presented in SI units. Unfortunately, this system is not yet universally used in the United States, where many engineers, as well as nontechnical people, still use the U.S. Customary system. We will therefore present examples and problems that are posed in either system of units.

When solving problems it is important to keep in mind that the SI system is an absolute one in which mass is a basic unit, so the mass of an object in kilograms will be the specified parameter. The weight at the earth's surface may then be obtained from $W = mg$. In contrast, force is the basic unit in the gravitational (U.S. Customary) system, so the weight in pounds (or ounces or tons) will be the specified parameter in this case. The mass is then found from $m = W/g$.

In general, we will solve problems in the same system of units as that used for the given information, and we shall not convert between different systems of units. Obviously, the units of any physical quantity can be changed with the aid of conversion factors. These conversion factors are numbers that are physically unit values, but not numerically, for example, 25.40 mm = 1 in. A brief set of factors are given in Appendix A.

EXAMPLE 2

Conversion factors may be used consecutively. As an illustration, use the fact that 1 in. ≡ 25.40 mm (the definition of an inch) to convert a speed of 60 miles/hour to units of (a) feet/second, and (b) meters/second.

Solution

Part a

The process of converting units may be regarded as one of canceling the units in the numerator and denominator by means of conversion factors until the desired units are obtained. The first part concerns a conversion of units in the U.S. Customary system only. Thus

$$60\,\frac{\text{mi}}{\text{hr}} = \left(60\,\frac{\text{mi}}{\text{hr}}\right)\left(\frac{1\text{ hr}}{3600\text{ s}}\right)\left(\frac{5280\text{ ft}}{1\text{ mi}}\right) = 88\text{ ft/s}$$

Incidentally, the above result is a conversion factor that we will employ frequently in *Dynamics*.

Part b

Here we must first convert the answer in part (a) to units of inches/second before we may apply the given SI conversion factor. Thus

$$60\,\frac{\text{mi}}{\text{hr}} = \left(88\,\frac{\text{ft}}{\text{s}}\right)\left(\frac{12\text{ in.}}{1\text{ ft}}\right)\left(\frac{25.40\text{ mm}}{1\text{ in.}}\right)\left(\frac{1\text{ m}}{1000\text{ mm}}\right) = 26.82\text{ m/s}$$

EXAMPLE 3

Using the facts that 1 in. equals 25.40 mm and that a body weighing 1 lb on the surface of the earth has a mass of 0.4536 kg, determine the factors for converting the British gravitational units of pounds force, slugs, mass, and density (slugs per cubic foot) into their SI equivalents.

Solution

The conversion factors between pounds and newtons is determined by using $W = mg$. Let us compute the weight in newtons of a 0.4536 kg body. Not knowing where the body was weighed, we use the average value $g = 9.807\ \text{m/s}^2$ to obtain

$$W = (0.4536\ \text{kg})(9.807\ \text{m/s}^2) = 4.448\ \text{N}$$

In other words,

$$1\ \text{lb} = 4.448\ \text{N}$$

Then, since 1 lb-s²/ft is 1 slug, we have

$$1\ \text{slug} = \left(1\ \frac{\text{lb-s}^2}{\text{ft}}\right)\left(\frac{4.448\ \text{N}}{1\ \text{lb}}\right)\left(\frac{1\ \text{ft}}{12\ \text{in.}}\right)\left(\frac{1\ \text{in.}}{25.4\ \text{mm}}\right)\left(\frac{1000\ \text{mm}}{1\ \text{m}}\right)$$

$$= 14.593\ \frac{\text{N-s}^2}{\text{m}} = 14.593\ \text{kg}$$

To convert the units of density, we use

$$1\ \frac{\text{slug}}{\text{ft}^3} = \left(1\ \frac{\text{slug}}{\text{ft}^3}\right)\left(\frac{14.593\ \text{kg}}{\text{slug}}\right)\left(\frac{1\ \text{ft}}{12\ \text{in.}}\right)^3\left[\left(\frac{1\ \text{in.}}{25.40\ \text{mm}}\right)\left(\frac{1000\ \text{mm}}{1\ \text{m}}\right)\right]^3$$

$$= 515.3\ \text{kg/m}^3$$

The differences between the values obtained here and those in the table in Appendix A are attributable to the fact that we were only using four significant figures for our calculations. The limitation of four significant figures is a result of the uncertainty of the location for the value of g.

HOMEWORK PROBLEMS

In the following problems, t is time, x is distance (length), v is speed (length/time), a is the magnitude of acceleration (length/time²), m is mass, and F is the magnitude of force.

1.13 Derive conversion factors for changing the following U.S. Customary units to their SI equivalents. (a) area: inch², (b) volume: foot³, (c) force: ounce, (d) pressure: pounds/inch².

1.14 Derive conversion factors for changing the following SI units to their U.S. Customary equivalents. (a) volume: m³, (b) speed: km/hr, (c) acceleration: m/s², (d) pressure: pascal.

1.15 Derive conversion factors for changing the following U.S. Customary units to their SI equivalents. (a) pressure: lb/ft², (b) speed: ft/s, (c) acceleration: mi/hr², (d) density: lb/ft³.

In Problems 1.16 and 1.17, c_1, c_2, etc., denote constants and θ is an angle in radians. Determine the dimensions of these constants if the formula is to be dimensionally correct.

1.16 (a) $a = c_1 v^2/x$, (b) $\frac{1}{2}mv^2 = c_1 x^2$, (c) $x = c_1 v + c_2 a^2$, (d) θ (degrees) $= c_1 \theta$ (radians).

1.17 (a) $d\theta/dt = c_1 + c_2 t + c_3 t^2$, (b) $F/x^2 = c_1 v\, dv/dx$, (c) $mc_1^2\, d^2\theta/dt^2 = Fx$, (d) $a = c_1 \cos(c_2 vx)$.

1.18 At a certain large university known for its athletic prowess, the following new set of physical units are in use. The basic length is the *touchdown* which is 100 yards, the basic time unit is the *dash* which is 9 s, and the basic force unit is the *golf ball*, which is 1.620 oz. Determine the conversion factors between these units and their SI equivalents.

1.19 At the university in Problem 1.18, the unit of mass is the *basket*, 1 basket being 1 golf ball-dash^2/touchdown. Determine the number of kilograms in one basket.

1.20 When a linear coil spring in the suspension system of an automobile is compressed by 2.5 in., it exerts a compressive force of 375 lb. Determine the stiffness of the spring in U.S. Customary and in SI units.

1.21 When a certain linear spring has a length of 180 mm the tension within it is 150 N. For a length of 160 mm the compressive force within the spring is 100 N. Determine the unstretched length and stiffness of the spring (a) in SI units, (b) in U.S. Customary units.

***1.22** A widely employed fluid mechanics equation for the pressure drop along a pipe of length l and diameter D is

$$P_1 - P_2 = f \frac{l}{D} \rho \frac{v^2}{2}$$

where P denotes pressure, ρ is the density of the fluid, v is its velocity, and f is the resistance coefficient, which is given by the expression

$$\sqrt{f} = 2.30 \log_{10}\left(\frac{Dv}{\nu}\sqrt{f}\right) - 0.91$$

Determine the dimensions of ν, the kinematic viscosity.

1.23 A restaurant buys 1000 lb of coffee at a standard locality and also a spring scale calibrated at that locality. Determine what the coffee weighs on the spring scale in a locality where (a) $g = 32.11$ ft/s^2, (b) $g = 9.791$ m/s^2.

1.24 The velocity v of a particle is known to depend upon the elapsed time according to $v = v_0 \ln(1 + t/t_0)$. It is known that $v = 50$ ft/s when $t = 4$ seconds, and that $v = 100$ m/s when $t = 12$ seconds. Determine the values of v_0 and t_0 in U.S. Customary units.

1.25 In Problem 1.24, determine the values of v_0 and t_0 in SI units.

CHAPTER 2

CONCURRENT FORCE SYSTEMS AND EQUILIBRIUM OF A PARTICLE

OBJECTIVES

1. To evaluate the components and direction angles of two- and three-dimensional vectors.

2. To determine the unit vector describing the direction of a vector.

3. To determine, in component form, the resultant of a set of concurrent forces.

4. To determine the properties of a set of concurrent forces required for them to have a specified resultant.

5. To evaluate the dot product of two vectors by using the component representation of each vector.

6. To compute the component of a vector in a specified direction and to write that component as a vector.

7. To draw a free body diagram of an object that is modeled as a particle.

8. To evaluate the forces required for static equilibrium of an object that is modeled as a particle.

A primary goal in *Statics* is to establish the basic laws governing the forces acting on bodies that are in a state of equilibrium. The concepts, techniques, and methods in this development are central to all engineering studies concerning the way in which physical systems respond to the forces acting on them. This chapter, which initiates our study, is devoted to the case of a particle, because a particle is the simplest model that can be used to represent a physical body.

A. BASIC CONCEPTS

By definition, a particle is a body that is considered to occupy a single point in space. Hence, the forces acting on a particle must be *concurrent* (intersect) at the particle. Consider the two forces $\bar{F}_1$ and $\bar{F}_2$ acting on the particle P shown in Figure 1. Forces $\bar{F}_1$ and $\bar{F}_2$, being vectors, add according to the parallelogram law. Their total effect on the particle is equivalent to the force $\bar{R}$ in Figure 1, where

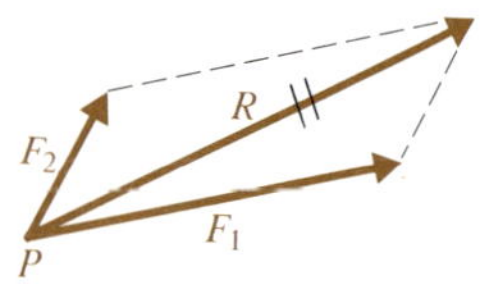

FIGURE 1

$$\bar{R} = \bar{F}_1 + \bar{F}_2$$

The vector $\bar{R}$ is called the *resultant* of $\bar{F}_1$ and $\bar{F}_2$. Note that we use a double slash on the arrow representing the resultant in order to distinguish it from an actual force acting on the particle.

If there were more than two forces acting on the particle the resultant of such a *system of forces* would be

$$\boxed{\bar{R} = \sum_{i=1}^{N} \bar{F}_i} \tag{1}$$

It is inherent to eq. (1) that the sum of vectors is independent of the order in which they are added. Further, note that the point of application of the resultant force is the point occupied by the particle.

Let us consider a number of ways in which the resultant force vector can be obtained. Clearly, as was done in Figure 1, we can use the parallelogram law to add the vectors consecutively. For example, in the case of three forces, we could first obtain the vector $\bar{R}_1 = \bar{F}_1 + \bar{F}_2$ and then find the total resultant according to

$$\bar{R} = (\bar{F}_1 + \bar{F}_2) + \bar{F}_3 = \bar{R}_1 + \bar{F}_3$$

as depicted in Figure 2a. This is known as the *associative law* of vectors.

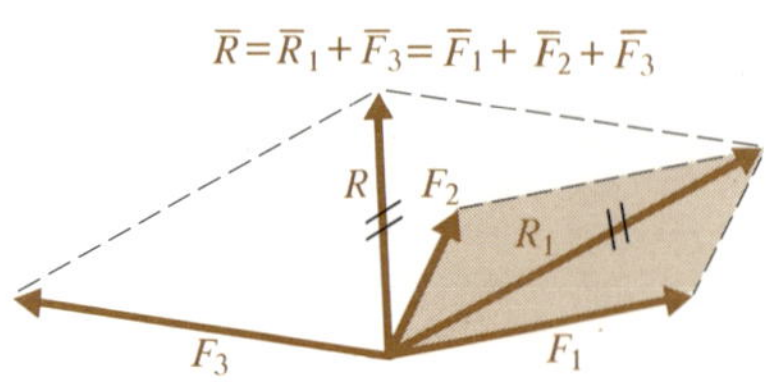

FIGURE 2a

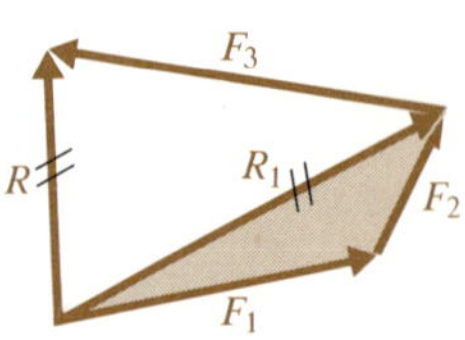

FIGURE 2b

Another view of the process for determining the result is obtained from the alternative rule for the addition of vector quantities. In this method the individual vectors in the sum are moved in space, with their orientations constant, until the tail of each vector in the sum coincides with the tip of the previous vector, as illustrated in Figure 2b. The resultant $\bar{R}$ then extends from the tail of the first vector ($\bar{F}_1$) to the tip of the last vector ($\bar{F}_3$). It is not necessary in this approach to actually determine the intermediate resultant force $\bar{R}_1$. In general, the resultant of a system of forces acting on a particle is the closing vector of a polygon whose other sides are composed of all the forces in the system.

Two major difficulties arise when using the approaches in Figure 2 to determine resultant vectors. Clearly, it is difficult to construct the appropriate diagrams for a system of three-dimensional vectors. Equally important, even in the case of planar force systems, the determination of the resultant usually involves the geometry and trigonometry of scalene (general) triangles.

A graphical solution in which the vector polygon is constructed to scale circumvents the need to use trigonometry. However, this approach places unnecessary limitations on the accuracy of the solution, and it still does not address the question of three-dimensional force systems.

Another procedure, which we shall use throughout the text, is to represent vectors by their components with respect to a set of mutually orthogonal (i.e., perpendicular) axes. These axes form an *xyz* coordinate system that we refer to as the *frame of reference*. This component approach for treating vectors has the significant advantage of simplifying the geometry and trigonometry to that of right triangles, while simultaneously providing a consistent approach for both planar and three-dimensional problems.

B. COMPONENTS OF A VECTOR

Before we can use the component approach for vector operations, such as addition, it is necessary for us to develop the ability to represent vectors in terms of their components with respect to an *arbitrarily chosen* rectangular Cartesian coordinate system *xyz*. A typical situation is depicted in Figure 3, where we wish to determine the components of a vector $\bar{A}$. The origin O of the coordinate system is chosen, for convenience, to coincide with the tail of the vector.

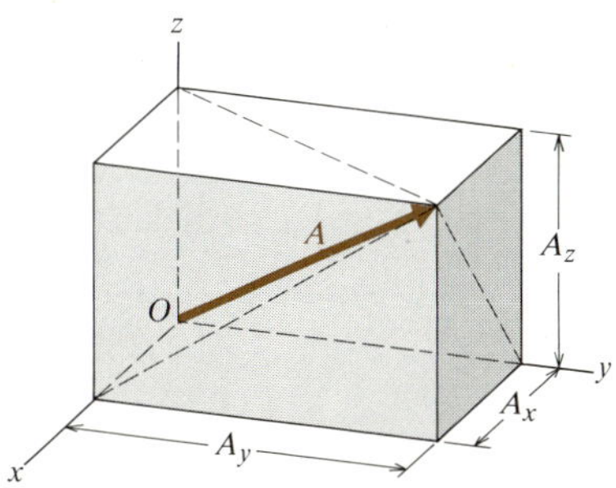

FIGURE 3

The dotted lines in Figure 3 are constructed to intersect the tip of the vector $\bar{A}$ and to intersect each coordinate axis at a right angle. From the intersections of these perpendiculars with the axes, we can form a rectangular parallelepiped, as depicted in the figure. The lengths A_x, A_y, and A_z are defined as the components of the vector $\bar{A}$ with respect to the x, y, and z coordinate axes, respectively. Because $\bar{A}$ is the diagonal of the parallelepiped, we find from the Pythagorean theorem that

$$|\bar{A}| = \sqrt{A_x^2 + A_y^2 + A_z^2} \tag{2}$$

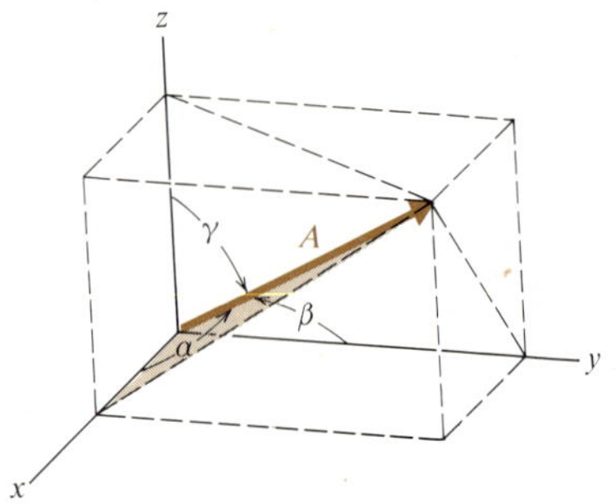

FIGURE 4

If we passed a plane through the vector $\bar{A}$ and the x axis, the angle that would appear between the two lines is α. This is depicted by the shaded plane in Figure 4. Clearly, $A_x = |\bar{A}| \cos \alpha$. More generally, we define the *direction angles* α, β, and γ in Figure 4 to be the angles between the vector $\bar{A}$ and the x, y, and z axes, respectively, when the tail of the vector is placed at the origin of the coordinate system. We may then write

$$A_x = |\bar{A}| \cos \alpha, \qquad A_y = |\bar{A}| \cos \beta, \qquad A_z = |\bar{A}| \cos \gamma \tag{3}$$

A component of a vector is negative if the projection intersects the negative coordinate axis. Thus, the direction cosines may be either positive or negative. Ambiguities in choosing the proper value for the direction angle are eliminated by requiring that

$$0 \le \alpha, \beta, \gamma \le 180^\circ$$

Although there are three direction angles associated with a vector, only two of them are independent quantities. The relationship between a set of direction angles is obtained by substituting eqs. (3) into eq. (2), which yields

$$\cos^2 \alpha + \cos^2 \beta + \cos^2 \gamma \equiv 1$$

The preceding discussion suggests that the properties of a free vector relative to a chosen coordinate system are completely described by its components. It therefore becomes important to develop a way to describe the orientation of a coordinate system. This is done by introducing the concept of a *unit vector*, which is a vector whose length is the dimensionless number one (hence, the word *unit*). It follows that the only information conveyed by a unit vector is its direction.

The orientation of the coordinate axes may be defined by introducing a set of unit vectors $\bar{i}$, $\bar{j}$, and $\bar{k}$ that are parallel to the x, y, and z axes, respectively, with the head of each unit vector pointing in the sense of the increasing coordinate value. These unit vectors are shown in black in Figure 5.

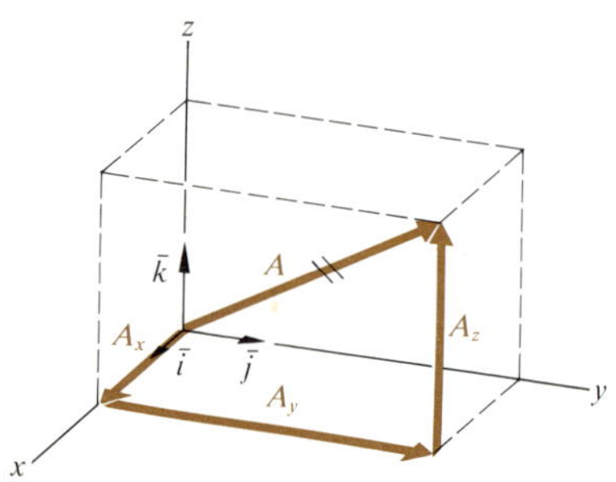

FIGURE 5

Figure 5 also depicts how the vector $\bar{A}$ may be constructed from its components. Starting from the tail of $\bar{A}$, a vector $A_x\bar{i}$ (shown in color) is first constructed. It is a vector of magnitude A_x that is parallel to the x axis, with the direction of this axis defined by $\bar{i}$. The tail of the vector $A_y\bar{j}$ (length A_y parallel to the y axis) is placed at the head of $A_x\bar{i}$ in order to add the two vectors. The vector $A_z\bar{k}$ is then added to that sum. The polygon rule for summing vectors then indicates that the sum of the three component vectors is the vector $\bar{A}$. Thus

$$\bar{A} = A_x\bar{i} + A_y\bar{j} + A_z\bar{k} \tag{4}$$

Recall that unit vectors are dimensionless. The physical units of a vector are the same as those of each of its components.

Equation (4) is the means whereby a vector may be described mathematically in terms of its components. It is clear that such a description is meaningful only if the corresponding coordinate axes and unit vectors are specified. This is demonstrated in the following example.

EXAMPLE 1

Represent the force $\bar{F}$ shown in the sketch in terms of its components with respect to the two coordinate systems xyz and $x'y'z'$.

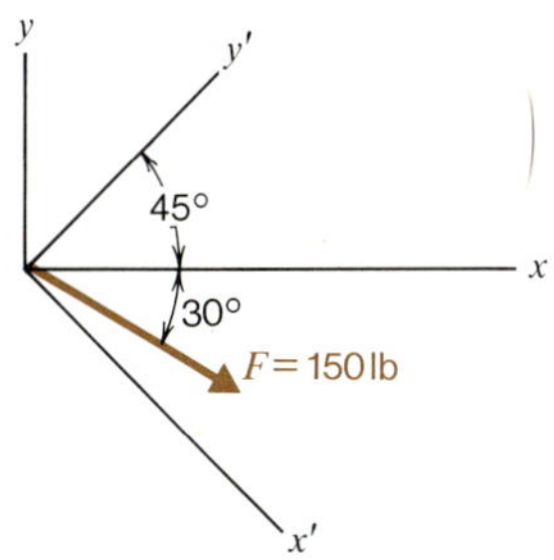

Solution

The force lies in the plane of the paper, so the direction angle between $\bar{F}$ and the z axis is 90°. Because $\cos 90° \equiv 0$, there is no component perpendicular to the paper. A sketch of the components of $\bar{F}$ with respect to the x and y axes shows that

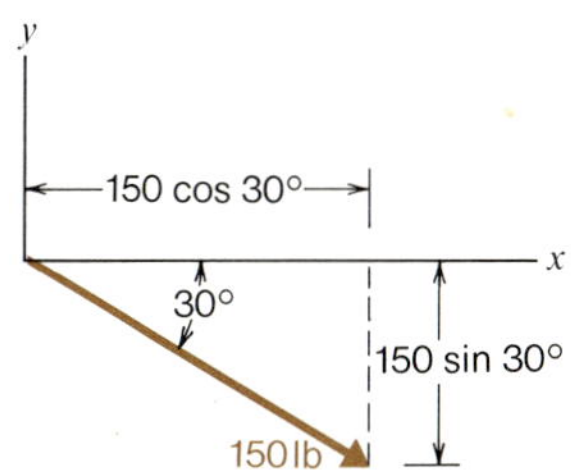

$$F_x = 150 \cos 30° = 129.90 \text{ lb}$$

$$F_y = -150 \sin 30° = -75.00 \text{ lb}$$

Note that F_y is negative because it is along the negative y axis.

Equation (4) then gives

$$\bar{F} = 129.90i - 75.00j \text{ lb}$$

where $\bar{i}, \bar{j}, \bar{k}$ are the unit vectors for the xyz coordinate system.

We now follow the same approach for the components with respect to $x'y'z'$, beginning with a sketch. From the sketch,

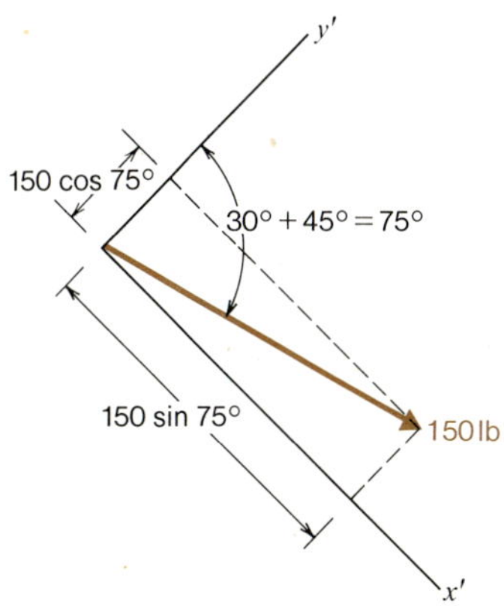

$$F_{x'} = 150 \sin 75° = 144.89 \text{ lb}$$

$$F_{y'} = 150 \cos 75° = 38.82 \text{ lb}$$

Thus

$$\bar{F} = 144.89\bar{i}' + 38.82\bar{j}' \text{ lb}$$

where $\bar{i}', \bar{j}', \bar{k}'$ are the unit vectors for the $x'y'z'$ axes.

Although the values of the components for the xyz axes are different from those for the $x'y'z'$ axes, both answers represent the same vector. A simple verification may be found by using eq. (2) to compute $|\bar{F}|$ for each result.

EXAMPLE 2

Determine the vector that locates a point P whose rectangular Cartesian coordinates are (400, 400, −200) mm with respect to the origin O. What are the direction angles of this vector?

Solution

A sketch shows that the desired *position vector* is $\bar{r}_{P/O}$, which is the position of point P with respect to point O (the notation should be read as r of P with respect to O). We see from the sketch that the components of the vector

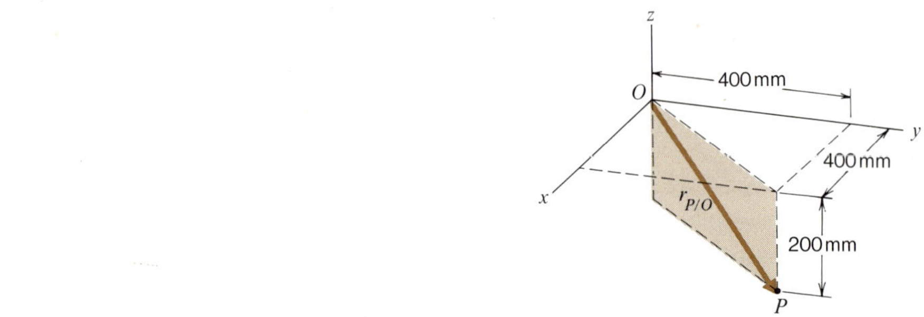

$\bar{r}_{P/O}$ in the x and y directions are both 400 mm, and the z component is −200 mm; the negative sign arises because $\bar{r}_{P/O}$ projects onto the negative z axis. Hence we write the vector as

$$\bar{r}_{P/O} = 400\bar{i} + 400\bar{j} - 200\bar{k} \text{ mm}$$

This expression illustrates the general result that the components of $\bar{r}_{P/O}$ are the coordinates of point P with respect to a coordinate system whose origin is point O.

We now find the direction cosines by calculating $|\bar{r}_{P/O}|$ and then applying eqs. (3), as follows.

$$|\bar{r}_{P/O}| = \sqrt{(400)^2 + (400)^2 + (200)^2} = 600 \text{ mm}$$

$$\cos\alpha = \frac{(r_{P/O})_x}{|\bar{r}_{P/O}|} = \frac{400}{600} = 0.6667$$

$$\cos\beta = \frac{(r_{P/O})_y}{|\bar{r}_{P/O}|} = \frac{400}{600} = 0.6667$$

$$\cos\gamma = \frac{(r_{P/O})_z}{|\bar{r}_{P/O}|} = \frac{-200}{600} = -0.3333$$

The values of the direction angles are restricted to the range from 0° to 180°, so

$$\alpha = \beta = \cos^{-1}(0.6667) = 48.19°$$

$$\gamma = \cos^{-1}(-0.3333) = 109.47°$$

Do you recall what it means when a direction angle is larger than 90°?

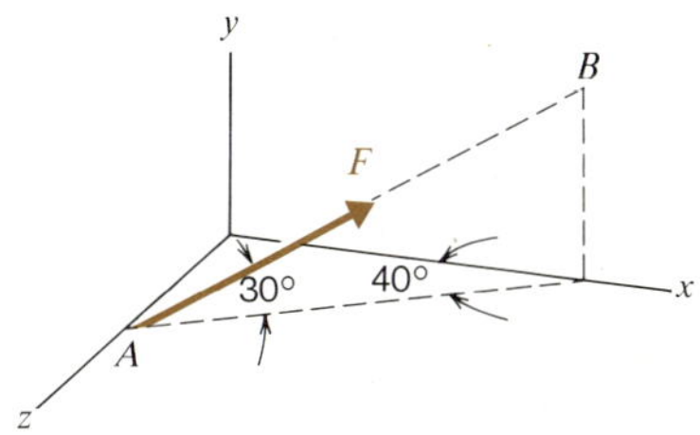

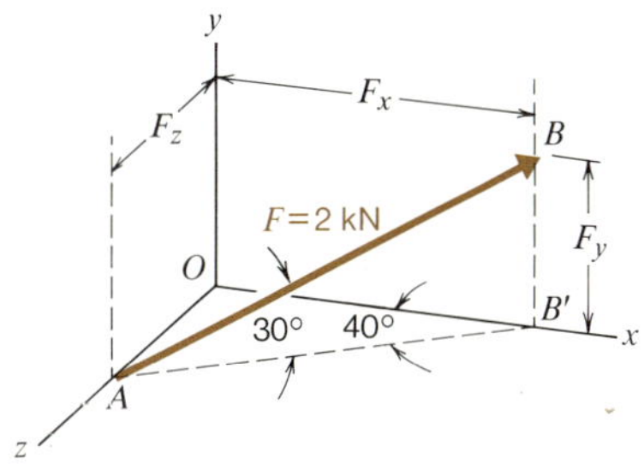

EXAMPLE 3

A force $\bar{F}$ whose magnitude is 2 kN is oriented from point A to point B, as depicted in the sketch. Write $\bar{F}$ as a vector in terms of the xyz coordinate system.

Solution

It is not necessary to place the tail of the vector at the origin before finding its components. We proceed by drawing the vector $\bar{F}$ to a scale that brings its head to point B. This sketch shows that the lengths of line segments OB', $B'B$, and OA are the absolute values of F_x, F_y, and F_z, respectively. Furthermore, if we pretend to start from the tail of $\bar{F}$ (point A) and proceed to the head as we follow these line segments, the movement will be in the negative z, positive x, and positive y directions, so $F_z < 0$, $F_x > 0$, $F_y > 0$. Finally, we note that the calculations of the required lengths are expedited by the fact that $AB'B$ and AOB' are both right triangles. Thus

$$F_y = +BB' = (2 \text{ kN}) \sin 30^\circ = 1 \text{ kN}$$

$$AB' = (2 \text{ kN}) \cos 30^\circ = 1.7321 \text{ kN}$$

$$F_x = +OB' = (AB') \cos 40^\circ = 1.3268 \text{ kN}$$

$$F_z = -OA = -(AB') \sin 40^\circ = -1.1133 \text{ kN}$$

Hence

$$\bar{F} = 1.0\bar{i} + 1.3268\bar{j} - 1.1133\bar{k} \text{ kN}$$

Additional Remarks

The preceding examples show that the description of a vector quantity according to eq. (4) fully describes the vector, provided we have a sketch of the coordinate system. Hence, it is acceptable to leave a solution for a vector quantity in the form of eq. (4), unless the magnitude and direction angles of the vector are specifically requested.

HOMEWORK PROBLEMS

2.1 The force $\bar{F}$ acting on the eyebolt has a magnitude of 160 N, and $\theta = 30^\circ$. Determine the components of $\bar{F}$ with respect to the xyz coordinate system shown and write the result as a vector.

2.2 The xyz components of the force $\bar{F}$ acting on the eyebolt are $F_x = 120$ lb and $F_y = 200$ lb. Determine the magnitude of $\bar{F}$ and the angle θ.

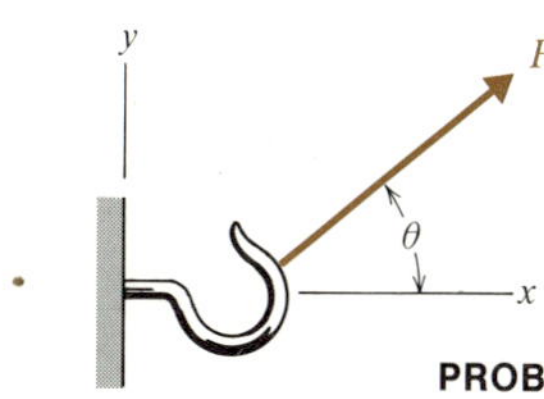

PROBS. 2.1 AND 2.2

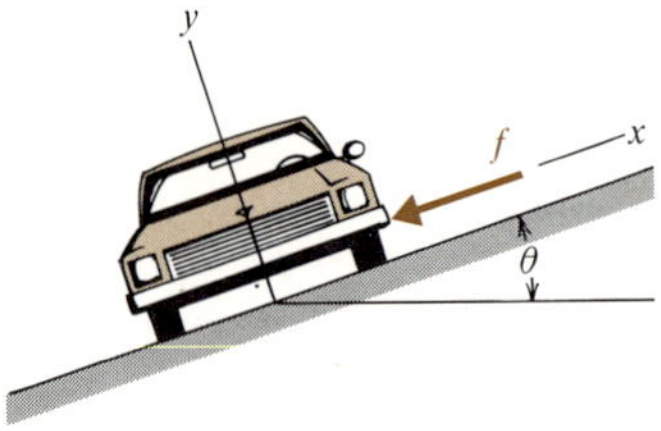

PROBS. 2.3 AND 2.4

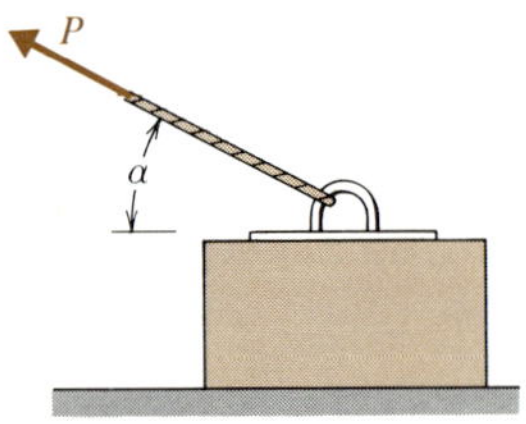

PROBS. 2.5 AND 2.6

2.3 The cross-sectional view shows a 800 kg automobile on a road that is banked at $\theta = 10°$. Determine the components of the automobile's weight force with respect to the xyz coordinate system.

2.4 The frictional force $\bar{f}$ acting on the automobile traversing the banked road, shown in cross section, has a magnitude of 500 lb. It is known that the vertical component of this force is 200 lb. Determine the angle θ at which the road is banked and the horizontal component of $\bar{f}$.

2.5 A tensile force $\bar{P}$ is applied to the free end of the cable. The horizontal component of the force exerted by the cable on the box is 300 N. Determine the magnitude and the vertical component of $\bar{P}$ for (a) $\alpha = 70°$, (b) $\alpha = 160°$.

2.6 A tensile force $\bar{P}$, whose magnitude is 400 N, is applied to the free end of the cable. It is known that the vertical component of $\bar{P}$ is twice as large as the horizontal component, and that the vertical component is upward. Determine the two possible values of the angle α, and the corresponding horizontal and vertical components of $\bar{P}$, for which the foregoing criteria are satisfied.

2.7 The 20 kN force is applied to the bar AB as shown. Determine the components of this force (a) horizontally and vertically, (b) parallel and perpendicular to the longitudinal axis of the bar.

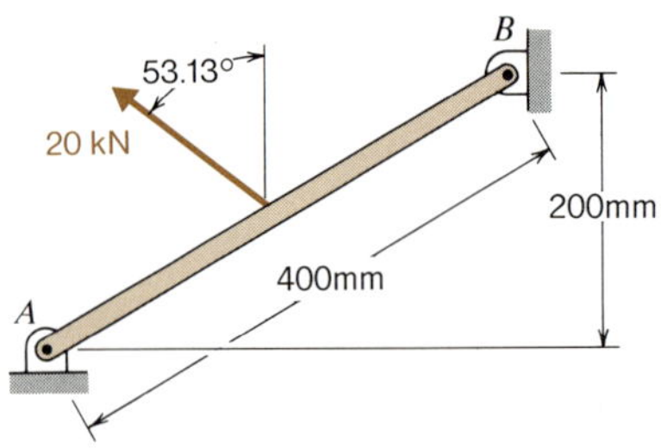

PROB. 2.7

2.8 The cart is pushed up the 15% grade (the slope is 15/100) by the 500 lb force. If $\theta = 30°$, determine the components of this force (a) parallel and perpendicular to the hill, (b) horizontally and vertically.

2.9 The cart is pushed up the 15% grade (the slope is 15/100) by the 500 lb force. The horizontal component of the 500 lb force is 400 lb. Determine (a) the angle θ, (b) the vertical component of the force, and (c) the components of the force parallel and perpendicular to the hill.

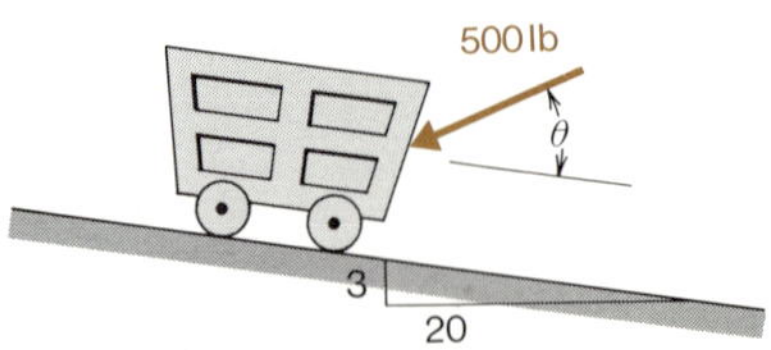

PROBS. 2.8 AND 2.9

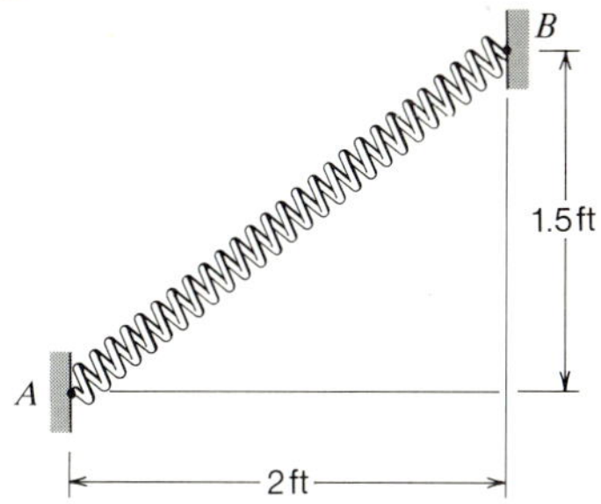

PROB. 2.10

2.10 A spring whose unstretched length is 2 ft and whose stiffness is 200 lb/ft is tied between points A and B. Determine the horizontal and vertical components of the force the spring exerts (a) on point A, (b) on point B.

2.11 The weight of collar A equals the vertical component of the spring force applied to the collar. The unstretched length of the spring is 600 mm and the stiffness of the spring is 150 N/m. If $x = 640$ mm, determine the mass of the collar.

***2.12** The weight of collar A equals the vertical component of the spring force applied to the collar. The unstretched length of the spring is 600 mm and the stiffness of the spring is 150 N/m. The mass of the collar is 2.5 kg. Determine the value of x corresponding to the given spring properties.

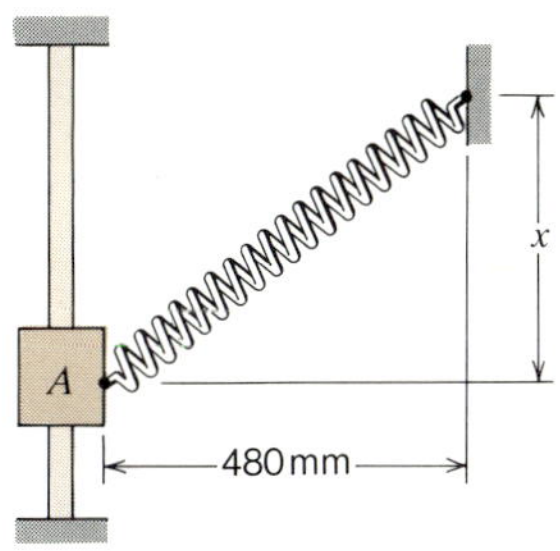

PROBS. 2.11 AND 2.12

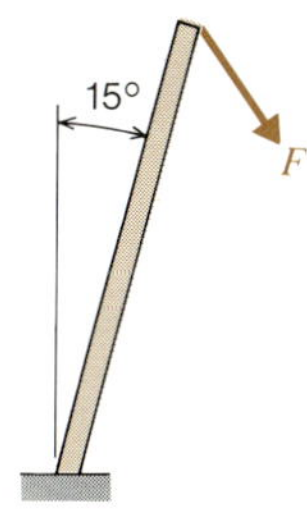

PROB. 2.13

***2.13** The force $\bar{F}$ acting on the inclined rod has horizontal and vertical components of 250 lb and 400 lb, respectively. Determine the components of $\bar{F}$ parallel and perpendicular to the longitudinal axis of the rod.

***2.14** The x and y axes shown form nonorthogonal Cartesian coordinates in the plane. It is desired to write the force $\bar{F}$ in the form $\bar{F} = F_x\bar{i} + F_y\bar{j}$. Determine the values of F_x and F_y. Are these components the perpendicular projections onto the corresponding axes?

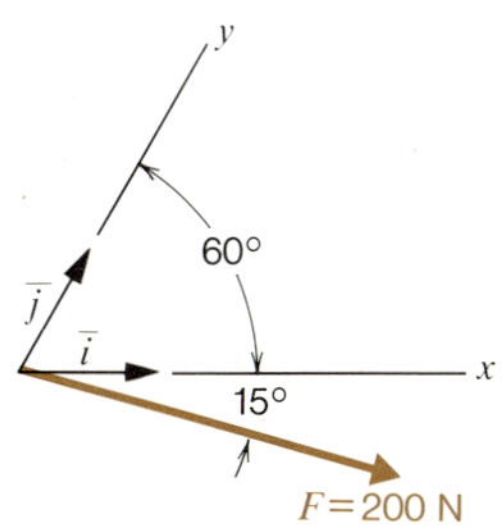

PROB. 2.14

* An asterisk denotes more difficult problems.

2.15 An observer on the ground at point A is looking at the top of a 200 m tall building at point B. (a) Determine the components of the position of point B with respect to point A and write the result as a vector. (b) Determine the direction cosines of this position vector with respect to the southerly and easterly directions.

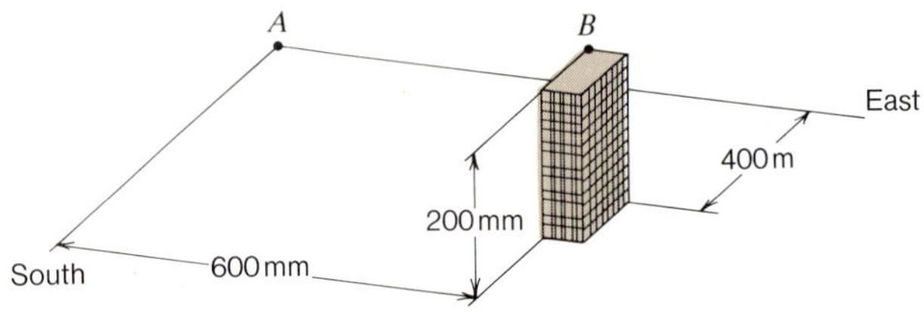

PROB. 2.15

2.16 The components of force $\bar{Q}$ with respect to a set of rectangular Cartesian axes xyz are $Q_x = 10$ lb, $Q_y = -15$ lb, $Q_z = -10$ lb. Determine the magnitude of $\bar{Q}$ and the direction angles of $\bar{Q}$ with respect to the x, y, and z axes.

2.17 The direction angles of a 90 kN force with respect to the x and y directions are $\alpha = 30°$, $\beta = 45°$. The vertical component of this force is in the positive z direction. Determine the components of this force in the x, y, and z directions.

2.18 Write the force $\bar{F}$ in terms of its components and determine the direction angles of this force with respect to the xyz reference frame.

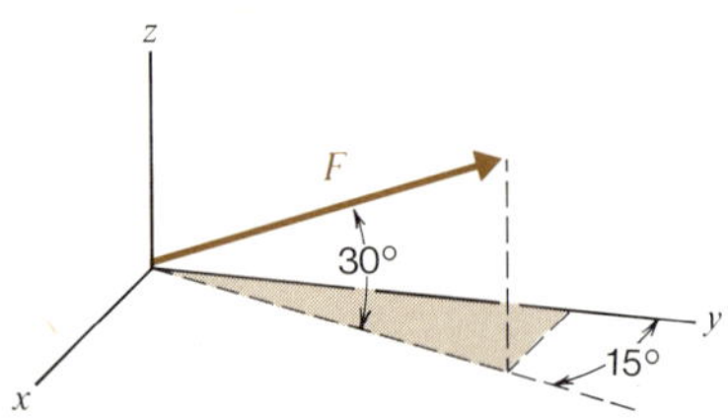

PROB. 2.18

2.19 A force $\bar{F}$ has a magnitude of 4 tons, and the direction angles between this force and the y and z axes are $\beta = 45°$, $\gamma = 120°$. It is also known that the x component of this force is negative. Determine the components of this force and write the result as a vector. Also, determine the direction angle between the force and the x axis.

2.20 A 1000 N force acts on the vertical pole shown. (a) Determine the components of the force with respect to the xyz coordinate system and write the force as a vector. (b) Determine the direction angles between the force and the xyz coordinate axes.

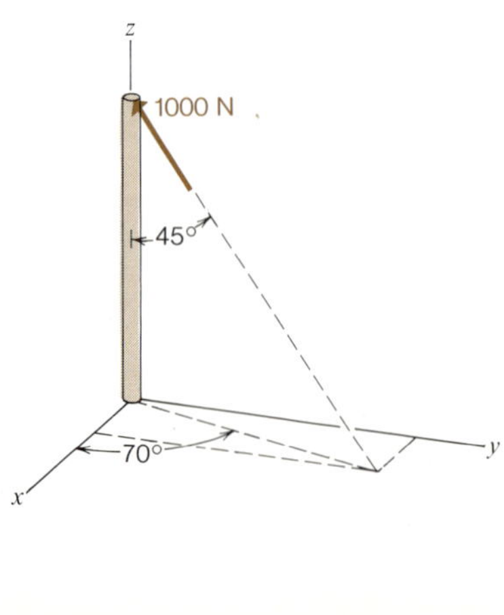

PROB. 2.20

2.21 The x, y, and z components of a 140 N force are in the proportion of $3{:}-2{:}6$. Determine these components and also the direction angles between the force and the coordinate axes.

***2.22** The direction angle between a force $\bar{A}$ and the x axis is $130°$, and $A_y = -2A_z$ with $A_z > 0$. Determine the direction angles between the force and the y and z axes.

C. GENERAL UNIT VECTORS

An asset of the component representation of vectors is that the vector operations take on the form of algebraic equations. For instance, if we wish to multiply a vector $\bar{A}$ by a scalar number c, by definition the result of this operation is a vector parallel to $\bar{A}$ whose magnitude is c times the magnitude of $\bar{A}$. It follows from eq. (4) and the distributive law that

$$c\bar{A} = cA_x\bar{i} + cA_y\bar{j} + cA_z\bar{k} \tag{5}$$

Equation (5) is particularly useful in the situation where we must describe a vector that is parallel to a line whose end points are specified. This situation is depicted in Figure 6, where a vector $\bar{A}$, having known magnitude $|\bar{A}|$, is formed such that it is parallel to the line from point Q to point P. The vector whose tail is at point Q and whose head is at point P is called the *position vector* $\bar{r}_{P/Q}$, where the notation should be read as the position of point P with respect to point Q. Thus, $\bar{A}$ is parallel to $\bar{r}_{P/Q}$.

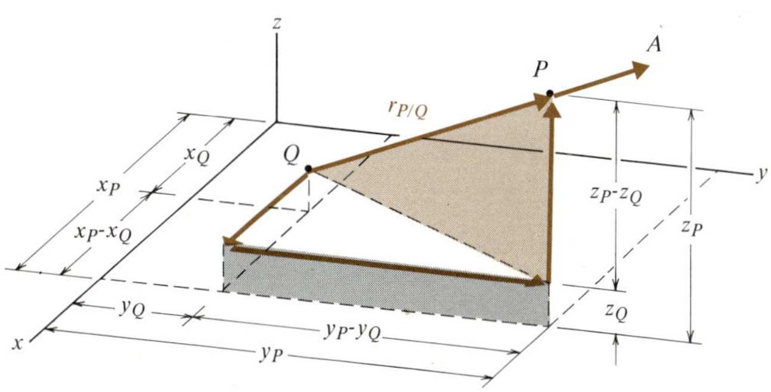

FIGURE 6

The coordinates of the end points of $\bar{r}_{P/Q}$ are denoted as (x_P, y_P, z_P) and (x_Q, y_Q, z_Q). We see from the figure that the components of $\bar{r}_{P/Q}$ can be written as the differences in the coordinates; that is

$$\bar{r}_{P/Q} = (x_P - x_Q)\bar{i} + (y_P - y_Q)\bar{j} + (z_P - z_Q)\bar{k} \tag{6}$$

Once $\bar{r}_{P/Q}$ is known, we may obtain a unit vector $\bar{e}_{P/Q}$ in the direction of $\bar{r}_{P/Q}$ (and hence $\bar{A}$) by writing

$$|\bar{r}_{P/Q}|\bar{e}_{P/Q} = \bar{r}_{P/Q}$$

$$\bar{e}_{P/Q} = \frac{\bar{r}_{P/Q}}{|\bar{r}_{P/Q}|}$$

Then, to obtain $\bar{A}$ we need only multiply the unit vector by the magnitude of $\bar{A}$, which gives

$$\boxed{\bar{A} = |\bar{A}|\bar{e}_{P/Q} = |\bar{A}|\frac{\bar{r}_{P/Q}}{|\bar{r}_{P/Q}|}} \tag{7}$$

EXAMPLE 4

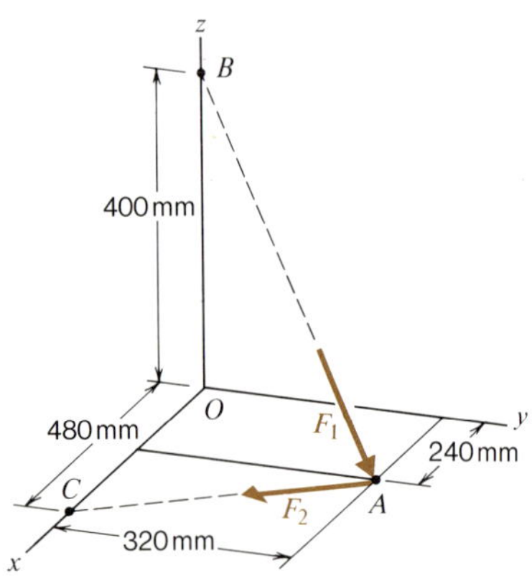

The magnitudes of the forces $\bar{F}_1$ and $\bar{F}_2$ shown in the sketch are 1.20 and 0.80 kN, respectively. Express these forces in terms of their components.

Solution

The forces $\bar{F}_1$ and $\bar{F}_2$ are directed along the lines from point B to point A, and point A to point C, respectively. A method for constructing the necessary position vectors $\bar{r}_{A/B}$ and $\bar{r}_{C/A}$ is to picture ourselves following the coordinate axes as we move from the tail to the tip of each vector. For instance, to move from point B to point A we must go 400 mm in the negative z direction, 240 mm in the positive x direction, and 320 mm in the positive y direction, so that

$$\bar{r}_{A/B} = 240\bar{i} + 320\bar{j} - 400\bar{k} \text{ mm}$$

Similarly,

$$\bar{r}_{C/A} = 240\bar{i} - 320\bar{j} \text{ mm}$$

From eq. (7) we then form

$$\bar{F}_1 = |\bar{F}_1|\bar{e}_{A/B}$$

$$= (1.2)\frac{\bar{r}_{A/B}}{|\bar{r}_{A/B}|} = (1.2)\frac{240\bar{i} + 320\bar{j} - 400\bar{k}}{\sqrt{(240)^2 + (320)^2 + (400)^2}}$$

$$= 0.5091\bar{i} + 0.6788\bar{j} - 0.8485\bar{k} \text{ kN}$$

$$\bar{F}_2 = (0.8)\frac{\bar{r}_{C/A}}{|\bar{r}_{C/A}|} = (0.8)\frac{240\bar{i} - 320\bar{j}}{\sqrt{(240)^2 + (320)^2}}$$

$$= 0.480\bar{i} - 0.640\bar{j} \text{ kN}$$

As an aside, note that $\bar{F}_2$, being a force in the xy plane, could have been obtained just as easily by determining the angle between the line AC and one of the coordinate axes and then using the method of Example 1. However, this method is awkward to employ in the case of the three-dimensional force $\bar{F}_1$.

HOMEWORK PROBLEMS

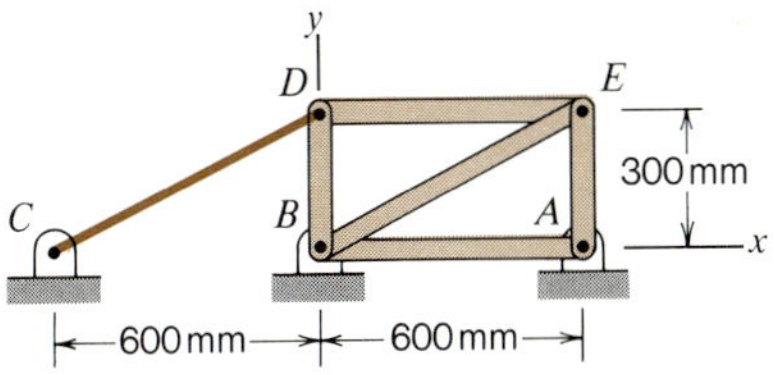

PROB. 2.23

2.23 The structure shown is pulled by cable CD. If the tensile force within the cable is 8000 N, determine the components with respect to the xyz axes of the force exerted by the cable on the structure at point D. Also, determine the unit vector parallel to this force.

2.24 The cabinet shown is braced by tensioned crosswires AC and BD. The tensile forces in both cables are 40 lb. Determine the x and y components of the unit vectors associated with the forces exerted by the cables on the cabinet at the four corners.

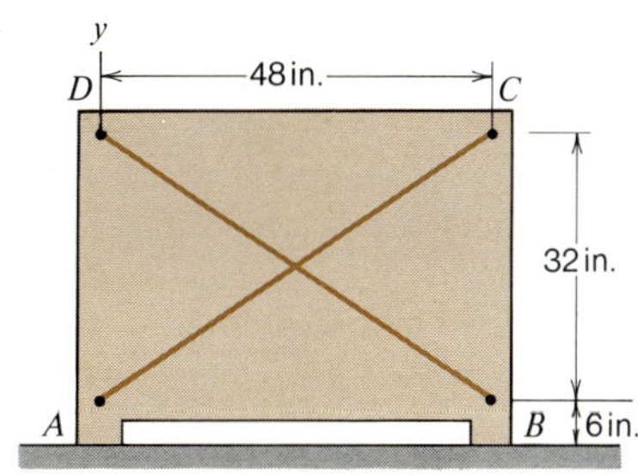

PROB. 2.24

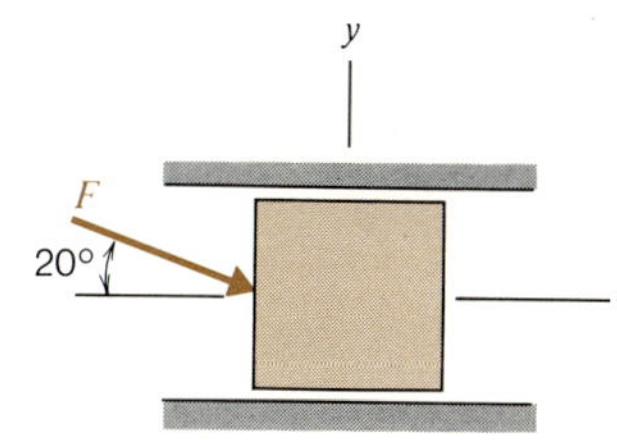

PROB. 2.25

2.25 The piston is pushed by the force $\bar{F}$. Write an expression for the unit vector in the direction of $\bar{F}$ in terms of the x and y axes shown.

2.26 Two cables are tied to eyebolt B. The tension in cable AB is 2 kN and the tension in cable BC is 3 kN. Determine the horizontal and vertical components of the force each cable exerts on the eyebolt.

2.27 The spring has a stiffness of 2 lb/in. and an unstretched length of 16 in. Determine the x and y components of the force exerted by the spring on rod AB, then write this force as a vector.

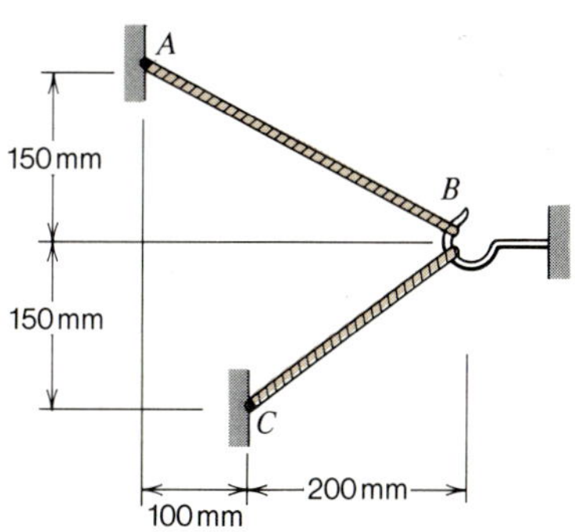

PROB. 2.26

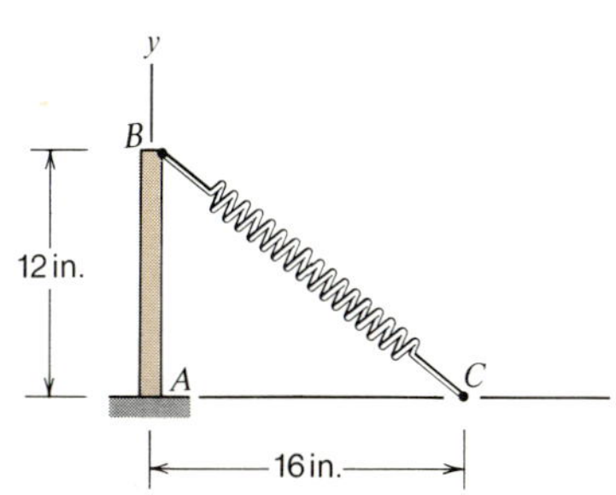

PROB. 2.27

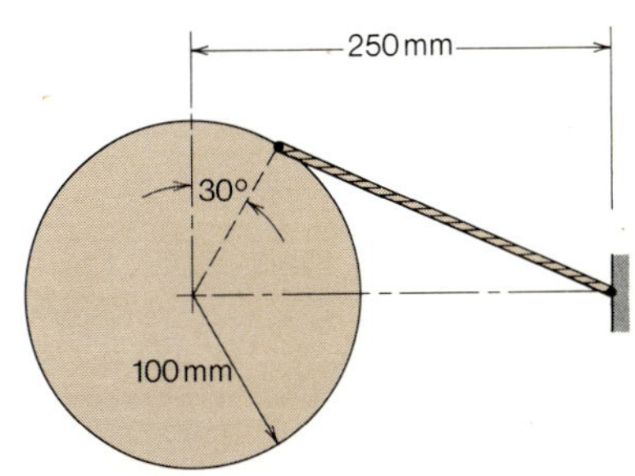

PROB. 2.28

2.28 The tension in the cable is 200 N. Find the horizontal and vertical components of the force exerted by the cable on the circular plate.

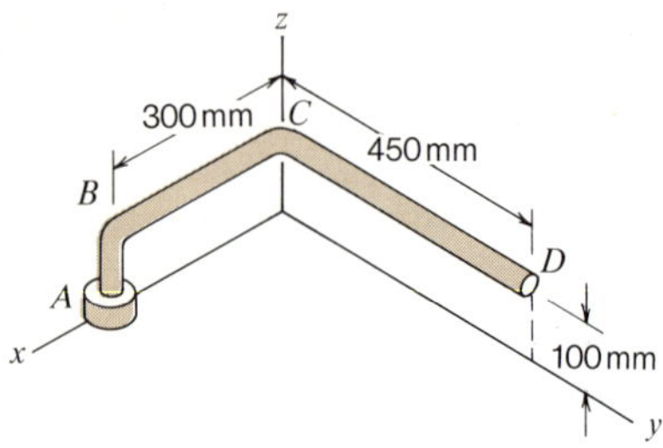

PROB. 2.29

2.29 Determine the unit vectors extending from point A to points B, C, and D of the bent bar shown.

2.30 The rectangular plate is supported in part by cables AB and CD. If the tensile force in each cable is 150 lb, determine the components of the force each cable exerts on the plate.

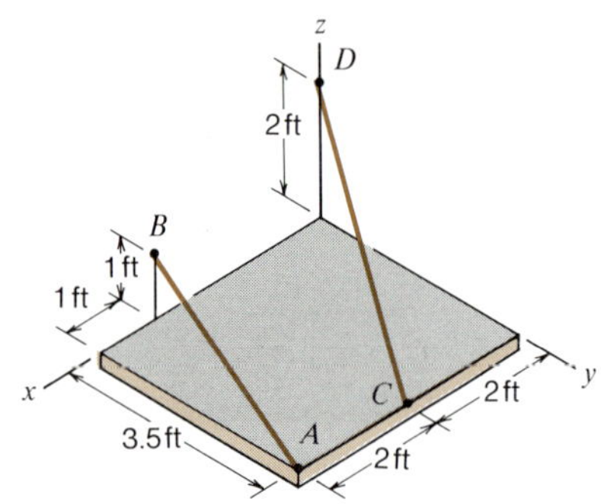

PROB. 2.30

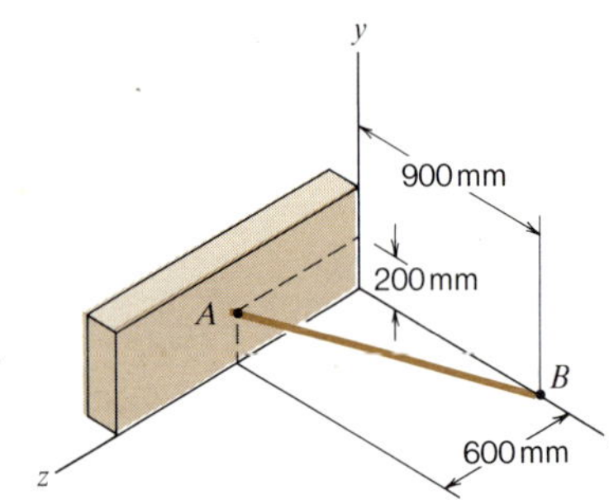

PROBS. 2.31 AND 2.32

2.31 The tension in cable AB is 1.5 kN. Determine the force exerted by the cable on the block. What are the direction cosines of this force with respect to the x, y, and z axes?

2.32 The block is being pulled at point A by cable AB. For what tensile force in the cable will the x component of the force exerted by the cable on the package be 200 N?

***2.33** The spring carries a tensile force of 800 N. Determine the force exerted by the spring on end C of the curved bar ABC.

***2.34** The spring has a stiffness of 0.7 kN/m and an unstretched length of 0.5 m. Determine the force exerted by the spring on end C of the curved bar ABC.

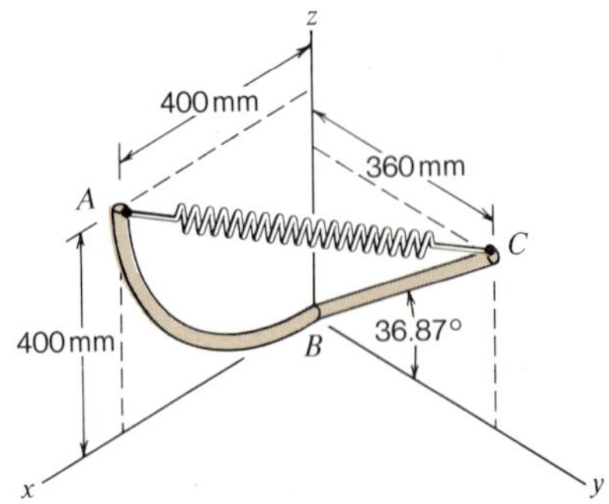

PROBS. 2.33 AND 2.34

D. DETERMINATION OF THE RESULTANT OF CONCURRENT FORCES

It was shown in eq. (1) that the resultant of a set of concurrent forces acting on a particle is the sum of the forces. Let us consider Figure 7, where two forces $\bar{F}_1$ and $\bar{F}_2$ are concurrent at the origin O. In order to evaluate the resultant of these forces, we move $\bar{F}_2$ parallel to itself so as to align the vectors "head to tail." This allows us to form the resultant according to the polygon rule for addition.

The components of the vectors $\bar{F}_1$, $\bar{F}_2$, and $\bar{R}$ forming the vector polygon are found in Figure 7 by projecting the tails and heads of these vectors onto the three orthogonal planes. This construction leads to the conclusion that the components of $\bar{R}$ are the sum of the corresponding components of $\bar{F}_1$ and $\bar{F}_2$.

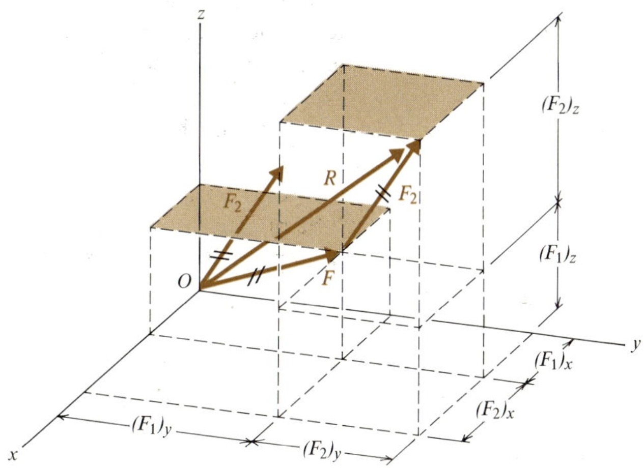

FIGURE 7

This conclusion can be extended to the case of the resultant of several forces by consecutively adding pairs of vectors, with the result that

$$\boxed{\bar{R} = R_x\bar{i} + R_y\bar{j} + R_z\bar{k} = \sum F_x\bar{i} + \sum F_y\bar{j} + \sum F_z\bar{k}} \tag{8}$$

where $\sum F_x$, $\sum F_y$, and $\sum F_z$ denote the sum of the force components in the x, y, and z directions, respectively. Also, remember that because $\bar{R}$ is the resultant of the system of forces acting on the particle at point O, the vector $\bar{R}$ must be applied at *the point of concurrency* of the force system.

Clearly, although eq. (8) was derived by considering the resultant of a set of forces, it is valid for any set of vectors. Thus, each component of the sum of a set of vectors is the sum of the corresponding components of the individual vectors.

The rule for the difference of two vectors $\bar{A}$ and $\bar{B}$ is a corollary to the foregoing development. For example, given the vector $\bar{B}$ we may write

$$-\bar{B} \equiv (-1)\bar{B} = -B_x\bar{i} - B_y\bar{j} - B_zk$$

Thus, the difference between $\bar{A}$ and $\bar{B}$ is

$$\bar{A} - \bar{B} = \bar{A} + (-\bar{B}) = (A_x - B_x)\bar{i} + (A_y - B_y)\bar{j} + (A_z - B_z)\bar{k} \tag{9}$$

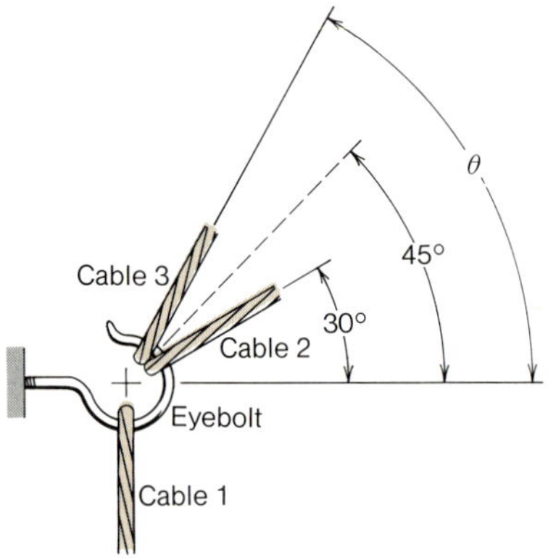

EXAMPLE 5

The tensile forces in three cables that exert a concurrent set of forces on an eyebolt are $F_1 = 6$ kN, $F_2 = 8$ kN, $F_3 = 10$ kN. Determine the angle θ of cable 3 for which the resultant of the forces the cables exert on the eyebolt is along the indicated dashed line. Also, write this resultant in vector form.

Solution

We know that cables pull on the object to which they are attached, so we sketch the forces applied by the cables as shown to the left. Also shown in the sketch is the coordinate system to be used for expressing components. Note that we chose one axis, the x axis, horizontal because all angles are referred to this line. The forces are all coplanar, so there are no components in the z direction.

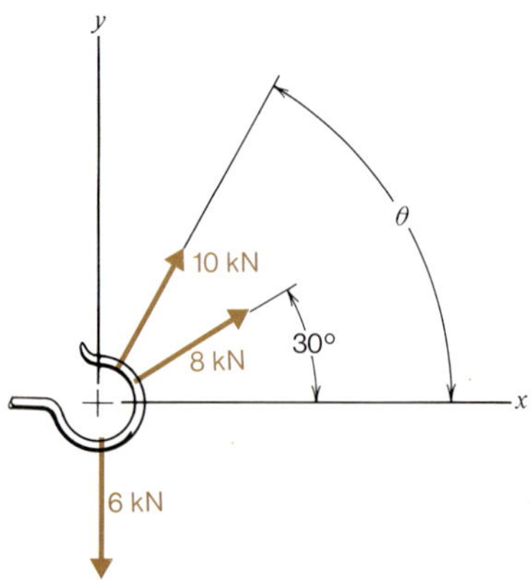

We determine the components of the forces by expressing each vector as the product of its magnitude and its unit vector. Cable 1 pulls downward, and cables 2 and 3 are at 30° and θ above the x axis, so

$$\bar{F}_1 = 6(-\bar{j}) = -6\bar{j} \text{ kN}$$

$$\bar{F}_2 = 8(\cos 30^\circ\,\bar{i} + \sin 30^\circ\,\bar{j}) = 6.928\bar{i} + 4.0\bar{j} \text{ kN}$$

$$\bar{F}_3 = 10(\cos\theta\,\bar{i} + \sin\theta\,\bar{j}) \text{ kN}$$

A requirement of the problem is that the unit vector for the resultant $\bar{R}$ must be along a line that is 45° above the x axis, which means that

$$\begin{aligned}\bar{R} &= R(\cos 45^\circ\,\bar{i} + \sin 45^\circ\,\bar{j})\\ &= 0.7071\,R(\bar{i} + \bar{j})\end{aligned}$$

Note that we assumed $\bar{R}$ to be up and to the right, that is, $R > 0$. A solution in which $R < 0$ would mean that $\bar{R}$ is down and to the left.

Now that we have employed all of the given information, the requirement that $\bar{R}$ is the resultant of $\bar{F}_1$, $\bar{F}_2$, and $\bar{F}_3$ is satisfied by adding like components. In general

$$\bar{R} = \bar{F}_1 + \bar{F}_2 + \bar{F}_3$$

$$\begin{aligned}0.7071\,R(\bar{i} + \bar{j}) &= -6\bar{j} + (6.928\bar{i} + 4.0\bar{j}) + 10(\cos\theta\,\bar{i} + 10\sin\theta\,\bar{j})\\ &= (6.928 + 10\cos\theta)\bar{i} + (-2 + 10\sin\theta)\bar{j}\end{aligned}$$

This equation expresses the fact that the vector on the left side of the equality sign must be the same as the vector on the right side. Because two vectors are identical only when they have the same components, we may equate the corresponding components on both sides of the equation. Thus

$$\bar{i}\text{ component:}\quad 0.7071\ R = 6.928 + 10\cos\theta$$

$$\bar{j}\text{ component:}\quad 0.7071\ R = -2 + 10\sin\theta$$

These are two simultaneous equations for the two unknowns R and θ. They may be solved by eliminating θ, as follows.

$$\cos\theta = 0.07071\ R - 0.6928$$

$$\sin\theta = 0.07071\ R + 0.2$$

$$\begin{aligned}\sin^2\theta + \cos^2\theta \equiv 1 &= (0.07071\ R - 0.6928)^2 + (0.07071\ R + 0.2)^2\\ &= 0.01\ R^2 - 0.06969\ R + 0.520\end{aligned}$$

This is a quadratic equation for R,

$$0.01\ R^2 - 0.06969\ R - 0.480 = 0$$

whose roots are

$$R = 11.240,\ -4.271\ \text{kN}$$

In order to find the corresponding values of θ, and also to check that neither value of R is an extraneous root, we substitute each value of R into the equations for $\sin\theta$ and $\cos\theta$. For $R = 11.240$, this gives

$$\cos\theta = 0.10198, \qquad \sin\theta = 0.9948$$

We see that θ is in the first quadrant, so

$$\theta = 84.15^\circ$$

The same procedure for $R = -4.271$ yields

$$\cos\theta = -0.9948, \qquad \sin\theta = -0.10198$$

$$\theta = 185.85^\circ$$

Therefore, there are two solutions to the problem.

$$\theta = 84.15^\circ, \qquad \bar{R} = 7.95(\bar{i} + \bar{j})\ \text{kN}$$

$$\theta = 185.85^\circ, \qquad \bar{R} = -3.02(\bar{i} + \bar{j})\ \text{kN}$$

EXAMPLE 6

A radio antenna is supported by three guy wires. The tensile force in cables AB, AC, and AD are 7 kilopounds (kips), 5 kips, and 8 kips, respectively. Determine the resultant force exerted on the antenna by these cables.

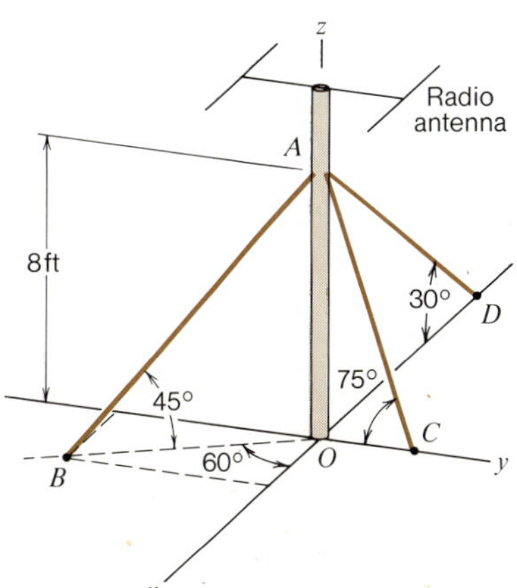

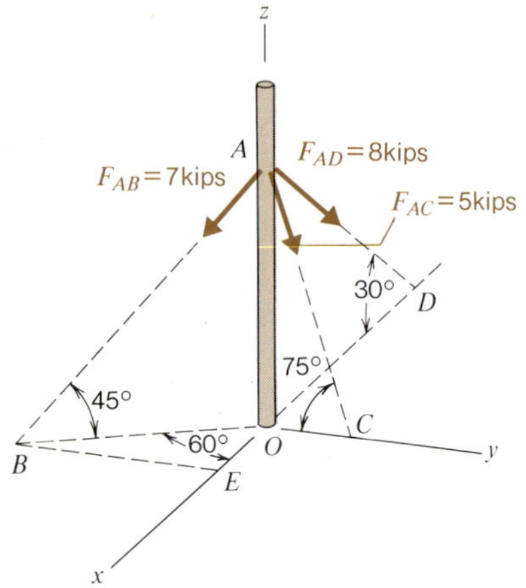

Solution

The first step is to draw a sketch showing the forces applied by the cables upon the mast; recall that cables have a pulling effect. Let us first write the vector expressions for $\bar{F}_{AC}$ and $\bar{F}_{AD}$, because they have the simplifying feature of coinciding with the coordinate planes. From the sketch we see that the components of $\bar{F}_{AD}$ are oriented along the negative x and negative z axes. Then, because the angle between the line AD and the x axis is 30°, it follows that

$$\begin{aligned}\bar{F}_{AD} &= 8(-\cos 30^\circ\,\bar{i} - \sin 30^\circ\,\bar{k})\\ &= -6.928\bar{i} - 4.0\bar{k} \text{ kips}\end{aligned}$$

In a similar manner we find that

$$\begin{aligned}\bar{F}_{AC} &= 5(\cos 75^\circ\,\bar{j} - \sin 75^\circ\,\bar{k})\\ &= 1.294\bar{j} - 4.830\bar{k} \text{ kips}\end{aligned}$$

The 8 ft dimension is not needed to describe $\bar{F}_{AB}$. Instead, we will employ the fact that the projections of line AB onto the coordinate axes are proportional to the components of the force in the guy wire. We may therefore find the components of $\bar{F}_{AB}$ by the letting the length $AB = |\bar{F}_{AB}| = 7$ kips. The lengths EO, BE, and AO are then the magnitudes of the x, y, and z components, respectively. We find these lengths by noting that AOB and OEB are both right triangles. Thus

$$\begin{aligned}AB &= 7 \text{ kips}\\ AO &= AB \sin 45^\circ = 4.950 \text{ kips}\\ OB &= AB \cos 45^\circ = 4.950 \text{ kips}\\ OE &= OB \cos 60^\circ = 2.475 \text{ kips}\\ BE &= OB \sin 60^\circ = 4.287 \text{ kips}\end{aligned}$$

The signs of the components are obtained by observing that a displacement from A to B, which is the sense of $\bar{F}_{AB}$, can be performed by moving along the negative z, positive x, and negative y axes, so

$$\bar{F}_{AB} = 2.475\bar{i} - 4.287\bar{j} - 4.950\bar{k} \text{ kips}$$

Finally, we determine the resultant by adding the corresponding components of the forces; that is

$$\begin{aligned}\bar{R} &= \bar{F}_{AB} + \bar{F}_{AC} + \bar{F}_{AB}\\ &= (2.475 + 0 - 6.928)\bar{i} + (-4.287 + 1.294 + 0)\bar{j}\\ &\quad + (-4.950 - 4.830 - 4.0)\bar{k}\\ &= -4.453\bar{i} - 2.993\bar{j} - 13.780\bar{k} \text{ kips}\end{aligned}$$

HOMEWORK PROBLEMS

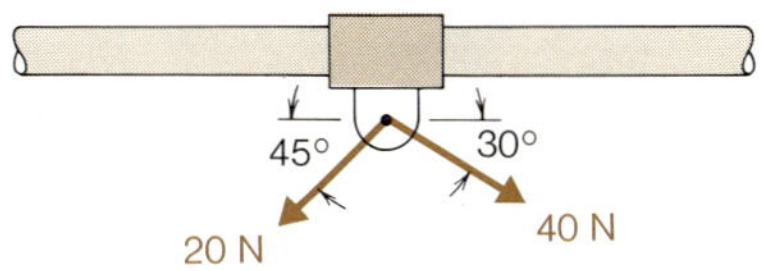

PROB. 2.35

2.35–2.37 Determine the magnitude and direction of the resultant of the two forces shown.

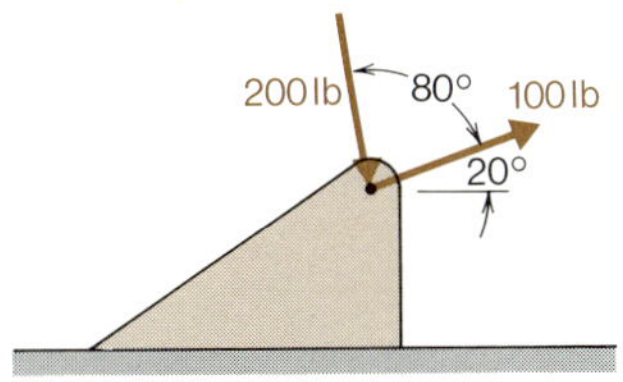

PROB. 2.36

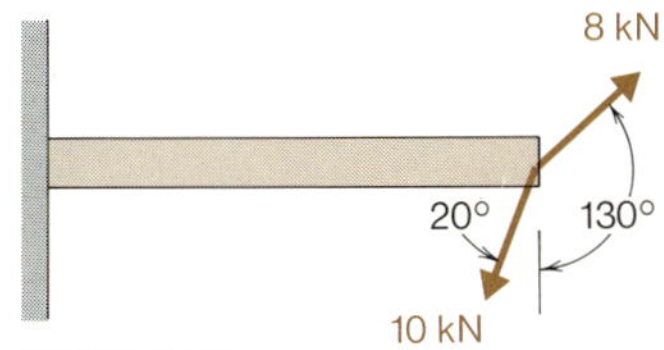

PROB. 2.37

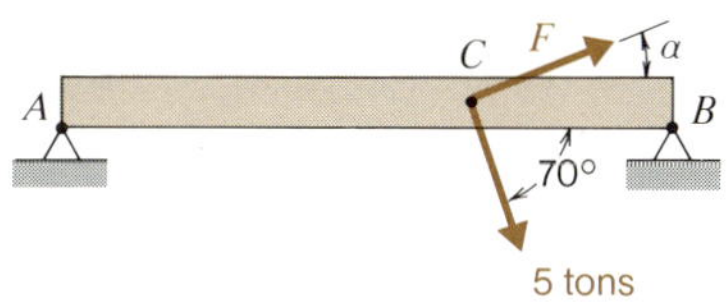

PROBS. 2.38, 2.39, AND 2.40

2.38 Two forces act at point C on the beam. Knowing that $F = 10$ tons and $\alpha = 40°$, determine the magnitude and angle relative to the beam of the resultant of these forces.

2.39 Two forces act at point C on the beam. If $\alpha = 40°$, determine the magnitude of the force $\bar{F}$ for which the resultant force is (a) horizontal, (b) vertical.

2.40 Two forces act at point C on the beam. If $F = 6$ tons, determine the angle α for which the resultant of the forces is horizontal. Also, write this resultant as a vector.

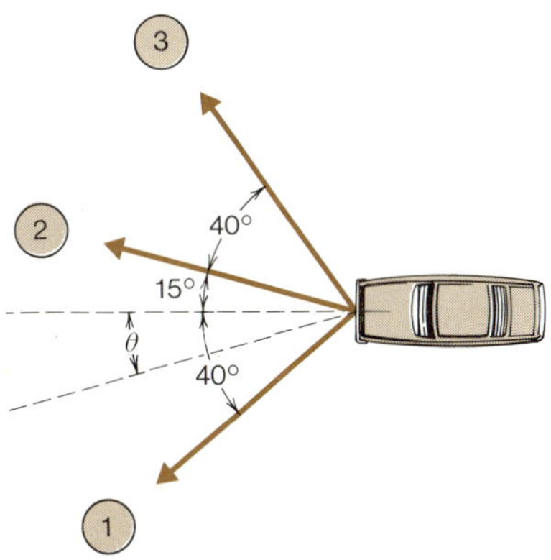

PROBS. 2.41 AND 2.42

2.41 Three people pull horizontally on cables attached to an automobile. If each person exerts a 200 N pull, determine the magnitude and the angle θ for the resultant force exerted on the vehicle.

***2.42** Three people pull horizontally on cables attached to an automobile. The tensile forces in cables 1 and 2 are 200 N. Determine the force applied to cable 3 for which the orientation of the resultant is (a) $\theta = 0$, (b) $\theta = 90°$.

2.43 It is desired that the resultant of the three forces acting on the roller guide be horizontal. If $F = 8$ kN, determine the corresponding angle β.

***2.44** Knowing that $\beta = 30°$, determine the magnitude of $\bar{F}$ for which the resultant of the force system shown has a magnitude of 5 kN.

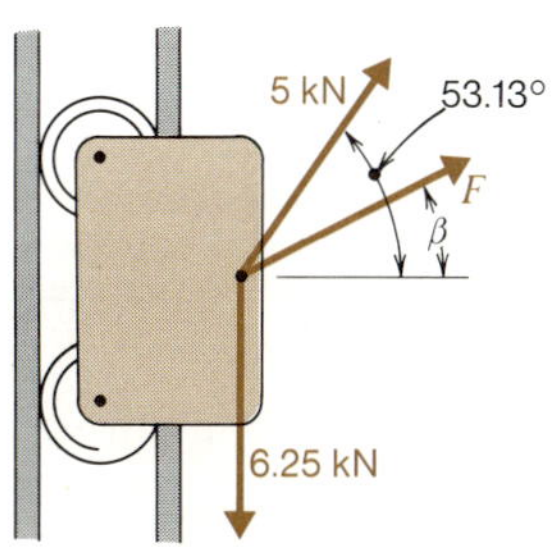

PROBS. 2.43 AND 2.44

***2.45** Determine the value of α required for the resultant of the three forces to be parallel to the inclined bar.

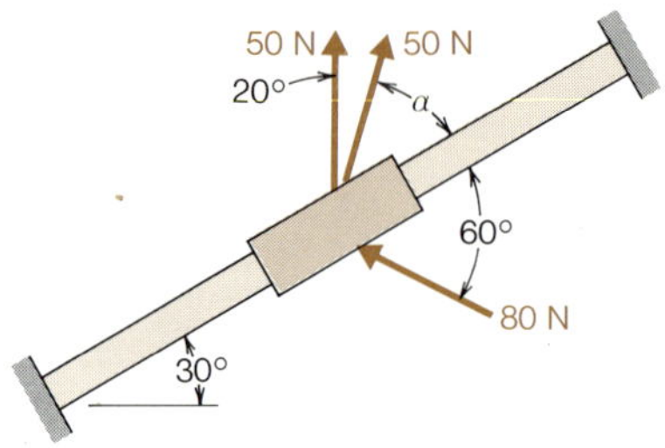

PROB. 2.45

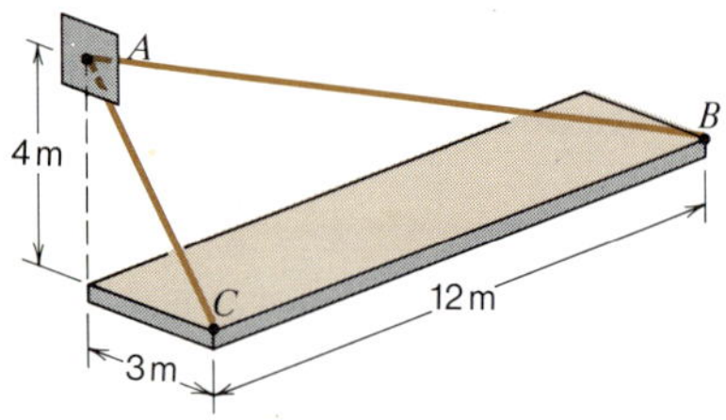

PROBS. 2.46 AND 2.47

2.46 The tensile force in cable AB is 2 kN, and that in cable AC is 3 kN. Determine the resultant force exerted by the cables on the support at point A.

2.47 The tensile force in cable AB is 2 kN. Determine the tensile force in cable AC for which the vertical component of the resultant force exerted by the cables on the support at A is 4 kN. Also, determine the corresponding resultant.

2.48 The square plate is suspended by four cables as shown. The tensile forces in cables AB and AC are 90 lb, and those in cables AD and AE are 60 lb. Determine the resultant force exerted by the cables on the support at point A.

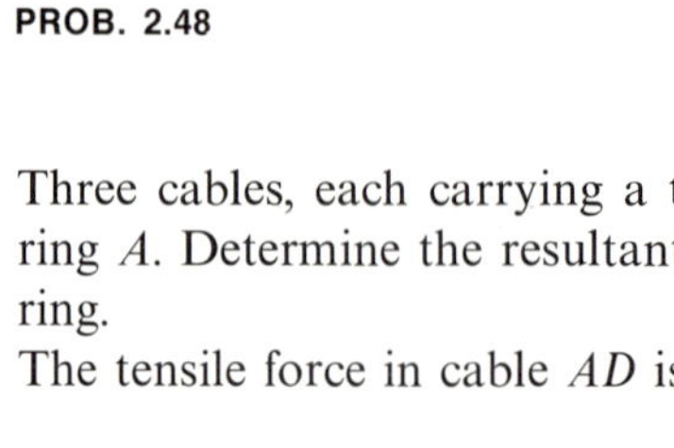

PROB. 2.48

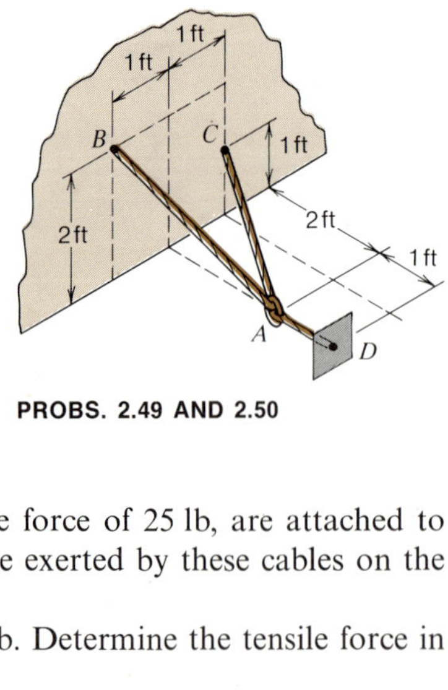

PROBS. 2.49 AND 2.50

2.49 Three cables, each carrying a tensile force of 25 lb, are attached to ring A. Determine the resultant force exerted by these cables on the ring.

***2.50** The tensile force in cable AD is 25 lb. Determine the tensile force in

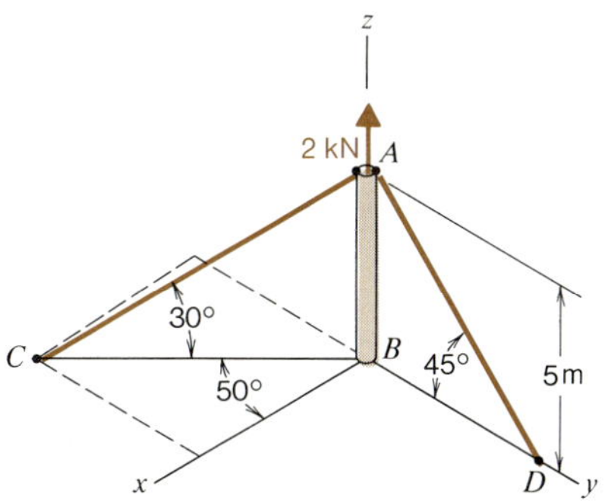

PROBS. 2.51 AND 2.52

the other two cables for which the resultant of the forces exerted by the cables on ring A is vertical. Also, determine this resultant.

2.51 An upward 2 kN force is applied at end A of the vertical pole, where two cables are also fastened. The tensile forces in these cables are $F_{AC} = 4$ kN, $F_{AD} = 5$ kN. Determine the resultant of the three forces applied to end A.

2.52 Two cables are attached to the vertical pole AB to steady it. An upward force of 2 kN is applied to the pole, with the effect that the resultant of the three forces acting at point A on the pole is parallel to the x axis. Determine the tensions in the two cables and the magnitude of the resultant.

2.53 Two cables and a 20 kip horizontal load act on end B of the pole. The tensions are 40 kips for cable BC and 30 kips for cable BD. Determine the resultant force acting on the pole at end B.

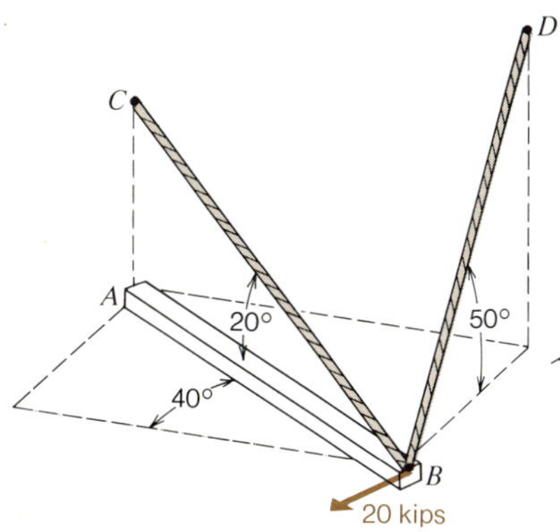

PROB. 2.53

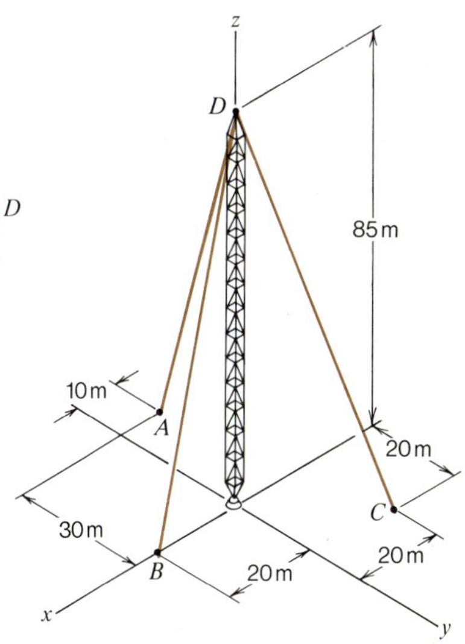

PROB. 2.54

2.54 In order to hold the television tower erect, the resultant loading exerted on end D must be 20 kN in the negative z direction. Determine the tensile force in each cable required to accomplish this.

2.55 Solve Problem 2.54 for the case where an additional force of 1 kN is applied to the tower at point D in the positive y direction.

E. DOT PRODUCT

The preceding sections described methods for determining the components of a vector in the directions of the axes of a rectangular Cartesian coordinate system. Each component is the product of the magnitude of the vector and the direction cosine, which is the cosine of the angle between the vector and the axis of interest. Such a product may be generalized in order to facilitate the determination of the components of a vector in a direction that is not a coordinate axis. The operation we will derive is called the *dot product* of two vectors.

Let us consider two free vectors $\bar{A}$ and $\bar{B}$, which have been positioned in Figure 8 such that their tails are coincident. The angle between these vectors is θ. When the tip of each vector is projected onto the other vector, we find that $|\bar{A}| \cos\theta$ is the component of $\bar{A}$ parallel to $\bar{B}$, whereas $|\bar{B}| \cos\theta$ is the component of $\bar{B}$ parallel to $\bar{A}$. Obviously, the value $|A|(|B| \cos\theta)$ is the same as the value $|\bar{B}|(|\bar{A}| \cos\theta)$. We write this product symbolically as $\bar{A} \cdot \bar{B}$, which should be read as $\bar{A}$ dot $\bar{B}$. Thus, the definition of a dot (or scalar) product is

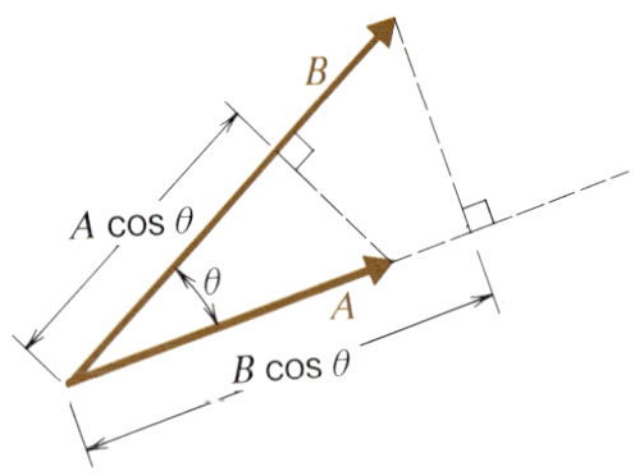

FIGURE 8

$$\bar{A} \cdot \bar{B} = \bar{B} \cdot \bar{A} \equiv |\bar{A}|\,|\bar{B}| \cos\theta \tag{10}$$

In words, the dot product is the magnitude of one vector times the component of the other vector parallel to the first one. The result is a *scalar*. Note that when $\theta > 90°$, the dot product is negative. Physically, this means that the component of $\bar{A}$ parallel to $\bar{B}$ is opposite the sense of $\bar{B}$ (or vice versa). When the two vectors are perpendicular to each other, eq. (10) gives

$$\bar{A} \cdot \bar{B} = |\bar{A}|\,|\bar{B}| \cos 90° = 0$$

If we calculate the dot product of a vector with itself, eq. (10) gives

$$\bar{A} \cdot \bar{A} = |\bar{A}|\,|\bar{A}| \cos 0 = |\bar{A}|^2$$

Two useful properties of the dot product are

$$p(\bar{A} \cdot \bar{B}) = (p\bar{A}) \cdot \bar{B} = \bar{A} \cdot (p\bar{B})$$
$$\bar{A} \cdot (\bar{B} + \bar{C}) = \bar{A} \cdot \bar{B} + \bar{A} \cdot \bar{C} \tag{11}$$

Both identities in eq. (11) are easily proven by use of the definition of the dot product, eq. (10).

The component representation of vectors may be employed in the calculation of a dot product. This is accomplished by first calculating the dot products of the unit vectors. For example, using eq. (10),

$$\bar{i} \cdot \bar{i} = (1)(1) \cos 0° = 1$$

Hence, for the unit vectors of a rectangular Cartesian coordinate system,

$$\bar{i} \cdot \bar{i} = \bar{j} \cdot \bar{j} = \bar{k} \cdot \bar{k} = 1 \tag{12a}$$

Any other combination of unit vectors leads to zero, for instance,

$$\bar{i} \cdot \bar{j} = (1)(1) \cos 90^\circ = 0$$

Therefore,

$$\bar{i} \cdot \bar{j} = \bar{j} \cdot \bar{i} = \bar{j} \cdot \bar{k} = \bar{k} \cdot \bar{j} = \bar{k} \cdot \bar{i} = \bar{i} \cdot \bar{k} = 0 \tag{12b}$$

In order to compute a dot product of two arbitrary vectors in terms of their components, we use eqs. (11) to write

$$\begin{aligned} \bar{A} \cdot \bar{B} &= (A_x\bar{i} + A_y\bar{j} + A_z\bar{k}) \cdot \bar{B} \\ &= A_x\bar{i} \cdot (B_x\bar{i} + B_y\bar{j} + B_z\bar{k}) + A_y\bar{j} \cdot (B_x\bar{i} + B_y\bar{j} + B_z\bar{k}) \\ &\quad + A_z\bar{k} \cdot (B_x\bar{i} + B_y\bar{j} + B_z\bar{k}) \end{aligned}$$

Then, in view of eqs. (12), this becomes

$$\boxed{\bar{A} \cdot \bar{B} = A_xB_x + A_yB_y + A_zB_z} \tag{13}$$

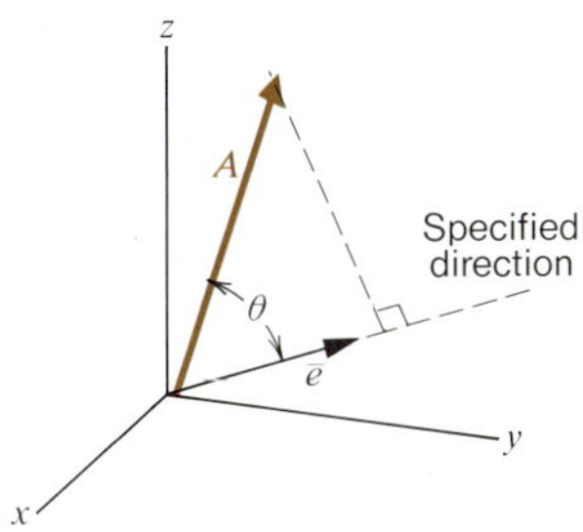

FIGURE 9

Our primary use for the dot product will be the determination of the component of a force parallel to a line that is not a coordinate axis. Let $\bar{e}$ be a unit vector parallel to the specified line, as is shown in Figure 9. We desire the component of the vector $\bar{A}$ in the specified direction. Using eq. (10) we have

$$\bar{A} \cdot \bar{e} = |\bar{A}|\,|\bar{e}| \cos \theta = |\bar{A}| \cos \theta \tag{14}$$

Equation (14) gives the scalar value of the desired component. A positive value means that the component is in the same direction as $\bar{e}$, whereas a negative value indicates the opposite direction.

In some situations, it is necessary to explicitly state the orientation of the vector $\bar{e}$. In such cases ambiguity is avoided by writing the component as a vector. Because any vector is the product of its magnitude and its unit vector, the *vector* representing the component of $\bar{A}$ along the specified line is

$$\boxed{\bar{A}_e = (\bar{A} \cdot \bar{e})\bar{e}} \tag{15}$$

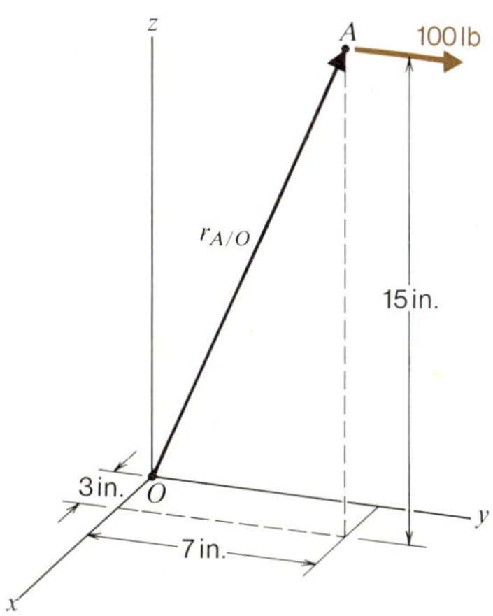

EXAMPLE 7

Determine the component of the 100 pound force in the direction OA. Write the component as a vector.

Solution

Let us begin by finding a unit vector $\bar{e}_{A/O}$. From the given coordinates we may write

$$\bar{e}_{A/O} = \frac{\bar{r}_{A/O}}{|\bar{r}_{A/O}|} = \frac{3\bar{i} + 7\bar{j} + 15\bar{k}}{\sqrt{3^2 + 7^2 + (15)^2}}$$
$$= 0.1783\bar{i} + 0.4161\bar{j} + 0.8917\bar{k}$$

Also,

$$\bar{F} = 100\bar{j} \text{ pounds}$$

Using eq. (14) we get

$$F_{OA} = \bar{F} \cdot \bar{e}_{A/O} = 100\bar{j} \cdot (0.1783\bar{i} + 0.4161\bar{j} + 0.8917\bar{k})$$
$$= 41.61 \text{ pounds}$$

Note that we did *not* have to find the angle between the force and line OA to use eq. (14).

Finally, eq. (15) yields

$$\bar{F}_{OA} = (\bar{F} \cdot \bar{e}_{A/O}) \cdot \bar{e}_{A/O} = 41.61\bar{e}_{A/O}$$
$$= 7.42\bar{i} + 17.31\bar{j} + 37.10\bar{k} \text{ pounds}$$

EXAMPLE 8

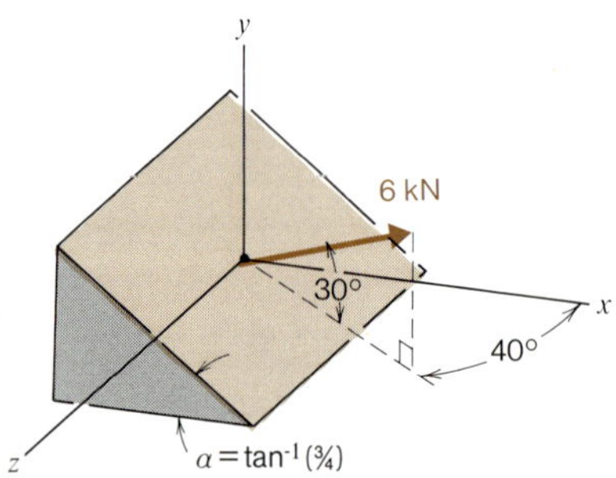

It is common practice in describing the resultant force acting on the surface of a body to refer to the component normal to the surface as a *tension force* if it pulls on the surface or as a *compression force* if it pushes on the surface. Determine the magnitude of the tension (or compression) force acting on the shaded rectangular area shown in the sketch, and then write this force as a vector.

Solution

We first express the 6 kN force in terms of its components. Note in the given sketch that the projection of the force onto the xz plane is 6(cos 30°) and that the angle between this projection and the x axis is 40°. Thus

$$\bar{F} = 6 \cos 30°(\cos 40° \, \bar{i} + \sin 40° \, \bar{k}) + 6 \sin 30° \, \bar{j}$$
$$= 3.980\bar{i} + 3.00\bar{j} + 3.340\bar{k} \text{ kN}$$

To determine the component of the force normal to the rectangular area, we will need the unit vector $\bar{e}_n$ normal to the surface. In this particular case $\bar{e}_n$ does not have a $\bar{k}$ component; that is, the normal vector is parallel to the xy plane, as shown in the sketch.

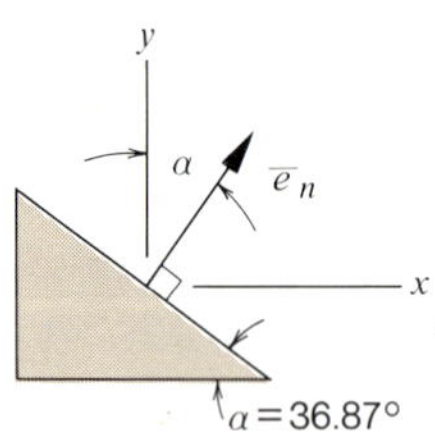

Because $\bar{e}_n$ is a unit vector, we have

$$\bar{e}_n = |\bar{e}_n| (\sin \alpha \bar{i} + \cos \alpha \, \bar{j}) = 0.60\bar{i} + 0.80\bar{j}$$

The component of the 6 kN force in the direction of $\bar{e}_n$ is found from the dot product to be

$$F_n = \bar{F} \cdot \bar{e}_n$$

$$\begin{aligned} F_n &= (3.980\bar{i} + 3.00\bar{j} + 3.340\bar{k}) \cdot (0.60\bar{i} + 0.80\bar{j}) \\ &= 3.980(0.60) + 3.00(0.80) + 0 \\ &= 4.788 \text{ kN} \end{aligned}$$

The dot product is positive, which means that the normal component of $\bar{F}$ is in the same direction as $\bar{e}_n$, which is outward from the surface. Therefore, $\bar{F}_n$ is a tensile force.

This force may be written in vector form by using eq. (15), which yields

$$\begin{aligned} \bar{F}_n = F_n\bar{e}_n &= 4.788(0.60\bar{i} + 0.80\bar{j}) \\ &= 2.873\bar{i} + 3.830\bar{j} \text{ kN} \end{aligned}$$

HOMEWORK PROBLEMS

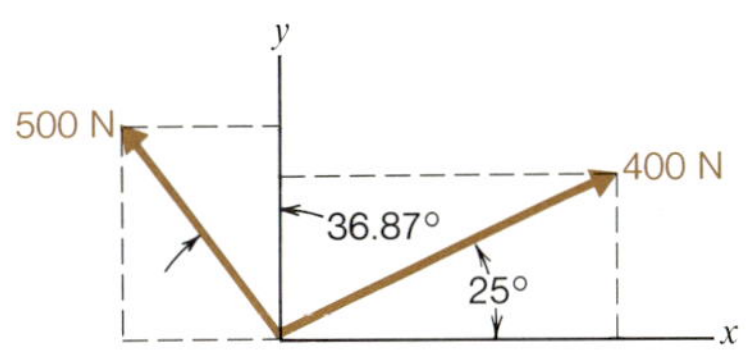

PROB. 2.56

2.56 Determine the dot product of the two forces shown by (a) using the formal definition of this product, (b) calculating the dot product of the component expressions for each force.

2.57 Use the dot product of $\bar{F}_1$ and $\bar{F}_2$ to derive the trigonometric identity for $\cos(\phi_2 - \phi_1)$.

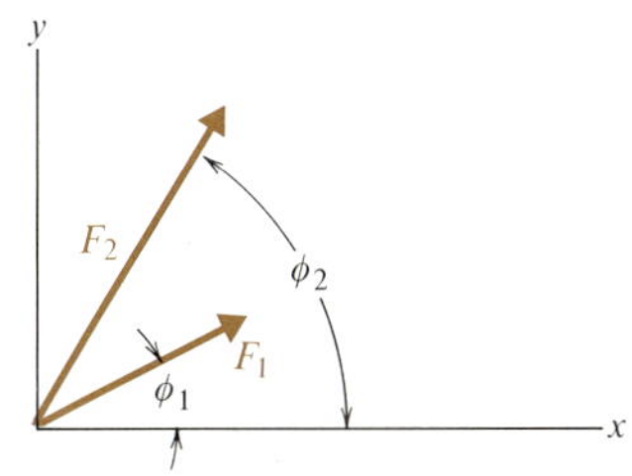

PROB. 2.57

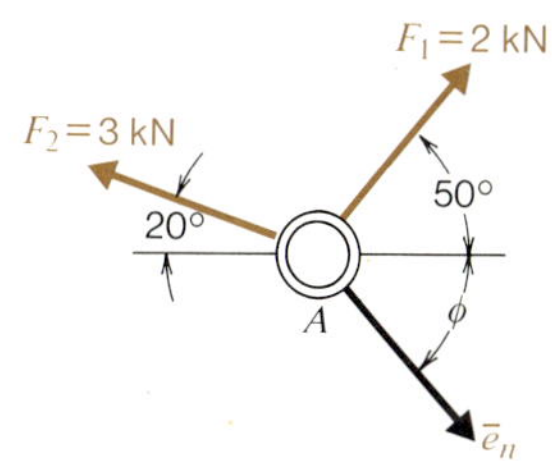

PROBS. 2.58 AND 2.59

2.58 Forces $\bar{F}_1$ and $\bar{F}_2$ are concurrent at ring A, and it is desired to determine the component R_n of their resultant in the direction of the unit vector $\bar{e}_n$ when $\phi = 40°$. Evaluate this component according to (a) $R_n = (\bar{F}_1 + \bar{F}_2) \cdot \bar{e}_n$, (b) $R_n = \bar{F}_1 \cdot \bar{e}_n + \bar{F}_2 \cdot \bar{e}_n$

2.59 The resultant of forces $\bar{F}_1$ and $\bar{F}_2$ is known to have a component of 0.5 kN in the direction of the unit vector $\bar{e}_n$. Determine the corresponding angle ϕ.

2.60 Prove that $\bar{A} \cdot (\bar{B} + \bar{C}) = \bar{A} \cdot \bar{B} + \bar{A} \cdot \bar{C}$ for the case where $\bar{A}$, $\bar{B}$, and $\bar{C}$ are coplanar.

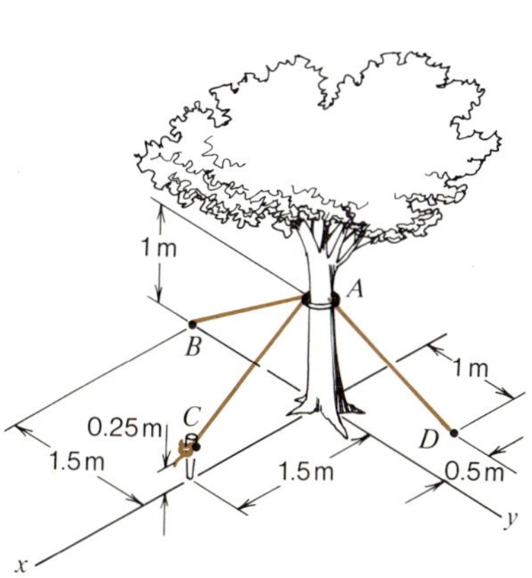

PROBS. 2.61, 2.62, AND 2.63

2.61 The tension in cable AB is 250 N. Determine the component in the direction of cable AC of the force exerted by cable AB on the tree. Express this component as a vector.

2.62 Determine the angle between cables AC and AD.

***2.63** Each cable bracing the tree carries a tension of 250 N. Determine the component of the resultant force exerted by all of the cables on the tree in the direction of cable AD. Express this component as a vector.

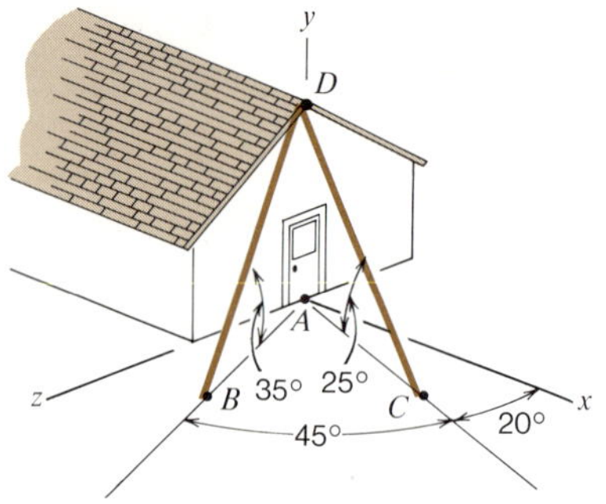

PROBS. 2.64, 2.65, AND 2.66

2.64 Cable CD has a tension of 500 lb. Determine the component in the direction of cable DB of the force this cable exerts on the building.

2.65 Determine the angle between the two cables that are attached to the building at point D.

***2.66** Cable BD has a tension of 200 lb, whereas the tension in cable CD is 100 lb. Determine the component parallel to line CD of the resultant force exerted by the cables on point D. Express this component as a vector.

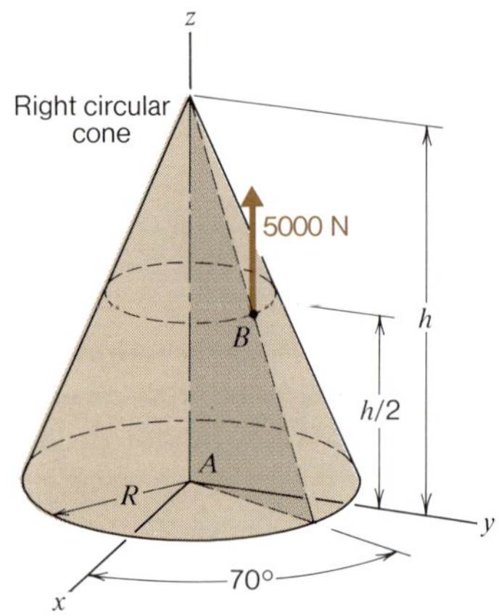

PROB. 2.67

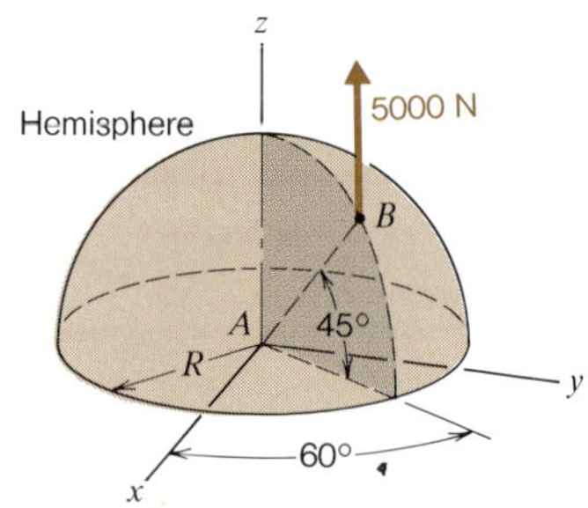

PROB. 2.68

2.67 and 2.68 A 5000 N vertical force is applied at point B on the surface shown. Determine the magnitude of the tension force produced by this force with respect to the tangent to the surface at point B. Then write this tension force as a vector with respect to the xyz reference frame shown.

F. EQUILIBRIUM OF A PARTICLE

1. Basic Equations

We have now developed techniques for evaluating and describing the resultant of a set of concurrent forces. An important application of these techniques is the study of the static equilibrium of a particle. We know from Newton's first law that a particle will remain at rest only if the resultant of all forces acting on the particle is zero. The resultant is found by summing the corresponding components of the individual forces. Thus, the condition of *static equilibrium* requires

$$\bar{R} \equiv \sum F_x \bar{i} + \sum F_y \bar{j} + \sum F_z \bar{k} = \bar{0}$$

The components of a zero vector are zero, from which we find the three scalar equations governing static equilibrium of a particle.

$$\boxed{\sum F_x = 0, \qquad \sum F_y = 0, \qquad \sum F_z = 0} \tag{16}$$

These equations state that in order for a particle to be in static equilibrium, the sum of all force components in each coordinate direction must be zero.

In thc special case of a planar system of forces, by definition there are no forces perpendicular to the plane. Equations (16) then reduce to the two scalar equations obtained by summing force components in that plane.

2. Free Body Diagrams

In general engineering practice, prior to writing the equations governing the behavior of a system, one models the essential features of the system. This model identifies the important characteristics of the system, and equally important, serves to remove extraneous information from the problem. In mechanics, we create such models by drawing *free body diagrams.* For particle systems these diagrams isolate each particle to which forces are applied and show *every* force exerted on the particle. With the free body diagram we also sketch an *xyz* coordinate reference system in order to minimize the chance for error in describing the force components, and also in order to simplify the task of checking our work.

In any problem it will be apparent that certain forces must be included in the free body diagram. Typical forces in this category are those exerted by cables, springs, and gravity (the weight force), and those that are given in the statement of the problem.

Forces that are less obvious, but equally important, are those that arise from the contact between the body and its surroundings. This contact means that the objects surrounding the body restrict the movement of the body. These restrictions, which are called constraint conditions, are imposed by means of *constraint forces.* Such forces are also called *reactions,* because they are the way in which the objects surrounding the body react to the tendency of the body to move. The topic of constraint forces will be considered in detail in the next chapter, where we study the equilibrium of rigid bodies. Nevertheless, let us now look at a simple example of a physical system with a constraint force.

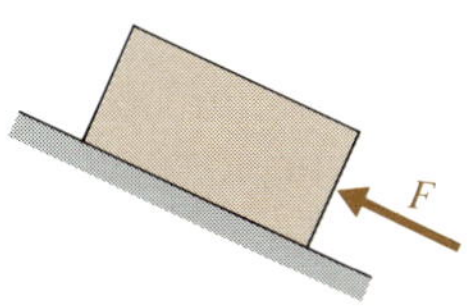

FIGURE 10a

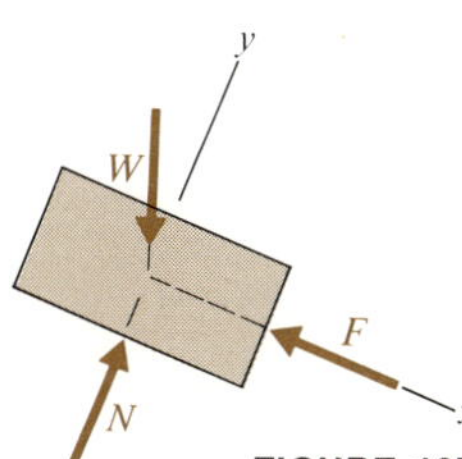

FIGURE 10b

Consider a box resting on a smooth (frictionless) inclined plane, as shown in Figure 10a. Clearly, the box cannot penetrate the solid surface of the incline. This is a constraint condition. The constraint condition corresponds to the constraint force, $\bar{N}$, normal to the plane, which is simply the interaction force between the box and the incline. In essence, in the free body diagram of Figure 10b we *remove* the plane and *account* for its presence by the interaction force $\bar{N}$. In addition, to complete the free body diagram, we show the given force $\bar{F}$ and the weight force $\bar{W}$. Only x and y axes are shown in the sketch, because we have a planar system of forces. The orientation of the axes may be chosen arbitrarily. In this case they were picked to facilitate the description of the components of the forces.

In the foregoing study the box was modeled as a particle. This means that we assumed that a system of concurrent forces exists. The shortcoming of such a model is that it cannot be used to study the effect of the point of application of $\bar{F}$ to the box. However, the two planar equilibrium equations that can be written for the box ($\sum F_x = 0, \sum F_y = 0$) will yield the values of F and N (in terms of W) required to have a state of static equilibrium.

It may be confusing to you that at times we can take the liberty of regarding physical systems occupying a region in space as particles. Throughout the text we will frequently make assumptions when modeling a system. These assumptions will not be loosely decided upon. Rather, they will be engineering decisions that result from our study of the various models that can be employed and the physical phenomena that these models can exhibit. Ultimately, however, the verification of an analytical study, and thus of all the assumptions it contains, comes from agreement with experimental results.

3. Problem Solving

At this juncture we can deduce a series of formal steps that will enable us to attack logically many problems relating to the equilibrium of bodies acted on by a system of concurrent forces. They are *not* items to be followed by rote; memorization does not lead to understanding. Rather, these steps should be used as a guideline to assist you in developing your own analytical abilities.

1. Draw a complete free body diagram. It should exhibit all constraint forces, all gravity forces, and all external forces acting on the system. When isolating one body from a larger system, be sure to account for the forces exerted by all other parts of the system that are in contact with the body of interest.
2. Choose an xyz reference frame that best fits the way in which the dimensions for the system are given. Show the axes in the free body diagram.
3. Evaluate the geometrical parameters necessary to describe all forces by their components. In planar problems, quantities such as angles that are not obvious from the given information should be shown in the free body diagram. For three-dimensional problems the forces should be written out in component form. Here, it may be necessary to use the concept of a general unit vector.
4. Write the equations of static equilibrium corresponding to the free body diagram drawn in step 1.
5. Count the number of scalar equations that result from step 4 and the number of unknowns contained in these equations. If there are more unknowns than equations, look for another free body diagram in the original system that will give information about one or more of the unknowns. Now repeat steps 1, 3, and 4. Eventually, a solvable set of equations will be obtained.
6. Solve the equations.

ILLUSTRATIVE PROBLEM 1

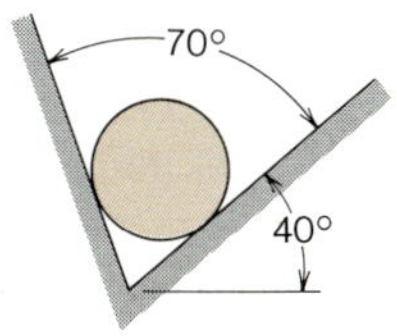

A 50 gram ball bearing rests in a smooth groove, as shown in the vertical cross section. Determine the forces acting on the bearing.

Solution

Step 1 We are interested in the forces acting on the ball bearing, so we draw a free body diagram of that object. Each wall of the groove exerts a normal force on the ball bearing, consistent with the constraint condition that the ball bearing cannot penetrate each surface. Note that the weight of the ball bearing in newtons is obtained by multiplying the mass in kilograms by g in meters/second.

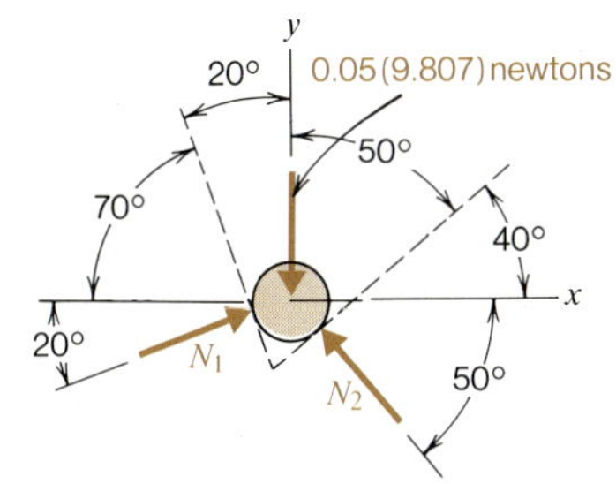

Steps 2 and 3 We have chosen the x and y axes to be horizontal and vertical because the angles to the forces from these directions are readily computed from the given geometrical information. The force system is planar, so the z axis is not relevant.

Step 4 The components of $\bar{N}_1$ and $\bar{N}_2$ are computed by projecting these forces onto the coordinate axes. Thus

$$\sum F_x = N_1 \cos 20^\circ - N_2 \cos 50^\circ = 0$$

$$\sum F_y = N_1 \sin 20^\circ + N_2 \sin 50^\circ - 0.05(9.807) = 0$$

Step 5 There are two equations in step 4 for the two unknown magnitudes N_1 and N_2, so we evaluate the trigonometric functions and solve the two simultaneous algebraic equations.

$$N_1 = 0.335 \text{ newtons}, \qquad N_2 = 0.490 \text{ newtons}$$

ILLUSTRATIVE PROBLEM 2

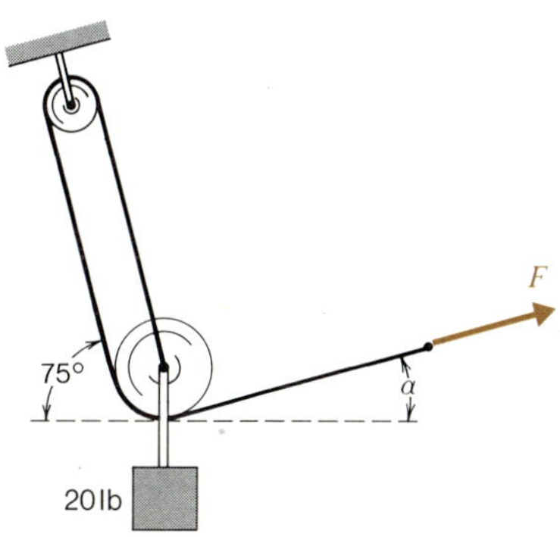

The 20 lb box is supported by a cable-pulley system as shown. Because the friction within the pulleys is negligible, the tension within the cable is unmodified as it passes over the pulley. Also, the pulleys have negligible mass. Determine the magnitude of the tensile force $\bar{F}$ and the angle α required for static equilibrium.

Solution

Step 1 The fact that the cable tension is constant along the length of a cable that passes over a frictionless pulley will be proven in the next chapter. The terminology used for referring to a pulley that is modeled as massless and frictionless is to say that it is *ideal*. Clearly, an ideal pulley is an approximation.

The gravity force in this problem acts on the block, whereas the lower pulley seems to be the focal point for the cables. We therefore draw free body diagrams of both bodies.

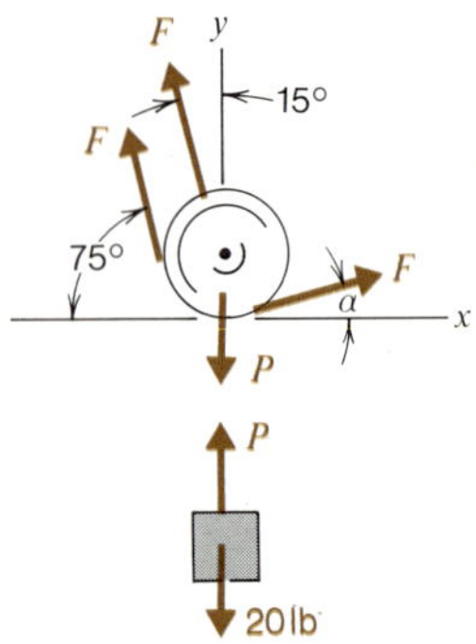

The easiest way to draw the diagram for the pulley is to imagine that we isolate the pulley by cutting through the cables. We then account for the cable by applying tensile forces to the cut ends. Modeling the pulley as ideal permits us to consider these tensile forces to be concurrent at the center of the pulley. In other words, we may consider the pulley to have a very small diameter. The given information tells us that the tensile force applied by each segment in the cable connecting the two pulleys is the same value F. Note that we are not considering equilibrium of the fixed pulley, because doing so would introduce into the problem another unknown force (the reaction holding that pulley to the ceiling).

Steps 2 and 3 The orientation of the cables is specified by their angle from the horizontal, so we choose the xy axes shown.

Step 4 We have two free body diagrams, so there are two sets of equilibrium equations. We begin with the box, because only one unknown force acts there. Thus,

$$\sum F_x \equiv 0, \qquad \sum F_y = P - 20 = 0$$

The equilibrium equations for the lower pulley are

$$\sum F_x = F \cos \alpha - 2F \cos 75^\circ = 0$$

$$\sum F_y = F \sin \alpha + 2F \sin 75^\circ - P = 0$$

Steps 5 and 6 We have one nontrivial equation for the box and two equations for the lower pulley, which may be solved first for P, and then for F and α. Thus

$$P = 20 \text{ lb}$$

$$\alpha = \cos^{-1}(2 \cos 75^\circ) = 58.83^\circ$$

$$F = P/(\sin \alpha + 2 \sin 75^\circ) = 7.18 \text{ lb}$$

ILLUSTRATIVE PROBLEM 3

A 500 kg crate is suspended by cables thare are joined at point A. Determine the tension in each of these cables. Points B, C and D are situated in the xy plane, and cable AE coincides with the vertical z axis.

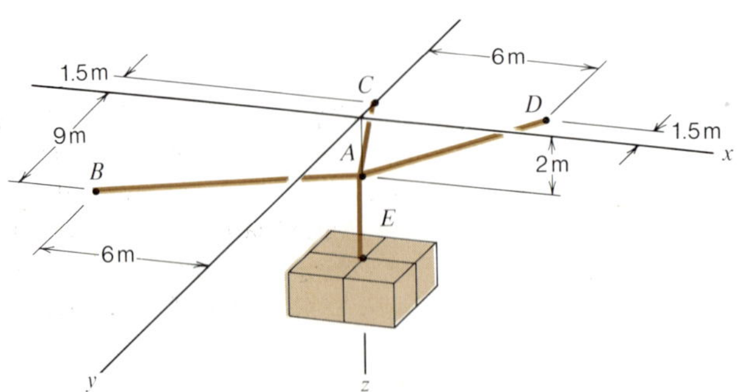

Solution

Step 1 We could draw free body diagrams of focal point A at which the cables meet, as well as of the crate, which is where the weight force is applied. However, we can make use of a shortcut that recalls a result in the previous illustrative problem. We know that the tension force in the cable AE must equal the weight of the crate, because those are the *only* forces acting on that body. A free body diagram of point A will then be sufficient to solve the problem.

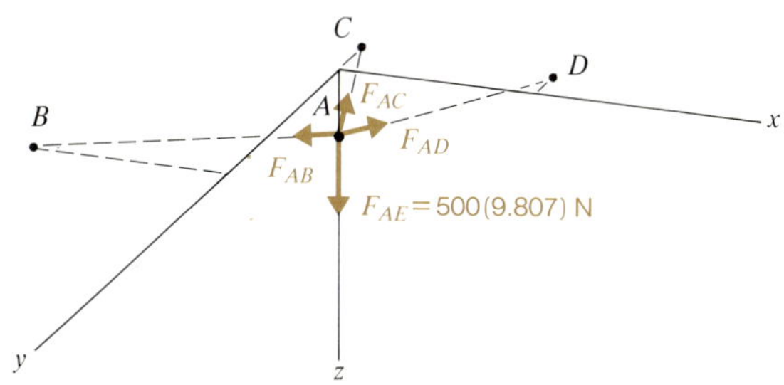

Step 2 The given xyz coordinate system is suitable for computations.

Step 3 Because the cables are oriented in three dimensions, it is best to express the forces in component form. The forces are oriented along lines from point A to points B, C, D, and E. Hence, from eq. (7) we have

$$\bar{F}_{AB} = F_{AB}\bar{e}_{B/A} = F_{AB}\frac{\bar{r}_{B/A}}{|\bar{r}_{B/A}|} = F_{AB}\frac{-6\bar{i} + 9\bar{j} - 2\bar{k}}{\sqrt{6^2 + 9^2 + 2^2}}$$

$$= F_{AB}(-\tfrac{6}{11}\bar{i} + \tfrac{9}{11}\bar{j} - \tfrac{2}{11}\bar{k})$$

$$\bar{F}_{AC} = F_{AC}\bar{e}_{C/A} = F_{AC}\frac{\bar{r}_{C/A}}{|\bar{r}_{C/A}|} = F_{AC}\frac{-1.5\bar{j} - 2\bar{k}}{\sqrt{(1.5)^2 + 2^2}}$$

$$= F_{AC}(-\tfrac{3}{5}\bar{j} - \tfrac{4}{5}\bar{k})$$

$$\bar{F}_{AD} = F_{AD}\frac{6\bar{i} - 1.5\bar{j} - 2\bar{k}}{\sqrt{6^2 + (1.5)^2 + 2^2}}$$

$$= F_{AD}(\tfrac{12}{13}\bar{i} - \tfrac{3}{13}\bar{j} - \tfrac{4}{13}\bar{k})$$

$$\bar{F}_{AE} = 4904\bar{k}\text{ N}$$

Step 4 For static equilibrium the sum of the force components at point A equals zero, therefore

$$\sum F_x = -\tfrac{6}{11}F_{AB} + \tfrac{12}{13}F_{AD} = 0$$

$$\sum F_y = \tfrac{9}{11}F_{AB} - \tfrac{3}{5}F_{AC} - \tfrac{3}{13}F_{AD} = 0$$

$$\sum F_z = -\tfrac{2}{11}F_{AB} - \tfrac{4}{5}F_{AC} - \tfrac{4}{13}F_{AD} + 4904 = 0$$

Steps 5 and 6 There are three equations for the three unknown cable tensions. We may solve the first two equations, in sequence, for F_{AD} and F_{AC} in terms of F_{AB}.

$$F_{AD} = \tfrac{13}{12}(\tfrac{6}{11})F_{AB} = 0.5909F_{AB}$$

$$\begin{aligned} F_{AC} &= \tfrac{5}{3}(\tfrac{9}{11}F_{AB} - \tfrac{3}{13}F_{AD}) \\ &= 1.6667[0.8182F_{AB} - 0.2308(0.5909)F_{AB}] \\ &= 1.1364F_{AB} \end{aligned}$$

The last equation for $\sum F_z$ then gives

$$\tfrac{2}{11}F_{AB} + \tfrac{4}{5}(1.1364)F_{AB} + \tfrac{4}{13}(0.5909)F_{AB} = 4904$$

$$F_{AB} = 3853 \text{ N}$$

The other forces are then found to be

$$F_{AD} = 0.5909(3853) = 2277 \text{ N}$$

$$F_{AC} = 1.1364(3853) = 4379 \text{ N}$$

HOMEWORK PROBLEMS

2.69 A 50 gram ball bearing rests in a smooth groove, as shown in the vertical cross section. Determine the forces acting on the bearing.

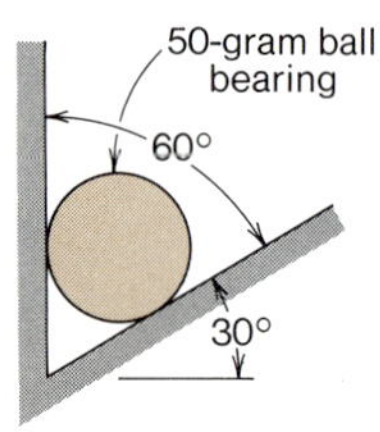

PROB. 2.69

2.70 and 2.71 A solid cylinder rests in a smooth groove, as shown in the vertical cross section. Determine the forces exerted on each cylinder.

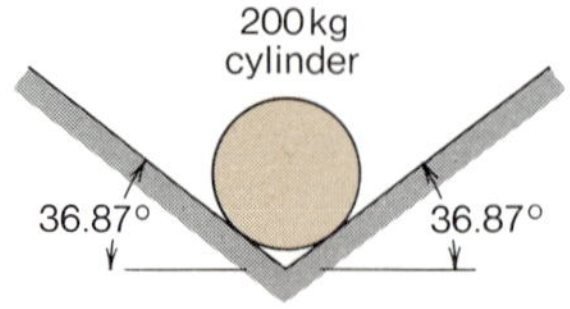

PROB. 2.70

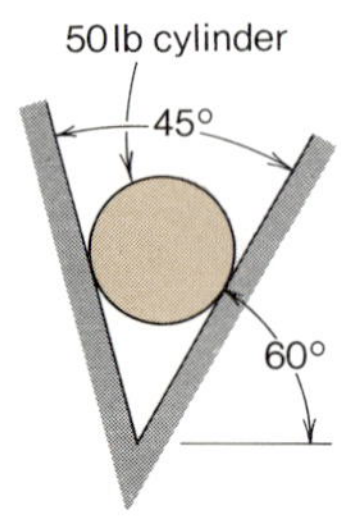

PROB. 2.71

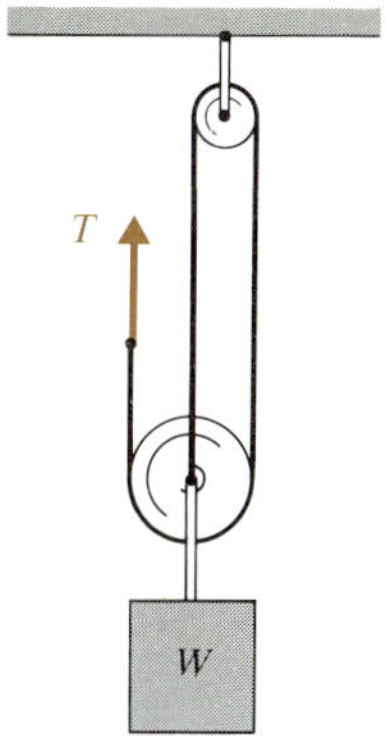

PROB. 2.72

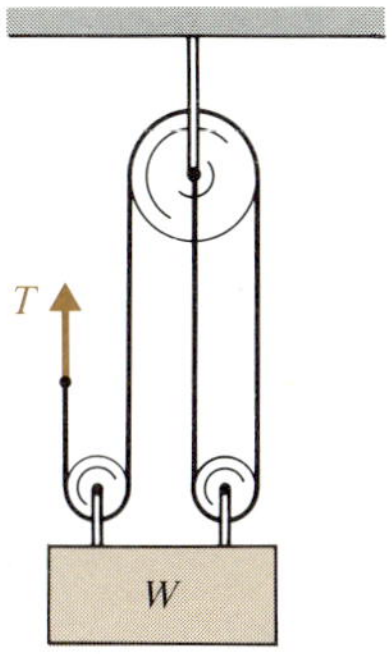

PROB. 2.73

2.72 and 2.73 Determine the tension T in the cable required for static equilibrium of the block whose weight is W.

2.74 Block A has a mass of 15 kg. Determine the mass of block B for static equilibrium. The pulleys are ideal.

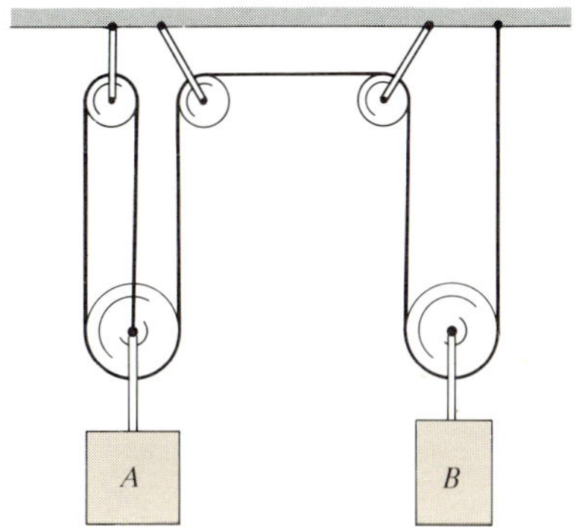

PROB. 2.74

PROB. 2.75

2.75 A 70 kg sign painter sitting on a 10 kg scaffold holds the free end of the cable. The pulleys have ideal properties. Determine the tension in the cable and the reaction force exerted between the painter and the scaffold.

2.76 If $\beta = 15^\circ$, determine the magnitude of the force $\bar{F}$ and the angle α for static equilibrium.

2.77 Solve Problem 2.76 for $\beta = -15^\circ$.

***2.78** Determine the smallest angle α for which the system will be in static equilibrium. Also find the corresponding values of the magnitude of the force F and the angle β. The pulleys are ideal.

2.79 and 2.80 Block A has a mass of 8 kg. Both inclines are smooth and the system is in static equilibrium. Determine the mass of block B.

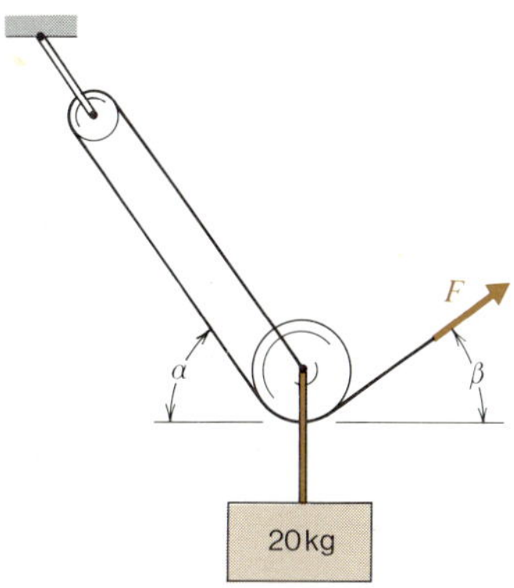

PROBS. 2.76 AND 2.78

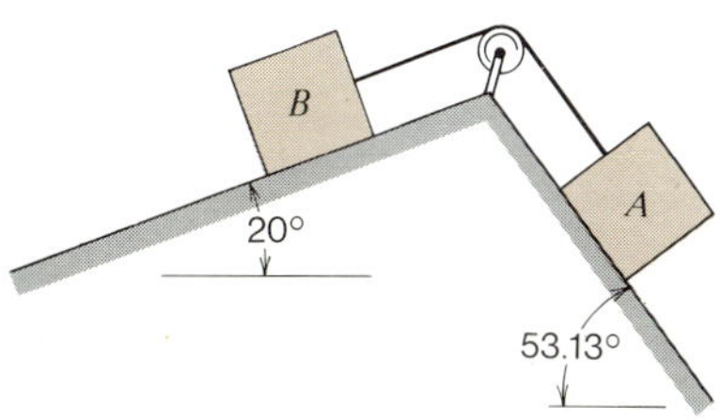

PROB. 2.79

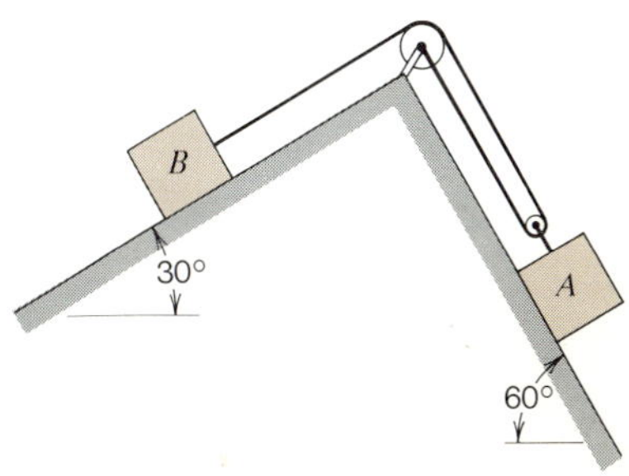

PROB. 2.80

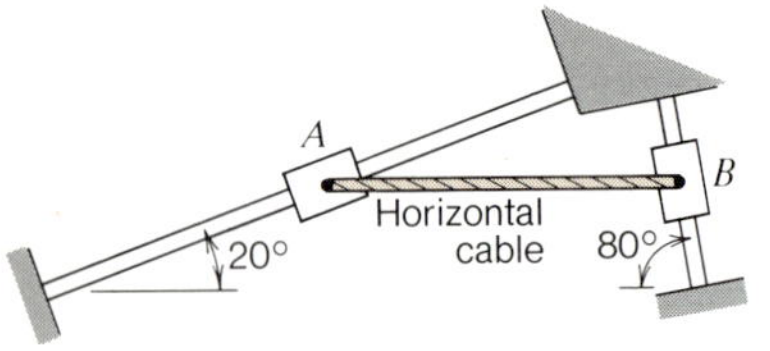

PROB. 2.81

***2.81** Both collars are at rest on the smooth guide bars. Determine the ratio of masses m_B/m_A.

2.82 The 50 kg cylinder A and the 100 kg cylinder B are held inside the crate shown in the vertical cross-sectional view. All surfaces of contact are smooth. Determine the contact forces exerted between the cylinders and between each cylinder and the sides of the crate.

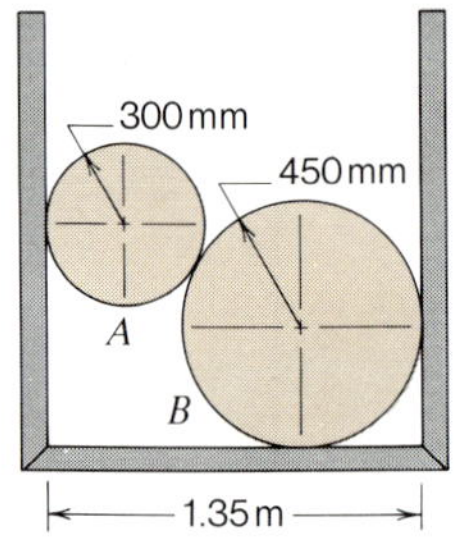

PROB. 2.82

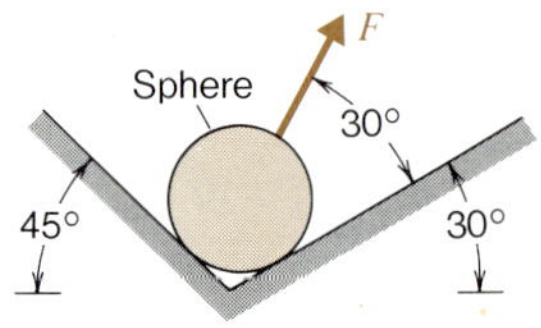

PROBS. 2.83 AND 2.84

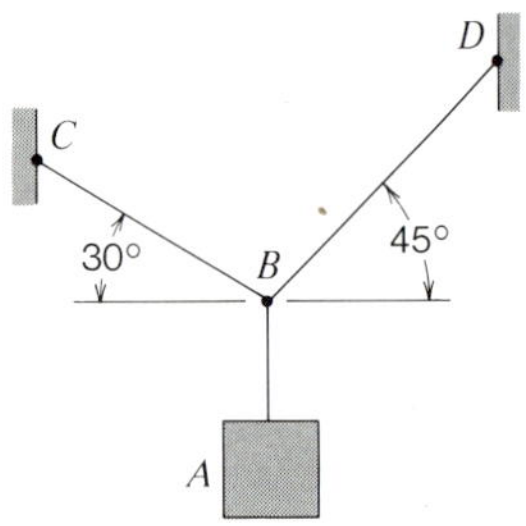

PROB. 2.85

2.83 A 5 lb steel sphere, which is supported in a smooth groove, has a force $F = 1$ lb applied to it, as shown in the vertical cross section. Determine the forces exerted between the sphere and the walls.

***2.84** A 10 kg sphere, which is supported in a smooth groove, has a force $\bar{F}$ applied to it as shown in the vertical cross section. Determine the maximum magnitude of this force for which the sphere will remain in contact with both walls.

2.85 Determine the tensions in cables BC and BD which support block A, whose mass is 5 kg.

2.86 Block A, whose weight is 90 lb, is supported in the position shown. Determine the weight of block B.

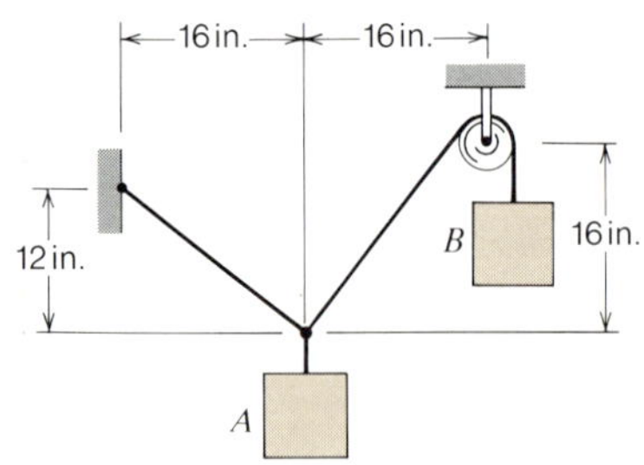

PROB. 2.86

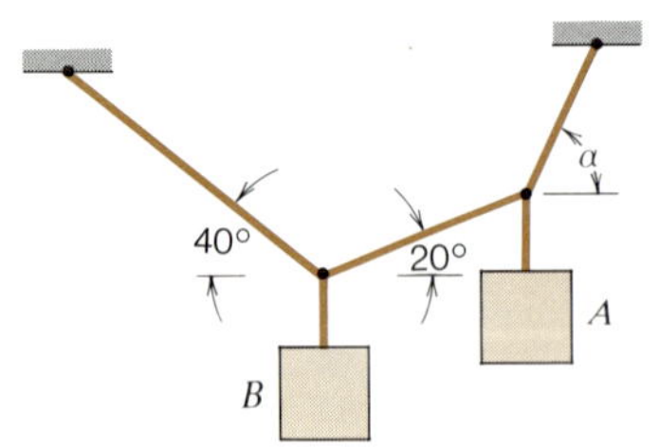

PROB. 2.87

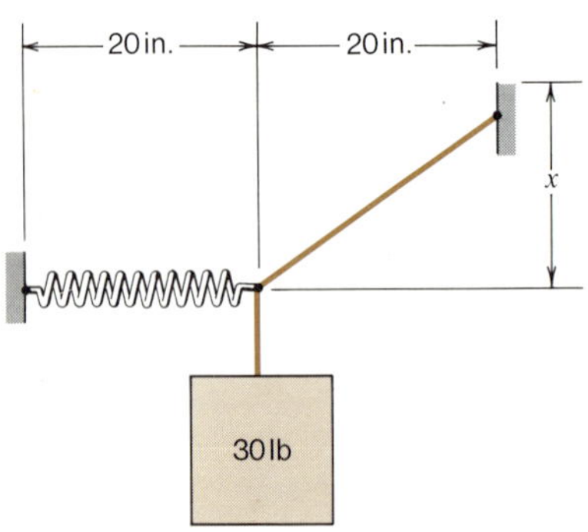

PROB. 2.88

2.87 The masses of blocks A and B are 20 and 40 kg, respectively. Determine the value of α and the tension forces in the cables for static equilibrium.

2.88 The spring has an unstretched length of 14 in. and a stiffness of 80 lb/ft. Determine the height x corresponding to static equilibrium in the position shown.

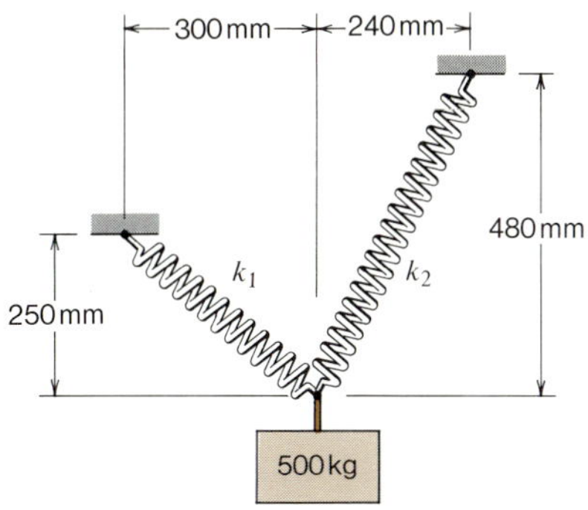

PROB. 2.89

*2.89 The stiffnesses of the springs are $k_1 = 20$ kN/m and $k_2 = 10$ kN/m. If the 500 kg block is suspended in static equilibrium in the position shown, determine the unstretched lengths of the springs.

2.90 The friction between the collar and its vertical guide is negligible. The lower spring has a stiffness $k_1 = 4$ N/m and an unstretched length $L_1 = 200$ mm, whereas the corresponding properties for the upper spring are $k_2 = 3$ N/m, $L_2 = 150$ mm. Determine the mass of the collar if $d = 80$ mm in the equilibrium position.

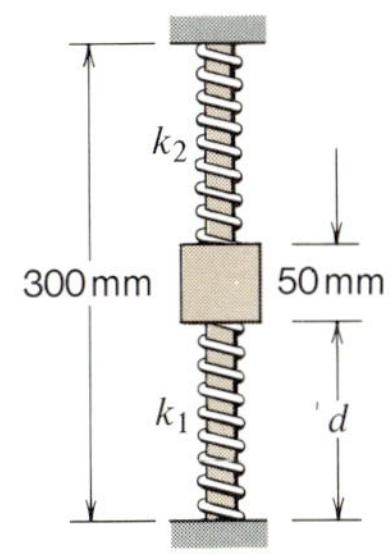

PROB. 2.90

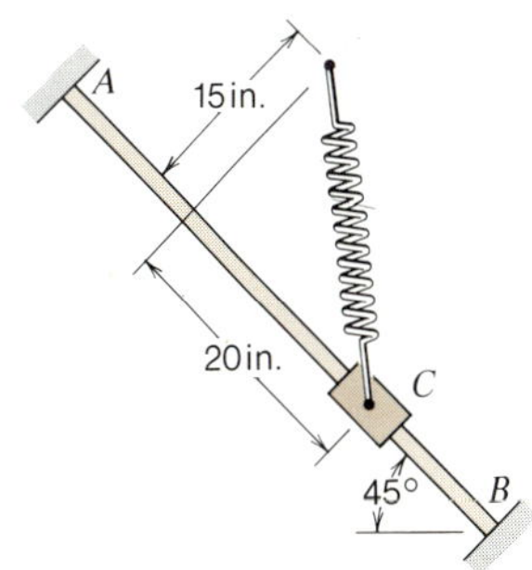

PROB. 2.91

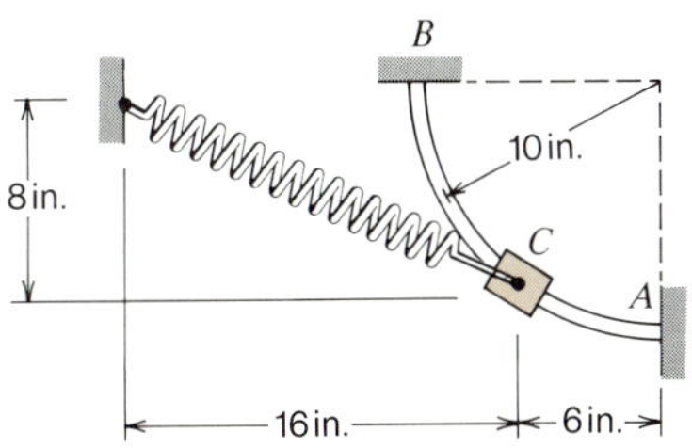

PROB. 2.92

*2.91 and 2.92 Collar C is supported by the smooth rod AB, and by a spring whose stiffness is 2 lb/in. and whose unstretched length is 15 in. The collar is in static equilibrium in the position shown. Determine the weight of the collar.

2.93 A 3000 lb automobile is suspended on a lightweight platform by four cables of equal length. Each cable carries the same tension because of symmetry. Determine this tension.

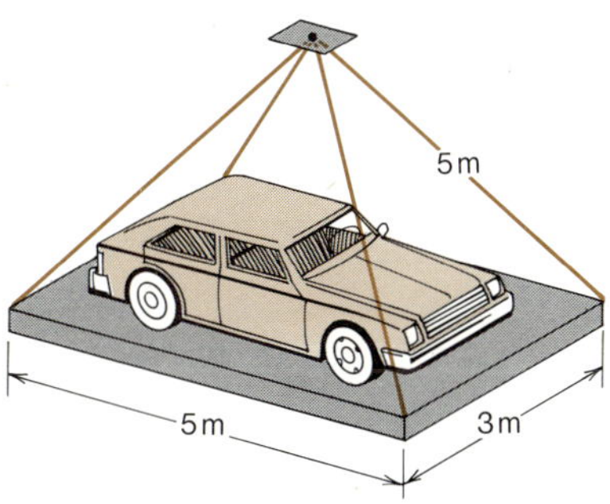

PROB. 2.93

2.94 The 100 kg box D is suspended from the ceiling by three cables whose ends A, B, and C form an equilateral triangle in the horizontal plane. Determine the tension in each cable.

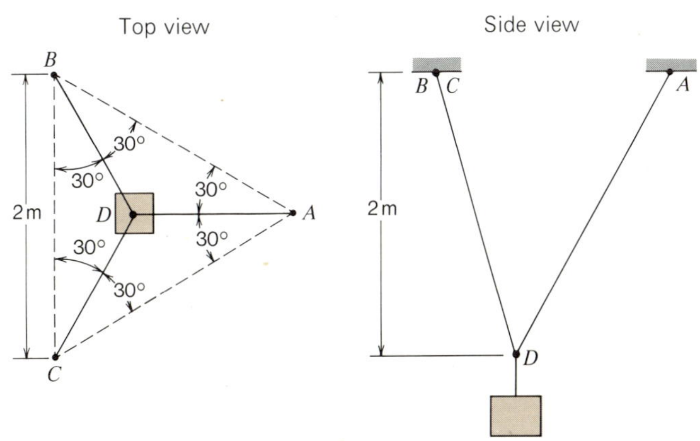

PROB. 2.94

2.95 and 2.96 A 15 kN horizontal force is applied to ring *A* where three cables are attached. Determine the tension in each cable.

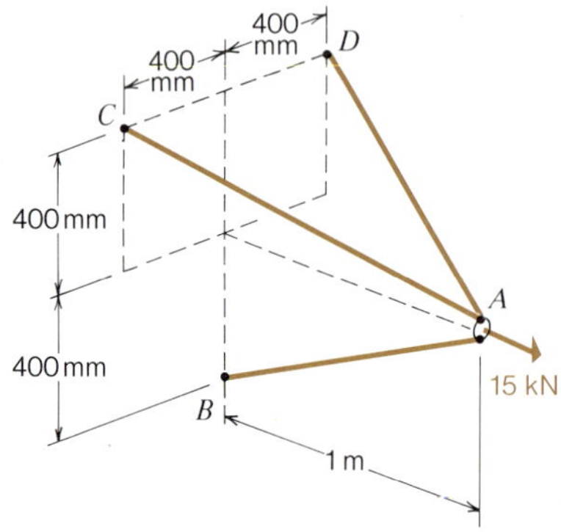

PROB. 2.95

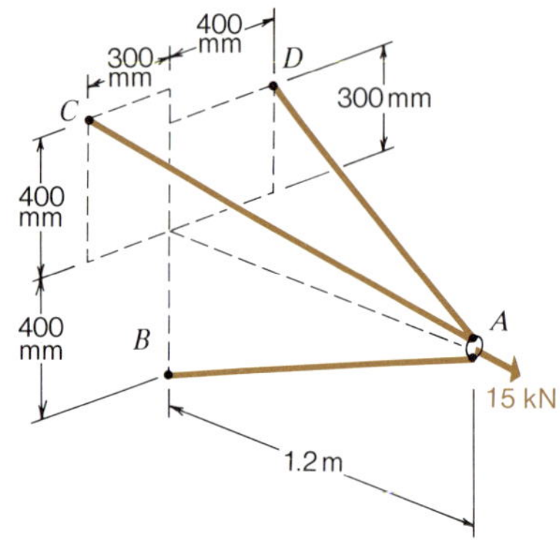

PROB. 2.96

***2.97** A 200 lb traffic light is suspended 12 ft above the ground by cables from three 16 ft tall poles located as shown. Determine the tension in each cable.

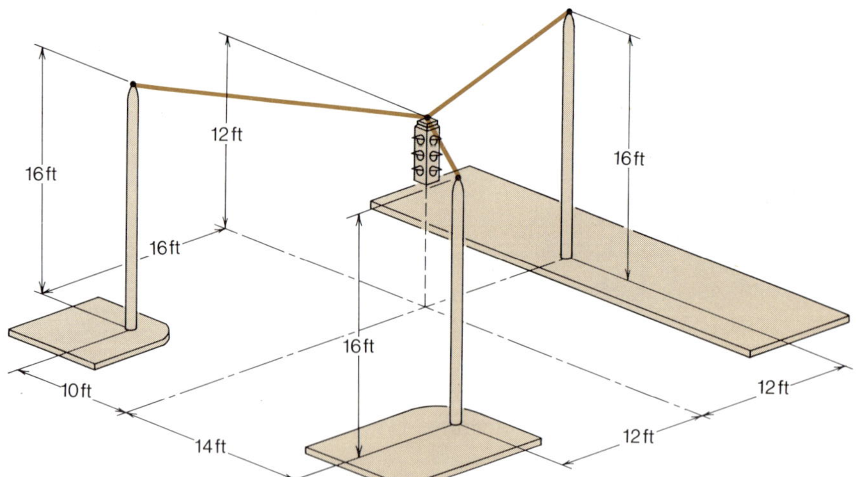

PROB. 2.97

2.98 A 1 kg ball bearing is pushed into the smooth right angle corner by the 20 N force shown, which passes through the center of the sphere. Determine the reaction forces exerted between the ball bearing and the vertical walls and horizontal floor.

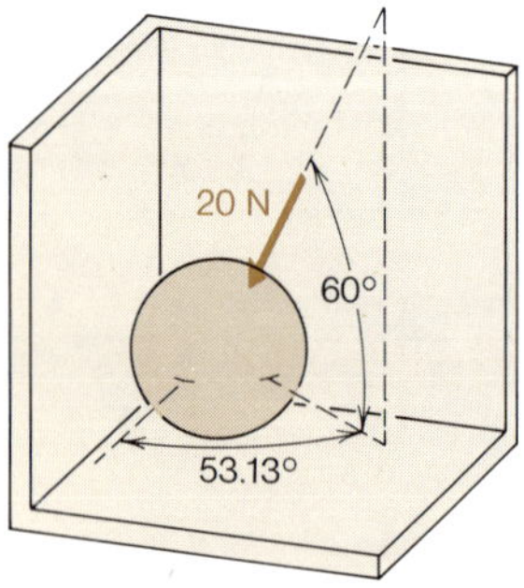

PROB. 2.98

***2.99** The diagram shows the side view and cross section of a V-block that is holding a steel sphere of negligible mass. The reaction between the sphere and the fixed vertical wall is 1000 N. Determine the reaction forces exerted between the sphere and the walls of the groove, and between the sphere and the inclined movable block.

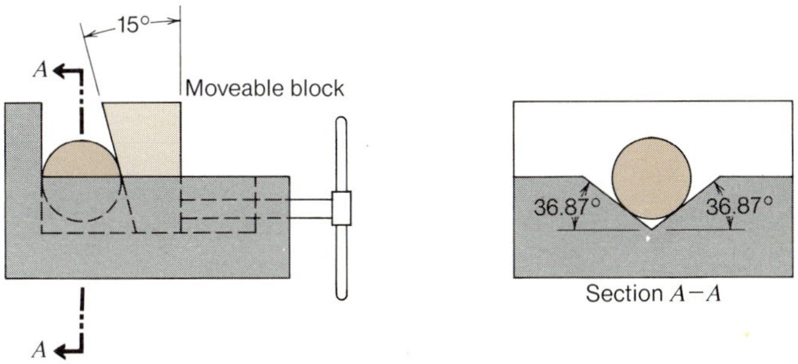

PROB. 2.99

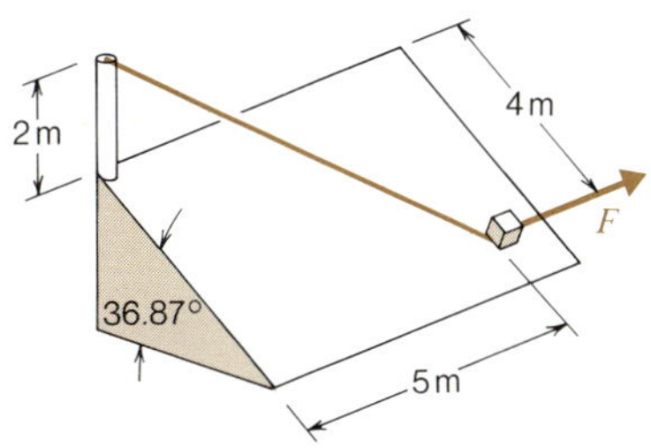

PROB. 2.100

2.100 A 100 kg crate is supported on the smooth inclined plane by the cable and the horizontal force $\bar{F}$ tangent to the incline, as shown. Determine the magnitude of $\bar{F}$ and the tension in the cable. Neglect the crate's dimensions.

***2.101** A 10 lb block is supported by three springs as shown. The stiffnesses of the springs are $k_1 = 5$ lb/in., $k_2 = 3$ lb/in., $k_3 = 4$ lb/in. Determine the unstretched length of each spring.

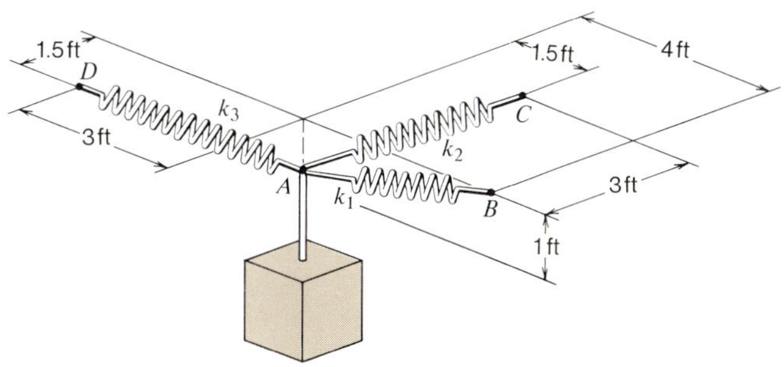

PROB. 2.101

CHAPTER 3

GENERAL FORCE SYSTEMS AND EQUILIBRIUM OF A RIGID BODY

OBJECTIVES

1. To compute the cross product of two vectors in terms of their components.

2. To compute the moment of a force about a point in a planar system by evaluating the lever arm of the force or the lever arms of its two components.

3. To compute the moment of a force about a point in a three-dimensional system by means of a cross product.

4. To compute the moment of a force about a line.

5. To resolve a set of given forces into a force-couple system acting at a specified point.

6. To evaluate the magnitude, direction, and location of a point on the line of action of the single force that is equivalent to a system of planar or parallel forces.

7. To inspect the supports of a rigid body in order to determine the nature of the reactions, and to use that information to draw a free body diagram.

8. To evaluate the unknown reactions holding a rigid body in equilibrium by solving the equations of static equilibrium.

9. To recognize situations of partial and improper constraint, as well as static indeterminacy, on the basis of the solvability of the equations of static equilibrium.

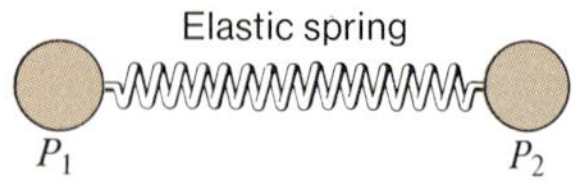

FIGURE 1a

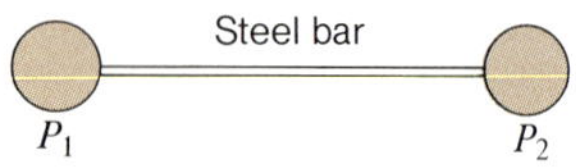

FIGURE 1b

A rigid body may be considered to be a collection of particles that are always at a fixed distance from each other. For example, consider the two-particle systems shown in Figure 1. Figure 1a shows two particles connected by an elastic spring. If the spring is extended or compressed, the particles will no longer be at their original distance from each other; hence the system is not a rigid body. In Figure 1b, if we neglect the secondary (small) effects of the deformation of the connecting steel bar, the particles are always at a fixed distance apart, and therefore the system is termed a rigid body.

The concept of a rigid body is a mathematical idealization, because there is no material that can remain undeformed under the influence of forces. Situations where the deformation of a body must be considered are treated extensively in the subject of the mechanics of materials, which builds upon the methods developed in this text.

Because a rigid body occupies more than a single point in space, it will generally be subjected to a system of nonconcurrent loads having a tendency to cause it to rotate. To describe this rotational tendency, we will develop the concept of the *moment of a force.* We shall see that the moment is dependent on the line of action of a force, as well as on its magnitude and direction.

The method for determining moments, in conjunction with the methods for describing forces that were developed in the previous chapter, are basic tools. Later in this chapter, they will enable us to study the static equilibrium of rigid bodies. This material is therefore the focal point in the study of Statics.

A. MOMENTS OF A FORCE ABOUT A LINE AND ABOUT A POINT

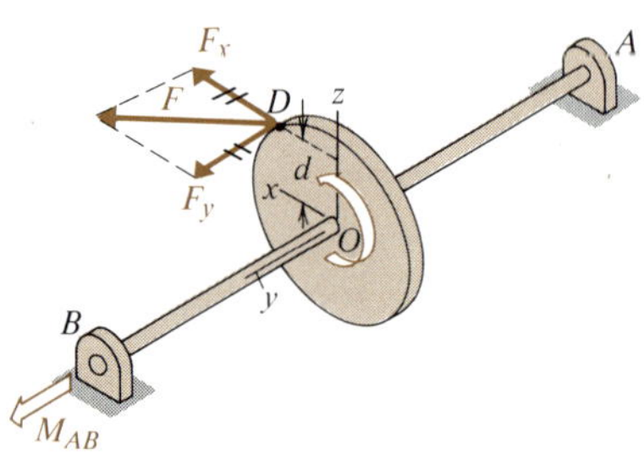

FIGURE 2

Let us begin this study by considering the system shown in Figure 2, where a force $\bar{F}$ is applied at point D on the perimeter of a disk mounted on a shaft. The y axis of the coordinate system is chosen parallel to the shaft AB, and the z axis is chosen to be perpendicular to $\bar{F}$. Because $\bar{F}$ is parallel to the xy plane, it may be replaced by its two component vectors $F_x\bar{i}$ and $F_y\bar{j}$, as shown.

The distance d is the perpendicular distance from the shaft to the line of action of the vector $F_x\bar{i}$. In the terminology of statics, d is the *lever arm* of the force $F_x\bar{i}$ about the axis AB.

For the situation of Figure 2, the component $F_x\bar{i}$ will cause the disk to rotate about the axis of the shaft in the sense of the curved arrow. The other component, $F_y\bar{j}$, causes no rotation about axis AB; it merely pulls on the disk parallel to that axis.

To quantify the rotational effect we define the concept of the *moment of a force about an axis.* Specifically, we define the moment $\bar{M}_{AB}$ of the force $\bar{F}$ about line AB such that its magnitude is the product of the component of the force $\bar{F}$ perpendicular to the line and the lever arm of the force (perpendicular distance) to the line. Thus

$$|\bar{M}_{AB}| = F_x d \tag{1}$$

Dimensionally, moment carries the physical units of force times distance. Throughout this text we shall consistently give the force unit of the moment first and then the length unit. This is done in order to avoid confusion with the units of energy, which also involve the product of force and length.

In addition to having a magnitude, the moment is a vector quantity having direction. This direction is associated with the sense of the rotational effect. We could use the curved line in Figure 2 to describe this direction by saying that the moment of $\bar{F}$ about line AB is counterclockwise as viewed from point B toward point A. Such descriptions prove to be inadequate in many situations.

Another way to describe the moment about a line is to employ the *right-hand rule*, as depicted in Figure 3. The fingers of the right hand are curled in the sense of the rotational effect of the moment about the axis. The unit vector parallel to the axis in the sense of the extended thumb is then defined as the direction of the moment. According to this description, the moment $\bar{M}_{AB}$ in Figure 2 is in the positive y direction, so

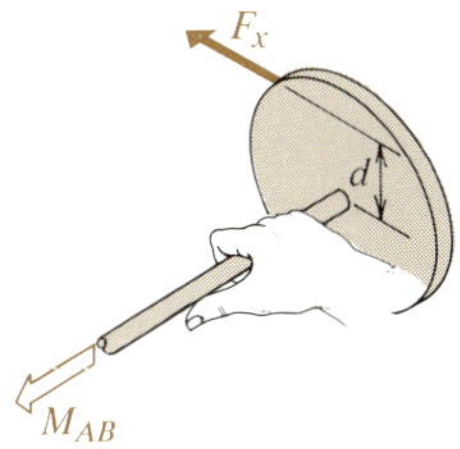

FIGURE 3

$$\bar{M}_{AB} = M_{AB}\bar{j} = F_x d\bar{j}$$

In order to avoid confusion between force and moment vectors, moment vectors will be depicted by double-lined arrows, as shown in Figure 3.

The concept of the moment of a force about a line will prove to be valuable in two-dimensional situations. For example, the moment $\bar{M}_{AB}$ in Figure 2 shows that the force component $F_x\bar{i}$ tends to produce a rotation in the xz plane about the y axis, whereas the component $F_y\bar{j}$ does not have any effect on rotation in this plane.

The overall rotational effect of $\bar{F}$ on the system can be described as the *moment of the force about a point*. In order to introduce this concept, suppose that instead of supporting the shaft in Figure 2 at ends A and B, there is only a single support at some point C on the shaft. This situation is depicted in Figure 4. We now want to determine the tendency of $\bar{F}$ to cause the system to rotate about point C. We indicate that the desired moment is about the point C by writing the quantity as $\bar{M}_C$, in contrast to writing a moment with a double subscript (as in Figure 3), which denotes a moment about an axis.

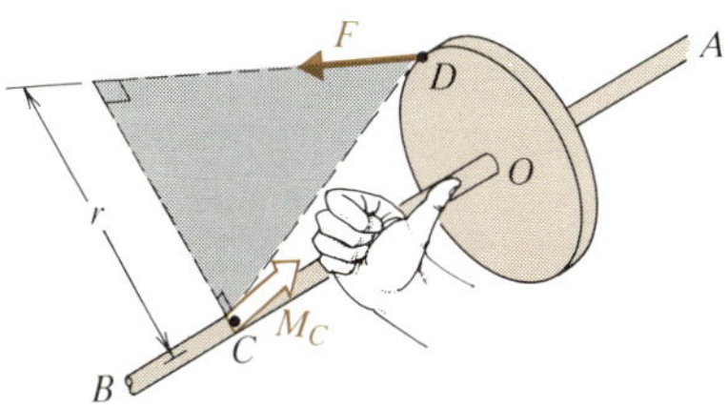

FIGURE 4

The distance r in Figure 4 is the perpendicular distance from the line of action of the force to point C. This distance is called the *lever arm of the force about point C*. The magnitude of $\bar{M}_C$ is defined to be the product of this lever arm and the magnitude of $\bar{F}$; that is

$$|\bar{M}_C| = |\bar{F}|r \tag{2}$$

The right-hand rule is again used to define the direction of the moment. To do this we form an analogy to the definition of the moment of the force about the axis of the shaft. The fingers of the right hand are curled in the sense of the rotation that would be imparted to the shaded plane formed by point C and the line of action of $\bar{F}$. The direction of $\bar{M}_C$ is then perpendicular to the shaded plane, in the sense of the extended thumb of the right hand.

The similarities in the definitions of the moment of a force about a line and the moment of a force about a point suggest that the two quantities are directly related. Later we shall study this relationship. Before doing so, we must increase our ability to describe the moment of a force. For example, in Figure 4 the determination of the lever arm r and the direction of $\bar{M}_C$ can be quite complicated. To avoid cumbersome computational procedures, we will now study the remaining vectorial tool we need, the cross product.

B. THE CROSS PRODUCT

The dot product of two vectors gives a scalar value involving the component of one vector parallel to the other vector. In contrast, the cross product gives a vector result in terms of the component of one vector perpendicular to the other vector. Because of the difference in the type of result of these two multiplication operations, the dot product is sometimes called the *scalar product*, and the cross product is sometimes called the *vector product*.

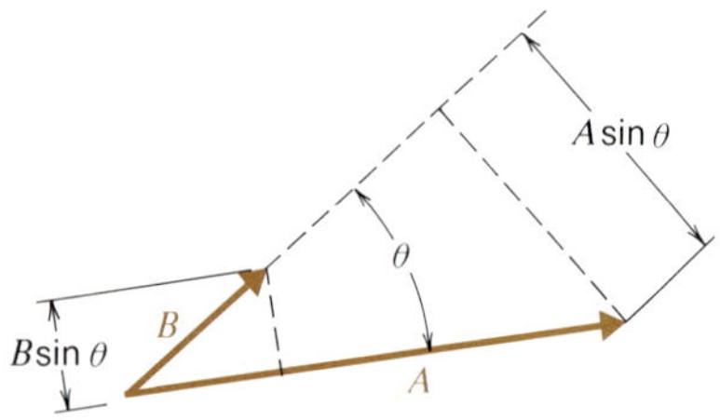

FIGURE 5

Consider the two vectors $\bar{A}$ and $\bar{B}$ shown in Figure 5. Without loss of generality, let the plane of the vectors coincide with the plane of the paper. With the angle between the vectors denoted as θ, the component of $\bar{A}$ perpendicular to $\bar{B}$ is $|\bar{A}| \sin\theta$, and the component $\bar{B}$ perpendicular to $\bar{A}$ is $|\bar{B}| \sin\theta$, as shown. The magnitude of the cross product $\bar{A} \times \bar{B}$ is defined to be the component of $\bar{B}$ perpendicular to $\bar{A}$ multiplied by the magnitude of $\bar{A}$; or vice versa.

$$|\bar{A} \times \bar{B}| = [|\bar{B}| \sin\theta]|\bar{A}| = [|\bar{A}| \sin\theta]|\bar{B}| = |\bar{A}|\,|\bar{B}| \sin\theta \tag{3}$$

It follows from this definition that $|\bar{A} \times \bar{B}| \equiv |\bar{B} \times \bar{A}|$ and that $\bar{A} \times \bar{B}$ is zero whenever $\bar{A}$ and $\bar{B}$ are parallel.

The direction of the cross product is defined to be normal to the plane formed by the two vectors. The sense of this normal is defined by employing the right-hand rule. We curl the fingers of the right hand about the common tail of the two vectors, from the first vector in the product to the second. The extended thumb is then oriented in the sense of the product.

According to this definition for the sense, $\bar{A} \times \bar{B}$ in Figure 5 is outward from the plane of the diagram, whereas $\bar{B} \times \bar{A}$ is inward. Because the order of multiplication does not affect the magnitude of the product given by eq. (3), we can conclude that the two products have equal magnitude and opposite sense. In other words,

$$\bar{B} \times \bar{A} = -\bar{A} \times \bar{B} \tag{4}$$

We will frequently employ the cross product operation in conjunction with a description of the vectors in terms of their components with respect to a coordinate system xyz. This requires consideration of algebraic rules. In view of the fact that the order in which the vectors appear in the cross product affects the result, as shown by eq. (4), we may wonder whether the other algebraic rules are different in form from those for the dot product. The answer is no. Specifically, we can prove that

$$p(\bar{A} \times \bar{B}) = (p\bar{A}) \times \bar{B} = \bar{A} \times (p\bar{B})$$

$$\bar{A} \times (\bar{B} + \bar{C}) = (\bar{A} \times \bar{B}) + (\bar{A} \times \bar{C}) \tag{5}$$
$$(\bar{A} + \bar{B}) \times \bar{C} = (\bar{A} \times \bar{C}) + (\bar{B} \times \bar{C})$$

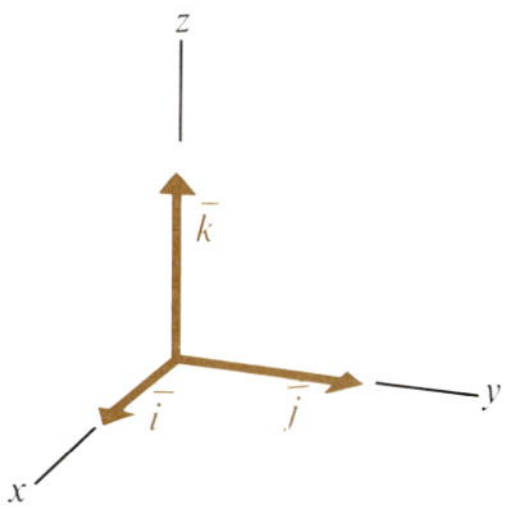

FIGURE 6

The evaluation of cross products in terms of components employs these laws in conjunction with the results for the cross products of the unit vectors. *It is essential that the unit vectors be defined in terms of a right-handed coordinate system*, in order to be consistent with the use of the right-hand rule in defining the cross product. Figure 6 shows a typical coordinate system.

As each of the unit vectors is obviously parallel to itself, it follows that

$$\bar{i} \times \bar{i} = \bar{j} \times \bar{j} = \bar{k} \times \bar{k} = \bar{0} \tag{6}$$

In contrast, the angle between any two different unit vectors is 90°, so

$$|\bar{i} \times \bar{j}| = |\bar{i}||\bar{j}| \sin 90^\circ = 1$$
$$= |\bar{j} \times \bar{k}| = |\bar{k} \times \bar{i}|$$

Also, the third unit vector is perpendicular to the coordinate plane of the other two unit vectors. The right-hand rule gives the appropriate sign. For example, when we curl our fingers from $\bar{j}$ to $\bar{i}$ in order to evaluate $\bar{j} \times \bar{i}$, our thumb is oriented opposite the sense of positive $\bar{k}$. The full set of relations are

$$\begin{aligned} \bar{i} \times \bar{j} &= -\bar{j} \times \bar{i} = \bar{k} \\ \bar{j} \times \bar{k} &= -\bar{k} \times \bar{j} = \bar{i} \\ \bar{k} \times \bar{i} &= -\bar{i} \times \bar{k} = \bar{j} \end{aligned} \tag{7}$$

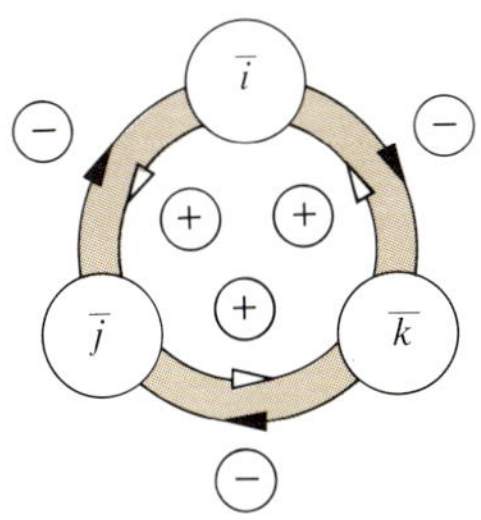

FIGURE 7

We will employs eqs. (7) frequently. We may remember them by considering the cross product of two unit vectors to be positive for positive alphabetical order and negative for negative alphabetical order, as depicted in Figure 7.

The cross product of two general vectors may now be determined by expressing them in terms of their components, and then employing eqs. (5) through (7). Thus

$$\begin{aligned} \bar{A} \times \bar{B} &= (A_x\bar{i} + A_y\bar{j} + A_z\bar{k}) \times \bar{B} \\ &= A_x[\bar{i} \times (B_x\bar{i} + B_y\bar{j} + B_z\bar{k})] + A_y[\bar{j} \times (B_x\bar{i} + B_y\bar{j} + B_z\bar{k})] \\ &\quad + A_z[\bar{k} \times (B_x\bar{i} + B_y\bar{j} + B_z\bar{k})] \\ &= A_x(B_y\bar{k} - B_z\bar{j}) + A_y(-B_x\bar{k} + B_z\bar{i}) + A_z(B_x\bar{j} - B_y\bar{i}) \end{aligned}$$

$$\boxed{\bar{A} \times \bar{B} = (A_yB_z - A_zB_y)\bar{i} + (A_zB_x - A_xB_z)\bar{j} + (A_xB_y - A_yB_x)\bar{k}} \tag{8}$$

It is noteworthy that in most problems one or both of the vectors will have some zero components, thereby simplifying the calculations.

An alternative computation of a cross product may be obtained by expanding a determinant. In such a formulation, the first row is the three unit vectors in alphabetical order, the second row is the corresponding components of the first vector, and the third row is the corresponding components of the second vector. For example,

$$\bar{A} \times \bar{B} = \begin{vmatrix} \bar{i} & \bar{j} & \bar{k} \\ A_x & A_y & A_z \\ B_x & B_y & B_z \end{vmatrix} \tag{9}$$

If you are familiar with determinants, you may use eq. (9) to perform your calculations. However, the method of eq. (8) will be used throughout this text.

EXAMPLE 1

Given the two vectors $\bar{A} = 3\bar{i} + 7\bar{j} + 4\bar{k}$ and $\bar{B} = 4\bar{i} + 3\bar{k}$, determine $\bar{A} \times \bar{B}$, $\bar{B} \times \bar{A}$ and the angle between the two vectors.

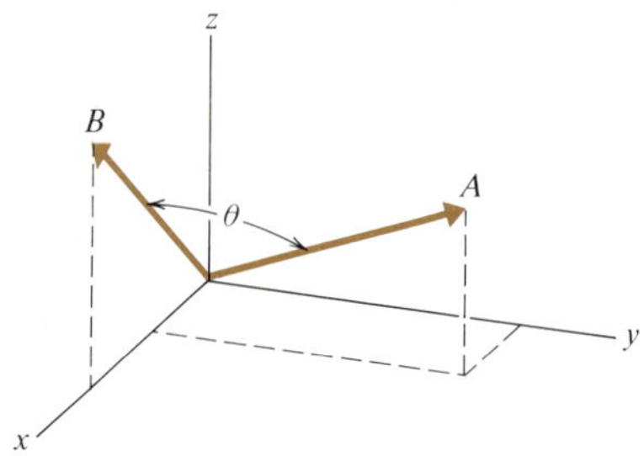

Solution

Following the procedure used to obtain eq. (8) yields

$$\begin{aligned} \bar{A} \times \bar{B} &= (3\bar{i} + 7\bar{j} + 4\bar{k}) \times (4\bar{i} + 3\bar{k}) \\ &= 3\bar{i} \times (4\bar{i} + 3\bar{k}) + 7\bar{j} \times (4\bar{i} + 3\bar{k}) + 4\bar{k} \times (4\bar{i} + 3\bar{k}) \\ &= 3\bar{i} \times 4\bar{i} + 3\bar{i} \times 3\bar{k} + 7\bar{j} \times 4\bar{i} + 7\bar{j} \times 3\bar{k} + 4\bar{k} \times 4\bar{i} + 4\bar{k} \times 3\bar{k} \\ &= -9\bar{j} - 28\bar{k} + 21\bar{i} + 16\bar{j} \\ &= 21\bar{i} + 7\bar{j} - 28\bar{k} \end{aligned}$$

$$\begin{aligned} \bar{B} \times \bar{A} &= (4\bar{i} + 3\bar{k}) \times (3\bar{i} + 7\bar{j} + 4\bar{k}) \\ &= 4\bar{i} \times (3\bar{i} + 7\bar{j} + 4\bar{k}) + 3\bar{k} \times (3\bar{i} + 7\bar{j} + 4\bar{k}) \\ &= 4\bar{i} \times 3\bar{i} + 4\bar{i} \times 7\bar{j} + 4\bar{i} \times 4\bar{k} + 3\bar{k} \times 3\bar{i} + 3\bar{k} \times 7\bar{j} + 3\bar{k} \times 4\bar{k} \\ &= 28\bar{k} - 16\bar{j} + 9\bar{j} - 21\bar{i} \\ &= -21\bar{i} - 7\bar{j} + 28\bar{k} \end{aligned}$$

Note that $\bar{A} \times \bar{B} = -\bar{B} \times \bar{A}$, as required by eq. (4).

In order to determine the angle θ between the two vectors, we use eq. (3), $|\bar{A} \times \bar{B}| = |\bar{A}|\,|\bar{B}| \sin\theta$, or

$$\sin\theta = \frac{|21\bar{i} + 7\bar{j} - 28\bar{k}|}{|3\bar{i} + 7\bar{j} + 4\bar{k}|\,|4\bar{i} + 3\bar{k}|} = 0.8298$$

Therefore

$$\theta = 56.08°$$

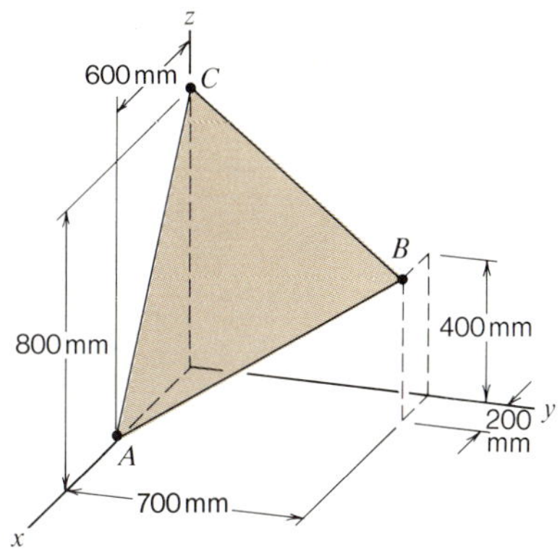

EXAMPLE 2

Three points A, B, and C are located in the sketch. Determine the angle ϕ between the lines AB and AC, the area of triangle ABC, and a unit vector normal to the plane of this triangle.

Solution

For the two vectors $\bar{A}$ and $\bar{B}$ in Figure 5, the component of one of these vectors perpendicular to the other is the altitude of the triangle formed by $\bar{A}$, $\bar{B}$, and the line connecting their heads. Thus, the area of the triangle formed by the vectors in Figure 5 is

$$\text{Area} = \tfrac{1}{2}|\bar{A} \times \bar{B}|$$

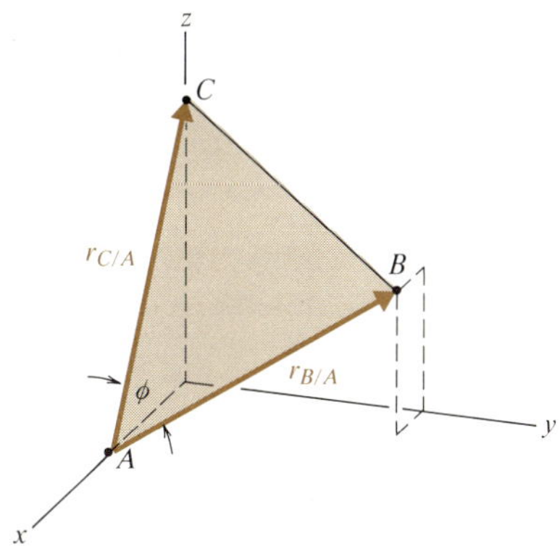

For the present problem, the angle ϕ may be obtained from the same cross product as that used to obtain the area if we employ the position vectors $\bar{r}_{B/A}$ and $\bar{r}_{C/A}$ shown in the sketch. Equation (3) then indicates that

$$\text{Area} = \tfrac{1}{2}|\bar{r}_{B/A} \times \bar{r}_{C/A}| \equiv \tfrac{1}{2}|\bar{r}_{B/A}|\,|\bar{r}_{C/A}| \sin\phi$$

We refer to the given sketch to determine the position vectors. A displacement from point A to point B is equivalent to displacements of 700 mm in the $+y$ direction, 400 mm in the $-x$ direction, and 400 mm in the $+z$ direction, so

$$\bar{r}_{B/A} = 0.70\bar{j} - 0.40\bar{i} + 0.40\bar{k}\ \text{m}$$

For $\bar{r}_{C/A}$ we follow the coordinate axes from point A to point C to find

$$\bar{r}_{C/A} = -0.60\bar{i} + 0.80\bar{k}\ \text{m}$$

The process of evaluating the vector product may be thought of as being similar to that for multiplying two scalar polynomials. Thus

$$\begin{aligned}
\bar{r}_{B/A} \times \bar{r}_{C/A} &= (0.7\bar{j} - 0.4\bar{i} + 0.4\bar{k}) \times (-0.6\bar{i} + 0.8\bar{k}) \\
&= -0.7(0.6)\bar{j} \times \bar{i} + 0.7(0.8)\bar{j} \times \bar{k} + 0.4(0.6)\bar{i} \times \bar{i} - 0.4(0.8)\bar{i} \times \bar{k} \\
&\quad - 0.4(0.6)\bar{k} \times \bar{i} + 0.4(0.8)\bar{k} \times \bar{k} \\
&= 0.42\bar{k} + 0.56\bar{i} + 0.32\bar{j} - 0.24\bar{j} \\
&= 0.56\bar{i} + 0.08\bar{j} + 0.42\bar{k}
\end{aligned}$$

Incidentally, as you become more familiar with the cross product, you will find that you can skip the second of the above computational steps. The area is therefore

$$\begin{aligned}
\text{Area} &= \tfrac{1}{2}|\bar{r}_{B/A} \times \bar{r}_{C/A}| \\
&= \tfrac{1}{2}[(0.56)^2 + (0.08)^2 + (0.42)^2]^{1/2} = 0.3523\ \text{m}^2
\end{aligned}$$

Next, we find $\sin\phi$ by solving the general relation for the area.

$$\sin\phi = \frac{|\bar{r}_{B/A} \times \bar{r}_{C/A}|}{|\bar{r}_{B/A}|\,|\bar{r}_{C/A}|} = \frac{2(\text{Area})}{|\bar{r}_{B/A}|\,|\bar{r}_{C/A}|}$$

The magnitudes of the position vectors are

$$|\bar{r}_{B/A}| = [(0.7)^2 + (0.4)^2 + (0.4)^2]^{1/2} = 0.90 \text{ m}$$
$$|\bar{r}_{C/A}| = [(0.6)^2 + (0.8)^2]^{1/2} = 1.00 \text{ m}$$

from which we find

$$\sin\phi = \frac{2(0.3523)}{0.90(1.00)}, \qquad \phi = 51.52^\circ$$

Finally, we divide $\bar{r}_{B/A} \times \bar{r}_{C/A}$ by its magnitude to obtain a *unit* vector normal to the plane of $\bar{r}_{B/A}$ and $\bar{r}_{C/A}$. This gives

$$\bar{e}_n = \frac{0.56\bar{i} + 0.08\bar{j} + 0.42\bar{k}}{2(0.3523)} = 0.7948\bar{i} + 0.1135\bar{j} + 0.5961\bar{k}$$

As all of the components of $\bar{e}_n$ are positive, we can conclude that $\bar{e}_n$ is directed outward from the origin.

HOMEWORK PROBLEMS

3.1 Compute $\bar{A} \times \bar{B}$ for the following vectors: (a) $\bar{A} = 5\bar{i} - \bar{j} - \bar{k}$, $\bar{B} = -4\bar{i} + 2\bar{j} + 4\bar{k}$, (b) $\bar{A} = 2\bar{i} + \bar{j}$, $\bar{B} = 3\bar{i} - 4\bar{k}$, (c) $\bar{A} = \bar{i} + 2\bar{j} + \bar{k}$, $\bar{B} = -3\bar{i} + 2\bar{j} - \bar{k}$

3.2 Given that $\bar{A} = \bar{i} + \bar{j} - \bar{k}$, $\bar{B} = \bar{i} - 2\bar{j} + 3\bar{k}$, and $\bar{C} = -\bar{i} + 2\bar{j} - 4\bar{k}$, compute (a) $(\bar{A} \times \bar{B}) \times \bar{C}$, (b) $\bar{A} \times (\bar{B} \times \bar{C})$.

3.3 For the vectors given in Problem 3.2, compute (a) $\bar{A} \cdot \bar{B} \times \bar{C}$, (b) $\bar{C} \cdot \bar{A} \times \bar{B}$.

3.4 Given that $\bar{A} = A_x\bar{i} - 5\bar{j} + 2\bar{k}$, $\bar{B} = -3\bar{i} + 2\bar{j} - B_z\bar{k}$, determine the values of A_x and B_z for which $\bar{A} \times \bar{B}$ is parallel (a) to the y axis, (b) to the x axis.

3.5 The product $\bar{A} \cdot \bar{B} \times \bar{C}$ of three arbitrary vectors $\bar{A}$, $\bar{B}$, and $\bar{C}$ is called the scalar triple product. Prove the following identities: (a) $\bar{A} \cdot \bar{B} \times \bar{C} \equiv \bar{A} \times \bar{B} \cdot \bar{C}$, (b) $\bar{A} \cdot \bar{B} \times \bar{C} \equiv \bar{B} \cdot \bar{C} \times \bar{A}$.

3.6 and 3.7 Determine the angle ϕ and the unit normal vector having a positive x component for triangle ABC.

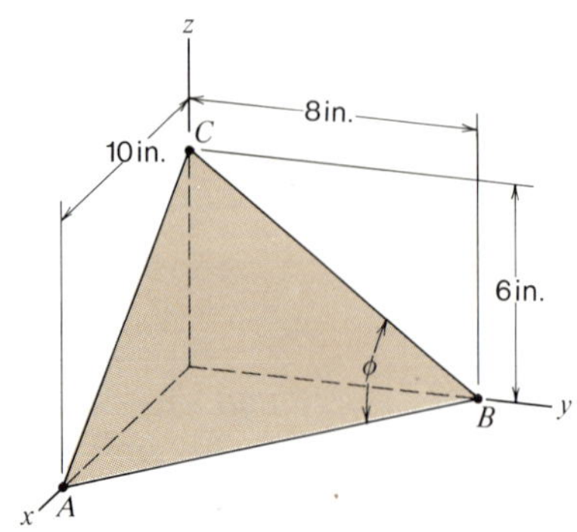

PROB. 3.6

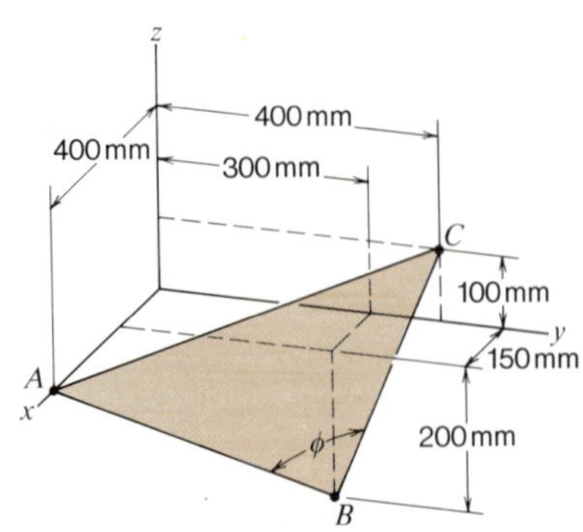

PROB. 3.7

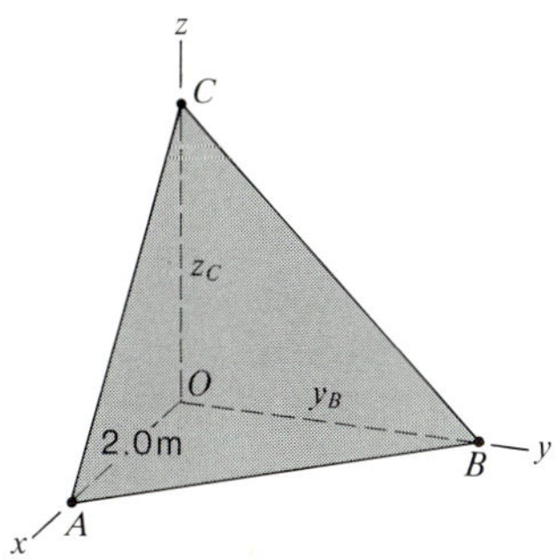

PROB. 3.9

*3.8 Forces $\bar{F}_1$ and $\bar{F}_2$ are both applied at point A. The force $\bar{F}_1 = 500\bar{i} + 400\bar{j} + z\bar{k}$ N is to form a 60° angle with force $\bar{F}_2 = -200\bar{i} + 100\bar{j} + 300\bar{k}$ N. Determine (a) the value of z, (b) the unit vector normal to the plane formed by $\bar{F}_1$ and $\bar{F}_2$.

*3.9 It is desired to pass a plane through point A whose unit normal vector is $0.4286\bar{i} - 0.8571\bar{j} + 0.2857\bar{k}$. Determine the values of y_B and z_C, thus locating where the plane intersects the corresponding coordinate axes.

*3.10 Line AB is perpendicular to line AC. Determine the value of x_B and the unit normal vector having a positive x component.

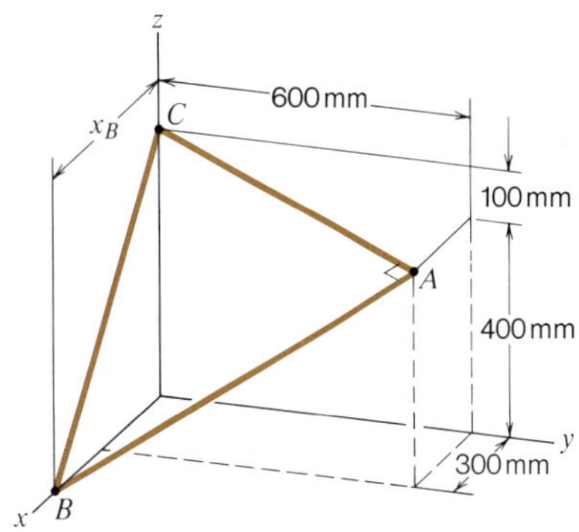

PROB. 3.10

C. COMPUTATION OF MOMENTS

1. Moment About a Point

Figure 8 depicts a general situation where a force $\bar{F}$ is applied along a designated line of action. We wish to determine the moment of this force about point C, which has been chosen arbitrarily.

The lever arm r is found by constructing the perpendicular projection from point C to the line of action of $\bar{F}$. By definition, the moment $\bar{M}_C$ in Figure 8 is $|\bar{F}|r$ in the direction that is normal to the plane containing point C and the force $\bar{F}$. The sense of $\bar{M}_C$ is shown in accord with the tendency of $\bar{F}$ to rotate the body about point C.

Figure 8 indicates that the length r is the component perpendicular to $\bar{F}$ of the position vector $\bar{r}_{P/C}$ from point C to *any* point P on the line of action of $\bar{F}$. It follows from the definition of a cross product, eq. (3), that the cross product of $\bar{F}$ and $\bar{r}_{P/C}$ has the magnitude of $\bar{M}_C$. Also, this cross product is perpendicular to the plane, as is $\bar{M}_C$. The proper order of multiplication of $\bar{F}$ and $\bar{r}_{P/C}$ is determined by picturing $\bar{F}$ to have its tail at point C. We then see that if the thumb of our right hand is oriented parallel to $\bar{M}_C$ in Figure 8, the fingers curl from $\bar{r}_{P/C}$ to $\bar{F}$. Thus

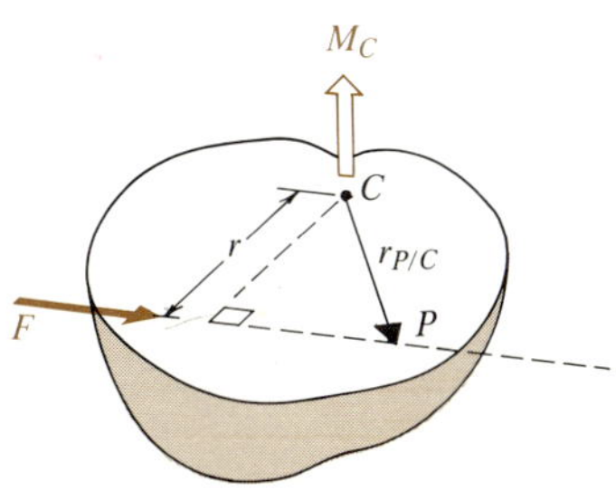

FIGURE 8

$$\bar{M}_C = \bar{r}_{P/C} \times \bar{F} \tag{10}$$

Equation (10) says that

> The moment of a force about a point C is obtained by crossing a position vector from point C to any point on the line of action of the force into the force vector.

This result gives us a rule for evaluating the moment of a force about a point in terms of vectors that are described in terms of their components. The methods for describing position and force vectors in terms of their components were developed in the previous chapter. Thus, the only new feature of the moment computation in this approach is the cross product.

Before we proceed to solve some examples, let us consider the planar case, in which the force and the reference point for the moment are both coincident with one of the coordinate planes. This is the situation in Figure 9, where we would like to compute the moment of force $\bar{F}$ about point C. Although the cross product specified in eq. (10) could be employed for this task, there are methods for planar problems such as this that are more direct. We shall describe three such methods here.

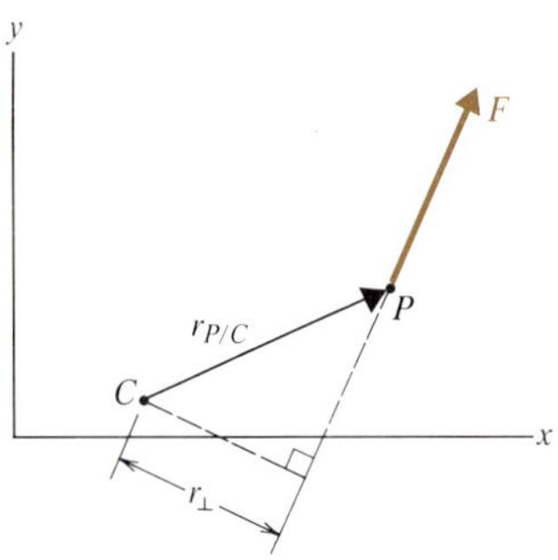

FIGURE 9

In Figure 9 the z axis is perpendicular to the plane, and $\bar{F}$ would tend to rotate a body that is pivoted at point C counterclockwise. One method for computing the moment of $\bar{F}$ about point C is based on the definition of a moment. Accordingly, this method requires that we compute the lever arm $r_\perp$ of $\bar{F}$ about point C. The magnitude of the moment is then $r_\perp|\bar{F}|$. It follows that the moment about point C is in the positive z direction, being given by

$$\bar{M}_C = r_\perp F\bar{k} \tag{11}$$

Note that in any planar situation, the moment is perpendicular to the plane in the sense of the right-hand rule for the rotational tendency of the force.

Another approach for evaluating the moment is often referred to as *Varignon's theorem*. Varignon was the mathematician who first demonstrated that a force may be replaced by its components in the evaluation of a moment. For us, this concept is merely a corollary of the algebraic rules for the cross product.

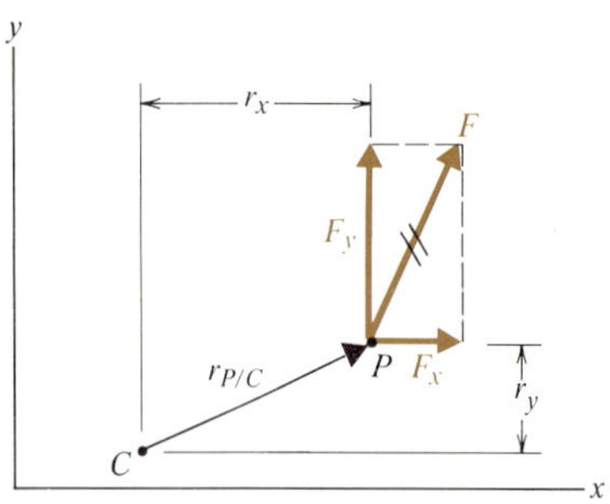

FIGURE 10

The basic quantities for this method are depicted in Figure 10. The lengths r_x and r_y are the distances in the x and y directions, respectively, between points C and P. Clearly, these are also the lever arms of the force components $F_y\bar{j}$ and $F_x\bar{i}$, respectively, about point C. The moment of each force component is parallel to the z axis for this planar situation, and we see that $F_y\bar{j}$ tends to cause a counterclockwise rotation about point C, whereas $F_x\bar{i}$ tends to cause a clockwise rotation. The total moment is the combined effect of the two components, so

$$\bar{M}_C = r_x F_y\bar{k} - r_y F_x\bar{k} \tag{12}$$

Even though this formula is different from the previous one, eq. (11), it gives the same value. An interesting aspect of this method is that it is obtained by replacing $\bar{r}_{P/C}$ and $\bar{F}$ by their components, as is obvious from Figure 10.

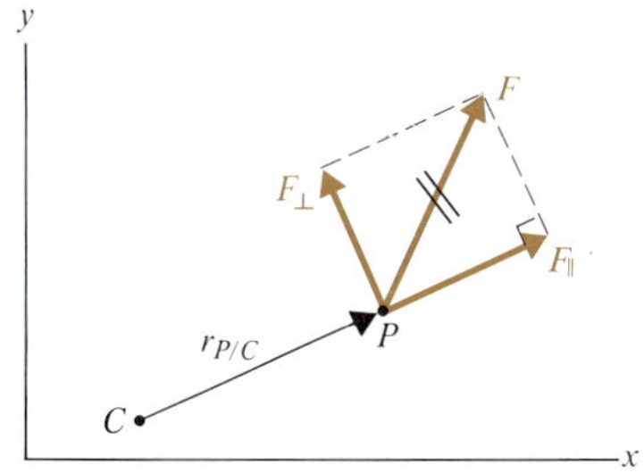

FIGURE 11

The same mathematical representation is employed in the cross product formulation. Thus, this approach uses the same information as the cross product in eq. (10).

The last method we shall present for evaluating moments in a planar case also replaces the force by its components. These components are now described as being parallel or perpendicular to the position vector, which is the situation in Figure 11.

Because the line of action of the component $F_{||}$ intersects point C, it exerts no moment about that point. On the other hand, the lever arm of the component $F_{\perp}$ is $|\bar{r}_{P/C}|$, and it tends to cause a counterclockwise rotation about point C. Thus

$$\bar{M}_C = |\bar{r}_{P/C}|F_{\perp}\bar{k} \tag{13}$$

In summary, there are three alternatives to the cross product, eq. (10), when the force and the point about which the moment is sought are situated in the same coordinate plane. In each case, lever arms are constructed and the signs of the moments are obtained by examining the rotational tendency. The direction of the moment may be described in terms of the unit vector perpendicular to the plane, or in words describing the overall rotational tendency.

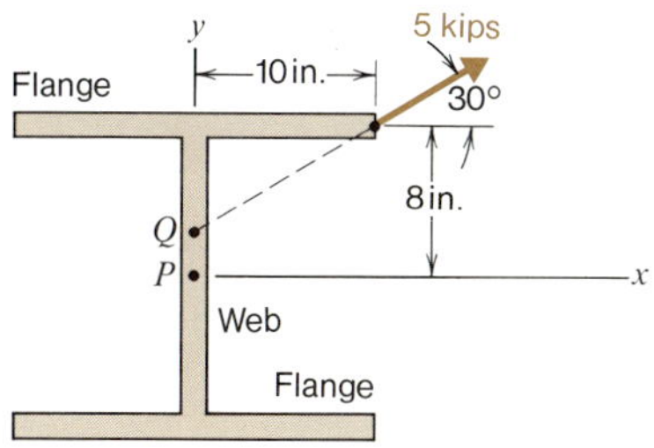

EXAMPLE 3

A 5 kip force is applied to the flange of the H beam as shown. Determine the moment of this force about the center C of the cross section by

(a) evaluating the lever arm of the force,
(b) considering horizontal and vertical force components acting at the flange point P,
(c) considering force components parallel and perpendicular to line PC acting at the flange point P,
(d) considering horizontal and vertical force components acting at the web point Q.

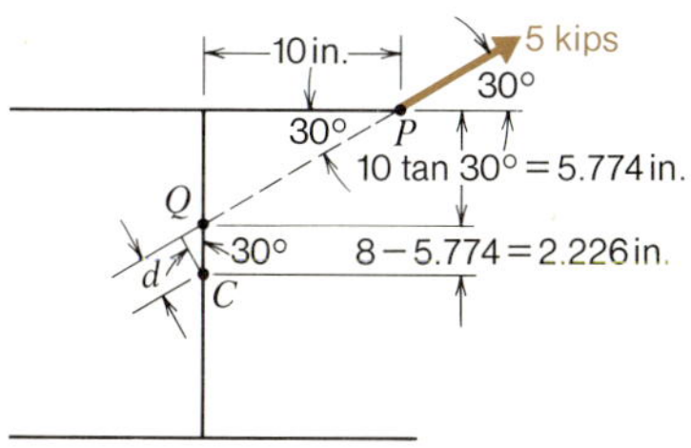

Solution

Method a

We first draw a sketch to evaluate the lever arm d. From this sketch

$$d = (2.226) \cos 30^\circ = 1.928 \text{ in.}$$

The force tends to rotate the beam clockwise about point C, so the moment may be described as

$$M_C = |\bar{F}|d = 5(1.928) = 9.640 \text{ kip-in. clockwise}$$

The equivalent vector form of this result is

$$\bar{M}_C = -9.64\bar{k} \text{ kip-in.}$$

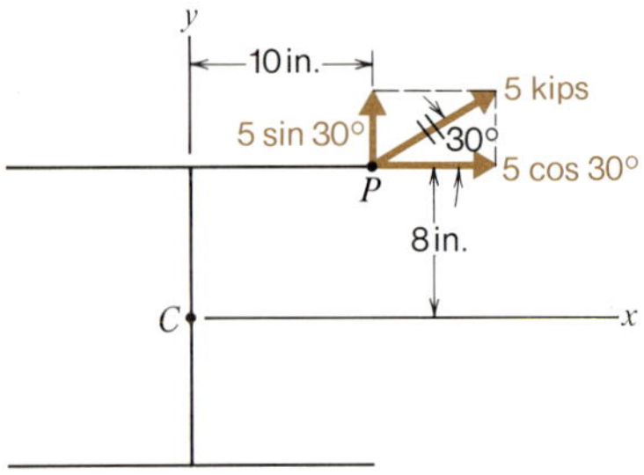

Method b

We begin by drawing a sketch showing the desired force components, as well as the given horizontal and vertical distances from point C to point P. Because the z axis is outward, counterclockwise moments are positive according to the right-hand rule. The total moment is found by multiplying each component by its lever arm and then adding the individual terms, consistent with the sign for each term. Thus

$$\bar{M}_C = 10(5 \sin 30^\circ)\bar{k} - 8(5 \cos 30^\circ)\bar{k} = -9.64\bar{k} \text{ kip-in.}$$

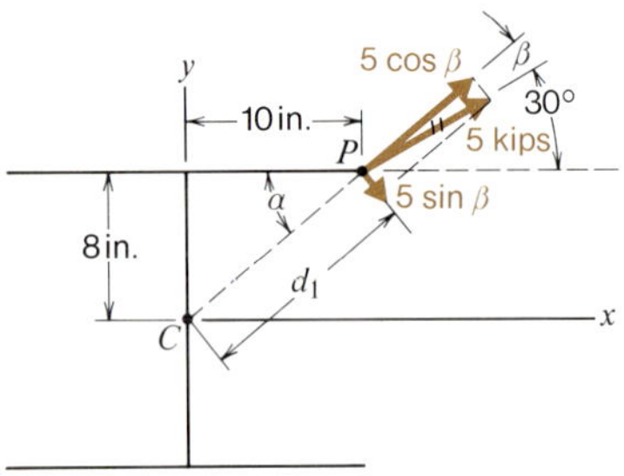

Method c

We begin this solution by drawing a sketch of the required force components and their lever arms. Only the force component perpendicular to line CP exerts a moment about point C. The angle β and the lever arm d_1 are computed as follows.

$$\beta + 30^\circ = \alpha \equiv \tan^{-1}\frac{8}{10} = 38.66^\circ$$

$$\beta = 8.66^\circ$$

$$d_1 = [8^2 + (10)^2]^{1/2} = 12.806 \text{ in.}$$

In accord with the right-hand rule associated with the z axis, counterclockwise moments are positive, so

$$\bar{M}_C = -d_1(5 \sin \beta)\bar{k} = -9.64\bar{k} \text{ kip-in.}$$

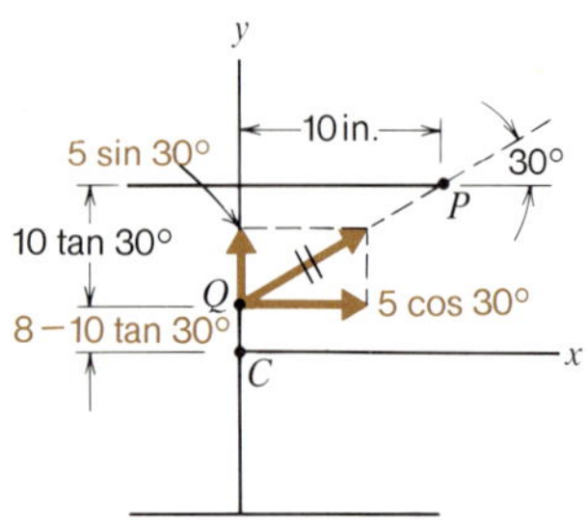

Method d

This approach is based on the principle of transmissibility, according to which a force may be applied anywhere along its line of action. The force components are the same as in Method b. The location of point Q is given in Method a. From the sketch we see that the vertical component of the force passes through point C, and that the horizontal component exerts a clockwise (negative) moment. Thus

$$\begin{aligned}\bar{M}_C &= -(8 - 10 \tan 30^\circ)(5 \cos 30^\circ)\bar{k} \\ &= -9.64\bar{k} \text{ kip-in.}\end{aligned}$$

Additional Remarks

The foregoing example describes four methods for evaluating a moment in a planar situation. The second method, which employs the force components and their lever arms, seems to be the most direct from the viewpoint of the geometrical computations required. This is a result of choosing the force components in harmony with the directions in which the dimensions are given. In other planar problems the points and forces may be specified in a different manner, in which case one of the other methods may be preferable.

EXAMPLE 4

A 10 kN force is applied as shown to the free end, point D, of the bent rod. Compute the moment of this force about point B.

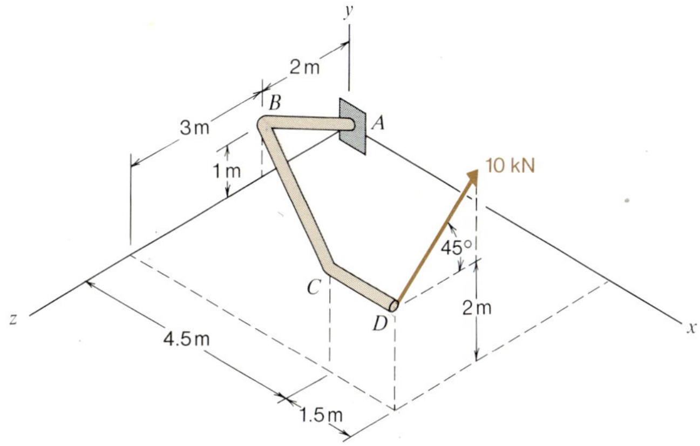

Solution

This is a three-dimensional problem. Therefore, we compute the moment about point B by constructing the position vector from this point to point D on the line of action of the force and then applying eq. (10). The diagram indicates that the force is parallel to the yz plane at an angle of 45° above the negative z direction. Thus

$$\bar{r}_{D/B} = 3\bar{k} + 6\bar{i} + \bar{j} \text{ m}$$

$$\bar{F} = 10(-\cos 45^\circ \bar{k} + \sin 45^\circ \bar{j}) = 7.071(\bar{j} - \bar{k}) \text{ kN}$$

$$\begin{aligned}\bar{M}_A &= \bar{r}_{D/B} \times \bar{F} = (3\bar{k} + 6\bar{i} + \bar{j}) \times 7.071(\bar{j} - \bar{k}) \\ &= 3(7.071)\bar{k} \times \bar{j} + 6(7.071)\bar{i} \times \bar{j} - 6(7.071)\bar{i} \times \bar{k} - (7.071)\bar{j} \times \bar{k} \\ &= -28.3\bar{i} + 42.4\bar{j} + 42.4\bar{k} \text{ kN-m}\end{aligned}$$

HOMEWORK PROBLEMS

3.11 A crate resting on the ground is acted upon by the 120 N force shown. Determine the moment of this force about corner A (a) by determining the lever arm of the force, (b) by replacing the force by its components acting at point B, (c) by replacing the force by its components acting at some other convenient point on its line of action.

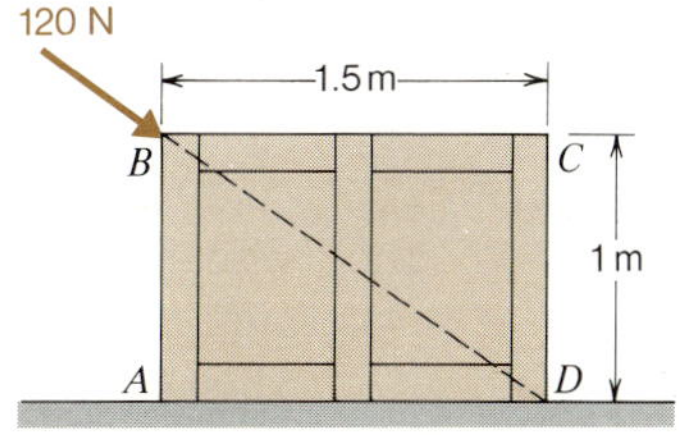

PROB. 3.11

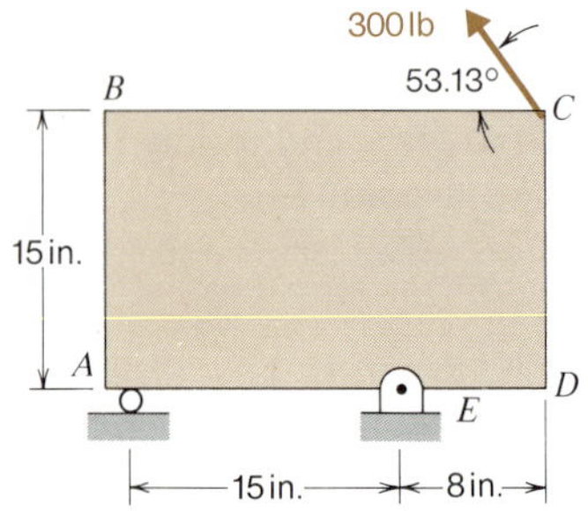

PROB. 3.12

3.12 A 300 lb force is applied to the rectangular plate at corner C. Determine the moment of this force about pin E (a) by computing its lever arm, (b) by replacing the force by its components acting at point C, (c) by replacing the force by its components acting at some other convenient point on its line of action.

3.13 Determine the moment of the 60 N force about the center O of the disk by (a) determining the lever arm of the force, (b) replacing the force by a convenient set of components.

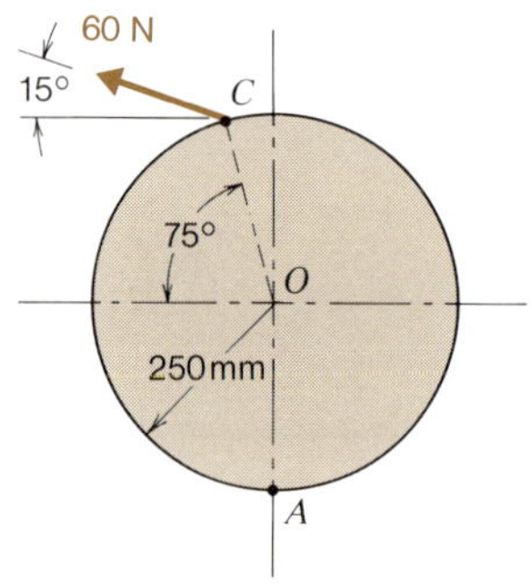

PROBS. 3.13 AND 3.14

3.14 Determine the moment of the 60 N force about point A on the perimeter of the disk by (a) determining the lever arm of the force, (b) replacing the force by a convenient set of components.

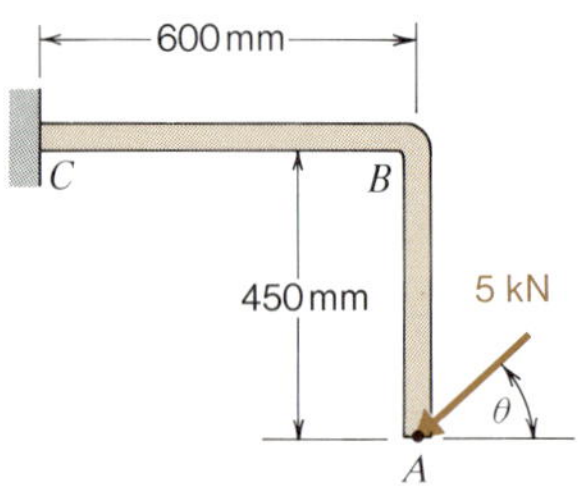

PROB. 3.15

3.15 A 5 kN force is applied to the bent arm at point A. Determine the moment of the force about point C (a) for $\theta = 0°$, (b) for $\theta = 30°$.

3.16 Control bar AB is inclined at 50° from the vertical. A 10 lb force is applied to the bar at end B. Determine the moment of this force about pin A (a) if the force is horizontal, (b) if $\phi = 120°$.

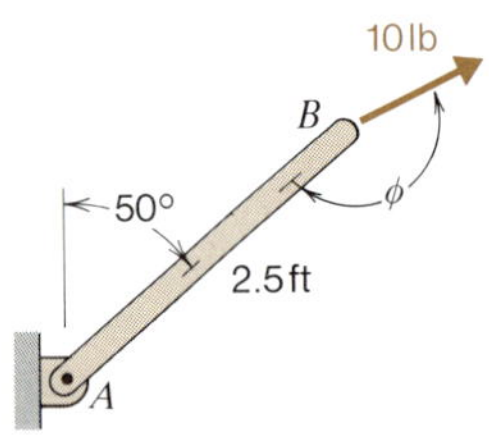

PROB. 3.16

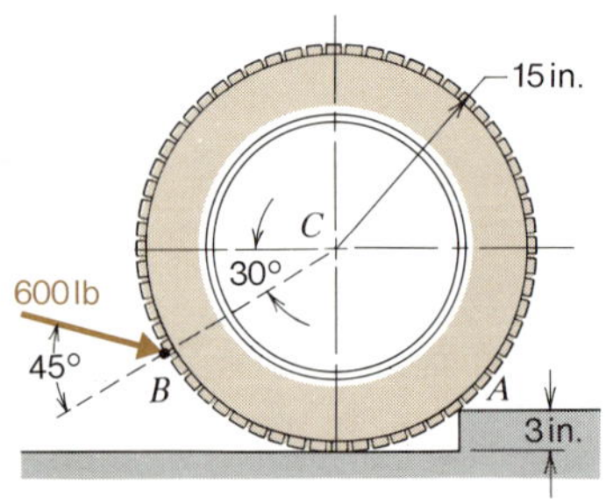

PROBS. 3.17 AND 3.18

3.17 A 600 lb force is applied to the wheel as shown. Determine the moment of this force about the center C.

3.18 A 600 lb force is applied to the wheel as shown. Determine the moment of this force about corner A on the curb.

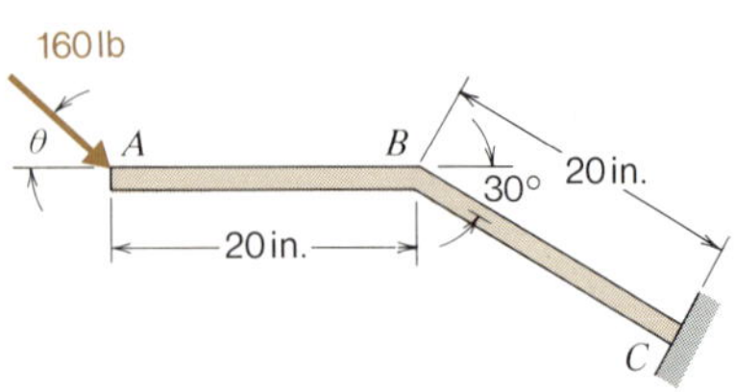

PROBS. 3.19 AND 3.20

3.19 Use two different methods to evaluate the moment of the 160 lb force about end C of the bent rod ABC when $\theta = 30°$.

3.20 Determine the moment of the 160 lb force about end C of bent rod ABC when $\theta = -30°$.

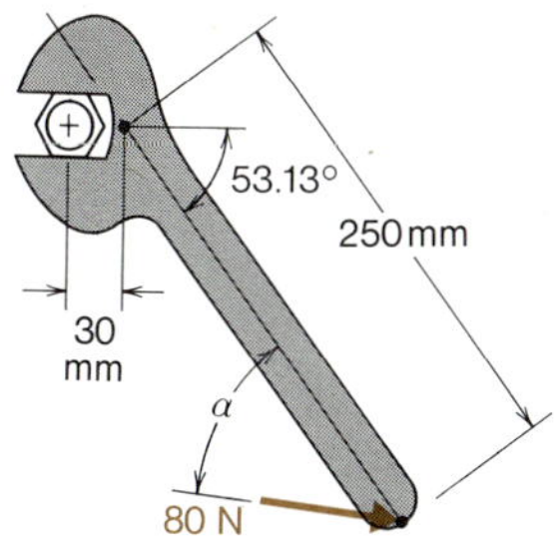

PROB. 3.21

3.21 An 80 N force is applied to the special purpose wrench as shown. Determine the moment of this force about the center of the bolt when $\alpha = 75°$.

3.22 For the system in Problem 3.15, determine the angle θ such that the 5 kN force applied at point A exerts the maximum moment about point C. What is the corresponding moment?

3.23 For the system in Problem 3.19, determine the angle θ for which the 160 lb force exerts the maximum moment about point C. What is the corresponding moment?

3.24 Determine the moment of the 2 kN force about (a) corner A, (b) corner B.

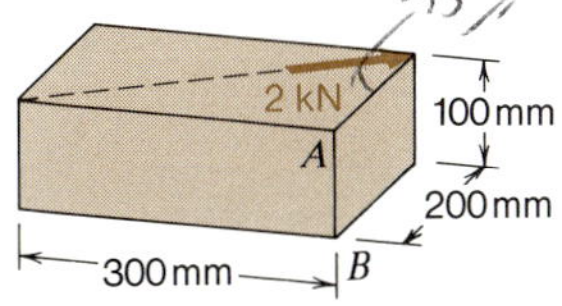

PROB. 3.24

3.25 A traffic light is steadied by cable AB. If the tension in this cable is 4 kN, determine (a) the moment about point D of the force this cable exerts on the pole, (b) the moment of this force about the base C.

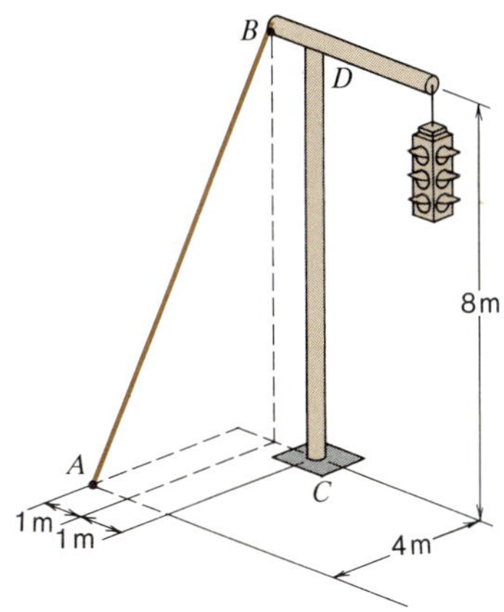

PROB. 3.25

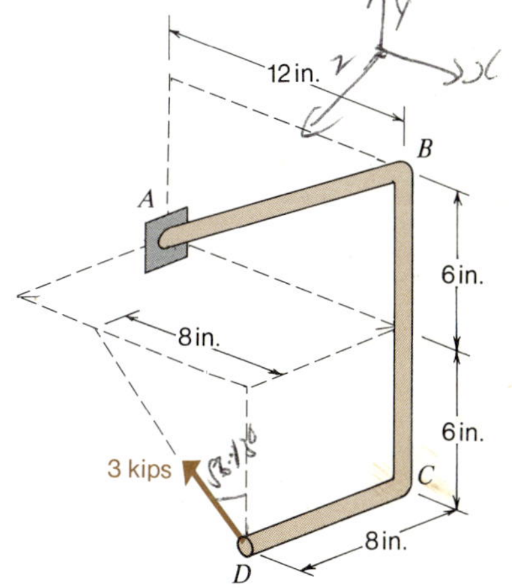

PROBS. 3.26 AND 3.27

3.26 Determine the moment of the 3 kip force about corner B of the bent bar.

***3.27** Determine the moment of the 3 kip force about end A of the bent bar.

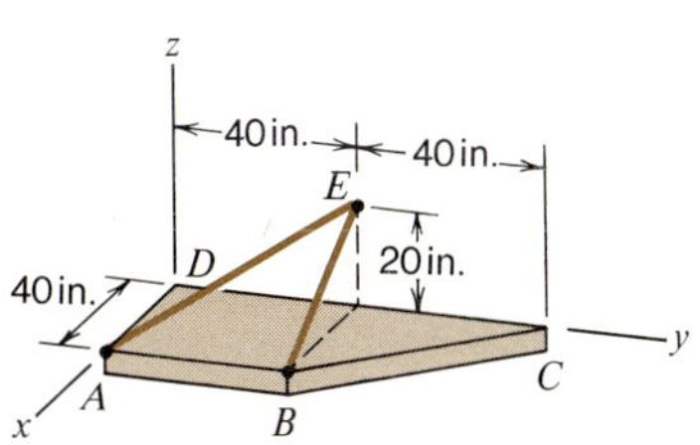

PROBS. 3.28 AND 3.29

3.28 The tensions in cables AE and BE are 400 lb and 500 lb, respectively. Determine the moment about corner D of the force each cable exerts on the trapezoidal plate.

3.29 The tensions in cables AE and BE are 400 lb and 500 lb, respectively. Determine the moment of the force each cable exerts on the trapezoidal plate about corner C.

3.30 A 1.5 kN force is applied to the pipe assembly as shown. Determine the moment of this force about (a) joint C, (b) end A.

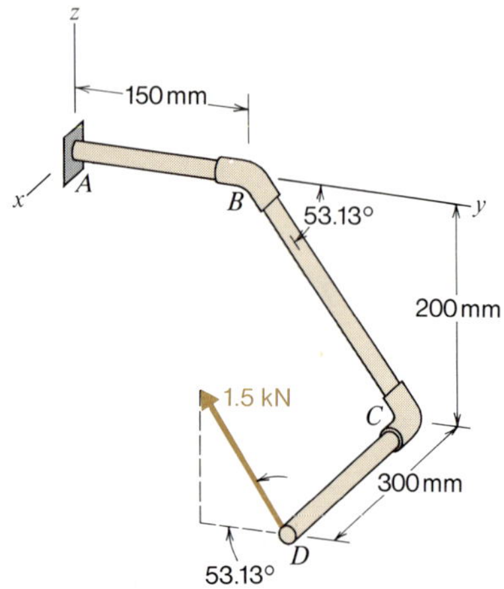

PROB. 3.30

3.31 Cable BC has a tension of 2 tons. Determine the moment about corner A of the force exerted by the cable on the box.

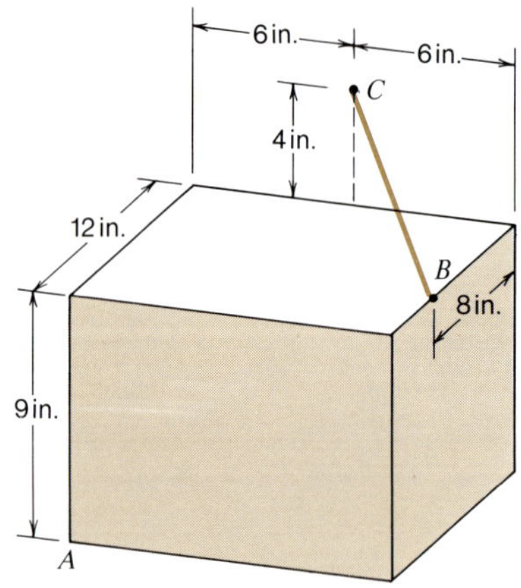

PROB. 3.31

***3.32** The rectangular sheet of plywood is held in position by the cable. The tension in the cable is 150 N. Determine the moment of the force exerted by the cable on the plywood about corner A.

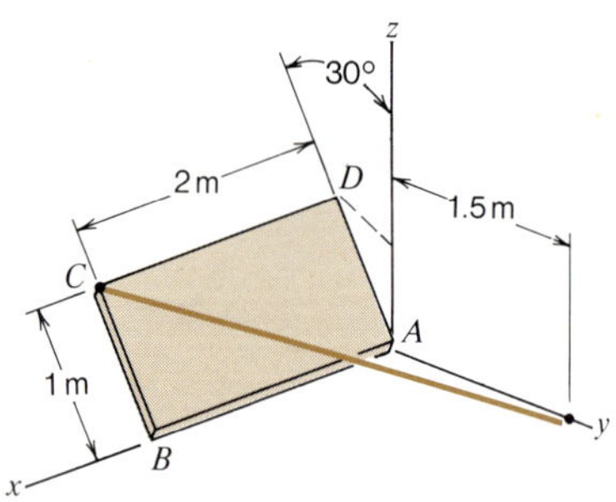

PROB. 3.32

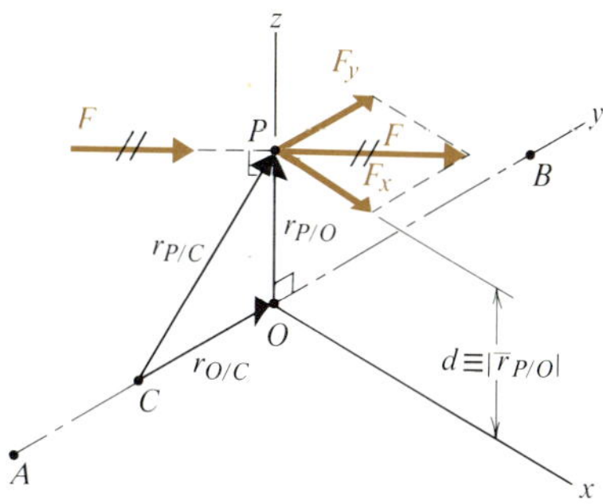

FIGURE 12

2. Moment About a Line

The moment computation in terms of a cross product, which is given by eq. (10), can be utilized to determine the moment of a force about a line. Consider the situation shown in Figure 12, where it is desired to determine the moment of the force $\bar{F}$ about the axis AB.

The first step is to construct the line OP, which is the line that perpendicularly intersects the line of action of $\bar{F}$ and the axis AB. By definition, the length OP is the lever arm of $\bar{F}$ about the axis AB. Once line OP is known, a convenient coordinate system may be defined by aligning the y axis with line AB and the z axis with line OP, as shown in Figure 12.

Because a force may be applied anywhere along its line of action, the force $\bar{F}$ in Figure 12 is replaced by its component forces $F_x\bar{i}$ and $F_y\bar{j}$ acting at point P. The component $F_y\bar{j}$ is parallel to line AB, so it exerts no moment about that line. The component $F_x\bar{i}$ is perpendicular to both lines AB and OP, so the right-hand rule and eq. (1) for the moment of a force about a line yield

$$\bar{M}_{AB} = F_x d\bar{j}$$

Now let us consider the moment of the force $\bar{F}$ about any point C located on line AB. Equation (10) states that

$$\bar{M}_C = \bar{r}_{P/C} \times \bar{F}$$

The position $\bar{r}_{P/C}$ may be written in component form by referring to Figure 12, which indicates that

$$\bar{r}_{P/C} = |\bar{r}_{O/C}|\bar{j} + d\bar{k}$$

The moment $\bar{M}_C$ is therefore

$$\begin{aligned}\bar{M}_C &= (|\bar{r}_{O/C}|\bar{j} + d\bar{k}) \times (F_x\bar{i} + F_y\bar{j}) \\ &= -|\bar{r}_{O/C}|F_x\bar{k} + d(F_x)\bar{j} - d(F_y)\bar{i}\end{aligned}$$

Notice that the y component of $\bar{M}_C$ is identical to $\bar{M}_{AB}$. Therefore, if $\bar{M}_C$ is known, then M_{AB} may be obtained by using the dot product with $\bar{j}$ to find the y component of $\bar{M}_C$. From this, M_{AB} may be written as a vector by multiplying the dot product by the unit vector $\bar{j}$. Thus we have the magnitude

$$M_{AB} = \bar{M}_C \cdot \bar{j}$$

and the vector moment

$$\bar{M}_{AB} = (\bar{M}_C \cdot \bar{j})\bar{j}$$

The reference frame for this derivation was intentionally chosen such that the y axis was coincident with the line of interest, AB. The relationship between the two types of moments in a situation where the line AB has an arbitrary orientation is found by replacing $\bar{j}$ by the unit vector $\bar{e}_{B/A}$ for the linc. Thus

$$\boxed{\bar{M}_{AB} = (\bar{M}_C \cdot \bar{e}_{B/A})\bar{e}_{B/A}} \tag{14}$$

This equation states that

> The magnitude of the moment of a force about a line is the component parallel to that line of the moment of the force about any point on that line. The direction of the moment is parallel to the line.

This theorem leads to several interesting observations. First, note that in all planar situations, the moment about a point is perpendicular to the plane. In this case the moment about a point is synonymous with the moment about the line that intersects the point and is perpendicular to the plane. For this reason, the moment about a point in a planar problem is often denoted as $\pm M_{Cz}$, with the sign of the value used to express the sense of the moment.

A related observation pertains to three-dimensional systems. Due to the relationship between the two types of moments, it follows that the components of the moment of a force about a point are the moments about the lines that are parallel to the respective coordinate axes and intersect the point. From this, we could develop another version of Varignon's theorem. In it, the moment about a point is obtained by considering the three planar situations that arise from looking down each coordinate axis. The difficulty with this approach is that the spatial perception question is often challenging. This frequently leads to errors in which the effects of some force components are overlooked. By contrast, the evaluation of moments by means of the vector component operations in eqs. (10) and (14) is a straightforward computation.

EXAMPLE 5

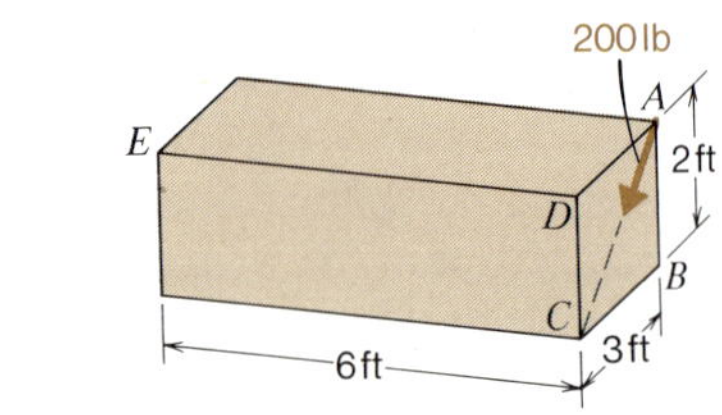

The 200 lb force is applied to the rectangular crate. Determine the moment of this force about diagonal BE.

Solution

We will need an xyz coordinate system in order to write the several vectors in component form. Axes parallel to each of the edges are chosen, in accord with the way the dimensions of the crate are specified.

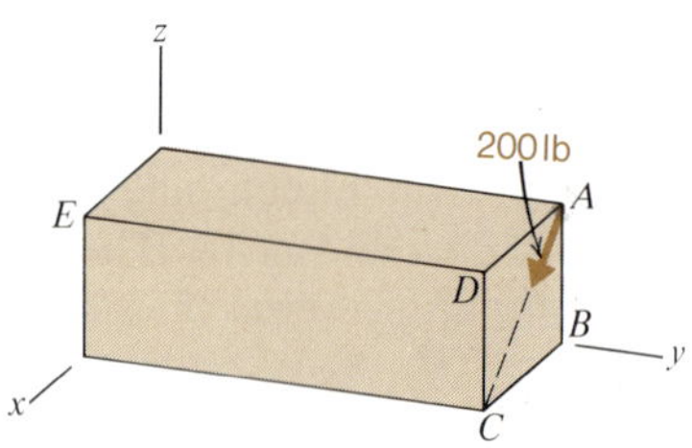

The force and point B coincide with a side of the box that is parallel to the xz plane. Thus, point B is a convenient choice for the computation of the moment of the force about a point on line BE. We therefore compute

$$\bar{r}_{A/B} = 2\bar{k} \text{ ft}$$

$$\bar{F} = 200\frac{\bar{r}_{C/A}}{|\bar{r}_{C/A}|} = 200\frac{3\bar{i} - 2\bar{k}}{[(3)^2 + (2)^2]^{1/2}} = 166.40\bar{i} - 110.94\bar{k} \text{ lb}$$

$$\bar{M}_B = \bar{r}_{A/B} \times \bar{F} = 332.8\bar{j} \text{ lb-ft}$$

The evaluation of the moment about line BE requires that we determine the unit vector for this line (in either sense). Thus

$$\bar{e}_{B/E} = \frac{\bar{r}_{B/E}}{|\bar{r}_{B/E}|} = \frac{-3\bar{i} + 6\bar{j} - 2\bar{k}}{[(3)^2 + (6)^2 + (2)^2]^{1/2}}$$
$$= -0.4286\bar{i} + 0.8571\bar{j} - 0.2857\bar{k}$$

The moment of the force about line BE may now be calculated.

$$\bar{M}_{BE} = (\bar{M}_B \cdot \bar{e}_{B/E})\bar{e}_{B/E} = (332.8)(0.8571)\bar{e}_{B/E}$$
$$= -122.2\bar{i} + 244.5\bar{j} - 81.5\bar{k} \text{ lb-ft}$$

HOMEWORK PROBLEMS

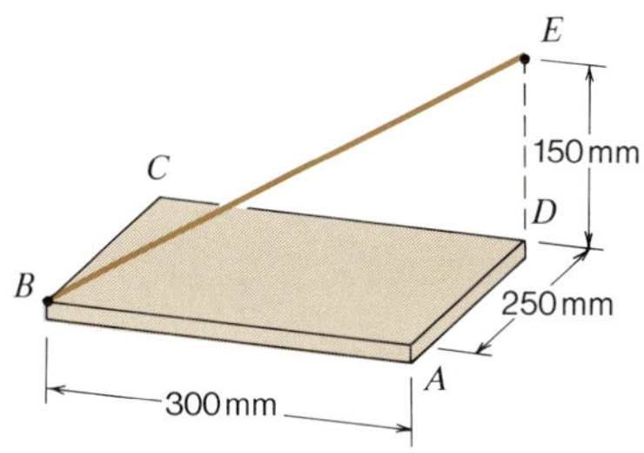

PROBS. 3.34 AND 3.35

3.33 For the system in Problem 3.25, determine the moment about the shaft CD of the force exerted by the cable on the pole. The tension in the cable is 4 kN.

3.34 The tension in cable BE is 100 N. Determine the moment about edge CD of the force exerted by the cable on the rectangular plate.

3.35 The tension in cable BE is 100 N. Determine the moment about diagonal AC of the force exerted by the cable on the rectangular plate.

3.36 A 20 lb force is applied at end A of the bent bar. Determine the moment of this force about segment CD.

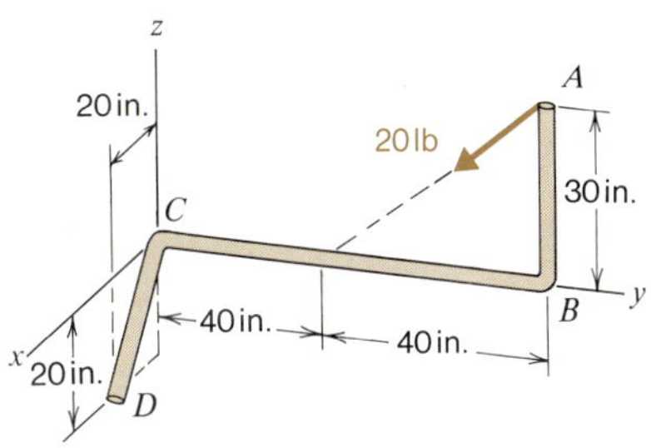

PROB. 3.36

3.37 A 50 lb force is applied to end D of the piping system. (a) Determine the moment of this force about pipe AB. (b) Briefly explain why the moments of this force about segments BC and CD are zero.

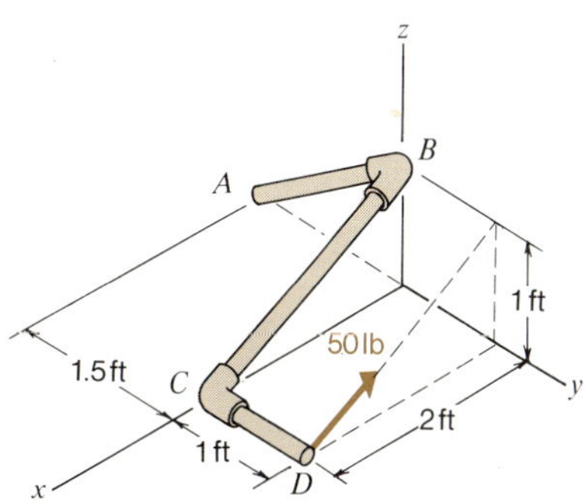

PROB. 3.37

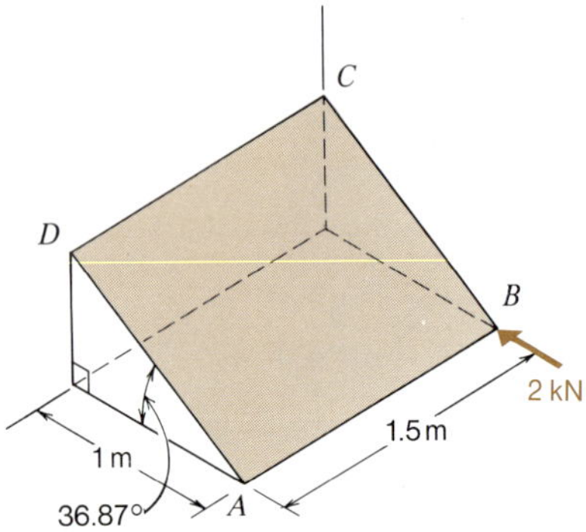

PROBS. 3.38 AND 3.39

3.38 Determine the moment about edge AD of the 2 kN force acting on the right triangular prism.

3.39 Determine the moment about diagonal AC of the 2 kN force acting on the right triangular prism.

3.40 For the system in Problem 3.26, determine the moment of the 3 kip force about bar segment AB.

3.41 Cable FG carries a tension force of 20 kN. Determine the moment about diagonal DE of the force the cable exerts on the box.

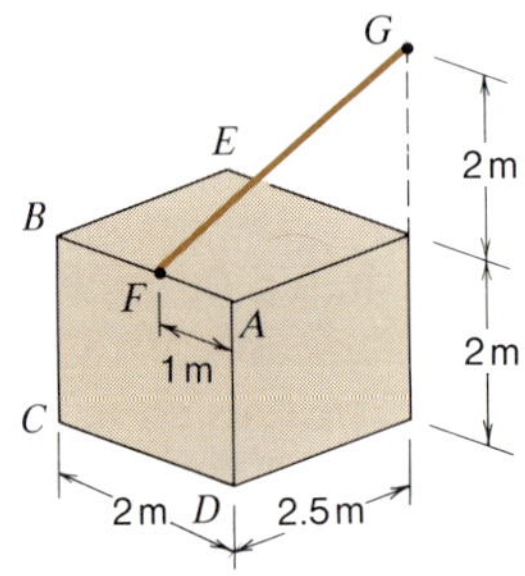

PROB. 3.41

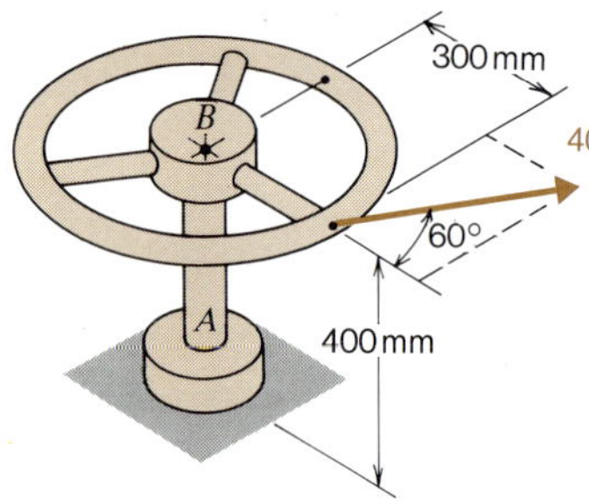

PROB. 3.42

***3.42** A horizontal 40 N force is applied to the valve wheel as shown. Determine the tendency of this force to cause the valve wheel to rotate about point A and about axis AB. Then use these results and the moment definitions to determine the lever arm of the force about point A and about axis AB.

***3.43** For the system in Problem 3.32, determine the moment of the force exerted by the cable on the plywood sheet about edge AD. The tension in the cable is 150 N.

D. FORCE COUPLES

We have developed procedures for characterizing the magnitude, direction, and line of action of a force. The sequence of topics that follow will use this knowledge as a basis for describing the total effect of a system of forces on a rigid body.

The most fundamental system of forces is formed by two forces that are equal in magnitude and opposite in direction. Such a system is said to form a *couple*. Figure 13 depicts a couple formed by a force $\bar{F}$ applied at point A and a force $-\bar{F}$ applied at point B. For convenience the diagram is drawn in the plane of the forces.

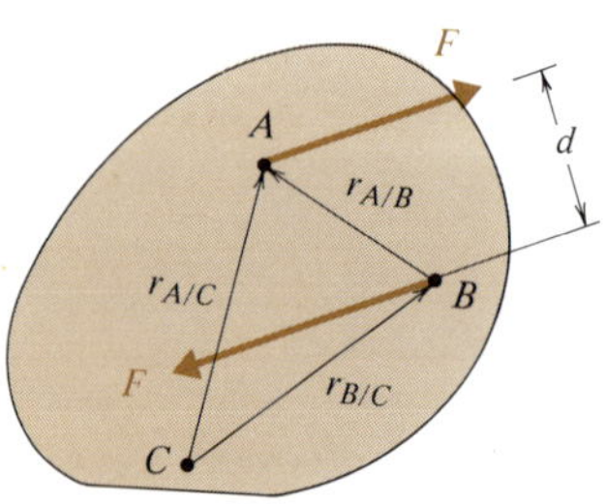

FIGURE 13

If the forces in this figure were collinear, we know from our study of concurrent force systems that $\bar{F}$ and $-\bar{F}$ would have a zero resultant, thereby canceling each other. In the illustrated situation the forces still do not push or pull on the body; they only exert a twisting effect. Let us compute the total moment these two forces exert about an arbitrary point C, which need not necessarily be coplanar with the forces. The position vectors shown in Figure 13 from point C to the forces yield

$$\bar{M}_C = \bar{r}_{A/C} \times \bar{F} + \bar{r}_{B/C} \times (-\bar{F}) = (\bar{r}_{A/C} - \bar{r}_{B/C}) \times \bar{F}$$

This expression may be simplified by applying the triangle law for the difference of two vectors, according to which Figure 13 indicates that

$$\bar{r}_{A/B} = \bar{r}_{A/C} - \bar{r}_{B/C}$$

The moment $\bar{M}_C$ is therefore

$$\boxed{\bar{M}_C = \bar{r}_{A/B} \times \bar{F}} \tag{15}$$

This result has several noteworthy features. First, we see that only the relative position of the points of application of the couple, as expressed by $\bar{r}_{A/B}$, affects the resultant moment. Thus, regardless of which point is selected for summing moments, a couple exerts the same moment. We call this resultant moment the *torque of the couple*, and denote it by $\bar{M}$. Furthermore, the component of $\bar{r}_{A/B}$ perpendicular to $\bar{F}$ is shown in Figure 13 to be the perpendicular distance d between the lines of action of the forces forming the couple. It follows that the magnitude of the torque is given by

$$\boxed{|\bar{M}| = |\bar{F}|\, d} \tag{16}$$

Finally, the direction of the torque is perpendicular to the plane formed by the forces, in the sense obtained from the right-hand rule when the fingers are curled in the sense of the combined rotational tendency of the forces.

From these observations we can conclude that different couples can have equivalent effects on a rigid body provided that the forces forming the couples are in parallel planes. *Equivalence* then requires that the products $|\bar{F}|d$ for the couples be the same, and that their torques have the same sense. Typical examples of equivalent couples are shown in Figure 14a for a planar system and Figure 14b for a three-dimensional system.

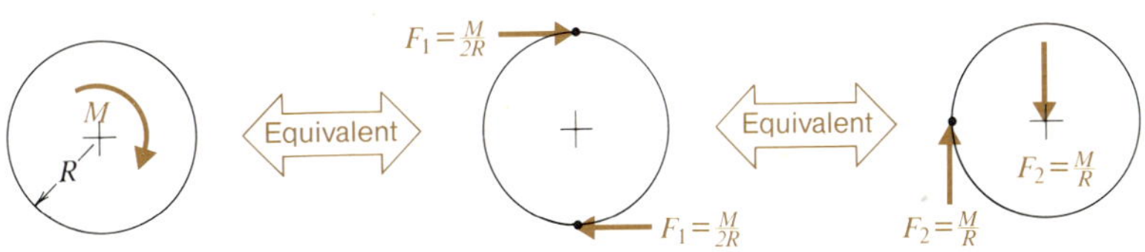

FIGURE 14a

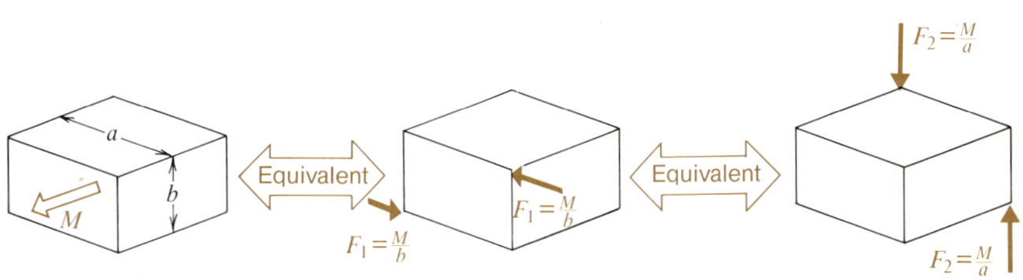

FIGURE 14b

The first of the representations of the couple in each of Figures 14 is the torque vector $\bar{M}$, rather than a pair of opposite force vectors. Recall that the moment of a couple is independent of the point chosen for computing moments. We therefore do not need to specify the point of application of the torque vector when we are concerned with the overall effect of a couple on a rigid body.

The fact that the torque vector is the only significant property of a couple leads to the conclusion that

> A set of couples acting on a rigid body is equivalent to a single couple whose torque is the sum of the moments, about an arbitrary point, of all forces forming the couples. The component of the torque of a couple parallel to a specific axis is the rotational tendency of the couple about that axis.

EXAMPLE 6

Determine the smallest forces applied at ends A and B of the beam shown that are equivalent to the given loading.

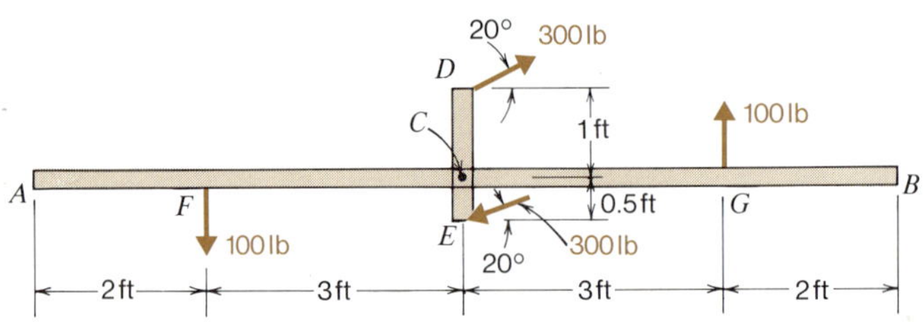

Solution

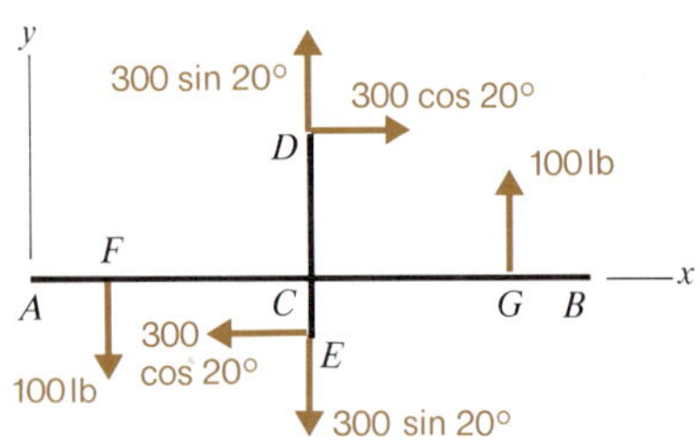

The planar force system consists of couples formed by the 100 lb forces and by the 300 lb forces. In order to find an equivalent couple, we must determine the total torque of the given couples. This torque may be evaluated by summing the moments of each force about any point; the geometry of the system suggests that we use point C.

The points of application of the forces are located by horizontal and vertical dimensions. Hence, we choose a coordinate system having horizontal and vertical axes. The coordinate system and force components are shown in the sketch.

The z axis is outward for a right-hand coordinate system, so counterclockwise moments are positive according to the right-hand rule. The vertical force components at points D and E pass through point C, making their lever arms about this point zero. Hence, the resultant torque is

$$\begin{aligned} M &= -300 \cos 20°(1.0) - 300 \cos 20°(0.5) + 100(3.0) + 100(3.0) \\ &= 177.14 \text{ lb-ft} \end{aligned}$$

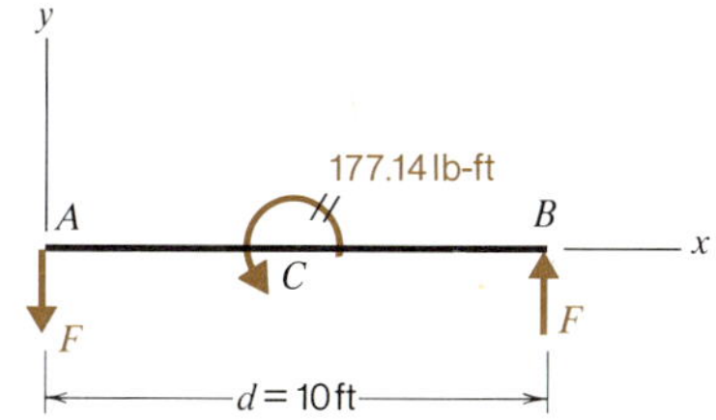

The problem now is to form a 177.14 lb-ft couple from the smallest forces $\bar{F}$ at point A and $-\bar{F}$ at point B. The magnitude of the torque of this couple is $|\bar{F}|d$, so the smallest $|\bar{F}|$ corresponds to the largest value for d. As shown in the sketch, this situation is obtained when $\bar{F}$ is perpendicular to the beam. Notice that the sense of the forces is determined from the requirement that they form a counterclockwise couple. From the sketch we then have

$$M = 177.14 = |\bar{F}|d = 10|\bar{F}|$$

$$|\bar{F}| = 17.714 \text{ lb}$$

EXAMPLE 7

A bus driver applies 60 N forces at points A and B on the steering wheel. These forces are situated in the vertical plane at an angle of 15° relative to the horizontal. What is the torque of this couple, and what portion of this torque has the effect of turning the steering wheel?

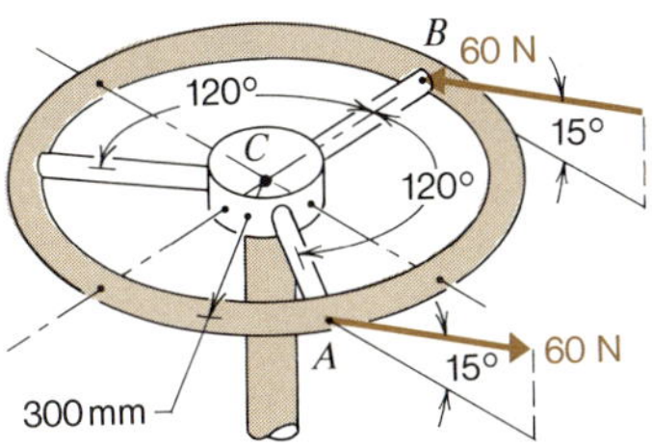

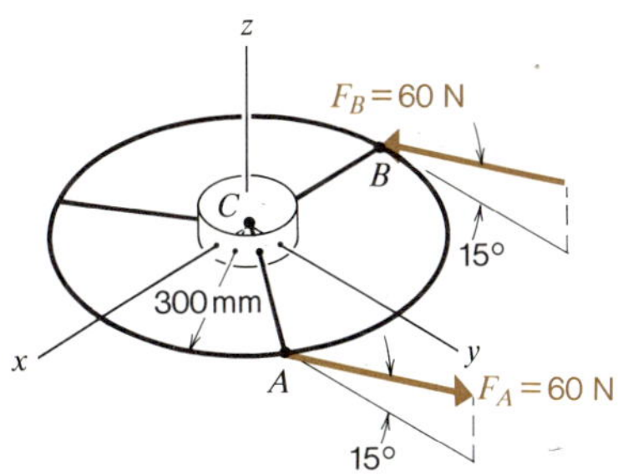

Solution

We will evaluate the torque of the couple by summing moments about the center point C, for the position vectors from this point to the points of application of the forces are easily described. The xyz reference frame shown in the sketch is chosen because points A, B, and C then all coincide with a coordinate plane, and also for the convenience it affords in describing force components.

The forces at points A and B are denoted as $\bar{F}_A$ and $\bar{F}_B$, so we have

$$\begin{aligned}\bar{F}_A &= -\bar{F}_B = 60(\cos 15°\,\bar{j} + \sin 15°\,\bar{k}) \\ &= 57.96\bar{j} + 15.53\bar{k} \text{ N}\end{aligned}$$

The position vectors from point C to the forces are

$$\bar{r}_{B/C} = -0.30\bar{i} \text{ m}$$

$$\begin{aligned}\bar{r}_{A/C} &= 0.30(\sin 30°\,\bar{i} + \cos 30°\,\bar{j}) \\ &= 0.15\bar{i} + 0.2598\bar{j} \text{ m}\end{aligned}$$

Hence, the torque of the couple is

$$\begin{aligned}\bar{M} &= \bar{r}_{A/C} \times \bar{F}_A + \bar{r}_{B/C} \times \bar{F}_B \\ &= (0.15\bar{i} + 0.2598\bar{j}) \times (57.96\bar{j} + 15.53\bar{k}) \\ &\quad + (-0.30\bar{i}) \times (-57.96\bar{j} - 15.53\bar{k}) \\ &= 0.15(57.96)\bar{k} + 0.15(15.53)(-\bar{j}) + 0.2598(15.53)(\bar{i}) \\ &\quad + 0.30(57.96)\bar{k} + 0.30(15.53)(-\bar{j}) \\ &= 4.03\bar{i} - 6.99\bar{j} + 26.08\bar{k} \text{ N-m}\end{aligned}$$

The steering wheel may rotate about the axis of the centerpost, which is the same as the z axis. Thus, the z component of $\bar{M}$ represents the tendency of the couple to rotate the steering wheel.

$$M_z = 26.08 \text{ N-m}$$

Furthermore, because M_z is positive, the driver is trying to turn the wheel counterclockwise looking down the z axis.

HOMEWORK PROBLEMS

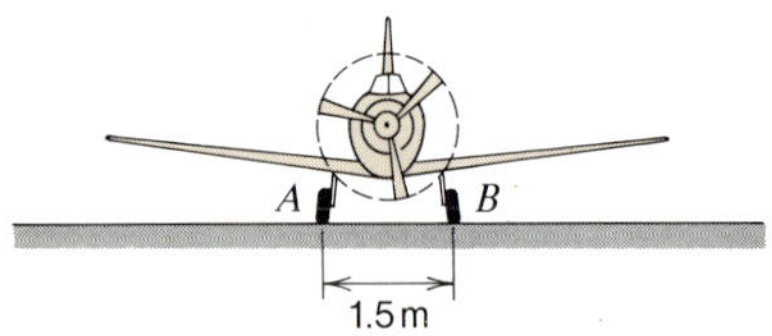

PROB. 3.44

3.44 The drive shaft from the engine applies a counterclockwise torque, as viewed from the front, of 1.8 kN-m to the propeller of the aircraft shown. Determine vertical reactions between the wheels and the ground at points A and B that are equivalent to this couple.

3.45 Determine the resultant torque of the two couples shown.

3.46 Determine the pair of vertical forces acting at ends A and B that are equivalent to the two couples shown.

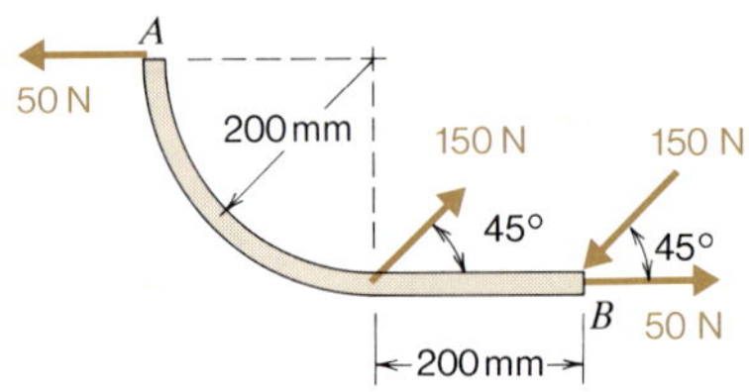

PROBS. 3.45 AND 3.46

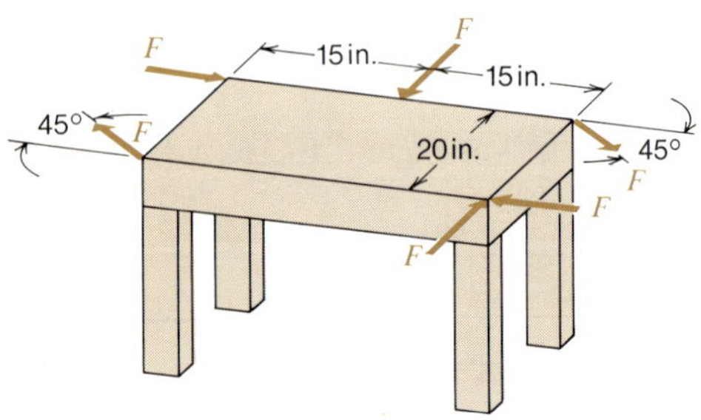

PROB. 3.47

3.47 Six forces in the plane of the table top have the same magnitude F. They exert an 80 lb-ft couple, which tends to cause the table to turn clockwise when viewed from above. Determine the value of F.

3.48 Two cables that pass over smooth pegs in the disk carry the tensions shown. Knowing that $F = 300$ N and that $\alpha = 30°$, determine the couple $\bar{M}$ that is equivalent to the force system exerted by the cables on the disk.

3.49 Two cables that pass over smooth pegs in the disk carry the tensions shown. The force system exerted by the cables on the disk should be equivalent to a clockwise couple whose magnitude is 500 N-m. Determine the minimum value of F and the corresponding angle α that will give the required couple.

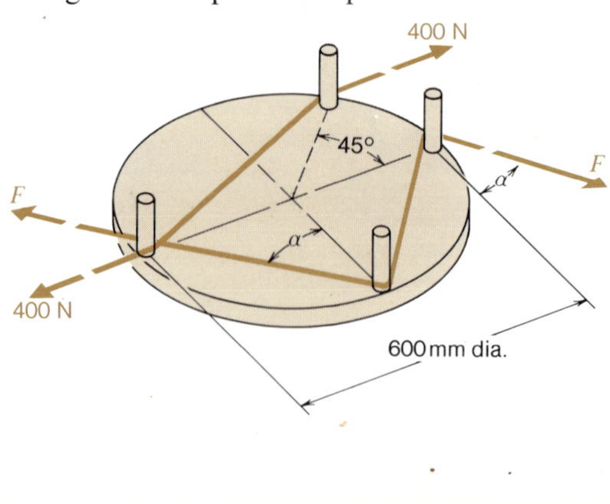

PROBS. 3.48 AND 3.49

3.50 The drive shaft applies a torque $M_1 = 5$ kN-m to the gear box and the driven shaft applies a torque $M_2 = 15$ kN-m, as shown. Determine the resultant couple acting on the gear box, specifying its magnitude and its direction angles with respect to the xyz coordinate system.

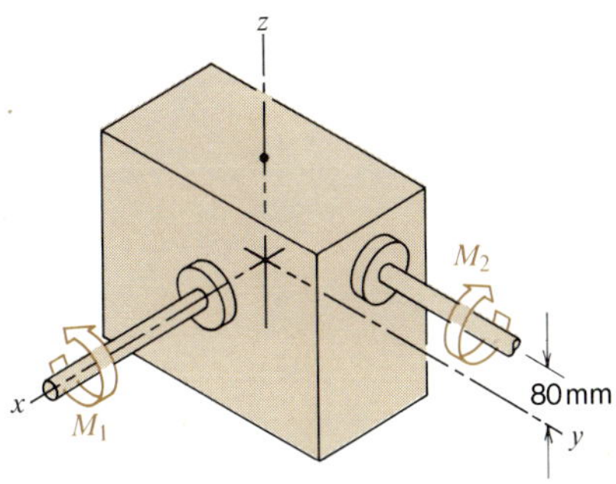

PROB. 3.50

3.51 Determine the couple that is equivalent to the effect of the four forces acting on the rectangular post.

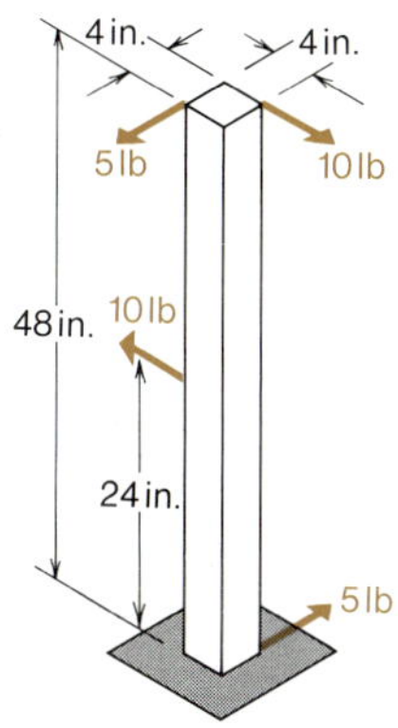

PROB. 3.51

3.52 Determine the couple that is equivalent to the effect of the four forces acting on the wide-flange beam.

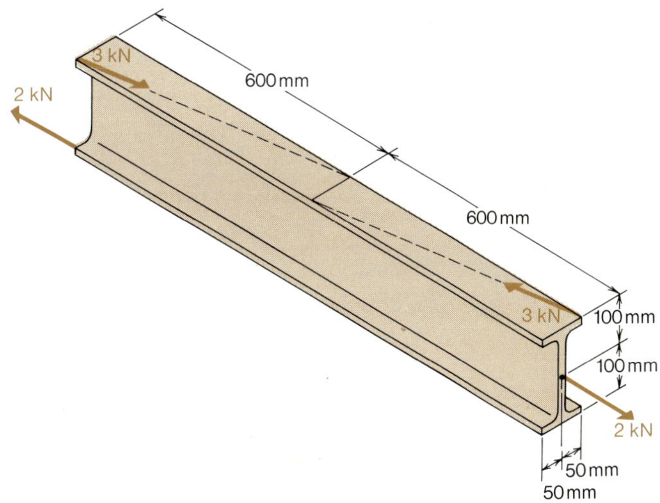

PROB. 3.52

***3.53** The four forces acting on the rectangular parallelepiped are equivalent to a couple $\bar{M}$. Determine the magnitude of $\bar{F}$, the angle α, and $\bar{M}$.

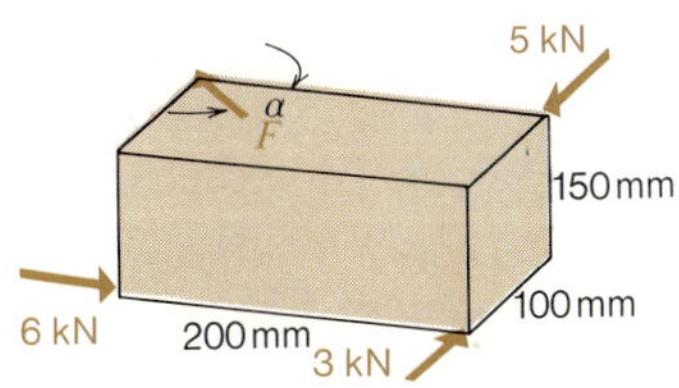

PROB. 3.53

*3.54 Two cables, each carrying a tension of 3 kips, are fastened to the crate. The effect of these cables and the force $\bar{F}$ acting at corner A is a couple $\bar{M}$. Determine $\bar{F}$ and $\bar{M}$.

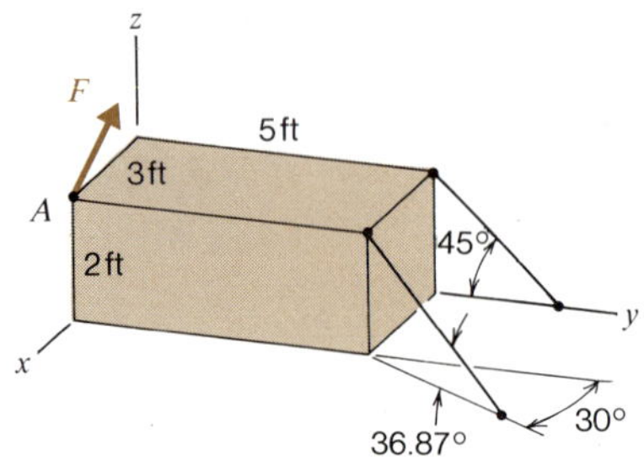

PROB. 3.54

E. RESULTANT OF A GENERAL SYSTEM OF FORCES

In the preceding section we showed that the torque of a couple represents the tendency of the couple to rotate a body. Our intuition tells us that a general system of forces, in which forces are applied at several points on a body, also tends to cause the body to rotate. We already know how to treat a system of concurrent forces. Therefore, let us begin the description of the effect of a general force system by considering how a force acting at a given point of application may be transferred to a more convenient point.

Consider a single force $\bar{F}$ applied to point P of a rigid body, as shown in Figure 15a. It is our desire to transfer $\bar{F}$ to the arbitrary point C. To achieve the transfer, we apply two forces $\bar{F}$ and $-\bar{F}$ at point C. These two forces are concurrent. Their resultant is zero, and therefore the force system in Figure 15b is equivalent to the force $\bar{F}$ in Figure 15a. However, the force $\bar{F}$ at point P and the force $-\bar{F}$ at point C form a couple $\bar{M}_C$ that is given by

$$\bar{M}_C = \bar{r}_{P/C} \times \bar{F} \qquad (17)$$

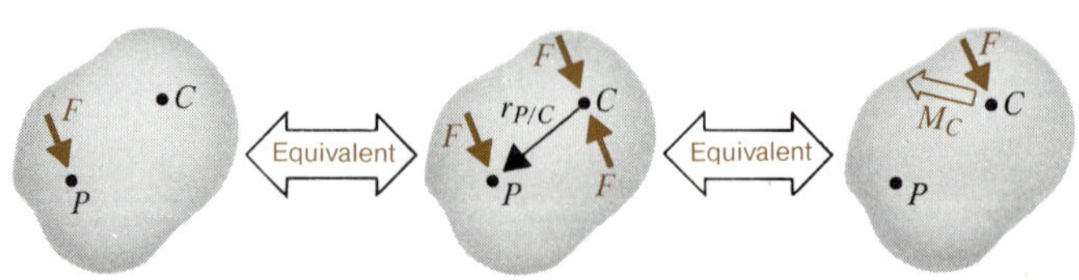

FIGURE 15a FIGURE 15b FIGURE 15c

This couple is simply the moment of the force $\bar{F}$ about point C. In other words, *the original force $\bar{F}$ is equivalent to the force $\bar{F}$ at point C and the couple $\bar{M}_C$, which is the moment of $\bar{F}$ about C.* This is called a *force-couple system.* The resulting equivalent force-couple system is shown in Figure 15c.

Now consider the problem of a general force system, such as the one shown in Figure 16a. Each force in the system may be transferred to a common point C, as is shown in Figure 16b, where $(\bar{M}_C)_i$ denotes the couple whose torque is the moment of force $\bar{F}_i$ about point C.

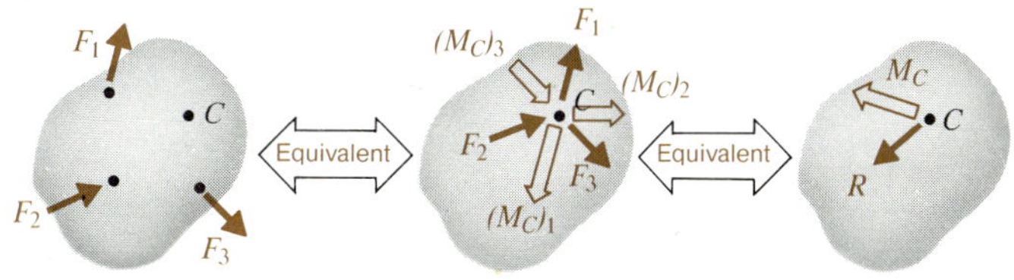

FIGURE 16a **FIGURE 16b** **FIGURE 16c**

The forces are now concurrent, so they may be summed to obtain their resultant $\bar{R}$. Further, because couples are additive, the individual couples $(\bar{M}_C)_i$ may be summed vectorially to obtain the resultant couple $\bar{M}_C$. Thus, $\bar{M}_C$ is the sum of the moments about point C of the forces in the original system. It then follows that the original system of forces is equivalent to the force-couple system in Figure 16c.

A general system of forces is equivalent to a force $\bar{R}$ applied at any point C, and a couple $\bar{M}_C$ provided that

$$\bar{R} = \sum \bar{F} \qquad \bar{M}_C = \sum \bar{M}_C \tag{18}$$

where $\sum \bar{F}$ denotes the sum of the forces in the original system and $\sum \bar{M}_C$ denotes the sum of the moments of the forces in the original system about point C.

Clearly, the resultant force $\bar{R}$ in the equivalent force-couple system is independent of the choice for point C; but the resultant couple $\bar{M}_C$ depends on which point we choose. This observation leads to a corollary of equation (18). *Two different force systems have equivalent effects on a rigid body if the forces in each system exert the same total moment about any point, as well as having identical sums.*

EXAMPLE 8

Determine the force-couple system acting at the center C of the stepped gear that is equivalent to the two forces shown.

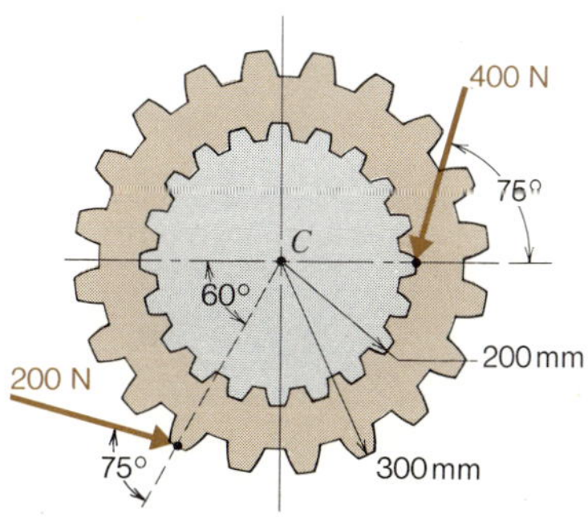

Solution

We first sketch the given forces, as well as an xyz coordinate system for evaluating force components. The resultant force $\bar{R}$ is the sum of the given forces, so

$$\begin{aligned}\bar{R} &= 400(-\cos 75°\,\bar{i} - \sin 75°\,\bar{j}) + 200(\cos 15°\,\bar{i} - \sin 15°\,\bar{j})\\ &= 90\bar{i} - 438\bar{j} \text{ newtons}\end{aligned}$$

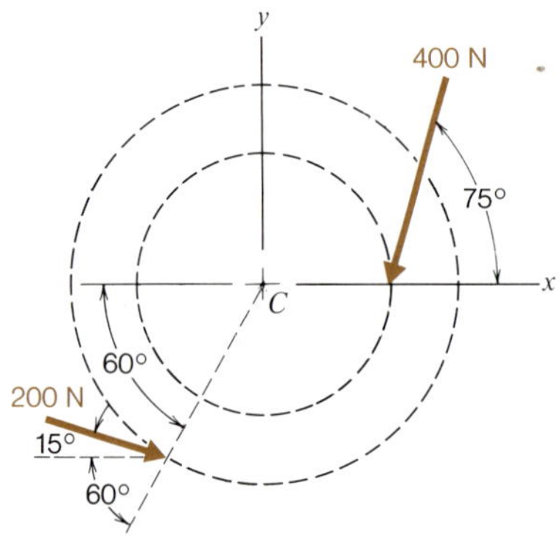

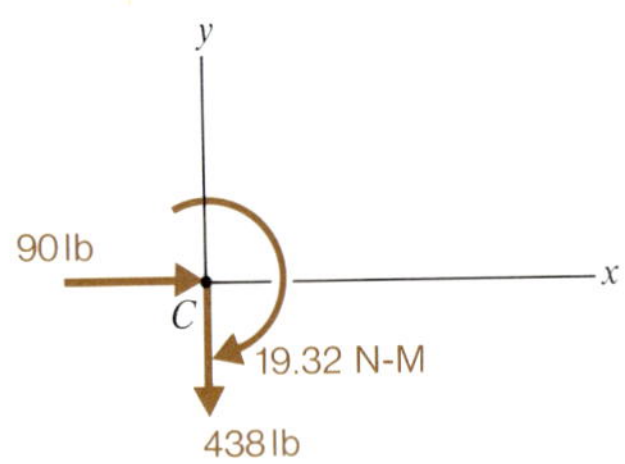

In the force-couple system we seek, $\bar{R}$ is applied at point C. The corresponding couple $\bar{M}_C$ is the sum of the moments about point C, which we compute by evaluating the component of each force perpendicular to the radial line from point C to the point of application of the force. Thus,

$$M_{Cz} = (200 \sin 75°)(0.3) - (400 \sin 75°)(0.2) = -19.32 \text{ N-m}$$

The equivalent system is illustrated to the left.

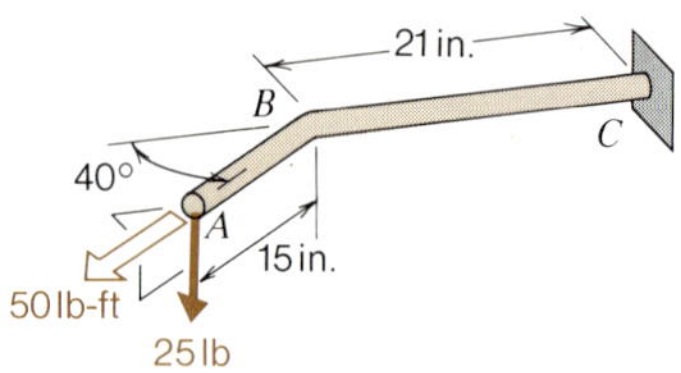

EXAMPLE 9

Replace the force-couple system acting at end A of the bent rod by an equivalent force-couple system acting at end C.

Solution

The xyz coordinate system in our sketch was chosen to fit the way in which the dimensions of the rod were given. There is only one force in the given system, so the resultant $\bar{R}$ in the system we seek is simply

$$\bar{R} = -25\bar{k} \text{ lb}$$

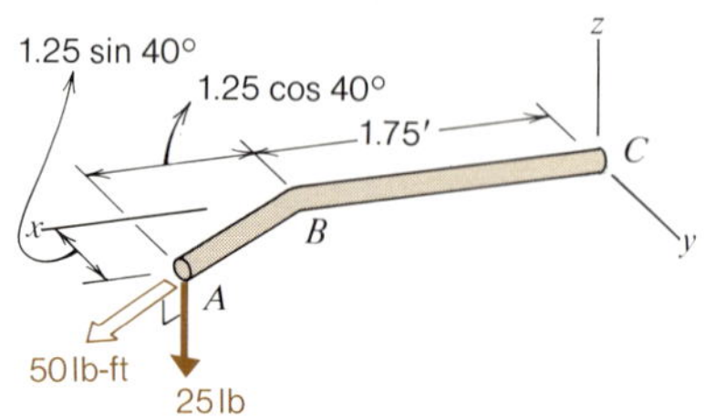

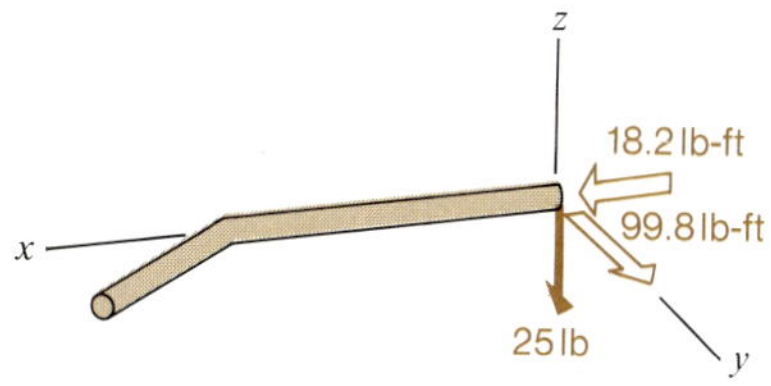

This force acts at end C. The corresponding couple $\bar{M}_C$ is the sum of the moments about C. The point of application of the 50 lb-ft couple is irrelevant to the moment it exerts, so

$$\begin{aligned}\bar{M}_C &= \bar{r}_{A/C} \times (-25\bar{k}) + 50(\cos 40^\circ \bar{i} + \sin 40^\circ \bar{j}) \\ &= [(1.75 + 1.25 \cos 40^\circ)\bar{i} + 1.25 \sin 40^\circ \bar{j}] \times (-25\bar{k}) + 38.30\bar{i} + 32.14\bar{j} \\ &= 18.2\bar{i} + 99.8\bar{j} \text{ lb-ft}\end{aligned}$$

This equivalent force couple system is sketched to the left.

HOMEWORK PROBLEMS

3.55 Replace the 200 N force that is applied to the tire wrench by an equivalent force-couple system at bolt A.

3.56 Replace the 200 N force that is applied to the tire wrench by an equivalent force-couple system at the center C of the tire.

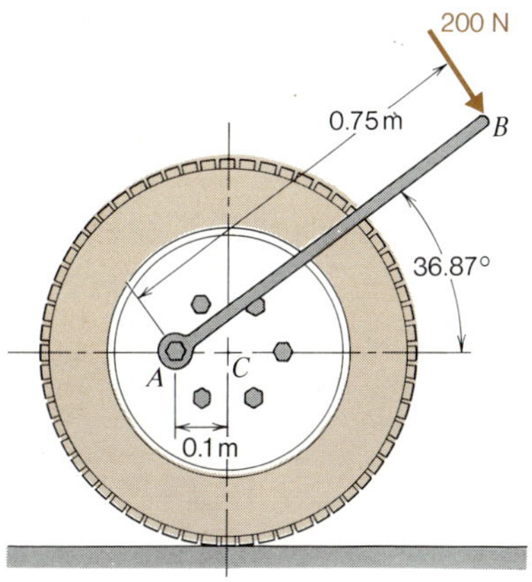

PROBS. 3.55 AND 3.56

3.57 and 3.58 Determine a force-couple system at end A of the curved bar that is equivalent to the loading shown.

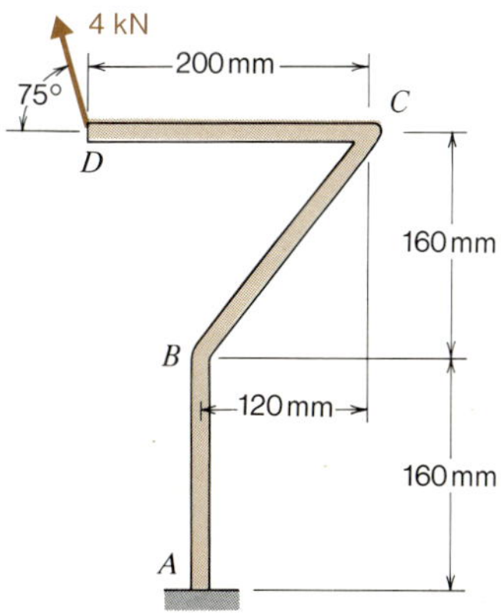

PROB. 3.57

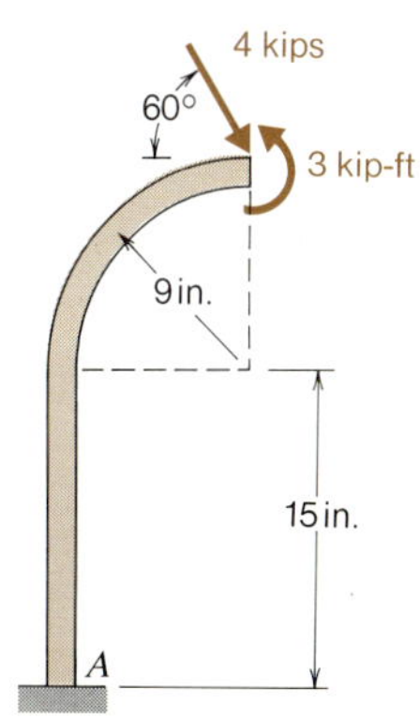

PROB. 3.58

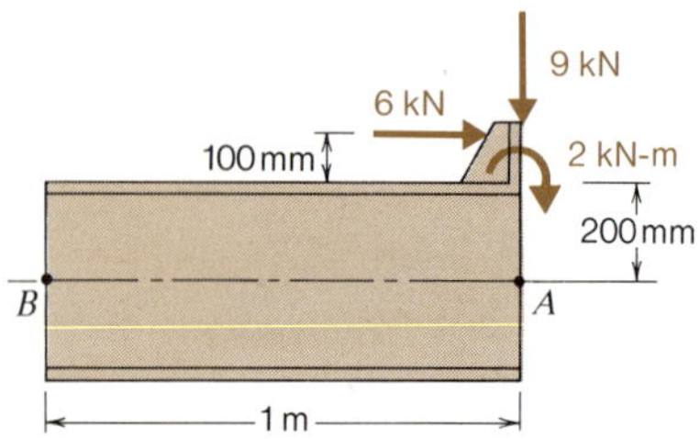

PROB. 3.59

3.59 The bracket is welded to the cantilever beam and is loaded as shown. Replace these loads by an equivalent force-couple system acting on the centerline of the beam at (a) point A, (b) point B.

3.60 A cable is passed over two ideal pulleys that are pinned to a block. The tension within the cable is 300 N. Determine an equivalent force-couple system acting (a) at pin A, (b) at pin B.

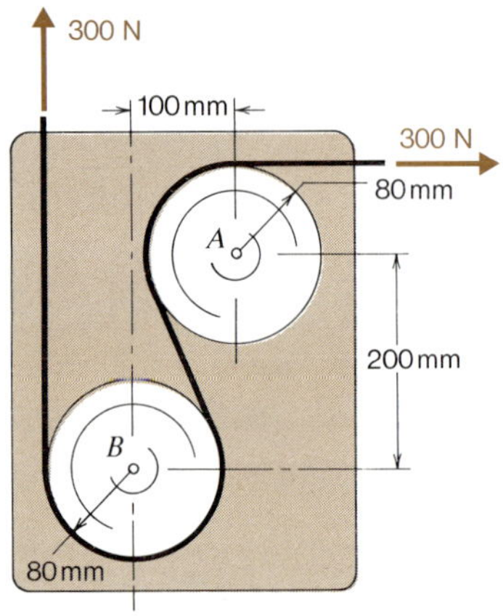

PROB. 3.60

3.61 and 3.62 Replace the loading shown by an equivalent force-couple system at the center of the disk.

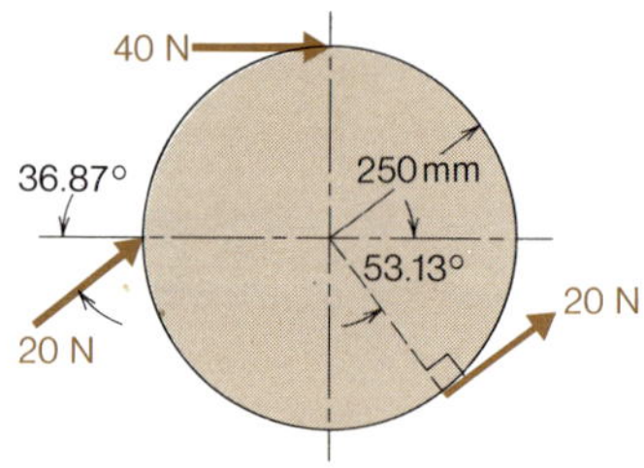

PROB. 3.61

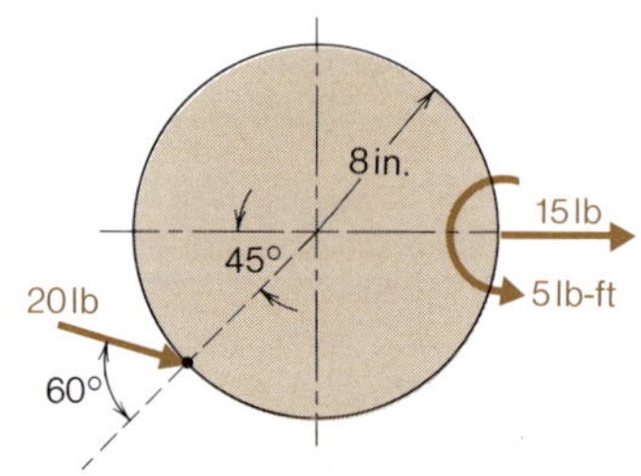

PROB. 3.62

3.63 The resultant of the force $\bar{F}$ and the other forces acting on the stepped gear is a single force acting at the center of the gear. Determine the magnitude of $\bar{F}$ and the corresponding resultant force.

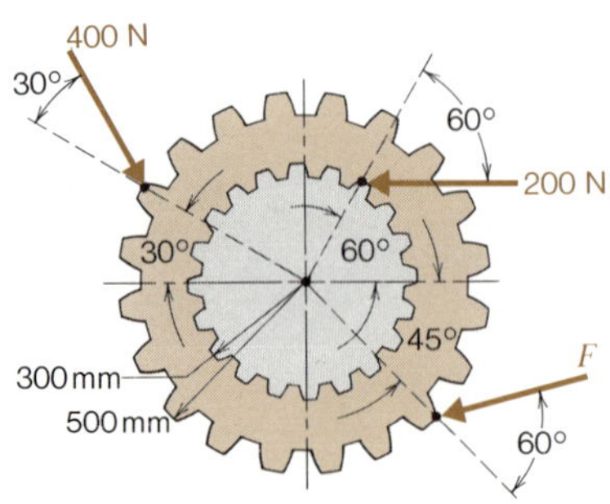

PROB. 3.63

3.64 The street light is supported by cable AC. The total weight of the structure is 400 lb acting at point G, and the 50 lb horizontal force represents the effect of the wind. The combined effect on the structure of gravity, wind, and the cable tension is equivalent to a single force acting on the base B. Determine the tension in the cable and the equivalent single force.

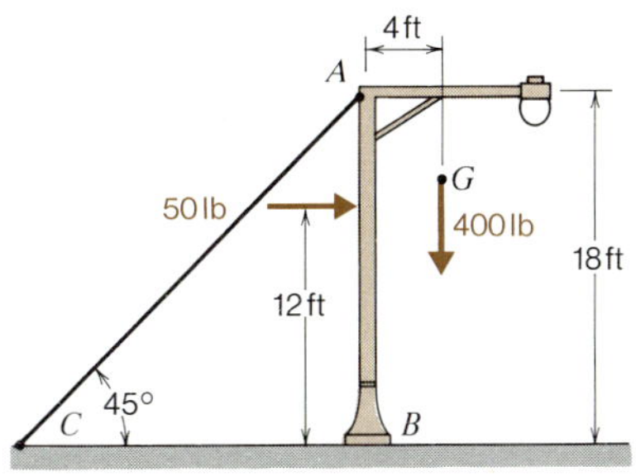

PROB. 3.64

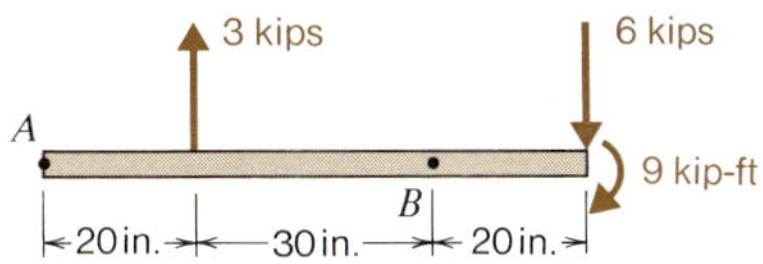

PROB. 3.65

3.65 Determine parallel forces acting at points A and B of the beam that are equivalent to the loading shown.

3.66 The shelf bracket carries the loads shown. Determine a horizontal force acting at point A and an arbitrary force at point B that are equivalent to this loading.

3.67 Determine parallel forces acting at points A and B of the shelf bracket that are equivalent to the loading shown.

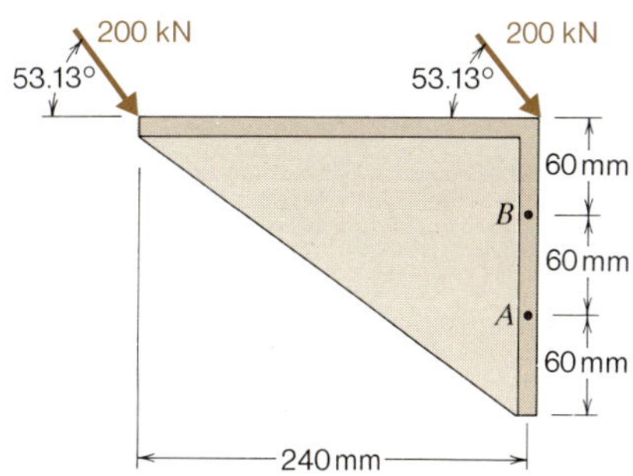

PROBS. 3.66 AND 3.67

3.68 A 250 kN force is applied to the bracket that is welded to the vertical column. Determine the equivalent force-couple system acting at point C on the centerline of the column.

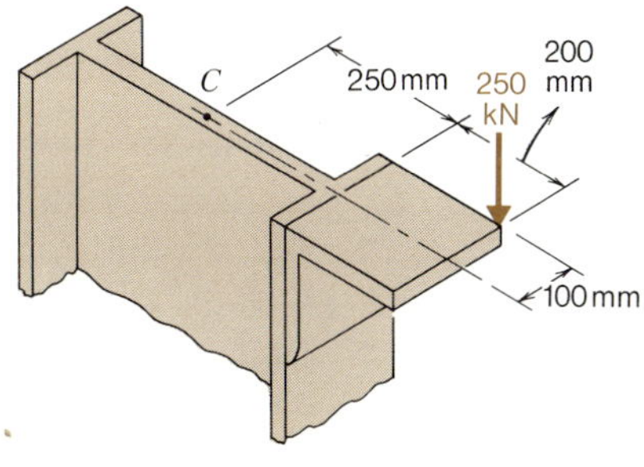

PROB. 3.68

3.69 A force and couple are applied at end A of the pipe assembly. Determine an equivalent force-couple system at end B.

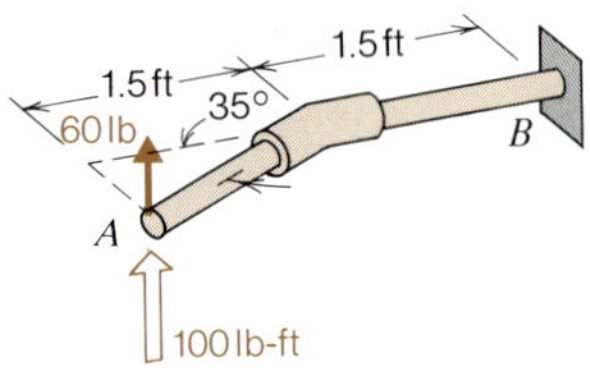

PROB. 3.69

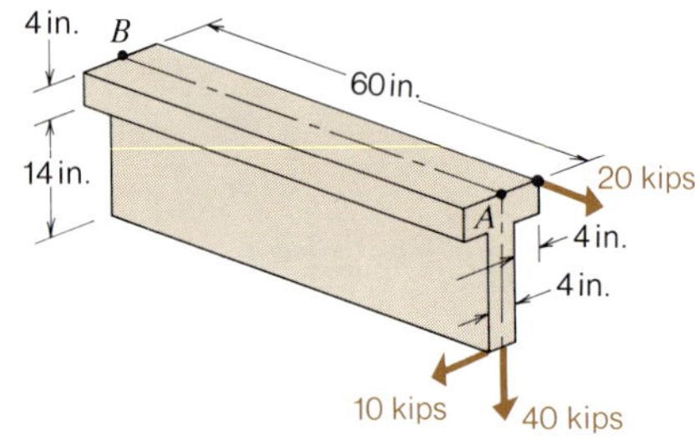

PROB. 3.70

3.70 The concrete T-beam is loaded as shown. Determine the equivalent force-couple system acting (a) at point A, (b) at point B.

3.71 A valve is tightened with the T-bar assembly by applying 75 N forces at points B and C as shown. (a) Replace these forces by an equivalent force-couple system acting at point A. (b) What portion of this equivalent system has the effect of turning the valve about axis AD?

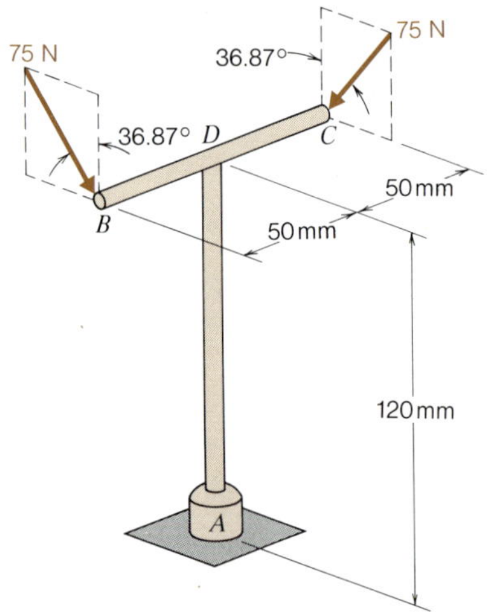

PROB. 3.71

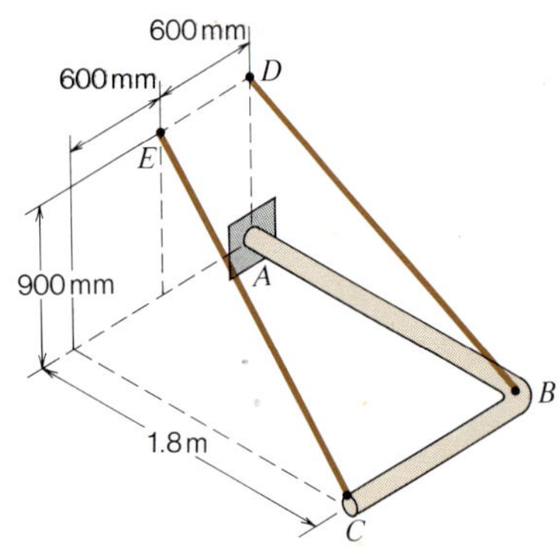

PROB. 3.72

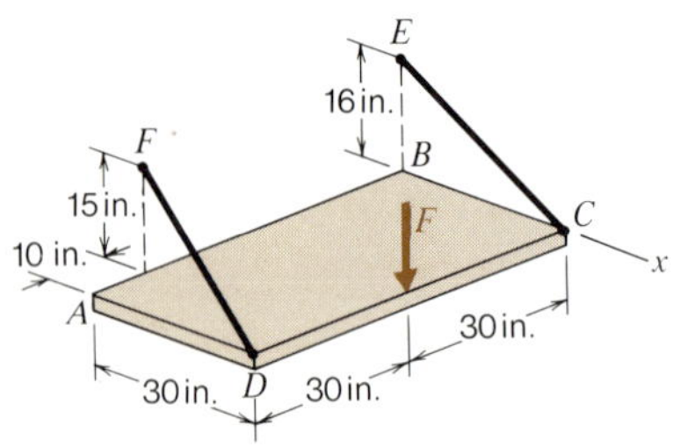

PROB. 3.73

***3.72** The tensions in cables BD and CE are 4 kN and 3 kN, respectively. Determine a force-couple system at end A of the bent bar that is equivalent to the forces exerted by the cables on the bar.

3.73 The tension in cable CE is 3400 lb and the tension in cable DF is 4200 lb. The force $\bar{F}$ is a downward load of 1000 lb. Determine the force-couple system acting at corner A that is equivalent to the effect on the plate of the cables and $\bar{F}$.

3.74 Replace the forces acting on the rectangular parallelepiped by an equivalent force-couple system (a) at corner A, (b) at corner B.

*3.75 In addition to the loads shown, a couple $\bar{M}$ is applied at corner B of the rectangular parallelepiped. The resultant of this loading is equivalent to a single force $\bar{F}$ at corner B. Determine $\bar{M}$ and $\bar{F}$.

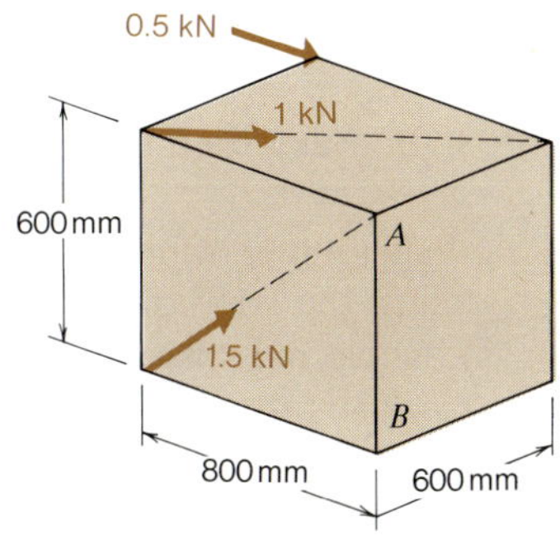

PROBS. 3.74 AND 3.75

F. SINGLE FORCE EQUIVALENTS

We have seen that a general force system can always be replaced by a force-couple system acting at an arbitrary point. The torque of the couple, which is the sum of the moments of the actual force system, depends on the point chosen for the force-couple representation. This result leads to the question of whether we can find a point at which the corresponding torque is zero. In such a case, we would be able to replace a system of forces by a *single force equivalent*. Such a representation greatly facilitates our understanding of the influence of force systems.

One situation where the sum of the moments of a force system is zero is the case of concurrent forces. We know that the single force equivalent in that case is the resultant force (that is, the sum of the forces), and that the line of action of this equivalent force intersects the point of concurrency. The task before us now is to determine whether single force equivalents exist in more general situations.

Suppose we try to represent the force system in Figure 17a by the single force $\bar{R}$ acting at some unknown point P, as depicted in Figure 17b. The first condition for equivalence is that

$$\bar{R} = \sum \bar{F} \tag{19}$$

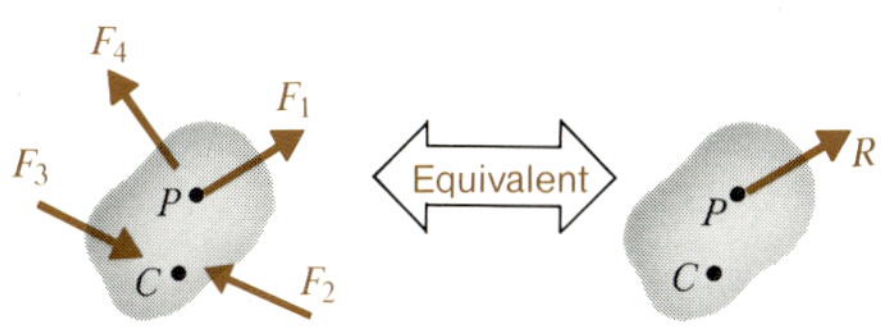

FIGURE 17a **FIGURE 17b**

The other condition pertains to the moment of the force system. Because the location of point P is not yet known, let us select a convenient, arbitrary point C and compute moments about this point for the alternative systems in Figures 17a and b. Equivalence then requires that

$$\bar{r}_{P/C} \times \bar{R} = \sum \bar{M}_C \tag{20}$$

We now note that the directions of both $\bar{R}$ and $\sum \bar{M}_C$ are determined by the given force system. However, the direction of the cross product is, by definition, perpendicular to both vectors in the product. This means that eq. (20) can be satisfied only when $\bar{R}$ is perpendicular to $\sum \bar{M}_C$. This situation occurs in two cases, in addition to the one where forces are concurrent. The two cases are presented below.

1. Planar Force Systems

When all forces coincide with a plane, their sum also lies in the plane, and their moment about any point in the plane is perpendicular to the plane. Thus, the condition for satisfying eq. (20) is fulfilled. In order to establish how we may ascertain the single force equivalent in this case, let us consider Figure 18a, where point C is any convenient point for the moment computation.

We first evaluate the components of the resultant $\bar{R}$ by summing the corresponding components of the original force system, as shown in Figure 18b.

$$R_x = \sum F_x, \qquad R_y = \sum F_y \tag{21}$$

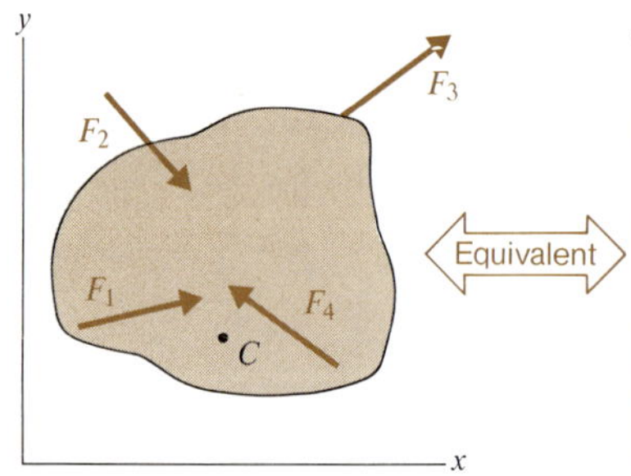

FIGURE 18a

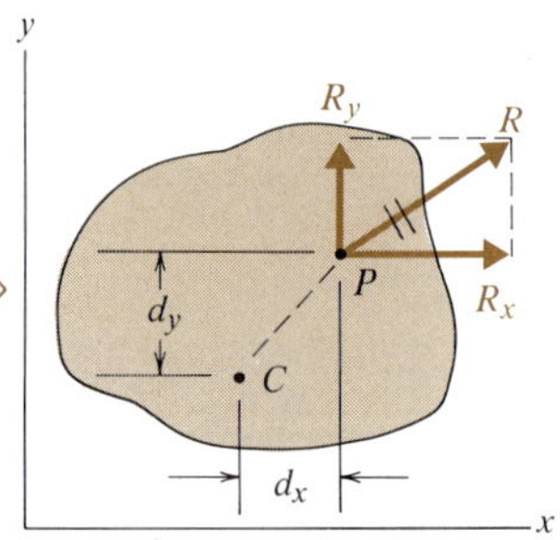

FIGURE 18b

We then determine the position of point P on the line of action of $\bar{R}$ by denoting the unknown components of $\bar{r}_{P/C}$ as d_x and d_y. Equivalence requires that the original forces and $\bar{R}$ exert the same moment about point C. Counterclockwise moments are positive for the chosen coordinate system, so

$$\sum M_{Cz} = d_x R_y - d_y R_x \tag{22}$$

The quantities R_x, R_y, and $\sum M_{Cz}$ are known values, so eq. (22) is the equation for the (straight) line of action.

2. Parallel Force Systems

When all forces are parallel to a given line, we align one of the coordinate axes with these forces, such as the z axis in the situation depicted in Figure 19a. The resultant $\bar{R}$ is then parallel to the z axis,

$$\boxed{\bar{R} = R\bar{k}, \qquad R = \sum F_z} \tag{23}$$

Because the forces in the original system exert no moment about the z axis, while their resultant $\bar{R}$ is parallel to the z axis, we have a case where $\bar{R}$ is perpendicular to $\bar{M}$.

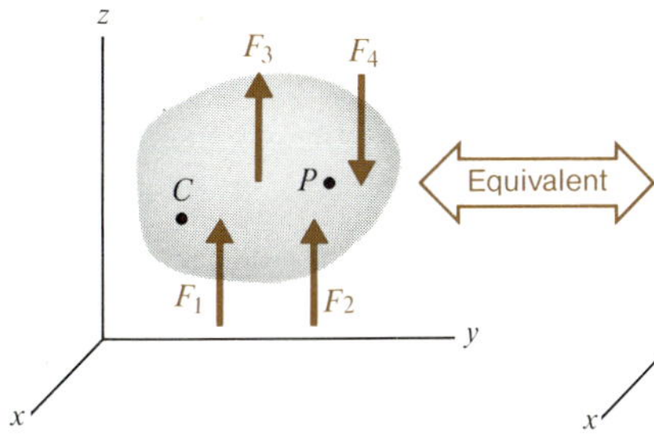

FIGURE 19a

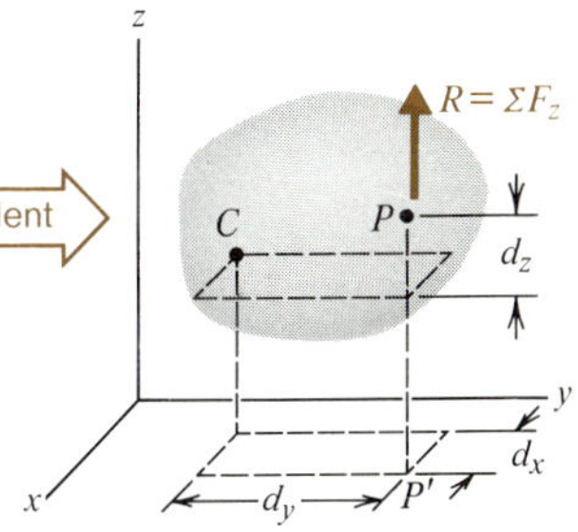

FIGURE 19b

We determine the location of point P on the line of action of $\bar{R}$ by equating the moments about point C of the two force systems in Figure 19. (Once again, point C is any convenient point.) The dimensions d_x, d_y, and d_z in Figure 19b are the unknown components of $\bar{r}_{P/C}$, so we require that

$$\sum \bar{M}_C = \bar{r}_{P/C} \times R\bar{k} \equiv (d_x\bar{i} + d_y\bar{j} + d_z\bar{k}) \times (R\bar{k})$$

The quantity $\sum \bar{M}_C$ represents the moment of the original force system about point C, so its components $\sum M_{Cx}$ and $\sum M_{Cy}$ are known values. Evaluating the cross product in the foregoing relation and equating like components then yields

$$\sum \bar{M}_C = \sum M_{Cx}\bar{i} + \sum M_{Cy}\bar{j} = d_yR\bar{i} - d_xR\bar{j}$$

$$\boxed{\sum M_{Cx} = d_yR, \qquad \sum M_{Cy} = -d_xR} \tag{24}$$

If the original force system does not form a couple, that is, if $R = \sum F_z \neq 0$, we may solve eq. (24) for d_x and d_y. This locates point P' where the line of action of $\bar{R}$ intersects the xy plane.

3. General System of Forces

For a general force system, there is no point C for which the torque $\bar{M}_C$ of the equivalent force-couple system is perpendicular to the resultant force $\bar{R}$. A slight simplification may be obtained by representing $\bar{M}_C$ by its components parallel and perpendicular to $\bar{R}$. We can then find a point P for which the torque $\bar{M}_P$ of the equivalent force-couple system is parallel to $\bar{R}$. Such a force system is called a *wrench*. The action of a wrench is typified by the manner in which we use a screwdriver, simultaneously pushing and twisting on the shaft on the screwdriver. The more general concept of an equivalent force-couple system is adequate for understanding the effect of the force system in this case. Therefore, we shall not pursue the procedure for evaluating the wrench.

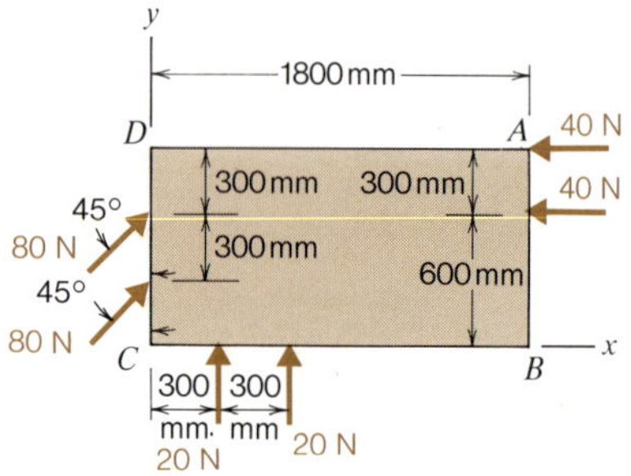

EXAMPLE 10

Three people exert horizontal forces on a rectangular table as shown. What single force has the same effect as these people, and where along edge BC should this force be applied?

Solution

The force system is planar, so it has a single equivalent force. This force is the sum of the given forces. We therefore sum force components using the coordinate system illustrated in the sketch. Note that we could first replace each pair of parallel forces by their single equivalent, but there is no advantage in doing that.

The force sum gives

$$R_x = \sum F_x = 2(80 \sin 45°) - 2(40) = 33.14 \text{ N}$$

$$R_y = \sum F_y = 2(80 \cos 45°) + 2(20) = 153.14 \text{ N}$$

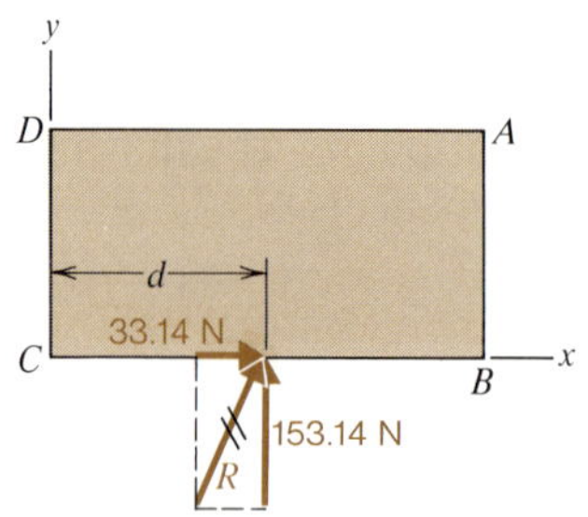

We now determine the point of application along edge BC with the aid of a sketch in which $\bar{R}$ is applied at an undetermined point along the edge. Then the moment of $\bar{R}$ about any point must equal the moment of the actual force system about that point.

The choice of the point for the moment sum is arbitrary; let us use point C. The z axis is outward, so counterclockwise moments are positive. The lever arms of the components of the original forces are found from the given diagram, and our sketch shows that only the y component of $\bar{R}$ exerts a moment about point C. Equating the moment of $\bar{R}$ about point C to that of the given forces then yields

$$\begin{aligned}\sum M_{Cz} &= 153.14(d) \\ &= -(80 \cos 45°)(0.3) - (80 \cos 45°)(0.6) \\ &\quad + 20(0.3) + 20(0.6) + 40(0.6) + 40(0.9) \\ d &= 0.1769 \text{ m}\end{aligned}$$

EXAMPLE 11

Two people are standing on the wing of a light aircraft, resulting in the loading shown. Determine the equivalent force and its point of application.

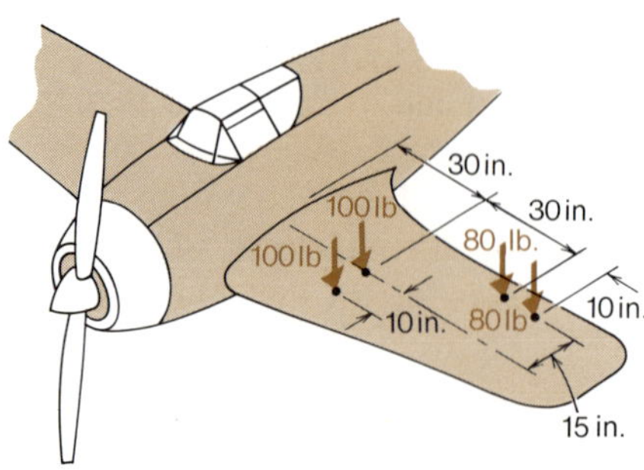

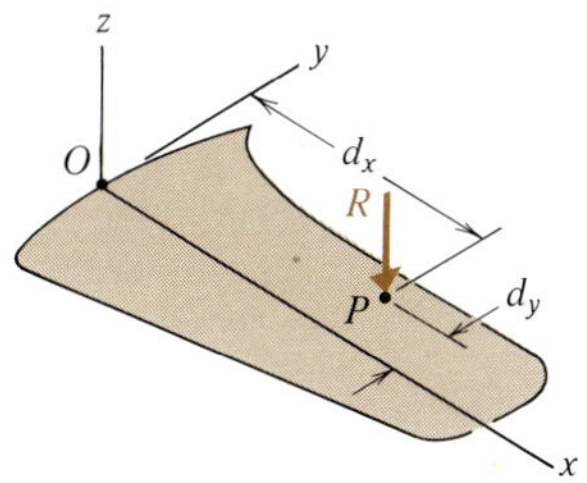

Solution

The given forces are parallel, so we draw a sketch of the resultant force $\bar{R}$ acting at point P in the xy plane. Note that the reference frame has been chosen such that the z axis is parallel to the given forces and the x and y axes align with the directions for the given dimensions.

The equivalent force $\bar{R}$ is the sum of the forces, so

$$\bar{R} = (-80 - 80 - 100 - 100)\bar{k} = -360\bar{k} \text{ lb}$$

The values d_x and d_y are found by equating moments. The origin O is a convenient point about which to sum moments. Thus

$$\begin{aligned}\sum \bar{M}_O &= (d_x\bar{i} + d_y\bar{j}) \times \bar{R} \equiv (d_x\bar{i} + d_y\bar{j}) \times (-360\bar{k}) \\ &= (30\bar{i}) \times (-100\bar{k}) + (30\bar{i} - 10\bar{j}) \times (-100\bar{k}) \\ &\quad + (60\bar{i} + 15\bar{j}) \times (-80\bar{k}) + (70\bar{i} + 15\bar{j}) \times (-80\bar{k}) \text{ lb-in.}\end{aligned}$$

We now evaluate the cross products, equate corresponding components, and solve the resulting algebraic equations. This yields

$$\bar{i} \text{ component:} \quad -360d_y = -1400$$

$$\bar{j} \text{ component:} \quad 360d_x = 16400$$

$$d_x = 45.6 \text{ in.,} \qquad d_y = 3.89 \text{ in.}$$

HOMEWORK PROBLEMS

3.76–3.78 Determine the single force that is equivalent to the loads acting on the cantilever beam. Specify the point of application of this resultant on the beam.

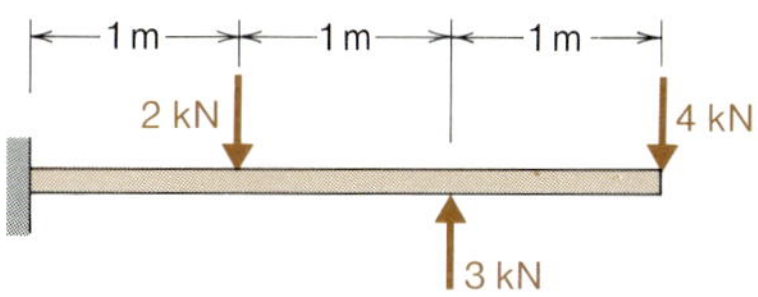

PROB. 3.76

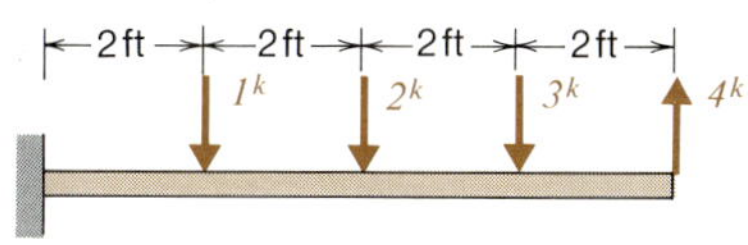

PROB. 3.77

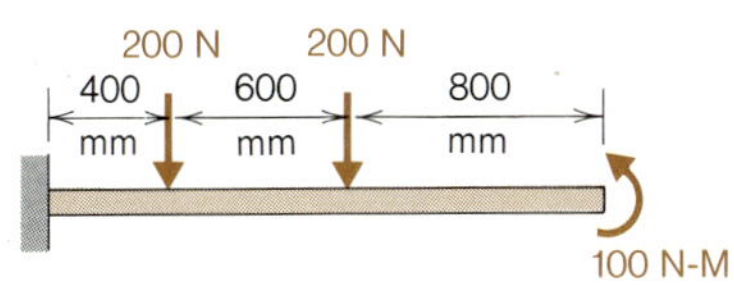

PROB. 3.78

3.79 Determine the distance d for which the loading on the bridge is equivalent to a single force acting at its midspan.

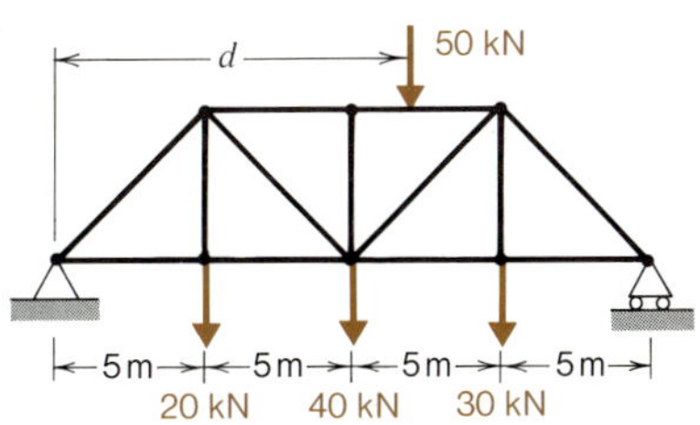

PROB. 3.79

3.80 In the tape guide shown, the recording tape that is tensioned at 10 N passes over the idler pulleys. Determine the single force that is equivalent to the forces exerted by the tape on the assembly. Where along the line AC does this force act?

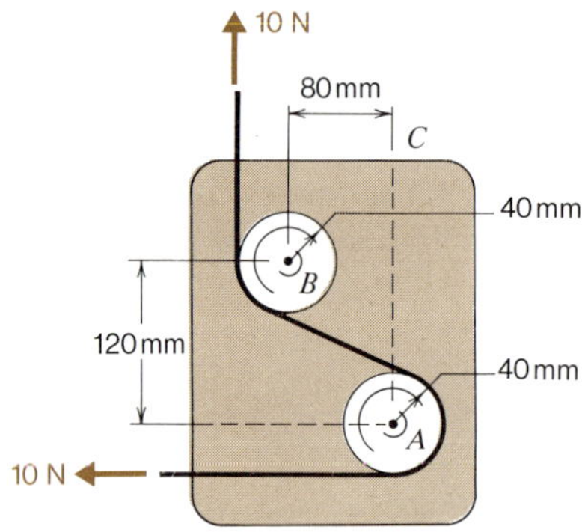

PROB. 3.80

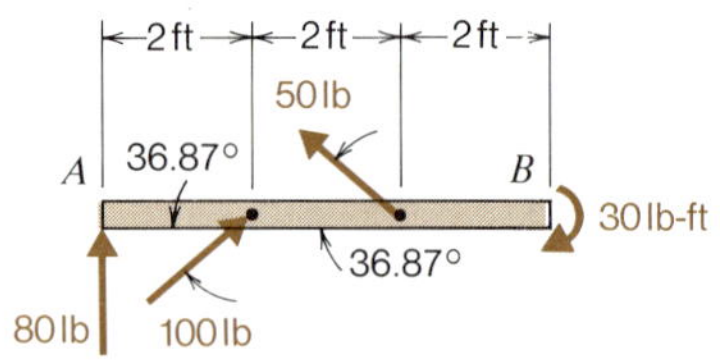

PROB. 3.81

3.81 Determine the single force that is equivalent to the loading shown. Give the magnitude, direction, and point of application along the beam of this equivalent force.

3.82 A street lamp supported by two cables is subjected to the loading shown. Determine the resultant of this system of forces and specify its point of application along line AB.

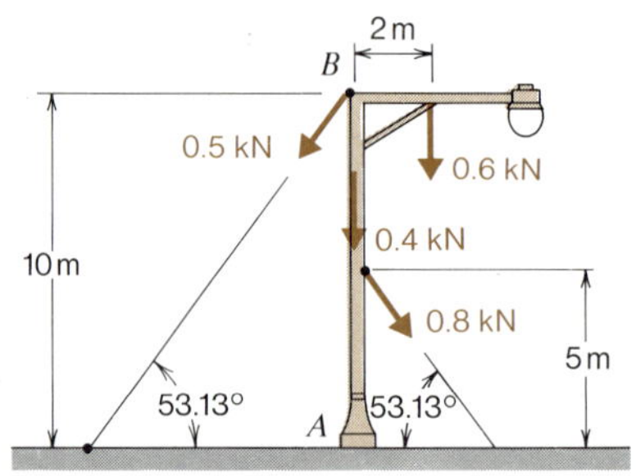

PROB. 3.82

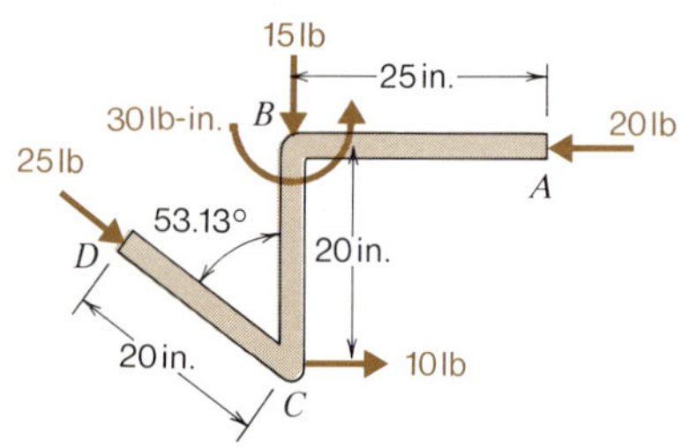

PROB. 3.83

3.83 Determine the magnitude, direction, and point of application along line AB of the force that is equivalent to the forces acting on the bent bar.

3.84 The truss structure is loaded as shown. Determine the magnitude of $\bar{F}$ and the angle α for which this system of forces is equivalent to a vertical force acting at point B.

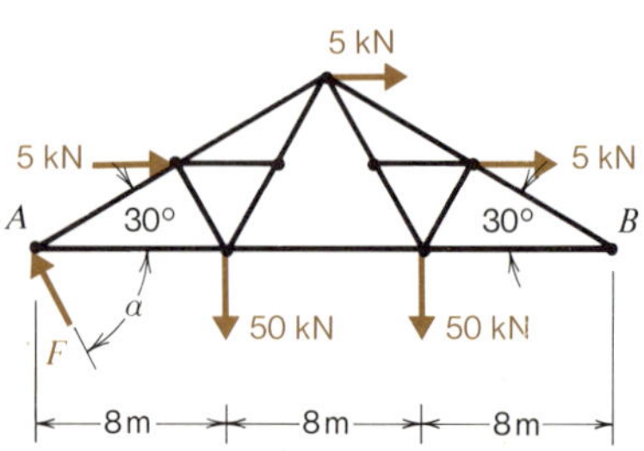

PROB. 3.84

3.85 A crate supported by two cables is pushed to one side by a person who exerts the 50 lb horizontal force. The weight of the crate is 400 lb, acting at point G. The tension in cable AC is 280 lb and the tension in cable BD is 240 lb. Determine the resultant of the forces acting on the crate, and specify its point of application along edge AB.

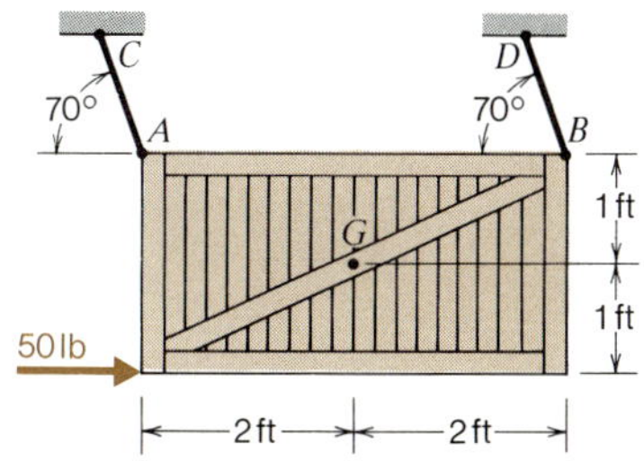

PROB. 3.85

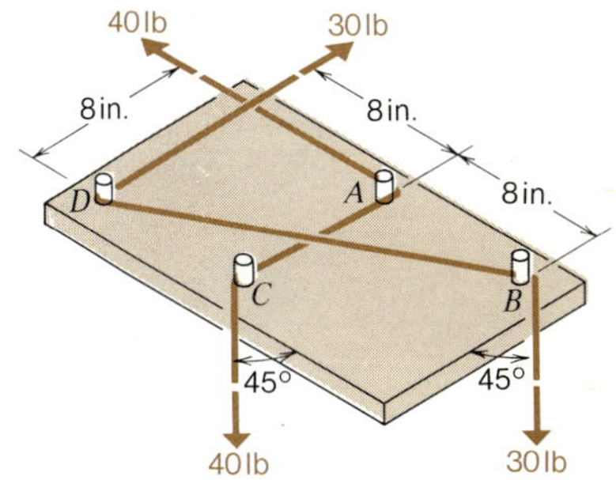

PROB. 3.86

***3.86** Two cords are pulled over smooth pegs embedded in the rectangular board. The tension in the cord passing over pegs A and C is 40 lb and the tension in the cord passing over pegs B and D is 30 lb. (a) Determine the magnitude and direction of the resultant force exerted on the board by the cords. (b) Determine the point of application of this resultant force along line AB. (c) Determine the point of application of the resultant force along line CD.

3.87 The tension in the cable is 10 kN and the weight of the uniform rectangular plate is 4 kN acting downward through point G. If $\alpha = 30°$, determine the force that is equivalent to the loading on the plate of the cable and gravity forces. Specify the point of application of this force along edge AB.

***3.88** The tension in the cable is 10 kN and the weight of the uniform rectangular plate is 14 kN acting downward through point G. The loading on the plate of the cable and gravity forces is equivalent to a single force acting at point A. Determine the angle α and the corresponding resultant force.

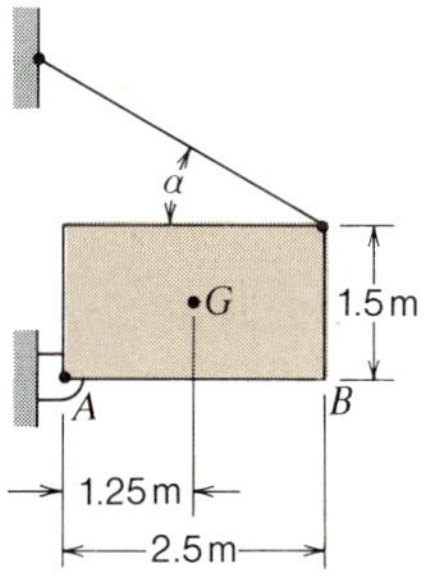

PROBS. 3.87 AND 3.88

3.89 A T-beam section is loaded as shown. Determine the resultant of these forces and its point of application on the top of the beam.

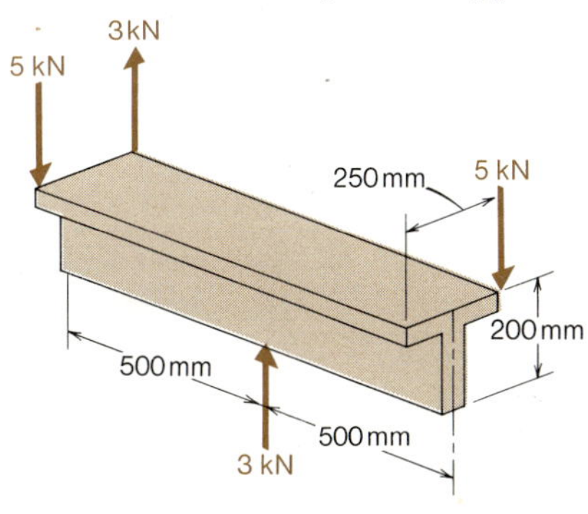

PROB. 3.89

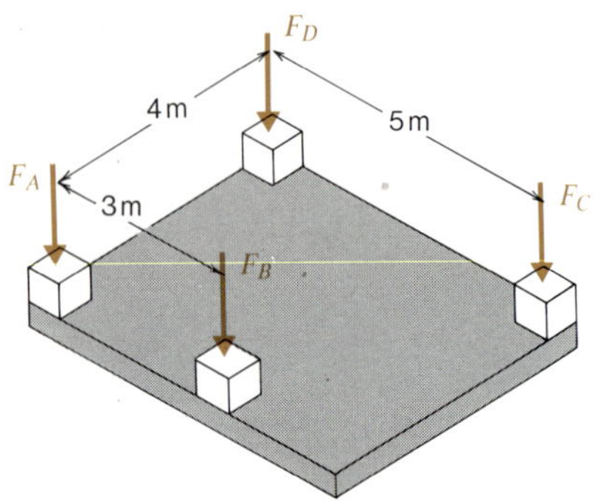

PROBS. 3.90 AND 3.91

3.90 Four columns are supported by the rectangular foundation slab. If the column loadings are $F_A = 15$ kN, $F_B = 18$ kN, $F_C = F_D = 12$ kN, determine the resultant column loading on the slab and its point of application.

3.91 Four columns are supported by the rectangular foundation slab. Two column loadings are known to be $F_A = F_C = 15$ kN. Determine the values of F_B and F_D for which the resultant column loading acts at the geometric center of the slab.

3.92 Three parallel columns equally spaced around the perimeter of the circular foundation slab carry the loads shown. Determine the resultant of this loading, and specify its point of application on the slab.

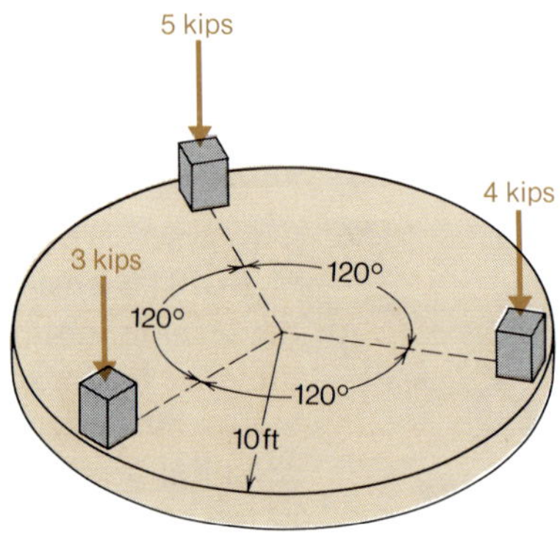

PROB. 3.92

***3.93** A semicircular rod embedded in a wall is loaded as shown. Determine the resultant of these forces and its point of application in the xy plane.

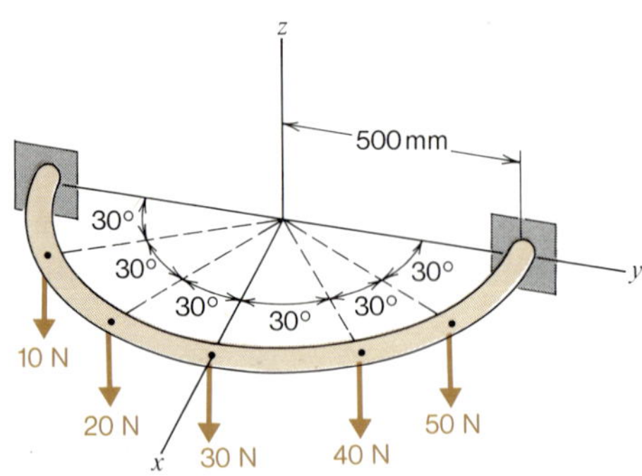

PROB. 3.93

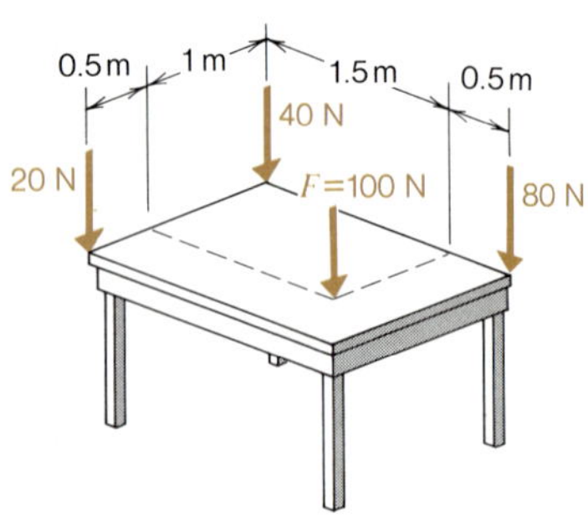

PROB. 3.94

***3.94** Three vertical forces are applied at corners of the table as shown. Determine where on the table a downward force $\bar{F}$, whose magnitude is 100 newtons, should be applied in order to make the resultant vertical force act at the geometric center of the table top.

G. EQUILIBRIUM OF RIGID BODIES

A fundamental application of the techniques for describing forces is the determination of the forces required to prevent a body from moving. This is the topic we will treat here. However, before we can investigate the relationship between the forces acting on a body in static equilibrium, we must learn to properly account for all forces.

1. Reactions and Free Body Diagrams

One aid in describing forces is the free body diagram. As we saw in Chapter 2, where we studied concurrent force systems, the free body diagram isolates the body of interest from its surroundings. In the diagram we show *all* forces being *exerted on the body*, bearing in mind that Newton's third law tells us that forces are the result of the interaction between bodies. The importance of a clear free body diagram cannot be overemphasized. The process of constructing the diagram aids us in understanding which parameters of the system are important. Equally important, the resulting diagram is a vital visual aid in making certain that we consider all forces during the solution process.

We will have no difficulty in depicting certain classes of forces in the diagram. Typical forces in this category are given forces, tensile forces exerted by cables, and the weight force. The latter, in particular, acts vertically downward, passing through the center of mass of the body. (Methods for determining this point are discussed in Chapter 6.)

Free body diagrams for rigid bodies are somewhat more difficult to construct than those we encountered in treating particles. The primary reason for this difficulty lies in the fact that the diagram accompanying the statement of the problem will usually depict the real system. That is, we will be shown the body (or bodies) of interest, as well as the means by which the body is supported and thus prevented from moving.

It is part of the task of constructing a free body diagram to examine the supports in order to deduce what types of force they apply to the body. It was mentioned in the previous chapter that these forces are sometimes called *constraint forces*, because they represent the way that the supports constrain (restrict) the body from moving. Another term for these forces is *reactions*, because they represent the way in which the supports react to the tendency of the body to move.

The method by which we establish the characteristics of the constraint forces is conveyed by their name. Recall that a force represents a pushing or pulling effect, whereas a couple represents a rotational effect. Thus, if a support restricts a point in a body from moving in a certain direction, the support will apply a reaction force to enforce this constraint. Similarly, a support that restricts rotation about an axis exerts a reaction couple parallel to that axis.

A primary part of the solution of an equilibrium problem is often the determination of the magnitude and orientation of the reaction forces and couples. Examination of a support will, in general, only reveal which components of the reactions are non-zero. The magnitude and sense of these components (e.g., left or right) are whatever is necessary to oppose the tendency of the other forces to cause the body to move. In other words, a free body diagram should depict *all possible* components of the reactions.

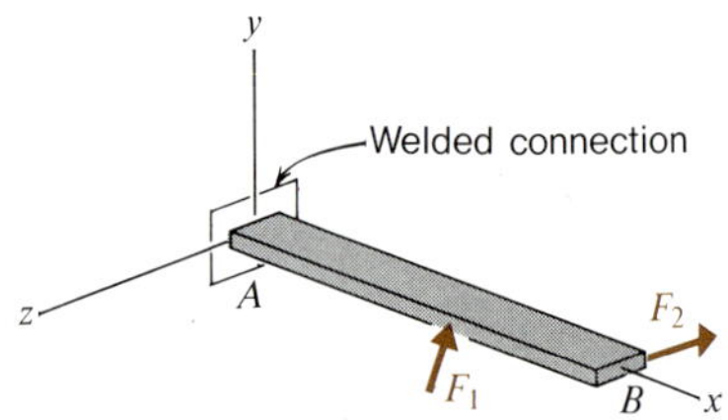

FIGURE 20

As an illustration of how these criteria are applied in the construction of a free body diagram, consider the welded connection between the rigid bar and the wall shown in Figure 20. The welding process in essence makes the materials of the joined bodies continuous (gluing has a similar effect), so that there can be no movement of one body relative to the other at their juncture. We assume that the wall in Figure 20 is rigid and fixed, so that end A of the bar is neither free to move in any direction nor to rotate about any axis.

The restriction against movement is imposed by a reaction force, and the restriction against rotation is imposed by a reaction couple. Therefore, the reaction force and couple at end A in Figure 20 have components parallel to each of the coordinate axes. We denote a reaction force at a specific point by the letter corresponding to that point, and a reaction couple by the symbol $\bar{M}$ followed by a subscript that is the letter for that point. Thus, the reaction at point A consists of a general force $\bar{A}$ having possible components A_x, A_y, and A_z, as well as a reaction couple $\bar{M}_A$ having possible components, M_{Ax}, M_{Ay}, and M_{Az}. The sense of the components may be assumed to be that of the positive coordinate axes; the fact that the sense of a reaction component is actually opposite the one we assume will be revealed by a negative sign in the problem solution. Figure 21 depicts the free body diagram of the rigid bar shown in Figure 20.

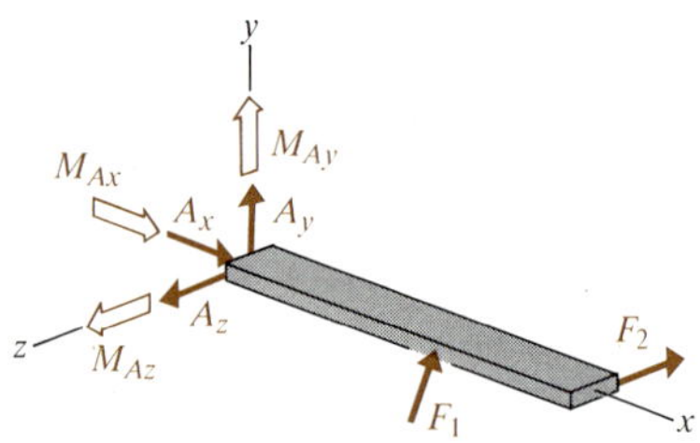

FIGURE 21

A welded joint represents the greatest degree of fixity (lack of freedom to move) that can be imposed by a connection. In fact, a welded support is usually described as a *fixed end*.

Whenever the type of support allows the body to move in a specific direction, or rotate about a specific axis, there will be no reaction force, or couple, in that direction. For example, suppose that a bar is supported at one end by a pin connection, such as that shown in Figure 22a. The pin acts like a miniature shaft. The end of the bar is fully constrained from moving in any direction by the pin and the brackets. The only possible way in which the bar may move is to rotate about the axis of the pin, which is the z axis in the figure. Thus, the reaction force and couple corresponding to this support are as shown in Figure 22b. Note that there is no couple in the z direction, because rotation about this axis is not restricted.

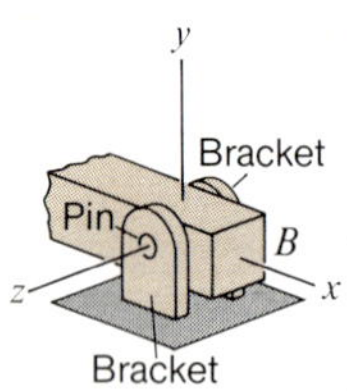

FIGURE 22a

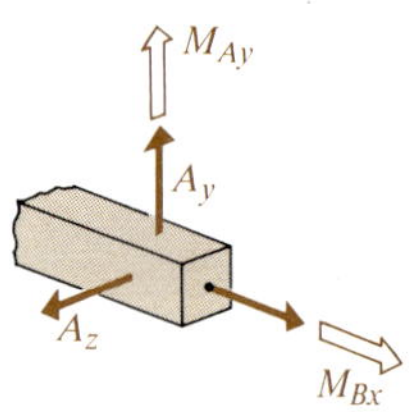

FIGURE 22b

The foregoing discussion implicitly assumed that the pin fits closely in the hole of the bar. Further, it was also assumed that the pin is rigid. This, of course, may not be the situation. For instance, a very common method of constructing a pin connection is to use a hinge, such as those found on doors. Anyone who has ever tried to support a door with just one hinge knows that it affords little constraint against rotation in any direction.

Thus, depending on specific knowledge of the type of construction, we may choose to retain or neglect the couples depicted in Figure 22b. We shall differentiate between the two alternatives by referring to the former condition as an ideal *pin*, and the latter one as a *hinge*, although this is not standard terminology.

When the loads that are applied to a body are situated in a common plane, we will usually be interested only in the reactions that restrict movement in that plane. This means that we may ignore reaction forces perpendicular to the plane and reaction couples whose torque vector lies in the plane (such couples prevent the body from rotating out of the plane). Thus, an ideal pin and a hinge in a planar situation such as the one in Figure 23a exert the same reactions, as depicted in Figure 23b. A comparison of Figures 23b and 22b shows that those reaction forces that tend to move the joint B out of the xy plane ($\bar{B}_z$, $\bar{M}_{Ax}$, $\bar{M}_{By}$) have been ignored in the free body diagram for the planar case.

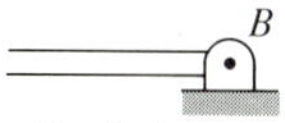

Physical support

FIGURE 23a

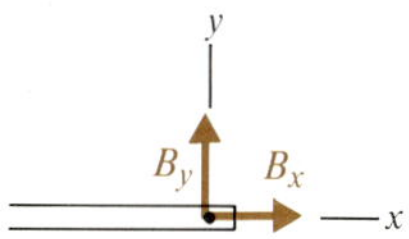

Reactions

FIGURE 23b

As an aid for our future studies, Table 1 depicts other common methods of connecting and supporting bodies, accompanied by a description of the nature of the corresponding restriction on the motion of the body and a sketch of the reaction force and couple components. This tabulation is presented in order to supplement, but not replace, our ability to examine the means of supporting a body. This is especially so because on some occasions we will encounter a support that is not depicted in Table 1.

TABLE 1
COMMON SUPPORTS AND REACTIONS

TYPE OF SUPPORT	MOTION RESTRICTION	REACTIONS
Cable 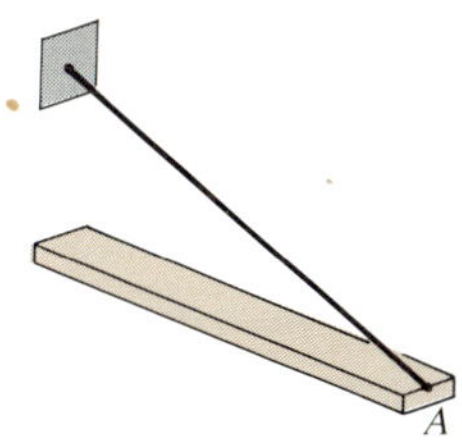	Attached point A cannot move outward, parallel to the cable. Free rotation.	$\bar{T}$ is a tensile force parallel to the cable 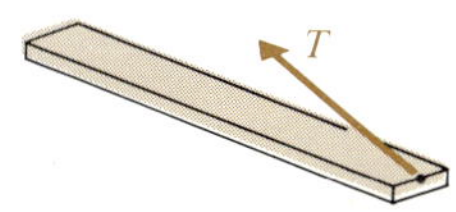
Short link 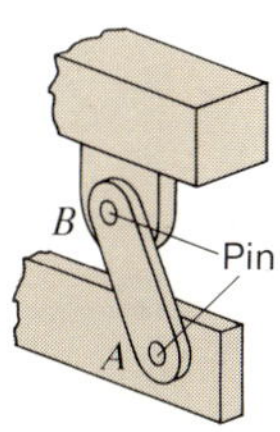	Attached point A cannot move parallel to the line AB of the link.	$\bar{F}_{AB}$ can be either a tensile or compressive force parallel to the link. 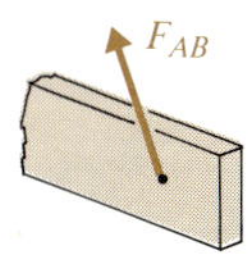

(*continued*)

TABLE 1 *(continued)*

TYPE OF SUPPORT	MOTION RESTRICTION	REACTIONS
Smooth surface 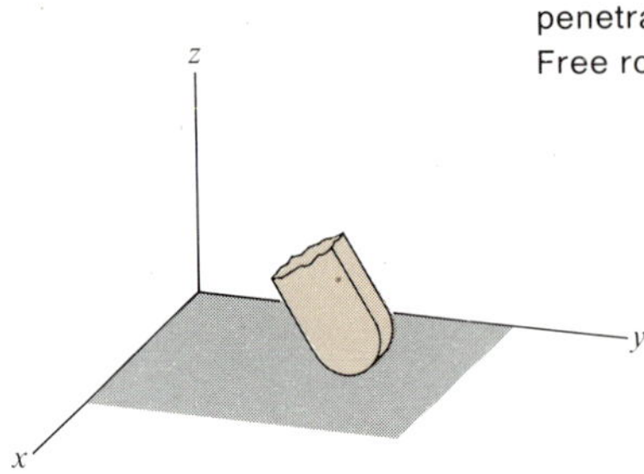	No movement that penetrates the surface. Free rotation.	$\bar{N}$ can only be normal upward 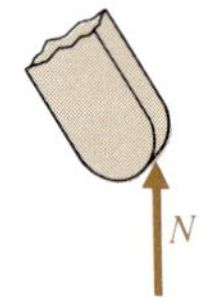
Rough surface or sharp edge 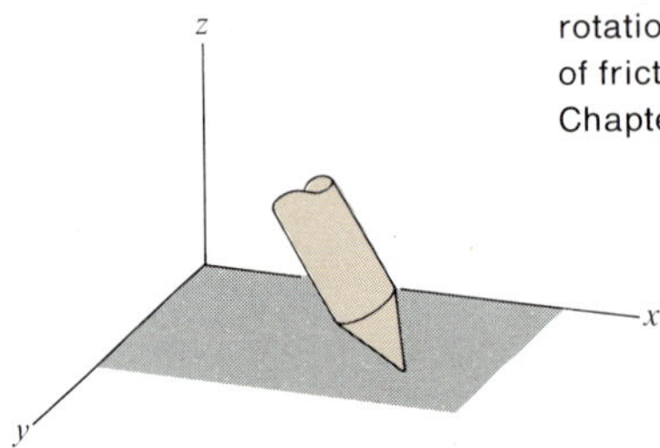	No movement that either penetrates the surface or is parallel to it. Free rotation. (For the effects of friction, see Chapter 5.)	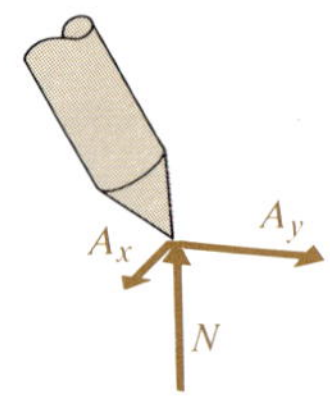
Small roller on a rough surface 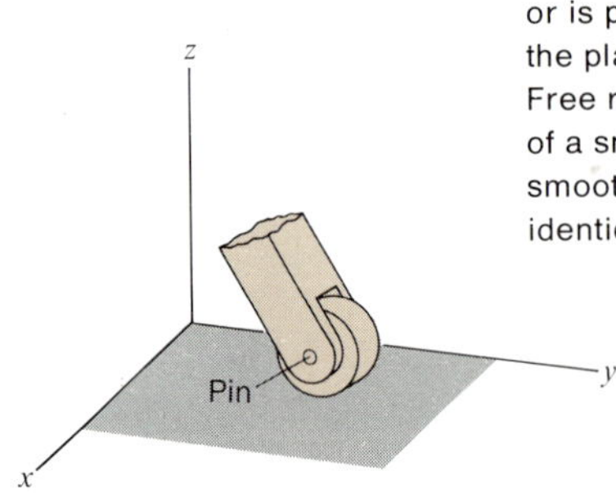	No movement that penetrates the surface or is perpendicular to the plane of the roller. Free rotation. The case of a small roller in a smooth groove is identical.	In the absence of a groove or friction, $\bar{A}_x = \bar{0}$. 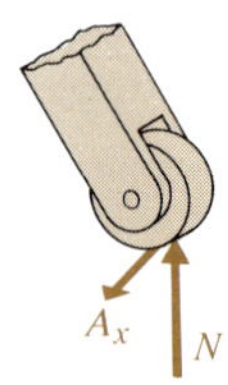
Collar on a smooth round bar 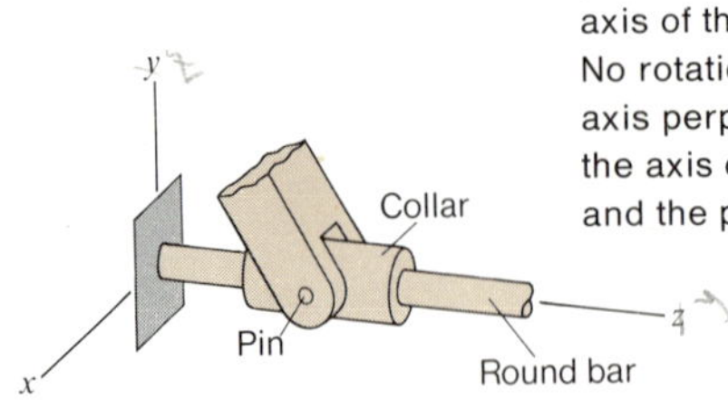	No movement perpendicular to the axis of the fixed bar. No rotation about an axis perpendicular to the axis of the fixed bar and the pin.	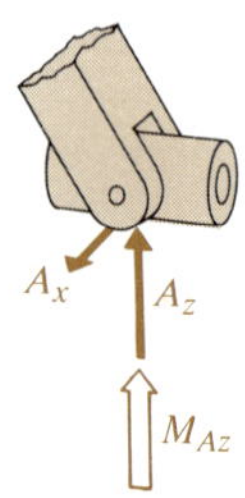

(continued)

TABLE 1 *(continued)*

TYPE OF SUPPORT	MOTION RESTRICTION	REACTIONS
Ball (and socket) joint 	No movement. Free rotation.	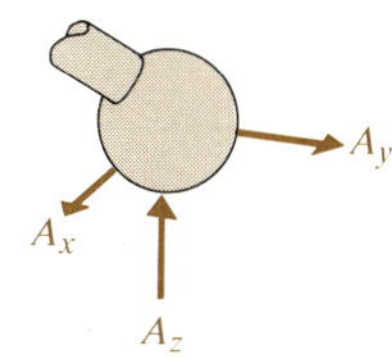
Bearings 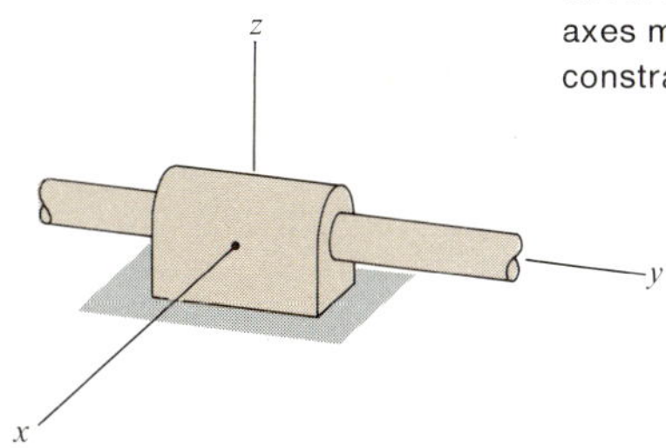	No movement in any direction perpendicular to the shaft. Depending on the design of the bearing, movement in the *y* direction or rotation about the *x* and *z* axes may also be constrained.	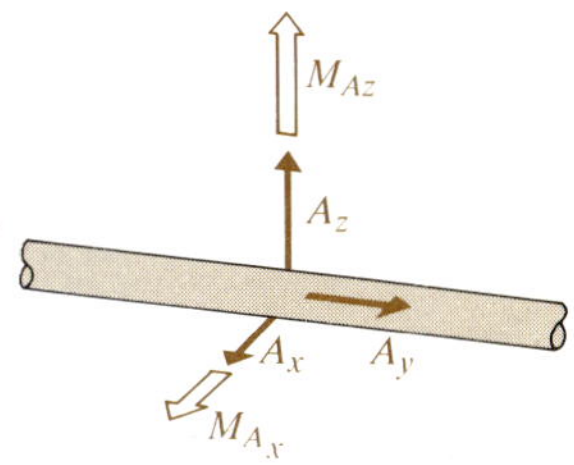 $\bar{A}_y = \bar{0}$ if the bearing does not carry thrust. $\bar{M}_{Ax} = \bar{M}_{Az} = \bar{0}$ if the bearing design does not restrict the corresponding rotations.

A question that arises in dealing with bodies that are supported by bearings, as depicted in Table 1, is whether or not to include the couples $\bar{M}_{Ax}$ and $\bar{M}_{Az}$ in drawing the free body diagram. The presence of these couples complicates the problem (frequently making it insoluble) by introducing additional reactions. Our approach in this text will be to ignore couple reactions for a bearing, unless it is specifically stated that there is a tight fit between the shaft and the bearing.

The following example will demonstrate how the concepts of constraints and constraint forces are implemented in drawing free body diagrams.

EXAMPLE 12

Draw free body diagrams for the systems shown. The centers of mass are denoted as points G.

a. A ladder

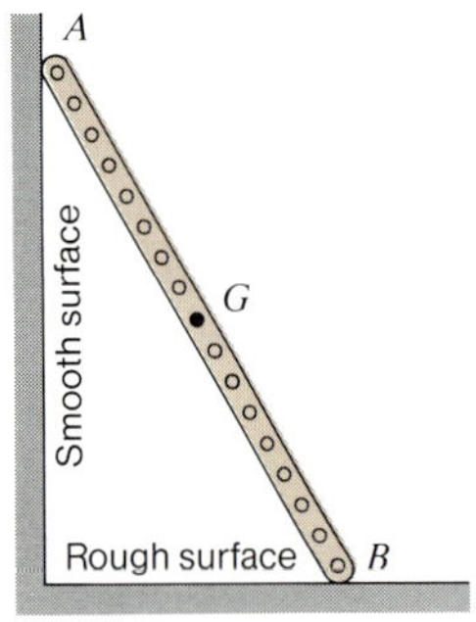

b. A disk attached to a block by a cable

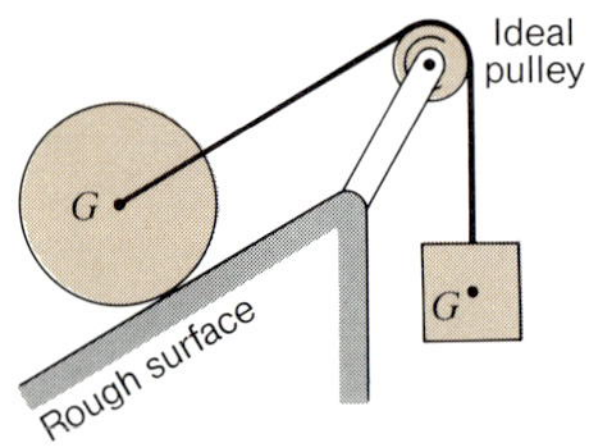

c. Bars AC and CD individually, and joined together (neglect gravity)

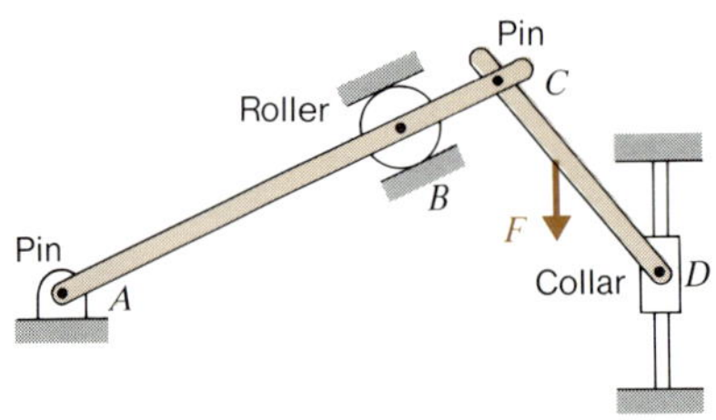

d. A door suspended by two hinges

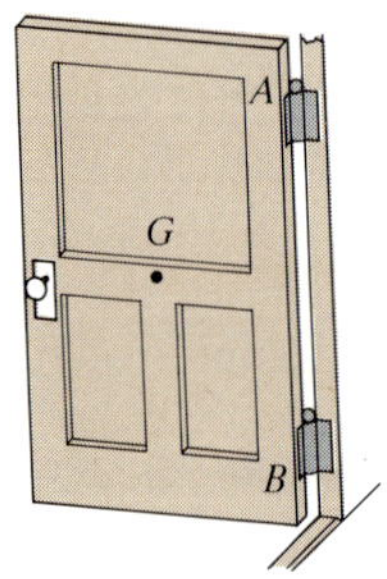

e. The entire truss structure (neglect gravity)

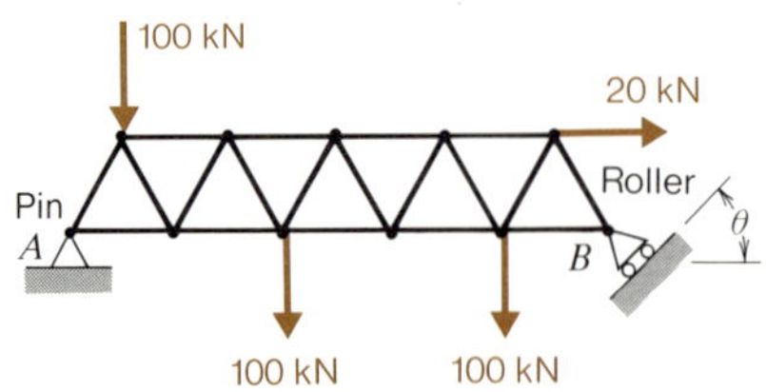

f. An antenna tower (neglect gravity)

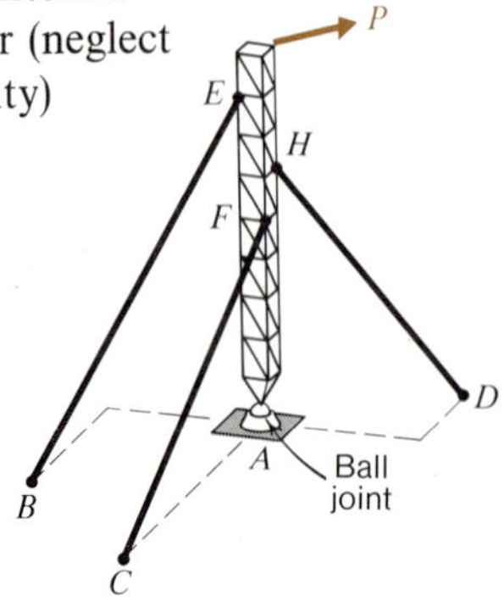

Solution

As an aid to including all forces, it is axiomatic that anything touching a body exerts a force on it.

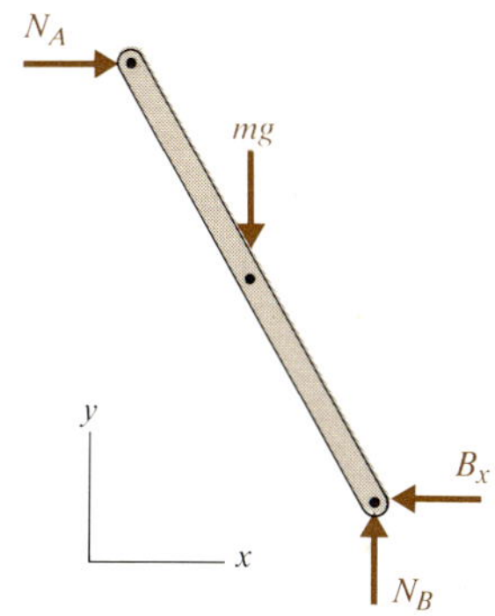

Part a

There is a force normal to the smooth wall at point A, whereas the rough surface exerts normal and tangential forces at point B. We show the weight force acting downward, passing through point G. The force $\bar{B}_x$ is shown pushing to the left because it opposes the way in which point B would tend to move. However, it is also correct to show $\bar{B}_x$ in the opposite direction, acting to the right, for we would then find a negative value for B_x. In contrast, it would make no sense to show $\bar{N}_A$ and $\bar{N}_B$ in opposite directions, because the surfaces can only push on the ladder.

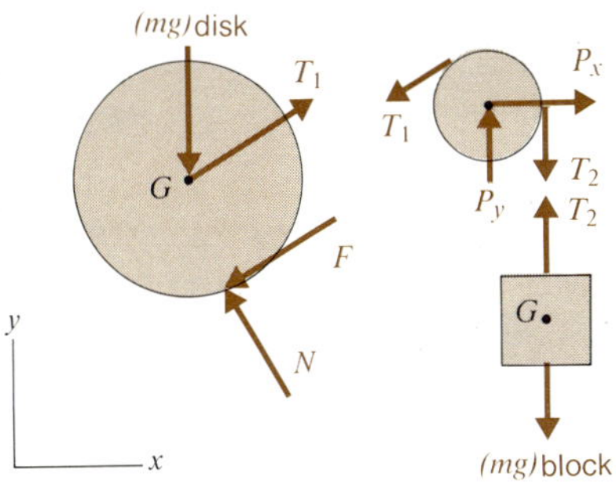

Part b

There are three connected bodies: the disk, the block, and the pulley. We shall draw free body diagrams of each. A useful technique to facilitate using Newton's third law is to *lay out the diagrams to resemble an assembly drawing*. The force $\bar{F}$ is the tangential component of the reaction between the disk and the rough surface. The forces $\bar{P}_x$ and $\bar{P}_y$ represent the reactions of the shaft on the pulley. The tensile forces T_1 and T_2 pull on the bodies they contact. We have already seen that for an ideal pulley, $T_1 = T_2$.

Part c

A pin joint exerts two reaction force components in a planar problem, whereas a roller or pin exerts no reaction force component in the direction in which motion is permitted. Further, the reactions exerted between the two bars at pin C must satisfy Newton's third law of action and reaction. These considerations yield the free body diagrams shown below for the individual bars. It should be noted that because roller B rides in a groove having two walls, the reaction at B may actually be in the opposite sense from the one shown. The same is true for each of the other reactions.

When we draw a free body of the two bars joined together, the reactions at pin B are internal to the system, so they do not appear in such a diagram. In effect, the diagram of the system is an overlay of the individual diagrams sketched below.

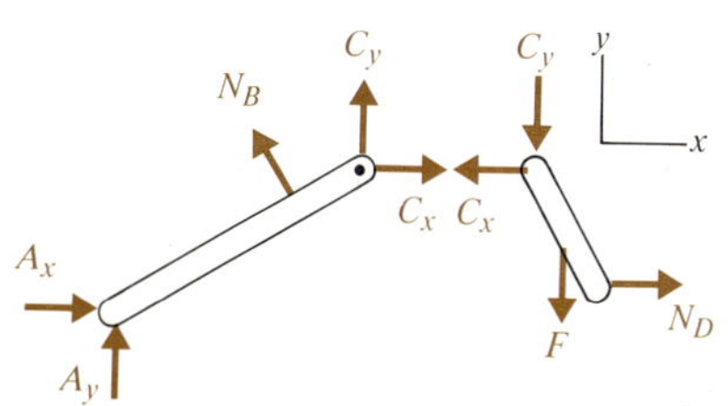

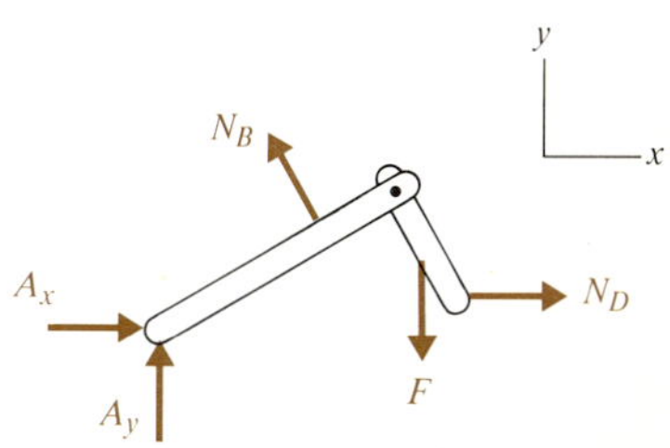

Part d

We agreed to neglect the couple reactions of hinges, hence the reactions at each hinge consist of three force components, as shown below.

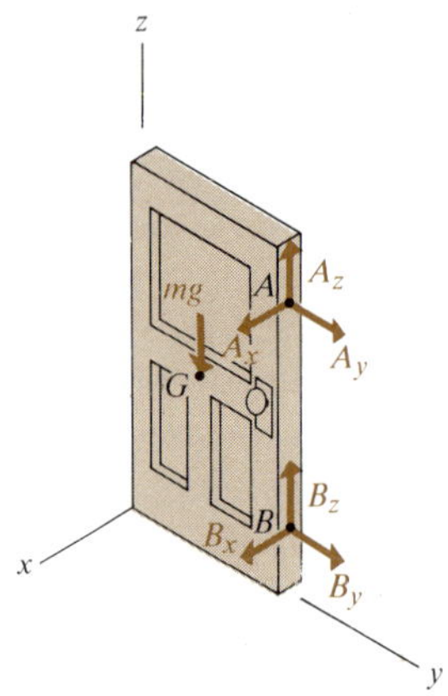

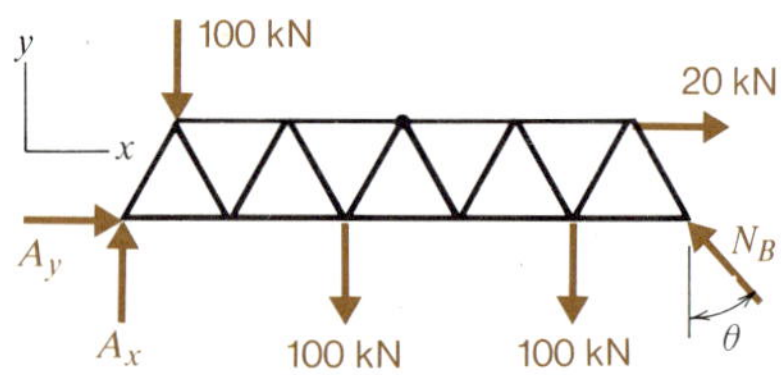

Part e

Note that the reaction of the roller is perpendicular to the inclined surface. For the type of roller shown, the reaction must be up and to the left, corresponding to the surface pushing up on the roller.

Part f

The cables exert tensile forces. The ball joint prevents movement of the base, but it does not restrict rotation. Thus, the reaction at joint A consists of a force having components A_x, A_y, and A_z.

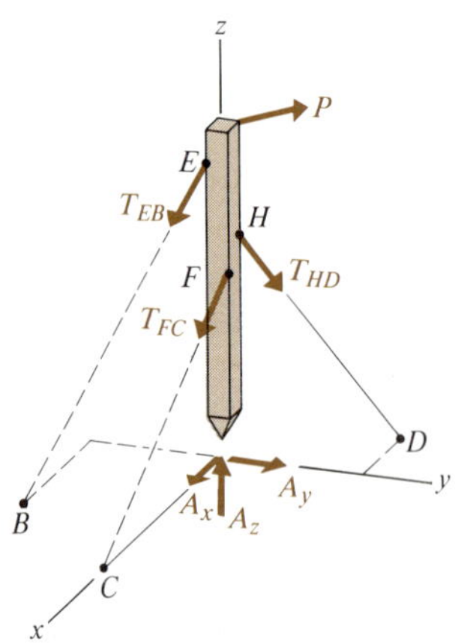

HOMEWORK PROBLEMS

Sketch free body diagrams for the systems shown. In each case the given forces are illustrated. In problems where the weight force is to be included, the location of the mass center, point G, is given. For simplicity, dimensions are omitted.

3.95 The traffic light

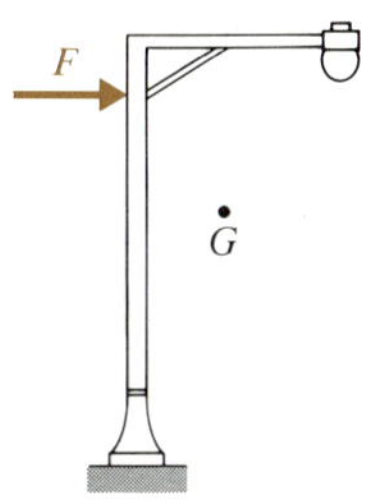

PROB. 3.95

3.96 The tire and wrench

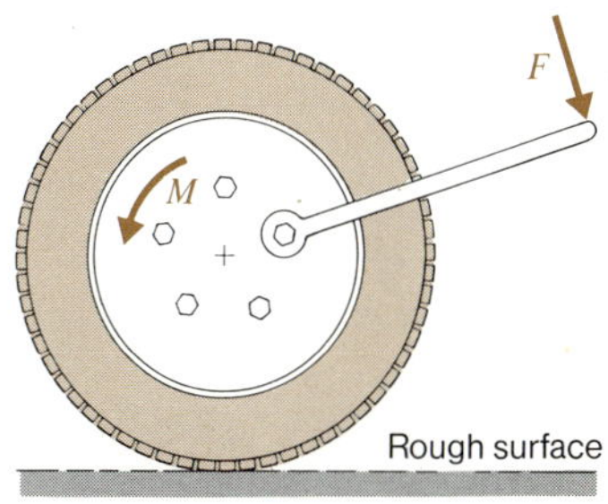

PROB. 3.96

3.97 The wrecking bar

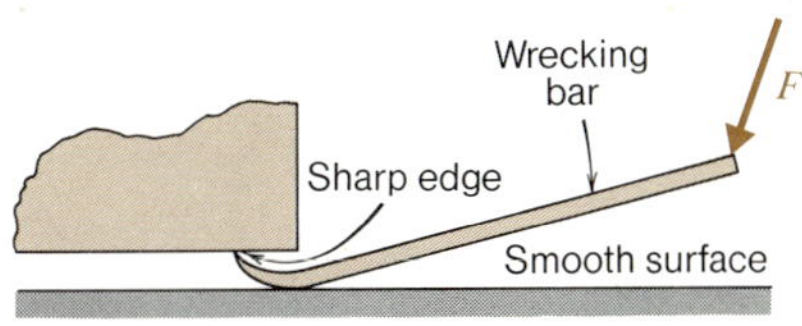

PROB. 3.97

3.98 The stepped disk

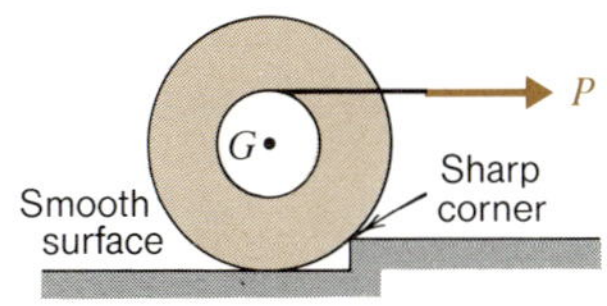

PROB. 3.98

3.99 The bent bar

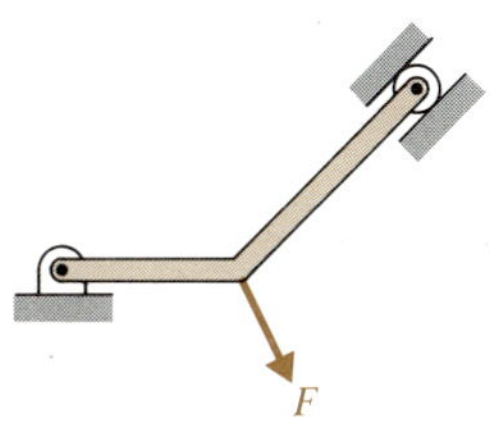

PROB. 3.99

3.100 The crate

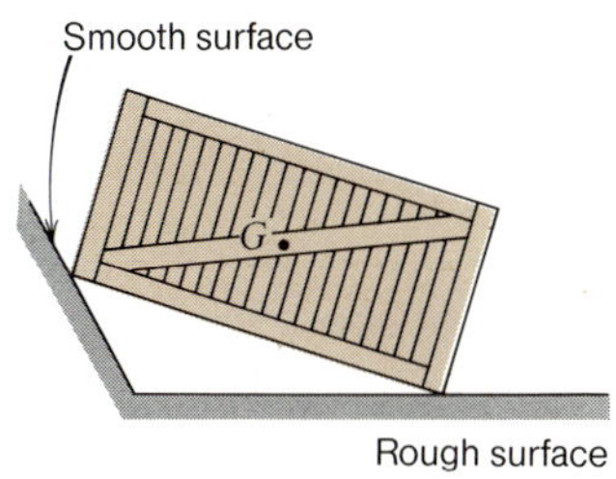

PROB. 3.100

3.101 The bar AB

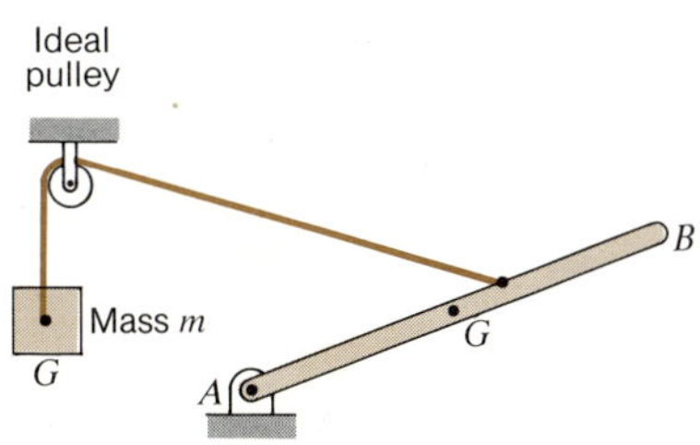

PROB. 3.101

3.102 The tape guide

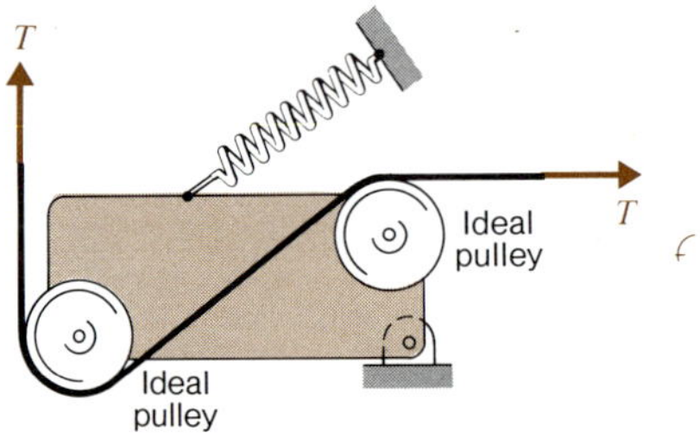

PROB. 3.102

3.103 The rectangular plate

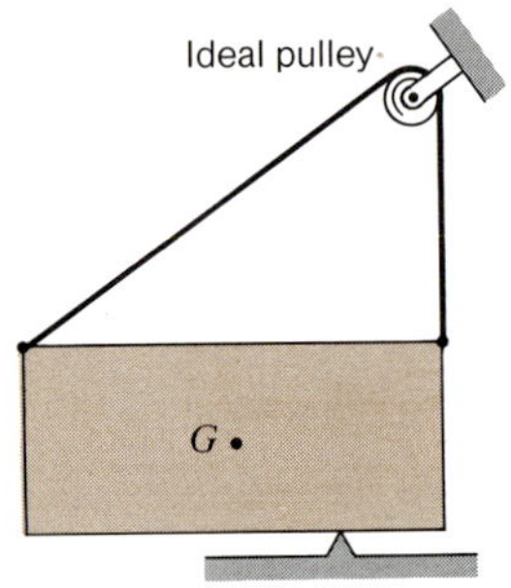

PROB. 3.103

3.104 Bars *AB* and *CD* individually

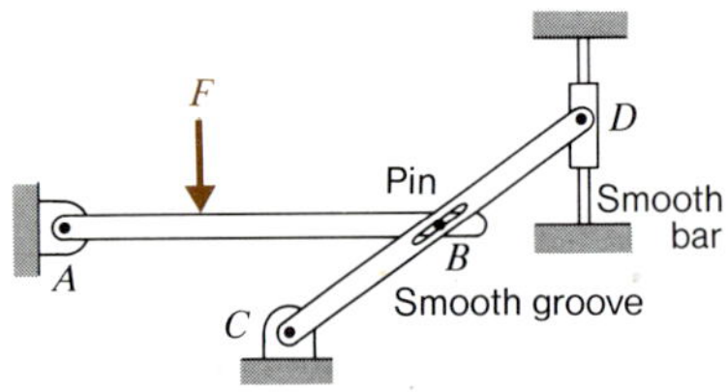

PROB. 3.104

3.105 The A-frame

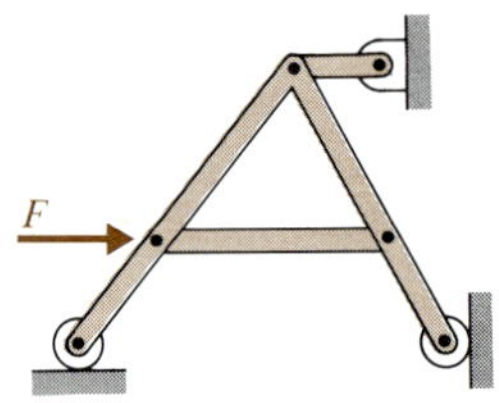

PROB. 3.105

3.106 The crane boom

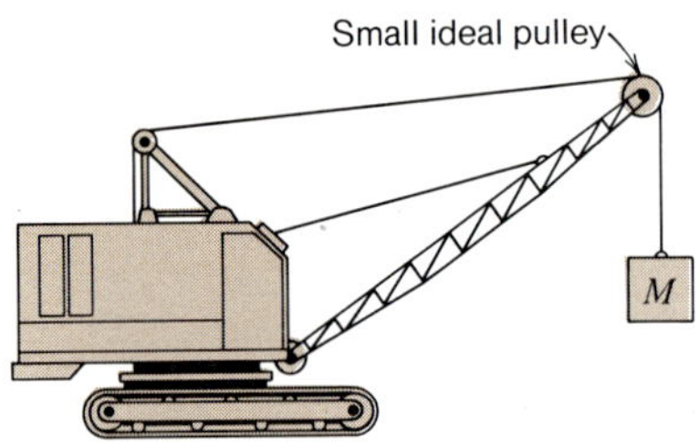

PROB. 3.106

3.107 Beam *AB*

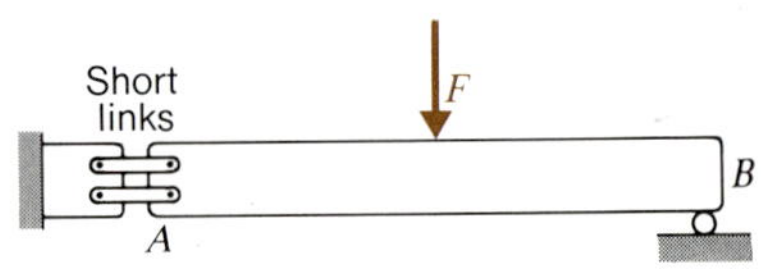

PROB. 3.107

3.108 The flywheel and bar *AB*, each individually

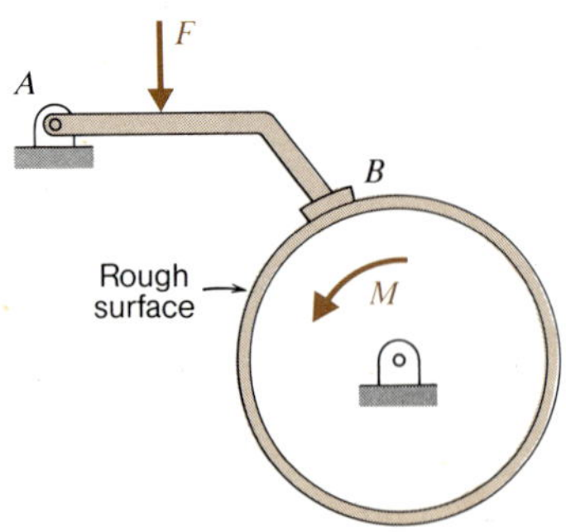

PROB. 3.108

3.109 The radio antenna

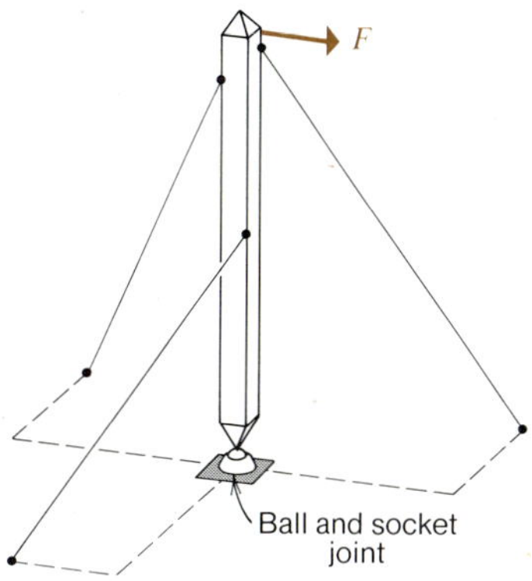

PROB. 3.109

3.110 The hatch cover

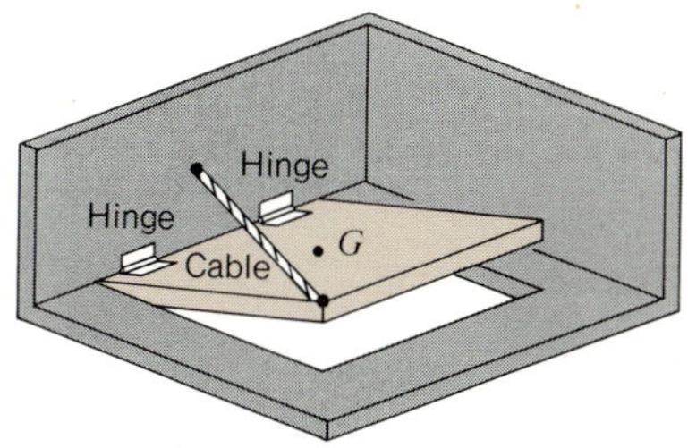

PROB. 3.110

3.111 The winch

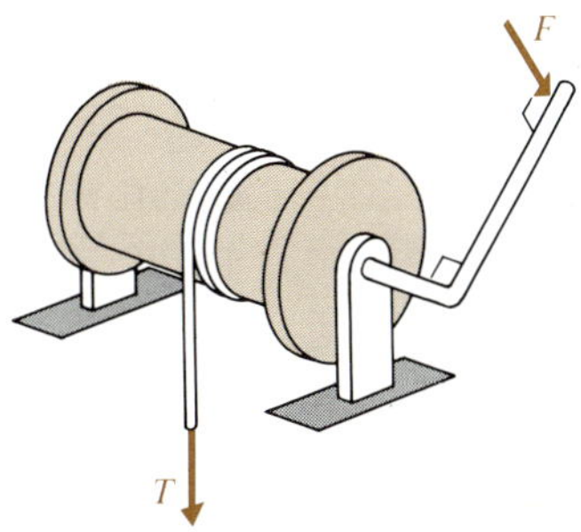

PROB. 3.111

3.112 The T-bar

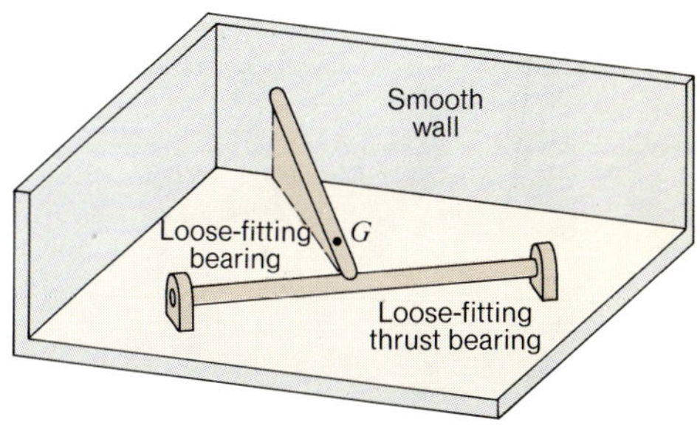

PROB. 3.112

3.113 The truss structure

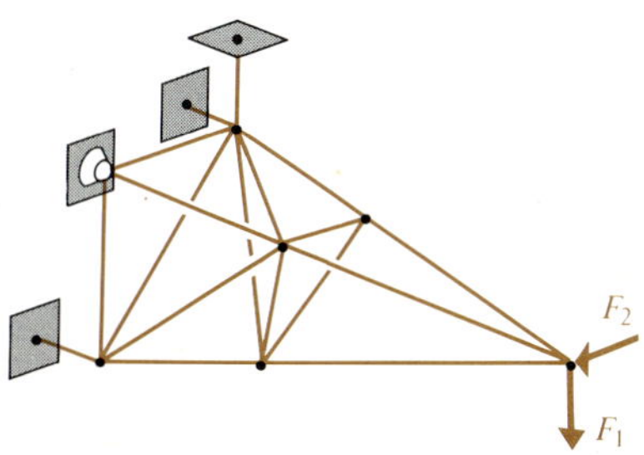

PROB. 3.113

3.114 The tubular guide for the flexible shaft

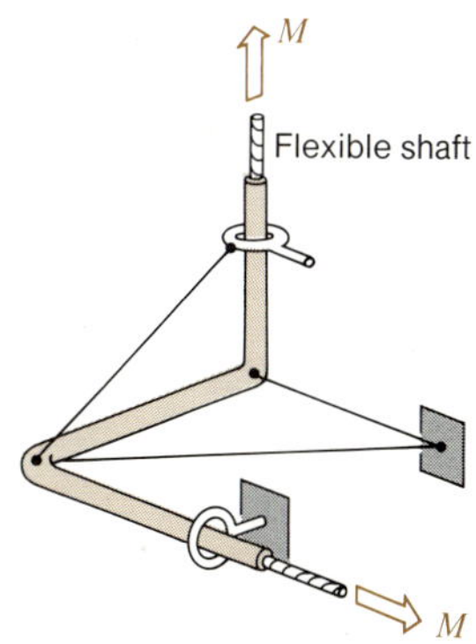

PROB. 3.114

2. Conditions for Static Equilibrium

We could develop a mathematical derivation of the criteria that must be satisfied if a body is to be in a state of static equilibrium. Such an approach would be a special case of the derivation of the equations of motion for a rigid body, which is detailed in the Dynamics text. In the present context, mathematical rigor does little to increase our understanding, so we will employ a heuristic line of reasoning.

Let us visualize replacing the actual forces acting on a body by an equivalent force-couple system acting at some convenient point C. The force $\bar{R}$, which is the sum of the actual forces, $\bar{R} = \sum \bar{F}_i$, describes the resultant pushing or pulling effect on point C. The couple $\bar{M}_C$ is the sum of the moments about point C of the actual forces, $\bar{M}_C = \sum \bar{r}_{i/C} \times \bar{F}_i$. This couple describes the resultant tendency of the actual forces to cause the body to rotate about point C. The concept of static equilibrium of a rigid body is that there is no tendency for the body to move. It follows that

> In order for a rigid body to be in a state of static equilibrium, the system of forces acting on the body must have both a zero resultant force and a zero resultant couple.

The most direct way of setting the resultant force and couple equal to zero is to actually compute the sum of the forces and the sum of the moments about any arbitrary point. We then equate the two resultant vectors to zero, thus obtaining

$$\begin{aligned} \sum \bar{F} &= \bar{0} \\ \sum \bar{M}_C &= \bar{0} \end{aligned} \tag{25}$$

In the case of a three-dimensional force system, the force and moment sums have three components each, so eqs. (25) are equivalent to the following six scalar *equations of static equilibrium*:

$$\begin{matrix} \sum F_x = 0 & \sum F_y = 0 & \sum F_z = 0 \\ \sum M_{Cx} = 0 & \sum M_{Cy} = 0 & \sum M_{Cz} = 0 \end{matrix} \tag{26}$$

The equilibrium equations for the case of a concurrent system of forces, that we derived in the previous chapter, are a special case of eqs. (26), as may be seen by letting point C be the point of concurrency.

A solution to the equations of static equilibrium yields the reactions and other unknown forces appearing in a free body diagram. The choice of the point C for summing moments is arbitrary. We shall generally choose point C to be along the line of action of at least one unknown reaction, in order to eliminate that unknown from the moment equation. This procedure simplifies the equations to be solved.

This ability to simplify the equilibrium equations suggests that we consider alternative formulations in which moments are summed about more than one point. Let us consider the case of a planar set of forces first.

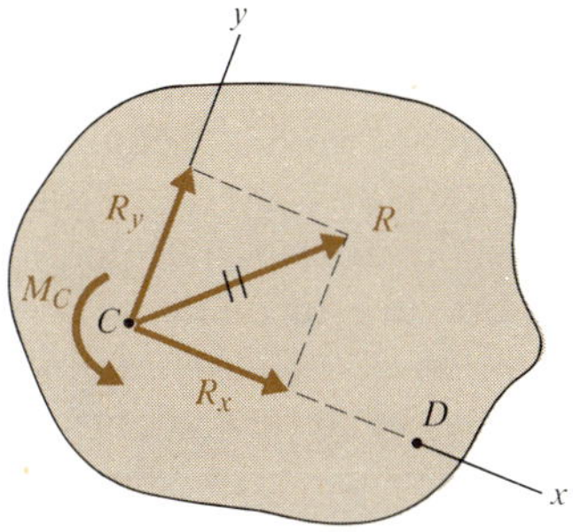

FIGURE 24

In Figure 24 we show a force $\bar{R}$ and couple $\bar{M}_C$ that are equivalent to an actual system of planar forces acting on the body. For this situation we will investigate the result of setting the moment sums about both points C and D to zero.

In order to simplify the notation, let us define x to be collinear with points C and D, and let y be the other axis in the plane of the force system. The components of $\bar{R}$ are then as depicted in the figure. Static equilibrium requires that R_x, R_y, and M_C all be zero.

Only the couple $\bar{M}_C$ exerts a moment about point C, whereas $\bar{M}_C$ and $R_y\bar{j}$ both exert moments about point D. It follows that the result of equating the moment sums about points C and D to zero will be that $M_C = 0$ and $R_y = 0$. An equation that then sets the remaining component R_x to zero may be obtained by summing forces in any direction other than that of the y axis. Alternatively, we may show that $R_x = 0$ by summing moments about a third point, provided that the point is not situated on the x axis.

We deduce from the foregoing that *when the system of forces is planar, there are only three nontrivial scalar equations of static equilibrium.* These equations may be obtained from either of three alternative formulations, as follows:

1. Equate to zero the moment about one point and the force sums in two directions. This is the approach of eqs. (26).
2. Equate to zero the moment sums about two points and the force sum in any direction that is not perpendicular to the line connecting the two chosen points.
3. Equate to zero the moment sums about three points that are not collinear.

We could present similar alternatives for the case of a three-dimensional system of forces. However, as we have already seen, the determination of moments in the three-dimensional case is somewhat lengthier than that for planar forces. Thus, the simplification in the resulting equilibrium equations would be offset by the additional computations required to obtain these equations. In order to have a unified approach for planar and three-dimensional forces, in both cases we will generally employ the formulation of eqs. (26), where the force and moment results are equated to zero. Hence, for a planar set of forces lying in the xy plane, the equilibrium equations we shall employ are

$$\sum F_x = 0, \qquad \sum F_y = 0, \qquad \sum M_{Cz} = 0 \tag{27}$$

3. Static Indeterminacy and Partial Constraints

The equations of static equilibrium are the conditions that must exist if a body is to remain at rest. The solvability of these equations is related to the manner in which the body is supported.

In the case of a general force system, there are six scalar equilibrium equations, which may be obtained by equating to zero the components of the resultant force sum and of the resultant moment sum. In the case of a planar force system, only three equations are nontrivial. Two considerations are involved in the comparison between the number of equilibrium equations to be solved and the number of unknown components associated with the reaction forces and couples. The first item is the adequacy of constraint, which pertains to whether the supports are adequate to prevent the body from moving. The second consideration is static determinacy, which addresses the question of whether the laws of static equilibrium are sufficient to evaluate the reactions.

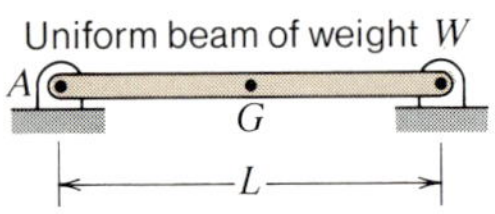

FIGURE 25a

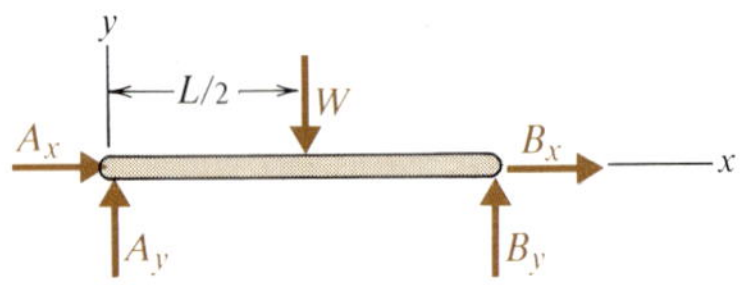

FIGURE 25b

We will discuss these ideas by considering the planar system of a horizontal beam loaded by its own weight. Suppose that such a beam is supported by pins at both ends, as depicted in Figure 25a. The corresponding free body diagram appears in Figure 25b.

We sum moments about point A in order to eliminate the reactions A_x and A_y with the result that the equilibrium equations are

$$\sum F_x = A_x + B_x = 0, \qquad \sum F_y = A_y + B_y - W = 0$$

$$\sum M_{Az} = B_y(L) - W\left(\frac{L}{2}\right) = 0$$

Solution of these equations yields $A_y = B_y = W/2$, $B_x = -A_x$. In other words, the laws of static equilibrium are adequate for finding the vertical reactions, but not the horizontal reactions. We might guess that $A_x = B_x = 0$, but that assumption would be incorrect.

The foregoing situation is one in which there are too many unknown reactions for the number of nontrivial equilibrium equations. We say that the system is *statically indeterminate*, because the laws of statics are insufficient for evaluating the forces required to prevent motion. Thus, the model of a rigid body is inadequate to analyze a statically indeterminate system. The unknown reactions, which are A_x and B_x for the beam in Figure 25a, can be determined only when we consider the manner in which the body deforms. Such questions are treated in the subject of the mechanics of materials.

By definition, static indeterminacy occurs when there are more reactions than necessary to prevent motion. Thus, we may say that static indeterminacy is synonymous with a state of *excessive constraint*. The equilibrium equations in this case may be solved for some of the reactions in terms of the extra ones. These extra reactions, which are either A_x or B_x for the beam in Figure 25, are *redundant* because the conditions of static equilibrium may be satisfied without them.

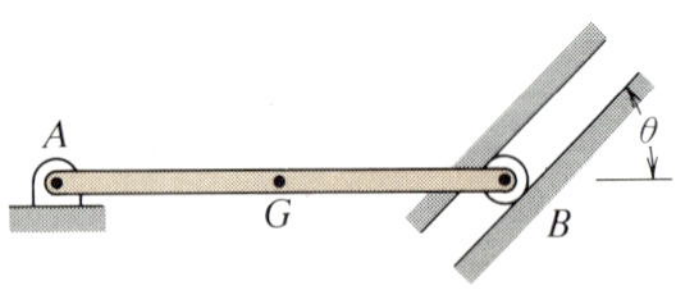

FIGURE 26a

Suppose that we remove the redundant reactions from a statically indeterminate system by modifying the supports. For example, let us change pin B into a roller, as depicted in Figure 26a.

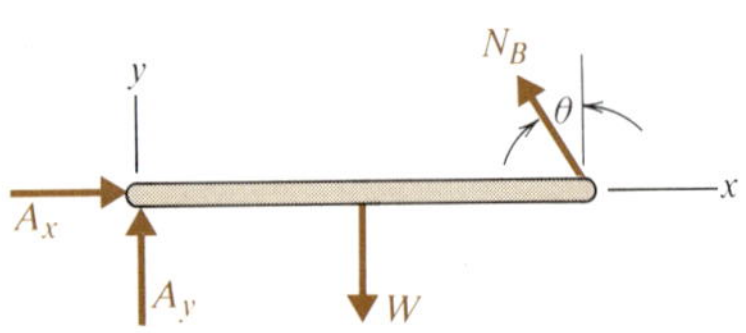

FIGURE 26b

The equilibrium equations obtained from the free body diagram in Figure 26b are

$$\sum F_x = A_x - N_B \sin\theta = 0, \qquad \sum F_y = A_y + N_B \cos\theta - W = 0$$

$$\sum M_{Az} = (N_B \cos\theta)L - W\left(\frac{L}{2}\right) = 0$$

If $\theta \neq 90°$, these equations may be solved for A_x, A_y, and N_B. The supports still prevent the system from moving, but the reactions may now be evaluated. We say that this is a *statically determinate* system, corresponding to supports that *properly constrain* the system.

The case where $\theta = 90°$ in Figure 26 is a special situation. In this case $\sum M_{Az} = 0$ cannot be satisfied ($\cos\theta = 0°$), and the force equilibrium equations only yield $A_y = W$, $A_x = -N_B$. Each equilibrium equation that cannot be satisfied corresponds to a type of motion that will occur: translational motion for a force equation, and rotational motion for a moment equation. (Precise definitions of these motion terms are given in the Dynamics text.) Thus, the fact that $\sum M_{Az} = 0$ cannot be satisfied when $\theta = 90°$ means that the supports permit the beam to rotate about point A. This is illustrated in Figure 27, where we see that the roller permits point B to move vertically, which is the direction in which this point would move if the beam were supported only by pin A.

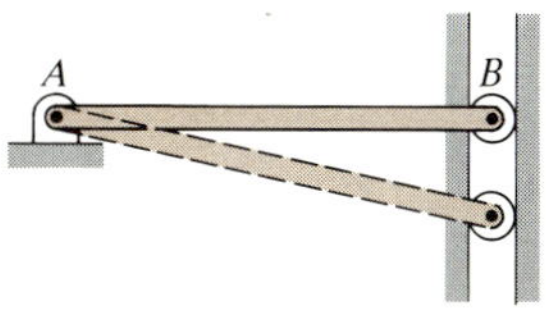

FIGURE 27

We say that the beam in Figure 27 is *improperly constrained*, because motion will occur even though the number of unknown reactions matches the number of nontrivial equilibrium equations. This condition arose because the manner in which the supports were arranged was incorrect for the prevention of motion.

It is interesting to note that when $\theta = 90°$ the reaction $\bar{N}_B$ in Figure 26b is equivalent to the horizontal reaction $B_x\bar{i}$ in Figure 25b. In other words, the modification of the indeterminate system in this case removed a reaction required for equilibrium, rather than the redundant reaction.

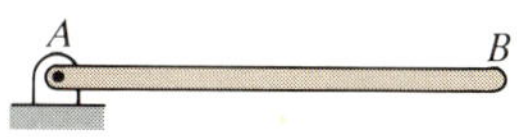

FIGURE 28a

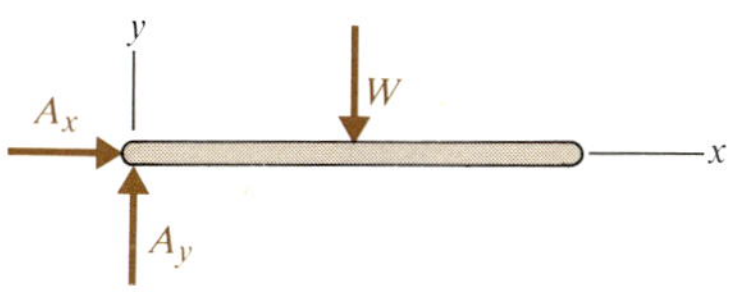

FIGURE 28b

Any modification of the supports in a statically determinate system that removes a reaction will result in a *partially constrained* system, that is, a system in which the supports cannot exert a sufficient number of reactions to prevent motion. As an example, let us suppose that the beam is supported only by pin A, as depicted in Figure 28a. The equilibrium equation $\sum M_{Az} = 0$ obtained from the free body diagram in Figure 28b cannot be satisfied because the weight W is the only force exerting a moment about point A. Thus, the beam will rotate about point A.

Confusion sometimes arises in distinguishing the conditions of partial and improper constraint. In a partially constrained situation, we can see by inspection that there are an insufficient number of reactions to satisfy all of the equilibrium equations. In contrast, the number of reactions matches the number of equilibrium equations in an improperly constrained system.

Suppose that the system of loads is such that there is no tendency for a possible motion to occur, as would happen if we applied a suitable upward force to the beams in Figures 27 and 28a. For this situation, the corresponding reactions in a partially constrained system are statically determinate, and they are statically indeterminate in an improperly constrained system. Partial constraints are inherent to dynamical systems, such as vehicles and machines, that are designed to be capable of motion.

Our main concern in this text is with statically determinate situations. Nevertheless, we should be able to discuss static determinacy and adequacy of constraints in planar and three-dimensional systems by considering the solvability of the equilibrium equations.

4. Problem Solving

The development of the ability to determine the forces acting on a body by applying the equations of static equilibrium is a primary goal of a course in statics. This capability is a prerequisite to the study of most of the topics in this text.

Problem solving is not an art, but rather a professional decision made by recognizing the proper keys. In what follows we present a logical way of thinking to identify the keys. It is not meant for memorization, but rather as a guide while working problems. Work enough problems *correctly* and the keys will become part of your thinking process.

1. Select the body (or bodies) to be isolated in a free body diagram (FBD) by examining where the forces of interest are applied. Draw a simple sketch of this body. Label any significant points if they are not already labeled.
2. Choose an *xyz* coordinate system for the description of the components of the force and position vectors. Usually, the coordinate axes should coincide with the directions in which the dimensions of the system are given. Show this coordinate system in the sketch of step 1.
3. Complete the FBD begun in step 1 by showing all forces acting on the body. In doing this, examine each support of the body to determine what types of reactions exist. The weight force, acting at the center of mass, should be included unless it is small compared to the other forces, or unless no information about it is given. If the dimensions describing a force do not appear in the given diagram, show the necessary dimensions in the FBD.
4. Choose a convenient point along the line of action of some unknown constraint force and compute the sum of the moments of the forces about this point. Equate the components of the moment sum to zero.
5. Equate to zero each component of the sum of the forces appearing in the FBD. (At times, this step may be omitted if it can be seen that the force sums involve unknown forces whose values are not desired, provided that these unknown forces have not appeared in the moment equation obtained in step 4.)
6. Solve the simultaneous equations that result from steps 4 and 5. Interpret the results, for example, by checking any initial assumptions.

Note that all of the simplifications of planar force systems, such as being able to sum force components directly and being able to compute moments without actually using the cross product, may be employed.

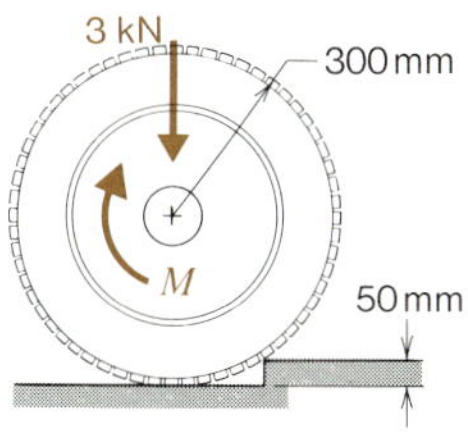

ILLUSTRATIVE PROBLEM 1

The rear axle of an automobile exerts a vertical 3 kN force and a couple $\bar{M}$ on the 20 kg tire, which is at rest. Determine the magnitude of $\bar{M}$ required to cause the tire to roll over the curb.

Solution

Steps 1 and 2 We isolate the tire in a free body diagram and select *xyz* axes that are horizontal and vertical because the vertical height of the curb is given in the sketch.

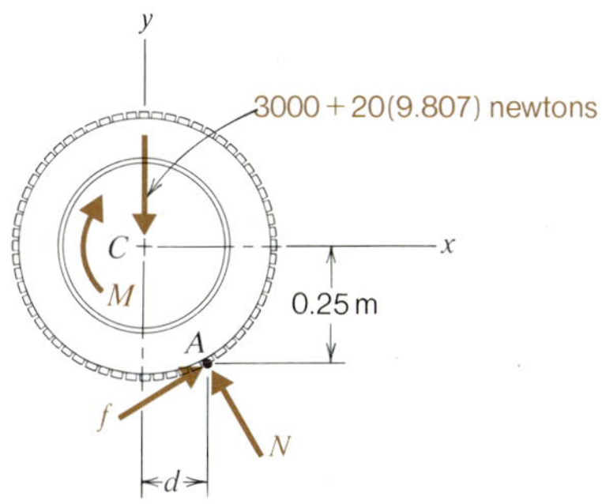

Step 3 In addition to the given forces, we show the weight force acting at the center of the tire. The curb is a sharp corner, so it exerts a general force on the tire. It is conventional to depict contact forces in terms of their tangent and normal components relative to the plane of contact. These are labeled $\bar{f}$ and $\bar{N}$ in the free body diagram.

In a general situation, there would also be a contact force between the ground and the tire. However, we want to study the case where the wheel is about to start rolling over the curb. At that instant, there is no contact with the ground, even though static equilibrium still exists.

Step 4 We select point *A* for the moment sum in order to eliminate the contact forces $\bar{f}$ and $\bar{N}$ from the moment equations. The lever arm *d* is

$$d = [(0.3)^2 - (0.25)^2]^{1/2} = 0.16583$$

so the moment equilibrium equation is

$$\sum M_{Az} = [3000 + 20(9.807)](0.16583) - M = 0$$

Step 5 The moment equation may be solved for *M*, so we omit the force summations.

Step 6 The solution of the equilibrium equation in step 4 is

$$M = 530 \text{ N-m}$$

This value is positive, so the clockwise direction in which it was shown is correct.

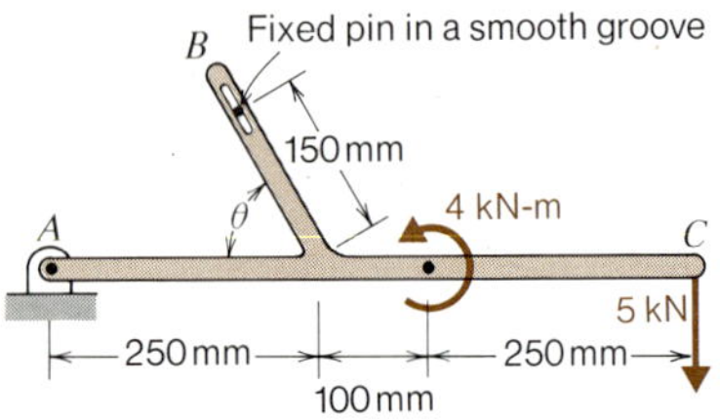

ILLUSTRATIVE PROBLEM 2

A 5 kN force and a 4 kN-m couple are applied to the forked bar. Determine the reaction forces at supports A and B when $\theta = 36.87°$.

Solution

Steps 1 and 2 We choose the forked bar for the FBD, because we wish to determine the forces required to support this body against the given set of loads. The x and y axes for this planar problem are selected as horizontal and vertical because the dimensions are given with respect to these directions.

Step 3 The reaction at pin B is normal to the surface of the smooth groove, so we have the following FBD.

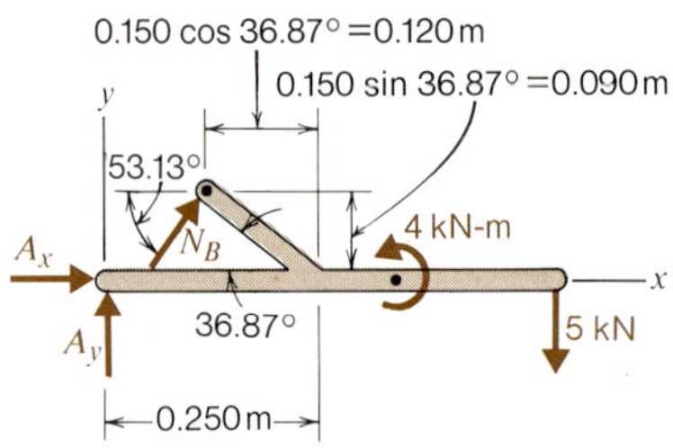

The dimensions shown in the FBD are needed to define the orientation and point of application of the normal force $\bar{N}_B$.

Step 4 We choose pin A for the moment sum in order to eliminate the unknown reactions at that pin. Thus, we have

$$\sum M_{Az} = -N_B \cos 53.13°(0.090) + N_B \sin 53.13°(0.250 - 0.120) + 4 - 5(0.600) = 0 \text{ kN-m}$$

Step 5 The equilibrium equations obtained by summing force components in the directions of the coordinate axes are

$$\sum F_x = N_B \cos 53.13° + A_x = 0 \text{ kN}$$

$$\sum F_y = N_B \sin 53.13° + A_y - 5 = 0 \text{ kN}$$

Step 6 We may now solve the moment equation for N_B and substitute this result into the force equations. Therefore

$$N_B[-0.6(0.9) + 0.8(0.13)] + 1.0 = 0 \quad \text{so} \quad N_B = -20 \text{ kN}$$

$$-20(0.60) + A_x = 0 \quad \text{so} \quad A_x = 12 \text{ kN}$$

$$-20(0.80) + A_y - 5 = 0 \quad \text{so} \quad A_y = 21 \text{ kN}$$

The negative value of N_B means that this force acts in the opposite direction from the one shown (guessed at) in the FBD. What does this mean in regard to which surface of the groove pin B is bearing against?

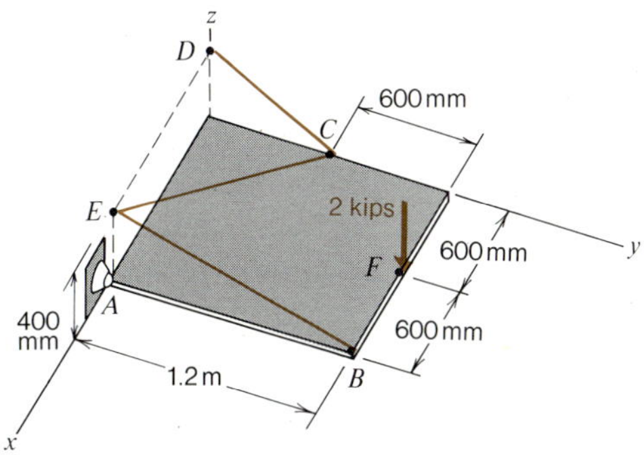

ILLUSTRATIVE PROBLEM 3

The thin square plate is loaded by a 2 kip force as shown. Determine the tension in each cable and the reactions exerted by the ball and socket joint at corner A.

Solution

Steps 1 and 2 The desired forces hold the plate in static equilibrium, so we draw a free body diagram of the plate. Two of the coordinate axes are aligned with the edges of the plate, consistent with the manner in which the dimensions are given.

Step 3 The ball joint prevents movement of corner A, so the reaction of that corner is a force having possible x, y, and z components. The cables exert tensile forces, and the weight of the plate is assumed to be negligible. We show these reactions and the 2 kip force in the free body diagram below.

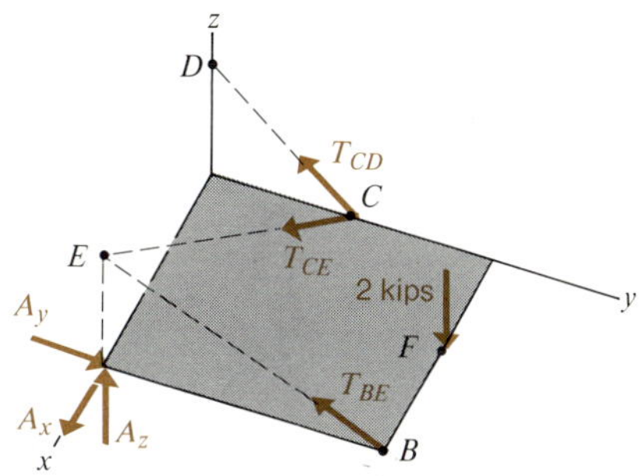

Step 4 We express the cable forces in terms of their components by writing their unit vectors.

$$\bar{T}_{BE} = T_{BE}\frac{\bar{r}_{E/B}}{|\bar{r}_{E/B}|} = T_{BE}\frac{-1.2\bar{j}+0.4\bar{k}}{[(1.2)^2+(0.4)^2]^{1/2}} = T_{BE}(-0.9487\bar{j}+0.3162\bar{k})$$

$$\bar{T}_{CD} = T_{CD}\frac{\bar{r}_{D/C}}{|\bar{r}_{D/C}|} = T_{CD}\frac{-0.6\bar{j}+0.4\bar{k}}{[(0.6)^2+(0.4)^2]^{1/2}} = T_{CD}(-0.8321\bar{j}+0.5547\bar{k})$$

$$\bar{T}_{CE} = T_{CE}\frac{\bar{r}_{E/C}}{|\bar{r}_{E/C}|} = T_{CE}\frac{1.2\bar{i}-0.6\bar{j}+0.4\bar{k}}{[(1.2)^2+(0.6)^2+(0.4)^2]^{1/2}} = T_{CE}(0.8571\bar{i}-0.4286\bar{j}+0.2857\bar{k})$$

We sum moments about corner A in order to eliminate the reactions at that corner from the moment equilibrium equations.

$$\begin{aligned}\sum \bar{M}_A &= \bar{r}_{B/A}\times\bar{T}_{BE}+\bar{r}_{C/A}\times\bar{T}_{CE}+\bar{r}_{C/A}\times\bar{T}_{CD}+\bar{r}_{F/A}\times(-2\bar{k})\\ &= 1.2\bar{j}\times T_{BE}(-0.9487\bar{j}+0.3162\bar{k})+(-1.2\bar{i}+0.6\bar{j})\times T_{CE}(0.8571\bar{i}\\ &\quad -0.4286\bar{j}+0.2857\bar{k})\\ &\quad +(-1.2\bar{i}+0.6\bar{j})\times T_{CD}(-0.8321\bar{j}+0.5547\bar{k})+(-0.6\bar{i}+1.2\bar{j})\times(-2\bar{k})\\ &= 0.3794T_{BE}\bar{i}+T_{CE}(0.1714\bar{i}+0.3428\bar{j})\\ &\quad +T_{CD}(0.3328\bar{i}+0.6656\bar{j}+0.9985\bar{k})-2.4\bar{i}-1.2\bar{j}=\bar{0}\end{aligned}$$

The scalar equations of moment equilibrium are obtained by collecting like components and setting them to zero.

$$\sum M_{Ax} = 0.3794T_{BE} + 0.1714T_{CE} + 0.3328T_{CD} - 2.4 = 0$$

$$\sum M_{Ay} = 0.3428T_{CE} + 0.6656T_{CD} - 1.2 = 0$$

$$\sum M_{Az} = 0.9985T_{CD} = 0$$

Step 5 The components of the cable forces are described by the vector expressions in step 4. The other force components are obvious from the free body diagram, so the force equilibrium equations are

$$\sum F_x = A_x + 0.8571T_{CE} = 0$$

$$\sum F_y = A_y - 0.9487T_{BE} - 0.8321T_{CD} - 0.4286T_{CE} = 0$$

$$\sum F_z = A_z + 0.3162T_{BE} + 0.5547T_{CD} + 0.2857T_{CE} - 2 = 0$$

Step 6 We first solve the moment equations in step 4 sequentially. Note that we are able to do so in this problem because we eliminated components of $\bar{F}_A$ from the equations by taking moments about point A.

$$T_{CD} = 0, \qquad T_{CE} = 3.50 \text{ kips}, \qquad T_{BE} = 4.74 \text{ kips}$$

Substituting these values into the force equations in step 5 yields

$$A_x = -3.00 \text{ kips}, \qquad A_y = 6.00 \text{ kips}, \qquad A_z = -0.50 \text{ kips}$$

Note that none of the cable tensions are negative, which must be the case, because cables can only pull on other bodies. The fact that $T_{CD} = 0$ means that it is not needed to support the given loading, but it would be important for other loadings on the plate.

ILLUSTRATIVE PROBLEM 4

An ideal pulley is pinned to end A of the bent bar ABC and a cable is wrapped around the pulley as shown. A tensile force of 200 lb is applied to the free end of the cable. (a) Prove that the tensile force within the cable is constant. (b) Determine the reactions at pin B and roller C. Is the roller bearing against the upper or lower surface?

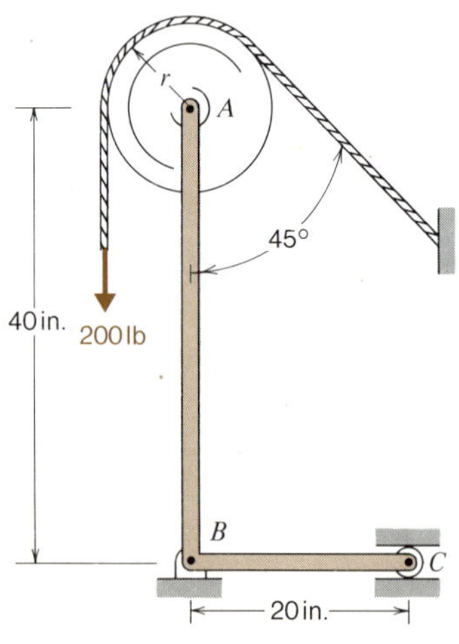

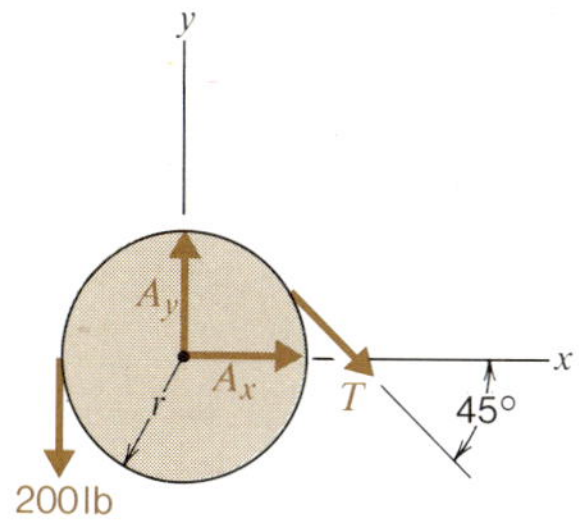

Solution

Up to now, we have accepted the fact that the tension in a cable that passes over an ideal pulley is constant. We shall now prove it.

Steps 1a and 2a In part (a) we are concerned with the interaction between the cable and the pulley, so we isolate the pulley in a free body diagram. Horizontal and vertical coordinate axes in this planar problem suit the directions in which dimensions are given.

Step 3a To the left of the pulley the tension within the cable is the 200 lb force applied to its end, and the tension on the right is considered to be some unknown value T. The pulley is also attached to a pin at its center, so there are two reaction force components at that point, as shown.

Step 4a Choosing pin A for the moment sum in order to eliminate the reactions at this location, the moment equilibrium equation is

$$\sum M_{Az} = -T(r) + 200(r) = 0 \text{ lb-in.}$$

Step 5a The force sums for the planar system of forces are

$$\sum F_x = T\cos 45^\circ + A_x = 0 \text{ lb}$$

$$\sum F_y = A_y - T\sin 45^\circ - 200 = 0 \text{ lb}$$

Step 6a The moment equation gives the desired result:

$$T = 200 \text{ lb}$$

If the tension were not constant, maintaining static equilibrium of the pulley would require a couple to resist the difference in the moments of the two tension forces.

The remaining force equilibrium equations give the values of A_x and A_y, which were not requested. Nevertheless, they are

$$A_x = -141.4 \text{ lb} \qquad A_y = 341.4 \text{ lb}$$

We will use these values later.

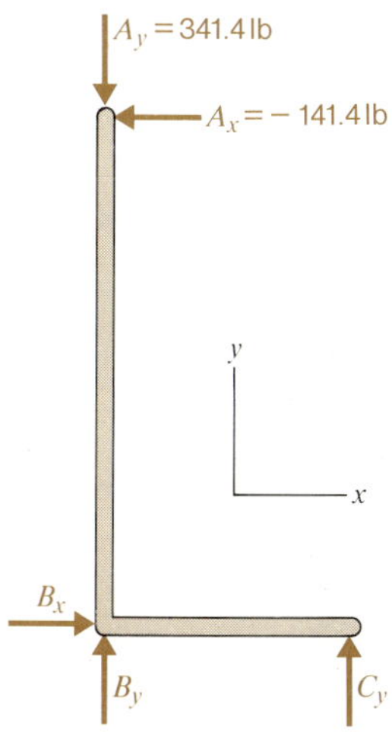

Steps 1b and 2b In order to determine the reactions at points B and C, we draw a free body diagram of the bent bar. The coordinate system utilized in part (a) will also suffice for this portion of the problem.

Step 3b Forces are exerted on the bent bar by the pin at point B, the roller at point C, and pin A, which holds the pulley. Hence, we have the free body diagram shown in the left margin.

Particular attention is drawn to the way in which the reaction forces at point A are entered into the free body diagram. We show $\bar{A}_x$ and $\bar{A}_y$ acting on the bar opposite from the directions in which they were assumed when we drew the free body diagram of the pulley, in accordance with Newton's third law. Then, after depicting the algebraic forces, we enter their numercial values with the same sign as they were calculated to have. Thus, we see that the pulley really pulls the bar to the right while pressing it down. This procedure is employed in order to prevent serious, and also common, sign errors. We will have more to say about this matter in the next chapter.

Step 4b Summing moments about pin B, we have

$$\sum M_{Bz} = C_y(20) + (-141.4)(40) = 0 \text{ lb-in.}$$

Step 5b The force sums are

$$\sum F_x = B_x - (-141.4) = 0 \text{ lb}$$

$$\sum F_y = B_y + C_y - 341.4 = 0 \text{ lb}$$

Step 6b The solutions of the equilibrium equations are readily found to be

$$C_y = 282.8 \text{ lb}, \qquad B_x = 141.4 \text{ lb}, \qquad B_y = 58.6 \text{ lb}$$

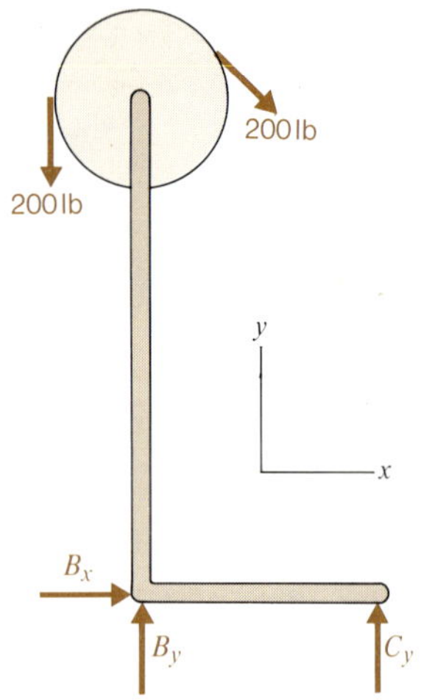

Because C_y is positive, it acts in the same sense as that shown in the FBD. Therefore, roller C is being pushed upward, which means that it is bearing on the lower surface.

As a final comment, let us consider how we would have approached the solution of part (b) if we were allowed to consider as an established fact that the tension within a cable is unchanged when the cable passes over an ideal pulley. In this case, we would isolate the assembly of bar ABC and the pulley in a free body diagram, as shown. That would leave only the reaction forces at points B and C to be determined. This approach eliminates the intermediate step of studying the interaction between the pulley and the bent bar.

ILLUSTRATIVE PROBLEM 5

The T-bar is supported by thrust bearing A, bearing B, and the smooth wall at end D. The bar fits loosely in both bearings. It is known that bar sections AB and CD each have a mass of 2 kg, and that their centers of mass are at their respective midpoints. Determine the reactions exerted by each bearing and by the wall.

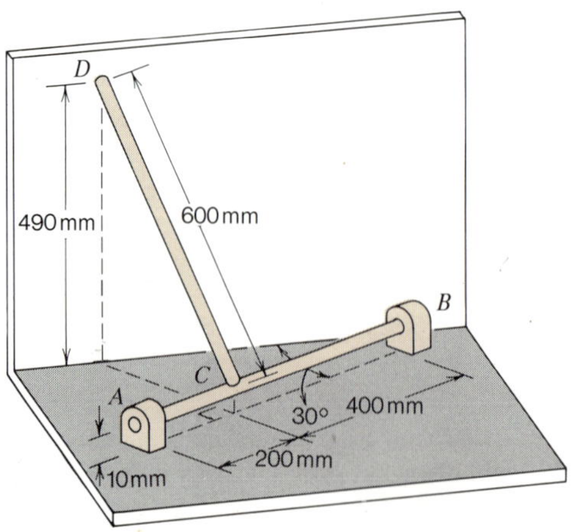

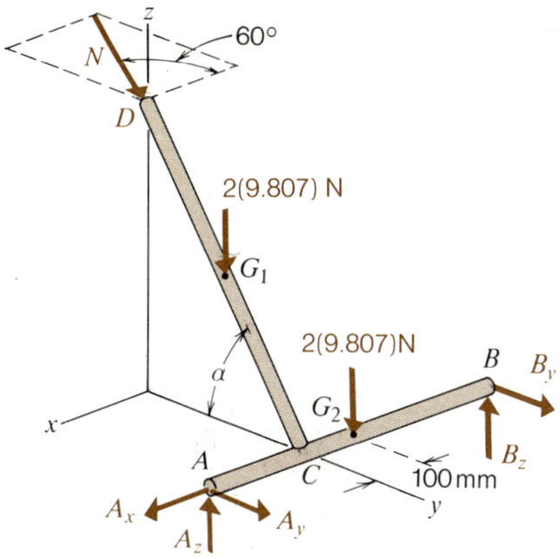

Solution

Steps 1 and 2 We isolate the T-bar in a free body diagram in order to evaluate the forces acting on it. The x axis is aligned parallel to segment AB and the z axis is vertical, in accord with the dimensioning of the system.

Step 3 The bearings are loose fitting, so they exert no couple reactions. Only bearing A exerts a thrust reaction parallel to segment AB. The reaction exerted by the wall is normal to the wall because the surface is smooth. This reaction is therefore horizontal, at a 60° angle from the x axis. Finally, we must remember to include the weight force of each segment in the free body diagram.

Step 4 We first write the normal reaction in terms of its components

$$\bar{N} = N(\cos 60°\,\bar{i} + \sin 60°\,\bar{j}) = N(0.50\bar{i} + 0.866\bar{j})$$

The angle of elevation α for segment CD is

$$\alpha = \sin^{-1}\left(\frac{0.48}{0.60}\right) = 53.13°$$

Moments will be summed about point A in order to eliminate the reactions at bearing A. For this, we require the position vectors for points G_1 and D.

$$\bar{r}_{D/A} = -0.2\bar{i} - 0.6\cos\alpha\bar{j} + 0.48\bar{k} = -0.2\bar{i} - 0.36\bar{j} + 0.48\bar{k} \text{ m}$$

$$\bar{r}_{G_1/A} = -0.2\bar{i} + 0.3(-\cos\alpha\bar{j} + \sin\alpha\bar{k}) = 0.2\bar{i} - 0.18\bar{j} + 0.24\bar{k} \text{ m}$$

Then

$$\begin{aligned}\sum \bar{M}_A &= \bar{r}_{B/A} \times (B_y\bar{j} + B_z\bar{k}) + \bar{r}_{D/A} \times \bar{N} + \bar{r}_{G_1/A} \times (-19.614\bar{k}) + \bar{r}_{G_2/A} \times (-19.614\bar{k}) \\ &= (-0.6\bar{i}) \times (B_y\bar{j} + B_z\bar{k}) + (-0.2\bar{i} - 0.36\bar{j} + 0.48\bar{k}) \times N(0.5\bar{i} + 0.866\bar{j}) \\ &\quad + (-0.2\bar{i} - 0.18\bar{j} + 0.24\bar{k}) \times (-19.614\bar{k}) + (-0.3\bar{i}) \times (-19.614\bar{k}) \\ &= 0.6B_z\bar{j} - 0.6B_y\bar{k} + N(-0.4157\bar{i} + 0.24\bar{j} + 0.0068\bar{k}) \\ &\quad + (3.531\bar{i} - 3.923\bar{j}) - 5.884\bar{j} = \bar{0}\end{aligned}$$

The component equations we obtain from this vector moment equation are

$$\sum M_{Ax} = -0.4157N + 3.531 = 0$$

$$\sum M_{Ay} = 0.6B_z + 0.24N - 3.923 - 5.884 = 0$$

$$\sum M_{Az} = -0.6B_y + 0.0068N = 0$$

Step 5 The force equilibrium equations are

$$\sum F_x = 0.5N + A_x = 0$$

$$\sum F_y = 0.866N + A_y + B_y = 0$$

$$\sum F_z = A_z + B_z - 2(19.614) = 0$$

Step 6 We first solve the moment equations in step 4.

$$N = 8.49 \text{ N}, \qquad B_z = 12.95 \text{ N}, \qquad B_y = 0.10 \text{ N}$$

The reactions at bearing A are then found by substituting these values into the force equations in step 5.

$$A_x = -4.25 \text{ N}, \quad A_y = -7.45 \text{ N}, \quad A_z = 26.28 \text{ N}$$

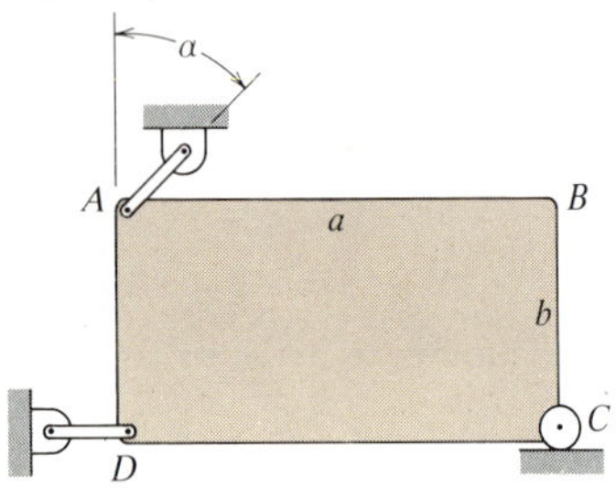

ILLUSTRATIVE PROBLEM 6

Determine the value of the angle α for the short link at point A that results in improper constraint of the rectangular plate.

Solution

Recall that the condition of partial constraint concerns equilibrium under a general loading. Hence, we will assume an arbitrary set of loads.

Steps 1 and 2 In order to study the nature of the constraints of the plate, we draw a free body diagram of the plate. The given dimensions suggest choosing coordinate axes parallel to the edges of the plate.

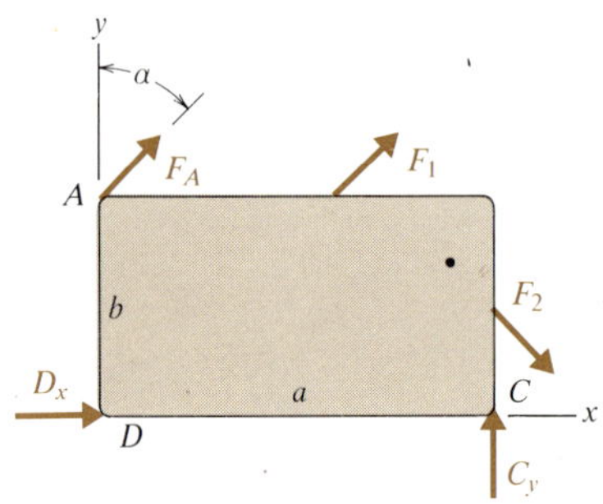

Step 3 Let two arbitrary forces $\bar{F}_1$ and $\bar{F}_2$ represent the general set of loads. The short links exert forces only along their axes, so the free body diagram is as shown.

Step 4 Select point C for the moment sum because the reaction forces $\bar{C}_y$ and $\bar{D}_x$ both intersect this point. The moment equilibrium equation is then

$$\sum M_{Cz} = -(F_A \sin \alpha)b - (F_A \cos \alpha)a + M_1 + M_2 = 0$$

where M_1 and M_2 are the moments of $\bar{F}_1$ and $\bar{F}_2$ respectively, about point C.

Step 5 The force equilibrium equations are

$$\sum F_x = D_x + F_A \sin \alpha + (F_1)_x + (F_2)_x = 0$$

$$\sum F_y = C_y + F_A \cos \alpha + (F_1)_y + (F_2)_y = 0$$

Step 6 The three equilibrium equations are now solved for the values of F_A, C_y, and D_x in terms of the angle α, yielding

$$F_A = \frac{M_1 + M_2}{b \sin \alpha + a \cos \alpha}$$

$$D_x = -(F_1)_x - (F_2)_x - \frac{(M_1 + M_2) \sin \alpha}{b \sin \alpha + a \cos \alpha}$$

$$C_y = -(F_1)_y - (F_2)_y - \frac{(M_1 + M_2)\cos\alpha}{b\sin\alpha + a\cos\alpha}$$

We now look for the condition of partial constraint, which is evidenced by an infinite value for at least one of the reactions. For this system, all three reactions are infinite when the denominator $b \sin\alpha + a\cos\alpha = 0$. Thus

$$\alpha = -\tan^{-1}\left(\frac{a}{b}\right)$$

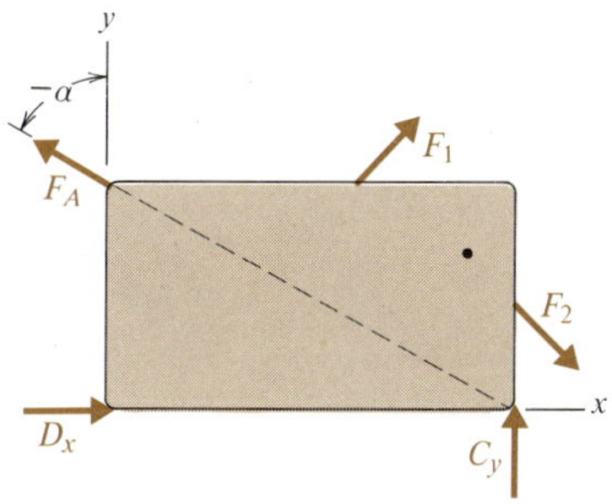

The reason that this value of α leads to partial constraint becomes obvious when we draw the corresponding free body diagram, as depicted to the left. In this situation all of the reactions intersect point C, so they cannot provide a moment to counteract the moment of the loading forces about point C.

In general, there are two ways of having improper constraint of a body for the case of a planar set of forces. As happened in this problem, all of the reaction forces may be concurrent. Alternatively, all of the reaction forces may be parallel. Such an arrangement permits movement perpendicular to the direction of the reactions.

HOMEWORK PROBLEMS

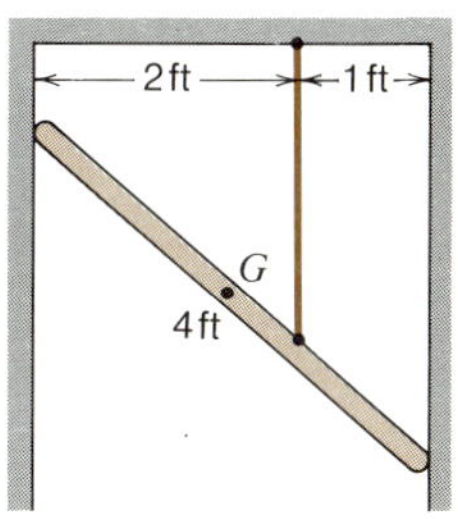

PROB. 3.115

3.115 The 10 lb, 4 ft long bar is suspended between two smooth walls as shown. The center of mass of the bar is at the midpoint G. Determine the reactions exerted by the walls on the bar.

3.116 The street light is loaded as shown. Determine the reaction at the fixed end A.

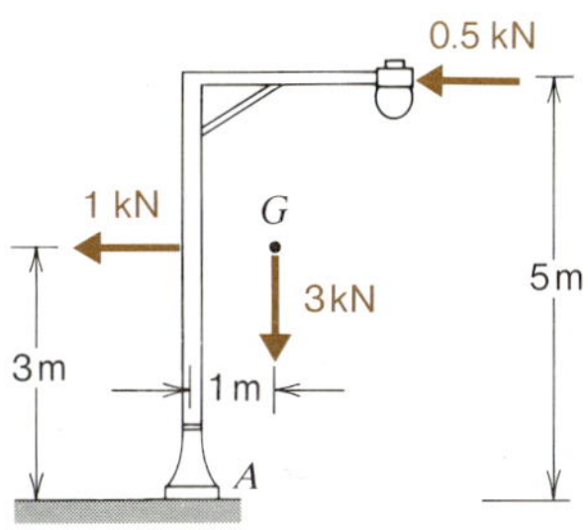

PROB. 3.116

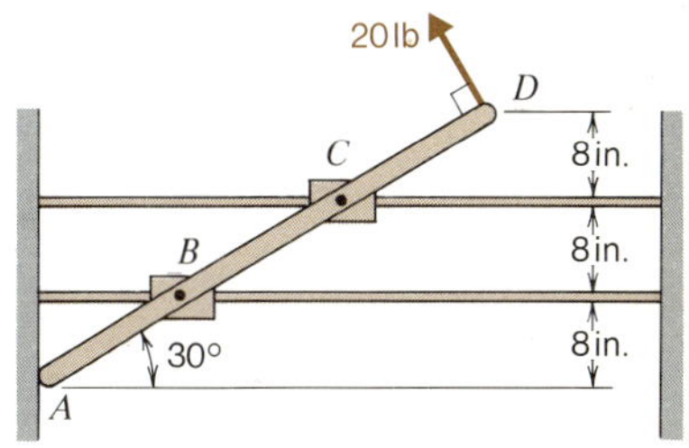

PROB. 3.117

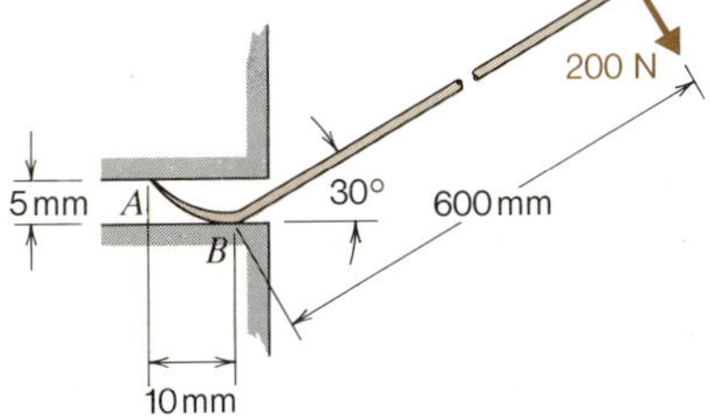

PROB. 3.118

3.117 The bar AD is supported by collars B and C, which ride on smooth horizontal rods. End A of the bar rests against the smooth vertical wall. Determine the reactions resulting from the 20 lb force shown.

3.118 A wrecking bar wedged into a crevice is loaded at its end by a 200 N force, as shown. Considering the surface of contact at point B to be smooth, determine the forces exerted on the bar at end A.

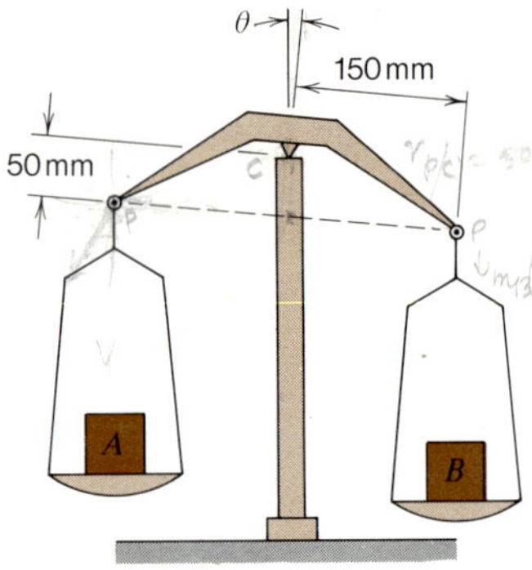

PROB. 3.119

3.119 The symmetrical balance beam scale is used to compare the unknown mass m_B to the standard mass m_A. The center of mass of the support arm AB coincides with the frictionless knife edge C. Determine m_B in the case where $m_A = 50$ grams and $\theta = 10°$.

3.120 For the balance beam in Problem 3.119, determine the ratio m_B/m_A in terms of the angle θ.

3.121 The lightweight pole is supported by two cables, and its lower end rests on a smooth horizontal surface. Determine the tension in each cable.

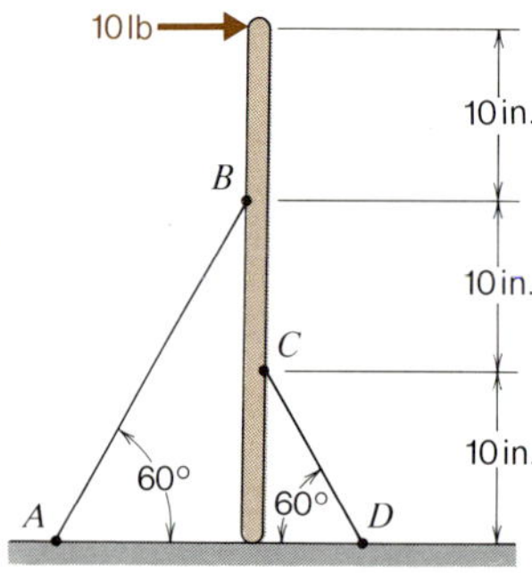

PROB. 3.121

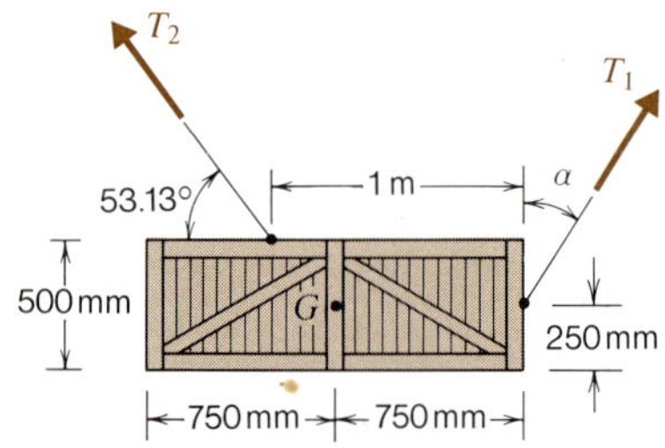

PROB. 3.122

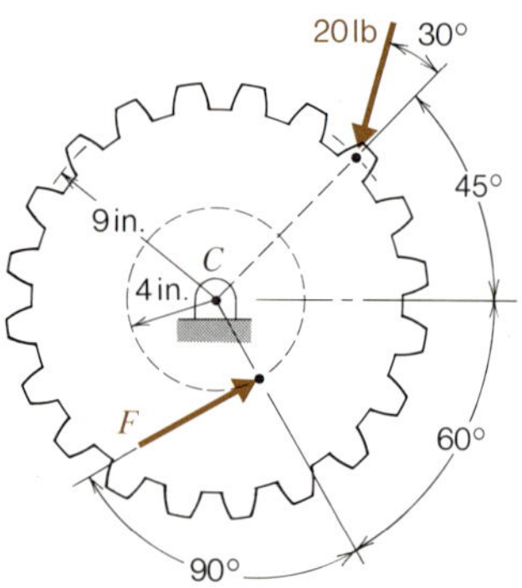

PROB. 3.123

3.122 Determine the tension in each cable and the angle α required to hold the 1000 kg crate in the position shown. Point G is the center of mass.

3.123 The stepped gear has a 20 lb force applied to its outer ring and a force $\bar{F}$ applied to its inner ring as shown. Determine the magnitude of $\bar{F}$ and the reaction at pin joint C if the gear is at rest.

3.124 A gear box mounted on a beam is subjected to two torques as shown. Determine the reactions at the supports of the beam.

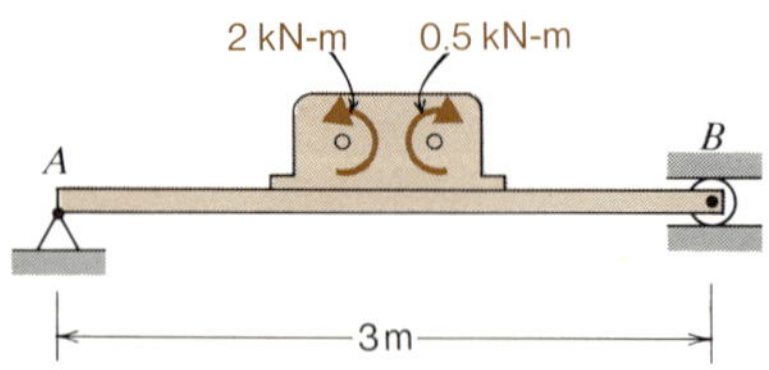

PROB. 3.124

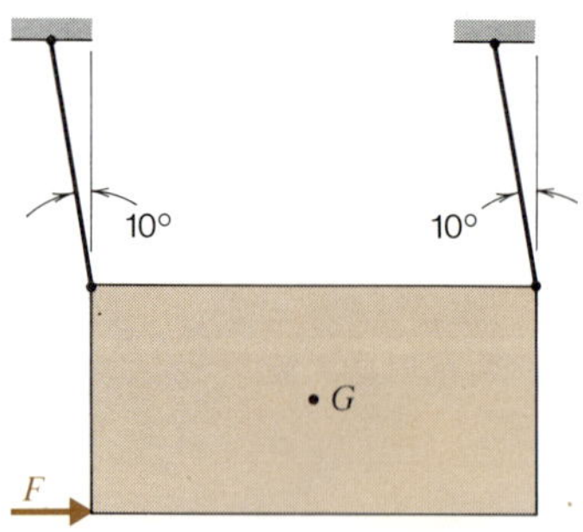

PROB. 3.125

3.125 Determine the magnitude of the force $\bar{F}$ and the tensions in the cables required to hold the 3 ton crate in the position shown. The crate's center of mass coincides with its geometric center G.

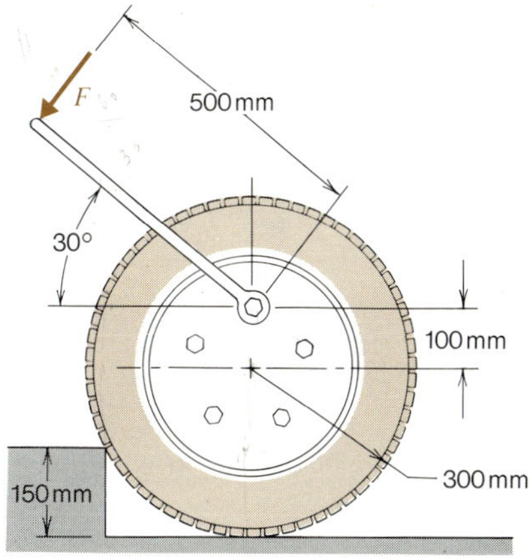

PROBS. 3.126 AND 3.127

3.126 The 20 kg tire rests on the smooth ground and the sharp curb. If the force applied to the tire wrench is $F = 20$ N in the direction shown, determine the force exerted on the tire by the curb.

3.127 Determine the maximum magnitude of the force $\overline{F}$ acting on the tire wrench for which the tire will not roll over the curb.

3.128 Determine the tension in the cable and the forces exerted by pin *A* on the lightweight beam.

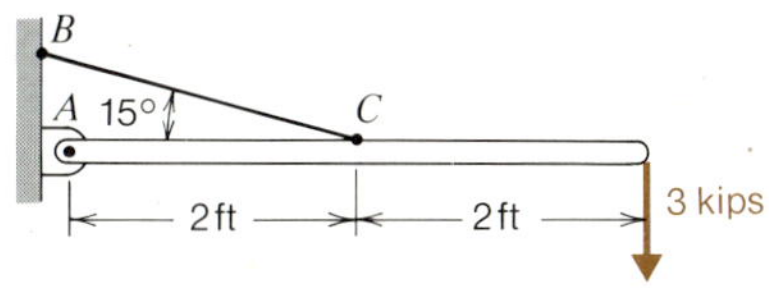

PROB. 3.128

3.129 Determine the reaction forces acting on the bent bar at ends *A* and *B*. Which surface of the groove does the roller bear against?

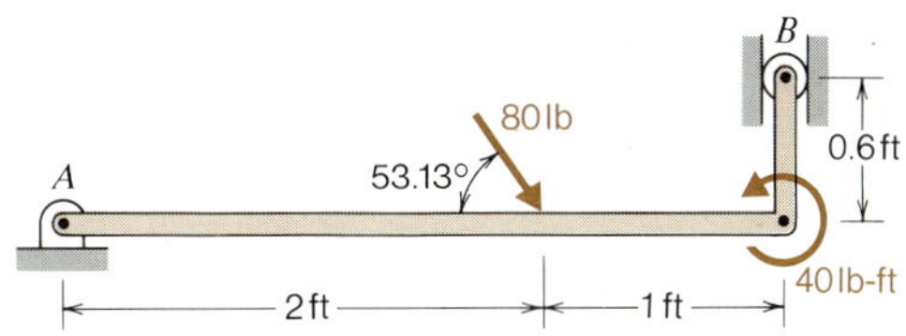

PROB. 3.129

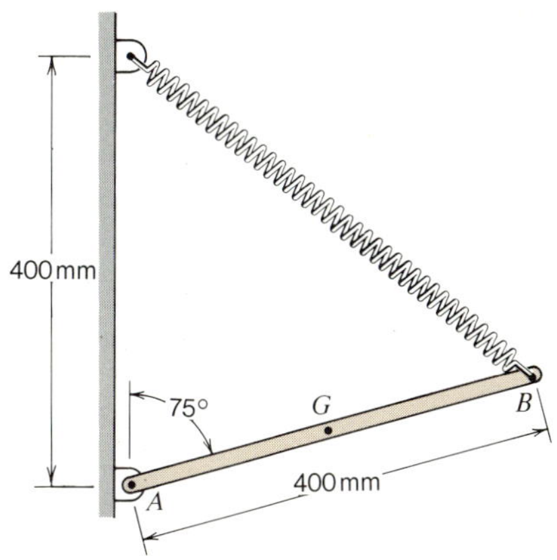

PROB. 3.130

***3.130** The stiffness of the spring is 2 kN/m and its unstretched length is 400 mm. Bar *AB* is in equilibrium in the position shown. Determine the mass of the bar, knowing that its center of mass is at the midpoint *G*.

3.131 Determine the forces acting on the triangular frame at points *A*, *B*, and *C*.

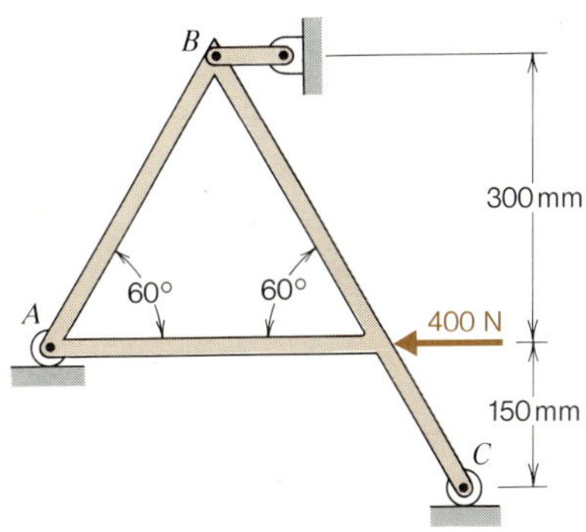

PROB. 3.131

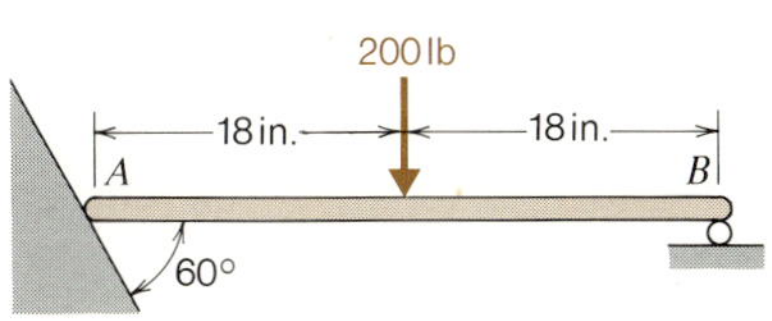

PROB. 3.132

3.132 End *A* of the lightweight beam shown is supported by a rough inclined surface. Determine the components of the reaction at end *A* tangent and normal to this surface.

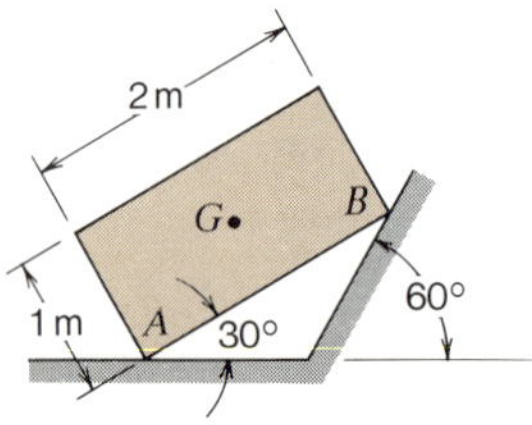

PROB. 3.133

***3.133** Corner A of the 1000 kg crate rests on the rough horizontal surface. The inclined surface supporting corner B is smooth. The center of mass of the crate is coincident with its geometric center G. Determine the forces acting on the crate at corners A and B.

3.134 The recording tape is wrapped over ideal pulleys A and B in the tape guide. The system is in a horizontal plane. Determine the force exerted by the spring at point C.

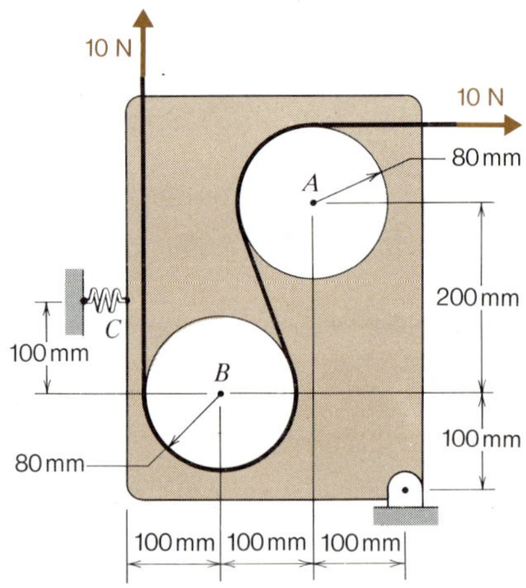

PROB. 3.134

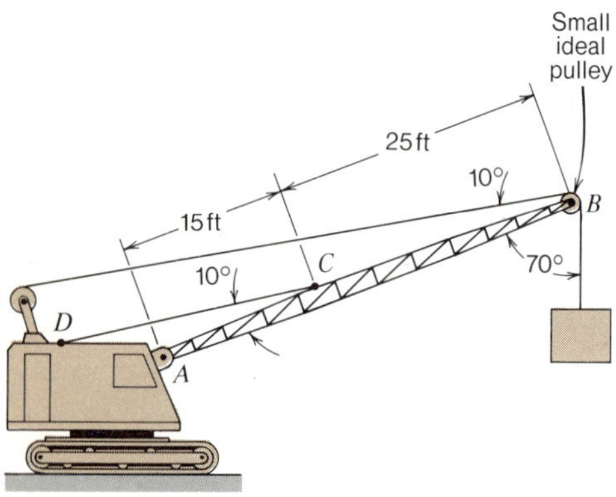

PROB. 3.135

3.135 The crane boom shown is supported by a pin at end A and cable CD. Determine the tension in cable CD required to hold the 10 ton crate in the position shown.

3.136 The 100 lb box is supported by a knife edge at point A and by a cable that passes over the ideal pulley B. The center of mass of the box is at the geometric center G. Determine the tension in the cable (a) when $d = 10$ in., (b) when $d = 28$ in.

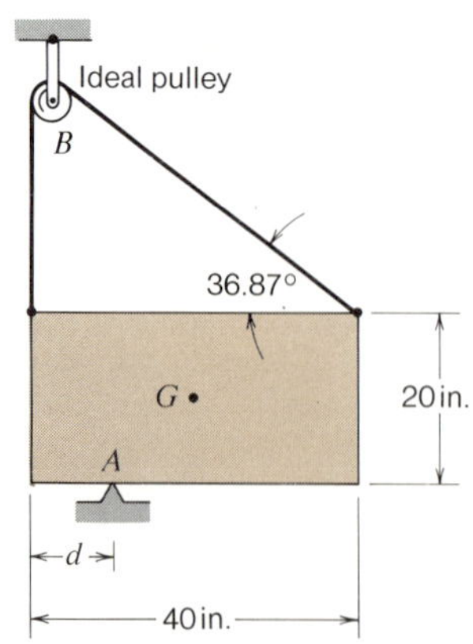

PROB. 3.136

3.137–3.139 Determine the reactions required to support the beam shown.

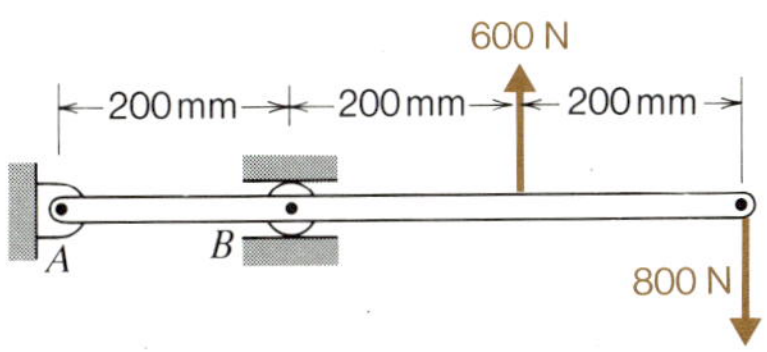

PROB. 3.137

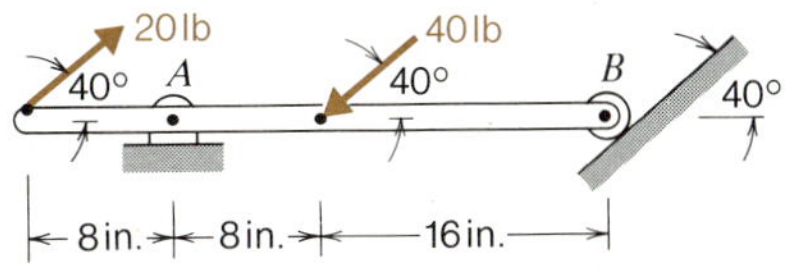

PROB. 3.138

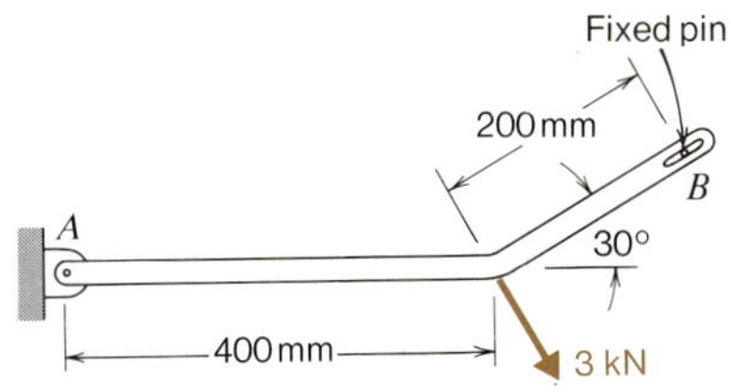

PROB. 3.139

3.140 The bent bar is supported by the fixed pin A and pin B, which fits into the smooth groove. Determine the reactions for the loading shown if $d = 600$ mm.

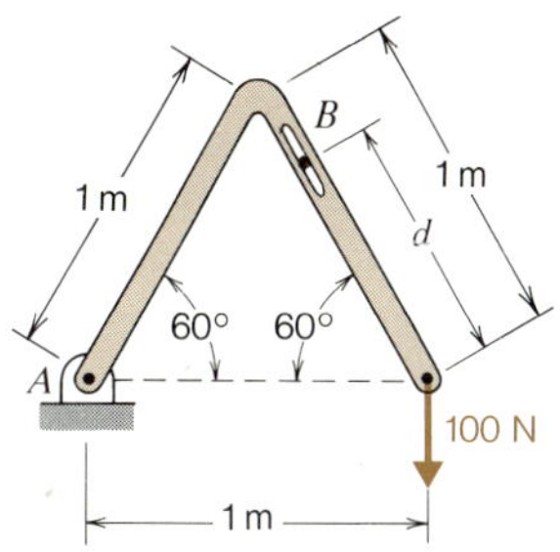

PROB. 3.140

3.141–3.143 Determine the reactions required to support the truss structure shown.

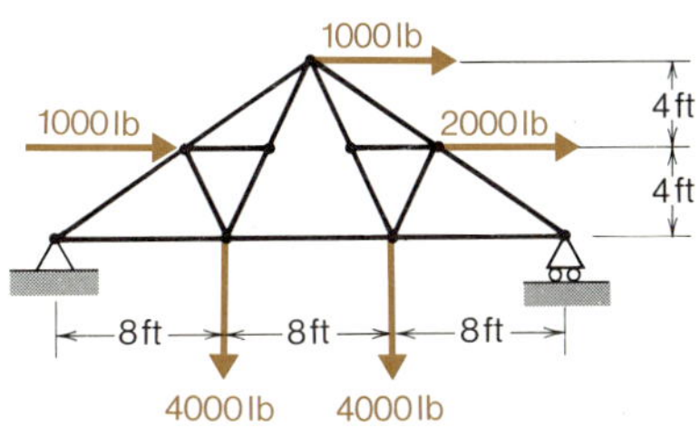

PROB. 3.141

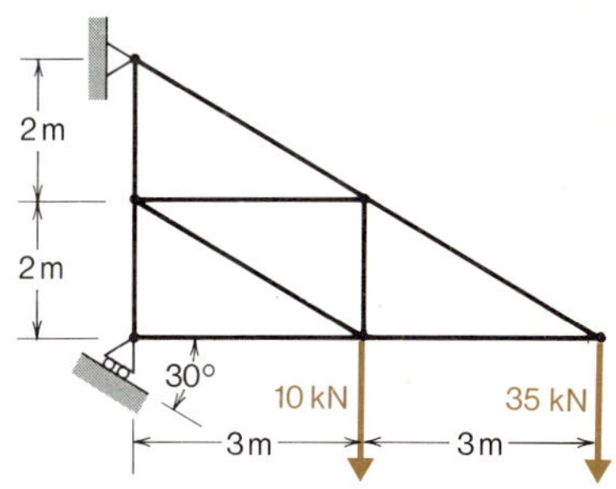

PROB. 3.142

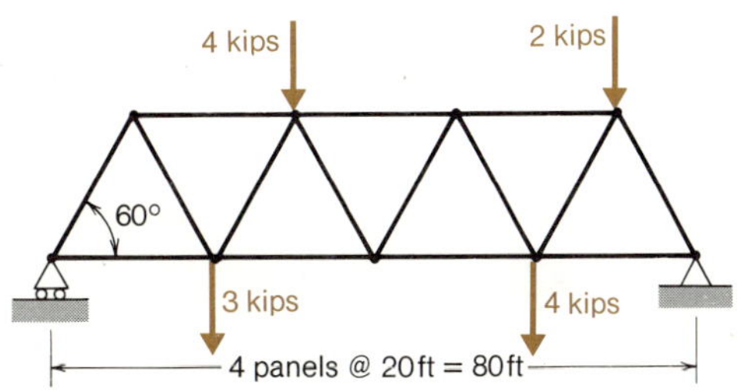

PROB. 3.143

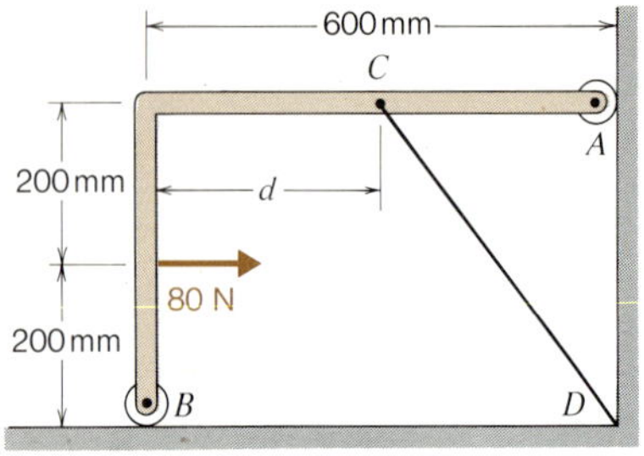

PROB. 3.144

3.144 The bent bar is supported by rollers at ends A and B and by cable CD. The distance $d = 300$ mm. Determine the reactions at the rollers and the tension in the cable.

3.145 The lightweight triangular plate is supported by three short links. Determine the reactive forces exerted on the plate by the links when the angle $\alpha = 60°$.

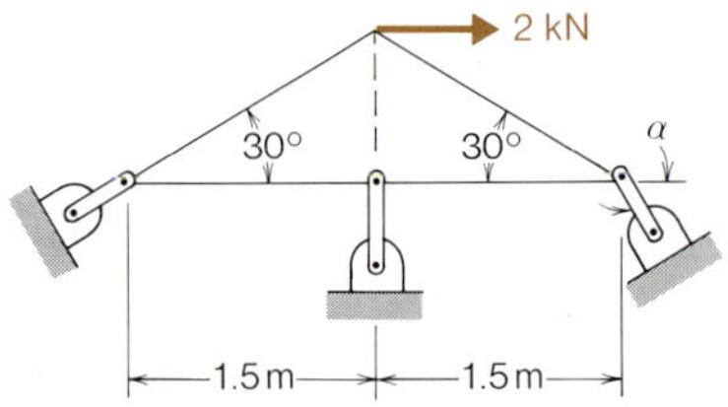

PROB. 3.145

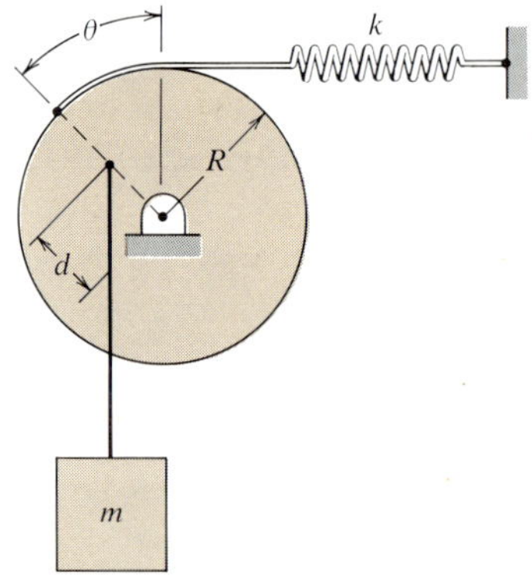

PROB. 3.146

3.146 A block of mass m is suspended from the cylindrical drum and a spring is attached to a cable that may wrap around the drum as shown. When $\theta = 0$ the spring is unstretched. Determine an expression relating θ to the other parameters.

***3.147** The lightweight beam is supported by a roller at end A and two short horizontal links at end B. Determine the reactions at ends A and B for the loading shown.

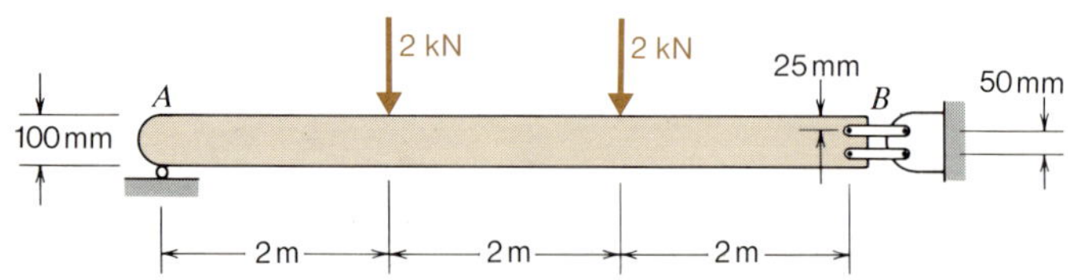

PROB. 3.147

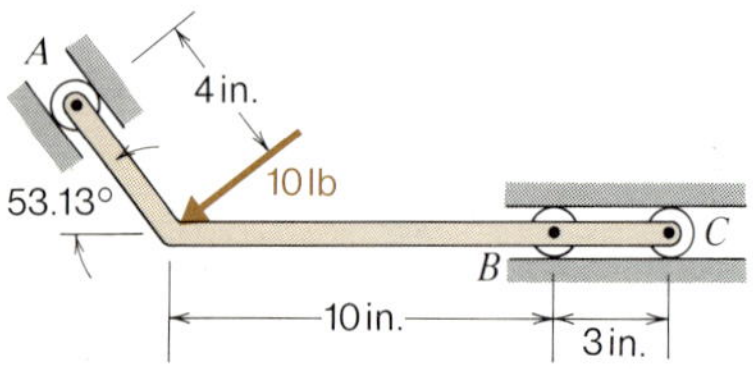

PROB. 3.148

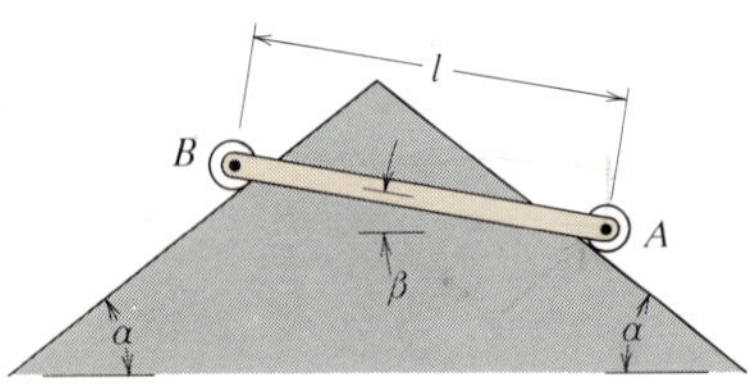

PROB. 3.149

3.148 Determine the force exerted on the bent bar by each roller. Which surface does each roller bear against?

3.149 Rollers A and B of masses m_A and m_B, respectively, are interconnected by a lightweight rod of length l and placed on the smooth surfaces shown. Knowing that in the static equilibrium position shown $\alpha = 30°$ and $\beta = 10°$, determine the ratio m_A/m_B.

***3.150** For the system in Problem 3.149, derive a general expression for the ratio m_A/m_B in terms of the angles α and β.

3.151 and 3.152 Derive a relation for the angle θ corresponding to equilibrium for the systems shown. In each case the bar has a mass m and its center of mass is at the midpoint G. All surfaces are smooth.

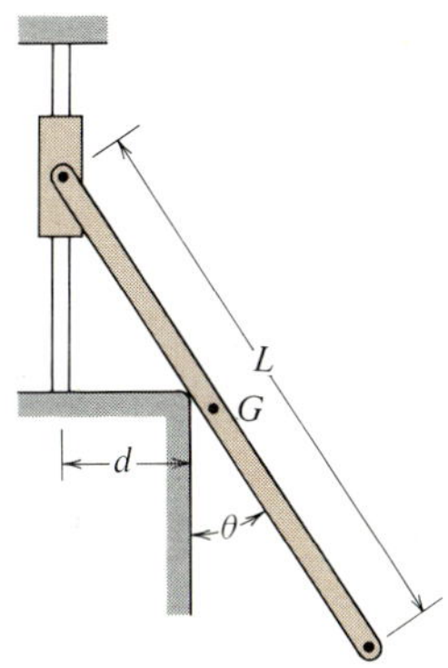

PROB. 3.151

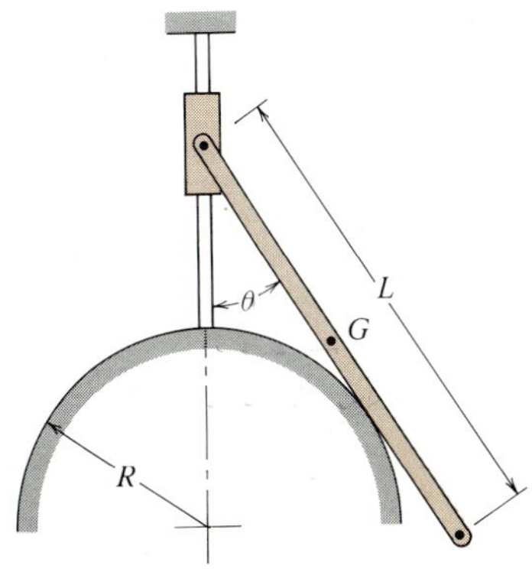

PROB. 3.152

***3.153** The uniform bar is supported by a cable that is attached to an anchored spring as shown. The spring is unstretched when the beam is in the horizontal position ($\theta = 0$). The spring stiffness is 2.5 kN/m and $l = 600$ mm. Determine the mass of the beam if $\theta = 30°$ in the static equilibrium position.

***3.154** Derive an expression for the mass of the uniform bar in terms of the spring stiffness k, the length l, and the angle θ in the static equilibrium position. The spring unstretched when $\theta = 0$.

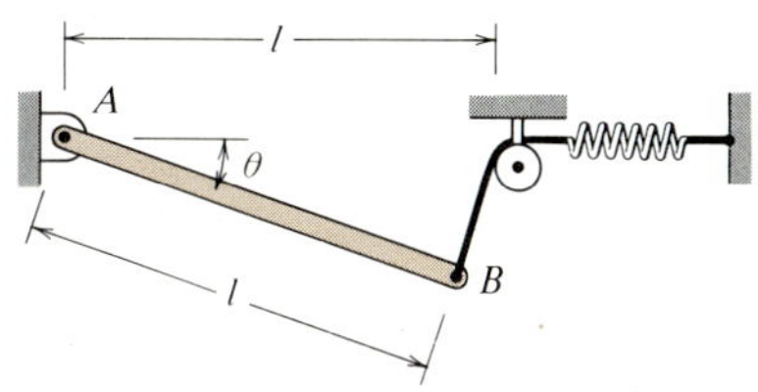

PROBS. 3.153 AND 3.154

3.155–3.158 Determine the value of the specified dimension in the problem listed below for which there is a situation of improper constraint.

3.155 The angle α in Problem 3.122.
3.156 The distance d in Problem 3.140.
3.157 The distance d in Problem 3.144.
3.158 The angle α in Problem 3.145.

3.159 In each case, determine whether (a) the plate is completely, partially, or improperly constrained, and (b) whether the reactions are statically determinate or indeterminate for cases of complete constraint. (*Hint:* Refer to the last paragraph of Illustrative Problem 6.)

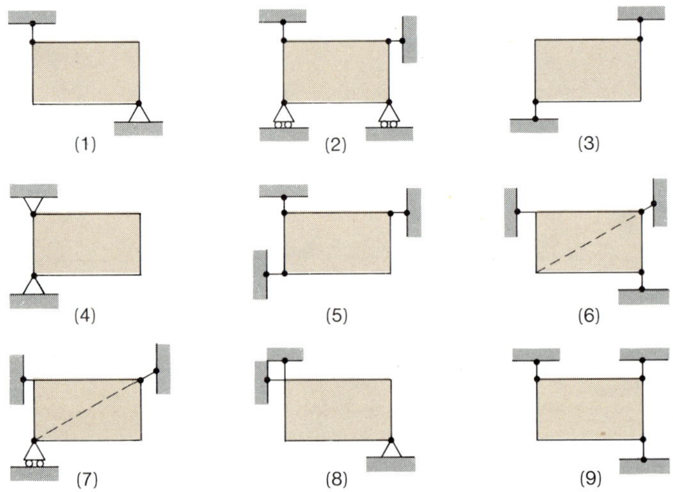

PROB. 3.159

3.160 Point C on top of the wide flange beam is subjected to the loads shown. Determine the reactions acting at point A on the axis of symmetry.

3.161 The bent bar, which is welded to the wall at end A, carries the loads shown. Determine the reactions.

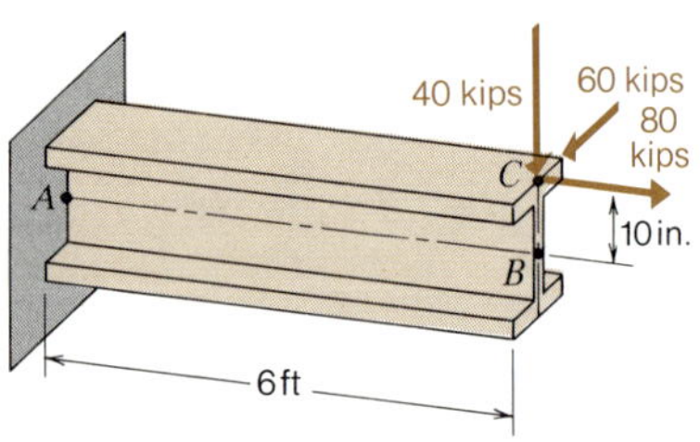

PROB. 3.160

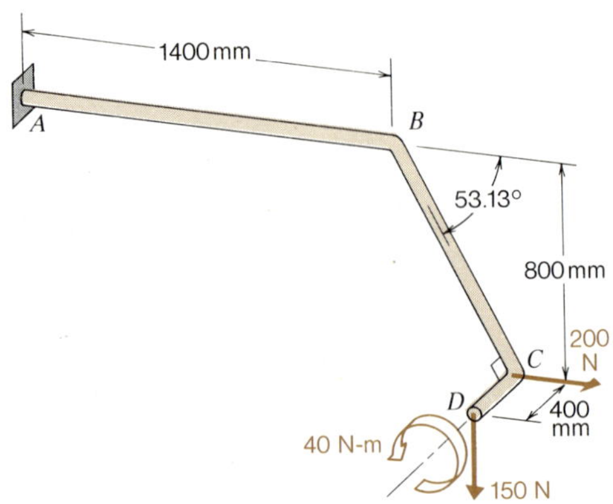

PROB. 3.161

3.162 The 10 kg horizontal flagpole is supported by a ball-and-socket joint at point A and two cables. Determine the tension in each cable when $d = 2$ m. The center of mass of the boom is at the midpoint.

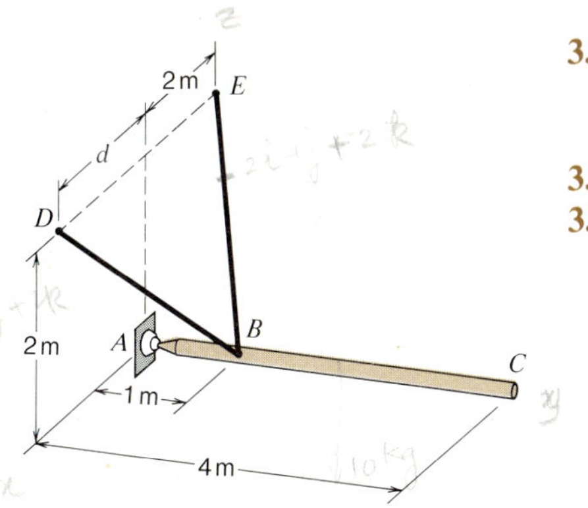

PROB. 3.162

3.163 Solve Problem 3.162 for $d = 1$ m.

3.164 The 20 kg package rests on the round table whose three legs are equally spaced along the perimeter. The mass of the table is 20 kg. Determine the force exerted on each leg by the smooth horizontal floor.

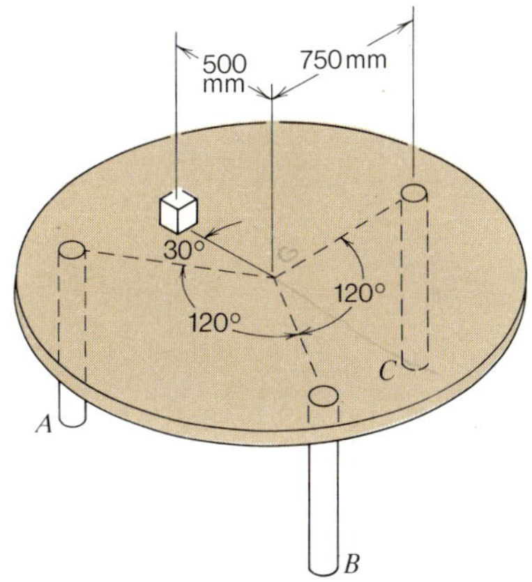

PROB. 3.164

3.165 The telephone pole is supported by a ball-and-socket joint at the bottom and three cables. Determine the tensions in the cables and the reaction exerted by the ball-and-socket joint for the loading shown.

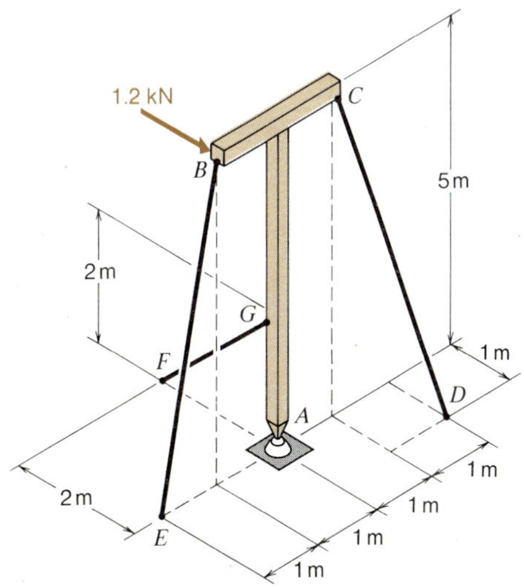

PROB. 3.165

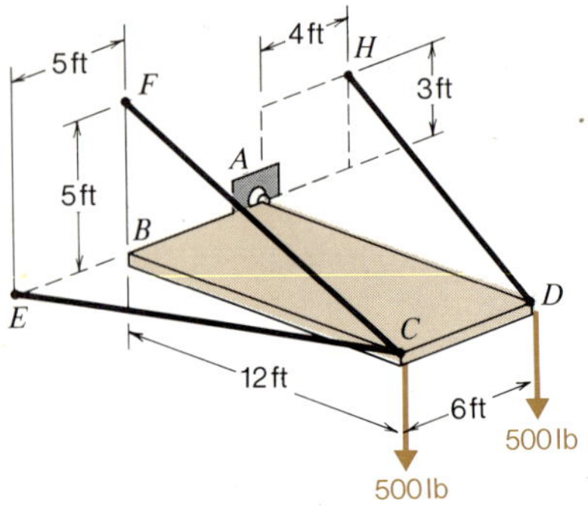

PROB. 3.166

3.166 The rectangular plate is attached to the wall by a ball-and-socket joint at corner A and held in the horizontal position by three cables. Determine the tensions in the cables and the reaction at corner A for the loading shown.

3.167 The horizontal boom is held in position at an angle of 60° with respect to the vertical wall by the ball-and-socket joint and the two cables. Determine the reaction at the ball-and-socket joint and the tension in each cable resulting from the vertical 50 kN loads.

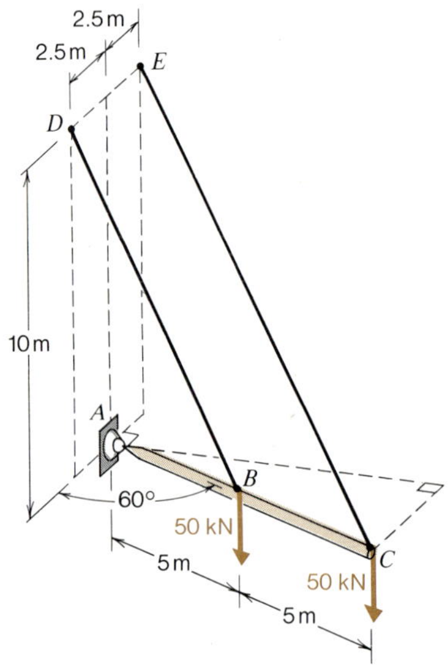

PROB. 3.167

3.168 The 10 kg rectangular trap door is held in the open position by a bar that exerts an axial force $\bar{F}$ on the door. Determine the magnitude of $\bar{F}$ and the reactions exerted on the door by the hinges at corners A and B. The center of mass of the door is its geometric center. Thrust is carried by hinge B only, and the hinges exert negligible couple reactions.

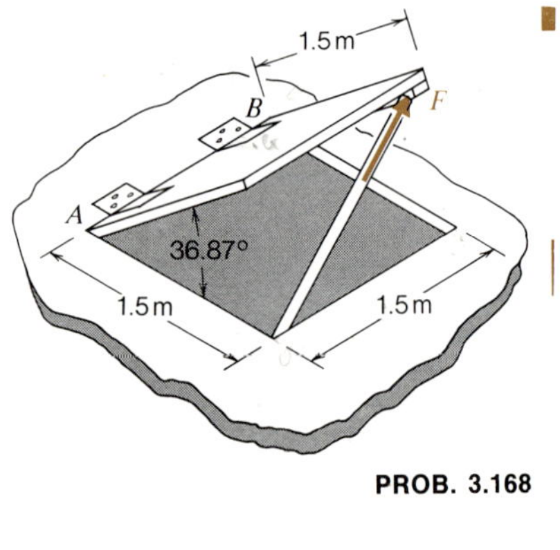

PROB. 3.168

3.169 A 2 ton crate is supported by a ball joint at point A, two short links at point B, and cable CD. The connections for the short links are also ball joints. The center of mass of the crate is at its geometric center. Determine the tensile force in the cable and the reaction forces of the links at point B.

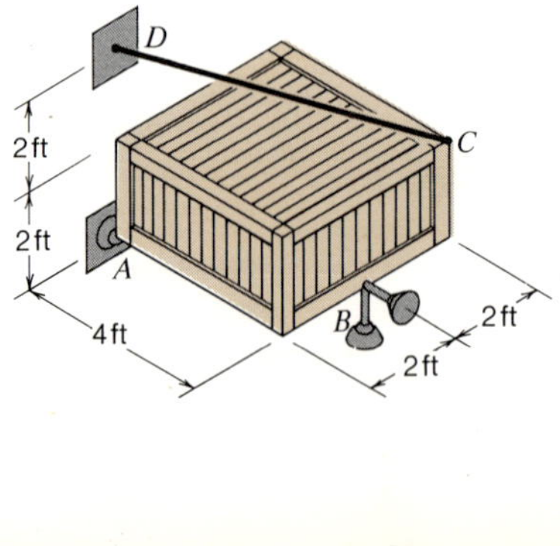

PROB. 3.169

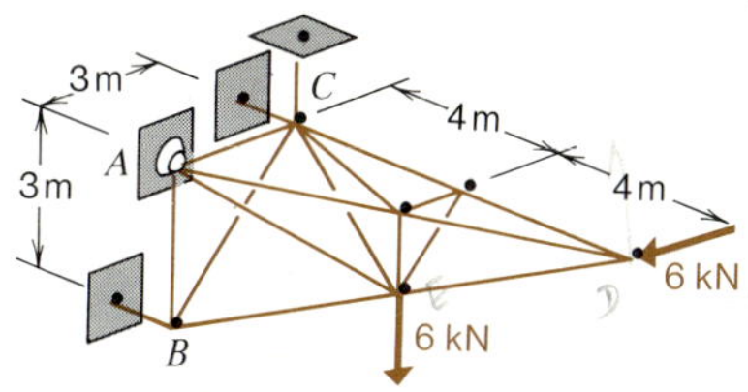

PROB. 3.170

3.170 The truss structure is supported by a ball-and-socket joint at A and short links at B and C. Determine the force exerted on the truss by the ball-and-socket joint.

3.171 Two identical rods of 2 kg mass each are welded at right angles and suspended by vertical wires. Section AB is horizontal and section BC is at 20° below the horizontal. Determine the tension in each wire. The center of gravity of each rod is at its midpoint.

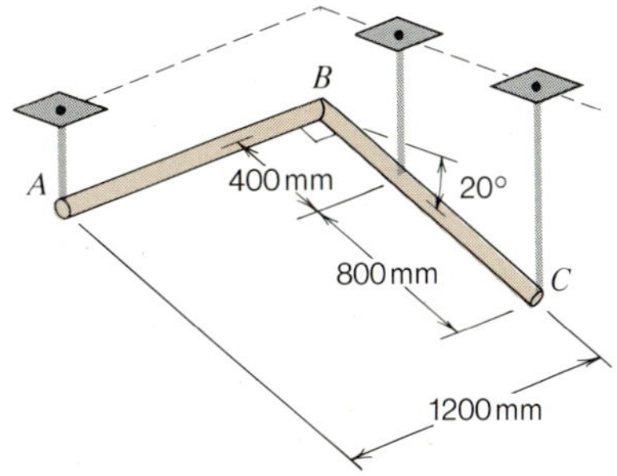

PROB. 3.171

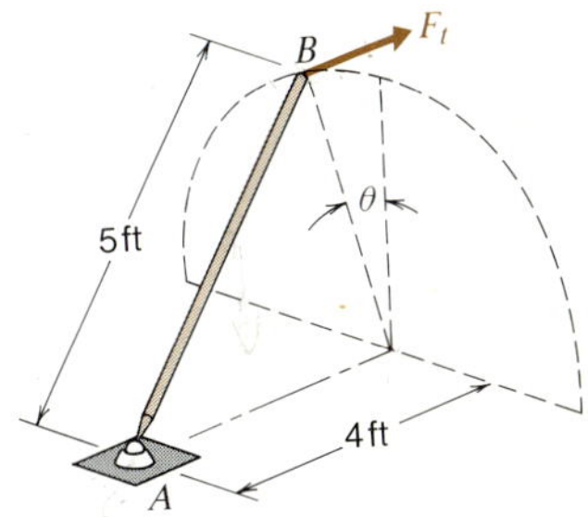

PROB. 3.172

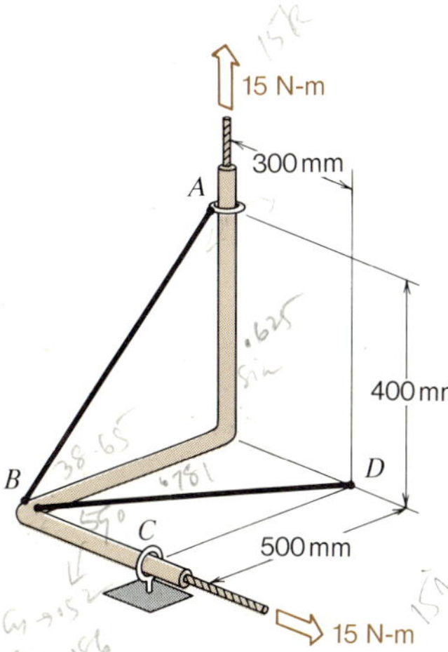

PROB. 3.174

3.172 The 5 lb bar is supported by a ball-and-socket joint at A and end B rests against a smooth wall. End B can be situated anywhere along the dashed semicircle depending on the value of the force $\bar{F}_t$, which is tangent to the circle. Determine $\bar{F}_t$ and the reactions at end A corresponding to a state of static equilibrium when $\theta = 53.13°$.

3.173 Solve Problem 3.172 for $\theta = 15°$.

3.174 A slender bent tube guides a flexible shaft that is transmitting a torque, with the result that the bent tube is loaded by the couples shown. The tube is supported by two smooth eyebolts and two cables. Determine the reactions exerted on the guide tube by the eyebolts.

***3.175** The lightweight bent bar is supported by bearings B and C, and end A rests against the smooth vertical wall. Determine all of the reactions resulting from the 50 lb vertical load that is applied to end D.

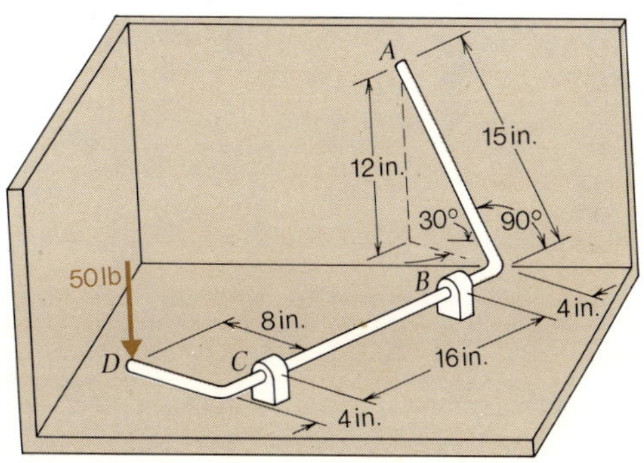

PROB. 3.175

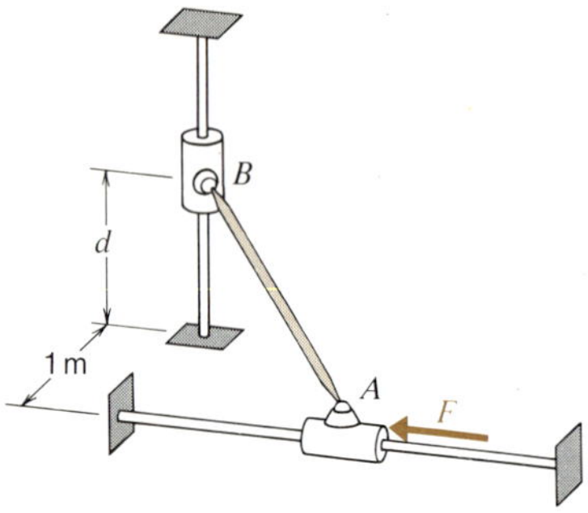

PROB. 3.176

***3.176** Ends A and B of the 5 kg rigid bar are attached by ball joints to lightweight collars that may slide over the smooth fixed rods shown. The center of mass of the bar is at its midpoint. Determine the horizontal force $\bar{F}$ applied to collar A that will result in static equilibrium in the position where $d = 1$ m. The length of the bar is 1.5 m.

***3.177** Solve Problem 3.160 for the position where $d = 0.5$ m.

***3.178** The 500 kg rectangular sign is held in position against the wall by loose fitting U-brackets at A and B that provide no support in the vertical direction. The center of mass of the plate is the geometric center. Determine the tension in each cable and the reactions exerted by the U-brackets.

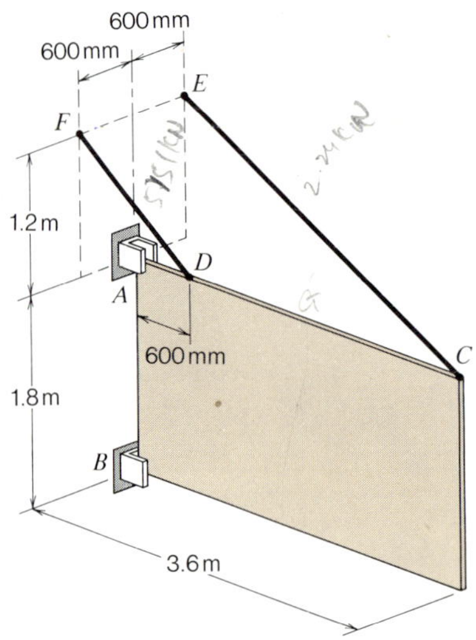

PROB. 3.178

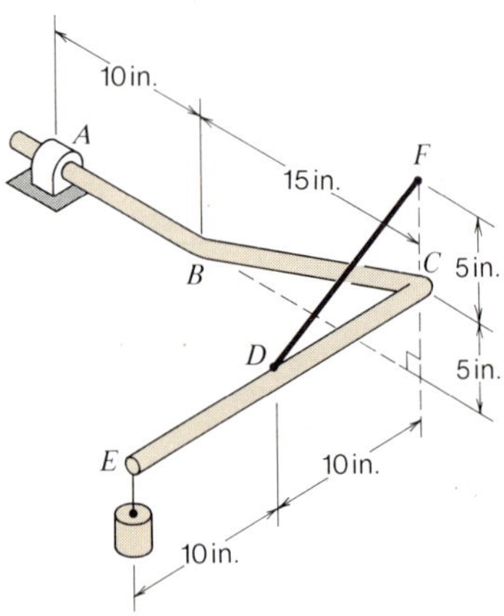

PROB. 3.179

***3.179** A 20 lb block is suspended from the bent bar at end E. The bar is supported by a thrust bearing at end A and by a cable. The thrust bearing is the tight fitting type whose construction only permits rotation of the rod about axis AB. Determine the tension in the cable and the reactions acting on the bar at end A.

***3.180** The bent bar is supported by three smooth eyebolts as shown. Is the system statically determinate and properly constrained?

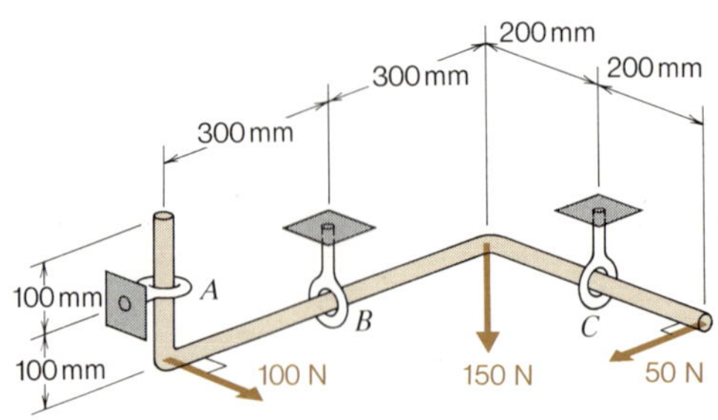

PROB. 3.180

CHAPTER 4

STRUCTURES

OBJECTIVES

1. To identify two force members in a structure.

2. To recognize both planar and three-dimensional structures that support loads in the manner of a truss.

3. To employ the method of joints to evaluate the axial force carried by each member in a truss.

4. To employ the method of sections in order to evaluate the axial forces carried by selected members in a truss, as well as to determine the reactions in a compound truss.

5. To identify when a structure acts like a frame or a machine, and then to draw free body diagrams of the entire structure and of each member.

6. To formulate and solve, in a logical sequence, the equilibrium equations for a frame or machine, in order to evaluate the forces acting on each member of the structure.

The methods for evaluating the forces acting on a rigid body were developed in the previous chapter. A few systems encountered there contained interconnected bodies. In those situations, it was sometimes necessary to investigate each body in the system, using the law of action and reaction.

The systems we will consider in this chapter share the feature of being composed of several interconnected bodies. These systems are *engineering structures*, which are designed for the purpose of safely resisting, transmitting, and, at times, modifying the various forces applied to them. The individual bodies forming a structure are called its *members*. In this chapter we will treat two categories of structures, trusses, and frames and machines. They differ in the manner in which their members transfer forces.

A. TRUSSES

1. Two-Force Members

A primary characteristic of a truss is that it is composed *solely* of two-force members, that is, members having forces applied at only two locations. This characteristic is important because the conditions of static equilibrium require that

> The forces acting at each of the two loaded points in a two-force member must be equal in magnitude, opposite in direction, and collinear.

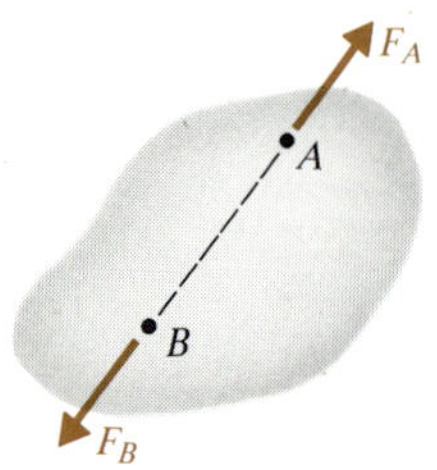

FIGURE 1

Such a force is illustrated in Figure 1.

We may readily verify that the force system in Figure 1 is the only way in which the body can be in static equilibrium when forces are applied at only points A and B. First, in order for there to be no moment about point A, the line of action of the force acting at B must intersect point A. Then, in order for the sum of the forces to be zero, the force acting at point A must be the negative of the one acting at point B. The force system in Figure 1 satisfies these criteria. Clearly, the case where both forces act in the opposite sense from that depicted in Figure 1 also satisfies the foregoing criteria.

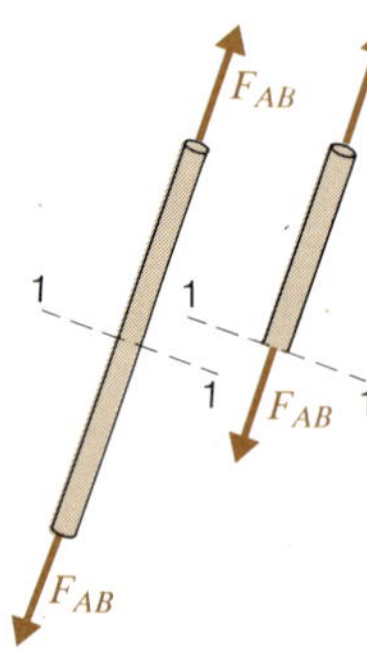

FIGURE 2a

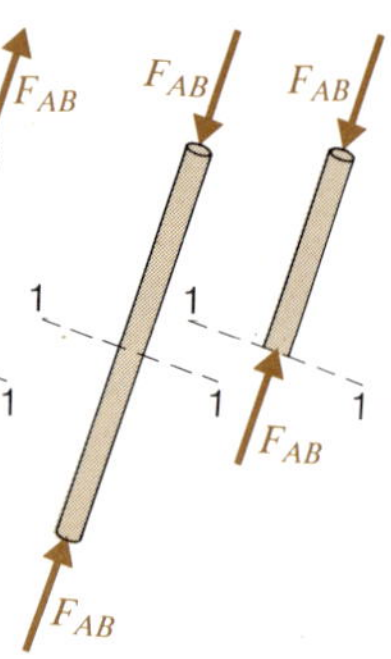

FIGURE 2b

A further restriction on the members of a truss is that they must be straight bars, such as the one illustrated in Figure 2a. The force system shown there is said to place the bar in a state of *tension*, because it tends to pull the bar apart, as if it were a cable. In fact, in this case the bar acts exactly like a cable, carrying a constant tension F_{AB} on every cross section, as shown in Figure 2a.

Note that the double subscript F_{AB} merely denotes the end points of the member. The order in which the subscript is written is unimportant.

The difference between the bar in Figure 2a and a cable is that the forces in the bar can also have the effect of pushing the ends together. This is the state of *compression*, illustrated in Figure 2b.

2. Characteristics of a Truss

A truss is a rigid framework of straight two-force members that are joined together at their ends. In order that the reactions at the ends of the members consist of forces, without couples, we model the connections of the members as pins in the case of planar trusses, or ball-and-socket joints in the case of three-dimensional trusses. The latter are usually referred to as *space trusses.* Figure 3 shows some typical planar trusses; those with a peak are generally used to support roofs, whereas the flat ones are generally used to support bridges. The dot in each diagram indicate the pin connections.

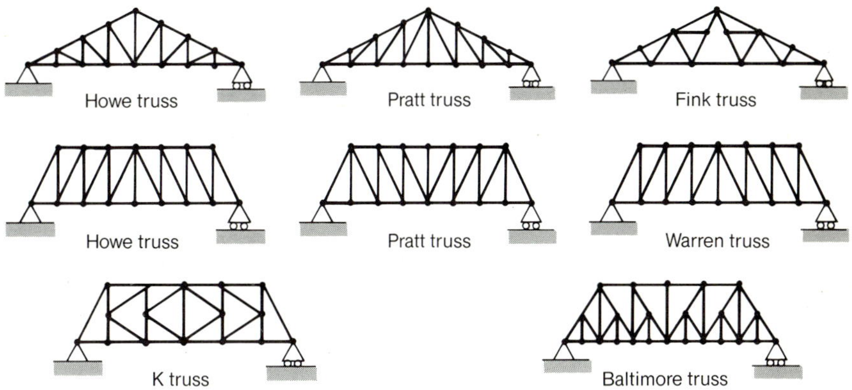

FIGURE 3

In an actual structure the members may be I-beams, channels, angle bars, or bars with other types of cross sections. The most common way of fastening members together is by means of welding, riveting, or bolting to gusset plates, as shown in Figure 4a. However, the model of a pin connection shown in Figure 4b is sufficiently accurate, provided that the centerlines of all members meeting at that connection are concurrent. (The proof of the validity of this approximation is beyond the scope of this book.) The terminology used to describe the model connection is to refer to it as a *joint*.

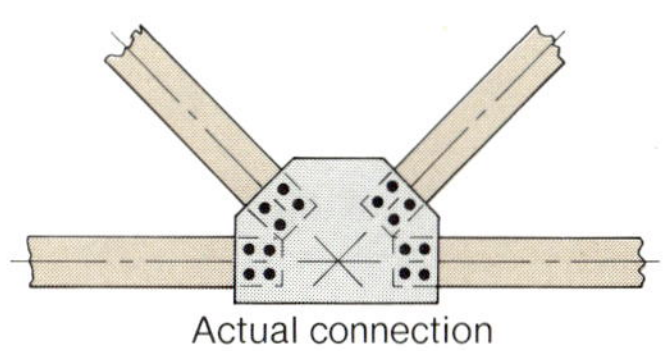

FIGURE 4a

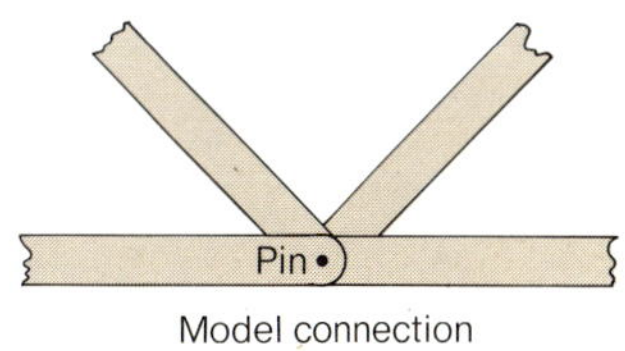

FIGURE 4b

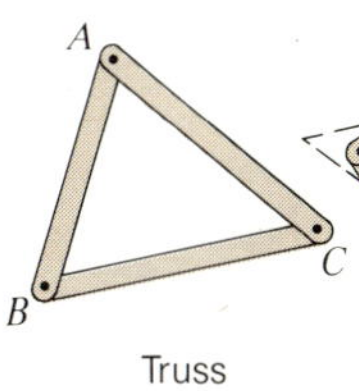

FIGURE 5a

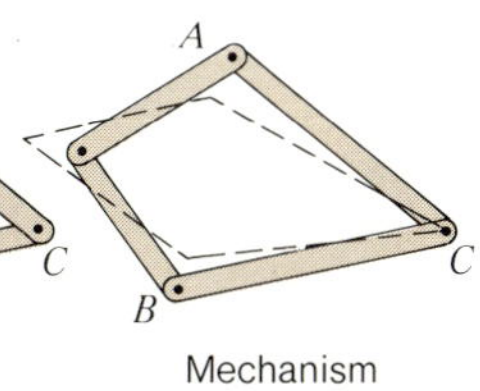

FIGURE 5b

It will be noted that the members in each of the trusses illustrated in Figure 3 are interconnected to form triangles. A comparison of the basic truss component in Figure 5a to the quadrilateral in Figure 5b shows that the triangular shape is essential to the formation of a rigid truss. If we try to form a polygon having more than three sides, the pin joints allow the bars to rotate relative to one another, thus forming a nonrigid structure called a *mechanism*.

Using the truss of Figure 5a as a basis, we can increase the size of the structure by attaching more members. If this addition results in the formation of additional basic triangular components, we have a *simple truss*, such as the planar one shown in Figure 6a and the space truss shown in Figure 6b.

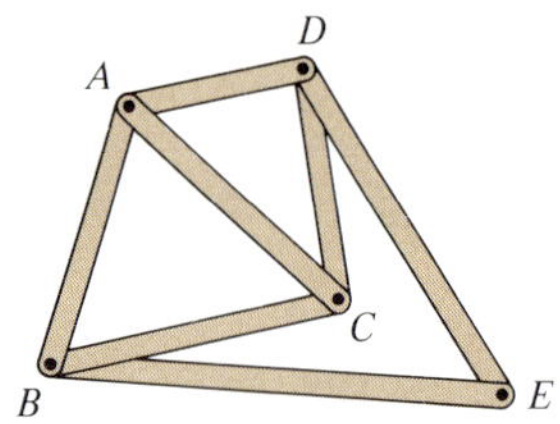

FIGURE 6a

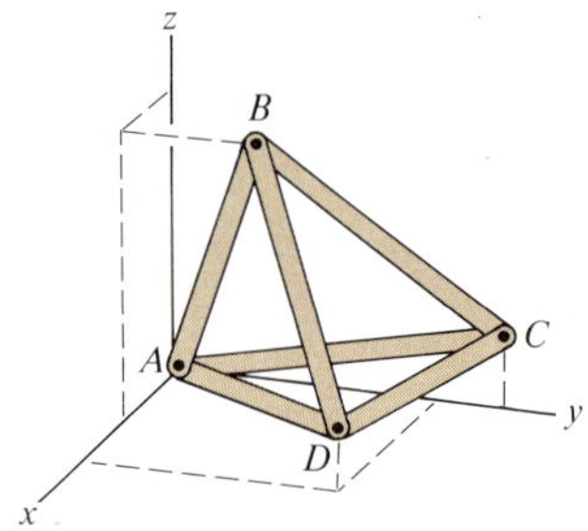

FIGURE 6b

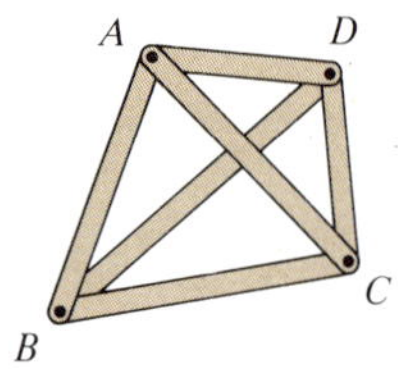

FIGURE 7

As we will see, simple trusses are statically determinate, allowing for the evaluation by the basic equilibrium equations of the forces being transmitted by each member. Although simple trusses will be the focus of our attention, not all trusses are simple. For example, the planar truss in Figure 7 was formed by adding three members to the basic truss component in Figure 5a. Not all of the members in this truss are needed to form the polygon $ABCD$; either of the diagonal members AD or BC could be removed to form a simple truss. In other words, this structure is excessively constrained; one of the members is *redundant*.

As was true for the determination of reactions, excessive constraint is synonymous with static indeterminacy. This means that the equations of static equilibrium are insufficient to evaluate the member forces, so the model of a rigid body is inadequate for the analysis. The evaluation of the member forces in a statically indeterminate truss requires the consideration of the small deformations of the members resulting from the loading.

There is a simple relationship between the number of joints and the number of members necessary to form a simple truss. Starting from the basic triangle in Figure 5a, where there are three members and three joints, we see that for a planar truss, such as that in Figure 6a, additional triangles are formed by adding two bars for every joint added. Thus, letting m be the number of members and j be the number of joints, we find that

$$m = 2j - 3 \qquad \text{for a simple planar truss} \tag{1a}$$

In the case of a simple space truss, we see that the requirement of rigid triangles leads to the formation of tetrahedrons, such as that in Figure 6b. Each additional joint now requires three additional members, so

$$m = 3j - 6 \qquad \text{for a simple space truss} \tag{1b}$$

Another characteristic of a truss pertains to the way in which it is loaded. Because the ends of the members are interconnected at the joints, it follows that, in order for there to be two force members, the external forces acting on the structure must be applied at the joints only. Any loading that is not applied at the joints requires that the member to which it is applied transfer more than just a tensile or compressive force. Hence, we have the following criteria for determining when a structure acts like a truss:

1. All members are straight.
2. The centerlines of all members that are connected at a joint must be concurrent at that joint.
3. The external forces acting on the structure must be applied at the joints.

A problem arises in considering the question of the weight of each member, because this force acts at the midpoint of the member and thus violates the second of the above criteria. In most cases the weight force can be neglected, because it will be small compared to the other forces acting on the structure.

There are two methods that may be employed to evaluate the member forces. We shall make it a standard practice when applying either method *to begin by first evaluating the reactions exerted by the supports,* so that these forces may be regarded as known when we determine the forces in the members. (There are truss problems where it is not necessary to first determine the reactions. We will discuss these cases later.)

3. Method of Joints

Our discussion of the method of joints begins by considering a single truss element, such as the one in Figure 8a. The forces $\bar{F}_A$, $\bar{F}_B$, and $\bar{F}_C$ represent known loads. In order to investigate the equilibrium of the structure, we consider each member of the structure individually in a free body diagram, as shown in Figure 8b.

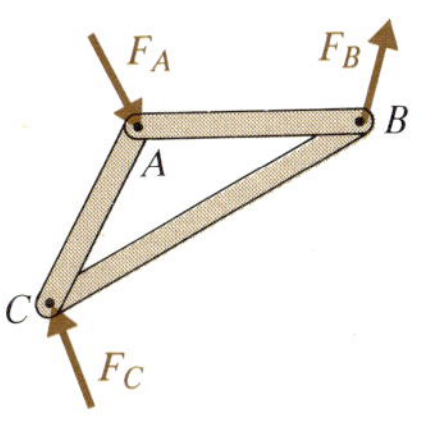

FIGURE 8a

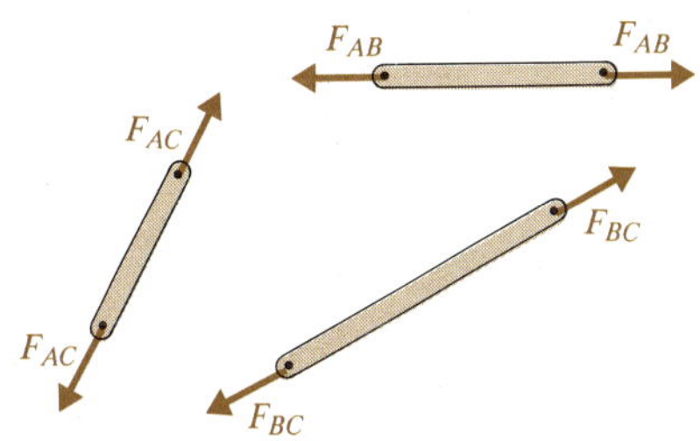

FIGURE 8b

Each truss member is loaded only at its ends. It follows from the concept of a two-force member that the forces acting at the ends of each truss member must be equal in magnitude, opposite in sense, and collinear. The axial forces depicted in Figure 8b were based on an assumption, for the sake of convenience, that each member is in tension.

In order to determine the axial force carried by each member, we recognize that the structure is formed from pin joints, as well as straight members. The pins forming the joints must also be in equilibrium, so we draw free body diagrams of each joint; hence the name *method of joints.*

Because we have assumed that each member is in tension, we know from Newton's third law that each member at a joint must pull away from that joint, toward the other end of the member. Thus, the free body diagrams of the members and joints of the truss in Figure 8a are as shown in Figure 9. Note that the external force acting at each joint must also be shown to complete the free body diagram.

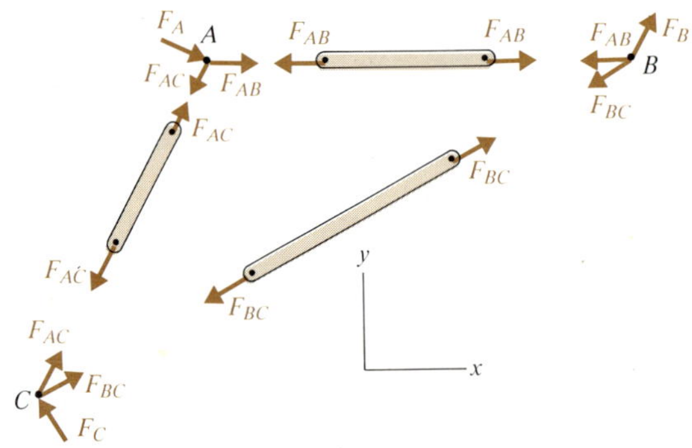

FIGURE 9

The unknown forces in the free body diagrams are those of the members of the truss. It is only necessary for us to consider equilibrium of the joints, because equilibrium of the members has been satisfied by the concept of a two-force member. The forces at each joint are concurrent, so each joint will be in equilibrium if the resultant force acting on it is zero. We formulate the equilibrium equations by choosing a convenient coordinate system, such as that shown in Figure 9, for this planar system. This gives $\sum F_x = 0$ and $\sum F_y = 0$ at each joint. In all, we may write six scalar equations for the planar system in Figure 9.

At this juncture it might seem that we have more equations than we need, in that the only unknowns are the three member forces. This is a subtle point. Recall that earlier we decided to begin by obtaining the reactions of the truss. This procedure enables us to say that the structure as a whole is in equilibrium. However, the entire structure will also be in equilibrium if the forces acting on each joint are equilibrated. In other words, we have three excess equations, corresponding to equilibrium of the entire structure. The extra equations may be employed to check the solution, as will be seen in the examples that follow.

We have assumed that all the members are in tension. This assumption is confirmed when the magnitude of the member force is found to be positive, whereas a negative value indicates a state of compression. Clearly, it is not necessary for us to regard tension as positive, but this convention allows for an orderly interpretation of the results. It is important for the design of a member to know whether the force being transmitted is tensile or compressive, because the design procedure for members depends entirely on this knowledge.

The foregoing discussion can be extended to treat general planar trusses, as well as space trusses. Briefly stated,

> The method of joints consists of drawing free body diagrams for all joints and then solving the equilibrium equations for the concurrent force systems acting at each joint.

In the case of a planar truss, we have $\sum F_x = 0$ and $\sum F_y = 0$ for each joint. Therefore, the j joints of a planar truss are governed by a total of $2j$ equilibrium equations. However, only $2j - 3$ of the joint equations are independent, because three equations were already used to evaluate the reactions.

We see that in order for the member forces in a planar truss to be statically determinate, there must be $2j - 3$ members. Notice that this is the number of members required for a simple planar truss, as given in eq. (1a). A planar truss having more than $2j - 3$ members is *internally indeterminate*, even though its reactions may be determinate. On the other hand, if there are fewer than $2j - 3$ members, the equilibrium equations for the joints cannot be satisfied. The truss is then not a rigid structure.

The foregoing are merely necessary conditions for having an internally determinate, rigid planar truss. It is possible to construct planar trusses that have j joints and $2j - 3$ members that are not rigid and statically determinate. The ultimate verification of these conditions rests in the solvability of the equilibrium equations for the joints.

In the case of a space truss we have $\sum F_x = 0$, $\sum F_y = 0$, and $\sum F_z = 0$ for each joint. Thus, a space truss having j joints will have $3j$ equilibrium equations for these joints. Deducting the six checking equations corresponding to the static equilibrium of the entire structure shows that the conditions of internal determinacy and rigidity require that there be $3j - 6$ members. This is the same number of members as that given by eq. (1b) for a simple space truss. A space truss with more than $3j - 6$ members is internally indeterminate, whereas a space truss having fewer than this number of members is not rigid.

EXAMPLE 1

Determine the force in each member of the truss shown using the method of joints.

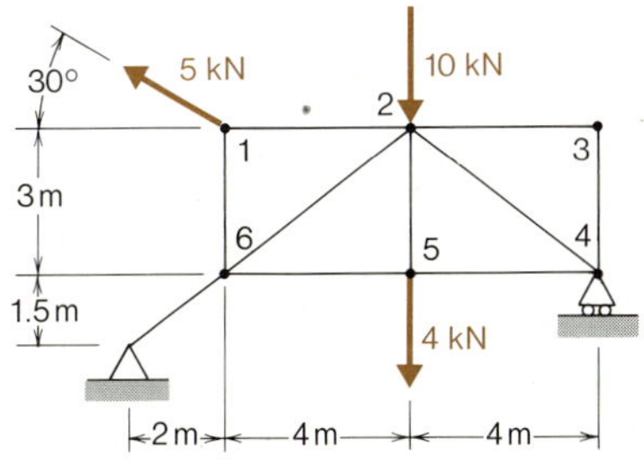

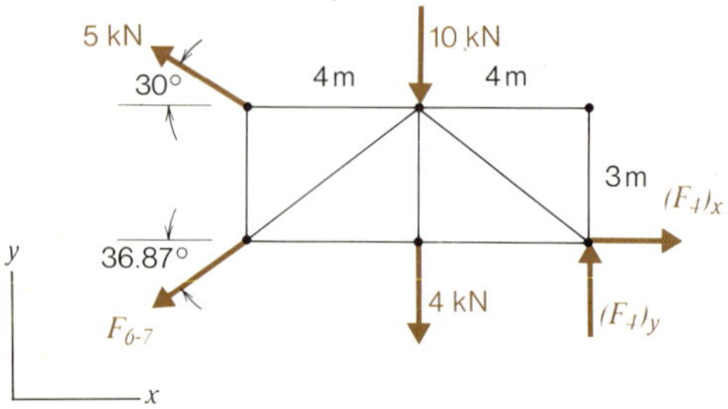

Solution

We begin by determining the reactions. Member 6–7 is a two-force member, so it can apply only an axial force to the rectangular portion of the truss. The other point of support of the truss is at pin joint 4, so we have the following free body diagram for the entire truss.

For consistency we have assumed that $\bar{F}_{6-7}$ is a tensile force. The dimensions of the truss are given in terms of horizontal and vertical distances, so the coordinate axes are aligned in these directions.

The z axis is outward for the coordinate system shown, making counterclockwise moments positive. In order to eliminate two reaction components, we choose joint 4 for the moment sum. The equilibrium equations are then

$$\sum M_{4z} = 10(4) + 4(4) + (F_{6-7} \sin 36.87°)8 - (5 \sin 30°)8 + (5 \cos 30°)3 = 0 \text{ kN-m}$$

$$\sum F_x = -F_{6-7} \cos 36.87° + (F_4)_x - 5 \cos 30° = 0$$

$$\sum F_y = -F_{6-7} \sin 36.87° + (F_4)_y - 4 - 10 + 5 \sin 30° = 0$$

We solve the moment equation for F_{6-7} and then determine $\bar{F}_4$ from the force sums. Thus

$$F_{6-7} = -10.206 \text{ kN}, \qquad (F_4)_y = 5.376 \text{ kN}, \qquad (F_4)_x = -3.835 \text{ kN}$$

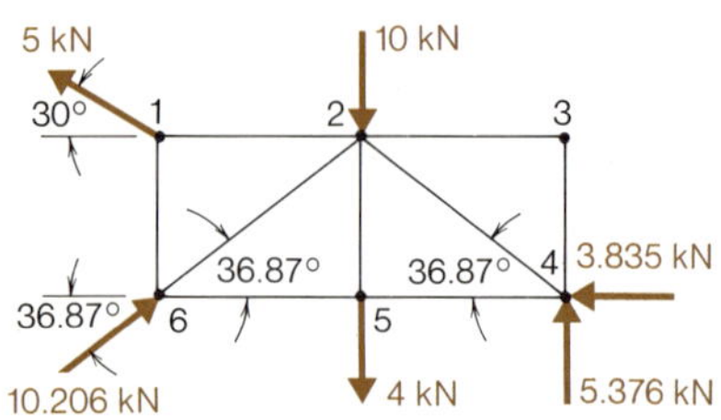

The equilibrium of each joint of the truss may now be considered. To assist in the calculations, we sketch the truss showing the angles between the members and the coordinate directions, and all loadings, including the reactions, in their correct orientation.

Note that the force $(\bar{F}_4)_x$ is shown pushing to the left, because its value was negative. Also note that the principle of transmissibility does not apply when finding the forces in the internal members of the structure, so each external force must be shown at the joint at which it acts.

This sketch can now be used to draw free body diagrams of each joint, using the convention of considering the members to be in tension. That is, we show each member pulling on the joint. Note that each joint is labeled in the free body diagrams.

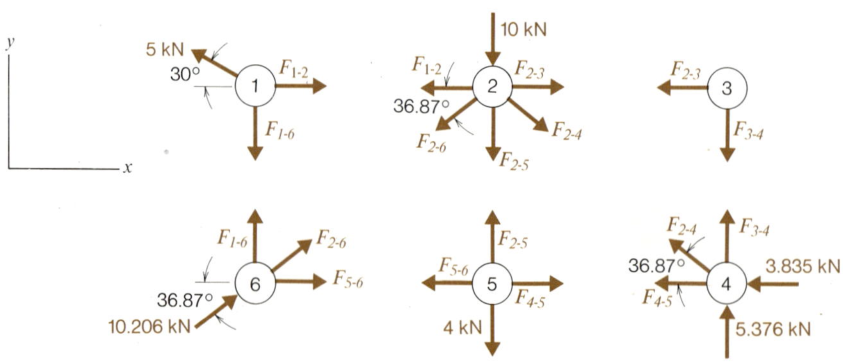

For computational purposes *it is best to consider joints connecting fewest members first*. This allows us to solve successively for the member forces rather than dealing with a large number of simultaneous equations. In other words, in planar problems we solve the equilibrium equations at joints that contain two or less unknowns first. After referring to the free body diagrams, the solution order we choose, by joints, is 1-3-4-5-6-2.

Joint 1:

$$\sum F_x = F_{1-2} - 5\cos 30^\circ = 0$$
$$\sum F_y = -F_{1-6} + 5\sin 30^\circ = 0$$
$$F_{1-2} = 4.330\ \text{kN}, \qquad F_{1-6} = 2.50\ \text{kN}$$

Joint 3:

$$\sum F_x = -F_{2-3} = 0 \qquad \sum F_y = -F_{3-4} = 0$$
$$F_{2-3} = F_{3-4} = 0$$

Joint 4:

$$\sum F_x = -F_{4-5} - F_{2-4}\cos 36.87^\circ - 3.835$$
$$\equiv -F_{4-5} - 0.8F_{2-4} - 3.835 = 0$$
$$\sum F_y = F_{3-4} + F_{2-4}\sin 36.87^\circ + 5.376$$
$$\equiv 0 + 0.6F_{2-4} + 5.376 = 0$$
$$F_{2-4} = -8.960\ \text{kN}, \qquad F_{4-5} = 3.333\ \text{kN}$$

Joint 5:

$$\sum F_x = F_{4-5} - F_{5-6} \equiv 3.333 - F_{5-6} = 0$$
$$\sum F_y = F_{2-5} - 4 = 0$$
$$F_{5-6} = 3.333\ \text{kN}, \qquad F_{2-5} = 4\ \text{kN}$$

Joint 6:

$$\sum F_x = F_{5-6} + F_{2-6}\cos 36.87^\circ + 10.206\cos 36.87^\circ$$
$$\equiv 3.333 + 0.8F_{2-6} + 8.165 = 0$$
$$\sum F_y = F_{1-6} + F_{2-6}\sin 36.87^\circ + 10.206\sin 36.87^\circ$$
$$\equiv 2.50 + 0.6F_{2-6} + 6.124 = 0$$
$$F_{2-6} = -14.373\ \text{kN}$$

Both equilibrium equations for joint 6 give the same value of F_{2-6} to the precision of our computations. This is the first check of the computations. The equilibrium equations for the remaining joint, number 2, give two more computational checks because the forces in all members are already known.

Joint 2:

$$\sum F_x = F_{2-3} + F_{2-4} \cos 36.87° - F_{2-6} \cos 36.87° - F_{1-2}$$
$$\equiv 0 + (-8.960)0.8 - (-14.373)0.8 - 4.330 \equiv 0.0004 \text{ kN}$$

$$\sum F_y = -(F_{2-6} + F_{2-4}) \sin 36.87° - F_{2-5} - 10$$
$$\equiv -(-14.373 - 8.960)0.6 - 4 - 10 \equiv -0.0002 \text{ kN}$$

The force sums for joint 2 are sufficiently close to zero in comparison to the precision of our computations to decide that the check equations are verified. As a result, we can be fairly certain that our calculations of the member forces, as well as the reactions, are correct. A member force that is negative indicates that the member is in compression.

Additional Remarks

When solving a problem by the method of joints, it may happen that you have chosen to consider the joints in the wrong order, so that the member forces for a certain joint cannot be found immediately (for example, more than two unknowns at a joint in a planar problem). If that should happen, proceed to the other joints until a sufficient number of forces have been calculated to allow the joint to be reconsidered.

Finally, recall that we began the solution by considering member 6–7 as a support of the rectangular truss. This allowed us to express the reaction on the left side of the truss as the axial force in that member. Suppose we had instead considered member 6–7 to be part of the truss, and ignored the fact that it is a two-force member. The corresponding reaction forces would have been those of pin 7, so we would have had four reaction components (two for pin 7 and two for pin 4). The equilibrium equations for the entire structure would have then been insoluble by themselves. In that case it would have been necessary to consider the equilibrium of the truss joints first, and then to use the results for the joints to calculate the reactions. Clearly, it is better to be able to recognize two-force members that support a structure.

EXAMPLE 2

Using the method of joints, determine the forces in the members of the space truss shown.

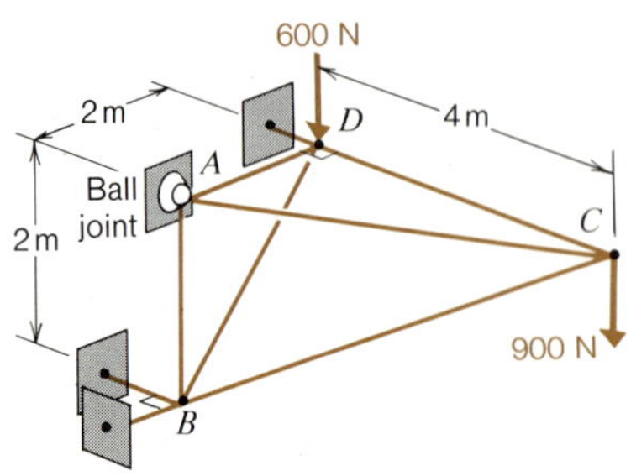

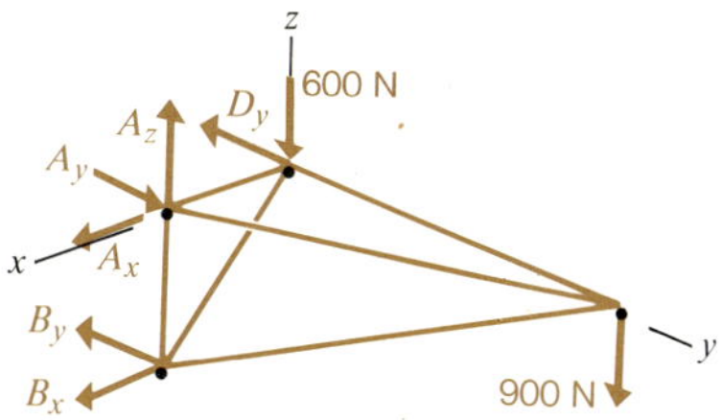

Solution

We first determine the reactions by drawing a free body diagram of the truss, noting that the short links at joints B and D can exert only axial forces. We have assumed in the free body diagram of the truss that these forces are tensile. As always, the xyz coordinate system is chosen to fit the directions in which dimensions are given.

In order to eliminate the three reaction components at the ball joint, point A is chosen for the moment sum. The equilibrium equations are then

$$\begin{aligned}\sum \bar{M}_A &= \bar{r}_{D/A} \times (-D_y\bar{j} - 600\bar{k}) + \bar{r}_{B/A} \times (B_x\bar{i} - B_y\bar{j}) + \bar{r}_{C/A} \times (-900\bar{k}) \\ &= (-2\bar{i}) \times (-D_y\bar{j} - 600\bar{k}) + (-2\bar{k}) \times (B_x\bar{i} - B_y\bar{j}) \\ &\quad + (-2\bar{i} + 4\bar{j}) \times (-900\bar{k}) \\ &= 2D_y\bar{k} - 1200\bar{j} - 2B_x\bar{j} - 2B_y\bar{i} - 1800\bar{j} - 3600\bar{i} = \bar{0} \text{ N-m}\end{aligned}$$

$$\sum \bar{F} = (A_x\bar{i} + A_y\bar{j} + A_z\bar{k}) - D_y\bar{j} + (B_x\bar{i} - B_y\bar{j}) - 900\bar{k} - 600\bar{k} = \bar{0} \text{ N}$$

Equating the components of the moment sum to zero yields

$$\bar{i} \text{ component:} \quad -2B_y - 3600 = 0 \quad \text{so} \quad B_y = -1800 \text{ N}$$

$$\bar{j} \text{ component:} \quad -2B_x - 3000 = 0 \quad \text{so} \quad B_x = -1500 \text{ N}$$

$$\bar{k} \text{ component:} \quad -2D_y = 0 \quad \text{so} \quad D_y = 0$$

Using these values of B_x, B_y, and D_y in equating the components of the force sum to zero yields

$$\bar{i} \text{ component:} \quad A_x + B_x = 0 \quad \text{so} \quad A_x = 1500 \text{ N}$$

$$\bar{j} \text{ component:} \quad A_y - D_y - B_y = 0 \quad \text{so} \quad A_y = -1800 \text{ N}$$

$$\bar{k} \text{ component:} \quad A_z - 1500 = 0 \quad \text{so} \quad A_z = 1500 \text{ N}$$

We now consider the equilibrium of the joints. The member forces are collinear with the corresponding members. In planar trusses we can obtain the components of the member forces directly from the free body diagram. For space trusses we obtain the force components by employing unit vectors parallel to each bar. Recall that in our notation a unit vector extends *to* the point denoted by the first subscript *from* the point denoted by the second subscript. The required unit vectors are

$$\bar{e}_{B/A} = -\bar{e}_{A/B} = \frac{\bar{r}_{B/A}}{|\bar{r}_{B/A}|} = -\bar{k}$$

$$\begin{aligned}\bar{e}_{C/A} = -\bar{e}_{A/C} &= \frac{\bar{r}_{C/A}}{|\bar{r}_{C/A}|} = \frac{-2\bar{i} + 4\bar{j}}{\sqrt{2^2 + 4^2}} \\ &= -0.4472\bar{i} + 0.8944\bar{j}\end{aligned}$$

$$\bar{e}_{D/A} = -\bar{e}_{A/D} = \frac{\bar{r}_{D/A}}{|\bar{r}_{D/A}|} = -\bar{i}$$

$$\bar{e}_{C/B} = -\bar{e}_{B/C} = \frac{\bar{r}_{C/B}}{|\bar{r}_{C/B}|} = \frac{-2\bar{i} + 4\bar{j} + 2\bar{k}}{\sqrt{2^2 + 4^2 + 2^2}}$$

$$= -0.4082\bar{i} + 0.8165\bar{j} + 0.4082\bar{k}$$

$$\bar{e}_{D/B} = -\bar{e}_{B/D} = \frac{-2\bar{i} + 2\bar{k}}{\sqrt{2^2 + 2^2}} = -0.7071\bar{i} + 0.7071\bar{k}$$

$$\bar{e}_{D/C} = -\bar{e}_{C/D} = -\bar{j}$$

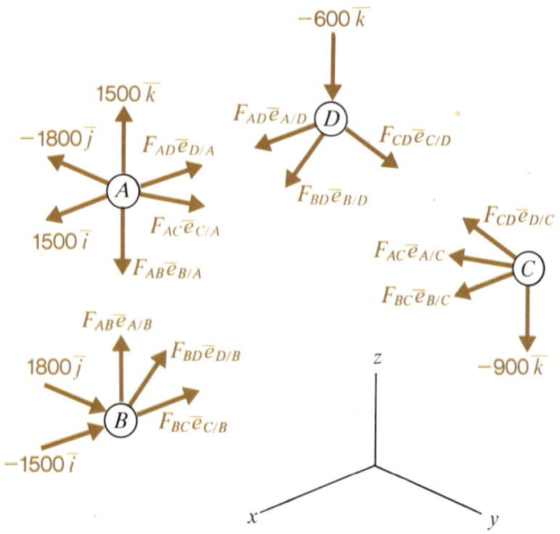

When drawing the free body diagrams for each joint, we assume that all members are in tension. Therefore, the axial force of each member at a joint is written as the product of the magnitude of the force and the unit vector from that joint to the other end of that member. For instance, the force that member BC exerts on joint B is $F_{BC}\bar{e}_{C/B}$, and the force that this member exerts on joint C is $F_{BC}\bar{e}_{B/C}$. Thus, we have the free body diagrams shown to the left.

Particular attention is called to the fact that the reaction forces are entered into the free body diagram in the direction in which they were calculated to act.

In the space truss we are analyzing, each joint connects only three members, so the joints may be considered in any order we desire. Starting with joint D, the equilibrium equations are

Joint D:

$$\sum \bar{F} = F_{AD}\bar{e}_{A/D} + F_{BD}\bar{e}_{B/D} + F_{CD}\bar{e}_{C/D} - 600\bar{k}$$
$$\equiv F_{AD}(+\bar{i}) + F_{BD}(+0.7071\bar{i} - 0.7071\bar{k})$$
$$+ F_{CD}(+\bar{j}) - 600\bar{k} = \bar{0}$$

Equating each component to zero yields

$$\bar{i} \text{ component:} \quad F_{AD} + 0.7071F_{BD} = 0$$
$$\bar{j} \text{ component:} \quad F_{CD} = 0$$
$$\bar{k} \text{ component:} \quad -0.7071F_{BD} - 600 = 0$$
$$F_{BD} = -849 \text{ N}, \qquad F_{AD} = 600 \text{ N}$$

Next, the equilibrium equations for joint C are

Joint C:

$$\sum \bar{F} = F_{AC}\bar{e}_{A/C} + F_{BC}\bar{e}_{B/C} + F_{CD}\bar{e}_{D/C} - 900\bar{k}$$
$$\equiv F_{AC}(0.4472\bar{i} - 0.8944\bar{j}) + F_{BC}(0.4082\bar{i}$$
$$- 0.8165\bar{j} - 0.4082\bar{k}) + (0)(-\bar{j}) - 900\bar{k} = \bar{0}$$

Equating like components to zero yields

$$\bar{i} \text{ component:} \quad 0.4472F_{AC} + 0.4082F_{BC} = 0$$
$$\bar{j} \text{ component:} \quad -0.8944F_{AC} - 0.8165F_{BC} = 0$$
$$\bar{k} \text{ component:} \quad -0.4082F_{BC} - 900 = 0$$
$$F_{BC} = -2205 \text{ N}, \qquad F_{AC} = 2013 \text{ N}$$

These values for F_{BC} and F_{AC} satisfy all three equations (the $\bar{j}$ equation is -2 times the $\bar{i}$ equation), so we have a validation of the calculations from the first excess equation.

The last unknown member force is found from the equilibrium equations for joint B, which is

Joint B:

$$\begin{aligned}\sum \bar{F} &= F_{AB}\bar{e}_{A/B} + F_{BC}\bar{e}_{C/B} + F_{BD}\bar{e}_{D/B} - 1500\bar{i} + 1800\bar{j} \\ &\equiv F_{AB}\bar{k} - 2205(-0.4082\bar{i} + 0.8165\bar{j} + 0.4082\bar{k}) \\ &\quad - 849(-0.7071\bar{i} + 0.7071\bar{k}) - 1500\bar{i} + 1800\bar{j} = \bar{0}\end{aligned}$$

In component form this yields

$$\bar{i} \text{ component:} \quad 900 + 600 - 1500 \equiv 0$$

$$\bar{j} \text{ component:} \quad -1800 + 1800 \equiv 0$$

$$\bar{k} \text{ component:} \quad F_{AB} - 900 - 600 = 0 \qquad F_{AB} = 1500 \text{ N}$$

The first two equations are checks on the calculations. With the forces in all members known, the component equations resulting from equilibrium of joint A will be three more checks. We leave it to you to perform the necessary computations for their verification. Once again, remember that if a member force is negative then the member is in compression.

HOMEWORK PROBLEMS

4.1–4.14 Determine the forces in the members of each truss shown. State whether each member is in tension or compression.

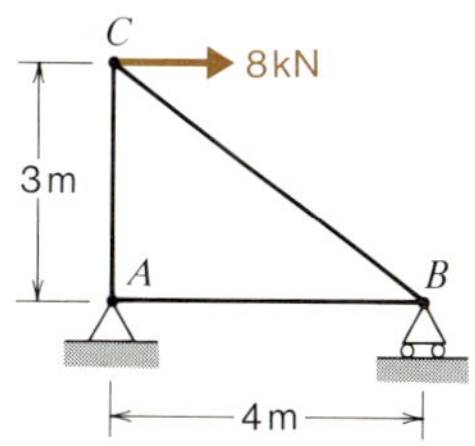

PROB. 4.1

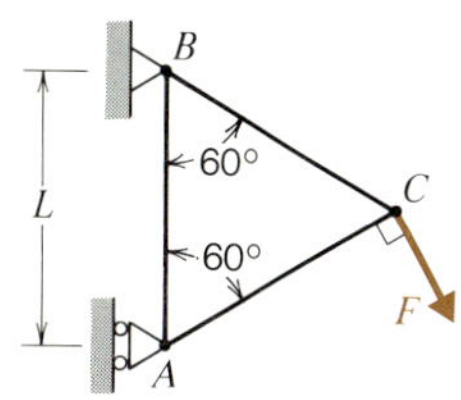

PROB. 4.2

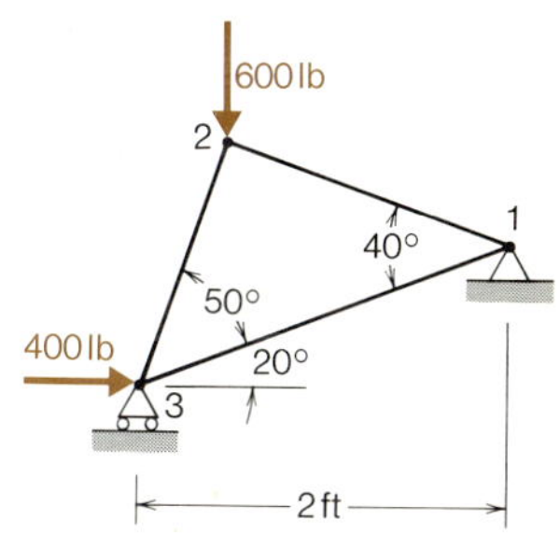

PROB. 4.3

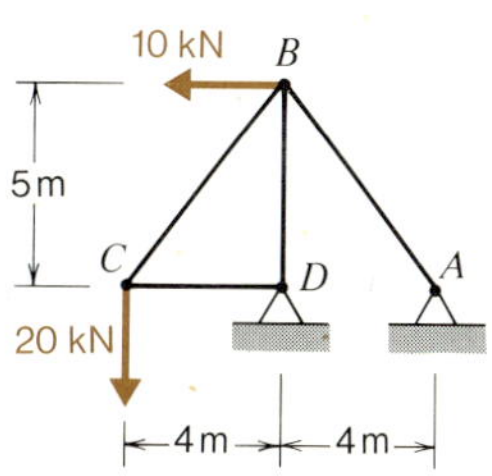

PROB. 4.4

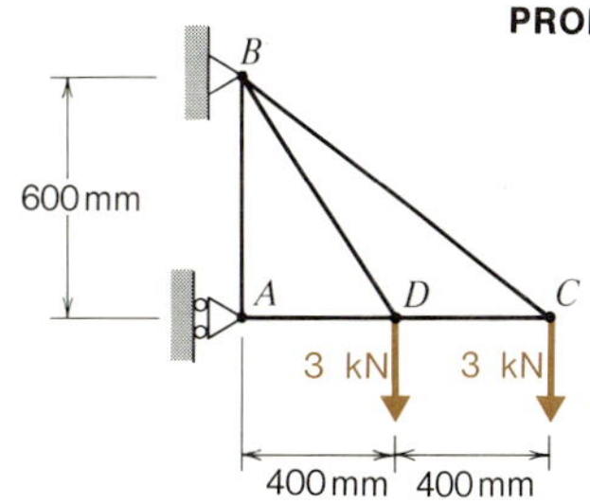

PROB. 4.5

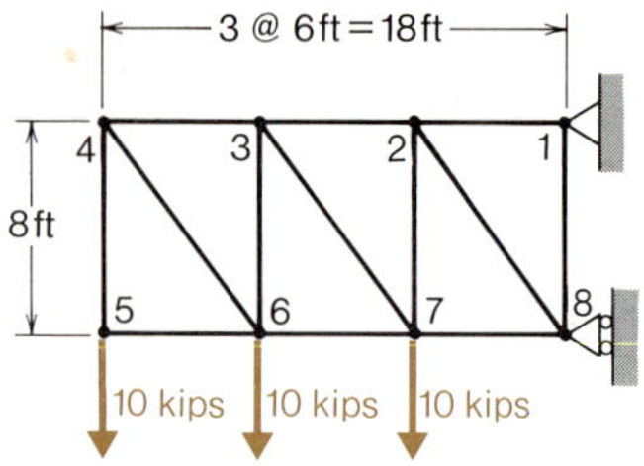

PROB. 4.6

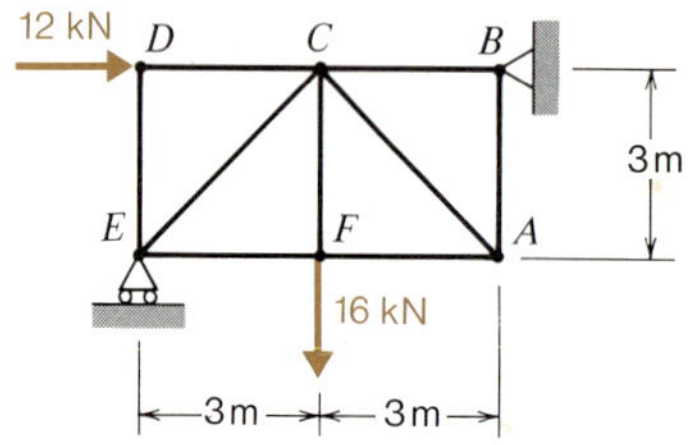

PROB. 4.7

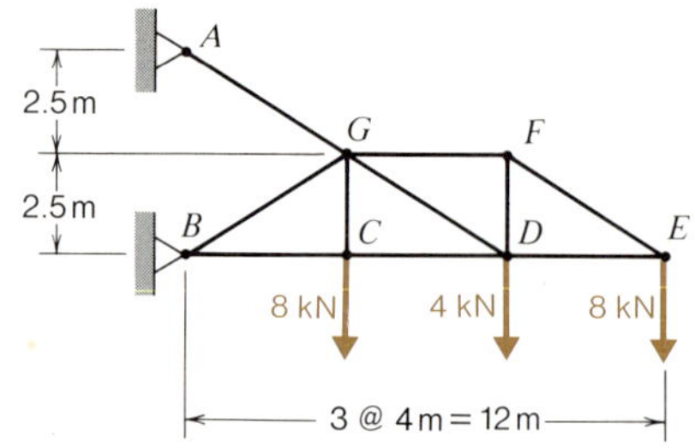

PROB. 4.8

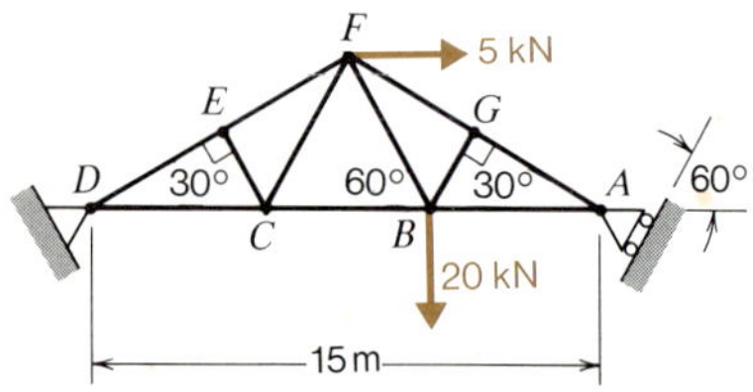

PROB. 4.9

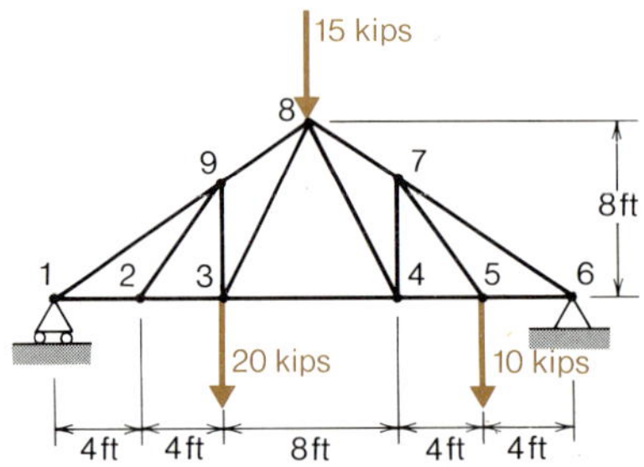

PROB. 4.10

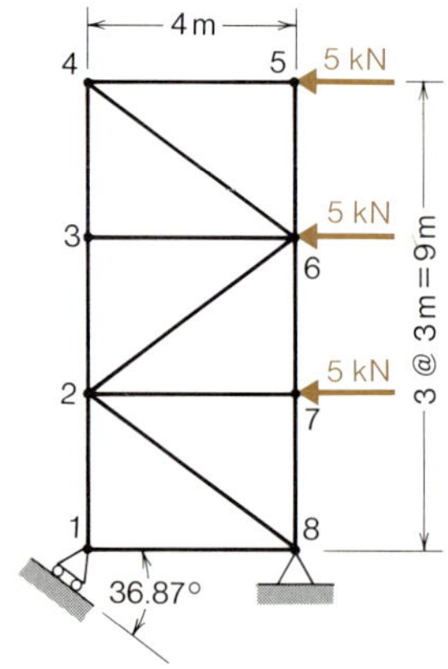

PROB. 4.11

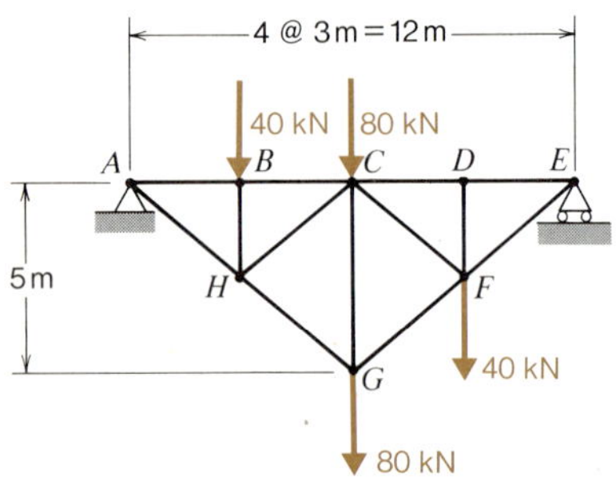

PROB. 4.12

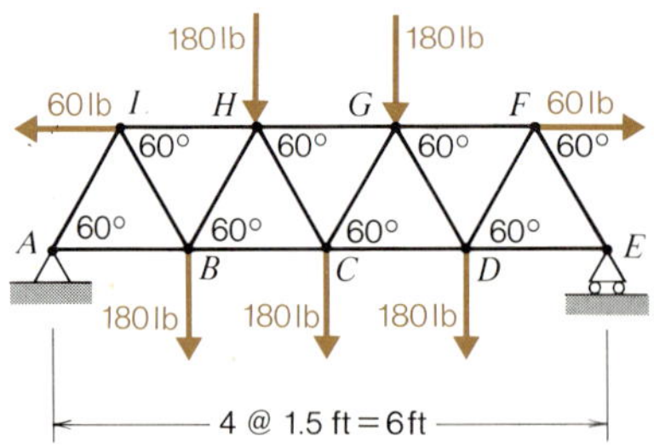

***PROB. 4.13**

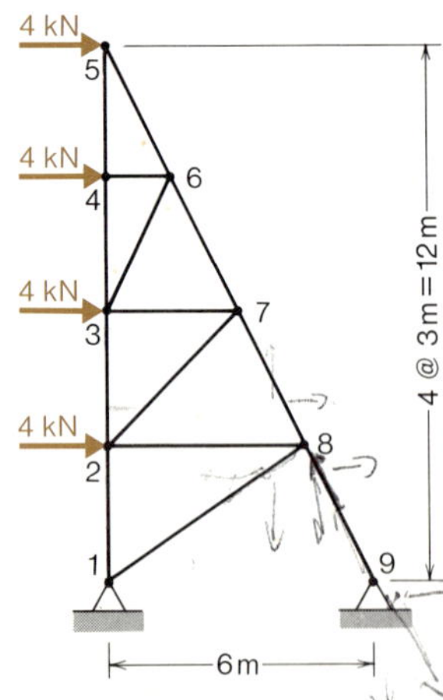

***PROB. 4.14**

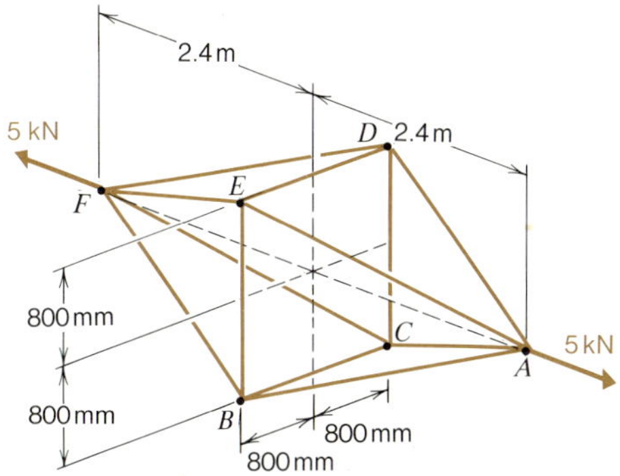

PROB. 4.15

4.15 Determine the forces in all members of the symmetrical space truss when it is subjected to the self-equilibrating set of loads shown. State whether the members are in tension or compression. *Hint:* Use symmetry to describe the members that are connected at pins A and F.

4.16 Determine the forces in the members of the space truss if $P = 30$ kN and $Q = 0$. Indicate whether the members are in tension or compression.

4.17 Solve Problem 4.16 if $P = 30$ kN and $Q = 20$ kN.

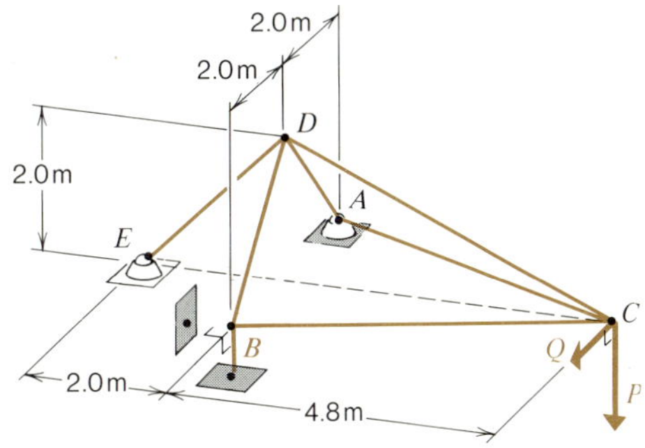

PROB. 4.16

4.18 Determine the forces in all members of the space truss if $P = 4$ kips and $Q = 0$. Indicate whether the members are in tension or compression.

4.19 Solve Problem 4.18 for $P = 4$ kips and $Q = 3$ kips.

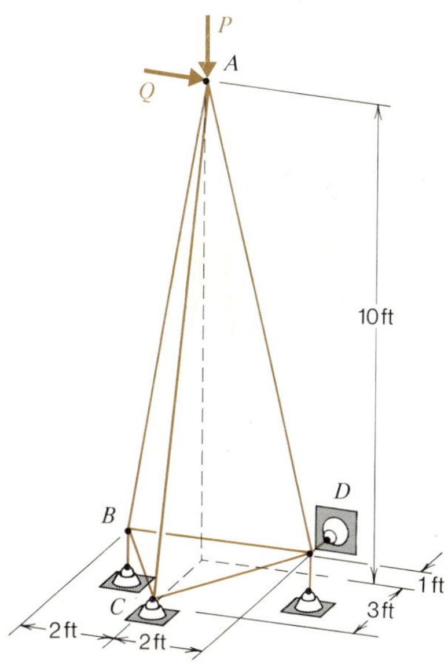

PROB. 4.18

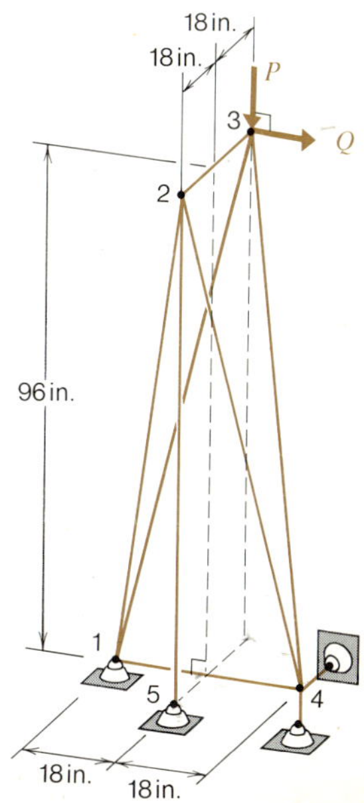

PROB. 4.20

***4.20** Determine the forces in all members of the space truss if $P = 400$ lb and $Q = 0$.

***4.21** Solve Problem 4.20 if $P = 400$ lb and $Q = 500$ lb.

*4.22 Determine the forces in all members of the space truss if $P = 12$ kN and $Q = 0$.

*4.23 Solve Problem 4.22 if $P = 12$ kN and $Q = 18$ kN.

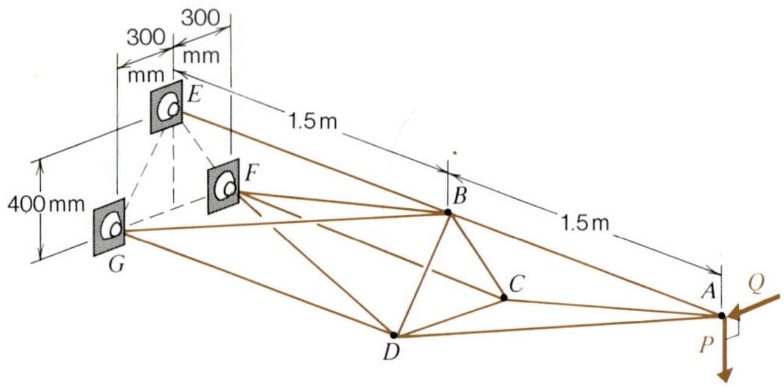

PROB. 4.22

4. Method of Sections

The primary objective of the method of joints is the determination of the forces in all members of a truss. To this end it is necessary to solve a large number of algebraic equations. Suppose that we are interested only in the forces in a few members of the truss. If we use the method of joints, we will have to determine the forces in many members that are of no interest to us. (Situations do arise where the desired member force can still be found from the equilibrium equations for one or two joints. However, such situations are not typical.)

The method of sections is intended specifically to be used for the determination of the forces in a few members of a truss. Equally important, in learning to use the method of sections we will strengthen our ability to disassemble a structure to create useful free body diagrams.

Recall that the method of joints was derived by breaking the truss up into its individual components of bars and pins. In the method of sections we decompose the truss into two segments, each of which remains intact. This decomposition is achieved by passing an imaginary *cutting plane* through the structure. For example, suppose that we wanted to know the force in member *CH* of the truss in Figure 10. In order to determine $\bar{F}_{CH}$ it must appear as an external force in a free body diagram. Hence, *the cutting plane must cut through the member of interest.*

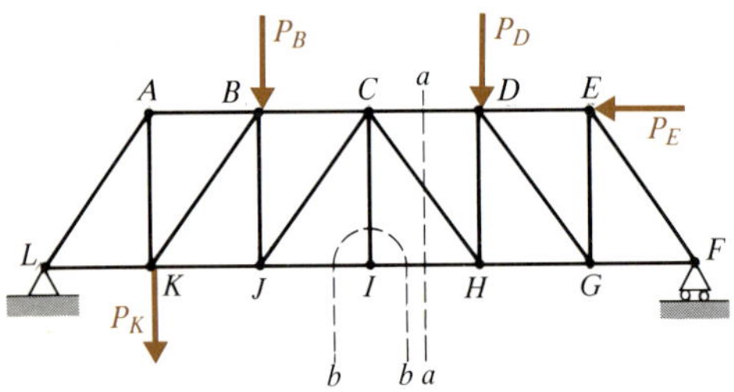

FIGURE 10

In the process of passing the cutting plane a–a through the structure, other members are also cut. In this manner the structure is decomposed into two parts, each of which must be in static equilibrium. The forces in the cut members are now external forces for these two truss segments. Once again, we assume the member forces to be tensile, so they pull on the isolated portions of the structure. The free body diagrams for the cut portions to the left and right of plane a–a are as shown in Figure 11.

Recalling that our standard procedure is to determine the reaction forces first, the only unknown forces in these free body diagrams are those

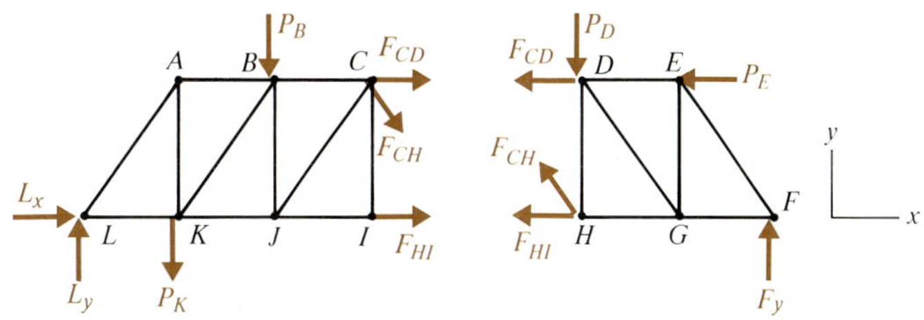

FIGURE 11

corresponding to the cut members: F_{CD}, F_{CH}, F_{HI}. Each isolated portion of the structure forms a rigid body carrying a planar force system, so we have three equations of static equilibrium for each body. The equations for *either* part will yield the values of the member forces being sought. Because the equilibrium of the entire structure was assured by the determination of the reactions exerted by the supports, the equilibrium equations for the remaining portion of the truss merely repeat the other equilibrium equations. The decision as to which portion of the cut truss should be equilibrated is arbitrary, although the portion having the smaller number of external forces and simpler geometry is usually the more convenient one to use for calculations.

The cutting plane need not be flat; it can be a curved plane. In fact, we could even use a curved plane to isolate a joint. For example, let us determine the force in member CI of the truss in Figure 10. Passing plane b–b through the truss (actually around the joint), we can isolate the joint in the free body diagram shown in Figure 12. Summing forces in the y direction shows that member IC is a zero force member.

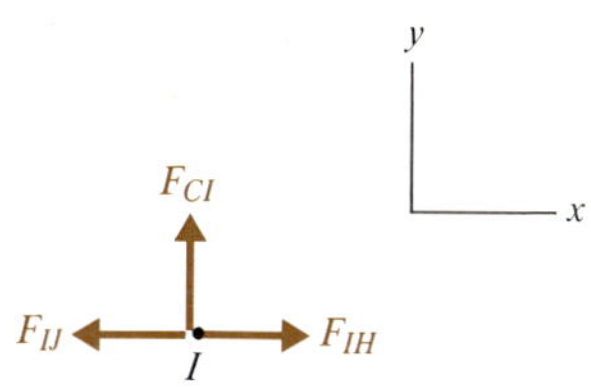

FIGURE 12

The major decision in using the method of sections involves the choice for the cutting plane. Let us first consider the case of a planar truss. If the cutting plane cuts through no more than three members, as was done in Figure 10, we may determine the member forces directly from the three equilibrium equations for an isolated portion of the structure. On the other hand, if a cutting plane cuts through four or more members, it will not be possible to determine all of the cut member forces using that section alone, although it might be possible to obtain values for a few of the forces.

The method of sections may also be applied to space trusses, in which case equilibrium of the isolated portion of the structure results in six scalar equations. If the cutting plane cuts through no more than six members, we will be able to determine all of the forces in the cut members. A cutting plane intersecting seven or more members will leave us unable to determine at least some of the member forces. In the examples that follow we shall concentrate on planar trusses, because the only true difference between these trusses and three-dimensional ones involves the techniques for forming the equilibrium equations.

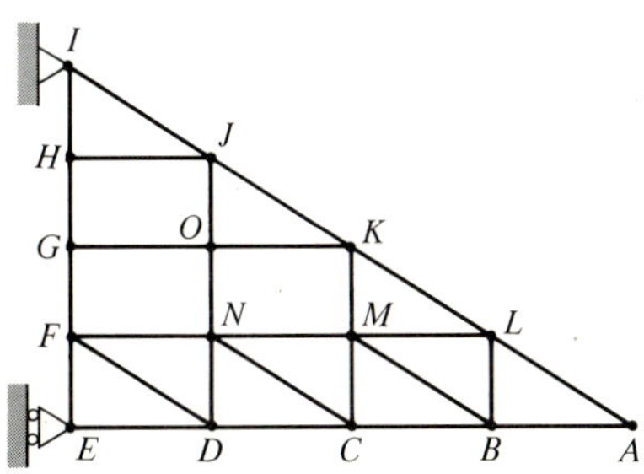

FIGURE 13

Finally, the discussion of the method of sections would be incomplete if we did not note that it is not universally applicable. In Figure 13 we have a simple truss that requires an excessive number of cutting planes to evaluate the force in an interior member, such as member NM. In such a case we resort to the method of joints, as we would if we wanted to know the forces in many or all of the members of any truss.

EXAMPLE 3

Determine the forces in members DE and DK of the truss.

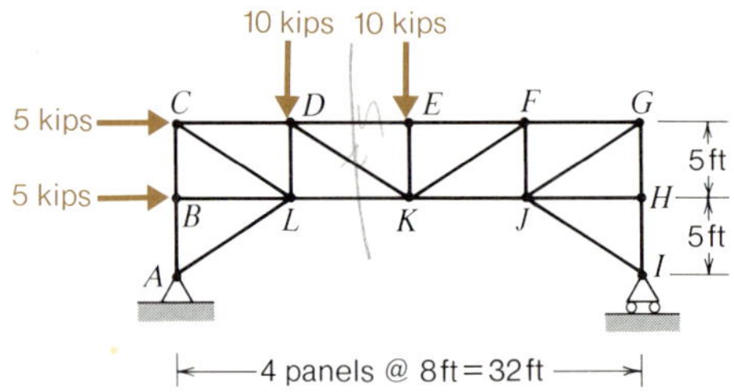

Solution

We first determine the reactions using a free body diagram of the entire truss, from which we find

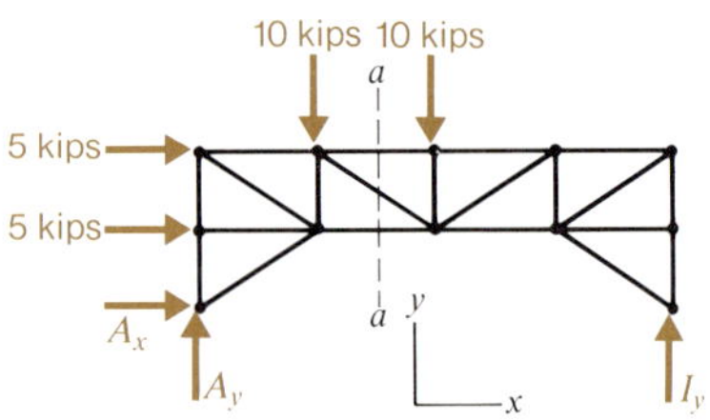

$$\sum M_{Az} = I_y(32) - 5(5) - 5(10) - 10(8) - 10(16) = 0$$

$$\sum F_y = A_x + 10 = 0, \qquad \sum F_y = A_y + I_y - 20 = 0$$

$$I_y = 9.844 \text{ kips}, \qquad A_y = 10.156 \text{ kips}, \qquad A_x = -10 \text{ kips}$$

The next step is to choose the cutting plane. This plane must cut through the member of interest, preferably while cutting through no more than three members altogether. The plane a–a in the first free body diagram is suitable for studying both members DE and DK.

We shall study the equilibrium of the truss segment to the right of cutting plane a–a because there are fewer external forces acting on the right segment than the left one. The free body diagram of the right section of the truss is shown in the margin. Remember that we assume that all members are in tension, so that the cut members are assumed to pull on the joints to which they are connected.

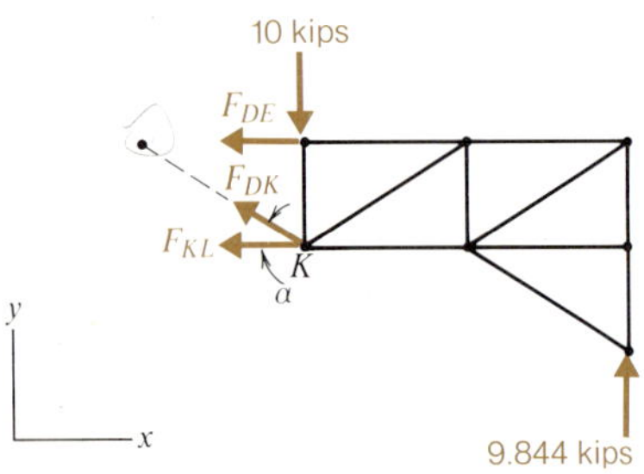

The angle α is found from the dimensions on the original sketch to be

$$\alpha = \tan\left(\frac{5}{8}\right) = 32.01^\circ$$

We find F_{DK} from a single equation by summing forces in the y direction for the section. The force F_{DE} is found directly by summing moments about joint K.

$$\sum F_y = F_{DK} \sin \alpha - 10 + 9.844 = 0$$

$$\sum M_{Kz} = F_{DE}(5) + 9.844(16) = 0$$

$$F_{DK} = 0.29 \text{ kips}, \qquad F_{DE} = -31.50 \text{ kips}$$

The fact that F_{DE} is negative means that it is in compression.

EXAMPLE 4

Determine the forces in members 9–12 and 7–10 of the truss shown.

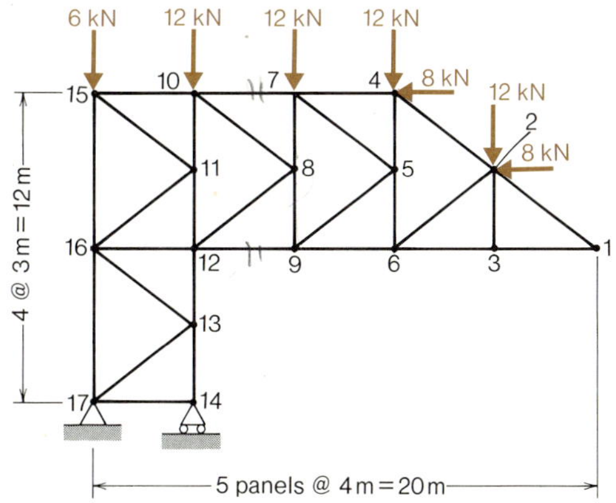

Solution

At the beginning of this chapter we stated that we would make it a standard practice to begin a truss analysis by determining the reactions at the supports. An exception arises here because the segment of the truss to the right of the cutting planes through members 9–12 and 7–10 does not contain any of the reactions. This will enable us to find the member forces without finding the reactions.

An inspection of the truss shows that any plane that cuts through members 9–12 and 7–10 also cuts through two or more additional members. Bearing in mind that there are only three equilibrium equations for a section, we note that members 7–8 and 8–9 are collinear. Therefore, a curved plane that cuts through members 7–10, 7–8, 8–9, and 9–12 will allow us to determine the desired truss member forces by means of moment equilibrium equations. This cutting plane a–a and the resulting free body diagram for the section to the right of the cutting plane is shown in the margin.

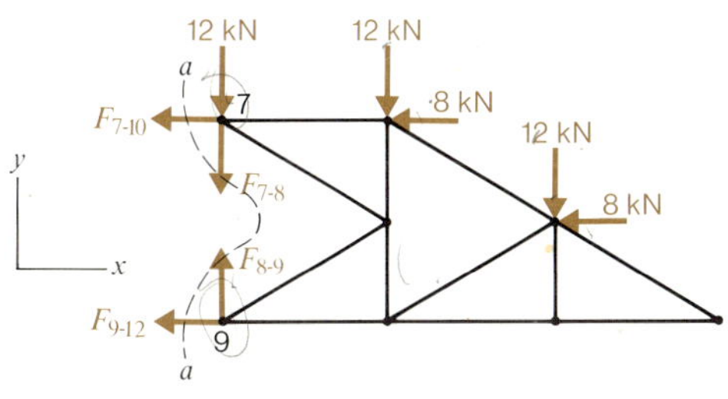

The force in member F_{7-10} is found by summing moments about joint 9. A moment sum about joint 7 will yield F_{9-12}. Thus

$$\sum M_{9z} = F_{7-10}(6) + 8(6) - 12(4) + 8(3) - 12(8) = 0$$

$$\sum M_{7z} = -F_{9-12}(6) - 12(4) - 8(3) - 12(8) = 0$$

which gives

$$F_{7-10} = 12 \text{ kN}, \qquad F_{9-12} = -28 \text{ kN}$$

EXAMPLE 5
Determine the tension in members 4–8 and 7–8 of the truss shown using the method of sections.

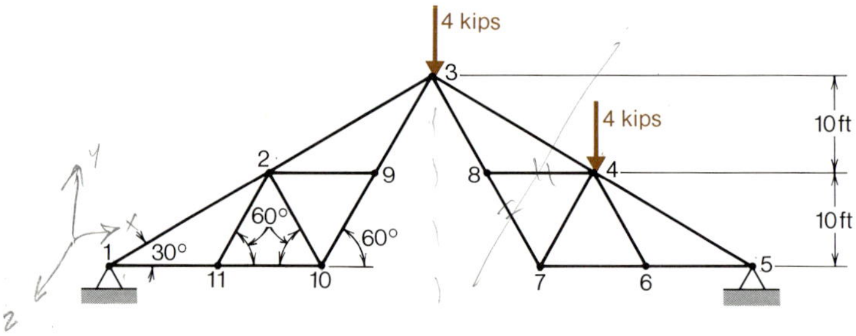

Solution
Perhaps the most notable feature of this truss is that it is supported by two pins, as shown by the reactions in the free body diagram. For equilibrium of the truss we have

$$\sum F_x = (F_1)_x + (F_5)_x = 0$$
$$\sum F_y = (F_1)_y + (F_5)_y - 4 - 4 = 0$$
$$\sum M_{1z} = (F_5)_y[4(10 \cot 30^\circ)] - 4[2(10 \cot 30^\circ)] - 4[3(10 \cot 30^\circ)] = 0$$
$$(F_5)_y = 5 \text{ kips}, \qquad (F_1)_y = 3 \text{ kips}, \qquad (F_5)_x = -(F_1)_x$$

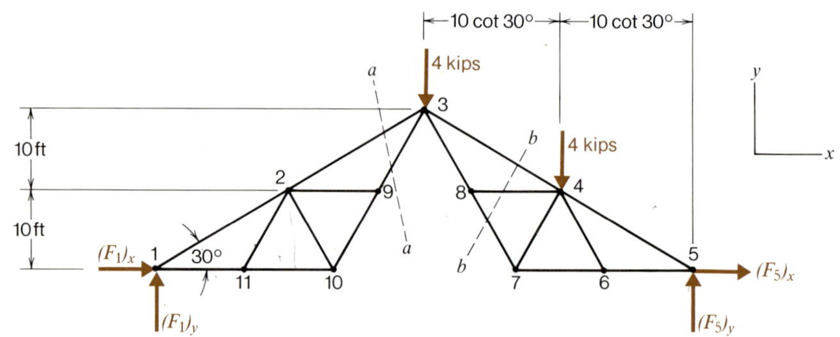

We cannot determine $(F_5)_x$ and $(F_1)_x$ from the foregoing equations. At first glance it might seem that these reactions are statically indeterminate. This is not so. Recall that an indeterminate structure has redundant supports that are not necessary for holding the system in static equilibrium. In the given truss, changing one of the pin supports to a roller will result in a nonrigid structure in which the roller support will move, allowing joint 3 to fall down. On this basis we surmise that the truss is statically determinate with the given supports.

The reason for the apparent ambiguity is that the truss is really composed of two simple trusses, each of which is rigid by itself. The left truss is supported at pin 1 and joint 3, and the right truss is supported at pin 5 and joint 3. In general, a truss formed from several simple trusses is called a *compound truss*. The Fink truss that was shown in Figure 3 is another compound truss. It resembles the truss in this problem, but it is rigid and statically determinate with a pin and roller support because of the presence of the additional horizontal member connecting the two portions.

From these observations, we conclude that the determination of the values of $(F_1)_x$ and $(F_5)_x$ requires that we consider the equilibrium of the simple trusses from which the compound truss is formed, as well as of the entire structure. For this consideration note that cutting plane a–a shown in the free body diagram of the entire truss cuts through only two members, although equilibrium of the sections formed by plane a–a will yield three additional equations. Hence, choosing to equilibrate the left section, we have the free body diagram shown in the left margin.

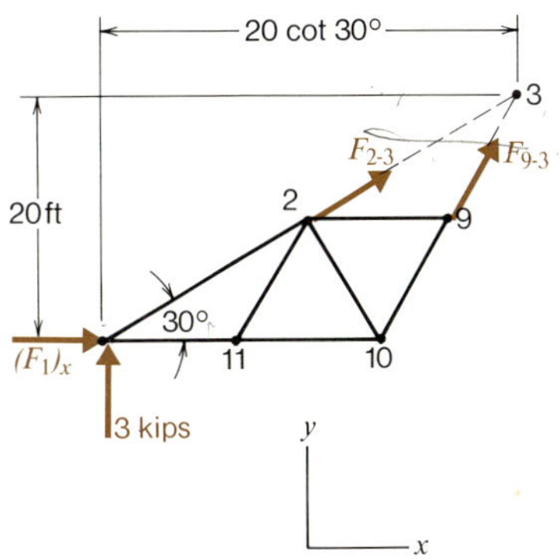

We obtain $(F_1)_x$ directly by summing moments about joint 3. Thus

$$\sum M_{3z} = (F_1)_x(20) - 3(20 \cot 30^\circ) = 0$$

This equation, in combination with the equilibrium relations for the entire structure, yields

$$(F_1)_x = 3 \cot 30^\circ = 5.196 \text{ kips}$$

$$(F_5)_x = -(F_1)_x = -5.196 \text{ kips}$$

Now that we know the reactions, a cutting plane can be chosen for determining the forces in members 4–8 and 7–8. We will use the cutting plane b–b shown in the free body diagram of the entire truss because it cuts through both members of interest while cutting through a total of only three members. Considering the portion of the truss to the right of the cutting plane for equilibrium leads to another free body diagram. We choose joint 4 for the moment sum in order to eliminate F_{4-3} and F_{4-8}. The equilibrium equations are

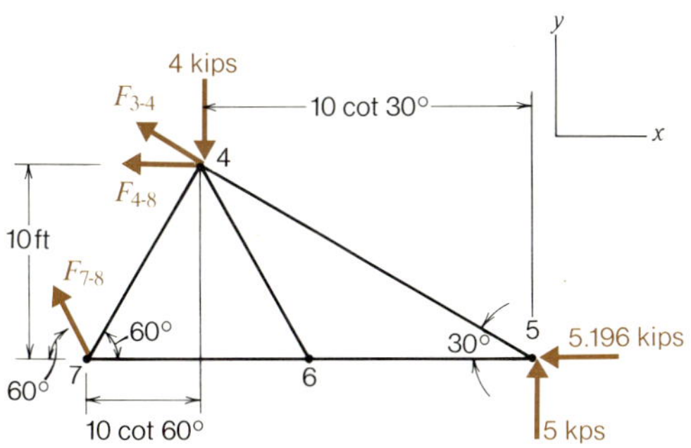

$$\sum M_{4z} = -F_{7-8} \sin 60^\circ(10 \cot 60^\circ) - F_{7-8} \cos 60^\circ(10) + 5(10 \cot 30^\circ) + (-5.196)(10) = 0$$

$$\sum F_x = -F_{7-8} \cos 60^\circ - F_{4-8} - F_{4-8} \cos 30^\circ - 5.196 = 0$$

$$\sum F_y = F_{3-4} \sin 30^\circ + F_{7-8} \sin 60^\circ + 5 - 4 = 0$$

$$F_{7-8} = 3.464 \text{ kips}, \qquad F_{3-4} = -8.00 \text{ kips}, \qquad F_{4-8} = 0 \text{ kips}$$

The value of F_{3-4} was obtained in the process of solving the equations, even though it was not requested.

Finally, note that member 4–8 could have been readily shown to be a zero force member by the method of joints; a free body diagram shows that it is the only force acting on joint 8 in the direction perpendicular to the line from joint 3 to joint 7. However, using the methods of joints to determine the force in member 7–8 requires that other joints be solved first.

HOMEWORK PROBLEMS

4.24 Determine the axial force in members B–C and B–F of the truss shown.

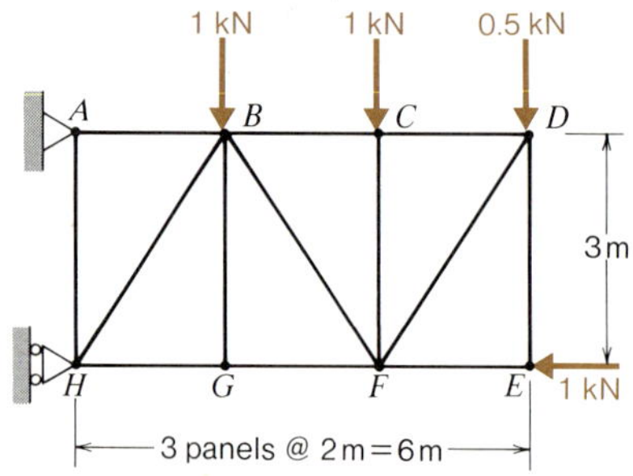

PROB. 4.24

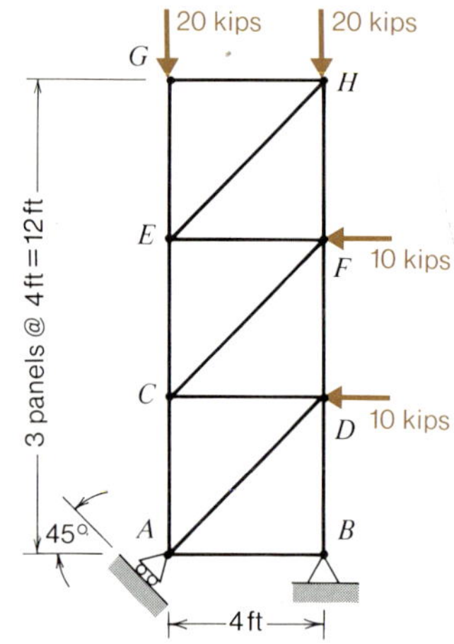

PROB. 4.25

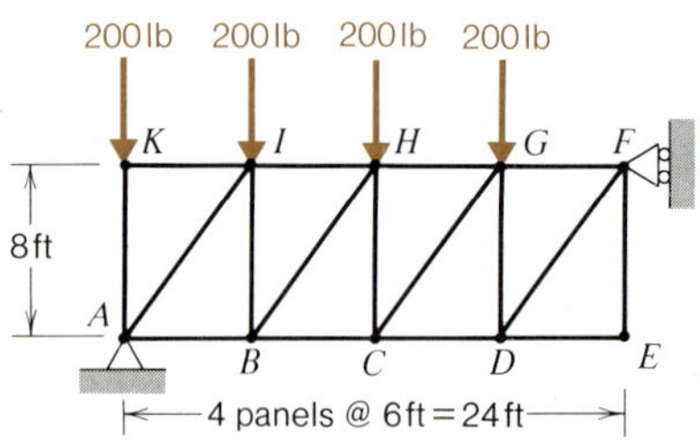

PROB. 4.27

4.25 Determine the axial force in members CE, CF, and GE of the truss shown.

4.26 Determine the axial force in members DE, CD, and CE of the truss in Problem 4.7.

4.27 Determine the axial force in members AK, BC, and CH of the truss shown.

4.28 Determine the axial force in member BF of the truss in Problem 4.9.

4.29 Determine the axial forces in members 3–4 and 4–7 of the truss shown.

4.30 Determine the axial force in members 2–8 and 8–9 of the truss shown.

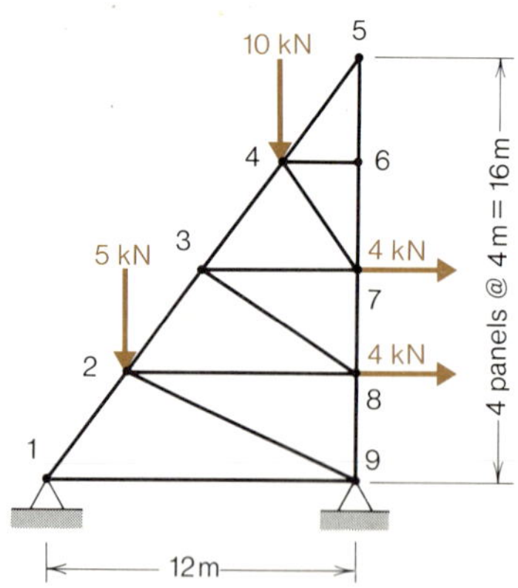

PROBS. 4.29 AND 4.30

4.31 Determine the force in members CF and FG of the truss in Problem 4.12.

4.32 Determine the axial force in members 3–4, 3–8, and 8–9 of the truss in Problem 4.10.

4.33 Determine the axial force in members *DE* and *DF* of the truss shown.

4.34 Find the axial force in members 5–11 and 11–12 of the truss shown.

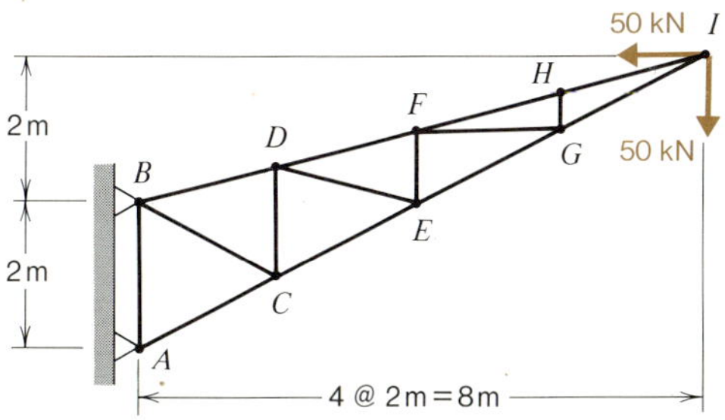

PROB. 4.33

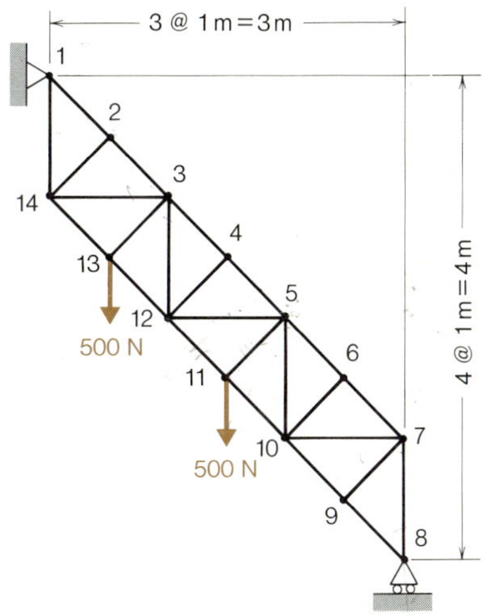

PROB. 4.34

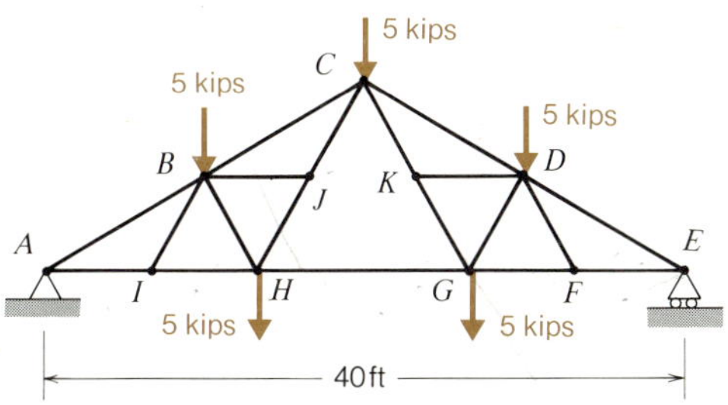

PROBS. 4.35 AND 4.36

4.35 The acute angles formed by the members in the truss shown are either 30° or 60°. Determine the force in member *CK* and *GH*.

4.36 The acute angles formed by the members in the truss shown are either 30° or 60°. Determine the axial force in members *BH* and *BJ*.

4.37 Determine the forces in members 4–9 and 8–9 of the truss shown.

***4.38** Determine the axial force in members 2–3 and 3–11 of the truss shown.

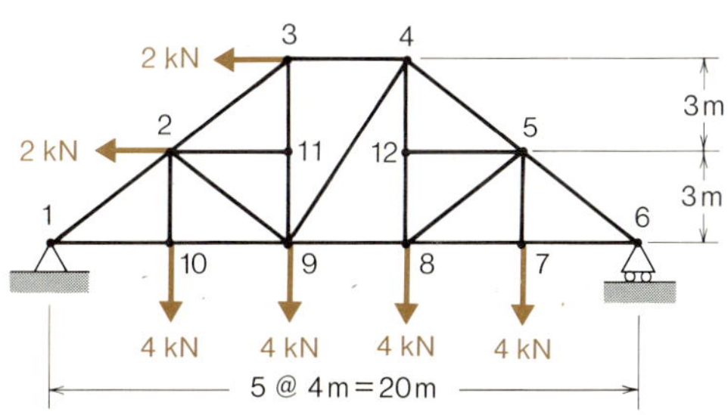

PROBS. 4.37 AND 4.38

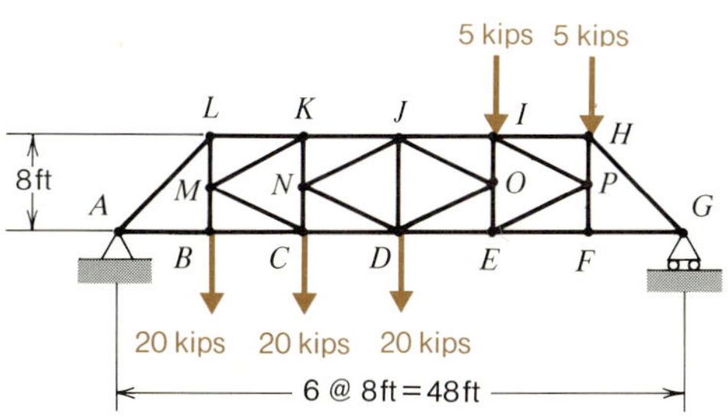

PROB. 4.39

4.39 Determine the axial force in member *I–J* of the truss shown.

4.40 Determine the axial forces in members 5–8 and 7–10 of the truss shown.

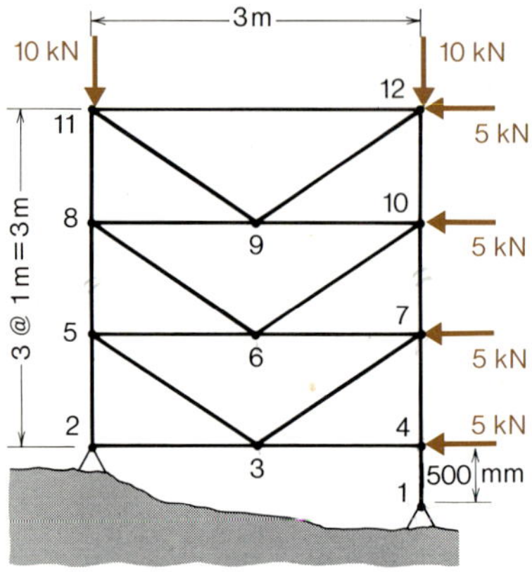

PROB. 4.40

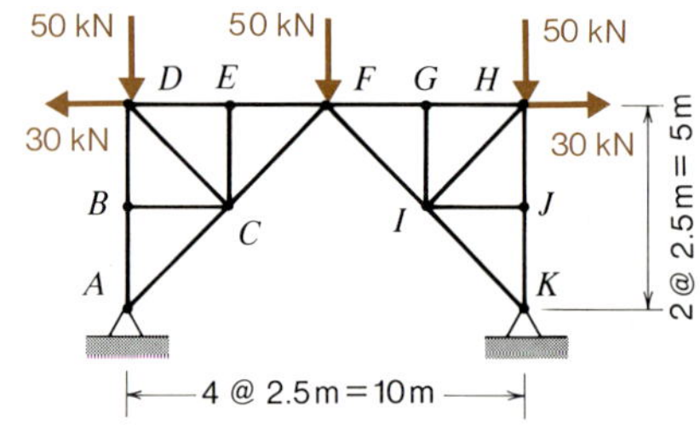

PROB. 4.41

4.41 Determine the axial force in members *EF* and *CF* of the truss shown.

4.42 Determine the axial force in members 5–6 and 6–15 of the truss shown.

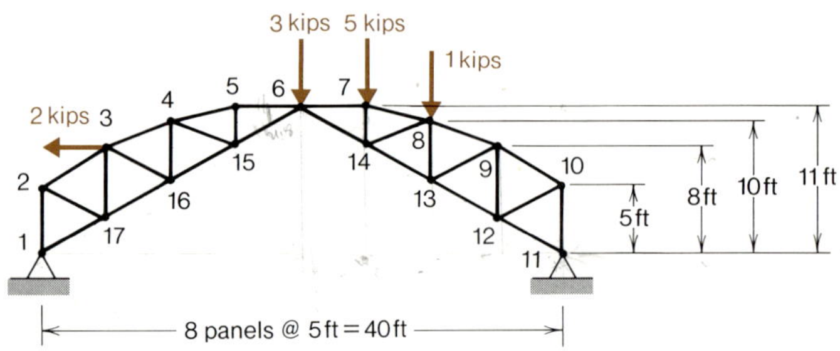

PROB. 4.42

4.43 Members 4–12 and 6–10 cross, but are not connected. Determine the force in these members.

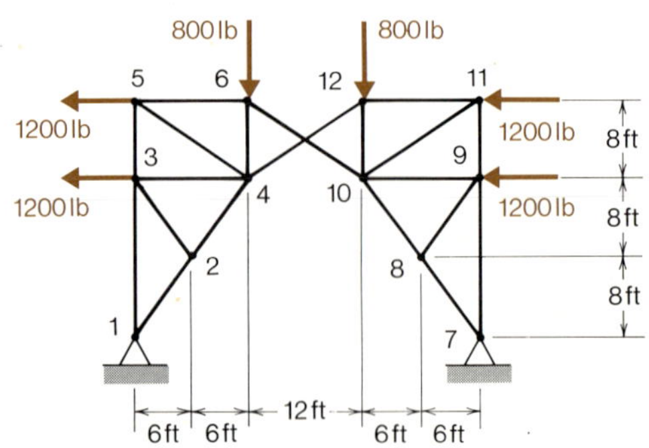

PROB. 4.43

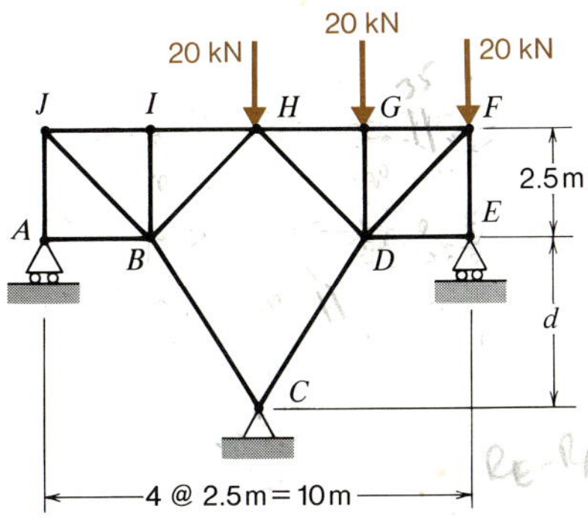

PROB. 4.44

***4.44** Determine the axial forces in members *DC* and *FG* of the truss shown when $d = 5$ m. Can these forces be found when $d = 2.5$ m? Explain your response.

***4.45** Determine the axial force in member 7–8 of the truss in Example 4.

***4.46** Determine the axial force in members 5–6 and 6–8 of the truss in Problem 4.40.

***4.47** The diagonal members in the truss shown exert negligible forces when required to be in compression because of a phenomenon called "buckling." Hence, they may be considered to sustain tensile axial forces only. Such members are called *counters*. These counters cross, but are not connected to each other. Determine the axial forces in counters *BF* and *CG* when $P = 40$ kN and $Q = 0$.

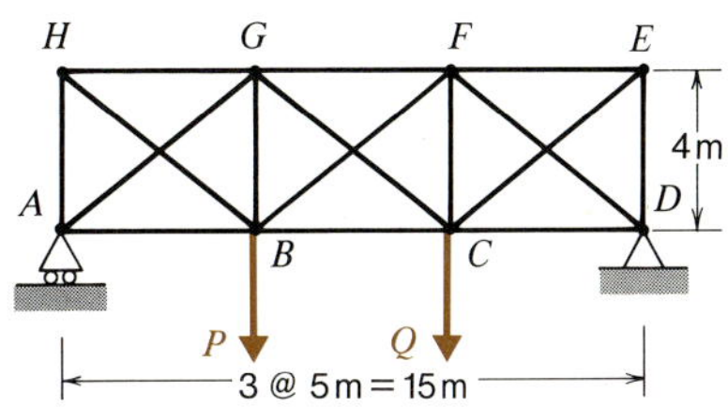

PROB. 4.47

***4.48** Solve Problem 4.47 for $P = 40$ kN and $Q = 80$ kN.

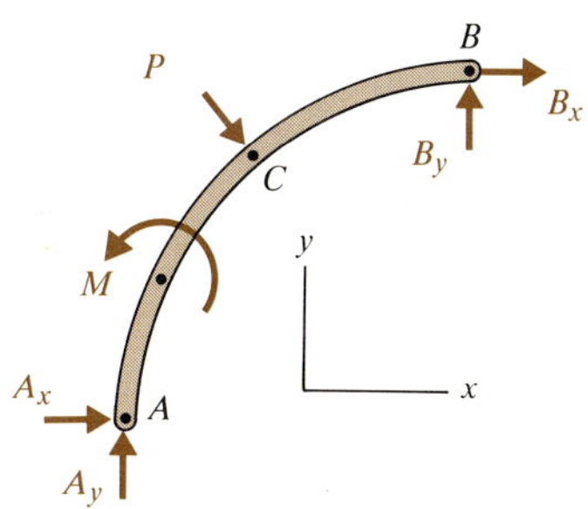

FIGURE 14

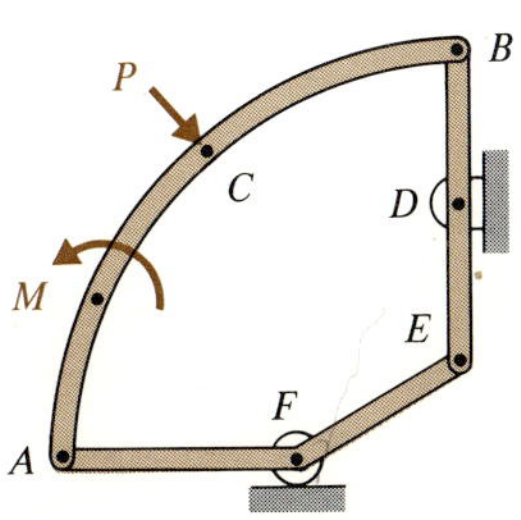

FIGURE 15

B. FRAMES AND MACHINES

A truss is a special structure. It has only straight members that are connected concurrently at its joints, and the loads it transmits are applied only at these joints. We will now explore how to determine the forces transmitted by the members of more general structures.

In this section we shall examine two types of general structures. They differ only in their application. *Frames* are rigid structures intended to support a given loading, whereas *machines* are nonrigid structures that transmit and modify a given set of input loading forces into another set of output forces. Our interest in machines in this text on Statics is with the relationship of input and output forces necessary for static equilibrium. We leave the question of the motion of a machine for the companion text on Dynamics. We shall not distinguish between the methods of solution for frames and machines.

Frames and machines share the characteristic that they contain *multi-force members*. Such members are loaded by general systems of forces. Consequently, we cannot investigate the static equilibrium of the structure on the basis of all members carrying only an axial force. For example, consider the curved member shown in Figure 14, that is pinned at ends *A* and *B* to other members of the structure in Figure 15. We do not know in advance how the reaction force components at end *A* compare to those at end *B*.

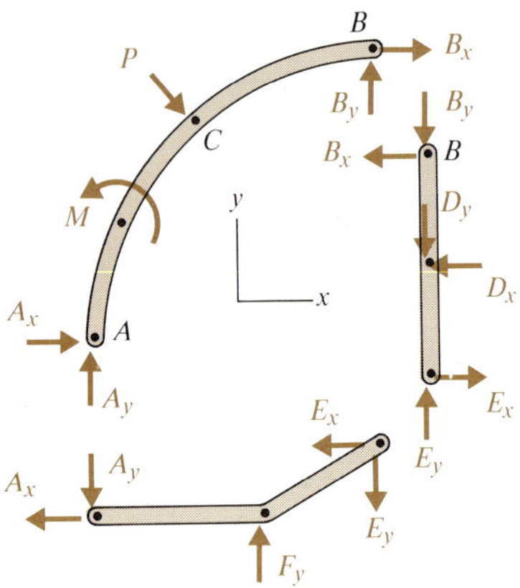

FIGURE 16

Rather, it is necessary for us to write the equations of static equilibrium for the member AB in order to relate these reactions to the given loading, which consists of a force $\bar{P}$ applied to point C and a couple $\bar{M}$.

It appears that the member in Figure 14 is statically indeterminate because there are four unknown reaction components, and only three equilibrium equations for the member. However, this is a member in a structure. Additional equations govern the static equilibrium of the other members of the structure, and of the structure as a whole. It is only after counting all such equations and the corresponding unknowns that we can judge the determinacy of the structure. For example, let us draw a free body diagram of each member in Figure 15, as shown in Figure 16. There are a total of nine unknowns in Figure 16. There are three bodies in this system, so we can write nine equations of static equilibrium. Thus, the frame is statically determinate.

From the foregoing, we deduce that the basic method for studying frames and machines is the process of simultaneously considering the static equilibrium (a) of the entire structure, and (b) of each of its members individually.

As always when dealing with bodies that are connected to each other, special care must be taken to satisfy Newton's third law of action and reaction when drawing free body diagrams. That is, when drawing free body diagrams, the forces exerted by one member on a second member must be equal, opposite, and collinear to the forces exerted by the second member on the first. As an example, consider the frame in Figure 17, whose weight is negligible compared to the force $\bar{F}_D$. If we draw the free body diagram for the structure and the two members ACD and BC, we arrive at the results shown in Figure 18, where the directions of the unknown forces are arbitrarily assumed using Newton's third law.

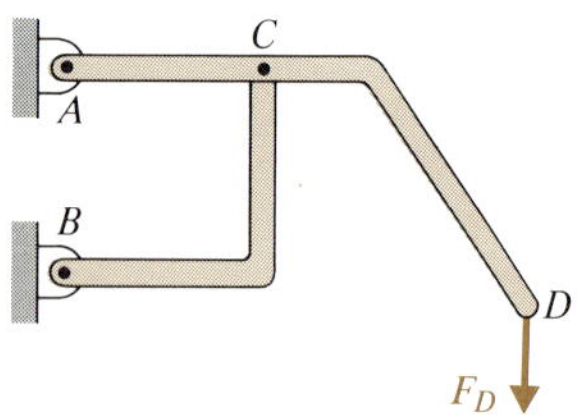

FIGURE 17

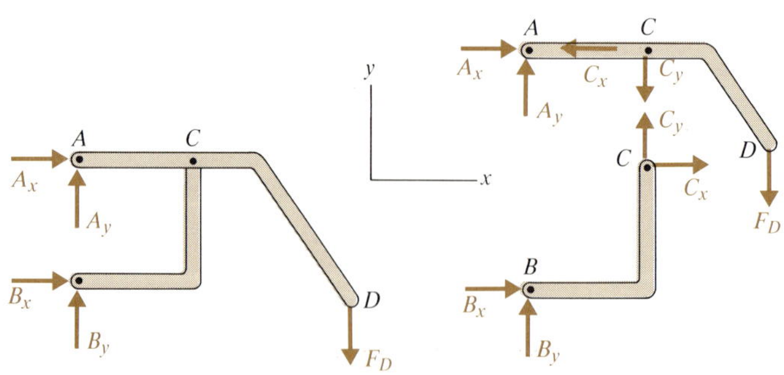

FIGURE 18

In these free body diagrams there are six unknown reaction components, for which we have nine equations of static equilibrium (three for each isolated body). As was true in our study of trusses, equilibrium of the entire truss is assured if the individual members are in equilibrium, so the three excess equations may be regarded as checks on the calculations.

We drew the free body diagrams of the individual members in Figure 18 by noting that their connection consists of a pin at joint *C*. We guessed the senses of the components of the pin force exerted by bar *ACD* on bar *BC* at point *C*. We then drew the reaction forces exerted by bar *BC* on bar *ACD consistently* with this initial guess. As usual, we will know that a direction has been improperly assumed if the force component is found to be negative.

> A useful check on whether the free body diagrams for individual members have been drawn consistent with Newton's third law is to superimpose the individual diagrams in order to form the original structure. If the reaction forces between the members cancel each other, with the result that the superposition exactly produces the free body diagram of the entire structure, then it can be concluded that the internal reaction forces are consistent.

We drew the free body diagrams in Figure 18 on the premise that we had not first carefully considered the nature of the members. However, a little analysis before we start sketching will save a good deal of computational effort. It will be noted that member *BC* is loaded only at its ends. Therefore, even though it is not a straight bar, it is a two-force member. In the discussion of two-force members at the beginning of this chapter, we noted that the two forces applied to curved, as well as straight, two-force members must be equal in magnitude, opposite in direction, and collinear with the line connecting their points of application. Thus, the observation that bar *BC* is a two-force member results in the alternative set of free body diagrams shown in Figure 19.

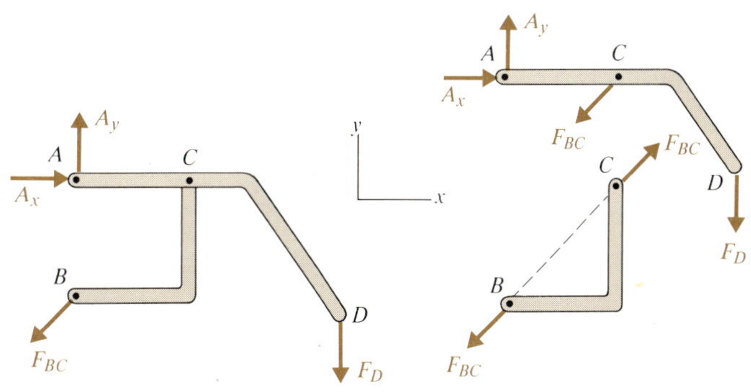

FIGURE 19

The computational savings previously alluded to result from the fact that equilibrium of member *BC* is assured, as well as from the fact that we now have only one unknown, F_{BC}, in place of the four unknown, B_x, B_y, C_x, and C_y in Figure 18. Nevertheless, it should be emphasized that the observation that certain members are two-force members is useful, but not crucial, to the solution process.

The foregoing discussion may be generalized in the following series of steps for determining the forces exerted between the members of a frame or a machine.

1. Inspect the structure to determine whether there are any two-force members.
2. Draw free body diagrams of the entire structure and of each member of the structure. In doing so, be certain to depict the forces acting on two-force members consistently, that is, equal in magnitude, opposite in direction, and collinear. An aid in satisfying Newton's third law in the depiction of the forces exerted between members is to position the free body diagrams of the individual members as one would draw an exploded assembly view of the structure.
3. Successively write and solve the equilibrium equations corresponding to the free body diagrams. The two-force members are automatically in equilibrium if their forces have been properly accounted for, so such members need not be considered. In following this step it is only necessary to formulate a sufficient number of equations to obtain the forces of interest. However, if all of the possible equilibrium equations are written, the excess equations may be used to check earlier computations.

These steps are equally valid for investigating planar and three-dimensional structures. For a given number of members, three-dimensional structures result in twice as many equilibrium equations as planar structures. The force and moment computations for three-dimensional structures are considerably more involved than those for planar structures. For these reasons, we shall restrict our attention to planar problems.

Finally, we note that it is possible for frames to be statically indeterminate. As always, this is manifested by an excess of unknown forces in comparison with the number of available equilibrium equations.

EXAMPLE 6

Bar AB is pinned at end B to a collar that may slide over the smooth bar CD. Determine the couple M_A required for static equilibrium of the system under the 1 kN loading shown. Also, determine the corresponding pin reactions at end A.

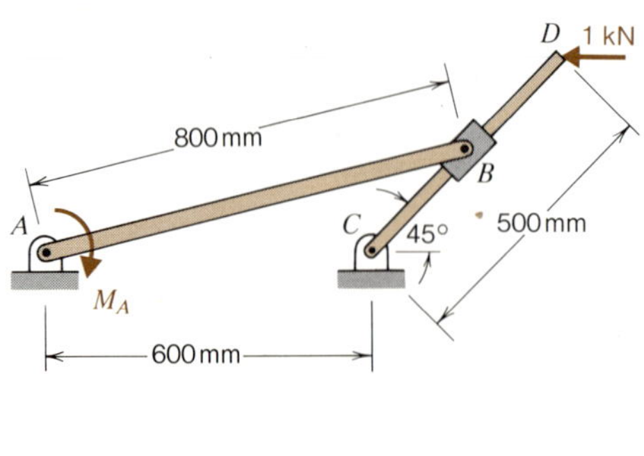

Solution

Recall that a couple is equivalent to a pair of oppositely directed forces having arbitrary points of application, from which it follows that member AB is *not* a two-force member. Consequently, we need free body diagrams of the entire structure, and of each member. We assume that the weight of each member is negligible in comparison to the loading, because no information about the mass of the members is given. Bar CD is smooth, so the reaction force exerted by the collar must be perpendicular to bar CD. The free body diagrams drawn below are based on the intuitive guess that the collar opposes the tendency of the 1 kN force to rotate bar CD about pin C.

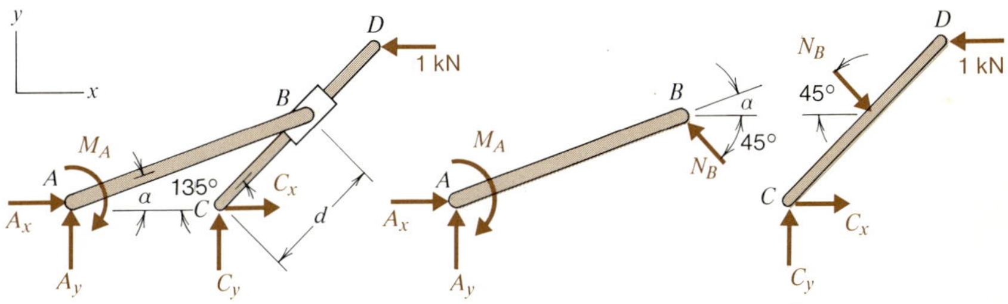

We will need the values of α and d to write the equilibrium equations, both of which may be found from the law of sines.

$$\frac{0.80}{\sin 135^\circ} = \frac{d}{\sin \alpha} = \frac{0.60}{\sin(180^\circ - 135^\circ - \alpha)}$$

Solving first for α, and then for d, yields

$$\sin(45^\circ - \alpha) = \frac{0.6}{0.8}\sin 135^\circ, \qquad \alpha = 12.97^\circ$$

$$d = \sin \alpha \frac{0.8}{\sin 135^\circ} = 0.2540 \text{ m}$$

The free body diagrams indicate that the values of M_A, A_x, and A_y may be determined from bar AB, after we find N_B. The value of N_B may be determined from bar CD by taking moments about point C.

$$\sum M_{Cz} = -N_B(d) + 1(0.5 \sin 45^\circ) = 0$$

$$N_B = 1.3921 \text{ kN}$$

The equilibrium equations for bar AB then yield

$$\sum M_{Az} = -M_A + N_B \sin(45^\circ + \alpha)(0.8) = 0$$

$$\sum F_x = A_x - N_B \cos 45^\circ = 0, \qquad \sum F_y = A_y + N_B \sin 45^\circ = 0$$

Substituting N_B and α into these equations gives

$$M_A = 0.944 \text{ kN-m}, \qquad A_x = -A_y = 0.984 \text{ kN}$$

It should be noted that the sequence of steps that we followed is not the only one possible. For example, we could have first obtained A_y (but not A_x) from equilibrium of the whole system. This value would then be used in equilibrating bar AB.

EXAMPLE 7

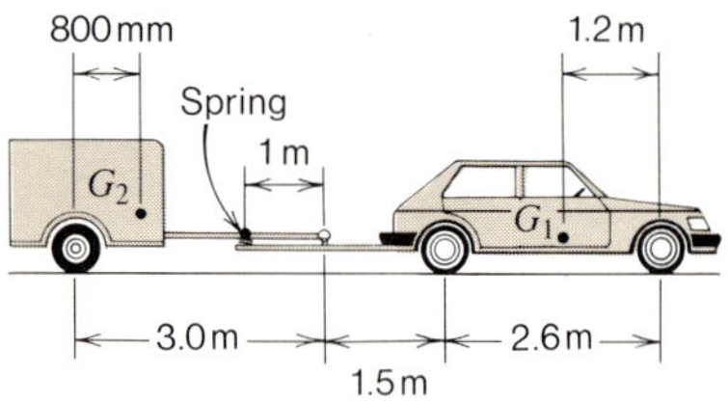

In the compensating trailer hitch shown, the spring is stretched to a tension force T in order to better distribute the loading of the 1000 kg trailer on the automobile. Compare the reactions between all tires and the ground when $T = 3$ kN to that obtained when $T = 0$. The centers of mass of the automobile and the trailer are at points G_1 and G_2, respectively.

Solution

The connection between the trailer and the hitch is a ball joint, so we have the following free body diagrams. Notice that there are no horizontal forces acting on the tires, which is consistent with the entire system being at rest on a level surface.

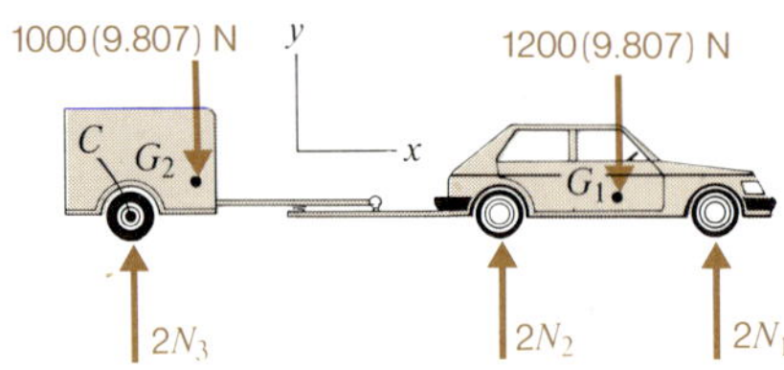

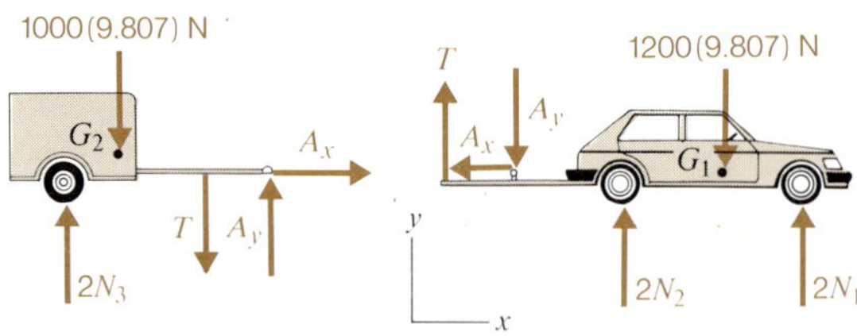

In this problem we are not interested in the reaction force components $\bar{A}_x$ and $\bar{A}_y$ at the ball joint, so we write only the equilibrium equations that do not involve these forces. Thus, letting point C be the center of the tire of the trailer, we have

Car and trailer:

$$\sum M_{Cz} = 2N_2(4.5) + 2N_1(7.1) - 1000(9.807)(0.8) - 1200(9.807)(5.9) = 0$$

$$\sum F_x \equiv 0$$

$$\sum F_y = 2N_1 + 2N_2 + 2N_3 - 1200(9.807) - 1000(9.807) = 0$$

Car:

$$\sum M_{Az} = -T(1.0) + 2N_2(1.5) + 2N_1(4.1) - 1200(9.807)(2.9) = 0$$

Trailer:

$$\sum M_{Az} = T(1.0) + 1000(9.807)(2.2) - 2N_3(3.0) = 0$$

In the four algebraic equations above there are three unknowns: N_1, N_2, and N_3, because T is a known parameter. Thus, there is one check equation. Solution of the equation for the trailer yields

$$N_3 = \frac{T}{6} + 3596 \text{ newtons}$$

and the simultaneous solution of the two remaining moment equations is

$$N_1 = 0.2885T + 2414 \text{ newtons}, \qquad N_2 = -0.4552T + 4777 \text{ newtons}$$

We now check these computations with the force sum in the y direction, which gives

$$\sum F_y = 2(0.2885T + 2414) + 2(-0.4552T + 4777) + 2\left(\frac{T}{6} + 3596\right) - 21573 \equiv 0$$

Hence, we conclude that the solutions are correct. Therefore, for $T = 3000$ newtons we find

$$N_1 = 3280, \qquad N_2 = 3411, \qquad N_3 = 4096 \text{ newtons}$$

and for $T = 0$ we find

$$N_1 = 2414, \qquad N_2 = 4777, \qquad N_3 = 3596 \text{ newtons}$$

These values show that an appreciable equalization of the forces on the front and rear wheels of the automobile is achieved through the action of the spring. Notice that it also increases the load on the wheels of the trailer.

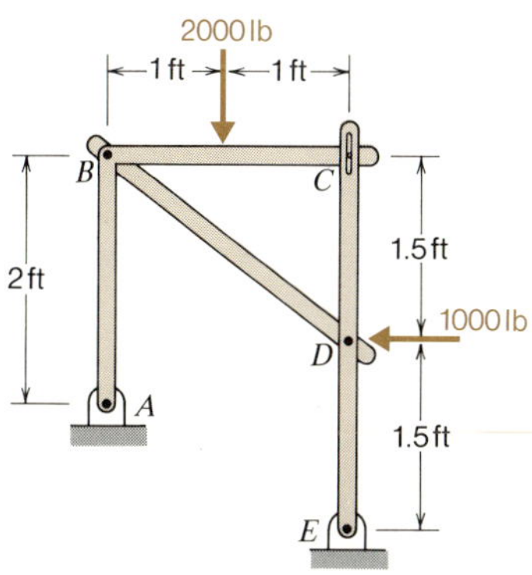

EXAMPLE 8

Determine the forces acting on member ABC at ends A and C. Neglect the weight of each member.

Solution

We are told that the weight of each member is negligible, so bar BD is a two-force member, which we shall assume to be in tension. We are not told to which member the 1000 lb force at joint D is applied. Such information is not important, because it can be proven that the *resultant force acting on each member is independent of how we choose to apportion the load between the members.* We shall choose to apply the 1000 lb force to member CDE, because applying it to the two-force member BD would require writing equilibrium equations for the member.

The pin at C is confined inside a smooth groove, so the reaction is normal to the walls of the groove. We assume that the pin forces acting on member ABC have positive components. The resulting free body diagrams appear below.

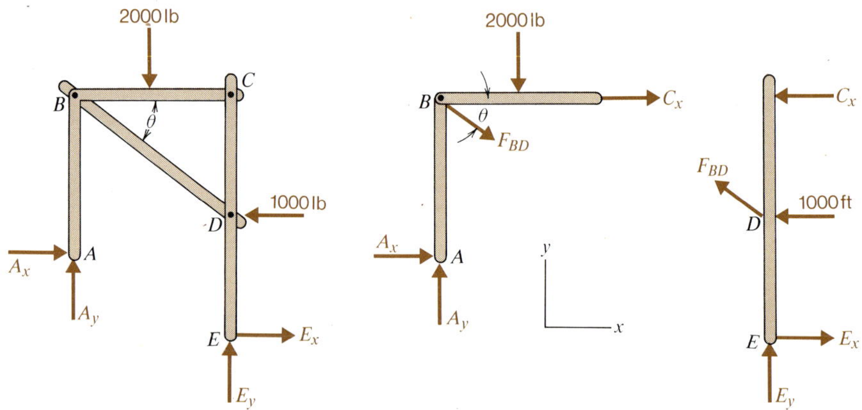

The angle θ is readily found to be

$$\theta = \tan^{-1}\left(\frac{1.5}{2}\right) = 36.87^\circ$$

The sequence of equilibrium equations to consider requires careful consideration, because each free body diagram has more than three unknown forces. Our procedure will be to sum moments about point E for the whole structure, in order to get an equation relating A_x and A_y. Equilibrium of bar ABC will then provide the required additional equations. Thus,

Entire structure:

$$\sum M_{Ez} = -A_x(1.0) - A_y(2.0) + 1000(1.5) + 2000(1.0) = 0$$

Member ABC

$$\sum M_{Bz} = A_x(2.0) - 2000(1.0) = 0$$

$$\sum F_x = A_x + F_{BD}\cos 36.87^\circ + C_x = 0$$

$$\sum F_y = A_y - F_{BD}\sin 36.87^\circ - 2000 = 0$$

The value of A_x is found from $\sum M_{Bz}$, and the other equations are then solved sequentially.

$$A_x = 1000 \text{ lb}, \qquad A_y = 1250 \text{ lb}, \qquad F_{BD} = -1250 \text{ lb}, \qquad C_x = 0 \text{ lb}$$

The forces acting on member ABC at its ends are therefore

$$\bar{F}_A = A_x\bar{i} + A_y\bar{j} = 1000\bar{i} + 1250\bar{j} \text{ lb}, \qquad \bar{F}_c = C_x\bar{i} = \bar{0}$$

In view of the fact that $\bar{F}_c = \bar{0}$, would this structure be usable without any connection at C?

Additional Remarks

You will note that it is necessary to account consistently for reactions whose sense is opposite to the one assumed in the free body diagram. Our procedure is to write equilibrium equations based on the assumed senses of the forces. Each computation is then performed by retaining the signs in the numerical values of the forces.

HOMEWORK PROBLEMS

4.49 and 4.50 Alternative free body diagrams for member *CD* in the frame are shown. Explain what, if anything is erroneous in each diagram.

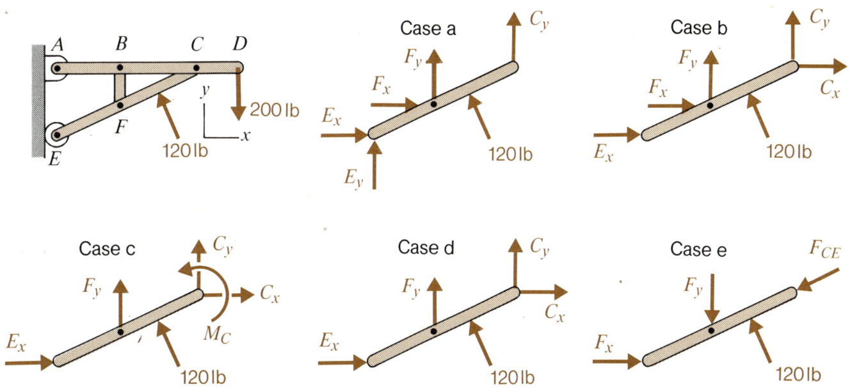

PROB. 4.49

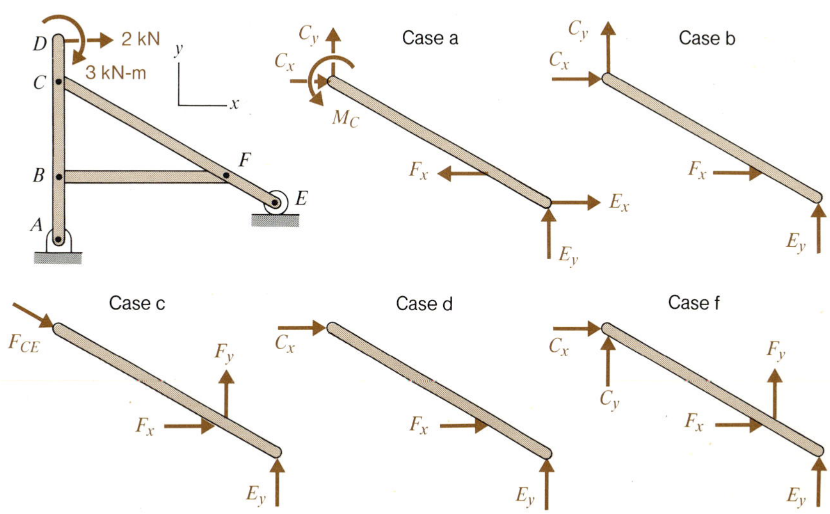

PROB. 4.50

4.51–4.53 Determine the reactions at pins A and C in the frames shown.

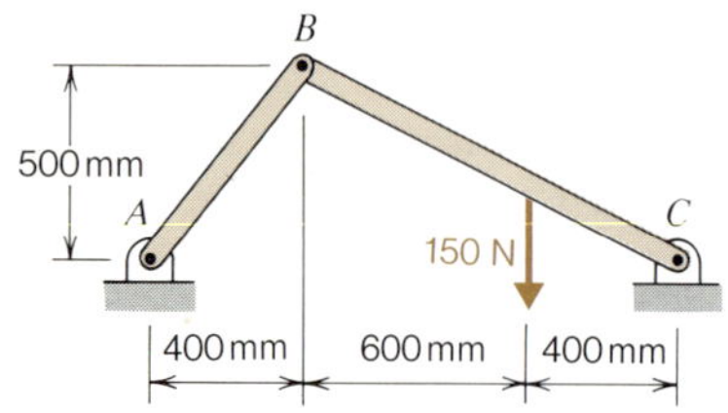

PROB. 4.51

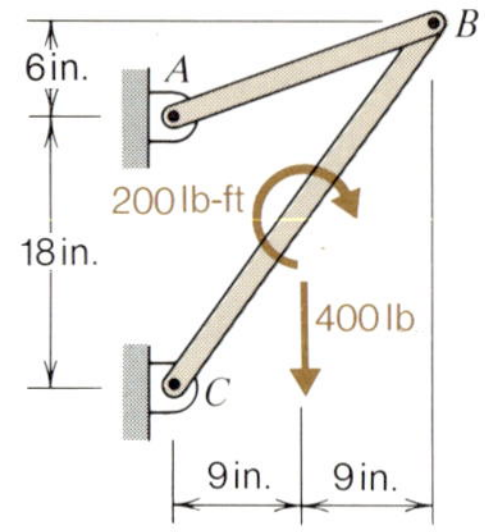

PROB. 4.52

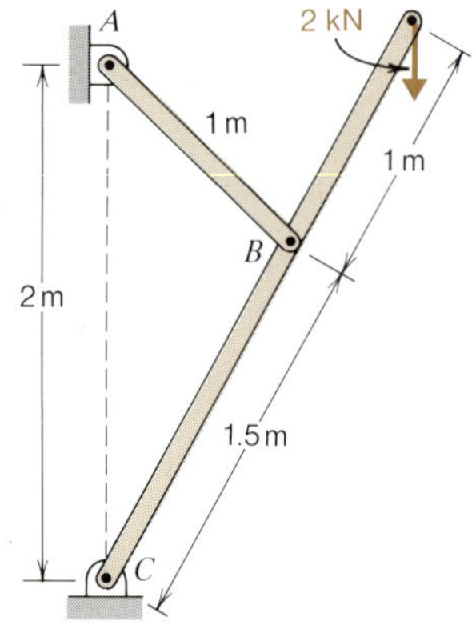

PROB. 4.53

4.54 A 500 lb-ft couple is applied to bar AB, as shown. This bar is pinned to collar B, which slides on the smooth bar CD. Determine the couple $\bar{M}_{CD}$ that must be applied to bar CD to maintain static equilibrium at $\theta = 90°$.

4.55 Solve Problem 4.54 for $\theta = 45°$.

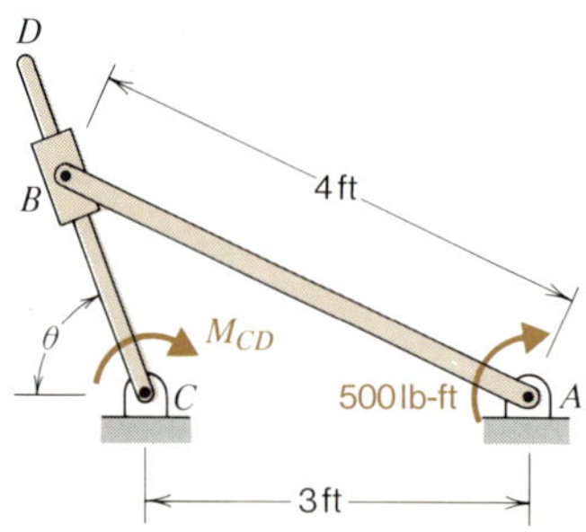

PROB. 4.54

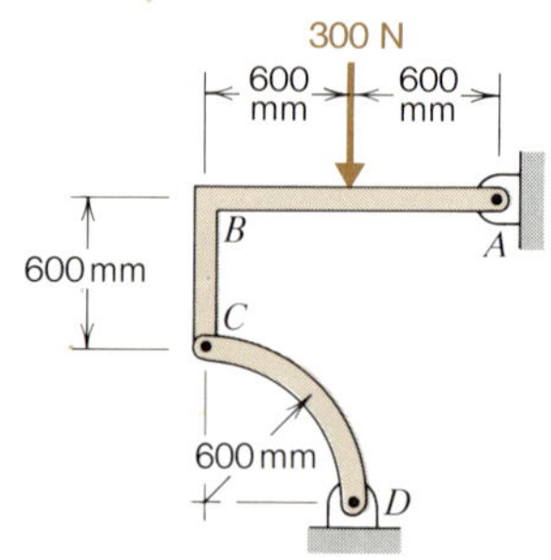

PROB. 4.56

4.56 Determine the reactions at pins A and C.

4.57 and 4.58 Determine the clamping force on the bolt and the force on pin A for the pliers shown.

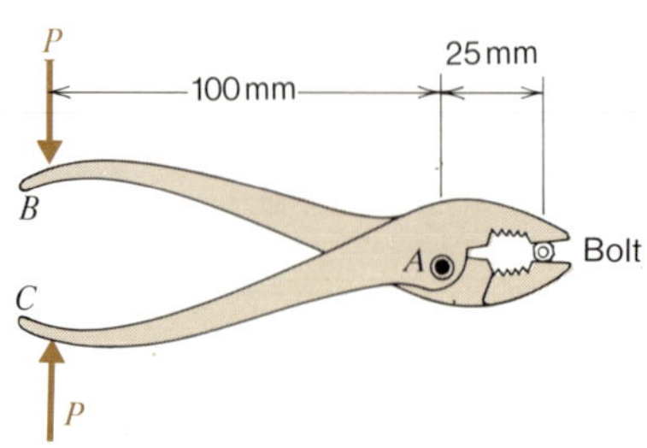

PROB. 4.57

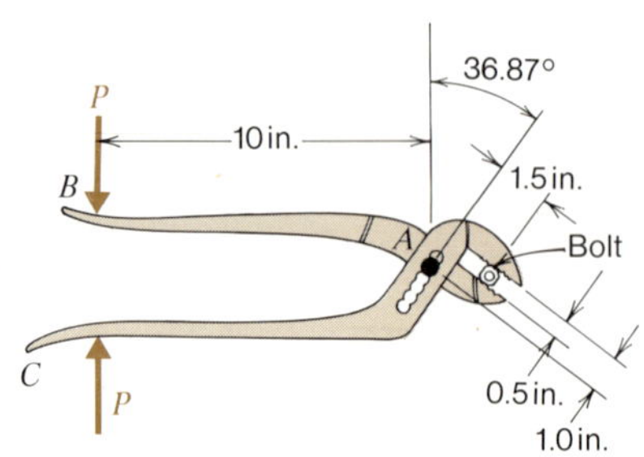

PROB. 4.58

4.59 and 4.60 Determine the reactions at ends A and B for each system of beams shown.

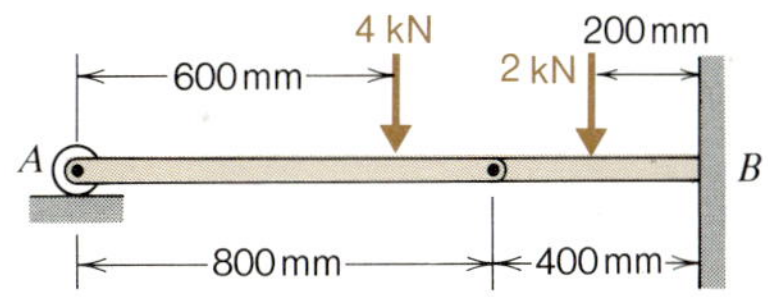

PROB. 4.59

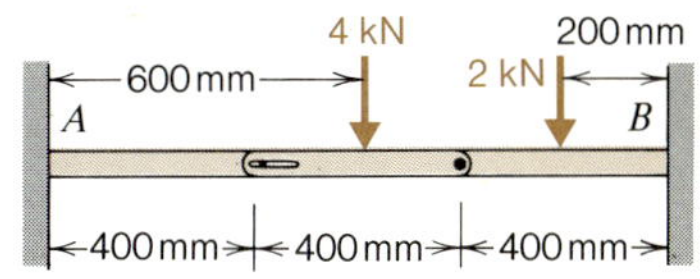

PROB. 4.60

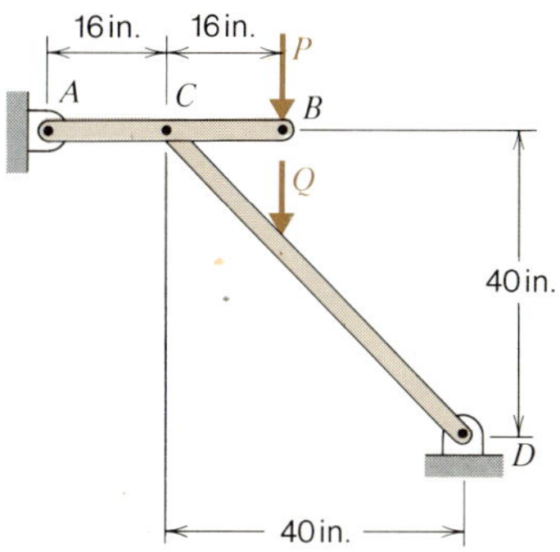

PROBS. 4.61 AND 4.62

4.61 Determine the force acting on member AB at pins A and C if $P = 0$ and $Q = 400$ lb.

4.62 Determine the force acting on member AB at pins A and C if $P = Q = 200$ lb.

4.63 Determine the forces acting on the smooth semicircular bar in the frame.

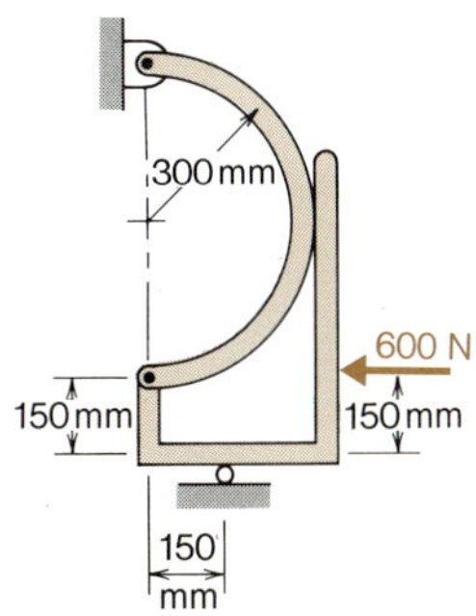

PROB. 4.63

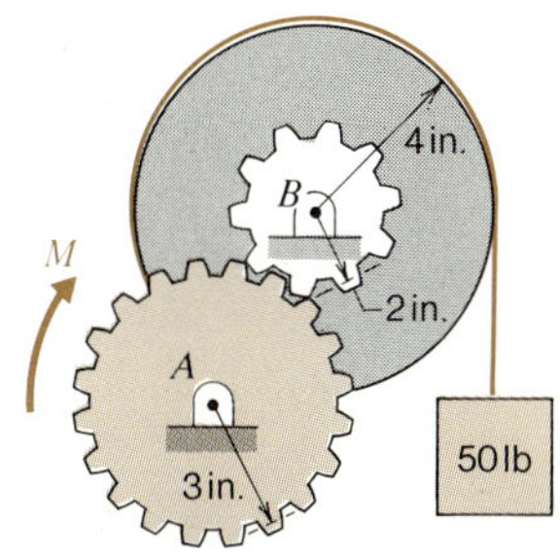

PROB. 4.64

4.64 Gear B is fastened to the drum. Determine the couple $\bar{M}$ that must be applied to gear A in order that the drum hold the 50 lb block in equilibrium.

4.65 A 500 N-m couple is applied to gear A, as shown. Stepped gear B, which transmits this couple to gear C, rides freely on its shaft. Determine the magnitude and sense of the couple $\bar{M}_C$ acting on gear C necessary for equilibrium.

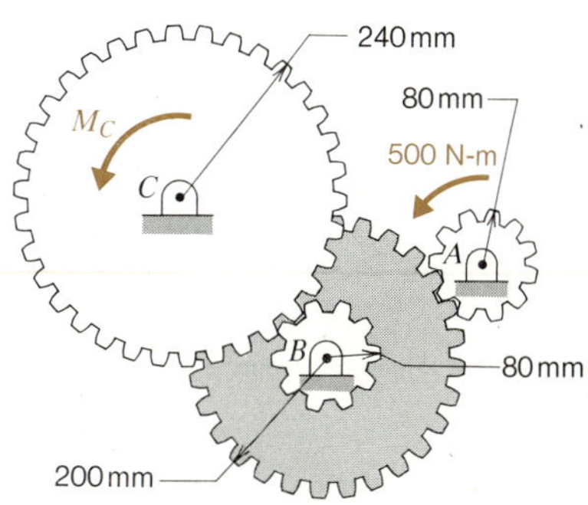

PROB. 4.65

4.66 Solve Problem 4.65 if a counterclockwise couple of 100 N-m is acting on the stepped gear B in addition to the 500 N-m couple on gear A.

4.67 Determine the force $\bar{F}$ that is required for static equilibrium.

4.68 Determine the tension in the cable required for equilibrium of the structure.

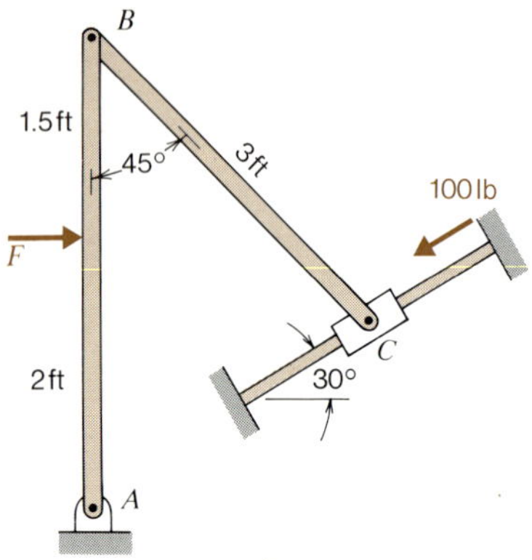

PROB. 4.67

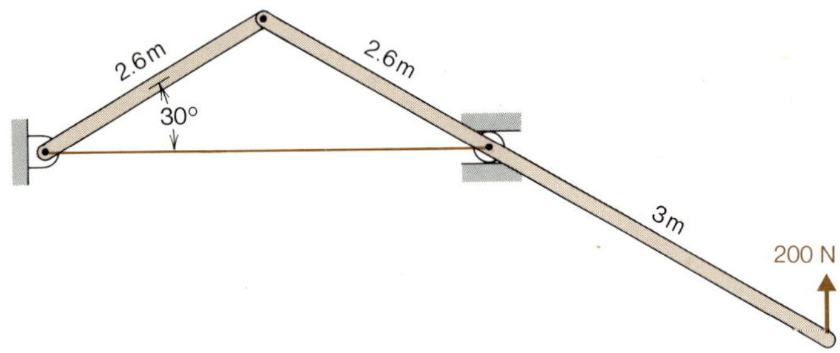

PROB. 4.68

4.69 The three-pin arch is subjected to the set of loads shown. Determine the reactions at pins A, B, and C.

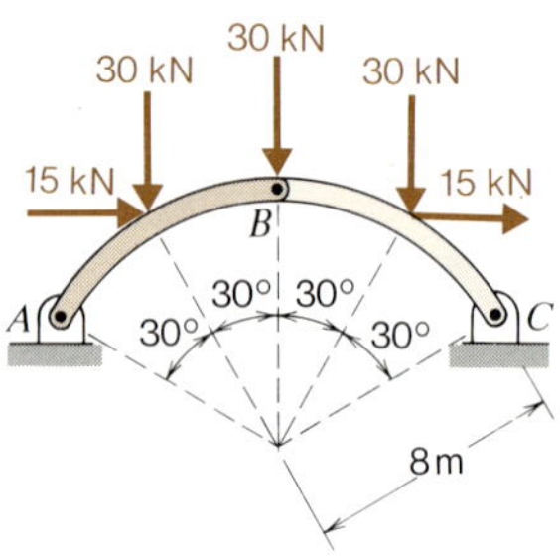

PROB. 4.69

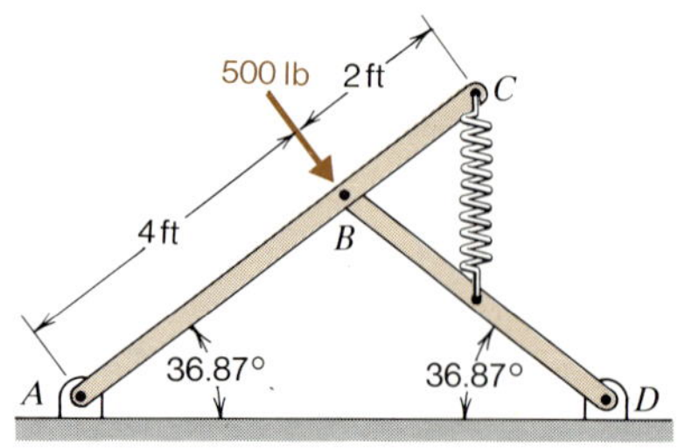

PROB. 4.70

4.70 The spring has a stiffness of 50 lb/in. and an unstretched length of 20 in. Determine the reactions at pins B and D.

4.71 Determine the force acting on member AC at end C.

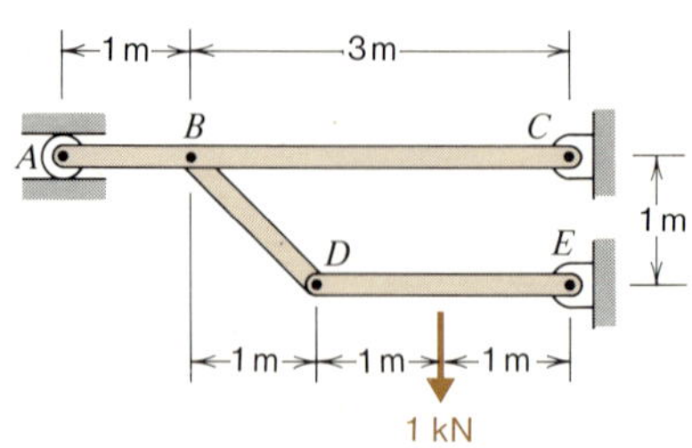

PROB. 4.71

4.72 Determine the forces acting on member AC if $P = 400$ N and $Q = 0$.
4.73 Determine the forces acting on member AC when $P = Q = 200$ N.

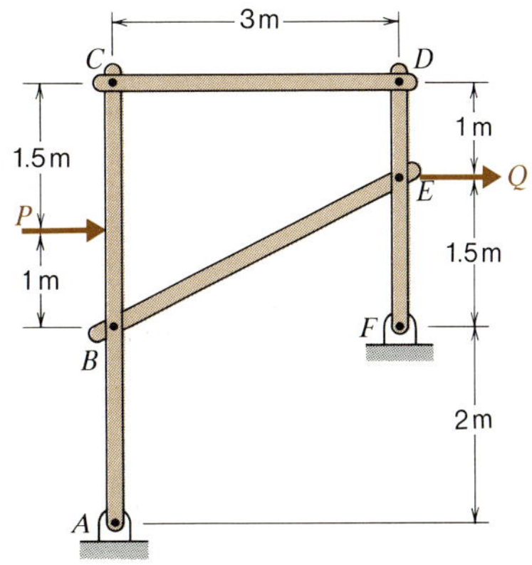

PROBS. 4.72 AND 4.73

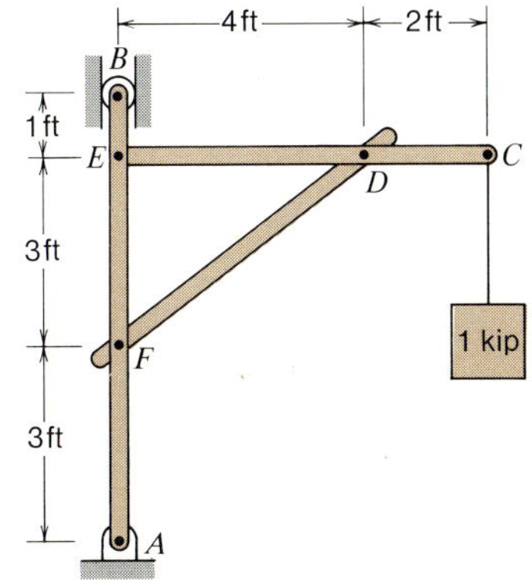

PROB. 4.74

4.74 The frame shown is made of uniform beams weighing 30 pounds per foot. Determine the forces exerted on bar AB at pins E and F.

4.75 and 4.76 Determine the forces acting on members AC and CE of the frame shown.

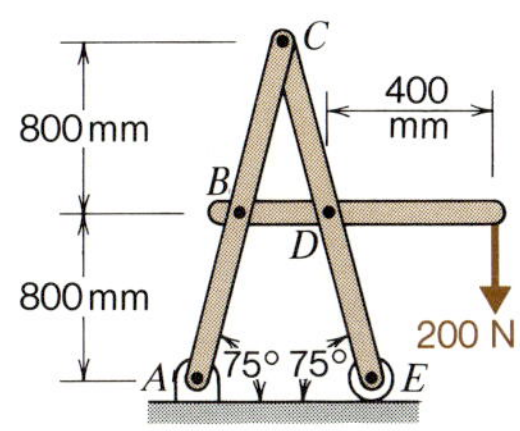

PROB. 4.75

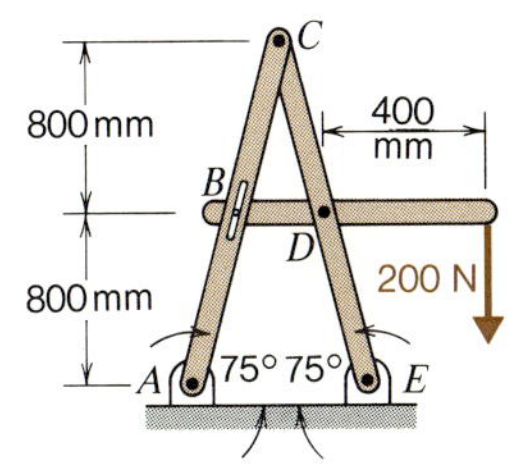

PROB. 4.76

4.77 Determine the reactions at pins A, B, and C when the frame supports a 100 kg package.

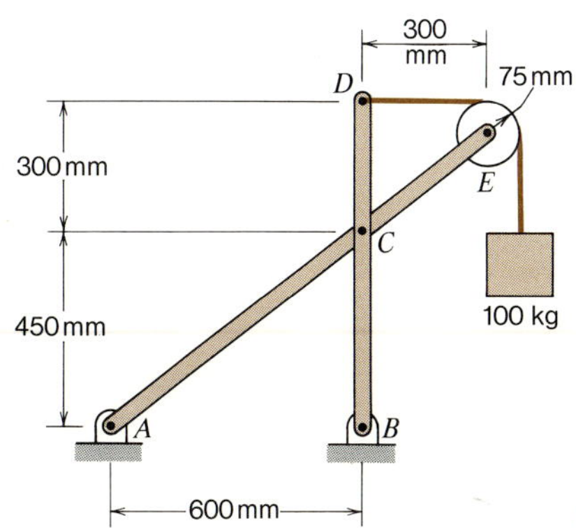

PROB. 4.77

4.78 The cylinder C weighs 50 lb and the weight of bars AB and EF is negligible. Determine the tension in cable AE.

4.79 Solve Problem 4.78 for the case where each bar weighs 30 lb

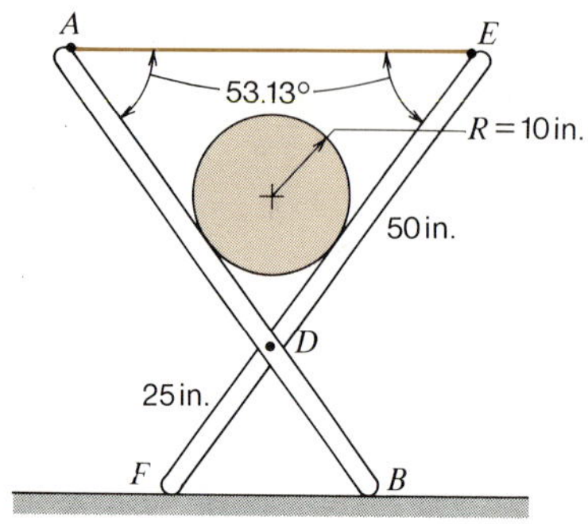

PROB. 4.78

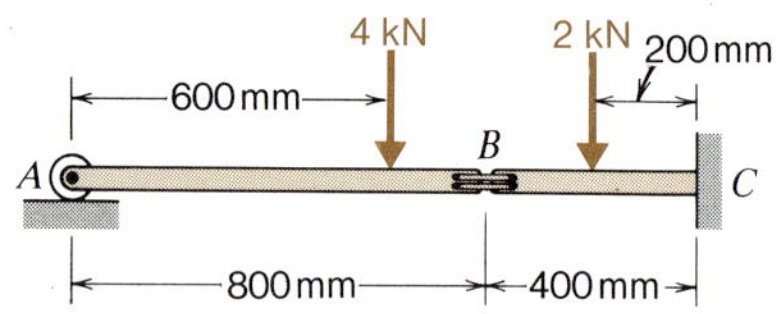

PROB. 4.80

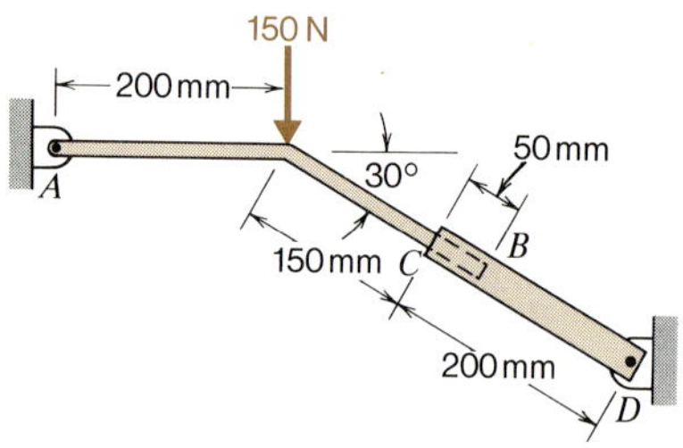

PROB. 4.81

***4.80** Determine the reaction acting on end C of the compound beam.

***4.81** Bar AB, having a circular cross section, fits closely in the smooth pipe CD. Determine the reactions at pins A and D.

4.82 A counterclockwise couple $\bar{M}_1$ is applied to the planetary gear system shown. The radius of the central gear is r_1 and that of the planetary gear is r_2. Determine the couples $\bar{M}_2$ and $\bar{M}_3$ acting on the spider and the outer gear, respectively, for the system to be in static equilibrium.

PROB. 4.82

4.83 Determine the force exerted on the bolt when the handle of the bolt cutter is gripped by the pair of 30 lb forces, as shown.

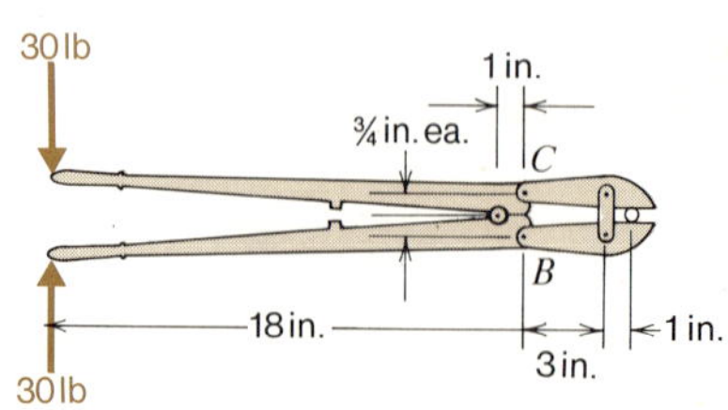

PROB. 4.83

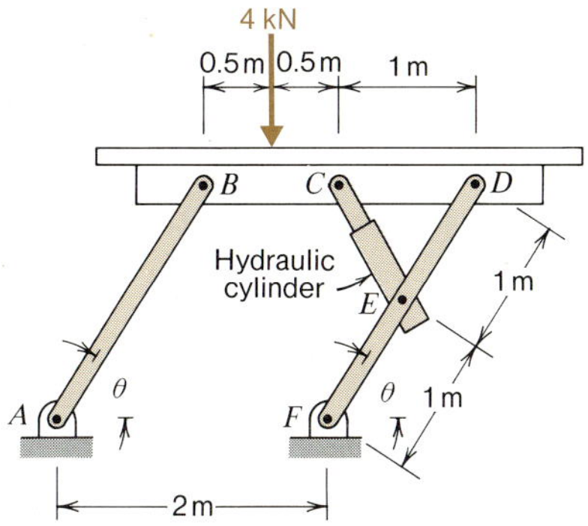

PROBS. 4.84 AND 4.85

4.84 Determine the forces acting on the platform *BCD* at pins *C* and *D* if the system is in equilibrium at $\theta = 30^\circ$.

***4.85** Determine the forces acting on the platform *BCD* at pins *C* and *D* in terms of the angle θ when the system is in static equilibrium.

4.86 Determine the reactions exerted on the frame at pins *A* and *G*.

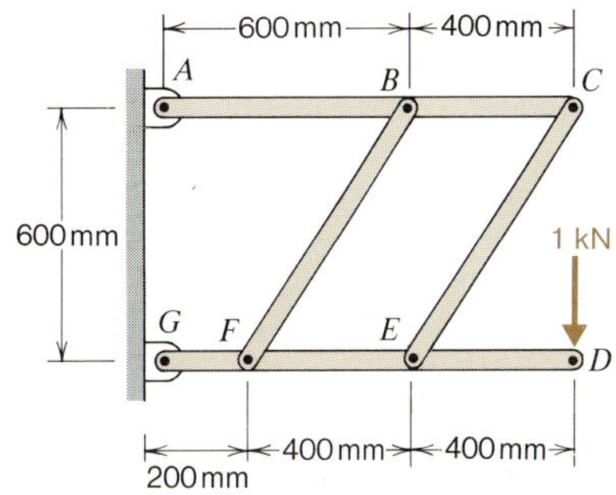

PROB. 4.86

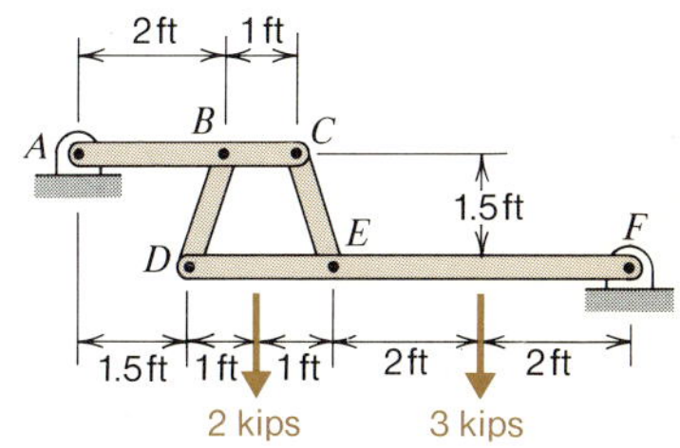

PROB. 4.87

***4.87** Determine the reactions at pins *D* and *E* for the loading shown.

***4.88** Determine the force $\bar{F}$ and the forces exerted by rack *BC* on gear *A* required for equilibrium when the gear is loaded by a 50 N-m couple, as shown.

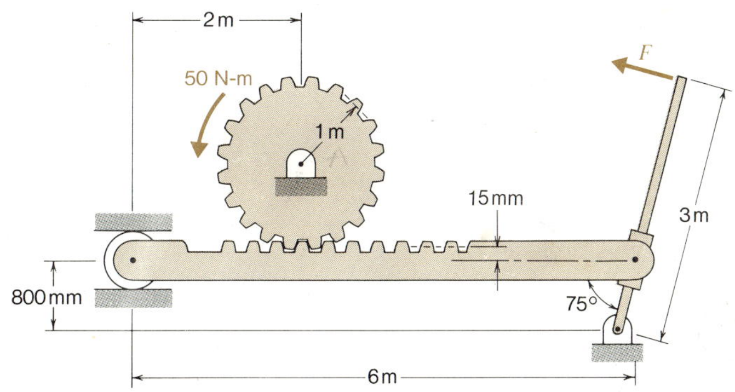

PROB. 4.88

***4.89** Determine the forces exerted on the bolt and on member *CD* when the locking pliers is gripped by the pair of 400 N forces, as shown.

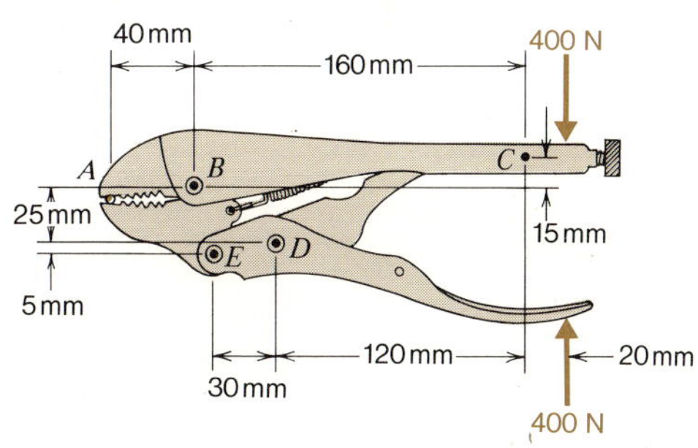

PROB. 4.89

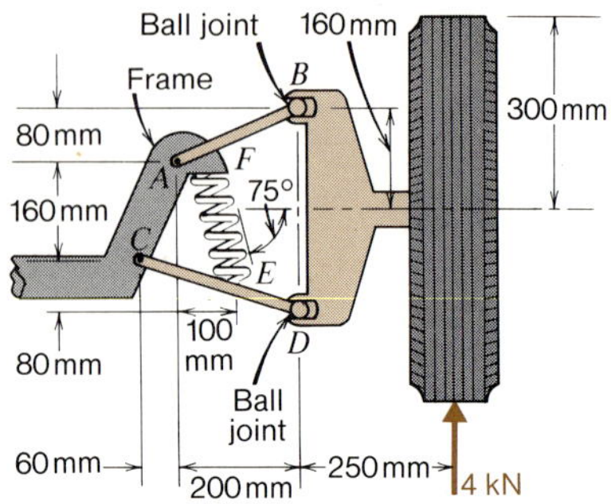

PROB. 4.90

*4.90 The suspension system shown transmits the 4 kN reaction between the tires and the ground to the frame of an automobile. Determine the force in the spring and the forces exerted by members AB and CD on the frame.

4.91 The spring has a stiffness of 5 kN/m and an unstretched length of 200 mm. If $\theta = 36.87°$ in the position of static equilibrium, determine the force $\bar{F}$.

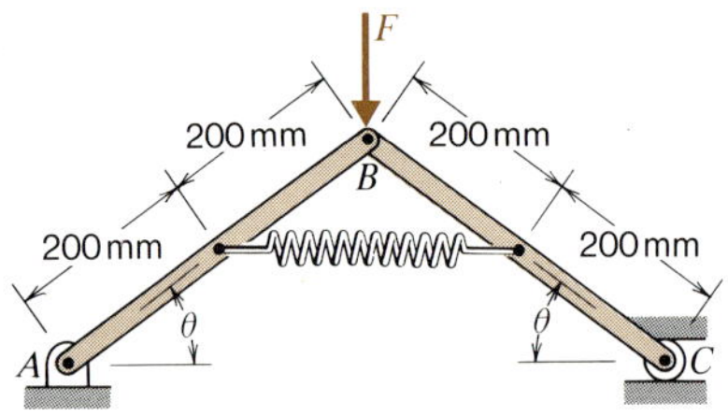

PROB. 4.91

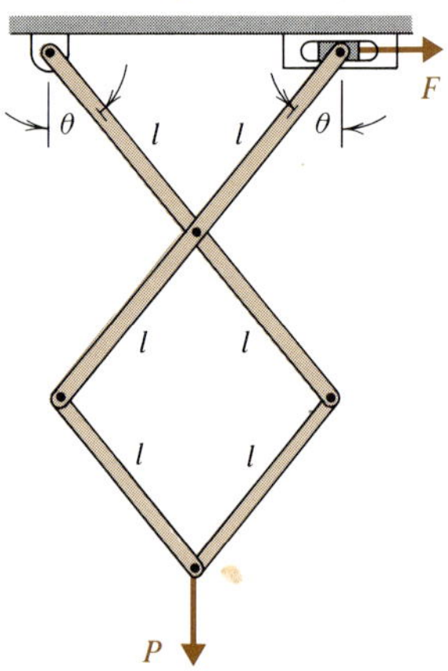

PROBS. 4.93 AND 4.94

*4.92 In Problem 4.91 derive an algebraic relation for the force $\bar{F}$ in terms of the angle θ in the static equilibrium position.

*4.93 Determine the magnitude of the horizontal force $\bar{F}$ in terms of the magnitude of $\bar{P}$ for static equilibrium in the position where $\theta = 60°$.

*4.94 Determine the magnitude of the horizontal force $\bar{F}$ in terms of the magnitude of $\bar{P}$ and the angle θ if the structure is in static equilibrium.

*4.95 The position of the scoop on the tractor is controlled by two parallel mechanisms, only one of which appears in the side view. Determine the forces acting on the main arm ADF when the scoop is carrying an 8 kN load in the position shown.

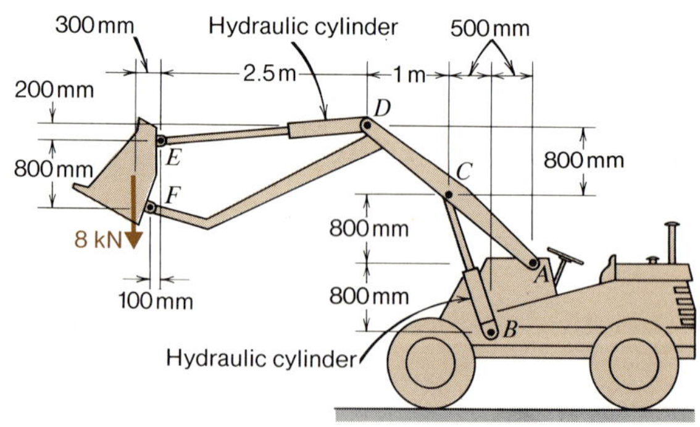

PROB. 4.95

CHAPTER 5

FRICTION

OBJECTIVES

1. To understand the principles of Coulomb (dry) friction.

2. To recognize situations involving friction in ordinary static equilibrium, impending motion, and sliding motion.

3. To evaluate the friction forces required to hold a system in static equilibrium.

4. To determine the properties of a system or its loads for which a system will be in a condition of impending motion.

5. To evaluate alternative types of impending motions, such as slipping and tipping, in order to determine which will occur.

6. To analyze the forces in machine systems with friction, such as wedges, screws, and belts.

Constructing free body diagrams often requires consideration of the reactive force exerted between two bodies that are in contact. Most of the situations we have treated thus far involved smooth contact surfaces, so that the reactions consisted of only force components normal to the plane of contact. In this chapter we shall consider the situation where the contact surfaces are rough. This results in the possibility of forces tangent to the plane of contact between the two bodies, whose effect is to oppose the movement of one surface relative to the other. Such forces are called *friction forces.*

Obviously, there is no surface that is absolutely smooth, but the concept of a system without friction is frequently a useful model to study. We refer to the condition of a smooth surface as an *ideal* model. The friction that we account for in a *real* model may originate from several sources. In general, friction forces can be classfied on the basis of whether the contacting surfaces are dry, or whether a fluid medium is involved.

Fluid friction forces arise partially, but not entirely, from the viscosity of fluids. Such forces are important phenomena in the design of conduits for conveying liquids, in the design of transportation systems such as airplanes, automobiles, and ships, and in the study of lubricants. The study of fluid friction requires knowledge of the mechanics of fluids and will not be treated here.

Dry friction is itself a complicated phenomenon that is still not fully understood. Current thinking is that it results from the microscopic irregularities present in all surfaces (hence the phrase: *rough surface*) and also from molecular attraction. Fortunately, elementary laws, first reported by Coulomb in 1781, are available for forming a simple model to describe dry friction. Coulomb's laws, which are the subject of the next section, should not be regarded as exact, because they possess certain anomalies. Nevertheless, they do provide accurate information in a broad range of applications that we will study in the following sections.

A negative connotation is usually associated with the word *friction,* because it gives rise to additional energy expenditures, and because it results in wear of surfaces in contact. This represents a limited view, because the existence of friction is very often useful, if not essential. For instance, how could we walk without friction?

A. DRY FRICTION

1. Coulomb's Laws

A primary feature of the friction forces exerted between surfaces in contact is that they oppose the direction in which the surfaces move, or would tend to move, relative to each other. In order to illustrate this, consider the block on the horizontal plane shown in Figure 1, where the magnitude of the horizontal force $\bar{P}$ is adjustable.

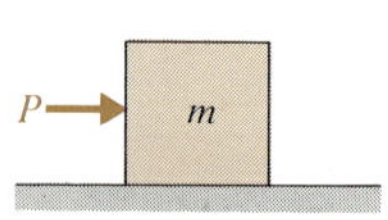

FIGURE 1

In the absence of friction, the block would certainly be pushed to the right by the force $\bar{P}$. Because friction opposes this tendency, the friction force $\bar{f}$ acting on the block is directed to the left, as shown in the free body diagram in Figure 2. The other forces acting on the block are the normal reaction $\bar{N}$ and the weight force mg. The friction force $\bar{f}$ and normal force $\bar{N}$ are the components of the total reaction force exerted by the ground on the block.

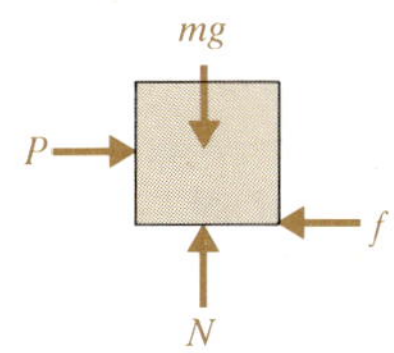

FIGURE 2

Let us now consider a simple experiment in which successively larger values of P, starting from zero, are applied to the block. When P is very small, the block does not move. The equations of static equilibrium corresponding to Figure 2 then indicate that $f = P$ and $N = mg$.

The block remains in static equilibrium, until a sufficiently large value of P results in movement of the block. Let us denote the largest force P for which the block does not move as P_{cr}. Once motion begins, the friction force acting on the block is still to the left. The magnitude of $\bar{f}$ may be evaluated by measuring the acceleration of the block and then applying Newton's second law.

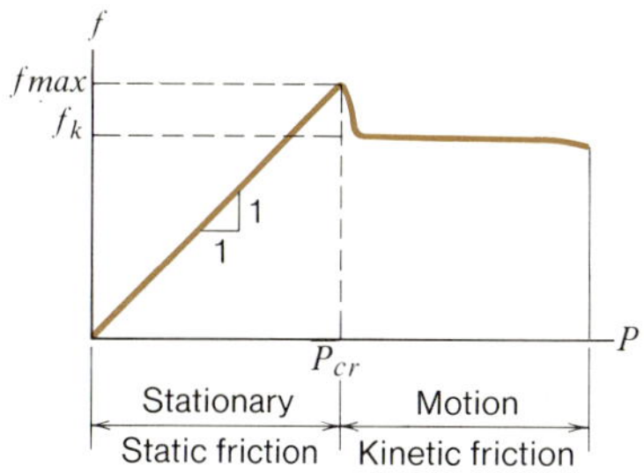

FIGURE 3

The magnitude of $\bar{f}$ corresponding to each imposed value of P in the foregoing experiment may be plotted in a graph, as shown in Figure 3. This graph shows that the largest friction force occurs when $P = P_{cr}$, which is the largest force that will not produce motion. Because $f = P$ for static equilibrium, it follows that $f_{max} = P_{cr}$.

The graph in Figure 3 also shows that if P is increased beyond P_{cr}, thereby causing the block to slide, then the friction force falls off rapidly to the value f_k. The magnitude of the friction force remains at this value for a wide range of values when $P > P_{cr}$. The terminology we shall use to describe the friction force when there is no sliding movement is to say that there is a *static friction force*. In the case where the contact surfaces are sliding over each other, the friction force is called a *kinetic friction force*.

An important property is found by comparing f to the magnitude N of the normal force. We know from the free body diagram, Figure 2, that $N = mg$, because even when the block is sliding, it has no acceleration in the vertical direction. For convenience, let us denote the ratio f_{max}/N as μ_s, and the ratio f_k/N as μ_k.

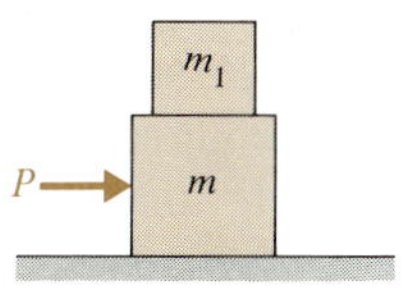

FIGURE 4

Suppose several experiments such as the one we have described here are performed. These experiments involve placing a block of mass m_1 on the original block, and then applying the horizontal force $\bar{P}$, as shown in Figure 4. A different value of m_1 is selected for each experiment. Each experiment would yield a graph showing f in terms of P. Each graph would resemble Figure 3, but the observed values of f_{max} (which is P_{cr}) and f_k would change from one experiment to the other.

Note that for the situation in Figure 4, the normal force in each experiment is $N = mg + m_1 g$. The surprising result of our series of experiments using different values of m_1 is that the values of μ_s $(\equiv f_{max}/N)$ and μ_k $(\equiv f_k/N)$ are constants. The name given to the parameter μ_s is the *coefficient of static friction*, and μ_k is called the *coefficient of kinetic friction*.

If we were to perform our experiments such that the force $\bar{P}$ tends to push the block to the left, we would find that the sense of the friction force $\bar{f}$ is reversed, but the scalar values μ_s and μ_k would not be affected. We shall use $|\bar{f}|$ to denote the magnitude of the friction force, in order to emphasize that μ_s and μ_k do not depend on the sense of the friction force.

The conclusion to be drawn from the experiments is that the static and kinetic friction forces are characterized by their coefficients μ_s and μ_k, respectively. The generalization of these observations are *Coulomb's laws of dry friction.*

(a) Static friction. In this case the pair of contact surfaces are not moving relative to one another. The friction and normal forces must then satisfy the following condition:

$$\boxed{|\bar{f}| \leq \mu_s N} \qquad (1)$$

As stated earlier, this equation governs only the magnitude of the friction force. The sense of the static friction force acting on a surface is opposite to the sense in which that surface tends to move relative to the other surface.

Equation (1) gives a relationship between the magnitudes of the friction and normal forces only when the equality sign pertains. That is the case of *impending motion*, in which the contacting surfaces are on the verge of sliding over each other.

(b) Kinetic friction. If two contacting surfaces are actually sliding over each other, then the friction and normal forces satisfy

$$\boxed{|\bar{f}| = \mu_k N} \qquad (2)$$

where μ_k is usually smaller than μ_s. In this situation, the friction force on each surface is opposite to the direction of the relative movement of the contacting surfaces.

A matter that should be obvious, but still deserves comment, is that there is a friction force only when the surfaces are in contact. Thus, Coulomb's laws as given by eqs. (1) and (2) are meaningful only when N is positive, corresponding to the surfaces pressing against each other.

The values of μ_s and μ_k are often considered to depend only on the types of materials in contact (for example, steel in contact with aluminum), and on the condition of each surface of contact (for example, highly polished or corroded). This is an approximation, for it gives rise to certain subtle differences with observed phenomena, such as the different tractive capabilities of two automobile tires of the same construction, but different dimensions. Further, as is exhibited by Figure 3, the value of μ_k is not truly a constant. A more refined experiment would show that, in reality, μ_k is slightly dependent on the relative velocity of the sliding surfaces.

Equations (1) and (2) are quite useful, even though they have limitations. Typical values of μ_s and μ_k are presented in Table 1. Note that there is no reason why the coefficients of friction should be less than unity, as exemplified by the value of μ_s given for contact between two cast iron surfaces. In actuality a broad range of values exists for each case given. The values shown in

TABLE 1

MATERIALS	μ_s	μ_k
Mild steel on mild steel	0.74	0.57
Aluminum on mild steel	0.61	0.47
Copper on mild steel	0.53	0.36
Cast iron on cast iron	1.10	0.15
Brake material on cast iron	0.40	0.30
Oak on cast iron	0.60	0.32
Leather on cast iron	0.60	0.56
Stone on cast iron	0.45	0.22
Rubber on metal	0.40	0.30
Rubber on wood	0.40	0.30
Rubber on pavement	0.90	0.80
Leather on wood	0.40	0.30
Glass on nickel	0.78	0.56

Table 1 merely give an idea of the general phenomena. For specific cases these values may be incorrect by more than 100%. Thus the tabulation should be employed only if values resulting from experiments on the actual system are unavailable.

An alternate way of representing the contact force exerted between two surfaces is to describe the resultant force that is formed by the normal and frictional forces. For example, consider the body in Figure 5, which is subjected to an arbitrary set of loads as it contacts a rough plane. The friction force $\bar{f}$ and the normal force $\bar{N}$ are the components of the resultant reaction $\bar{R}$, which is inclined at an angle ϕ from the normal to the plane of contact. Clearly

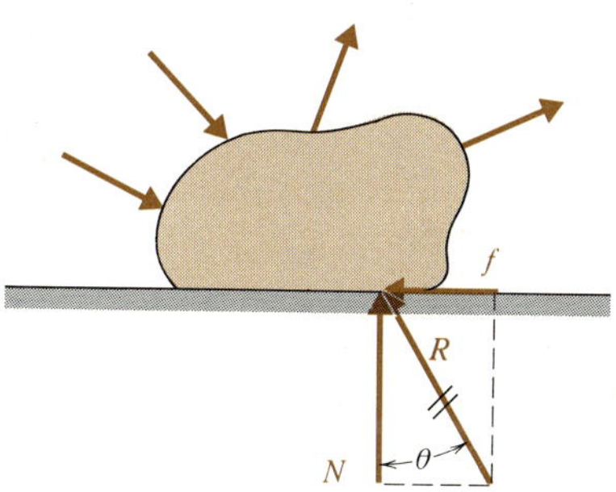

FIGURE 5

$$\phi = \tan^{-1}\left(\frac{|\bar{f}|}{N}\right)$$

In the case of static friction, eq. (1) then indicates that

$$\phi \leq \tan^{-1}(\mu_s)$$

This means that when there is no sliding, the angle ϕ for the total reaction $\bar{R}$ is smaller than or equal to the angle $\tan^{-1}(\mu_s)$. Impending motion corresponds to the equality sign, so the largest possible value of ϕ is

$$\boxed{\phi_s \equiv \tan^{-1}(\mu_s)} \tag{3}$$

The angle ϕ_s is called the *angle of static friction.* A similar relation is obtained in the case of kinetic friction, for which eq. (2) gives

$$\boxed{\phi_k = \tan^{-1}(\mu_k)} \tag{4}$$

The value ϕ_k is the *angle of kinetic friction.*

The physical significance of ϕ_s and ϕ_k may be demonstrated by considering a block that is placed on an inclined plane. It can be shown that ϕ_s is the largest angle of elevation for which the block will not slide down the incline. Similarly, ϕ_k is the angle of elevation for which the block will continue to slide down the incline at a constant speed once it is set in motion.

Equations (3) and (4) represent alternative ways of specifying the coefficient of friction. However, our approach to Statics involves treating reaction forces in terms of their components rather than their magnitude and direction. For this reason, we will not use the concept of the angle of friction when we solve problems.

The primary interest in a course in Statics is in systems that are at rest, which corresponds to the case of static friction. Thus, the friction and normal forces must satisfy all of the static equilibrium equations, as well as eq. (1). There are two types of problems that generally arise in this regard.

The first involves systems for which it must be determined whether equilibrium is possible under a specified set of external loads. In this case, the friction force $\bar{f}$ and normal force $\bar{N}$ are treated as unknown reactions that are to be found from the equilibrium equations. The value of $|\bar{f}|$ represents the amount of friction required for static equilibrium, whereas the value $\mu_s N$ represents the maximum possible value that can be attained. Thus, substituting the forces $\bar{f}$ and $\bar{N}$ obtained from the equilibrium equations into eq. (1) is a test of whether the surfaces can develop sufficient friction to support the external loads.

The second type of static equilibrium problem involving friction corresponds to the case of impending motion. Such problems are identifiable by wording that requests the maximum or minimum value of some parameter for which the system will not move. In this case the equations of static equilibrium must be satisfied simultaneously with the special case of eq. (1), where $|\bar{f}| = \mu_s N$. Care must be taken here to be sure that $\bar{f}$ opposes the tendency of the other forces to cause motion.

EXAMPLE 1

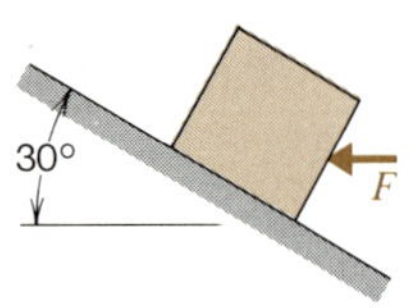

The coefficients of friction between the inclined plane and the 2.5 kg block are $\mu_s = 0.25$ and $\mu_k = 0.20$. The block is initially at rest when the horizontal force $\bar{F}$ is applied to it. Determine (a) the friction force when $F = 20$ N, (b) the friction force when $F = 5$ N, (c) the minimum magnitude of $\bar{F}$ for which the block will slide uphill.

Solution

In parts (a) and (b) we cannot tell by inspection whether there is sliding or not, for this depends on the comparison of the friction and normal reactions. We therefore will assume that there is no sliding, and compare the values of the friction force $\bar{f}$ and normal force $\bar{N}$ obtained from the equations of static equilibrium. On the other hand, in part (c) we are interested in the condition where uphill motion is impending, so for that case, we set $|\bar{f}| = \mu_s|\bar{N}|$. Note that in parts (a) and (b) we can assume a direction for f, because a negative sign in the solution for the magnitude of $\bar{f}$ will indicate that the direction was assumed incorrectly. However, in part (c) the friction

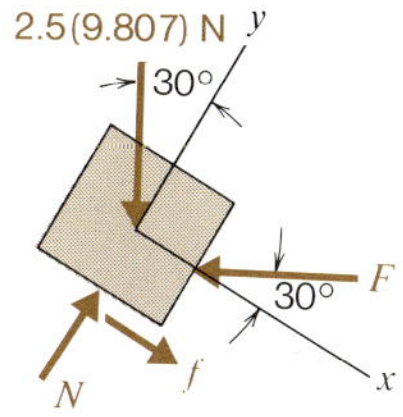

force must be opposite the direction of the impending movement of the block relative to the hill, so $\bar{f}$ *must* be depicted as being downhill.

For parts (a) and (b) let us assume that the friction force is down the incline, so the free body diagram for these cases is as shown. Because this problem does not involve consideration of the lines of action of the forces, we need only consider the equations for force equilibrium. These are

$$\sum F_x = -F\cos 30° + f + 2.5(9.807)\sin 30° = 0 \text{ newtons}$$

$$\sum F_y = N - F\sin 30° - 2.5(9.807)\cos 30° = 0 \text{ newtons}$$

Thus

$$f = 0.866F - 12.26$$

$$N = 0.50F + 21.23$$

Substituting the given values of the applied force F in parts (a) and (b) yields

$$\text{part (a):}\quad f_a = 5.06 \qquad N_a = 31.23 \text{ newtons}$$

$$\text{part (b):}\quad f_b = -7.93 \qquad N_b = 23.73 \text{ newtons}$$

The negative value of f_b tells us that the friction force is up the incline, which means that the tendency of the block is to move downward.

We cannot accept the preceding values of f until they have been checked against the maximum static friction force possible in each case, $|\bar{f}| = \mu_s N$. Thus

$$\text{part (a):}\quad \mu_s N_a = 0.25(31.23) = 7.81 \text{ newtons} > |\bar{f}_a|$$

$$\text{part (b):}\quad \mu_s N_b = 0.25(23.73) = 5.93 \text{ newtons} < |\bar{f}_b|$$

We observe that eq. (2) is satisfied in part (a), but not in part (b). Thus, in part (a) there is no sliding, and

$$\bar{f} = f_a\bar{i} = 5.06\bar{i} \text{ newtons}$$

In part (b) the contact surfaces cannot sustain a friction force sufficient to prevent sliding. The direction in which the sliding will occur can be deduced to be down the plane, because that is the tendency indicated by the result for f_b. Therefore

$$\bar{f} = -\mu_k N_b \bar{i} = -4.75\bar{i} \text{ newtons}$$

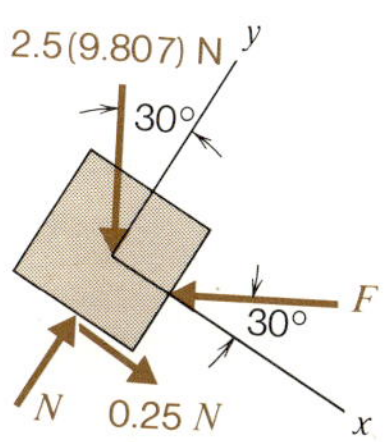

To solve part (c) we draw a new free body diagram, because in this situation of impending motion, $\bar{f}$ is $\mu_s N$ down the plane. The equilibrium equations are now

$$\sum F_x = -F\cos 30° + 0.25N + 2.5(9.807)\sin 30° = 0 \text{ newtons}$$

$$\sum F_y = N - F\sin 30° - 2.5(9.807)\cos 30° = 0 \text{ newtons}$$

Solving these equations simultaneously yields

$$F = 23.7 \text{ newtons}$$

HOMEWORK PROBLEMS

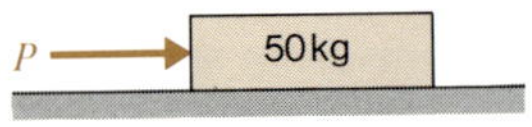

PROB. 5.1

5.1 Determine the horizontal force $\overline{P}$ that must be exerted on the 50 kg block to cause the block to move. The coefficient of static friction is 0.30.

5.2 The coefficient of static friction is $\mu_s = 0.4$. Determine the maximum angle θ for which the 20 lb block will not slide down the incline.

5.3 The coefficients of friction between the 20 lb block and the inclined plane are $\mu_s = 0.20$ and $\mu_k = 0.15$. Determine the friction force acting on the block if (a) $\theta = 5°$, (b) $\theta = 20°$.

PROBS. 5.2 AND 5.3

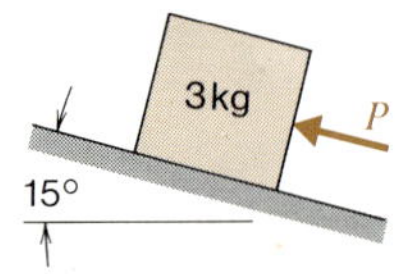

PROBS. 5.4 AND 5.5

5.4 The coefficients of friction between the 3 kg block and the inclined plane are $\mu_s = 0.50$, $\mu_k = 0.35$. Determine the friction force acting on the block if (a) $P = 4$ newtons, (b) $P = 25$ newtons.

5.5 The coefficient of static friction between the 3 kg block and the inclined plane is $\mu_s = 0.60$. Determine the largest force $\overline{P}$ for which the block will not slide up the incline.

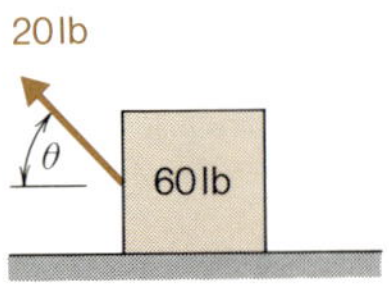

PROB. 5.6

5.6 The coefficient of static friction between the 60 lb block and the horizontal surface is 0.20. Determine the smallest angle of elevation θ of the 20 lb force for which the block will not slide.

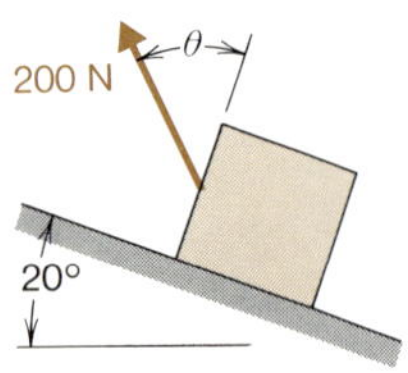

PROBS. 5.7 AND 5.8

5.7 The mass of the block is 25 kg and $\theta = 60°$. If the block is at rest, determine (a) the friction force acting on the block, (b) the corresponding minimum value of μ_s for which static equilibrium is possible.

5.8 The coefficient of static friction between the block and the inclined surface is 0.75. If the largest angle θ for which the block does not slide is 30°, determine the mass of the block.

2. Systems with Friction

Before we consider some examples involving engineering systems containing friction forces, let us review the important results of the previous section, in particular, the procedures for setting up and solving friction problems. We have seen that there are basically two classes of statics problems involving friction. In one situation the external loads are specified, and we must evaluate the friction forces. We address such problems by assuming first that there is no sliding. The friction force is depicted in the free body diagram as being tangent to the contact plane in any sense we choose. The equilibrium equations yield the magnitude and actual sense of the friction force required for static equilibrium. Then the initial assumption of static friction is verified if the results for the friction and normal forces satisfy $|\overline{f}| \le \mu_s N$. When this criterion is not satisfied, it may be concluded that the given loads cause the surfaces to slide over each other. The magnitude of the friction force for the sliding motion is $\mu_k N$. This force is in the same direction as the static friction force found from the initial assumption of no sliding, because the static friction force indicates the direction in which sliding is tending to occur.

The second general class of friction problems involves situations where we must evaluate some property of the system, or of the external loads, that will cause motion to be impending. The magnitude of the friction force is then $\mu_s N$, which is the maximum friction force that precedes sliding. The sense of this friction force must be chosen to be opposite the direction of the impending relative movement. The equations of static equilibrium yield the desired information about the system. Occasionally, it might be necessary to consider both possible senses for the impending motion, particularly if the direction of impending motion cannot be predicted by inspection.

The preceding discussion focuses on only one aspect of a system, specifically, the reaction force generated between two rough surfaces in contact. In general, the system to be studied may be anything from a single particle to a system of rigid bodies. The equations of static equilibrium to be employed are those that are appropriate to the system. Thus, the basic methods for studying static equilibrium developed in earlier chapters are still required for a successful solution of a static friction problem.

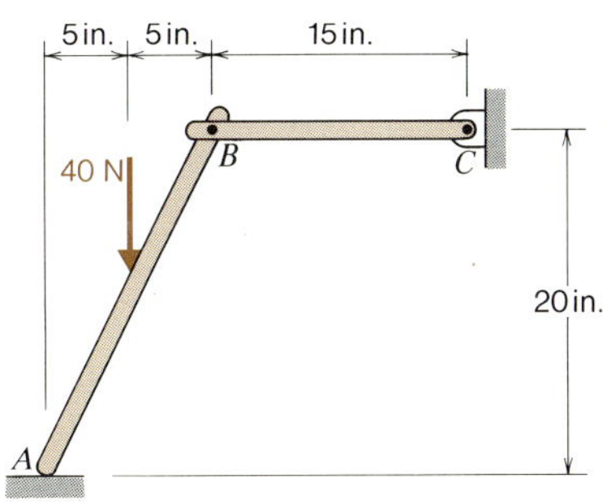

EXAMPLE 2

A vertical 40 N load is applied to the midpoint of bar AB. The bars have negligible mass. Determine (a) the friction and normal forces acting on bar AB at end A, (b) the minimum coefficient of static friction at end A required for static equilibrium.

Solution

The system is a structural frame, in which member BC is a two-force member. Because this member can exert only an axial force, which we shall assume to be compression, a free body diagram of member AB will be sufficient. In addition to the axial force at end B, the other forces acting on bar AB are the 40 N load and the friction and normal forces at end A. Our task in this problem is to evaluate the forces required for equilibrium, so we are free to assume the friction force to be in either sense. As shown in the free body diagram, we have assumed $\bar{f}$ to act to the right, because that is opposite to the sense in which end A would move if there were no friction.

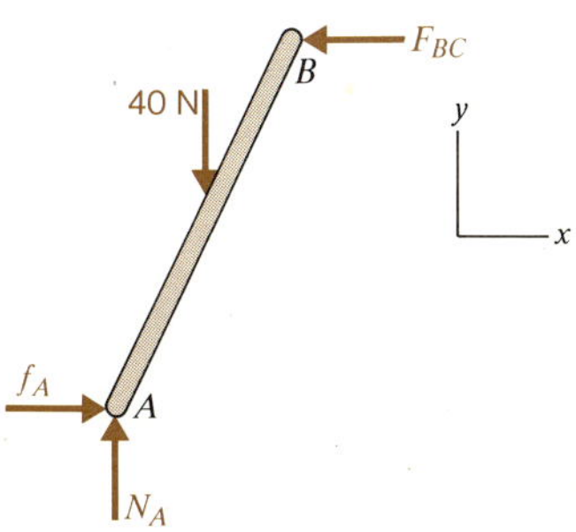

The equilibrium equations for bar AB are

$$\sum M_{Az} = 20F_{BC} - 40(5) = 0$$

$$\sum F_x = f_A - F_{BC} = 0 \qquad \sum F_y = N_A - 40 = 0$$

$$F_{BC} = f_A = 10 \text{ lb}, \qquad N_A = 40 \text{ lb}$$

This is the solution to part (a). To solve part (b) we use the fact that the minimum μ_s corresponds to impending motion. Thus, we set $f_A = \mu_s N_A$ to find

$$(\mu_s)_{\min} = \frac{f_A}{N_A} = \frac{10}{40} = 0.25$$

EXAMPLE 3

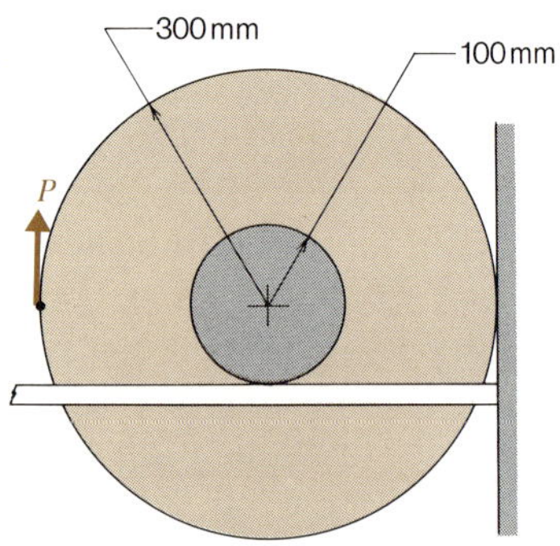

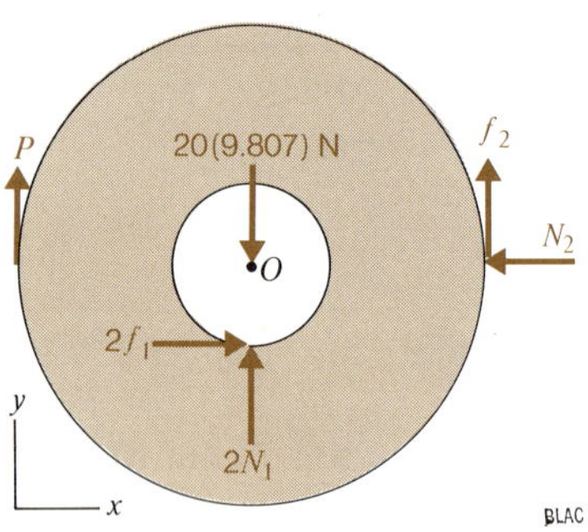

A vertical force P acts at the outer perimeter of a 20 kg stepped disk that is supported by two horizontal rails as it rests against a vertical wall. (Only one rail is visible in the side view.) The coefficient of static friction between all contacting surfaces is 0.40. Determine the largest force $\bar{P}$ that will not move the disk.

Solution

Two types of impending motion are possible. A large force $\bar{P}$ might cause the disk to rotate clockwise in place, or it might lift the disk off the horizontal rails. In the former case, all friction forces will be $\mu_s N$ opposing the clockwise rotation; the latter motion is characterized by the friction and normal forces at the horizontal rail being zero. The free body diagram we have sketched includes both possibilities. (The factor 2 for the $\bar{f}_1$ and $\bar{N}_1$ reactions arises because there are two rails.)

The equilibrium equations for the disk are

$$\sum M_{Oz} = -P(0.3) + f_2(0.3) + 2f_1(0.10) = 0$$

$$\sum F_x = 2f_1 - N_2 = 0$$

$$\sum F_y = 2N_1 + f_2 + P - 196.14 = 0$$

There are five unknowns in these three equations. Two more equations are obtained from the assumption of impending slipping by setting

$$f_1 = \mu_s N_1 \qquad \text{and} \qquad f_2 = \mu_s N_2 \qquad (\mu_s = 0.4)$$

Now the static equilibrium equations become

$$\sum M_{Oz} = -0.3P + 0.12N_2 + 0.08N_1 = 0$$

$$\sum F_x = 0.8N_1 - N_2 = 0$$

$$\sum F_y = 2N_1 + 0.4N_2 + P - 196.14 = 0$$

Simultaneous solution of these equations yields

$N_1 = 67.47$ newtons, $\quad N_2 = 53.98$ newtons, $\quad P = 39.61$ newtons

We obtained positive values for N_1 and N_2, which means that we were correct in assuming that slippage is the impending motion. A negative value for either normal force would indicate that the disk will lose contact at the corresponding surface. In particular, if N_1 turned out to be negative, then we would go back to the equations of static equilibrium and set N_1 and f_1 equal to zero and solve for P, which then would be a lower value than the one obtained above.

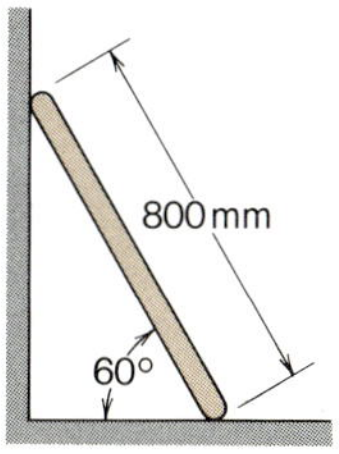

EXAMPLE 4

The coefficient of static friction between the uniform 10 kg bar and the surfaces it contacts is 0.25. Can the bar remain at rest in the position shown?

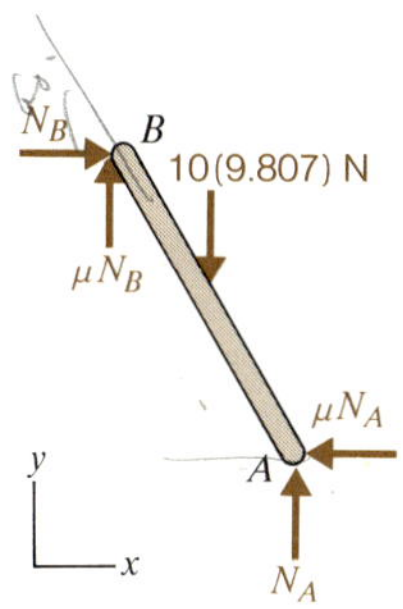

Solution

We draw a free body diagram of the bar, with the friction force at each end being μN, in the sense opposite the direction in which that end would move if there were no friction. We could attempt to find the friction and normal forces acting at each end of the bar, and then check the static friction condition, $|f| \leq \mu_s N$. However, this approach will not succeed because, as shown by the free body diagram, these forces constitute four unknowns that are related by only three equilibrium equations. We must therefore resort to an alternative approach.

We shall evaluate the coefficient of static friction μ for which the bar is in a state of impending motion. This value represents the minimum value of μ_s for which static equilibrium is possible. The equilibrium equations corresponding to the free body diagram are

$$\sum M_{Az} = -N_B(0.8 \sin 60°) - \mu N_B(0.8 \cos 60°) + 98.07(0.4 \cos 60°) = 0$$

$$\sum F_x = N_B - \mu N_A = 0$$

$$\sum F_y = N_A + \mu N_B - 98.07 = 0$$

These equations are solved as follows.

$$\sum M_{Az}\colon N_B = \frac{19.614}{(0.6928 + 0.4\mu)}, \qquad \sum F_x\colon N_A = \frac{19.614}{\mu(0.6928 + 0.4\mu)}$$

$$\sum F_y\colon \frac{19.614}{(0.6928 + 0.4\mu)}\left[\frac{1}{\mu} + \mu\right] = 98.07$$

Simplifying the last equation yields

$$1 + \mu^2 = \frac{98.07}{19.614}\mu(0.6928 + 0.4\mu)$$

$$\mu^2 + 3.464\mu - 1 = 0$$

The positive root of this quadratic equation is

$$\mu = \tfrac{1}{2}\{-3.464 + [(3.464)^2 + 4]^{1/2}\} = 0.268$$

The negative root has no physical meaning.

The actual coefficient of static friction, $\mu_s = 0.25$, is smaller than this minimum value required for equilibrium. We therefore conclude that sufficient friction cannot be developed to resist the tendency of the weight force to move the bar. Thus, the bar cannot remain at rest in the given position.

HOMEWORK PROBLEMS

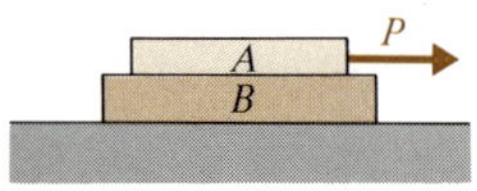

PROB. 5.9

5.9 Determine the force $\bar{P}$ required to cause motion in the system shown. At which surfaces will sliding occur? The coefficient of static friction between the 2 kg block A and the 8 kg block B is $\mu_s = 0.2$ and the coefficient of static friction between block B and the horizontal plane is $\mu_s = 0.1$.

5.10 The coefficient of static friction between the person and the floor is 0.50, and the coefficient of static friction between the crate and the floor is 0.25. Determine the largest mass for the crate that the person can move by pulling on the cable if the person's mass is 70 kg.

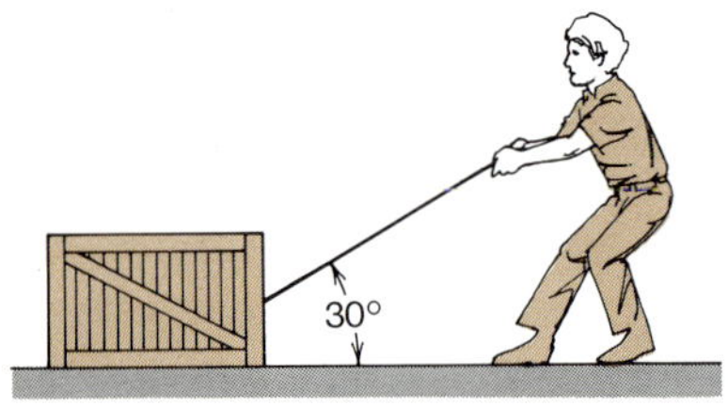

PROB. 5.10

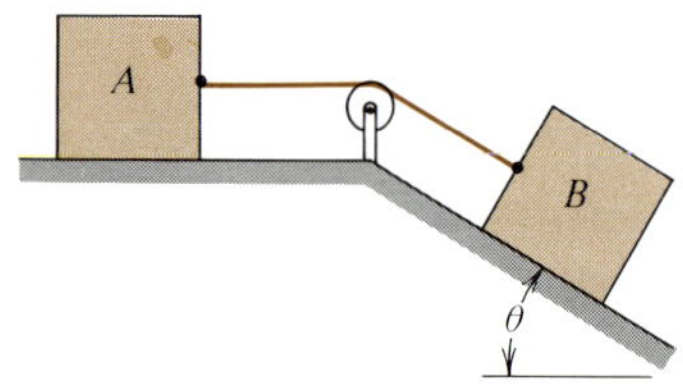

PROB. 5.11

5.11 The coefficient of static friction between each block and the surface it contacts is 0.2. Block A weighs 4 lb and block B weighs 6 lbs. Determine the largest angle θ of the incline for which the blocks will remain at rest.

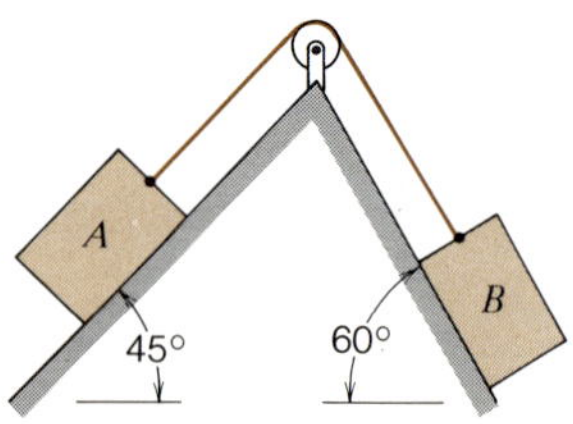

PROB. 5.12

5.12 The masses of blocks A and B are 5 and 10 kg, respectively. The coefficient of static friction between block A and its incline is 0.20. Determine the minimum coefficient of static friction between block B and its incline required to maintain equilibrium. If this coefficient is not sufficiently large, in which direction will the blocks slide?

5.13 Solve Problem 5.12 if block A has a mass of 10 kg and block B has a mass of 5 kg.

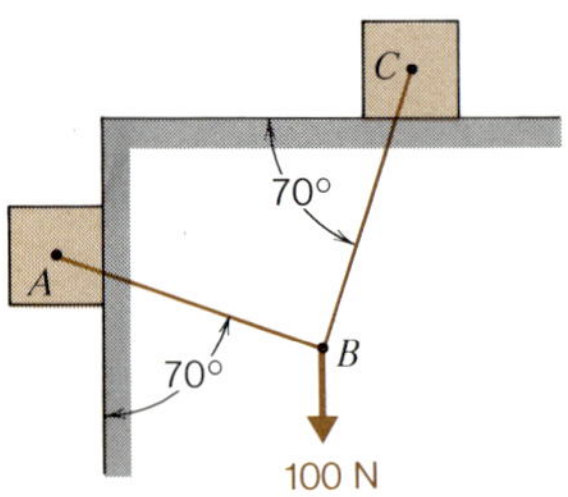

PROB. 5.14

5.14 The masses of blocks A and C are 10 and 20 kg, respectively. These blocks are interconnected by cords AB and BC. Knowing that the system is in equilibrium, determine (a) the friction force acting on each block, (b) the minimum allowable coefficient of static friction between each block and its surface for static equilibrium.

5.15 Two 3 kg collars are connected by a cord and are in equilibrium in the position shown. The coefficient of static friction between collar A and its inclined guide is 0.30. Determine the minimum allowable value of the coefficient of static friction between collar B and its guide.

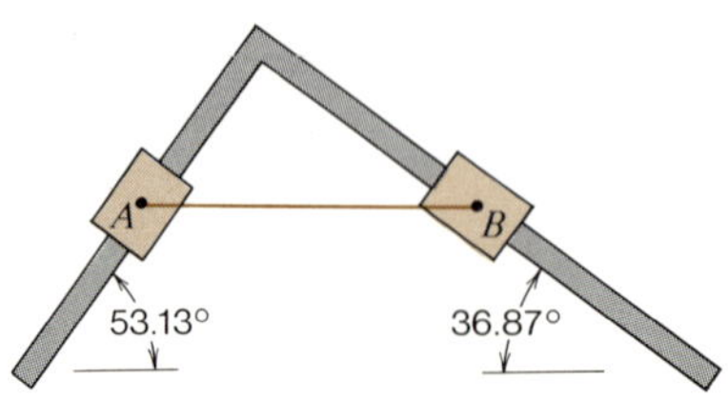

PROB. 5.15

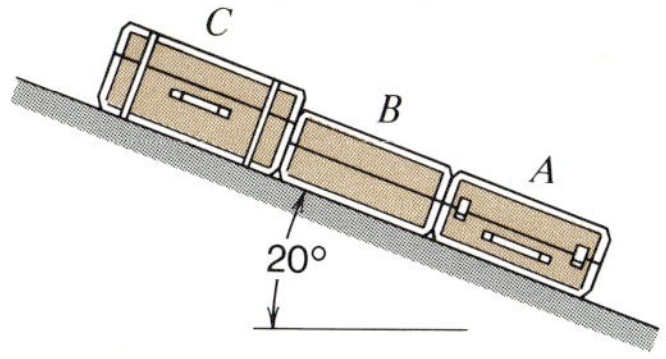

PROB. 5.16

***5.16** Three suitcases A, B, and C are placed on a chute. Their weights are $w_A = w_B = 30$ lb, $w_C = 45$ lb. The coefficients of static friction between the suitcases and the chute are $\mu_A = \mu_B = 0.40$, $\mu_C = 0.20$. Determine which, if any, of the suitcases will slide down the chute. *Hint:* Successively consider the possibility of impending motion of suitcase A alone, then of both suitcases A and B, and finally of all three suitcases.

***5.17** A 10 lb block resting on a plane inclined at 20° above the horizontal is subjected to a horizontal force $\bar{P}$ parallel to the plane, as shown. It is observed that the block will start to move if the magnitude of $\bar{P}$ exceeds 3 lb. Determine (a) the coefficient of static friction between the block and the inclined plane, (b) the direction in which the block starts to slide if the magnitude of $\bar{P}$ exceeds 3 lb.

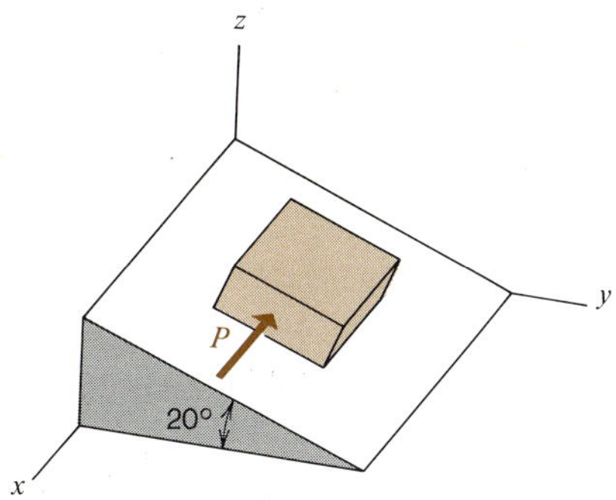

PROB. 5.17

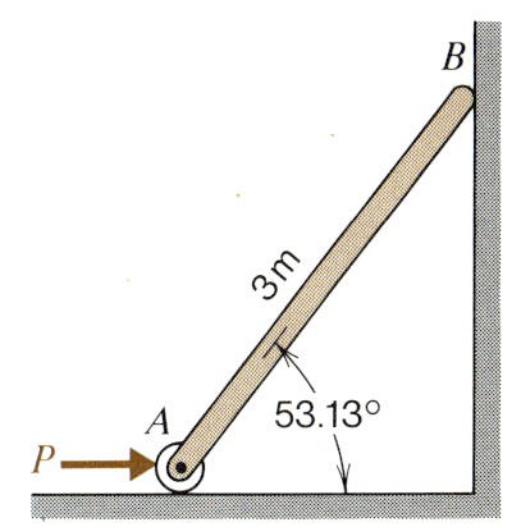

PROBS. 5.18 AND 5.19

5.18 The uniform 80 kg bar rests against the rough vertical wall. The horizontal force $\bar{P}$ is 400 newtons acting to the right. Determine (a) the friction force at end B, and (b) the minimum coefficient of static friction between the bar and the wall for which static equilibrium is possible.

5.19 The coefficient of static friction between the 80 kg uniform bar and the rough vertical wall is 0.3. Determine the smallest horizontal force $\bar{P}$ at end A for which static equilibrium is possible.

5.20 The uniform 20 lb bar is supported by a roller on the inclined wall and by the rough horizontal surface. The bar is in static equilibrium and $\theta = 30°$. Determine (a) the friction force acting on end A, and (b) the minimum coefficient of static friction between the bar and the horizontal surface.

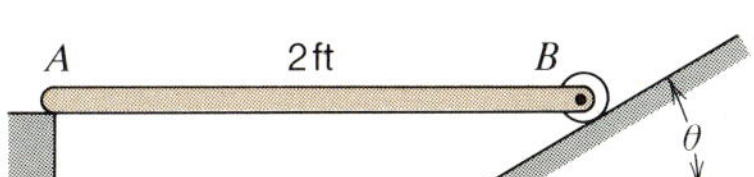

PROBS. 5.20 AND 5.21

5.21 The coefficient of static friction between the uniform 20 lb bar and the horizontal surface at end A is 0.40. The inclined surface at end B is smooth. Determine the maximum angle θ for which static equilibrium is possible.

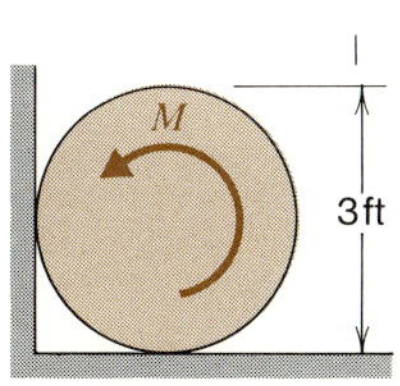

PROB. 5.22

5.22 The 3 ft diameter cylinder weighs 500 lb. The coefficient of friction between the cylinder and both surfaces is $\mu_s = 0.32$. Determine the maximum magnitude of the counterclockwise couple $\bar{M}$ for which sliding will not occur.

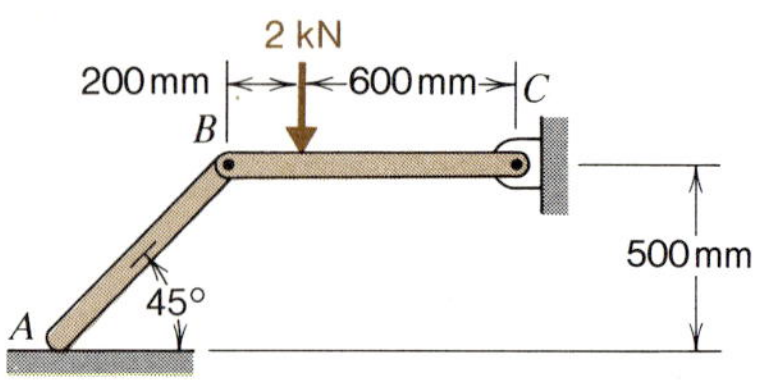

PROB. 5.23

5.23 A 2 kN load is applied to the linkage, as shown. Knowing that the system is in equilibrium, determine (a) the friction force acting on bar AB at end A, (b) the minimum value of the coefficient of static friction for which static equilibrium is possible. The mass of the bars is negligible.

5.24 The semicylinder of mass m and radius R is in a condition of impending motion. The coefficient of static friction between the semicylinder and the inclined planes are the identical values μ, and point G is the center of mass of the semicylinder. Determine μ.

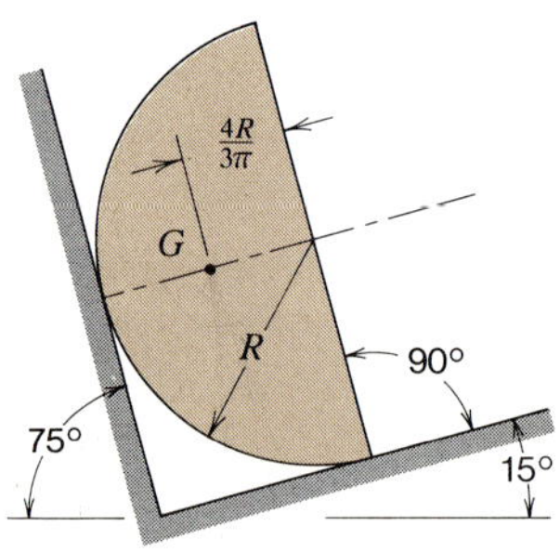

PROB. 5.24

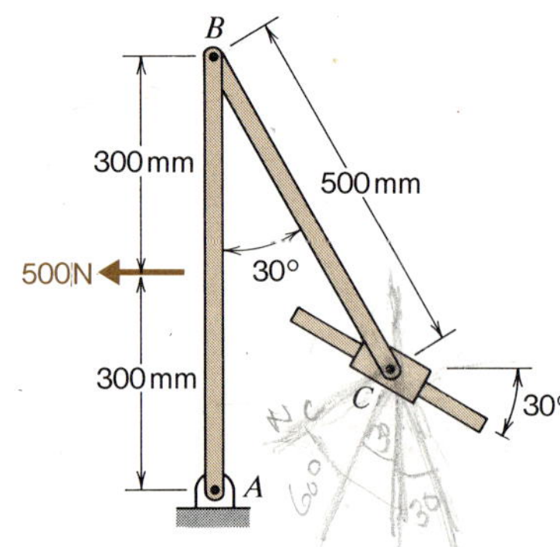

PROB. 5.25

5.25 The coefficient of static friction between collar C and the inclined bar is 0.60. The masses of the bars and the collar are negligible, and the system is in static equilibrium. Determine (a) the friction and normal forces acting on the collar, and (b) the minimum possible coefficient of static friction between the collar and the guide for which the system will remain in static equilibrium.

***5.26** Solve Problem 5.25 if the mass of each bar is 20 kg.

5.27 A cable is wrapped around the 50 kg stepped drum and a tensile force $\bar{P}$ is applied to the free end of the cable, as shown. The drum is in static equilibrium. Determine (a) the magnitude of $\bar{P}$ and of the friction force acting on the drum, (b) the direction in which the point on the drum in contact with the inclined surface is tending to move, and (c) the minimum value of the coefficient of static friction for which static equilibrium is possible.

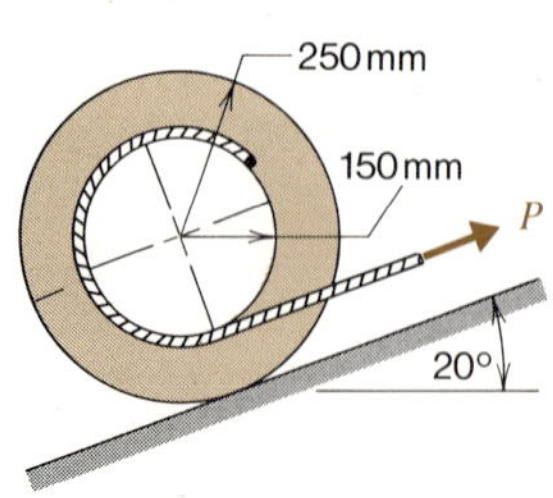

PROB. 5.27

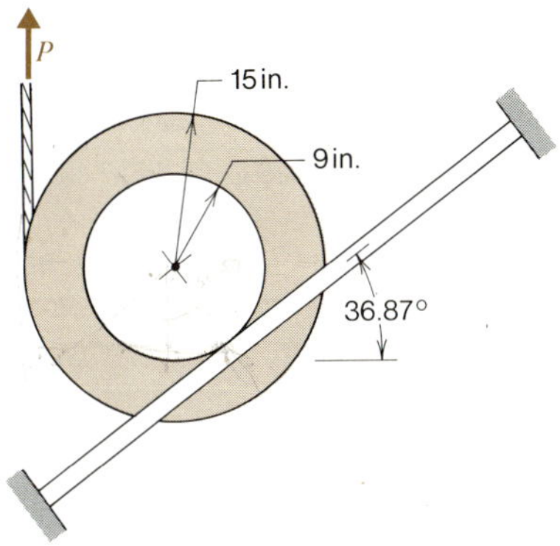

PROB. 5.28

5.28 The 1000 lb stepped drum is held in static equilibrium by the tensile force $\bar{P}$ applied to the cable. The drum is supported on its inner radius by two rails (only one is visible in the side view shown). Determine (a) the magnitude of $\bar{P}$, (b) the friction force exerted by each rail on the drum, and (c) the minimum allowable coefficient of static friction between the drum and a rail.

5.29 The coefficients of static friction between the ladder and the wall, and the ladder and the ground, are 0.2 and 0.3, respectively. Determine the minimum horizontal distance d between the center of mass G of the 60 kg worker and the wall for which the ladder will not slide. The mass of the ladder is negligible.

5.30 Solve Problem 5.29 if the mass of the ladder is 10 kg.

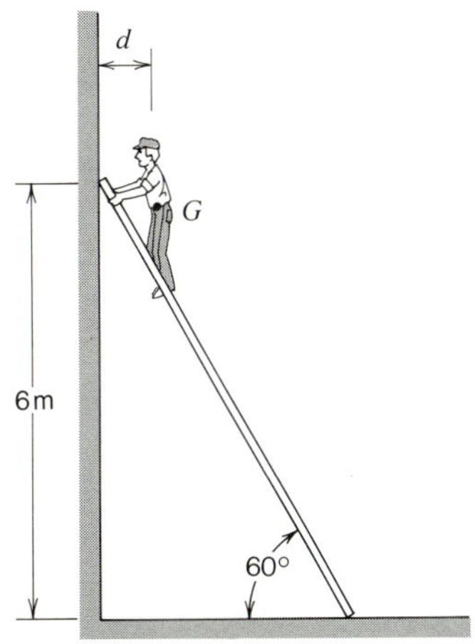

PROB. 5.29

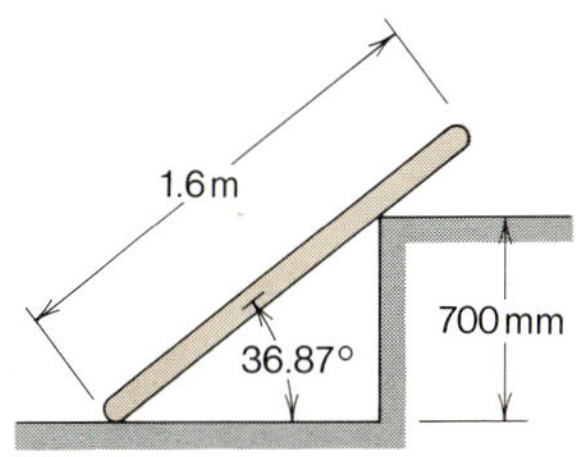

PROB. 5.31

5.31 The coefficient of friction between the uniform 30 kg bar and both surface is $\mu_s = 0.20$. Is the system in static equilibrium?

5.32 The coefficient of friction between the uniform 5 lb bar and both inclined walls is 0.30. Is the bar in static equilibrium in the horizontal position shown?

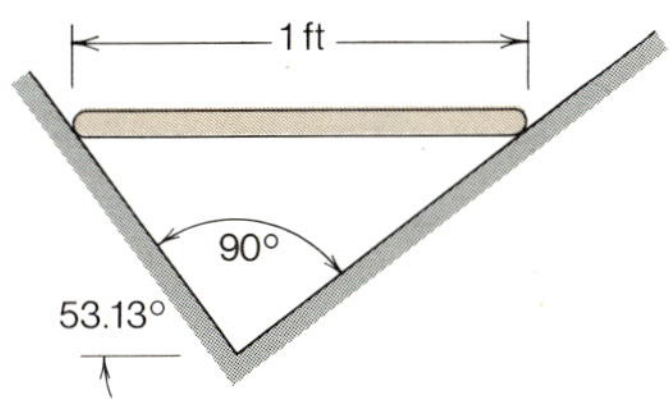

PROB. 5.32

5.33 Point G is the center of mass of the 8 kg semicylinder. If the system is in static equilibrium, determine (a) the angle θ, and (b) the minimum allowable coefficient of static friction between the semicylinder and the ground.

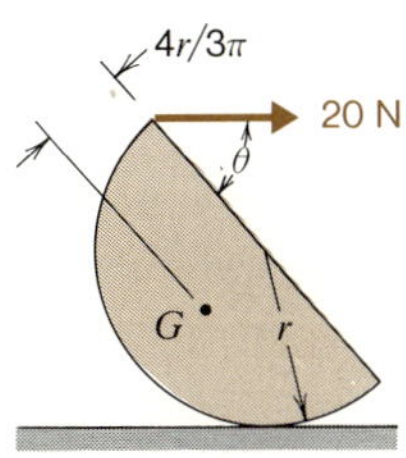

PROB. 5.33

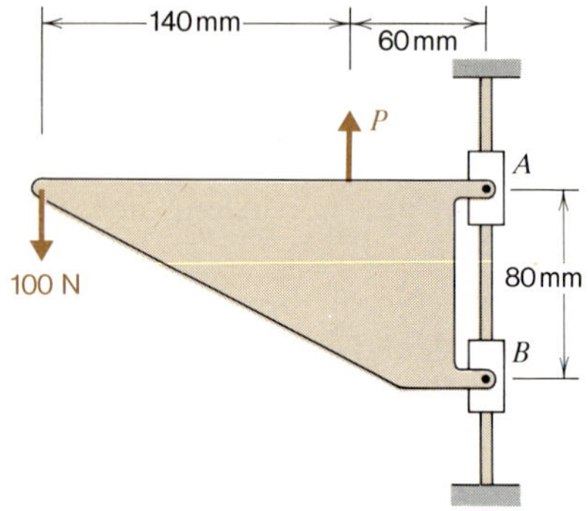

PROBS. 5.34 AND 5.35

5.34 The adjustable bracket is supported on the vertical guide by collars at A and B. The coefficient of static friction between each collar and the guide is 0.20, and mass of the bracket is negligible. Determine the magnitude of the vertical force $\bar{P}$ required to move the bracket upward from rest.

5.35 In Problem 5.34, determine the magnitude and sense of the vertical force $\bar{P}$ required to move the bracket downward from the rest position.

5.36 The 200 lb-in. couple is applied to rod AB in order to resist the 70 lb force applied to piston C. Knowing that the linkage is in static equilibrium at the position where $\theta = 90°$, determine (a) the friction force acting on the piston, and (b) the minimum coefficient of static friction between the piston and its cylinder for which this situation is possible.

5.37 Solve Problem 5.36 for the position where $\theta = 135°$.

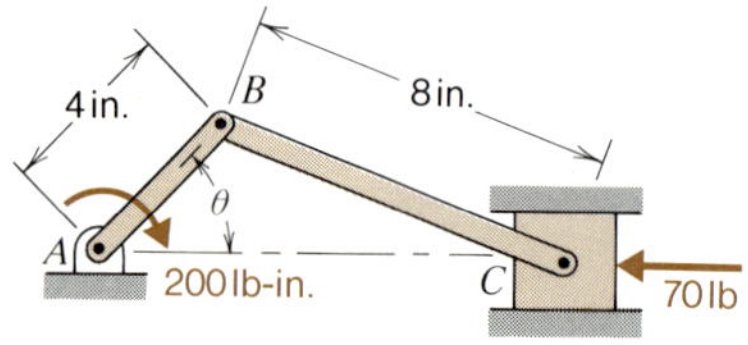

PROBS. 5.36 AND 5.37

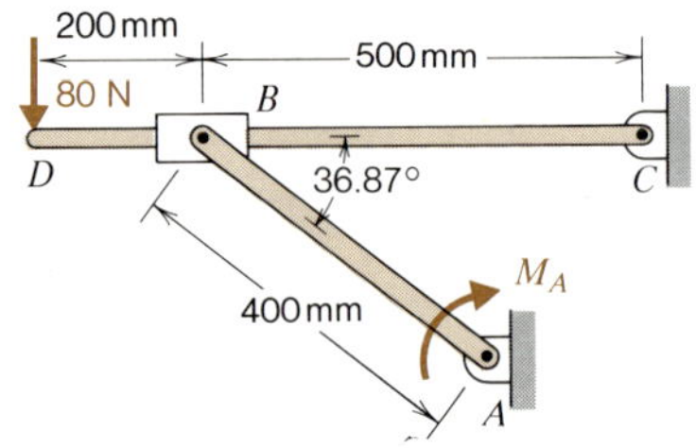

PROBS. 5.38 AND 5.39

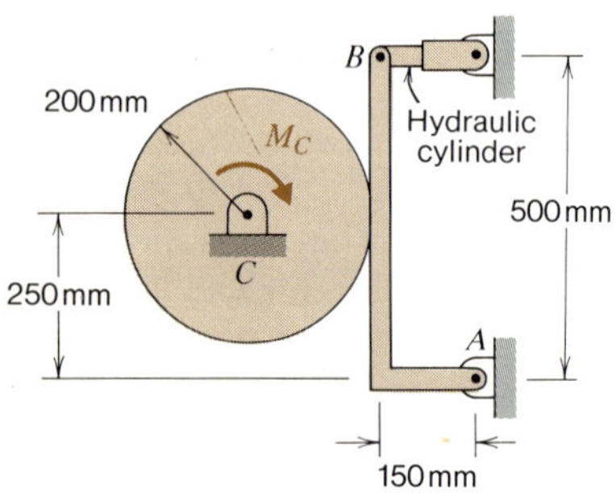

PROBS. 5.40 AND 5.41

5.38 The coefficient of static friction between collar B and bar CD is 0.25. The bars have negligible mass. Determine the magnitude of the couple $\bar{M}_A$ for which bar AB will rotate clockwise from rest.

5.39 In Problem 5.38, determine the magnitude and sense of the couple $\bar{M}_A$ for which bar AB will rotate counterclockwise from rest.

5.40 A couple $\bar{M}_C$ is applied to the brake drum C. Determine the smallest force exerted by the hydraulic cylinder on brake arm AB for which the brake drum will not rotate if $\bar{M}_C$ is 300 N-m clockwise. The coefficient of static friction between the drum and the arm is 0.75.

5.41 Solve Problem 5.40 if $\bar{M}_C$ is 300 N-m counterclockwise.

*5.42 The coefficient of static friction between the 2 kg bar and both surfaces it contacts is 0.20. Determine the largest and smallest values of the angle θ for which the bar is in static equilibrium.

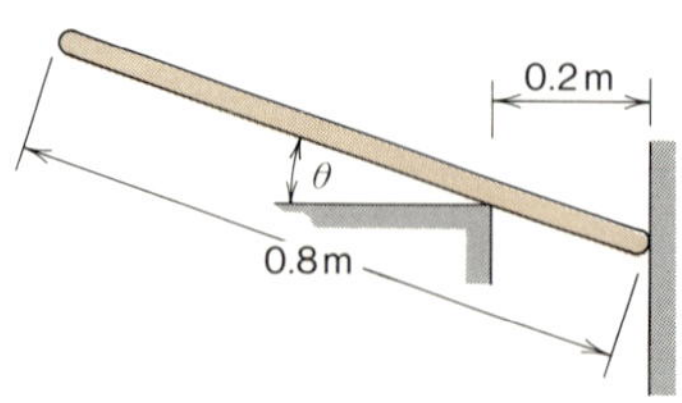

PROB. 5.42

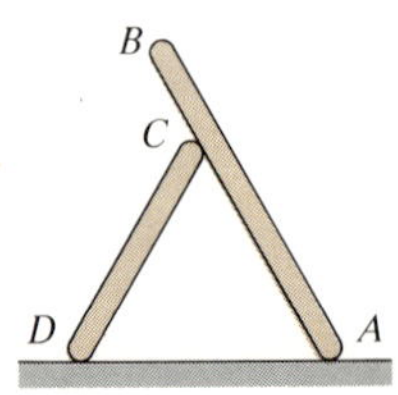

PROB. 5.43

*5.43 The 3 lb bar AB is supported by the 2 lb bar CD. Determine the minimum values of the coefficients of static friction between each pair of contact surfaces for which static equilibrium is possible.

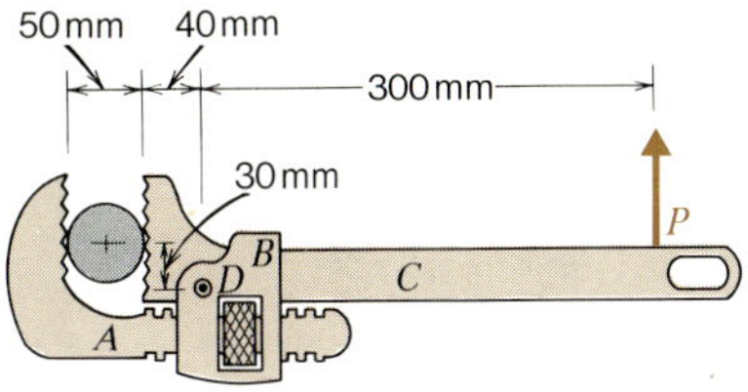

PROB. 5.44

*5.44 A 50 mm diameter rod is being gripped by the Stilson wrench shown. Members A and B may be regarded as a single rigid body connected to member C only by pin D. It is observed that, regardless of the magnitude of the force $\bar{P}$, the wrench does not slip over the rod (the wrench is said to be *self-locking*). Determine the minimum coefficients of static friction at both points of contact between the wrench and the pipe.

*5.45 A 5 ft long bar is connected to the floor by a ball-and-socket joint that is 4 ft from the vertical wall supporting the other end of the bar. The bar weighs 20 lb and the coefficient of static friction between the bar and the wall is 0.35. Determine the maximum value of the angle θ for which static equilibrium is possible. *Hint:* Because of the support at end A, end B tends to move tangent to the dashed circle.

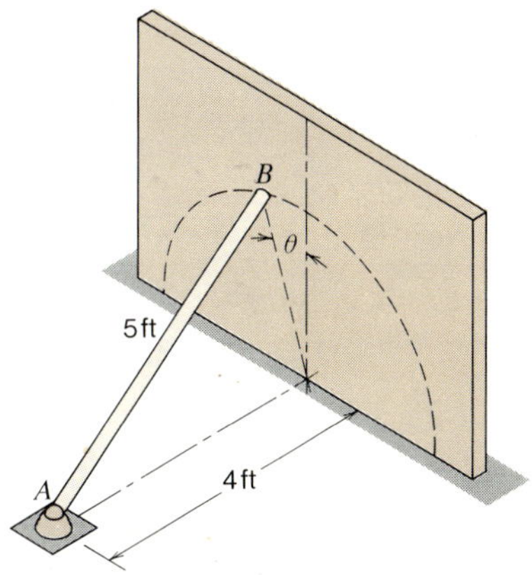

PROB. 5.45

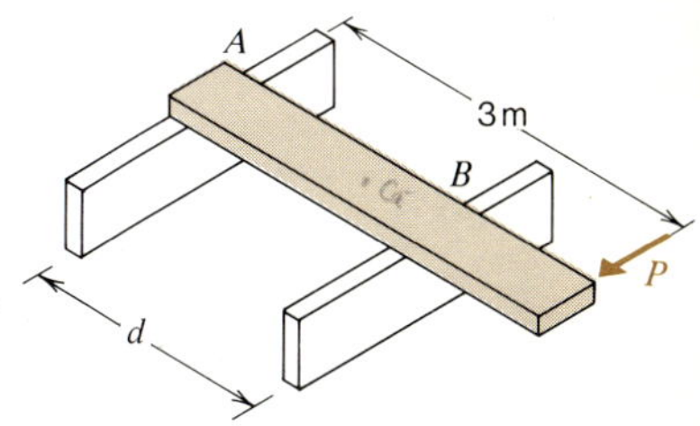

PROB. 5.46

5.46 A 30 kg plank is resting on two horizontal joists, perpendicular to the axis of the joists in the horizontal plane. The coefficient of static friction between the plank and the joists is 0.40. Determine the force $\bar{P}$, parallel to the joists, required to move the plank when $d = 2$ m. *Hint:* The friction forces acting on the plank may be considered to be parallel to the joists.

3. Sliding or Tipping

The impending motion we have considered thus far originated from the tendency of the pair of contacting surfaces to slide over each other. Another type of impending motion that is possible is the tendency of the external forces acting on a body to overturn (that is, tip over) the body. This type of motion arises whenever the frictional and normal reactions cannot exert a moment that is sufficient to balance the moment of the external forces.

As an example, consider the horizontal force that the person in Figure 6a must apply to the file cabinet in order to move the cabinet. If the person exerts a sufficiently large force close to the ground, as in Figure 6b, the

FIGURE 6a FIGURE 6b FIGURE 6c

cabinet will tend to slide. In contrast, pushing on the cabinet at a high location may result in tipping the cabinet over, as in Figure 6c.

A free body diagram of the file cabinet showing the pushing force $\bar{P}$ at a height h above the ground appears in Figure 7. Impending motion corresponds to the largest force $\bar{P}$ for which static equilibrium is possible. Summing moments about the point O at which the friction and normal forces act leads to the following equations of static equilibrium.

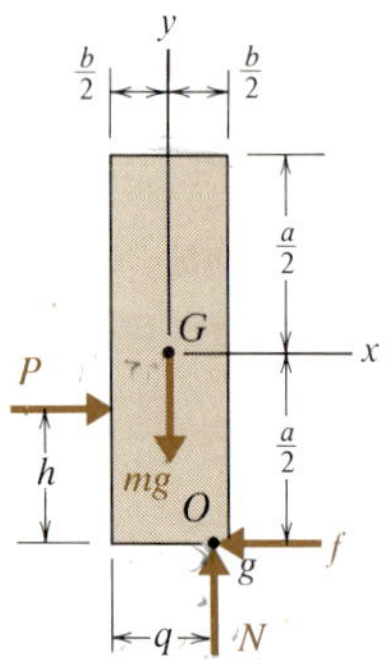

FIGURE 7

$$\sum F_x = P - f = 0, \qquad \sum F_y = N - mg = 0$$

$$\sum M_{Oz} = mg(q - b/2) - Ph = 0$$

The force sums indicate that $f = P$ and $N = mg$, and the moment equation yields a relation for the distance q locating the line of action of the normal force, specifically

$$q = \frac{b}{2} + \frac{Ph}{mg}$$

It follows that increasing P from a small value increases the distance q and the magnitude of the friction force $\bar{f}$ required for static equilibrium.

Clearly, if the cabinet starts to tip over, its lower right edge will be the only location where there is contact with the ground; this corresponds to $q = b$. If the value of P is such that the moment equilibrium equation yields $q = b$, then tipping is impending. On the other hand, if the value of P leads to $P = f = \mu_s N$, then slipping is impending.

The foregoing are the two possible types of impending motion. The impending motion that will actually occur is the one corresponding to the smaller force P. Note that the coefficient of static friction does not appear in the moment equation used to determine the critical value of P for tipping. Thus, the case of simultaneous tipping and slipping is a very special one.

The preceding discussion of the filing cabinet is illustrative of the general procedure we shall employ to study whether impending motion of an object corresponds to tipping or slipping. We begin by drawing a free body diagram that depicts the situation prior to either state of impending motion. Thus, the normal forces depict a general state of contact, rather than the one corresponding to impending overturning. Also, the friction force on each contact surface is considered to be independent of the normal force. The resulting free body diagram is used to obtain a set of general static equilibrium equations.

Impending tipping is investigated by considering the normal forces to act at the edges that would be in contact if the body were to actually overturn. The static equilibrium equations for this case yield the parameters that will cause tipping.

Impending slipping is studied as another special case of the general equilibrium equations. The friction forces at the surfaces where slip is impending are considered to be $f = \mu_s N$. Substitution of these relations into the equilibrium equations yields the parameters for slipping. The critical state is the one corresponding to the parameters that will occur first (for example, the smallest force or the shallowest angle of elevation for an incline).

EXAMPLE 5

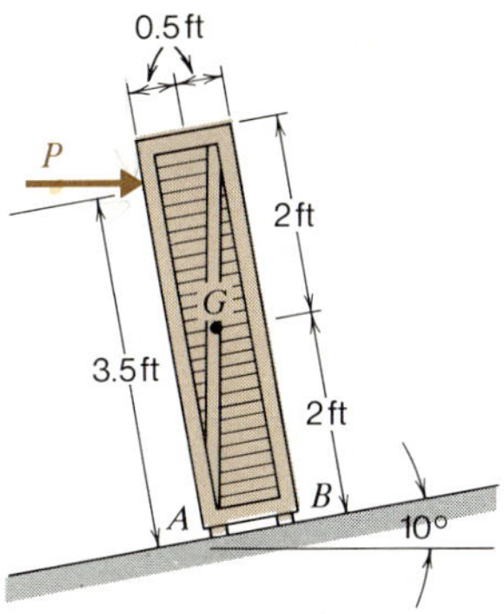

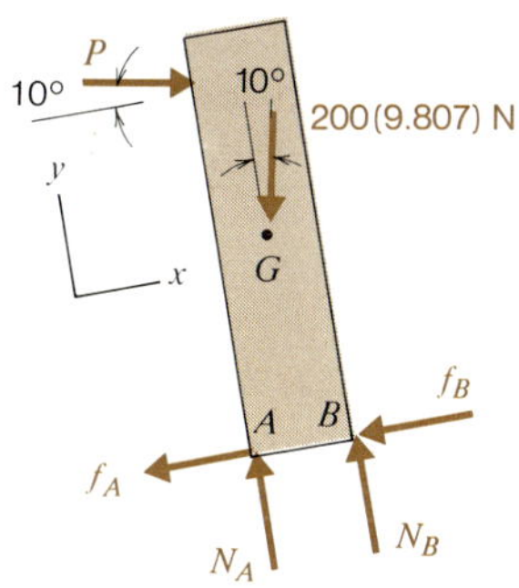

The 200 kg crate, whose center of mass is point G, is at rest on the hill. The crate is supported by small skids at points A and B. The coefficient of static friction between the incline and the skids of 0.25. Determine the largest horizontal force $\bar{P}$ acting in the sense shown for which the crate will not move.

Solution

We start by drawing a free body diagram of the crate, for which we assume that neither tipping nor slipping is impending. Accordingly, the friction and normal forces at each skid are denoted as f and N, respectively. The equilibrium equations corresponding to this free body diagram are

$$\sum M_{Az} = N_B(1.0) - 1961.4 \cos 10^\circ(0.5) + 1961.4 \sin 10^\circ(2.0) - P \cos 10^\circ(3.5) = 0$$

$$\sum F_x = -f_A - f_B - 1961.4 \sin 10^\circ + P \cos 10^\circ = 0$$

$$\sum F_y = N_A + N_B - 1961.4 \cos 10^\circ - P \sin 10^\circ = 0$$

In order to study impending tipping, we note that a very large force $\bar{P}$ might cause the crate to overturn such that only skid B contacts the incline. Thus, we set $f_A = N_A = 0$ in the above equilibrium equations. We then find that

$$\sum F_y\colon N_B = 0.17365P + 1931.6$$

Placing this result into the moment equation yields

$$\sum M_{Az} = (0.17365P + 1931.6) - 965.8 + 681.2 - 3.447P = 0$$

$$P = 503 \text{ newtons}$$

The case of impending slipping is investigated by setting $f_A = \mu_s N_A$ and $f_B = \mu_s N_B$, with $\mu_s = 0.25$ at each skid. This gives

$$\sum F_y\colon N_A + N_B = 0.17365P + 1931.6$$

$$\begin{aligned}\sum F_x &= -0.25(N_A + N_B) - 340.6 + 0.9848P \\ &= -0.25(0.17365P + 1931.6) - 340.6 + 0.9848P = 0\end{aligned}$$

$$P = 875 \text{ newtons}$$

The value of P for tipping is smaller than that required to slide the crate. Thus, we conclude that the crate will tip and

$$P_{max} = 503 \text{ newtons}$$

Additional Remarks

It will be noted that we solved the case of impending slipping without employing the moment equilibrium equation. This is a consequence of the coefficients of static friction at skids A and B being equal. If these coefficients were different, we would have needed $\sum F_y$ and $\sum M_{Az}$ to find the individual values of N_A and N_B.

An interesting aside to this problem is its similarity to the problem posed in part (c) of Example 1. It can be seen that concern about the possibility of impending tipping, which always requires that we consider moment equilibrium, complicates the solution of problems involving impending motion.

HOMEWORK PROBLEMS

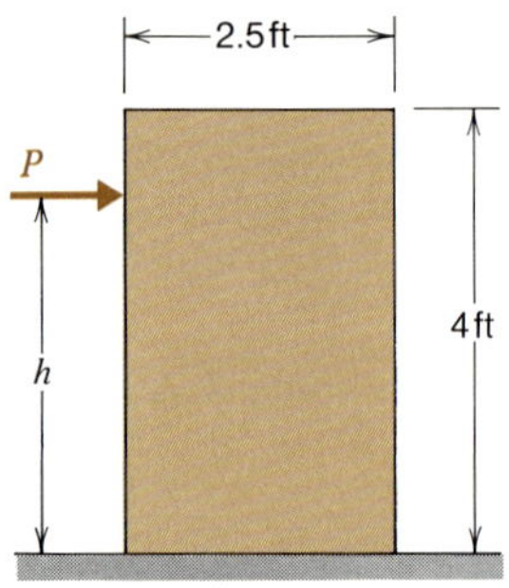

PROBS. 5.47 AND 5.48

5.47 The coefficient of static friction between the 1 ton granite block and the floor is 0.35, and $h = 2.5$ ft. Determine the largest horizontal force $\bar{P}$ that will not move the block.

5.48 The coefficient of static friction between the 1 ton granite block and the floor is 0.35. Determine the largest distance h for the horizontal force $\bar{P}$ for which slipping will occur before tipping.

5.49 A 60 kg refrigerator, having center of mass G, is mounted on four casters. The coefficient of static friction between a locked caster and the floor is 0.50, and the frictional resistance of a rolling caster is negligible. Determine the magnitude of the horizontal force $\bar{P}$ required to move the refrigerator when $h = 1.0$ m if all casters are locked.

5.50 Solve Problem 5.49 if only casters B are locked.

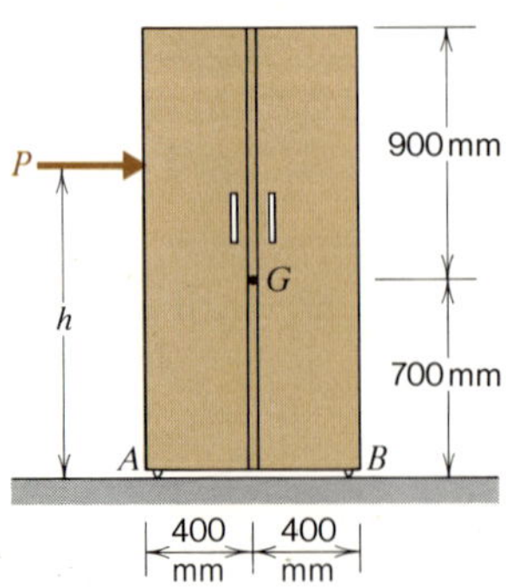

PROB. 5.49

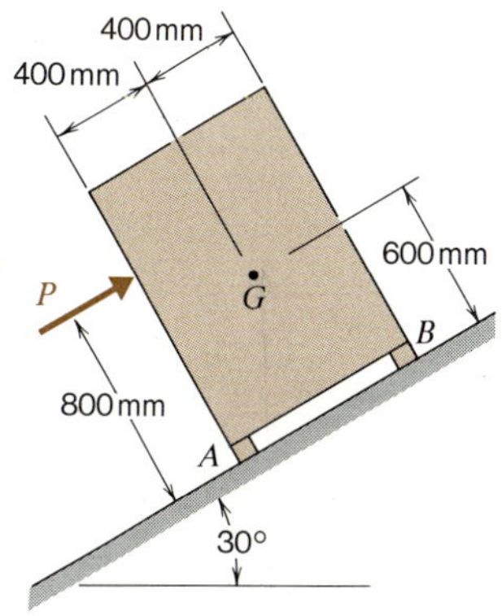

PROB. 5.52

5.51 In Problem 5.49, determine the largest value of h for which the refrigerator will not tip (a) if all casters are locked, (b) if only casters B are locked. (c) Explain why it is impossible to tip the refrigerator over with the force $\overline{P}$ when only casters A are locked.

5.52 A 60 kg cabinet whose center of mass is point G is supported by skids along edges A and B. The coefficients of static friction between the skids and the incline are $\mu_A = \mu_B = 0.75$. Determine the largest force $\overline{P}$, parallel to the incline, for which the cabinet will not move.

5.53 Solve Problem 5.52 if $\mu_A = 0.5$ and $\mu_B = 0.90$.

5.54 Determine the magnitude and sense of the force $\overline{P}$ for which the cabinet in Problem 5.52 is in a state of impending motion down the incline.

5.55 The center of mass of the 120 lb table is point G. The table is supported by four casters, each of which is locked. The coefficient of static friction between a locked caster and the incline is 0.60. Determine the magnitude and sense of the horizontal force $\overline{P}$ for which the table is in a state of impending motion down the incline.

5.56 Solve Problem 5.55 if only caster A are locked. The friction between the unlocked casters B and the incline is negligible.

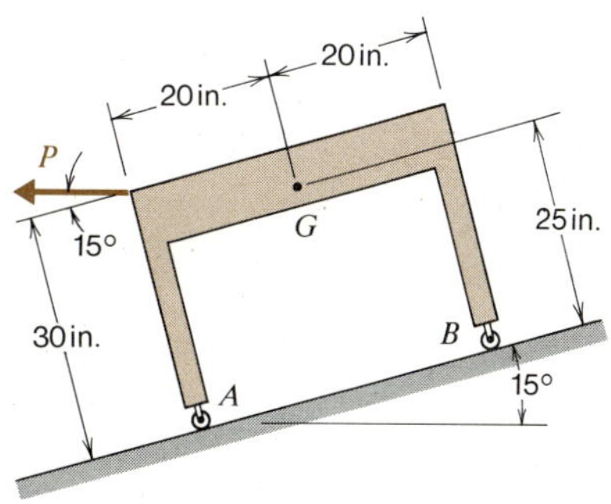

PROB. 5.55

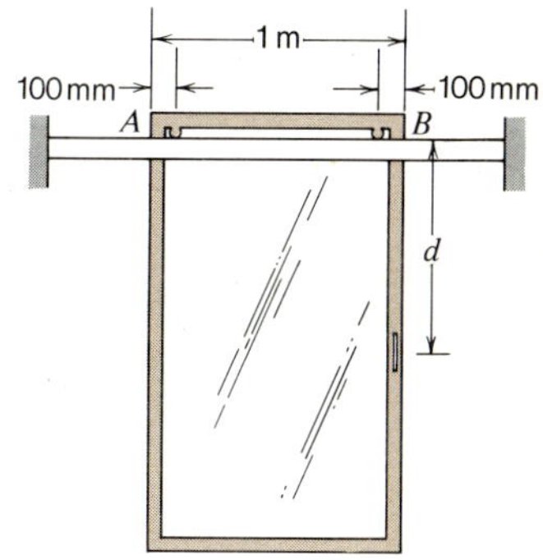

PROB. 5.57

5.57 The 25 kg door is mounted on the horizontal rail by means of runners A and B. The coefficients of friction for these runners are $\mu_A = \mu_B = 0.15$. The door handle is pulled to the right to open the door. Determine (a) the maximum distance d for the door handle for which the door will not tip when it is opened, (b) the force required to open the door if d equals the value found in part (a), and (c) the force required to open the door if d equals one-half the value found in part (a).

*5.58 Solve Problem 5.57 if $\mu_A = 0.20$, $\mu_B = 0.10$.

5.59 The 8 lb thin semicylindrical shell is to be towed to the left. The coefficient of static friction between the shell and the surface is 0.35. Determine the largest angle α for which the cable tension T will cause the shell to slide to the left without tipping. What is the corresponding value of T?

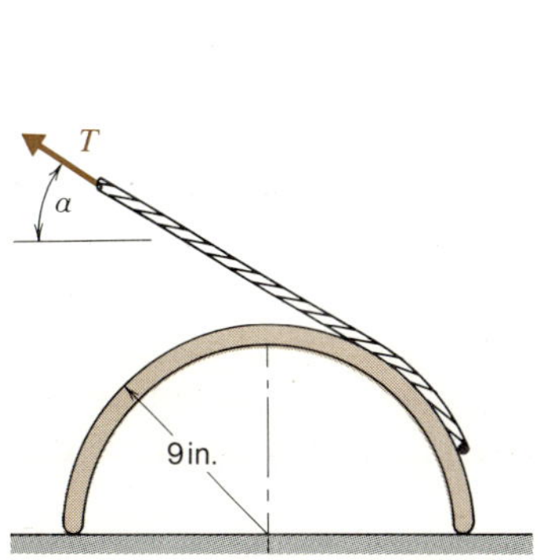

PROB. 5.59

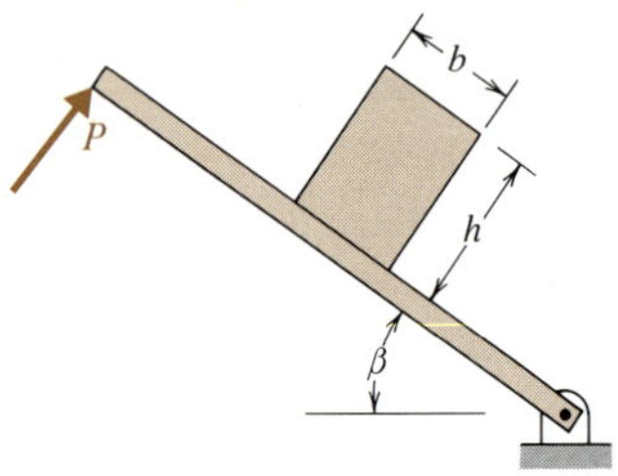

PROB. 5.60

5.60 The angle β of the incline is very gradually increased until the block of mass m moves. Develop formulas that indicate whether the block slides or tips in terms of the coefficient of friction μ, the ratio b/h, and the angle β at which movement occurs.

***5.61** The bar of mass m is supported in the vertical plane by pins A and B, which are separated by a distance d. The coefficient of static friction between each pin and the bar is 0.70. Determine the largest value of d for which static equilibrium is possible.

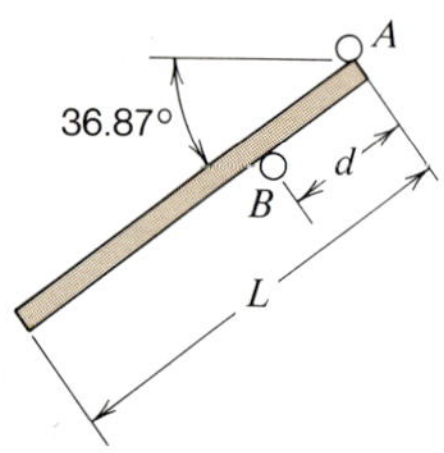

PROB. 5.61

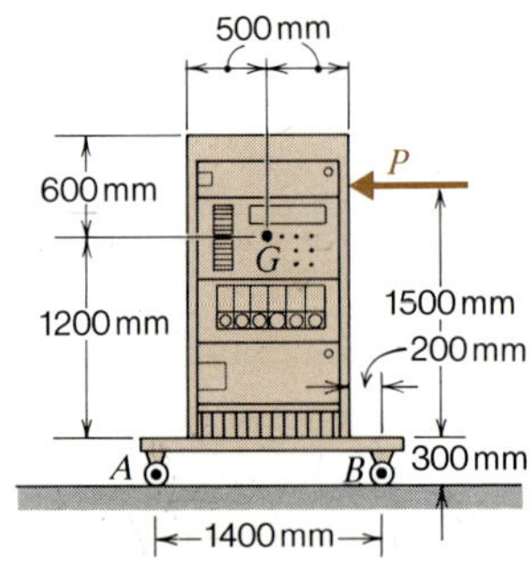

PROB. 5.62

***5.62** A 500 kg electronic computer, having center of mass G, is placed on the 20 kg dolly. The casters A of the dolly are locked. The coefficients of friction between the computer and the dolly, and between the casters and the floor, are 0.50 and 0.30, respectively. Determine the maximum force $\bar{P}$ that can be applied without causing the computer to move.

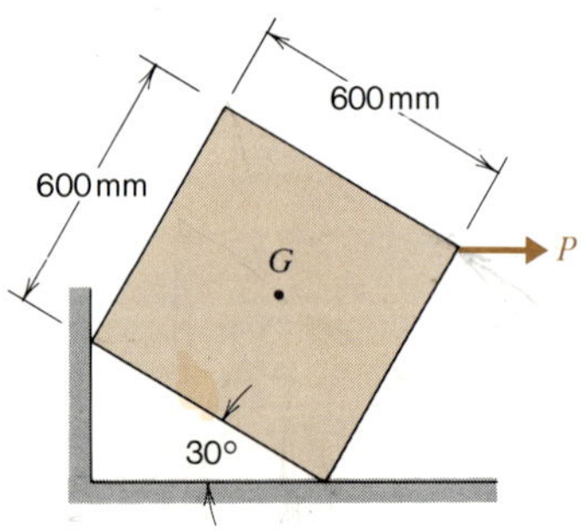

PROB. 5.63

***5.63** The 80 kg block rests against the vertical wall in the position shown. Determine the largest horizontal force $\bar{P}$ for which the block will not move if the coefficient of friction between the ground and the block is 0.50, and the vertical wall is smooth.

***5.64** Solve Problem 5.63 for the case where the coefficient of friction between the block and each surface it contacts is 0.5.

B. BASIC MACHINES HAVING FRICTION

There are a multitude of simple mechanical devices, that we refer to as basic machines, that can be used to either move or prevent the motion of physical objects. These devices can magnify or convert the input forces applied to them into a different set of output forces that are applied to other systems. For example, a wheel fixed to an axle has the capability of transforming a torque exerted by its axle into a forward force that can propel a vehicle.

In Chapter 4, we developed the basic techniques for investigating the equilibrium of machines. The common feature of the machines we shall consider here is the important, and sometimes useful, effect of friction.

1. Wedges

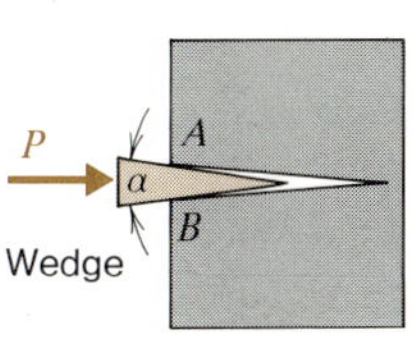

FIGURE 8a

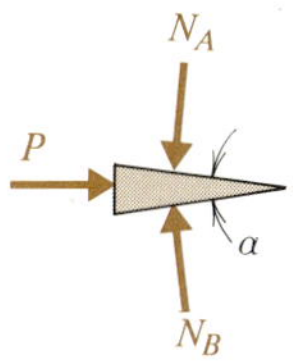

FIGURE 8b

A wedge is merely a block having two flat surfaces that form a small angle relative to each other. Consider the triangular wedge in Figure 8a that is being pushed into a crack in a large body by the force $\bar{P}$.

In the situation where the surfaces of contact between the wedge and the large body are smooth, the free body diagram of the wedge is as shown in Figure 8b. It is not difficult to see that, if the angle α is sufficiently small, the magnitudes of the normal forces $\bar{N}_A$ and $\bar{N}_B$ will be much larger than P. Thus, this wedge could be used to enlarge the crack. An example of this type of action is the head of an axe.

However, Figure 8b is not a correct representation of the action of the wedge, because friction cannot be neglected. To study the effect of friction we must distinguish between the situation where the tendency is for the wedge to be forced further into the crack, as shown in the free body diagram in Figure 9a, and the case where the tendency is for the wedge to be squeezed out of the crack, as shown in the free body diagram in Figure 9b.

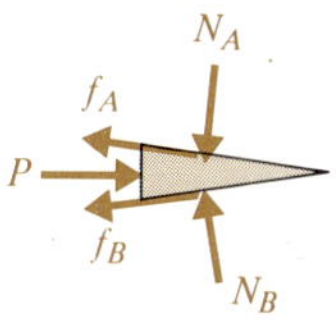

Wedge being forced in

FIGURE 9a

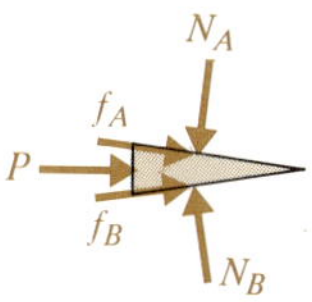

Wedge being squeezed out

FIGURE 9b

For a given force $\bar{P}$, there are four unknown forces acting on the wedge in each case illustrated in Figure 9. Because we are not concerned here with the points of application of the forces acting on the wedge, there are only two scalar equations of force equilibrium. We can conclude then that the free body diagrams in Figure 9 represent statically indeterminate systems having two redundant forces.

Fortunately, our interest in wedges lies in the study of impending motion. One problem we wish to investigate is how the wedge can be forced further into the crack. This is the case of impending inward motion. It may be studied by using the free body diagram of Figure 9a with $f = \mu_s N$ for each friction force. For such a problem the friction forces are no longer independent unknowns, so the system becomes a statically determinate one.

Another problem we wish to investigate is how the wedge can be pulled out of the crack. In this case we have impending outward motion, so we use the free body diagram of Figure 9b, setting $f = \mu_s N$ for each friction force. Once again, the friction forces are no longer independent unknowns, so the system is a statically determinate one.

The preceding discussion enables us to observe that the existence of friction has both good and bad effects in wedges. The bad effect arises because the friction force for impending inward motion, Figure 9a, opposes the input force $\bar{P}$. Therefore, the existence of friction means that a larger force $\bar{P}$ is needed to drive the wedge in. The beneficial effect of friction is associated with removing the wedge from the crack. If the coefficients of friction are very small, impending outward motion of the wedge, Figure 9b, still corresponds to an inward force $\bar{P}$. In other words, it will be necessary to hold the wedge in position. In contrast, if the coefficients of static friction are sufficiently large, for impending outward motion a force $\bar{P}$ in the opposite sense of the one depicted in the free body diagram (that is, a pulling force) will be required to remove the wedge. This means that the wedge will stay in place even if $\bar{P}$ is not applied to it. For this situation, the wedge is said to be *self-locking*.

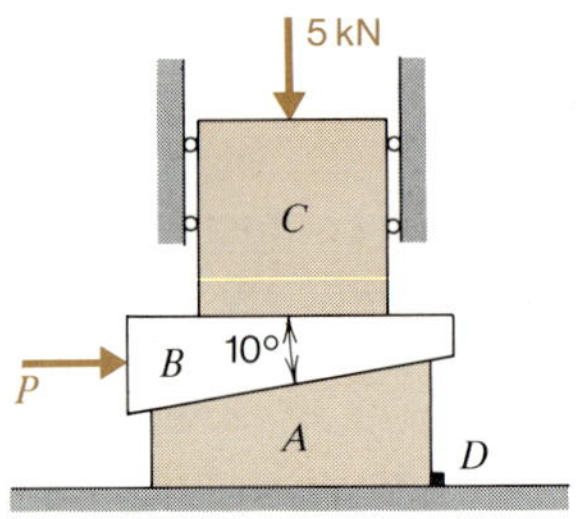

EXAMPLE 6

Wedges A and B are used to raise block C against the 5 kN load that is acting on it. The coefficient of friction between wedge B and all surfaces in contact with it is 0.10. Cleat D holds wedge A in position. (a) Determine the minimum force $\bar{P}$ required to raise block C. (b) Determine if the wedge is self-locking.

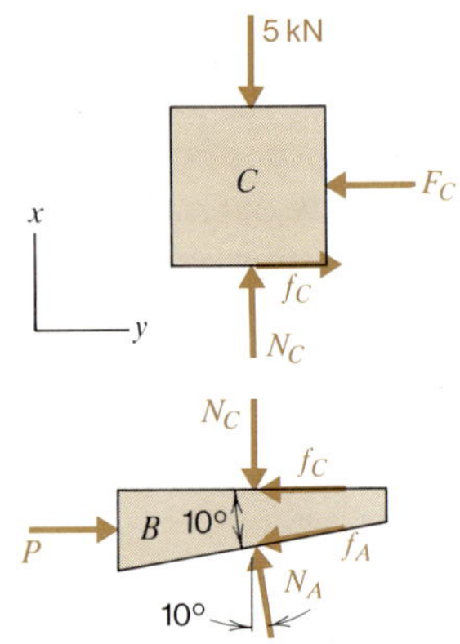

Solution

The first consideration here is which free body diagrams to draw. In view of the presence of cleat D, the wedge A is in effect part of the ground, so there is no reason to consider this wedge. On the other hand, wedge B and block C are individual bodies whose equilibrium must be assured. Thus, we will draw free body diagrams of these bodies. In part (a) we are interested in impending motion of wedge B to the right, whereas to investigate self-locking, we want to consider impending motion of wedge B to the left. In order to have one set of free body diagrams for both cases, we denote the friction forces acting on this wedge as $\bar{f}_C$ and $\bar{f}_A$. We then set $\bar{f} = \pm\mu_s N$, with the sign chosen to match the type of impending motion. For the purpose of drawing free body diagrams, $\bar{f}_C$ and $\bar{f}_A$ have been depicted in the diagrams as though block B was tending to move to the right. Note that the forces exerted by the wedge on block C must satisfy Newton's third law.

The equilibrium equations for these bodies are

$$\textbf{Wedge B:}\quad \sum F_x = P - N_A \sin 10^\circ - f_C - f_A \cos 10^\circ = 0$$

$$\sum F_y = N_A \cos 10^\circ - f_A \sin 10^\circ - N_C = 0$$

$$\textbf{Block C:}\quad \sum F_y = N_C - 5 = 0 \text{ kN}$$

We are not interested in $\bar{F}_C$, so we have disregarded $\sum F_x$ for block C. First, to solve part (a) we set

$$f_C = +0.10N_C \qquad f_A = +0.10N_A$$

Substitution of these expressions and $N_C = 5$ kN (obtained from equilibrium of block C) into the first two equilibrium equations yields

$$P - 0.1736N_A - (0.50) - 0.9848(0.10N_A) = 0$$

$$0.9848N_A - 0.1736(0.10N_A) - 5 = 0$$

$$N_A = 5.168 \text{ kN}, \qquad P = 1.906 \text{ kN}$$

We now turn our attention to part (b). In order to evaluate if the wedge is self-locking, we will determine whether the force $\bar{P}$ should be to the right or left when motion is impending to the left. In this case, we have

$$f_A = -\mu_s N_A \qquad f_C = -\mu_s N_C$$

The equilibrium equation for block C still gives $N_C = 5\text{ kN}$, so the equilibrium equations for the wedge now take the form

$$P - 0.1736N_A - (-0.50) - 0.9848(-0.10N_A) = 0$$

$$0.9848N_A - 0.1736(-0.10N_A) - 5 = 0$$

$$N_A = 4.989\text{ kN}, \qquad P = -0.125\text{ kN}$$

The fact that P is negative means that the force P must pull on the wedge to the left to move the wedge in that direction. Thus, the wedge is self-locking.

HOMEWORK PROBLEMS

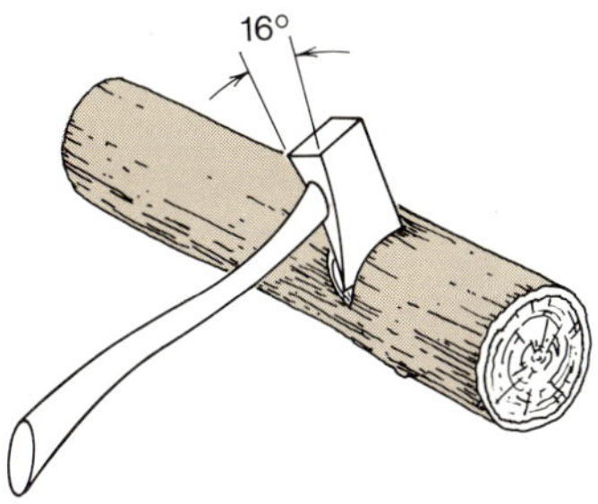

PROB. 5.65

5.65 The faces of the blade of the axe form a 16° angle. Determine the minimum coefficient of static friction between the blade and the log for which the axe is self-locking.

5.66 and 5.67 Determine the largest horizontal force $\bar{P}$ that may be applied to the lightweight wedge A without raising block B. The coefficient of state friction between all surfaces is 0.15.

5.68 and 5.69 Determine the magnitude and sense of the horizontal force $\bar{P}$ for which there is impending motion to the right of the lightweight wedge A. Is the wedge self-locking? The coefficient of static friction between all surfaces is 0.15.

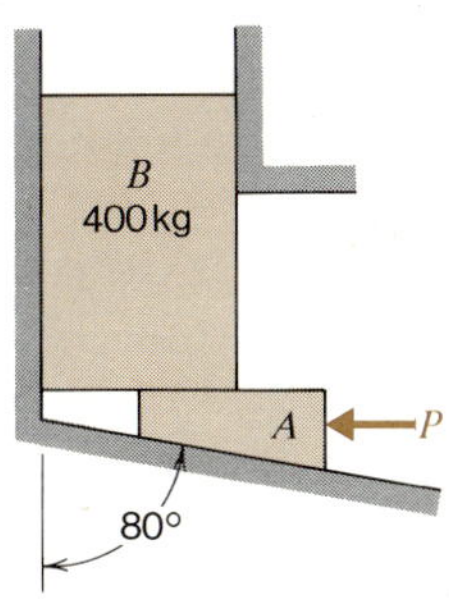

PROBS. 5.66 AND 5.68

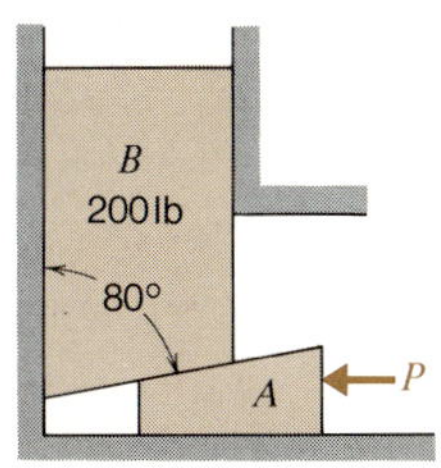

PROBS. 5.67 AND 5.69

5.70 and 5.71 Determine the downward force $\bar{P}$, applied to the 20 kg block, required to move the 500 kg block. The coefficient of static friction between all surfaces is 0.15.

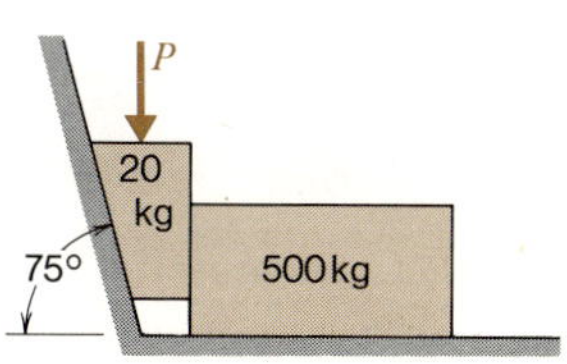

PROB. 5.70

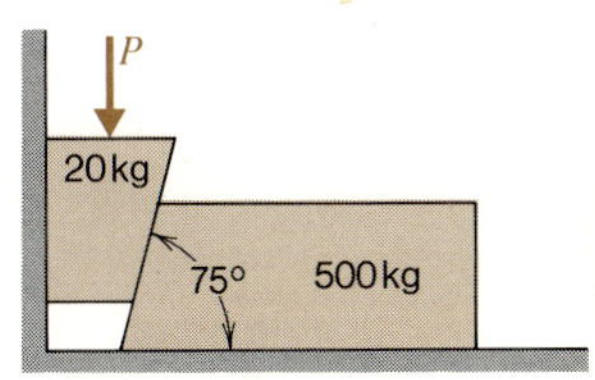

PROB. 5.71

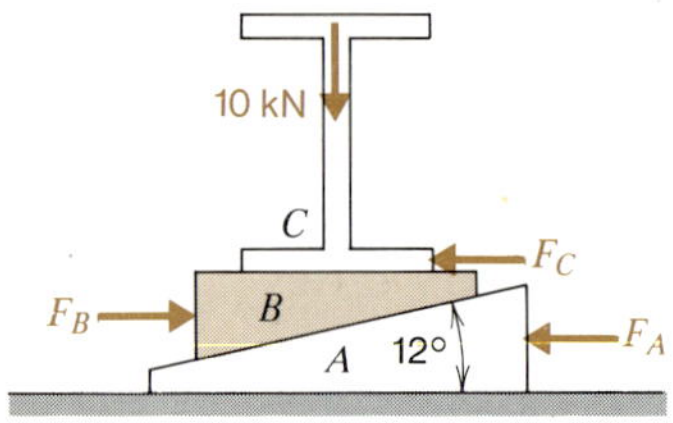

PROB. 5.72

5.72 The H-beam is to be adjusted into its final position by applying the force $\bar{F}_B$ to wedge B. The reaction on the end of the beam is 10 kN. The coefficient of static friction between all surfaces is 0.20. (a) What is the minimum force $\bar{F}_B$ required to raise the beam? (b) What are the corresponding forces $\bar{F}_C$ and $\bar{F}_A$ required to prevent the beam and lower wedge from shifting?

5.73 and 5.74 Determine the smallest horizontal force $\bar{P}$ applied to the wedge that will raise beam AB. The masses of the wedge and the beam are negligible, and the coefficient of static friction for both sides of the wedge is 0.20.

5.75 and 5.76 The masses of the wedge and beam AB are negligible, and the coefficient of static friction for both sides of the wedge is 0.20. Is the wedge self-locking?

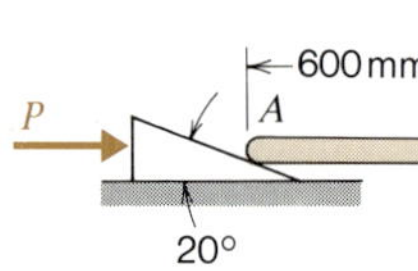

PROBS. 5.73 AND 5.75

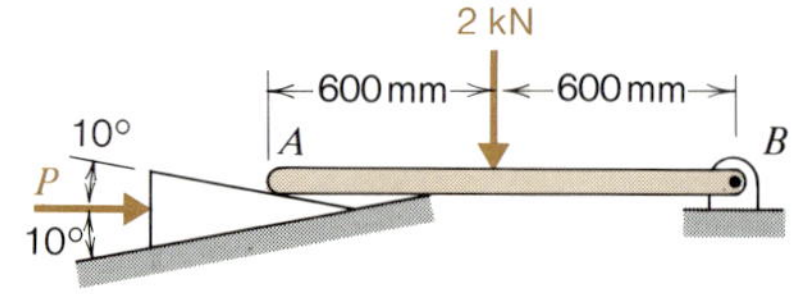

PROBS. 5.74 AND 5.76

*5.77 A 2000 kg lathe having center of mass G is to be leveled by means of a wedge at corner B. The coefficient of static friction between the wedge and the surfaces it contacts is 0.10. Determine (a) the minimum horizontal force $\bar{P}$ required to raise corner B of the lathe, and (b) the minimum coefficient of static friction between corner A of the lathe and the floor for which the lathe will not slide when the force $\bar{P}$ in part (a) is applied to the wedge. The lathe may be considered to be in the horizontal position.

*5.78 Point G is the center of mass of the 2000 kg lathe. The coefficient of static friction for both sides of the wedge at corner B is μ_B, and the coefficient of static friction between corner A and the ground is μ_A. It is desired that the lathe remain in the position shown when the horizontal force $\bar{P}$ is removed from the wedge. Determine the minimum allowable values of μ_B and μ_A.

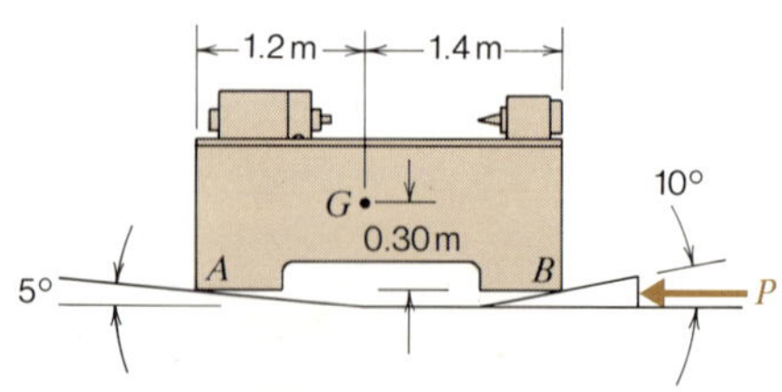

PROBS. 5.77 AND 5.78

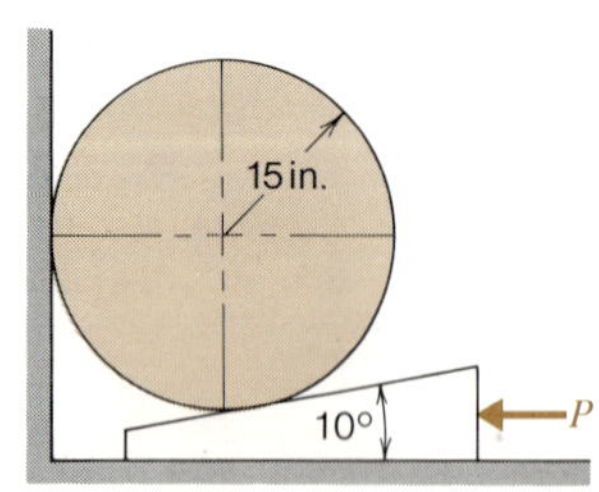

PROB. 5.79

*5.79 Determine the minimum force $\bar{P}$ required to lift the 500 lb drum shown. The mass of the wedge is negligible and the coefficient of friction between all surfaces is 0.25. Which surfaces are on the verge of sliding for this force $\bar{P}$?

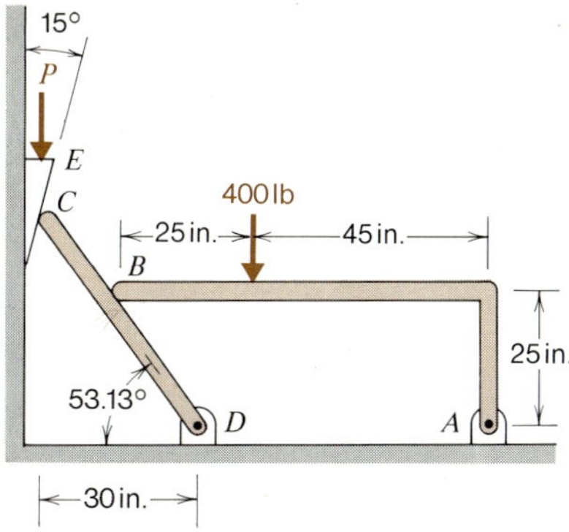

PROB. 5.80

***5.80** A vertical force $\bar{P}$ is to be applied to wedge E in order to push end C of bar CD to the right. The coefficient of static friction between all surfaces is 0.15. Determine the minimum force $\bar{P}$ that will achieve this goal.

***5.81** Determine if the system in Problem 5.80 is self-locking.

---

2. Square Threaded Screws

A screw is a basic machine for transforming a couple applied about an axis into an axial force that can be used to move objects or to fasten objects together. Familiar examples of screw-type machines are an automobile bumper jack (to move objects) and a bolt and nut set (to fasten objects).

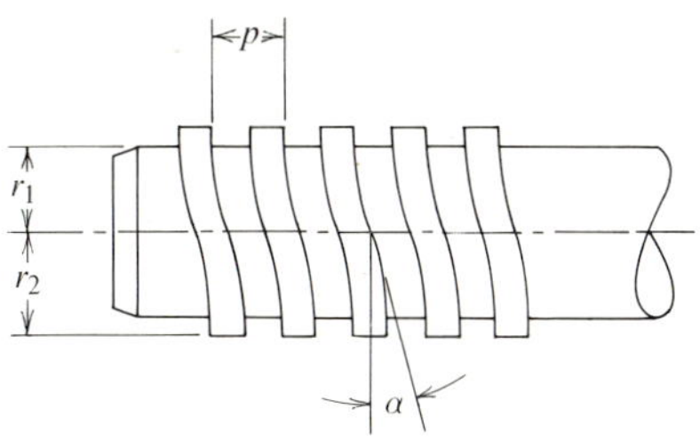

FIGURE 10

There are several common shapes for the threads of a screw. We will limit our efforts here to the analysis of square threads, because the analysis of such threads is quite similar to that of wedges. A typical square-threaded screw is illustrated in Figure 10, where the dimensions r_1 and r_2 are the inner and outer radii of the thread. The axial distance p, over which the thread pattern repeats itself, is called the *pitch.*

A screw thread is helical, which means that its edges may be visualized as being formed by wrapping a right triangle around a cylinder. The *lead angle* α for the tangent to the screw thread in Figure 10 is one of the angles of such a triangle.

Our analysis of the forces acting on the screw thread will employ the mean (that is, average) radius of the thread

$$r = \tfrac{1}{2}(r_1 + r_2) \tag{5}$$

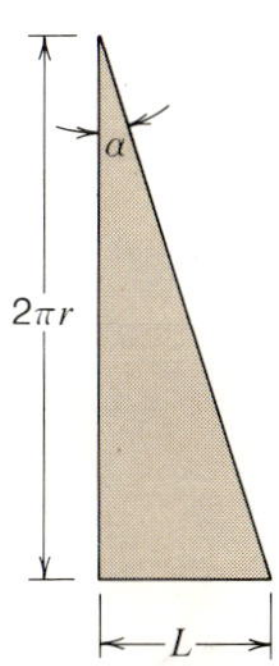

FIGURE 11

The triangle forming the corresponding helix is as shown in Figure 11, where the side of the triangle adjacent to the lead angle α is the circumference of a cylinder whose radius is r, that is, $2\pi r$. The dimension L in Figure 11 is called the lead. It represents the distance the screw will advance (or retreat) in one revolution. Screws usually have a single thread that is wrapped continously around their axis. Such a screw is said to be *single-threaded.* Screws that have two or three parallel threads are termed *double- or triple-*

threaded, respectively. Hence, a distance equal to the lead L, measured along the axis of the screw, will contain n threads in an *n-threaded-screw*. It follows that the lead L is related to the pitch p, which is the spacing between threads, by

$$L = np \tag{6}$$

Figure 11 then indicates that

$$\tan \alpha = \frac{L}{2\pi r} = \frac{np}{2\pi r} \tag{7}$$

Now that the geometrical relations describing the threads have been derived, the equilibrium equations for a screw may be formulated. Figure 12 depicts a square threaded screw that carries a couple $\bar{M}$ that would advance the screw in opposition to an axial load $\bar{F}$.

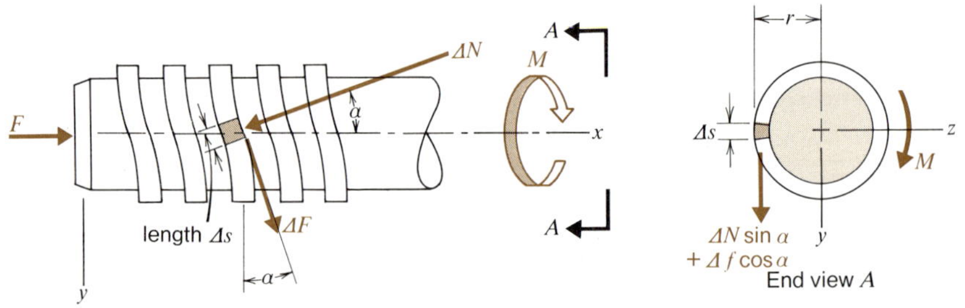

FIGURE 12

The axial force pushes the screw into contact with the mating screw threads on the adjoining body. As a result, a normal force $\Delta\bar{N}$ and frictional force $\Delta\bar{f}$ are developed on the load-bearing surface of each segment of the thread, such as the segment Δs in Figure 12. For each segment Δs, such as the one in Figure 12, there is a matching segment on the opposite side of the screw carrying a comparable set of forces. For the coordinate system in Figure 12, the only equations of static equilibrium not identically satisfied by such a force system are $\sum F_x = 0$ and $\sum M_x = 0$.

The axial component of the forces acting on segment Δs is

$$\Delta F_x = \Delta f \sin \alpha - \Delta N \cos \alpha \tag{8}$$

The end view in Figure 12 indicates that the moment of the forces $\Delta\bar{N}$ and $\Delta\bar{f}$ about the x axis is obtained by multiplying the y component of these forces by the mean radius r for the bearing surface of the thread. Thus,

$$\Delta M_x = (\Delta N \sin \alpha + \Delta f \cos \alpha)r \tag{9}$$

When the individual contributions of each segment are added together, we find the total normal and frictional force components N and f, respectively. The equilibrium equations for the whole screw are then found to be

$$\sum F_x = F + f \sin \alpha - N \cos \alpha = 0 \tag{10}$$

$$\sum M_x = -M + (N \sin \alpha + f \cos \alpha)r = 0 \tag{11}$$

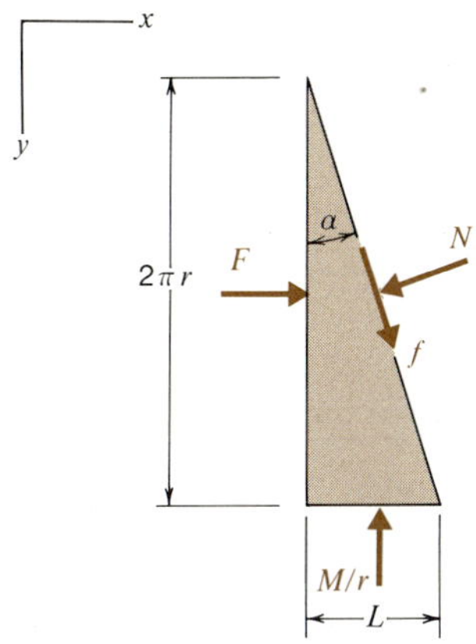

FIGURE 13

It is not necessary to memorize these relations, because they may be obtained for any problem by considering an analogy involving a wedge in the shape of the triangle in Figures 11 and 12. The orientation of the forces $\Delta\bar{N}$ and $\Delta\bar{f}$ applied to each thread segment Δs in Figure 12 suggests that the inclined face of the wedge should carry the normal force $\bar{N}$ and friction force $\bar{f}$. The axial force $\bar{F}$ in this analogy becomes the normal force acting on the side of the triangle whose length is the circumference $2\pi r$. There is no load acting perpendicular to $\bar{F}$ at the left end of the screw, so we consider the left face of the wedge to be smooth.

Finally, the couple $\bar{M}$ in Figure 12 would advance the screw if the other forces were not present. We therefore regard M/r to be the magnitude of the normal force acting on the face whose length is L. A positive value of M has been defined to correspond to a couple that is tending to advance the screw, so the force M/r is considered to be one that is pushing the wedge inward. The resulting analogy is shown in Figure 13. It is readily verified that eqs. (10) and (11) are identical to the equilibrium equations $\sum F_x = 0$ and $\sum F_y = 0$ for the wedge.

A primary concern in designing a screw is the relationship between the magnitudes of the axial force $\bar{F}$ and the torsional couple $\bar{M}$. Let us consider the axial force $\bar{F}$ to be given. Then the two static equilibrium equations are insufficient to evaluate the three unknowns M, N, and f. This is the usual result for wedges having friction. They are statically indeterminate in the general condition.

Fortunately, the cases of impending motion are the most important. For example, we might wish to evaluate the couple $\bar{M}$ required to advance the screw. In Figure 13 this corresponds to impending upward motion of the wedge. The sense of the friction force $\bar{f}$ depicted in this figure is consistent with this tendency, so we have

$$\boxed{f = +\mu_s N \qquad \text{for impending advancement}} \tag{12a}$$

Another consideration is the couple $\bar{M}$ required to retract the screw, which means that the wedge in Figure 13 will be tending to move downward. In this case the friction force $\bar{f}$ will be in the opposite sense from the one shown, so that

$$\boxed{f = -\mu_s N \qquad \text{for impending retraction}} \tag{12b}$$

With the aid of eq. (12a), the equilibrium equations for the wedge may by solved for the largest couple $\bar{M}$ that may be applied to the screw without causing it to advance against the axial force $\bar{F}$. Similarly, eq. (12b) and the equilibrium equations for the wedge yield the couple $\bar{M}$ corresponding to the impending retraction of the screw.

This analogy to a wedge indicates that a square-threaded screw might be self-locking. Figure 14 depicts the forces acting on a wedge when retraction of a screw is impending. The friction and normal forces are equivalent to a resultant force $\bar{R}$ at an angle ϕ_s from the normal force, where $\tan\phi_s = \mu_s$. Figure 14 illustrates the situation when $\phi_s < \alpha$. In this case, the y component

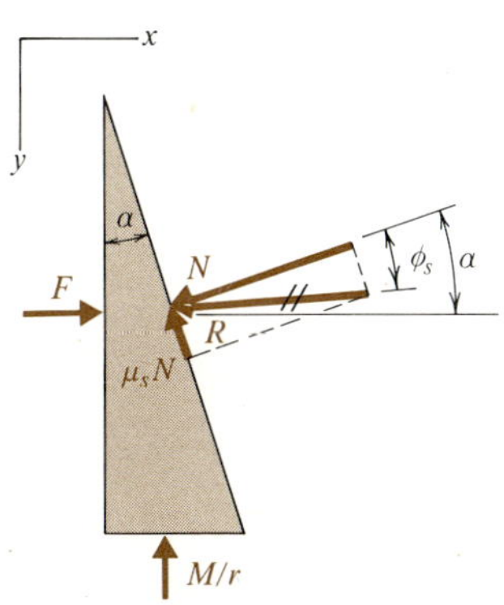

FIGURE 14

of $\bar{R}$ tends to cause the wedge to withdraw (move downward), so a force M/r in the sense shown is required for static equilibrium.

When $\phi_s > \alpha$, the y component of $\bar{R}$ will tend to push the wedge further inward. A force M/r in the opposite sense from the one shown is required to obtain the state of impending motion. This means that the screw can be retracted only if $M < 0$, which corresponds to a couple that is actually tending to retract the screw. Therefore,

> For a self-locking bolt having square single threads:
>
> $$\alpha < \phi_s \equiv \tan^{-1}(\mu_s)$$

Our discussions thus far have considered situations where it is desired to advance or retract a screw from a state of rest. In the situation where the screw is rotating, the threads are sliding over each other. If the screw is rotating slowly, so that inertial effects are negligible, the wedge analogy is still valid. The only modification that is needed to treat a slowly rotating screw is to replace μ_s by μ_k, the coefficient of kinetic friction.

EXAMPLE 7

A bolt having square single threads with a mean diameter of 12 mm and a pitch of 4 mm is screwed into a fixed metal block. The coefficients of static and kinetic friction between the block and the bolt are 0.35 and 0.27, respectively. Friction between the head of the bolt and the block is negligible. Determine the axial force in the bolt if (a) 50 N-m is the largest couple that a socket wrench may apply without causing the bolt to turn, (b) 50 N-m is the couple that a socket wrench must apply to turn the bolt slowly. (c) For the axial force found in case (a), determine the magnitude and sense of the couple for which retraction of the screw is impending.

Solution

The mean radius is one-half the mean diameter, and the lead equals the pitch for a single-pitched screw ($n = 1$), so

$$r = 0.006 \text{ m}, \qquad L = 0.004 \text{ m}$$

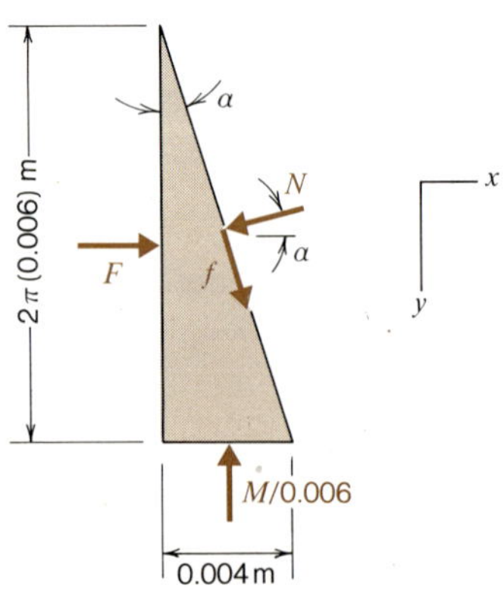

The corresponding wedge analogy is sketched to the left, from which the lead angle is found to be

$$\alpha = \tan^{-1}\left[\frac{0.004}{2\pi(0.006)}\right] = 6.057°$$

The general equilibrium equations for the wedge are

$$\sum F_x = F - N\cos\alpha + f\sin\alpha \equiv F - 0.9944N + 0.1055f = 0$$

$$\sum F_y = -\frac{M}{0.006} + N\sin\alpha + f\cos\alpha \equiv -\frac{M}{0.006} + 0.1055N + 0.9944f = 0$$

Case (a) corresponds to impending advancement of the wedge, so we set $f = +\mu_s N \equiv 0.35N$. Substituting this expression and $M = 50$ N-m into the equilibrium equations then yields

$$\sum F_x\colon F = [0.9944 - 0.1055(0.35)]N$$

$$\sum F_y\colon [0.1055 + 0.9944(0.35)]N = \frac{50}{0.006}$$

$$N = 18.373 \text{ kN}, \qquad F = 17.592 \text{ kN}$$

Case (b) is treated by considering the wedge to be moving inward slowly, so $f = +\mu_k N = 0.27N$. The equilibrium equations then yield

$$\sum F_x\colon F = [0.9944 - 0.1055(0.27)]N$$

$$\sum F_y\colon [0.1055 + 0.9944(0.27)]N = \frac{50}{0.006}$$

$$N = 22.28 \text{ kN}, \qquad F = 21.52 \text{ kN}$$

Case (c) corresponds to impending withdrawal of the wedge, so $f = -0.35N$. When the value of F in part (a) is substituted into the equilibrium equations, it is found that

$$\sum F_x\colon (17.592) = [0.9944 + 0.1055(0.35)]N$$

$$\sum F_y\colon \frac{M}{0.006} = [0.1055 - 0.9944(0.35)]N$$

$$N = 17.06 \text{ kN}, \qquad M = -4.14 \text{ kN-m}$$

Because M is negative, the couple required to retract the screw must be in the sense of the rotation for a retraction. It may be concluded that the screw is self-locking.

Additional Remarks

The difference between the results for cases (a) and (b) has an important, subtle consequence in the general problem of tightening screws to a predetermined clamping force by means of a torque wrench. (A torque wrench has a dial that indicates the moment being applied to the screw.) Clearly, it is easy to see when a screw is rotating, and almost impossible to sense by hand that the screw is on the verge of rotating. Therefore, torque specifications corresponding to a desired axial force in a screw usually require that the screw be turned with the wrench until the wrench indicates the specified torque. After achieving this adjustment, it would require a considerably higher torque to start the screw rotating again to tighten it further.

HOMEWORK PROBLEMS

5.82 The C clamp holds two pieces of wood together. The clamp has a single square thread with a pitch of 3 mm and a mean diameter of 18 mm. The coefficient of static friction is 0.20. The clamping force is 800 N. Determine the torque required to (a) tighten the clamp further, (b) loosen the clamp.

5.83 Solve Problem 5.82 if the screw is double-threaded.

PROB. 5.82

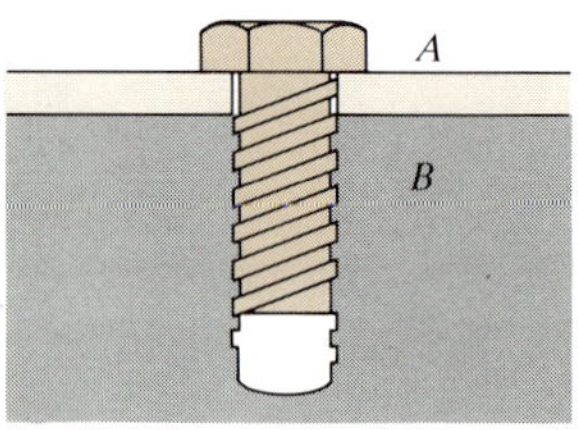

PROB. 5.84

5.84 The metal plate A shown in the diagram is fastened to block B by the single-threaded screw, whose square threads have a pitch of 1/16 in. The mean radius of the threads is 1/4 in. The coefficients of static and kinetic friction for the threads are 0.30 and 0.25, respectively, and friction between plate A and the head of the screw is negligible. The screw is tightened by means of a torque wrench. Determine the clamping force on plate A if the torque wrench reads 40 lb-ft (a) when the screw is about to turn, (b) when the screw is turning.

5.85 The automatic screwdriver shown operates by pushing the handle A downward with a force $\bar{P}$ to turn the threaded shaft B, which holds the bit. The square threads have a pitch of 100 mm, and the mean radius of the thread is 15 mm. The coefficients of static and kinetic friction for the threads are 0.15 and 0.12, respectively. If shaft B is single-threaded, determine the force $\bar{P}$ required to tighten the wood screw to 2 N-m in one slow continuous motion from the position where the screw is first started in the hole.

5.86 The turnbuckle is used to keep the cable taut. Its eyebolts have square single threads whose mean radius and pitch are 0.25 in. and 0.10 in., respectively. The coefficient of static friction between the threaded sleeve and the eyebolts is 0.40. Determine the torque that must be applied to the sleeve to tighten the turnbuckle further if it has already been adjusted to produce a tensile force of 1000 lb in the cable. Eyebolt A has a left-hand thread and eyebolt B has a right-hand thread.

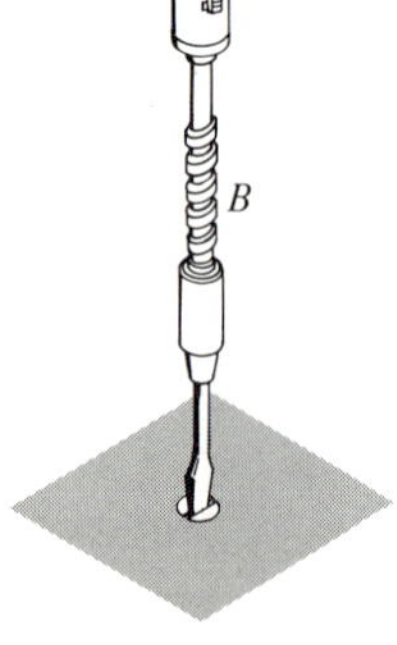

PROB. 5.85

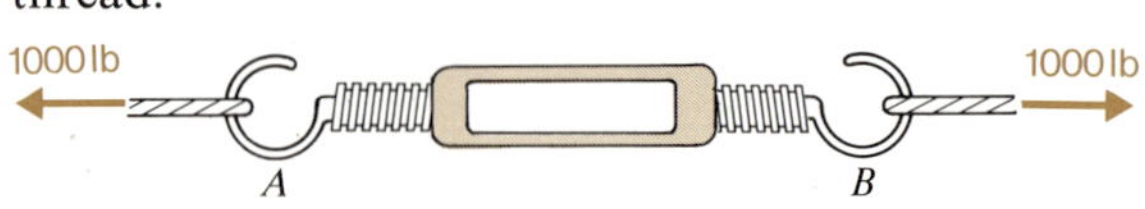

PROB. 5.86

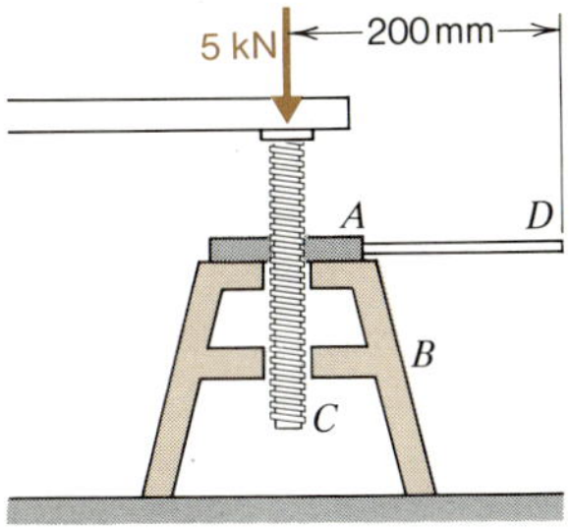

PROB. 5.88

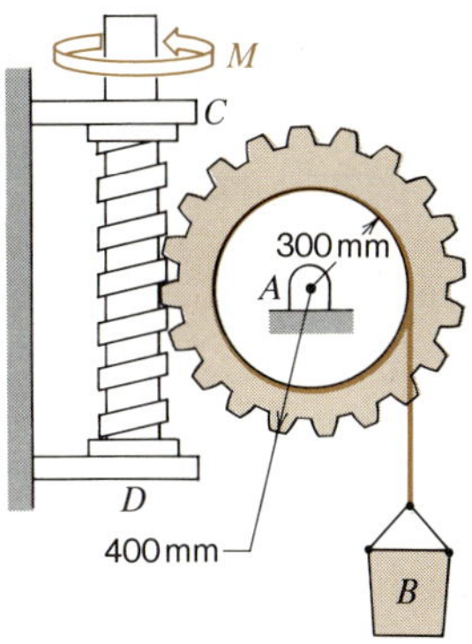

PROB. 5.90

5.87 Determine the couple that must be applied to the sleeve in Problem 5.86 in order to loosen the cables.

5.88 The jack shown in cross section consists of the threaded collar A, which bears on the frame B as it rides on screw C. The collar may be rotated by turning bar AD. The screw itself is prevented from rotating by the object being supported. The single square thread has inner and outer radii of 20 mm and 25 mm, and its pitch is 10 mm. The coefficient of friction for the screw threads is 0.40. The friction between collar A and the frame is negligible. Determine the smallest force applied perpendicular to the diagram at end D of bar AD that will raise an object that applies a 5 kN axial force to the end of the screw.

5.89 For the jack in Problem 5.88, determine the smallest force applied perpendicular to the diagram at end D of bar AD that will lower an object that applies a 5 kN axial force to the end of the screw.

***5.90** Worm gear CD having a single square thread of 50 mm mean radius and 10 mm pitch meshes with gear A. Gear A is rigidly attached to the drum that holds the cable suspending the 100 kg bucket B. The coefficient of static friction between the threads of the worm gear and the teeth of gear A is 0.20. Determine the torque $\bar{M}$ that must be applied to the shaft CD to lift the bucket slowly.

***5.91** Use the wedge analogy and eqs. (6) and (7) to derive expressions for the ratio of the torque M to the axial load F, that is, M/F, in terms of the mean radius, pitch, number of threads, and coefficient of static friction when a square-threaded screw is in a state of (a) impending advancement, (b) impending retraction.

5.92 The scissors jack shown supports an automobile exerting an 800 lb load. The screw has a double square thread whose pitch is 1/8 in. The mean radius of the thread is 3/4 in. Knowing that the coefficient of static friction is 0.33, determine the couple $\bar{M}_{AB}$ that should be applied to the screw (a) to raise the automobile further, (b) to lower the automobile. The screw fits loosely in collar A. Friction between the shoulder in the screw and this collar is negligible.

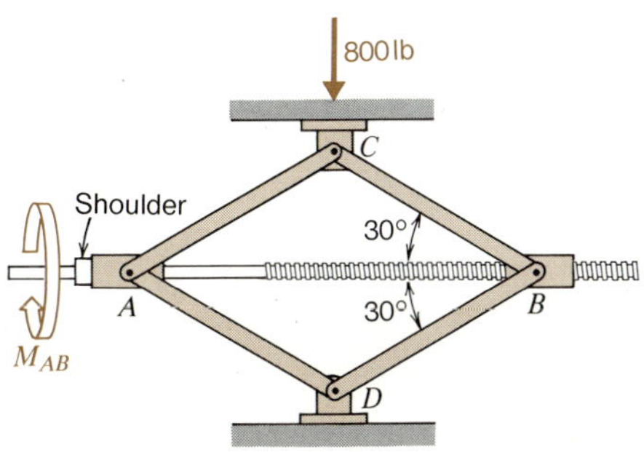

PROB. 5.92

3. Belt Friction

Cables that are guided over pulleys are effective devices for transferring forces between bodies. When the surface of a pulley is smooth, any difference between the cable tensions on either side of the pulley would cause the cable to slip. Sometimes, this would be a desirable result. However, there are situations where friction is required to sustain different tensions in each segment of the cable. This is the case when cables and pulleys are employed to transfer a torque from one pulley to another. For example, belt drives are utilized in internal combustion engines to drive auxiliary equipment, such as an electric generator.

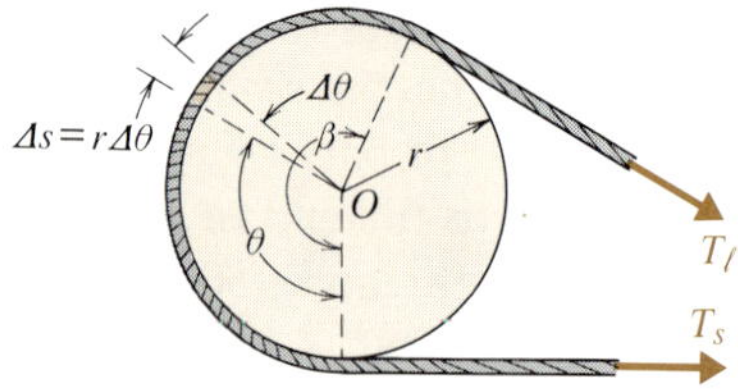

FIGURE 15

Let us consider a cable that passes over a fixed pulley as shown in Figure 15, where the angle β is the *angle of wrap* between the pulley and the cable. Without loss of generality, we will consider the cable to be tending to turn the pulley clockwise. This means that $T_l > T_s$, where the subscripts l and s correspond to the words large and small.

It then follows that the cable exerts a moment $\bar{M}_0$ about the center of the pulley whose magnitude is given by

$$M_0 = (T_l - T_s)r \tag{13}$$

and whose sense is the same as that of the moment of $\bar{T}_l$ about center O.

We wish to explore the relationship between T_l and T_s in the case where sliding between the cable and the pulley is impending. To do this, let us isolate a small element of cable Δs that is in contact with the pulley in the situation where the tendency of the cable is to rotate the pulley clockwise. The contact between the cable and the pulley leads to a frictional force Δf and normal force ΔN. The element Δs tends to move tangent to the pulley, in the positive x direction as shown in Figure 16. Therefore, when sliding is impending, the friction force acting on the element is $\Delta f = \mu_s \Delta N$ in the negative x direction. To complete the free body diagram, we have the tension force T at the position θ on one side of the element and the tension force $T + \Delta T$ at the position $\theta + \Delta\theta$ on the other side of the element. The x axis is chosen to be tangent to the pulley at the midpoint of the element, so it follows that the angle between either tensile force and the x axis is $\Delta\theta/2$.

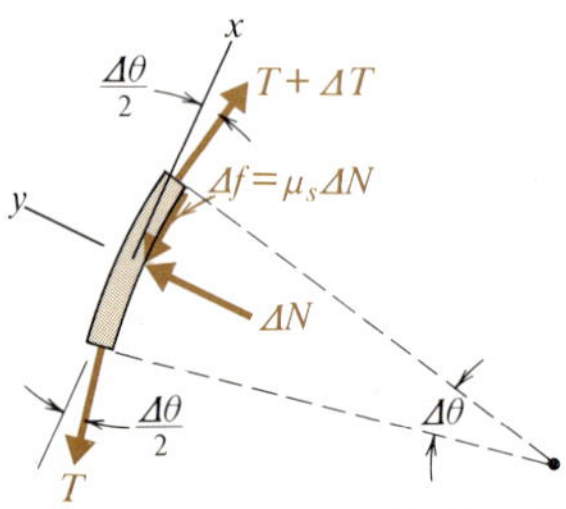

FIGURE 16

The equations for force equilibrium of the element are then found to be

$$\sum F_x = -T\cos\frac{\Delta\theta}{2} + (T + \Delta T)\cos\frac{\Delta\theta}{2} - \mu_s \Delta N = 0$$

$$\sum F_y = -T\sin\frac{\Delta\theta}{2} - (T + \Delta T)\sin\frac{\Delta\theta}{2} + \Delta N = 0$$

These reduce to

$$\Delta T\cos\frac{\Delta\theta}{2} = \mu_s \Delta N \tag{14}$$

$$(2T + \Delta T)\sin\frac{\Delta\theta}{2} = \Delta N \tag{15}$$

Let us now consider what happens when we isolate smaller and smaller elements, which means that the value of $\Delta\theta$ is reduced. In the limit of an infinitesimal element, $\Delta\theta$ approaches $d\theta$. The trigonometric terms in eqs. (14) and (15) become

$$\cos\frac{d\theta}{2} = 1 \qquad \sin\frac{d\theta}{2} = \frac{d\theta}{2}$$

Higher order differentials are negligible when the incremental values become infinitesimal, so eqs. (14) and (15) reduce to

$$dT = \mu_s\, dN \qquad \text{and} \qquad T\, d\theta = dN \tag{16}$$

Our interest is with the tension force, not the normal force, so eliminating dN yields

$$\frac{dT}{T} = \mu_s\, d\theta \tag{17}$$

Equation (17) may be integrated. Figure 15 indicates that the tension at $\theta = 0$ is T_s, while that at $\theta = \beta$ is T_l. Thus

$$\int_{T_s}^{T_l} \frac{dT}{T} = \int_0^\beta \mu_s\, d\theta$$

$$\ln T_l - \ln T_s \equiv \ln\frac{T_l}{T_s} = \mu_s \beta$$

A more useful form for this result is

$$\boxed{\frac{T_l}{T_s} = e^{\mu_s \beta}} \tag{18}$$

Let us review the meaning of each term in eq. (18). The angle β, measured in radians, is the angle of wrap of the cable around the pulley. It is possible for β to be larger than 2π. For instance, if there are n full turns of the cable around the pulley, then $\beta = 2\pi n$. The symbols T_l and T_s represent the larger and the smaller of the tension forces on either side of the pulley; the tendency of the cable is to slide toward the side where the tension is T_l.

Note that eq. (18) describes the situation where a cable is about to slide over *any fixed cylindrical object*, such as a binding post. It may also be employed to describe a cable that is wrapped around a rotating cylindrical object when there is no slipping of the cable, provided that inertial effects are negligible. Further, eq. (18) may be modified to describe the situation where the cable is actually sliding over the cylindrical object. In this case, the symbol μ_s should be replaced by the symbol μ_k, the coefficient of kinetic friction.

Finally, note that eq. (18) was derived for the case of sliding or impending motion. It is not valid for the general, statically indeterminate case. A further restriction on eq. (18) is that it is valid only if contact between the cable and the fixed object is along a cylindrical surface. It must be modified if we wish to analyze a V-belt and its pulley. This modification is the subject of Homework Problem 5.107.

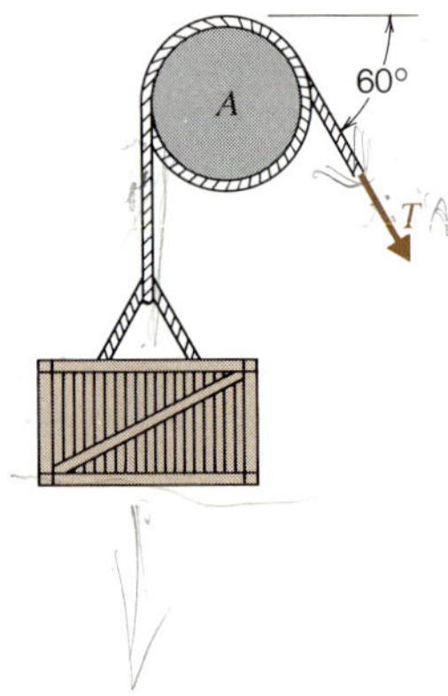

EXAMPLE 8

The tensile force T is applied to the free end of a cable that is wrapped around the horizontal binding post A for one complete turn and then attached to the 60 lb crate. The coefficient of static friction between the post and the cable is 0.30. Determine (a) the minimum tension T for which the crate will not descend, (b) the maximum tension T for which the crate will not rise.

Solution

The tensile force in the vertical portion of the cable is obviously the weight of the crate, 60 lb. When the crate is to be lowered, the cable is tending to move in the direction that the crate is pulling it, so we have $T_l = 60$ lb. We then seek the value $T = T_s$ for impending slipping of the cable.

Conversely, when the crate is to be raised, the cable is about to move in the direction in which its free end is being pulled. In that case T corresponds to T_l and $T_s = 60$ lb.

The angle of wrap is determined from the diagram and the given information. Adding the 150° of contact depicted in the diagram to the one additional turn specified in the problem, we have

$$\beta = 150^\circ\left(\frac{\pi}{180^\circ}\right) + 2\pi = 8.901 \text{ rad}$$

Equation (18) for part (a) then yields

$$\frac{60}{T} = e^{(0.30)(8.901)} = 14.444$$

$$T = 4.15 \text{ lb}$$

Similarly, for part (b) we have

$$\frac{T}{60} = e^{(0.30)(8.901)}$$

$$T = 866 \text{ lb}$$

We can see that it is easy for a person to lower the crate and quite difficult for a person to raise it. This is a consequence of the fact that the exponential function increases rapidly with an increase in its argument.

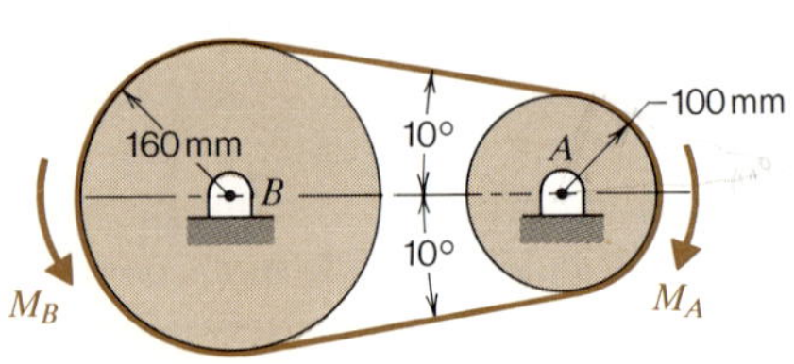

EXAMPLE 9

The coefficient of static friction between the flat belt and each pulley is 0.40. It is known that the maximum torque M_A that a motor can apply to pulley A without causing the belt to slip over either pulley is 20 N-m. For this torque determine (a) the tension in the belt on either side of pulley A, (b) the torque M_B on pulley B for equilibrium, and (c) the pulley at which slip is impending.

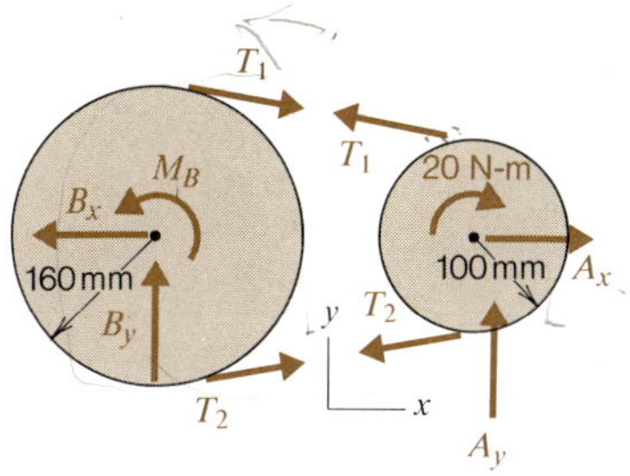

Solution

There are two rigid bodies, specifically the pulleys, whose equilibrium must be assured. A free body diagram of each pulley will show the tensile force exerted by each segment of the belt, as well as the reaction forces exerted by each shaft in order to resist the pulling effect of the belt. Only the equations of moment equilibrium about the center of each pulley are required, because the values of the reaction forces are not being sought. Thus,

$$\text{pulley } A: \quad \sum M_{Az} = -20 + (T_1 - T_2)0.10$$

$$\text{pulley } B: \quad \sum M_{Bz} = M_B + (T_2 - T_1)0.16$$

These equations allow us to solve part (b) immediately, for they show that

$$T_1 - T_2 = \frac{20}{0.10} = \frac{M_B}{0.16}$$

$$M_B = 32 \text{ N-m}$$

The above leaves one equilibrium equation involving the two unknowns T_1 and T_2. Determining T_1 and T_2 individually requires another equation. It is known from the given information that slip is impending, so eq. (18) may be employed. The angles of wrap for the two pulleys are found from the given diagram to be

$$\beta_A = 160\left(\frac{\pi}{180}\right) = 2.793 \text{ rad} \qquad \beta_B = 200\left(\frac{\pi}{180}\right) = 3.491 \text{ rad}$$

The pulley at which slip is impending is not specified, so we shall consider both possibilities. The moment equilibrium equations indicate that $T_1 - T_2 > 0$, so T_1 corresponds to T_l. If slip at pulley A is impending, then eq. (18) for $\mu_s = 0.4$ gives

$$\frac{T_1}{T_2} = e^{(0.40)(2.793)}; \qquad T_1 = 3.056\, T_2$$

In the case where slip is impending at pulley B, eq. (18) indicates that

$$\frac{T_1}{T_2} = e^{(0.40)(3.491)}; \qquad T_1 = 4.040\, T_2$$

In order to determine which of these alternatives is appropriate, consider the situation prior to impending slip. When the torque $M_A = 0$, the tensions T_1 and T_2 are identical. Increasing M_A results in an increase of the ratio T_1/T_2. We have found that if this ratio attains the value 3.056, slip occurs at pulley A. Because T_1/T_2 is higher for impending motion at pulley B than it is at pulley A, we conclude that

slip is impending at pulley A

We now substitute $T_1/T_2 = 3.056$ into the moment equation for pulley A to find

$$T_1 - T_2 \equiv (3.056 - 1)\, T_2 = \frac{20}{0.10}$$

$$T_2 = 97.3 \text{ N} \qquad \text{for slip at pulley } A$$

HOMEWORK PROBLEMS

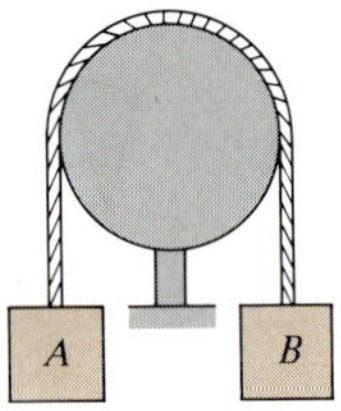

PROB. 5.93

5.93 The rope connecting the 3 kg block A with block B passes over a fixed cylinder for which the coefficient of static friction is 0.30. Determine the largest and smallest masses of block B for which static equilibrium is possible.

5.94 A ship is secured by wrapping a rope around the capstan. A dock worker can apply a 200 N force to counteract a 7 kN force by the ship. Determine the number of complete turns of the rope about the capstan required to keep the rope from slipping if the coefficient of static friction is 0.25.

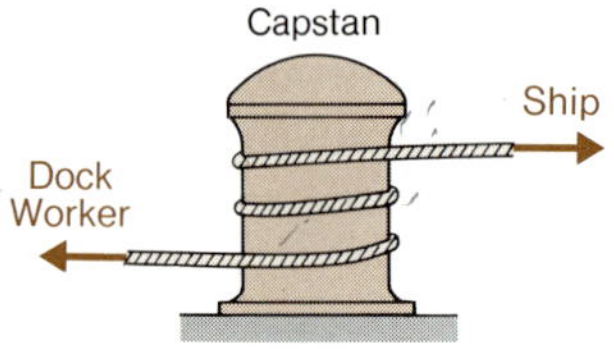

PROB. 5.94

PROB. 5.95

5.95 A 160 lb person stands on a 20 lb scaffold attached to one end of a rope. The rope passes over a fixed cylinder and the free end of the cable is grasped by the person. The coefficient of static friction between the cable and the cylinder is 0.50. Determine the force the person must apply to (a) lower the scaffold, (b) raise the scaffold.

5.96 A cable is wrapped around a fixed cylinder through an angle β. If the tension T on one end of the cable is constant, show that $T = (T_1 T_2)^{1/2}$, where T_1 and T_2 are the maximum and minimum values of the tension on the other end of the rope for which the cable will not slip.

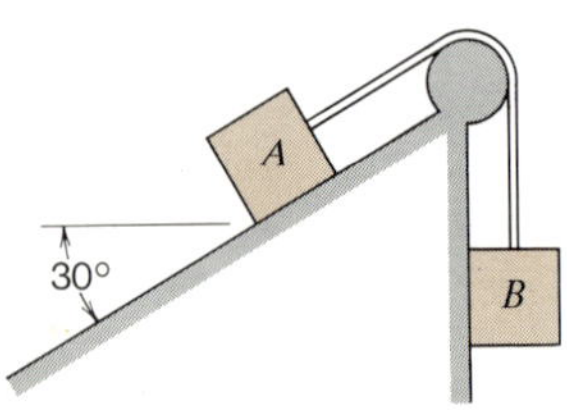

PROB. 5.97

5.97 The 5 kg block A is at rest on the 30° smooth incline. The coefficient of static friction between the cable and the fixed circular guide is 0.50. Determine the largest mass for block B for which the system is in static equilibrium. The incline is smooth.

5.98 Solve Problem 5.97 if the coefficient of static friction between block A and the incline is 0.25.

*5.99 Three rough pegs protrude from the 5 lb block A, which slides in smooth vertical guides. Determine the tension T at the lower end of the cable for which block A is in static equilibrium in a condition of impending motion. The coefficient of static friction between the cable and the pegs is 0.60.

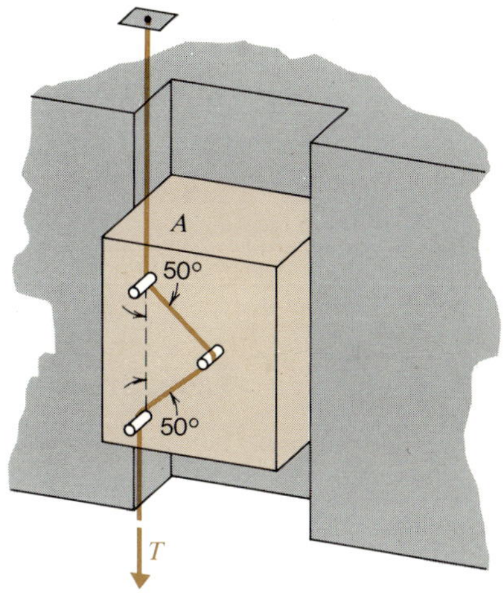

PROB. 5.99

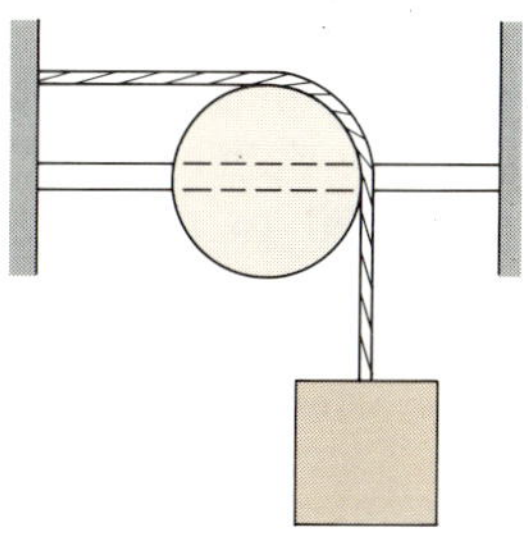

PROB. 5.100

5.100 The coefficient of static friction between the disk and the horizontal guide is μ_1, and the coefficient of static friction between the disk and the cable is 0.4. Determine the smallest value of μ_1 for which the suspended block of mass m will not move.

5.101 A 50 lb force is applied to a band brake that restrains flywheel A against the counterclockwise couple $\bar{M}_A$. The coefficients of static and kinetic friction between the belt of the brake and the flywheel are 0.40 and 0.30, respectively. Determine the couple $\bar{M}_A$ required to (a) hold the flywheel at rest in a condition of impending slip, (b) rotate the flywheel in the direction of the couple at a constant rate of rotation.

*5.102 Solve Problem 5.101 for the case where the couple $\bar{M}_A$ is clockwise.

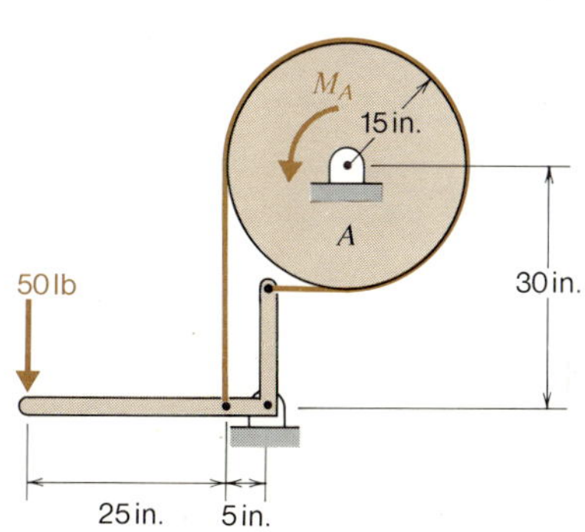

PROB. 5.101

5.103 and 5.104 Determine the maximum couple $\bar{M}_A$ that the motor may apply to pulley A without exceeding the maximum allowable belt tension of 2500 N. Also determine the corresponding couple $\bar{M}_B$ exerted on pulley B by its drive shaft. Where will slipping first occur? The coefficient of static friction between the belt and the pulleys is 0.35 and the pulleys are rotating at a constant rate.

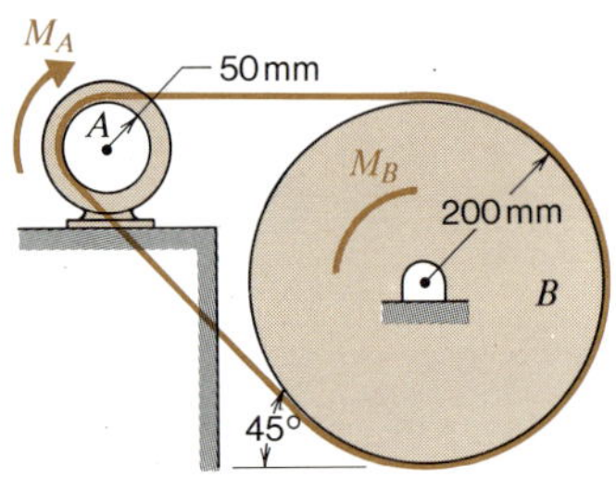

PROB. 5.103

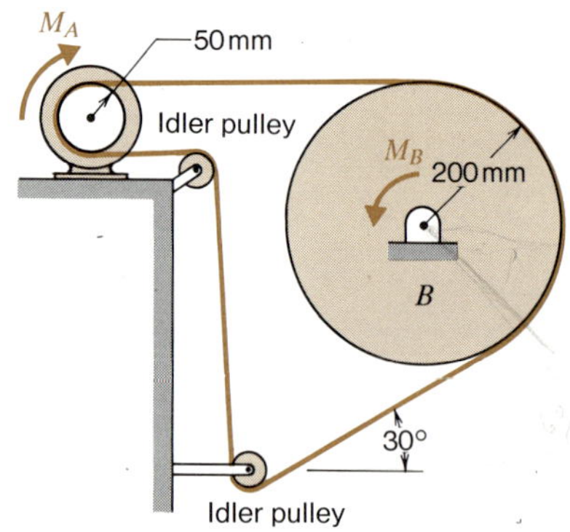

PROB. 5.104

***5.105** The belt drive shown is called a Rockwood mount. In it the 40 kg motor, whose center of mass coincides with its shaft, is mounted on a pivoted platform. Determine the maximum clockwise couple $\bar{M}_A$ the motor may apply to the drive pulley A, and also determine the corresponding couple $\bar{M}_B$ applied to pulley B by its shaft. The coefficient of friction between the belt and the pulley is 0.50. Neglect the mass of the platform.

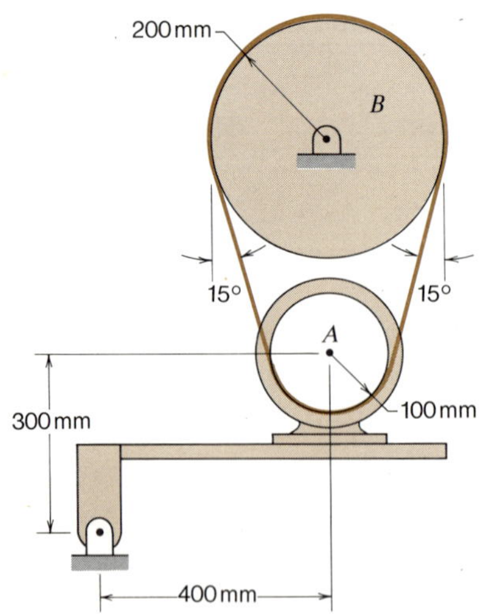

PROB. 5.105

***5.106** Solve Problem 5.105 in the case where the motor applies a counterclockwise couple to pulley A.

*5.107 A V-belt and its pulley are in contact along the nonparallel sides of the belt, as shown in the cross sectional view A–A. Contact is not made along a cylindrical surface, so eq. (18) is not valid. Isolate an element ds of the V-belt and follow similar steps to those employed in deriving eq. (18) to show that, for impending slip,

$$\frac{T_l}{T_s} = e^{\mu_s \beta/\sin(\alpha/2)}$$

Hint: Determine the radial component of the resultant of the normal forces exerted by the pulley on the sides of an element ds of the V-belt.

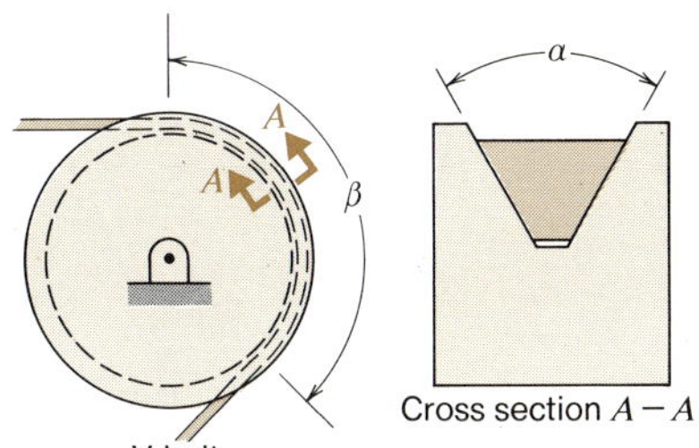

PROB. 5.107

4. Rolling Friction

Perhaps the simplest and most important of the basic machines is the wheel. However, it is not without reason that we have made consideration of the resistance to rolling of a wheel the last item in this chapter, for it is a very complicated phenomenon to analyze.

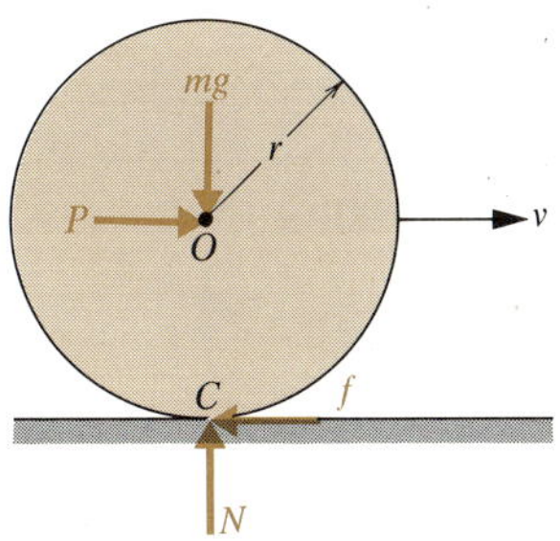

FIGURE 17

Consider a homogeneous disk of mass m that is rolling to the right along a flat horizontal surface at a constant speed v as a result of a horizontal force $\bar{P}$ that is applied to its center O. If the disk and the surface are rigid, the only reaction forces possible are friction and normal forces exerted by the surface on the disk at the point of contact C. These are shown in the free body diagram, Figure 17. The disk is rolling along at a constant speed, so we may apply the equations of static equilibrium. By summing moments about center O, we find that $\bar{f} = \bar{0}$. Force equilibrium in the horizontal direction then requires that $\bar{P} = \bar{O}$. This leads to a paradox, for it says that a wheel will roll along a horizontal plane at a constant speed without being pushed. However, we know that there must be resistance, for all freely rolling bodies slow down.

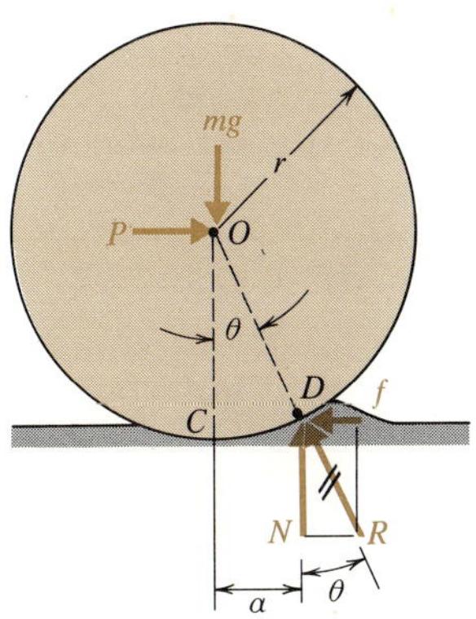

FIGURE 18

To resolve the paradox, we must abandon the model of rigid bodies. As we have mentioned on other occasions, a concentrated force is a modeling assumption, for in reality all materials deform. This results in the distribution of the contact force over a certain amount of surface area. Let us assume that the disk we are considering is much less deformable than the surface it is rolling on. Then, Figure 18 gives an exaggerated, but qualitatively accurate, picture of the deformation of the surface as the disk rolls to the right.

The deformation pattern of the surface moves ahead with the disk, in a fashion similar to the bow wave preceding a boat in water. Thus, the phenomenon of rolling resistance bears little resemblance to the concept of dry friction described by Coulomb's laws, for which the model of rigid bodies are sufficient.

As shown in Figure 18, the creation of an area of contact between the disk and the surface means that the resultant contact force $\bar{R}$ exerted between these bodies need no longer be applied at the lowest point C. Instead, it may be applied at point D, which is an angle θ ahead of point C. Now the equilibrium equation obtained by summing moments about the center of the disk requires that the line of action of the resultant force $\bar{R}$ be the line OD.

The vertical component of the reaction $\bar{R}$ in Figure 18 may be interpreted as the reaction force $\bar{N}$ normal to the undeformed surface of the floor, and the horizontal component of $\bar{R}$ may be interpreted as the friction force $\bar{f}$. Thus, we have

$$f = R \sin \theta \qquad N = R \cos \theta \tag{19}$$

It is clear from the equation for force equilibrium in the horizontal direction that the horizontal force $\bar{P}$ required to move the wheel forward at a constant velocity is no longer zero. We have resolved the paradox caused by assuming that the floor is rigid.

We noted earlier that Figure 18 gives an exaggerated picture of the deformation pattern. In most instances, the area of contact, and therefore the angle θ, is very small. It proves more convenient in such cases to replace the angle θ by the horizontal distance a locating the point D where the resultant force acts. It can be seen from Figure 18 that

$$\sin \theta = \frac{a}{r} \qquad \cos \theta = \left(1 - \frac{a^2}{r^2}\right)^{1/2}$$

Substituting these expressions into eqs. (19) and eliminating the magnitude R of the resultant force yields

$$f = \frac{\sin \theta}{\cos \theta} N = \frac{a/r}{(1 - a^2/r^2)^{1/2}} N$$

Then, because a/r is much smaller than unity, this reduces to

$$\boxed{f = \frac{a}{r} N} \tag{20}$$

The distance a is called the *coefficient of rolling resistance*, and it has units of length. Early investigators, such as Coulomb, regarded it as a parameter that is independent of the magnitude of the normal force $\bar{N}$, and also independent of the radius r. If this were true, a/r would be a factor of proportionality between the friction and normal forces, similar to the coefficients of dry friction. A further implication of eq. (20) in that case is that the rolling resistance on a wheel is inversely proportional to its radius.

Typical experimental values for the magnitude of the coefficient of rolling resistance a that have been reported are: for mild steel on mild steel, $a = 0.18$ to 0.38 mm: for hardened steel on hardened steel, $a = 0.005$ to 0.012 mm.

The foregoing analysis is contradicted by recent experimental and theoretical work. Although differing among each other, the investigators found that a is slightly dependent on the magnitude of $\bar{N}$ and strongly dependent on the radius r. In view of this disagreement, eq. (20) should be regarded as useful for discussion and preliminary analysis purposes only. We will therefore not pursue the application of this formulation to solve problems.

CHAPTER 6

DISTRIBUTED FORCES

OBJECTIVES

1. To evaluate the mass and the location of the center of mass of a composite body.
2. To locate the centroid of volumes, planar areas, curved surfaces, and curved lines.
3. To recognize situations where the task of finding the center of mass may be simplified by using centroidal concepts.
4. To locate the center of mass or the centroid by using integral calculus.
5. To evaluate the resultant of a given line load or pressure distribution, and to evaluate the reactions acting on a body that carries such a load.
6. To evaluate the hydrostatic pressure loading acting on a body that is immersed in a liquid.
7. To evaluate the resultant of a hydrostatic loading and the reactions acting on a body that carries such a load.

In this chapter we will focus on situations where forces are spread out over a region of surface area or volume of a body. Such forces are called *distributed forces.* The preceding chapters treated forces that act at a point, that is, *concentrated forces.* Such forces do not exist in reality, for any material will deform when a force is applied to it, resulting in the distribution of force over a small amount of contact area. Thus, when we consider a force to be concentrated, we are assuming that the contact area is sufficiently small to represent it as a point.

Our objective here will be to determine how to represent distributed forces by equivalent concentrated forces in situations where the region of application of the distributed force system is not negligible.

A. GRAVITATIONAL FORCE

A common distributed force system is the gravitational attraction exerted on a body by the planet earth. The *weight* of a body is the magnitude of the resultant of the gravity force on each particle of the body. The point of application of this resultant force is the *center of mass* or, equivalently, the *center of gravity.* Previously, we regarded the location of the center of mass as a given quantity. Now the task is to learn how to determine its location.

1. Center of Mass and Geometric Centroid of a Composite Body

Suppose that a body is fabricated from one substance. In that case the *mass density* ρ, whose dimensions are mass per unit volume, does not change from point to point within the body. Such a body is said to be *homogeneous.* The center of mass of a homogeneous body coincides with its *geometric centroid*, which we may think of as the average location of the volume of the body.

A useful analogy for understanding the difference between the center of mass and the centroid of a *nonhomogeneous* body is to think of the scores achieved by a group of students on an examination. One measure of the performance of the group is the median, which is the grade that was equaled or exceeded by half of the students. This is analogous to the centroid. Another measure of the group performance is the mean (i.e., average) grade, which is analogous to the center of mass. In general, these two grades will not be identical.

The fact that the center of mass and the centroid of a homogeneous body are coincident is important because it is standard practice to tabulate the volume and centroid location of a variety of basic geometrical shapes. Appendix B of this text is typical of such tabulations.

The mass of a homogeneous body having one of the tabulated basic shapes is found by multiplying the volume by the density,

$$\boxed{m = \rho \mathcal{V}} \tag{1}$$

The center of mass of such a body is found by referring to the tabulation to locate the centroid.

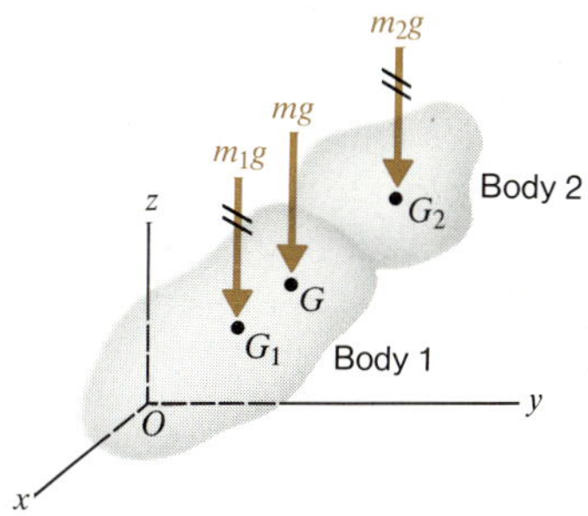

FIGURE 1

A slightly more complicated situation arises if the body in which we are interested can be pictured as being formed from two or more homogeneous bodies having fundamental shapes. Such an object is called a *composite body*. Figure 1 shows a composite body that is formed from two elements whose centers of mass G_1 and G_2 are known.

The total gravitational force, mg, acts at the center of mass G of the composite body. It is equivalent to the sum of the weight forces $m_1 g$ and $m_2 g$ acting at the centers of mass G_1 and G_2 of the individual elements. To quantify this result let us use the xyz coordinate system appearing in Figure 1. The first requirement for equivalence is that the two force systems have the same sum. After summing forces in the negative z direction and canceling the common factor g, we find that

$$m = m_1 + m_2 \tag{2}$$

This is the obvious fact that the total mass is the sum of the masses of the elements.

The other requirement for equivalence is that the two force systems exert the same moment about any point, which we choose to be the origin O in Figure 1. The components of the position vector from point O to each center of mass are the x, y, and z coordinates of that point, which we denote as $(x_G)_i$, $(y_G)_i$, and $(z_G)_i$. Hence, for moment equivalence we write

$$\sum \bar{M}_O = \bar{r}_G \times (-mg\bar{k}) \equiv (\bar{r}_G)_1 \times (-m_1 g\bar{k}) + (\bar{r}_G)_2 \times (-m_2 g\bar{k})$$

$$\begin{aligned}\sum \bar{M}_O = {} & (x_G\bar{i} + y_G\bar{j} + z_G\bar{k}) \times (-mg\bar{k}) \\ & \equiv [(x_G)_1\bar{i} + (y_G)_1\bar{j} + (z_G)_1\bar{k}] \times (-m_1 g\bar{k}) \\ & \quad + [(x_G)_2\bar{i} + (y_G)_2\bar{j} + (z_G)_2\bar{k}] \times (-m_2 g\bar{k})\end{aligned}$$

After the cross products are evaluated and corresponding components are equated, cancelation of the common factor g yields

$$m x_G = m_1 (x_G)_1 + m_2 (x_G)_2 \tag{3a}$$

$$m y_G = m_1 (y_G)_1 + m_2 (y_G)_2 \tag{3b}$$

We did not obtain an equation governing z_G in this case because gravity was parallel to the z axis. A relationship for z_G may be obtained by imagining that we rotate the composite body and reference frame in Figure 1, such that the gravitational forces act in the negative x direction. We would then find that

$$m z_G = m_1 (z_G)_1 + m_2 (z_G)_2 \tag{3c}$$

Equations (3) are said to be the *first moments of mass* with respect to the x, y, and z coordinates, respectively. We may solve these equations for the coordinates of the center of mass G of the composite body, after first determining the mass m from eq. (2).

The extension of eqs. (2) and (3) to the case of a composite body formed from more than two elements is obtained by adding the contribution of each element to the mass equation, and to the first moment of mass equation.

A standard procedure used to account for a vacant region in a body is to consider the void to be one of the elements of the composite shape. We then subtract the contributions of this element to the mass and first moment of mass in the calculations. For example, if a body whose mass is initially m_1 has a hole drilled in it, so that a mass m_2 is removed, eq. (3a) yields

$$m x_G = m_1 (x_G)_1 - m_2 (x_G)_2$$

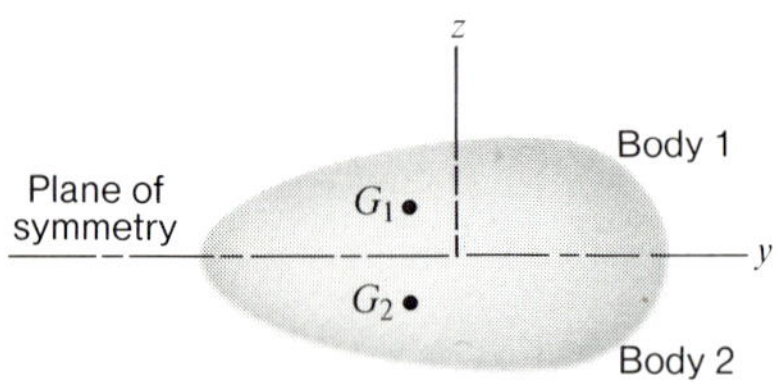

FIGURE 2

In general, it is necessary for us to solve each of eqs. (3) for the coordinates of point G. However, there are occasions when we may use the property of symmetry to reduce the number of computations. (We use the term *symmetry* to describe the mass distribution, not merely the shape of the body.) Figure 2 depicts a cross section of a body that is symmetric with respect to the xy plane.

We may consider this body to be a composite of the upper half and the mirror-image lower half. By virtue of the symmetry property, the coordinates of the centers of mass must be related by $(x_G)_2 = (x_G)_1$, $(y_G)_2 = (y_G)_1$, $(z_G)_2 = -(z_G)_1$. The last relation when substituted into eq. (3c) yields $z_G = 0$. Our choice of labels for the plane of symmetry was arbitrary, so we may conclude that

> The center of mass of a body having a plane of symmetry is situated somewhere on that plane.

Whenever a body is not homogeneous, first moments of mass are required for the determination of the location of the center of mass. A slight simplification is obtained for homogeneous bodies by using eq. (1) to factor out the density, ρ, from eqs. (2) and (3).

$$\boxed{\mathscr{V} = \mathscr{V}_1 + \mathscr{V}_2} \tag{4}$$

Therefore eqs. (3) become

$$\boxed{\begin{aligned} \mathscr{V} x_G &= \mathscr{V}_1 (x_G)_1 + \mathscr{V}_2 (x_G)_2 \\ \mathscr{V} y_G &= \mathscr{V}_1 (y_G)_1 + \mathscr{V}_2 (y_G)_2 \\ \mathscr{V} z_G &= \mathscr{V}_1 (z_G)_1 + \mathscr{V}_2 (z_G)_2 \end{aligned}} \tag{5}$$

Equations (5) are the *first moments of volume*. The point G whose coordinates they describe is the *geometric centroid* of the composite body, regardless of whether or not the body is homogeneous. When the body is homogeneous, this point is also the center of mass.

The modifications required in these equations to treat composite bodies formed from more than two basic shapes or containing holes are precisely the same as those discussed earlier.

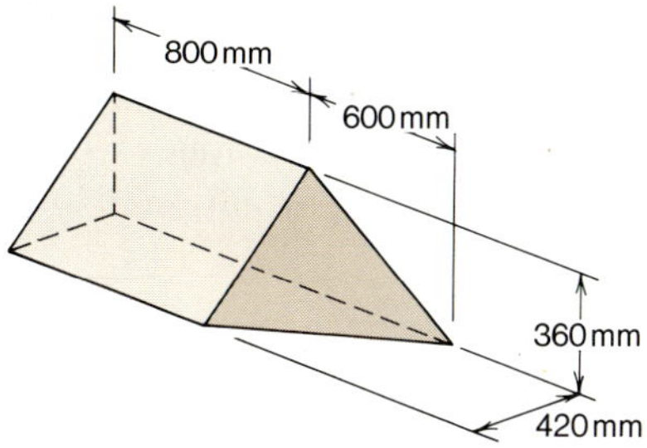

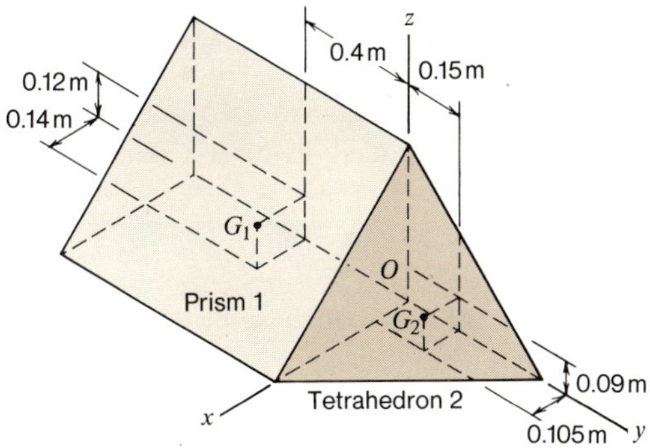

EXAMPLE 1

The object is made of concrete. Determine its mass and the location of its center of mass.

Solution

We see from Appendix B that this body is a combination of a triangular prism, which we label 1, and an orthogonal tetrahedron, which we label 2. We draw a sketch of these elements in order to detail the dimensions locating their centers of mass, as obtained from the Appendix. The sketch also shows the xyz coordinate system we have chosen. The axes align with the directions for the dimensions, and the origin is located at a convenient point.

The body is homogeneous, so we shall use first moments of volume, followed by eq. (1) to find the mass. Appendix D lists density in terms of the density of water, from which we find

$$\rho_{\text{conc}} = 2.4\rho_{\text{water}} = 2.4(10^3)\ \text{kg/m}^3$$

The next step is to evaluate the volume and the coordinates of the center of mass of each portion. Then we shall employ eqs. (4) and (5).

Prism 1:

$$\mathscr{V}_1 = \tfrac{1}{2}(0.8)(0.42)(0.36) = 0.06048\ \text{m}^3$$

$$(x_G)_1, (y_G)_1, (z_G)_1 = 0.14, -0.4, 0.12$$

Tetrahedron 2:

$$\mathscr{V}_2 = \tfrac{1}{6}(0.6)(0.42)(0.36) = 0.01512\ \text{m}^3$$

$$(x_G)_2, (y_G)_2, (z_G)_2 = 0.105, 0.15, 0.09$$

Composite:

$$\mathscr{V} = \mathscr{V}_1 + \mathscr{V}_2 = 0.07560\ \text{m}^3$$

$$0.07560(x_G) = 0.06048(0.14) + 0.01512(0.105)$$

$$0.07560(y_G) = 0.06048(-0.40) + 0.01512(0.15)$$

$$0.07560(z_G) = 0.06048(0.12) + 0.01512(0.09)$$

$$x_G, y_G, z_G = 0.1330, -0.290, 0.1140\ \text{meters}$$

Finally, we evaluate the mass.

$$m = \rho_{\text{conc}}\mathscr{V} = 2.4(10^3)(0.07560) = 181.4\ \text{kg}$$

Additional Remarks

A frequent source of error in these problems is to ignore the signs of the coordinates of the various centers of mass. Such difficulties can be avoided by drawing a sketch carefully and labeling the important points with equal care.

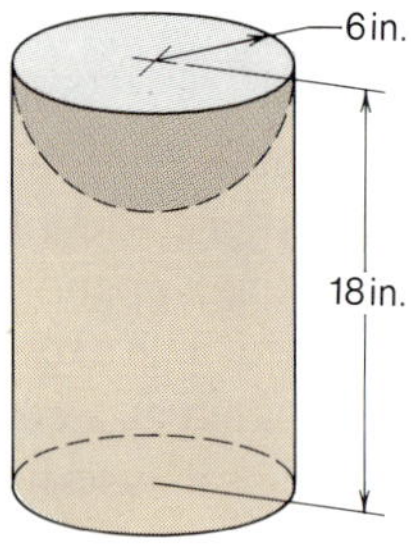

EXAMPLE 2

A hemispherical cavity at the top of a steel cylinder is filled with aluminum, as shown. Determine the mass and location of the center of mass of this body.

Solution

There are a number of ways in which we may decompose this object into a set of elements. One of the more direct approaches is to say that, in effect, steel is removed within the cavity and replaced by aluminum. In other words, a hemisphere whose density is $\rho_{\text{steel}} - \rho_{\text{al}}$ is removed from the complete steel cylinder. Thus

$$m = \rho_{\text{steel}}\mathscr{V}_{\text{cyl}} - (\rho_{\text{steel}} - \rho_{\text{alum}})\mathscr{V}_{\text{hemis}}$$

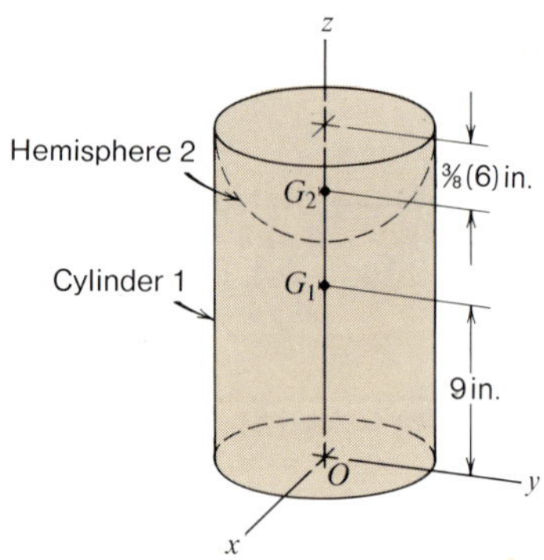

We use Appendix B to determine the center of mass of these elements and show the location of these points in a sketch.

The coordinate system we have chosen has the z axis aligned with the centerline of the cylinder, with the origin located on the bottom face of the cylinder. The body is symmetric about the z axis, so the center of mass must be situated on this axis. Thus, by symmetry

$$x_G = y_G = 0$$

The body is composed of two different materials, so we must evaluate first moments of mass. The densities of steel and aluminum are calculated from Appendix D.

$$\rho_{\text{steel}} = 7.84\rho_{\text{water}} = 7.84\,\frac{62.4\text{ lb/ft}^3}{g} = 15.207\text{ lb-s}^2/\text{ft}^4$$

$$\rho_{\text{alum}} = 2.69\rho_{\text{water}} = 2.69\,\frac{62.4\text{ lb/ft}^3}{g} = 5.218\text{ lb-s}^2/\text{ft}^4$$

We now tabulate the mass and the z coordinate of the center of mass of each basic body. Remember that the units of density and volume must be consistent.

Cylinder 1:

$$m_1 = \rho_{\text{steel}}\mathscr{V}_1 = 15.207[\pi(\tfrac{6}{12})^2(\tfrac{18}{12})] = 17.915\text{ lb-s}^2/\text{ft}$$

$$(z_G)_1 = \tfrac{9}{12} = 0.75\text{ ft}$$

Hemisphere 2:

$$m_2 = (\rho_{\text{steel}} - \rho_{\text{alum}})\mathscr{V}_2 = (15.207 - 5.218)\left[\tfrac{2}{3}\pi\left(\tfrac{6}{12}\right)^3\right] = 2.615 \text{ lb-s}^2/\text{ft}$$

$$(z_G)_2 = \tfrac{18}{12} - \tfrac{3}{8}\left(\tfrac{6}{12}\right) = 1.3125 \text{ ft}$$

Because the hemisphere is removed from the cylinder, we subtract the properties of the hemisphere when we form the composite. Thus,

$$m = m_1 - m_2 = 15.30 \text{ lb-s}^2/\text{ft}$$

$$15.30(z_G) = 17.915(0.75) - 2.615(1.3125)$$

$$z_G = 0.654 \text{ ft}$$

HOMEWORK PROBLEMS

6.1–6.12 Evaluate the mass and determine the location of the center of mass of a body having the shape and composition specified below.

6.1 Steel

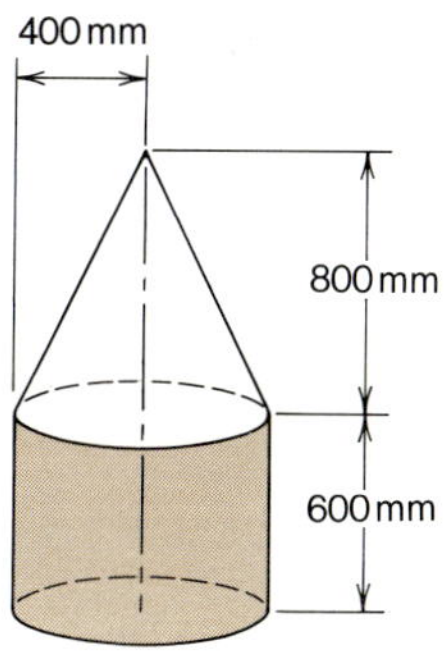

PROB. 6.1

6.2 Aluminum

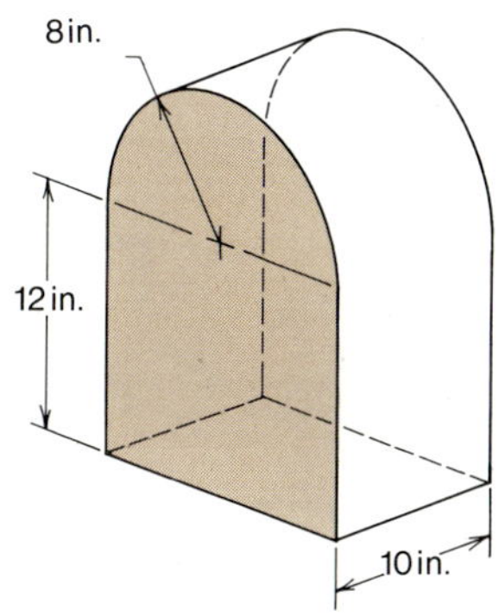

PROB. 6.2

6.3 Lead

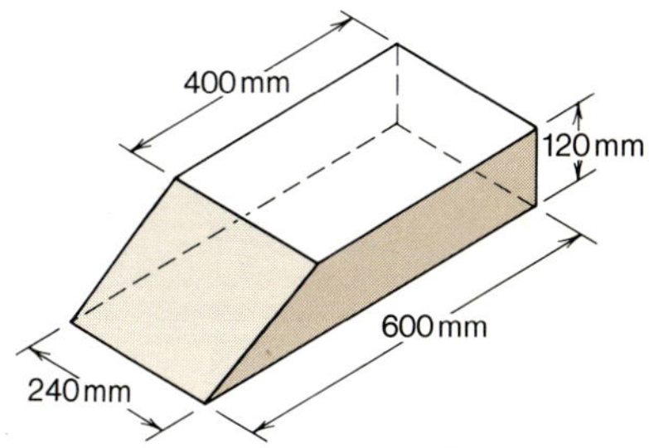

PROB. 6.3

6.4 Glass

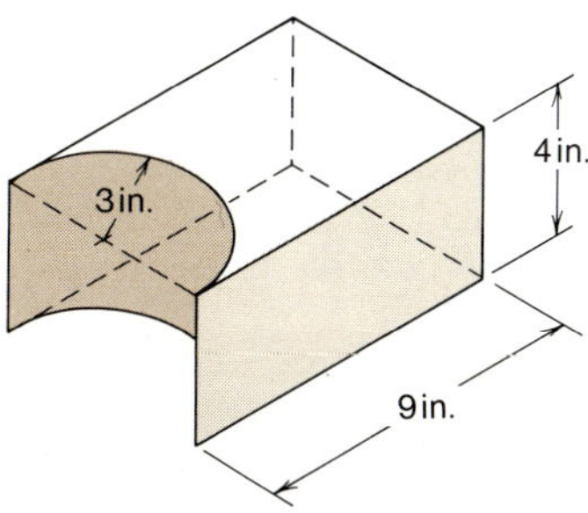

PROB. 6.4

6.5 Concrete

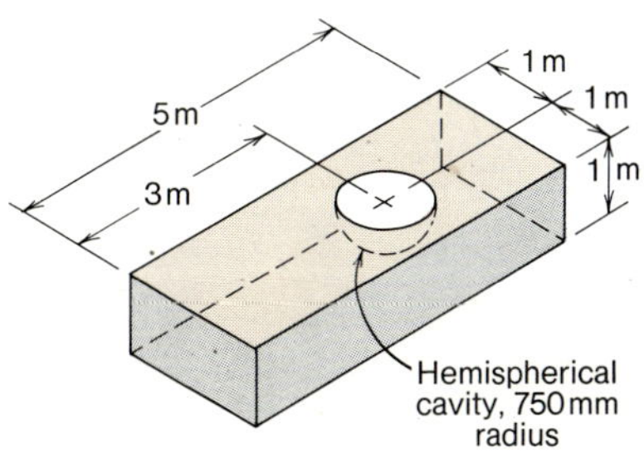

PROB. 6.5

6.6 Cast iron

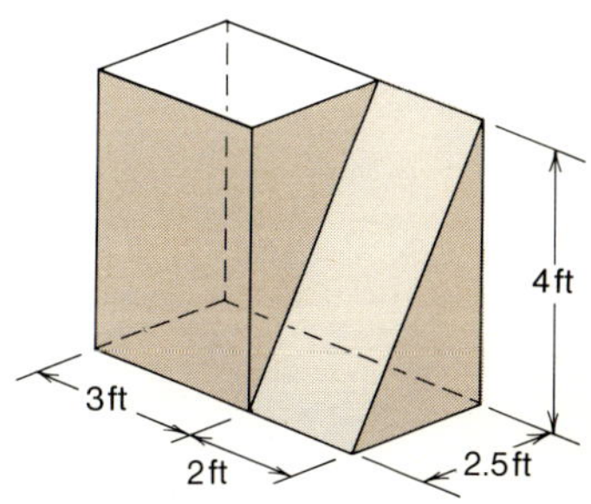

PROB. 6.6

6.7 Steel

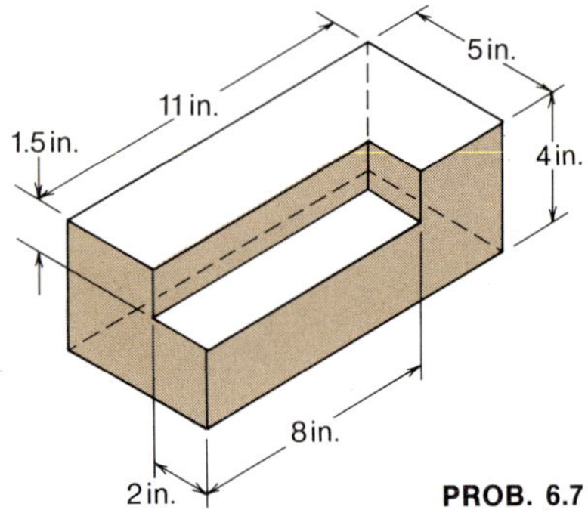

PROB. 6.7

6.8 Glass

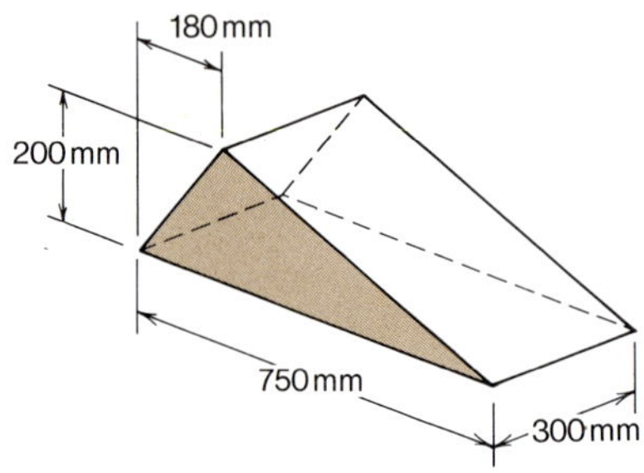

PROB. 6.8

6.9 Copper

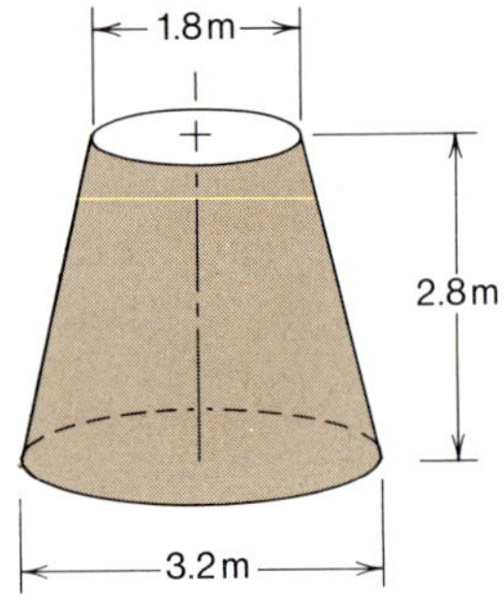

PROB. 6.9

6.10 Hard oak wood

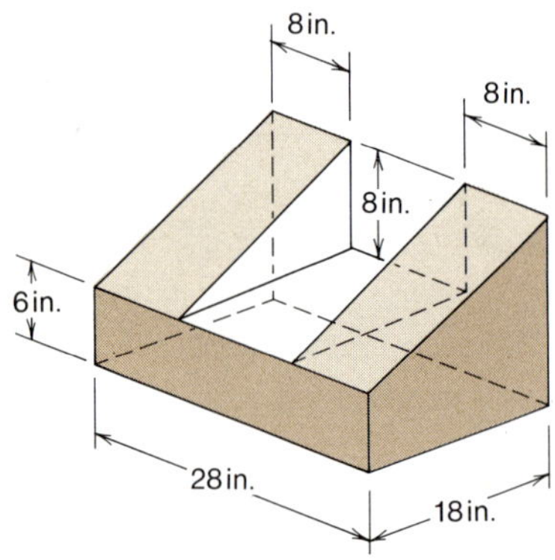

PROB. 6.10

6.11 Arbitrary material (density $= \rho$)

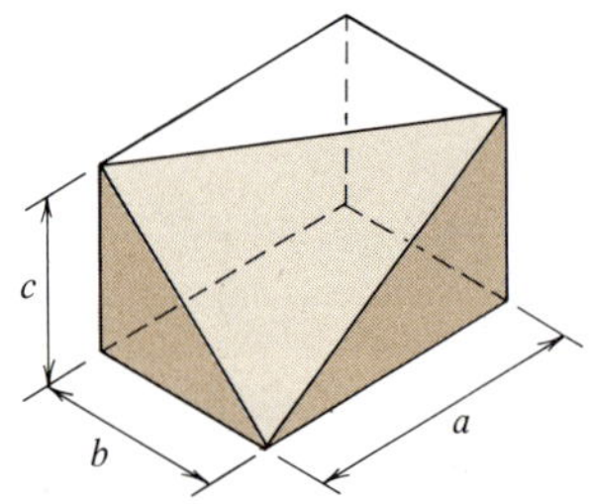

PROB. 6.11

***6.12** Concrete

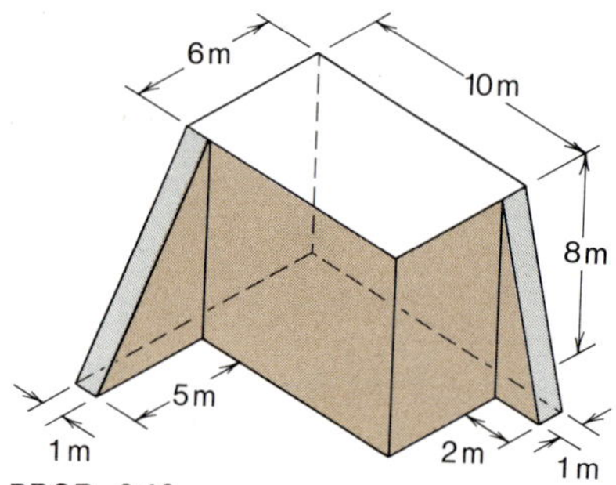

PROB. 6.12

6.13 and 6.14 The aluminum disk shown has a steel inset whose faces are flush with the faces of the disk. The thickness of the disk is as indicated. Determine the location of the center of mass of this body.

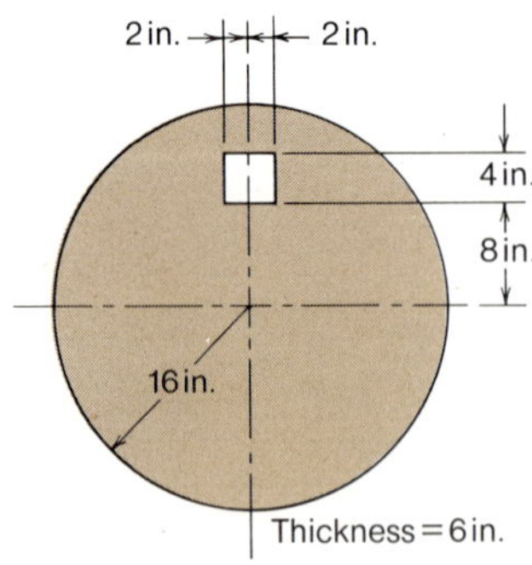

PROB. 6.13

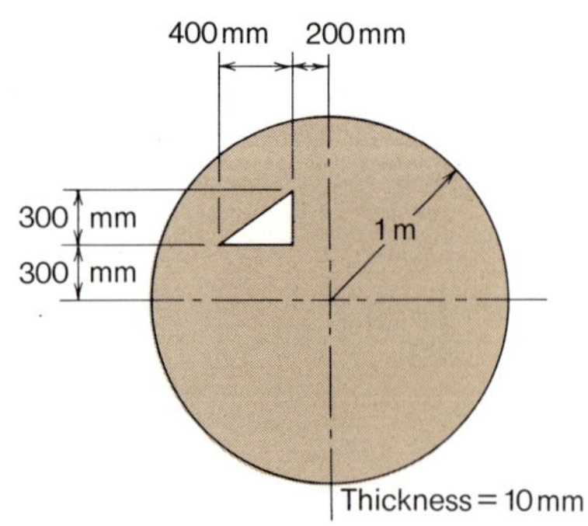

PROB. 6.14

6.15 Determine the mass and the location of the center of mass of the body shown in Problem 6.1 if the conical cap is made of steel and the cylindrical bottom is made of lead.

6.16 Determine the mass and the location of the center of mass of the concrete body shown in Problem 6.5 if the hemispherical cavity is filled with mercury.

6.17 Determine the mass and the location of the center of mass of the body shown in Problem 6.6 if the triangular prismatic element is composed of cast iron and the rectangular parallelepiped portion is composed of aluminum.

***6.18** Determine the mass and the location of the center of mass of the body shown in Problem 6.12 if the block is composed of concrete and the triangular prismatic elements are composed of steel.

***6.19 and 6.20** · The cylindrical steel rod fits tightly in a hole in the aluminum block. The end of the rod that is not visible in the sketch is flush with the corresponding face of the block. Determine the mass and the location of the center of mass of the body.

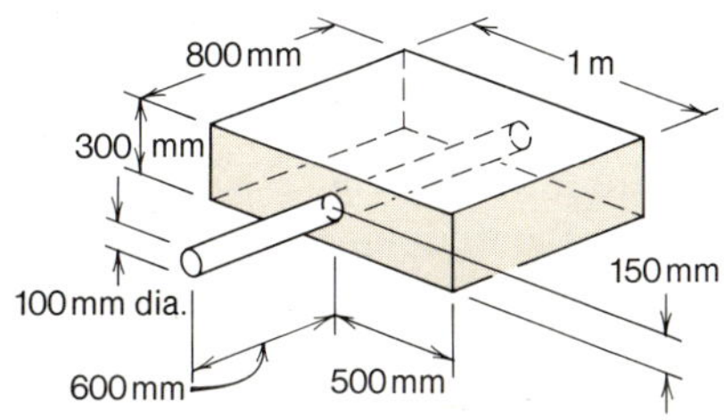

PROB. 6.19

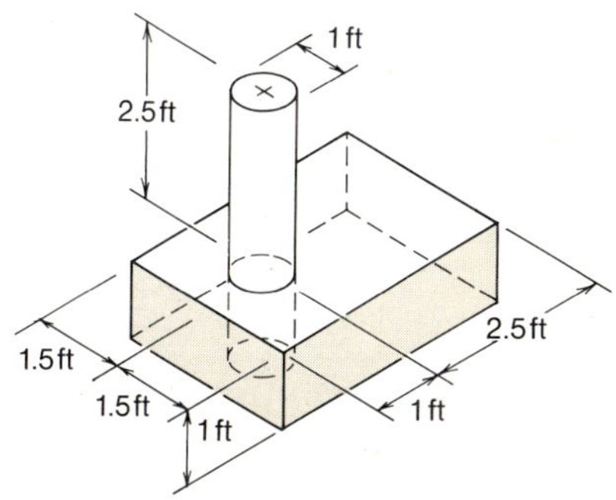

PROB. 6.20

---

2. Centroids of Composite Surfaces and Lines

A type of element that occurs frequently is one whose thickness is small compared to its other dimensions. It is permissible in this case to consider the mass to be concentrated along the surface of the body. Such a body is said to be a *shell*—a familiar example is an egg shell. A shell that is flat is usually called a *plate*.

A typical composite shell, formed from two elements having surface area $\mathscr{A}_1$ and $\mathscr{A}_2$, is depicted in Figure 3. At any location on the shell, we measure the thickness h perpendicular to the local tangent plane. If h is constant, and if the shell is homogeneous, then the mass per unit surface area of the shell is the constant value ρh. Equation (1), for mass, then gives for each element of the composite body

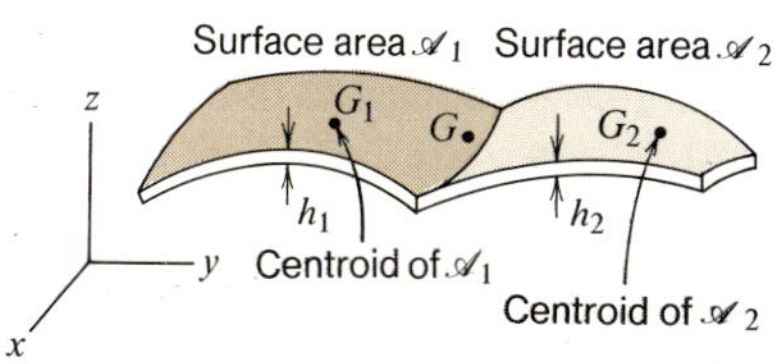

FIGURE 3

$$\boxed{m_i = \rho h \mathscr{A}_i} \tag{6}$$

For the case of constant thickness, eqs. (2) and (3) reduce to

$$\mathscr{A} = \mathscr{A}_1 + \mathscr{A}_2 \tag{7}$$

$$\begin{aligned} \mathscr{A} x_G &= \mathscr{A}_1(x_G)_1 + \mathscr{A}_2(x_G)_2 \\ \mathscr{A} y_G &= \mathscr{A}_1(y_G)_1 + \mathscr{A}_2(y_G)_2 \\ \mathscr{A} z_G &= \mathscr{A}_1(z_G)_1 + \mathscr{A}_2(z_G)_2 \end{aligned} \tag{8}$$

Equations (8) are the *first moments of surface area*. Their solution gives the location of the centroid G of the composite surface. This point is coincident with the location of the center of mass of a homogeneous, constant thickness shell whose shape is the composite surface. If the shell is not flat, so that the areas are not coplanar, the centroid most likely will not be situated on the surface of the composite. Rather, point G will be a theoretically determined point in space.

Suppose that a body has a constant thickness h (not necessarily small) in a direction perpendicular to two parallel flat-end surfaces. In geometry, this type of body is called a *cylinder*. First moments of area are also useful for locating the center of mass of a homogeneous cylinder.

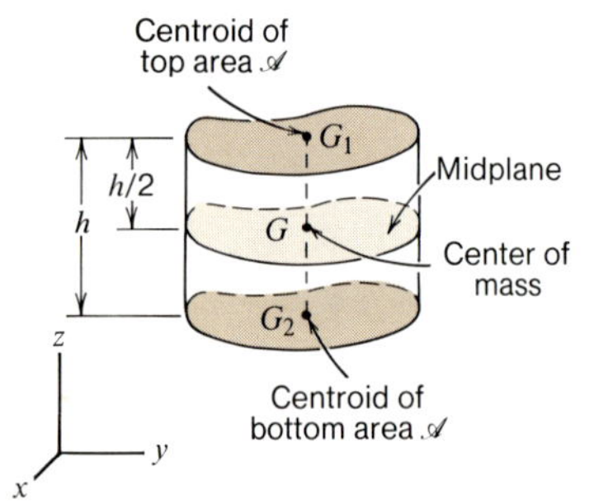

FIGURE 4

Figure 4 shows an arbitrary cylinder, for which the xy plane is parallel to the end surfaces whose area is $\mathscr{A}$. The cylinder is symmetrical with respect to the midplane that cuts the cylinder in half, so the center of mass G must be situated along this midplane. This defines the value of z_G. The values of x_G and y_G may be determined by noting that point G is also situated along the straight line, parallel to the z axis, that intersects the centroids G_1 and G_2 of the upper and lower end faces. Thus, points G, G_1, and G_2 must have the same x and y coordinates. It follows that the task of locating the center of mass of a cylinder reduces to an evaluation of the location of the centroid of one of the parallel end surfaces.

Another object that is frequently encountered in engineering systems is a *slender bar*. A pictorial description of a bar requires that a line, called the *axis* of the bar, be specified first. Then, at each point along the axis, a cross section having area $\mathscr{A}_c$ is constructed perpendicular to the axis.

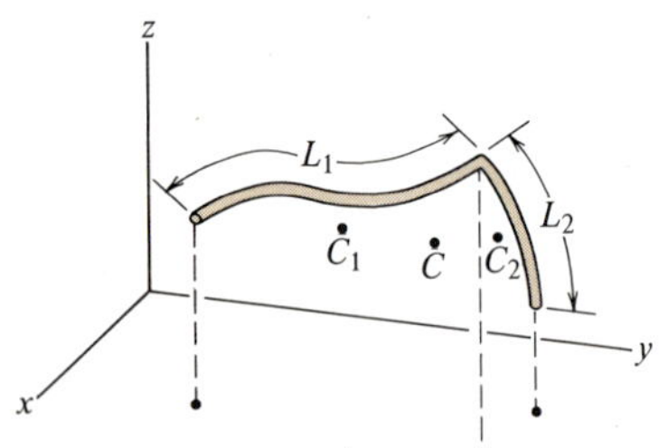

FIGURE 5

The bar's axis need *not* be a straight line. The property of slenderness means that the dimensions defining the cross-sectional area $\mathscr{A}_c$ are small compared to the arc length of the axis. It is permissible then to consider the mass of the bar to be concentrated along the axis. A typical composite bar is depicted in Figure 5.

When the cross section has a constant area $\mathscr{A}_c$ and the composite bar is homogeneous, the mass per unit arc length along the axis of the bar is the constant value $\rho\mathscr{A}_c$. In this case, eq. (1) for mass gives, for each element of the composite bar,

$$m_i = \rho\mathscr{A}_c L_i \tag{9}$$

where L_i is the arc length of the axis of each portion of the composite bar. For the case of constant cross-sectional area, $\mathscr{A}_c$, eqs. (2) and (3) reduce to

$$L = L_1 + L_2 \tag{10}$$

$$\begin{aligned} Lx_c &= L_1(x_c)_1 + L_2(x_c)_2 \\ Ly_c &= L_1(y_c)_1 + L_2(y_c)_2 \\ Lz_c &= L_1(z_c)_1 + L_2(z_c)_2 \end{aligned} \tag{11}$$

Equations (11) are the *first moments of arc length*; they define the centroid of the curved line that forms the axis of the slender bar. When the bar is homogeneous with a constant cross-sectional area $\mathscr{A}_c$, then this centroid coincides with the center of mass of the bar. It will be noted from Figure 5 that the centroid will probably not be situated on the bar.

Appendix C in this text gives a tabulation of the centroids of several fundamental planar areas and curves. This tabulation features a number of shapes whose three-dimensional analog does not appear in Appendix C. Thus, the concept of first moments of area or arc length gives us increased versatility in locating the center of mass of a composite body.

EXAMPLE 3

Determine the location of the centroid of the trapezoidal area.

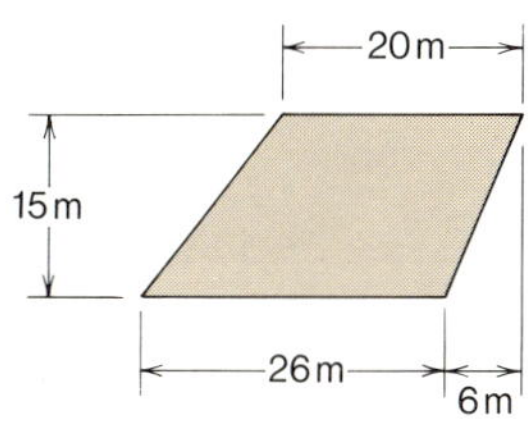

Solution

A trapezoid may be treated as a composite body. One way of doing so is to form two scalene triangles by drawing a line from any corner to the diagonally opposite corner. However, we note from Appendix C that the centroid of a triangle is always at a distance of 1/3 of the height from the base. The consequence of this fact is that the values of x_G and y_G for a right triangle are very easy to obtain. Thus, we shall consider the trapezoid to be a composite of two right triangles and a rectangle. The sketch shows the location of the centroids of these elements, as well as the *xyz* coordinate system we shall use.

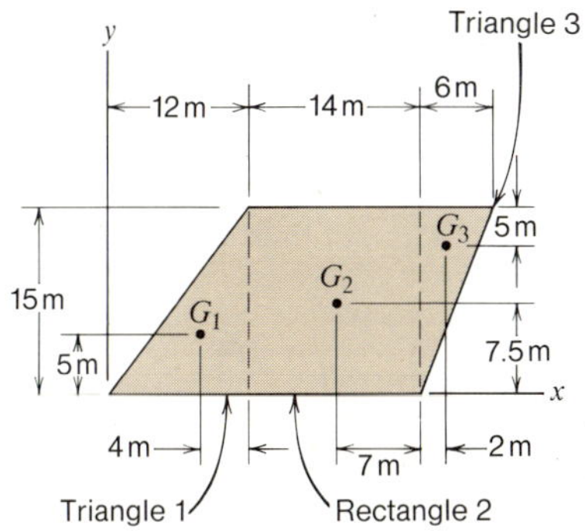

We now evaluate the area and the location of the centroid of each portion of the composite.

Triangle 1:

$$\mathscr{A}_1 = \tfrac{1}{2}(12)(15) = 90 \text{ m}^2, \qquad (x_G)_1 = 8 \text{ m}, \qquad (y_G)_1 = 5 \text{ m}$$

Rectangle 2:

$$\mathscr{A}_2 = (14)(15) = 210 \text{ m}^2, \qquad (x_G)_2 = 19 \text{ m}, \qquad (y_G)_2 = 7.5 \text{ m}$$

Triangle 3:

$$\mathscr{A}_3 = \tfrac{1}{2}(6)(15) = 45 \text{ m}^2, \qquad (x_G)_3 = 28 \text{ m}, \qquad (y_G)_3 = 10 \text{ m}$$

The composite is the sum of these elements, so

$$\mathscr{A} = \mathscr{A}_1 + \mathscr{A}_2 + \mathscr{A}_3 = 345 \text{ m}^2$$

$$345x_G = \mathscr{A}_1(x_G)_1 + \mathscr{A}_2(x_G)_2 + \mathscr{A}_3(x_G)_3 = 90(8) + 210(19) + 45(28)$$

$$345y_G = \mathscr{A}_1(y_G)_1 + \mathscr{A}_2(y_G)_2 + \mathscr{A}_3(y_G)_3 = 90(5) + 210(7.5) + 45(10)$$

$$x_G = 17.30 \text{ m}, \qquad y_G = 7.17 \text{ m}$$

EXAMPLE 4

The copper wire has a circular cross section whose radius is 5 mm. Determine the mass and the location of the center of mass of the wire.

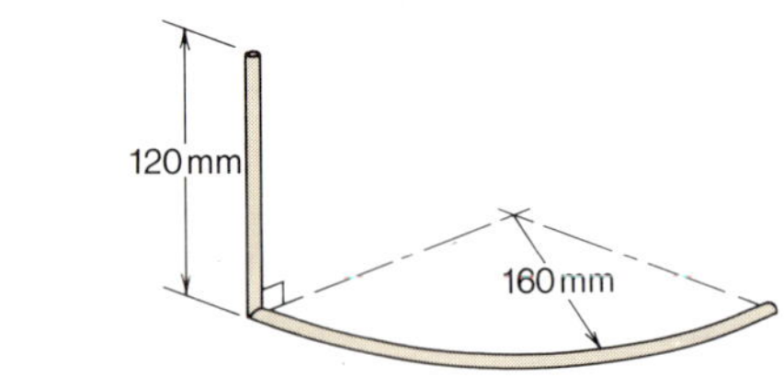

Solution

The cross-sectional diameter of the wire is small compared to its length, so we may consider the wire to be a curved line. The wire is a combination of a quarter circular arc 1 and a straight segment 2. The listing for a quarter circular arc in Appendix C suggests that we choose the *xyz* coordinate system pictured in the sketch.

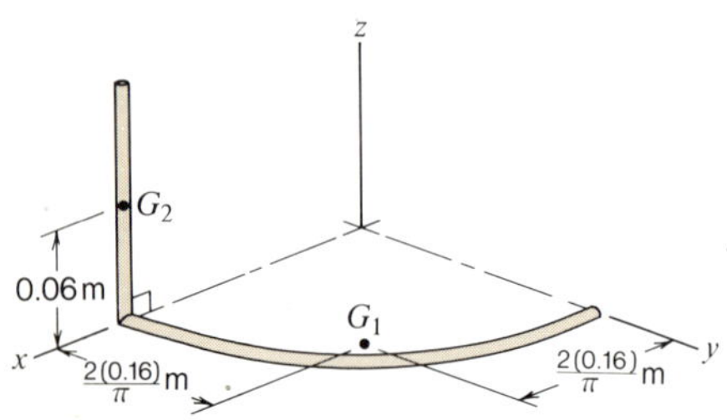

The mass per unit length is

$$\rho_{\text{copper}}\mathscr{A}_c = [8.91(10^3) \text{ kg/m}^3]\pi(0.005 \text{ m})^2 = 0.6998 \text{ kg/m}$$

We shall locate the center of mass by using first moments of arc length and then find the mass by using the total arc length. Specifically,

Circular arc 1:

$$L_1 = 0.16\left(\frac{\pi}{2}\right) = 0.2513 \text{ m}$$

$$(x_G)_1, (y_G)_1, (z_G)_1 = \frac{2(0.16)}{\pi}, \frac{2(0.16)}{\pi}, 0$$

Straight line 2:

$$L_2 = 0.12 \text{ m}; \qquad (x_G)_2, (y_G)_2, (z_G)_2 = 0.16, 0, 0.16 \text{ m}$$

Composite:

$$L = L_1 + L_2 = 0.3713 \text{ m}$$

$$0.3713x_G = L_1(x_G)_1 + L_2(x_G)_2 = 0.2513\left(\frac{0.32}{\pi}\right) + 0.12(0.16)$$

$$0.3713y_G = L_1(y_G)_1 + L_2(y_G)_2 = 0.2513\left(\frac{0.32}{\pi}\right) + 0.12(0)$$

$$0.3713z_G = L_1(z_G)_1 + L_2(z_G)_2 = 0.2513(0) + 0.12(0.06)$$

$$x_G = 0.1206 \text{ m}, \qquad y_G = 0.0689 \text{ m}, \qquad z_G = 0.0194 \text{ m}$$

$$m = \rho_{\text{copper}}\mathscr{A}_c L = (0.6998)(0.3713) = 0.2598 \text{ kg}$$

EXAMPLE 5

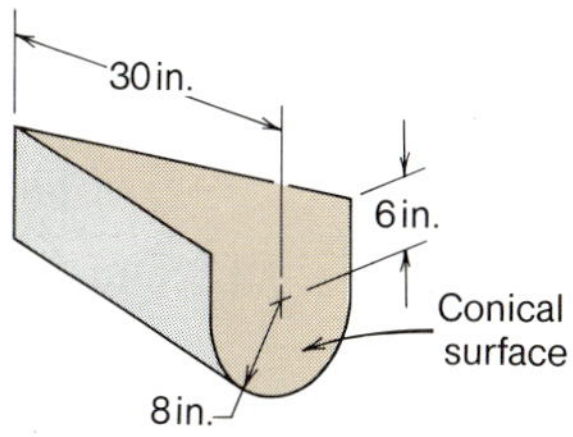

A 1/16 inch thick sheet of aluminum is bent into the shape shown. Determine the mass and the location of the center of mass of this body.

Solution

The thickness is quite small compared to the other dimensions of the body. Therefore we shall consider this composite to be a thin-walled body composed of a semiconical shell (body 1) and two rectangular plates (bodies 2 and 3). Our sketch shows side and top views of the shell.

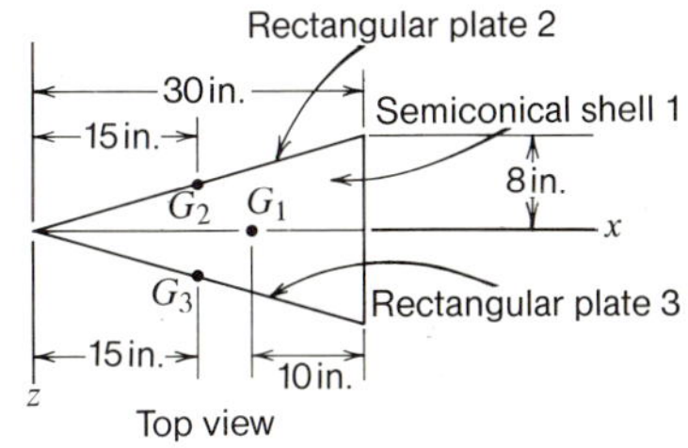

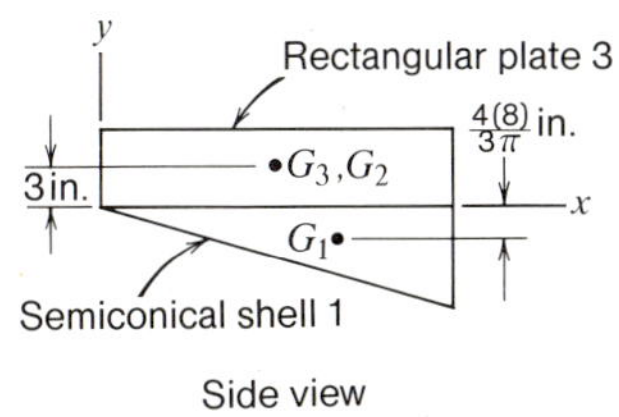

The xyz coordinate system we have chosen corresponds to the directions for the dimensions locating point G_1 as tabulated in Appendix B. The coordinates of points G_2 and G_3 are found by noting that the centroid of a rectangle is at its midpoint. The body is symmetric with respect to the xy plane, so

$$z_G = 0$$

The mass per unit surface area is found with the aid of Appendix D. Because the mass unit is a derived unit in U.S. Customary units, we convert the density to a consistent set of units, as follows.

$$\rho_{\text{alum}} h = \left[2.69\left(\frac{62.4\ \text{lb/ft}^3}{32.17\ \text{ft/s}^2}\right)\left(\frac{1\ \text{ft}}{12\ \text{in.}}\right)^4\right]\left(\frac{1}{16}\ \text{in.}\right)$$
$$= 1.5727(10^{-5})\ \text{lb-s}^2/\text{in.}^3$$

We will determine the values of x_G and y_G by using first moments of area and then compute the mass from the total surface area. Thus,

Semiconical shell 1:

$$\mathscr{A}_1 = \frac{\pi}{2}(8)[(8)^2 + (30)^2]^{1/2} = 390.17\ \text{in.}^2$$

$$(x_G)_1, (y_G)_1 = 20, -\frac{32}{3\pi} \text{ in.}$$

Rectangular plates 2 and 3:

$$\mathscr{A}_2 = \mathscr{A}_3 = 6[(30)^2 + (8)^2]^{1/2} = 186.29 \text{ in.}^2$$

$$(x_G)_2, (y_G)_2 = (x_G)_3, (y_G)_3 = 15, 3 \text{ in.}$$

Composite:

$$\mathscr{A} = \mathscr{A}_1 + \mathscr{A}_2 + \mathscr{A}_3 = 762.8 \text{ in.}^2$$

$$762.8x_G = \mathscr{A}_1(x_G)_1 + \mathscr{A}_2(x_G)_2 + \mathscr{A}_3(x_G)_3 = 390.2(20) + 2(186.3)(15)$$

$$762.8y_G = \mathscr{A}_1(y_G)_1 + \mathscr{A}_2(y_G)_2 + \mathscr{A}_3(y_G)_3 = 390.2\left(-\frac{32}{3\pi}\right) + 2(186.3)(3)$$

$$x_G = 17.56 \text{ in.}, \qquad y_G = -0.271 \text{ in.}$$

$$m = \rho_{\text{alum}} h\mathscr{A} = 1.5727(10^{-5})(762.8) = 0.01200 \text{ lb-s}^2/\text{in.}$$

HOMEWORK PROBLEMS

6.21–6.23 Determine the location of the centroid of the area shown.

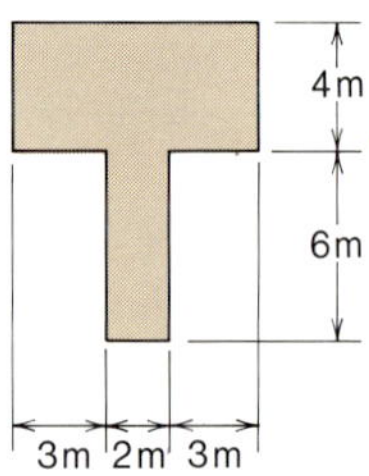

PROB. 6.21

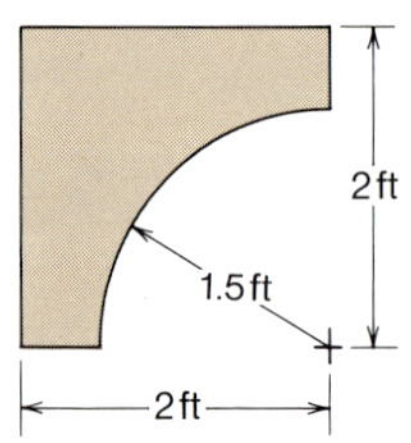

PROB. 6.22

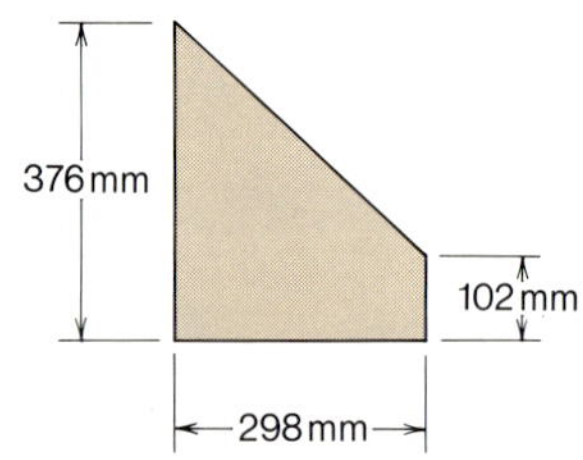

PROB. 6.23

6.24 and 6.25 The cross section of a standard beam is as shown. Determine the area and the location of the centroid of the cross section. It is permissible to neglect the fillets (i.e., the rounding of interior and exterior corners).

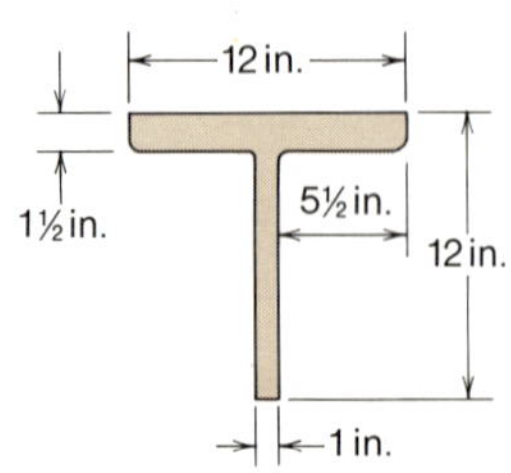

PROB. 6.24

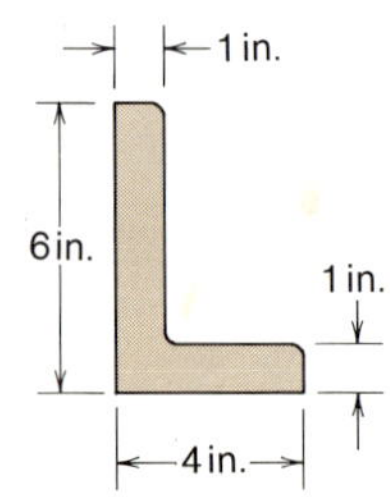

PROB. 6.25

6.26–6.29 Determine the mass and the center of mass of the homogeneous plate described below.

6.26 Steel, 20 mm thick

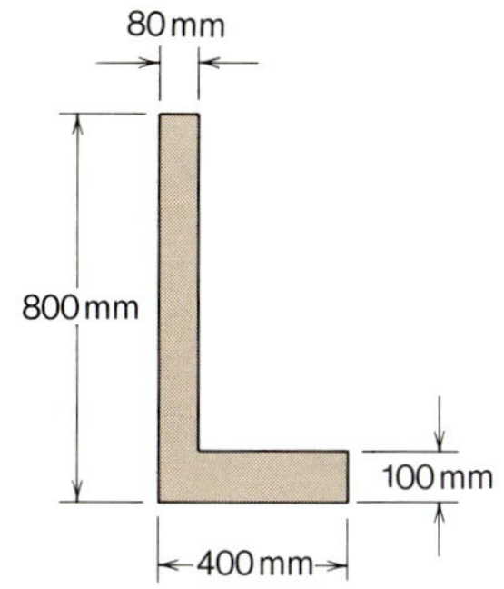

PROB. 6.26

6.27 Cast iron, 3/4 in. thick

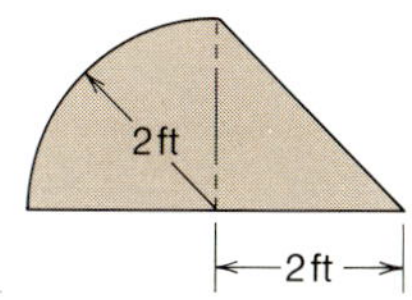

PROB. 6.27

6.28 Glass, 8 mm thick

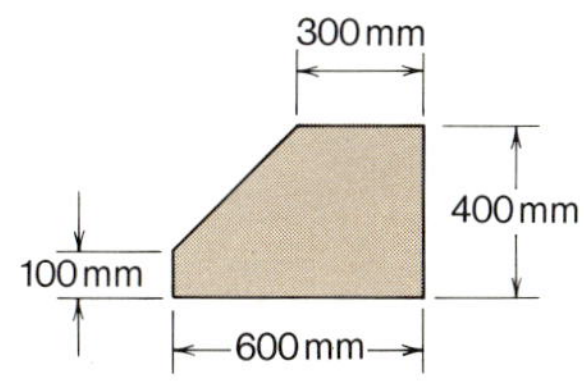

PROB. 6.28

***6.29** Aluminum, 10 mm thick

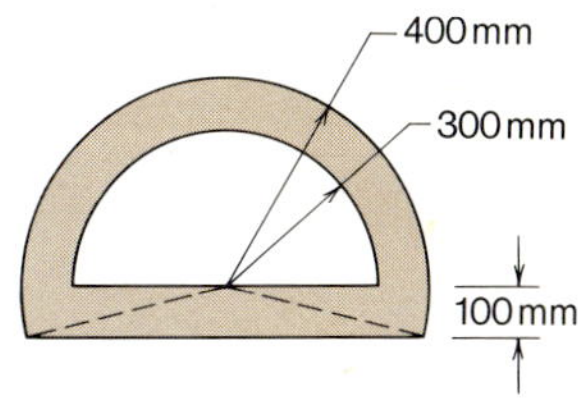

PROB. 6.29

6.30–6.35 Determine the mass and the location of the center of mass of the wire described below.

6.30 Copper, 2 mm dia.

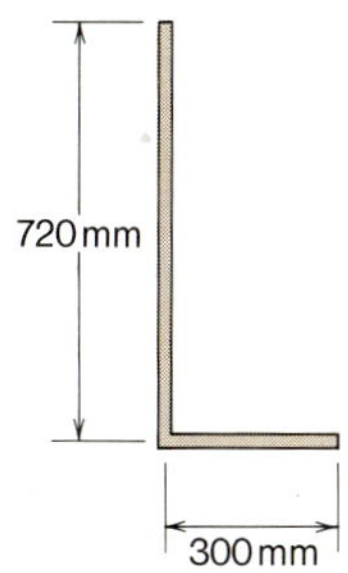

PROB. 6.30

6.31 Aluminum, 1/16 in. dia.

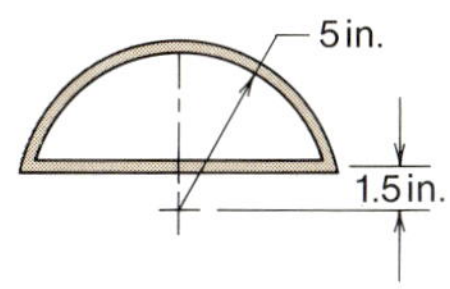

PROB. 6.31

6.32 Steel, 6 mm dia.

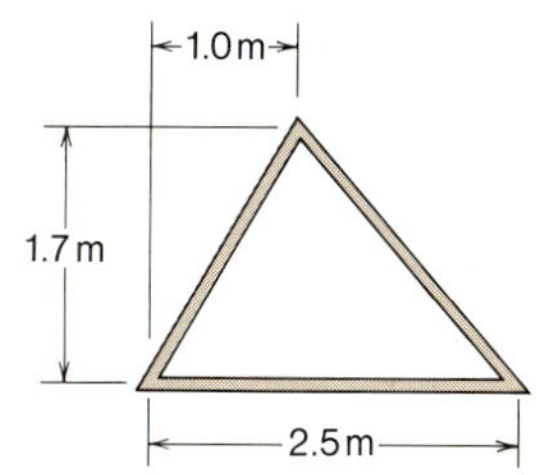

PROB. 6.32

6.33 Aluminum, 1/4 in. dia.

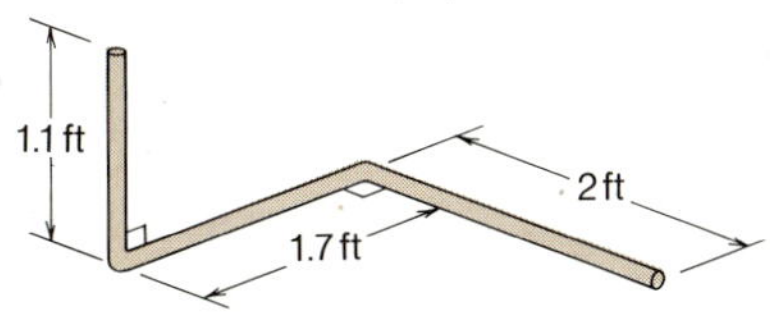

PROB. 6.33

6.34 Copper, 3 mm dia.

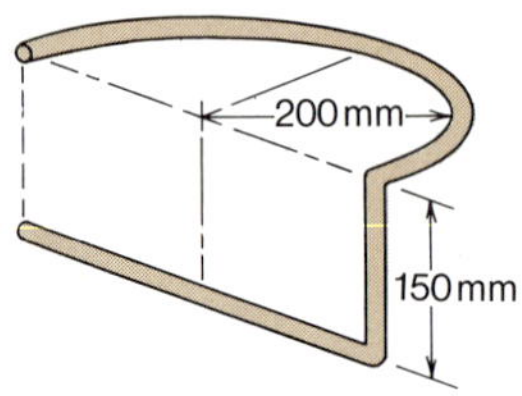

PROB. 6.34

6.35 Steel, 10 mm dia.

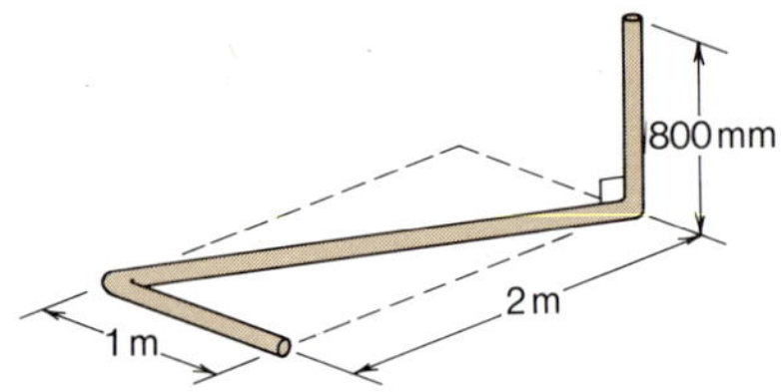

PROB. 6.35

6.36–6.39 The container shown is made out of sheet steel having the thickness specified. Determine the mass and the location of the center of mass of the container.

6.36 Thickness = 25 mm

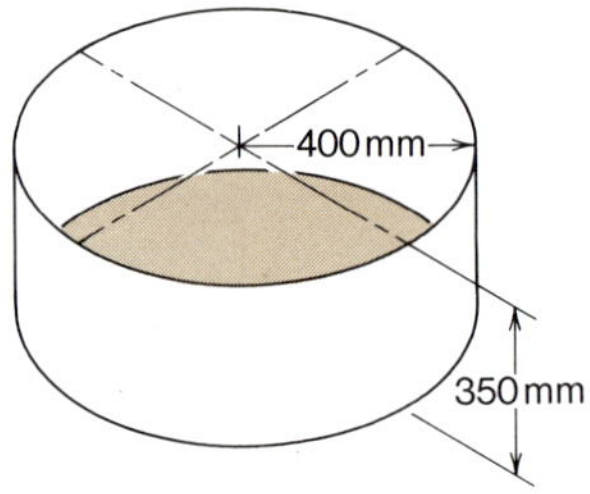

PROB. 6.36

6.37 Thickness = 10 mm

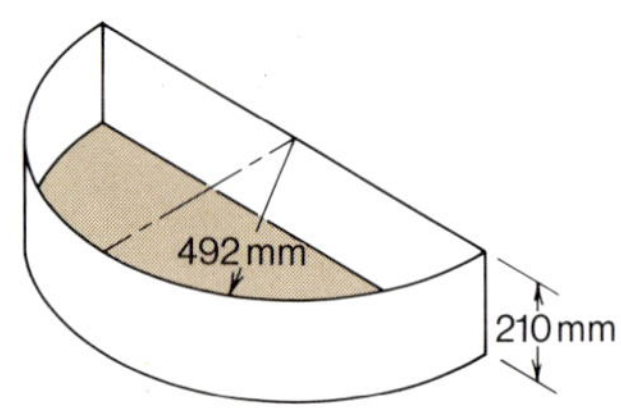

PROB. 6.37

6.38 Thickness = 1/4 in.

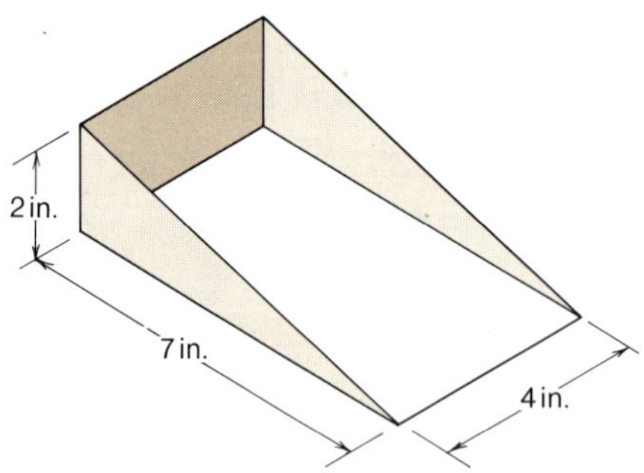

PROB. 6.38

6.39 Thickness = 5 mm

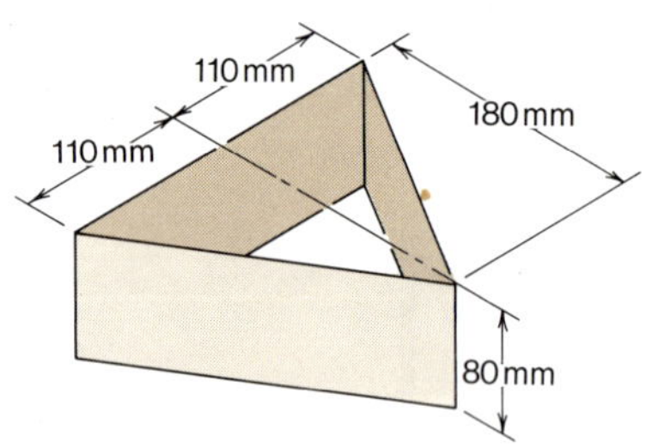

PROB. 6.39

*3. Center of Mass and Centroid by Integration

At this juncture, we have discussed only the techniques for locating the center of mass of a composite body. We will now develop methods for locating the center of mass by means of integral calculus. This will enable us to treat bodies whose shape does not appear in a tabulation. Indeed, it will allow us to learn how the geometric and mass properties given in tabulations are derived.

Let us consider a body that has N elements. We examine this body to determine whether it is homogeneous, and whether its shape can be described in terms of surface area or arc length, rather than volume. Based on such a consideration, we choose one of the sets of first moment equations. Clearly, one need only be acquainted with the mass equations; the other first moments are simplifications. Let us denote the quantity we have chosen (either mass m, volume $\mathscr{V}$, area $\mathscr{A}$, or arc length L) by the symbol Q (meaning "quantity").

Because the body has N elements, its total quantity is

$$Q = \sum_{i=1}^{N} Q_i$$

The first moment with respect to the x coordinate is representative of all three coordinates; it has the general form

$$Qx_G = \sum_{i=1}^{N} (x_G)_i Q_i$$

Suppose that we represent the body by infinitesimal elements, each of whose quantity is dQ. An infinite number of elements is then required to represent the body. The summation of an infinite number of infinitesimal quantities is the process of integration. We denote the region occupied by the body as $\mathscr{R}$, so we find

$$\boxed{Q = \int_{\mathscr{R}} dQ} \tag{12}$$

and

$$\boxed{\begin{aligned} Qx_G &= \int_{\mathscr{R}} x_c \, dQ \\ Qy_G &= \int_{\mathscr{R}} y_c \, dQ \\ Qz_G &= \int_{\mathscr{R}} z_c \, dQ \end{aligned}} \tag{13}$$

In eqs. (13), x_c, y_c, and z_c are the coordinates of the *centroid* of an element dQ. The symbol $\int_{\mathscr{R}}$ merely indicates that we need to account for every element dQ in the region $\mathscr{R}$.

Evaluating these integrals requires that we describe the element dQ and the position of points in the body. Let us recall two methods from analytical geometry for locating a point. One is to write the rectangular coordinates x, y, z of the point. The other method is to specify the point in *cylindrical coordinates*. In the latter, we choose one variable to be the distance relative to one of the coordinate planes. For example, in Figure 6 the distance z locates the distance of point P above the xy plane. We then locate the position of point P parallel to the xy plane by using the polar coordinates (R, ϕ). The set of numbers (R, ϕ, z) are the cylindrical coordinates of point P.

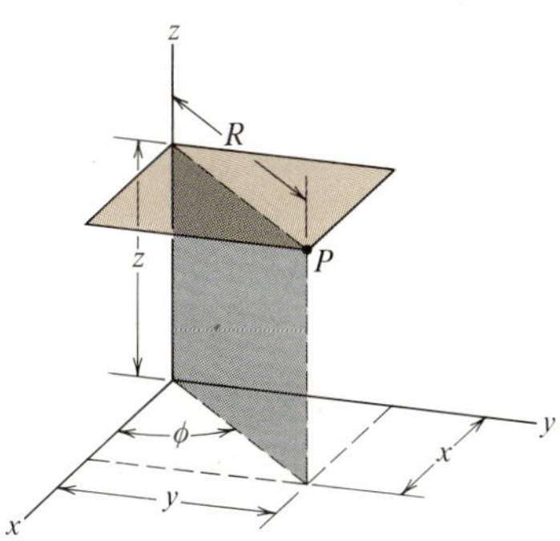

FIGURE 6

The integrals in the first moments, eqs. (13), may be formulated in terms of either the Cartesian or cylindrical coordinates. The use of the latter requires that we determine the *coordinate transformation*, which is the relationship between the Cartesian and the cylindrical coordinates. The transformation corresponding to the coordinates defined in Figure 6 is

$$x = R\cos\phi, \qquad y = R\sin\phi, \qquad z = z$$

The decision to use either Cartesian or cylindrical coordinates is usually not arbitrary. We select the set of coordinates that most easily permits us to relate the limits of the integrals to the boundaries of the body.

(a) Single Integrals. The simplest manner in which one may integrate over a region in space is to form a first-order differential element. The value of dQ then corresponds to the portion of the body bounded by two infinitesimally different values of a selected spatial coordinate. The formulations of the integrals in the first moments requires that we express dQ, as well as x_c, y_c, and z_c, in terms of only this chosen coordinate.

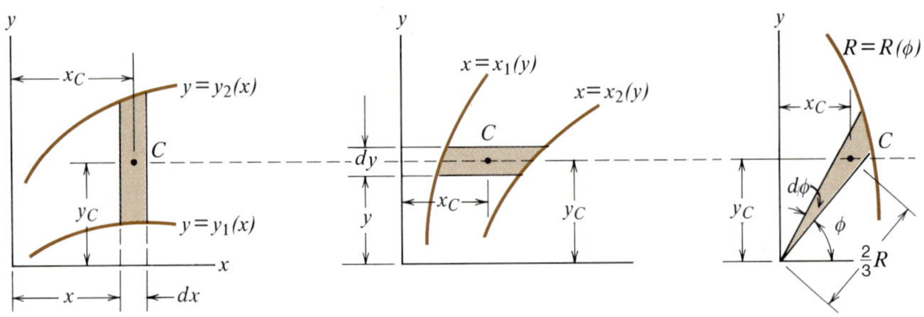

$d\mathscr{A}=[y_2(x)-y_1(x)]dx$, $x_C=x$, $y_C=\frac{1}{2}[y_2(x)+y_1(x)]$

$d\mathscr{A}=[x_2(y)-x_1(y)]dy$, $x_C=\frac{1}{2}[x_2(y)+x_1(y)]$, $y_C=y$

$d\mathscr{A}=\frac{1}{2}R^2d\phi$, $x_C=\frac{2}{3}R(\phi)\cos\phi$, $y_C=\frac{2}{3}R(\phi)\sin\phi$

FIGURE 7

Typical first-order elements of planar area are shown in Figure 7. In each case the expressions for dA, x_c, and y_c are obtained by neglecting higher order differentials. Hence, the element in Cartesian coordinates is a rectangle and the element in polar coordinates is an isosceles triangle whose equal sides are $R(\phi)$. In Figure 7 note that we employed functional notation. For example, $R(\phi)$ does not indicate R times ϕ, but rather that R is a function of the angle ϕ. More important, it means that R may be written in terms of ϕ for the purpose of integration.

The element of arc length of a planar curve is also expressible as a first-order differential, as indicated in Figure 8. It will be noted that because these elements reduce to a point as $dL \to 0$, the values of x_c and y_c are those of the point.

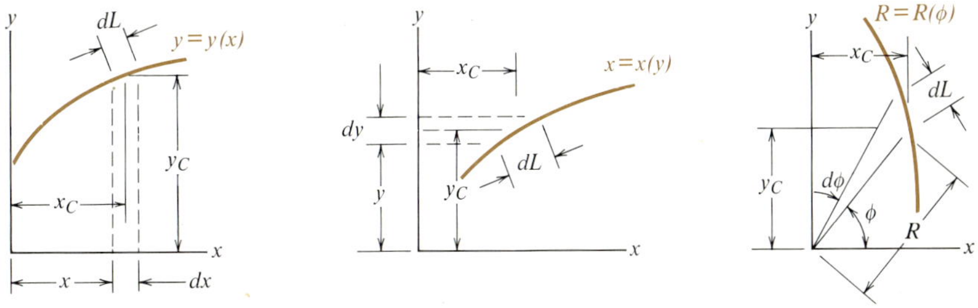

$$dL = \left[1 + \left(\frac{dy}{dx}\right)^2\right]^{1/2} dx \qquad dL = \left[1 + \left(\frac{dx}{dy}\right)^2\right]^{1/2} dy \qquad dL = \left[\left(\frac{dR}{d\phi}\right)^2 + R^2\right]^{1/2} d\phi$$

$$x_C = x,\ y_C = y(x) \qquad x_C = x(y),\ y_C = y \qquad x_C = R(\phi)\cos\phi,\ y_C = R(\phi)\sin\phi$$

FIGURE 8

In general, it is cumbersome to utilize first-order elements to describe three-dimensional bodies. However, they are useful for treating bodies of revolution, that is, a body that possesses axial symmetry. Figure 9a depicts a body of revolution, and Figure 9b depicts a shell of revolution. The surface in both cases may be represented in functional form as $R = R(z)$. Because of axial symmetry, the center of mass is on the z axis.

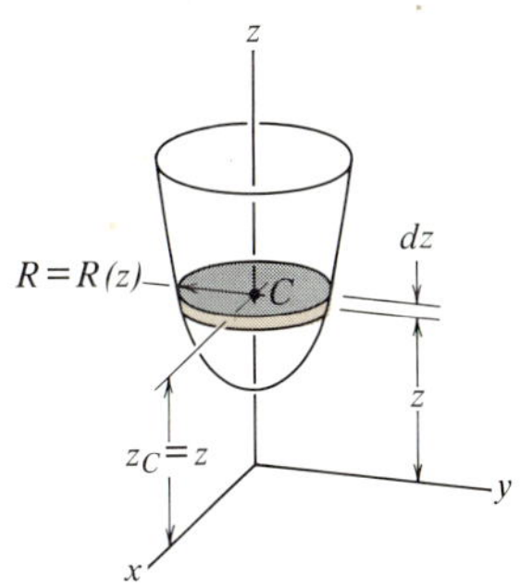

$$d\mathscr{V} = \pi[R(z)]^2 dz$$

FIGURE 9a

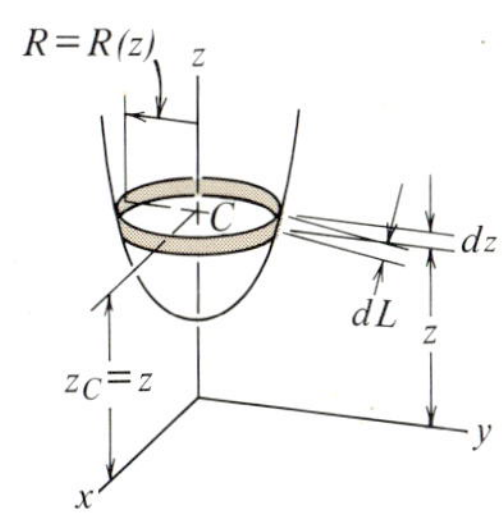

$$d\mathscr{A} = 2\pi R(z)\,dL = 2\pi R(z)\left[1 + \left(\frac{dR}{dz}\right)^2\right]^{1/2} dz$$

FIGURE 9b

The disk element in Figure 9a may be employed for first moments of volume. It may also be applied to first moments of mass for nonhomogeneous bodies, provided that the body's density depends only on the z coordinate. This is because when $\rho = \rho(z)$, the density at all points in the disk at each z is constant, so that

$$dm = \rho(z)\, d\mathscr{V} = \pi\rho(z)[R(z)]^2\, dz$$

The following examples are typical of situations where first-order differential elements are useful.

EXAMPLE 6

Determine the location of the centroid of the shaded area.

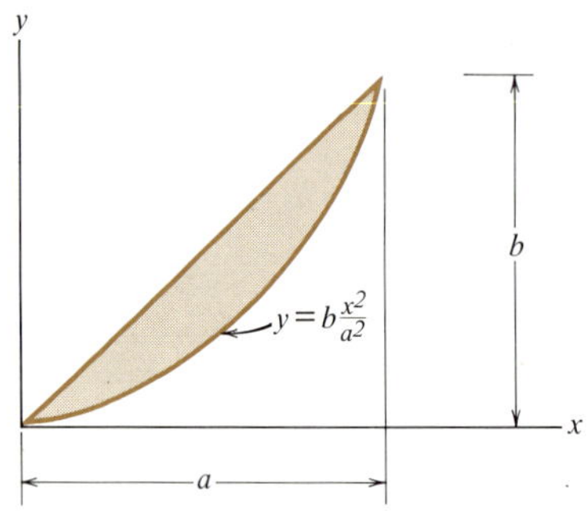

Solution

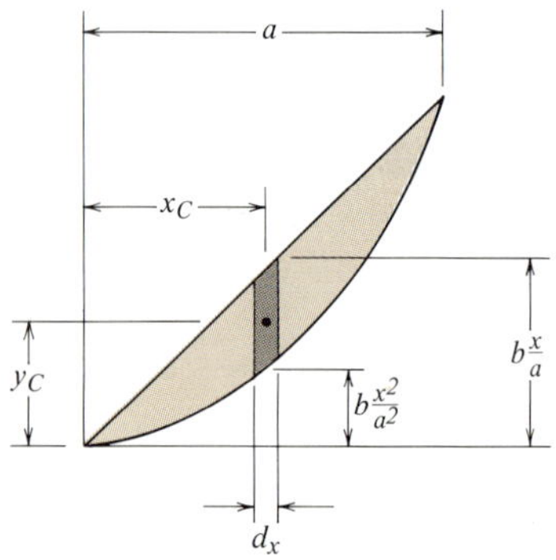

We shall employ the vertical differential element shown in the sketch in the margin. The height of the upper edge of the strip is obtained from the equation for a straight line.

The properties of this element are

$$dA = \left(\frac{bx}{a} - \frac{bx^2}{a^2}\right)dx; \qquad x_c = x, \qquad y_c = \frac{1}{2}\left(b\frac{x}{a} + b\frac{x^2}{a^2}\right)$$

These elements are situated in the region $0 \leq x \leq a$, so the integrals for the first moment of area are

$$\mathscr{A} = \int_0^a \left(\frac{bx}{a} - \frac{bx^2}{a^2}\right)dx = \frac{1}{2}\frac{b}{a}(a^2) - \frac{1}{3}\frac{b}{a^2}(a^3) = \frac{1}{6}ab$$

$$\mathscr{A}x_G = \int_0^a x\left(\frac{bx}{a} - \frac{bx^2}{a^2}\right)dx = \frac{1}{3}\frac{b}{a}(a^3) - \frac{1}{4}\frac{b}{a^2}(a^4) = \frac{1}{12}a^2b$$

$$\mathscr{A}y_G = \int_0^a \frac{1}{2}\left(\frac{bx}{a} + \frac{bx^2}{a^2}\right)\left(\frac{bx}{a} - \frac{bx^2}{a^2}\right)dx = \frac{1}{2}\int_0^a \left(\frac{b^2x^2}{a^2} - \frac{b^2x^4}{a^4}\right)dx$$

$$= \frac{1}{2}\left[\frac{1}{3}\frac{b^2}{a^2}(a^3) - \frac{1}{5}\frac{b^2}{a^4}(a^5)\right] = \frac{1}{15}ab^2$$

These equations yield

$$x_G = \frac{a^2b/12}{ab/6} = \frac{1}{2}a, \qquad y_G = \frac{ab^2/15}{ab/6} = \frac{2}{5}b$$

EXAMPLE 7

Determine by integration the x, y coordinates of the centroid of the circular arc.

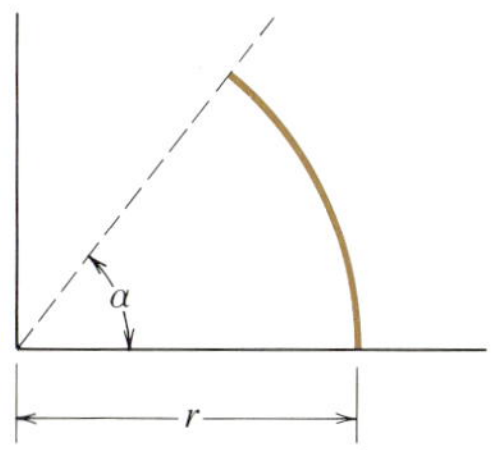

Solution

Polar coordinates (i.e., cylindrical coordinates in the xy plane) are useful for this problem. These coordinates and the element dL are depicted in the sketch, from which we find

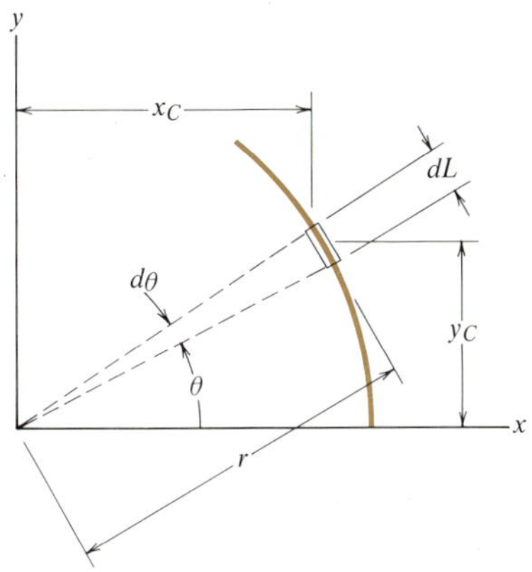

$$dL = r\,d\theta, \qquad x_c = r\cos\theta, \qquad y_c = r\sin\theta$$

The elements are situated in $0 \le \theta \le \alpha$, so the integrals for the first moment are

$$L = \int_0^\alpha dL = \int_0^\alpha r\,d\theta = r\alpha$$

$$Lx_G = \int_0^\alpha x_c\,dL = \int_0^\alpha r^2\cos\theta\,d\theta = r^2\sin\alpha$$

$$Ly_G = \int_0^\alpha y_c\,dL = \int_0^\alpha r^2\sin\theta\,d\theta = r^2(1-\cos\alpha)$$

Thus

$$x_G = r\frac{\sin\alpha}{\alpha}, \qquad y_G = \frac{r(1-\cos\alpha)}{\alpha}$$

EXAMPLE 8

The density of the hemisphere varies with distance h from the base according to $\rho = \rho_0(2 - h/a)$. Determine the mass and location of the center of mass of the hemisphere in terms of ρ_0 and a.

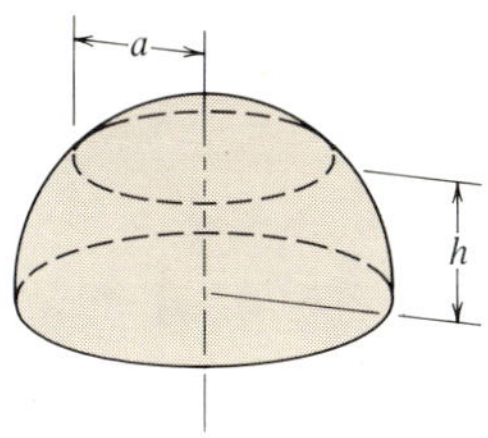

Solution

We must employ first moments of mass because the density is not constant. We align the z axis with the axis of symmetry, and place the origin at the base, so that $z = h$. The differential element consists of a disk whose radius is $R(z)$ and whose thickness is dz. Because all points on the surface of the sphere are at the radius a from the origin, it must be that

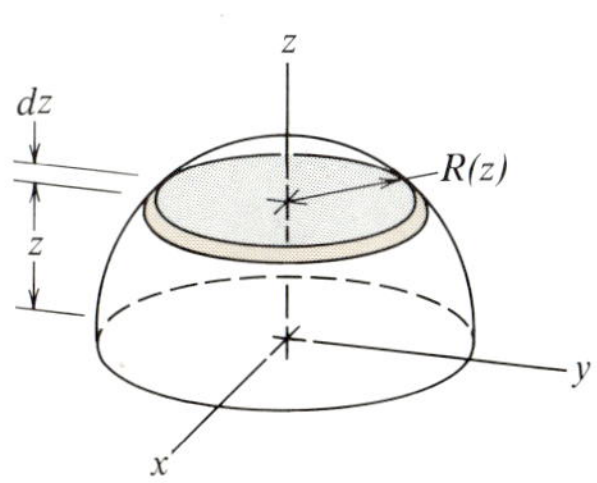

$$x^2 + y^2 + z^2 = a^2$$

However, $R(z)$ is the distance from a point on the surface to the z axis, so

$$R(z) = (x^2 + y^2)^{1/2} = (a^2 - z^2)^{1/2}$$

It follows that

$$dm = \rho(z)\,d\mathscr{V} = \left[\rho_0\left(2 - \frac{z}{a}\right)\right]\pi[R(z)]^2\,dz$$

$$= \rho_0\pi\left(2a^2 - az - 2z^2 + \frac{z^3}{a}\right)dz$$

These elements are situated in $0 \le z \le a$, so

$$m = \int_0^a dm = \rho_0 \pi \int_0^a \left(2a^2 - az - 2z^2 + \frac{z^3}{a} \right) dz = \frac{13}{12} \pi \rho_0 a^3$$

Because of symmetry

$$x_G = y_G = 0$$

Each point in the element is at a distance z from the xy plane, so $z = z_c$. Therefore

$$m z_G \equiv \frac{13}{12} \pi \rho_0 a^3 z_G = \int_0^a z_c \, dm$$

$$= \rho_0 \pi \int_0^a z \left(2a^2 - az - 2z^2 + \frac{z^3}{a} \right) dz = \frac{11}{30} \pi \rho_0 a^4$$

$$z_G = \frac{11/30}{13/12} a = \frac{22}{65} a$$

HOMEWORK PROBLEMS

6.40–6.47 Determine the location of the centroid of each shaded area shown.

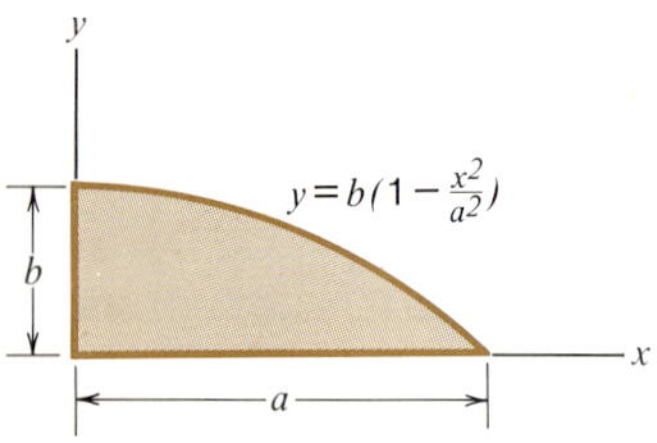

PROB. 6.40

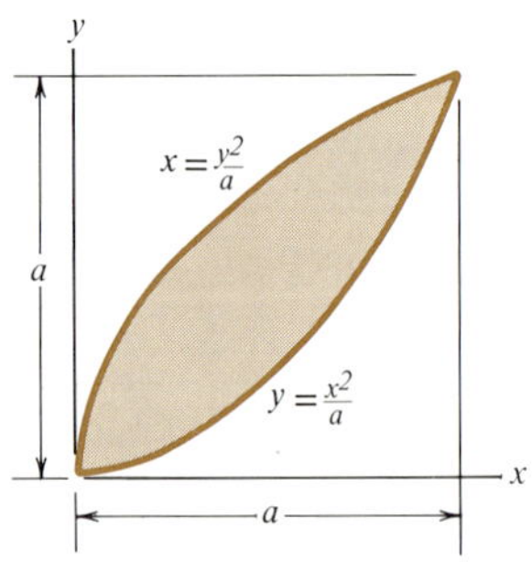

PROB. 6.41

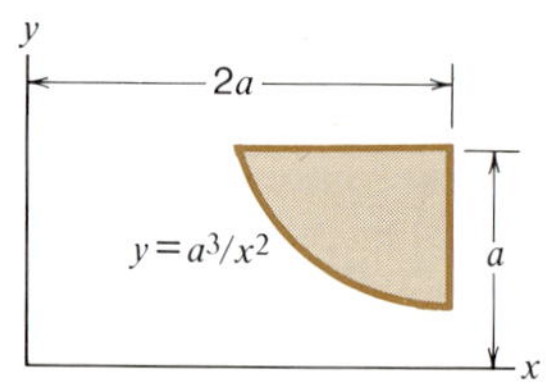

PROB. 6.42

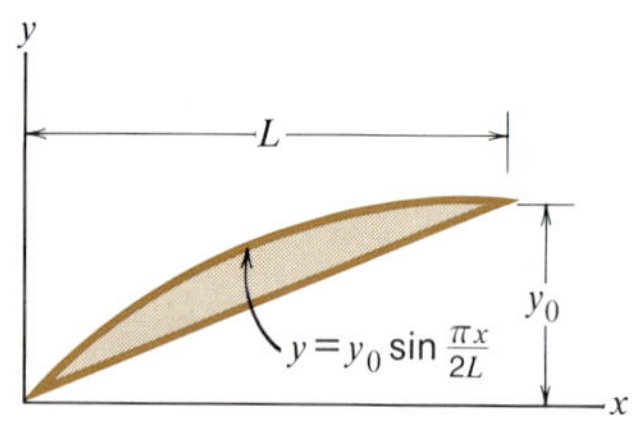

PROB. 6.43

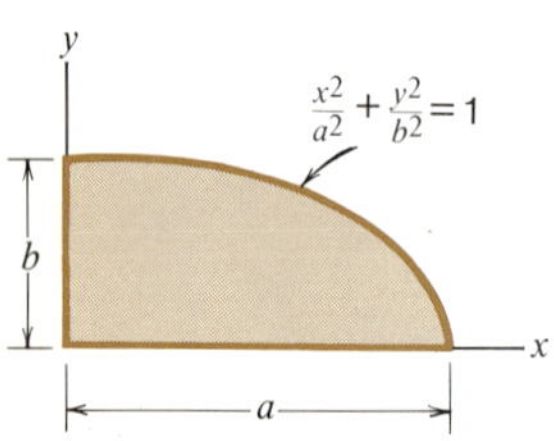

PROB. 6.44

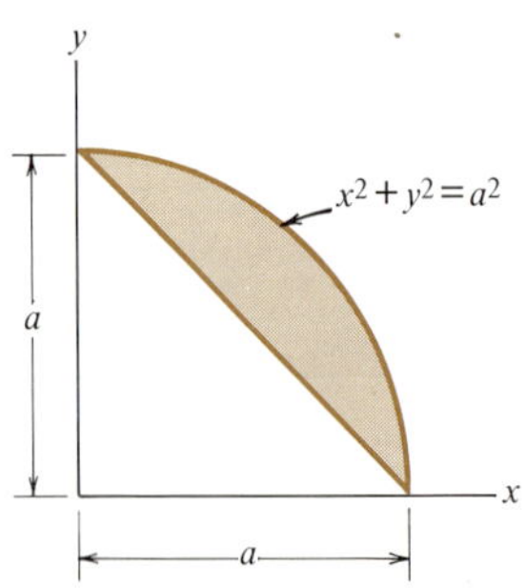

PROB. 6.45

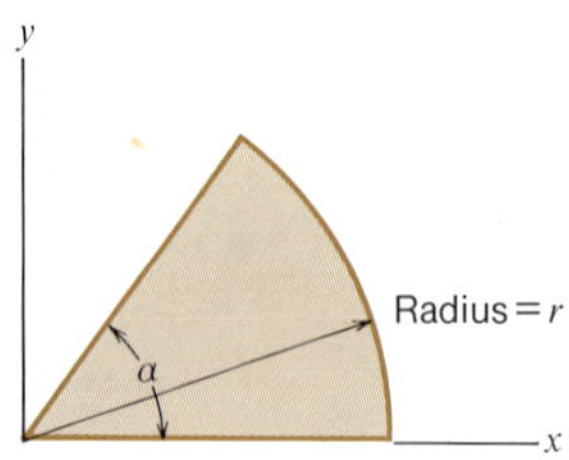

PROB. 6.46

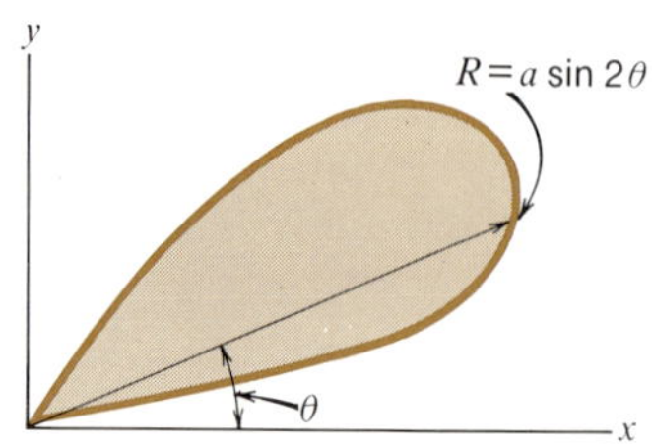

PROB. 6.47

6.48–6.50 Determine the location of the centroid of each arc shown.

6.48 Circular arc

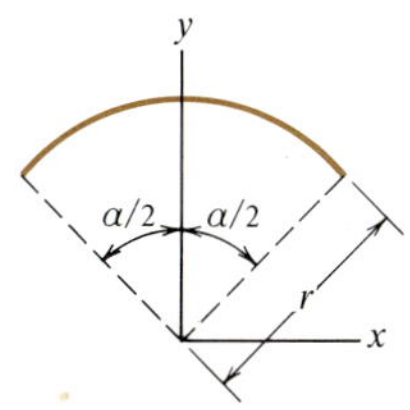

PROB. 6.48

6.49 Spiral arc

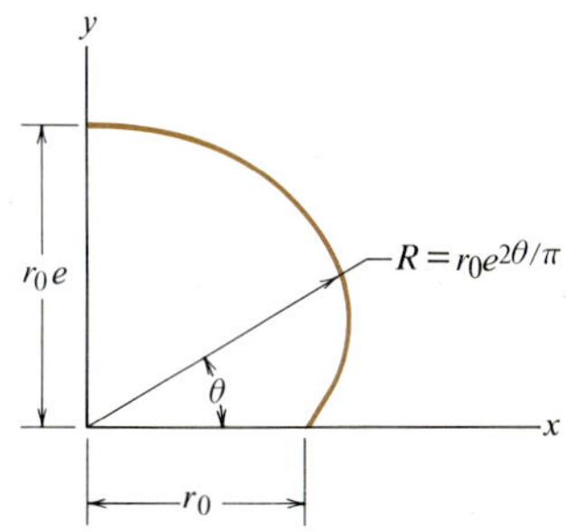

PROB. 6.49

***6.50** Parabolic arc

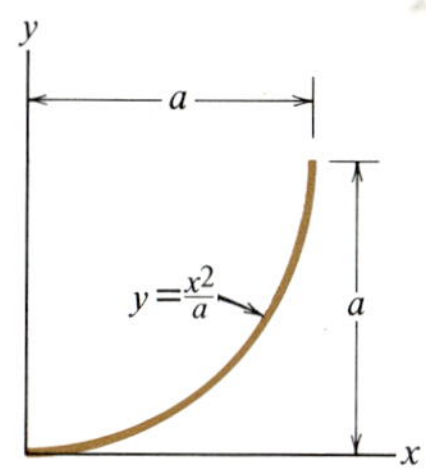

PROB. 6.50

6.51–6.53 Determine the location of the centroid of each body shown.

6.51 Cone

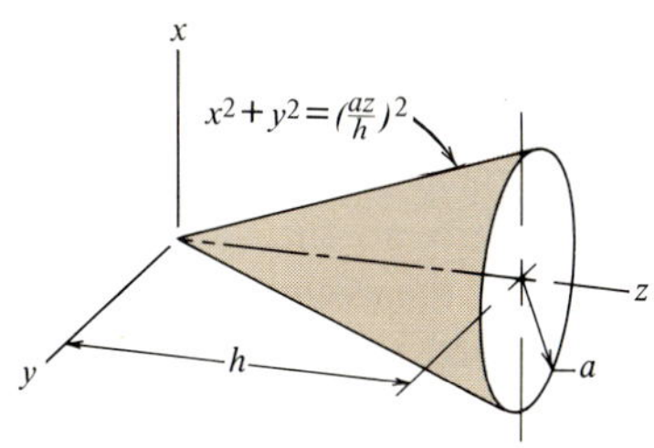

PROB. 6.51

6.52 Paraboloid of revolution

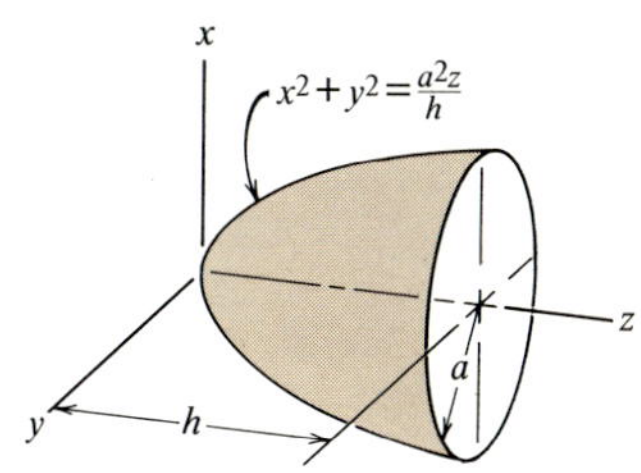

PROB. 6.52

6.53 Ellipsoid of revolution

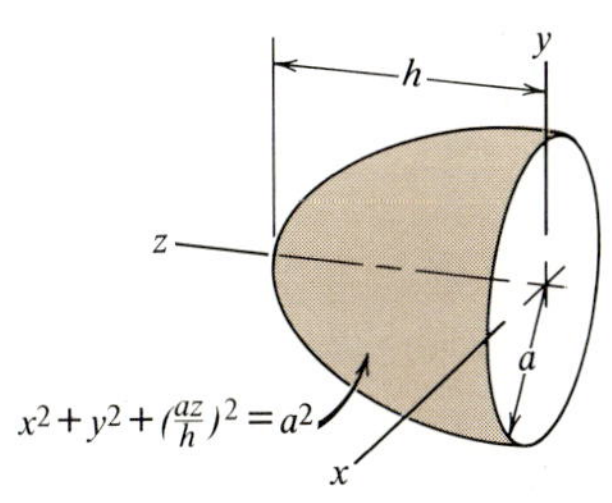

PROB. 6.53

***6.54** Determine the mass and the location of the center of mass of the cone in Problem 6.51 if the density varies according to $\rho = \rho_0(3 - x/h)$.

6.55 and 6.56 Determine the location of the center of mass of each shell shown.

***6.55** Conical shell

***6.56** Hemispherical shell

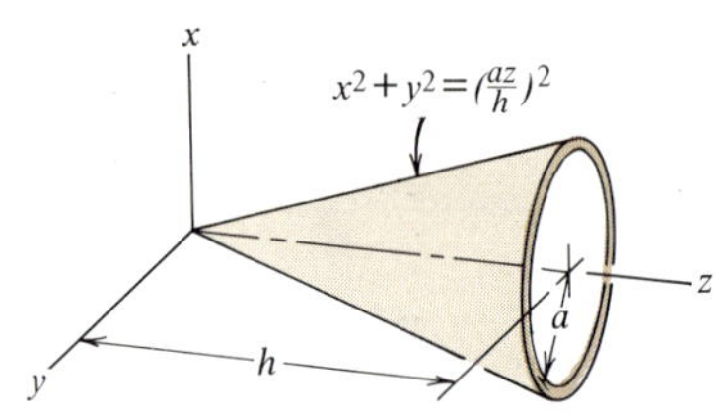

PROB. 6.55

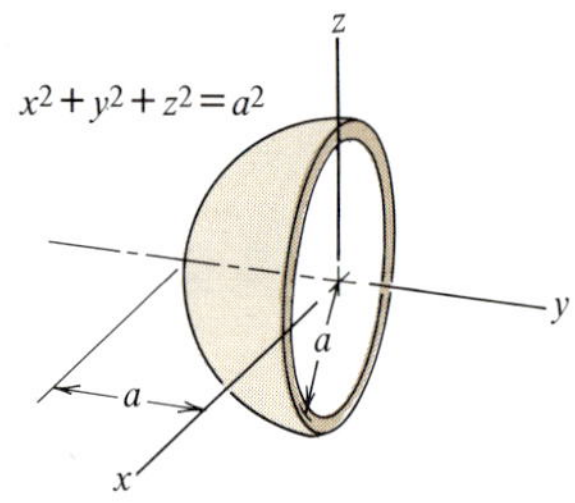

PROB. 6.56

(b) Multiple Integrals. Situations occur in which the concept of first-order differential elements is not suitable for computations. One reason for this might be that it is difficult to express x_c, y_c or z_c because the element does not have a familiar shape. Another reason, in the case of nonhomogeneous bodies, might be that the density is not constant across any first-order element.

The most general way in which one can integrate over a region in space is to employ the highest order (in other words, the smallest) differential element possible. Such an element corresponds to the region bounded by differential increments in each of the spatial coordinates for the body. Typical elements that arise in this formulation are depicted in Figure 10. Each of these elements reduce to a point in the limit as their sides approach zero, so the values of x_c, y_c, and z_c for these elements are the coordinates of these elements in this limit. The volume elements in Figure 10 may be used to evaluate the properties of a planar area, lying in the xy plane, by replacing dz by 1.

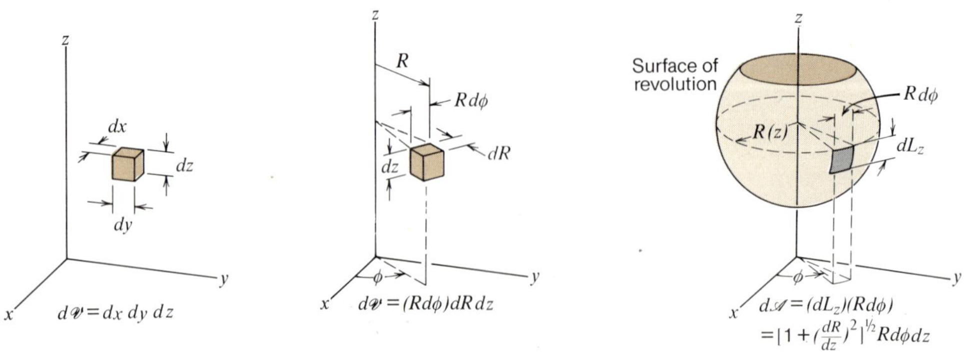

FIGURE 10

In general, our understanding of the sequence and limits for multiple integrals is greatly increased by drawing a view of a typical cross section of the body. If the problem is formulated in terms of the cylindrical coordinates (R, ϕ, z), this cross section will usually be at a typical value of z. Such an approach is equally useful for analyzing volumes and curved shells.

It should be noted that the only curved shells we have discussed form a segment of a surface of revolution. Also, we have discussed only curved lines that are situated in a plane. Comparable treatments of more general curved lines and shells are topics in advanced calculus.

EXAMPLE 9

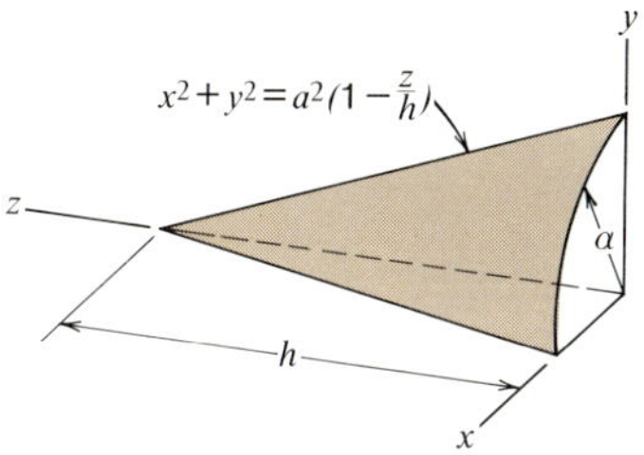

Determine the location of the center of mass of the homogeneous quarter of a cone.

Solution

The body is symmetric with respect to a plane that bisects the right angle between the xz and yz plane, so that

$$x_G = y_G$$

The body is homogeneous, so it is sufficient to evaluate first moments of volume.

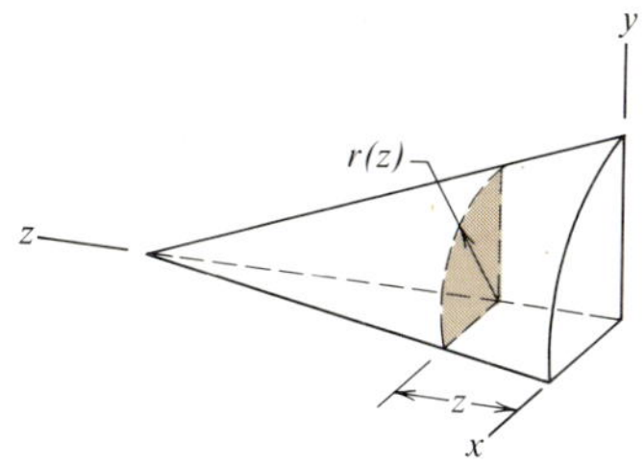

The integration scheme we will use is to consider all differential elements $d\mathscr{V}$ lying on the shaded plane at a distance z from the xy plane. This plane, as seen when viewed downward along the z axis, is depicted in the next sketch. All such planes may be accounted for by integrating over $0 \le z \le h$.

The radius $r(z)$ of this section is found from our sketch of the cross section, which shows that for any points on the circular perimeter

$$(x^2 + y^2)^{1/2} = r(z)$$

When this expression is substituted into the equation for the conical surface, it is found that

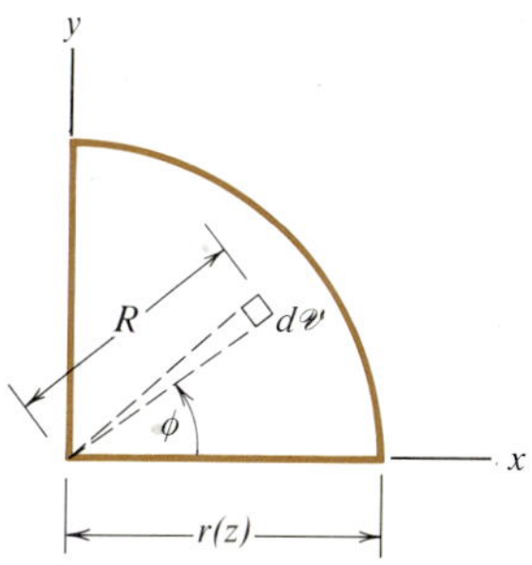

$$r(z) = a\left(1 - \frac{z}{h}\right)$$

The element of volume we shall employ is the one for cylindrical coordinates

$$d\mathscr{V} = R\,d\phi\,dR\,dz$$

and the Cartesian coordinates of the point associated with this element are

$$x_c = R\cos\phi, \qquad y_c = R\sin\phi, \qquad z_c = z$$

We may account for all elements on the cross section by integrating over $0 \le R \le r(z)$, $0 \le \phi \le \pi/2$ (the order of integration is arbitrary). Thus

$$\mathscr{V} = \int_0^h \int_0^{a(1-z/h)} \int_0^{\pi/2} d\mathscr{V} = \int_0^h \int_0^{a(1-z/h)} \int_0^{\pi/2} R\,d\phi\,dR\,dz$$

$$= \int_0^h \int_0^{a(1-z/h)} \frac{\pi}{2} R\,dR\,dz = \frac{\pi}{2}\int_0^h \frac{1}{2}a^2\left(1-\frac{z}{h}\right)^2 dz = \frac{\pi}{12}a^2h$$

$$\mathscr{V} x_G = \int_0^h \int_0^{a(1-z/h)} \int_0^{\pi/2} x_c\,d\mathscr{V} = \int_0^h \int_0^{a(1-z/h)} \int_0^{\pi/2} R\cos\phi\,R\,d\phi\,dR\,dz$$

$$= \int_0^h \int_0^{a(1-z/h)} R^2\,dR\,dz = \int_0^h \frac{1}{3}a^3\left(1-\frac{z}{h}\right)^3 dz = \frac{a^3h}{12}$$

$$\mathscr{V} z_G = \int_0^h \int_0^{a(1-z/h)} \int_0^{\pi/2} z_c\,d\mathscr{V} = \int_0^h \int_0^{a(1-z/h)} \int_0^{\pi/2} zR\,d\phi\,dR\,dz$$

$$= \int_0^h z\left(\frac{\pi}{4}\right)a^2\left(1-\frac{z}{h}\right)^2 dz = \frac{\pi}{48}a^2h^2$$

$$x_G = \frac{a^3h/12}{\pi a^2h/12} = \frac{a}{\pi} = y_G, \qquad z_G = \frac{\pi a^2h^2/48}{\pi a^2h/12} = \frac{h}{4}$$

EXAMPLE 10

The density of the rectangular parallelepiped varies according to

$$\rho = \rho_0\left(1 + \frac{x}{a}\frac{y}{b}\right)$$

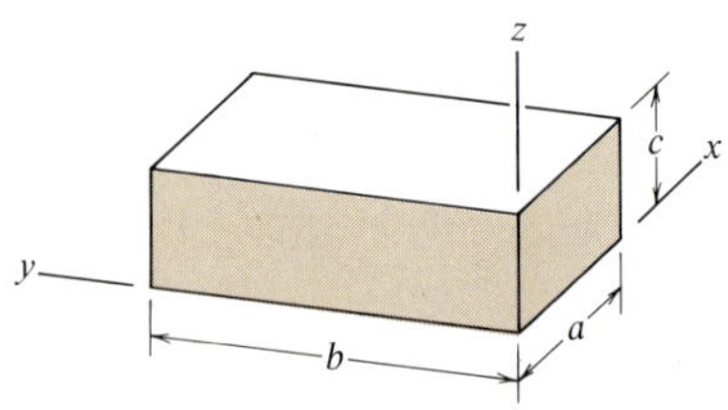

Determine the mass and the x coordinate of the center of mass of this body.

Solution

The given xyz coordinate system readily describes the shape of the body, as well as the dependence of the density. The mass of an element is

$$dm = \rho\,d\mathscr{V} = \rho_0\left(1 + \frac{x}{a}\frac{y}{b}\right)dx\,dy\,dz$$

The region occupied by the parallelepiped is $0 \le x \le a, 0 \le y \le b, 0 \le z \le c$. Integrating the mass elements in this region yields

$$m = \int_0^c \int_0^b \int_0^a \rho_0\left(1 + \frac{xy}{ab}\right)dx\,dy\,dz = \rho_0 \int_0^c \int_0^b \left(a + \frac{ay}{2b}\right)dy\,dz$$

$$= \rho_0 \int_0^c (ab + \tfrac{1}{4}ab)\,dz = \tfrac{5}{4}\rho_0 abc$$

The value of x_G is found by forming the first moment of mass.

$$mx_G = \int_0^c \int_0^b \int_0^a x(\rho_0)\left(1 + \frac{xy}{ab}\right)dx\,dy\,dz$$

$$= \rho_0 \int_0^c \int_0^b \left(\frac{a^2}{2} + \frac{a^2y}{3b}\right)dy\,dz = \rho_0 \int_0^c \left(\frac{a^2b}{2} + \frac{a^2b}{6}\right)dz = \frac{2}{3}\rho_0 a^2bc$$

The result of substituting the results for m is

$$x_G = (\tfrac{2}{3}\rho_0 a^2bc)/(\tfrac{5}{4}\rho_0 abc) = \tfrac{8}{15}a$$

HOMEWORK PROBLEMS

6.57–6.62 Determine the location of the center of mass of the homogeneous body shown.

6.57 Triangular prism

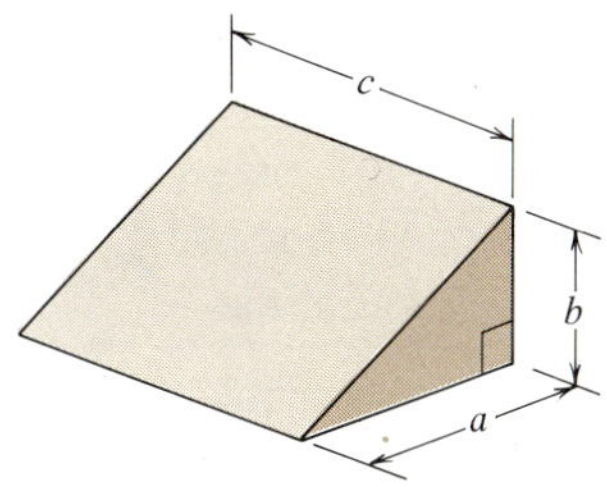

PROB. 6.57

6.58 Orthogonal tetrahedron

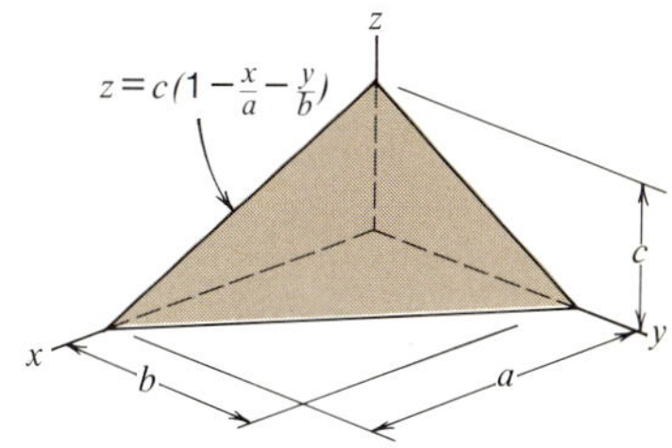

PROB. 6.58

6.59 Hyperbolic paraboloid

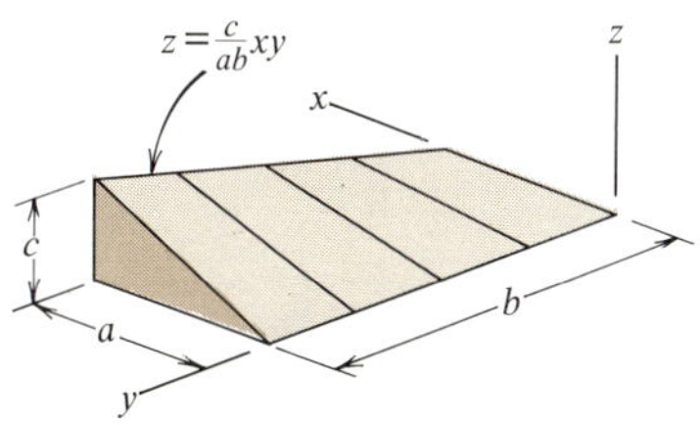

PROB. 6.59

***6.60** Hemispherical segment

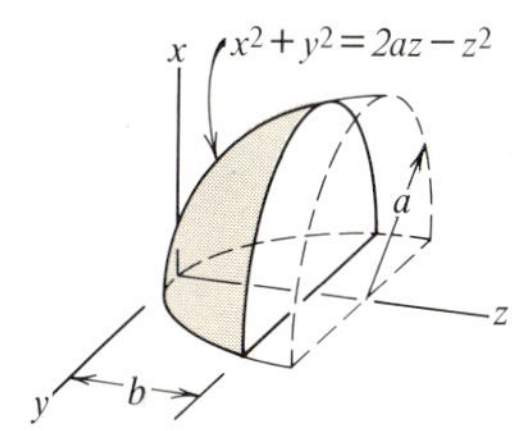

PROB. 6.60

6.61 Exponential horn segment

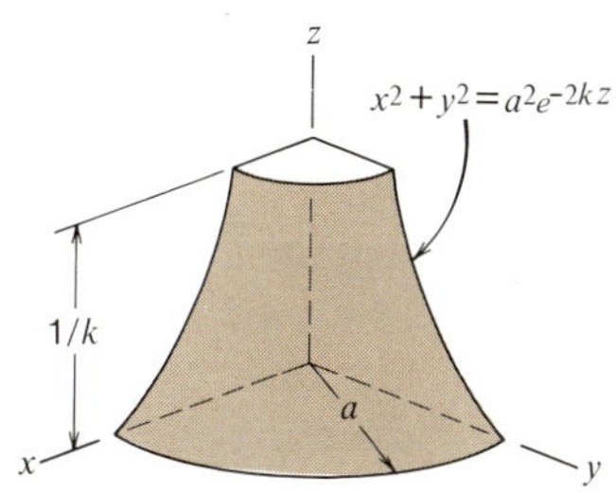

PROB. 6.61

***6.62** Truncated cylinder

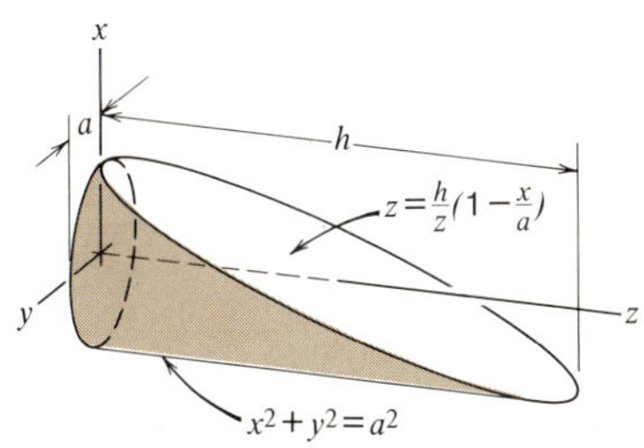

PROB. 6.62

***6.63** The density of the orthogonal tetrahedron in Problem 6.58 varies according to $\rho = \rho_0(1 + z/2c)$. Determine the mass and the location of the center of mass of this body.

***6.64** The density of the body in Problem 6.61 varies with the radial distance R from the z axis according to $\rho = \rho_0(1 + R/a)$. Determine the mass and the location of the center of mass of this body.

6.65–6.67 Determine the location of the center of mass of the homogeneous, constant-thickness shell.

6.65 Semiconical shell

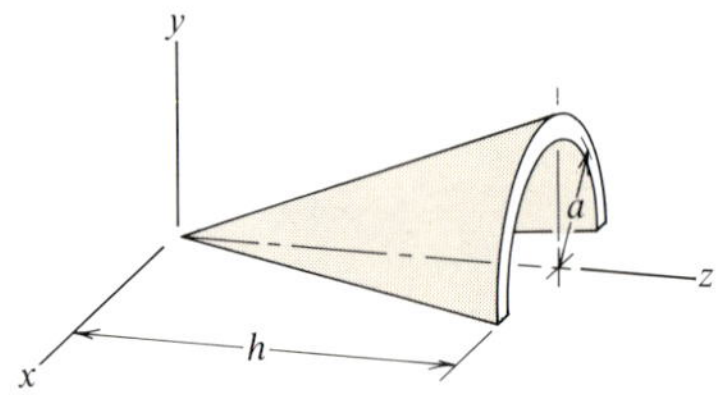

PROB. 6.65

6.66 Spherical shell segment

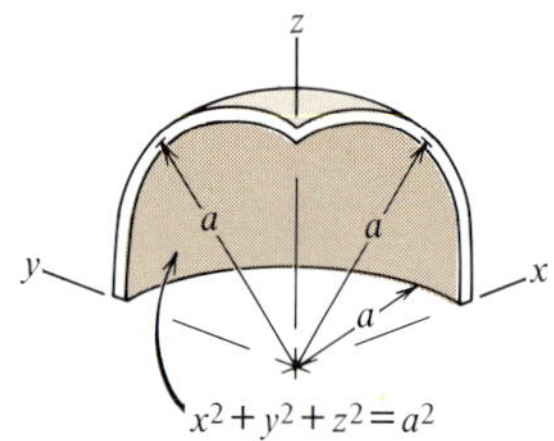

PROB. 6.66

***6.67** Truncated cylindrical shell segment. *Hint:* Use cylindrical coordinates (R, ϕ, z).

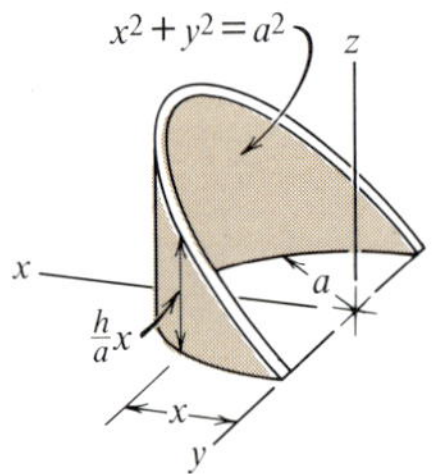

PROB. 6.67

B. SURFACE LOADINGS

The gravitational forces we considered in Section A are examples of distributed force systems that are applied over the space occupied by a body; they are termed *body forces.* Another important type of force system is one that acts over the surface area of a body; this is the case of a *surface force.* Typically, distributed surface forces can result from the interaction of two bodies that are in contact or from pressure exerted by fluids, such as air and water.

Surface forces may be categorized according to the manner in which they are distributed. A *line load* has the property of being distributed over a line, whereas an *area load* is a force that is spread out over a surface area. The particular cases of line and area loads that we shall study here arise frequently in engineering practice. They share the common characteristic of acting perpendicular to the surface to which they are applied.

1. Line Loads

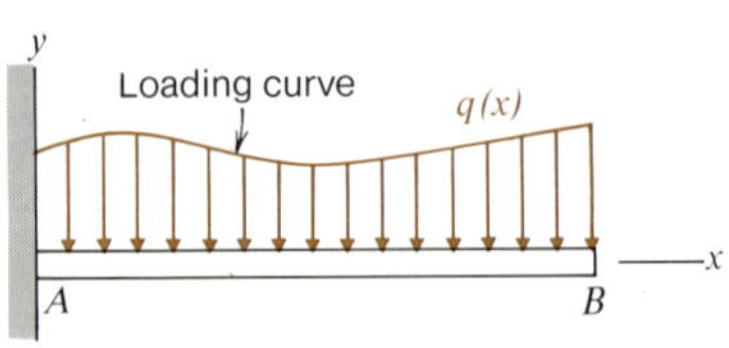

FIGURE 11

Our concern in this section is the situation where a force acting perpendicular to a straight line is distributed along the line. This line load is denoted by the vector $\bar{q}$. *The magnitude q of this loading has units of force per unit length* measured along the line that is loaded. Such a loading is typified by Figure 11, where the horizontal beam AB carries a vertical line load.

Since the load is variable, the magnitude $q(x)$ may vary from point to point along the beam. This dependency is defined by letting x be the horizontal distance along the beam from some convenient origin such as end A in Figure 11. The function $q(x)$ then defines the magnitude of the line load at each location along the beam.

A pictorial representation of the line load is obtained by plotting $q(x)$ as a function of x along the beam, and then using arrows at several locations along the beam to indicate the sense of the loading. The resulting curve, $y = q(x)$, is called the *loading curve.*

Let us consider the effect of the line load acting over a portion of a beam in the interval from x to $x + dx$. Because dx is infinitesimal, q is essentially constant in this interval. The line load on this portion of a beam is therefore equivalent to a concentrated force $q(x)\,dx$ acting at position x. This is illustrated in Figure 12a.

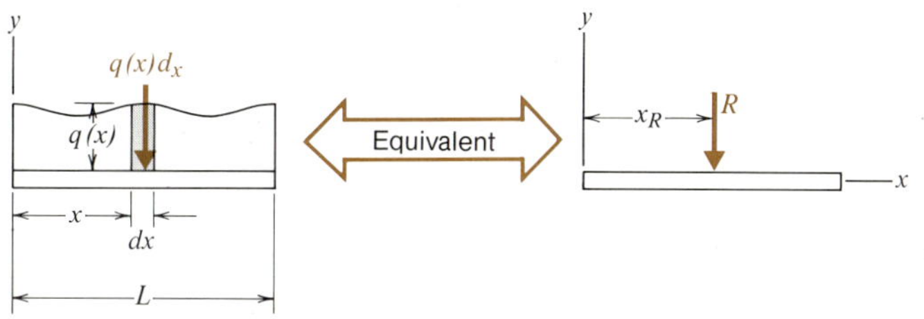

FIGURE 12a **FIGURE 12b**

When we consider the interval dx at each location of the beam, we obtain an infinite number of infinitesimal forces $q(x)\,dx$. Each of these forces is parallel to the y axis. We know from our study in Chapter 3 that such a system of forces is equivalent to a single concentrated force $\bar{R}$ acting parallel to the y axis. This force is depicted in Figure 12b.

The first requirement for equivalence is that the magnitude of $\bar{R}$ must be identical to the sum of the magnitudes of the infinitesimal forces. A summation of infinitesimal quantities is an integral, so we find that

$$R = \int_0^L q(x)\,dx \tag{14}$$

The other requirement for equivalence is that $\bar{R}$ and the distributed force system must exert the same moment about any point. A convenient point for the moment computation is the origin, $x = 0$. We see in Figure 12 that the lever arm of $\bar{R}$ is x_R, and the lever arm of the infinitesimal force $q(x)\,dx$ is the value of x for this force. In this case, the moments in both systems of forces are clockwise. Using an integral to sum the moment of each infinitesimal force results in

$$Rx_R = \int_0^L xq(x)\,dx \tag{15}$$

If R is found to be non-zero from eq. (14), then eq. (15) may be solved for the corresponding value of x_R. The case where $R \equiv 0$ corresponds to a line load that exerts a couple.

The pictorial representation of a line load by means of a loading curve provides a useful interpretation of eqs. (14) and (15). The differential element of area under the loading curve in an interval from x to $x + dx$ is the force $q(x)\,dx$; this is the area of the shaded strip in Figure 12a. Thus, the integral in eq. (14) gives the area under the loading curve, and the integral in eq. (15) is the first moment of the area under the loading curve with respect to the x coordinate. This means that

> The resultant $\bar{R}$ of a line load $\bar{q}$ acting perpendicular to a straight line is the area under the loading curve, in the sense of positive q. This resultant force intersects the centroid of the area under the loading curve.

The importance of this analogy is that it enables us to use the area properties tabulated in Appendix C to evaluate the resultant of a line load whenever the loading curve forms a basic shape or a composite of basic shapes. In this regard, it is convenient to note that the assumed direction of positive q is a downward loading, as shown in Figures 11 and 12. Hence, negative q constitutes an upward loading. Symbolically, a negative loading curve is shown below the line carrying the load, pointing upward. This is illustrated in Example 12.

EXAMPLE 11

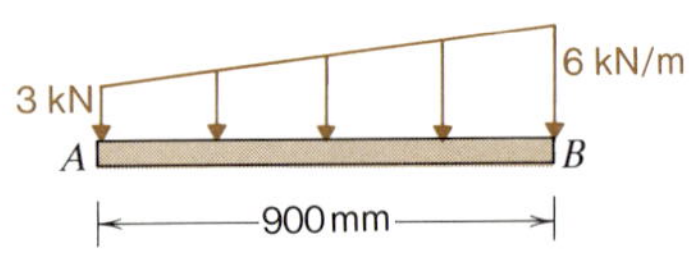

Determine the resultant of the trapezoidal line load and its point of application along the beam.

Solution

Because the properties of a trapezoid do not appear in Appendix C, we will treat the area formed by the loading curve as a composite shape. There are several such composites that may be formed. One that often proves to be useful is to consider the trapezoid to be a combination of two triangles that are formed by one of the diagonals. This is illustrated in the sketch.

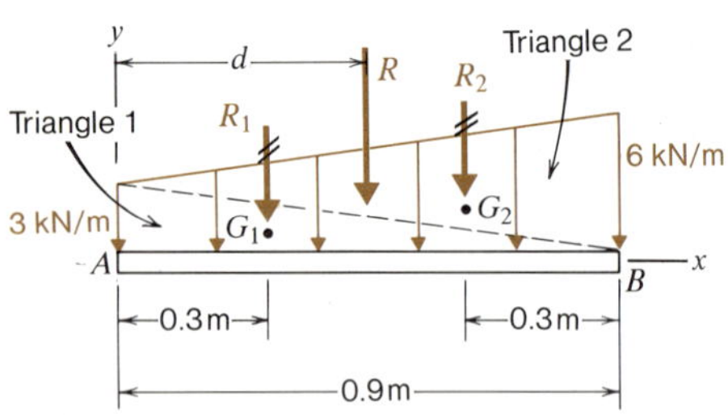

There is a resultant force $\bar{R}_i$ for each triangle. The point of application of each force along the beam, which is shown in the sketch, is readily found from the fact that the centroid of any triangle is at a distance of 1/3 of the height from the base. (Note that the vertical position of the centroids is irrelevant to line loads.) The individual resultant forces are depicted in the sketch with two hatchmarks in order to indicate that their combined effect is the total resultant $\bar{R}$.

The magnitudes of R_1 and R_2 are the areas of the corresponding triangles.

$$R_1 = \frac{1}{2}\left(3\,\frac{\text{kN}}{\text{m}}\right)(0.9\text{m}) = 1.35\text{ kN}, \qquad R_2 = \frac{1}{2}\left(6\,\frac{\text{kN}}{\text{m}}\right)(0.9\text{m}) = 2.70\text{ kN}$$

It follows that

$$R = \sum R_i = R_1 + R_2 = 4.05\text{ kN}$$

The dimension d locating $\bar{R}$ is found by equating the moment of $\bar{R}$ to the moment of $\bar{R}_1$ and $\bar{R}_2$. Summing moments about end A yields

$$\sum M_{Az} = -Rd = -R_1(0.3) - R_2(0.6)$$

$$4.05d = 1.35(0.3) + 2.70(0.6)$$

$$d = 0.50 \text{ meter}$$

EXAMPLE 12

The beam carries the line load and the concentrated force shown. Determine the reactions at supports A and B.

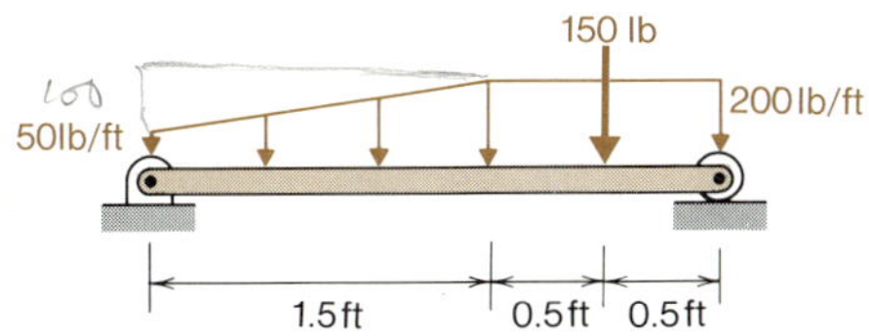

Solution

The loading curve forms a composite of areas that are tabulated in Appendix C. If we treat it as a combination of positive areas only, we will need to consider three basic shapes. However, we can regard the area under the loading curve to be a rectangle of height 200 lb/ft, from which a triangle is removed. Removal of an area means that the area is negative, so we depict the triangular area as an upward line load in the sketch. We construct this sketch as a free body diagram of the beam, in order to determine the reactions acting on the beam.

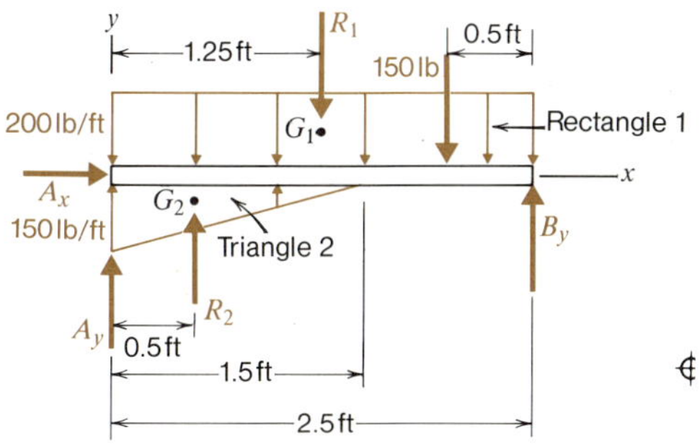

The magnitude of each resultant R_i is the corresponding area, so

$$R_1 = \left(200 \frac{\text{lb}}{\text{ft}}\right)(2.5 \text{ ft}) = 500 \text{ lb}, \qquad R_2 = \frac{1}{2}\left(150 \frac{\text{lb}}{\text{ft}}\right)(1.5 \text{ ft}) = 112.5 \text{ lb}$$

These resultants act through the centroid of each area, as shown in the free body diagram.

We may now write the equilibrium equations for the beam, which are

$$\sum M_{Az} = B_y(2.5) - 150(2.0) - R_1(1.25) + R_2(0.5)$$

$$\equiv 2.5B_y - 300 - 500(1.25) + 112.5(0.5) = 0$$

$$\sum F_x = A_x = 0$$

$$\sum F_y = A_y + B_y - 150 - R_1 + R_2$$

$$\equiv A_y + B_y - 150 - 500 + 112.5 = 0$$

The solution of these equations is

$$B_y = 347.5 \text{ lb}, \qquad A_x = 0, \qquad A_y = 190 \text{ lb}$$

It will be noted that we did not have to find the resultant of the given distributed load. Thus, it was not necessary to replace the individual resultants $\bar{R}_1$ and $\bar{R}_2$ by a total resultant $\bar{R}$, as was done in the previous example.

HOMEWORK PROBLEMS

6.68–6.79 Determine the resultant of the distributed loading and the point of application along the beam of this resultant.

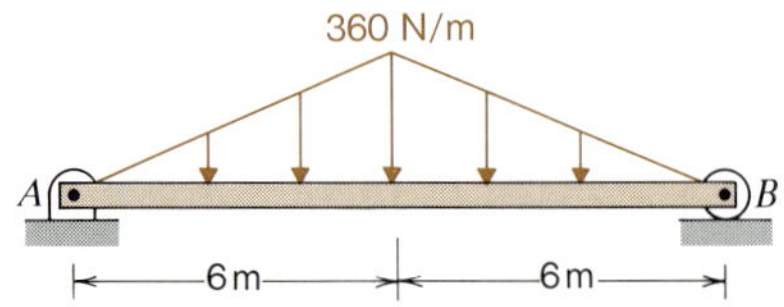

PROBS. 6.68 AND 6.80

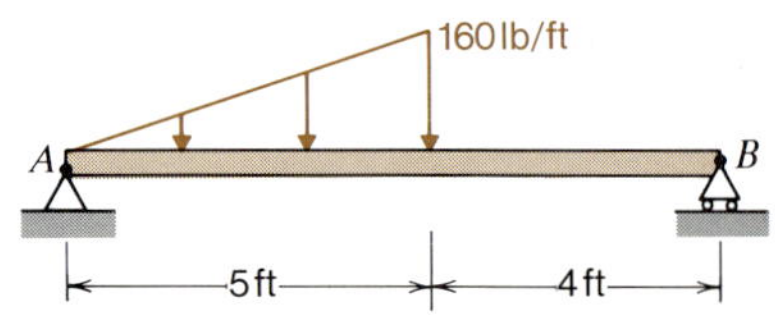

PROBS. 6.69 AND 6.81

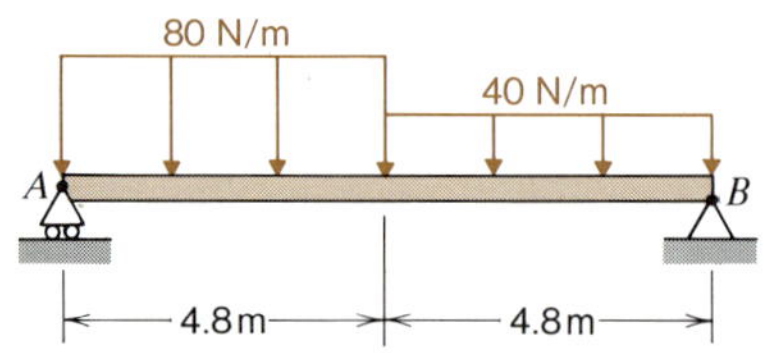

PROBS. 6.70 AND 6.82

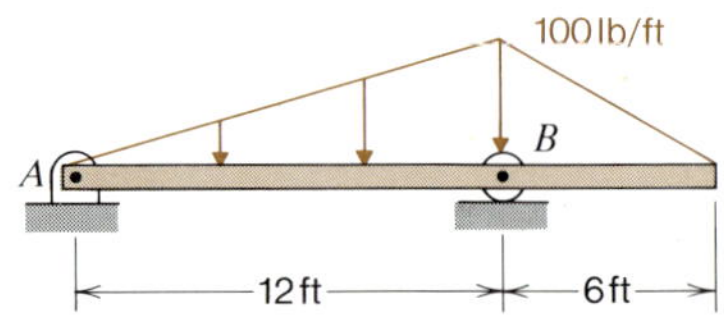

PROBS. 6.71 AND 6.83

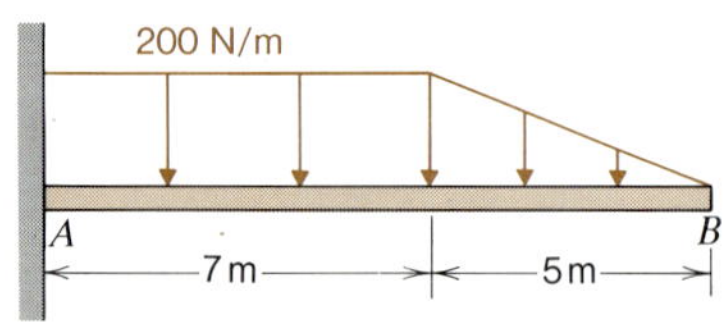

PROBS. 6.72 AND 6.84

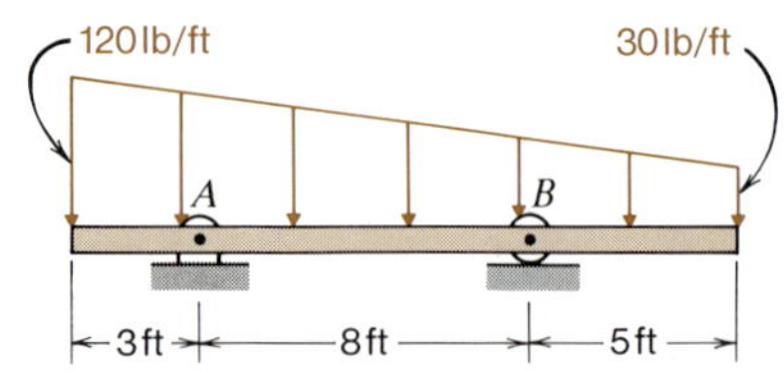

PROBS. 6.73 AND 6.85

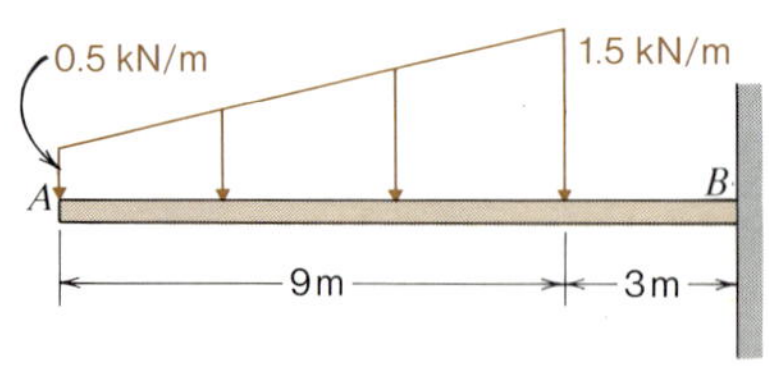

PROBS. 6.74 AND 6.86

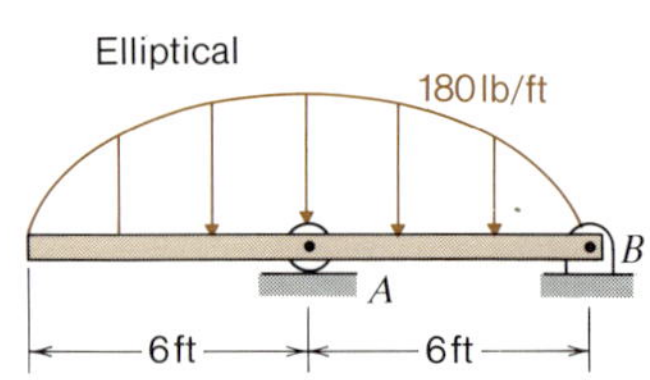

PROBS. 6.75 AND 6.87

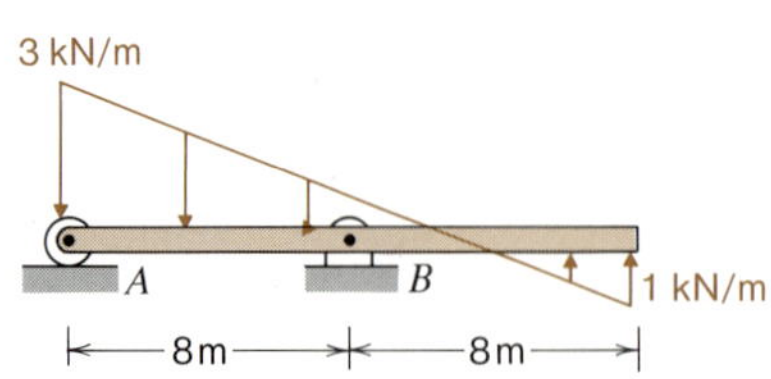

PROBS. 6.76 AND 6.88

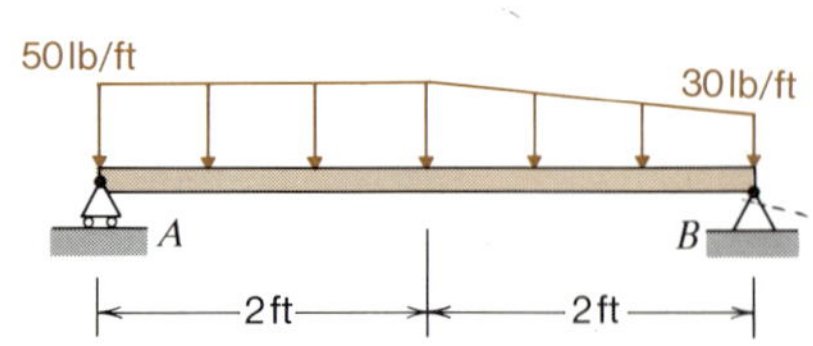

PROBS. 6.77 AND 6.89

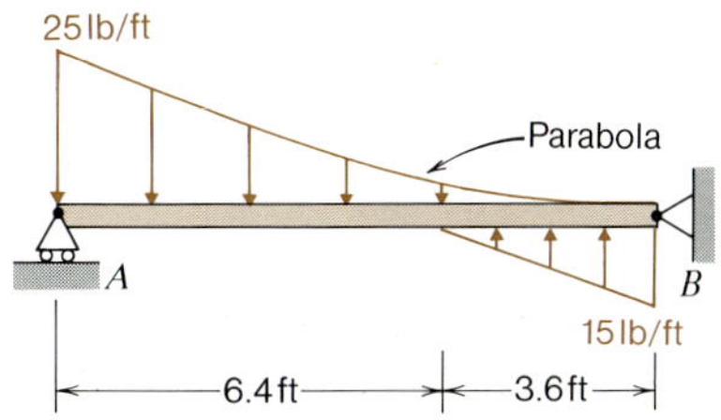

PROBS. 6.78 AND 6.90

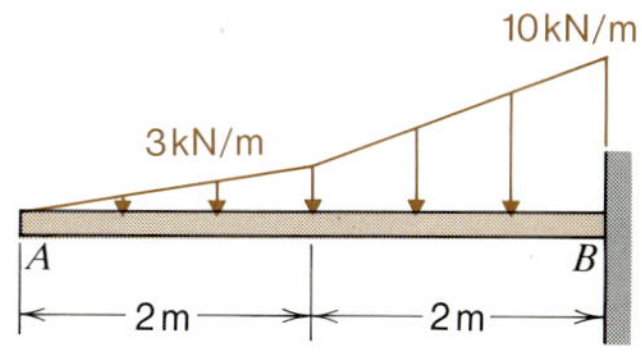

PROBS. 6.79 AND 6.91

6.80–6.91 Determine the reactions at the supports of the beam shown.

6.92–6.95 Determine by integration the resultant of the line load and the point of application along the beam of this resultant.

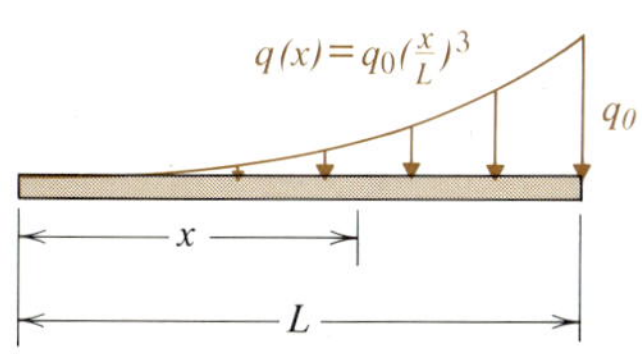

PROB. 6.92

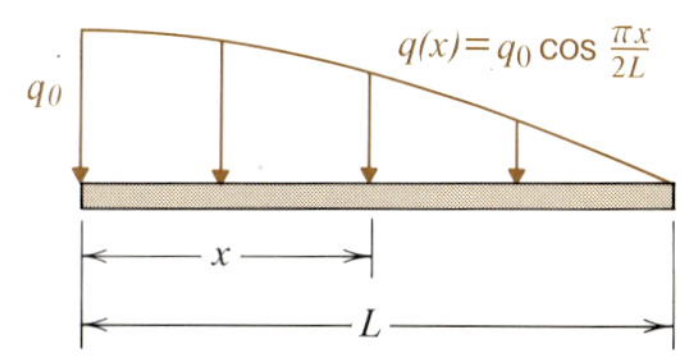

PROB. 6.93

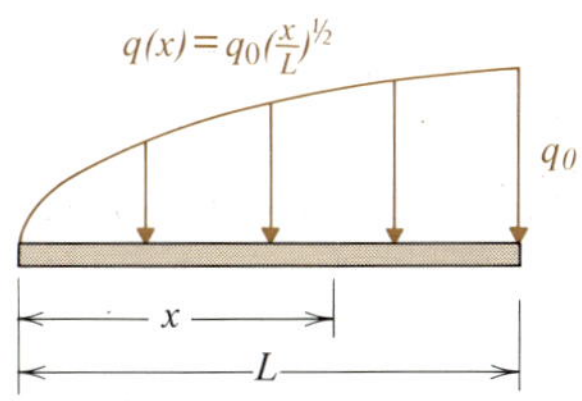

PROB. 6.94

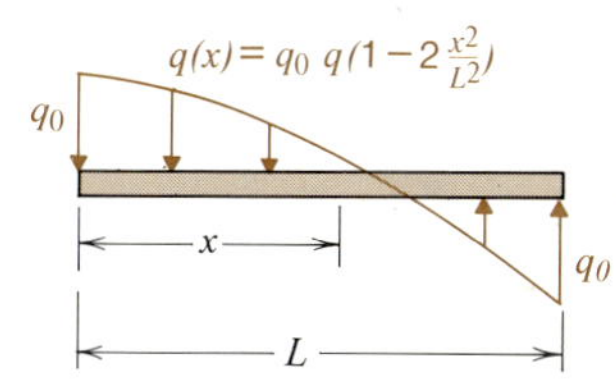

PROB. 6.95

***6.96 and *6.97** The resultant of the loading is a couple. Determine the ratio q_2/q_1.

***6.98 and *6.99** The resultant of the loading acts at the left end of the beam. Determine the ratio q_2/q_1.

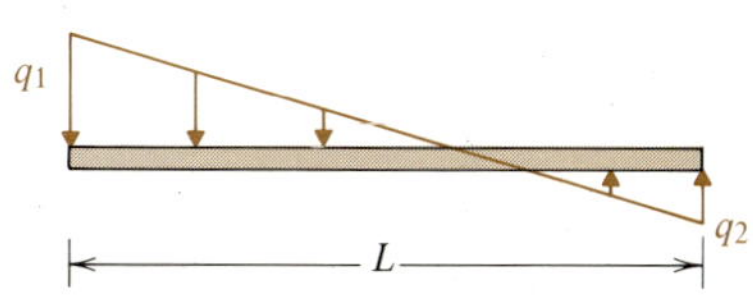

PROBS. 6.96 AND 6.98

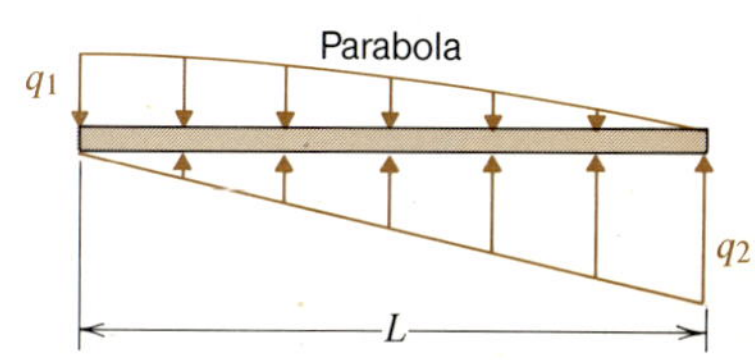

PROBS. 6.97 AND 6.99

2. Pressure Distributions

A distributed force system that acts normal (i.e., perpendicular) to a surface area is called *pressure*. Pressure has the dimensions of force per unit area. It is defined to be positive when it pushes down on the surface. Negative pressure means a suction effect on the surface. For simplicity, we will consider only the case where the surface is flat.

Without loss of generality, the surface will be considered to be coincident with the xy plane. A *pressure distribution* acting on a flat surface may be depicted pictorially by using arrows parallel to the z axis, with the length of each arrow being proportional to the pressure at the point where the arrow intersects the surface. A typical pressure distribution is shown in Figure 13. It can be seen that this pictorial representation forms an imaginary body that we shall call the *pressure space*. The height of the pressure space above each point on the surface is $p(x, y)$, which is the value of the pressure at that point.

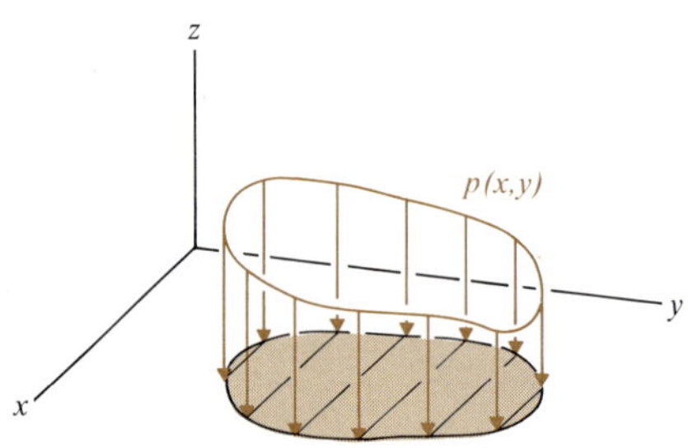

FIGURE 13

The effect of the pressure distribution may be described by considering an infinitesimal element of area $d\mathscr{A}$ surrounding a point (x, y). The pressure acting on $d\mathscr{A}$ is essentially constant at $p(x, y)$. Thus, the portion of the pressure distribution acting on the area $d\mathscr{A}$ is equivalent to an infinitesimal concentrated force $p(x, y)d\mathscr{A}$ parallel to the z axis. This force acts at point (x, y) on the surface. One such force is shown in Figure 14a.

The total effect of the pressure distribution acting at each point of the area $\mathscr{A}$ is an infinite number of parallel infinitesimal forces like the one appearing in Figure 14a. A set of parallel forces may be replaced by a single

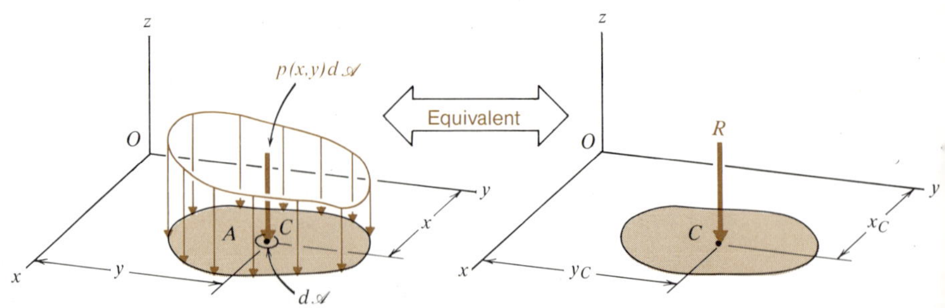

FIGURE 14a **FIGURE 14b**

equivalent force. Therefore, the pressure distribution in Figure 14a is equivalent to a concentrated force $\bar{R}$, normal to the surface and acting at some point C, as shown in Figure 14b.

The summation of each infinitesimal force becomes an integral over the surface area, so

$$R = \int_{\mathscr{A}} p(x, y)\, d\mathscr{A} \tag{16}$$

The coordinates x_c and y_c required to specify where $\bar{R}$ is applied to the surface are obtained by equating the moments of the two equivalent force systems in Figure 14. We choose origin O for the moment sum, and add the moment of each infinitesimal force by means of an integral. This yields

$$\sum \bar{M}_O = (x_c\bar{i} + y_c\bar{j}) \times (-R\bar{k}) = \int_{\mathscr{A}} (x\bar{i} + y\bar{j}) \times \{-p(x, y)\, d\mathscr{A}\bar{k}\}$$

The result of carrying out the cross products and equating matching components is

$$\begin{aligned} Rx_c &= \int_{\mathscr{A}} xp(x, y)\, d\mathscr{A} \\ Ry_c &= \int_{\mathscr{A}} yp(x, y)\, d\mathscr{A} \end{aligned} \tag{17}$$

The point on the loaded surface whose coordinates are the values x_c and y_c is called the *center of pressure*.

As we did for line loads, we may develop an analogy that is useful for evaluating the resultant. Recall that the height of the pressure space above the xy plane is $p(x, y)$. It follows that $p(x, y)\, d\mathscr{A}$ is a differential element of volume for the pressure space. In this interpretation, eq. (16) gives the volume of the pressure space, and eqs. (17) represent the first moments of volume with respect to the x and y coordinates. Hence

> The magnitude of the resultant of a pressure distribution acting on a planar surface is the volume of the pressure space. The direction of this resultant force is perpendicular to the surface, in the sense of the pushing effect for positive pressure. The line of action of the resultant force intersects the geometric centroid of the pressure space.

A negative pressure should be regarded as a negative contribution to the pressure space.

The importance of the analogy between the resultant $\bar{R}$ and the pressure space is that it will frequently enable us to make use of tabulated geometric properties when determining the resultant force. This occurs whenever the pressure space can be represented in terms of the fundamental shapes given in Appendix B.

In some situations a pressure distribution may be simplified to an equivalent line load. Of particular importance is the case where a pressure distribution acting on a rectangular area does not vary across the width of the rectangle. The intensity of the equivalent line load is $q = pb$, where b is the width of the rectangle in the direction of constant pressure. This line load is taken to act along the centerline of the rectangular area, as illustrated in Figure 15 for a pressure distribution that depends only on the x coordinate.

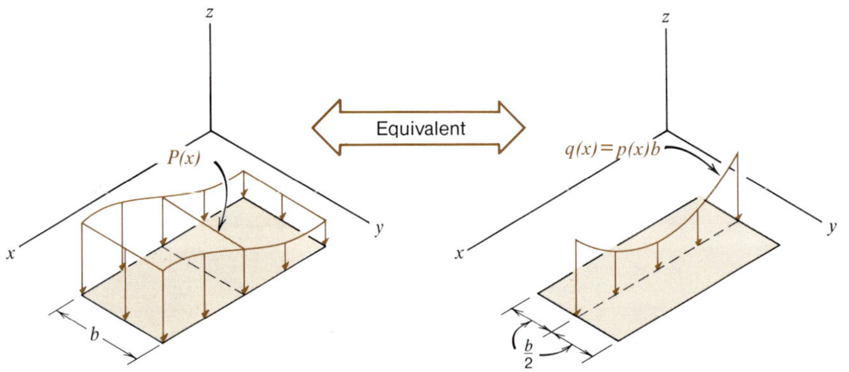

FIGURE 15

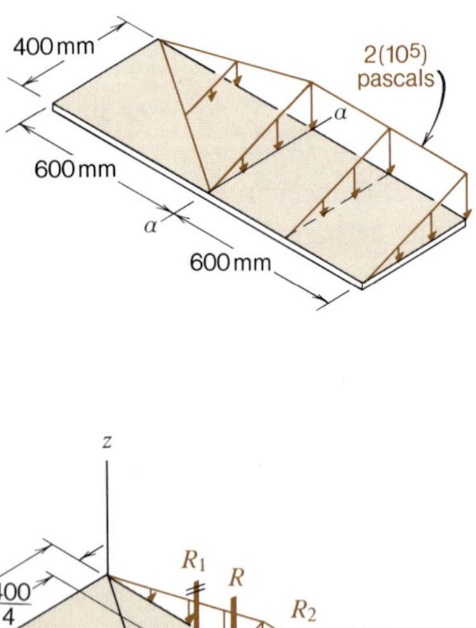

EXAMPLE 13

The pressure distribution on a rectangular plate is as shown, with a maximum value of $2(10^5)$ pascals. Determine the resultant of this pressure distribution and the corresponding center of pressure.

Solution

Although the pressure is applied to a rectangular area, it is not constant in the direction of either edge. Therefore, we do not seek an equivalent line load. The given diagram depicts the pressure space to be an orthogonal tetrahedron to the left of line a–a and a prism to the right of line a–a. Recognizing these basic shapes means that we may employ the analogy between the resultant and the geometric properties of the pressure space.

We begin with a sketch that shows the total resultant $\bar{R}$ and the resultants $\bar{R}_i$ $(i = 1, 2)$ for the basic shapes forming the pressure space. The forces $\bar{R}_i$ intersect the centroids of their spaces, so the sketch also shows the location of the centroid of each basic shape in the xy plane, as found in Appendix B.

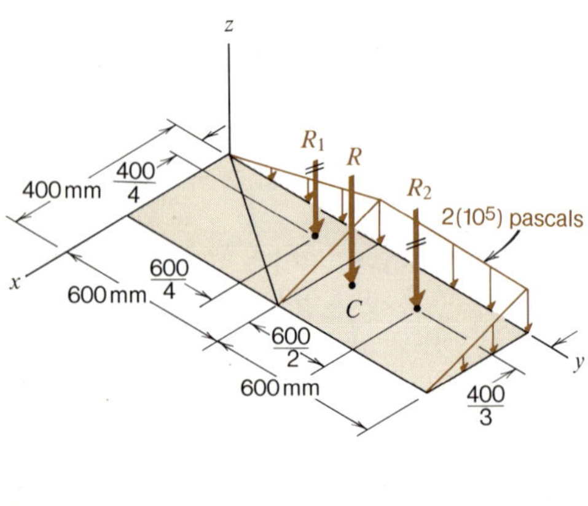

The values of R_1 and R_2 are calculated as the volume of the corresponding pressure spaces. By definition 1.0 pascal $\equiv$ 1.0 N/m², so we have

$$R_1 = \tfrac{1}{6}(0.4)(0.6)2(10^5)\text{ N} = 8.00\text{ kN}$$

$$R_2 = \tfrac{1}{2}(0.4)(0.6)2(10^5)\text{ N} = 24.00\text{ kN}$$

$$R = R_1 + R_2 = 32.0\text{ kN}$$

The center of pressure, being the point where the resultant $\bar{R}$ is applied to the plate, is determined by equating the moment of $\bar{R}$ to the total moment of $\bar{R}_1$ and $\bar{R}_2$. Referring to the sketch for the position vectors, we then have

$$\begin{aligned}(x_c\bar{i} + y_c\bar{j}) \times \bar{R} &= [0.10\bar{i} + (0.60 - 0.15)\bar{j}] \times \bar{R}_1 \\ &\quad + [0.133\bar{i} + (0.60 + 0.30)\bar{j}] \times \bar{R}_2 \text{ N-m} \\ (x_c\bar{i} + y_c\bar{j}) \times (-32.0\bar{k}) &= (0.10\bar{i} + 0.45\bar{j}) \times (-8.0\bar{k}) \\ &\quad + (0.133\bar{i} + 0.90\bar{j}) \times (-24.0\bar{k}) \\ -32.0y_c\bar{i} + 32.0x_c\bar{j} &= -25.2\bar{i} + 4.0\bar{j}\end{aligned}$$

Equating corresponding components then yields

$$x_c = 0.125 \text{ m} \qquad y_c = 0.788 \text{ m}$$

HOMEWORK PROBLEMS

6.100–6.105 The indicated uniform wind pressure is distributed over the shaded area of the traffic sign. Determine the resultant force exerted by the wind and the corresponding center of pressure.

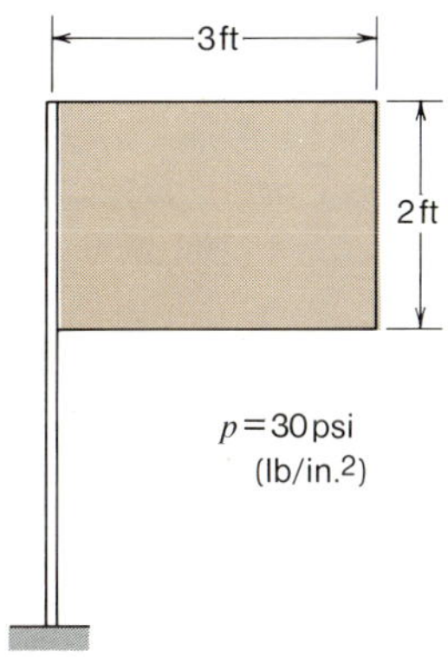

PROB. 6.100

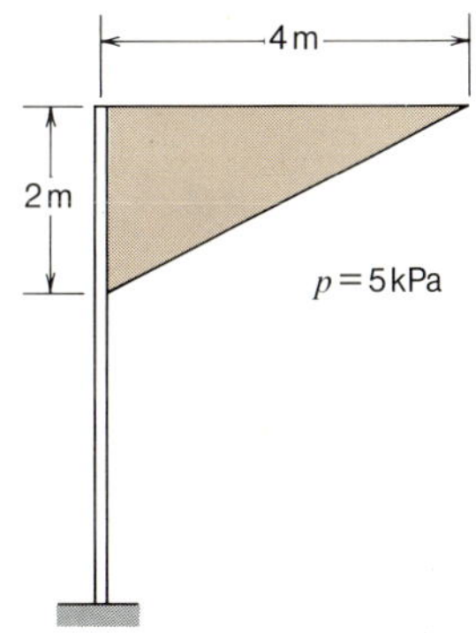

PROB. 6.101

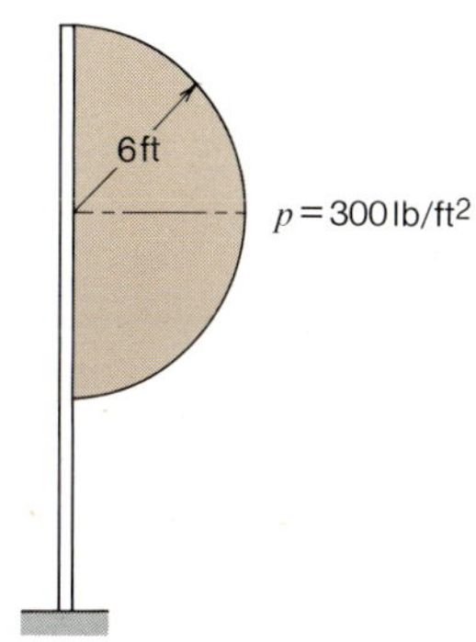

PROB. 6.102

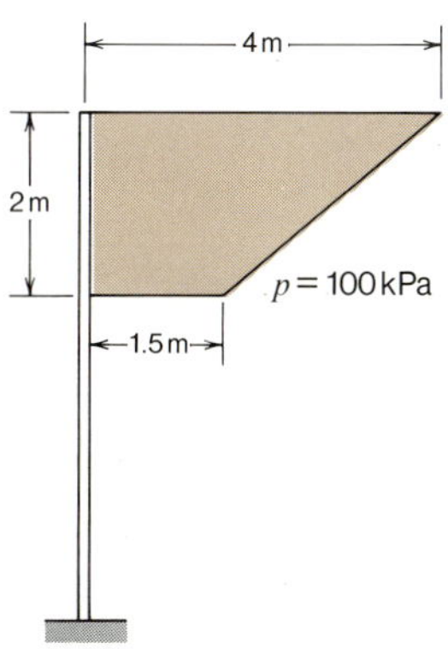

PROB. 6.103

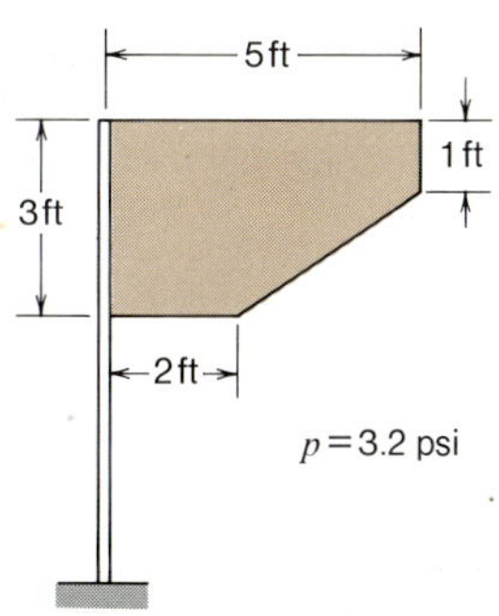

PROB. 6.104

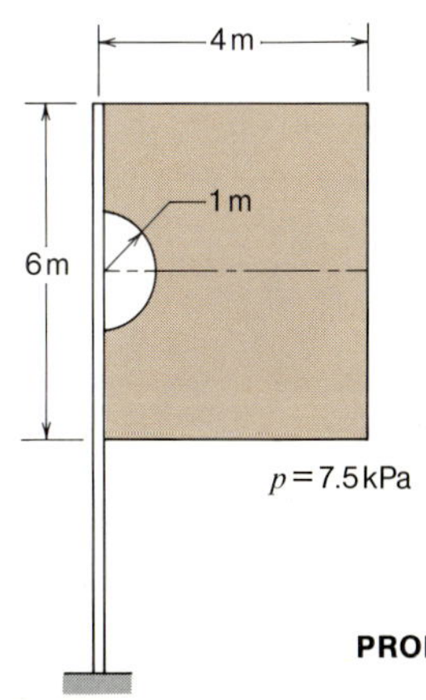

PROB. 6.105

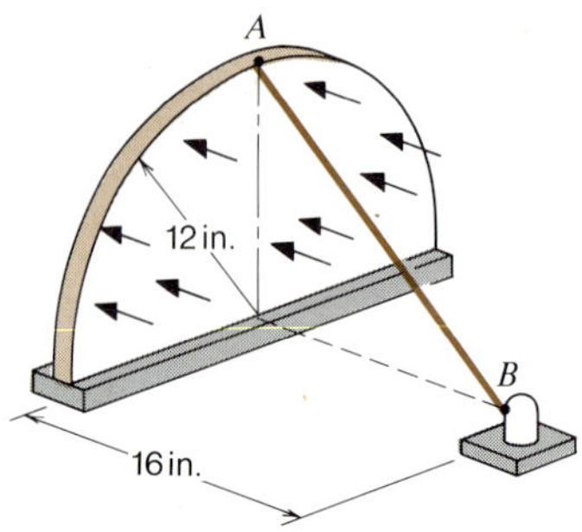

PROB. 6.106

6.106 The semicircular plate is supported in a wind tunnel by a groove along its lower edge and by cable AB. The lateral pressure on the plate is 40 psi. Determine (a) the resultant wind force on the plate, (b) the tension in cable AB. The wind is blowing from the right.

6.107–6.110 The snow load on a roof is as shown. Determine the resultant force exerted by the snow on the roof. Specify the point of application of this force.

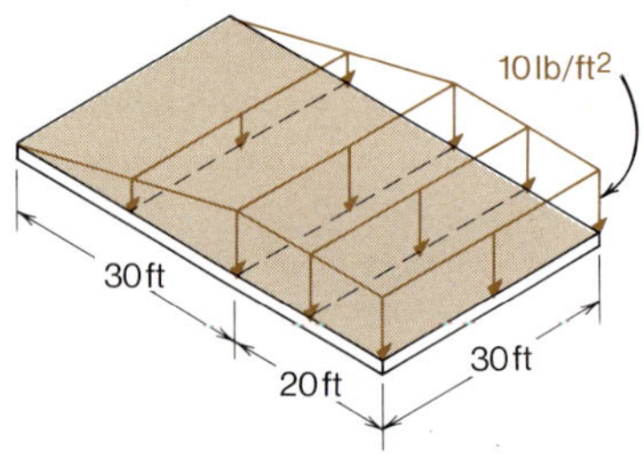

PROB. 6.107

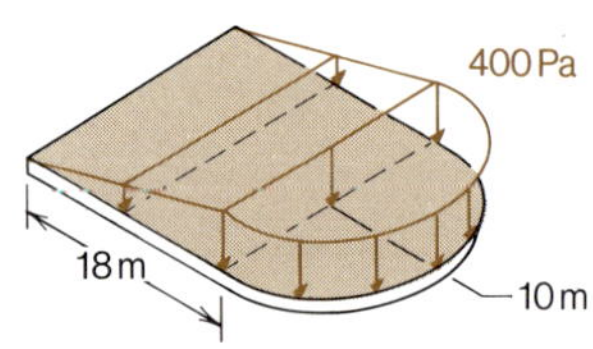

PROB. 6.108

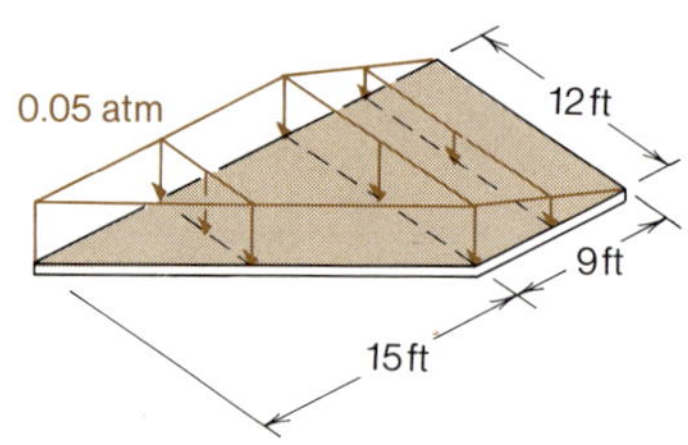

PROB. 6.109

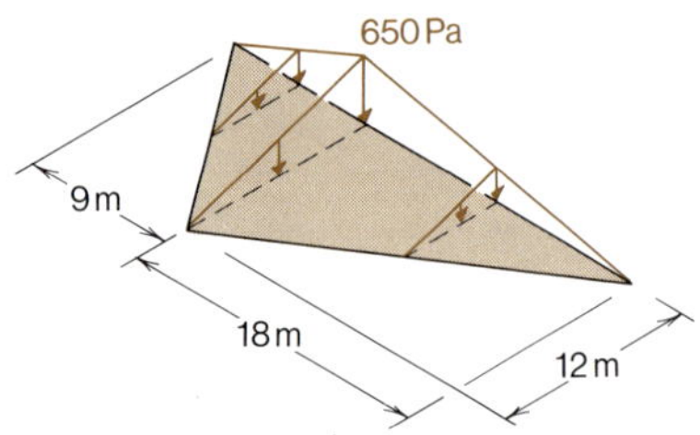

PROB. 6.110

***6.111 and *6.112** Determine by integration the resultant and the center of pressure for the given pressure distribution.

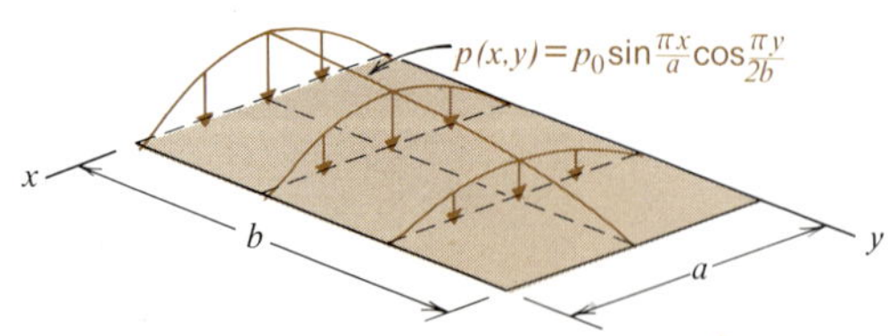

PROB. 6.111

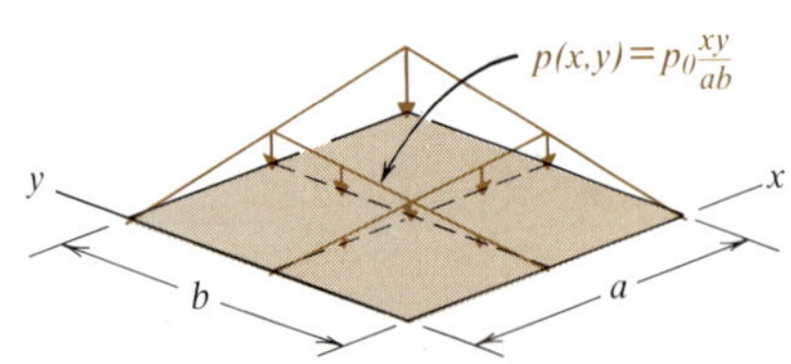

PROB. 6.112

C. STATICS OF LIQUIDS

A common source of distributed loads is a fluid, such as oil, water, or air. The distinguishing characteristic of a *fluid* is that, if it is at rest, it can apply only a pressure loading to another stationary body.

The density of a *gas* changes radically with large changes in pressure. In contrast, the density of a *liquid* changes slightly with pressure, so a liquid may be modeled as a fluid of constant density. In other words, a liquid is essentially *incompressible*. Here we shall study only liquids, for the study of gases requires the consideration of the temperature of the material.

When a liquid is placed in an open container, it forms a horizontal surface, called the *free surface*. The pressure in a liquid is related directly to the vertical distance below the free surface. To develop this relationship, consider the element of liquid shown as a rectangular parallelepiped in Figure 16. Note that the xy plane is chosen to coincide with the free surface.

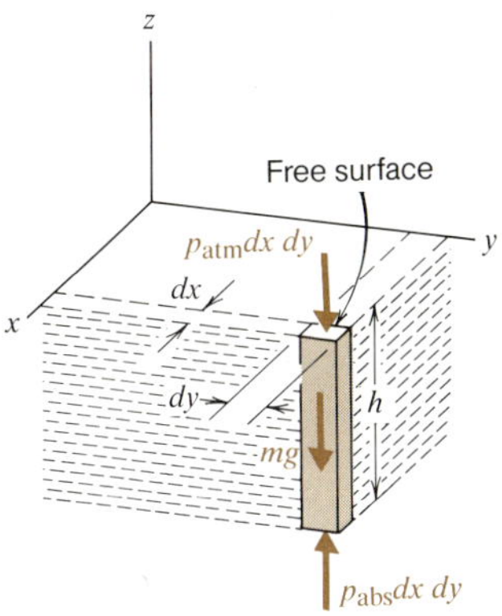

FIGURE 16

The pressure at the upper face of the element depicted in Figure 16 is the atmospheric pressure, p_{atm}. It results from the atmosphere pushing on the free surface; for standard conditions, $p_{atm} = 1\text{ atm} = 1.0133(10^5)\text{ N/m}^2 = 101.33\text{ kPa} = 14.7\text{ psi}$. The total, or absolute, pressure acting on the lower face is denoted as p_{abs}.

Only the vertical forces are depicted in Figure 16. The resultant force acting on the upper and lower faces is the corresponding pressure multiplied by the area, $dx\,dy$, of these faces. Because a fluid exerts only pressure forces, there are no vertical force components acting on the vertical faces. The other vertical force acting on the element we have isolated is the weight mg. For static equilibrium in the vertical direction, it must be that

$$\sum F_z = p_{abs}\,dx\,dy - p_{atm}\,dx\,dy - mg = 0$$

The mass m of the element is the product of the density, ρ, of the fluid and the volume, $h\,dx\,dy$, of the element. When this mass is substituted into the foregoing equilibrium equation, it is found that

$$p_{abs} = p_{atm} + \rho g h$$

In most systems the important parameter is the increase of pressure from the atmospheric value, because the condition of a body after it is constructed defines the reference state of zero loading. The difference $p_{abs} - p_{atm}$ is defined as the *gage pressure*, p. (The word *gage* is not misspelled!) Thus, the gage pressure is given by

$$\boxed{p = \rho g h} \qquad (18)$$

Equation (18) defines a *hydrostatic pressure distribution*.

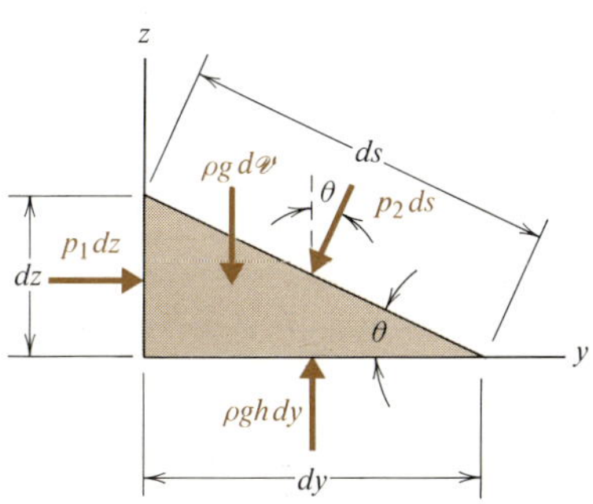

FIGURE 17

As derived, the pressure given by eq. (18) is that acting on a horizontal plane located a distance h below the free surface. We will now prove that this pressure is exerted equally on surfaces having any orientation at a depth h below the free surface. This proof is accomplished by considering the small prismatic element of fluid dm in Figure 17. Its width in the horizontal direction perpendicular to the plane of the diagram is unity.

We denote the pressures on the inclined and vertical faces of the element by the unknown values p_1 and p_2. The force on each face depicted in the figure is obtained by multiplying the pressure on that face by the area of the face (remembering that the width is unity).

The equations of equilibrium for the element of fluid are

$$\sum F_y = p_1\,dz - p_2\,ds\sin\theta = 0$$

$$\sum F_z = \rho gh\,dy - p_2\,ds\cos\theta - \rho g\,d\mathcal{V} = 0$$

where ρg is the *specific weight* (units of force per unit volume) of the fluid. The weight of the element $\rho g\,d\mathcal{V} = \rho g\,dy\,dz/2$, being a second-order differential, is negligible when compared to the other terms in the z force sum. The $\sum F_z$ equation then reduces to

$$\sum F_z = \rho gh\,dy - p_2\,ds\cos\theta = 0$$

From Figure 17 we note that $ds\sin\theta = dz$ and $ds\cos\theta = dy$. Thus the equilibrium equations yield

$$p_1 = p_2 = \rho gh$$

This result was obtained for an arbitrary value of θ, so we may conclude that

> The gage pressure at a point in a liquid is proportional to the depth h of that point below the free surface, according to $p = \rho gh$. This pressure is exerted equally on all surfaces at that depth, regardless of their orientation.

The second sentence in this statement is known as Pascal's law. It should be noted that in the U.S. Customary system of units it is common to prescribe the specific weight ρg of a material, whereas in the SI system the density ρ of a material is the preferred prescribed parameter.

Now that the relationship governing the pressure in a liquid has been established, we may proceed to the study of the resultant force exerted by a liquid on another object. We will consider three special types of surfaces that occur frequently in engineering practice. The case of an arbitrary surface is left for texts on fluid mechanics.

1. Rectangular Surfaces

Consider the situation where two edges of a submerged rectangular surface are horizontal. Any line in the rectangle parallel to these edges is situated at a constant depth h, so the hydrostatic pressure is constant along this line. As a result, the pressure distribution acting on the rectangular surface may be converted into an equivalent line load. This is done by multiplying the pressure by the horizontal width b of the rectangle.

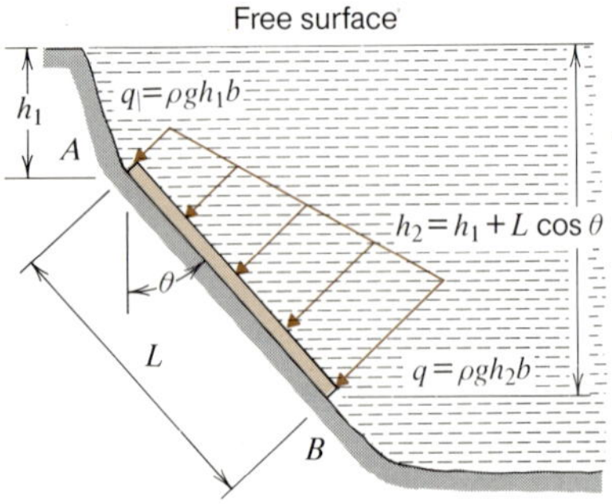

FIGURE 18

A typical case of the pressure on a rectangular plate of length L that is exposed to a liquid on only one side is shown in side view in Figure 18, where b is the width of the surface perpendicular to the plane of the figure.

At the upper edge A the depth is h_1. Here, the gage pressure is ρgh_1, and the line load is $q = pb \equiv \rho gh_1 b$. At the lower edge B the depth is h_2, so the line load at this location is $q = \rho gh_2 b$. Because the distance along the line

AB is linearly related to the depth, the line load is trapezoidal, as illustrated. Obvious exceptions occur when edge A is at the free surface, in which case $h_1 = 0$ and the loading is a triangular area; and when the plate is horizontal ($\theta = 90°$), for then $h_1 = h_2$ and the loading function forms a rectangle.

EXAMPLE 14

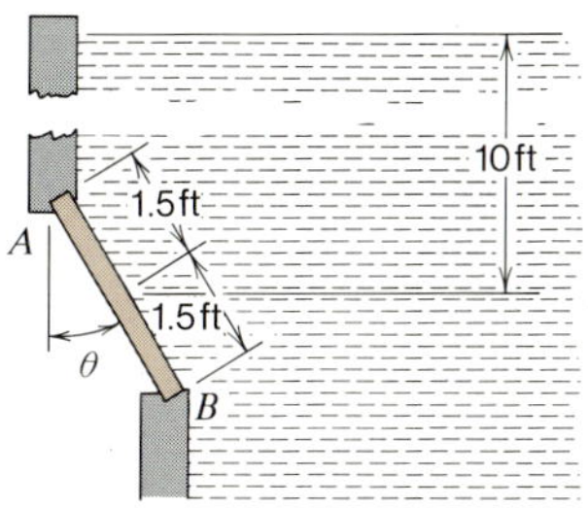

The center of the rectangular window in an aquarium is 10 ft below the surface. The horizontal width of the window is 2 ft. Determine the resultant force exerted by the water on the window and the location of the center of pressure in terms of the angle θ.

Solution

We denote the pressure at the upper and lower edges of the window as p_A and p_B, respectively. The intensity of the line load at each location is obtained by multiplying the pressure by 2 ft, which is the horizontal width of the window. The line load acting on the window is shown in the sketch.

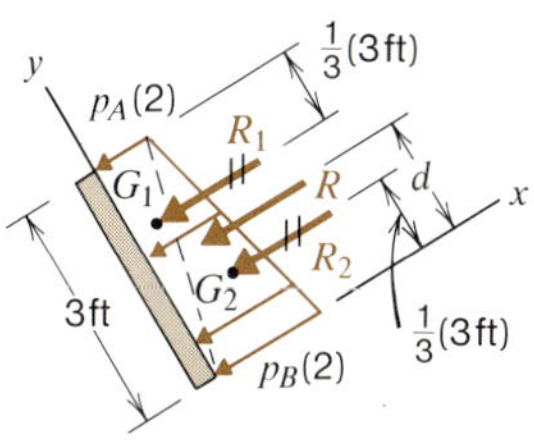

Edge A is 1.5 cos θ meters in the vertical direction above the center, and edge B is the same vertical distance below the center. The specific weight of water is $\rho g = 62.4\ \text{lb/ft}^3$, so

$$p_A = \rho g h_A = 62.4(10 - 1.5 \cos \theta)\ \text{lb/ft}^2$$

$$p_B = \rho g h_B = 62.4(10 + 1.5 \cos \theta)\ \text{lb/ft}^2$$

We shall treat the trapezoidal line load as a composite of two triangular loads, as indicated in the sketch. The resultant force for each triangle is the area of the triangle, acting perpendicular to the window. Thus

$$R_1 = \tfrac{1}{2}(2\, p_A)(3) = 187.2(10 - 1.5 \cos \theta)\ \text{lb}$$

$$R_2 = \tfrac{1}{2}(2\, p_B)(3) = 187.2(10 + 1.5 \cos \theta)\ \text{lb}$$

These two forces are equivalent to the total resultant $\bar{R}$, so

$$R = R_1 + R_2 = 187.2(20) = 3744\ \text{lb}$$

The center of pressure is the location along the window where $\bar{R}$ acts. The dimension d locating this point is found by equating the moments of $\bar{R}_1$ and $\bar{R}_2$ to the moment of $\bar{R}$ about some convenient point, which we choose to be point B. Recall that $\bar{R}_1$ and $\bar{R}_2$ intersect the centroids of the corresponding triangles. Thus

$$\sum M_{Bz} = R(d) = R_1(2) + R_2(1)$$

$$3744d = 187.2(10 - 1.5 \cos \theta)2 + 187.2(10 + 1.5 \cos \theta)(1)$$

$$\equiv 5616 - 280.8 \cos \theta$$

$$d = 1.5 - 0.075 \cos \theta\ \text{ft}$$

Additional Remarks

It will be noted that the resultant force is independent of the angle of inclination θ. Also, the distance d locating the center of pressure is always less than 1.5 ft (assuming that $\theta < 90°$), so the center of pressure is lower than the center of the window. These results will be shown in a later section to be true for any flat surface.

EXAMPLE 15

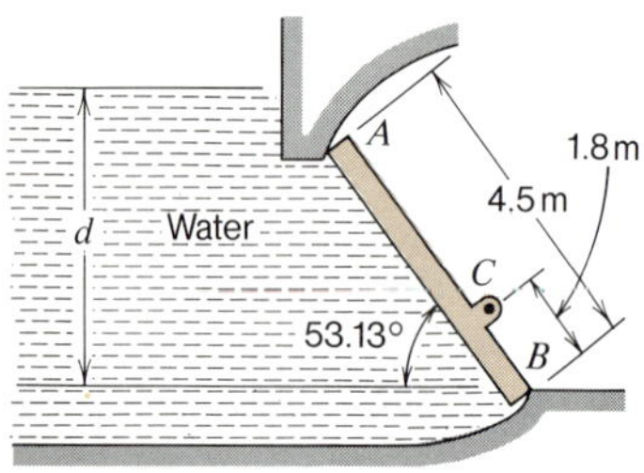

The automatic valve shown in cross section consists of a 4.5 by 3 m rectangular plate AB that can rotate about a horizontal shaft passing through point C. Neglecting the weight of this plate, determine the depth of water d in the reservoir for which the valve will open.

Solution

We shall draw a free body diagram of the valve plate in order to investigate the effect of the pressure distribution on the static equilibrium of the system. There is no contact at edges A and B when the valve is about to open, so there are no reaction forces at either location.

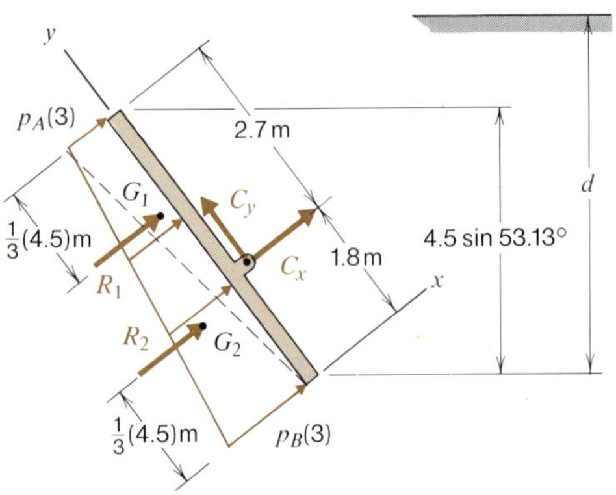

The value of the line load at the upper and lower edges is the pressure at each location multiplied by the 3 m width. The depth at edge B is the unknown quantity d, and edge A is 4.5 sin 53.13° = 3.6 m higher in the vertical direction than edge B. We will keep the specific weight of water in algebraic form, so

$$p_B = \rho g d, \qquad p_A = \rho g(d - 3.6)\ \text{N/m}^2$$

We treat the trapezoidal line load as the composite of two triangles; the resultant force for each is the corresponding triangular area.

$$R_1 = \tfrac{1}{2}[p_A(3)](4.5) = \rho g(6.75)(d - 3.6)$$

$$R_2 = \tfrac{1}{2}[p_B(3)](4.5) = \rho g(6.75)d$$

Each resultant acts at the centroid of the corresponding triangle.

There is no need to combine $\bar{R}_1$ and $\bar{R}_2$ into a single resultant force, because we are concerned only with the static equilibrium of the valve plate. The reactions at the shaft C are of no interest, so we shall sum moments about that point.

$$\sum M_{Cz} = -R_1\left(2.7 - \frac{4.5}{3}\right) + R_2\left(1.8 - \frac{4.5}{3}\right) = 0$$

Substituting the above expressions for R_1 and R_2 yields

$$-\rho g(6.75)(d - 3.6)(1.2) + \rho g(6.75)d(0.3) = 0$$

$$d = 4.80\ \text{m}$$

Before we interpret this result, we note that $d = 4.8$ m means that the free surface is higher than edge A. If the value of d was found to be less than

3.6 m, which is the vertical distance from edge B to edge A, we would have to redo the solution. There would then be only a triangular line load over that portion of the valve plate that is immersed in the fluid.

The value $d = 4.8$ m that we obtained represents the condition where the valve plate is on the verge of rotating. When $d < 4.8$ m, the moment of R_2 about point C is larger than the moment of R_1 (this may be seen by considering the case where the free surface is at edge A). Thus, when $d <$ 4.8 meters, the water pressure tends to cause the valve plate to rotate counterclockwise. The supports at edges A and B do not permit such rotation, so $d = 4.8$ meters is the maximum depth for which the gate remains closed. It will be noted that the value of ρg was not needed, which means that the valve would open at a depth of 4.8 m for any liquid.

HOMEWORK PROBLEMS

6.113 and 6.114 Determine the hydrostatic force acting on the 10 ft section of seawall shown when the water is at flood stage and locate the corresponding center of pressure. The density of seawater is 2.5% greater than that of fresh water.

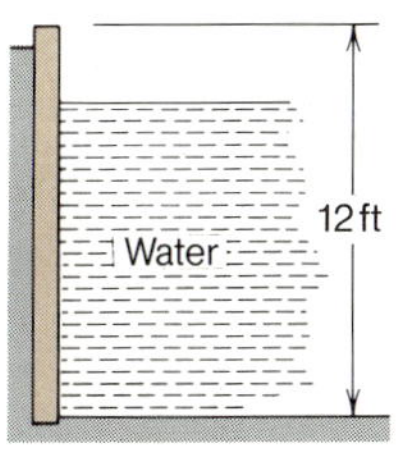

PROB. 6.113

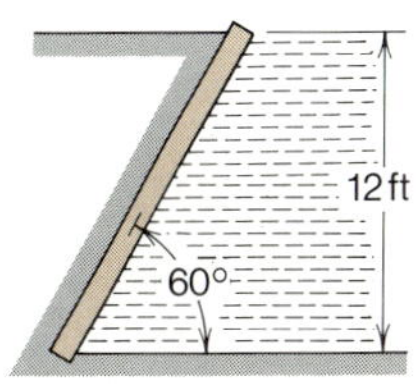

PROB. 6.114

6.115 The fish tank holds water to a depth of 1.5 m. For each face, including the bottom, determine the resultant force exerted by the water. Also determine the point of application of each resultant force (the center of pressure).

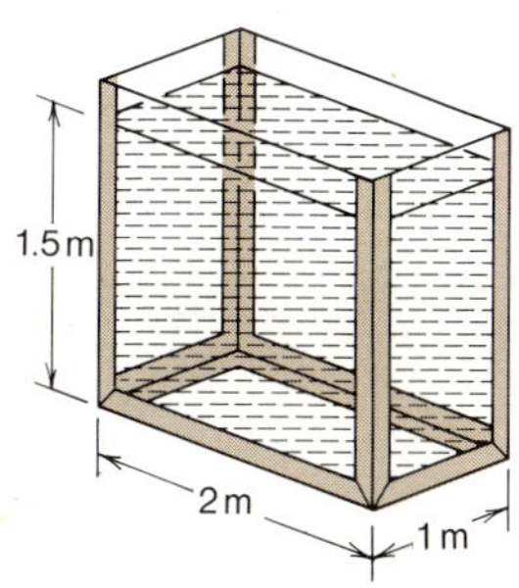

PROB. 6.115

6.116 and 6.117 The horizontal width of the gate at the bottom of a retaining wall is 6 ft. Determine the resultant force exerted by the water on the gate and the corresponding center of pressure.

6.118 and 6.119 The horizontal width of the sliding gate at the bottom of a retaining wall is 6 ft. If the couple reactions at the upper and lower edges are negligible, determine the forces exerted by the grooves to resist the water pressure.

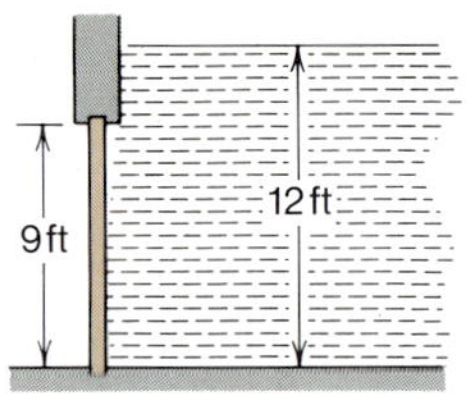

PROBS. 6.116 AND 6.118

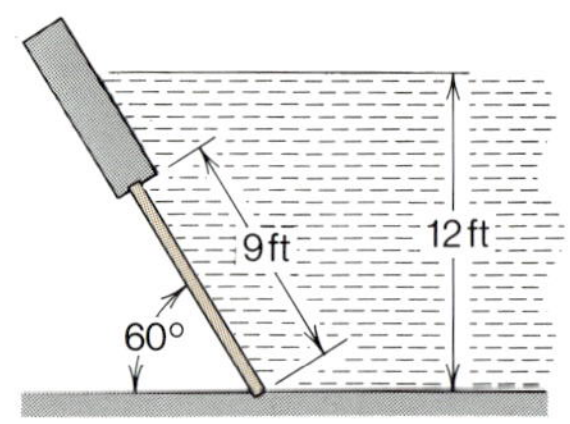

PROBS. 6.117 AND 6.119

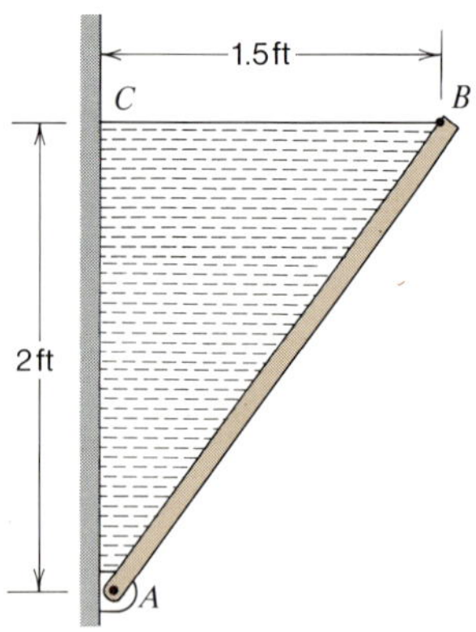

PROB. 6.120

6.120 The lower edge of panel AB in the triangular trough is supported by a series of pins. The upper edge is supported by a total of 20 cables that are equally spaced along the 40 ft length of the trough. Determine the tension in a cable when the trough is filled to the top with water.

6.121 and 6.122 The 750 mm wide gate, which may pivot about the horizontal shaft A, is held in the closed position by the preloaded spring. Determine the compressive force in the spring for which the gate will open when the depth of the oil is $d = 1.5$ m.

6.123 Solve Problem 6.121 for $d = 3$ m.

6.124 Solve Problem 6.122 for $d = 3$ m.

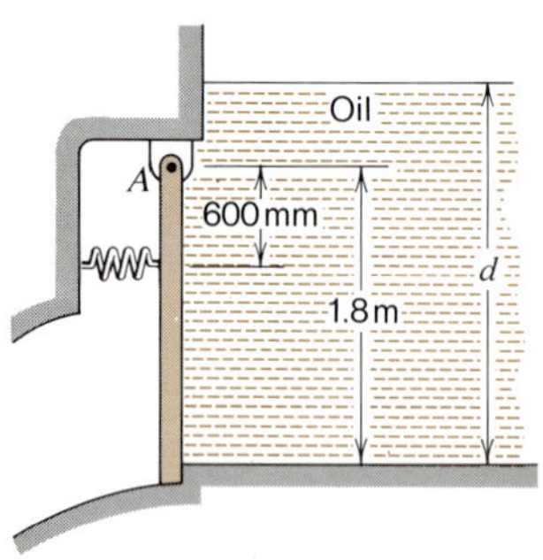

PROB. 6.121

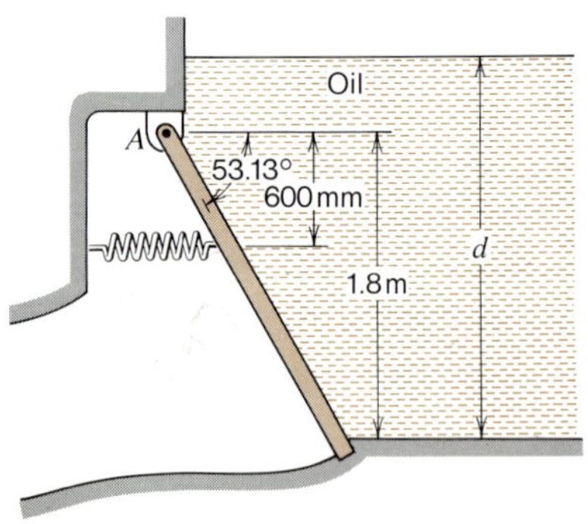

PROB. 6.122

6.125 and 6.126 The cross section of a 60 ft long section of concrete formwork is shown. The left upright panel is attached to the anchored panel on the right by 40 equally spaced tie rods. Determine the tension in the tie rod when concrete, having a density of 150 lb/ft^3, is in its liquid state. Assume that the lower edge of the upright panel, edge A, is simply supported.

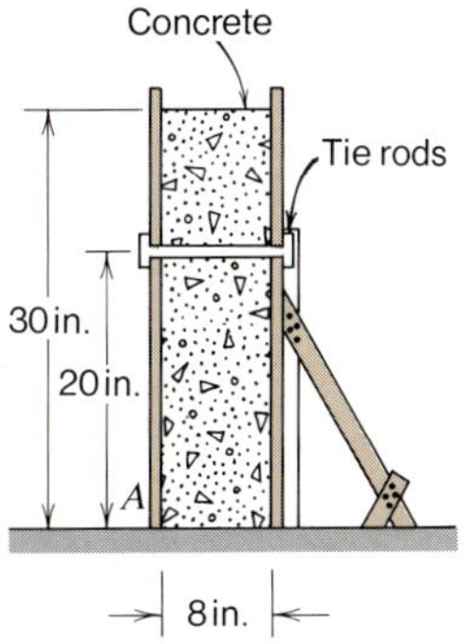

PROB. 6.125

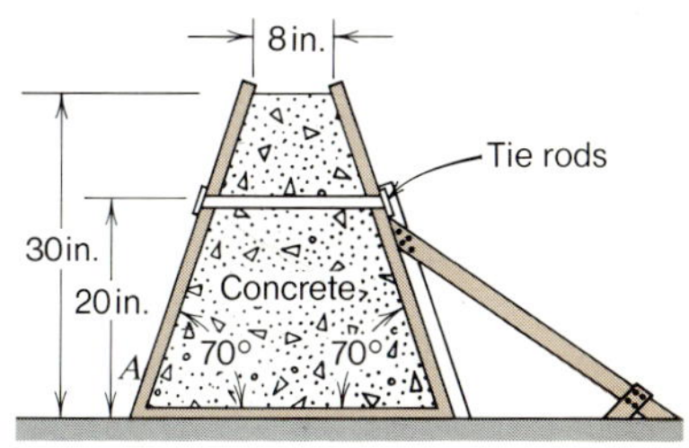

PROB. 6.126

***6.127** Panel AB in the 2 m long tank separates a pool of water from a pool of mercury. Determine the depth d of mercury for which neither cable is slack. Also determine the corresponding magnitude of the resultant forces exerted by the mercury and the water on the panel.

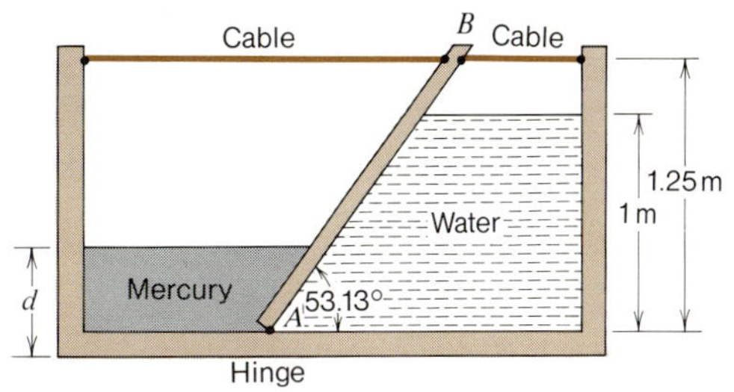

PROB. 6.127

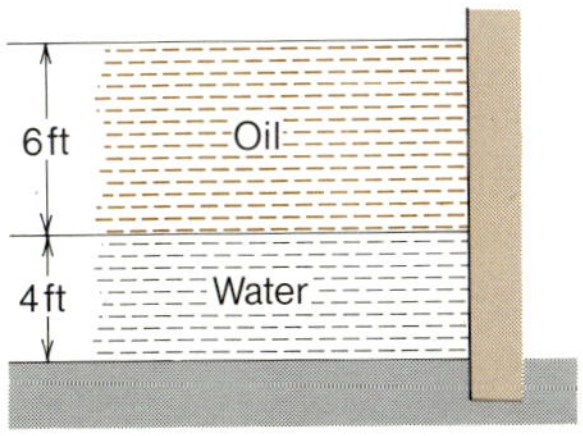

PROB. 6.128

***6.128** The vertical wall, whose horizontal width is 8 ft, is exposed to a layer of oil and a layer of water. Determine the resultant of the combined loading of oil and water and the corresponding location of the center of pressure. *Hint:* The hydrostatic pressure distributions in the oil and the water are proportional to the corresponding densities, and the pressure does not change as the oil–water interface is crossed.

*2. Cylindrical Surface

In this section we shall describe the hydrostatic force exerted on a curved surface that is a sector of a horizontal (arbitrary) cylinder. A side view of such a surface, whose cross section in the vertical plane is the curve AB, appears in Figure 19. The horizontal width (perpendicular to the plane of the diagram) of this surface is denoted as b.

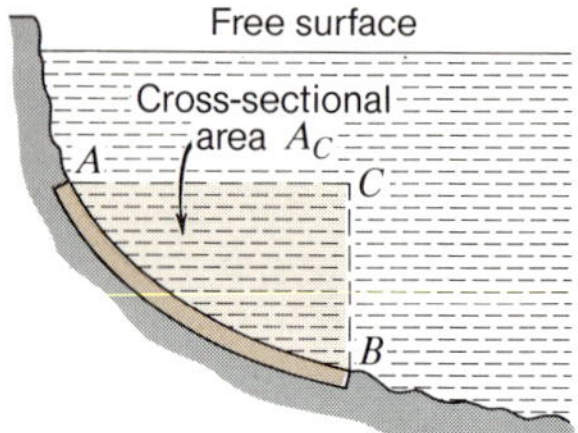

FIGURE 19

Because the surface is curved, the direction of the pressure force exerted by the liquid is not constant. We have not presented methods for treating nonparallel distributed force systems. Instead, we shall develop a method for this problem that avoids having to deal directly with the hydrostatic pressure exerted by the liquid on the curved surface.

This approach will include a portion of the liquid in the free body diagram. Let us pass horizontal and vertical planes through the upper and lower edges of the surface AB, as shown with dashed lines in Figure 19. In this manner a cylindrical body of liquid having cross-sectional area $\mathscr{A}_c$ is formed. This portion of the liquid is isolated from the surrounding liquid in Figure 20.

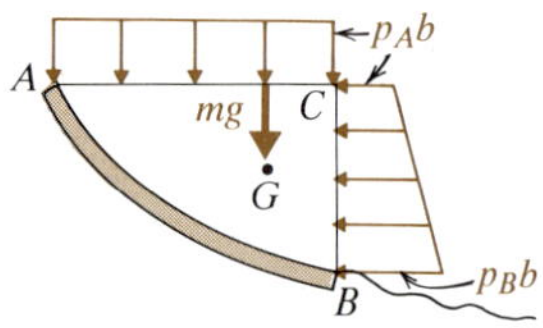

FIGURE 20

We shall now study the loading that is applied to the combination of the curved surface and the cylindrical body of liquid. All points along the horizontal surface AC are at the same depth as edge A, so the pressure distribution on this surface is equivalent to a rectangular line load p_Ab. The pressure acting on the vertical surface BC increases from p_A at edge C to p_B at edge B. This pressure distribution is equivalent to a trapezoidal line load that varies from p_Ab at the top to p_Bb at the bottom.

In addition to these line loads, another force that must be considered is the weight of the cylindrical body of liquid. As viewed in Figure 20, the center of mass G coincides with the centroid of the cross-sectional area $\mathscr{A}_c$. The volume is the area $\mathscr{A}_c$ multiplied by the width, so

$$m = \rho \mathscr{V} = \rho \mathscr{A}_c b$$

The line loads and weight force in Figure 20 *may be replaced by an equivalent force-couple system* acting at a specific point. If desired, the point may be selected to be the one for which the couple vanishes. Because the cylindrical body of liquid is in static equilibrium, *this equivalent system is the resultant of the pressure distribution acting on the curved surface AB.*

The task in some problems is to evaluate the reactions exerted by the supports of the curved surface. In such cases it is not necessary to determine the resultant of the pressure distribution. The line loads and weight of the cylindrical body of liquid may be treated directly as loads in a combined free body diagram of the curved surface and the body of liquid.

EXAMPLE 16

The cylindrical surface shown in cross section is formed from the quarter-circular arc AB. The horizontal width is b. Determine a force-couple system at the center, C, of the arc that is equivalent to the force exerted by the liquid of density ρ on the cylindrical surface.

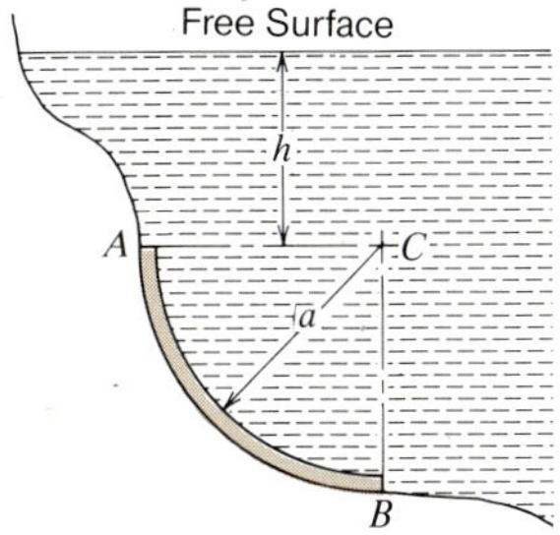

Solution

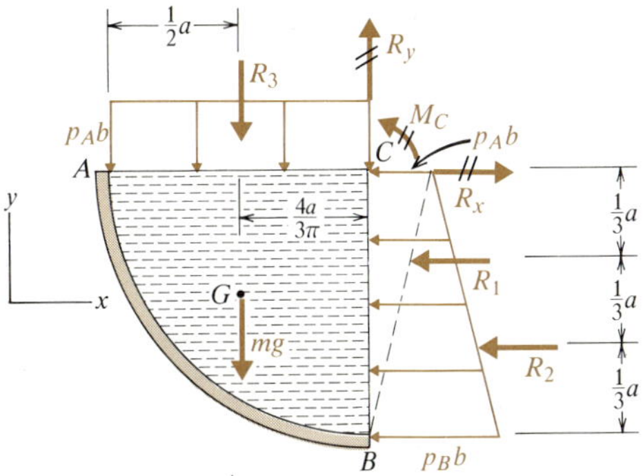

The cylindrical body of liquid we need to consider is a quarter-circular sector when viewed from the side. The sketch shows this body in contact with curved surface AB. The force components R_x and R_y, and the couple $\bar{M}_C$, which act at point C, are marked by double slashes in order to indicate that they are equivalent to the line loads and weight force shown. The depth at edges A and B are h and $h + a$, respectively. Thus

$$p_A = \rho gh, \qquad p_B = \rho g(h + a)$$

The trapezoidal line load is treated as a composite of two triangles, each of whose resultants intersects the centroid of the triangle. Thus

$$R_1 = \tfrac{1}{2}(p_Ab)(a) = \tfrac{1}{2}\rho ghab$$

$$R_2 = \tfrac{1}{2}(p_Bb)(a) = \tfrac{1}{2}\rho g(h + a)ab$$

$$R_3 = (p_Ab)(a) = \rho ghab$$

The cross-sectional area of the body of liquid is $\pi a^2/4$, so its mass is

$$m = \rho \mathscr{A}_c b = \frac{\pi}{4}\rho a^2 b$$

The location of the centroid of the cross-sectional area is listed in Appendix C.

For equivalence of the forces, it must be that

$$\sum F_x = R_x = -R_1 - R_2 \equiv -\tfrac{1}{2}\rho gab(2h + a)$$

$$\sum F_y = R_y = -R_3 - mg = -\rho gab\left(h + \frac{\pi}{4}a\right)$$

Point C is chosen for the evaluation of moment equivalence.

$$\sum M_{Cz} = M_C = R_3\left(\frac{a}{2}\right) - R_1\left(\frac{a}{3}\right) - R_2\left(\frac{2a}{3}\right) + mg\left(\frac{4a}{3\pi}\right)$$

$$M_C \equiv (\rho ghab)\frac{a}{2} - (\tfrac{1}{2}\rho ghab)\frac{a}{3} - [\tfrac{1}{2}\rho g(h + a)ab]\frac{2a}{3}$$

$$+ \frac{\pi}{4}\rho g a^2 b\left(\frac{4a}{3\pi}\right) \equiv 0$$

These results indicate that the pressure distribution is equivalent to a single force

$$\bar{R} = \rho gab\left[-\left(h + \frac{a}{2}\right)\bar{i} - \left(h + \frac{\pi}{4}a\right)\bar{j}\right]$$

which acts at the center of the circle.

Additional Remarks

The fact that there is no couple in the resultant acting at point C could have been predicted independently. It arises because the pressure acting on the circular arc AB is perpendicular to the tangent to the circular arc. Thus, the pressure forces that are distributed along the arc are all concentric at the center C; they form a system of concurrent forces. Clearly, a concentric system of forces does not exert a moment about the point of concentricity.

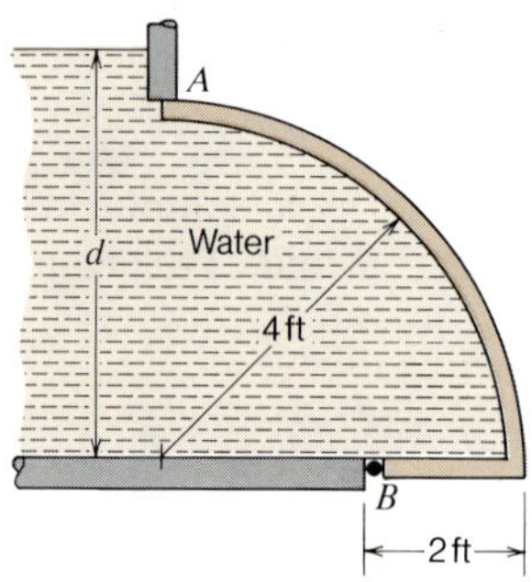

EXAMPLE 17

The 3 ft wide bent plate is supported by a horizontal shaft at edge B, and it bears against the bottom of the wall at edge A. Determine the reaction at edge A for a depth of water $d = 8$ ft. The weight of the plate is negligible.

Solution

The cylindrical body of water we shall consider is formed by a vertical plane through edge A and a horizontal plane through edge B. We need to find the reactions acting on the bent plate, so we draw a free body diagram of the combination of the plate and the body of water.

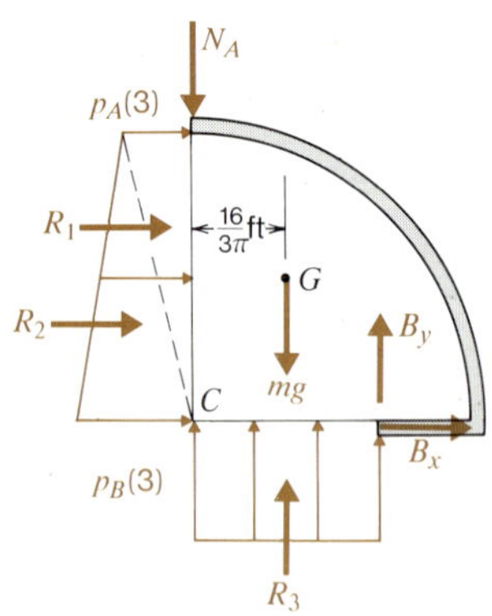

The line load along segment BC of the horizontal surface is the reaction corresponding to the pressure p_B exerted by the water on the level floor. This line load extends only over the portion of the cylindrical body of water that is not in contact with the bent plate. When $d = 8$ ft, the depth at edge A is 4 ft, and the depth at edge B is 8 ft, so

$$p_A = \rho g(4), \qquad p_B = \rho g(8)$$

The trapezoidal line load is treated as a composite of two triangles, whose resultants are

$$R_1 = \tfrac{1}{2}p_A(3)(4) = 24\rho g, \qquad R_2 = \tfrac{1}{2}p_B(3)(4) = 48\rho g$$

$$R_3 = p_B(3)(2) = 48\rho g$$

The weight of the body of water is

$$mg = \rho \mathscr{A}_c(3)g = \rho g\left(\frac{\pi}{4}\right)(4)^2(3) = 37.70\rho g$$

This force acts at the centroid of the quarter circle.

Because we only wish to find $\bar{N}_A$, we sum moments about the shaft B. Thus

$$\sum M_{Bz} = -R_1\left(\frac{8}{3}\right) - R_2\left(\frac{4}{3}\right) - R_3(1) + mg\left(2 - \frac{16}{3\pi}\right) + N_A(2) = 0 \text{ lb-ft.}$$

$$2N_A = 24\rho g\left(\frac{8}{3}\right) + 48\rho g\left(\frac{4}{3}\right) + 48\rho g(1) - 37.70\rho g(0.3023)$$

$$N_A = 82.30\rho g = 5.136 \text{ kips}$$

where the value of ρg in U.S Customary units is 62.4 lb/ft^3. The value of this reaction demonstrates the enormous forces required to contain large bodies of water.

HOMEWORK PROBLEMS

6.129 and 6.130 A viewing window in an aquarium is 2.5 m long with the cross section shown. Determine the force acting at the center O that is equivalent to the pressure distribution exerted by the water on the window.

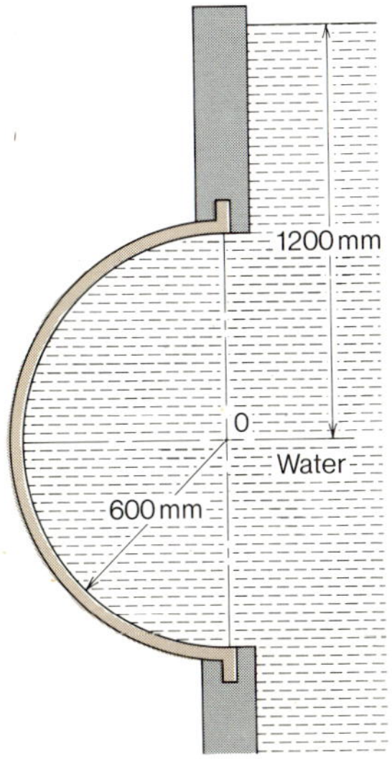

PROB. 6.129

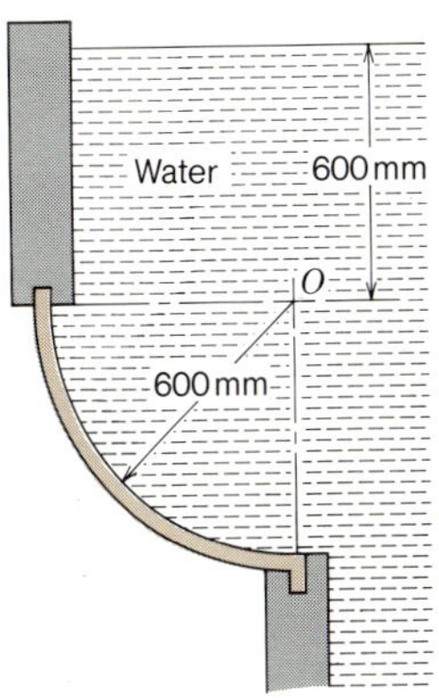

PROB. 6.130

6.131–6.134 Determine the magnitude and direction of the force that is equivalent to the pressure distribution exerted by the water on the curved surface AB.

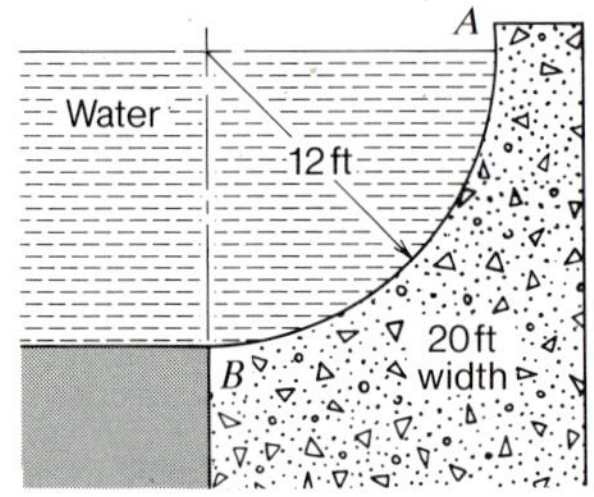

PROB. 6.131

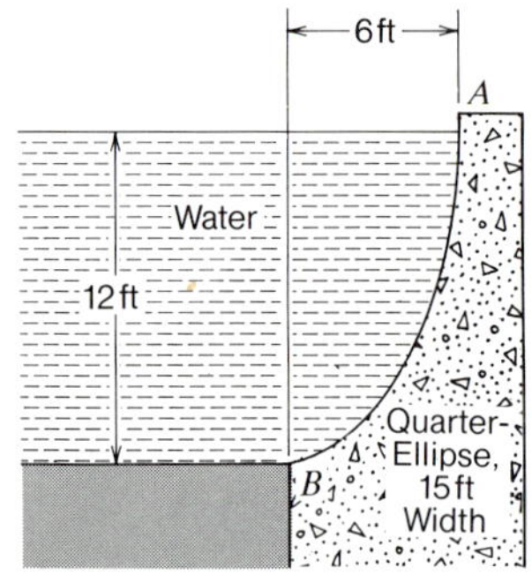

PROB. 6.132

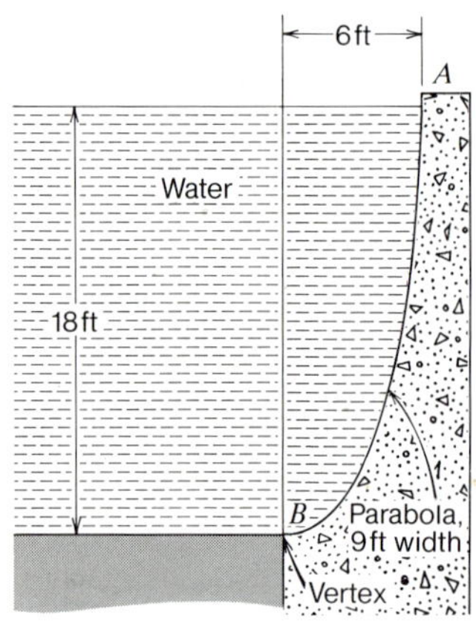

PROB. 6.133

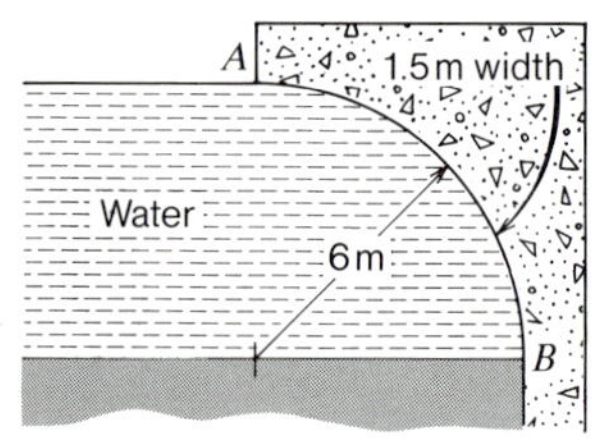

PROB. 6.134

***6.135 and *6.136** A window in the hull of a boat has the cross section shown. The window is 3 ft wide. Determine a force-couple system at edge A that is equivalent to the pressure distribution exerted by the water on the window.

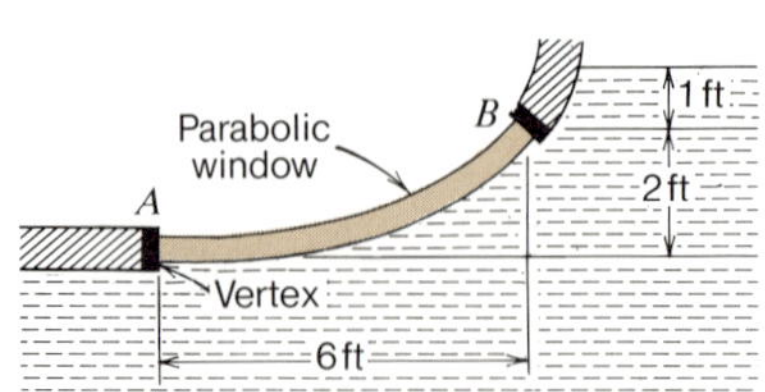

PROB. 6.135

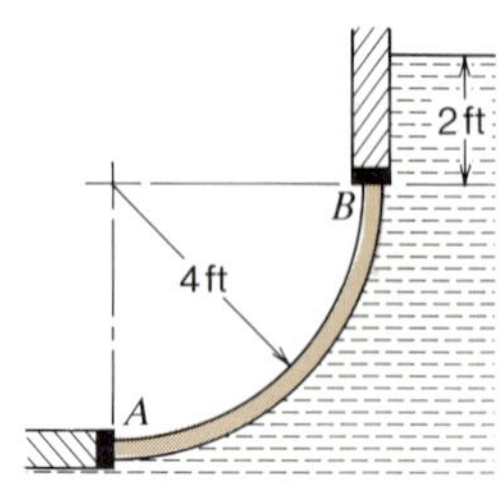

PROB. 6.136

6.137–6.139 A water conduit, whose weight is negligible, has the cross section shown. The conduit is supported by hinges along the lower edge A and a total of 20 cables at edge B that are equally spaced along the length of the conduit. Determine the tension in each cable when the conduit is filled to the top.

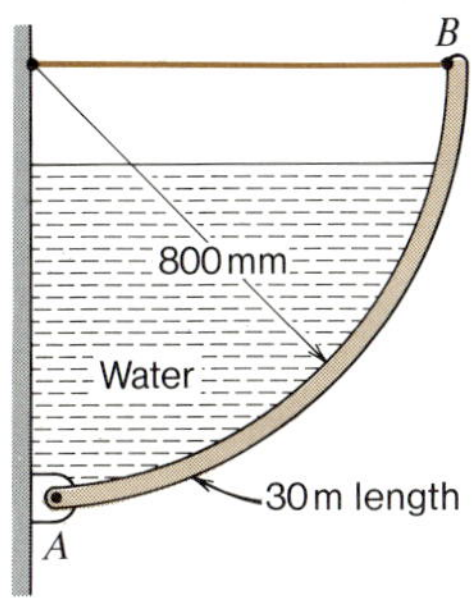

PROB. 6.137

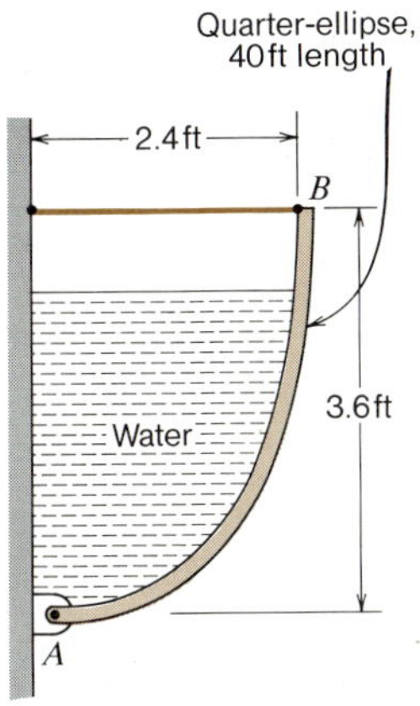

PROB. 6.138

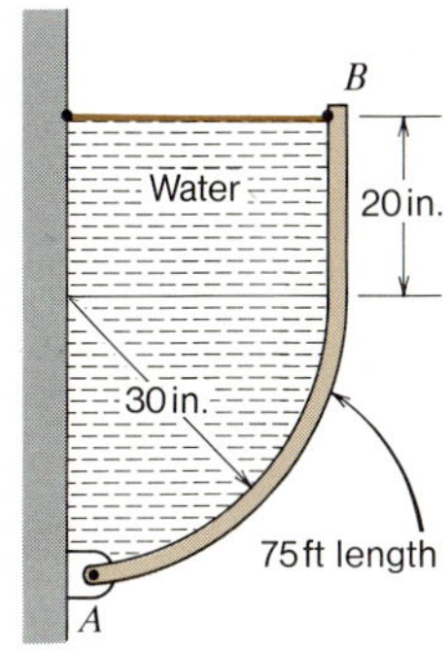

PROB. 6.139

6.140 and 6.141 A bent plate 750 mm long is used as a gate in a dam. Determine the horizontal force exerted by the hydraulic cylinder required to open the gate when the depth of oil is $d = 1.5$ meters.

6.142 Solve Problem 6.140 when $d = 2.5$ m

6.143 Solve Problem 6.141 when $d = 2.5$ m.

6.144 and 6.145 A panel in a water tank is supported by a horizontal shaft along its upper edge A, and its lower edge B bears against the bottom edge of the tank. The horizontal width of the panel is 4 ft. Determine the reactions at edges A and B when $d = 3$ ft.

6.146 Solve Problem 6.144 when $d = 5$ ft.

6.147 Solve Problem 6.145 when $d = 5$ ft.

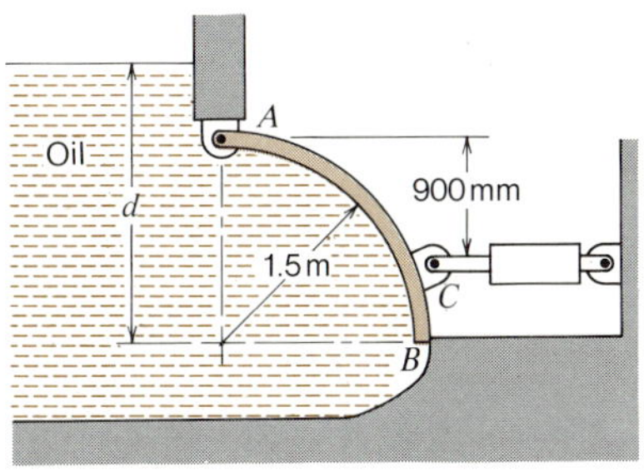

PROB. 6.140

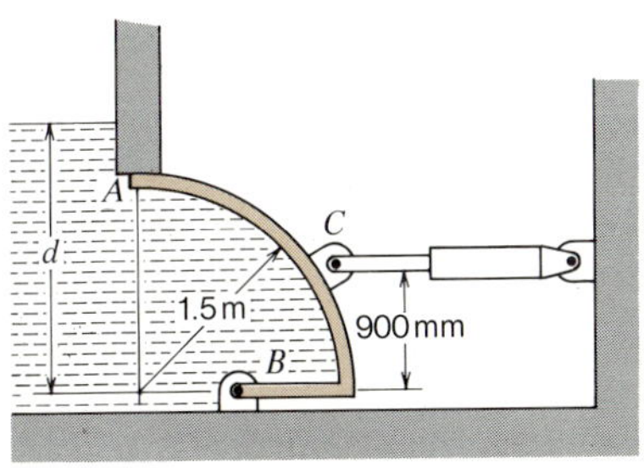

PROB. 6.141

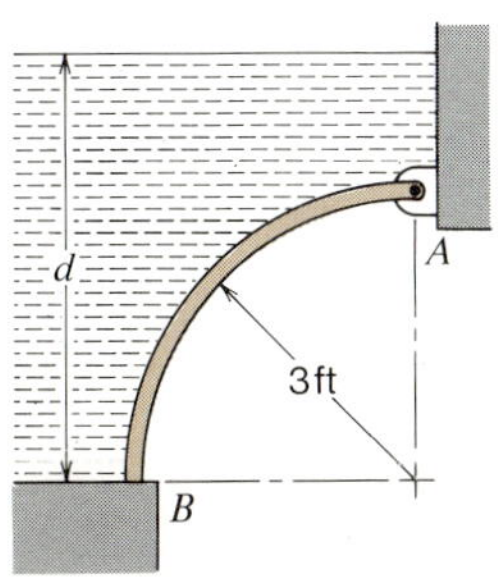

PROB. 6.144

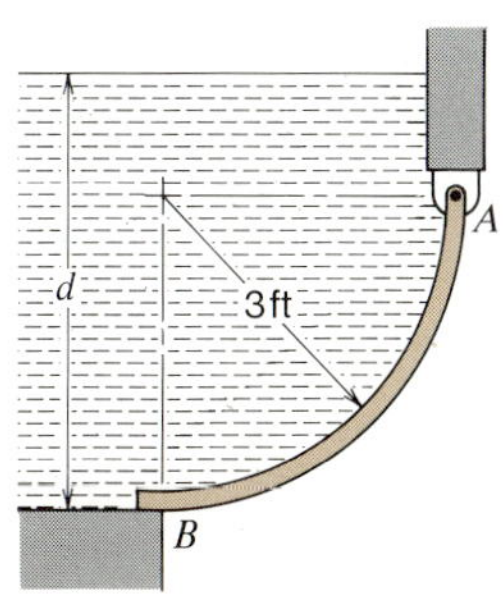

PROB. 6.145

3. Arbitrary Planar Surfaces

We investigated the hydrostatic pressure distribution acting on a rectangular surface by converting it into an equivalent line load. Such a simplification is not applicable to the pressure distribution acting on a planar surface of arbitrary shape.

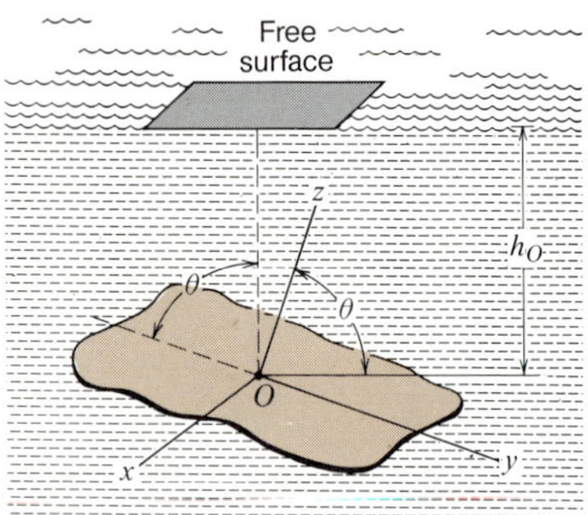

FIGURE 21

Consider the situation in Figure 21, where the gray rectangle represents the free surface and the colored region is a flat surface that is inclined at an angle θ from the vertical. The coordinate axes are selected such that the xy plane coincides with the plane of the inclined surface, with the x axis horizontal. This means that the yz plane is vertical, and that the angle between the negative y axis and a vertical line is θ, as shown.

Because the x axis is horizontal, all points in the surface at a constant value of y are at the same depth. Let h_O denote the depth of the origin O, whose location will be specified later. Then, the depth of other points is given by

$$h = h_O + y \cos \theta$$

It follows that the pressure distribution is

$$p = \rho g h = \rho g (h_O + y \cos \theta) \tag{19}$$

To describe the resultant of this pressure distribution we could resort to the analogy with the geometric properties of the pressure space, as developed in Section B.2 of this chapter. Except in rare situations, the pressure space resulting from eq. (19) will not be easily represented in terms of the basic shapes in Appendix B. We therefore shall employ eqs. (16) and (17) found in Section B.2. A differential element of area for the surface is $d\mathscr{A}$, so eq. (16) gives

$$R = \int_{\mathscr{A}} p \, d\mathscr{A} = \rho g h_O \int_{\mathscr{A}} d\mathscr{A} + \rho g \cos \theta \int_{\mathscr{A}} y \, d\mathscr{A}$$

Clearly, the first integral on the right side gives the area $\mathscr{A}$, and the second integral is the first moment of area with respect to the y coordinate. In order to eliminate the latter term, we now *locate the origin O* at the centroid of the surface area $\mathscr{A}$. The result is

$$\boxed{R = \rho g h_O \mathscr{A}} \tag{20}$$

This resultant force is perpendicular to the plane in the negative z direction. An interpretation of eq. (20) is to say that $\rho g h_O$, where h_O is the depth of the centroid of the area, is the *average pressure* acting on the area.

Now that the resultant force has been determined, we may proceed to locate the center of pressure. Note that $\bar{R}$ does not act at the centroid O of the surface area $\mathscr{A}$. Equations (17) give

$$R x_R \equiv \rho g h_O \mathscr{A} x_R = \int_{\mathscr{A}} x p \, d\mathscr{A}$$
$$= \rho g h_O \int_{\mathscr{A}} x \, d\mathscr{A} + \rho g \cos \theta \int_{\mathscr{A}} x y \, d\mathscr{A}$$

$$Ry_R \equiv \rho g h_O \mathscr{A} y_R = \int_{\mathscr{A}} yp\, d\mathscr{A}$$
$$= \rho g h_O \int_{\mathscr{A}} y\, d\mathscr{A} + \rho g \cos\theta \int_{\mathscr{A}} y^2\, d\mathscr{A}$$

As a result of the fact that the origin is located at the centroid, the first moments of area in the foregoing equations vanish. Let us define the following symbols:

$$I_x = \int_{\mathscr{A}} y^2\, d\mathscr{A} \qquad I_{xy} = \int_{\mathscr{A}} xy\, d\mathscr{A} \tag{21}$$

The expressions for x_R and y_R may then be written as

$$x_R = \frac{I_{xy}\cos\theta}{h_O \mathscr{A}}, \qquad y_R = \frac{I_x \cos\theta}{h_O \mathscr{A}} \tag{22}$$

The quantity I_x is called the *second moment of area about the x axis* or, alternatively, the *area moment of inertia about the x axis.* The reason for associating it with the x axis is that y is the distance of an element of area $d\mathscr{A}$ from the x axis. The value of y^2 can never be negative, so the integrand for I_x is never negative, and the value of I_x will always be a positive quantity. Hence, the center of pressure is never higher than the centroid of the surface area.

The parameter I_{xy} is the *area product of inertia.* This quantity is studied in detail in Chapter 8. Briefly, I_{xy} is a measure of the asymmetry of the area $\mathscr{A}$. If the area is symmetrical with respect to either the x or y axes, then $I_{xy} = 0$. In that case eqs. (22) give $x_R = 0$, which means that the center of pressure is situated on the y axis. In all of the problems we shall consider in this section the area will have this type of symmetry.

Values of the area moments of inertia for some basic geometric shapes about centroidal axes are given in Appendix C. Methods for calculating this property for geometric shapes not appearing in Appendix C are described in Chapter 8.

Before we apply eqs. (20) and (22), let us make some fundamental observations regarding them. The fact that the pressure at the centroid of the area is the average for the entire area is logical. What is surprising is that the angle of inclination θ does not affect the magnitude of the resultant. The expression for y_R shows that, for a given area $\mathscr{A}$, the center of pressure moves upward (y_R decreases) if the depth h_O or the angle of inclination θ is increased. These tendencies are a result of the fact that an increase of either of these parameters increases the ratio of the pressure of the highest point on the area to the pressure at the lowest point. It is also interesting to note that similar expressions arise in the study of mechanics of materials when the state of stress in beams carrying axial and transverse loads is studied.

EXAMPLE 18

The 1 m diameter circular plate allows access to the interior of the spherical water tank when the tank is empty. Determine the resultant water force exerted on the plate and the location of the center of pressure when the water is full, but a free surface is still present.

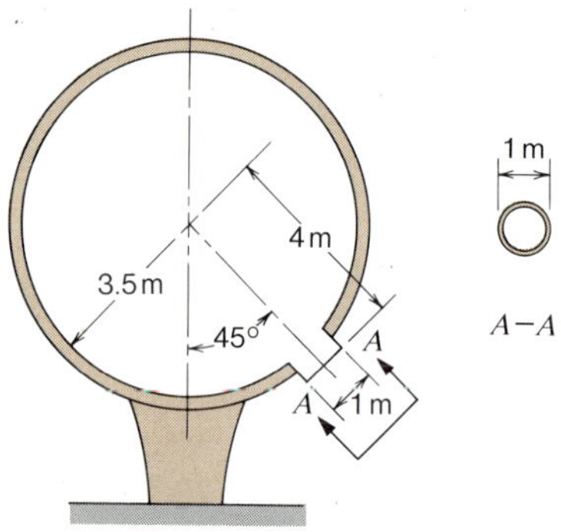

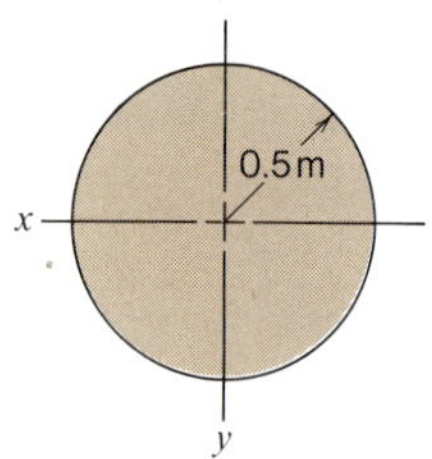

Solution

We need a coordinate system xyz for the circular plate. In order to employ eqs. (20) and (22), we let x and y coincide with the plane of the plate, and align the x axis horizontally. The origin is at the centroid of the plate, which is the center of the circle. The coordinate system is depicted in the sketch.

From the diagram accompanying the statement of the problem, we calculate that when the free surface is at the top of the tank, the depth of the center of the plate is

$$h_O = 3.5 + 4.0 \cos 45^\circ = 6.328 \text{ m}$$

The angle of inclination of the plate is seen to be $\theta = 45^\circ$. From Appendix C we have

$$\mathscr{A} = \pi(0.5)^2 = 0.7854 \text{ m}^2$$

$$I_x = \frac{\pi}{4}(0.5)^4 = 0.04909 \text{ m}^4$$

Also, the plate is symmetric about both the x and y axes, so we know that the center of pressure is situated on the y axis. (Symmetry about one axis is sufficient for this.) The density of water is $\rho = 1000 \text{ kg/m}^3$, so eq. (20) gives

$$R = 1000(9.807)(6.328)(0.7854) = 48.76 \text{ kN}$$

For the location of the center of pressure, equation (22) gives

$$y_R = \frac{0.04909(\cos 45^\circ)}{6.328(0.7854)} = 6.98(10^{-3}) \text{ m}$$

Hence, the center of pressure is 6.98 mm below the center of the 500 mm radius plate.

HOMEWORK PROBLEMS

6.148–6.151 A window in an aquarium is situated on a vertical wall. Determine the resultant force exerted by the water on the window and locate the center of pressure if (a) $d = 1$ ft, (b) $d = 10$ ft.

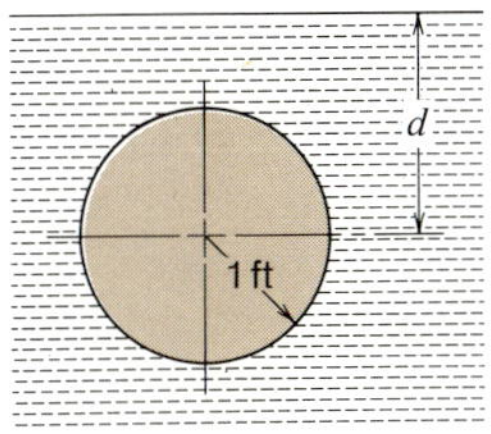

PROB. 6.148

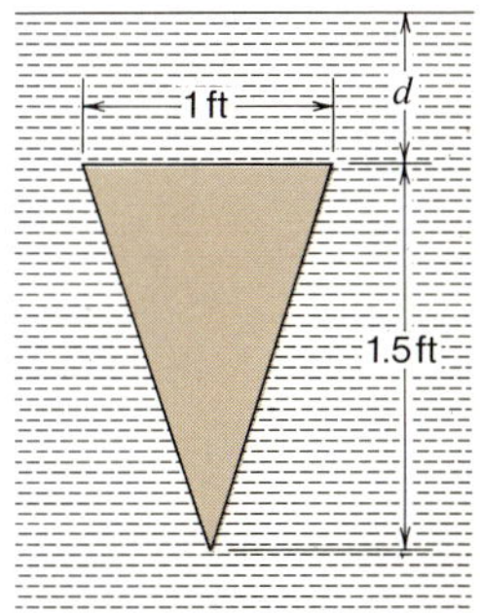

PROB. 6.149

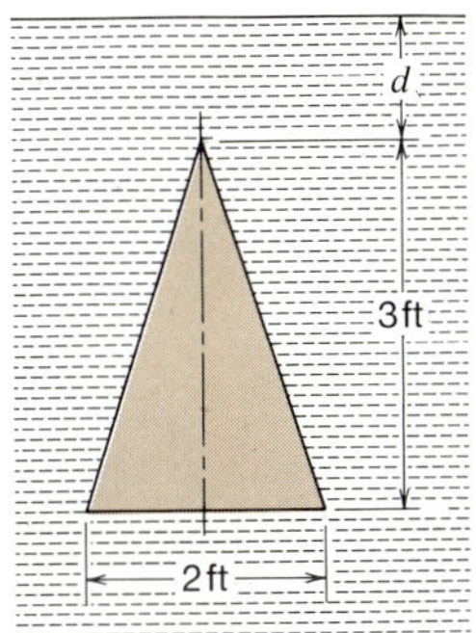

PROB. 6.150

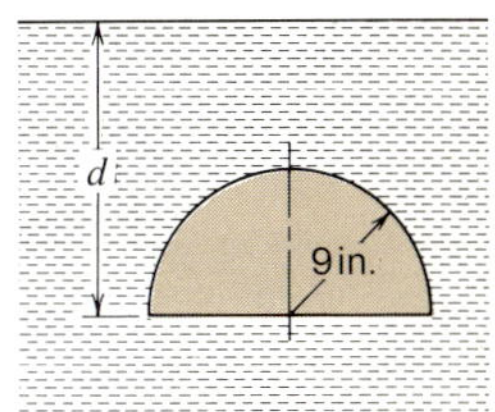

PROB. 6.151

6.152–6.154 The flat plate shown fits into a matching opening in the vertical wall of a tank of oil. The density of the oil is 800 kg/m^3. The plate is hinged about the horizontal axis AB. Determine the force $\bar{P}$ applied at point C required to keep the plate in place.

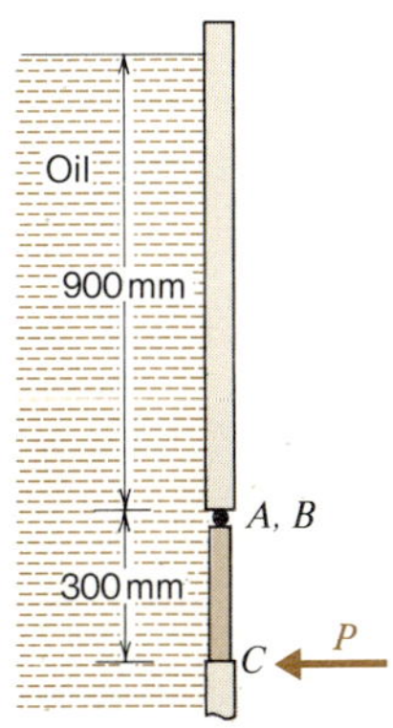

PROBS. 6.152–6.154

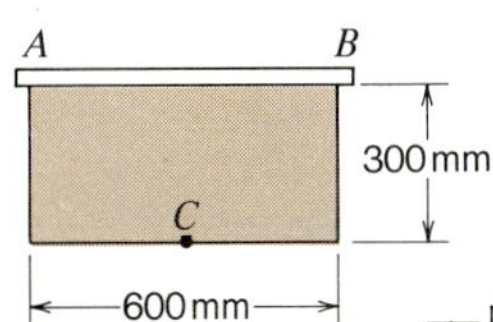

PROB. 6.152

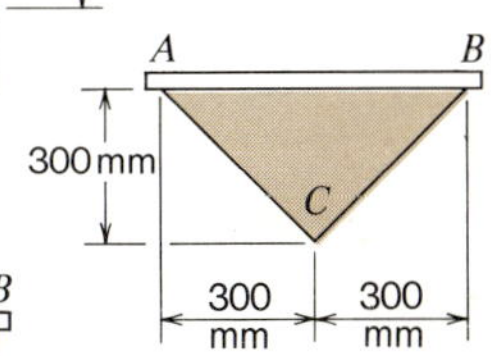

PROB. 6.153

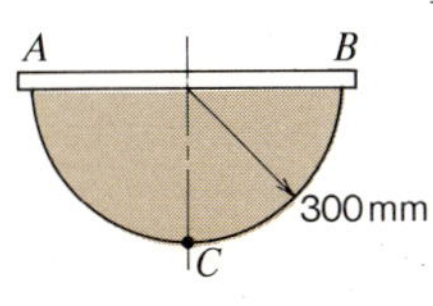

PROB. 6.154

***6.155 and 6.156** The end of a water trough is inclined at 36.87° from the vertical. The end plate shown is attached to the walls of the trough by bolts at points *A*, *B*, and *C*. Determine the tension in the bolts when the trough is filled to the top.

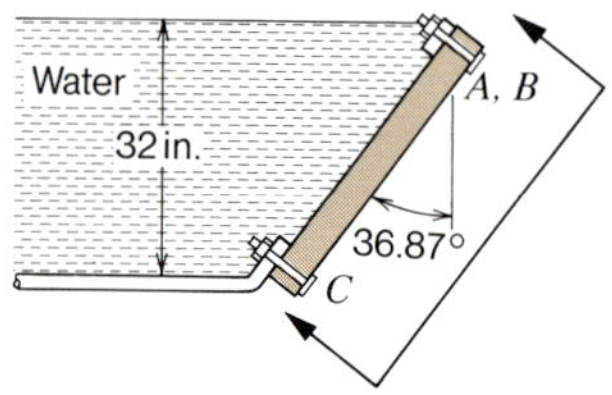

PROBS. 6.155 AND 6.156

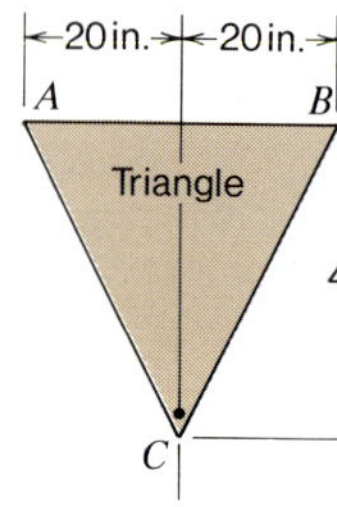

PROB. 6.155

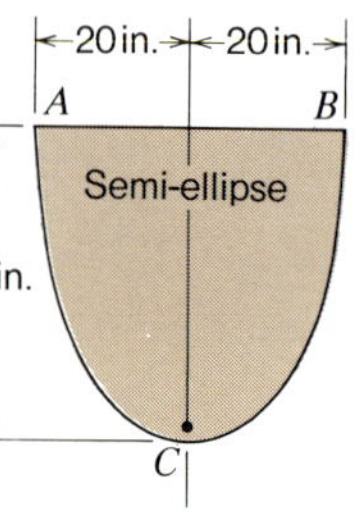

PROB. 6.156

CHAPTER 7

INTERNAL FORCE ANALYSIS

OBJECTIVES

1. To draw a free body diagram that depicts the internal reaction at a cross section of a structural element in terms of axial and shear forces, as well as bending and torsional moments.

2. To evaluate the internal reaction at a specified cross section of a structural element.

3. To graph the position dependence of the shear force and bending moment internal to a straight beam.

4. To evaluate the tensile force and sag of a cable that is loaded by a series of concentrated forces or by a uniform distributed load.

5. To employ the catenary equation for the deflection of a cable sagging under its own weight, and to evaluate the tensile force in such a cable.

Chapter 4 focused on structures and machines whose interconnected members support and modify the loads applied to the system. The development in that chapter was devoted to the determination of the reactions acting on the members. Our goal here is to describe how the forces applied at one location of a member are transmitted internally to another location in that member.

Knowledge of the distribution and flow of internal forces is essential to the design of the members of a structure. In this chapter, we will discuss two common types of structural members: (A) slender bars that function as beams, and (B) cables that carry loads that are applied along their length.

A. BEAMS

1. Basic Definition

The primary physical characteristic of a bar that functions as a beam is that it is slender. The property of slenderness means that the line connecting the centroids of the cross sections of the bar is long compared to the dimensions of the cross sections. This centroidal axis serves as a reference line for the bar. A beam is said to be straight or curved in accordance with the shape of its centroidal axis.

The property of slenderness is not sufficient to establish that a bar acts like a beam, although all beams must have that property. The second item of importance is the type of internal forces the bar transmits. To investigate this matter let us pass an imaginary cutting plane through an arbitrary bar and consider the force system exerted by one portion of the cut bar on the other.

In the most general situation the internal forces will be equivalent to a force-couple system. The point of reference that we choose for this force-couple system is the centroid of the cross-sectional area. The reason for choosing the centroid is that many formulas derived in mechanics of materials to describe the load-carrying ability of bars require such a resolution for the internal forces. Thus, considering the cut bar in Figure 1, the internal forces exerted on the cross-sectional surface to the right of the cutting plane by the portion of the bar to the left of the plane are equivalent to a resultant force $\bar{R}$ acting at the centroid C and a couple $\bar{M}_C$.

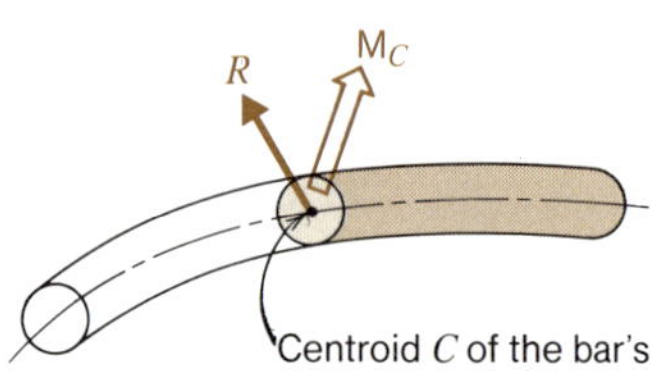

FIGURE 1

Although they are not depicted in Figure 1, it follows from Newton's third law that the internal forces exerted by the portion of the bar to the right of the cutting plane on the exposed cross section to the left of the cutting plane consist of a resultant force $-\bar{R}$, applied at the centroid, and a couple $-\bar{M}_C$.

We shall describe the internal force $\bar{R}$ and couple $\bar{M}_C$ in terms of their components. For this, an xyz coordinate system is chosen such that the centroid C is the origin, and the positive x axis is directed outward from the exposed cross section. It follows that the yz plane is coincident with the cross section, and that the x axis is tangent to the centroidal axis of the bar at that cross section. The cross section appearing in Figure 1 is shown in an expanded view in Figure 2.

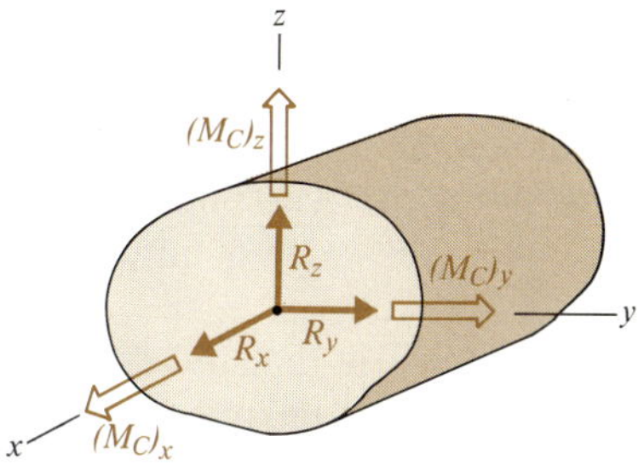

FIGURE 2

Although the basic method we shall present for evaluating the internal forces is equally valid for straight and curved beams, here we shall investigate only straight beams. This leads to a significant simplification in the formulation, because it will not be necessary to account for change in the orientation of the *xyz* coordinate system from one cross section to another.

Now that a convenient coordinate system has been defined, let us consider further the components of the internal force and moment. In Figure 2 the force components R_y and R_z, which are parallel to the cross section, are called the *shear forces*. As was true when we treated straight two-force members, the component R_x, normal to the cross section, is called a *tensile* or *compressive* force, according to whether R_x is in the same or opposite sense of the x axis. The components of the internal moment parallel to the cross section, that is, $(M_C)_y$ and $(M_C)_z$, are called *bending moments* because they have the effect of tending to bend the bar. Finally, the component $(M_C)_x$ normal to the cross section is called the *torsional moment* because it has the effect of tending to twist the bar about its centroidal axis.

The primary factor in deciding whether a bar acts like a beam is that

> In a beam the shear forces and bending moments are significant in comparison to the axial (tensile or compressive) force and the torsional moment.

Clearly, the foregoing definition excludes cables and the straight members of a truss, for both transmit only axial forces. It also excludes a straight bar, frequently called a torsion bar, whose only loads are torsional moments.

The majority of multiforce members in the frames and machines we analyzed in Chapter 4 are beams. Beams of all descriptions may be found in the structural framework of buildings, automobiles, trucks, and airplanes. They are crucial to many industrial machines and parts of many hand tools, such as the handles of pliers and wrenches, which function like beams. A beam is one of the fundamental structural elements for a designer.

It frequently proves convenient to classify straight beams according to the manner in which they are supported. Some typical types are displayed

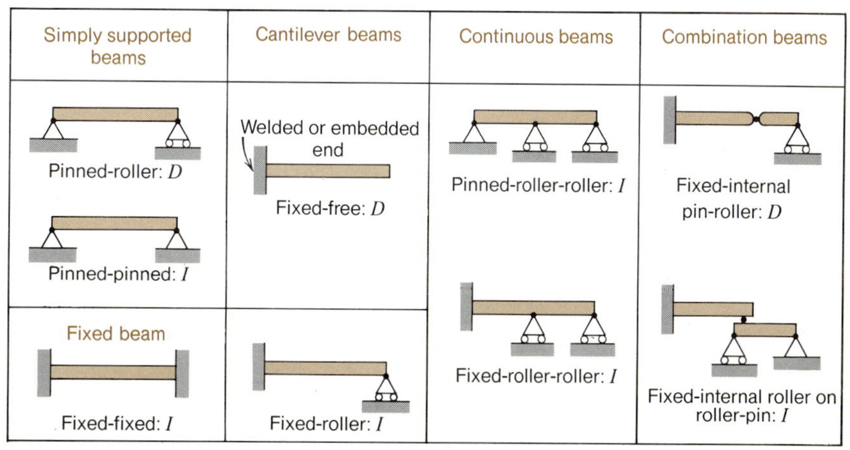

FIGURE 3

in Figure 3. They have been depicted in the horizontal position solely for convenience; in practice, beams may have any orientation in space.

All except one of the beams exhibited in Figure 3 have two names. The more descriptive name explicitly states the type of supports, whereas the other name (heading each entry) is the one more commonly employed by engineers.

Another feature of the beams in Figure 3 is that in several cases, the supports exert more reactions that the number of available static equilibrium equations. These are statically indeterminate systems whose analysis requires consideration of the deformation of the beam caused by the beam loading. Note that each system in Figure 3 is labeled D or I to indicate that it is statically determinate or statically indeterminate, respectively. Statically indeterminate systems are treated in courses on mechanics of materials and structures.

2. Determination of the Internal Forces

A beam may be subjected to concentrated or distributed loads, or to a combination of both. The reactions exerted by the supports of a statically determinate beam, which is the only kind we shall consider, may be determined by the methods previously developed in Chapters 3, 4 and 6. After the reactions have been determined, they become additional (known) loads for an investigation of the internal forces.

Suppose that we wish to know the internal forces acting on a cross section at some specific location along the centroidal axis of a beam. In order to have these internal forces appear in a free body diagram, we employ an imaginary cutting plane that is perpendicular to the beam's centroidal axis. In this manner, the beam is (fictitiously) separated into two segments, with the cross section of interest being one end of each segment.

Let us consider the straight beam shown in Figure 4a, where the illustrated loading system includes the reactions. Let us determine the internal forces acting at location C on the beam. We use the cutting plane a–a to sever the beam into two segments at the cross section of interest.

Figure 4b shows a free body diagram for each segment. In addition to the external loads that act on each side of the cutting plane, these free body diagrams show the internal force $\bar{R}$ and the internal couple $\bar{M}$ acting at

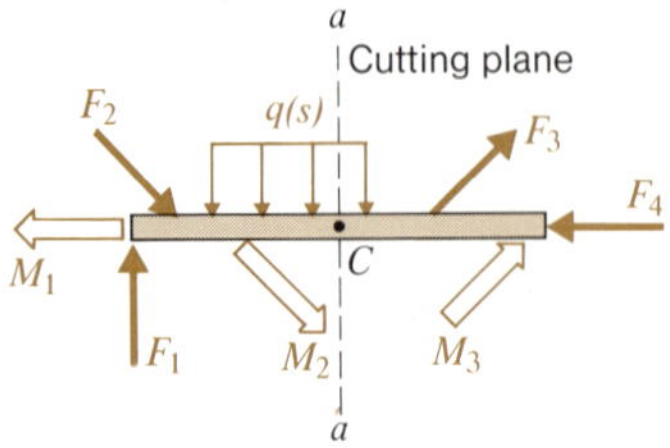

FIGURE 4a

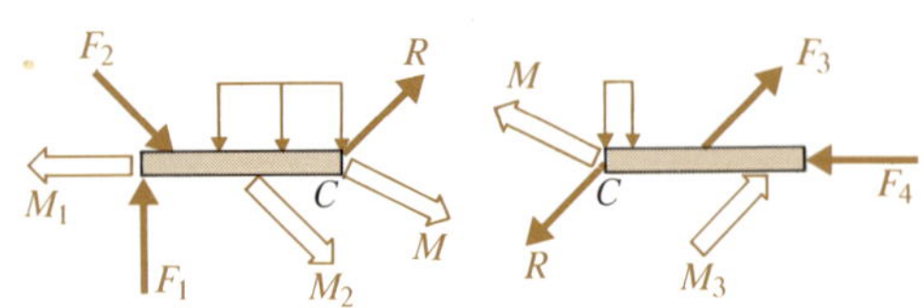

FIGURE 4b

the centroid of the cross section C. This force-couple system is merely the force system required to keep each segment of the cut beam in static equilibrium. Because the entire beam is in static equilibrium under the system of external loads in Figure 4a, it must be that equilibrium of either segment of the cut beam assures static equilibrium of the other segment. Thus, the equations of static equilibrium for either segment of the cut beam may be used to determine $\bar{R}$ and $\bar{M}$.

It is advantageous to formulate the equilibrium equations for one of the segments of the cut beam in terms of a moment sum about the centroid of the cross section C. In this way the unknown force $\bar{R}$ is eliminated from the moment equation. Thus, the process of determining the internal force-couple system acting on a specific cross section of a beam can be summarized as follows.

> First isolate in a free body diagram the portion of the beam and of the loading system to *either* side of the cutting plane for that cross section. Then formulate the equilibrium equations for the chosen portion of the beam by summing moments about the centroid of the cross section and summing forces for the isolated portion.

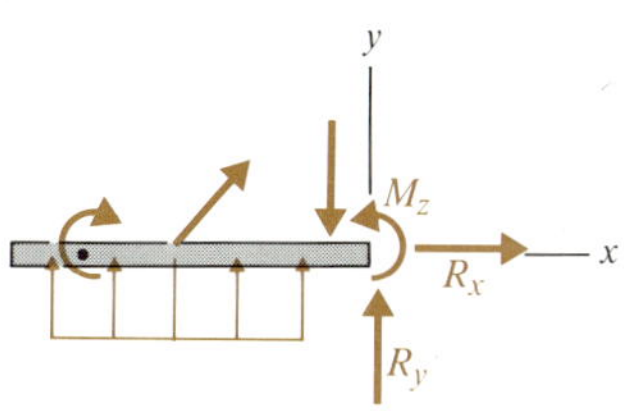

FIGURE 5

The components of the internal force should be identified as axial or shear forces, and the components of the internal couple should be identified as bending or torsional moments. Such an identification is readily performed for a straight beam by recalling that the x axis should be the outward normal for the exposed cross section, so that R_x is the axial force and M_x is the torsional moment.

In the case of a planar system of external loads, the internal force-couple system will also be planar. This means that it will consist of only an axial force R_x, a shear force R_y, and a bending moment M_z. These are depicted in Figure 5 for the segment to the left of a typical beam after being cut.

EXAMPLE 1

Determine the internal forces acting on the cross section of bar ABC at location E. Neglect the weight of the bar.

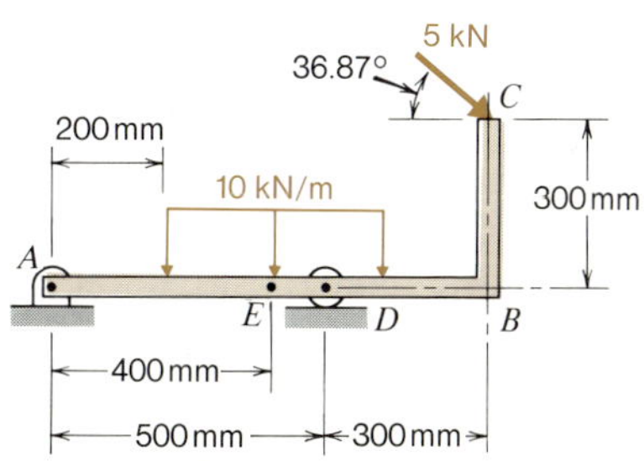

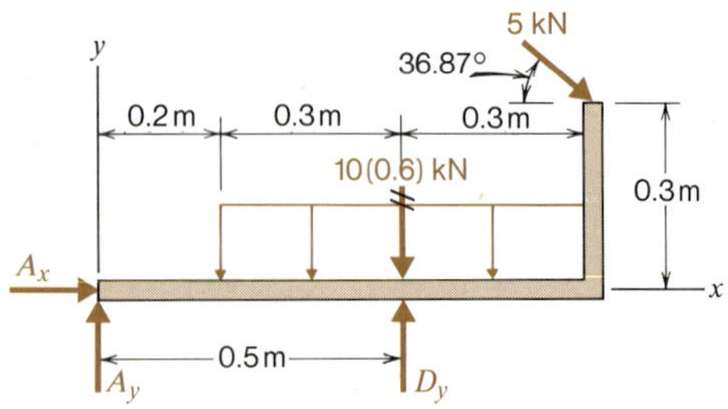

Solution

We first determine the reaction forces. For this, we draw a free body diagram replacing the 10 kN/m distributed load by its resultant acting at the centroid of the area of the loading function. The equilibrium equations for the entire beam are formulated in terms of any convenient coordinate system, such as the one shown. A moment sum about point A yields

$$\sum M_{Az} = D_y(0.5) - 6.0(0.5) - (5 \cos 36.87°)(0.3) - (5 \sin 36.87°)(0.8) = 0$$

$$\sum F_x = A_x + 5 \cos 36.87° = 0$$

$$\sum F_y = A_y + D_y - 6.0 - 5 \sin 36.87° = 0$$

$$D_y = 13.2 \text{ kN}, \qquad A_x = -4 \text{ kN}, \qquad A_y = -4.2 \text{ kN}$$

Now that the reactions have been found, we pass a cutting plane through point E, perpendicular to the centroidal axis of the bar. Either portion of the cut bar may be chosen for consideration; we will choose the left end because there are fewer forces on that side. In the free body diagram for the isolated portion chosen, the reaction forces at pin A are depicted in the direction in which they were calculated to act.

The xyz coordinate system in the free body diagram of the beam segment is chosen such that the x axis is the outward normal for the exposed face. Because this is a planar problem, the internal force system consists of only an axial force R_x, shear force R_y, and bending moment M_z.

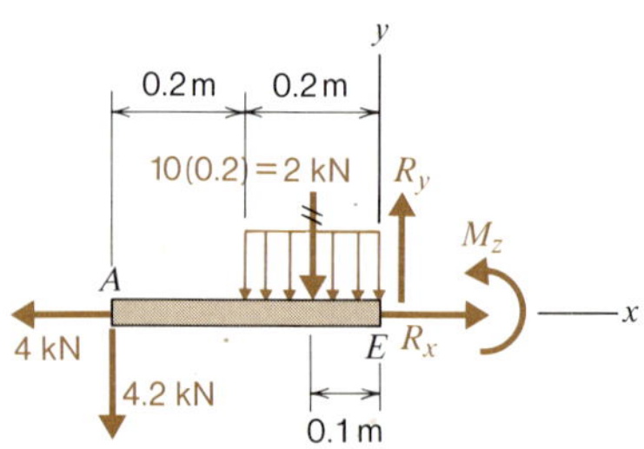

The free body diagram contains only that portion of the distributed load acting on the isolated section. The static equilibrium equations are

$$\sum M_{Ez} = 2(0.1) + 4.2(0.4) + M_z = 0$$

$$\sum F_x = R_x - 4.0 = 0, \qquad \sum F_y = -2.0 - 4.2 + R_y = 0$$

which give

$$M_z = -1.88 \text{ kN-m}, \qquad R_x = 4.0 \text{ kN}, \qquad R_y = 6.2 \text{ kN}$$

These results may be interpreted to mean that at point E the beam carries a tensile force (positive R_x) of 4.0 kN, a vertical shear force of 6.2 kN, and a bending moment of 1.88 kN-m clockwise about the axis perpendicular to the plane of the paper.

EXAMPLE 2

Determine the axial force and the magnitude of the shear force, of the bending moment, and of the torsional moment in bar AB at location C.

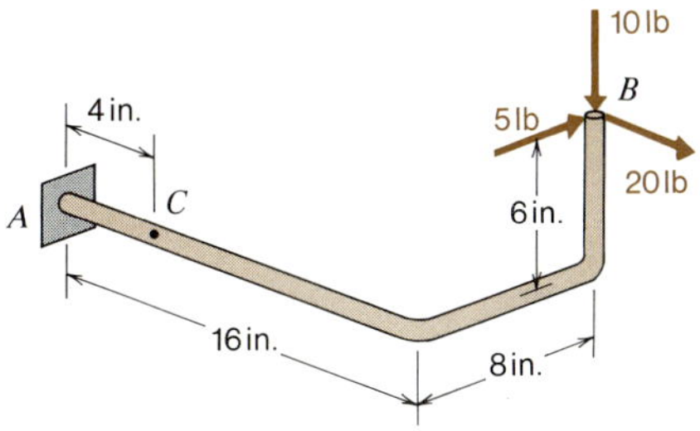

Solution

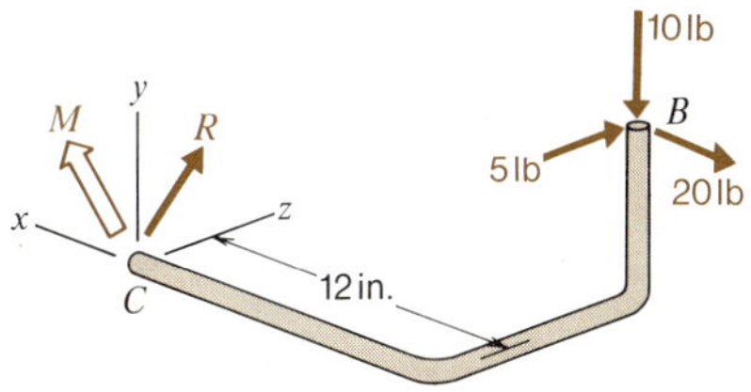

All of the reactions are situated to the left of a vertical cutting plane through the bar at point C. Hence, using the segment to the right of that cutting plane makes it possible to evaluate the internal forces without having to first find the reactions. A free body diagram of this segment shows a general force-couple system $\bar{R}$ and $\bar{M}$ acting on the cross section. The x axis is selected to be the outward normal for the exposed cross section, and the orientations of the other axes match the directions in which the dimensions are given.

The equilibrium equations for the cut segment are formulated in terms of a moment sum about the centroid of the cross section C. Thus

$$\sum \bar{F} = \bar{R} - 20\bar{i} - 10\bar{j} + 5\bar{k} = \bar{0}$$

$$\sum \bar{M}_C = \bar{M} + \bar{r}_{B/C} \times (-20\bar{i} - 10\bar{j} + 5\bar{k})$$

$$\equiv \bar{M} + (-12\bar{i} + 6\bar{j} + 8\bar{k}) \times (-20\bar{i} - 10\bar{j} + 5\bar{k}) = \bar{0}$$

These equations yield

$$\bar{R} = 20\bar{i} + 10\bar{j} - 5\bar{k} \text{ lb}, \qquad \bar{M} = -110\bar{i} + 100\bar{j} - 240\bar{k} \text{ lb-in.}$$

The axial force is the x component of $\bar{R}$, and both the y and z force components represent shear forces. The magnitude of the shear force is the resultant of these components, so

$$\text{axial force} = R_x = 20 \text{ lb}, \qquad \text{shear force} = [R_y^2 + R_z^2]^{1/2} = 11.18 \text{ lb}$$

Similarly, the x component of $\bar{M}$ is the torsional moment, and the magnitude of the bending moment is the resultant of the y and z components.

$$\text{torsional moment} = |M_x| = 110 \text{ lb-in.}$$

$$\text{bending moment} = [M_y^2 + M_z^2]^{1/2} = 260 \text{ lb-in.}$$

HOMEWORK PROBLEMS

7.1–7.16 Determine the axial and shear forces and the bending moment acting on the cross section at point C of the beam shown.

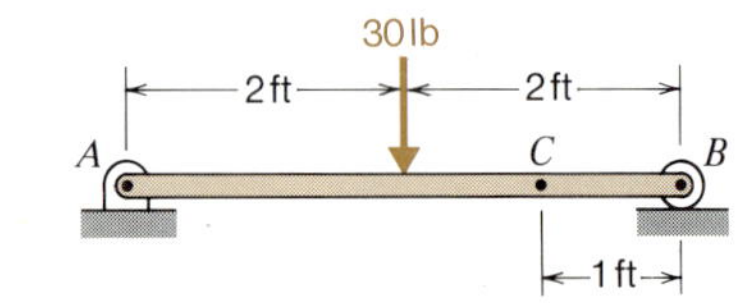

PROB. 7.1

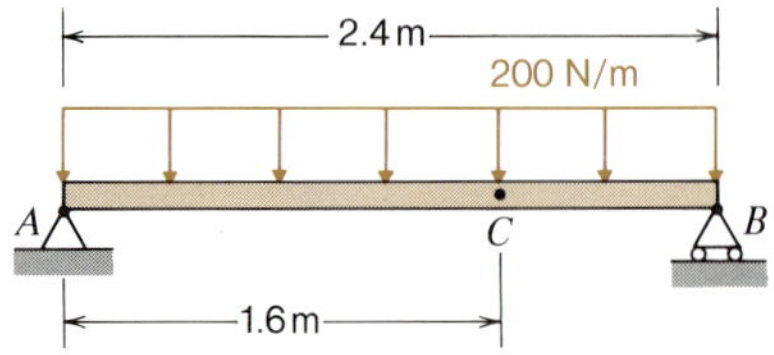

PROB. 7.2

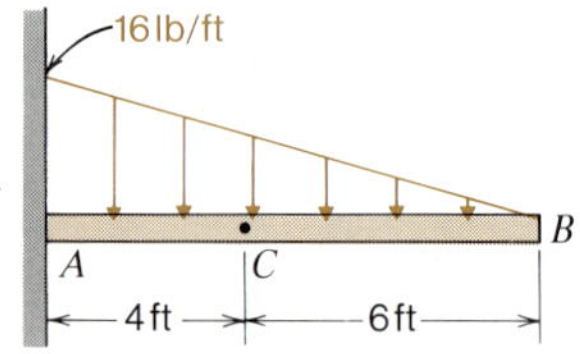

PROB. 7.3

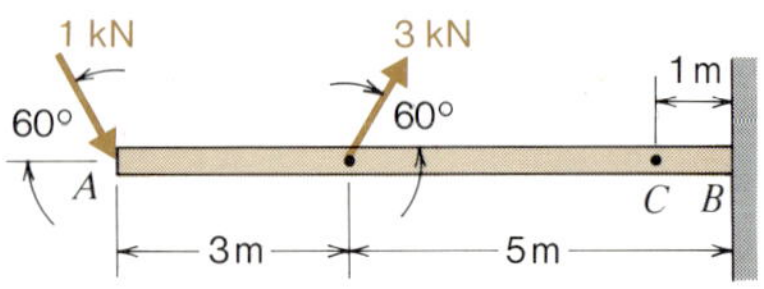

PROB. 7.4

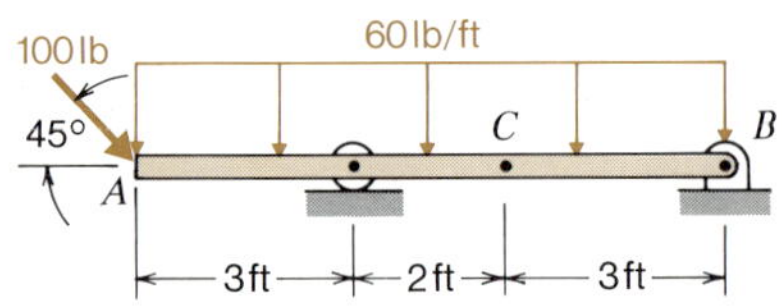

PROB. 7.5

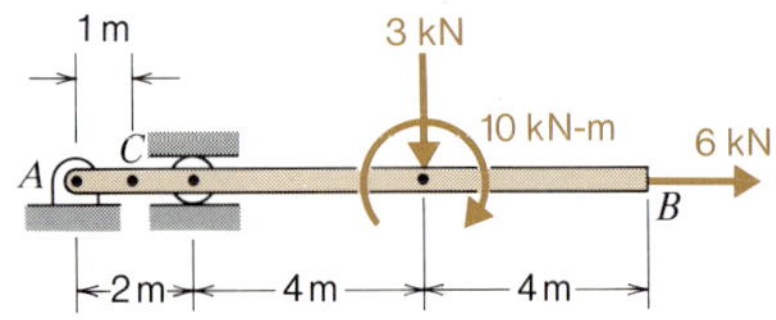

PROB. 7.6

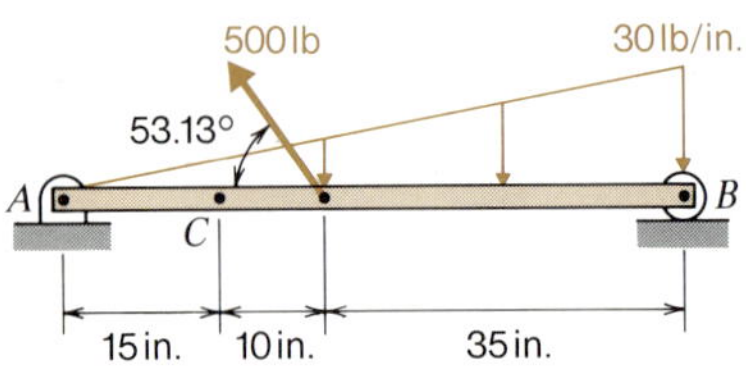

PROB. 7.7

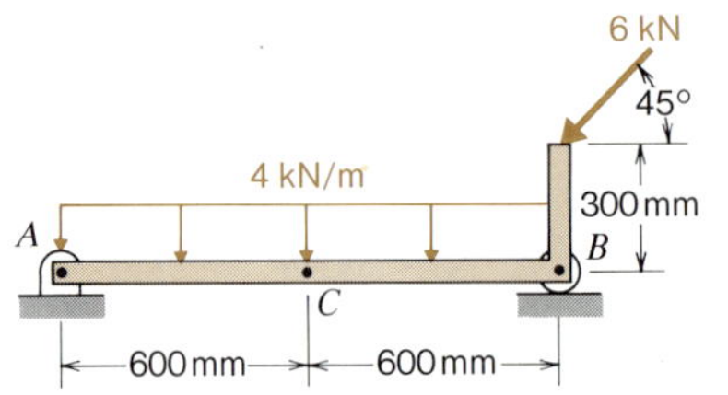

PROB. 7.8

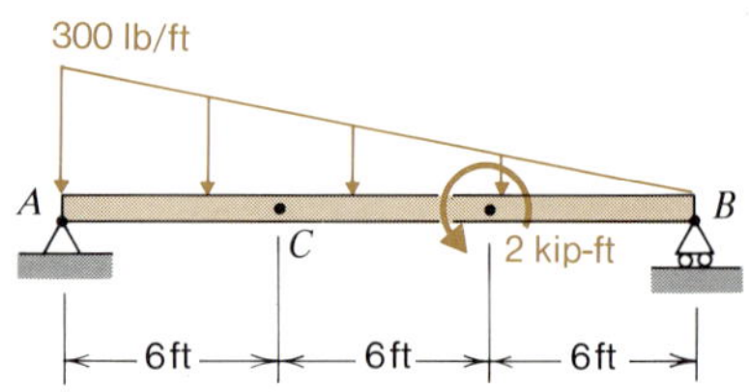

PROB. 7.9

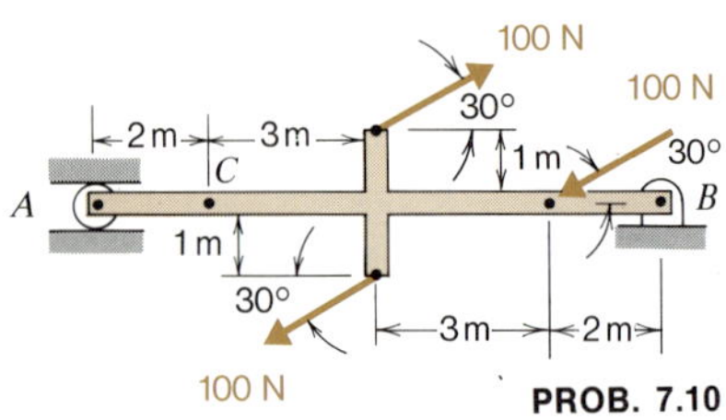

PROB. 7.10

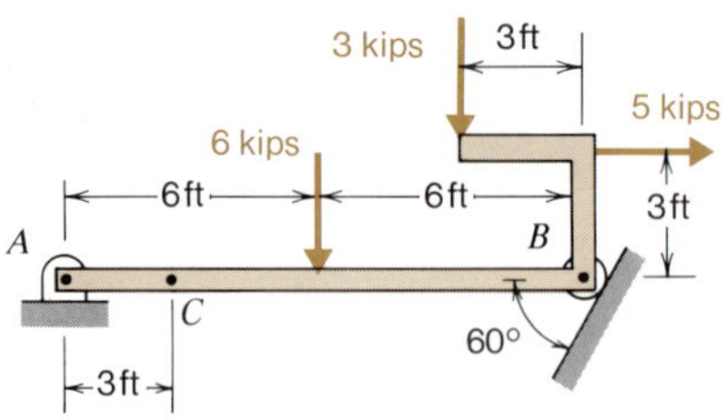

PROB. 7.11

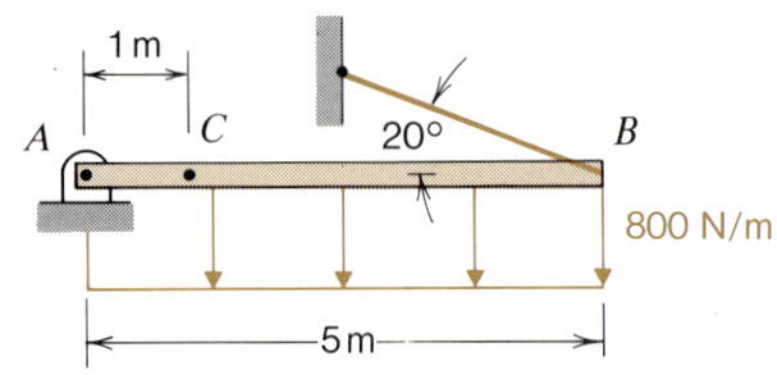

PROB. 7.12

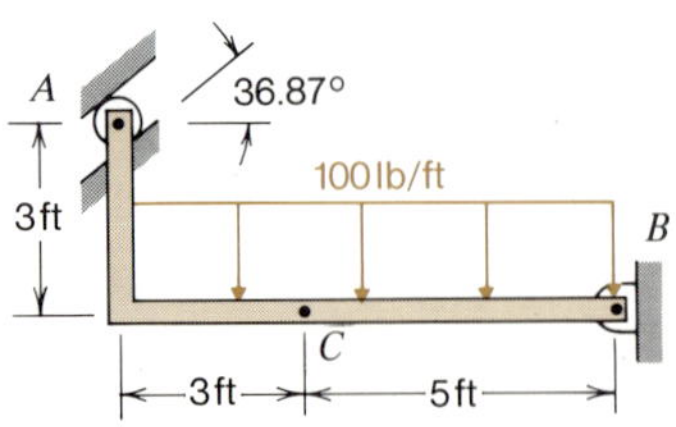

PROB. 7.13

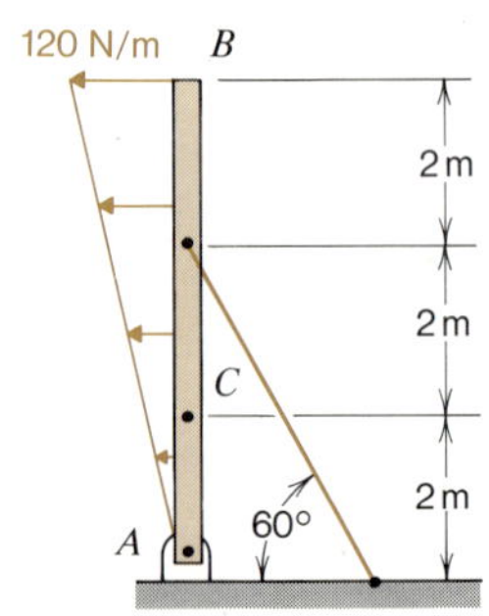

PROB. 7.14

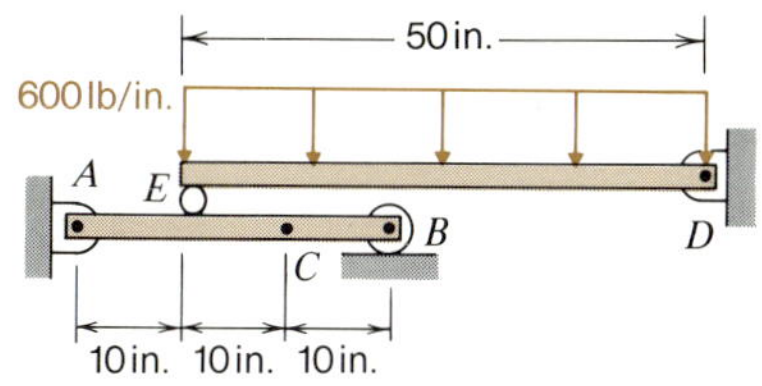

*PROB. 7.15

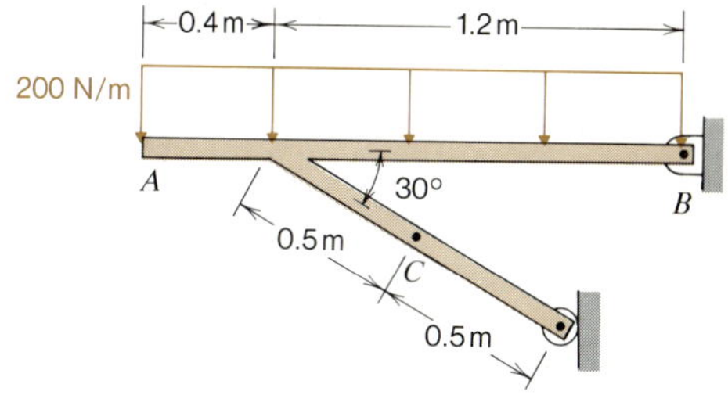

*PROB. 7.16

7.17–7.20 Determine the axial force and the magnitude of the shear force, of the bending moment, and of the torsional moment acting on the cross section at location C of the bent bar.

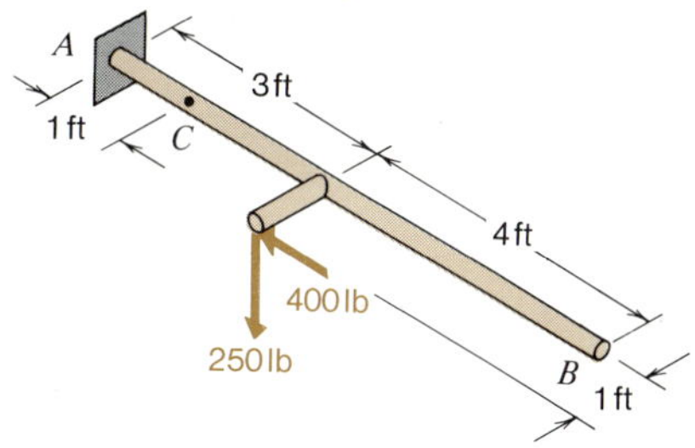

PROB. 7.17

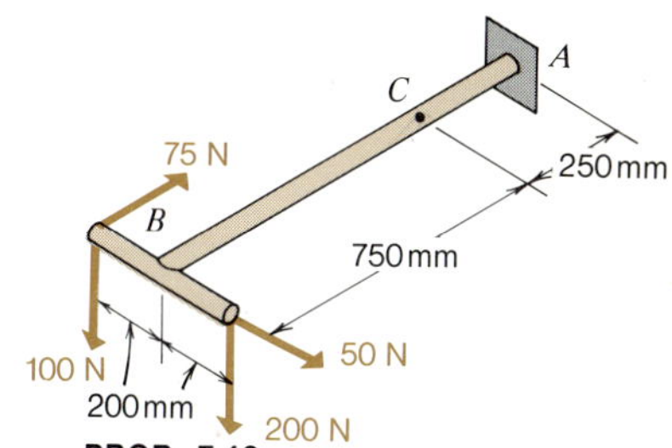

PROB. 7.18

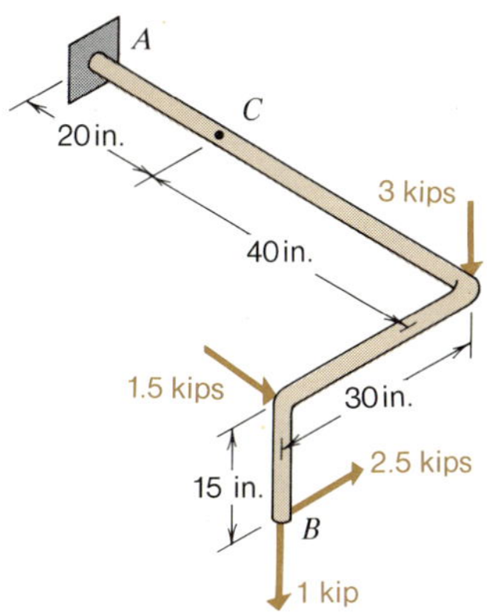

*PROB. 7.19

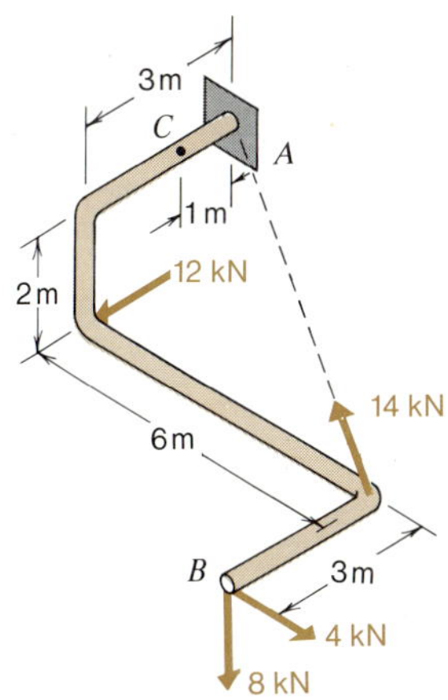

*PROB. 7.20

3. Shear and Bending Moment Diagrams

The design of a structural member requires knowledge of the internal forces acting at each of its cross sections. Such information can be obtained by systematically applying the concepts discussed in the previous section. We will develop more general techniques in the following sections. For simplicity, only the case of straight beams carrying a planar system of transverse forces and couples will be analyzed.

A typical loading is depicted in Figure 6a. In order to be able to describe all of the cross sections with the same xyz coordinate system, the origin is located at one end of the beam and the x axis is aligned with the centroidal axis of the beam. The location of a particular cross section may then be defined by specifying its position x relative to the origin.

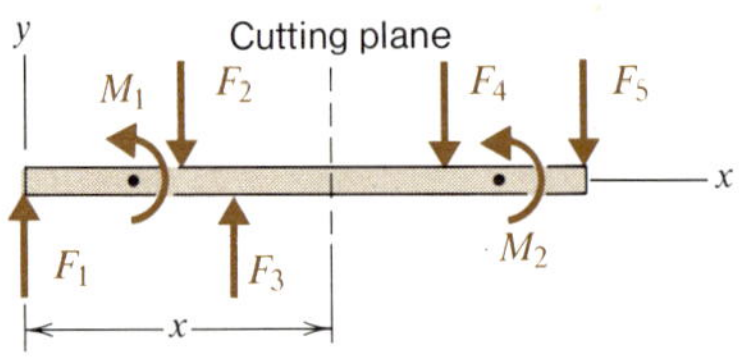

FIGURE 6a

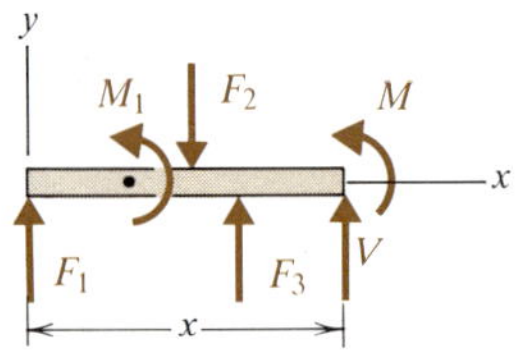

FIGURE 6b

The internal forces on a cross section at position x are found by cutting the beam at that location. The equations of static equilibrium for either segment of the cut beam will yield the internal forces. We choose the left segment, depicted in Figure 6b. Because all the external forces act transversely to the beam, in a common plane, the internal forces consist only of a shear force and a bending moment. These parameters shall be denoted by V and M, respectively.

An investigation of the internal forces on all cross sections of the beam reduces to a determination of the dependence of V and M on the position x. The functions $V(x)$ and $M(x)$ are usually depicted graphically in *shear and bending moment diagrams*. Such diagrams require nonambiguous definitions for positive shear and positive bending moment, both in order to avoid using inconsistent signs in the computations and in order to enable someone else to interpret correctly the diagrams we draw. These definitions are called *sign conventions*. (For example, we have already been using a sign convention by saying that an internal force normal to a cross section is positive if it is a tensile force.)

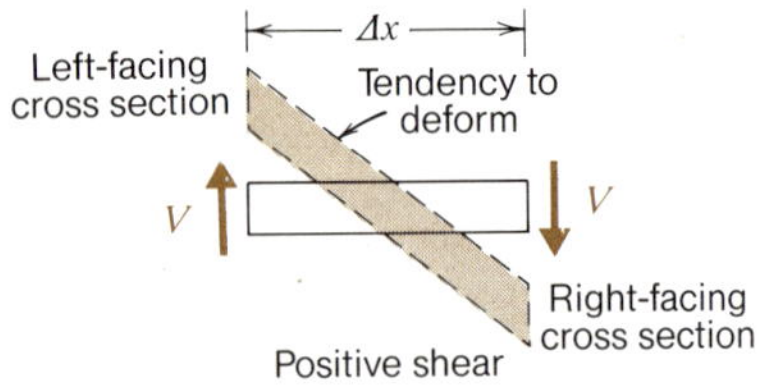

FIGURE 7a

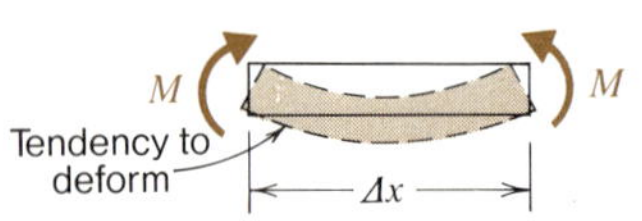

FIGURE 7b

The sign convention we will use here is obtained by considering the type of deformation a section of the beam of length Δx would undergo if it carried only a shear force or only a bending moment. These situations are depicted in Figures 7a and 7b, respectively.

For convenience, we will refer to a cross section whose outward normal faces left as a *left-facing cross section*, and a cross section whose outward normal faces right as a *right-facing cross section*. Figure 7a shows that in our sign convention,

A shear force is positive if it pushes the left-facing cross section upward, or equivalently if it pushes the right-facing cross section downward.

Alternatively, we can look upon this definition as: *Positive shear causes a clockwise rotation of the beam element being analyzed.*

Figure 7b shows that we will consider *positive bending moment* to be one that has the effect of bending the middle of the bar downward, that is, one that *causes the centroidal axis to assume a shape that is convex downward.* This means that in our sign convention,

A bending moment is positive if it is clockwise on the left-facing cross section, or equivalently, if it is counterclockwise on the right-facing cross section.

At this time, it is necessary that you memorize these sign conventions, although they will become quite natural to use after you have had some experience with them.

The deformation effects in Figure 7 have been greatly exaggerated for the sake of clarity. In reality, the deformation of a beam is usually quite small. Thus, it is not a contradiction to consider the beam to be rigid when equations of static equilibrium are formulated.

Now that the sign conventions for shear and bending moment have been established, the task of constructing shear and bending moment diagrams may be addressed. The most direct procedure is to cut the beam into two segments by means of a cutting plane at position x. The equilibrium equations for either segment yield functions for the shear V and bending moment M in terms of x. The functions $V(x)$ and $M(x)$ are then plotted to obtain the desired diagrams.

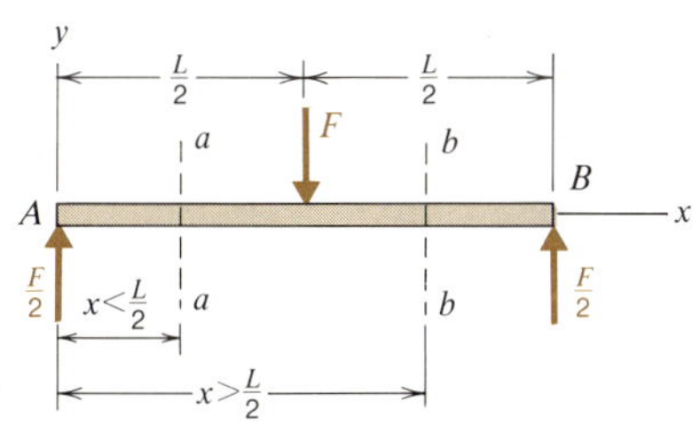

FIGURE 8

A key element of this procedure is that a variety of cutting planes must be considered. As the value of x is increased from zero, corresponding to cross sections at increasing distances from the origin, the free body diagram will be modified by the addition or deletion of external concentrated forces or couples. As an example, consider a lightweight simply supported beam loaded by a downward force $\bar{F}$ at its midpoint. After the reactions are evaluated, the free body diagram of the entire beam is as shown in Figure 8.

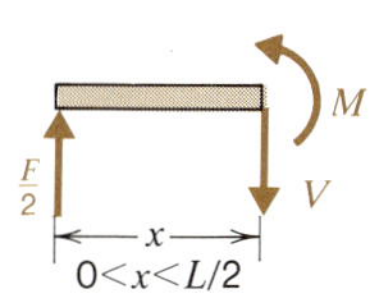

FIGURE 9a

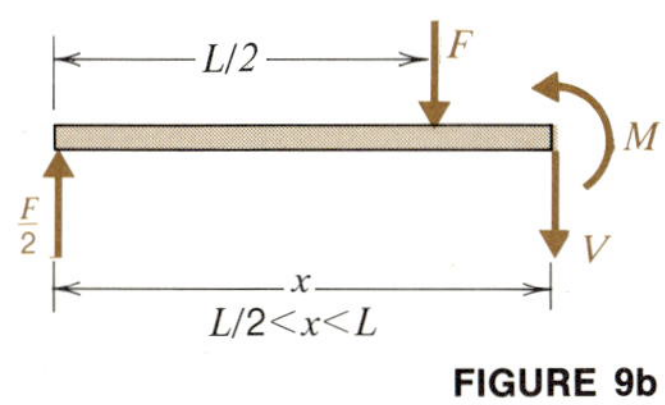

FIGURE 9b

Suppose we choose to evaluate $V(x)$ and $M(x)$ by using the beam segment to the left of the cutting plane at location x. It can be seen that for cutting plane a–a the only external force acting on the left segment is the reaction at end A. When the cutting plane is to the right of the midpoint, plane b–b, the force $\bar{F}$ is an additional force acting on the left segment. Hence, free body diagrams for both segments are needed to completely describe the shear and moment distributions. These appear in Figures 9a and 9b. It is important to note that in each diagram V and M are depicted in the positive direction, as defined by our sign convention.

The static equilibrium equations for each free body diagram give

$$0 < x < \frac{L}{2}: \qquad V = \frac{F}{2} \qquad M = \frac{F}{2}x$$

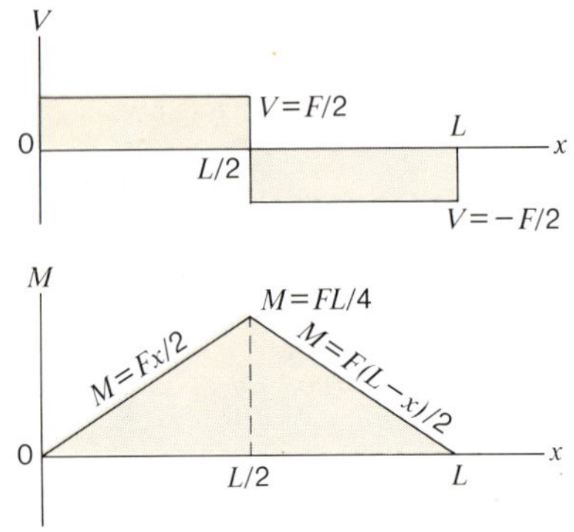

FIGURE 10

$$\frac{L}{2} < x < L: \qquad V = -\frac{F}{2} \qquad M = \frac{F}{2}x - F\left(x - \frac{L}{2}\right) = \frac{F}{2}(L - x)$$

These functions are now plotted to yield the shear and bending moment diagrams shown in Figure 10. (Plotting techniques are discussed in the Examples that follow.)

The correctness of these diagrams may be partially verified by checking that the values of the shear and bending moments at the ends match the force-couple systems acting at these locations. For instance, from Figure 8 we see that there is an upward force $F/2$ and no couple acting at the left end of the beam. According to our sign convention, a positive shear is upward on a left-facing cross section, which is the situation for the left end. Therefore, the external force-couple system acting at the left end is equivalent to a positive shear force $F/2$ and no bending moment. These values are confirmed by the shear and bending moment diagrams in Figure 10.

The terminology we use to describe the foregoing check is to say that we have verified that the shear and bending moment diagrams *close* at the left end. The check for closure at both ends is very important, even though it may seem obvious. We leave it to you to verify closure at the right end of the diagrams in Figure 10.

A noteworthy feature of the diagrams in Figure 10 is that at the midpoint, where the force is applied, the value of the shear force and the derivative (slope) of the bending moment function are discontinuous. A point of discontinuity in a function or its derivative represents a mathematical *singularity*, because it is an undefined situation. In addition to concentrated forces, other types of discontinuities arise at locations where couples are applied to a beam, and where a distributed loading function changes.

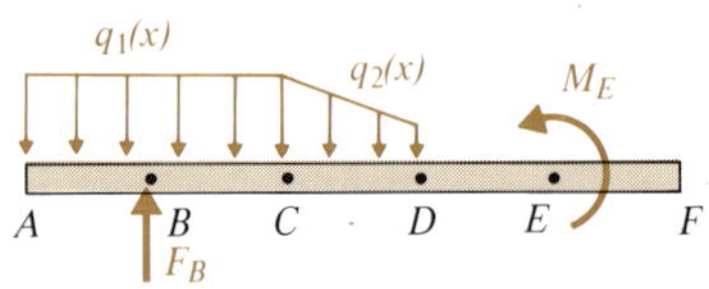

FIGURE 11

For example, consider the beam shown in Figure 11. Discontinuities occur at point B, where the concentrated force $\bar{F}_B$ is applied; at point C, where the distributed load function changes from $q_1(x)$ to $q_2(x)$; at point D, where the distributed load changes from q_2 to zero; and at point E, where the couple M_E is applied. To obtain the shear and bending moment diagrams for this beam, we would need to use cutting planes that pass between adjacent points of discontinuity. Thus, a total of five different cases are required to analyze the beam.

EXAMPLE 3

Sketch the shear and bending moment diagrams for the simply supported beam shown.

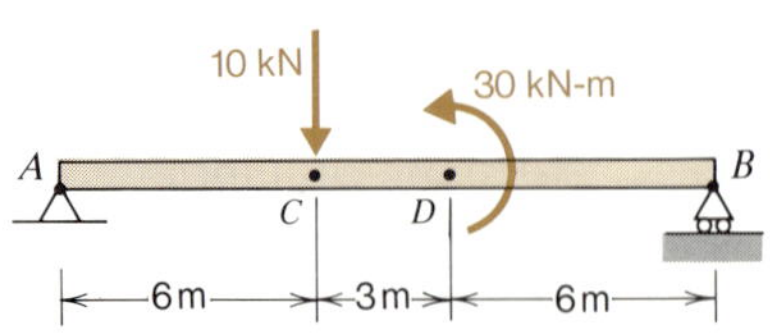

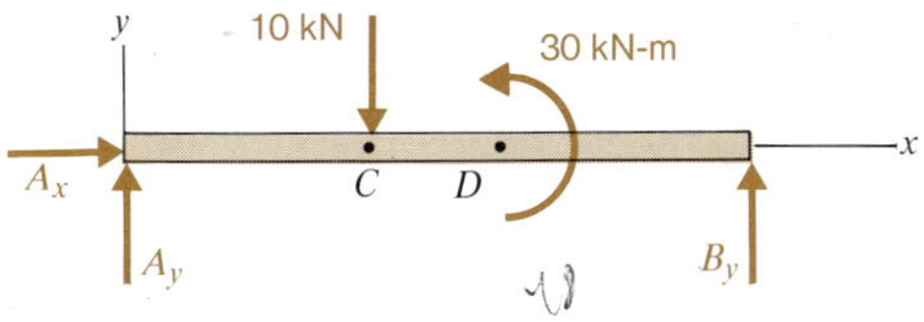

Solution

The first step in any analysis of internal forces is the determination of the reactions. Thus, we draw a free body diagram of the beam and solve the equilibrium equations.

$$\sum M_{Az} = 30 - 10(6) + B_y(15) = 0 \text{ kN-m}$$
$$\sum F_x = A_x = 0$$
$$\sum F_y = A_y + B_y - 10 = 0 \text{ kN}$$
$$B_y = 2 \text{ kN}, \qquad A_x = 0, \qquad A_y = 8 \text{ kN}$$

We next ascertain where discontinuities occur in the loading. There is a concentrated force applied at point C and a couple applied at point D. This means that we will require three cutting planes, one between the left end A and point C, the second between points C and D, and the last between point D and the right end B.

It is best to perform the computation in an orderly fashion. Experience has shown that students are less prone to errors if they progress across the beam in one direction, usually the direction of increasing values of x. Measuring x from end A, we will isolate the section of the beam to the left of each of the three aforementioned cutting planes.

When we draw the free body diagrams, we depict the shear and bending moment in the positive direction according to the appropriate sign convention. On a right-facing cross section we consider downward shear and counterclockwise bending moment positive, so the free body diagrams and corresponding computations of V and M are as shown below.

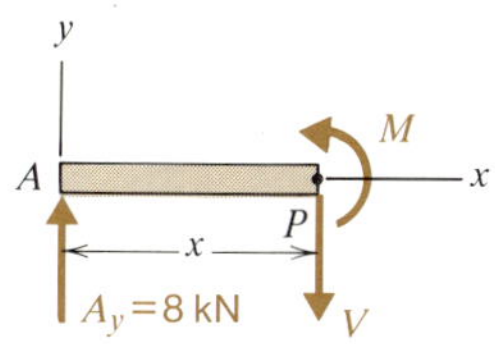

Region: $0 < x < 6$ m

$$\sum M_{Pz} = M - 8(x) = 0, \qquad \sum F_y = -V + 8 = 0$$
$$V = 8 \text{ kN}, \qquad M = 8x \text{ kN-m}$$

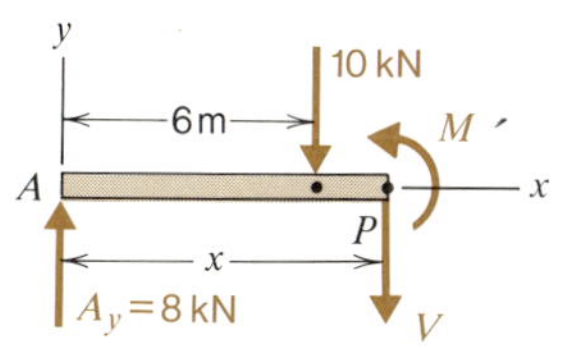

Region: $6 < x < 9$ m

$$\sum M_{Pz} = M + 10(x - 6) - 8(x) = 0, \qquad \sum F_y = -V - 10 + 8 = 0$$
$$V = -2\text{kN}, \qquad M = 60 - 2x \text{ kN-m}$$

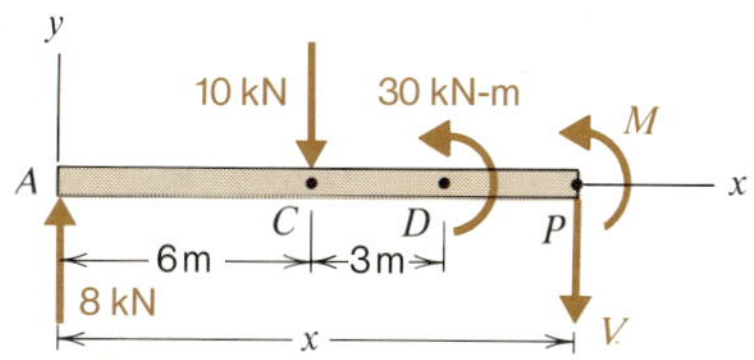

Region: $9 < x < 15$ m

$$\sum M_{Pz} = M + 30 + 10(x - 6) - 8(x) = 0$$
$$\sum F_y = -V - 10 + 8 = 0$$
$$V = -2 \text{ kN}, \qquad M = 30 - 2x \text{ kN-m}$$

We now have sufficient information to draw the shear and bending moment diagrams. It is not necessary to actually plot these diagrams point by point. The expressions for V and M in each region may be evaluated at the end points of that region; the values at these points may then be connected by the curve corresponding to the equations for V and M in the region. For

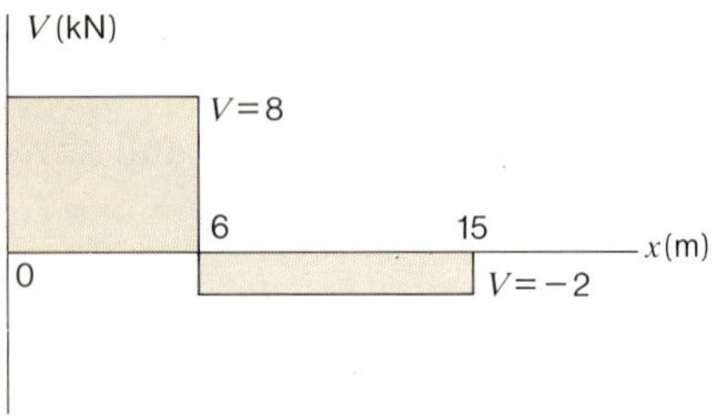

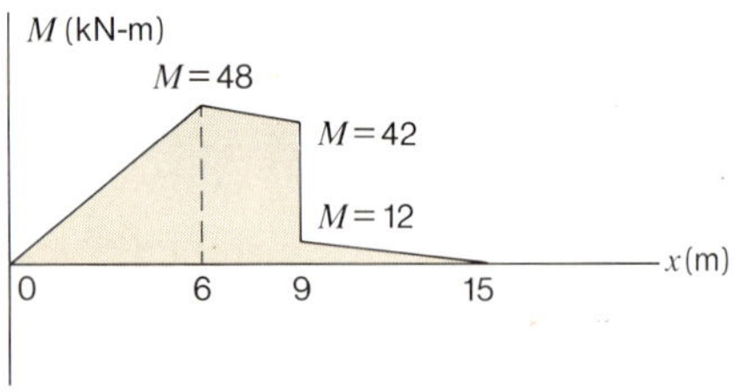

example, in the range $0 < x < 6$ m, we calculate and plot the values of V and M at $x = 0$ and $x = 6$. The value of V is constant, so it is represented by a horizontal line. The value of M is linearly dependent on the value of x, so it is represented by a straight sloping line.

The result of the plotting is the shear and bending moment diagrams shown. Notice that the singularity introduced by a couple loading is such that the value of the bending moment is discontinuous, but there is no discontinuity in the shear at that point. Also, notice that the singularity introduced by a concentrated force is such that the value of the shear is discontinuous, but there is no discontinuity in the bending moment at that point.

Closure of the shear and bending moment diagrams is verified by referring to the free body diagram of the entire beam. At end A we see that because $A_y = 8$ kN upward and there is no couple, V should be positive 8 kN (end A is a left-facing cross section) and M should be zero. At end B we see that because $B_y = 2$ kN upward and there is no couple, V should be negative 2 kN (right-facing cross section) and M should be zero. These values are all confirmed by, and consistent with, the diagrams we have drawn.

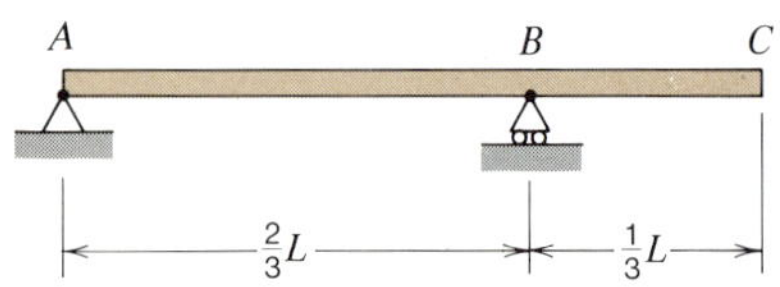

EXAMPLE 4

Sketch the shear and bending moment diagrams for the horizontal overhanging beam shown. The beam has constant weight per unit length w.

Solution

The first step in the solution is the determination of the reactions. The beam carries a constant distributed loading w, as shown in the free body diagram. The slash marks on the resultant $\bar{R}$ of the line load are used to denote that $\bar{R}$ is not an actual concentrated force. The equilibrium equations for the beam yield

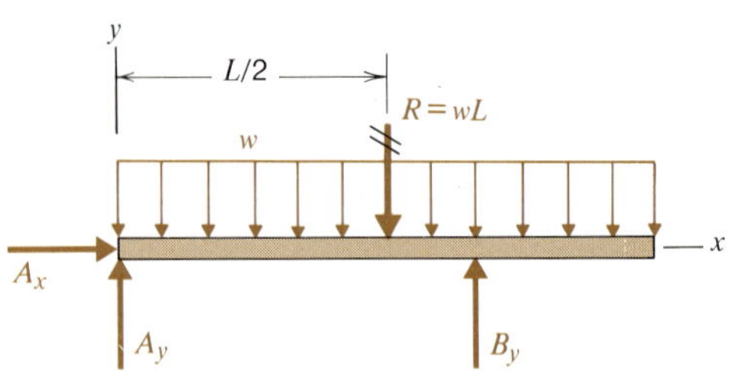

$$\sum M_{Az} = B_y\left(\frac{2L}{3}\right) - wL\left(\frac{L}{2}\right) = 0$$

$$\sum F_x = A_x = 0, \qquad \sum F_y = A_y + B_y - wL = 0$$

$$A_x = 0, \qquad A_y = \tfrac{1}{4}wL, \qquad B_y = \tfrac{3}{4}wL$$

We next locate the discontinuities. The distributed loading is the same along the whole length of the beam, so the only change in the loading occurs when the cutting plane passes point B, where a reaction force is applied. Thus, we must analyze two regions, one to the left and one to the right of point B. We measure x from the left end A and determine the internal forces by isolating the section of the beam to the left of the cutting plane. The required free body diagrams and equilibrium equations are given below. Notice that because we are concerned with the shear and bending moment acting on the right-facing cross section, the shear V is depicted as positive downward and the bending moment M is depicted as positive counterclockwise.

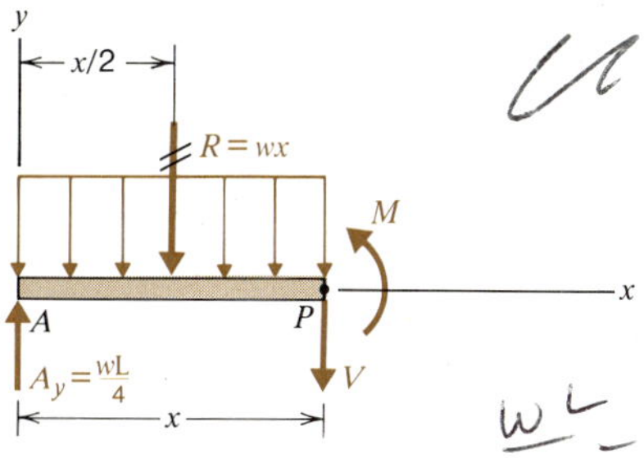

Region: $0 < x < 2L/3$

$$\sum M_{Pz} = M + wx\left(\frac{x}{2}\right) - \frac{wL}{4}(x) = 0$$

$$\sum F_y = -V - wx + \frac{wL}{4} = 0$$

$$V = w\left(\frac{L}{4} - x\right), \qquad M = -\frac{w}{4}(2x^2 - Lx)$$

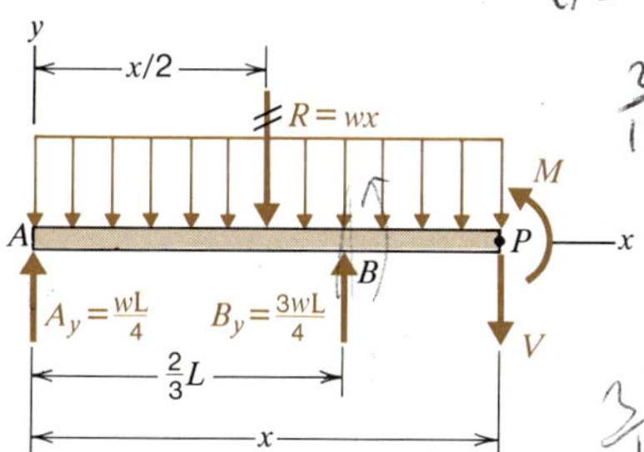

Region: $2L/3 < x < L$

$$\sum M_{Pz} = M - \frac{3wL}{4}\left(x - \frac{2L}{3}\right) + wx\left(\frac{x}{2}\right) - \frac{wL}{4}(x) = 0$$

$$\sum F_y = -V + \frac{3wL}{4} + \frac{wL}{4} - wx = 0$$

$$V = w(L - x), \qquad M = -w\left(\frac{x^2}{2} - Lx + \frac{1}{2}L^2\right)$$

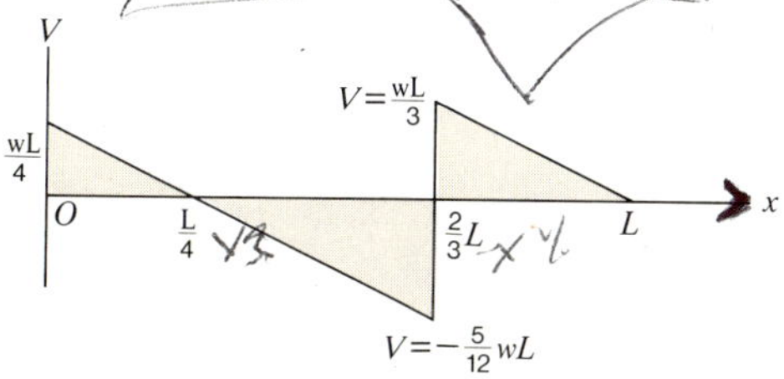

The shear diagram is sketched by noting that both expressions for V are linear in x, so the curves for V in both ranges are straight lines. Evaluating each function for V at the limits of its region of validity and plotting the corresponding points yields the shear diagram shown.

The bending moment diagram is sketched by noting that both expressions for M are quadratic polynomials in which the coefficient of the x^2 term is negative. This means that the curves representing M in either range are parabolas that are concave downward. We may determine the location of the horizontal tangent of each parabola by setting $dM/dx = 0$. This yields

Region: $0 < x < 2L/3$

$$\frac{dM}{dx} = -\frac{w}{4}(4x - L) = 0 \qquad \text{so} \qquad x = L/4$$

Region: $2L/3 < x < L$

$$\frac{dM}{dx} = -w(x - L) = 0 \qquad \text{so} \qquad x = L$$

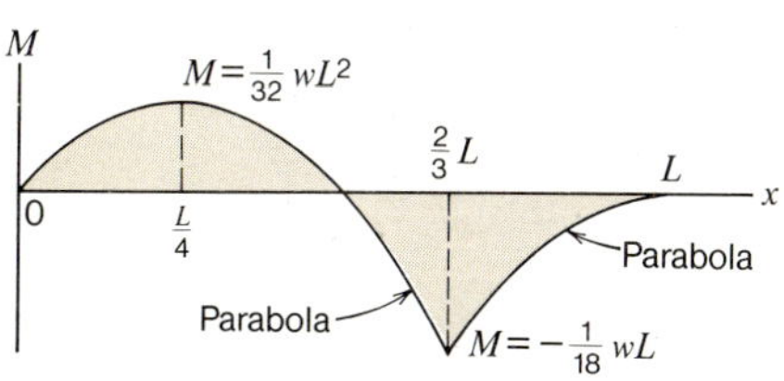

The foregoing knowledge of the shape of the bending moment curve, together with the plotted points representing the values of each bending moment function at the limits of its range of validity, allow us to draw the bending moment diagram. Can you verify that the shear and bending moment diagrams close?

It is interesting to note that the bending moment has its extreme values ($dM/dx = 0$) at the locations $x = L/4$ and $x = L$ where the shear is zero. This is not a coincidence, as will be proven in the next subsection.

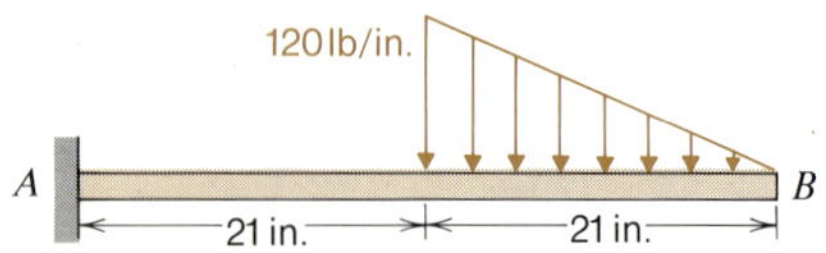

EXAMPLE 5

Sketch the shear and bending moment diagrams for the cantilever beam carrying the triangular distributed load shown.

Solution

We begin with a free body diagram of the entire beam, which we use to determine the reactions at the fixed end A. In the free body diagram the force $\bar{R}$ is the resultant of the distributed load.

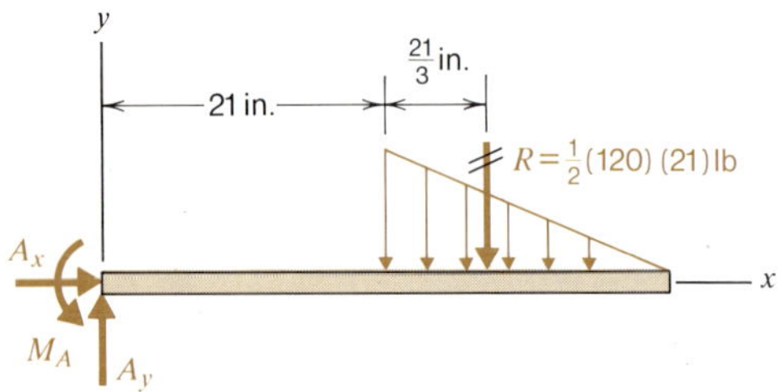

$$\sum M_A = M_A - 1260(21 + 7) = 0$$

$$\sum F_x = A_x = 0, \qquad \sum F_y = A_y - 1260 = 0$$

$$A_x = 0, \qquad A_y = 1260 \text{ lb}, \qquad M_A = 35{,}280 \text{ lb-in.}$$

The only discontinuity in the loading occurs at the midpoint, where the distributed load begins. We place the origin of xyz at the left end, and consider cutting planes at progressively larger values of x. For the region $0 < x < 21$ in., where the cutting plane is to the left of the distributed loading, the free body diagram and equilibrium equations for the left beam segment are as described below.

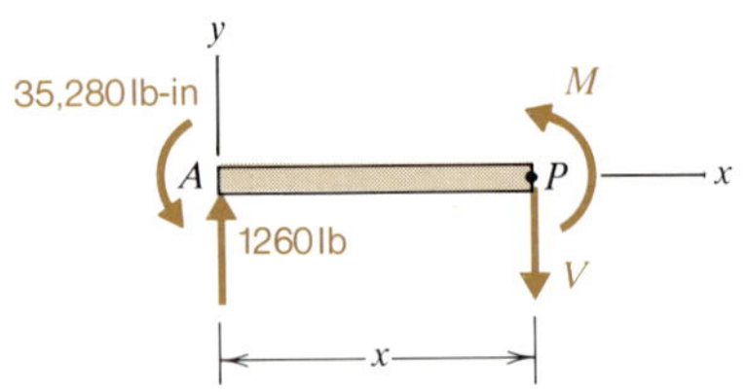

Region: $0 < x < 21$ in.

$$\sum M_{Pz} = M + 35{,}280 - 1260(x) = 0, \qquad \sum F_y = -V - 1260 = 0$$

$$V = 1260 \text{ lb}, \qquad M = 1260x - 35{,}280 \text{ lb-in.}$$

When we consider cutting planes in the region $21 < x < 42$, we find that there is a trapezoidal line load acting on the beam segment to the left of the cutting plane. (Recall that the free body diagram for the isolated beam segment contains only those external loads that actually act on that segment.)

An alternative for this situation that is simpler than describing the trapezoidal load is to consider the beam segment to the right of the cutting plane. As shown in the free body diagram, this segment carries only a triangular line load. The cross section of this segment that is exposed by the cutting plane faces left, so positive V is upward and positive M is clockwise. Note that because x measures positions from the left end, the length of the beam segment is $42 - x$ inches.

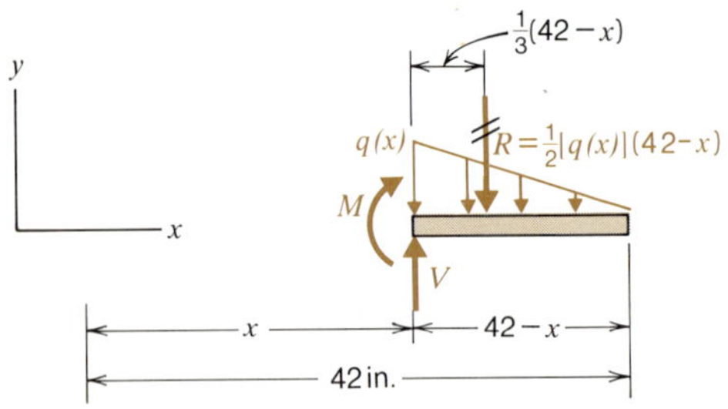

Region: $21 < x < 42$ in.

$$\sum M_{Pz} = -M - R_1[\tfrac{1}{3}(42 - x)] = 0, \qquad \sum F_y = V - R_1 = 0$$

$$V = R_1 = \tfrac{1}{2}q(x)(42 - x)$$

$$M = -\tfrac{1}{3}R_1(42 - x) = -\tfrac{1}{6}q(x)(42 - x)^2$$

The determination of V and M in terms of x now requires that we determine the expression for the loading function $q(x)$. This is obtained by equating the slope of the straight loading curve in the given sketch to the

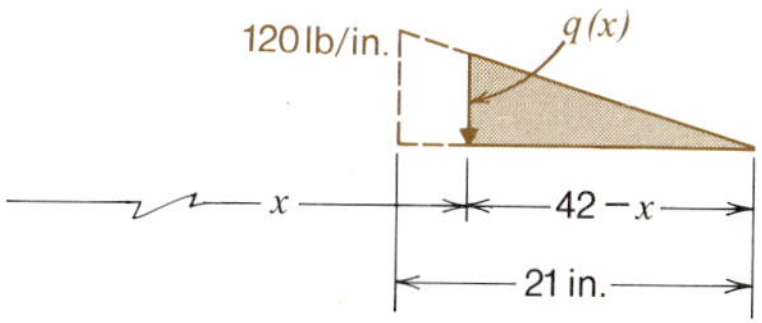

slope found from the free body diagram of the beam segment under consideration. Thus, from the sketch in the margin

$$\frac{q(x)}{42 - x} = \frac{120}{21}, \qquad q(x) = 120\left(2 - \frac{x}{21}\right) \text{ lb/in.}$$

The last set of equilibrium equations now take the form

Region: $21 < x < 42$ in.

$$V = \frac{1}{2}\left[120\left(2 - \frac{x}{21}\right)\right](42 - x) = 1260\left(2 - \frac{x}{21}\right)^2 \text{ lb}$$

$$M = -\frac{1}{6}\left[120\left(2 - \frac{x}{21}\right)\right](42 - x)^2 = -8820\left(2 - \frac{x}{21}\right)^3 \text{ lb-in.}$$

The beam's shear and bending moment diagrams may now be sketched. For $0 < x < 21$ in., V is constant and M increases linearly with x, so both functions plot as straight lines. In the region $21 < x < 42$ in., V is a quadratic polynomial and M is a cubic polynomial. The curve for V is readily identified as an upward-curving parabola having its lowest point at $x = 42$ in. where $V = 0$. Similarly, we see that the curve for M is a downward-curving cubic whose maximum value is zero at $x = 42$ in.

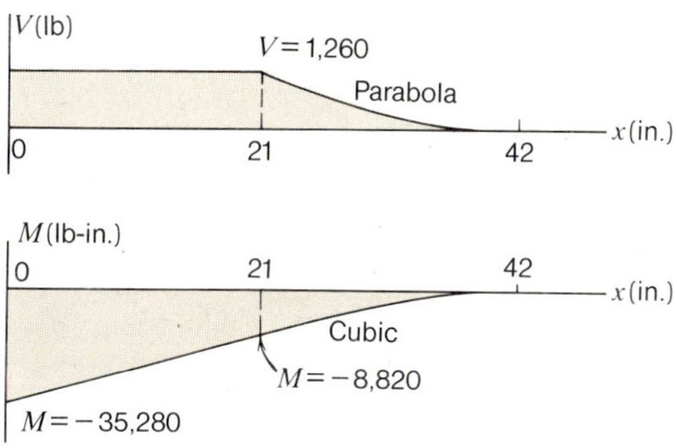

We evaluate the expressions for M and V in the region $0 < x < 21$ at $x = 0$ and $x = 21$, and the expressions for M and V in the region $21 < x < 42$ at $x = 21$. Then, by plotting these points and connecting them by the appropriate curves, we obtain the accompanying diagrams.

These diagrams close at both ends, because there is no force-couple system acting at the right end B, and the reactions at end A, which is a left-facing cross section, are equivalent to a positive (upward) shear and a negative (counterclockwise) bending moment.

HOMEWORK PROBLEMS

7.21–7.32 Draw shear and bending moment diagrams for the beam shown. Give all critical values.

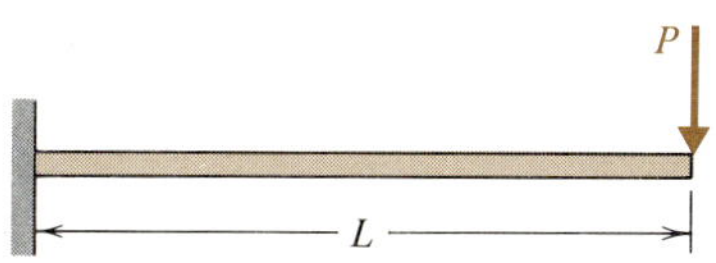

PROB. 7.21

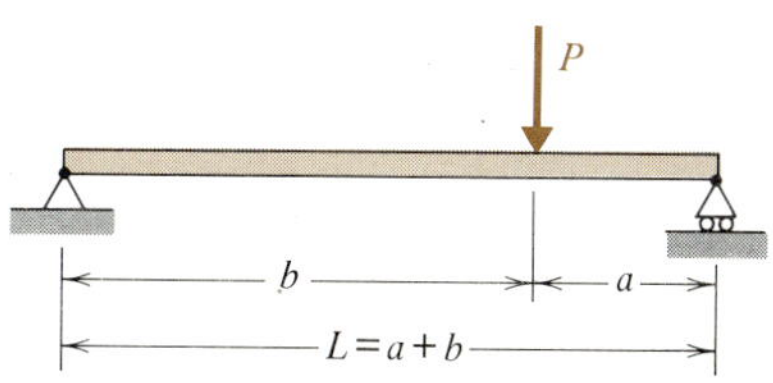

PROB. 7.22

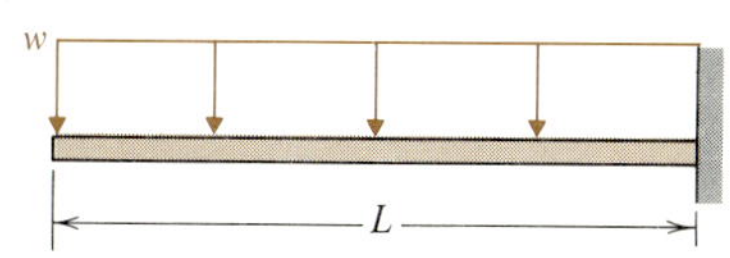

PROB. 7.23

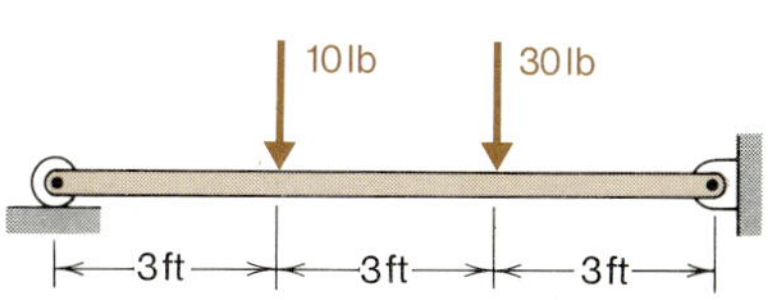

PROB. 7.24

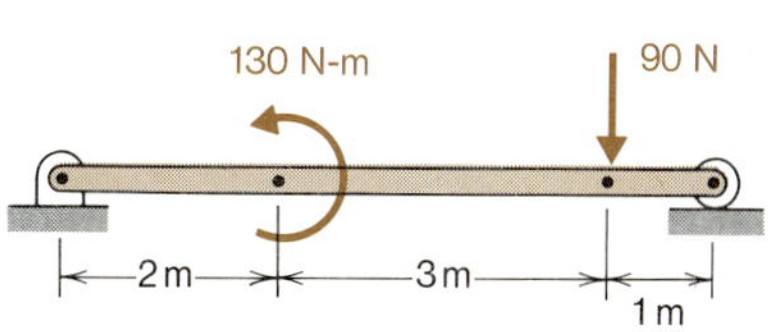

PROB. 7.25

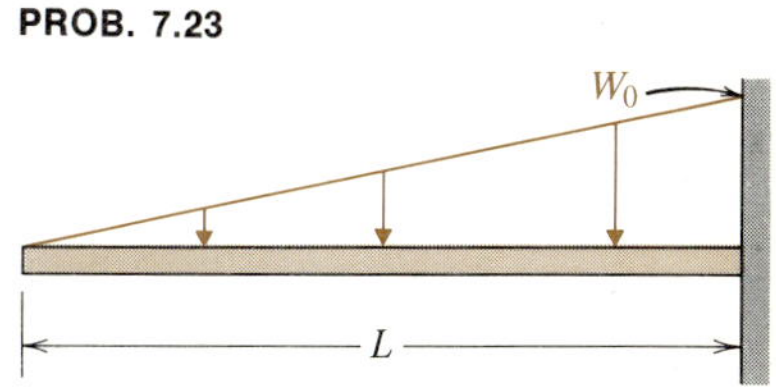

PROB. 7.26

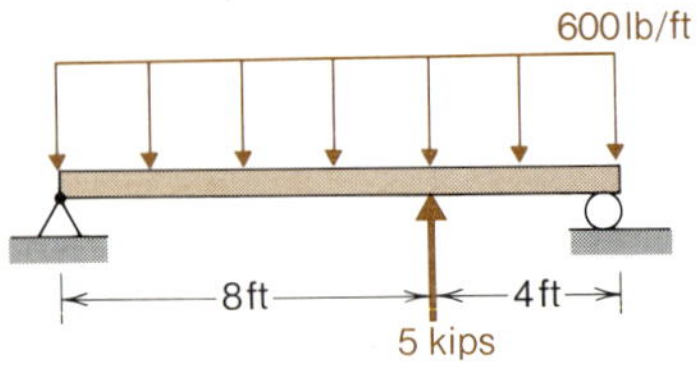

PROB. 7.27

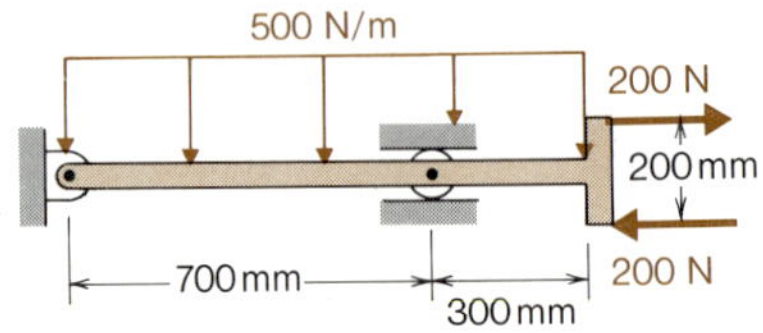

PROB. 7.28

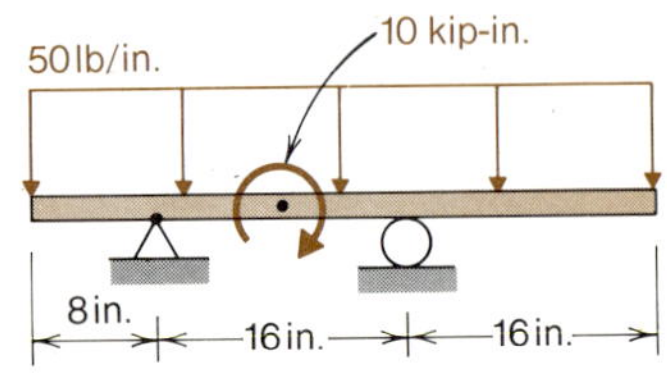

PROB. 7.29

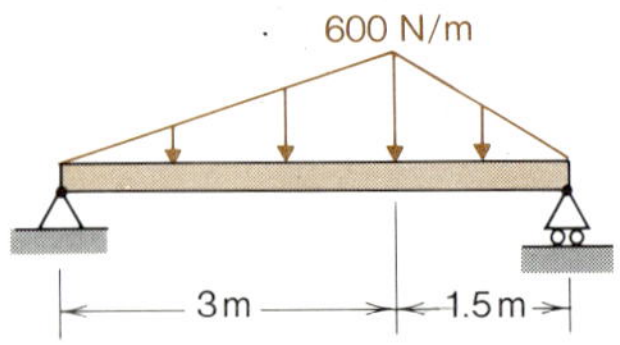

PROB. 7.30

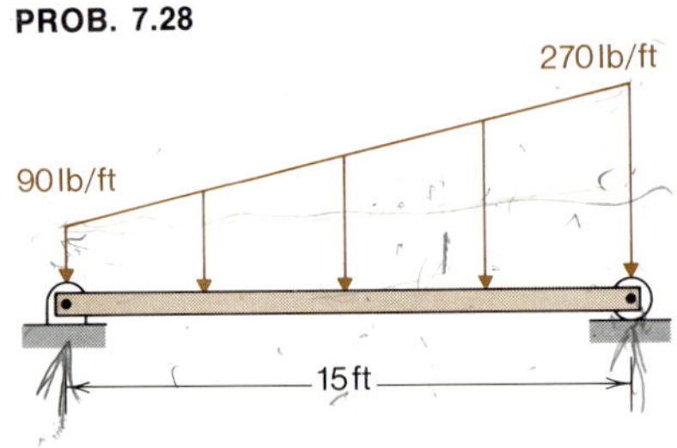

PROB. 7.31

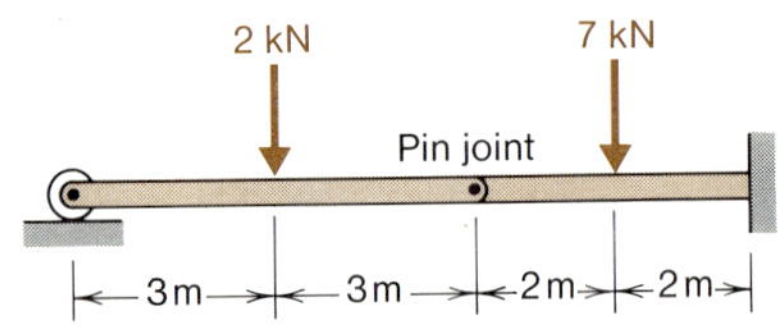

PROB. 7.32

*4. Relations Between Bending Moment, Shear, and Distributed Load

After having studied a few examples and solved some homework problems, it should be apparent to you that although the theory for the determination of the shear and bending moment diagrams is conceptually straightforward, in practice the computations can get quite involved. In this section we will employ some general relations between the internal forces and the loading on a beam. The deeper understanding that this study will bring will help to substantially reduce the amount of calculations involved in the determination of shear and bending moment diagrams.

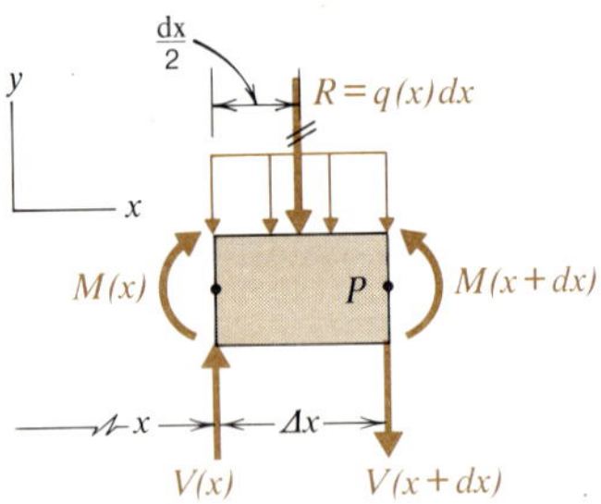

FIGURE 12

General relations between the internal force system and the external loading may be obtained by isolating an infinitesimal segment of length dx of a beam. The segment in Figure 12 carries a line load $q(x)$, but no concentrated force or couple loading.

The forces acting on this beam segment are the shear and bending moment on the left-facing cross section at position x, and on the right-facing cross section at position $x + dx$. These force-couple systems, which are depicted in their positive sense in Figure 12, need not be the same on both cross sections. The free body diagram for the segment is completed by replacing the line load by its equivalent concentrated force resultant $\bar{R}$. Because dx is infinitesimal, the change in the value of q from x to $x + dx$ may be ignored. Thus, the line load is considered to be rectangular, and $R = q(x)\,dx$ acts at the midpoint.

The equilibrium equations for the beam segment are

$$\sum M_{Pz} = M(x + dx) - M(x) - V(x)\,dx + q(x)\,dx\,\frac{dx}{2} = 0$$

$$\sum F_x \equiv 0, \qquad \sum F_y = V(x) - V(x + dx) - q(x)\,dx = 0$$

We divide each of the nontrivial equations above by dx to find

$$\frac{V(x+dx)-V(x)}{dx} = -q(x), \qquad \frac{M(x+dx)-M(x)}{dx} = V(x) + q(x)\frac{dx}{2}$$

Because dx is infinitesimal, these relations become

$$\frac{dV}{dx} = -q(x) \tag{1a}$$

$$\frac{dM}{dx} = V(x) \tag{1b}$$

Equations (1) are called *differential equations* because they equate the derivative of one variable to another function. These equations are useful for plotting the shear and bending moment diagrams. For example, eq. (1b) tells us that the bending moment has a horizontal tangent at the location where the shear is zero. (We noted this fact in Example 4.)

Another use for eqs. (1) is associated with the determination of the shear and bending moment resulting from a given loading. This calls for an integral representation of the relations. Suppose that we know the shear and bending moment at some value x_A and wish to determine the value of these quantities at some other value x_B. Equations (1) may be used for this determination by multiplying each equation by dx and integrating between the limits corresponding to the two points. This gives

$$\int_{V_A}^{V_B} dV = V_B - V_A = -\int_{x_A}^{x_B} q(x)\,dx$$
$$\int_{M_A}^{M_B} dM = M_B - M_A = \int_{x_A}^{x_B} V(x)\,dx$$

Rearranging terms yields

$$V_B = V_A - \int_{x_A}^{x_B} q(x)\,dx \tag{2a}$$

$$M_B = M_A + \int_{x_A}^{x_B} V(x)\,dx \tag{2b}$$

From the fact that an integral of a function is the area under the curve representing that function, eq. (2a) can be seen to be equivalent to the statement that

> The shear force at x_B is the shear force at x_A *minus* the area under the distributed load curve in the region between x_A and x_B.

Similarly, eq. (2b) states that

> The bending moment at x_B is the bending moment at x_A *plus* the area under the shear curve in the region between x_A and x_B.

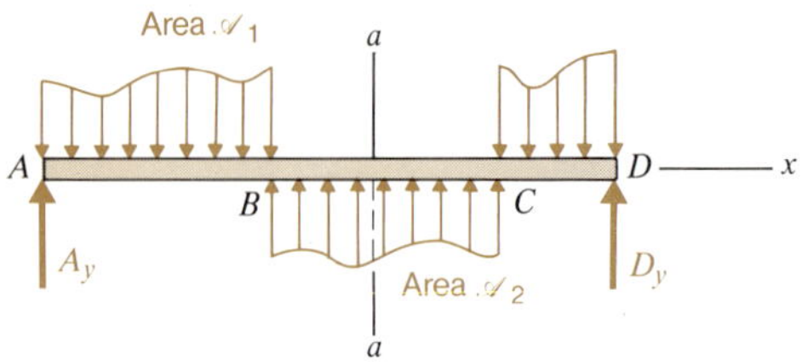

FIGURE 13

Our interest in this text is the quantitative determination of the internal forces at a few key locations along a beam, for example, where the bending moment and shear assume extreme values, as well as the determination of a pictorial representation of the way in which the internal forces vary with location. We can use the foregoing statements to achieve this goal. Consider the situation depicted in Figure 13, where a beam is carrying a distributed load and concentrated forces (reactions) at its ends.

According to our sign conventions, the vertical force $\bar{A}_y$ at end A is equivalent to a positive shear force. According to eq. (2a), the shear force at the location of the cutting plane a–a appearing in the figure is given by

$$V_1 = V_A - (\mathscr{A}_1 - \mathscr{A}_2) = A_y - \mathscr{A}_1 + \mathscr{A}_2$$

where $\mathscr{A}_1$ and $\mathscr{A}_2$ are areas formed by the loading function, as indicated in Figure 13. In regard to the area $\mathscr{A}_2$, it is important to realize that between points B and C, the load, and thus the area, is negative because we considered downward loads to be positive in deriving eqs. (2).

Let us determine the shear diagram for the beam in Figure 13. We begin by recalling that discontinuous changes in the value of shear occur only at locations where there are concentrated transverse forces. Thus, the curves representing the shear distribution in the regions bounded by the singular points B and C are continuous. We may use eq. (2a) to obtain the value of shear at any critical point, particularly the ends of the beam and the singular points B and C. Then, the character of the shear curves connecting the plotted points may be explored by using equation (1a), $dV/dx = -q$, to ascertain how dV/dx (the slope of the shear curve) and V depend on x.

The foregoing procedure for constructing the shear diagram, and a similar procedure for bending moment diagrams, will be demonstrated in the next two examples. Before proceeding, however, we shall consider the cases of loadings by concentrated forces and couples in order to gain further insight into the discontinuities they cause in the shear and bending moment diagrams.

Figures 14a and 14b depict an infinitesimal beam segment of length dx whose midpoint is point C, where a concentrated load is applied. If any distributed load is present, its effect on such a segment is negligible because the resultant $q(x)\,dx$ is infinitesimal, whereas the concentrated loads are finite. The cases of a concentrated force $\bar{F}_C$ or couple $\bar{M}_C$ are depicted in Figures 14a and 14b, respectively.

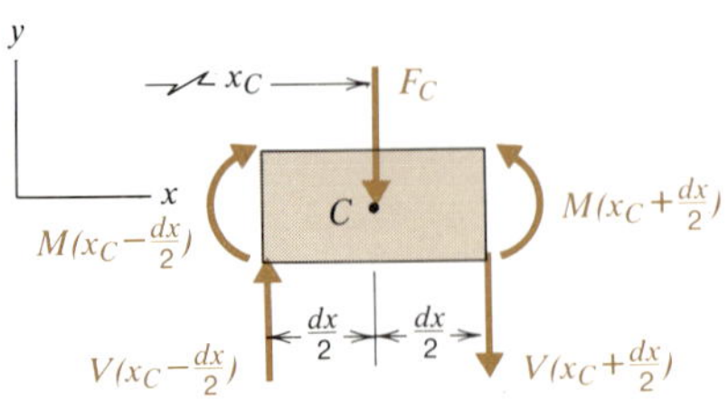

FIGURE 14a

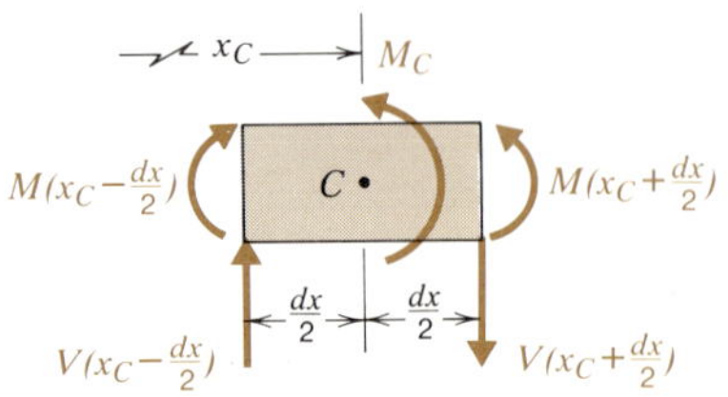

FIGURE 14b

Equilibrium of the element in Figure 14a requires that

$$\sum M_{Cz} = M\left(x_C + \frac{dx}{2}\right) - M\left(x_C - \frac{dx}{2}\right) - \left[V\left(x_C - \frac{dx}{2}\right) + V\left(x_C + \frac{dx}{2}\right)\right]\left(\frac{dx}{2}\right) = 0 \tag{3a}$$

$$\sum F_y = V\left(x_C - \frac{dx}{2}\right) - V\left(x_C + \frac{dx}{2}\right) - F_C = 0 \tag{3b}$$

The presence of the finite term F_C in the force equation means that the equations cannot be divided by dx. Instead, we note that because dx is infinitesimal, $x_C - dx/2$ is the position of the cross section immediately to the left of point C, and $x_C + dx/2$ is the position of the cross section immediately to the right of point C. We denote these positions as x_C^- and x_C^+, respectively. Equations (3) then become

$$\sum M_{Cz} = M(x_C^+) - M(x_C^-) - [V(x_C^-) + V(x_C^+)]\left(\frac{dx}{2}\right) = 0$$

$$\sum F_y = V(x_C^-) - V(x_C^+) - F_C = 0$$

Terms of order dx are negligible, so these equations reduce to

$$V(x_C^+) = V(x_C^-) - F_C \tag{4a}$$

$$M(x_C^+) = M(x_C^-) \tag{4b}$$

In words, eqs. (4) state that

> The value of the bending moment is continuous at the singular point of a downward concentrated force, but the shear is discontinuous, being less to the right of the point of application of the force by an amount equal to the magnitude of the force.

Note that in the case where the concentrated force is upward, it is necessary to regard F_C as a negative quantity in eq. (4a).

Similar steps can be followed to investigate the equilibrium of the beam element in Figure 14b. Such an investigation show that

$$V(x_C^+) = V(x_C^-) \tag{5a}$$

$$M(x_C^+) = M(x_C^-) - M_C \tag{5b}$$

In words, eqs. (5) state that

> The value of the shear is continuous at the singular point of a counterclockwise couple, but the bending moment is discontinuous, being less to the right of the point of application of the couple by an amount equal to the magnitude of the couple.

Note that a clockwise couple is treated by regarding M_C as a negative quantity.

Referring back to any of the shear and bending moment diagrams previously determined, particularly in Examples 3 and 4, will show that eqs. (4) and (5) validate the results given there.

In the context of our general goal of determining shear and moment diagrams, we see that eqs. (2) may be used to relate shears or bending moments within the regions lying between points where concentrated forces and couples are applied, whereas eqs. (4) and (5) may be used to go from one side of these singular points to the other. In this manner, the values of shear and bending moment at the singular points and other critical points may be determined. After these values are plotted, the characteristics of the curves connecting the plotted values in each region bounded by the singular points can be investigated with the aid of eqs. (1).

The remarkable feature of this method for constructing the shear and bending moment diagrams is that once the reactions have been obtained, it is no longer necessary to write the equations of static equilibrium, as is demonstrated in the following examples.

EXAMPLE 6

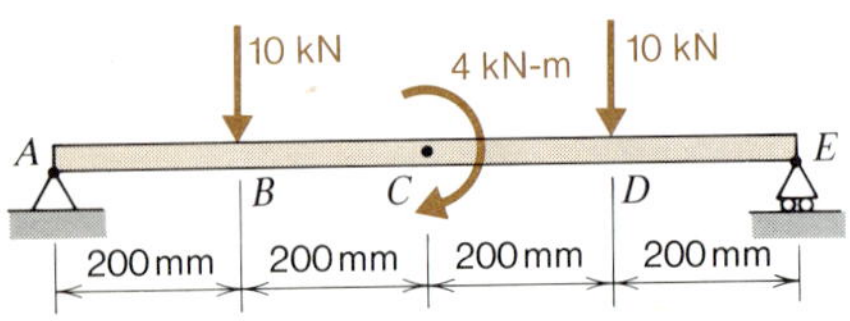

The simply supported beam carries the loading shown. Use the derived relations between load, shear, and bending moment to sketch the shear and bending moment diagrams. Give all critical values.

Solution

We begin by drawing a free body diagram of the entire beam in order to determine the reactions. The corresponding equilibrium equations are

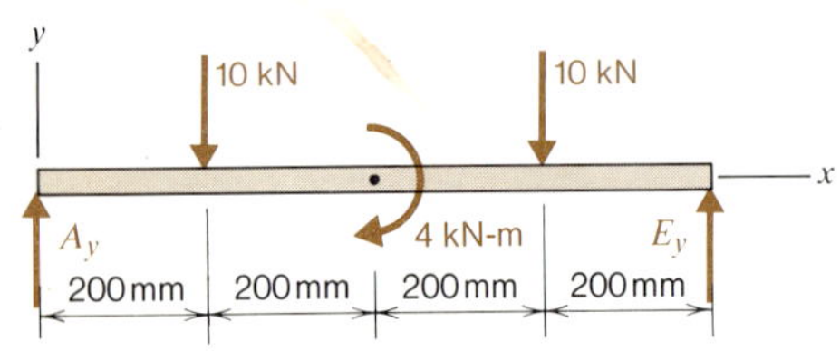

$$\sum M_{Az} = -10(0.20) - 4 - 10(0.60) + E_y(0.80) = 0$$

$$\sum F_x \equiv 0, \qquad \sum F_y = A_y + E_y - 20 = 0$$

$$E_y = 15 \text{ kN}, \qquad A_y = 5 \text{ kN}$$

We now draw the loading diagram by repeating the free body diagram of the beam, with the unknown reactions replaced by their calculated values. Singularities are readily identified to exist at points B, C, and D.

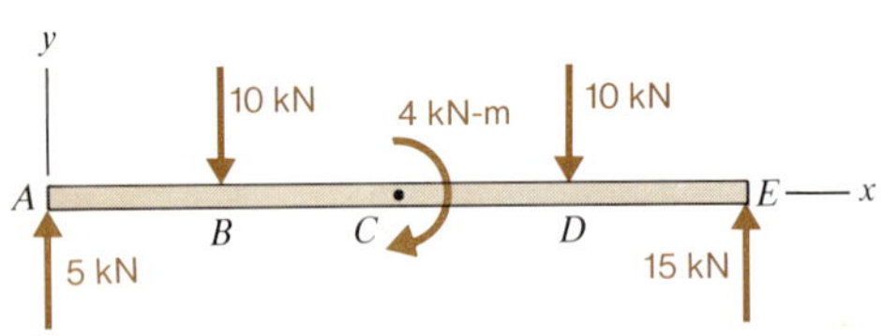

From the load diagram we may construct the shear diagram. At point A, the force is equivalent to a positive shear (upward on a left-facing cross section), $V = 5$ kN. From eq. (4a) we observe that the concentrated *downward* forces result in discontinuous *decreases* of the shear by 10 kN each as points B and D are passed from left to right when sketching the shear diagram. Further, with respect to the couple, we observe from eq. (5a) that the shear is continuous at point C. Finally, the shear is constant between these singular points because there is no line load, $dV/dx = 0$. These observations lead to the shear diagram shown to the left. As a check, we note that the shear diagram closes at end E, because the upward 15 kN force at this end is equivalent to a negative shear.

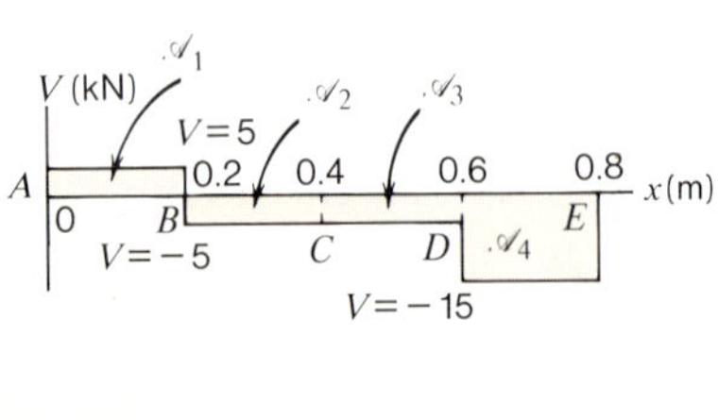

We now use the shear diagram to construct the moment diagram. From the load diagram, we see that there is no moment at end A. From eq. (4b) we note that the bending moment is continuous at points B and D, where there are concentrated forces, and from eq. (5b) we observe that the *clockwise* 4 kN-m couple at point C results in a discontinuous *increase* in bending moment in that amount as point C is passed from left to right in the diagram.

The values of bending moment at the singularties may then be calculated by using eq. (2b) to proceed from one singularity to the next. In terms of the areas indicated in the shear diagram, this equation gives

$$M(x_B = 0.20) = M(x_A = 0) + \mathscr{A}_1 = 0 + 5(0.20) = 1.0 \text{ kN-m}$$

$$M(x_C = 0.40^-) = M(x_B = 0.20) - \mathscr{A}_2 = 1.0 - 5(0.20) = 0$$

$$M(x_C = 0.40^+) = M(x_C = 0.40^-) + 4 = 4.0 \text{ kN-m}$$

$$M(x_D = 0.60) = M(x_C = 0.40^+) - \mathscr{A}_3 = 4.0 - 5(0.20) = 3.0 \text{ kN-m}$$

$$M(x_E = 0.80) = M(x_D = 0.60) - \mathscr{A}_4 = 3.0 - 15(0.2) = 0$$

Note that the bending moment diagram closes at end E.

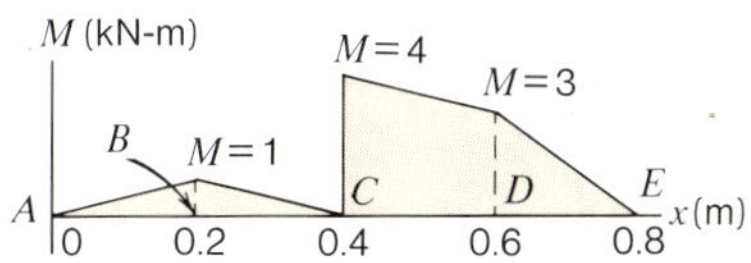

To plot these values we determine the type of curves connecting the plotted points. From eq. (1b), $dM/dx = V$. Because V is constant between each pair of singular points, we see that each curve has constant slope. In other words, the curves are straight lines. Thus, we obtain the bending moment diagram shown to the left.

EXAMPLE 7

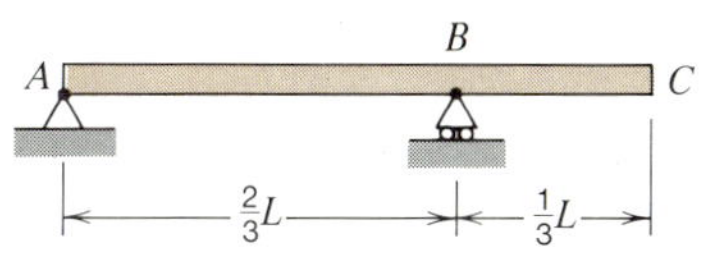

The overhanging beam of weight w per unit length studied in Example 4 is depicted again in the sketch. Use the relations between distributed load, shear, and bending moment to sketch the shear and bending moment diagrams. Give all cirtical values.

Solution

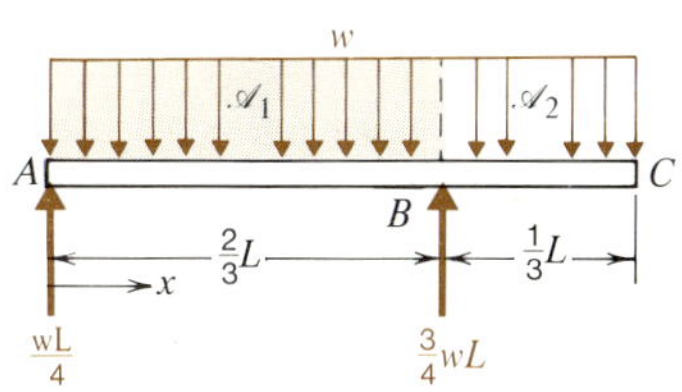

The necessary preliminary step of determining the reactions was performed in Example 4, so it is not repeated here. After the reactions are determined the loading diagram is drawn by showing all loads, including the reactions, in terms of their known magnitude and sense. This diagram indicates that the only singularity occurs at point B, where there is a concentrated upward force.

The shear diagram is obtained from the loading diagram. The force $wL/4$ at end A is equivalent to a positive shear. From eq. (4a) we find that the concentrated *upward* force at point B results in a discontinuous *increase* in shear by $3wL/4$ as this point is passed in the diagram from left to right. Then, the shear at the points of discontinuity are calculated by using eq. (2a) and the areas indicated in the loading diagram, as follows.

$$V\left(x_B = \frac{2L^-}{3}\right) = V(x_A = 0) - \mathscr{A}_1 = \frac{1}{4}wL - w\left(\frac{2L}{3}\right) = -\frac{5}{12}wL$$

$$V\left(x_B = \frac{2L^+}{3}\right) = V\left(x_B = \frac{2L^-}{3}\right) - F_B = -\frac{5}{12}wL - \left(-\frac{3}{4}wL\right) = \frac{1}{3}wL$$

$$V(x_C = L) = V\left(x_B = \frac{2L^+}{3}\right) - \mathscr{A}_2 = \frac{1}{3}wL - w\left(\frac{L}{3}\right) = 0$$

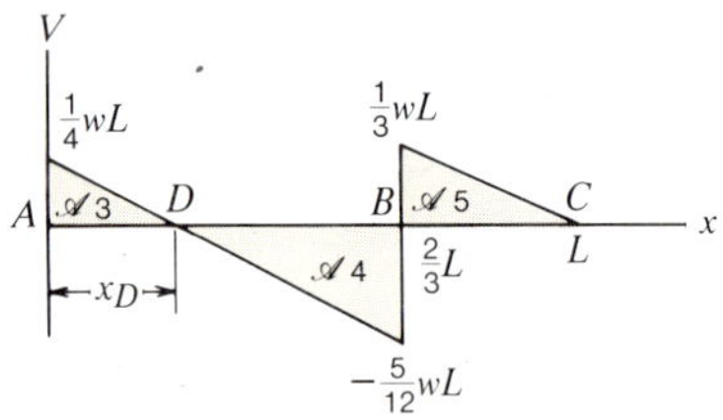

As a check, we note that the shear closes at end C, because no force is applied at this end.

These points are now plotted and eq. (1a), $dV/dx = -q$, is used to ascertain the types of curves connecting the plotted points. Because q is constant, it follows that V is linearly dependent on x (integration of a constant). In other words, the shear curves are straight lines. Further, because q is positive, these straight lines are sloping downward. The shear diagram resulting from these considerations is as shown.

The determination of the bending moment proceeds in a similar manner, using the areas indicated in the shear diagram to evaluate the integrals arising from eq. (2b). The determination of these areas requires that we first obtain the distance x_D locating point D where $V = 0$. The proportionality law for similar triangles yields

$$\frac{x_D}{(\frac{1}{4}wL)} = \frac{(2L/3)}{(\frac{1}{4}wL + \frac{5}{12}wL)} \qquad \text{so} \qquad x_D = \frac{L}{4}$$

The area $\mathscr{A}_4$ is negative because the shear is negative. Also, the beam does not carry any couple loads, so the moment diagram is continuous. Thus, the computations of the bending moment at the singular points proceed as follows, starting at $x_A = 0$, where we know that $M = 0$.

$$M\left(x_D = \frac{L}{4}\right) = M(x_A = 0) + \mathscr{A}_3 = 0 + \frac{1}{2}\left(\frac{1}{4}wL\right)\left(\frac{L}{4}\right) = \frac{1}{32}wL^2$$

$$M\left(x_B = \frac{2L}{3}\right) = M\left(x_D = \frac{L}{4}\right) - \mathscr{A}_4$$

$$= \frac{1}{32}wL^2 - \frac{1}{2}\left(\frac{5}{12}wL\right)\left(\frac{2L}{3} - \frac{L}{4}\right) = -\frac{1}{18}wL^2$$

$$M(x_C = L) = M\left(x_B = \frac{2L}{3}\right) + \mathscr{A}_5$$

$$= -\frac{1}{18}wL^2 + \frac{1}{2}\left(\frac{1}{3}wL\right)\left(\frac{L}{3}\right) = 0$$

Notice that the bending moment closes at end C because there is no couple at that location.

The bending moment diagram is obtained by first plotting the values of bending moment that were determined. To draw the curves connecting the plotted points, we recall eq. (1b), $dM/dx = V$. The shear curves are straight lines between the singularities, so V is linearly dependent on x. It then follows that M is dependent on x^2 (integrating x), so the moment curves are parabolas.

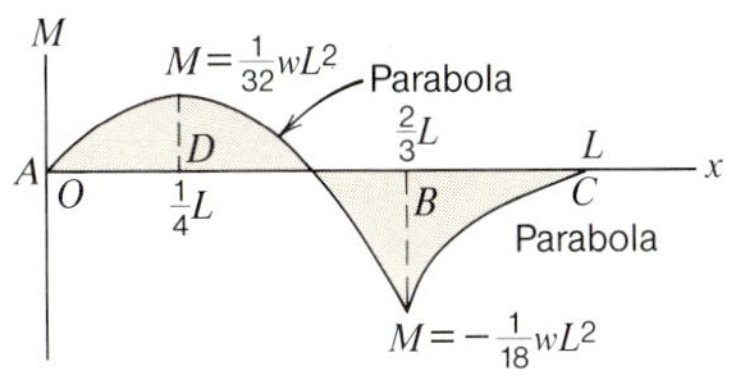

The nature of these parabolas is investigated by using $dM/dx = V$ to study their slope. Proceeding first from point A to point D, we see that the slope of the parabola decreases from a positive value at $x_A = 0$ to zero at $x_D = L/4$ (where $V = 0$), and then becomes large negatively for increasing values of x beyond x_D. This can be true only if the parabola between points A and B curves downward with a horizontal tangent at x_D.

Similar reasoning may be applied to the region between points B and C. There is a discontinuity in the shear diagram, so the parabolic curve depicting the bending moment in this region is different from that in the region between points A and B. Because the slope $dM/dx = V$ decreases from a positive value at x_B to zero at x_C, we may conclude that the parabola curves downward with a horizontal tangent at x_C. These considerations lead to the bending moment diagram shown to the left. Clearly, the diagrams we obtained here are identical to those obtained in Example 4.

Additional Remarks

The advantages of the procedure we have just presented become evident when we note that the majority of the effort in the solution of the examples was devoted to explanations, not calculations. By evaluating the areas under a few curves, we can quickly and accurately obtain the required diagrams. It is no longer necessary to study the free body diagrams corresponding to different cutting planes. Needless to say, the formulas for the areas of some elementary shapes given in Appendix C may prove to be of value when using this method.

HOMEWORK PROBLEMS

7.33–7.44 Using the relationships between the transverse load, the shear, and the bending moment, sketch shear and bending moment diagrams for the beams indicated below. Give all critical values and show all computations.

Homework Problem	7.33	7.34	7.35	7.36	7.37	7.38
Beam in Problem	7.21	7.23	7.22	7.24	7.25	7.27
Homework Problem	7.39	7.40	7.41	*7.42	*7.43	*7.44
Beam in Problem	7.26	7.28	7.29	7.31	7.32	7.30

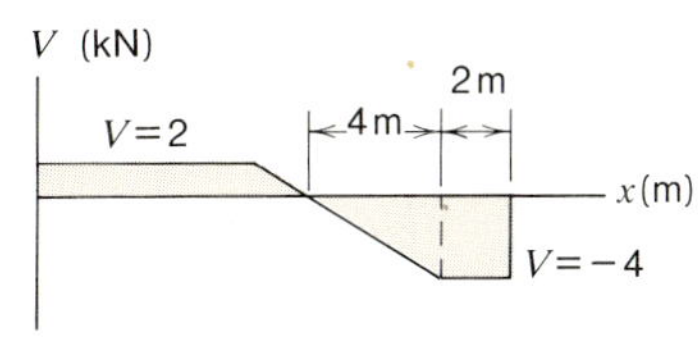

PROB. 7.45

***7.45 and *7.46** The shear diagram for a simply supported beam is partially known, as depicted in the accompanying figure. It is also known that no couples are applied to the beam. Determine (a) the distributed and concentrated forces acting on the beam, (b) the length of the beam and all distances required to specify where the loads are applied, and (c) the magnitude and location of the maximum bending moment in the beam.

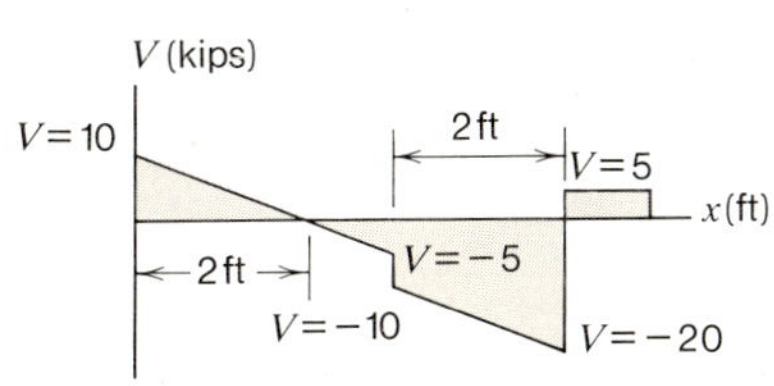

PROB. 7.46

B. FLEXIBLE CABLES

Like a beam, a *flexible cable* is a slender body. The primary difference between these two types of structural components is that the ratio of the length to the largest cross-sectional dimension for a cable (of the order of 1000 or more) is much larger than that for a beam. Because of this, a cable offers negligible resistance to bending and it bows out if one tries to apply a compressive force to it. Thus, the only internal force of any significance that it may withstand is an axial tension force.

Flexible cables are used in a variety of ways in structures and machines. They are a main element in such systems as suspension bridges, conveyor belts, and alpine cable cars. Their behavior is important to the design of electric and telephone transmission lines.

One use of a cable that we have already treated is to tie it between two points having known positions as a means of interconnecting these points. In this usage the tensile force is usually so large that we may consider the cable to be taut; that is, the tensile force is collinear with the line connecting the end points.

Our concern in this chapter is with situations where cables carry transverse loads that are sufficiently large to make it unacceptable to consider the cable to be stretched in a straight line between its ends. We then say that the cable *sags*.

1. Discrete Transverse Loads

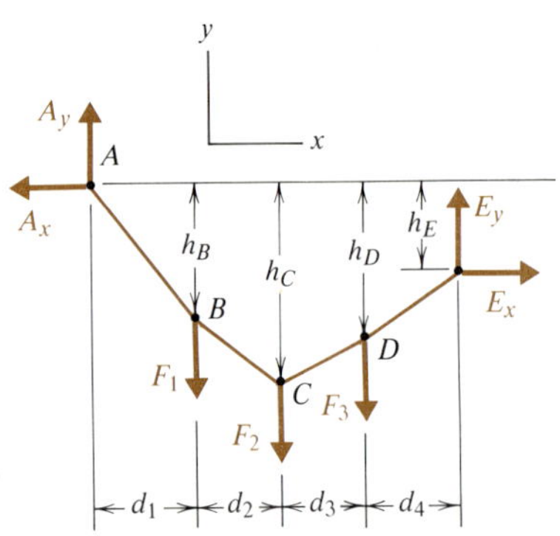

FIGURE 15

Here we will consider the case where a cable supports a set of transverse loads at discrete points along its length. Such a situation is depicted in Figure 15, which shows the free body diagram of a cable that is anchored to pins at ends A and E as it supports a series of transverse loads $\bar{F}_i$. The lines of action of the vertical loads are specified by the dimensions d_1, d_2, d_3, and d_4, which we will regard as known quantities.

If the loads are much larger than the weight of the cable, we may regard the segments of the cable between the points of application of the loads to be taut. Each segment then carries a constant tensile force, although the tension may vary from segment to segment. Thus, the analysis of the forces within the cable in Figure 15 reduces to the determination of the tensile forces T_{AB}, T_{BC}, T_{CD}, and T_{DE}.

Let us assume that the location of ends A and E of the cable are known, so that h_E is a known dimension. Then, additional information we seek is the vertical distances h_B, h_C, and h_D. These distances represent the *sag* of the cable. Including the four unknown reaction components, the analysis of this cable requires that we determine 11 unknowns (4 tensions, 4 reactions, and 3 sag distances).

Because the cable is functioning as a series of straight two-force members, it might seem that the problem can be solved by the same methods as those used for trusses. This is true, but there is one new feature here that complicates the solution. Specifically, in a truss the location of all joints is known, whereas the locations of joints B, C, and D for the cable are part of the information we seek.

Let us investigate the number of available equations. There are five joints (points A through E). Because the forces at each joint are collinear, there are two equilibrium equations for each joint, making a total of 10 available equilibrium equations. In practice, we usually consider three of the equilibrium equations for the joints as checks. These equations are replaced by the three equilibrium equations corresponding to the free body diagram of the entire structure, Figure 15. Nevertheless, there are only 10 independent equations of static equilibrium, and there are 11 unknowns; we need one more equation.

The cable is *not* statically indeterminate. Rather, the difficulty lies in the fact that we have not fully specified the geometry of the system.

One class of problems we can pursue pertains to situations where we know the length of the cable. In that case we would use the Pythagorean theorem to express the length of each segment of the cable in terms of the sags h_B, h_C, and h_D. The required additional equation is then obtained by equating the sum of the individual segment lengths to the total length.

A complication arises in a problem of this type, because the Pythagorean theorem gives rise to terms in the equation for the length that are the square roots of the unknown sags. Thus, although the formulation of the required equations is fairly straightforward, the solution of these equations is laborious.

A simpler situation occurs when one of the sags is specified. This, of course, reduces by one the number of unknowns to be determined, so that the static equilibrium equations are sufficient. There is a simple procedure available for formulating the solution for this class of problems.

For example, suppose that we know h_C in Figure 15. We employ the method of sections for trusses, using a cutting plane that cuts the cable to one side of point C. Choosing a cutting plane to the left of this point, the corresponding free body diagram for the portion of the cable to the left of the plane is shown in Figure 16.

When we sum moments about point C, we obtain an equation relating A_y and A_x. In conjunction with the three equilibrium equations for the entire system, we can determine all of the reactions. Further, the force equilibrium equations for the section in Figure 16 give the tension T_{BC} and the angle of elevation θ_{BC} of cable BC.

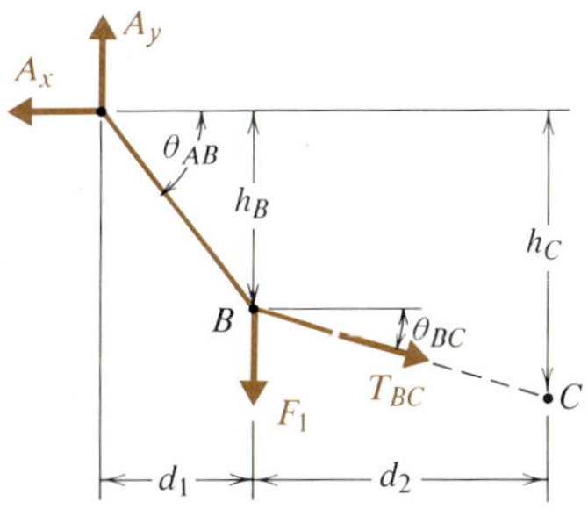

FIGURE 16

Once the reactions are known, the method of joints may be used to determine the tension and angle of elevation of each cable segment. Finally, the sag of each point can be determined from the laws of trigonometry.

EXAMPLE 8

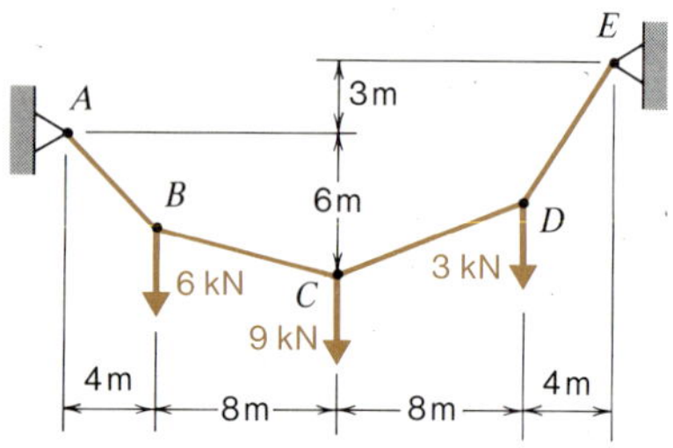

The cable carries three transverse loads, as shown. Determine (a) the elevations of joints B and D, (b) the maximum tension, and (c) the maximum angle of elevation.

Solution

The free body diagram and equilibrium equations for the entire system are as given below.

$$\sum M_{Az} = 24B_y - 3B_x - 6(4) - 9(12) - 3(20) = 0 \text{ kN-m}$$
$$\sum F_x = -A_x + B_x = 0 \text{ kN}$$
$$\sum F_y = A_y + B_y - 18 = 0 \text{ kN}$$

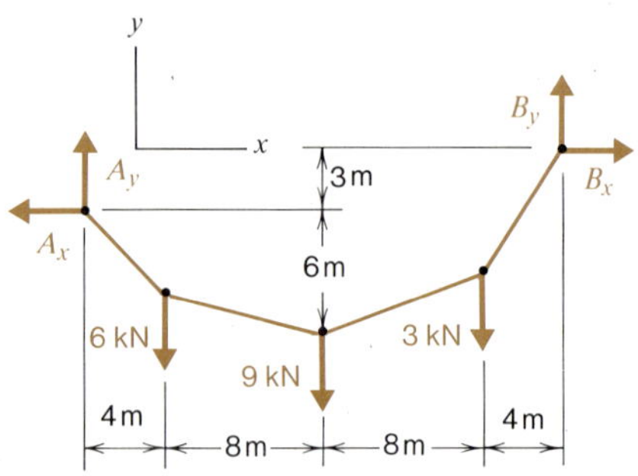

These equations are not sufficient to solve for the reactions. We now pass a cutting plane through the cable near point C, which is the point whose sag is given. A free body diagram for the portion of the cable on either side of the cutting plane leads to additional equilibrium equations. The free body diagram shown is for the left portion, and the corresponding equilibrium equations are

$$\sum M_{Cz} = A_x(6) - A_y(12) + 6(8) = 0$$
$$\sum F_x = T_{BC} \cos \theta_{BC} - A_x = 0$$
$$\sum F_y = A_y - T_{BC} \sin \theta_{BC} - 6 = 0$$

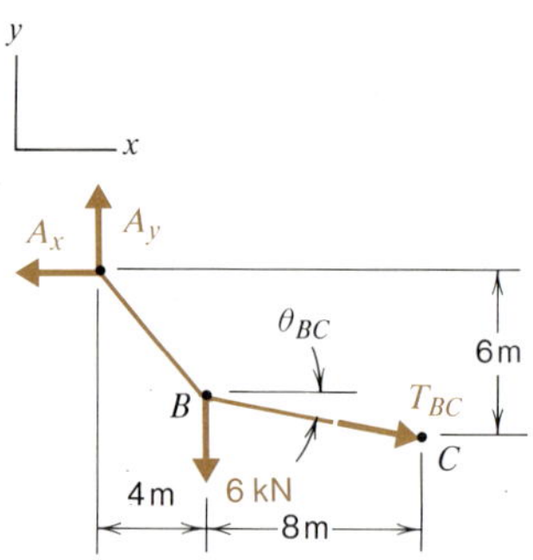

The simultaneous solutions of the moment equation for the section and the equations for the entire system are

$$A_x = B_x = 9.60 \text{ kN}, \qquad A_y = 8.80 \text{ kN}, \qquad B_y = 9.20 \text{ kN}$$

The force equations for the section of cable then become

$$T_{BC} \cos \theta_{BC} = A_x = 9.60$$
$$T_{BC} \sin \theta_{BC} = A_y - 6 = 2.80$$

The solutions of these equations are

$$\theta_{BC} = \tan^{-1}\left(\frac{2.80}{9.60}\right) = 16.26°$$

$$T_{BC} = \frac{9.60}{\cos 16.26°} = 10.00 \text{ kN}$$

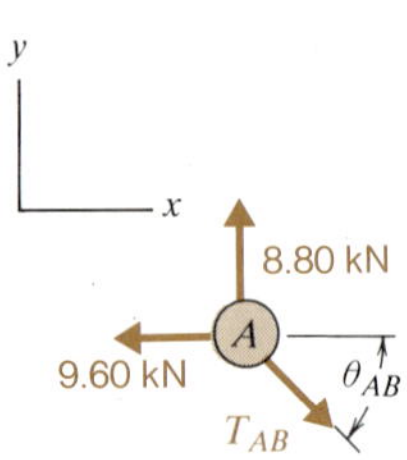

We now proceed to investigate the joints, starting at either end. Considering pin A first, the free body diagram and equilibrium equations are as follows.

$$\sum F_x = T_{AB} \cos \theta_{AB} - 9.60 = 0$$
$$\sum F_y = 8.80 - T_{AB} \sin \theta_{AB} = 0$$

These equations yields

$$\theta_{AB} = 42.51°, \qquad T_{AB} = 13.023 \text{ kN}$$

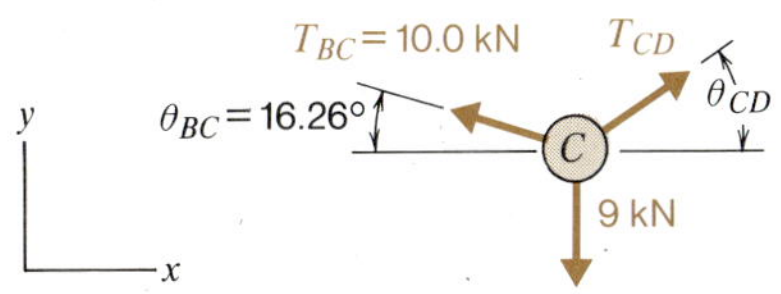

Because we now know the tension and angle of elevation of segments AB and BC, the equilibrium equations for joint B only give checks on the computations. We skip this step and proceed to joint C. The free body diagram of this joint yields the following equilibrium equations, in which it is assumed that point D is higher than point C.

$$\sum F_x = T_{CD} \cos \theta_{CD} - 10.00 \cos 16.26° = 0$$

$$\sum F_y = T_{CD} \sin \theta_{CD} + 10.00 \sin 16.26° - 9 = 0$$

$$\theta_{CD} = 32.86°, \qquad T_{CD} = 11.428 \text{ kN}$$

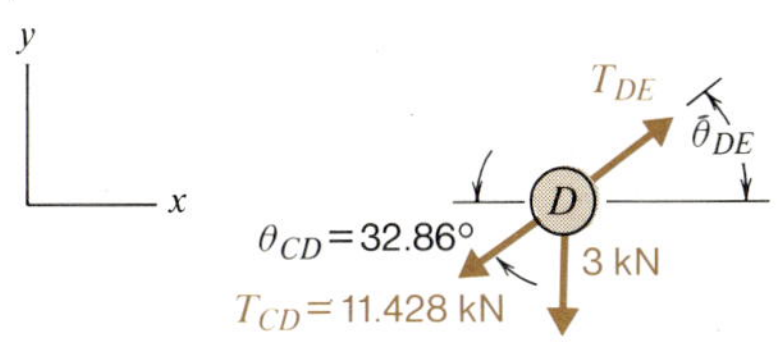

Finally, for joint D we have

$$\sum F_x = T_{DE} \cos \theta_{DE} - 11.428 \cos 32.86° = 0$$

$$\sum F_y = T_{DE} \sin \theta_{DE} - 11.428 \sin 32.86° - 3 = 0$$

$$\theta_{DE} = 43.78°, \qquad T_{DE} = 13.297 \text{ kN}$$

We have found all of the tensions and all of the angles of elevation, so we omit consideration of the equilibrium of joint E, which would only serve as a check on the calculations. From these results we may answer parts (b) and (c), for we see that the maximum tension and angle of elevation are those for segment DE, being

$$\theta_{\max} = \theta_{DE} = 43.8°, \qquad T_{\max} = T_{DE} = 13.30 \text{ kN}$$

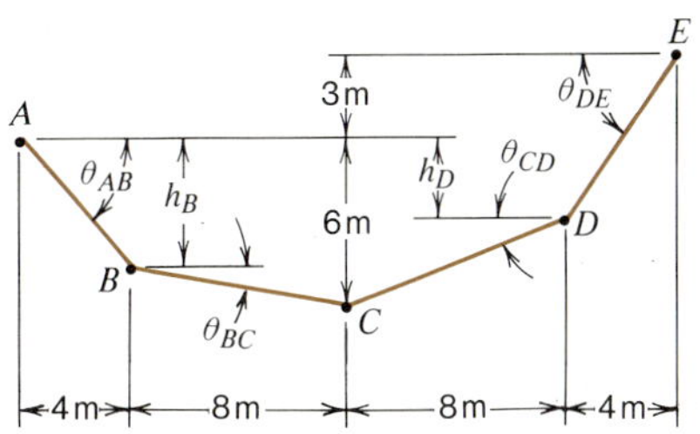

We should not be surprised that the tension is maximum in the cable having the steepest slope, for equilibrium requires that the horizontal component $T_i \cos \theta_i$ of the tension in the ith segment be a constant equal to A_x $(= -E_x)$.

To determine the sags from the angle of elevation we draw a sketch of the geometry. From this sketch we see that

$$h_B = 4 \tan \theta_{AB} = 3.667 \text{ m}$$

and

$$h_D + 3 = 4 \tan \theta_{DE} \qquad \text{so} \qquad h_D = 0.833 \text{ m}$$

The sketch also suggests another check, in addition to the excess equilibrium equations, on the computations. Specifically, we may ascertain whether, starting from either pin A or E, the sag h_C of point C actually is found to be 6 m. Starting from pin A, we have

$$h_C = h_B + 8 \tan \theta_{BC} = 6.000 \text{ m}$$

and starting from pin E, we find

$$h_C = h_D + 8 \tan \theta_{CD} = 6.001 \text{ m}$$

Because both values of h_C are in agreement with the given value, the check validates the results.

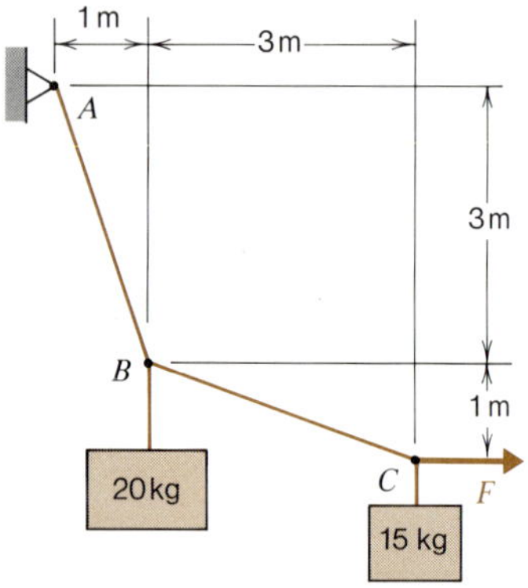

PROB. 7.47

HOMEWORK PROBLEMS

7.47 Two blocks are supported by the lightweight cable ABC, as shown. Determine (a) the magnitude of the horizontal force $\bar{F}$, and (b) the corresponding sag h_B.

7.48–7.51 A cable supports two vertical loads. Determine (a) the sag h_C, (b) the maximum tension, and (c) the maximum angle of elevation.

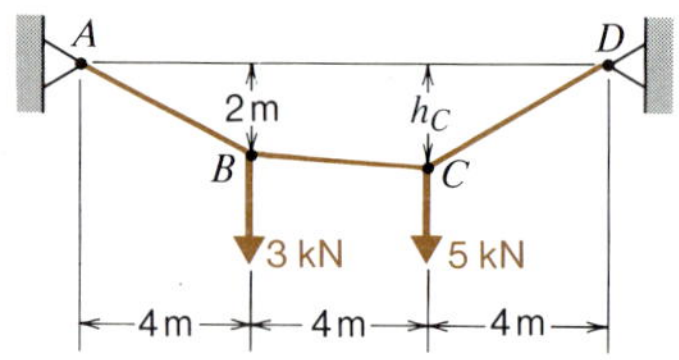

PROB. 7.48

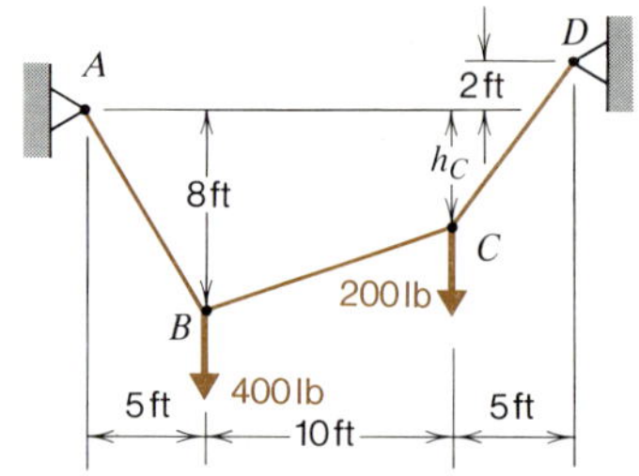

PROB. 7.49

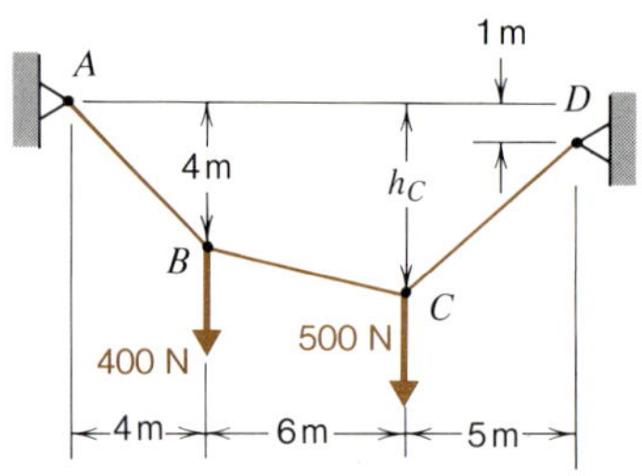

PROB. 7.50

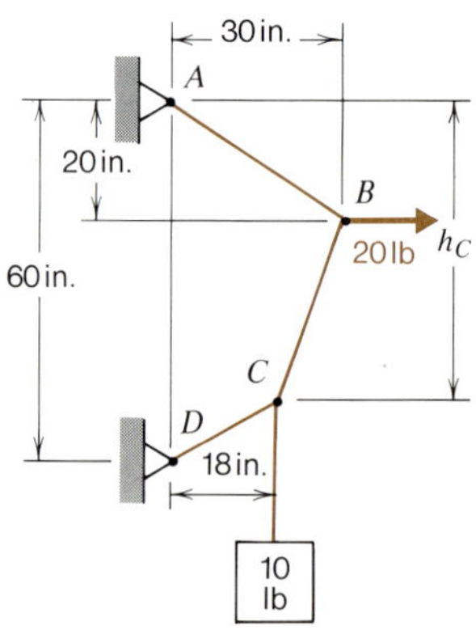

PROB. 7.51

7.52 and *7.53 A cable supports the three transverse loads shown. Determine (a) the maximum tension, (b) the sags h_B and h_D, and (c) the length of the cable.

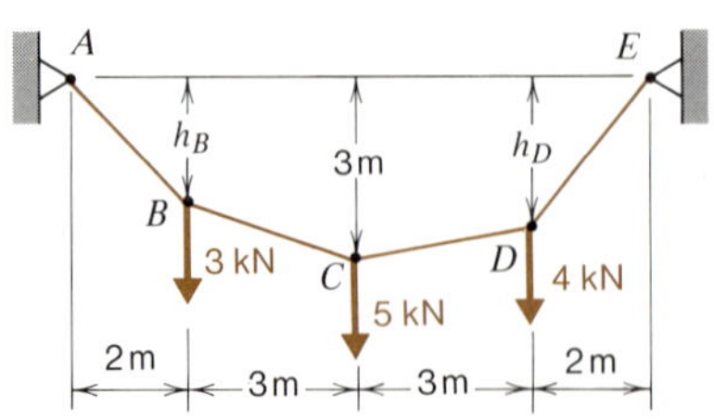

PROB. 7.52

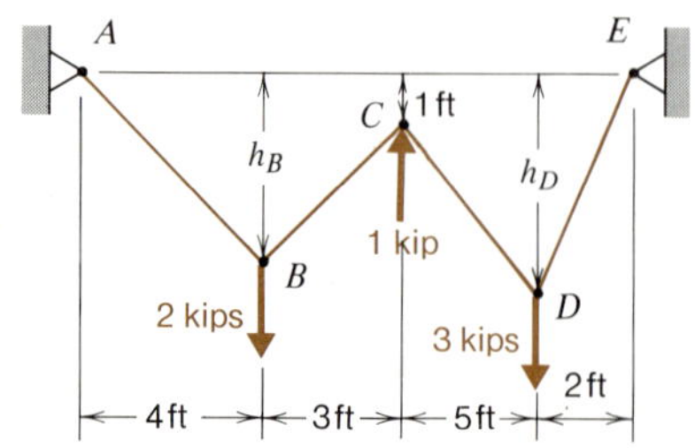

PROB. 7.53

***7.54** A set of horizontal and vertical forces are applied to cable *ABCDE*. Determine (a) the distances d_B and d_D, (b) the maximum tension, and (c) the maximum angle of elevation.

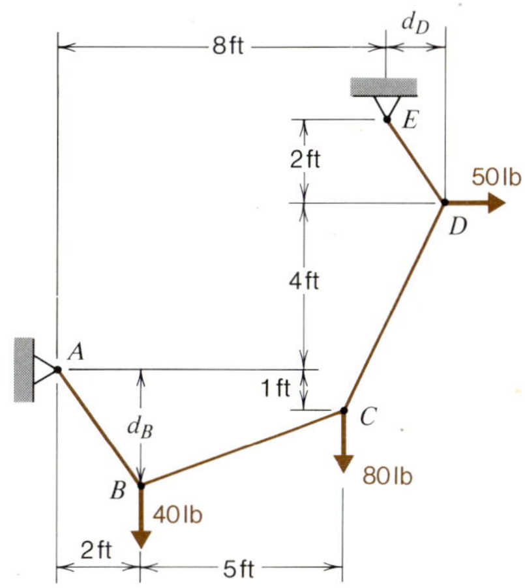

PROB. 7.54

2. Distributed Loads—Parabolic Sag

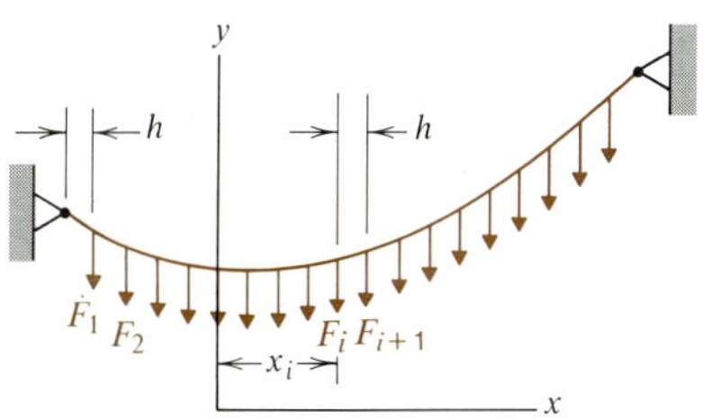

FIGURE 17

Consider a cable supporting many discrete loads that are applied at equally spaced intervals a distance h apart in the horizontal direction. We wish to determine how the cable sags under such a loading distribution. This situation is depicted in Figure 17. It was indicated in the previous section that the solution of this problem requires additional information, such as the sag at some point along the cable or the total length of the cable. Assuming that such information is given, we may determine the sag by the tedious procedure of considering the equilibrium equations for each discrete point where a load is applied, as outlined in the preceding section.

With the advent of high-speed electronic computers, such an approach for the determination of the sag is not unreasonable. However, engineers in earlier times did not have such computational aids, so they developed an approximate, but more convenient, method for determining the sag. This method was based on replacing the discrete loads in Figure 17 by a continuous distributed load $q(x)$, where q has units of force per unit length. The functional dependence of $q(x)$ is defined such that at point i along the cable, having the horizontal coordinate x_i as shown in the figure, $q(x_i) = F_i/h$. In other words, $q(x)$ is the average line load.

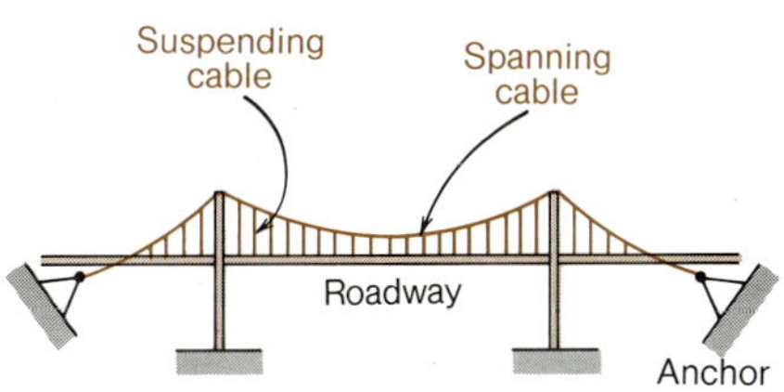

FIGURE 18

The case where $q(x)$ is a constant value, q_0, is most usual. This situation arises in the design of a suspension bridge, such as the one depicted in Figure 18, where a number of vertical cables are suspended from the spanning cables to support the roadway. If we neglect the weight of the cables and the stiffness of the roadway, we may treat the spanning cables as though they were loaded only by the constant weight per unit length of the roadway.

Equations pertaining to the sag of the cable may be obtained in general terms, independent of the precise details of the distributed load $q(x)$. It is convenient to select an xyz coordinate system whose origin is at the point where the cable has a horizontal tangent, that is, at the lowest point of the cable.

Let us consider a finite section of the cable of length x in the horizontal direction, having one end at the origin and the other end at an arbitrary point P. The free body diagram for such an element is shown in Figure 19.

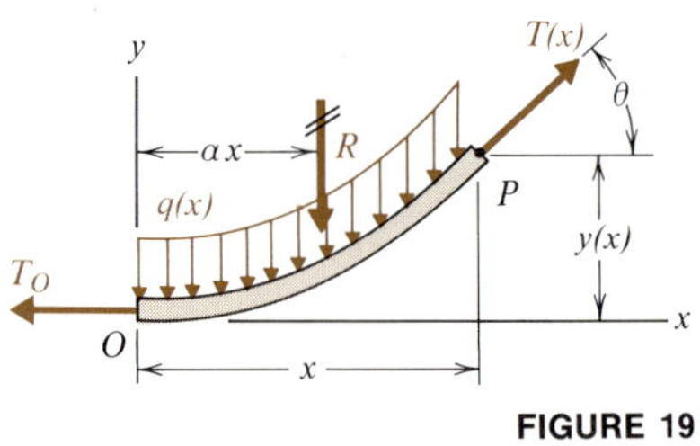

FIGURE 19

The force $\bar{R}$ shown in the figure is the resultant of the distributed load. Its magnitude and the distance αx defining its line of action are known if the loading function $q(x)$ is known. For example, for a uniform load q_0, $R = q_0 x$ and $\alpha = \frac{1}{2}$. The other forces acting on the element are the tensile forces tangent to the cable at each end. The term T_O denotes the tension at the horizontal tangent, point O.

Figure 19 shows that the sag of the cable is defined by the function $y(x)$. To determine this function, as well as the tension within the cable, we formulate the equations of equilibrium for the element. Choosing point P on the right for the moment sum, we obtain

$$\sum M_{Pz} = R(x - \alpha x) - T_O y = 0$$

$$\sum F_x = T \cos\theta - T_O = 0, \qquad \sum F_y = T \sin\theta - R = 0$$

The equation for $\sum F_x$ indicates that the sag of the cable is such that the tensile force T always has a constant horizontal component T_O. The value of T_O will be an important parameter in the solution of problems. The solution of the equilibrium equations in terms of T_O and R is

$$T = (T_O^2 + R^2)^{1/2} \tag{6a}$$

$$\tan\theta = \frac{R}{T_O} \tag{6b}$$

$$y = \frac{(1 - \alpha)xR}{T_O} \tag{6c}$$

An important feature of these equations is that they are not independent because the slope of the cable is $\tan\theta$, so

$$\tan\theta = \frac{dy}{dx} = \frac{R}{T_O} \tag{7}$$

When the loading $q(x)$ is specified, so that the resultant R may be evaluated, eq. (7) can be solved for y in terms of x. This solution would identically satisfy eq. (6c), which is a verification of the assumption that the cable carries a tensile force. It follows that only eqs. (6a) and (6c) are needed for the evaluation of the sag. Equation (7) is useful for the evaluation of the slope.

In order to proceed further, we must define the load function. As noted earlier, our primary interest is with the case of a uniform load. Setting

$q(x) = q_0$, $R = q_0 x$, and $\alpha = \frac{1}{2}$ in eqs. (6a) and (6c) yields

$$y = \frac{q_0}{2T_O} x^2 \tag{8a}$$

$$T = (T_O^2 + q_0^2 x^2)^{1/2} \tag{8b}$$

The equation for y shows that a cable carrying a *uniform horizontal load* sags into a *parabolic* shape. It is readily verified that the equation for $\tan\theta$ obtained from eqs. (7) and (8) is identical to that given by eq. (6b), confirming the correctness of the analysis.

As was true in the case of a cable carrying concentrated loads, there are two standard types of problems we can analyze, assuming that the locations of the ends of the cable are known. In one, we know the sag of some point on the cable. Equations (8) will prove sufficient for such cases.

The second type of problem is the one where we know the length of the cable. To solve problems of this type we need an expression relating the length of the cable to the other parameters for the system. This expression is obtained by evaluating the arc length of the parabola formed by the cable.

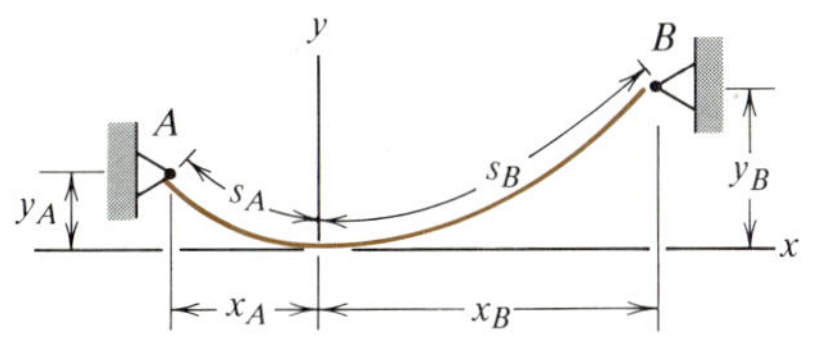

FIGURE 20

Consider the situation depicted in Figure 20. The total length of the cable is the sum of the arc lengths s_A and s_B measured from the origin to each end. Both quantities are arc lengths along the parabola measured from the origin, so we need only consider how an expression for one arc length is obtained. In courses on calculus, it is shown that the arc length s_B may be obtained by evaluating the following integral.

$$s_B = \int_0^{x_B} \left[1 + \left(\frac{dy}{dx}\right)^2\right]^{1/2} dx$$

Upon substitution for y from eq. (8a), the integral becomes

$$s_B = \int_0^{x_B} \left[1 + \frac{q_0^2}{T_O^2} x^2\right]^{1/2} dx \tag{9}$$

An exact solution for this integral may be found using a table of integrals; however, the resulting expression is cumbersome for calculations. A more convenient expression may be obtained by expanding the integrand in eq. (9) in an infinite series using the binomial theorem, which is

$$(1 + Z)^n = 1 + nZ + \frac{n(n-1)}{2!} Z^2 + \frac{n(n-1)(n-2)}{3!} Z^3 + \cdots$$

With this expression, eq. (9) becomes

$$s_B = \int_0^{x_B} \left(1 + \frac{q_0^2 x^2}{2T_O^2} - \frac{q_0^4 x^4}{8T_O^4} + \cdots\right) dx$$

$$= x_B\left(1 + \frac{q_0^2 x_B^2}{6T_O^2} - \frac{q_0^4 x_B^4}{40T_O^4} + \cdots\right)$$

Finally, using eq. (8) to replace q_0/T_O by the elevation y_B at end B yields

$$s_B = x_B\left[1 + \frac{2}{3}\left(\frac{y_B}{x_B}\right)^2 - \frac{2}{5}\left(\frac{y_B}{x_B}\right)^4 + \cdots\right] \tag{10}$$

The infinite series in eq. (10) is convergent when $y_B/x_B < 0.5$. If y_B/x_B is substantially lower than this value, as it is in many engineering applications, then the convergence is very rapid and the listed terms provide sufficient accuracy. On the other hand, if y_B/x_B exceeds 0.5, the exact integral of eq. (9) is required.

EXAMPLE 9

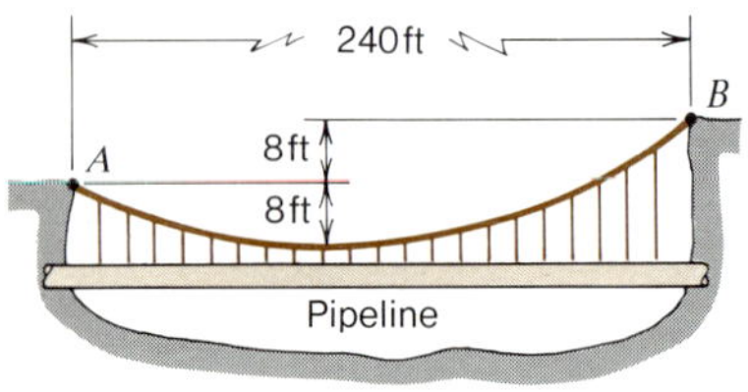

A cable supporting a pipeline weighing 1 kip/ft is to be suspended between pins A and B, whose positions are as shown. The design calls for the low point on the cable to be 8 ft below the elevation of pin A. Determine (a) the maximum tension in the cable, and (b) the required length of the cable.

Solution

In this problem the location of the horizontal tangent (the low point) is not completely given. This point is the origin from which the position of all points on the cable are measured, so its location must be found. This evaluation is achieved with the aid of a sketch of the geometry and the coordinate system.

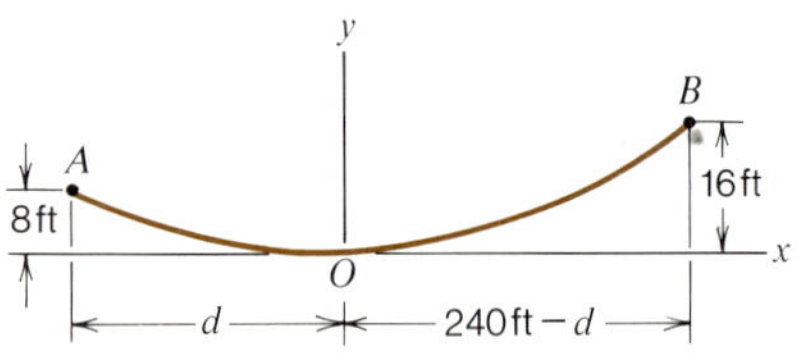

Let d be the horizontal distance of the origin O from pin A. We see from the sketch that the coordinates of point A are ($x_A = -d$, $y_A = 8$), and those for point B are ($x_B = 240 - d$, $y_B = 16$). Both of these points lie on the parabola defined by eq. (8a). Therefore, we have

$$y_A = 8 = \frac{q_0}{2T_O}(x_A)^2 = \frac{q_0}{2T_O}d^2$$

$$y_B = 16 = \frac{q_0}{2T_O}(x_B)^2 = \frac{q_0}{2T_O}(240 - d)^2$$

The distributed load is the weight of the pipeline, $q_0 = 1000$ lb/ft. Hence, the foregoing represents two equations whose simultaneous solution gives the values of d and T_O. We solve for q_0/T_O from the first equation.

$$\frac{q_0}{T_O} = \frac{16}{d^2}$$

The second equation then becomes a quadratic in d. Specifically, it gives

$$16 = \left(\frac{8}{d^2}\right)(240 - d)^2$$

$$d^2 + 480d - 5.76(10^4) = 0$$

$$d = 99.41, \quad -579.4 \text{ ft}$$

The negative value is an extraneous root for this problem, so $d = 99.41$ ft. Substituting the value of q_0 and d, we then find T_O.

$$T_O = q_0\left(\frac{d^2}{16}\right) = 1000\frac{(99.41)^2}{16} = 6.177(10^5)\text{ lb}$$

These are the parameters required to obtain the information requested. From eq. (8b) we see that the tension is a maximum at the point that is farthest from the horizontal tangency. This is at pin B because the distance d is less than half the horizontal distance between pins A and B. Thus, using eq. (8b) with $x_B = 240 - d = 140.59$ ft yields

$$T_{\max} = (T_O^2 + q_0^2 x_B^2)^{1/2} = [(6.177)^2(10^{10}) + (10)^6(140.59)^2]^{1/2}$$
$$= 6.33(10^5)\text{ lb}$$

To determine the length of the cable we note that y_A/x_A and y_B/x_B are both extremely small. We therefore may employ eq. (10). This yields

$$s_A = x_A\left[1 + \frac{2}{3}\left(\frac{y_A}{x_A}\right)^2 - \frac{2}{5}\left(\frac{y_A}{x_A}\right)^4\right]$$
$$= 99.41\left[1 + \frac{2}{3}\left(\frac{8}{99.41}\right)^2 - \frac{2}{5}\left(\frac{8}{99.41}\right)^4\right]$$
$$= 99.41 + 0.4275$$
$$s_B = 140.59\left[1 + \frac{2}{3}\left(\frac{16}{140.59}\right)^2 - \frac{2}{5}\left(\frac{16}{140.59}\right)^4\right]$$
$$= 140.59 + 1.2045$$
$$s = s_A + s_B = 240 + 1.632 = 241.63\text{ ft}$$

These results show that because of the relatively small amount by which the cable sags, the tension within the cable is much larger than the total weight of the pipeline, (1000 lb/ft)(240 ft), and the required length of the cable is only 1.63 ft longer than the horizontal distance spanned.

HOMEWORK PROBLEMS

7.55 The span of a suspension bridge between two towers of equal height is 1000 m and the sag of its cables at midspan is 100 m. The roadway, having a mass of $20(10^3)$ kg/m, is supported equally by two suspension cables. Determine (a) the maximum and minimum tensions in a cable, and (b) the length of a cable between the towers.

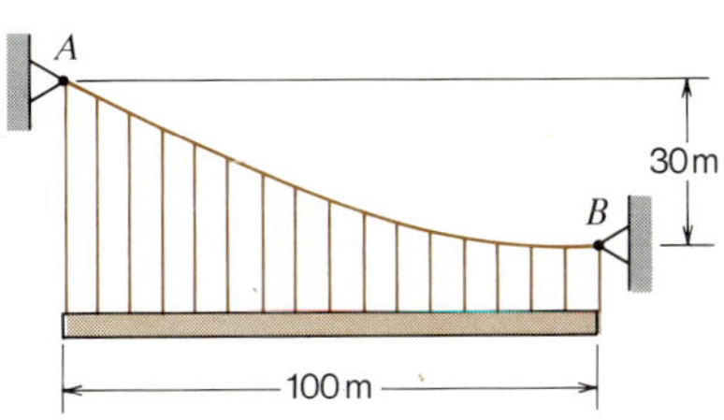

PROB. 7.56

7.56 Cable AB supports a 200kN total load that is distributed uniformly along the horizontal. The slope of the cable is zero at the lower end. Determine (a) the maximum and minimum tensions in the cable, and (b) the location of the point on the cable where the tension is the average of the values determined in part (a).

7.57 A cable spanning a distance of 300 ft between pins located at the same elevation supports a uniform load of 3 lb/ft. The maximum tension in the cable is 2000 lb. Determine (a) the maximum sag, (b) the slope of the cable at its supports, and (c) the length of the cable.

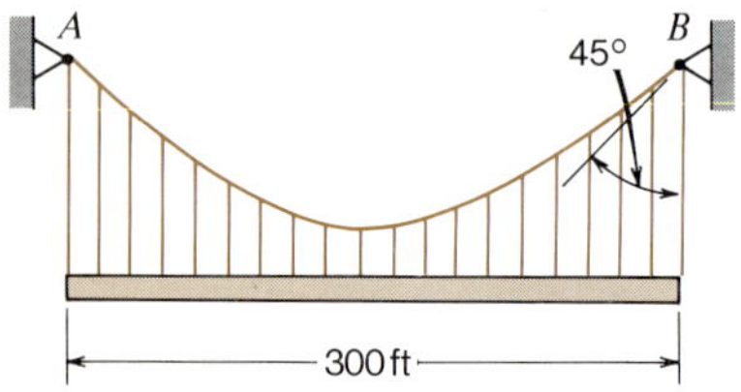

PROB. 7.58

7.58 Cable AB, carrying a uniform load of 500 lb/ft, has a span of 300 ft. Knowing that the tangent to the cable at an end is at a 45° angle from the vertical, determine the maximum sag and the maximum tension in the cable.

7.59 A span of a suspension bridge between two towers of equal height is 4000 ft. The sag of each of its four cables at midspan is 200 ft. The maximum permissible tension in each cable is 8000 kips. Determine the maximum uniform distributed line load that is permissible.

7.60 Cables AB and BC supporting a uniform horizontal load of 750 lb/ft are both pinned to tower B, as shown. Determine the resultant force exerted by both cables on the tower.

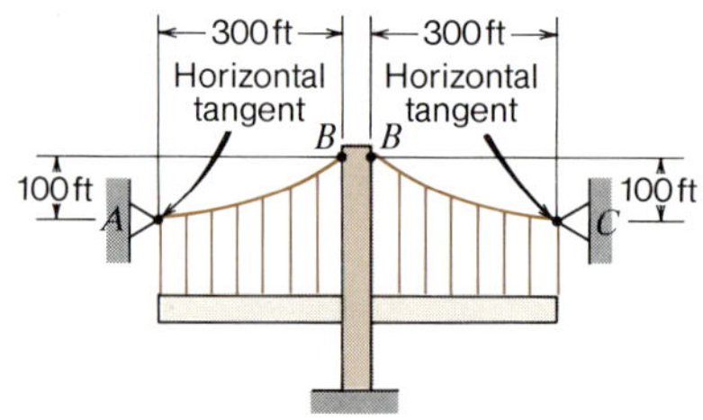

PROB. 7.60

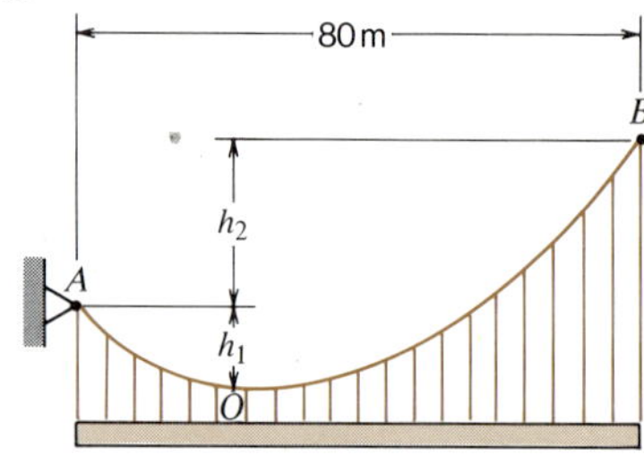

PROBS. 7.61 AND 7.62

7.61 A cable spanning a distance of 80 meters between pins A and B carries a uniform horizontal load of 400 N/m. It is known that the lowest point on the cable is 20 m horizontally from end A, and that the sag of this point is $h_1 = 0.50$ m. Determine (a) the minimum tension in the cable, (b) the elevation h_2 of pin B above pin A, and (c) the length of the cable.

7.62 A cable spanning a distance of 80 m between pins A and B carries a uniform horizontal load of 400 N/m. It is known that the maximum sag below point A is $h_1 = 2$ m, and that the elevation of pin B above pin A is $h_2 = 4$ m. Determine (a) the minimum and maximum tensions in the cable, and (b) the length of the cable.

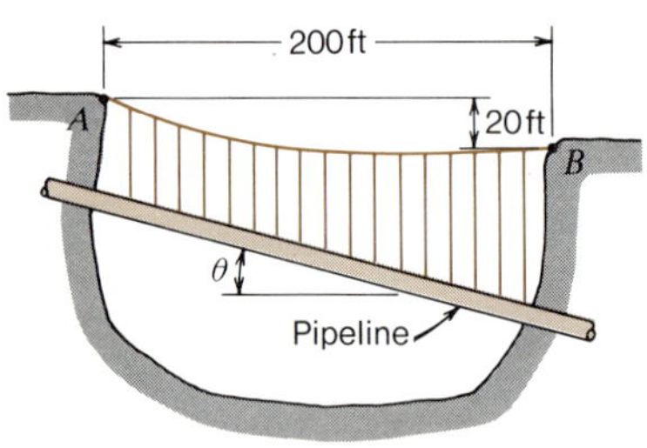

PROB. 7.63

7.63 The pipeline, whose mass per unit length is 200 lb/ft is oriented horizontally ($\theta = 0$) as it passes over the gorge. The ends of the cable are located as shown. It is known that at its lowest point the cable is 10 ft below pin B. Determine (a) the maximum tension within the cable, and (b) the length of the cable.

7.64 Solve Problem 7.63 if the pipeline is inclined at $\theta = 20°$.

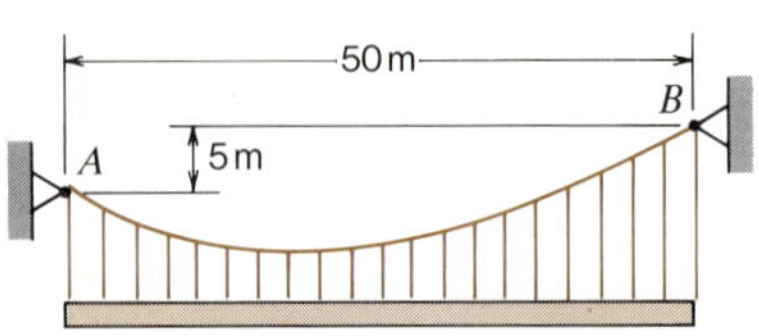

PROB. 7.65

7.65 A cable spanning the distance between pins A and B supports a uniform load of 200 N/m, distributed along the horizontal. Knowing that the lowest point on the cable is 10 m horizontally from pin A, determine the lowest elevation of the cable below pin A and the length of the cable.

7.66 Solve Problem 7.65 if, instead of knowing the horizontal location of the low point of the cable, it is known that the minimum tension in the cable is 20 kN.

7.67 A cable carrying a uniform horizontal load of 800 lb/ft spans the 40 ft distance between pins A and B. Pin B is 20 ft higher than pin A, and the cable at midspan is 9 ft higher than pin A, as shown. Determine (a) the maximum tension in the cable, and (b) the length of the cable.

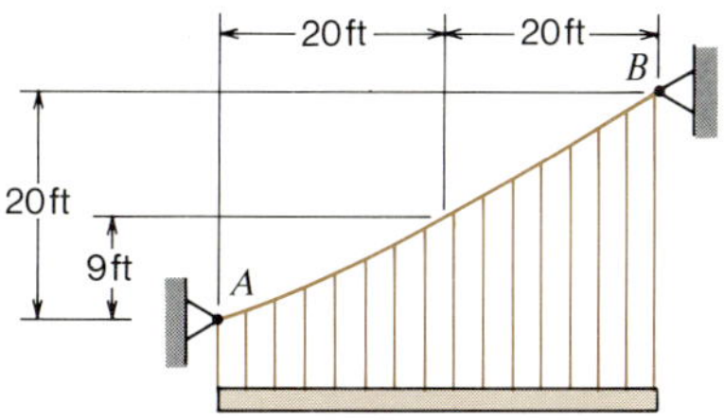

PROB. 7.67

3. Catenary

Frequently, we are interested in the sag of a cable under its own weight. We shall only consider the case of a uniform cable having a weight per unit length w_0. Then the gravitational force acting on an element of cable of infinitesimal arc length ds is $w_0\,ds$, as depicted in Figure 21.

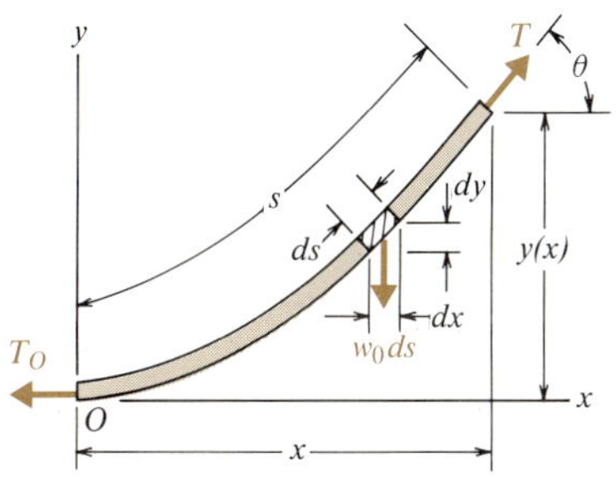

FIGURE 21

The situation in Figure 21 is a special case of a cable carrying an arbitrary distributed load, described by Figure 19 and eqs. (6). Let us therefore consider the resultant of the distributed gravity force acting on the section of cable. Because w_0 is constant, it follows that $w_0 s$ is the resultant gravitational force or, equivalently, the weight of the section of cable. Hence, $R = w_0 s$. By definition, the weight force acts at the center of mass, which is the centroid of the curve formed by the cable. This creates a problem, because the location of the center of mass depends on the shape of the curve formed by the cable in Figure 21, and this shape is what we wish to determine. Therefore, we cannot locate the line of action of the resultant force $\bar{R}$, which means that the parameter α in eq. (6c) is unknown. It follows that we cannot determine the function $y(x)$ describing the sag of the cable solely by solving the equilibrium equations for the element.

To resolve this difficulty, we recall that eqs. (6b) and (6c) are not independent because eq. (7), $\tan\theta = dy/dx$, must be satisfied. Equation (6b) does not depend on the parameter α, so we will use that equation in place of eq. (6c). Substituting $R = w_0 s$ into eqs. (6a) and (6b) gives

$$T = [T_O^2 + w_0^2 s^2]^{1/2} \tag{11a}$$

$$\tan\theta = \frac{dy}{dx} = \frac{w_0}{T_O}\,s \tag{11b}$$

Equation (11b) is a *differential equation*. It relates the differentials dy and dx of the infinitesimal element shown in Figure 21. This figure also shows that, by the Pythagorean theorem,

$$(ds)^2 = (dx)^2 + (dy)^2$$

Because s appears in the right side of eq. (11b), we use the foregoing relationship to eliminate dy by writing

$$\frac{dy}{dx} = \frac{[(ds)^2 - (dx)^2]^{1/2}}{dx} = \frac{w_0}{T_O} s$$

$$\left[\left(\frac{ds}{dx}\right)^2 - 1\right]^{1/2} = \frac{w_0}{T_O} s$$

$$\frac{ds}{dx} = 1 + \left(\frac{w_0}{T_O}\right)^2 s^2$$

Solving for dx yields

$$dx = \frac{ds}{1 + (w_0/T_O)^2 s^2} \tag{12}$$

This last equation may be integrated with the aid of a table of integrals. As shown in Figure 21, s is measured from the origin, so $s = 0$ when $x = 0$. Thus, a definite integral of eq. (12) gives

$$\int_0^x dx = \int_0^s \frac{ds}{[1 + (w_0/T_O)^2 s^2]^{1/2}}$$

$$x = \frac{T_O}{w_0}\left[\sinh^{-1}\left(\frac{w_0}{T_O} s\right) - \sinh^{-1}(0)\right] \tag{13}$$

The function $v = \sinh u$ is the hyperbolic sine of u, and the function $u = \sinh^{-1} v$ is its inverse, called the arc hyperbolic sine. There is some similarity between the hyperbolic and trigonometric functions, as a result of their definitions in terms of complex numbers. For our purposes here, we will need the following properties of the hyperbolic sine and cosine.

$$\sinh u \equiv \tfrac{1}{2}(e^u - e^{-u}), \qquad \cosh u \equiv \tfrac{1}{2}(e^u + e^{-u}) \tag{14a}$$

$$\frac{d}{du} \sinh u = \cosh u, \qquad \frac{d}{du} \cosh u = \sinh u \tag{14b}$$

$$\cosh^2 u - \sinh^2 u = 1 \tag{14c}$$

From the definitions, eqs. (14a), we see that

$$\sinh(-u) = -\sinh u, \qquad \cosh(-u) = \cosh u \tag{14d}$$

$$\sinh(0) = 0, \qquad \cosh(0) = 1 \tag{14e}$$

Values of the hyperbolic functions may be found in standard numerical tables, as well as by evaluating them directly. They are keyboard functions on many electronic calculators.

We may now return to the evaluation of the sag. With the aid of eq. (14e), we find from eq. (13) that

$$s = \frac{T_O}{w_0} \sinh\left(\frac{w_0 x}{T_O}\right) \tag{15}$$

This result is important because it provides a relationship between the horizontal distance and the length of the cable, analogous to eq. (10) for a cable supporting a uniform horizontal load.

Next, we determine the slope of the cable at any horizontal position x by substituting eq. (15) into eq. (11b). This yields

$$\frac{dy}{dx} = \sinh\left(\frac{w_0 x}{T_O}\right)$$

We obtain the sag y from this equation by multiplying by dx and then forming a definite integral. The cable intersects the origin, so $y = 0$ at $x = 0$. It follows that

$$\int_0^y dy = \int_0^x \sinh\left(\frac{w_0 x}{T_O}\right) dx = \frac{T_O}{w_0}\left[\cosh\left(\frac{w_0 x}{T_O}\right) - \cosh(0)\right]$$

$$y = \frac{T_O}{w_0}\left[\cosh\left(\frac{w_0 x}{T_O}\right) - 1\right] \tag{16}$$

The curve formed by plotting $y(x)$ is called a *catenary*.

In addition to the curve formed by the sagging cable, we also need to know the tension within the cable. The required expression is found by substituting eq. (15) for s into eq. (11a), and then using the identity given by eq. (14c). This yields

$$T = \left[T_O^2 + w_0^2\left(\frac{T_O}{w_0}\right)^2 \sinh^2\left(\frac{w_0 x}{T_O}\right)\right]$$

$$T = T_O \cosh\left(\frac{w_0 x}{T_O}\right) \tag{17}$$

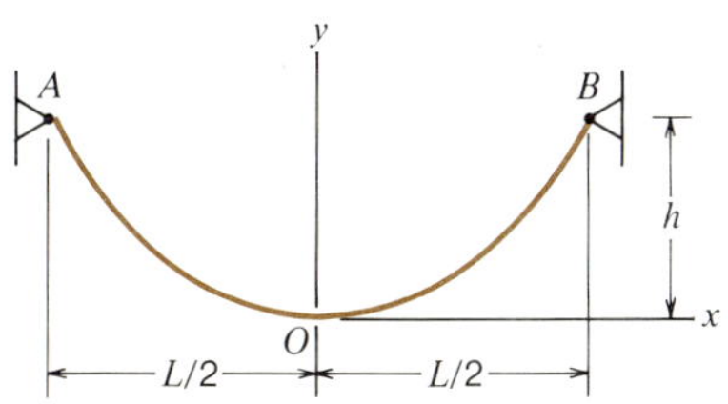

FIGURE 22

Equations (15)–(17) are sufficient for solving problems. Assuming that we know the location of the pins holding the ends of the cable, a key parameter to be determined is the constant T_O, which is the constant horizontal component of the tension. Let us consider what is involved in such a determination for a cable whose pins are at the same elevation, as illustrated in Figure 22.

By symmetry, the origin O, which is located at the horizontal tangent, must be located midway in the span, as shown. In terms of the maximum sag h, the coordinates of the end points A and B of the catenary are ($x_A = -L/2$, $y_A = h$) and ($x_B = L/2$, $y_B = h$). Substituting these coordinates into eq. (16) gives

$$h = \frac{T_O}{w_0}\left[\cosh\left(\frac{w_0 L}{2T_O}\right) - 1\right]$$

Two *nondimensional* combinations of the parameters may be identified in the foregoing, specifically, h/L and w_0L/T_O. In terms of these quantities, we then find

$$\frac{h}{L}\left(\frac{w_0L}{T_O}\right) = \cosh\left(\frac{w_0L}{2T_O}\right) - 1 \qquad (18)$$

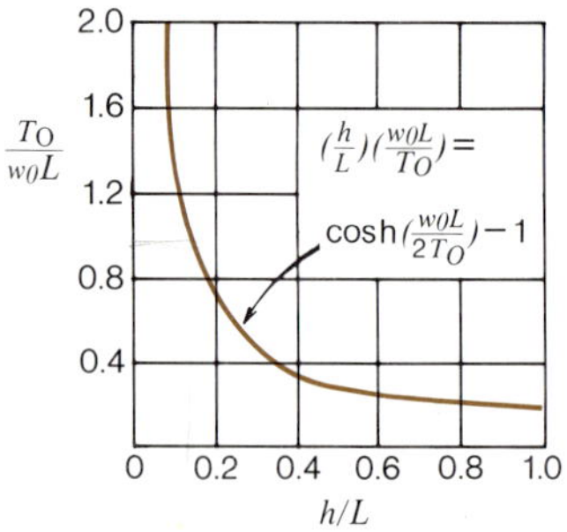

FIGURE 23

Although this equation cannot be solved in closed-form for w_0L/T_O in terms of h/L, we can evaluate h/L as a function of w_0L/T_O; a graph of the result is shown in Figure 23. This graph may be used by itself to find the value of w_0L/T_O corresponding to a specified value of h/L. When more accurate values are needed, the graph provides a starting point for a trial-and-error solution of eq. (18).

In closing, it should be noted that there is no resemblance between the formulas obtained here and those obtained in the previous section for a cable carrying a uniform horizontal load. This is attributable to the fact that the weight of a cable is distributed uniformly along its length, not along the horizontal. However, in the case where the tension T_O is sufficiently large to make the sag ratio very small ($h/L < 0.10$), the cable is almost oriented horizontally. In such a case we may obtain an approximate solution by considering the weight to be a uniform horizontal load, with $q_0 = w_0$ in the formulas for a uniform load in the previous section.

EXAMPLE 10

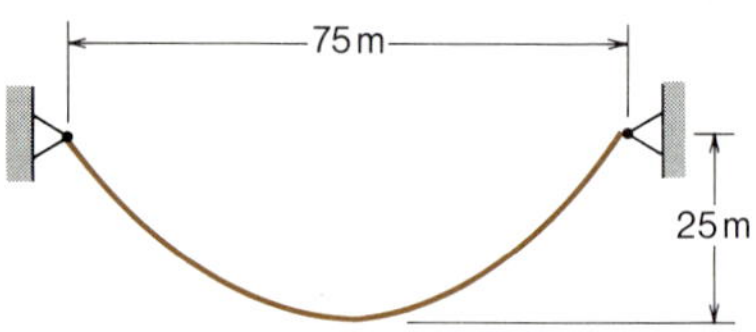

A uniform cable whose mass per unit length is 10 kg/m hangs as shown. Determine the maximum tension in the cable and its length.

Solution

From the given information we calculate the weight per unit length and the ratio of maximum sag to span

$$w_0 = (10 \text{ kg/m})(9.807 \text{ m/sec}^2) = 98.07 \text{ N/m}$$

$$\frac{h}{L} = \frac{25}{75} = \frac{1}{3}$$

The other important parameter for an analysis is T_O/w_0L. The value of T_O/w_0L that we read from the graph given in Figure 23 corresponding to $h/L = 1/3$ is

$$\frac{T_O}{w_0L} = 0.42$$

Substitution of the value of w_0 and L will yield T_O. However, let us assume that we want more accuracy. We may follow a trial-and-error procedure by using the estimate of T_O/w_0L to evaluate the difference between the left- and right-hand sides of eq. (18). The estimate is then improved by modifying the value of T_O/w_0L to reduce this difference. This computation may be performed in tabular form, as indicated below, where

$$\text{LHS} = \left(\frac{h}{L}\right)\left(\frac{w_0L}{T_O}\right) = \left(\frac{1}{3}\right)\left(\frac{w_0L}{T_O}\right)$$

$$\text{RHS} = \cosh\left[\frac{1}{2}\left(\frac{w_0L}{T_O}\right)\right] - 1$$

T_O/w_0L	LHS	RHS	LHS − RHS
0.42	0.7937	0.7964	−0.0027
0.419	0.7955	0.8006	−0.0051
0.421	0.7918	0.7921	−0.0003
0.4211	0.7916	0.7917	−0.0001
0.4212	0.7914	0.7913	0.0001

The second trial value of T_O/w_0L in the tabulation was based on a guess that the initial value was too high. This guess was shown to be wrong by the fact that there was a larger difference between the values of LHS and RHS corresponding to the second estimate than there was for the initial estimate. The succeeding values of T_O/w_0L were chosen in accord with the trend indicated by the first two trial values. The last two trials indicate that the value T_O/w_0L is somewhere between 0.4211 and 0.4212. From the former value, we calculate that

$$T_O = 0.4211w_0L = 0.4211(98.07)(75) = 3097 \text{ N}$$

The tension in the cable is given by eq. (17). This equation shows that T has its maximum value at the location where x is a maximum, which is one of the ends. Setting $x = L/2 = 37.5$ m, we find

$$T_{\max} = T|_{x=L/2} = 3097 \cosh\left[\frac{(98.07)(37.5)}{3097}\right] = 5549 \text{ N}$$

The total length of the cable is twice the arc length from the lowest point on the cable, which is where the origin is located. Therefore, from eq. (15) we find

$$s_{\text{total}} = 2s|_{x=L/2} = 2\left(\frac{3097}{98.07}\right)\sinh\left[\frac{(98.07)(37.5)}{3097}\right] = 93.9 \text{ m}$$

HOMEWORK PROBLEMS

7.68 A uniform 0.4 kg/m cable is attached to collar A, which may slide along the smooth horizontal bar. It is observed that a horizontal force $P = 10$ N is required to hold the collar in place at $d = 900$ mm. Determine (a) the sag h, and (b) the length of the cable.

7.69 A uniform 0.4 kg/m cable is attached to collar A, which may slide along the smooth horizontal bar. A horizontal force $\bar{P}$ holds the collar at $d = 600$ mm, and the sag $h = 360$ mm. Determine the magnitude of $\bar{P}$.

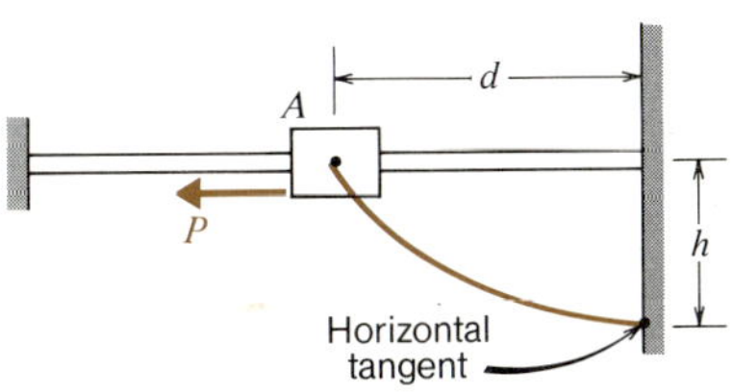

PROBS. 7.68 AND 7.69

7.70 A uniform 0.5 lb/ft cable is suspended between two supports at equal elevation. The span is 250 ft and the sag is 100 ft. Determine the minimum and maximum tensions in the cable and the length of the cable.

***7.71** A uniform 0.2 kg/m cable spans a distance of 50 m between two supports at equal elevations. The sag is 2.5 m. (a) Determine the maximum and minimum tension in the cable. (b) Compare the values found in part (a) to the approximate values obtained by considering the 0.2 kg/m mass per unit length of the cable to be distributed along the horizontal.

7.72 A 0.5 kg/m cable is suspended between two buildings with its ends at equal elevations. The length of the cable is 24 m and the maximum sag is 10 m. Determine (a) the horizontal distance spanned by the cable, and (b) the minimum and maximum tensions. *Hint:* Formulate expressions for the length and the maximum sag of the cable and then use identity (14c).

7.73 An electrical transmission cable having a weight per unit length of 2 lb/ft is suspended between, and is pinned to, a series of towers. If $L = 400$ ft and $h = 5$ ft, determine the horizontal and vertical components of the resultant force exerted by segment AB on tower A.

***7.74** The maximum permissible tension in a 6 kg/m electrical transmission cable is 50 kN. If $L = 80$ m, determine the maximum length of cable that can be used and the corresponding tension. *Hint:* There are two values of T_O that give a maximum tension of 50 kN.

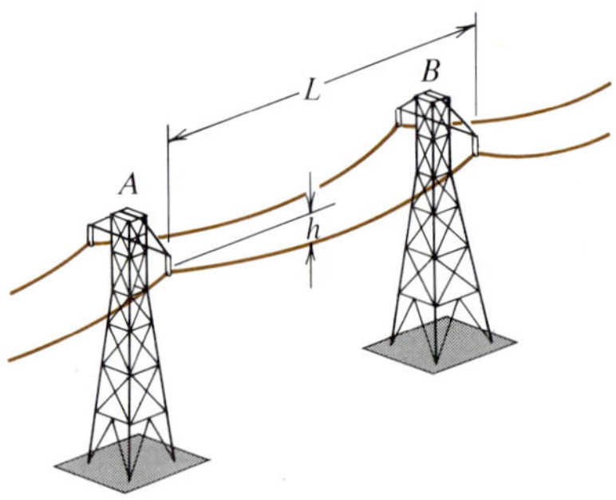

PROBS. 7.73 AND 7.74

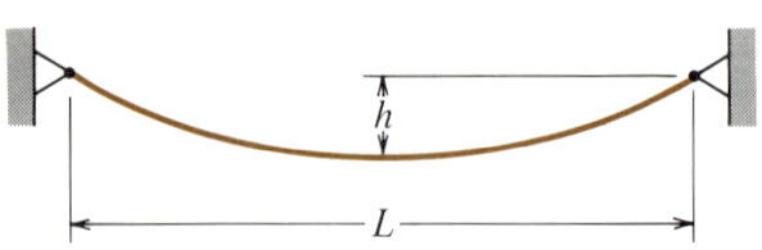

PROBS. 7.75–7.77

7.75 A 50 ft long cable weighing 2 lb/ft spans a distance L between two pins at equal elevation. Knowing that the maximum sag $h = 0.2L$, determine (a) the distance L, and (b) the horizontal and vertical components of the reaction force at one of the pin supports.

7.76 Determine the sag to span ratio h/L of a cable whose total weight equals the minimum tension in the cable.

7.77 Determine the sag to span ratio h/L of a cable whose total weight equals the maximum tension in the cable.

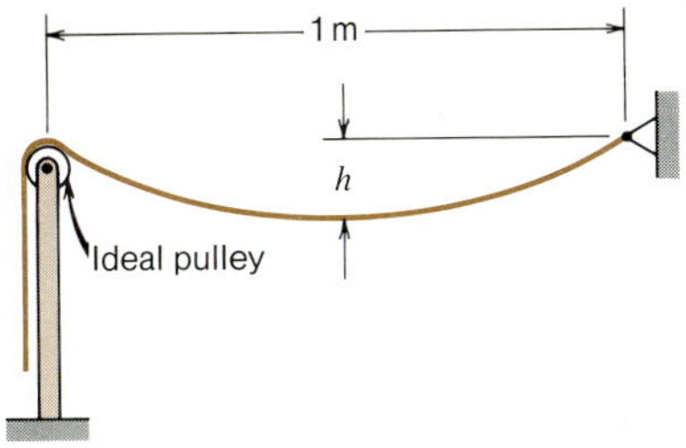

PROB. 7.78

*7.78 A uniform 2 m long cable having a total mass of 2 kg is suspended as shown. Determine the *two* values of h for which the cable is in equilibrium.

7.79 The mass per unit length of the cable is 0.4 kg/m. Determine the *two* values of the sag h for which the maximum tension in the cable is 300 N.

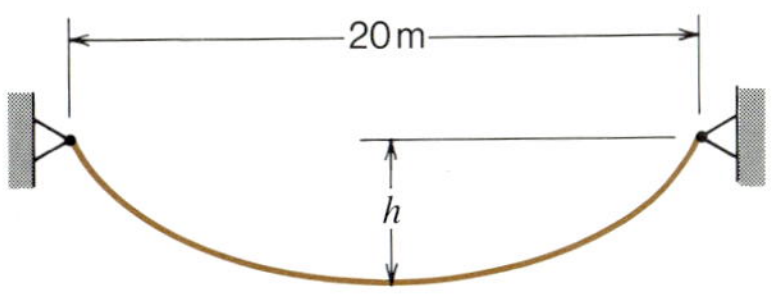

PROB. 7.79

CHAPTER 8

AREA MOMENTS AND PRODUCTS OF INERTIA

OBJECTIVES

1. To use mathematical integration to evaluate the area moments and product of inertia.
2. To recognize principal axes on the basis of symmetry.
3. To use the parallel axis theorems to transfer area inertia properties between parallel coordinate systems.
4. To evaluate the inertia properties of composite areas by using tabulated properties for basic shapes.
5. To construct a Mohr's circle describing the relationship between the area inertia properties for rotated coordinate systems.

Two important properties of a planar figure that were treated in Chapter 6 are the area and the location of the centroid. Additional numerical properties that characterize the size and shape of a planar figure are its *area moments* and *product of inertia*. These parameters occur in a variety of applications, such as hydrostatic pressure loadings on flat surfaces (see Section C3 in Chapter 6), stress distributions in elastic beams, and the rotational inertia of rigid bodies.

A. BASIC DEFINITIONS

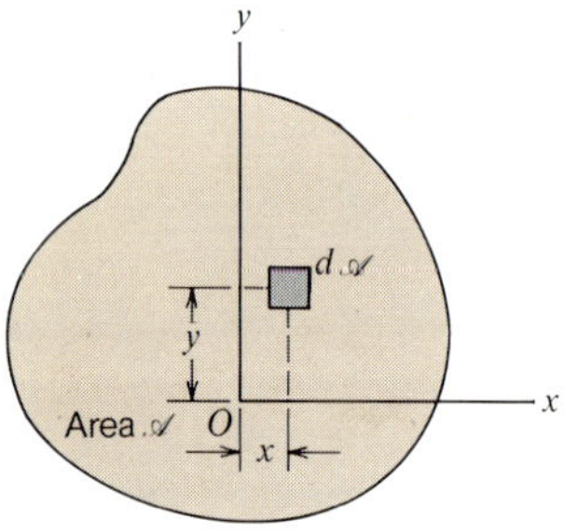

FIGURE 1

Consider the area $\mathscr{A}$ in Figure 1. The location of an infinitesimal element of area $d\mathscr{A}$ may be described in terms of the x and y coordinates relative to an xyz coordinate system whose origin is an arbitrary point O.

A *moment of inertia* provides a measure of the distribution of the elements of area about an axis. For example, the element $d\mathscr{A}$ in Figure 1 is at a distance $|y|$ from the x axis. Instead of the absolute value of y, the measure of distance that occurs in the moment of inertia formula is y^2. The weighted contribution of an element of area is $y^2\, d\mathscr{A}$. Integrating this quantity over the entire area then yields the moment of inertia about the x axis.

Recall that $y\, d\mathscr{A}$ is the integrand in the first moment of area, which is used to locate the centroid (i.e., the average location) of the elements $d\mathscr{A}$ from the origin. For the *area moments of inertia*, in general, the area $d\mathscr{A}$ of an element is multiplied by a quadratic product of the coordinates of the element (e.g., y^2). Such terms are then integrated to account for all elements in the body. Because an area is multiplied by a second power of the coordinates, the area moments of inertia are sometimes called *second moments of area*. The moment of inertia about the x and y axes, which are denoted as I_x and I_y, respectively, are accordingly defined as

$$I_x = \int_{\mathscr{A}} y^2\, d\mathscr{A} \qquad I_y = \int_{\mathscr{A}} x^2\, d\mathscr{A} \tag{1}$$

The phrase *moment of inertia* about the x axis is a result of the similarity between the properties of an area as defined above and the properties of the mass distribution of a rigid body. Thus, although the word *area* is frequently omitted when referring to the moments of inertia, it is acceptable to do so only when there is no possibility of confusion with the mass properties of a rigid body.

The values of I_x and I_y are always positive, because their integrands are always positive or zero. It follows, then, that if two areas have the same shape (for instance, circular, or rectangular in a fixed proportion), the larger area will have the larger values for the moments of inertia. Also, for two different shapes having the same area, the one whose area is distributed farther from an axis will have the larger value for the moment of inertia about that axis.

The parameters we have just defined convey information about the distribution of area about the x and y axes. A third moment of inertia is concerned with the distribution of area about the z axis. Referring to Figure 1, we see that the z axis is perpendicular to the plane of the area, intersecting the origin O. Hence, the distance from this axis to an area element $d\mathscr{A}$ is $(x^2 + y^2)^{1/2}$. Squaring this distance and multiplying it by $d\mathscr{A}$ yields the integral for the (area) *polar moment of inertia* J_z.

$$J_z = \int_{\mathscr{A}} (x^2 + y^2)\, d\mathscr{A} \tag{2}$$

The reason for using the symbol J_z rather than I_z for the polar moment of inertia is that it is conventional to do so.

The polar moment of inertia is not an independent quantity. To see this, it is only necessary to express eq. (2) as a sum of integrals rather than an integral of a sum. Employing eqs. (1) then leads to the identity

$$J_z = I_x + I_y \tag{3}$$

A common way to describe the moments of inertia of a shape about an axis is to give the *radius of gyration* k about that axis. This quantity is simply the square root of the ratio of the corresponding moment of inertia to the area $\mathscr{A}$. That is,

$$k_x = \sqrt{\frac{I_x}{\mathscr{A}}}, \qquad k_y = \sqrt{\frac{I_y}{\mathscr{A}}}, \qquad k_z = \sqrt{\frac{I_z}{\mathscr{A}}} \tag{4}$$

A radius of gyration has units of length, because the area moments of inertia have units of the fourth power of length. The radii of gyration have a simple interpretation. Suppose that we have a thin rectangular strip whose area $\mathscr{A}$ is the same as that for the general shape of interest. Let us orient this strip parallel to the x axis at a distance equal to the value of k_x, as shown in Figure 2a. Because all of the elements of area are then essentially at the distance k_x from the x axis, the moment of inertia of this strip about the x axis is $k_x^2 \mathscr{A} \equiv I_x$. Similarly, the moment of inertia about the y axis of the strip in Figure 2b is identical to I_y for the general shape, and the polar moment of inertia about the z axis for the thin ring in Figure 2c is identical to J_z for the general shape. In view of eq. (3), the radii of gyration are related by

$$k_z^2 = k_x^2 + k_y^2 \tag{5}$$

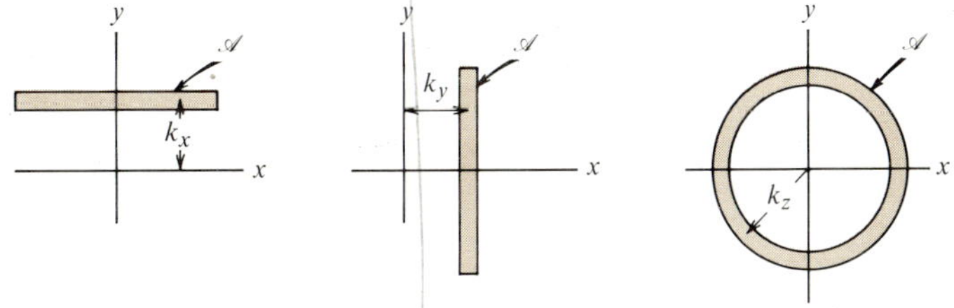

FIGURE 2a FIGURE 2b FIGURE 2c

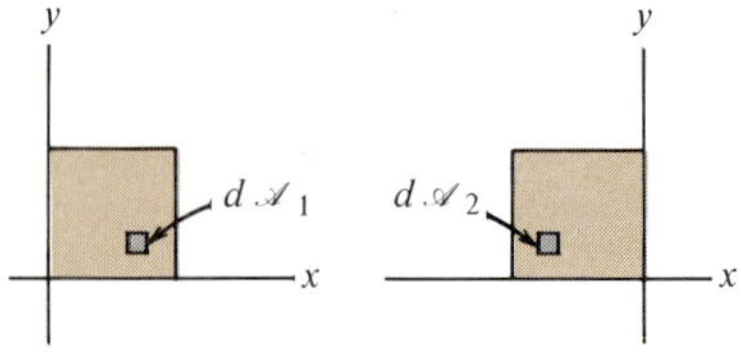

FIGURE 3

It might seem that the moments of inertia are the only parameters needed to characterize the distribution of area about an axis. This is not so. Referring to the two squares shown in Figure 3 shows that for every element of area $d\mathscr{A}_1$ in the square on the left, there is a companion element $d\mathscr{A}_2$ in the right square. Both of these elements are situated at the same distances from each of the coordinate axes, so both squares have identical moments of inertia. Yet clearly, these squares differ, in the sense that they are located in different coordinate quadrants.

The manner in which the area of a planar figure is situated in the coordinate quadrants is described by the area *product of inertia*. This quantity is defined as

$$I_{xy} = I_{yx} = \int_{\mathscr{A}} xy\, d\mathscr{A} \tag{6}$$

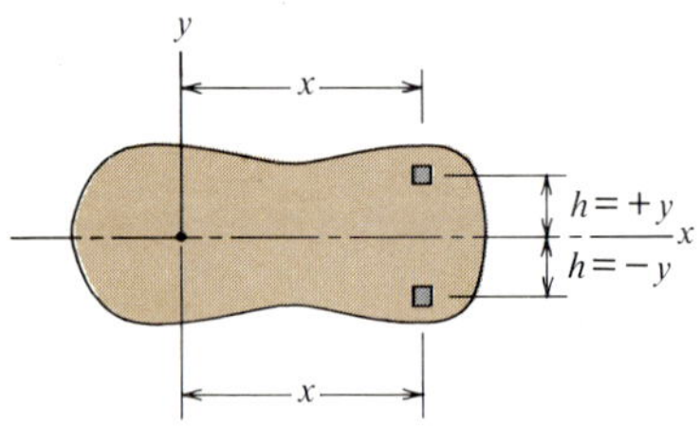

FIGURE 4

Because xy is positive if an element of area is located in the first or third quadrant, and negative in the second or fourth quadrant, we conclude that positive I_{xy} means that area predominates in the first and/or third quadrants. Obviously, when I_{xy} is negative, the area predominates in the second and/or fourth quadrants.

In the important case where a shape is symmetric about one of the coordinate axes, the value of xy for an element to one side of the axis of symmetry is canceled by the value of xy for the mirror-image element to the other side. This is demonstrated in Figure 4 for a shape that is symmetric about the x axis. Thus

> Whenever a planar shape has an axis of symmetry that is either the x or y axis, then $I_{xy} \equiv 0$.

In general, when the x and y coordinate axes give a zero product of inertia, we say that they are *principal axes*. All shapes, not just symmetric ones, have principal axes. Later in this chapter we will see how to locate the principal axes for an arbitrary shape.

For the remainder of this chapter we will consider various techniques for evaluating the area moments and products of inertia. We could go on to define higher moments of area. However, such quantities do not find application in the field of mechanics.

B. INTEGRATION TECHNIQUES

Because moments and product of inertia are defined in terms of integrals, the obvious way to determine the properties is to evaluate the integrals. Doing so requires that we describe the position of points contained within the closed curve forming the area $\mathscr{A}$.

Various techniques for integrating over an area were discussed in Chapter 6 in the course of locating the centroid of a rigid body. The approach we shall employ here is to form the smallest possible differential element of area. Recall that the elements of a planar area and their location may be

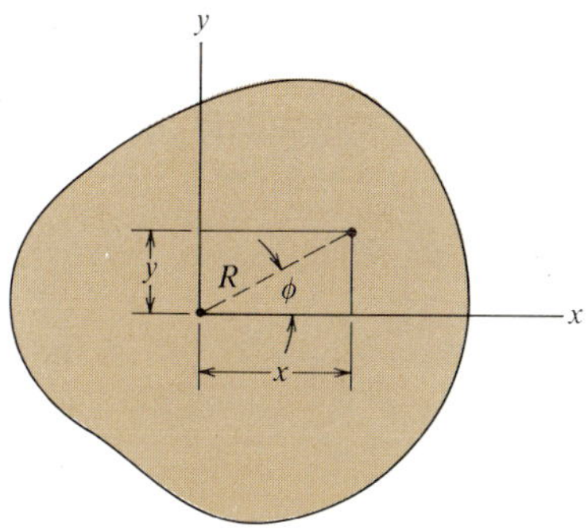

FIGURE 5

described in terms of either Cartesian or polar coordinates, depending on the manner in which the boundary of the area is specified.

For the planar area shown in Figure 5, a possible choice for polar coordinates are the radial distance R and angle θ relative to the x axis. These parameters are related to the Cartesian coordinates by

$$x = R\cos\phi \qquad y = R\sin\phi$$

It should be noted that other choices for the polar coordinates are possible, for instance, by measuring the angle ϕ from a different axis.

The differential element $d\mathscr{A}$ that we shall employ is a second-order differential of area. It corresponds to the area bounded by two values of each coordinate that differ by an infinitesimal amount. The basic element we shall select will be either of the two types depicted in Figure 6. When cubic and higher differentials are neglected, it is found that

$$\text{Cartesian coordinates—Figure 6a:} \qquad d\mathscr{A} = dx\,dz \tag{7a}$$

$$\text{Cylindrical coordinates—Figure 6b:} \qquad d\mathscr{A} = R\,d\phi\,dR \tag{7b}$$

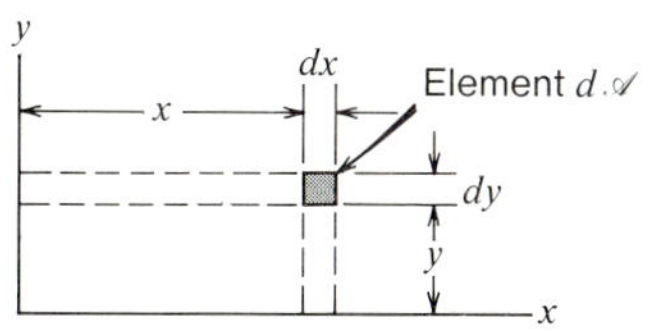

FIGURE 6a

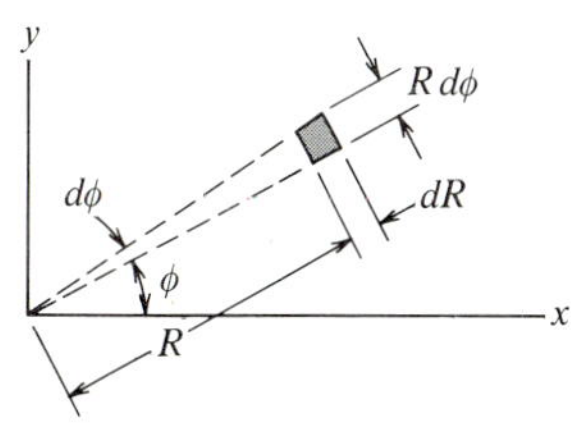

FIGURE 6b

The following examples will demonstrate that the methods for formulating the limits of an integral extending over all of the area are the same as those we employed in locating centers of mass by integration.

EXAMPLE 1

Calculate the moments and product of inertia about the axes of the given coordinate system for the parabolic sector shown.

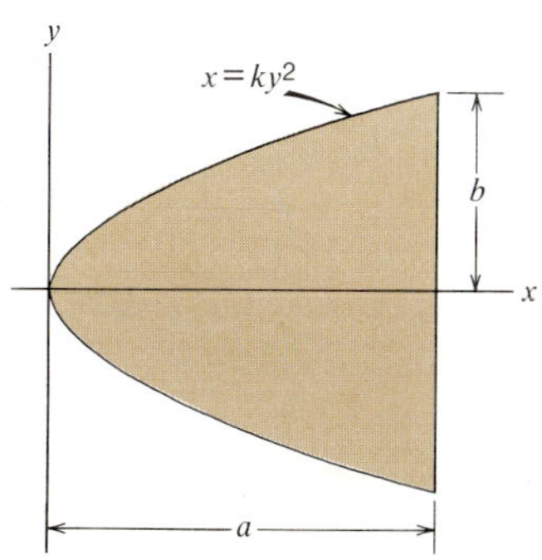

Solution

Cartesian coordinates will be used to formulate the solution, because the given parabola is defined in Cartesian coordinates. The constant k is determined by noting that the point (a, b) is on the parabola. From $x = ky^2$, we find that

$$a = kb^2, \qquad \text{or} \qquad k = a/b^2$$

The element of area $d\mathscr{A}$ for the Cartesian coordinates is shown in the following sketch.

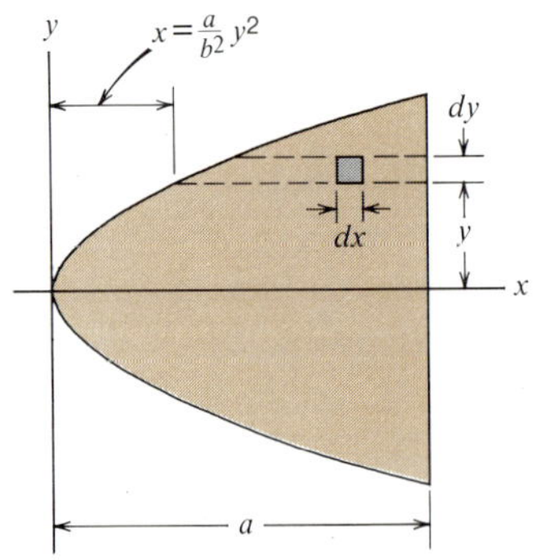

An integration over the entire area may be achieved by integrating first over the range of all x variables or all y variables, thus forming horizontal or vertical strips, respectively. As indicated in the preceding sketch, the equation for the parabola gives the value of x on the parabola for each value of y. Therefore, we will form horizontal strips by integrating x over the range $ay^2/b^2 \leq x \leq a$. All horizontal strips are then accounted for by integrating y over the range $-b \leq y \leq b$.

The parabolic sector is symmetrical about the x axis, so

$$I_{xy} = 0$$

Also, because the polar moment of inertia J_z may be obtained from eq. (3), it is only necessary for us to evaluate the integrals in eqs. (1). These are

$$I_x = \int_{-b}^{b} \int_{ay^2/b^2}^{a} y^2 \, dx \, dy = \int_{-b}^{b} y^2 \left(a - \frac{ay^2}{b^2} \right) dy = \frac{4}{15} ab^3$$

$$I_y = \int_{-b}^{b} \int_{ay^2/b^2}^{a} x^2 \, dx \, dy = \frac{1}{3} \int_{-b}^{b} \left(a^3 - \frac{a^3 y^6}{b^6} \right) dy = \frac{4}{7} a^3 b$$

From eq. (3) we then find

$$J_z = I_x + I_y = \frac{4}{105} ab(7b^2 + 15a^2)$$

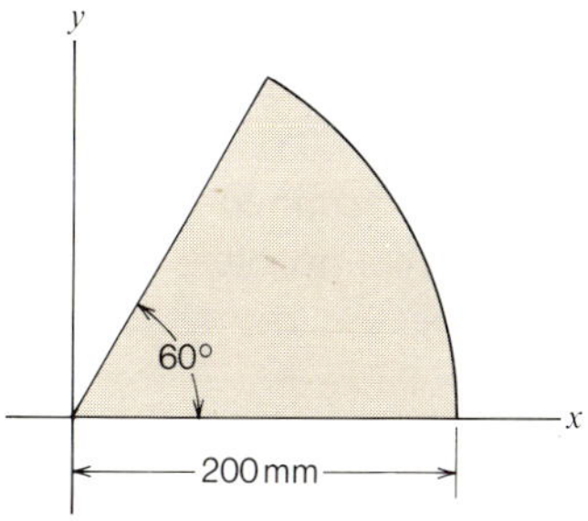

EXAMPLE 2

Calculate the moments and product of inertia of the circular sector shown with respect to the xyz coordinate system. Also, determine the corresponding radii of gyration.

Solution

It is logical to employ polar coordinates in this problem, because points on the arc are at a constant radial distance from the origin, and equally important, points on the straight sides are at constant polar angles ϕ from the coordinate axes.

The polar coordinates depicted in the sketch are related by

$$x = R\cos\phi, \qquad y = R\sin\phi$$

and the element of area is $d\mathscr{A} = R\,d\phi\,dR$. As indicated in the sketch, the sequence of integrations we choose will form a differential sector by integrating over $0 \le R \le 0.20$ m. All of these sectors will then be accounted for by integrating over $0 \le \phi \le \pi/3$ radians.

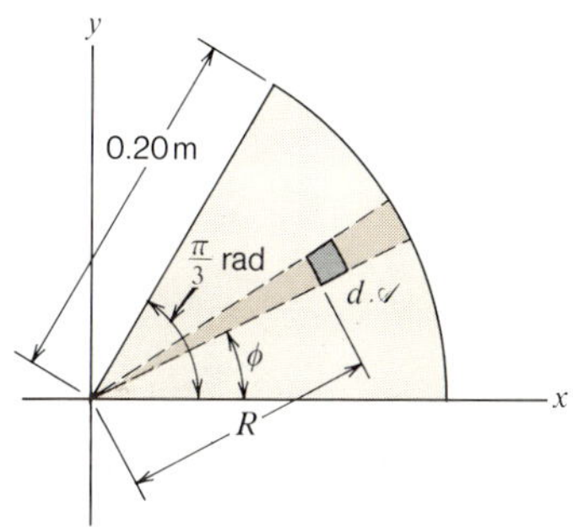

The differentials in the expression for $d\mathscr{A}$ are rearranged below to match the order of integration. The moments of inertia are

$$I_x = \int_0^{\pi/3}\int_0^{0.2} y^2 R\,dR\,d\phi = \int_0^{\pi/3}\int_0^{0.2} R^3\sin^2\phi\,dR\,d\phi$$

$$= \tfrac{1}{4}(0.2)^4\int_0^{\pi/3}\sin^2\phi\,d\phi = 1.228(10^{-4})\text{ m}^4$$

$$I_y = \int_0^{\pi/3}\int_0^{0.2} x^2 R\,dR\,d\phi = \int_0^{\pi/3}\int_0^{0.2} R^3\cos^2\phi\,dR\,d\phi$$

$$= \tfrac{1}{4}(0.2)^4\int_0^{\pi/3}\cos^2\phi\,d\phi = 2.960(10^{-4})\text{ m}^4$$

$$I_{xy} = \int_0^{\pi/3}\int_0^{0.2} xyR\,dR\,d\phi = \int_0^{\pi/3}\int_0^{0.2} R^3\cos\phi\sin\phi\,dR\,d\phi$$

$$= \tfrac{1}{4}(0.2)^4\int_0^{\pi/3}\cos\phi\sin\phi\,d\phi = 1.50(10^{-4})\text{ m}^4$$

Equation (3) then gives

$$J_z = I_x + I_y = 4.188(10^{-4})\ \text{m}^4$$

The radii of gyration may now be determined using their definitions, eqs. (4). For this we need the area $\mathscr{A}$. In this problem it is not necessary to evaluate an integral to determine $\mathscr{A}$, because the sector is one-sixth of a full circle. Thus

$$\mathscr{A} = \tfrac{1}{6}\pi(0.2)^2 = 2.094(10^{-2})\ \text{m}^2$$

Hence, the radii of gyration are

$$k_x = \sqrt{\frac{1.228(10^{-4})}{2.094(10^{-2})}} = 0.0766\ \text{m}, \qquad k_y = \sqrt{\frac{2.960(10^{-4})}{2.094(10^{-2})}} = 0.1189\ \text{m}$$

$$k_z = \sqrt{\frac{4.188(10^{-4})}{2.094(10^{-2})}} = 0.1414\ \text{m}$$

HOMEWORK PROBLEMS

8.1–8.4 Use integration to evaluate the area moment of inertia I_x of the shaded area shown.

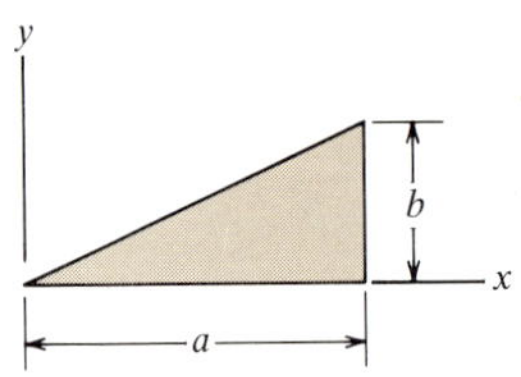

PROBS. 8.1 AND 8.5

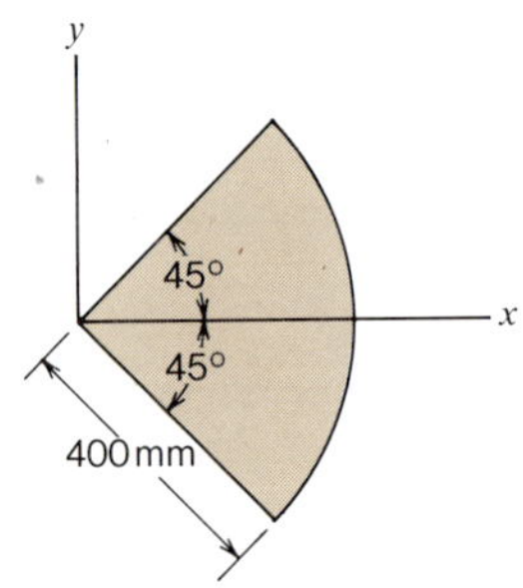

PROBS. 8.2 AND 8.6

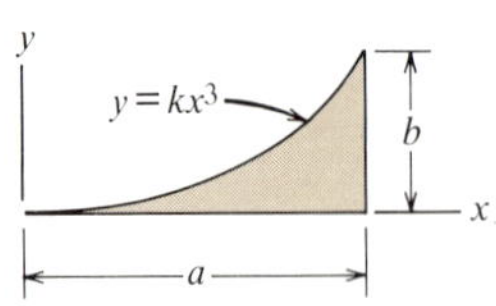

PROBS. 8.3 AND 8.7

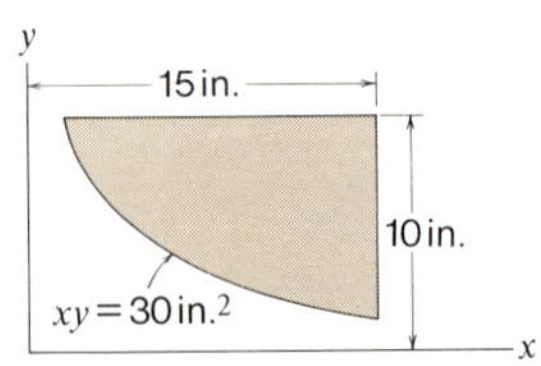

PROBS. 8.4 AND 8.8

8.5–8.8 Use integration to evaluate the area moment of inertia I_y and the area product of inertia I_{xy} of the shaded area shown.

8.9–8.12 Integrate over the shaded area shown to evaluate the polar moment of inertia J_z and the radius of gyration k_z of the shaded area shown.

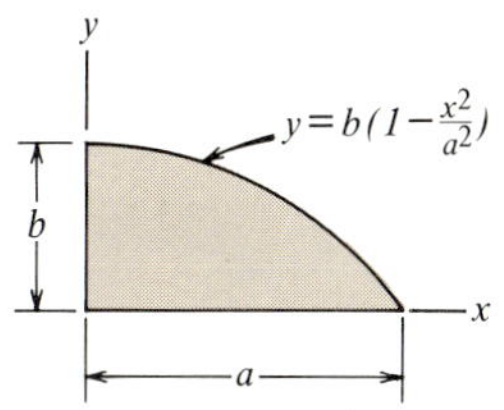

PROBS. 8.9 AND 8.13

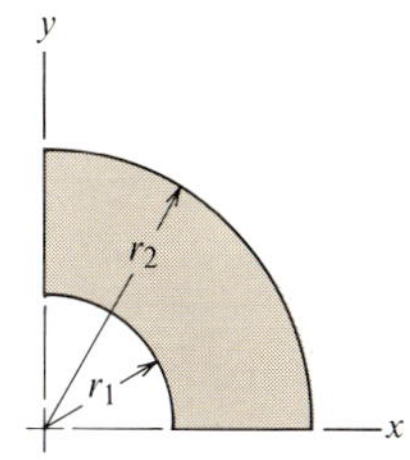

PROBS. 8.10 AND 8.14

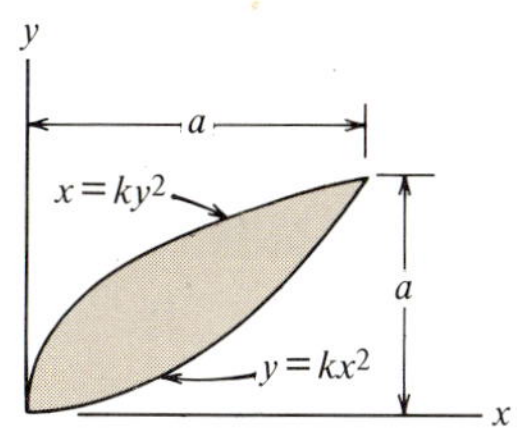

PROBS. 8.11 AND 8.15

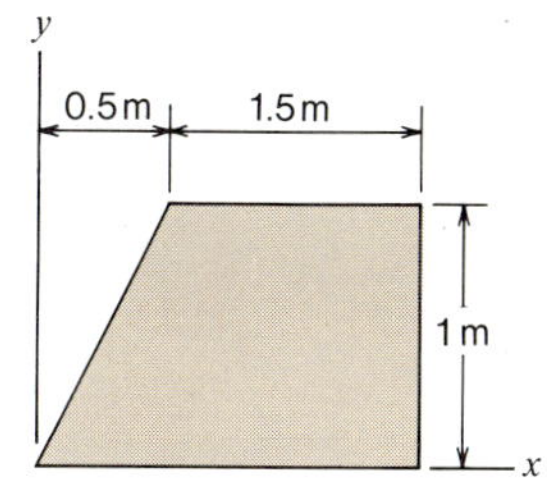

PROBS. 8.12 AND 8.16

8.13–8.16 By integration, evaluate the area moments of inertia I_x and I_y and the area product of inertia I_{xy} of the shaded area shown.

8.17–8.19 By integration, evaluate the radii of gyration k_x, k_y, and k_z of the shaded area shown.

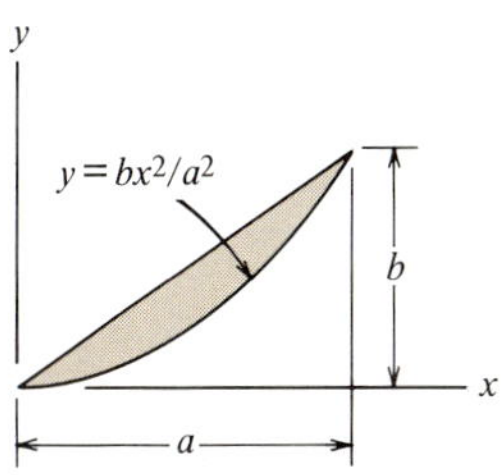

PROB. 8.17

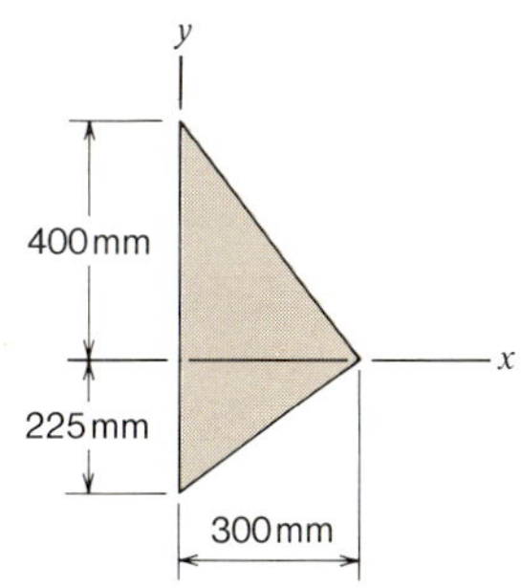

PROB. 8.18

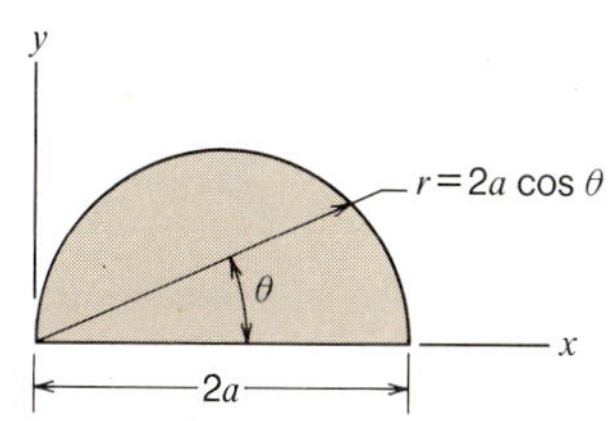

PROB. 8.19

C. PARALLEL AXIS THEOREMS

Suppose that the area moments and products of inertia of an area are known with respect to one set of xyz coordinate axes. It would be useful to have algebraic formulas that would give the values of these parameters with respect to other coordinate systems. Otherwise, a new set of integrations would be required for each coordinate system. In this section we will develop the relationship between the moments and products of inertia for two mutually parallel coordinate systems.

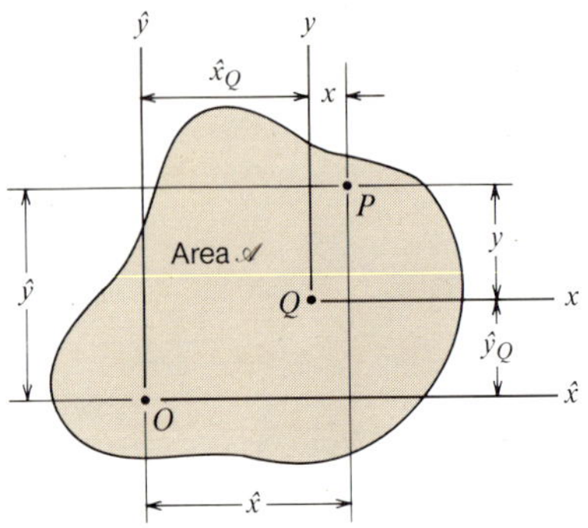

FIGURE 7

Consider the two coordinate systems xyz and $\hat{x}\hat{y}\hat{z}$ used to describe the arbitrary shape in Figure 7. As shown in the figure, the coordinates of the origin Q of the xyz coordinate system with respect to the $\hat{x}\hat{y}\hat{z}$ system are $(\hat{x}_Q, \hat{y}_Q, 0)$. Note that the distance between the x and $\hat{x}$ axis is $|\hat{y}_Q|$ and the distance between the y and $\hat{y}$ axes is $|\hat{x}_Q|$, and that the signs for these quantities are chosen in accord with the quadrant of the $\hat{x}\hat{y}\hat{z}$ system in which the origin Q is located. From Figure 7, we see that the coordinates of an arbitrary point P are related by

$$x = \hat{x} - \hat{x}_Q, \qquad y = \hat{y} - \hat{y}_Q \tag{8}$$

Let us begin by determining I_x. From its definition and eqs. (8), we find that

$$\begin{aligned} I_x &= \int_{\mathscr{A}} y^2 \, d\mathscr{A} = \int_{\mathscr{A}} (\hat{y} - \hat{y}_Q)^2 \, d\mathscr{A} \\ &= \int_{\mathscr{A}} \hat{y}^2 \, d\mathscr{A} - 2\hat{y}_Q \int_{\mathscr{A}} \hat{y} \, d\mathscr{A} + (\hat{y}_Q)^2 \int_{\mathscr{A}} d\mathscr{A} \end{aligned}$$

The last of the integrals on the right side of the preceding equation is readily identified to give the area $\mathscr{A}$, and the first integral is seen to be the definition of $I_{\hat{x}}$. Recall that the integral in the second term is the first moment of area with respect to the $\hat{y}$ coordinate. Hence, *by choosing the origin O of $\hat{x}\hat{y}\hat{z}$ to be the centroid of the area, we can make this integral vanish.*

It is not difficult to show that similar steps for I_y and J_z yield the following set of *parallel axis theorems* for converting the moments of inertia for one coordinate system into those of the other.

$$\boxed{\begin{aligned} I_x &= I_{\hat{x}} + \mathscr{A}(\hat{y}_Q)^2 \\ I_y &= I_{\hat{y}} + \mathscr{A}(\hat{x}_Q)^2 \\ J_z &= J_{\hat{z}} + \mathscr{A}[(\hat{x}_Q)^2 + (\hat{y}_Q)^2] \end{aligned}} \tag{9}$$

Two important facts to bear in mind when applying eqs. (9) are that they are valid only if the origin of the $\hat{x}\hat{y}\hat{z}$ coordinate system coincides with the centroid of the area, and that $(\hat{x}_Q, \hat{y}_Q)$ are the coordinates of the origin Q of the xyz system measured with respect to the $\hat{x}\hat{y}\hat{z}$ axes.

The interpretation of eqs. (9) is fairly direct. As we noted earlier, $|\hat{y}_Q|$ and $|\hat{x}_Q|$ represent the distances between the parallel pairs of x and y axes, respectively. Similarly, it may be seen by applying the Pythagorean theorem to Figure 7 that $[(\hat{x}_Q)^2 + (\hat{y}_Q)^2]^{1/2}$ is the distance between the parallel pair of z axes. Thus, the area moment of inertia about an axis that does not intersect the center of mass is always larger than the value for a parallel centroidal axis. The difference between the corresponding values is the area multiplied by the square of the distance between the axes.

The coordinate transformations of eq. (8) may also be applied to derive a parallel axis theorem for the area product of inertia, I_{xy}. Specifically,

$$I_{xy} = \int_{\mathscr{A}} xy \, d\mathscr{A} = \int_{\mathscr{A}} (\hat{x} - \hat{x}_Q)(\hat{y} - \hat{y}_Q) \, d\mathscr{A}$$

$$I_{xy} = \int_{\mathscr{A}} \hat{x}\hat{y}\, d\mathscr{A} - \hat{x}_Q \int_{\mathscr{A}} \hat{y}\, d\mathscr{A} - \hat{y}_Q \int_{\mathscr{A}} \hat{x}\, d\mathscr{A} + \hat{x}_Q\hat{y}_Q \int_{\mathscr{A}} d\mathscr{A}$$

The first integral is the definition of $I_{\hat{x}\hat{y}}$. The second and third integrals are zero because the $\hat{x}$ and $\hat{y}$ axes pass through the centroid of the area. Hence, the parallel axis theorem for the product of inertia is

$$\boxed{I_{xy} = I_{\hat{x}\hat{y}} + \mathscr{A}\hat{x}_Q\hat{y}_Q} \tag{10}$$

Notice that in eqs. (9) the sign of $\hat{x}_Q$ and $\hat{y}_Q$ are not significant, because those parallel axis theorems depend only on the distance between corresponding axes. However, particular attention must be paid to the sign of $\hat{x}$ and $\hat{y}$ when applying eq. (10). Again we remind you that

1. the centroid of the area $\mathscr{A}$ is the origin of the $\hat{x}\hat{y}\hat{z}$ coordinate system,
2. point Q is the origin of the parallel xyz coordinate system,
3. $(\hat{x}_Q, \hat{y}_Q, 0)$ are the coordinates of point Q measured from the $\hat{x}\hat{y}\hat{z}$ system.

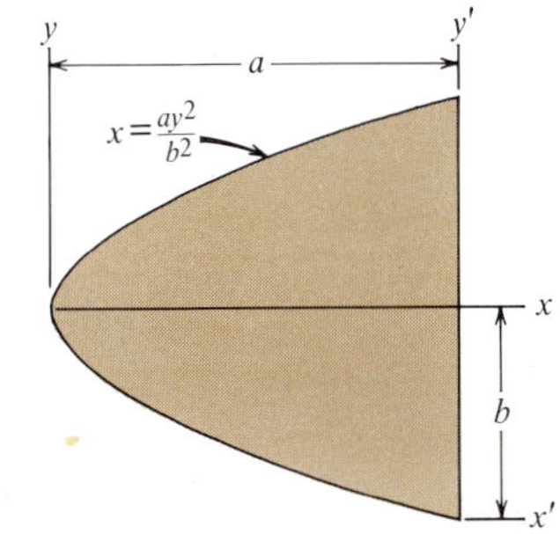

EXAMPLE 3

The area moments and product of inertia of the parabolic sector with respect to the xyz coordinate system were determined in Example 1. Determine these parameters for the $x'y'z'$ system shown.

Solution

It is important to remember that the parallel axis theorems are valid only when one of the coordinate systems is centroidal. Thus, in addition to the given systems xyz and $x'y'z'$, a parallel set of centroidal axes $\hat{x}\hat{y}\hat{z}$ will be needed. This means that the location of the centroid of the parabolic sector, as well as its area, must be determined before the parallel axis theorems are employed.

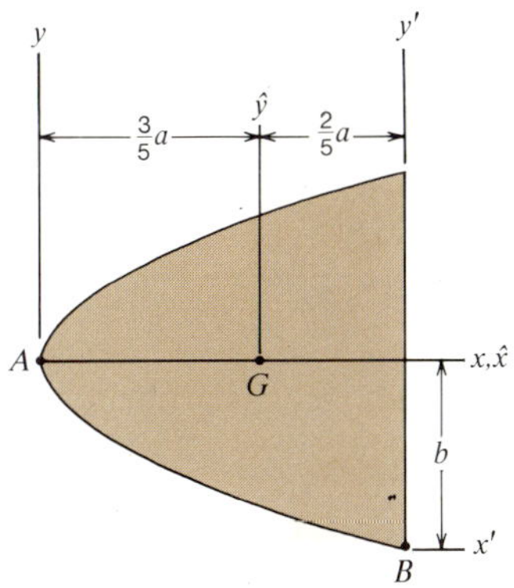

The x axis cuts the given parabolic sector into two semiparabolic sections. Because of symmetry, each of these shapes has the same x coordinate for its centroid, and the y coordinate of the full sector is zero. Thus, the centroidal coordinate system $\hat{x}\hat{y}\hat{z}$ is as shown in the sketch. Doubling the area of a semiparabolic section as given in Appendix C yields

$$\mathscr{A} = \tfrac{4}{3}ab$$

The coordinates of the origin A of the xyz system measured from the $\hat{x}\hat{y}\hat{z}$ axes are $(-3a/5, 0, 0)$. The parallel axis theorems may now be solved for $I_{\hat{x}}$, $I_{\hat{y}}$, and $I_{\hat{x}\hat{y}}$ in terms of the noncentroidal parameters evaluated in Example 1. This gives

$$I_{\hat{x}} = I_x - \mathscr{A}(\hat{y}_A)^2 = \tfrac{4}{15}ab^3 - (\tfrac{4}{3}ab)(0) = \tfrac{4}{15}ab^3$$

$$I_{\hat{y}} = I_y - \mathscr{A}(\hat{x}_A)^2 = \tfrac{4}{7}a^3b - (\tfrac{4}{3}ab)(-\tfrac{3}{5}a)^2 = \tfrac{16}{175}a^3b$$

$$I_{\hat{x}\hat{y}} = I_{xy} - \mathscr{A}\hat{x}_A\hat{y}_A = 0 - (\tfrac{4}{3}ab)(-\tfrac{3}{5}a)(0) = 0$$

Note that it is not necessary to apply the parallel axis theorem for $J_{\hat{z}}$ because we will be able to use eq. (3) to determine $J_{z'}$.

The coordinates of the origin B of the $x'y'z'$ coordinate system with respect to $\hat{x}\hat{y}\hat{z}$ are $(2a/5, -b, 0)$. We now apply the parallel axis theorems again in order to transfer from the $\hat{x}\hat{y}\hat{z}$ system to the $x'y'z'$ system. This yields

$$I_{x'} = I_{\hat{x}} + \mathscr{A}(\hat{y}_B)^2 = \tfrac{4}{15}ab^3 + (\tfrac{4}{3}ab)(-b)^2 = \tfrac{8}{5}ab^3$$

$$I_{y'} = I_{\hat{y}} + \mathscr{A}(x_B)^2 = \tfrac{16}{175}a^3b + (\tfrac{4}{3}ab)(\tfrac{2}{5}a)^2 = \tfrac{32}{105}a^3b$$

$$I_{x'y'} = I_{\hat{x}\hat{y}} + \mathscr{A}\hat{x}_B\hat{y}_B = 0 + (\tfrac{4}{3}ab)(\tfrac{2}{5}a)(-b) = -\tfrac{8}{15}a^2b^2$$

We then find that

$$J_{z'} = I_{x'} + I_{y'} = \tfrac{8}{105}ab(21b^2 + 4a^2)$$

HOMEWORK PROBLEMS

8.20–8.28 Determine the moment of inertia I_x, the polar moment of inertia J_z, and the product of inertia I_{xy} for the shaded area and corresponding coordinate system shown. The properties in Appendix C may be employed where necessary.

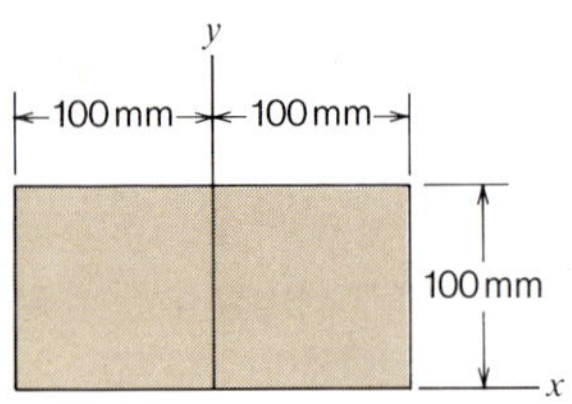

PROB. 8.20

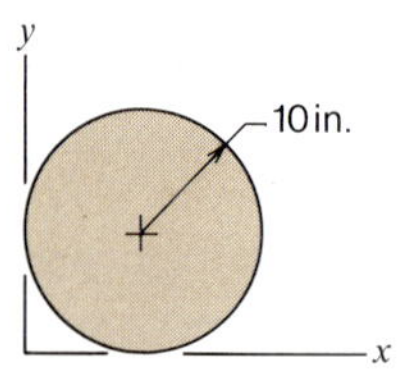

PROB. 8.21

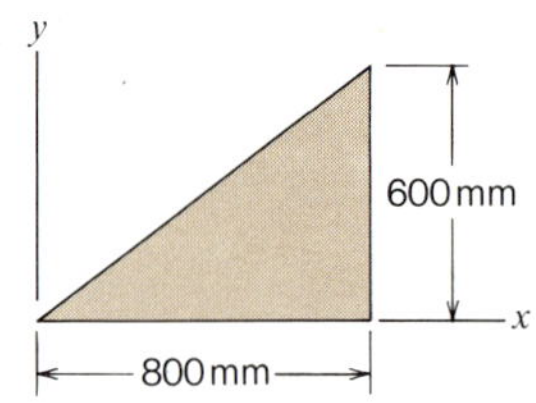

PROB. 8.22

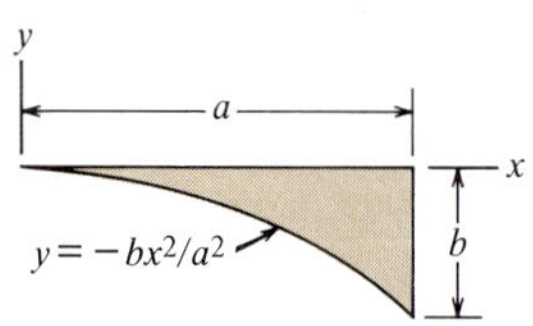

PROB. 8.23

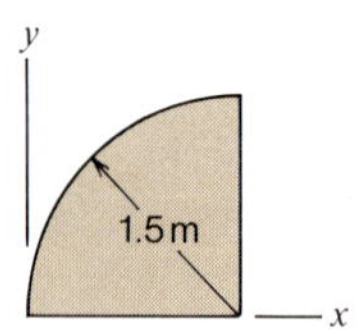

PROB. 8.24

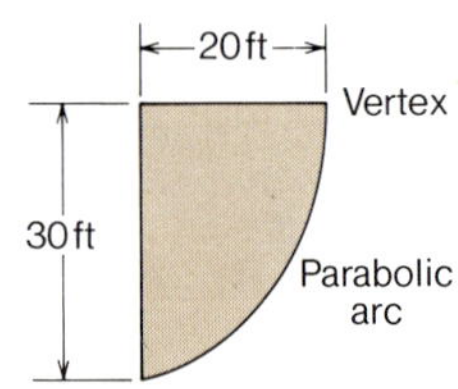

PROB. 8.25

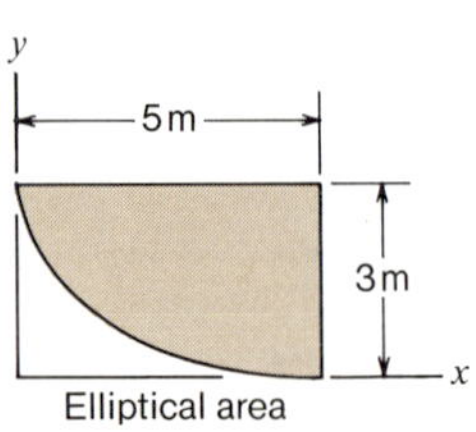

PROB. 8.26

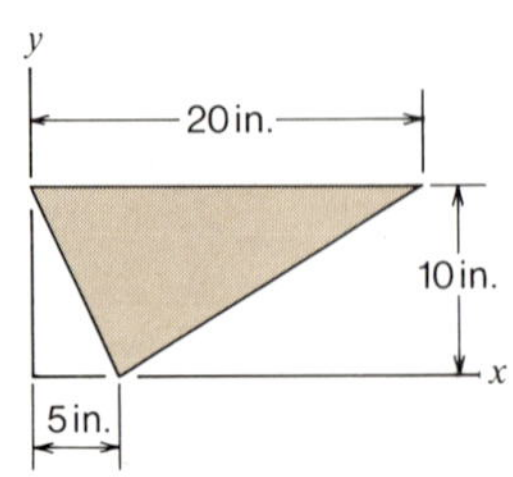

PROB. 8.27

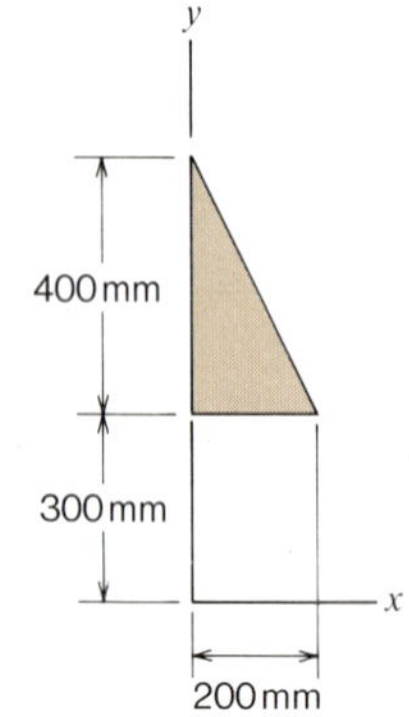

PROB. 8.28

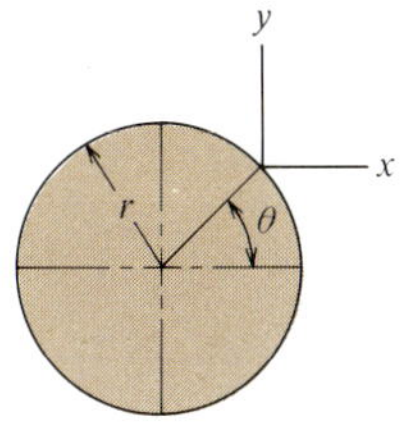

PROB. 8.29

8.29 The origin of the xyz axes is situated on the perimeter of the circle, and it is known that $I_x = 0.20\ \text{m}^4$, $J_z = 0.40\ \text{m}^4$. Determine the radius r and all possible values of the angle θ locating the origin of xyz.

8.30 The origin of the xyz axes is situated on edge AB of the rectangle, and it is known that $I_y = 20\ \text{in.}^4$, $I_{xy} = -5\ \text{in.}^4$. Determine the width a and the dimension c locating the origin.

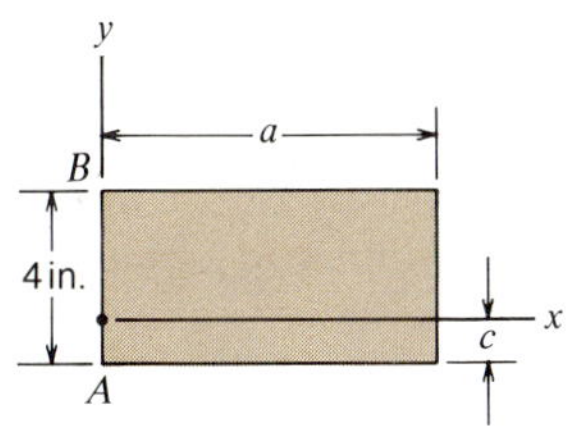

PROB. 8.30

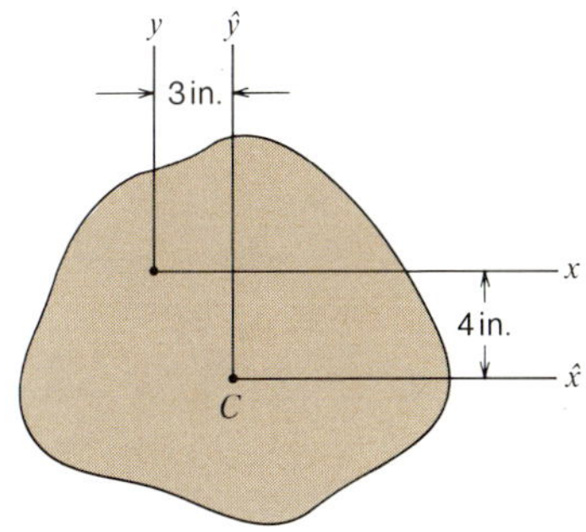

PROB. 8.31

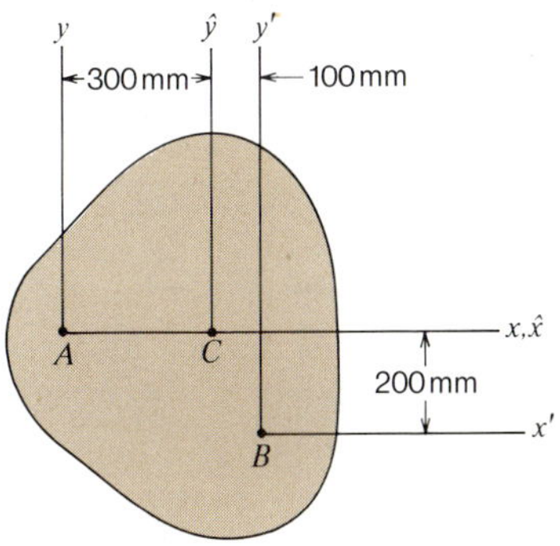

PROB. 8.32

8.31 Point C is the centroid of the shaded area. It is known that for the xyz coordinate system shown, $I_x = 1230\ \text{in.}^4$ and $J_z = 2580\ \text{in.}^4$, and for the centroidal $\hat{x}\hat{y}\hat{z}$ coordinate system $J_{\hat{z}} = 1830\ \text{in.}^4$. Determine (a) the area $\mathscr{A}$, (b) the values of $I_{\hat{x}}$ and $I_{\hat{y}}$.

8.32 Point C is the centroid of the shaded area shown. Because of symmetry it is known that the centroidal $\hat{x}\hat{y}\hat{z}$ axes are principal axes, so that the product of inertia with respect to these axes is zero. It is also known that for the other coordinate systems shown, $I_y = 8.50(10^{-2})\ \text{m}^4$, $I_{x'} = 3.25(10^{-2})\ \text{m}^4$, and $I_{y'} = 6.50(10^{-2})\ \text{m}^4$. Determine (a) the area $\mathscr{A}$, (b) the moment of inertia $I_{\hat{x}}$, (c) the products of inertia I_{xy} and $I_{x'y'}$.

D. COMPOSITE AREAS

The area moments and products of inertia with respect to a centroidal coordinate system have been tabulated in Appendix C for a variety of common shapes. The properties of other shapes often may be evaluated with the aid of this tabulation by making use of the concept of a composite area.

Figure 8 depicts an area that is formed from two basic shapes having areas $\mathscr{A}_1$ and $\mathscr{A}_2$. In the definitions of the inertia properties of an area, integrals extending over the total area $\mathscr{A}$ may be expressed as the sum of integrals extending over the individual areas $\mathscr{A}_1$ and $\mathscr{A}_2$. The following identities are then found.

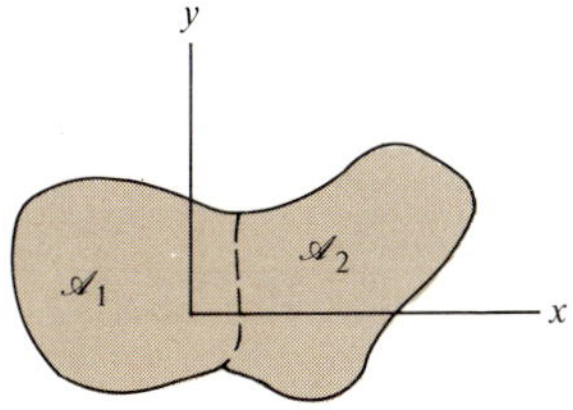

FIGURE 8

$$
\begin{aligned}
I_x &= (I_x)_1 + (I_x)_2 \\
I_y &= (I_y)_1 + (I_y)_2 \\
J_z &= (J_z)_1 + (J_z)_2 \\
I_{xy} &= (I_{xy})_1 + (I_{xy})_2
\end{aligned}
\tag{11}
$$

Equations (11) state that the moments and products of inertia for the composite are the sum of the corresponding values for the individual areas. The important aspect of this summation is that *the properties of the individual shapes must all be expressed with respect to the same xyz coordinate system before they may be added.* Because the properties of the basic shapes in Appendix C are for centroidal axes, it will usually be necessary to employ the parallel axis theorems in order to obtain the properties of each shape with respect to a common coordinate system.

EXAMPLE 4

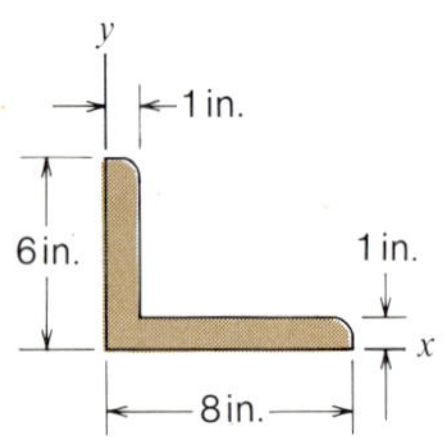

The cross section of an angle iron used in the construction of steel structures is shown. Neglecting the fillets (rounded-off edges), determine the values of I_x, I_y, and I_{xy} for the coordinate system shown.

Solution

When the fillets are neglected, the cross section is a composite of the rectangles formed by each leg. These individual shapes and their parallel centroidal axes are depicted in the sketch, where Q is the origin of the given coordinate system.

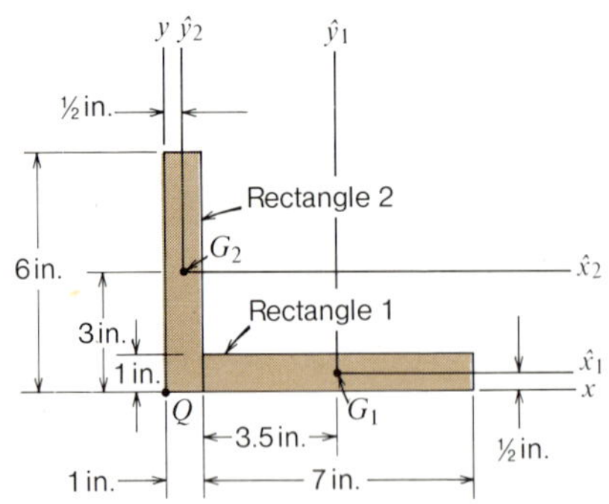

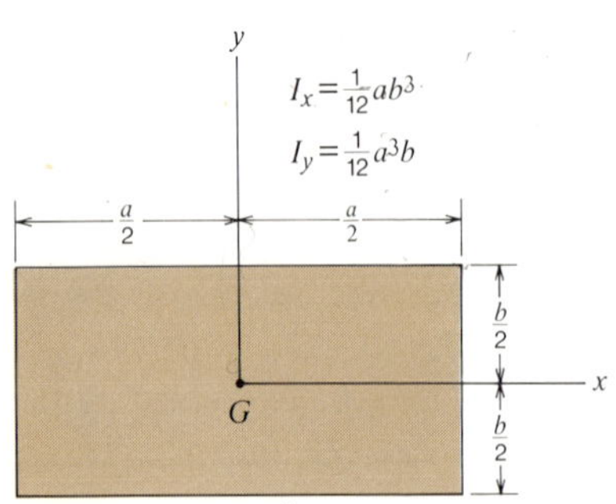

The inertia properties for a rectangle are given in Appendix C; they are repeated in the second sketch for convenience. Consistent with the procedure for composite shapes developed in the treatment of centroids, we first express the properties of each individual rectangle. Note that the coordinates of the origin Q with respect to the origin of each of the centroidal axes are needed in order to transfer the centroidal inertia properties to xyz.

Rectangle 1: $A_1 = 7(1) = 7 \text{ in.}^2, (\hat{x}_Q, \hat{y}_Q)_1 = (-4.5, -0.5) \text{ in.}$

$$(I_x)_1 = (I_{\hat{x}})_1 + A_1(-0.5)^2 = \tfrac{1}{12}(7)(1)^3 + 7(-0.5)^2 = 2.33 \text{ in.}^4$$

$$(I_y)_1 = (I_{\hat{y}}) + A_1(-4.5)^2 = \tfrac{1}{12}(7)^3(1) + 7(-4.5)^2 = 170.33 \text{ in.}^4$$

$$(I_{xy})_1 = (I_{\hat{x}\hat{y}})_1 + A_1(-4.5)(-0.5) = 0 + 7(-4.5)(-0.5) = 15.76 \text{ in.}^4$$

Rectangle 2: $A_2 = (1)(6) = 6 \text{ in.}^2, (\hat{x}_Q, \hat{y}_Q)_2 = (-0.5, -3) \text{ in.}$

$$(I_x)_2 = (I_{\hat{x}})_2 + A_2(-3)^2 = \tfrac{1}{12}(1)(6)^3 + 6(-3)^2 = 72.00 \text{ in.}^4$$

$$(I_y)_2 = (I_{\hat{y}})_2 + A_2(-0.5)^2 = \tfrac{1}{12}(1)^3(6) + 6(-0.5)^2 = 2.00 \text{ in.}^4$$

$$(I_{xy})_2 = (I_{\hat{x}\hat{y}})_2 + A_2(-0.5)(-3) = 0 + 6(-0.5)(-3) = 9.00 \text{ in.}^4$$

The properties for the composite may now be determined directly from eqs. (11).

$$I_x = (I_x)_1 + (I_x)_2 = 74.33 \text{ in.}^4$$

$$I_y = (I_y)_1 + (I_y)_2 = 172.33 \text{ in.}^4$$

$$I_{xy} = (I_{xy})_1 + (I_{xy})_2 = 24.76 \text{ in.}^4$$

Additional Remarks

It is often necessary to determine the inertia properties of a composite shape with respect to axes whose origin is the centroid of the composite. This may be achieved by first evaluating the properties of the composite with respect to a convenient noncentroidal coordinate system, such as *xyz* in this problem. Such inertia properties may be transferred to desired centroidal axes by again applying the parallel axis theorems to the composite.

HOMEWORK PROBLEMS

8.33–8.36 Determine the moment of inertia I_x, and the radius of gyration k_x of the shaded area shown.

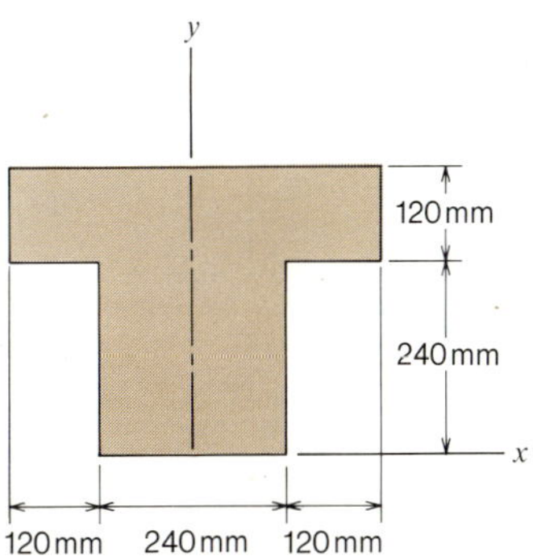

PROBS. 8.33 AND 8.37

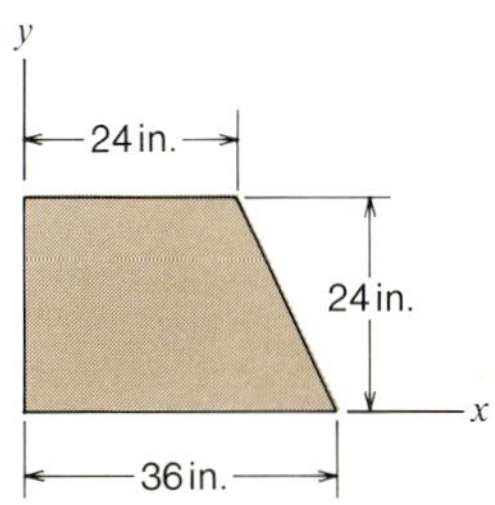

PROBS. 8.34 AND 8.38

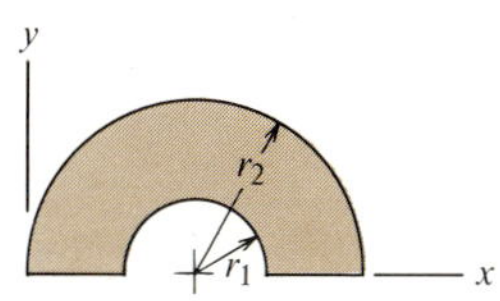

PROBS. 8.35 AND 8.39

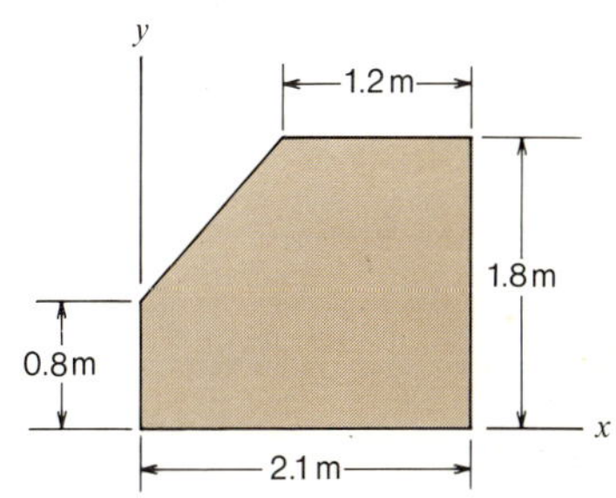

PROBS. 8.36 AND 8.40

8.37–8.40 Determine the polar moment of inertia J_z, and the product of inertia I_{xy} of the shaded area shown.

8.41–8.44 Point G is the centroid of the shaded area shown. Determine the distance y_G locating this point, and the corresponding value of I_x.

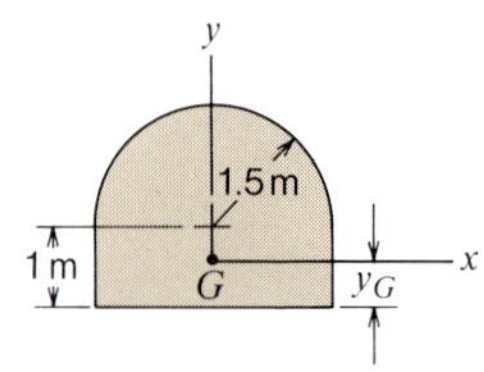

PROB. 8.41

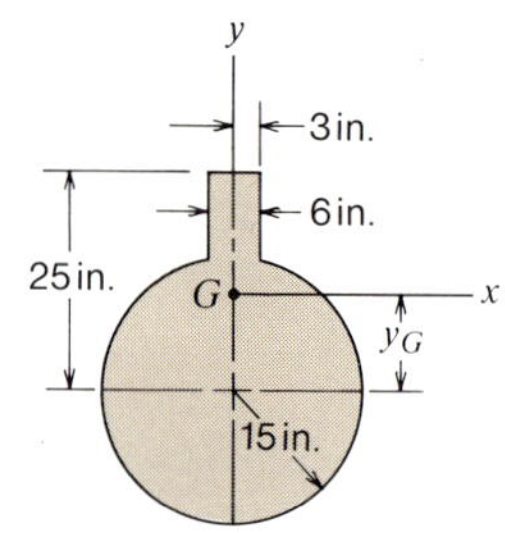

PROB. 8.42

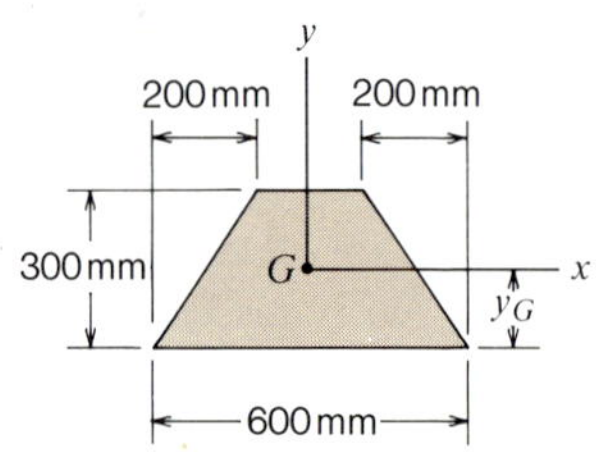

PROB. 8.43

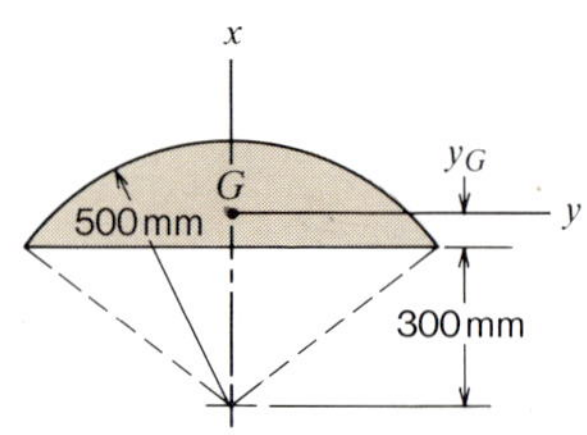

***PROB. 8.44**

8.45–8.48 The shaded area depicts the cross section of a structural beam for which point G is the centroid. Determine the location of this point and the values of I_x, I_y, J_z, and I_{xy} for the centroidal xyz coordinate system.

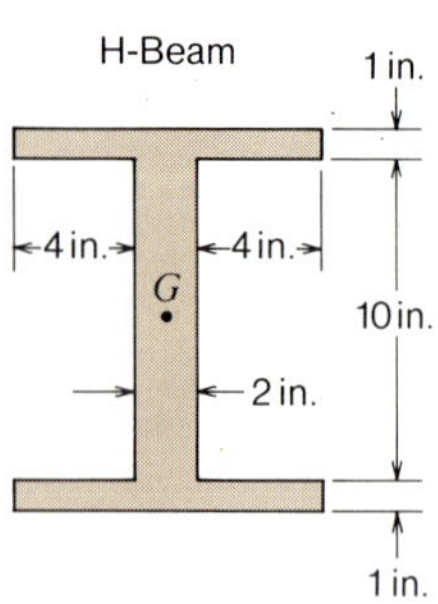

PROB. 8.45

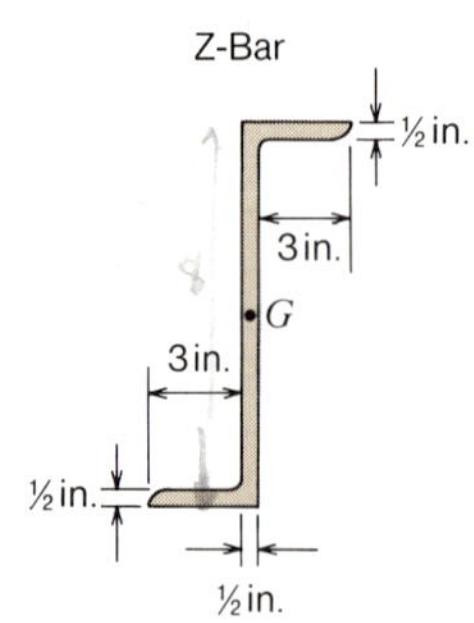

PROB. 8.46

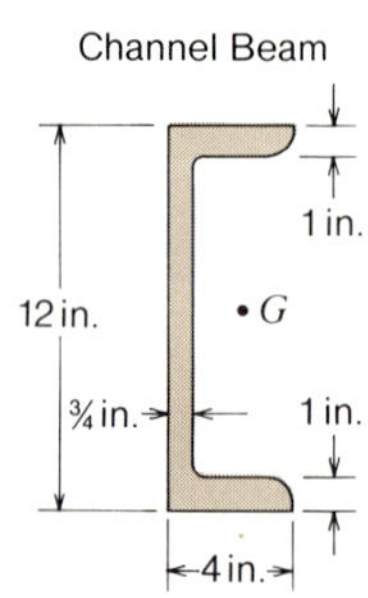

PROB. 8.47

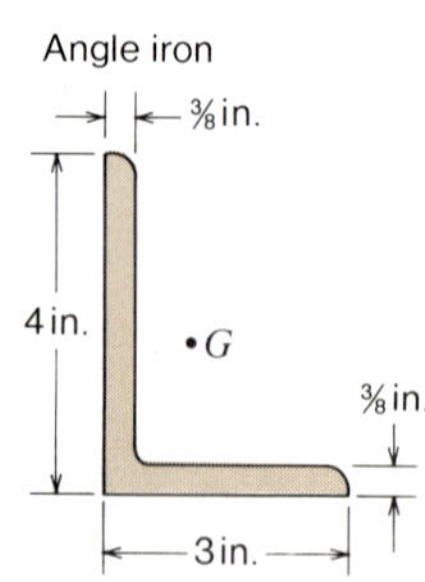

PROB. 8.48

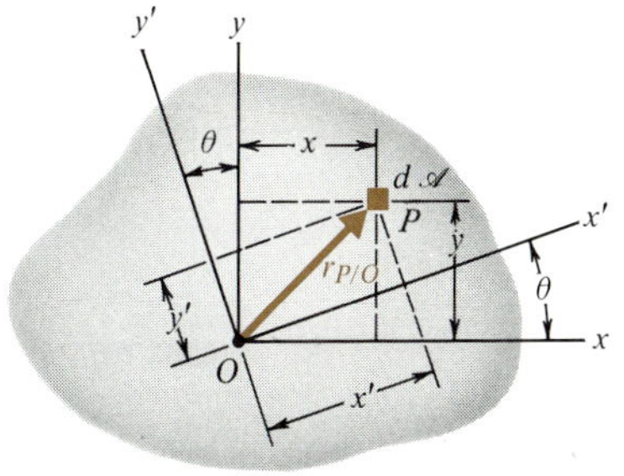

FIGURE 9

E. ROTATION TRANSFORMATION—MOHR'S CIRCLE

The parallel axis theorems enable us to determine the moments and product of inertia corresponding to one coordinate system in terms of the properties for another parallel coordinate system. In the following development we will derive the relationship between the inertia properties for two coordinate systems having identical origins, but different orientations. A typical situation is shown in Figure 9, where the xyz system is the one for which the moments and products of inertia are known.

We need expressions for the (x', y') coordinates in terms of the (x, y) values. These could be obtained by using trigonometry to measure various lengths in Figure 9. An alternative method employs the fact that the position vector $\bar{r}_{P/O}$ may be described in terms of either coordinate system. Thus we write

$$\bar{r}_{P/O} = x'\bar{i}' + y'\bar{j}' = x\bar{i} + y\bar{j}$$

In order to match the two expressions for $\bar{r}_{P/O}$, the unit vectors $\bar{i}$ and $\bar{j}$ will now be expressed in terms of their $x'y'z'$ components.

$$\bar{i} = (\cos\theta)\bar{i}' - (\sin\theta)\bar{j}', \qquad \bar{j} = (\sin\theta)\bar{i}' + (\cos\theta)\bar{j}'$$

Substitution of these expressions into the equations for $\bar{r}_{P/O}$ yields

$$\bar{r}_{P/O} = x'\bar{i}' + y'\bar{j}' = x[(\cos\theta)\bar{i}' - (\sin\theta)\bar{j}'] + y[(\sin\theta)\bar{i}' + (\cos\theta)\bar{j}']$$

Equating corresponding components gives

$$x' = x\cos\theta + y\sin\theta, \qquad y' = -x\sin\theta + y\cos\theta \tag{12}$$

We now employ this transformation in the expression for $I_{x'}$, as follows.

$$\begin{aligned} I_{x'} &\equiv \int_{\mathscr{A}} (y')^2\,d\mathscr{A} = \int_{\mathscr{A}} (-x\sin\theta + y\cos\theta)^2\,d\mathscr{A} \\ &= \cos^2\theta \int_{\mathscr{A}} y^2\,d\mathscr{A} - 2\cos\theta\sin\theta \int_{\mathscr{A}} xy\,d\mathscr{A} + \sin^2\theta \int_{\mathscr{A}} x^2\,d\mathscr{A} \end{aligned}$$

The three integrals on the right side of this expression are I_x, I_{xy}, and I_y, respectively. When similar steps are followed for $I_{y'}$ and $I_{x'y'}$, we obtain the following set of *rotation transformations* for the area moments of inertia.

$$\boxed{\begin{aligned} I_{x'} &= I_x\cos^2\theta + I_y\sin^2\theta - 2\,I_{xy}\sin\theta\cos\theta \\ I_{y'} &= I_x\sin^2\theta + I_y\cos^2\theta + 2\,I_{xy}\sin\theta\cos\theta \\ I_{x'y'} &= I_{xy}(\cos^2\theta - \sin^2\theta) + (I_x - I_y)\sin\theta\cos\theta \end{aligned}} \tag{13}$$

Notice that a relation for the polar moment of inertia is unnecessary because the z' and z axes coincide, so that $J_{z'} \equiv J_z$. This leads to the interesting result that

$$J_z = I_{x'} + I_{y'} \equiv I_x + I_y$$

Other quantities have rotation transformations in the same form as eqs. (13). Three that are of particular importance in the field of mechanics are stress and strain in materials, and the mass moments of inertia for moving bodies.

Equations (13) are easy enough to employ. Nevertheless, by performing a few simple manipulations we may obtain a pictorial representation of the transformation that makes it unnecessary to remember formulas. To do this, we employ the following trigonometric identities.

$$\cos^2\theta = \tfrac{1}{2}(1+\cos 2\theta), \qquad \sin^2\theta = \tfrac{1}{2}(1-\cos 2\theta), \qquad \sin\theta\cos\theta = \tfrac{1}{2}\sin 2\theta$$

Substitution of these identities into eqs. (13) then yields

$$I_{x'} = \tfrac{1}{2}(I_x + I_y) + \tfrac{1}{2}(I_x - I_y)\cos 2\theta - I_{xy}\sin 2\theta \tag{14a}$$

$$I_{y'} = \tfrac{1}{2}(I_x + I_y) - \tfrac{1}{2}(I_x - I_y)\cos 2\theta + I_{xy}\sin 2\theta \tag{14b}$$

$$I_{x'y'} = I_{xy}\cos 2\theta + \tfrac{1}{2}(I_x - I_y)\sin 2\theta \tag{14c}$$

We next transpose the terms $\frac{1}{2}(I_x + I_y)$ from the right side to the left in eqs. (14a) and (14b). When either eq. (14a) or (14b) is squared, and the result is then added to the square of eq. (14c), it is found that

$$\left[I_{p'} - \left(\frac{I_x + I_y}{2}\right)\right]^2 + (I_{x'y'})^2 = \left(\frac{I_x - I_y}{2}\right)^2 + (I_{xy})^2 \tag{15}$$

where the symbol p' represents either x' or y', depending on which equation was squared.

Now recall that we began this discussion by considering the parameters I_x, I_y, and I_{xy} corresponding to the xyz system to be known values. Therefore, eq. (15) can be regarded as a relationship between $I_{x'}$ and $I_{x'y'}$, or alternatively, between $I_{y'}$ and $I_{x'y'}$. In terms of a graph with $I_{p'}$ plotted along the abscissa and $I_{x'y'}$ plotted along the ordinate, eq. (15) represents a circle of radius R that is centered on the abscissa at a distance d from the origin, where

$$R = \left[\left(\frac{I_x - I_y}{2}\right)^2 + (I_{xy})^2\right]^{1/2} \qquad \text{and} \qquad d = \tfrac{1}{2}(I_x + I_y) \tag{16}$$

It is interesting to note that because $J_z \equiv I_x + I_y$, the value of d is one-half the polar moment of inertia.

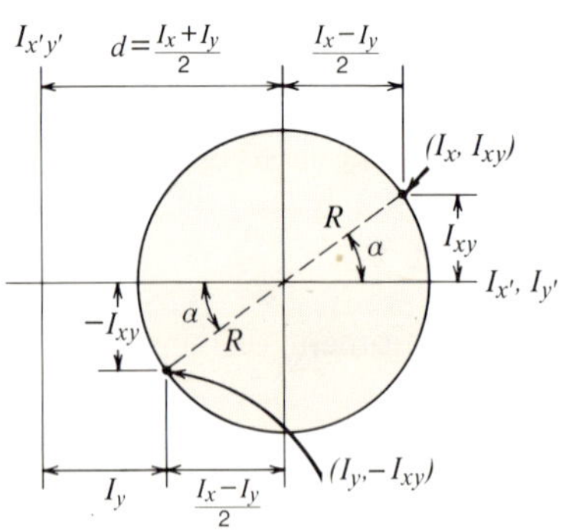

FIGURE 10

The parameters R and d are readily determined from a geometric construction by plotting the points (I_x, I_{xy}) and $(I_y, -I_{xy})$ and then connecting these points by a straight line. This construction and the resulting circle are illustrated in Figure 10.

When $\theta = 0°$, the x and x' axes in Figure 9 are coincident, as are the y and y' axes. Hence, the points we have plotted in Figure 10 may be considered to represent $I_{x'}$, $I_{y'}$, and $I_{x'y'}$ when $\theta = 0$. The question we must now consider is which points on the circle correspond to the values $(I_{x'}, I_{x'y'})$, and $(I_{y'}, I_{x'y'})$. In order to answer this question, we use Figure 10 to derive the following relations for the angle α.

$$R\cos\alpha = \tfrac{1}{2}(I_x - I_y), \qquad R\sin\alpha = I_{xy} \tag{17}$$

These equations and eq. (16) for d are now used to eliminate the terms $I_x + I_y$, $I_x - I_y$, and I_{xy} from eqs. (14). With the aid of the trigonometric identities for the sine and cosine of the sum of two angles, it is found that

$$I_{x'} = d + R(\cos\alpha\cos 2\theta - \sin\alpha\sin 2\theta) \equiv d + R\cos(\alpha + 2\theta) \quad \text{(18a)}$$

$$I_{y'} = d - R(\cos\alpha\cos 2\theta - \sin\alpha\sin 2\theta) \equiv d - R\cos(\alpha + 2\theta) \quad \text{(18b)}$$

$$I_{x'y'} = R(\sin\alpha\cos 2\theta + \cos\alpha\sin 2\theta) \equiv R\sin(\alpha + 2\theta) \quad \text{(18c)}$$

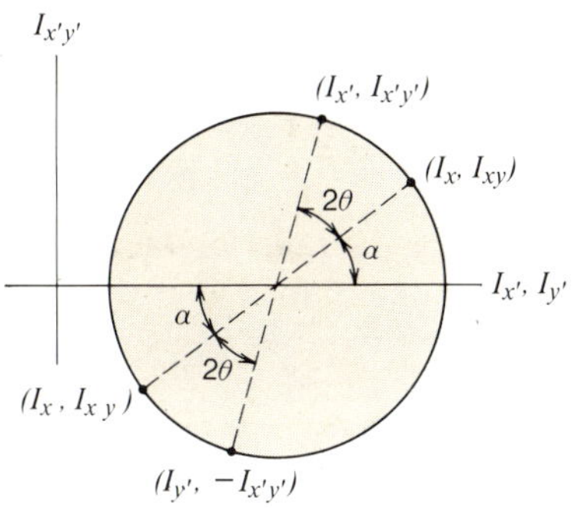

FIGURE 11

As illustrated in Figure 11, these relations show that the point $(I_{x'}, I_{x'y'})$ is obtained by rotating the radial line that intersects the point (I_x, I_{xy}) counterclockwise through an angle 2θ. A similar statement applies to the point $(I_{y'}, -I_{x'y'})$. It is significant that the angle θ is the angle of rotation, measured counterclockwise, from the xyz coordinate system (whose inertia properties are known) to the $x'y'z'$ system. Thus the inertia properties are obtained by rotating the diametral line by twice the angle between the x and x' (or y and y') axes. The sense of the rotation of this line is the same as the sense of the rotation from xyz to $x'y'z'$.

This pictorial method for representing the rotation transformations of eqs. (13) is known as *Mohr's circle* after the engineer Otto Mohr. In summation, the Mohr's circle for area moments of inertia is employed as follows to determine the inertia properties for the $x'y'z'$ coordinate system.

1. Using the abscissa to plot moments of inertia and the ordinate for the product of inertia, plot the two points (I_x, I_{xy}) and $(I_y, -I_{xy})$.
2. Connect the two points plotted in the previous step. The length of this line is the diameter of the Mohr's circle, and the point of intersection of this line with the abscissa is the center of the circle. Sketch the corresponding circle.
3. Draw a diameter in the Mohr's circle that is at an angle 2θ from the line in step 2, in the same sense as the angle of rotation θ from the x axis to the x' axis.
4. The point of intersection of the diameter just drawn with the Mohr's circle, adjacent to the point (I_x, I_{xy}) in the sense of the rotation from the x axis to the x' axis, has the coordinates $(I_{x'}, I_{x'y'})$. The point of intersection adjacent to the point $(I_y, -I_{xy})$ in the sense of the rotation is the point $(I_{y'}, -I_{x'y'})$.

In the construction of Mohr's circle, the points must be plotted consistently with the sign of I_{xy}. Also, there is no restriction that I_x be larger than I_y, even though this was the case in the derivation. Mohr's circle could be employed to obtain a graphical solution. However, we will employ it solely as a pictorial aid that makes it unnecessary to remember eqs. (13).

An important aspect of the rotation transformation is exhibited directly by Mohr's circle. Specifically, referring to Figure 11, notice that the two points where the circle intersects the abscissa correspond to orientations of the $x'y'z'$ system for which $I_{x'y'} = 0$. A coordinate system for which the product of inertia is zero is called a set of *principal axes*, and the corresponding moments of inertia are the *principal* values. Earlier, in the discussion of shapes having an axis of symmetry, we saw that any coordinate system containing this axis is a set of principal axes. We now see that all planar figures have a set of principal axes for any choice of the origin.

The angle of rotation $\tilde{\theta}$ from the xyz axes having known properties to the principal axes is easily determined from Figure 10 or Figure 11. Recalling that the angle of rotation is one-half the angle in Mohr's circle, and considering counterclockwise rotations to be positive rotations, we see that $2\tilde{\theta} = -\alpha$. (The alternative $2\tilde{\theta} = 180° - \alpha$ gives the same coordinate system as the foregoing, except that the coordinate axes are labeled differently.) It is then found from eqs. (17) that

$$\tilde{\theta} = -\tfrac{1}{2}\tan^{-1}\left(\frac{2I_{xy}}{I_x - I_y}\right) \tag{19}$$

In practice, there is no need to remember eq. (19), for the magnitude and sense of the angle of rotation $\tilde{\theta}$ are easily determined after the Mohr's circle has been sketched.

Additional features of the rotation transformation are apparent from the Mohr's circle in Figure 11. We see that the principal moments of inertia are the maximum and minimum values of I_x and I_y that can be obtained in a rotation of the coordinate axes. Also, the axes for which the product of inertia is a maximum are at an angle of 45° ($2\theta = 90°$) from the principal axes.

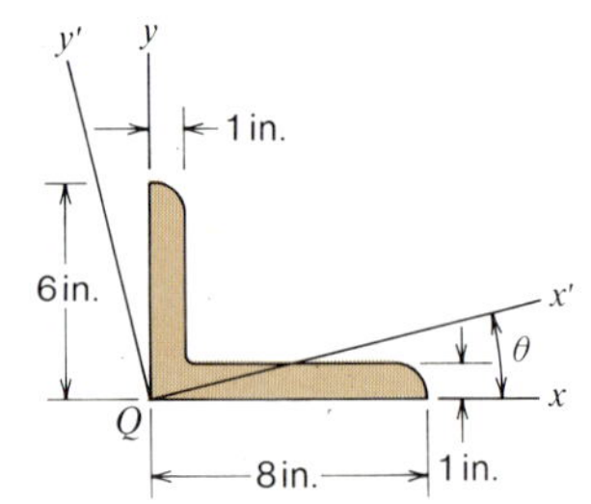

EXAMPLE 5

The moments and product of inertia of the cross section of an angle iron with respect to a given set of xyz axes were found in Example 4 to be $I_x = 74.33$ in.^{4}, $I_y = 172.33$ in.^{4}, $I_{xy} = 24.76$ in.^{4}. Determine (a) the values $I_{x'}$, $I_{y'}$, and $I_{x'y'}$ corresponding to an $x'y'z'$ system rotated about corner Q by an angle $\theta = 40°$, and (b) the orientation relative to the xyz axes of the principal axes whose origin is corner Q; also determine the principal moments of inertia.

Solution

The points $(I_x, I_{xy}) = (74.33, 24.76)$ and $(I_y, -I_{xy}) = (172.33, -24.76)$ are plotted; they are labeled as points P_x, P_y for future reference. The distance d locating the center of the circle, the radius R and the angle α for the diameter P_xP_y are calculated to be

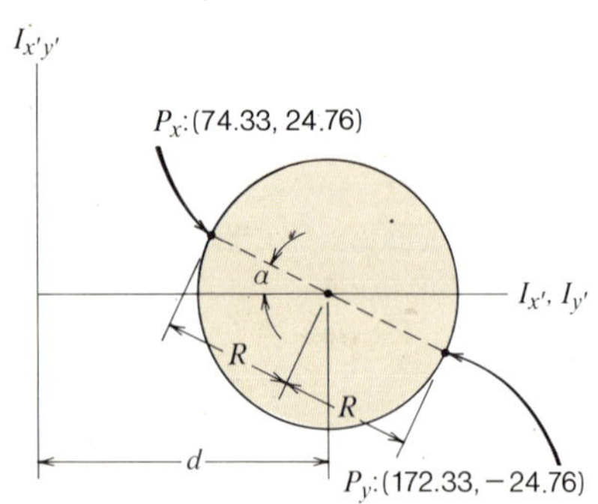

$$d = \tfrac{1}{2}(74.33 + 172.33) = 123.33$$

$$R = \left[\left(\frac{172.33 - 74.33}{2}\right)^2 + (24.76)^2\right]^{1/2} = 54.90$$

$$\alpha = \sin^{-1}\left(\frac{24.76}{54.90}\right) = 26.80°$$

The inertia properties for part (a) correspond to a set of axes that are rotated by $\theta = 40°$ counterclockwise. The corresponding diameter in the

Mohr's circle is rotated counterclockwise from line P_xP_y by $2\theta = 80°$. This diameter is labeled as Q_1Q_2. The principal axes for part (b) correspond to the points S_1 and S_2 at which the Mohr's circle intersects the abscissa. This represents a rotation of the diameter P_xP_y counterclockwise by the angle α. These diameters are illustrated in the second sketch.

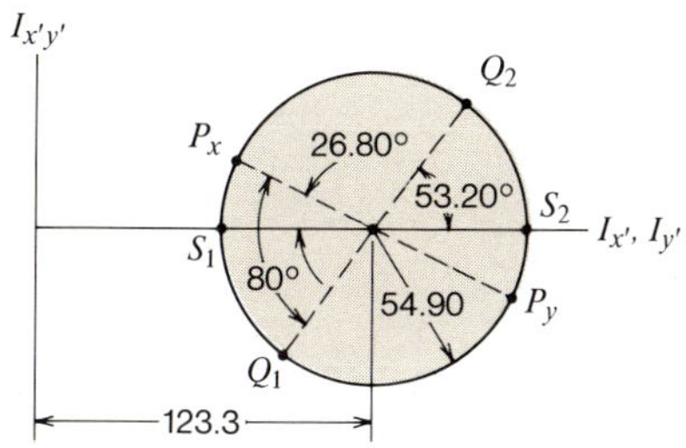

The solution of part (a) is obtained by noting that in a counterclockwise rotation of P_xP_y, point P_x goes to point Q_1, so point Q_1 has the coordinates $(I_{x'}, I_{x'y'})$. Similarly, point P_y goes to point Q_2, so the coordinates of point Q_2 are $(I_{y'}, -I_{x'y'})$. The corresponding distances are computed from trigonometry, in accord with eqs. (10). The result is

$$I_{x'} = 123.33 - 54.90 \cos(53.20°) = 90.44 \text{ in.}^4$$

$$I_{y'} = 123.33 + 54.90 \cos(53.20°) = 156.22 \text{ in.}^4$$

$$I_{x'y'} = -54.90 \sin(53.20°) = -43.96 \text{ in.}^4$$

For part (b), point P_x goes to point S_1 and point P_y goes to point S_2. Hence, points S_1 and S_2 have the coordinates $(I_{\bar{x}}, 0)$ and $(I_{\bar{y}}, 0)$, respectively, where $I_{\bar{x}}$ and $I_{\bar{y}}$ are the principal moments of inertia. Thus, the principal axes are rotated from xyz by

$$\tilde{\theta} = \tfrac{1}{2}(26.80°) = 13.4° \text{ counterclockwise}$$

and the principal moments of inertia are

$$I_{\bar{x}} = 123.33 - 54.90 = 68.43 \text{ in.}^4,$$

$$I_{\bar{y}} = 123.33 + 54.90 = 178.23 \text{ in.}^4$$

The resulting principal axes are shown in the adjacent sketch. Notice that their orientation could not be guessed on the basis of the geometry of the cross section.

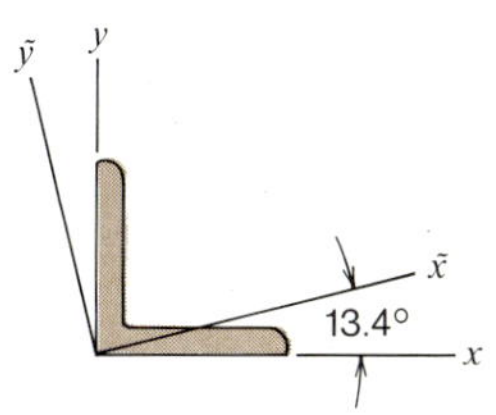

HOMEWORK PROBLEMS

8.49–8.52 Point A is the centroid of the shaded area. Determine the moments of inertia $I_{x'}$ and $I_{y'}$, and the product of inertia $I_{x'y'}$ for $x'y'z'$ axes at $\theta = 30°$. The properties in Appendix C may be used where necessary.

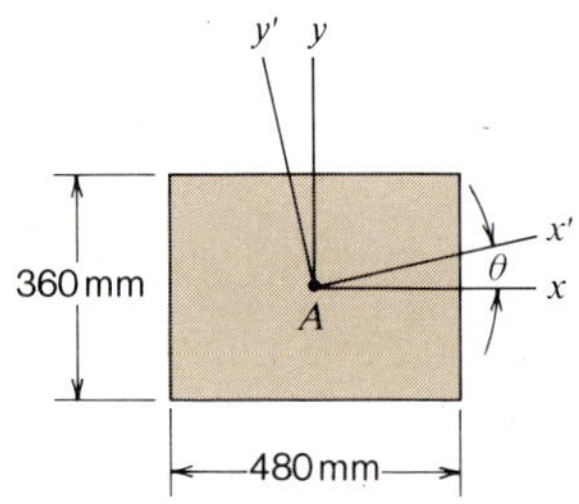

PROBS. 8.49 AND 8.53

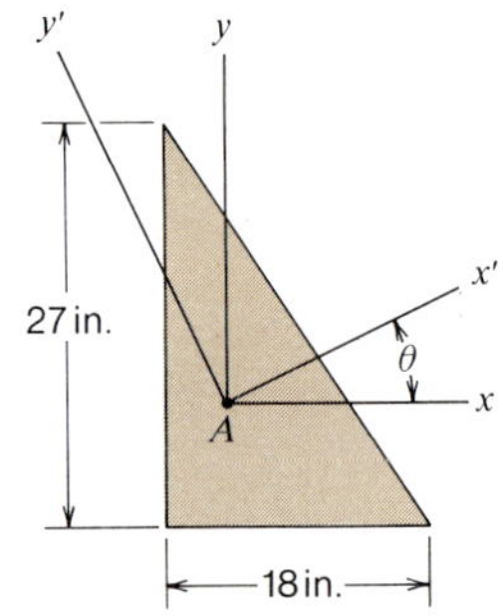

PROBS. 8.50 AND 8.54

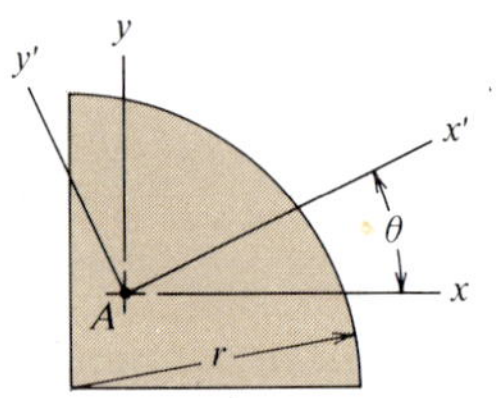

PROBS. 8.51 AND 8.55

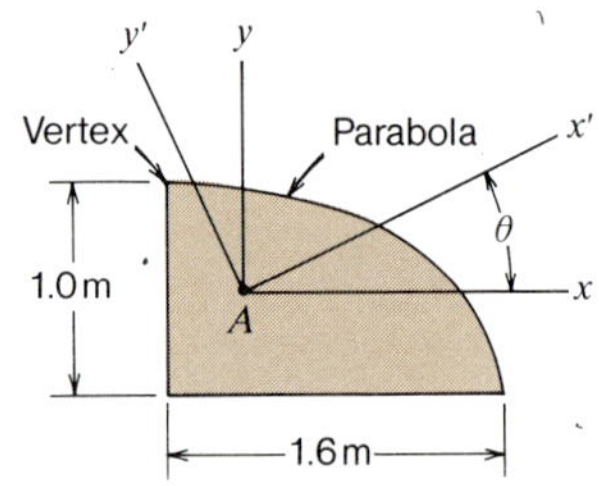

PROBS. 8.52 AND 8.56

8.53–8.56 Point A is the centroid of the shaded area. Determine the moments of inertia $I_{x'}$ and $I_{y'}$ and the product of inertia $I_{x'y'}$ for $x'y'z'$ axes at (a) $\theta = -36.87°$, (b) $\theta = 135°$. The properties in Appendix C may be used where necessary.

8.57–8.59 Determine the moments of inertia $I_{x'}$ and $I_{y'}$, and the product of inertia $I_{x'y'}$ of the shaded area for the coordinate system $x'y'z'$ having origin A. The properties in Appendix C may be used where necessary.

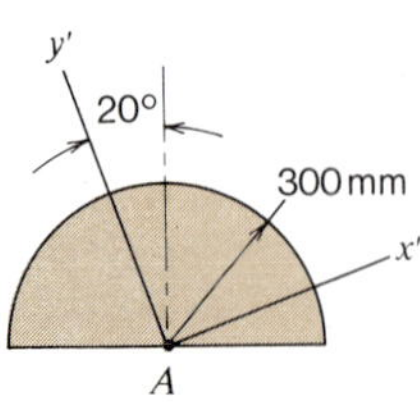

PROB. 8.57

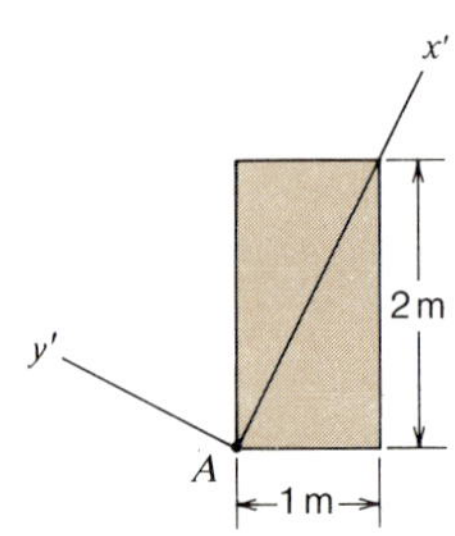

PROB. 8.58

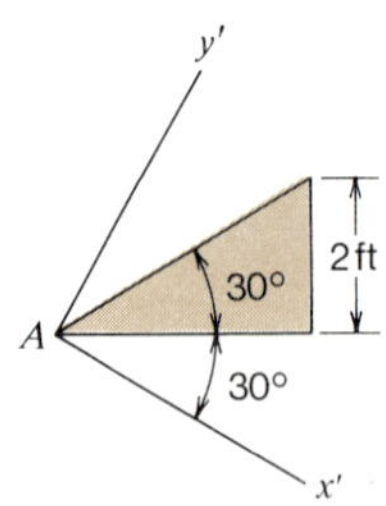

***PROB. 8.59**

8.60–8.64 For the shape in the problem listed below, determine the orientation of the principal axes having origin A, and also determine the corresponding principal moments of inertia.

Homework Problem	8.60	8.61	8.62	8.63*	8.64*
Shape in Problem	8.50	8.51	8.52	8.58	8.59

8.65–8.69 For the shape in the problem listed below, determine the orientation of the axes having origin A for which the product of inertia has its maximum value. Also determine the moments and product of inertia for this coordinate system.

Homework Problem	8.65	8.66	8.67	8.68*	8.69*
Shape in Problem	8.51	8.49	8.52	8.58	8.59

8.70 Letting $\tilde{x}\tilde{y}\tilde{z}$ denote a set of principal axes for an arbitrary area and xyz be another set of axes having the same origin, prove that

$$I_{xy} = \sqrt{I_x I_y - I_{\tilde{x}} I_{\tilde{y}}}$$

CHAPTER 9

VIRTUAL WORK AND ENERGY METHODS

OBJECTIVES

1. To be able to select a generalized coordinate describing the position of the system.

2. To relate the displacement of points in a virtual movement to the increment in the generalized coordinate.

3. To evaluate the work done by a system of forces in a virtual displacement.

4. To use the principal of virtual work to evaluate either the static equilibrium position of a partially constrained system, or a selected reaction in a fully constrained system.

5. To describe the potential energy of a spring or a gravitational force in terms of the generalized coordinate.

6. To evaluate the static equilibrium of a system containing generalized coordinates by using the principle of virtual work and energy.

7. To evaluate the stability of a static equilibrium position by examining the potential energy.

The methods developed thus far for studying the static equilibrium of systems require that we isolate each of the rigid bodies forming the system. The equations for force and moment equilibrium then provide the basis for determining all unknown forces, provided that the system is statically determinate. In such analyses the interaction forces exerted between bodies in a system must be accounted for, even when knowledge of these forces is not of interest.

Unlike those methods, formulations employing the *principle of virtual work* provide a systems viewpoint, in which interconnected rigid bodies are considered as an ensemble. The important aspect of the interconnection between individual bodies will be the motion constraint (restriction) it imposes. The principle of virtual work will make it possible to evaluate the reaction forces supporting the system, without detailed consideration of the system's internal forces.

The starting point for a treatment of this topic is the concept of the work done by a force. For simplicity, we shall limit ourselves to planar problems, although the resulting principles are equally applicable to three-dimensional problems.

A. WORK DONE BY A FORCE

The concept of the *work done* in moving an object is that the force producing the movement is multiplied by the distance the object moves. In order to develop an accurate definition, let us consider the work done by a force $\bar{F}$ applied to a point P when the point follows a fixed path S in space. A typical situation is shown in Figure 1.

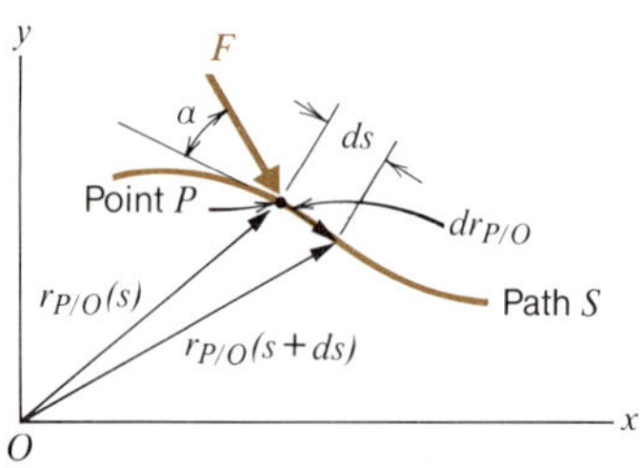

FIGURE 1

When point P moves an infinitesimal arc length ds, the position vector $\bar{r}_{P/O}$ of that point changes by a small amount $d\bar{r}_{P/O}$. Because the arc is infinitesimal, the magnitude of $d\bar{r}_{P/O}$ is ds and the direction of $d\bar{r}_{P/O}$ is tangent to the path, as illustrated in the figure.

The result of this differential displacement is that the force does a differential amount of work dU, which is defined as

$$dU \equiv \bar{F} \cdot d\bar{r}_{P/O} \equiv |\bar{F}|\, ds \cos\alpha \tag{1}$$

where, as shown in Figure 1, α is the angle between $\bar{F}$ and the local tangent.

The work dU given by eq. (1) may be pictured in either of two ways. We can say that dU is the component of force tangent to the path S multiplied by the distance ds traveled along the path, as depicted in Figure 2a. Alternatively, we can say that dU is the component of the displacement vector $d\bar{r}_{P/O}$ parallel to the force $\bar{F}$ multiplied by the magnitude of this force, as shown in Figure 2b.

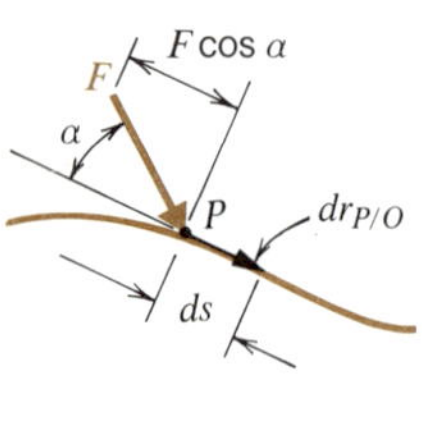

FIGURE 2a

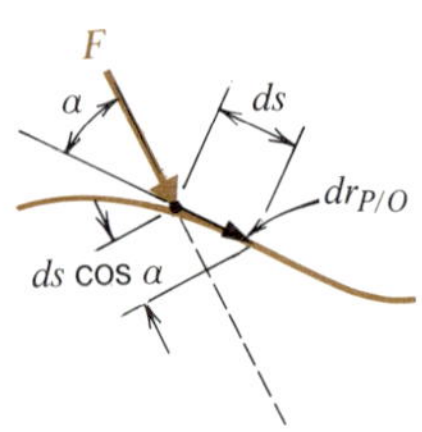

FIGURE 2b

Some special features of the concept of work can be found by studying eq. (1). The value of the work dU may be positive or negative, depending on whether α is less than, or greater than 90°, respectively. The case where dU is negative means that the force opposes the movement, because in that case the component of force parallel to the path is opposite the direction of movement along the path. Also, when the force is perpendicular to the displacement, that is, when $\alpha = 90°$, the work done by the force is zero. Finally, note that we can state that the work done is the force times the distance traveled only when $\alpha = 0$, that is, only when the force is parallel to the displacement.

So far, we have considered only the work done by a single force. Because work is a scalar quantity, the total work done by a system of forces acting on a body may be obtained by algebraically adding the work done by each force. Let dU_i denote the work done by the ith force in a set of N forces; then

$$dU = \sum_{i=1}^{N} dU_i \tag{2}$$

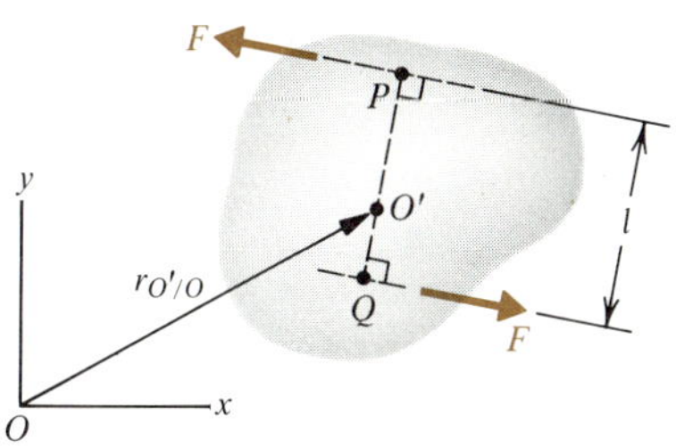

FIGURE 3

Equation (2) expresses the means by which we may determine the infinitesimal work done by a system of forces. In the special case where two forces form a couple, it may be specialized even further. A couple $\bar{F}$ and $-\bar{F}$ acting on an arbitrary rigid body is depicted in Figure 3. The moment of this couple is $M = Fl$ counterclockwise, or vectorially, $\bar{M} = Fl\bar{k}$. We now wish to determine the work done by this couple.

The most general infinitesimal change in the position of a rigid body can be shown to consist of a superposition of an infinitesimal displacement of an arbitrarily chosen point O', such as the point shown in Figure 3, and an infinitesimal rotation about point O'. (Proofs of this statement are contained in Chapters 13 and 15 of the companion text on Dynamics.) In order to simplify the following discussion, we draw a line through point O' that is perpendicular to the lines of action of the forces. Points P and Q, which are intersections of this construction line with the lines of action of the forces, are considered to be the points of application of the forces.

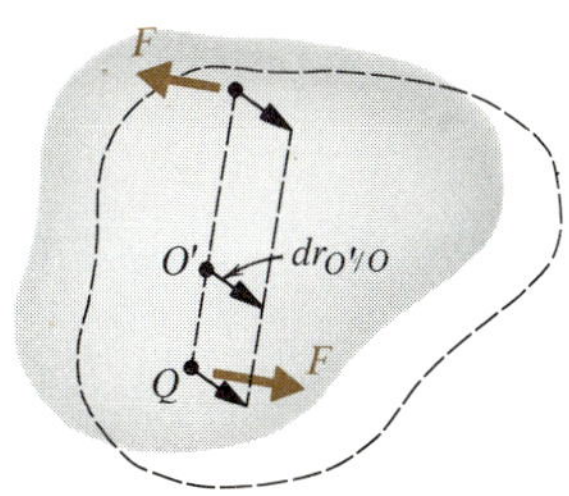

FIGURE 4a

Let us consider the infinitesimal displacement $d\bar{r}_{O'/O}$ of point O' separately from the infinitesimal rotation $d\theta$ about point O'. The first type of movement is illustrated in Figure 4a and the second is shown in Figure 4b. The total work done by the couple will then be the sum of the work done in each type of movement.

For the situation in Figure 4a, which is a translational displacement, points O', P, and Q all undergo the same displacement $d\bar{r}_{O'/O}$. Hence, the work done by the couple is

$$dU_{\text{trans}} = \bar{F} \cdot d\bar{r}_{O'/O} + (-\bar{F}) \cdot d\bar{r}_{O'/O} \equiv 0$$

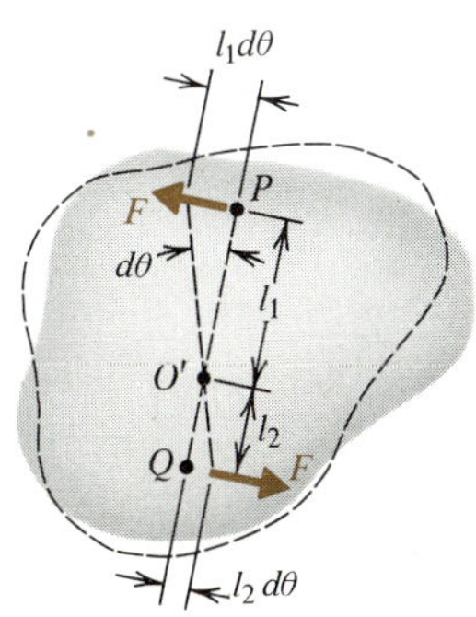

FIGURE 4b

thus demonstrating that a couple does no work when a body translates. This makes sense, because we have seen that a couple does not exert a pushing effect on a rigid body.

In the rotational movement in Figure 4b, point P moves in the direction of $\bar{F}$ and point Q moves in the direction of $-\bar{F}$, so each force does positive work. Also, because the angle of rotation is infinitesimal, each displacement is parallel to the corresponding force, and its magnitude equals the arc length. The total work is the sum of the work done by each force, so

$$dU_{\text{rot}} = F(l_1\, d\theta) + F(l_2\, d\theta) = F(l_1 + l_2)\, d\theta$$

where l_1 and l_2 are the distances from point O' to points P and Q, respectively.

It can be seen from Figure 4b that the lever arm l for the couple is $l = l_1 + l_2$. It then follows that the total work done by the couple in the infinitesimal translation and rotation is

$$dU = dU_{\text{rot}} + dU_{\text{trans}} = Fl\, d\theta$$

$$\boxed{dU = |\bar{M}|\, d\theta} \qquad (3)$$

Note that the rotation $d\theta$ was considered to be in the same sense as the moment of the couple, that is, counterclockwise, so the work done is positive. In situations where the rotation is in the opposite sense from the moment, the work done is negative, indicating that the couple is opposing the rotation. In general, the best way to avoid a sign error is to choose the signs of moments and rotations according to the right-hand rule with respect to the z axis normal to the plane.

An important matter requiring consideration is the units for work. Equation (1) indicates that work has the dimensions of force times length. This is consistent with eq. (3), because the angle of rotation must be measured in radians. In order to avoid confusing moment and work, we generally describe a quantity of work by giving its length unit first and then its force unit, which is the opposite order from that used to describe a moment. Thus, work is expressed in units of ft-lb or in.-lb in U.S. Customary units. The SI system eliminates this ambiguity by defining the unit of work to be a *joule* (J), where 1 J is defined as 1 m-N.

B. VIRTUAL MOVEMENT

Because static systems are at rest, and a force does no work if the point at which it is applied does not move, it is logical to ask how the concept of work can be used to solve statics problems. To answer this question we resort to the concept of a *virtual movement*. The word *virtual* is a classic term; what is really meant is *fictitious* or *imaginary*. In a virtual movement we imagine that the system moves away from its static equilibrium position, and all forces acting on the system are considered to maintain their magnitude and direction. Also, the points at which the forces are applied to the *bodies* in the system are unchanged. For example, in the case of a fully constrained structure, in which a virtual movement can be obtained only by imagining that a physical support is removed, the force corresponding to this support is still considered to act on the system.

In order to indicate that we are not considering a real displacement of the system, we use the Greek symbol δ (delta) instead of d to denote a differential virtual quantity. Thus, the *virtual displacement* of a point P resulting from the virtual movement we give the system is denoted as $\delta\bar{r}_{P/O}$ and the *virtual work* δU of a force $\bar{F}$ applied at this point is

$$\delta U = \bar{F} \cdot \delta\bar{r}_{P/O} \tag{4}$$

We shall now introduce some terminology necessary for a quantified description of the virtual displacements $\delta\bar{r}_{P/O}$ of all points at which forces act. Geometric parameters, such as angles of elevation or position coordinates of points, change when a system is given a virtual movement. The incremental change of some of these parameters may be related directly to the change in other parameters. The minimum number of geometric parameters that must be specified in order to locate the new position of the system is defined as the *number of degrees of freedom.* Those parameters that are specified in order to define the new position are said to be *generalized coordinates.*

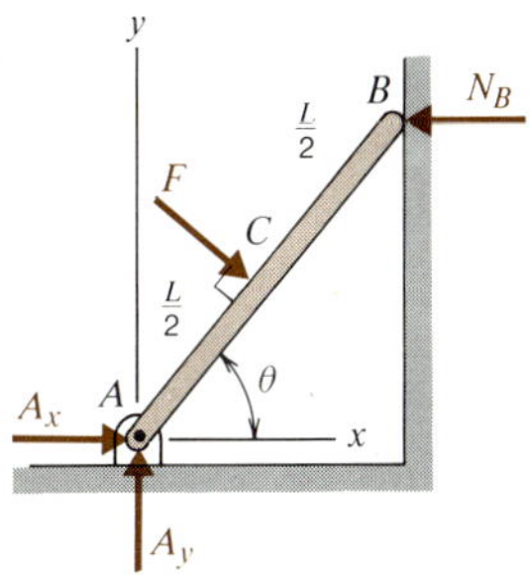

FIGURE 5

A simple system that is useful for illustrating these ideas is the inclined bar in Figure 5, which is loaded by a force $\bar{F}$ perpendicular to the bar. A virtual movement of this bar may be produced by imagining what would happen if the wall at end B were removed, allowing bar AB to rotate. The position of the bar after the virtual movement may be described completely in terms of the change, $\delta\theta$, in the angle of elevation θ. Thus, θ may be used as the generalized coordinate, which means that the virtual movement has one degree of freedom.

Because the virtual movement is a figment of our imagination, we may consider the bar to rotate in either direction. In order to minimize the opportunity for making sign errors, it will be *standard practice to consider the generalized coordinate to increase in magnitude.* Consequentially, the virtual movement depicted in Figure 6 shows a counterclockwise rotation corresponding to an increase of the angle θ by $\delta\theta$.

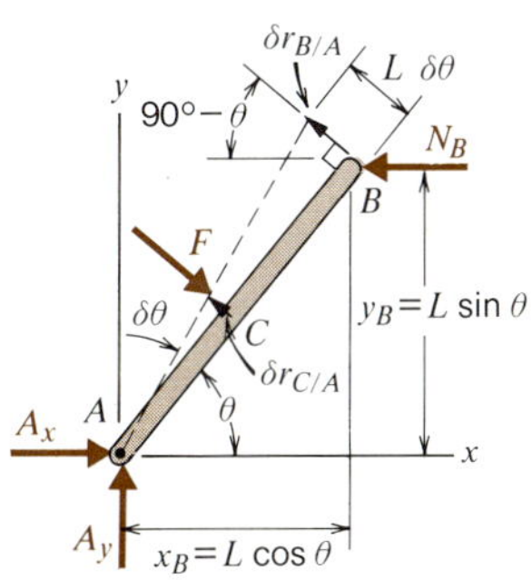

FIGURE 6

As a result of the virtual rotation, the forces acting on the bar do work. The computation of this virtual work requires that the virtual displacement of the point of application of each force be related to the value of $\delta\theta$. There are basically two ways of achieving this. One, called the *geometric approach,* involves applying the laws of geometry and trigonometry to the diagram of the virtual displacement, which is what is shown in Figure 6.

The virtual movement depicted in Figure 6 is particularly suitable for the geometric approach. The movement depicted there is a pure planar rotation. That is, considering the points of the body lying in the plane, one point remains fixed as the other points follow circular arcs centered at the fixed point. We are dealing with an infinitesimal angle of rotation $\delta\theta$, so a point at a radial distance r from this fixed point will displace a distance $r\,\delta\theta$. Furthermore, because of the smallness of $\delta\theta$, the direction of the virtual displacement of a point will be perpendicular to the radial line from the fixed point, in the sense of the rotation.

These general observations enable us to construct directly the required expressions for the virtual displacements. For instance, point B is at a distance L from the fixed point A. Thus, the virtual displacement $\delta\bar{r}_{B/A}$ is

$L\,\delta\theta$ perpendicular to the radial line AB. From the fact that the virtual rotation is counterclockwise, it follows that point B displaces upward and to the left. By referring to Figure 6 for the appropriate direction cosines, we find that

$$\begin{aligned}\delta\bar{r}_{B/A} &= L\,\delta\theta[-\cos(90^\circ - \theta)\bar{i} + \sin(90^\circ - \theta)\bar{j}] \\ &= L\,\delta\theta[-\sin\theta\,\bar{i} + \cos\theta\,\bar{j}]\end{aligned} \tag{5}$$

Clearly, comparable expressions for the virtual displacements of other points may also be obtained by this method.

Although the geometric approach is suitable for situations where a body undergoes a pure planar rotation, it is more difficult to apply when a body undergoes the general type of movement that we discussed earlier. Some of the concepts of motion analysis, developed in a dynamics course, provide useful aids in applying the geometric approach.

An alternative that is suitable for our purposes is called the *analytical method.* In this approach, the virtual displacements of points are obtained by differentiating general expressions for position vectors. The analytical method requires an xyz coordinate system whose origin remains fixed when the system is given the virtual movement.

The origin of the coordinate system in Figure 6 has been placed at pin A, for clearly it is a convenient reference point for discussing a virtual rotation of the bar about pin A. In the analytical method, the position of the point of application of each force doing work must be expressed algebraically in terms of the generalized coordinate. The reactions at pin A in Figure 6 are applied at a fixed point, so they do no work. Therefore, we need to describe only the positions of points B and C.

$$\bar{r}_{B/A} = L\cos\theta\,\bar{i} + L\sin\theta\,\bar{j}, \qquad \bar{r}_{C/A} = \tfrac{1}{2}\bar{r}_{B/A} \tag{6}$$

A differential increment in a generalized coordinate produces an infinitesimal change in the position coordinates. The position change may be evaluated by the chain rule for differentiation. The xyz coordinate axes are fixed, so $\bar{i}$ and $\bar{j}$ are constants. Therefore, the virtual displacements of points B and C are found from eqs. (6) to be

$$\begin{aligned}\delta\bar{r}_{B/A} &= \left(\frac{d}{d\theta}\bar{r}_{B/A}\right)\delta\theta = \left[\frac{d}{d\theta}(L\cos\theta)\bar{i} + \frac{d}{d\theta}(L\sin\theta)\bar{j}\right]\delta\theta \\ &= -L\,\delta\theta\sin\theta\,\bar{i} + L\,\delta\theta\cos\theta\,\bar{j} \\ \delta\bar{r}_{C/A} &= \tfrac{1}{2}\delta\bar{r}_{B/A} = -\tfrac{1}{2}L\,\delta\theta\sin\theta\,\bar{i} + \tfrac{1}{2}L\,\delta\theta\cos\theta\,\bar{j}\end{aligned} \tag{7}$$

The analytical approach is verified by the fact that eqs. (5) and (7) give identical expressions for $\delta\bar{r}_{B/A}$. Note that if it had been desired to describe the virtual displacements when θ has a specific numerical value, this value may be employed in the analytical approach only *after* the algebraic expressions have been obtained.

Once the virtual displacements have been described (by either the geometrical or analytical method), the forces are expressed in terms of their vector components. The virtual work is then obtained from eq. (4). For the case of the virtual movement in Figure 6, this procedure yields

$$\delta U = (-N_B\bar{i}) \cdot \delta\bar{r}_{B/A} + (F \sin\theta\,\bar{i} - F\cos\theta\,\bar{j}) \cdot \delta\bar{r}_{C/A}$$
$$= (-N_B)(-L\,\delta\theta \sin\theta) + (F\sin\theta)\left(-\frac{L}{2}\,\delta\theta\sin\theta\right)$$
$$+ (-F\cos\theta)\left(\frac{L}{2}\,\delta\theta\cos\theta\right)$$
$$= \left[N_B L\sin\theta - \frac{FL}{2}(\sin^2\theta + \cos^2\theta)\right]\delta\theta$$
$$= \left[N_B L\sin\theta - \frac{FL}{2}\right]\delta\theta \tag{8}$$

In the foregoing we discussed one type of virtual movement of the bar in Figure 5, specifically, the virtual rotation produced when we imagine that the support at end B is removed. Another type of virtual movement occurs if we pretend that the pin support at point A is changed to a roller, thus allowing for a virtual displacement of point A horizontally. As shown in Figure 7, this results in a vertical virtual displacement of end B.

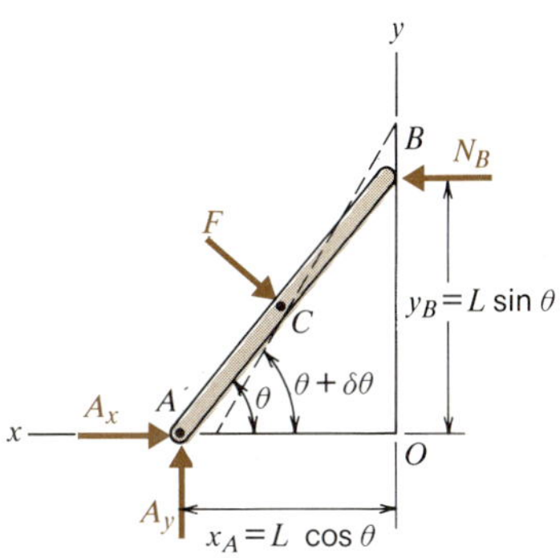

FIGURE 7

The position of the bar after the virtual movement in Figure 7 takes place may be established if the increment $\delta\theta$ in the angle of elevation is stipulated. Thus, θ may be employed once again as the generalized coordinate.

The coordinate system in Figure 7 is different from that used in Figure 6, because we need a set of axes having an origin that is fixed during the virtual movement. However, the steps in computing the virtual displacement analytically follow the prescribed order. First, the position of the points where forces are applied is expressed in terms of the generalized coordinate θ. This gives

$$\begin{aligned}\bar{r}_{B/O} &= y_B\bar{j} = L\sin\theta\,\bar{j}\\ \bar{r}_{C/O} &= \tfrac{1}{2}x_A\bar{i} + \tfrac{1}{2}y_B\bar{j} = \tfrac{1}{2}L\cos\theta\,\bar{i} + \tfrac{1}{2}L\sin\theta\,\bar{j}\\ \bar{r}_{A/O} &= x_A\bar{i} = L\cos\theta\,\bar{i}\end{aligned} \tag{9}$$

Next, the virtual displacements are found by using the chain rule to compute the differential changes. Thus,

$$\begin{aligned}\delta\bar{r}_{B/O} &= \frac{d\bar{r}_{B/O}}{d\theta}\,\delta\theta = L\,\delta\theta\cos\theta\,\bar{j}\\ \delta\bar{r}_{C/O} &= \frac{d\bar{r}_{C/O}}{d\theta}\,\delta\theta = \tfrac{1}{2}L(-\sin\theta\,\bar{i} + \cos\theta\,\bar{j})\,\delta\theta\\ \delta\bar{r}_{A/O} &= \frac{d\bar{r}_{A/O}}{d\theta}\,\delta\theta = -L\,\delta\theta\sin\theta\,\bar{i}\end{aligned} \tag{10}$$

The corresponding virtual work is then found from eq. (4) to be

$$\begin{aligned}\delta U &= (N_B\bar{i})\cdot\delta\bar{r}_{B/O} + (-F\sin\theta\,\bar{i} - F\cos\theta\,\bar{j})\cdot\delta\bar{r}_{C/O} \\ &\quad + (-A_x\bar{i} + A_y\bar{j})\cdot\delta\bar{r}_{A/O} \\ &= 0 + \tfrac{1}{2}FL\,\delta\theta(\sin^2\theta - \cos^2\theta) + A_xL\,\delta\theta\sin\theta \\ &= (-\tfrac{1}{2}FL\cos 2\theta + A_xL\sin\theta)\,\delta\theta \end{aligned} \tag{11}$$

In this case, the fact that $\bar{N}_B$ does no work could have been anticipated without calculation, because $\bar{N}_B$ is perpendicular to the virtual displacement given to point B.

A feature of the concept of virtual movement that sometimes proves troublesome for students is that the choice of the geometric parameter to use as the generalized coordinate is arbitrary. For instance, the distance x_A locating end A of the bar could have been employed as the generalized coordinate for the virtual movement in Figure 7. The position of the other points may be described in terms of x_A with the aid of the Pythagorean theorem, which yields $y_B = (L^2 - x_A^2)^{1/2}$. The position vectors are then

$$\begin{aligned}\bar{r}_{B/O} &= y_B\bar{j} = (L^2 - x_A^2)^{1/2}\bar{j} \\ \bar{r}_{C/O} &= \tfrac{1}{2}x_A\bar{j} + \tfrac{1}{2}y_B\bar{j} = \tfrac{1}{2}x_A\bar{i} + \tfrac{1}{2}(L^2 - x_A^2)^{1/2}\bar{j} \\ \bar{r}_{A/O} &= x_A\bar{i}\end{aligned} \tag{12}$$

The corresponding virtual displacements are

$$\begin{aligned}\delta\bar{r}_{B/O} &= \frac{d\bar{r}_{B/O}}{dx_A}\,\delta x_A = -\frac{x_A\,\delta x_A}{(L^2 - x_A^2)^{1/2}} \\ \delta\bar{r}_{C/O} &= \frac{d\bar{r}_{C/O}}{dx_A}\,\delta x_A = \frac{1}{2}\,\delta x_A\bar{i} - \frac{1}{2}\frac{x_A\,\delta x_A}{(L^2 - x_A^2)^{1/2}}\bar{j} \\ \delta\bar{r}_{A/O} &= \delta x_A\bar{i}\end{aligned} \tag{13}$$

A subtle aspect of the concept of a generalized coordinate is that the force components must also be described in terms of this quantity. For this, we note from Figure 7 that

$$\cos\theta = \frac{x_A}{L}, \qquad \sin\theta = \frac{y_B}{L} = \frac{(L^2 - x_A^2)^{1/2}}{L}$$

The virtual work is then

$$\begin{aligned}\delta U &= N_B\bar{i}\cdot\delta\bar{r}_{B/O} + \left[-F\frac{(L^2 - x_A^2)^{1/2}}{L}\bar{i} - F\frac{x_A}{L}\bar{j}\right]\cdot\delta\bar{r}_{C/O} \\ &\quad + (A_x\bar{i} + A_y\bar{j})\cdot\delta\bar{r}_{A/O} \\ &= \frac{1}{2}\frac{F}{L}\left[-(L^2 - x_A^2)^{1/2} + \frac{x_A^2}{(L^2 - x_A^2)^{1/2}}\right]\delta x_A - A_x\,\delta x_A\end{aligned}$$

It is clear that this expression for δU is different from eq. (11), even though it describes the same type of virtual movement. In general, the choice of the generalized coordinate will affect the resulting expression for the virtual work.

An essential observation regarding the two cases we have considered is that when a virtual movement was produced by imagining that a support is removed, the reaction force associated with the support does virtual work. This is a general result of the fact that the geometric constraint condition imposed by the support is violated in the virtual movement. In contrast, the reaction forces corresponding to supports that are unmodified during a virtual movement do no work.

The two types of virtual movement discussed above were produced by removing one support. After removal of the support, the system had one degree of freedom, so the virtual movement could be described in terms of a single generalized coordinate. It follows that if we consider removing more than one support at a time, more than one generalized coordinate would be required. This added complication is unnecessary in statics, so it will not be pursued further.

The bar in Figure 5 is a fully constrained system, because its supports fully prevent movement. A primary goal of a statics analysis of such systems is the determination of the reaction forces. There are systems that are only partially constrained, that is, that move unless some unknown force (or forces) has the necessary value for static equilibrium. For our purposes, such systems may be analyzed from the same viewpoint as that used for fully constrained systems. The generalized coordinate is then the geometric parameter required to describe the virtual movement of the system consistent with the existing supports. The unknown external force required to prevent the system from moving, in effect, is then an additional constraint force.

We have seen how the concept of a virtual movement leads to the virtual work of the forces acting on the system. In the following sections we shall develop the principles that enable us to use this concept to solve statics problems. However, because the determination of virtual displacements is the key to evaluating the virtual work, the following examples and homework problems are provided to gain practice in this technique.

EXAMPLE 1

A block of mass m is suspended from peg B on the drum. The spring, whose stiffness is k, is unstretched when $\theta = 0$. Compute the virtual work done by the weight of the block and the spring in a virtual rotation of the drum for an arbitrary value of θ.

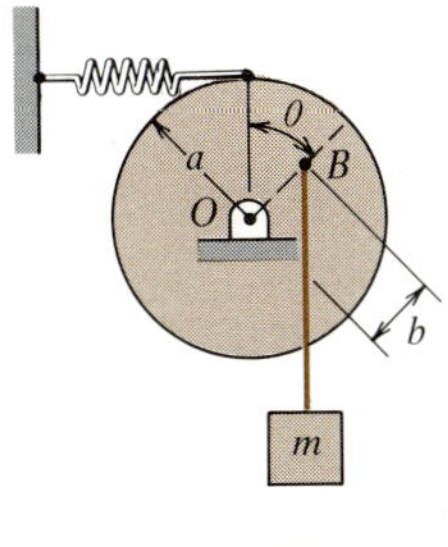

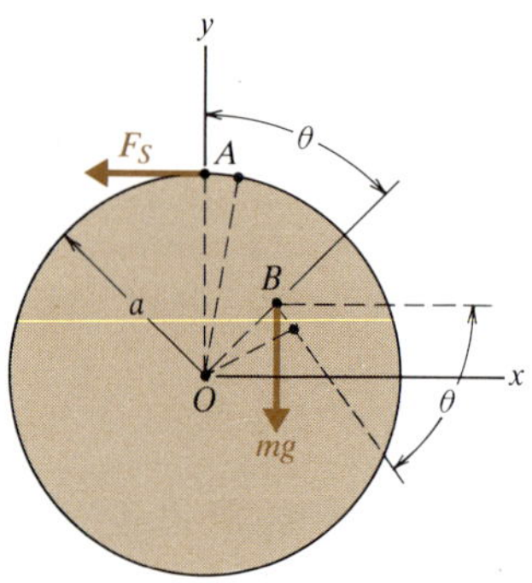

Solution

A useful initial step is to sketch the displacement of the points of application of the forces resulting from the virtual displacement. The virtual movement consists of a pure rotation about point O, in which each point moves in an infinitesimal circular arc. The geometric approach is particularly suitable to the description of pure rotations, because the displacement of a point is merely the radial distance to the center of rotation multiplied by the infinitesimal angle of rotation. The direction of each displacement is perpendicular to the radial line. These displacements are depicted in the sketch, from which it is found that

$$\delta\bar{r}_{A/O} = a\,\delta\theta\,\bar{i}, \qquad \delta\bar{r}_{B/O} = b\,\delta\theta(\cos\theta\,\bar{i} - \sin\theta\,\bar{j})$$

Because the spring is unstretched when $\theta = 0$, the arc length $a\theta$ is the elongation, so

$$\bar{F}_s = k(a\theta)(-\bar{i})$$

The virtual work is therefore

$$\begin{aligned}\delta U &= \bar{F}_s \cdot \delta\bar{r}_{A/O} + (-mg\bar{j}) \cdot \delta\bar{r}_{B/O} \\ &= -ka^2\,\delta\theta + mgb\sin\theta\,\delta\theta\end{aligned}$$

EXAMPLE 2

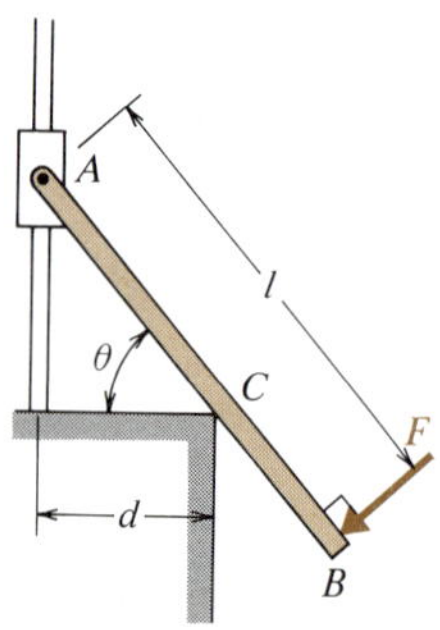

The force F acts perpendicular to bar AB. Determine the virtual work done by $\bar{F}$ in a virtual movement of the bar that does not violate the constraints of collar A and corner C.

Solution

Collar A prevents movement perpendicular to the guide, and corner C prevents the point of contact from moving downward, perpendicular to the axis of the bar. The angle θ is a convenient generalized coordinate for the bar. Thus, the generalized movement depicted in the sketch shows the result of increasing this angle by an amount $\delta\theta$, consistent with the constraints on the movement.

The movement of the bar is not a simple rotation, so the analytical method is employed. Point D is a convenient fixed point, so it is selected as the origin. The virtual work done by $\bar{F}$ requires an expression for the virtual displacement of point B, where the force is applied. This is obtained by describing the position of point B in terms of the generalized coordinate, which is found from the sketch to be

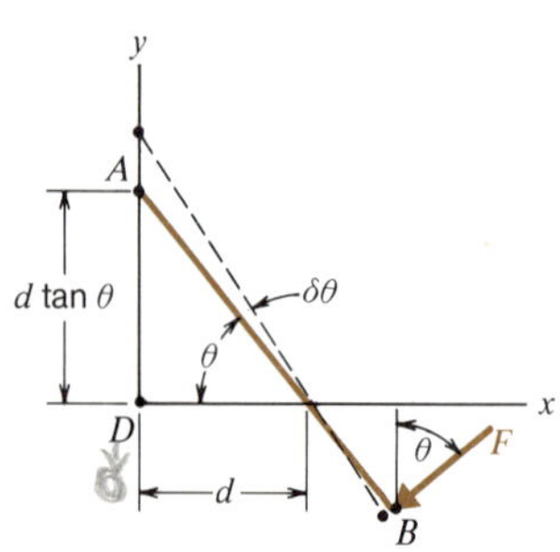

$$\bar{r}_{B/O} = l\cos\theta\,i + (d\tan\theta - l\sin\theta)\bar{j}$$

The virtual displacement is then

$$\delta\bar{r}_{B/O} = \frac{d\bar{r}_{B/O}}{d\theta}\,\delta\theta = \left[-l\sin\theta\,\bar{i} + \left(\frac{d}{\cos^2\theta} - l\cos\theta\right)\bar{j}\right]\delta\theta$$

The force components are evaluated from the sketch,

$$\bar{F} = -F \sin \theta \bar{i} - F \cos \theta \bar{j}$$

and the corresponding virtual work is found to be

$$\delta U = \bar{F} \cdot \delta \bar{r}_{B/O} = (-F \sin \theta \bar{i} - F \cos \theta \bar{j}) \cdot \left[-l \sin \theta \bar{i} + \left(\frac{d}{\cos^2 \theta} - l \cos \theta \right) \bar{j} \right] \delta\theta$$

$$= \left[Fl \sin^2 \theta - \frac{Fd}{\cos \theta} + Fl \cos^2 \theta \right] \delta\theta$$

$$= F\left(l - \frac{d}{\cos \theta} \right) \delta\theta$$

EXAMPLE 3

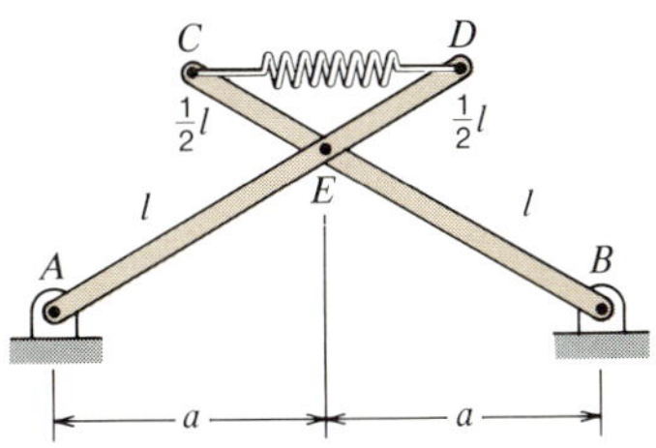

The spring has a stiffness k and an unstretched length $\frac{3}{4}a$. The frame is given a virtual movement that violates only the constraint of pin support B against horizontal movement. Determine the virtual work done by the spring in terms of the distance a.

Solution

As shown in the sketch, the prescribed virtual movement results in point A remaining fixed, and the points E on each bar move by identical amounts. It is desired to evaluate the virtual work in terms of the parameter a. Thus, a is chosen as the generalized coordinate for the system.

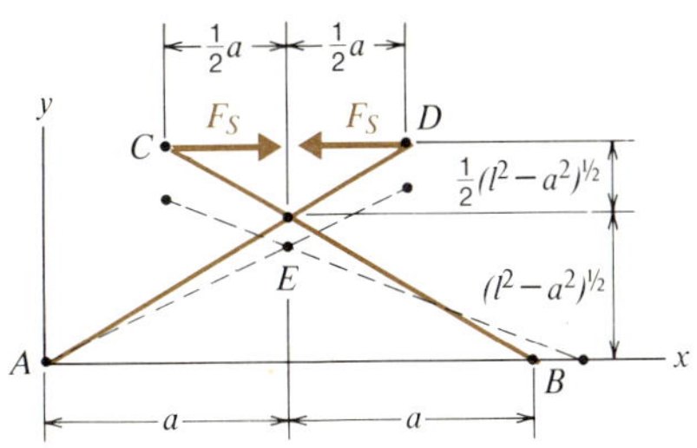

The analytical method is selected for the description of the virtual displacements, because bar CB executes a general type of movement. Pin A is a convenient choice for the fixed origin. The dimensions shown in the sketch, which are obtained by the Pythagorean theorem and the law of similar triangles, yield the following expressions for the positions of points C and D at which the spring force acts.

$$\bar{r}_{C/A} = \tfrac{1}{2}a\bar{i} + \tfrac{3}{2}(l^2 - a^2)^{1/2}\bar{j}, \qquad \bar{r}_{D/A} = \tfrac{3}{2}a\bar{i} + \tfrac{3}{2}(l^2 - a^2)^{1/2}\bar{j},$$

The corresponding virtual displacements are

$$\delta\bar{r}_{C/A} = \frac{d\bar{r}_{C/A}}{da}\,\delta a = [\tfrac{1}{2}\bar{i} - \tfrac{3}{2}a(l^2 - a^2)^{-1/2}\bar{j}]\,\delta a$$

$$\delta\bar{r}_{D/A} = \frac{d\bar{r}_{D/A}}{da}\,\delta a = [\tfrac{3}{2}\bar{i} - \tfrac{3}{2}a(l^2 - a^2)^{-1/2}\bar{j}]\,\delta a$$

The elongation of the spring is the distance a between points C and D minus the unstretched length $\frac{3}{4}a$. Thus, $F_s = k(\frac{1}{4}a)$. The virtual work done by the spring is then

$$\delta U = F_s\bar{i} \cdot \delta\bar{r}_{C/A} + (-F_s\bar{i}) \cdot \delta\bar{r}_{D/A}$$
$$= (\tfrac{1}{4}ka)(\tfrac{1}{2}\,\delta a) + (-\tfrac{1}{4}ka)(\tfrac{3}{2}\,\delta a) = -\tfrac{1}{4}ka\,\delta a$$

HOMEWORK PROBLEMS

9.1 and 9.2 The spring that is wrapped around the outer perimeter of the drum is unstressed when $\theta = 0$. Use the geometric method to determine the virtual work done by the spring and the weight of the block in a virtual rotation of the drum about pivot O for an arbitrary value of θ.

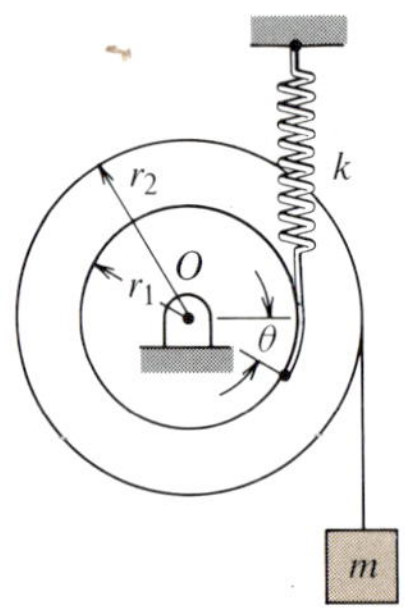

PROB. 9.1

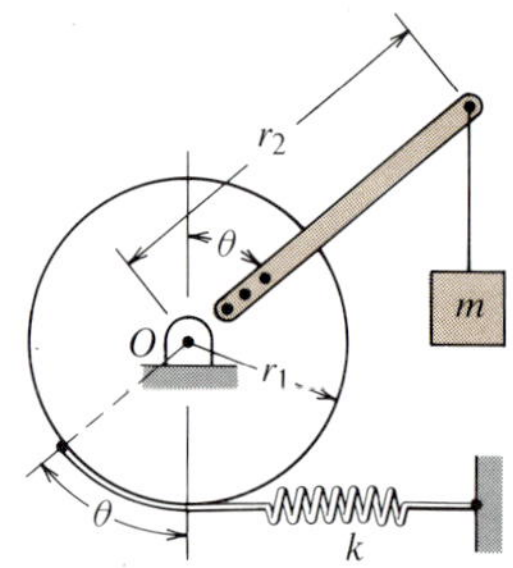

PROB. 9.2

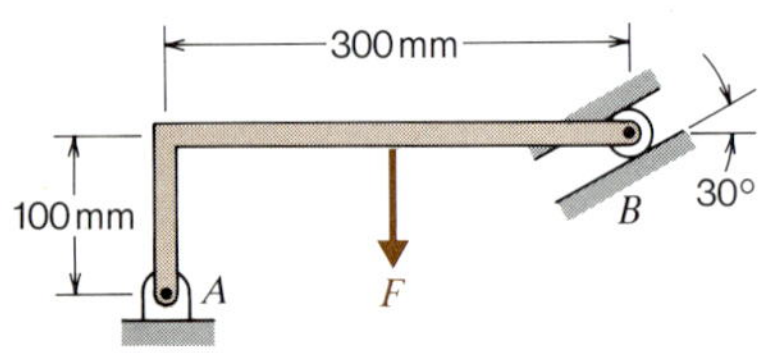

PROB. 9.3

9.3 Determine the work done by the force $\bar{F}$ and the reaction at end B of the curved bar in a virtual movement that violates the constraint of roller B.

9.4 Bar AB is pinned to collar C, which is guided by the vertical rod. Determine the work done by the force $\bar{F}$ at end B in a virtual movement that does not violate the constraints of the collar and the horizontal surface.

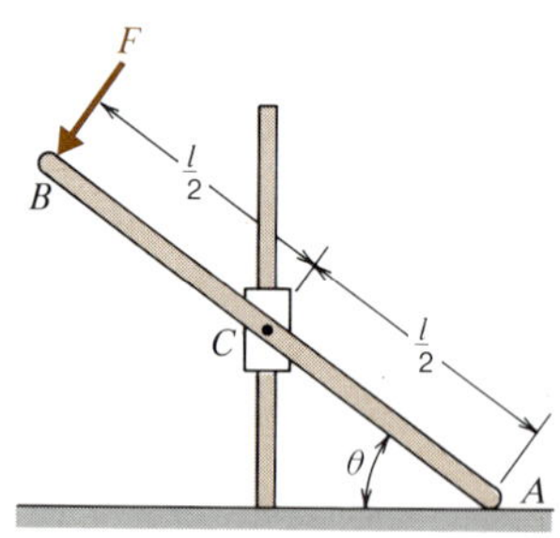

PROB. 9.4

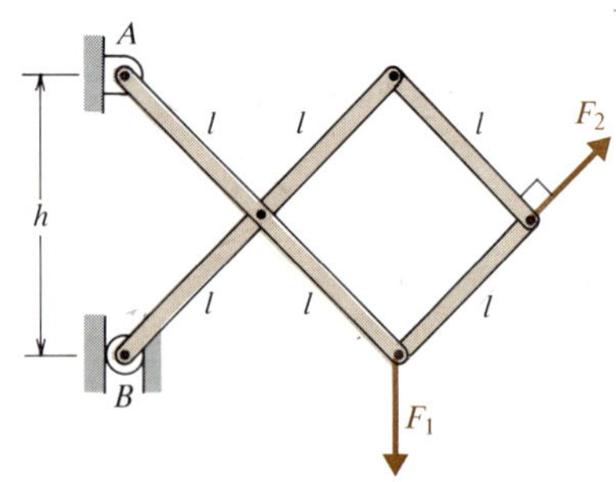

PROBS. 9.5 AND 9.6

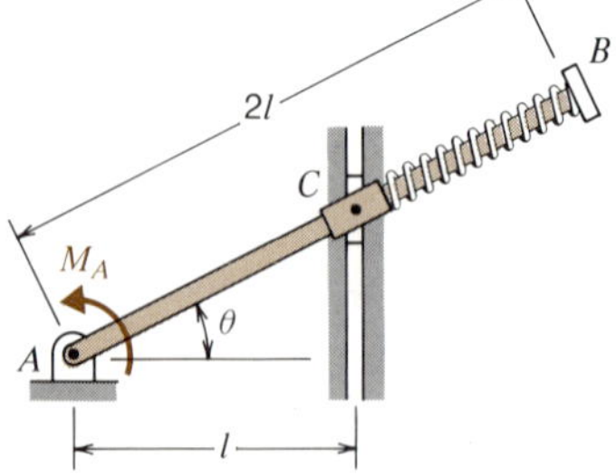

PROB. 9.7

9.5 Determine the work done by the force $\bar{F}_1$ in a virtual movement that does not violate the constraints of pin A and roller B.

9.6 Determine the work done by the force $\bar{F}_2$ in a virtual movement that does not violate the constraints of pin A and collar B.

9.7 Collar C, which rides on bar AB, is pinned to a block that is guided by the straight groove. The spring, whose stiffness is k, is unstressed when $\theta = 0$. Determine the virtual work done by the spring in a virtual rotation of bar AB about pin A.

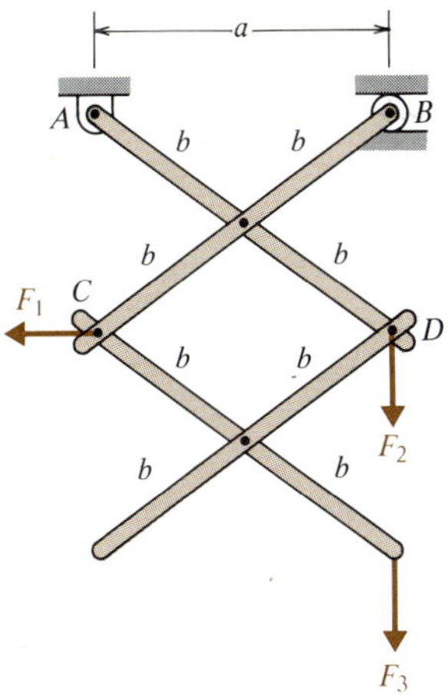

PROB. 9.8

9.8 Determine the virtual work done by the loads $\bar{F}_1$, $\bar{F}_2$, and $\bar{F}_3$ as a result of a virtual movement in which all pin joints remain intact as roller B is displaced horizontally.

***9.9** Determine the work done by the force $\bar{F}$ at end B of the bar in a virtual movement that does not violate the constraint of collar A or the cylindrical surface.

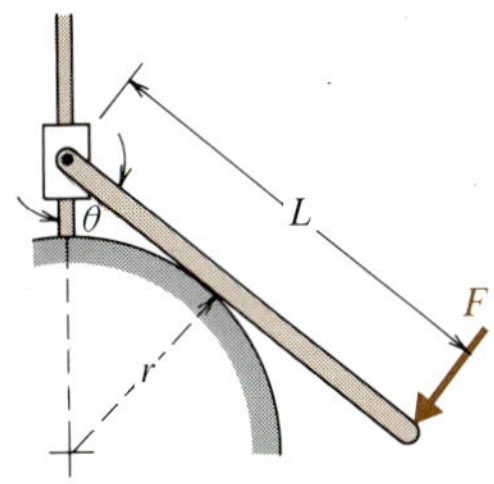

PROB. 9.9

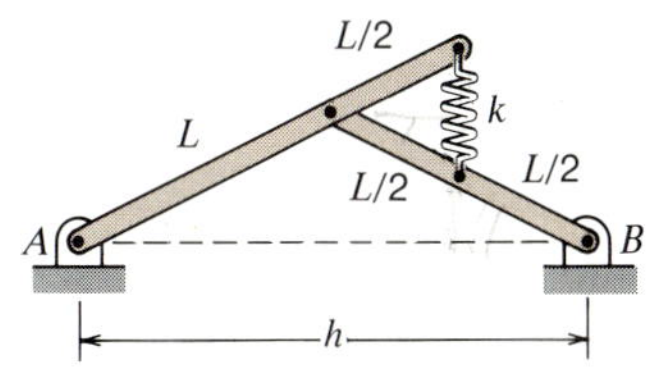

PROB. 9.10

***9.10** The spring has stiffness k and unstretched length $L/2$. Determine the work done by the spring in a virtual movement that violates the constraint of pin B against horizontal movement in the case where $h = 1.6L$.

C. PRINCIPLE OF VIRTUAL WORK

In the preceding section we saw that it is possible to choose a virtual movement such that only selected unknown forces acting on the system do work. Therefore, if we could determine what the total amount of virtual work should equal, we would be able to form a relationship for the unknown forces doing work in the virtual movement. In order to achieve this goal, consider the body shown in Figure 8a, which is in equilibrium under the loading of an arbitrary planar system of forces.

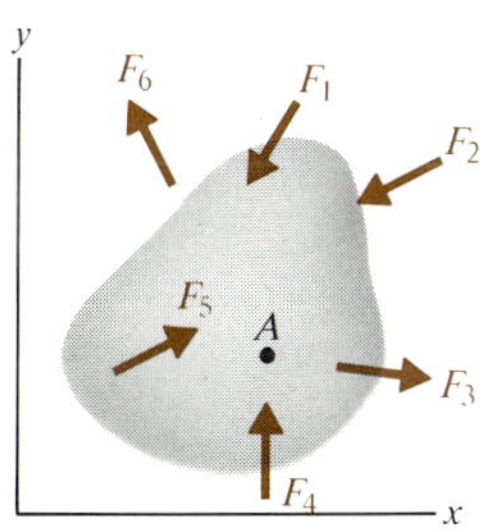

FIGURE 8a

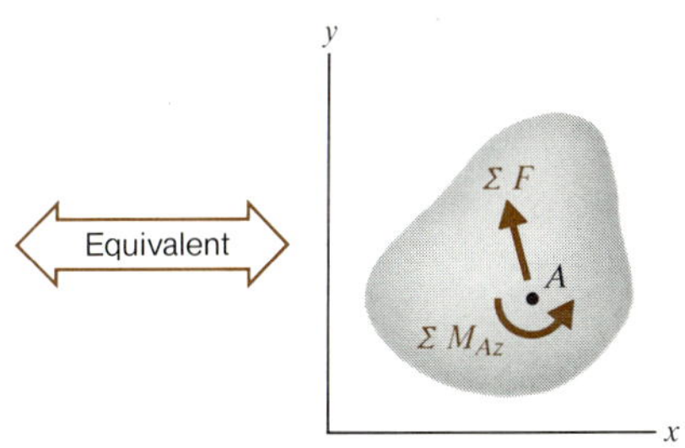

FIGURE 8b

We learned in Chapter 3 that the system of forces in Figure 8a is equivalent to the force-couple system acting at an arbitrary point A shown in Figure 8b. In this figure $\sum \bar{F}$ denotes the sum of all forces acting on the body, and $\sum M_{Az}$ denotes the sum of the moments of all forces about point A.

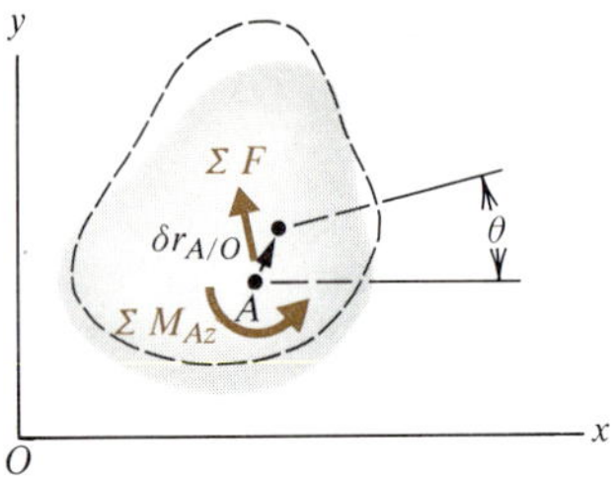

FIGURE 9

Recall that the most general infinitesimal movement of a body is a superposition of a translation following a point plus a rotation about that point. When point A of the body in Figure 8 is selected as the point, a general virtual movement is as shown in Figure 9. In the translation the couple does no work, and in the rotation the force does no work. Hence, the total virtual work done by the forces acting on the body is

$$\delta U = \sum \bar{F} \cdot \delta \bar{r}_{A/O} + \sum M_{Az} \delta\theta \tag{14}$$

This is a general expression for the virtual work done by an arbitrary set of forces. However, we have specified that the body is in equilibrium. The equilibrium conditions require that $\sum \bar{F} = \bar{0}$ and $\sum M_{Az} = 0$. It then follows from eq. (14) that

$$\boxed{\delta U = 0} \tag{15}$$

This simple equation is the *principle of virtual work*. As derived, it applies only to an isolated rigid body. However, the principle is also valid for systems of interconnected rigid bodies, because the total virtual work done by the forces acting on the system is the sum of the virtual work done in moving each body in the system, which is zero. Hence, the principle of virtual work may be stated as

> The virtual work δU done by all forces acting on any system in static equilibrium is zero.

The key feature of this principle is related to the fact that the type of virtual movement given to the system is at our discretion. A movement that violates only one constraint condition results in the appearance of only the associated constraint force (out of the full set of unknown reactions), in the virtual work δU. Setting $\delta U = 0$ then yields a single scalar equation for the unknown force.

For example, consider the bar of Figure 5, which is shown again in Figure 10. Recall that $\bar{F}$ is a given loading. When we wish to find $\bar{N}_B$ we imagine that the vertical wall is removed, leaving the pin at point A intact. This leads to a virtual rotation about point A, in which the reactions at end A do no work. The resulting virtual work was found in eq. (8). Setting that expression for $\delta U = 0$ yields

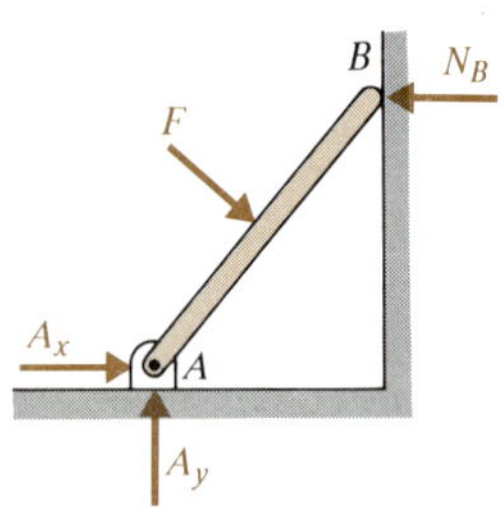

$$\delta U = \left(N_B L \sin\theta - \frac{FL}{2} \right) \delta\theta = 0$$

FIGURE 10

Because $\delta\theta$ is an arbitrary nonzero quantity, it must be true that

$$N_B = \frac{F}{2 \sin\theta}$$

Should we wish to determine the reaction $\bar{A}_x$, we would imagine that the constraint of pin A against horizontal movement is removed. We would then maintain the constraint of pin A against vertical movement, as well as thc constraint against horizontal motion of end B imposed by the wall. In the resulting virtual movement forces $\bar{A}_y$ and $\bar{N}_B$ do no work. The virtual work resulting from this type of virtual movement was found in eq. (11). Setting that expression for $\delta U = 0$ yields

$$\delta U = (-\tfrac{1}{2}FL\cos 2\theta + A_x L\sin\theta)\,\delta\theta = 0$$

$$A_x = \frac{F\cos 2\theta}{2\sin\theta}$$

Clearly, the preceding requires a large effort to solve some simple problems. The values of N_B and A_x for the bar could have been determined from the equations of static equilibrium. In fact, summing moments about pin A would yield the value of N_B in a single equation, and a moment sum about the point of intersection of the horizontal line through end B and the vertical line through pin A would yield the value of A_x in a single equation. Nevertheless, even for this simple system, the principle of virtual work gives us a new viewpoint of the influence of the forces acting on the system.

The significant area of application of the principle of virtual work is for systems consisting of interconnected rigid bodies. To demonstrate this, consider the situation depicted in Figure 11a, where two arbitrary bodies are connected by a frictionless pin at point A.

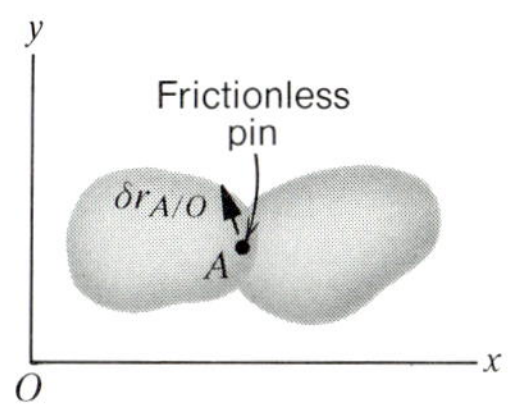

FIGURE 11a

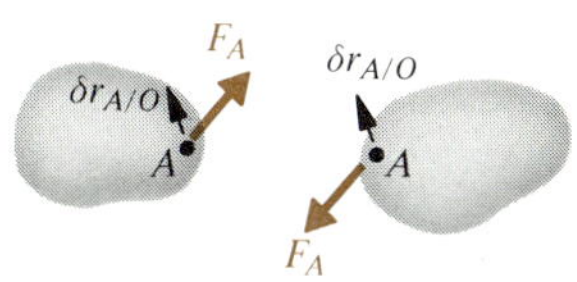

FIGURE 11b

According to Newton's third law, the interaction forces exerted between the two bodies are $\bar{F}_A$ and $-\bar{F}_A$, as shown in Figure 11b. Thus, in a virtual movement of the two bodies in which the bodies remain connected, so that the displacement of their common point of connection is the vector $\delta\bar{r}_{A/O}$, the virtual work of $\bar{F}_A$ and $-\bar{F}_A$ is

$$\delta U_{\text{pin}} = \bar{F}_A \cdot \delta\bar{r}_{A/O} + (-\bar{F}_A)\cdot\delta\bar{r}_{A/O} \equiv 0$$

Thus, frictionless pins do no work in a virtual movement in which the connected bodies remain joined.

The reactions associated with most types of connections between bodies are determined on the basis that friction is negligible. With this assumption, we create *ideal* models of the system, for which the internal reaction forces are constraint forces. For example, the reactions acting at pin A on each body in Figure 11b are the constraint forces required to ensure that the displacements of point A on each body are identical. In general, constraint forces do no work in virtual movements that are consistent with their constraint. This permits us to conclude that

> The internal reactions in an ideal multibody system do no work in a virtual movement that is consistent with the interconnections between the bodies.

Further insight into this statement, as well as an understanding of the effects of friction, may be gained by considering the reaction forces corresponding to the collar in Figure 12a. Let us assume that friction is significant in treating the tendency of the collar to slide over bar 1. The friction and normal components of the reaction acting on each bar at this connection are readily described in terms of coordinate axes for which y is tangent and x is perpendicular to the bar over which the collar may slide, as shown in Figure 12b.

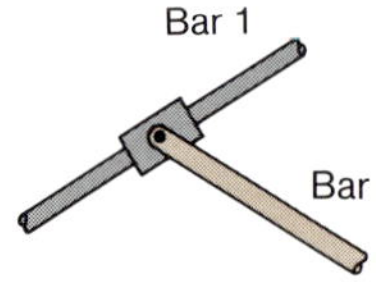

FIGURE 12a

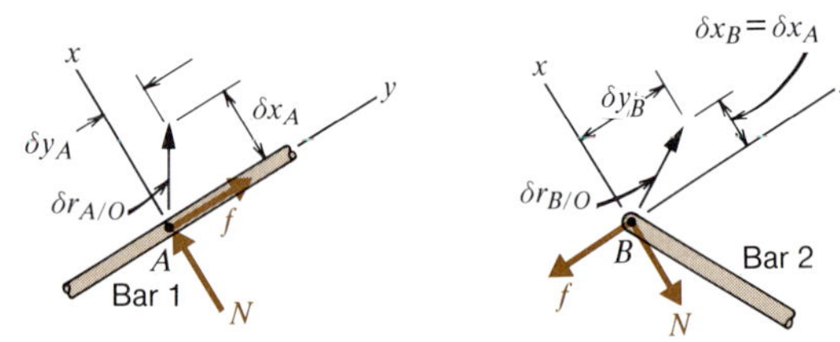

FIGURE 12b

The virtual displacement of points A and B, which are the points of application on each bar of the friction and normal forces, are considered individually. The sliding motion that is allowed by the collar means that points A and B may move by different amounts δy_A and δy_B in the y direction. The constraint condition associated with the collar is that points A and B must displace by identical amounts $\delta x_A \equiv \delta x_B$ in the x direction. From this constraint condition, the total virtual work for the situation in Figure 12b is readily computed to be

$$\begin{aligned} \delta U_{\text{collar}} &= (N\bar{i} + f\bar{j}) \cdot \delta\bar{r}_{A/O} + (-N\bar{i} - f\bar{j}) \cdot \delta\bar{r}_{B/O} \\ &= (N\bar{i} + f\bar{j}) \cdot (\delta x_A \bar{i} + \delta y_A \bar{j}) + (-N\bar{i} - f\bar{j}) \cdot (\delta x_A \bar{i} + \delta y_B \bar{j}) \\ &= f(\delta y_A - \delta y_B) \end{aligned} \tag{16}$$

It will be noted that the normal reaction $\bar{N}$ does no work. This is a consequence of the fact that it enforces the constraint that points A and B execute the same displacement in the x direction. The friction reaction f is not a constraint force, because the collar does not restrict the movements of points A and B in the y direction. This is seen in eq. (16) by the fact that $\delta y_A - \delta y_B$, which is the movement in the y direction of point A relative to point B, does not vanish. From eq. (16), it is also clear that an idealized model of the collar, in which $f \equiv 0$, leads to no virtual work being done by the internal reaction forces. This is in agreement with the earlier general statement regarding internal reactions in ideal systems.

We may conclude from the foregoing observations that the principle of virtual work, eq. (15), is suitable for studying ideal systems. The internal reactions in ideal systems do not hinder a virtual movement that leaves intact the connections between bodies. Because these reactions do not appear in the virtual work, eq. (15) will yield a single equation for the unknown force required for static equilibrium.

The arguments for an ideal system are contrasted by the situation for a nonideal system, in which unknown internal friction forces do virtual work. In the nonideal case, eq. (15) does not lead to a single relation for a single unknown. Problems involving nonideal systems are solved more easily by returning to the static equilibrium methods that were developed in the earlier chapters.

Other types of internal forces also do work. Of particular importance in this regard is a spring, which does work whenever the virtual movement has the effect of changing the length of the spring. This does not represent an obstacle to the application of the principle of virtual work, because a spring force is known when the elongation is known.

D. PROBLEM SOLVING

The preceding developments may be synthesized into a formal approach for solving problems by means of the principle of virtual work.

1. Draw a free body diagram of the entire system in the position of interest as an aid in accounting for all forces. General reaction forces, such as those exerted by pin supports, may be broken up into any convenient set of components, for instance horizontal and vertical.
2. Draw another sketch showing the position of the system before any virtual movement, and also showing the supports of the system.
3. Determine the type of virtual movement desired. For fully constrained systems, this is done by violating only the constraint corresponding to the reaction of interest. For partially constrained systems the virtual movement should be the type the system would undergo if the external loads were not those required for static equilibrium. Show the position of the system after the virtual movement in the sketch of step 2.
4. Choose an *xyz* coordinate system whose origin is located at a point that does not move during the virtual displacement. As always, the coordinate axes should have orientations that suit the geometry of the system. Show this coordinate system both in the free body diagram of step 1 and the sketch of step 2.
5. Choose any convenient geometric parameter whose value changes during the virtual movement as the generalized coordinate.
6. Evaluate the virtual displacement of the points of application of the forces. In general, follow the analytical approach for this determination by first expressing the position vectors of the points in terms of an arbitrary value of the generalized coordinate. The required virtual displacements are then the differential changes in the position vectors resulting from an infinitesimal increase of the generalized coordinate. If the generalized coordinate has a specific value in the

equilibrium position, substitute that value *after* evaluating the derivatives.

Optionally, in situations where a body is given a pure planar rotation in a virtual movement, the geometrical method may prove easy to apply. The virtual displacement of a point is then the product of the angle of the virtual rotation, in radians, and the radial distance from the point to the fixed point, perpendicular to the radial line in the sense of the rotation.

7. Express the forces doing work in terms of their components, using the free body diagram as an aid. Compute the total virtual work δU as the sum of the dot product of each force and the virtual displacement of its point of application. Equate δU to zero.
8. Solve the equation $\delta U = 0$ for the unknown force.

Clearly, the key step among those outlined above is step 6. In the illustrative problems that follow we will favor the analytical viewpoint for constructing virtual displacements because it is useful for all types of virtual movement. Equally important, because this method is more mathematical, it requires less insight into the nature of the virtual movement of each body and therefore reduces the chance of making sign errors.

ILLUSTRATIVE PROBLEM 1

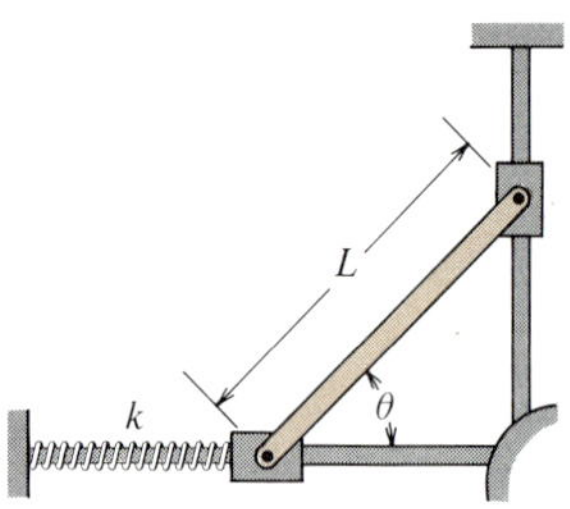

The slender bar of mass m is at an angle of elevation θ in the static equilibrium position. The spring is unstretched in the position where $\theta = 60°$ Derive an expression for the stiffness k of the spring by using the principle of virtual work.

Solution

Step 1 The spring force is depicted in the free body diagram as a compressive force $k\Delta$, where Δ is the amount by which the spring is shortened in comparison to its unstretched length at $\theta = 60°$. Thus

$$\Delta = L \cos \theta - L \cos 60°$$

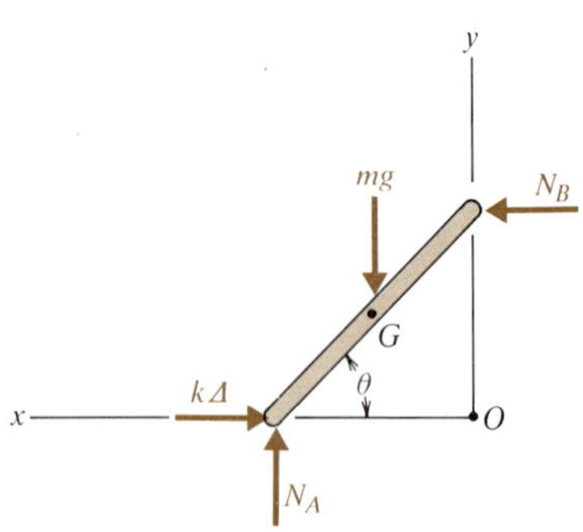

Steps 2 and 3 An expression for k that does not involve the reactions N_A and N_B may be obtained by giving the bar a virtual movement in which the constraints of the collars are not violated. Then $\bar{N}_A$ and $\bar{N}_B$ do no work. The position of the bar prior to the virtual movement is illustrated by the solid line.

Step 4 The coordinate system shown in the free body diagram and in the sketch of the virtual movement is aligned with the axes of the guides for the collars.

Step 5 The position of bar AB in the virtual movement is conveniently described in terms of the angle of elevation θ.

Step 6 The only forces doing work are the gravity force at the midpoint G and the spring force at end A. The virtual movement consists of a general motion in which no point on the bar remains fixed. Hence, the analytical method for virtual displace-

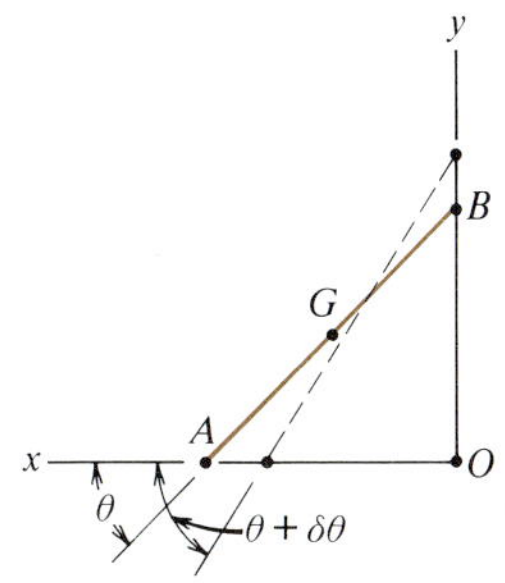

ments is chosen. The position of point G is obtained from the fact that it is the midpoint of the bar.

$$\bar{r}_{A/O} = L\cos\theta\,\bar{i}, \qquad \bar{r}_{G/O} = \frac{L}{2}\cos\theta\,\bar{i} + \frac{L}{2}\sin\theta\,\bar{j}$$

Then

$$\delta\bar{r}_{A/O} = \frac{d\bar{r}_{A/O}}{d\theta}\,\delta\theta = (-L\sin\theta\,\bar{i})\,\delta\theta$$

$$\delta\bar{r}_{G/O} = \frac{d\bar{r}_{G/O}}{d\theta}\,\delta\theta = \left(-\frac{L}{2}\sin\theta\,\bar{i} + \frac{L}{2}\cos\theta\,\bar{j}\right)\delta\theta$$

Step 7 The expression for Δ in Step 1 permits the virtual work to be written as

$$\begin{aligned}\delta U &= (-k\Delta\bar{i})\cdot\delta\bar{r}_{A/O} + (-mg\bar{j})\cdot\delta\bar{r}_{G/O}\\ &= -k(L\cos\theta - L\cos 60°)\bar{i}\cdot(-L\sin\theta\,\bar{i})\,\delta\theta - mg\bar{j}\cdot\left(-\frac{L}{2}\sin\theta\,\bar{i} + \frac{L}{2}\cos\theta\,\bar{j}\right)\delta\theta\\ &= kL^2(\cos\theta - 0.5)(\sin\theta)\,\delta\theta - mg\frac{L}{2}(\cos\theta)\,\delta\theta\end{aligned}$$

Cancellation of the non-zero common factor $\delta\theta$ yields

$$k = \frac{mg\cos\theta}{L\sin\theta(2\cos\theta - 1)}$$

ILLUSTRATIVE PROBLEM 2

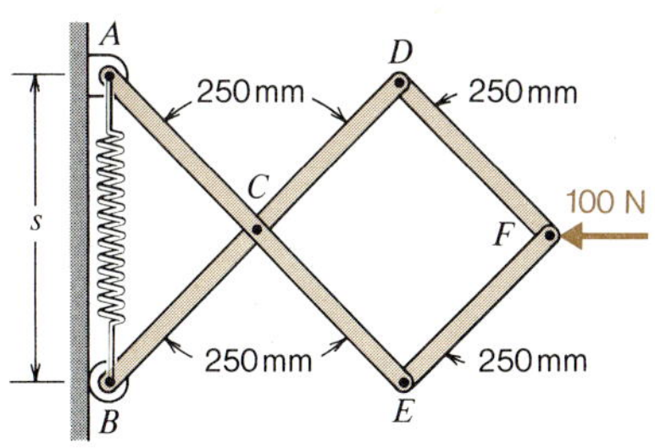

The parallelogram frame is loaded by a horizontal 100 N force. The unstretched length of the spring is 350 mm. Determine the required stiffness k of the spring in terms of the distance s in the static equilibrium position.

Solution

Step 1 The spring is extended by $s - 0.35$ m, so the tension in the spring is $k(s - 0.35)$. Thus, the free body diagram of the entire frame is as shown.

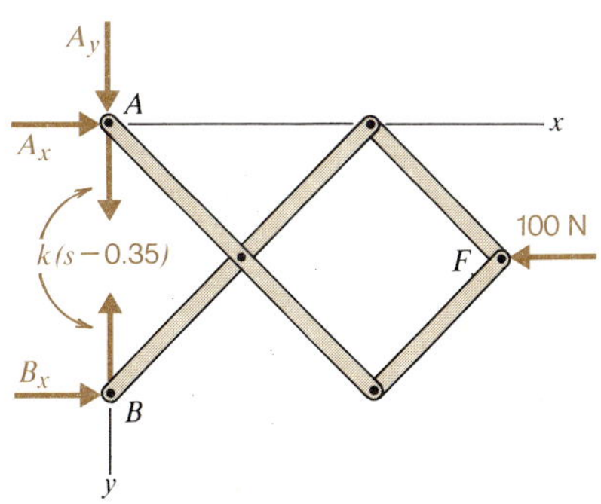

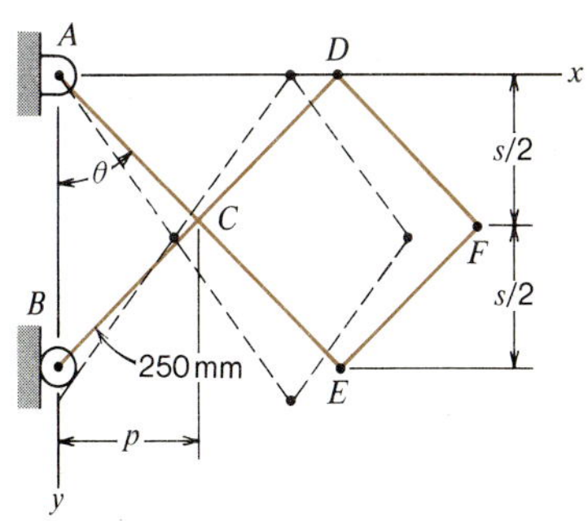

Step 2 The solid lines in the second sketch depict the frame before any virtual movement.

Step 3 The frame is a partially constrained system, because the distance s depends on the loads being carried by the frame. Thus the virtual displacement we give to the system is one in which the value of s is changed, with the constraints of the pins and the roller maintained. As depicted by the dashed lines in the sketch of step 2, the result of this virtual movement is that roller B moves vertically.

Step 4 The sketch of the virtual displacement shows that pin A is fixed, so it is the point chosen as the origin. The y axis shown in the free body diagram (and in the sketch of the geometry) is chosen downward for convenience.

Step 5 A suitable choice for the generalized coordinate is the distance s or the angle θ shown in the sketch of step 2. Locations of points in the frame are easily described in terms of either of these variable parameters. We choose s solely because the value of s in the equilibrium position is given.

Step 6 Point A is fixed, so the reaction forces there do no work. Hence, we need to construct only the general position vectors for points B and F. The Pythagorean theorem is used to express the distance p in terms of the generalized coordinate s. It is found that

$$\bar{r}_{B/A} = s\bar{j}$$

$$\bar{r}_{F/A} = 3p\bar{i} + \frac{s}{2}\bar{j} = 3\left[(0.25)^2 - \left(\frac{s}{2}\right)^2\right]^{1/2}\bar{i} + \frac{s}{2}\bar{j}$$

The virtual displacements are the differential increases in the position vectors resulting from an infinitesimal increase δs. Therefore

$$\delta\bar{r}_{B/A} = \frac{d\bar{r}_{B/A}}{ds}\,\delta s = (\bar{j})\,\delta s$$

$$\delta\bar{r}_{F/A} = \frac{d\bar{r}_{F/A}}{ds}\,\delta s = \left\{3\left(\frac{1}{2}\right)\left[(0.25)^2 - \left(\frac{s}{2}\right)^2\right]^{-1/2}(-s)\bar{i} + \frac{1}{2}\bar{j}\right\}\delta s$$

Step 7 The reaction force $\bar{B}_x$ does no work because it is horizontal and point B displaces vertically (or from another viewpoint, because the constraint imposed by the roller has not been violated). Thus, the principle of virtual work gives

$$\begin{aligned}\delta U &= [-(s - 0.35)k\bar{j}]\cdot\delta\bar{r}_{B/A} + (-100\bar{i})\cdot\delta\bar{r}_{F/A} \\ &= -(s - 0.35)k\,\delta s + (-100)\left(-\frac{3}{2}s\right)\left(0.0625 - \frac{s^2}{4}\right)^{-1/2}\delta s \\ &= \left[-(s - 0.35)k + \frac{150s}{(0.0625 - s^2/4)^{1/2}}\right]\delta s = 0 \text{ joules}\end{aligned}$$

Step 8 Canceling the common factor δs in the preceding expression yields

$$k = \frac{300s}{(s - 0.35)(0.250 - s^2)^{1/2}} \text{ N/m}$$

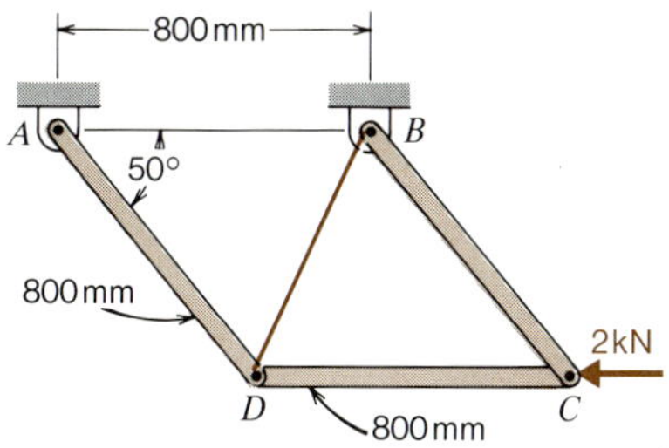

ILLUSTRATIVE PROBLEM 3

The lightweight frame supports the horizontal 2 kN load. Determine the tension in cable BD.

Solution

Step 1 In order to determine the cable tension, it must appear as an external force in the free body diagram. This is achieved by passing a cutting plane through the cable. The orientation of the cable is obtained by noting that sides AB and AD in triangle ABD are equal, so the angles opposite these sides must also be equal.

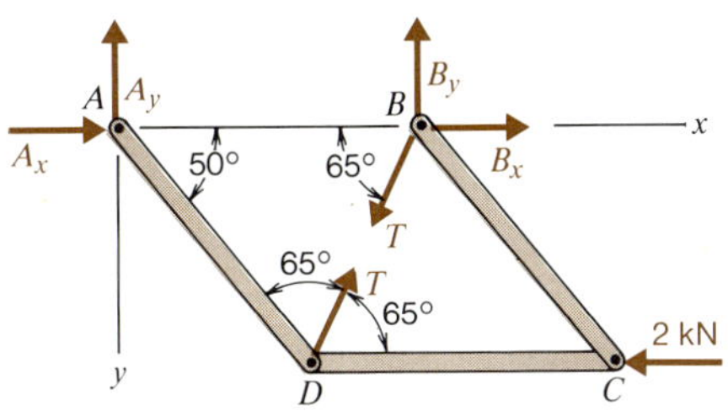

Steps 2 and 3 In order to obtain an expression for the cable tension T that does not involve the reactions at pins A and B, we consider a virtual movement that does not violate the constraints of these pins. The positions of the frame before and after the virtual movement are depicted by solid and dashed lines, respectively, in the sketch.

Step 4 Pin A is a convenient fixed origin for the xyz reference system. The axes are depicted in both the free body diagram and the sketch of the position of the frame.

Step 5 The shape of the parallelogram $ABCD$ is conveniently described in terms of the acute angle θ that its sides form. The static equilibrium position corresponds to $\theta = 50°$, as shown in the sketch in step 2.

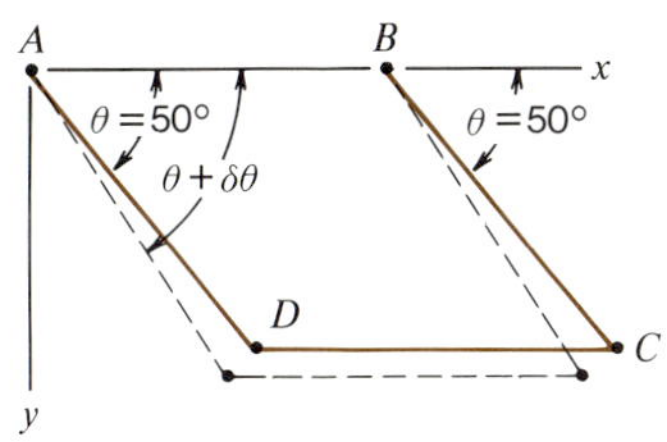

Step 6 Both bars AD and BC execute virtual rotations about their fixed ends, so the geometric method for describing virtual displacements could be employed. However, we will use the analytical method in order to gain more experience in its application. This approach is begun by writing expressions for positions in terms of an *algebraic value* of the generalized coordinate. Only the forces at pins C and D do work, so

$$\bar{r}_{D/A} = 0.8 \cos \theta \bar{i} + 0.8 \sin \theta \bar{j}$$

$$\bar{r}_{C/A} = (0.8 \cos \theta + 0.8)\bar{i} + 0.8 \sin \theta \bar{j}$$

The virtual displacement in the position where $\theta \doteq 50°$ is then found by evaluating the derivatives for this angle. Thus

$$\delta \bar{r}_{D/A} = \left. \frac{d\bar{r}_{D/A}}{d\theta} \right|_{\theta=50°} \delta\theta = \left[-0.8 \sin \theta \bar{i} + 0.8 \cos \theta \bar{j} \right] \Big|_{\theta=50°} \delta\theta$$

$$= (-0.6128\bar{i} + 0.5142\bar{j})\,\delta\theta$$

$$\delta \bar{r}_{C/A} = \left. \frac{d\bar{r}_{C/A}}{d\theta} \right|_{\theta=50°} \delta\theta \equiv \delta \bar{r}_{D/A}$$

Step 7 The orientation of the cable force acting on pin D is shown in the free body diagram. From this, it is found that

$$\delta U = (T \cos 65° \bar{i} - T \sin 65° \bar{j}) \cdot \delta \bar{r}_{D/A} + (-2\bar{i}) \cdot \delta \bar{r}_{C/A}$$

$$= [T \cos 65°(-0.6128) - T \sin 65°(0.5142)]\,\delta\theta - 2(-0.6128)\,\delta\theta$$

$$= 0 \text{ joules}$$

Canceling the common factor $\delta\theta$ results in

$$T(-0.7250) + 1.2256 = 0$$

$$T = 1.69 \text{ kN}$$

ILLUSTRATIVE PROBLEM 4

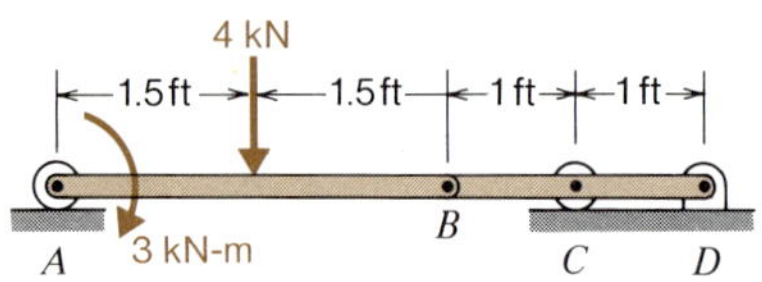

Beam AB, which is pinned at end B to beam BD, carries the 4 kN force and 3 kN-m torque shown. Determine the reaction at roller C.

Solution

Step 1 The internal forces at pin B do not appear in a free body diagram of the entire system.

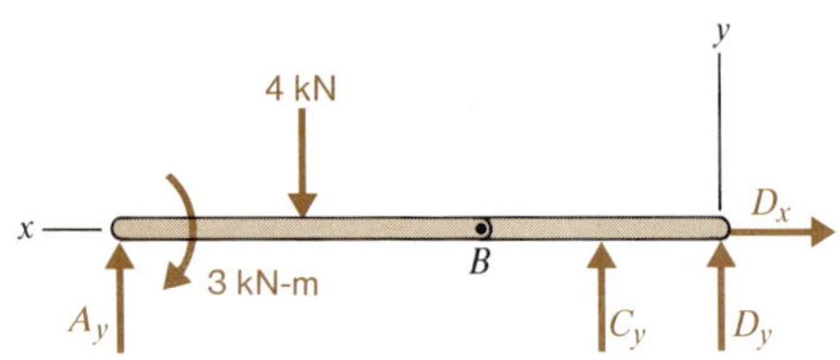

Steps 2 and 3 In order to relate the reaction C_y to the loading, the system will be given a virtual movement that violates only the constraint of roller C against vertical movement. In this way the other reactions will do no work. The position of the system before and after the selected virtual movement is illustrated by solid and dashed lines, respectively, in the second sketch. Point E in this sketch is the point of application of the 4 kN force.

Step 4 Because pin D remains fixed in the virtual movement, it is chosen as the origin of the xyz coordinate system, as shown in both sketches.

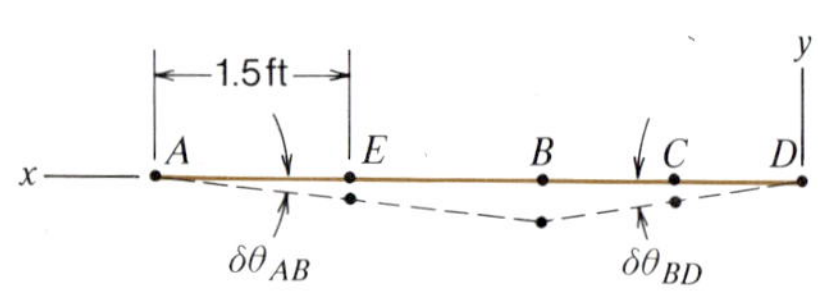

Step 5 The sketch constructed in steps 2 and 3 indicates that the position of the system after the virtual movement may be described in terms of the virtual rotation $\delta\theta_{BD}$. Using this angle, the position of pin B and roller A may be determined. Hence, θ_{BD} is chosen as the generalized coordinate.

Step 6 The description of the virtual displacements by the analytical method requires expressions for the position vectors in terms of an arbitrary (i.e., non-zero) value of the generalized coordinate θ_{BD}. The geometric approach is easier to apply here. Because the angle $\delta\theta_{BD}$ is infinitesimal, the virtual rotation of bar BD from its horizontal position results in each point on this bar displacing an infinitesimal distance in the vertical direction. The fact that the horizontal displacement of pin B is negligible means that the horizontal movement of the other points on bar AB may also be neglected. Hence, bar AB executes a virtual rotation about roller A when the bars are horizontal. The distance that each point displaces is therefore the angle of rotation times the distance from the point of rotation, so

$$\delta\bar{r}_{C/D} = (1\,\delta\theta_{BD})(-\bar{j}), \qquad \delta\bar{r}_{B/D} = (2\,\delta\theta_{BD})(-\bar{j}) = (3\,\delta\theta_{AB})(-\bar{j})$$

$$\delta\bar{r}_{E/A} = (1.5\,\delta\theta_{AB})(-\bar{j})$$

That $\delta\bar{r}_{B/D}$ may be expressed in terms of either $\delta\theta_{BD}$ or $\delta\theta_{AB}$ results from the constraint of pin B, which has not been violated in the virtual movement. It follows from the expressions for $\delta\bar{r}_{B/D}$ that

$$\delta\theta_{AB} = \tfrac{2}{3}\,\delta\theta_{BD}$$

Step 7 The virtual work done by a couple is the moment of the couple multiplied by the virtual angle of rotation in the sense of the couple, so

$$\begin{aligned}\delta U &= 3\,\delta\theta_{AB} + (-4\bar{j})\cdot\delta\bar{r}_{E/A} + C_y\bar{j}\cdot\delta\bar{r}_{C/D}\\ &= 3\,\delta\theta_{AB} + 4(1.5\,\delta\theta_{AB}) - C_y\,\delta\theta_{BD}\\ &= (9)(\tfrac{2}{3}\,\delta\theta_{BD}) - C_y\,\delta\theta_{BD} = 0\end{aligned}$$

$$C_y = 6 \text{ kips (upward)}$$

Additional Remarks

An overview of the illustrative problems we have solved indicates that each system was characterized by relatively simple geometric relations for the position of the system. For example, the frames featured isosceles triangles and parallelograms. This simplicity made it possible to describe the virtual displacements in a straightforward manner. When the geometry is more complicated, as is the case in a frame whose members have different lengths, the task of describing the virtual displacement is more cumbersome. Such systems are often more readily studied by the conventional force and moment equilibrium methods employed in the earlier chapters.

HOMEWORK PROBLEMS

Note: *Each of the following problems should be solved by the principle of virtual work.*

9.11 For the system in Problem 9.1, determine the angle θ in the static equilibrium position.

9.12 For the system in Problem 9.2, determine the angle θ in the static equilibrium position.

9.13 Bar AB of mass m is supported at an angle of inclination θ. Determine the tension in cable BC.

9.14 (a) Derive a relation between the mass m of bar AB, the mass M of the suspended block, and the angle θ. (b) Solve this relation for the angle θ when $M = m/4$. (c) Determine the largest value of M for which end B will not touch the pulley.

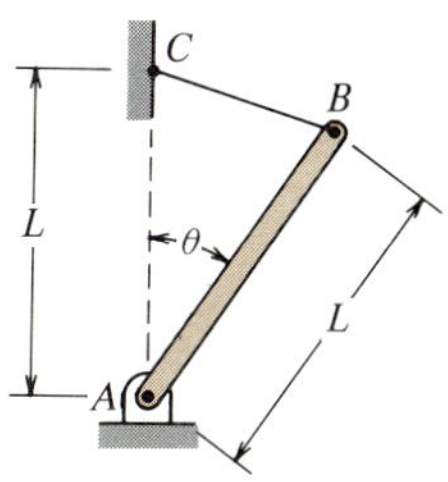

PROB. 9.13

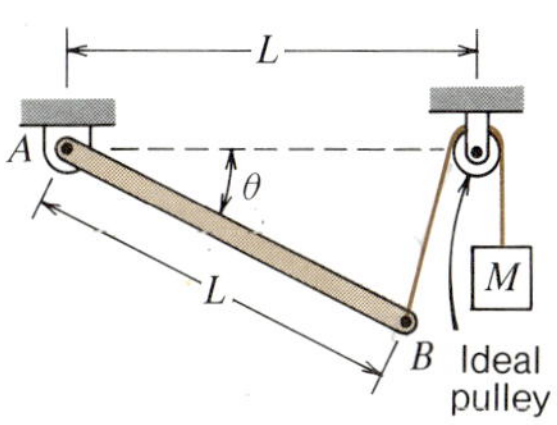

PROB. 9.14

9.15 Bar AB, whose mass is m, is pinned to collar A, which rides on the vertical rod. The spring has a stiffness k and unstretched length l. Determine the value of k for static equilibrium at $\theta = 60^\circ$.

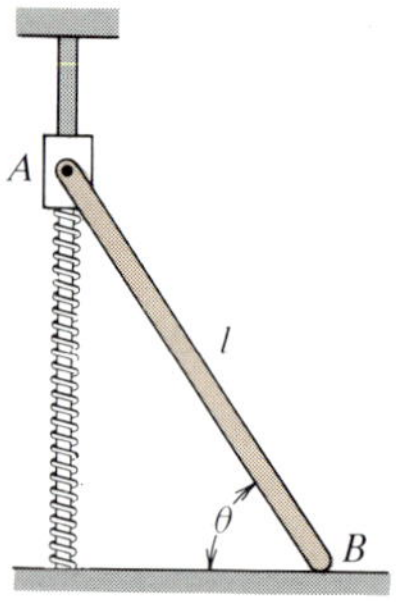

PROB. 9.15

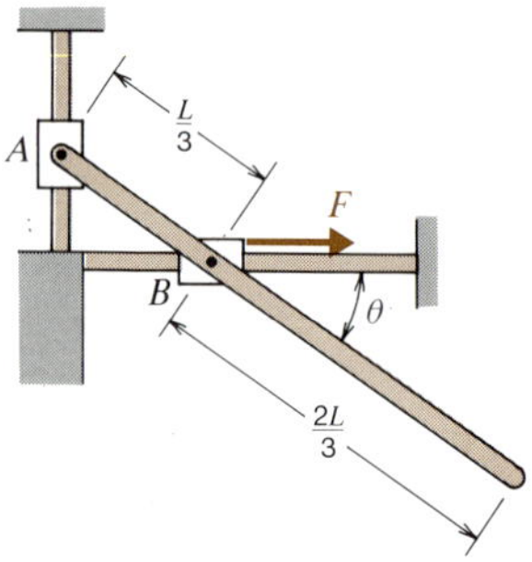

PROB. 9.16

9.16 Determine in terms of the angle of elevation θ the horizontal force $\bar{F}$ required to hold the bar of mass m in static equilibrium.

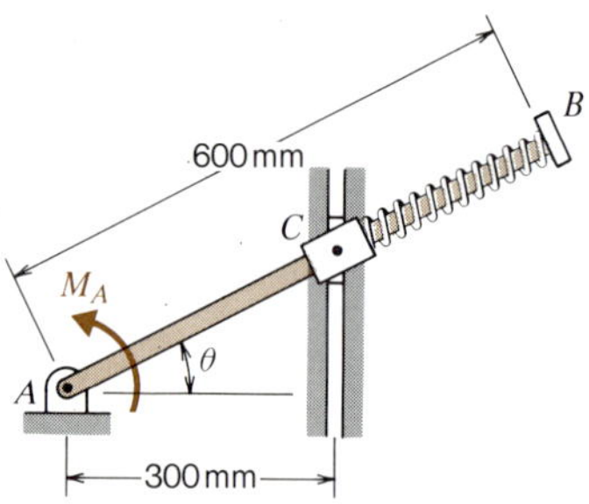

PROB. 9.17

9.17 Collar C, which rides on the smooth bar AB, is pinned to a block that is guided by the smooth straight groove. The spring is unstressed when $\theta = 0$ and its stiffness is 5 kN/m. Determine the magnitude of the couple $\bar{M}_A$ required to maintain equilibrium in terms of θ.

9.18 Bar AB, having a mass m, is pinned to the collar, which rides on the vertical rod. Friction is negligible. Determine the value of θ for static equilibrium.

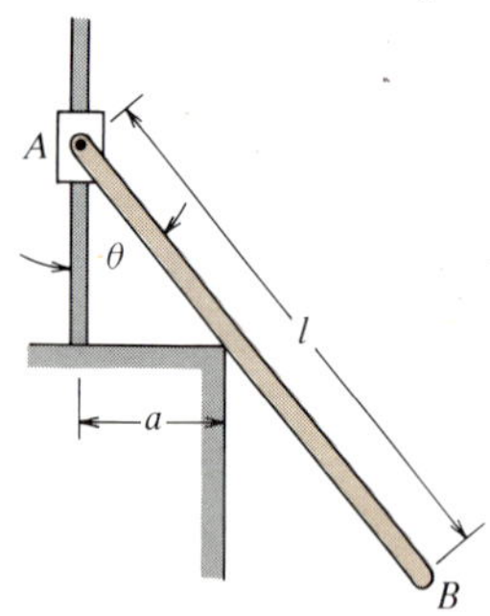

PROB. 9.18

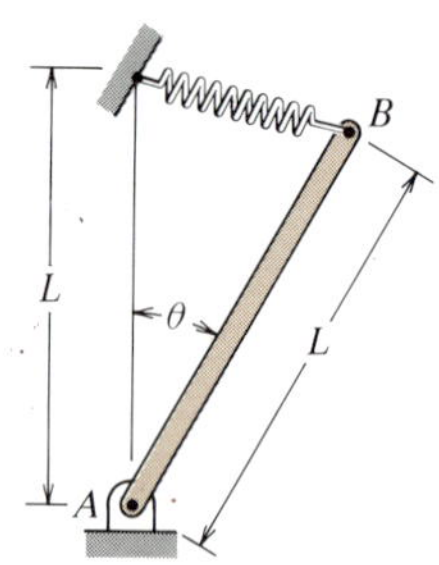

PROB. 9.19

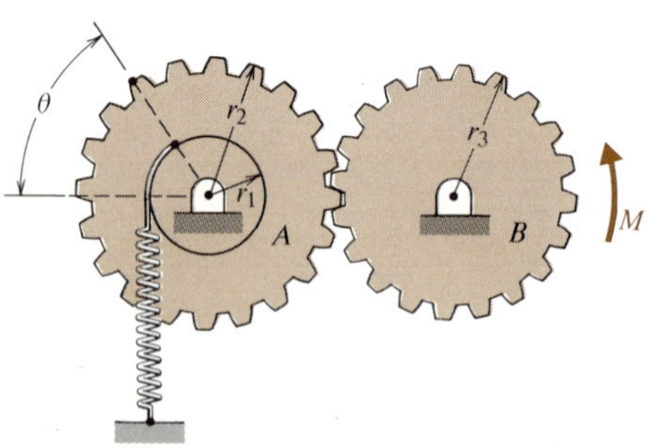

PROB. 9.20

9.19 The slender bar of mass m is held in static equilibrium at an angle of inclination θ by a spring whose stiffness is k and whose unstretched length is $L/2$. Derive an equation governing θ.

9.20 The spring of stiffness k that is wrapped around the inner drum of gear A is unstretched when $\theta = 0$. Derive an expression for the couple M, applied to gear B, required for static equilibrium at a specified angle of rotation θ. *Hint:* A virtual movement of the gears that is consistent with the constraint of the gear teeth causes the pair of contacting teeth to displace the same distance circumferentially.

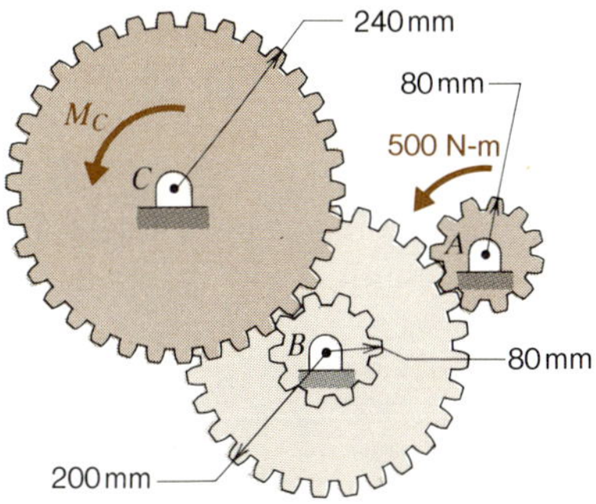

PROB. 9.21

9.21 A 500 N-m couple is applied to gear A, as shown. Gear B, which transmits this couple to gear C, rides freely on its shaft. Determine the magnitude and sense of the couple $\bar{M}_C$ acting on gear C necessary for equilibrium. See the hint for Problem 9.20.

9.22 Solve Problem 9.21 if a counterclockwise couple of 100 N-m is acting on the stepped gear B in addition to the couples on gears A and C.

9.23 and 9.24 The bars in the frame have negligible mass. Determine the ratio F_2/F_1 required for static equilibrium in terms of the angle θ.

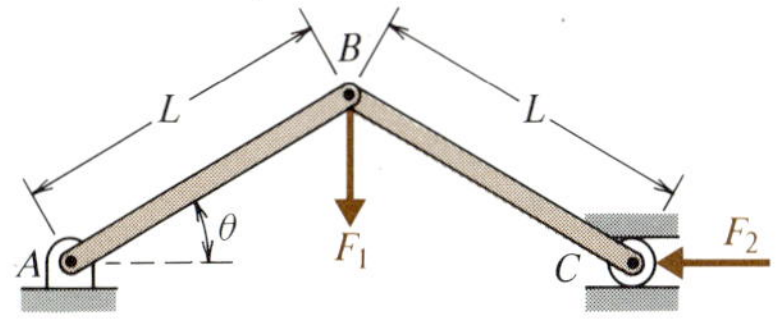

PROB. 9.23

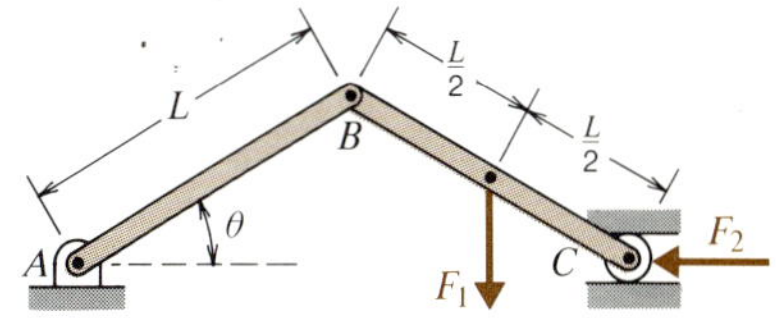

PROB. 9.24

9.25 and 9.26 Determine the tension in cable AB in terms of the distance a and the magnitude of the force $\bar{F}$. The mass of the bars is negligible.

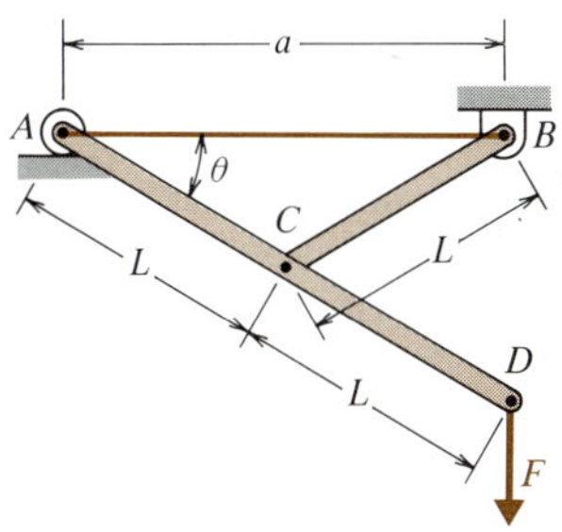

PROB. 9.25

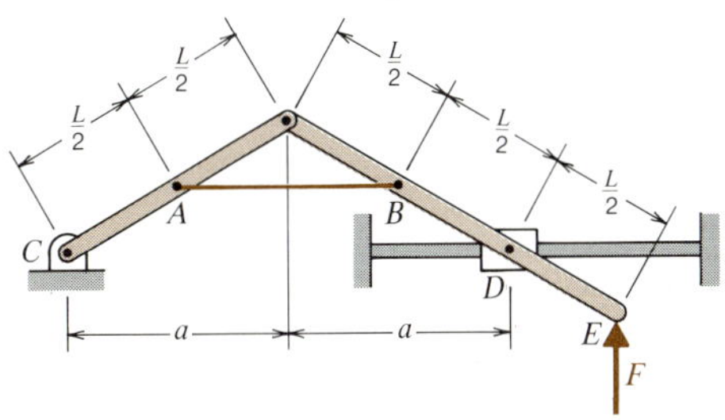

PROB. 9.26

***9.27** A 40 lb force is applied to arm ABC of the press punch shown in cross section. Determine the force the punch applies to the piece of sheet metal in the position shown.

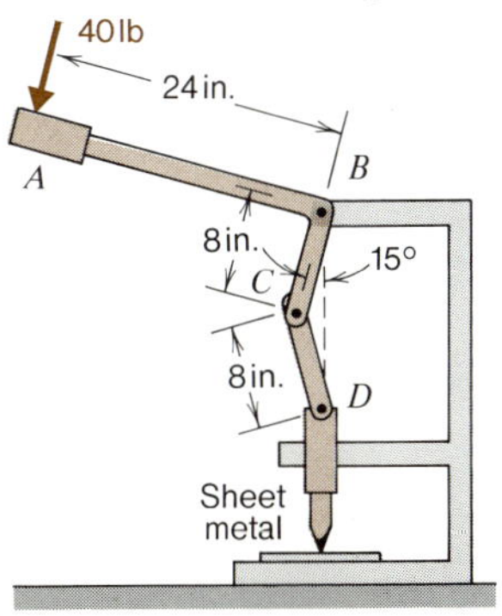

PROB. 9.27

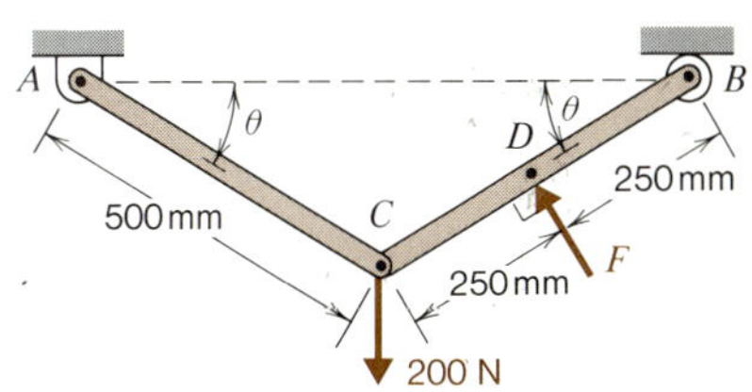

PROB. 9.28

***9.28** Determine the magnitude of the force $\bar{F}$ required for static equilibrium of the frame when $\theta = 20°$. The mass of the bars is negligible.

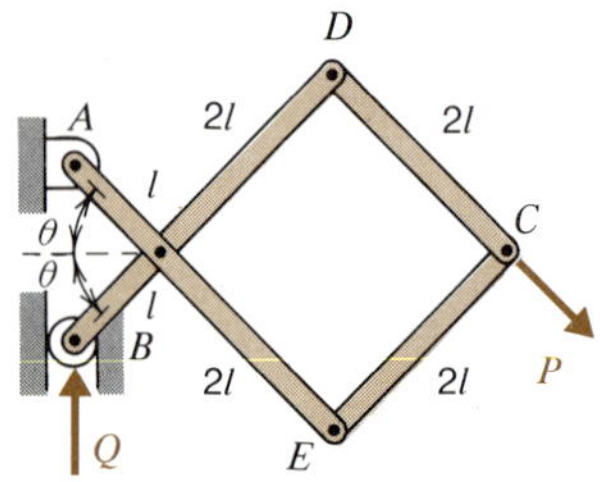

PROB. 9.29

9.29 The load $\bar{P}$ acts parallel to bar CD. Determine the magnitude and sense of the vertical force $\bar{Q}$ acting on roller B required for static equilibrium. The bars have negligible mass.

9.30 and 9.31 The spring is unstretched when $\theta = 60°$. The mass of the bars is negligible. Derive an expression for the spring stiffness k in terms of the magnitude of the load $\bar{P}$, the length l, and the angle θ at static equilibrium.

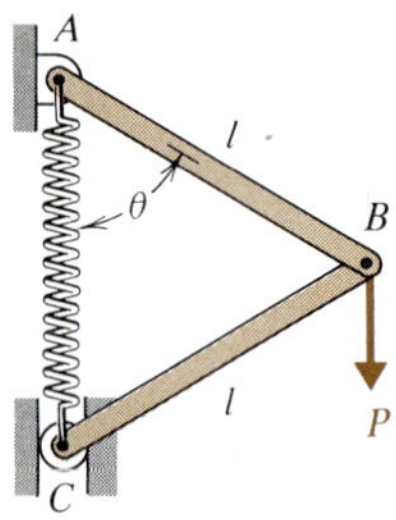

PROB. 9.30

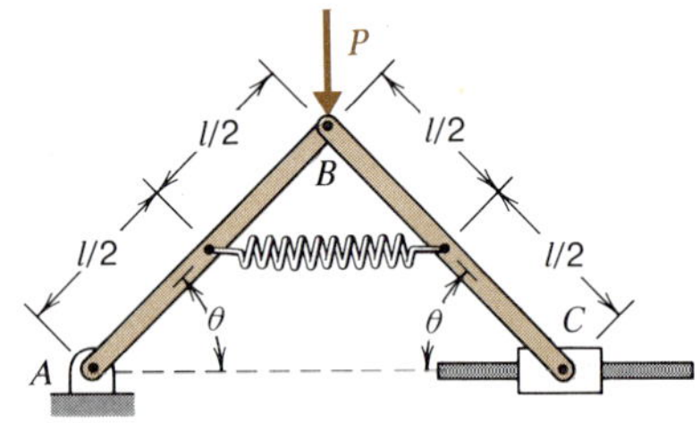

PROB. 9.31

9.32 Lightweight bars AB and CD support the 400 lb block. The spring is unstretched when $\theta = 90°$. Derive an equation relating the spring stiffness to the angle θ for static equilibrium in the tilted position illustrated.

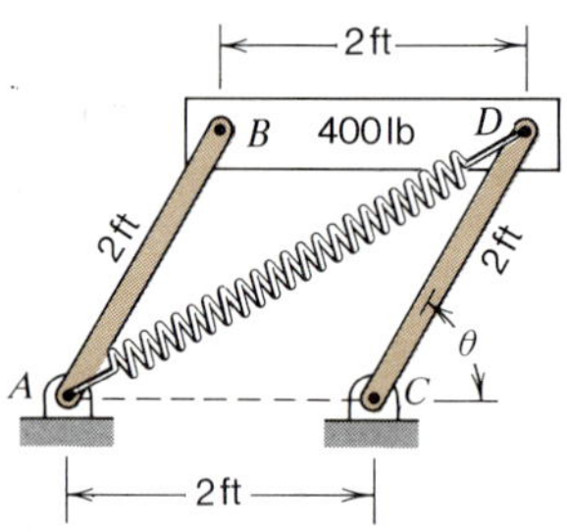

PROB. 9.32

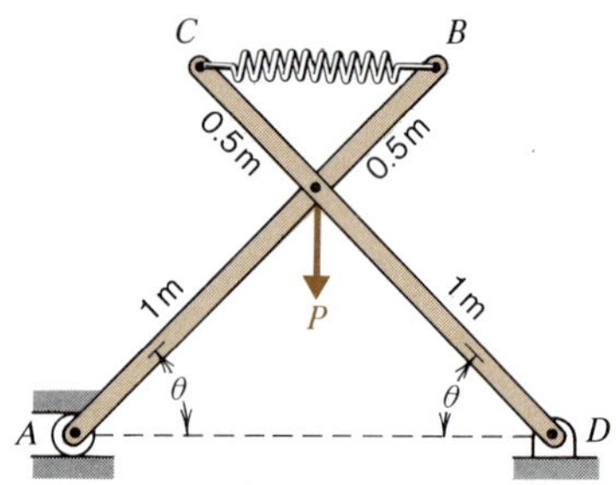

PROB. 9.33

9.33 The spring has a stiffness of 2 kN/m and an unstretched length of 700 mm. Determine the magnitude of the load $\bar{P}$ required for static equilibrium at $\theta = 36.87°$.

***9.34** A 500 kg crate rests on the platform, which is supported by the mechanism shown. Determine the force in the hydraulic cylinder as a function of θ.

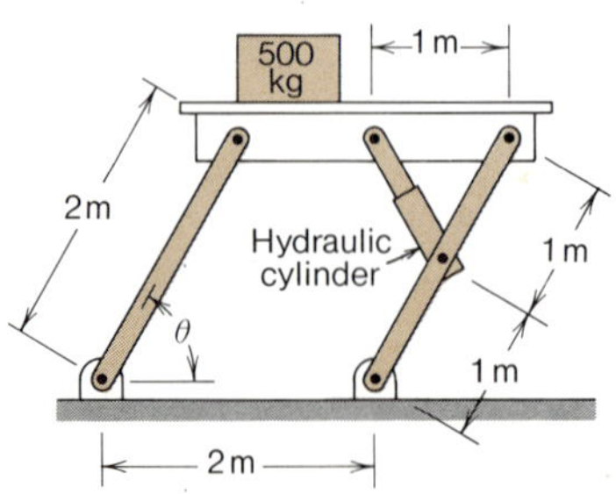

PROB. 9.34

***9.35** The lightweight frame is in equilibrium in the position shown. The unstretched length of the spring is 800 mm. Determine the stiffness of the spring.

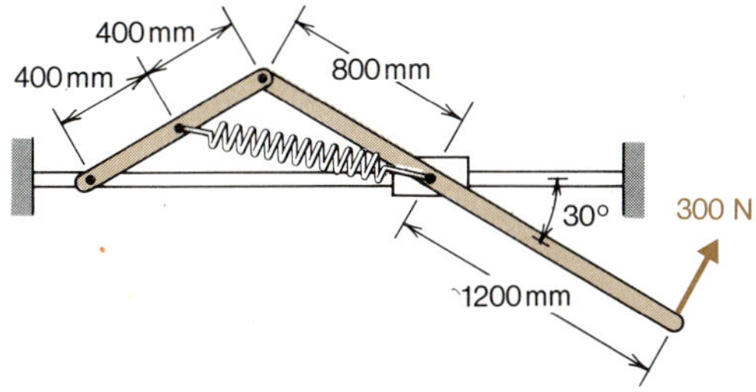

PROB. 9.35

9.36 Determine the reaction at roller B when the system of lightweight beams carries the set of loads shown.

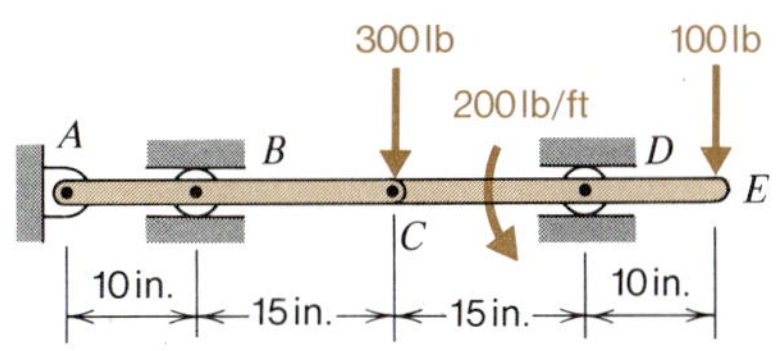

PROBS. 9.36 AND 9.37

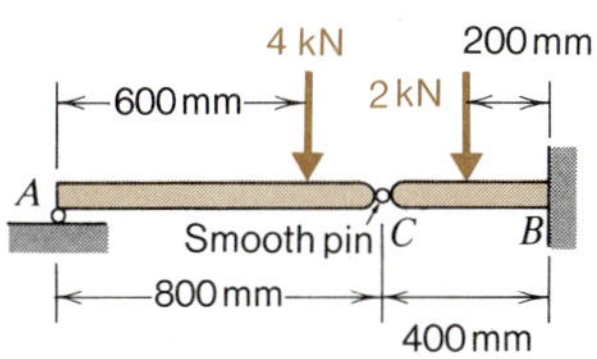

PROB. 9.38

9.37 Determine the vertical reaction at pin A when the system of lightweight beams carries the set of loads shown.

9.38 Determine the couple reaction at the fixed end B of the interconnected, lightweight beams for the set of loads shown.

***9.39** Determine the reaction at roller C of the system of lightweight beams shown.

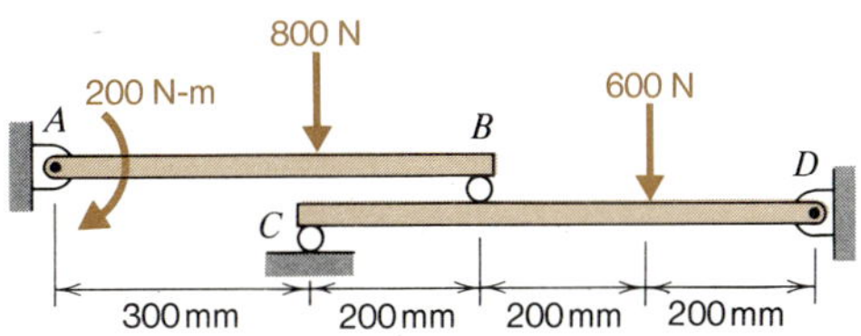

PROB. 9.39

***9.40** Determine the reaction at roller C resulting from the set of loads on the lightweight system of beams shown.

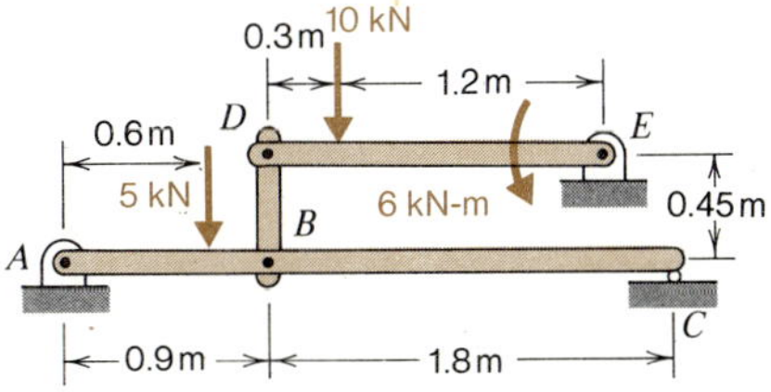

PROB. 9.40

E. PRINCIPLE OF VIRTUAL WORK AND POTENTIAL ENERGY

The development of the principle of virtual work requires consideration of the work done by a force in an infinitesimal displacement of its points of application. In contrast, the study of the actual motion of systems, which is the subject of dynamics, leads to the work done in finite movement of points. One feature of such a study, the concept of *potential energy*, is also relevant to statics, for it will greatly simplify the determination of the virtual work done by certain types of forces.

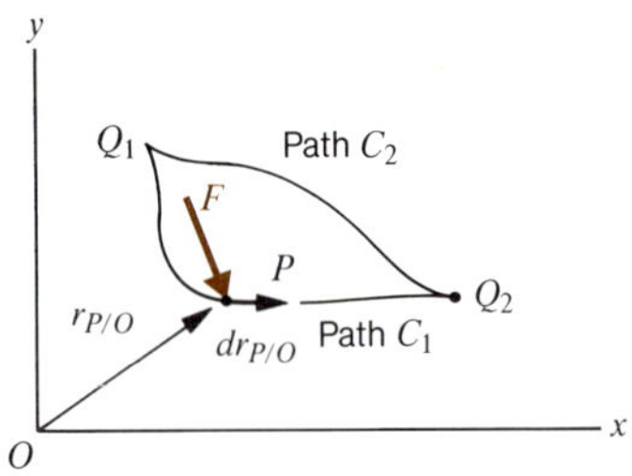

FIGURE 13

Figure 13 illustrates the basic situation to be studied. The force $\bar{F}$ is applied to point P, which may move from the initial position Q_1 to the final position Q_2 along either of two arbitrary paths, C_1 or C_2.

When P moves from one position to another along either path, the position vector changes by an amount $d\bar{r}_{P/O}$. In general, $d\bar{r}_{P/O}$ is different for each location along the path. The infinitesimal work corresponding to this infinitesimal displacement is $dU = \bar{F} \cdot d\bar{r}_{P/O}$. The total work in moving from the initial to final position is the sum of the work for each portion of the movement, so

$$U_{1\to 2} = \int_{r_1}^{r_2} \bar{F} \cdot d\bar{r}_{P/O} \tag{17}$$

where r_1 and r_2 denote the coordinates of points Q_1 and Q_2, respectively.

In general, both the magnitude and direction of $\bar{F}$ will depend on the location of point P. Furthermore, the angle between $\bar{F}$ and the infinitesimal displacement vector $d\bar{r}_{P/O}$ will depend on the orientation of the tangent to the path. We can therefore expect that the value of the work obtained from eq. (17) will depend on whether path C_1 or path C_2 has been followed in the finite displacement.

Our interest in statics, however, is in the exceptional situation where the work done does not depend on the path followed. In that case it must be true that the work $U_{1\to 2}$ is a function only of the locations of the initial point Q_1 and the final point Q_2. The locations of points Q_1 and Q_2 appear only as the limits of the integral for $U_{1\to 2}$. The work done in this situation can depend only on the difference of a function V evaluated at the initial and final positions. For later convenience, this difference is written as the initial value of V less the final value of V. Hence,

$$\boxed{U_{1\to 2} = V_1 - V_2} \tag{18}$$

where V_1 is the value of this function at position Q_1 and V_2 is the value of the function at position Q_2.

The function V is called the *potential energy* of the force $\bar{F}$ and the force is said to be *conservative*. To see why the terminology "conservative" is used, suppose that point P were to follow a closed path in which it goes from position Q_1 to Q_2 along path C_1, and then returns to position Q_1 along path C_2. Equation (18) then indicates that the work in the return along path C_2 is

$$U_{2\to 1} = V_2 - V_1 = -U_{1\to 2} \tag{19}$$

This means that the total work done in following the closed path is zero, because

$$U_{1\to 2\to 1} = U_{1\to 2} + U_{2\to 1} = 0 \tag{20}$$

Equations (19) and (20) show that the work done by the force $\bar{F}$ when point P moves from one position to another is not lost. Rather, it is stored (thus, the word conservative) in the system and may be used to return point P back to its original position. In the special case where $V_2 = 0$, we have $U_{1\to 2} = V_1$. Thus the potential energy may be regarded as the stored ability of the force to do work in order to return point P to the position of zero potential energy.

Two types of conservative forces that especially interest us are the weight force acting on a body at the surface of the earth and the force of a spring. Let us consider the weight force first. Figure 14 shows a body whose center of mass G moves from position Q_1 to position Q_2. Note that for the purpose of computing the potential energy of the weight force shown, we need not consider what causes the movement.

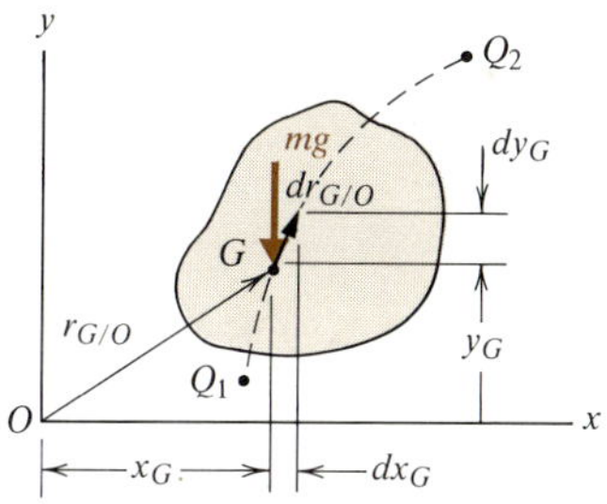

FIGURE 14

To describe the vectors shown in Figure 14, we shall utilize an xyz coordinate system whose y axis is vertical and whose origin O is located at an arbitrary position, as illustrated. When the coordinates of point G are denoted as (x_G, y_G, z_G), an infinitesimal displacement $d\bar{r}_{G/O}$ can be seen from Figure 14 to be equivalent to a displacement dx_G to the right and dy_G upward. That is,

$$d\bar{r}_{G/O} = dx_G\bar{i} + dy_G\bar{j}$$

The gravity force acts downward, so the work done is

$$\begin{aligned} U_{1\to 2} &= \int_{r_1}^{r_2} (-mg\bar{j}) \cdot d\bar{r}_{G/O} = \int_{y_1}^{y_2} (-mg\, dy_G) \\ &= -mgy_2 + mgy_1 \end{aligned} \tag{21}$$

where y_1 and y_2 denote the y coordinates of points Q_1 and Q_2, respectively.

The path followed by the center of mass in the derivation of eq. (21) was arbitrary, so it must be that gravity is a conservative force. By comparing the specific result of eq. (21) to the general result for a conservative force, eq. (18), we can see that

$$\boxed{V_{\text{gravity}} = mgy_G} \tag{22}$$

where y_G represents the vertical height of the center of mass G above the origin.

An alternative way of regarding the distance y_G is to think of the x axis in Figure 14 as defining a reference elevation, called the *datum.* Then y_G is the elevation of the center of mass above the datum. When point G is below the datum, the value of y_G is negative. This means that the gravity force will oppose any movement that brings the center of mass back to the datum.

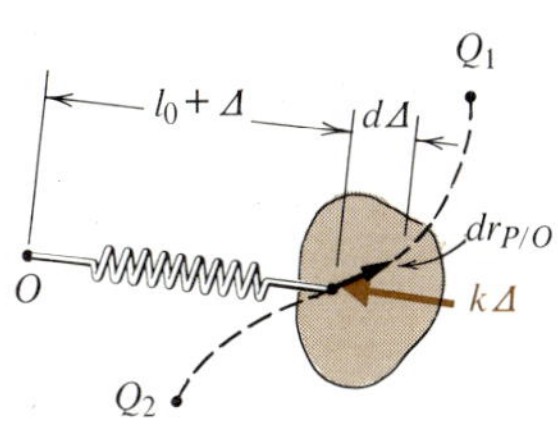

FIGURE 15

Let us now consider the force exerted by a spring. In Figure 15 there is a spring anchored at the fixed point O and attached at its other end to point P in a moving body. The path followed by point P as it moves from the initial position Q_1 to the final position Q_2 is arbitrary.

The spring in Figure 15 has an extension Δ beyond the unstretched length l_O, so the spring force applied to the body is $k\Delta$, directed from point P to point O. The infinitesimal displacement $d\bar{r}_{P/O}$ may be resolved into components perpendicular and parallel to the line OP. The portion of the movement that is perpendicular to line OP is also perpendicular to the spring force. No work is done in this type of movement.

On the other hand, suppose that the component of $d\bar{r}_{P/O}$ parallel to line OP is directed outward, so that the length of the spring is increased by an amount $d\Delta$. The spring force is in the opposite sense from this portion of the movement, so the work is negative in the amount $-k\Delta\, d\Delta$. The total work done by the spring in a movement from position Q_1 to position Q_2 is therefore

$$U_{1\to 2} = \int_{r_1}^{r_2} (-k\Delta\, d\Delta) = -\tfrac{1}{2}k\Delta_2^2 + \tfrac{1}{2}k\Delta_1^2 \tag{23}$$

where Δ_1 and Δ_2 denote the extension of the spring at positions Q_1 and Q_2, respectively.

Equation (23) was obtained without knowledge of the path followed by point P, so the spring force is conservative. Comparison of eq. (23) with the general result of eq. (18) then shows that

$$\boxed{V_{\text{spring}} = \tfrac{1}{2}k\Delta^2} \tag{24}$$

Although this result was obtained by considering a spring that is extended, it is equally valid for springs that can sustain a compressive force $k\Delta$ due to a decrease Δ in its length. Consequently, the potential energy of a spring is positive regardless of whether it is elongated or shortened. This means that a spring will do positive work when the system returns to the position where the spring is undeformed.

We could go on to consider other types of conservative forces, but eqs. (22) and (24) will suffice for the systems we wish to consider. An important aspect of the potential energy of a force is that it is a scalar, so the potential energy of each of several conservative forces is additive as a scalar.

Let us now return to the reason for considering the concept of potential energy in a course on statics. Suppose that a generalized coordinate, labeled q, has been chosen to describe the virtual movement of a system. By definition, the position of all points may be described in terms of q. This means that the total potential energy V of all forces is a function of q; that is, $V = V(q)$. Then the virtual work done by the conservative forces in an infinitesimal displacement δq of the generalized coordinate is found from eq. (18) to be

$$\delta U^{(\text{cons})} = V(q) - V(q + \delta q) = -\frac{dV}{dq}\,\delta q \tag{25}$$

We now separate the total virtual work δU into the portion $\delta U^{(\mathrm{cons})}$ attributable to the set of conservative forces acting on the system, and the portion $\delta U^{(\mathrm{nc})}$ attributable to the nonconservative forces acting on the system. This can be done because work is a scalar function. Then, upon application of eq. (25), the principle of virtual work becomes

$$\delta U = \delta U^{(\mathrm{nc})} + \delta U^{(\mathrm{cons})} = \delta U^{(\mathrm{nc})} - \frac{dV}{dq}\,\delta q = 0$$

$$\boxed{\delta U^{(\mathrm{nc})} = \frac{dV}{dq}\,\delta q} \tag{26}$$

Equation (26) is the principle of virtual work and energy. In words, it states that

> In the static equilibrium position, the virtual work done by the nonconservative forces equals the virtual change in the potential energy of the conservative forces.

The usefulness of this principle is that we no longer have to evaluate the virtual displacements of the points of application of conservative forces. All that is necessary to account for the conservative forces is that we express their potential energy V in terms of the generalized coordinate q and then evaluate the derivative dV/dq. Thus, eq. (26) is merely the result of a refinement in the method of computing the work done by a certain class of forces. Of course, all forces of uncertain nature should be regarded as nonconservative.

In the special situation where all of the forces doing work are conservative, in which case we say that we have a *conservative system*, eq. (26) reduces to

$$\boxed{\frac{dV}{dq} = 0} \tag{27}$$

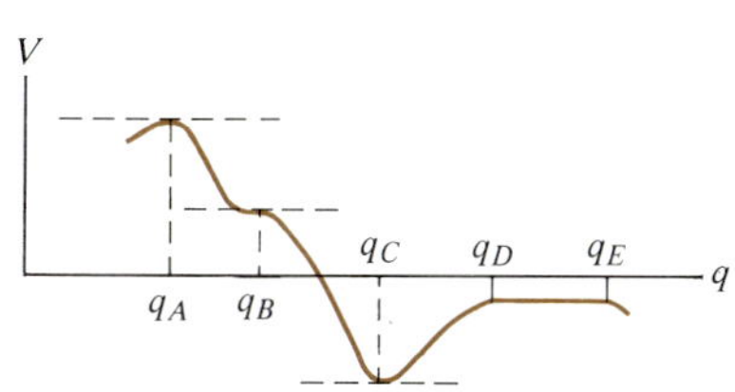

FIGURE 16

This is known as the *principle of stationary potential energy* for a one-degree-of-freedom system, because it states that in the static equilibrium position of a conservative system, the potential energy is either a maximum or minimum value, or possibly a horizontal inflection point. A typical potential energy function for a conservative system is illustrated in Figure 16. The values q_A, q_B, q_C, and the region between q_D and q_E all correspond to values of the generalized coordinate for which the system is in static equilibrium.

EXAMPLE 4

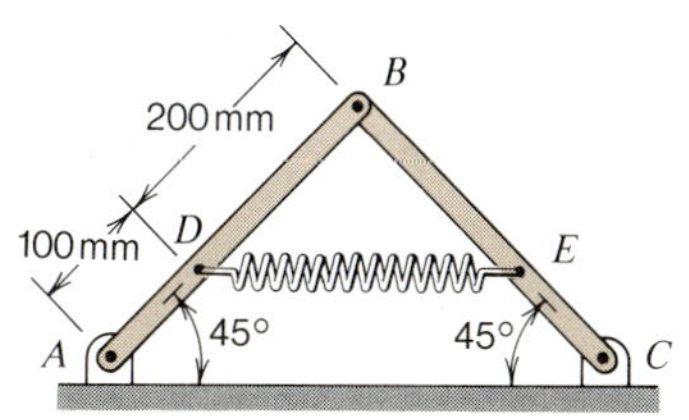

Bars AB and BC each have a mass of 5 kg. The spring has a stiffness of 500 N/m and an unstretched length of 200 mm. Determine the horizontal component of the reaction at pin C.

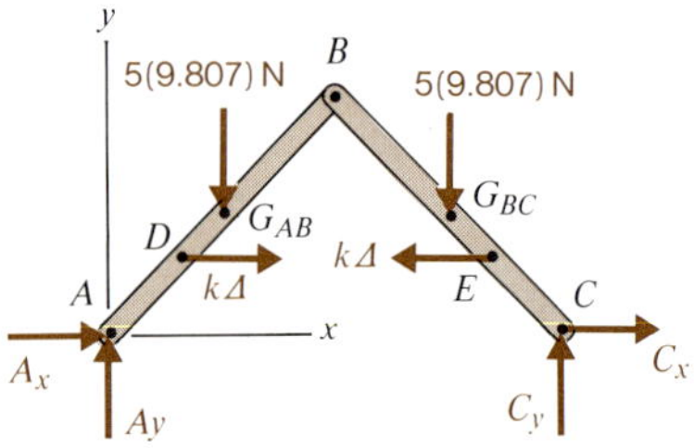

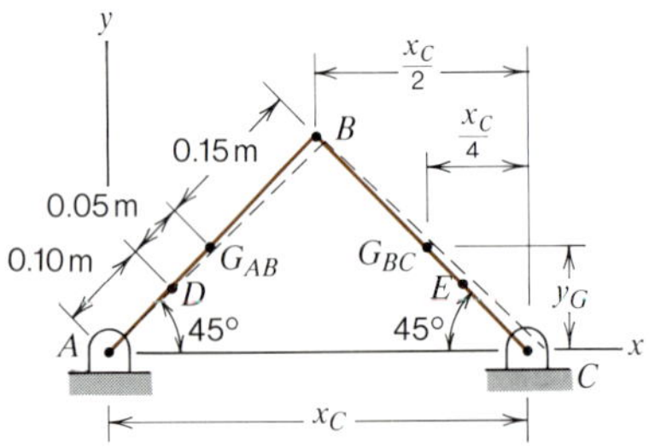

Solution

Aside from the fact that it is no longer necessary to compute the virtual displacement of the points of application of conservative forces, the basic method for solving problems that was developed in the last section is essentially unmodified by the utilization of the concept of potential energy. We begin with a free body diagram of the system.

We have been requested to determine the reaction C_x. Therefore, the virtual movement we give the system is one that violates the constraint of pin C against horizontal movement. The resulting virtual movement is pictured in the geometric sketch. Note that the reactions at pin A do no work because pin A is fixed in the virtual movement, and the vertical reaction at pin C does no work because the virtual displacement of point C is perpendicular to this force. Point A, which is the only point on the frame that is fixed in the virtual movement, is selected as the origin for the coordinate system.

A suitable choice for the generalized coordinate is either the angle θ describing the angle of elevation of the bars, or the horizontal distance x_C. For the sake of variety from previous problems, let us use the latter. The only nonconservative force doing work in the virtual displacement is the reaction C_x. Thus, the only virtual displacement we need to formulate is that of point C. The position vector of this point in terms of x_C is simply

$$\bar{r}_{C/A} = x_C \bar{i}$$

so

$$\delta \bar{r}_{C/A} = \frac{d\bar{r}_{C/A}}{dx_C}\,\delta x_C = \bar{i}\,\delta x_C$$

It then follows that the work $\delta U^{(\text{nc})}$ is

$$\delta U^{(\text{nc})} = C_x \bar{i} \cdot \delta \bar{r}_{C/A} = C_x\,\delta x_C$$

We now formulate the potential energy of the conservative weight and spring forces. The x axis is a convenient datum for the potential energy of gravity, as indicated in the free body diagram.

The total potential energy is the sum of the energy associated with each conservative force, so

$$V = 2(5)(9.807)y_G + \tfrac{1}{2}(500)\Delta^2$$

where y_G is the elevation of the centers of mass and Δ is the elongation of the spring.

The application of eq. (26) requires that we express V in terms of the generalized coordinate $q \equiv x_C$. The sketch of the geometry indicates that the horizontal distance from pin C to the center of mass of bar BC is $x_C/4$, so the Pythagorean theorem yields

$$y_G = [(0.15)^2 - (x_C/4)^2]^{1/2} \text{ m}$$

The spring extension Δ is the difference between the length of the spring l_{DE} in the stretched position and the unstretched length of 0.20 m. The sketch indicates that triangles BDE and BAC are similar, so

$$\frac{l_{DE}}{0.20} = \frac{x_C}{0.30} \qquad \text{or} \qquad l_{DE} = 0.6667x_c$$

It follows that

$$V = 98.07(0.0225 - x_C^2/16)^{1/2} + 250(0.6667x_C - 0.20)^2$$

The work-energy principle, eq. (26), now yields

$$\delta U^{(\text{nc})} = \frac{dV}{dx_C}\,\delta x_C$$

$$C_x\,\delta x_C = \big[\tfrac{1}{2}(98.07)(0.0225 - x_C^2/16)^{-1/2}(-x_C/8) + 250(2)(0.6667x_C - 0.20)(0.6667)\big]\,\delta x_C \text{ joules}$$

The value of x_C for the frame prior to the virtual movement is

$$x_C = 2(0.30)\cos 45^\circ = 0.4243 \text{ m}$$

This value is substituted into the work-energy equation, and the factor δx_C is canceled, with the result that

$$C_x = 3.10 \text{ N}$$

EXAMPLE 5

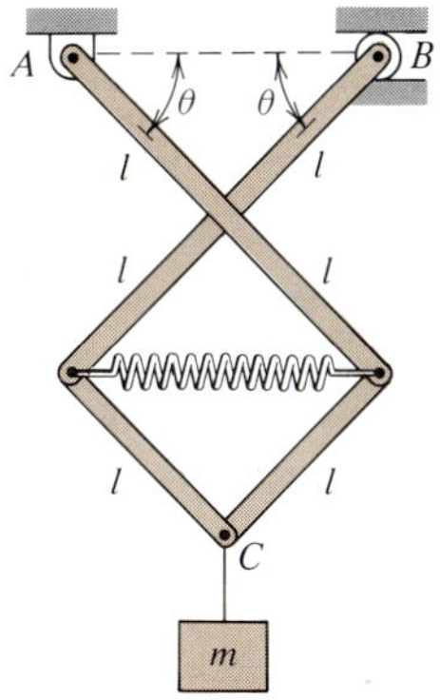

A block of mass m is suspended from the lightweight frame. The spring, whose stiffness is k, is unstressed when $\theta = 36.87°$. Derive an expression for k in terms of the value of θ for the static equilibrium position.

Solution

As always, the solution begins with a free body diagram of the system and a separate sketch describing the geometry of the system. In order to obtain an expression for k that does not involve the reactions at pin A and roller B, the frame is given a virtual movement that is consistent with the constraints of the supports. The only forces doing work in this movement are those of the spring and gravity, both of which are conservative. Therefore, eq. (27) may be employed, and it will not be necessary to determine virtual displacements of points.

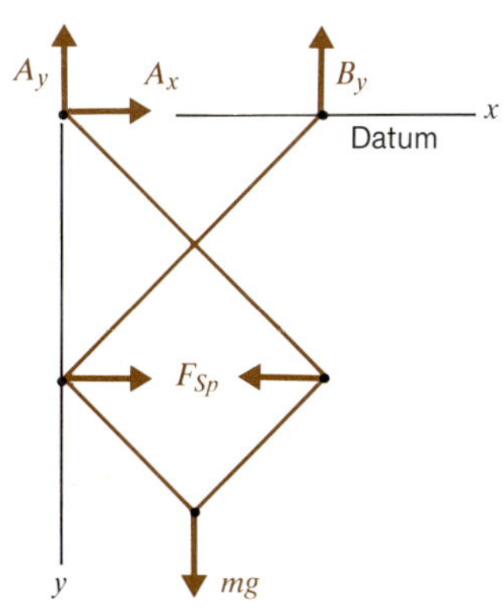

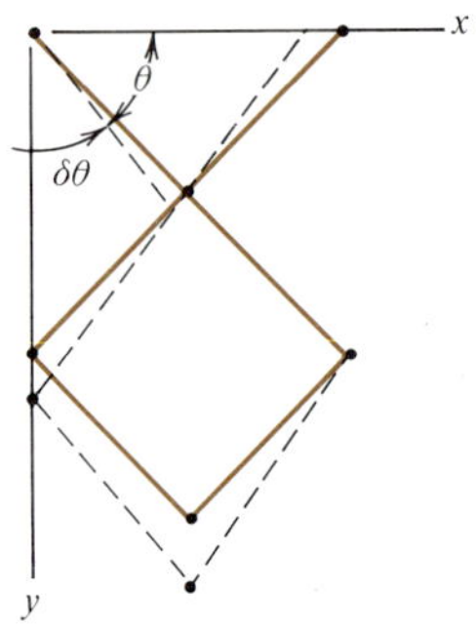

The sketch of the geometry shows that the virtual movement is characterized by a change in the angle θ, so θ is chosen as the generalized coordinate. Pin A, which is fixed, is chosen as the origin of the xyz axes, and the x axis is selected as the datum for the determination of the potential energy of the gravity force. Pin C is below the datum, so the height h_c *above* the datum is

$$h_c = -3l \sin \theta$$

The length of the spring is $2l \cos \theta$. Because the spring is unstretched when $\theta = 36.87°$, we may write

$$\Delta = 2l \cos \theta - 2l \cos 36.87° = l(2 \cos \theta - 1.6)$$

The expression for the total potential energy in terms of the generalized coordinate is therefore

$$\begin{aligned} V &= V_{\text{gravity}} + V_{\text{spring}} = mgh_c + \tfrac{1}{2}k\Delta^2 \\ &= -mg(3l \sin \theta) + \tfrac{1}{2}kl^2(2 \cos \theta - 1.6)^2 \end{aligned}$$

Equation (27) then indicates that for static equilibrium

$$\frac{dV}{d\theta} = -3mgl \cos \theta + kl^2(2 \cos \theta - 1.6)(-2 \sin \theta) = 0$$

$$k = \frac{3mg}{l}\left(\frac{\cot \theta}{1.6 - 2 \cos \theta}\right)$$

HOMEWORK PROBLEMS

9.41 Solve Problem 9.15 by using the concept of potential energy.

9.42 Solve Problem 9.19 by using the concept of potential energy.

9.43 Two identical springs of stiffness k are anchored at points A and B, and joined together at point C in order to support a block of mass m. In the position where $\theta = 0$, the tension in each spring is T_0. Derive an expression for the value of m required to have static equilibrium at a specified value of θ.

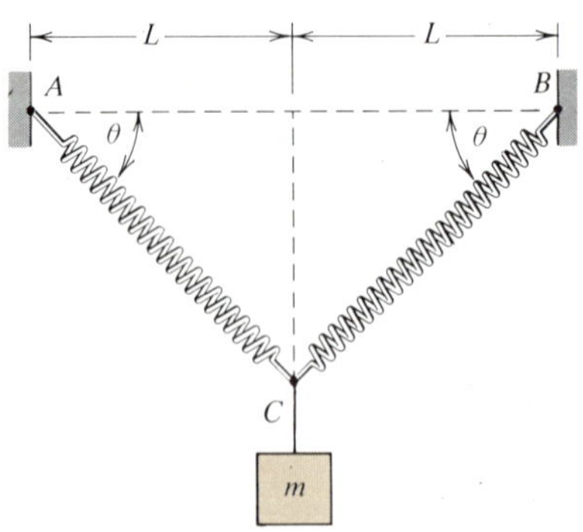

PROB. 9.43

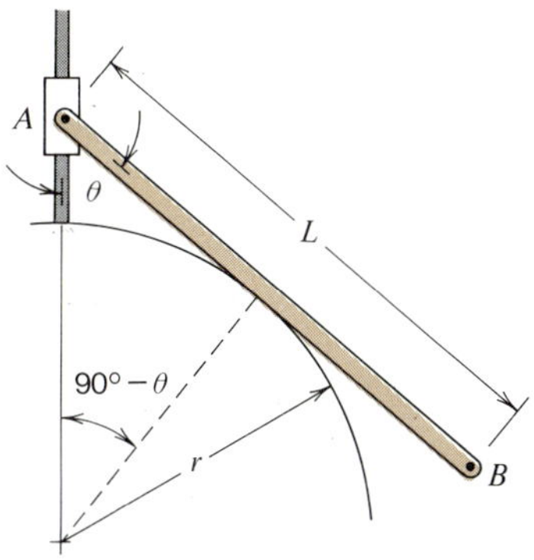

PROB. 9.44

9.44 Bar AB of mass m is supported by a lightweight collar A, which rides on a vertical guide bar. Friction is negligible. Derive an expression for the angle θ in the static equilibrium position.

9.45 The frame consists of two bars of mass m that are situated in the vertical plane. The spring has a stiffness k and an unstretched length l. Derive an expression in terms of k, m, and l for the angle θ in the static equilibrium position.

9.46 Solve Problem 9.31 by using the concept of potential energy.

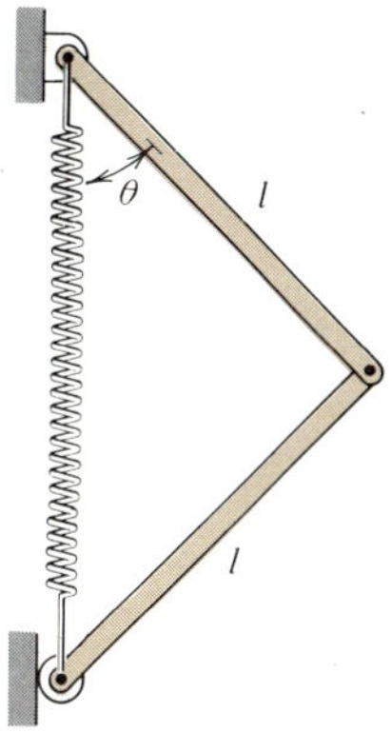

PROB. 9.45

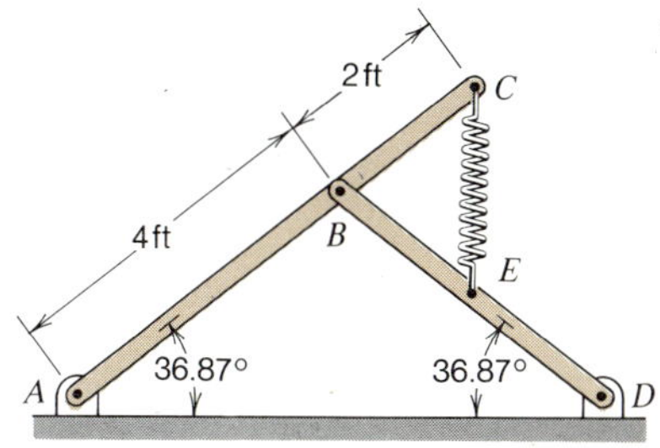

PROB. 9.47

9.47 The spring has a stiffness of 50 lb/in. and an unstretched length of 20 in. Neglecting the mass of the bars, determine the horizontal component of the reaction at pin support D.

9.48 Solve Problem 9.32 by using the concept of potential energy.

9.49 and 9.50 In the lightweight frame shown, the identical springs of stiffness k are unstretched when $\theta = 45°$. Determine the relationship between the magnitude of the force $\bar{P}$ and the value of θ in the static equilibrium positions.

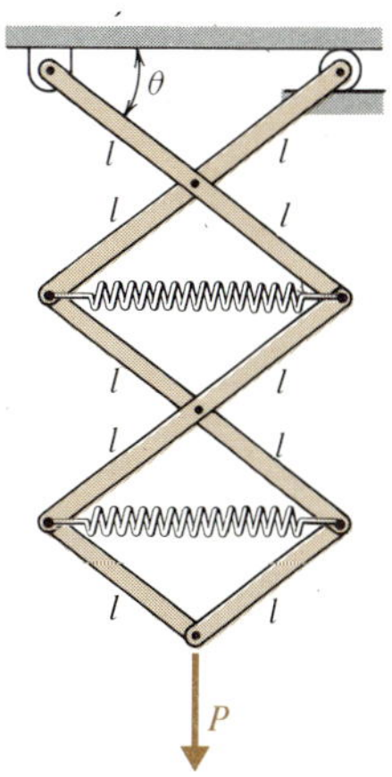

PROB. 9.49

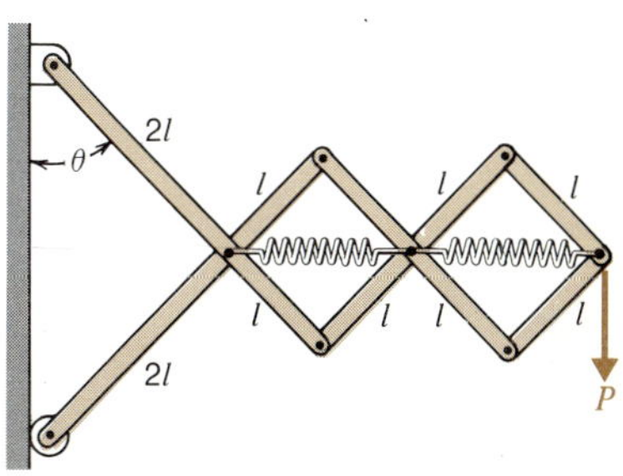

PROB. 9.50

*9.51 Collar C, which rides on the smooth bar BD, is pinned to the ground. The spring, whose stiffness is 30 lb/in., is unstressed when $\theta = 30°$. Determine an expression for the vertical force $\bar{F}$ at point B required for equilibrium in terms of the angle θ.

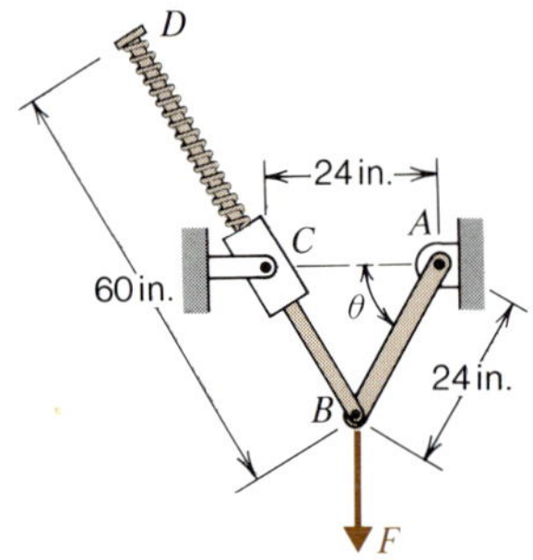

PROB. 9.51

*F. STABILITY OF THE EQUILIBRIUM POSITION

It is often found that more than one static equilibrium position is possible in a partially constrained system, that is, in a system whose supports do not fully prevent movement. For example, it is theoretically possible for a bar that is free to pivot about one end in the vertical plane to balance in the upright position. Proving that this position is not realizable in practice requires an investigation of the stability of the equilibrium position.

A very simple situation illustrating the question of the stability of the equilibrium position is the case of a particle placed on a smooth surface whose vertical profile is shown in Figure 17. At positions A, B, and C the tangent to the hill is horizontal. It then follows that each of these locations is a theoretical static equilibrium for the block. Let us consider each of these.

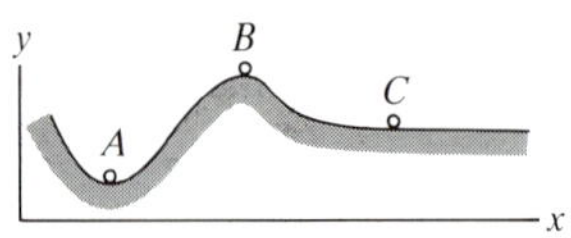

FIGURE 17

Position A is the lowest point in a valley. If it is moved away from this position by a slight disturbance and then released, the particle will return to the low point. This is said to be a *stable* equilibrium position. Point B is the top of a hill. There, a small disturbance away from the top causes the particle to fall down the hill. This is an *unstable* equilibrium position. At location C the particle will remain at any point along the level portion of the hill, because there is neither a tendency to move away nor to return to location C after a disturbance. This is the condition of *neutral* equilibrium.

Two observations regarding the situation in Figure 17 are significant. First, notice that in each case the only constraint acting on the particle is that of the hill, which exerts a normal force on the particle in order to prevent movement of the particle in the normal direction. In the movement satisfying this constraint, that is, in a displacement of the particle tangent to the hill, the only force doing work is gravity. Thus, the particle on the hill forms a partially constrained system that is conservative with respect to virtual movements satisfying the existing constraints. The second observation is that when the x axis is chosen as the datum for the potential energy, then the potential energy of the system is proportional to the coordinate y; that is, $V = mgy$. If we now choose the horizontal coordinate as the generalized coordinate, a graph of the potential energy as a function of the generalized coordinate has the same shape as the vertical profile of the hill shown in Figure 17. This suggests that the type of stability of a partially constrained conservative system can be determined by the properties of its potential energy function.

We are treating only systems with one degree of freedom, so there is one generalized coordinate, denoted q, which describes the position of the partially constrained system. Suppose that Figure 16, which is repeated here as Figure 18, represents the potential energy of such a system. When $q = q_A$, where the potential energy is a maximum (analogous to a particle at the peak of a hill), there is unstable equilibrium. If the system is given a slight disturbance from this location, the conservative forces will act to move the system farther away from q_A. At location C, where the potential energy is minimum (a particle at the bottom of a valley), the conservative forces act to return the system to this location after a disturbance. This is a stable equilibrium position. Thus we have the following criteria for conservative systems.

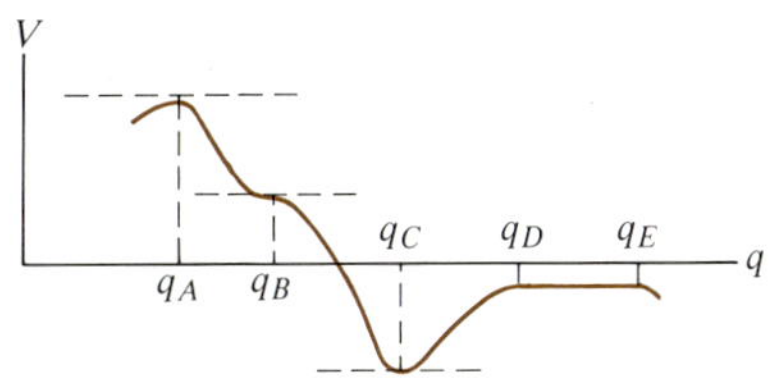

FIGURE 18

$$\left.\frac{dV}{dq}\right|_{q=q_0} = 0 \quad \text{and} \quad \left.\frac{d^2V}{dq^2}\right|_{q=q_0} > 0: \quad q_0 \text{ represents a stable equilibrium position}$$
$$\left.\frac{dV}{dq}\right|_{q=q_0} = 0 \quad \text{and} \quad \left.\frac{d^2V}{dq^2}\right|_{q=q_0} < 0: \quad q_0 \text{ represents an unstable equilibrium position} \tag{28}$$

There are several possibilities in the case where the second derivative of V is zero. For example, all locations in a region where the potential energy is constant, such as the one between q_D and q_E in Figure 18, are possible equilibrium positions ($dV/dq \equiv 0$). Thus, there is no tendency of the system to either move away from, or return to, a specific location in this region. This means that the system is in *neutral equilibrium*.

The fact that $d^2V/dq^2 = 0$ at an equilibrium position might also mean a horizontal inflection point, as is the case for location q_B in Figure 18. At such locations, a change in the generalized coordinate in one sense will result in a decrease in the potential energy. It must therefore be concluded that an inflection point represents unstable equilibrium.

It is also possible that the potential energy of an equilibrium position is actually a maximum or a minimum, even though $d^2V/dq^2 = 0$. This matter depends on the higher derivatives of V. One way of considering each possibility at a location where $d^2V/dq^2 = 0$ is to actually compute V at values of the generalized coordinate that are slightly smaller and larger than at the equilibrium position. A comparison of these values to the potential energy at the equilibrium position will indicate how the potential energy varies in that vicinity.

It should be noted that the stability theorems developed here apply only to conservative systems. The stability of partially constrained systems in which nonconservative forces do work is a more complicated question, being concerned with the dynamics of the system. Thus, before applying eqs. (28) it must be ascertained that all forces doing work are conservative.

EXAMPLE 6

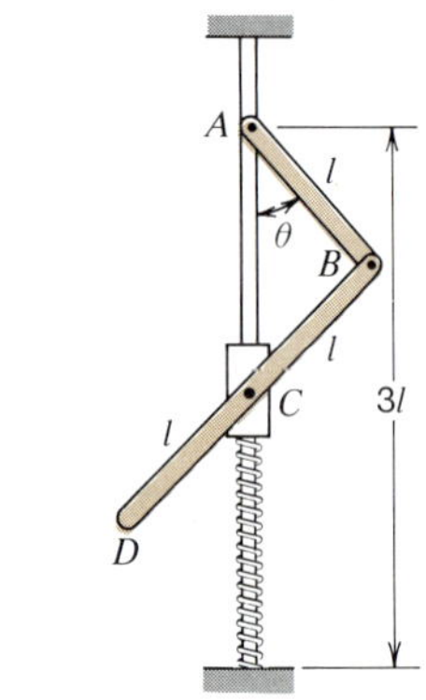

The mass of bars AB and BD are m and $2m$, respectively. The spring supporting collar C has a stiffness k and an unstretched length $3l$. Derive an expression for the angle θ at equilibrium, and evaluate the stability of each equilibrium position in terms of k, m, and l.

Solution

The first step is to draw a free body diagram of the frame and a diagram of the virtual movement corresponding to an increase of the generalized coordinate, which we select as θ. These diagrams indicate that the only forces doing work are the gravitational and spring forces. Hence, the system is conservative.

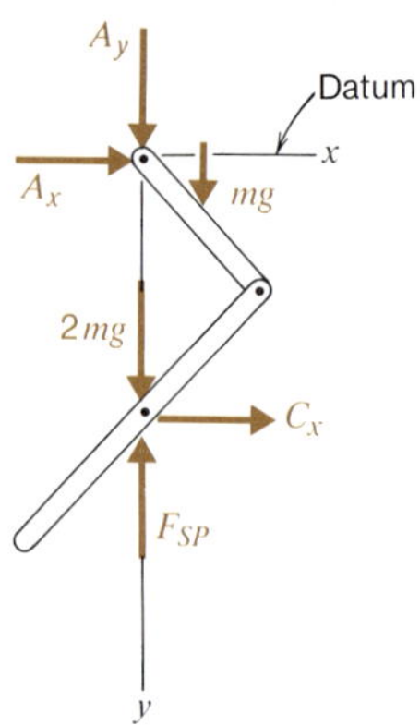

Pin A is chosen as the origin for the xyz axes because it is fixed in the virtual movement. The x axis is defined as the datum for gravity. The center of mass of each bar is below this datum, and the geometrical sketch indicates that the deformation of the spring is $\Delta = 2l\cos\theta$. Thus, the total potential energy is

$$V = mg\left(-\frac{l}{2}\cos\theta\right) + 2mg(-2l\cos\theta) + \tfrac{1}{2}k(2l\cos\theta)^2$$

$$= -\tfrac{9}{2}mgl\cos\theta + 2kl^2\cos^2\theta$$

The derivatives of V required for the stability investigation are

$$\frac{dV}{d\theta} = \frac{9}{2}mgl\sin\theta - 4kl^2\cos\theta\sin\theta$$

$$\frac{d^2V}{d\theta^2} = \frac{9}{2}mgl\cos\theta + 4kl^2(\sin^2\theta - \cos^2\theta)$$

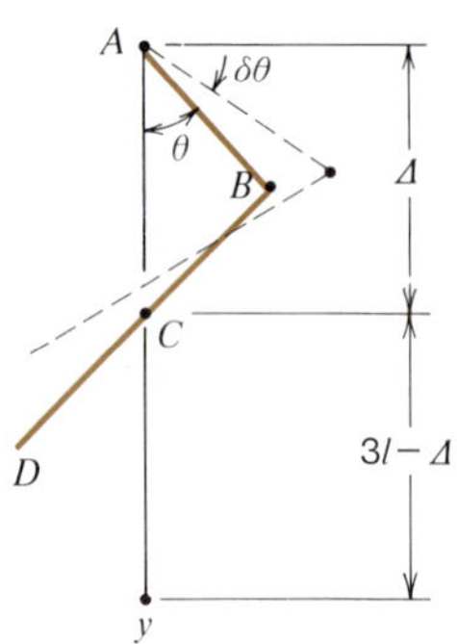

The equilibrium positions satisfy $dV/d\theta = 0$. This yields

$$\sin\theta = 0 \qquad \text{or} \qquad \cos\theta = \frac{9}{8}\frac{mg}{kl}$$

We consider each position separately. The case $\sin\theta = 0$ clearly means $\theta = 0$. It is then found that

$$\left.\frac{d^2V}{d\theta^2}\right|_{\theta=0} = \frac{9}{2}mgl - 4kl^2$$

This position is stable when the second derivative is positive. Thus

$$\text{if} \quad k < \frac{9}{8}\frac{mg}{l}, \qquad \text{then} \quad \theta = 0 \quad \text{is stable}$$

The non-zero position for θ is investigated by using the identity $\sin^2\theta \equiv 1 - \cos^2\theta$. Thus

$$\left.\frac{d^2V}{d\theta^2}\right|_{\cos\theta=(9/8)(mg/kl)} = \frac{9}{2}mgl\left(\frac{9}{8}\frac{mg}{kl}\right) + 4kl^2\left[1 - 2\left(\frac{9}{8}\frac{mg}{kl}\right)^2\right]$$

$$= 4kl^2 - \frac{81}{16}\frac{(mg)^2}{k}$$

This position is stable when the second derivative is positive, which yields

$$4kl^2 > \frac{81}{16}\frac{(mg)^2}{k} \qquad \text{for stability}$$

Thus

$$\text{if} \quad k > \frac{9}{8}\frac{mg}{l}, \qquad \text{then} \quad \theta = \cos^{-1}\left(\frac{9}{8}\frac{mg}{kl}\right) \quad \text{is stable}$$

It will be noted that the range of values of k for which the non-zero value of θ is stable is the same as the range for which $\theta = 0$ is unstable. Consequently, the frame is collapsed in the position $\theta = 0$, unless the spring is sufficiently stiff.

HOMEWORK PROBLEMS

9.52 Blocks A and B of mass m, resting against the smooth wall and floor, are connected by a bar of mass $2m$. The spring attached to block A has a stiffness k and is unstressed when the bar is vertical. (a) Determine the equilibrium positions of the bar. (b) Determine the range of values of k for which the equilibrium positions found in part (a) are stable.

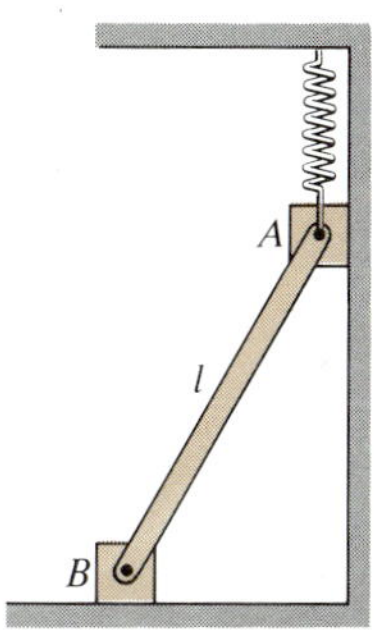

PROB. 9.52

9.53 Determine the range of values of the stiffness k of the spring in Problem 9.45 for which the frame is in stable equilibrium with the bars vertical.

9.54 In each case illustrated, a bar of mass m is supported by two identical bars of negligible mass. For each system determine the type of stability that exists for the equilibrium position shown.

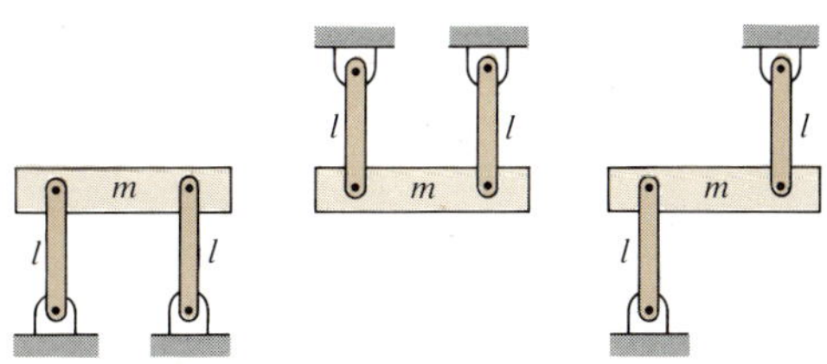

(a) (b) (c)

PROB. 9.54

9.55 The spring is unstretched when $\theta = 60^\circ$, and it is known that $\theta = 30^\circ$ is a static equilibrium position. Each bar has a mass m. Determine the stiffness k in terms of m and l, and describe the stability of this position.

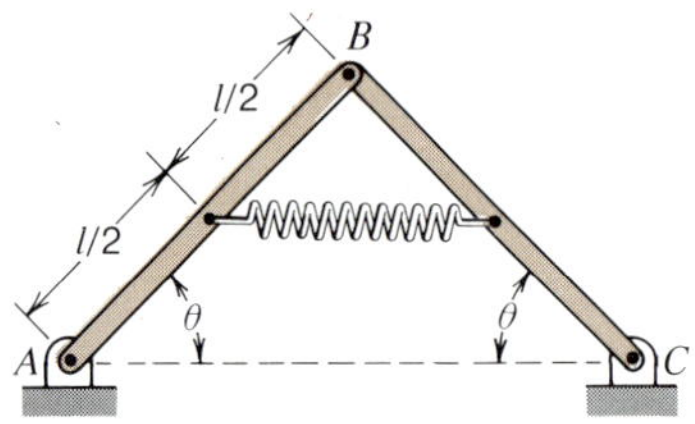

PROB. 9.55

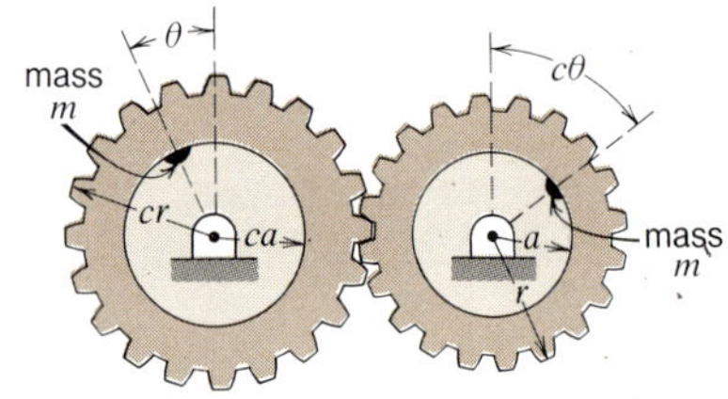

PROB. 9.56

9.56 Two gears of radii r and cr that coincide with the vertical plane carry identical masses m at radii a and ca, as shown. For the case $c = 1$, determine the angle of rotation θ of the left gear for which the system is in static equilibrium. Then describe the stability of each equilibrium position.

9.57 Solve Problem 9.56 for the case where $c = 2$.

***9.58** The lamp assembly, whose center of mass is G, has a mass of 2 kg. The bars supporting the assembly have negligible mass. The unstretched length of the spring is $2l$, and the system is in static equilibrium when $\theta = 120^\circ$. Determine the stiffness k of the spring, and then evaluate the stability of the equilibrium position.

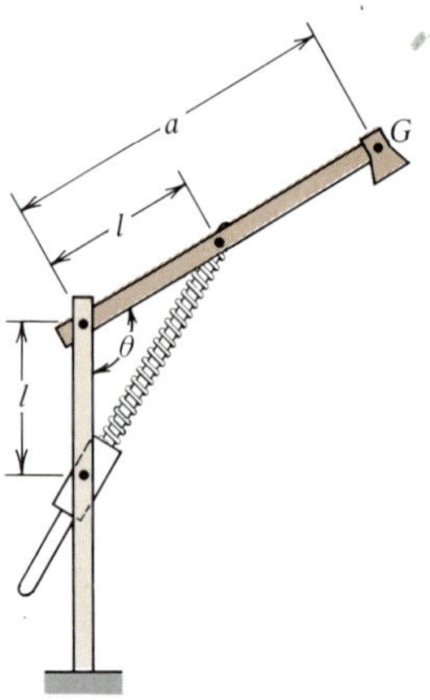

PROB. 9.58

9.59 Lightweight bars AB and CD support bar BC whose mass is m. The identical cross-bracing springs have an unstretched length of $0.75l$. Determine the range of values of k for which the static equilibrium position $\theta = 0$ is stable.

***9.60** Lightweight bars AB and CD support the 50 kg bar BC. The distance $l = 800$ mm and the unstretched length of the identical cross-bracing springs is 600 mm. Determine the stiffness k of each spring required for static equilibrium at $\theta = 53.13^\circ$. Then describe the stability of this equilibrium position.

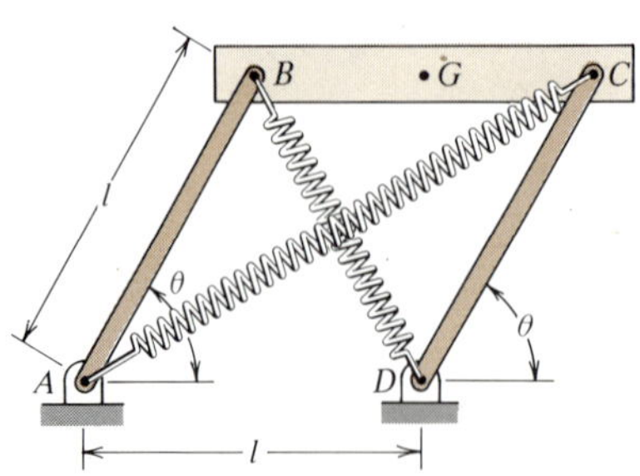

PROBS. 9.59 AND 9.60

***9.61** The sphere at end A has a mass m, and the bars have negligible mass. The spring is unstressed in the position corresponding to $\theta = 0$. Derive an expression for the stiffness k of the spring that is required for static equilibrium at a specified value of θ. Then determine the stability of the corresponding equilibrium position.

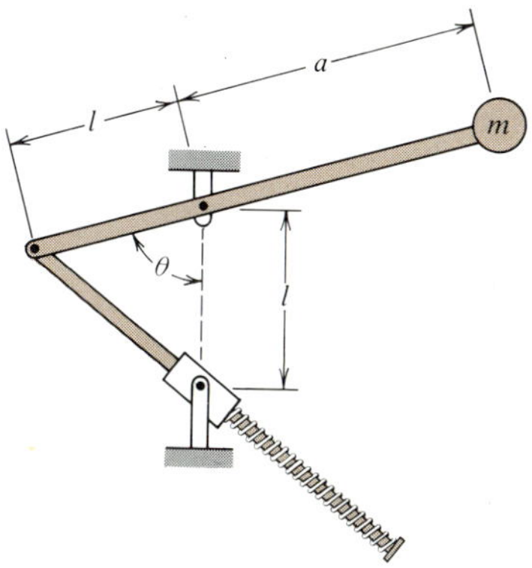

PROB. 9.61

***9.62** The lightweight T-bar $ABCD$ supports a block of mass m. In the position shown, the identical springs have equal tensions T_0. Determine the range of values of the stiffness k of each spring for which the illustrated position is a stable equilibrium position. *Hint:* Consider the angle of rotation of the T-bar from the vertical position to be sufficiently small to consider points A and C to move only vertically.

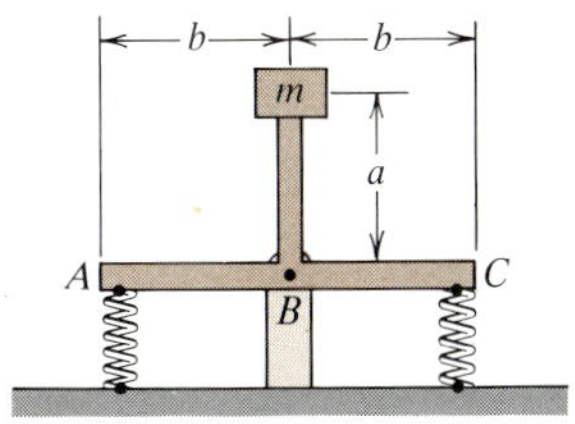

PROB. 9.62

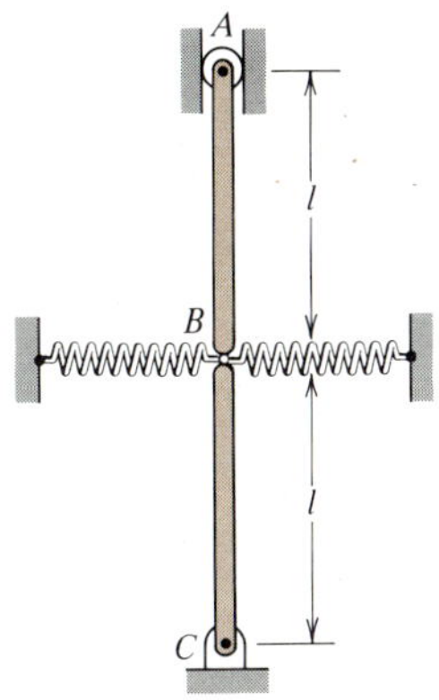

PROB. 9.63

***9.63** Bars AB and BC each have a mass m. When the bars are in the vertical position shown, the identical springs have a tension P. Determine the range of values of the stiffness k of each spring for which the vertical position is a stable equilibrium position. *Hint:* Consider an infinitesimal rotation of bar BC from a nonvertical initial angle θ. Also, consider θ to be sufficiently small to express the spring deformation as $l\theta$.

CHAPTER 10

FUNDAMENTAL CONCEPTS

OBJECTIVES

1. To review the concept of creating particle, rigid body, and deformable body models.

2. To understand the definitions and physical meaning of velocity and acceleration.

3. To review Newton's laws of particle motion.

4. To review the SI and U.S. Customary systems of units.

5. To review the basic operations of vector algebra in terms of components.

6. To develop the capability to transform the description of a vector from one set of coordinate axes to a related set.

7. To understand the basic theorems describing computations in vector calculus.

An investigation of the *dynamics* of a physical system, such as a machine, reveals the relationship between the forces acting on the system and the motion of that system. Many of us develop an intuitive understanding of this subject through our daily experiences. The study of dynamics will provide quantitative predictions of the relationship between forces and motion. It will also reveal situations where our intuition is wrong.

The methods we employ to describe forces were developed in a course on statics. The primary new feature here will be that the systems are moving.

A. BASIC DEFINITIONS

The investigation of the dynamic response of a physical system can generally be broken into two parts, one treating the subject of *kinematics*, and the other the subject of *kinetics*. A kinematical study is one that deals with methods for describing the motion of a system without regard to the role of the forces causing the motion. A kinetics study considers the relationship of the forces to the kinematical variables representing the motion of the system.

A key phrase used above is *physical system*. By it we mean a collection of basic atomic elements that combine to form bodies. In the field of mechanics, for the convenience of problem solving, we tend to categorize these collections of elements into three broad groupings: particles, rigid bodies, and deformable bodies.

PARTICLE A body whose dimensions are negligible is said to be a particle. It follows, then, that a particle occupies only a single point in space.

RIGID BODY A body occupying more than one point in space is said to be rigid if all of the constituent elements of matter within the body are always at fixed distances from each other.

DEFORMABLE BODY A body is said to be deformable if its constituent elements of matter experience changes in their distances from each other that are significant to the problem being investigated.

Clearly there is a certain amount of ambiguity in these definitions. For instance, is the group of molecules that compose a gas a system of particles or a deformable body? Another basic question we could ask is whether any body can correctly be considered to be rigid, because all real materials deform when forces are exerted on them.

There are no absolute answers to these questions, for the *model* of the system we form depends on what knowledge we wish to gain about the response of the system.

B. THE MODELING PROCESS

Our general approach in studying a physical system begins by constructing a conceptual model of the system. That is, we consider its components to be either particles, rigid bodies, or deformable bodies. We then employ our knowledge of kinematics to describe the motion of the chosen model. The response of the system is determined by relating the results of the kinematical study to the force concepts of kinetics. The general approach to the subject of dynamics can be summarized in the conceptual diagram given in Figure 1.

The dotted paths in Figure 1 illustrate a key feature of the modeling process. That is, to be sure of the validity of the model initially chosen to represent the system, it is necessary that the model display any relevant phenomena that are observed experimentally. If not, the model must be improved. In part, modeling is an educated trial-and-error procedure. That is, the modeling process is an art based on prior experience and physical intuition. On the other hand, the modeling process is also a science based on a knowledge of the purpose of the physical system and of the available analytical methods to study the phenomena exhibited by the system.

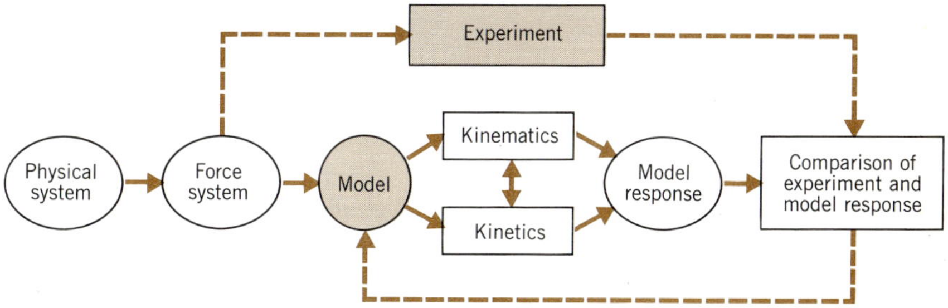

FIGURE 1

As an example, consider the case of the flight of a golf ball. We might tend to model the ball as a particle occupying a single point in space. As will be seen in our studies, there are some situations where such an approach results in a successful analysis; that is, the theoretical and experimental results are in good agreement. On the other hand, considering the golf ball to be a particle does not explain why the ball hooks, slices, and otherwise deviates from a straight course when the wind is calm. The latter phenomena stem from the aerodynamic forces produced by the spin of the golf ball. An accurate study of a golf ball's flight requires that it be modeled as a rigid body, as opposed to a simple particle. Carrying these thoughts one step further, if the problem was one of determining how a golf club imparts energy to a ball, we would quickly find that the correct model would be one where the golf ball (and club) must be modeled as deformable bodies. From the foregoing example we see that the model chosen to represent the physical system is a crucial element in the solution process.

This book does not treat the dynamics of deformable bodies. However, many of the principles and techniques that we develop here in studying particles and rigid bodies are also employed in the study of deformable bodies.

C. FOUNDATIONS OF NEWTONIAN MECHANICS

An investigation of dynamics requires a quantified description of motion. As you probably recall from your previous courses in elementary physics, the three fundamental kinematical quantities used to describe the motion of a point are position, velocity, and acceleration. These basic terms are defined with respect to a frame of reference that describes the three-dimensional space within which the motion occurs. This frame of reference can be visualized as a set of rectangular Cartesian axes xyz.

The laws of motion stated by Sir Isaac Newton (1642–1727) in his historic work *Principia* (1687) are the foundation of the study of dynamics. To employ these laws we must postulate that the frame of reference is *absolute*, which means that it is *fixed*, that is, stationary, in space. In most situations, the motion of the earth is negligible in comparison to the motion of the object of interest. The absolute reference frame in such cases may be considered to be attached to the earth.

Although considering the earth to be fixed is not exactly correct, the errors that result from this assumption are shown in Chapter 15 to be negligible for most systems. Indeed, the assumptions associated with the modeling process usually introduce errors having a greater magnitude.

The notion that anything, such as a reference frame, is fixed in space conflicts with the concepts of the theory of relativity, which tell us that nothing is stationary. Fortunately, it can be shown that the laws of the theory of relativity reduce to the simpler laws of Newtonian mechanics in the special case where bodies are moving much more slowly than the speed of light. This is certainly the case for most systems of engineering significance.

Let us now return to the basic kinematical quantities. The location of a point P within an arbitrarily chosen xyz reference frame cannot be specified until we know both the length and the orientation of the line from the origin O to point P. For example, the location of point P with respect to the origin O in Figure 2 is described by the *position vector* $\bar{r}_{P/O}$. (The notation should be read as "r of P with respect to O").

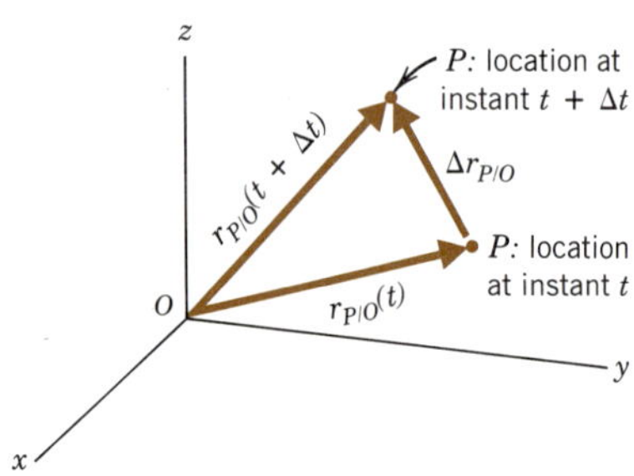

FIGURE 2

You will note that Figure 2 indicates that as point P moves in space its position vector changes, so that $\bar{r}_{P/O}$ is a function of time. The vector $\Delta\bar{r}_{P/O}$ represents the change in the vector $\bar{r}_{P/O}$ in the time interval from t to $t + \Delta t$. According to the rules for adding vectors diagrammatically, from Figure 2 we may write

$$\bar{r}_{P/O}(t) + \Delta\bar{r}_{P/O} = \bar{r}_{P/O}(t + \Delta t)$$

so that

$$\Delta\bar{r}_{P/O} = \bar{r}_{P/O}(t + \Delta t) - \bar{r}_{P/O}(t)$$

Division of $\Delta\bar{r}_{P/O}$ by Δt yields a quantity that resembles the time derivative of a scalar function

$$\frac{df}{dt} \equiv \lim_{\Delta t \to 0} \frac{\Delta f}{\Delta t} = \lim_{\Delta t \to 0} \frac{f(t + \Delta t) - f(t)}{\Delta t}$$

A similar definition is used for differentiation of a vector function, where the time rate of change of the position vector $\bar{r}_{P/O}$ is called the *velocity* $\bar{v}_P$. This is the second of the three basic kinematical quantities. Thus

$$\bar{v}_P \equiv \frac{d}{dt}\bar{r}_{P/O} \equiv \lim_{\Delta t \to 0} \frac{\Delta \bar{r}_{P/O}}{\Delta t} \equiv \lim_{\Delta t \to 0} \frac{\bar{r}_{P/O}(t + \Delta t) - \bar{r}_{P/O}(t)}{\Delta t} \tag{1}$$

The notation for velocity has been simplified slightly by not indicating that $\bar{v}_P$ is defined with respect to point O. This is permissible because the velocity with respect to all fixed points is the same, as will be proven in Chapter 11, Section E.

The velocity $\bar{v}_P$, being a vector, has magnitude and direction. The magnitude is defined as the *speed* v_P: $v_P = |\bar{v}_P|$. The speed expresses a rate of motion (e.g., 50 km/hr) and is a scalar quantity. When we are requested to determine how fast a point is moving, it is synonymous with being asked to find the speed. The direction of $\bar{v}_P$ tells us in which direction the point P is traveling. We shall have more to say about speed and velocity in the next chapter.

The third of the basic kinematical quantities, acceleration, is defined in a manner that is similar to eq. (1) for velocity. The *acceleration* $\bar{a}_P$ is the time rate of change of the velocity. This means

$$\bar{a}_P = \frac{d}{dt}\bar{v}_P = \lim_{\Delta t \to 0} \frac{\bar{v}_P(t + \Delta t) - \bar{v}_P(t)}{\Delta t} \tag{2}$$

The kinematical quantities of position and velocity are pieces of information that our eyes and brain are attuned to sense. Were it not for Newton's laws of motion, it would not be obvious that we should also be interested in acceleration. Because Newton's laws are the basis for our future studies we restate them here, in modern language.

FIRST LAW A particle will remain at rest or move with constant speed along a straight line, unless it is acted upon by a resultant force.

SECOND LAW When a resultant force is exerted on a particle, the acceleration of that particle is parallel to the direction of the force and the magnitude of the acceleration is proportional to the magnitude of the force.

THIRD LAW Each force exerted upon a body is the result of an interaction with another body, the forces of action and reaction being equal in magnitude, opposite in direction, and collinear.

In our studies we shall find that the first law is included in the second; it merely emphasizes that the state of motion of a particle can be changed only by the action of forces. The third law helps to define what forces are acting on a body.

The first and third laws are the only ones needed to solve problems in statics. The added feature in dynamics is the application of the second law. This law may be reworded to state that the force $\bar{F}$ acting on a particle is proportional to the acceleration $\bar{a}_P$ of the particle. As you probably know, the constant of proportionality in this relation is the mass m of the particle. Thus, we have the familiar equation

$$\bar{F} = m\bar{a}_P \tag{3}$$

Before we can use this equation to solve kinetics problems, we must be sure that the units chosen to describe each term are mutually consistent. In the next section we consider this topic.

D. SYSTEMS OF UNITS

Acceleration, mass, and force are dimensional quantities whose magnitude is defined with respect to a set of basic units. Because they are related by Newton's second law, eq. (3), the units chosen for the measurement of these quantities must satisfy the *law of dimensional homogeneity*. This means that the quantities on either side of the equality sign must have the same dimension—in colloquial language, "apples cannot be equated to oranges."

Let us denote the dimensions of length, time, mass, and force as L, T, M, and F, respectively. The magnitude of position has dimensions of length, so eqs. (1) and (2) mean that the dimensions of acceleration are L/T^2. Dimensional homogeneity of eq. (3) requires that

$$F = M(L/T^2)$$

This equation states that the dimensions F, M, L, and T are interrelated. Thus, the basic units chosen to measure three of these dimensions may be chosen arbitrarily, whereas the units for the fourth dimension must be consistent with this relationship.

Historically, the units for measuring length and time are well defined and, through the efforts of international standards committees, universally accepted. This leaves us with the choice of defining either a standard unit of mass or of force, and then deriving the unit for the undefined quantity. For this, we apply Newton's second law to relate the weight w and acceleration g of a body that is falling freely in a vacuum at the earth's surface.

$$w = mg \tag{4}$$

Mass is a basic property of a body that does not vary with location. Thus, a system of units is said to be *absolute* if a unit of mass is defined according to some standard and the unit of force is obtained from eq. (4). In contrast, a system of units that employs force as a standard unit is said to be *gravitational*.

Presently, the systems of units in widest usage are the metric absolute and U.S. Customary (British gravitational) systems. In both systems, the value for the derived dimension will depend on the value of g, whose value varies slightly for different locations on the earth's surface. For most problems we shall use

$$g = 9.807 \text{ m/s}^2 \text{ (metric)}$$

$$g = 32.17 \text{ ft/s}^2 \text{ (U.S. Customary)}$$

$$g = 386.0 \text{ in./s}^2 \text{ (U.S. Customary)}$$

These are average values and will be examined in Example 5, Chapter 15.

In Table 1 we summarize the commonly used systems of units and indicate how the derived units are obtained. The basic units in each system are defined with respect to standard bodies or with respect to physical phenomena. For example, a kilogram is the mass of a particular bar of platinum alloy. On the other hand, a meter is defined as 1,650,763.73 wavelengths in a vaccum of the orange-red line of krypton-86. The derived quantities in the table are the starred values.

The choice for systems of units presented in Table 1 is further complicated by the fact that certain multiples and submultiples of the basic units in each system are given their own names, such as the mile, which is 5280 ft, or the dyne, which is one hundred-thousandth of a newton. To simplify this situation, the SI (Standard International) system is now being adopted worldwide.

TABLE 1
SYSTEMS OF UNITS ***DERIVED QUANTITIES**

SYSTEM	LENGTH	TIME	MASS	FORCE
Metric absolute	Meter (m)	Second (s)	Kilogram (kg)	Newton (N)* = kg-m/s^2
Metric gravitational	Meter (m)	Second (s)	Metric slug* = kg-s^2/m	Kilogram (kg)
British absolute	Foot (ft)	Second (s)	Pound (lb)	Poundal (pdl)* = lb-ft/s^2
British gravitational (U.S. Customary)	Foot (ft)	Second (s)	Slug* = lb-s^2/ft	Pound (lb)

Essentially, the SI system is the metric absolute system with only one unit being used to describe each basic quantity. The SI units for length, time, mass, and force are meters, seconds, kilograms, and newtons, respectively. Decimal multiples and submultiples of the basic units are indicated by the appropriate prefixes. Thus, in the SI system, lengths of 0.2 meters and 200 millimeters are identical and correct. On the other hand, $2(10^{-3})$ newtons, or equivalently 2 millinewtons, is the correct description for a force whose magnitude is 200 dynes. Lists of the SI units and of the preferred prefixes may be found in Appendix A.

Many of the examples and problems in this text will be presented in SI units. Unfortunately, this system is not yet universally used in the United States, where the U.S. Customary system has traditionally been employed. Both systems of units will therefore be featured in the examples and problems presented in this text. In general, we shall solve problems in the same system of units as that used for the given information; we shall not convert between different systems of units.

Obviously, the units of any physical quantity can be changed with the aid of conversion factors. These conversion factors are numbers that describe the equivalence between the measurement of the same quantity using two different systems of units, such as 25.40 mm = 1 in. From this equivalence equation, we may form a ratio to cancel the original system of units of a quantity in favor of the desired system. For example,

$$8 \text{ in.} = 8 \text{ in.}\left(\frac{25.40 \text{ mm}}{1 \text{ in.}}\right) = 203.2 \text{ mm}$$

A brief set of conversion factors is given in Appendix A.

Conversion factors may be employed in a sequential cancellation of units. For instance, to change from a speed in miles per hour to a speed in meters per second, we can compute

$$1\,\frac{\text{mi}}{\text{hr}}\left(\frac{1 \text{ hr}}{3600 \text{ s}}\right)\left(\frac{5280 \text{ ft}}{1 \text{ mi}}\right)\left(\frac{12 \text{ in.}}{1 \text{ ft}}\right)\left(\frac{25.40 \text{ mm}}{\text{in.}}\right)\left(\frac{1 \text{ m}}{1000 \text{ mm}}\right) = 0.4470 \text{ m/s}$$

Incidentally, a conversion factor that is contained in the preceding equation is that 1 mi/hr = (88/60) ft/s. Our main use for this factor will be to convert selected kinematical data into units that are consistent with the law of dimensional homogeneity, as given in Table 1.

It is important to recognize that equations relating different variables are often stated in a dimensional form. For example, the statement: "$v = 40t$ where v is m/s and t is seconds" is a dimensional equation. In such situations the numerical coefficients must be regarded as having dimensions, in order that the law of dimensional homogeneity be satisfied. This means that changing the units for the variables in a dimensional equation necessitates a corresponding change in the units of the coefficients.

EXAMPLE 1

A standard 1 kg body is known to weigh 2.205 lb. Use the fact that 1 in. = 25.40 mm as the only other conversion factor between U.S. Customary and SI units to convert the density unit of 1 slug/ft^3 into its SI equivalent.

Solution

Recall that 1 slug is defined as 1 lb-s^2/ft. A conversion factor for mass may be obtained from the given information with the aid of Newton's second law. It is not stated where the body was weighed, so we employ $g = 32.17$ ft/s^2 to obtain

$$m = 1 \text{ kg} = \frac{w}{g} = \frac{2.205 \text{ lb}}{32.17 \text{ ft/s}^2} = 0.06854 \text{ slugs}$$

Then

$$1\,\frac{\text{slug}}{\text{ft}^3} = \left(\frac{1 \text{ slug}}{\text{ft}^3}\right)\left(\frac{1 \text{ kg}}{0.06854 \text{ slugs}}\right)\left[\left(\frac{1 \text{ ft}}{12 \text{ in.}}\right)\left(\frac{1 \text{ in.}}{25.40 \text{ mm}}\right)\left(\frac{1000 \text{ mm}}{1 \text{ meter}}\right)\right]^3$$

$$= 515.2 \text{ kg/m}^3$$

The differences between the values obtained here and those in the table in Appendix A are attributable to the fact that we were using only four significant figures for our calculations, which is the maximum allowed by our choice for the value of g.

EXAMPLE 2

A 15 lb body is observed to accelerate at 400 in./s^2. Determine the magnitude of the resultant force acting on this body.

Solution

A body is described in the U.S. Customary system in terms of its weight in pounds, whereas its *mass in kilograms is the specified parameter in SI units*. (The unit of pounds mass is employed in the U.S.-British absolute system. We will not use it in this text.)

We may find the mass of the body from $w = mg$. For this we employ $g = 386$ in./s^2 in order to match the units of the given acceleration. Thus

$$m = \frac{w}{g} = \frac{15}{386} = 0.03886 \text{ lb-s}^2/\text{in.}$$

Note that the mass unit is *not* a slug ($\equiv 1$ lb-s^2/ft). This is a consequence of employing a gravitational system of units, in which the mass unit depends on the length unit selected.

The force required to accelerate the body at 400 in./s^2 may now be found from Newton's second law

$$F = ma = (0.03886 \text{ lb-s}^2/\text{in.})(400 \text{ in./s}^2)$$
$$= 15.54 \text{ lb}$$

Additional Remarks

Both of these examples are illustrative of the general question of the number of significant figures to be employed in a calculation. Each example involved the value of g, which has only three significant figures when the location on the earth's surface is not specified.

Another factor in this regard is the matter of the technique of computation. For example, some people write out intermediate results appearing on their calculator, retaining three or four significant figures, and then reenter these values when they are required. In contrast, others exploit the full internal accuracy of their calculator by storing all intermediate results. A comparison of the results obtained by either technique might show substantial discrepancies in the third significant figure.

For these reasons, we will regard intermediate results to have four place accuracy, but we will round off final results to three significant figures. In any case, the first two digits of an answer obtained by any computational scheme should certainly agree!

HOMEWORK PROBLEMS

The symbols employed in the following problems are: t for time, x for distance, v for speed, a for the magnitude of acceleration, m for mass, and F for the magnitude of force.

10.1 Use the conversion factors for length and force given in Appendix A to derive conversion factors for changing the following U.S. Customary units into their SI equivalents. (a) area: 1 ft^2, (b) acceleration: 1 mi/hr^2, (c) pressure: 1 lb/in.^2

10.2 Use the conversion factors for length and mass given in Appendix A to derive conversion factors for changing the following SI units into their U.S. Customary unit equivalents. (a) speed: 1 km/hr, (b) volume: 1 liter, (c) mass per unit surface area: 1 kg/m^2.

10.3 Use the facts that a 1 kg body weighs 2.205 lb, and that 25.40 mm = 1 in. to convert $F = 60$ newtons into units of pounds.

10.4 Use the facts that a 1 kg body weighs 2.205 lb, and that 1 m = 3.281 ft, to convert m = 25 $\text{lb-s}^2\text{/in.}$ into kilogram units.

10.5 The power developed by a force is Fv. In the SI system the basic unit of power is a watt (W), a watt being the power developed by a 1 N force when it moves a body at a speed of 1 m/s. In the U.S. Customary system the basic unit of power is a horsepower, a horsepower being the power developed by a 550 lb force when it moves a body at a speed of 1 ft/s. Determine (a) how many kilowatts there are in a horsepower, and (b) how many horsepower there are in a kilowatt.

10.6 A 50 kg body is pushed by a 600 newton force. Determine the acceleration of this body is units of (a) m/s^2, (b) km/hr/s, (c) ft/s^2.

10.7 A restaurant buys 1000 lb of coffee at a standard locality and also a spring scale calibrated at that locality. Determine what the coffee weighs on the spring scale in a locality where (a) $g = 32.20\ \text{ft/s}^2$, (b) $g = 9.800\ \text{m/s}^2$.

10.8 In each of the following formulas c_1, c_2, etc., denote constants and θ is an angle in radians. Determine the dimensions of these constants if the formula is to be dimensionally correct. (a) $a = c_1 v^2/x$, (b) $\frac{1}{2}mv^2 = \frac{1}{2}c_1 x^2$, (c) $x = c_1 x_0^2 + c_2 v_0 + c_3 t^2$, (d) θ (degrees) $= c_1 \theta$ (radians), (e) $(d\theta/dt)^2 = c_1 x + c_2 t$, (f) $t = c_1 \sqrt{x}$, (g) $mc_1^2\, d^2\theta/dt^2 = Fv$.

10.9 At a certain large university known for its athletic prowess, the following new set of physical units are in use. The basic length unit is the *touchdown*, which is 100 yd, the basic time unit is the *dash*, which is 9 s, and the basic force unit is the *golf ball*, which is 1.620 oz. Determine the conversion factors between these units and their SI equivalents.

10.10 At the university in Problem 10.9 the unit of mass is the *basket*, one basket being one golf ball-dash^2/touchdown. Determine the number of kilograms in one basket.

*10.11 The relation between v and x in a certain motion is known to be $v = 50 \sin(\pi x^2)$, where v is km/hr and x is km. (a) Evaluate the dimensionality of this equation, and from that result determine the units of the coefficients "50" and "π." (b) Determine the corresponding relation in which v is m/s and x is meters. (c) Determine the corresponding relation in which v is ft/s and x is feet.

* Problems preceded by an asterisk may be expected to present a greater degree of difficulty for the typical student. Refer to the preface for a discussion of the ordering of homework problems in a set.

E. VECTORS

Many of the quantities that occur in the study of dynamics behave as vectors, which means that they have both magnitude and direction. This section reviews the basic operations of vector algebra, which were developed in the subject of statics. Some basic theorems of vector calculus will also be presented. They form the foundation for studying the kinematical relationships between position, velocity, and acceleration.

1. Vector Algebra

A vector is displayed graphically by means of an arrow whose orientation shows the direction of the vector. The length of the arrow is proportional to the magnitude of the vector. This is the scheme employed in Figure 3.

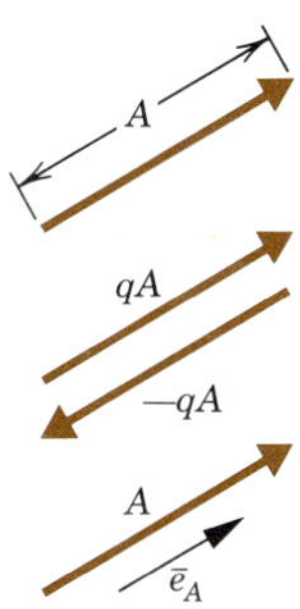

FIGURE 3

(a) Multiplication by a Scalar If q is a scalar quantity, then the product $q\bar{A}$ is defined to have the property that

$$|q\bar{A}| = |q|\,|\bar{A}|$$

The resulting vector $q\bar{A}$ is parallel to $\bar{A}$, in the same sense as $\bar{A}$ if q is positive, and in the opposite sense if q is negative.

The concept of a unit vector is an important application of the multiplication by a scalar. Suppose that we multiply $\bar{A}$ by $(1/|\bar{A}|)$. The result of this operation is a vector $\bar{e}_A$, parallel to $\bar{A}$, whose magnitude is the *dimensionless* number one. We call $\bar{e}_A$ a *unit vector*. Thus

$$\bar{e}_A = \frac{\bar{A}}{|\bar{A}|} \tag{5}$$

As you may recall from statics, this equation proves to be very useful for representing vector quantities.

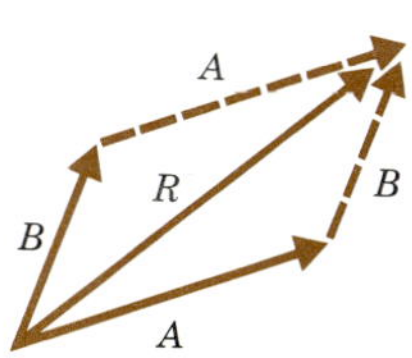

FIGURE 4

(b) Addition of Vectors If two quantities having magnitude and direction are actually vectors, they must add according to the *parallelogram law*. In this law, the tails of two vectors $\bar{A}$ and $\bar{B}$ are made to coincide, in order to form two adjacent sides in a parallelogram, as shown in Figure 4. The resultant $\bar{R} = \bar{A} + \bar{B}$ is formed by the diagonal that intersects the tails of $\bar{A}$ and $\bar{B}$.

An equivalent rule for adding vectors may also be obtained from Figure 4. Specifically, consider the triangle formed by making the tail of one vector coincide with the head of the other. The resultant is the third side of the triangle.

The difference of two vectors can be obtained as a corollary of the addition rule. Subtracting $\bar{B}$ from $\bar{A}$ is the same as adding $-\bar{B}$ to $\bar{A}$, where $-\bar{B} \equiv (-1)\bar{B}$ is a vector having the same magnitude as $\bar{B}$, but opposite direction. Thus

$$\bar{C} = \bar{A} - \bar{B} \equiv \bar{A} + (-\bar{B})$$

The difference $\bar{C}$ could be depicted in Figure 4 as the vector extending along the diagonal from the tip of $\bar{B}$ to the tip of $\bar{A}$.

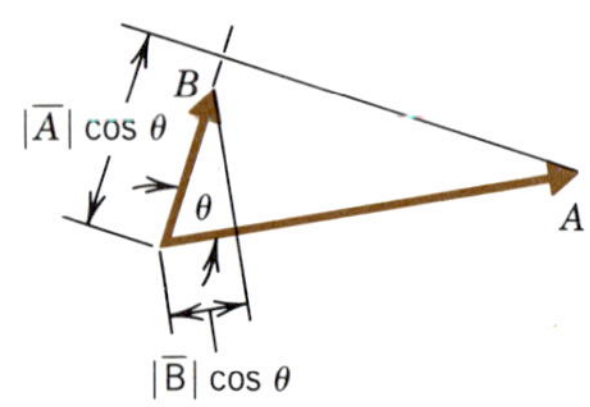

FIGURE 5

(c) Dot Product of Vectors The product $\bar{A} \cdot \bar{B}$ is defined to be the product of the magnitude of one vector with the component of the other vector in the direction of the first. From Figure 5 we see that this definition can be written mathematically as

$$\bar{A} \cdot \bar{B} = \bar{B} \cdot \bar{A} \equiv |\bar{A}|\,|\bar{B}| \cos \theta$$

where θ is the angle between the vectors $\bar{A}$ and $\bar{B}$. Notice that the dot product will be negative if $\theta > 90°$, which means that the projection of $\bar{A}$ onto $\bar{B}$ is opposite the sense of $\bar{B}$, or vice versa. Also, from this definition, $\bar{A} \cdot \bar{A} = |\bar{A}|^2$.

(d) Cross Product of Vectors The definition of the cross product $\bar{A} \times \bar{B}$ is based on the fact that two intersecting vectors $\bar{A}$ and $\bar{B}$ define a flat plane. The cross product is defined to be a vector normal (perpendicular) to the plane of the two vectors. To specify the sense of this normal we use the "right-hand rule," in accordance with which we curl the fingers of our right hand from the first vector to the second and define the sense of the result to be in the direction of the oustretched thumb, as shown in Figure 6.

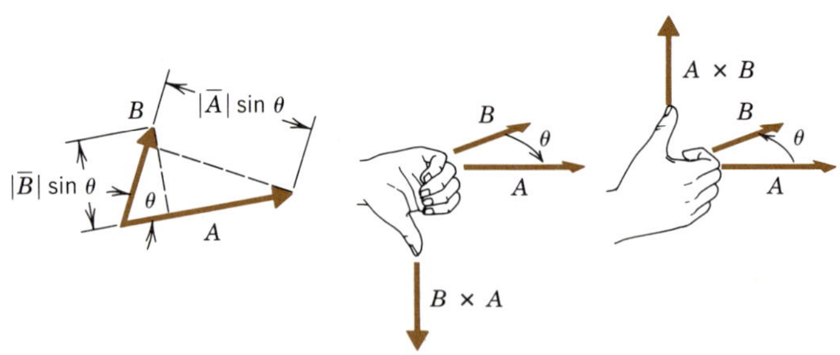

FIGURE 6

The definition of the magnitude of the cross product is

$$|\bar{A} \times \bar{B}| \equiv |\bar{A}|\,|\bar{B}| \sin \theta$$

From Figure 6 it can be seen that the definition involves the component of one vector normal to the other vector.

Note that $\bar{B} \times \bar{A}$ and $\bar{A} \times \bar{B}$ have the same magnitude, but they are oriented in opposite directions, as shown in Figure 6. It follows that

$$\bar{B} \times \bar{A} = -\bar{A} \times \bar{B}$$

(e) Theorems of Vector Algebra The definition of the basic operations leads to rules for combining these operations that resemble those of scalar algebra. Thus

$$q(\bar{A} + \bar{B}) = q\bar{A} + q\bar{B}$$

$$q(\bar{A} \cdot \bar{B}) = (q\bar{A}) \cdot \bar{B} = \bar{A} \cdot (q\bar{B})$$

$$q(\bar{A} \times \bar{B}) = (q\bar{A}) \times \bar{B} = \bar{A} \times (q\bar{B})$$

$$\bar{C} \cdot (\bar{A} + \bar{B}) = (\bar{C} \cdot \bar{A}) + (\bar{C} \cdot \bar{B})$$

$$\bar{C} \times (\bar{A} + \bar{B}) = \bar{C} \times \bar{A} + \bar{C} \times \bar{B}$$

(f) Rectangular Components Describing vectors diagrammatically by using arrows is an approach that has two serious shortcomings. First, three-dimensional vectors are difficult to represent in a diagram. Second, even for two-dimensional vectors, the addition of vectors according to the parallelogram law requires extensive use of the trigonometry of scalene triangles. To circumvent these difficulties, throughout the text we shall employ the alternative approach of representing vectors by their components with respect to a set of orthogonal (mutually perpendicular) directions.

In order to describe the components of a vector $\bar{A}$ with respect to a rectangular Cartesian coordinate system xyz, we place the origin of the coordinate axes at the tail of the vector. We then drop perpendiculars (indicated by dashed lines in Figure 7) from the tip of the vector onto the three coordinate axes. From the intersections of these perpendiculars with the axes, we form the rectangular parallelipiped depicted in the figure. The lengths A_x, A_y, and A_z are the components of the vector $\bar{A}$ with respect to the xyz coordinate system.

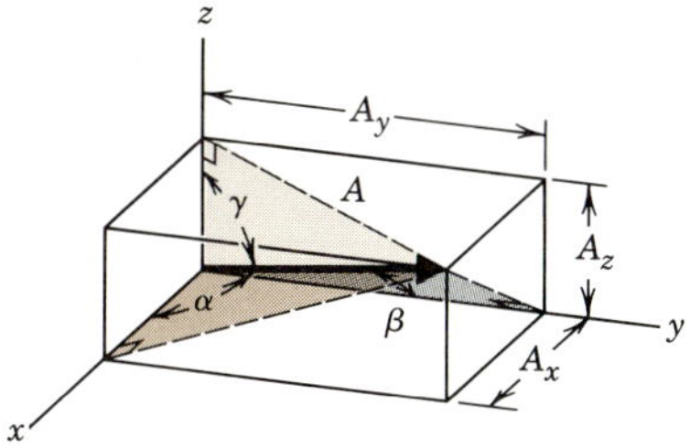

FIGURE 7

In terms of the angles α, β, γ between the vector $\bar{A}$ and the respective coordinate axes, we have

$$A_x = |\bar{A}| \cos \alpha \qquad A_y = |\bar{A}| \cos \beta \qquad A_z = |\bar{A}| \cos \gamma \tag{6}$$

The angles α, β, y are called the *direction angles*; the cosines of the direction angles are the *direction cosines*. Because $\bar{A}$ is a diagonal for the parallelipiped, it follows that

$$|\bar{A}| = [A_x^2 + A_y^2 + A_z^2]^{1/2} \tag{7}$$

Substituting equations (6) into equation (7) yields

$$\cos^2 \alpha + \cos^2 \beta + \cos^2 \gamma = 1$$

which shows that only two of the direction angles are independent quantities.

From the foregoing we see that the components of a vector completely describe the properties of a vector, provided that we know which coordinate system is associated with the components. To specify the coordinate system we use the unit vectors $\bar{i}, \bar{j}$ and $\bar{k}$, which are oriented in the directions of the

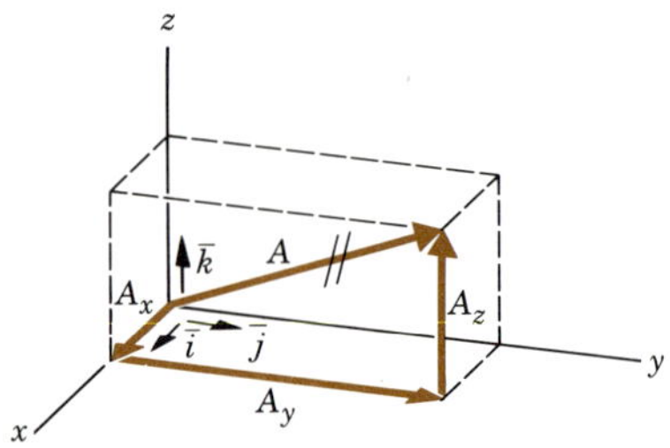

FIGURE 8

positive x, y, and z axes, respectively. Then, for example, eq. (5) shows that $A_x\bar{i}$ is a vector parallel to the x axis whose magnitude is the component A_x. Figure 8, which illustrates the three component vectors, also shows that the vector sum of these components is $\bar{A}$. In other words,

$$\bar{A} = A_x\bar{i} + A_y\bar{j} + A_z\bar{k} \tag{8}$$

This is the *vector component representation* of $\bar{A}$.

An important result of representing vectors by their components is interpretation of the equality of vectors: Two vectors can be equal only when their components are equal. In other words,

$$\bar{A} = \bar{B} \quad \text{means} \quad A_x = B_x, \qquad A_y = B_y, \qquad A_z = B_z \tag{9}$$

Equation (9) is particularly useful for our studies, as it shows that a vector equation can be subdivided into the three scalar equations obtained by equating corresponding components from each side of the equation.

In general, the various vector operations are more easily performed in terms of components than from their basic definitions. To obain the unit vector $\bar{e}_A$, we substitute eqs. (7) and (8) into eq. (5), with the result that

$$\bar{e}_A = \frac{\bar{A}}{|\bar{A}|} = \frac{A_x\bar{i} + A_y\bar{j} + A_z\bar{k}}{\sqrt{A_x^2 + A_y^2 + A_z^2}} \tag{10a}$$

Equation (10*a*) is the form used when the components of the vector $\bar{A}$ are known. When the direction angles are known, it is more convenient to substitute eqs. (6) and (8) into eq. (5). This yields

$$\bar{e}_A = \frac{\bar{A}}{|\bar{A}|} = \cos\alpha\,\bar{i} + \cos\beta\,\bar{j} + \cos\gamma\,\bar{k} \tag{10b}$$

EXAMPLE 3

The bent rod *ABC* is supported by cable *CD*, which carries a tensile force of 4.5 kN. Determine the vector component representation of the force exerted by the cable on the rod.

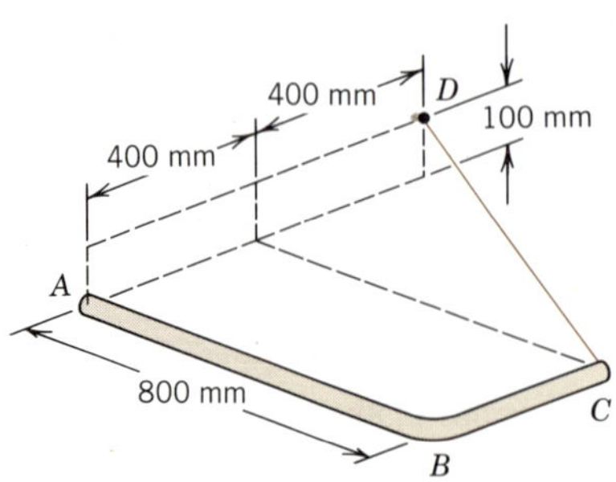

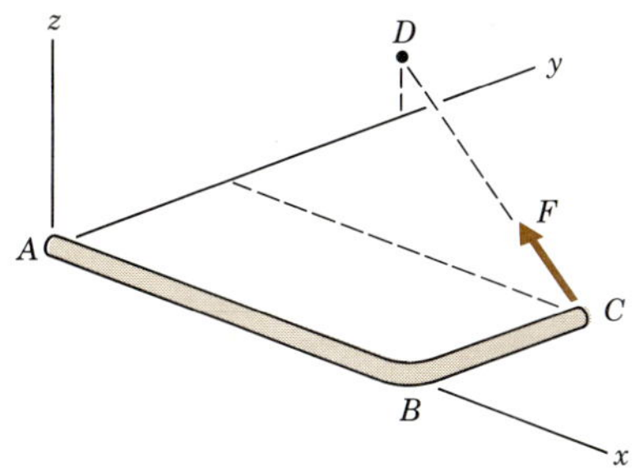

Solution

As shown in the second sketch, the tensile force exerted by the cable on the rod is in the same direction as $\bar{r}_{D/C}$, and its magnitude is 4.5 kN. Thus

$$\bar{F} = |\bar{F}|\bar{e}_{D/C} = 4.5 \frac{\bar{r}_{D/C}}{|\bar{r}_{D/C}|} \text{ kN}$$

The coordinate axes in the sketch match the directions for the given dimensions. The vector $\bar{r}_{D/C}$ is constructed by considering a movement from point C to point D that follows the coordinate axes. Then

$$\bar{F} = (4.5)\frac{-0.8\bar{i} + 0.4\bar{j} + 0.1\bar{k}}{[(0.8)^2 + (0.4)^2 + (0.1)^2]^{1/2}}$$
$$= -4\bar{i} + 2\bar{j} + 0.5\bar{k} \text{ kN}$$

(g) Computations Using Vector Components The operations occurring in a vector equation are readily performed by employing the basic theorems of vector algebra.

The computation of the sum or difference of two vectors is straightforward, for

$$\bar{A} \pm \bar{B} = (A_x \pm B_x)\bar{i} + (A_y \pm B_y)\bar{j} + (A_z \pm B_z)\bar{k} \tag{11}$$

By induction, eq. (11) can be extended to treat more than two vectors.

The dot product is obtained by first evaluating the products of the unit vectors using the definition of the dot product. For example,

$$\bar{i}\cdot\bar{i} = |\bar{i}|\,|\bar{i}| \cos 0 = 1$$

Accordingly,

$$\begin{aligned} &\bar{i}\cdot\bar{i} = \bar{j}\cdot\bar{j} = \bar{k}\cdot\bar{k} = 1 \\ &\bar{i}\cdot\bar{j} = \bar{j}\cdot\bar{i} = \bar{j}\cdot\bar{k} = \bar{k}\cdot\bar{j} = \bar{k}\cdot\bar{i} = \bar{i}\cdot\bar{k} = 0 \end{aligned} \tag{12}$$

The dot product of two general vectors then follows from the distributive law.

$$\begin{aligned} \bar{A}\cdot\bar{B} &= (A_x\bar{i} + A_y\bar{j} + A_z\bar{k})\cdot\bar{B} \\ &= A_x[\bar{i}\cdot(B_x\bar{i} + B_y\bar{j} + B_z\bar{k})] \\ &\quad + A_y[\bar{j}\cdot(B_x\bar{i} + B_y\bar{j} + B_z\bar{k})] \\ &\quad + A_z[\bar{k}\cdot(B_x\bar{i} + B_y\bar{j} + B_z\bar{k})] \\ &= A_xB_x + A_yB_y + A_zB_z \end{aligned} \tag{13}$$

A similar approach is employed to develop the cross product in component form. The definition of the cross product, when applied to the unit vectors yields

$$\begin{aligned} &\bar{i}\times\bar{i} = \bar{j}\times\bar{j} = \bar{k}\times\bar{k} = \bar{0} \\ &\bar{i}\times\bar{j} = -\bar{j}\times\bar{i} = \bar{k} \end{aligned} \tag{14}$$

$$\bar{j} \times \bar{k} = -\bar{k} \times \bar{j} = \bar{i}$$
$$\bar{k} \times \bar{i} = -\bar{i} \times \bar{k} = \bar{j}$$

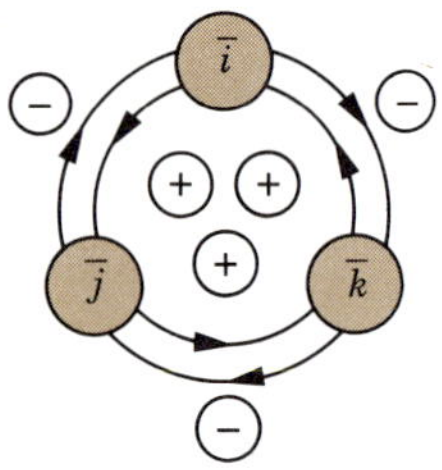

FIGURE 9

A useful device for remembering the cross product of different unit vectors is to consider positive alphabetical order to be a progression from $\bar{i}$ to $\bar{j}$ to $\bar{k}$ back to $\bar{i}$. This corresponds to the counterclockwise sequence shown in Figure 9. Positive alphabetical progression (*ccw*) yields the third unit vector with a positive sign, whereas a negative progression (*cw*) yields the third unit vector with a negative sign.

The distributive law for cross products yields the following sequence of operations for evaluating the product of two general vectors.

$$\begin{aligned}\bar{A} \times \bar{B} &= (A_x\bar{i} + A_y\bar{j} + A_z\bar{k}) \times \bar{B} \\ &= A_x[\bar{i} \times (B_x\bar{i} + B_y\bar{j} + B_z\bar{k}] \\ &\quad + A_y[\bar{j} \times (B_x\bar{i} + B_y\bar{j} + B_z\bar{k})] \\ &\quad + A_z[\bar{k} \times (B_x\bar{i} + B_y\bar{j} + B_z\bar{k})] \\ &= A_x(B_y\bar{k} - B_z\bar{j}) + A_y(-B_x\bar{k} + B_z\bar{i}) + A_z(B_x\bar{j} - B_y\bar{i}) \\ &= (A_yB_z - A_zB_y)\bar{i} + (A_zB_x - A_xB_z)\bar{j} + (A_xB_y - A_yB_x)\bar{k} \qquad (15a)\end{aligned}$$

An alternative way of computing the cross product, which involves the evaluation of a determinant, is a convenient device for expediting calculations. This method gives

$$\bar{A} \times \bar{B} = \begin{vmatrix} \bar{i} & \bar{j} & \bar{k} \\ A_x & A_y & A_z \\ B_x & B_y & B_z \end{vmatrix} \qquad (15b)$$

To illustrate the various vector operations, let us now solve a problem that reviews basic computations in statics.

EXAMPLE 4

Consider the 4.5 kN force $\bar{F}$ exerted by the cable on rod *ABC* in Example 3. Determine the moment of this force about point *A* and about the x axis.

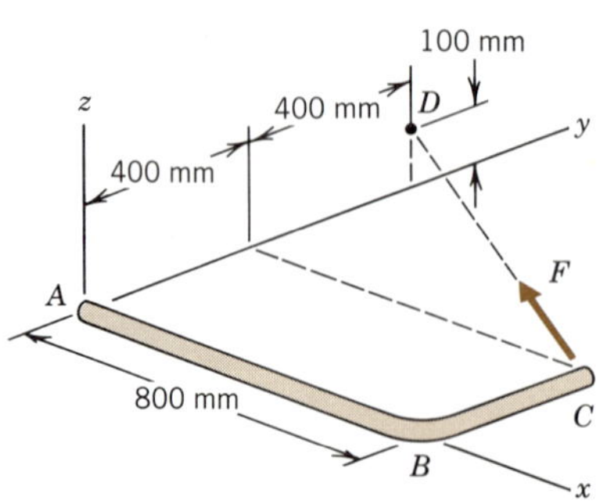

Solution

It was found in Example 3 that

$$\bar{F} = -4\bar{i} + 2\bar{j} + 0.5\bar{k} \text{ kN}$$

The moment about point A is obtained from a cross product with a position vector from point A to any point on the line of action of the force. Point C is convenient for this purpose. This gives

$$\begin{aligned}\bar{M}_A = \bar{r}_{C/A} \times \bar{F} &= (0.8\bar{i} + 0.4\bar{j}) \times (-4\bar{i} + 2\bar{j} + 0.5\bar{k}) \\ &= (0.8)(2)\bar{k} + (0.8)(0.5)(-\bar{j}) + (0.4)(-4)(-\bar{k}) + (0.4)(0.5)(\bar{i}) \\ &= 0.20\bar{i} - 0.40\bar{j} + 3.20\bar{k} \text{ kN-m}\end{aligned}$$

Because point A is situated on the x axis, the moment about the x axis is the component of $\bar{M}_A$ parallel to the axis, directed in the sense of that component. Thus

$$\bar{M}_{Ax} = (\bar{M}_A \cdot \bar{i})\bar{i} = 0.20\bar{i} \text{ kN-m}$$

(h) Equivalent Components A few situations will arise in later chapters that involve the representation of a vector in terms of its components with respect to two coordinate systems. Figure 10 depicts the components of an arbitrary two-dimensional vector $\bar{A}$ with respect to both the xyz and XYZ axes. One case that will arise corresponds to knowing the components of $\bar{A}$ with respect to xyz, and wishing to evaluate the components with respect to XYZ. In another case, one component with respect to each coordinate system is known, and it is necessary to determine the other components.

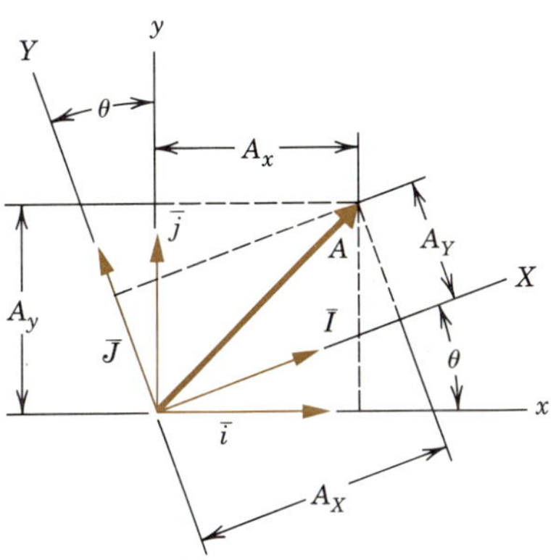

FIGURE 10

Both cases may be addressed by the same general approach. It begins by expressing the unit vectors for one coordinate system in terms of their components with respect to the other system. The choice is arbitrary; let us use xyz components for Figure 10. The angle between the X and x axes in Figure 10 is θ, with the X axis lying in the first quadrant of xy. Similarly, the Y axis is in the second quadrant of xy at an angle θ from the y axis. Thus

$$\bar{I} = \cos\theta\,\bar{i} + \sin\theta\,\bar{j}, \qquad \bar{J} = -\sin\theta\,\bar{i} + \cos\theta\,\bar{j}$$

The vector $\bar{A}$ may be written in terms of its components with respect to either set of axes, so

$$\bar{A} = A_x\bar{i} + A_y\bar{j} = A_X\bar{I} + A_Y\bar{J}$$

Substitution of the expressions for $\bar{I}$ and $\bar{J}$ then gives

$$\begin{aligned}\bar{A} &= A_x\bar{i} + A_y\bar{j} \\ &= A_X(\cos\theta\,\bar{i} + \sin\theta\,\bar{j}) + A_Y(-\sin\theta\,\bar{i} + \cos\theta\,\bar{j})\end{aligned}$$

These two different representations of $\bar{A}$ are in terms of xy components. Equations (9) tell us that like components must match in such situations, which yields

$$\bar{i}\text{ component:}\quad A_x = A_X\cos\theta - A_Y\sin\theta$$

$$\bar{j}\text{ component:}\quad A_y = A_X\sin\theta + A_Y\cos\theta$$

These two equations may be solved for two components if the other two are known.

EXAMPLE 5

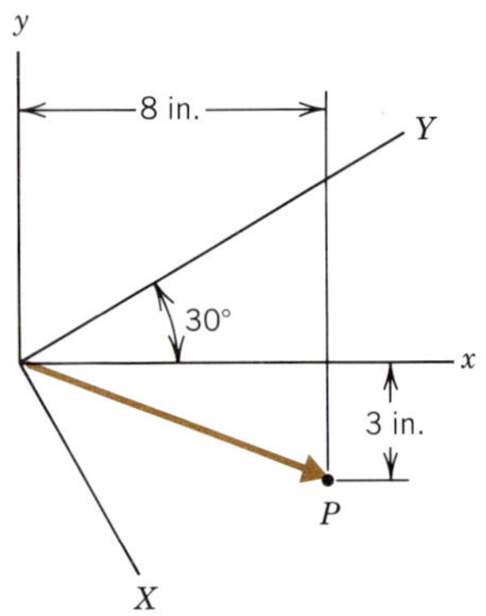

Express the position vector $\bar{r}_{P/O}$ in terms of its components with respect to the XYZ coordinate axes.

Solution

The given diagram indicates that

$$\bar{r}_{P/O} = 8\bar{i} - 3\bar{j} \text{ in.}$$

Because we wish to determine the XY components, it is convenient to represent $\bar{i}$ and $\bar{j}$ in terms of the unit vectors $\bar{I}$ and $\bar{J}$. The unit vector $\bar{i}$ is in the first quadrant of XY at 30° from the Y axis, and $\bar{j}$ is 60° from the Y axis in the second quadrant, so

$$\bar{i} = \sin 30^\circ\, \bar{I} + \cos 30^\circ\, \bar{J} = 0.50\bar{I} + 0.8660\bar{J}$$

$$\bar{j} = -\sin 60^\circ\, \bar{I} + \cos 60^\circ\, \bar{J} = -0.8660\bar{I} + 0.50\bar{J}$$

Then

$$\bar{r}_{P/O} = 8(0.50\bar{I} + 0.8660\bar{J}) - 3(-0.8660\bar{I} + 0.50\bar{J})$$

$$\bar{r}_{P/O} = 6.60\bar{I} + 5.43\bar{J} \text{ in.}$$

HOMEWORK PROBLEMS

10.12–10.15 Write the position vector $\bar{r}_{B/A}$ in the sketch in terms of xyz components.

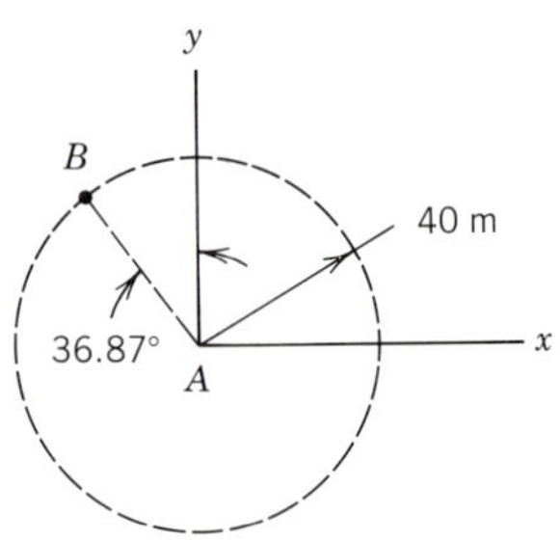

PROB. 10.12

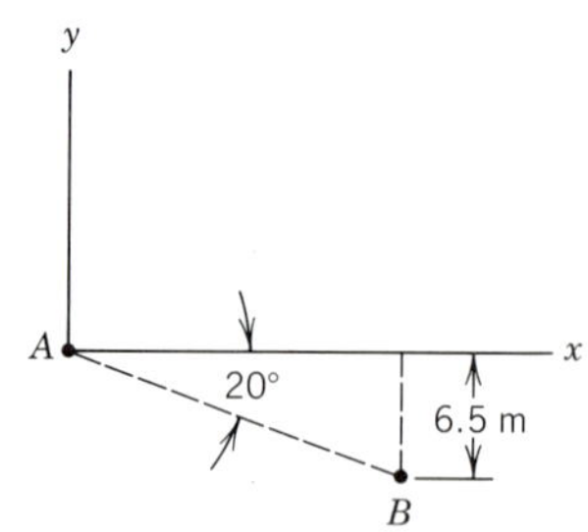

PROB. 10.13

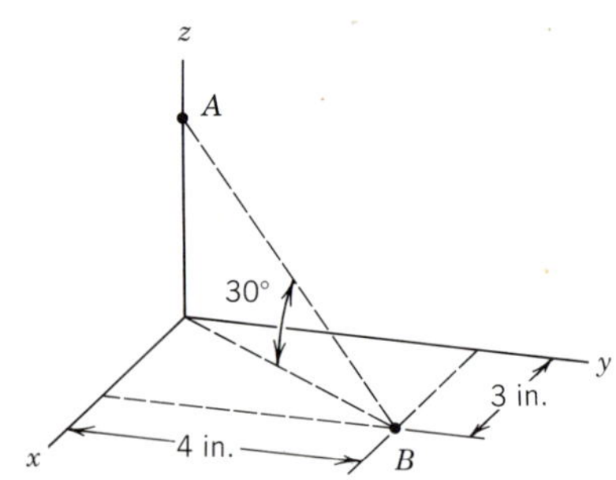

PROBS. 10.14 AND 10.16

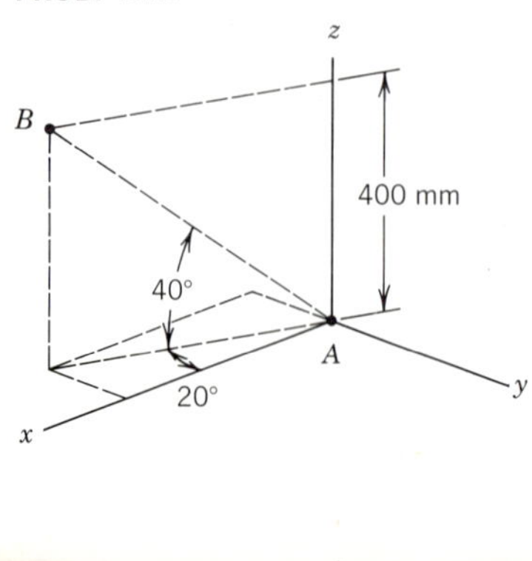

PROBS. 10.15 AND 10.17

10.16 and 10.17 Determine the xyz components of the unit vector $\bar{e}_{B/A}$ extending from point A to point B in the sketch. Then evaluate the direction angles between this vector and the coordinate axes.

10.18 Given the vectors $\bar{A} = \bar{i} + 2\bar{j}$, $\bar{B} = 4\bar{i} - 3\bar{j} - \bar{k}$, and $\bar{C} = -2\bar{i} - 5\bar{j} + \bar{k}$, evaluate (a) $\bar{B} \cdot \bar{C}$, (b) $\bar{A} \cdot (\bar{B} + \bar{C})$, (c) $(\bar{A} - 2\bar{B}) \cdot \bar{C}$.

10.19 For the vectors in Problem 10.18 evaluate (a) $\bar{A} \times \bar{B}$, (b) $\bar{B} \times \bar{C}$, (c) $\bar{B} \times (2\bar{A} + \bar{C})$.

10.20 For the vectors in Problem 10.18 evaluate (a) $(\bar{A} \times \bar{B}) \times \bar{C}$, (b) $\bar{A} \times (\bar{B} \times \bar{C})$, (c) $\bar{A} \cdot (\bar{B} \times \bar{C})$.

10.21 A force $\bar{F}$ of magnitude 200 N is applied at point A. Determine the moment of $\bar{F}$ about corner O.

10.22 A force $\bar{R}$ of magnitude 300 N is applied at point B. Determine the moment of $\bar{R}$ about corner C.

10.23 Each cable has a tension of 800 lb. Determine the moment about corner A of the forces exerted by these cables on plate $ABCD$.

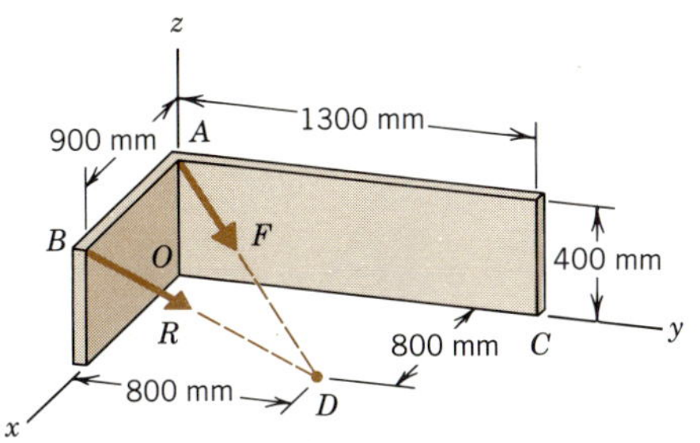

PROBS. 10.21 AND 10.22

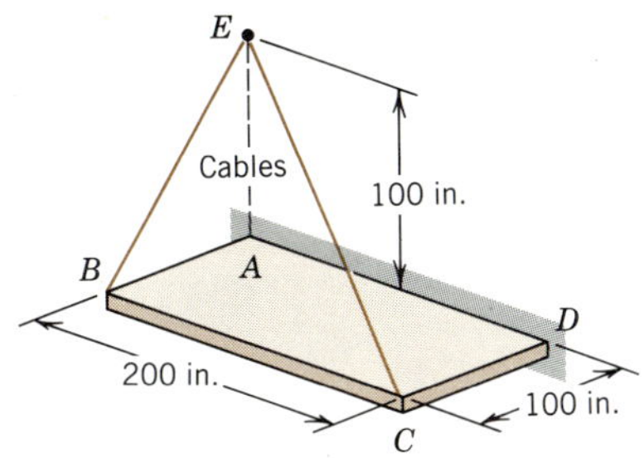

PROBS. 10.23 AND 10.24

10.24 Each of the cables supporting the 500 lb flat plate has a tension of 800 lb. The weight force acts at the centroid of the plate. Determine the force-couple system acting at point D that is equivalent to the forces exerted by gravity and the supporting cables for the plate.

10.25 The 40 kg uniform boom is supported by cables AC and BD and a ball and socket joint at O. Determine the tension in each cable.

10.26 The center of mass of the 5 ton crate coincides with the geometric centroid of the volume. A 6000 lb force $\bar{F}$ acting parallel to the diagonal line BC pushes the crate at point C. Determine (a) the force-couple system acting at point A that is equivalent to the combined effect of gravity and $\bar{F}$, and (b) the moment of these forces about the edge AE.

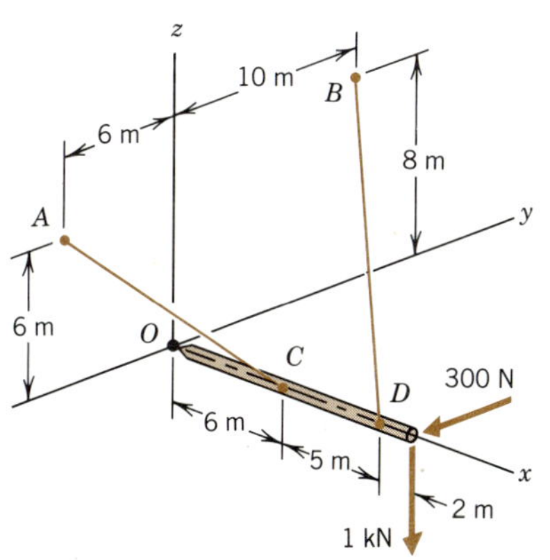

PROB. 10.25

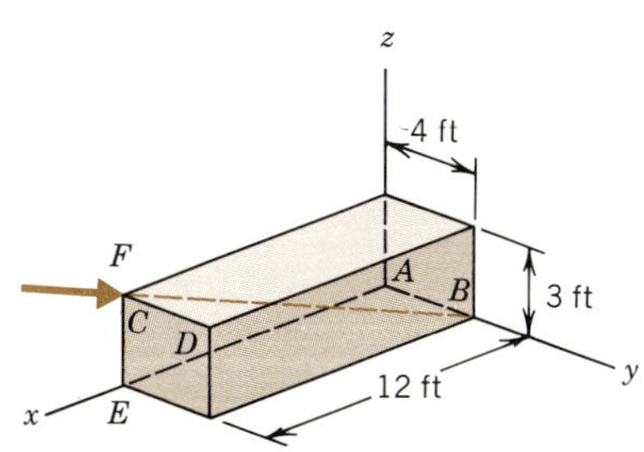

PROB. 10.26

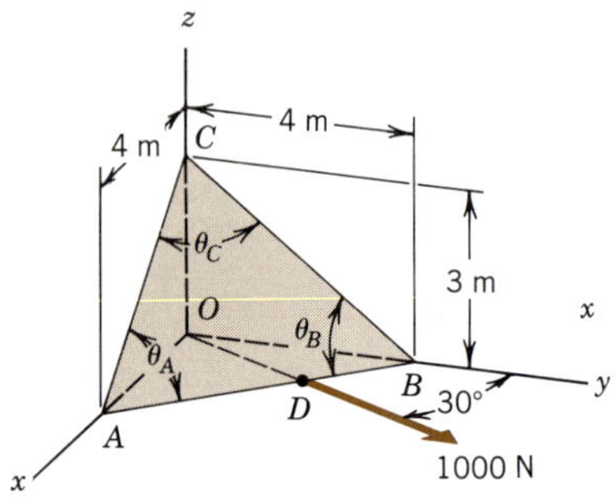

PROBS. 10.27 AND 10.28

***10.27** Determine the three vertex angles θ_A, θ_B, and θ_C for the triangle ABC shown. Also determine the unit vector normal to the plane of the triangle that is oriented outward from the origin.

***10.28** A 1000 N force parallel to the xy plane is applied at point D along edge AB of the triangle, as shown. Resolve this force into two force vectors, one perpendicular to the plane of triangle ABC, the other parallel to that plane.

10.29–10.31 Write the position vector $\bar{r}$ shown in the sketch in terms of its XY components.

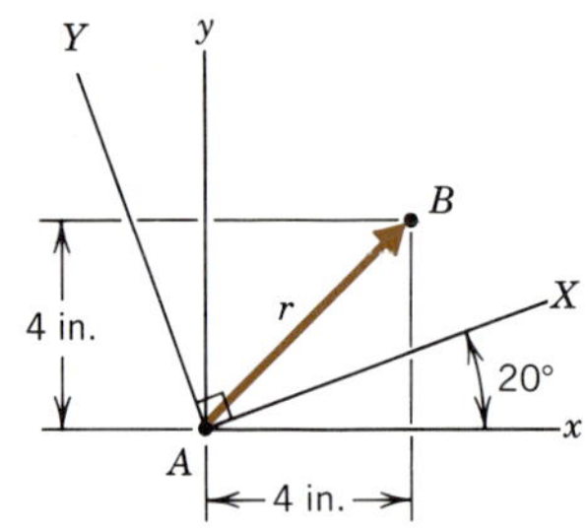

PROB. 10.29

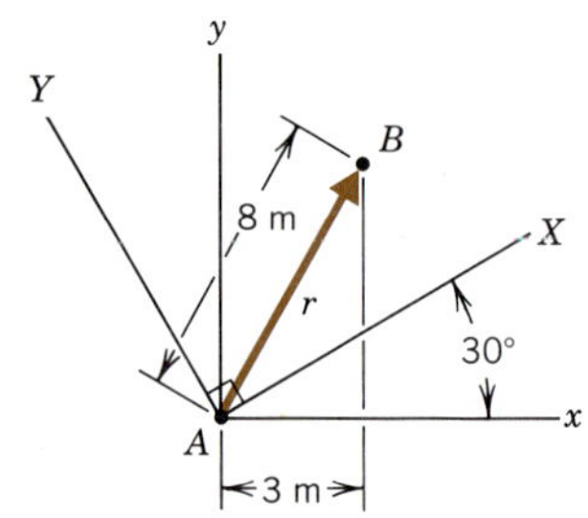

PROB. 10.30

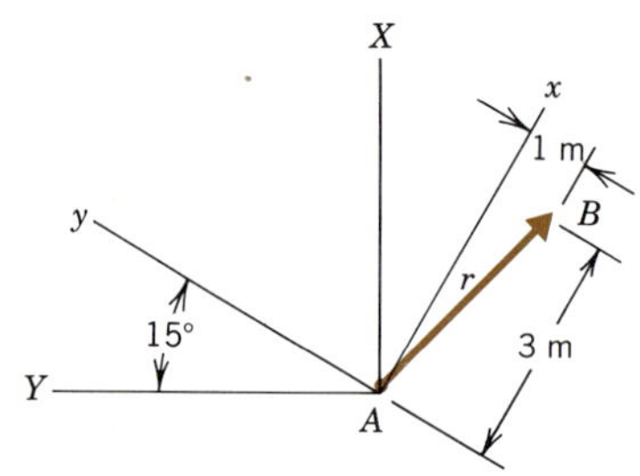

PROB. 10.31

***10.32 and *10.33** The position vector $\bar{r}$ depicted in the sketch has components $r_x = 5$ ft and $r_Y = 2$ ft. Determine the components r_y and r_X.

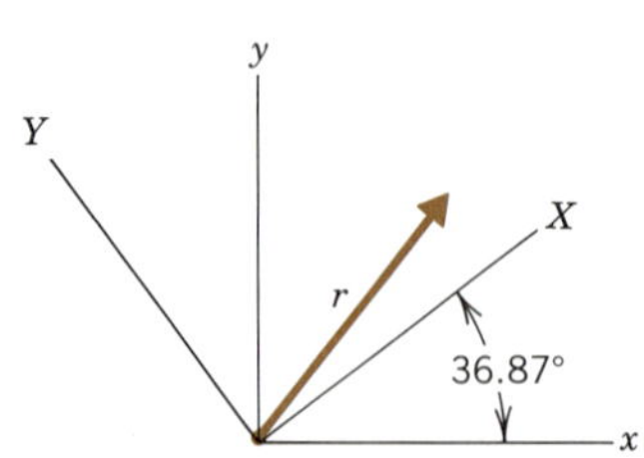

PROB. 10.32

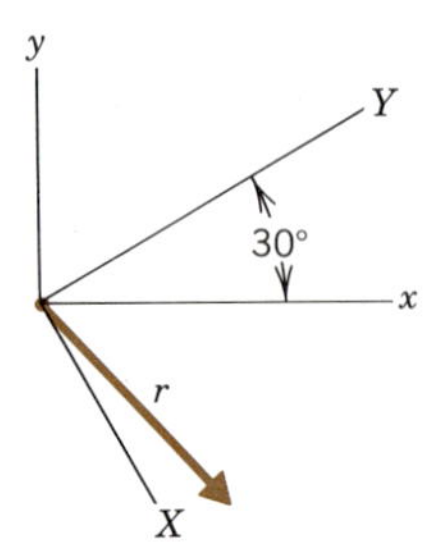

PROB. 10.33

2. Vector Calculus

(a) Time Derivatives Derivatives with respect to time are of primary importance in dynamics. Hence, in the following discussion we shall use time t as the independent variable. The derivative of any vector $\bar{A}(t)$ that depends on time is defined to be

$$\frac{d\bar{A}}{dt} = \lim_{\Delta t \to 0} \frac{\bar{A}(t + \Delta t) - \bar{A}(t)}{\Delta t} \tag{16}$$

The following example is intended to give you a basic appreciation of the significance of eq. (16), and to emphasize the meaning of the basic concept of velocity.

EXAMPLE 6

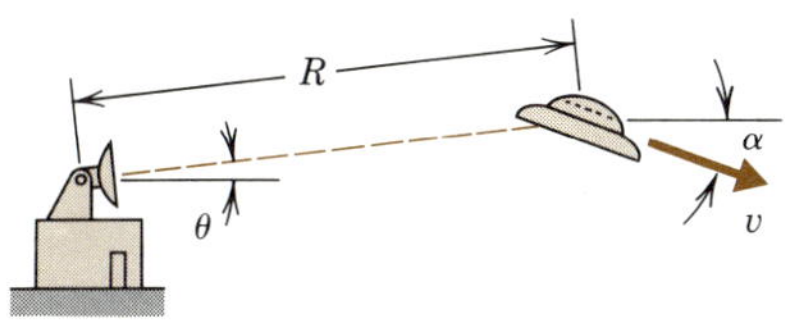

A radar station tracks a UFO by measuring the distance R to it, and the angle of elevation θ. At a certain instant the UFO is sighted at $R = 4100$ m, $\theta = 33.7°$. After an interval of 0.15 s the UFO is at $R = 4240$ m and $\theta = 29.3°$. What is the approximate speed of the UFO and the approximate angle α at which it is diving or climbing?

Solution

The speed is the magnitude of the velocity $\bar{v}_P$, and α defines the direction of the velocity. The velocity may not be found by differentiation because the functions defining R and θ are unknown. We may approximate the velocity by using the given values of R and θ to locate the position of the UFO at times t and $t + \Delta t$, where $\Delta t = 0.15$ s. Then

$$\bar{v}_P \simeq \frac{\bar{r}_{P/O}(t + \Delta t) - \bar{r}_{P/O}(t)}{\Delta t}$$

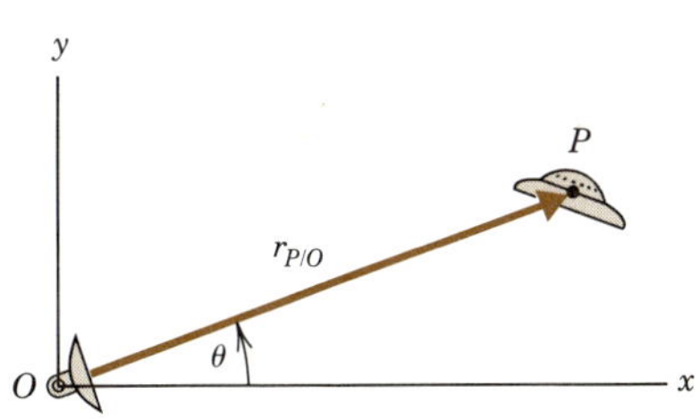

For general values of R and θ the vector $\bar{r}_{P/O}$ going from the radar station to the UFO is shown in the diagram to be

$$\bar{r}_{P/O} = R(\cos\theta\,\bar{I} + \sin\theta\,\bar{J})$$

Thus

$$\begin{aligned}\bar{r}_{P/O}(t) &= 4100(\cos 33.7°\,\bar{i} + \sin 33.7°\,\bar{j}) \\ &= 3411\bar{i} + 2275\bar{j}\text{ m}\end{aligned}$$

$$\begin{aligned}\bar{r}_{P/O}(t + \Delta t) &= 4240(\cos 29.3°\,\bar{i} + \sin 29.3°\,\bar{j}) \\ &= 3698\bar{i} + 2075\bar{j}\text{ m}\end{aligned}$$

$$\begin{aligned}\bar{v}_P &= \frac{1}{0.15}[(3698 - 3411)\bar{i} + (2075 - 2275)\bar{j}] \\ &= 1913\bar{i} - 1333\bar{j}\text{ m/s}\end{aligned}$$

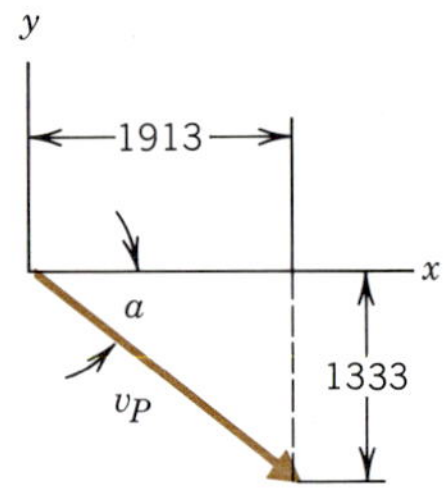

A sketch of this vector indicates that the magnitude and direction of $\bar{v}_P$ are given by

$$v_P = |\bar{v}_P| = [(1913)^2 + (1333)^2]^{1/2} = 2332 \text{ m/s}$$

$$\alpha = \tan^{-1}\frac{1333}{1913} = 34.9°$$

Thus the speed of the UFO is 2332 m/s and it is diving at an angle of 34.9° from the horizontal.

(b) Basic Properties The similarity of the definitions of the derivatives of scalar and vector variables has the important consequence that the basic theorems of vector calculus are directly analogous to those for scalar calculus. These theorems are presented below; the corresponding proofs follow the same steps as those for scalar variables and are therefore not presented here. Let $\bar{A}$ and $\bar{B}$ be two vector functions and let c be a scalar function. Then

$$\begin{aligned} \frac{d}{dt}(\bar{A} \pm \bar{B}) &= \frac{d\bar{A}}{dt} \pm \frac{d\bar{B}}{dt} \\ \frac{d}{dt}(c\bar{A}) &= \frac{dc}{dt}\bar{A} + c\frac{d\bar{A}}{dt} \\ \frac{d}{dt}(\bar{A} \cdot \bar{B}) &= \frac{d\bar{A}}{dt} \cdot \bar{B} + \bar{A} \cdot \frac{d\bar{B}}{dt} \\ \frac{d}{dt}(\bar{A} \times \bar{B}) &= \frac{d\bar{A}}{dt} \times \bar{B} + \bar{A} \times \frac{d\bar{B}}{dt} \end{aligned} \tag{17}$$

Note that in evaluating the derivative of a cross product, the order of multiplication must be maintained, because the cross product is not commutative ($\bar{A} \times \bar{B} \neq \bar{B} \times \bar{A}$).

The first and second identities in eqs. (17) will be of particular use to us. They provide the basis for differentiating the component representation of a vector.

In some situations a vector $\bar{A}$ will be known in terms of a parameter α, whose value in turn depends on t. The rate of change of $\bar{A}$ may then be evaluated by applying the *chain rule* for differentiation, as follows.

$$\frac{d\bar{A}}{dt} = \frac{d\bar{A}}{d\alpha}\frac{d\alpha}{dt} \tag{18}$$

Next, let us consider the process of integration, which also has the same basis for vectors as for scalars. In dynamics, we are generally concerned with definite integrals. Let t_0 denote the initial time. Then

$$\bar{A} = \frac{d\bar{B}}{dt} \quad \text{means} \quad \bar{B}(t) = \bar{B}(t_0) + \int_{t_0}^{t} \bar{A}(t)\,dt \tag{19}$$

Equations (17)–(19) will be employed extensively in the next chapter.

In order to simplify writing derivatives, we shall hereafter denote a derivative with respect to time by a dot over the quantity. For example, the basic definitions, eqs. (1) and (2), may now be rewritten as

$$\bar{v}_P \equiv \dot{\bar{r}}_{P/O} \qquad \bar{a}_P = \dot{\bar{v}}_P = \ddot{\bar{r}}_{P/O}$$

As can be seen, two dots over a quantity denotes the second derivative.

In closing this section we call attention to an important fact that follows from the foregoing development. Consider the representation of a vector in terms of its unit vector, as given by eq. (5). Then

$$\frac{d\bar{A}}{dt} = \frac{d}{dt}(|\bar{A}|\bar{e}_A) = \left[\frac{d}{dt}|\bar{A}|\right]\bar{e}_A + |\bar{A}|\frac{d}{dt}\bar{e}_A$$

from which we see that a vector is constant only if both its magnitude is constant ($d|\bar{A}|/dt = 0$) and its orientation is constant ($d\bar{e}_A/dt = \bar{0}$).

F. PROBLEM SOLVING

The goal of a course in dynamics is the development of an understanding of the laws governing the motion of physical systems, with a primary emphasis on the application of this knowledge to solving problems. In general, engineering is the process of applying established physical laws to the solution of practical problems. Therefore, a common methodology has evolved for solving problems, although in each area the physical laws are different. The pertinent steps are detailed below.

1. Read (or formulate) the problem. Draw a sketch of the system.
2. Show any known information in the sketch, or else list it. Also, list the information that is to be determined from the solution.
3. Choose a model to represent the system. For this model, write the equations corresponding to all physical principles relating the given and desired information.
4. Substitute the given information into these equations.
5. Count the number of equations and number of unknowns. An excess of unknowns usually means that either some given information was not used or a principle governing the model was not considered. In contrast, an excess of equations usually means that improper assumptions were made, or a principle employed in step 3 was incorrectly formulated, or, possibly, that such a principle is not truly applicable to the problem.
6. Solve the equations in step 5.

These steps are very general. At crucial points in the text we will present detailed steps for problem solving. These will be followed by a series of Illustrative Problems. The purpose of these problems is to demonstrate the applications of the procedural steps to a broad class of problems. This is in contrast to Examples, such as the six we have already studied, whose purpose is to demonstrate specific principles or techniques.

HOMEWORK PROBLEMS

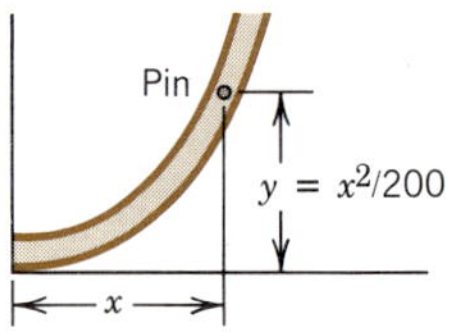

PROB. 10.34

10.34 A pin in a machine follows a parabolic groove defined by $y = x^2/200$, where x and y are measured in millimeters. When first observed, the pin is at $x = 235.2$ mm, and 2 ms later it is at $x = 219.4$ mm. Determine the approximate velocity of the pin.

10.35 A radar station tracks an airplane. At one sighting the airplane is at $R = 3500$ m, $\theta = 55.2°$. If this airplane is traveling at 1836 km/hr at a climb angle of 30°, determine what the values of R and θ will be when it is sighted 0.50 s later.

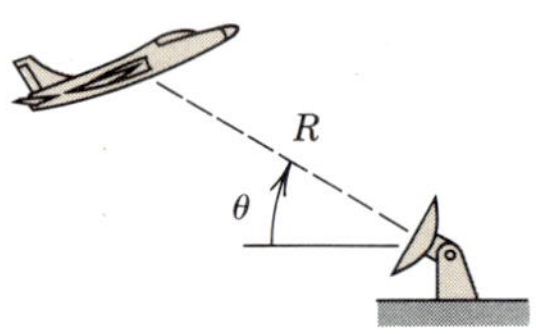

PROB. 10.35

10.36 At a third sighting of the UFO in Example 6, 0.10 s after the second sighting, the UFO is at $R = 4450$ m, $\theta = 25.6°$. Determine the approximation of (a) the velocity of the UFO during this second interval, and (b) the acceleration of the UFO. Would you say that it is under control?

10.37 An automobile is observed while traversing a level circular turn of 500 m radius. An observer at the center O of the curve initially sees the vehicle at a position where $\theta = 30°$ east of south. One second later, the car is at the position where $\theta = 33°$ east of south. What are the speed and the direction in which the vehicle is moving, approximately?

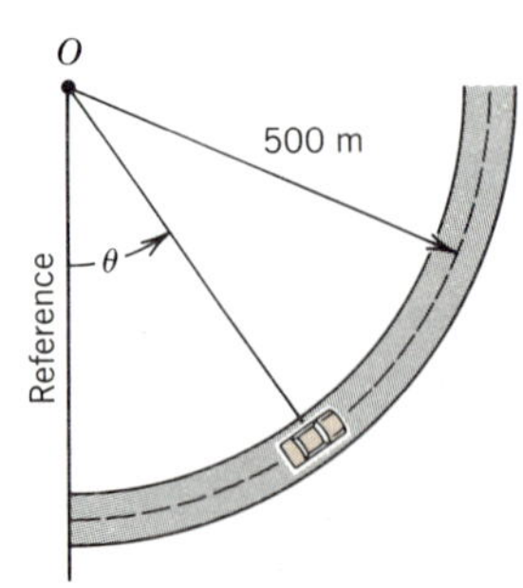

PROBS. 10.37 AND 10.38

10.38 Two seconds after the first sighting, a third observation of the automobile in Problem 10.37 places it at $\theta = 36^\circ$ east of south. Determine an approximation to the acceleration of the vehicle.

***10.39** In order to track an airplane whose flight path does not pass over the station, a radar station measures the azimuth angle ϕ, as well as the angle of elevation θ and the distance R, as shown. Two sightings of the airplane 1.6 s apart place the object first at $R = 22{,}050$ ft, $\theta = 46.1^\circ$, $\phi = 32.2^\circ$, and then at $R = 21{,}800$ ft, $\theta = 47.3^\circ$, $\phi = 32.5^\circ$. Determine its speed and describe the direction in which it is moving.

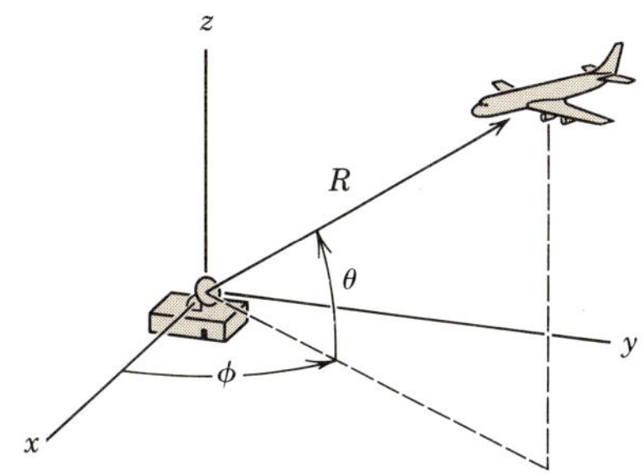

PROB. 10.39

CHAPTER 11

KINEMATICS OF A PARTICLE

OBJECTIVES

1. To be able to use the constant acceleration formulas that describe the position and velocity of a point in rectilinear motion.

2. To understand the definitions of the basic parameters for path variables, Cartesian coordinates, and cylindrical coordinates, and from that to identify when each description is most suitable for the solution of a given problem.

3. To be able to use the kinematical formulas in path variables, Cartesian coordinates, and cylindrical coordinates, with the aid of the chain rule for differentiation where necessary.

4. To develop the ability to convert velocities and accelerations from one kinematical description to another, and in so doing, to be able to use two kinematical descriptions jointly.

5. To relate the relative position, velocity, and acceleration of two points to their absolute motions.

6. To relate the motion of bodies that are interconnected by cables and pulleys.

The kinematics concepts to be developed here relate the position, velocity, and acceleration of a particle. The model of a particle, which considers a body to occupy a single point, is rudimentary, but it has many significant applications. Equally important, the ability to describe the motion of points is a prerequisite for the study of more complex models.

A. BASIC KINEMATICAL PROPERTIES

One of the principal properties of the motion of a particle is its *path*, which is the line (curved or straight) the particle follows as it travels through space. We depict the path of a particle by tracing the point occupied by the particle at each instant of time, thereby forming the locus of these points. A major topic for consideration in this chapter is the relationship between the particle path and the kinematical quantities of position, velocity, and acceleration, whose definitions we shall repeat here from Chapter 10.

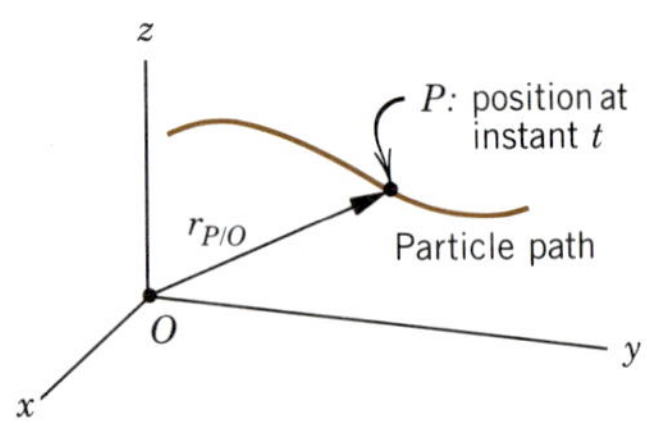

FIGURE 1

The xyz coordinate system in Figure 1, whose origin is point O, serves as a fixed frame of reference. The vector $\bar{r}_{P/O}$ is the position of point P relative to point O at time t. By definition, the absolute velocity of point P, which will be denoted as $\bar{v}$, is the time rate of change of position, so

$$\bar{v} = \frac{d}{dt}\bar{r}_{P/O} \equiv \dot{\bar{r}}_{P/O} \tag{1}$$

It will be noted that the symbol $\bar{v}$ appears without a subscript. We will prove later that it is not necessary to retain the subscript O because the absolute velocity is the same regardless of where the origin O is located. Also, a subscript corresponding to the label for the moving point will be employed only in situations involving more than one moving point.

Now that we have a mathematical definition for velocity, a similar approach may be followed in defining acceleration. Acceleration is the time rate of change of the velocity, so that

$$\bar{a} \equiv \dot{\bar{v}} \equiv \ddot{\bar{r}}_{P/O} \tag{2}$$

The three vectors $\bar{r}_{P/O}$, $\bar{v}$, and $\bar{a}$, can be described by their components with respect to the xyz coordinate system shown in Figure 1. This coordinate system can also be used to write the equations describing the particle's path. These and other uses of a coordinate system will be discussed in the following sections.

B. KINEMATICAL RELATIONSHIPS

There are several methods for describing the basic kinematical properties of a particle. The differences between them are associated with the parameters that are employed to describe the position of a point.

1. Comparison of Fixed Coordinate Systems and Path Variables

One possible way of specifying the position is to give the rectangular Cartesian coordinates of the particle as functions of time, that is, to give $x(t)$, $y(t)$, and $z(t)$. Such functions could be evaluated at a specific instant in order to find the corresponding values of x, y, and z. A plot of the values of x, y, and z for each instant would yield a trace of the path followed by the particle.

Now consider an alternative approach. Suppose that we are given the particle's path. The position of the particle along its path could be defined by the arc length that the particle has traveled from its initial position. The arc length is called a *path variable.* An example of such a description is to be told that a particle follows a circular path in the xy plane and that at $t = t_1$ it has traveled one-quarter of the circumference of the circle from its initial position in a clockwise direction. As we can see in Figure 2, this locates the position of the particle, provided that we know its initial position.

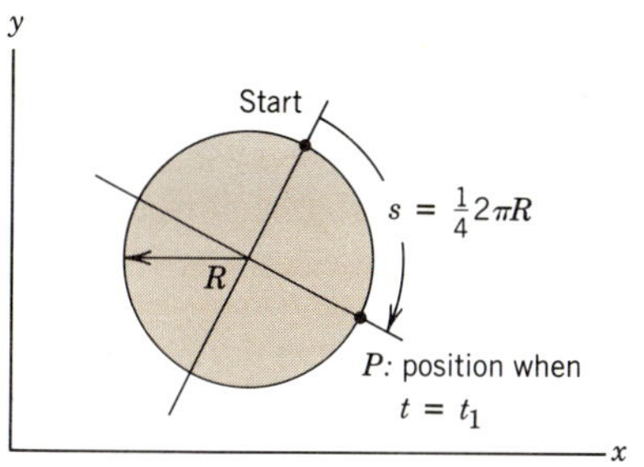

FIGURE 2

The fundamental difference between the Cartesian coordinate description and path variables is that in Cartesian coordinates the specification of the position at any instant is dependent on the choice of coordinate axes, and, as we shall see, is independent of knowledge of the path. The same is true for the velocity and acceleration vectors. Hence in the Cartesian coordinate approach, and in the cylindrical coordinate description which will be developed later, the choice of coordinate system will change the basic variables. On the other hand, in the path variable approach all quantities, including position, are defined with respect to the path. In this method the description is essentially independent of the choice of coordinate system; the main purpose of the coordinate system in this case is to describe the path.

2. Straight Path—Constant Acceleration

The situation where the path of a point is a straight line is called *rectilinear motion.* Such a motion can be described in terms of any of the approaches discussed in the preceding section. Here we shall restrict our remarks to Cartesian coordinates.

Consider a point moving along a straight path. Without loss of generality, the x axis is chosen to coincide with the path, as shown in Figure 3. The location of the origin O represents the reference position for distance. The length s is the distance from the origin to point P, so

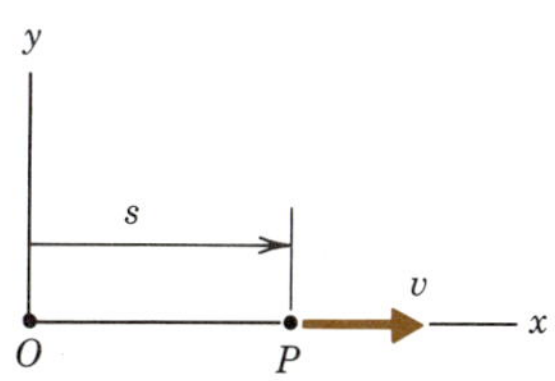

FIGURE 3

$$\bar{r}_{P/O} = s\bar{i}$$

The unit vector $\bar{i}$ has a fixed orientation, so eqs. (1) and (2) indicate that

$$\bar{v} = \dot{s}\,\bar{i}, \qquad \bar{a} = \ddot{s}\,\bar{i}$$

Because all of the kinematical vectors are parallel to the x axis, we may omit the unit vector $\bar{i}$, and note that a negative scalar value for each variable represents a vector that is oriented in the sense of $-\bar{i}$. The scalar form of these relations is therefore

$$v = \dot{s}, \qquad a = \dot{v} = \ddot{s} \tag{3}$$

These formulas yield the velocity and acceleration when the position is known as a function of time. The process of evaluating the velocity and position when the acceleration is given requires that the values of v and s at the start of motion be specified. These values are called *initial conditions.* Without loss of generality, the initial time may be defined as $t = 0$, so the initial conditions are $v = v_0$ and $s = s_0$ when $t = 0$.

The case of variable acceleration will be treated in Chapter 12. The treatment here will be restricted to the case of constant acceleration, which occurs frequently in practice. The task of finding v and s when the value of a is given involves integrations of eq. (3). For this, the derivatives are converted into integrals by multiplying each term in eq. (3) by dt. Thus

$$dv = a\,dt, \qquad ds = v\,dt$$

These differentials may be integrated directly because a is a constant. It is convenient to use a definite integral in which the lower limits are the initial values of v, s, and t. Let us first obtain the velocity

$$\int_{v_0}^{v} dv = a \int_0^t dt \Rightarrow v - v_0 = at$$

$$v = v_0 + at \tag{4}$$

This result is now employed to obtain s. Because v_0 and a are constants, it is found that

$$\int_{s_0}^{s} ds = \int_0^t v\,dt \equiv \int_0^t (v_0 + at)\,dt$$

$$s = s_0 + v_0 t + \tfrac{1}{2}at^2 \tag{5}$$

Equations (4) and (5) are probably familiar to you from your physics courses, but it must be emphasized that *they are valid only when a is constant.*

It is often necessary to obtain an explicit relationship between the velocity v and position s. Rather than algebraically eliminating t from eqs. (4) and (5), the chain rule for differentiation may be employed. Consider v to be a function of s; then

$$a = \frac{dv}{dt} = \frac{dv}{ds}\frac{ds}{dt} = \frac{dv}{ds}v$$

In order to convert this into differential form, both sides are multiplied by ds. A definite integral containing the initial conditions as the lower limits yields

$$a \int_{s_0}^{s} ds = \int_{v_0}^{v} v\,dv \Rightarrow a(s - s_0) = \tfrac{1}{2}(v^2 - v_0^2)$$

$$v^2 = v_0^2 + 2a(s - s_0) \tag{6}$$

Equations (3)–(6) are the fundamental scalar relationships required to solve rectilinear motion problems in which the acceleration is constant.

EXAMPLE 1

A ball is thrown vertically upward with a speed of 20 m/s. Determine the maximum height the ball would attain if air resistance were negligible, and also determine the elapsed time to that position.

Solution

Let s be the height of the ball above the ground. The maximum value of s occurs when the ball comes to rest, immediately prior to reversing its direction of motion. The acceleration is the free fall value g directed downward, opposite the sense in which s is measured. Thus, $a = -g = -9.807 \text{ m/s}^2$. The height s_{max} for $v = 0$ may be obtained from eq. (6), with $v_0 = 20$ m/s, and $s_0 = 0$. Thus

$$0 = (20)^2 + 2(-9.807)s_{max} \Rightarrow s_{max} = 20.4 \text{ m}$$

The corresponding time may be obtained by setting $v = 0$ in eq. (4).

$$0 = 20 + (-9.807)t \Rightarrow t = 2.04 \text{ s}$$

EXAMPLE 2

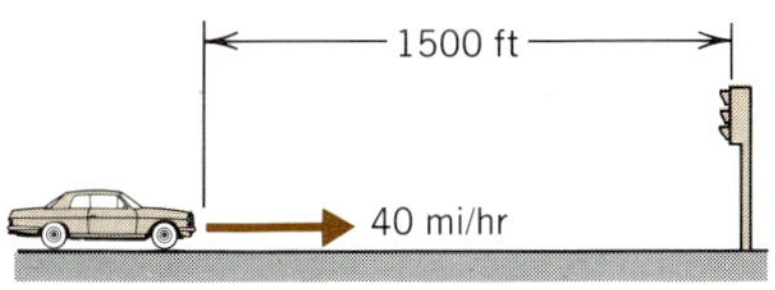

The driver of an automobile traveling 40 mi/hr sees a traffic light 1500 ft ahead turn red. The light will be red for 15 s, and then green for 30 s. (a) What constant acceleration will enable the automobile to pass the light at the instant it turns green, and what is the corresponding speed when it passes the light? (b) What constant acceleration will enable the automobile to pass the traffic light with a speed of 10 mi/hr? Will the light be red or green at that instant?

Solution

The given initial speed is first converted into a consistent set of units.

$$v_0 = 40 \text{ mi/hr}\left(\frac{88 \text{ ft/s}}{60 \text{ mi/hr}}\right) = 58.67 \text{ ft/s}$$

Part (a) requires the acceleration for $s = 1500$ ft when $t = 15$ seconds. This matches the parameters in eq. (5), which relates s and t.

$$1500 = 58.67(15) + \tfrac{1}{2}a(15)^2 \Rightarrow a = 5.511 \text{ ft/s}^2$$

The corresponding value of v may be found from eq. (4) by using this value of a.

$$v = 58.67 + 5.511(15) = 141.3 \text{ ft/s}$$

In part (b), it is given that the final speed is

$$v = 10(\tfrac{88}{60}) = 14.67 \text{ ft/s}$$

In this case it is necessary to determine the value of a for $s = 1500$ ft when $v = 14.67$ ft/s. Equation (6) is convenient to use in this case because time is not involved explicitly.

$$(14.67)^2 = (58.67)^2 + 2a(1500) \Rightarrow a = -1.0757 \text{ ft/s}^2$$

The question of whether the light is red or green may be answered by determining the time for which eq. (4) is valid.

$$14.67 = 58.67 + (-1.0757)t \Rightarrow t = 40.90 \text{ s}$$

The time is 25.90 seconds greater than the time when the light first turns green, which is within the green interval.

HOMEWORK PROBLEMS

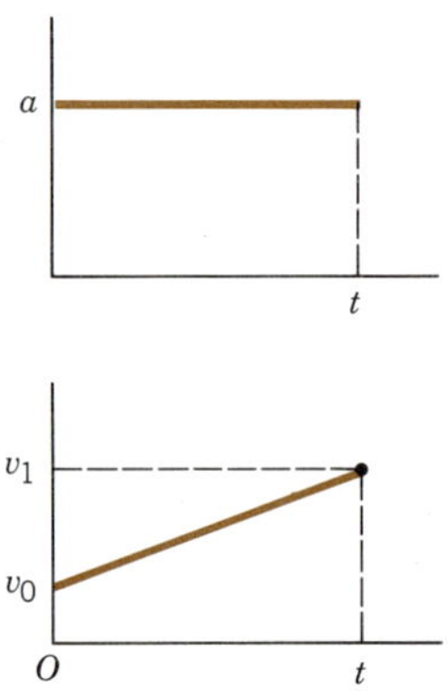

PROB. 11.1

11.1 A plane is flown along a straight path with constant acceleration. Its initial speed is v_0, and its final speed is v_1. If distance is measured from the position when $t = 0$, show that the distance traveled is the product of the average velocity $(v_0 + v_1)/2$ and the time t.

11.2 A ball is tossed vertically upward with speed v_0. Neglecting air resistance, determine the elapsed time for it to fall to its point of origin.

11.3 A rock is thrown downward from a bridge with a speed of 5 m/s and hits the water 4 s later. Determine (a) the height of the bridge, and (b) the speed with which the rock hits the water.

11.4 A way to simulate the effect of a head-on collision of two cars, each traveling at 50 mi/hr, is to bring to rest, in a distance of 10 ft, one car having an initial speed of 100 mi/hr. Determine the corresponding constant rate of deceleration, and the elapsed time for stopping the car.

11.5 A helicopter that is ascending at a constant rate of 3 m/s drops a buoy into the sea from an altitude of 100 m above sea level. Determine the altitude of the helicopter when the buoy strikes the water. Neglect air resistance.

11.6 A particle in rectilinear motion with an initial velocity of 90 ft/s to the right, decelerates at the rate of 20 ft/s^2 until it has a velocity of 30 ft/s to the left. Determine (a) the elapsed time, (b) the distance between the initial and final positions, and (c) the total distance traveled by the particle.

11.7 Derive eq. (6) in this chapter directly from eqs. (4) and (5).

11.8 A locomotive has a constant acceleration along a straight east–west track. The locomotive is 1 km east of a marker initially, and 8 min later it is 2 km west of that marker. It is also known that the locomotive came to rest momentarily 5 min after the initial time. Determine the velocity of the locomotive at the initial and final positions.

11.9 An automobile in rectilinear motion has a constant acceleration such that it is 12 ft, 28 ft, and 60 ft from a reference position when $t = 0$, 1 s, and 2 s, respectively. Determine (a) the initial velocity of the automobile, and (b) the distance from the reference position when $t = 8$ s.

11.10 A motorist in an automobile traveling 72 km/hr at a distance of 60 m behind a truck going at the same speed wishes to pass the truck. In order to accomplish this maneuver safely, the automobile must accelerate and reach a position 60 m in front of the truck in no more than 30 s. (a) Determine the minimum constant acceleration of the automobile that is safe. (b) Determine the corresponding speed of the car at the end of the passing maneuver.

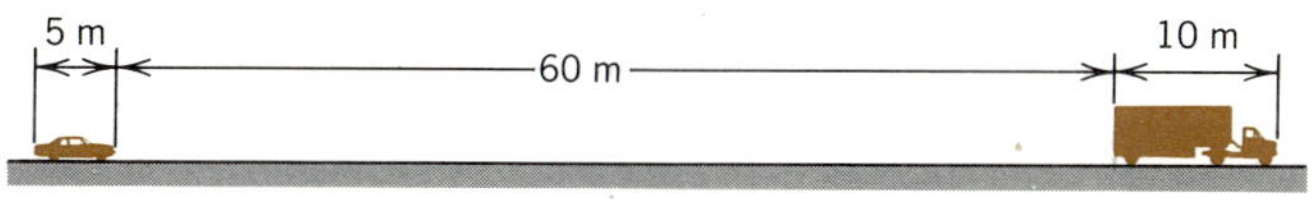

PROB. 11.10

11.11 A race car starts from rest and travels a straight path for 2000 m, where it comes to rest. The maximum rate of acceleration of the car is $0.7g$ and its deceleration is limited to $1.0g$. Determine (a) the maximum velocity attained by the car, and (b) the least possible time t required to cover the distance.

***11.12** A subway car has a maximum acceleration of 4 ft/s^2, a maximum speed of 45 mph, and a maximum deceleration rate of 8 ft/s^2. Determine the minimum time required to start from rest at one station and come to rest at the next station 3 miles away.

---

3. Velocity in Path Variables

The position along a straight path was described in the previous section by the distance traveled along the path from a reference position. The same parameter is employed to describe curvilinear motion in terms of path variables, but the distance traveled is now the arc length measured along the path. This type of measurement is typified by the odometer reading on an automobile, which indicates the elapsed milage along the driven path.

In Figure 4 the location of point P on a specified path is given by the arc length s measured from reference point A. Thus, the position vector $\bar{r}_{P/O}$ is a function of s, $\bar{r}_{P/O} \equiv \bar{r}_{P/O}(s)$. In turn, the arc length s is time dependent, so $s = s(t)$.

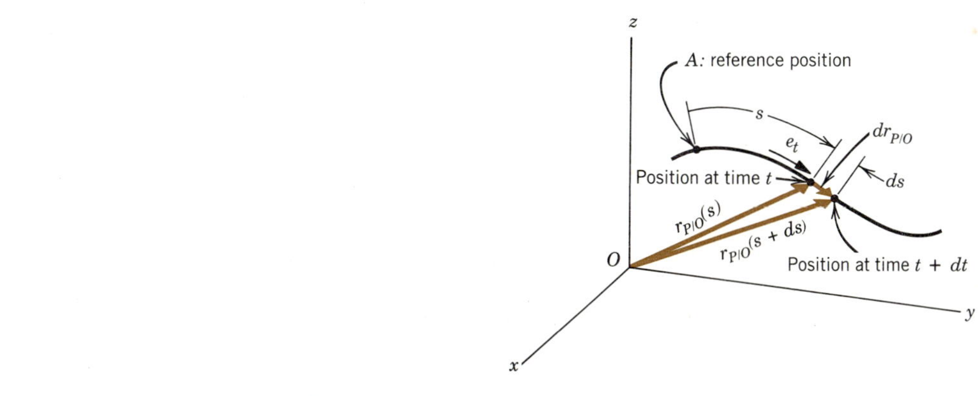

FIGURE 4

It follows that the velocity may be obtained by using the chain rule to differentiate the position with respect to time.

$$\bar{v} = \frac{d\bar{r}_{P/O}}{dt} = \frac{d\bar{r}_{P/O}}{ds}\,\dot{s} \tag{7}$$

The quantity $\dot{s}$ is a number representing the time rate of change of the distance traveled. In order to understand the term $d\bar{r}_{P/O}/ds$ in eq. (7), let us consider the position of point P at a differential arc length ds further along the path.

First, the fact that $d\bar{r}_{P/O}$ is indeed the increase in the position is verified in Figure 4 by the vector triangle, which indicates that

$$\bar{r}_{P/O}(s + ds) = \bar{r}_{P/O}(s) + d\bar{r}_{P/O}$$

Because the two positions in Figure 4 are separated only by an infinitesimal distance, the chord connecting these positions is the same as the tangent. It follows that $d\bar{r}_{P/O}$, which forms this chord, is tangent to the path, and that the magnitude of $d\bar{r}_{P/O}$ is the arc length ds. Furthermore, $d\bar{r}_{P/O}$ is in the sense of a progression along the path in the direction of increasing s.

These observations lead to the conclusion that

$$d\bar{r}_t \tag{8}$$

where

> $\bar{e}_t$ is a unit vector that is coincident with the tangent to the path at arc length s. The sense of $\bar{e}_t$ is the sense of a displacement in the direction of increasing s.

When eq. (8) is divided by ds, we obtain

$$\frac{d\bar{r}_{P/O}}{ds} = \bar{e}_t$$

Thus, the term $d\bar{r}_{P/O}/ds$ in eq. (7) defines the direction of the velocity, and $\dot{s}$ is the magnitude of the velocity. By definition, the latter quantity is the speed, so

$$\boxed{v = |\bar{v}| = \dot{s}, \qquad \bar{v} = v\,\bar{e}_t} \tag{9}$$

Equations (9) state that the velocity is tangent to the path and that the speed, which is the magnitude of the velocity, is the time rate of change of the arc length traveled along the path. The sign of $\dot{s}$ defines the sense of the velocity. For example, if $\dot{s}$ is negative (i.e., if s is decreasing), the velocity is in the sense of a return to the reference position, A.

One significant aspect of eq. (9) is that the relation for the speed is identical to the one obtained for rectilinear motion. We also see that v is independent of the choice of the fixed origin O. The new feature of motion along a curved path is that the orientation of $\bar{e}_t$ depends on the position of the particle along the path.

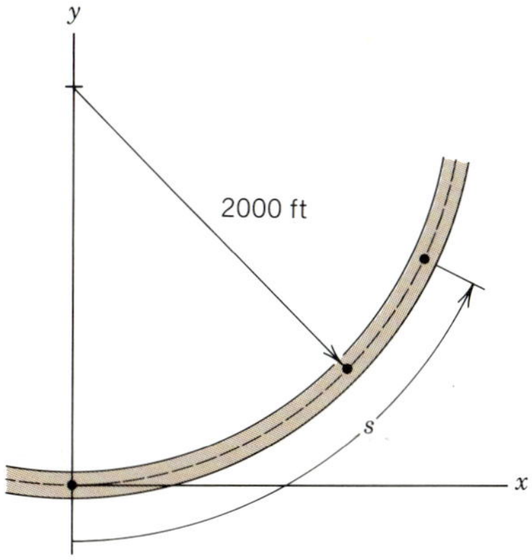

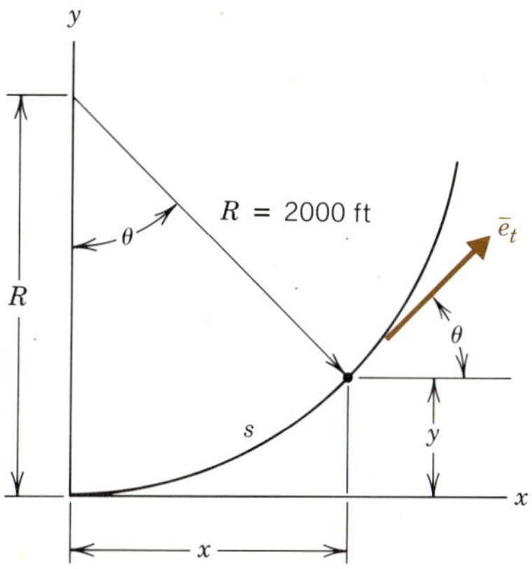

EXAMPLE 3

A locomotive follows a level, 2000 ft radius curve such that the distance it has traveled from the start is $s = 40t + t^2$ (s is in feet and t is in seconds). Determine the position and velocity of the train in terms of xy components for $t = 10$ s.

Solution

The position relative to the path and the speed are found by evaluating s and $\dot{s}$ at $t = 10$ s.

$$s = (40t + t^2)\big|_{t=10} = 500 \text{ ft}$$

$$v = \dot{s} = (40 + 2t)\big|_{t=10} = 60 \text{ ft/s}^2$$

The unit vector $\bar{e}_t$ is shown in the sketch as the tangent to the path at this value of s. The angle θ defining the orientation of $\bar{e}_t$ is found from the fact that s is a circular arc that subtends the angle θ. Then the coordinates and the components of $\bar{e}_t$ are readily obtained in terms of θ.

$$\theta = \frac{s}{R} = \frac{500}{2000} = 0.25 \text{ rad} = 14.32^\circ$$

$$x = R \sin\theta = 494.8 \text{ ft}, \qquad y = R - R\cos\theta = 62.2 \text{ ft}$$

$$\bar{e}_t = (1)(\cos\theta\,\bar{i} + \sin\theta\,\bar{j}) = 0.9689\bar{i} + 0.2474\bar{j}$$

The values of x and y are the components of the position vector. The velocity, which is $\bar{v} = v\bar{e}_t$ in path variables, is found in xy components by using the above expression for $\bar{e}_t$.

$$\bar{r} = x\bar{i} + y\bar{j} = 494.8\bar{i} + 62.2\bar{j} \text{ ft}$$

$$v = 60\bar{e}_t = 60(0.9689\bar{i} + 0.2474\bar{j})$$
$$= 58.1\bar{i} + 14.8\bar{j} \text{ ft/s}$$

HOMEWORK PROBLEMS

11.13 The small sphere at the end of the 40 in. cable has a speed of 4 ft/s in the position shown. Determine the horizontal and vertical components of the velocity in this position.

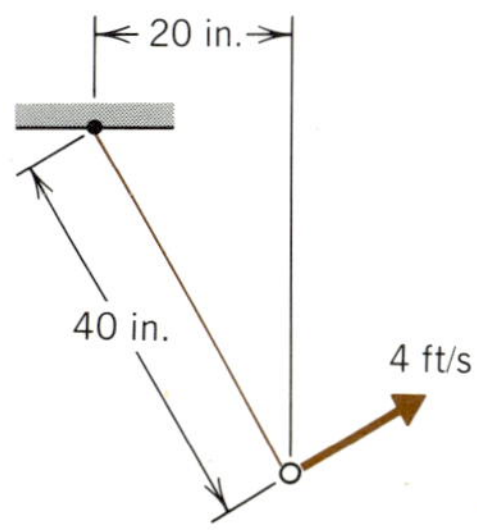

PROB. 11.13

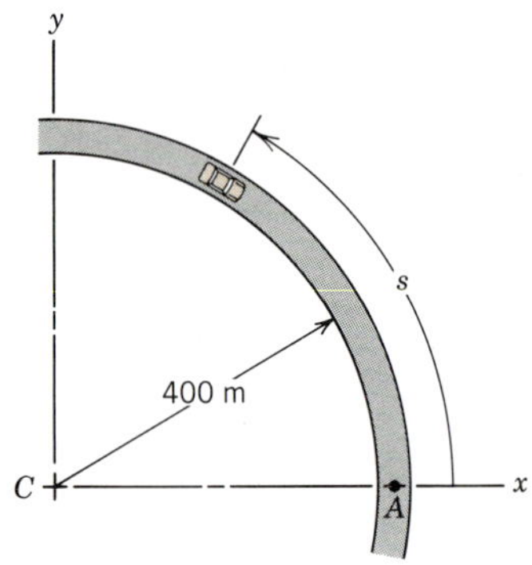

PROB. 11.14

11.14 An automobile moves with a constant speed of 110 km/hr along the 400 m radius road. Determine its position and velocity in terms of xy components after the vehicle has traveled 100 m along the road from point A.

11.15 The upper portion of the roller coaster is a segment of a circular arc of 100 ft radius. The distance a car travels from the top is given by $s = 20t + 8t^2$ (s is in feet and t is in seconds). Determine the horizontal and vertical components of the velocity of the car when $t = 1$ s.

11.16 In Problem 11.15, if the tangent at point B is at an angle of 36.87° from the horizontal, as shown, determine the velocity of the car at at this position.

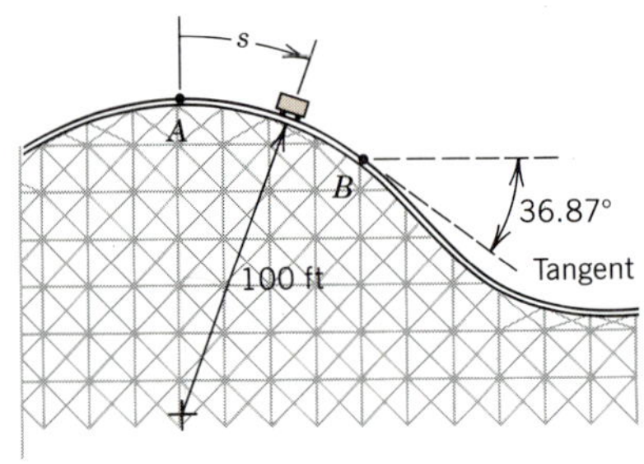

PROBS. 11.15 AND 11.16

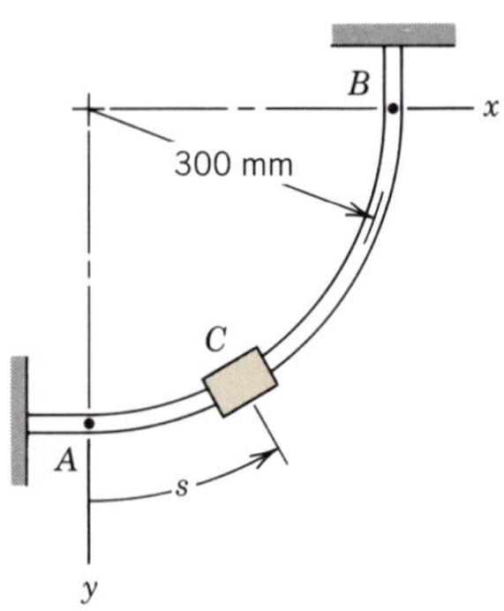

PROBS. 11.17 AND 11.18

11.17 The distance s traveled by collar C from position A is given by $s = 400t \exp(-0.5t)$ (s is in millimeters and t is in seconds). Determine the position and velocity in terms of xy components when $t = 0.5$ s.

11.18 For the collar in Problem 11.17 determine the position and velocity in terms of xy coordinates for the instant when the speed is maximum.

4. Acceleration in Path Variables

The speed obviously might not be constant, but it is equally important to note that $\bar{e}_t$ also changes with time. This is a consequence of the fact that $\bar{e}_t$ depends on the arc length s locating the moving point, while s depends on time. Evaluation of the acceleration by differentiating eq. (9) therefore requires the chain rule for $\dot{\bar{e}}_t$. Thus

$$\bar{a} = \dot{\bar{v}} = \dot{v}\,\bar{e}_t + v\,\dot{\bar{e}}_t = \dot{v}\,\bar{e}_t + v\left(\frac{d\bar{e}_t}{ds}\,\dot{s}\right)$$
$$= \dot{v}\,\bar{e}_t + v^2\,\frac{d\bar{e}_t}{ds} \tag{10}$$

The only new term in eq. (10) is $d\bar{e}_t/ds$. Let us consider this term for the case of *planar motion*, which means that the curved path lies in a plane.

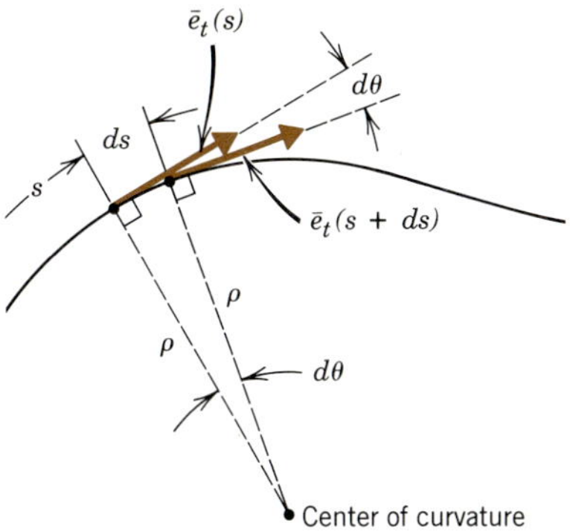

FIGURE 5

Figure 5 depicts the tangent $\bar{e}_t$ obtained from two positions that are separated by an arc length ds along such a path. Once again, this separation distance is depicted by a finite length in Figure 5 for the sake of clarity.

The geometrical construction in Figure 5 depicts perpendiculars to the tangent vectors at the two neighboring positions. The point of intersection of these perpendiculars is called the *center of curvature*, and the distance from either position to the center of curvature is ρ, the *radius of curvature*. In essense, what this construction does is to define a circular arc of radius ρ that best approximates the segment of the actual path between the two neighboring positions.

As indicated in Figure 5, the angle between the two perpendiculars is also the angle between the two tangent unit vectors. From the formula for a circular arc length, this angle is found to be

$$d\theta = \frac{ds}{\rho} \tag{11}$$

By definition, $d\bar{e}_t$ is the increase in $\bar{e}_t$, according to

$$\bar{e}_t(s + ds) = \bar{e}_t(s) + d\bar{e}_t$$

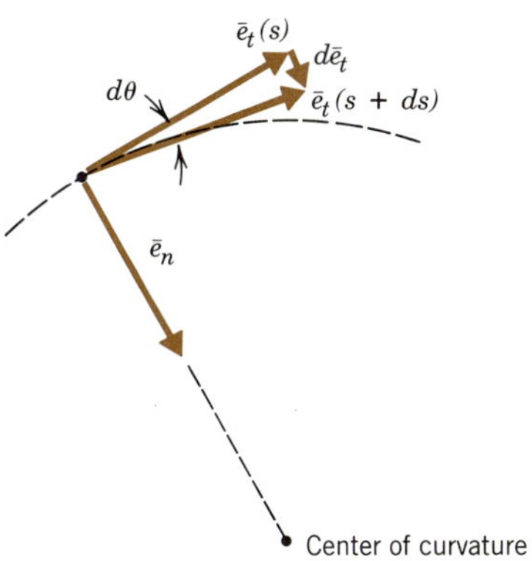

FIGURE 6

This vector sum is depicted in Figure 6 by making the tails of the unit vectors in Figure 5 coincide.

The fact that $\bar{e}_t$ is a unit vector means that its length is one. Then because $d\theta$ is infinitesimally small, it is permissible to regard $d\bar{e}_t$ in Figure 6 to form the arc of a unit radius circle that subtends an angle $d\theta$. This, together with eq. (11), yields

$$|d\bar{e}_t| = (1)\, d\theta = \frac{ds}{\rho}$$

Furthermore, the smallness of $d\theta$ means that $d\bar{e}_t$ is essentially perpendicular to $\bar{e}_t(s)$. Finally, a comparison of Figures 5 and 6 indicates that $d\bar{e}_t$ is oriented toward the center of curvature. It follows that

$$d\bar{e}_t = |d\bar{e}_t|\bar{e}_n = \frac{ds}{\rho}\,\bar{e}_n \tag{12}$$

where $\bar{e}_n$ is the *unit normal vector*, the second fundamental vector for path variables. Its definition is that

> $\bar{e}_n$ is normal (i.e., perpendicular) to the tangent vector $\bar{e}_t$ at arc length s. It is oriented toward the center of curvature.

Substitution of eq. (12) into eq. (10) results in cancellation of the factor ds, so the equation for acceleration in path variables is

$$\boxed{\bar{a} = \dot{v}\,\bar{e}_t + \frac{v^2}{\rho}\,\bar{e}_n} \tag{13}$$

This result means that the acceleration has two components. The *tangential acceleration*, $a_t = \dot{v}$, is the rate of change of the speed. It will be oriented either forward or backward along the path, depending on whether the speed is increasing or decreasing. The *normal acceleration*, $a_n = v^2/\rho$, is called the *centripetal acceleration*. It is *always* directed toward the center of curvature because ρ is a (positive) distance and v^2 is a positive number.

The magnitude of the acceleration is the resultant of the tangential and centripetal accelerations. Because $\bar{e}_t$ and $\bar{e}_n$ are perpendicular unit vectors this resultant is obtained from the sum of the squares of the components.

$$|\bar{a}| = \left[\dot{v}^2 + \left(\frac{v^2}{\rho}\right)^2\right]^{1/2} \tag{14}$$

Recall that a vector is not constant if either its magnitude or direction varies. The tangential acceleration accounts for the changing magnitude of the velocity. It was the only term encountered in the study of rectilinear motion. The centripetal acceleration is the consequence of the change in the direction of the velocity.

As was true in rectilinear motion, the relationships between s, v, and $\dot{v}$ may involve a situation where either s or $\dot{v}$ is given. If $s(t)$ is known, successive differentiations will yield the other parameters. The general case where the tangential acceleration, $\dot{v}$, is the given function is more complicated and is deferred to the next chapter. The important case where the tangential acceleration is constant, however, can be treated here. All that is necessary to obtain relations for the speed and distance traveled in this situation is to replace the parameter a in eqs. (4)–(6) by $\dot{v}$. This modification is required because the given value of $\dot{v}$ is only one component of acceleration when the path is curved, as indicated by eq. (14).

EXAMPLE 4

Shown are the path of a particle and representations of the velocity and acceleration of the particle at that instant in time. Indicate whether each of the conditions shown is possible or not.

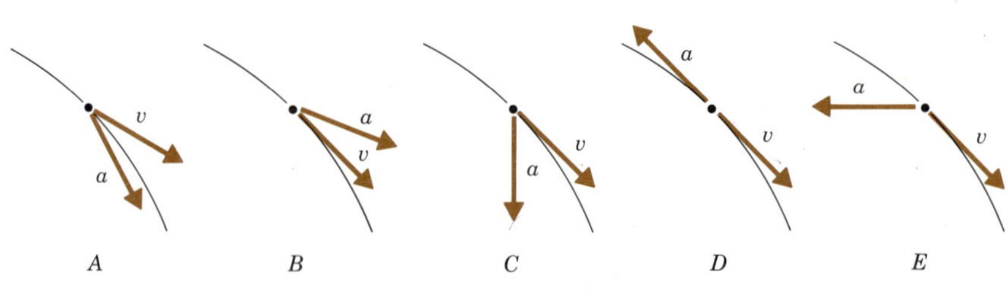

Solution

Figure A: The velocity is not tangent to the path, as is required by eq. (9). This is therefore not a possible situation.

Figure B: The velocity is consistent because it is tangent to the path. However, the component of acceleration perpendicular to the tangent is not oriented inward toward the center of curvature. This is not possible, according to eq. (13).

Figure C: The velocity is tangent to the path and the normal acceleration component is oriented inward toward the center of curvature. This is a possible situation.

Figure D: Both the velocity and acceleration here are tangent to the path. This means that the normal component of acceleration is zero. However, $a_n \equiv v^2/\rho = 0$ only if $v = 0$ or $\rho = \infty$. The speed is not zero because the figure indicates that there is a velocity vector, and $\rho = \infty$ occurs for a straight path or at an inflection point, neither of which is the case here. It follows that this is not a possible situation.

Figure E: The velocity is tangent to the path and the normal component of acceleration is toward the center of curvature, so this situation is possible. The fact that the component of acceleration tangent to the path is oriented in the opposite sense from the velocity means that $\dot{v}$ has the opposite sign from that of v. In other words, the speed is decreasing.

EXAMPLE 5

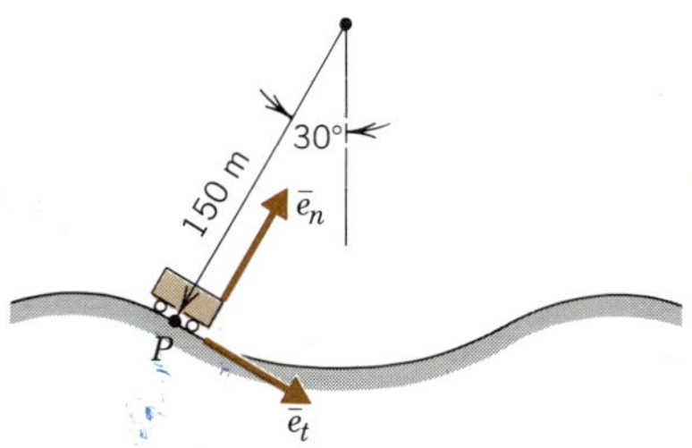

The diagram shown is a portion of a roller coaster track. It consists of a circle in a vertical plane of radius 150 m. Find the magnitude of the total acceleration of the roller coaster if, at the time shown, the car is moving at 50 km/hr and is gaining speed at the rate of $0.4g$ ($1g = 9.807$ m/s^2).

Solution

The given information is clearly in the form of path variables (path, speed, and rate of increase of speed). Therefore select the unit vectors at P as shown. In general

$$\bar{a} = \dot{v}\,\bar{e}_t + \frac{v^2}{\rho}\bar{e}_n$$

The given information is

$$v = 50{,}000 \text{ m/hr}\left(\frac{1}{3600}\frac{\text{s}}{\text{hr}}\right) = 13.889 \text{ m/s}$$

$$\dot{v} = 0.4(9.807) = 3.923 \text{ m/s}^2$$

so eq. (13) gives

$$\bar{a} = 3.923\bar{e}_t + \frac{(13.889)^2}{150}\bar{e}_n = 3.923\bar{e}_t + 1.286\bar{e}_n \text{ m/s}^2$$

Then the magnitude of $\bar{a}$ is

$$|\bar{a}| = \sqrt{(3.923)^2 + (1.286)^2} = 4.13 \text{ m/s}^2$$

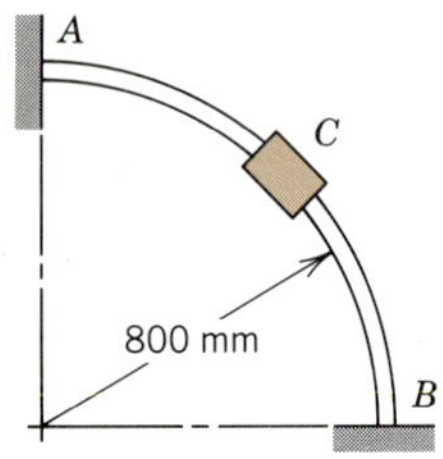

EXAMPLE 6

Collar C starts from rest at position A and gains speed at the constant rate of 4 m/s^2. Determine the acceleration of the collar after it has traveled 200 mm along the guide rod. Draw a sketch of this acceleration.

Solution

It is given that $\dot{v} = 4$ m/s^2 and that the radius of curvature of the circular path is $\rho = 0.8$ m. The evaluation of $\bar{a}$ according to eq. (13) requires that we define the unit vectors $\bar{e}_t$ and $\bar{e}_n$, and determine the value of v at the desired position $s = 0.2$ m.

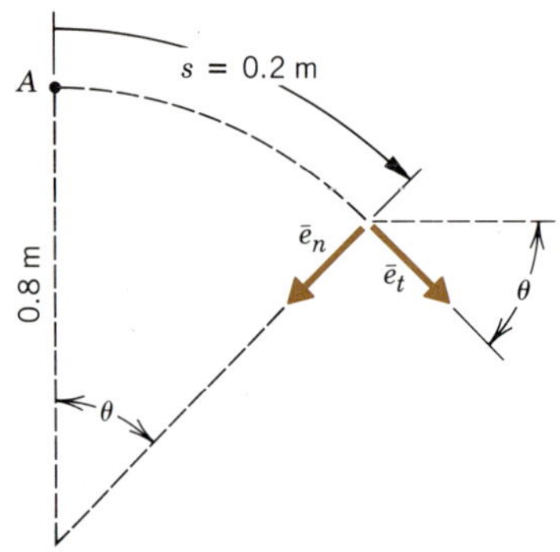

As shown in the sketch, the orientation of the unit vectors are defined by the angle θ subtended by the arc

$$\theta = \frac{s}{R} = \frac{0.2}{0.8} = 0.25 \text{ rad} = 14.32^\circ$$

The value of v is obtained from the fact that $\dot{v}$ is constant, and that initially $v = 0$. Thus eq. (6) indicates that at $s = 0.2$ m,

$$v^2 = 2\dot{v}s = 2(4)(0.2) = 1.6 \text{ m}^2/\text{s}^2$$

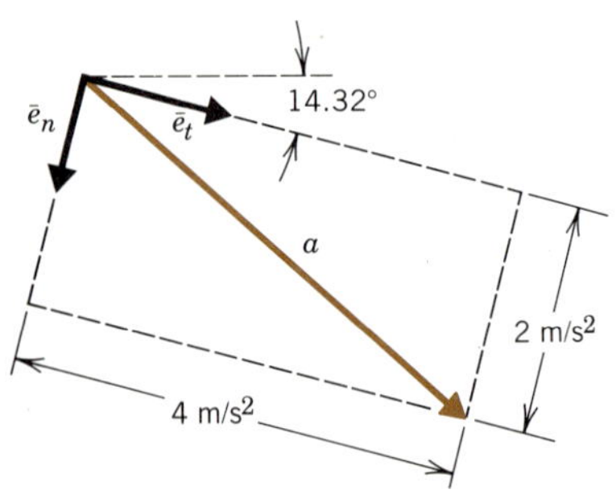

Equation (12) then yields

$$\bar{a} = 4\bar{e}_t + \frac{1.6}{0.8}\bar{e}_n = 4\bar{e}_t + 2\bar{e}_n \text{ m/s}^2$$

Finally this result is depicted in a sketch by drawing a vector that extends 4 units in the direction of $\bar{e}_t$ and 2 units in the direction of $\bar{e}_n$. *If a sketch were not requested, a sufficient answer would be to write the vector in component form, and to draw a sketch of the unit vectors.*

HOMEWORK PROBLEMS

11.19 A bicyclist has a constant speed of 18 mi/hr in a left turn. What is the minimum radius of this turn for which the acceleration will not exceed 5 ft/s^2?

11.20 An automobile traveling at a constant speed of 60 km/hr on a 40 m radius curve has a certain acceleration. What rate of change of speed will instantaneously increase the magnitude of that acceleration by 50%?

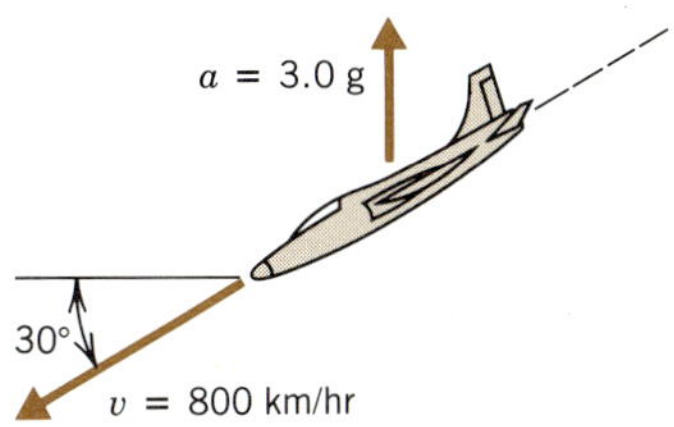

PROB. 11.21

11.21 At a particular instant an airplane is diving at a 30° angle with a speed of 800 km/hr. Its instantaneous acceleration is $3.0g$ upward. Determine (a) the rate of increase of the speed, (b) the unit vector normal to the path of the airplane, and (c) the radius of curvature of the path.

11.22 A freight train heading east enters a 1500 m radius curve to the north. The arc length traveled along the curve is given by $s = 80t - t^2$ (s is in meters and t is in seconds). What is the speed, the heading in degrees with respect to the easterly direction, and the acceleration of the train after it has traveled 1200 m?

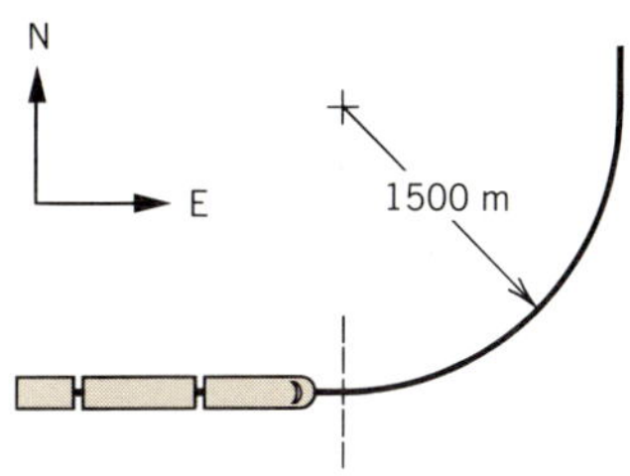

PROB. 11.22

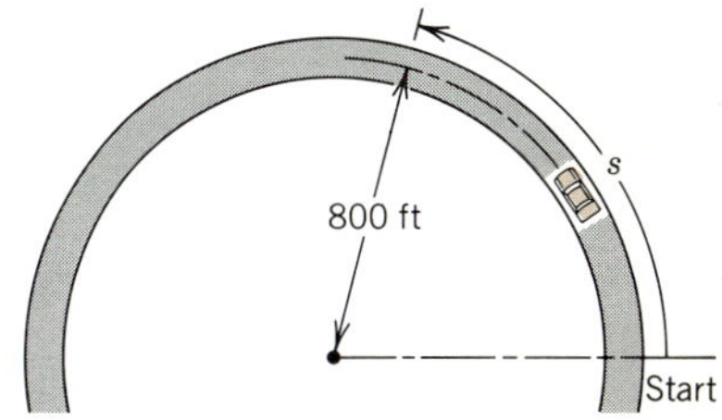

PROB. 11.23

11.23 The motion of an automobile along a 800 ft radius track is defined by $s = 6t^2 - 0.2t^3$, where t is measured in seconds and s is in feet. Determine the distance traveled, the speed, and the acceleration of the vehicle when $t = 5$ s.

11.24 Determine the distance traveled, the speed, and the acceleration of the automobile in Problem 11.23 for the instant when its speed is maximum.

11.25 An automobile gains speed at a constant rate as it travels between points A and B along a road whose vertical profile is as shown. The speed at point A is 40 km/hr and the speed at point B is 80 km/hr. Determine the magnitude of the acceleration at point B, where the radius of curvature is 180 m.

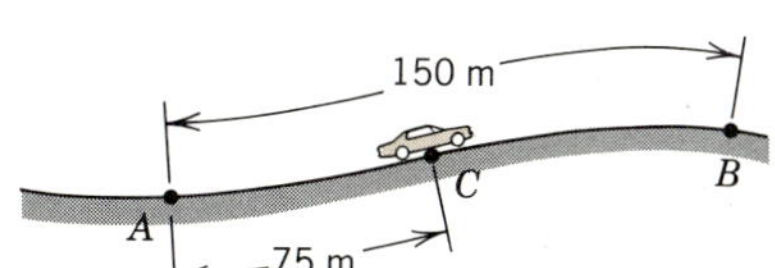

PROB. 11.25

11.26 The magnitude of the acceleration of the automobile in Problem 11.25 at point C is $0.5g$. Determine the radius of curvature of the road at this location.

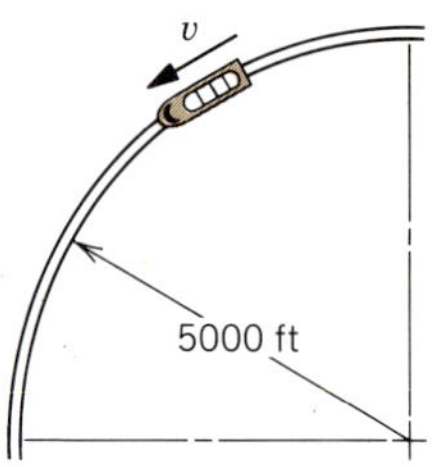

PROB. 11.27

11.27 A monorail train is traveling at a speed of 150 mi/hr along a 5000 ft radius curve when the brakes are applied. For passenger comfort the magnitude of the acceleration of the train cannot exceed 16 ft/s^2. Determine (a) the maximum rate at which the speed may be decreased and (b) the distance required to stop the train for the deceleration rate found in part (a).

11.28 The constant speed of the particle spinning in a circular horizontal path at the end of a cable of length l is known to depend on the constant height h illustrated in the diagram according to $v^2 = v_0^2 + 2gh$. For what value of v_0 will the acceleration have a magnitude of $2v_0^2/l$ when the height is $0.8l$?

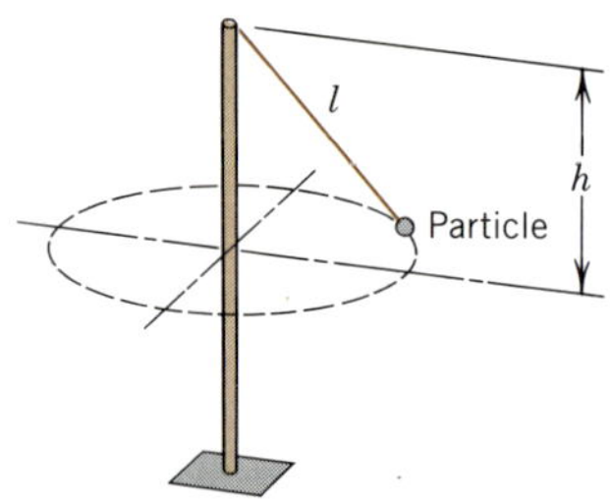

PROB. 11.28

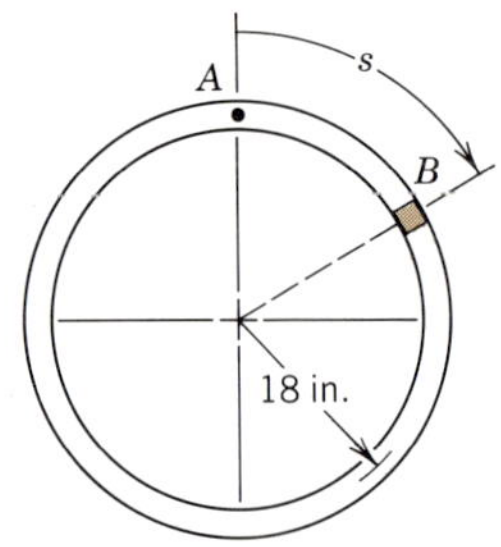

PROB. 11.29

11.29 Block B moves through the hollow tube such that the distance s from top position A is given by $s = -t + 3t^2 - 2t^4$, where s is in feet and t is in seconds. Determine the acceleration of the block when $t = 0.3$ s and draw a sketch of the result.

***11.30** Determine the position, velocity, and acceleration of the block in Problem 11.29 at the instant when the speed is maximum. Draw a sketch of the result.

***11.31** Determine the position, velocity, and acceleration of the block in Problem 11.29 when (a) $t = 0.25$ s, (b) $t = 0.75$ s. Draw a sketch of both results and compare them.

***11.32** The arc length s for the pendulum in a vertical plane, shown in the diagram, is given by $s = (l/10)\sin(\sqrt{g/l}\,t)$. Determine the acceleration of the pendulum, and show the result in a sketch for (a) $s = 0$, (b) $s = l/10$, and (c) $s = l/20$.

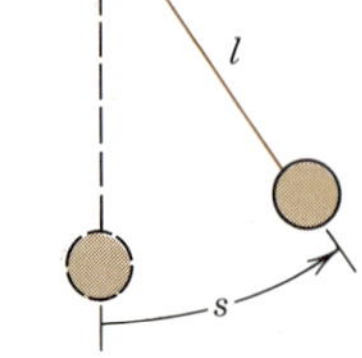

PROB. 11.32

5. Evaluation of the Path Variable Parameters

In the special case where the path is circular, the radius of curvature, the orientation of the unit vectors, and the arc length s are readily obtained from familiar geometrical properties. We shall present the process of determining these parameters for a general three-dimensional path in the next section. Here, we will consider the case of a planar path described by the functional equation $y = f(x)$. This is illustrated in Figure 7.

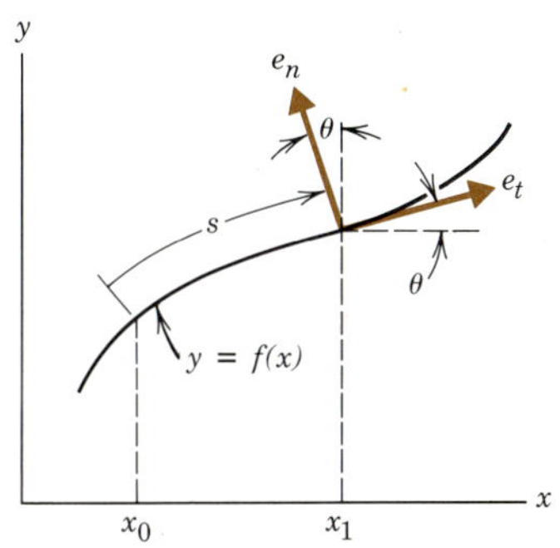

FIGURE 7

The vector $\bar{e}_t$ is tangent to the path, and the slope of the tangent is dy/dx. Thus the angle θ defining $\bar{e}_t$ and $\bar{e}_n$ is given by

$$\theta = \tan^{-1}\left(\frac{dy}{dx}\right) \tag{15}$$

An element of arc length ds is the hypotenuse of a right triangle whose other sides are dx and dy, so $(ds)^2 = (dx)^2 + (dy)^2$

$$ds = \left[1 + \left(\frac{dy}{dx}\right)^2\right]^{1/2} dx \tag{16a}$$

The total arc length between the starting position x_0 and the general position x_1 is therefore

$$s = \int_{x_0}^{x_1} \left[1 + \left(\frac{dy}{dx}\right)^2\right]^{1/2} dx \tag{16b}$$

Finally, an expression for the radius of curvature is obtained by first differentiating eq. (15) in order to form $d\theta$.

$$d\theta = \left(\frac{d}{dx}\theta\right)dx = \left[\frac{1}{1 + \left(\frac{dy}{dx}\right)^2}\frac{d^2y}{dx^2}\right]dx$$

Substituting this equation and eq. (16a) into eq. (11) then yields

$$\frac{1}{\rho} = \frac{|d^2y/dx^2|}{[1 + (dy/dx)^2]^{3/2}} \tag{17}$$

where the absolute value of the second derivative is taken in order to ensure that ρ is positive. As an aside, it should be noted that it is sometimes easier to obtain derivatives by considering the path to be defined in functional form by $x = g(y)$. Such situations can be treated by interchanging the symbols x and y in the formulas.

Once the components of $\bar{a}$ are evaluated with the aid of this expression for ρ, the equation for θ may be employed to depict the unit vectors. Indeed, the angle θ can be used to determine the xy components of $\bar{e}_t$ and $\bar{e}_n$. These in turn can be used to determine the xy components of velocity and acceleration.

One use of the formula for the arc length s, eqs. (16), is to treat situations where the speed of a point is known in terms of the distance s, in other words, where $v(s)$ is given. This may be accomplished using the chain rule to evaluate the tangential acceleration, as follows.

$$\dot{v} = \frac{dv}{ds}\dot{s} = v\frac{dv}{ds} \tag{18}$$

In conclusion, recall that the path variable formulation addresses situations where the path is known, and the motion along the path is described in terms of the distance s and/or the speed v. We shall see that if such information is not available, then there are, in general, better ways to proceed.

EXAMPLE 7

An airplane moves along a path in the vertical plane given by $y = 500 \sin(\pi x/4000)$, where the units of x and y are feet. At position B, the airplane is moving at 300 mi/hr, and it is slowing at 50 ft/s^2. Determine the velocity and acceleration at this position and show the results in a sketch. Then express the velocity and acceleration in terms of xy components.

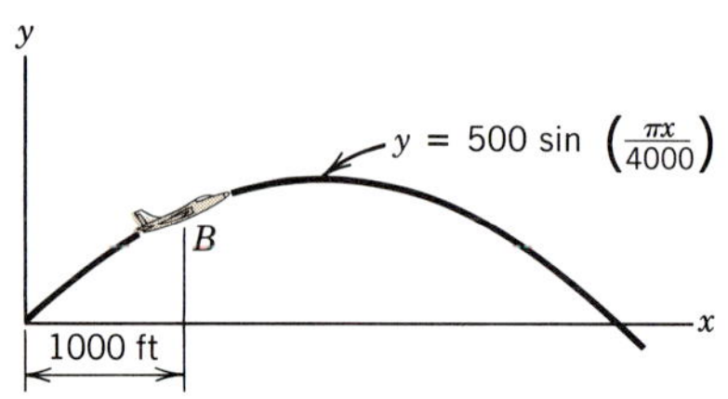

Solution

The values of v and $\dot{v}$ are given

$$v = 300\left(\tfrac{88}{60}\right) = 440 \text{ ft/s}, \qquad \dot{v} = -50 \text{ ft/s}^2$$

We will use eq. (17) to evaluate the radius of curvature. Then, with v, $\dot{v}$, and ρ known, we may compute the acceleration components at position B.

The derivatives of the equation for the path are evaluated at $x = 1000$ ft, which corresponds to position B.

$$\left(\frac{dy}{dx}\right)_B = 500\left(\frac{\pi}{4000}\right)\cos\left[\frac{\pi(1000)}{4000}\right] = 0.2777$$

$$\left|\left(\frac{d^2y}{dx^2}\right)_B\right| = \left|-500\left(\frac{\pi}{4000}\right)^2 \sin\left[\frac{\pi(1000)}{4000}\right]\right| = 2.181(10^{-4})$$

$$\frac{1}{\rho_B} = \frac{2.181(10^{-4})}{[1 + (0.2777)^2]^{3/2}} = 1.9510(10^{-4})$$

The velocity and acceleration in tangent and normal components are therefore

$$\bar{v} = v\,\bar{e}_t = 440\bar{e}_t \text{ ft/s}$$

$$\bar{a} = \dot{v}\,\bar{e}_t + \frac{v^2}{\rho}\bar{e}_n = -50\bar{e}_t + 37.77\bar{e}_n \text{ ft/s}^2$$

In a problem that does not stipulate the form of the results, these expressions would describe the velocity and acceleration, provided that they were accompanied by a sketch of $\bar{e}_t$ and $\bar{e}_n$. The tangent vector is in the sense of the movement, and the normal vector is perpendicular to $\bar{e}_t$, downward because the path is curving down. (The sense of $\bar{e}_n$ may be determined mathematically from the fact that $d^2y/dx^2 < 0$.) The angle θ defining the orientation of the unit vectors is

$$\theta = \tan^{-1}\left(\frac{dy}{dx}\right)_B = 15.52^\circ$$

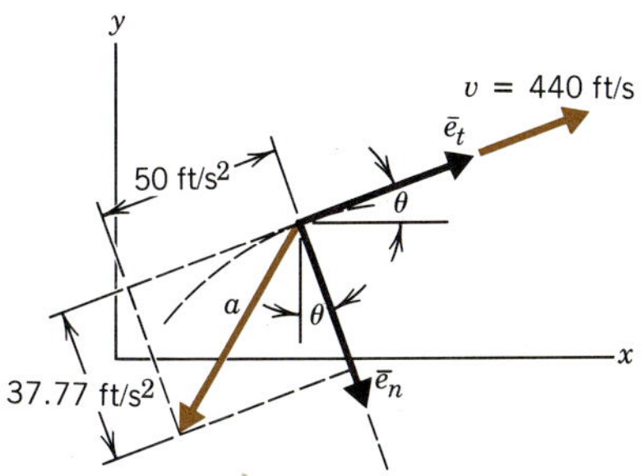

The velocity and acceleration are now depicted in the sketch by plotting their components along $\bar{e}_t$ and $\bar{e}_n$, with the result shown.

The final task in this problem is to convert $\bar{v}$ and $\bar{a}$ into xy components. We begin by expressing the unit vectors in terms of xy components.

$$\bar{e}_t = \cos\theta\,\bar{i} + \sin\theta\,\bar{j} = 0.9635\bar{i} + 0.2676\bar{j}$$

$$\bar{e}_n = \sin\theta\,\bar{i} - \cos\theta\,\bar{j} = 0.2676\bar{i} - 0.9635\bar{j}$$

Replacement of $\bar{e}_t$ and $\bar{e}_n$ in the earlier results for $\bar{v}$ and $\bar{a}$ then yields the desired result.

$$\bar{v} = 440(0.9635\bar{i} + 0.2676\bar{j}) = 424\bar{i} + 118\bar{j}\ \text{ft/s}$$

$$\bar{a} = -50(0.9635\bar{i} + 0.2676\bar{j}) + 37.77(0.2676\bar{i} - 0.9635\bar{j})$$
$$= -38.1\bar{i} - 49.8\bar{j}\ \text{ft/s}^2$$

EXAMPLE 8

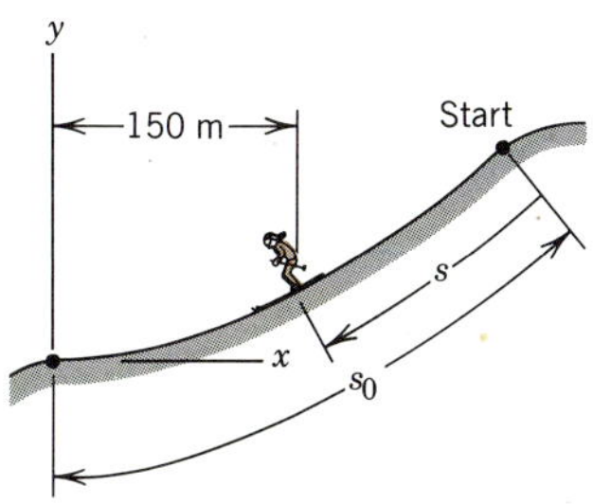

A ski jump is a curve in the vertical plane defined by $y = x^2/200$, where x and y are the horizontal and vertical coordinates of a point relative to the bottom of the ski jump, measured in meters. In one trial a skier began her run at $x = 150$ m and her speed was observed to vary according to $v = 30\sin(\pi s/2s_0)$ m/s, where s is the distance she has traveled and s_0 is the arc length from the start to bottom. Determine the magnitude of the acceleration of the skier when she is at the bottom of the ski jump.

Solution

The path variable approach is suitable because the path is given and the speed is known in terms of s.

$$\bar{a} = \dot{v}\,\bar{e}_t + \frac{v^2}{\rho}\bar{e}_n$$

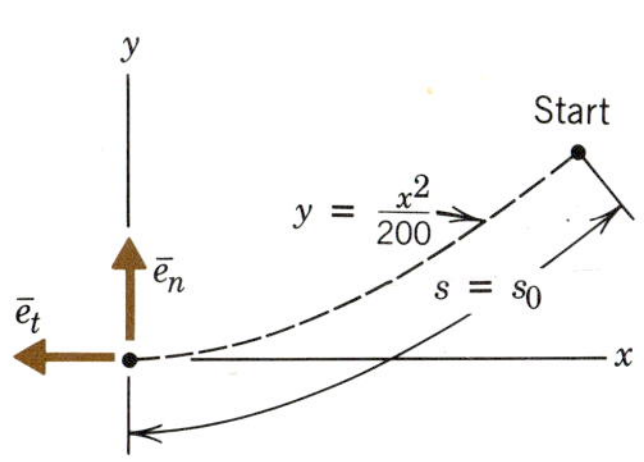

The sketch of the path shows the unit vectors at the bottom, where $s = s_0$. The value of s_0 is obtained from the formula for arc length.

$$s_0 = \int_0^{150}\left[1 + \left(\frac{dy}{dx}\right)^2\right]^{1/2} dx = \int_0^{150}\left[1 + \left(\frac{x}{100}\right)^2\right]^{1/2} dx$$
$$= 0.005\{x((100)^2 + x^2)^{1/2} + (100)^2\ln[x + ((100)^2 + x^2)^{1/2}]\}\Big|_0^{150}$$
$$= 194.95\ \text{m}$$

The foregoing integral was evaluated using a table of integrals; such tabulations are often useful for computing the arc length.

Now that the value of s_0 is known, the value of v may be found by substitution into the given velocity function. Computation of the value of $\dot{v}$ requires use of the chain rule.

$$v = 30 \sin\left(\frac{\pi s}{2s_0}\right)\bigg|_{s=s_0} = 30 \text{ m/s}$$

$$\dot{v} = v\frac{dv}{ds}\bigg|_{s=s_0} = 30\left[30\left(\frac{\pi}{2s_0}\right)\cos\left(\frac{\pi s}{2s_0}\right)\right]\bigg|_{s=s_0} = 0$$

Note that $\dot{v} = 0$ at $s = s_0$ because the value of v is a maximum at that position.

We now need to determine the radius of curvature ρ. In order to use eq. (17), the values of dy/dx and d^2y/dx^2 are calculated at the bottom position, $x = 0$.

$$\frac{dy}{dx}\bigg|_{x=0} = \frac{x}{100}\bigg|_{x=0} = 0, \qquad \frac{d^2y}{dx^2}\bigg|_{x=0} = \frac{1}{100} = 0.01$$

Then

$$\frac{1}{\rho_0} = \frac{0.01}{[1+0]^{3/2}} = 0.01$$

The resulting acceleration is

$$\bar{a} = 0\bar{e}_t + (30)^2(0.01)\bar{e}_n = 9\bar{e}_n \text{ m/s}^2$$

$$|\bar{a}| = 9 \text{ m/s}^2$$

HOMEWORK PROBLEMS

11.33 A toy slot car follows a 4 ft radius track. The speed of the toy is related to the distance traveled, s, according to $v = 2\sqrt{s}$ (s is in feet and v is in feet per second). Determine the acceleration (a) when $s = 1$ ft, (b) when $s = 6.25$ ft. Draw a sketch of the results.

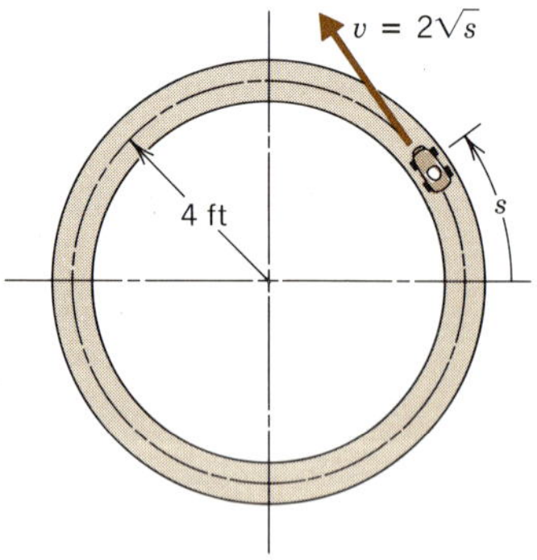

PROB. 11.33

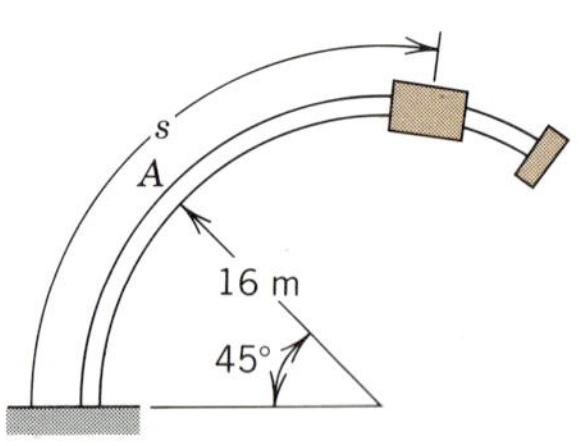

PROB. 11.34

11.34 A collar follows a 16 m radius circular guide. The speed of the collar is related to the arc length s by $v = 10s$, where s is measured in meters and v has units of meters per second. Determine (a) the value of $\dot{v}$ when the collar reaches position A, (b) the acceleration $\bar{a}$ of the collar at position A. Sketch these results.

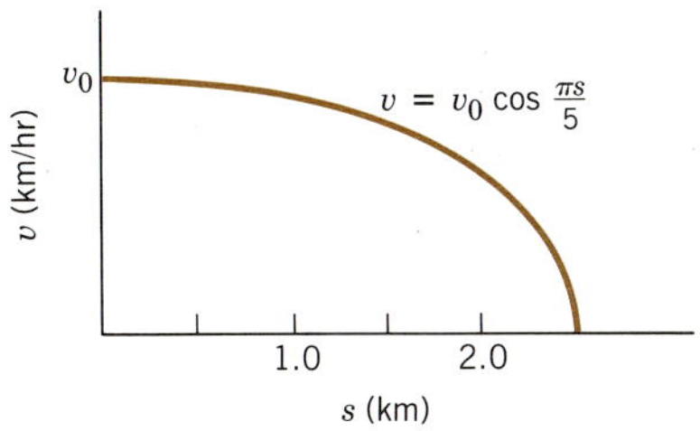

PROB. 11.35

11.35 The computerized controller of a high-speed passenger train slows the train according to the given graph. A train is on a 10 km curve with an initial speed $v_0 = 360$ km/hr when the brakes are applied. Determine the magnitude of the acceleration when $s = 1.5$ km.

11.36 For Problem 11.35 determine the position where $\dot{v}$ has its maximum magnitude.

11.37 An automobile moves with a constant speed of 90 km/hr along a highway whose vertical elevation y in meters is given by $y = 2\cos(\pi x/100)$, where x is the horizontal distance traveled, also in meters. Determine the acceleration of the automobile and sketch the result for the position $x = 100$ m.

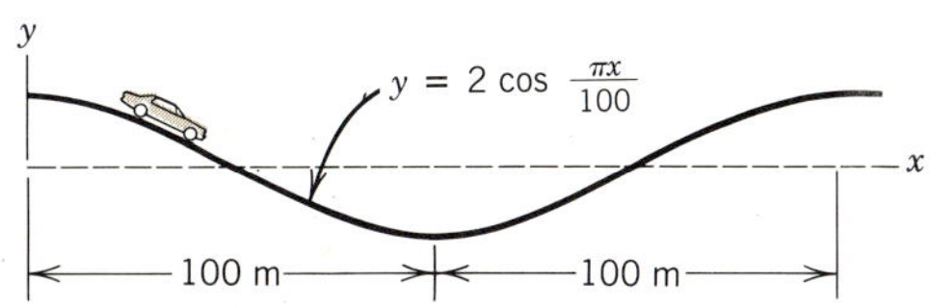

PROB. 11.37

11.38 The brakes of the automobile in Problem 11.37 are suddenly applied at $x = 200$ m, causing the speed to decrease instantaneously at the rate of $0.5g$. Determine and sketch the resulting acceleration at this position.

***11.39** The throttle control of the automobile in Problem 11.37 is suddenly changed at $x = 125$ m, instantaneously causing the speed to increase at 3 m/s^2. Determine the xy components of the velocity and acceleration at this position.

11.40 Arm AB rotates at a rate such that the pin follows the parabolic groove $y = 4 - x^2$(x and y are in feet) at a constant speed of 3 ft/s. Determine the horizontal and vertical components of the velocity and acceleration of the pin when $x = 2$ ft.

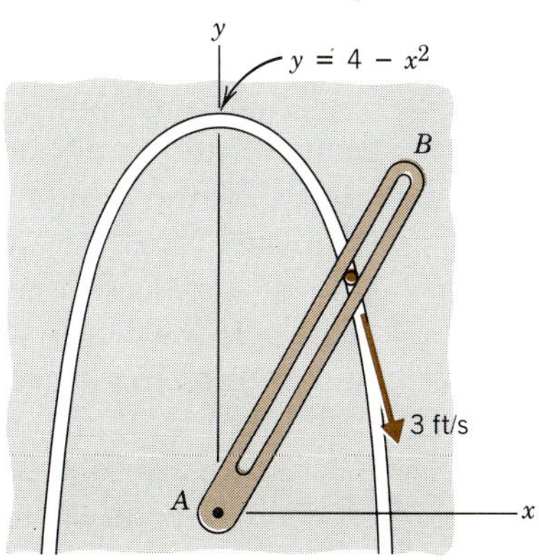

PROB. 11.40

***11.41** Determine the magnitude of the acceleration of the skier in Example 8 when she is a horizontal distance of 75 m relative to the bottom.

***11.42** Determine the horizontal and vertical components of the velocity and acceleration of the skier in Example 8 when she is at a horizontal distance of 50 m relative to the bottom.

---

*6. Path Variables in Three-Dimensional Motion

The path variable formula for acceleration, eq. (13), indicates that the acceleration of a particle consists of tangent and normal components. Although this result was derived for a planar curve, it is also valid for motion along a three-dimensional curvilinear path. The application of the formula in this case requires a description of the tangent and normal directions.

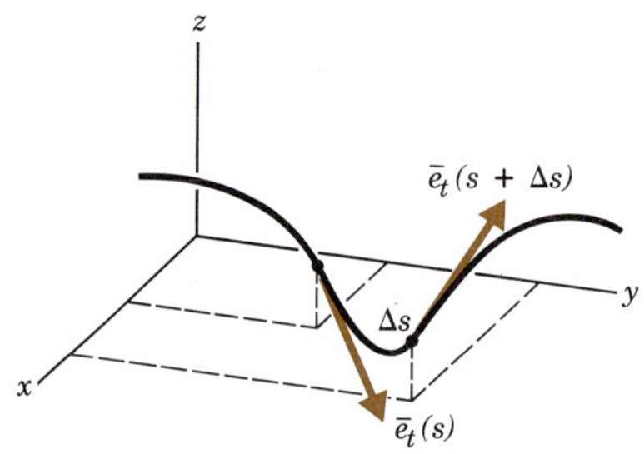

FIGURE 8

A general three-dimensional path is shown in Figure 8. As Δs becomes infinitesmal, but non-zero, it can be proved that $\bar{e}_t(s)$ and $\bar{e}_t(s + \Delta s)$ lie in a common plane, but are not collinear. This plane is the *osculating plane.* To get some idea of the meaning of the osculating plane, we suggest the following. Bend a wire clothes hanger into an arbitrary shape and pick out some point on the hanger. Put the hanger into contact with a flat table (as shown in Figure 9) and rotate the wire until you find the orientation for which a small region of the wire in the vicinity of the point seems to be fully in contact with the table. You have found the osculating plane for the chosen point on the wire. Note that the center of curvature for the infinitesmal arc length *ds* containing the chosen point also lies in the osculating plane. If the wire is bent into a planar curve, the osculating plane for such a curve is the plane of the wire. In the special case where the wire forms a straight line, *any* plane that contains the line is the osculating plane.

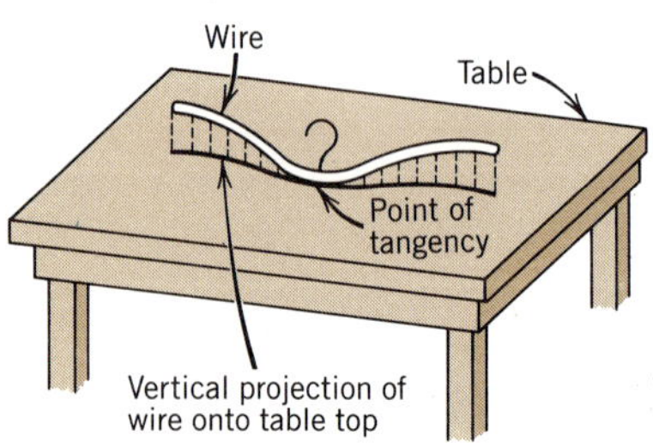

FIGURE 9

In summary, at an arbitrary point on a particle's path, the unit vector $\bar{e}_n$ is perpendicular with the tangent vector $\bar{e}_t$, lies in the osculating plane, and is directed toward the center of curvature.

When we think of the Cartesian coordinates *xyz*, we associate the unit vectors $\bar{i}$, $\bar{j}$, $\bar{k}$ to the system. Similarly, for the path-dependent variables $\bar{e}_t$ and $\bar{e}_n$ we can define a unit vector perpendicular to the osculating plane (the plane of $\bar{e}_t$ and $\bar{e}_n$) in order to complete the triad. That direction is called the *binormal* and is defined as

$$\bar{e}_b \equiv \bar{e}_t \times \bar{e}_n \tag{19}$$

(Clearly, this is directly analogous to $\bar{i} \times \bar{j} = \bar{k}$.) Thus the vector triad is $\bar{e}_t, \bar{e}_n, \bar{e}_b$.

The importance of $\bar{e}_b$ for particle motion is that when one considers Newton's second law, $\bar{F} = m\bar{a}$, from eq. (13) we see that the component of $\bar{a}$ in the direction of $\bar{e}_b$ is zero. This means that we have an equation of statics to call upon in problem solving.

In general, the determination of the path variable quantities for three-dimensional motion involves more mathematical manipulations than that required for path variables in two-dimensional motion. This is illustrated in the following example.

EXAMPLE 9

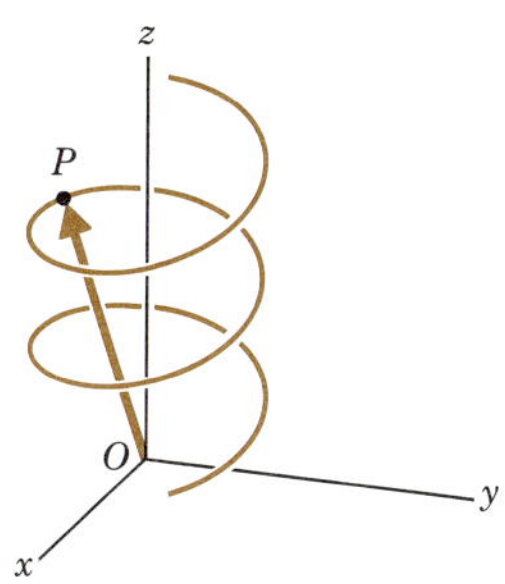

A common way of defining a three-dimensional path is in parametric form. A helix is defined in parametric form by

$$x = A\cos\alpha, \qquad y = A\sin\alpha, \qquad z = C\alpha$$

where α is the parameter. Determine the tangent, normal, and binormal unit vectors and the radius of curvature for an arbitrary value α.

Solution

Equations (8), (12), and (19) are the basic relations for finding $\bar{e}_t$, $\bar{e}_n$, and $\bar{e}_b$ in this problem. They yield

$$\bar{e}_t = \frac{d\bar{r}_{P/O}}{ds}, \qquad \bar{e}_n = \rho\frac{d\bar{e}_t}{ds}, \qquad \bar{e}_b = \bar{e}_t \times \bar{e}_n$$

Point P has coordinates (x, y, z) so

$$\begin{aligned}\bar{r}_{P/O} &= x\bar{i} + y\bar{j} + z\bar{k} \\ &= A\cos\alpha\,\bar{i} + A\sin\alpha\,\bar{j} + C\alpha\bar{k}\end{aligned}$$

Differentiation of $\bar{r}_{P/O}$ with respect to the arc length s requires the chain rule because $\bar{r}_{P/O}$ is known as a function of α.

$$\begin{aligned}\bar{e}_t &= \frac{d\bar{r}_{P/O}}{ds} = \frac{d\bar{r}_{P/O}}{d\alpha}\frac{d\alpha}{ds} \\ &= \frac{d\alpha}{ds}[-A\sin\alpha\,\bar{i} + A\cos\alpha\,\bar{j} + C\bar{k}]\end{aligned}$$

We do not know $d\alpha/ds$. To determine this quantity we use the fact that $\bar{e}_t$ must be a unit vector, so

$$\begin{aligned}|\bar{e}_t| = 1 &= \frac{d\alpha}{ds}[(-A\sin\alpha)^2 + (A\cos\alpha)^2 + C^2]^{1/2} \\ &= \frac{d\alpha}{ds}[A^2 + C^2]^{1/2}\end{aligned}$$

$$\frac{d\alpha}{ds} = (A^2 + C^2)^{-1/2}$$

so that

$$\bar{e}_t = (A^2 + C^2)^{-1/2}(-A \sin \alpha \bar{i} + A \cos \alpha \bar{j} + C\bar{k})$$

The normal vector $\bar{e}_n$ may now be determined by using the chain rule to differentiate $\bar{e}_t$.

$$\bar{e}_n = \rho \frac{d\bar{e}_t}{ds} = \rho \frac{d\bar{e}_t}{d\alpha}\frac{d\alpha}{ds}$$

$$= \rho(A^2 + C^2)^{-1/2}(-A \cos \alpha \bar{i} - A \sin \alpha \bar{j})(A^2 + C^2)^{-1/2}$$

$$= -\rho(A^2 + C^2)^{-1}(A \cos \alpha \bar{i} + A \sin \alpha \bar{j})$$

However, $\bar{e}_n$ is also a unit vector.

$$|\bar{e}_n| = 1 = \rho(A^2 + C^2)^{-1}[(A \cos \alpha)^2 + (A \sin \alpha)^2]^{1/2}$$

$$\equiv \rho(A^2 + C^2)^{-1}A$$

This enables us to find the radius of curvature and $\bar{e}_n$.

$$\rho = \frac{A^2 + C^2}{A}$$

$$\bar{e}_n = -\frac{1}{A}(A \cos \alpha \bar{i} + A \sin \alpha \bar{j})$$

Finally, $\bar{e}_b$ is constructed such that $\bar{e}_t$, $\bar{e}_n$, and $\bar{e}_b$ are mutually perpendicular.

$$\bar{e}_b = \bar{e}_t \times \bar{e}_n$$

$$= -\frac{1}{A}(A^2 + C^2)^{-1/2}(-A \sin \alpha \bar{i} + A \cos \alpha \bar{j} + C\bar{k})$$

$$\times (A \cos \alpha \bar{i} + A \sin \alpha \bar{j})$$

$$= (A^2 + C^2)^{-1/2}(C \sin \alpha \bar{i} - C \cos \alpha \bar{j} + A\bar{k})$$

HOMEWORK PROBLEMS

11.43 A conical helix is defined in parametric form by

$$x = 40\alpha \cos \pi\alpha \qquad y = 40\alpha \sin \pi\alpha \qquad z = 30\alpha$$

where x, y, and z have units of millimeters. Determine the tangent, normal, and binormal unit vectors when $z = 18$ mm.

11.44 In the position where $\alpha = 3.25$ a particle following the conical helix of Problem 11.43 has a speed of 50 m/s and is gaining speed at 10 m/s^2. Determine its corresponding position, velocity, and acceleration.

11.45 The elevation z of a hyperbolic paraboloid is related to the horizontal position coordinates x and y by $z = -0.1xy$ where all lengths are in feet. A sphere rolling on this surface moves such that $x = 5 \sin \alpha$,

$y = 5 \cos \alpha$, where x and y are in feet and α is a parameter. Determine the tangent, normal, and binormal unit vectors for the path of the sphere at $x = 2.5$ ft.

11.46 The sphere in Problem 11.45 has a speed of 10 ft/s, and it is slowing at 5 ft/s^2 when it is at $x = 4$ ft. Determine the corresponding position, velocity, and acceleration.

---

7. Rectangular Cartesian Coordinates

The underlying principle for the Cartesian coordinate description of the kinematical relationships is to express the components of vector quantities in a fixed xyz coordinate system. The unit vectors $\bar{i}, \bar{j}, \bar{k}$ describe the positive x, y, and z directions.

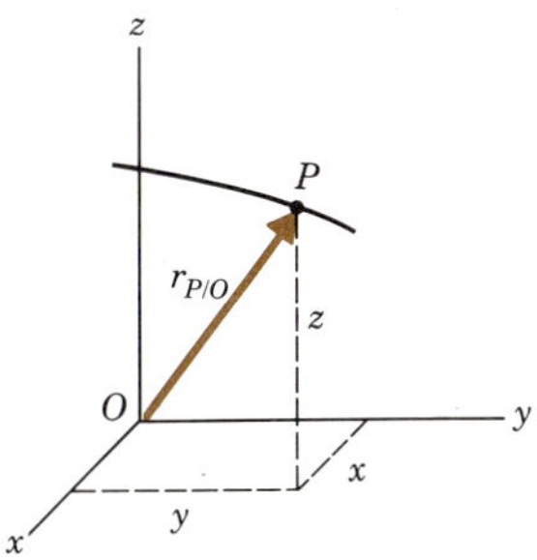

FIGURE 10a

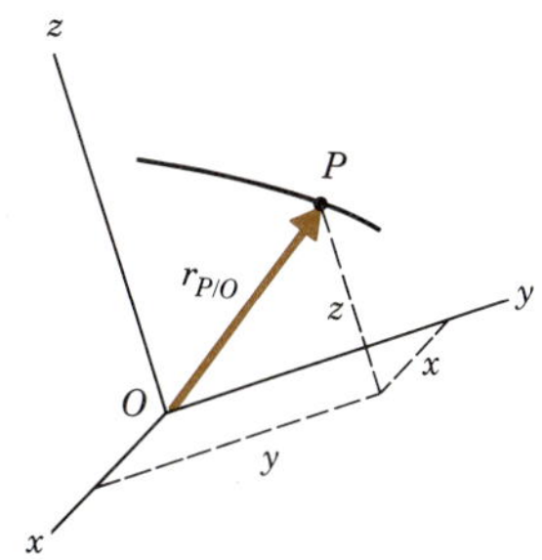

FIGURE 10b

Figure 10a shows a point P whose coordinates are (x, y, z). The components of the position vector $\bar{r}_{P/O}$ parallel to the $\bar{i}$, $\bar{j}$, and $\bar{k}$ unit vectors are

$$\boxed{\bar{r}_{P/O} = x\bar{i} + y\bar{j} + z\bar{k}} \tag{20}$$

In Figure 10b we selected a Cartesian coordinate system having the same origin O but different orientation from the one shown in Figure 10a. Note that the values of x, y, and z are different, as are, of course, the $\bar{i}$, $\bar{j}$, and $\bar{k}$ directions. Nevertheless, the resulting vector $\bar{r}_{P/O}$ represents the same physical position in space. This emphasizes the statement that a description in component form of any vector is meaningful only if the choice of unit vectors is specified.

When the particle moves, the values of the x, y, and z components change with time. The functions $x(t)$, $y(t)$, and $z(t)$ define the path of the particle in parametric form, with t being the parameter. For instance, $x = y = 0$, $z = -\frac{1}{2}t^2$ defines a path that is the z axis.

The velocity is found by differentiating the position vector with respect to time. Because xyz is a fixed system, $\bar{i}$, $\bar{j}$, and $\bar{k}$ have constant magnitude and direction, so that $\dot{\bar{i}}, = \dot{\bar{j}} = \dot{\bar{k}} = \bar{0}$. Because x, y, and z are the only variables in the expression for $\bar{r}_{P/O}$,

$$\bar{v} = \dot{x}\bar{i} + \dot{y}\bar{j} + \dot{z}\bar{k}$$

Thus, each component of velocity is the rate of change of the corresponding position coordinate.

$$\bar{v} = v_x\bar{i} + v_y\bar{j} + v_z\bar{k}$$
$$v_x = \dot{x}, \qquad v_y = \dot{y}, \qquad v_z = \dot{z} \tag{21}$$

The speed may be obtained from eq. (21) by evaluating the magnitude of the velocity

$$v = |\bar{v}| = [v_x^2 + v_y^2 + v_z^2]^{1/2}$$

An expression for acceleration is derived from the velocity by differentiation. For this, the unit vectors are again treated as constants, with the result that

$$\bar{a} = \dot{v}_x\bar{i} + \dot{v}_y\bar{j} + \dot{v}_z\bar{k} \tag{22}$$

An important feature of the Cartesian coordinate formulas is the relationship between corresponding components of position, velocity, and acceleration. Each equation is derived by differentiating the preceding one with respect to time. This is reminiscent of the relationship between arc length, speed, and acceleration in path variables. One facet of this similarity is that it permits the application of the constant acceleration formulae. For example, *if* the x component of acceleration is constant, $\dot{v}_x = a_x$ (= constant), then

$$v_x = (v_x)_0 + a_x t, \qquad x = x_0 + (v_x)_0 t + \tfrac{1}{2}a_x t^2$$
$$v_x^2 = (v_x)_0^2 + 2a_x(x - x_0)$$

(Be sure not to use these relations when a_x is *not* constant.)

The similarity with path variables also provides a different viewpoint for the motion. Recall that the unit vectors for Cartesian coordinates have constant orientation. Thus the motion may be pictured as a superposition of rectilinear motions in each of the three coordinate directions. It follows that the Cartesian coordinate formulation is useful when the motion of a point is described with respect to fixed directions, such as horizontal and vertical.

EXAMPLE 10

The coordinates of a particle in planar motion are $x = 40t^4$, $y = 200t^2 - 800$, where x and y are in millimeters and t is in seconds. Determine the position, velocity, and acceleration of the particle when $t = 2$ s, and identify its path.

Solution

The coordinates are given as a function of t, so the velocity and acceleration components may be found by successive differentiations.

$$x = 40t^4, \qquad y = 200t^2 - 800 \text{ mm}$$
$$v_x = \dot{x} = 160t^3, \qquad v_y = \dot{y} = 400t \text{ mm/s}$$
$$\dot{v}_x = 480t^2, \qquad \dot{v}_y = 400 \text{ mm/s}^2$$

We evaluate these terms at $t = 2$, and then write the components in vector form

$$\bar{r} = 0.64\bar{i} \text{ m}$$
$$\bar{v} = v_x\bar{i} + v_y\bar{j} = 1.28\bar{i} + 0.8\bar{j} \text{ m/s}$$
$$\bar{a} = \dot{v}_x\bar{i} + \dot{v}_y\bar{j} = 1.92\bar{i} + 0.4\bar{j} \text{ m/s}^2$$

The path followed by the particle may be determined by finding an equation relating x and y. This requires elimination of t, which may be achieved by solving either the x or y equation for t, and substituting that result into the other equation. For example,

$$t = \left(\frac{x}{40}\right)^{1/4} \Rightarrow y = 200\left[\left(\frac{x}{40}\right)^{1/4}\right]^2 - 800 = 200\left(\frac{x}{40}\right)^{1/2} - 800$$

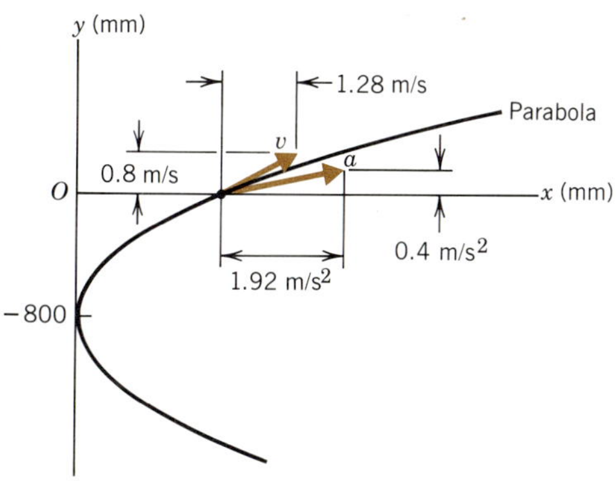

A simpler form of this relation is

$$x = 40\left(\frac{y}{200} + 4\right)^2$$

As shown in the sketch, this is a parabola. The sketch also depicts the position, velocity, and acceleration of the particle when $t = 2$ s.

Notice that the velocity is depicted as the tangent to the path, and the acceleration component perpendicular to the tangent is in the sense of the curvature. These results are consistent with the path variable formulas.

EXAMPLE 11

The circular rod guiding collar C is defined by $x^2 + y^2 = R^2$. Collar C is made to move such that its horizontal velocity component, v_x, is constant (as long as $|x| < R$). Determine the velocity and acceleration at a general position.

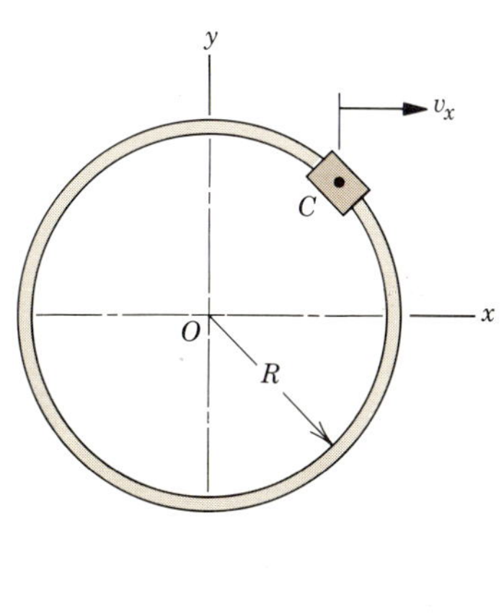

Solution

Cartesian coordinates are suitable for this problem because the movement along the given path is described in terms of the velocity component v_x. Specifically, v_x is a given constant and $\dot{v}_x = 0$.

Evaluation of the velocity and acceleration requires the determination of $v_y \equiv \dot{y}$ and $\dot{v}_y$. The relationship between x and y is known, as is the value $\dot{x} = v_x$. This suggests that derivatives be found by employing the chain rule. First we solve for y in terms of x.

$$y = (R^2 - x^2)^{1/2}$$

Then, with $\dot{v}_x = 0$

$$v_y = \dot{y} = \frac{d}{dt}(R^2 - x^2)^{1/2} = \frac{1}{2}\frac{(-2x\dot{x})}{(R^2 - x^2)^{1/2}} = -\frac{xv_x}{(R^2 - x^2)^{1/2}}$$

$$\dot{v}_y - \frac{d}{dt}\left[-\frac{xv_x}{(R^2 - x^2)^{1/2}}\right]$$

$$= -\frac{\dot{x}v_x}{(R^2 - x^2)^{1/2}} + \frac{1}{2}\frac{xv_x(-2x\dot{x})}{(R^2 - x^2)^{3/2}}$$

$$\equiv -\frac{R^2}{(R^2 - x^2)^{3/2}}v_x^2$$

Thus the velocity and acceleration are

$$\bar{v} = v_x\bar{i} + v_y\bar{j} = v_x\bar{i} - \frac{x}{(R^2 - x^2)^{1/2}}v_x\bar{j}$$

$$\bar{a} = \dot{v}_x\bar{i} + \dot{v}_y\bar{j} = -\frac{R^2}{(R^2 - x^2)^{3/2}}v_x^2\bar{j}$$

As an aside, v_y and $\dot{v}_y$ could have been evaluated by an alternative procedure. The equation for x in terms of t could have been obtained by using the constant speed relationship. Substitution of that into $y = (R^2 - x^2)^{1/2}$ then gives an equation for $y(t)$, whose derivatives give $\bar{v}$ and $\bar{a}$ in terms of time t.

HOMEWORK PROBLEMS

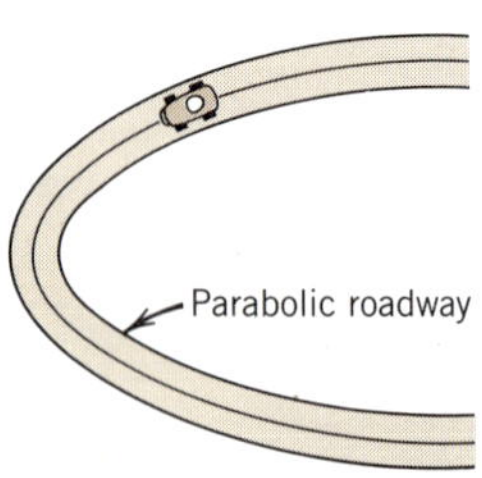

PROB. 11.47

11.47 A toy race car follows a level road defined by $y = x^2/0.2$, where x and y and measured in feet. When first observed, the car is at $x = 0.2552$ ft and 2 ms later it is at $x = 0.2394$ ft. Determine the approximate velocity of the car.

11.48 The planar motion of a particle is defined by $x = t^2 + 8t$ and $y = t^3/3 - 4$ where x and y are in meters and t is in seconds. Determine the magnitude and angle relative to the x axis of the velocity and acceleration when $t = 3$ seconds.

11.49 The planar motion of a particle is defined by $x = 4t - 3$, $y = \sqrt{24t - 12}$, where x and y are expressed in meters and t is in seconds.

Determine the velocity and acceleration of the particle (a) when $t = 1$ s, and (b) when $t = 3$ s. Show the results in a sketch of the path.

11.50 The coordinates of a particle in planar motion are defined by $x = e^{t/2}$, $y = 16e^{-t/2}$, where x and y are measured in feet and t is in seconds. Show that the path is a hyperbola and determine the velocity and acceleration when $t = 2$ s. Show the solution in a sketch.

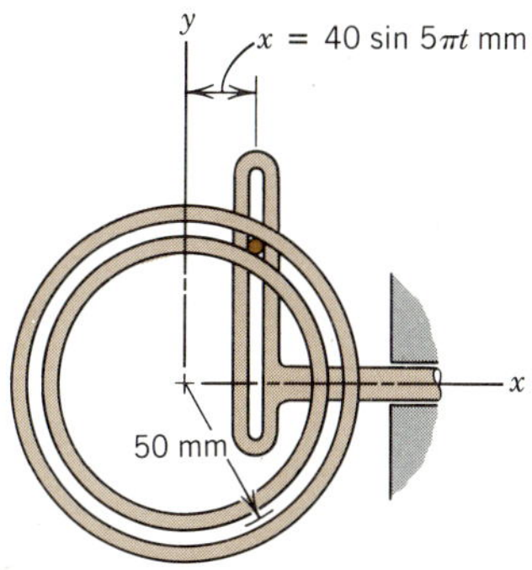

PROB. 11.51

11.51 The (x, y) coordinates of a sliding pin connecting the two grooved members of the machine shown in the diagram simultaneously satisfy the relations $x^2 + y^2 = 2500$, $x = 40 \sin 5\pi t$, where x and y have units of millimeters and t is in seconds. Determine the velocity and acceleration of the pin when $t = 0$.

11.52 Solve Problem 11.51 for $t = \frac{1}{30}$ s.

11.53 A ball is thrown horizontally off the top of a building and follows a parabolic path in the vertical plane defined by $z = 0$, $y = -cx^2$, where y is its elevation measured from the top of the building and x is the horizontal distance traveled by the ball. The horizontal component of velocity, v_x, is constant. Find the acceleration of the particle in terms of v_x and c.

11.54 Block B, which is tied to peg A by an elastic cable, slides on a flat surface. The motion is defined by $x = C \cos pt$, $y = D \sin pt$. Show that, at any instant, the acceleration is directed toward the origin and that its magnitude is proportional to the distance from peg A to the block.

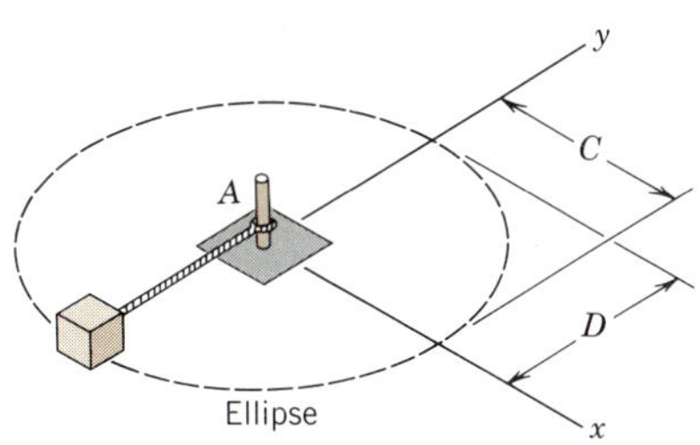

PROB. 11.54

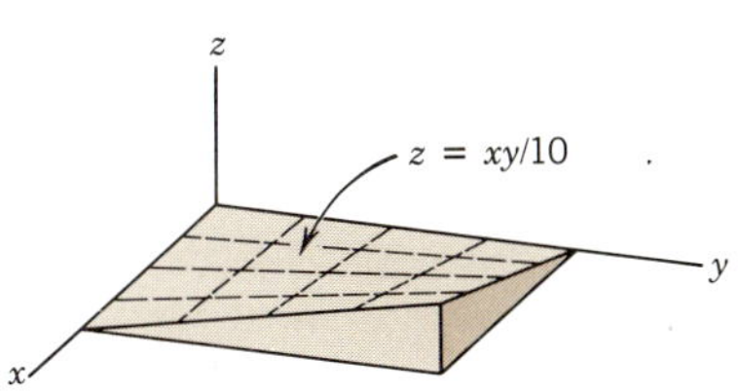

PROB. 11.55

11.55 The surface shown has the shape of a hyperbolic paraboloid $z = xy/10$, where the length unit is meters. A ball rolling on the roof is observed to have the motion $x = 10 - 40t$, $y = 10 - 20t^2$, where t is in seconds. Determine the velocity and acceleration of the ball when it reaches the edge where $x = 0$.

11.56 An electron in a cathode-ray tube follows the path

$$y = 0.9x, \qquad z = (x^2 - 2500)^{1/2}$$

where x, y, and z are measured in millimeters. The value of x varies with time t according to $x = 1.5(10^6)t^4$, where t is in seconds. Determine the velocity and acceleration of the electron when $x = 150$ mm.

8. Cylindrical Coordinates

For Cartesian coordinates we selected three independent quantities, namely the x, y, and z coordinates of a particle, to uniquely determine its position in the chosen reference frame. Then, in terms of the unit vectors of the coordinate system, we obtained position, velocity, and acceleration formulas for the particle. In this section we follow a similar procedure using a different set of quantities to define the position of a particle.

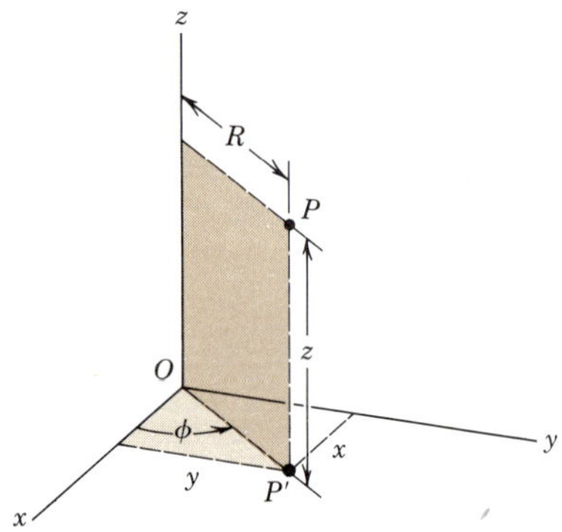

FIGURE 11

In order to define the position of point P with respect to the xyz coordinate system in Figure 11, we construct line PP' that intersects point P and is perpendicular to the xy plane. For notation, we denote the length of line OP' as R and the angle between the x axis and line OP' as ϕ; the length of line PP' is the Cartesian coordinate z.

The values (R, ϕ, z) are the *cylindrical coordinates* of point P. The terminology used to discuss them is to say that R is the *radial distance*, that ϕ is the *transverse angle*, and that z is the *axial distance*. The name "cylindrical coordinates" originates from the fact that if the value of R is specified, point P will be situated somewhere on the surface of a cylinder of radius R. The parameter R can be a function of time, hence the cylinder need *not* be circular. Note that for the special case of planar motion in the xy plane, that is $z \equiv 0$, the remaining coordinates (R, ϕ) are *polar coordinates*.

The cylindrical coordinates locate point P even if the path is unknown. This is illustrated by the fact that there is a general coordinate transformation between the cylindrical and Cartesian coordinates. In order to avoid ambiguity in this transformation, the transverse angle ϕ is restricted to the range $0 \le \phi < 2\pi$. Then because x and y are two sides of the right triangle in Figure 11 whose hypotenuse is R, it follows that

$$x = R \cos \phi, \qquad y = R \sin \phi$$

so that

$$R = (x^2 + y^2)^{1/2}, \qquad \phi = \tan^{-1}\left(\frac{y}{x}\right)$$

The three cylindrical coordinates have a set of three corresponding unit vectors, as did the Cartesian coordinates. The *radial direction*, which is described by the unit vector $\bar{e}_R$, is the direction in which point P would move if its radial distance R were increased. Similarly, the *transverse direction* is described by the unit vector $\bar{e}_\phi$; it is the direction in which point P would move if its transverse angle ϕ were increased. Because R is measured perpendicular to the z axis, both $\bar{e}_R$ and $\bar{e}_\phi$ are parallel to the xy plane. They are depicted in Figure 12 as viewed looking down the z axis. An important feature of these unit vectors is that their orientation changes as the angle ϕ changes.

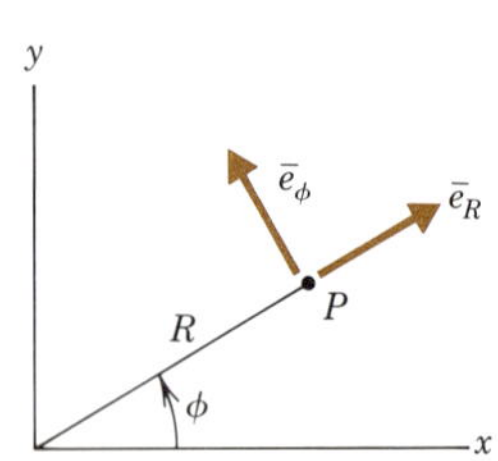

FIGURE 12

The third unit vector $\bar{k}$ is the *axial direction*, corresponding to the axial distance z. It never changes direction; thus it is the same as the (constant) unit vector that is employed in Cartesian coordinates.

All three unit vectors appear in Figure 13, along with the position vector $\bar{r}_{P/O}$. It will be noted that the unit vectors are mutually orthogonal, with $\bar{e}_R \times \bar{e}_\phi = \bar{k}$.

The three unit vectors $\bar{e}_R$, $\bar{e}_\phi$, and $\bar{k}$ may be used to express any vector in terms of its radial, transverse, and axial components. We will now develop such a description for the position, velocity, and acceleration of a point.

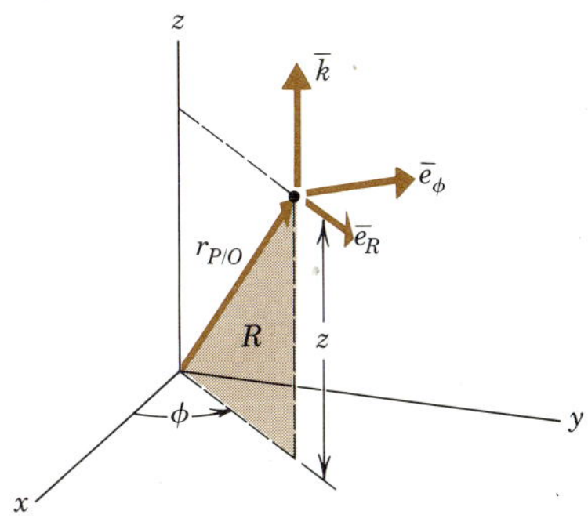

FIGURE 13

Figure 13 indicates that the radial component of $\bar{r}_{P/O}$ is R, and the axial component is z. This position vector has no transverse component because it lies in the plane formed by $\bar{e}_R$ and $\bar{k}$. Therefore

$$\boxed{\bar{r}_{P/O} = R\,\bar{e}_R + z\,\bar{k}} \tag{23}$$

Previously we showed that the values of R, ϕ, and z were required to define the position. Then why doesn't the value of ϕ appear in this expression for $\bar{r}_{P/O}$? The answer is that the equation for $\bar{r}_{P/O}$ really does depend on the value of ϕ, but not explicitly, because ϕ must be known in order to define (locate) the direction of $\bar{e}_R$.

An expression for the velocity $\bar{v}$ is obtained by differentiating $\bar{r}_{P/O}$ in eq. (23). For this, recall that the orientation of $\bar{e}_R$ depends on ϕ, so the chain rule is needed for $\dot{\bar{e}}_R$. In contrast, $\dot{\bar{k}} \equiv \bar{0}$ because $\bar{k}$ is constant. Thus

$$\bar{v} = \dot{R}\,\bar{e}_R + R\,\dot{\bar{e}}_R + \dot{z}\,\bar{k} = \dot{R}\,\bar{e}_R + R\dot{\phi}\,\frac{d\bar{e}_R}{d\phi} + \dot{z}\,\bar{k}$$

The terms $\dot{R}, \dot{\phi}$, and $\dot{z}$ are recognizable as the rates of change of the cylindrical coordinates, but the term $d\bar{e}_R/d\phi$ requires further consideration.

Let us look down the z axis in order to see how $\bar{e}_R$ changes when ϕ is increased by an infinitesimal amount $d\phi$. This is the situation in Figure 14.

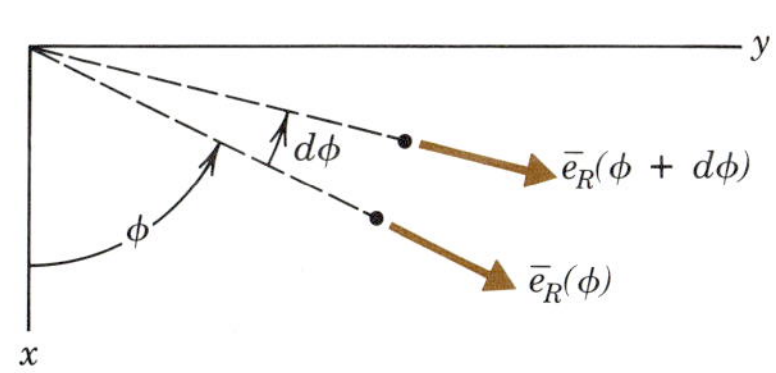

FIGURE 14a

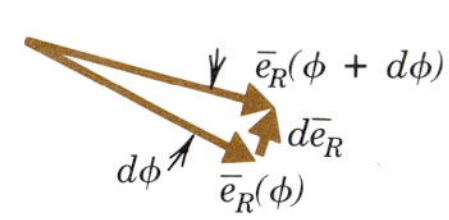

FIGURE 14b

The difference between the two $\bar{e}_R$ vectors is the vector $d\bar{e}_R$ shown in Figure 14b, that is

$$d\bar{e}_R \equiv \bar{e}_R(\phi + d\phi) - \bar{e}_R(\phi)$$

The length of both $\bar{e}_R$ vectors is unity, by definition. The length of $d\bar{e}_R$ is determined by calculating the arc length, which is permissible because $d\phi$ is very small.

$$|d\bar{e}_R| = |\bar{e}_R|\,d\phi \equiv d\phi$$

The direction of $d\bar{e}_R$ is perpendicular to $\bar{e}_R$, again because of the smallness of $d\phi$. The sense of this perpendicular is that of the increase in ϕ. In other words, $d\bar{e}_R$ is in the direction of $\bar{e}_\phi$. In conjunction with the result for the magnitude of $d\bar{e}_R$, it follows that

$$\frac{d\bar{e}_R}{d\phi} = \bar{e}_\phi \tag{24}$$

Substitution of this expression into the earlier expression for $\bar{v}_P$ then yields

$$\bar{v} = \dot{R}\,\bar{e}_R + R\dot{\phi}\bar{e}_\phi + \dot{z}\,\bar{k} \tag{25}$$

Equation (25) is the velocity equation in cyclindrical coordinates. The radial component $v_R = \dot{R}$ occurs when the particle moves in the radial direction. The axial component $v_z = \dot{z}$ occurs when the particle moves in the axial direction. The transverse component $v_\phi = R\dot{\phi}$ occurs when ϕ is not a constant. We can see that, for a given value of $\dot{\phi}$, the particle will move faster for larger values of R. The quantity $\dot{\phi}$ is termed the *angular speed.*

In order to find the acceleration formula we differentiate the velocity, eq. (25). The chain rule is employed once again to differentiate the unit vectors.

$$\bar{a} = \ddot{R}\,\bar{e}_R + \dot{R}\,\dot{\bar{e}}_R + \dot{R}\,\dot{\phi}\bar{e}_\phi + R\ddot{\phi}\bar{e}_\phi + R\dot{\phi}\dot{\bar{e}}_\phi + \ddot{z}\,\bar{k}$$

$$\equiv \ddot{R}\,\bar{e}_R + R\dot{\phi}\frac{d\bar{e}_R}{d\phi} + \dot{R}\dot{\phi}\bar{e}_\phi + R\ddot{\phi}\bar{e}_\phi + R\dot{\phi}^2\frac{d\bar{e}_\phi}{d\phi} + \ddot{z}\,\bar{k}$$

The term $d\bar{e}_R/d\phi$ is given in eq. (24). The same procedure used to obtain eq. (24) may be followed to evaluate $d\bar{e}_\phi/d\phi$. This would yield

$$\frac{d\bar{e}_\phi}{d\phi} = -\bar{e}_R \tag{26}$$

Substitution of both derivatives into the acceleration expression, followed by grouping of like components, then gives the following general relation.

$$\bar{a} = (\ddot{R} - R\dot{\phi}^2)\bar{e}_R + (R\ddot{\phi} + 2\dot{R}\dot{\phi})\bar{e}_\phi + \ddot{z}\bar{k} \tag{27}$$

Let us look at this expression for acceleration by considering motion along some simple paths. If a particle moves along a radial line, then only R will change with time so that eq. (27) reduces to $\bar{a} = \ddot{R}\bar{e}_R$. If a particle moves along a line parallel to the z axis, only z changes with time and eq. (27) reduces to $\bar{a} = \ddot{z}\bar{k}$. If a particle moves on a circular path of constant radius R parallel to the xy plane, then eq. (27) reduces to

$$\bar{a} = -R\dot{\phi}^2\bar{e}_R + R\ddot{\phi}\bar{e}_\phi$$

The first term on the right side of this equation is perpendicular to the circular path, so it represents the centripetal acceleration. The second term is the tangential component of acceleration because $\bar{e}_\phi$ is perpendicular to the radius.

For a general motion, the radial acceleration in eq. (27) combines the effect of varying radial distance in rectilinear motion and the centripetal acceleration in motion along a curved path. One of the terms in the transverse acceleration in eq. (27) did not arise for either of the simpler paths. The portion $R\ddot{\phi}$, which is the tangential acceleration for the circular path, is obviously associated with angular acceleration. However, the term $2\dot{R}\dot{\phi}$ is an interaction effect that occurs only when R and ϕ both vary. It is called the *Coriolis acceleration* effect, so named after G. Coriolis (1792–1843), who first disclosed its existence.

One indication that cylindrical coordinates should be employed in a problem is that the motion is described in terms of the radial distance from a fixed axis. Another clue is that the angle or rate of rotation about a fixed axis is significant. Once a velocity or acceleration has been evaluated, it is adequate to express the result in terms of its radial, transverse, and axial components, provided that the corresponding unit vectors are shown in a sketch.

EXAMPLE 12

A locomotive approaches the pivot point O of the turntable at a constant speed of 1.5 m/s as the turntable rotates at a constant rate of 2 rev/min. Determine the velocity and acceleration of the locomotive at the instant shown. Draw a sketch of each result.

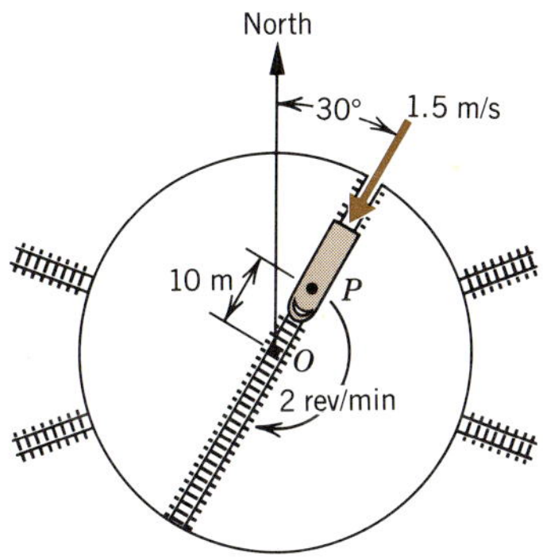

Solution

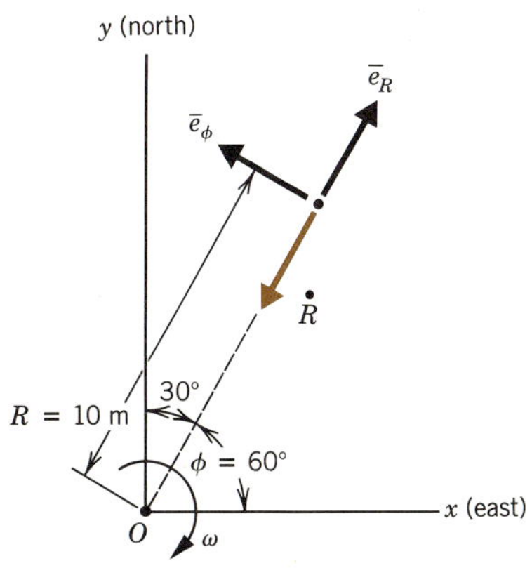

The motion is described in terms of a rate of rotation about an axis through point O, and a rate of approach to that axis. Also, the motion occurs in the horizontal plane. This suits the cylindrical coordinate formulation, with the axial motion identically zero—in other words, polar coordinates.

The accompanying sketch depicts R, ϕ, and the corresponding unit vectors at the given instant. Note that $\bar{e}_R$ always points outward from the pivot, and that $\bar{e}_\phi$ is perpendicular to $\bar{e}_R$ in the sense of increasing ϕ.

The general expressions for $\bar{v}$ and $\bar{a}$ are

$$\bar{v} = \dot{R}\,\bar{e}_R + R\dot{\phi}\bar{e}_\phi$$

$$\bar{a} = (\ddot{R} - R\dot{\phi}^2)\bar{e}_R + (R\ddot{\phi} + 2\dot{R}\dot{\phi})\bar{e}_\phi$$

The given angular speed is the rate of change of ϕ. The angular unit for this rate must be radians, not revolutions, and ϕ is decreasing for clockwise rotation, so $\dot{\phi} < 0$. Thus

$$\dot{\phi} = -2\,\frac{\text{rev}}{\text{min}}\left(\frac{2\pi\ \text{rad}}{\text{rev}}\right)\left(\frac{1\ \text{min}}{60\ \text{s}}\right) = -0.2094\ \text{rad/s}$$

The radial distance decreases when the locomotive approaches the pivot, which means that $\dot{R} < 0$, and the given speed is the magnitude of $\dot{R}$. Thus, at this instant,

$$R = 10\ \text{m}, \qquad \dot{R} = -1.5\ \text{m/s}$$

Both $\dot{R}$ and $\dot{\phi}$ are constant, so $\ddot{R} = \ddot{\phi} = 0$. These values are substituted into the formulas for $\bar{v}$ and $\bar{a}$, with the result that

$$\bar{v} = -1.5\bar{e}_R - 2.09\bar{e}_\phi\ \text{m/s}$$

$$\bar{a} = -0.439\bar{e}_R + 0.628\bar{e}_\phi\ \text{m/s}^2$$

These vectors are depicted in the sketch by constructing their components along the unit vectors.

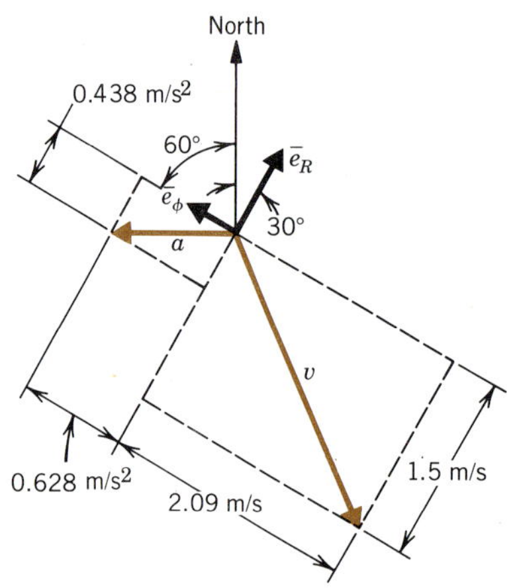

EXAMPLE 13

Points on a spirally wound helical coil satisfy $R = k_1\theta^2$, $z = k_2\theta$, where k_1 and k_2 are constants and θ is the transverse angle shown. It is observed that a collar that slides over the coil rotates about the z axis such that $\theta = k_3 t^2$. Determine the velocity and acceleration of the collar at any instant in terms of k_1, k_2, and k_3.

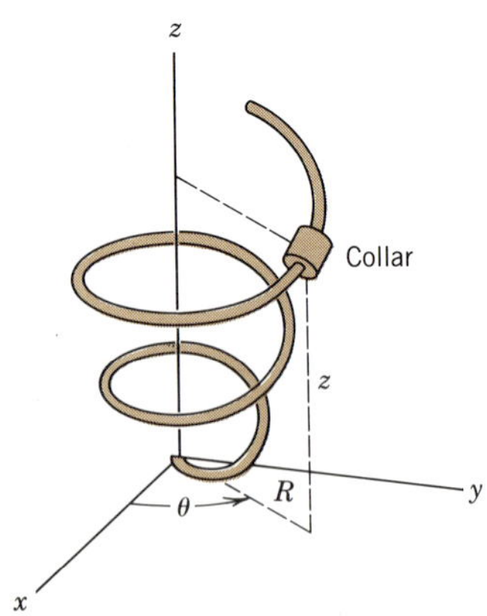

Solution

The path of the collar is given as equations for the radial and axial distances relative to the z axis, and the movement along the collar is defined in terms of the angle θ for the radial line. These are the parameters occurring in cylindrical coordinates (with ϕ being replaced by θ). The corresponding formulas for velocity and acceleration require derivatives of R, θ, and z. These may be obtained by substituting the given expression for $\theta(t)$ into the equations for R and z.

$$\theta = k_3 t^2 \qquad \dot{\theta} = 2k_3 t \qquad \ddot{\theta} = 2k_3$$
$$R = k_1\theta^2 = k_1 k_3^2 t^4 \qquad \dot{R} = 4k_1 k_3^2 t^3 \qquad \ddot{R} = 12 k_1 k_3^2 t^2$$
$$z = k_2\theta = k_2 k_3 t^2 \qquad \dot{z} = 2k_2 k_3 t \qquad \ddot{z} = 2k_2 k_3$$

Then

$$\bar{v} = \dot{R}\bar{e}_R + R\dot{\theta}\bar{e}_\theta + \dot{z}\bar{k}$$
$$= 4k_1 k_3^2 t^3 \bar{e}_R + 2k_1 k_3^3 t^5 \bar{e}_\theta + 2k_2 k_3 t \bar{k}$$

$$\bar{a} = (\ddot{R} - R\dot{\theta}^2)\bar{e}_R + (R\ddot{\theta} + 2\dot{R}\dot{\theta})\bar{e}_\theta + \ddot{z}\bar{k}$$
$$= 4k_1 k_3^2 t^2 (3 - k_3^2 t^4)\bar{e}_R + 18 k_1 k_3^3 t^4 \bar{e}_\theta + 2k_2 k_3 \bar{k}$$

EXAMPLE 14

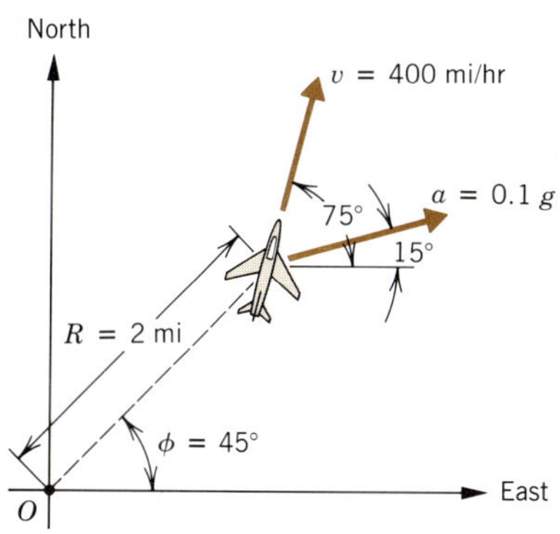

The position, velocity, and acceleration of an aircraft in level flight are as shown. Determine the values of $\dot{R}$, $\ddot{R}$, $\dot{\phi}$ and $\ddot{\phi}$ at this instant.

Solution

The given velocity and acceleration could be resolved into any set of components. We choose cylindrical components in order to obtain information about R and ϕ, which are polar coordinates for the planar motion.

We begin by considering the velocity because it is a lower-order derivative than the acceleration. The sketch shows the polar coordinates, the unit vectors, and the velocity at the given instant. In terms of radial and transverse components, the given velocity is

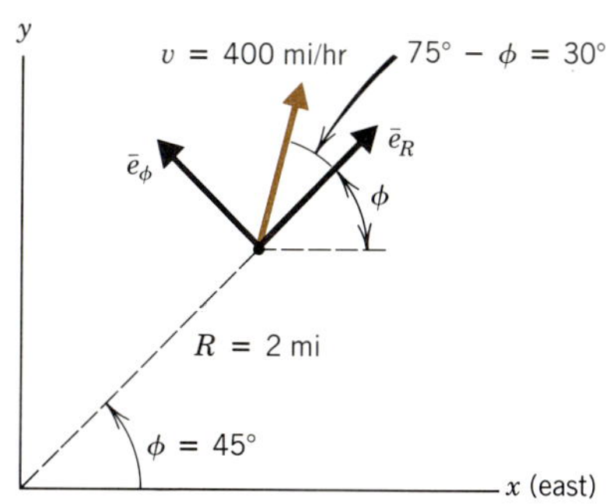

$$\bar{v} = 400\left(\tfrac{88}{60}\right)(\cos 30^\circ \,\bar{e}_R + \sin 30^\circ \,\bar{e}_\phi)$$
$$= 508.1\bar{e}_R + 293.3\bar{e}_\phi \text{ ft/s}$$

In general, the velocity equation is

$$\bar{v} = \dot{R}\bar{e}_R + R\dot{\phi}\bar{e}_\phi$$

The only way in which these alternative descriptions of the velocity vector can be the same is if the corresponding components match. Hence

$$\bar{e}_R \text{ component:} \quad 508.1 = \dot{R}$$

$$\bar{e}_\phi \text{ component:} \quad 293.3 = R\dot{\phi}$$

The value of R at this instant is

$$R = (2 \text{ mi})\left(\frac{5280 \text{ ft}}{\text{mi}}\right) = 10{,}560 \text{ ft}$$

so that

$$\dot{R} = 508.1 \text{ ft/s}$$

$$\dot{\phi} = \frac{293.3}{10560} = 0.027777 \text{ rad/s}$$

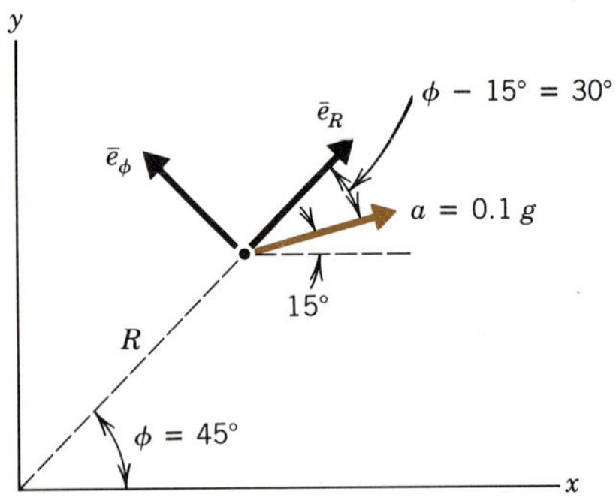

Next, the same approach is applied to the acceleration. The sketch of the given acceleration is used to represent $\bar{a}$ in component form. For this, we use $g = 32.17$ ft/s^2, because the unit for R is feet.

$$\bar{a} = (0.1)(32.17)(\cos 30^\circ \bar{e}_R - \sin 30^\circ \bar{e}_\phi)$$
$$= 2.786\bar{e}_R - 1.6085\bar{e}_\phi \text{ ft/s}$$

The acceleration formula gives

$$\bar{a} = (\ddot{R} - R\dot{\phi}^2)\bar{e}_R + (R\ddot{\phi} + 2\dot{R}\dot{\phi})\bar{e}_\phi$$

Substituting the values of R, $\dot{R}$, and $\dot{\phi}$, which have already been determined, into this general expression for $\bar{a}$, and equating corresponding components in the two acceleration equations then yields

$$\bar{e}_R \text{ component:} \quad 2.786 = \ddot{R} - 10{,}560(0.027777)^2$$

$$\bar{e}_\phi \text{ component:} \quad -1.6085 = 10{,}560\ddot{\phi} + 2(508.1)(0.027777)$$

The solution of these equations is

$$\ddot{R} = 10.93 \text{ ft/s}^2, \qquad \ddot{\phi} = -2.82(10^{-3}) \text{ rad/s}^2$$

HOMEWORK PROBLEMS

11.57 The planar motion of a particle is defined by the relations $R = 3t^2 - t^3$ and $\phi = 2t^2$, where R is expressed in inches, t in seconds, and ϕ in radians. Determine the velocity and acceleration of the particle when $t = 1$ s.

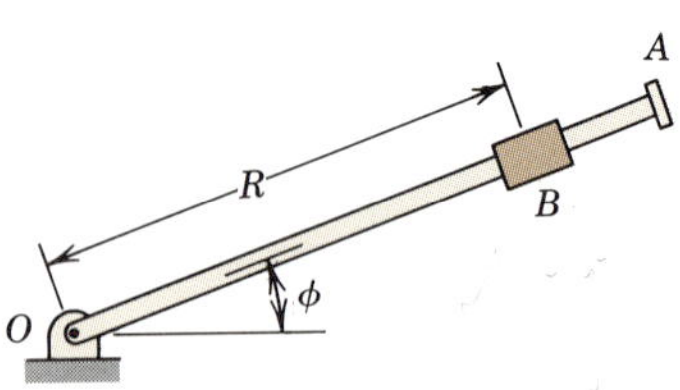

PROBS. 11.58 AND 11.59

11.58 The rotation of arm OA about point O is defined by the relation $\phi = 0.15t^2$, where ϕ is expressed in radians and t in seconds. Block B slides along the arm in such a way that its distance from point O is $r = 1 - 0.20t^2$, where r is expressed in meters and t in seconds. Determine the velocity and the acceleration of block B after the arm OA has rotated through 30°.

11.59 The rotation of arm OA is defined by the relation $\phi = 0.2t^3$, where ϕ is expressed in radians and t in seconds. Block B slides along the arm in such a way that its distance from point O is $R = 3 - 0.2\phi$, where R is expressed in feet and t in seconds. When $\phi = 0.67$ radians determine: (a) the velocity of block B, (b) the speed of block B, and (c) the acceleration of block B.

11.60 The rotating arm AB guides pin P through the fixed spiral, which is defined by $R = 4\phi$, where R has units of inches and ϕ has units of radians. The arm has a constant counterclockwise angular speed of 16 rad/s. Determine the velocity and acceleration of the pin when $\phi = 45°$.

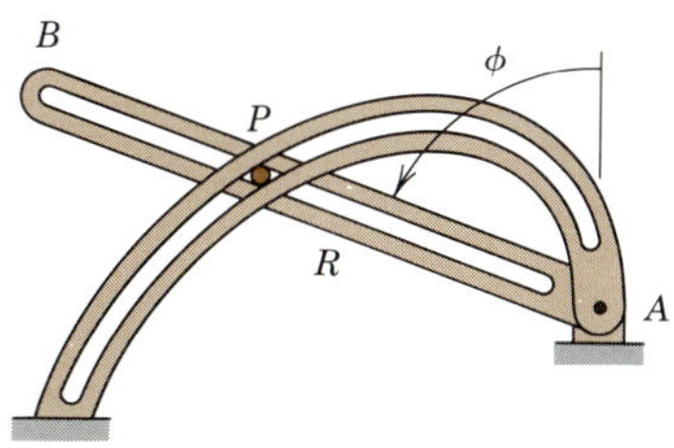

PROB. 11.60

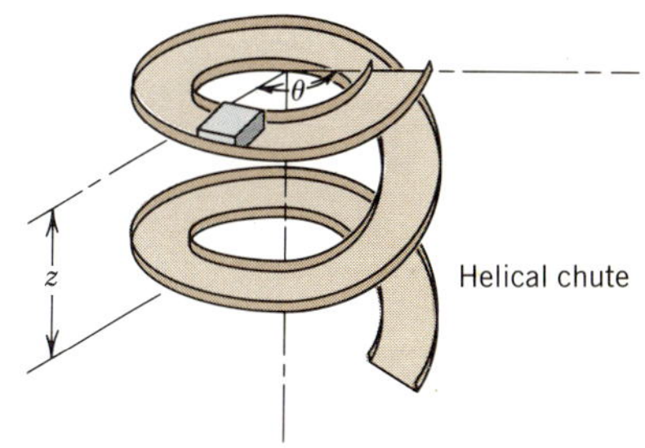

PROB. 11.61

11.61 A package is descending a helical chute that is defined by $R = 0.8$ m, $z = 0.4\theta/2\pi$ m, where θ is measured in radians. The motion of the package is defined by $\theta = 0.5t + t^2$, where t is in seconds. Determine the acceleration of the package when $t = 1$ s.

11.62 The cylindrical coordinates (R, ϕ, z) of a particle are given by $R = 10t^2$, $\phi = 5t^3$, $z = 2t^4$, where R and z are expressed in meters, ϕ is in radians, and t is in seconds. (a) Determine the expressions for the velocity and acceleration of the particle. (b) Draw a sketch of the path.

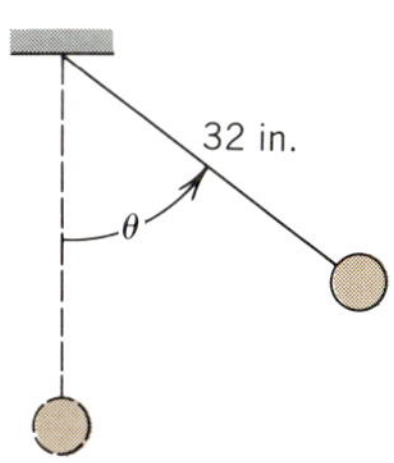

PROB. 11.63

11.63 The pendulum shown is moving to the right and slowing down. When $\theta = 15°$ it has a speed of 16 in./s and the magnitude of its acceleration is 20 in./s^2. Determine the angular speed $\dot{\theta}$ and angular acceleration $\ddot{\theta}$ at this instant.

11.64 At the instant shown, the telescoping passenger ramp is 45 m long and is being extended at the constant rate of 0.5 m/s. What is the constant rotation rate $\dot{\phi}$ about pivot A that will produce an acceleration of 1 m/s^2 at end B at this instant? What is the corresponding speed of end B at this instant?

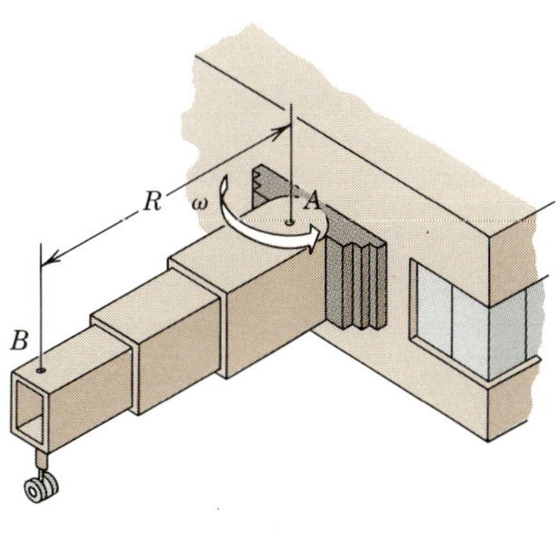

PROB. 11.64

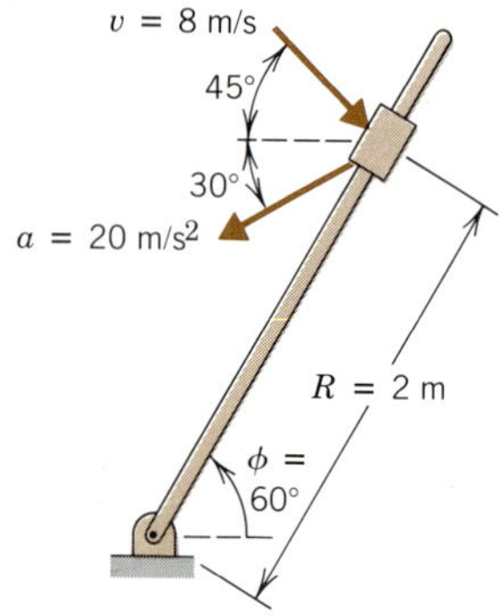

PROB. 11.65

11.65 The sliding block C is known to have the instantaneous velocity and acceleration illustrated in the diagram. Determine the corresponding values of $\dot{R}$, $\ddot{R}$, $\dot{\phi}$, and $\ddot{\phi}$ in this position.

11.66 The piston supporting block B is being extended as it pivots about point A. At the instant shown, the speed $v = 10$ ft/s and $\alpha = 20°$. Also, the acceleration is $a = 20$ ft/s^2 and $\beta = 40°$. Determine the corresponding values of $\dot{R}$, $\ddot{R}$, $\dot{\phi}$, and $\ddot{\phi}$.

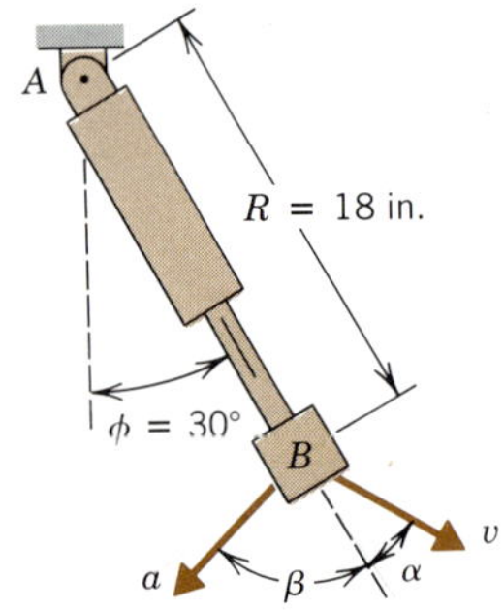

PROBS. 11.66 AND 11.67

***11.67** The piston supporting block B is being extended as it pivots about point A. At the instant shown $\dot{R} = -10$ ft/s and $\ddot{R} = 20$ ft/s^2. It is also known that the velocity is horizontal and that the acceleration is vertical. Determine the corresponding values of $\dot{\phi}$ and $\ddot{\phi}$.

***11.68** The crane shown has a 30 m boom AB that rotates about axis CD. At the instant in time when $\theta = 53.13°$, $\dot{\theta} = 0.10$ rad/s, $\ddot{\theta} = -0.20$ rad/s^2, $\dot{\phi} = 0.20$ rad/s, and $\ddot{\phi} = 0.10$ rad/s^2. Determine the acceleration of point B at this instant. *Hint:* Express the radial and axial distances relative to axis CD in terms of θ, and then evaluate the derivatives of of these distances.

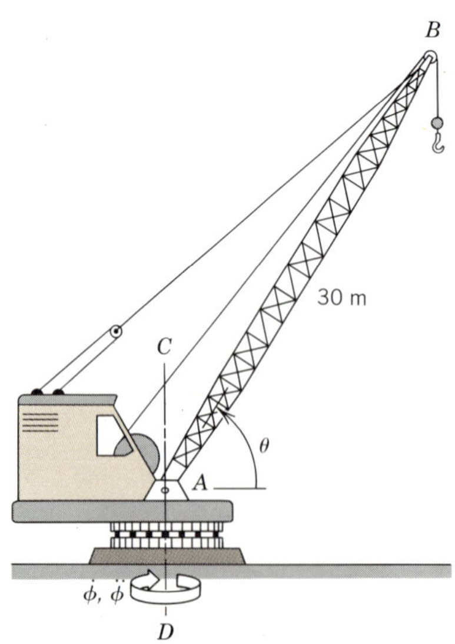

PROB. 11.68

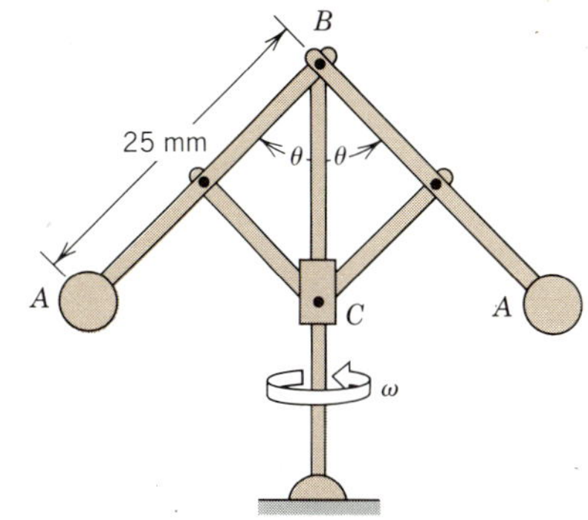

PROB. 11.69

***11.69** A governor consists of two heavy balls A on the end of 25 mm rods that are hinged to the vertical shaft at B. The position of the collar C on the vertical shaft may be adjusted to change the angle θ. The entire device rotates at a constant rate $\omega = 100$ rpm about the vertical. The angular speed $\dot{\theta}$ is a constant 0.2 rad/s. Determine the velocity and acceleration of each ball for $\theta = 30°$. See the hint for Problem 11.68.

C. JOINT KINEMATICAL DESCRIPTIONS

Thus far, we have solved kinematics problems by employing only one of the three kinematical descriptions: path variables, Cartesian coordinates, or cylindrical coordinates. In a few situations we converted a velocity or acceleration into Cartesian components after it was evaluated by one of the other descriptions.

In general, there are two types of problems in which we may anticipate that it will be suitable to employ two kinematical formulations. In the first, the given information permits evaluation of the velocity and/or the acceleration in terms of one formulation. However, the parameters that are to be evaluated correspond to a different formulation. In the second type of problem, the motion is specified in terms of parameters that arise in two different formulations. The primary task in this case is to evaluate the velocity and/or the acceleration. The technique we will develop here is equally valid for both situations.

Consider the general planar path in Figure 15. The unit vectors for each kinematical description at a typical location are illustrated in the figure.

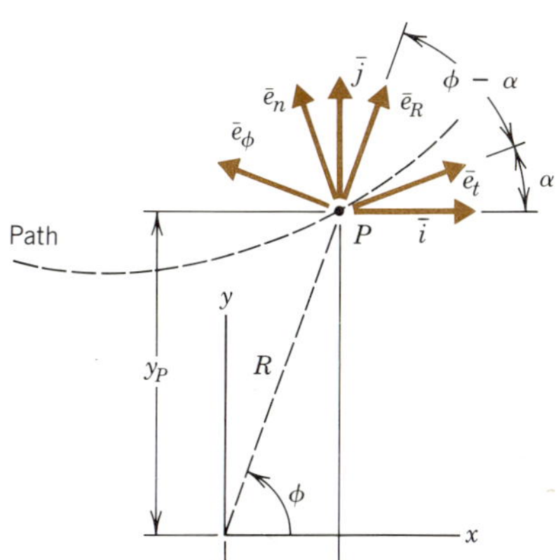

FIGURE 15

Let us develop the desired approach by considering a specific case in which the path and the angular speed $\dot{\phi}$ are specified; the problem is to determine the corresponding velocity.

One possible formulation involves writing the equation for the path in terms of polar coordinates. However, such an equation might not be known, or it might be very difficult to determine if the geometry is complicated.

These difficulties can be avoided by matching the cylindrical coordinate and path variable formulas for velocity. Specifically, the alternative velocity descriptions are

$$\bar{v} = \dot{R}\,\bar{e}_R + R\dot{\phi}\bar{e}_\phi = v\,\bar{e}_t \qquad (28)$$

For any position along the path the radial distance R, transverse angle ϕ, and angle α for the tangent in Figure 15 may be determined from the laws of geometry. This, in turn, defines the unit vectors in eq. (28) for the velocity. Because $\dot{\phi}$ is a given parameter, the only unknowns in these equations are $\dot{R}$ and v.

The process of equating the two equations involves matching like components. This may be done by evaluating the components of one set of unit vectors with respect to the other set of vectors. (Such an operation was developed in Section E.1.h of Chapter 10). For instance, Figure 15 indicates that the polar coordinate description of $\bar{e}_t$ is

$$\bar{e}_t = \cos(\phi - \alpha)\bar{e}_R - \sin(\phi - \alpha)\bar{e}_\phi$$

After substitution of this expression, eq. (28) becomes

$$\bar{v} = \dot{R}\bar{e}_R + R\dot{\phi}\bar{e}_\phi = v[\cos(\phi - \alpha)\bar{e}_R - \sin(\phi - \alpha)\bar{e}_\phi]$$

Equating like components yields

$$\bar{e}_R \text{ component:} \quad \dot{R} = v\cos(\phi - \alpha)$$

$$\bar{e}_\phi \text{ component:} \quad R\dot{\phi} = -v\sin(\phi - \alpha)$$

These two equations may be solved for $\dot{R}$ and $\dot{\phi}$ because the other parameters are known. The resulting solutions are the quantities required to complete eq. (28) for the velocity.

One corollary of the foregoing procedure is that the same approach is applicable to acceleration problems. However, doing so usually requires that the velocity problem be solved first. This is to be expected, just as it is necessary to obtain the first derivative of a function before the second derivative can be evaluated.

In summary, two kinematical descriptions may be employed jointly according to the following procedure:

1. Write the velocity in terms of the formula for each type of description. Substitute into these formulas any parameters that are given or can be found from geometry.
2. Express one set of unit vectors in terms of their components with respect to the other set. Substitute these representations into the velocity formulas.
3. Match like components of velocity, and solve the resulting equations for the unknowns.
4. If the problem pertains to acceleration, repeat steps 1–3 using the acceleration formulas and the results of the velocity solution.

One important aspect of the foregoing development is that the choice of kinematical descriptions depends on how the motion is defined and on the type of information to be determined. This is true even for rectilinear motion or motion along a circular path. Figure 16 shows some typical formulations for these simple paths. Note that the Cartesian and cylindrical

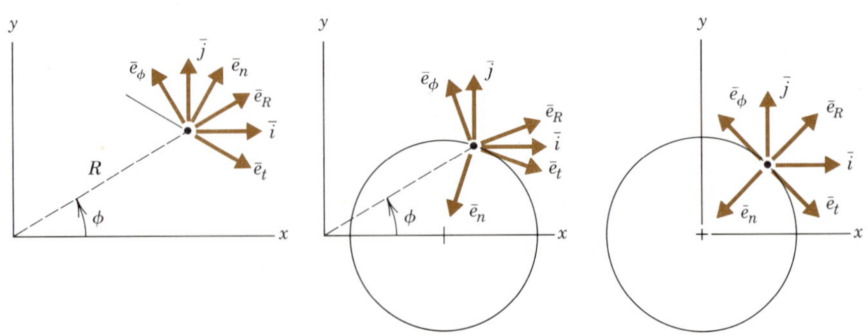

FIGURE 16

coordinate unit vectors depend on how the corresponding coordinate systems are defined.

EXAMPLE 15

The launching of a space shuttle is tracked by a radar antenna at point A. The rocket has a constant acceleration of $2g$ starting from rest. Determine the rate of rotation ω of the antenna when the space shuttle is at an altitude of 1000 m.

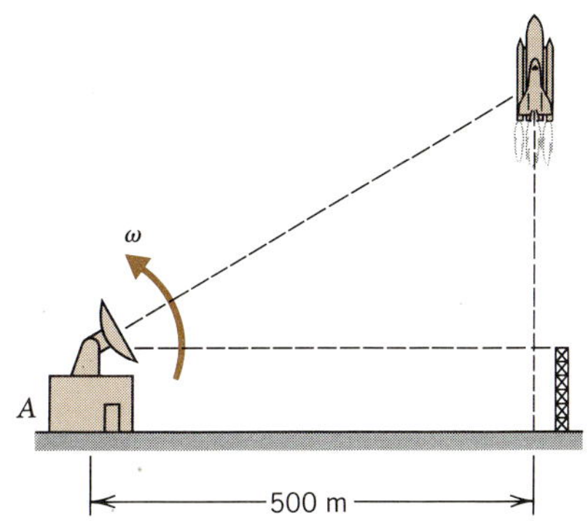

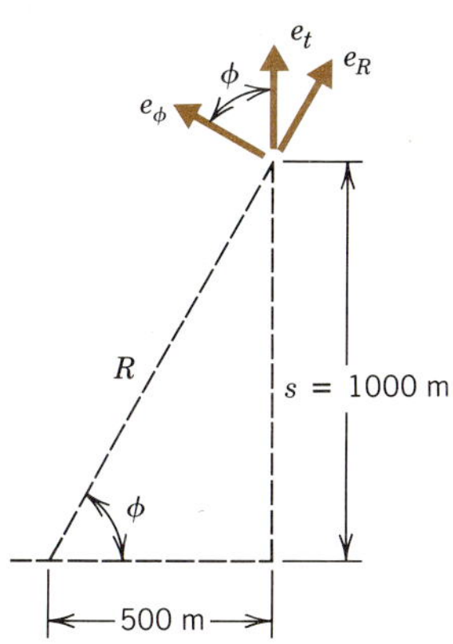

Solution

The space shuttle is executing rectilinear motion at a constant acceleration, so its velocity may be evaluated at any position. The desired value ω is the angular velocity of the line from the antenna to the shuttle. This suggests a formulation involving path variables and cylindrical coordinates. The basic parameters for each description are shown in the sketch for an altitude of 1000 m.

The formulas for velocity are

$$\bar{v} = v\,\bar{e}_t = \dot{R}\,\bar{e}_R + R\dot{\phi}\bar{e}_\phi$$

where $\dot{\phi} = \omega$ for this problem. The value of R and the angle ϕ, which is needed to define $\bar{e}_R$ and $\bar{e}_\phi$, are found from the s etch to be

$$R = [(1000)^2 + (500)^2]^{1/2} = 1118.0 \text{ m}$$

$$\phi = \tan^{-1}\left(\frac{1000}{500}\right) = 63.43^\circ$$

The speed v may be obtained from the rectilinear motion formulas. For zero initial speed, the speed after traveling 1000 m is

$$v^2 = 2\dot{v}(s - s_0) = 2(2)(9.807)(1000)$$

$$v = 198.06 \text{ m/s}$$

The unit vector $\bar{e}_t$ is indicated in the sketch to be at an angle ϕ from $\bar{e}_\phi$, so its cylindrical coordinate components are

$$\bar{e}_t = \sin\phi\,\bar{e}_R + \cos\phi\,\bar{e}_\phi = 0.8944\bar{e}_R + 0.4473\bar{e}_\phi$$

Substitution of v, R, and $\bar{e}_t$ into the velocity formulas, followed by matching the components, yields

$$\begin{aligned}\bar{v} &= 198.06(0.9844\bar{e}_R + 0.4473\bar{e}_\phi)\\ &= \dot{R}\bar{e}_R + 1118.0\omega\bar{e}_\phi\end{aligned}$$

$$\bar{e}_\phi \text{ component:} \quad 198.06(0.4473) = 1118\omega$$

$$\omega = 0.0792 \text{ rad/s}$$

Note that the radial components of velocity in each description were not needed here.

EXAMPLE 16

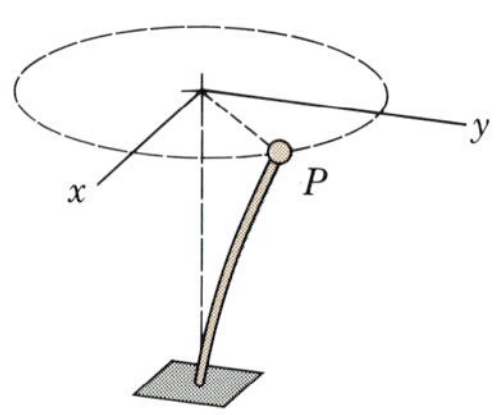

It is observed that the small sphere P at the end of a flexible rod moves in a horizontal plane such that $x = 3\cos(2t)$, $y = 4\sin(2t)$, where x and y are in inches and t is in seconds. Determine the speed, the rate of change of the speed, and the radius of curvature of the path when $t = \pi/8$ s.

Solution

The motion is fully specified in terms of Cartesian coordinates, but the values of the path variable parameters are requested. These are therefore the two descriptions to be employed. First, the speed is evaluated by considering the velocity formulas.

$$\bar{v} = v\bar{e}_t = \dot{x}\bar{i} + \dot{y}\bar{j}$$

Here the values of $\dot{x}$ and $\dot{y}$ may be evaluated, and v and $\bar{e}_t$ are the unknowns. In calculating the trigonometric functions, the radian mode must be used. Thus

$$\dot{x} = -3(2)\sin\left(2\frac{\pi}{8}\right) = -4.243 \text{ in./s}$$

$$\dot{y} = 4(2)\cos\left(2\frac{\pi}{8}\right) = 5.657 \text{ in./s}$$

These values are now substituted into the formulas. Then the fact that v is the magnitude of the velocity yields both the speed and the tangent vector.

$$\bar{v} = v\bar{e}_t = -4.243\bar{i} + 5.657\bar{j}$$

$$v = |\bar{v}| = [(4.243)^2 + (5.657)^2]^{1/2} = 7.071 \text{ in./s}$$

$$\bar{e}_t = \frac{\bar{v}}{v} = -0.600\bar{i} + 0.800\bar{j}$$

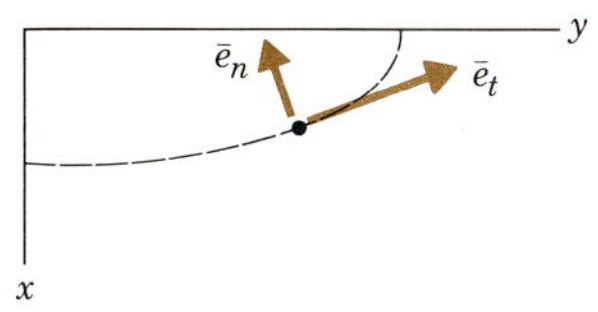

The orientation of this vector is shown in the sketch.

The values of $\dot{v}$ and ρ, both of which are acceleration parameters, may now be found from the acceleration formulas.

$$\bar{a} = \dot{v}\,\bar{e}_t + \frac{v^2}{\rho}\,\bar{e}_n = \ddot{x}\,\bar{i} + \ddot{y}\,\bar{j}$$

The Cartesian acceleration components when $t = \pi/8$ are

$$\ddot{x} = -3(2)^2 \cos\left(2\frac{\pi}{8}\right) = -8.485 \text{ in./s}^2$$

$$\ddot{y} = -4(2)^2 \sin\left(2\frac{\pi}{8}\right) = -11.314 \text{ in./s}^2$$

These values, and the result for v, are substituted into the formulas.

$$\bar{a} = \dot{v}\,\bar{e}_t + \frac{(7.071)^2}{\rho}\,\bar{e}_n = -8.485\bar{i} - 11.314\bar{j}$$

The value of $\bar{e}_t$ was determined earlier, which means that this relation has three unknowns.

In order to proceed further we need to use the fact that $\bar{e}_t$ and $\bar{e}_n$ are perpendicular unit vectors, so $\bar{e}_t \cdot \bar{e}_n = 0$. This permits evaluation of the tangential component of acceleration, $\dot{v}$, by means of a dot product.

$$\begin{aligned} \bar{a} \cdot \bar{e}_t = \dot{v} &= (-8.485\bar{i} - 11.314\bar{j}) \cdot \bar{e}_t \\ &= (-8.485\bar{i} - 11.314\bar{j}) \cdot (-0.60\bar{i} + 0.80\bar{j}) \\ \dot{v} &= -3.960 \text{ in./s}^2 \end{aligned}$$

The radius of curvature may now be determined by expressing the magnitude of the acceleration in terms of its tangent and normal components, as well as its xy components.

$$|\bar{a}| = \left[\dot{v}^2 + \frac{(7.071)^4}{\rho^2}\right]^{1/2} = [(8.485)^2 + (11.314)^2]^{1/2}$$

$$\frac{(7.071)^4}{\rho^2} = (8.485)^2 + (11.314)^2 - (3.960)^2$$

$$\rho = 3.68 \text{ in.}$$

It is interesting to note in closing that the path of sphere P is an ellipse which satisfies

$$\left(\frac{x}{3}\right)^2 + \left(\frac{y}{4}\right)^2 = 1$$

We have determined the radius of curvature of the path without using eq. (17) for ρ!

EXAMPLE 17

Consider the system in Example 11, in which collar C slides along the circular rod of radius R such that the horizontal component of velocity v_x is constant. Determine the velocity and acceleration at a general position by using two kinematical descriptions.

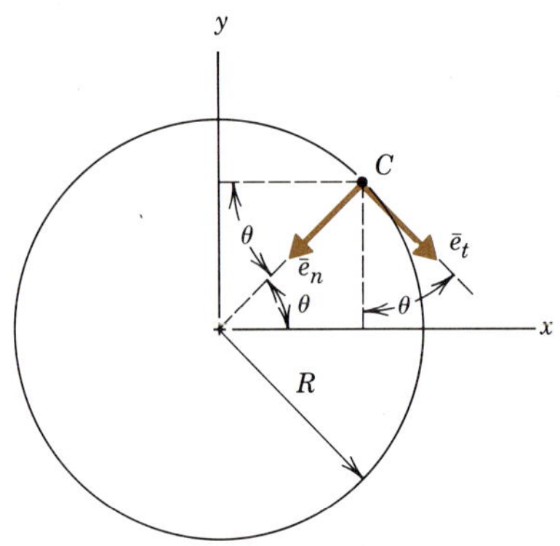

Solution

The path is given, and the movement along the path is defined by a velocity component in the (fixed) horizontal direction. This suggests using path variables and Cartesian coordinates. The parameters for each are shown in the sketch, with the collar at an arbitrary position on its path.

The velocity formulas are

$$\bar{v} = v\,\bar{e}_t = \dot{x}\,\bar{i} + \dot{y}\,\bar{j}$$

where $\dot{x} = v_x$. The tangent vector $\bar{e}_t$ is found from the sketch in terms of the angle θ locating point C.

$$\bar{e}_t = \sin\theta\,\bar{i} - \cos\theta\,\bar{j}$$

Thus

$$\bar{v} = v(\sin\theta\,\bar{i} - \cos\theta\,\bar{j}) = v_x\bar{i} + \dot{y}\bar{j}$$

Matching like components yields

$$\bar{i}\text{ component:}\quad v\sin\theta = v_x$$

$$\bar{j}\text{ component:}\quad -v\cos\theta = \dot{y}$$

$$v = v_x/\sin\theta, \qquad \dot{y} = -v_x\cot\theta$$

so

$$\bar{v} = v_x\bar{i} - v_x\cot\theta\,\bar{j}$$

We begin the same procedure for acceleration by writing

$$\bar{a} = \dot{v}\bar{e}_t + \frac{v^2}{\rho}\,\bar{e}_n = \dot{v}_x\bar{i} + \dot{v}_y\bar{j}$$

Because v_x is constant, $\dot{v}_x = 0$. The radius of curvature for the circular path is $\rho = R$, and the normal vector $\bar{e}_n$ is found from the sketch to be

$$\bar{e}_n = -\cos\theta\,\bar{i} - \sin\theta\,\bar{j}$$

Substituting these expressions and the earlier results for $\bar{e}_t$ and v yields

$$\bar{a} = \dot{v}(\sin\theta\bar{i} - \cos\theta\bar{j}) + \frac{v_x^2}{R\sin^2\theta}(-\cos\theta\,\bar{i} - \sin\theta\,\bar{j})$$

$$= \dot{v}_y\bar{j}$$

Equating like components yields

$$\bar{i}\text{ component:}\quad \dot{v}\sin\theta - \frac{v_x^2}{R\sin^2\theta}\cos\theta = 0$$

$$\bar{j} \text{ component:} \quad -\dot{v}\cos\theta - \frac{v_x^2}{R\sin\theta} = \dot{v}_y$$

Thus

$$\dot{v} = v_x^2 \frac{\cos\theta}{R\sin^3\theta}$$

$$\dot{v}_y = -\frac{v_x^2}{R}\left[\frac{\cos^2\theta}{\sin^3\theta} + \frac{1}{\sin\theta}\right] \equiv -\frac{v_x^2}{R}\frac{1}{\sin^3\theta}$$

$$\bar{a} = -\frac{v_x^2}{R}\frac{1}{\sin^3\theta}\bar{j}$$

These results may be checked by substituting $x = R\cos\theta$ into the results of Example 11. Such confirmation is illustrative of the fact that there are often several ways to set up kinematics problems and to display results.

HOMEWORK PROBLEMS

11.70 When the automobile is in the position shown, a photographer at position O observes that he must rotate his camera clockwise at 1.2 rad/s in order to track the automobile. Determine the speed v of the automobile in this position.

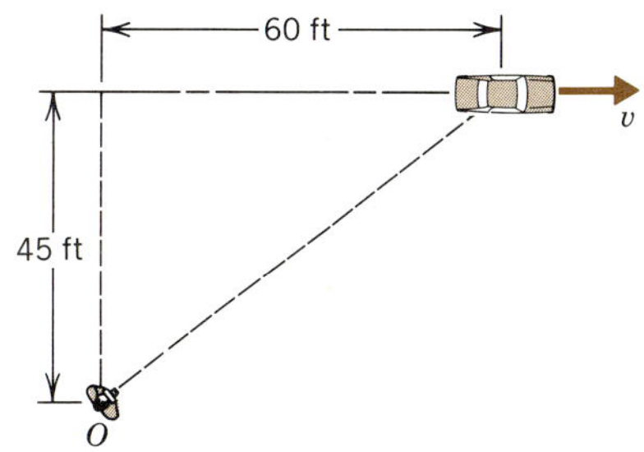

PROB. 11.70

11.71 The photographer in Problem 11.70 must decrease the rate of rotation of his camera at the rate of 0.4 rad/s^2 in order to track the automobile in the given position. Determine the corresponding acceleration of the automobile.

11.72 The motion of a ball that is thrown horizontally with an initial speed of 20 m/s is defined by $x = 20t$, $y = -4.903t^2$, where x and y are in meters and t is in seconds. Determine the speed, the rate of change of the speed, and the radius of curvature of the path when $t = 0$.

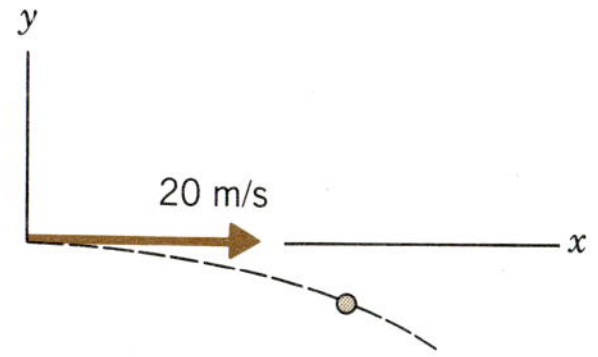

PROB. 11.72

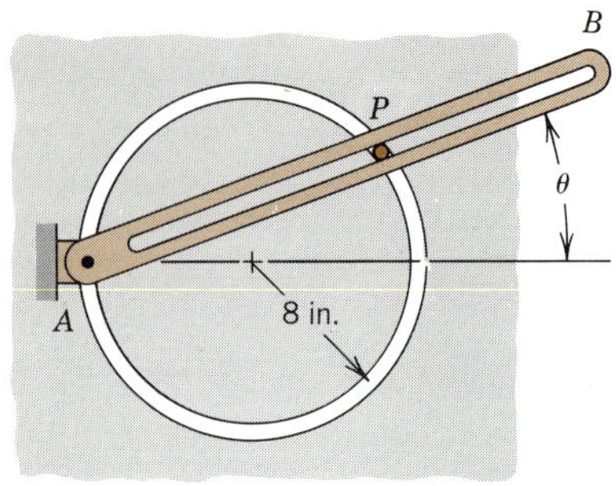

PROB. 11.74

11.73 Solve Problem 11.72 for $t = 2$ s.

11.74 Bar AB rotates clockwise at a constant angular speed of 4 rad/s as it guides pin P through the fixed circular slot. Determine the velocity of pin P when $\theta = 45^\circ$.

11.75 Determine the acceleration of pin P in Problem 11.74 when $\theta = 45^\circ$.

11.76 Solve Problem 11.74 for $\theta = 30^\circ$.

11.77 The telescoping passenger ramp is being extended at the constant rate of 2 m/s as it rotates about pivot A at the constant rate $\omega = 0.02$ rad/s. For the instant when $R = 50$ m, determine the speed and the rate of change of the speed of end B.

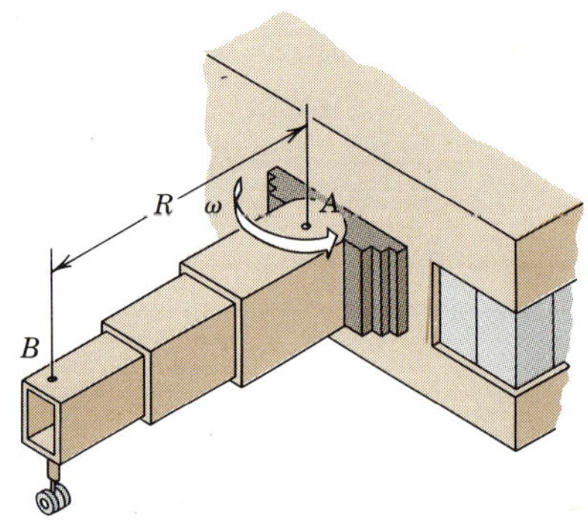

PROB. 11.77

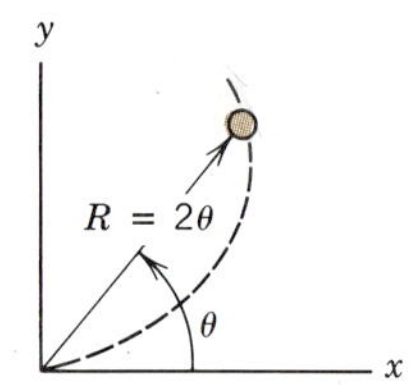

PROB. 11.78

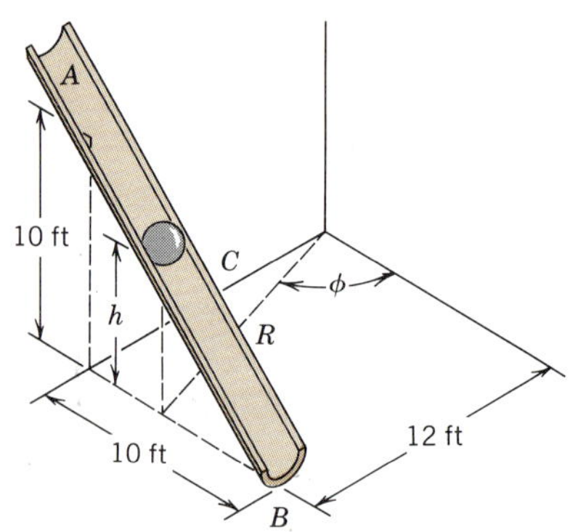

PROB. 11.79

11.78 An electron in a cyclotron follows a spiral in the horizontal plane defined in terms of polar coordinates by $R = 2\theta$ (mm), where θ is in radians. The electron rotates about the vertical axis at a constant rate of $6(10^4)$ rad/s. Determine the tangential and normal components of acceleration, and the radius of curvature of the path when $R = 3$ m.

***11.79** The ball rolls down the ramp with a constant acceleration of 20 ft/s^2, and its speed at position A is 5 ft/s. Determine the values of $\dot{R}$, $\dot{\phi}$, and $\dot{h}$ at position A.

***11.80** For the ball in Problem 11.79, determine the values of $\ddot{R}$ and $\ddot{\phi}$ at position A.

***11.81** Solve Problem 11.79 when the ball is at position B.

D. PROBLEM SOLVING

By now you should have a reasonable feel for the meaning of the variables and unit vectors that appear in the three kinematical descriptions presented. Problem solving using these descriptions (i.e., picking the right one) is not an art, but rather a decision made by recognizing the proper keys. In what follows we present a logical way of thinking to identify the keys. It is not meant for memorization, but rather as a guide while working problems. Work enough problems *correctly* and the key will become part of your thinking process.

1. Read the problem carefully and express all quantitative information in equation form, at the same time converting the information to a consistent set of units. For instance, if a problem states "After the particle has traveled 700 ft, it has a speed of 30 mi/hr," you should write down

$$\text{when } s = 700 \text{ ft},$$

$$v = 30\,\frac{\text{mi}}{\text{hr}}\left(\frac{88 \text{ ft/s}}{60 \text{ mi/hr}}\right) = 44 \text{ ft/s}$$

2. If the path is defined but no diagram is provided, make a sketch of the path.
3. Choose path variables, Cartesian coordinates, or cylindrical coordinates. Make this choice based on which formulation most closely fits the information given in the problem and, equally important, based on the solution variables whose values are sought. Select the coordinate parameters and unit vectors and show them in the sketch. Now, in terms of these coordinates, write general geometric relationships for the lengths and angles.
4. Write down the formulas in the kinematical description you have chosen that relate the given information to the variables of the solution. After studying the formulas and the given information, from the sketch write any additional information needed to relate the system's geometry.
5. Solve the equations you have developed.

In the illustrative problems that follow, we shall refer to these steps by their appropriate numbers.

ILLUSTRATIVE PROBLEM 1

A helicopter descends at a speed of 10 m/s with an acceleration of $0.5g$. The main rotor blades are 5 m long, oriented horizontally, and are rotating at a constant rate of 450 rpm. What is the speed and magnitude of the acceleration of a point on the tip of a blade?

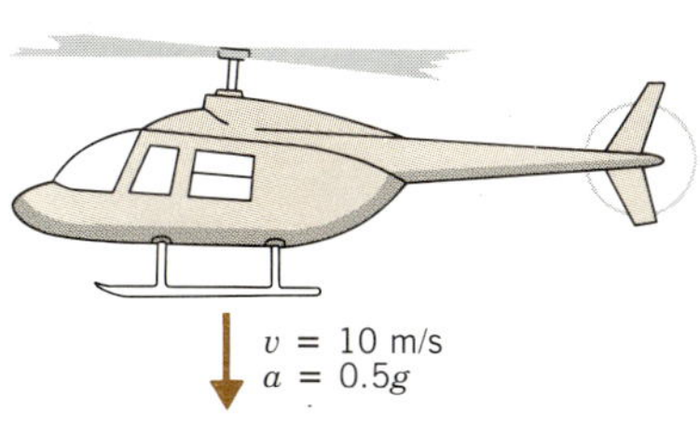

Solution

Step 1 Given $v_z = 10$ m/s; $a_z = 0.5g[(9.807 \text{ m/s}^2)/1g] = 4.904 \text{ m/s}^2$, both down, $\omega_{\text{rotor}} = 450$ rpm $(2\pi$ rad/rev) (1 min/60 s) $= 15\pi$ rad/s, $\dot{\omega} = 0$ (Note that the angular units for rotation must be radians.) Find $|\bar{v}|$ and $|\bar{a}|$.

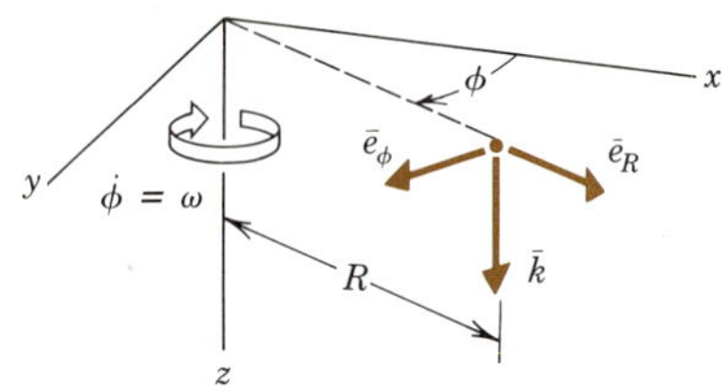

Steps 2 and 3 The motion of the axis of the rotor shaft is the same as that of the helicopter. Therefore, we know the rate of rotation of a radial line from the rotor axis to a blade tip and the motion of the tip parallel to the rotor shaft axis. This is ideal for cylindrical coordinates. If we measure z positive downward, then $\bar{k}$ is parallel to the velocity and acceleration of the helicopter. In order to produce a right-handed system we measure positive ϕ in the direction of rotation.

Step 4

$$\bar{v} = \dot{R}\,\bar{e}_R + R\dot{\phi}\bar{e}_\phi + \dot{z}\,\bar{k}$$

$$|\bar{v}| = \sqrt{\bar{v}\cdot\bar{v}}$$

$$\bar{a} = (\ddot{R} - R\dot{\phi}^2)\bar{e}_R + (R\ddot{\phi} + 2\dot{R}\dot{\phi})\bar{e}_\phi + \ddot{z}\,\bar{k}$$

$$|\bar{a}| = \sqrt{\bar{a}\cdot\bar{a}}$$

Step 5

$$R = 5\text{ m} \qquad \dot{R} = \ddot{R} = 0$$

$$\dot{\phi} = 15\pi\text{ rad/s} \qquad \ddot{\phi} = 0$$

$$\dot{z} = 10\text{ m/s} \qquad \ddot{z} = 4.904\text{ m/s}^2$$

$$\bar{v} = 5(15\pi)\bar{e}_\phi + 10\bar{k}$$

$$|\bar{v}| = [(75\pi)^2 + (10)^2]^{1/2} = 235.8\text{ m/s}$$

$$\bar{a} = -5(15\pi)^2\bar{e}_R + 4.904\,\bar{k}$$

$$|\bar{a}| = [(1125\pi^2)^2 + (4.904)^2]^{1/2} = 1.110(10^4)\text{ m/s}^2$$

ILLUSTRATIVE PROBLEM 2

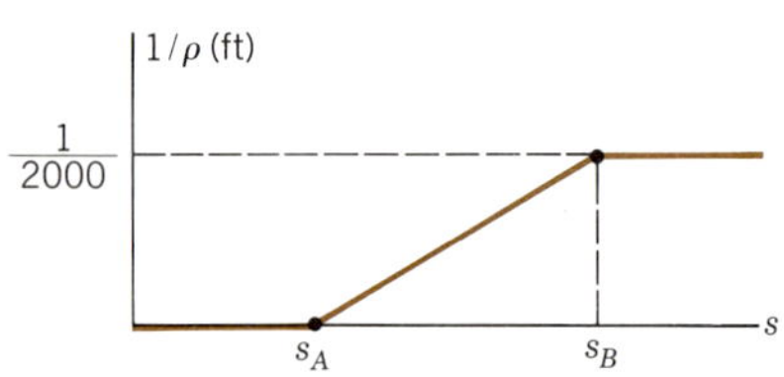

The transition from a straight track to a 2000 ft radius turn for a rail road train is achieved by increasing the radius of curvature gradually, as shown in the graph. A train reaches position A with a speed of 90 mph. At this position it begins to slow down at the constant rate of $0.1g$. If the magnitude of the acceleration at position B is $0.2g$, determine the distance traveled between positions A and B.

Solution

Step 1 Given that $\rho_A = 0$, $\rho_B = 2000$ ft, $v_A = 90(88/60)$ ft/s, constant $\dot{v} = -0.1g \equiv -3.217$ ft/s^2, $|\bar{a}|_B = 0.2(32.17)$ ft/s^2; find $s_B - s_A$.

Steps 2 and 3 A sketch of the path is not required, because only the magnitude of the acceleration is involved. We choose path variables because the movement along the path is given in terms of the speed.

Step 4 The acceleration at position B in path variables is

$$\bar{a} = \dot{v}\,\bar{e}_t + \frac{v_B^2}{\rho_B}\,\bar{e}_n$$

In this formula only the value of v_B at position B is not given. This quantity is related to the distance $s_B - s_A$ through the constant acceleration formula

$$v_B^2 = v_A^2 + 2\dot{v}(s_B - s_A)$$

Step 5 The actual value of v_B may be found from the fact that $|\bar{a}| = 0.2g$ at position B. Thus

$$|\bar{a}|_B = 0.2(32.17) = \left[\dot{v}^2 + \left(\frac{v_B^2}{\rho_B}\right)^2\right]^{1/2}$$

$$= \left[(3.217)^2 + \frac{v_B^4}{(2000)^2}\right]^{1/2}$$

$$(6.434)^2 = (3.217)^2 + \frac{v_B^4}{(2000)^2}$$

$$v_B = 105.57 \text{ ft/s}$$

The constant velocity equation then yields

$$(105.57)^2 = \left[90\left(\frac{88}{60}\right)\right]^2 + 2(-3.217)(s_B - s_A)$$

$$s_B - s_A = 976 \text{ ft}$$

ILLUSTRATIVE PROBLEM 3

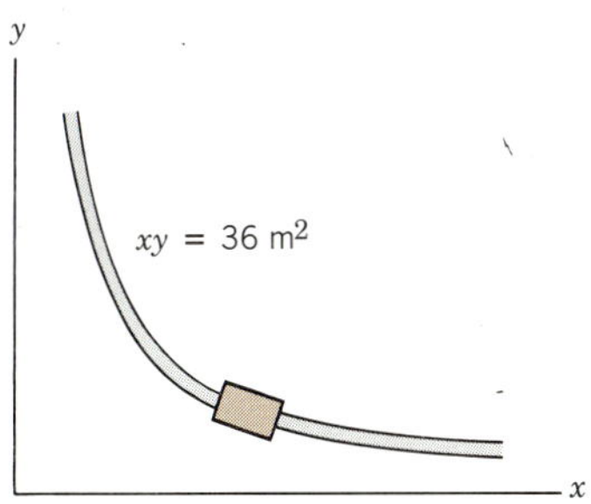

A slider moves in the vertical plane along a wire that is bent into the hyperbola $xy = 36 \text{ m}^2$. The vertical component of velocity of the slider is constant at 10 m/s downward. Determine the horizontal component of the acceleration of the slider when it is at $x = 9$ m.

Solution

Step 1 Given that $xy = 36 \text{ m}^2$, $v_y = -10$ m/s, $\dot{v}_y = 0$. Find a_x when $x = 9$ m.

Steps 2 and 3 Use Cartesian coordinates in terms of the given xy system because the velocity component v_y is given. The given sketch shows the coordinate directions.

Step 4 The acceleration formula is

$$\bar{a} = \dot{v}_x\bar{i} + \dot{v}_y\bar{j}$$

where $v_x = \dot{x}$ and $v_y = \dot{y} = -10$ m/s, $\ddot{v}_y = 0$.

Step 5 We calculate $\dot{x}$ and $\ddot{x}$ using the chain rule.

$$x = \frac{36}{y}$$

$$\dot{x} = -\frac{36}{y^2}\dot{y} = \frac{360}{y^2}$$

$$\ddot{x} = -\frac{360(2)}{y^3}\dot{y} = \frac{7200}{y^3}$$

The value of y corresponding to $x = 9$ m is

$$y = \frac{36}{9} = 4 \text{ m}$$

from which it is found that

$$\ddot{x} = \frac{7200}{4^3} = 112.5 \text{ m/s}^2$$

$$\bar{a} = 112.5\bar{i} \text{ m/s}^2$$

ILLUSTRATIVE PROBLEM 4

Consider the slider in Illustrative Problem 3, which is guided by a wire that is bent in the shape of the hyperbola $xy = 36 \text{ m}^2$. In the present case, the slider is moving from left to right at a constant speed of 10 m/s. Determine the horizontal and vertical components of the acceleration of the slider when it is at $x = 9$ m.

Solution

The difference between this situation and the one in the previous problem is that here the movement along the path is described in terms of the speed.

Step 1 Given that $xy = 36 \text{ m}^2$, $v = 10$ m/s to the right, $\dot{v} = 0$. Find a_x and a_y when $x = 9$ m.

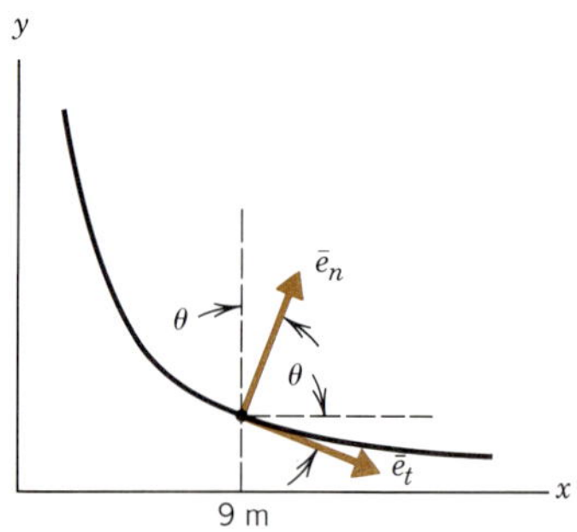

Steps 2 and 3 Path variables are chosen for the evaluation of the acceleration, because the speed and path are known. The x and y components of $\bar{a}$ may then be evaluated by converting the components. The vector $\bar{e}_t$ in the sketch is depicted in the sense of the movement. The path and the angle θ for the unit vectors is defined by

$$y = \frac{36}{x}, \quad \theta = \tan^{-1}\left(-\frac{dy}{dx}\right)$$

Step 4 The formulas for the acceleration and the radius of curvature are

$$\bar{a} = \dot{v}\,\bar{e}_t + \frac{v^2}{\rho}\,\bar{e}_n = 0\bar{e}_t + \frac{10^2}{\rho}\,\bar{e}_n$$

$$\frac{1}{\rho} = \frac{|d^2y/dx^2|}{[1 + (dy/dx)^2]^{3/2}}$$

Step 5 The equation for the path yields

$$y = \frac{36}{x}, \qquad \frac{dy}{dx} = -\frac{36}{x^2}, \qquad \frac{d^2y}{dx^2} = \frac{72}{x^3}$$

$$\theta = \tan^{-1}\left(\frac{36}{x^2}\right), \qquad \frac{1}{\rho} = \frac{72/x^3}{[1 + (36/x^2)]^{3/2}}$$

It follows that at $x = 9$ m,

$$\theta = \tan^{-1}\left(\frac{36}{9^2}\right) = 23.96^\circ$$

$$\frac{1}{\rho} = \frac{72/9^3}{[1 + (36/9^2)^2]^{3/2}} = 0.07537 \text{ m}^{-1}$$

$$\bar{a} = 10^2(0.07537)\bar{e}_n = 7.537\bar{e}_n \text{ m/s}^2$$

The horizontal and vertical components of this acceleration are found by using the sketch to describe the x and y components of $\bar{e}_n$.

$$\bar{e}_n = \sin\theta\,\bar{i} + \cos\theta\,\bar{j} = 0.4061\bar{i} + 0.9138\bar{j}$$

$$\bar{a} = 7.537(0.4061\bar{i} + 0.9138\bar{j}) = 3.06\bar{i} + 6.89\bar{j} \text{ m/s}^2$$

The differences between the methods of solution for this and the previous problem, as well as the difference in their results, are noteworthy. This is a consequence of the differences in the manner in which the movement along the path is specified.

ILLUSTRATIVE PROBLEM 5

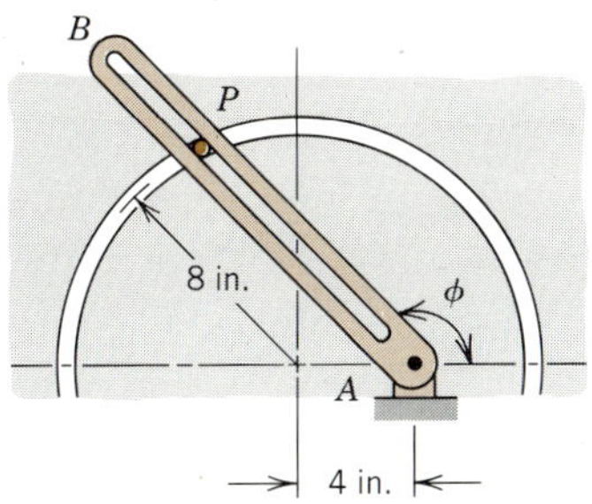

Arm AB rotates clockwise at the constant rate of 5 rad/s. Determine the velocity and acceleration of pin P when $\phi = 90°$.

Solution

Step 1 Given that $\dot{\phi} = -5$ rad/s, $\ddot{\phi} = 0$. Find $\bar{v}$ and $\bar{a}$ when $\phi = 90°$.

Steps 2 and 3 The movement along the path is described in terms of the rate of rotation of the radial line. This suggests the use of cylindrical coordinates. Evaluating the radial distance from pivot A to pin P in terms of the angle ϕ at a general position requires the law of cosines. We will follow the alternative of also employing path variables, which is possible because the path is known. The unit vectors for both formulations are shown in the sketch for the position $\phi = 90°$.

ρ = 8 in.
$R = (8^2 - 4^2)^{1/2} = 6.928$ in.
$\theta = \cos^{-1}\left(\frac{4}{8}\right) = 60°$

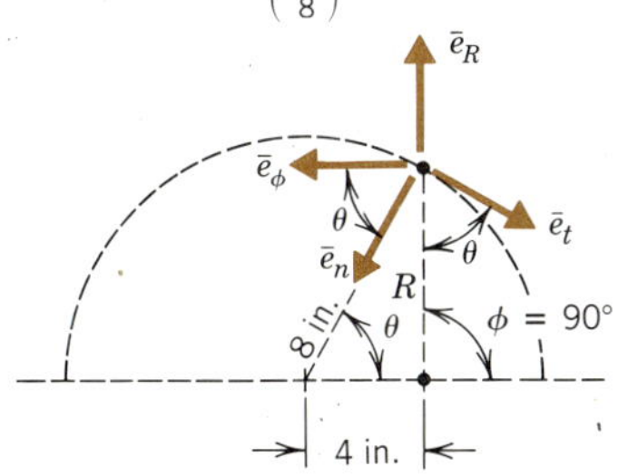

Step 4

$$\bar{v} = \dot{R}\,\bar{e}_R + R\dot{\phi}\bar{e}_\phi = v\,\bar{e}_t$$

$$\bar{a} = (\ddot{R} - R\dot{\phi}^2)\bar{e}_R + (R\ddot{\phi} + 2\dot{R}\dot{\phi})\bar{e}_\phi = \dot{v}\,\bar{e}_t + \frac{v^2}{\rho}\bar{e}_n$$

Step 5 Matching the two kinematical descriptions requires expressions for the unit vectors of one description with respect to the other. Thus

$$\bar{e}_t = -\cos\theta\,\bar{e}_R - \sin\theta\,\bar{e}_\phi = -0.50\bar{e}_R - 0.8660\bar{e}_\phi$$

$$\bar{e}_n = -\sin\theta\,\bar{e}_R + \cos\theta\,\bar{e}_\phi = -0.8660\bar{e}_R + 0.50\bar{e}_\phi$$

Substitution of $\bar{e}_t$, R, and ϕ into the velocity equations, followed by equating like components, yields

$$\bar{v} = \dot{R}\,\bar{e}_R + 6.928(-5)\bar{e}_\phi = v(-0.50\bar{e}_R - 0.8660\bar{e}_\phi)$$

$$\bar{e}_R \text{ component:} \quad \dot{R} = -0.50v$$

$$\bar{e}_\phi \text{ component:} \quad -34.64 = -0.8660v$$

$$\bar{v} = 40 \text{ in./s}, \qquad \dot{R} = -20 \text{ in./s}$$

$$\bar{v} - -20\bar{e}_R - 34.6\bar{e}_\phi = 40\bar{e}_t \text{ in./s}$$

The same procedure is now applied to the acceleration. The results for v and $\dot{R}$, as well as the other parameters, are substituted into the acceleration formulas. Thus

$$\bar{a} = [\ddot{R} - 6.928(-5)^2]\bar{e}_R + [0 + 2(-20)(-5)]\bar{e}_\phi$$

$$= \dot{v}(-0.50\bar{e}_R - 0.8660\bar{e}_\phi) + \frac{(40)^2}{8}(-0.8660\bar{e}_R + 0.50\bar{e}_\phi)$$

$$\bar{e}_R \text{ component:} \quad \ddot{R} - 173.20 = -0.5\dot{v} - 173.20$$

$$\bar{e}_\phi \text{ component:} \quad 200 = -0.8660\dot{v} + 100$$

$$\dot{v} = -115.47 \text{ in./s}^2, \qquad \ddot{R} = 57.74 \text{ in./s}^2$$

$$\bar{a} = -115.5\bar{e}_R + 200\bar{e}_\phi = -115.5\bar{e}_t + 200\bar{e}_n \text{ in./s}^2$$

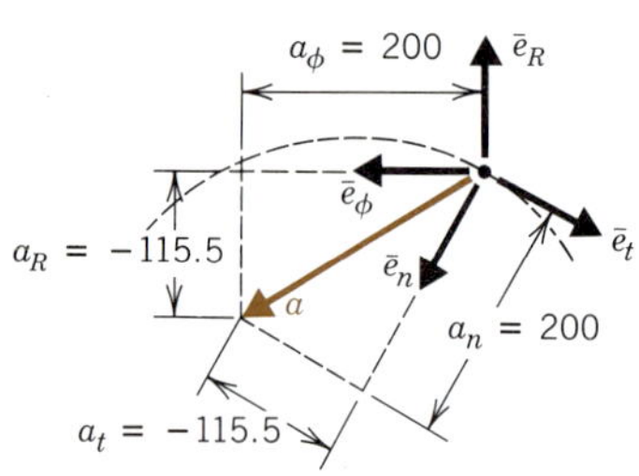

The similarity between these two descriptions of the resulting acceleration is surprising. The sketch to the left shows the acceleration in terms of both sets of components.

HOMEWORK PROBLEMS

11.82 Five seconds after it is released, a ball rolling along an incline is observed to have a speed of 20 m/s and to have traveled 62.5 m from the release point. Determine the constant rate of acceleration and the initial speed of the ball.

11.83 The distance (in feet) traveled by a box on a straight conveyor belt is given by $s = 20t^3 - 5t^4$, where t is in seconds. Determine the acceleration at the instant (a) when s is maximum, (b) when the speed is maximum.

11.84 Determine the maximum acceleration of the box in Problem 11.83 and the instant at which it occurs.

11.85 Solve Problem 11.83 for the case where the radius of curvature of the conveyor belt is 50 ft.

11.86 An automobile whose initial speed is 45 km/hr accelerates to 55 km/hr in a time interval of 2 s. The road is a horizontal curve whose radius is 250 m. Determine the magnitude of the acceleration of the automobile one second after it began to gain speed.

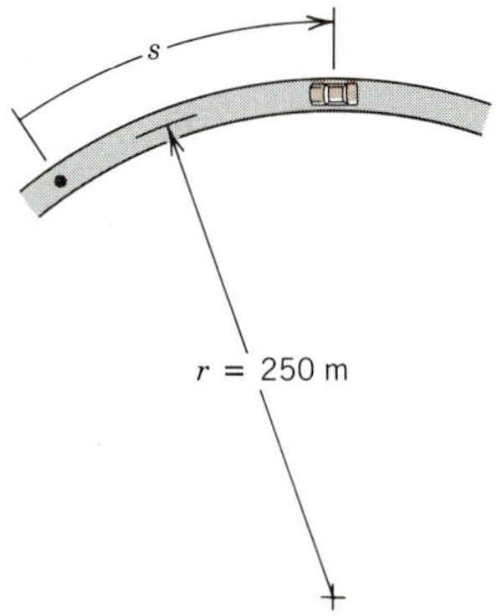

PROB. 11.86

11.87 A stunt rider on a motorcycle passes the bottom position A on the circular loop with a speed of 80 km/hr. The speed begins to decrease at this position at the constant rate of $0.3g$. Determine the acceleration of the motorcyclist's head one second after passing position A. Draw a sketch showing the position and acceleration at this instant.

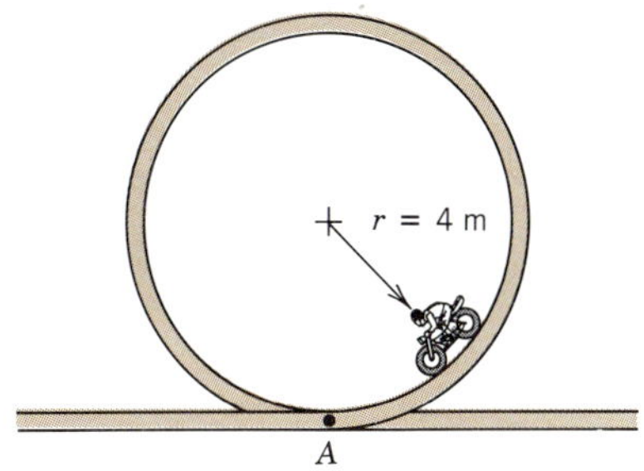

PROB. 11.87

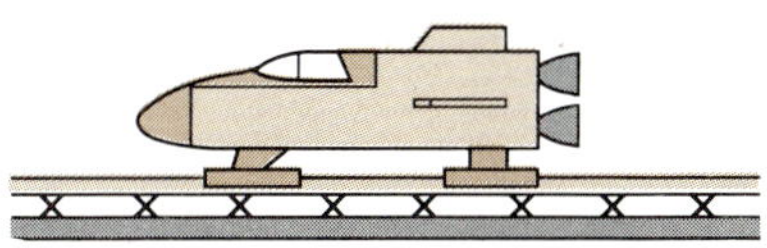

PROB. 11.88

11.88 The speed of a rocket sled on a straight track is $v = 2100s - 600s^2$, where s is the distance traveled in kilometers and v is in kilometers per hour. Determine the magnitude of the acceleration at $s = 1$ km.

11.89 Solve Problem 11.88 for a test track that is a circular arc whose radius is 10 km.

11.90 The motion of a particle is defined by $x = (1 + t)^2$ and $y = 4(1 + t)^{-2}$, where the coordinates are meters and time is in seconds. (a) Determine the path of the particle. (b) Find the velocity and acceleration when $t = 2$ s.

11.91 Two bars are welded to a nut that advances along a bolt whose pitch (the distance the bolt advances in one revolution) is 4 mm. The 80 mm bar–nut system rotates at the constant rate of 30 rpm. What are the velocity and acceleration of a point at the free end of one of the bars?

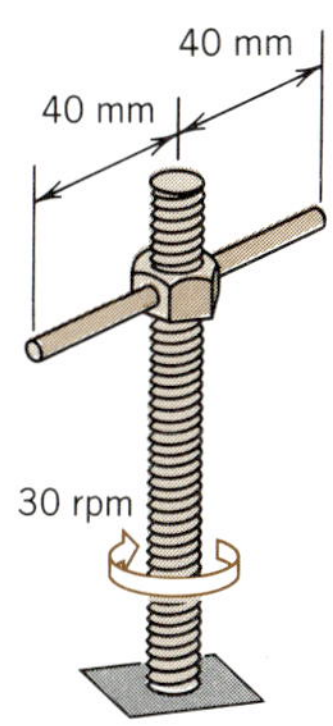

PROB. 11.91

11.92 A pipe rotates about the vertical axis at 30 rpm as water flows through it at 20 m/s. Determine the velocity and the magnitude of the acceleration of the particle of fluid that is at $R = 50$ mm from the vertical axis.

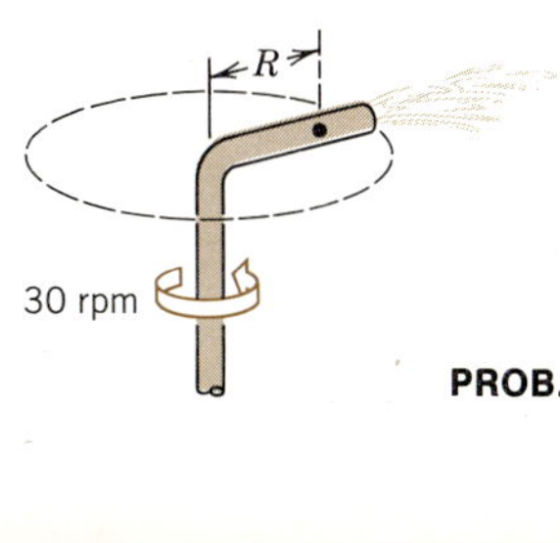

PROB. 11.92

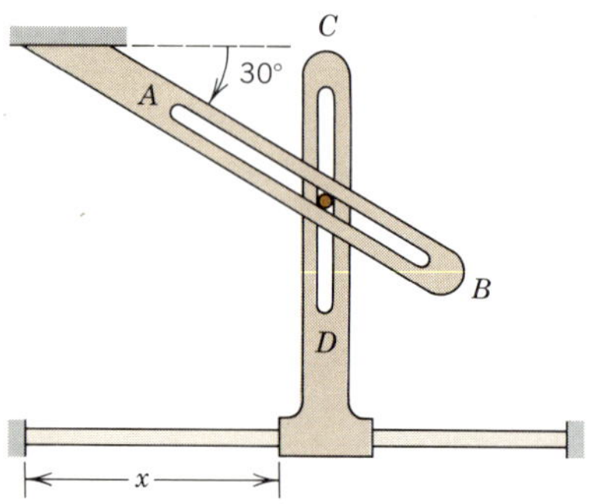

PROB. 11.94

*11.93 Determine the speed and the rate of change of the speed of the particle of fluid in Problem 11.92.

11.94 The pin is guided through the fixed straight groove AB by arm CD, which moves horizontally to the right according to $X = 200 \times (1 - e^{-t/2})$ mm. Determine the speed and acceleration of the pin in the position where $X = 100$ mm.

11.95 A satellite is moving at a constant speed in a circular orbit at a distance R from the center of the earth. The magnitude of its acceleration is $g(R_e/R)^2$, where $R_e = 6370$ km is the radius of the earth and $g = 9.807\ \text{m/s}^2$. Determine the value of R for which the satellite is in a synchronous orbit, which is an orbit that requires 1 day for a complete revolution. Also determine the corresponding speed of the satellite. (Consider a day to be 24 hr.)

11.96 An automobile follows a circular track at a constant speed of 52 km/hr. An observer is situated at position O. For the instant shown, determine $\dot{R}$ and $\dot{\theta}$.

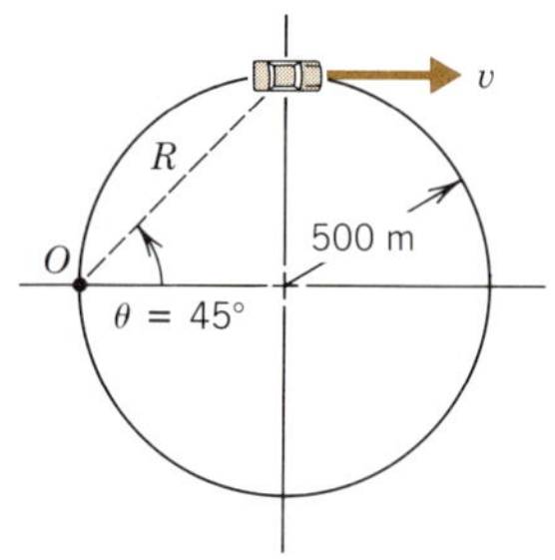

PROB. 11.96

11.97 Determine the values of $\ddot{R}$ and $\ddot{\theta}$ when the automobile in Problem 11.96 is in the illustrated position.

11.98 The diagram shows an elliptical curve forming a portion of a toy race track. A car at position A is slowing at the rate of $3g$ and the magnitude of its acceleration is $4.5g$. (a) Determine the speed of the car at this position. (b) Draw a sketch of the acceleration at this position.

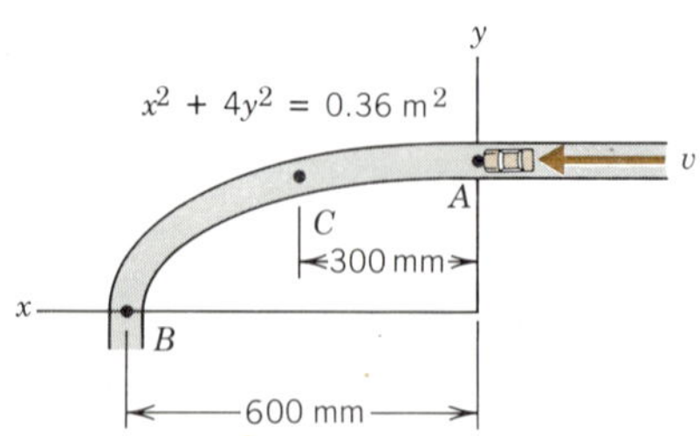

PROB. 11.98

11.99 Solve Problem 11.98 for the case where the given information corresponds to position C.

11.100 The parameter x is the increase in the length of a spring whose unstretched length is l. The speed of a block that is attached to this spring is

$$v = \pm[v_0^2 - k_1x^2 - \tfrac{1}{2}k_2x^4]^{1/2}$$

where v_0, k_1, and k_2 are constants, and the sign of v indicates whether x is increasing or decreasing. (a) Derive an expression for the acceleration of the block in terms of x. (b) Determine the value of x at which the speed is largest. What are the speed and acceleration at this position?

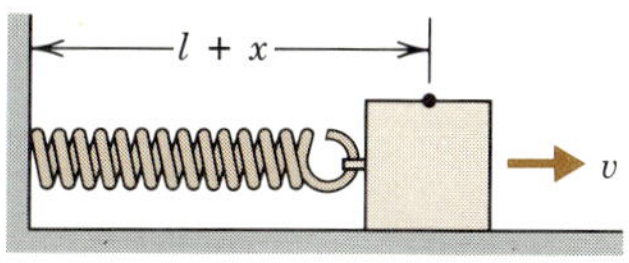

PROB. 11.100

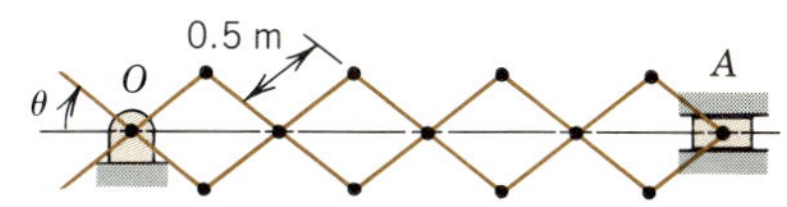

PROB. 11.101

11.101 A *lazy tongs* consisting of bars 1 m long is pivoted at a fixed point O. The pin A is constrained to move along OA. Find the velocity and acceleration of pin A when $\theta = 50°$, if $\dot{\theta}$ is a constant 3 rad/s clockwise. *Hint:* Express the distance from point O to point A in terms of the angle θ.

11.102 The rectangular Cartesian coordinates of a particle are given by $x = 200 \sin \pi t/10$, $y = 200 \cos \pi t/10$, $z = 400 \cos \pi t/20$, where x, y, and z are expressed in millimeters and t is in seconds. Determine the velocity and acceleration when $t = 2$ s, and show that the path lies on the surface of a circular cylinder.

11.103 The collar slides along the bent rod ABC such that the distance traveled is given by $s = 30t + 5t^2$, where s has units of inches and t is in seconds. Determine the acceleration of the collar as a function of t. Draw a graph showing the results for each acceleration component.

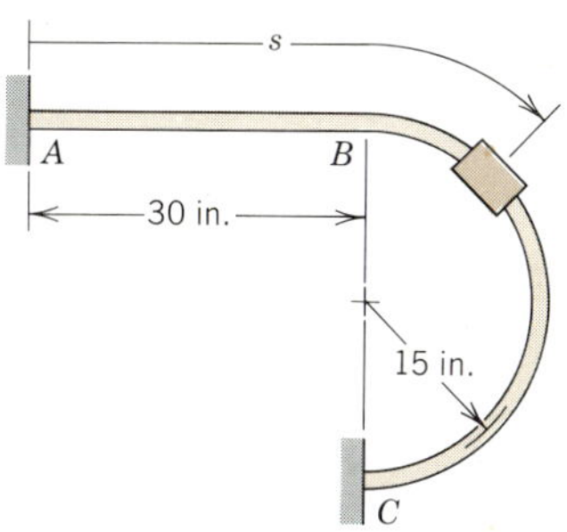

PROB. 11.103

11.104 Pin P slides inside the circular slot clockwise at a constant speed of 1 m/s, causing arm AB to rotate. Determine the angular velocity $\dot{\phi}$ and angular acceleration $\ddot{\phi}$ for the position $\theta = 0$.

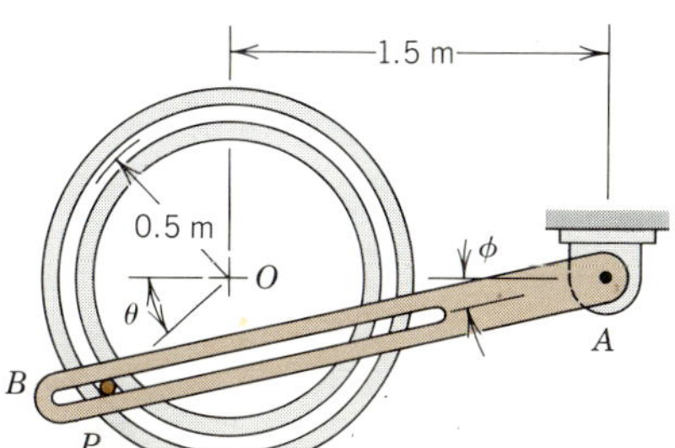

PROB. 11.104

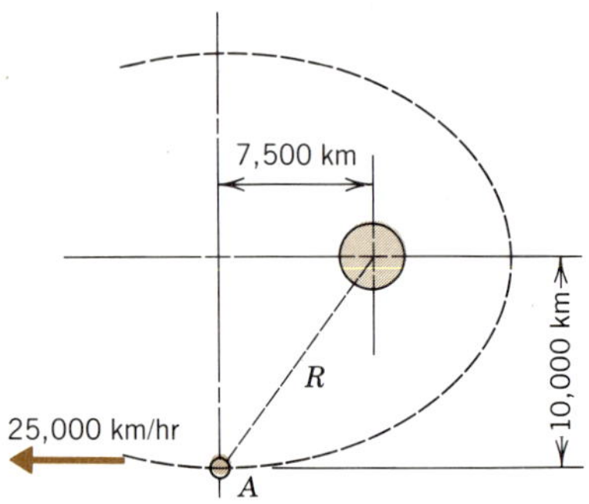

PROB. 11.106

11.105 Solve Problem 11.104 for the position $\theta = 90^\circ$.

11.106 The acceleration of a satellite in an elliptical orbit has a magnitude of $g(R_e/R)^2$, where $g = 9.807$ m/s^2 and $R_e = 6370$ km is the radius of the earth. The acceleration is always directed toward the center of the earth. At point A the satellite has a speed of 25,000 km/hr. What is the rate of change of the speed and the radius of curvature of the path at this location?

11.107 A race car on a 450 m radius curve is being tracked by a hand-held radar set at position A. When the car is at position B, its speed is 120 km/hr. For this instant, determine the radial speed $\dot{R}$ and the angular speed $\dot{\phi}$ at which the radar set is being rotated. What speed is indicated by the radar device in this situation?

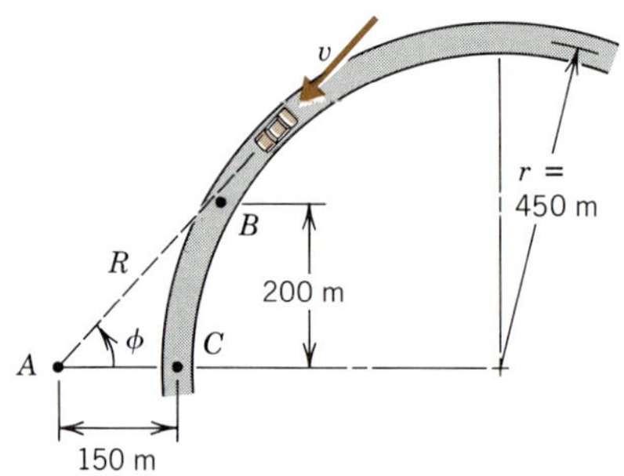

PROB. 11.107

***11.108** The speed of the race car in Problem 11.107 is constant. Determine the values of $\ddot{R}$ and $\ddot{\phi}$ when the automobile is at position C.

11.109 A conveyor belt changes from a straight line to a circular arc of 90 in. radius by means of a transition curve. The radius of curvature as a function of the distance s along the conveyor belt is given in the graph. For the case of a constant belt speed of 5 ft/s, sketch graphs showing the magnitudes of each component of the acceleration of a package as a function of its distance s.

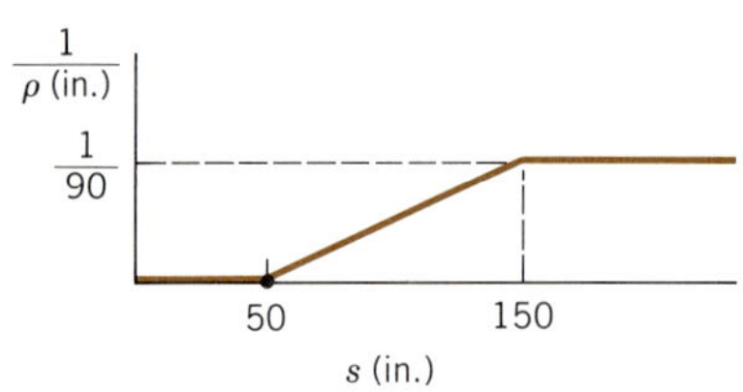

PROB. 11.109

***11.110** Solve Problem 11.109 if the conveyor belt moves with a constant speed of 2 ft/s until a package reaches $s = 50$ in., at which time it gains speed at the constant rate of 20 in./s^2.

11.111 A portion of a machine consists of a block that is free to slide along a rotating rod. The block is tied by means of an inextensible string that is wrapped around a fixed drum, as shown. The length L of the string shown varies according to $L = 5 + 0.3\theta$ (L in meters and θ in radians). If θ varies according to the formula $\theta = (\pi/6)(3 - 2t^3)$, what is the acceleration of point A on the block when $t = 1$ s?

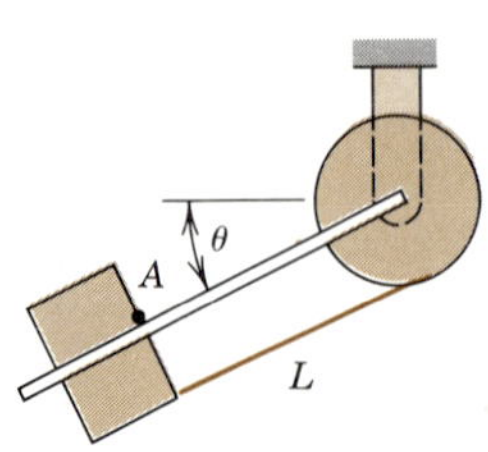

PROB. 11.111

11.112 Tube AB rotates clockwise at a constant rate of 120 rev/min in the vertical plane. The piston contracts such that the distance R decreases at the constant rate of 10 m/s. Determine the horizontal and vertical components of the velocity and acceleration of end C of the piston at the instant when $R = 600$ mm and $\theta = 30°$.

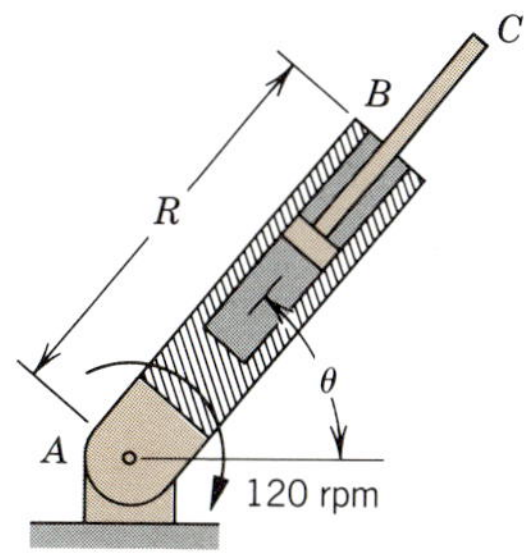

PROB. 11.112

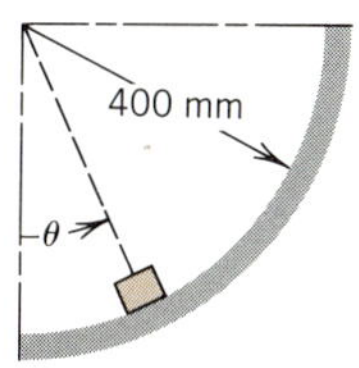

PROB. 11.113

11.113 A block is pushed up the circular hill; its horizontal component of velocity is constant at 200 mm/s. Determine the velocity and acceleration of the block when $\theta = 53.13°$.

11.114 A cycloidal path is defined in parametric form by $x = 0.5(\gamma - \sin\gamma)$ ft, $y = -0.5\cos\gamma$ ft, where γ is the parameter in units of radians. In a certain case it is known that the motion of a particle following this path is given by $\gamma = 4t^2$. Determine the velocity and acceleration of the particle when $t = 2$ s. Draw a sketch of the results.

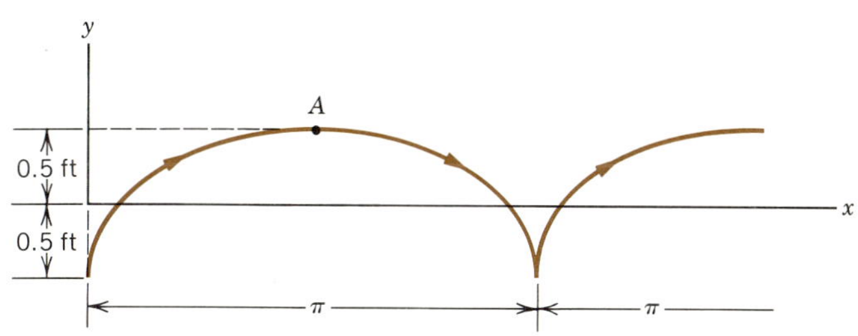

PROB. 11.114

11.115 For the motion in Problem 11.114, determine the velocity and acceleration when the particle is at position A. Draw a sketch of the results.

***11.116** For the motion in Problem 11.114 determine the speed, the rate of increase of the speed, and the radius of curvature of the path when $t = 3$ s.

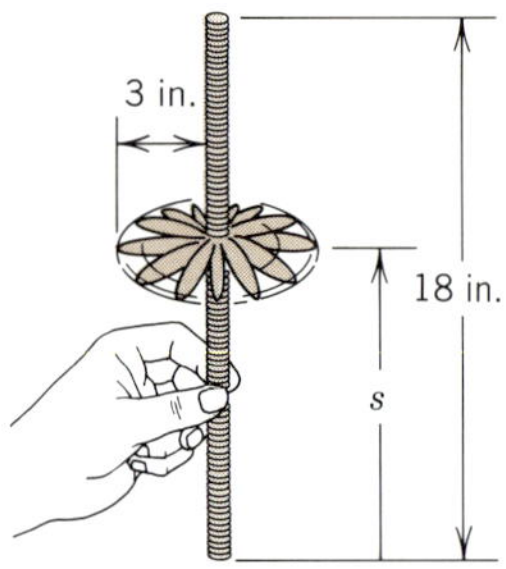

PROB. 11.117

11.117 The propeller on the toy shown is pushed along a threaded shaft so that it flies into the air while rotating. The pitch of the screw results in one revolution of the propeller for an advancement of 0.5 in. The distance s from the bottom of the shaft to the hub is given by $s = 3 + 1.5t^3$ where s is in inches and t is in seconds. (a) How long does it take for the hub to reach the top of the shaft? (b) What is the magnitude of the acceleration of the tip of a propeller blade at the instant before the hub leaves the shaft?

11.118 A hydraulic piston inside arm AB causes the length of the arm to increase at the constant rate of 2.5 m/s, thereby moving collar C along the circular guide. Determine the velocity and acceleration of the collar when $\theta = 90°$.

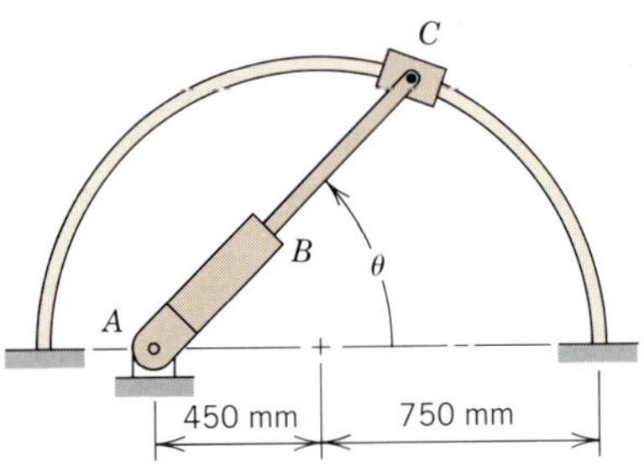

PROB. 11.118

***11.119** Solve Problem 11.118 when collar C is at $\theta = 53.13°$.

11.120 A pin slides in two yokes as shown. Yoke A has an instantaneous velocity of 50 mm/s to the right and an acceleration of 20 mm/s^2 to the left and yoke B has a constant downward velocity of 10 mm/s. For the position shown, determine the velocity and acceleration of the pin.

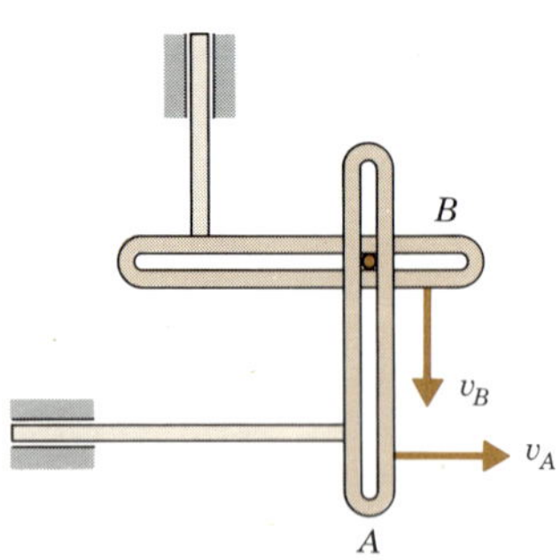

PROB. 11.120

***11.121** Determine the speed, the rate of change of the speed, and the radius of curvature of the path in Problem 11.120. Show the location of the center of curvature in a sketch.

11.122 For sufficiently large values of the radial distance R from the center line of a *vortex* in a fluid, the cylindrical components of the velocity of a particle of fluid are $v_R = 0$, $v_\phi = A/R$, $v_z = 0$, where A is a constant. Determine an expression for the acceleration of a particle. What is the path of a particle of fluid?

***11.123** An important type of flow in the study of fluids is the *doublet*, for which the velocity components in cylindrical coordinates are $v_R = (A \cos \phi)/R^2$, $v_\phi = (A \sin \phi)/R^2$, $v_z = 0$, where A is a constant. Determine an expression for the acceleration of a particle.

***11.124** Along a portion of a railroad line the transition curve from a straight track to a curve is defined by $y = x^2/2000$, where x and y are in meters. Measuring the distance s from the position where the locomotive is at $x = 0$, the motion of the train is defined by $s = t^2/10$, where s is the arc length in meters and t is in seconds. What is the speed, the rate of change of the speed, and the magnitude of the acceleration of the locomotive when $x = 1000$ m?

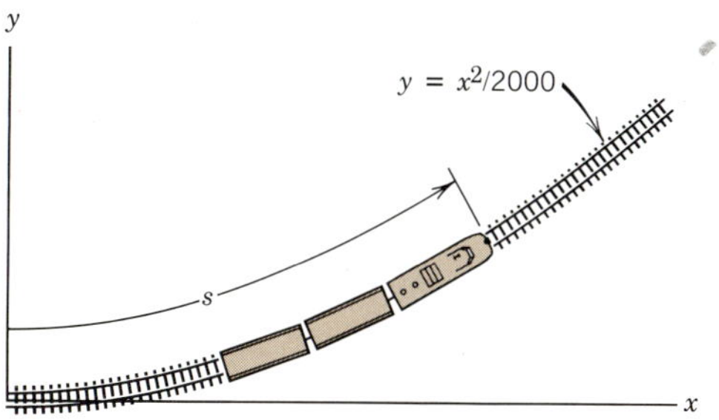

PROB. 11.124

***11.125** The amusement park ride rotates about its vertical axis at the constant rate of 10 rpm as the arms swing in and out. At the instant when $\theta = 30^\circ$ the arms are swinging outward at $\dot{\theta} = 0.3$ rad/s and this rate is decreasing at 0.1 rad/s^2. Determine the acceleration of the cabin at this instant.

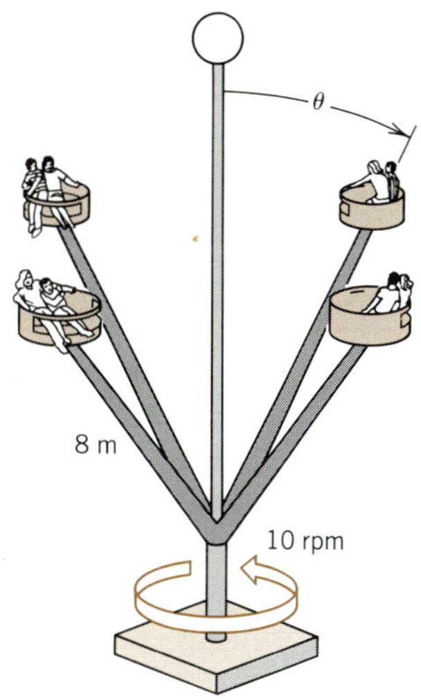

PROB. 11.125

E. RELATIVE MOTION

1. Formulation of Kinematical Relationships

The kinematical equations developed so far describe how we see the motion of a particle when our own position is a fixed point in space, that is, when the origin of the reference frame is a fixed point. Clearly, if we were to view

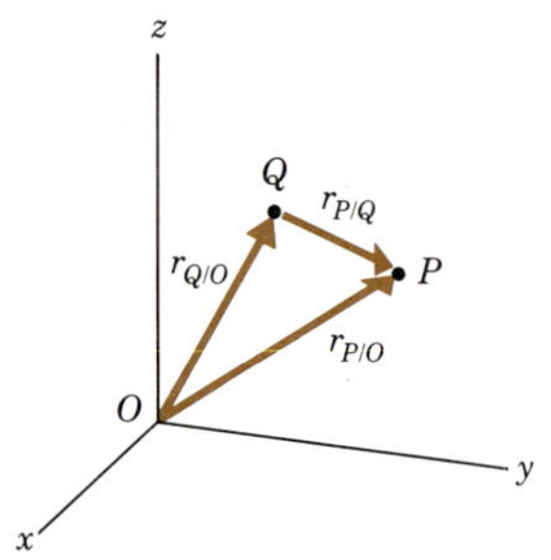

FIGURE 17

the same particle motion from the window of a moving vehicle, our observation would be modified by our motion. A general situation such as this is illustrated in Figure 17.

The *xyz* coordinate system is a fixed frame of reference, whose origin is point O. Points P and Q are both moving with respect to this reference frame. In the situation to be addressed here, the motion of point P will be described indirectly. For example, the position $\bar{r}_{P/O}$ may be established if the position of point Q, $\bar{r}_{Q/O}$, and the position of point P with respect to point Q, $\bar{r}_{P/Q}$, are both known. The relationship between these positions is found from the vector triangle in Figure 17 to be

$$\boxed{\bar{r}_{P/O} = \bar{r}_{Q/O} + \bar{r}_{P/Q}} \tag{29}$$

Motion parameters that we describe with respect to a fixed point are called *absolute* quantities, whereas those that are referred to a moving point are called *relative* quantities. Thus, eq. (29) states that the absolute position of point P is the sum of the absolute position of point Q and the relative position of point P with respect to point Q.

The next kinematical quantity of interest is velocity. The time derivative of the relative position vector $\bar{r}_{P/Q}$ is the relative velocity $\bar{v}_{P/Q}$ of point P with respect to point Q. That is,

$$\frac{d}{dt}(\bar{r}_{P/Q}) = \dot{\bar{r}}_{P/Q} = \bar{v}_{P/Q} \tag{30}$$

Differentiating eq. (29) yields the following relationship between the absolute and relative velocities:

$$\dot{\bar{r}}_{P/O} = \dot{\bar{r}}_{Q/O} + \dot{\bar{r}}_{P/Q}$$

$$\boxed{\bar{v}_P = \bar{v}_Q + \bar{v}_{P/Q}} \tag{31}$$

In a similar manner we can reason that the acceleration of point P as seen by the observer at Q must be

$$\ddot{\bar{r}}_{P/Q} = \dot{\bar{v}}_{P/Q} = \bar{a}_{P/Q} \tag{32}$$

Then, by differentiating eq. (31), we find that

$$\boxed{\bar{a}_P = \bar{a}_Q + \bar{a}_{P/Q}} \tag{33}$$

Equations (29), (31), and (33) all have similar form. They tell us that the absolute motion of some point P may be obtained by adding to the absolute motion of another point Q the motion of point P with respect to point Q. A different viewpoint is obtained by solving these equations for the relative motion, which gives

$$\boxed{\begin{aligned} \bar{r}_{P/Q} &= \bar{r}_P - \bar{r}_Q \\ \bar{v}_{P/Q} &= \bar{v}_P - \bar{v}_Q \\ \bar{a}_{P/Q} &= \bar{a}_P - \bar{a}_Q \end{aligned}} \tag{34}$$

In other words, the motion of point P with respect to point Q is the difference between the absolute motions of points P and Q.

A variety of ways in which the relative motion relations may be employed will be developed in the examples that follow. One immediate result arises in the special case where point Q is also a fixed point, for then $\bar{v}_Q = \bar{a}_Q = \bar{0}$. This means that

$$\bar{v}_P = \bar{v}_{P/Q} \quad \text{and} \quad \bar{a}_P = \bar{a}_{P/Q}$$

thus proving the earlier statements that the absolute velocity and acceleration are independent of the location of the fixed point of observation.

EXAMPLE 18

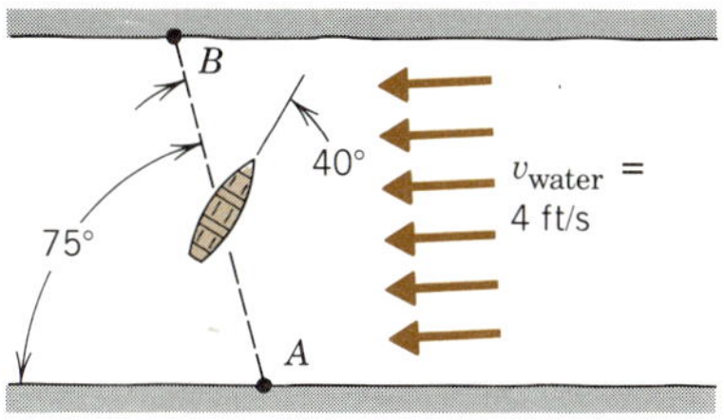

A rowboat is following a straight path from point A to point B as it travels across a river that is flowing at a constant speed of 4 ft/s as shown. The boat has a speed of 6 ft/s with respect to the river, and its axis is oriented 40° upstream from line AB. Determine the absolute speed of the rowboat and the speed at which the river is flowing.

Solution

The rowboat moves relative to the water in the direction of its axis because the shape of its hull prevents sideways slippage. If the river were not flowing, the velocity of the boat relative to the water would also be its absolute velocity. When the river is flowing, the boat must face in a transverse direction relative to the desired direction, in order to compensate for the downstream movement of the river.

The relative velocity eq. (31) may be written in terms of this situation as

$$\bar{v}_{\text{boat}} = \bar{v}_{\text{water}} + \bar{v}_{\text{boat/water}}$$

These vectors will be described in component form with the aid of a coordinate system that is aligned with the river, as shown in the sketch.

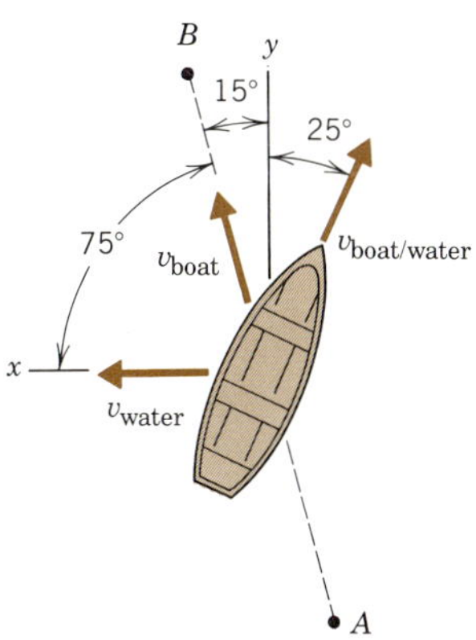

The orientations of the velocity vectors depicted in this sketch correspond to the given information that the river is flowing downstream, that the boat is proceeding from point A to point B, and that the axis of the boat is oriented 40° upstream relative to line AB. In conjunction with the fact that $|\bar{v}_{\text{boat/water}}| = 6$ ft/s, it follows that

$$\begin{aligned}\bar{v}_{\text{boat}} &= v_{\text{boat}}(\sin 15^\circ\,\bar{i} + \cos 15^\circ\,\bar{j}) \\ &= v_{\text{boat}}(0.2588\bar{i} + 0.9659\bar{j})\end{aligned}$$

$$\bar{v}_{\text{water}} = v_{\text{water}}\bar{i}$$

$$\begin{aligned}\bar{v}_{\text{boat/water}} &= 6(-\sin 25^\circ\,\bar{i} + \cos 25^\circ\,\bar{j}) \\ &= -2.536\bar{i} + 5.438\bar{j} \text{ ft/s}\end{aligned}$$

The relative velocity equation then requires that

$$v_{\text{boat}}(0.2588\bar{i} + 0.9659\bar{j}) = v_{\text{water}}\bar{i} - 2.536\bar{i} + 5.438\bar{j}$$

$$\bar{i} \text{ component:} \quad 0.2588 v_{\text{boat}} = v_{\text{water}} - 2.536$$

$$\bar{j} \text{ component:} \quad 0.9659 v_{\text{boat}} = 5.438$$

$$v_{\text{boat}} = 5.63 \text{ ft/s}, \qquad v_{\text{water}} = 3.99 \text{ ft/s}$$

EXAMPLE 19

At the instant depicted in the diagram, car A on the east-west overpass has a constant speed of 36 km/hr, and car B has a constant speed of 72 km/hr along the highway. The radius of curvature of the highway is 100 m. Determine the velocity and acceleration of car A relative to car B.

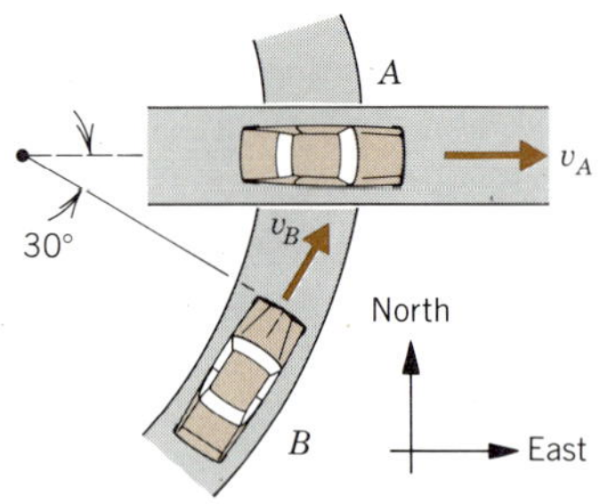

Solution

This problem requires the determination of the motion of car A relative to car B, in a situation where the absolute motion of each is specified. Because the speed and path for each car are given, we describe the absolute velocities and accelerations by using path variables. Then the relative motion will be evaluated by expressing the absolute parameters in terms of their components with respect to the xyz coordinate system shown.

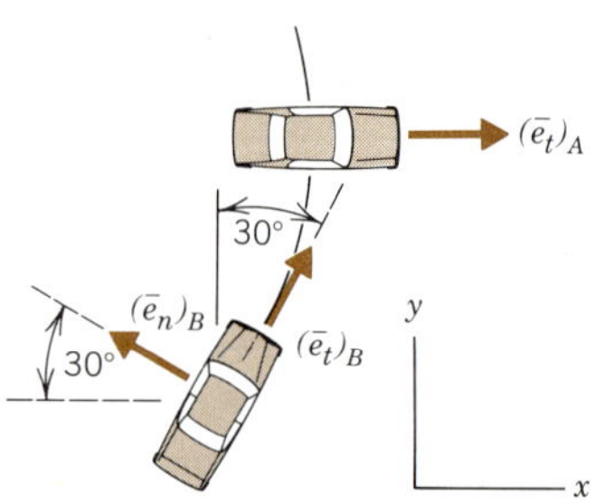

It is given that $v_A = 36$ km/hr $= 10$ m/s, $\dot{v}_A = 0$, $v_B = 72$ km/hr $= 20$ m/s, $\dot{v}_B = 0$, and $\rho_B = 100$ m. Thus

$$\bar{v}_A = 10(\bar{e}_t)_A \qquad \bar{v}_B = 20(\bar{e}_t)_B \text{ m/s}^2$$

$$\bar{a}_A = \bar{0} \qquad \bar{a}_B = \frac{(20)^2}{100}(\bar{e}_n)_B = 4(\bar{e}_n)_B \text{ m/s}^2$$

In order to represent these quantities in terms of their xy components, we transform their path variable unit vectors. The sketch indicates that

$$(\bar{e}_t)_A = \bar{i}, \qquad (\bar{e}_t)_B = \sin 30^\circ \bar{i} + \cos 30^\circ \bar{j}$$

$$(\bar{e}_n)_B = -\cos 30^\circ \bar{i} + \sin 30^\circ \bar{j}$$

Thus

$$\bar{v}_A = 10\bar{i}, \qquad \bar{v}_B = 20(0.50\bar{i} + 0.8660\bar{j}) = 10.0\bar{i} + 17.32\bar{j}$$

$$\bar{a}_A = \bar{0}, \qquad \bar{a}_B = 4(-0.8660\bar{i} + 0.50\bar{j}) = -3.464\bar{i} + 2\bar{j}$$

Finally, from eqs. (34) we find that

$$\bar{v}_{A/B} = \bar{v}_A - \bar{v}_B = -17.32 \text{ m/s}^2$$

and

$$\bar{a}_{A/B} = \bar{a}_A - \bar{a}_B = 3.464\bar{i} - 2.0\bar{j} \text{ m/s}^2$$

Thus we see that with respect to car B, at this instant car A seems to be moving south at 17.32 m/s, with acceleration components of 3.464 m/s^2 east and 2.0 m/s^2 south.

EXAMPLE 20

At the instant shown yoke A has a speed of 20 m/s up and to the right, and yoke B is moving at 5 m/s to the left. Determine the velocity of pin P.

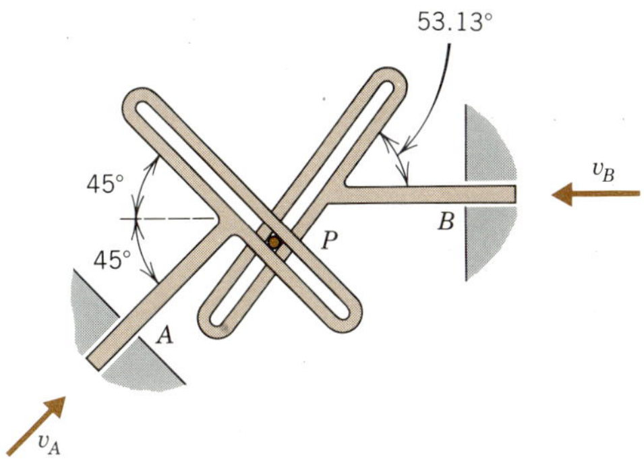

Solution

The motion of pin P may be visualized by regarding the pin to move parallel to the slot in each yoke as the yokes move. This suggests that there are two relative motions. The velocity of pin P will be determined by making these motions agree.

We will assume that pin P is moving upward relative to each slot. The sketch therefore depicts each relative velocity to be parallel to each groove, upward. The illustrated xyz coordinate system is convenient for describing the various velocities.

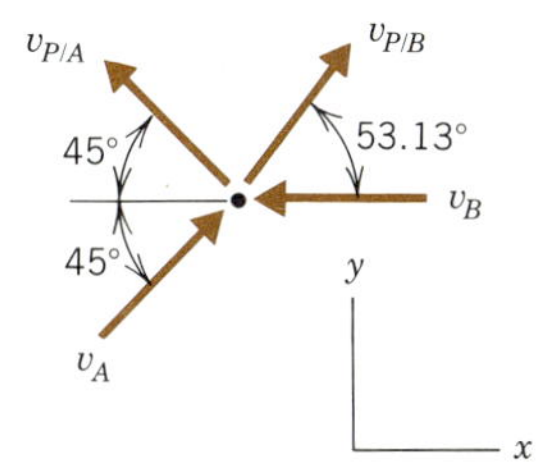

For the given values of v_A and v_B, the absolute and relative velocities are known from the sketch to be

$$\bar{v}_A = 20(\cos 45^\circ \bar{i} + \sin 45^\circ \bar{j}) = 14.142(\bar{i} + \bar{j}) \text{ m/s}$$

$$\bar{v}_{P/A} = v_{P/A}(-\cos 45^\circ \bar{i} + \sin 45^\circ \bar{j}) = v_{P/A}(0.7071)(-\bar{i} + \bar{j})$$

$$\bar{v}_B = -5\bar{i} \text{ m/s}$$

$$\bar{v}_{P/B} = v_{P/B}(\cos 53.13^\circ \bar{i} + \sin 53.13 \bar{j}) = v_{P/A}(0.6\bar{i} + 0.8\bar{j})$$

The relative velocity equation (31) indicates that

$$\bar{v}_P = \bar{v}_A + \bar{v}_{P/A} = \bar{v}_B + \bar{v}_{P/B}$$

The result of substituting the expressions for the velocities is that

$$\begin{aligned}\bar{v}_P &= 14.142(\bar{i} + \bar{j}) + v_{P/A}(0.7071)(-\bar{i} + \bar{j}) \\ &= -5\bar{i} + v_{P/B}(0.6\bar{i} + 0.8\bar{j})\end{aligned}$$

Matching corresponding components then yields

$$\bar{i}\text{ component:}\quad 14.142 - 0.7071v_{P/A} = -5 + 0.6v_{P/B}$$

$$\bar{j}\text{ component:}\quad 14.142 + 0.7071v_{P/A} = 0.8v_{P/B}$$

These equations are solved for $v_{P/A}$ and $v_{P/B}$, and the results used to evaluate $\bar{v}_P$.

$$v_{P/B} = 23.77\text{ m/s},\qquad v_{P/A} = 6.90\text{ m/s}$$

$$\bar{v}_P = -5\bar{i} + 23.77(0.6\bar{i} + 0.8\bar{j})$$
$$= 9.26\bar{i} + 19.02\bar{j}\text{ m/s}$$

Notice that $v_{P/A}$ and $v_{P/B}$ are positive, which confirms the assumption that pin P is moving upward relative to both slots.

HOMEWORK PROBLEMS

11.126 The pilot of an airplane notes that the velocity with respect to the air is 300 km/hr due north. At a ground-based radar station the plane is observed to be flying at 310 km/hr in a direction 7° west of north. Determine the speed and heading of the air flow.

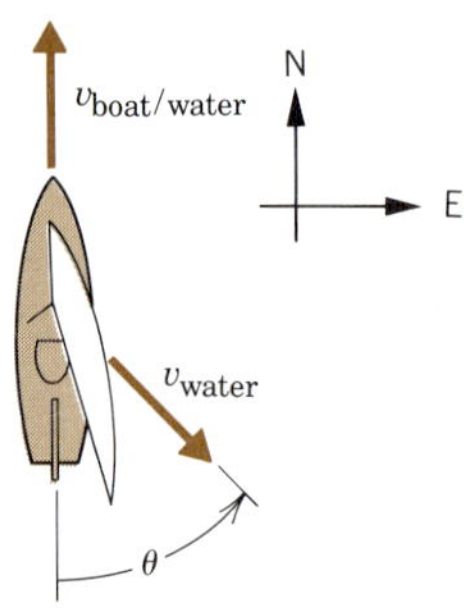

PROBS. 11.127 AND 11.128

11.127 The sailboat is moving at a speed of 10 ft/s relative to the water with its keel oriented northward. The water is flowing at 5 ft/s at a heading $\theta = 60°$. Determine the absolute speed and heading of the boat.

11.128 The sailboat is moving at a speed of 5 ft/s relative to the water with its keel oriented northward. The water is flowing at $\theta = 36.87°$, with the result that the sailboat seems to be drifting eastward. Determine the speed of the water and the speed of the boat.

11.129 The nose of a small airplane is oriented 20° south of west and the airplane is traveling west. Knowing that the wind is from the southwest at 110 km/hr, determine the airspeed and groundspeed of the airplane.

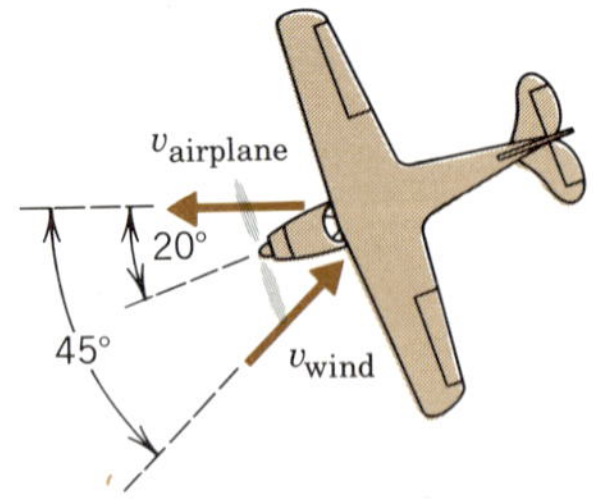

PROB. 11.129

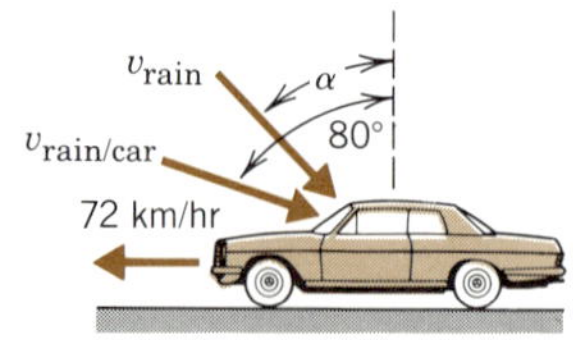

PROB. 11.130

11.130 Passengers in an automobile traveling along a level road at 72 km/hr think that raindrops are falling at 80° from the vertical. The absolute speed of a raindrop is known to be 30 km/hr. Determine the angle α at which an observer on the ground sees the raindrops falling.

11.131 An airplane that can fly at v km/hr in still air is to travel to a point $\bar{r} = R(\cos\theta\,\bar{i} + \sin\theta\,\bar{j})$ km from its starting point at the origin in a wind of velocity $v_w\bar{i}$ km/hr. Determine in terms of v, R, θ, and v_w: (a) the direction in which the plane should be headed, (b) the ground speed v_0 of the airplane, (c) the time it will take to make the flight, (d) the wind speed v_{max} for which it becomes impossible for the flight to be made in a straight-line path.

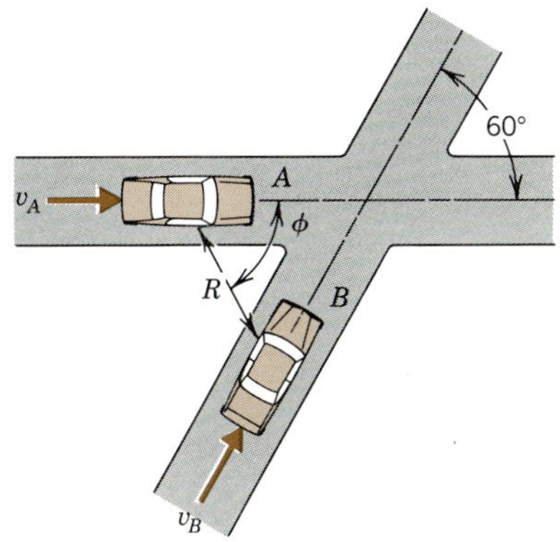

PROB. 11.132

11.132 Automobile A is moving at 90 km/hr and accelerating at 2 m/s^2 as it approaches the intersection. Automobile B is moving at 60 km/hr and decelerating at 4 m/s^2. Determine the velocity and acceleration of automobile B with respect to automobile A.

***11.133** In Problem 11.132 determine the rate of change of the separation distance R and of the relative angle ϕ of automobile B with respect to automobile A if $R = 150$ m and $\phi = 45°$ at the instant shown. *Hint:* Describe the velocity $\bar{v}_{B/A}$ in terms of cylindrical coordinates.

11.134 When bicyclists A and B are in the positions shown, $v_A = 10$ mi/hr, $\dot{v}_A = 0$, $v_B = 15$ mi/hr, and $\dot{v}_B = -8$ ft/s^2. Determine the corresponding velocity and acceleration of bicycle A with respect to bicycle B.

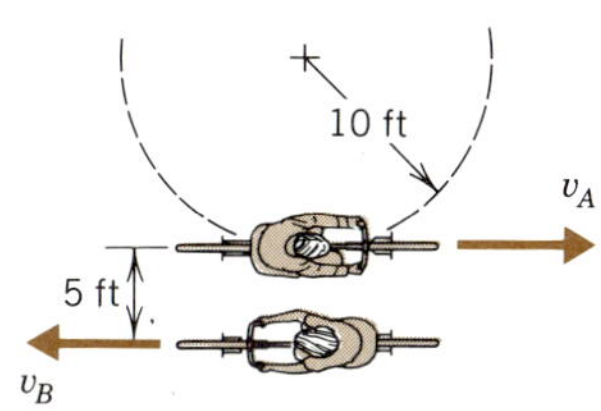

PROB. 11.134

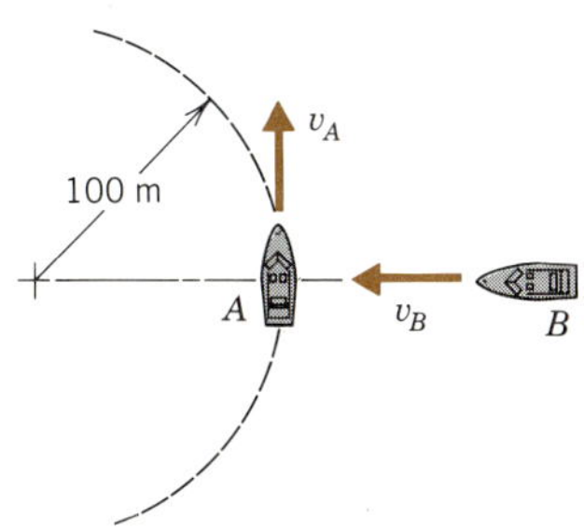

PROB. 11.135

11.135 Motorboat A travels at a constant speed of 36 km/hr. At the instant shown, motorboat B is moving at 50 km/hr and is slowing down at the rate of 6 m/s^2. Determine the velocity and acceleration of boat B as observed from boat A.

11.136 At the instant shown, automobile A has a constant speed of 25 m/s, and automobile B has a speed of 30 m/s and is gaining speed at the rate of 5 m/s^2. Determine the corresponding velocity and acceleration of automobile A with respect to automobile B.

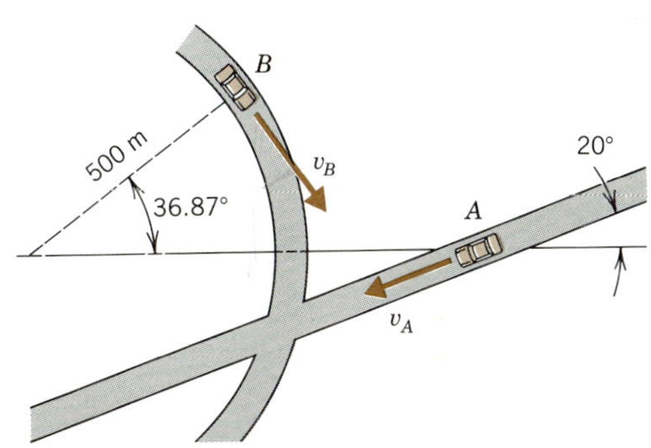

PROB. 11.136

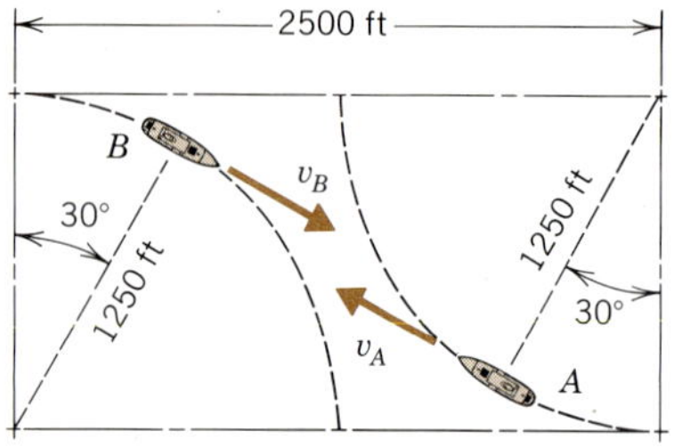

PROB. 11.137

11.137 In order to avoid a collision, ships A and B maneuver in circular paths. At the instant shown, ship A has a constant speed of 9 ft/s, and ship B has a constant speed of 5 ft/s. Determine the corresponding position, velocity, and acceleration of ship B with respect to ship A.

11.138 Solve Problem 11.137 if the speed of ship A at the given instant is decreasing at 1 ft/s^2.

11.139 Two automobiles are traveling in opposite directions along a divided highway. Both vehicles have a speed of 55 mi/hr. Determine the velocity and acceleration of vehicle A with respect to vehicle B when the angle of separation $\phi = 0°$.

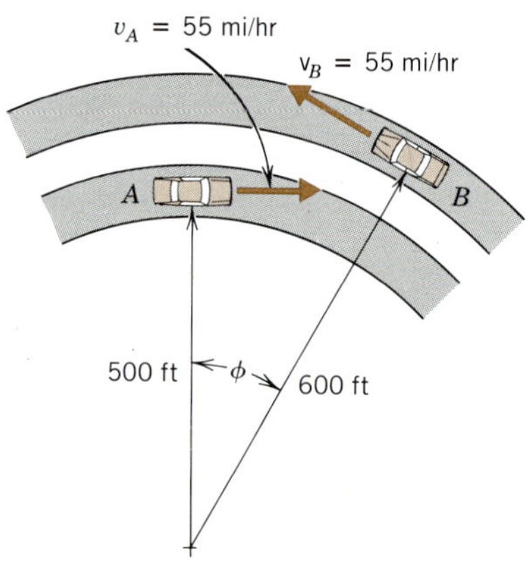

PROB. 11.139

11.140 Solve Problem 11.139 for $\phi = 30°$.

11.141 Pin P slides within the fixed slot A, as it is made to move by yoke B. Determine the velocity of the pin in terms of the speed v_B of the yoke.

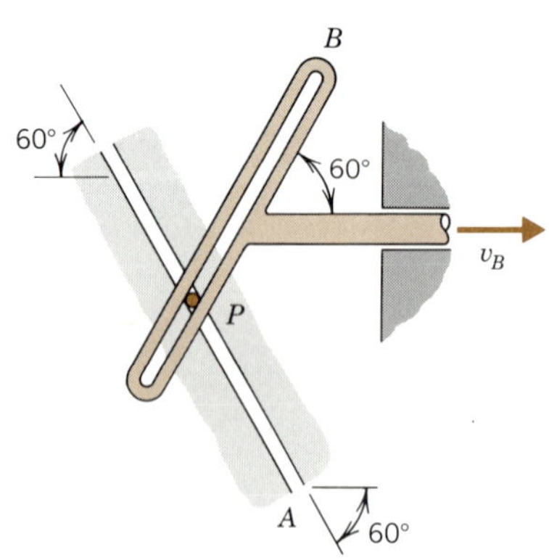

PROB. 11.141

11.142 At the instant shown yoke A is moving to the right at 40 mm/s, and yoke B is moving upward at 120 mm/s. Determine the velocity of pin P.

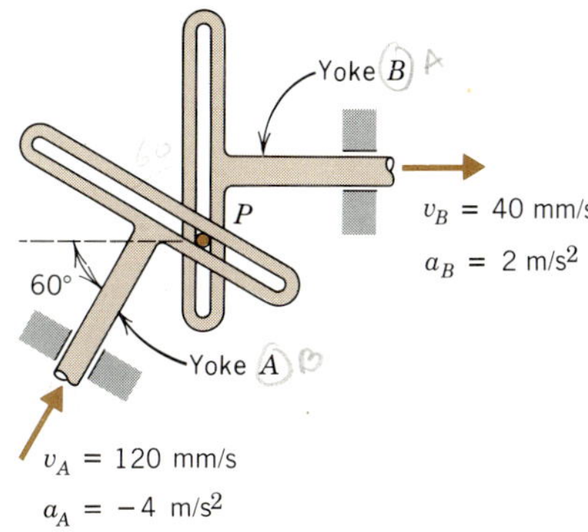

PROB. 11.142

11.143 In Problem 11.142 yoke A is slowing at the rate of 4 m/s^2 and yoke B is speeding up at 2 m/s^2. Determine the acceleration of pin P.

***11.144** Pin C slides within the fixed vertical slot A and the slot in yoke B. Yoke B is moving to the left at the constant speed of 5 ft/s. Determine the velocity of pin C in the position $\theta = 15°$.

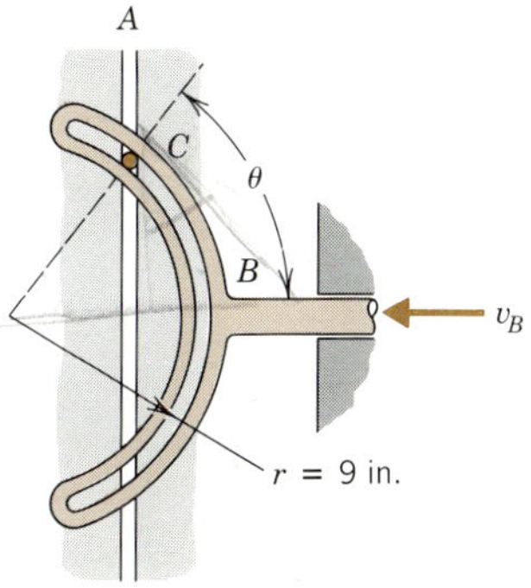

PROB. 11.144

***11.145** Determine the velocity and acceleration of pin C in Problem 11.144 for $\theta = 36.87°$.

***11.146** A pin at the end of the 300 mm long arm AB follows the circular groove in arm CD, which is moving to the right at a constant speed of 60 m/s. For the instant when $\phi = 45°$, determine the velocity of the pin.

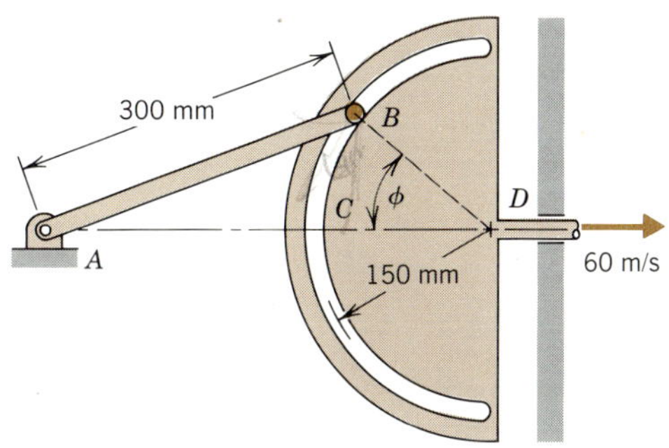

PROB. 11.146

2. Cable–Pulley Systems

When cables and pulleys are utilized to connect bodies, we are concerned with the motion of more than one point, just as we were in the preceding study of relative motion. In rigid body mechanics, cables are generally modeled as inextensible, massless bodies. The condition that the length of the cable is constant leads to a relationship between the motion of the bodies in the system.

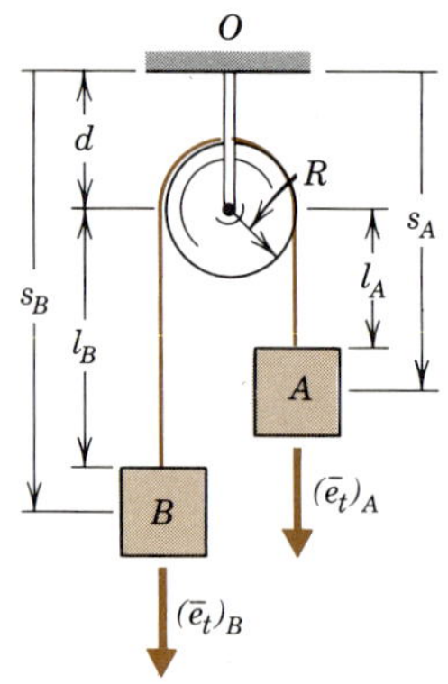

FIGURE 18

Consider blocks A and B in Figure 18, which are connected by an inextensible cable. The distances s_A and s_B locate the blocks with respect to the fixed reference point O. The tangent unit vector describing the rectilinear motion of each block is defined in the figure in the direction of increasing s, consistent with the derivation of the path variable equations. The corresponding velocity and acceleration formulas are

$$\bar{v}_A = v_A(\bar{e}_t)_A, \qquad \bar{v}_B = v_B(\bar{e}_t)_B$$

$$\bar{a}_A = \dot{v}_A(\bar{e}_t)_A, \qquad \bar{a}_B = \dot{v}_B(\bar{e}_t)_B$$

where

$$v_A = \dot{s}_A, \qquad v_B = \dot{s}_B$$

The relationship between v_A and v_B, and therefore between $\dot{v}_A$ and $\dot{v}_B$, is obtained by constructing the length of the cable. For the situation in Figure 18, this length is

$$l = l_A + \pi R + l_B$$

where πR is the length of the cable in contact with the pulley. From Figure 18 we may write

$$l_A = s_A - C_A \quad \text{and} \quad l_B = s_B - C_B$$

where C_A and C_B are constants whose magnitudes depend on the fixed distance d and the dimensions of the masses. Replacing l_A and l_B in the equation for l gives

$$l = s_A - C_A + \pi R + s_B - C_B$$

or

$$s_A + s_B = C$$

where C is a combined single constant whose value is not important.

This equation states that a change in s_A must be compensated by the opposite change in s_B. Differentiating this relation then yields

$$\dot{s}_A + \dot{s}_B = \dot{C} \equiv 0$$

In other words $\dot{s}_B = -\dot{s}_A$, so that

$$v_B = -v_A, \qquad \dot{v}_B = -\dot{v}_A$$

The interpretation of these relations is that if block A is given a certain velocity or acceleration, either up or down, then block B will execute the same motion in the opposite direction.

This approach may be generalized for the case of any cable–pulley system. First, we write path variable formulas for each moving body in terms of the distance along the path from a convenient fixed point. Then we describe the length of each portion of the cable in terms of the path variable distances. Now an equation for the cable length is formed, which ignores any constant lengths, and the sum is equated to a constant. Differentiation of the sum then yields relations for the speed and the rates of change of the speeds.

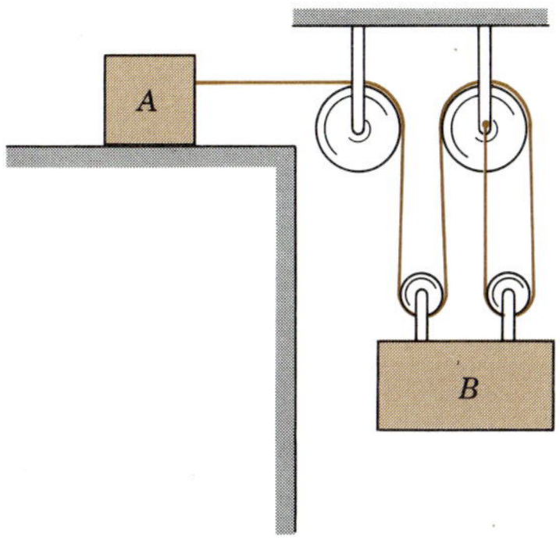

EXAMPLE 21
Block A is moving to the left at 6 ft/s and decelerating at the rate of 12 ft/s^2. Determine the velocity and acceleration of block B.

Solution
We begin with a sketch that shows the distances s_A and s_B locating the position of each block. A convenient point of reference for these distances is point O at one of the fixed pulleys. The tangent vector for each block is defined in the sketch to be in the sense of increasing s for that block.

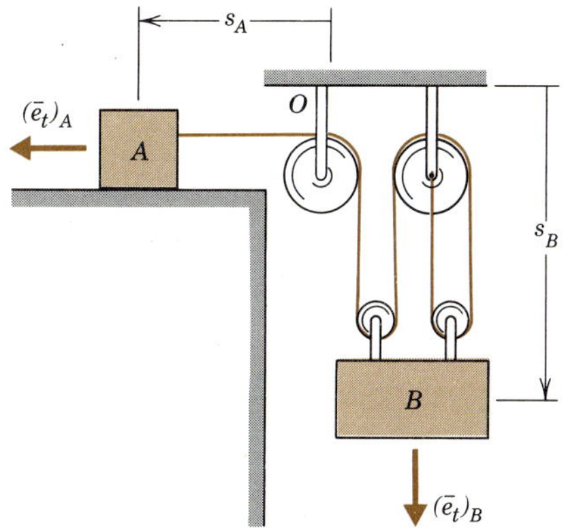

Each block is in rectilinear motion, so

$$\bar{v}_A = v_A(\bar{e}_t)_A, \qquad \bar{a}_A = \dot{v}_A(\bar{e}_t)_A$$

$$\bar{v}_B = v_B(\bar{e}_t)_B, \qquad \bar{a}_B = \dot{v}_B(\bar{e}_t)_B$$

For a movement of block A to the left at the given speed, it must be that

$$v_A = \dot{s}_A = 6 \text{ ft/s}$$

This speed is decreasing, so $\dot{v}_A < 0$. Thus

$$\dot{v}_A = -12 \text{ ft/s}^2$$

The corresponding information about block B is obtained by expressing the length of the cable in terms of s_A and s_B. There are four vertical segments of cable and one horizontal segment, so

$$4s_B + s_A = C$$

Differentiating this equation yields

$$4\dot{s}_B + \dot{s}_A \equiv 4v_B + v_A = 0$$

$$v_B = -\tfrac{1}{4}v_A, \qquad \dot{v}_B = -\tfrac{1}{4}\dot{v}_A$$

Substitution of the given values of v_A and $\dot{v}_A$ then results in

$$v_B = -1.5 \text{ ft/s}, \qquad \dot{v}_B = 3 \text{ ft/s}^2$$

$$\bar{v}_B = -1.5(\bar{e}_t)_B \text{ ft/s}, \qquad \bar{a}_B = 3(\bar{e}_t)_B \text{ ft/s}^2$$

In words, block B is moving upward at 1.5 ft/s, and slowing down at 3 ft/s^2.

HOMEWORK PROBLEMS

11.147–11.149 Determine the velocity and acceleration of block B corresponding to the given motion of block A.

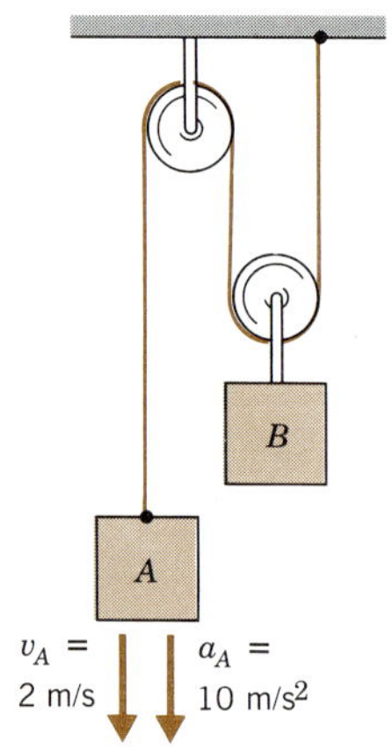

PROB. 11.147

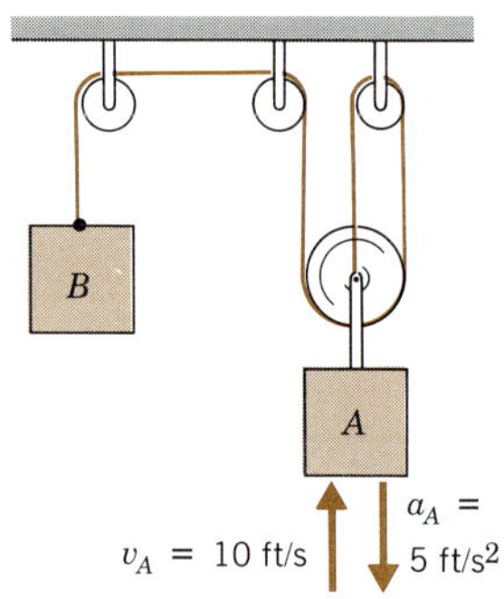

PROB. 11.148

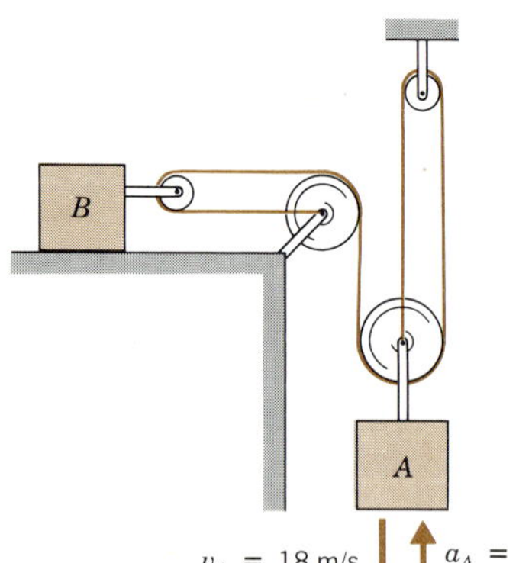

PROB. 11.149

11.150 Block A is given an acceleration of $0.5g$ to the right. Determine (a) the acceleration of block B and (b) the acceleration of block B with respect to block A.

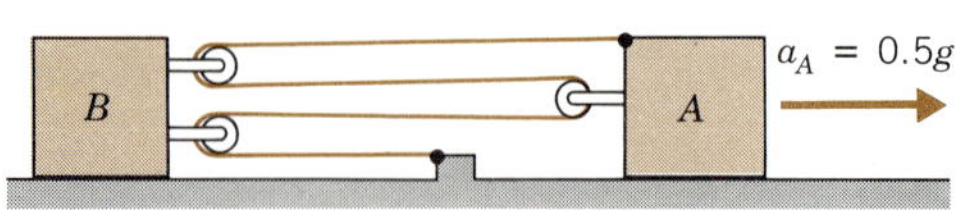

PROB. 11.150

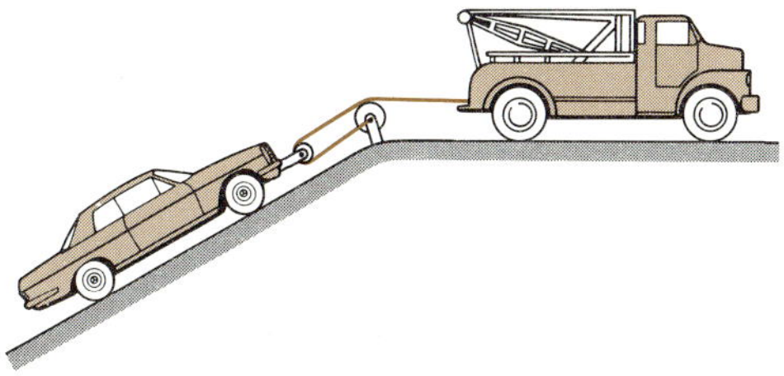

PROB. 11.151

11.151 The tow truck A pulls the automobile B out of a ravine by driving forward. Starting from the position shown, the distance the truck moves is given by $s = t^2/2 - t^3/100$, where s is measured in meters and t in seconds. Determine how far the automobile is pulled after 10 s and its velocity at the end of this time interval.

11.152 The electrically powered winch pulls the cable upward at the constant speed of 1 m/s. What is the velocity of the platform?

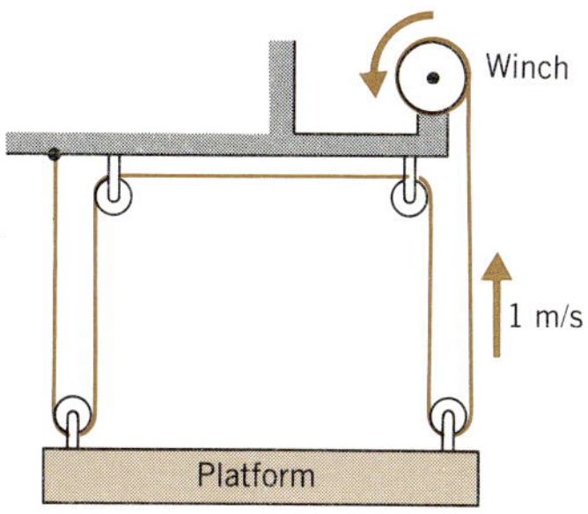

PROB. 11.152

11.153 Block B is moving downward with a speed v_B. What is the velocity of block A?

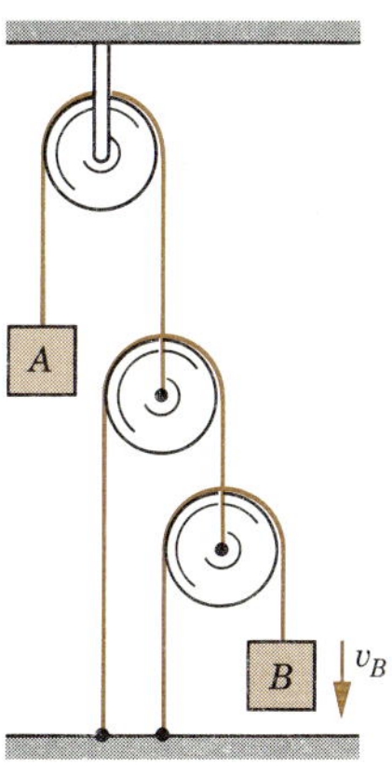

PROB. 11.153

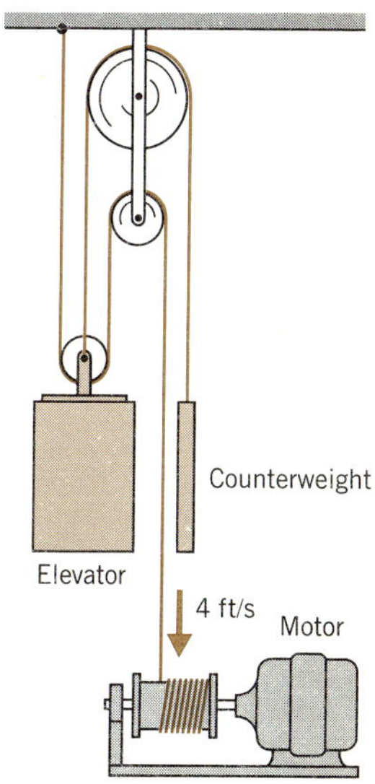

PROB. 11.154

11.154 The motor pulls the cable supporting the elevator downward at the constant rate of 4 ft/s. Determine (a) the velocity of the counterweight, and (b) the distance the counterweight moves during a 5 s interval.

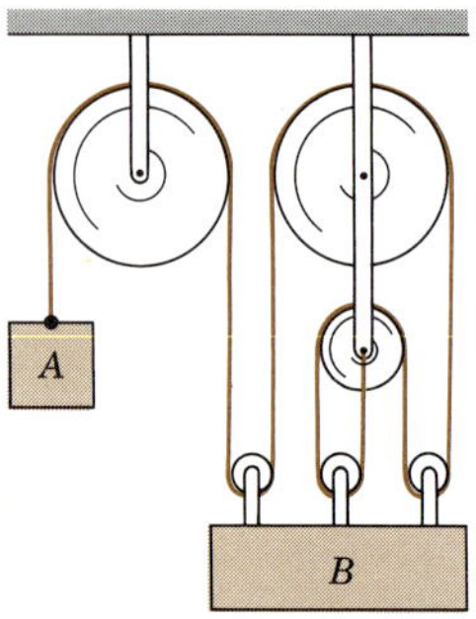

PROB. 11.155

***11.155** Block A is moving downward at 840 mm/s with respect to block B. Determine the absolute velocities of blocks A and B.

***11.156** Two alternative designs for an elevator are shown in the diagram. In design A the pulley attached to the motor winds the cable in with an absolute speed v. In design B the speed v with which the cable is wound in is with respect to the elevator. Compare the velocity of the elevator in each design.

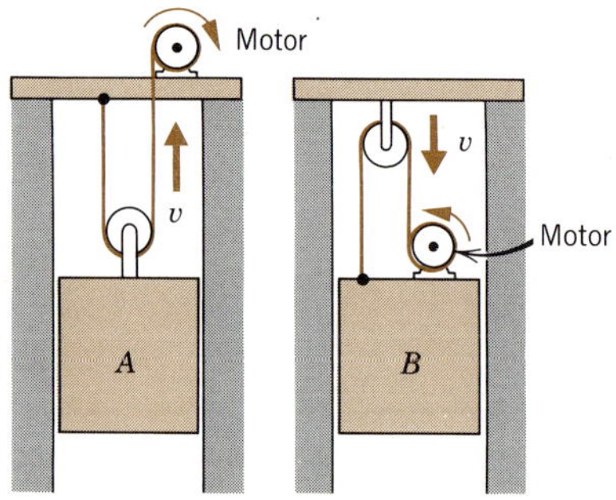

PROB. 11.156

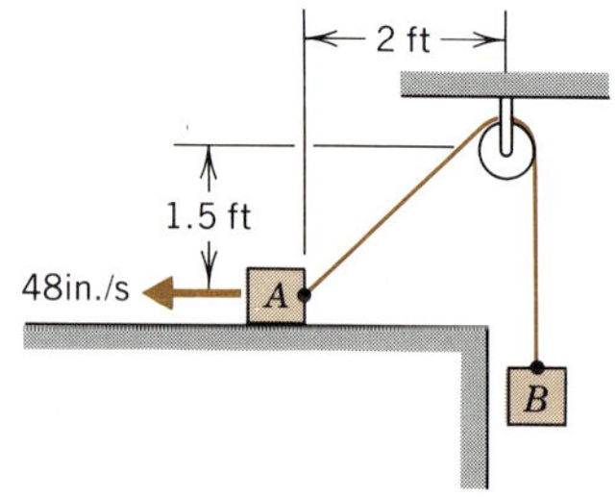

PROB. 11.157

***11.157** Block A has a constant velocity of 48 in./s to the left. For the position shown, determine the velocity of block B. (*Note:* The radius of the pulley may be neglected.)

***11.158** Block A is falling at 3 m/s in the illustrated position. Determine the corresponding velocity of block B. (*Note:* The radius of each pulley may be neglected.)

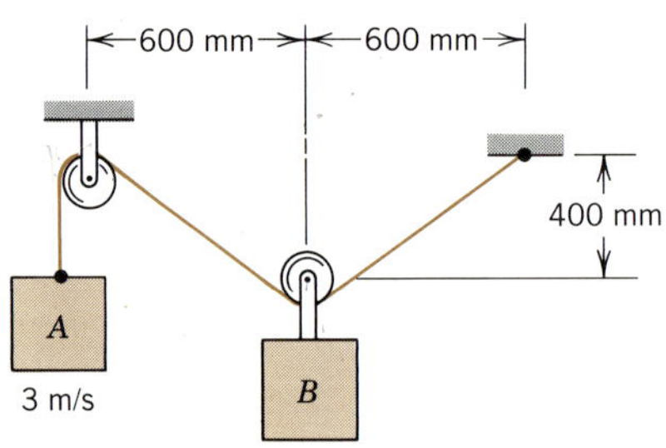

PROB. 11.158

CHAPTER 12

KINETICS OF A PARTICLE

OBJECTIVES

1. To sketch a free body diagram of an object that is modeled as a particle.

2. To obtain the equations of motion for a particle by using Newton's second law in conjunction with the kinematical description appropriate to the motion and the force system.

3. To solve the differential equation(s) describing the rectilinear motion of a particle in the case where variables may be separated.

4. To recognize similarities between the equations of rectilinear and curvilinear motion, in order to facilitate the use of constant acceleration formulas and the process of separating variables when evaluating the curvilinear motion of a particle.

5. To evaluate the work done by forces that depend on position, both by integration and by means of the potential energy of conservative forces.

6. To employ the work-energy principle in order to relate the velocity of a particle at two successive positions.

7. To employ the linear impulse-momentum principles in order to relate the velocity of a particle at two successive instants, with emphasis on impulsive force and central impact problems.

8. To employ the principle of conservation of angular momentum in order to relate transverse velocity components and angular velocities when there is a central force system acting on a particle.

9. To be able to recognize when Newton's second law and its first integrals should be jointly employed to evaluate the motion of a particle.

In this chapter we shall apply our knowledge of the kinematics of a particle to determine how the forces exerted on the particle affect its motion. The foundation for this study is Newton's second law of motion.

The resultant force, $\sum \bar{F}$, acting on a particle is related to the acceleration $\bar{a}$ of that particle by

$$\sum \bar{F} = m\bar{a} \tag{1}$$

We employ this relationship by expressing the forces and accelerations in terms of their components with respect to either Cartesian coordinates, cylindrical coordinates, or path variables. As one of our objectives, we shall learn how to recognize which of these coordinate descriptions is best suited for the formulation and solution of particular classes of dynamics problems.

Before continuing, it will prove beneficial to place Newton's two other laws of motion in proper perspective. The first law is embodied in the second law, reaffirming the fact that $\bar{a} = \bar{0}$ when $\sum \bar{F} = \bar{0}$. In effect, by stating this law separately, we are considering the study of statics to be independent of the study of dynamics. The third law proves most useful when we model physical systems as interacting, but independent, bodies. For such instances the action and reaction will be needed to represent the interaction forces correctly.

The foregoing brings to mind the concept of a free body diagram, in which we isolate a body from its surroundings, showing all forces exerted on the body of interest. This diagram serves two purposes. Drawing the diagram helps us to understand how the forces affect the body, and the resulting diagram aids us to account consistently for all the forces. The free body diagram is a primary tool in solving problems.

An important feature that arises in analyzing the motion of a system concerns the restrictions, called *constraints*, that are often placed on the motion. In the case of particle motion, the most common constraint is that the particle must follow a specific path. Such constraint conditions can be thought of as being imposed by the application of *reaction* forces. For example, consider the particle in Figure 1a, which is moving under the influence of a given force $\bar{F}_1$ along a smooth curved path in a *horizontal plane*. The free body diagram of the particle at a general point P must show the reaction force $\bar{N}$ that is normal to the path. This force has the effect of restraining the velocity of the particle to be tangent to the path. Without this force the particle would follow a different path from the prescribed one. Thus, Figure 1b is the free body diagram for this problem.

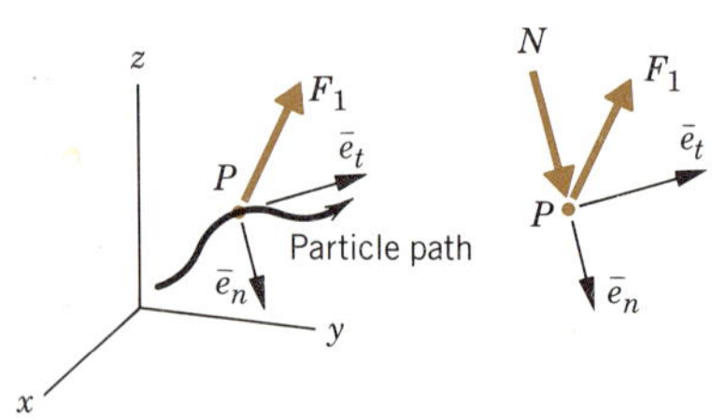

FIGURE 1

For ease in reading Figure 1, we have accentuated the arrows depicting forces in order to distinguish them from the arrows that describe unit vectors. Also note that the forces normal to the plane of motion, in this case the vertical forces, are not shown in Figure 1b. Included in these forces is the constraint force necessary to prevent motion (due to gravity) perpendicular to the plane. If we were interested in this force, we would draw a free body diagram showing the system as it is seen when viewed horizontally.

In general, an examination of the free body diagram and the given information will enable us to place the particular system into broad classes of problems. One class involves situations where the motion of the physical system is given and we seek the forces producing this motion. For a particle, this reduces to a kinematical study using the methods in Chapter 11 to determine $\bar{a}$. Then a free body diagram and $\sum \bar{F} = m\bar{a}$ will yield the necessary information about the forces.

In the second class of problems, the forces exerted on a physical system are given and we seek to determine the system's motion. For the case of a particle, we may determine its acceleration from $\bar{a} = \sum \bar{F}/m$. It is then necessary to follow an integration process (the opposite of differentiation) to determine the velocity and position vectors. A third class of problems combines some features of the two foregoing situations.

In this chapter, we shall study how to determine the curvilinear motion of a particle under the action of a given force system. However, in order to understand this process clearly, we will first consider rectilinear motion. The discussion of this motion in the previous chapter treated only constant acceleration, which corresponds to forces that have a constant effect. The situations that we will consider here have greater generality.

A. EQUATIONS OF RECTILINEAR MOTION

We begin the study of the motion of a particle along a straight path by drawing a free body diagram that shows *all* forces acting on the particle. To complete the sketch we also need a set of unit vectors, which of course involves a decision as to which kinematical formulation we should select to describe the motion. As always, this decision depends on how the information given in the problem is presented.

Let us assume that the given forces are functions of the path variable quantities, which are the distance s (traveled along the path), the speed v, and the time t. It is consistent in this case to formulate the problem in terms of path variables. We would then depict the unit vectors $\bar{e}_t$, $\bar{e}_n$, and $\bar{e}_b$ in the free body diagram. Recall that the binormal direction is $\bar{e}_b \equiv \bar{e}_t \times \bar{e}_n$, that is, the binormal is perpendicular to the plane formed by $\bar{e}_t$ and $\bar{e}_n$.

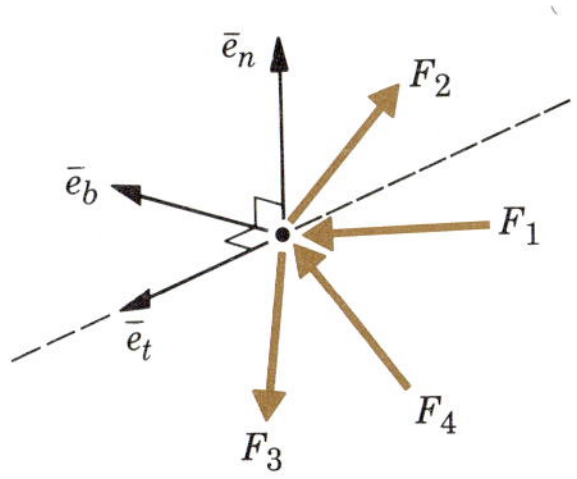

FIGURE 2

The general free body diagram for a particle in motion along a straight line appears in Figure 2. The vectors $\bar{F}_1$, $\bar{F}_2$, etc., denote the individual forces acting on the particle.

We now wish to apply Newton's second law to this problem. To do this we resolve all vectors into their components with respect to the directions of the unit vectors. The resultant force $\sum \bar{F}$ is the sum of the vectors $\bar{F}_i$,

$$\sum \bar{F} = \sum \bar{F}_i \equiv (\sum F_t)\bar{e}_t + (\sum F_n)\bar{e}_n + (\sum F_b)\bar{e}_b \tag{2}$$

where $\sum F_t$, $\sum F_n$, and $\sum F_b$ are the sum of the force components in the tangent and two normal directions, respectively. From kinematics we know that for rectilinear motion

$$\bar{a} = \dot{v}\bar{e}_t \tag{3}$$

Thus, because two vectors are equal only when their components are equal, the vector relation $\sum \bar{F} = m\bar{a}$ reduces to the scalar equations

$$\sum F_t = m\dot{v}$$
$$\sum F_n = 0 \qquad (4)$$
$$\sum F_b = 0$$

Equations (4) are the basic equations for rectilinear motion. The components perpendicular to the direction of motion express the fact that there must be a balance of forces in these directions, whereas the tangential component relates the forces to the rate of increase of the speed.

Equations (4) may be used to determine the forces for a known value of $\dot{v}$, or to determine $\dot{v}$ for a known set of forces. In the first case we have algebraic equations for the force components that can be solved by conventional methods. Here we shall concentrate on the latter case, in that it gives rise to an interesting situation. Let us assume that we know the functional dependence of the forces. Then, when we substitute the functions describing these forces into eq. (4), we obtain a *differential equation*, that is, an equation relating the derivative of a function to other functions. The solution of this equation will tell us how the speed is related to the elapsed time and to the distance traveled and how the distance traveled depends on the time. These data are defined by the functions $v(t)$, $v(s)$, and $s(t)$.

It will soon be obvious to you that we cannot determine these characteristics of the motion unless we know what state the particle was in prior to the application of the forces. Remember, we require a force to alter a state of motion. To define the initial state, we need *initial conditions*, which are the time t_0 when the motion was initiated, the distance s_0 describing the initial location of the particle, and the initial speed v_0.

B. SEPARATION OF VARIABLES

In general, there are many methods, both analytical and computerized, for solving the differential equation of motion obtained from equation (4). An effective approach is the method of separating variables. This method, which is applicable in situations where the differential equation contains only two variables, regroups the terms into integrable combinations. In the sections that follow, we will describe several common situations for which this method is convenient for problem solving.

1. General Procedure

The situations that we will consider may be discussed in terms of a general equation. Let β denote the independent variable (i.e., the derivatives in the equation are with respect to β), and let w denote the dependent variable (i.e., w depends on β, $w = w(\beta)$). Suppose that the differential equation of motion has the form

$$\frac{dw}{d\beta} = f(\beta)g(w) \qquad (5)$$

where $f(\beta)$ and $g(w)$ denote arbitrary functions.

The expression "separating variables" refers to the process of manipulating this equation in order to have terms containing only w on one side of the equation, and terms containing only β on the other side. For the situation in eq. (5) this may be achieved by considering the derivative to be a ratio of the differentials dw and $d\beta$. Thus, eq. (5) may be rewritten in the equivalent form

$$\frac{dw}{g(w)} = f(\beta)\,d\beta \tag{6}$$

The significance of eq. (6) is that each term is a differential quantity that may be integrated. *Definite integrals* are useful in dynamics studies. The *lower limit* in each integral will correspond to the *initial value* of the variable being integrated; let these be denoted as β_0 and w_0 for the case of eq. (6). The *upper limit* will be the *general value* of the variable. Integrating eq. (6) then yields

$$\int_{w_0}^{w} \frac{dw}{g(w)} = \int_{\beta_0}^{\beta} f(\beta)\,d\beta \tag{7}$$

These integrals may be evaluated, when the functions $f(\beta)$ and $g(w)$ are specified. The result is an *algebraic relation* for the value of w corresponding to the value of β.

2. Acceleration as a Function of Time

This case arises when it is known how the forces causing a body to move vary with time. Let $a(t)$ denote the acceleration function. Then the equation of motion is

$$\dot{v} \equiv \frac{dv}{dt} = a(t) \tag{8}$$

The dependent variable in this case is v. Separating variables according to the procedure in the previous section yields $dv = a(t)dt$. Integrating

$$\boxed{\int_{v_0}^{v} dv \equiv v - v_0 = \int_{t_0}^{t} a(t)\,dt} \tag{9}$$

After the integral is evaluated for a specified $a(t)$, it is a simple matter to solve the equation for v as a function of t. Note that the initial time t_0 is usually, but not always, $t_0 = 0$.

EXAMPLE 1

Starting from rest when $t = 0$, a sprinter accelerates at the rate $a = kt^{1/2}$, where k is a constant. If the speed of the runner when $t = 1$ second is 20 ft/s, determine k.

Solution

We will evaluate the relationship between v and t in terms of k, and then determine k by setting $v = 20$ ft/s when $t = 1$ s. The given initial conditions are $v = 0$ when $t = 0$, so

$$\dot{v} = \frac{dv}{dt} = kt^{1/2}$$

$$\int_0^v dv = \int_0^t kt^{1/2}\,dt \Rightarrow v = \tfrac{2}{3}kt^{3/2}$$

Substituting the final value of v yields

$$20 \text{ ft/s} = \tfrac{2}{3}k\,(1 \text{ s})^{3/2}$$

$$k = 30 \text{ ft/s}^{5/2}$$

(Note that the units of k could have been obtained from the given equation for the acceleration.)

EXAMPLE 2

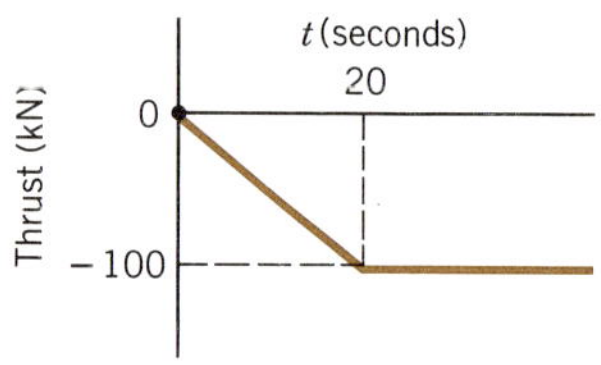

A 25,000 kg airplane lands at a speed of 200 km/hr, at which time the engines begin to produce the reverse thrust T described in the graph. Draw the corresponding graph showing the speed of the aircraft as a function of time.

Solution

In order to relate the thrust to the acceleration we draw a free body diagram that shows the reverse thrust acting opposite the direction $\bar{e}_t$ of the rectilinear motion.

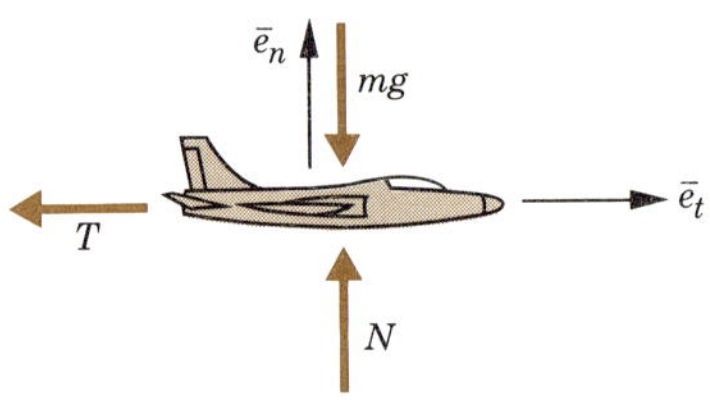

The force equation for the tangential direction is

$$\sum F_t = -T = m\dot{v} \Rightarrow \dot{v} = -\frac{T}{m}$$

The graph shows that the thrust is given by different functions for times before and after 20 s.

$$T = \begin{cases} -\frac{100}{20}t \text{ kN}; & 0 \le t \le 20 \text{ s} \\ -100 \text{ kN}; & t \ge 20 \text{ s} \end{cases}$$

It is therefore necessary to evaluate separately the motion for each time interval.

The initial instant for the first interval is $t_0 = 0$, and the initial speed is given to be

$$v_0 = \frac{200}{3.6} = 55.56 \text{ m/s}$$

The mass $m = 25{,}000$ kg, so the interval $0 \leq t \leq 20$ is described by

$$\dot{v} = \frac{dv}{dt} = -\frac{T}{m} = -\frac{(5t)(10^3)(\text{N})}{25(10^3)(\text{kg})}$$

$$\int_{55.65}^{v} dv = \int_0^t (-0.2t)\,dt$$

$$v - 55.56 = -0.1t^2 \text{ m/s}$$

We see from this that plotting v versus t will produce a parabola. The speed at the end of this interval, $t = 20$ s, is

$$v_1 = 55.56 - 0.1(20)^2 = 15.56 \text{ m/s}$$

Because $v_1 > 0$, more than 20 s are required to bring the aircraft to rest. It is therefore necessary to consider the second time interval, for which the initial time is $t = 20$ s. The speed at that instant is the value v_1 given below. Thus for $t > 20$,

$$\dot{v} = \frac{dv}{dt} = -\frac{T}{m} = -\frac{100(10^3)(\text{N})}{25(10^3)(\text{kg})} = -4 \text{ m/s}^2$$

$$\int_{15.56}^{v} dv = \int_{20}^{t} (-4)\,dt$$

$$v - 15.56 = -4(t - 20)$$

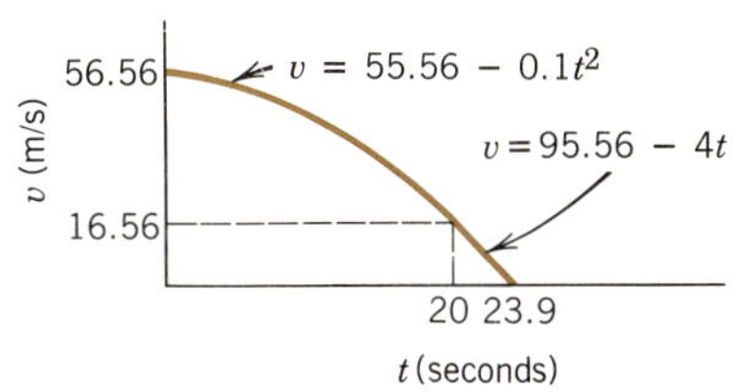

This result indicates that the plot of v versus t for the second time interval is a straight line. The airplane stops when $v = 0$. Thus, from above

$$t = \frac{15.56 + 80}{4} = 23.9 \text{ s}$$

Plotting the values of v when $t = 0$, 20, and 23.9 yields the accompanying graph.

HOMEWORK PROBLEMS

12.1 and 12.2 A block on the ground is initially at rest when it is given the acceleration shown. Determine the velocity of the block when $t = 5$ seconds and $t = 15$ seconds.

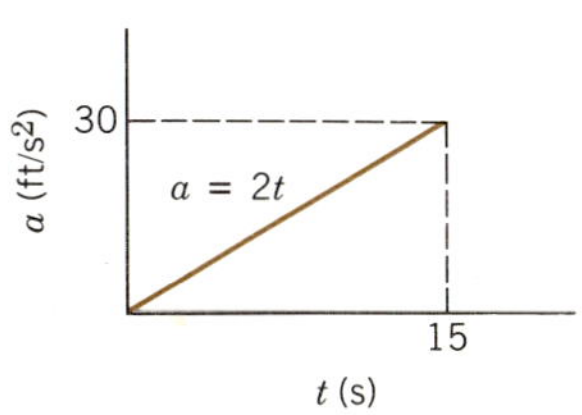

PROB. 12.1

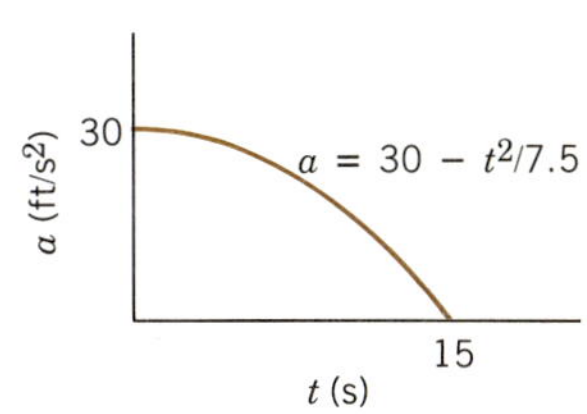

PROB. 12.2

12.3 A 500 gram iron block is pulled to the right by an electromagnetic force $F = 50 - 4t$, where F is in newtons and t is in seconds. The block is initially moving to the left at 20 m/s. Determine the time required to bring the block to rest.

12.4 The block in Problem 12.3 is initially at rest when it is subjected to the given force. Determine the maximum speed obtained by the block, and the corresponding elapsed time.

12.5 The acceleration of a particle starting from rest along a straight line is known to follow the relation

$$a = k_n t^n$$

where k_n is a constant that depends on the nonnegative integer n. It is also known that the distance $s = l$ when $t = \tau$ seconds. Determine (a) an expression for k_n in terms of l, τ, and n, and (b) the speed of the particle when $t = \tau$. Does this speed increase or decrease with increasing values of n?

12.6 A 2 kg ball in vertical motion is falling under the influence of gravity at a speed of 10 m/s when it is subjected to an additional downward force $F = 10t$, where F is in newtons and t is in seconds. Determine the time at which the sphere will attain a speed 40 m/s.

12.7 The thrust T of a toy rocket that follows a straight vertical path is shown in the diagram. The toy weighs 2 lb and rests on its platform until the thrust is sufficient to launch it. Determine the maximum speed of the toy.

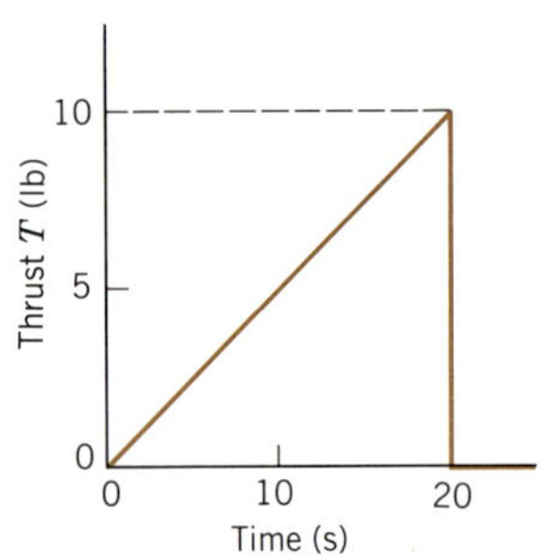

PROB. 12.7

12.8–12.10 An automobile accelerates from rest along a level road according to the graph shown. Determine the velocity of the automobile when $t = 10$ and 20 s, and sketch the velocity as a function of time.

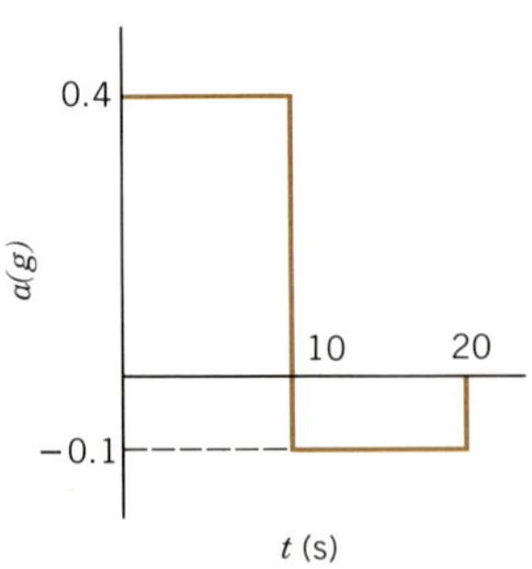

PROB. 12.8

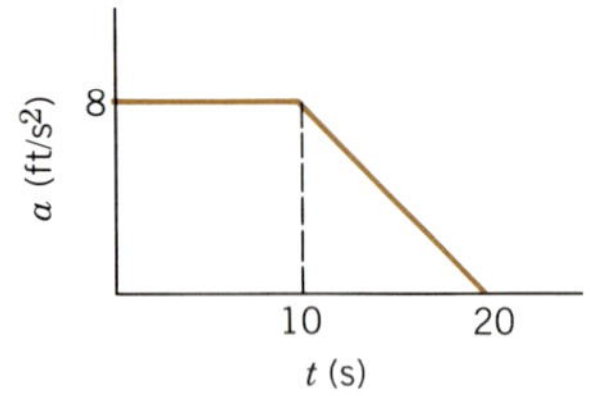

PROB. 12.9

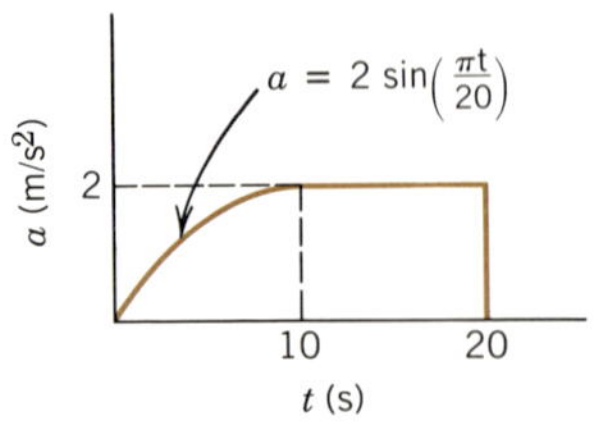

PROB. 12.10

12.11 Experimental measurement of the contact force F exerted between a 48 gram golf ball and a club indicates that the force-time plot is as shown. However, because of an error the scale factor for the force was lost, so the value of F_{max} is unknown. If the velocity of the ball after the impact was 70 m/s, determine the value of F_{max}.

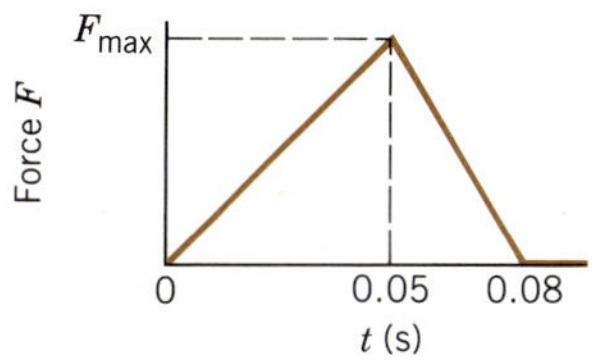

PROB. 12.11

---

3. Acceleration as a Function of Position

Often, the data provided for a situation is not explicitly measured in terms of time. For example, the tangential component of force can be measured in terms of the position of a body. Newton's second law, eqs. (4), then states that the acceleration depends on the distance s that the body has moved along its path from some reference position.

Let $a(s)$ denote the acceleration function, so $\dot{v} = a(s)$. We do not wish to use $\dot{v} = dv/dt$ to separate variables, for that would explicitly introduce the variable t. Instead, we employ the chain rule for differentiation, because the acceleration and velocity depend on s. Hence,

$$\dot{v} = \frac{dv}{ds}\frac{ds}{dt} \equiv v\frac{dv}{ds} = a(s) \tag{10}$$

Separating variables is achieved by multiplying by ds. The lower limits in the integrals are the speed v_0 at the initial position s_0.

$$v\,dv = a(s)\,ds$$

$$\boxed{\int_{v_0}^{v} v\,dv \equiv \tfrac{1}{2}(v^2 - v_0^2) = \int_{s_0}^{s} a(s)\,ds} \tag{11}$$

An important consideration here is the matter of positive direction. The chain rule, eq. (11), uses $ds/dt = +v$; in other words, positive v is in the sense of increasing s. This means that

> The tangential direction $\bar{e}_t$, which is used for defining positive vector components, should be shown in the free body diagram to be in the sense of increasing s.

EXAMPLE 3
The 0.5 kg block is released from rest, and slides down the smooth 30° incline. In its inital position, the spring, whose stiffness is 30 N/m, is elongated by 400 mm. Determine the speed of the block when it has moved 200 mm down the incline.

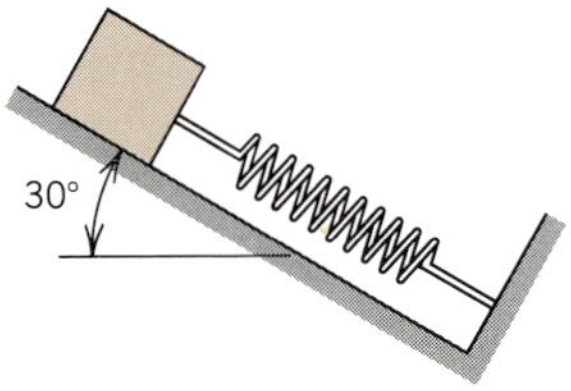

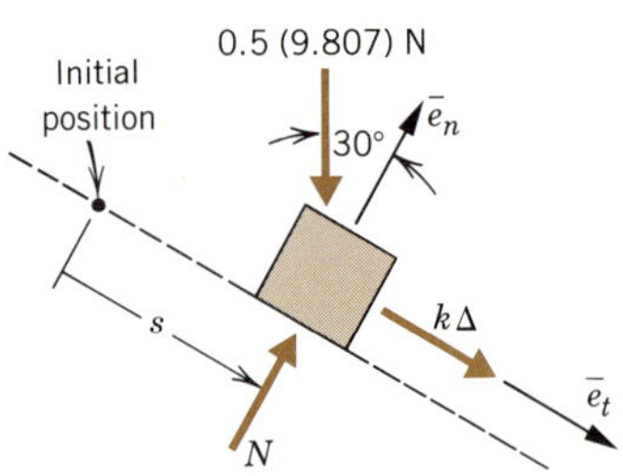

Solution
The magnitude of the spring force is $k\Delta$, where k is the stiffness and Δ is the elongation. This force is considered in the free body diagram to pull the block down the hill, because it is given that the spring is elongated initially.

The position of the block is measured in terms of the distance s from the initial location, and $\bar{e}_t$ is defined in the direction of increasing s. In order to describe the elongation Δ, we note that moving the block a distance s will *shorten* the spring. Then, because $\Delta = 0.4$ m in the initial position, the general relation is

$$\Delta = 0.4 - s \text{ m}$$

Now that the spring force has been characterized, the equations of motion may be formulated. Using the unit vectors in the free body diagram to define positive vector components yields

$$\sum F_t = k\Delta + 0.5(9.807) \sin 30^\circ$$
$$\equiv 30(0.4 - s) + 2.452 = 0.5\dot{v}$$
$$\sum F_n = N - 0.5(9.807) \cos 30^\circ = 0$$

The normal equation merely gives the force $\bar{N}$, which we do not need to evaluate. The tangential equation clearly shows that the acceleration is a function of s, so we employ the chain rule to separate variables. The given initial conditions are that $v_0 = 0$ when $s = 0$, which leads to

$$\dot{v} = v\frac{dv}{ds} = \frac{30}{0.5}(0.4 - s) + \frac{2.452}{0.5}$$
$$\int_0^v v\,dv = \int_0^s [28.904 - 60s]\,ds$$
$$\tfrac{1}{2}v^2 = 28.904s - 30s^2$$

The velocity at $s = 0.2$ m is readily calculated to be

$$v = [2\,(28.904)(0.2) - 2(30)(0.2)^2]^{1/2} = 3.03 \text{ m/s}$$

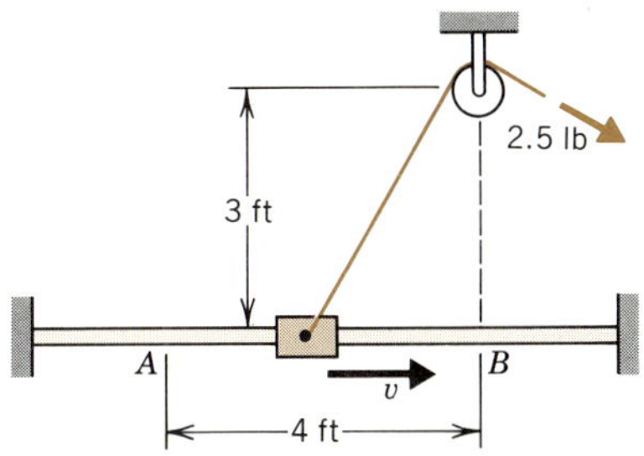

EXAMPLE 4

The 2 lb collar is pulled along the smooth horizontal bar by a cable that carries a 2.5 lb tensile force. The collar is initially at rest in position A. Determine the speed of the collar as it passes position B.

Solution

The free body diagram indicates that the only force having unknown magnitude is the reaction $\bar{N}$. The other forces have constant magnitude. However, the component of the resultant force in the tangential direction is not constant because the angle θ for the cable is variable.

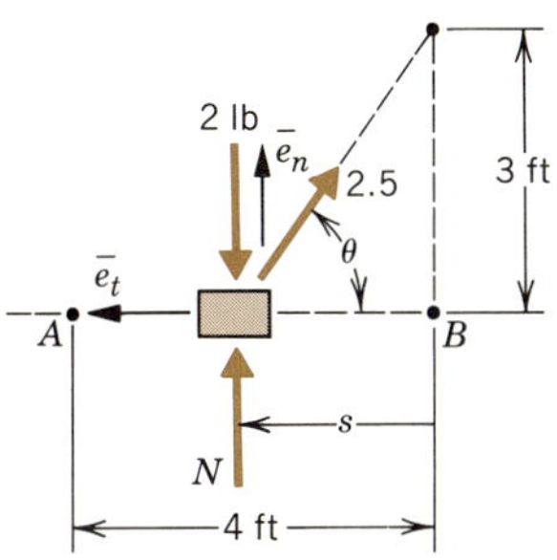

The angle θ is readily related to the distance s from point B, so this distance is selected as the path variable parameter. The unit vector $\bar{e}_t$ is defined in the sense of increasing s. The equations of motion for this system are

$$\sum F_t = -2.5 \cos\theta = \frac{2}{32.17}\dot{v}$$

$$\sum F_n = N - 2 + 2.5 \sin\theta = 0$$

The normal equation merely defines the reaction $\bar{N}$. In order to proceed further with the tangential force equation, we use the right triangle in the free body diagram to write

$$\cos\theta = \frac{s}{(s^2 + 3^2)^{1/2}}$$

Thus

$$\dot{v} = v\frac{dv}{ds} = \frac{32.17}{2}(-2.5)\frac{s}{(s^2 + 3^2)^{1/2}}$$

The initial position is point A, where $s_0 = 4$ ft, and the corresponding initial speed is $v_0 = 0$. Separating variables by multiplying by ds yields

$$\int_0^v v\,dv = \int_4^s \left[-\frac{40.21s}{(s^2 + 9)^{1/2}}\right] ds$$

$$\tfrac{1}{2}v^2 = -40.21(s^2 + 9)^{1/2}\Big|_4^s = 40.21[5 - (s^2 + 9)^{1/2}]$$

(A table of integrals will often be useful for the less familiar integrals.) Position B corresponds to $s = 0$, from which it follows that

$$v_B = [2(40.21)(5 - 3)]^{1/2} = 12.68 \text{ ft/s}$$

Additional Remarks

You might have noticed in Examples 3 and 4 that the normal force equation was not needed for evaluating the motion. However, if dry friction were present, then a friction force $\mu_k N$ would act in the opposite direction from the velocity. The magnitude of the reaction force $\bar{N}$ would then be obtained from the normal force equation.

HOMEWORK PROBLEMS

12.12 The traction force acting on a 900 kg automobile varies with the distance s traveled along a level road according to $F = 0.6 \sin(\pi s/2000)$, where s is in meters and F is in kilonewtons. The automobile starts from rest. Determine the speed of the automobile at $s = 1500$ m.

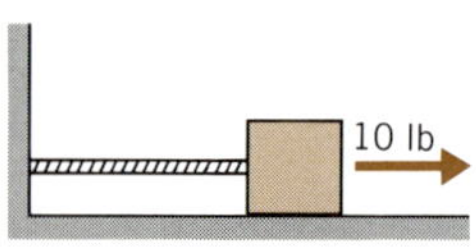

PROB. 12.13

12.13 The 10 lb force pulls the 5 lb block to the right. The tensile force exerted by the rubber band in opposition to this movement is $F = 12s + 8s^3$, where F is in pounds and s is the distance traveled in feet. Determine the speed of the block when $s = 9$ inches.

12.14 Solve Example 3 if the coefficient of kinetic friction between the ground and the block is $\mu_k = 0.2$.

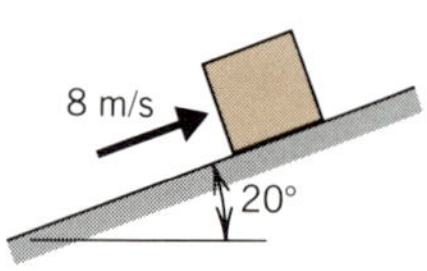

PROB. 12.15

12.15 The 1.2 kg block is given an initial speed of 8 m/s up the 20° incline. The coefficient of kinetic friction is $\mu_k = 0.15$. How far will the block slide up the incline before it stops?

12.16 The coefficient of static friction between the 500 lb crate and the flat bed trailer is 0.3. The truck has an initial speed of 30 mi/hr. Determine the shortest distance in which the truck may be stopped without causing the crate to slide.

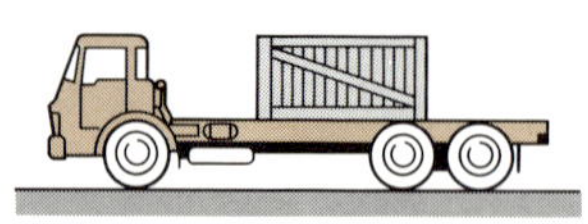

PROB. 12.16

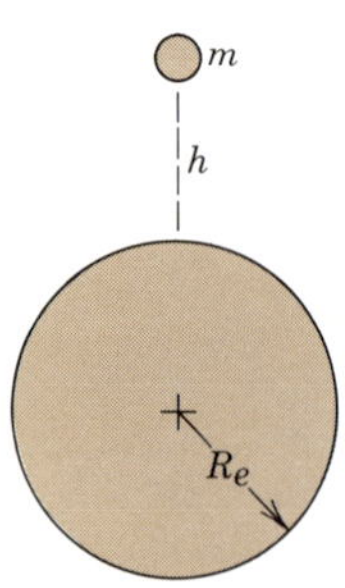

PROB. 12.17

12.17 The inverse square law for the gravitational attraction exerted on a particle of mass m is

$$F = \frac{mg}{(1 + h/R_e)^2}$$

where $g = 32.17$ ft/s^2, h is the altitude of the particle above the surface of the earth, and $R_e = 2.091(10^7)$ ft is the mean radius of the earth. In a hypothetical situation, a particle is released from rest at a height $h = 20$ mi. Neglecting air resistance, determine the velocity of the particle when it hits the earth. Compare this result to the value obtained by considering the gravity force to be simply mg.

12.18 A steel block weighing 800 lb is to be picked up by an electromagnet whose attractive force in pounds is given by the inverse square formula $F = 3(10^4)/s^2$ lb, where s is the distance in feet. The block was initially at $s = 5$ ft. Determine the velocity with which the block strikes the electromagnet.

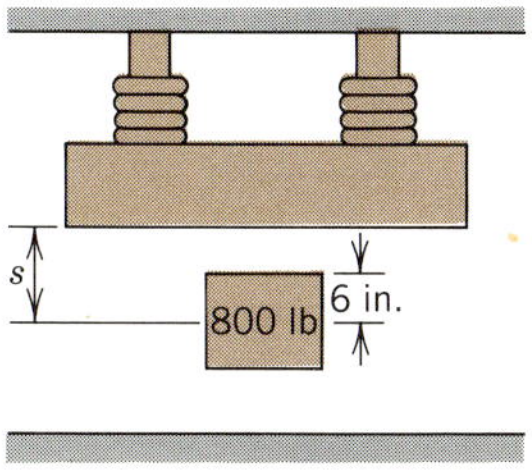

PROB. 12.18

12.19 The 10 kg block is pulled upward from the rest position shown by a constant cable tension of 800 N. Determine the velocity of the block when it reaches the elevation of the pulley if friction is negligible.

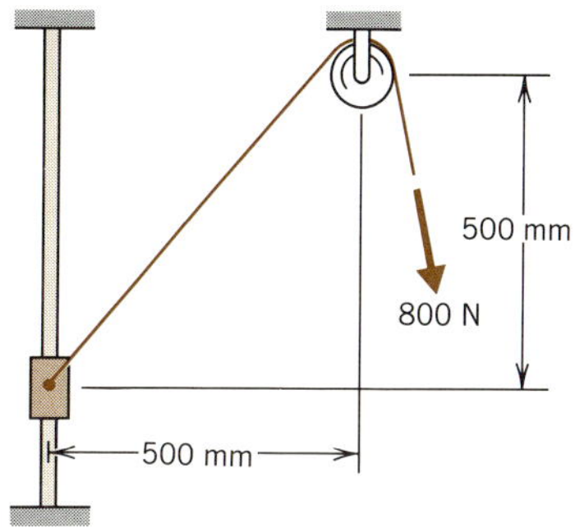

PROB. 12.19

***12.20** Solve Problem 12.19 if the coefficient of kinetic friction between the block and the bar is $\mu_k = 0.16$.

12.21 Solve Example 4 if the coefficient of kinetic friction between the collar and the block is $\mu_k = 0.1$.

***12.22** A 2 lb collar slides over a horizontal rod under the action of a linear spring whose stiffness is 20 lb/ft. There is no force in the spring when the collar is at point A. In a certain experiment the collar is released from rest at point B. Determine the velocity of the collar when it passes point A. Friction is negligible.

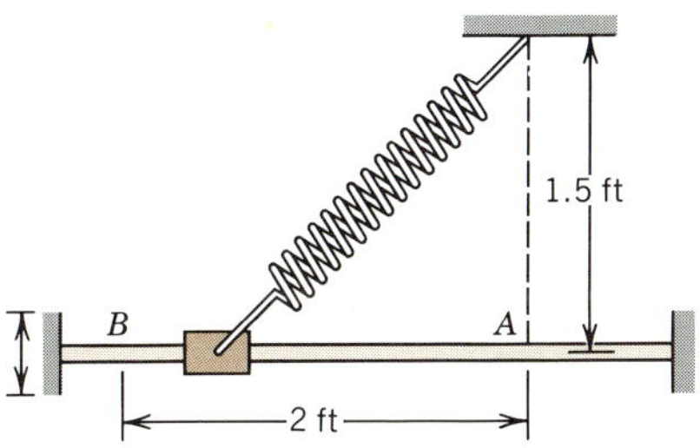

PROB. 12.22

4. Acceleration as a Function of Velocity

Friction forces are always dependent on the velocity because they are in the sense that is opposite to the movement of the body. In addition, their magnitude is often dependent on the speed of the body, most commonly in the case of the frictional resistance of a fluid. Other situations such as this arise in electromechanical systems. Summing forces in the tangential direction in these situations yields an acceleration that is a function of speed, $\dot{v} = a(v)$.

In this case the acceleration function does not contain an independent variable, so we may consider v to depend on either t or s. Considering v to depend on t yields

$$\dot{v} = \frac{dv}{dt} = a(v) \tag{12}$$

In order to separate variables here we multiply by dt and divide by $a(v)$. Definite integration, with the initial conditions v_0 and t_0 as the lower limits, yields an equation relating the speed v to the elapsed time.

$$\int_{v_0}^{v} \frac{dv}{a(v)} = \int_{t_0}^{t} dt = t - t_0 \tag{13}$$

Instead of using eq. (12) we may consider v to depend on s. The chain rule for differentiation then gives

$$\dot{v} = v\frac{dv}{ds} = a(v) \tag{14}$$

Separating variables here requires multiplying by ds and dividing by $a(v)$. The lower limits for the integrals are then initial position s_0 and initial speed v_0. The result is now an equation relating the speed to the distance traveled.

$$\int_{v_0}^{v} \frac{v\,dv}{a(v)} = \int_{s_0}^{s} ds \equiv s - s_0 \tag{15}$$

The value of v corresponding to a specified t or s may be found by solving eq. (13) or (15). The choice as to which formulation should be employed depends on the type of information given, as well as that which must be found in the solution.

EXAMPLE 5

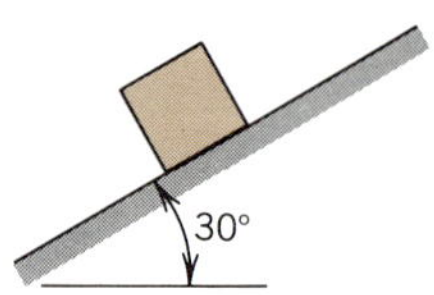

The viscous friction between a 30° inclined plane coated by a layer of lubricant and a 10 lb block is given by $f = 0.25v$, where f is expressed in pounds and v is in feet per second. The block is released from rest. Determine the terminal speed and the time required to attain one half of the terminal speed.

Solution

The *terminal speed* is the constant speed that would be attained eventually if the motion were to continue for a long time. This is the speed at which the force causing the motion (gravity for this system) has the same tangential component as the friction force opposing the motion. Because there is no resultant tangential force, the terminal speed may be found by setting $\dot{v} = 0$.

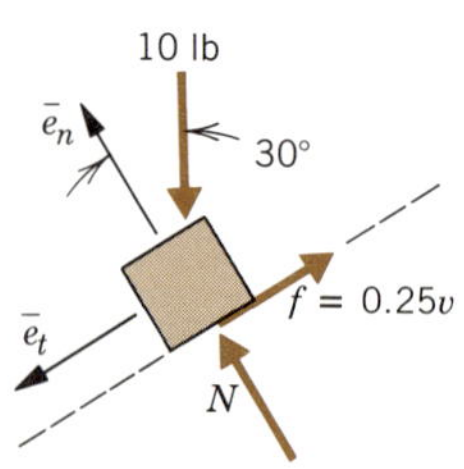

The acceleration $\dot{v}$ is found by using the free body diagram to form the components of $\sum \bar{F} = m\bar{a}$. Note that the friction force $\bar{f}$ is opposite to the sense of the motion, which is defined by $\bar{e}_t$. The tangential force equation is

$$\sum F_t = 10 \sin 30^\circ - 0.25v = \frac{10}{32.17}\dot{v}$$

The normal force equation is omitted because it describes only the normal reaction.

The terminal speed is found by letting $\dot{v} = 0$, which yields

$$v_{\text{term}} = \frac{10 \sin 30^\circ}{0.25} = 20 \text{ ft/s}$$

It is now necessary to determine the time required to attain $v = 0.5v_{\text{term}} \equiv 10$ ft/s. In order to relate v and t, we write $\dot{v} = dv/dt$ and separate variables. The result for initial conditions $v_0 = 0$ when $t = 0$ is

$$\dot{v} = \frac{dv}{dt} = \frac{32.17}{10}(10 \sin 30^\circ - 0.25v)$$

$$\int_0^v \frac{dv}{(16.085 - 0.8043v)} = \int_0^t dt = t$$

$$t = -\frac{1}{0.8043} \ln(16.085 - 0.8043v)\bigg|_0^v$$

Because the difference of natural logarithms of two quantities is the natural logarithm of the quotient, this result may be written as

$$t = \frac{1}{0.8043} \ln\left(\frac{16.085}{16.085 - 0.8043v}\right)$$

For $v = 10$ ft/s, this yields

$$t = \frac{1}{0.8043} \ln\left(\frac{16.085}{8.042}\right) = 0.862 \text{ s}$$

One check of the equation relating v and t is to verify the terminal speed by letting $t \to \infty$. In order for t to be infinite, the denominator in the natural logarithm term must also be infinite. The value v_{term} giving this condition agrees with the earlier value.

EXAMPLE 6

The block in Example 5 is given an initial speed of twice the terminal speed, $v_0 = 2v_{\text{term}}$. Determine how far the block will have traveled when its speed is 5% larger than the terminal value.

Solution

The acceleration as a function of velocity, as well as the terminal speed, was found in the previous example. Here we need to relate speed to distance traveled, so we use the chain rule.

$$\dot{v} = v\frac{dv}{ds} = \frac{32.17}{10}(10 \sin 30^\circ - 0.25v)$$

The initial conditions are that $v_0 = 2v_{\text{term}} = 40$ ft/s when $s_0 = 0$, so separating variables and integrating yields

$$\int_{40}^v \frac{v\,dv}{(16.085 - 0.8043v)} = \int_0^s ds$$

$$s = \frac{1}{(0.8043)^2}\left[(16.085 - 0.8043v) - 16.085\ln(16.085 - 0.8043v)\right]\Big|_{v=40}^{v}$$

$$= 1.2433(40 - v) + 24.86\ln\left[\frac{16.085 - 0.8043(40)}{16.085 - 0.8043v}\right]$$

The value of s when $v = 1.05\,v_{term} = 21$ ft/s is computed to be

$$s = 98.1 \text{ ft}$$

HOMEWORK PROBLEMS

PROB. 12.23

12.23 The resistance of the water to the motion of a 500 lb speedboat is given by $f = 0.20v$, where f is expressed in pounds and v is in feet per second. The speedboat's engine is turned off when the boat attains a speed of 60 mi/hr. Determine the time required to reduce the speed of the boat to 10 mi/hr.

12.24 For the speedboat in Problem 12.23, determine how far the boat will coast while its speed is reduced from 60 mi/hr to 10 mi/hr.

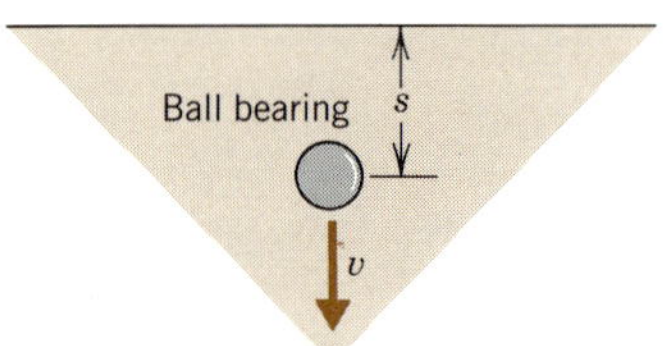

PROBS. 12.25 AND 12.26

12.25 The frictional resistance on a 2 oz ball bearing falling through oil is given by $f = 0.03v$, where f is expressed in pounds and v is in feet per second. The ball bearing is released from rest. Determine the terminal speed of the ball bearing and the time required to attain 75% of that speed.

12.26 The frictional resistance on a ball bearing falling through oil is known to be proportional to the speed of the ball bearing, $f = kv$. It is also known that the terminal speed is 800 mm/s. The ball bearing is released from rest. Determine the distance the ball bearing will descend before it attains 90% of its terminal speed.

12.27 The ball bearing in Problem 12.26 is dropped into the oil with an initial speed of 2 m/s. How long will it take to reduce the speed of the ball bearing to 1 m/s?

12.28 The air resistance on a 200 gram ball in free flight is given by $f = 2(10^{-4})v^2$, where f is in newtons and v is in meters per second. If the ball is dropped from rest 500 m above the ground, determine the speed at which it hits the ground. What percentage of the terminal speed is the result?

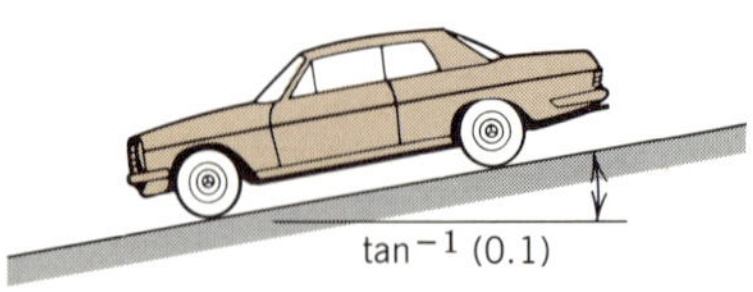

PROB. 12.29

12.29 The brakes of an automobile whose mass is 1500 kg are released and the automobile rolls down a 10% grade. The air resistance is kv^2, where $k = 2.5$ N-s^2/m^2, and the rolling resistance of the tires is $0.01N$, where N is the resultant normal force exerted by the ground. Determine the speed of the automobile after it has rolled 100 m.

12.30 Determine the speed of the automobile in Problem 12.29 20 s after it began to roll. What percentage of the terminal speed is the result?

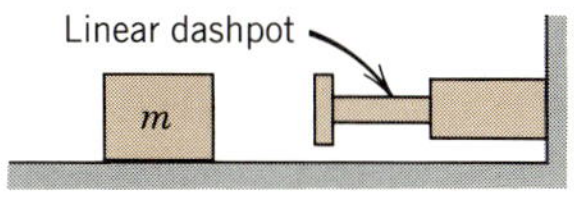

PROB. 12.31

12.31 A block of mass m is sliding on a smooth horizontal surface at a speed v, when it hits a shock-absorbing device called a *linear dashpot*. The dashpot slows the block by exerting a retarding force $f = kv$, where v is the speed of the block and k is a constant. Derive expressions for (a) v as a function of elapsed time t, and (b) v as a function of the distance s the block moves after it contacts the dashpot.

12.32 A 200 gram ball is thrown vertically upward from the ground with a speed of 25 m/s. The air resistance is $f = 2(10^{-4})v^2$, where f is in newtons and v is in meters per second. Determine the maximum height attained by the ball.

*12.33 In Problem 12.32, determine the speed with which the ball returns to the ground and the total elapsed time from the instant when the ball is released to the instant when it returns to the ground.

*12.34 The resistance of the water to the motion of a $5(10^4)$ ton ship is known to be proportional to the square of the speed. It is also known that the propellers must produce a thrust of a quarter of a million pounds to give the ship a speed of 40 ft/s. The ship is traveling at 40 ft/s when the engine is turned off and the ship slows to a speed of 10 ft/s. Determine (a) the distance the ship travels, and (b) the elapsed time.

---

*5. Relating Speed and Distance Traveled

In the preceding sections we solved the differential equation resulting from summing forces in the tangential direction. The result in each case was an expression relating the speed v to the elapsed time t, or to the distance traveled along the path s. A complete description of the motion requires the evaluation of an explicit relationship between s and t. Because $v \equiv \dot{s}$, this task gives rise to another differential equation.

Suppose that the function $v(t)$ has been obtained, and let s_0 be the initial position at time t_0. Separating variables in this case yields

$$v = \frac{ds}{dt} = v(t)$$

$$\boxed{\int_{s_0}^{s} ds \equiv s - s_0 = \int_{t_0}^{t} v(t)\,dt} \tag{16}$$

A similar approach is used when $v(s)$ is known. However, separating variables now requires division by $v(s)$. Thus

$$v = \frac{ds}{dt} = v(s)$$

$$\boxed{\int_{s_0}^{s} \frac{ds}{v(s)} = \int_{t_0}^{t} dt \equiv t - t_0} \tag{17}$$

The primary difference between eqs. (16) and (17) is that the first gives the value of s at a specified time t, whereas the second equation yields the time t required to reach a specified distance s. Regardless of which of these relations is derived, the other may be obtained by an algebraic solution.

EXAMPLE 7

Determine how far the airplane in Example 2 travels from its landing point until it comes to rest.

Solution

The velocity as a function of time was evaluated in Example 2. That result defined different expressions for the time intervals $0 \leq t \leq 20$ seconds and $t > 20$ s. These expressions were

$$v = 55.56 - 0.1t^2 \text{ m/s}; \qquad 0 \leq t \leq 20$$

$$v = 95.56 - 4t \text{ m/s}; \qquad t > 20$$

It was found from these results that $v = 0$ when $t = 23.9$ s, so now we must determine the distance traveled in both time intervals.

For the first time interval, the initial conditions are $s_0 = 0$ when $t = 0$. Separating variables using $ds/dt = v(t)$ yields

$$v = \frac{ds}{dt} = 55.56 - 0.1t^2$$

$$\int_0^s ds = \int_0^t (55.56 - 0.1t^2)\,dt$$

$$s = 55.56t - 0.03333t^3$$

The value of s at the end of this time interval is

$$s(t = 20) = 844.6 \text{ m}$$

The second time interval may now be studied, with the position when $t = 20$ being the initial conditions. Thus

$$v = \frac{ds}{dt} = 95.56 - 4t$$

$$\int_{844.6}^s ds = \int_{20}^t (95.56 - 4t)\,dt$$

$$s - 844.46 = 95.56(t - 20) - 2(t^2 - (20)^2)$$

This yields

$$s(t = 23.9) = 875 \text{ m}$$

EXAMPLE 8

It is desired to accelerate an automobile from rest along a straight road such that its speed v is related to the distance s it travels by $v = ks^{2/3}$, where k is a constant. Determine how the acceleration must vary with time in order to achieve this motion.

Solution

Because $v(s)$ is known, the acceleration may be found from the chain rule.

$$a = \dot{v} = v\frac{dv}{ds} = (ks^{2/3})(\tfrac{2}{3}ks^{-1/3}) = \tfrac{2}{3}k^2s^{1/3}$$

This gives a in terms of s. In order to find a in terms of t, it is necessary to evaluate s as a function of t. For this we use the given function for $v(s)$ as follows.

$$v = \frac{ds}{dt} = ks^{2/3}$$

The initial condition is $s = 0$ when $t = 0$, so separating variables yields

$$\int_0^s \frac{ds}{ks^{2/3}} = \int_0^t dt$$

$$\frac{3}{k}s^{1/3} = t; \qquad s^{1/3} = \frac{1}{3}kt$$

Substituting this result into the earlier expression for a yields

$$a = \tfrac{2}{3}k^2[(\tfrac{1}{3}kt)] = \tfrac{2}{9}k^3t$$

HOMEWORK PROBLEMS

12.35 An advanced spacecraft is designed to have a speed of $v = 4(10^{-5})t^2$ (v is in feet per second and t is in seconds) until it reaches 100,000 ft/s. How far will it travel from its position when $t = 0$ until it attains its final speed?

12.36 The velocity of a collar at the end of a spring is known to be $v = 250 \sin(5t)$, where t is in seconds and v is in millimeters per second. Sketch the distance traveled as a function of time, and use that graph to determine the largest distance the collar will travel from its initial position and the elapsed time for that movement.

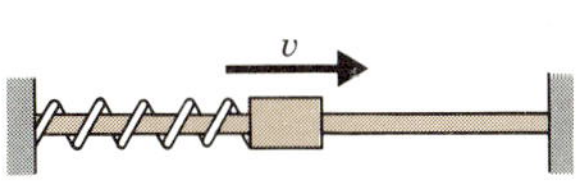

PROB. 12.36

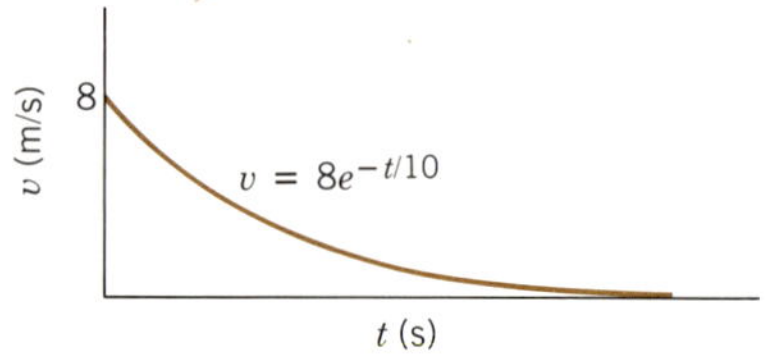

PROB. 12.37

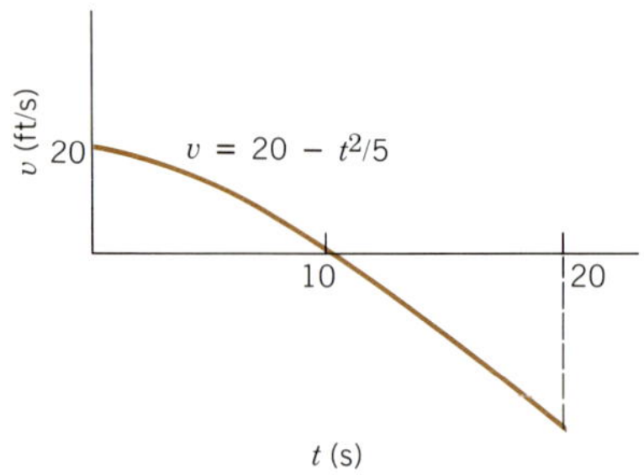

PROB. 12.38

12.37–12.40 A particle located at $s = 0$ when $t = 0$ moves along a straight path with the velocity shown. Sketch the distance traveled as a function of time for $t \le 20$ s.

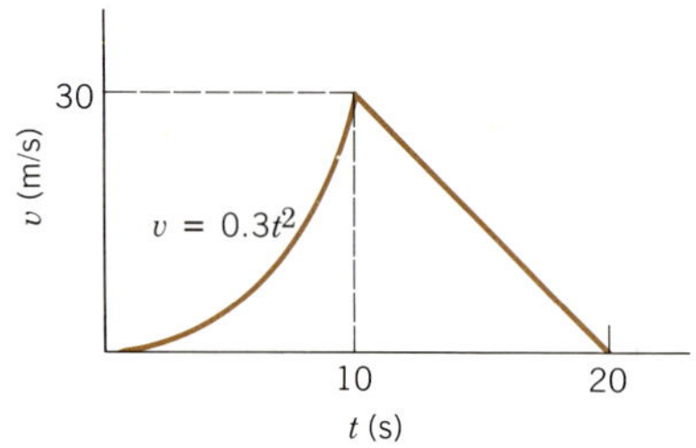

PROB. 12.39

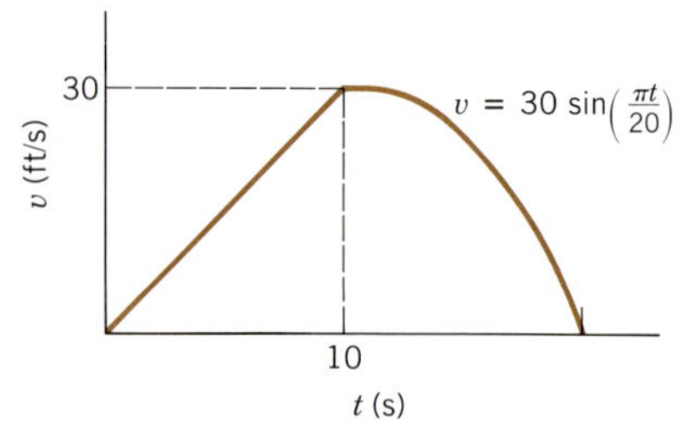

PROB. 12.40

12.41 A block at $s = 4$ m moves toward the origin with a speed of $v = -6s$ m/s. Determine the time to reach (a) $s = 1$ m, (b) $s = 0.25$ m. What is the acceleration of the block at each position?

12.42 The velocity of a small sphere that has been dropped into a viscous fluid is $v = 60\exp(-2s)$, where s is the depth in feet and v is in feet per second. Determine the time required for the sphere to reach a depth of 2 ft, and also find its acceleration at that position.

C. EQUATIONS OF CURVILINEAR MOTION

The study of the motion of a particle as it follows a curved path has much in common with the case of rectilinear motion. With the aid of a free body diagram, the scalar equations of motion are obtained from the components of Newton's second law, eq. (1). The kinematical description and corresponding unit vectors employed in this formulation should be selected to match the way in which the motion and forces are specified. This will become apparent in the developments that follow.

1. Path Variables

The path variable formulation describes the position along the path in terms of the arc length s. The tangent, normal, and binormal directions are used to describe the components of vector quantities. The binormal direction $\bar{e}_b$ has no acceleration component because it is defined to be perpendicular to the (instantaneous) plane of motion. The acceleration components in path variables were found in Chapter 11 to be

$$\bar{a} = \dot{v}\bar{e}_t + \frac{v^2}{\rho}\bar{e}_n + 0\bar{e}_b$$

where

$$\bar{v} = \dot{s}\bar{e}_t$$

As with the acceleration, the resultant force may be described in terms of path variables. Its component in each direction is obtained by summing the components of the individual forces appearing in the free body diagram. Formulating $\sum \bar{F} = m\bar{a}$ by matching components then yields the following three scalar equations.

$$\sum F_t = ma_t \equiv m\dot{v}$$
$$\sum F_n = ma_n \equiv m\frac{v^2}{\rho} \tag{18}$$
$$\sum F_b = ma_b \equiv 0$$

It is significant that, in comparison to eq. (4) for rectilinear motion, there is only one significant change due to curvature of the path. Specifically, the resultant force must now have a non-zero normal component because there is a centripetal component of acceleration.

In situations where $\sum F_t$ is known as a function of t, s, or v, the problem may give rise to a differential equation, as it did in our study of rectilinear motion. In other cases, the tangential force equation may be needed only for an algebraic relation between $\sum F_t$ and $\dot{v}$.

Finally, note that there is a *static* force balance in the binormal direction. Situations requiring consideration of this direction generally arise when a reaction force has a component parallel to $\bar{e}_b$, as well as parallel to $\bar{e}_t$ or $\bar{e}_n$.

EXAMPLE 9

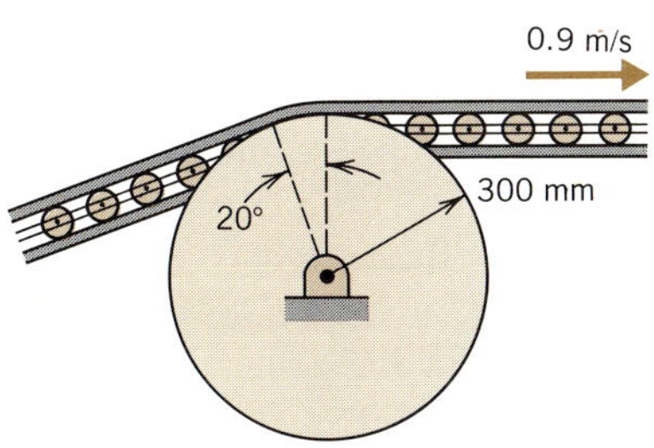

The conveyor belt moves with a constant speed of 0.9 m/s up the incline before it passes over the 300 mm radius idler wheel. The coefficient of static friction between a package and the belt is $\mu_k = 0.4$. Determine the maximum deceleration rate that can be imparted to the belt without causing a package that is at 20° from the vertical to slip.

Solution

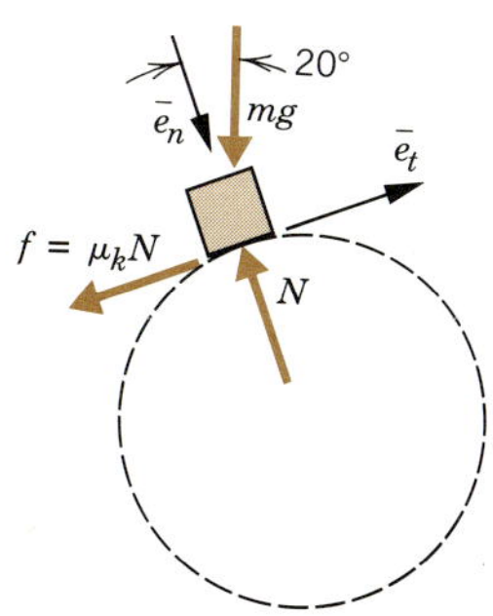

The friction and normal forces are tangent and normal to the plane of contact, respectively. The maximum deceleration rate corresponds to $f = \mu_k N$, with the direction of f being opposite the tendency of the package to continue its upward motion, as shown in the free body diagram. The other force depicted in the diagram is the weight mg.

The path over the idler wheel is known to be a circle having a 300 mm radius. Hence, a formulation in terms of path variables is convenient to evaluate $\dot{v}$. The acceleration corresponding to the given speed is therefore

$$\bar{a} = \dot{v}\bar{e}_t + \frac{(0.9)^2}{0.3}\bar{e}_n = \dot{v}\bar{e}_t + 2.7\bar{e}_n \text{ m/s}^2$$

where the unit vectors are depicted in the free body diagram. The corresponding equations of motion are

$$\sum F_t = -mg \sin 20^\circ - \mu_k N = ma_t = m\dot{v}$$

$$\sum F_n = mg \cos 20^\circ - N = ma_n = m(2.7)$$

With $g = 9.807$ m/s^2 and $\mu_k = 0.4$, the solution of these equations is

$$N = m(g \cos 20^\circ - 2.7) = 6.516m$$

$$\dot{v} = -g \sin 20^\circ - \mu_k \frac{N}{m} = -5.96 \text{ m/s}^2$$

Note that $\dot{v}$ is smaller than it would be for a straight 20° incline because the centripetal acceleration lowers the magnitude of N.

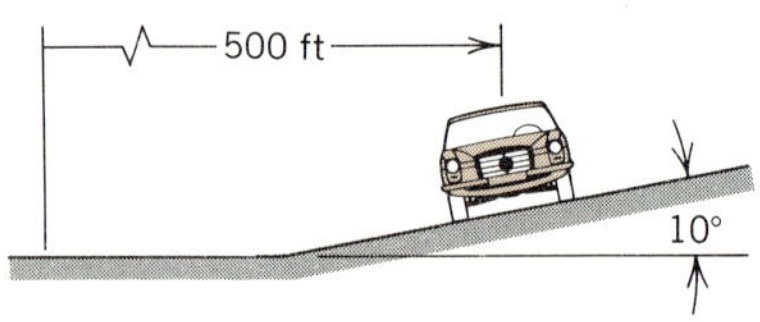

EXAMPLE 10

A 3000 lb automobile is traveling at 54 m/hr along a level highway whose radius of curvature is 500 ft. The highway is banked at 10° from the horizontal. Determine the total normal and frictional forces exerted between the tires and the road when the automobile remains in its lane. What is the minimum coefficient of static friction required to prevent slippage?

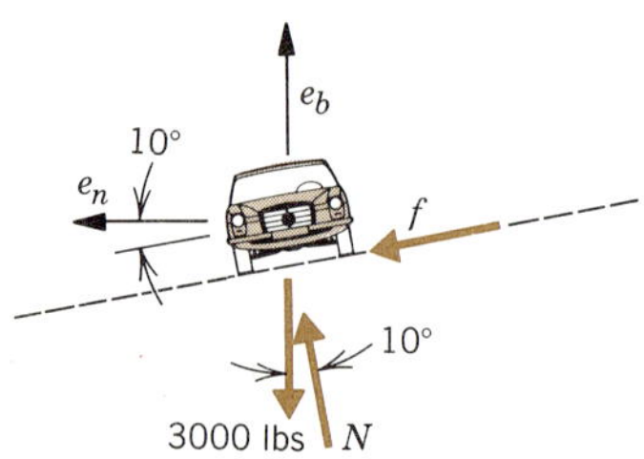

Solution

The friction force can be regarded as having two components parallel to the road surface. The first, which is the force $\bar{f}$ tangent to the incline in the free body diagram, keeps the vehicle in its lane. This force is depicted in the diagram assuming that the automobile would move to an outer lane if friction were not present. The second friction force would be tangent to the path, and therefore perpendicular to the plane of the diagram. In this problem, the tangential friction force is zero because the speed is constant.

It is important to note that the path of the automobile is a 500 ft radius circle ($\rho = 500$ ft), which is situated in the *horizontal* plane. Thus, $\bar{e}_n$ is horizontal, rather than parallel, to the incline. The unit vector $\bar{e}_b$, which is vertical, is needed because the friction and normal forces have vertical *and* horizontal components. The speed is constant at 54(88/60) = 79.2 ft/s, so the acceleration is

$$\bar{a} = \frac{(79.2)^2}{500} \bar{e}_n = 12.545\bar{e}_n \text{ ft/s}^2$$

Summing force components, with $g = 32.17$ ft/s^2, yields

$$\sum F_n = N \sin 10^\circ + f \cos 10^\circ = ma_n = \frac{3000}{32.17}(12.545)$$

$$\sum F_b = N \cos 10^\circ - f \sin 10^\circ - 3000 = 0$$

The foregoing are simultaneous equations for f and N. Their solution is

$$N = 3158 \text{ lb}, \qquad f = 631 \text{ lb}$$

Because f is positive, our initial assumption that the vehicle would slide outward in the absence of friction was correct.

In order to calculate the minimum allowable μ_s, we recall that $\mu_s N$ is the largest friction force that can exist between two surfaces. That is $f \leq \mu_s N$. Thus, if the automobile in this problem is about to skid, equality holds, and

$$(\mu_s)_{\min} = \frac{631}{3158} = 0.20$$

HOMEWORK PROBLEMS

12.43 The following diagrams show the velocity $\bar{v}$, the acceleration $\bar{a}$, and the resultant force $\bar{F}$ for a particle moving along the indicated path. Some of these diagrams violate the principles of dynamics. Indicate which are incorrect and explain why.

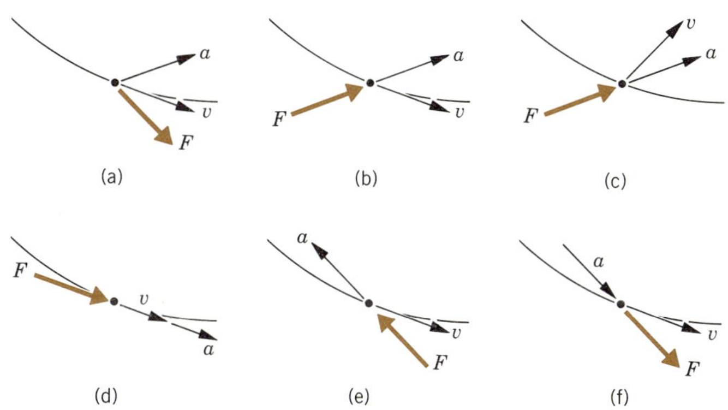

PROB. 12.43

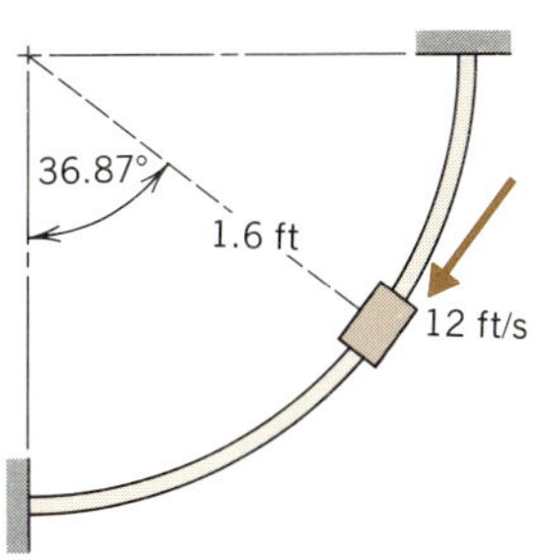

PROB. 12.44

12.44 The 10 oz collar is sliding down the smooth curved rod with an instantaneous speed of 12 ft/s in the position shown. Determine the force exerted by the rod on the collar and the acceleration of the collar in this position.

12.45 A 50 gram sphere is initially held by two cables in the position shown. The left cable is cut. Determine the tension in the right cable before and immediately after the cable is cut. Also determine the initial acceleration of the sphere.

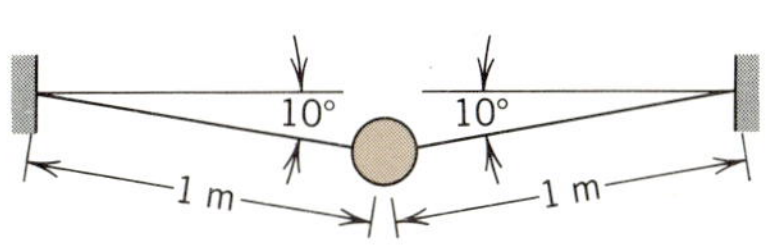

PROB. 12.45

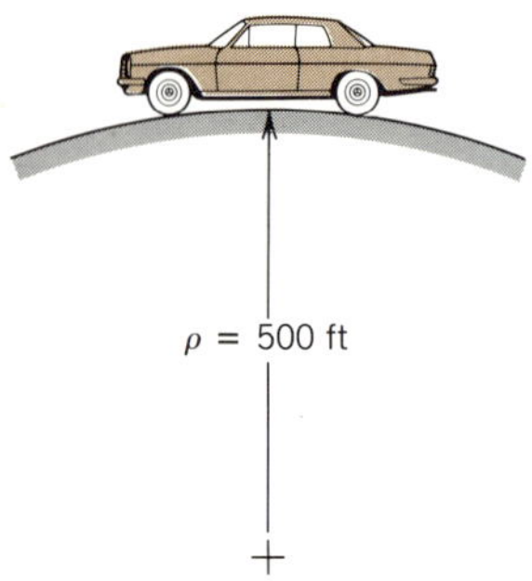

PROB. 12.46

12.46 An automobile is traveling at a constant speed over a hill. At the top the radius of curvature is 500 ft. (a) Determine the maximum speed the automobile can have in this position and still remain in contact with the road. (b) If the automobile is traveling at 60 mi/hr and the brakes are suddenly applied at the top of the hill, determine the maximum possible deceleration rate. The coefficient of friction between each tire and the ground is 0.70.

12.47 The longitudinal axis of the guided missile is aligned tangent to the path at an angle of 35° from the vertical. The missile's thrust is 500,000 lb and its weight is 160,000 lb. If the speed is $v = 5000$ ft/s, determine the rate of change of the speed of the rocket, and the radius of curvature of the path.

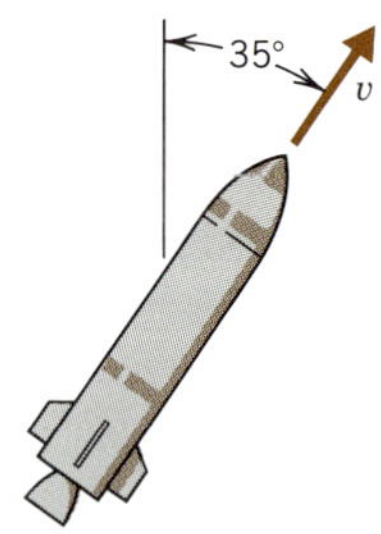

PROB. 12.47

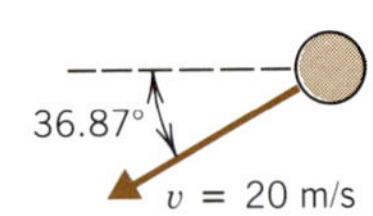

PROB. 12.48

12.48 A 150 gram ball is falling at the velocity shown. Air resistance is negligible. For this instant, determine the rate of change of the speed and the radius of curvature of the path.

12.49 Solve Problem 12.48 if an air resistance force of 0.5 N is present.

12.50 A small sphere of mass m is attached to a cord of length l. The ball rotates in a horizontal circle for which the angle between the cord and the vertical axis is θ. Derive expressions for the tension in the cord and the constant speed of the ball in terms of l and θ.

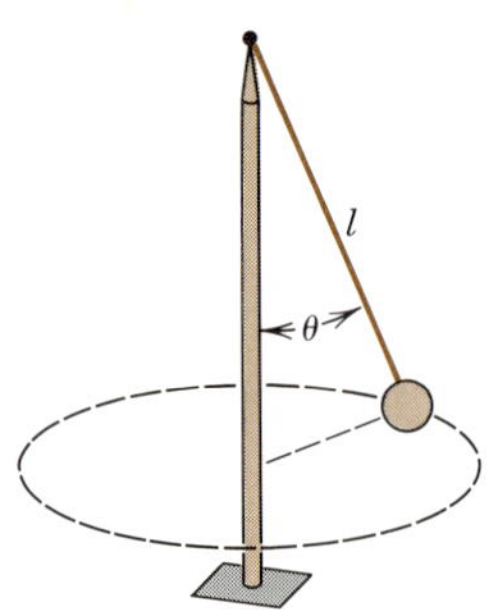

PROB. 12.50

12.51 The driver of a miniature go-cart wants to perform a stunt by driving along a circular track that is banked at 90°, so that the surface of the track is vertical. The top speed of the car is 45 mi/hr and the coefficient of friction between the tires and the ground is 0.8. What is the maximum radius r that the track can have?

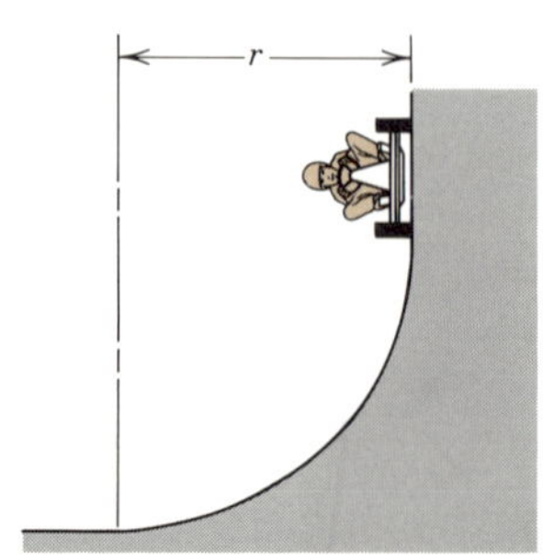

PROB. 12.51

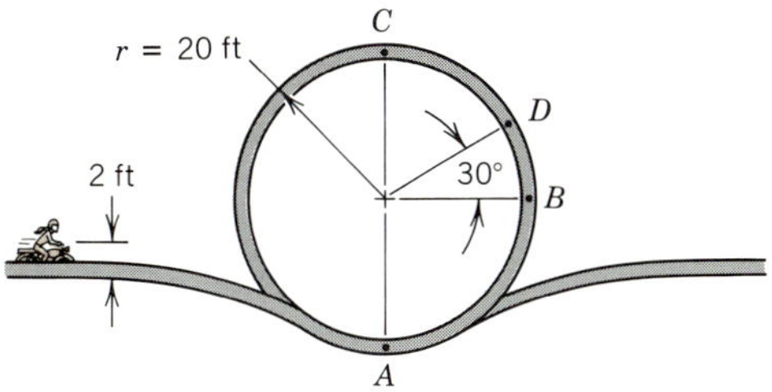

PROB. 12.52

12.52 The center of mass of a stunt rider and her motorcycle is 2 ft above the road surface. If the motorcycle maintains a constant speed of 30 mi/hr in the vertical loop, determine the resultant normal and traction forces exerted on the wheels by the track at positions A and B. The total weight of the rider and the motorcycle is 750 lb.

12.53 Solve Problem 12.52 for positions C and D.

12.54 The speed at which an automobile can traverse a banked curve without friction is called the *rated speed.* Determine the rated speed of the highway curve shown. Is this speed dependent on the mass m?

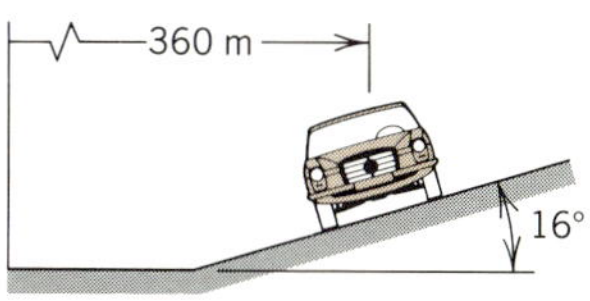

PROBS. 12.54

12.55 The automobile is traveling at 45 m/s along the banked curve in the horizontal plane. Determine the magnitude and sense of the friction force. The mass of the automobile is 1100 kg.

12.56 Solve Problem 12.55 when the speed is 10 m/s.

---

2. Rectangular Cartesian Coordinates

Rectangular Cartesian coordinates describe vectors in terms of components in fixed directions, for example, horizontal and vertical. To determine whether this formulation is suitable for a kinetics problem requires that we consider the way in which the forces, as well as the kinematical parameters for the system, are defined.

Let (x, y, z) denote the coordinates of a particle with respect to a rectangular Cartesian coordinate system whose orientation has been specified in a sketch. The acceleration of the particle can be denoted by

$$\bar{a} = \dot{v}_x \bar{i} + \dot{v}_y \bar{j} + \dot{v}_z \bar{k} = \ddot{x}\bar{i} + \ddot{y}\bar{j} + \ddot{z}\bar{k}$$

Matching like components of Newton's second law in this case yields

$$\begin{aligned} \sum F_x &= ma_x \equiv m\dot{v}_x \equiv m\ddot{x} \\ \sum F_y &= ma_y \equiv m\dot{v}_y \equiv m\ddot{y} \\ \sum F_z &= ma_z \equiv m\dot{v}_z \equiv m\ddot{z} \end{aligned} \tag{19}$$

Each of these equations provides an algebraic relation for the force required to achieve a specified movement in that direction. Alternatively, if the force component is known, the result is an expression for the corresponding acceleration component. Knowledge of the acceleration can also make it possible to evaluate the velocity and position of the particle, as will be apparent in the next example.

In the common case of planar motion in the xy plane, $a_z \equiv 0$. The z equation describes a static force balance in that case. It need be considered only when the reaction forces preventing movement out of the plane are of interest.

EXAMPLE 11

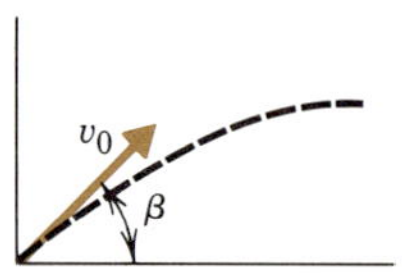

A particle is thrown upward with an initial velocity v_0 at an angle of elevation α. Determine the position of this object t seconds after it is thrown. Air resistance is negligible.

Solution

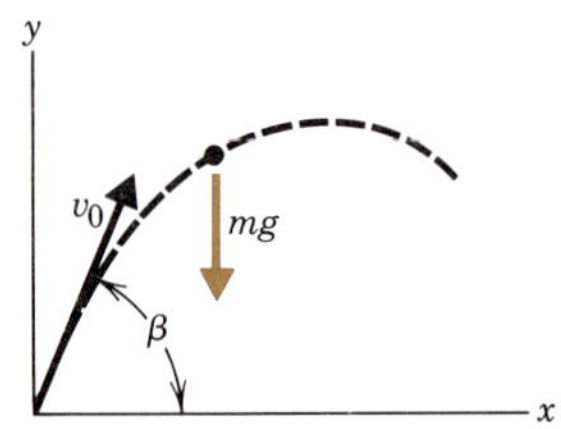

The object is in free fall after it is thrown. As shown in the free body diagram, the weight of the object is the only force acting on the particle. Cartesian coordinates are suitable because the gravitational force is in the (constant) vertical direction. For convenience, the origin of the xyz coordinate system has been chosen to coincide with the particle's initial location, and the xy plane is defined to be the vertical plane containing the initial velocity $\bar{v}_0$.

This is a case of planar motion, because the initial velocity is in the xy plane and there is no force to move the particle out of this plane. The acceleration is

$$\bar{a} = \dot{v}_x \bar{i} + \dot{v}_y \bar{j}$$

where

$$v_x = \dot{x}, \qquad v_y = \dot{y}$$

Using the free body diagram to obtain the force components then yields

$$\sum F_x = 0 = ma_x = m\dot{v}_x$$
$$\sum F_y = -mg = ma_y = m\dot{v}_y$$

The acceleration components are found from these equations to be

$$\dot{v}_x = 0, \qquad \dot{v}_y = -g$$

The accelerations in the horizontal and vertical directions are both constant. In order to employ the constant acceleration formulas, the initial conditions must be described. It is known that the object is at the origin initially, and that its initial velocity is $\bar{v}_0$. Thus, for $t = 0$

$$x_0 = y_0 = z_0 = 0$$
$$(v_x)_0 = v_0 \cos\beta, \qquad (v_y)_0 = v_0 \sin\beta$$

The motion in each direction may be evaluated individually as tabulated below.

x direction: $\dot{v}_x \equiv dv_x/dt = 0$

$$\int_{v_0 \cos\beta}^{v_x} dv_x = 0, \qquad v_x = v_0 \cos\beta \equiv dx/dt$$

$$\int_0^t v_0 \cos\beta \, dt = \int_0^x dx, \qquad x = (v_0 \cos\beta)t$$

***y* direction:** $\dot{v}_y = -g \equiv dv_y/dt$

$$\int_{v_0 \sin\beta}^{v_y} dv_y = \int_0^t -g\,dt, \qquad v_y = -gt + v_0 \sin\beta \equiv \frac{dy}{dt}$$

$$\int_0^t (-gt + v_0 \sin\beta)\,dt = \int_0^y dy, \qquad y = -\tfrac{1}{2}gt^2 + (v_0 \sin\beta)t$$

It will be noted that the object moves with a constant horizontal component of velocity. In contrast, the vertical component of velocity is initially upward, but it is decreasing. For large enough t, this velocity component will be downward. (Clearly, $v_y = 0$ represents the instant when y is maximum.)

EXAMPLE 12

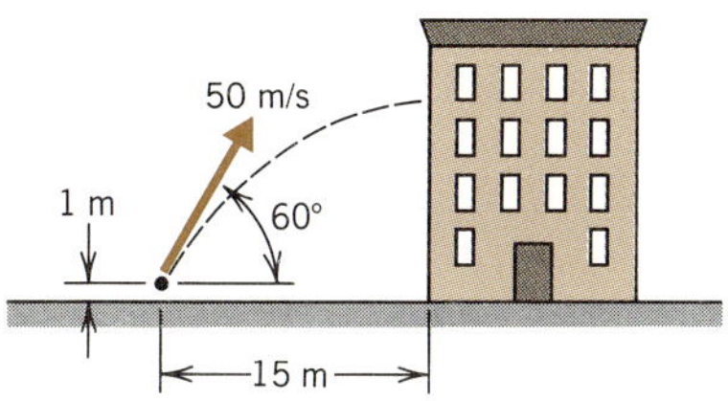

Firefighters are holding the nozzle of a hose 1 m above the ground and 15 m from the wall. Water exits from the nozzle with a velocity of 50 m/s at an angle of 60° above the horizontal, and air resistance is negligible. Use the results of Example 11 to determine where the stream of water hits the wall. Also determine the velocity with which the water strikes the wall.

Solution

As in the previous example, it is convenient to select a Cartesian coordinate system to describe the trajectory of a particle of water, with its origin situated at the initial location. Its xy plane is defined to be the vertical plane containing the initial velocity. This is depicted in the sketch.

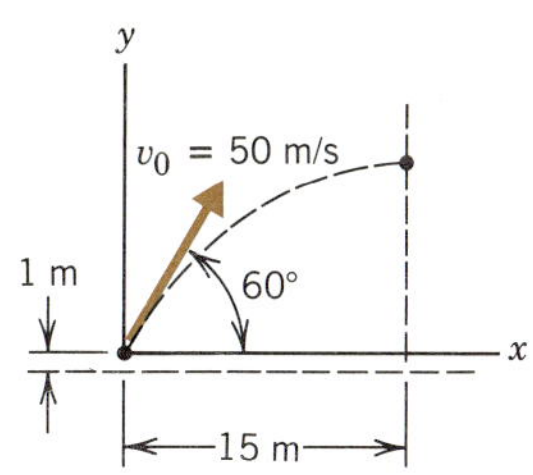

It is obvious from the sketch that the water hits the wall at $x = 15$ m. The task here is to determine the corresponding value of y. For this, we note that the initial velocity is

$$\bar{v}_0 = 50(\cos 60^\circ \bar{i} + \sin 60^\circ \bar{j}) = 25\bar{i} + 43.30\bar{j} \text{ m/s}$$

There is no acceleration in the x direction, so v_x remains unchanged from its initial value. Thus

$$v_x = 25 \text{ m/s}, \qquad x = 25t \text{ m}$$

The motion in the y direction consists of a constant downward acceleration at $g = 9.807$ m/s^2, starting from an initial upward velocity of 43.30 m/s. The constant acceleration formulas yield

$$v_y = -9.807t + 43.30 \text{ m/s}$$

$$y = -\tfrac{1}{2}(9.807)t^2 + 43.30t \text{ m}$$

The time required for the water to travel from the nozzle to the wall is is found by setting $x = 15$ m. This yields

$$15 = 25t \Rightarrow t = 0.60 \text{ s}$$

$$y|_{t=0.6} = -\tfrac{1}{2}(9.807)(0.6)^2 + 43.30(0.6) = 24.2 \text{ m}$$

The velocity with which the water hits the wall is found by substituting $t = 0.6$ s.

$$v_x = 25 \text{ m/s}, \qquad v_y = -9.807(0.6) + 43.30 = 37.4 \text{ m/s}$$

$$\bar{v} = 25\bar{i} + 37.4\bar{j} \text{ m/s}$$

EXAMPLE 13

The horizontal force $\bar{F}$ moves the 3 kg package up a hill whose profile is $y = x^2/4$, where x and y are in meters. It is known that the horizontal component of the velocity of the package is constant at 5 m/s. Determine how the magnitude of $\bar{F}$ must depend on x in order to achieve this motion. Friction is negligible.

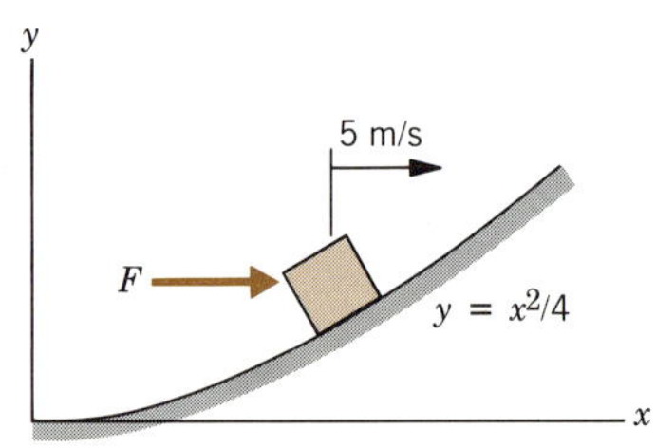

Solution

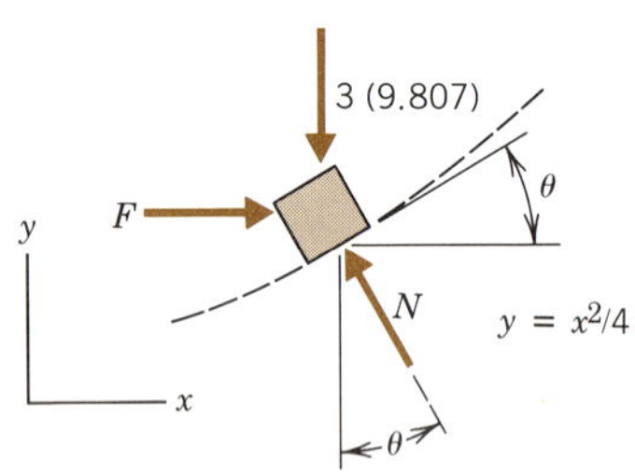

The contact force $\bar{N}$ in the free body diagram is perpendicular to the tangent of the path, with the orientation of this tangent being defined by θ. The movement along the path was specified by giving $v_x = 5$ m/s, $\dot{v}_x = 0$. It is consistent to employ Cartesian coordinates oriented horizontally and vertically, as shown in the diagram,

When $z \equiv 0$, the Cartesian coordinate formula for acceleration is

$$\bar{a} = \dot{v}_x\bar{i} + \dot{v}_y\bar{j}; \qquad v_x = \dot{x}, \qquad v_y = \dot{y}$$

In general, x and y are related by the equation of the path, $y = x^2/4$. Differentiating this relation successively yields $\dot{y}$ and $\ddot{y}$ because it is known that $v_x = \dot{x} = 5$ m/s and $\dot{v}_x = 0$.

$$\dot{y} = \frac{d}{dt}(\tfrac{1}{4}x^2) = \tfrac{1}{2}x\dot{x} = 2.5x \text{ m/s}$$

$$\ddot{y} = \frac{d}{dt}(\dot{y}) = 2.5\dot{x} = 12.5 \text{ m/s}^2$$

Thus, the acceleration is constant in the upward direction.

$$\bar{a} = 12.5\bar{j} \text{ m/s}^2$$

Another kinematical variable that will be needed is the angle θ for the tangent to the path. This may be obtained from the slope formula,

$$\theta = \tan^{-1}\frac{dy}{dx} = \tan^{-1}(\tfrac{1}{2}x)$$

(Occasionally, it is easier to describe θ by using the velocity components to construct $\bar{e}_t$.)

The equations of motion may now be formulated. The free body diagram indicates that

$$\sum F_x = F - N\sin\theta = ma_x = 0$$

$$\sum F_y = N\cos\theta - 3(9.807) = ma_y = 3(12.5)$$

Because θ is determined by the value of x, there are two equations for two unknowns.

Sequential solution of the equations of motion yields

$$N = 66.92/\cos\theta \text{ N}$$

$$F = N\sin\theta = 66.92\tan\theta = 33.46x \text{ N}$$

HOMEWORK PROBLEMS

12.57 A ball is thrown horizontally at 90 ft/s from a cliff 100 ft above the ground. Determine the elapsed time for the ball to hit the ground and the corresponding horizontal distance traveled by the ball.

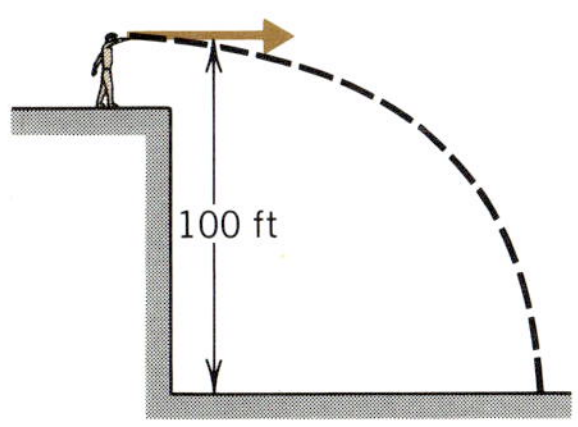

PROB. 12.57

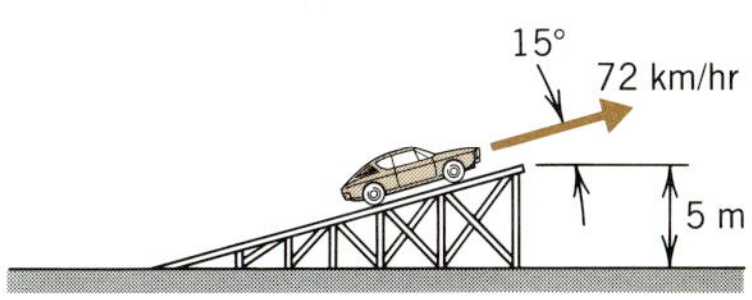

PROB. 12.58

12.58 An automobile is driven as a stunt off the ramp at a speed of 72 km/hr. Determine the maximum height the automobile achieves above the ground and the corresponding elapsed time.

12.59 Determine where the automobile in Problem 12.58 will hit the ground. What is its velocity immediately prior to hitting the ground?

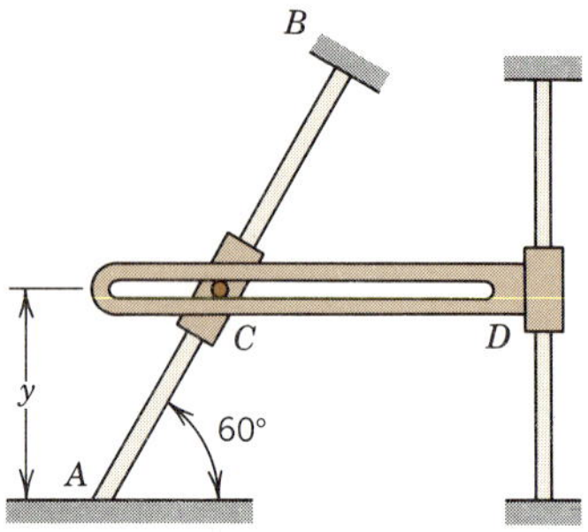

PROBS. 12.60 AND 12.61

12.60 Collar C, whose mass is 200 grams, is pulled up the fixed rod AB by the slotted guide bar D. The vertical position of the collar is $y = 4t + 2t^3$ m. Determine the forces acting on the collar when $t = 2$ s.

12.61 Collar C, which weighs 12 oz, is constrained to move along the fixed rod AB. Bar D starts from $y = 0$ and moves upward with a constant acceleration of $3g$. Evaluate the forces acting on the collar. Do these forces change with elapsed time?

12.62 The 2 lb collar C is pulled to the left along rod AB, whose shape in the horizontal plane is defined by $xy = 3\ \text{ft}^2$. Arm D, which produces this motion, has a constant speed of 20 ft/s. Evaluate each force acting on the collar as it passes the position $x = 1$ ft.

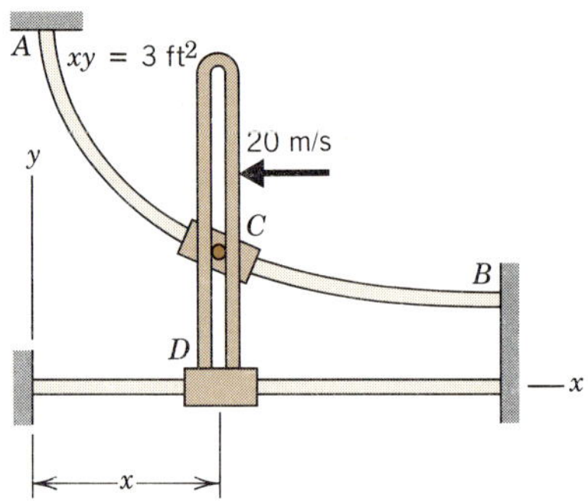

PROB. 12.62

12.63 A 1 oz steel ball bearing in free flight passes through a magnetic field that exerts a constant horizontal pull of 2 oz in the x direction. Initially, the ball bearing is at a point 1 ft above the ground and its velocity is $20\bar{j}$ ft/s. Determine (a) the location of point B where the ball bearing hits the xy plane, and (b) the velocity of the ball bearing when it reaches point B.

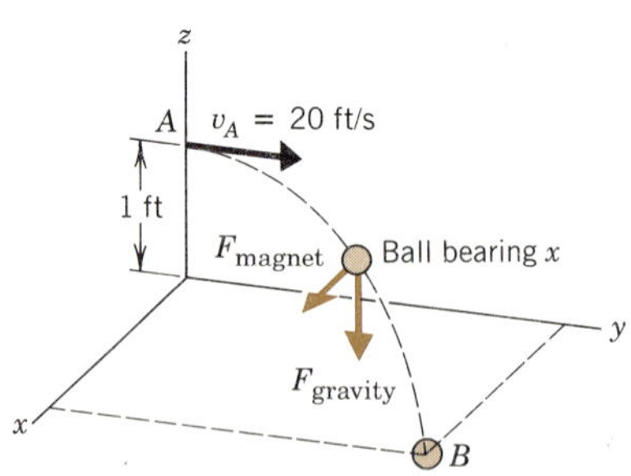

PROB. 12.63

12.64 Use the equation derived in Example 11 to show that the path of the particle in free flight is a parabola. Then derive an expression for the velocity of the particle in terms of the horizontal distance it has traveled.

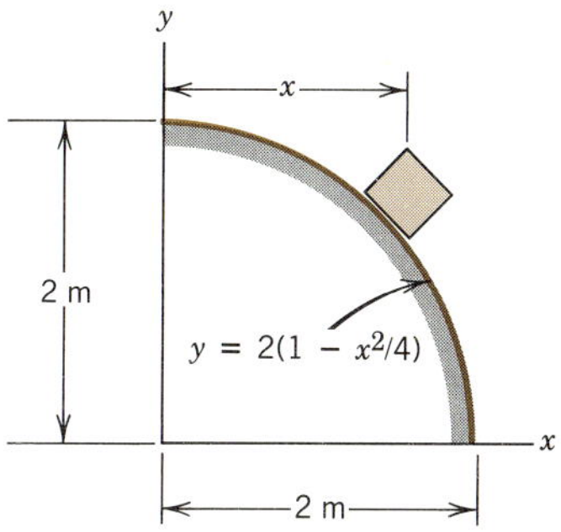

PROB. 12.65

12.65 When the block, which is sliding down the curved hill, reaches $x = 2$ m, its horizontal velocity component is $v_x = 5$ m/s, and its horizontal acceleration component is $\dot{v}_x = -2$ m/s^2. Evaluate the forces exerted by the hill on the block in this position.

***12.66** A 200 gram pin is pulled through a circular groove by the grooved vertical arm, which is moving to the right at the steady speed of 20 m/s. Determine the normal force exerted by each groove on the pin in the position where $\theta = 36.87°$.

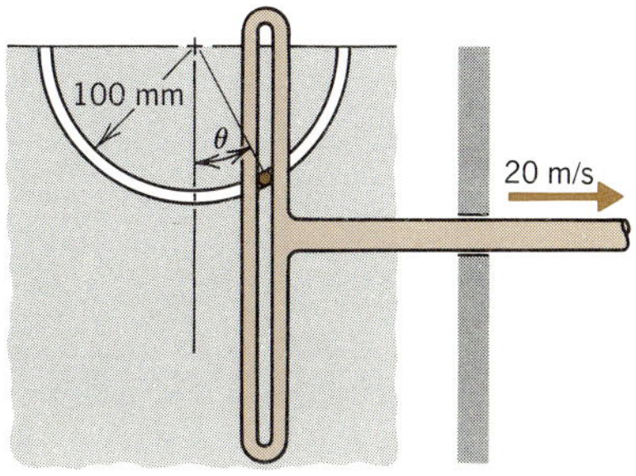

PROB. 12.66

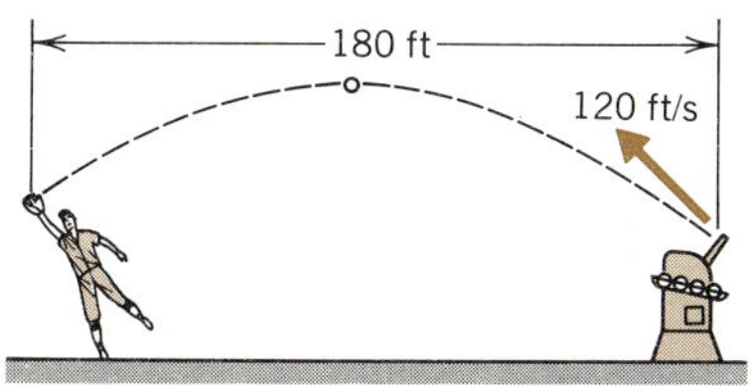

PROB. 12.67

***12.67** A pitching machine can throw baseballs at any angle of elevation with a speed of 120 ft/s. A ball player standing 180 ft away catches the ball at the same height above the level ground as that from which it was thrown. Neglecting air resistance, determine (a) the two possible angles of elevation for the throw, and (b) the final speed of the ball and the time duration of the flight for both possible angles.

---

3. Cylindrical Coordinates

The distinctive feature of cylindrical coordinates is that they employ a fixed axial direction, $\bar{k}$, perpendicular to a plane. Vectors are described relative to this plane in terms of their components with respect to the (non-fixed) radial and transverse directions, $\bar{e}_R$ and $\bar{e}_\phi$, respectively.

The acceleration in cylindrical coordinates is

$$\bar{a} = (\ddot{R} - R\dot{\phi}^2)\bar{e}_R + (R\ddot{\phi} + 2\dot{R}\dot{\phi})\bar{e}_\phi + \ddot{z}\bar{k}$$

The scalar equations of motion that are obtained by matching like components of $\sum \bar{F}$ and $m\bar{a}$ are

$$\begin{aligned} \sum F_R &= ma_R \equiv m(\ddot{R} - R\dot{\phi}^2) \\ \sum F_\phi &= ma_\phi = m(R\ddot{\phi} + 2\dot{R}\dot{\phi}) \\ \sum F_z &= ma_z = m\ddot{z} \end{aligned} \tag{20}$$

The definition of cylindrical coordinates, like that for Cartesian coordinates, does not require knowledge of the path. Key considerations indicating the suitability of cylindrical coordinates are that the forces are described with respect to a fixed axis or point, or that the angle of rotation and radial distance to a fixed axis (or point) are the significant kinematical quantities.

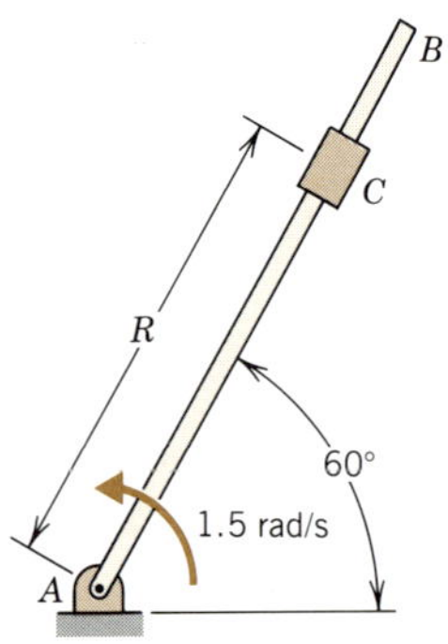

EXAMPLE 14

The smooth bar AB rotates in the vertical plane at the constant rate of 1.5 rad/s. At the instant shown, the 400 gram collar C is at $R = 800$ mm, and it is sliding inward relative to the bar at 3 m/s. For this instant, determine the force exerted by bar AB on the collar, and the acceleration term $\ddot{R}$.

Solution

We begin with a free body diagram of the collar, which depicts the gravitational force and the normal force $\bar{N}$ exerted by the rod. This is a planar problem involving rotation about the fixed point A, so only the cylindrical coordinates (R, ϕ) are needed. Note that $\bar{e}_\phi$ is oriented in the sense of increasing ϕ.

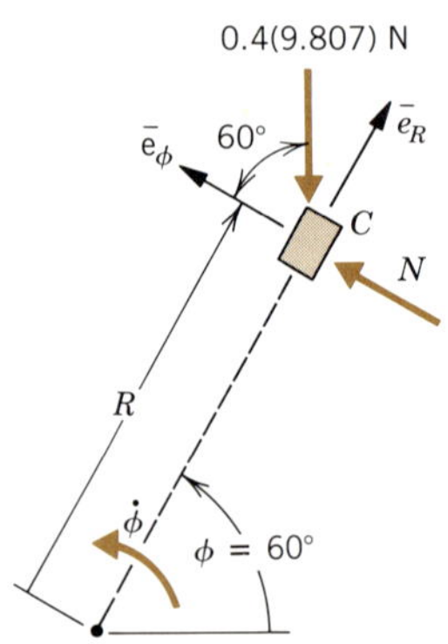

The given information is that at the instant of interest $\phi = 60°$, $\dot{\phi} = 1.5$ rad/s, $\ddot{\phi} = 0$, $R = 0.8$ m, and $\dot{R} = -3$ m/s (the negative sign means that R is decreasing for inward motion). The acceleration is therefore

$$\begin{aligned}\bar{a} &= [\ddot{R} - 0.8(1.5)^2]\bar{e}_R + [0 + 2(-3)(1.5)]\bar{e}_\phi \\ &= (\ddot{R} - 1.80)\bar{e}_R - 9\bar{e}_\phi \text{ m/s}^2\end{aligned}$$

Equating the components of the forces in the free body diagram to the matching components of $m\bar{a}$ yields

$$\sum F_R = -0.4(9.807)\sin 60° = 0.4a_R = 0.4(\ddot{R} - 1.8)$$

$$\sum F_\phi = N - 0.4(9.807)\cos 60° = 0.4a_\phi = 0.4(-9)$$

$$\ddot{R} = -6.69 \text{ m/s}^2, \qquad N = -1.64 \text{ N}$$

The result for $\ddot{R}$ is the acceleration of the collar *with respect to rod AB* at this instant. The fact that N is negative merely means that it is acting in the opposite sense from the one appearing in the free body diagram.

EXAMPLE 15

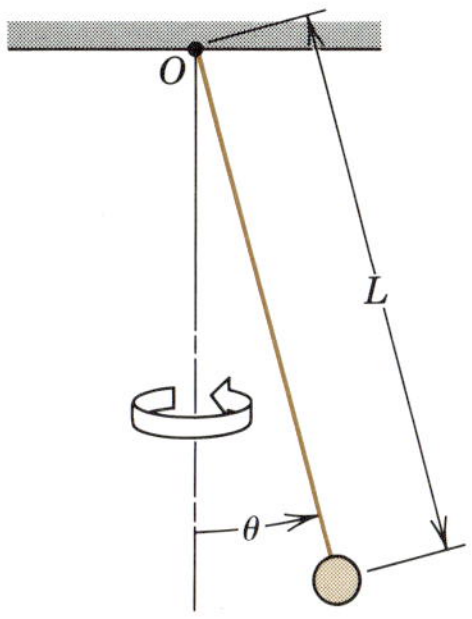

A small sphere is suspended by a cable from pivot O in the ceiling. The sphere requires 0.5 s to complete one revolution about the vertical axis with $\theta = 15°$. What is the length L?

Solution

In the free body diagram, the gravitational force is labeled as mg because the mass of the sphere is not given. (We will use $g = 32.17$ ft/s^2, although it is equally convenient to use SI units here.) The tensile force exerted by the cable is always oriented toward the fixed point O. We will assume that air resistance is negligible.

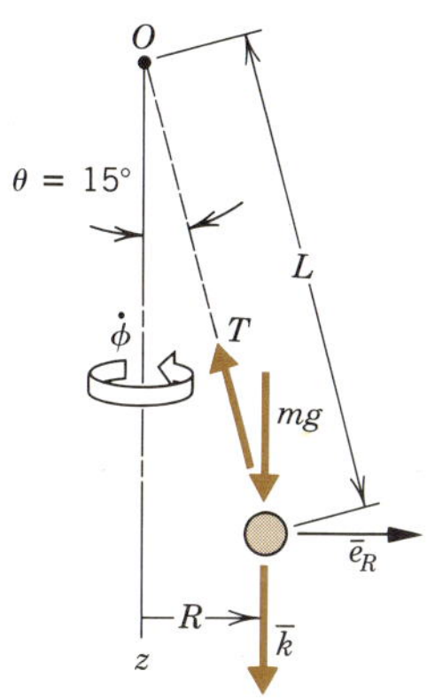

The cylindrical coordinate formulation will be used because the rate of rotation about the fixed vertical axis, which is denoted as z, is given. Although z is measured downward, its sense will not be important. Note that $\bar{e}_R$ is perpendicular to the z axis, by definition. It is *not* parallel to the line from the pivot to the sphere.

A constant value of $\theta = 15°$ means that

$$R = L \sin 15° = 0.2588L, \qquad \dot{R} = \ddot{R} = 0$$

$$z = L \cos 15° = 0.9659L, \qquad \dot{z} = \ddot{z} = 0$$

It is given that one revolution about the z axis requires 0.5 s, so the angular speed is

$$\dot{\phi} = \left(\frac{1 \text{ rev}}{0.5 \text{ s}}\right)\left(\frac{2\pi \text{ rad}}{1 \text{ rev}}\right) = 12.5664 \text{ rad/s}$$

This value is constant, so $\ddot{\phi} = 0$.

The acceleration formula gives

$$\bar{a} = [0 - (0.2588L)(12.5664)^2]\bar{e}_R + 0\bar{e}_\phi + 0\bar{k} = -40.87L\bar{e}_R$$

from which the components of Newton's second law are found to be

$$\sum F_R = -T \sin 15^\circ = ma_R = -m(40.87L)$$

$$\sum F_\phi \equiv 0$$

$$\sum F_z = mg - T \cos 15^\circ = ma_z = 0$$

Eliminating T from the above relations yields (after cancellation of the factor m)

$$L = \frac{g \tan 15^\circ}{40.87} = 0.211 \text{ ft}$$

It will be noted that the equation for $\sum F_\phi$ provided no information. This was a consequence of correctly assuming that $\ddot{\phi}$ is zero, which is the case when friction is negligible.

HOMEWORK PROBLEMS

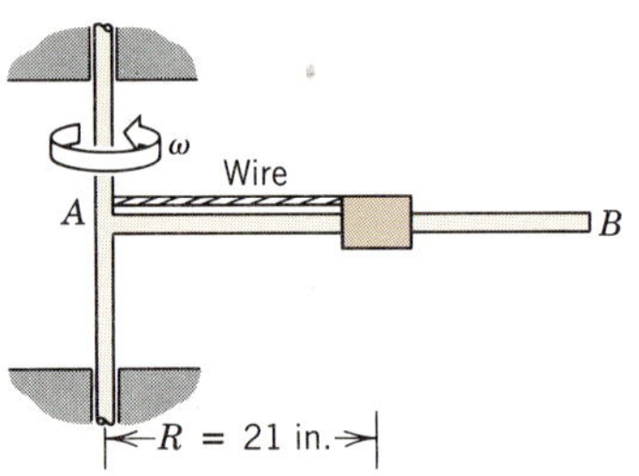

PROBS. 12.68 AND 12.69

12.68 Bar AB rotates at a constant rate in the horizontal plane. The wire holding the 2 lb collar has a tensile strength of 50 lb. What rate of rotation (rev/min) will result in breaking the wire?

12.69 Bar AB is rotating at the constant rate of 40 rev/min when the wire breaks. Determine the value of $\ddot{R}$ at the *instant after* the wire breaks.

12.70 Rod AB rotates in the horizontal plane about pivot A. When $\omega = 0$, the distance d to the inner end of the spring is 900 mm. The stiffness of the spring is 200 N/m, and the collar has a mass of 5 kg. Determine the value of d for a constant rotation rate of 45 rev/min.

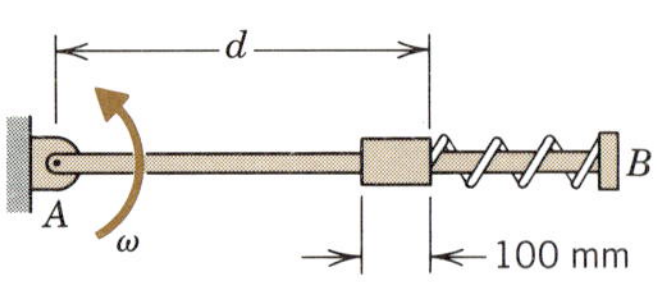

PROB. 12.70

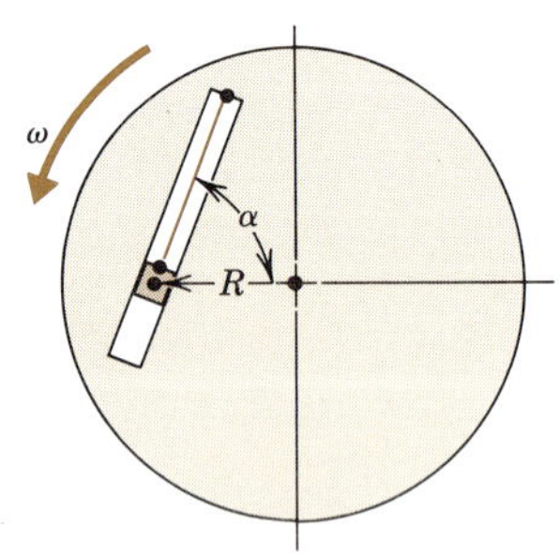

PROB. 12.71

12.71 A block is held in position within a groove in a rotating disk by a cable. Derive expressions for the tension force in the cable and the normal force exerted by the groove on the block in terms of the constant angular speed ω of the disk and the dimensions R and α shown. The motion occurs in the horizontal plane.

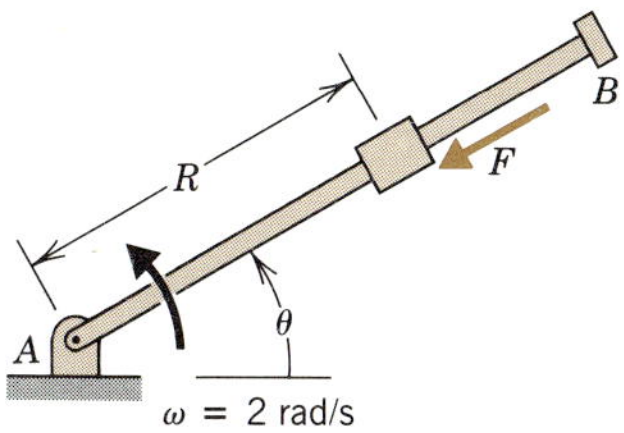

PROBS. 12.72 AND 12.73

12.72 The 2 kg block slides over the rigid arm AB, which is rotating in the vertical plane at 2 rad/s counterclockwise. A radial force F gives the block a constant inward speed of 1 m/s. When $\theta = 30^\circ$, the radial distance R is 400 mm. Determine the corresponding value of F if friction is negligible.

12.73 The 2 kg block slides over the rigid arm AB, which is rotating in the vertical plane at 2 rad/s counterclockwise and slowing at the rate of 1.5 rad/s^2. A radial force F gives the block a constant inward speed of 1 m/s. When $\theta = 30^\circ$, the radial distance R is 400 mm. Determine the corresponding value of F if the coefficient of kinetic friction between the collar and arm AB is 0.10.

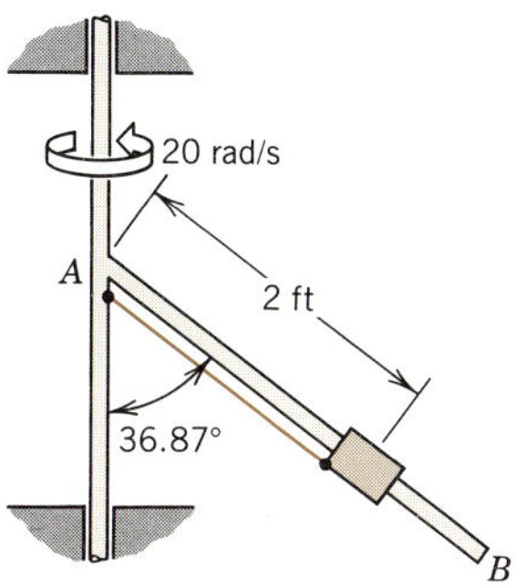

PROB. 12.74

12.74 The 4 oz collar is held in place by the cable as the system rotates about the vertical axis at 20 rad/s. Bar AB is smooth. Determine the tension in the cable and the force exerted on the collar by bar AB.

12.75 Passengers on an amusement park ride sit in seats that are suspended by 25 ft long cables from the horizontal arm AB, as shown. Arm AB rotates about the vertical axis at a constant angular speed ω. If $\theta = 36.87^\circ$, determine the corresponding value of ω.

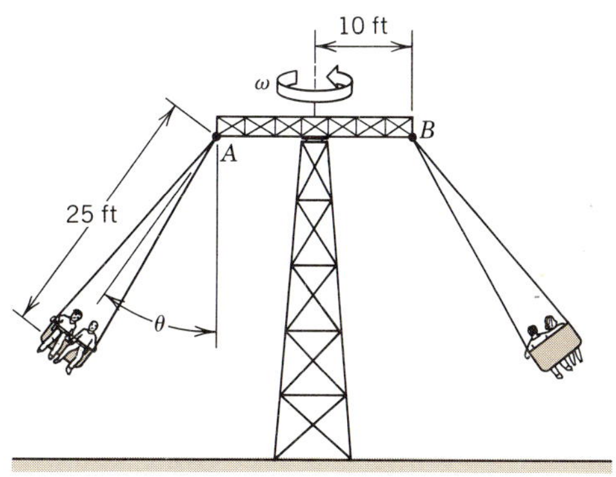

PROB. 12.75

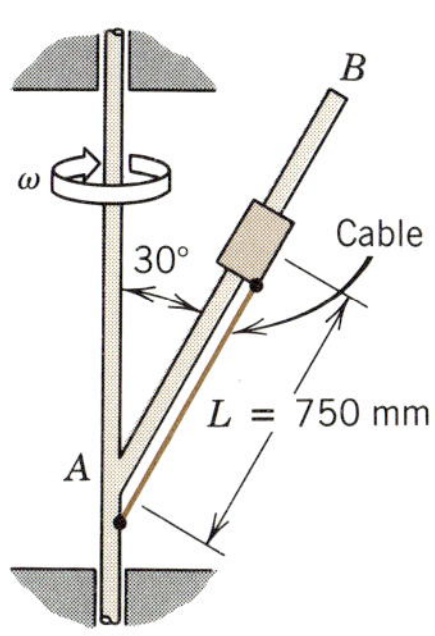

PROBS. 12.76 AND 12.77

12.76 The smooth bar AB rotates about the vertical axis at a constant angular speed ω as it supports the 500 gram collar. Determine the minimum value of ω for which the cable is taut.

*12.77 The smooth bar AB supporting the 500 gram collar rotates about the vertical axis at the constant rate of 10 rad/s. The cable, which is taut initially, suddenly breaks. Determine the value of $\ddot{L}$ at the instant after the break occurs.

D. PROBLEM SOLVING

Each of the preceding sections has featured the formulation of the equations of motion according to a specified kinematical description. In this section we will be concerned with the decision as to which description should be employed. Situations will be addressed where several of the techniques will be combined to solve problems. For example, evaluating the acceleration might involve the joint utilization of two kinematical formulations, as was done in the previous chapter; or a curvilinear motion problem might yield a differential equation, for which the techniques presented for rectilinear motion are useful.

The following steps are intended to synthesize the developments presented thus far. They are a logical framework for solving problems. It is the organization of these steps, rather than their memorization, to which you should devote your attention.

1. Draw a free body diagram. If several bodies are interconnected, draw a free body diagram of each body, being sure to satisfy Newton's third law of action and reaction. As an aid in constructing these diagrams, remember that each constraint (restriction) upon the motion of the body can be imposed only by a reaction force.
2. Write down or depict in the free body diagram all given information about the forces and kinematical variables. List the quantities required for solution.
3. Choose a coordinate system for describing vector components, and show the associated unit vectors in the free body diagram. This choice should be based on suiting the kinematical quantities and the forces associated with the system.
4. Following the kinematical methods in Chapter 11, express the acceleration of each body in terms of the coordinate system chosen in step 3.
5. Sum the force components in the direction of each unit vector and and equate each term to the corresponding component of $m\bar{a}$. Check that the number of algebraic and differential equations match the number of unknowns. Common reasons for these two numbers being unequal are a force that was not properly accounted for, a given quantity that was not used, or a kinematical constraint that was not considered.
6. Solve the equations of motion.

ILLUSTRATIVE PROBLEM 1

The coefficient of kinetic friction between the 2 kg block A and the incline is $\mu_k = 0.15$. Block B has a mass of 5 kg. The system is released from rest. Determine the velocity of block A after it has moved 1 m.

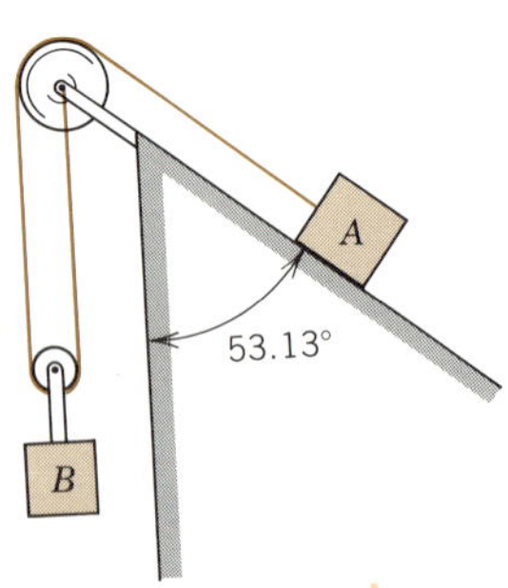

Solution

Steps 1 and 2 Because blocks A and B are individual bodies, we must draw a free body diagram of each. We will assume that the pulleys are ideal (i.e., massless and frictionless), which means that the tension in each segment of the cable is T. In turn, this means that the resultant cable force acting on block B is $2T$. In addition to the weights, we must also consider the friction force. A comparison between the masses of the blocks seems to indicate that block A is tending to slide up the incline, which

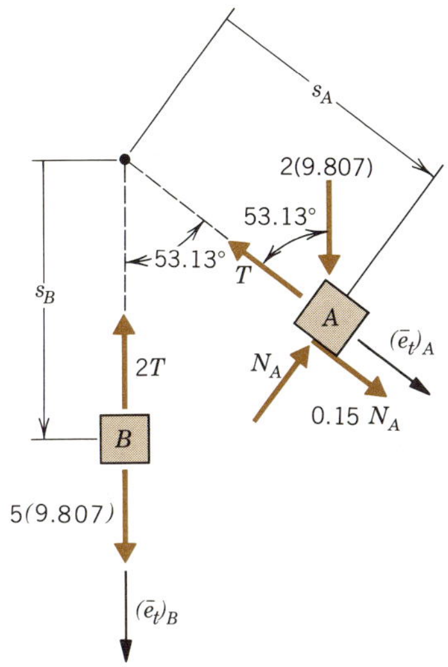

corresponds to the friction force on block A acting down the incline. (If this guess is wrong, a quantity that we assume to be positive will be found to be negative.) The free body diagrams are therefore as shown.

Step 3 The kinematical method for describing cable-pulley systems that was developed in Chapter 11, Section E.2, is useful here. Accordingly, we let s_A and s_B denote the distances of the block from the upper pulley, which is a convenient fixed reference point. The tangent vector for each block is then defined to be in the direction of increasing s, as depicted.

Step 4 Each block is in rectilinear motion, so

$$\bar{a}_A = \dot{v}_A(\bar{e}_t)_A; \qquad v_A = \dot{s}_A$$

$$a_B = \dot{v}_B(\bar{e}_t)_B; \qquad v_B = \dot{s}_B$$

The fact that the cable is inextensible gives a relation between v_A and v_B. Thus

$$s_A + 2s_B = \text{const}$$

$$\dot{s}_A + 2\dot{s}_B \equiv v_A + 2v_B = 0 \Rightarrow v_B = -0.5v_A$$

$$\bar{a}_A = \dot{v}_A(\bar{e}_t)_A \qquad \bar{a}_B = -0.5\dot{v}_A(\bar{e}_t)_B$$

Step 5 The equations of motion for block A are

$$\sum F_t = 0.15N_A + 2(9.807)\cos 53.13^\circ - T = (ma_t)_A = 2\dot{v}_A$$

$$\sum F_n = N_A - 2(9.807)\sin 53.13^\circ = (ma_n)_A = 0$$

Only the tangential force equation for block B is useful. It is

$$\sum F_t = 5(9.807) - 2T = (ma_t)_B = 5(-0.5\dot{v}_A)$$

Step 6 The three force equations have three unknowns, T, N_A, and $\dot{v}_A$. Their simultaneous solution is

$$N_A = 15.69 \text{ N}, \qquad T = 20.52 \text{ N}, \qquad \dot{v}_A = -3.199 \text{ m/s}^2$$

The negative value for $\dot{v}_A$ means that block A is accelerating up the plane. Hence our initial assumption is correct. The acceleration is constant. Therefore, the speed after block A has traveled 1 m starting from rest can be obtained by the constant acceleration equation, with $\Delta s_A = -1$ m corresponding to a reduction in s_A during the motion.

$$v_A^2 = 0 + 2\dot{v}_A(\Delta s_A) = 2(-3.199)(-1) \Rightarrow v_A = 2.53 \text{ m/s}$$

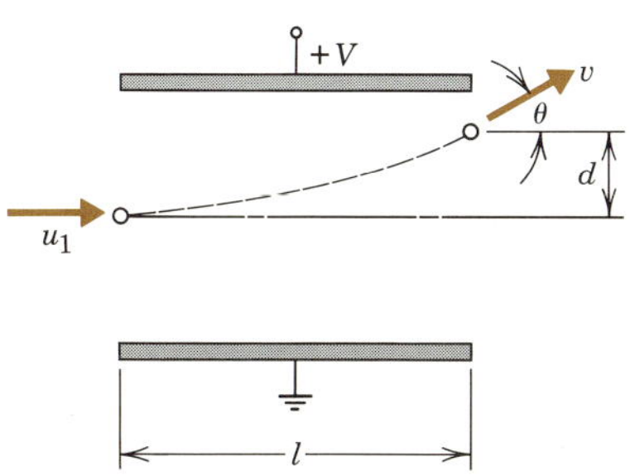

ILLUSTRATIVE PROBLEM 2

An electron is injected into the field between charged electrical plates with horizontal velocity u_1 when $t = 0$. The electron is attracted to the positively charged plate by a force eV/b, where e is the charge of an electron, V is the voltage difference, and b is the plate separation. Consider the case where $V = ct$. Determine the deflection d and the corresponding velocity $\bar{u}_2$ at the location where the electron exits the field. The gravitational force acting on the electron is negligible.

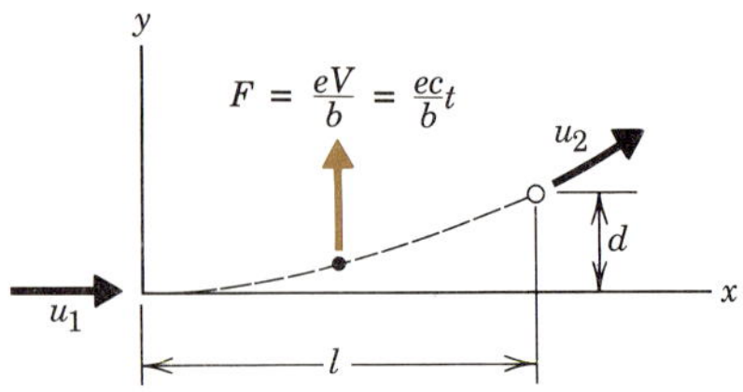

Solution

Step 1 The attraction force is the only appreciable force on the electron. It is shown in the free body diagram to act in the fixed direction toward the positive plate.

Step 2 Given the velocity $\bar{u}_1$ at the initial position depicted in the free body diagram, determine the deflection d and final velocity $\bar{u}_2$.

Step 3 Cartesian coordinates are the logical kinematical description because the force acts in a fixed direction and the path is not known. We have chosen the xy plane to coincide with the plane formed by the force and the initial velocity $\bar{u}_1$.

Step 4 There is no force to cause movement in the z direction, so the motion occurs in the xy plane. The corresponding kinematical formulas are

$$\bar{r} = x\bar{i} + y\bar{j}$$

$$\bar{v} = \dot{x}\bar{i} + \dot{y}\bar{j}; \qquad v_x = \dot{x}, \qquad v_y = \dot{y}$$

$$\bar{a} = \dot{v}_x\bar{i} + \dot{v}_y\bar{j}$$

Step 5 The component form of Newton's second law is

$$\sum F_x = 0 = ma_x = m\dot{v}_x$$

$$\sum F_y = \frac{ec}{b}t = m\dot{v}_y$$

These are two equations for determining the acceleration components.

Step 6 The equation for $\sum F_x$ indicates that $a_x = 0$. This means that v_x does not change from its initial value u_1. Furthermore, the initial position is $x = 0$, so

$$v_x = u_1, \qquad x = u_1 t$$

The equation for $\sum F_y$ yields

$$\dot{v}_y = \frac{ec}{bm}t$$

This is the same situation as the one where acceleration in rectilinear motion depends on time. The process of separating variables may be employed here. The initial conditions are $v_y = 0$ and $y = 0$ when $t = 0$.

$$\dot{v}_y = \frac{dv_y}{dt} = \frac{ec}{bm}t$$

$$\int_0^{v_y} dv_y = \int_0^t \frac{ec}{bm}t\,dt \Rightarrow v_y = \frac{1}{2}\frac{ec}{bm}t^2 = \frac{dy}{dt}$$

$$\int_0^y dy = \int_0^t \frac{1}{2}\frac{ec}{bm}t^2\,dt \Rightarrow y = \frac{1}{6}\frac{ec}{bm}t^3$$

The foregoing describes the motion in general. We evaluate the final position by setting $x = l$. Thus

$$l = x_f = u_1 t_f \Rightarrow t_f = u_1/l$$

$$d = y_f = \frac{1}{6}\frac{ec}{bm} t_f^3 = \frac{1}{6}\frac{ec}{bm}\frac{u_1^3}{l^3}$$

$$\bar{u}_2 = (v_x)_f \bar{i} + (v_y)_f \bar{j} = u_1 \bar{i} + \frac{1}{2}\frac{ec}{bm}\left(\frac{u_1}{l}\right)^2 \bar{j}$$

The only difference between this motion and that of a projectile in free fall is that here the vertical acceleration is not constant.

ILLUSTRATIVE PROBLEM 3

The rate of rotation of arm AB is constant at 450 rpm, and the arm is smooth. The cable holding the 12 oz collar on the rotating arm suddenly breaks, with the result that the collar slides outward. Determine the speed of the collar and the force exerted by arm AB on the collar immediately before the collar hits the bumper at end B.

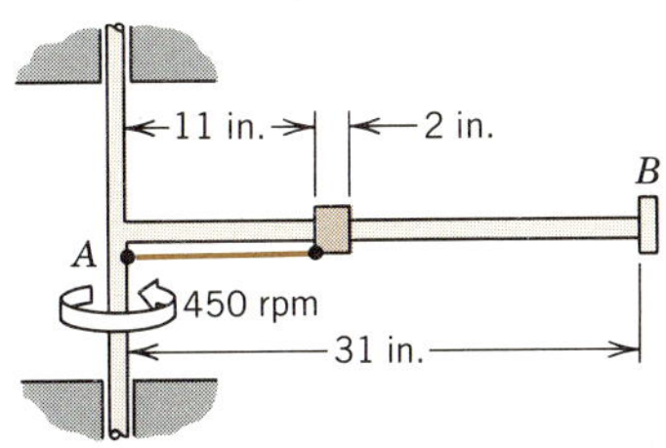

Solution

Step 1 Arm AB exerts normal reactions on the collar in both the horizontal and vertical plane. These are depicted in the free body diagram, which is drawn in two views.

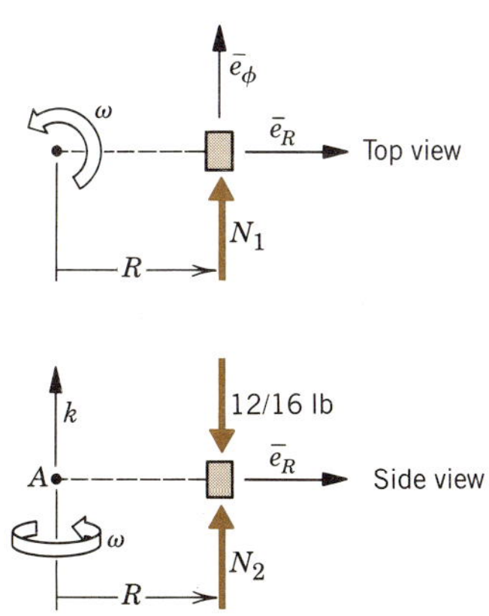

Step 2 It is given that $\omega = 450(2\pi/60)$ rad/s and that $\dot{\omega} = 0$. It is also known that the distance R from the axis of rotation to the center of mass of the collar is $R_1 = 12$ in. at the initial position and $R_2 = 30$ in. for the final position. At the instant the cable breaks, the collar is not moving outward, so $\dot{R}_1 = 0$. Find v, N_1, and N_2.

Step 3 It is convenient to employ the cylindrical coordinates defined in the free body diagram, primarily because the distance from a fixed axis is important and the rate of rotation is given.

Step 4 In cylindrical coordinates the velocity and acceleration are

$$\bar{v} = \dot{R}\bar{e}_R + R\dot{\phi}\bar{e}_\phi + \dot{z}\bar{k}$$
$$\bar{a} = (\ddot{R} - R\dot{\phi}^2)\bar{e}_R + (R\ddot{\phi} + 2\dot{R}\dot{\phi})\bar{e}_\phi + \ddot{z}\bar{k}$$
$$\equiv [\ddot{R} - (15\pi)^2 R]\bar{e}_R + [0 + 2\dot{R}(15\pi)]\bar{e}_\phi + 0\bar{k}$$

Step 5 The force components in the free body diagram give

$$\sum F_R = 0 = ma_R = \frac{12/16}{386}[\ddot{R} - 225\pi^2 R]$$

$$\sum F_\phi = N_1 = ma_\phi = \frac{12/16}{386}[30\pi\dot{R}]$$

$$\sum F_z = N_2 - \frac{12}{16} = 0$$

These three equations have three unknowns, N_1, N_2, and R. Note that the length unit has been defined to be inches, hence the choice for g.

Step 6 The values of the reaction N_1 and the speed both require the value of $\dot{R}$ when $R = 30$ in. The radial force equation yields a differential equation for R, specifically

$$\ddot{R} = 225\pi^2 R$$

Even though the symbol here is R, this differential equation suggests the case where acceleration ($\ddot{R}$) is dependent on position (R). Solving the differential equation by means of separating variables is achieved with the aid of the chain rule for differentiation. The initial conditions are that $\dot{R} = 0$ when $R = 12$ in.

$$\ddot{R} \equiv \frac{d}{dt}(\dot{R}) = \left[\frac{d}{dR}(\dot{R})\right]\frac{dR}{dt} = \dot{R}\frac{d\dot{R}}{dR} = 225\pi^2 R$$

$$\int_0^{\dot{R}} \dot{R}\,d\dot{R} = 225\pi^2 \int_{12}^{R} R\,dR \Rightarrow \tfrac{1}{2}\dot{R}^2 = \tfrac{1}{2}(225\pi^2)[R^2 - (12)^2]$$

Thus, at $R_2 = 30$ in.,

$$\dot{R}_2 = 1295.8 \text{ in./s}$$

The speed is therefore

$$v_2 = [v_R^2 + v_\phi^2]^{1/2} = [\dot{R}_2^2 + R_2^2\dot{\phi}^2]^{1/2}$$
$$= [(1295.8)^2 + (30)^2(15\pi)^2]^{1/2} = 1918 \text{ in./s}$$

and the reaction force components are found from the equations of motion to be

$$N_1 = 237 \text{ lb}, \qquad N_2 = 0.75 \text{ lb}$$

The large value for N_1 is a consequence of holding ω constant as the value of R increases.

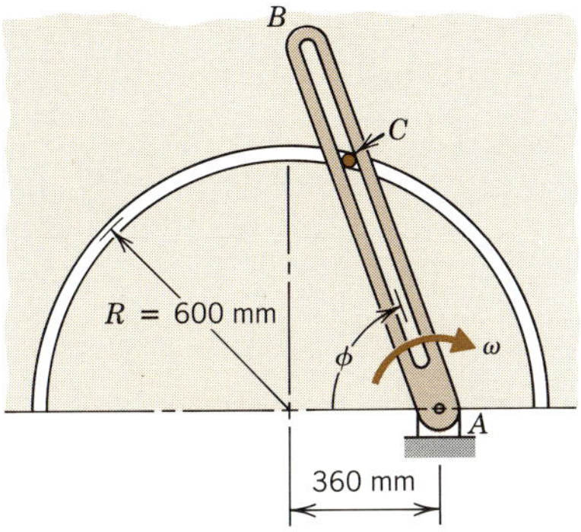

ILLUSTRATIVE PROBLEM 4

Arm AB rotates clockwise in the horizontal plane at the constant rate of 90 rpm, thereby moving the 400 gram pin C along the fixed circular groove. Determine the force exerted on pin C by arm AB when $\phi = 90°$.

Solution

Step 1 The only forces acting on pin C in the horizontal plane are contact forces $\bar{N}_1$ and $\bar{N}_2$, which are perpendicular to the walls of each groove, as shown in the free body diagram. Note that the diagram depicts the situation in the position of interest, $\phi = 90°$.

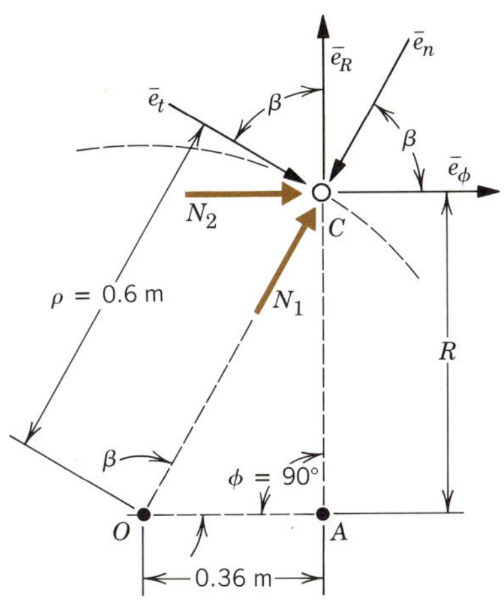

Step 2 Given that $m = 0.4$ kg, $\omega = \dot{\phi} = 90(2\pi/60)$ rad/s, and $\dot{\omega} = 0$. Find N_2.

Step 3 The movement along the path is defined by the rate of rotation of a radial line. This suggests the polar coordinates shown in the free body diagram. However, the properties of the path are more readily described in terms of path variables, with the tangent and normal vectors depicted in the diagram. Having important information given in this dual manner suggests the joint kinematical formulation that was developed in Section C of Chapter 11.

Step 4 The kinematical analysis begins with the position parameters. From right triangle OAC we obtain

$$\beta = \cos^{-1}\left(\frac{0.36}{0.6}\right) = 53.13°, \qquad R = [(0.6)^2 - (0.36)^2]^{1/2} = 0.48 \text{ m}$$

We may now express $\bar{e}_t$ and $\bar{e}_n$ in terms of polar components. The free body diagram gives the orientation of these vectors, which leads to

$$\bar{e}_t = -\cos\beta\,\bar{e}_R + \sin\beta\,\bar{e}_\phi = -0.6\bar{e}_R + 0.8\bar{e}_\phi$$

$$\bar{e}_n = -\sin\beta\,\bar{e}_R - \cos\beta\,\bar{e}_\phi = -0.8\bar{e}_R - 0.6\bar{e}_\phi$$

The joint descriptions for the velocity are

$$\bar{v} = \dot{R}\bar{e}_R + R\dot{\phi}\bar{e}_\phi = \dot{R}\bar{e}_R + 0.48(3\pi)\bar{e}_\phi$$
$$= v\bar{e}_t = v(-0.6\bar{e}_R + 0.8\bar{e}_\phi)$$

Equating like components we find

$$\bar{e}_R\colon\quad \dot{R} = -0.6v; \qquad \bar{e}_\phi\colon\quad 0.48(3\pi) = 0.8v$$
$$v = 1.8\pi \text{ m/s} \qquad \dot{R} = -1.08\pi \text{ m/s}$$

The equations for acceleration in the two formulations are

$$\bar{a} = (\ddot{R} - R\dot{\phi}^2)\bar{e}_R + (R\ddot{\phi} + 2\dot{R}\dot{\phi})\bar{e}_\phi$$
$$= [\ddot{R} - 0.48(3\pi)^2]\bar{e}_R + 2(-1.08\pi)(3\pi)\bar{e}_\phi$$
$$= \dot{v}\,\bar{e}_t + \frac{v^2}{\rho}\,\bar{e}_n$$
$$= \dot{v}(-0.6\bar{e}_R + 0.8\bar{e}_\phi) + \frac{(1.8\pi)^2}{0.6}(-0.8\bar{e}_R - 0.6\bar{e}_\phi)$$

Matching corresponding components yields

$$\bar{e}_R\colon\quad \ddot{R} - 4.32\pi^2 = -0.6\dot{v} - 4.32\pi^2$$
$$\bar{e}_\phi\colon\quad -6.48\pi^2 = 0.8\dot{v} - 3.24\pi^2$$
$$\dot{v} = -4.05\pi^2 \text{ m/s}^2, \qquad \ddot{R} = 2.43\pi^2 \text{ m/s}^2$$

The acceleration is now found to be

$$\bar{a} = -1.89\pi^2\bar{e}_R - 6.48\pi^2\bar{e}_\phi \text{ m/s}^2$$
$$= -4.05\pi^2\bar{e}_t + 5.40\pi^2\bar{e}_n \text{ m/s}^2$$

Steps 5 and 6 Because the acceleration has been found in terms of both path variables and polar coordinates, we may formulate Newton's second law with either. The free body diagram indicates that $\bar{N}_1$, which we do not seek, exerts no component in the $\bar{e}_t$ direction. We therefore write

$$\sum F_t = N_2 \sin\beta \equiv 0.8N_2 = ma_t = 0.4(-4.05\pi^2)$$
$$N_2 = -19.99\bar{e}_\phi \text{ lb}$$

Note that the minus sign means that our initial guess for the direction of $\bar{N}_2$ in the free body diagram was incorrect.

HOMEWORK PROBLEMS

12.78 A 3000 lb automobile accelerates at a constant rate from rest to 60 mi/hr in 10 s. The rolling resistance attributable to the undriven wheels is 20 lb. What is the tractive force exerted on the drive wheels by the ground during the acceleration? Air resistance is negligible.

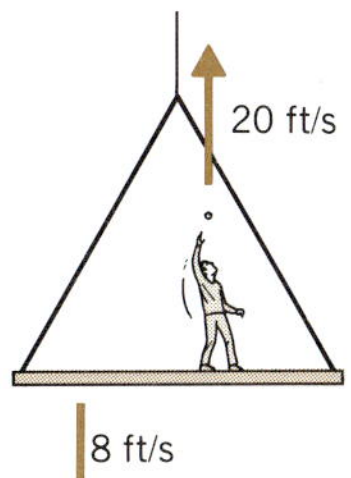

PROB. 12.79

12.79 A person throws a ball upward at 20 ft/s relative to an open platform, which is descending at the rate of 8 ft/s. Determine the elapsed time and distance traveled by the platform until the ball is caught.

12.80 A 3 kg block sliding down the inclined rod is brought to a stop by a cord that is held horizontally. The tension in the cord is 100 N and the coefficient of friction between the block and the rod is 0.4. The block was initially moving at 2 m/s. Determine the distance the block slides before coming to rest and the corresponding elapsed time.

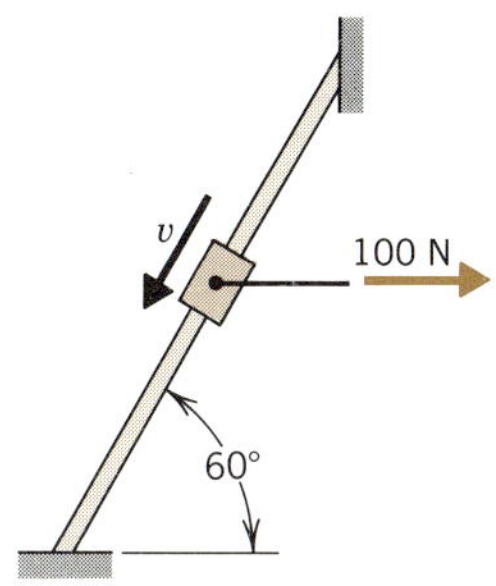

PROB. 12.80

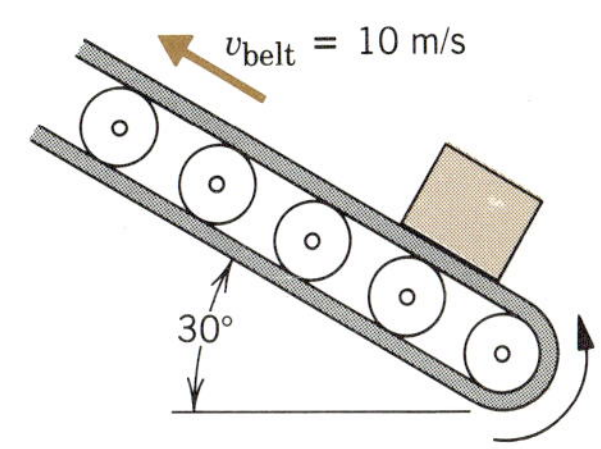

PROB. 12.81

12.81 A package is at rest when it is placed on a conveyor belt that is moving up a 30° incline at a constant speed of 10 m/s. The coefficient of kinetic friction between the package and the belt is 0.75. Determine how far the package will move up the incline before it ceases to slip relative to the conveyor belt. Also determine the corresponding elapsed time. *Hint*: Slipping ceases when the belt and the package have the same velocity.

PROB. 12.82

12.82 The tractor accelerates at $0.3g$ as it pulls the sled. The coefficient of kinetic friction between the sled and the ground is $\mu_k = 0.6$. Determine the tension in the cable and the net traction force exerted between the tractor and the ground.

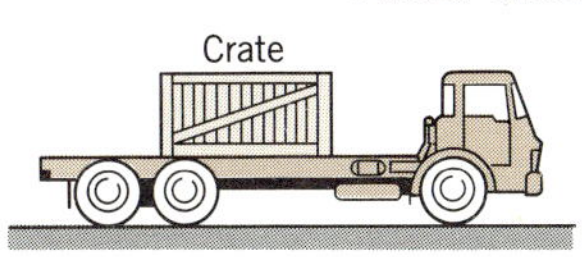

PROB. 12.83

12.83 A crate weighing 1 ton rests on the flat bed of a 5 ton truck traveling at 55 mi/hr. The coefficient of static friction between the crate and the truck is $\mu_s = 0.50$. Determine the minimum length of time in which the truck may be brought to rest without causing the crate to shift relative to the truck. What is the total braking force that the tires must produce for this deceleration?

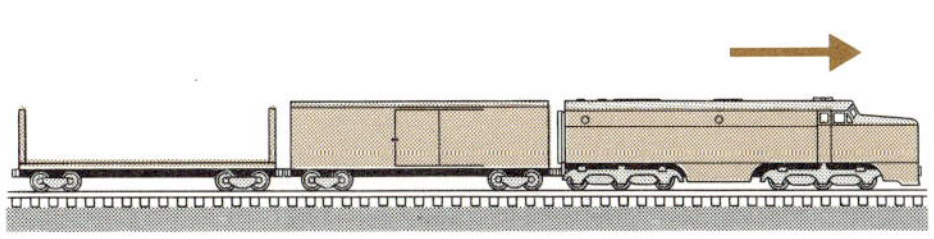
PROB. 12.84

12.84 A locomotive pulls a two-car train with a constant acceleration from rest to 20 m/s in 1 min along a level track. The mass of each car is 10^4 kg and the mass of the locomotive is $2(10^4)$ kg. The frictional resistance on each car and the locomotive attributable to the undriven wheels is 1 kN each. Determine the traction force of the locomotive and the force exerted within the couplers between the cars during the acceleration.

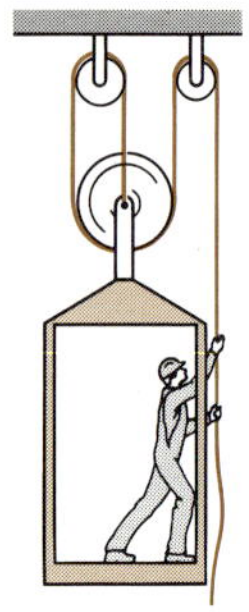

PROB. 12.85

12.85 The 70 kg sign painter applies a tensile force of 220 N to the free end of the cable. Determine (a) the acceleration of the painter, and (b) the force exerted between the painter and the platform.

12.86 Each block has a mass of 20 kg. What force $\bar{F}$ must be applied to block A to accelerate block B at 6 m/s^2? What is the corresponding tension force in the cable? Friction is negligible. The mass of the platform is negligible.

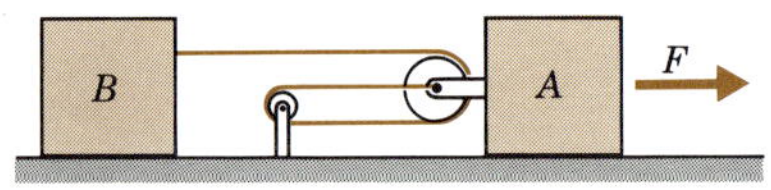

PROB. 12.86

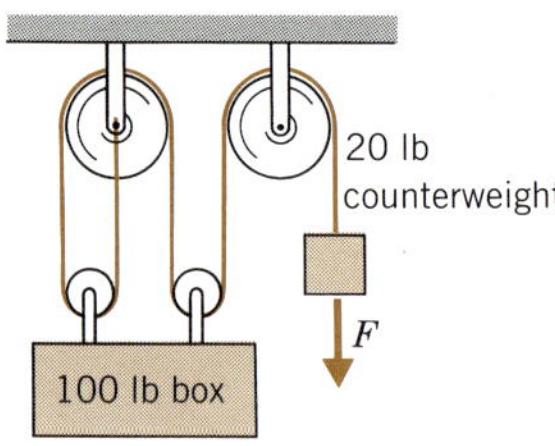

PROB. 12.87

12.87 The 100 lb box was initially descending at 8 ft/s. What constant force $\bar{F}$ must be applied to the 20 lb counterweight to bring the box to rest in 4 s? How far does the counterweight move in this time interval?

12.88 and 12.89 Blocks A and B, whose weights are 15 and 20 lb, respectively, are initially at rest. Determine the velocity of these blocks 5 s after they are released. All surfaces are smooth.

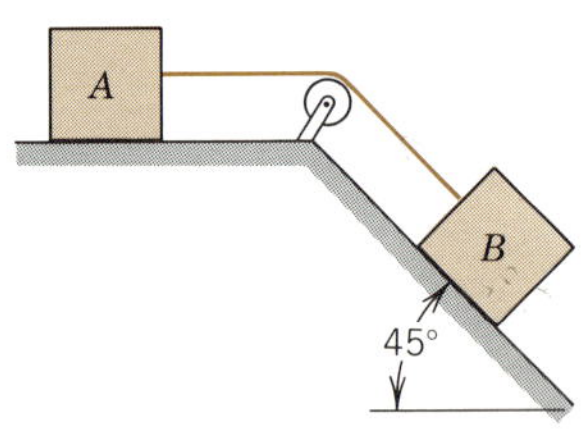

PROB. 12.88

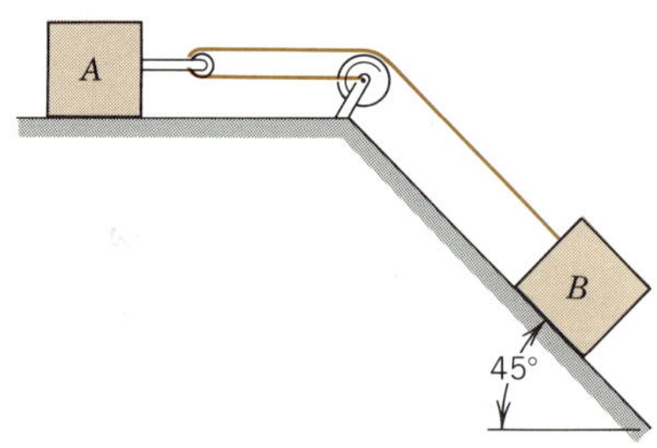

PROB. 12.89

***12.90** Solve Problem 12.89 if the coefficient of kinetic friction between all contacting surfaces is $\mu_k = 0.10$.

***12.91** A 600 lb motorboat is pointed upstream in a river flowing at 6 ft/s and the thrust of the outboard motor is 20 lb, with the result that the boat holds its position relative to the riverbank. The frictional force of the water is proportional to the speed of the boat relative to the water. Determine the true speed of the boat 30 s after the engine is turned off.

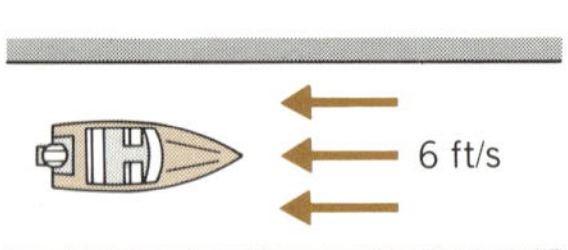

PROB. 12.91

***12.92** Determine the distance the boat in Problem 12.91 will drift before it attains a speed of 3 ft/s.

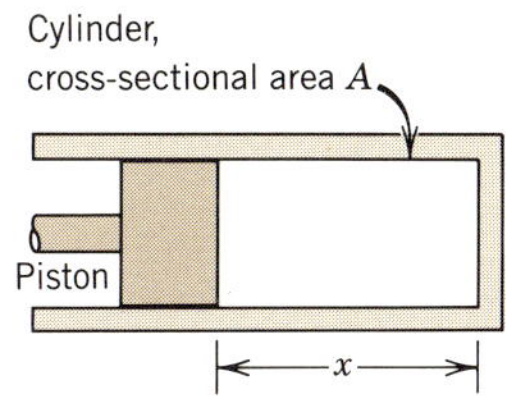

PROB. 12.93

***12.93** The 0.5 kg piston slides within a cylinder whose cross-sectional area $A = 5000\ \text{mm}^2$. The gas pressure p within the cylinder is inversely proportional to the volume of gas Ax, where x is as shown. Also, when $x = 50$ mm, the pressure p equals the constant atmospheric pressure $p_0 = 1.0133(10^5)$ pascal. If the piston passes the position $x = 50$ mm with a velocity to the right of 500 mm/s, determine the distance the piston travels before coming to rest.

12.94 Two cords are attached to a 1 kg sphere that revolves in a horizontal circle of 500 mm radius at a constant speed of 2.5 m/s. Determine the tension force in each cord.

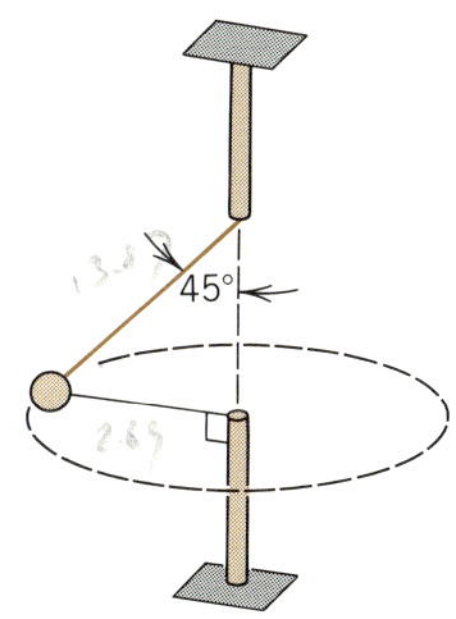

PROB. 12.94

12.95 What is the greatest speed of the sphere in Problem 12.94 for which the lower cable is not slack?

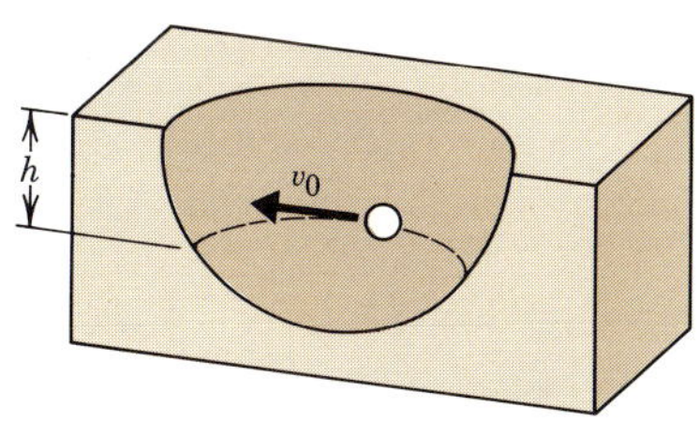

PROB. 12.96

12.96 A small ball bearing follows a horizontal circle as it rolls at a constant speed v on the interior of a hemispherical bowl of radius r. Neglecting friction, determine the relationship between v and the depth h of the ball below the rim of the bowl.

12.97 The 300 gram collar is pulled by a spring along the curved rod that lies in the horizontal plane. The spring has a stiffness of 150 N/m, and its unstretched length is 350 mm. The collar has a speed of 10 m/s in the position shown, and the guide rod is smooth. Determine the acceleration of the collar and the force exerted by the rod on the collar in the position shown.

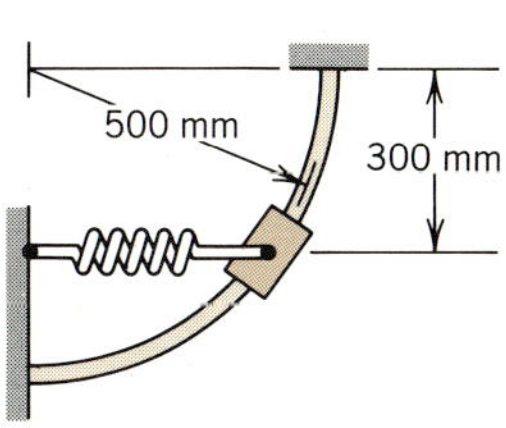

PROB. 12.97

12.98 Solve Problem 12.97 for the case where the coefficient of kinetic friction between the collar and the rod is $\mu_k = 0.40$.

12.99 An automobile is slowed at a constant rate from 108 km/hr to 72 km/hr in a distance of 600 m as the vehicle follows a level curve of 500 m radius. What are the magnitude and angle relative to the road of the horizontal force exerted between the automobile and the ground at the initial position?

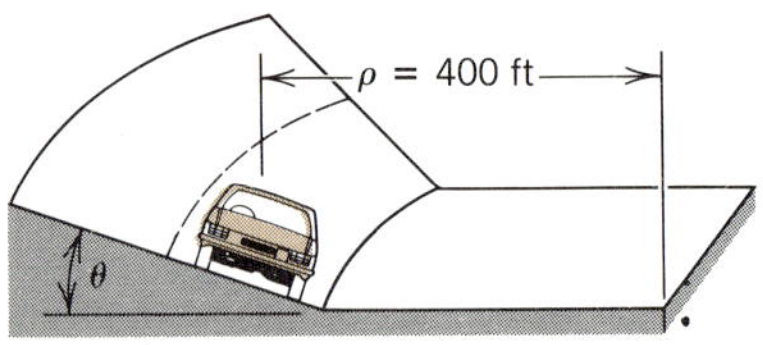

PROB. 12.100

12.100 A 2400 lb automobile has a speed of 60 mi/hr along a curved horizontal roadway, whose radius of curvature is 400 ft. For the case where the road is unbanked ($\theta = 0$), determine the resultant friction force acting on the vehicle when it begins to slow down at $0.3g$. What is the corresponding minimum allowable coefficient of static friction, μ_s?

***12.101** Solve Problem 12.100 for the case where the road is banked at $\theta = 10°$.

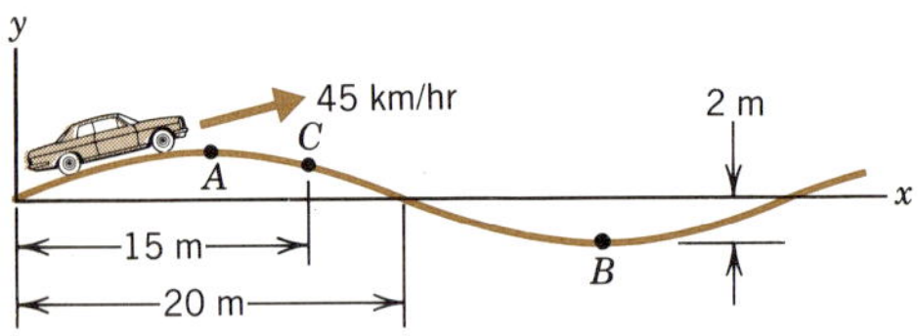

PROB. 12.102

12.102 The vertical elevation of a straight road is defined by $y = 2\sin(\pi x/20)$, where x and y are measured in meters. Determine the maximum rate at which an automobile traveling at 45 km/hr can decelerate at (a) point A and (b) point B. The coefficient of friction between the tires and the ground is 0.60.

***12.103** Determine the maximum rate at which the automobile in Problem 12.102 can decelerate at position C.

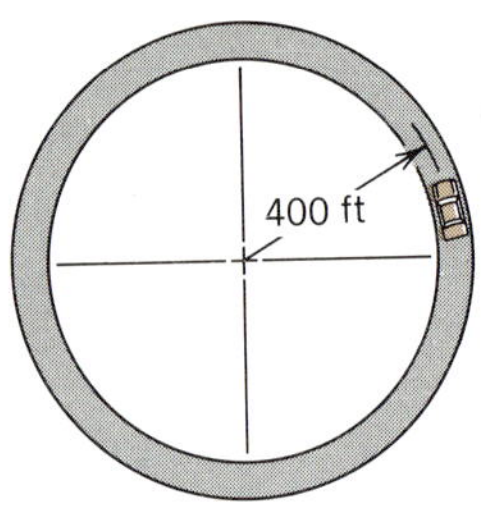

PROB. 12.104

***12.104** A 2400 lb automobile on a circular track having a 400 ft radius of curvature starts from rest. It gains speed at the rate of $5\sin(0.1t)$ ft/s^2, where t is in seconds. The track is level and unbanked. Determine the magnitude of the resultant traction force exerted between the tires and the ground when $t = 15$ s.

***12.105** A 20 kg block of ice slides within a chute. Section AB forms a level 90° turn of 5 m radius. The viscous friction force between the ice and the metal chute is given by $f = 0.2v$ N, where v is the speed in meters per second. At position A the ice block is entering the turn at a speed of 10 m/s. Determine (a) the speed of the ice block at position B, and (b) the normal force exerted by the side wall of the chute at position B.

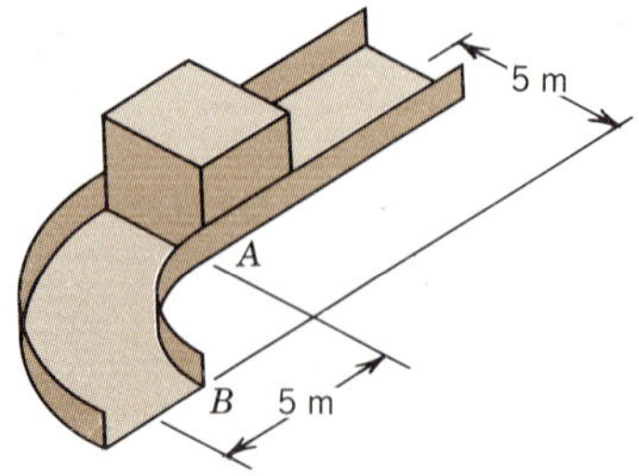

PROB. 12.105

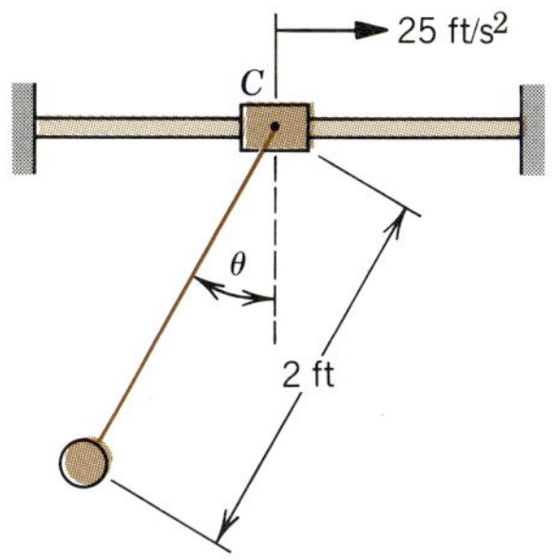

PROB. 12.106

12.106 The 6 oz sphere is tied to a cable hung from collar C. This collar accelerates to the right at the constant rate of 25 ft/s^2. Determine the constant angle of inclination θ.

12.107 A rider on an open platform that is descending at the constant speed of 3 m/s throws a ball. Relative to the platform, the ball's initial velocity is horizontal at 12 m/s. The ground is 10 m below the location where the ball is thrown. (a) Where does the ball hit the ground? (b) Determine the time required for the ball to hit the ground.

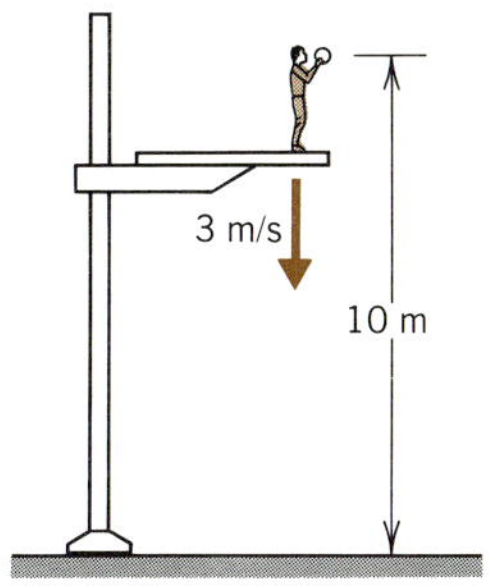

PROB. 12.107

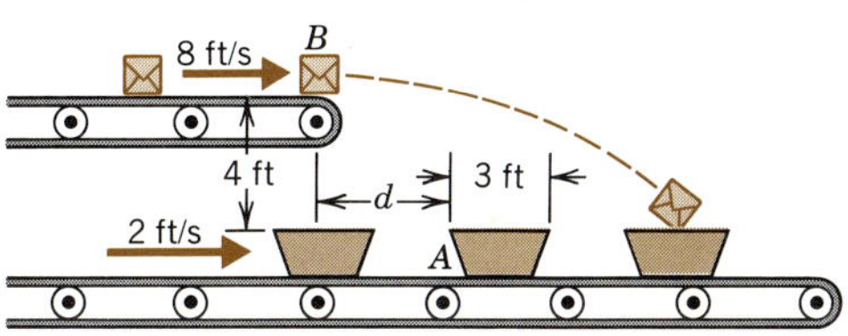

PROB. 12.108

12.108 Mail parcels are projected from a horizontal conveyor belt moving at 8 ft/s and are to be caught by bins running on a parallel conveyor belt 4 ft below. The speed of this conveyor belt is 2 ft/s. In order to catch package B in bin A, what is the smallest distance d that bin A should be ahead of package B when the package is projected from the upper conveyor belt?

12.109 The longitudinal axis of a 1000 lb guided missile remains horizontal as the engine produces a constant thrust of 2000 lb. If the missile was launched from an aircraft traveling at 900 mi/hr in level flight at an altitude of 1000 ft, determine the elapsed time before it hits the ground, and the corresponding distance it travels horizontally.

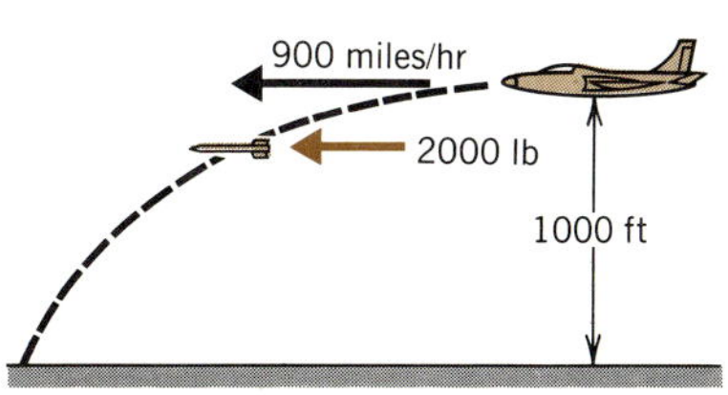

PROB. 12.109

12.110 In Problem 12.109, determine the velocity of the missile at the instant before it hits the ground.

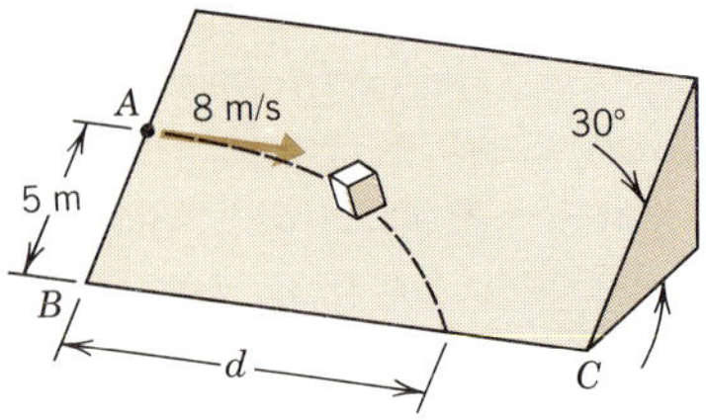

PROB. 12.111

12.111 A 600 gram steel block slides on a smooth 30° incline. The block was released from position A with an initial velocity of 8 m/s directed parallel to the horizontal edge BC. Determine the distance d at which the block crosses edge BC.

12.112 A person exerts the 30 lb horizontal force in order to push the 20 lb cart up the curved hill. In the position shown, the horizontal component of the velocity of the cart is 5 ft/s. Determine the acceleration of the cart in this position.

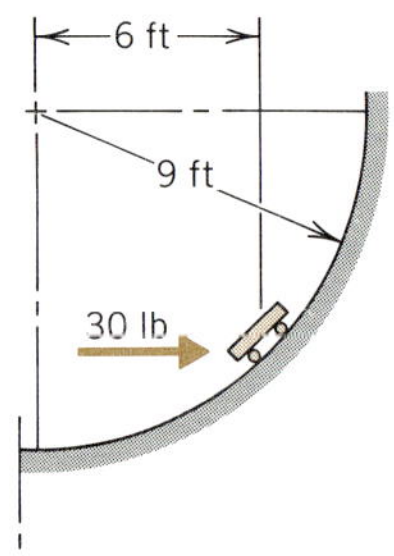

PROB. 12.112

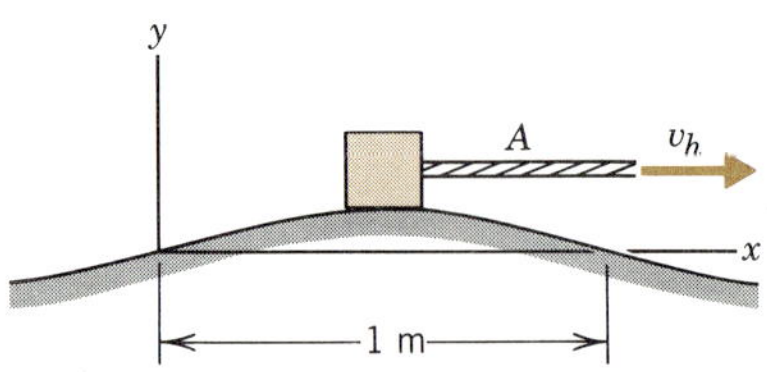

PROB. 12.113

12.113 The 800 gram block is pulled by a cable over a wavy surface whose elevation in the vertical plane is $y = 0.1\sin(\pi x)$ where x and y are in meters. The cable A, which is always oriented horizontally, imparts a constant horizontal velocity component v_h to the block. Determine the maximum value of v_h for which the block will remain in contact with the surface at $x = 0.5$ m.

12.114 Determine the maximum value of v_h in Problem 12.113 for which the block will remain in contact with the surface at $x = 0.25$ m.

12.115 A ball is thrown down a 15° incline with an initial velocity of 50 ft/s at 30° above the horizontal. Determine the distance d at which the ball returns to the incline, and also determine the corresponding elapsed time. *Hint:* Measure position parallel and perpendicular to the incline.

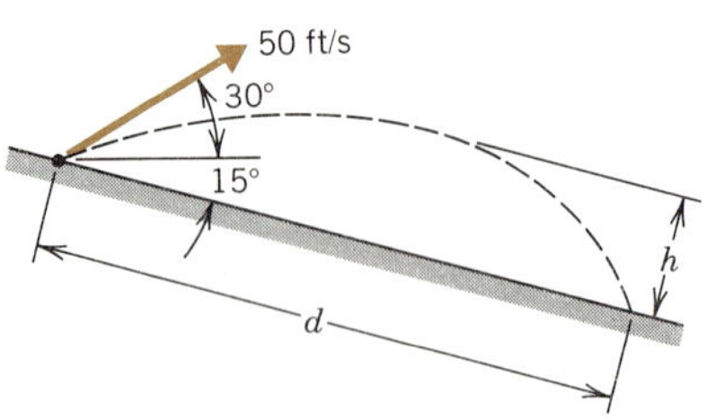

PROB. 12.115

12.116 In Problem 12.115, determine the maximum height h, perpendicular to the incline, that is attained by the ball.

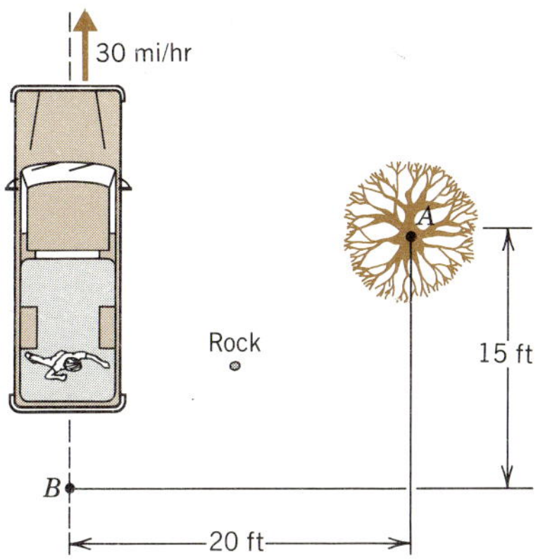

PROB 12.117

***12.117** A person standing on the bed of a pickup truck traveling at 30 mi/hr on a straight road throws a rock horizontally, aiming the throw perpendicular to the velocity of the truck, with the result that the rock hits a tree at point A, as shown in the top view. Knowing that the rock was thrown at position B, 15 ft before the location of the tree, determine (a) the true initial velocity of the rock, (b) the location of the rider when the rock hits the tree, and (c) the distance the rock has fallen when it hits the tree.

12.118 The speed governor consists of a plunger that moves outward as the angular speed increases until it contacts switch S. When $\omega = 0$, the plunger is in the position shown and the force in the spring is 80 N. The stiffness of the spring is 400 N/m. The 250 gram mass of the plunger may be considered to be concentrated at point G. Determine the constant value of ω for which the plunger will contact the switch. The system lies in the horizontal plane.

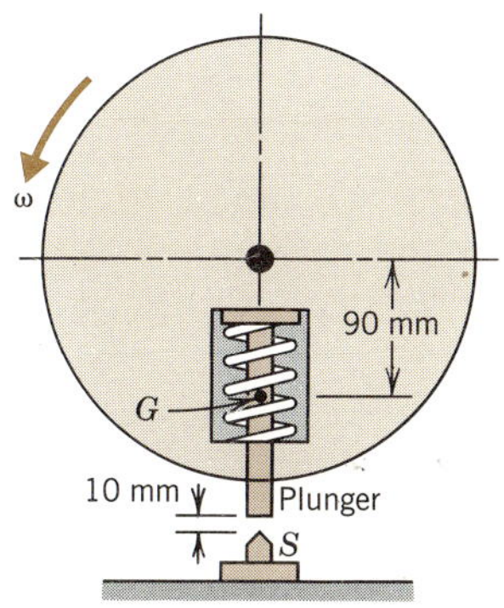

PROB. 12.118

12.119 and 12.120 Hollow plastic spheres weighing 2 oz are tested within the centrifuge device whose cross section is shown. Determine the normal forces exerted between a sphere and each of the smooth walls for a constant rate of rotation of 3600 rpm.

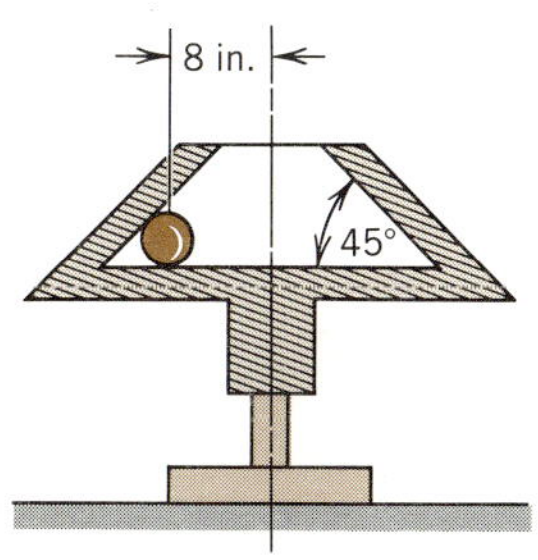

PROB. 12.119

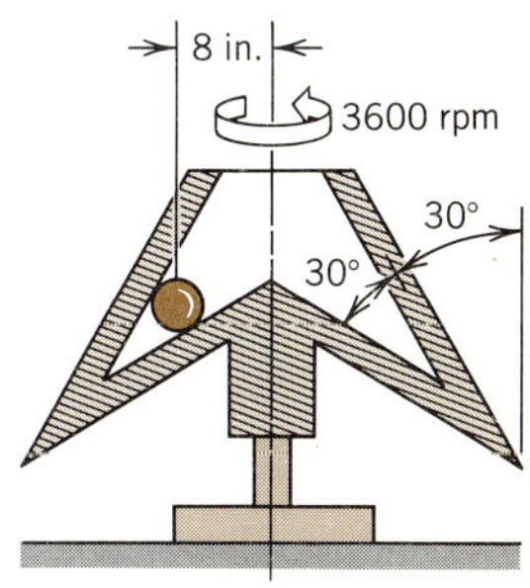

PROB. 12.120

PROB. 12.121

12.121 A 1.5 kg block sliding on a smooth horizontal table is tied to a cable that passes through hole O. At the instant shown, $R = 800$ mm, the velocity of the block is $v = 10$ m/s at $\beta = 90°$, and the cable tension is 100 N. Determine the value of $\ddot{R}$ at this instant.

12.122 Solve Problem 12.121 for the case where $\beta = 60°$.

12.123 A satellite in orbit above the earth has the instantaneous position and velocity shown. The inverse square law for gravitation is $F = mgR_e^2/(R_e + y)^2$, where $R_e = 6730$ km is the mean radius of the earth. Determine the value of $\ddot{y}$ at this position.

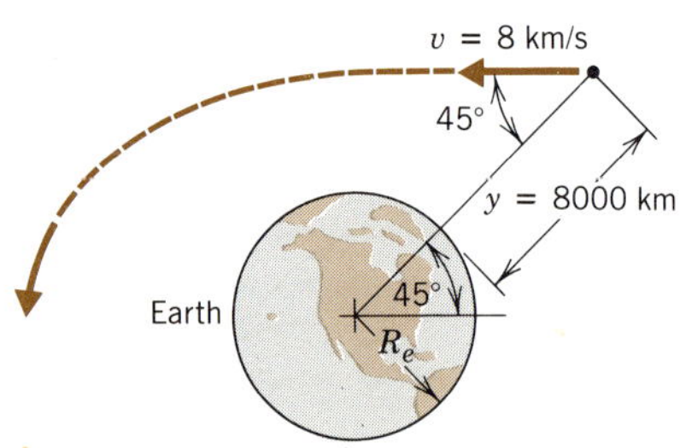

PROB. 12.123

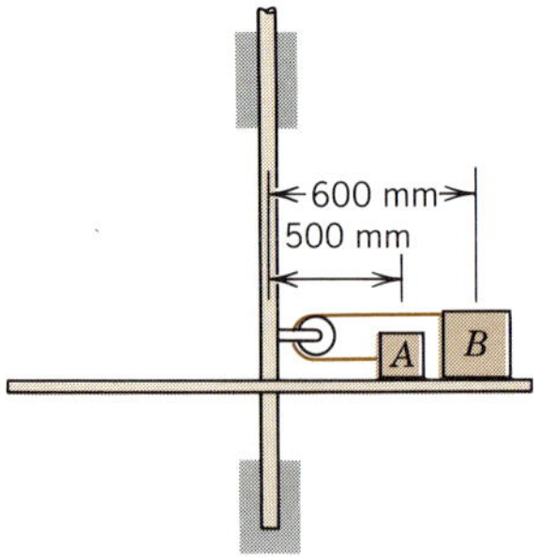

PROB. 12.124

12.124 Two blocks rest on a turntable that is rotating about the vertical axis at a constant angular speed ω. The blocks are connected by an inextensible cable. The masses of blocks A and B are 1.0 and 1.5 kg, respectively, and the coefficient of friction is 0.30. Determine the maximum angular speed for which the blocks will not slide. Which way will the blocks slide if ω exceeds this value?

12.125 Arm AB, which rotates in the horizontal plane at the constant rate of 10 rad/s, pushes collar C through the straight groove. The mass of collar C is 0.2 kg. Determine the force exerted by arm AB on the collar when $\phi = 53.13°$.

12.126 Arm AB rotates clockwise at the constant rate of 180 rpm in the vertical plane. This causes the 400 gram pin C to follow the fixed groove defined by $R = 0.5 + \cos^2 \phi$, where R is in meters. Determine the forces exerted on the pin when $\phi = 90°$.

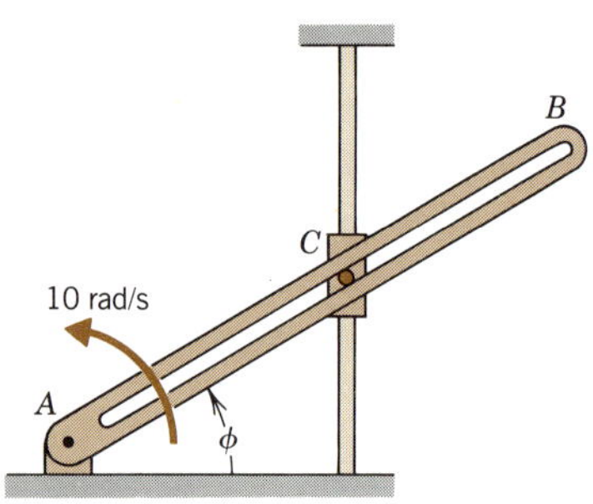

PROB. 12.125

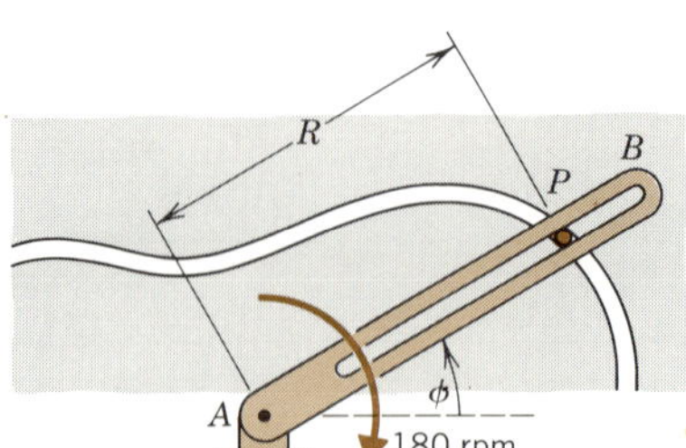

PROB. 12.126

***12.127** Solve Problem 12.126 for $\phi = 45°$.

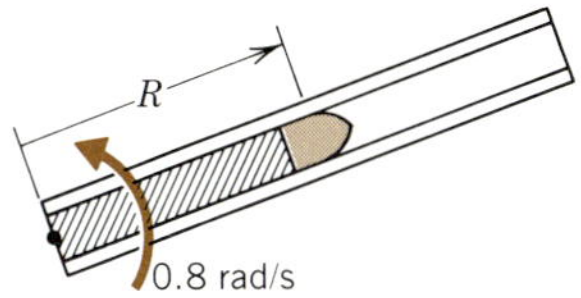

PROB. 12.128

12.128 The compressed air gun fires a 4 oz pellet by exerting a constant force of 10 lb. The pellet accelerates from rest at $R = 3$ in., while the barrel rotates in the horizontal plane at the constant rate of 0.8 rad/s. Determine (a) the acceleration of the pellet with respect to the barrel at $R = 15$ in., and (b) the force exerted by the barrel on the bullet at that position.

***12.129** Collar A, whose mass is 0.5 kg, is pushed outward along bar BC by the spring, while bar BC rotates about the vertical axis at the constant rate of 60 rpm. The spring has a stiffness of 50 N/m and an unstretched length of 300 mm. The collar is released from rest with respect to the bar at $R = 200$ mm. Determine the speed of the collar at $R = 300$ mm.

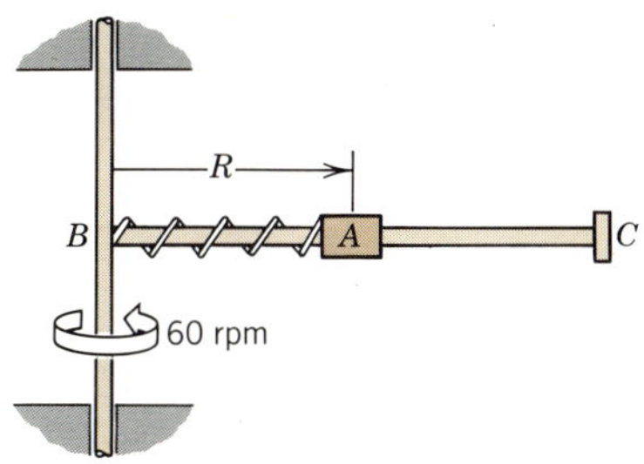

PROB. 12.129

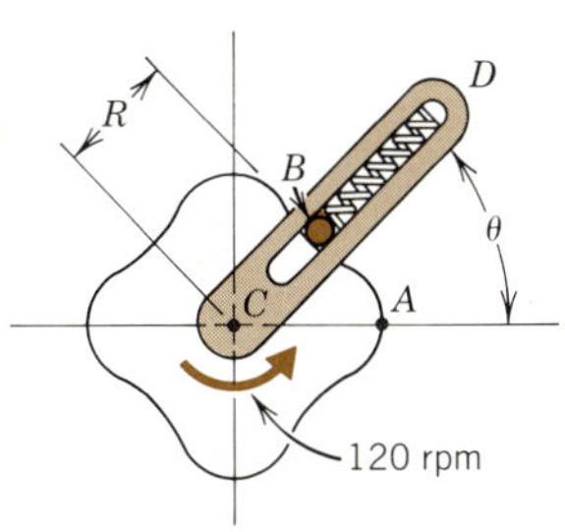

PROB. 12.130

12.130 The arm CD rotates counterclockwise in a horizontal plane at the constant rate of 120 rpm as it guides an 8 oz pin over the fixed camshaft. The equation of the camshaft is $R = 10 + 2 \cos 4\theta$, where R is in inches and θ is in radians. The minimum compressive force in the spring is 9 lb and the spring stiffness is 3 lb/in. Determine the force exerted by the cam on the pin when $\theta = 0°$.

***12.131** The free end of the cable is pulled in at a constant speed of 40 in./s, thereby pulling the 500 gram collar to the left. Determine the cable tension when $x = 20$ in.

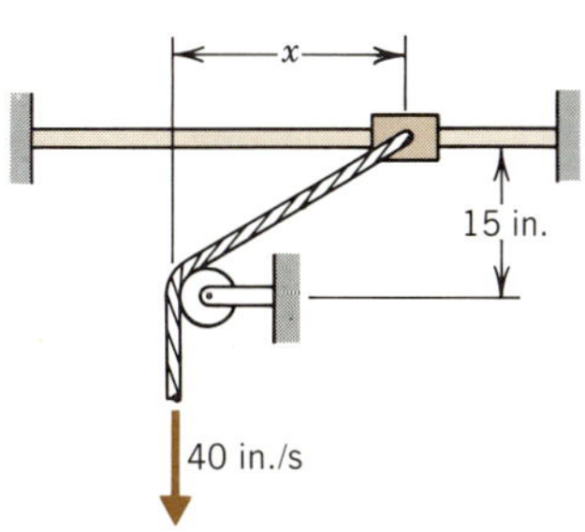

PROB. 12.131

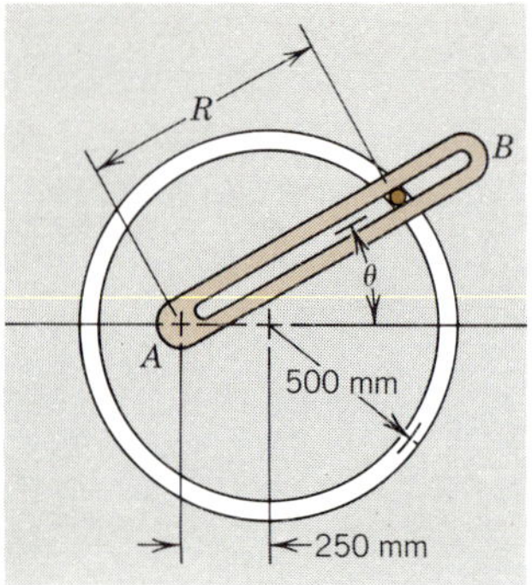

PROB. 12.132

*12.132 The eccentric arm AB rotates clockwise at the steady rate of 180 rpm, causing the 100 gram pin to follow the fixed circular groove whose radius is 500 mm. The system lies in the horizontal plane and friction is negligible. Determine the force exerted between arm AB and the pin when $\theta = 90°$.

*12.133 The hydraulic piston is being extended at the constant rate of 8 ft/s when collar B is in the position shown on the circular rod. The collar weighs 2 lb, and the system is situated in the horizontal plane. Determine the force $\bar{F}$ acting along line AB that the piston must exert on the collar in order to achieve this motion.

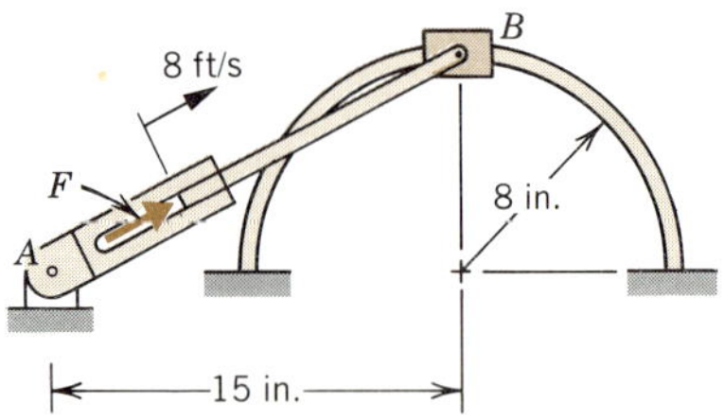

PROB. 12.133

E. NEWTON'S EQUATION OF MOTION—SYSTEM OF PARTICLES

In all of the examples and problems discussed thus far in this text we have modeled some large bodies, such as automobiles and airplanes, as though they were a single particle. In order to understand the significance of such a model, let us consider a general system of particles.

A primary property of such a system is the position of its center of mass. This point is defined by studying the resultant of the gravitational force acting on each of the N particles in Figure 3a. Because these forces are all parallel to the z direction, the magnitude of their resultant $\bar{w}$ is the sum of their individual magnitudes, $m_i g$. This is the obvious fact that the resultant is the total weight

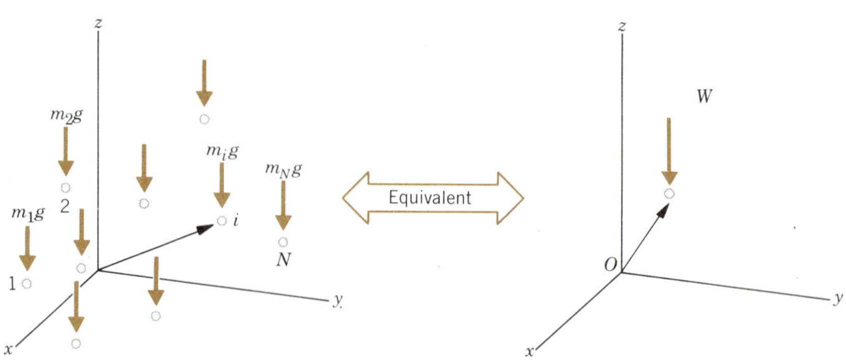

FIGURE 3

$$w = mg \qquad \text{and} \qquad m = \sum_{i=1}^{w} m_i$$

The line of action of $\bar{w}$ must intersect the center of mass of the system of particles, by definition. It was shown in Statics that the line of action of an equivalent force is obtained by equating the moments of the equivalent force systems in Figure 3a. This yields

$$\sum \bar{M}_O = \sum_{i=1}^{N} \bar{r}_{i/O} \times (-m_i g\bar{k}) \equiv \bar{r}_{G/O} \times (-mg\bar{k})$$

The foregoing equality must also be true in the case where gravity is not acting in the z direction. For instance, if the gravity field were aligned with the y axis, then in the foregoing equation we would replace $\bar{k}$ by $\bar{j}$. This observation essentially permits us to cancel the cross product and unit vectors. The result is a general vector representation of the *first moment of mass* with respect to point O, as follows.

$$\boxed{m\bar{r}_{G/O} = \sum_{i=1}^{N} m_i \bar{r}_{i/O}} \tag{21}$$

Computations employing this relation are usually performed in component form. However, the vector form, eq. (21), is more useful for the discussion here.

Before we consider Newton's laws for the system of particles, let us categorize the individual forces acting on each particle in the system according to their source. We denote by $\bar{F}_i$ the resultant force on particle i from sources external to the system. In addition to the external forces, there will be interaction forces $\bar{f}_{ij}\,(i \neq j)$ on each particle i due to every other particle j in the system.

Examples of external forces are gravity and contact forces exerted between the earth and the particle being studied. Interaction forces can arise from particles within the system that are in contact, and from the gravitational attractions between particles. A simple case is illustrated in Figure 4 for a system of three particles. Notice that each pair of interaction forces satisfies Newton's third law, so that $\bar{f}_{ij} = -\bar{f}_{ji}$.

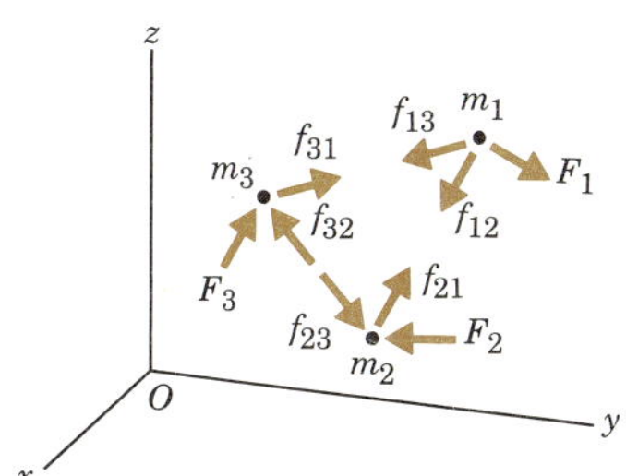

FIGURE 4

Let us now extend Figure 4 to the case where N particles are contained in the system. Consider the arbitrary ith particle from that group. Newton's second law for that particle requires that we sum the external force $\bar{F}_i$ and each internal force $\bar{f}_{ij}$ acting on the particle, so that

$$\bar{F}_i + \sum_{\substack{j=1 \\ i \neq j}}^{N} \bar{f}_{ij} = m_i \bar{a}_i = m_i \ddot{\bar{r}}_{i/O} \tag{22}$$

Let us now form the resultant of the whole set of forces acting on the system. This resultant is obtained by adding the forces acting on each particle. As was observed earlier for the system in Figure 4, the forces $\bar{f}_{ij}$

and $\bar{f}_{ji}$ associated with any two particles i and j cancel each other. Therefore, when we sum eq. (22) for each particle, we find

$$\sum_{i=1}^{N} \bar{F}_i = \sum_{i=1}^{N} m_i \ddot{\bar{r}}_{i/O} = \frac{d^2}{dt^2} \sum_{i=1}^{N} m_i \bar{r}_{i/O}$$

This expression gains greater significance when we note that the summation in the right side defines the center of mass of the system according to eq. (21). Hence

$$\sum_{i=1}^{N} \bar{F}_i = \frac{d^2}{dt^2}(m\bar{r}_{G/O}) = m\ddot{\bar{r}}_{G/O}$$

$$\boxed{\sum_{i=1}^{N} \bar{F}_i = m\bar{a}_G} \tag{23}$$

In words, eq. (23) states that the acceleration of the center of mass is affected only by the total external force, and, equally important, it is unaffected by the interaction between the particles of the system.

The primary reason for presenting eq. (23) at this point in our studies is the direct analogy between this equation and Newton's second law. This shows that *when we model the collection of atoms composing a physical body as a single particle, we are studying the motion of the center of mass of the body.*

Equation (23) is essential to the study of kinetics of rigid bodies, which we will begin in Chapter 14. We will also use it later in this chapter to investigate phenomena involving interacting particles.

F. WORK–ENERGY RELATIONS

We have seen that the application of Newton's laws often results in a differential equation relating some of the kinematical parameters describing the motion of a particle. The work–energy principle represents a standard integral of such differential equations. Basically, it relates the motion of a particle at two different positions along its path. Although the work–energy principle circumvents the need to solve a differential equation, it is not suitable for all situations. Therefore, it complements, but does not replace the formulation of the equations of motion.

1. Work Done by Forces

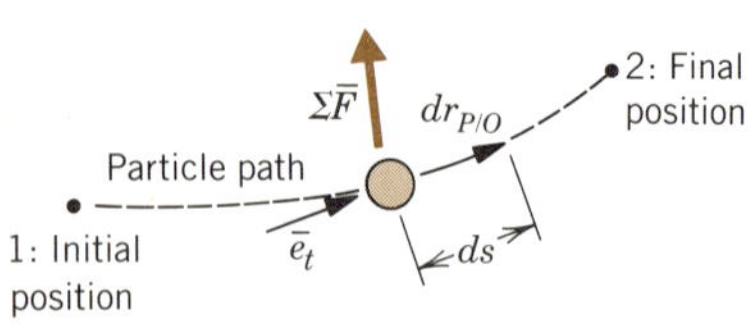

FIGURE 5

Consider the particle in Figure 5 as it travels an infinitesimal arc length ds along its path under the action of the resultant force $\sum \bar{F}$. The infinitesimal amount of work dU done by this resultant force in moving the particle is defined to be the product of the component of the force in the direction of motion and the infinitesimal distance traveled.

$$dU = \sum F_t \, ds$$

Thus, the total work done by the resultant force in moving the particle from point 1 to point 2, as indicated in Figure 5, is

$$U_{1\to 2} \equiv \int_{s_1}^{s_2} \sum F_t\, ds \tag{24}$$

The evaluation of this expression requires that we express the resultant force in terms of its tangent and normal components.

A more general definition of the work done can be obtained by referring to the infinitesimal change in the position vector of the particle $d\bar{r}_{P/O}$, as illustrated in Figure 5. Specifically, we have

$$U_{1\to 2} = \int_{r_1}^{r_2} \sum \bar{F} \cdot d\bar{r}_{P/O} \tag{25}$$

where r_1 and r_2 denote the coordinates of the initial and final positions of point P measured from an arbitrary fixed origin O.

Note that work is the product of a force and a length quantity. The unit of work in U.S. Customary units is usually foot-pound. (The length unit is listed first in order to avoid ambiguity with the units for torque: pound-feet.) This confusion does not arise in SI units, where work is measured in joules (1 J ≡ 1 m-N).

EXAMPLE 16

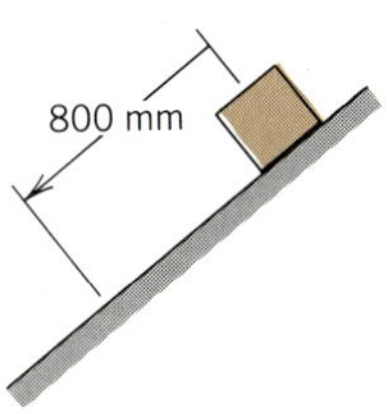

A 2 kg block slides 800 mm down a 40° incline. The coefficient of kinetic friction is 0.25. Determine the work done by the forces acting on the block.

Solution

The free body diagram shows that the forces acting on the block are gravity and friction. Path variables with the tangent vector oriented downward are suitable for this rectilinear motion.

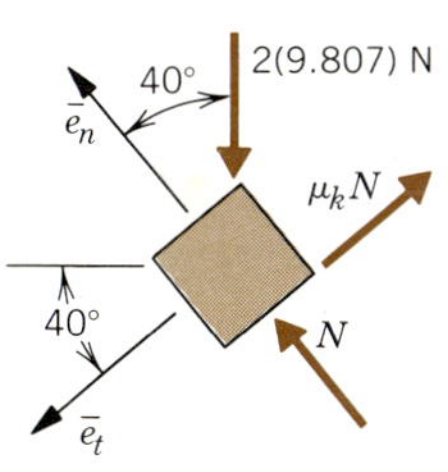

The normal force $\bar{N}$ does no work because it has no tangential component. However, we must determine its magnitude in order to account for the friction force. There is no acceleration in the normal direction, so

$$\sum F_n = N - 2(9.807)\cos 40^\circ = ma_n = 0$$

$$N = 15.03 \text{ N}$$

The tangential component of the resultant force may now be determined.

$$\sum F_t = 2(9.807)\sin 40^\circ - 0.25(15.03) = 8.850\ \text{N}$$

The work done in a 0.80 m movement of the block is then found to be

$$U_{1\to 2} = \int_0^{0.80} (8.850)\, ds = 8.850 \int_0^{0.80} ds$$
$$= 7.08\ \text{J}$$

Note that, in this problem, $\sum F_t$ was a constant value, so the computation of the work reduced to the simple relation, "component of force in the direction of displacement multiplied by the distance traveled."

EXAMPLE 17

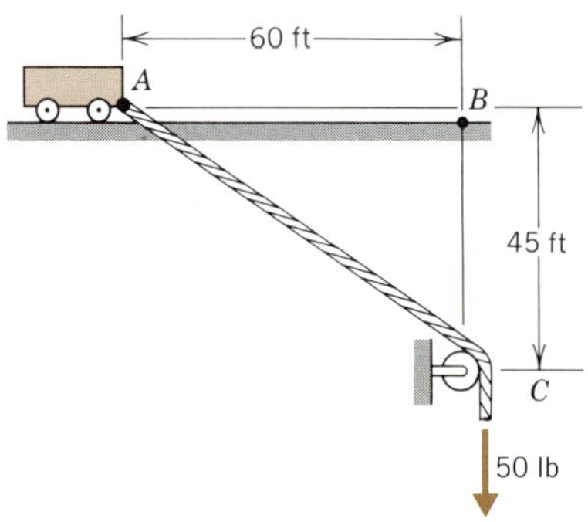

A 100 lb cart is pulled along a horizontal rail by the cable, which carries a tension of 50 lb. The cart moves from position A to position B. Determine the work done by the forces acting on the cart in this movement.

Solution

The free body diagram shows that the only force acting on the cart in the tangential direction is the cable tension, which is 50 lb. In contrast to the previous problem, the tangential component is not constant because the angle β for the cable depends on the position of the cart.

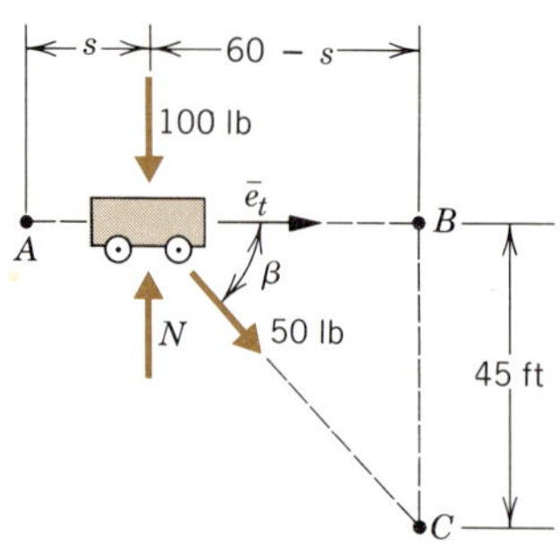

We have defined the path variables such that s measures the distance from the initial position A, and $\bar{e}_t$ is in the sense of increasing s. The work is then found by using the dashed triangle to express β.

$$\sum F_t = 50\cos\beta = 50\,\frac{60 - s}{[(60 - s)^2 + (45)^2]^{1/2}}$$

$$U_{A\to B} = \int_0^{60} 50\,\frac{(60 - s)}{[(60 - s)^2 + (45)^2]^{1/2}}\, ds$$

$$= -50[(60 - s)^2 + (45)^2]^{1/2}\Big|_{s=0}^{s=60} = 1500\ \text{ft-lb}$$

This system typifies a common situation in which there is a simpler method of employing the basic definitions, eqs. (24) or (25), for computing the work done. Here the work done by the cable is the same as the work done by the tensile force acting at the free end of the cable. This force is constant, so we need to compute only the distance the free end is pulled.

The length from the pulley C to the cart for the initial position A and final position B are found from the free body diagram to be

$$l_A = [(45)^2 + (60)^2]^{1/2} = 75 \text{ ft}, \qquad l_B = 45 \text{ ft}$$

The cable length does not change, so the free end of the cable moves downward by

$$\Delta l = l_A - l_B = 30 \text{ ft}$$

This movement of the free end is in the same sense as the 50 lb tensile force acting on the end, so

$$U_{A\to B} = 50\,\Delta l = 1500 \text{ ft-lb}$$

The value obtained by both methods is obviously identical, although the methods are quite different.

EXAMPLE 18

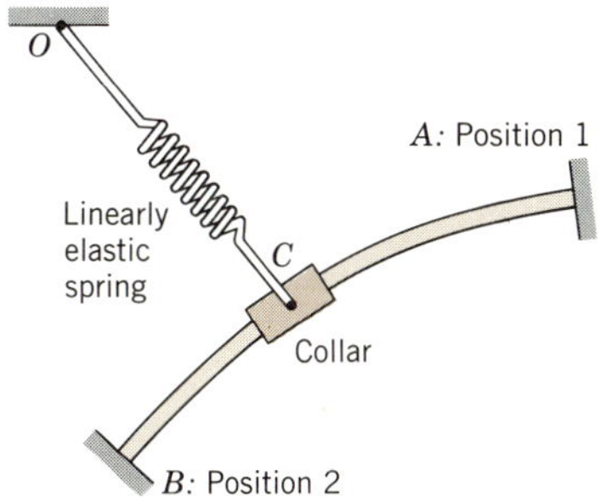

One end of a linear elastic spring whose unstretched length is L_0 is attached to a fixed point O, and the other end is attached to collar C, which slides on a smooth curved bar, as shown in the diagram. The spring and collar are coplanar. Determine the work done by the spring force when the collar C is moved from point A to point B.

Solution

The force in the spring depends on the distance from the fixed end O. Cylindrical coordinates having origin O are convenient for this situation. Note that the radial distance R to the collar is merely the sum of the unstretched length L_0 and the elongation Δ.

$$R = L_0 + \Delta$$

Also the spring force is collinear with line OC, so it has only a radial component.

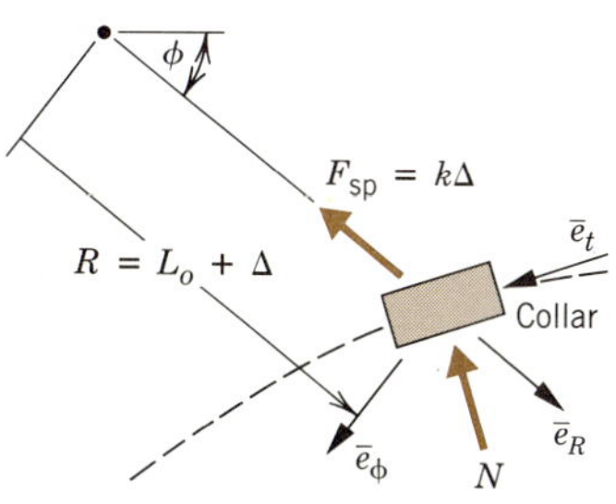

The vector relation for work, eq. (25), is convenient when cylindrical coordinates are used. For this planar situation, we have

$$\bar{F}_{sp} = -k\Delta\bar{e}_R$$

$$d\bar{r}_{P/O} = \bar{v}_P\,dt = (\dot{R}\bar{e}_R + R\dot{\phi}\bar{e}_\phi)\,dt$$

$$= dR\,\bar{e}_R + R\,d\phi\bar{e}_\phi$$

$$U_{1\to 2} = \int_{r_1}^{r_2} \bar{F}_{sp}\cdot d\bar{r}_{P/O} = -\int_{r_1}^{r_2} k\Delta\,dR$$

Now note that increasing R by dR increases Δ by the same amount, so $dR = d\Delta$. Thus

$$U_{1\to 2} = -\int_{\Delta_1}^{\Delta_2} k\Delta\,d\Delta = \tfrac{1}{2}k(\Delta_1^2 - \Delta_2^2)$$

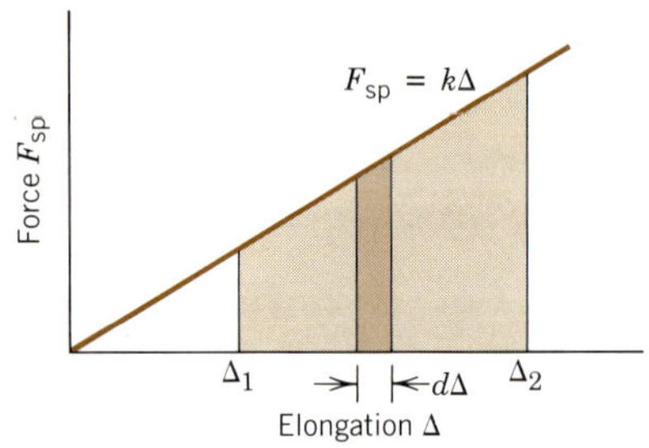

We can gain added insight to the meaning of this result by examining a graph of the force exerted by the spring as a function of its elongation, as shown in the margin.

When the motion of the particle results in increasing the elongation of the spring from Δ_1 to Δ_2, the force applied by the spring is opposite the motion of the particle, so that

$$U_{1\to 2} = -\int_{\Delta_1}^{\Delta_2} F_{sp}\,d\Delta$$

This integral is the lightly shaded trapezoidal area in the sketch. Hence

$$U_{1\to 2} = -\tfrac{1}{2}[(F_{sp})_1 + (F_{sp})_2](\Delta_2 - \Delta_1)$$

$$= -\tfrac{1}{2}[k\Delta_1 + k\Delta_2](\Delta_2 - \Delta_1) = \tfrac{1}{2}k(\Delta_1^2 - \Delta_2^2)$$

This expression, which is the same as the one obtained earlier, is valid for any linear spring.

Additional Remarks

Our experience in these examples leads to a few observations that will prove to be useful both here and in the studies that follow.

1. The work done by forces is a *scalar quantity*. It may be evaluated either by forming the resultant of the force system or else by adding the work done by each force in the system.
2. Any force that acts normal to the direction of motion, such as a reaction, does no work. (It has no component in the direction of motion.)
3. When the tangential component of a force is in the opposite sense from the movement, the work *must* be negative.
4. Evaluation of the work done by a force according to eqs. (24) or (25) *requires that the force depend only on the position.* When the force depends on the other kinematical properties, such as speed, the work can be computed only after these properties are known (i.e., after the problem has been solved by another method).

HOMEWORK PROBLEMS

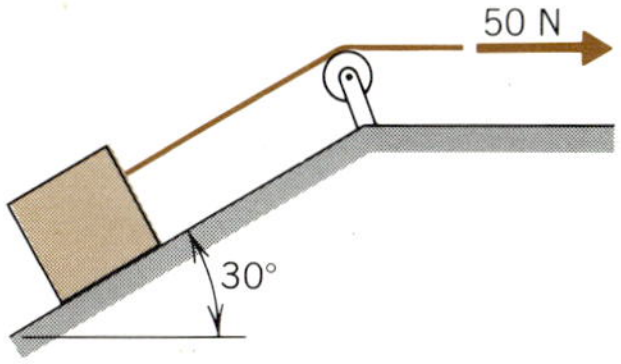

PROB. 12.135

12.134 A 10 lb block slides 3 ft down a 20° incline after it is released. The coefficient of static friction is $\mu_k = 0.3$. Determine the work done by the forces acting on the block in this motion.

12.135 The 10 kg package is pulled 2 m up the incline by a cable that carries a 50 N tension. Determine the work done by the forces acting on the package in this movement.

12.136 A horizontal 40 N force pushes the 2 kg collar 1.20 m down the inclined rod. The coefficient of kinetic friction between the collar and the rod is 0.40. Determine the work done by the forces acting on the collar.

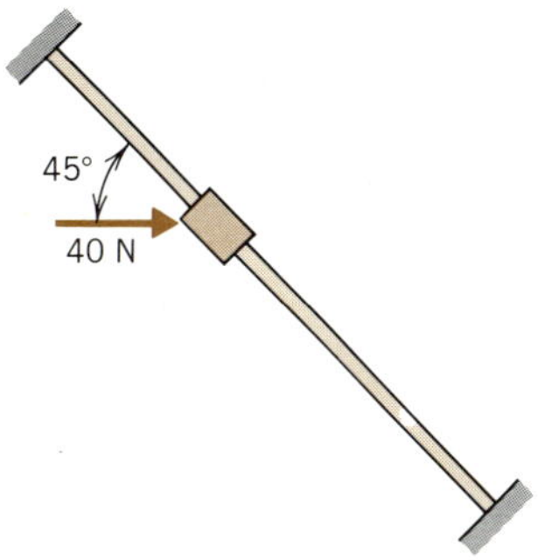

PROB. 12.136

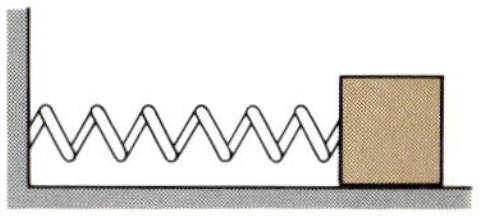

PROB. 12.137

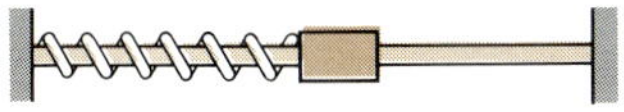

PROB. 12.138

12.137 A 5 lb block sliding on a level surface ($\mu_k = 0.8$) is brought to rest by a spring whose stiffness is 20 lb/ft. Determine the work done by the forces acting on the block in the course of compressing the spring 8 in.

12.138 A spring whose stiffness is 2 kN/m is initially compressed 250 mm when the collar is released. The coefficient of friction between the collar and the rod is 0.2. Determine the work done by the forces acting on the 3 kg collar in a movement of 200 mm from release.

***12.139** As the 50 kg package descends the incline, it is caught by a bumper (spring) at point A. The stiffness of the spring, which is initially held at 50 N compression by the two tensioned cables, is 100 N/m. The coefficient of friction is 0.20. Determine the total work done by the forces acting on the package during a movement of the block 100 mm from point A.

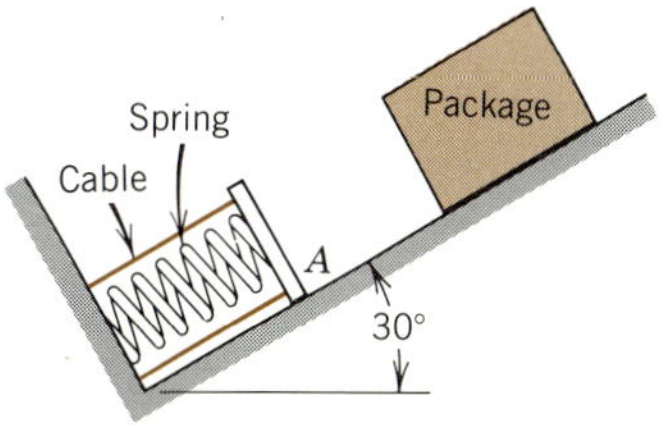

PROB. 12.139

12.140 The 2 lb sphere is released from the position $\theta = 45°$. Determine the work done by the gravitational force acting on the sphere during the descent of the sphere to $\theta = 0$.

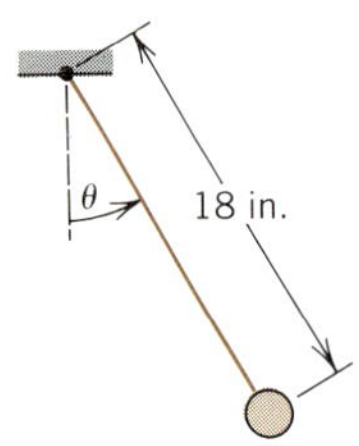

PROB. 12.140

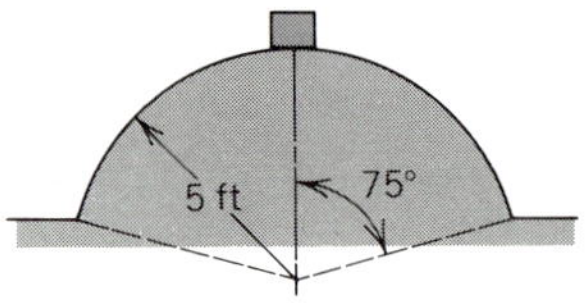

PROB. 12.141

12.141 The 200 gram block is pushed off the top of the hill. Determine the work done by gravity during the descent of the block to the bottom of the hill.

12.142 A cable carrying a constant tension of 10 N is fastened to the 800 gram collar. The collar moves from position A to position B along the smooth vertical rod. Determine the work done by the forces acting on the collar in this movement.

12.143 A constant tension force of 100 lb at the end of the cable raises the suspended 40 lb crate 1.75 ft from the position shown. Determine the total work done in raising the crate.

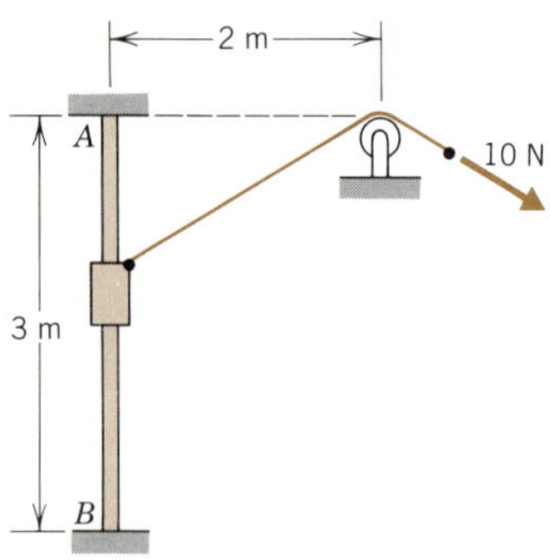

PROB. 12.142

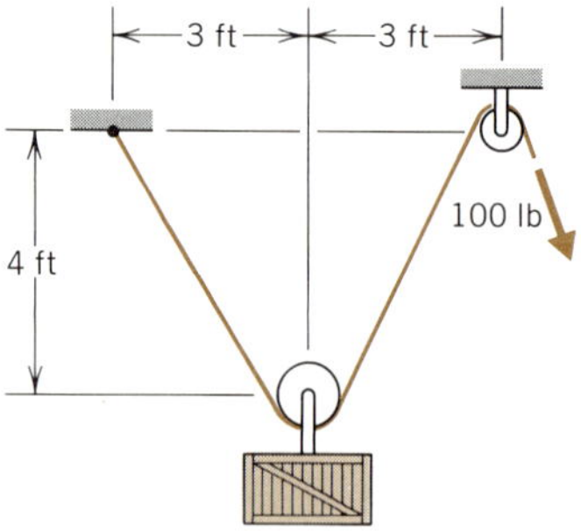

PROB. 12.143

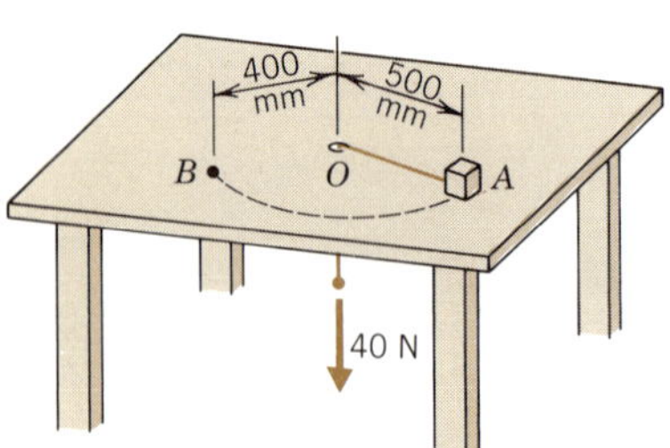

PROB. 12.144

12.144 A 2 kg block is tied to a cable as it slides on a smooth table in the horizontal plane. The cable is passed through a hole in the table at point O, and a 40 N force is applied to its free end. This force causes the block to spiral inward, from point A where $R = 500$ mm, to point B where $R = 400$ mm. Determine the work done by the cable (a) by considering the force applied to the block, (b) by considering the force applied to the free end of the cable.

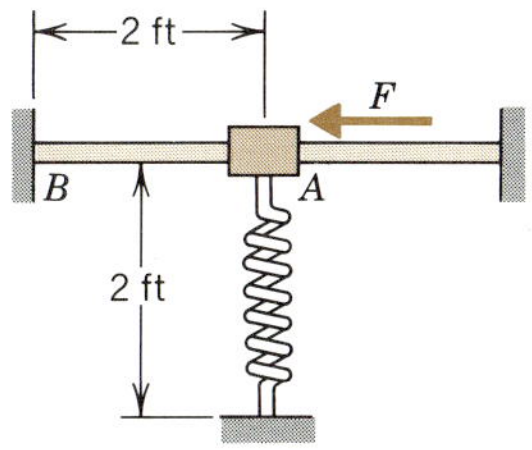

PROB. 12.145

***12.145** A spring whose stiffness is 200 lb/ft and whose unstretched length is 2 ft is fastened to the collar. The collar is pushed from position A to position B along the smooth horizontal rod. Determine the magnitude of $\bar{F}$ if the total work done by the forces acting on the collar is zero.

***12.146** Solve Problem 12.145 if the unstretched length of the spring is 18 in. and all other parameters are unchanged.

12.147 A 0.5 kg collar is pushed along the rigid bar by the constant 10 N force shown. Determine the work done by all the forces acting on the block in moving the block from position A to position B. The z axis is vertical, and friction is negligible.

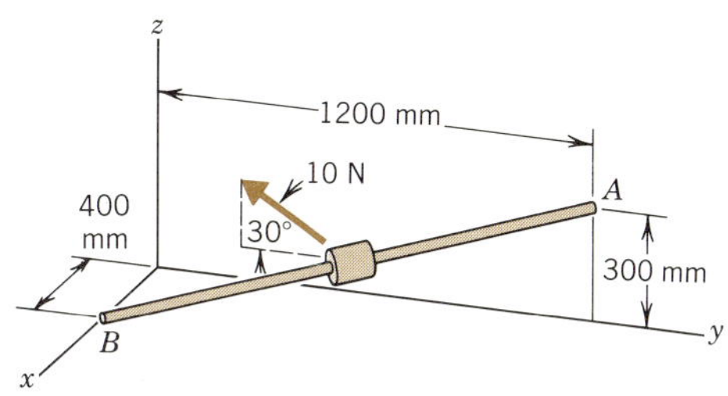

PROB. 12.147

---

2. Potential Energy and Conservative Forces

In Example 18 we evaluated the work done by a spring and noted that an important aspect of this work is that it is not necessary to know the path. For instance, the work done by the spring force acting on the collar in Figure 6 does not depend on which of the alternative paths C_1 or C_2 the collar follows. The only parameters that are needed are the initial deformation Δ_1, the final deformation Δ_2, and the spring stiffness k.

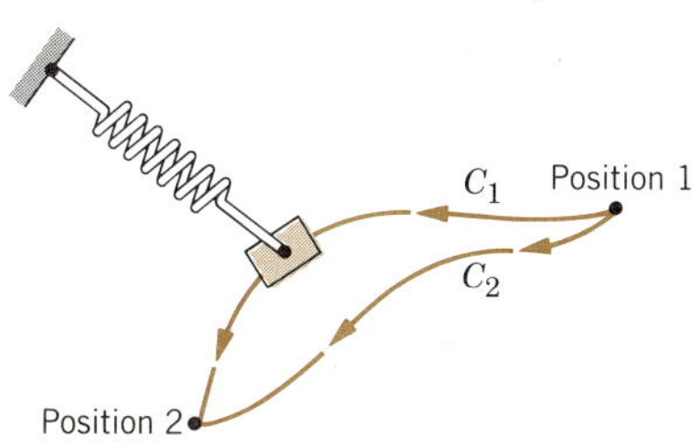

FIGURE 6

There are other forces that have the property that the work they do is independent of the path connecting the initial and final positions of their point of application. They are called *conservative forces*. The property of independence of path means that the work done by a conservative force may be obtained from a function, denoted as V, that depends only on the position of the point at which the force is applied. The work done may be proven to be the difference between the values of this function in the initial and final positions, which here are denoted as V_1 and V_2, respectively.

$$\boxed{U_{1\to 2} = V_1 - V_2} \qquad \text{for conservative forces} \qquad (26)$$

The function V is called the *potential energy* of the force. An important property of the potential energy is the *datum*, which is the location where the potential energy is zero. When the final position is the datum, $V_2 = 0$; it follows from eq. (26) that

$$U_{1\to 2} = V_1$$

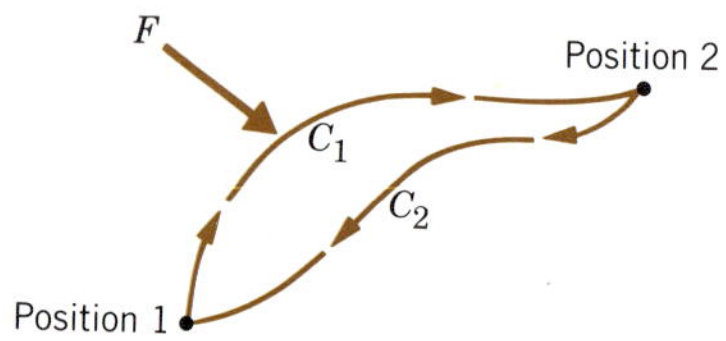

FIGURE 7

This shows that the potential energy is the amount of work that a force can do to move its point of application from the initial position to the position of zero potential energy.

Let us now consider the conservative force $\bar{F}$ in Figure 7 as its point of application follows the closed path shown. This path is formed by following the curve C_1 from position 1 to position 2, and then by following the curve C_2 back to position 1. The work done in each portion of the motion is obtained from eq. (26), with the initial potential energy in each case written first,

$$U_{1\to 2} = V_1 - V_2, \qquad U_{2\to 1} = V_2 - V_1 \equiv -U_{1\to 2}$$

The total work is the sum of the two parts,

$$U_{1\to 2\to 1} = U_{1\to 2} + U_{2\to 1} \equiv 0$$

In other words, a conservative force does no work when its point of application follows a closed path. (This property is often used to demonstrate that a force, such as friction, is not conservative.)

The conservative forces in which we shall be most interested are due to springs and gravity. Our earlier investigation of force exerted by a spring yielded an expression for the work done. It may be compared to the general relation between work and potential energy, eq. (26).

$$(U_{1\to 2})_{\text{spring}} = (V_{\text{spring}})_1 - (V_{\text{spring}})_2 = \tfrac{1}{2}k(\Delta_1^2 - \Delta_2^2)$$

This can be true for arbitrary positions 1 and 2 only if the potential energy at a typical position is

$$\boxed{V_{\text{spring}} = \tfrac{1}{2}k\Delta^2} \tag{27}$$

where $\Delta = l - l_0$ is the elongation of the spring from its length l_0.

A similar approach may be employed to determine the potential energy of a gravity force. When the motion occurs close to the earth's surface the gravity force on a body is mg vertically downward. We regard it as acting at the center of mass G of that body. This is the situation depicted in Figure 8.

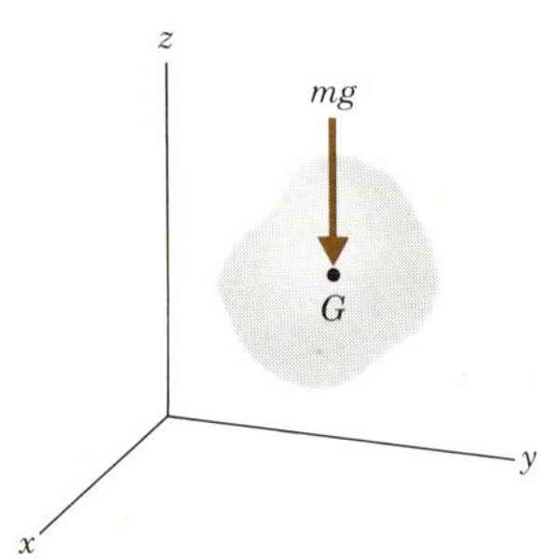

FIGURE 8

In order to evaluate the work done by the gravity force in a general movement, we will employ the xyz coordinate system in Figure 8, whose z axis is oriented vertically. Then

$$d\bar{r}_{G/O} \equiv \bar{v}_G\,dt = dx\,\bar{i} + dy\,\bar{j} + dz\,\bar{k}$$

$$\bar{F}_{\text{gravity}} = -mg\bar{k}$$

$$(U_{1\to 2})_{\text{gravity}} = \int_{r_1}^{r_2} \bar{F}_{\text{gravity}} \cdot d\bar{r}_{G/O} = \int_{z_1}^{z_2} (-mg\,dz)$$

$$= -mg(z_2 - z_1)$$

Comparing this to the general eq. (26) yields

$$\boxed{V_{\text{gravity}} = mgz} \tag{28}$$

where z is the vertical position of the center of mass.

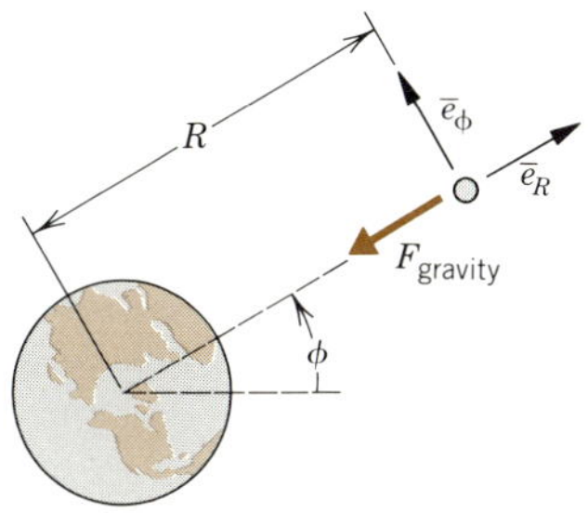

FIGURE 9

Note that the location of the xy plane is arbitrary. Because $z \equiv 0$ on this plane, it represents the datum elevation for potential energy of gravity. That is, at the datum $V_{gravity} = 0$. The case where the center of mass descends below the datum is treated by making z negative.

Sometimes the motion of the center of mass is not close to the earth's surface, or it is necessary to account for the curvature of the earth. As shown in Figure 9, this situation may be treated by using cylindrical coordinates whose origin is the center of mass of the earth. The infinitesimal displacement in this case is

$$d\bar{r}_{P/O} = \bar{v}_p\, dt = dR\,\bar{e}_R + R\,d\phi\bar{e}_\phi + dz\,\bar{k}$$

The gravity force is obtained from the inverse square law.

$$\boxed{\bar{F}_{gravity} = \frac{GMm}{R^2}(-\bar{e}_R)} \tag{29}$$

where G is the universal gravitational constant, M is the mass of the earth, and R is the distance from the body to the center of the earth. The value of GM may be estimated by noting that $F_{gravity} = mg$ at the earth's surface. Substituting for $F_{gravity}$ in the inverse square law yields

$$GM = gR_e^2$$

where $R_e = 3960$ mi $= 6370$ km is the average radius of the earth.

The work done by this force may now be determined.

$$U_{1\to 2} = \int_{r_1}^{r_2} \bar{F}_{gravity} \cdot d\bar{r}_{P/O} = \int_{R_1}^{R_2} \left(-mg\frac{R_e^2}{R^2}\,dR\right)$$

$$= mgR_e^2\left(\frac{1}{R_2} - \frac{1}{R_1}\right)$$

Comparing this to eq. (26) indicates that

$$\boxed{V_{gravity} = -mg\frac{R_e^2}{R}} \tag{30}$$

This expression is obviously different from eq. (28) for motion near the earth's surface. Part of the difference results from accounting for the change in the magnitude of the gravity force. Less obvious is the fact that the datum associated with each expression is different. The datum for gravitational potential energy ($V = 0$) is $R = \infty$, which is the location where the gravitational force is zero. A difference in the definition of the datum is not important because it only adds a constant to the value of V. Such a constant drops out when we compute the work by evaluating the change in the potential energy between the initial and final positions.

A more mathematical treatment of potential energy can be established by proving that a conservative force is defined by the *gradient* of its potential energy. For rectangular Cartesian coordinates this operation is given by

$$\bar{F} = -\bar{\nabla} V \equiv -\frac{\partial V}{\partial x}\bar{i} - \frac{\partial V}{\partial y}\bar{j} - \frac{\partial V}{\partial z}\bar{k}$$

where the symbol $\bar{\nabla}$ is called the "grad" or "del" operator. The gradient formulation has many useful applications in the study of mechanics, particularly fluid mechanics.

EXAMPLE 19

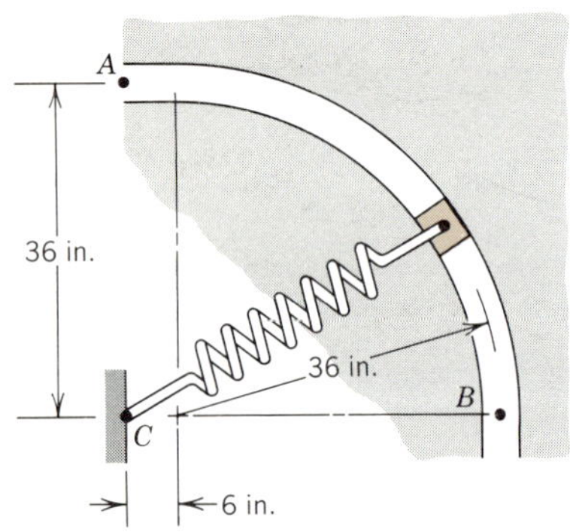

The 8 oz block slides inside the smooth groove, which lies in the vertical plane. The spring has an unstretched length of 30 in. and a stiffness of 2 lb/in. Determine the work done by the forces acting on the block in a movement from position A to position B.

Solution

The free body diagram shows that the only forces doing work (i.e., having a component in the tangential direction) are the spring force and gravity. Both are conservative forces, so the work they do may be evaluated by determining their potential energies.

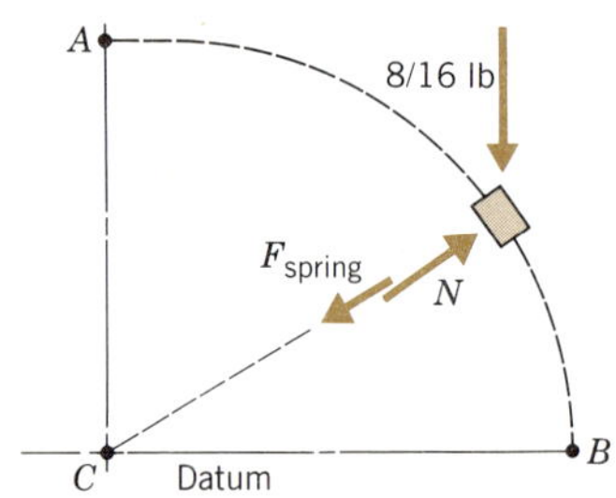

The datum that is defined in the diagram is for the (constant) gravity force. The total potential energy at each position is the sum of the individual terms.

$$V = V_{\text{gravity}} + V_{\text{spring}} = mgz + \tfrac{1}{2}k\Delta^2$$

At position A, the block is 30 in. above the (gravitational) datum and the spring is 36 in. long. We subtract the unstretched length from the latter to find $\Delta_A = 36 - 30 = 6$ in., which gives

$$V_A = \tfrac{8}{16}(36) + \tfrac{1}{2}(2)(6)^2 = 54 \text{ in.-lb}$$

Position B is at the datum elevation, so $z_B = 0$. The spring is 42 in. long in that position, so $\Delta_B = 42 - 30 = 12$ in. Thus

$$V_B = \tfrac{8}{16}(0) + \tfrac{1}{2}(2)(12)^2 = 144 \text{ in.-lb}$$

The work done in going from position A to position B is found from the change in the potential energy

$$U_{A\rightarrow B} = V_A - V_B = -90 \text{ lb-in.}$$

EXAMPLE 20
A 400 kg ball is to be raised from the earth's surface. Determine the work done by gravity according to the inverse square law, and determine the error in the constant force approximation for increases in elevation of (a) 10 km and (b) 10,000 km.

Solution
Let us begin with the inverse square law relation. The initial distance R from the center of mass is $R_1 = R_e = 6.370(10^6)$ m. Equation (30) indicates that

$$V_1 = -0.4(9.807)6.370(10^6) = -2.498824(10^7)\text{ J}$$

(We have used more precision than usual in order to avoid numerical inaccuracies later.)

The change in the elevation is the same as the increase in the radial distance. Thus, in case (a)

$$R_2 = R_e + 10\text{ km} = 6.380(10^6)\text{ m}$$

$$V_2 = -0.4(9.807)\frac{(6.370)^2(10^{12})}{6.380(10^6)} = -2.494907(10^7)\text{ J}$$

Equation (26) for work then yields

$$(U_{1\to 2})_a = V_1 - V_2 = -3.917(10^4)\text{ J}$$

The same procedure may be followed for case (b).

$$R_2 = R_e + 10{,}000\text{ km} = 1.6370(10^7)\text{ m}$$

$$V_2 = -(0.4)(9.807)\frac{(6.370)^2(10^{12})}{1.6370(10^7)} = -9.723584(10^6)\text{ J}$$

$$(U_{1\to 2})_b = V_1 - V_2 = -1.5225(10^7)\text{ J}$$

The computations for the constant force approximation, eq. (28), are easier. The datum is arbitrary; we shall set it at the earth's surface. This is the initial position, so $z_1 = 0$, which gives $V_1 = 0$. The elevation increase is the new value of z in each case. Case (a) is then

$$z_2 = 10\text{ km} = 10^4\text{ m}$$

$$V_2 = 0.4(9.807)10^4 = 3.923(10^4)\text{ J}$$

$$(U_{1\to 2})_a = V_1 - V_2 = -3.923(10^4)\text{ J}$$

These computations for case (b) yield

$$z_2 = 10{,}000\text{ km} = 10^7\text{ m}$$

$$V_2 = 0.4(9.807)(10^7) = 3.923(10^7)\text{ J}$$

$$(U_{1\to 2})_b = V_1 - V_2 = -3.923(10^7)\text{ J}$$

A comparison of these results with the earlier values indicates that the constant force approximation is 0.2% too large in case (a), where the motion is fairly close to the earth. This is contrasted by an excess prediction of 157% for case (b), where the change in the distance from the center of the earth is substantial.

HOMEWORK PROBLEMS

12.148 Solve Problem 12.137 by using potential energy.

12.149 Solve Problem 12.140 by using potential energy.

12.150 Solve Problem 12.139 by using potential energy.

12.151 Derive an expression for the potential energy $V_{gravity}$ of the gravitational force acting on the sphere in terms of the mass m, the length l, and the angle θ. Plot the results for $-180^\circ \leq \theta \leq 180^\circ$. What is the significance of the locations where $V_{gravity}$ is a maximum or a minimum?

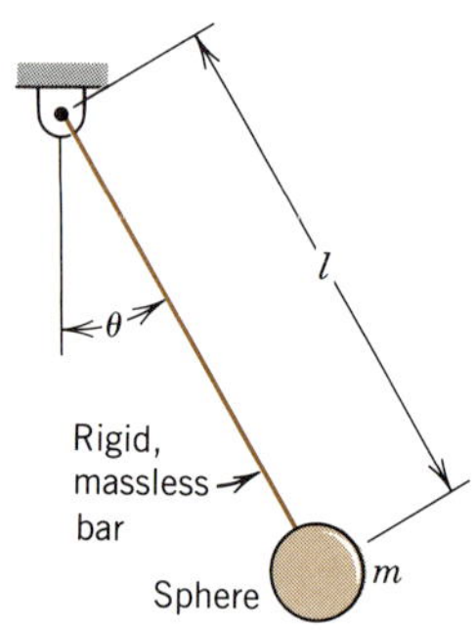

PROB. 12.151

12.152 A 4000 lb satellite in elliptical orbit about the earth has a minimum altitude of 100 miles and a maximum altitude of 400 miles above the earth's surface. Determine the work done by gravity when the satellite moves from the highest to the lowest altitude.

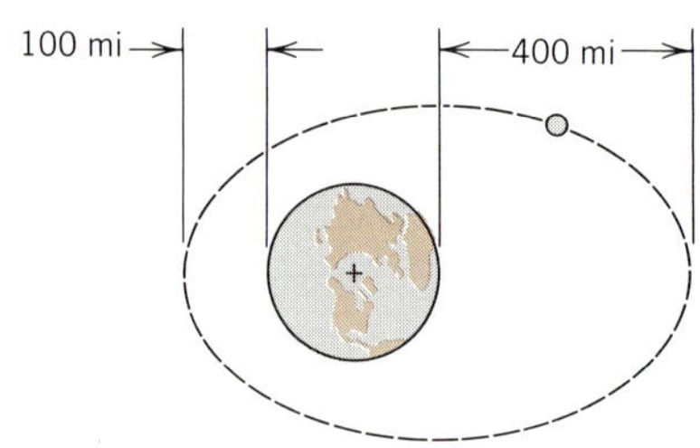

PROB. 12.152

12.153 A 5 kg package falls 8 km after being ejected from an airplane. Determine the work done by gravity during the fall according to (a) the inverse square law, (b) the constant force approximation.

12.154 Solve Problem 12.145 by using potential energy.

12.155 A spring whose stiffness is 50 lb/ft and whose unstretched length is 28 in. is used to slow the 12 oz slider. Determine the work done by the forces acting on the slider during a movement from position A to position B. The groove is smooth.

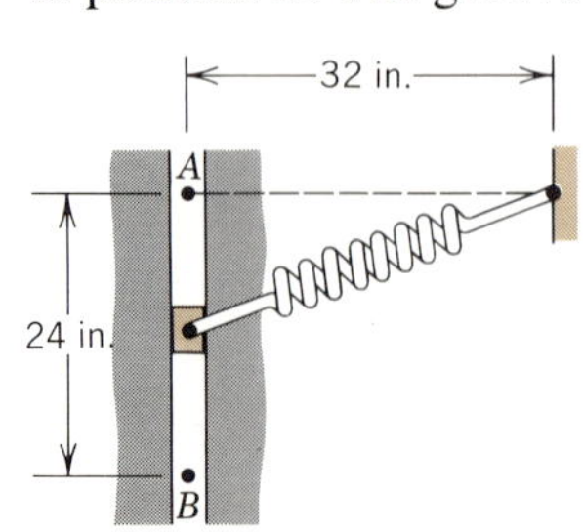

PROB. 12.155

12.156 The elevation of a 200 kg object above the surface of the earth is increased by 1 km. Determine the work done by the gravitational force acting on this body (a) if the object was initially at the surface of the earth, (b) if the object was initially 6370 km above the earth's surface.

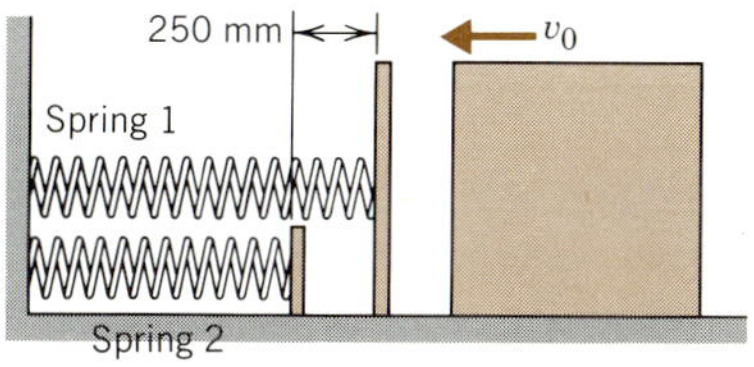

PROB. 12.157

12.157 The variable stiffness bumper consists of spring 2, which is parallel to, but 250 mm shorter than, spring 1. The stiffness of each spring individually is 2 kN/m. A sliding block compresses the bumper by 500 mm. Determine the work done by the bumper force exerted on the block during (a) the movement of the block 250 mm beyond the point of first contact, (b) the total movement of the block.

***12.158** A 1 kg collar slides in a vertical plane over the smooth inclined guide under the action of a 100 N tangential force and an elastic spring whose stiffness is 200 N/m. The spring's unstretched length is 400 mm. Determine the total work done by all the forces acting on the collar in moving the collar from point A to point B.

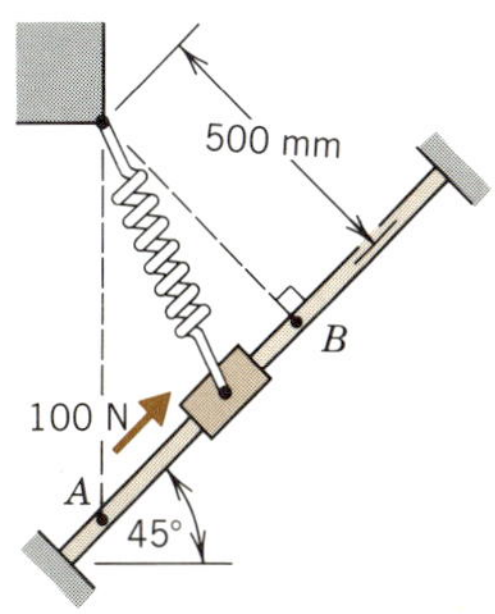

PROB. 12.158

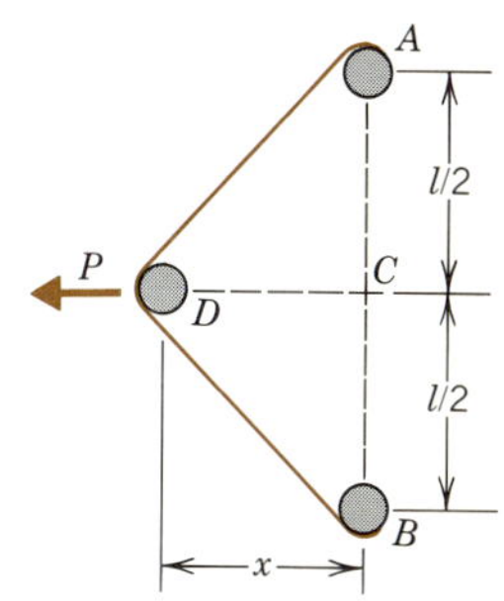

PROB. 12.159

***12.159** A slingshot is formed by stretching a rubber band horizontally between pegs A and B with an initial tension force $\bar{F}_0$. An object is pulled by the horizontal force $\bar{P}$ and held at point D. The stiffness of the rubber band is k. (a) Derive an expression for the potential energy V_{rb} in terms of k, F_0, and the dimensions x and l shown. (b) How much work is done by the force $\bar{P}$ in moving the object from point C to point D? (c) When the rubber band is released and the object is projected past point C, how much work does the rubber band do?

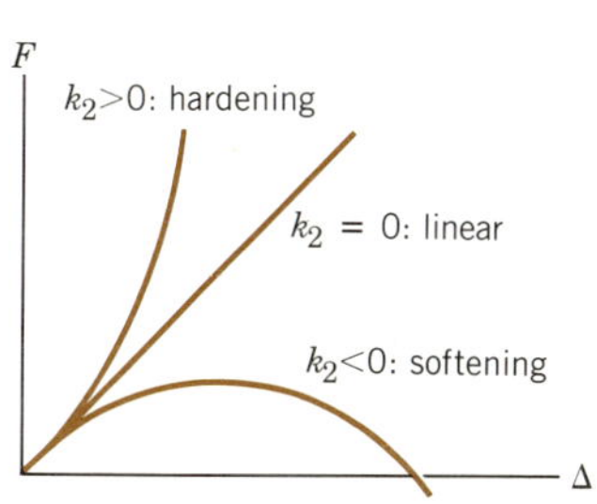

PROB. 12.160

***12.160** The tensile force in a nonlinear spring is given by $F = k_1\Delta + k_2\Delta^3$, where Δ is the elongation. As shown, the spring is said to be of the hardening or softening type according to whether k_2 is positive or negative. (a) Derive an expression for the potential energy V of such a spring. (b) For the case where it is of the softening type, determine the values of Δ where $F = 0$ and where $V = 0$. Why don't the two coincide?

3. General Principle of Work and Energy

The tangential component of force, $\sum F_t$, occurs in Newton's second law, as well as in the definition of work. This is the basis for relating work to kinematical variables. Specifically, substituting $\sum F_t = m\dot{v}$ into the formula for work, eq. (24), yields the following identity

$$U_{1\to 2} = \int_{s_1}^{s_2} \sum F_t \, ds = \int_{s_1}^{s_2} m\dot{v}\, ds$$

In order to proceed further let us consider the speed v to be a function of s, so that $\dot{v} = v(dv/ds)$ according to the chain rule for differentiation. The above equation for the work then becomes

$$U_{1\to 2} = \int_{s_1}^{s_2} mv\left(\frac{dv}{ds}\right) ds = \int_{v_1}^{v_2} mv\, dv$$

$$= \tfrac{1}{2}m(v_2^2 - v_1^2) \tag{31}$$

In words, the work done by the forces acting on a particle is the amount by which the quantity $\frac{1}{2}mv^2$ changes from the initial to the final position. This quantity is the *kinetic energy*, which we will denote by the symbol T.

$$\boxed{T = \tfrac{1}{2}mv^2 \equiv \tfrac{1}{2}m\bar{v}\cdot\bar{v}} \tag{32}$$

Equation (31) for the work may then be rewritten as

$$T_2 - T_1 = U_{1\to 2} \tag{33}$$

Using this relation to solve problems requires that, for each problem, we compute the work done by all forces. The computations can be considerably reduced if we use the concept of potential energy to evaluate the work done by conservative forces. Let us decompose the total work $U_{1\to 2}$ into the contributions of the conservative forces, denoted as $U_{1\to 2}^{(c)}$, and the nonconservative forces, denoted as $U_{1\to 2}^{(nc)}$. Equation (33) then becomes

$$T_2 - T_1 = U_{1\to 2}^{(c)} + U_{1\to 2}^{(nc)}$$

Now let *V be the total potential energy*, that is, the sum of the potential energy associated with each conservative force. From eq. (26) we then have

$$T_2 - T_1 = V_1 - V_2 + U_{1\to 2}^{(nc)}$$

Rearranging the terms in this expression yields

$$\boxed{T_2 + V_2 = T_1 + V_1 + U_{1\to 2}^{(nc)}} \tag{34}$$

This relation expresses the *work–energy principle.* It may be used to relate the speed at two positions. As indicated by their names, T and V are forms of energy. They both describe mechanical systems, so they are called mechanical energies. We may interpret the above work–energy relation to mean that the total mechanical energy in the final position is the sum of the energy in the initial position and the work done by nonconservative forces.

This principle should not be confused with the first law of thermodynamics. That law states that the sum of all forms of energy is constant. The work done by nonconservative forces may be interpreted in a thermodynamic sense to represent a conversion between mechanical energy and other forms of energy. For example, the friction force on a sliding body always opposes the sliding motion, so the work done by the friction force is negative. The work–energy relation indicates that the total mechanical energy will decrease in the motion. Mechanical energy that is dissipated in this manner is manifested mainly in the production of heat.

In the case where all of the forces are conservative, $U_{1\to 2}^{(nc)} \equiv 0$. This results in the *conservation of mechanical energy*

$$T_2 + V_2 = T_1 + V_1$$

There is no need to consider this as a separate principle, because it is merely a special case of the work–energy relation, eq. (32).

Before we employ eq. (32) to solve problems, we must note that it provides a useful relation between the speed and the force system only if the work can be evaluated. Our earlier discussion of work indicates that this is possible when the forces depend on position, but *not* when they depend on the elapsed time or the speed of the particle.

EXAMPLE 21

A 50 ton railroad car rolling down the 10° incline is to be stopped by the bumper. The spring whose stiffness is 10^4 lb/in. is precompressed 3 in. due to the mounting bracket. Determine the speed v_0 with which the car may contact the bumper if the maximum additional compression of the bumper is to be 1 ft.

Solution

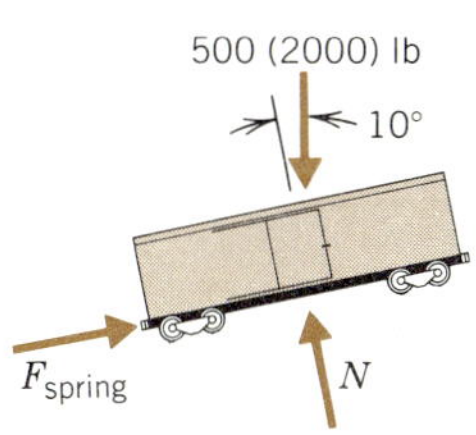

A free body diagram of the railroad car shows that the forces doing work are due to gravity and the spring. Both of these forces are conservative. It is easier in this problem to compute the work done by gravity directly, because the component of the force in the direction of motion is constant. (In general, it is acceptable to ignore the conservative aspect of a force, but it is *never* correct to describe a force in terms of its potential energy when its work is computed directly.)

The railroad car moves 1 ft, and the tangential component of the gravity force is $10^5 \sin 10°$, so

$$U_{1\to 2}^{(nc)} = (10^5 \sin 10°)(1) = 1.7365(10^4) \text{ ft-lb}$$

The spring is initially shortened by 3/12 ft; so $\Delta_1 = 0.25$ ft. The additional shortening of the spring is 1 ft, so $\Delta_2 = 1.25$ ft. The potential energy in the two positions is therefore

$$V_1 = \tfrac{1}{2}(10^4)(12)(0.25)^2 = 0.375(10^4) \text{ ft-lb}$$

$$V_2 = \tfrac{1}{2}(10^4)(12)(1.25)^2 = 9.375(10^4) \text{ ft-lb}$$

The kinetic energy in the initial position for a speed v_0 is

$$T_1 = \frac{1}{2}\frac{10^5}{32.17}v_0^2 = 1.5542(10^3)v_0^2$$

The speed is zero when the compression of the spring is maximum, so $T_2 = 0$. The work–energy equation then gives

$$T_2 + V_2 = T_1 + V_1 + U_{1\to 2}^{(\mathrm{nc})}$$

$$0 + 9.375(10^4) = 1.5542(10^3)v_0^2 + 0.375(10^4) + 1.7365(10^4)$$

$$v_0 = 6.84 \text{ ft/s}$$

EXAMPLE 22

A simple pendulum of mass m and length l is given an initial velocity v_0 to the right at its lowest point. Determine the minimum value of v_0 necessary to cause the mass to follow a full circle if the mass is connected to the pivot A by a cable.

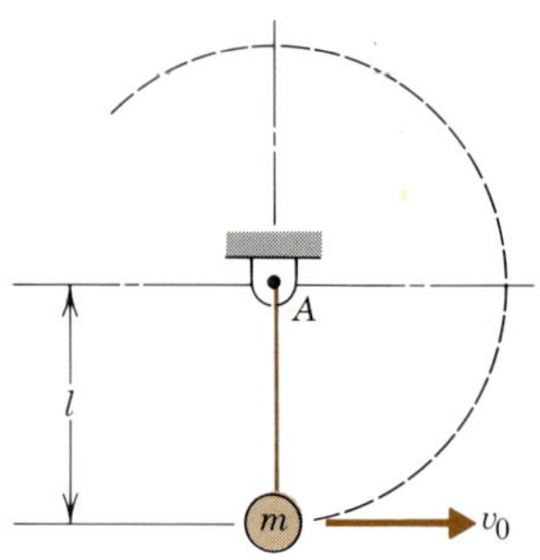

Solution

The speed is given in the initial position. The cord must be taut at the top of the path in order for the mass to follow a full circle. The free body diagram for the top position will be used to relate the speed v_2 at this position to the cable tension F.

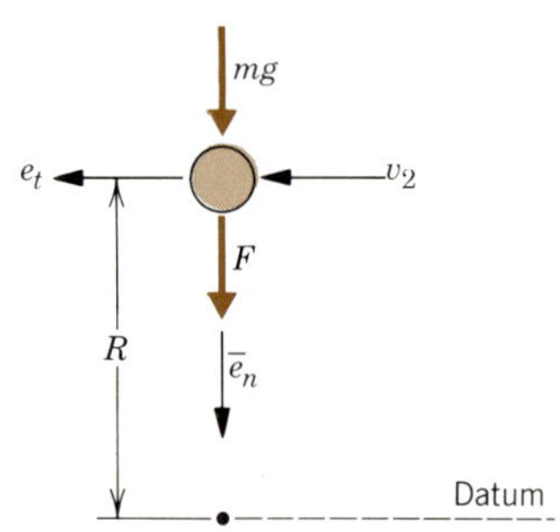

The normal component of $\sum \bar{F} = m\bar{a}$ gives

$$F + mg = m\frac{v_2^2}{l}$$

The smallest value of v_2^2 corresponds to $F = 0$, so the desired speed at the top is

$$v_2^2 = gl$$

The kinetic energies at the initial and final positions are therefore

$$T_1 = \tfrac{1}{2}mv_0^2, \qquad T_2 = \tfrac{1}{2}mv_2^2 = \tfrac{1}{2}mgl$$

The work–energy equation relates v_2 to v_0. The only force doing work is gravity, which is conservative. Hence $U^{(nc)}_{1\to 2} = 0$. For a gravitational datum located at the pivot, the initial height is $-l$ and the final height is $+l$. The corresponding potential energies are

$$V_1 = mg(-l), \qquad V_2 = mgl$$

The work–energy equation for a conservative system now gives

$$T_2 + V_2 = T_1 + V_1$$

$$\tfrac{1}{2}mgl + mgl = +\tfrac{1}{2}mv_0^2 - mgl$$

$$v_0 = \left(5\frac{g}{l}\right)^{1/2}$$

EXAMPLE 23

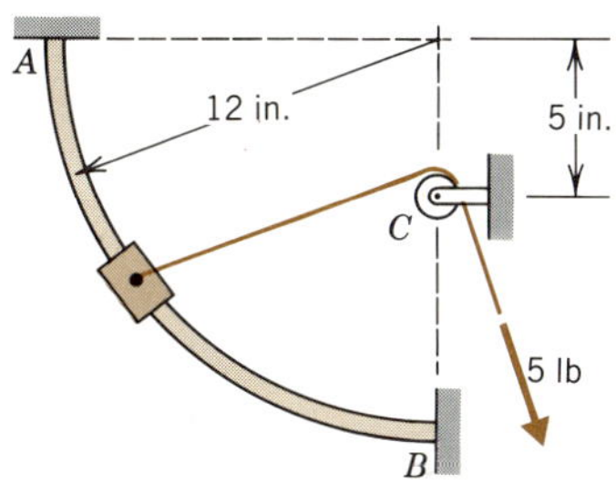

The 4 oz collar is pulled by the cable from position A to position B along the curved bar, which is in the horizontal plane. The speed of the collar at these positions are $v_A = 6$ ft/s, and $v_B = 10$ ft/s. The tension in the cable is constant at 5 lb. Determine the average friction acting on the collar between these positions.

Solution

The average value of a force may be obtained if the work done by that force can be evaluated. This average is the work divided by the distance traveled. The distance traveled by the collar from position A to position B is $(\pi/2)(12)$ in., so the work done by the friction force depicted in the free body diagram may be written as

$$U^{(f)}_{1\to 2} = -f_{av}(6\pi) \text{ in.-lb}$$

where the negative sign is chosen because friction opposes the movement.

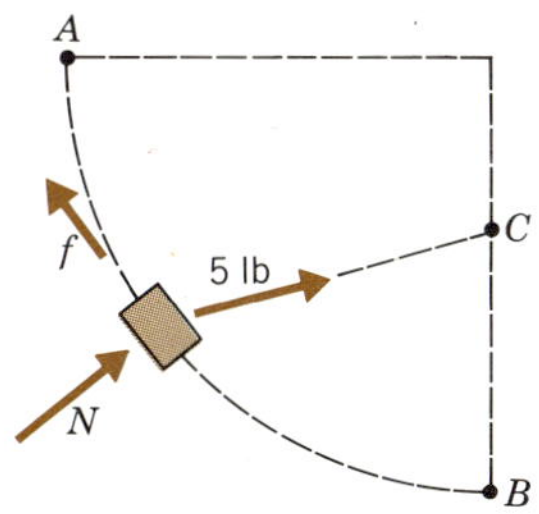

The free body diagram also depicts the other forces acting on the collar. In addition to friction, the only force doing work is the cable tension. The work done by this force is determined by computing the distance that the free end of the cable is pulled. The length of cable from the collar to the pulley at each position is

$$l_{AC} = [(5)^2 + (12)^2]^{1/2} = 13 \text{ in.}, \qquad l_{BC} = 12 - 5 = 7 \text{ in.}$$

The amount by which this length decreases is the outward movement of the free end, so

$$U_{1\to 2}^{(\text{cable})} = 5(6) = 30 \text{ in.-lb}$$

The kinetic energy at both positions may be computed from the given speeds.

$$T_1 = \frac{1}{2}\left[\frac{4/16}{32.17(12)}\right][6(12)]^2 = 1.6786 \text{ in.-lb}$$

$$T_2 = \frac{1}{2}\left[\frac{4/16}{32.17(12)}\right][10(12)]^2 = 4.663 \text{ in.-lb}$$

The work–energy relation then gives an equation for f_{av}.

$$T_2 = T_1 + U_{1\to 2}^{(\text{cable})} + U_{1\to 2}^{(f)}$$

$$4.663 = 1.6786 + 30 - 6\pi f_{av}$$

$$f_{av} = 1.433 \text{ lb}$$

Additional Remarks

As a closure for this discussion of the principle of work and energy, let us consider a problem that is similar to the last example. Suppose that the coefficient of friction μ_k were given and it was necessary to determine the speed in the final position. We could not use the work–energy principle for that task. This limitation arises because the normal force $\bar{N}$ in the curvilinear motion depends on the speed v, through $\sum F_n = ma_n$. The friction force, being $f = \mu_k N$, would also be dependent on v. We would then be unable to evaluate the work done by friction, because we do not know how v changes with distance traveled.

It often happens that the first integrals of $\sum \bar{F} = m\bar{a}$ (i.e., energy or momentum principles) lead to circuitous reasoning. In such situations, the only recourse is to formulate the differential equation of motion for the system using the techniques developed earlier in this chapter.

HOMEWORK PROBLEMS

12.161 A 3000 lb automobile descends a 5% grade at 45 mi/hr. The coefficient of friction between the tires and the ground is 0.60. What is the minimum distance required to reduce the speed (a) to 15 mi/hr, (b) to zero? Neglect the resistance of the air.

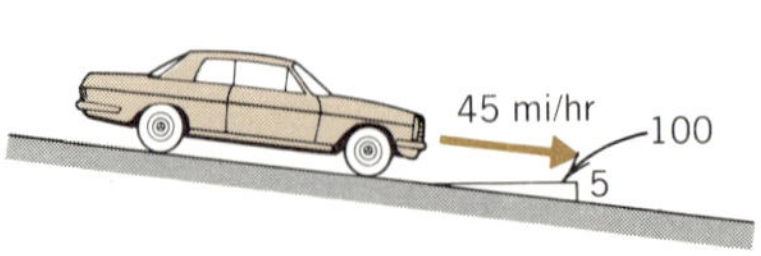

PROB. 12.161

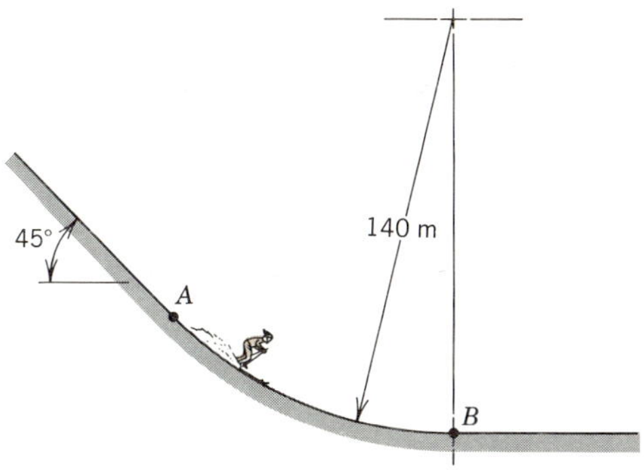

PROB. 12.162

12.162 A 70 kg skier descending a 45° slope has a speed of 80 km/hr at point A where the circular transition curve begins. Determine the speed of the skier at the bottom point B. Friction is negligible.

12.163 Solve Problem 12.162 if the average friction force is 25 N.

12.164 Determine the minimum height h at which the 200 gram toy car may be released to successfully traverse the hoop. Friction is negligible.

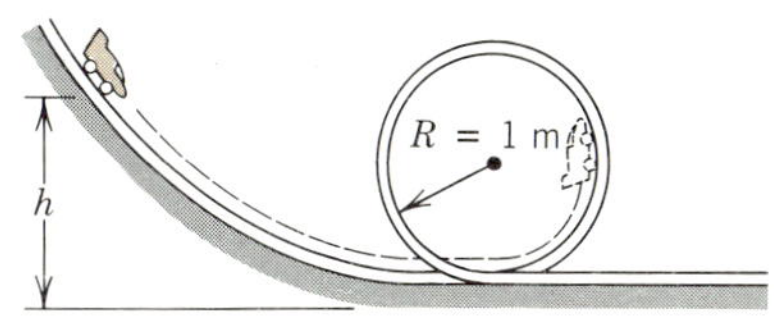

PROB. 12.164

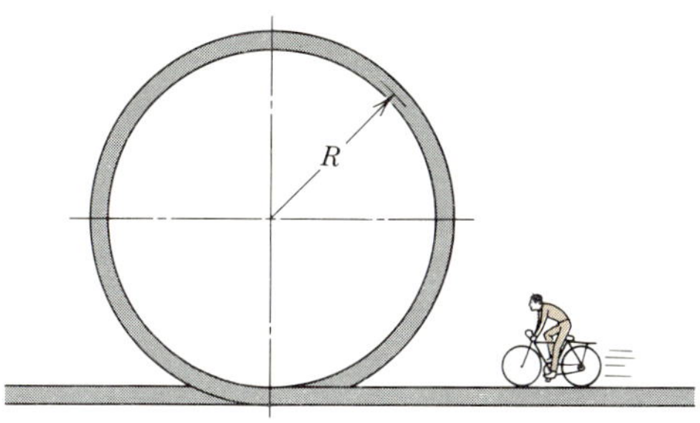

PROB. 12.165

12.165 The maximum speed of a bicyclist on level ground is 10 m/s. What is the largest radius R of a vertical loop that the rider may safely negotiate without pedaling the bicycle? The center of mass of the bicyclist and bicycle is 1 m above the ground.

12.166 Solve Example 22 in the case where the mass is connected to pivot A by a lightweight, rigid rod.

12.167 The 10 kg package sliding down the ramp has a speed of 10 m/s at position A, and its speed at position B is 9 m/s. Determine the average friction force acting on the package between these positions.

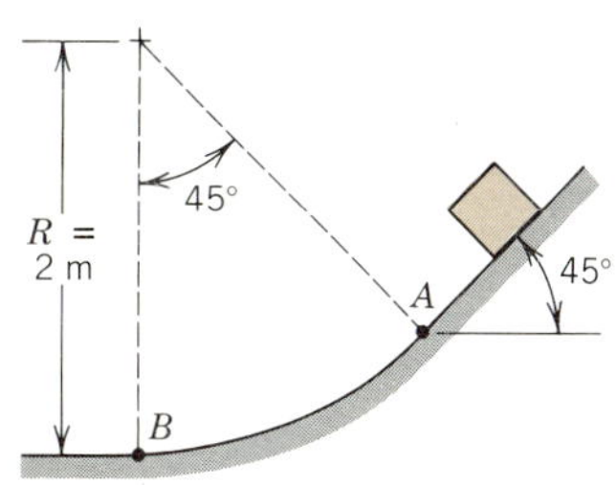

PROB. 12.167

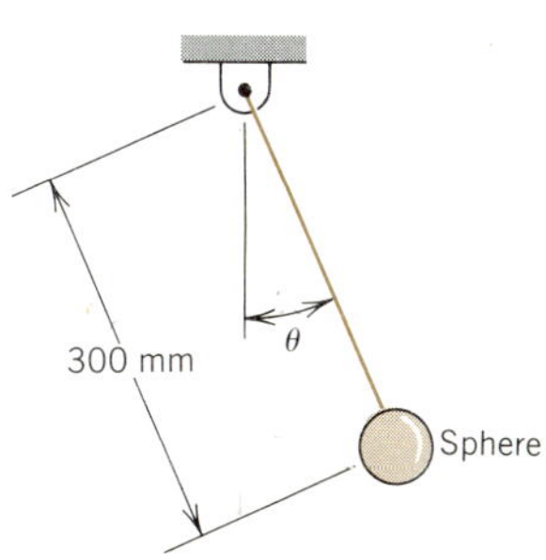

PROB. 12.168

12.168 At $\theta = 15°$ the pendulum is returning to its static equilibrium position at 2.0 m/s. Determine (a) the speed of the sphere when it passes the vertical position, (b) the maximum value of the angle θ during the motion.

12.169 A 10 kg package slides down from point A and is brought to rest at point B by a spring whose stiffness is 5 kN/m. Neglecting friction, determine the maximum compression of the spring.

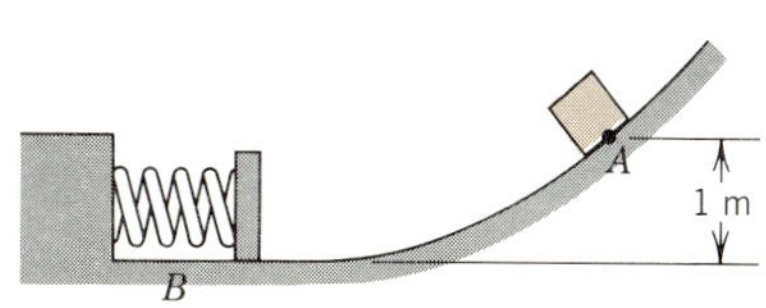

PROB. 12.169

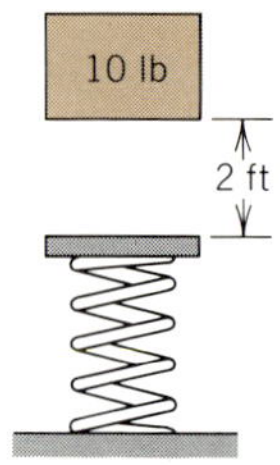

PROB. 12.170

12.170 A 10 lb block is dropped from a height of 2 ft onto a spring whose stiffness is 2 lb/in. Determine the maximum deformation of the spring and the maximum compressive force it sustains.

12.171 What constant tension force $\bar{F}$ must be applied to the end of the cable to give the 20 lb block a downward velocity of 2 ft/s after the block descends 4 ft from rest?

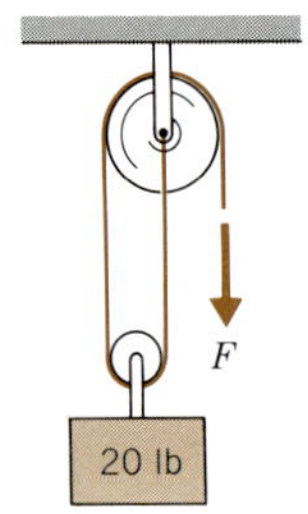

PROB. 12.171

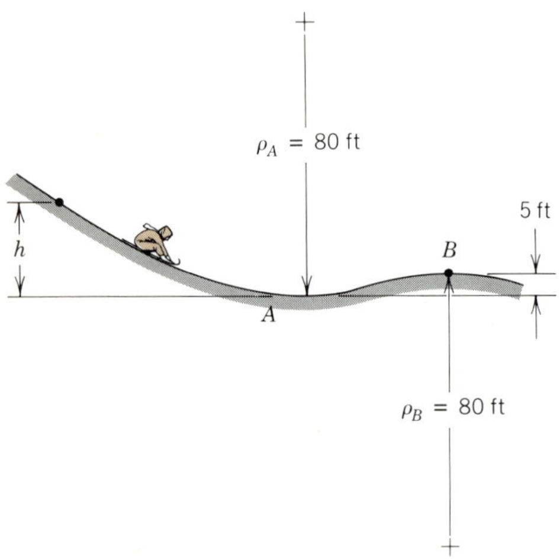

PROB. 12.172

12.172 A 125 lb rider on a 10 lb toboggan starts from $h = 25$ ft. Determine her velocity and the force she exerts on the toboggan at (a) the low point A, (b) the peak B. Neglect friction and air resistance.

12.173 For safety, it is necessary that the toboggan in Problem 12.172 always remain in contact with the track. What is the maximum allowable height h from which the rider may start?

12.174 In the position shown, the spring, whose stiffness is 4 kN/m, is 100 mm shorter than its unstretched length. The 250 gram collar is not attached to the spring. Determine the speed with which the collar passes the position where the spring is unstretched. The rod is smooth.

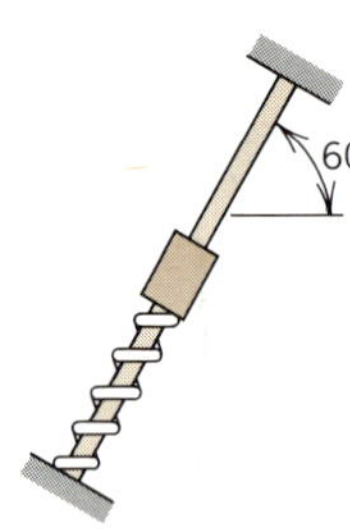

PROB. 12.174

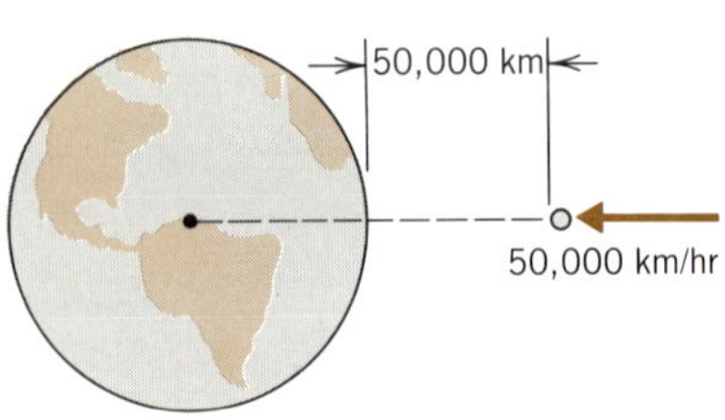

PROB. 12.176

12.175 Determine the maximum distance the collar in Problem 12.174 travels up the inclined rod.

12.176 A 10 kg meteorite is heading directly toward the center of the earth. At an altitude of 50,000 km from the center of the earth it has a speed of 50,000 km/hr. Determine the speed of the meteorite when it is 100 km above the earth's surface.

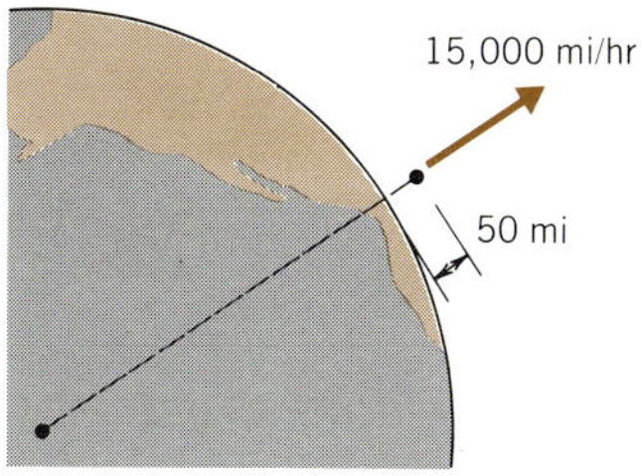

PROB. 12.177

12.177 An exploratory rocket is traveling directly away from the center of the earth, with its engine off. It has a speed of 15,000 mi/hr at an altitude of 50 mi above the earth's surface. Determine the maximum altitude attained by the rocket.

12.178 The 30 lb crate is initially at rest at position A when a 50 lb tensile force is applied to the free end of the cable. Determine the speed of the crate after it has been raised 1.75 ft.

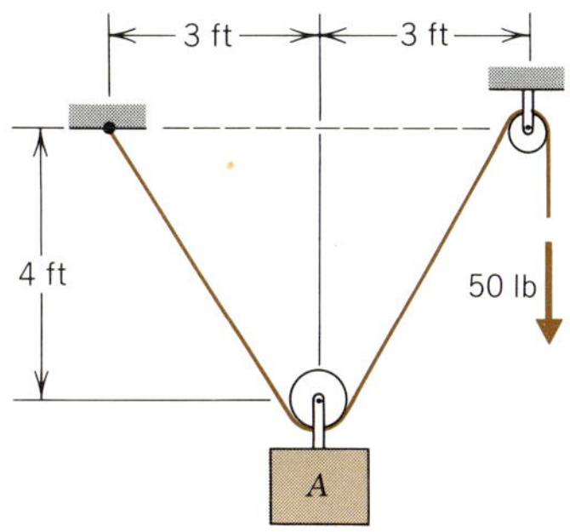

PROB. 12.178

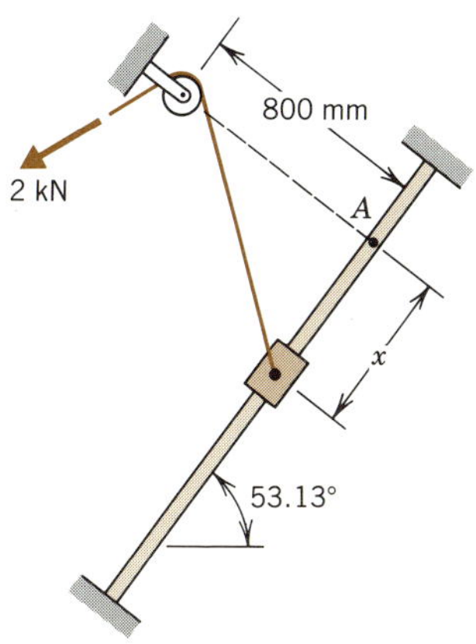

PROB. 12.179

12.179 The 2 kg collar is moving downward at a speed v_0 when it passes position A. It is fastened to the cable whose tension is constant at 2 kN. The maximum distance attained by the collar is $x = 600$ mm. Determine v_0. Friction is negligible.

12.180 The 8 oz collar is pulled down the curved rod ABC by the spring, whose stiffness is 2 lb/in. and whose unstretched length is 15 in. The collar is released from rest at position A. For position B, determine (a) the speed of the collar, and (b) the contact force exerted by the curved rod on the collar.

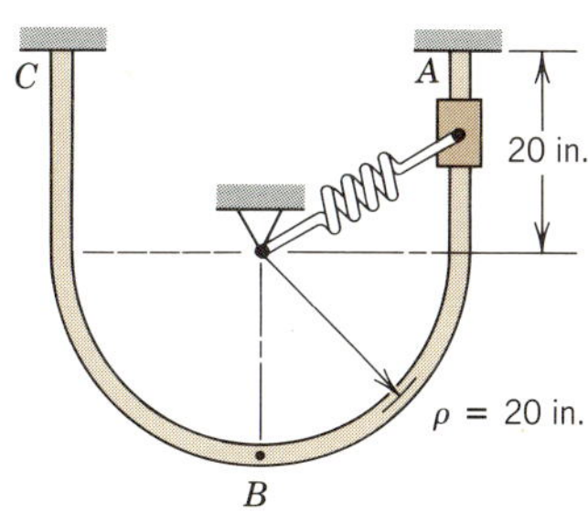

PROB. 12.180

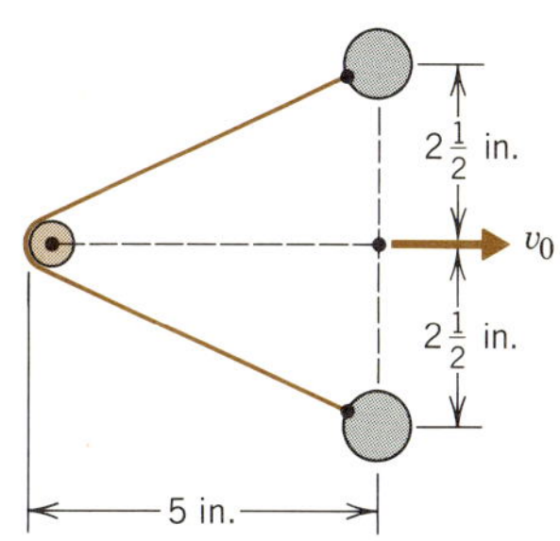

PROB. 12.181

12.181 The slingshot consists of a rubber band whose stiffness is 200 lb/ft, and whose unstretched length is 3 in. A 2 oz pellet is held in the position shown and then released. Determine the speed v_0 with which the pellet leaves the slingshot.

12.182 The railroad car in Example 21 hits the bumper at an initial speed $v_0 = 5$ ft/s. Determine the maximum compressive force developed in the spring.

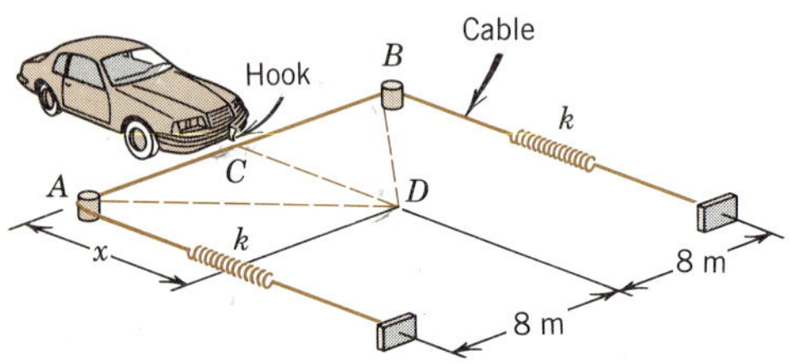

PROB. 12.183

12.183 The arresting gear slows an automobile in a crash test by catching the hook at midpoint C on the cable. The cable carries a tension of 15 kN when the hook is not engaged, and each spring has a stiffness of 30 kN/m. A 1200 kg automobile comes to rest with the hook at position D, where $x = 6$ m. Determine the speed at which the hook first contacted the cable at point C. The vehicle rolls freely.

***12.184** The automobile in Problem 12.183 engages the cable at a speed of 50 km/hr. Determine the distance x locating the position D of the hook when the automobile comes to rest.

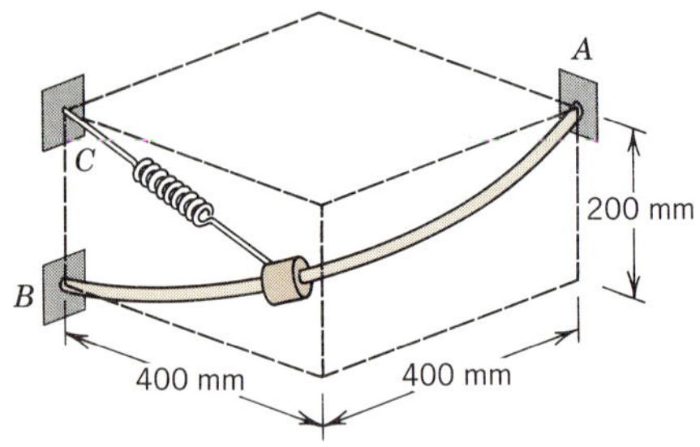

PROB. 12.185

***12.185** A 2 kg collar slides over the rigid guide AB. The spring has a stiffness of 5 N/m and an unstretched length of 100 mm. If the collar is released from rest at point A, what is its speed when it reaches point B?

***12.186** An elastic cord whose stiffness is 24 lb/ft and whose unstretched length is 16 in. is attached to a 2 lb collar. The collar rides on the circular guide $ABCD$ in a horizontal plane. When the collar is at point A it has a velocity of 10 ft/s to the left. Determine (a) the maximum speed of the collar, (b) whether the collar reaches point D, and if so, its speed at this location. *Hint:* Recall that a cord cannot sustain a compressive force.

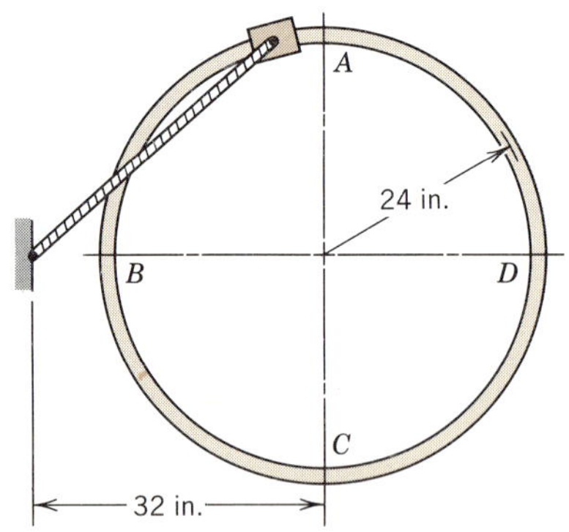

PROB. 12.186

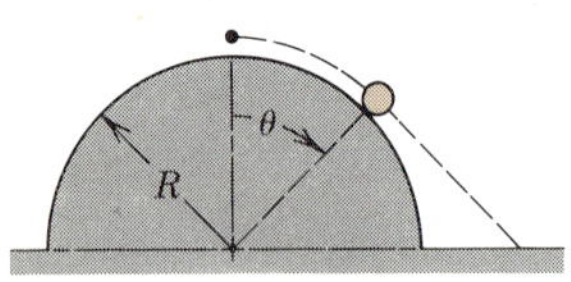

PROB. 12.187

***12.187** A small sphere falls down the surface of a smooth semicircular cylinder. It starts from the top with a negligible initial speed. Determine the angle θ at which the ball bearing first ceases to be in contact with the surface. *Hint:* The speed at angle θ is related to θ by the condition that the contact force is zero.

4. Work and Energy for a System of Particles

The work done by a force is a scalar quantity, as are the potential and kinetic energies. It follows that the work–energy relation for a system of N particles is obtained by summing the corresponding equations for each particle. Let V denote the *total* potential energy for all conservative forces acting on a system, and let T denote the *total* kinetic energy of all of the particles of the system. That is,

$$V = \sum V_i \qquad \text{and} \qquad T = \sum_{i=1}^{N} (\tfrac{1}{2} m_i v_i^2) \tag{35}$$

Summing the work–energy equation for each particle shows that the same form of the equation as developed in the previous section governs the energy of the system. In other words, eq. (34) is still valid.

$$\boxed{T_2 + V_2 = T_1 + V_1 + U_{1\to 2}^{(\mathrm{nc})}} \tag{34}$$

where $U_{1\to 2}^{(\mathrm{nc})}$ is now the *total* work done by all nonconservative forces.

This principle will prove to be useful later when we study rigid bodies. Its suitability for the current topic of a discrete system of particles is more limited, but it provides a single equation relating the motion of successive positions. This suggests that eq. (34) should be employed in situations where the motion may be described in terms of one parameter. A typical situation of this class of problems arises in cable–pulley systems.

EXAMPLE 24

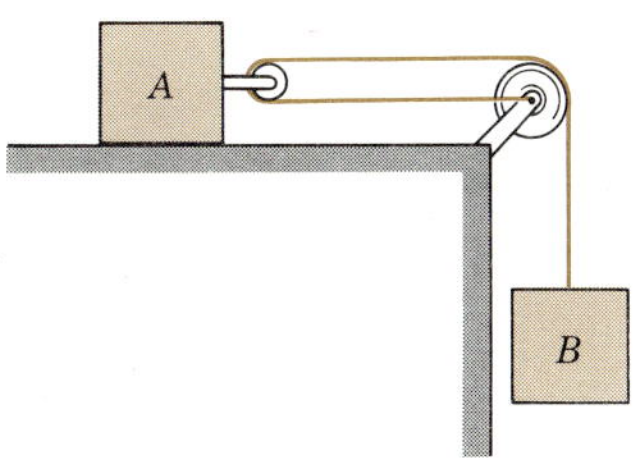

The system shown in the sketch is released from rest. Determine the velocity of each block when block B has descended 1.5 m. The masses of the blocks are $m_A = 12$ kg, $m_B = 6$ kg, and the coefficient of kinetic friction between block A and the horizontal surface is 0.20.

Solution

The first step is to draw a free body diagram of each block, with the positions measured according to the procedure presented in Chapter 11 for cable–pulley systems.

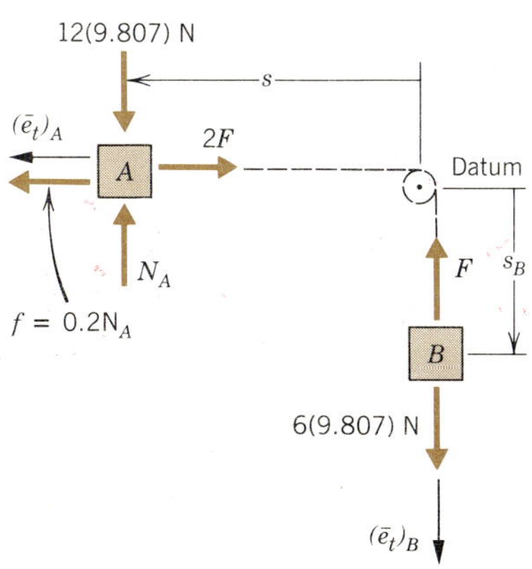

Constructing the length of the cable yields

$$2s_A + s_B = C, \qquad 2v_A + v_B = 0$$

These relations tell us that when s_B increases by 1.5 m in the descent of block B, then s_A will decrease by 0.75 m, that is, block A moves 0.75 m to the right. Similarly when block B is moving downward at speed v_B, then block A is moving to the right at speed $v_B/2$.

The free body diagrams indicate that the forces doing work are gravity, which is conservative, friction, and the cable force. Because block A has no acceleration in the vertical direction, static equilibrium yields $N_A =$ 12(9.807) N. From this, the friction force is seen to be constant, so the work done by this force is

$$U_{1\to 2}^{(f)} = -0.2[12(9.807)](0.75) = -17.653 \text{ J}$$

We could express the work done by the cable, and discover that $U_{1\to2}^{(\text{cable})} = 0$. Rather, we will pursue a general argument for describing internal forces. Let us consider the subsystem formed by the cables and pulleys; it has ideal properties (massless, inextensible, and frictionless). This means that this subsystem has no mechanical energy, and no work is done on it. It follows that it can do no work in moving the blocks, so

$$U_{1\to2}^{(\text{cable})} = 0$$

For the gravitational datum appearing in the free body diagram,

$$V_1 = 6(9.807)(-s_B), \qquad V_2 = 6(9.807)(-s_B - 1.5)$$

The system is initially at rest, and $v_A = -\frac{1}{2}v_B$ in the final position. The corresponding kinetic energies are

$$T_1 = 0, \qquad T_2 = \tfrac{1}{2}(6)v_B^2 + \tfrac{1}{2}(12)(\tfrac{1}{2}v_B)^2 = 4.5v_B^2$$

The work–energy equation for the system yields

$$T_2 + V_2 = T_1 + V_1 + U_{1\to2}^{(f)} + U_{1\to2}^{(\text{cable})}$$

$$4.5v_B^2 + 58.84(-s_B - 1.5) = 58.84(-s_B) - 17.653$$

$$v_B = 3.96 \text{ m/s downward}$$

$$v_A = \left|-\tfrac{1}{2}v_B\right| = 1.98 \text{ m/s to the right}$$

Additional Remarks

The reasoning that was used in this example to discuss the work done by the cable is valid for other connections between bodies. All that is required is that the connection have the ideal properties of being rigid, massless, and frictionless. Note that this is often descriptive of a rod connecting two particles, but it is not valid for a connection using a spring. (Why not?)

HOMEWORK PROBLEMS

12.188 Determine the force $\bar{F}$ that must be applied to the 20 lb block A to make the 20 lb block B attain a speed of 8 ft/s after moving 4 ft to the right from rest. Neglect friction.

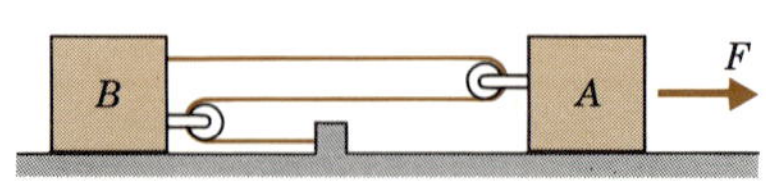

PROB. 12.188

12.189 Solve Problem 12.188 if the coefficient of kinetic friction between the blocks and the ground is 0.20.

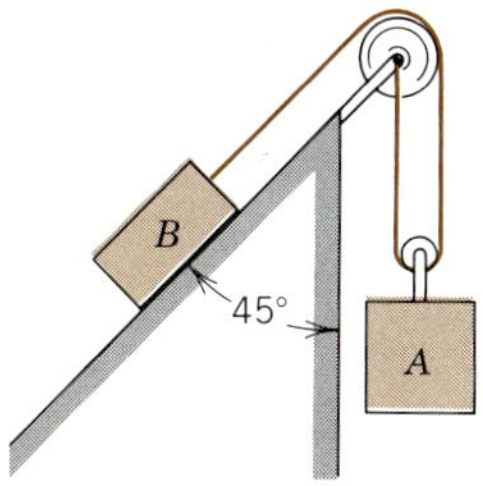

PROB. 12.190

12.190 When released from rest, block A attains a speed of 1 m/s after rising 2 m. The coefficient of friction between block B and the incline is 0.40, and the mass of block B is 5 kg. Determine the mass of block A.

12.191 The spring whose stiffness is 800 N/m is initially elongated 160 mm. The masses of the blocks are $m_A = 1$ kg and $m_B = 0.5$ kg. Determine the velocity of block A at the position where the spring has no force. Friction is negligible.

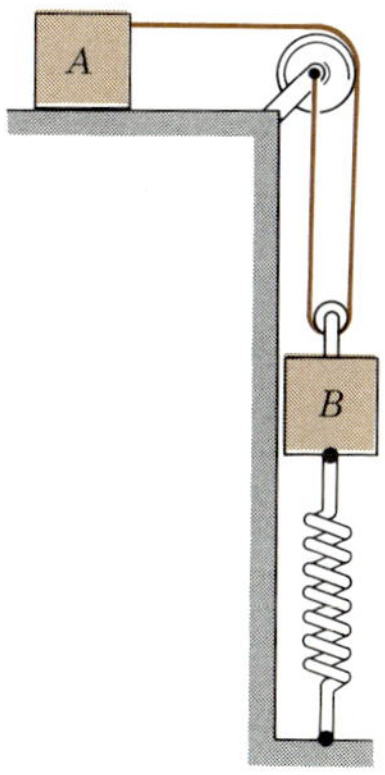

PROB. 12.191

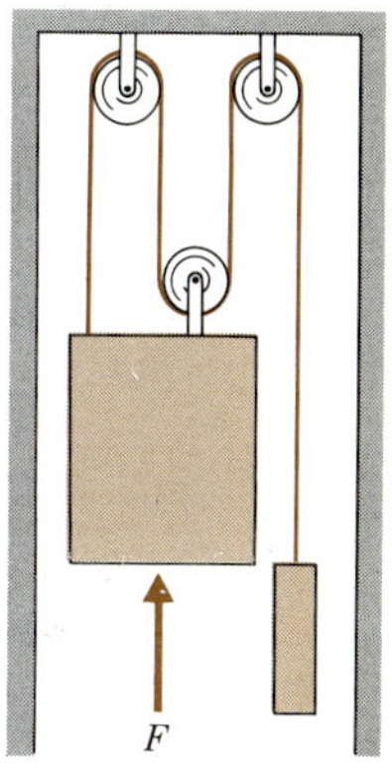

PROB. 12.192

12.192 The elevator and its passengers have a combined weight of 2400 lb, and the counterweight weighs 600 lb. The elevator is moving upward at 10 ft/s after traveling 25 ft from its rest position. Determine the force $\bar{F}$.

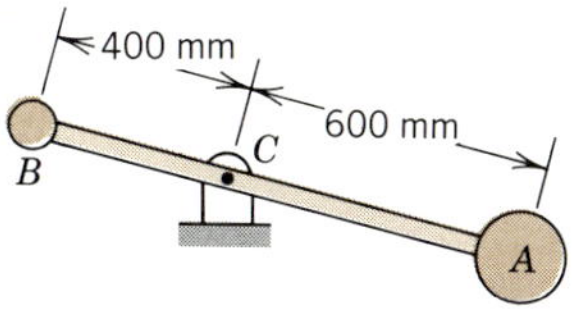

PROB. 12.193

12.193 Two spheres A and B are attached to opposite ends of the rigid bar. The masses of the spheres are $m_A = 2$ kg, $m_B = 1$ kg, and the mass of the bar is negligible. The bar is released from rest in the horizontal position. Determine the angular speed of the bar when it passes the vertical position.

12.194 Spheres A and B, which weigh 2 lb each, are attached to the lightweight rigid bar. The elastic cable AC has a stiffness of 5 lb/in. and an unstretched length of 12 in. If the bar is released from the horizontal position shown, what is its angular speed when it passes the vertical position?

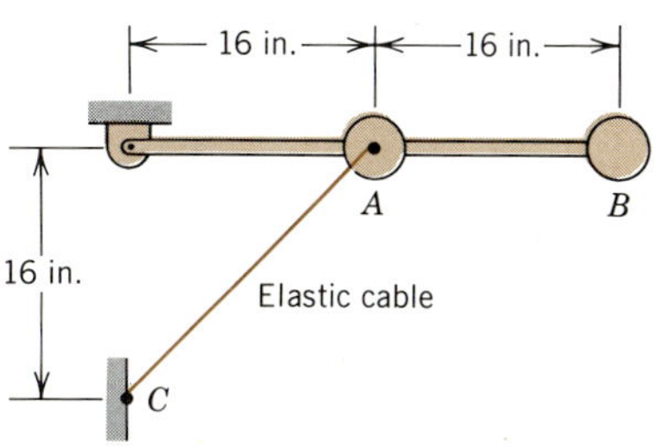

PROB. 12.194

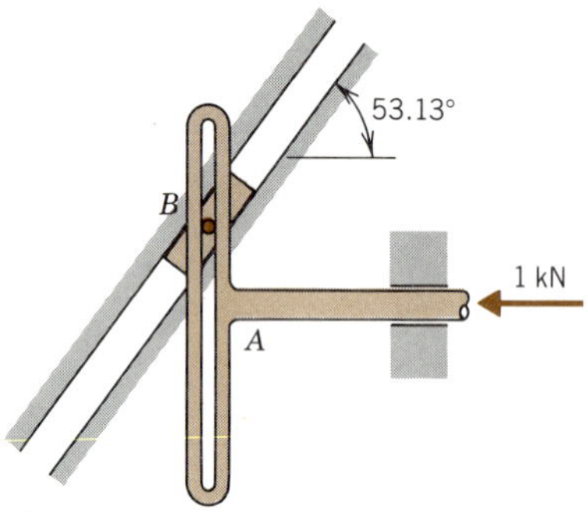

PROB. 12.195

12.195 The 5 kg piston B is pulled through the groove by the 10 kg slotted arm A. The system lies in the horizontal plane, and friction is negligible. Determine the speed of the piston after it has traveled 500 mm along the groove.

12.196 Sliders A and B are connected by a rigid bar of negligible mass in a vertical plane. If block A is released from rest in the position shown, determine its velocity after it falls (a) 2 ft, (b) 4 ft. Each block weighs 2 lb, and friction is negligible.

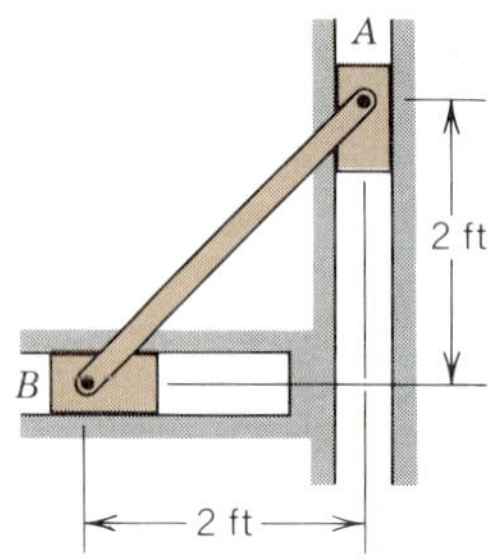

PROB. 12.196

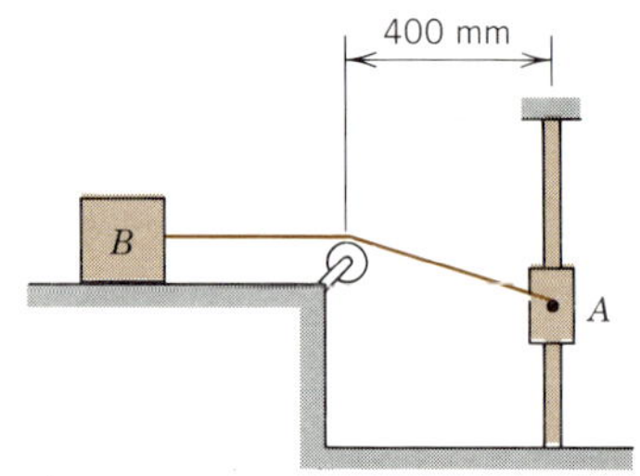

PROB. 12.197

***12.197** Block A is released from rest at the same elevation as block B with the cable taut. Neglecting friction, determine the velocity of block B after block A has fallen 300 mm. The mass of each block is 0.6 kg.

G. POWER

The concept of power is directly related to the energy concepts we have just studied. Basically, the *mechanical power* supplied to a system by a force is the time rate at which this force does work; that is,

$$\boxed{\text{power} \equiv \frac{dU}{dt}} \tag{36}$$

A more specific relation for power is obtained by considering the definition of the work performed by a force in an infinitesimal displacement

$$dU = \bar{F} \cdot d\bar{r}_{P/O}$$

Dividing this expression by dt yields the rate of change of U.

$$\frac{dU}{dt} = \bar{F} \cdot \frac{d\bar{r}_{P/O}}{dt}$$

$$\boxed{\text{power} = \bar{F} \cdot \bar{v}} \tag{37}$$

The unit of power in the SI system is a watt (W), which is equal to 1 J/s, which is 1 N-m/s. In the U.S. Customary system the basic unit is a horsepower (hp), one horsepower being defined as 550 ft-lb/s. (One horsepower is 776 W.)

For a different viewpoint of the power developed by a force, let us consider the basic work-energy principle, eq. (34). It may be rewritten as

$$U_{1\to 2}^{(\mathrm{nc})} = (T_2 - T_1) + (V_2 - V_1)$$

In the situation where position 2 is different from position 1 by a differential amount, this equation becomes

$$dU^{(\mathrm{nc})} = dT + dV$$

Dividing this result by dt gives

$$\boxed{\text{power}^{(\mathrm{nc})} = \frac{d}{dt}(T + V)} \qquad (38)$$

We see that the total power supplied to a system by a nonconservative force equals the rate of increase of the potential and kinetic energy of the system.

A common type of nonconservative force is friction, which generally does negative work. Thus, the power associated with a friction force is negative, representing a dissipation of energy essentially in the form of heat.

EXAMPLE 25

An automobile weighs 2500 lbs, and the air resistance is $0.012v^2$ lb, where v is in ft/s. The power developed by the automobile's electric motor at any speed is constant at 80 hp, of which 25% is dissipated in the drive train. Rolling resistance is negligible. Determine (a) the maximum speed of the vehicle, (b) the acceleration of the vehicle at one-half the maximum speed, and (c) the power dissipated by air resistance at one-half the maximum speed.

Solution

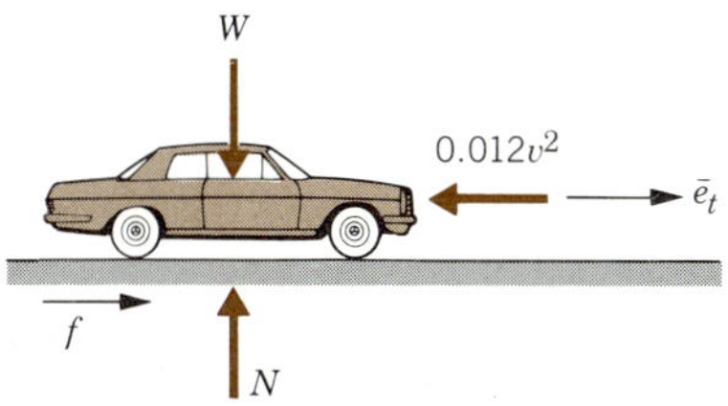

A free body diagram of the automobile must account for the friction force $\bar{f}$ that acts on the drive wheels. This force, which is due to the torque applied to these wheels by the drive train, pushes the automobile forward. The power produced by the engine to propel the car is 75% (80 hp) = 60 hp. This is the power imparted to the vehicle by the friction force $\bar{f}$. Thus

$$\text{power} = (60\text{ hp})\left(\frac{550\text{ ft-lb/s}}{1\text{ hp}}\right) = fv$$

$$f = \frac{33{,}000}{v}\text{ lb}$$

Summing forces in the tangential direction yields an equation for the acceleration.

$$\sum F_t = \frac{33{,}000}{v} - 0.012v^2 = \frac{2500}{32.17}\dot{v}$$

The maximum speed is defined by the condition $\dot{v} = 0$. This gives

$$\frac{33{,}000}{v_{max}} - 0.012v_{max}^2 = 0 \Rightarrow v_{max} = 140.10 \text{ ft/s}$$

The acceleration when $v = \frac{1}{2}v_{max} = 70.05$ ft/s is readily found to be

$$\dot{v} = \frac{32.17}{2500}\left[\frac{33{,}000}{70.05} - 0.012(70.05)^2\right] = 5.30 \text{ ft/s}$$

Finally, the power dissipated by air resistance is the product of the resistance force and the speed. For a speed of 70.05 ft/s, this is

$$P_{dis} = (0.012v^2)v = 0.012(70.05)^3 = 4125 \text{ ft-lb/s}$$

$$= 4125 \text{ ft-lb/s}\left(\frac{1 \text{ hp}}{550 \text{ ft-lb/s}}\right) = 7.5 \text{ hp}$$

In words, from the 80 hp developed by the engine, 20 hp is lost in the drive train and 7.5 hp is lost due to air resistance. The remainder, 52.5 hp, is being used to increase the mechanical energy $T + V$. The percentage of the power supply that is used to increase the mechanical energy is called the *mechanical efficiency*. The mechanical efficiency of this automobile at $\frac{1}{2}v_{max}$ is

$$\text{efficiency} = \frac{52.5}{80}(100\%) = 65.6\%$$

What is the efficiency when the speed is v_{max}?

HOMEWORK PROBLEMS

PROB. 12.198

12.198 A 30,000 kg truck has a constant speed of 40 km/hr up a 10% grade. The air resistance is $F = 40.0v^2$, where F is in newtons and v is the speed of the vehicle in meters per second. It is known that 80% of the power produced by the engine is actually used to propel the truck. Determine the power produced by the engine.

12.199 In Problem 12.198, what acceleration will the truck have when it reaches level ground if its engine continues to produce the same power?

12.200 Four identical diesel-electric locomotives weighing 30 tons each accelerate a train from rest. There are 150 freight cars in the train. The weight of each car is 60 tons, and the rolling resistance on each car is 0.2% of its weight. Air resistance is negligible. Each locomotive is producing 2000 hp of usable mechanical power. Determine (a) the maximum speed of the train, (b) the acceleration at a speed of 60 ft/s, and (c) the power dissipated by friction at a speed of 60 ft/s.

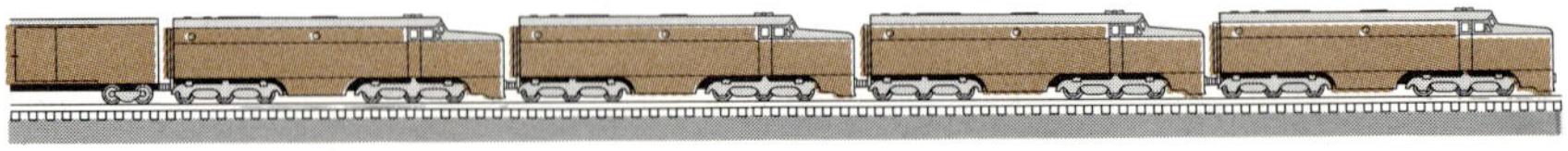

PROB. 12.200

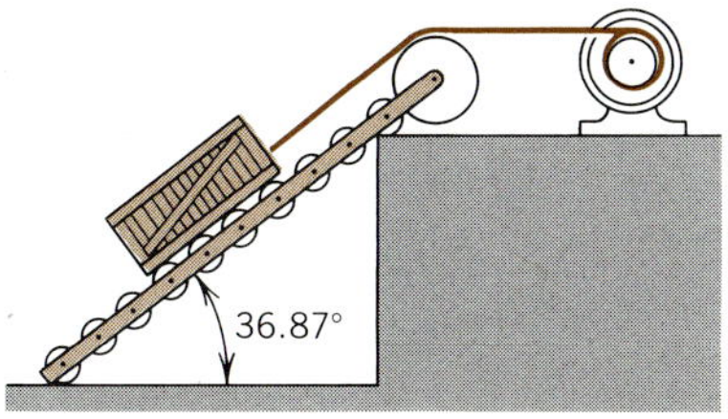

PROBS. 12.201 AND 12.202

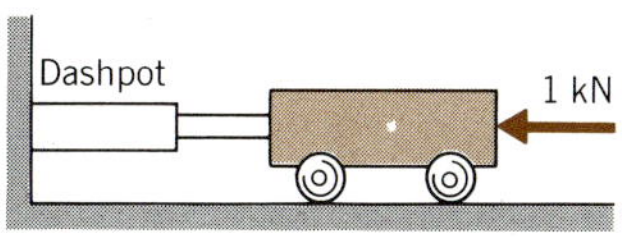

PROB. 12.203

12.201 An electrically powered winch is utilized to move the 50 kg crate from rest at the base of the ramp to 5 m/s in a distance of 10 m with a constant acceleration. Friction is negligible. Determine the maximum horsepower produced by the motor during this motion.

12.202 An electrically powered winch develops a constant usable power of 1 kW to move the 50 kg crate up the inclined ramp. Friction between the crate and the ramp is negligible. Determine (a) the maximum speed v_{max} for the crate, assuming that the ramp was very long, and (b) the acceleration of the crate when $v = \frac{1}{2}v_{max}$.

12.203 The shock absorber consists of a dashpot, which resists the movement of the 100 kg cart with a force of $25v$ newtons, where v is in meters per second. A constant force of 1 kN is applied to the cart. Determine (a) the maximum speed v_{max} that the block could attain if its movement was unlimited, and (b) the power dissipated by the dashpot when $v = v_{max}$.

12.204 From Problem 12.203 determine (a) the acceleration of the block when $v = \frac{1}{2}v_{max}$, and (b) the power dissipated by the dashpot when $v = \frac{1}{2}v_{max}$.

12.205 The electric motor raises block A at the constant speed $v_A = 15$ ft/s. Each block weighs 4 lb. Determine the power required of the motor.

***12.206** The motor is switched on when block A is descending at 15 ft/s, immediately producing 0.5 hp to slow the system. Determine the deceleration rate of block A when the motor is turned on. Each block weighs 4 lb.

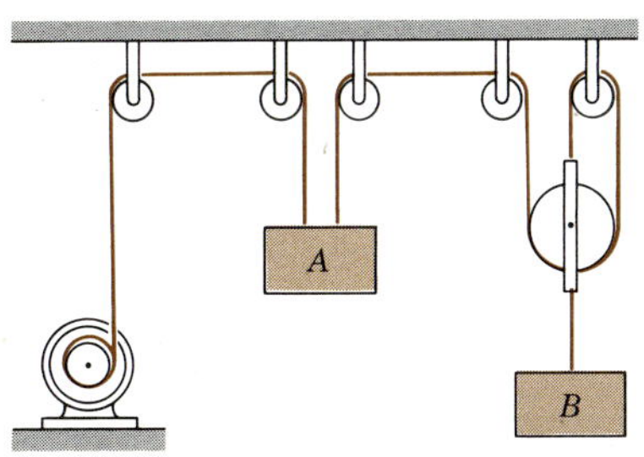

PROBS. 12.205 AND 12.206

***12.207** When a motorboat exceeds 3 m/s, its hull begins to *plane* over the water. It is known that the water resistance is then given by $F_w = k_1/v + k_2 v^2$, where v is the speed of the boat. It is also known that 120 kW is required to propel the boat at 15 m/s, and that a speed of 5 m/s requires the minimum power output from the engine. (a) Determine the values of the coefficients k_1 and k_2. (b) At what speed is the water resistance a minimum? Why isn't this speed the same as that for which the power is a minimum?

***12.208** Determine the minimum time in which the automobile in Example 25 can attain a speed of 60 mi/hr.

***12.209** The crate in Problem 12.202 starts from rest. Determine the distance it will have traveled up the incline when its speed is 2 m/s.

H. LINEAR IMPULSE-MOMENTUM RELATIONS

The momentum principles that we shall study are additional standard integrals of the differential equations of motion. In contrast to the work–energy principle, *momentum concepts are suitable for particle motion when the forces depend only on time.*

The derivation begins with Newton's second law, $\sum \bar{F} = m\bar{a}$. When the forces depend only on t, the variables in the vector equation may be separated by writing $\bar{a} \equiv \dot{\bar{v}} \equiv d\bar{v}/dt$. Thus

$$m\, d\bar{v} = \sum \bar{F}\, dt$$

Integrating over an arbitrary time interval from t_1 to t_2 yields

$$m \int_{v_1}^{v_2} d\bar{v} = \int_{t_1}^{t_2} \sum \bar{F}\, dt$$

$$\boxed{m\bar{v}_2 = m\bar{v}_1 + \int_{t_1}^{t_2} \sum \bar{F}\, dt} \tag{39}$$

where $\bar{v}_1$ and $\bar{v}_2$ are the initial and final velocities of the particle.

The quantity $m\bar{v}$ is the *linear momentum* of the particle. The integral in the right side of eq. (39) is called the *linear impulse* of the force resultant. In words, eq. (39) states that the change in the linear momentum equals the linear impulse. Notice that this is a vector relation representing three scalar equations.

Let us now consider how to apply eq. (39) in conjunction with the several kinematical descriptions. In the case of rectangular Cartesian coordinates, the unit vectors are constants. This permits us to evaluate the impulse by integrating the components of the resultant force. The momentum principle then becomes,

$$\boxed{m\bar{v}_2 = m\bar{v}_1 + \left[\int_{t_1}^{t_2} \sum F_x\, dt\right]\bar{i} + \left[\int_{t_1}^{t_2} \sum F_y\, dt\right]\bar{j} + \left[\int_{t_1}^{t_2} \sum F_z\, dt\right]\bar{k}} \tag{40}$$

Three scalar equations may be obtained from eq. (40) by matching like components.

Let us now consider what is involved in evaluating the linear impulse in situations where we are using either path variables or cylindrical coordinates. The unit vectors in these coordinate descriptions vary, so they cannot be treated as constants in the integral for the impulse. For example, in the case of path variables we have

$$\int_{t_1}^{t_2} \sum \bar{F}\, dt = \int_{t_1}^{t_2} \left(\sum F_t \bar{e}_t\right) dt + \int_{t_1}^{t_2} \left(\sum F_n \bar{e}_n\right) dt$$

For a general motion, the unit vectors $\bar{e}_t$ and $\bar{e}_n$ depend on the current position. Because this position is unknown, the impulse cannot be evaluated. Therefore, we shall make it a practice to formulate the linear impulse-momentum in terms of rectangular Cartesian coordinates.

It will be noted that the principle of linear impulse and momentum is merely a formalization of our earlier procedure for separating variables when force is a function of time. Nevertheless, the impulse-momentum principle often provides an insight that we might overlook were we concerned solely with solving the differential equation.

One such situation occurs when we are interested only in the average value of the forces in a time interval, because in that case the linear impulse is

$$\int_{t_1}^{t_2} \sum \bar{F}\,dt = (\sum \bar{F}_{\text{av}})(t_2 - t_1) = (\sum \bar{F}_{\text{av}})\,\Delta t \tag{41}$$

Other applications of the impulse-momentum principles are for systems of interacting particles and for impulsive forces, such as those exerted between colliding bodies. These are the next topics we shall study.

1. Interacting Particles

Let us begin with the equation for the acceleration of the center of mass G of a system of N particles, eq. (23),

$$\sum_{i=1}^{N} \bar{F}_i = m\bar{a}_G = m\frac{d\bar{v}_G}{dt}, \qquad m = \sum_{i=1}^{N} m$$

where $\bar{F}_i$ is the external force acting on particle i. (Recall that the forces that are internal to the system do not affect the motion of the center of mass.) The variables in this equation may be separated with the result that

$$m(\bar{v}_G)_2 - m(\bar{v}_G)_1 = \int_{t_1}^{t_2} \sum_{i=1}^{N} \bar{F}_i\,dt \equiv \sum_{i=1}^{N} \int_{t_1}^{t_2} \bar{F}_i\,dt \tag{42}$$

The velocity of the center of mass may be related to the velocities of the particles. In order to accomplish this we recall that the position of the center of mass may be obtained from the first moment of mass.

$$m\bar{r}_{G/O} = \sum_{i=1}^{N} m_i\bar{r}_{i/O}$$

Differentiating with respect to time yields

$$m\bar{v}_G = \sum_{i=1}^{N} m_i\bar{v}_i$$

Substituting this equation into eq. (42) we find that

$$\sum_{i=1}^{N} m_i(\bar{v}_i)_2 = \sum_{i=1}^{N} m_i(\bar{v}_i)_1 + \sum_{i=1}^{N} \int_{t_1}^{t_2} \bar{F}_i\,dt \tag{43}$$

Equation (43) states that the increase in the total momentum of a system during a time interval is equal to the resultant impulse of the external forces.

Occasionally, we will encounter systems in which the resultant impulse of the external forces is zero. Also, there are situations where the external forces are insignificant in comparison to the forces exerted within the system. In either case, eq. (43) becomes

$$\sum_{i=1}^{N} m_i(\bar{v}_i)_2 = \sum_{i=1}^{N} m_i(\bar{v}_i)_1$$

which means that the total momentum of the system is constant during the time interval of interest. This is the principle of *conservation of linear momentum* for a system.

The primary feature of eq. (43) is that it enables us to study the overall motion of a system without considering the system's internal forces. In contrast, when an internal force is to be considered, the momentum equation for the particle it acts on is required. Such a case is illustrated in the next example.

EXAMPLE 26

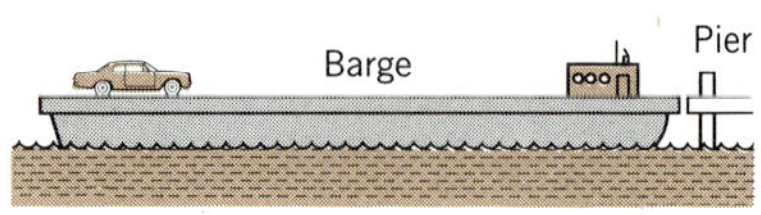

The 1500 kg automobile starts from rest and attains a speed of 5 m/s relative to the 6000 kg barge in 5 s. Because someone forgot to attach the mooring cable, the barge moves away from the pier. Determine (a) the velocity of the barge after 5 s, (b) the average traction force exerted by the tires on the barge, and (c) the distance the barge moves in this 5 s interval.

Solution

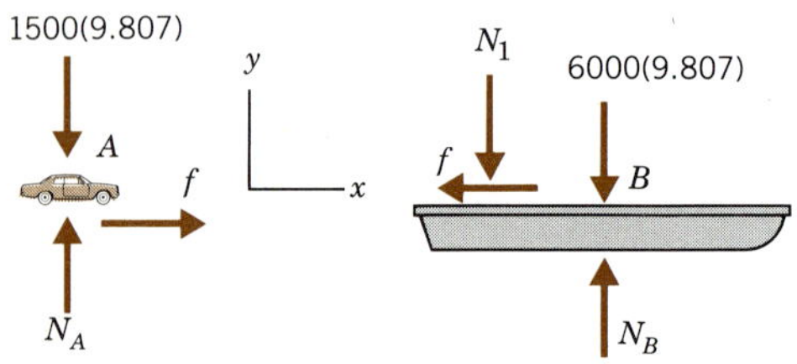

The first step is to draw free body diagrams of the automobile and of the barge, for which the water resistance is neglected. The force $\bar{f}$ is the traction force, and is oriented to the right on the automobile because that is the direction in which the automobile accelerates.

Prior to formulating the momentum equations, let us compile the given information. The system is at rest initially, hence $(\bar{v}_A)_1 = (\bar{v}_B)_1 = \bar{0}$ when $t = 0$. Also the velocity of the automobile with respect to the barge is $5\bar{i}$ m/s when $t = 5$ s. The barge is stated to move to the left, so $(\bar{v}_B)_2 = -v_B\bar{i}$, where v_B is to be determined. To relate $\bar{v}_A$ and $\bar{v}_B$ we formulate a relative velocity equation.

$$(\bar{v}_A)_2 = (\bar{v}_B)_2 + (\bar{v}_{A/B})_2 = (5 - v_B)\bar{i} \text{ m/s}$$

The information in part (a) does not involve the interaction forces $\bar{N}_1$ and $\bar{f}$. This suggests that the impulse-momentum equation for the entire system has to be formulated. Because the motion is in the x direction, we need only formulate that component. The horizontal force $\bar{f}$ is internal to the system, so we have a case of conservation of linear momentum. This means that

$$[m_A(v_{Ax})_2 + m_B(v_{Bx})_2] = [m_A(v_{Ax})_1 + m_B(v_{Bx})_1]$$

$$1500(5 - v_B) + 6000(-v_B) = 0$$

$$v_B = 1 \text{ m/s}, \qquad \bar{v}_B = -1\bar{i} \text{ m/s}$$

The average traction force may now be evaluated by considering the barge alone, in order to make $\bar{f}$ appear as an external force. The x component of the impulse-momentum equation for the 5 s interval in this case is

$$m_B(v_{Bx})_2 = m_B(v_{Bx})_1 - f(5)$$

$$6000(-1) = 0 - 5f, \qquad f = 1200 \text{ N}$$

It should be noted that the signs of the components in both of the above impulse-momentum relations were selected in accord with the xyz system defined in the free body diagram.

In order to determine the distance the barge moves, we shall assume that the barge has a constant acceleration, so

$$(v_{Bx})_2 = (v_{Bx})_1 + \dot{v}_{Bx}\,\Delta t, \; -1 = \dot{v}_{Bx}(5)$$

$$\dot{v}_{Bx} = -0.2 \text{ m/s}$$

Here the negative sign means that the acceleration is to the left. The distance traveled is then

$$\Delta x = (v_{Bx})_1\,\Delta t + \tfrac{1}{2}\dot{v}_{Bx}(\Delta t)^2 = \tfrac{1}{2}(-0.2)(5)^2$$
$$= -2.5 \text{ m}$$

The negative sign for Δx arises because the value of x decreases in the motion.

HOMEWORK PROBLEMS

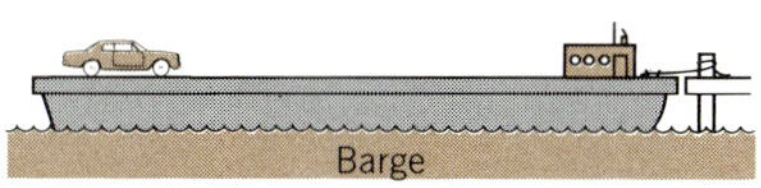

PROB. 12.210

12.210 The 2500 lb automobile accelerates from rest and attains a speed of 20 ft/s relative to the 5000 lb barge in 2 s. Determine the average force in the mooring cable that is required to hold the barge stationary and the average traction force exerted by the tires on the barge.

12.211 With the aid of a catapult, a 20 ton aircraft is launched in an easterly direction from the deck of an aircraft carrier. The wind is calm and the aircraft has an airspeed of 150 mi/hr when launched, attained after being accelerated for 4 s. If the engines provide 40,000 lb of thrust, determine the force provided by the catapult (a) if the aircraft carrier is at rest, and (b) if the aircraft carrier is heading east at a constant speed of 30 mi/hr.

12.212 The electrically operated winch pulls the 200 kg crate to the left over the smooth deck of the cart. The total mass of the cart is 100 kg. Three seconds after starting from rest, the crate has a speed of 2 m/s relative to the cart. For this time interval, determine (a) the average cable tension, (b) the final velocity of the cart, and (c) the distance the cart moves.

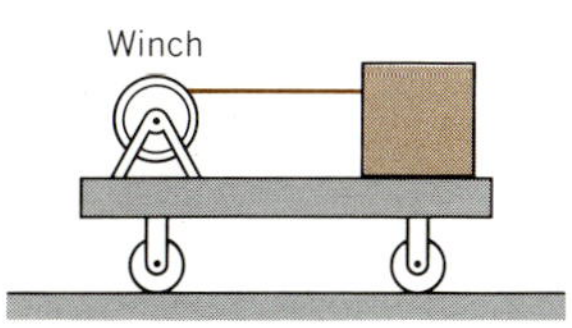

PROB. 12.212

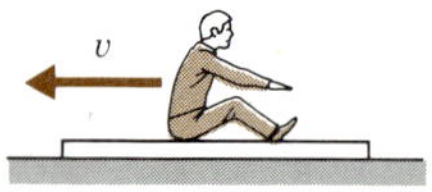

PROB. 12.213

*12.213 A 40 kg boy sitting on a 10 kg board rocks backward from rest. The horizontal velocity of his center of mass is v after 0.25 s has elapsed. The coefficients of static friction are 0.6 between the boy and the board and 0.5 between the board and the ground. Determine the minimum value of v such that (a) the board will slide over the ground, (b) the boy will slide over the board. (c) Which type of sliding occurs first?

12.214 An 8000 lb tractor towing a 40,000 lb trailer starts from rest and accelerates along level ground to a speed of 30 mi/hr in 20 s. The frictional resistances on the tractor and the trailer attributable to the undriven wheels are 200 lb each. Determine the average traction forces produced by the drive wheels of the tractor during the acceleration and determine the average pull exerted by the tractor on the trailer. Neglect air resistance.

PROB. 12.214

12.215 A three-car commuter train slows from an initial speed of 72 km/hr on a level track. Each car has a mass of (10^4) kg, and the coefficient of friction between the wheels and the track is 0.20. Determine the minimum amount of time required to stop the train and the corresponding force within the couplers if the brakes in all cars are functioning. Air resistance is negligible.

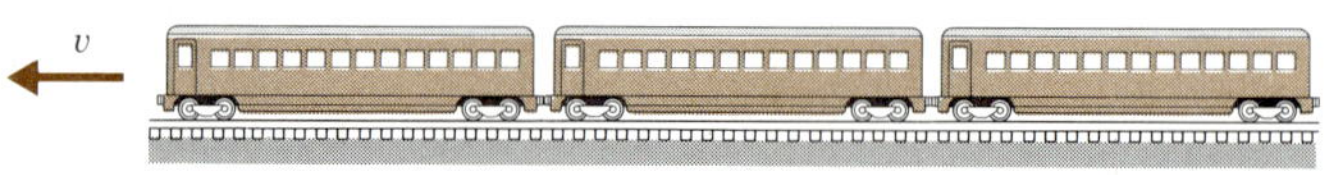

PROB. 12.215

12.216 Solve Problem 12.215 in the case where the brakes in the middle car fail.

12.217 As the 3000 lb automobile descends the hill, the total friction force $\bar{f}$ exerted between the tires and the ground varies as shown, where a negative value represents braking. The initial speed of the car was 60 mi/hr. Determine the speed of the car (a) after 15 s, and (b) after 30 s. Neglect air resistance.

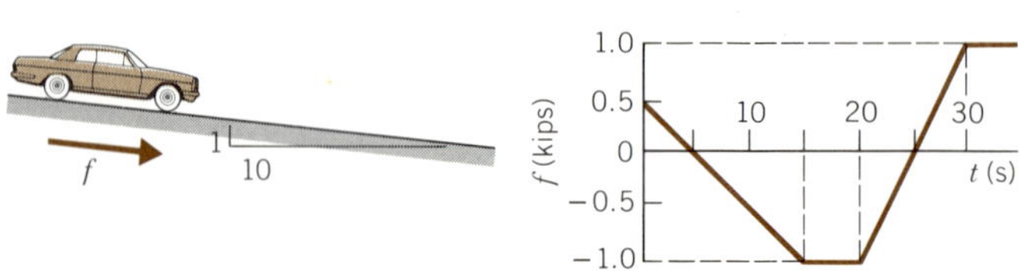

PROB. 12.217

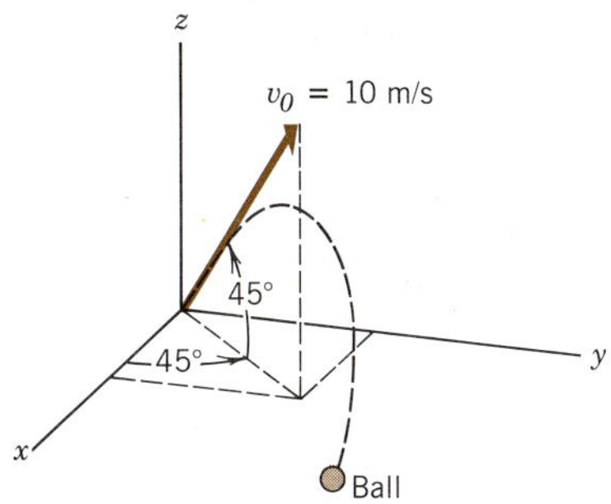

PROB. 12.218

***12.218** A 0.5 kg ball is thrown at 10 m/s as shown. The xy plane is horizontal. The ball is subjected to a constant 1 N force from a wind that is blowing in the negative y direction. How long will it be before the ball is traveling parallel to the xz plane? Determine the velocity of the ball at this time.

***12.219** The 20 lb block slides down the smooth ramp of the 100 lb cart. The system was at rest initially. Determine (a) the velocity of the block and of the cart when the box reaches the bottom of the ramp, and (b) the distance the cart shifts during this motion.

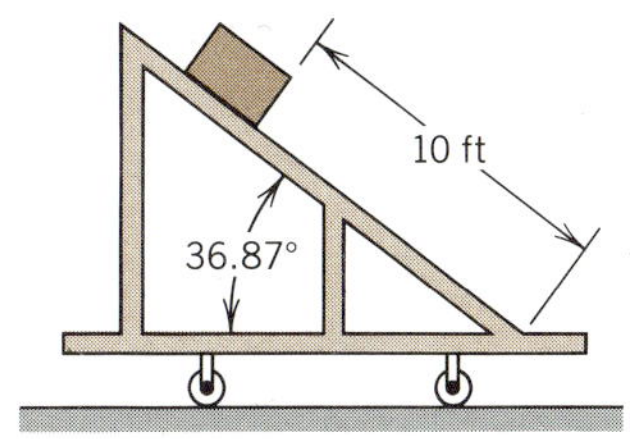

PROB. 12.219

---

2. Impulsive Forces

The term *impulsive force* describes any force that acts on a body over a very small time interval, during which it imparts a large acceleration (many g's). Because the change in velocity is the time integral of acceleration, an impulsive force can significantly change the velocity of a body, even though it acts over a small time interval. Typical situations involving impulsive forces occur in explosions and in collisions.

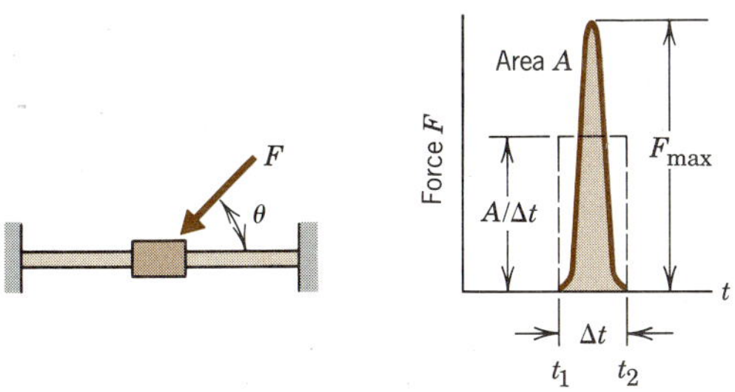

FIGURE 10

Let us consider the situation in Figure 10a where a collar is hit by an impulsive force $\bar{F}$ whose direction angle θ is constant. The measurement of the impulsive force $\bar{F}$ versus time in Figure 10b is difficult to obtain. It is often sufficient to obtain an order of magnitude estimate of F_{max} by using the average force. For this, we determine the impulse $A \equiv \int F\,dt$ by computing the shaded area under the curve in Figure 10b. The average force having the same impulse is then

$$F_{av} = \frac{A}{\Delta t}$$

where $\Delta t = t_2 - t_1$.

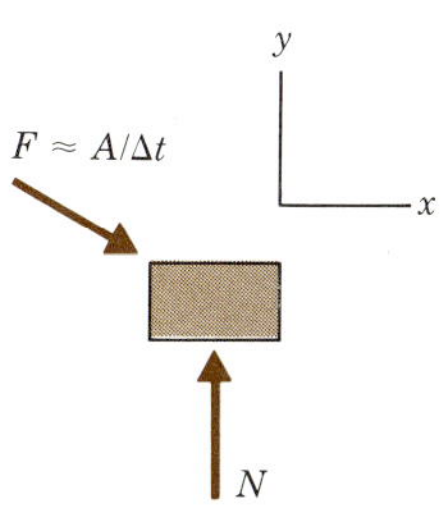

FIGURE 11

If the rod guiding the collar is smooth, the free body diagram during the interval of interest is as shown in Figure 11. Note that in the free body diagram we omitted the gravitational force. The impulsive force, F_{av}, is very large, so the influence of the gravitational force is negligible during the small time interval Δt. In contrast, the reaction force $\bar{N}$ is not negligible. This is because it must be large in order to resist the tendency of the vertical component of $\bar{F}$ to accelerate the collar in the y direction.

The final velocity $\bar{v}_2$ and an estimate of the reaction $\bar{N}$ may be obtained from the impulse-momentum equation for the collar.

If we were to observe the collar by eye, we would think that the change from $\bar{v}_1$ to $\bar{v}_2$ occurs instantaneously. This is not the actual situation, for an instantaneous change in velocity requires an infinite acceleration, hence an infinite force. Nevertheless, it is a simplification that proves to be reasonably descriptive.

A related approximation is that although the velocity changes suddenly, the change in position is negligible. This follows from the fact that the velocities are not exceptionally large, whereas the elapsed time is very small. The foregoing approximations are exemplified by the collision of a falling ball and a hard surface. The ball seems to rebound instantly, with the rebound starting at the location where the contact was first made.

We shall now outline a general procedure for treating impulsive forces that follows from the preceding discussion.

1. Impulsive forces are described in terms of an average value giving the same impulse as the actual force.
2. Nonimpulsive forces, such as gravity, are ignored during the impulsive effect.
3. The linear-impulse momentum principle is used to study the unknown impulsive forces and the velocity change during the time interval of the impulse.
4. Velocity is considered to change instantaneously.
5. Position does not change during the impulsive effect.

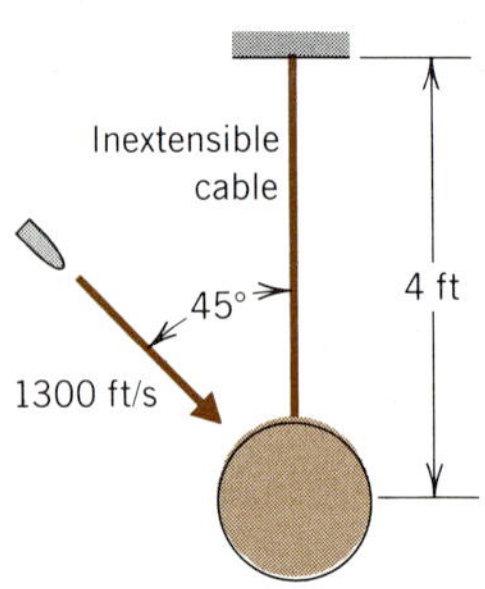

EXAMPLE 27

A 2 oz bullet traveling at 1300 ft/s in the direction shown hits and is embedded in the 8 lb sphere. The impact is completed in 0.05 s. Determine (a) the average increase in the tension of the cable during the impact, (b) the speed of the sphere immediately after the impact, and (c) the average force acting on the bullet during the impact.

Solution

The impulsive forces acting during the impact are the interaction force $\bar{F}_b$ exerted between the bullet and the sphere, and also the cable tension $\bar{F}_c$, which prevents the sphere from moving downward. The free body diagrams that we have drawn ignore the nonimpulsive weight forces. They also depict the cable tension as always being vertical, because movement during the impact is negligible.

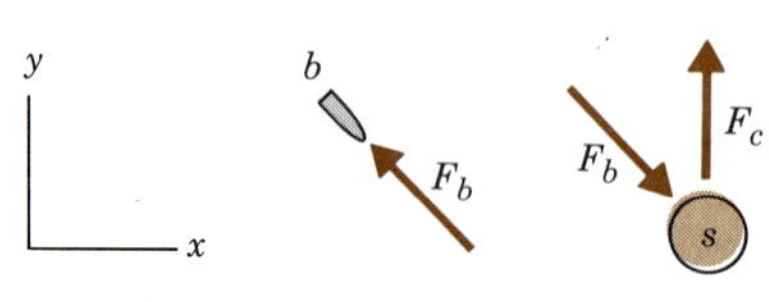

The initial velocities of the bullet and the sphere are given to be

$$(\bar{v}_b)_1 = 1300(\sin 45^\circ\,\bar{\imath} - \cos 45^\circ\,\bar{\jmath}) = 919.2(\bar{\imath} - \bar{\jmath})\ \text{ft/s}$$

$$(\bar{v}_s)_1 = \bar{0}$$

After the impact has been completed, the sphere and the bullet move as though they are one body. The cable permits movement in only the horizontal direction for the position of the impact, so

$$(\bar{v}_b)_2 = (\bar{v}_s)_2 = v_2\bar{\imath}$$

In order to eliminate the internal force $\bar{F}_b$ from the evaluation of the motion, we formulate the impulse-momentum relation for the entire system. The force $\bar{F}_c$ is external, so

$$\frac{2/16}{32.17}(v_2\bar{\imath}) + \frac{8}{32.17}(v_2\bar{\imath}) = \frac{2/16}{32.17}(919.2)(\bar{\imath} - \bar{\jmath}) + (F_c\bar{\jmath})\Delta t$$

$$i \text{ component:}\quad 0.2526v_2 = 3.572$$

$$j \text{ component:}\quad 0 = -3.572 + 0.05F_c$$

$$v_2 = 14.141 \text{ ft/s}, \qquad F_c = 71.4 \text{ lb}$$

The average force acting on the bullet may now be found from the impulse momentum equation for the bullet, which is

$$\frac{2/16}{32.17}(14.141\bar{\imath}) = \frac{2/16}{32.17}(919.2)(\bar{\imath} - \bar{\jmath}) + \bar{F}_b(0.05)$$

$$\bar{F}_b = -70.3\bar{\imath} + 71.4\bar{\jmath}$$

Note that $\bar{F}_b$ is *not* in the opposite sense from the initial velocity of the bullet. Why not?

HOMEWORK PROBLEMS

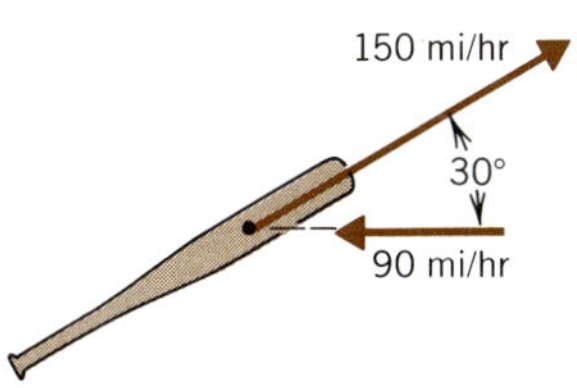

PROB. 12.220

12.220 The initial and final velocities of a 5 oz baseball being hit by a bat are shown. Knowing that the ball and bat are in contact for 0.05 s, determine the average force exerted between the ball and the bat.

12.221 A 50 gram bullet traveling at 500 m/s is fully embedded in the 5 kg block 0.015 s after first contact. The block is then slowed by the spring, whose stiffness is 100 N/mm. The horizontal surface is smooth. Determine (a) the velocity of the block immediately after the impact, (b) the average force acting on the bullet during the impact, and (c) the maximum distance the spring is compressed.

PROB. 12.221

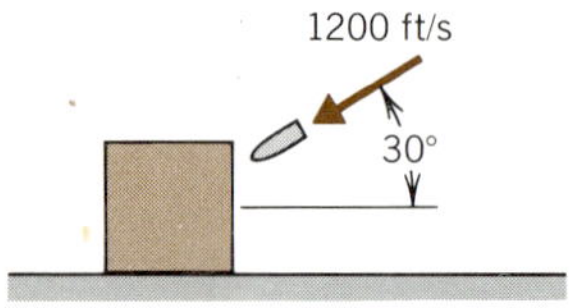

PROB. 12.222

12.222 The initial velocity of the 1 oz bullet is 1200 ft/s, and the 2 lb block is initially at rest. The bullet is fully embedded in the block 0.004 s after first contact. If the ground is smooth, determine (a) the velocity of the block immediately after impact, and (b) the average force exerted on the bullet during the impact.

12.223 A cable is fastened to the 20 kg block in order to restrain the block during the impact of the 150 gram bullet. The bullet is fully embedded in the block 0.01 s after first contact, and the ground is smooth. If the initial velocity of the bullet is $v_1 = 800$ m/s in the direction shown, determine (a) the average tensile force in the cable during the impact, and (b) the corresponding average force acting on the bullet.

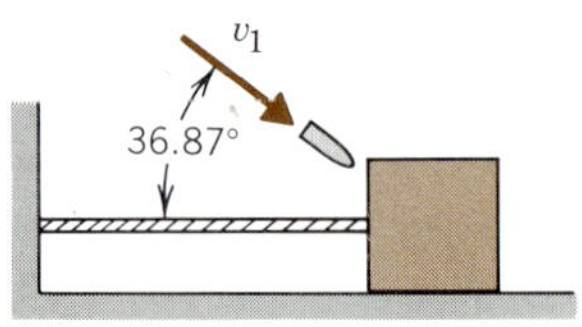

PROB. 12.223

12.224 The cable in Problem 12.223 will break whenever the tensile force in it is 12 kN. Determine the smallest initial speed v_1 that will cause the cable to break.

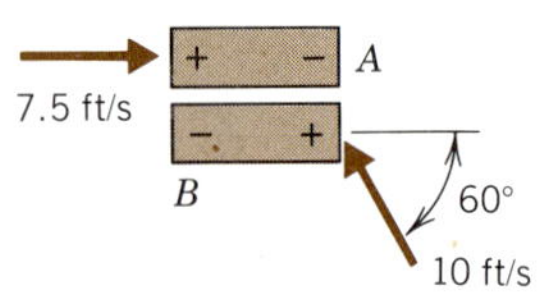

PROB. 12.225

12.225 Bar magnets A and B, weighing 1 lb and 1.5 lbs, respectively, become attached after they contact with the initial speeds shown. The horizontal surface over which the magnets slide is smooth. Determine the velocity of the magnets after their impact.

12.226 The velocities of a 20 kg block of wood and an 80 gram bullet immediately before they collide are shown. The bullet is embedded in the block 0.02 s after the first contact. Determine the final velocity of the block of wood and the average force exerted on the block by the bullet during the collison.

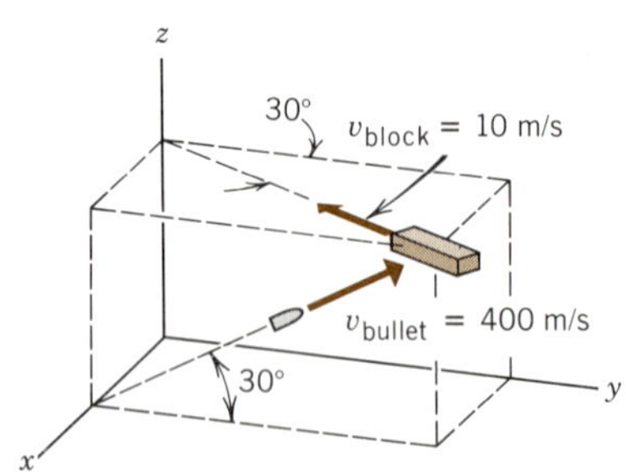

PROB. 12.226

12.227 Different weight bullets are fired from a rifle with the same kinetic energy of 2500 J. Each bullet emerges from the barrel 20 ms after being fired. Determine the average force required to hold the rifle stationary for a (a) 25 gram bullet, and (b) 50 gram bullet.

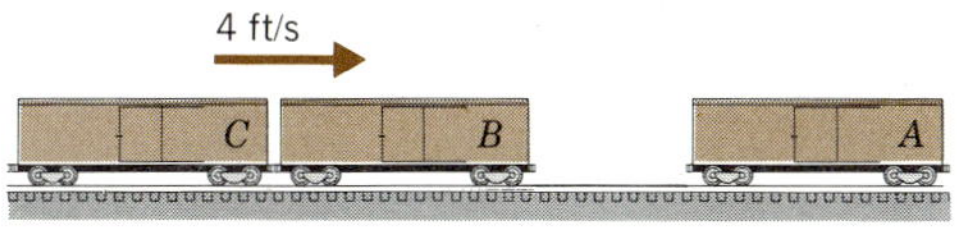

PROB. 12.228

12.228 Three identical 30 ton freight cars are to be coupled. The coupling process is completed in 0.50 s. Car A is at rest and the coupled cars B and C are traveling at 4 ft/s. Determine the speed of the cars after the coupling and the average force in each of the two couplers during the coupling.

***12.229** The sketch is a schematic view of a cannon and its recoil mechanism. Twelve milliseconds (12 ms) after being fired, the shell emerges with the absolute velocity shown. The mass of the barrel and its movable support is 600 kg. Also, the spring has a stiffness of 1000 kN/m. Determine (a) the average increase in the reaction force on the rails during the firing, (b) the recoil velocity of the barrel, and (c) the maximum distance that the spring is compressed.

PROB. 12.229

---

3. Central Impact

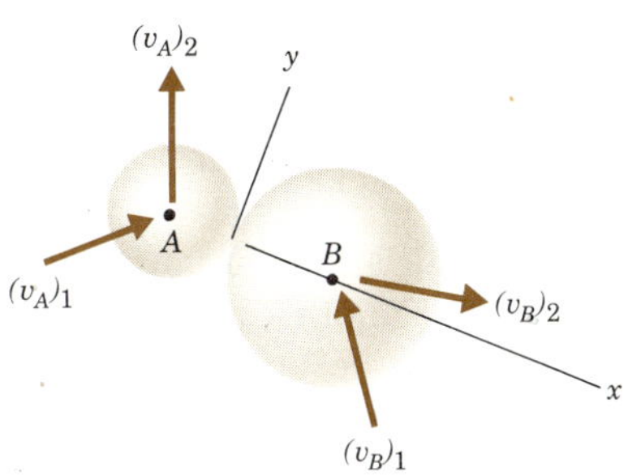

FIGURE 12a

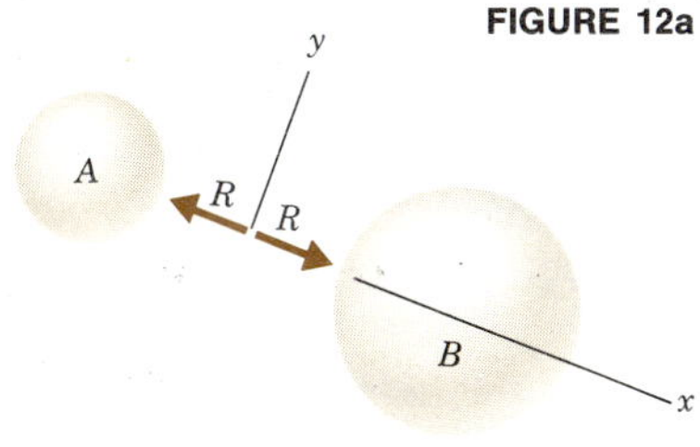

FIGURE 12b

A common way in which impulsive forces are generated is through the collision of bodies. Figure 12a depicts a general condition of a collision (*impact*) of two spheres, where $(\bar{v}_A)_1$ and $(\bar{v}_B)_1$ are the velocities of the spheres immediately before impact, and $(\bar{v}_A)_2$ and $(\bar{v}_B)_2$ are the velocities immediately after the impact. To study this system we choose a coordinate system xyz such that the x axis is normal to the plane of contact (the yz plane) of the spheres, as shown in the figure.

The situation depicted in Figure 12a is called *central impact*. This is the case when the centers of mass of both bodies and the point of contact are all situated along a common line normal to the plane of contact. The case where these three points are not all situated along the normal is called *eccentric impact*. The latter is a topic in rigid body kinetics, which is treated in Chapter 14.

We assume that the spheres are perfectly smooth, so that the only interaction force between the bodies is the normal force $\bar{R}$ illustrated in Figure 12b. The bodies first contact at time t_1 and then begin to deform. After attaining their maximum deformation, at time t_{max}, the bodies begin to regain their shapes, until they lose contact at time t_2. To a certain extent the collision process resembles the sudden compression and release of a spring, but it is much more complex.

In order to avoid the necessity of studying the deformation characteristics of the bodies, let us consider the results of a number of experiments in which we measure the final velocities of the two spheres for many different initial velocities. We find in these experiments that the ratio of the initial and final relative velocity components in the x direction (normal to the plane of contact) is essentially constant for a range of velocities of engineering significance. In order for the particles to collide and then separate, we must have

$$(v_{Ax})_1 > (v_{Bx})_1$$

$$(v_{Ax})_2 \leq (v_{Bx})_2$$

Therefore, the experiments tell us that

$$\boxed{\varepsilon = -\frac{(v_{Bx})_2 - (v_{Ax})_2}{(v_{Bx})_1 - (v_{Ax})_1}} \tag{44}$$

where ε is a positive constant called the *coefficient of restitution.*

Equation (44) also has meaning in regard to the interaction force $\bar{R}$. Let us define the period of deformation by $t_1 \leq t \leq t_{\max}$ and the period of restitution by $t_{\max} \leq t \leq t_2$. We can show that

$$\varepsilon = \frac{\int_{t_{\max}}^{t_2} R\,dt}{\int_{t_1}^{t_{\max}} R\,dt} \tag{45}$$

In words, the ratio of the impulses of the contact force in the two time intervals is the coefficient of restitution.

A collision generally involves a loss of mechanical energy, the energy normally being expended in the form of heat and sound. Only when $\varepsilon = 1$, which is the case of a *perfectly elastic impact*, is there conservation of mechanical energy. The case $\varepsilon = 0$, which is a *perfectly plastic impact*, represents a maximum loss of energy.

Equation (44) represents only one relationship for the velocity components $(v_{Ax})_2$ and $(v_{Bx})_2$. Clearly we require another relationship to determine the velocities of both bodies after the collision. Toward this end, we use the free body diagrams of Figure 12b. The force $\bar{R}$ which is the only significant force, is internal to the system formed by the two bodies. This means that the linear momentum of the system is conserved. The x component of the momentum equation gives

$$m_A(v_{Ax})_2 + m_B(v_{Bx})_2 = m_A(v_{Ax})_1 + m_B(v_{Bx})_1 \tag{46}$$

Solving this equation simultaneously with eq. (44) yields the x components of the velocities.

The components of velocity parallel to the plane of contact are determined by noting (in the free body diagram, Figure 12b) that no impulsive force acts in either the y or z direction. It follows that the corresponding velocity components are constant.

$$\begin{aligned} (v_{Ay})_2 &= (v_{Ay})_1 \qquad & (v_{By})_2 &= (v_{By})_1 \\ (v_{Az})_2 &= (v_{Az})_1 \qquad & (v_{Bz})_2 &= (v_{Bz})_1 \end{aligned} \tag{47}$$

We should note the limitations of the foregoing development. Equation (44), for the coefficient of restitution, is applicable to all cases of central impact. On the other hand, eqs. (46) and (47) were obtained for the special case where the impulsive interaction force $\bar{R}$, shown in Figure 12b, was the only force. In some problems the bodies will be restrained by impulsive forces that may invalidate either eq. (46) or eq. (47). Situations where this occurs are featured in a few of the homework problems that follow.

Finally, consider the case when a small body collides with the earth. In such a situation the enormous mass of the earth results in a negligible change in the earth's velocity. Therefore, to study such problems we consider the earth to be fixed, and disregard the system momentum principle. The velocity of the small body is found by using eqs. (44) and (47), with $\bar{v}_A \equiv \bar{0}$ for a fixed earth.

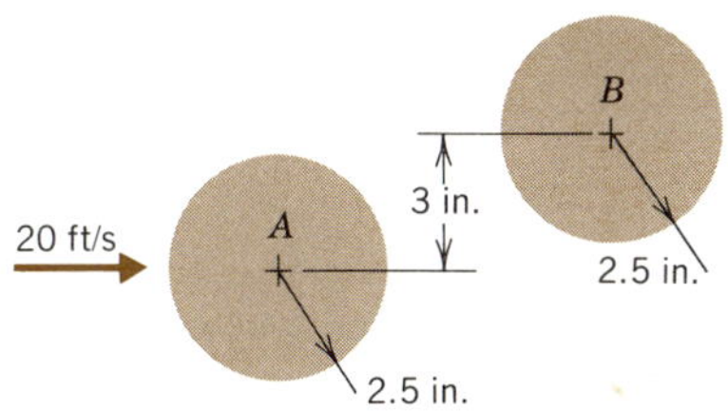

EXAMPLE 28

Disk A on an air hockey game board collides with disk B, which is initially at rest. The disks are of identical mass and the coefficient of restitution is ε. Determine (a) the velocities with which the disks rebound from the collision, and (b) the percentage of the initial kinetic energy that is dissipated in the collision.

Solution

The study of the collision process requires that we first establish the contact plane. Therefore, we make a sketch that shows the two disks in contact. Note that consistency with eq. (45) for ε requires that the x axis be perpendicular to the contact plane.

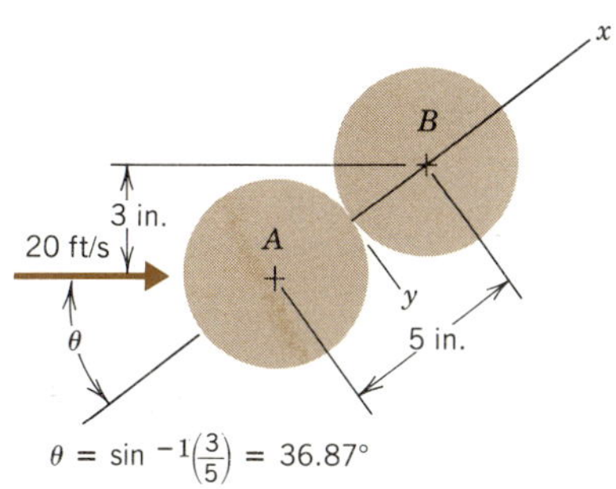

The initial velocities are

$$(\bar{v}_B)_1 = \bar{0}, \qquad (\bar{v}_A)_1 = 20(\cos\theta\,\bar{i} + \sin\theta\,\bar{j}) = 16\bar{i} + 12\bar{j} \text{ ft/s}$$

Because the contact force is in the x direction, the velocity components in the y direction are unchanged. Hence

$$(v_{By})_2 = 0, \qquad (v_{Ay})_2 = 12$$

There are no impulsive external forces acting on the system. The motion in the x direction is therefore governed by the equation for the coefficient of restitution and by conservation of linear momentum for the system.

$$\varepsilon = -\frac{(v_{Bx})_2 - (v_{Ax})_2}{0 - 16}$$

$$m(v_{Ax})_2 + m(v_{Bx})_2 = m(16) + 0$$

Solving these equations simultaneously gives

$$(v_{Ax})_2 = 8(1 - \varepsilon), \qquad (v_{Bx})_2 = 8(1 + \varepsilon)$$

The final velocity vectors are formed by combining the x and y components. The results are written below, and also shown in the sketch.

$$(\bar{v}_A)_2 = 8(1 - \varepsilon)\bar{i} + 12\bar{j} \text{ ft/s}, \qquad (\bar{v}_B)_2 = 8(1 + \varepsilon)\bar{i} \text{ ft/s}$$

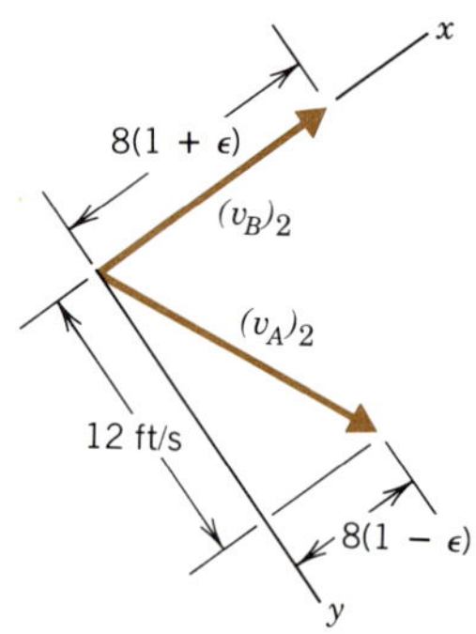

Now that the velocities have been found, the kinetic energy of the system may be evaluated.

$$T_1 = \tfrac{1}{2}m_A(v_A)_1^2 + \tfrac{1}{2}m_B(v_B)_1^2 = \tfrac{1}{2}m(20)^2 = \tfrac{1}{2}m(400)$$

$$\begin{aligned} T_2 &= \tfrac{1}{2}m_A(v_A)_2^2 + \tfrac{1}{2}m_B(v_B)_2^2 \\ &= \tfrac{1}{2}m[8^2(1 - \varepsilon)^2 + (12)^2 + 8^2(1 + \varepsilon)^2] \\ &= \tfrac{1}{2}m[272 + 128\varepsilon^2] \end{aligned}$$

The percent decrease in kinetic energy is

$$\begin{aligned} \Delta T(\%) &= \frac{T_2 - T_1}{T_1}(100) = \frac{272 + 128\varepsilon^2 - 400}{400}(100) \\ &= -32(1 - \varepsilon^2)\% \end{aligned}$$

From this result we can deduce that 32% of the kinetic energy is dissipated in a perfectly plastic collision and no energy is lost in a perfectly elastic collision.

In closing, we call attention to a common misconception about a perfectly plastic collision, which is that the two bodies are joined after the collision. This need not be so, as can be seen in this problem by setting $\varepsilon = 0$ in $(\bar{v}_A)_2$ and $(\bar{v}_B)_2$.

EXAMPLE 29

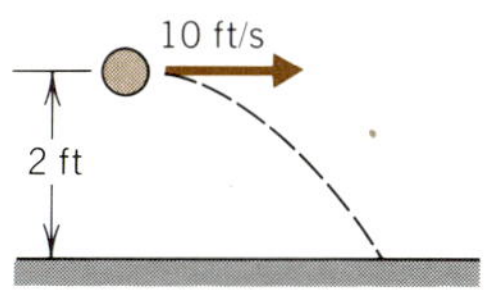

A ball bearing is moving horizontally at 10 ft/s when it is 2 ft above the ground. Determine the magnitude and direction of the velocity of the bearing immediately before and immediately after it collides with the surface; the coefficient of restitution is 0.70.

Solution

A useful first step is to define the xyz coordinate system for the contact plane in a sketch. The information we seek is v_1 and θ_1 for the instant of first contact, and v_2 and θ_2 for the rebound.

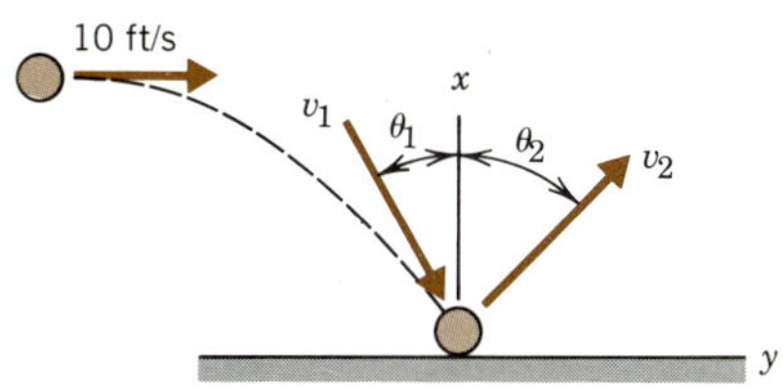

The first time interval in this problem features free fall of the sphere, where the only force acting on the sphere is its weight. Hence, there is no acceleration in the y direction and the acceleration in the x direction is $\dot{v}_x = -g$. For a fall of 2 ft, the constant acceleration formulas give

$$(v_1)_y = 10 \text{ ft/s}, \qquad (v_1)_x^2 = 2g(2) = 128.68 \text{ ft}^2/\text{s}^2$$

$$(v_1)_x = -11.344 \text{ ft/s}$$

The negative sign was assigned to $(v_1)_x$ because we know that it is in the negative x direction. Our sketch defines the components of $\bar{v}$ in terms of θ_1, from which it follows that

$$v_1 = [(v_1)_x^2 + (v_1)_y^2]^{1/2} = 15.122 \text{ ft/s}$$

$$\theta_1 = \tan^{-1}\left[\frac{-(v_1)_y}{(v_1)_x}\right] = 41.40^\circ$$

The horizontal surface, being part of the ground, is fixed, so we set the velocity of this body to zero. The relation for ε yields

$$0.7 = -\left[\frac{(v_x)_2 - 0}{(v_x)_1 - 0}\right] = -\frac{(v_x)_2}{(-11.344)}, \qquad (v_x)_2 = 7.941 \text{ ft/s}$$

There are no impulsive forces acting on the sphere in the y direction, so

$$(v_y)_2 = (v_y)_1 = 10 \text{ ft/s}$$

The speed v_2 and the angle θ_2 are then found with the aid of the sketch.

$$v_2 = [(v_2)_x^2 + (v_2)_y^2]^{1/2} = 12.77 \text{ ft/s}$$

$$\theta_2 = \tan^{-1}\frac{(v_2)_y}{(v_2)_x} = 51.55°$$

Additional Remarks

The results of both of the foregoing examples should help you avoid some common misconceptions. The work–energy principle cannot be used to relate the initial and final velocities in a collison (unless $\varepsilon = 1$), because the amount of mechanical energy lost in the collision is not known in advance. Also, the angle of incidence does not equal the angle of reflection when a small body hits the ground (again, unless $\varepsilon = 1$).

HOMEWORK PROBLEMS

PROB. 12.230

12.230 Two automobiles collide head-on. Vehicle A has a mass of 1200 kg and an initial speed of 3 m/s, and vehicle B has a mass of 1500 kg and an initial speed of 2 m/s. The coefficient of restitution is 0.20. Determine (a) the velocities of the two vehicles following the collision, and (b) the amount of mechanical energy dissipated by the collision.

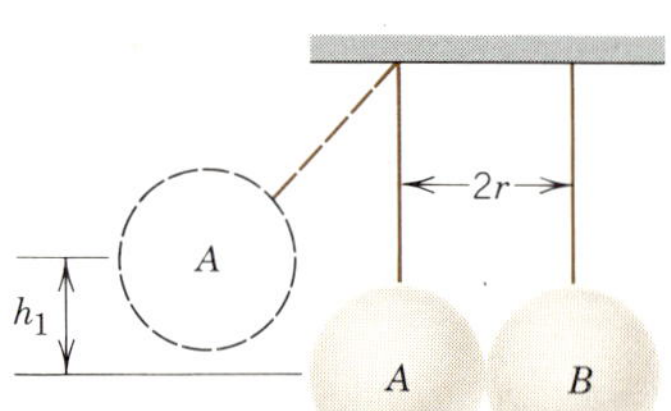

PROB. 12.231

12.231 Spheres A and B have identical mass m and radius r. The coefficient of restitution is 0.80. If sphere A is released from a height h_1, determine the velocity of each sphere after they collide.

12.232 The velocities of the 2 lb block A and the 1 lb disk B on a smooth horizontal table are $v_A = 0$, $v_B = 20$ ft/s. The coefficient of restitution is 0.40. Determine the velocities after they collide.

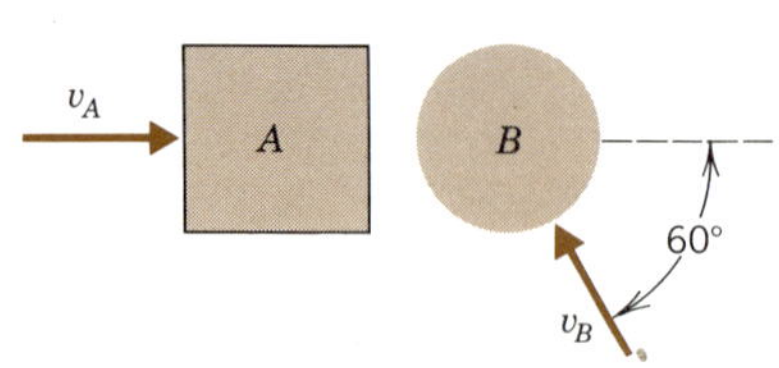

PROB. 12.232

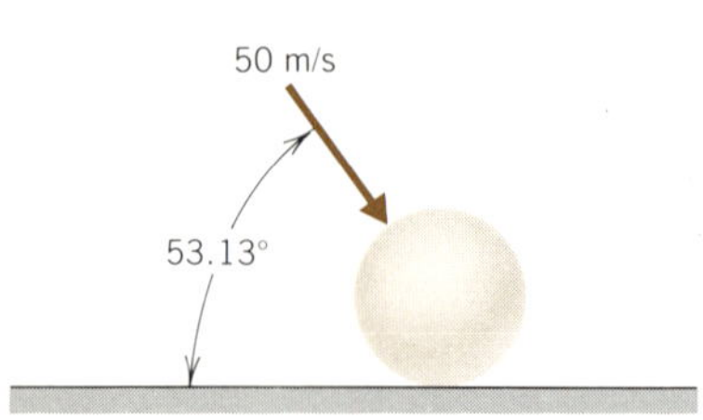

PROB. 12.235

12.233 Solve Problem 12.232 for the case of a perfectly plastic collision.

12.234 Solve Problem 12.232 for the case where $v_A = 10$ ft/s and $v_B = 20$ ft/s.

12.235 The initial velocity of a 400 gram ball is 50 m/s in the direction shown. The ball remains in contact with the ground for 25 ms, and the coefficient of restitution is 0.5. Determine (a) the rebound velocity of the ball, and (b) the average impact force acting on the ball during the collision.

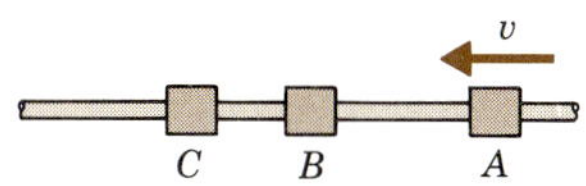

PROB. 12.236

*12.236 Three identical collars slide over a smooth horizontal rod. Initially, block A is moving to the left at speed v and the other blocks are at rest. Determine how many collisions will occur and the velocity of each block after all collisions are completed in the case where the collisions are (a) perfectly elastic, and (b) perfectly plastic.

*12.237 Solve Problem 12.236 in the case where $\varepsilon = 0.50$.

12.238 Billiard ball A has a speed of 8 m/s, and billiard ball B is at rest. The coefficient of restitution is 0.9. It is required that the velocity $(\bar{v}_B)_2$ of billiard ball B after the collision be in the direction shown. Determine (a) the offset distance d, and (b) the value $(v_B)_2$.

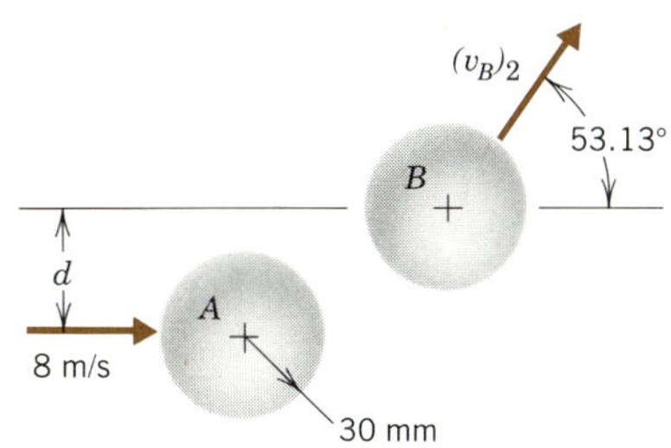

PROB. 12.238

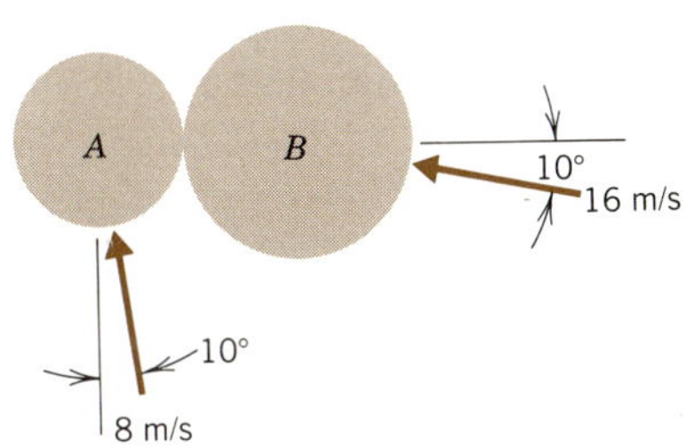

PROB. 12.239

12.239 Disks A and B, having masses of 0.8 and 1.0 kg, collide with the initial velocities shown. Determine their velocities after they collide if the collision is perfectly plastic.

12.240 Solve Problem 12.239 for the case where $\varepsilon = 0.80$.

12.241 The rolling ball hits the bump with a speed of 40 ft/s. The coefficient of restitution is 0.3. Determine the rebound velocity of the ball.

PROB. 12.241

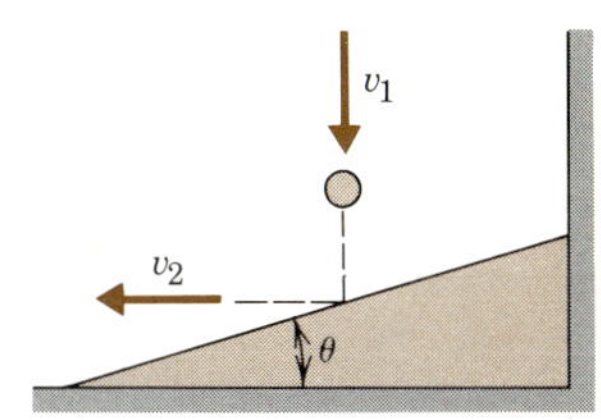

PROB. 12.242

12.242 A ball is dropped vertically onto the inclined plane, hitting it at a speed v_1. It is desired to have the ball rebound horizontally. Derive expressions for the required angle θ for the incline and the corresponding rebound speed v_2 in terms of the coefficient of restitution ε.

12.243 Two disks sliding over a horizontal surface collide in the position shown. Their weights are $w_A = 15$ oz, $w_B = 3$ oz, and the coefficient of restitution is 0.80. As shown in the diagram, the initial velocity of disk B and the final velocity of disk A are unknown. Determine these unknown quantities.

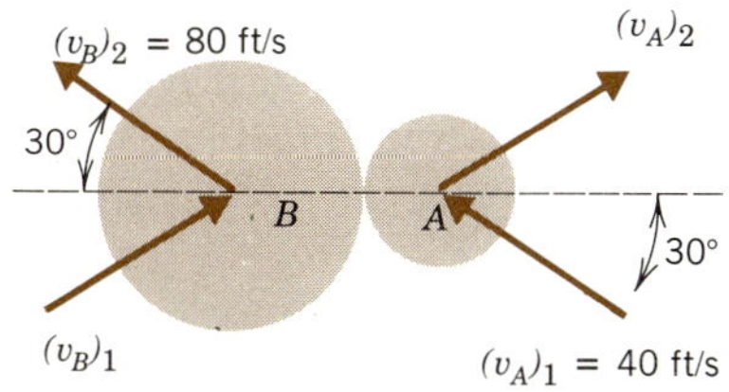

PROB. 12.243

12.244 Show that, in the absence of friction, a ball rolling on a horizontal plane will emerge from a corner such that its final velocity $\bar{v}_2$ is $-\varepsilon\bar{v}_1$, where $\bar{v}_1$ is the initial velocity. Also explain what would happen in the case where $\varepsilon = 0$.

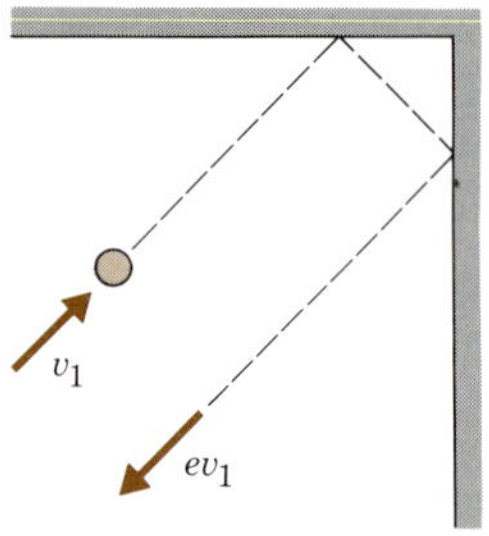

PROB. 12.244

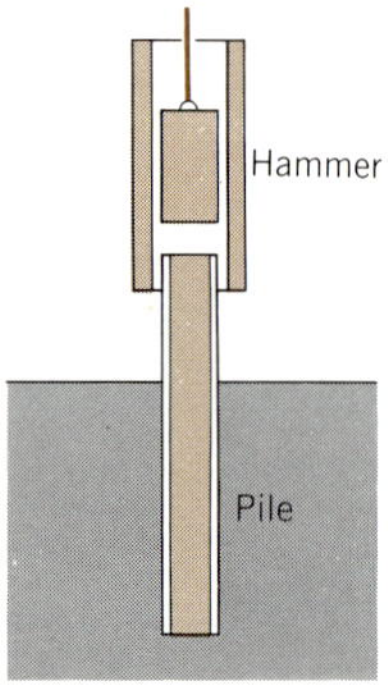

PROB. 12.245

12.245 The 900 kg hammer of a pile driver falls from a height of 2 m above a 150 kg pile. The collision between the hammer and the pile is perfectly plastic, driving the pile 150 mm into the ground. Determine the average resistance of the ground to the motion of the pile.

12.246 A ball bearing is thrown horizontally at a speed v_1 at a distance h_1 above the ground. Derive expressions for the distances d and h_2 locating the highest point attained by the ball bearing after it hits the ground. Let the coefficient of restitution be ε.

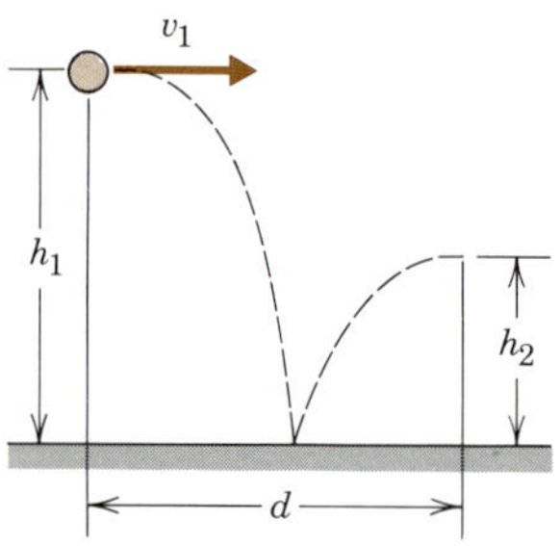

PROB. 12.246

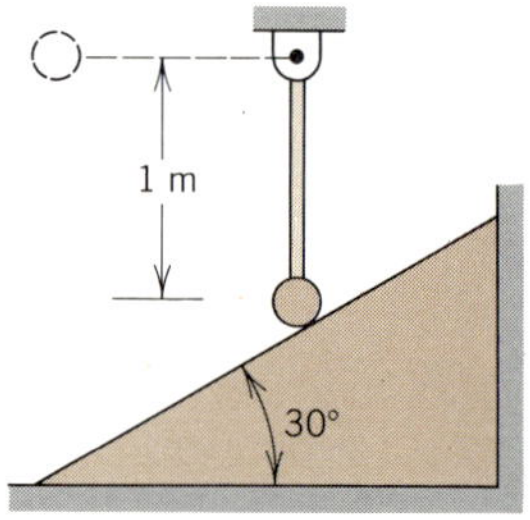

PROB. 12.247

***12.247** The pendulum is released from rest in the horizontal position and strikes the inclined surface when it is in the vertical position. The coefficient of restitution is 0.6. The connection between the sphere and the pivot is a rigid, massless bar. Determine the rebound velocity of the sphere. *Hint:* The bar exerts an impulsive axial force.

***12.248** A 4 oz ball bearing hits the 2 lb steel wedge with a horizontal velocity of 20 ft/s. Assuming that all surfaces are smooth, determine the velocities of both bodies after the collision. The coefficient of restitution is 0.75. *Hint:* The reaction force exerted by the ground on the wedge is impulsive.

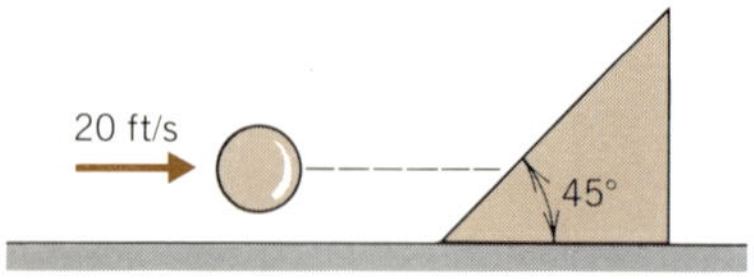

PROB. 12.248

I. ANGULAR IMPULSE-MOMENTUM RELATIONS

The relationships to be developed in this section are additional standard integrals of the differential equations of motion involving the linear momentum of a particle. In contrast to the linear impulse-momentum principles, however, the result obtained here will be relationships useful for systems that are rotating about an axis or a point. In addition to their applications to particle systems, the principles will be fundamental to further studies, such as the kinetics of rigid bodies. As was done previously, we shall first study the motion of a single particle, and then consider a system of particles.

1. Single Particle

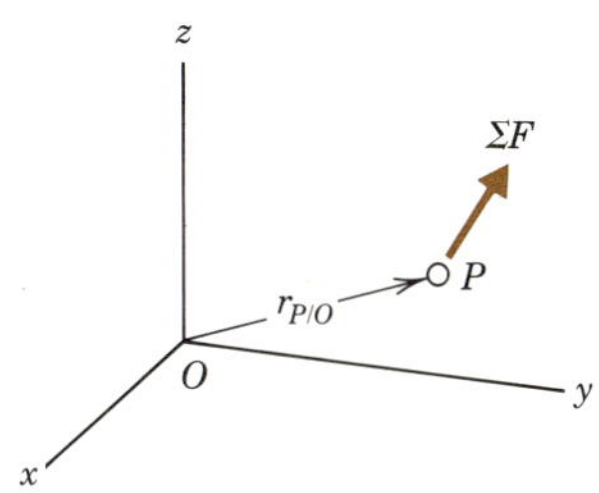

FIGURE 13

The focus of attention here shall be the moment of forces. Figure 13 depicts the resultant force $\sum \bar{F}$ acting on a particle P of mass m. Newton's second law, $\sum \bar{F} = m\bar{a}$, leads to the following relationship between the moment about point O and the kinematical parameters.

$$\sum \bar{M}_O \equiv \bar{r}_{P/O} \times \sum \bar{F} = \bar{r}_{P/O} \times m\bar{a}$$

A more descriptive equation may be obtained by using the fact that $\bar{a} = \dot{\bar{v}}$, and the rule for the time derivative of a cross product.

$$\frac{d}{dt}(\bar{a} \times \bar{b}) \equiv \dot{\bar{a}} \times \bar{b} + \bar{a} \times \dot{\bar{b}}$$

This yields

$$\sum \bar{M}_O = \bar{r}_{P/O} \times m\dot{\bar{v}} \equiv \frac{d}{dt}(\bar{r}_{P/O} \times m\bar{v}) - \dot{\bar{r}}_{P/O} \times m\bar{v}$$

Now note that $\dot{\bar{r}}_{P/O} \equiv \bar{v}$, which means that $\dot{\bar{r}}_{P/O}$ is parallel to $m\bar{v}$. Hence, the last cross product is identically zero.

The quantity $\bar{r}_{P/O} \times m\bar{v}$ remaining in the moment equation is called the *angular momentum* of the particle with respect to point O. It will be denoted for brevity as $\bar{H}_O$. Thus, the moment equation may be written as

$$\boxed{\begin{aligned} \bar{H}_O &= \bar{r}_{P/O} \times m\bar{v} \\ \bar{M}_O &= \dot{\bar{H}}_O \end{aligned}} \tag{48}$$

The definition of $\bar{H}_O$ is equivalent to saying that the angular momentum is the moment of the linear momentum about point O.

When the moment $\bar{M}_O$ is known as a function of time, it is possible to integrate the relation between $\bar{H}_O$ and $\bar{M}_O$. This gives

$$\boxed{(\bar{H}_O)_2 = (\bar{H}_O)_1 + \int_{t_1}^{t_2} \bar{M}_O \, dt} \tag{49}$$

Note the similarity to eq. (39) for linear momentum. Hence it makes sense to call $\int \bar{M}_O \, dt$ the *angular impulse*. In situations where the resultant force exerts no moment about the origin, eq. (49) tells us that $(\bar{H}_O)_1 = (\bar{H}_O)_2$, which is the statement of the principle of *conservation of angular momentum*.

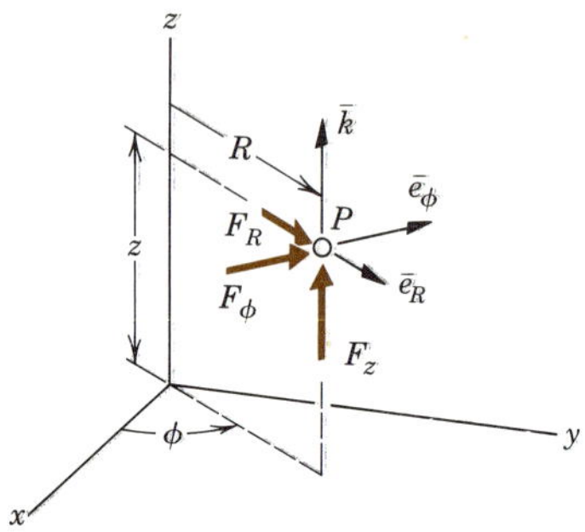

FIGURE 14

The moment exerted by a force is usually associated with a rotational tendency. For this reason, let us consider a general situation in which a particle is rotating about a fixed axis. Such an axis is the z axis for a set of cylindrical coordinates, depicted in Figure 14. The vectors $F_R\bar{e}_R$, $F_\phi\bar{e}_\phi$, and $F_z\bar{k}$ in this figure are the cylindrical components of the resultant force acting on particle P.

We shall focus our attention on the axial component of the angular impulse-momentum relation, because the particle is rotating about that axis. Figure 14 shows that $F_R\bar{e}_R$ intersects the z axis and $F_z\bar{k}$ is parallel to it. Hence, only $F_\phi\bar{e}_\phi$ exerts a moment about the z axis. The lever arm of this force is R, so the z component of the moment is

$$\sum M_{Oz} \equiv (\bar{r}_{P/O} \times \sum \bar{F})_z = RF_\phi \tag{50}$$

The moment of the linear momentum $m\bar{v}$ may be described in a similar manner. Only the transverse component mv_ϕ has a moment about the z axis. Thus

$$\boxed{H_{Oz} \equiv (\bar{r}_{P/O} \times m\bar{v})_z = R(mv_\phi)} \tag{51}$$

The integral form of the angular impulse-momentum relation, eq. (49), may be written in explicit form with the aid of the foregoing expressions for $\sum M_{Oz}$ and H_{Oz}. Specifically,

$$\boxed{mR_2(v_\phi)_2 = mR_1(v_\phi)_1 + \int_{t_1}^{t_2} RF_\phi\,dt} \tag{52}$$

It is unusual to know how R and F_ϕ vary with time, except in the special case of a *central force system.* Such a set of forces has the property of having no transverse component, $F_\phi \equiv 0$. Equation (52) then describes the *conservation of angular momentum* about the z axis, because it states that $H_{Oz} = mRv_\phi$ does not change.

The motion of satellites around the earth, and of the planets around the sun, are important cases of central force motion. The conservation of angular momentum is a primary consideration in deriving the equations describing orbital motion.

In order to obtain a different viewpoint for the angular impulse-momentum relation, recall the kinematical formulas for cylindrical coordinates.

$$v_\phi = R\dot{\phi}, \qquad a_\phi = R\ddot{\phi} + 2\dot{R}\dot{\phi}$$

The transverse component of Newton's second law is

$$F_\phi = ma_\phi$$

so the moment about the z axis must satisfy

$$\sum M_{Oz} = RF_\phi = R(ma_\phi) = m(R^2\ddot{\phi} + 2R\dot{R}\dot{\phi})$$

$$\equiv m\frac{d}{dt}(R^2\dot{\phi}) \equiv \frac{d}{dt}(mRv_\phi)$$

$$\equiv \frac{d}{dt} H_{Oz}$$

We may conclude from this that if a particle rotating about the z axis has a transverse force, the angular impulse-momentum principle provides the same information as the transverse component of Newton's second law. This is not true for central force motion. In that case, the conservation of angular momentum, given by eq. (52) with $F_\phi \equiv 0$, provides a useful relation for v_ϕ at different positions.

A corollary of the foregoing observation regarding $F_\phi = ma_\phi$ is that the angular impulse-momentum relation is independent of the work–energy relation. The latter may be traced back to $\sum F_t = ma_t$. Most problems involving central force motion require both relations for their solution, as is demonstrated in the next example.

EXAMPLE 30

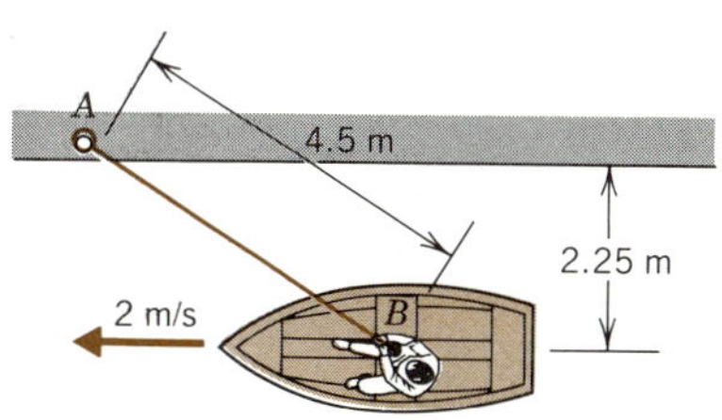

A rider at position B in a rowboat pulls on the cable, which is attached to the mooring post A. The cable tension is constant at 100 N. The total mass of the boat and its rider is 120 kg, and the center of mass coincides with point B. Determine the minimum distance to which the rider will approach point A when the initial motion is as shown.

Solution

A free body diagram of the boat shows that the only horizontal force is the cable tension, which always intersects point A. Because there is no force component perpendicular to line AB, this is a case of central force motion about a vertical axis through point A. The vertical axis through point A is the axial direction for the cylindrical coordinates defined in the free body diagram.

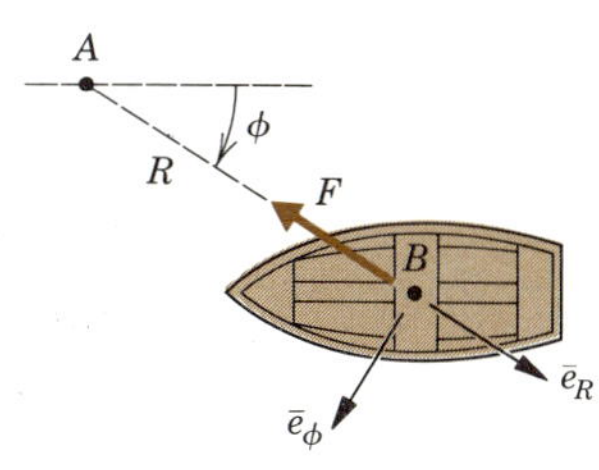

The initial position is given as

$$R_1 = 4.5 \text{ m}, \qquad \phi_1 = \sin^{-1}\left(\frac{2.25}{4.5}\right) = 30^\circ$$

Because $\bar{v}_1$ is parallel to the dock, we find that

$$(v_\phi)_1 = v_1 \sin 30^\circ = 1.0 \text{ m/s}$$

The final position is specified to be the one where R has its minimum value. For a minimum, $\dot{R} = 0$, but $\dot{R} = v_R$. This means that the velocity must be in the transverse direction when R is a minimum. (The same is true when R is a maximum.) Consequently, we have

$$(v_\phi)_2 = v_2$$

One relation for v_2 is obtained by using the fact that angular momentum is conserved in the central force motion. The radial distance and transverse velocity are known in the initial position. Equation (52) with $F_\phi = 0$ gives

$$R_2(v_\phi)_2 \equiv R_2 v_2 = R_1(v_\phi)_1 = 4.5(1.0) \text{ m}^2/\text{s}$$

The work–energy principle may also be employed to relate R_2 and v_2. The cable is pulled inward by the distance $(4.5 - R_2)$ m, and the cable tension is 100 N. The corresponding work is

$$U_{1\to 2}^{(\text{nc})} = 100(4.5 - R_2) \text{ J}$$

For an initial speed of 2 m/s, we then find that

$$\tfrac{1}{2}(120)v_2^2 = \tfrac{1}{2}(120)2^2 + 100(4.5 - R_2)$$

A single equation for R_2 may be obtained by solving the angular momentum equation for v_2.

$$v_2 = \frac{4.5}{R_2}$$

Placing v_2 into the work–energy equation yields

$$60\left(\frac{4.5}{R_2}\right)^2 = 240 + 450 - 100R_2$$

$$100R_2^3 - 690R_2^2 + 1215 = 0$$

This cubic equation may be solved by trial-and-error, or by numerical algorithms that are available on computers and some programmable calculators. The three roots are

$$R_2 = -1.2230, \quad 1.500, \quad 6.623$$

The first and last of these roots have no physical meaning. Therefore

$$R_{\text{min}} = 1.50 \text{ m}$$

HOMEWORK PROBLEMS

12.249 Determine the velocity components of the rowboat in Example 30 when the distance from the mooring post to position B on the boat is 3 m.

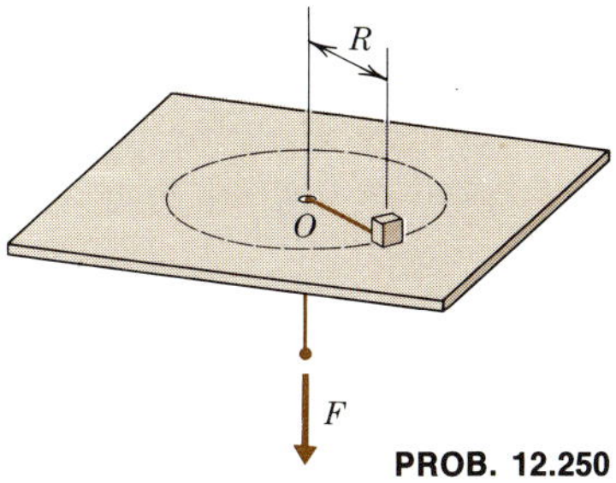

PROB. 12.250

12.250 The 2 lb block slides over the smooth horizontal surface under the restraint of the cable that passes through a hole at point O. The block was initially moving at 30 ft/s in a circular path whose radius is $R = 2$ ft, when the cable tension was decreased suddenly to a constant value of 5 lb. Determine the velocity components of the block when $R = 3$ ft.

12.251 Determine the maximum radial distance R for the motion of the block in Problem 12.250.

12.252 An artificial satellite makes its closest approach to the surface of the earth (perigee) in an elliptical orbit. It has an altitude of 400 km and a speed of 33,400 km/hr. Determine the altitude and speed of of the satellite when it is farthest from the earth (apogee).

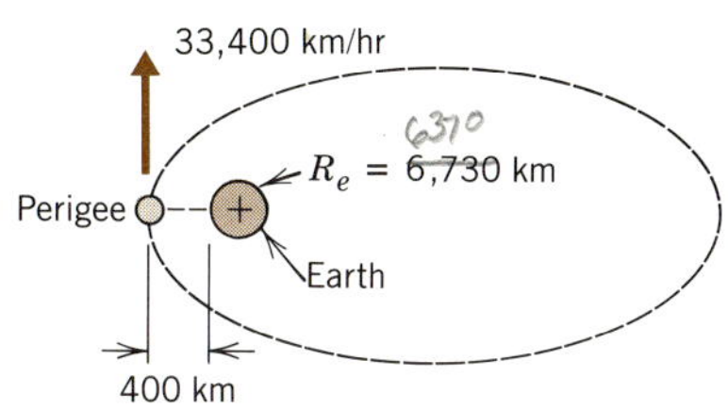

PROB. 12.252

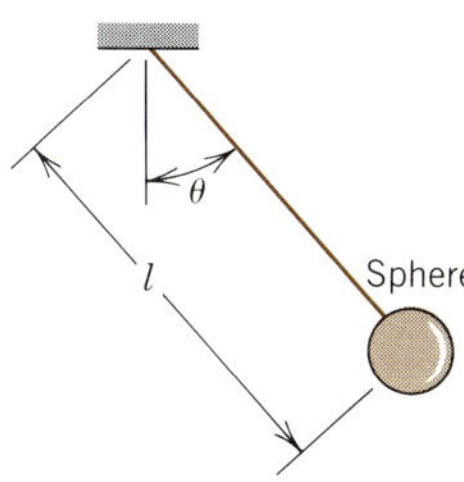

PROB. 12.253

12.253 Use the relation $M_{Oz} = \dot{H}_{Oz}$ to derive the differential equation of motion governing the angle θ.

12.254 An elastic cable, whose stiffness is 5 lb/ft and whose unstretched length is 2 ft, is fastened to a fixed pin at A and to block B. The block, which weighs 12 oz, is given the initial motion shown. Determine the velocity components of the block in the position where the cable becomes slack.

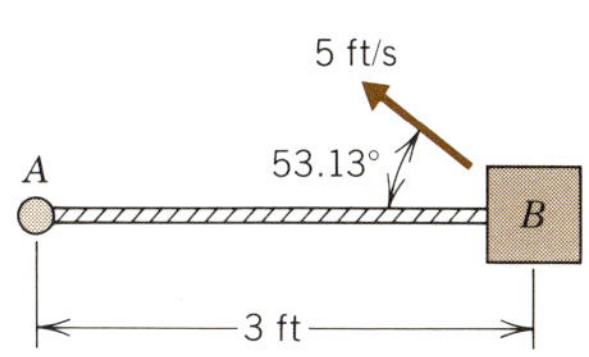

PROB.12.254

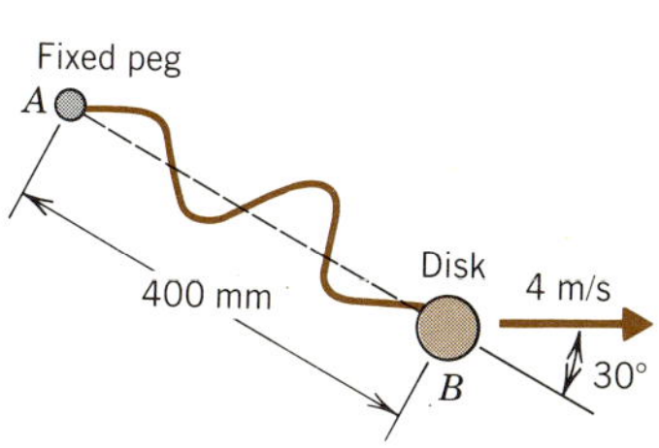

PROB. 12.255

12.255 A 200 gram disk on a smooth horizontal plane is tied to a rubber band whose opposite end is tied to a peg at point A. The disk is set into motion at point B, as shown. The stiffness of the rubber band is 8 N/m and the rubber band is 600 mm long when it is slack. Determine the components of the velocity of the disk when it is 1 m from point A.

12.256 Determine the maximum distance between the disk and point A for the motion in Problem 12.255. *Hint:* The equation for R_{max} will be a quartic polynomial.

***12.257** A ball bearing is projected tangentially along the rim of the cone shown at a speed v_1. At its lowest point the ball bearing is 12 in. above the apex of the cone. Determine (a) the initial speed v_1 of the ball bearing, and (b) the speed of the ball bearing at is lowest point.

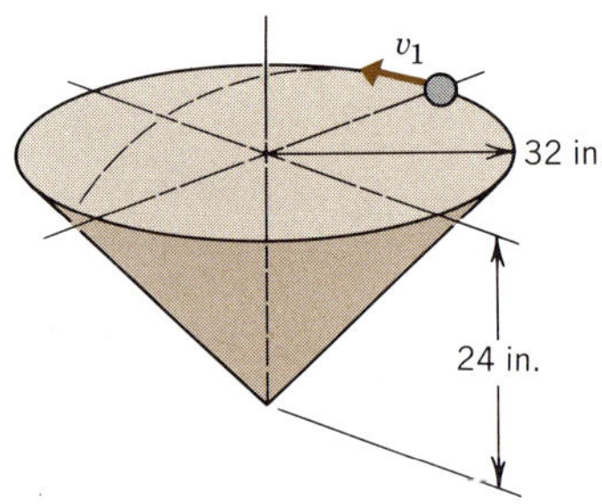

PROB. 12.257

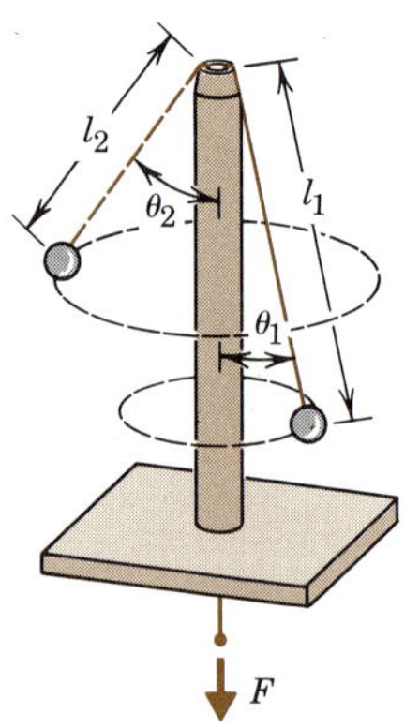

PROB. 12.258

***12.258** An inextensible cable is passed through the hollow vertical post and then tied to a 0.5 kg sphere. The sphere is made to revolve in a horizontal circle for which $l_1 = 2$ m and $\theta_1 = 30°$. The cable is then drawn in very slowly until the sphere follows a horizontal circle for which $\theta_2 = 60°$. Determine (a) the initial speed v_1 of the sphere, (b) the final speed v_2, and (c) the amount of work done by the cable in pulling the sphere inward. *Hint:* Newton's second law must be satisfied for the circular motions at θ_1 and θ_2.

***12.259** A particle slides down the smooth interior surface of a hemisphere of radius r. It was released from the rim with a velocity $\bar{v}_1$ directed at an angle of 30° with the horizontal, as shown. Determine the speed v_1 for which the minimum elevation of the particle below the rim is $0.8r$.

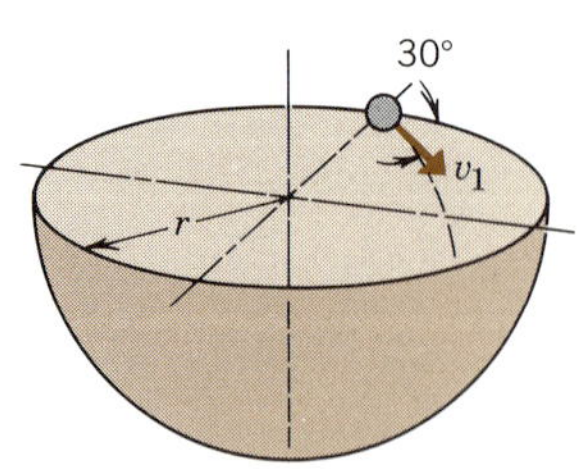

PROB. 12.259

---

2. System of Particles

Our objective here is to extend the angular impulse-momentum principle for a single particle to systems of interacting particles. At the same time,

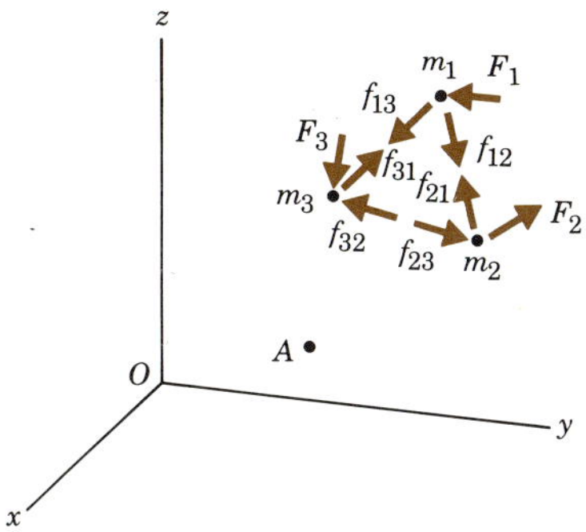

FIGURE 15

we will consider the possibility of summing moments about a moving point, which we will denote as point A. In order to begin this discussion, consider Figure 15, which shows the forces acting on a typical three particle system.

Two types of forces appear in this figure. The force $\bar{F}_i$ represents the external force on particle i, whereas $\bar{f}_{ij}$ is the internal force exerted on particle i by particle j. Newton's third law tells us that for every $\bar{f}_{ij}$, there is a collinear force $\bar{f}_{ji} = -\bar{f}_{ij}$ acting on particle j.

The foundation of our study is Newton's second law. The resultant force acting on a specified particle i is the sum of $\bar{F}_i$ and all of the $\bar{f}_{ij}$. Thus

$$\bar{F}_i + \sum_{\substack{j=1 \\ j \neq i}}^{N} \bar{f}_{ij} = m\bar{a}_i$$

Let us now compute the total moment about point A for all of these resultant forces. The moment obtained from each pair of interaction forces $\bar{f}_{ij}$ and $\bar{f}_{ji}$ is zero, because the only difference between these forces is their opposite senses. In words, *only the external forces affect the total moment.* (This fact was basic to our study of statics.) The moment of each external force is $\bar{r}_{i/A} \times \bar{F}_i$, where $\bar{r}_{i/A}$ is defined in Figure 16 to be the position of particle i with respect to point A. Thus, the total moment is

$$\sum \bar{M}_A = \sum_{i=1}^{N} \bar{r}_{i/A} \times \bar{F}_i = \sum_{i=1}^{N} \bar{r}_{i/A} \times m_i\bar{a}_i \tag{53}$$

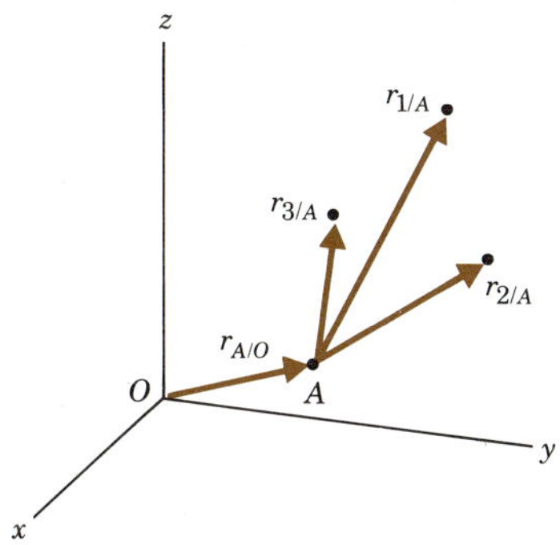

FIGURE 16

The occurrence of the relative position $\bar{r}_{i/A}$ in eq. (53) suggests that we can introduce the kinematics of relative motion.

$$\bar{r}_{i/O} = \bar{r}_{A/O} + \bar{r}_{i/A}, \qquad \bar{v}_i = \bar{v}_A + \bar{v}_{i/A}, \qquad \bar{a}_i = \bar{a}_A + \bar{a}_{i/A}$$

Substituting these relations into eq. (53) and using the identity for the derivative of a cross product (see Section I.1) then yields

$$\begin{aligned}
\sum \bar{M}_A &= \sum_{i=1}^{N} [\bar{r}_{i/A} \times m_i(\bar{a}_A + \bar{a}_{i/A})] \\
&= \sum_{i=1}^{N} (\bar{r}_{i/A} \times m_i\bar{a}_A) + \sum_{i=1}^{N} \left(\bar{r}_{i/A} \times m_i \frac{d}{dt}\bar{v}_{i/A}\right) \\
&= \left(\sum_{i=1}^{N} m_i\bar{r}_{i/A}\right) \times \bar{a}_A + \sum_{i=1}^{N} \left[\frac{d}{dt}(\bar{r}_{i/A} \times m_i\bar{v}_{i/A}) - \left(\frac{d}{dt}\bar{r}_{i/A} \times m_i\bar{v}_{i/A}\right)\right]
\end{aligned} \tag{54}$$

The first simplification in the foregoing is that $(d/dt)\bar{r}_{i/A} \equiv \bar{v}_{i/A}$, so the last cross product term is identically zero. Another important observation is that the first term on the right-hand side is the first moment of mass with respect to point A. It locates the center of mass G of the system of particles, according to

$$m\bar{r}_{G/A} = \sum_{i=1}^{N} m_i\bar{r}_{i/A}$$

Equation (54) for the total moment therefore reduces to

$$\sum \bar{M}_A = \bar{r}_{G/A} \times m\bar{a}_A + \dot{\bar{H}}_A \tag{55}$$

where

$$\bar{H}_A = \sum_{i=1}^{N} \bar{r}_{i/A} \times m_i\bar{v}_{i/A} \tag{56}$$

The quantity $\bar{H}_A$ defined above is called the *angular momentum of the system of particles with respect to point A*. One way of interpreting $\bar{H}_A$ is to think of $m_i\bar{v}_{i/A}$ as the momentum of particle i relative to point A. Then $\bar{r}_{i/A} \times m_i\bar{v}_{i/A}$ is the moment about point A of the relative momentum of the ith particle, and $\bar{H}_A$ is the sum for all particles.

Another noteworthy feature of eq. (56) is that $\bar{H}_A = \bar{0}$ whenever each particle has the same velocity as point A ($\bar{v}_{i/A} \equiv \bar{0}$). In the case where point A is fixed, we have $\bar{v}_{i/A} \equiv \bar{v}_i$. The relative angular momentum then reduces to the sum of the (absolute) angular momentum of each particle, which is defined in the first of eqs. (48).

A significant difference between eq. (55) and the simple relation $\bar{M}_O = (d/dt)\bar{H}_O$ for a single particle is that, if point A is arbitrarily chosen, it is necessary to determine the acceleration of this point in addition to determining the rate of change of the angular momentum $\bar{H}_A$.

It is easier to formulate the moment equation when it has the following simple form

$$\sum \bar{M}_A = \dot{\bar{H}}_A \tag{57}$$

In order for this to occur, point A *must* fit one of three special cases that reduces the term $\bar{r}_{G/A} \times m\bar{a}_A$ to zero.

1. If point A is fixed, or has constant velocity, then A is an inertial reference point and $\bar{a}_A \equiv \bar{0}$.
2. If point A is the center of mass, then $\bar{r}_{G/A} \equiv \bar{0}$.
3. If point A is accelerating directly toward or away from the center of mass, then $\bar{a}_A$ is parallel to $\bar{r}_{G/A}$ and their cross product vanishes.

Restricting ourselves to formulating the angular momentum with respect to one of these special points in no way limits the scope of the problems we can investigate.

We now complete the derivation of angular momentum principles by considering the situation where the moment of the forces is a known function of time. For such a case eq. (57) can be integrated to obtain

$$(\bar{H}_A)_2 = (\bar{H}_A)_1 + \int_{t_1}^{t_2} \sum \bar{M}_A \, dt \tag{58}$$

Equations (57) and (58) are derivative and integral forms, respectively, of the angular impulse-momentum principle. As was true for a single particle, the concept of angular momentum for a system is useful when the system is rotating about an axis through a fixed point. For example, if the external forces exert no moment about the axis of rotation, then eq. (58) states that the *corresponding* component of $\bar{H}_A$ about the fixed axis is constant.

Another important use of the concept of angular momentum is for treating the motion of rigid bodies. Equation (57) will be employed in Chapters 14 and 16 as one of the fundamental equations of motion for rigid bodies.

EXAMPLE 31

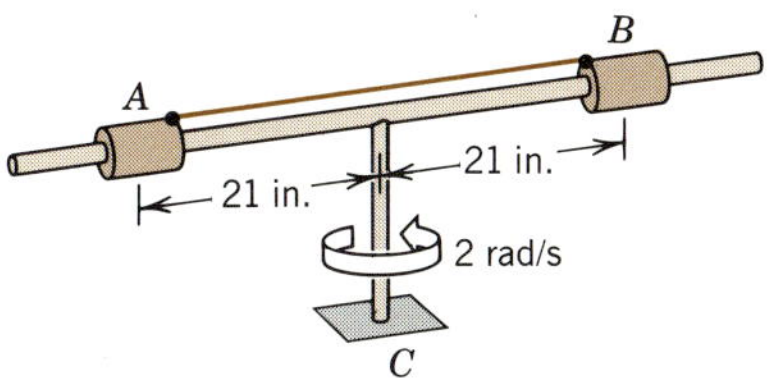

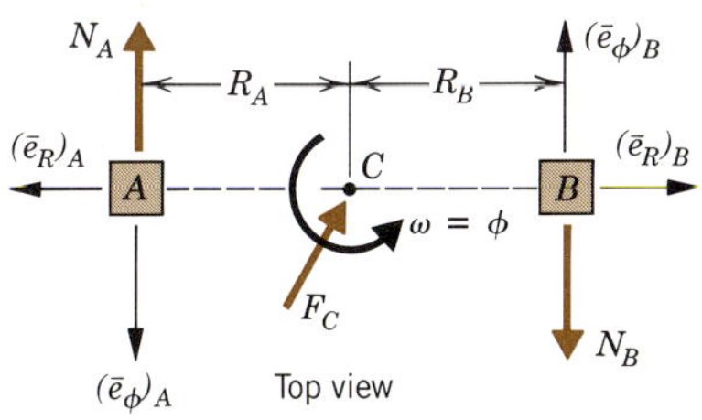

The 1.2 lb block A and the 1.5 lb block B are tied together by a cable. These blocks may slide over the crossbar, whose mass is negligible, while the entire system rotates freely about the pivot C. The blocks are not sliding in the initial motion shown. For the position where the blocks have slid 3 in. over the crossbar, determine (a) the angular velocity of the crossbar, and (b) the velocity components of block A.

Solution

It is logical to anticipate that block B, which is heavier, will move outward. (If this guess is wrong, we will find that our solution leads to an impossible situation.) The free body diagrams show that the only forces acting on each block in the horizontal plane of motion are the normal force exerted by the crossbar and the cable tension. In order to eliminate these forces, we shall include the crossbar and cable in the system. The only external force acting on this system in the horizontal plane is the reaction $\bar{F}_C$ at the pivot.

The blocks are tied together, so the outward motion of block B is matched by the same inward movement of block A.

$$\Delta R_A = -\Delta R_B, \qquad \dot{R}_A = -\dot{R}_B$$

The kinematical relations for the initial and final motions are therefore

$$R_{A1} = R_{B1} = 21 \text{ in.}, \qquad R_{B2} = 24 \text{ in.}, \qquad R_{A2} = 18 \text{ in.}$$

$$\bar{v}_{B1} = R_{B1}\omega_1(\bar{e}_\phi)_B = 42(\bar{e}_\phi)_B \text{ in./s}$$

$$\bar{v}_{A1} = R_{A1}\omega_1(\bar{e}_\phi)_A = 42(\bar{e}_\phi)_A \text{ in./s}$$

$$\bar{v}_{B2} = \dot{R}_B(\bar{e}_R)_B + R_{B2}\dot{\phi}_2(\bar{e}_\phi)_B = \dot{R}_B(\bar{e}_R)_B + 24\omega_2(\bar{e}_\phi)_B \text{ in./s}$$

$$\bar{v}_{A2} = \dot{R}_A(\bar{e}_R)_A + R_{A2}\dot{\phi}_2(\bar{e}_\phi)_A = -\dot{R}_B(\bar{e}_R)_A + 18\omega_2(\bar{e}_\phi)_A \text{ in./s}$$

The angular momentum of the system about the z axis is conserved because the external forces exert no moment about that axis. We formulate H_{Cz} for the system with the aid of eq. (51) for each particle. Thus

$$(H_{Cz})_1 = m_A R_{A1}(v_{\phi A})_1 + m_B R_{B1}(v_{\phi B})_1$$

$$= \frac{1.2}{g}(21)(42) + \frac{1.5}{g}(21)(42) = 6.169 \text{ lb-in.-s}$$

$$(H_{Cz})_2 = m_A R_{A2}(v_{\phi A})_2 + m_B R_{B2}(v_{\phi B})_2$$

$$= \frac{1.2}{g}(18)(18\omega_2) + \frac{1.5}{g}(24)(24\omega_2) = 3.246\omega_2$$

$$(H_{Cz})_2 = (H_{Cz})_1 \Rightarrow \omega_2 = \frac{6.169}{3.246} = 1.9005 \text{ rad/s}$$

This value of ω_2 yields the transverse velocity of block A. In order to determine the radial velocity, as well as to check our assumption that block B moves outward, we note that the work–energy principle may be employed. The cable tension does no work on the system because the cable is rigid, massless, and frictionless. Similar reasoning applies to the normal forces, because the crossbar has ideal properties. Thus

$$U_{1\to 2}^{(\text{nc})} = 0$$

The motion occurs in the horizontal plane, so $V \equiv 0$. The kinetic energy is formed from *both* velocity components for each block, according to

$$T = \tfrac{1}{2} m_A \bar{v}_A \cdot \bar{v}_A + \tfrac{1}{2} m_B \bar{v}_B \cdot \bar{v}_B$$

$$T_1 = \frac{1}{2}\frac{1.2}{g}(42)^2 + \frac{1}{2}\frac{1.5}{g}(42)^2 = 6.169 \text{ in.-lb}$$

$$T_2 = \frac{1}{2}\frac{1.2}{g}[\dot{R}_B^2 + (18\omega_2)^2] + \frac{1}{2}\frac{1.5}{g}[\dot{R}_B^2 + (24\omega_2)^2]$$

$$= 3.497(10^{-3})\dot{R}_B^2 + 5.862 \text{ in.-lb}$$

Because the work and potential energy functions are zero, $T_2 = T_1$.

$$3.497(10^{-3})\dot{R}_B^2 + 5.862 = 6.169, \qquad \dot{R}_B = 9.37 \text{ in./s}$$

$$\bar{v}_{A2} = -\dot{R}_B(\bar{e}_R)_A + 18\omega_2(\bar{e}_\phi)_A = -9.37(\bar{e}_R)_A + 34.2(\bar{e}_\phi)_A \text{ in./s}$$

Finally we note that our assumption about the sense of the sliding motion is confirmed by the fact that $\dot{R}_B$ is a real number. If we had assumed that block B slid inward we would have found that $\dot{R}_B^2 < 0$ (try it!), which is not possible.

HOMEWORK PROBLEMS

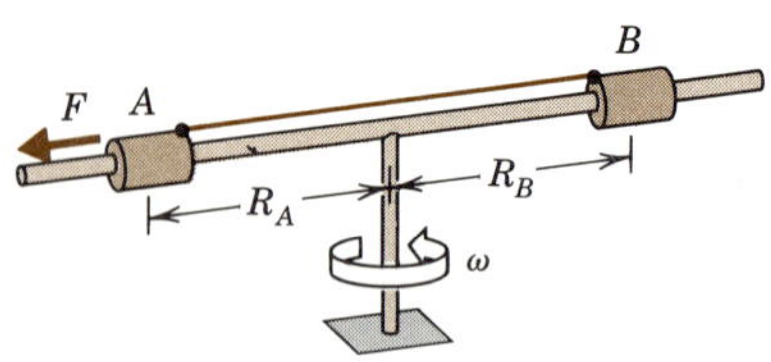

PROB. 12.260

12.260 The identical 250 grams collars may slide over the massless crossbar, which rotates freely about the vertical axis. The collars are initially stationary with respect to the crossbar, corresponding to $\omega = 8$ rad/s, $R_A = R_B = 500$ mm. A variable radial force $\bar{F}$ is to be applied to collar A, such that $R_A = 900$ mm is the maximum value for the ensuing motion. Determine (a) the speed of each collar when $R_A = 900$ mm, and (b) the average magnitude of $\bar{F}$.

12.261 In Problem 12.260, the radial force $\bar{F}$ is constant at 32 N. Determine the components of the velocity of collar A when $R_A = 600$ mm.

12.262 Collars A and B, whose masses are 500 grams each, are not sliding over their guides in the initial motion shown. The guide bars are massless, and there is negligible friction. For the position where collar A has slid 400 mm outward, determine (a) the angular speed ω, and (b) the speed of collar B.

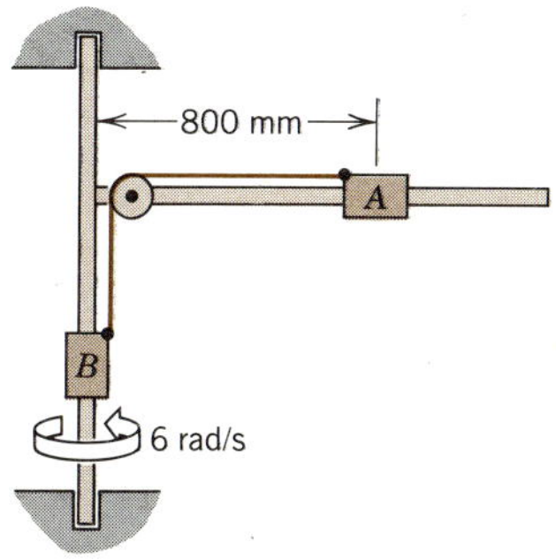

PROB. 12.262

12.263 Solve Example 31 for the case where the blocks are initially at $R_A = 24$ in., $R_B = 18$ in., and all other parameters are as given.

12.264 The device shown is a flyball speed governor that spins about the vertical axis. The angle θ for each small 400 gram sphere may be changed by pushing rod AB upward. The mass of each rod in this linkage is neligible. In one case, the system was initially rotating at 1200 rpm with θ constant at 60°. A force $\bar{F}$ caused θ to decrease, very slowly, to the constant value $\theta = 30°$. Determine (a) the angular speed in the final position, and (b) the average value of $\bar{F}$ required to change θ.

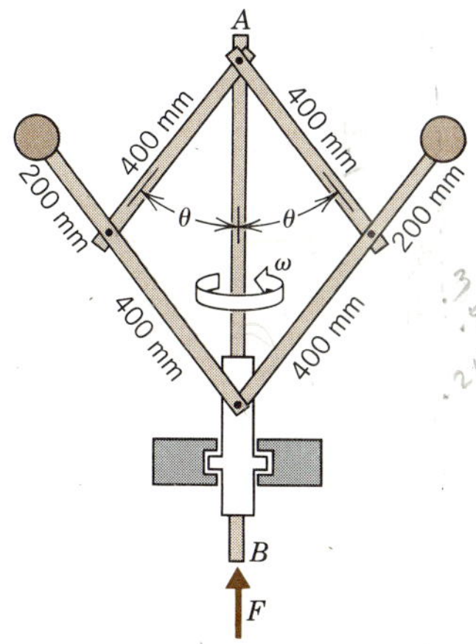

PROB. 12.264

12.265 Two identical 1 kg disks of negligible diameter are fastened to the rigid bar AC, which is free to pivot in the vertical plane about point C. The system is at rest when disk A is hit by a 100 gram bullet that is traveling at 600 m/s as shown. The bullet becomes embedded in the disk. Determine the angular velocity of the system immediately after the collision is completed.

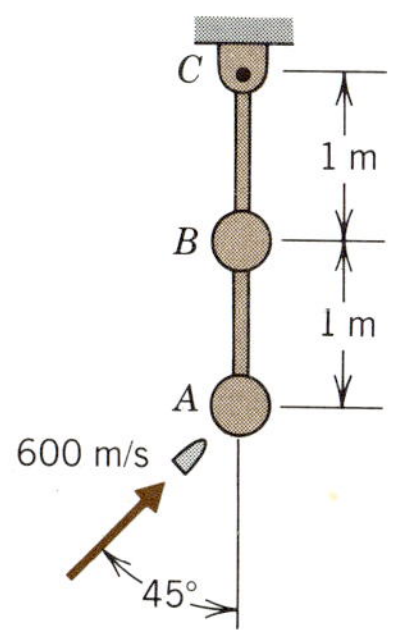

PROB. 12.265

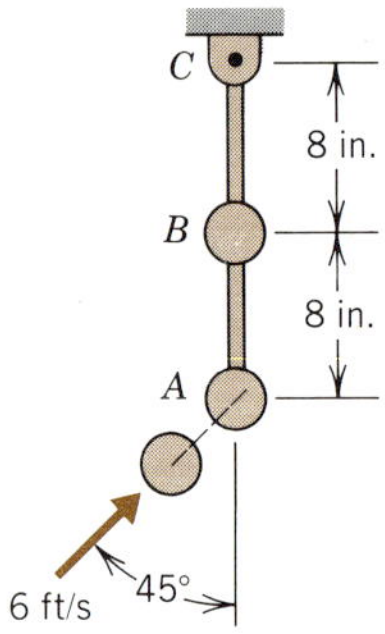

PROB. 12.266

***12.266** Two identical 1 lb disks of negligible diameter are fastened to the rigid bar AC, which is free to pivot in the vertical plane about point C. The system is at rest when disk A is hit by another 1 lb disk that is traveling at 6 ft/s, as shown. The coefficient of restitution for the collision is 0.50. Determine the angular velocity of bar AC after the collision is completed.

12.267 and 12.268 The assembly shown swings in the vertical plane about pin A. The mass of each sphere is m. Use eq. (57) to derive the differential equation governing the angle θ.

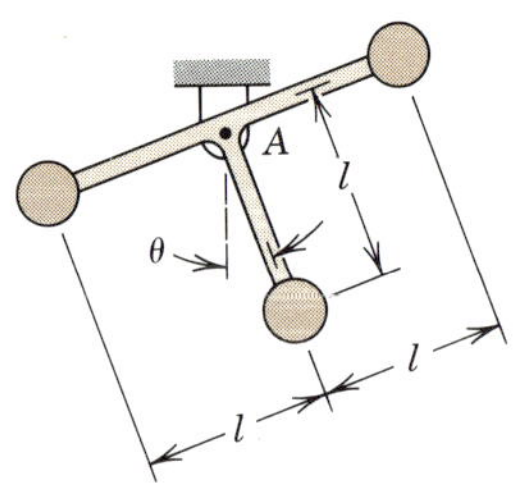

PROB. 12.267

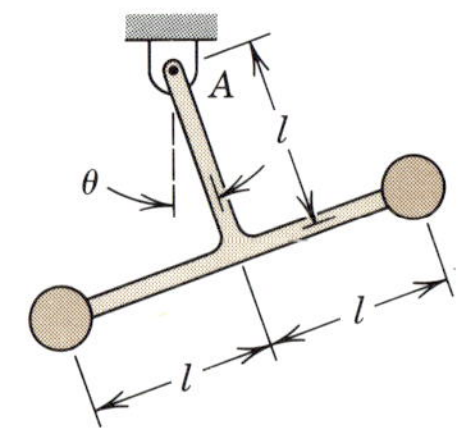

PROB. 12.268

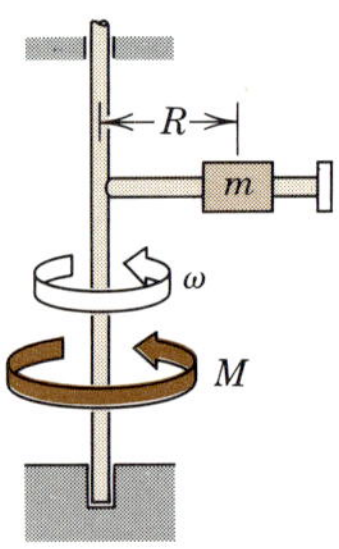

PROB. 12.269

***12.269** The collar of mass m can slide over the smooth rigid crossbar as the bar rotates about the vertical axis. The mass of the bar is negligible. A constant counterclockwise torque $\bar{M}$ is applied to the bar, causing it to rotate, and also causing the collar to slide along it. Use eq. (57) and Newton's second law to derive two differential equations governing the unknown variables ω and R.

***12.270** A constant clockwise couple $\bar{M}$ is applied to the bar AB, causing the system to rotate in the vertical plane about the pin at point A. This causes the block of mass m to slide along the bar. Bar AB is smooth and massless. Derive the differential equations of motion for the radial distance R and the angle θ by using eq. (57) and Newton's second law.

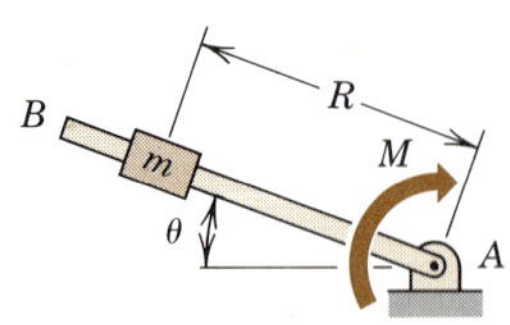

PROB. 12.270

J. SUMMARY OF PARTICLE DYNAMICS

The fundamental tool for a complete investigation of the motion of particle systems is Newton's second law expressed in terms of the appropriate kinematical description. Here we concentrated on path variables, rectangular

Cartesian coordinates, and cylindrical coordinates for the kinematical descriptions. These methods are sufficient for problems where only a force or an acceleration value are to be determined.

In some problems we must relate the initial and final velocity parameters of the system. One method for achieving this is to derive and solve the differential equation of motion by formulating Newton's second law for a general position. There are many techniques for solving differential equations of motion, the most elementary, where applicable, being the method of separating variables.

This approach for relating the initial and final velocities is complemented by the principles of work and energy, linear impulse and momentum, and angular impulse and momentum. All of these principles are standard solutions of the differential equations of motion. The key features to look for in applying these principles are summarized below.

1. Energy considerations are generally useful when the forces depend only on the position of the particle.
2. Linear momentum is suitable to situations where the rectangular components of the forces are functions of time, particularly for impulsive forces.
3. Angular momentum is suitable to situations where bodies are rotating about a specific axis.

It should be noted that these principles may be utilized jointly to solve a problem, or individually to treat separate intervals in the motion of a body.

One topic featuring most of the principles of particle dynamics is orbital motion, where a body moves under the influence of the gravitational attraction of a larger body. This subject is currently receiving much attention because of its obvious relevance to space flight. The remainder of this chapter provides an introduction to the study of orbital motion.

*K. ORBITAL MOTION

Newton's gravitational law postulates that the force of attraction between two bodies is a function of the distance between them. In our earlier studies we investigated the ballistic motion of a particle for the case where the particle was close to the earth's surface. Thus, the distance from the earth's surface to the particle was small compared to the radius of the earth. This allowed us to approximate gravity as a constant force, thereby simplifying the formulation and solution of the equations of motion.

Our aim here is to account for the varying magnitude and direction of the gravitational force acting on a particle that is located sufficiently far from the attracting body's mass center to make the force a variable quantity. For the sake of simplicity we shall consider only the case where the particle is much less massive than the larger body, as is the case for an artificial satellite that orbits the earth. This makes it permissible to consider the larger body to form a fixed (inertial) reference frame for the motion.

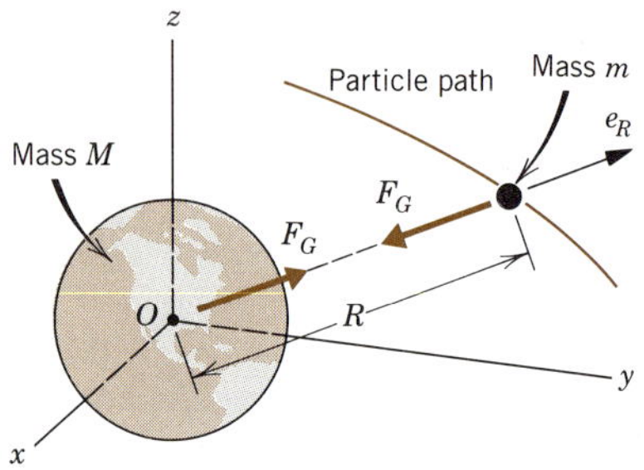

FIGURE 17

1. The Gravitational Attraction Force

In Figure 17 the xyz coordinate system describing the fixed reference frame has its origin at the center of mass O of the larger body. According to Newton's third law, the gravitational attraction forces exerted on each body by the other are equal in magnitude, opposite in direction, and collinear with the line connecting the centers of the two masses. The radial direction $\bar{e}_R$ is defined in Figure 17 to represent this line. The other two unit vectors corresponding to a set of cylindrical coordinates will be defined later.

Because the resultant force $\bar{F}_G$ acting on the particle is always directed toward the fixed point O, we have

$$\bar{F}_G = -F_G\bar{e}_R \quad \text{acting on the particle}$$

In general, any situation where the resultant force acting on a particle is always oriented toward a fixed point is called *central force motion.*

General relations for central force motion may be specialized to gravitational attraction by using Newton's inverse square law of gravitation,

$$\boxed{\bar{F}_G = -\frac{GMm}{R^2}\bar{e}_R} \tag{59}$$

where M and m are the mass of the large fixed body and of the particle, respectively.

As we saw in Section F.2 of this chapter, the value of GM for motion around the earth (the *intensity* of the gravity force) may be estimated by setting $F_G = mg$ when the particle is at the surface of the earth, that is, when $R = R_e = 3960$ miles $= 6370$ km. Then

$$\boxed{\begin{aligned} GM = gR_e^2 &= 3.979(10^{14})\ \text{m}^3/\text{s}^2 \\ &= 1.4604(10^{16})\ \text{ft}^3/\text{s}^2 \end{aligned}} \tag{60}$$

Situations where the large body is not the earth may be studied by using for the universal gravitational constant

$$\begin{aligned} G &= 6.673(10^{-11})\ \text{m}^3/(\text{kg-s}^2) \\ &= 3.442(10^{-8})\ \text{ft}^4/(\text{lb-s}^4) \end{aligned}$$

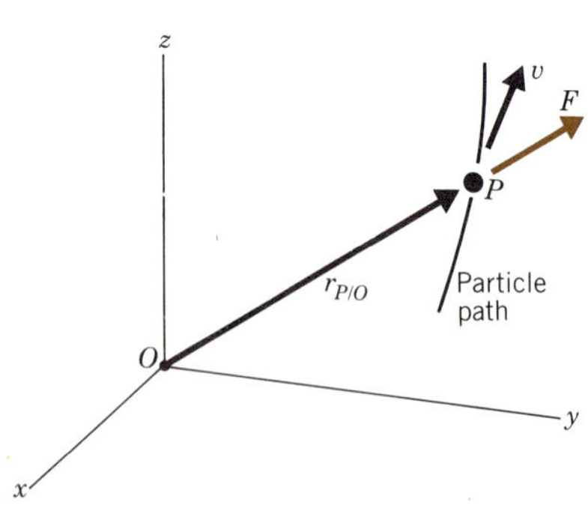

FIGURE 18

2. Central Force Motion

In this section we shall consider the motion of a particle under the influence of a *general* central force $\bar{F}$, whose magnitude may be a variable quantity. In Figure 18, point O is the fixed point through which this force always acts, and point P is the location of the particle.

Let the initial position and velocity of the particle be arbitrary and known. Given these initial conditions, we wish to determine the path the particle follows and the relationships that exist among its kinematical variables.

We begin by noting that in central force motion the moment of $\bar{F}$ about the origin O is zero. Hence, the angular impulse-momentum principle for a single particle, formulated about the fixed point O, yields

$$\sum \bar{M}_O \equiv \bar{0} \equiv \dot{\bar{H}}_O = \frac{d}{dt}(\bar{r}_{P/O} \times m\bar{v})$$

It follows that the angular momentum $\bar{H}_O$ is constant throughout the motion. For convenience let us divide $\bar{H}_O$ by the constant mass m to obtain

$$\bar{h}_O = \frac{\bar{H}_O}{m} = \bar{r}_{P/O} \times \bar{v} \equiv \text{constant} \tag{61}$$

where $\bar{h}_O$ is called the angular momentum per unit mass.

In order for the vector $\bar{h}_O$ in eq. (61) to be constant, both its magnitude and direction must be constant. In view of the definition of the direction of the cross product, this constant vector is normal to the plane formed by $\bar{r}_{P/O}$ and $\bar{v}$. This means that $\bar{h}_O$ can have a constant orientation only if the plane of $\bar{r}_{P/O}$ and $\bar{v}$ is constant, so

> A particle moving under the influence of a central force is in planar motion.

The radial direction has been defined to be collinear with $\bar{r}_{P/O}$, such that $\bar{e}_r = \bar{r}_{P/O}/|\bar{r}_{P/O}|$. The polar angle ϕ, measured from the x axis, is used to define the orientation of the radial line in the plane of motion; and $\bar{k}$ is perpendicular to the plane. This is the situation depicted in Figure 19.

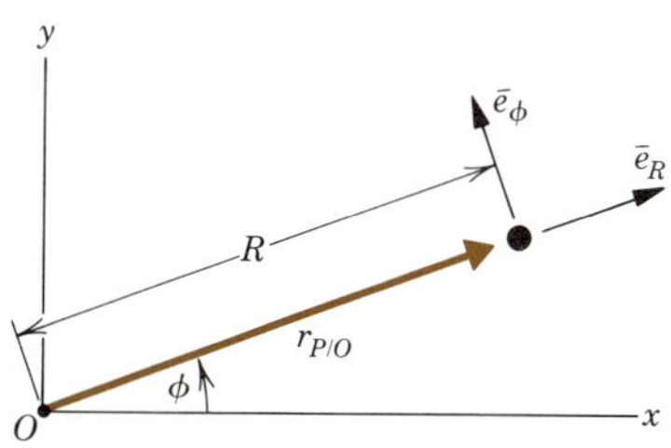

FIGURE 19

Setting $z \equiv 0$ for planar motion, the kinematical formulas for the position and velocity of particle P in terms of cylindrical coordinates are

$$\bar{r}_{P/O} = R\bar{e}_R, \qquad \bar{v} = \dot{R}\bar{e}_R + R\dot{\phi}\bar{e}_\phi$$

Substituting these expressions into eq. (61) yields

$$\bar{h}_O = R\bar{e}_R \times (\dot{R}\bar{e}_R + R\dot{\phi}\bar{e}_\phi) = R^2\dot{\phi}\bar{k}$$

from which we note that $\bar{h}_O$ has a single component in the $\bar{k}$ direction, which is in agreement with the earlier discussion. Equally important, from the initial conditions we can evaluate the magnitude h_O of this vector. The requirement that the value of h_O be constant for the motion provides an important general relation.

$$\boxed{h_O \equiv R^2\dot{\phi} \equiv Rv_\phi = \text{constant}} \tag{62}$$

This result has a geometric interpretation. Consider the position of a particle at two instants that are separated by an infinitesimal interval dt, as shown in Figure 20. Let us compute the shaded area dA, which is the area swept out by the position vector in the time interval from t to $t + dt$. As shown in the figure, the base of the triangle is R and the height is $R\,d\phi$, so that

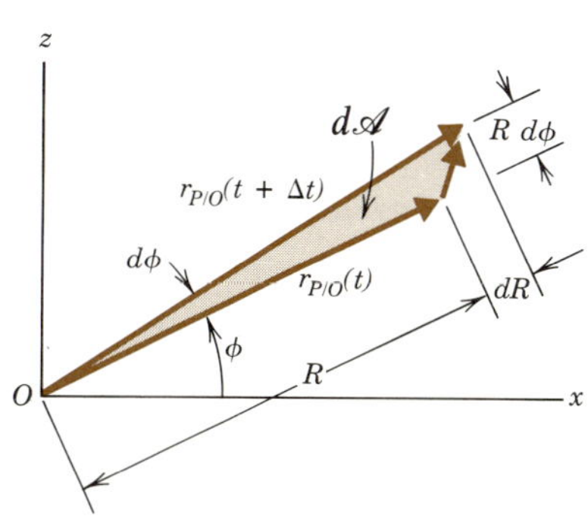

FIGURE 20

$$d\mathscr{A} = \tfrac{1}{2}R(R\,d\phi)$$

Dividing by dt and comparing the result to eq. (62) gives

$$\frac{d\mathscr{A}}{dt} = \frac{1}{2}R^2\frac{d\phi}{dt} = \frac{1}{2}R^2\dot{\phi} = \frac{1}{2}h_O \tag{63}$$

The quantity $d\mathscr{A}/dt$ is called the *areal velocity*. It is the rate at which the position vector sweeps out area. Because h_O is a constant, we see that the areal velocity is constant in central force motion.

We now wish to determine the planar path of a particle undergoing central force motion. The central force $\bar{F}$ has only one component (in the radial direction) so from Newton's second law we find the scalar equations of motion to be

$$\begin{aligned} \sum F_R &= F = ma_R = m(\ddot{R} - R\dot{\phi}^2) \\ \sum F_\phi &= 0 = ma_\phi = m(R\ddot{\phi} + 2\dot{R}\dot{\phi} \\ \sum F_z &= 0 \end{aligned} \tag{64}$$

Equation (62) may now be used to eliminate $\dot{\phi}$ from the two nontrival equations of motion. Solving for $\dot{\phi}$ we have

$$\dot{\phi} = \frac{h_O}{R^2}$$

This relation may be differentiated with respect to time, with the result that

$$\ddot{\phi} = -\frac{2h_O}{R^3}\dot{R}$$

Substituting these expressions for $\dot{\phi}$ and $\ddot{\phi}$ into eqs. (64) gives

$$F = m\left[\ddot{R} - R\left(\frac{h_O}{R^2}\right)^2\right]$$

$$0 \equiv m\left[R\left(-\frac{2h_O}{R^3}\dot{R}\right) + 2\dot{R}\left(\frac{h_O}{R^2}\right)\right]$$

Note that the second equation (the transverse force component) is also trivial. This is a consequence of the fact that the angular impulse-momentum principle for a single particle is equivalent to the transverse component of Newton's second law. All that remains is a single scalar equation of motion, the solution to which defines the particle's path.

$$\ddot{R} - \frac{h_O^2}{R^3} = \frac{F}{m} \tag{65}$$

Equation (65) is a nonlinear differential equation. To facilitate its solution, let us transform the equation from its present form, relating R and t, to a simpler one relating R and ϕ. Evaluating R when $\dot{R}$ is a function of ϕ requires the chain rule.

$$\dot{R} = \frac{d}{dt}(R) = \frac{dR}{d\phi}\dot{\phi}$$

Using eq. (62) to eliminate $\dot{\phi}$ leads to a general relation between $\dot{R}$ and R, which may then be differentiated to relate $\ddot{R}$ and R. Specifically

$$\dot{R} = \frac{dR}{d\phi}\dot{\phi} = \frac{h_O}{R^2}\frac{dR}{d\phi}$$

$$\ddot{R} \equiv \frac{d}{dt}(\dot{R}) = \frac{d\dot{R}}{d\phi}\dot{\phi} = \frac{h_O}{R^2}\frac{d}{d\phi}\left(\frac{h_O}{R^2}\frac{dR}{d\phi}\right)$$

It is useful to introduce a new variable $q = 1/R$, so that

$$\frac{dR}{d\phi} = \frac{d}{d\phi}\left(\frac{1}{q}\right) = -\frac{1}{q^2}\frac{dq}{d\phi}$$

The expression for $\ddot{R}$ is transformed by this change of variables to

$$\ddot{R} = h_O q^2 \frac{d}{d\phi}\left[h_O q^2\left(-\frac{1}{q^2}\frac{dq}{d\phi}\right)\right] = h_O q^2 \frac{d}{d\phi}\left(-h_O\frac{dq}{d\phi}\right) = -h_O^2 q^2 \frac{d^2 q}{d\phi^2}$$

The result of substituting these expressions for R and $\ddot{R}$ into eq. (65) is

$$\boxed{\frac{d^2 q}{d\phi^2} + q = -\frac{F}{m h_O^2 q^2}, \qquad q = \frac{1}{R}} \tag{66}$$

Equation (66) is the general form of the equation for the path of a particle undergoing central force motion. To proceed further we must define the functional dependence of the central force.

3. Gravitational Central Force Motion—Conic Sections

In order to study the central force motion associated with gravitational attraction, the radial component of $\bar{F}_G$ in eq. (59) is substituted for the general force component F. Making this substitution in eq. (66) requires that the relation $1/R = q$ be used. The differential equation for the path is then found to be

$$\frac{d^2 q}{d\phi^2} + q = -\frac{1}{m h_O^2 q^2}(-GMmq^2) = \frac{GM}{h_O^2} \tag{67}$$

This is a linear, nonhomogeneous, second-order differential equation with constant coefficients. Its solution is a principal topic in a first course on differential equations. It is readily verified by direct substitution that the general solution of eq. (67) is

$$\boxed{q = \frac{1}{R} = C\cos(\phi - \beta) + \frac{GM}{h_O^2}} \tag{68}$$

This is the equation for the *trajectory* of a particle undergoing gravitational central force motion. The constants C and β are constants of integration that, as we shall see, are determined from the initial conditions.

Although it is not obvious, eq. (68) is the equation in polar coordinates of the curve formed from a conic section. For this reason, we now will review briefly some properties of a conic section. To construct the conic section of

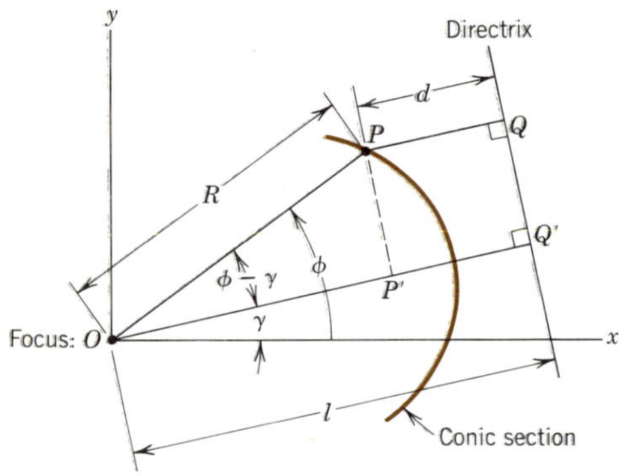

FIGURE 21

Figure 21 we first choose a fixed point O, called the *focus*, and a fixed line, called the *directrix*. The line OQ', perpendicular to the directrix and passing through the focus, is the axis of symmetry of the conic section.

A conic section has the property that the radial distance R from the focus to any point P on the curve is in a constant ratio to the perpendicular distance d from the directrix to point P, that is

$$e = \frac{R}{d} \tag{69}$$

where the constant e is called the *eccentricity*; it is always greater than or equal to zero.

For a specific choice of the focus and directrix, we can determine the the constant angle γ and the constant distance l shown in Figure 21 by constructing the line OQ'. Hence, let us express the distance d in terms of these parameters. From Figure 21 we see that

$$l = |\bar{r}_{P'/O}| + |\bar{r}_{Q'/P'}| = R\cos(\phi - \gamma) + d$$

Solving this expression for d and substituting the result into eq. (69) yields

$$e = \frac{R}{l - R\cos(\phi - \gamma)}$$

$$\frac{1}{R} = \frac{1}{el}[1 + e\cos(\phi - \gamma)] \tag{70}$$

Comparing this expression for R to eq. (68) for the gravitational motion gives

$$\boxed{\gamma = \beta, \qquad \frac{1}{l} = C, \qquad \frac{1}{el} = \frac{GM}{h_O^2}} \tag{71}$$

Thus, once the constants h_O, β, and C are evaluated from the initial conditions for the motion, we can determine the eccentricity e, the distance l from the focus to the directrix, and the angle γ between the axis of symmetry of the conic section and the x axis. These computations fully define the path of the particle.

Before we perform this determination, note that a conic section can be categorized by the value of the eccentricity. We have the following possibilities:

When $e = 0$ the path is a circle.
When $0 \leq e < 1$ the path is an ellipse.
When $e = 1$ the path is a parabola.
When $e > 1$ the path is a hyperbola.

The ellipse is the only closed conic section (considering the circle as a special ellipse). It is of particular interest to us because many space vehicles, as well as the planets, follow elliptical *orbits*. Basic properties of an elliptical

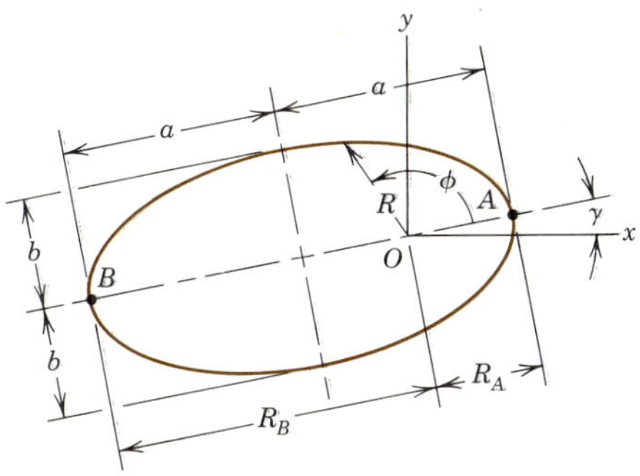

FIGURE 22

path are the lengths of the semimajor and semiminor axes, which are the values a and b shown in Figure 22. From this figure we see that the minimum and maximum values of R occur when the particle is at point A ($\phi = \gamma$) and when it is at point B ($\phi = 180^\circ + \gamma$), and also that $2a = R_A + R_B$. Solving eq. (70) for the values R_A and R_B at these two values of ϕ yields

$$\frac{1}{R_A} = \frac{1}{el}(1 + e), \qquad \frac{1}{R_B} = \frac{1}{el}(1 - e)$$

Now, using eq. (71) to eliminate the term el gives

$$R_A = \frac{h_O^2}{GM}\left(\frac{1}{1+e}\right), \qquad R_B = \frac{h_O^2}{GM}\left(\frac{1}{1-e}\right)$$

Hence

$$a = \frac{1}{2}(R_A + R_B) = \frac{h_O^2}{GM}\left(\frac{1}{1-e^2}\right)$$
$$R_A = a(1 - e) \qquad R_B = a(1 + e) \tag{72}$$

The value of b can then be shown to be

$$b = a(1 - e^2)^{1/2} \tag{73}$$

In the case of orbital motion about the earth, point A, where R is a minimum, is called the *perigee*, and point B, where R is a maximum, is called the *apogee*.

4. Initial Conditions

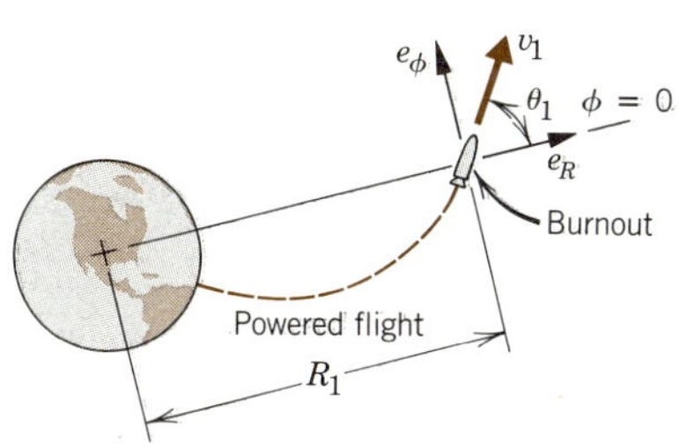

FIGURE 23

To aid in our understanding of the role of the initial conditions, consider the problem of placing a satellite into orbit, as shown in Figure 23. At the completion of the powered portion of the flight, the space vehicle is at a radial distance R_1 and has a velocity $\bar{v}_1$ directed at an angle θ_1 from the radial line. The values of R_1, v_1, and θ_1 are the initial conditions. As shown in Figure 23, the initial orientation of the radial line is defined to be $\phi = 0$. Then the initial conditions may be stated as

$$R = R_1, \qquad v_R = v_1 \cos \theta_1, \qquad v_\phi = v_1 \sin \phi_1 \quad \text{when } \phi = 0 \tag{74}$$

Evaluating the conic section parameters for the orbit requires that we first determine the constants h_O, C, and β defining the path in eq. (68). The value of h_O is readily found from the definition

$$h_O = Rv_\phi = R_1 v_1 \sin \theta_1 \tag{75}$$

Another relation for the constants is found by substituting $R = R_1$ when $\phi = \phi_1$ in the equation for the trajectory. Thus

$$\frac{1}{R_1} = C \cos \beta + \frac{GM}{h_O^2}$$

Because h_O is known from eq. (75), the foregoing is one equation relating C and β. The second equation is obtained by requiring that the tangent to the path at the initial position be aligned along the initial velocity. For this, we use the kinematical property that

$$\cot\theta_1 = \frac{(v_R)_1}{(v_\phi)_1} = \left(\frac{dR/dt}{R\,d\phi/dt}\right)_1 = \frac{1}{R_1}\left(\frac{dR}{d\phi}\right)_1 \tag{76}$$

Differentiating eq. (68) with respect to ϕ in order to express $dR/d\phi$ yields

$$-\frac{1}{R^2}\frac{dR}{d\phi} = -C\sin(\phi - \beta)$$

Evaluating this relation at the initial position and then substituting eq. (76) yields

$$-\frac{1}{R_1}\cot\theta_1 = C\sin\beta$$

In summary, the values of C and β may be determined by solving

$$C\cos\beta = \frac{1}{R_1} - \frac{GM}{h_O^2}$$
$$C\sin\beta = -\frac{1}{R_1}\cot\theta_1 \tag{77}$$

Special care must be taken in the solution of these relations to place the value of β in the correct quadrant.

Once the value of C is computed, we may determine the properties of the resulting trajectory. In particular, we are interested in the eccentricity. We find from eqs. (71) that

$$e = \frac{h_O^2}{GM}C \tag{78}$$

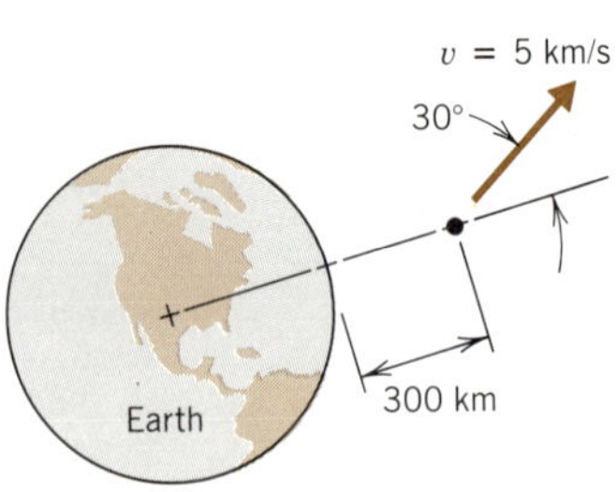

EXAMPLE 32

A satellite is placed into its orbit at an altitude of 300 km above the earth's surface at a speed of 5 km/s and an angle $\theta = 30^\circ$, as illustrated. Determine the equation for its trajectory, and show that its orbit is elliptical. Also determine the perigee and apogee distances. Is the launch successful?

Solution

Remember that the radial distance R is the distance from the focus of the central force. Taking the earth's radius to be 6370 km, the given initial conditions are that, when $\phi = 0$,

$$R_1 = 6370 + 300 = 6670 \text{ km} = 6.670(10^6) \text{ m}$$

$$v_1 = 5 \text{ km/s} = 5000 \text{ m/s}, \qquad \theta_1 = 30^\circ$$

We use these initial values to determine the parameters h_O, C, and β. First, eq. (75) gives

$$h_O = 6.670(10^6)(5)(10^3) \sin 30^\circ = 1.6675(10^{10}) \text{ m}^2\text{/s}$$

Substituting the foregoing parameters into eqs. (77) yields

$$C \cos \beta = \frac{1}{6.670(10^6)} - \frac{3.979(10^{14})}{(1.6675)^2(10^{20})} = -1.2811(10^{-6})$$

$$C \sin \beta = -\frac{1}{6.670(10^6)} \cot 30^\circ = -2.597(10^{-7})$$

Both $\sin \beta$ and $\cos \beta$ are negative, so β is situated in the third quadrant. Then

$$\tan \beta \equiv \frac{\sin \beta}{\cos \beta} = \frac{0.2597}{1.2811} \Rightarrow \beta = -168.54^\circ$$

$$C = -\frac{1.2811(10^{-6})}{\cos \beta} = 1.3071(10^{-6}) \text{ 1/m}$$

The corresponding eccentricity is found from eq. (78) to be

$$e = \frac{(1.6675)^2(10^{20})}{3.979(10^{14})}(1.3071)(10^{-6}) = 0.9134$$

Because $e < 1$, the trajectory is an ellipse.

Equation (68) for the trajectory then gives

$$\frac{1}{R} = C \cos(\phi - \beta) + \frac{GM}{h_O^2}$$
$$= 1.3071(10^{-6}) \cos(\phi + 168.54^\circ) + 1.4310(10^{-6}) \text{ 1/m}$$

The perigee and apogee distances are the minimum and maximum values for R. Setting the cosine term first equal to $+1$ and then to -1 yields

$$\frac{1}{R_{\text{perigee}}} = (1.3071 + 1.4310)10^{-6} \Rightarrow R_{\text{perigee}} = 3.652(10^5) \text{ m} = 365 \text{ km}$$

$$\frac{1}{R_{\text{apogee}}} = (-1.3071 + 1.4310)10^{-6} \Rightarrow R_{\text{apogee}} = 8.071(10^6) \text{ m} = 8071 \text{ km}$$

The fact that the perigee distance is smaller than the radius of the earth means that the satellite will crash into the earth before it reaches its perigee; the launch was unsuccessful.

HOMEWORK PROBLEMS

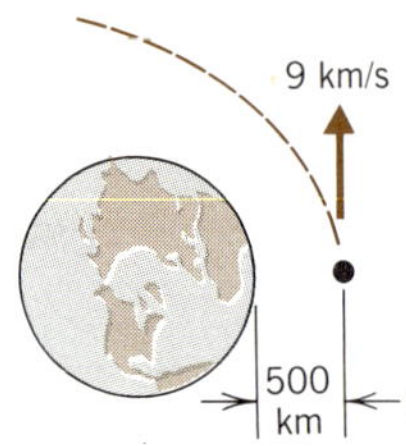

PROB. 12.271

12.271 A satelite is launched at an altitude of 500 km above the earth's surface with a velocity of 9 km/s parallel to the earth's surface. Determine the eccentricity of the orbit and the maximum and minimum altitudes for the satellite's trajectory.

12.272 The altitude of a satellite at apogee is 4040 miles and at perigee it is 540 miles. Determine the equation of the trajectory and the velocity of the satellite at apogee and at perigee.

12.273 A satellite in circular orbit at an altitude of 1130 km is to be given a new orbit. To do this the engines are fired tangent to the orbit, imparting an additional velocity of 4 km/s to the satellite. Determine the equation for the new orbit and the eccentricity of the orbit. Is the orbit open or closed?

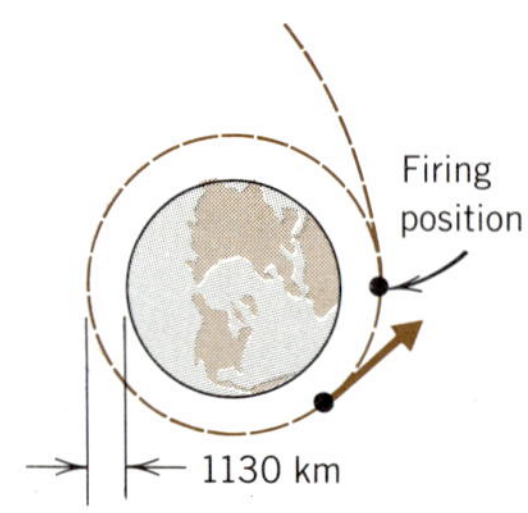

PROB. 12.273

12.274 Solve Problem 12.273 if the engines are aligned radially, giving the satellite a radial velocity component of 4 km/s outward.

***12.275** A satellite in circular orbit at an altitude of 240 miles is to be moved to a higher coplanar circular orbit at an altitude of 6040 miles. To accomplish this the engines are fired tangentially at point P_1. This increases the speed by Δv_1, placing the satellite into an elliptical orbit. (This orbit is called the Hohmann transfer ellipse.) At point P_2 the rockets are again fired tangentially, increasing the speed by Δv_2, in order to place the satellite in the desired circular orbit. Determine Δv_1 and Δv_2.

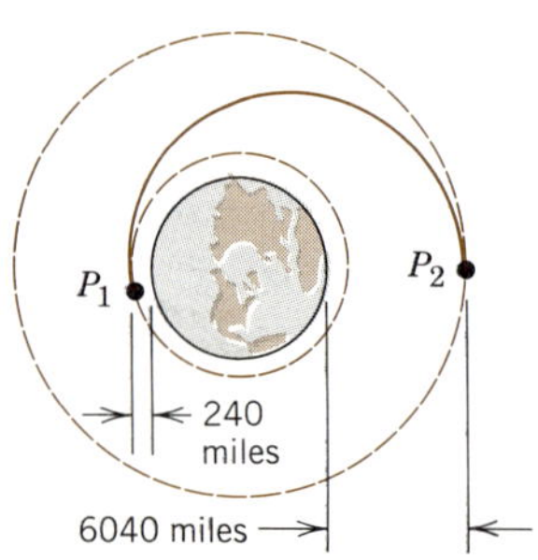

PROB. 12.275

12.276 A satellite is in a circular orbit at an altitude of 630 km when retrorockets are fired tangent to the orbit, slowing the rocket by 5000 km/hr. Determine the angle ϕ locating the point S where the statellite will splash down. Neglect air resistance.

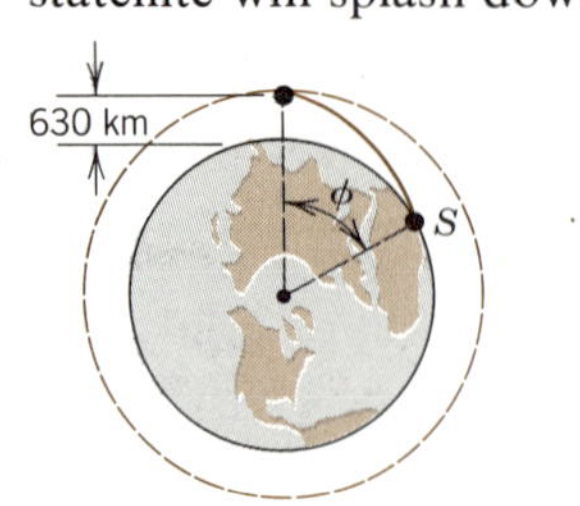

PROB. 12.276

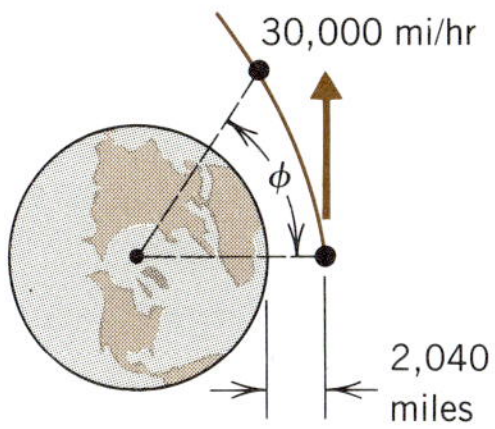

PROB. 12.277

12.277 At an altitude of 2040 miles above the earth's surface, a space probe is launched into a hyperbolic trajectory, giving it a velocity of $3(10^4)$ mi/hr tangent to the earth's surface. Determine the largest value of the angle ϕ that is possible for this trajectory.

12.278 At an altitude of 30,000 km above the surface of the earth, a meteorite has a velocity of 15 km/s at $\theta = 120°$. Determine (a) the eccentricity of the trajectory of the meteorite, (b) the minimum altitude for the trajectory, (c) the angle ϕ at the position of closest approach. Will the meteorite enter the earth's atmosphere? (Consider the atmosphere to extend to an altitude of 40 miles.)

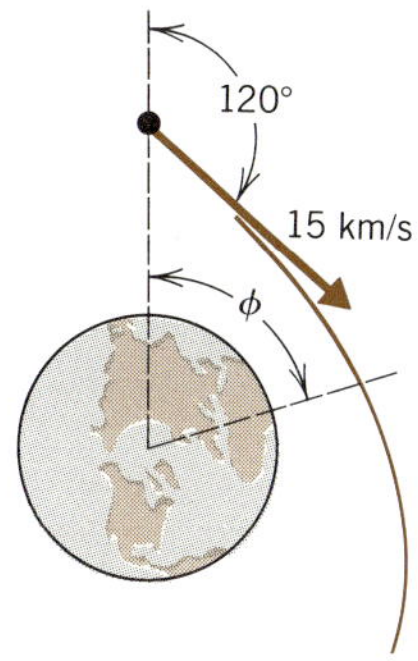

PROB. 12.278

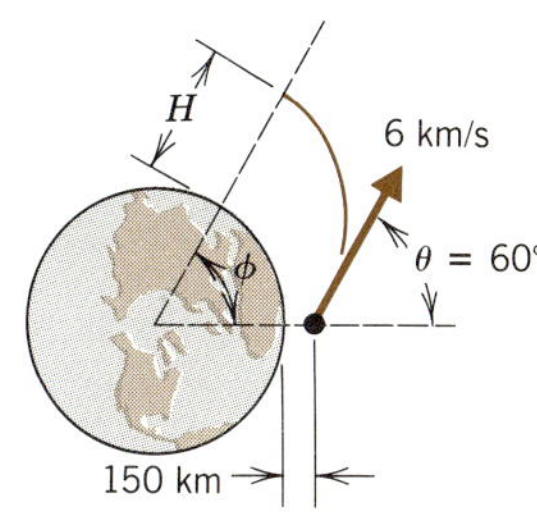

PROB. 12.279

12.279 The engine of a missile is shut off at an altitude of 150 km, at which time the velocity is 6 km/s at an angle $\theta = 60°$. Determine the equation for the trajectory, the maximum altitude H attained by the missile, and the polar angle ϕ at this location.

12.280 Determine the angle ϕ defining the location where the missile in Problem 12.279 returns to the surface of the earth. Air resistance is negligible.

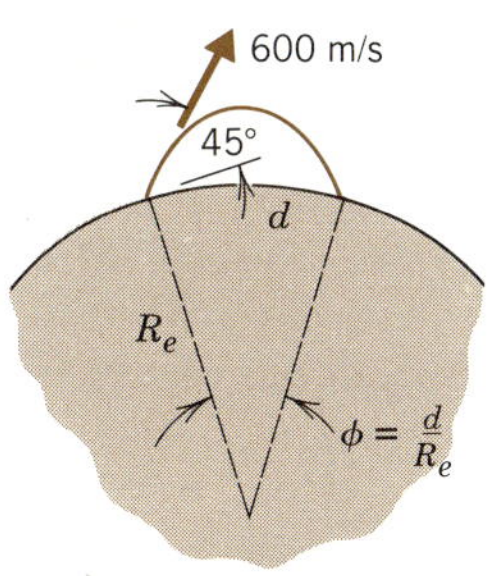

PROB. 12.281

***12.281** A projectile is fired from the surface of the earth at 600 m/s at an angle of elevation of 45°. (a) Using the orbit equations, determine the distance d locating where the projectile returns to the earth. (b) Compare the result of part (a) to that obtained when the earth is considered flat and the gravity force is considered constant.

---

5. Energy

In the preceding section we saw how the trajectory of a particle is determined. The evaluation of the velocity of the particle when it is at a specific position along its trajectory requires further development. There are many ways in which we may proceed. Here we choose the work–energy principles, for aside from being a direct approach, it is a good tool for gaining further insight into the orbital motion problem.

When the gravitational force is defined by the inverse square law, eq. (59), the potential energy of this conservative force was shown in Section F.2 of this chapter to be

$$V_{\text{gravity}} = -\frac{GMm}{R}$$

The only force acting on the particle is gravity, so mechanical energy is conserved,

$$T + V_{\text{gravity}} = \text{constant}$$

Let us denote the constant energy per unit mass as E.

$$E \equiv \frac{1}{m}(T + V_{\text{gravity}}) = \frac{1}{2}v^2 - \frac{GM}{R} \tag{79}$$

We may use eq. (79) to compute the value of E corresponding to given initial conditions. Equating this value of E to that at a different position then yields a relation for the speed v in terms of the value of R at the new position. The radial component of velocity v_R may then be determined by noting that the speed is the magnitude of the velocity. Thus

$$v^2 = v_R^2 + v_\phi^2$$

The transverse velocity component is known from eq. (75) for h_O, so we find that

$$\begin{aligned} v_\phi &= \frac{h_O}{R} \\ v_R &= \pm\left[v^2 - \left(\frac{h_O}{R}\right)^2\right]^{1/2} \end{aligned} \tag{80}$$

Equations (79) and (80) therefore define the velocity components as a function of R.

The energy parameter E may also be used to evaluate the trajectory properties. For this, eq. (76) and the first of eqs. (80) are used to eliminate the parameters R_1 and θ_1 from eqs. (77) for C and β. This yields

$$C\cos\beta = \frac{(v_\phi)_1}{h_O} - \frac{GM}{h_O^2}$$

$$C\sin\beta = -\frac{(v_\phi)_1}{h_O}\frac{(v_R)_1}{(v_\phi)_1} \equiv -\frac{(v_R)_1}{h_O}$$

In order to solve for C, each of the above equations is squared and they are added together.

$$\begin{aligned} C^2(\cos^2\beta + \sin^2\beta) \equiv C^2 &= \left[\frac{(v_\phi)_1}{h_O} - \frac{GM}{h_O^2}\right]^2 + \left[\frac{(v_R)_1}{h_O}\right]^2 \\ &\equiv \frac{1}{h_O^2}\left[(v_R)_1^2 + (v_\phi)_1^2 - \frac{2(v_\phi)_1 GM}{h_O} + \left(\frac{GM}{h_O}\right)^2\right] \\ &\equiv \frac{1}{h_O^2}\left[v_1^2 - \frac{2GM}{R_1} + \left(\frac{GM}{h_O}\right)^2\right] \end{aligned}$$

Referring back to eq. (79) enables us to rewrite this relation as

$$C = \frac{1}{h_O}\left[2E + \left(\frac{GM}{h_O}\right)^2\right]^{1/2} \tag{81}$$

Substitution of this expression into eq. (78) yields a very useful relation for the eccentricity.

$$e = \left[2E\left(\frac{h_O}{GM}\right)^2 + 1\right]^{1/2} \tag{82}$$

Recalling the types of trajectories associated with the different ranges of values of the eccentricity, we see that:

Circular trajectory: $e = 0$, so $E = -0.5(GM/h_O)^2$
Elliptical trajectory: $0 \le e < 1$, so $-0.5(GM/h_O)^2 \le E < 0$
Parabolic trajectory: $e = 1$, so $E = 0$
Hyperbolic trajectory: $e > 1$, so $E > 0$

Thus, the energy level E is a direct indicator of the type of trajectory.

An important consequence of the foregoing is the concept of the *escape speed*, which is the minimum speed that a body must have in order to be in an open trajectory that extends to an infinite value of R. Because the parabola is the open trajectory having the lowest energy level, we set $E = 0$ to find v_{escape}. This yields

$$v_{\text{escape}} = \left(\frac{2GM}{R}\right)^{1/2} \tag{83}$$

Note that the type of trajectory depends only on the energy given to the body initially; the direction of the initial velocity has no effect on this consideration.

EXAMPLE 33

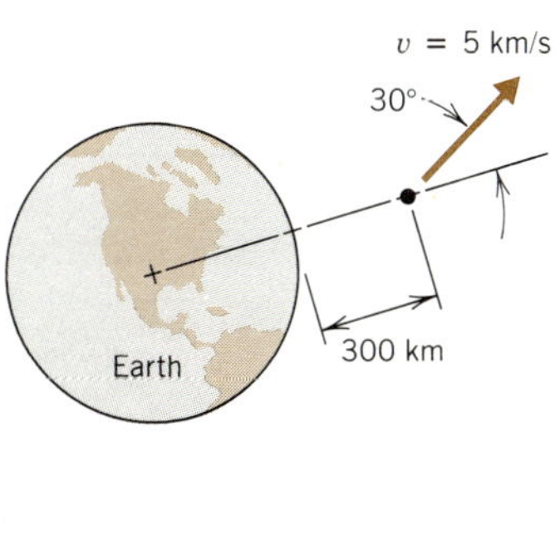

The system and initial conditions presented in Example 32 are repeated in the accompanying diagram. Once again, determine the perigee and apogee distance, but now use the energy and angular momentum equations derived this section.

Solution

The given initial conditions allow us to compute the energy parameter E and angular momentum parameter h_O, as follows.

$$E = \frac{1}{2}v_1^2 - \frac{GM}{R_1} = \frac{1}{2}[5(10^3)]^2 - \frac{3.979(10^{14})}{(6370 + 300)(10^3)}$$

$$= -4.716(10^7)\ \text{m}^2/\text{s}^2$$

$$h_0 = R_1(v_\phi)_1 = R_1 v_1 \sin\theta_1 = (6.670)(10^6)5(10^3)\sin 30^\circ$$
$$= 1.6675(10^{10})\ \mathrm{m^2/s}$$

Equations (79) and (80) then provide the following relations for the motion at a general position

$$\frac{1}{2}v^2 - \frac{3.979(10^{14})}{R} = -4.716(10^7)\ \mathrm{m^2/s^2}$$

$$v_\phi = \frac{1.6675(10^{10})}{R}\ \mathrm{m/s}, \qquad v_R = (v^2 - v_\phi^2)^{1/2}$$

At the perigee position the value of R is a minimum, while R has its maximum value at its apogee. Therefore $\dot{R} \equiv v_R = 0$ at both positions. Setting $v_R = 0$ in the above relation for v_R leads to the conclusion that $v_\phi = v$ at perigee and apogee. (This result could have been obtained by noting that the tangent to the path is perpendicular to the radial line at perigee and apogee.) Substituting the expression for v_ϕ $(= v)$ into the energy equation gives

$$-4.716(10^7) = \frac{1}{2}\left[\frac{1.6675(10^{10})}{R}\right]^2 - \frac{3.979(10^{14})}{R}$$

This expression simplifies to a quadratic equation, called the *absidial quadratic*, whose roots are the perigee and apogee distances.

$$4.716R^2 - 3.979(10^7)R + 1.3903(10^{13}) = 0$$

$$R_{\text{perigee}} = 3.65(10^5)\ \mathrm{m} = 365\ \mathrm{km}$$

$$R_{\text{apogee}} = 8.072(10^6)\ \mathrm{m} = 8072\ \mathrm{km}$$

To check the accuracy of our calculations, note that these values are identical to those obtained in Example 32.

As an aside, it should be mentioned that the multitude of relations that have been derived often means that there is a choice of methods for solving a problem. No single approach is, in general, optimal.

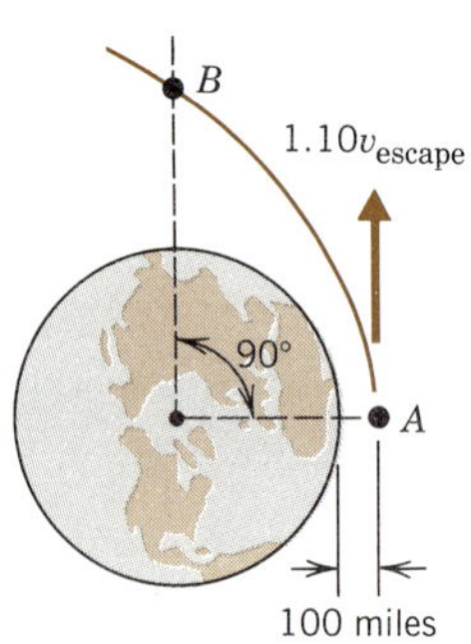

EXAMPLE 34

A space probe is injected into orbit at position A (altitude of 100 miles) with a velocity that is perpendicular to the radial line and is 10% greater than the local escape velocity. Determine the altitude and velocity of the probe at position B, where the radial line from the center of the earth has swept out a 90° angle.

Solution

Because the given initial velocity is perpendicular to the radial line, $v_R = 0$ initially. Equation (83) then leads to the following initial conditions for the position where $\phi = 0$:

$$R_1 = (100 + 3960)5280 = 2.144(10^7)\ \text{ft}$$

$$v_1 = (v_\phi)_1 = 1.10 v_{\text{escape}} = 1.10 \left(\frac{2GM}{R_1}\right)^{1/2}$$

$$= 1.10 \left[\frac{2(1.4606)(10^{16})}{2.144(10^7)}\right]^{1/2} = 4.060(10^4)\ \text{ft/s}$$

The corresponding energy and angular momentum parameters are found to be

$$E = \frac{1}{2} v_1^2 - \frac{GM}{R_1} = 1.4306(10^8)\ \text{ft}^2/\text{s}^2$$

$$h_O = R_1 (v_\phi)_1 = 8.704(10^{11})\ \text{ft}^2/\text{s}$$

To determine the distance for a specific value of ϕ we must first determine the coefficients C and β appearing in eq. (68) for the trajectory. Because the probe is launched perpendicularly to the radial line ($\theta_1 = 90°$), the launch point A is on the axis of symmetry and $\beta = 0$. [This is readily verified from the second of eqs. (77).] The value of C may be computed from eq. (78), after the eccentricity is found from eq. (82). Thus

$$e = \left[2(1.4306)(10^8)\left(\frac{8.704}{1.4604}\right)^2 (10^{-10}) + 1\right]^{1/2} = 1.4200$$

$$C = \frac{GM}{h_O^2} e = \frac{1.4604(10^{16})}{(8.704)^2(10^{22})}(1.4200) = 2.737(10^{-8})\ 1/\text{ft}$$

Equation (68) for the trajectory then becomes, with $\beta = 0$,

$$\frac{1}{R} = C \cos\phi + \frac{GM}{h_O^2} = 2.737(10^{-8}) \cos\phi + 1.9276(10^{-8})$$

To find the radial distance of point B we substitute $\phi = 90°$. This gives

$$\frac{1}{R_B} = 1.9276(10^{-8}), \qquad R_B = 5.187(10^7)\ \text{ft} = 9825\ \text{miles}$$

so

$$\text{altitude} = 9825 - 3960 = 5865\ \text{miles}$$

We can now determine the velocity at this location by employing eqs. (79) and (80). The former equation yields

$$E = 1.4306(10^8)\ \text{ft}^2/\text{s}^2 = \frac{1}{2} v_B^2 - \frac{GM}{R_B} = \frac{1}{2} v_B^2 - \frac{1.4606(10^{16})}{5.187(10^7)}$$

$$v_B = 2.914(10^4)\ \text{ft/s}$$

The velocity components obtained from eqs. (80) are then found to be

$$(v_\phi)_B = \frac{h_O}{R_B} = 1.6781(10^4)\ \text{ft/s}$$

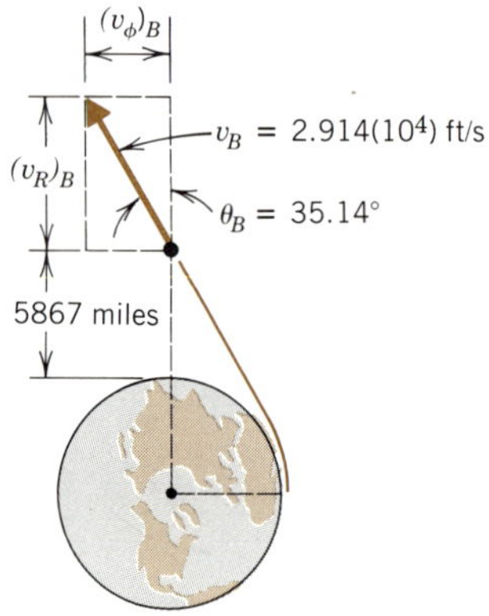

$$(v_R)_B = [v_B^2 - (v_\phi)_B^2]^{1/2} = 2.383(10^4)\text{ ft/s}$$

The motion at position B is depicted in the sketch. The angle θ_B describing the orientation of $\bar{v}_B$ relative to the radial line is

$$\theta_B = \sin^{-1}\left[\frac{(v_\phi)_B}{v_B}\right] = 35.16^\circ$$

Because $e > 1$, the trajectory is a hyperbola.

HOMEWORK PROBLEMS

12.282 An earth satellite has a speed of 5.5 mi/s at its minimum altitude of 400 miles. Determine (a) its angular momentum per unit mass, (b) its energy per unit mass, (c) that its trajectory is an ellipse, (d) its speed at apogee, and (e) its altitude at apogee.

12.283 A weather satellite is placed into orbit at an altitude of 330 km above the earth's surface, moving parallel to the earth's surface. Determine (a) the maximum initial velocity that will keep the satellite in the earth's gravitational field. (b) The initial speed that will cause the orbit to be a circle. (c) The minimum initial speed for a successful orbit. (The perigee for this orbit is the radius of the earth, so it is called the *grazing orbit*). Determine the eccentricity and energy per unit mass for such an orbit. (d) Determine the initial speed that will result in a second (different) orbit of the same eccentricity as that of part (c), with the launch position now being the perigee. Find its energy per unit mass and altitude at apogee.

12.284 A meteorite at a distance of 60,000 km from the center of the earth has a velocity of 50,000 km/hr at $\theta = 165^\circ$. Determine the velocity of the meteorite at its closest approach to the earth.

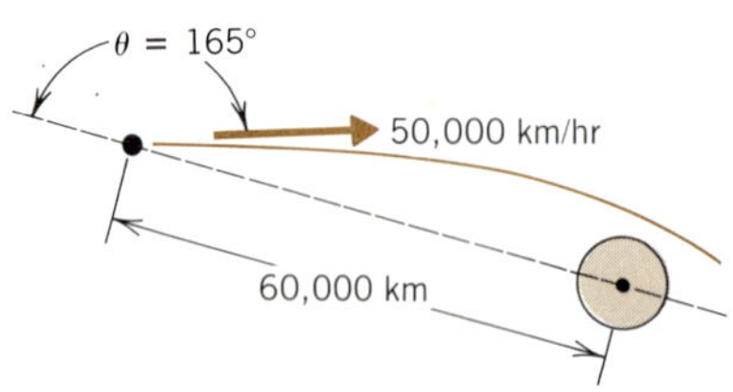

PROB. 12.284

***12.285** Determine the velocity of the meteorite in Problem 12.284 at a distance of 25,000 km from the center of the earth.

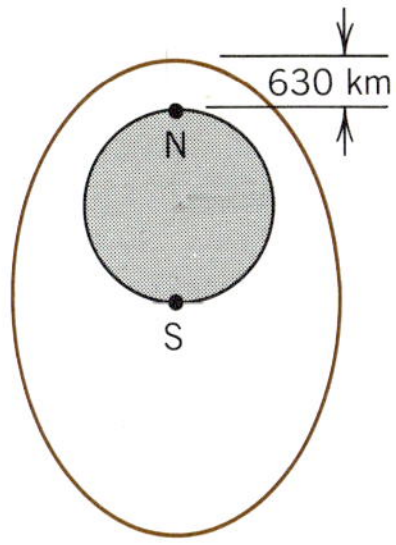

PROB. 12.286

12.286 A satellite is in a polar orbit of eccentricity 0.40 for which perigee occurs over the north pole at an altitude of 630 km. Determine the velocity of the satellite as it passes over (a) the south pole, (b) the equator.

12.287 A space probe is launched into a parabolic trajectory by shutting down the engines at an altitude of 90 miles, at which point the velocity is tangent to the earth's surface. Determine the velocity of the probe when it reaches an altitude of 10,000 miles.

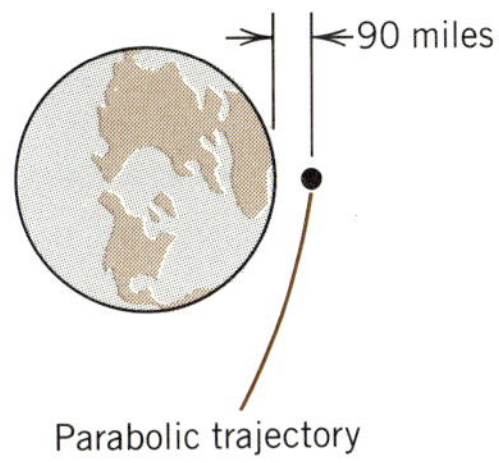

PROB. 12.287

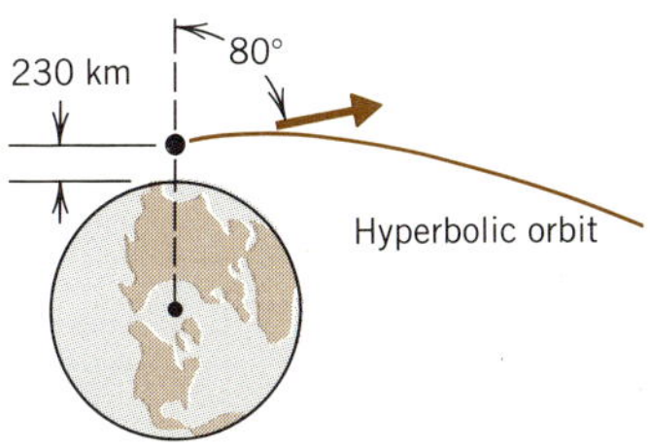

PROB. 12.288

***12.288** A space probe is following a hyperbolic trajectory of eccentricity 1.50. The engines were shut down at an altitude of 230 km, at which point the velocity was 80° from the outward radial direction. Determine the velocity of the space probe when it is (a) 1000 km above the earth's surface, (b) 10,000 km above the earth's surface.

12.289 At its closest approach to the earth, a satellite is at an altitude of 530 km and has a speed of 36,000 km/hr. Determine the velocity of the satellite when $\phi = 60^\circ$.

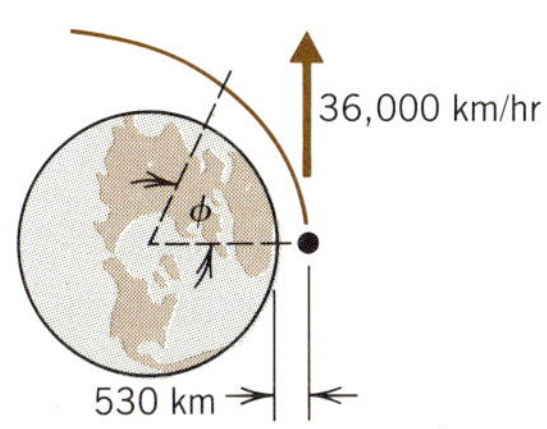

PROB. 12.289

12.290 Solve Problem 12.289 for $\phi = 120^\circ$.

---

6. The Period of an Orbit

Our study of the dynamics of gravitational central force motion is almost completed. We have seen how the position parameters R and ϕ are related through the equations of the trajectory. We have also seen how the velocity at any position along the trajectory may be evaluated. An important parameter we have not yet discussed is time.

In general, the easiest way to bring time into the problem is to use the fact that the angular momentum is constant. In the first of eqs. (80) we substitute $\dot{\phi} = v_\phi/R$. This yields

$$\dot{\phi} \equiv \frac{d\phi}{dt} = \frac{h_O}{R^2}$$

Eliminating R between this relation and eq. (68) leads to a first-order differential equation for $\phi(t)$.

$$\frac{d\phi}{dt} = h_O \left[C\cos(\phi - \beta) + \frac{GM}{h_O^2} \right]^2$$

where the values of C and β are known from the initial conditions. This equation is solved by separating variables and integrating, with the result that

$$t = \frac{1}{h_O} \int_O^\phi \frac{d\phi}{[C\cos(\phi - \beta) + GM/h_O^2]^2} \tag{84}$$

The integral may be evaluated with the aid of a table of integrals, but the result is too lengthy to present here.

A very important time parameter is the *period* of an elliptical orbit, that is, the length of time required for a particle to return to its initial position, thus completing one orbit. For this determination we recall eq. (63), which states that for any central force motion $\dot{\mathscr{A}} = h_O/2$, where $\mathscr{A}$ is the area swept out by the radial line. Because h_O is a constant, integrating the expression for $\dot{\mathscr{A}}$ shows that $\mathscr{A}$ increases proportionately to the elapsed time t, that is,

$$\boxed{\mathscr{A} = \tfrac{1}{2} h_O t} \tag{85}$$

For an elliptical trajectory, the area swept out in a full period τ is the area of the ellipse, which is

$$\mathscr{A} = \pi ab$$

where a and b are the lengths of the semimajor and semiminor axes, respectively. Using the foregoing area and setting $t = \tau$ in eq. (85) yields

$$\tau = \frac{2\pi ab}{h_O}$$

At this juncture it is interesting to note the work of the early astronomer Johannes Kepler (1571–1630), who painstakingly studied our solar system. He postulated the following three laws governing planetary motion:

1. The orbit of the planets are ellipses for which the sun is one focus.
2. The areal velocity is constant.
3. The square of the period of the elliptical orbit is in a constant ratio to the cube of its semimajor axis.

We have already proven the first two laws, and the third law may be obtained by using eqs. (72) and (73) to express τ in terms of a. This gives

$$\tau = \frac{2\pi ab}{h_O} = \frac{2\pi a}{h_O} a(1 - e^2)^{1/2} = \frac{2\pi a^2}{h_O}\left[\frac{h_O^2}{GMa}\right]^{1/2}$$

$$\boxed{\tau = \frac{2\pi}{(GM)^{1/2}} a^{3/2}} \tag{86}$$

The remarkable aspect of Kepler's laws is that, although we obtained them as mathematical consequences of Newton's laws of gravity and motion, Kepler's work preceded Newton; they were the result of careful observation and empiricism.

EXAMPLE 35

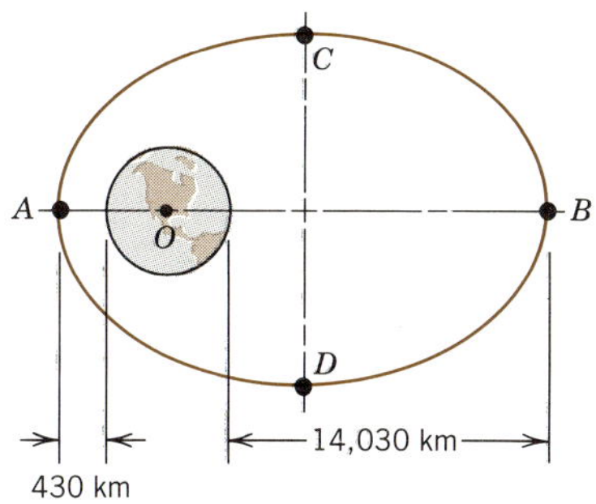

The perigee and apogee altitudes for an artificial satellite are 430 km and 14,030 km, respectively. Determine the time required for the satellite to travel from point C on the semiminor axis of its elliptical orbit, past the perigee A, and then on to point D, which is the other point on the semiminor axis. Compare this result to the period of the orbit.

Solution

This problem could be solved by integrating eq. (84), but a shortcut, based on the concept of the areal velocity, is available. This is so because the sectorial area bounded by the arc CAD and the radial lines OC and OD can be evaluated. In this approach we first determine the eccentricity e and the lengths a and b of the semiaxes of the ellipse.

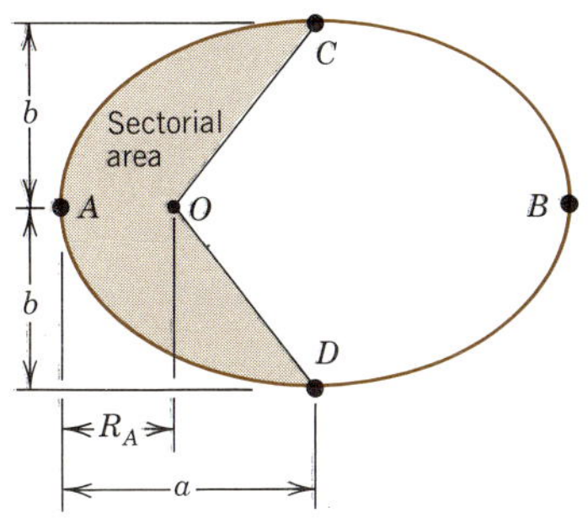

It is given that

$$R_A = 6{,}370 + 430 = 6{,}800 \text{ km} = 6.80(10^6) \text{ m}$$

$$R_B = 6{,}370 + 14{,}030 = 20{,}400 \text{ km} = 2.04(10^7) \text{ m}$$

From eqs. (72) and (73) we find that

$$a = \tfrac{1}{2}(R_A + R_B) = 1.360(10^7) \text{ m}$$

$$e = \frac{R_B}{a} - 1 = 0.5$$

$$b = a(1 - e^2)^{1/2} = 1.1778(10^7)\ \text{m}$$

The first of eqs. (72) also provides the value of h_O. Thus

$$h_O = [GMa(1 - e^2)]^{1/2} = [(3.979)(10^{14})(1.360)(10^7)(0.75)]^{1/2}$$
$$= 6.371(10^{10})\ \text{m}^2/\text{s}$$

We find $\mathscr{A}$ by deducting the triangular area OCD from the area of the semi-ellipse. Thus

$$\mathscr{A} = \tfrac{1}{2}\pi ab - \tfrac{1}{2}(2b)(a - R_A) = 1.7152(10^{14})\ \text{m}^2$$

Denoting the desired time as t^*, eq. (84) yields

$$t^* = \frac{2\mathscr{A}}{h_O} = 5384\ \text{s} = 1.496\ \text{hr}$$

In contrast, the orbital period is

$$\tau = \frac{2(\pi ab)}{h_O} = 15{,}797\ \text{s} = 4.388\ \text{hr}$$

Thus the time t^* is 32% less than $\tau/2$. Why isn't $t^* = \tau/2$?

EXAMPLE 36

The engines of the launch vehicle burn out at the correct altitude to inject a weather satellite into a stationary equatorial orbit about the earth. The satellite also has the proper speed at burnout for a stationary orbit, but due to a malfunction, the velocity at burnout is only at an angle of 80° with the radial line. Determine the eccentricity and period of the orbit.

Solution

We must first determine the initial conditions by studying the desired stationary circular orbit, which is an orbit whose period is 1 day. For a circular orbit at radius R_{st} and speed v_{st}, Newton's second law for the radial direction gives

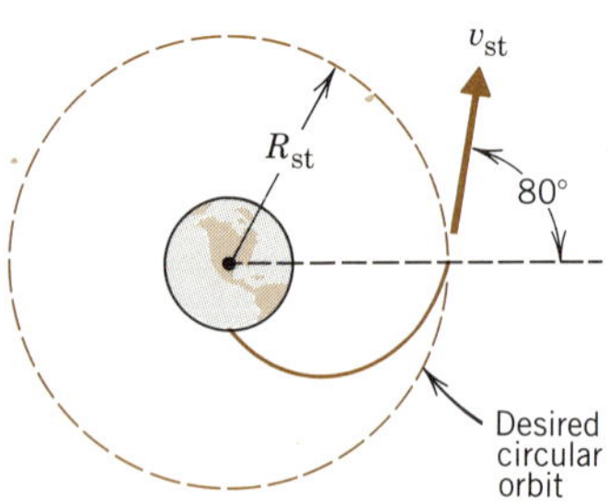

$$\sum F_R = -\frac{GMm}{R_{st}^2} = ma_R = -m\frac{v_{st}^2}{R_{st}}$$

$$v_{st} = \left(\frac{GM}{R_{st}}\right)^{1/2}$$

Furthermore, for a circular orbit $a \equiv R_{st}$ and for a stationary orbit $\tau = 23.93$ hr. The value of R_{st} may therefore be obtained by solving eq. (86). For this we will use feet as the length unit. Thus

$$(R_{st})^{3/2} = \frac{\tau(GM)^{1/2}}{2\pi} = \frac{23.93(3600)}{2\pi}[1.4604(10^{16})]^{1/2}$$

$$R_{st} = 1.4002(10^8)\text{ ft}$$

$$v_{st} = \left[\frac{1.4604(10^{16})}{1.4002(10^8)}\right]^{1/2} = 1.0213(10^4)\text{ ft/s}$$

These values are also the initial distance and speed for the actual orbit. Due to the 80° launch angle, the initial transverse speed is

$$(v_\phi)_1 = v_{st}\sin 80^\circ = 1.0057(10^4)\text{ ft/s}$$

Thus the value of h_O for the orbit is

$$h_O = R_1(v_\phi)_1 = 1.4002(10^8)(1.0057)(10^4)$$
$$= 1.4083(10^{12})\text{ ft}^2/\text{s}$$

and the corresponding value of E is

$$E = \frac{1}{2}(v_1)^2 - \frac{GM}{R_1} = \frac{1}{2}(1.0213)^2(10^8) - \frac{1.4604(10^{16})}{1.4002(10^8)}$$
$$= -5.215(10^7)\text{ ft}^2/\text{s}$$

We may now compute the eccentricity from eq. (82). This yields

$$e = \left[2(-5.215)(10^7)\left(\frac{1.4083}{1.4604}\right)^2(10^{-8}) + 1\right]^{1/2} = 0.17365$$

The orbit is elliptical because $e < 1$. The semimajor axis of this orbit is found from the first of eqs. (72) to be

$$a = \frac{h_O^2}{GM(1-e^2)} = \frac{(1.4083)^2(10^{24})}{(1.4604)(10^{16})[1-(0.17365)^2]}$$
$$= 1.4002(10^8)\text{ ft}$$

Finally, the period is found from eq. (86) to be

$$\tau = \frac{2\pi}{(1.4604)^{1/2}(10^8)}[(1.4002)(10^8)]^{3/2}$$
$$= 8.615(10^4)\text{ s} = 23.93\text{ hr}$$

The fact that a is the same as R_{st}, thus giving the same period for the stationary and elliptical orbits, is not coincidental. If eq. (82) were to be used to eliminate the eccentricity from the relation for a in eq. (72), the result would be an expression for a that depends only on the energy level E. This means that the period is a function only of the magnitude of the initial velocity, not its direction. Nevertheless, the elliptical orbit in this problem is not stationary. Why?

HOMEWORK PROBLEMS

12.291 The altitude of an artificial satellite at apogee is 1230 km and at perigee it is 330 km. Determine the period of the orbit.

12.292 A satellite is in an elliptical orbit around the earth whose eccentricity is 0.90. The perigee altitude is 540 miles. Determine the period of the orbit.

12.293 The eccentricity of the earth's orbit is 0.0167 and its mean distance from the sun (the average of the perigee and apogee distance) is $1.495(10^8)$ km. Taking the earth's period as 365.3 days, determine (a) the angular momentum per unit mass for the earth, and (b) the maximum and minimum distances from the earth to the sun. The mass of the sun is approximately $3.33(10^5)$ larger than the mass of the earth.

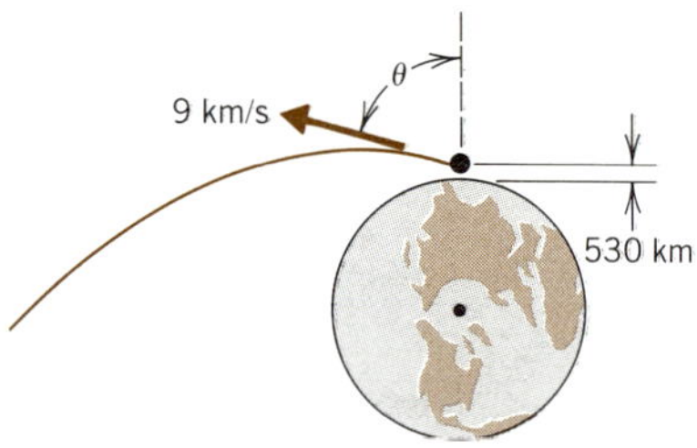

PROB. 12.294

12.294 When the engines of the launch vehicle of a satellite are shut off, the altitude is 530 km and the velocity is 9 km/s at $\theta = 90°$ from the outward radial line. Determine the period of the resulting orbit.

12.295 Solve Problem 12.294 if $\theta = 85°$ for the launch and all other parameters are unchanged.

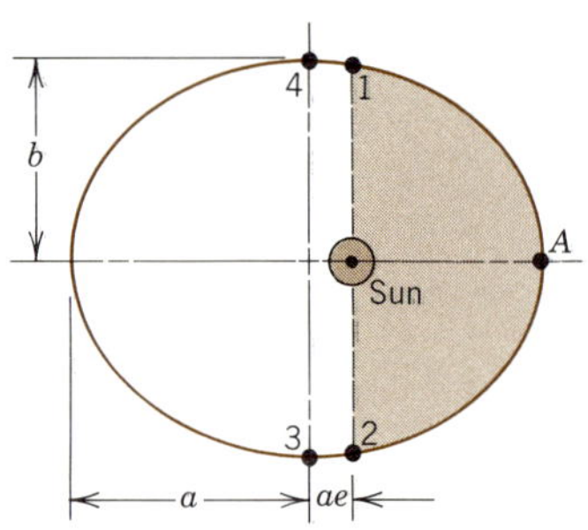

PROB. 12.296

***12.296** Determine the approximate time required for the earth to traverse from one end of the latus rectum of its orbit (point 1), through perihelion (point A) to the other end of the latus rectum (point 2) sweeping out the shaded area shown. The eccentricity of the earth's orbit is 0.0167. *Hint:* Approximate the shape of area 1-2-3-4 as a trapezoid. The mass of the sun is $3.328(10^5)$ times that of the earth.

***12.297** A satellite is in a circular orbit at an altitude of 1000 miles. The engines are fired tangentially at point A, placing the satellite into an elliptical orbit whose apogee is at point B, in order to place it in a new circular orbit at an altitude of 10,000 miles. Determine the length of time after the firing required for the satellite to reach apogee B on the transfer orbit.

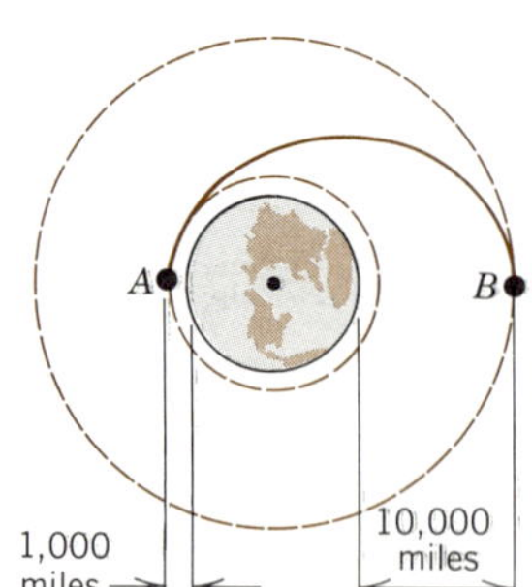

PROB. 12.297

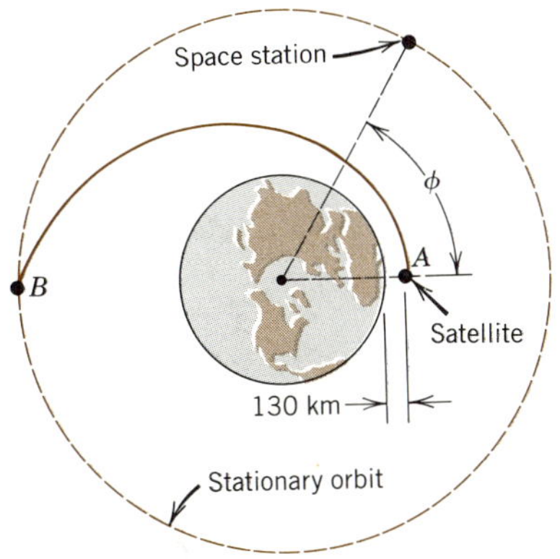

PROB. 12.298

***12.298** A satellite is destined for a rendezvous with a space station in a stationary orbit. The engines of the launch vehicle are shut down at point A, where the altitude is 130 km and the velocity is perpendicular to the radial line from the center of the earth. At what angle ϕ should the space station be when the engines of the satellite are turned off if rendezvous is to be accomplished at point B?

CHAPTER 13

KINEMATICS OF A RIGID BODY IN PLANAR MOTION

OBJECTIVES

1. To evaluate the velocity and acceleration of a point on a rigid body when the motion of another point and the angular motion of the body are given.

2. To identify points on a rigid body whose movement is constrained, and to write the corresponding constraint equations governing the velocity and acceleration of such points.

3. To use a rigid body's constraint equations to evaluate the velocity and acceleration of points in the body, as well as the angular velocity and angular acceleration of that body.

4. To employ the concept of an instantaneous center of zero velocity to relate the velocity of points in a rigid body to the angular velocity of the body.

5. To relate the angular motions of two bodies that roll over each other without slipping, and to use such relations to determine the motions of points on either body.

6. To determine the angular velocity and angular acceleration of each body in a linkage system, and to determine the velocity and acceleration of points on each of the links.

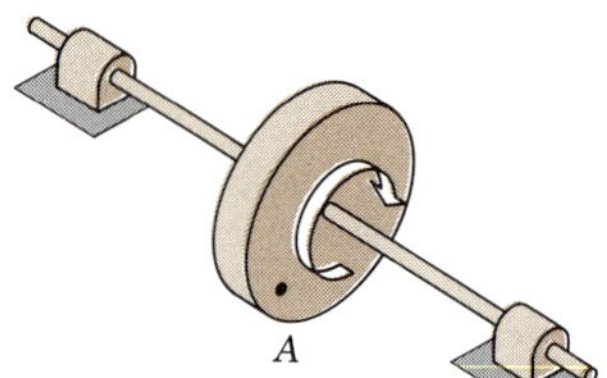

FIGURE 1

Now that we have completed a study of the dynamics of particles, we turn our attention to rigid bodies. Our aim in this chapter is to describe the movement of a rigid body that is executing *planar motion.* This type of motion occurs when the path of each point lies wholly in a plane, with the planar paths for all points in the body being parallel to each other. A simple example is a wheel mounted on a rotating shaft. The plane of the motion of any point in that system (say point A in Figure 1) is perpendicular to the axis of the shaft.

A. BASIC CONSIDERATIONS

1. The Rigid Body Model

The restriction to planar motion substantially simplifies the kinematical description in comparison to a general three-dimensional motion. (The latter is treated in Chapter 15.) One of the attributes of this simplification is that it will be possible to depict the motion by means of a planar diagram. Without loss of generality, we will always let the xy plane coincide with the plane of motion.

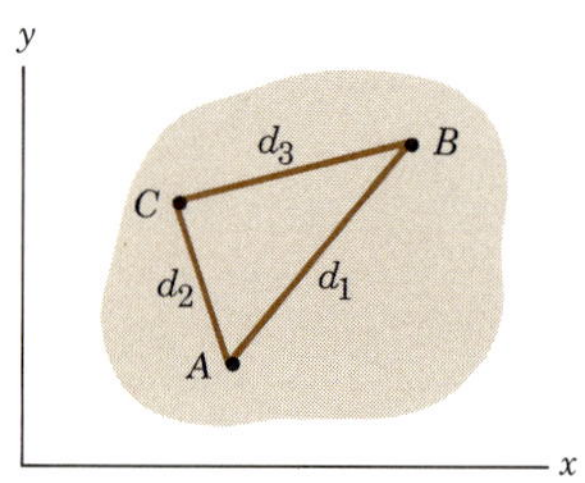

FIGURE 2

By definition, a rigid body is a system of particles that are always at a fixed distance from each other. This geometric restriction means that the movement of any point in the body relative to any other is limited. In order to discuss this matter, let us consider three arbitrarily selected points in the rigid body appearing in Figure 2.

The definition of rigidity requires that the distances d_1, d_2, and d_3 between points A, B, and C be constants. The consequence of this condition is that the shape of triangle ABC does not change, regardless of where or how the body moves. Preservation of the shape of this triangle, or any other geometric pattern formed by points in a rigid body, will enable us to form kinematical relations to describe the motion of the body.

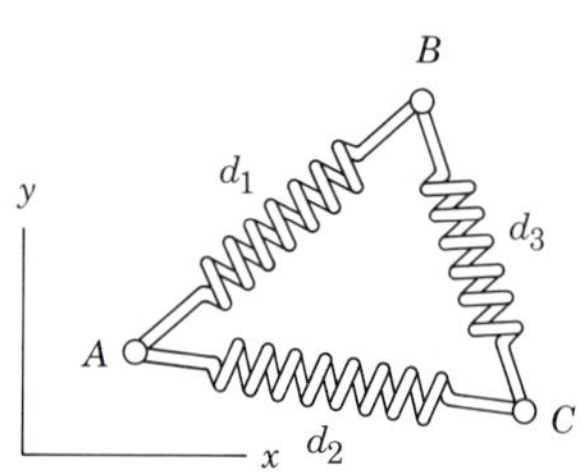

FIGURE 3

The existence of such kinematical relations is contrasted by the situation where particles in a system are not rigidly connected. For example, the three particles in Figure 3 are connected by springs. The distances d_1, d_2, and d_3 are variables because springs are deformable. The deformation of the springs depends on their axial forces. Hence, the shape of triangle ABC, or, equivalently, the relative position between the particles, can be ascertained only by studying the kinetics of the system.

The system in Figure 3 could be considered to be a rigid body if the stiffnesses of the springs were very large. In that case an axial force in the spring would produce a negligible deformation. In effect, we would then be saying that the springs act like rigid bodies.

From the foregoing we can surmise that the concept of a rigid body is a model that we create as an approximation. All materials deform when forces are applied to them. When we use a rigid body model, we are assuming that the movement of points due to deformation is negligible. The subject of the dynamics of deformable bodies, which is an advanced topic, is founded on many of the concepts developed in the study of rigid bodies.

2. Position

The triangle depicted in Figure 2 is useful for characterizing the kinematics of rigid body motion. Let us consider this triangle to be scribed onto a body. Thus, the length of each side is known, and the angles θ_A, θ_B, and θ_C formed by the sides, as depicted in Figure 4, are also known.

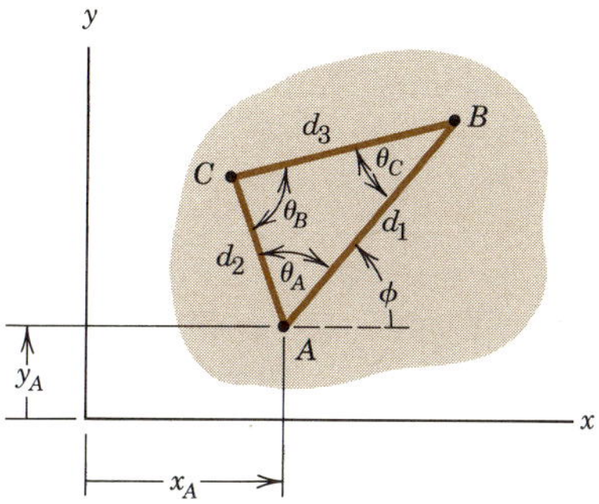

FIGURE 4

Let us now consider what parameters are needed to describe where triangle ABC is situated in space. At any instant, it would be sufficient to know the coordinates (x_A, y_A) locating point A, and the angle ϕ between the x axis and side AB. From this information and the known length d_1, the position of point B can be constructed. Because the triangle retains its shape, the other sides and point C may then be located without any additional information.

The same reasoning applies if the position of either point B or C, or the angular position of any other line, had been given. Because the points A, B, and C were arbitrarily selected, we may conclude that

> The position of a rigid body at any instant is defined by the absolute position of a point in the body and the angular orientation of any line in the body.

3. Representation of the Motion—Chasle's Theorem

Our knowledge of the information required to locate the position of a rigid body permits us to describe how the position changes with time. Figure 5 depicts the position of triangle ABC at an instant subsequent to the one in Figure 4. The original position of the triangle is indicated by the dashed lines in Figure 5.

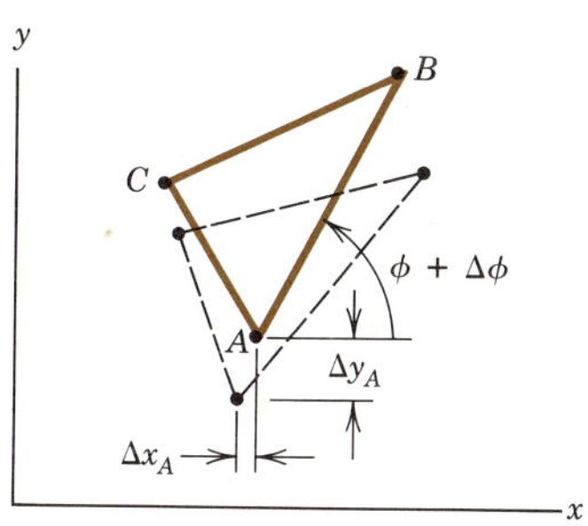

FIGURE 5

The distances Δx_A and Δy_A describe the movement of point A, and $\Delta\phi$ describes the angle of rotation of line AB. By the definition of a rigid body we know that the angles between all sides of the triangle are constant. It follows then that each line undergoes the same rotation. Hence

> Angular motion is an overall property of the motion of the body. It is the same regardless of which points in the body are being discussed.

Now, suppose that the movement of point A is specified. In addition to this movement, points B and C may move *relative to point A*. This relative

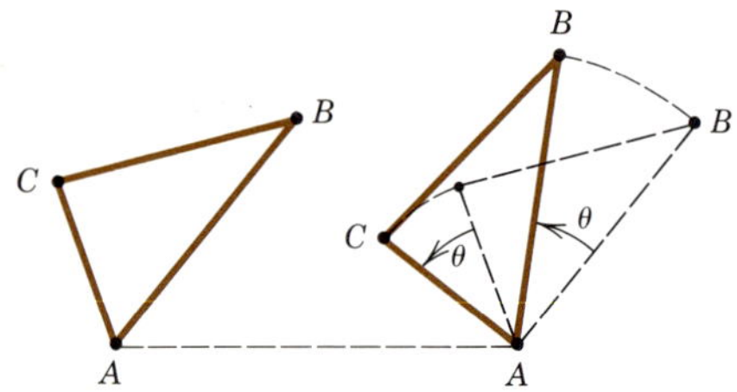

FIGURE 6

motion must be such that the distances d_1 and d_2 remain constant. In other words, each point moves in a circular path relative to point A. This situation is depicted in Figure 6, where A undergoes a motion to the right, and lines AB and AC rotate such that points B and C undergo additional motions *relative* to A. Furthermore, because the angular motion is an overall property, the radial lines from point A to points B and C undergo the same rotation, θ. The arbitrariness of the choice for points A, B, and C leads to the conclusion that

> The motion of a rigid body consists of a superposition of two movements. The first part consists of a movement of all points following the motion of an arbitrarily selected point in the body. The second movement is a rotation about the selected point in which all lines rotate by the same amount.

This statement is known as *Chasle's theorem.*

Special terms are used to describe the motion of rigid bodies. One simple type of motion occurs when every line in the body retains its original orientation—that is, there is no rotation. This is called a *translation.* Chasle's theorem indicates that all points in a translating body undergo the same movement. Figure 7 shows a typical translation. Notice that the points in the body may follow a curved path. It is a common misconception to think that translation is equivalent to rectilinear motion of each point in the body.

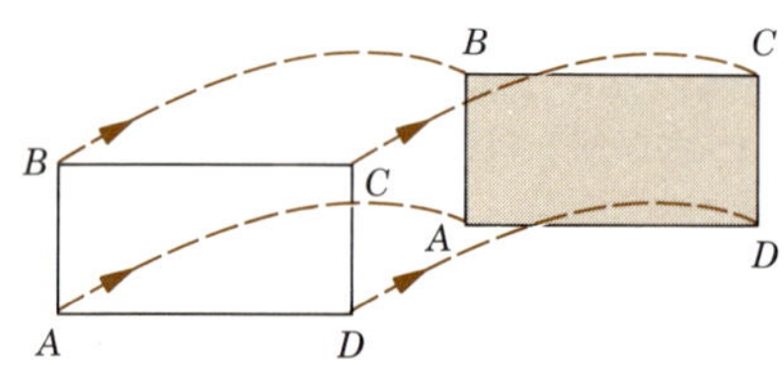

FIGURE 7

Another simple type of motion occurs when one point in the body is fixed in space. Chasle's theorem states that the motion of the body may be considered to consist solely of a rotation about the fixed point; this is called a *pure rotation.* Thus the body in Figure 8 is executing a pure rotation about point A, with the axis of rotation being the line that intersects point A perpendicular to the plane of motion.

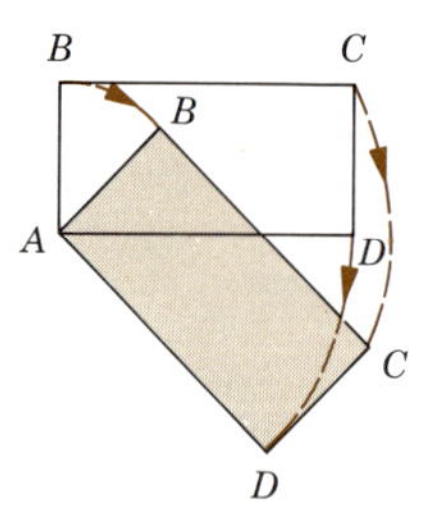

FIGURE 8

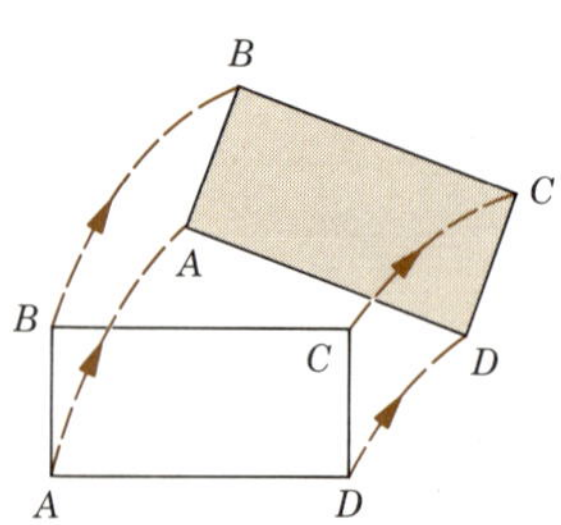

FIGURE 9

Chasle's theorem may be reworded to state that the motion of a rigid body is the superposition of a translation of an arbitrary point (on the body or its extension) and a pure rotation about that point. In a pure rotation the superposition keeps some point fixed. In contrast, no point remains fixed in a *general motion.* This was the case shown in Figure 6. Figure 9 also depicts a general case of planar motion.

The qualitative insight provided by Chasle's theorem will enable us to form algebraic relations for determining the velocities and accelerations of points on a rigid body. These kinematical properties are the topic of the next section.

B. KINEMATICAL EQUATIONS OF PLANAR MOTION

The important kinematic variables for rigid body motion are the same as those for particle motion, namely, position, velocity, and acceleration. Here, we will derive equations for these quantities that analytically express the superposition principle given by Chasle's theorem.

1. Basic Formulas

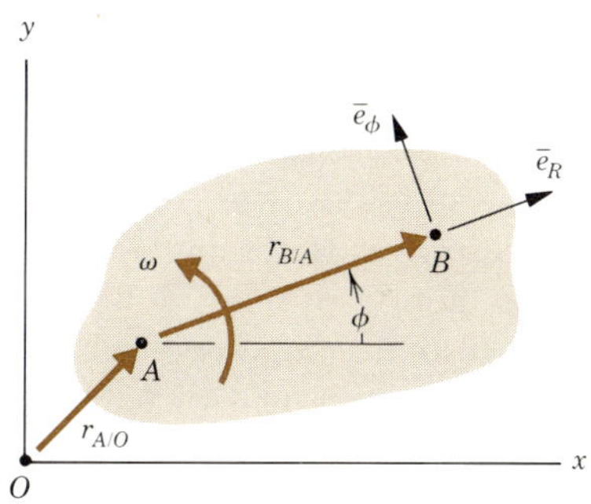

FIGURE 10

Suppose that the motion of point A in the rigid body shown in Figure 10 is known, as is the angular rotation of the body, ω. Point B is an arbitrarily chosen point whose motion we seek.

The discussion of Chasle's theorem indicated that point B moves in a circle relative to point A. Cylindrical coordinates having their origin at point A are useful for describing this relative motion. These are depicted in Figure 10. Point B is at a distance $r_{B/A}$ from point A, directed in the radial direction. Thus

$$\bar{r}_{B/A} = r_{B/A}\bar{e}_R \tag{1}$$

The radial line from A to B is rotating at the rate ω, that is,

$$\frac{d}{dt}(\phi) \equiv \dot{\phi} \equiv \omega \tag{2}$$

Note that ϕ describes the orientation of a specific radial line. However, ω is independent of the choice for points A and B, because Chasle's theorem states that the rotation is the same for all lines in a rigid body. The absolute value of ω is called the *angular speed* of the rigid body.

Because point B is moving in a circle relative to point A, the cylindrical coordinate formulas for velocity and acceleration give

$$\begin{aligned} \bar{v}_{B/A} &= r_{B/A}\omega\bar{e}_\phi \\ \bar{a}_{B/A} &= -r_{B/A}\omega^2\bar{e}_R + r_{B/A}\dot{\omega}\bar{e}_\phi \end{aligned} \tag{3}$$

The difficulty with these relations is that it will often be necessary to describe the kinematical variables in terms of their Cartesian (i.e., x and y) components. This type of operation was performed in Chapter 11, where $\bar{e}_R$ and $\bar{e}_\phi$ were expressed in terms of $(\bar{i}, \bar{j})$ components, and then substituted into the cylindrical component expressions. Such an operation is tedious, particularly if it must be performed often.

In order to circumvent this difficulty, we note that, for the unit vectors defined in Figure 10, $\bar{e}_\phi \equiv \bar{k} \times \bar{e}_R$. This fact, in conjunction with eq. (1), permits us to rewrite eqs. (3) as

$$\begin{aligned} \bar{v}_{B/A} &= r_{B/A}\omega(\bar{k} \times \bar{e}_R) \equiv \bar{\omega} \times \bar{r}_{B/A} \\ \bar{a}_{B/A} &= -r_{B/A}\omega^2\bar{e}_R + r_{B/A}\dot{\omega}(\bar{k} \times \bar{e}_\phi) \\ &\equiv -\omega^2\bar{r}_{B/A} + \bar{\alpha} \times \bar{r}_{B/A} \end{aligned} \tag{4}$$

where

$$\bar{\omega} = \omega \bar{k}, \qquad \bar{\alpha} = \dot{\omega} \bar{k} \tag{5}$$

The vector $\bar{\omega}$ is the *angular velocity* of the body, and $\bar{\alpha}$ is the *angular acceleration.* Both vectors are perpendicular to the plane of motion, so they are parallel to the axis for the rotational portion of the motion. As usual, the angular unit for $\bar{\omega}$ and $\bar{\alpha}$ must be radians.

The orientation of these angular motion vectors *must* be made to *agree with the right-hand rule.* The fingers of the right hand are curled in the sense (clockwise or counterclockwise) of the rotation of the body. The direction of $\bar{\omega}$ (either $+k$ or $-\bar{k}$) is then found by comparing the sense of the extended thumb to the sense of the positive z axis. The sense of $\bar{\alpha}$ is found by comparing it to $\bar{\omega}$. Both vectors have the same or opposite orientation, according to whether the rate of rotation is increasing or decreasing, respectively.

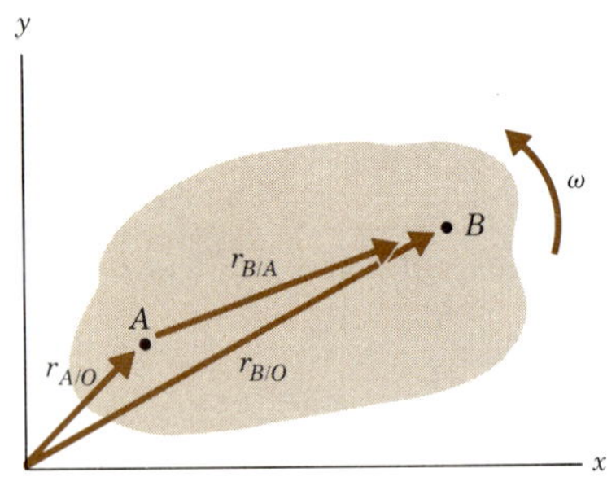

FIGURE 11

Equations (4) describe the relative motion of point B with respect to point A. The relationship between the absolute motions of points A and B is readily formed by referring to the position vectors depicted in Figure 11.

$$\bar{r}_{B/O} = \bar{r}_{A/O} + \bar{r}_{B/A} \tag{6}$$

$$\bar{v}_B = \bar{v}_A + \bar{v}_{B/A} = \bar{v}_A + \bar{\omega} \times \bar{r}_{B/A} \tag{7}$$

$$\bar{a}_B = \bar{a}_A + \bar{a}_{B/A} = \bar{a}_A + \bar{\alpha} \times \bar{r}_{B/A} - \omega^2 \bar{r}_{B/A} \tag{8}$$

The utility of writing the kinematical relations in the foregoing form is that they are readily formulated in terms of Cartesian components. Doing so requires that we select a convenient xyz coordinate system, with the z axis perpendicular to the plane of motion. Then all vectors must be expressed in terms the chosen coordinate system. The resulting eqs. (6)–(8) describe the position, velocity, and acceleration of a point in terms of x and y components.

EXAMPLE 1

At the instant shown collar B is moving down the guide rod at a speed of 20 ft/s, and speeding up at 10 ft/s^2. The bar is swinging counterclockwise at 4 rad/s, and slowing at 0.5 rad/s^2. Determine the horizontal and vertical components of velocity and acceleration of end A at this instant.

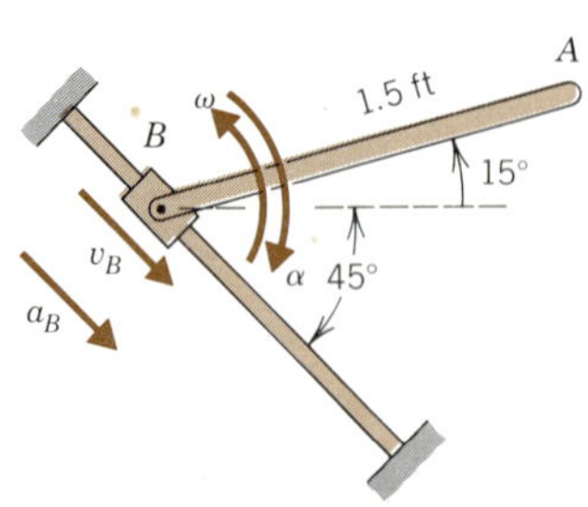

Solution

The xyz coordinate system selected in the sketch agrees with the type of components that have been requested. Equally important is the fact that it agrees with the configuration of the system

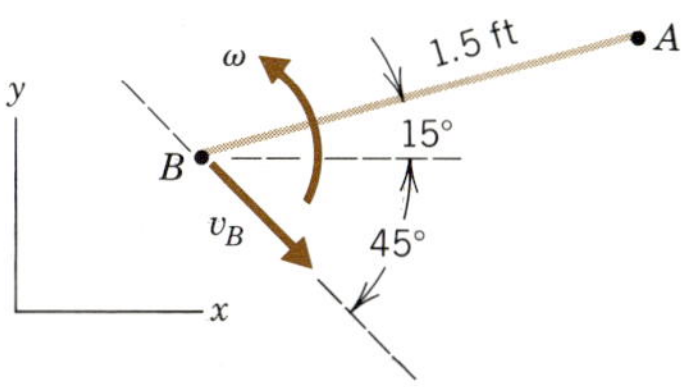

The given information is that

$$v_B = 20 \text{ ft/s}, \qquad \dot{v}_B = 10 \text{ ft/s}^2$$

$$\omega = 4 \text{ rad/s}, \qquad \dot{\omega} = -0.5 \text{ rad/s}^2$$

Point B is following a straight path, so

$$\bar{v}_B = v_B \bar{e}_t = 20(\cos 45^\circ \bar{i} - \sin 45^\circ \bar{j}) = 14.142\bar{i} - 14.142\bar{j} \text{ ft/s}$$

$$\bar{a}_B = \dot{v}_B \bar{e}_t = 10(\cos 45^\circ \bar{i} - \sin 45^\circ \bar{j}) = 7.071\bar{i} - 7.071\bar{j} \text{ ft/s}^2$$

The rotational speed of link AB is in the positive z direction according to the right-hand rule. Hence

$$\bar{\omega} = \omega\bar{k} = 4\bar{k} \text{ rad/s}, \qquad \bar{\alpha} = \dot{\omega}\bar{k} = -0.5\bar{k} \text{ rad/s}^2$$

Note that $\bar{\alpha}$ is directed oppositely from $\bar{\omega}$, as it should be when the rate of rotation is decreasing.

The movement of point B is specified. Hence, we shall consider this to be the reference point for the relative motion. The position of point A, whose motion we seek, is found from the sketch to be

$$\bar{r}_{A/B} = 1.5(\cos 15^\circ \bar{i} + \sin 15^\circ \bar{j}) = 1.4489\bar{i} + 0.3882\bar{j} \text{ ft}$$

Equation (7) then yields

$$\begin{aligned}\bar{v}_A &= \bar{v}_B + \bar{\omega} \times \bar{r}_{A/B} \\ &= (14.142\bar{i} - 14.142\bar{j}) + 4\bar{k} \times (1.4489\bar{i} + 0.3882\bar{j}) \\ &= 12.59\bar{i} - 8.35\bar{j} \text{ ft/s}\end{aligned}$$

Similarly, eq. (8) gives

$$\begin{aligned}\bar{a}_A &= \bar{a}_B + \bar{\alpha} \times \bar{r}_{A/B} - \omega^2 \bar{r}_{A/B} \\ &= (7.071\bar{i} - 7.071\bar{j}) + (-0.5\bar{k}) \times (1.4489\bar{i} + 0.3882\bar{j}) \\ &\quad - 4^2(1.4489\bar{i} + 0.3882\bar{j}) \\ &= -15.92\bar{i} - 14.01\bar{j} \text{ ft/s}^2\end{aligned}$$

HOMEWORK PROBLEMS

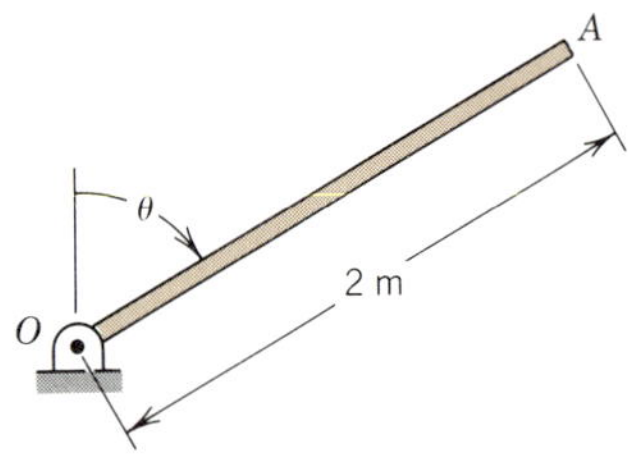

PROB. 13.1

13.1 Bar OA rotates in a plane about the fixed point O. At the instant where $\theta = 30^\circ$, the angular velocity is 4 rad/s clockwise and the angular acceleration is 6 rad/s^2 counterclockwise. Find (a) the velocity and (b) the acceleration of point A at this instant in terms of horizontal and vertical components.

13.2 The angle of rotation of the reel of a tape recorder is defined by $\theta = 1.2t + 0.01 \sin 30\pi t$, where θ is measured in radians and t in seconds. Determine the horizontal and vertical components of the acceleration of point A on the rim of the reel when (a) $t = 0$, (b) $t = 1/60$ s, and (c) $t = 1/45$ s.

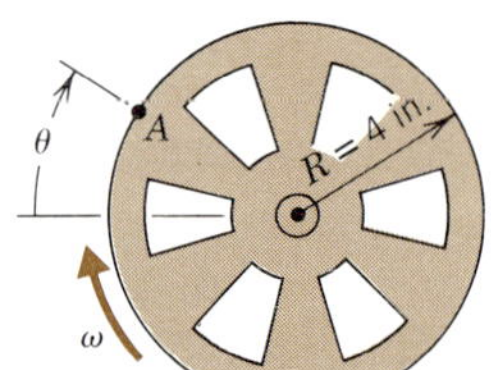

PROB. 13.2

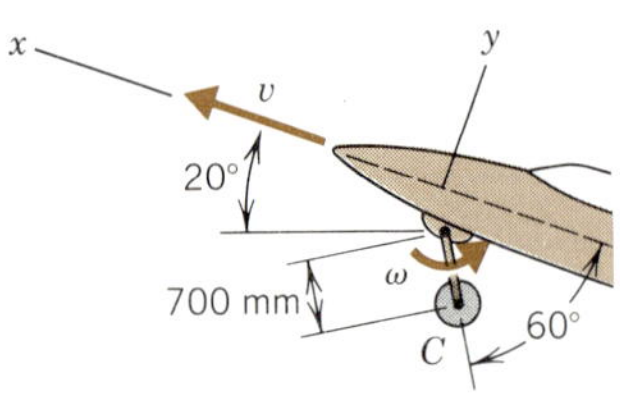

PROB. 13.3

13.3 The landing gear of an aircraft is being retracted with an angular speed of 2 rad/s and an angular acceleration of 5 rad/s^2 in the position shown. The aircraft has an instantaneous speed of 100 m/s, which is increasing at 5 m/s^2. The climb angle is constant at 20°. Determine the components of $\bar{v}_C$ and $\bar{a}_C$ with respect to the x and y axes given in the sketch.

13.4 Determine the horizontal and vertical components of the velocity of point C on the landing gear in Problem 13.3.

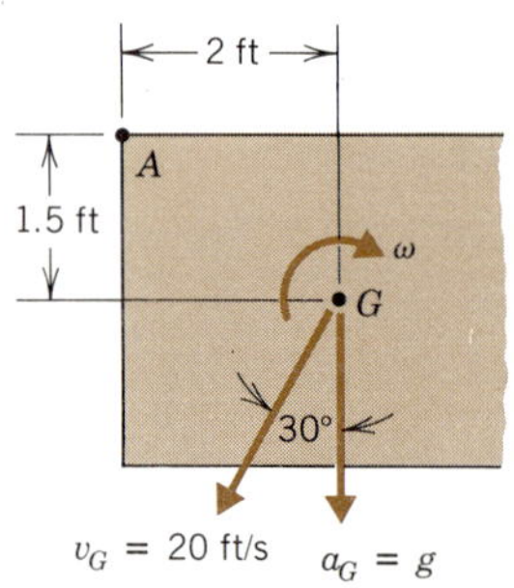

PROB. 13.5

13.5 The motion of the center of mass G of a freely falling box is as shown. The angular speed, which is 2 rad/s at this instant, is increasing at 5 rad/s^2. Determine the velocity and acceleration of corner A.

13.6 Collar A moves to the right at a constant speed of 8 m/s. In the position shown the angular speed of bar AB is $\omega = 8$ rad/s, and this value is decreasing at 50 rad/s^2. Determine the velocity and acceleration of end B.

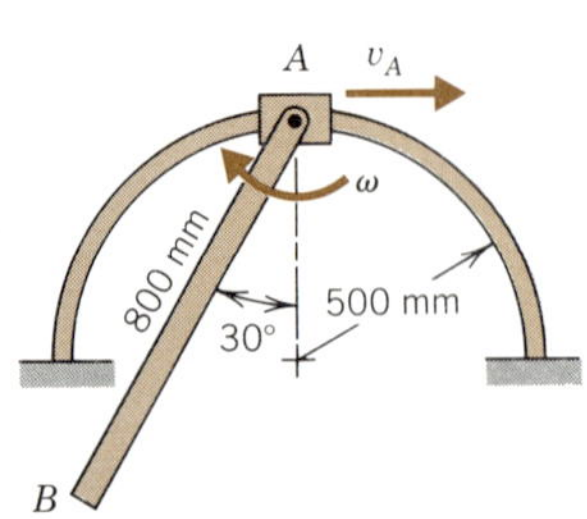

PROB. 13.6

PROB. 13.7

13.7 The truck shown in the sketch is at rest. The tailgate of the truck is lowered such that the vertical component of the velocity of its free end is constant at 5 m/s. Determine the angular velocity and angular acceleration of the tailgate for the instant when $\theta = \tan^{-1}(3/4)$.

13.8 Bar AB rotates counterclockwise at the constant rate of 90 rev/min. In the position shown, arm BC has an angular velocity of 10 rad/s and an angular acceleration of 20 rad/s^2, both clockwise. Determine the velocity of end C at this instant.

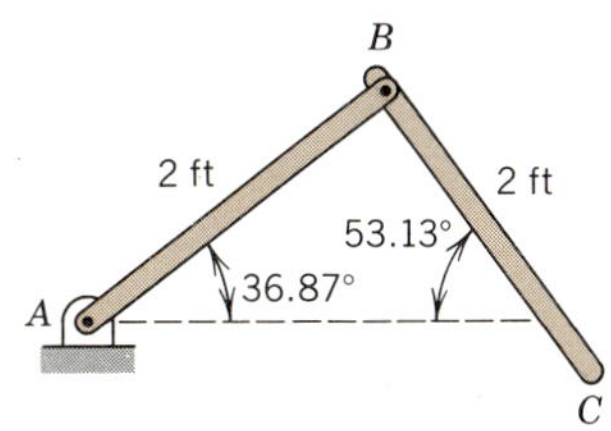

PROB. 13.8

13.9 Determine the acceleration of end C of bar BC in Problem 13.8.

***13.10** Bars AB and BC rotate in the vertical plane such that the angles γ and β are identical. Bar AB is rotating clockwise at the constant rate of 20 rev/min. Determine, for the instant when $\gamma = 36.87°$, (a) the angular velocity of bar BC, and (b) the velocity of end C.

***13.11** Consider the system of bars in Problem 13.10. For the instant when $\gamma = 36.87°$, determine (a) the angular acceleration of bar BC, and (b) the acceleration of end C.

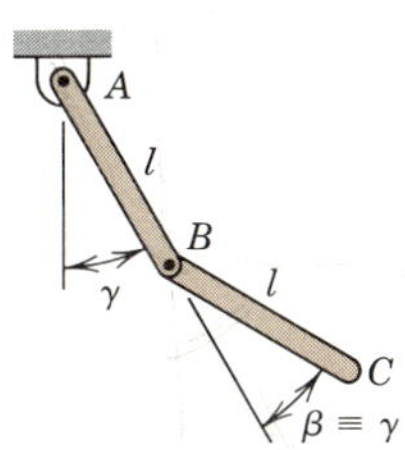

PROB. 13.10

2. Constrained Motion

Equations (7) and (8) were employed in the previous section in situations where the motion of a point and the angular motion of the body are specified. The unknown quantities in those cases were the velocity and acceleration of another point. This is a small part of the task in most practical problems.

The more typical situation is one in which the angular motion is not given. Instead, information about the movement of more than one point is known. In order to ascertain what kind of information might be provided, let us first consider the velocity equation.

$$\bar{v}_B = \bar{v}_A + \bar{\omega} \times \bar{r}_{B/A} \tag{9}$$

When points A and B are specified, $\bar{r}_{B/A}$ is a known vector. For planar motion, $\bar{v}_A$ and $\bar{v}_B$ have two components each, and $\bar{\omega}$ has one unknown component ($\bar{\omega}$ must be parallel to $\bar{k}$). This gives a total of five scalar variables at any instant. Equation (9) has two components. Matching like components on either side of the equality sign will yield *two scalar equations* for the aforementioned five variables.

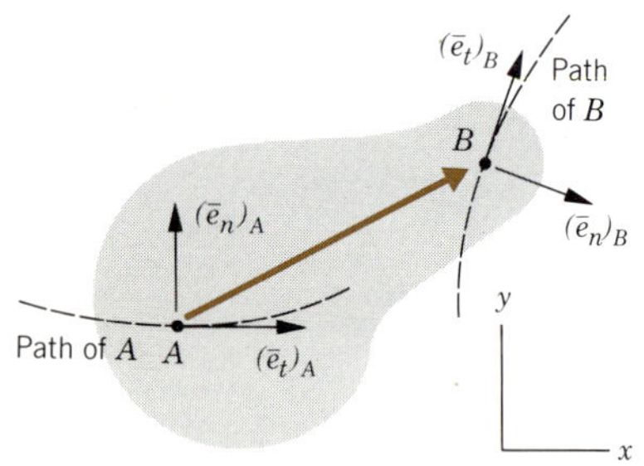

FIGURE 12

Let us start by considering a system containing only one rigid body. A typical problem in this case is to specify that two points on the body follow known paths. Such specifications are called constraint conditions, because they constrain (i.e., restrict) the freedom of points to move. A typical situation is depicted in Figure 12.

The path variable unit vectors for points A and B are known, because the instantaneous position of the points is known. The velocities may therefore be written as

$$\bar{v}_A = v_A(\bar{e}_t)_A, \qquad \bar{v}_B = v_B(\bar{e}_t)_B \tag{10}$$

These *constraint equations* are the mathematical statement of how the constraint conditions affect the velocities.

By constraining points A and B to follow specific paths, the speeds v_A and v_B are the only unknowns in their velocity. We have now reduced the total number of unknown variables from five to three: v_A, v_B, and ω.

It is necessary that one of these three quantities be specified. Doing so establishes the overall rate of movement of the rigid body. The scalar equations obtained from eq. (9), after substitution of the constraint eqs. (10), may then be solved for the two unknown variables.

The same reasoning applies for the acceleration. The basic relation is

$$\bar{a}_B = \bar{a}_A + \bar{\alpha} \times \bar{r}_{B/A} - \omega^2 \bar{r}_{B/A} \tag{11}$$

The constraint that both points follow their specified paths allows us to write their accelerations as

$$\bar{a}_A = \dot{v}_A(\bar{e}_t)_A + \frac{v_A^2}{\rho_A}(\bar{e}_n)_A, \qquad \bar{a}_B = \dot{v}_B(\bar{e}_t)_B + \frac{v_B^2}{\rho_B}(\bar{e}_n)_B \tag{12}$$

where the radii of curvature, like the unit vectors, are known in terms of the instantaneous positions.

Our standard practice is to solve first for all velocity variables. (This is much like evaluating the first derivative of a function prior to evaluating the second derivative.) Then only $\dot{v}_A$ and $\dot{v}_B$ remain as possible unknowns in the constraint eqs. (12) for $\bar{a}_A$ and $\bar{a}_B$. As we found for the velocities, constraining a point to follow a specific path leaves only one possible unknown in the acceleration of that point.

The angular speed ω is known from the solution to the velocity problem. Thus the only other possible unknown arising in eq. (11) is $\dot{\omega}$ (recall that $\bar{\alpha} \equiv \dot{\omega}\bar{k}$). When eqs. (12) are substituted into eq. (11), matching like components yields two scalar equations. It follows that one of the variables $\dot{v}_A$, $\dot{v}_B$ or $\dot{\omega}$ is required to establish the overall acceleration rate. The other two variables may then be determined by solving the component equations.

A final item to bear in mind is that the general procedure is to relate the motion of only the constrained points. Doing so allows us to determine fully the motion of one of these points, as well as the angular motion of the rigid body. The motion of other points in the body may then be determined directly from the basic eqs. (7) and (8).

EXAMPLE 2

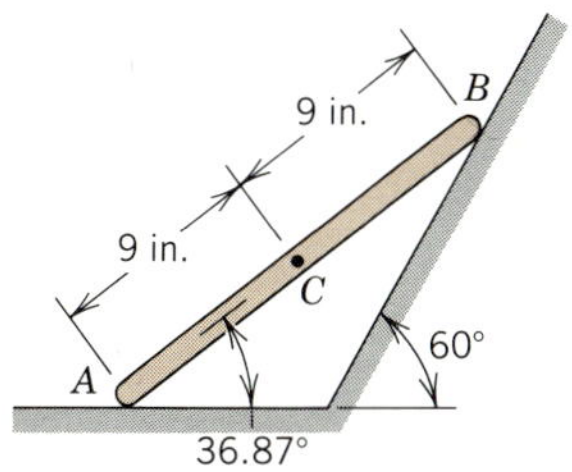

In the position shown end A of the bar is moving to the left at 5 ft/s. Determine the velocity of end B and of midpoint C of the bar.

Solution

The velocity of point A is given. Point B is another constrained point because it must follow the incline. The sketch depicts the unit vectors describing the straight path of each point for this instant, as well as the xyz coordinate axes we will employ. Note that the movement of point C is not guided. Because its path is not immediately known, we will consider it later.

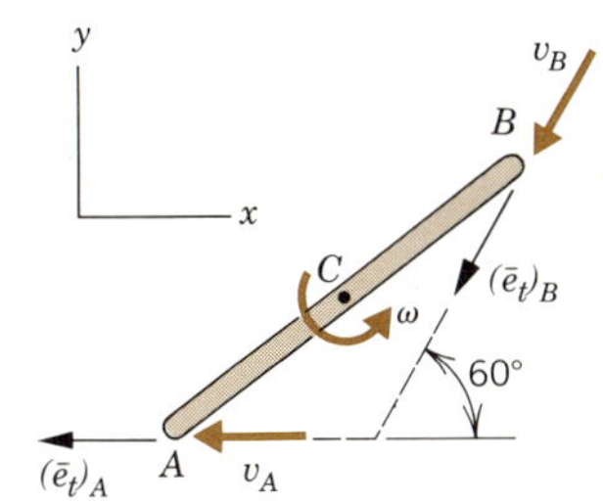

The sketch indicates that we are assuming that point B is moving down and that $\bar{\omega}$ is counterclockwise. It is difficult to guess the actual sense of these movements, but doing so is unimportant. An incorrect guess will merely result in a negative sign for the corresponding rate variable.

The constrained points A and B are related by

$$\bar{v}_B = \bar{v}_A + \bar{\omega} \times \bar{r}_{B/A}$$

In order to match components in this equation, all vectors must be described in terms of xyz components. The sketch allows us to write the constraint conditions for velocity as

$$\bar{v}_A = v_A(\bar{e}_t)_A = 5(-\bar{i}) \text{ ft/s}$$

$$\bar{v}_B = v_B(\bar{e}_t)_B = v_B(-\cos 60^\circ \bar{i} - \sin 60^\circ \bar{j}) = v_B(-0.50\bar{i} - 0.8660\bar{j})$$

For planar motion in the assumed sense of the rotation

$$\bar{\omega} = \omega\bar{k}$$

The position $\bar{r}_{B/A}$ may be obtained from the given sketch.

$$\bar{r}_{B/A} = \tfrac{18}{12}(\cos 36.87^\circ \bar{i} + \sin 36.87^\circ \bar{j}) = 1.20\bar{i} + 0.90\bar{j} \text{ ft}$$

We now substitute all of these vectors into the velocity equation, and then match like components.

$$v_B(-0.5\bar{i} - 0.8666\bar{j}) = -5\bar{i} + \omega\bar{k} \times (1.20\bar{i} + 0.90\bar{j})$$
$$= -5\bar{i} - 0.9\omega\bar{i} + 1.20\omega\bar{j}$$

$$\bar{i}\text{ component:}\quad -0.5v_B = -5 - 0.9\omega$$

$$\bar{j}\text{ component:}\quad -0.866v_B = 1.2\omega$$

$$v_B = 4.350\text{ ft/s},\qquad \omega = -3.139\text{ rad/s}$$

$$\bar{v}_B = 4.350(-0.5\bar{i} - 0.8660\bar{j}) = -2.175\bar{i} - 3.767\bar{j}\text{ ft/s}$$

$$\bar{\omega} = -3.139\bar{k}\text{ rad/s}$$

Now that $\bar{\omega}$ is known, $\bar{v}_C$ may be computed from

$$\bar{v}_C = \bar{v}_A + \bar{\omega} \times \bar{r}_{C/A}$$

where $\bar{v}_A$ was given and $\bar{r}_{C/A} = \frac{1}{2}\bar{r}_{B/A}$. Thus

$$\bar{v}_C = -5\bar{i} + (-3.139\bar{k}) \times (0.60\bar{i} + 0.45\bar{j})$$
$$= -3.59\bar{i} - 1.88\bar{j}\text{ ft/s}$$

EXAMPLE 3

Determine the angular acceleration of the bar in Example 2 if the speed of end A is decreasing at 10 ft/s^2.

Solution

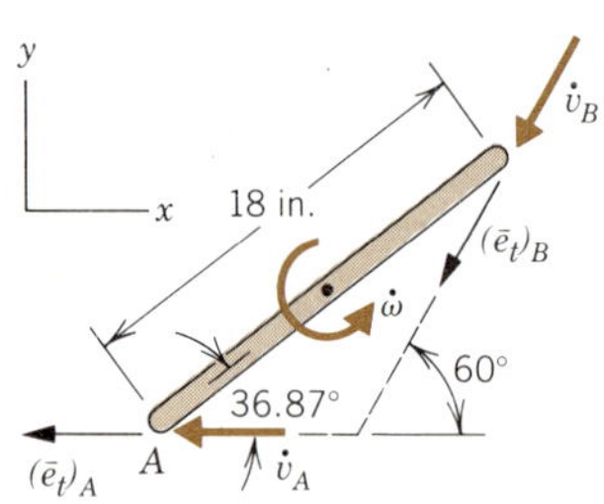

For convenience, we redraw the sketch of the system that was employed in Example 2. The angular and linear velocities were found there, so we proceed directly to the acceleration equation.

The motion of end A is given, and end B is constrained to follow the incline. The acceleration of these constrained points is related by

$$\bar{a}_B = \bar{a}_A + \bar{\alpha} \times \bar{r}_{B/A} - \omega^2\bar{r}_{B/A}$$

Both ends are following straight paths, so

$$\bar{a}_A = \dot{v}_A(\bar{e}_t)_A = (-10)(-\bar{i})\text{ ft/s}^2$$

$$\bar{a}_B = \dot{v}_B(\bar{e}_t)_B = \dot{v}_B(-0.5\bar{i} - 0.866\bar{j})$$

The values of ω and $\bar{r}_{B/A}$ were found in Example 2 to be

$$\omega = -3.139\text{ rad/s},\qquad \bar{r}_{B/A} = 1.20\bar{i} + 0.90\bar{j}\text{ ft}$$

Positive ω is considered to represent counterclockwise rotation, so

$$\bar{\alpha} = \dot{\omega}\bar{k}$$

Substitution of the foregoing expressions into the equation relating $\bar{a}_B$ and $\bar{a}_A$ gives

$$\dot{v}_B(-0.5\bar{i} - 0.866\bar{j}) = 10\bar{i} + \dot{\omega}\bar{k} \times (1.2\bar{i} + 0.9\bar{j}) - (-3.139)^2(1.2\bar{i} + 0.9\bar{j})$$
$$= 10\bar{i} - 0.9\dot{\omega}\bar{i} + 1.2\dot{\omega}\bar{j} - 11.824\bar{i} - 8.868\bar{j}$$

We now match like components in this equation and solve the resulting scalar equations.

$$\bar{i}\text{ component:}\quad -0.5\dot{v}_B = 10 - 0.9\dot{\omega} - 11.824$$

$$\bar{j}\text{ component:}\quad -0.866\dot{v}_B = 1.2\dot{\omega} - 8.868$$

$$\dot{v}_B = 7.373\text{ ft/s}^2, \qquad \dot{\omega} = 2.069\text{ rad/s}^2$$

$$\bar{\alpha} = 2.07\bar{k}\text{ rad/s}^2$$

EXAMPLE 4

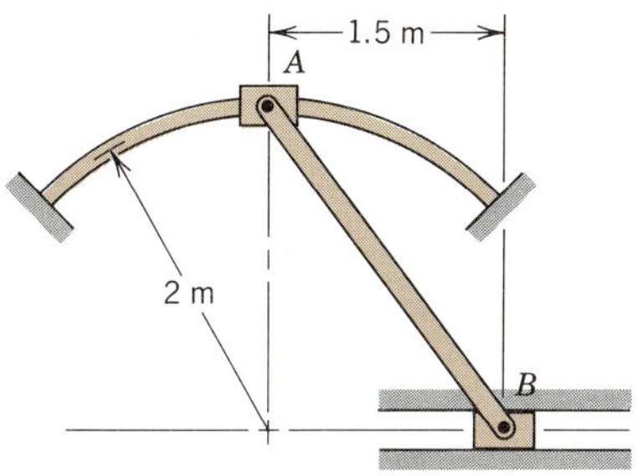

Collar A is moving to the right at the constant speed of 10 m/s. Determine the angular velocity and angular acceleration of rod AB in the position shown.

Solution

The movement of collar A is given and the path of collar B is known. These are the constrained points whose motion we will relate. The sketch shows the path variable unit vectors at the specified position.

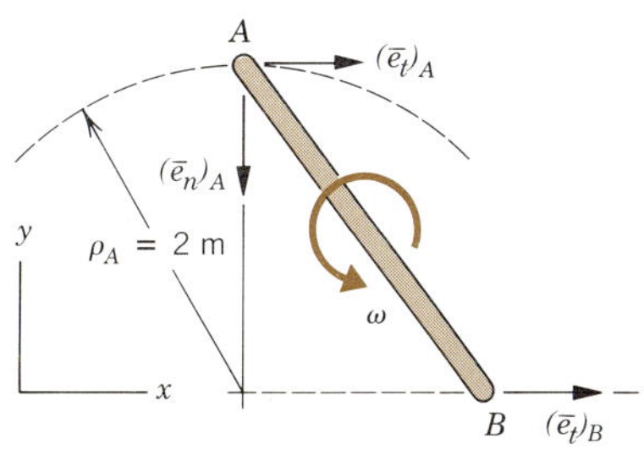

The velocity relationship is

$$\bar{v}_B = \bar{v}_A + \bar{\omega} \times \bar{r}_{B/A}$$

The speed of collar A is given, and the speed of collar B is denoted as v_B. We assume that the bar is rotating counterclockwise, and we use the path variable formulas to describe the velocities of the collars. Thus

$$\bar{v}_A = 10(\bar{e}_t)_A = 10\bar{i}\text{ m/s}, \qquad \bar{v}_B = v_B(\bar{e}_t)_B = v_B\bar{i}$$

$$\bar{\omega} = \omega\bar{k}, \qquad \bar{r}_{B/A} = 1.5\bar{i} - 2.0\bar{j}\text{ m}$$

Substituting these vectors into the equation relating $\bar{v}_A$ and $\bar{v}_B$ yields

$$v_B\bar{i} = 10\bar{i} + \omega\bar{k} \times (1.5\bar{i} - 2.0\bar{j}) = 10\bar{i} + 2\omega\bar{i} + 1.5\omega\bar{j}$$

$$\bar{i}\text{ component:}\quad v_B = 10 + 2\omega$$

$$\bar{j}\text{ component:}\quad 0 = 1.5\omega$$

The solution of the equations is

$$\bar{\omega} = \bar{0}, \qquad v_B = 10\text{ m/s}$$

The result that there is no angular velocity can be explained physically. Consider the motion of the bar as a superposition of a horizontal translation following end A and a rotation about end A. End B moves horizontally in the translation, which agrees with its constraint condition. In contrast, at this instant there can be no rotation about end A, because that would cause a movement of end B in a nonhorizontal direction.

Even though the angular velocity is zero and collar A has a constant speed, there is no reason to expect that the angular acceleration is zero. Indeed, we should expect $\bar{\alpha} \neq \bar{0}$, because collar A has a centripetal acceleration.

When $\bar{\omega} = \bar{0}$, the acceleration equation reduces to

$$\bar{a}_B = \bar{a}_A + \bar{\alpha} \times \bar{r}_{B/A}$$

For planar motion with the collars following their constrained motion, we have

$$\bar{a}_A = \dot{v}_A(\bar{e}_t)_A + \frac{v_A^2}{\rho_A}(\bar{e}_n)_A = \frac{10^2}{2}(-\bar{j})\ \text{m/s}^2$$

$$\bar{a}_B = \dot{v}_B(\bar{e}_t)_B = \dot{v}_B\bar{i}, \qquad \bar{\alpha} = \alpha\bar{k}$$

The acceleration relation then gives

$$\dot{v}_B\bar{i} = -50\bar{j} + \alpha\bar{k} \times (1.5\bar{i} - 2.0\bar{j}) = -50\bar{j} + 2\alpha\bar{i} + 1.5\alpha\bar{j}$$

$$\bar{i}\ \text{component:} \quad \dot{v}_B = 2\alpha$$

$$\bar{j}\ \text{component:} \quad 0 = -50 + 1.5\alpha$$

The angular acceleration is found by solving for α above.

$$\bar{\alpha} = \alpha\bar{k} = 33.3\bar{k}\ \text{rad/s}^2$$

HOMEWORK PROBLEMS

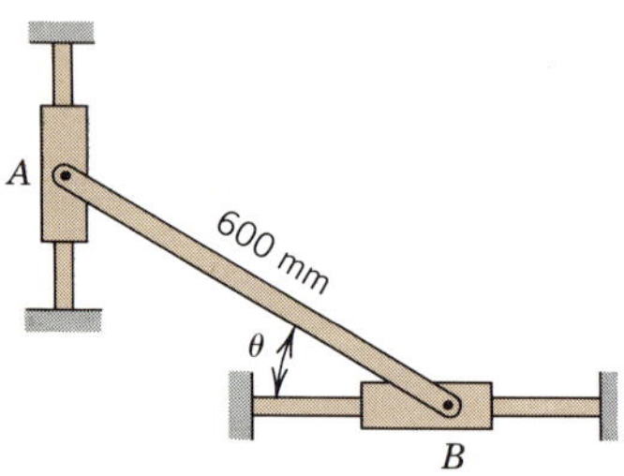

PROBS. 13.12 AND 13.13

13.12 At the instant when $\theta = 36.87^\circ$, collar A is moving downward at 25 m/s. Determine the velocity of collar B at this instant.

13.13 Collar A moves upward at a constant speed of 25 m/s. For the instant when $\theta = 60^\circ$, determine (a) the velocity of collar B, and (b) the acceleration of collar B.

13.14 When bar AB passes the horizontal position, slider B is moving upward through the slot at 4 ft/s, and it is slowing at the rate of 8 ft/s^2. Determine the corresponding (a) velocity of slider A, and (b) acceleration of slider A.

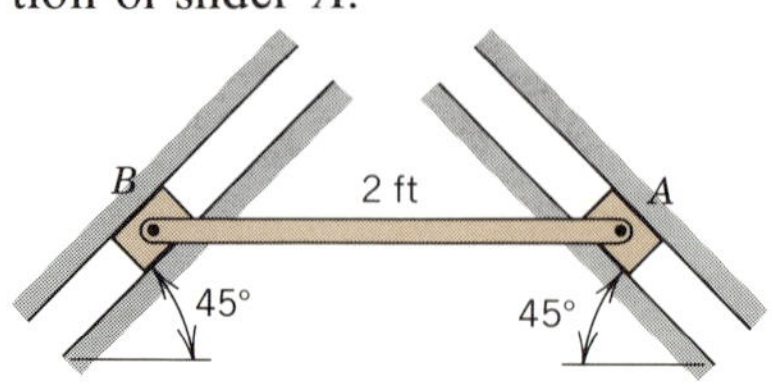

PROB. 13.14

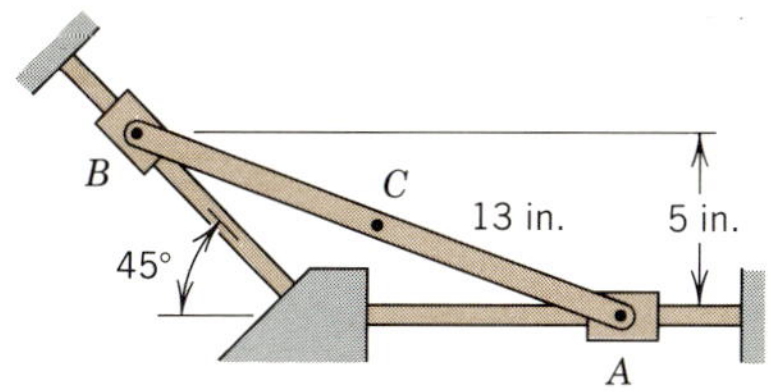

PROB. 13.15

13.15 In the position shown collar A at the end of the 13 in. rod is moving to the left at 300 in./s and is gaining speed at the rate of 100 in./s^2. Determine the corresponding velocities of (a) collar B and (b) midpoint C.

13.16 For the system in Problem 13.15, determine the accelerations of (a) collar B and (b) midpoint C.

13.17 Observation of the motion of the circular disk reveals the following information about the instantaneous velocity of points A, B, and C: $v_{Ax} = 20$m/s, $v_{By} = -15$ m/s, $v_{Cx} = 25$ m/s. Determine the angular velocity of the disk and the velocity of the midpoint O.

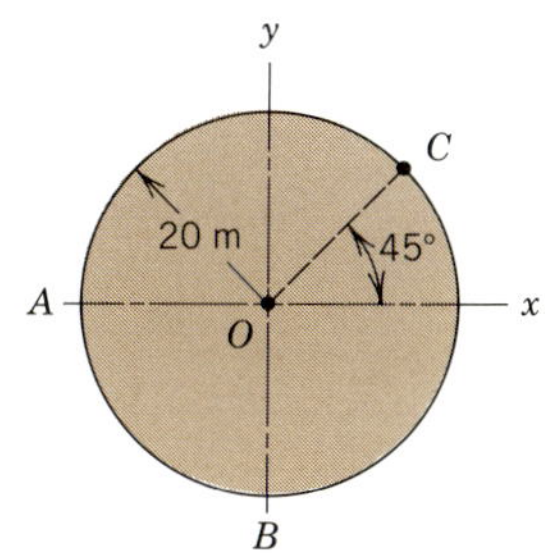

PROB. 13.17

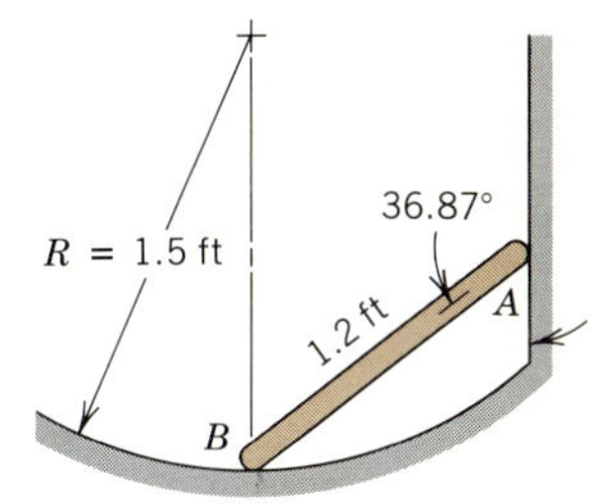

PROBS. 13.18 AND 13.19

13.18 End B of the rod is moving to the right at a constant speed of 10 ft/s. Determine the velocity of end A for the instant shown.

***13.19** In the position shown end A of the rod is moving upward at 8 ft/s and slowing at 4 ft/s^2. Determine the velocity and acceleration of end B at this instant.

13.20 The rigid link ABC is a portion of the control linkage in an automotive carburetor. At the instant shown, the control rod BE has a velocity of 100 mm/s and an acceleration of 40 mm/s^2, both to the left. Determine the velocity and acceleration of point C at this instant.

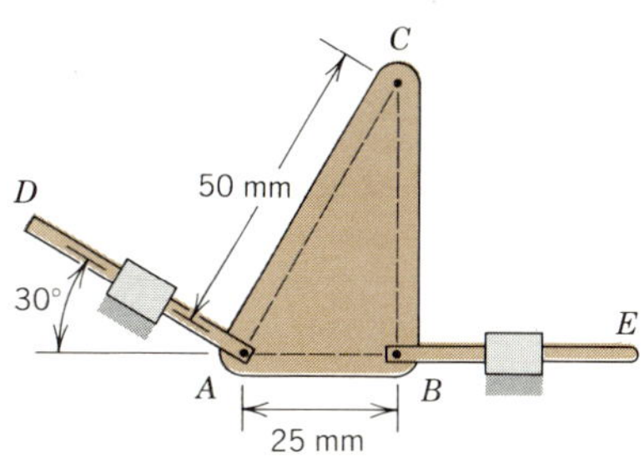

PROB. 13.20

13.21 In the position shown collar B is moving to the right and speeding up. Its horizontal velocity component is 25 ft/s, and its horizontal acceleration component is 10 ft/s^2. Determine the velocity of collar A in the position shown.

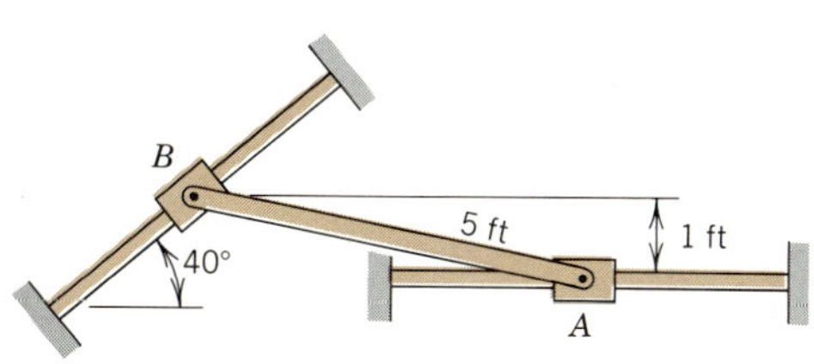

PROB. 13.21

***13.22** Determine the acceleration of collar A in Problem 13.21 at the instant shown.

***13.23** The box has just begun to move in the position shown, so the angular velocity is zero. The angular acceleration is known to be 3 rad/s^2 counterclockwise. Determine the corresponding accelerations of corners A and C. *Hint:* Align the x and y axes parallel to the edges of the box.

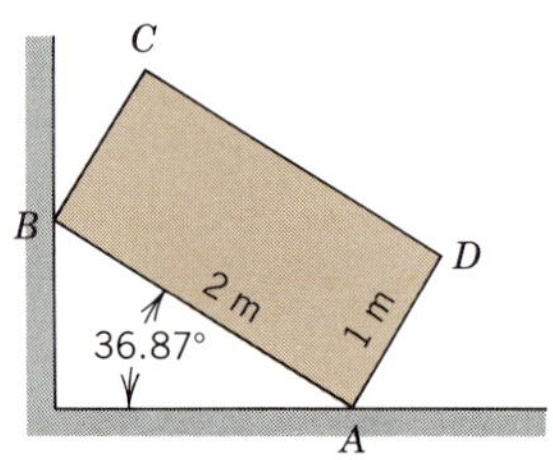

PROB. 13.23

***13.24** Solve Problem 13.23 if the angular velocity in the given position is 10 rad/s counterclockwise.

C. APPLICATION OF THE KINEMATICAL EQUATIONS

In this section we will consider three special applications of the kinematical equations. The results of these discussions will prove to be of great value in formulating many problems.

1. Instantaneous Center of Zero Velocity

We will demonstrate here that a rigid body undergoing a general planar motion has a point whose velocity is zero. Using this point as the reference point in Chasle's theorem will enable us to simplify the determination of the velocities of other points of the body.

Consider the rigid body of Figure 13. Let us assume that we have measured the velocities at A and B, hence $\bar{v}_A$ and $\bar{v}_B$ are known. We then can relate the velocity of either point A or B to any other point C. From eq. (7) we have

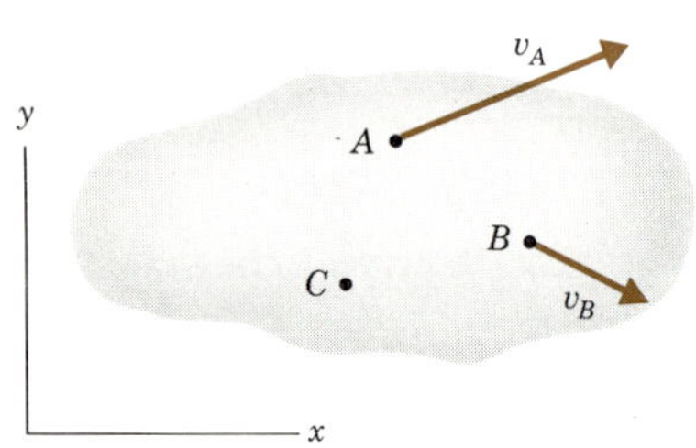

FIGURE 13

$$\bar{v}_A = \bar{v}_C + \bar{\omega} \times \bar{r}_{A/C} \tag{13}$$

$$\bar{v}_B = \bar{v}_C + \bar{\omega} \times \bar{r}_{B/C} \tag{14}$$

The problem we wish to explore is contained in two questions. What are the conditions for $\bar{v}_C = \bar{0}$, and how can point C be found? From eqs. (13) and (14) we see that for $\bar{v}_C = \bar{0}$,

$$\bar{v}_A = \bar{\omega} \times \bar{r}_{A/C}, \qquad \bar{v}_B = \bar{\omega} \times \bar{r}_{B/C} \tag{15}$$

The definition of the cross product states that $\bar{\omega} \times \bar{r}$ is perpendicular to both $\bar{\omega}$ and $\bar{r}$. Because $\bar{\omega}$ is perpendicular to the plane of motion, it must be that $\bar{\omega} \times \bar{r}$ is perpendicular to $\bar{r}$, and lies in the xy plane.

In other words, eqs. (15) indicate that the position vector $\bar{r}_{A/C}$ from the point of zero velocity to point A must be perpendicular to $\bar{v}_A$. Of course, a similar statement is true for point B. When we construct the perpendiculars to $\bar{v}_A$ and $\bar{v}_B$, they will intersect at point C, which is the only point common to both perpendiculars.

In Figure 14 we have shown this construction for the rigid body motion in Figure 13. The important conclusion is that the velocity formulas in eq. (15) resemble those for a body in pure rotation. It follows that, for the purposes of computing *velocity only*, we may consider the body to be rotating about point C.

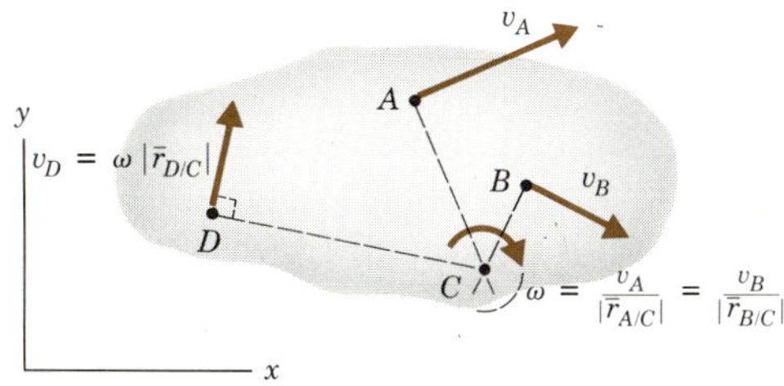

FIGURE 14

Let D be *any* point in the body. It must be that $\bar{v}_D$ is perpendicular to $\bar{r}_{D/C}$ in the sense of the rotation of the body; the speed of this point is

$$v_D = \omega|\bar{r}_{D/C}| \qquad \text{when } C \text{ is the instant center} \tag{16}$$

The construction of v_D, as well as the relationship between ω and the velocity of the other points, is depicted in Figure 14.

Point C has zero velocity only at the instant for which we have drawn the diagram, because at a later instant the location of points A and B, as well as the velocities of these points, will have changed. The velocity formulas, eqs. (15), resemble those for a body in pure rotation about point C. Therefore, we call point C an *instantaneous center of zero velocity*. This is frequently shortened to *instant center*.

These statements are true for a general planar motion of a rigid body. If the body is in pure rotational motion, the instant center is the fixed center of rotation, point O in Figure 15a. On the other hand, in a pure translation, all points in the body have the same velocity. Then $\bar{v}_A$ is parallel to $\bar{v}_B$, so that the perpendiculars to $\bar{v}_A$ and $\bar{v}_B$ do not intersect. Thus there is no instant center for a body in pure translation, as depicted in Figure 15b. Also note that frequently the instant center is not situated on the body. In such cases we can, if we wish, think of the instant center as lying on an imaginary extension of the body, as depicted in Figure 15c. Finally, we emphasize that although the instantaneous center can be located graphically, it can be located more exactly by trigonometry and geometry.

An obvious question to ask now is, can we use the instant center for determining accelerations? With the exception of pure rotation of a rigid body, the instant center is not fixed. Therefore the various points of the rigid body do not follow circular paths, and hence in general we *cannot* use the acceleration equations for circular motion about the instant center. Our approach for acceleration problems is to use the vector eq. (8).

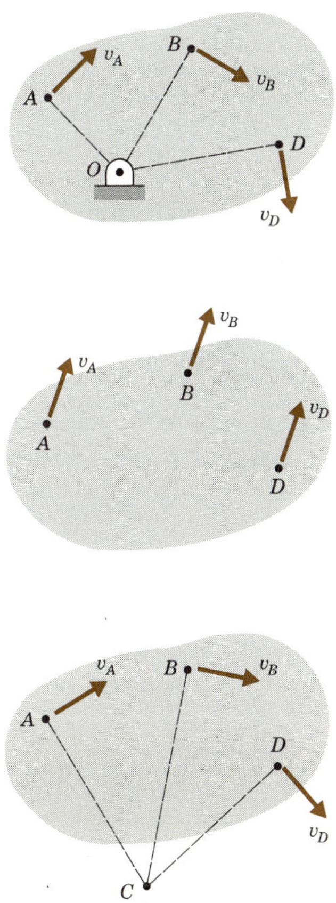

FIGURE 15

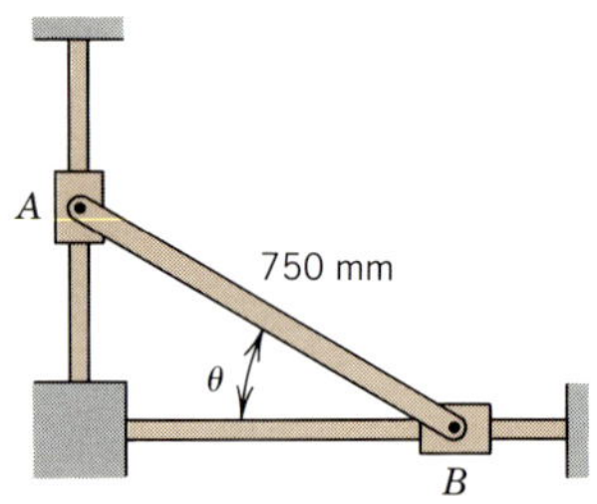

EXAMPLE 5

When $\theta = 30^\circ$, collar A is moving downward at 250 mm/s. Determine the corresponding velocity of collar B.

Solution

The first step is to make a sketch of the bar in the position when $\theta = 30^\circ$. In this sketch we draw an arrow representing $\bar{v}_A$, which is downward. We also know that $\bar{v}_B$ is horizontal, because that is the direction of the tangent to the path followed by collar B. We can see that collar B must move to the right when collar A moves down, so we depict $\bar{v}_B$ directed to the right. (The sense of an unknown velocity will not always be this obvious. In such cases a negative value for the speed will indicate that the actual sense of the velocity is opposite the assumed direction.)

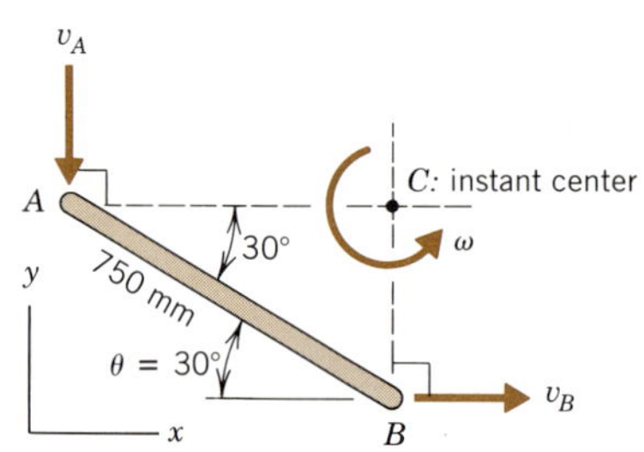

In order to locate the instant center, we draw dashed lines through points A and B, perpendicular to the velocity vector at each point. The point of intersection of these perpendiculars is the instant center C.

At this instant the bar seems to be rotating about point C, as depicted in the sketch. The sense of $\bar{\omega}$ shown in the sketch is consistent with the direction of the velocity vectors at points A and B.

For circular motion of each point centered at point C, it must be that the speed is the angular speed multiplied by the radial distance.

$$v_A = \omega|\bar{r}_{A/C}| \qquad v_B = \omega|\bar{r}_{B/C}|$$

It is given that $v_A = 0.25$ m/s, and the distances are found by solving triangle ABC. Thus

$$|\bar{r}_{A/C}| = 0.75 \cos 30^\circ = 0.6495 \text{ m}, \qquad |\bar{r}_{B/C}| = 0.75 \sin 30^\circ = 0.3750 \text{ m}$$

$$v_A = 0.25 = \omega(0.6495) \Rightarrow \omega = 0.3849 \text{ rad/s}$$

$$v_B = \omega(0.3750) = 0.1443 \text{ m/s}$$

The speed v_B is positive, in accord with our initial guess. The corresponding velocity is

$$\bar{v}_B = v_B(\bar{e}_t)_B = 0.1443\bar{i} \text{ m/s}$$

EXAMPLE 6

Use the method of instant centers to solve Example 2, in which the bar at the position shown is moving to the left at 5 ft/s. The velocities of points *B* and *C* are the desired results.

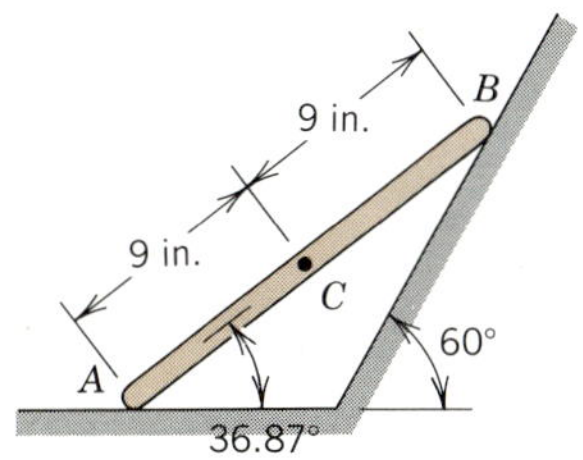

Solution

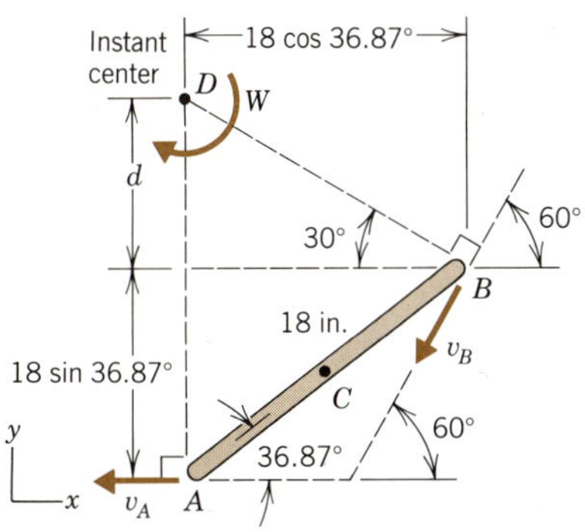

The sketch of the bar in the position of interest shows that $\bar{v}_A$ and $\bar{v}_B$ are parallel to the contacting surfaces. The lines through the constrained end points perpendicular to the velocities at these points intersect at the instant center *D*. Note that the movement of midpoint *C* is not constrained by other bodies, so we defer considering that point.

The lengths given in the sketch are the horizontal and vertical projections of the 18 in. length of bar *AB*, from which the vertical distance *d* is found to be

$$d = (18 \cos 36.87^\circ) \tan 30^\circ = 8.314 \text{ in.}$$

It is given that $v_A = 5(12)$ in./s, so setting the speed of each constrained point equal to ω times the radial distance gives

$$v_A = 60 = \omega|\bar{r}_{A/D}| = \omega(18 \sin 36.87^\circ + d) = \omega(19.114)$$

$$v_B = \omega|\bar{r}_{B/D}| = \omega\left(\frac{18 \cos 36.87^\circ}{\cos 30^\circ}\right) = 16.628\omega$$

$$\omega = 3.139 \text{ rad/s}, \qquad v_B = 52.20 \text{ in./s}$$

The value of v_B is positive, which confirms our assumption that v_B is downward. The corresponding velocity is

$$\bar{v}_B = v_B(\bar{e}_t)_B = 52.20(-\cos 60^\circ \bar{i} - \sin 60^\circ \bar{j})$$
$$= -26.1\bar{i} - 45.2\bar{j} \text{ in./s}$$

Now that ω is known, the velocity of point *C* may be determined. We could obtain v_C by multiplying ω by the radial distance from the instant center, that is, $v_C = \omega|\bar{r}_{C/D}|$. However, determining $|\bar{r}_{C/D}|$ involves the trigonometry of a scalene triangle. Expressing the direction of $\bar{v}_C$ is another complication.

An easier approach is to obtain the vector $\bar{v}_C$ directly from

$$\bar{v}_C = \bar{v}_D + \bar{\omega} \times \bar{r}_{C/D} \quad \text{where} \quad \bar{v}_D \equiv \bar{0}$$

The position $\bar{r}_{C/D}$ may be found from the instant center diagram by noting that point C is the midpoint of the bar. Hence it is situated midway both horizontally and vertically.

$$\begin{aligned}\bar{r}_{C/D} &= \tfrac{1}{2}(18 \cos 36.87°)\bar{i} - [d + \tfrac{1}{2}(18 \sin 36.87°)]\bar{j} \\ &= 7.20\bar{i} - 13.714\bar{j} \text{ in.}\end{aligned}$$

The angular rotation is shown in the diagram to be clockwise, which corresponds to the negative z direction by the right-hand rule.

$$\bar{\omega} = \omega(-\bar{k}) = -3.139\bar{k} \text{ rad/s}$$

This yields

$$\begin{aligned}\bar{v}_C &= (-3.139\bar{k}) \times (7.20\bar{i} - 13.714\bar{j}) \\ &= -43.0\bar{i} - 22.6\bar{j} \text{ in./s}\end{aligned}$$

The results obtained here are identical to those in Example 2, except for the difference in length units.

Additional Remarks

Examples 5 and 6 demonstrate that the instant center method becomes more involved in trigonometric calculations as the complexity of the geometry increases. Furthermore, a comparison of Examples 2 and 6, which solved the same problem, indicates that geometrical complexity is less of a difficulty for the vector equation method. Both methods are useful. However, it must be emphasized that the instant center method can be used only to relate velocities, whereas the vector equation method is suitable for both velocity and acceleration.

HOMEWORK PROBLEMS

13.25 Use the instant center method to solve Problem 13.12.

13.26 Two people are moving a table. Knowing that $\bar{v}_A$ is 15 in./s parallel to the diagonal line AC and that $\bar{v}_C$ is parallel to edge BC, as shown, determine the velocity that the person at corner B imparts to the table.

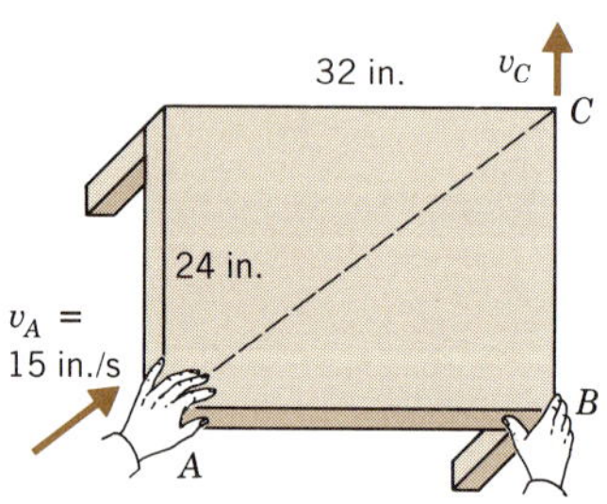

PROB. 13.26

13.27 In the position shown $v_B = 18$ m/s. Determine the corresponding (a) angular velocity of the bar, and (b) velocity of midpoint C.

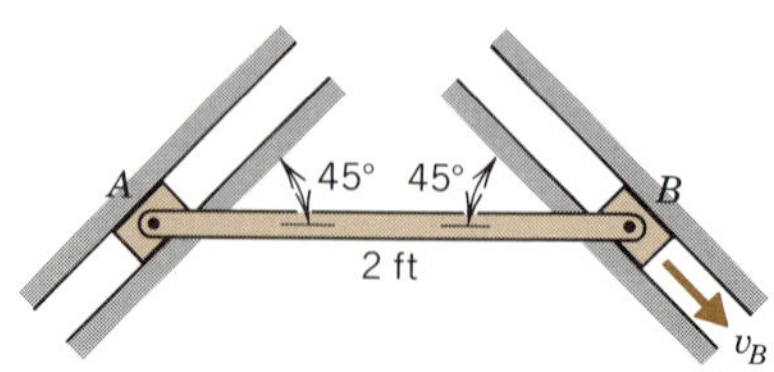

PROB. 13.27

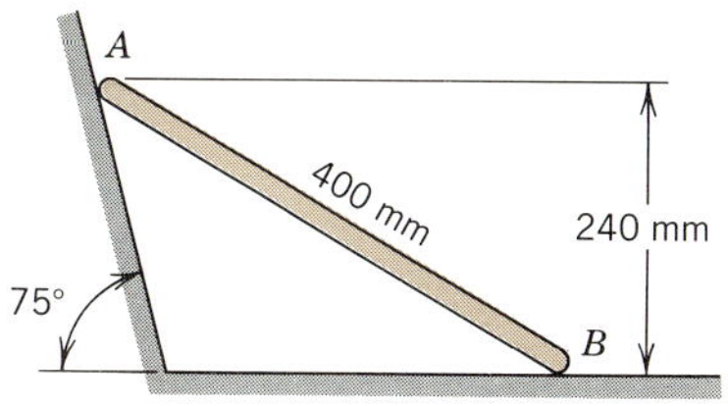

PROB. 13.28

13.28 End B of the bar is moving to the left at a speed of 16 m/s. Determine the angular velocity of the bar and the velocity of end A in the position shown.

13.29 Slider A moves to the left at a constant speed of 20 ft/s. For the instant when $\phi = 30°$, determine the angular velocity of the rod and the velocity of collar B.

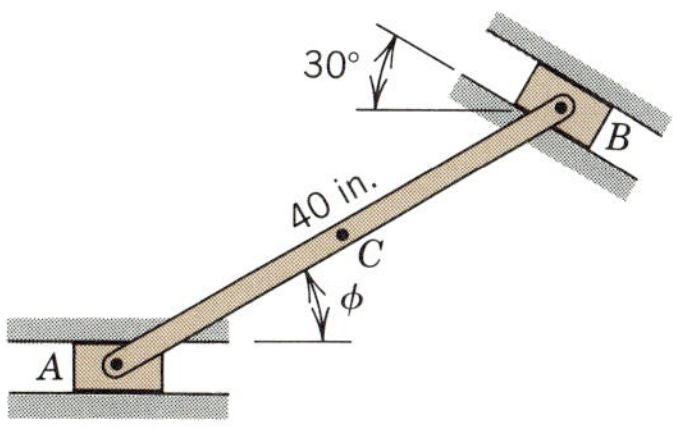

PROBS. 13.29–13.31

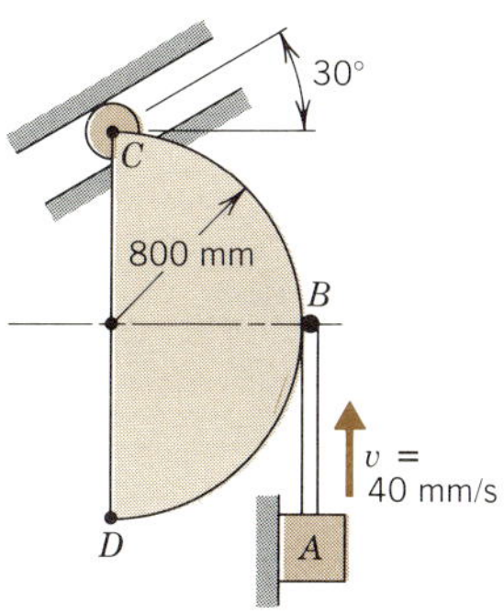

PROB. 13.32

13.30 Slider A moves to the left at a constant speed of 20 ft/s. Determine the velocity of midpoint C when the bar is at $\phi = 30°$.

13.31 Slider A moves to the left at a constant speed of 30 ft/s. Determine the velocity of midpoint C when the bar is at $\phi = 45°$.

13.32 The hydraulic piston at A is being extended at the rate of 40 mm/s. Determine the angular velocity of the semicircular plate and the velocity of point D at the instant shown.

13.33 Pushrod AB is moving to the right at a speed of 5 m/s. For the position shown determine (a) the velocity of pushrod CD, and (b) the velocity of end E of the angle bracket.

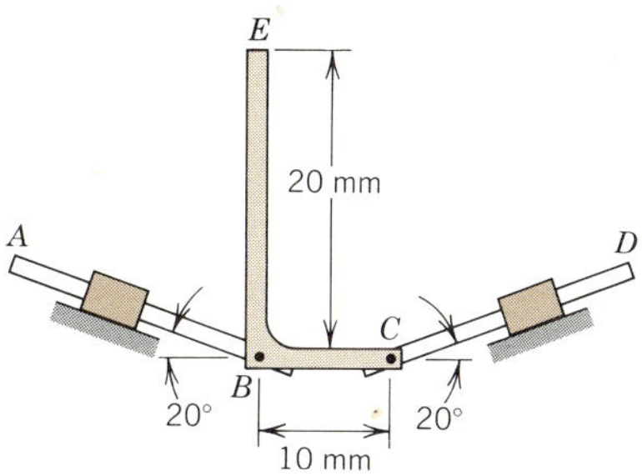

PROB. 13.33

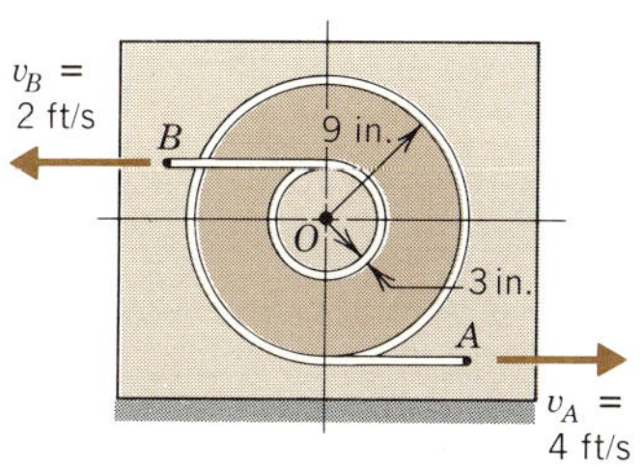

13.34 The stepped pulley is pinned at its center O to the sliding block. Two cords are wrapped around the pulley and pulled with constant velocities, as shown in the diagram. Determine (a) the location of the instant center of the pulley, (b) the angular velocity of the pulley, and (c) the velocity of the block.

PROB. 13.34

***13.35** The piston actuates the motion of the T-bar by giving follower C a horizontal velocity component v. Roller B has a velocity of 10 m/s upward when $\theta = 45^\circ$. Determine v.

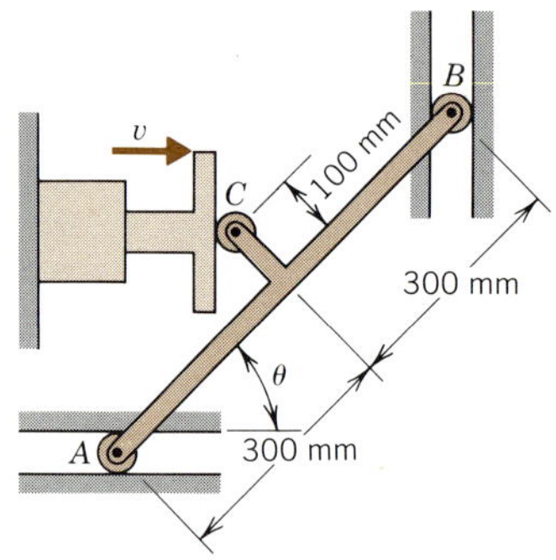

PROB. 13.35

***13.36** In Problem 13.35, roller B is moving upward at 8 m/s when $\theta = 53.13^\circ$. Determine the corresponding horizontal velocity component v.

---

2. Rolling Without Slipping

A common constraint on the motion of a body is to require that it roll without slipping. This condition pertains to the motion of surfaces that contact each other at a single location. A general situation of this kind involving two curved surfaces is depicted in Figure 16.

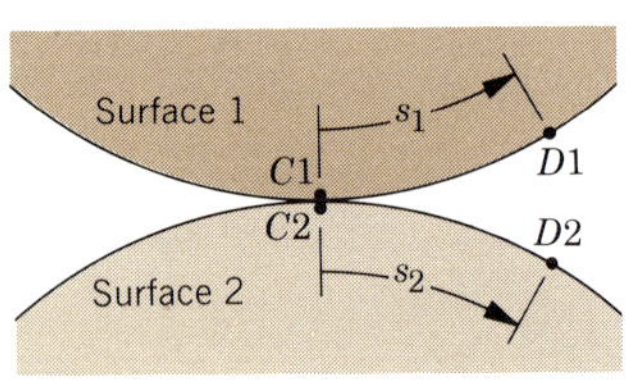

FIGURE 16

A *rolling* motion of these two surfaces is one in which the point of contact continuously changes. In Figure 16 points $C1$ and $C2$ are the points of contact on each surface at some instant, and points $D1$ and $D2$ are the corresponding points at a later instant. The arc lengths s_1 and s_2 measured along each surface are kinematically unrelated when there is slippage. In contrast, the condition that there is *no slipping* means that the arc lengths along each surface are equal.

$$s_1 \equiv s_2$$

This constraint condition may be used to relate the change in the position of rigid bodies as they roll without slipping. The relationship between the velocities of contacting pairs of points may be described with the aid of a coordinate system whose axes T and N in the plane of motion are tangent and normal, respectively, to the plane of tangency. This is depicted in Figure 17.

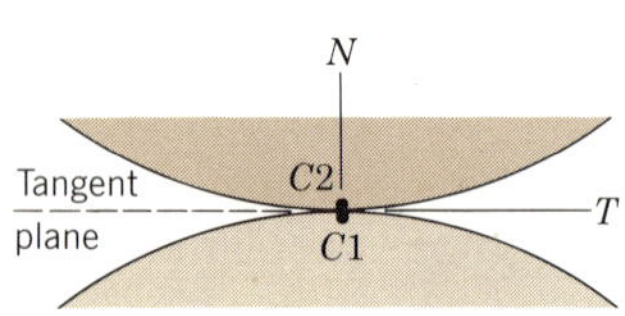

FIGURE 17

Rolling requires that the surfaces remain in contact. In that rigid bodies cannot penetrate each other, the contacting points must be moving at the same rate perpendicular to the tangent plane, $(v_{C1})_N = (v_{C2})_N$. Furthermore, if there is no slipping, then points $C1$ and $C2$ must also be moving at the same rate in the tangential direction, $(v_{C1})_T = (v_{C2})_T$. Because the corresponding components of $\bar{v}_{C1}$ and $\bar{v}_{C2}$ in both directions are equal, it follows that

$$\boxed{\bar{v}_{C1} = \bar{v}_{C2}} \tag{17}$$

Equation (17) is the constraint condition for velocities. The absence of slipping also constrains the accelerations. In the course of rolling, the points of contact $C1$ and $C2$ accelerate away from each other, so the $\bar{N}$ components of $\bar{a}_{C1}$ and $\bar{a}_{C2}$ will not be the same. However, points $C1$ and $C2$ must follow each other in the direction of the tangent plane. Thus the condition of no slip requires that

$$\boxed{(a_{C1})_T = (a_{C2})_T} \tag{18}$$

The most common engineering systems featuring rolling without slipping involve gears and wheels. The gear teeth prevent slipping, and friction tends to oppose slipping for wheels. The circular shape of these systems substantially simplifies the application of the constraint conditions.

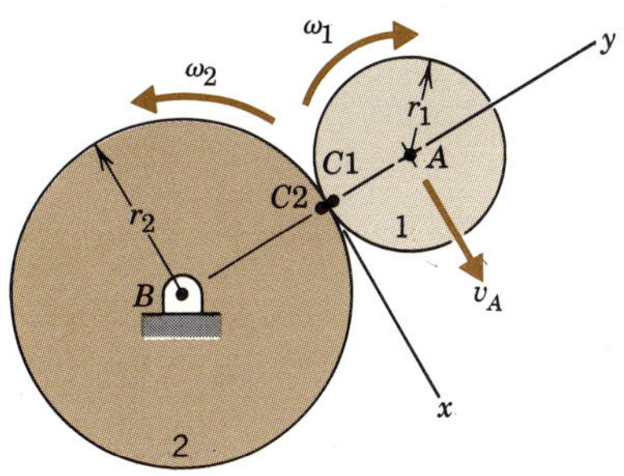

FIGURE 18

A general situation involving two disks is illustrated in Figure 18. Disk 2 is in pure rotation about its center B, and disk 1 is rolling without slipping over disk 2. The simplifications resulting from the circular geometry are:

1. The center A of gear 1 is constrained to follow a circular path. Therefore the motion of point A is fully known when the values of v_A and $\dot{v}_A$ have been determined.
2. The geometry of the system is constant. This means that the *algebraic form* of the kinematical relation between the speed v_A and the angular speeds ω_1 and ω_2 does not change. Such a relation may be differentiated in order to relate the acceleration rates $\dot{v}_A$, $\dot{\omega}_1$, and $\dot{\omega}_2$.

Items (1) and (2) are also valid for other cases of rolling without slipping, such as a disk rolling over the interior of a circular surface or a wheel rolling over a flat surface. In the latter case, the path of the center of the wheel is a straight line.

An important illustration of these ideas is a wheel rolling over the ground. The general features of this motion are treated in the next example.

EXAMPLE 7

The wheel is rolling over the ground without slipping. Its center is moving to the right with a speed v that is changing at the rate $\dot{v}$. Determine (a) the angular velocity and angular acceleration of the wheel, and (b) the velocity and acceleration of points A, B, and C on the perimeter of the wheel.

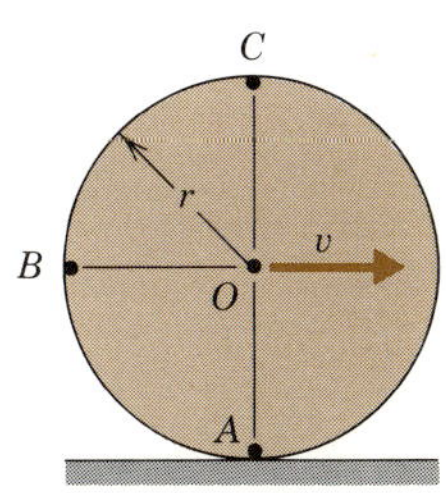

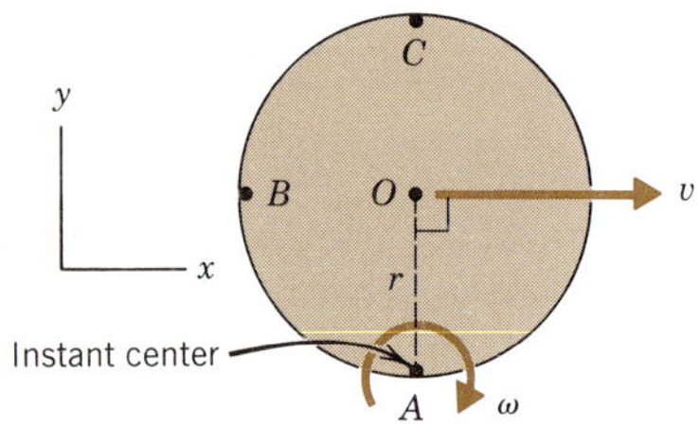

Solution

There is no slippage between the ground and the wheel, so the contact point A has no velocity, $\bar{v}_A = \bar{0}$. Point A is therefore the instant center, as shown in the sketch. Regarding the wheel to be rotating clockwise about the instant center at the angular speed ω enables us to write

$$v = \omega r \Rightarrow \bar{\omega} = \omega(-\bar{k}) = -\frac{v}{r}\bar{k}$$

At this juncture, we could evaluate the velocities of the other points on the perimeter by referring them to the instant center. However, accelerations are also needed, so it is preferable to proceed directly to the evaluation of $\bar{\alpha}$. For this, we note that the instant center diagram gives $v = \omega r$ for any position of the wheel. It is therefore a general relation that can be differentiated. Thus

$$\dot{v} = \dot{\omega} r \Rightarrow \bar{\alpha} = \dot{\omega}(-\bar{k}) = -\frac{\dot{v}}{r}\bar{k}$$

The kinematical vector equations involve the same procedure for velocities and accelerations. Hence, we shall employ that method to study points A, B, and C. We know that point O is following a straight path parallel to the x axis, so

$$\bar{v}_O = v\bar{i} \qquad \text{and} \qquad \bar{a}_O = \dot{v}\bar{i}$$

Let P denote an arbitrary point. With $\bar{\omega}$ and $\bar{\alpha}$ known, we have, in general,

$$\bar{v}_P = \bar{v}_O + \bar{\omega} \times \bar{r}_{P/O} = v\bar{i} + \left(-\frac{v}{r}\bar{k}\right) \times \bar{r}_{P/O}$$

$$\bar{a}_P = \bar{a}_O + \bar{\alpha} \times \bar{r}_{P/O} - \omega^2 \bar{r}_{P/O} = \dot{v}\bar{i} + \left(-\frac{\dot{v}}{r}\bar{k}\right) \times \bar{r}_{P/O} - \left(\frac{v^2}{r^2}\right)\bar{r}_{P/O}$$

We know that $\bar{v}_A = \bar{0}$, but calculating $\bar{v}_A$ from the above relation serves as a check. Setting $\bar{r}_{A/O} = -r\bar{j}$ yields

$$\bar{v}_A = v\bar{i} + \left(-\frac{v}{r}\bar{k}\right) \times (-r\bar{j}) \equiv \bar{0}$$

$$\bar{a}_A = \dot{v}\bar{i} + \left(-\frac{\dot{v}}{r}\bar{k}\right) \times (-r\bar{j}) - \left(\frac{v}{r}\right)^2(-r\bar{j}) = \frac{v^2}{r}\bar{j}$$

Another check obtained from the above is that $\bar{a}_A$ is perpendicular to the contact plane. This satisfies the no slip constraint on acceleration, eq. (17).

For points B and C, we have $\bar{r}_{B/O} = -r\bar{i}$, and $\bar{r}_{C/O} = r\bar{j}$. The results obtained from the above relations for $\bar{v}_P$ and $\bar{a}_P$ are

$$\bar{v}_B = v\bar{i} + v\bar{j}, \qquad \bar{a}_B = \left(\dot{v} + \frac{v^2}{r}\right)\bar{i} + \dot{v}\bar{j}$$

$$\bar{v}_C = 2v\bar{i}, \qquad \bar{a}_C = 2\dot{v}\bar{i} - \frac{v^2}{r}\bar{j}$$

Additional Remarks

Further consideration of the wheel in Example 7 provides additional insight into rolling without slipping. In the solution we related the motion of other points to that of the center O. According to Chasle's theorem, this is equivalent to regarding the motion of the wheel to be a superposition of a pure translation following point O and a pure rotation about point O. The superposition of velocity and acceleration are sketched in Figure 19.

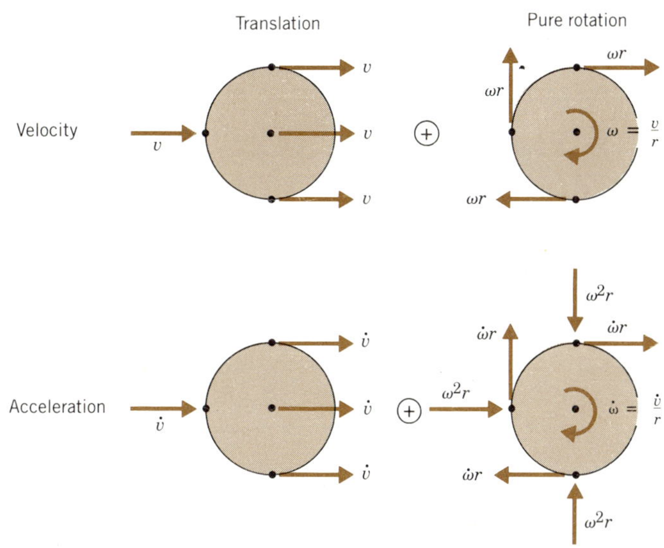

FIGURE 19

These superpositions agree with the results we obtained. Comparing the velocity and acceleration of the contact point in the translational and rotational parts of the motion shows that the two effects cancel each other parallel to the (horizontal) tangent plane, as required by the constraint equations.

Another item of interest for a rolling wheel is the path followed by a point on its perimeter, such as point A in the previous example. Do you remember that this path is a *cycloid*? The cusps in the cycloidal path in Figure 20 are the points in space where point A is in contact with the ground. The distance between cusps is the circumference of the wheel, because the arc lengths measured along each surface must be equal.

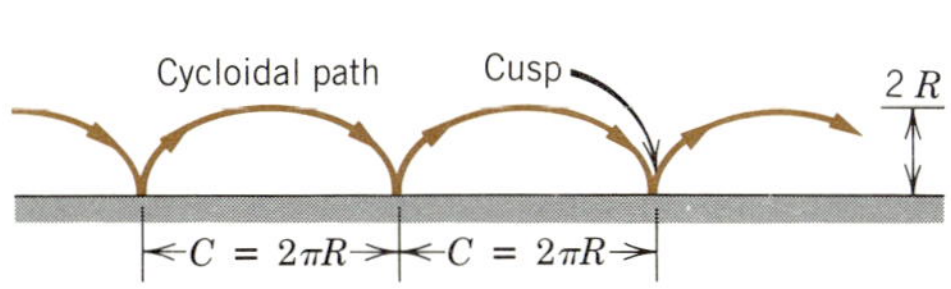

FIGURE 20

When point A approaches a cusp in its path, its velocity is downward and its speed is decreasing. After passing the cusp the velocity has reversed direction and the speed increases. Thus at the cusp, point A has zero velocity and its acceleration is upward, as we already determined in the solution to Example 7.

EXAMPLE 8

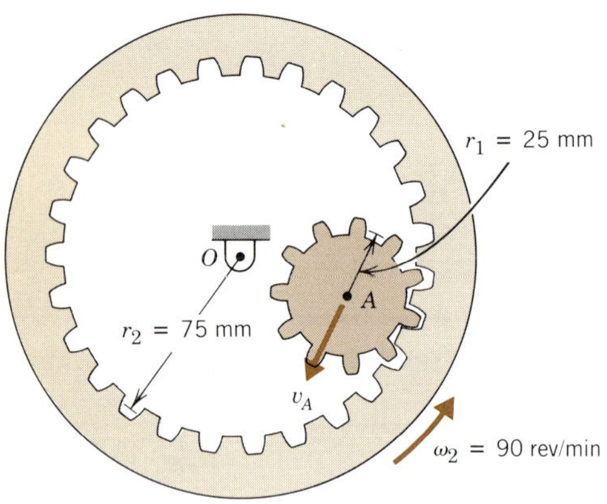

The ring gear is rotating counterclockwise about pin O at the constant rate $\omega_2 = 900$ rev/min. The center A of the inner gear is descending such that $v_A = 10$ m/s, $\dot{v}_A = 5$ m/s^2. Determine the angular velocity and angular acceleration of the inner gear.

Solution

In order to avoid errors in units, express ω_2 in radians per second. Thus

$$\omega_2 = 90\left(\frac{2\pi}{60}\right) = 30\pi \text{ rad/s}$$

No point on the inner gear is an obvious instant center. For this reason, we will obtain the algebraic relation between v_A and the angular speed by the vector equation method.

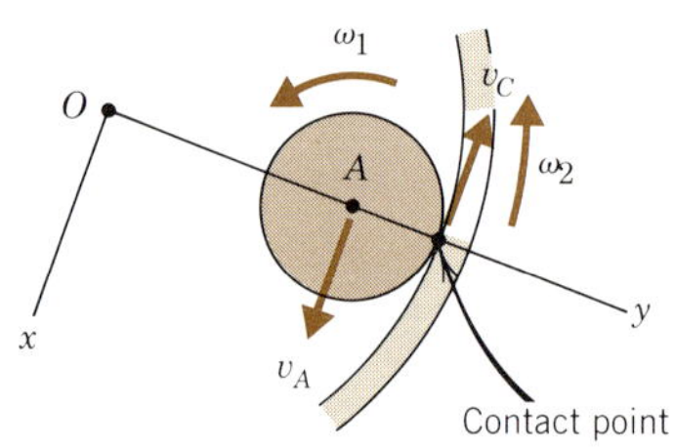

In the sketch the y axis is selected to align with the center of each gear. Doing so makes the x axis parallel to the plane of tangency.

The constrained points on the smaller gear are the point of contact C and the center A. In algebraic terms, we know that $v_C = \omega_2 r_2$, in the negative x direction, because the outer gear is in a pure rotation about point O. We also know that point A follows a circular path, so $\bar{v}_A = v_A \bar{i}$. We assume that the angular speed of gear 1 is counterclockwise, so $\bar{\omega}_1 = \omega_1 \bar{k}$. The velocities of the two constrained points are therefore related by

$$\bar{v}_C = \bar{v}_A + \bar{\omega}_1 \times \bar{r}_{C/A}$$

$$\omega_2 r_2(-\bar{i}) = v_A \bar{i} + \omega_1 \bar{k} \times (r_1 \bar{j}) = (v_A - \omega_1 r_1)\bar{i}$$

$$\bar{i}\text{ component:}\quad -\omega_2 r_2 = v_A - \omega_1 r_1 \Rightarrow \omega_1 = \frac{1}{r_1}(v_A + \omega_2 r_2)$$

The foregoing equation is always valid for this system, so it may be differentiated. Thus

$$\dot{\omega}_1 = \frac{1}{r_1}(\dot{v}_A + \dot{\omega}_2 r_2)$$

The given rates for this system then yield

$$\omega_1 = \frac{1}{0.025}[10 + 30\pi(0.075)] = 682.7 \text{ rad/s}$$

$$\dot{\omega}_1 = \frac{1}{0.025}[5 + (0)] = 200 \text{ rad/s}^2$$

$$\bar{\omega}_1 = \omega_1 \bar{k} = 683\bar{k} \text{ rad/s}$$

$$\bar{\alpha}_1 = \dot{\omega}_1 \bar{k} = 200\bar{k} \text{ rad/s}^2$$

HOMEWORK PROBLEMS

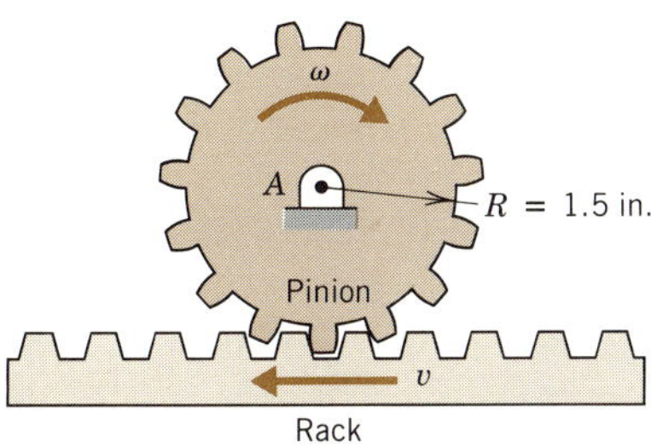

PROB. 13.37

13.37 A portion of the steering gear of an automobile consists of the rack and pinion set shown in the sketch. The pinion rotates about a fixed axis passing through point A, and the rack is constrained to translate only. If the pinion is rotating at 0.5 rev/s clockwise and this rate is increasing at 4 rev/s^2, determine the velocity and acceleration of the rack.

13.38 The wheel rolls without slipping to the left. The speed of its center is constant at $v_O = 900$ mm/s. Determine the velocity and acceleration of points A, B, and C.

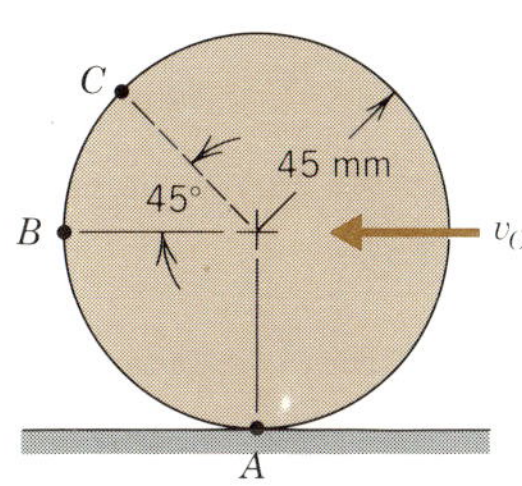

PROB. 13.38

13.39 Solve Problem 13.38 if the given value of v_O is increasing at 18 m/s^2.

13.40 The cylindrical tank is rolling backward without slipping on the flatbed of a truck. The center of the tank has a velocity of 10 ft/s to the left, and the truck has a forward speed of 30 ft/s. Determine the angular velocity of the tank.

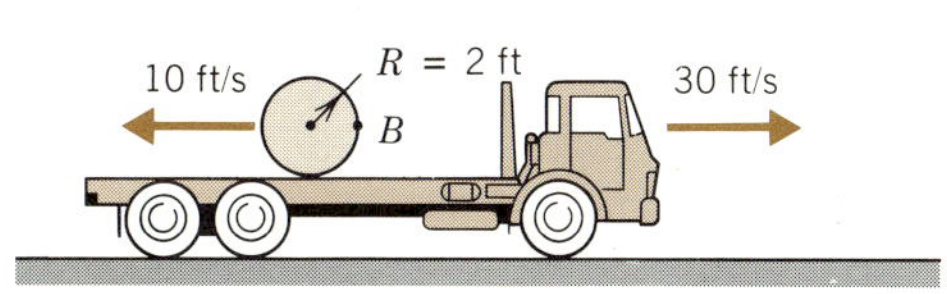

PROB. 13.40

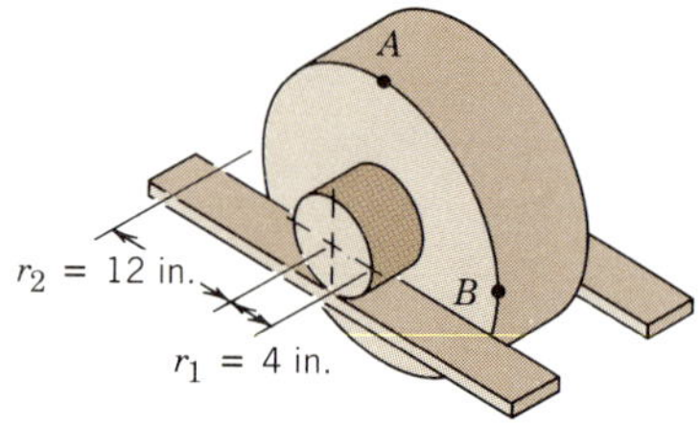

PROB. 13.42

13.41 Determine the velocity of point B on the perimeter of the cylindrical tank in Problem 13.40.

13.42 The stepped disk rolls without slipping over a set of rails. The center of the disk is moving to the right at a constant speed of 4 ft/s. Determine the velocity of points A and B.

13.43 Determine the accelerations of points A and B on the stepped disk in Problem 13.42.

13.44 A box is placed on rollers and given an acceleration of 4 ft/s^2 to the right. There is no slippage. At a specific instant it is known that the total acceleration of the point on the perimeter of a roller that is in contact with the crate is 5 ft/s^2. Determine the velocity and acceleration of the centers of the rollers at this instant.

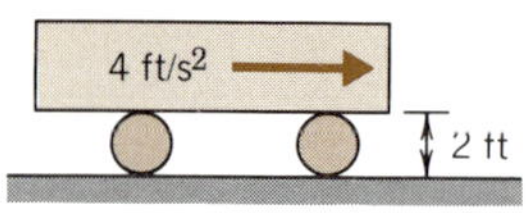

PROB. 13.44

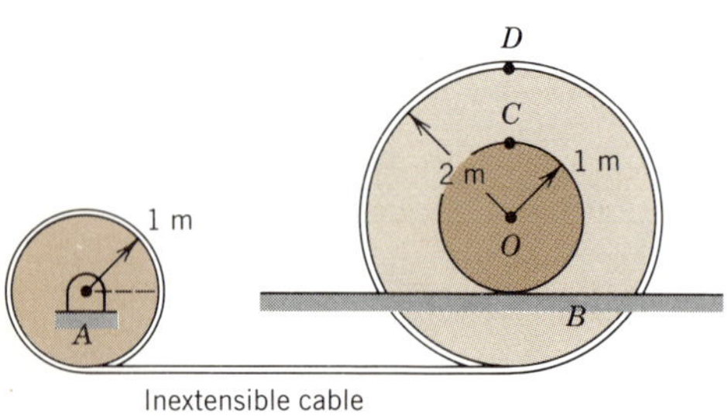

PROBS. 13.45 AND 13.46

13.45 The stepped drum rolls over the horizontal rail without slipping. Its angular velocity and angular acceleration are 5 rad/s and 20 rad/s^2, both clockwise. Determine the angular velocity and angular acceleration of drum A.

13.46 Drum A rotates clockwise at the constant rate of 240 rev/min. The stepped drum B rolls without slipping over the horizontal rail. Determine the velocity and acceleration of points O, C, and D.

13.47 A hoist consists of a motor driven gear A that rotates the larger stepped gear B. Determine (a) the gear ratio ω_B/ω_A, and (b) the velocity of the block when $\omega_A = 300$ rev/min.

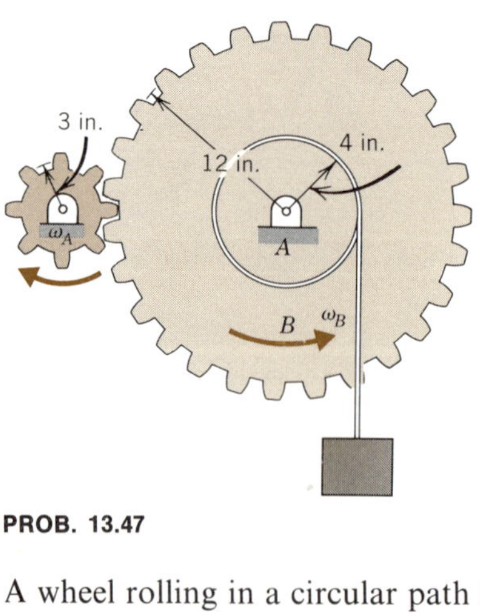

PROB. 13.47

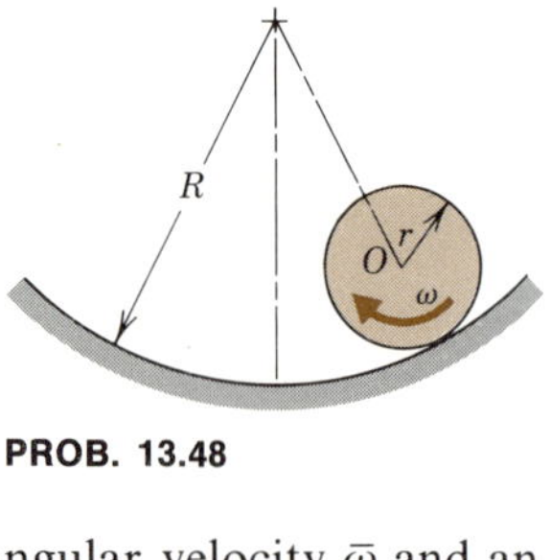

PROB. 13.48

13.48 A wheel rolling in a circular path has an angular velocity $\bar{\omega}$ and an angular acceleration $\bar{\alpha}$, both clockwise. There is no slipping at the contact point. Determine the velocity and acceleration of the center point O.

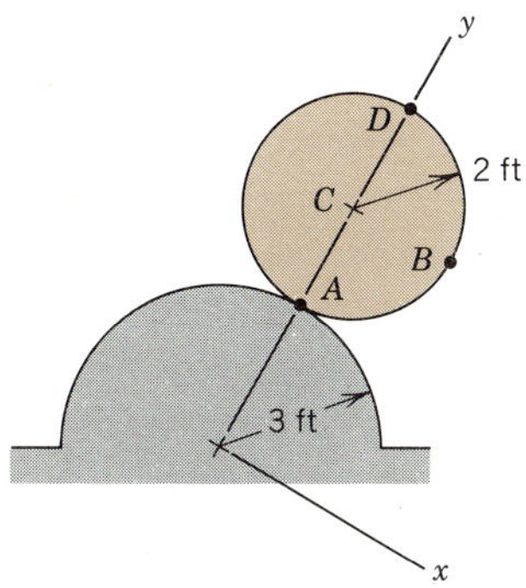

PROB. 13.49

13.49 The wheel is rolling without slipping down the semicircular hill. The speed of the center C of the wheel is 5 ft/s. Determine (a) the angular velocity of the wheel, and (b) the velocity of points A, B, and D on the wheel.

***13.50** The center C of the wheel in Problem 13.49 is gaining speed at the rate of 10 ft/s^2. Determine (a) the angular acceleration of the wheel, and (b) the accelerations of points A, B, and D on the wheel.

13.51 Racks A and B are both moving to the right with speeds of 20 m/s and 8 m/s, respectively. Determine the velocity of (a) point C and (b) point D.

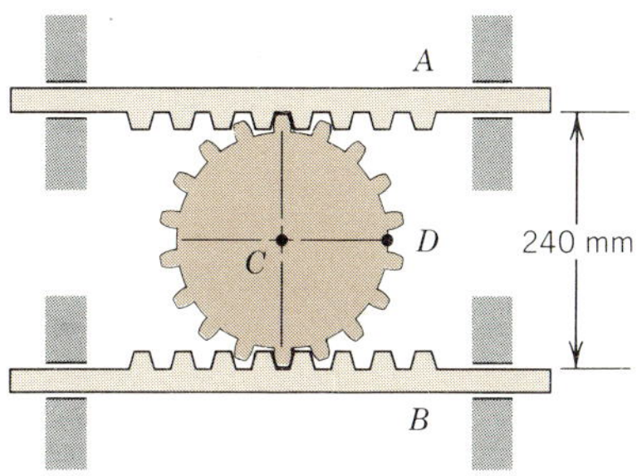

PROBS. 13.51 AND 13.52

***13.52** Rack A is moving to the right at a constant speed of 16 m/s, and rack B is moving to the left with a speed of 8 m/s, which is decreasing at 5 m/s^2. Determine (a) the angular velocity and the angular acceleration of the pinion gear, and (b) the velocity and acceleration of points C and D.

---

3. Mechanical Linkages

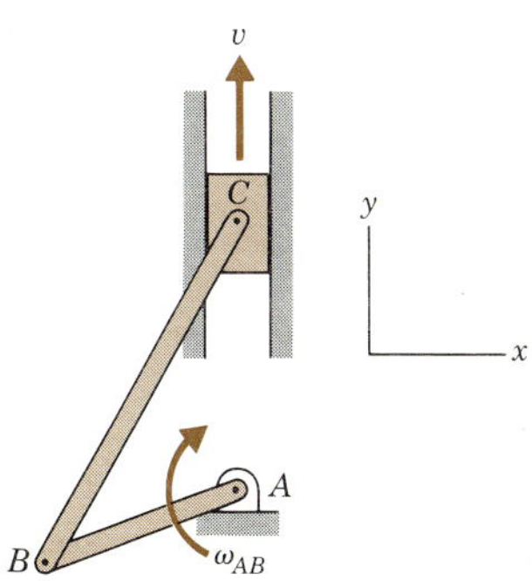

FIGURE 21

A mechanical linkage is a mechanism for transferring and transforming motion. For example, the crankshaft and connecting rod in Figure 21 transform the rectilinear motion of the piston into the rotatory motion of the crankshaft.

The connection between the bodies forming a linkage imposes constraints on the way in which each link may move relative to its neighbor. Only the case of pin connections will be considered here; other types of connections are treated in Chapter 15. The constraint imposed by a pin is straightforward. A pin is a point of commonality between adjacent links. The pinned point on one link must have the same position, velocity, and acceleration as the corresponding point on the neighboring link. For example, points B on bars AB and BC in Figure 21 must have the same velocity.

In a typical situation the motion of one of the bodies forming the linkage is specified. Let us suppose that the motion of the piston in Figure 21 is given. The task is to determine the corresponding angular velocity and angular acceleration of the crankshaft and of the connecting rod when the linkage is in a specified position.

The method of solution here is essentially the same as the one we have already employed, in that we focus on the constrained points and formulate relative motion equations. *The velocities are studied first*, because we need the angular velocities to calculate accelerations. We know the velocity of pin C because it follows the translating piston. Another constrained point is pin A, which is at rest. The remaining constrained point is pin B, which joins the two bars.

We may obtain expressions for $\bar{v}_B$ by proceeding through the linkage from either end. When we start from end A, and set $\bar{v}_A = \bar{0}$, we find that

$$\bar{v}_B = \bar{\omega}_{AB} \times \bar{r}_{B/A} \tag{19}$$

On the other hand, proceeding from piston C to pin B yields

$$\bar{v}_B = \bar{v}_C + \bar{\omega}_{BC} \times \bar{r}_{B/C} \tag{20}$$

Note that we cannot use the relative velocity equation to directly express the motion between constrained points A and C. They are on two separate bodies, which have two distinct angular velocities.

Let us consider a counterclockwise rotation to be positive, rather than worrying about its actual sense. Hence, for planar motion with the coordinate system defined in Figure 21, we set $\bar{\omega}_{AB} = \omega_{AB}\bar{k}$ and $\bar{\omega}_{BC} = \omega_{BC}\bar{k}$. (A negative value for either of the unknown angular speeds will mean that the corresponding rotation is clockwise.)

The only unknowns in eqs. (19) and (20) for $\bar{v}_B$ are the values of ω_{AB} and ω_{BC}, respectively. Matching the x and y components of the right side of these equations will yield two scalar equations for these two unknowns. After these equations are solved for the angular velocity, the velocity of any point in the system may be determined.

The acceleration problem may be solved in the same manner. We form two expressions for the acceleration of pin B. These expressions are obtained by proceeding through the linkage from either end. Specifically, with $\bar{a}_A = \bar{0}$ for the fixed pin,

$$\begin{aligned} \bar{a}_B &= \bar{\alpha}_{AB} \times \bar{r}_{B/A} - \omega_{AB}^2\bar{r}_{B/A} \\ \bar{a}_B &= \bar{a}_C + \bar{\alpha}_{BC} \times \bar{r}_{B/C} - \omega_{BC}^2\bar{r}_{B/C} \end{aligned} \tag{21}$$

Planar motion requires that $\bar{\alpha}_{AB} = \alpha_{AB}\bar{k}$ and $\bar{\alpha}_{BC} = \alpha_{BC}\bar{k}$. The value of $\bar{a}_C$ is known because the piston's motion is given, and both angular speeds were determined by solving the velocity problem. Hence, the angular acceleration rates are the only unknowns in the alternative expressions for $\bar{a}_B$. Matching like components yields the two scalar equations necessary to determine α_{AB} and α_{BC}. These values may then be used to determine the acceleration of any other point on the linkage.

We close this discussion by noting that the instant center method is an alternative for solving the velocity problem. As we will see in the third of the following examples, the constraints on the velocity of each pin make it possible to locate the instant center for each body on the linkage. Such knowledge yields relations between the velocity of each pin and the angular velocity of each link.

Of course, the method of instant centers is not valid for accelerations. If it is necessary to determine accelerations in a linkage, it will probably be preferable to employ the vector equation approach throughout the analysis.

EXAMPLE 9

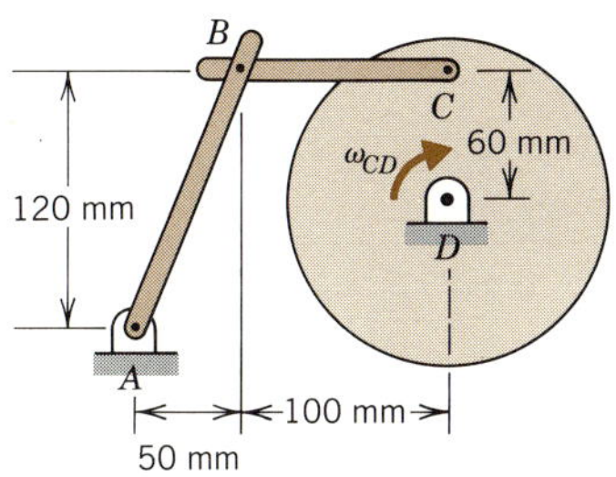

The flywheel is rotating clockwise at 40 rev/min. Determine the angular velocity of bar AB for the position shown.

Solution

We will demonstrate here how the vector equation method is employed. The first step is to draw a sketch of the linkage that depicts the xyz coordinate system. The primary consideration in choosing the orientation of the axes is convenience in writing the position vectors. We therefore align x and y in the directions of the given dimensions.

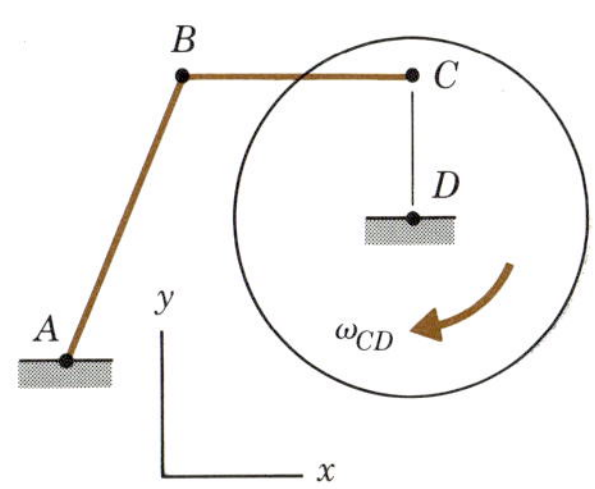

Ends A and D of the linkage are fixed pins, so $\bar{v}_A = \bar{v}_D = \bar{0}$. The angular velocity of the flywheel is given (it must be converted to rad/s), and the angular velocity of each link is perpendicular to the plane of motion. Thus

$$\bar{\omega}_{CD} = 40\left(\frac{2\pi}{60}\right)(-\bar{k}) = -4.189\bar{k} \text{ rad/s}$$

$$\bar{\omega}_{AB} = \omega_{AB}\bar{k}, \qquad \bar{\omega}_{BC} = \omega_{BC}\bar{k}$$

The velocity of pin C is defined by the (known) motion of the flywheel, and the velocity of pin B is not known. Hence, we proceed through the linkage from each end to pin B. Because $\bar{v}_A = \bar{0}$, starting from pin A yields

$$\bar{v}_B = \bar{\omega}_{AB} \times \bar{r}_{B/A}$$

When we start from pin D, we first use the fact that pin C is a point on the flywheel. Setting $\bar{v}_D = \bar{0}$ then gives

$$\bar{v}_C = \bar{\omega}_{CD} \times \bar{r}_{C/D}$$

We may now proceed to pin B by considering pin C to be a point on bar BC.

$$\bar{v}_B = \bar{v}_C + \bar{\omega}_{BC} \times \bar{r}_{B/C} \equiv \bar{\omega}_{CD} \times \bar{r}_{C/D} + \bar{\omega}_{BC} \times \bar{r}_{B/C}$$

The next step is to evaluate the position vectors, which are found from the given sketch to be

$$\bar{r}_{B/A} = 0.05\bar{i} + 0.12\bar{j} \text{ m}, \qquad \bar{r}_{C/D} = 0.06\bar{j} \text{ m}, \qquad \bar{r}_{B/C} = -0.10\bar{i} \text{ m}$$

Substituting the angular velocities and position vectors into the velocity equations then gives

$$\bar{v}_B = \omega_{AB}\bar{k} \times (0.05\bar{i} + 0.12\bar{j}) = -0.12\omega_{AB}\bar{i} + 0.05\omega_{AB}\bar{j}$$

$$\bar{v}_B = (-4.189\bar{k}) \times (0.06\bar{j}) + \omega_{BC}\bar{k} \times (-0.10\bar{i})$$
$$= 0.2513\bar{i} - 0.10\omega_{BC}\bar{j}$$

Equating like components of $\bar{v}_B$ yields

$$\bar{i} \text{ component:} \quad -0.12\omega_{AB} = 0.2513$$

$$\bar{j} \text{ component:} \quad 0.05\,\omega_{AB} = -0.10\omega_{BC}$$

$$\omega_{AB} = -2.094, \qquad \omega_{BC} = 1.047$$

$$\bar{\omega}_{AB} = -2.09\bar{k} \text{ rad/s}$$

EXAMPLE 10

The angular speed of the flywheel in Example 9 is constant. Determine the angular acceleration of bar *AB* in the position given there.

Solution

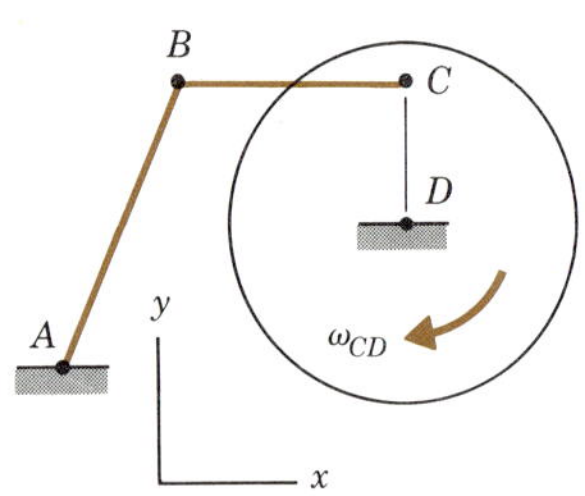

We follow the same approach as that used in the previous example to solve the velocity problem. Thus we proceed from pin *A* to pin *B* using the fact that they are both points in bar *AB*.

$$\bar{a}_B = \bar{a}_A + \bar{\alpha}_{AB} \times \bar{r}_{B/A} - \omega_{AB}^2\bar{r}_{B/A}$$

In order to reach pin *B* starting from end *D*, we need to use the fact that the intermediate pin *C* is a point on the flywheel and on bar *BC*. Thus

$$\bar{a}_C = \bar{a}_D + \bar{\alpha}_{CD} \times \bar{r}_{C/D} - \omega_{CD}^2\bar{r}_{C/D}$$

$$\bar{a}_B = \bar{a}_C + \bar{\alpha}_{BC} \times \bar{r}_{B/C} - \omega_{BC}^2\bar{r}_{B/C}$$

Note that pins *A* and *D* are fixed, so $\bar{a}_A = \bar{a}_D = \bar{0}$.

Solving Example 9 is consistent with the general requirement that the velocity problem be solved first. The angular speed of each bar is known from that solution. One advantage of having used the vector equation method in the velocity problem is that the position vectors required for the acceleration problem have already been formulated. The only new vectors required are the angular accelerations. Because ω_{CD} is constant and the motion is planar,

$$\bar{\alpha}_{CD} = \bar{0}, \qquad \bar{\alpha}_{AB} = \dot{\omega}_{AB}\bar{k}, \qquad \bar{\alpha}_{BC} = \dot{\omega}_{BC}\bar{k}$$

We now substitute each vector into the acceleration equations. The relation between pins *B* and *A* becomes

$$\bar{a}_B = \bar{0} + \dot{\omega}_{AB}\bar{k} \times (0.05\bar{i} + 0.12\bar{j}) - (-2.094)^2(0.05\bar{i} + 0.12\bar{j})$$
$$= -0.12\dot{\omega}_{AB}\bar{i} + 0.05\dot{\omega}_{AB}\bar{j} - 0.2192\bar{i} - 0.5262\bar{j} \text{ m/s}^2$$

The equations relating pin B to pin D yield

$$\bar{a}_C = \bar{0} + \bar{0} - (-4.189)^2(0.06\bar{j}) = -1.0529\bar{j} \text{ m/s}^2$$
$$\bar{a}_B = -1.0529\bar{j} + \dot{\omega}_{BC}\bar{k} \times (-0.10\bar{i}) - (1.047)^2(-0.10\bar{i})$$
$$= -1.0529\bar{j} - 0.10\dot{\omega}_{BC}\bar{j} + 0.1096\bar{i} \text{ m/s}^2$$

Matching like components in the two expressions for $\bar{a}_B$ then gives

$$\bar{i} \text{ component:} \quad -0.12\dot{\omega}_{AB} - 0.2192 = 0.1096$$
$$\bar{j} \text{ component:} \quad 0.05\dot{\omega}_{AB} - 0.5262 = -1.0529 - 0.10\dot{\omega}_{BC}$$
$$\dot{\omega}_{AB} = -2.74, \qquad \dot{\omega}_{BC} = -3.90$$
$$\bar{\alpha}_{AB} = -2.74\bar{k} \text{ rad/s}^2$$

EXAMPLE 11

Bar BC is rotating counterclockwise at 5 rad/s when the linkage is in the position shown. Determine the corresponding velocity of collar A.

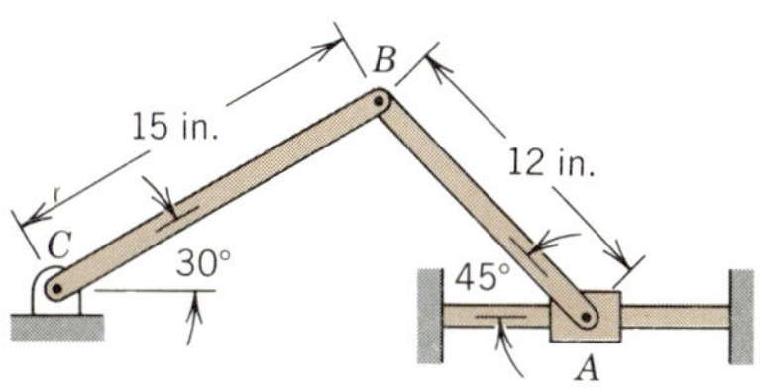

Solution

We will demonstrate here how the instant center method may be employed to solve the velocity problem for a linkage. The first step is to make a sketch of the linkage in the desired position. This sketch displays the known information about the ends of the linkage, specifically, that pin C is fixed and that pin A moves horizontally. (We have assumed that $\bar{v}_A$ is to the left.) The sketch also depicts the given angular velocity of bar BC.

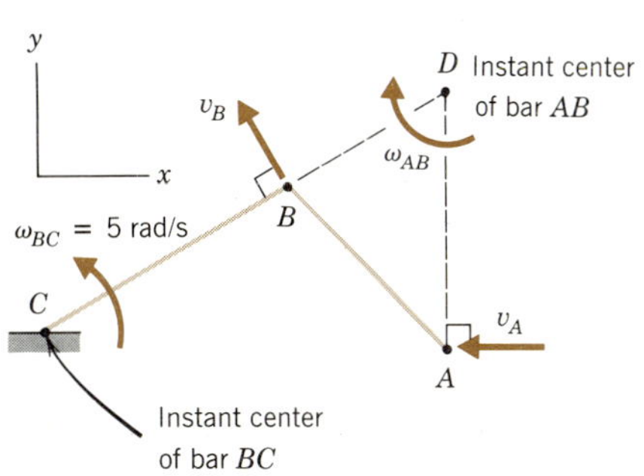

Because $\bar{v}_C = \bar{0}$, point C is the instant center of bar BC. The velocity of pin B therefore must be perpendicular to line BC, in the upward sense shown.

We have now depicted in the sketch the direction in which the ends of bar AB move. This information enables us to locate the instant center of bar AB by drawing dashed lines through each point perpendicular to the velocity of that point. The intersection, point D, is the instant center.

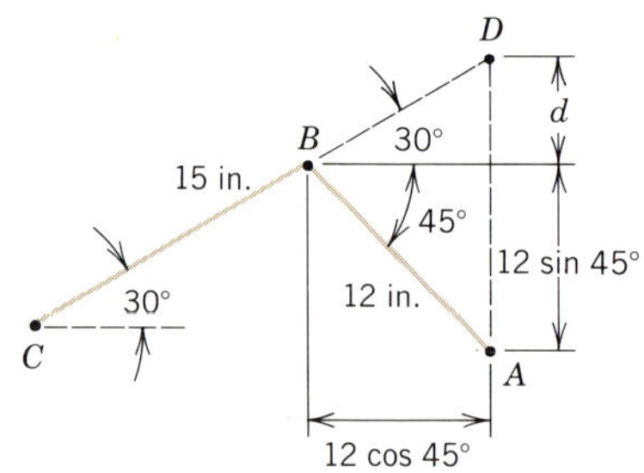

It will be noted that $\bar{\omega}_{AB}$ is depicted clockwise in the instant center diagram. This results from knowing that v_B is upward. Considering each point to be in circular motion about an instant center then gives

$$v_B = \omega_{BC}|\bar{r}_{B/C}| = \omega_{AB}|\bar{r}_{B/D}|$$

$$v_A = -\omega_{AB}|\bar{r}_{A/D}|$$

The distance $|\bar{r}_{B/C}| = 15$ in., but the other distance must be evaluated. Several trigonometric formulations are possible; the one we have depicted in the sketch uses horizontal and vertical distances. We find from the sketch that

$$|\bar{r}_{B/D}| = \frac{12 \cos 45^\circ}{\cos 30^\circ} = 9.798 \text{ in.}$$

$$d = |\bar{r}_{B/D}| \sin 30^\circ = 4.899 \text{ in.,} \qquad |\bar{r}_{A/D}| = d + 12 \sin 45^\circ = 13.384 \text{ in.}$$

The velocity relations then yield

$$v_B = 5(15) = \omega_{AB}(9.798) \Rightarrow \omega_{AB} = 7.655 \text{ rad/s}$$

$$v_A = \omega_{AB}(13.384) = 102.45$$

$$\bar{v}_A = -v_A\bar{i} = -102.4\bar{i} \text{ in./s}$$

HOMEWORK PROBLEMS

13.53 and 13.54 Bar AB is rotating counterclockwise at the rate of 600 rev/min. Determine the angular velocities of the other bars for the position shown.

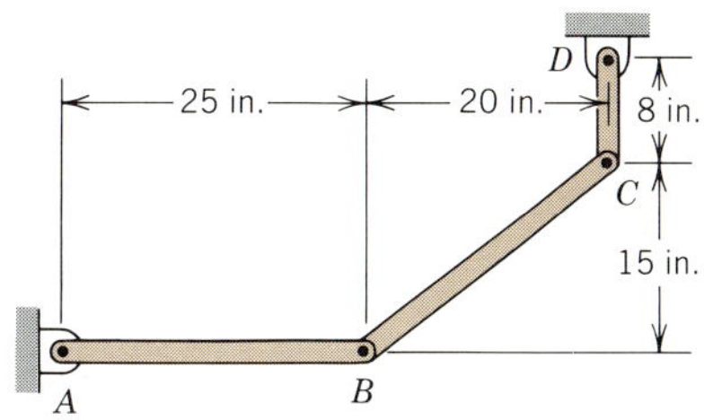

PROB. 13.53

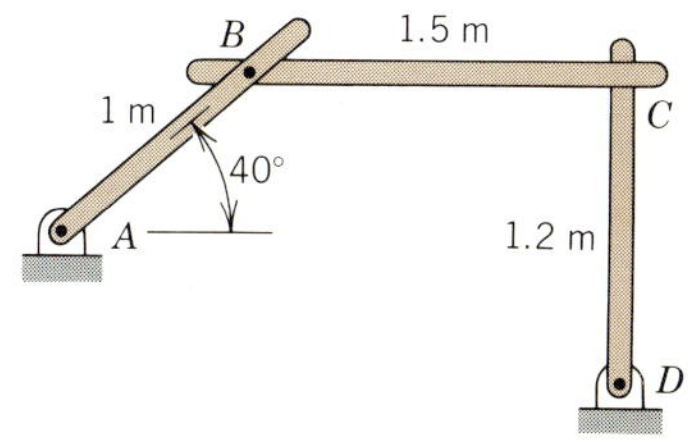

PROB. 13.54

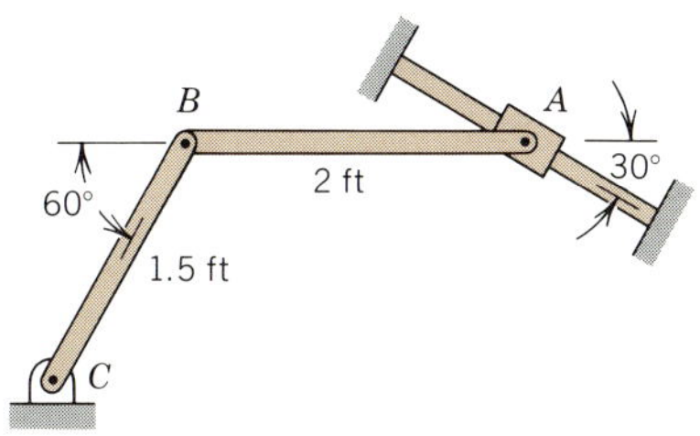

PROB. 13.55

13.55 Collar A is moving to the left at 50 ft/s in the position shown. Determine the angular velocity of bar BC.

13.56 Slider A has a downward velocity of 600 mm/s. Determine the angular velocity of rod BC and the velocity of slider C for the position shown.

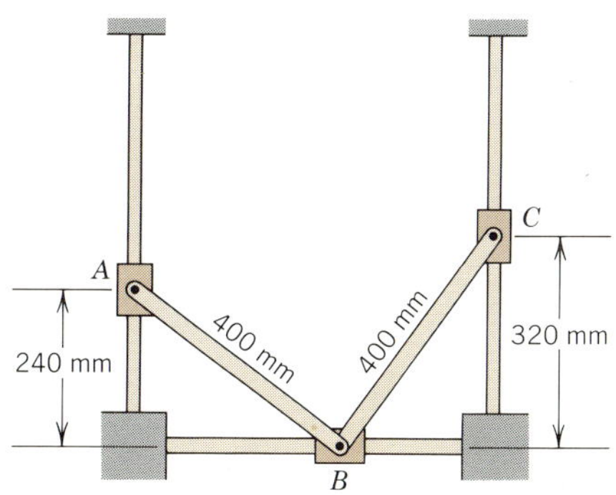

PROB. 13.56

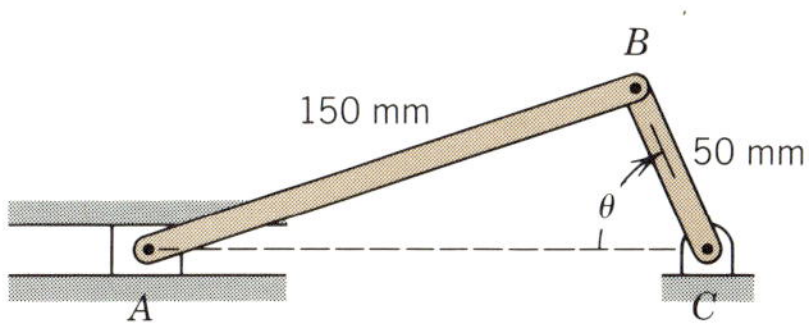

PROB. 13.57

13.57 Bar BC is rotating clockwise at 1800 rev/min in the position shown. Determine the corresponding velocity of piston A when $\theta = 36.87°$.

***13.58** Solve Problem 13.57 for $\theta = 120°$.

13.59 A hydraulically driven piston moves point C to the right with a constant acceleration of 4.5 m/s². At the instant depicted in the diagram, point C has a velocity of 15 m/s, also to the right. Determine the angular velocity and angular acceleration of both rods at this instant.

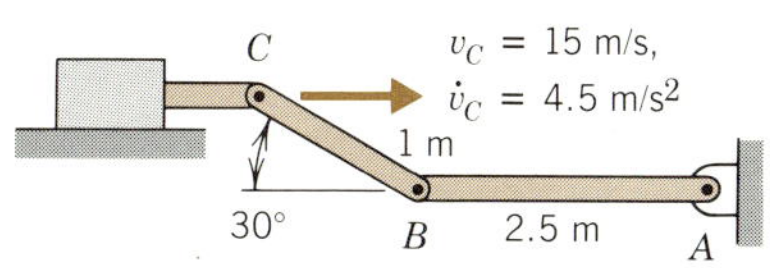

PROB. 13.59

13.60 and 13.61 Bar AB rotates counterclockwise at the constant rate of 10 rad/s. Determine the angular velocity and angular acceleration of the other bars for the position shown.

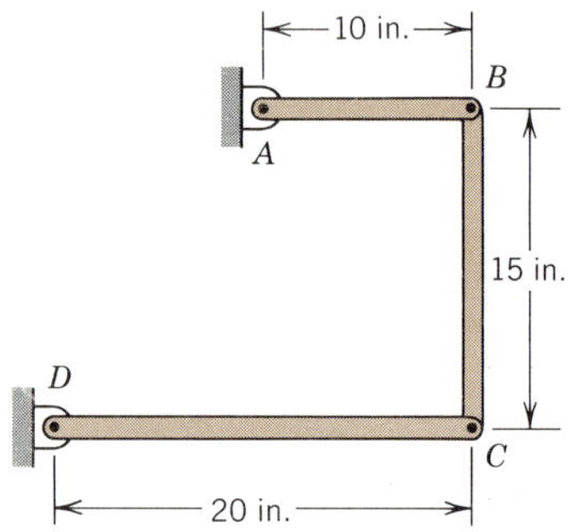

PROB. 13.60

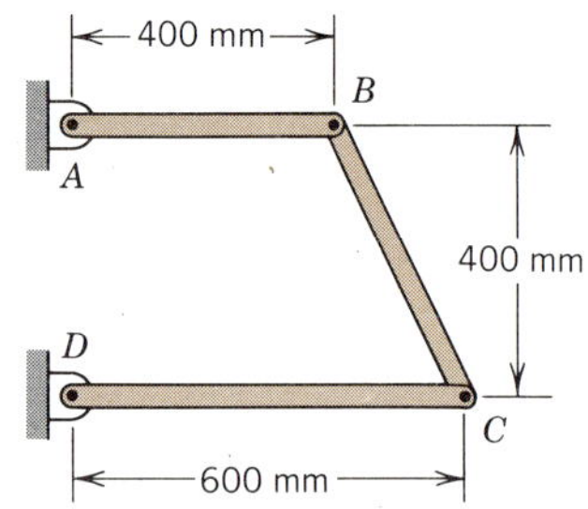

PROB. 13.61

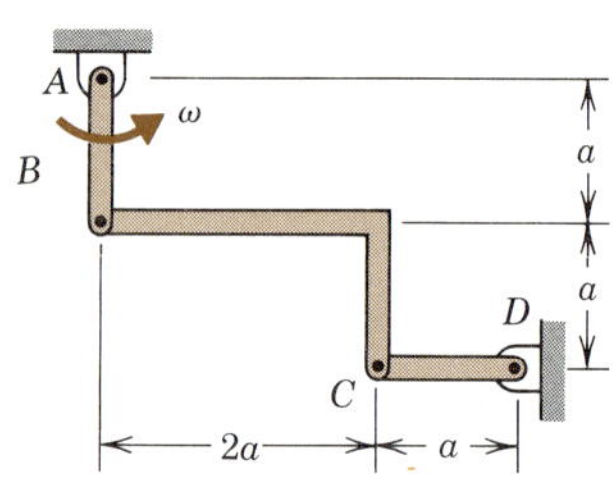

PROB. 13.62

13.62 In the position shown, the angular velocity of bar AB is the constant value $\bar{\omega}$. Determine the corresponding angular velocities and angular accelerations of bars BC and CD.

13.63 Bar AB in Problem 13.54 is slowing at 5 rad/s. Determine the corresponding angular accelerations of the other bars.

13.64 The hydraulic piston pushes pin C to the right. The system is at rest in the position shown when the piston is suddenly given an acceleration $\dot{v}_C = 50$ ft/s². Determine the corresponding angular acceleration of bar AB.

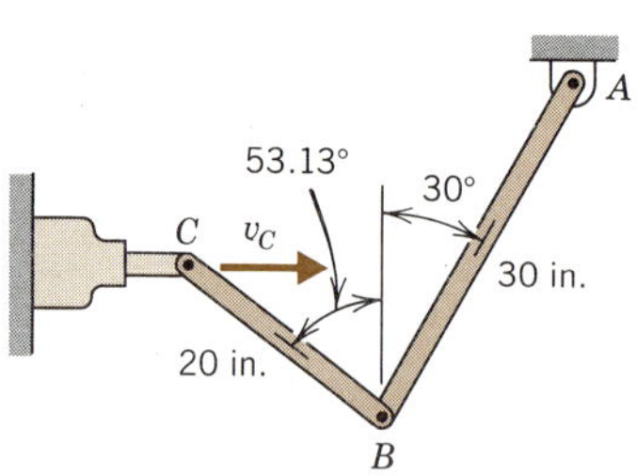

PROB. 13.64

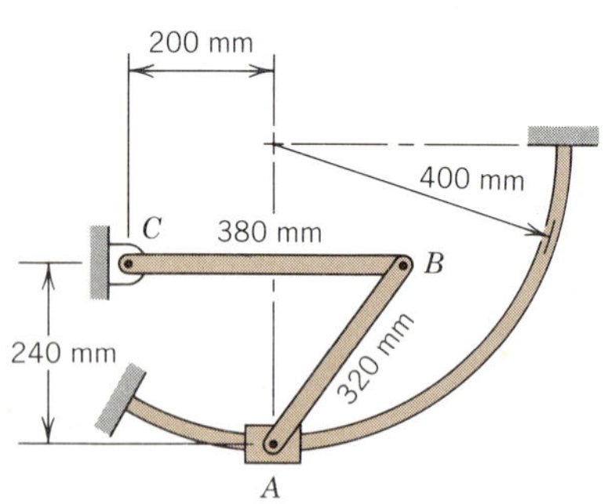

PROB. 13.67

13.65 Solve Problem 13.64 if $v_C = 10$ ft/s and $\dot{v}_C = -20$ ft/s² in the given position.

***13.66** The rate of rotation of bar BC in Problem 13.57 is increasing at 10 rev/min/s. Determine the acceleration of the piston.

***13.67** Bar BC rotates counterclockwise at the constant rate of 90 rev/min. Determine the acceleration of collar A when the system is in the position shown.

D. PROBLEM SOLVING

The basic kinematical relations were derived at the beginning of this chapter, and then applied in a series of specialized studies. A general problem in planar kinematics might combine various features of the topics in each section. For example, a linkage can contain a gear that is rolling without slipping. It is useful to synthesize the various developments into a unified approach for problem solving.

The angular and linear velocities must be determined in any situation, even one in which we are only requested to determine accelerations. Thus the overall approach begins by addressing the evaluation of velocities. Common to all studies is the following.

1. Draw a simple sketch of the system in the position of interest. Choose an xyz coordinate system such that x and y, which are in the plane of motion, make it as easy as possible to express the position of points in the system. For instance, if horizontal and vertical dimensions are given, choose x and y such that one is horizontal and the other is vertical.
2. Delineate the given information in the sketch, or express it in equation form. List the quantities that are requested in the problem statement.

The Velocity Problem

3. Write down the equations relating the velocities of each of the constrained points of a body to the angular velocity of the body. Do this for each body. Typical constrained points to look for are
   a. fixed points
   b. points pinned to another body
   c. points required to follow a specific path
   d. points of contact for bodies that roll
4. Write down the mathematical restrictions imposed on *all* constrained points. For instance, if a point A must follow a straight line in the x direction, write down $\bar{v}_A = v_A \bar{i}$.
5. Express all position vectors and angular velocity vectors by their xyz components. Remember that for planar motion $\bar{\omega} = \pm\omega\bar{k}$.
6. Substitute the constraint conditions from step 4 and the position vectors and angular velocity vectors from step 5 into the velocity equations from step 3. Calculate the cross products.
7. Collect components in the x and y directions. Count the number of scalar equations you have. Count the number of scalar unknowns in these equations. If the two numbers are not equal, you have probably forgotten to use a given piece of information or forgotten to utilize a constraint condition that the system must obey.
8. Solve the equations.
9. After determining the velocity of any point on the body and the angular velocity of the body, the velocity of unconstrained points may be found from the relative velocity equation.

Alternative to Steps 3–9 The method of instant centers may be used in lieu of the vector equation approach. This method relates the velocity of points in a body to the corresponding angular velocity by regarding the body to be executing a pure rotation about the instant center.

The Acceleration Problem

Steps 10–16 If accelerations must be determined, repeat steps 3 through 9, replacing the word *velocity* by the word *acceleration*. In these steps consider the solutions for velocities as given information. Remember that instant centers are not valid for determining acceleration.

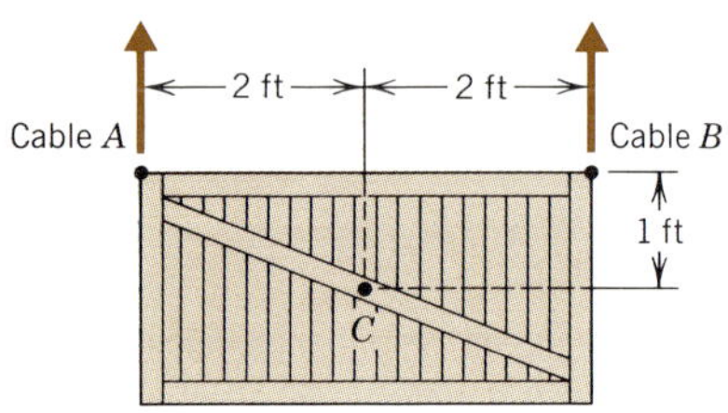

ILLUSTRATIVE PROBLEM 1

A cabinet was being lowered by two workers holding cables A and B. They were letting the cables out at the same rate of 3 ft/s downward with the cabinet maintained in the horizontal orientation shown. The workers tightened their grips suddenly. The rate of descent immediately began to slow at 5 ft/s^2 for cable A and 1 ft/s^2 for cable B. Determine the acceleration of the centroid C and the angular acceleration of the cabinet at the instant immediately after the workers tightened their grips.

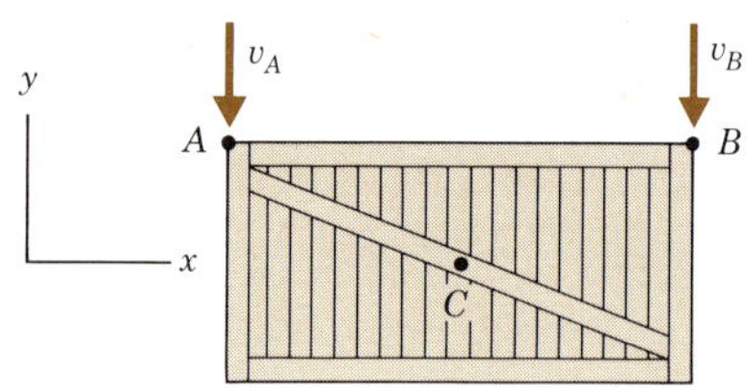

Solution

Step 1 The points labeled in the sketch are corners A and B where the cables are attached, and centroid C.

Step 2 The sketch depicts the fact that corners A and B descend at the rate that the cables are let out. It is given that $v_A = v_B = 3$ ft/s, $\dot{v}_A = -5$ ft/s^2, and $\dot{v}_B = -1$ ft/s^2 (the negative signs correspond to decreasing speeds).

Steps 3–9 Points A and B have the same velocity, so there is no relative velocity $\bar{v}_{B/A}$. This situation arises only when there is no angular velocity, therefore

$$\bar{\omega} = \bar{0}$$

(The same result can also be found by the instant center method.)

The instant in which we are interested is the one at which the cabinet has just begun to rotate, so there will be an angular acceleration. An analogous situation in the motion of particles is one in which a particle at rest is given an acceleration.

Step 10 The constrained points are points A and B whose accelerations are given. The path of point C is not prescribed, so we will consider its acceleration at a later step.

Because $\bar{\omega} = \bar{0}$ at the instant of interest, the acceleration equation relating the constrained points reduces to

$$\bar{a}_B = \bar{a}_A + \bar{\alpha} \times \bar{r}_{B/A}$$

Step 11 The motion imposed on the constrained points is that

$$\bar{a}_A = \dot{v}_A(-\bar{j}) = 5\bar{j} \text{ ft/s}^2$$

$$\bar{a}_B = \dot{v}_B(-\bar{j}) = 1\bar{j} \text{ ft/s}^2$$

Step 12 Set $\bar{\alpha} = \dot{\omega}\bar{k}$ for planar motion. The original picture of the cabinet gives

$$\bar{r}_{B/A} = 4\bar{i} \text{ ft/s}$$

Steps 13–15 Substitute all vectors into the relation between $\bar{a}_A$ and $\bar{a}_B$ in Step 10. Then match like components

$$1\bar{j} = 5\bar{j} + \dot{\omega}\bar{k} \times (4\bar{i}) \equiv 5\bar{j} + 4\dot{\omega}\bar{j}$$

$$\bar{j} \text{ component:} \quad 1 = 5 + 4\dot{\omega}$$

We have one scalar equation and one unknown

$$\dot{\omega} = -1 \Rightarrow \bar{\alpha} = -1\bar{k} \text{ rad/s}^2$$

Step 16 The acceleration of point C may be related to either $\bar{a}_A$ or $\bar{a}_B$. The former gives

$$\bar{a}_C = \bar{a}_A + \bar{\alpha} \times \bar{r}_{C/A} - \omega^2\bar{r}_{C/A}$$
$$= 5\bar{j} + (-1\bar{k}) \times (2\bar{i} - \bar{j}) - \bar{0} = -1\bar{i} + 3\bar{j} \text{ ft/s}^2$$

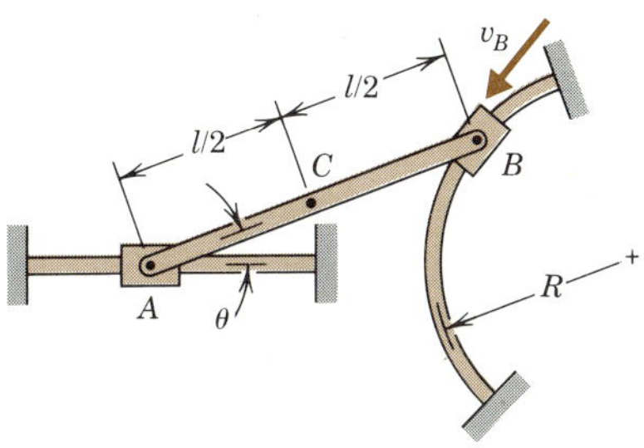

ILLUSTRATIVE PROBLEM 2

When $\theta = 0$, slider B is moving downward along a curved bar with a speed v_B and gaining speed at a rate $\dot{v}_B$. Determine the corresponding velocities and accelerations of slider A and midpoint C of bar AB in terms of v_B, $\dot{v}_B$, l, and R.

Solution

Step 1 The sketch shows the bar at $\theta = 0$, as well as the path followed by each end.

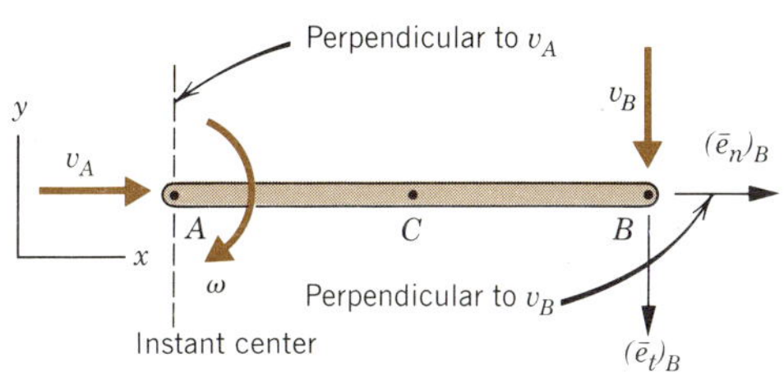

Step 2 The given velocity $\bar{v}_B$ is shown in the sketch. We need to evaluate $\bar{v}_A$, $\bar{v}_C$, $\bar{a}_A$, and $\bar{a}_C$.

Steps 3–9 The geometry of this system is not complicated, so the instant center method will be employed for velocities. The velocity of each slider is tangent to the path it follows, as indicated in the sketch. The perpendiculars to each velocity intersect at point A. Hence point A is the instant center, which gives

$$\bar{v}_A = \bar{0}$$

$$v_B = \omega|\bar{r}_{B/A}| = \omega l, \qquad v_C = \omega|\bar{r}_{C/A}| = \omega\frac{1}{2}$$

We solve for ω and v_C in terms of v_B, and use the senses for both rates indicated by the instant center diagram, with the result that

$$\bar{\omega} = \frac{v_B}{l}(-\bar{k}), \qquad \bar{v}_C = \frac{1}{2}v_B(-\bar{j})$$

Step 10 The constrained points are A and B, which must follow the guides. We therefore write

$$\bar{a}_A = \bar{a}_B + \bar{\alpha} \times \bar{r}_{A/B} - \omega^2\bar{r}_{A/B}$$

Step 11 The speed v_B is changing at the given rate $\dot{v}_B$, as slider B follows a circular path of radius R. The unit vectors for this path appear in the instant center diagram, from which we find that

$$\bar{a}_B = \dot{v}_B(\bar{e}_t)_B + \frac{v_B^2}{R}(\bar{e}_n)_B = \dot{v}_B(-\bar{j}) + \frac{v_B^2}{R}(\bar{i})$$

Slider A follows a straight path parallel to the x axis, so

$$\bar{a}_A = \dot{v}_A\bar{i}$$

Step 12 Set $\bar{\alpha} = \dot{\omega}\bar{k}$ and note that $\bar{r}_{A/B} = -l\bar{i}$.

Step 13 Substitute all vectors and the expression for ω found in the velocity solution into the acceleration equation

$$\dot{v}_A\bar{i} = -\dot{v}_B\bar{j} + \frac{v_B^2}{R}\bar{i} + \dot{\omega}\bar{k} \times (-l\bar{i}) - \left(\frac{v_B}{l}\right)^2(-l\bar{i})$$

$$= -\dot{v}_B\bar{j} + \frac{v_B^2}{R}\bar{i} - \dot{\omega}l\bar{j} + \frac{v_B^2}{l}\bar{i}$$

Steps 14 and 15 Match like components and solve.

$$\bar{i} \text{ component:} \quad \dot{v}_A = \frac{v_B^2}{R} + \frac{v_B^2}{l}$$

$$\bar{j} \text{ component:} \quad 0 = -\dot{v}_B - \dot{\omega}l \Rightarrow \dot{\omega} = -\frac{\dot{v}_B}{l}$$

$$\bar{a}_A = v_B^2\left(\frac{1}{R} + \frac{1}{l}\right)\bar{i}$$

Step 16 The acceleration of point C may now be found from

$$\bar{a}_C = \bar{a}_A + \bar{\alpha} \times \bar{r}_{C/A} - \omega^2\bar{r}_{C/A}$$

$$= v_B^2\left(\frac{1}{R} + \frac{1}{l}\right)\bar{i} + \left(-\frac{\dot{v}_B}{l}\bar{k}\right) \times \frac{l}{2}\bar{i} - \left(\frac{v_B}{l}\right)^2\frac{l}{2}\bar{i}$$

$$\bar{a}_C = v_B^2\left(\frac{1}{R} + \frac{1}{2l}\right)\bar{i} - \frac{1}{2}\dot{v}_B\bar{j}$$

ILLUSTRATIVE PROBLEM 3

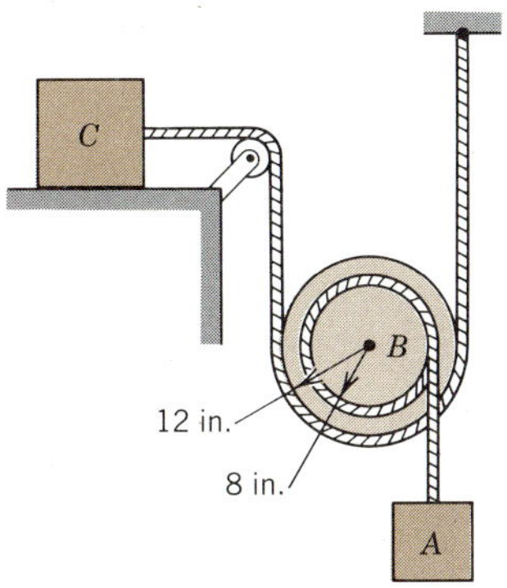

The cable suspending block A is wrapped around the inner drum of the stepped pulley B. The pulley is suspended by the cable, which passes over the outer drum and is then connected to block C, which is moving to the right at 42 in./s. There is no slipping between the stepped pulley and the cables. Determine (a) the angular velocity of the pulley, (b) the velocity of block A, and (c) the rate at which the cable suspending block A is being wrapped or unwrapped around the stepped pulley.

Solution

Step 1 The sketch of the system identifies the points of tangency between the cable and the stepped pulley. We will employ the no slip condition for these points.

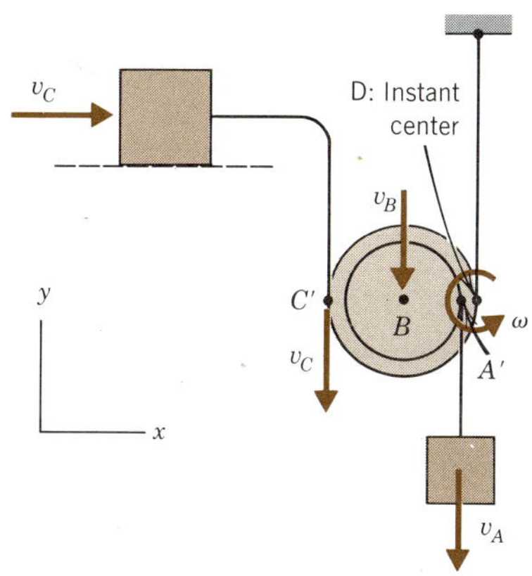

Step 2 It is given that $v_C = 42$ in./s. We need to find $\bar{\omega}$ of the pulley, $\bar{v}_A$, and the rate at which the length of cable from the pulley to block A is changing. (Note that $\bar{v}_A$ could be obtained by the method in Chapter 11, in which we construct the cable length.)

Steps 3–8 The overall constraint on the motion is that there is no slipping between the cables and the pulley. Thus each cable and the pulley have the same velocity at the point at which they are tangent. (The straight segments of the cable act like flat surfaces over which the pulley rolls.) The method of instant centers is convenient for evaluating the velocities of the points of tangency because these points are arranged along the same diameter.

We use the sketch in step 1 for depicting the instant center. The right segment of the cable is at rest, because it is attached to the ceiling. Thus $\bar{v}_D = \bar{0}$, which means that point D is the instant center of the pulley. Similar reasoning indicates that the speeds of points A' and C' are the speeds of the corresponding blocks. Hence considering each point on the pulley to be following a circular path around point D gives

$$v_{C'} \equiv v_C = 42 \text{ in./s} = \omega|\bar{r}_{C'/D}| = \omega(24)$$

$$v_B = \omega|\bar{r}_{B/D}| = \omega(12)$$

$$v_A = \omega|\bar{r}_{A/D}| = \omega(4)$$

The solution of these equations is

$$\omega = 1.750 \text{ rad/s}, \qquad v_B = 21 \text{ in./s}, \qquad v_A = 7 \text{ in./s}$$

As shown in the instant center diagram, the angular velocity is counterclockwise and the velocity of each point is perpendicular to the line to the instant center, that is, downward. Thus

$$\bar{\omega} = 1.750\bar{k} \text{ rad/s}$$

$$v_A = -7\bar{j} \text{ in./s}$$

Step 9 The rate of change of the length of cable between the pulley and block A is a relative velocity. Specifically, if block A and the center B of the pulley had the same velocity, then the cable length would be constant.

Let l denote the suspended length. Then

$$\dot{l} = |\bar{v}_{A/B}| = |\bar{v}_A - \bar{v}_B| = v_A - v_B$$

The values of v_A and v_B were found by solving the instant center equations. Thus

$$\dot{l} = 7 - 21 = -14 \text{ in./s}$$

The fact that $\dot{l}$ is negative means that the cable is being wound up.

It should be noted that the relative velocity $\bar{v}_{A/B}$ is associated with the angular rotation, that is, $\bar{v}_{A/B} = \bar{\omega} \times \bar{r}_{A/B}$. The cable winds up because the pulley is rotating counterclockwise.

ILLUSTRATIVE PROBLEM 4

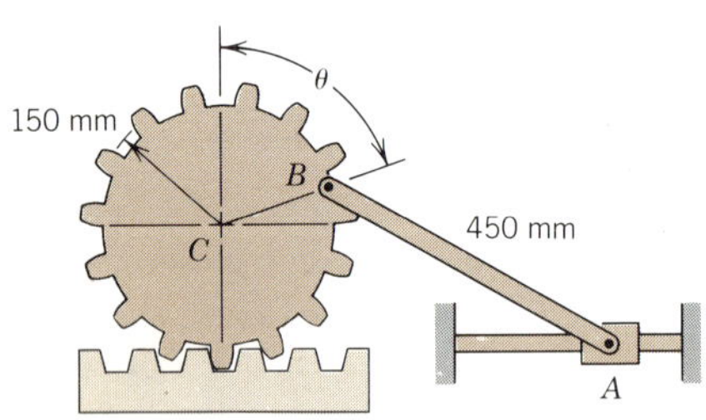

Collar A moves to the right at a constant speed of 20 m/s. Determine the velocity and acceleration of the center C of the gear when $\theta = 90°$.

Solution

Step 1 The sketch depicts the system when $=90°$.

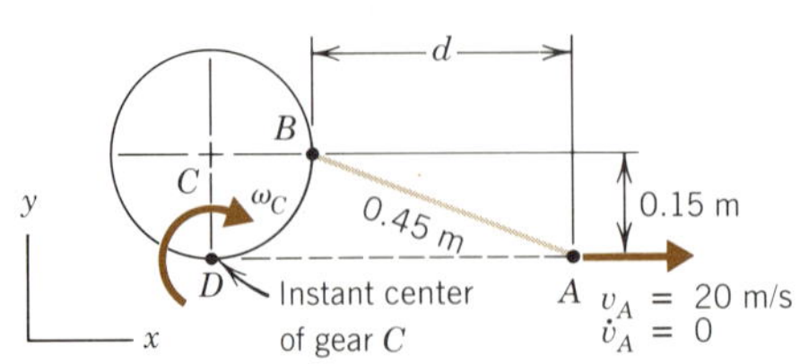

Step 2 The given motion of collar A is shown in the sketch. We need to evaluate $\bar{v}_C$ and $\bar{a}_C$.

Steps 3 and 4 This system is a linkage in which one of the links is the gear, which rolls without slipping over the horizontal rack. In addition to point A, whose velocity is given, other constrained points are (a) pin B, which connects the bar and the gear, (b) contact point D, which does not slip relative to the rack, and (c) center C, which follows a horizontal path, so $\bar{v}_C = v_C\bar{i}$ and $\bar{a}_C = a_C\bar{i}$.

This problem requires evaluation of accelerations, so we shall use the vector equation approach for relating $\bar{v}_B$ to the angular velocities $\bar{\omega}_C$ of the gear and $\bar{\omega}_{AB}$ of the bar. The center of a rolling wheel is the primary point of reference. Hence we write

$$\begin{aligned}\bar{v}_B &= \bar{v}_A + \bar{\omega}_{AB} \times \bar{r}_{B/A} \\ &= \bar{v}_C + \bar{\omega}_C \times \bar{r}_{B/C}\end{aligned}$$

Also, the rack is stationary, so point D is the instant center for the gear, $v_D \equiv 0$. We guess that the gear will rotate clockwise, and $|\bar{r}_{C/D}| = 0.15$ m. This gives

$$v_C = \omega_C|\bar{r}_{C/D}| \Rightarrow \bar{v}_C = 0.15\omega_C\bar{i}$$

Step 5 The position vectors appearing in the foregoing velocity relations are

$$\bar{r}_{B/A} = -d i + 0.15\bar{j} \equiv -[(0.45)^2 - (0.15)^2]^{1/2}\bar{i} + 0.15\bar{j} = -0.4243\bar{i} + 0.15\bar{j} \text{ m}$$

$$\bar{r}_{B/C} = 0.15\bar{i} \text{ m}$$

We have already assumed that $\bar{\omega}_C$ is clockwise. Let us assume the same sense for $\bar{\omega}_{AB}$. Thus

$$\bar{\omega}_C = \omega_C(-\bar{k}), \qquad \bar{\omega}_{AB} = \omega_{AB}(-\bar{k})$$

Step 6 Substituting the given value for $\bar{v}_A$, the constraint equation for $\bar{v}_C$, and the position and angular velocity vectors into the basic velocity relations yields

$$\begin{aligned}\bar{v}_B &= 20\bar{i} + (-\omega_{AB}\bar{k}) \times (-0.4243\bar{i} + 0.15\bar{j}) \\ &= 0.15\omega_C\bar{i} + (-\omega_C\bar{k}) \times (0.15\bar{i})\end{aligned}$$

Steps 7 and 8 Matching like components of $\bar{v}_B$ yields

$$\bar{i} \text{ component:} \quad 20 + 0.15\omega_{AB} = 0.15\omega_C$$

$$\bar{j} \text{ component:} \quad 0.4243\omega_{AB} = -0.15\omega_C$$

We have two equations for the two unknown angular speeds. Their solution is

$$\omega_{AB} = -34.83 \text{ rad/s}, \qquad \omega_C = 98.52 \text{ rad/s}$$

Then the constraint equation for $\bar{v}_C$ gives

$$\bar{v}_C = 0.15(98.52)\bar{i} = 14.78\bar{i} \text{ m/s}$$

Steps 10 and 11 These are the analogs for acceleration of steps 3 and 4. Describing the acceleration of pin B in terms of the bodies it connects gives

$$\begin{aligned}\bar{a}_B &= \bar{a}_A + \bar{\alpha}_{AB} \times \bar{r}_{B/A} - \omega_{AB}^2\bar{r}_{B/A} \\ &= \bar{a}_C + \bar{\alpha}_C \times \bar{r}_{B/C} - \omega_C^2\bar{r}_{B/C}\end{aligned}$$

where $\bar{a}_A = \bar{0}$ for a constant v_A along a horizontal path.

The relation $v_C = \omega_C|\bar{r}_{C/D}|$ is valid at any instant because the point at which the gear contacts the rack is always the instant center. This relation may therefore be differentiated to find $\dot{v}_C = \dot{\omega}_C|\bar{r}_{C/D}|$. When used in conjunction with the fact that point C follows a straight horizontal path, this relation yields

$$\bar{a}_C = \dot{v}_C\bar{i} = 0.15\dot{\omega}_C\bar{i}$$

Step 12 The position vectors were expressed earlier. Increasing angular speeds, that is, $\dot{\omega}_C > 0$ and $\dot{\omega}_{AB} > 0$, correspond to clockwise angular accelerations according to our initial assumptions. Hence, let

$$\bar{\alpha}_C = \dot{\omega}_C(-\bar{k}), \qquad \bar{\alpha}_{AB} = \dot{\omega}_{AB}(-\bar{k})$$

Step 13 The expressions for each vector, as well as the values of ω_{AB} and ω_C that were found in step 8, are substituted into the alternative expressions for $\bar{a}_B$

$$\bar{a}_B = \bar{0} + (-\dot{\omega}_{AB}\bar{k}) \times (-0.4243\bar{i} + 0.15\bar{j}) - (34.83)^2(-0.4243\bar{i} + 0.15\bar{j})$$
$$= 0.15\dot{\omega}_C\bar{i} + (-\dot{\omega}_C\bar{k}) \times (0.15\bar{i}) - (98.52)^2(0.15\bar{i})$$

Steps 14 and 15 Matching like components of $\bar{a}_B$ yields two equations for the two unknowns, $\dot{\omega}_C$ and $\dot{\omega}_{AB}$. These equations may therefore be solved.

$$\bar{i} \text{ component:} \quad 0.15\dot{\omega}_{AB} + (34.83)^2(0.4243) = 0.15\dot{\omega}_C - (98.52)^2(0.15)$$
$$\bar{j} \text{ component:} \quad 0.4243\dot{\omega}_{AB} - (34.83)^2(0.15) = -0.15\dot{\omega}_C$$
$$\dot{\omega}_{AB} = -3114 \text{ rad/s}^2, \qquad \dot{\omega}_C = 10{,}022 \text{ rad/s}^2$$

The constraint equation for $\bar{a}_C$ then gives

$$\bar{a}_C = 0.15(10{,}022)\bar{i} = 1503\bar{i} \text{ m/s}^2$$

The large value of $\bar{a}_C$ is a consequence of the high rates of rotation ω_{AB} and ω_C, which produce very rapid changes in the geometrical arrangement of the constrained points.

HOMEWORK PROBLEMS

13.68 Block B is moving to the right at a constant speed of 600 mm/s and end A of the bar is held stationary, $v_A = 0$. Determine the angular velocity and angular acceleration of the bar in the position where $\theta = 30°$.

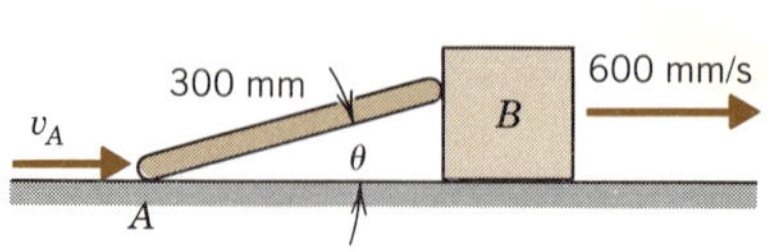

PROB. 13.68

13.69 Solve Problem 13.68 if end A of the bar is moving to the right at a constant speed of 900 mm/s and the other parameters are unchanged.

13.70 Determine the speed v_A and the angular velocity of each pulley in terms of the speed v_B for the upward motion of the block. There is no slipping between the cables and the pulleys.

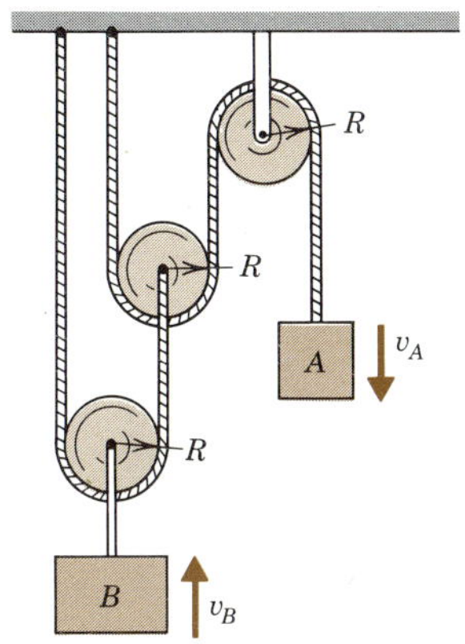

PROB. 13.70

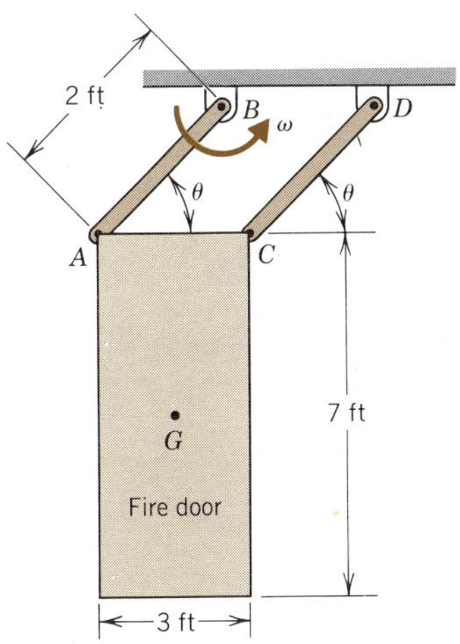

PROBS. 13.71 AND 13.72

13.71 When $\theta = 30^\circ$ the rotation of bar AB is known to be $\omega = 1.2$ rad/s, $\dot{\omega} = 1.5$ rad/s^2. Determine the corresponding velocity and acceleration of the centroid G of the fire door.

13.72 When $\theta = 36.87^\circ$, the centroid G of the fire door has a speed of 4 ft/s, and the magnitude of its acceleration is 15 ft/s^2. Determine ω and $\dot{\omega}$ for bar AB.

13.73 The flywheel has a constant clockwise angular velocity $\bar{\omega}$. For the instant shown, determine the velocity and acceleration of (a) pin C and (b) midpoint E.

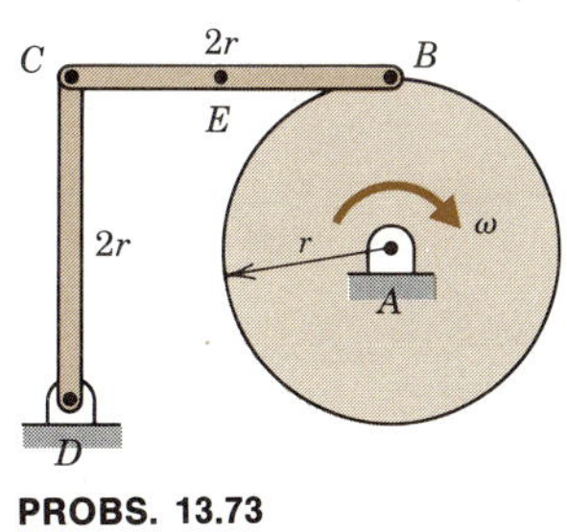

PROBS. 13.73

***13.74** Determine the velocity of pin C and of midpoint E when the flywheel in Problem 13.73 has rotated 180° from the position shown.

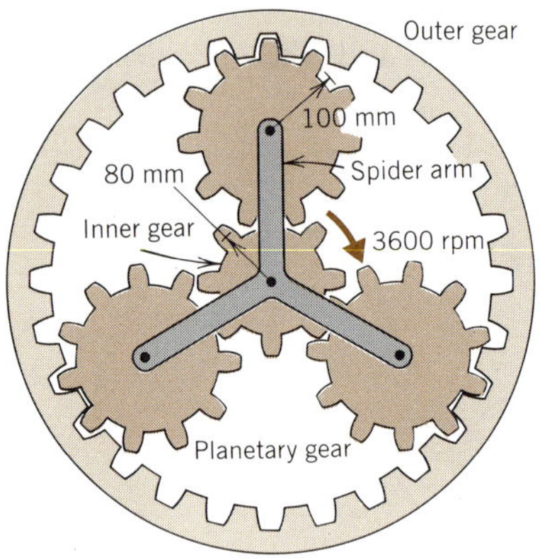

PROB. 13.75

13.75 A motor rotates the inner gear in the planetary gear transmission at a rate of 3600 rev/min clockwise. Determine the angular velocity of the outer gear when the spider arms are held fixed.

13.76 The outer gear in Problem 13.75 is held fixed. Determine the angular velocity of the spider arms.

13.77 At the instant shown, the elevator is descending at the rate of 4 m/s and is slowing down at 0.5 m/s^2. The cable does not slip over the pulley. Determine (a) the angular velocity and angular acceleration of the large upper pulley, and (b) the velocity and acceleration of the counterweight.

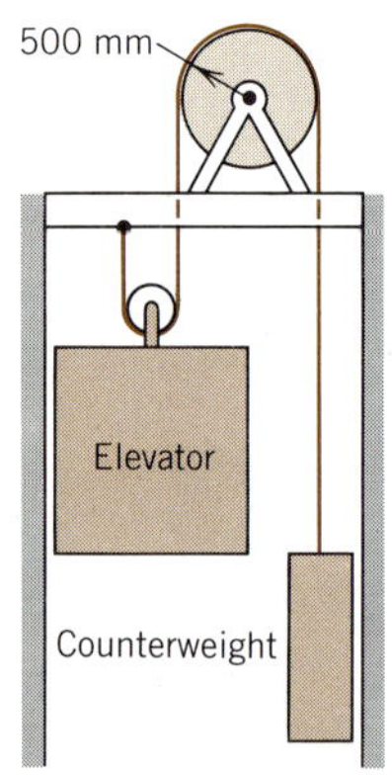

PROB. 13.77

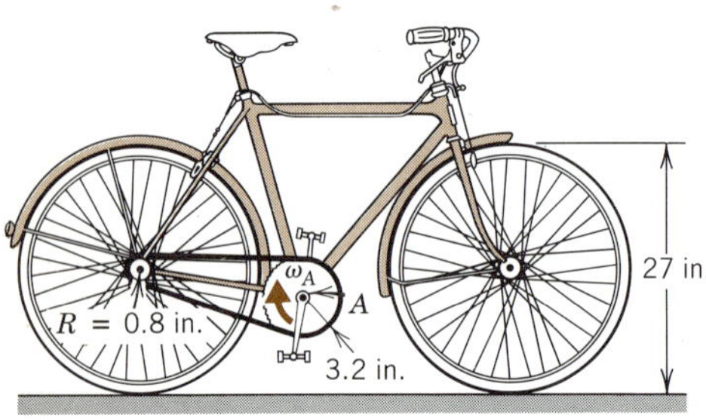

PROB. 13.78

13.78 The pedals rotate gear A on the bike at $\omega_A = 1.2$ rev/s. Determine the speed of the bicycle.

13.79 In the position shown bar AB is rotating at a constant rate of 120 rev/min. Determine the velocity of the midpoint G of bar BC at this instant.

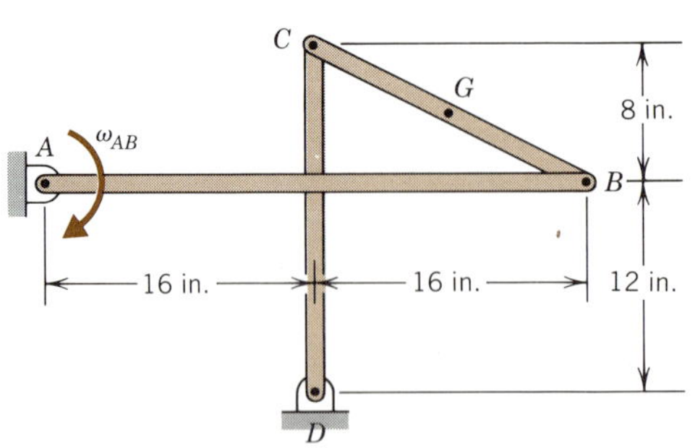

PROB. 13.79

13.80 Determine the angular acceleration of each bar in Problem 13.79.

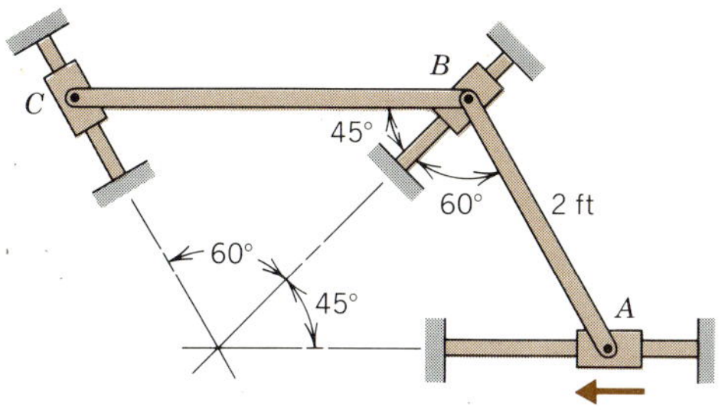

PROB. 13.81

***13.81** Collar A is moving to the left at 20 ft/s when the linkage is in the position shown. Determine the corresponding velocities of collars B and C.

13.82 Bar AB is rotating clockwise at 450 rev/min when the linkage is in the position shown. Determine the corresponding angular velocities of the other bars.

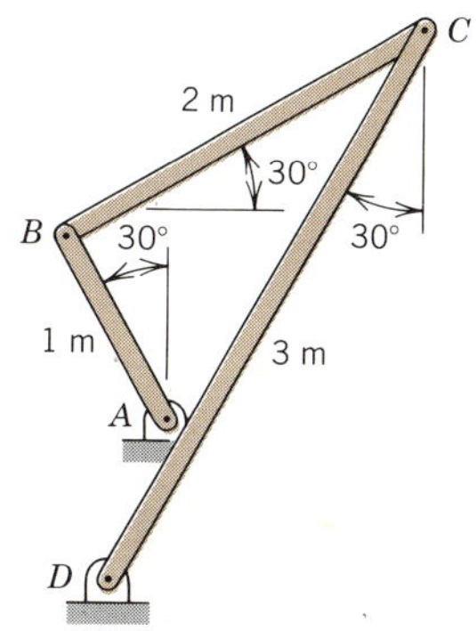

PROBS. 13.82 AND 13.83

13.83 Bar AB is rotating clockwise at 450 rev/min and is slowing at the rate of 50 rev/min/s when the linkage is in the given position. Determine the corresponding angular acceleration of the other bars.

13.84 The semicircular cylinder rolls without slipping. It is released from rest at $\theta = 0$. The angular acceleration at that instant is 2 rad/s^2 counterclockwise. Determine the corresponding acceleration of the centroid G.

***13.85** At the instant when $\theta = 36.87°$, the semicircular cylinder is rolling without slipping at an angular velocity of 2 rad/s. The rate of rotation is increasing at 1 rad/s^2. Determine the corresponding velocity and acceleration of the centroid G.

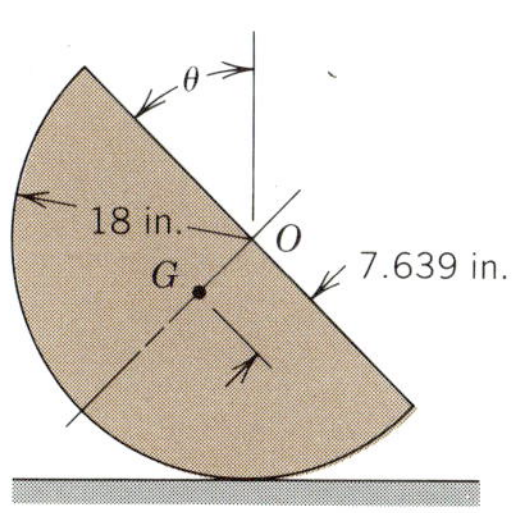

PROBS. 13.84 AND 13.85

13.86 The disk rotates clockwise at 5000 rev/min. Determine the velocity and acceleration of piston C when $\theta = 30°$.

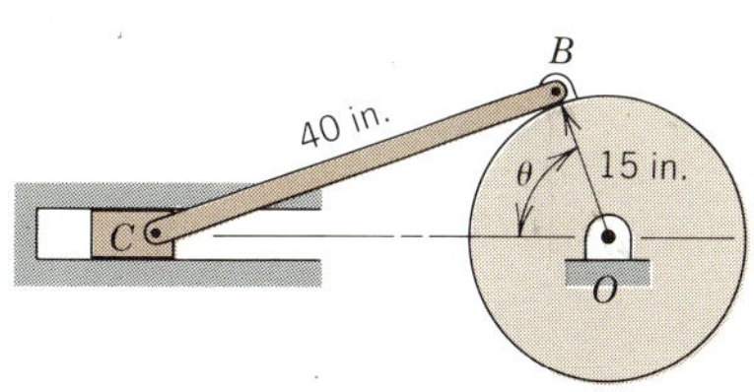

PROB. 13.86

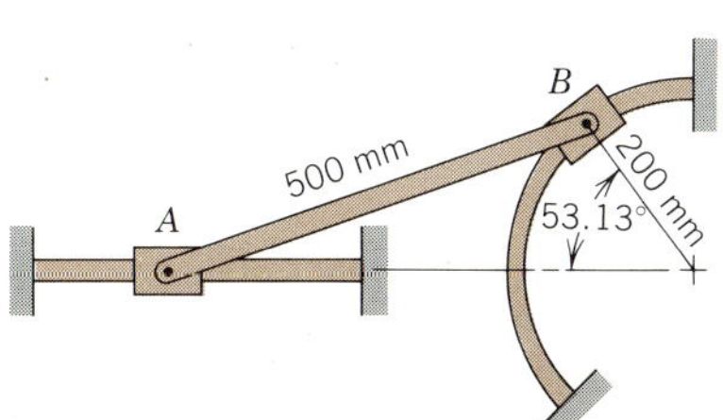

PROB. 13.87

***13.87** Collar A is moving to the right at a constant speed of 40 m/s. For the instant shown, determine the velocity and acceleration of collar B.

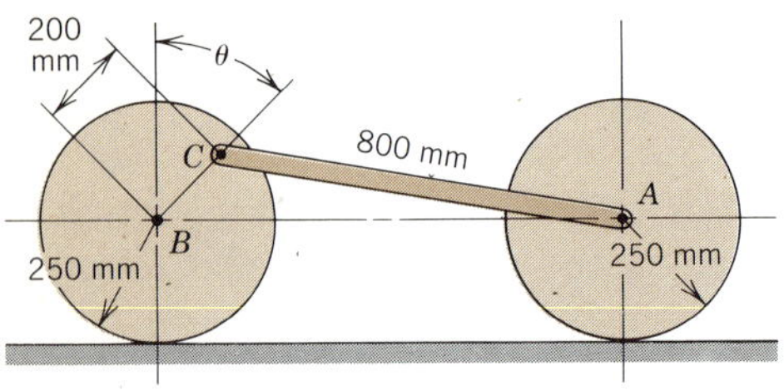

PROB. 13.89

13.88 In the linkage studied in Illustrative Problem 4, collar A is moving to the right at 2 m/s and gaining speed at 20 m/s^2 when $\theta = 0$. Determine the corresponding velocity and acceleration of the center C of the gear.

13.89 Two wheels that are interconnected by rod AC roll along the ground without slipping. Point A has a constant velocity of 5 m/s to the right. Determine the velocity and acceleration of point B in the position where $\theta = 0$.

13.90 For the system of Problem 13.89, determine the velocity and acceleration of wheel B in the position where $\theta = 90°$.

***13.91** Arm AB is rotating counterclockwise at the constant rate of 10 rad/s. The large inner gear is stationary. Determine the velocity and acceleration of (a) center B and (b) perimeter point C.

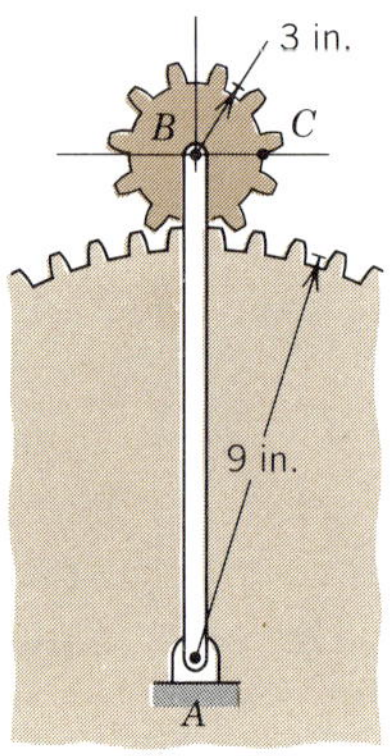

PROB. 13.91

CHAPTER 14

KINETICS OF A RIGID BODY IN PLANAR MOTION

OBJECTIVES

1. To use the method of integration to locate the center of mass of a body, and to determine the moment of inertia of a body about a specified axis.

2. To employ the parallel axis theorem to evaluate intertia properties in one coordinate system by using the properties for a parallel coordinate system (a) for a single rigid body and (b) for composite bodies.

3. To formulate the basic equations of motion, which requires (a) using free body diagrams as an aid in summing force components and moments; (b) identifying the appropriate point for summing moments, based on the type of motion; (c) using the kinematical constraint conditions to relate the angular acceleration and the acceleration of the center of mass.

4. When appropriate, to employ auxiliary conditions, such as Coulomb's friction equations, in order to make the equations of motion solvable.

5. To solve the equations of motion, and to interpret the results to be sure that they agree with assumptions that led to the solution.

6. To employ the work–energy principle to relate the velocity of a rigid body at two different positions. This involves (a) expressing the kinetic energy of a rigid body in terms of the velocity of a point or the angular velocity of the body, (b) using potential energy to account for the work done by conservative forces, and (c) evaluating the work done by non-conservative forces.

7. To employ the angular and linear impulse momentum equations to relate the motion of a rigid body at two different time instants, (a) when time is explicit in the problem, and (b) when the problem involves impulsive forces.

In this chapter we shall establish relationships between force systems acting on a rigid body and the motion of the body. We will restrict our attention to the broad class of problems where the significant forces are parallel to the plane of motion, and to rigid bodies that have a plane of symmetry that is parallel to the plane of motion. Problems that do not fit one or both of these specifications will be discussed in Chapter 16, where three-dimensional effects are studied.

A. KINETICS PRINCIPLES

1. Basic Considerations

In essence, a rigid body is a continuum of particles of infinitesimal mass that are in contact with their neighboring particles. It is obviously not possible to describe individually the motion of each of these particles. Fortunately, the constraint that the particles maintain a fixed geometry with respect to one another simplifies the treatment. The only kinematical parameters that are needed to describe a rigid body are those that occur in Chasle's theorem, specifically, the movement of any point in the body and the angle of rotation of the body.

Thus we will employ kinetics principles for a particle system that pertain to the motion of a point and the rotation about an axis. The principles we select for this purpose match our physical understanding of the properties of a force system.

First of all, we know that the resultant of a force system is associated with a net pushing or pulling effect. The resultant, $\sum \bar{F}$, acting on a general system of particles is related to the acceleration of the center of mass G by

$$\boxed{\sum \bar{F} = m\bar{a}_G} \tag{1}$$

where m is the total mass of the system.

Although this equation is similar to Newton's second law for a single particle, there are some important differences. For example, the resultant, $\sum \bar{F}$, is formed from the external forces acting on the system, in contrast to the full set of forces that affect the acceleration of a particle.

Another difference between eq. (1) and Newton's second law is that a particle occupies only a single point in space. In a rigid body, the location of the center of mass, as well as the mass itself, must be known.

In a general system of particles, the center of mass shifts relative to the particles as the arrangement of the particles varies with time. However, in a rigid body the center of mass, G, is a point in the body. In other words, eq. (1) is the principle we shall use to investigate the motion of a particular point, the center of mass, in a rigid body.

Let us now consider the rotation of a rigid body. For this, we use our intuitive understanding that the tendency of a force system to cause a body to rotate is measured by the moment of the resultant force. Because the moment of a force occurs in the angular impulse-momentum principle, it is consistent to use this principle to study the angular motion of a rigid body.

Equation (1) is the time derivative of the *linear* impulse-momentum principle for a system of particles. Similarly, we use the derivative form of the *angular* impulse-momentum principle, that is

$$\sum \bar{M}_A = \frac{d}{dt} \bar{H}_A \tag{2}$$

As was true for the resultant force, the resultant moment $\sum \bar{M}_A$ accounts for only the external forces; each internal force cancels the moment of its reaction when we treat the entire system.

An important restriction on eq. (2) is that it is valid only if point A is

1. the center of mass, or
2. a fixed point, or
3. a point accelerating directly toward or away from the center of mass.

This restriction will play an important role in the developments to follow.

Clearly, eq. (2) is not usable in its current form—we need to relate $\bar{H}_A$ to the angular velocity. This is the topic of the next section.

2. Angular Momentum

The basic expression for the angular momentum $\bar{H}_A$ was developed in Chapter 12. Forming it involves the linear momentum of each particle i relative to point A (see Figure 1), $m_i\bar{v}_{i/A}$. Then $\bar{H}_A$ is the sum of the moments about point A of each relative momentum.

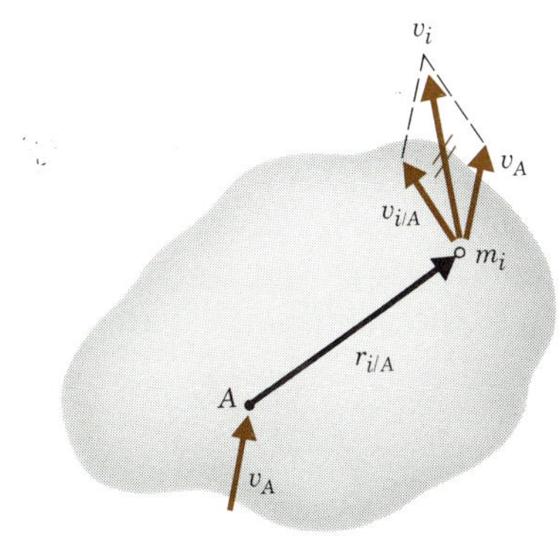

FIGURE 1

$$\bar{H}_A = \sum_{i=1}^{N} \bar{r}_{i/A} \times m_i\bar{v}_{i/A} \tag{3}$$

We will use the general rigid body in Figure 1 to obtain the special form of $\bar{H}_A$ associated with rigid body motion. The kinematical equations governing rigid body motion relate the velocity of each particle to that of point A.

It is important to note that using such a formulation places a subtle restriction on point A. Specifically, *point A must be attached to the body.* The foregoing restriction is important when we consider the points A for which eq. (2) is valid. For example, all points that are part of the inertial reference frame are fixed. However, it is not valid to use such a point when formulating $\sum \bar{M}_A = \dot{\bar{H}}_A$ for a rigid body in general motion, because no point in such a body is fixed. The center of mass G presents the opposite situation. It obviously is a valid point in the rigid body about which to formulate the angular momentum, regardless of the type of motion the body undergoes.

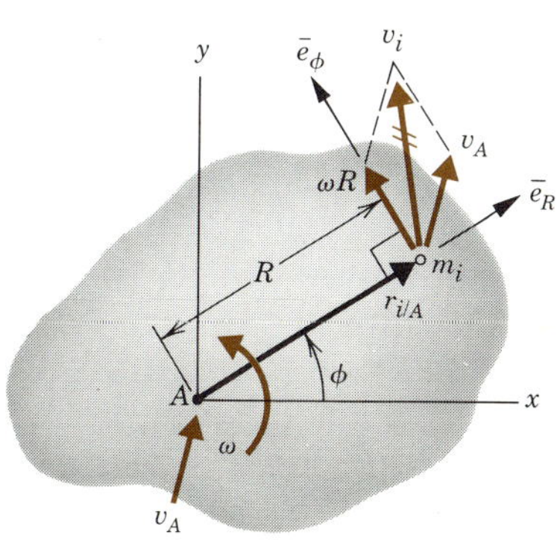

FIGURE 2

Figure 2 depicts the superposition of velocities that is obtained from Chasle's theorem. We consider the motion of particle m_i to be produced by a translation following point A and a rotation about point A. For convenience, let R denote the radial distance from point A to m_i, that is, R is the magnitude of the relative position vector

$$\bar{r}_{i/A} = R\bar{e}_R \tag{4}$$

As shown in Figure 2, the rotational velocity is ωR, perpendicular to $\bar{r}_{i/A}$. The velocity $\bar{v}_i$ of the particle is the vector sum of this rotational velocity $\omega R e_\phi$ and the translational velocity $\bar{v}_A$. It follows that

$$\bar{v}_{i/A} \equiv \bar{v}_i - \bar{v}_A = \omega R \bar{e}_\phi \tag{5}$$

The relative linear momentum is obtained by multiplying $\bar{v}_{i/A}$ by the mass m_i. Because of the perpendicular orientation of $\bar{v}_{i/A}$, the lever arm of this vector about point A is the radial distance R. Both $\bar{r}_{i/A}$ and $m_i\bar{v}_{i/A}$ lie in the plane, so the right-hand rule shows that the moment is perpendicular to the plane in the same sense as that of $\bar{\omega}$. Hence

$$\bar{r}_{i/A} \times m_i\bar{v}_{i/A} = R(m_i\omega R)\bar{k} = m_i R^2 \bar{\omega} \tag{6}$$

We must now sum this term for each particle in the body. In order to perform such an operation we consider the rigid body of Figure 2 to be composed of an infinite number of infinitesimal masses $dm = m_i$. Then the summation becomes an integral over all of the mass elements dm in the region $\mathscr{V}$ occupied by the body. In conjunction with eq. (6), this changes eq. (3) into

$$\bar{H}_A = \int_{\mathscr{V}} \bar{r}_{i/A} \times (dm)\bar{v}_{i/A} = \int_{\mathscr{V}} R^2 \bar{\omega}\, dm = \bar{\omega} \int_{\mathscr{V}} R^2\, dm$$

The angular velocity $\bar{\omega}$ was factored out of the integral because it is a function of time only. The remaining integral is a basic inertial property called the *moment of inertia*. Note that it is a *constant* because the radial distance R to each mass element cannot change due to the body's rigidity. We will use the notation I_{zz} to denote the moment of inertia. Thus

$$\boxed{\bar{H}_A = I_{zz}\bar{\omega} \quad \text{for planar motion}} \tag{7}$$

where

$$\boxed{I_{zz} = \int_{\mathscr{V}} R^2\, dm} \tag{8}$$

Care must be taken to avoid confusing I_{zz} with the second moment of area, which is sometimes called the area moment of inertia. We will consider the relation between the mass and area parameters in a later section.

Let us make it our standard procedure to

> Place the origin of the xyz coordinate system at the point A selected for the moment equation.

Such a coordinate system is depicted in Figure 2.

The subscript z in I_{zz} serves two purposes. It reminds us that the z axis is aligned perpendicular to the plane of motion. It also indicates that R is the perpendicular distance from the z axis to the mass element dm. (The only reason for the double subscript z is consistency. It agrees with the notation that arises in the kinetics of rigid bodies in three-dimensional motion, as we will see in Chapter 16.)

B. INERTIA PROPERTIES

1. The Meaning of Moment of Inertia

The developments in the previous sections indicate that the mass properties that influence the planar motion of a rigid body are

1. its mass m,
2. the location of its center of mass G within the body,
3. its moment of inertia I_{zz} about a specified z axis perpendicular to the plane of motion.

The first two items were discussed in Statics. Two approaches were used there. One involved integration of infinitesimal mass elements, and the other treated a rigid body as a composite of basic shapes. We will evaluate the moment of inertia by comparable methods.

Before we actually determine the inertia properties of a body, let us consider the significance of the moment of inertia. The value of I_{zz} conveys information regarding the distribution of the mass about the z axis. As an example, we have selected the z axis for the thin ring in Figure 3 to be the axis of symmetry. Each element of mass is essentially situated along the perimeter. The distance R from the z axis to each element is equal to the radius r of the ring, so this distance may be factored out of the integral.

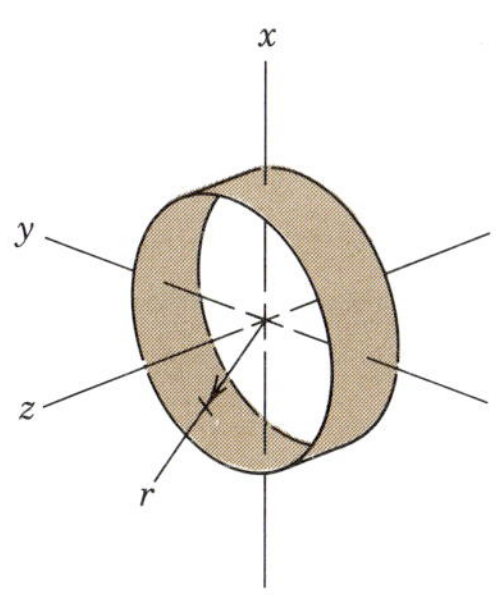

FIGURE 3

$$I_{zz} = \int_{\mathscr{V}} R^2\, dm = r^2 \int_{\mathscr{V}} dm = mr^2$$

Clearly, if two rings have the same mass, then the one with the larger radius has the larger moment of inertia.

Suppose we divide the value of I_{zz} for any body by the mass m of that body, and then take the square root of that quotient. The result is called the *radius of gyration*, k_z, of the body about the z axis,

$$\boxed{k_z = \sqrt{\frac{I_{zz}}{m}}} \tag{9}$$

Because $k_z = r$ for the ring in Figure 3, the radius of gyration of any body is the radius of a ring having the same inertia properties as the body under consideration. Note that the integrand in the definition of I_{zz}, eq. (8), is always positive, so that I_{zz} will always be positive. Hence k_z is a real number. Also, the value of k_z must be less than or equal to the largest distance R from the z axis to a mass element. This fact is useful for identifying computational errors in the evaluation of I_{zz}.

It sometimes happens that the radius of gyration of a body is known. The corresponding value of I_{zz} is readily obtained in that case by solving eq. (9). The following sections address situations where the inertia properties are less obvious.

2. Inertia Properties by Integration

Consider a situation where a body has a shape that is not familiar to us. In that case the mass and the location of the center of mass, as well as the moment of inertia, must be determined. The same integration procedure may be followed for all three tasks.

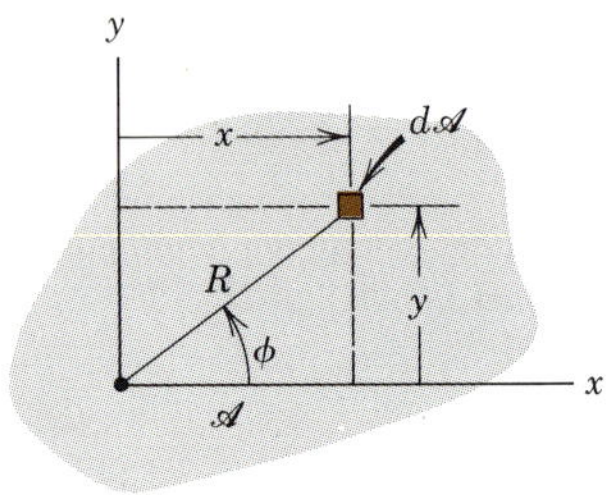

FIGURE 4

The area $\mathscr{A}$ in Figure 4 represents the outline in the xy plane of a rigid body. We will restrict our attention to bodies whose density depends on the position in the xy plane, which includes homogeneous (constant density) bodies. Then the mass element dm may be described in terms of the area element $d\mathscr{A}$.

Denote as h the thickness of the body in the z direction at the location of the mass element. (In general, h is *not* constant.) Then the infinitesimal volume of the mass element is $d\mathscr{V} = h\,d\mathscr{A}$. The area $d\mathscr{A}$ essentially occupies a single point in the xy plane, and we are allowing for no variation of density in the z direction. It follows that the density throughout the mass element is the single value ρ, which leads to

$$dm = \rho\, d\mathscr{V} = \rho h\, d\mathscr{A} \tag{10}$$

The mass m is the sum of the mass elements. Thus

$$\boxed{m = \int_{\mathscr{A}} \rho h\, d\mathscr{A}} \tag{11}$$

The coordinates (x_G, y_G) of the center of mass are defined by the first moments of mass with respect to x and y. The coordinates of the mass element in Figure 4 are (x, y), so

$$\boxed{\begin{aligned} mx_G &= \int_{\mathscr{V}} x\, dm \equiv \int_{\mathscr{A}} x(\rho h\, d\mathscr{A}) \\ my_G &= \int_{\mathscr{V}} y\, dm \equiv \int_{\mathscr{A}} y(\rho h\, d\mathscr{A}) \end{aligned}} \tag{12}$$

The distance R from the z axis to the mass element may also be expressed in terms of x and y. This merely involves applying the Pythagorean theorem to Figure 4. Thus the definition of the moment of inertia becomes

$$\boxed{I_{zz} = \int_{\mathscr{A}} R^2(\rho h\, d\mathscr{A}) \equiv \int_{\mathscr{A}} (x^2 + y^2)\rho h\, d\mathscr{A}} \tag{13}$$

As an aside, we should note that the moment of inertia involves quadratic powers of x and y. In contrast, eqs. (12) involve first powers. It is for this reason that I_{zz} is often called a *second moment of mass.*

This name brings to mind a confusion that sometimes arises. There is an area property called the *second moment of area*, or the *area moment of inertia.* We shall use the symbol J_z to denote this parameter. Its definition is (see Chapter 9 in Statics)

$$J_z = \int_{\mathscr{A}} R^2\, d\mathscr{A} \tag{14}$$

Compare eqs. (13) and (14). If the body is flat (constant thickness h) and homogeneous (constant density ρ), then ρh may be factored out of the integral for I_{zz}. This yields

$$\boxed{I_{zz} = \rho h J_z \quad (\text{constant } \rho h)} \tag{15}$$

This is the only case where either I_{zz} or J_z may be found by knowing the other parameter.

Let us now focus on the primary topic of this section, the computation of inertia properties by integration. Equations (11)–(13) require that we account for every element of area contained within the area $\mathscr{A}$. The integrals over $\mathscr{A}$ may be formulated in terms of either rectangular Cartesian coordinates (x, y) or polar coordinates (R, ϕ). In general we use the set of coordinates that most conveniently describe the shape of the boundary.

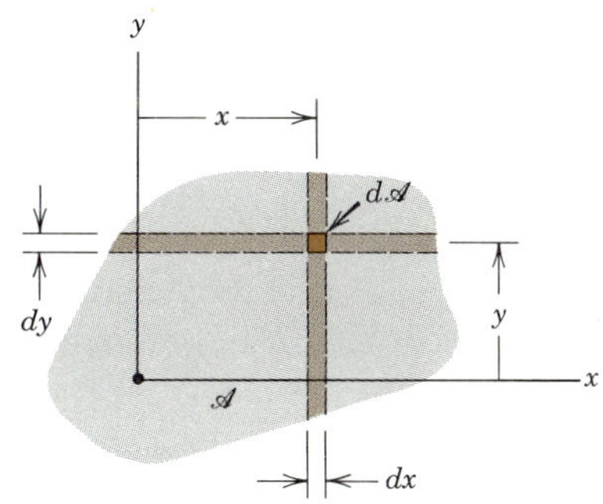

FIGURE 5

In the Cartesian coordinate system shown in Figure 5, the area element is

$$d\mathscr{A} = dx\,dy$$

An integration over $\mathscr{A}$ in this case begins by integrating first over either x or y, with the other coordinate held fixed. This forms either vertical strips (integrate over y) or horizontal strips (integrate over x). Then an integration over the other coordinate accounts for all strips. Note that the Cartesian coordinate formulation also requires that the density ρ and thickness h be expressed in terms of their x and y dependence whenever they are not constant.

The integration scheme for polar coordinates is depicted in Figure 6.

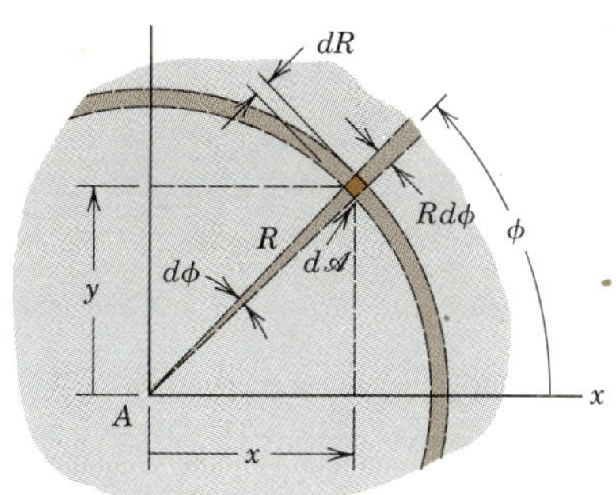

FIGURE 6

We see there that the area $d\mathscr{A}$ is the region formed by incrementing the radial distance by dR and the polar angle by $d\phi$. The arc length associated with the latter is $R\,d\phi$ (when second order differentials are neglected). Thus

$$d\mathscr{A} = R\,d\phi\,dR$$

The integration sequence now forms a sector when the range of values of R is integrated first, or a circular ring if the first integration is over the range of ϕ. (Which of these sequences is more suitable depends on the shape of the body.)

An important aspect of using cylindrical coordinates is that each term in an integrand must be expressed in terms of its dependence on R and ϕ. In addition to expressing ρ and h in such form, when they are not constant, this requires that the first moments of mass use the coordinate transformation from (x, y) to (R, ϕ), specifically

$$x = R\cos\phi, \qquad y = R\sin\phi$$

The following examples display a few highlights of the integration techniques for evaluating inertia properties. More detail on this subject is available in any calculus textbook that treats multiple integration.

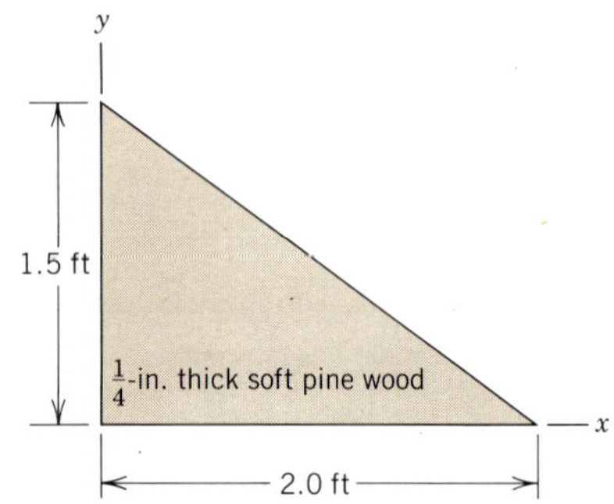

EXAMPLE 1

A 3/4 in. sheet of soft pine wood is cut into the triangular shape shown in the diagram. Using the methods of integration, locate the center of mass of the sheet and then determine the value of I_{zz} for the coordinate axes shown.

Solution

The starting point for an integration over an area is the description of the boundary. Two sides of the triangle coincide with the coordinate axes, and the hypotenuse is defined by the straight line formula, $y = mx + b$. The constant m is the slope, and b is the y intercept, which gives

$$y = -\frac{1.5}{2}x + 1.5 = 1.5 - 0.75x \text{ ft}$$

It is obvious from the foregoing that the integrals should be formulated in terms of (x, y) coordinates. As the sketch indicates, we shall first form a

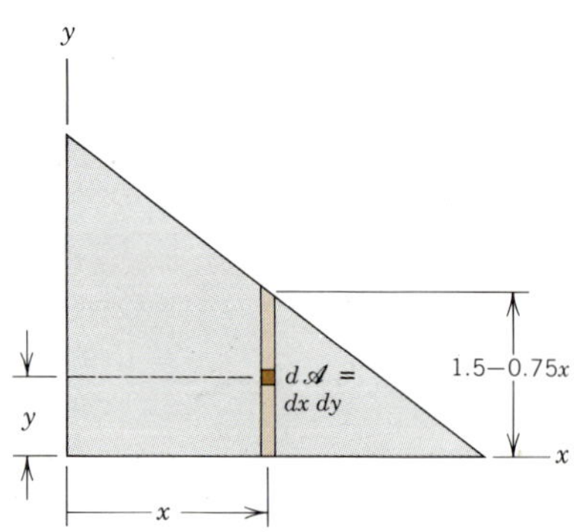

vertical differential strip by integrating in the y direction from $y = 0$ to $y = 1.5 - 0.75x$. Then all strips will be summed by integrating from $x = 0$ to $x = 2$ ft. The thickness of the sheet is constant at $h = (3/4)/12$ ft, and the density of soft pine obtained from Appendix D is

$$\rho = 0.48\left(\frac{62.4 \text{ lb/ft}^3}{32.17 \text{ ft/s}^2}\right) = 0.9311 \text{ lb-s}^2/\text{ft}^4$$

Both ρ and h are constant factors, so

$$m = \rho h \int_0^2 \int_0^{1.5-0.75x} dy\,dx = \rho h \int_0^2 (1.5 - 0.75x)\,dx$$

$$= 0.08729 \text{ lb-s}^2/\text{ft}$$

The first moments of mass are evaluated by the same procedure.

$$mx_G = \rho h \int_0^2 \int_0^{1.5-0.75x} x\,dy\,dx = \rho h \int_0^2 x(1.5 - 0.75x)\,dx$$

$$= 0.05819 \text{ lb-s}^2$$

$$my_G = \rho h \int_0^2 \int_0^{1.5-0.75x} y\,dy\,dx = \rho h \int_0^2 \tfrac{1}{2}(1.5 - 0.75x)^2\,dx$$

$$= 0.04364 \text{ lb-s}^2$$

Substituting for m then yields

$$x_G = 0.667 \text{ ft}, \qquad y_G = 0.500 \text{ ft}$$

The value of I_{zz} is obtained by using the Cartesian coordinate form in eq. (13), but the integration sequence is unchanged.

$$I_{zz} = \rho h \int_0^2 \int_0^{(1.50-0.75x)} (x^2 + y^2)\,dy\,dx$$

$$= \rho h \int_0^2 [x^2(1.5 - 0.75x) + \tfrac{1}{3}(1.5 - 0.75x)^3]\,dx$$

$$= 0.09093 \text{ lb-s}^2\text{-ft}$$

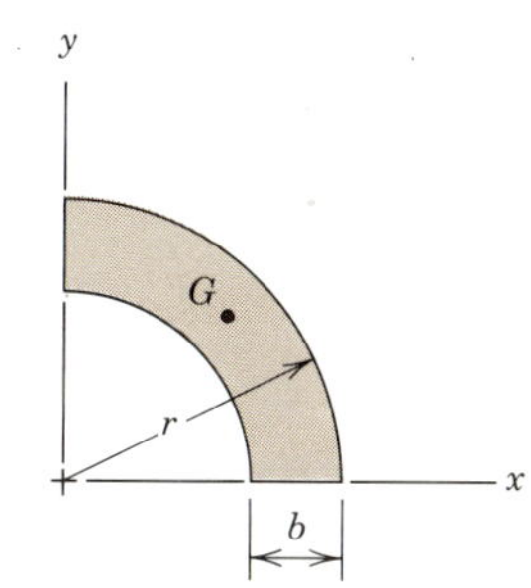

EXAMPLE 2

The homogeneous body shown has a constant thickness h perpendicular to the plane. Derive expressions for the mass m, the coordinates (x_G, y_G) of the center of mass, and the moment of inertia I_{zz} in terms of the outer radius r and the width b. Then evaluate the limiting case of a wire by letting the width b approach zero while the combination ρhb approaches the mass per unit length σ of the wire.

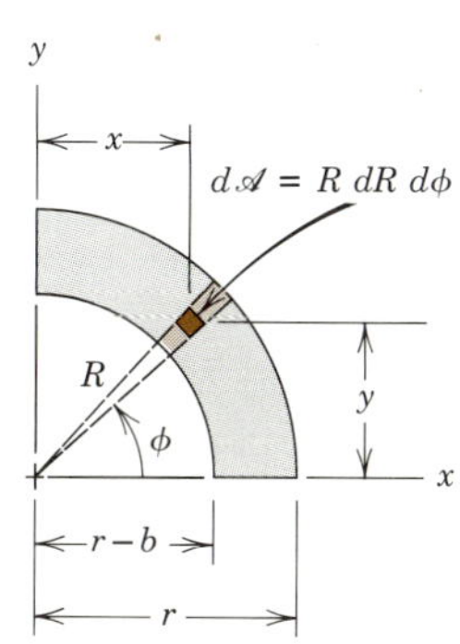

Solution

Two edges of the shape are circles centered on the origin of the coordinate axes and the other two edges are aligned with radial lines. This situation is ideal for using polar coordinates. Because each edge corresponds to a constant value of one of the R, ϕ coordinates, the sequence of integrations is arbitrary. As indicated in the sketch, we will integrate first over $r - b \leq R \leq r$, and then integrate over $0 \leq \phi \leq \pi/2$.

The area of the shaded element shown in the sketch is $d\mathscr{A} = R\,d\phi\,dR$, and the factor ρh is constant. Thus the mass is

$$m = \rho h \int_0^{\pi/2} \int_{r-b}^{r} R\,d\phi\,dR = \frac{1}{2}\rho h\left(\frac{\pi}{2}\right)[r^2 - (r - b)^2]$$

$$= \frac{\pi}{4}\rho h(2rb - b^2)$$

This body is symmetric with respect to a line at $\phi = 45°$. Hence, the center of mass must be situated on this line. It follows that $x_G \equiv y_G$, so we need to evaluate only the first moment of mass with respect to x. The transformation is seen in the sketch to be $x = R\cos\phi$. Thus the corresponding first moment is

$$mx_G = \rho h \int_0^{\pi/2} \int_{r-b}^{r} (R\cos\phi) R\,d\phi\,dR$$

$$= \frac{1}{3}\rho h\left(\sin\frac{\pi}{2}\right)[r^3 - (r-b)^3] = \frac{1}{3}\rho h(3r^2 b - 3rb^2 + b^3)$$

$$x_G = \frac{(mx_G)}{m} = \frac{4}{3\pi}\frac{3r^2 - 3rb + b^2}{2r - b} = y_G$$

The integration for I_{zz} follows the same procedure

$$I_{zz} = \rho h \int_0^{\pi/2} \int_{r-b}^{r} R^2(R\,d\phi\,dR) = \frac{1}{4}\rho h\left(\frac{\pi}{2}\right)[r^4 - (r-b)^4]$$

$$= \frac{\pi}{8}\rho h(4r^3 b - 6r^2 b^2 + 4rb^3 - b^4)$$

The foregoing are general expressions. The limiting case of a wire is obtained by replacing ρhb by σ and equating to zero any term that still contains b. Thus

$$\lim_{b\to 0} m = \lim_{b\to 0}\left[\frac{\pi}{4}(\rho hb)(2r - b)\right] = \frac{\pi}{2} r\sigma$$

$$\lim_{b\to 0} x_G = \lim_{b\to 0}\left[\frac{4}{3\pi}\frac{3r^2 - 3rb + b^2}{2r - b}\right] = \frac{2}{\pi} r = y_G$$

$$\lim_{b\to 0} I_{zz} = \lim_{b\to 0}\left[\frac{\pi}{8}(\rho hb)(4r^3 - 6r^2 b + 4rb^2 - b^3)\right] = \frac{\pi}{2} r^3\sigma$$

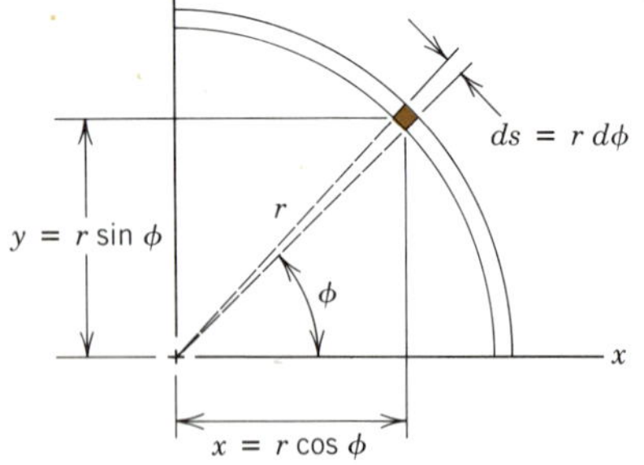

If we were only interested in obtaining the properties of a wire, we could have formulated the mass element dm in a different manner. The mass can be considered to be concentrated on a circle of radius r. As shown in the sketch, the mass element would then be the mass contained in an infinitesimal arc length $ds = r\,d\phi$. The corresponding mass is $dm = \sigma\,ds$. Then

$$m = \int_l \sigma\,ds = \sigma\int_0^{\pi/2} r\,d\phi = \frac{\pi}{2}\sigma r$$

$$mx_G = \int_l x(\sigma\,ds) = \sigma\int_0^{\pi/2}(r\cos\phi)(r\,d\phi) = \sigma r^2$$

$$x_G = y_G = \frac{\sigma r^2}{m} = \frac{2r}{\pi}$$

$$I_{zz} = \int_l r^2(\sigma\, ds) = \sigma \int_0^{\pi/2} r^2(r\, d\phi) = \frac{\pi}{2}\sigma r^3$$

EXAMPLE 3
Determine, by integration, the moment of inertia of a homogeneous sphere of radius a about an axis passing through the center of the sphere.

Solution
The axis of interest, which we denote as z, must intersect the center of the sphere. In order to make use of the symmetry of this body, we place the origin at the center. The shape of a sphere makes it logical to use polar coordinates to locate the differential mass elements. The significant variables are shown in the diagram below.

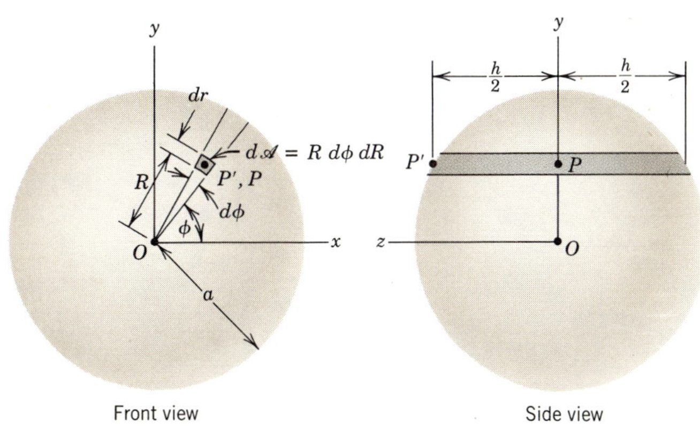

In this sketch the point P locates where the mass element intersects the xy plane and point P' locates where the mass element intersects the surface of the sphere. The length of the line OP is $R = \sqrt{x^2 + y^2}$, and the length of the line OP' is the radius a. Thus, because the mass element is perpendicular to the xy plane, the Pythagorean theorem gives

$$\frac{h}{2} = \sqrt{a^2 - R^2}$$

In order to obtain the contribution of all mass elements, we first integrate over the range of ϕ from 0 to 2π, and then integrate over the range of R from 0 to a. Because ρ is a constant, we have

$$I_{zz} = \rho \int_0^a \int_0^{2\pi} R^2 h(R\, d\phi\, dR) = 2\rho \int_0^a \int_0^{2\pi} R^3 \sqrt{a^2 - R^2}\, d\phi\, dR$$

$$= 2\rho(2\pi) \int_0^a R^3 \sqrt{a^2 - R^2}\, dR = \tfrac{8}{15}\pi\rho a^5$$

The last integral was evaluated with the aid of a table of integrals. In order to eliminate ρ, we note that the mass of a sphere is

$$m = \rho \mathscr{V} = \frac{4}{3}\pi\rho a^3 \Rightarrow \rho = \frac{3}{4\pi}\frac{m}{a^3}$$

Placing ρ into I_{zz} yields

$$I_{zz} = \tfrac{2}{5}ma^2$$

HOMEWORK PROBLEMS

14.1–14.4 Use the method of integration to determine I_{zz} in terms of the density ρ of each body shown.

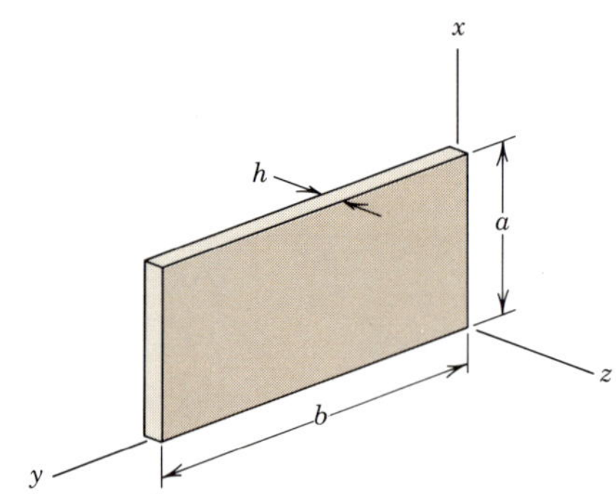

$I_{zz} = \frac{1}{3}\rho abh(a^2 + b^2)$

PROB. 14.1

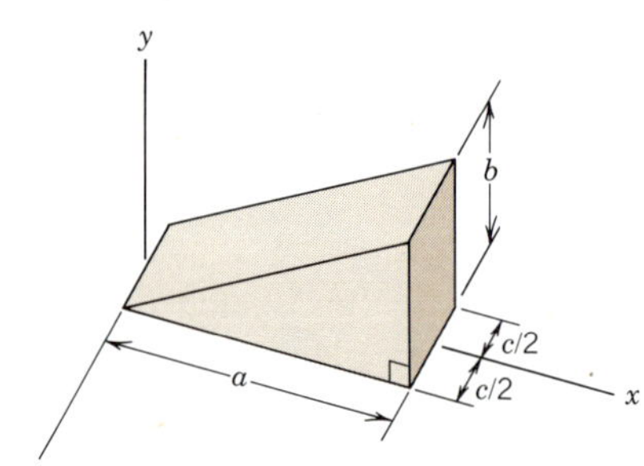

$I_{zz} = \frac{1}{12}\rho abc(3a^2 + b^2)$

PROB. 14.2

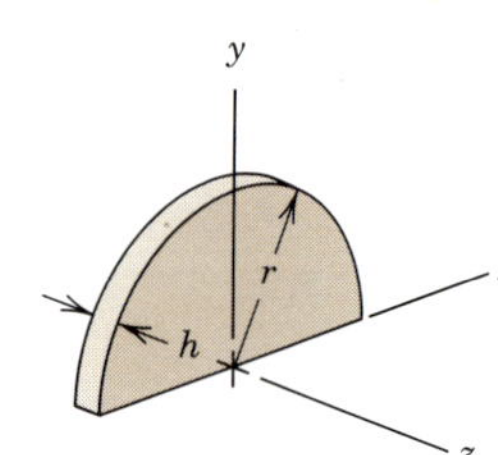

$I_{zz} = \frac{1}{4}\rho\pi r^4 h$

PROB. 14.3

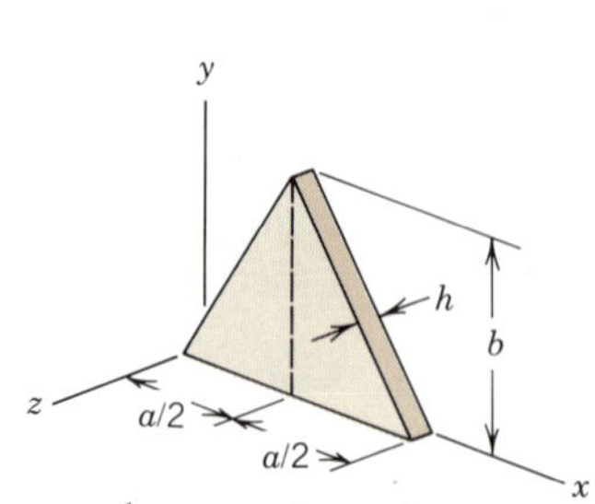

$I_{zz} = \frac{1}{48}\rho hab(7a^2 + 4b^2)$

PROB. 14.4

14.5–14.8 Each homogeneous body shown has mass m. Use the method of integration to determine the location of the center of mass and the value of I_{zz}.

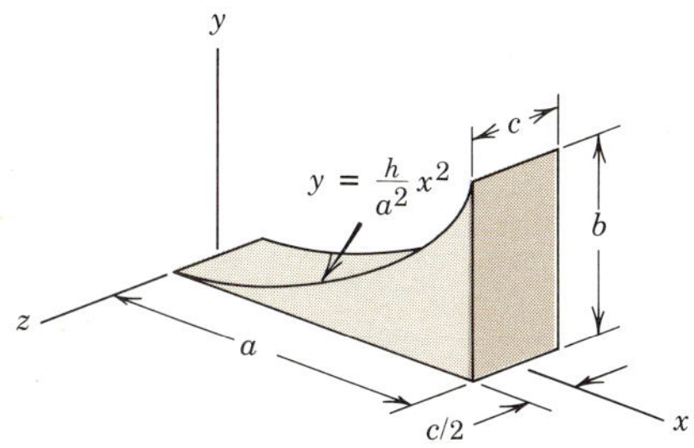

PROB. 14.5

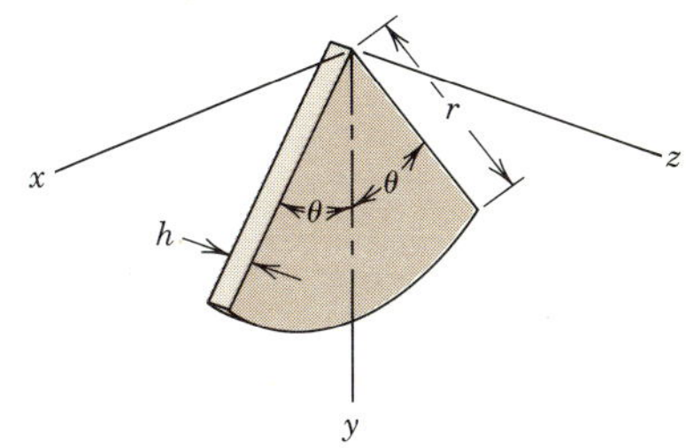

PROB. 14.6

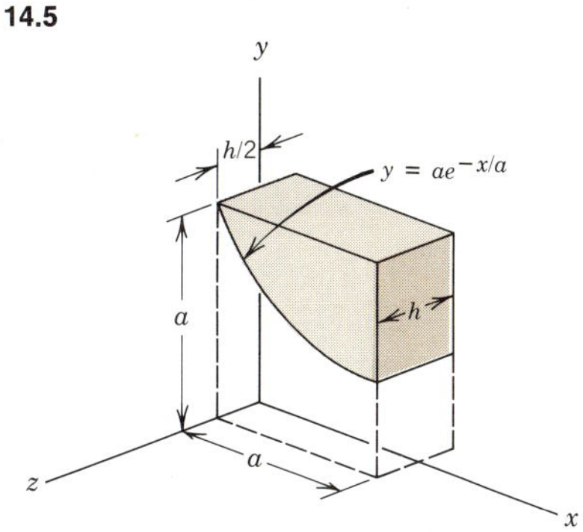

PROB. 14.7

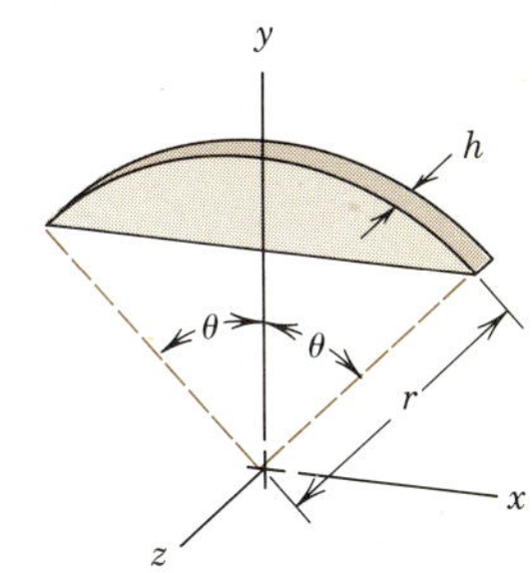

PROB. 14.8

14.9 and 14.10 The wire shown has a mass per unit length σ. Use the method of integration to determine the mass, the location of the center of mass, and I_{zz}.

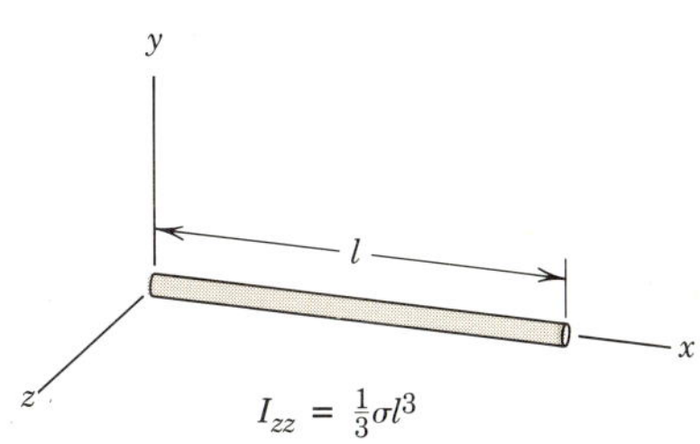

PROB. 14.9

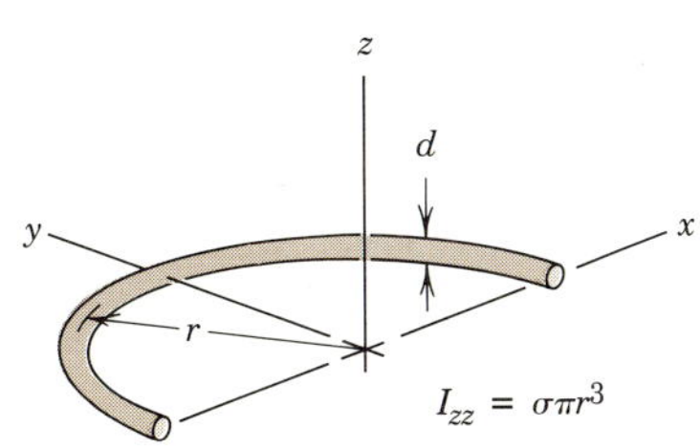

PROB. 14.10

***14.11 and *14.12** The body shown has a mass m. Use the method of integration to determine the height of the center of mass above the xy plane, as well as the value of I_{zz}.

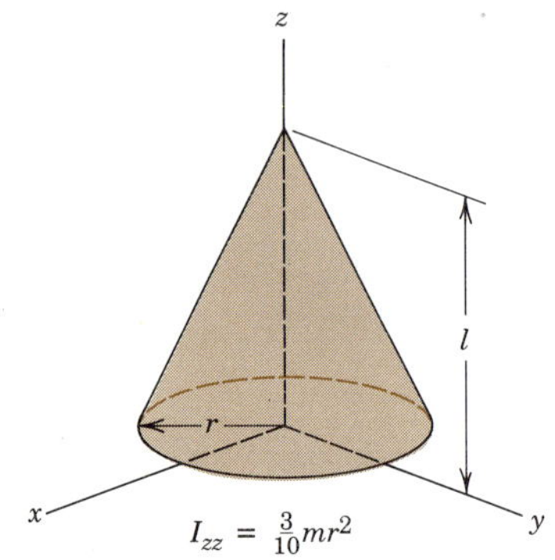

PROB. 14.11

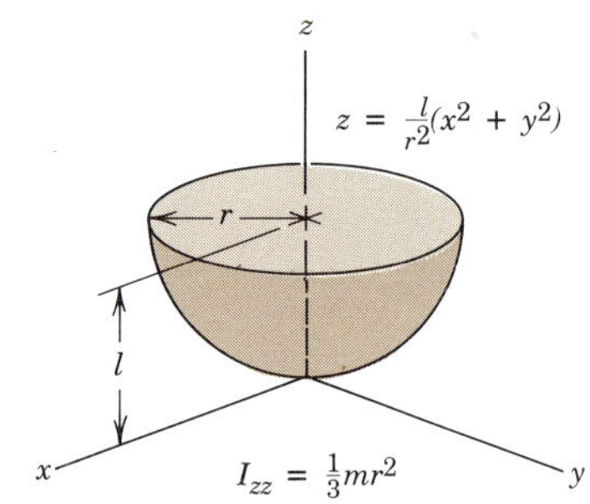

PROB. 14.12

---

3. The Parallel Axis Theorem

Each body that we have examined thus far can be termed an elementary shape. Such shapes occur frequently in engineering systems. For that reason it is useful to tabulate their inertia properties after they have been found by the method of integration. A standard table of centroidal locations and moments of inertia is given in Appendix B. In problem solving it often happens that we need the moment of inertia for a noncentroidal axis.

Suppose that we know $I_{\hat{z}\hat{z}}$, where $\hat{x}\hat{y}\hat{z}$ is a coordinate system whose origin is the center of mass. This is the situation depicted in Figure 7. It is our desire to use our knowledge of $I_{\hat{z}\hat{z}}$ to determine the moment of inertia I_{zz} about a different axis. In a planar motion the only moments of inertia that are needed are for axes that are perpendicular to the plane. Hence, we are interested in the case where the $\hat{z}$ and z axes are parallel. We have selected the x and y axes in Figure 7 to be parallel to the corresponding $\hat{x}$ and $\hat{y}$ axes as a convenience.

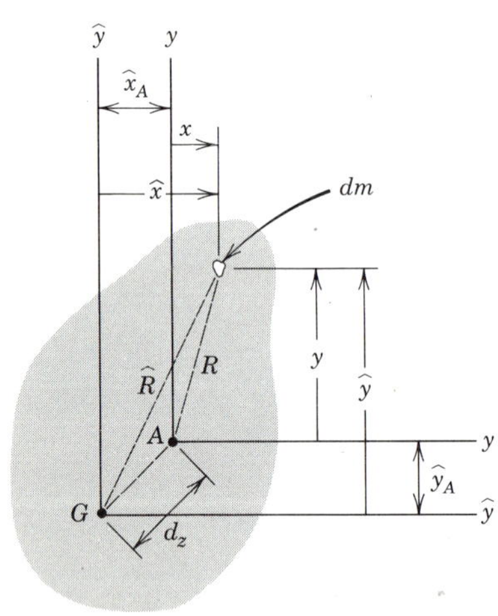

FIGURE 7

Let us assume that the moment of inertia with respect to the centroidal $\hat{x}\hat{y}\hat{z}$ system is known. We therefore locate the new origin, at A, by referring to the $\hat{x}\hat{y}\hat{z}$ coordinates measured from G. Thus

> $(\hat{x}_A, \hat{y}_A, 0)$ are the coordinates of point A with respect to the $\hat{x}\hat{y}\hat{z}$ coordinate system whose origin is the center of mass.

The moments of inertia with respect to each axis are

$$I_{\hat{z}\hat{z}} = \int_{\mathscr{V}} \hat{R}^2\, dm \equiv \int_{\mathscr{V}} (\hat{x}^2 + \hat{y}^2)\, dm$$

$$I_{zz} = \int_{\mathscr{V}} R^2\, dm = \int_{\mathscr{V}} (x^2 + y^2)\, dm$$

The situation depicted in Figure 7 corresponds to positive x_A and y_A.

Inspection of the figure reveals that the position coordinates of dm with respect to the parallel coordinate systems are related by

$$x = \hat{x} - \hat{x}_A, \qquad y = \hat{y} - \hat{y}_A$$

We substitute this relation into I_{zz} and use the fact that $\hat{x}_A$ and $\hat{y}_A$ are constants. This gives

$$\begin{aligned} I_{zz} &= \int_{\mathscr{V}} (x^2 + y^2)\,dm = \int_{\mathscr{V}} [(\hat{x} - \hat{x}_A)^2 + (\hat{y} - \hat{y}_A)^2]\,dm \\ &= \int_{\mathscr{V}} (\hat{x}^2 + \hat{y}^2)\,dm - 2\hat{x}_A \int_{\mathscr{V}} \hat{x}\,dm - 2\hat{y}_A \int_{\mathscr{V}} \hat{y}\,dm + [(\hat{x}_A)^2 + (\hat{y}_A)^2] \int_{\mathscr{V}} dm \end{aligned} \tag{16}$$

Because the origin of the $\hat{x}\hat{y}\hat{z}$ system is the center of mass, first moments of mass with respect to this coordinate system are zero. Therefore, the second and third integrals in eq. (16) for I_{zz} both vanish. Further, the first integral is the definition of $I_{\hat{z}\hat{z}}$, and the last integral is the mass m. Equation (16) then reduces to

$$\boxed{I_{zz} = I_{\hat{z}\hat{z}} + m[(\hat{x}_A)^2 + (\hat{y}_A)^2]} \tag{17a}$$

This is one form of the *parallel axis theorem*. An alternate, and more familiar, form is one featuring the distance d_z separating the pair of z axes in Figure 7. The result of using the Pythagorean theorem for d_z is

$$\boxed{I_{zz} = I_{\hat{z}\hat{z}} + m(d_z)^2} \tag{17b}$$

The application of either of eqs. (17) is straightforward, provided that you remember that the $\hat{z}$ *axis must pass through the center of mass*. If it does not, then the first moments of mass do not vanish and eqs. (17) are not valid.

HOMEWORK PROBLEMS

14.13 and 14.14 Use the value of I_{zz} given in the problem listed to determine the moment of inertia about a parallel axis which intersects the point at $y = r$.

14.13 Problem 14.11

14.14 Problem 14.12

14.15–14.19 For the body in the problem listed below, use the value of I_{zz} given there to determine the moment of inertia about a parallel axis that intersects the center of mass. The centroidal locations tabulated in Appendix B may be employed when necessary.

Problem Number	14.15	14.16	14.17	14.18	14.19
Body	14.1	14.2	14.3	14.9	14.10

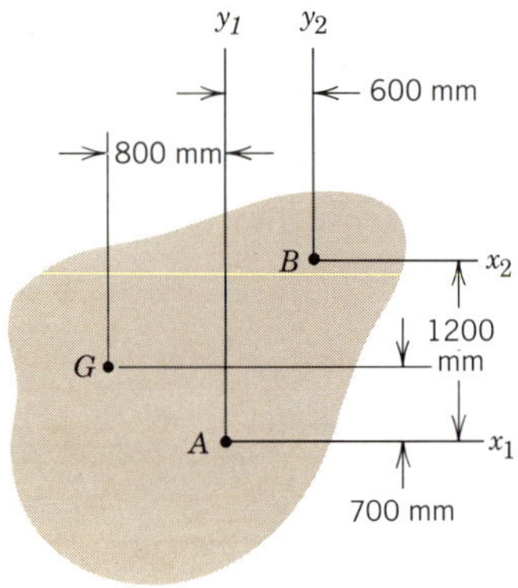

PROB. 14.20

14.20 Two mutually parallel coordinate systems are employed for the homogeneous body shown, whose center of mass is point G. It is known that $m = 2.5$ kg and that $I_{z_1 z_1} = 15$ kg–m^2. Determine $I_{z_2 z_2}$.

***14.21** The distances d_1 and d_2 and the angles θ and ϕ locate points A and B with respect to the centroidal axes $\hat{x}\hat{y}\hat{z}$. The mass of the body is m and the moment of inertia about an axis parallel to $\hat{z}$ intersecting point A is I_A. Use the law of cosines to determine the moment of inertia I_B about an axis parallel to $\hat{z}$ that intersects point B. The expression for I_B may contain only the parameters m, I_A, d_1, d_2, θ, and ϕ. What simplification in this relation results when $\theta = 90°$?

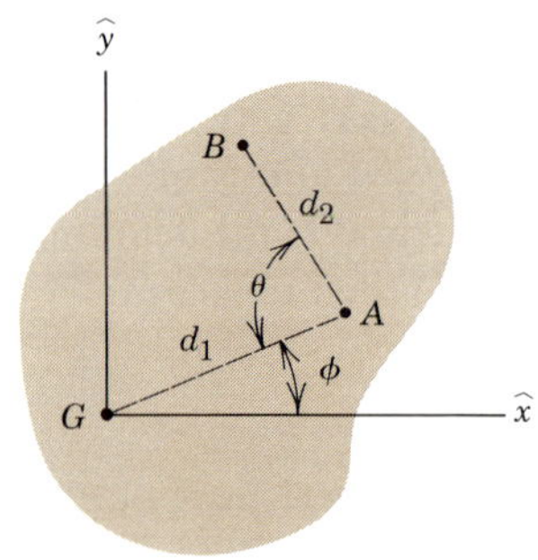

PROB. 14.21

4. Composite Bodies

One situation where the inertia properties in Appendix B are particularly useful is the treatment of bodies whose shape is a composite of several basic shapes. Such a situation is shown in Figure 8, where the composite body is a combination of two individual bodies having volumes $\mathcal{V}_1$ and $\mathcal{V}_2$.

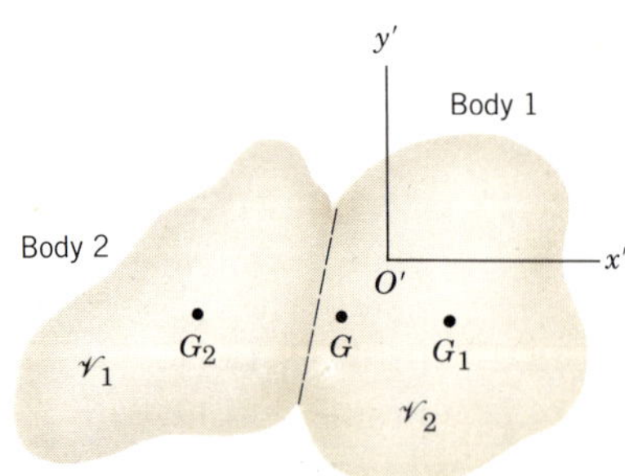

FIGURE 8

The tabulation in Appendix B describes homogeneous bodies only. Hence we will limit our attention to those cases where the individual bodies are homogeneous, although they may have differing densities ρ_1 and ρ_2. Because each part is homogeneous, the mass of each is the product of its density and its volume. It follows that the total mass is the sum of the masses of the individual bodies, so

$$\boxed{\begin{gathered} m_1 = \rho_1 \mathscr{V}_1, \qquad m_2 = \rho_2 \mathscr{V}_2 \\ m = m_1 + m_2 \end{gathered}} \tag{18}$$

The weight forces $m_i g$ acting on each body i at its center of mass G_i must exert the same moment as the total weight force mg acting at the center of mass G of the composite. This reasoning is the basis of the first moments of mass, which yield relations for the coordinates of point G. The coordinate system employed for formulating the first moments may have *arbitrary* origin and orientation, as does the $x'y'z'$ coordinate system appearing in Figure 8.

In practice we try to give the coordinate axes the same orientation as that used to dimension the body. The origin O' is usually located at a meaningful point in the system, such as a corner or the center of mass of one of the individual bodies.

Equating the first moments of mass with respect to x' and y' for the composite and for the individual bodies yields

$$\boxed{\begin{gathered} m x'_G = m_1 x'_{G_1} + m_2 x'_{G_2} \\ m y'_G = m_1 y'_{G_1} + m_2 y'_{G_2} \end{gathered}} \tag{19}$$

The values of x'_G and y'_G obtained from these relations describe where the center of mass of the composite is located within the body.

The only inertia property for planar motion that now remains to be evaluated is the moment of inertia I_{zz}. The z axis for this quantity may be different from the z' axis, which arises in the determination of the center of mass. In order to evaluate I_{zz} we return to its definition, eq. (8). The region $\mathscr{V}$ occupied by the composite is the sum of the regions $\mathscr{V}_1$ and $\mathscr{V}_2$ occupied by the individual bodies. Thus

$$I_{zz} = \int_{\mathscr{V}_1 + \mathscr{V}_2} R^2\, dm = \int_{\mathscr{V}_1} R_1^2\, dm + \int_{\mathscr{V}_2} R_2^2\, dm$$

Furthermore, the integrals on the right side are the definitions of the moments of inertia for the individual bodies about the z axis. It then follows that

$$\boxed{I_{zz} = (I_{zz})_1 + (I_{zz})_2} \tag{20}$$

This relation states that the moments of inertia of the individual bodies are additive, *provided that they are for the same z axis.*

When we solve kinetics problems, the z axis in eq. (20) will be the one that intersects the point we have selected for the moment sum. The moments of inertia for the fundamental geometric shapes in Appendix B may not be given about an axis that coincides with this z axis. In such cases it will be necessary first to apply the parallel axis theorem (17) to determine the value of I_{zz} for each component of the composite body before applying eq. (20).

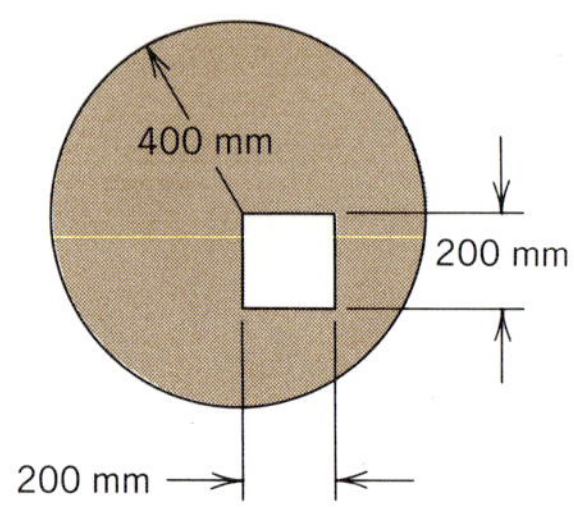

EXAMPLE 4

The aluminum body shown in the sketch has a thickness of 80 mm. Determine the mass of the body, the location of its center of mass, and its moment of inertia about an axis perpendicular to the plane of the diagram through the center of mass.

Solution

The body has a composite shape formed by removing a square parallelepiped (call it body 2) from a circular cylinder (call it body 1). This decomposition is depicted in a sketch that also shows the locations of the centers of mass of the individual bodies. The $x'y'z'$ coordinate system we will employ for formulating the first moment is aligned parallel to the edges of the rectangular cutout. The center of the circle is a convenient choice for the origin of $x'y'z'$.

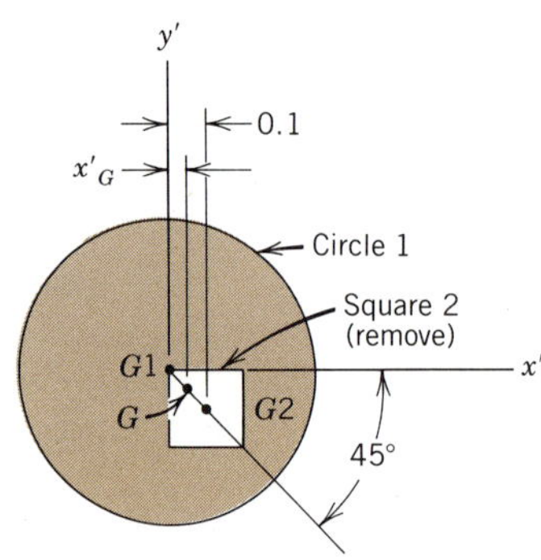

The body is symmetric about a line that is 45° below the x axis through the center G_1. The center of mass must be situated on the line. We therefore may reduce the amount of calculations by setting $y_G = -x_G$.

It is convenient to evaluate composite bodies by a systematic procedure. We will begin by expressing the information given in Appendix B for each individual body. Specifically, we evaluate each mass, the (x', y') coordinates of each center of mass, and the moment of inertia about a centroidal $\hat{z}$ axis for each mass. For this, we find from Appendix D that the density of aluminum is $\rho = 2690 \text{ kg/m}^3$, from which we find:

Circular Cylinder 1:

$$m_1 = \rho \mathcal{V}_1 = 2690(\pi)(0.4)^2(0.08) = 108.17 \text{ kg}$$

$$x'_{G_1} = y'_{G_1} = 0$$

$$(I_{\hat{z}\hat{z}})_1 = \tfrac{1}{2} m_1 r^2 = \tfrac{1}{2}(108.17)(0.4)^2 = 8.654 \text{ kg-m}^2$$

Rectangular Parallelepiped 2:

$$m_2 = \rho \mathcal{V}_2 = 2690(0.2)^2(0.80) = 8.608 \text{ kg}$$

$$x'_{G_2} = 0.10 \text{ m}, \qquad y'_{G_2} = -0.10 \text{ m}$$

$$(I_{\hat{z}\hat{z}})_2 = \tfrac{1}{12}m_2(a^2 + b^2) = \tfrac{1}{12}(8.608)[(0.2)^2 + (0.2)^2] = 0.0574 \text{ kg-m}^2$$

Next we use these properties to evaluate the mass and first moment of mass for the composite. Thus

$$m = m_1 - m_2 = 99.56 \text{ kg}$$

$$mx'_G = m_1 x_{G_1} - m_2 x_{G_2}$$

$$x'_G = \frac{1}{99.56}[0 - 8.608(0.10)] = -0.008646 \text{ m} \equiv -y'_G$$

Note that x'_G is negative, which means that the center of mass is to the left of the center of the circle.

We may now evaluate the moment of inertia of the individual bodies about the z axis through point G. Note that $(I_{\hat{z}\hat{z}})_1$ and $(I_{\hat{z}\hat{z}})_2$ are centroidal values, so the md_z^2 term is *additive* in each case. The distances $(d_z)_1$ and $(d_z)_2$ between points G_1 and G, and points G_2 and G, respectively, are found from the sketch with the aid of the Pythagorean theorem. Thus

$$(d_z)_1^2 = 2(x'_G)^2 = 1.4950(10^{-4}) \text{ m}^2$$

$$(d_z)_2^2 = 2(0.10 - x'_G)^2 = 2(0.10 + 0.00865)^2 = 2.361(10^{-2}) \text{ m}^2$$

$$(I_{zz})_1 = (I_{\hat{z}\hat{z}})_1 + m_1(d_z)_1^2 = 8.670 \text{ kg-m}^2$$

$$(I_{zz})_2 = (I_{\hat{z}\hat{z}})_2 + m_2(d_z)_2^2 = 0.261 \text{ kg-m}^2$$

Both values of I_{zz} are for the same axis. In this case, their difference is the value for the composite body.

$$I_{zz} = (I_{zz})_1 - (I_{zz})_2 = 8.41 \text{ kg-m}^2$$

HOMEWORK PROBLEMS

14.22–14.25 Determine the moment of inertia about the z axis shown using the inertia properties in Appendix B.

14.22 Aluminum cylinder and cone

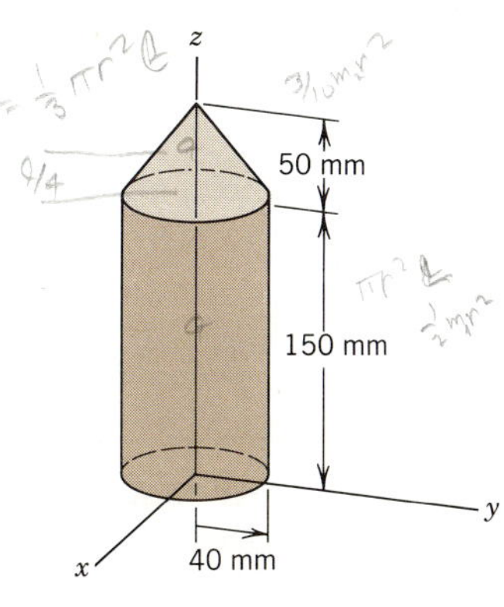

PROB. 14.22

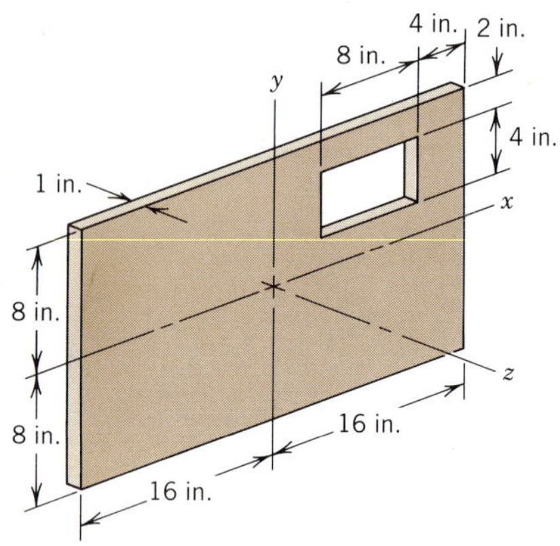

PROB. 14.23

14.23 Steel plate
14.24 Cast iron plate

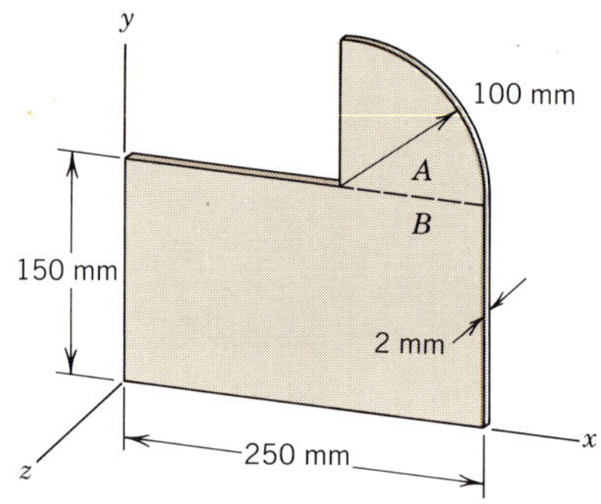

PROB. 14.24

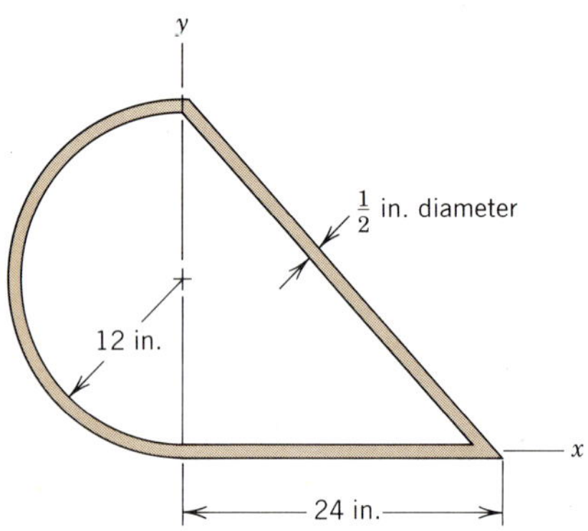

PROB. 14.25

14.25 Copper wire loop

14.26 Determine I_{zz} for the steel plate in Problem 14.23 in the case where the rectangular cutout is filled with lead flush with the surfaces of the plate.

14.27 Determine I_{zz} for the plate in Problem 14.24 if the rectangular section is composed of aluminum and the quarter circular panel is composed of steel.

14.28–14.30 For the body in the problem listed below, determine the mass, the location of the center of mass G, and the moment of inertia $I_{\hat{z}\hat{z}}$, where the $\hat{z}$ axis intersects point G and is parallel to the given z axis.

Problem Number	14.28	14.29	14.30
Body	14.23	14.24	14.25

C. THE GENERAL EQUATIONS OF PLANAR MOTION

In the previous section we provided the foundation for evaluating the inertia properties relevant to planar motion. The mass and location of the center of mass G are required to formulate the equation for translational motion of a rigid body, eq. (1), and the moment of inertia I_{zz} is needed to formulate the angular momentum of a rigid body, eq. (7).

Recall that I_{zz} is a constant parameter because the body is rigid. Thus the only variable in $\bar{H}_A$ is the angular velocity $\bar{\omega}$. The time derivative of $\bar{\omega}$ is the angular acceleration $\bar{\alpha}$. Therefore, taking the time derivative of $\bar{H}_A$ in eq. (7) and substituting that result into the moment eq. (2) yields

$$\sum \bar{M}_A = \frac{d}{dt}\bar{H}_A = \frac{d}{dt}(I_{zz}\bar{\omega}) = I_{zz}\bar{\alpha} \tag{21}$$

For planar motion $\bar{\alpha}$ is perpendicular to the plane. Without loss of generality, we will consider the motion to take place in the xy plane and write

$$\bar{\omega} = \omega\bar{k}, \qquad \bar{\alpha} \equiv \dot{\bar{\omega}} = \alpha\bar{k} \tag{22}$$

where

A positive value of α corresponds to an angular acceleration that is in the sense of the positive z axis according to the right-hand rule.

Because $\bar{\alpha}$ is in the $\bar{k}$ direction, it follows from eq. (21) that only the z component of $\sum \bar{M}_A$, or equivalently the moment about the z axis intersecting point A, is significant for planar motion. Therefore, when using eq. (21), the result is a scalar relation between the moment exerted by the force system and the angular acceleration α. Accordingly, it is useful to rewrite the basic equation for $\bar{a}_G$ as two scalar equations, one for each of the two planar components. The result is the following set of *three scalar equations of motion* governing the translation and rotation of a rigid body in planar motion.

$$\begin{aligned} \sum F_x &= m(a_G)_x \\ \sum F_y &= m(a_G)_y \\ \sum M_{Az} &= I_{zz}\alpha \end{aligned} \tag{23}$$

The first step in applying eqs. (23) is selecting the point A for formulating the moment equation. The criterion for selecting this point is that it be one of the allowable points discussed earlier. The specific choice for the three types of rigid body motions (pure translation, pure rotation, and general motion) are discussed in the next section.

The task of evaluating the force and moment sums in a dynamic system requires that we draw a free body diagram. The computation of the force sums from such a diagram should be familiar to you because it was treated in Chapter 12. However, it is useful to review the computation of moment for a planar force system.

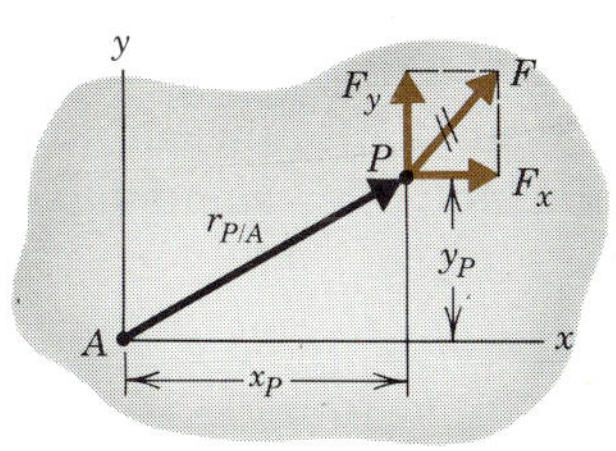

FIGURE 9

Figure 9 shows an xyz coordinate system whose origin A is the point about which we wish to sum moments. Any force may be replaced by its components, as we have done for the force $\bar{F}$ in Figure 9. The distances x_P and y_P locate the point of application P of this force in the directions of the coordinate axes.

The distances x_P and y_P are also the lever arms of the force components F_y and F_x, respectively. The moment of each component is its lever arm multiplied by the *magnitude* of the component. The *sign* of each moment is deduced by considering the tendency of that component to rotate the body. The right-hand rule is then used to compare that rotational tendency to the sense of the positive z axis. Thus, F_y in Figure 9 exerts a positive moment because it would cause a counterclockwise (positive z) rotation about point A. In contrast, F_x would tend to cause a clockwise rotation about point A, so its moment is negative. The moment of $\bar{F}$ is the sum of the moments of its components, so

$$\boxed{M_{Az} = x_P F_y - y_P F_x} \tag{24}$$

The foregoing is the primary method we will employ. The cross product form, $\bar{M}_A = \bar{r}_{P/A} \times \bar{F}$, is fully equivalent to it. Another method is to compute the lever arm of the entire force $\bar{F}$ directly. You may use the approach with which you are most familiar.

Although we will employ the basic equations of planar motion (23) to study moving bodies, they also are important for static systems. Obviously, $\bar{a}_G \equiv \bar{0}$ and $\bar{\alpha} \equiv \bar{0}$ for a body that is at rest. Equations (23) in this case reduce to the equations of equilibrium, which were axioms for the Statics text. Equations (23) also prove the principle of transmissibility, which stated that a force acting on a rigid body may be applied anywhere along its line of action, with the same overall effect. This is proven here by the fact that the moment of a force depends only on its magnitude, direction, and line of action.

D. THE PHYSICAL SIGNIFICANCE OF THE EQUATIONS OF MOTION

Our purpose here is to consider the equations of motion from the viewpoint of providing physical insight to the equations. This overview will help us to solve problems because it will emphasize the need to be careful in selecting the point for the moment sum. This discussion of eqs. (23) will build on the concepts developed in the study of the statics of rigid bodies.

According to the principle of equivalence of force systems, the effect on a rigid body of an arbitrary force system can be represented by an equivalent force-couple system acting at any point P. In this equivalent system, the force that is applied at point P is the resultant of the forces acting on the body, and the torque of the couple is the total moment of all the forces about point P.

Figure 10a shows a rigid body subjected to an arbitrary planar force system, where $\hat{x}\hat{y}\hat{z}$ is the coordinate system whose origin is the center of mass G. We use the center of mass G as the point P for the equivalent force-couple system because it is always a valid point for summing moments in the

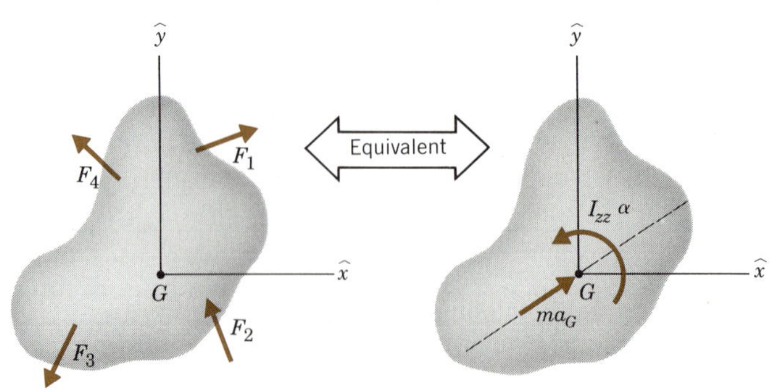

FIGURE 10a **FIGURE 10b**

formulation of the dynamics problem. Thus, eqs. (23) tell us that the force system in Figure 10a is equivalent to a force $m\bar{a}_G$ applied at point G and a couple $I_{zz}\bar{\alpha}$, as illustrated in Figure 10b. Using this concept of equivalence, let us examine each of the three types of rigid body motion: pure translation, pure rotation, and general motion.

1. Pure Translational Motion

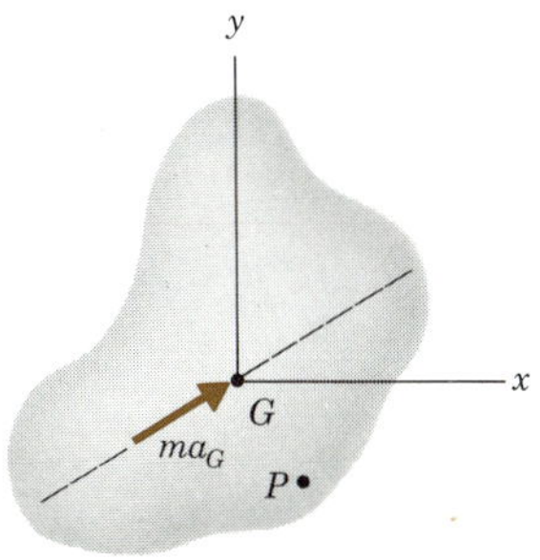

FIGURE 11

By definition, a translating rigid body never changes its angular orientation, so $\bar{\omega} \equiv \bar{0}$ and $\bar{\alpha} \equiv \bar{0}$. In this case the equivalent force system in Figure 10b reduces to that shown in Figure 11. Note that each point in a translating body has the same acceleration, so the motion is fully defined by the resultant force equations. The moment equation enables us to evaluate the properties that an unknown force must have in order to prevent rotation.

Suppose that we were to try to formulate the moment equation by computing the moment about any point other than the center of mass, for example point P in Figure 11. Then in general the moment of the actual force system about point P would have to be equated to the moment of the equivalent force $m\bar{a}_G$ about that point. This is an unnecessary complication of the moment equation, particularly if the acceleration $\bar{a}_G$ is not known.

An exceptional case arises if the point P lies on the line of action of the equivalent force $m\bar{a}_G$, shown as a dashed line in Figure 11. However, there is much less chance for error if we

> Always sum moments about the center of mass when a body is in pure translational motion.

The equations of motion in this case reduce to

$$\sum F_x = m(a_G)_x, \qquad \sum F_y = m(a_G)_y, \qquad \sum M_{Gz} = 0 \tag{25}$$

EXAMPLE 5

A person is trying to balance a stick of mass m and length l in a position where the stick is tilted at an angle θ from the vertical. Determine the acceleration of this person's hand and the force the person exerts on the stick.

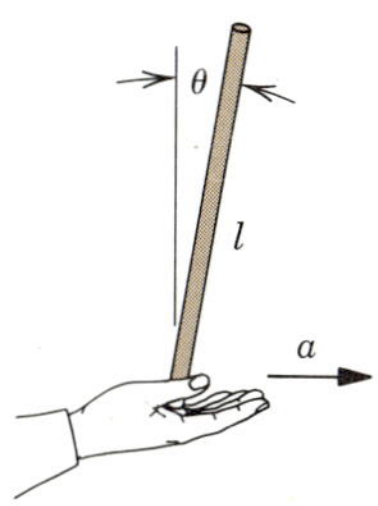

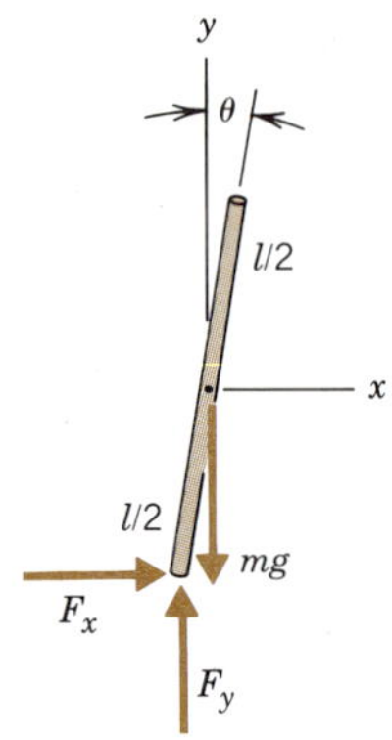

Solution

We begin by constructing a free body diagram that depicts the gravitational force, as well as the horizontal and vertical forces exerted by the person. The stick is translating because the angle of inclination is constant. We place the origin of the xyz coordinate system at the center of mass in order to remember that the moment equation will be formulated about that point.

All points in a translating body have the same acceleration, so $\bar{a}_G = a\bar{i}$. The equations of motion then yield

$$\sum F_x = F_x = m(a_G)_x = ma$$

$$\sum F_y = F_y - mg = m(a_G)_y = 0$$

$$\sum M_{Gz} = F_x\left(\frac{l}{2}\cos\theta\right) - F_y\left(\frac{l}{2}\sin\theta\right) = 0$$

$$F_y = mg, \qquad F_x = mg\tan\theta, \qquad a = g\tan\theta$$

These results merely state that the resultant reaction at the bottom of the stick must intersect the center of mass. (Why?) The magnitude of this reaction must be such that the vertical component counteracts gravity and the horizontal component produces the acceleration.

---

2. Pure Rotational Motion

A body undergoing pure rotation in a plane has an axis perpendicular to the plane of motion that is fixed. The center of mass in this case follows a circular path about the axis of rotation. We denote the radius of this circle as R_G, and define a set of cylindrical coordinates whose origin is point O on the fixed axis. These parameters appear in Figure 12.

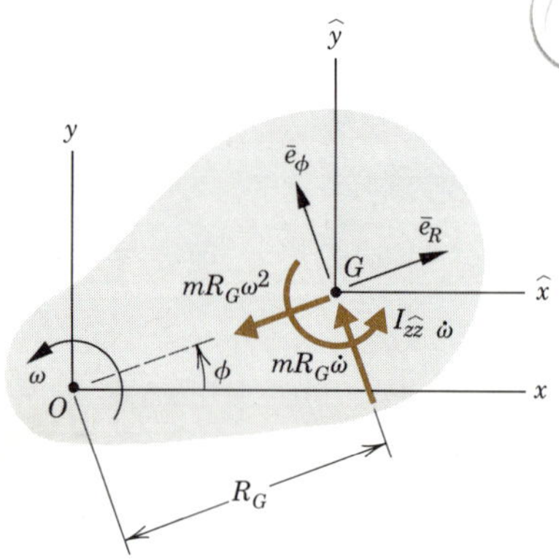

FIGURE 12

The acceleration of the center of mass in this formulation is

$$\bar{a}_G = -R_G\omega^2\bar{e}_R + R_G\dot{\omega}\bar{e}_\phi$$

Multiplying each component by m leads to the representation of the equivalent force-couple system depicted in Figure 12.

Note that the equivalent force $-mR\omega^2\bar{e}_R$ exerts no moment about point O, and that the distance R_G is the lever arm of the equivalent force $mR_G\dot{\omega}\bar{e}_\phi$. Therefore, when we compute $\sum \bar{M}_O$ for the actual force system and equate this to the sum of the moments about point O of the equivalent force system, we find that

$$\sum M_{Oz} = I_{\hat{z}\hat{z}}\dot{\omega} + mR_G\dot{\omega}(R_G) = (I_{\hat{z}\hat{z}} + mR_G^2)\dot{\omega}$$

This may be simplified further by using the parallel axis theorem and noting that R_G is the distance between the centroidal $\hat{z}$ axis and the noncentroidal z axis. Thus the moment sum around point O reduces to

$$\sum M_{Oz} = I_{zz}\alpha$$

where $\alpha = \dot{\omega}$ for the situation in Figure 12.

The foregoing demonstrates why the moment eq. (23) is valid when the point selected for that moment is fixed. Suppose we had summed moments about an arbitrary point. Then the equivalent force $-mR_G\omega^2\bar{e}_R$ would have entered into the moment (unless the point was along line OG). Also the lever arm of the equivalent force $mR_G\dot{\omega}$ would not, in general, be R_G, so that the parallel axis theorem could not have been employed.

The question of whether we should compute the moment of the force system about point O or point G is easily answered when we consider the forces acting on the system. In order to have a fixed axis of rotation, there must be constraint forces applied to this axis, such as the reaction forces produced by a pin joint. It follows that in order to eliminate these unknown reactions from the moment equation, we shall

> Always sum moments about the fixed point O when a body is in pure rotational motion.

The corresponding equations of motion are

$$\sum F_x = m(a_G)_x, \qquad \sum F_y = m(a_G)_y, \qquad \sum M_{Oz} = I_{zz}\alpha \tag{26}$$

where I_{zz} is about an axis through point O.

EXAMPLE 6

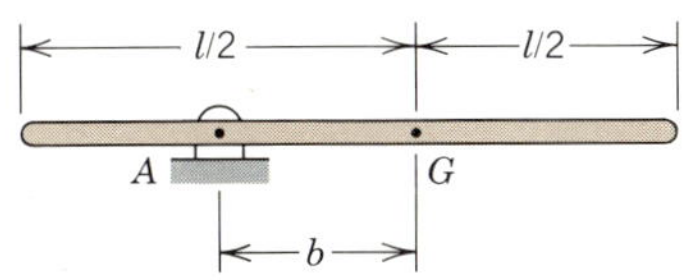

The bar of mass m is released from rest in the position shown, causing it to swing in the vertical plane. Determine the angular acceleration of the bar and the force exerted by pin A at the instant after release.

Solution

Pin A restricts movement horizontally and vertically by applying reaction forces in those directions. These forces, as well as the weight force, are shown in the free body diagram. Note that the origin of the xyz coordinate system

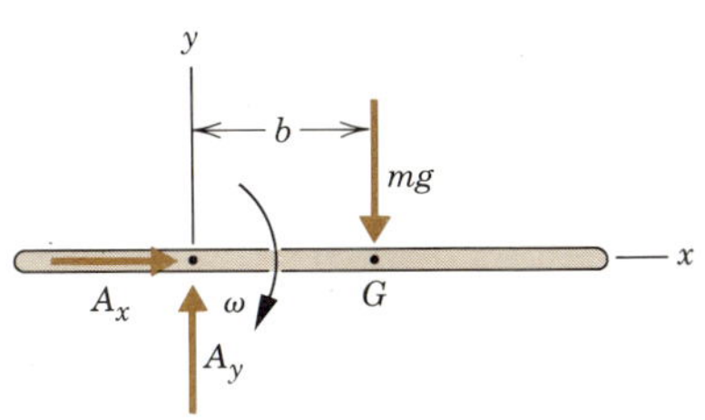

is placed at the fixed point A, which we will use to formulate the moment equation. The moment of inertia about the z axis is found from Appendix B and the parallel axis theorem to be

$$I_{zz} = \tfrac{1}{12}ml^2 + mb^2$$

The rotation will be in the clockwise direction, so $\bar{\omega} = \omega(-\bar{k})$ and $\bar{\alpha} = \dot{\omega}(-\bar{k})$. Note that $\omega = 0$ at the instant after release, although $\dot{\omega} \neq 0$. The acceleration of the center of mass is related to the angular motion by

$$\overline{a}_G = \overline{a}_O + \overline{\alpha} \times \overline{r}_{G/O} - \omega^2 \overline{r}_{G/O}$$
$$= \overline{0} + (-\dot{\omega}\overline{k}) \times (b\overline{i}) - \overline{0} = -\dot{\omega}b\overline{j}$$

(We have used the kinematical vector relation in order to obtain $\overline{a}_G$ in terms of its x and y components.)

We may now formulate the equations of motion. Positive terms in the moment equation are counterclockwise according to the right-hand rule because the z axis is outward. Thus

$$\sum F_x = A_x = m(a_G)_x = 0$$
$$\sum F_y = A_y - mg = m(a_G)_y \equiv m(-\dot{\omega}b)$$
$$\sum M_{Oz} = -mg(b) = I_{zz}\alpha \equiv m(\tfrac{1}{12}l^2 + b^2)(-\dot{\omega})$$
$$\dot{\omega} = \left(\frac{12b}{l^2 + 12b^2}\right)g$$

$$A_x = 0; \qquad A_y = mg - m\left(\frac{12b^2}{l^2 + 12b^2}\right)g \equiv \frac{l^2}{l^2 + 12b^2}mg$$

You might have guessed at the beginning that $A_x = 0$, but the result for A_y could not have been foreseen. You might have guessed that $A_y = mg$, but that is not the case because the center of mass is accelerating downward.

---

3. General Motion

For this type of motion, which is a superposition of translation and rotational effects, no point in the body is fixed. Also, whether or not there is any point in the body whose acceleration is directed toward or away from the center of mass depends on the geometry and constraints of the system. This can be illustrated by the simple problem of a circular wheel rolling without slipping down an incline.

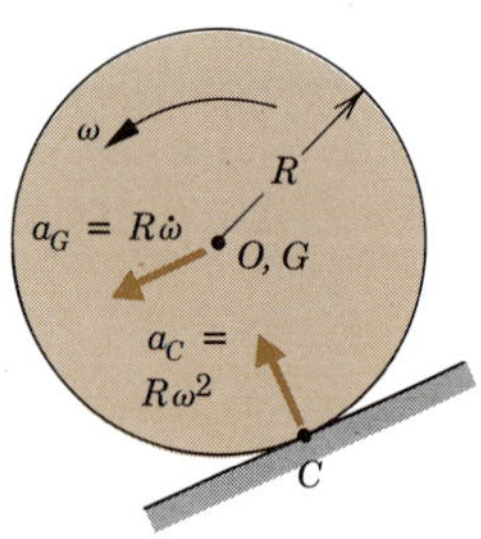

FIGURE 13a

If the wheel is balanced so that the center of mass G coincides with the geometric center O, then we have the kinematical situation illustrated in Figure 13a. For the no slip condition, point C of the body, which is in contact with the incline, must accelerate directly toward point O. Thus, if points O and G coincide, then point C is an allowable point for applying eq. (23). In contrast, if the wheel is unbalanced, so that the center of mass is at a distance l from the geometric center O, we then have the kinematical situation shown in Figure 13b. From this figure we conclude that the contact point C does not accelerate directly toward point G, except for the special cases that arise at instants when points G, O, and C are collinear. By writing a relative acceleration equation between points G and O, the three components of acceleration shown at point G can be identified.

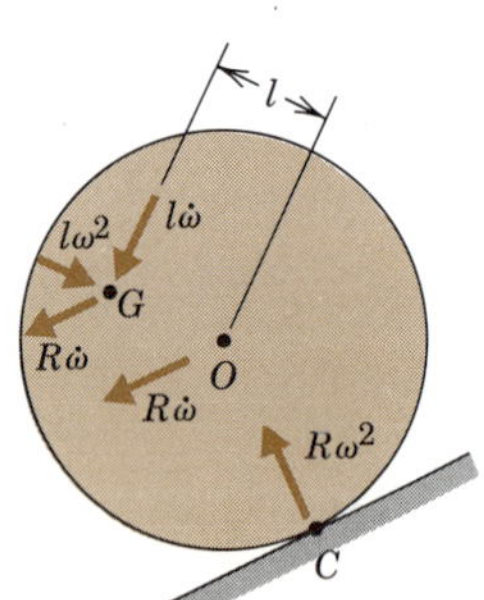

FIGURE 13b

In order to have an approach that is valid for all cases, we will

> Always sum moments about the center of mass G when a body is in general motion.

The equations of motion in this case are

$$\sum F_x = m(a_G)_z, \qquad \sum F_y = m(a_G)_y, \qquad \sum M_{Gz} = I_{zz}\alpha \tag{27}$$

where I_{zz} is about an axis through the center of mass G.

EXAMPLE 7

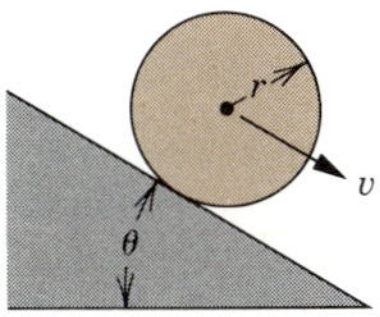

The homogeneous wheel is rolling down the incline with an instantaneous speed v. There is no slipping. Determine the acceleration of the center of the wheel and the forces exerted on the wheel by the incline.

Solution

The forces exerted by the incline are a normal component $\bar{N}$ and frictional component $\bar{f}$. The sense of $\bar{f}$ in the free body diagram is based on an intuitive guess; a negative magnitude for $\bar{f}$ would indicate that this guess was wrong.

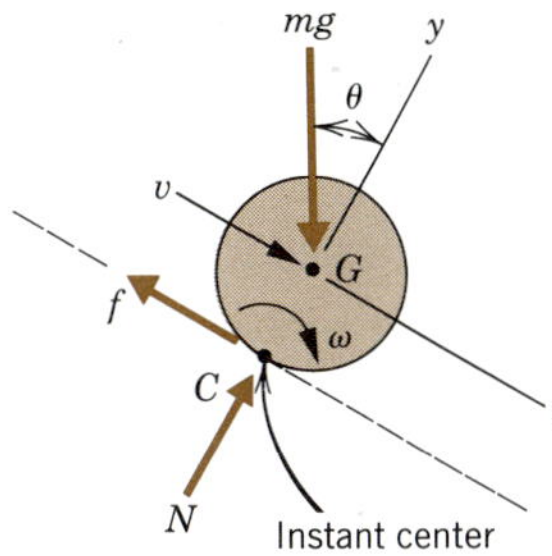

The origin of the xyz coordinate system is placed at the center of mass, which is the geometrical center for a homogeneous wheel. Note that the x axis has been aligned parallel to the incline in order to expedite the kinematical description as well as the force sums. The moment of inertia is found from Appendix B to be

$$I_{zz} = \tfrac{1}{2}mr^2$$

The rotation is related to the motion of the center of mass by the no slip condition. The contact point C is the instant center of zero velocity, so

$$v = \omega r \quad \text{and} \quad \dot{v} = \dot{\omega} r$$

For the clockwise rotation appearing in the diagram, the angular acceleration is

$$\bar{\alpha} = \dot{\omega}(-\bar{k}) = -\frac{\dot{v}}{r}\bar{k}$$

Also, the center of mass is following a straight path so

$$\bar{a}_G = \dot{v}\bar{i}$$

Now that the kinematical relations have been developed, we may formulate the equations of motion. Thus

$$\sum F_x = mg \sin \theta - f = m(a_G)_x = m\dot{v}$$

$$\sum F_y = N - mg \cos \theta = m(a_G)_y = 0$$

$$\sum M_{Gz} = -f(r) = I_{zz}\alpha = \tfrac{1}{2}mr^2\left(-\frac{\dot{v}}{r}\right)$$

Simultaneous solution of the equations for $\sum F_x$ and $\sum M_{Gz}$ yields

$$\dot{v} = \tfrac{2}{3}g \sin \theta, \qquad f = \tfrac{1}{3}mg \sin \theta$$

$$N = mg \cos \theta$$

An interesting aspect of this result is that $\dot{v} = g \sin \theta$ in the case of a *smooth* incline ($f \equiv 0$). Friction causes the wheel to roll, thereby slowing the forward acceleration. Also note that the result for $\dot{v}$ is independent of r. This means that regardless of how small a rolling object is, it is never correct to model it as a particle—we must account for its rotation by modeling it as a rigid body.

E. PROBLEM SOLVING

The formulation of the equations of motion is a synthesis of the kinematical methods developed in Chapter 13 and the methods for treating force systems that were developed in Statics. This is evidenced by Examples 5–7. The reason for not yet presenting any homework problems involving the equations of motion is that it is necessary to have a methodical approach in order to avoid errors. The sequence of steps we present here is essentially the same as that used to solve Examples 5–7.

1. Determine the mass and location of the center of mass G of each body in the system for which such information is not given.
2. Draw a free body diagram of each individual body in the system, making use of Newton's law of action and reaction where appropriate. Remember that each constraint on the motion of a system can only be imposed by a corresponding reaction force or moment. The free body diagram(s) should also show the location of the center of mass, even when the gravitational force is perpendicular to the plane of motion.
3. Select an xyz reference frame for each body. For bodies in translation or general motion, the origin A of this frame should be the center of mass of the body, whereas for bodies that are rotating about a fixed axis the origin should be situated on that axis. The x and y axes should be parallel to the plane of motion and their orientation should, where possible, fit the geometry of the physical system. In problems where there is more than one body, the xyz reference frame for each body should be parallel to each other, making it easier to equate various vectors.
4. Check that all known values for forces appear in the free body diagram(s). Write down any other given information in equation

form, using the xyz frame to describe vector components. Also, list the quantities requested in the problem statement.

5. Calculate I_{zz} for any body that has rotational motion.
6. Following the methods developed in Chapter 13 for treating the kinematics of rigid bodies in planar motion, relate the angular acceleration and the x and y components of the acceleration of the center of mass. Be certain to satisfy all of the constraints imposed on the motion of the system.
7. Form the scalar equations of motion: $\sum F_x = m(a_G)_x$, $\sum F_y = m(a_G)_y$, and $\sum M_{Az} = I_{zz}\alpha$, for each body. Now substitute the kinematical relations developed in step 6 into the scalar equations of motion.
8. Count the number of scalar equations and the number of scalar unknowns to be determined. If the two numbers do not match, you probably have either improperly employed the law of action and reaction or else forgotten a reaction force or the corresponding kinematical constraint condition.
9. Solve the scalar equations of motion. Then check that the solution satisfies the initial assumptions. For example, if you have assumed that two surfaces do not slip relative to each other, check to be sure that the friction force $\bar{f}$ and normal force $\bar{N}$ satisfy the condition $|\bar{f}| < \mu_s|\bar{N}|$.

The following six illustrative problems demonstrate the application of the foregoing steps. They have been grouped in pairs treating pure translation, pure rotation, and general motion.

ILLUSTRATIVE PROBLEM 1

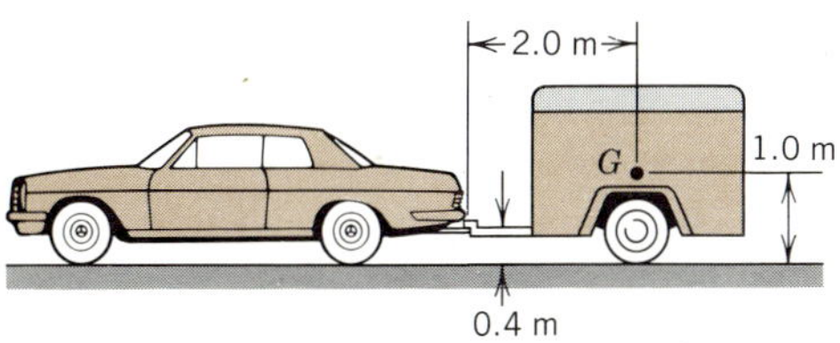

The automobile shown accelerates at a constant rate on a level road. It starts from rest and reaches the speed of 70 km/hr in a distance of 200 m. Determine the reactive force at the hitch of the 400 kg trailer during the motion. Neglect friction.

Solution

Step 1 The location of the center of mass G of the trailer is given.

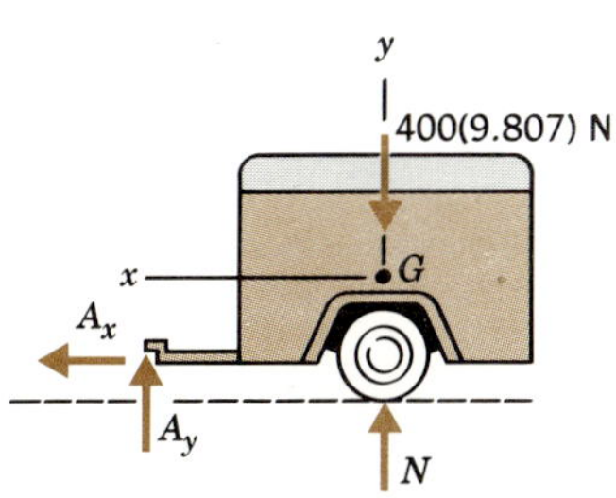

Steps 2 and 3 Because the reactive force at the hitch can be determined only when it appears as an external force, we isolate the trailer in a free body diagram. Note that we do not consider the automobile, because we have no information about this body. The hitch is a pin, so it exerts horizontal and vertical reactions. The origin of the coordinate system is placed at the center of mass because the trailer is in pure translation. The x axis is oriented to the left for convenience.

Step 4 It is given that $v_1 = 0$ and $v_2 = (70/3.6)$ m/s for a distance $\Delta s = 200$ m with constant $\dot{v}$. We must evaluate A_x and A_y.

Step 5 The value of I_{zz} is not needed for a pure translation motion.

Step 6 There is no rotation, so $\omega = \alpha = 0$. The value of $\dot{v}$ is found from the constant acceleration formula

$$v_2^2 = v_1^2 + 2\dot{v}\,\Delta s$$

Substituting the given values yields

$$\left(\frac{70}{3.6}\right)^2 = 2\dot{v}(200) \Rightarrow \dot{v} = 0.9452 \text{ m/s}^2$$

For a translation, all points have the same acceleration so

$$\bar{a}_G = \dot{v}\bar{i} = 0.9452\bar{i} \text{ m/s}^2$$

Steps 7–9

$$\sum F_x = A_x = m(a_G)_x = 400(0.9452)$$

$$\sum F_y = -400(9.807) + A_y + N = m(a_G)_y = 0$$

$$\sum M_{Gz} = (2A_y + 0.6A_x) = 0$$

$$A_x = 378 \text{ N}, \qquad A_y = -113 \text{ N}$$

The negative value of A_y means that the front of the trailer is tending to lift upward.

ILLUSTRATIVE PROBLEM 2

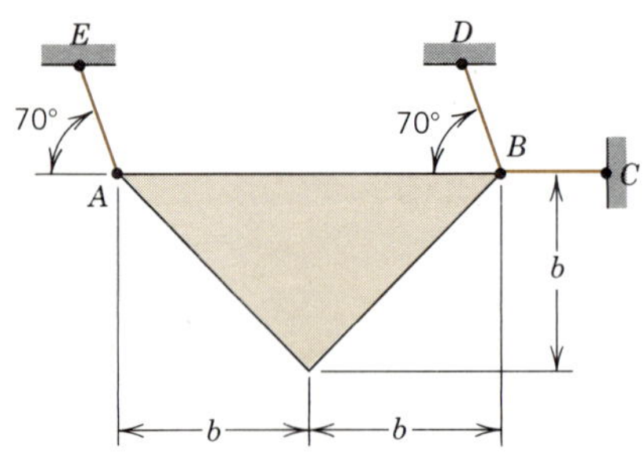

The uniform plate of mass m is released from its rest position in a vertical plane by cutting cable BC. Determine the forces in cables BD and AE and the acceleration of the center of mass at the instant after motion begins.

Solution

Step 1 The center of mass G for the triangular shape is along the centerline at a distance $b/3$ below the top.

Step 2 The free body diagram shows the two cable tensions remaining after cable BC is cut. Note that the tension forces are in their original orientations because the position of the plate cannot change instantaneously after the cable is cut.

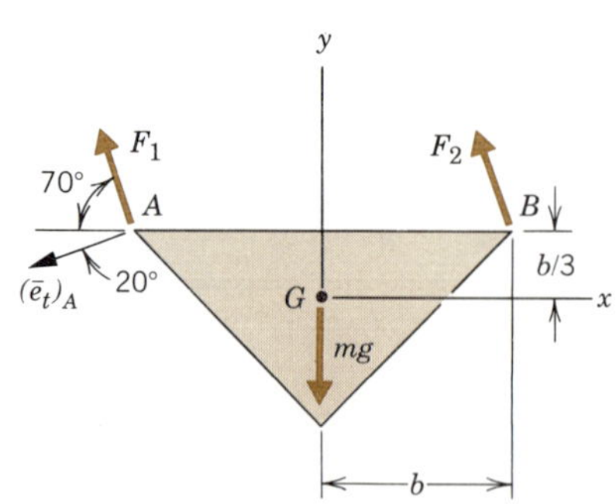

Step 3 The quadrilateral $ABDE$ is a parallelogram, so side AB is al zontal. From this we deduce that this is a case of pure translation. The orig. is placed at the center of mass for a translational motion. The coordinate axes have been oriented horizontally and vertically because those are the directions in which the dimensions of the plate are given.

Step 4 Given that $\bar{v} = \bar{0}$ initially, find F_1, F_2, and $\bar{a}_G$ at that instant.

Step 5 The value of I_{zz} is not needed when a body is in pure translation.

Step 6 Set $\bar{\omega} = \bar{\alpha} = \bar{0}$ for translational motion. The constrained points are A and B, where the cables are attached. There are several methods for describing the effect of the constraints on $\bar{a}_G$. The one we shall employ uses the fact that point A follows a circular path centered at point E (assuming that cable AE is not slack). The tangential direction for this path is depicted in the sketch. Because $\bar{a}_G = \bar{a}_A = \bar{a}_B$ when $\bar{\omega} = \bar{\alpha} = \bar{0}$, and $v = 0$ at the instant of release, we have

$$\bar{a}_G = \bar{a}_A = \dot{v}(\bar{e}_t)_A = \dot{v}(-\cos 20^\circ \bar{i} - \sin 20^\circ \bar{j})$$

Step 7

$$\sum F_x = -(F_1 + F_2)\cos 70^\circ = m(a_G)_x = m(-\dot{v}\cos 20^\circ)$$

$$\sum F_y = (F_1 + F_2)\sin 70^\circ - mg = m(a_G)_y = m(-\dot{v}\sin 20^\circ)$$

$$\sum M_{Gz} = F_1 \cos 70^\circ\left(\frac{b}{3}\right) - F_1 \sin 70^\circ (b) + F_2 \cos 70^\circ\left(\frac{b}{3}\right)$$

$$+ F_2 \sin 70^\circ (b) = I_{zz}\alpha = 0$$

Steps 8 and 9 We have three equations for the three unknowns F_1, F_2, and $\dot{v}$. They are solved simultaneously, with the result that

$$\dot{v} = 0.3420g, \qquad F_1 = 0.5269\ mg, \qquad F_2 = 0.4128\ mg$$

Once again, the results emphasize that we cannot depend on our intuition in dynamics. Many students set $F_1 = F_2$ in the free body diagram because they think the *dynamic* system is symmetrical. However, the lack of symmetry is evident in the moment equation.

ILLUSTRATIVE PROBLEM 3

The 100 lb block is released from rest. Compute the angular velocity of the 60 lb uniform drum and the tension in the cable 5 s after release.

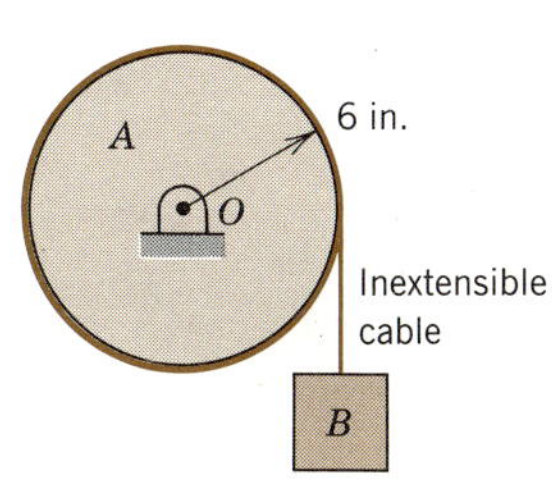

Solution

Step 1 The block and the drum are individual bodies, so we must construct a free body diagram of each. We will model the block as a particle and assume that the center of mass of the drum coincides with the shaft supporting the drum.

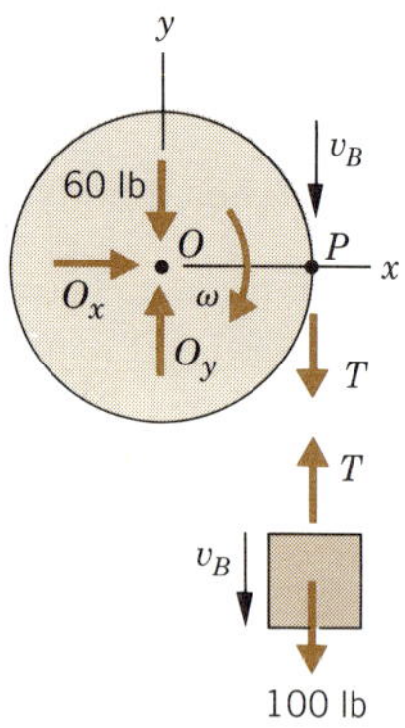

Steps 2 and 3 The shaft can apply horizontal and vertical reactions to support the drum, resulting in the free body diagram shown. Note that the tension T does not equal the weight of the block because the block is accelerating. The origin of the xyz coordinate system for the drum is placed at the fixed point O.

Step 4 Given that $v_B = \omega = 0$ initially, find ω and T when $t = 5$ s.

Step 5 The moment of inertia of the block is not required because we are treating it as a particle. We consider the drum to be a homogeneous cylinder. Appendix B then gives

$$I_{zz} = \frac{1}{2}mr^2 = \frac{1}{2}\frac{60}{32.17}\left(\frac{6}{12}\right)^2 = 0.2331 \text{ lb-s}^2\text{-ft}$$

Step 6 The cable is not slipping over the drum, so the velocity at the point of tangency, P, must be $\bar{v}_B$. However, the drum is in pure rotation about point O. For the directions of motion shown in the free body diagrams, it follows that

$$\bar{\alpha} = \dot{\omega}(-\bar{k})$$

$$v_B = \omega(r) = 0.5\omega, \qquad \dot{v}_B = 0.5\dot{\omega}$$

$$\bar{a}_B = \dot{v}_B(-\bar{j}) = -0.5\dot{\omega}\bar{j}$$

Step 7 The force sum for the drum will involve the unknown reactions at point A. Because we are not interested in these forces, we sum moments for the drum only.

$$\sum M_{Oz} = -T(0.5) = I_{zz}\alpha = 0.2331(-\dot{\omega})$$

For the block, we sum forces in the y direction and use the expression for $\bar{a}_B$ determined in step 6.

$$\sum F_y = T - 100 = m(a_B)_y = \frac{100}{32.17}(-0.5\dot{\omega})$$

Step 8 We have two scalar equations for the two scalar unknowns T and $\dot{\omega}$.

Step 9 The solution of the equations of motion in step 7 is

$$\dot{\omega} = 49.49 \text{ rad/s}^2, \qquad T = 23.07 \text{ lb}$$

The values of T and $\dot{\omega}$ are constant. We obtain the desired angular velocity by applying the formula for constant acceleration. Hence, when $t = 5$ s, we have

$$\omega_2 = \omega_1 + \dot{\omega}t = 0 + 49.49(5)$$

$$\bar{\omega}_2 = -247\bar{k} \text{ rad/s}$$

We should note that the value of $\dot{\omega}$ is substantially smaller than it would be if a 100 lb tensile force was applied directly at point P. The effect of the block is to add inertial resistance, as well as to introduce the 100 lb gravity force.

ILLUSTRATIVE PROBLEM 4

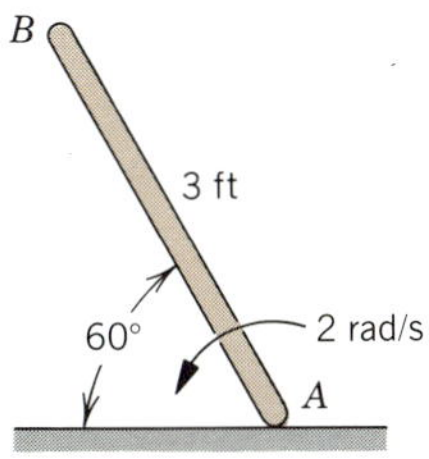

In the position shown the 6 lb rod is falling at 2 rad/s, and end A is not sliding over the ground. Determine the corresponding angular acceleration of the rod and the minimum coefficient of static friction between end A and the ground.

Solution

Step 1 From symmetry the center of mass of the rod is at its midpoint.

Step 2 The reaction at end A consists of a normal (vertical) force $\bar{N}$ and a friction (horizontal) force $\bar{f}$. We have guessed the sense of the latter to be to the right.

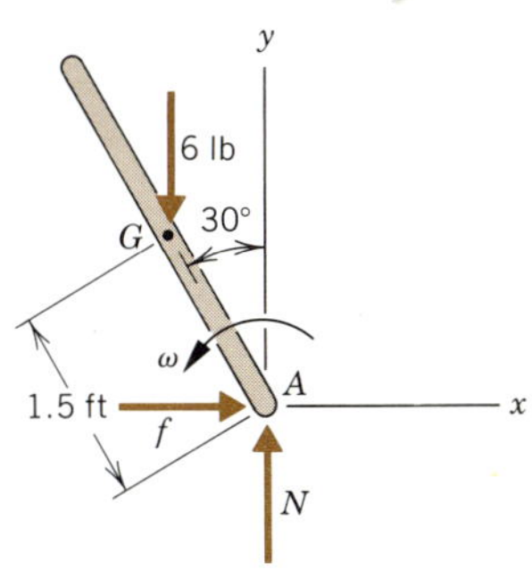

Step 3 We model a rod as having zero width. In this approximation, the same point at end A is always in contact with the ground. The condition of no slipping means that end A is also not moving horizontally. We conclude that end A is a fixed point in the pure rotation of the rod. Accordingly, we place the origin of xyz at this point.

Step 4 Given that $\omega = 2$ rad/s in the illustrated position, find $\dot{\omega}$ and the minimum μ_s for no sliding at end A.

Step 5 We use Appendix B and the parallel axis theorem to compute

$$I_{zz} = \frac{1}{12}ml^2 + m\left(\frac{l}{2}\right)^2 = \frac{1}{3}ml^2 = \frac{1}{3}\frac{6}{32.17}(3)^2 = 0.5595 \text{ lb-s}^2\text{-ft}$$

Step 6 For the sense of the rotation given

$$\bar{\omega} = 2\bar{k} \text{ rad/s}, \qquad \bar{\alpha} = \dot{\omega}\bar{k}$$

The acceleration of the center of mass in this pure rotation ($\bar{a}_A \equiv \bar{0}$) is

$$\bar{a}_G = \bar{\alpha} \times \bar{r}_{G/A} - \omega^2 \bar{r}_{G/A}$$

For the instant of interest, we find

$$\bar{r}_{G/A} = 1.5(-\sin 30^\circ \bar{i} + \cos 30^\circ \bar{j}) = -0.75\bar{i} + 1.2990\bar{j} \text{ ft}$$

$$\bar{a}_G = (\dot{\omega}\bar{k}) \times (-0.75\bar{i} + 1.2990\bar{j}) - (2)^2(-0.75\bar{i} + 1.2990\bar{j})$$
$$= (-1.2990\dot{\omega} + 3)\bar{i} + (-0.75\dot{\omega} - 5.196)\bar{j} \text{ ft/s}^2$$

Step 7 We form the force equations using the components of $\bar{a}_G$ found above. Thus

$$\sum F_x = f = m(a_G)_x = \frac{6}{32.17}(-1.2990\dot{\omega} + 3)$$

$$\sum F_y = N - 6 = m(a_G)_y = \frac{6}{32.17}(-0.75\dot{\omega} - 5.196)$$

$$\sum M_{Oz} = 6(1.5 \sin 30^\circ) = I_{zz}\alpha = 0.5595(\dot{\omega})$$

Steps 8 and 9 The three unknowns in the three equations of motion are $\dot{\omega}$, N, and f. Solving these equations yields

$$\dot{\omega} = 8.043 \text{ rad/s}^2 \Rightarrow \bar{\alpha} = 8.04\bar{k} \text{ rad/s}^2$$

$$N = 3.906 \text{ lb}, \qquad f = -1.389 \text{ lb}$$

The minus sign obtained in the solution for f indicates that we guessed wrong as to which way the rod would slip. The minimum allowable μ_s that prevents sliding is the one for which $|\bar{f}| = \mu_s N$. Hence.

$$(\mu_s)_{\min} = \frac{|\bar{f}|}{N} = \frac{1.389}{3.906} = 0.356$$

ILLUSTRATIVE PROBLEM 5

The 30 kg bar is released from rest in the position shown. The wall and the floor are smooth. Determine the reactions at ends A and B, and the angular acceleration of the bar at the instant after release.

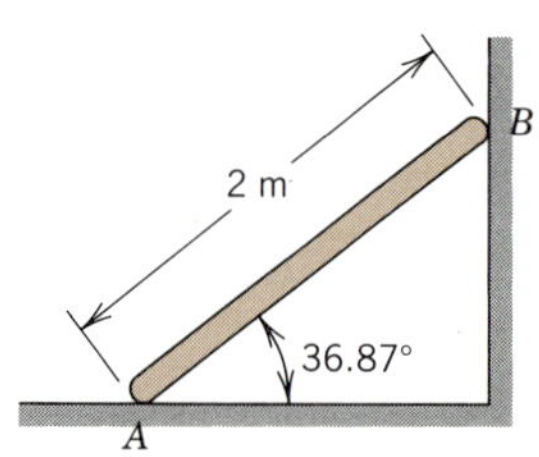

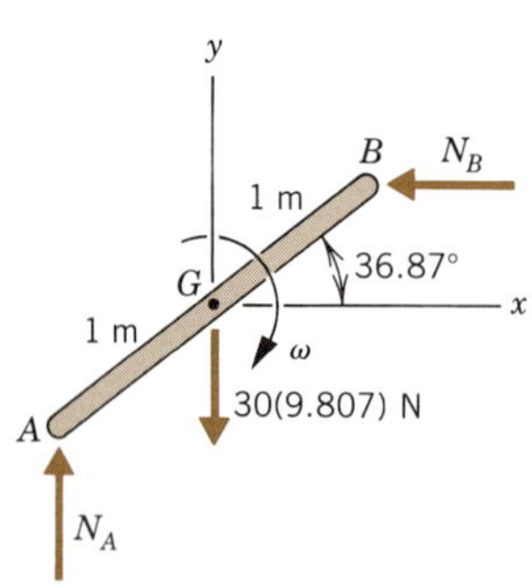

Solution

Step 1 The location of the center of mass G is known by symmetry.

Steps 2 and 3 As shown in the free body diagram, the surfaces contacting the bar exert only normal forces because they are smooth. The origin of xyz is placed at point G because the bar is in a general motion.

Step 4 Given that $\bar{v}_G = \bar{\omega} = \bar{0}$ for the position in the free body diagram, determine the corresponding $\bar{a}_G$, N_A, and N_B.

Step 5 Appendix B gives

$$I_{zz} = \tfrac{1}{12}ml^2 = \tfrac{1}{12}(30)(2)^2 = 10 \text{ kg-m}^2$$

Step 6 The acceleration $\bar{a}_G$ and the angular acceleration $\bar{\alpha}$ are related because the ends A and B must follow the surfaces they contact. We will employ the vector equation method of Chapter 13 to develop this relation.

First we note that at this instant $v_A = v_B = 0$ and $\omega = 0$. We therefore proceed to the acceleration formulation, employing the vector method to relate the acceleration of the constrained points A and B. Setting $\omega = 0$ gives

$$\bar{a}_B = \bar{a}_A + \bar{\alpha} \times \bar{r}_{B/A}$$

Both points follow straight paths, so

$$\bar{a}_A = \dot{v}_A(-\bar{i}), \qquad \bar{a}_B = \dot{v}_B(-\bar{j})$$

The position $\bar{r}_{B/A}$ may be found from the free body diagram, where it may also be seen that we have assumed a clockwise rotation. Thus

$$\bar{\omega} = \omega(-\bar{k}), \qquad \bar{\alpha} = \dot{\omega}(-\bar{k})$$

$$\bar{r}_{B/A} = 2(\cos 36.87^\circ\, \bar{i} + \sin 36.87^\circ\, \bar{j}) = 1.6\bar{i} + 1.2\bar{j}$$

The acceleration equation then gives

$$-\dot{v}_B\bar{j} = -\dot{v}_A\bar{i} + (-\dot{\omega}\bar{k}) \times (1.6\bar{i} + 1.2\bar{j})$$

$$\bar{i}\text{ component:}\quad 0 = -\dot{v}_A + 1.2\dot{\omega} \Rightarrow \dot{v}_A = 1.2\dot{\omega}$$

$$\bar{j}\text{ component:}\quad -\dot{v}_B = -1.6\dot{\omega} \Rightarrow \dot{v}_B = 1.6\dot{\omega}$$

These relations show that the only kinematical unknown is $\dot{\omega}$. With $\bar{a}_A$ known in terms of $\dot{\omega}$, we may now determine the acceleration of point G. Thus, for $\bar{\omega} = \bar{0}$ we have

$$\bar{a}_A = -\dot{v}_A\bar{i} = -1.2\dot{\omega}\bar{i}$$

$$\begin{aligned}\bar{a}_G &= \bar{a}_A + \bar{\alpha} \times \bar{r}_{G/A} = -1.2\dot{\omega}\bar{i} + (-\dot{\omega}\bar{k}) \times (0.8\bar{i} + 0.6\bar{j})\\ &= -0.6\dot{\omega}\bar{i} - 0.8\dot{\omega}\bar{j}\end{aligned}$$

Step 7 The components of $\bar{a}_G$ were related to $\dot{\omega}$ in the previous step. Substituting those expressions as we form each force equation yields

$$\sum F_x = -N_B = m(a_G)_x = 30(-0.6\dot{\omega})$$

$$\sum F_y = N_A - 30(9.807) = m(a_G)_y = 30(-0.8\dot{\omega})$$

$$\begin{aligned}\sum M_{Gz} &= -N_A(1 \cos 36.87^\circ) + N_B(1 \sin 36.87^\circ)\\ &= I_{zz}\alpha = 10(-\dot{\omega})\end{aligned}$$

Steps 8 and 9 The solution of the three equations for the three unknowns is

$$\dot{\omega} = 5.884 \text{ rad/s}^2 \Rightarrow \bar{\alpha} = -5.88\bar{k} \text{ rad/s}^2$$

$$N_A = 153.0 \text{ N}, \qquad N_B = 105.9 \text{ N}$$

ILLUSTRATIVE PROBLEM 6

The diagram shows the cross section of an 80 lb spool of wire. The radius of gyration of the spool is 10 in. Determine the acceleration of the mass center G for the cases where the coefficients of friction are (a) $\mu_s = 0.25$, $\mu_k = 0.20$, and (b) $\mu_s = 0.15$, $\mu_k = 0.10$.

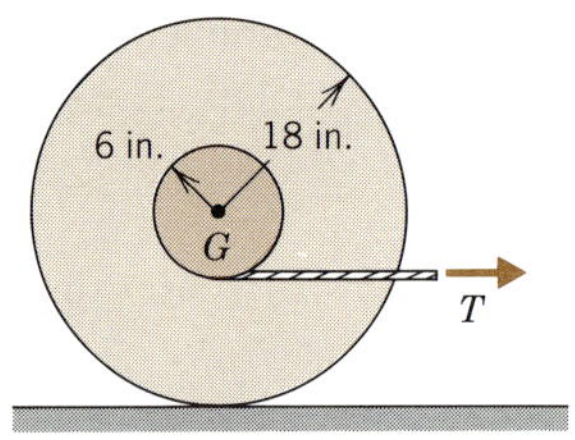

Solution

Step 1 The location of the center of mass is given.

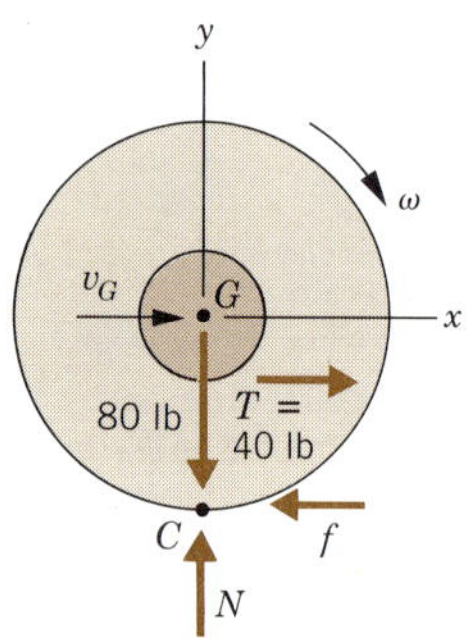

Steps 2 and 3 A situation involving rolling was treated in Example 7. The new feature here is that no information is provided as to whether the spool slips for either set of friction coefficients. We therefore consider the friction force $\bar{f}$ in the free body diagram as an unknown.

The drum is in general motion regardless of the situation for slipping. Thus the origin of the xyz coordinate system is placed at the center of mass G.

Steps 4 and 5 The given value of T is shown in the free body diagram, and the radius of gyration is given as 10 in. The corresponding moment of inertia is

$$I_{zz} = mk_z^2 = \frac{80}{32.17}\left(\frac{10}{12}\right)^2 = 1.7269 \text{ lb-s}^2\text{-ft}$$

We must evaluate $\bar{a}_G$ when (a) $\mu_s = 0.25$, $\mu_k = 0.20$, and (b) $\mu_s = 0.15$, $\mu_k = 0.10$.

Step 6 The center of mass follows a horizontal path. We cannot be sure of the sense of the motion, nor of the sense of the rotation; our guesses are shown in the free body diagram. The corresponding accelerations are

$$\bar{a}_G = \dot{v}_G\bar{i}, \qquad \bar{\alpha} = \dot{\omega}(-\bar{k})$$

If there is no slipping, then point C is the instant center for the wheel and

$$v_G = \omega r = \omega\left(\frac{18}{12}\right) \text{ ft/s}, \qquad \dot{v}_G = \dot{\omega}\left(\frac{18}{12}\right) \text{ ft/s}^2$$

However, we will hold this last relation in reserve because we do not know whether the assumption of no slipping is valid.

Step 7

$$\sum F_x = 40 - f = m(a_G)_x = \frac{80}{32.17}\dot{v}_G$$

$$\sum F_y = N - 80 = m(a_G)_y = 0$$

$$\sum M_{Gz} = 40\left(\frac{6}{12}\right) - f\left(\frac{18}{12}\right) = I_{zz}\alpha = 1.7269(-\dot{\omega})$$

Step 8 The three equations of motion have four unknowns: f, N, $\dot{v}_G$, and $\dot{\omega}$. For the fourth equation we must consider the question of slipping. For this we reason that, if slipping is occurring in this system, it is a result of a transition from a condition of no slip prior to the application of the cable force $\bar{T}$. This argument indicates that we should first assume the no slip condition, $\dot{v}_G = 1.5\omega$, given in step 6. It will be necessary to check this assumption by computing the corresponding value of $|\bar{f}|$, which may not exceed the maximum value $\mu_s N$ for static friction.

Step 9 Substitution of $\dot{\omega} = \dot{v}_G/1.5$ into the equations in step 7 yields

$$40 - f = 2.487\dot{v}_G, \qquad N - 80 = 0$$

$$20 - 1.5f = -1.7269\left(\frac{\dot{v}_G}{1.5}\right)$$

$$N = 80 \text{ lb}, \qquad \dot{v}_G = 8.194 \text{ ft/s}^2, \qquad f = 19.62 \text{ lb}$$

Case (a)

The maximum possible friction force when $\mu_s = 0.25$ is

$$f_{\max} = \mu_s N = 0.25(80) = 20 \text{ lb}$$

We conclude that our guess was correct because the contacting surfaces can sustain the friction force $f = 19.62$ lb required for the spool not to slip. Thus

$$\bar{a}_G = \dot{v}_G\bar{i} = 8.19\bar{i} \text{ ft/s}^2$$

Case (b)

The maximum possible friction force when $\mu_s = 0.15$ is

$$f_{\max} = \mu_s N = 0.15(80) = 12 \text{ lb}$$

Our assumption of no slip is incorrect here, because the friction force $f = 19.62$ lb cannot be developed by the contacting surfaces.

We conclude that the contacting surfaces are sliding over each other in this case. Hence the friction force is $\mu_k N$. The sense of this force is the same as the sense of $\bar{f}$ required to prevent slipping. Thus, because f was positive for the no slip case, we set $f = +\mu_k N = 0.1N$ here. Substituting this into the basic equations of motion in step 7 yields

$$40 - 0.1N = 2.487\dot{v}_G, \qquad N - 80 = 0$$

$$20 - 1.5(0.1N) = -1.7269\,\dot{\omega}$$

$$N = 80 \text{ lb}, \qquad \dot{v}_G = 12.87 \text{ ft/s}, \qquad \dot{\omega} = -4.633 \text{ rad/s}^2$$

$$\bar{a}_G = \dot{v}_G\bar{i} = 12.9\bar{i} \text{ ft/s}^2$$

It is readily verified that $\dot{v}_G$ and $\dot{\omega}$ are kinematically unrelated in this case of rolling and slipping; specifically, $\dot{v}_G \neq 1.5\dot{\omega}$. Indeed, even the sense of the rotation here is opposite the sense in case (a).

HOMEWORK PROBLEMS

a. Pure Translational Motion

PROB. 14.31

14.31 The truck is accelerating at $\dot{v} = 1 \text{ m/s}^2$ as the tailgate is held in the position shown by two cables (only one is shown in the side view). The tailgate has a mass of 15 kg. What is the tension force in each cable?

14.32 Determine the deceleration rate $\dot{v}$ of the truck in Problem 14.31 for which the cables holding the tailgate become slack.

14.33 A homogeneous cylinder of mass m and radius r is pulled along a rough surface as shown by a truck whose acceleration is a. If the cylinder is sliding over the ground without rotating, determine the coefficient of friction and the force the truck applies to the two cables (only one is shown in the side view).

PROB. 14.33

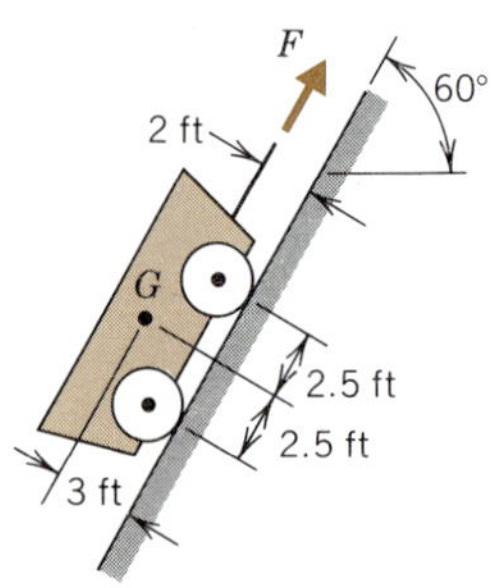

PROB. 14.34

14.34 A cable tension $F = 2.0$ tons is used to pull the 2 ton coal car up the ramp. The mass and rolling resistance of the wheels are negligible. Determine the acceleration of the car and the corresponding reaction forces exerted by the ground.

14.35 For the coal car in Problem 14.34, determine the maximum value of the cable tension F for which both sets of wheels remain in contact with the ramp. Also determine the corresponding acceleration.

14.36 A cyclist moving at a speed of 15 mph applies the front wheel brakes only. The cyclist wishes to avoid being thrown over the handle bars. Determine the shortest safe stopping distance for which the rear wheel will not lose contact with the ground. Point G is the center of mass of the bicycle and the cyclist.

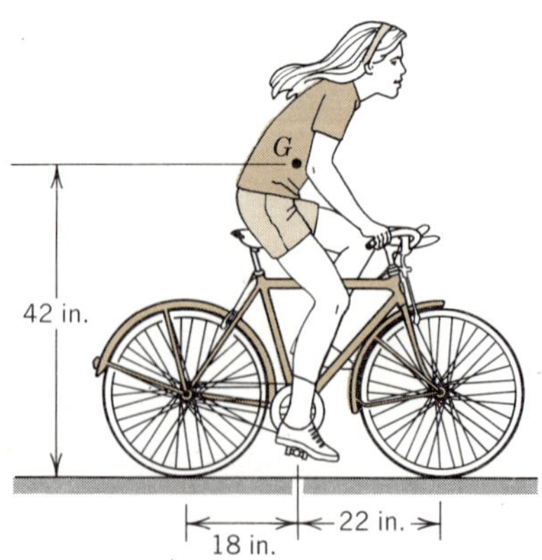

PROB. 14.36

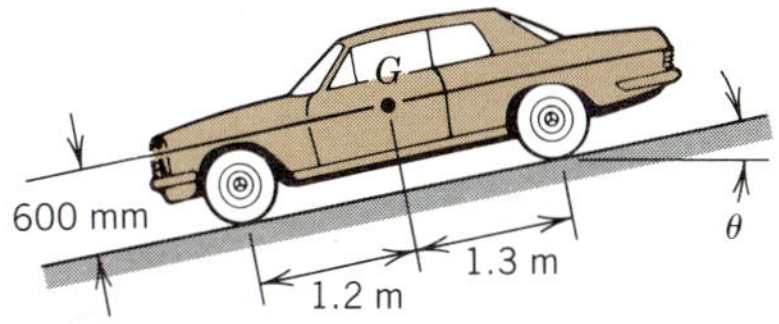

PROBS. 14.37 AND 14.38

14.37 An automobile coasts freely down a hill inclined at $\theta = 10^\circ$. Determine (a) the normal reactions between the road and tires, and (b) the acceleration of the 1500 kg vehicle. The mass of the wheels is negligible, as is rolling resistance.

14.38 The automobile is accelerated on a level road ($\theta = 0$) such that each driven wheel is on the verge of skidding. The coefficient of static friction between each tire and the ground is 0.65. The mass of each wheel is negligible, as is rolling resistance for an undriven wheel. Determine the acceleration if the vehicle has rear-wheel drive.

14.39 Solve Problem 14.38 for the case where the automobile has (a) front-wheel drive, and (b) four-wheel drive.

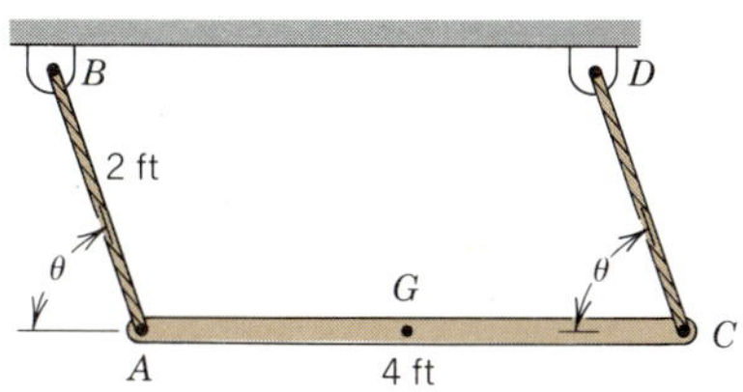

PROB. 14.40

14.40 The 400 lb rod, which is suspended by cables AB and CD, is released from rest in the position where $\theta = 30^\circ$. Determine the acceleration of the center of mass G in that position and the corresponding tension in each cable.

14.41 Solve Problem 14.40 for the position $\theta = 90^\circ$ if the center of mass has a speed of 10 ft/s in that position.

***14.42** Solve Problem 14.40 for the position $\theta = 60^\circ$ if the center of mass has a speed of 6 ft/s in that position.

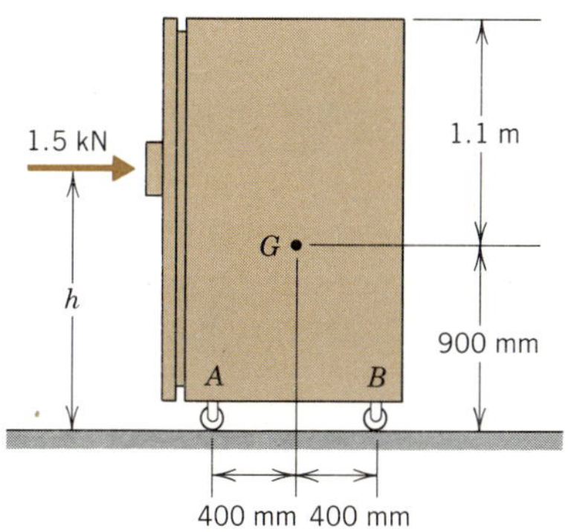

PROB. 14.43

14.43 The 120 kg chest is pushed to the right by the 1.5 kN horizontal force. The sets of casters A and B are unlocked, and their rolling resistance is negligible. Determine (a) the largest height h for which the chest will not tip over, and (b) the corresponding acceleration of the chest.

***14.44** Solve Problem 14.43 in the case where both sets of casters are locked. The coefficients of friction between each caster and the ground are $\mu_s = 0.40$, $\mu_k = 0.30$.

14.45 A truck traveling at a speed of 45 mi/hr brakes to a stop. Determine the shortest stopping distance for the truck such that the 200 lb crate will neither tilt nor slip. The coefficient of static friction between the crate and the truck is 0.25.

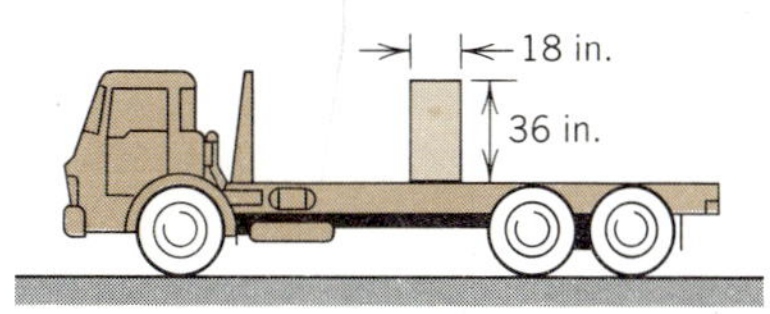

PROB. 14.45

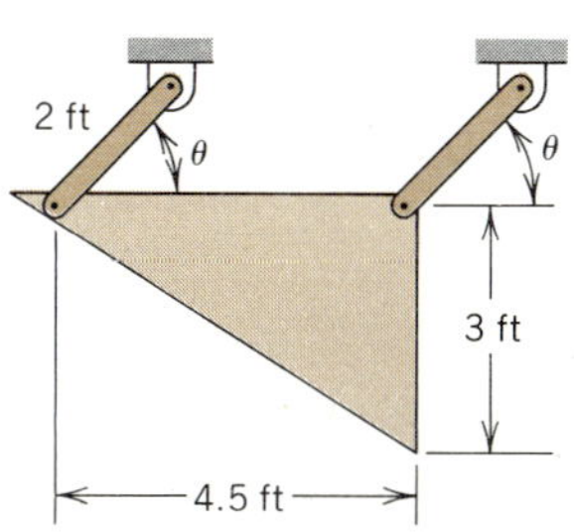

PROB. 14.46

14.46 The 50 lb triangular plate is released from rest at $\theta = 36.87^\circ$. The weights of links AB and CD are negligible. For this position determine the tension in each link and the acceleration of the center of mass.

***14.47** Solve Problem 14.46 for the case where the center of mass has a speed of 16 ft/s in the position where $\theta = 53.13^\circ$.

b. Pure Rotational Motion

14.48 The 8 lb platter of a record turntable has a radius of gyration of 5 in. What torque is required to accelerate the platter from rest to an angular speed of $33\frac{1}{3}$ rev/min in a time interval of one second?

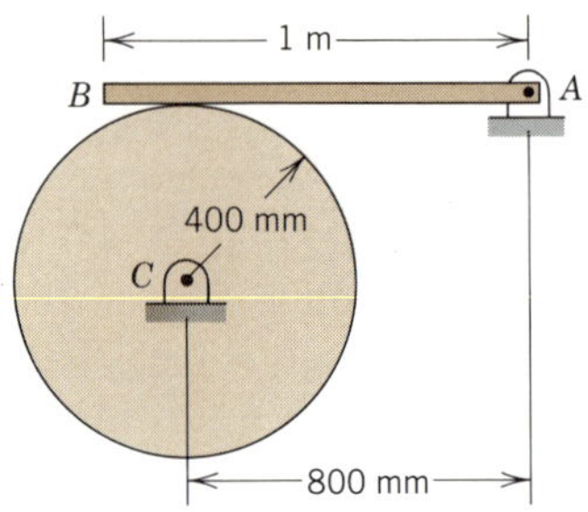

PROB. 14.49

14.49 The radius of gyration of the 100 kg flywheel is 400 mm. The flywheel was rotating counterclockwise at angular speed ω when the 50 kg beam AB was brought into contact with the rim. The coefficient of kinetic friction is 0.25. Determine the deceleration rate $-\dot{\omega}$.

14.50 Determine the angular acceleration of the 100 kg pulley (a) if a 400 N force is applied to the end of the cable, (b) if a block whose weight is 400 N is suspended from the end of the cable.

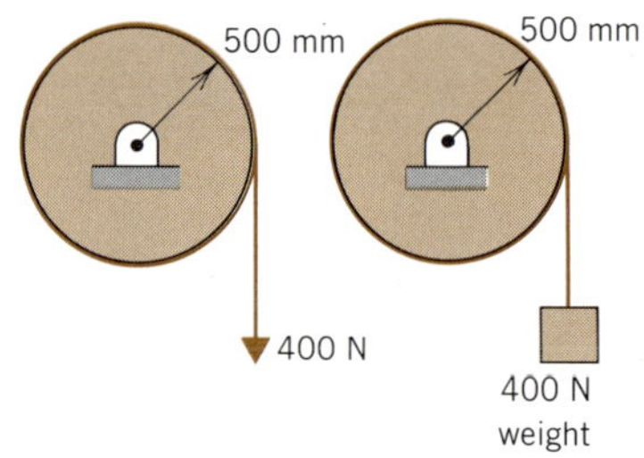

PROB. 14.50

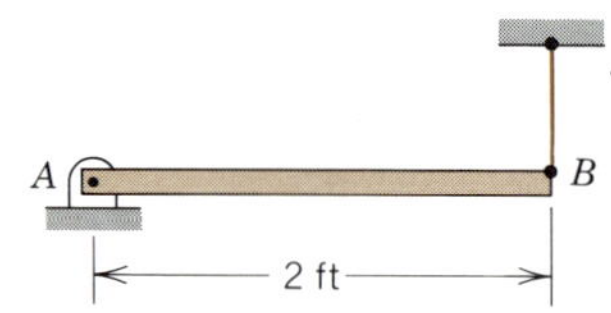

PROB. 14.51

14.51 The homogeneous bar weighs 15 lb. The cord at end B breaks. Determine the angular acceleration of the bar and the reaction at the support A at this instant.

14.52 Solve Problem 14.51 for the case where the support at end A exerts a constant frictional couple of 2 lb-ft opposing the motion of the bar.

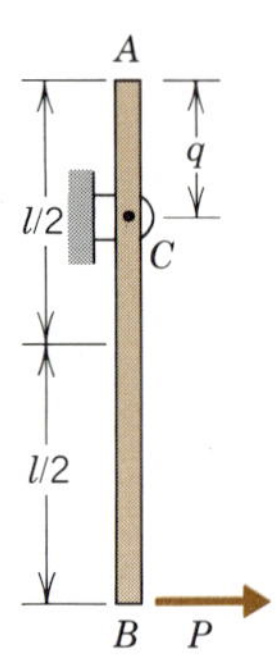

PROBS. 14.53 AND 14.54

14.53 A horizontal force $\bar{P}$ is applied at end B of the uniform rod of mass m. The bar is initially at rest in the position shown. If the distance $q = L/4$, determine (a) the angular acceleration of the rod in this position, (b) the components of the reaction force at pin C for this position.

14.54 A horizontal force $\bar{P}$ is applied at end B of the uniform rod of mass m. The bar is at rest in the position shown. Determine the value of q for which the horizontal component of the reaction force at pin C is zero in this position, and also determine the corresponding angular acceleration.

14.55 Gear A weighs 5 lb and its radius of gyration is 5 in. Gear B weighs 15 lb and its radius of gyration is 7 in. Determine the angular acceleration of these gears for the given couple on gear A.

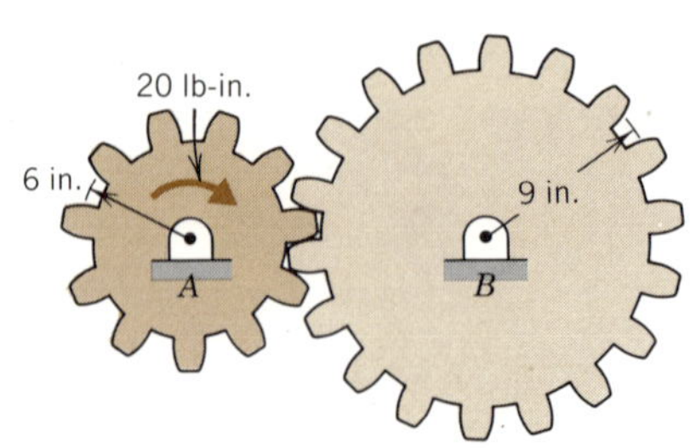

PROB. 14.55

PROB. 14.56

14.56 A dynamometer is used to simulate the inertial resistance of an automobile. To do this the vehicle is held stationary and the drive wheels cause the drum of radius R to rotate clockwise. Letting m_a denote the mass of the automobile, derive an expression for the moment of inertia of the drum about its axis of rotation for which the tire traction required to impart an angular acceleration to the drum is the same as that required to actually accelerate the vehicle on level ground at $\dot{v} = R\dot{\omega}$. The mass of the tires is negligible.

14.57 A force $\bar{F}$ is applied to arm AD in order to slow the 800 kg flywheel whose radius of gyration is 160 mm. The coefficient of friction between the brake shoe D and the rim is 0.60. Determine the force $\bar{F}$ required to stop the flywheel after 30 revolutions for an initial angular velocity of 600 rev/min clockwise.

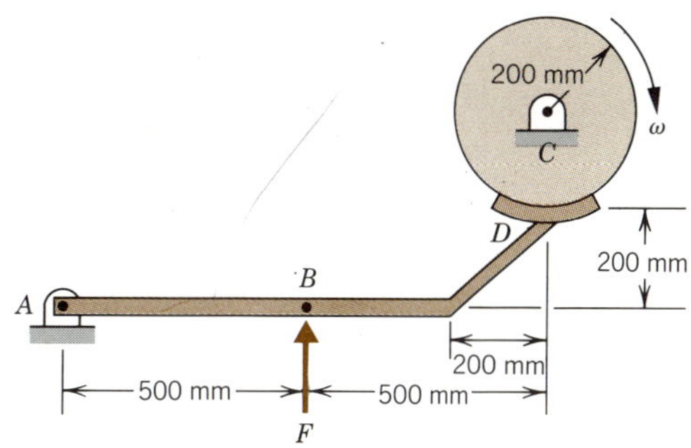

PROB. 14.57

14.58 Solve Problem 14.57 for an initial angular velocity of 600 rev/min counterclockwise.

14.59 The 30 lb double pulley A has a 15 in. radius of gyration. Block A weighs 40 lb and block B weighs 20 lb. Determine the angular acceleration of the pulley.

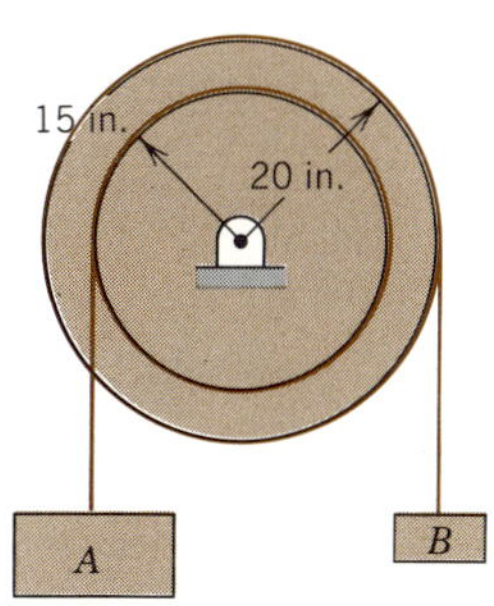

PROB. 14.59

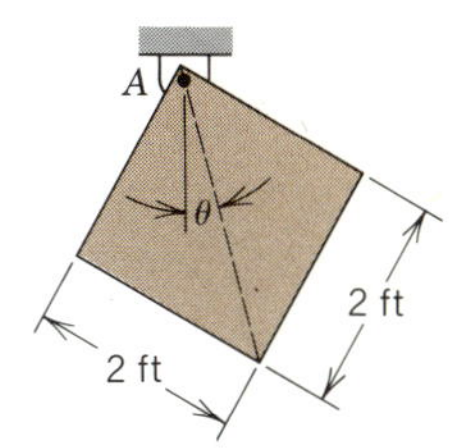

PROB. 14.60

14.60 The square plate weighs 25 lb. Its angular velocity is 30 rad/s counterclockwise when $\theta = 10°$. Determine the corresponding angular acceleration of the plate and the reaction force at pin A.

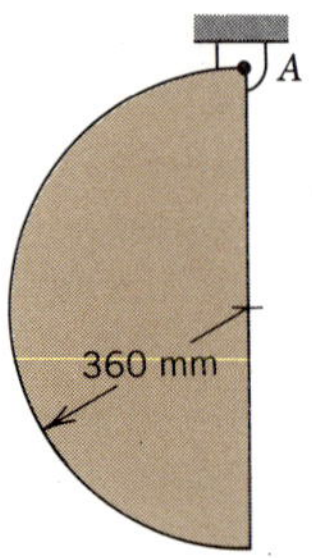

PROB. 14.61

14.61 The homogeneous semicircular cylinder whose mass is 4 kg is released from rest with its flat side aligned vertically, as shown. Determine the angular acceleration of the cylinder and the reaction forces at pin A at the instant after release.

***14.62** Solve Problem 14.61 if the semicircular cylinder has an angular velocity of 5 rad/s in the position shown.

14.63 Disks A and B have identical mass m. Initially, disk A is rotating clockwise at a rate of 300 rev/min and disk B is at rest. Disk B is then gently released and brought into contact with disk A. The coefficient of friction between the two disks is 0.25, and both disks rotate freely about their shafts. Determine the time required for the disks to attain the same angular speed and thus cease to slip over each other.

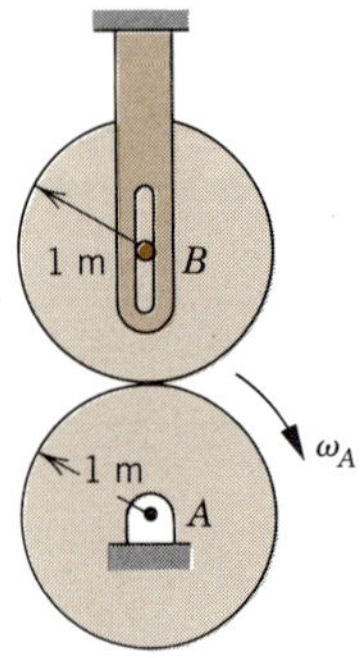

PROB. 14.63

14.64 In Problem 14.63, a clockwise couple $\bar{M}$ is applied to disk A to maintain its angular speed at 300 rev/min. Determine (a) the required value of M, and (b) the time required for the disks to cease to slip over each other.

***14.65** The equilateral triangular plate of mass m pivots about pin A. (a) Derive a differential equation of motion for the angle θ for the triangular plate. (b) The angle ϕ for the simple pendulum formed from a cable and a particle of mass m shown obeys the same equation of motion as that found in part (a). What is the length l of the cable?

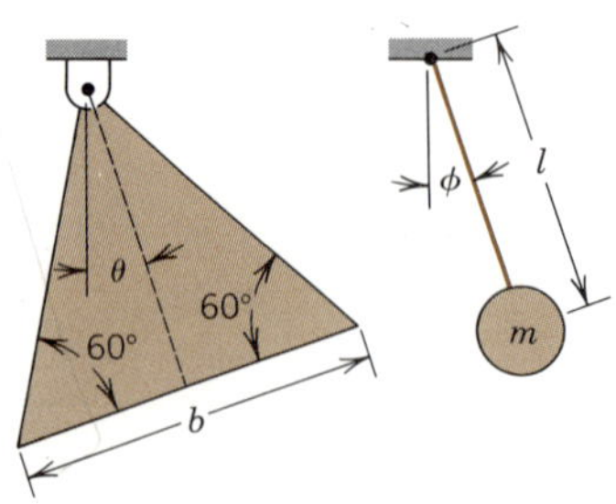

PROB. 14.65

***14.66** A 2 ton homogeneous block is suspended by cables OA and OC, which are attached to the stationary crane at point O. The system is released from rest at $\theta = 70^\circ$. Determine the cable tensions and the angular acceleration $\dot{\omega}$ of the block at the instant after release.

***14.67** Solve Problem 14.66 for the case where $\omega = 1.5$ rad/s when $\theta = 80^\circ$.

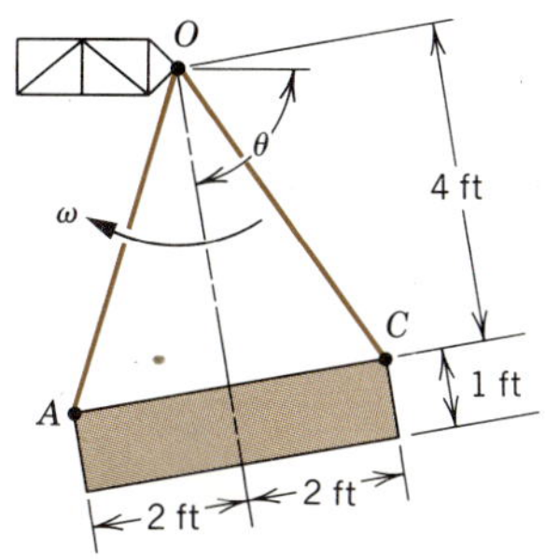

PROB. 14.66

c. General Motion

14.68 The diagram shows a 1 ton beam on a smooth horizontal surface. The beam is initially at rest when tensions F_A and F_B are applied to the horizontal cables. Determine the values of F_A and F_B required to impart accelerations $\dot{v}_A = 8$ ft/s^2, $\dot{v}_B = 16$ ft/s^2 to the points at which the cables are attached.

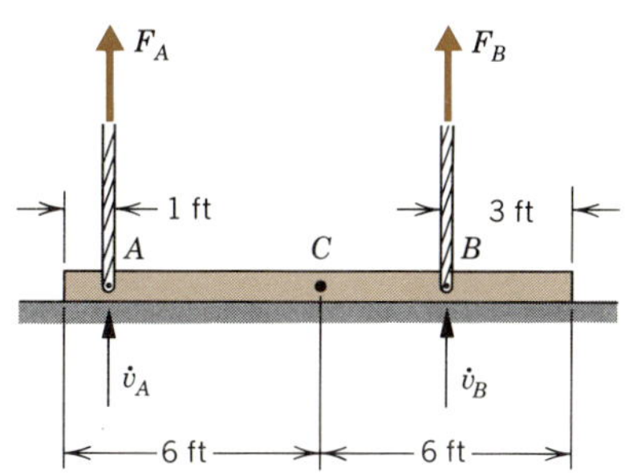

PROB. 14.68

14.69 The beam in Problem 14.68 is subjected to cable tensions $F_A = 400$ lb, $F_B = 600$ lb. Determine (a) the angular acceleration and the acceleration of the center of mass C, and (b) the accelerations of the attachment points A and B.

14.70 In order to correct the course of the missile, the thrust of engine A will be increased. The mass of the missile is (10^5) kg and each engine has a thrust of $2(10^3)$ kN when the increase is to occur. It is desired to rotate the missile by 1° in 1 s. Determine (a) the required increase in the thrust of engine A, and (b) the acceleration of the center of mass G immediately after the thrust of engine A is increased. The missile may be approximated as a thin uniform bar 60 m long.

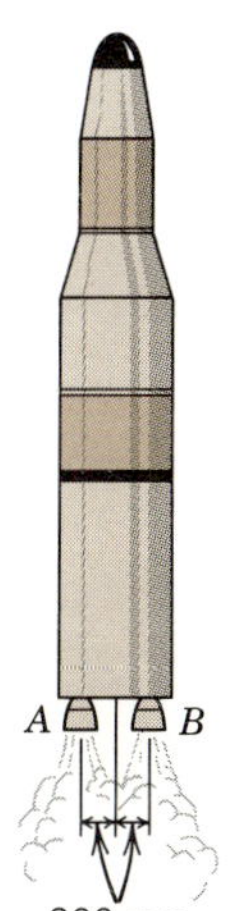

PROB. 14.70

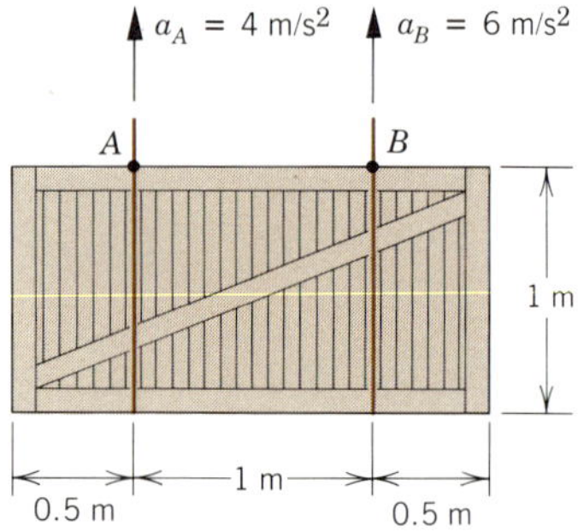

PROB. 14.71

14.71 The 5000 kg crate is at rest in the position shown when cables A and B are given upward accelerations of 4 m/s^2 and 6 m/s^2, respectively. Determine the tension in each cable. The crate may be regarded as a homogeneous body.

14.72 The 50 lb force is applied horizontally ($\theta = 0$) to the roller, whose weight is 100 lb. There is no slipping between the roller and the ground. Determine (a) the acceleration of the center of the roller, (b) the friction force, and (c) the minimum coefficient of friction for which slipping does not occur.

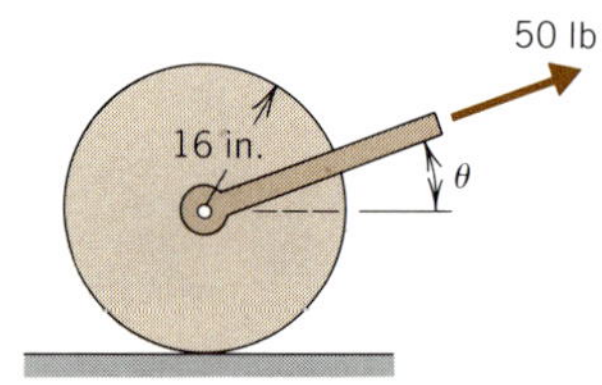

PROB. 14.72

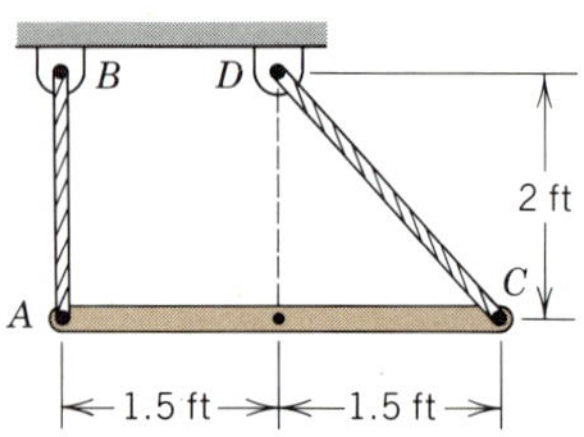

PROB. 14.74

14.73 Solve Problem 14.72 for the case where $\theta = 30°$.

14.74 The 5 lb bar is suspended by cables AB and CD. It is released from rest in the horizontal position shown. Determine the acceleration of the center of mass and the tension in each cable at the instant after release.

14.75 The coefficients of friction between the 3 kg wheel and the ground are $\mu_s = 0.30$, $\mu_k = 0.25$. Determine the acceleration of the center of the wheel and the angular acceleration of the wheel.

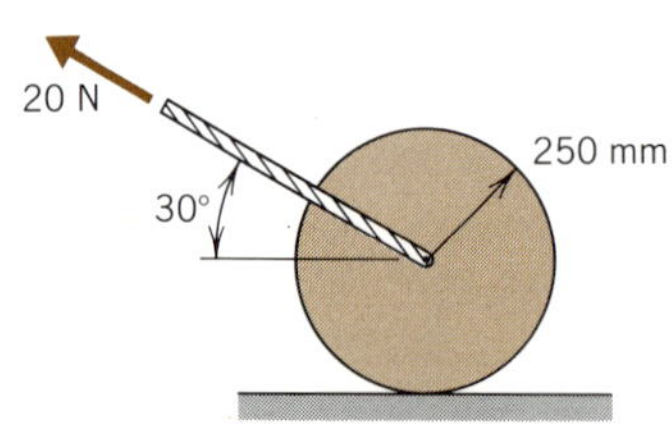

PROB. 14.75

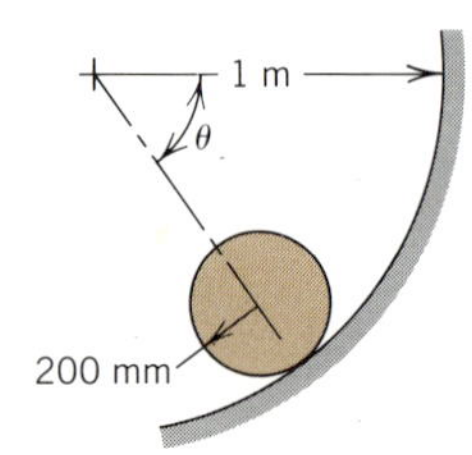

PROB. 14.76

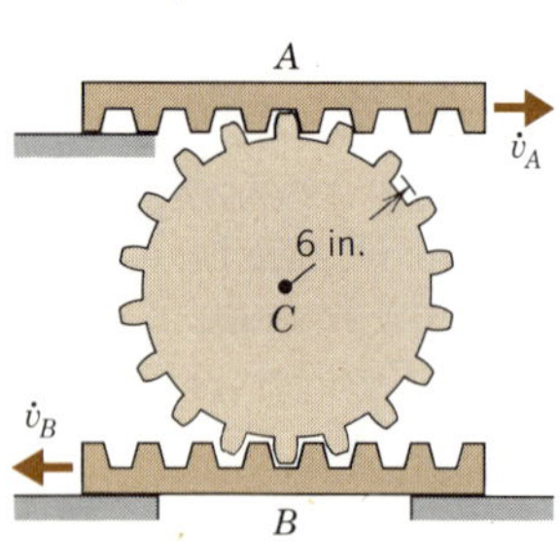

PROB. 14.77

14.76 The 2.5 kg homogeneous cylinder rolls without slipping down the curved surface. In the position where $\theta = 36.87°$, the speed of the mass center is 4 m/s. Determine the acceleration of the cylinder in this position and the components of the reaction forces acting on the cylinder.

14.77 The acceleration of racks A and B are $\dot{v}_A = 30$ ft/s^2 to the right and $\dot{v}_B = 10$ ft/s^2 to the left. Determine the horizontal components of the gear tooth forces exerted by each rack on the 20 lb gear C. The radius of gyration of this gear about its center is 4 in.

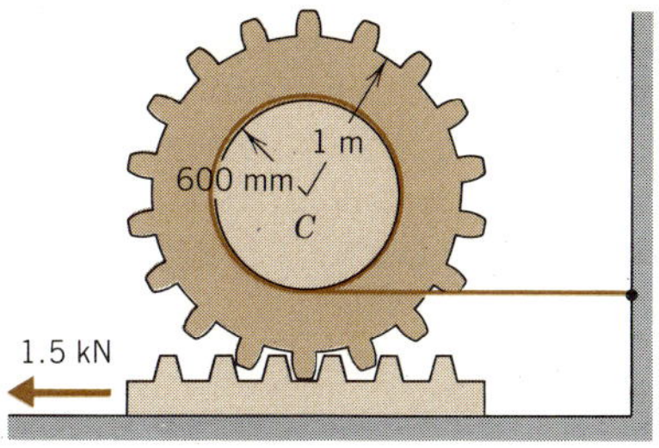

PROB. 14.78

14.78 The 1.5 kN horizontal force is applied to the rack, whose mass is negligible. Also, the ground is smooth. The gear and drum assembly has a mass of 300 kg and a radius of gyration of 500 mm. Determine the acceleration of the center C and of the rack.

14.79 Solve Problem 14.78 for the case where the mass of the rack is 150 kg and all other parameters are as given there.

14.80 The 4 lb stepped pulley in the sketch is pinned at its center O to the 8 lb sliding block. The tension forces in the cables are $T_A = 10$ lb, $T_B = 5$ lb. The radius of gyration of the pulley is 6 in., and friction is negligible. Determine the angular acceleration of the pulley and the acceleration of the block.

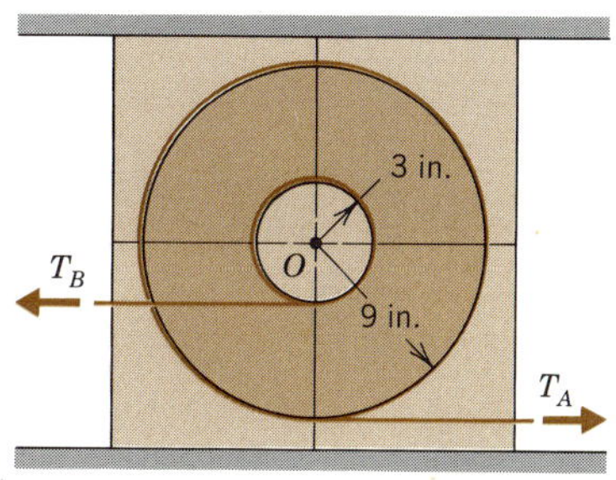

PROB. 14.80

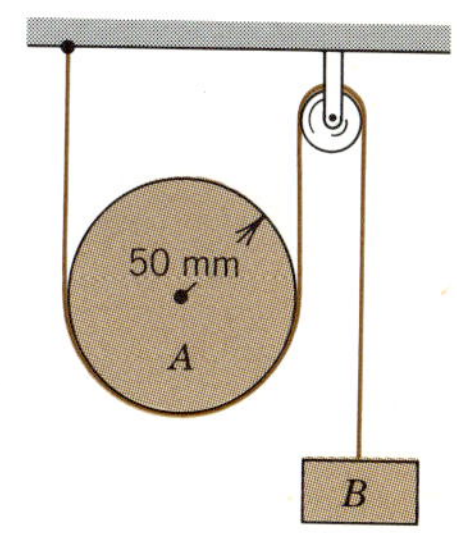

PROB. 14.81

14.81 The 30 kg wheel A has a radius of gyration of 40 mm. Determine the acceleration of the center of the wheel when the 20 kg counterweight B is released from rest.

14.82 The semicylinder of mass m and radius R is released from rest in the position shown. Determine (a) the minimum value of the coefficient of friction for which the semicylinder will roll without slipping, and (b) the corresponding initial angular acceleration.

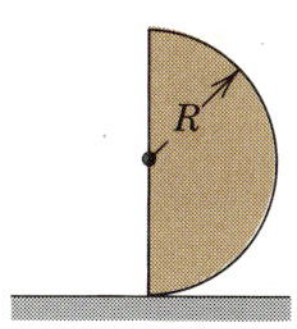

PROB. 14.82

14.83 Gear A is stationary. Gear B weighs 10 lb and has a radius of gyration of 5 in. The weight of bar AB is negligible. If the system is released from rest in the position shown, determine (a) the initial angular acceleration of the bar, and (b) the initial acceleration of the mass center of gear B.

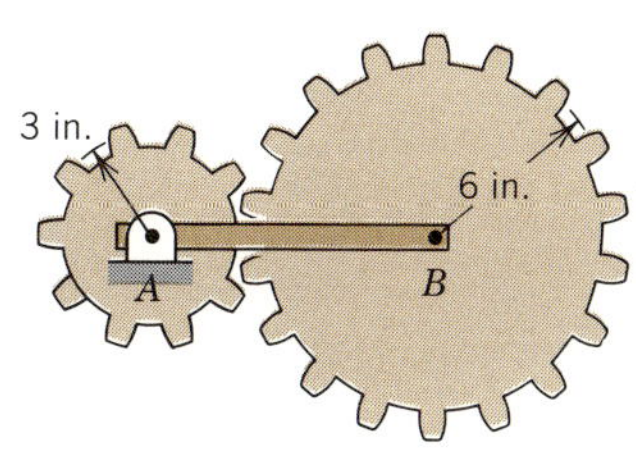

PROB. 14.83

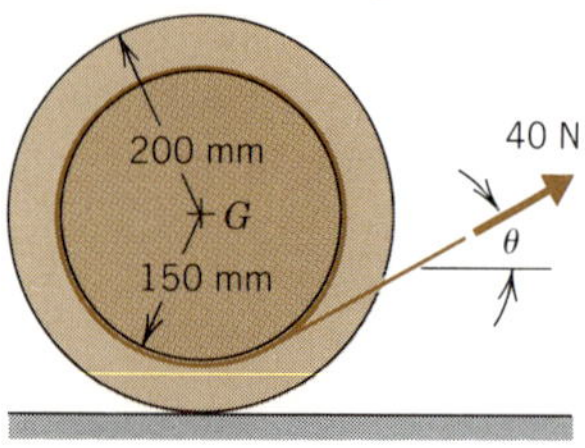

PROB. 14.84

14.84 The stepped pulley has a mass of 10 kg and a radius of gyration of 180 mm, and the coefficients of friction between the pulley and the horizontal surface are $\mu_s = 0.40$, $\mu_k = 0.35$. The angle θ is held constant as the end of the cable is pulled by a 40 N force. Determine the acceleration of the center of mass G and the angular acceleration of the disk if $\theta = 90°$.

14.85 Solve Problem 14.84 for the case where $\theta = 75°$.

14.86 A couple $\bar{M}_G$ is applied to the 20 lb bar AB, such that end A has a constant speed of 18 in./s. Determine $\bar{M}_G$ and the tension in cables AC and BD when the bar is in the horizontal position shown.

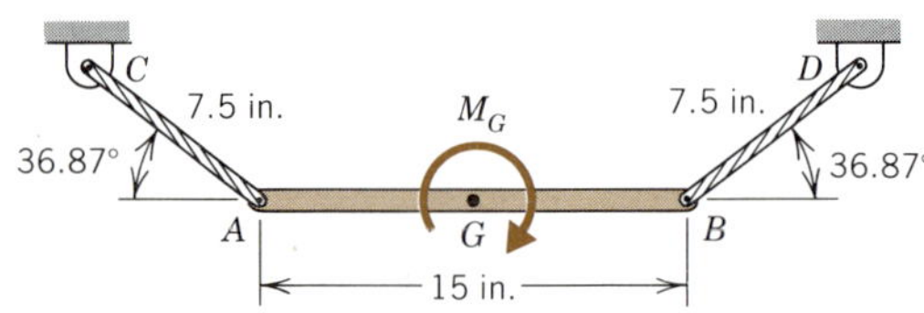

PROB. 14.86

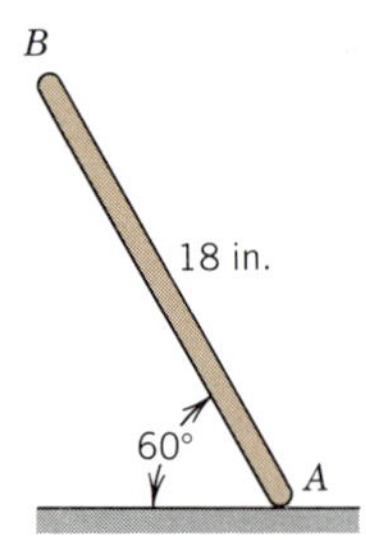

PROB. 14.87

14.87 The 2.5 lb bar is released from rest in the position shown. The ground is smooth, so the horizontal movement of end A is not constrained. Determine the acceleration of the bar's center of mass and its angular acceleration at the instant after release.

***14.88** The 50 kg homogeneous plate is at rest when the cable suspending corner B is cut. Determine the acceleration of the plate's center of mass and the tension in the cable at corner A at the instant motion begins.

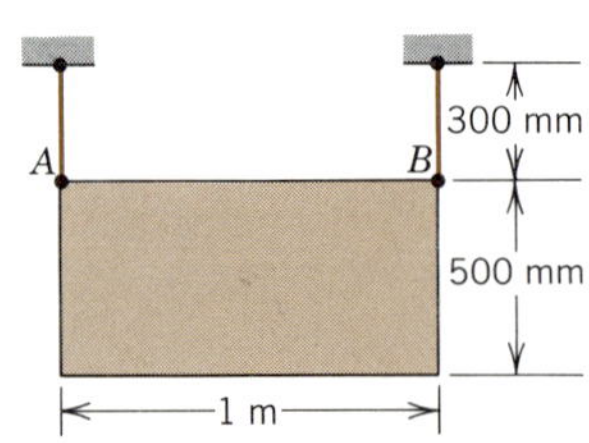

PROB. 14.88

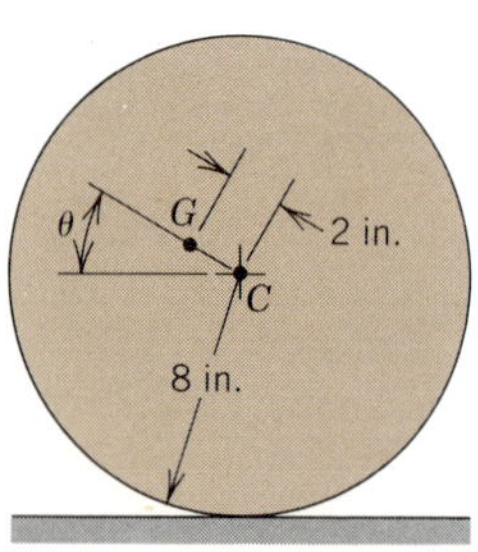

PROB. 14.89

***14.89** A 5 lb wheel of 8 in. radius was imperfectly manufactured, so its mass center G is 2 in. from its geometric center C, as shown. The radius of gyration with respect to an axis through the center C is 7 in. As a result of the imperfection, the velocity of point C varies when the wheel rolls. In the position where $\theta = 0$, point C is moving to the left at 2 in./s. The wheel is rolling without slipping. Determine (a) the corresponding acceleration of the wheel's center C, and (b) the minimum coefficient of static friction for which slipping does not occur in this position.

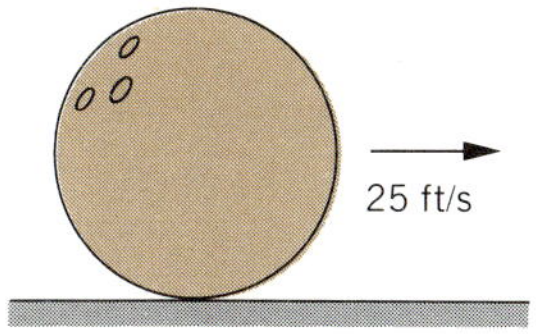

PROB. 14.90

14.90 When projected onto the floor, the 16 lb bowling ball is moving horizontally at 25 ft/s and is not rotating. The coefficient of kinetic friction is 0.20. Determine (a) the speed of the ball when it begins to roll on the ground without slipping, and (b) the distance traveled for the slipping phase of the motion. Regard the bowling ball as a homogeneous 1 ft diameter sphere. *Hint:* Slipping will cease when the velocity at the point of contact with the ground is zero.

14.91 When the thin hoop of radius R and weight W was placed on the floor, its center O was at rest and it was rotating at an angular velocity $\bar{\omega}_1$. Letting the coefficient of kinetic friction be μ_k, derive expressions for (a) the speed of point O at the instant when the hoop ceases to slip relative to the ground, and (b) the time during which slipping occurs.

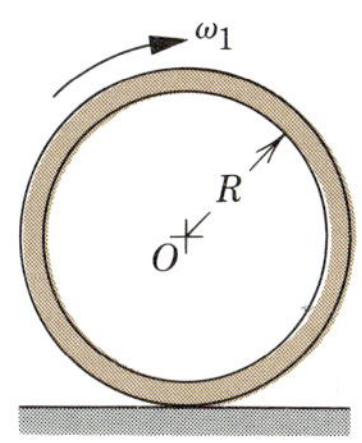

PROB. 14.91

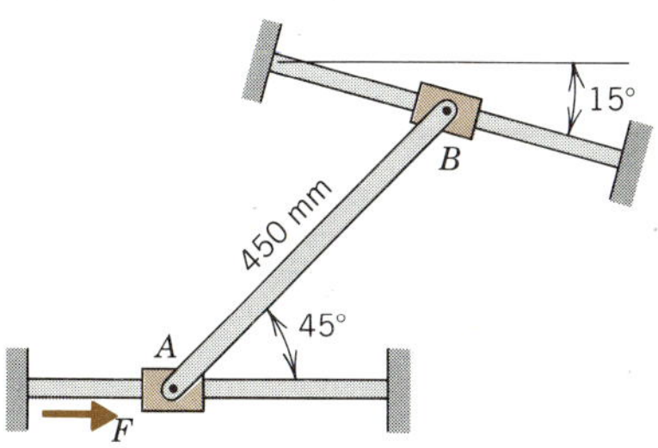

PROBS. 14.92 AND 14.93

***14.92** Bar AB has a mass of 2 kg, and collars A and B are massless and frictionless. The horizontal force $\bar{F}$ gives collar A a constant velocity of 10 m/s. For the position shown, determine $\bar{F}$ and the reactions at ends A and B.

***14.93** Bar AB has a mass of 2 kg, and collars A and B are massless and frictionless. The horizontal force is $F = 40$ N and the bar has just begun to move in the position shown. Determine the corresponding angular acceleration of the bar and the reactions at ends A and B.

***14.94** Solve Problem 14.93 in the case where collar A is moving to the right at 2 m/s in the specified position.

***14.95** The 500 lb box is released from rest in the position shown. The wall and the ground are smooth. Determine (a) the angular acceleration of the box, and (b) the reactions at corners A and B at the instant after release.

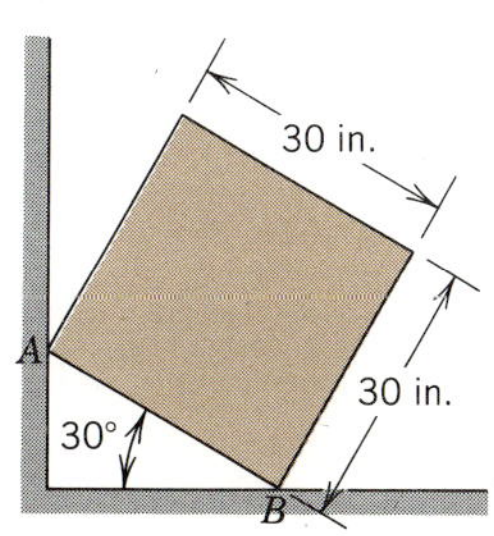

PROB. 14.95

F. PRINCIPLE OF WORK AND ENERGY

Our earlier study of particle motion demonstrated that energy and momentum principles greatly facilitate solving certain classes of problems, such as those involving conservative or impulsive forces. These principles are, in essence, special solutions of the differential equations that arise when using Newton's second law.

In this section we will adapt the energy principle for a system of particles, derived in Chapter 12, to the special case where the particles form a rigid body in planar motion. As in the previous sections, we restrict the study to the case of bodies that have a plane of symmetry parallel to the plane of motion, reserving the more general case for Chapter 16 on three-dimensional motion.

1. Kinetic Energy

In the symbolism for the work–energy formulation for a system of particles, T is the total kinetic energy of the system, V is the potential energy of all conservative forces acting on the system, and $U_{1\to 2}^{(\text{nc})}$ is the total work done by the nonconservative forces acting on the system. Then the principle of work and energy is described by the following scalar equation.

$$\boxed{T_2 + V_2 = T_1 + V_1 + U_{1\to 2}^{(\text{nc})}} \qquad (28)$$

where the subscripts 1 and 2, respectively, denote the initial and final position.

The first task in applying eq. (28) to rigid body motion is to describe the kinetic energy. In Figure 14 a rigid body is modeled as a continuum of infinitesimal elements by setting $m_i = dm$. The origin of the xyz coordinate system is an *arbitrary* point A in the body.

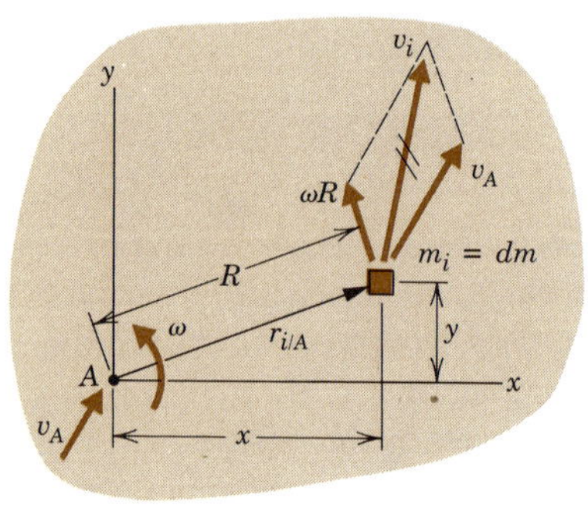

FIGURE 14

Chasle's theorem states that the motion of the body is a translation following point A and a rotation about point A. The superposition of velocities depicted in Figure 14 may be written as

$$\bar{v}_i = \bar{v}_A + \bar{\omega} \times \bar{r}_{i/A}$$

The kinetic energy of mass element i is $\frac{1}{2}(dm)v_i^2 \equiv \frac{1}{2}\bar{v}_i \cdot \bar{v}_i\, dm$. This is a differential quantity. Summing differentials is the process of integration, so

$$T = \int_{\mathscr{V}} \tfrac{1}{2}\bar{v}_i \cdot \bar{v}_i\, dm \equiv \tfrac{1}{2} \int_{\mathscr{V}} (\bar{v}_A + \bar{\omega} \times \bar{r}_{i/A}) \cdot (\bar{v}_A + \bar{\omega} \times \bar{r}_{i/A})\, dm$$

where $\mathscr{V}$ denotes the region occupied by the body. Next, we apply the rules of vector algebra. The variables $\bar{v}_A$ and $\bar{\omega}$ do not depend on which mass element is under consideration, so they may be factored out of the integrals. Thus

$$T = \tfrac{1}{2} \int_{\mathscr{V}} \bar{v}_A \cdot \bar{v}_A\, dm + \tfrac{1}{2} \int_{\mathscr{V}} [\bar{v}_A \cdot (\bar{\omega} \times \bar{r}_{i/A}) + (\bar{\omega} \times \bar{r}_{i/A}) \cdot \bar{v}_A]\, dm$$
$$+ \tfrac{1}{2} \int_{\mathscr{V}} (\bar{\omega} \times \bar{r}_{i/A}) \cdot (\bar{\omega} \times \bar{r}_{i/A})\, dm$$
$$= \tfrac{1}{2}v_A^2 \int_{\mathscr{V}} dm + v_A \cdot \left(\bar{\omega} \times \int_{\mathscr{V}} \bar{r}_{i/A}\, dm\right) + \tfrac{1}{2} \int_{\mathscr{V}} |\bar{\omega} \times \bar{r}_{i/A}|^2\, dm \qquad (29)$$

The first integral is obviously the mass. Also, the rotational velocity term $\bar{\omega} \times \bar{r}_{i/A}$ may be seen in Figure 14 to have a magnitude of ωR, where R is the distance from point A to the mass element. Thus

$$\int_{\mathscr{V}} |\bar{\omega} \times \bar{r}_{i/A}|^2 \, dm = \int_{\mathscr{V}} \omega^2 R^2 \, dm = \omega^2 \int_{\mathscr{V}} R^2 \, dm = \omega^2 I_{zz}$$

From the foregoing we may rewrite eq. (29) as

$$T = \tfrac{1}{2} m v_A^2 + \bar{v}_A \cdot \bar{\omega} \times \int_{\mathscr{V}} \bar{r}_{i/A} \, dm + \tfrac{1}{2} I_{zz} \omega^2 \tag{30}$$

This relation for kinetic energy requires the evaluation of the integral appearing in the middle term. In order to discuss this integral, we note from Figure 14 that the components of $\bar{r}_{i/A}$ are the x and y coordinates of the point occupied by dm. In other words, the integral in eq. (30) is the vector form of the first moment of mass.

$$m\bar{r}_{G/A} = \int_{\mathscr{V}} \bar{r}_{i/A} \, dm$$

Equation (30) now reduces to

$$T = \tfrac{1}{2} m v_A^2 + m\bar{v}_A \cdot (\bar{\omega} \times \bar{r}_{G/A}) + \tfrac{1}{2} I_{zz} \omega^2 \tag{31}$$

This is the formula for kinetic energy when the origin A is arbitrarily selected. In principle, we could use eq. (31) because it contains parameters that we are familiar with. However, evaluating the middle term is an unnecessary complication.

The formula and the corresponding computations are simplified substantially when the middle term is identically zero. One situation where this is *always* true arises when we select the origin to be the center of mass G, so that $\bar{r}_{G/A} \equiv \bar{0}$. We then find

$$\boxed{T = \tfrac{1}{2} m v_G^2 + \tfrac{1}{2} I_{zz} \omega^2; \quad z \text{ axis intersects } G} \tag{32}$$

Equation (32) indicates that there are two contributions to the total kinetic energy. The term $\frac{1}{2} m v_G^2$ is the *kinetic energy of translation*, which is associated with the motion of the center of mass G. The term $\frac{1}{2} I_{zz} \omega^2$ is the *kinetic energy of rotation* about the center of mass. The latter term is zero for a body in pure translation ($\omega \equiv 0$), as well as for a particle ($I_{zz} \equiv 0$).

There is another situation in which eq. (31) simplifies. Suppose that we place the origin A at the fixed point for a body in pure rotation, or else at the instant center for a body in general motion. In either case, $\bar{v}_A = \bar{0}$. A fixed point in a pure rotation is obviously an instant center as well, so we may write eq. (31) in the alternative form

$$\boxed{T = \tfrac{1}{2} I_{zz} \omega^2; \quad z \text{ axis intersects the instant center}} \tag{33}$$

From the viewpoint of this formula, a body has only rotational kinetic energy about its instant center.

In practice, we always employ eq. (33) to treat bodies in pure rotation. It is reasonable to employ it also to treat general motion provided that we have identified the instant center during the kinematical analysis. Clearly, eq. (33) is not valid for pure translation. In contrast, eq. (32) is valid for *all* cases of planar motion.

2. Work Done by Forces

Now that we have derived the formula for the kinetic energy of a rigid body, let us examine the other terms in the work–energy principle, eq. (28). The potential energy V is determined in the same manner as was done for particles. Each conservative force may be considered individually, as their potential energies are additive.

We may also use the earlier procedure for the nonconservative forces in which we trace the path of the point of application of each force and then compute the work of each force individually. However, in some situations the points of application may not have simple motions. Let us derive an alternative method for computing the work term for such cases.

Consider the force $\bar{F}$ in Figure 15, which is applied at point P in the

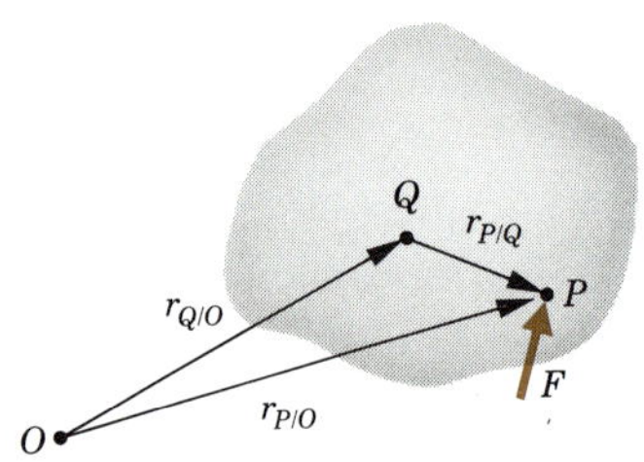

FIGURE 15

rigid body. The work done by the force $\bar{F}$ in changing the position of point P with respect to any fixed point O is

$$U_{1\to 2} = \int_{(r_{P/O})_1}^{(r_{P/O})_2} \bar{F} \cdot d\bar{r}_{P/O} \tag{34}$$

where the limits in the integral denote the coordinates of the initial and final positions of point P.

Suppose that the path of point P is unknown, but the path of another point, Q, is known. (This would be the case if point Q was a constrained point.) The infinitesimal displacement $d\bar{r}_{P/O}$ corresponds to a displacement $d\bar{r}_{Q/O}$ of point Q and a rotation $\overline{d\theta}$ of the body. The relation between these motions may be obtained from the kinematical equation for the velocity of points in a rigid body.

We develop this relationship by noting that $d\bar{r}_{P/O}$ is the position change resulting from a velocity $\bar{v}_P$ during the interval dt. Similar reasoning applies to $d\bar{r}_{Q/O}$ and $\overline{d\theta}$. Thus

$$\begin{aligned} d\bar{r}_{P/O} &\equiv \bar{v}_P\,dt = (\bar{v}_Q + \bar{\omega} \times \bar{r}_{P/Q})\,dt \\ &\equiv d\bar{r}_{Q/O} + \overline{d\theta} \times \bar{r}_{P/Q} \end{aligned}$$

This expression is a restatement of Chasle's theorem in terms of infinitesimal movements.

Substitution of $d\bar{r}_{P/O}$ into eq. (34) for the work yields

$$U_{1\to 2} = \int_{(r_{Q/O})_1}^{(r_{Q/O})_2} \bar{F} \cdot d\bar{r}_{Q/O} + \int_{\theta_1}^{\theta_2} \bar{F} \cdot (\overline{d\theta} \times \bar{r}_{P/Q}) \tag{35}$$

The first term is recognizable as the work that $\bar{F}$ would do if it were applied at point Q.

In order to investigate the second term in eq. (35), consider the situation in Figure 16, where l is the separation distance between points P and Q, $l = |\bar{r}_{P/Q}|$. Also, β is the angle of the force $\bar{F}$ relative to the line PQ. We find from this sketch that $\overline{d\theta} \times \bar{r}_{P/Q} = l\,d\theta$ perpendicular to line PQ. Then

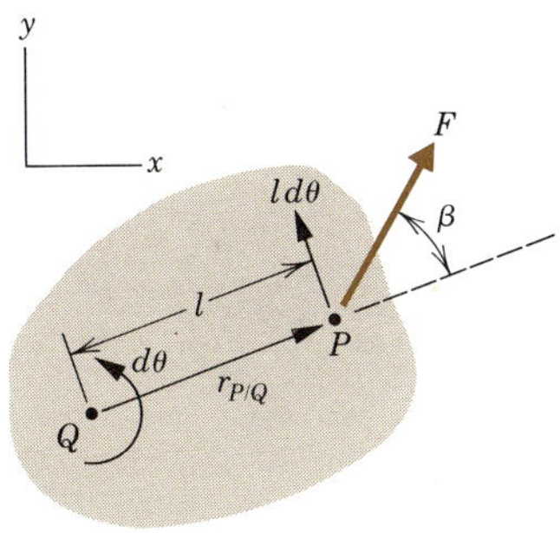

FIGURE 16

$$\bar{F} \cdot (\overline{d\theta} \times \bar{r}_{P/Q}) = (F \sin \beta)(l\,d\theta) \tag{36}$$

where $F \sin \beta$ is the component of $\bar{F}$ perpendicular to line PQ. Changing the order of multiplication gives a new viewpoint to this term. Particularly, note that the moment M_{Qz} of $\bar{F}$ about point Q in Figure 16 is

$$M_{Qz} = l(F \sin \beta)$$

so that eq. (36) is the product of M_{Qz} and $d\theta$. It follows that eq. (35) for the work done by $\bar{F}$ may be rewritten as

$$\boxed{U_{1\to 2} = \int_{(Q/O)_1}^{(r_{Q/O})_2} \bar{F} \cdot d\bar{r}_{Q/O} + \int_{\theta_1}^{\theta_2} M_{Qz}\,d\theta} \tag{37}$$

Equation (37) reduces to the more familiar definition of work in eq. (34) when P and Q are the same point.

The explanation of the new formula for work lies in the concept of equivalent force systems, which was developed in Statics. Applying the force $\bar{F}$ at point P is equivalent to applying the force-couple system $\bar{F}$ and M_{Qz} at point Q. This equivalency is depicted in Figure 17. Hence, the first integral in eq. (37) is the work done by the equivalent force in the translation of the rigid body following point Q, while the second term is the work done by the equivalent moment in the rotation of the body about point Q.

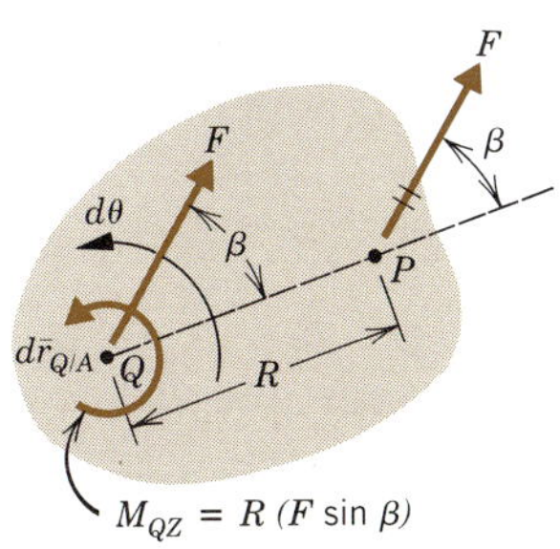

FIGURE 17

Note that the situation depicted in Figures 16 and 17 corresponds to the case where the couple and the rotation are in the same sense. The work done when the couple is in the *opposite* sense from the rotation leads to *negative work* (just as a force opposing the movement of its point of application does negative work.)

When the work done by a system of nonconservative forces needs to be computed, each force can be treated individually according to eq. (37). Alternatively, the entire system of forces can be replaced by a single force-couple system at point Q. The force and moment terms in eq. (37) are then the resultants of the force system.

The second alternative leads to an important general statement. Specifically, the internal forces in a rigid body *never* do work. This results because the force and moment exerted by one particle on another are canceled by the reaction applied by the second particle to the first.

An important physical system featuring the application of eq. (37) is a wheel that rolls without slipping. In Figure 18a, a disk has rolled along a flat surface an infinitesimal distance dx_A from its initial position. Points C' and C are the initial and the current points of contact with the flat surface, respectively. Thus, $d\theta$ is the angle of rotation.

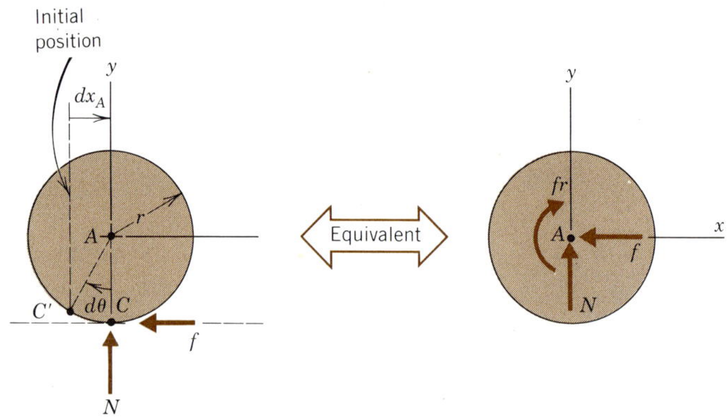

FIGURE 18a **FIGURE 18b**

It is difficult to describe the work done by the contact forces in Figure 18a, because the point on the body to which they are applied changes as the wheel rolls. We therefore replace these forces by the equivalent force-couple system in Figure 18b. The reason for selecting the center A for this representation is that the path followed by this point is a horizontal line. The couple $M_{Az} = -fr$ shown in Figure 18b is the moment of the forces in Figure 18a about point A.

The sense of the rotation $d\theta$ is about the negative z axis according to the right-hand rule. Also the movement of point A is horizontal. Thus, the infinitesimal work dU corresponding to the infinitesimal movement is found from eq. (37) to be

$$dU = -f(dx_G) + M_{Az}(-d\theta) = -f\,dx_G + fr\,d\theta \tag{38}$$

This is a general relation that does not account for the no slip condition. In order to describe the effect of the no slip constraint, note that the angular speed ω and the speed of point A are related by $v_A = \omega r$ when there is no slipping. This kinematical relation may be expressed in differential form as

$$dx_G \equiv v_A\,dt = (\omega r)\,dt = \left(\frac{d\theta}{dt}\,r\right)dt = (d\theta)r$$

Substituting the no slip condition into eq. (38) gives

$$dU = -f(r\,d\theta) + fr\,d\theta \equiv 0$$

It is not surprising that the normal reaction acting on the rolling wheel does no work. However, you might not have anticipated that the friction force also does no work. The general notion of friction is that it slows any movement. When surfaces slide over each other, this retarding effect is manifested as an energy dissipation corresponding to a negative value for the work done by friction. In contrast, when there is rolling without slipping, the retarding effect is that any translation also produces a rotation. Then some of the work done by the loads acting on the rolling body is used to create the angular kinetic energy, rather than going entirely to produce translational kinetic energy.

It is a general fact that

> The contact forces causing a wheel to roll without slipping never do work.

Knowing this substantially expedites the treatment of rolling without slipping, because there is no longer a need to replace the friction and normal forces by an equivalent force-couple system.

As a closure, let us summarize the concepts discussed in this section. We have found that the kinetic energy may be evaluated by either eq. (32) or eq. (33). The work–energy principle, eq. (28), therefore relates the linear and angular velocity of a body at two positions. The work done by a conservative force, such as gravity or a spring force, may be described by using the corresponding potential energy, (see Chapter 12). Equation (37) is needed to treat nonconservative forces.

The use of eq. (28) is limited by the fact that it is a single scalar equation. Thus, it yields the velocity of a body at a specified position only when that motion can be described in terms of a single rate variable, such as the velocity of a point or the angular velocity.

Finally, recall that it is sometimes impossible to evaluate the work until the motion of the system has been determined. This is the case when the magnitude of a force is dependent on elapsed time or velocity. Such situations cannot be addressed by employing the work–energy principle.

EXAMPLE 8

When $\theta = 36.87^\circ$ the velocity of collar A is v_A, directed as shown. The mass of bar AB is m and the mass of the sliders is negligible. The system lies in the vertical plane. (a) Derive an expression for the corresponding kinetic energy using eq. (33). (b) Solve part (a) by using eq. (33). (c) If the bar was at rest when $\theta = 0$, determine v_A.

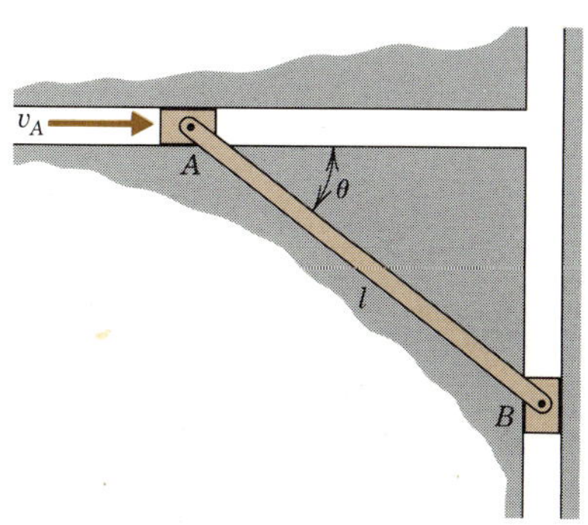

Solution

In either formulation it is necessary to relate v_A to the angular speed ω. We will formulate this relation by the method of instant centers. This will provide the information required for part (b). The construction of the perpendiculars to the velocity of each collar is depicted in the sketch of the bar at $\theta = 36.87°$; the intersection of these lines is the instant center C.

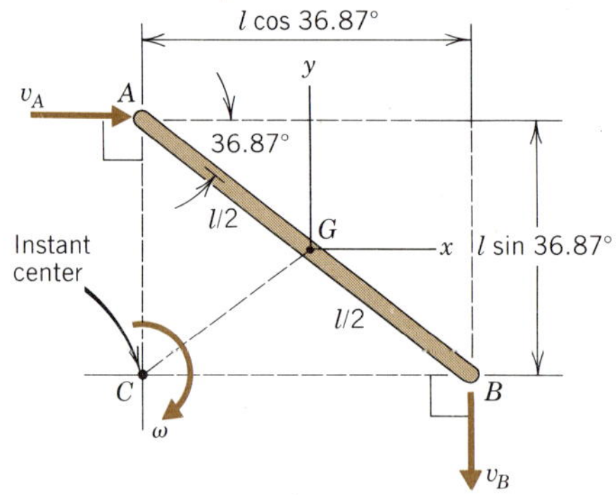

Part a

Equation (32) features the speed of the center of mass and the angular velocity of the bar. These must be related to v_A. The corresponding instant center equations are

$$v_A = \omega r_{A/C}, \qquad v_G = \omega r_{G/C}$$

A trigonometric analysis of the instant center diagram yields

$$r_{A/C} = l \sin 36.87° = 0.6l, \qquad r_{G/C} = \frac{l}{2}$$

$$\omega = \frac{v_A}{r_{A/C}} = 1.6667\,\frac{v_A}{l}, \qquad v_G = \left(1.6667\,\frac{v_A}{l}\right)\left(\frac{l}{2}\right) = 0.8333 v_A$$

We have placed the origin of the coordinate system at point G in order to employ eq. (32). The corresponding moment of inertia is given in Appendix B as

$$I_{zz} = \tfrac{1}{12} ml^2$$

Thus eq. (32) gives

$$T = \frac{1}{2} m(0.8333 v_A)^2 + \frac{1}{2}\left(\frac{1}{12} ml^2\right)\left(1.6667\,\frac{v_A}{l}\right)^2$$
$$= 0.4630 m v_A^2$$

Part b

Only the relationship between v_A and ω is needed for eq. (33). This was found in the instant center analysis to be $\omega = 1.6667\, v_A/l$. However, there is a trade-off for not having to relate v_G to ω. We must now compute the moment of inertia about an axis parallel to the z axis intersecting the instant center. The distance in the plane of motion between the center of mass and the instant center was found earlier to be $r_{G/C} = l/2$, so the parallel axis theorem for the desired z axis gives

$$I_{zz} = \frac{1}{12}ml^2 + m\left(\frac{l}{2}\right)^2 = \frac{1}{3}ml^2$$

Then eq. (33) becomes

$$T = \frac{1}{2}\left(\frac{1}{3}ml^2\right)\left(1.6667\frac{v_A}{l}\right)^2 = 0.4630mv_A^2$$

Part c

The bar in the initial position is at rest, and we must find the velocity in a second position. This is the type of situation that suggests using the work–energy principle. The purpose of a free body diagram in such a formulation is to assist in accounting for the work done by all forces. Note that *the free body diagram should show a general position.*

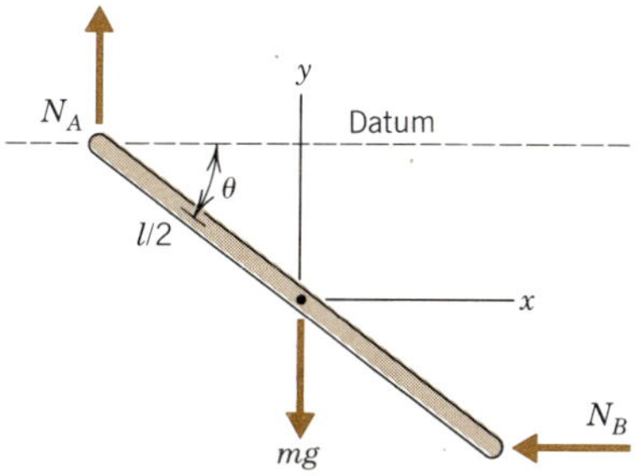

Gravity is conservative. It is useful in general to place the datum for the gravitational potential energy at an elevation that is geometrically significant. We have selected the elevation of slider A. When $\theta = 0$, the center of mass is at the datum. In the final position it is a distance $(l/2)$ sin 36.87° below the datum. Thus

$$V_1 = 0, \qquad V_2 = mg\left(-\frac{l}{2}\sin 36.87^\circ\right) = -0.3mgl$$

The reaction forces $\bar{N}_A$ and $\bar{N}_B$ are always perpendicular to the direction in which their points of application move. Hence $U_{1\to 2}^{(nc)} \equiv 0$. The body was initially at rest when $\theta = 0$, and the final kinetic energy was found in both parts (a) and (b). Thus

$$T_1 = 0, \qquad T_2 = 0.4630mv_A^2$$

Equation (28) therefore yields

$$0.4630mv_A^2 - 0.3mgl = 0 + 0 + 0$$

$$v_A = 0.805\,(gl)^{1/2}$$

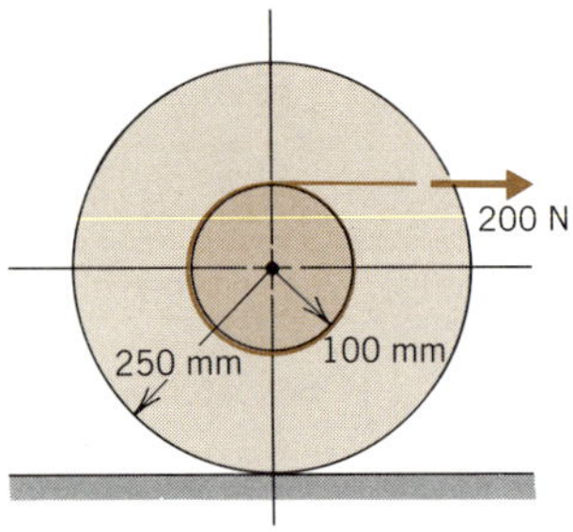

EXAMPLE 9

A 200 N force is applied to the free end of a cable wrapped around the drum of the stepped disk, as shown. The disk has a mass of 40 kg and a radius of gyration of 150 mm. Knowing that the disk starts from rest and rolls without slipping, determine its speed after moving 5 m.

Solution

We begin by constructing a free body diagram of the stepped disk. Note that we have placed the origin of xyz at the center of mass because we will employ eq. (32) for the kinetic energy associated with this general motion.

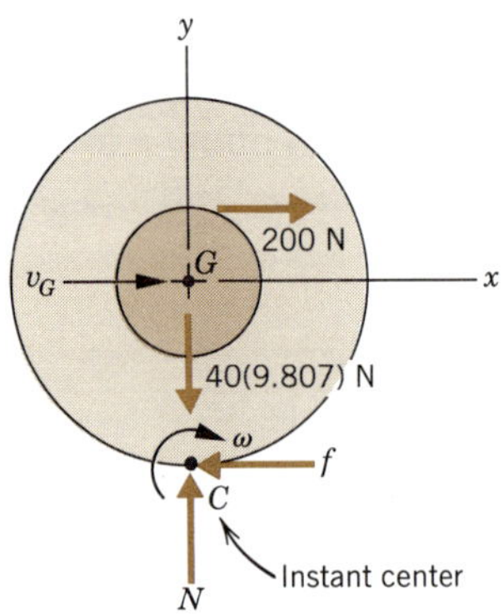

A key factor in employing the work–energy principle is being able to evaluate the work term. Hence this is where we begin the analysis. The x axis is a convenient datum for the gravitational potential energy. Because the center of mass is always situated at that elevation, set

$$V \equiv 0$$

There is no slipping at the contact point C. Earlier we learned that for rolling without slipping the friction force as well as the normal force does no work. Because the point on the drum that is tangent to the cable is changing continuously, we shall replace the cable tension by a force $\bar{F}$ and couple M_{Gz} at the mass center G. Computing the moment of the cable tension about point G gives

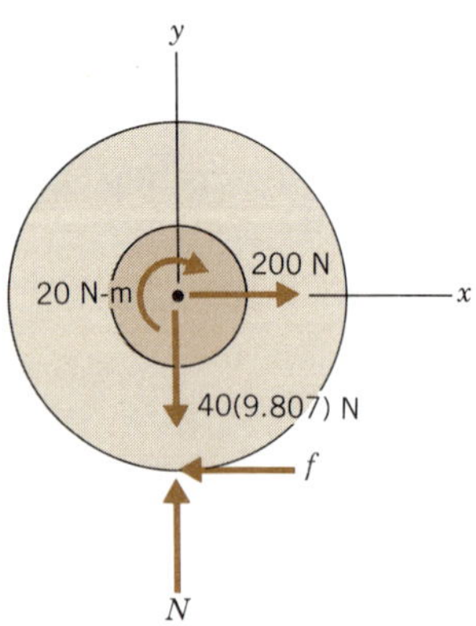

$$F_x = 200 \text{ N}, \qquad F_y = 0$$

$$M_{Gz} = -200(0.1) = -20 \text{ N-m}$$

These forces are shown on the equivalent free body diagram.

The infinitesimal displacement of the center G is dx_G in the x direction, and the rotation is $d\theta$ clockwise. Also, because point C is the instant center, $v_G = 0.25\omega$, so that $dx_G = 0.25\, d\theta$. Thus eq. (37) gives

$$U_{1\to 2}^{(\text{nc})} = \int_0^{\Delta x_G} F_x\, dx_G + \int_0^{\Delta y_G} F_y\, dy_G + \int_0^{\Delta\theta} M_{Gz}(-d\theta)$$

$$U_{1\to 2}^{(nc)} = \int_0^{\Delta x_G} 200\,dx_G + \int_0^{\Delta x_G} (-20)\left(-\frac{dx_G}{0.25}\right)$$

$$= \int_0^{\Delta x_G} (200 + 80)dx_G = 280\,\Delta x_G$$

where $\Delta x_G = 5$ m is the distance that point G moves.

The last quantity requiring evaluation is the kinetic energy. The disk was initially at rest, so $T_1 = 0$. For the final kinetic energy we employ eq. (32). Setting $\omega = v_G/0.25$ for no slipping gives

$$T_2 = \frac{1}{2}mv_G^2 + \frac{1}{2}I_{zz}\left(\frac{v_G}{0.25}\right)^2$$

The given radius of gyration corresponds to the z axis through point G, and $m = 40$ kg. Thus

$$T_2 = \frac{1}{2}(40)v_G^2 + \frac{1}{2}[40(0.15)^2]\left(\frac{v_G}{0.25}\right)^2 = 27.20v_G^2$$

We may now substitute the individual terms into the work–energy principle, eq. (28). Thus

$$T_2 + V_2 = T_1 + V_1 + U_{1\to 2}^{(nc)}$$

$$27.20v_G^2 + 0 = 0 + 0 + 280(5)$$

$$v_G = 7.17 \text{ m/s}$$

EXAMPLE 10

The stiffness of the spring is 30 lb/in., and its unstretched length is 10 in. Each bar weighs 5 lb, and the linkage lies in the horizontal plane. The system is released from rest at $\theta = 30^\circ$. Determine the angular speed of bar AB when $\theta = 60^\circ$. The mass of collar C is negligible.

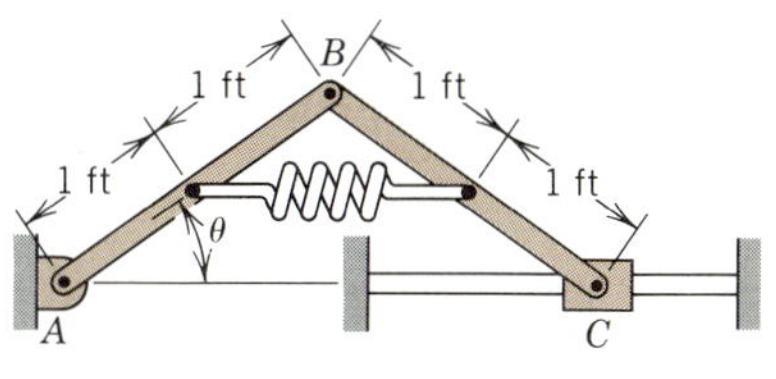

Solution

The first step is to draw a free body diagram of the linkage in order to discuss the work done by the forces. Note that pin B is assumed to be ideal, so it neither stores nor dissipates energy. Hence we need not consider the forces exerted at this pin.

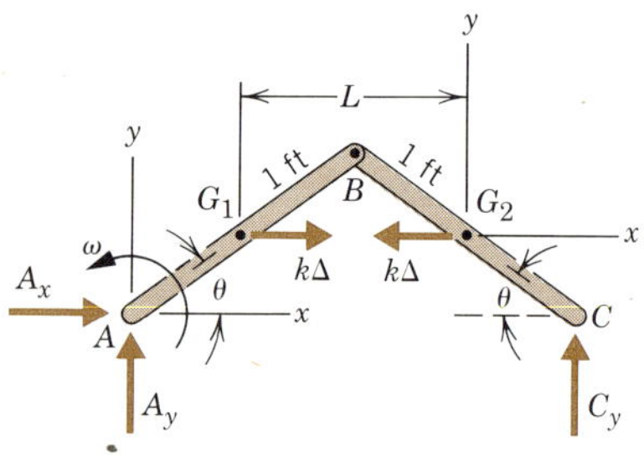

There are two bodies in this linkage. We shall use a different coordinate system to describe each. Bar AB is in pure rotation about pin A, so we place one origin at point A. Equation (33) then describes T_{AB}. In contrast, bar BC is in general motion. We will employ eq. (32) for T_{BC}, with the second origin placed at the center of mass G_2 of bar BC.

Point A is fixed, and end C follows a horizontal path. The reactions A_x, A_y, and C_y therefore do not work, $U_{1\to 2}^{(\mathrm{nc})} = 0$. The spring forces $k\Delta$ are conservative; hence the potential energy of the spring is $V = \frac{1}{2}k\Delta^2$. The unstretched length of the spring is given as 10/12 ft, and the free body diagram shows that the length of the spring in a general position is $L = 2(1 \cos\theta)$ ft. Thus

$$V = \frac{1}{2}k\left(2\cos\theta - \frac{10}{12}\right)^2$$

Setting $\theta_1 = 30^\circ$ and $\theta_2 = 60^\circ$ then yields

$$V_1 = \frac{1}{2}\left(30\,\frac{\text{lb}}{\text{in.}}\right)\left(\frac{12\text{ in.}}{1\text{ ft}}\right)\left(2\cos 30^\circ - \frac{10}{12}\right)^2 = 145.38\text{ ft-lb}$$

$$V_2 = \frac{1}{2}(30)(12)\left(2\cos 60^\circ - \frac{10}{12}\right)^2 = 5\text{ ft-lb}$$

The last step is to evaluate the kinetic energy. The system is at rest in the initial position, so $T_1 = 0$. For the final position, eqs. (32) and (33) for each bar give

$$T_{AB} = \tfrac{1}{2}I_{z_1z_1}\omega_{AB}^2, \qquad T_{BC} = \tfrac{1}{2}mv_{G_2}^2 + \tfrac{1}{2}I_{z_2z_2}\omega_{BC}^2$$

These kinetic energies are scalar quantities, so

$$T_2 = T_{AB} + T_{BC}$$

The kinematical variable we wish to determine is ω_{AB}. We must therefore express v_{G_2} and ω_{BC} in terms of ω_{AB} at the position $\theta = 60^\circ$.

In a general linkage this would involve a full linkage analysis. However, the symmetry of this linkage system leads to an acceptable shortcut. The bars form equal angles with the horizontal, so the rotation of bar AB counterclockwise must equal the rotation of bar BC clockwise. Hence

$$\bar{\omega}_{AB} = \omega_{AB}\bar{k}, \qquad \bar{\omega}_{BC} = \omega_{AB}(-\bar{k})$$

With $\bar{\omega}_{BC}$ now known in terms of $\bar{\omega}_{AB}$, we may express $\bar{v}_B$ in terms of $\bar{\omega}_{AB}$ and use that result to construct $\bar{v}_{G_2}$. Thus

$$\bar{v}_B = \bar{\omega}_{AB} \times \bar{r}_{B/A}, \qquad \bar{v}_{G_2} = \bar{v}_B + \bar{\omega}_{BC} \times \bar{r}_{G_2/B}$$

When $\theta = 60^\circ$ the position vectors are

$$\bar{r}_{B/A} = 2(\cos 60^\circ\,\bar{i} + \sin 60^\circ\,\bar{j}) = 1.0\bar{i} + 1.7321\bar{j}\text{ ft}$$

$$\bar{r}_{G_2/B} = 1(\cos 60^\circ\,\bar{i} - \sin 60^\circ\,\bar{j}) = 0.5\bar{i} - 0.8660\bar{j}\text{ ft}$$

so the velocity equations give

$$\bar{v}_B = \omega_{AB}\bar{k} \times (1.0\bar{i} + 1.7321\bar{j}) = -1.7321\omega_{AB}\bar{i} + \omega_{AB}\bar{j}$$

$$\bar{v}_{G_2} = (-1.7321\omega_{AB}\bar{i} + \omega_{AB}\bar{j}) + (-\omega_{AB}\bar{k}) \times (0.5\bar{i} - 0.866\bar{j})$$
$$= -2.598\omega_{AB}\bar{i} + 0.5\omega_{AB}\bar{j}$$

$$v_{G_2}^2 \equiv \bar{v}_{G_2} \cdot \bar{v}_{G_2} = (2.598\omega_{AB})^2 + (0.5\omega_{AB})^2 = 7.0\omega_{AB}^2$$

The only parameters remaining to be evaluated are the mass and the moments of inertia. The value of $I_{z_2z_2}$ is given by Appendix B, but computing $I_{z_1z_1}$ requires using the parallel axis theorem.

$$m = \frac{5}{32.17} = 0.15542 \text{ lb-s}^2/\text{ft}$$

$$I_{z_2z_2} = \tfrac{1}{12}m(2)^2 = 0.05181 \text{ lb-s}^2\text{-ft}$$

$$I_{z_1z_1} = \tfrac{1}{12}m(2)^2 + m(1)^2 = 0.2072 \text{ lb-s}^2\text{-ft}$$

Substituting the expressions for $\bar{\omega}_{BC}$, $v_{G_2}^2$ and the moments of inertia into the kinetic energy expressions yields

$$T_{AB} = \tfrac{1}{2}(0.2072)\omega_{AB}^2 = 0.1036\omega_{AB}^2$$

$$T_{BC} = \tfrac{1}{2}(0.15542)(7.0\omega_{AB}^2) + \tfrac{1}{2}(0.05181)\omega_{AB}^2 = 0.5699\omega_{AB}^2$$

$$T_2 = T_{AB} + T_{BC} = 0.6735\omega_{AB}^2$$

We have now determined each term in the work energy principle, eq. (28). We then find that

$$T_2 + V_2 = T_1 + V_1 + U_{1\to 2}^{(\text{nc})}$$

$$0.6735\omega_{AB}^2 + 5 = 0 + 145.38 + 0$$

$$\omega_{AB} = 14.44 \text{ rad/s}$$

HOMEWORK PROBLEMS

Kinetic Energy Problems

14.96 In the position shown, the 5 kg uniform rod is rotating counterclockwise with an angular speed of 120 rev/min. Determine the kinetic energy of the bar (a) using eq. (32), (b) using eq. (33).

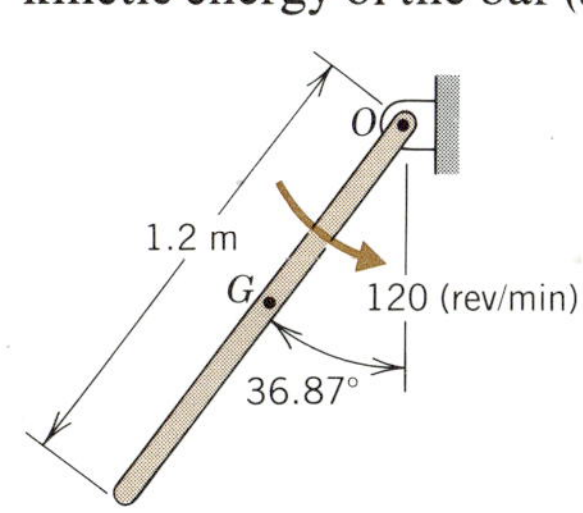

PROB. 14.96

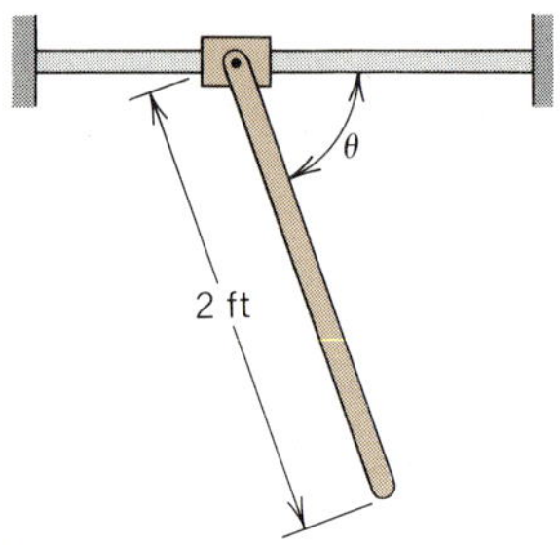

PROB. 14.97

14.97 When $\theta = 90°$, the 40 lb bar is swinging counterclockwise at 10 rad/s as the collar moves to the right at 20 ft/s. Determine the corresponding kinetic energy of the bar.

14.98 Solve Problem 14.97 if the parameters given there are for the position $\theta = 30°$.

14.99 The mass of each of the uniform bars is m and that of the flywheel disk is $8m$. In the position shown the disk is rotating at angular speed ω counterclockwise. Determine the kinetic energy of each bar and of the disk.

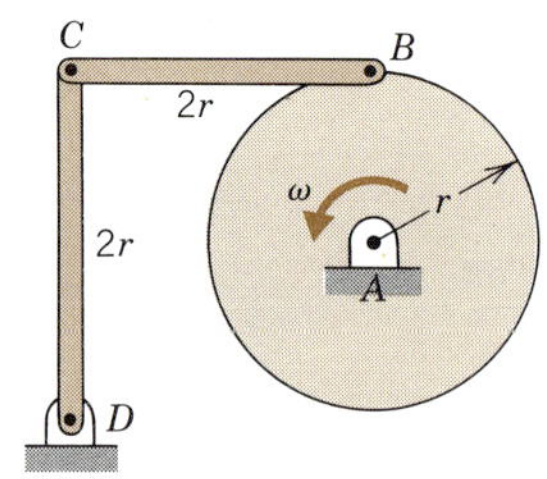

PROB. 14.99

14.100 The 50 lb pulley has a radius of gyration of 12 in. and is rotating at $\omega = 240$ rev/min clockwise. Determine the total kinetic energy of the system.

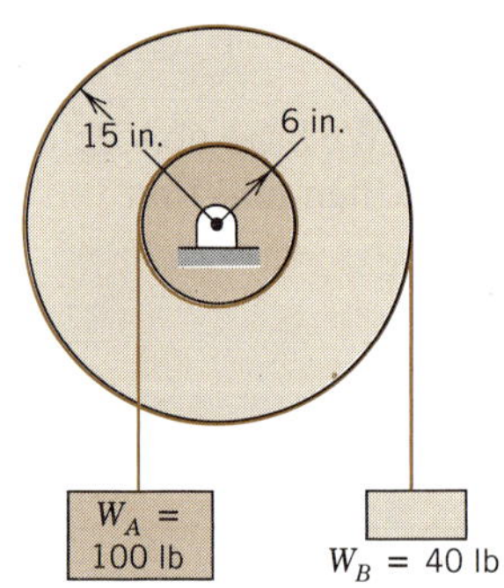

PROB. 14.100

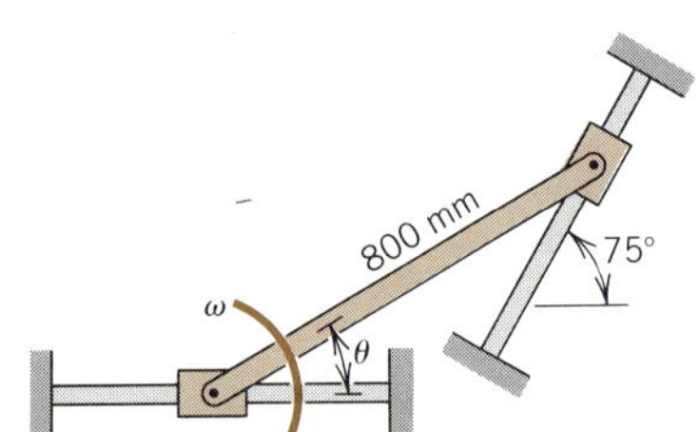

PROB. 14.101

14.101 Describe the kinetic energy of the 6 kg bar in terms of its angular speed ω when the bar is at $\theta = 0°$.

***14.102** Solve Problem 14.101 for $\theta = 30°$.

Work-Energy Problems

14.103 Bar AB has a mass of 60 kg and the mass of links AC and BD is negligible. The system is initially at rest in a vertical plane at $\theta = 15°$. Determine the angular speed of bar AC when $\theta = 90°$

14.104 Solve Problem 14.103 for the case where the mass of bars AC and BD are 15 kg each.

14.105 A 2000 kg turbine engine has a radius of gyration of 1 m. When shut down while operating at 1500 rev/min, it requires 4000 rev to come to rest. Determine the average magnitude of the frictional torque caused by the bearings and air resistance.

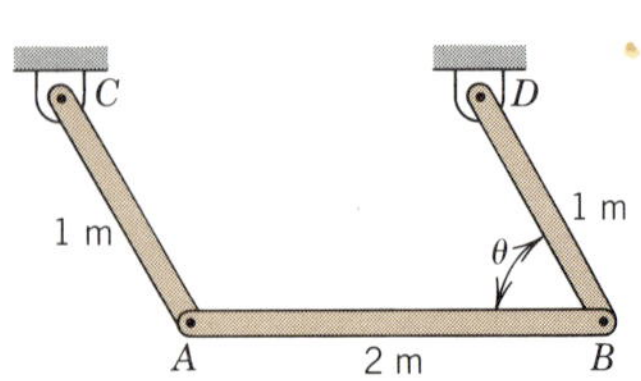

PROB. 14.103

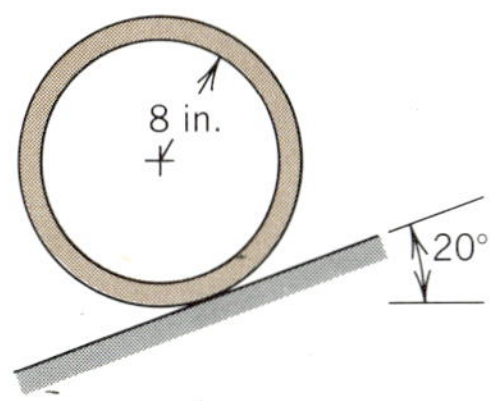

PROB. 14.106

14.106 The ring is released from rest. Determine the speed of the ring after it has rolled 8 ft down the incline.

14.107 The 10 kg cylinder B is placed in contact with a conveyor belt traveling at 40 m/s. The cylinder is initially at rest and the coefficient of kinetic friction between the cylinder and belt is 0.25. Determine the number of revolutions required for the perimeter of the cylinder to attain a speed of 40 m/s, which is the condition at which slipping ceases. The mass of link AB is negligible.

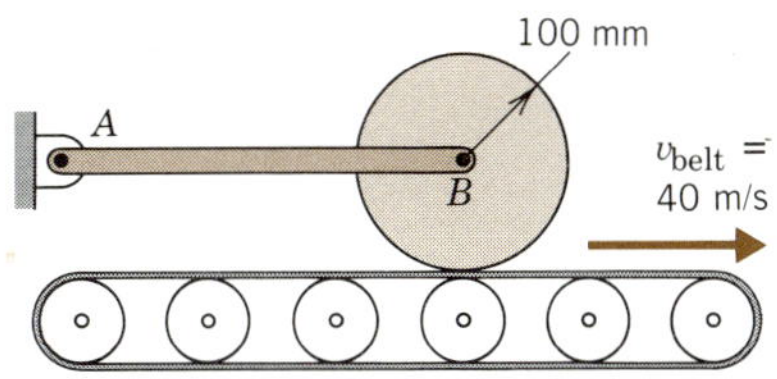

PROB. 14.107

14.108 Gear A weighs 300 lb and has a radius of gyration of 15 in., and rack B weighs 200 lb. Determine the horizontal force $\bar{F}$ acting on the rack for which gear A accelerates from rest to an angular speed of 120 rev/min after 20 rev.

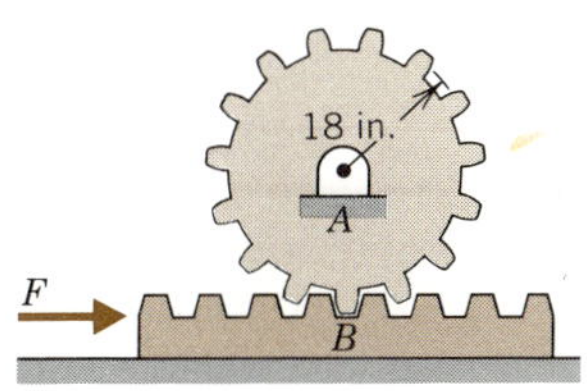

PROB. 14.108

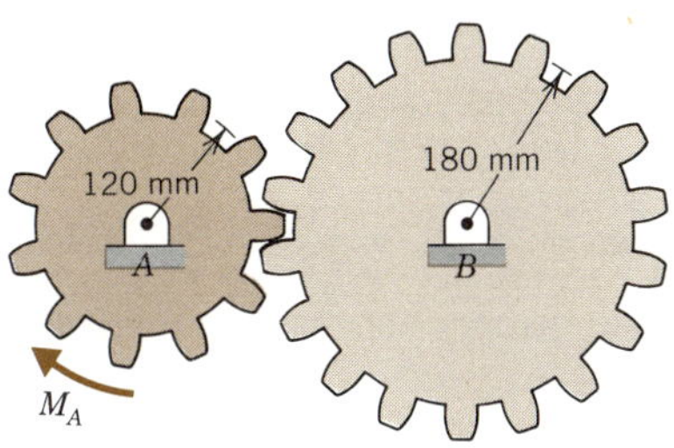

PROB. 14.109

14.109 Gear A has a mass of 2 kg and a radius of gyration of 100 mm, and gear B has a mass of 4 kg and a radius of gyration of 150 mm. Determine the couple $\bar{M}_A$ required to make gear A rotate at 1800 rev/min after 10 revolutions starting from rest.

14.110 The 2 lb gear A and the 5 lb geared drum B have radii of gyration of 4 in. and 5 in., respectively. Determine the torque $\bar{M}_A$ required to give the 1 lb block C a speed of 10 ft/s after it has been raised 2 ft.

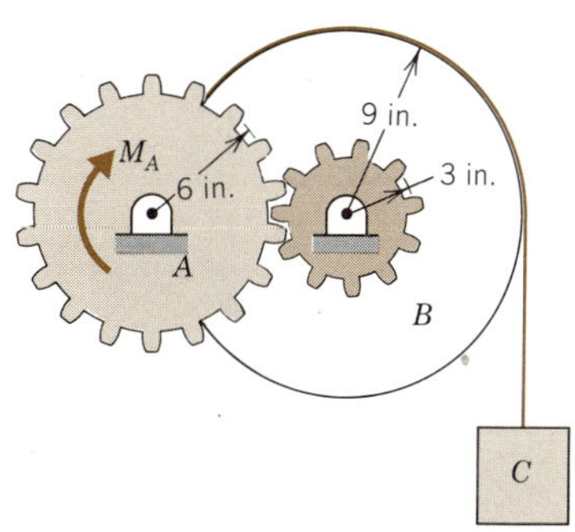

PROB. 14.110

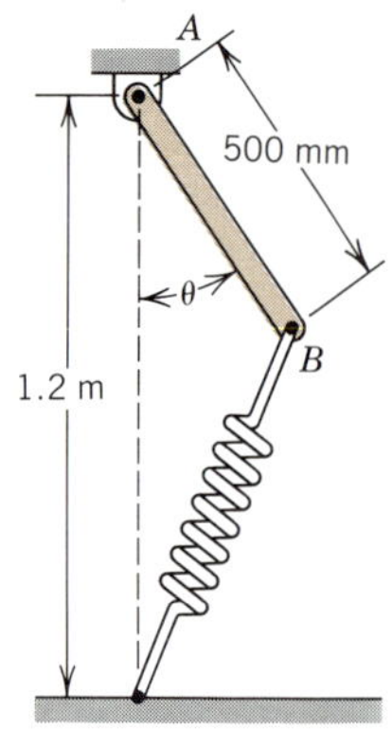

PROBS. 14.111 AND 14.112

14.111 The 2 kg bar is released from rest in the horizontal position, $\theta = 90^\circ$. The stiffness of the spring is 20 N/m and its unstretched length is 600 mm. Determine the angular speed of the bar when $\theta = 0$.

14.112 A couple $\bar{M}_A$ acting at end A of the bar AB in Problem 14.111 results in an angular speed of 5 rad/s when the bar reaches $\theta = 0$ after being released from rest at $\theta = 90^\circ$. Determine $\bar{M}_A$.

14.113 The 20 lb bar moves in the vertical plane under the constraint of sliders A and B. The spring, whose stiffness is 50 lb/ft, is unstressed in the position $\theta = 0$. If the bar is released from rest at $\theta = 38^\circ$, determine the speed of slider A when $\theta = 0$. The weight of each slider is negligible.

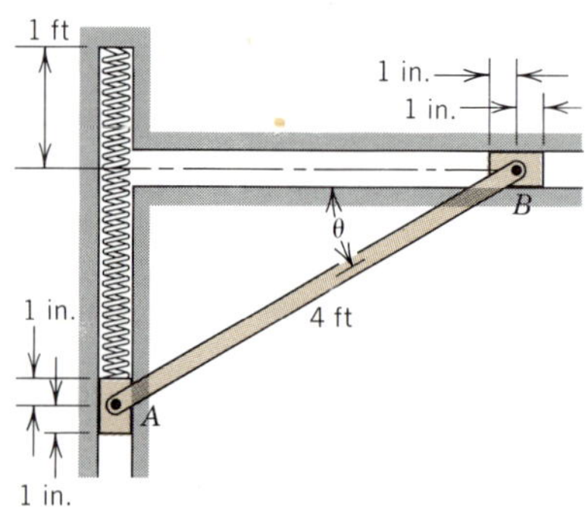

PROB. 14.113

14.114 Solve Problem 14.113 for the case where each slider weighs 2 lb.

14.115 A 40 lb disk is raised by the belt. There is no slipping. Determine the tension force F applied to the belt for which the disk will have an upward velocity of 16 ft/s after it has moved upward 3 ft from its initial rest position.

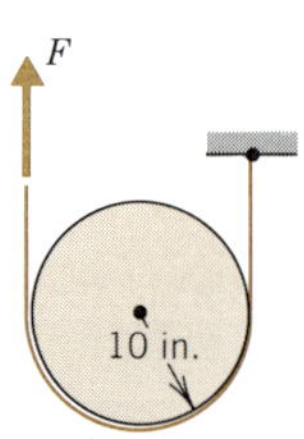

PROB. 14.115

14.116 The 100 lb geared drum is attached to the wall by a spring-loaded cable. The stiffness of the spring is 200 lb/ft, and the radius of gyration of the drum is 15 in. If the drum is released from rest in the position where the spring has a tensile force of 150 lb, determine the speed of the center of the drum when the cable becomes slack.

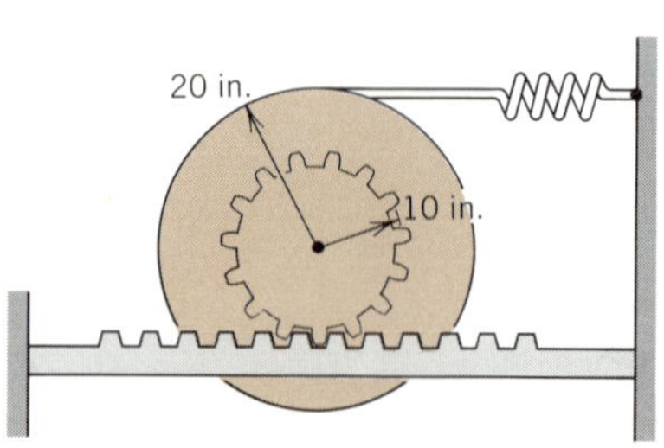

PROB. 14.116

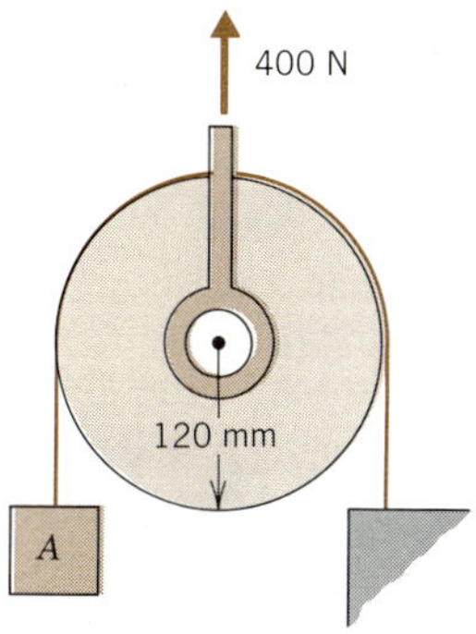

PROB. 14.117

14.117 The 5 kg wheel of radius of gyration 80 mm is hoisted by the 400 N force. Assuming that the cord does not slip on the wheel, determine the distance the 10 kg mass A must be raised, starting from rest, to attain a speed of 15 m/s.

14.118 A tensile force of 100 N is applied vertically to the outer drum of the stepped disk, which rolls without slipping between the inner drum and the horizontal rails, only one of which is shown. The mass of the disk is 20 kg and its radius of gyration is 300 mm. Determine the velocity of the disk after it has rolled 5 m from its initial rest position.

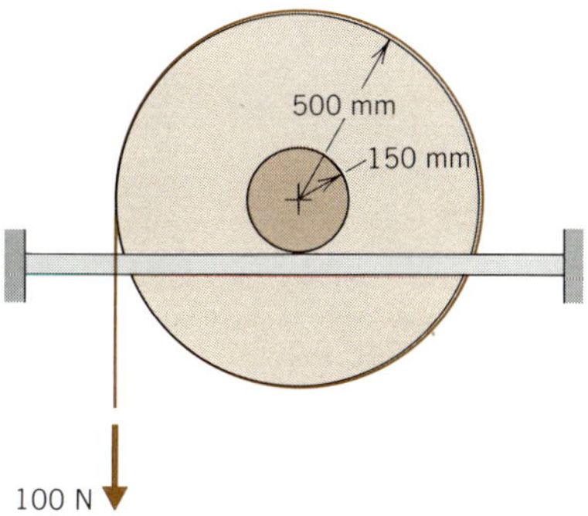

PROB. 14.118

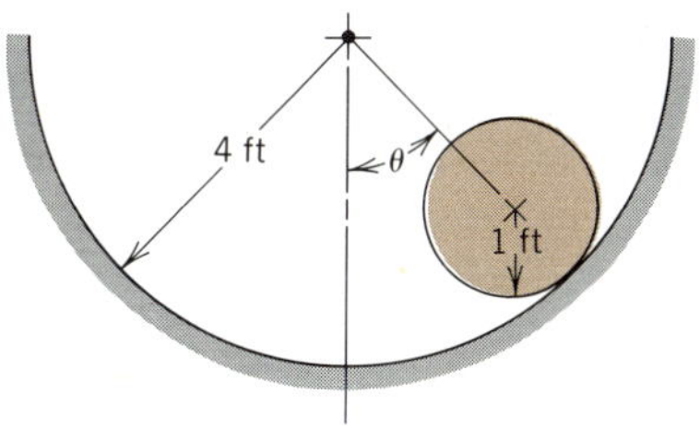

PROB. 14.119

14.119 A 15 lb sphere of 1 ft radius rolls without slipping inside a curved surface of 4 ft radius. Knowing that the sphere is released from rest at $\theta = 60^\circ$, determine (a) the velocity of the center of the sphere as it passes $\theta = 0^\circ$, and (b) the magnitude of the normal reaction at that instant.

14.120 The linkage is released from rest at $\theta = 75^\circ$. Each bar has a mass of 800 grams. Determine the angular speed of bar AB for the vertical position $\theta = 0$. The mass of collar C is negligible.

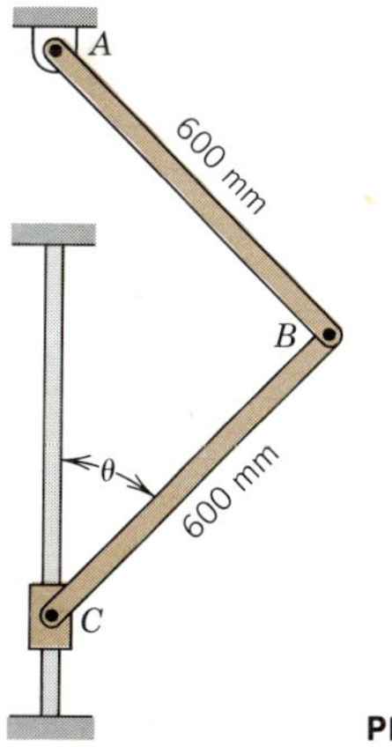

PROB. 14.120

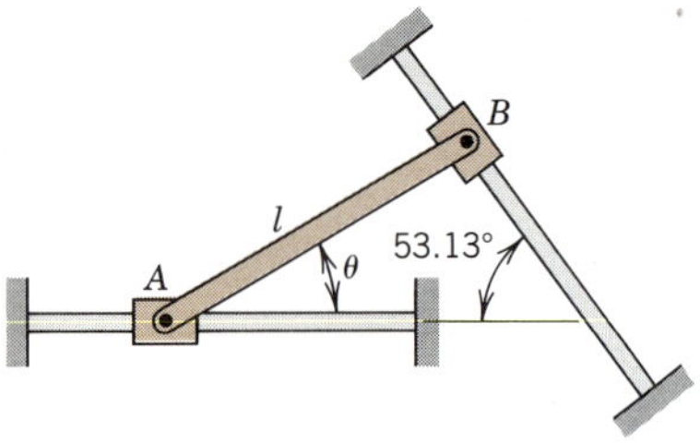

PROB. 14.122

14.121 Solve Problem 14.120 for the final position $\theta = 36.87°$.

***14.122** Bar AB of mass m and length l is released from rest in the vertical plane at $\theta = 36.87°$. Derive an expression for its angular speed when it falls to the horizontal position $\theta = 0$. The collars ride on smooth rods.

***14.123** Solve Problem 14.122 for the final position $\theta = -10°$.

14.124 Each bar has a mass of 5 kg and the stiffness of the spring is 2 kN/m. The unstretched length of the spring is 400 mm, and the linkage lies in the horizontal plane. Determine the constant force $\bar{F}$ acting on piston C for which bar AB will come to rest at $\theta = 0$ after being given an initial counterclockwise angular velocity of 8 rad/s when $\theta = 60°$.

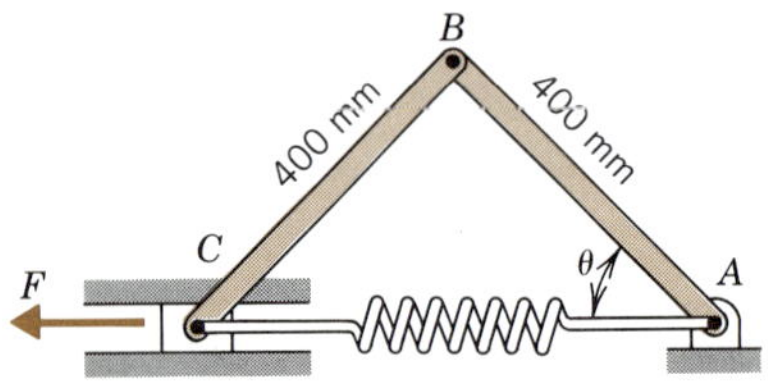

PROB. 14.124

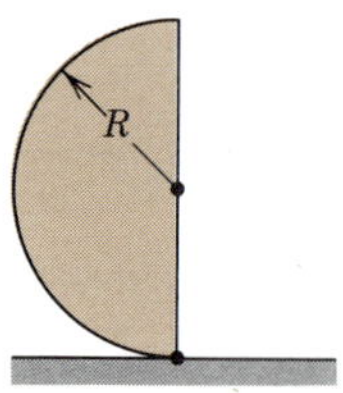

PROB. 14.125

14.125 The solid homogeneous semicylinder is released from rest in the position shown. Determine its angular speed when its mass center is at its lowest point. Assume that slipping does not occur.

***14.126** The mechanism shown lies in a horizontal plane. Bar AB weighs 10 lb and gear B weighs 5 lb and has a 3 in. radius of gyration. The spring has a stiffness of 20 lb/ft and an unstretched length of 12 in. If the system was at rest at $\theta = 0$ when gear B was given a very slight push to the right, determine the speed of gear B when $x = 2$ in. The weight of collar A is negligible.

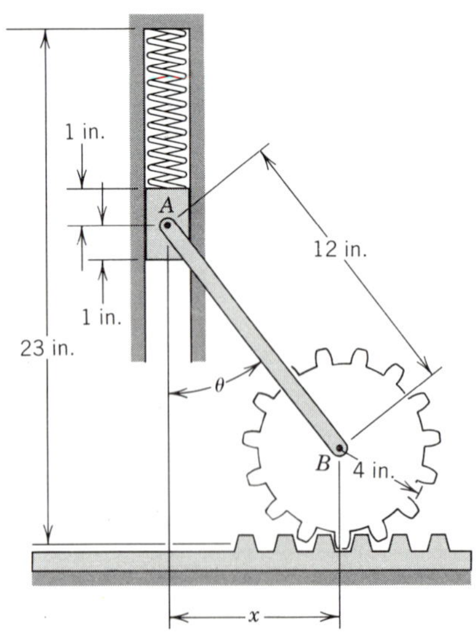

PROB. 14.126

***14.127** For the mechanism in Problem 14.126, determine the maximum speed of gear B and corresponding value of x.

G. IMPULSE AND MOMENTUM

1. Basic Relations

The linear and angular momentum principles were developed in Chapter 12 for an arbitrary system of particles. In general, they treat situations in which the forces are known as functions of time. The basic equations of motion for a rigid body were shown earlier to be the time derivatives of the impulse-momentum equations. Applying the principles for a general system of particles to the case of a rigid body necessitated only that we relate the angular momentum to the angular velocity of the body. This relation is given by eq. (7).

When the forces acting on a body are known as functions of time, we may integrate the basic equations of motion. The force equations then yield a relation between the linear momentum $m\bar{v}_G$ and the impulse of the resultant force, $\int \sum \bar{F}\, dt$. Similarly, integrating the moment equation yields a relation between the angular momentum $I_{zz}\bar{\omega}$ and the impulse of the resultant moment, $\int \sum \bar{M}_A\, dt$.

As usual, we employ definite integrals for this step. The component form of the principles is then found to be

$$\begin{aligned} m(v_{Gx})_2 &= m(v_{Gx})_1 + \int_{t_1}^{t_2} \sum F_x\, dt \\ m(v_{Gy})_2 &= m(v_{Gy})_1 + \int_{t_1}^{t_2} \sum F_y\, dt \\ I_{zz}\omega_2 &= I_{zz}\omega_1 + \int_{t_1}^{t_2} \sum M_{Az}\, dt \end{aligned} \tag{39}$$

Of special note is the fact that these relations use the coordinate axes to assign the signs to force and velocity components. This is done for the angular momentum equation using the right-hand rule with respect to the z axis, which is perpendicular to the plane of motion. Also note that the origin A of the xyz coordinate system must fit the earlier specifications regarding allowable points for summing moments.

One situation where the momentum equations are particularly useful is for impulsive forces. In Chapter 12 we saw that such forces are very large, but act over a very short time interval. The major effect of impulsive forces is described by their impulse, rather than their actual variation with time. This simplification results because we regard the velocity to change instantaneously, while the position does not change. Another simplification that arises in this case is that we may neglect nonimpulsive forces during the time interval when the impulsive forces are acting. A summary of the general procedure for treating impulsive forces is given in Chapter 12.

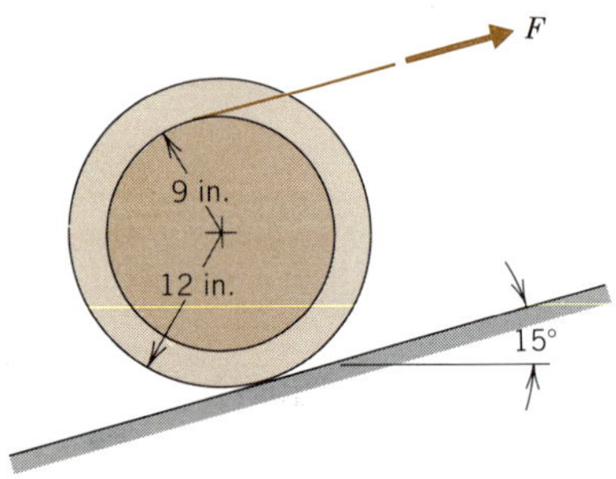

EXAMPLE 11

A tension $F = 10 + 5t$ lb is applied to the cable, which is wrapped around the drum of the stepped disk. The disk, which weighs 20 lb and has a radius of gyration of 8 in., was initially at rest. Determine the velocity of the disk when $t = 2$ s, as well as the average friction force during that interval. Assume that the disk rolls without slipping.

Solution

The impulse-momentum equations will be employed here because we wish to determine the motion at a specified time. The free body diagram and xyz coordinate system are constructed as shown, using the usual considerations

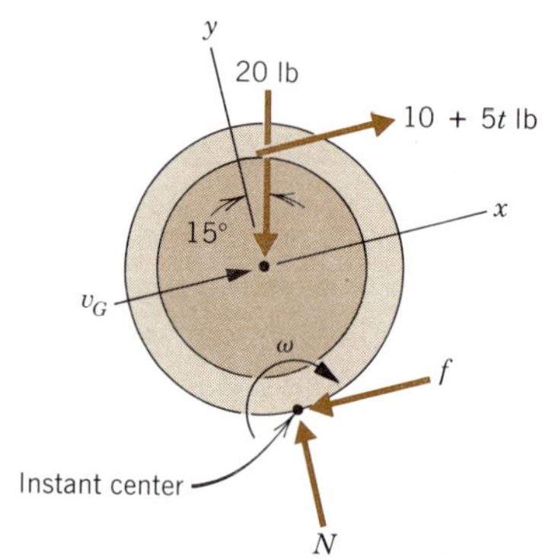

for a system in general motion. The mass and the moment of inertia about the z axis are found to be

$$m = \frac{20}{32.17(12)} = 0.05181 \text{ lb-s}^2/\text{in.}$$

$$I_{zz} = mk_z^2 = 0.05181(8)^2 = 3.316 \text{ lb-s}^2\text{-in.}$$

The next step is to identify which properties of the motion are kinematically determined. There is no slipping at the contact point C, so $v_G = r\omega = 12\omega$ in./s. The sense of the movement depicted in the free body diagram then yields

$$\bar{v}_G = v_G \bar{i}, \qquad \bar{\omega} = -\frac{v_G}{12}\bar{k}$$

The disk is at rest when $t = 0$, so $(v_G)_1 = \omega_1 = 0$. The signs of all vector components in the linear and angular momentum equations are assigned by referring to the positive sense of the respective axes. Hence writing these equations for the 2 s time interval gives

$$x \text{ linear momentum:} \quad 0.05181(v_G)_2 = 0 + \int_0^2 (10 + 5t - 20 \sin 15^\circ - f)\, dt$$

$$z \text{ angular momentum:} \quad 3.316\left(-\frac{v_G}{12}\right) = 0 + \int_0^2 [-(10+5t)9 - (f)12]\,dt$$

Note that the y momentum equation is not needed because the normal reaction $\bar{N}$ is not of interest.

Both momentum equations involve the impulse of the friction force $\bar{f}$. We define the average value of $\bar{f}$ as the constant force that exerts the same impulse. Thus

$$f_{av}(2) = \int_0^2 f\,dt$$

When the other integrals in the momentum equations are evaluated, we obtain

$$0.05181(v_G)_2 = (10 - 20\sin 15^\circ)2 + \tfrac{1}{2}(5)(2)^2 - 2f_{av}$$

$$-\frac{3.316}{12}v_G = -90(2) - \tfrac{1}{2}(45)(2)^2 - 12(2f_{av})$$

Solving these equations simultaneously yields

$$(v_G)_2 = 563.2 \text{ in./s}, \quad \Rightarrow \quad (\bar{v}_G)_2 = 563\bar{i} \text{ in./s}$$

$$f_{av} = -4.766 \text{ lb}, \quad \Rightarrow \quad \bar{f}_{av} = -4.77\bar{i} \text{ lb}$$

EXAMPLE 12

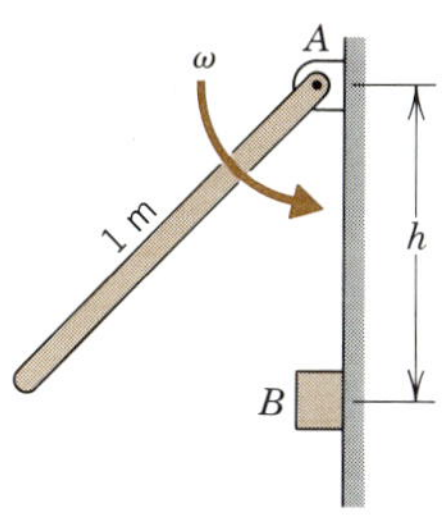

The 18 kg bar was rotating at $\omega = 20$ rad/s immediately before it struck bumper B. The bar remains in contact with the bumper for 17.5 ms, after which it rebounds with an angular speed of 3 rad/s. The distance h locating the bumper is such that the reaction at pivot A is minimized during the collision. Determine h and the corresponding average force exerted by the bumper.

Solution

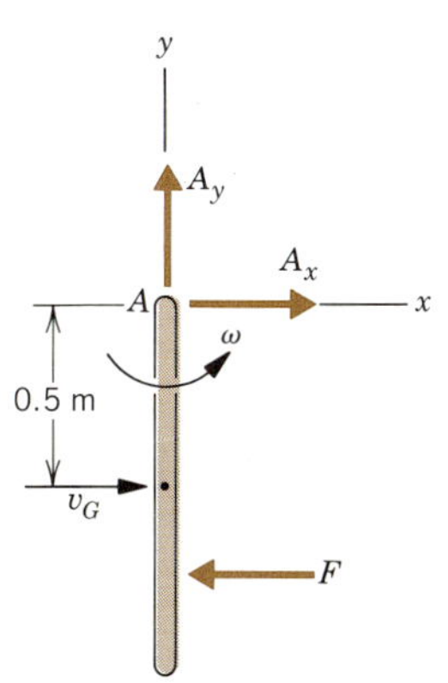

The free body diagram shows the bar in the vertical position because that is the position at which the impact occurs. Note that the weight force is ignored because we expect it to be much smaller than the impulsive force exerted by the bumper. Also, the origin of xyz is placed at the pivot because the bar is in pure rotation about that point.

The moment of inertia about the z axis is

$$I_{zz} = \tfrac{1}{12}(18)(1)^2 + (18)(0.5)^2 = 6.0 \text{ kg-m}^2$$

As shown in the free body diagram, a counterclockwise rotation about pivot A corresponds to a movement of the center of mass G in the positive x direction. Thus

$$\bar{\omega} = \omega\bar{k}, \qquad \bar{v}_G = 0.5\omega\bar{i}$$

The initial and final values of ω are given as $\omega_1 = 20$ rad/s, $\omega_2 = -3$ rad/s.

We now formulate the momentum equations, with all forces treated as average (i.e., constant) values, and h used as a parameter.

$$x \text{ linear momentum:} \quad 18(v_{Gx})_2 = 18(v_{Gx})_1 + \int_0^{\Delta t} (A_x - F)\,dt$$

$$18(0.5)(-3) = 18(0.5)(20) + (A_x - F)\,\Delta t$$

$$y \text{ linear momentum:} \quad 0 = 0 + \int_0^{\Delta t} A_y\,dt = A_y\,\Delta t$$

$$z \text{ angular momentum:} \quad 6\omega_2 = 6\omega_1 + \int_0^{\Delta t} (-Fh)\,dt$$

$$6(-3) = 6(20) - Fh\,\Delta t$$

where $\Delta = 0.0175$ s for the collision. We find from these equations that

$$F = \frac{7886}{h}\text{ N}, \qquad A_x = \frac{7886}{h} - 11829\text{ N}, \qquad A_y = 0\text{ N}$$

The minimum magnitude of the reaction at A corresponds to $A_x = 0$, because $A_y = 0$ (when the gravity force is neglected). Thus

$$h = \frac{7886}{11829} = 0.667\text{ m}$$

The corresponding value of the contact force at the bumper is

$$F = 11.83\text{ kN}$$

The location on a rotating body where the reaction to an impulsive force is minimized is called the *center of percussion.*

HOMEWORK PROBLEMS

14.128 The 600 kg turbine of a jet engine is rotating at 4000 rev/min when the engine is shut down. The radius of gyration of the turbine is 700 mm. Friction in the bearings produces a torque of 500 N-m. Determine the time required for the engine to come to rest.

14.129 A torque $M = 500t^2$ lb-ft, where t is elapsed time in seconds, is applied to the 1000 lb disk. Frictional bearing resistance produces a torque of 100 lb-ft. Determine the angular speed of the disk when $t = 3$ s.

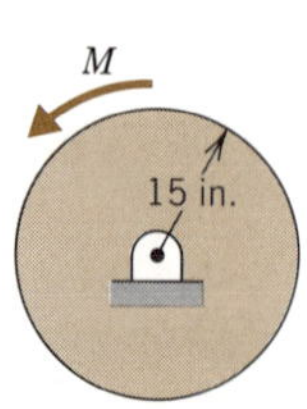

PROB. 14.129

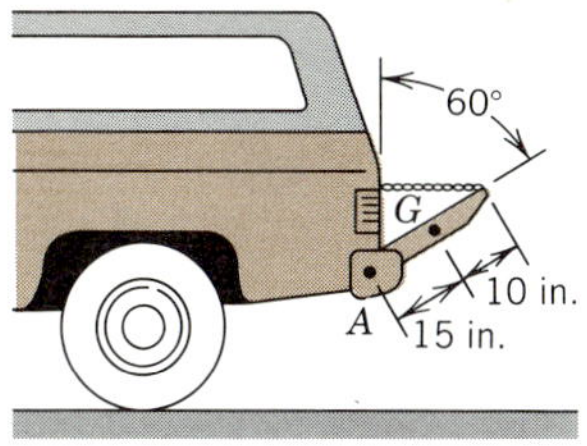

PROB. 14.130

14.130 The 10 lb tailgate is held in position by two cables, only one of which is visible in the side view. The speed of the truck varies as $v = 2.5t^{3/2}$ ft/s where t is the elapsed time in seconds. Determine the average cable tension and the average reaction at hinge A during the first 10 s of elapsed time.

14.131 A horizontal force $F = 50 + 3t$ N, where t is seconds, is applied to the 50 kg sliding door. The casters A and B roll freely. Determine (a) the velocity of the door when $t = 4$ s, (b) the average reactions at A and B for the 4 s time interval in part (a).

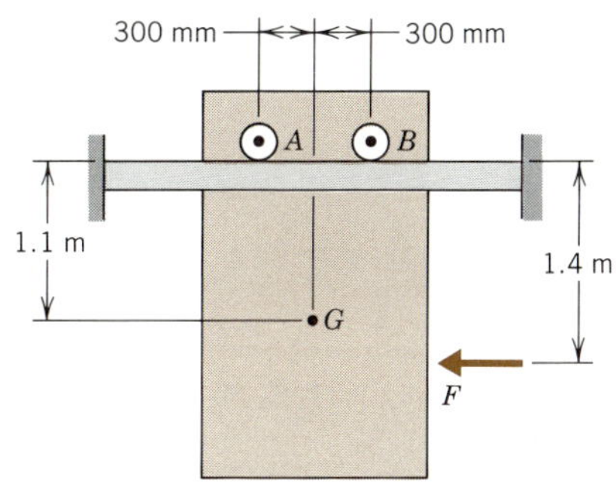

PROB. 14.131

14.132 Solve Problem 14.131 if both casters are locked. The coefficient of sliding friction between the guides and the rail is 0.10.

14.133 The 8 lb stepped disk has a 5 in. radius of gyration. The disk is released from rest when an upward force $F = 8 + 2t$ lb is applied to the end of the cable. Determine the velocity of the center of the disk and the angular velocity when $t = 4$ s.

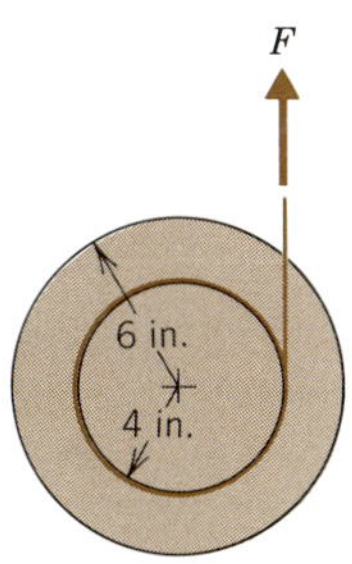

PROB. 14.133

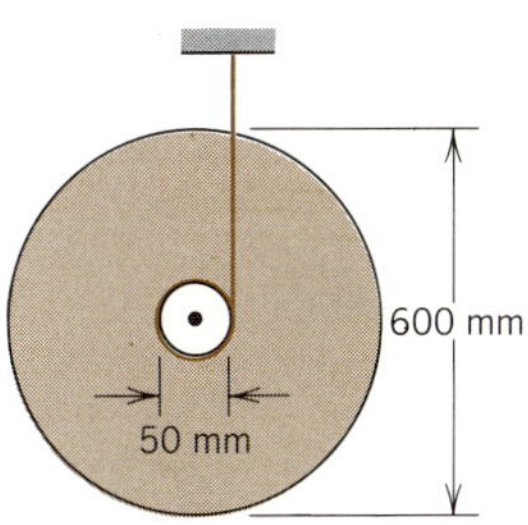

PROB. 14.134

14.134 The spool shown in cross section consists of two 600 mm diameter drums having 4 kg mass each, connected by the 50 mm shaft whose inertia is negligible. Determine the velocity of the spool 5 s after it is released from rest.

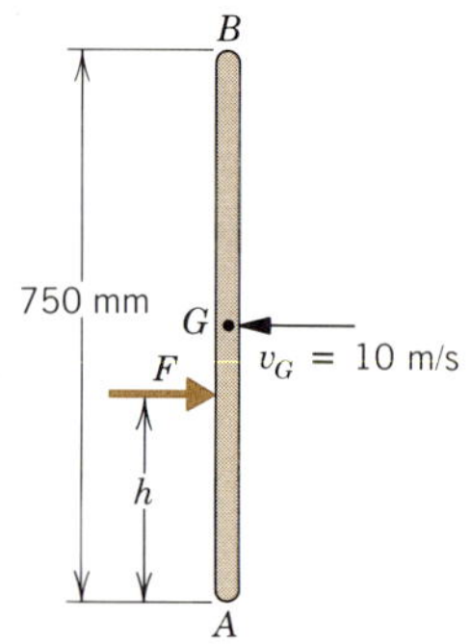

PROB. 14.135

14.135 The 2 kg bar is initially translating at 10 m/s, as shown, when it is subjected to the impulsive force $\bar{F}$, whose average magnitude over a 0.10 s interval is 500 N. If $h = 250$ mm, determine the velocity of the center of mass and the angular velocity of the bar immediately after the impact.

14.136 Immediately after the application of the impulsive force $\bar{F}$ in Problem 14.135, point A is observed to be the instantaneous center of rotation. What are the corresponding values of h and of the angular velocity after the impact?

14.137 A downward force $F = 4t$ N (t is in seconds) is applied to block A, which is suspended from pulley B. The block has a mass of 2 kg, and the pulley has a mass of 1.5 kg and a radius of gyration of 200 mm. The system was at rest when $t = 0$. Determine (a) the average tension during the first 5 s of motion, and (b) the angular speed of the pulley when $t = 5$ s.

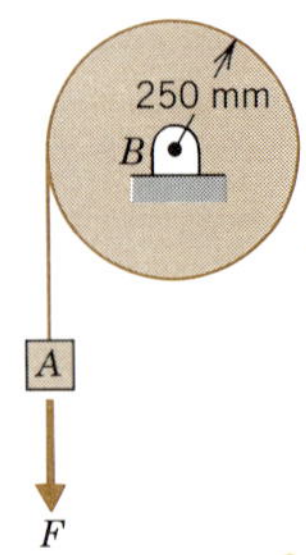

PROB. 14.137

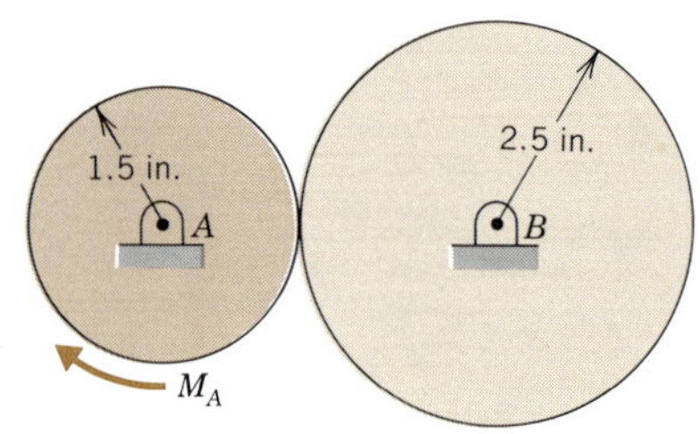

PROB. 14.138

14.138 Gear A weighs 1 oz and has a radius of gyration of 0.75 in. The corresponding parameters for gear B are 1.5 oz and 1.0 in. Determine the constant torque $\bar{M}_A$ acting on gear A that will cause gear A to rotate at 16,000 rev/m after 2 s, starting from rest.

***14.139** A torque $M_A = k(t^{0.8})$ lb-in. is applied to gear A in Problem 14.138. Gear B was initially rotating at 6000 rev/min. Determine the constant k required to bring the gear train to rest when $t = 10$ s.

14.140 The 10 kg semicircular plate strikes corner C with an initial angular velocity of 15 rad/s counterclockwise. It rebounds from the collision with an angular speed of 9 rad/s, after being in contact with corner C for 12 ms. Determine the average force exerted by corner C and the average reaction at pin A in the case where $h = 750$ mm.

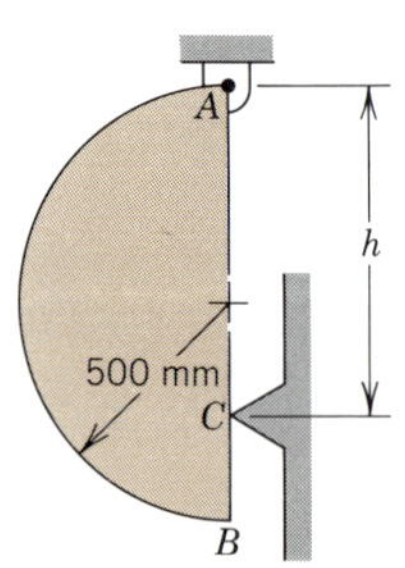

PROB. 14.140

14.141 Determine the distance h in Problem 14.140 for which corner C is at the center of percussion for the semicircular plate.

14.142 A 120 lb ice skater spins with knees bent and arms outstretched at an angular speed of 12 rev/min. In this position the skater has an 18 in. radius of gyration about the axis of rotation. When standing the skater's radius of gyration is 6 in. and the mass center rises by 1 ft. Neglecting friction from the ice, determine (a) the angular speed of the skater in the upright position, (b) the work done by the skater's muscles in going from the kneeling to the upright position.

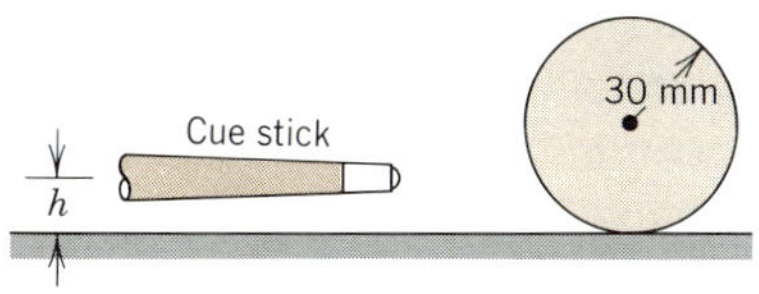

PROBS. 14.143 AND 14.144

14.143 The 300 gram billiard ball of 30 mm radius is struck by a cue stick that exerts an average force of 600 N horizontally over a 0.005 s interval. If $h = 10$ mm, determine the velocity of the center of the ball and the angular velocity of the ball immediately after the impact.

14.144 The 300 gram billiard ball of 30 mm radius is struck by a cue stick that exerts an average force of 600 N horizontally over a 0.005 s interval. Immediately after being hit, the billiard ball rolls without slipping. Determine (a) the height h for the cue stick, and (b) the velocity of the ball after the impact.

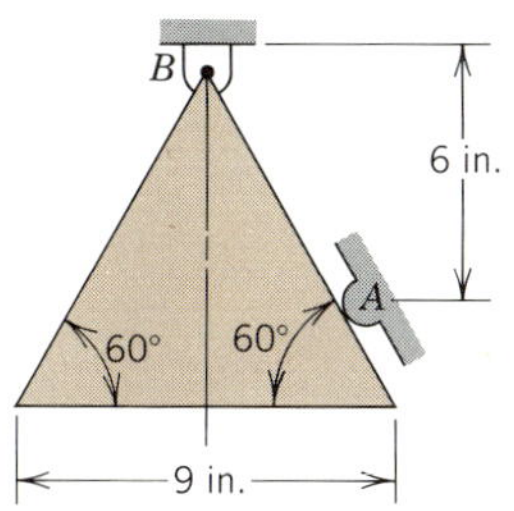

PROB. 14.145

14.145 The 2 lb equilateral plate is rotating at 5 rad/s when it hits projection A with the plate's centerline oriented vertically. The collision has a duration of 8 ms, after which the plate rebounds at 1 rad/s. Determine the average contact force at projection A and the average reaction at pivot B.

***14.146** The 20 kg flywheel A, whose radius of gyration is 250 mm, and the 40 kg disk B are connected by means of a belt that does not slip over the contacting surfaces. The flywheel is driven by a torque $M = 10t$ where M is in newton-meters and t is in seconds. Determine the time required for the disk to attain an angular velocity of 100 rad/s starting from rest.

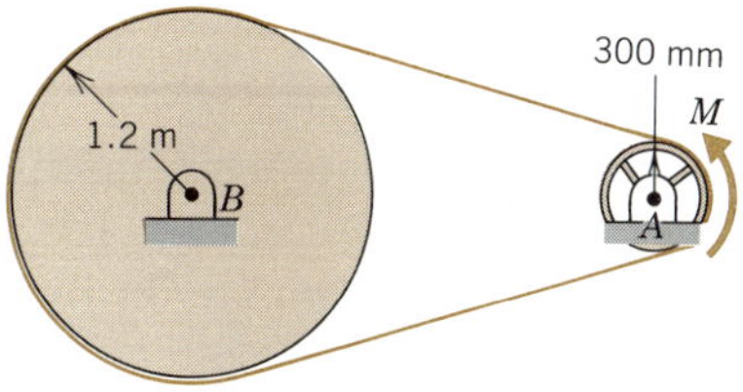

PROB. 14.146

***14.147** Two cables are wrapped securely around the inner drum of the stepped disk. The disk weighs 150 lb and its radius of gyration is 16 in. The coefficients of friction between the drum and the horizontal surface are $\mu_s = 0.40$, $\mu_k = 0.30$. A tensile force $F = 10t$ lb, where t is elapsed time in seconds, is applied to the cable. Determine (a) the value of t at which the disk begins to skid over the ground, and (b) the velocity of the center of the cylinder 2 s after skidding begins.

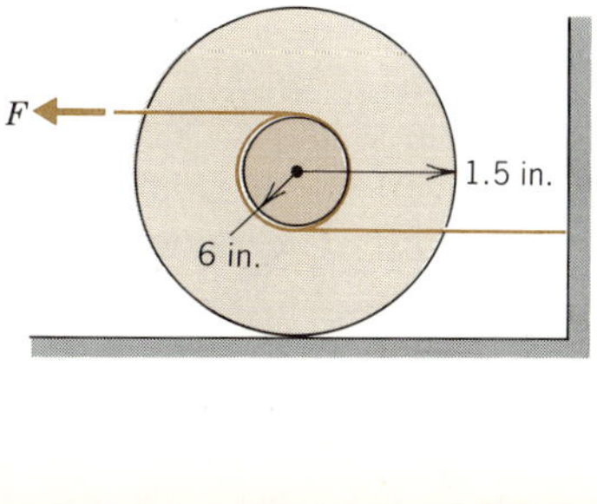

PROB. 14.147

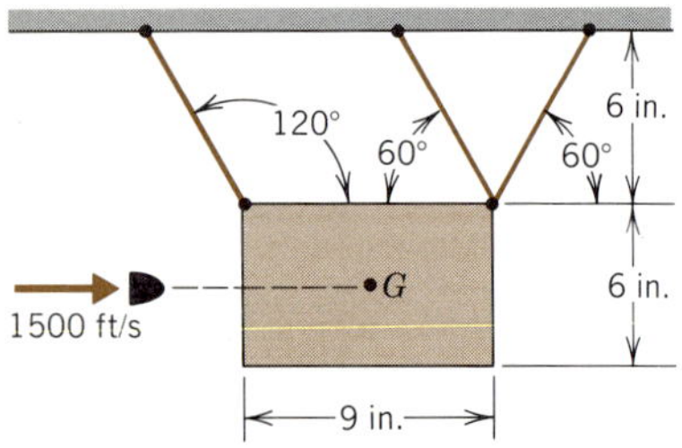

PROB. 14.148

*14.148 The 20 lb wooden block is initially at rest when it is struck by a 2 oz bullet that is traveling at 1500 ft/s in the direction shown. The bullet becomes embedded, coming to rest with respect to the block 0.004 s after the initial contact. Determine (a) the velocity of the center of mass of the block immediately after the impact, and (b) the average tensile force within each cable during the impact.

14.149 A 50 gram bullet traveling at 2000 km/hr in the direction shown hits the 10 kg square plate. The bullet becomes embedded, coming to rest within the plate 0.002 s after the initial contact. Determine (a) the angular velocity of the plate immediately after the impact, and (b) the impulsive reaction at pin A.

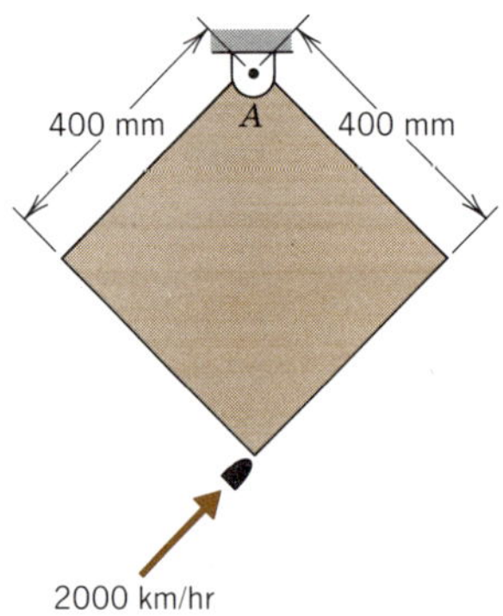

PROB. 14.149

*2. Eccentric Impact

The situation in Example 12 involved a collision between bodies in which the normal to the plane of contact did not coincide with the line connecting the mass centers of the two bodies. This is called an *eccentric impact*. In that example the motion of the system after the collision was given. Our aim here is to learn how to determine the motion of a body when it rebounds from a collision.

When we studied the collision of particles in Chapter 12 (for which the line of impact coincides with the line connecting the mass centers), we were able to solve problems by using the concept of conservation of linear momentum in conjunction with the coefficient of restitution, ε, which was a measure of the loss of energy during the collision. In essence, we recognized that particles are not really rigid and that energy can be lost due to deformation during impact. The problem for rigid bodies is quite similar, but it is complicated slightly by the fact that the velocity at the point of impact is not the same as the velocity of the mass center.

Consider the general type of collision depicted in Figure 19. Points A and B are the centers of mass of the colliding bodies. These are always allowable points for summing moments, so each is selected as the origin for an xyz coordinate system. For convenience in treating the contact force and the velocity of each body, these coordinate systems have been selected

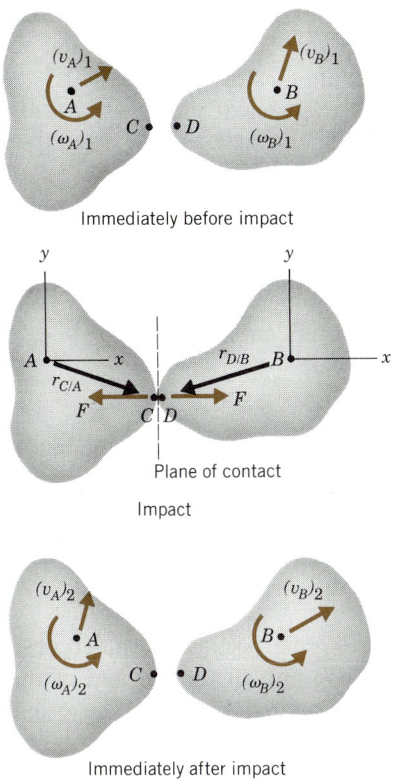

FIGURE 19

to have parallel orientations, with the y axes aligned parallel to the plane of contact (the mutual plane of tangency). In the following, we shall discuss the procedures for determining the velocity of each mass center, $\bar{v}_A$ and $\bar{v}_B$, and the angular velocity of each body $\bar{\omega}_A$ and $\bar{\omega}_B$, at the end of the collisions.

The actual collision process is quite complicated. Fortunately, we may formulate an empirical relationship based on experimental results. Recall that in the case of colliding particles, the coefficient of restitution ε was related to the relative velocity components perpendicular to the plane of contact. Specifically, it was the ratio of these components before and after the collision. The same type of ratio is found in the case of rigid body motion, with one important modification: the relative velocity is for the points at which the bodies first contact each other.

The points of contact are labeled C and D in Figure 19. The corresponding equation for ε is

$$\varepsilon = \frac{[(v_{D/C})_x]_2}{[(v_{C/D})_x]_1}$$

This relationship has greater meaning when we realize that the denominator is the rate at which the contacting surfaces were approaching each other before the collision, while the numerator is the rate at which these surfaces are separating after the collision.

The relation for the coefficient of restitution may be written in a more useful form by recalling that a relative velocity is the difference of two absolute velocities. Thus

$$\varepsilon = -\left[\frac{(v_{Dx})_2 - (v_{Cx})_2}{(v_{Dx})_1 - (v_{Cx})_1}\right] \tag{40}$$

An important limitation of this equation is that it applies only to collisions where friction is negligible. Also, it is accurate only for a limited range of relative velocities.

Equation (40) relates the velocities of the points of contacts on each body. However, we are interested in the velocities of the centers of mass. We therefore employ the relative velocity relations for each of the rigid bodies:

$$\bar{v}_C = \bar{v}_A + \bar{\omega}_A \times \bar{r}_{C/A}, \qquad \bar{v}_D = \bar{v}_B + \bar{\omega}_B \times \bar{r}_{D/B} \tag{41}$$

When we use eq. (41) to evaluate the x components of $\bar{v}_C$ and $\bar{v}_D$, eq. (40) then represents one relation for the unknowns. Other relationships are obtained from the linear and/or angular impulse-momentum principle, which, of course, involves the impulse of the contact force $\bar{F}$ exerted between the two bodies. As shown in Figure 19, this force is directed along the normal to the plane of contact.

To see how these equations are used, suppose that the bodies in Figure 19 are unconstrained in their motion, so that the contact force $\bar{F}$ is the only impulsive force acting on the body. In such a situation, we apply the linear and angular momentum principles *for each body*, the latter with respect to the centers of mass. There are x and y components of linear momentum and a z component of angular momentum. This gives three scalar equations of momentum for each body. Thus, after using equation (40), we have a total of seven scalar equations. The seven scalar unknowns are the x and y components of $\bar{v}_A$ and $\bar{v}_B$, the z components of $\bar{\omega}_A$ and $\bar{\omega}_B$, and the impulse of the contact force $\bar{F}$. Hence, we have a well-posed problem.

In contrast, there are additional impulsive reaction forces when a body rotates about a fixed axis, as we saw in Example 12. We can treat such a situation by formulating the angular impulse-momentum equation with respect to a point on the fixed axis. By doing so, the impulse of the reaction forces appear only in the linear momentum equations for the body.

EXAMPLE 13

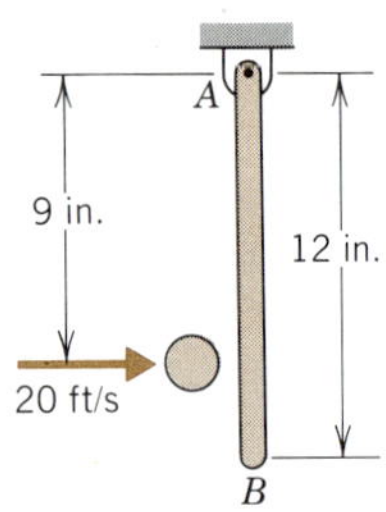

The 2 lb bar AB is at rest when it is hit by the 1 lb sphere, which is traveling horizontally at 20 ft/s. The coefficient of restitution is 0.6. Determine the the velocity of the sphere and the angular velocity of the bar immediately after the collision.

Solution

The sphere may be modeled as a particle because it undergoes a direct central impact. We draw free body diagrams of the bar and of the sphere, in which the nonimpulsive weight forces are omitted. The origin of the coordinate system for the bar is placed at the fixed point A in order to eliminate the moment of the impulsive reaction force exerted by the pivot. Note that the y axis is aligned parallel to the plane of contact, consistent with eq. (40).

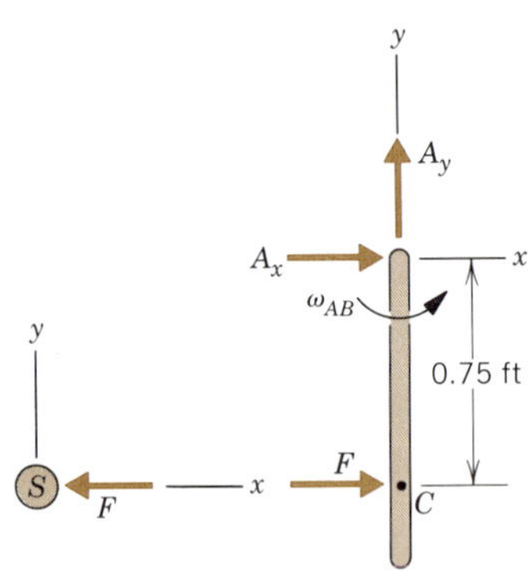

The masses and moment of inertia are

$$m_s = \frac{1}{32.17} = 0.03108 \text{ lb-s}^2/\text{ft}, \qquad m_{AB} = \frac{2}{32.17} = 0.06217 \text{ lb-s}^2/\text{ft}$$

$$I_{zz} = \frac{1}{12}(0.06217)(1)^2 + (0.06217)(0.5)^2 = 0.02072 \text{ lb-s}^2\text{-ft}$$

The initial conditions are

$$(\omega_{AB})_1 = 0, \qquad (\bar{v}_s)_1 = 20\bar{i} \text{ ft/s}$$

We are not interested in the reactions A_x and A_y, so we consider only the angular impulse-momentum equation for the bar. Also, the sphere will have no final velocity component in the y direction, because there is no force acting in that direction.

$$(\bar{v}_s)_2 = v_s\bar{i}$$

Therefore we need only consider the x component of linear momentum of the sphere.

angular momentum of the bar:

$$0.02072\omega_2 = 0 + \int_0^{\Delta t} F(0.75)\,dt = 0.75F_{av}\,\Delta t$$

linear momentum of the sphere:

$$0.03108(v_s)_2 = 0.03108(20) + \int_0^{\Delta t} (-F)\,dt = 0.6217 - F_{av}\,\Delta t$$

The additional equation required to solve for $(v_s)_2$, ω_2, and $(F_{av}\,\Delta t)$ is found from the coefficient of restitution. For pure rotation about pin A, the velocity of the point of contact after the collision is

$$(\bar{v}_C)_2 = (\bar{\omega}_{AB})_2 \times \bar{r}_{C/A} = \omega_2\bar{k} \times (-0.75\bar{j}) = 0.75\omega_2\bar{i}$$

Setting $(\bar{v}_C)_1 = \bar{0}$ in eq. (40) then gives

$$\varepsilon = 0.6 = -\left[\frac{(v_{Cx})_2 - (v_{sx})_2}{(v_{Cx})_1 - (v_{sx})_1}\right] = -\left[\frac{0.75\omega_2 - (v_s)_2}{0 - 20}\right]$$

$$0.75\omega_2 - (v_s)_2 = 12$$

Solving this relation simultaneously with the momentum equations yields

$$(v_s)_2 = 2.64 \quad \Rightarrow \quad (\bar{v}_s)_2 = 2.64\bar{i} \text{ ft/s}$$

$$\omega_2 = 19.52 \quad \Rightarrow \quad (\bar{\omega}_{AB})_2 = 19.52\bar{k} \text{ rad/s}$$

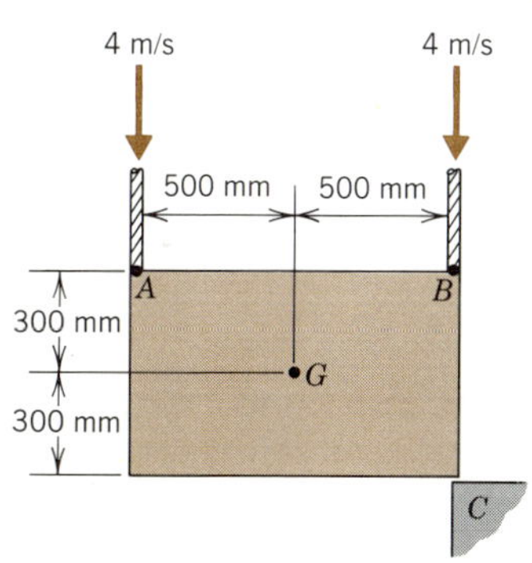

EXAMPLE 14

The 500 kg crate is translating downward with the horizontal orientation shown when it accidentally strikes the ledge at a speed of 4 m/s. The radius of gyration of the crate about its center of mass G is 300 mm. The coefficient of restitution for the collision is 0.40, and the time duration is 15 ms. Determine the velocity of the crate's center of mass and its angular velocity after the collision. Also, determine the average force exerted by the ledge during the collision.

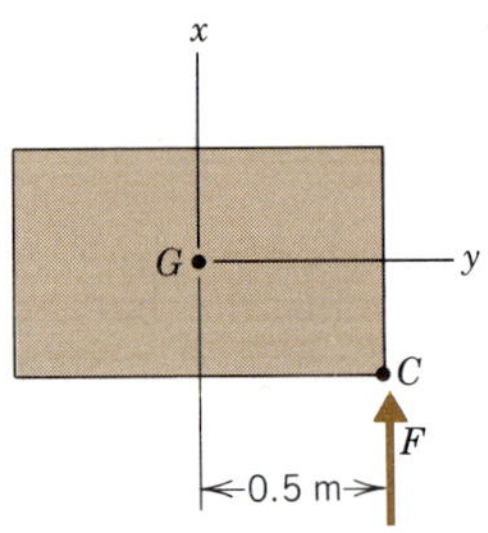

Solution

The weight force is not impulsive. Thus the free body diagram need only to account for the contact force $\bar{F}$. Consistency with eq. (40) for ε requires that we align the x axis perpendicular to the plane of contact, that is, vertical. The origin of xyz is at the center of mass because the crate will be in general motion after the collision. The moment of inertia corresponding to the given radius of gyration is

$$I_{zz} = 500(0.3)^2 = 45 \text{ kg-m}^2$$

The initial conditions are known to be

$$(\bar{v}_G)_1 = -4\bar{i} \text{ m/s}, \qquad \bar{\omega}_1 = \bar{0}$$

We consider F to be the average value. Then the impulse-momentum equations for the 15 ms time interval are:

$$x \text{ linear momentum:} \quad 500(v_{Gx})_2 = 500(-4) + F(0.015)$$

$$y \text{ linear momentum:} \quad 500(v_{Gy})_2 = 0 + 0$$

$$z \text{ angular momentum:} \quad 45\omega_2 = 0 + (-0.5F)(0.015)$$

The foregoing represent three equations for four unknowns, although the y equation is readily solved. Note that we do not consider the momentum of the earth because it is massive and assumed fixed.

For the fourth equation, we employ the coefficient of restitution. This requires consideration of the component of $\bar{v}_C$ in the x direction. Initially, the crate is translating, so $(v_{Cx})_1 = -4$ m/s. For the final velocity, we have

$$(\bar{v}_C)_2 = (\bar{v}_G)_2 + \bar{\omega}_2 \times \bar{r}_{C/G} = (\bar{v}_G)_2 + \omega_2\bar{k} \times (-0.3\bar{i} + 0.5\bar{j})$$

$$\bar{i} \text{ component:} \quad (v_{Cx})_2 = (v_{Gx})_2 - 0.5\omega_2$$

Substituting $(v_{Cx})_1$ and $(v_{Cx})_2$ into eq. (40) and setting the velocity of the earth to zero then yields

$$\varepsilon = 0.4 = -\frac{(v_{Cx})_2}{(v_{Cx})_1} = -\frac{[(v_{Gx})_2 - 0.5\omega_2]}{(-4)}$$

$$(v_{Gx})_2 - 0.5\omega_2 = 1.6$$

This relation and the momentum equations may now be solved simultaneously. The result is

$$(v_{Gx})_2 = -2.52 \text{ m/s}, \qquad (v_{Gy})_2 = 0 \Rightarrow (\bar{v}_G)_2 = -2.52\bar{i} \text{ m/s}$$

$$\omega_2 = -8.24 \text{ rad/s}, \quad \Rightarrow \quad \bar{\omega}_2 = -8.24\bar{k} \text{ rad/s}$$

$$\bar{F} = 49.4\bar{i} \text{ kN}$$

Note that a negative ω_2 corresponds to a counterclockwise rotation for the axes we have used. Thus the center of mass continues to move downward, but it has slowed due to the contact force $\bar{F}$. This force also causes the rotation.

HOMEWORK PROBLEMS

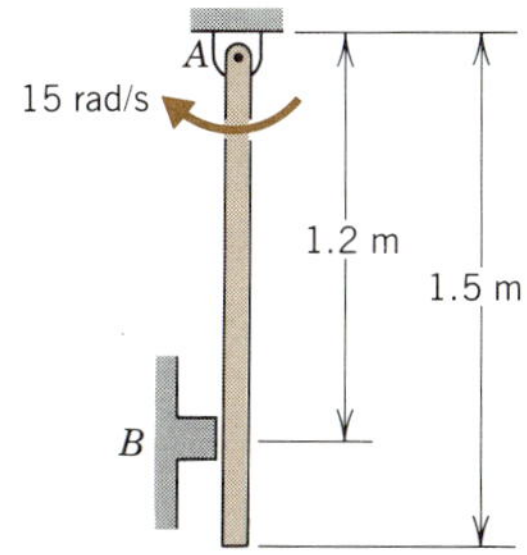

PROB. 14.150

14.150 The 50 kg bar strikes the bumper B at an angular speed of 15 rad/s. The coefficient of restitution is 0.5, and the duration of the impact is 12 ms. Determine the angular speed with which the bar rebounds from the collision, and also evaluate the average force exerted by the bumper.

14.151 The 2 kg slender rod is hit by a 150 gram ball traveling at 20 m/s in the direction shown. The coefficient of restitution is 0.80. Determine the angular velocity of the bar and the velocity of the ball immediately after the collision.

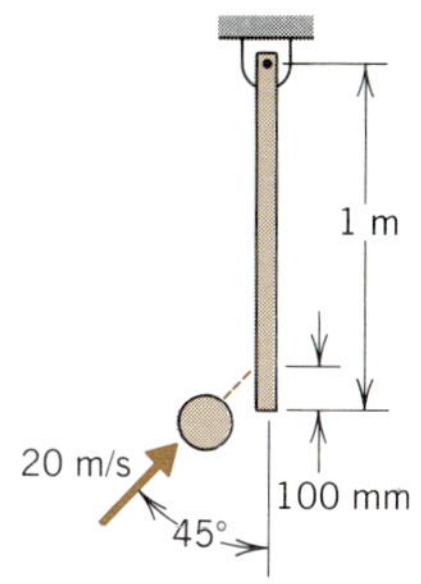

PROB. 14.151

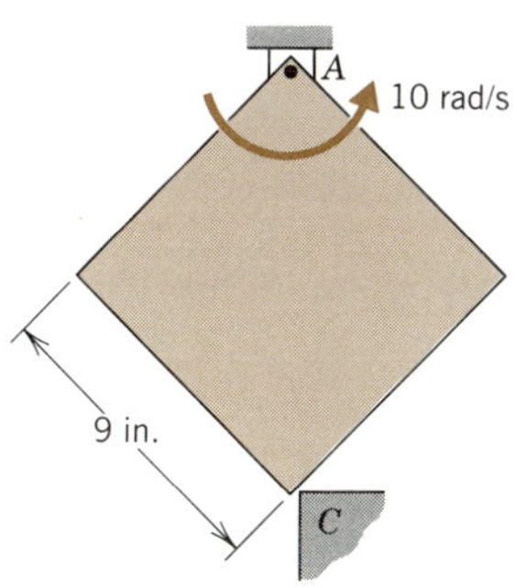

PROB. 14.152

14.152 The 12 lb square plate strikes corner C at an angular speed of 10 rad/s. The coefficient of restitution if 0.4 and the duration of the impact is 9 ms. Determine (a) the angular velocity of the plate after the collision, (b) the average force exerted by the corner C, and (c) the average reaction force at the pivot A.

14.153 The 100 lb semicylinder was translating downward at 50 ft/s when it contacted the ledge at corner B, as shown. Determine the velocity of the center of mass G and the angular velocity of the plate after the collision.

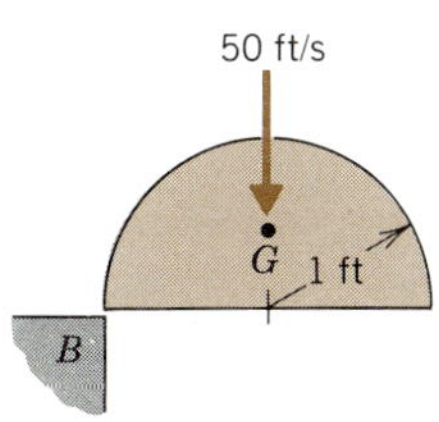

PROB. 14.153

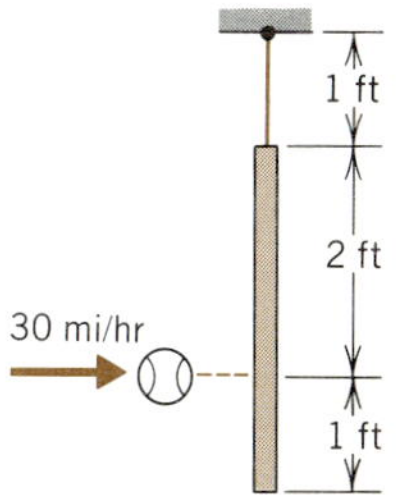

PROB. 14.154

14.154 A 5.50 oz baseball traveling horizontally at 30 mi/hr hits the 4 lb wooden bar. The coefficient of restitution for the collision is 0.50. Determine the velocity of both ends of the bar immediately after the collision.

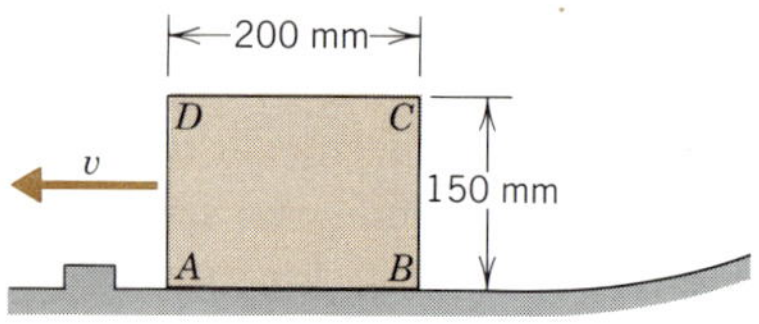

PROB. 14.155

14.155 A 10 kg package on an assembly line slides along the curved chute in the vertical plane until it hits a small cleat at point A. If the speed of the package just before it hits the cleat is $v = 5$ m/s, determine the angular velocity of the package immediately after the collision. Assume that the impact is perfectly plastic.

14.156 What is the maximum speed v at which the package in Problem 14.155 can hit the cleat without tipping over onto edge AD?

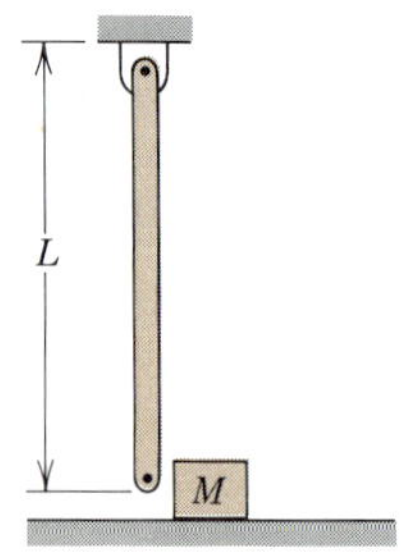

PROB. 14.157

14.157 The bar, whose mass is m, is rotating at angular speed ω_1 when it strikes the stationary block of mass M. The block rebounds from the collision at speed v_2, whereas the bar is at rest after the collision. Derive expressions for the coefficient of restitution ε and the speed v_2 in terms of ω_1, m, M, and L.

14.158 and 14.159 The 2 lb wooden stick is translating at the velocity shown when it hits the wall with the orientation depicted. The coefficient of restitution is 0.3. Determine the velocity of the stick's center of mass and its angular velocity after the impact.

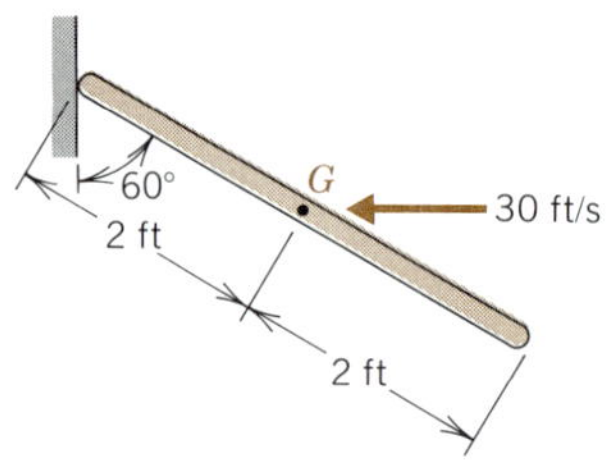

PROB. 14.158

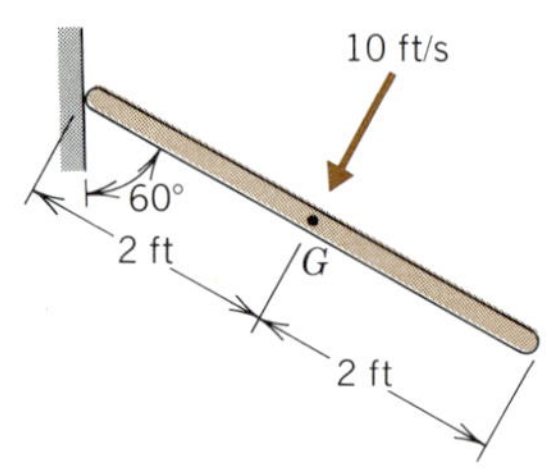

PROB. 14.159

***14.160** The 500 gram homogeneous sphere is rolling without slipping down the incline at 10 m/s when it hits the horizontal surface. Assuming that the collision with the horizontal surface is perfectly plastic, determine the velocity of the center of the sphere and its angular velocity immediately after the impact. Does the sphere roll without slipping immediately after the impact?

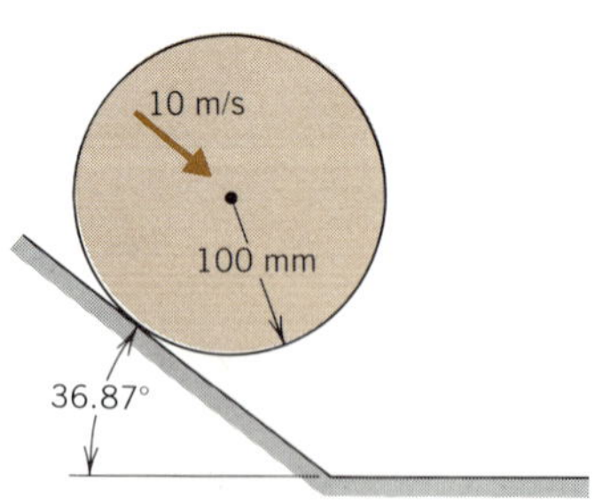

PROB. 14.160

***14.161** Solve Problem 14.160 if the coefficient of restitution for the collision with the horizontal surface is 0.90.

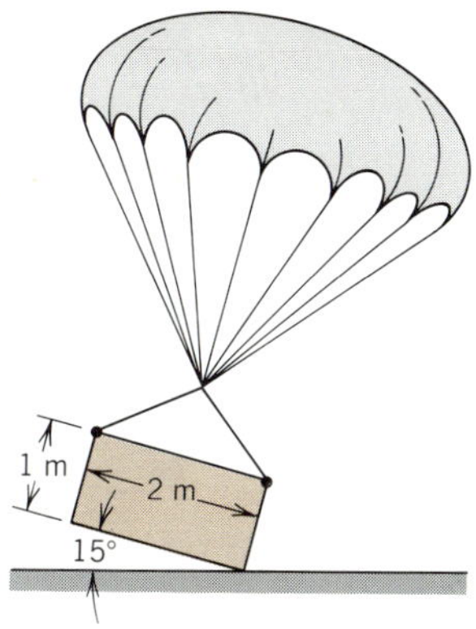

PROB. 14.162

***14.162** A 1000 kg crate is dropped by parachute, hitting the ground in the position shown. At the instant of impact the crate is in rectilinear translation, downward, with a speed of 20 m/s. Knowing that the coefficient of restitution is 0.20, determine the motion of the crate at the end of the impact. The radius of gyration of the crate about its center of mass G is 0.80 m.

***14.163** The batter is swinging the bat at a 5.5 oz baseball that is traveling horizontally at 100 ft/s. At the instant of impact shown, the batter's hands are moving forward at 15 ft/s and rotating the bat at 10 rad/s. The bat weighs 2 lb and its radius of gyration about the mass center is 16 in. The coefficient of restitution is 0.60. Assuming that the force exerted by the batter is negligible during the collision, determine the velocity of the ball after the impact.

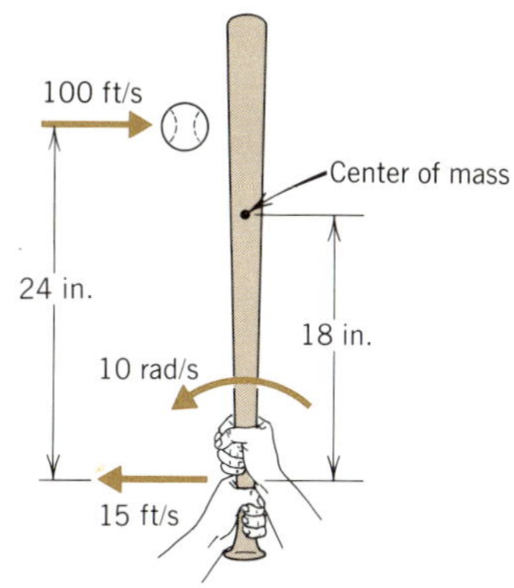

PROB. 14.163

CHAPTER 15

KINEMATICS IN A MOVING REFERENCE FRAME

OBJECTIVES

1. To select a moving reference frame that is suitable for describing the angular motion of bodies and the linear motion of points.

2. To evaluate the angular velocity and angular acceleration of a moving reference frame.

3. To determine the velocity and acceleration of a specified point. This involves (a) determining the terms describing the motion of the moving reference frame; (b) determining the terms describing the motion of the point with respect to the moving reference frame; (c) combining the individual terms in the relative velocity and relative acceleration equations.

4. To satisfy the restrictions on the motion of constrained points in order to determine unknown linear or angular speeds, and the rates of change of these speeds.

Picture the following situation. You are in the Space Shuttle when you observe an interesting phenomenon. You wish to describe the location of this phenomenon to someone on the earth who will observe it through a telescope. Considering the fact that your vehicle is moving with respect to the earth, what information should you convey to the observer on the earth?

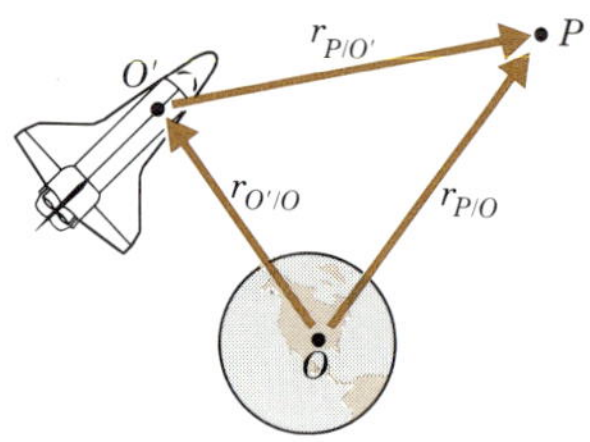

FIGURE 1

For the time being, let us consider the earth to be fixed. Then as shown in Figure 1, the observer on the earth at point O needs to know the absolute (true) position, $\bar{r}_{P/O}$, of the phenomenon at point P. However, you are observing the position of point P relative to your present location O', as given by $\bar{r}_{P/O'}$. The vector triangle depicted in Figure 1 indicates that the observer on the earth can deduce $\bar{r}_{P/O}$ by following your position, $\bar{r}_{O'/O}$. Then

$$\bar{r}_{P/O} = \bar{r}_{O'/O} + \bar{r}_{P/O'} \tag{1}$$

This equation is exactly like the one we developed in Chapter 11 to describe relative motion. The new feature here is the way in which the relative position vector will be described. In the postulated situation, where you are in the Space Shuttle, you would most likely use your vehicle as your frame of reference for describing $\bar{r}_{P/O'}$. For example, you could state the distances that point P is ahead of, to the right of, and above the vehicle. However, these directions will change as the vehicle moves. It therefore is is necessary for you to describe to the ground observer information regarding the orientation of your vehicle.

Whereas the preceding example discussed a system of advanced technology, we could have made the same arguments with respect to the moving parts of a machine. In either case the new task before us is clear: to describe the motion of a particle with respect to a moving reference frame. The kinematical tools that will emerge from the following developments are essential to the solution of three-dimensional dynamics problems. Equally important, when applied to two-dimensional problems, in many cases they simplify a previously complex analysis.

A. REFERENCE FRAMES AND ANGULAR VELOCITY

It will be necessary to distinguish between kinematical quantities that are observed from a fixed reference frame and those that are described with respect to a moving reference frame. We shall do this by standardizing the notation. In Figure 2 the fixed reference frame, whose origin is point O, is labeled as XYZ. Its units vectors are $\bar{I}$, $\bar{J}$, $\bar{K}$. Thus

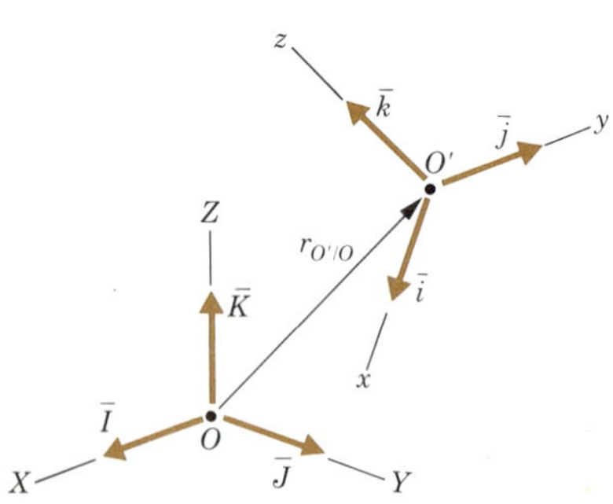

FIGURE 2

> Capital letters will denote the coordinate axes and unit vectors for a fixed reference frame.

The coordinate system xyz in Figure 2 has the moving origin O' and units vectors $\bar{i}, \bar{j}, \bar{k}$. In other words

Lower-case letters will denote coordinate axes and unit vectors for a moving reference frame.

Describing the motion of the origin O' of the moving coordinate system presents no new difficulties, because various kinematical techniques for determining such a description were developed in the previous chapters.

Now consider a point P whose position coordinates (x, y, z) with respect to the moving coordinate system are known. This is the situation depicted in Figure 3. The position vector $\bar{r}_{P/O'}$ may be written in terms of the known coordinates as

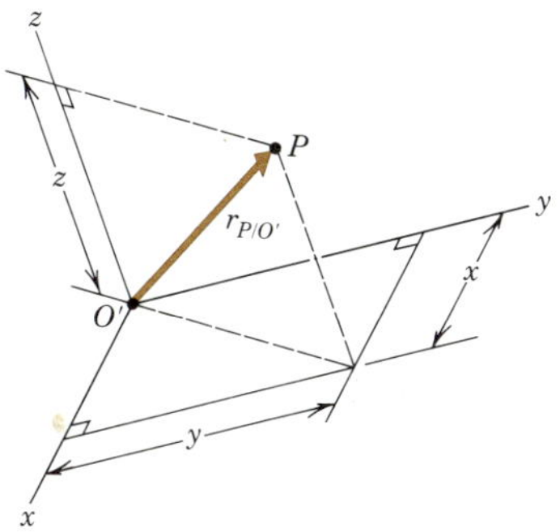

FIGURE 3

$$\bar{r}_{P/O'} = x\bar{i} + y\bar{j} + z\bar{k} \tag{2}$$

Determining the rate at which this position vector changes is a major element in computing the velocity of point P. Even when the x, y, z coordinates are constant, the rotation of the coordinate axes will result in a time rate of change in $\bar{r}_{P/O'}$. We therefore need to develop a way of describing the rotation of the xyz coordinate axes.

Let us consider an infinitesimal rotation that a moving system undergoes during the infinitesimal time interval from t to $t + dt$. The resulting rotation may be described in terms of the infinitesimal rotations $d\theta_x$, $d\theta_y$, and $d\theta_z$, where $d\theta_x$ is a rotation about an axis that is parallel to the x axis at time t, and so on. (By an infinitesimal rotation we mean that the magnitude of the angle of rotation in radians squared is negligible when compared to the angle itself.)

We have already described rotation in planar motion as a vector by using the right-hand rule relative to the axis of rotation. *Intuitively*, it makes sense to also treat the three rotations, $d\theta_x$, $d\theta_y$, $d\theta_z$, in vector form. In this case the total infinitesimal rotation $\overline{d\theta}$ can be obtained as a vector addition of the individual rotations.

$$\overline{d\theta} = d\theta_x\bar{i} + d\theta_y\bar{j} + d\theta_z\bar{k} \tag{3}$$

The basic definition of angular velocity is that $\bar{\omega}$ is the rate of change of the angle of rotation directed in accordance with the right-hand rule. Therefore

$$\bar{\omega} \equiv \frac{\overline{d\theta}}{dt} = \frac{d\theta_x}{dt}\bar{i} + \frac{d\theta_y}{dt}\bar{j} + \frac{d\theta_z}{dt}\bar{k} = \omega_x\bar{i} + \omega_y\bar{j} + \omega_z\bar{k} \tag{4}$$

In order to understand this result let us write $\bar{\omega}$ in terms of its unit vector as $\bar{\omega} = \omega\bar{e}_\omega$. This form is reminiscent of $\bar{\omega} = \omega\bar{k}$ for planar motion. The unit vector $\bar{e}_\omega$ for the three-dimensional rotation defines the orientation of the axis of rotation of $\bar{\omega}$ at the instant t. The magnitude ω is the corresponding rate of rotation. In planar motion, only the value of ω was a variable. A three-dimensional motion can also feature an axis of rotation whose orientation varies.

Treating the rotations $d\theta_x$, $d\theta_y$, and $d\theta_z$ as vectors contains an important subtlety. It assumes that the resulting orientation of the xyz coordinate system is independent of the sequence in which the rotations occur. That assumption permitted us to write the resulting rotation $\overline{d\theta}$ in eq. (3) without indicating the sequence of rotations. The fact that infinitesimal rotations add as vectors is proven rigorously in Section A.2. In Section A.1 we demonstrate that in the case of finite (large) rotations, the final orientation *does* depend on the order of rotation.

Sections A.1 and A.2 are presented for those who desire a rigorous treatment. If you choose, without loss of continuity, you can accept eqs. (3) and (4) on the basis of the foregoing discussion and proceed directly to Section B.

1. Finite Rotations

It is a simple matter to demonstrate by example that, when we deal with finite rotations, we must know the order in which the rotations are imposed. Figures 4a and 4b illustrate what happens when rectangle $OABC$, which initially lies in the xy plane, is rotated first about the x axis by $-90°$ (Figure 4a), and then about the z axis by 90° (Figure 4b). The result of the two rotations is that the rectangle lies in the yz plane.

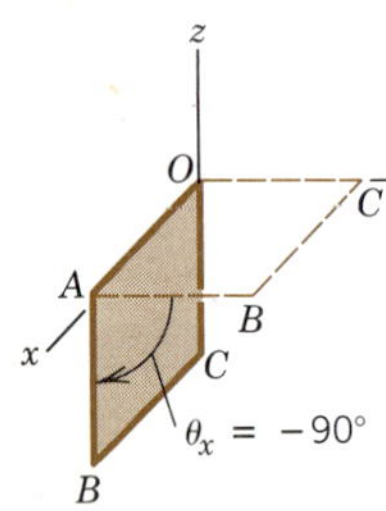

FIGURE 4a

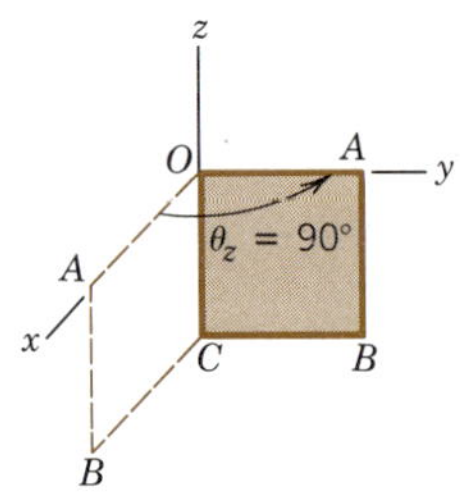

FIGURE 4b

Alternatively, the rotation about the z axis could have been imposed first, followed by the rotation about the x axis. This is illustrated in Figures 5a and 5b. In this case the position of the rectangle is the xz plane. Therefore we can conclude that, in general, the order of rotation is important when considering finite rotations.

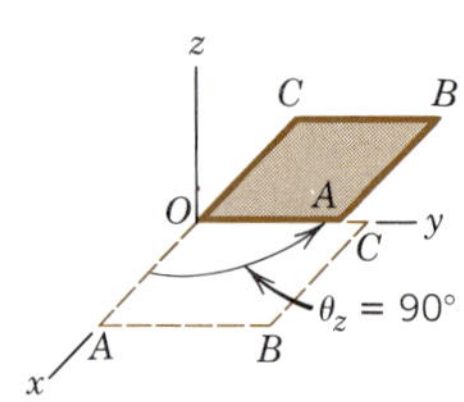

FIGURE 5a

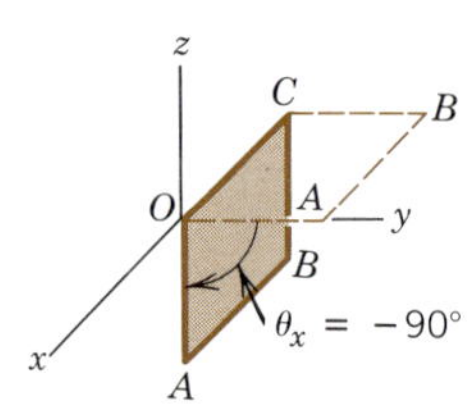

FIGURE 5b

2. Infinitesimal Rotations and Unit Vectors

Let us now consider the case of infinitesimal rotations by considering how an arbitrary unit vector

$$\bar{e} = e_x\bar{i} + e_y\bar{j} + e_z\bar{k}$$

is changed by successive rotations $d\theta_x$, $d\theta_y$, and $d\theta_z$. Figure 6 is a view looking down the x axis at the result of the rotation $d\theta_x$. In this rotation the vector $\bar{e}$ changes by the amount $d\bar{e}_1$ to become the vector $\bar{e}_1$. The vectors $(\bar{e})_\perp$ and $(\bar{e}_1)_\perp$ in Figure 6 are the components of $\bar{e}$ and $\bar{e}_1$, respectively, parallel to the yz plane. Thus

$$(\bar{e})_\perp = e_y\bar{j} + e_z\bar{k} = \bar{e} - e_x\bar{i} \tag{5}$$

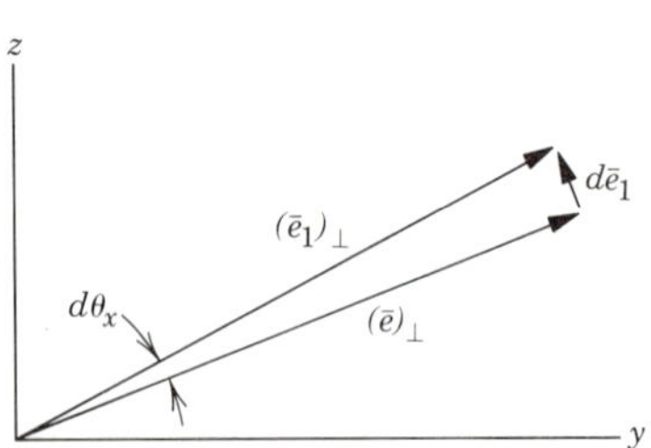

FIGURE 6

In the limit as the value of $d\theta_x$ goes to zero, we see from Figure 6 that $d\bar{e}_1$, which is parallel to the yz plane, becomes perpendicular to $(\bar{e})_\perp$. Furthermore, because the chord of an infinitesimal circular sector is equal to the arc length, we have

$$|d\bar{e}_1| = |(\bar{e})_\perp|\, d\theta_x$$

Thus, in vector form,

$$d\bar{e}_1 = (d\theta_x \bar{i}) \times (\bar{e})_\perp$$

Now, in view of eq. (5), we have

$$d\bar{e}_1 = (d\theta_x \bar{i}) \times (\bar{e} - e_x \bar{i}) \equiv (d\theta_x \bar{i}) \times \bar{e}$$

The vector triangle in Figure 6 indicates that $(\bar{e}_1)_\perp$ is the sum of $\bar{e}_\perp$ and $d\bar{e}_1$. The component of $\bar{e}$ in the direction of the x axis does not change in a rotation about that axis, so it follows that

$$\bar{e}_1 = \bar{e}_\perp + d\bar{e}_1 + e_x \bar{i} = \bar{e} + (d\theta_x \bar{i}) \times \bar{e} \tag{6}$$

We shall now operate on $\bar{e}_1$ by performing the rotation $d\theta_y$. This results in $\bar{e}_1$ changing by $d\bar{e}_2$ to become $\bar{e}_2$. By following similar steps to those used in obtaining equation (6), it is easy to show that

$$\bar{e}_2 = \bar{e}_1 + d\bar{e}_2 = \bar{e}_1 + (d\theta_y \bar{j}) \times \bar{e}_1 \tag{7}$$

Finally, with the rotation $d\theta_z$ the vector $\bar{e}_2$ changes by $d\bar{e}_3$ to become the vector $\bar{e}_3$, according to

$$\bar{e}_3 = \bar{e}_2 + d\bar{e}_3 = \bar{e}_2 + (d\theta_z \bar{k}) \times \bar{e}_2 \tag{8}$$

Equations (6)–(8) detail how a unit vector $\bar{e}$ becomes a vector $\bar{e}_3$. When we substitute eq. (6) into eq. (7) and then substitute eq. (7) into eq. (8), we obtain an expression for $\bar{e}_3$ in terms of $\bar{e}$. The first of these two steps gives

$$\bar{e}_2 = \bar{e} + (d\theta_x \bar{i}) \times \bar{e} + (d\theta_y \bar{j}) \times [\bar{e} + (d\theta_x \bar{i}) \times \bar{e}]$$

The product of infinitesimal quantities is negligible in comparison to the quantities themselves. Ignoring the term in the above expression for $\bar{e}_2$ whose magnitude is $d\theta_x\, d\theta_y$ gives

$$\begin{aligned}\bar{e}_2 &= \bar{e} + d\theta_x \bar{i} \times \bar{e} + d\theta_y \bar{j} \times \bar{e}\\ &= \bar{e} + (d\theta_x \bar{i} + d\theta_y \bar{j}) \times \bar{e}\end{aligned}$$

Substituting this expression for $\bar{e}_2$ into eq. (8) and dropping second-order differentials yields

$$\begin{aligned}\bar{e}_3 &= \bar{e} + (d\theta_x \bar{i} + d\theta_y \bar{j}) \times \bar{e} + d\theta_z \bar{k} \times \bar{e}\\ &= \bar{e} + (d\theta_x \bar{i} + d\theta_y \bar{j} + d\theta_z \bar{k}) \times \bar{e}\\ &= \bar{e} + \overline{d\theta} \times \bar{e}\end{aligned}$$

This equation tells us that the total infinitesimal change in the unit vector $\bar{e}$ is

$$d\bar{e} = \bar{e}_3 - \bar{e} = \overline{d\theta} \times \bar{e} \tag{9}$$

From this we can conclude that only the resultant rotation vector $\overline{d\theta}$ is significant to infinitesimal rotations.

B. RELATIVE VELOCITY—DIFFERENTIATION IN A MOVING REFERENCE FRAME

In Section A.2 we proved that the change of a unit vector $\bar{e}$ due to an arbitrary infinitesimal rotation $\overline{d\theta}$ is given by

$$d\bar{e} = \overline{d\theta} \times \bar{e} \tag{10}$$

Equation (10) states that in an infinitesimal rotation $\overline{d\theta}$ the unit vector $\bar{e}$ rotates by an angle $\overline{d\theta}$ about an instantaneous axis of rotation that is parallel to the direction of $\overline{d\theta}$.

It follows from eq. (10) that, as a unit vector rotates, its rate of change is given by

$$\frac{d\bar{e}}{dt} \equiv \dot{\bar{e}} = \frac{\overline{d\theta}}{dt} \times \bar{e}$$

$$\boxed{\dot{\bar{e}} = \bar{\omega} \times \bar{e}} \tag{11}$$

Equation (11) provides a method for describing the rotational characteristics of a moving reference frame. We shall now utilize this result to study the motion of a point P. In Figure 3 we selected an xyz system that is rotating and translating. The origin of the xyz system is at point O', and the vector $\bar{r}_{P/O'}$ locates the position of point P within this system. The velocity $\bar{v}_{P/O'}$ of point P with respect to origin O' accounts for the total rate of change of the position vector, according to

$$\bar{v}_{P/O'} \equiv \frac{d}{dt}(\bar{r}_{P/O'})$$

Differentiating the component form of $\bar{r}_{P/O'}$, eq. (2), shows that

$$\bar{v}_{P/O'} = \dot{x}\bar{i} + x\dot{\bar{i}} + \dot{y}\bar{j} + y\dot{\bar{j}} + \dot{z}\bar{k} + z\dot{\bar{k}} \tag{12}$$

In order to evaluate the derivatives of the unit vectors that occur in this relation

$$\boxed{\text{Let } \bar{\omega} \text{ be the angular velocity of the } xyz \text{ reference frame.}}$$

Because $\bar{i}, \bar{j}, \bar{k}$ are attached to this reference frame, $\bar{\omega}$ is the angular velocity for these unit vectors. It follows from eq. (10) that the derivative of these unit vectors is

$$\dot{\bar{i}} = \bar{\omega} \times \bar{i} \qquad \dot{\bar{j}} = \bar{\omega} \times \bar{j} \qquad \dot{\bar{k}} = \bar{\omega} \times \bar{k}$$

With these results eq. (11) becomes

$$\begin{aligned}\bar{v}_{P/O'} &= \frac{d}{dt}(\bar{r}_{P/O'}) = \dot{x}\bar{i} + \dot{y}\bar{j} + \dot{z}\bar{k} + x(\bar{\omega} \times \bar{i}) + y(\bar{\omega} \times \bar{j}) + z(\bar{\omega} \times \bar{k}) \\ &= \dot{x}\bar{i} + \dot{y}\bar{j} + \dot{z}\bar{k} + \bar{\omega} \times (x\bar{i} + y\bar{j} + z\bar{k}) \\ &= (\dot{x}\bar{i} + \dot{y}\bar{j} + \dot{z}\bar{k}) + \bar{\omega} \times \bar{r}_{P/O'}\end{aligned}$$

We shall now define a new process of differentiation of a vector using the symbol $\delta(\ )/\delta t$. This differentiation process accounts only for the rate of change of the scalar components of a vector. It ignores the fact that the unit vectors for the components are not constant. In terms of this definition,

$$\frac{\delta}{\delta t}(\bar{r}_{P/O'}) = \dot{x}\bar{i} + \dot{y}\bar{i} + \dot{z}\bar{k} \tag{13}$$

Therefore, $\delta(\bar{r}_{P/O'})/\delta t$ is the rate of change of $\bar{r}_{P/O'}$ that someone would observe if that person did not realize that the reference frame was rotating. In other words, *$\delta(\bar{r}_{P/O'})/\delta t$ is the velocity of point P relative to the moving reference frame.* For notational convenience let us denote this term by

$$\frac{\delta}{\delta t}(\bar{r}_{P/O'}) \equiv (\bar{v}_{P/O'})_{\text{rel}}$$

This allows us to write the velocity of P with respect to O' as

$$\bar{v}_{P/O'} = \frac{d}{dt}(\bar{r}_{P/O'}) = (\bar{v}_{P/O'})_{\text{rel}} + \bar{\omega} \times \bar{r}_{P/O'} \tag{14}$$

In words, this equation states that the velocity of point P with respect to the origin of the moving reference frame xyz (point O') is the velocity of point P relative to xyz, $(\bar{v}_{P/O'})_{\text{rel}}$, plus an additional term $\bar{\omega} \times \bar{r}_{P/O'}$ that accounts for the velocity caused by the angular motion $\bar{\omega}$ of the moving reference system.

Equation (14) describes the rate of change of a position vector. We could similarly use the $\delta/\delta t$ operation to evaluate the derivative of any vector $\bar{A}$. This would only necessitate replacing the x, y, z components of $\bar{r}_{P/O'}$ by the A_x, A_y, A_z components of $\bar{A}$. Thus

$$\boxed{\frac{d\bar{A}}{dt} = \frac{\delta\bar{A}}{\delta t} + \bar{\omega} \times \bar{A}} \tag{15}$$

Note that eq. (11), which we will use extensively later, is a special case of eq. (15). The simpler form of eq. (11) results because the unit vector $\bar{e}$ has constant components.

C. ABSOLUTE VELOCITY

Let us now return to the example in Figure 1. That situation may be generalized as follows. The motion of a point P is known as seen from a moving reference frame xyz whose origin is point O'. We wish to employ this knowledge to determine the motion of point P as seen from a fixed reference frame XYZ having origin O. We have already established that the positions are related by eq. (1).

$$\boxed{\bar{r}_{P/O} = \bar{r}_{O'/O} + \bar{r}_{P/O'}} \tag{1}$$

We now wish to describe the velocity of point P with respect to the fixed reference frame. We shall refer to this quantity as the *absolute velocity* $\bar{v}_P$. In view of the basic definition of velocity, we have

$$\bar{v}_P = \frac{d}{dt}(\bar{r}_{P/O})$$

so that differentiating eq. (1) gives

$$\bar{v}_P = \frac{d}{dt}(\bar{r}_{O'/O}) + \frac{d}{dt}(\bar{r}_{P/O'}) = \bar{v}_{O'} + \bar{v}_{P/O'}$$

By using eq. (14) we obtain the final result,

$$\boxed{\bar{v}_P = \bar{v}_{O'} + (\bar{v}_{P/O'})_{\text{rel}} + \bar{\omega} \times \bar{r}_{P/O'}} \tag{16}$$

This relation states that the absolute velocity of point P equals the absolute velocity of the origin of the moving reference frame plus the velocity of point P relative to the moving reference frame plus the cross product of the angular velocity $\bar{\omega}$ of the moving reference frame and the relative position vector $\bar{r}_{P/O'}$. In the simple case of a reference frame that is only translating, $\bar{\omega} \equiv \bar{0}$ and $\bar{v}_{P/O'} = (\bar{v}_{P/O'})_{\text{rel}}$. Equation (16) then becomes the relative motion equation that was studied in Chapter 11, specifically

$$\bar{v}_P = \bar{v}_{O'} + \bar{v}_{P/O'}$$

Next, consider the case where point P is on a rigid body. Let us attach the xyz reference system to the body. By doing this *the angular velocity of the reference frame is the angular velocity of the body*, $\bar{\omega} = \bar{\omega}_{\text{body}}$, because the reference frame and the body undergo the same rotation. Also, the distances between points in a rigid body cannot change, so the (x, y, z) coordinates of point P are constants. In other words, point P is not moving with respect to xyz, thus $(\bar{v}_{P/O'})_{\text{rel}} \equiv \bar{0}$. Equation (16) then gives

$$\bar{v}_P = \bar{v}_{O'} + \bar{\omega}_{\text{body}} \times \bar{r}_{P/O'} \tag{17}$$

This result, of course, is the same as the velocity formula that was obtained for rigid bodies in planar motion (Chapter 13). The new feature is that we have now shown the result to be true for three-dimensional motion, where the axis of rotation of the body can have a variable orientation.

Before we consider some examples that utilize eq. (16), let us discuss criteria to be employed in the selection of a moving reference frame. Basically, three decisions regarding the xyz frame have to be made.

1. Where should its origin O' be located?
2. What angular velocity should it have?
3. What orientation should it have at the instant of interest?

These questions can be answered by studying eq. (16). First, we note that the absolute velocity of point O' is required; hence the origin should have a fairly simple type of motion. For example, a fixed point or a point following a planar path would be ideal. The decision regarding the angular velocity $\bar{\omega}$ to be ascribed to the moving frame is based on the fact that another term we need to compute is the velocity of point P relative to the moving reference frame. Hence we look for a reference frame whose rotation makes it comparatively easy to determine $(\bar{v}_{P/O'})_{\text{rel}}$.

Finally, to select the instantaneous orientation of the xyz frame, we note that the components of several vectors are required. These components must *all* be with respect to the same coordinate system in order to perform the operations of vector addition and multiplication. The components can be with respect to either the fixed unit vectors $\bar{I}, \bar{J}, \bar{K}$, or the moving unit vectors $\bar{i}, \bar{j}, \bar{k}$. As a general rule, dynamicists tend to use $\bar{i}, \bar{j}, \bar{k}$ components. We shall therefore select the orientation of the xyz frame to be such that the components of the vectors are easy to describe.

EXAMPLE 1

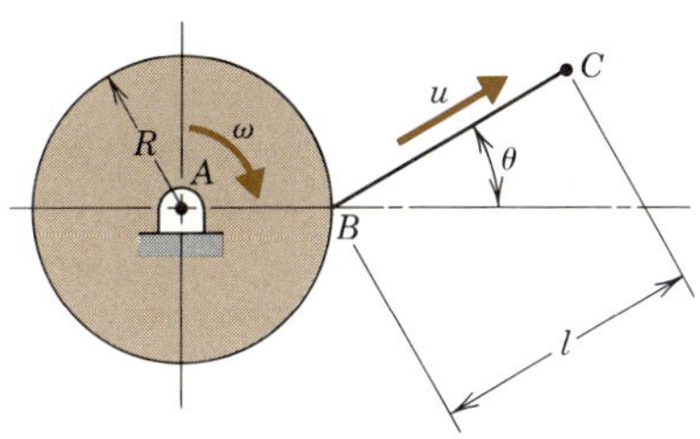

Turbine rotors and impeller pumps, which have fluid flowing over the vanes or blades, lead to important kinematical studies. Let us idealize such a problem. A disk of radius R rotates with constant angular velocity $\bar{\omega}$ about a fixed axis. A straight vane of length l is welded rigidly to the disk, and the angle between the vane and a radial line is θ, as shown in the sketch. If a particle of fluid slides outward along the vane at a speed u relative to the vane, determine the velocity of the particle as it flies off the end of the vane.

Solution

The disk and blade BC are in pure rotation about an axis perpendicular to the plane of the diagram through point A. It is consistent to select a fixed XYZ coordinate to have its origin at point A with the Z axis aligned with the axis of rotation.

In order to select the position and orientation of the moving xyz reference frame, let us describe the motion of the fluid particle in words. The particle moves along the vane BC, as the vane rotates about the Z axis. We therefore let the xyz reference frame rotate with the disk and the vane. In that case the motion of the particle *relative to xyz* reduces to a rectilinear motion parallel to the vane. There are several possibilities for the origin of xyz: (a) on the axis of rotation, that is, point A, (b) at the junction of the disk and the vane, that is, point B, (c) at the end of the vane, that is, point C on the vane. Let us study each of these possibilities.

Approach a

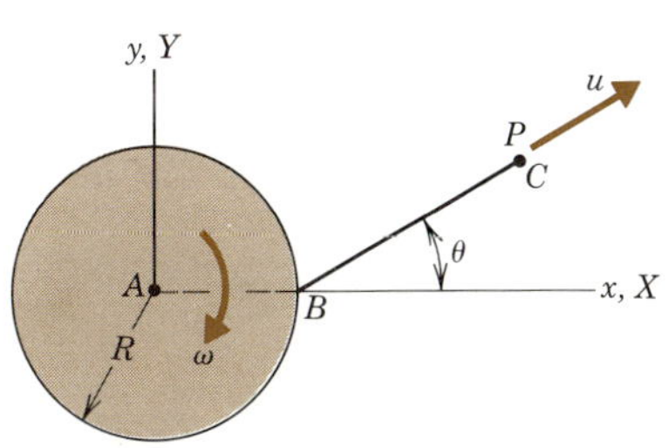

Select xyz to have origin O' at point A and attach xyz to the disk. We orient the x axis along line AB, and select the y axis to also lie in the plane of motion. This orientation, which is shown in the sketch, suits the geometry of this system. For the origins we have selected, the relative velocity equation becomes

$$\bar{v}_P = \bar{v}_A + (\bar{v}_{P/A})_{\text{rel}} + \bar{\omega} \times \bar{r}_{P/A}$$

A prime consideration in any relative motion problem is describing the angular motion of xyz. Hence this is where we begin. The reference frame has the angular velocity of the disk (it is attached to this body). Thus

$$\bar{\omega} = \omega(-\bar{K}) = \omega(-\bar{k})$$

Note that we have expressed $\bar{\omega}$ in terms of both XYZ and xyz components. This is done in order to anticipate a step in the acceleration problem, which we shall consider in the next section.

We now describe the other terms in the velocity equation. Point A is fixed, so $\bar{v}_A = \bar{0}$. The relative velocity of particle P is along line BC. We use the sketch to construct this velocity and the position vector.

$$(\bar{v}_{P/A})_{\text{rel}} = u(\cos\theta\,\bar{i} + \sin\theta\,\bar{j})$$

$$\bar{r}_{P/A} = (R + l\cos\theta)\bar{i} + l\sin\theta\,\bar{j}$$

Substituting the individual terms into the velocity equation then yields

$$\begin{aligned}\bar{v}_P &= \bar{0} + u(\cos\theta\,\bar{i} + \sin\theta\,\bar{j}) + (-\omega\bar{k}) \times [(R + l\cos\theta)\bar{i} + l\sin\theta\,\bar{j}] \\ &= (u\cos\theta + \omega l\sin\theta)\bar{i} + (-\omega R + u\sin\theta - \omega l\cos\theta)\bar{j}\end{aligned}$$

Approach b

The origin of xyz is now placed at point B, with xyz rotating with the disk. The orientation used in the previous approach fits the present case also, as shown in the sketch. The relative velocity equation is now

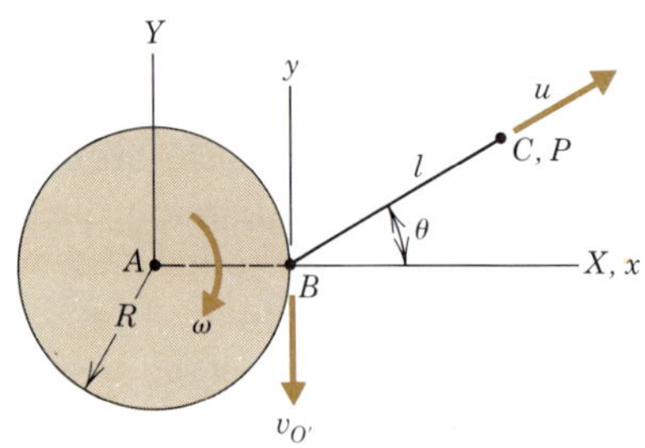

$$\bar{v}_P = \bar{v}_B + (\bar{v}_{P/B})_{\text{rel}} + \bar{\omega} \times \bar{r}_{P/B}$$

The angular motion of this xyz reference frame and the relative velocity of the particle P are the same as in approach (a). Thus

$$\bar{\omega} = (-\omega\bar{k}), \qquad (\bar{v}_{P/B})_{\text{rel}} = u(\cos\theta\,\bar{i} + \sin\theta\,\bar{j})$$

The position vector is

$$\bar{r}_{P/B} = l(\cos\theta\,\bar{i} + \sin\theta\,\bar{j})$$

In this approach origin B is moving. Its velocity may be derived in one of two ways. We can either note that point B is following a circular path centered at point A, so that its velocity is ωR tangent to the path, or alternatively that point B is a point on the rigid disk, in which case we can employ the rigid body eq. (17). Let us follow the latter approach. Then

$$\bar{v}_B = \bar{v}_A + \bar{\omega}_{\text{disk}} \times \bar{r}_{B/A} = \bar{0} + (-\omega\bar{k}) \times R\bar{i} = -\omega R\bar{j}$$

The equation for $\bar{v}_P$ then gives

$$\bar{v}_P = -\omega R\bar{j} + u(\cos\theta\,\bar{i} + \sin\theta\,\bar{j}) + (-\omega\bar{k}) \times l(\cos\theta\,\bar{i} + \sin\theta\,\bar{j})$$
$$= (u\cos\theta + \omega l\sin\theta)\bar{i} + (-\omega R + u\sin\theta - \omega l\cos\theta)\bar{j}$$

Because the xyz system selected in approach (b) is parallel to the one used in approach (a) the components of $\bar{v}_P$ should be identical for each approach. A comparison shows this to be true.

Approach c

The origin of xyz is now situated at point C. The angular motion of xyz is the same as that of the disk, and the earlier orientation of the axis is still useful. This is shown in the sketch. The velocity equation is now

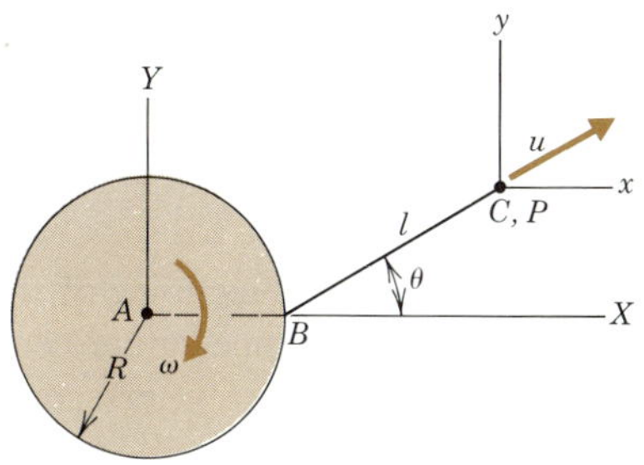

$$\bar{v}_P = \bar{v}_C + (\bar{v}_{P/C})_{\text{rel}} + \bar{\omega} \times \bar{r}_{P/C}$$

where $\bar{\omega}$ and $(\bar{v}_{P/C})_{\text{rel}}$ are the same as in the earlier approaches,

$$\bar{\omega} = -\omega\bar{k}, \qquad (\bar{v}_{P/C})_{\text{rel}} = u(\cos\theta\,\bar{i} + \sin\theta\,\bar{j})$$

Because the fluid particle is at the origin C at this instant, $\bar{r}_{P/C} = \bar{0}$.

In order to determine the velocity of point C, we consider it to be a point on a rigid disk (with the attached vane). Then

$$\bar{v}_C = \bar{v}_A + \bar{\omega}_{\text{disk}} \times \bar{r}_{C/A} = \bar{0} + (-\omega\bar{k}) \times [(R + l\cos\theta)\bar{i} + l\sin\theta\,\bar{j}]$$
$$= \omega l\sin\theta\,\bar{i} - \omega(R + l\cos\theta)\bar{j}$$

The velocity $\bar{v}_P$ is then found to be

$$\bar{v}_P = \omega l\sin\theta\,\bar{i} - \omega(R + l\cos\theta)\bar{j} + u(\cos\theta\,\bar{i} + \sin\theta\,\bar{j}) + \bar{0}$$
$$= (u\cos\theta + \omega l\sin\theta)\bar{i} + (-\omega R + u\sin\theta - \omega l\cos\theta)\bar{j}$$

Additional Remarks

Let us summarize the foregoing by commenting on the advantages and disadvantages of each approach. In approach (c) the method we used for determining the velocity of the origin may not have been obvious to you, and the alternative of using the concepts of circular motion leads to some complications due to the geometry of the system.

The solution was obtained fairly directly in both approaches (a) and (b). However, of the two, in approach (b) the origin O' was a moving point whose velocity was easy to compute, and each term in the velocity equation had components that were very simple to write. This turns out to be true in general. That is, if there is a point intermediate to the moving point P and the fixed point O, and if the kinematical description of this point is fairly direct, we shall use such a point as the origin of the moving reference system.

Students often try to solve a problem such as Example 1 by letting the origin of xyz move along with the point P of interest; the reference frame is no longer attached to the disk. In that case the relative velocity and the position vector are both zero. The only non-zero term remaining in the right side of the velocity equation is the velocity of the origin, $\bar{v}_P \equiv \bar{v}_{O'}$. In other words, "The answer is the answer." The moving reference frame does not aid solving for $\bar{v}_P$ in this case. We should *never* let the origin of the moving reference frame follow the point in which we are interested.

EXAMPLE 2

The motor spins the flywheel at $\omega_1 = 1800$ rev/min, as the turntable rotates about the vertical axis at $\omega_2 = 100$ rev/min. Determine the velocity of point B on the perimeter of the flywheel at the instant shown.

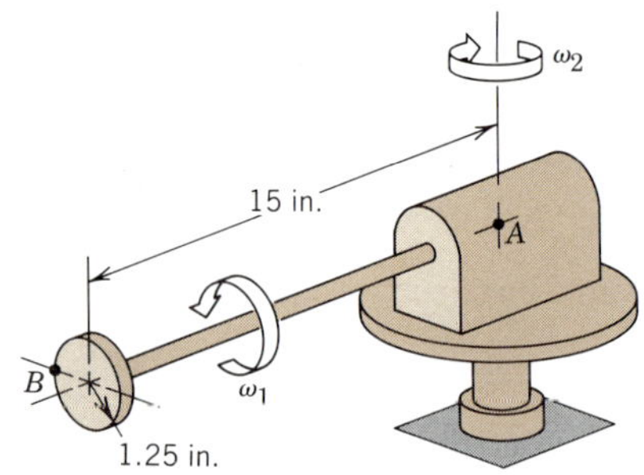

Solution

In the previous example we discussed the matter of the choice of the origin of the moving reference frame. Based on that experience, either center C of the flywheel or point A, where the axis of the motor intersects the axis of the turntable, are suitable for the origin of xyz. We shall select point C to be this origin. The velocity equation therefore is

$$\bar{v}_B = \bar{v}_C + (\bar{v}_{B/C})_{\text{rel}} + \bar{\omega} \times \bar{r}_{B/C}$$

Two possibilities arise for the angular motion to be ascribed to xyz. We could attach the reference frame to the motor or to the flywheel. We will consider both approaches. For the sake of clarity when you solve a problem, you *must explicitly state* the body to which the moving reference frame is attached.

Approach a

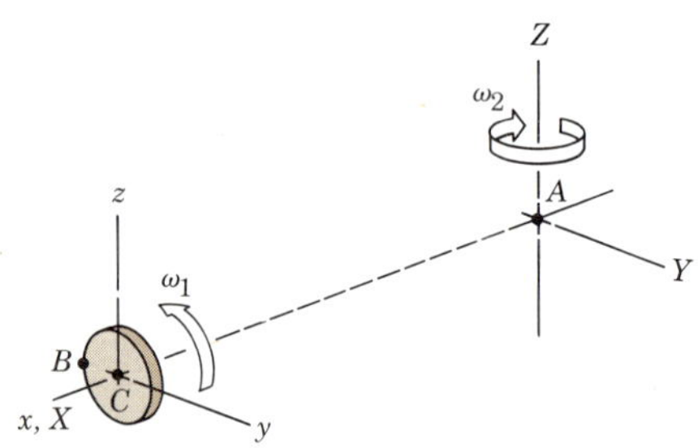

Let xyz be attached to the motor. The orientation of xyz shown in the sketch is chosen primarily for its convenience in describing the two axes of rotation. Notice that the z and Z axes are always parallel. However, the x axis is coincident with the fixed X axis only at the instant shown.

The first item to be considered is the angular velocity of xyz. Although we have not yet discussed the problem of accelerations, it will be seen later that there are two separate uses for $\bar{\omega}$. One use, which fits the aim of this problem, is to substitute the $\bar{\omega}$ vector into the velocity equation in terms of its components with respect to the xyz axes, in order to perform the necessary computations. A further use is to have an expression for the $\bar{\omega}$ vector that is valid at any arbitrary instant. That expression can then be differentiated with respect to time in order to determine the angular acceleration $\bar{\alpha}$.

In order to be consistent with the procedures we will develop for evaluating accelerations, we first describe $\bar{\omega}$ in a general manner. For this, we note that xyz is rotating at the rate ω_2 about the vertical axis. Both the Z and z axes are always vertical. Thus according to the right-hand rule

$$\bar{\omega} = \omega_2(-\bar{K}) = \omega_2(-\bar{k})$$

The given value of ω_2 must be converted to rad/s, so that

$$\bar{\omega} = -100\left(\frac{2\pi}{60}\right)\bar{k} = -10.472\bar{k} \text{ rad/s}$$

The velocity of the origin C is readily obtained by considering it to be a point on the motor. Thus

$$\bar{v}_C = \bar{v}_A + \bar{\omega}_{\text{motor}} \times \bar{r}_{C/A} = \bar{\omega} \times \bar{r}_{C/A}$$
$$= -10.472\bar{k} \times 15\bar{i} = -157.08\bar{j} \text{ in./s}$$

In order to describe the relative motion, let us pretend that the motor, to which xyz is attached, is not moving. This still leaves the rotation of the flywheel. Thus point B is following a 1.25 in. circular path relative to xyz at an angular speed ω_1. At the instant depicted in the sketch, this gives

$$(\bar{v}_{B/C})_{\text{rel}} = 1.25\omega_1(-\bar{k}) = -1.25(1800)\left(\frac{2\pi}{60}\right)\bar{k} = -235.6\bar{k} \text{ in./s}$$

$$\bar{r}_{B/C} = -1.25\bar{j} \text{ in./s}$$

Thus the velocity equation reduces to

$$\bar{v}_B = -157.1\bar{j} - 235.6\bar{k} + (-10.472\bar{k}) \times (-1.25\bar{j})$$
$$= 13.1\bar{i} - 157\bar{j} - 236\bar{k} \text{ in./s}$$

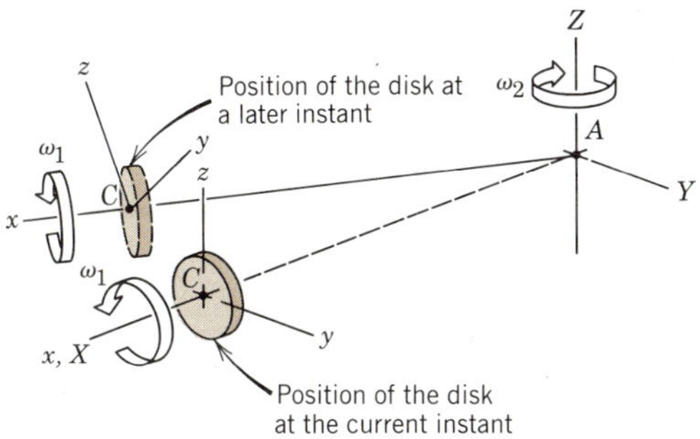

Appraoch b

Now xyz is attached to the flywheel. We will employ the orientation of the axes depicted in the sketch for the previous solution. However, in the present case the axes of xyz are parallel to XYZ only at this instant. Observe that as the flywheel and motor rotate, the x axis will remain aligned with the motor shaft, and the z axis will rotate away from the vertical.

The total rotation of the flywheel may be described by saying that it rotates about the (moving) x axis at ω_1 as this axis rotates about the (fixed) Z axis at ω_2. Thus a general description of $\bar{\omega}$ found by the right-hand rule is

$$\bar{\omega} = \omega_1(\bar{i}) + \omega_2(-\bar{K})$$

Substituting the values of ω_1 and ω_2, and the fact that $\bar{K}$ is parallel to the z axis at this instant then gives

$$\bar{\omega} = 1800\left(\frac{2\pi}{60}\right)(\bar{i}) + 100\left(\frac{2\pi}{60}\right)(-\bar{k})$$
$$= 188.50\bar{i} - 10.472\bar{k} \text{ rad/s}$$

The method for finding $\bar{v}_C$ in the previous approach is valid here also. Thus

$$\bar{v}_C = -157.08\bar{j} \text{ in./s}$$

The relative motion in this approach is particularly simple to describe. If we pretend that xyz is not moving, the flywheel stops. Thus

$$(\bar{v}_{B/C})_{\text{rel}} = \bar{0}, \qquad \bar{r}_{B/C} = -1.25\bar{j} \text{ in.}$$

Then

$$\bar{v}_C = -157.08\bar{j} + (188.50\bar{i} - 10.472\bar{k}) \times (-1.25\bar{j})$$
$$= -13.1\bar{i} - 157\bar{j} + 236\bar{k} \text{ in./s}$$

Additional Remarks

Let us assess the comparative advantages of the two approaches. In approach (a) the angular velocity of the reference frame (rotation about one axis) was easier to describe and understand than the angular velocity of the xyz axes in approach (b) (rotation about two axes). However, in approach (b) the motion relative to the moving reference system was much simpler to formulate. You should be familiar with both approaches. Indeed, when we treat the kinetics of rigid bodies in three-dimensional motion it will be necessary to describe the angular motion of bodies that rotate about more than one axis.

HOMEWORK PROBLEMS

15.1 Rod ABC rotates clockwise at 10 rad/s as the slider moves outward at 1 m/s. Determine the velocity of the slider when it is at (a) position B, and (b) position C.

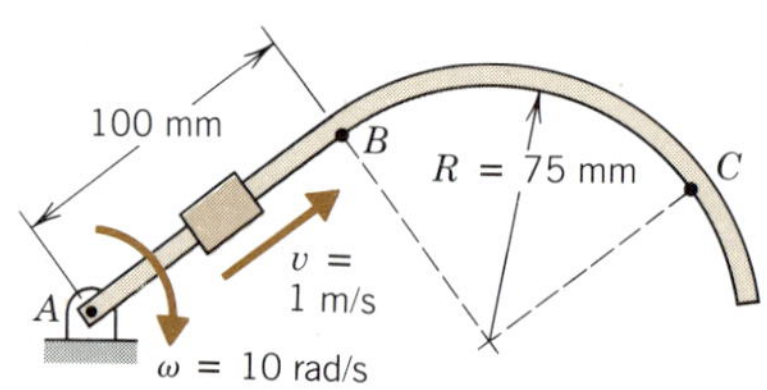

PROB. 15.1

15.2 Slider A is moving through the groove at a constant speed $u = 10$ in./s relative to the rotating disk. Determine the velocity of slider A for the position shown.

15.3 Slider B is moving through the groove at a constant speed $u = 10$ in./s relative to the rotating disk. Determine the velocity of slider B for the position shown.

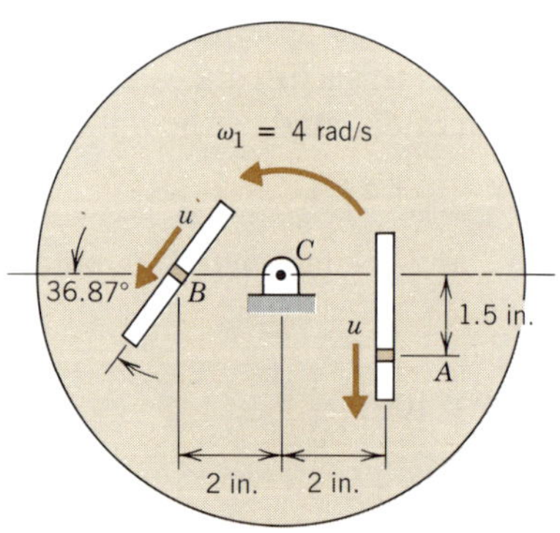

PROBS. 15.2 AND 15.3

15.4 The circular hoop is rotating about the vertical axis at 20 rad/s as the collar moves relative to the hoop at a constant speed of 4 m/s. Determine the velocity of the slider (a) when $\theta = 0°$, (b) when $\theta = 90°$.

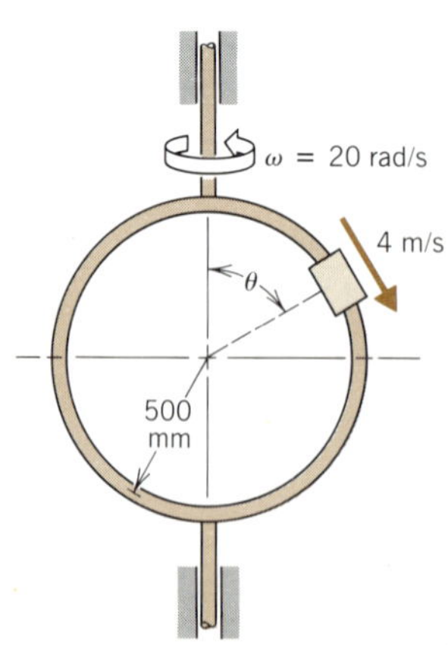

PROB. 15.4

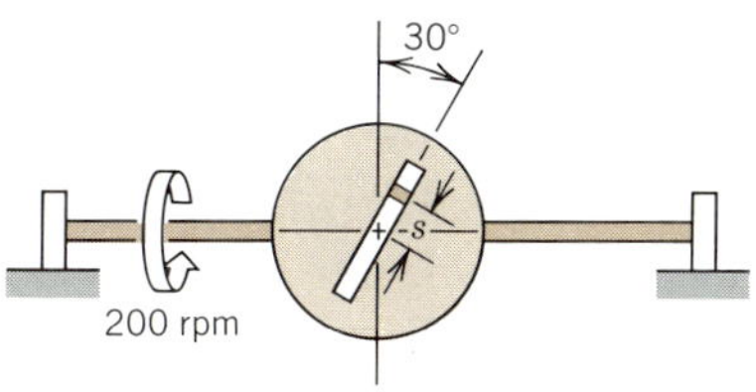

PROB. 15.6

15.5 Solve Problem 15.4 for the instant when the collar is at (a) $\theta = 30°$, and (b) $\theta = 120°$.

15.6 The slider oscillates within the groove of the disk, such that the distance $s = 20 \sin 10\pi t$, where s is measured in millimeters and t is in seconds. The disk is fixed to the horizontal axis, which is rotating at 200 rpm. Determine the velocity of the slider when $t = 30$ ms.

15.7 A passenger in the front seat of automobile A follows automobile B by observing the separation distance R and the angle ϕ, which is measured with respect to the longitudinal axis of automobile A. In the position shown, automobile A has a speed of 45 mi/hr, $\phi = 45°$, and $\dot{\phi} = -0.2$ rad/s. Determine the speed of automobile B.

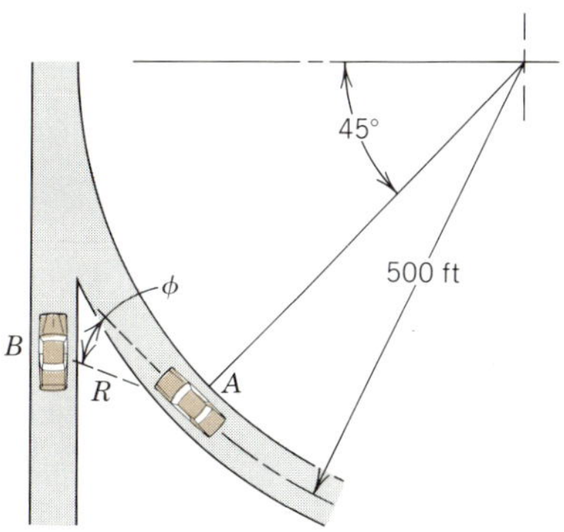

PROB. 15.7

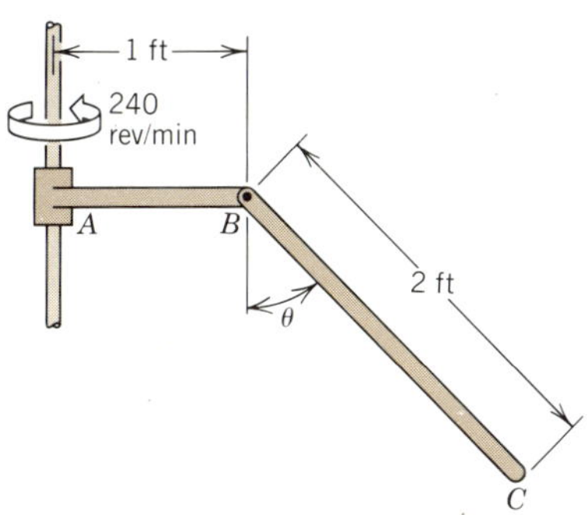

PROB. 15.8

15.8 Bar BC is pinned at end B to bar AB, which rotates about the vertical axis at 240 rev/min. At the instant when $\theta = 90°$, this angle is increasing at the rate of 5 rad/s. Determine the velocity of end C at this instant.

15.9 The angle θ in Problem 15.8 is decreasing at 3 rad/s when $\theta = 36.87°$, Determine the velocity of end C at this instant.

15.10 The radar antenna rotates about the fixed vertical axis at 2 rev/min as the angle θ oscillates at $\theta = \pi/6 + (\pi/3) \sin \pi t$, where θ is in units of radians and t is in seconds. Determine the velocity of the probe P when $t = 0.25$ s.

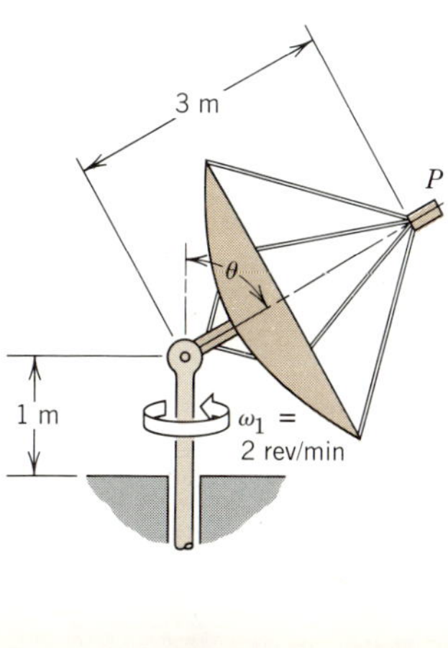

PROB. 15.10

15.11 In Problem 15.10 determine the angular velocity of the antenna when $t = 0.75$ s. Then use that result to determine the corresponding velocity of the probe P.

15.12 A small electric motor pivots about the horizontal pivot A. The disk that is attached to the shaft has an angular speed of 900 rev/min, and the motor is rotating downward at 5 rad/s. Determine the angular velocity of the disk.

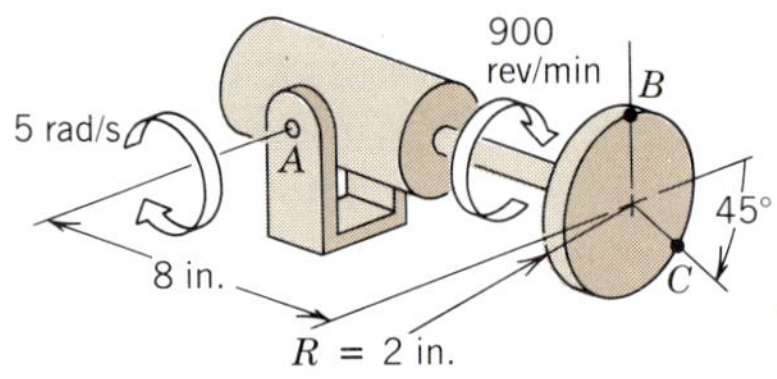

PROB. 15.12

15.13 Determine the velocity of point C on the disk in Problem 15.12.

D. ABSOLUTE ACCELERATION

Our next task is to develop a formula relating the accelerations of a point relative to fixed and moving reference frames. For this we use the definition that acceleration is the time derivative of velocity. Differentiating eq. (16) yields

$$\bar{a}_P = \frac{d}{dt}(\bar{v}_P) = \frac{d}{dt}(\bar{v}_{O'}) + \frac{d}{dt}(\bar{v}_{P/O'})_{\text{rel}} + \frac{d}{dt}(\bar{\omega} \times \bar{r}_{P/O'})$$

$$= \bar{a}_{O'} + \frac{d}{dt}(\bar{v}_{P/O'})_{\text{rel}} + \frac{d\bar{\omega}}{dt} \times \bar{r}_{P/O'} + \bar{\omega} \times \frac{d}{dt}(\bar{r}_{P/O'}) \tag{18}$$

The term $d\bar{\omega}/dt$ is the angular acceleration, just as it was in planar motion. Indeed, it is denoted by the same symbol

$$\boxed{\bar{\alpha} = \frac{d\bar{\omega}}{dt}} \tag{19}$$

The last term in eq. (18) contains the factor $d(\bar{r}_{P/O'})/dt$. We have encountered this derivative previously; it is given by eq. (14). Differentiating $(\bar{v}_{P/O'})_{\text{rel}}$ in eq. (18) presents us with a similar problem, because both vectors describe what an observer in the moving reference system is witnessing. This suggests that we should again employ the concept of a derivative with respect to the moving reference system, as defined by eq. (15). This step gives

$$\frac{d}{dt}[(\bar{v}_{P/O'})_{\text{rel}}] = \frac{\delta}{\delta t}[(\bar{v}_{P/O'})_{\text{rel}}] + \bar{\omega} \times (\bar{v}_{P/O'})_{\text{rel}} \tag{20}$$

The first term in the right-hand side of eq. (20) expresses how an observer in the moving reference system would see the relative velocity change. In other words, it is the relative acceleration,

$$(\bar{a}_{P/O'})_{\text{rel}} \equiv \frac{\delta}{\delta t}\left[(\bar{v}_{P/O'})_{\text{rel}}\right] \equiv \frac{\delta^2}{\delta t^2}(\bar{r}_{P/O'})_{\text{rel}} \tag{21}$$

Substitution of eqs. (14), (19), (20), and (21) into eq. (18) yields the final form of the acceleration equation in relative motion.

$$\boxed{\bar{a}_P = \bar{a}_{O'} + (\bar{a}_{P/O'})_{\text{rel}} + 2\bar{\omega} \times (\bar{v}_{P/O'})_{\text{rel}} + \bar{\alpha} \times \bar{r}_{P/O'} + \bar{\omega} \times (\bar{\omega} \times \bar{r}_{P/O'})} \tag{22}$$

Two special cases arise when applying eq. (22). Suppose that the xyz frame is in pure translation, in which case $\bar{\omega}$ and $\bar{\alpha}$ are both zero. This leads to the relative acceleration equation in Chapter 11.

$$\bar{a}_P = \bar{a}_{O'} + (\bar{a}_{P/O'})_{\text{rel}} \qquad (\bar{a}_{P/O'})_{\text{rel}} \equiv \bar{a}_{P/O'}$$

The other special case occurs when point P is a point in a rigid body and xyz is fixed to the body. In this case there is no motion relative to xyz, so eq. (22) reduces to

$$\bar{a}_P = \bar{a}_{O'} + \bar{\alpha}_{\text{body}} \times \bar{r}_{P/O'} + \bar{\omega}_{\text{body}} \times (\bar{\omega}_{\text{body}} \times \bar{r}_{P/O'}) \tag{23}$$

In order to gain further understanding of eq. (22), let us consider its terms individually. Both $\bar{a}_{O'}$ and $(\bar{a}_{P/O'})_{\text{rel}}$ are self-explanatory. The term $2\bar{\omega} \times (\bar{v}_{P/O'})_{\text{rel}}$ does not occur in the simpler cases of translating reference frame and of rigid body motion. If we had naively tried to construct the general equation (22) on the basis of the foregoing special cases, we surely would have omitted this important effect, which represents the interaction of the motion relative to the xyz frame and the rotation of xyz. We refer to $2\bar{\omega} \times (\bar{v}_{P/O'})_{\text{rel}}$ as the Coriolis acceleration. Note its similarity to the transverse acceleration term $2\dot{R}\dot{\phi}\bar{e}_\phi$ in cylindrical coordinates (Chapter 11).

Next observe that $\bar{\alpha} \neq 0$ whenever there is any change in the angular velocity. The increase in the velocity of point P resulting from such a change is described by the term $\bar{\alpha} \times \bar{r}_{P/O'}$. A primary new feature of three-dimensional kinematics is that the axis of rotation need not have a fixed orientation. Hence $\bar{\alpha}$ will generally not be parallel to $\bar{\omega}$, and $\bar{\alpha} \times \bar{r}_{P/O'}$ will not represent solely a transverse acceleration.

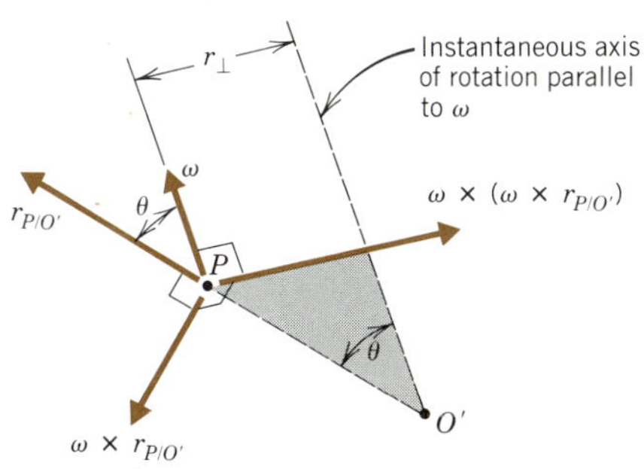

FIGURE 7

Figure 7 depicts the last term in eq. (22). In accord with the rules for a cross product, the tails of $\bar{\omega}$ and $\bar{r}_{P/O}$ are made to coincide. (It is useful to select point P as the point of coincidence.) The angle between these vectors is denoted as θ. Then $\bar{\omega} \times \bar{r}_{P/O'}$ has the magnitude $|\omega|\,|\bar{r}_{P/O'}|\sin\theta$ by definition. Further, this cross product is perpendicular to the shaded plane formed by $\bar{\omega}$ and $\bar{r}_{P/O'}$. It follows that $\bar{\omega} \times \bar{r}_{P/O'}$ is perpendicular to $\bar{\omega}$, so

$$|\bar{\omega} \times (\bar{\omega} \times \bar{r}_{P/O'})| = |\bar{\omega}|^2 |\bar{r}_{P/O'}| \sin\theta$$

Now note that $|\bar{r}_{P/O}| \sin\theta \equiv r_\perp$, that is, the perpendicular distance from point P to the instantaneous axis of rotation. Also, Figure 7 shows that the result of the second cross product is always oriented from point P toward the axis of rotation. In other words $\bar{\omega} \times (\bar{\omega} \times \bar{r}_{P/O'})$ is the centripetal acceleration of point P, corresponding to rotation of the moving reference frame about an axis through point O'. In planar motion, the position vectors we construct are perpendicular to $\bar{\omega}$. Only in that special case does $\bar{\omega} \times (\bar{\omega} \times \bar{r}_{P/O'})$ reduce to the simpler term, $-\omega^2 \bar{r}_{P/O}$.

We may conclude from the foregoing that eq. (23) is the generalization of the rigid body acceleration equation in Chapter 13. It still describes the superposition of translational motion following point O' and a pure rotation about point O'. Furthermore, recall that eq. (17) for rigid body accelerations is the same as the one derived in Chapter 13. It then follows that Chasle's theorem, which we developed heuristically for planar motion, has now been proven to be descriptive of all motions of a rigid body.

Our earlier investigations of velocity led to criteria regarding the selection of the moving reference frame. Those conclusions also apply for the determination of acceleration. The primary new feature for such an analysis is the determination of the angular acceleration $\bar{\alpha}$.

The following three-step standard technique will assist in the task of determining both the angular velocity and angular acceleration vectors.

1. Express $\bar{\omega}$ in general terms. This is achieved by first using the unit vectors of either the fixed or moving coordinate system to describe the various axes of rotation, whichever gives a constant description. Thus, if an axis of rotation is fixed in space we describe it in terms of the fixed $\bar{I}\,\bar{J}\,\bar{K}$, whereas if an axis of rotation has varying orientation, we describe it in terms of the moving $\bar{i}\,\bar{j}\,\bar{k}$. Equally important, we use algebraic variables to denote the rates of rotation.
2. Differentiate $\bar{\omega}$ to obtain a general expression for $\bar{\alpha}$, using $\dot{\bar{e}} = \bar{\omega} \times \bar{e}$ to evaluate the derivatives of $\bar{i}$, $\bar{j}$, and $\bar{k}$.
3. The instantaneous values of $\bar{\omega}$ and $\bar{\alpha}$ are now determined by substituting values of the rates of rotation and expressing the unit vectors of XYZ in terms of their xyz equivalents.

These steps will be a key feature of the following examples.

EXAMPLE 3

The turbine rotor of Example 1 is shown again in the accompanying sketch. Find the acceleration of a particle of fluid immediately prior to the instant when it flies off the end of the vane if the rotation rate ω and relative speed u are both constant.

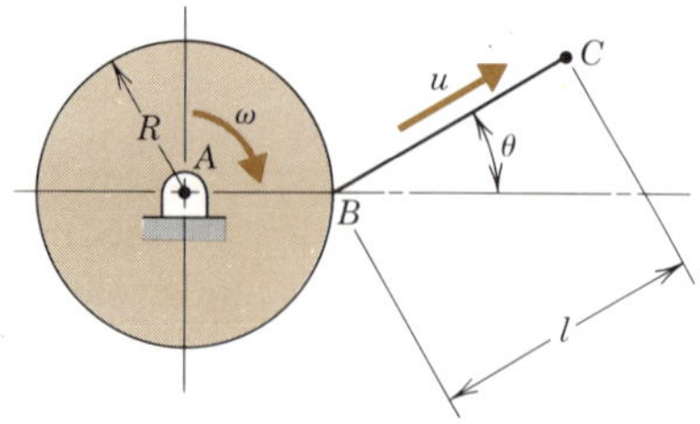

Solution

In Example 1 we found that attaching the moving xyz to the disk and the blade expedited the description of the relative motion term. Also, the task of describing the motion of the origin was simplified when the origin was placed at the pivot A. Hence this is the moving reference frame we choose to employ here. The orientation shown in the sketch is also the same as

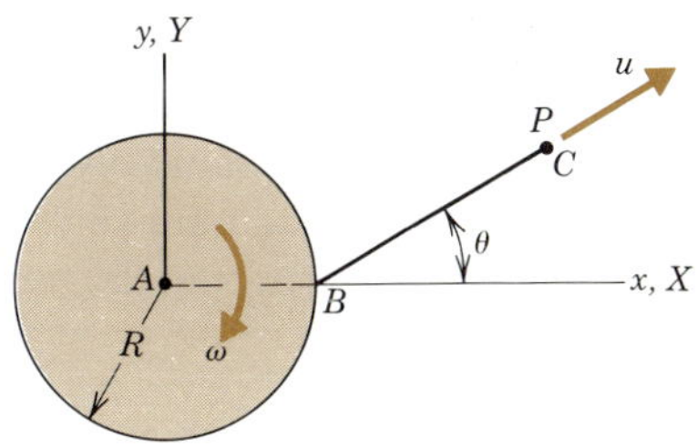

the one we selected earlier. The acceleration eq. (22) then becomes

$$\bar{a}_P = \bar{a}_A + (\bar{a}_{P/A})_{\text{rel}} + 2\bar{\omega} \times (\bar{v}_{P/A})_{\text{rel}} + \bar{\alpha} \times \bar{r}_{P/A} + \bar{\omega} \times (\bar{\omega} \times \bar{r}_{P/A})$$

Because we have selected the origin of xyz to coincide with the fixed point A, we have $\bar{a}_A = \bar{0}$. Also, following our method for describing the angular motion, the general expressions for $\bar{\omega}$ and $\bar{\alpha}$ are

$$\bar{\omega} = \omega(-\bar{K}): \quad \text{always}$$

$$\bar{\alpha} = \frac{d\bar{\omega}}{dt} = \dot{\omega}(-\bar{K}) + \omega(-\dot{\bar{K}}) = -\dot{\omega}\bar{K}$$

The rate ω is constant and $\bar{K} \equiv \bar{k}$, so that

$$\bar{\omega} = -\omega\bar{k}, \qquad \bar{\alpha} = \bar{0}$$

For the evaluation of the relative motion, we visualize the situation where the reference frame is not moving. Then the particle P would be in in rectilinear motion along line BC. Because the speed u in this relative motion is constant, we have

$$\bar{r}_{P/A} = (R + l\cos\theta)\bar{i} + l\sin\theta\,\bar{j}$$

$$(\bar{v}_{P/A})_{\text{rel}} = u\bar{e}_{C/B} = u(\cos\theta\,\bar{i} + \sin\theta\,\bar{j})$$

$$(\bar{a}_{P/A})_{\text{rel}} = \bar{0}$$

We now substitute the individual terms into the acceleration equation.

$$\begin{aligned}\bar{a}_P &= \bar{0} + \bar{0} + 2(-\omega\bar{k}) \times u(\cos\theta\,\bar{i} + \sin\theta\,\bar{j}) \\ &\quad + \bar{0} + (-\omega\bar{k}) \times \{(-\omega\bar{k}) \times [(R + l\cos\theta)\bar{i} + l\sin\theta\,\bar{j}]\} \\ &= -2\omega u\cos\theta\,\bar{j} + 2\omega u\sin\theta\,\bar{i} + (-\omega\bar{k}) \times [-\omega(R + l\cos\theta)\bar{j} + \omega l\sin\theta\,\bar{i}] \\ &= [2\omega u\sin\theta - \omega^2(R + l\cos\theta)]\bar{i} - (2\omega u\cos\theta + \omega^2 l\sin\theta)\bar{j}\end{aligned}$$

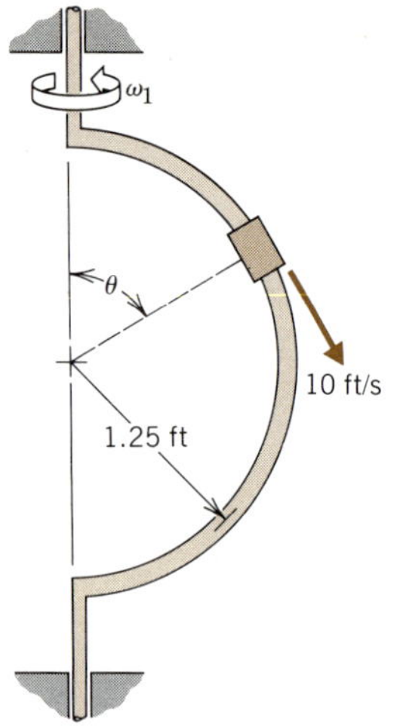

EXAMPLE 4

The collar slides along the curved bar at the constant rate of 10 ft/s. When the collar is at $\theta = 60°$, the rotation about the vertical axis is $\omega_1 = 5$ rad/s and $\dot{\omega}_1 = -2$ rad/s^2. Determine the corresponding acceleration of the collar.

Solution

The collar is moving *relative* to the bar in a circular path as the bar rotates. Thus we fix xyz to the bar, with its origin at the center O. For convenience, let XYZ have the same origin. The orientations of the coordinate systems shown in the sketch are suitable for describing the angular motion, as well as the vector describing the position of point P. The corresponding equation for $\bar{a}_P$ is

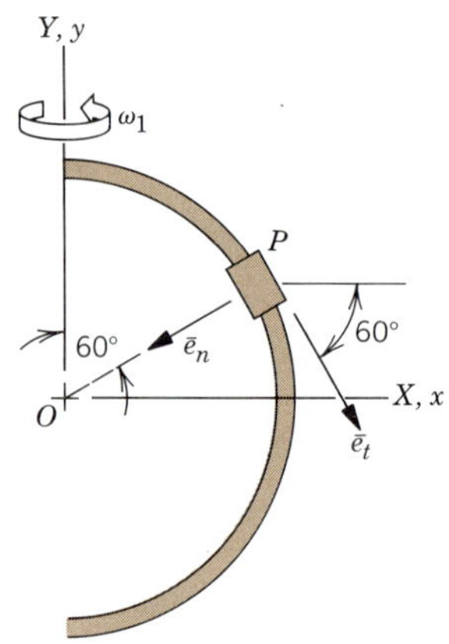

$$\bar{a}_P = \bar{a}_O + (\bar{a}_{P/O})_{\text{rel}} + 2\bar{\omega} \times (\bar{v}_{P/O})_{\text{rel}} + \bar{\alpha} \times \bar{r}_{P/O} + \bar{\omega} \times (\bar{\omega} \times \bar{r}_{P/O})$$

The guide bar is rotating about the fixed Y axis at ω_1, so the general expressions for $\bar{\omega}$ and $\bar{\alpha}$ are

$$\bar{\omega} = \omega_1 \bar{J}, \qquad \bar{\alpha} = \frac{d\bar{\omega}}{dt} = \dot{\omega}_1 \bar{J}: \quad \text{always}$$

The sketch shows that $\bar{J} = \bar{j}$. Substituting the given values of ω_1 and $\dot{\omega}_1$ then gives

$$\bar{\omega} = 5\bar{j} \text{ rad/s}, \qquad \bar{\alpha} = -2\bar{j} \text{ rad/s}$$

The origin is fixed, so $\bar{a}_O = \bar{0}$. In order to describe the relative motion, we pretend that xyz is not moving. This stops the guide bar. The sketch shows that the collar then follows a 1.25 ft radius circle at a constant speed of 10 ft/s relative to the bar. Thus

$$\bar{r}_{P/O} = 1.25(\sin 60° \, \bar{i} + \cos 60° \, \bar{j}) = 1.0825\bar{i} + 0.625\bar{j} \text{ ft}$$

$$(\bar{v}_{P/O})_{\text{rel}} = v_{\text{rel}}\bar{e}_t = 10(\cos 60° \, \bar{i} - \sin 60° \, \bar{j}) = 5.0\bar{i} - 8.66\bar{j} \text{ ft/s}$$

$$(\bar{a}_{P/O})_{\text{rel}} = \dot{v}_{\text{rel}}\bar{e}_t + \frac{v_{\text{rel}}^2}{\rho}\bar{e}_n = \bar{0} + \frac{10^2}{1.25}(-\sin 60°\,\bar{i} - \cos 60°\,\bar{j})$$
$$= -69.28\bar{i} - 40.0\bar{j} \text{ ft/s}^2$$

The result of substituting the individual terms into the acceleration equation is

$$\bar{a}_P = \bar{0} + (-69.28\bar{i} - 40.0\bar{j}) + 2(5\bar{j}) \times (5.0\bar{i} - 8.66\bar{j})$$
$$+ (-2\bar{j}) \times (1.0825\bar{i} + 0.625\bar{j})$$
$$+ 5\bar{j} \times [5\bar{j} \times (1.0825\bar{i} + 0.625\bar{j})]$$
$$= -96.3\bar{i} - 40.0\bar{j} - 47.8\bar{k} \text{ ft/s}^2$$

EXAMPLE 5

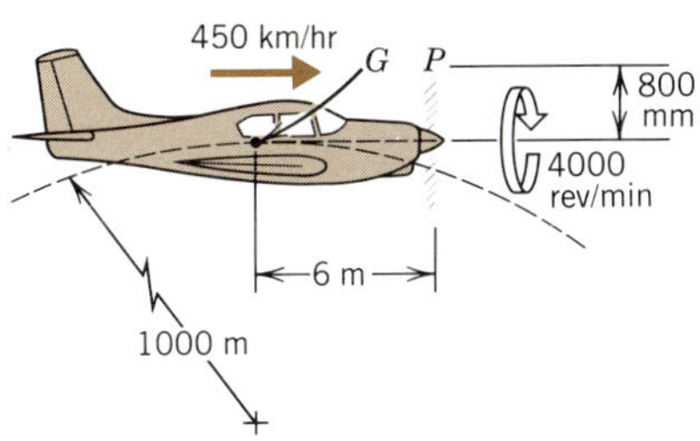

The center of mass G of the airplane is following a vertical circle of 1000 m radius at a constant speed of 450 km/hr. The propeller rotates at the constant rate of 4000 rev/min about the motor shaft, which always remains tangent to the path of G. Determine the acceleration of tip P of the propeller for the position shown.

Solution

The motion of point G is given. The sketch also indicates that point G is on the axis of the propeller shaft, so it may be considered to be a (fictitious) extension of the propeller. These considerations suggest that we fix xyz to the propeller, with its origin at point G. We may now write

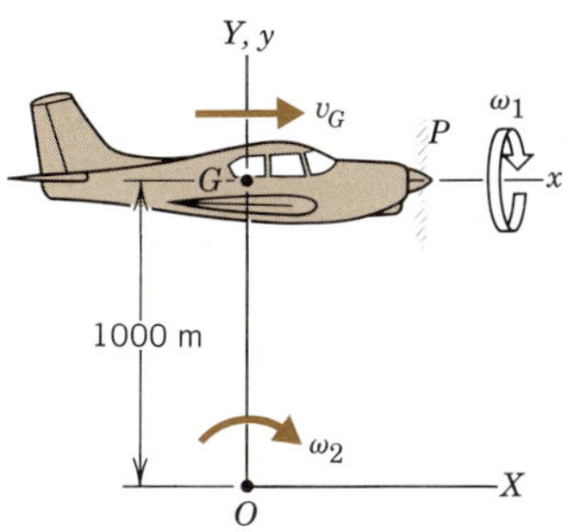

$$\bar{a}_P = \bar{a}_G + (\bar{a}_{P/G})_{\text{rel}} + 2\bar{\omega} \times (\bar{v}_{P/G})_{\text{rel}} + \bar{\alpha} \times \bar{r}_{P/G} + \bar{\omega} \times (\bar{\omega} \times \bar{r}_{P/G})$$

The orientations chosen for the xyz and XYZ axes (shown in the sketch) assist in describing the axes of rotation and the position vector $\bar{r}_{P/G}$. The origin of XYZ is placed at the center of curvature of the airplane's path because it is a convenient fixed point.

Two rotations are indicated in the sketch. The rate ω_1 is the rate at which the motor spins the propeller about the negative x axis. The rate ω_2 accounts for the fact that the motor is also rotating about the (horizontal) negative Z axis due to the circular path of the airplane. Hence the angular velocity is

$$\bar{\omega} = \omega_1(-\bar{i}) + \omega_2(-\bar{K}): \quad \text{always}$$

This general expression may be differentiated by using $\dot{\bar{i}} \equiv \bar{\omega} \times \bar{i}$. Thus

$$\bar{\alpha} = \frac{d\bar{\omega}}{dt} = -\dot{\omega}_1\bar{i} - \omega_1\dot{\bar{i}} - \dot{\omega}_2\bar{K}$$
$$= -\dot{\omega}_1\bar{i} - \omega_1(\bar{\omega} \times \bar{i}) - \dot{\omega}_2\bar{K}$$

It was given that $\omega_1 = 4000(2\pi/60)$ rad/s, and $\dot{\omega}_1 = 0$. For circular motion of point G,

$$\omega_2 = \frac{v_G}{R_G} = \frac{450/3.6}{1000} = 0.125 \text{ rad/s}$$

$$\dot{\omega}_2 = \frac{\dot{v}_G}{R_G} = 0$$

Also at the given instant $\bar{K} = \bar{k}$, so

$$\bar{\omega} = -418.9\bar{i} - 0.125\bar{k} \text{ rad/s}: \quad \text{instantaneously}$$
$$\bar{\alpha} = \bar{0} - 418.9(-418.9\bar{i} - 0.125\bar{k}) \times \bar{i} - \bar{0}$$
$$= 52.36\bar{j} \text{ rad/s}^2: \quad \text{instantaneously}$$

Since this is the first time that we have determined $\bar{\alpha}$ by this method, let us consider its physical meaning. The rotation $\omega_2(-\bar{K})$ is constant in both magnitude and direction, so it does not give an angular acceleration term. However, the rotation $\omega_1(-\bar{i})$ is about the shaft of the propeller. This shaft is rotating with the rest of the airplane at the rate $\omega_2(-\bar{K})$. Thus the angular velocity component $\omega_1(-\bar{i})$ is itself rotating in a manner that causes its head (negative x) to move upward, in the direction of the y axis. This is the direction of the $\bar{\alpha}$ vector we have constructed.

For the acceleration of point G, we note that for circular motion with constant v_G,

$$\bar{a}_G = \frac{v_G^2}{\rho_G}(\bar{e}_n)_G = \frac{(450/3.6)^2}{1000}(-\bar{j}) = -15.625\,\bar{j} \text{ m/s}^2$$

Point P is not moving relative to xyz, because it is a point on the rigid body to which xyz is attached. Thus

$$\bar{r}_{P/G} = 6\bar{i} + 0.8\bar{j} \text{ m}$$

$$(\bar{v}_{P/G})_{\text{rel}} = (\bar{a}_{P/G})_{\text{rel}} = \bar{0}$$

Then

$$\bar{a}_P = -15.625\bar{j} + \bar{0} + \bar{0} + 52.36\bar{j} \times (6\bar{i} + 0.8\bar{j})$$
$$+ (-418.9\bar{i} - 0.125\bar{k}) \times [(-418.9\bar{i} - 0.125\bar{k}) \times (6\bar{i} + 0.8\bar{j})]$$
$$= -0.1\bar{i} - 140{,}397\bar{j} - 0.02\bar{k} \text{ m/s}^2$$

It is interesting to compare this result to the value of $\bar{a}_P$ in level flight. When the speed of the airplane is constant, there is only a centripetal acceleration as the blade spins around at ω_1. The tip P is 0.8 m from the axis of rotation, which gives $|\bar{a}_P| = 0.8\omega_1^2 = 140{,}368 \text{ m/s}^2$. This is the predominant contribution to the result for $\bar{a}_P$ we obtained above.

EXAMPLE 6

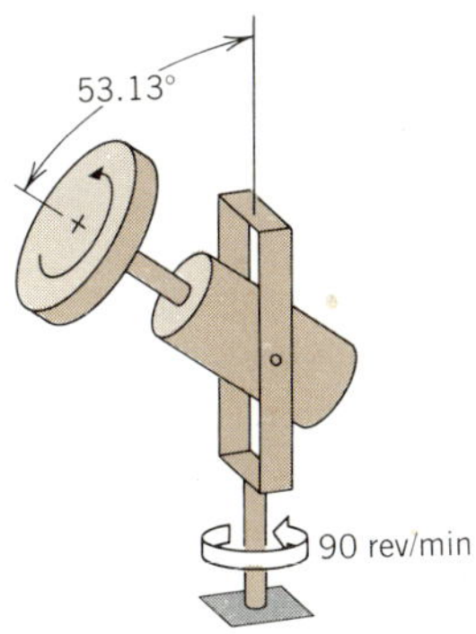

The motor is rotating at the constant rate of 90 rev/min about a vertical axis with its shaft oriented 53.13° from the vertical. The motor is rotating the disk at 15,000 rev/min, when a brake is applied to decrease that rate at 100 rad/s². Determine the angular velocity and angular acceleration of the disk at the instant after the brake is applied.

Solution

We *must* attach xyz to the disk in order that the reference frame and the body have the same motion. The choice for the origin is not important, because this problem does not involve the acceleration of points. The sketch, which has been drawn as a side view, shows that the orientation of the axes of rotation is the prime consideration in selecting the orientations of xyz and XYZ.

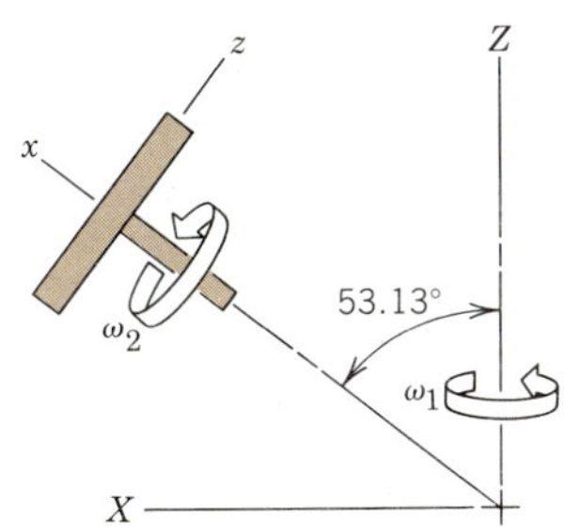

The rotation ω_1 is about the fixed Z axis, and the spin rate ω_2 is about the motor shaft which is the (moving) x axis. It follows that the angular velocity is

$$\bar{\omega} = \omega_1 \bar{K} + \omega_2 \bar{i}: \quad \text{always}$$

The rate ω_1 and the direction $\bar{K}$ are both constant. Then differentiating $\bar{\omega}$ with the aid of eq. (11) gives

$$\bar{\alpha} = \frac{d\bar{\omega}}{dt} = \dot{\omega}_1 \bar{K} + \dot{\omega}_2 \bar{i} + \omega_2 \dot{\bar{i}} = \dot{\omega}_2 \bar{i} + \omega_2 (\bar{\omega} \times \bar{i})$$

In order to evaluate cross products, we must express $\bar{K}$ in terms of its $(\bar{i}, \bar{j}, \bar{k})$ components. For the instant depicted in the sketch,

$$\bar{K} = \cos 53.13° \, \bar{i} + \sin 53.13° \, \bar{k} = 0.6\bar{i} + 0.8\bar{k}$$

Next we convert the given values of ω_1 and ω_2 to radian units, and note that $\dot{\omega}_2 < 0$ because ω_2 is decreasing. Thus

$$\bar{\omega} = 90\left(\frac{2\pi}{60}\right)(0.6\bar{i} + 0.8\bar{k}) + 15{,}000\left(\frac{2\pi}{60}\right)\bar{i}$$
$$= 1576.5\bar{i} + 7.540\bar{k} \text{ rad/s}$$

When we substitute $\bar{\omega}$ and the given information into the earlier expression for $\bar{\alpha}$, we find

$$\bar{\alpha} = -100\bar{i} + 15{,}000\left(\frac{2\pi}{60}\right)(1576.5\bar{i} + 7.540\bar{k}) \times \bar{i}$$
$$= -100\bar{i} + 11{,}844\bar{j} \text{ rad/s}^2$$

The first term in $\bar{\alpha}$ is associated with decreasing the spin rate ω_2. The second term arises because the angular velocity component $\omega_2\bar{i}$ representing the rotation about the axis of the motor is itself rotating about the vertical axis at ω. This causes the tip of $\omega_2\bar{i}$ to move out of the plane of the sketch. The outward direction in this sketch is $+\bar{j}$ because xyz is a right-handed system.

HOMEWORK PROBLEMS

15.14 Slider A moves through the groove at the constant speed $u = 20$ in./s relative to the plate. The angular speed of the plate is constant. Determine the acceleration of slider A in the given position.

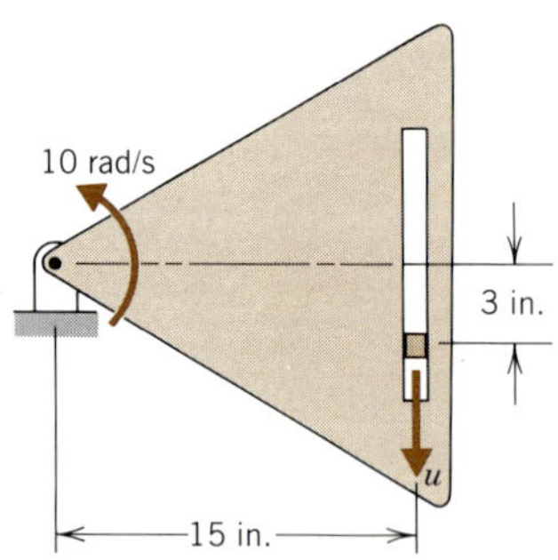

PROB. 15.14

15.15 Rod ABC rotates clockwise at 10 rad/s as the slider moves outward at 1 m/s. Determine the acceleration of the slider when it reaches position C.

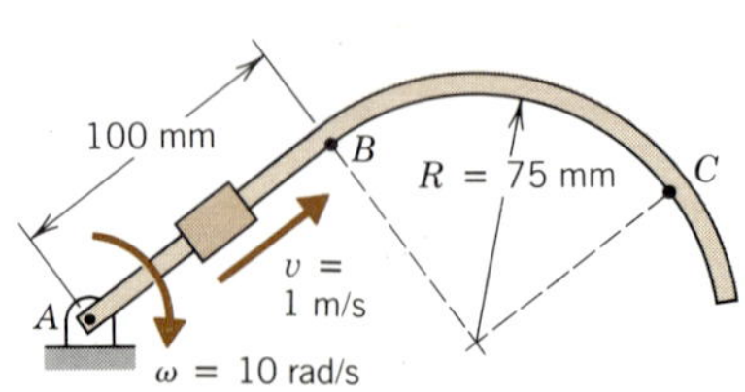

PROB. 15.15

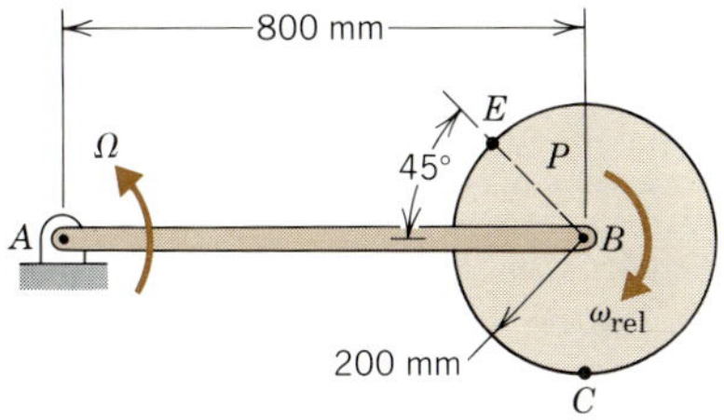

PROB. 15.16

15.16 The disk P rotates relative to arm AB at $\omega_{rel} = 600$ rev/min as the arm rotates at $\Omega = 300$ rev/min. Determine the acceleration of point C.

15.17 Determine the acceleration of point E in Problem 15.16.

15.18 The circular hoop is rotating about the vertical axis at the constant rate $\Omega = 8$ rad/s. The block is moving at a constant speed of 6 ft/s relative to the hoop. Determine the acceleration of the slider at the instant when $\theta = 90°$.

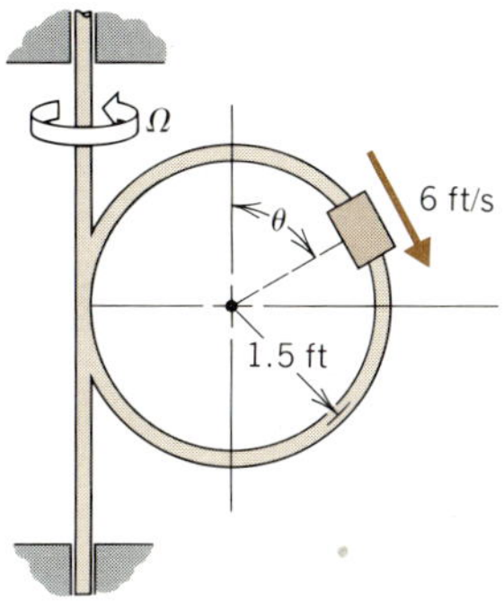

PROB. 15.18

15.19 Solve Problem 15.18 for the position $\theta = 53.13°$ if $\Omega = 8$ rad/s and $\dot{\Omega} = 5$ rad/s^2.

15.20 A centrifuge is being used to study the tolerance of human beings to motion. It consists of a main arm that rotates about the A–A axis at an angular speed ω_1, as shown in the sketch, and a cockpit that may rotate with respect to the main arm at an angular speed ω_2 about the C–C axis. The head of the person being tested is stationary with respect to the cockpit seat. Consider a test where the rotation rates are constant at $\omega_1 = 60$ rad/min, $\omega_2 = 90$ rad/min. Determine the acceleration of the rider's eyes for the position shown.

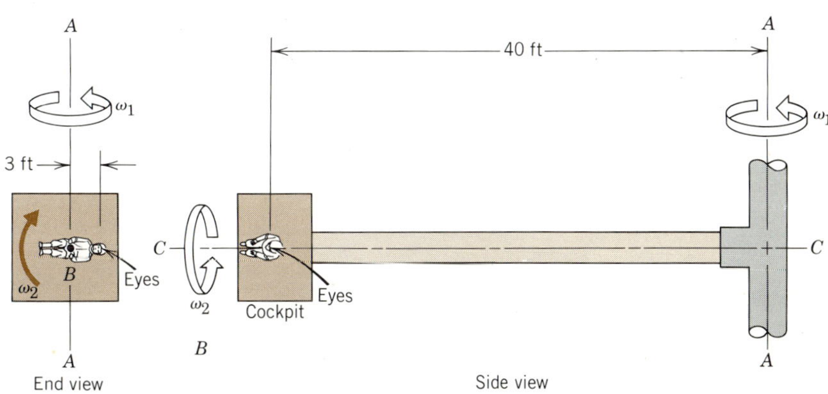

PROB. 15.20

15.21 Bar BC is pinned at end B to bar AB, which rotates about the vertical axis at 240 rev/min. Bar BC rotates at the constant rate $\dot{\theta} = 5$ rad/s relative to AB. Determine the acceleration of end C at the instant when $\dot{\theta} = 90°$.

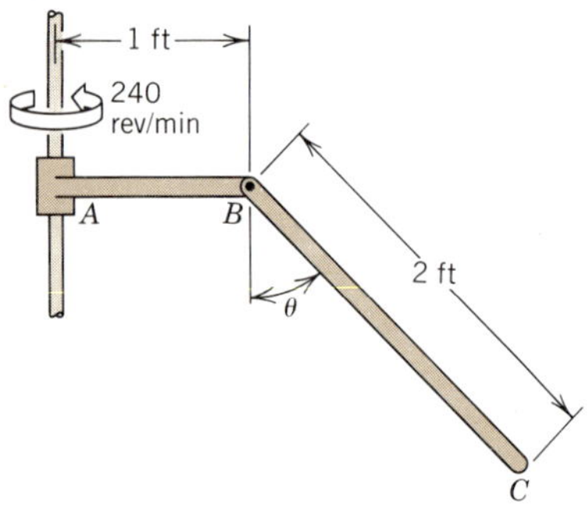

PROB. 15.21

15.22 The disk rotates at a constant rate $\omega_1 = 20$ rad/s with respect to arm AB as the arm rotates about the vertical axis at $\omega_2 = 5$ rad/s. Determine the acceleration of point C.

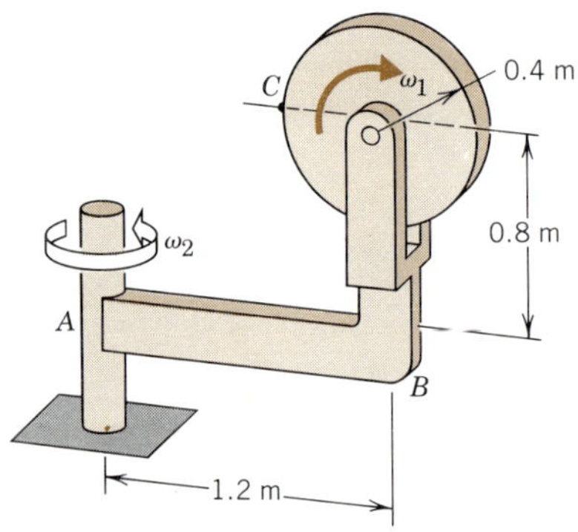

PROB. 15.22

15.23 In Problem 15.22, $\omega_1 = 25$ rad/s, $\dot{\omega}_1 = 50$ rad/s^2, $\omega_2 = 5$ rad/s, and $\dot{\omega}_2 = -10$ rad/s^2. Determine the angular acceleration of the disk.

15.24 The amusement park ride consists of two cars that are fixed to supporting arms, which may rotate about a horizontal pivot at point A. This pivot is attached to the vertical shaft AB, which rotates at angular speed Ω. At the instant of interest $\theta = 30°$, $\dot{\theta} = 0.5$ rad/s, $\ddot{\theta} = -1$ rad/s^2, and Ω is constant at 15 rev/min. In terms of components with respect to the xyz coordinate axes shown, determine the acceleration of point P.

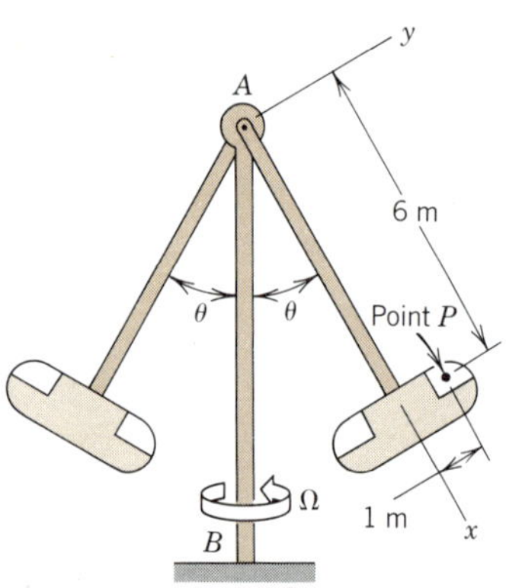

PROB. 15.24

15.25 The radar antenna rotates about the fixed vertical axis at 2 rev/min as the angle θ oscillates at $\theta = \pi/6 + (\pi/3) \sin \pi t$, where θ is in units of radians and t is in seconds. Determine the angular acceleration of the antenna when $t = 0.25$ s.

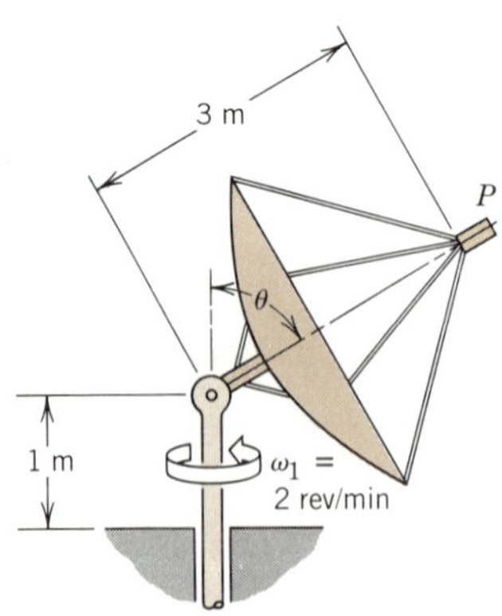

PROB. 15.25

***15.26** Determine the acceleration of the probe P in Problem 15.25 when $t = 0.25$ s.

***15.27** The football is thrown in a wobbling manner. It is spinning about its axis of symmetry at 80 rev/min as this axis rotates about the horizontal axis A–A at 40 rev/min. The angle between the axis of symmetry and axis A–A is constant at 15°. Determine the angular acceleration of the football in the position shown.

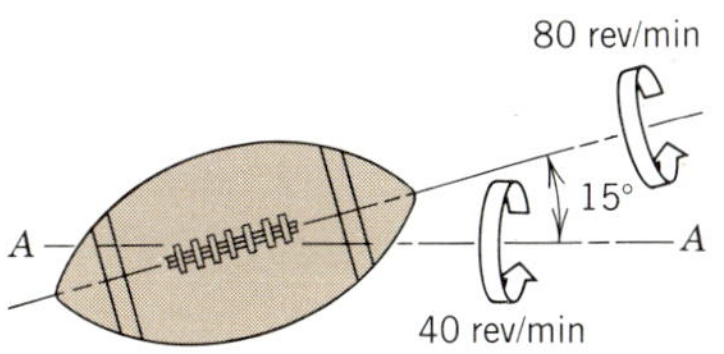

PROB. 15.27

E. THE EARTH AS A MOVING REFERENCE FRAME

The principles we have developed in this and previous chapters have all depended on the existence of a fixed point that could serve as the origin of a reference system. According to Einstein's theory of relativity, no point in the universe is truly fixed. Nevertheless, in most engineering problems concerning phenomena close to the earth's surface, we may consider the center of the earth to be fixed.

In this section we will study the effect of the rotation of the earth about an axis that contains this fixed point of reference and passes through the north and south poles. Specifically, let us determine the effect of the earth's rotation on the motion of the particle P shown in Figure 8.

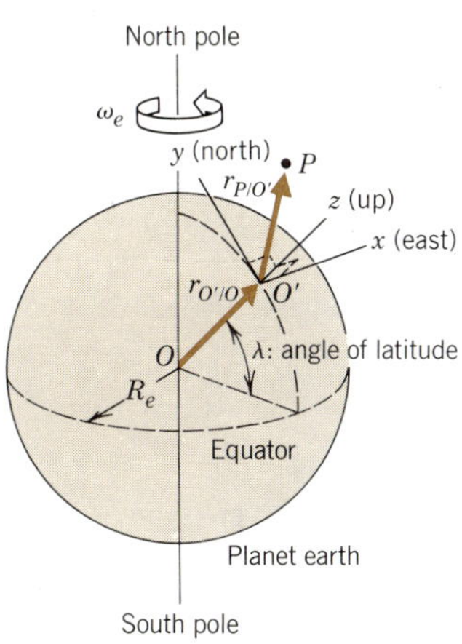

FIGURE 8

We will consider an elementary model for the earth. In this model the earth is regarded as a homogeneous, perfect sphere of radius $R_e = 6370$ km rotating about its fixed polar axis in a counterclockwise direction as seen by a fixed observer looking down at the North Pole. This rate of rotation is 1 rev/day, so that

$$\omega_e = \frac{2\pi \text{ rad}}{23.934 \text{ hr}} \times \frac{1 \text{ hr}}{3600 \text{ s}} = 7.292(10^{-5}) \text{ rad/s}$$

Therefore, our model ignores such effects as deviation from a spherical shape, local variation in the mass density of the earth, and the very slight wobble (nutation) of the polar axis.

Let the resultant force on the particle of Figure 8 be $\bar{F}$. From Newton's second law we know that the absolute acceleration of the particle is

$$\bar{a}_P = \frac{\bar{F}}{m}$$

To describe the acceleration we use an xyz reference frame that is attached to the earth with origin O' at the earth's surface. The relative motion equation (22) then gives

$$\bar{a}_P = \frac{\bar{F}}{m} = \bar{a}_{O'} + (\bar{a}_{P/O'})_{\text{rel}} + \bar{\alpha} \times \bar{r}_{P/O'} + 2\bar{\omega} \times (\bar{v}_{P/O'})_{\text{rel}} + \bar{\omega} \times (\bar{\omega} \times \bar{r}_{P/O'}) \quad (24)$$

The angular velocity $\bar{\omega}$ of the xyz axes is the same as that of the earth. Thus $\bar{\omega}$ is parallel to the polar axis, with components in the y and z directions only

$$\bar{\omega} = \omega_e(\cos \lambda\, \bar{j} + \sin \lambda\, \bar{k}) \quad (25)$$

Further, because we regard both the polar axis and the rate of rotation to be invariant, it follows that

$$\bar{\alpha} = \bar{0}$$

To determine $\bar{a}_{O'}$ we note that point O' is a point on the earth's surface. It therefore follows a circular path in a plane perpendicular to the polar axis. To obtain an explicit expression for $\bar{a}_{O'}$ we use eq. (23) for rigid body motion, which yields

$$\begin{aligned} \bar{a}_{O'} &= \bar{a}_O + \bar{\alpha} \times \bar{r}_{O'/O} + \bar{\omega} \times (\bar{\omega} \times \bar{r}_{O'/O}) \\ &= \bar{\omega} \times (\bar{\omega} \times \bar{r}_{O'/O}) \end{aligned}$$

Compare this result to the last term in equation (24), $\bar{\omega} \times (\bar{\omega} \times \bar{r}_{P/O'})$. For a particle near the earth's surface.

$$|\bar{r}_{P/O'}| \ll |\bar{r}_{O'/O}| = R_e = 6370 \text{ km} \quad (26)$$

Hence the term $\bar{\omega} \times (\bar{\omega} \times \bar{r}_{P/O'})$ is negligible compared to $\bar{a}_{O'}$.

With the foregoing in mind, equation (24) becomes

$$\frac{\bar{F}}{m} = (\bar{a}_{P/O'})_{\text{rel}} + 2\bar{\omega} \times (\bar{v}_{P/O'})_{\text{rel}} + \bar{\omega} \times (\bar{\omega} \times \bar{r}_{O'/O}) \quad (27)$$

where the kinematical quantities for an observer on the earth at point O' are

$$\begin{aligned} \bar{r}_{P/O'} &= x\bar{i} + y\bar{j} + z\bar{k} \\ (\bar{v}_{P/O'})_{\text{rel}} &= \dot{x}\bar{i} + \dot{y}\bar{j} + \dot{z}\bar{k} \quad (28) \\ (\bar{a}_{P/O'})_{\text{rel}} &= \ddot{x}\bar{i} + \ddot{y}\bar{j} + \ddot{z}\bar{k} \end{aligned}$$

In the examples that follow we shall consider two cases where eq. (27) has special significance.

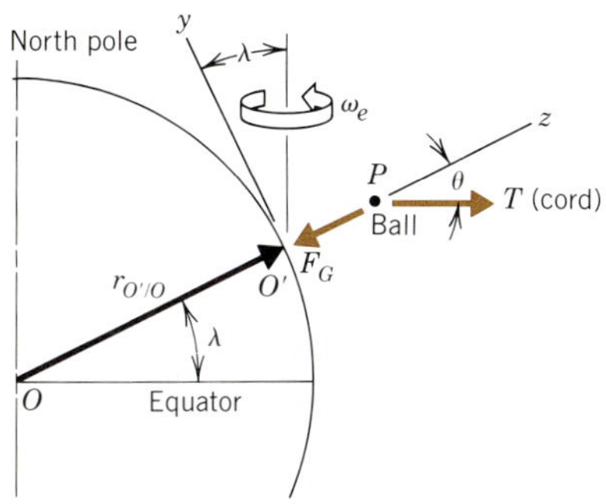

EXAMPLE 7

A ball is suspended near the earth's surface from a cord and is brought to rest relative to the earth. Determine the force in the cord and the direction in which the cord is oriented.

Solution

We first draw a free body diagram of the ball that also shows the xyz frame fixed to the surface of the earth. Note that we have given the cord an arbitrary orientation, rather than assuming it to be oriented along the radial line from the center of the earth. Also, we do not assume that the gravitational force $\bar{F}_G$ is the weight. The reasons for doing this will be obvious in the work that follows.

The magnitude of $\bar{F}_G$ is given by the universal law of gravitation,

$$|\bar{F}_G| = \frac{GM_e m}{R_e^2}$$

Therefore, from the free body diagram we see that the force sum is

$$\bar{F} = \left(T\cos\theta - \frac{GM_e m}{R_e^2}\right)\bar{k} - T\sin\theta j$$

The ball is constrained to have no motion relative to the earth, so eq. (27) reduces to

$$\frac{\bar{F}}{m} = \bar{\omega} \times (\bar{\omega} \times \bar{r}_{O'/O})$$

As we have seen in the general discussion, $\bar{\omega} = \omega_e(\cos\lambda\,\bar{j} + \sin\lambda\,\bar{k})$ and $\bar{r}_{O'/O} = R_e\bar{k}$. The centripetal acceleration is therefore

$$\bar{\omega} \times (\bar{\omega} \times \bar{r}_{O'/O}) = R_e\omega_e^2\cos\lambda(\sin\lambda\,\bar{j} - \cos\lambda\,\bar{k})$$

Notice that the computation shows that the centripetal acceleration is perpendicular to the polar axis, as it should be.

By equating this term to $\bar{F}/m$, we have the two scalar equations

$$\frac{T}{m}\cos\theta - \frac{GM_e}{R_e^2} = -R_e\omega_e^2\cos^2\lambda$$

$$-\frac{T}{m}\sin\theta = R_e\omega_e^2\sin\lambda\cos\lambda$$

The identities $(T\cos\theta)^2 + (T\sin\theta)^2 \equiv T^2$ and $(T\sin\theta)/(T\cos\theta) \equiv \tan\theta$ then yield

$$T = m\left[\left(\frac{GM_e}{R_e^2}\right)^2 - R_e\omega_e^2\left(2\frac{GM_e}{R_e^2} - R_e\omega_e^2\right)\cos^2\lambda\right]^{1/2}$$

$$\tan\theta = -\frac{R_e\omega_e^2\sin\lambda\cos\lambda}{(GM_e/R_e^2) - R_e\omega_e^2\cos^2\lambda}$$

Let us now interpret this solution. In the sense of a statics problem, we would say that the tension T equals the weight mg, where g is the *local* free fall acceleration. We would also say that the cord is oriented vertically. From the results we see that the apparent weight of the body is not solely the force of gravity and that the vertical direction is not truly along the radial line from the center of the earth. However, from the solution for θ we can see that $\theta \equiv 0$ at $\lambda = 0, \pm 90°$, (i.e., at the equator and the north and south poles). Furthermore, because

$$\frac{GM}{R_e^2} = \frac{6.670(10^{-11})(5.976)(10^{24})}{(6.370)^2(10^6)^2} = 9.823 \text{ m/s}^2$$

$$R_e\omega_e^2 = (6.370)(10^6)(7.292)^2(10^{-5})^2 = 3.388(10^{-2}) \text{ m/s}^2$$

we can show that the maximum value $|\theta| = 0.099°$ occurs at $\lambda = \pm 44.90°$. This value of θ is sufficiently small to neglect it in engineering problems. Similarly, if we let $T = mg$, we find that the value of g increases steadily from $g = 9.786 \text{ m/s}^2$ at the equator to $g = 9.823 \text{ m/s}^2$ at the poles. This fluctuation is not significant to the usual accuracy required for engineering problems, and hence it has become common practice to use $g = 9.807 \text{ m/s}^2$ as an average. However, for completeness, we note that there is an *international gravity formula* that accounts for both the rotation of the earth and the deviation of the earth from a spherical shape. It is

$$g = 9.780(1 + 0.005288 \sin^2 \lambda - 0.000006 \sin^2 2\lambda) \text{ m/s}^2$$

EXAMPLE 8

A low-pressure area forms within a fluid, such as an open drain in a tank of water, or a low-pressure atmospheric weather system. Assuming that the fluid is constrained to move solely in a horizontal plane, describe the resulting fluid motion.

Solution

In the previous example we saw that the term $\bar{\omega} \times (\bar{\omega} \times \bar{r}_{O'/O})$ represents a small correction to the weight force that is included in the value of g and the sense of the local vertical. The relative and Coriolis acceleration terms are the primary contributions in eq. (27) for treating motion relative to the earth.

$$\frac{\bar{F}}{m} = (\bar{a}_{P/O'})_{\text{rel}} + 2\bar{\omega} \times (\bar{v}_{P/O'})_{\text{rel}}$$

In this problem the motion is constrained to the horizontal (x, y) plane. We are therefore interested only in the horizontal components of the acceleration, which depends on the vertical (z) component of $\bar{\omega}$, $\omega_z = \omega_e \sin \lambda$. Then the relative acceleration is given by

$$(\bar{a}_{P/O'})_{\text{rel}} = \frac{\bar{F}_{\text{hor}}}{m} - 2(\omega_z \bar{k}) \times (\bar{v}_{P/O'})_{\text{rel}}$$

The term $\bar{F}_{\text{hor}}$ denotes the horizontal component of the resultant force acting on a particle of water. It is due to the pressure difference, so it is directed toward the center of the low-pressure area.

We now introduce some qualitative reasoning. As a particle of fluid starts from rest relative to the earth, the pressure difference causes the particle to accelerate toward the center of the low-pressure area. Therefore, $(\bar{v}_{P/O'})_{\text{rel}}$ initially is also toward the center of the low-pressure area. The sketch shows this initial relative velocity, denoted as v_{rel}, for several fluid particles.

The corresponding relative acceleration $\bar{a}_{\text{rel}}$ may be obtained by subtracting the Coriolis acceleration $2(\omega_z\bar{k}) \times \bar{v}_{\text{rel}}$ from the absolute acceleration $\bar{F}_{\text{hor}}/m$. This construction is shown in the sketch, for motion in the Northern Hemisphere ($\omega_z\bar{k}$ is counterclockwise in this hemisphere.)

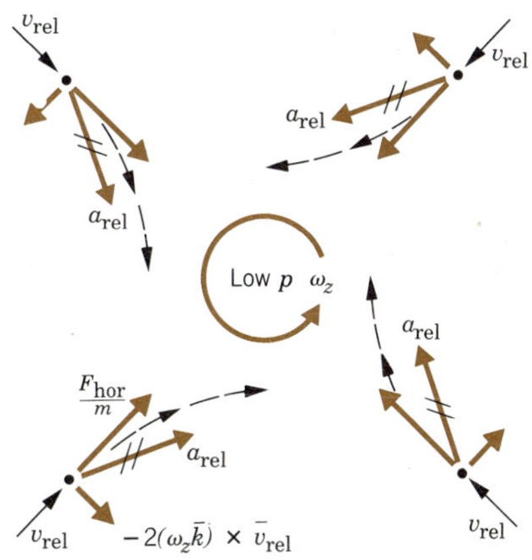

We see from the sketch that the relative acceleration is to the right of the absolute acceleration, when viewed looking inward. This acceleration produces velocities, and hence tangent vectors, which also deviate to the right of the radial line. The ensuing path of a fluid particle is depicted by the series of dashed arrows.

We see that the Coriolis acceleration effect tends to induce a counterclockwise motion relative to the earth's surface, when viewed looking downward. This effect reverses in the Southern Hemisphere, because ω_z is negative there. We conclude that the fluid motion will consist of an inward movement, plus a counterclockwise rotation in the Northern Hemisphere or a clockwise rotation in the Southern Hemisphere.

Knowledge of this effect is essential to the understanding of meteorological and oceanographic phenomena, such as hurricanes, whirlpools, and ocean currents. However, the preceding discussion does not provide a full description because we have not considered the forces that are generated within the fluid by its motion.

Additional Remarks

The technique we have employed in this section may be generalized. It often happens that we wish to determine the relative motion when the absolute motion is known. Suppose that $\bar{v}_P$ and $\bar{a}_P$, as well as $\bar{\omega}$ and $\bar{\alpha}$, are known. Then $(\bar{v}_{P/O'})_{\text{rel}}$ may be found using eq. (16). This result, in conjunction with the other parameters, then makes it possible to determine $(\bar{a}_{P/O'})_{\text{rel}}$ from eq. (22).

HOMEWORK PROBLEMS

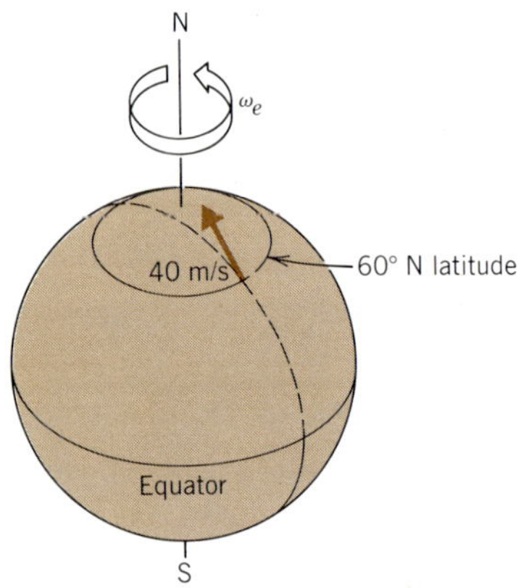

PROB. 15.28

15.28 A 1500 kg automobile is traveling north at 40 m/s along a straight road. It is in the Northern Hemisphere, at 60° latitude. Determine the reaction force exerted by the ground on the vehicle in terms of north, east, and vertical components.

15.29 The automobile in Problem 15.28 is moving west. Determine the corresponding reaction forces.

15.30 A ball bearing rolls on a frictionless horizontal plane located at latitude 45° north. It was given an initial velocity of 50 ft/s east. Determine the acceleration of the ball bearing with respect to the earth at the instant it was released.

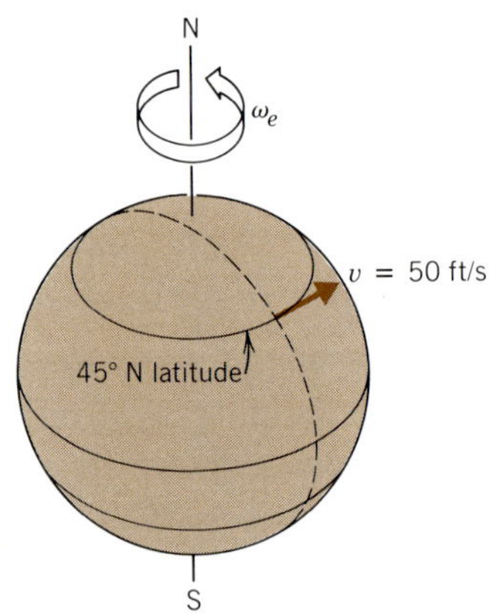

PROB. 15.30

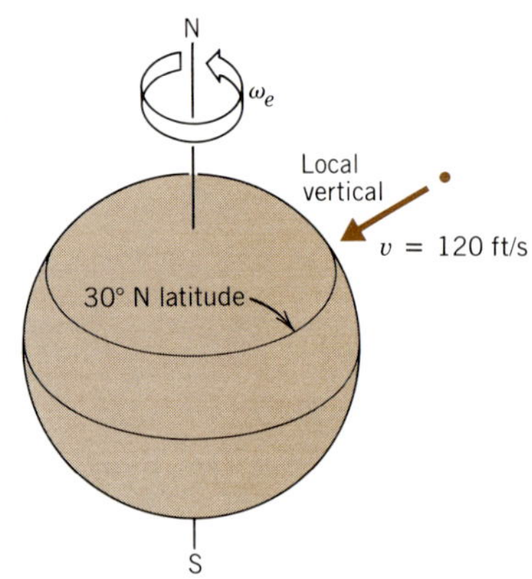

PROB. 15.31

15.31 An object at 75° north latitude is falling freely. Its velocity with respect to the earth is 120 ft/s vertically downward. Air resistance is negligible. Determine the acceleration of the object with respect to the earth.

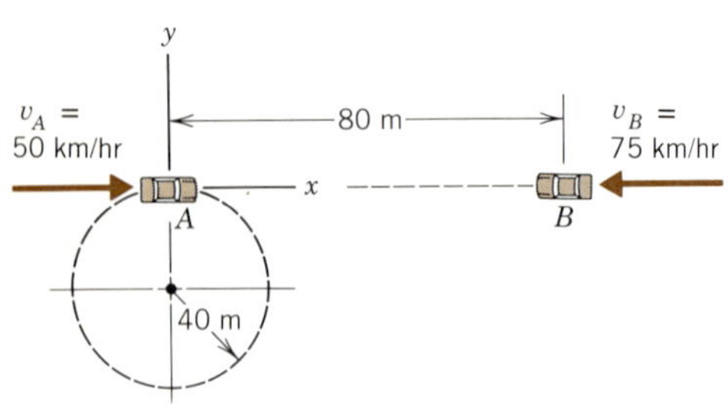

PROB. 15.32

15.32 Automobile A is executing a 40 m radius turn at a constant speed of 50 km/hr. Automobile B has a constant speed of 75 km/hr along a straight path. The reference frame is attached to automobile A. For the instant where the vehicles are in the positions shown, determine the velocity of automobile B with respect to automobile A as seen from the xyz axes.

*15.33 In Problem 15.32, determine the acceleration of automobile B with respect to automobile A as seen from the xyz axes.

15.34 A rocket is launched to the south. Its initial velocity is 2000 ft/s relative to the earth at an angle of elevation of 30° above the horizontal. The location is at latitude 45° north. Neglecting air resistance, determine the initial acceleration of the rocket with respect to the earth.

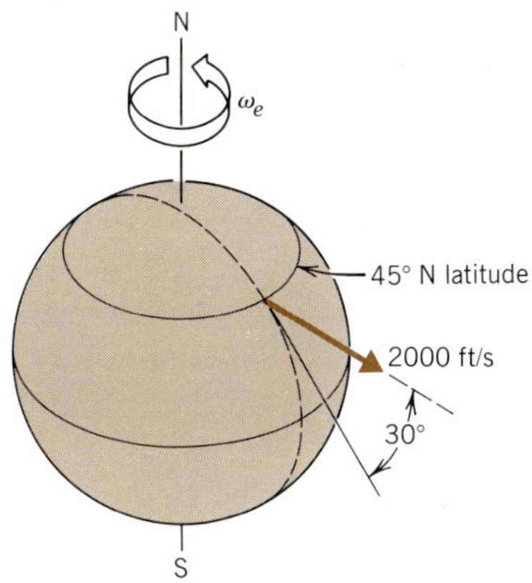

PROB. 15.34

*15.35 A block of mass m, released from rest, slides down a vertical pole without opposition from friction. Derive expressions for the horizontal components of the contact force exerted by the pole on the block in terms of the time t that the block has been falling and the angle of latitude λ.

F. PROBLEM SOLVING

Following the approach of previous chapters, we now present a systematic procedure for the determination of expressions for linear and angular velocity and acceleration. These expressions may be the only results required in a problem, or they may represent the kinematical input needed for the study of the kinetics of the system.

1. Sketch the system, showing the axes of rotation and, symbolically, the angular speeds about these axes. This sketch should also show all significant points, including points whose motion is desired and all points that are constrained.
2. Show your choice for the moving reference frame xyz in the sketch. The general criteria for this step are summarized as follows:
   a. Let the origin O' of xyz be a point in the system that is executing a simple motion, such as a point following a known planar path.
   b. Let the xyz axes have as many rotational velocities as you can easily comprehend, in order to expedite the description of $(\bar{v}_{P/O'})_{\text{rel}}$

and $(\bar{a}_{P/O'})_{\text{rel}}$. In the situation where it is desired to study the angular motion of a specific body, attach xyz to that body.

c. Choose the instantaneous orientation of the xyz frame to give a simple description for the geometry of the system, in particular, the description of the axes of rotation and of the relative position vectors.

When some aspect of the rotation of the system is described with respect to a fixed observer, select a fixed reference frame XYZ such that its origin O is a convenient fixed point and one of the axes is parallel to the fixed axis of rotation. (For an example of a situation where the fixed XYZ system is not required, see Illustrative Problem 3.)

3. Following the method described on page 760, obtain a general description for the angular velocity $\bar{\omega}$ of the xyz frame. If the problem involves acceleration, differentiate this expression to obtain a general expression for $\bar{\alpha}\,(=\dot{\bar{\omega}})$. The instantaneous expressions for $\bar{\omega}$ and $\bar{\alpha}$ are then obtained by replacing the components of the $(\bar{I}, \bar{J}, \bar{K})$ unit vectors by the $(\bar{i}, \bar{j}, \bar{k})$ representation of the xyz frame. In problems where constraint conditions must be satisfied, it will prove useful to leave all velocity and acceleration rates in algebraic (symbolic) form.

Velocity Problem

4. If any angular speeds were not given, relate the velocities of all constrained points to the rates of rotation using the relative velocity equation (16). The rates of rotation are obtained by satisfying the constraints on the velocities of these points. Then substitute all rotation rates into the expressions for $\bar{\omega}$ and $\bar{\alpha}$.

5. If the problem requires the velocity of any unconstrained points, use eq. (16) for these determinations.

Acceleration Problem

6 and 7. Using the relative acceleration eq. (22), repeat the procedures of steps 4 and 5 in terms of the acceleration variables and the restrictions on the accelerations of the constrained points.

ILLUSTRATIVE PROBLEM 1

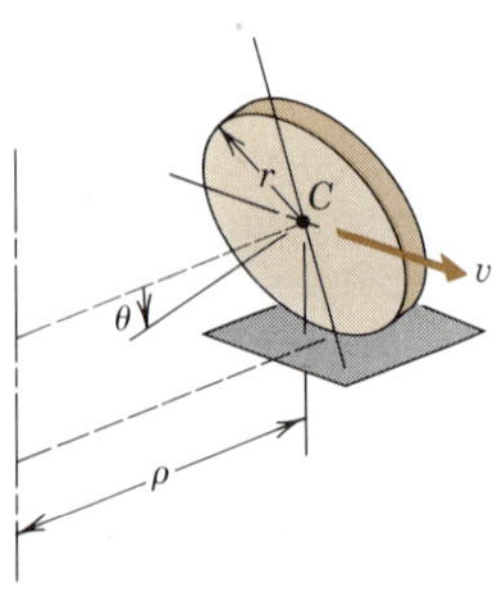

The disk is rolling without slipping over the ground. The centroid C has a constant speed v as it follows a circular path of radius ρ. The disk is banked, such that its axis is always at an angle θ from the horizontal, as shown. Determine the angular velocity and angular acceleration of the disk.

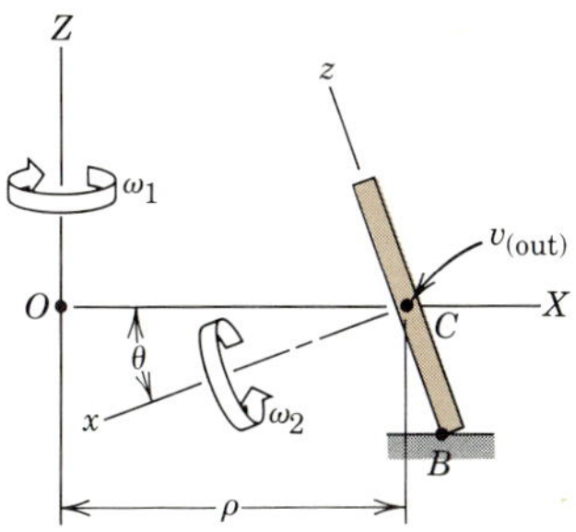

Solution

Step 1 The best view for depicting the rotation of the disk is a front view in which the disk is seen on edge. The sketch indicates that the motion of the disk consists of two rotations. One is about a vertical axis at angular speed ω_1. The other is about an axis perpendicular to the plane of the disk and intersecting point C; the angular speed for this rotation is ω_2. One of the points denoted in the sketch is the point of contact B. This point is constrained to have no velocity because of the no slip condition.

Step 2 Select the centroid C as the origin for xyz because its path and speed are given. Attach xyz to the disk in order that $\bar{\omega}$ and $\bar{\alpha}$ describe the rotation of that body. The x axis is selected to coincide with an axis of the disk in order to describe the rotation $\bar{\omega}_2$. The only significant feature of the stationary XYZ system is that the Z axis is selected to coincide with the (fixed) vertical axis of rotation, thus facilitating the description of $\bar{\omega}_1$.

Step 3 The overall rotation is a combination of the angular rates ω_1 about the Z axis and ω_2 about the x axis. For the assumed sense of each rotation, $\bar{\omega}$ is

$$\bar{\omega} = \omega_1(-\bar{K}) + \omega_2(-\bar{i}): \quad \text{always}$$

Differentiating this gives

$$\bar{\alpha} = \dot{\bar{\omega}} = -\dot{\omega}_1\bar{K} - \dot{\omega}_2\bar{i} - \omega_2\dot{\bar{i}} = -\dot{\omega}_1\bar{K} - \dot{\omega}_2\bar{i} - \omega_2(\bar{\omega} \times \bar{i})$$

We might anticipate that ω_1 and ω_2 are constants, and therefore set $\dot{\omega}_1 = \dot{\omega}_2 = 0$. Instead we will derive this fact.

There are some constrained points whose motion we will need to consider. We therefore retain ω_1 and ω_2 as parameters. The expression for the xyz components of $\bar{K}$ are found from the sketch to be

$$\bar{K} = -\sin\theta\,\bar{i} + \cos\theta\,\bar{k}$$

This leads to expressions for the instantaneous $\bar{\omega}$ and $\bar{\alpha}$.

$$\bar{\omega} = -\omega_1(-\sin\theta\,\bar{i} + \cos\theta\,\bar{k}) - \omega_2\bar{i} = (\omega_1\sin\theta - \omega_2)\bar{i} - \omega_1\cos\theta\,\bar{k}: \quad \text{instantaneously}$$

$$\begin{aligned}\bar{\alpha} &= -\dot{\omega}_1(-\sin\theta\,\bar{i} + \cos\theta\,\bar{k}) - \dot{\omega}_2\bar{i} - \omega_2[(\omega_1\sin\theta - \omega_2)\bar{i} - \omega_1\cos\theta\,\bar{k}] \times \bar{i} \\ &= (\dot{\omega}_1\sin\theta - \dot{\omega}_2)\bar{i} + \omega_1\omega_2\cos\theta\,\bar{j} - \dot{\omega}_1\cos\theta\,\bar{k}: \quad \text{instantaneously}\end{aligned}$$

Step 4 The centroid C and the point of contact B are both constrained. For circular motion of point C about the vertical axis at the speed v, we have

$$\omega_1 = \frac{v}{r_{C/O}} = \frac{v}{\rho}.$$

Also we know that $\bar{v}_B = \bar{0}$ because there is no slipping relative to the ground. However, points B and C are both points on the disk. Thus

$$\bar{v}_B = \bar{0} = \bar{v}_C + \bar{\omega} \times \bar{r}_{B/C}$$

At the instant depicted in the sketch, the y axis is horizontal and the z axis coincides with the line through points B and C. Then at this instant

$$\bar{v}_C = v\bar{j}, \qquad \bar{r}_{B/C} = -r\bar{k}$$

Substituting these expressions and $\bar{\omega}$ into the equation for $\bar{v}_B$ then gives

$$\bar{0} = v\bar{j} + [(\omega_1 \sin\theta - \omega_2)\bar{i} - \omega_1 \cos\theta\,\bar{k}] \times (-r\bar{k})$$
$$= [v + (\omega_1 \sin\theta - \omega_2)r]\bar{j}$$
$$\bar{j} \text{ component:} \quad v + (\omega_1 \sin\theta - \omega_2)r = 0$$

We have already found that $\omega_1 = v/\rho$. Solving the foregoing relation for ω_2 therefore yields

$$\omega_2 = \omega_1 \sin\theta + \frac{v}{r} = \frac{v}{\rho}\sin\theta + \frac{v}{r}$$

We now note that the expressions for ω_1 and ω_2 in terms of v are general results because they were derived for an arbitrary instant. Then because v is constant, we find

$$\dot{\omega}_1 = \frac{\dot{v}}{\rho} = 0, \qquad \dot{\omega}_2 = \frac{\dot{v}}{\rho}\sin\theta + \frac{\dot{v}}{r} = 0$$

The angular velocity and angular acceleration corresponding to these rotation rates are

$$\bar{\omega} = \left[\frac{v}{\rho}\sin\theta - \left(\frac{v}{\rho}\sin\theta + \frac{v}{r}\right)\right]\bar{i} - \frac{v}{\rho}\cos\theta\,\bar{k} = -\frac{v}{r}\bar{i} - \frac{v}{\rho}\cos\theta\,\bar{k}$$

$$\bar{\alpha} = \left(\frac{v}{\rho}\right)\left(\frac{v}{\rho}\sin\theta + \frac{v}{r}\right)\cos\theta\,\bar{j} = \left(\frac{v^2}{\rho^2}\sin\theta\cos\theta + \frac{v^2}{\rho r}\cos\theta\right)\bar{j}$$

Additional Remarks

There is a tendency for students to try to solve problems such as this by using *planar motion* concepts, specifically, by saying that $v = \omega_2 r$ based on considering point B to be the instant center for no slipping. This gives $\omega_2 = v/r$, which is only one term in the derived expression for ω_2. Such reasoning is erroneous because the rotation about the axis perpendicular to the disk is due, in part, to the rotation ω_1 about the vertical axis. The intuitive reasoning could have been correct only if the x axis were oriented horizontally.

ILLUSTRATIVE PROBLEM 2

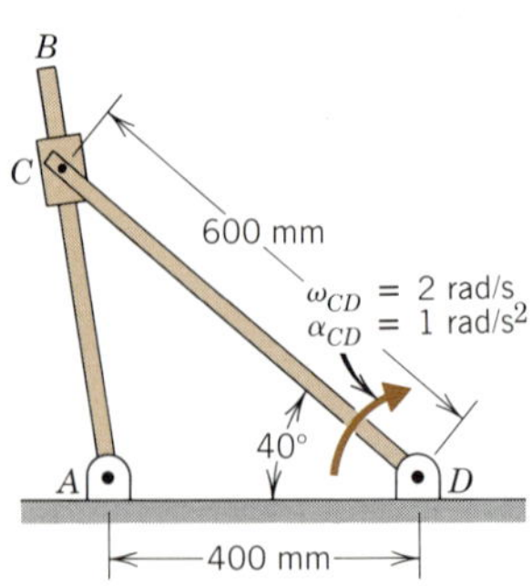

The linkage arm CD rotates clockwise about the pivot D with an angular velocity of 2 rad/s and an angular acceleration of 1 rad/s^2. The collar at C is pinned to rod CD and slides along rod AB. For the position shown, determine the angular velocity and angular acceleration of rod AB.

Solution

This problem is similar to the planar linkage problems of Chapter 13, where all members were pinned together. The new feature of this problem is that the collar allows for motion of point C *relative* to bar AB. We will employ a moving reference frame to describe this motion. On the other hand, the collar is pinned to rod CD, so the rigid body motion formulas will be applied to describe the motion of this rod.

Step 1

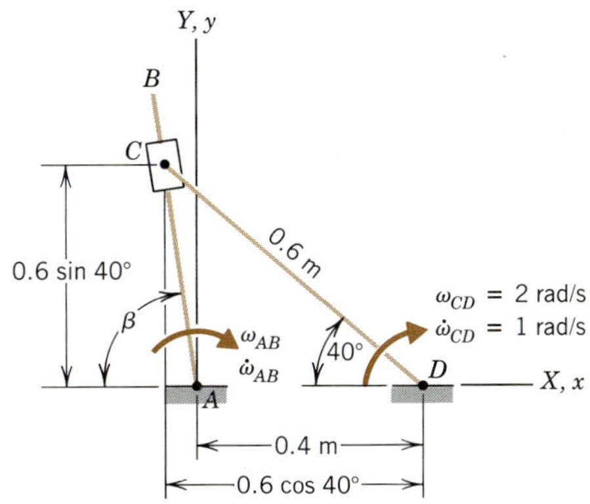

Step 2 The fixed reference frame is selected such that X and Y are situated in the plane of motion. This is the only significant feature of XYZ. In order to describe the motion relative to rod AB, we attach the moving xyz frame to this rod with its origin also at the fixed point A. At the instant of interest the x axis is selected to coincide with line AD; this expedites the writing of the position vectors. Both the fixed and moving reference frames are shown in the sketch in step 1.

Step 3 Because the xyz frame is attached to rod AB, and because this is a planar motion problem, the rotation of this reference frame is given by

$$\bar{\omega} = \omega_{AB}(-\bar{K}) = -\omega_{AB}\bar{k}$$

$$\bar{\alpha} = \dot{\omega}_{AB}(-\bar{K}) = -\alpha_{AB}\bar{k}$$

The values ω_{AB} and α_{AB} are the quantities we must determine.

Step 4 The constrained points in this problem are the fixed points A and D, and point C on the collar. We can relate points A and C using the relative motion equation (16). Thus

$$\bar{v}_C = \bar{v}_A + (\bar{v}_{C/A})_{\text{rel}} + \bar{\omega} \times \bar{r}_{C/A}$$

However, point C is also a point on rod CD. Its velocity is therefore given by the rigid body equation (17), which becomes

$$\bar{v}_C = \bar{v}_D + \bar{\omega}_{CD} \times \bar{r}_{C/D}$$

Points A and D are fixed, so $\bar{v}_A = \bar{v}_D = \bar{0}$. Also, we know from the given information that

$$\bar{\omega}_{CD} = 2(-\bar{k}) \text{ rad/s}$$

The two expressions for $\bar{v}_C$ will be equated, but we must first describe the individual terms in these expressions. Toward this end, in our sketch we describe the instantaneous position of collar C, from which we find

$$\bar{r}_{C/A} = -(0.60 \cos 40^\circ - 0.40)\bar{i} + 0.60 \sin 40^\circ \, \bar{j} = -0.0596\bar{i} + 0.3857\bar{j} \text{ m}$$

$$\bar{r}_{C/D} = -0.60 \cos 40^\circ \, \bar{i} + 0.60 \sin 40^\circ \, \bar{j} = -0.4596\bar{i} + 0.3857\bar{j} \text{ m}$$

$$\beta = \tan^{-1}\left(\frac{0.6 \sin 40^\circ}{0.6 \cos 40^\circ - 0.4}\right) = 81.21^\circ$$

The angle β is needed to describe $(\bar{v}_{C/A})_{\text{rel}}$. Because the collar is free to slide along rod AB, this relative velocity must be parallel to the rod. The magnitude of this term, denoted by v_{rel}, is an unknown quantity. Assuming the relative velocity to be outward then gives

$$\begin{aligned}(\bar{v}_{C/A})_{\text{rel}} &= v_{\text{rel}}(-\cos \beta \, \bar{i} + \sin \beta \, \bar{j}) \\ &= v_{\text{rel}}(-0.1528\bar{i} + 0.9883\bar{j})\end{aligned}$$

The result of substituting the appropriate terms into the first expression for $\bar{v}_C$ is

$$\begin{aligned}\bar{v}_C &= \bar{0} + v_{\text{rel}}(-0.1528\bar{i} + 0.9883\bar{j}) + (-\omega_{AB}\bar{k}) \times (-0.0596\bar{i} + 0.3857\bar{j}) \\ &= (-0.1528v_{\text{rel}} + 0.3857\omega_{AB})i + (0.9883v_{\text{rel}} + 0.0596\omega_{AB})\bar{j} \text{ m/s}\end{aligned}$$

Similarly, the second expression for $\bar{v}_C$ gives

$$\begin{aligned}\bar{v}_C &= \bar{0} + (-2\bar{k}) \times (-0.4596\bar{i} + 0.3857\bar{j}) \\ &= 0.7714\bar{i} + 0.9192\bar{j} \text{ m/s}\end{aligned}$$

By equating the individual components of $\bar{v}_C$, we obtain the following two simultaneous equations.

$$\bar{i} \text{ component:} \quad (v_C)_x = -0.1528v_{\text{rel}} + 0.3857\omega_{AB} = 0.7714$$

$$\bar{j} \text{ component:} \quad (v_C)_y = 0.9883v_{\text{rel}} + 0.0596\omega_{AB} = 0.9192$$

The solution of these equation is

$$v_{\text{rel}} = 0.7906 \text{ m/s}, \qquad \omega_{AB} = 2.313 \text{ rad/s}$$

$$\bar{\omega}_{AB} = \omega_{AB}(-\bar{k}) = -2.313\bar{k} \text{ rad/s}$$

Step 5 Not applicable.

Step 6 We repeat the procedures of step 4 using the acceleration variables. Thus,

$$\begin{aligned}\bar{a}_C &= \bar{a}_A + (\bar{a}_{C/A})_{\text{rel}} + 2\bar{\omega} \times (\bar{v}_{C/A})_{\text{rel}} + \bar{\alpha} \times \bar{r}_{C/A} + \bar{\omega} \times (\bar{\omega} \times \bar{r}_{C/A}) \\ &= \bar{a}_D + \bar{\alpha}_{CD} \times \bar{r}_{C/D} + \bar{\omega}_{C/D} \times (\bar{\omega}_{C/D} \times \bar{r}_{C/D})\end{aligned}$$

Note that the value of v_{rel} in step 4 is needed to construct $(\bar{v}_{C/A})_{\text{rel}}$.

$$(\bar{v}_{C/A})_{\text{rel}} = 0.7906(-0.1528\bar{i} + 0.9883\bar{j}) = -0.1208\bar{i} + 0.7814\bar{j} \text{ m/s}$$

Points A and D are fixed, so $\bar{a}_A = \bar{a}_D = \bar{0}$. From the given information we obtain

$$\bar{\alpha}_{CD} = 1(-\bar{k}) \text{ rad/s}^2$$

The other terms, with the exception of the relative acceleration, are defined in steps 3 and 4. For the relative acceleration we note that because the xyz frame is attached to rod AB, $(\bar{a}_{C/A})_{\text{rel}}$ must account for the sliding motion of the collar along the rod. Thus, letting $|(\bar{a}_{C/A})_{\text{rel}}| \equiv \dot{v}_{\text{rel}}$, we have

$$(\bar{a}_{C/A})_{\text{rel}} = \dot{v}_{\text{rel}}(-\cos\beta\,\bar{i} + \sin\beta\,\bar{j})$$
$$= \dot{v}_{\text{rel}}(-0.1528\bar{i} + 0.9883\bar{j})$$

When we substitute the individual terms into the first expression for $\bar{a}_C$, we obtain

$$\begin{aligned}
\bar{a}_C &= \bar{0} + \dot{v}_{\text{rel}}(-0.1528\bar{i} + 0.9883\bar{j}) \\
&\quad + 2(-2.313\bar{k}) \times (-0.1208\bar{i} + 0.7814\bar{j}) + (-\alpha_{AB}\bar{k}) \times (-0.0596\bar{i} + 0.3857\bar{j}) \\
&\quad + (-2.313\bar{k}) \times [(-2.313\bar{k}) \times (-0.0596\bar{i} + 0.3857\bar{j})] \\
&= (-0.1528\dot{v}_{\text{rel}} + 3.615 + 0.3857\alpha_{AB} + 0.3189)\bar{i} \\
&\quad + (0.9883\dot{v}_{\text{rel}} + 0.5588 + 0.0596\alpha_{AB} - 2.073)\bar{j}\ \text{m/s}^2
\end{aligned}$$

The second expression for $\bar{a}_C$ gives

$$\begin{aligned}
\bar{a}_C &= \bar{0} + (-\bar{k}) \times (-0.4596\bar{i} + 0.3857\bar{j}) \\
&\quad + (-2\bar{k}) \times [(-2\bar{k}) \times (-0.4596\bar{i} + 0.3857\bar{j})] \\
&= 2.224\bar{i} - 1.083\bar{j}\ \text{m/s}^2
\end{aligned}$$

The two scalar equations obtained by equating the two expressions for $\bar{a}_C$ are

$$\bar{i}\ \text{component:}\quad (a_C)_x = -0.1528\dot{v}_{\text{rel}} + 3.934 + 0.3857\dot{\omega}_{AB} = 2.224$$

$$\bar{j}\ \text{component:}\quad (a_C)_y = 0.9883\dot{v}_{\text{rel}} - 1.514 + 0.0596\dot{\omega}_{AB} = -1.083$$

The simultaneous solution of these equations is

$$\dot{v}_{\text{rel}} = 0.6871\ \text{m/s}^2, \qquad \dot{\omega}_{AB} = -4.161\ \text{rad/s}^2$$
$$\bar{\alpha}_{AB} = \dot{\omega}_{AB}(-\bar{k}) = 4.161\bar{k}\ \text{rad/s}^2$$

In closing, let us note the physical significance of the method of the solution. The given rotation values for rod *CD* defined the absolute velocity and acceleration of collar *C*. The relative motion equations ensure that collar *C* and rod *AB* have the same velocity and acceleration components perpendicular to line *AB*. The velocity and acceleration components parallel to line *AB* serve to define the relative velocity and relative acceleration components.

ILLUSTRATIVE PROBLEM 3

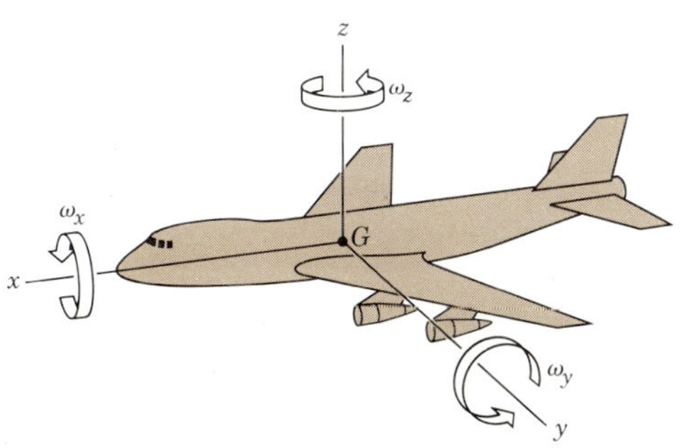

The rotations of an airplane about a set of xyz axes attached to the airplane are termed the roll rate ω_x, the pitching rate ω_y, and the yaw rate ω_z. A passenger on a jumbo jet 30 m behind and 2 m above the center of mass *G* of the airplane is walking aft at the constant rate of 1 m/s. Determine the difference between the acceleration that this passenger experiences and the acceleration that the center of mass of the airplane undergoes, if the roll rate is 0.5 rad/s, the pitching rate is -0.3 rad/s, and the yaw rate is 0.1 rad/s. All rotation rates are constant.

Solution

Step 1 The diagram accompanying the statement of the problem will suffice for the solution.

Step 2 We shall use the moving xyz frame given in the problem, which is attached to the aircraft. Because all rotation rates are referred to the axes of this coordinate system, and because we are not concerned with the absolute acceleration of the origin G, we do not need a fixed reference frame XYZ.

Step 3 The roll, pitch, and yaw rates are with respect to the x, y, and z axes, respectively. Hence the general descriptions of $\bar{\alpha}$ and $\bar{\omega}$ are

$$\bar{\omega} = \omega_x\bar{i} + \omega_y\bar{j} + \omega_z\bar{k}: \quad \text{always}$$

$$\begin{aligned}\bar{\alpha} &= \dot{\omega}_x\bar{i} + \omega_x\dot{\bar{i}} + \dot{\omega}_y\bar{j} + \omega_y\dot{\bar{j}} + \dot{\omega}_z\bar{k} + \omega_z\dot{\bar{k}} \\ &= \dot{\omega}_x\bar{i} + \dot{\omega}_y\bar{j} + \dot{\omega}_z\bar{k} + \omega_x(\bar{\omega} \times \bar{i}) + \omega_y(\bar{\omega} \times \bar{j}) + \omega_z(\bar{\omega} \times \bar{k}) \\ &= \dot{\omega}_x\bar{i} + \dot{\omega}_y\bar{j} + \dot{\omega}_z\bar{k} + \bar{\omega} \times (\omega_x\bar{i} + \omega_y\bar{j} + \omega_z\bar{k}) \\ &= \dot{\omega}_x\bar{i} + \dot{\omega}_y\bar{j} + \dot{\omega}_z\bar{k} + \bar{\omega} \times \bar{\omega} \\ &\equiv \dot{\omega}_x\bar{i} + \dot{\omega}_y\bar{j} + \dot{\omega}_z\bar{k}: \quad \text{always}\end{aligned}$$

This result for angular acceleration is interesting, because it states that the components of angular acceleration of the airplane are the derivatives of the components of the angular velocity. The fact that this is generally true when we fix the moving reference system to a body will have great significance for our treatment of the equations of motion of a rigid body in the next chapter.

Notice that the above expressions for $\bar{\omega}$ and $\bar{\alpha}$ already give the components with respect to xyz. Thus finding the instantaneous $\bar{\omega}$ and $\bar{\alpha}$ requires only substitution of the given rates of rotation.

$$\bar{\omega} = 0.5\bar{i} - 0.3\bar{j} + 0.1\bar{k} \text{ rad/s}$$

The rates of rotation are constant, so $\bar{\alpha} = \bar{0}$.

Steps 4–6 These steps are not applicable, for there are no constraint conditions to satisfy and we do not seek the absolute velocity of any points.

Step 7 We are interested in the value of $(\bar{a}_P - \bar{a}_G) \equiv a_{P/G}$, where point P is the position of the passenger. This is obtained from equation (22), namely,

$$\bar{a}_P - \bar{a}_G = (\bar{a}_{P/G})_{\text{rel}} + 2\bar{\omega} \times (\bar{v}_{P/G})_{\text{rel}} + \bar{\alpha} \times \bar{r}_{P/G} + \bar{\omega} \times (\bar{\omega} \times \bar{r}_{P/G})$$

From the information given in the problem and the choice of coordinate axes in the sketch, we may write,

$$\bar{r}_{P/G} = -30\bar{i} + 2\bar{k} \text{ m}$$

Also, from the given information we know that the passenger has a constant speed of 1 m/s in the negative x direction relative to the airplane. Thus

$$(\bar{v}_{P/G})_{\text{rel}} = -1.0\bar{i} \text{ m/s}, \qquad (\bar{a}_{P/G})_{\text{rel}} = \bar{0}$$

The solution is then found by substituting the values of $\bar{\omega}$ and $\bar{\alpha}$ in step 3, and the relative motion variables determined above into the acceleration equation. This step gives

$$\begin{aligned}\bar{a}_P - \bar{a}_G &= \bar{0} + 2(0.5\bar{i} - 0.3\bar{j} + 0.1\bar{k}) \times (-1.0\bar{i}) \\ &\quad + \bar{0} + (0.5\bar{i} - 0.3\bar{j} + 0.1\bar{k}) \times [(0.5\bar{i} - 0.3\bar{j} + 0.1\bar{k}) \times (-30\bar{i} + 2\bar{k})] \\ &= 3.10\bar{i} + 4.24\bar{j} - 2.78\bar{k} \text{ m/s}^2\end{aligned}$$

The magnitude of this result is $0.61g$, which exceeds the value desired in normal operation of passenger aircraft. However, atmospheric disturbances can cause rotation rates comparable to those considered in this problem

---

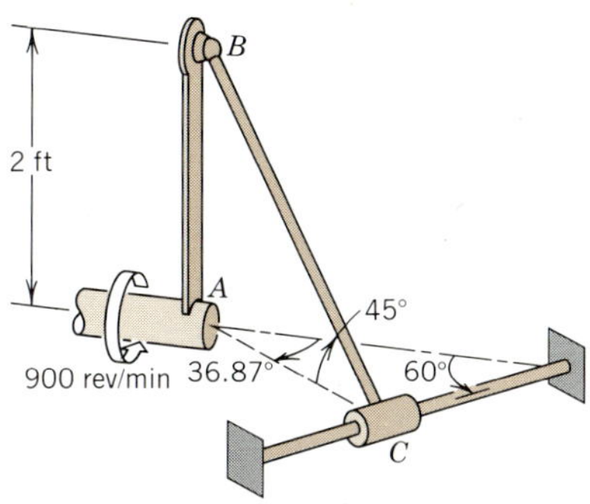

ILLUSTRATIVE PROBLEM 4

Crank *AB* rotates about the horizontal shaft at 900 rpm. Bar *BC* is connected to the crank and to collar *C* by ball joints. Determine the angular velocity of bar *BC* and the velocity of collar *C* in the position shown.

Solution

This system is the three-dimensional analog of the planar linkages we studied in Chapter 13. The similarities in the analyses will be apparent as we proceed.

Step 1

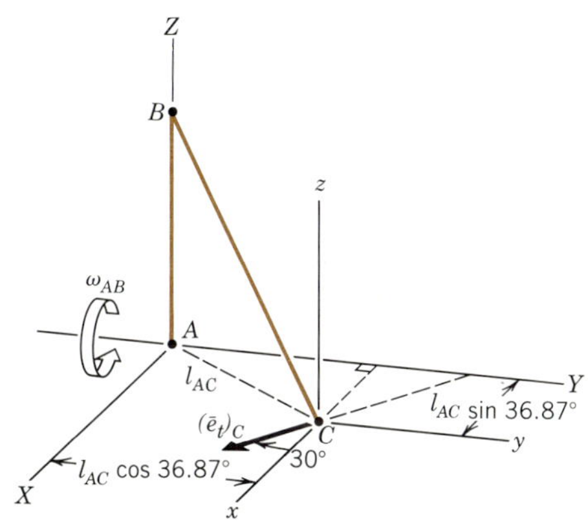

Step 2 Bar *AB* is rotating about a fixed axis, so we will describe its angular velocity by using the fixed *XYZ* reference frame. The moving reference frame *xyz* is attached to bar *BC* in order to describe the three-dimensional motion of that body. Point *C* is selected as the origin of *xyz* because the path of that point is a straight line. The orientation of *xyz* shown in the sketch is suitable for describing the motion of point *C*, as well as the position vectors.

Step 3 Only velocities are to be determined. We therefore omit finding angular accelerations. Bar *AB* is rotating about the *X* axis so

$$\bar{\omega}_{AB} = \omega_{AB}\bar{J} = 900\left(\frac{2\pi}{60}\right)\bar{j} = 94.25\bar{j} \text{ rad/s}$$

The nature of the rotation of bar *BC* is not obvious, so we let the angular velocity of this body be an unknown general vector

$$\bar{\omega} = \bar{\omega}_{BC} = \omega_x\bar{i} + \omega_y\bar{j} + \omega_z\bar{k}$$

Step 4 Point B is constrained to follow both bars. Thus

$$\bar{v}_B = \bar{v}_A + \bar{\omega}_{AB} \times \bar{r}_{B/A}$$
$$= \bar{v}_C + \bar{\omega}_{BC} \times \bar{r}_{B/C}$$

Point A is fixed ($\bar{v}_A = \bar{0}$) and point C must follow a guide that lies in the xy plane. Thus

$$\bar{v}_C = v_C(\bar{e}_t)_C = v_C(\cos 30^\circ \bar{i} - \sin 30^\circ \bar{j})$$
$$= v_C(0.866\bar{i} - 0.50\bar{j})$$

The position vector $\bar{r}_{B/C}$ involves the length l_{AC}, which we find by noting that BAC is a 45° right triangle. Thus

$$l_{AC} = \frac{l_{AB}}{\tan 45^\circ} = l_{AB} = 2 \text{ ft}$$

$$\bar{r}_{B/C} = -l_{AC} \sin 36.87^\circ \bar{i} - l_{AC} \cos 36.87^\circ \bar{j} + 2\bar{k}$$
$$= -1.20\bar{i} - 1.60\bar{j} + 2\bar{k} \text{ feet}$$

$$\bar{r}_{B/A} = 2\bar{k} \text{ feet}$$

Substituting all vectors into the equations for $\bar{v}_B$ then yields

$$\bar{v}_B = \bar{0} + (94.25\bar{j}) \times 2\bar{k} = 188.50\bar{i} \text{ ft/s}$$
$$= v_C(0.866\bar{i} - 0.50\bar{j}) + (\omega_x\bar{i} + \omega_y\bar{j} + \omega_z\bar{k}) \times (-1.2\bar{i} - 1.6\bar{j} + 2\bar{k})$$
$$= v_C(0.866\bar{i} - 0.50\bar{j}) + (2\omega_y + 1.6\omega_z)\bar{i} + (-1.2\omega_z - 2\omega_x)\bar{j} + (-1.6\omega_x + 1.2\omega_y)\bar{k}$$

Matching components in these alternative expressions for $\bar{v}_B$ yields

$$\bar{i} \text{ component:} \quad (v_B)_x = 188.50 = 0.866v_C + 2\omega_y + 1.6\omega_z$$

$$\bar{j} \text{ component:} \quad (v_B)_y = 0 = -0.5v_C - 1.2\omega_z - 2\omega_x$$

$$\bar{k} \text{ component:} \quad (v_B)_z = 0 = -1.6\omega_x + 1.2\omega_y$$

Clearly the foregoing are only three equations for the four unknowns ω_x, ω_y, ω_z, and v_C. This shortage results from the fact that bar BC could be spinning about its own axis without any effect on the velocity of points. Nothing is causing such a spin. We will set it to zero by requiring that $\bar{\omega}_{BC}$ have no component parallel to line BC. Thus

$$\bar{\omega}_{BC} \cdot \bar{e}_{B/C} = (\omega_x\bar{i} + \omega_y\bar{j} + \omega_z\bar{k}) \cdot \frac{\bar{r}_{B/C}}{|\bar{r}_{B/C}|}$$

$$= \frac{1}{|\bar{r}_{B/C}|}(-1.2\omega_x - 1.6\omega_y + 2\omega_z) = 0$$

This is the fourth equation, which completes the formulation. The simultaneous solution of these equations is

$$\omega_x = -118.2 \qquad \omega_y = -157.6 \qquad \omega_z = -197.0 \qquad v_C = 946$$

$$\bar{\omega}_{BC} = -118.2\bar{i} - 157.6\bar{j} - 197.0\bar{k} \text{ rad/s}$$

$$\bar{v}_C = 946(0.866\bar{i} - 0.50\bar{j}) = 819\bar{i} - 473\bar{j} \text{ ft/s}$$

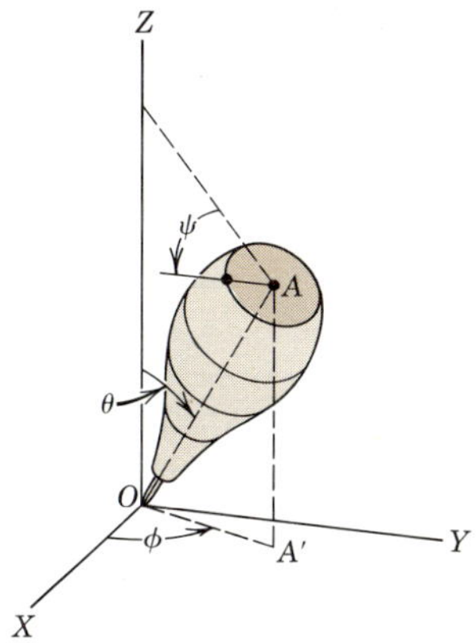

ILLUSTRATIVE PROBLEM 5

In order to define the orientation of a spinning top, a set of *Eulerian angles* is defined. The *precession angle* ϕ describes the angle of rotation about the fixed Z axis of the centerline OA of the top, and the *nutation angle* θ describes the angle between the Z axis and the centerline. Then the *spin angle* ψ defines the rotation of the top about its centerline. Treating these angles as arbitrary functions of time, obtain expressions for the angular velocity and angular acceleration of the top.

Solution

As we shall see, the new feature of this problem is that the three angles give rise to rotations about three axes, which will cause us to employ two moving reference systems.

Step 1 To simplify the sketch we draw only the centerline OA and the line AB, which are sufficient for illustrating the Eulerian angles.

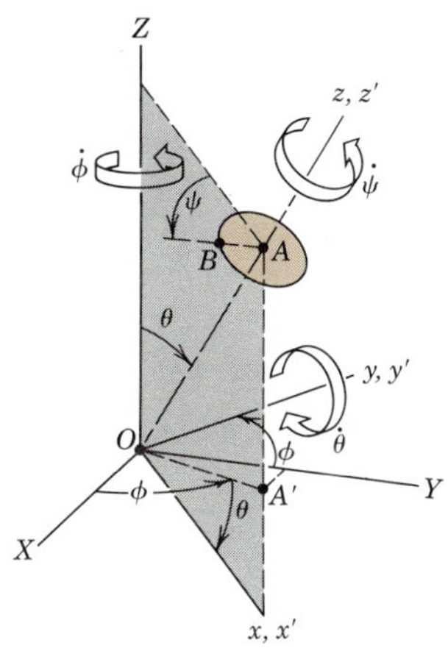

Step 2 The fixed reference frame XYZ is given, and the moving xyz system must be attached to the top in order that the xyz axes and the top have the same angular motion. Point O is a convenient origin for xyz. A useful orientation is to let the z axis coincide with the centerline OA, and to select the x axis such that the Z axis coincides with the xz plane. This plane, which is shaded in the sketch, also contains the line OA' formed by projecting point A onto the XY plane.

Step 3 The orientation of xyz changes if any of the three angles θ, ϕ, or ψ change. Thus xyz has possible rotations about all three axes. The precession angle ϕ describes the orientation of the shaded plane with respect to the XZ plane, so $\dot{\phi}$ represents a precessional rotation about the Z axis. The spin angle ψ defines the rotation of line AB about the centerline OA so $\dot{\psi}$ corresponds to a spinning rotation about the z axis.

Now consider the nutation angle θ. As shown in the sketch, θ describes the orientation of the centerline *in the shaded plane.* Thus $\dot{\theta}$ represents a rotation about an axis perpendicular to the shaded plane. At the instant shown in the sketch, the y axis is such a perpendicular. However, any change in the spin angle ψ will cause the y axis to rotate away from the perpendicularity we desire. In other words, the nutational rotation $\dot{\theta}$ is *always* about an axis y' that does not undergo the spinning motion. For this reason we introduce a second moving reference frame $x'y'z'$; it undergoes only the precessional and nutational motions. As shown in the sketch, the xyz and $x'y'z'$ axes are coincident at the instant under consideration. However, only the z and z' axes always coincide. According to the right-hand rule, the nutation rate $\dot{\theta}$ is always about the positive y' axis.

Let us denote the angular velocity of xyz as $\bar{\omega}$ and the angular velocity of $x'y'z'$ as $\bar{\Omega}$. The general expressions are

$$\bar{\omega} = \dot{\phi}\bar{K} + \dot{\theta}\bar{j}' + \dot{\psi}\bar{k}: \quad \text{always}$$

$$\bar{\Omega} = \dot{\phi}\bar{K} + \dot{\theta}\bar{j}': \quad \text{always}$$

Because xyz is fixed to the top, $\bar{\omega}$ is the angular velocity of that body. The angular acceleration of the top is then

$$\bar{\alpha} = \dot{\bar{\omega}} = \ddot{\phi}\bar{K} + \ddot{\theta}\bar{j}' + \dot{\theta}\,\dot{\bar{j}}' + \ddot{\psi}\bar{k} + \dot{\psi}\dot{\bar{k}}$$

The unit vector $\bar{j}'$ is fixed to the $x'y'z'$ system, which is rotating at the angular speed $\bar{\Omega}$, whereas the unit vector $\bar{k}$ is fixed to the xyz system, which is rotating at the angular speed $\bar{\omega}$. Hence

$$\bar{\alpha} = \ddot{\phi}\bar{K} + \ddot{\theta}\bar{j}' + \dot{\theta}(\bar{\Omega} \times \bar{j}') + \ddot{\psi}\bar{k} + \dot{\psi}(\bar{\omega} \times \bar{k})$$

For the last step we obtain the instantaneous angular velocity and angular acceleration vectors by expressing the unit vectors in terms of their xyz components. The sketch in step 1 shows that the Z axis coincides with the xy plane, with θ being the angle backward from z to Z. Thus

$$\bar{K} = -\sin\theta\,\bar{i} + \cos\theta\,\bar{k}$$

Also, the y and y' axes are coincident at this instant, so $\bar{j}' = \bar{j}$. The general expressions therefore become

$$\begin{aligned}
\bar{\omega} &= \dot{\phi}(-\sin\theta\,\bar{i} + \cos\theta\,\bar{k}) + \dot{\theta}\bar{j} + \dot{\psi}\bar{k} \\
&= -\dot{\phi}\sin\theta\,\bar{i} + \dot{\theta}\bar{j} + (\dot{\phi}\cos\theta + \dot{\psi})\bar{k}
\end{aligned}$$

$$\begin{aligned}
\bar{\Omega} &= -\dot{\phi}\sin\theta\,\bar{i} + \dot{\theta}\bar{j} + \dot{\phi}\cos\theta\,\bar{k} \\
\bar{\alpha} &= \ddot{\phi}(-\sin\theta\,\bar{i} + \cos\theta\,\bar{k}) + \ddot{\theta}\bar{j} + \dot{\theta}(-\dot{\phi}\sin\theta\,\bar{i} + \dot{\theta}\bar{j} + \dot{\phi}\cos\theta\,\bar{k}) \times \bar{j} \\
&\quad + \ddot{\psi}\bar{k} + \dot{\psi}[-\dot{\phi}\sin\theta\,\bar{i} + \dot{\theta}\bar{j} + (\dot{\phi}\cos\theta + \dot{\psi})\bar{k}] \times \bar{k} \\
&= (-\ddot{\phi}\sin\theta - \dot{\phi}\dot{\theta}\cos\theta + \dot{\theta}\dot{\psi})\bar{i} + (\ddot{\theta} + \dot{\phi}\dot{\psi}\sin\theta)\bar{j} \\
&\quad + (\ddot{\psi} + \ddot{\phi}\cos\theta - \dot{\phi}\dot{\theta}\sin\theta)\bar{k}
\end{aligned}$$

The foregoing expressions are always valid, for the expressions were evaluated at a general instant (the angles are arbitrary). Equations such as these are widely employed for advanced studies of the motion of rigid bodies, especially for computerized studies.

HOMEWORK PROBLEMS

15.36 The disk is welded to the shaft such that the centerline of the disk forms a 30° angle with the shaft. The shaft is rotating at 3000 rev/min and slowing at 50 rev/min/s. Determine the velocity and acceleration of point B on the disk.

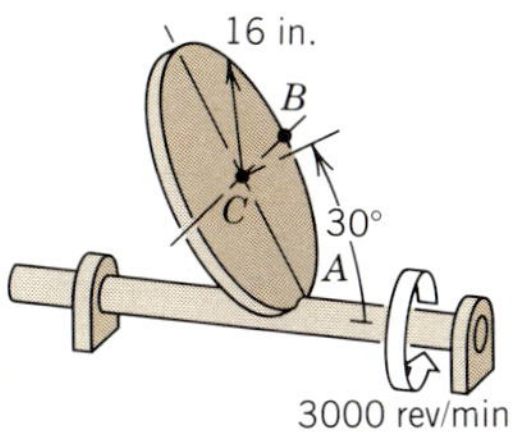

PROB. 15.36

15.37 The curved bar rotates about the inclined axis AB at the constant rate of 10 rad/s. When collar C is in the position shown, it is moving at $u = 5$ m/s with respect to the curved bar. Determine the velocity of the collar in terms of its components with respect to the xyz coordinate system shown.

15.38 The relative speed u in Problem 15.37 is increasing at $\dot{u} = 2\ \text{m/s}^2$ when the collar is at the given position. Determine the corresponding xyz components of the acceleration of the collar.

15.39 An insect crawls along the exterior surface of the cylindrical hoop shown at a constant rate of 8 in./s as the hoop rotates about the fixed horizontal axis at the constant rate of 0.25 rad/s. Determine the xyz components of the acceleration of the insect when it is at the position shown.

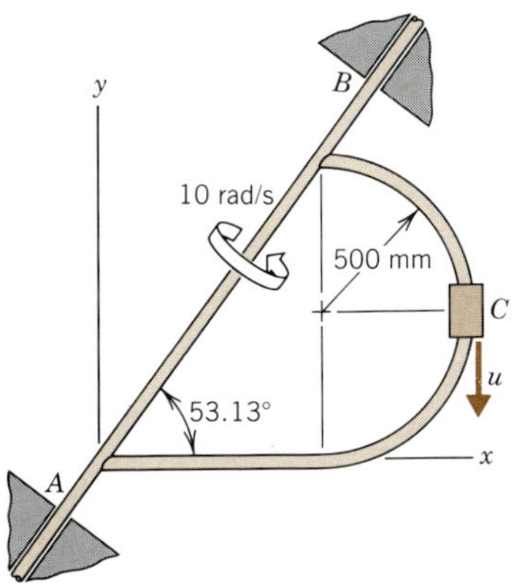

PROB. 15.37

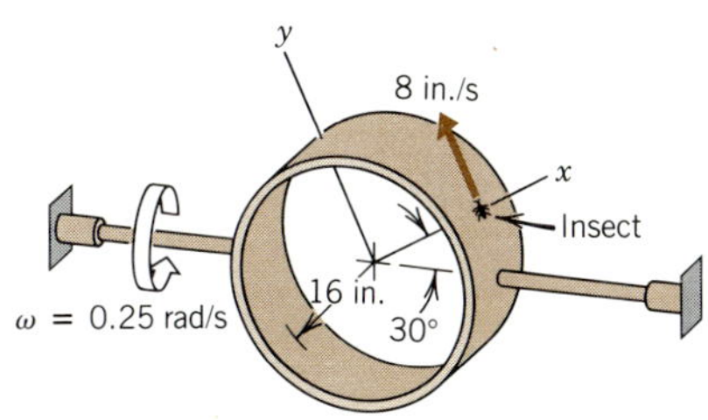

PROB. 15.39

15.40 A person who is stationed at point C on the perimeter of the rotating platform is observing a baseball in free flight. At the instant shown, the ball is at point B and is moving horizontally with a speed of 40 ft/s. The angular velocity of the platform is 2 rad/s clockwise. For this instant, determine (a) the velocity of the baseball as seen by this person, (b) the acceleration of the baseball as seen by this person. Draw a sketch of the results.

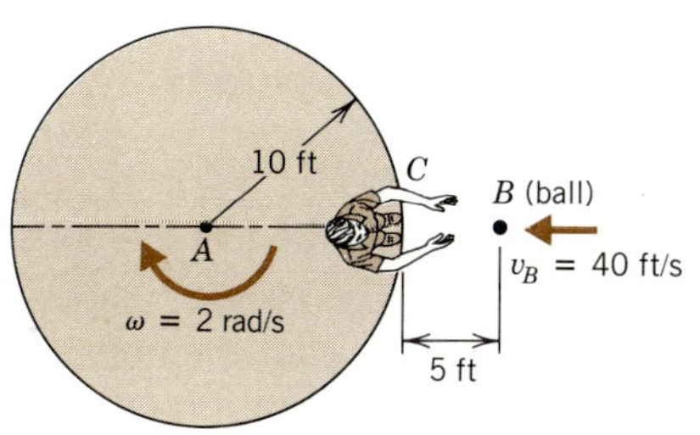

PROB. 15.40

15.41 Two automobiles are traveling in opposite directions along a curve in a divided highway. Both vehicles are traveling at a constant speed of 100 km/hr. For the position where $\theta = 0$, determine (a) the absolute velocity and acceleration of car B relative to car A, (b) the velocity and acceleration of car B relative to a passenger in car A who is unaware of the motion of that vehicle.

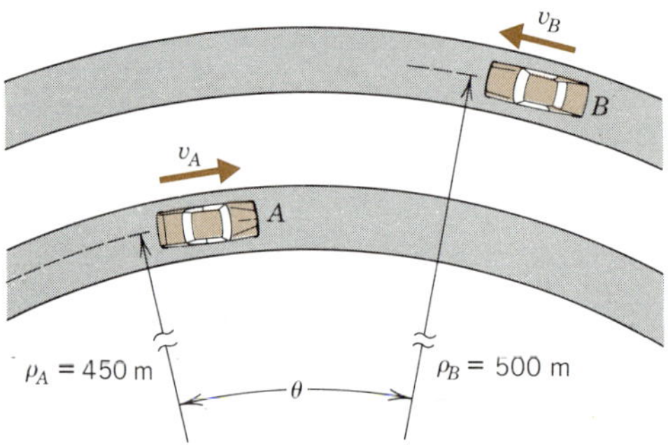

PROB. 15.41

***15.42** Solve Problem 15.41 for the position where $\theta = 15°$.

15.43 Water flows through the garden sprinkler at a speed of 15 m/s relative to an arm. The water that emerges at the nozzle is observed to move radially outward. Determine the angular velocity $\bar{\omega}$ of the sprinkler arms and the velocity of a particle of water when it emerges from a nozzle.

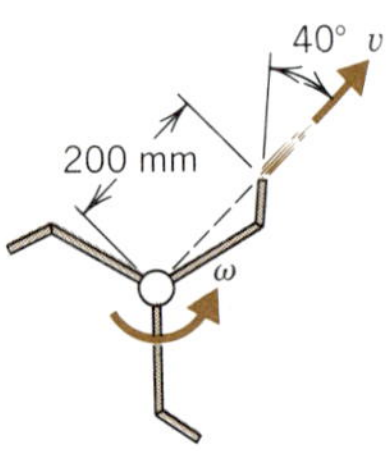

PROB. 15.43

15.44 The 200 gram bullet has a constant speed *relative to the rifle barrel* of 2700 km/hr when it is in the position shown. In order to follow the target, the rifle is being rotated about the vertical axis at the constant rate $\Omega = 0.05$ rad/s. The angle of inclination is constant at $\theta = 60°$. Determine (a) the acceleration of the bullet, and (b) the y and z components of the force exerted by the rifle barrel on the bullet.

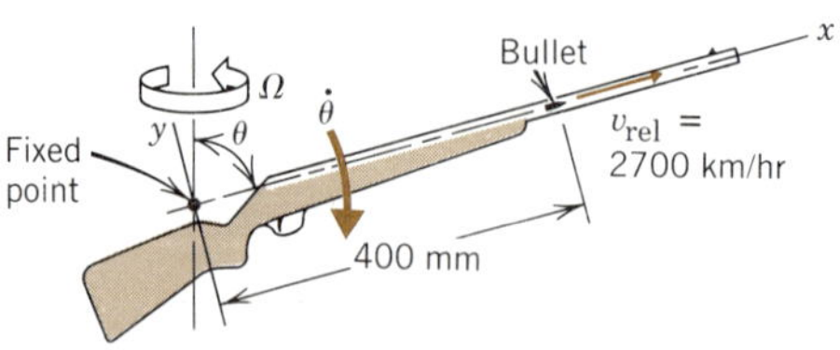

PROB. 15.44

15.45 Solve Problem 15.44 for the situation where the angle of inclination is increasing at the constant rate $\dot{\theta} = 0.2$ rad/s when $\theta = 75°$.

15.46 The disk has a constant angular velocity of 80 rad/s about its axis as the fork that supports the disk rotates about the horizontal axis at 10 rad/s. Determine the velocity and acceleration (a) of point B, (b) of point C.

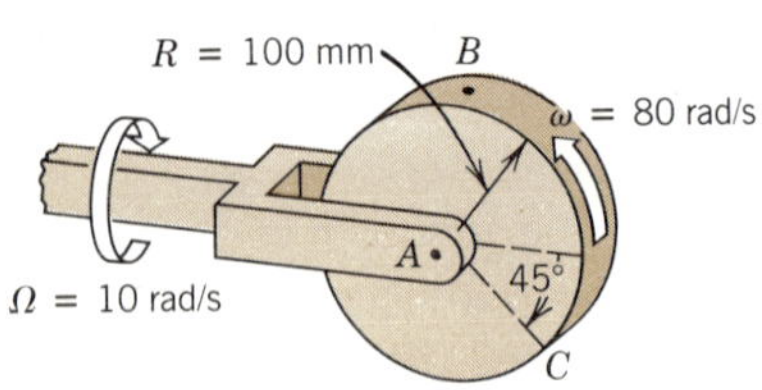

PROB. 15.46

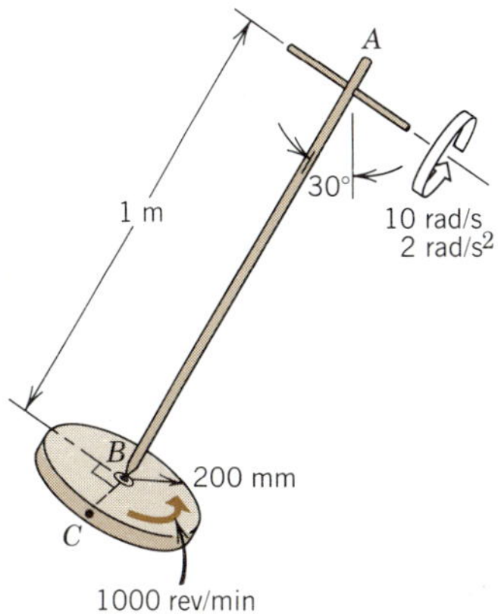

PROB. 15.47

15.47 The disk is rotating about its axis at a constant rate of 1000 rev/min as arm AB pivots about a horizontal axis. In the position shown, arm AB is inclined 30° from the vertical and has an angular speed of 10 rad/s, which is increasing at the rate of 2 rad/s^2. Determine the angular velocity and angular acceleration of the disk.

15.48 Determine the velocity and acceleration of point C of the disk in Problem 15.47.

15.49 Rod AC is attached to collar A and passes through pivoted collar B. Collar A moves upward with a constant velocity of 30 ft/s. For the instant when the bar is in the position shown, determine (a) the angular velocity of rod AC, and (b) the velocity of the point on the rod in contact with the pivoted collar B.

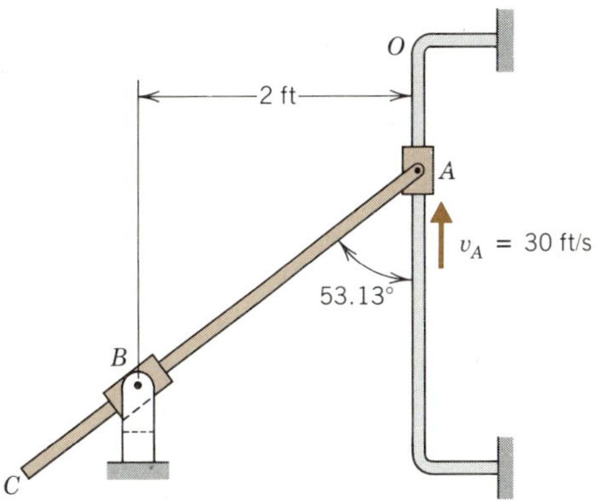

PROB. 15.49

15.50 Determine the angular acceleration of the bar in Problem 15.49 for the given instant.

15.51 Collar B is pinned to bar AB and slides over bar CD. In the position shown, bar CD is rotating clockwise at 50 rad/s. Determine the angular velocity of bar AB at this instant.

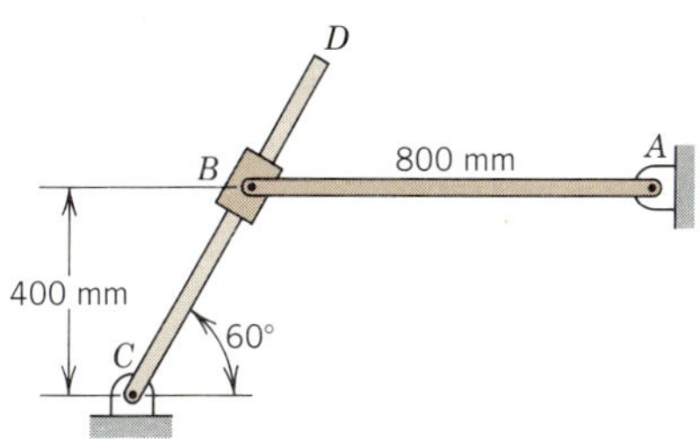

PROB. 15.51

15.52 Gear B rotates freely about its shaft, which has a constant angular speed of 10 rad/s about the vertical axis. Gear A is stationary. Determine (a) the angular velocity and angular acceleration of gear B, and (b) the velocity and acceleration of point C.

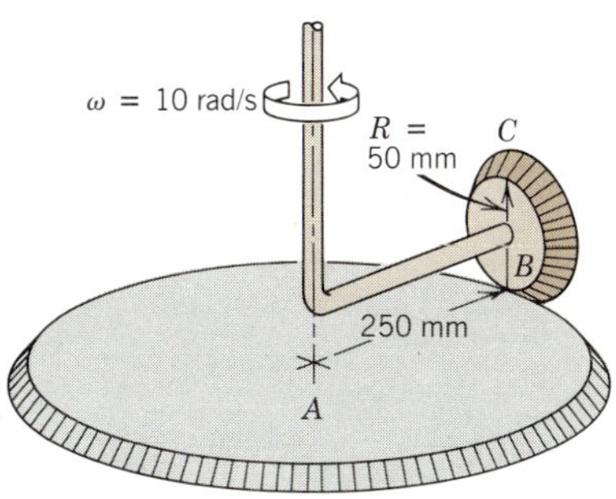

PROB. 15.52

***15.53** Solve Problem 15.52 if the rate of rotation ω_1 is increasing at 20 rad/s^2.

15.54 In the position shown, the aircraft is executing a diving turn to the right by pitching its nose down at $\omega_p = 0.5$ rad/s and rolling at $\omega_r = 0.3$ rad/s. The roll rate is decreasing at 0.6 rad/s^2. Determine the velocity and acceleration of point A relative to the center of mass G.

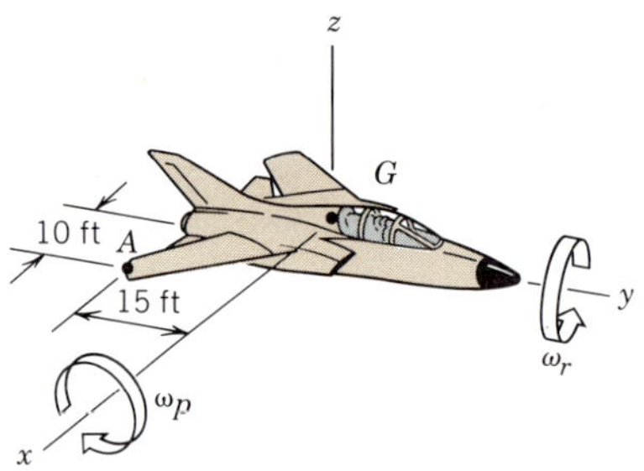

PROB. 15.54

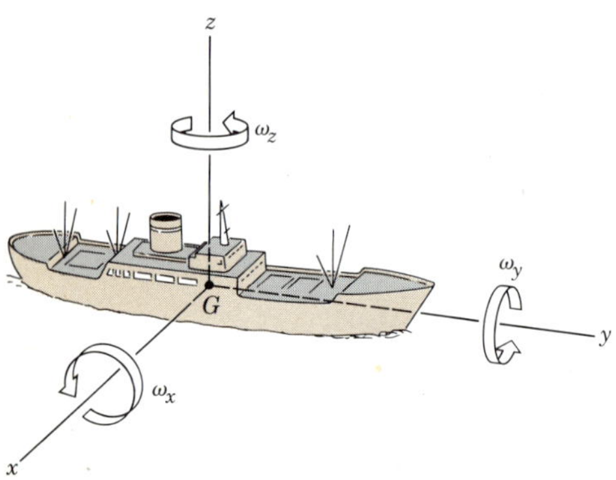

PROB. 15.55

15.55 The center of mass G of a large ship has a constant forward velocity of 6 m/s. Due to the action of the ocean waves, however, it is pitching, rolling, and yawing about the x, y, and z axes, respectively, which are attached to the ship. At this instant these axes have the horizontal and vertical orientations they would have in a calm sea but $\omega_x = 0.08$ rad/s, $\omega_y = 0.04$ rad/s, $\omega_z = 0.02$ rad/s. Also, all rotation rates have their maximum value at this instant. Determine the absolute velocity and acceleration of the bow of the ship, which is 120 m forward and 15 m above point G.

15.56 In Problem 15.55, an elevator on the ship is 40 m behind, 10 m above, and 5 m to the right (looking forward) of point G. The elevator is moving upward at the constant rate of 4 m/s relative to the ship. Determine the absolute velocity and acceleration of the elevator.

15.57 and 15.58 Sliders A and B are connected to bar AB by ball joints. Slider B has a speed of 3 m/s to the right. Determine the velocity of slider A in the position shown.

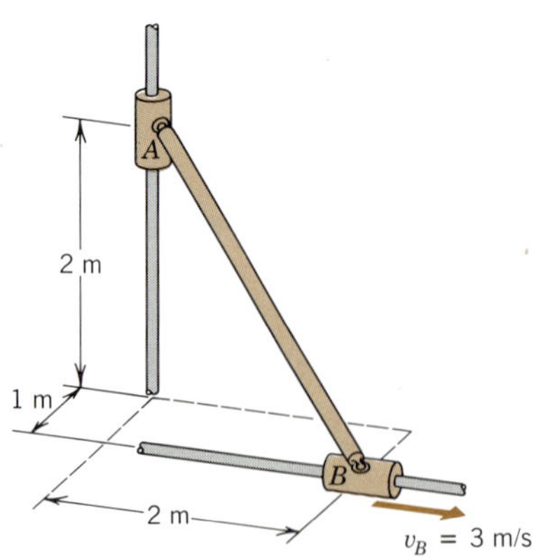

PROB. 15.57

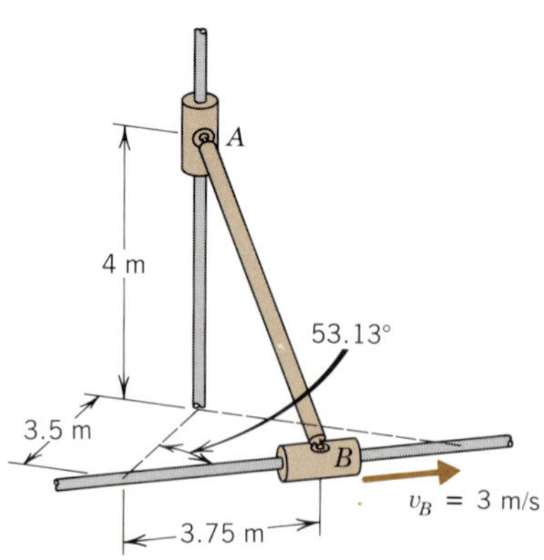

PROB. 15.58

***15.59** The speed of slider B in Problem 15.57 is constant. What is the corresponding acceleration of slider A? *Hint:* After $\bar{\omega}_{AB}$ has been found, let $\bar{\alpha}_{AB} = \alpha_x \bar{i} + \alpha_y \bar{j} + \alpha_z \bar{k}$ and eliminate the spinning portion of $\bar{\alpha}_{AB}$ by setting $\bar{\alpha}_{AB} \cdot \bar{e}_{B/A} = 0$.

15.60 The machine shown in side view rotates about the vertical axis at an angular speed ω_1, which is increasing at $\dot{\omega}_1$. The disk rotates at angular speed ω_2 about the shaft AB, such that there is no slipping relative to the horizontal surface C. (a) Derive expressions for ω_2 and $\dot{\omega}_2$ in terms of ω_1 and $\dot{\omega}_1$, and (b) determine the angular velocity and angular acceleration of the disk.

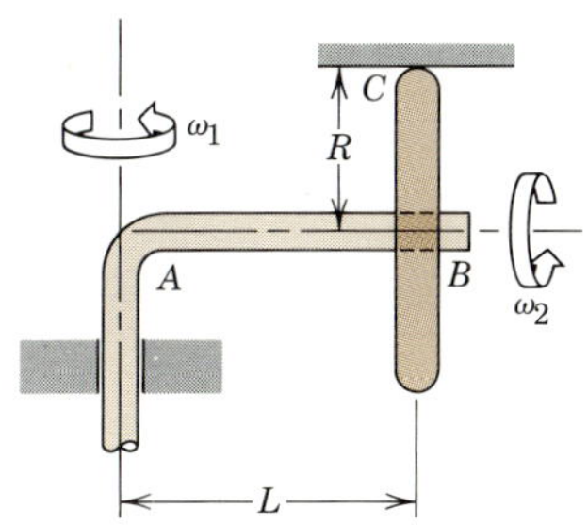

PROB. 15.60

15.61 The radar-tracking antenna has sensors that can measure the radial distance r to an airplane along the axis OA, as well as the orientation of this axis in terms of the angle of elevation θ and the angle of rotation ϕ about the fixed vertical axis. At the instant of interest $\theta = 0$, $\dot{\theta} = 0.2$ rad/s, $\ddot{\theta} = 0$, $\phi = 60°$, $\dot{\phi} = 0.5$ rad/s and $\ddot{\phi} = 0$. Note that the x axis is collinear with OA, and the y axis lies in the vertical plane. Determine the angular velocity and angular acceleration of the antenna at this instant.

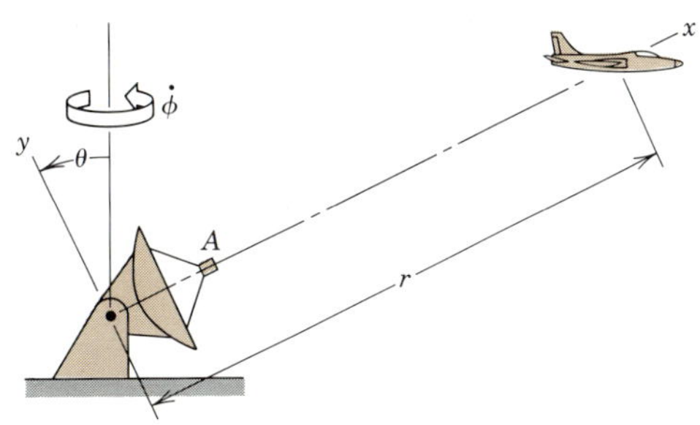

PROB. 15.61

15.62 At the instant when the antenna in Problem 15.61 is at $\theta = 15°$, $\phi = 90°$, the rates of rotation are $\dot{\theta} = 0.15$ rad/s, $\dot{\phi} = 0.3$ rad/s, $\ddot{\theta} = 0$, $\ddot{\phi} = 0.08$ rad/s^2. The position of the aircraft at that instant is

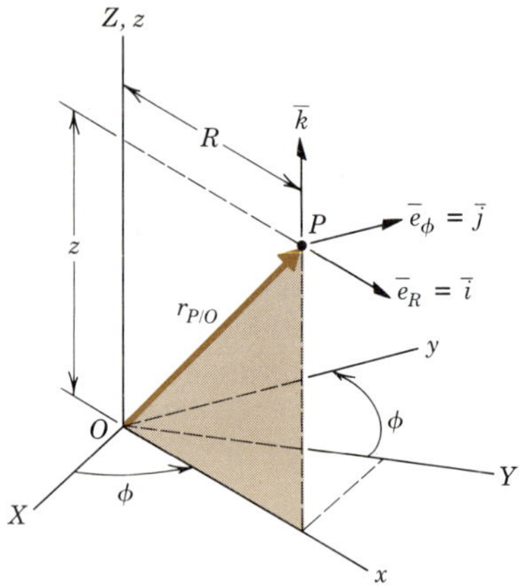

PROB. 15.63

given by $r = 1000$ ft, $\dot{r} = 600$ ft/s, $\ddot{r} = 10$ ft/s^2. Determine the corresponding velocity and acceleration of the aircraft in terms of components with respect to the xyz coordinate system shown.

15.63 Use the moving reference frame shown to derive expressions for the position, velocity, and acceleration of point P in terms of the cylindrical coordinates (R, ϕ, z) and the corresponding unit vectors $(\bar{e}_R, \bar{e}_\phi, \bar{k})$.

15.64 Spherical coordinates are another useful set of position coordinates. In this system the location of a point is defined by the distance r from the origin, the angle θ from the Z axis, and the angle ϕ from the X axis. Use the moving xyz reference frame shown to derive expressions for the position, velocity, and acceleration in terms of the spherical coordinates (r, ϕ, θ) and the corresponding orthogonal unit vectors $(\bar{e}_r, \bar{e}_\phi, \bar{e}_\theta)$.

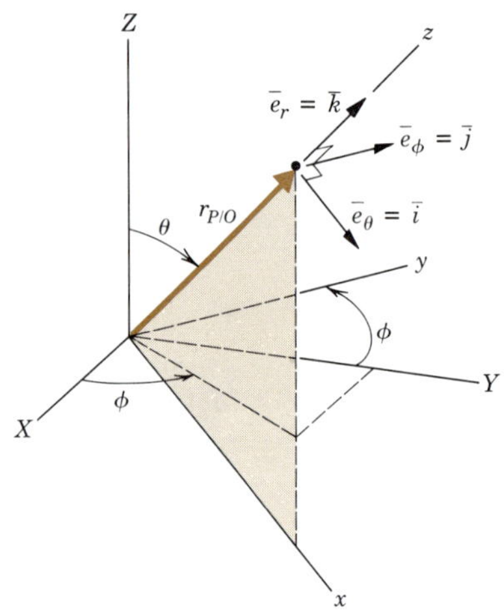

PROB. 15.64

***15.65** A motorcyclist banks the vehicle into a turn at an angle of 30° from the vertical and follows a horizontal 60 m radius curve at 90 km/hr. Each wheel has a diameter of 800 mm. Determine the angular velocity and angular acceleration of a wheel.

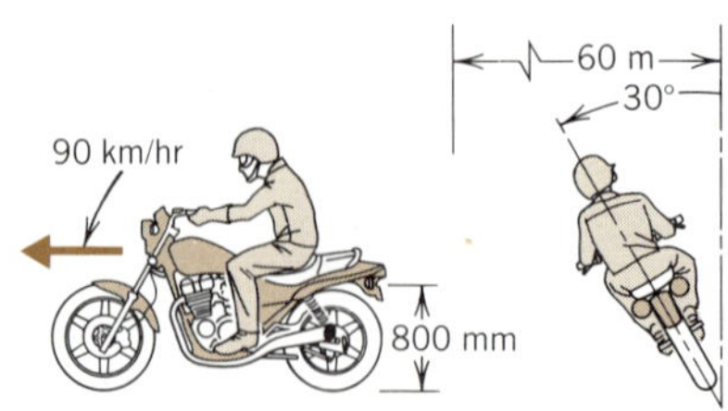

PROB. 15.65

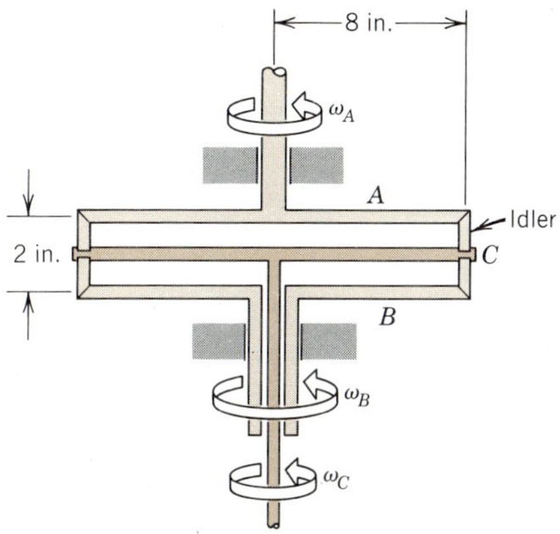

PROB. 15.66

15.66 The gear differential shown in cross section consists of parallel horizontal gears that rotate at $\omega_A = 2400$ rev/min, $\omega_B = 800$ rev/min. Determine (a) the angular speed ω_C of the crossbar connecting the idler gears, and (b) the angular velocity and angular acceleration of idler gear C.

15.67 Water enters the radial flow turbine from the interior and then flows along a turbine blade at a constant speed of 150 ft/s relative to a blade. At its tip, the tangent to a blade forms an angle of 53.13° with respect to the radial line, and the radius of curvature of a blade at that location is 5 in. The turbine is turning at a constant rate of 50 rad/s. Determine the velocity and acceleration of a particle of water at the tip of a turbine blade. Show the results in a sketch.

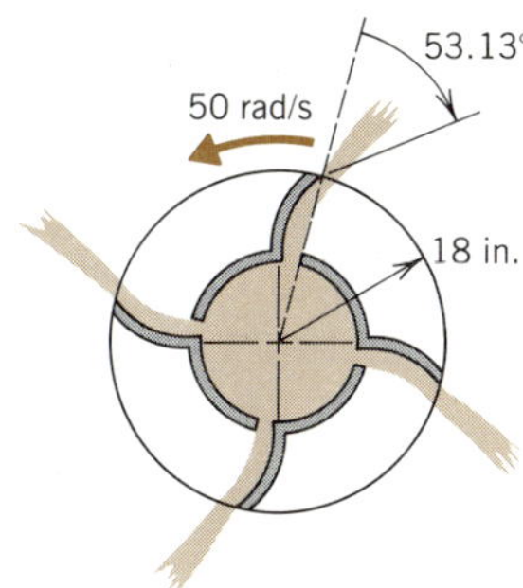

PROB. 15.67

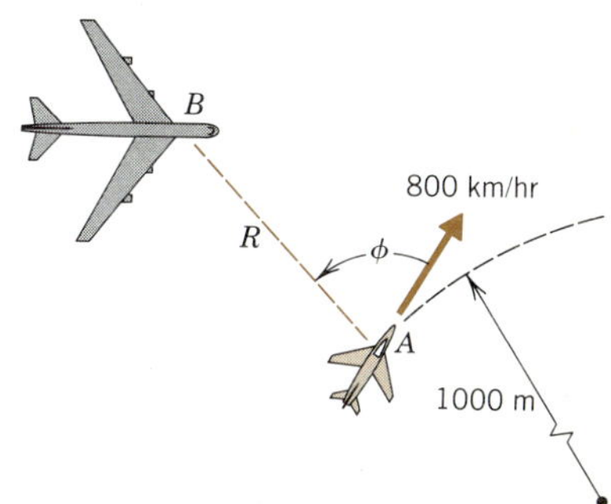

PROB. 15.68

***15.68** The radar equipment on board aircraft A follows aircraft B by measuring the distance R and angle ϕ with respect to the aircraft's longitudinal axis. Both aircraft are in level flight at the same altitude. Aircraft A is in a tight horizontal turn having a radius of 1000 m, and its speed is constant at 800 km/hr. At the instant shown, radar instruments indicate that $R = 500$ m, $\phi = 75°$, $\dot{\phi} = -0.1$ rad/s, $\dot{R} = 80$ m/s, $\ddot{R} = 60$ m/s^2, $\ddot{\phi} = 0.04$ rad/s^2. Determine the absolute velocity and absolute acceleration of aircraft B. Show the results in a sketch. *Hint:* Use polar coordinates to describe the motion of aircraft B with respect to a moving reference frame attached to airplane A.

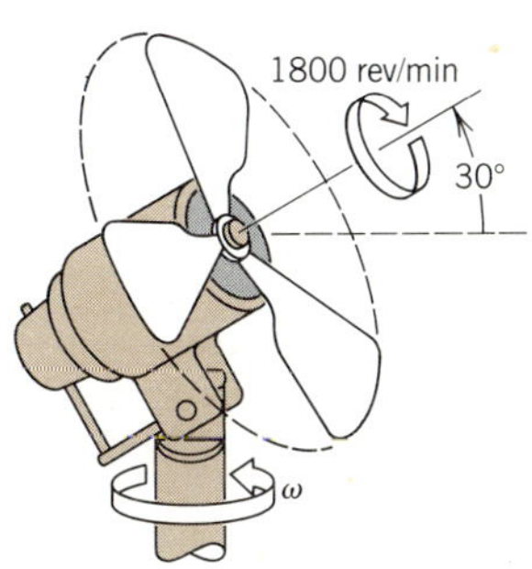

PROB. 15.69

15.69 An oscillating electric fan is aimed at 30° above the horizontal. The rotation of the fan about the vertical axis is defined by $\omega_1 = \sin \pi t$, where ω_1 is in radians per second and t is in seconds. The fan blades are spinning at 1800 rev/min. Determine the angular velocity and angular acceleration of the blades when $t = 2$ s.

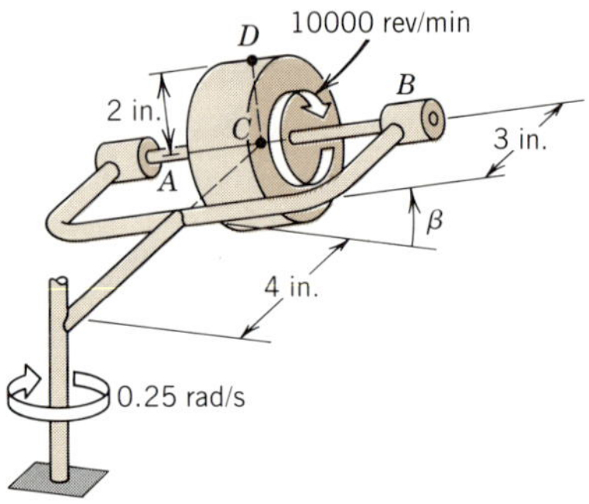

PROB. 15.70

15.70 The gyroscopic turn indicator consists of a flywheel that spins about axis AB at 10,000 rev/min, as the assembly rotates about the vertical axis at the constant rate of 0.25 rad/s. The angle of tilt is constant at $\beta = 20°$. Determine (a) the angular velocity and angular acceleration of the disk, and (b) the velocity and acceleration of point D on the flywheel.

***15.71** The motor is mounted on a bracket that is rotating about the vertical shaft at 10 rev/min. The angular speed of the disk about the axis of the motor shaft is 300 rev/min, and the centerline of the shaft is inclined at a constant 36.87° angle from the horizontal. Determine (a) the angular velocity and the angular acceleration of the disk, and (b) the velocity and acceleration of point C on the disk.

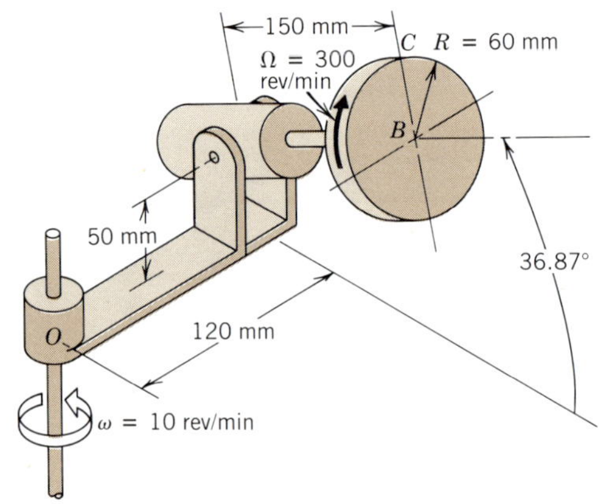

PROB. 15.71

15.72 Both bars slide through holes in block E, with the result that the angle between the bars is constant at 45°. Bar AB has a constant counterclockwise angular velocity of 10 rad/s. Determine the corresponding velocity of block E in the position shown. *Hint:* Both bars rotate by the same amount because the angle between them is constant.

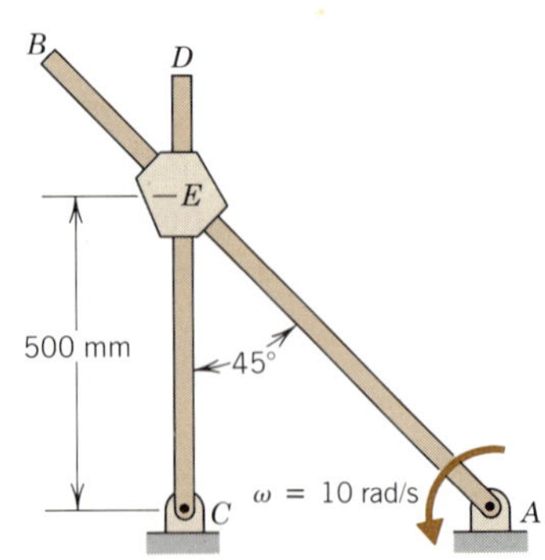

PROB. 15.72

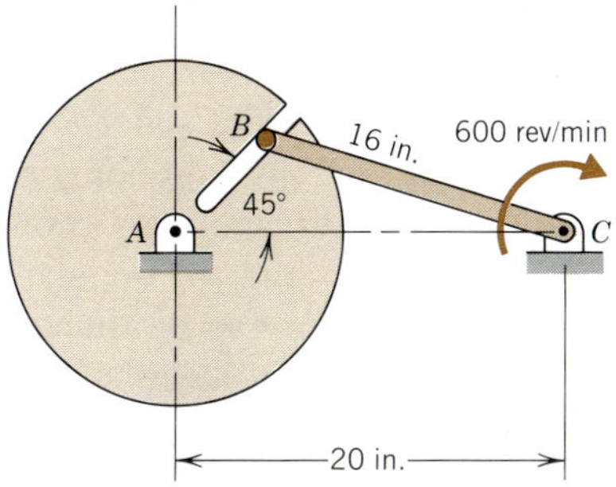

PROB. 15.74

15.73 Determine the acceleration of block E in Problem 15.72.

15.74 Pin B, which is embedded in the rotating rod BC, slides within the groove in disk A. In the position shown, rod BC is rotating clockwise at 600 rev/min. Determine the corresponding angular velocity of disk A and the speed of pin B relative to the groove.

***15.75** Slider A is moving downward at the constant rate of 600 mm/s. Collar C is pinned to bar CD and slides over bar AB. Determine the angular velocities of both bars in the position shown.

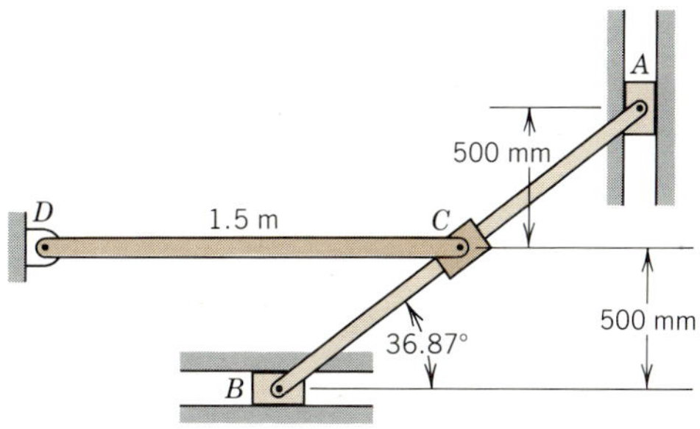

PROB. 15.75

15.76 The cone rolls without slipping over the horizontal plane, such that its centerline makes one revolution about the vertical axis every 2 s. Determine (a) the rate ω_1 at which the cone spins about its centerline, and (b) the angular velocity and angular acceleration of the cone.

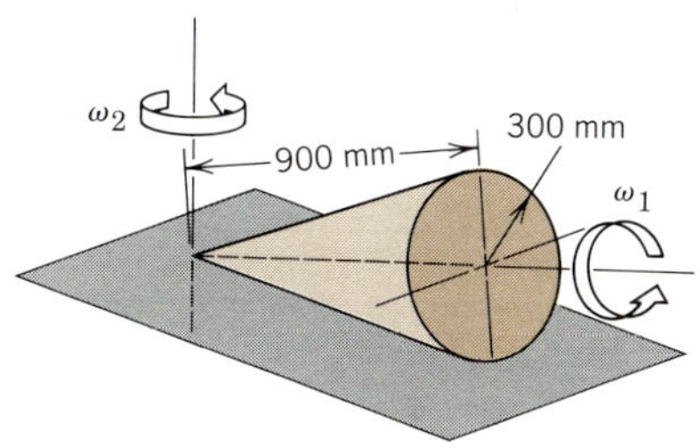

PROB. 15.76

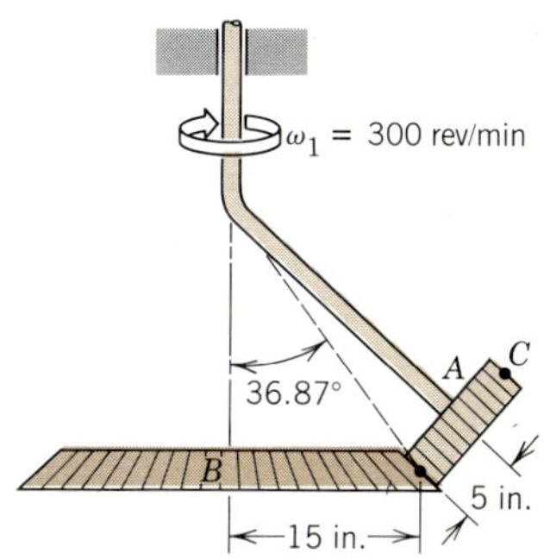

PROB. 15.77

***15.77** The shaft of gear A rotates about the fixed vertical axis at a constant rate of 300 rev/min. This gear rotates freely about its shaft as it meshes with gear B, which is stationary. Determine the angular velocity and angular acceleration of gear A.

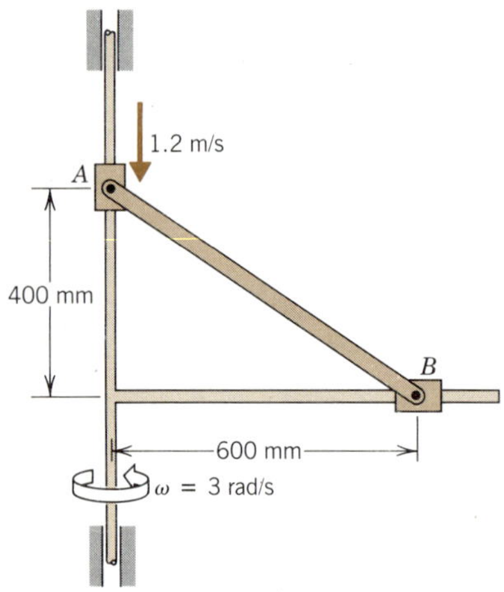

PROB. 15.78

*15.78 Sliders A and B are pinned to bar AB. Slider A has a constant downward velocity of 1.2 m/s as the T bar assembly rotates about the vertical axis at 3 rad/s. Determine the velocity of slider B.

*15.79 The sketch shows a universal joint of the type commonly employed in the drive train of automobiles. Considering ω_1 as a known value, determine the corresponding value of ω_2 and the angular velocity of the crossbar in the position where arm CD of the crossbar is horizontal and arm AB coincides with the vertical plane, as shown. *Hint:* Attach a moving reference frame to the crossbar with the origin at stationary point O.

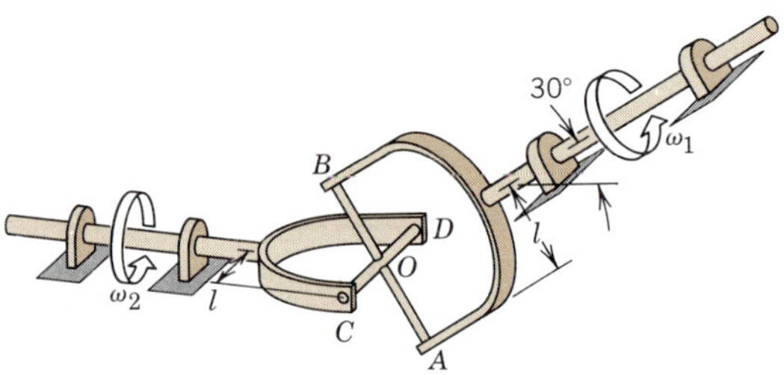

PROB. 15.79

*15.80 Collar C and bars AB and BC are interconnected by ball joints. Bar AB is rotating about the fixed point A at the constant rate of 20 rad/s. Determine the velocity of collar C for the position shown.

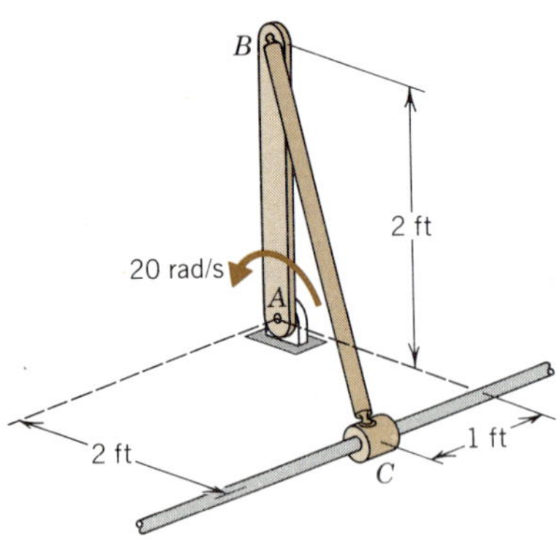

PROB. 15.80

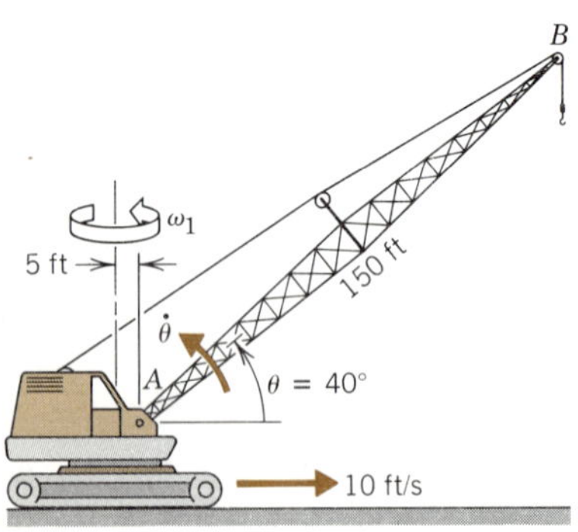

PROB. 15.82

*15.81 Determine the acceleration of collar C in Problem 15.80.

*15.82 At the instant shown, the crane is moving forward at a constant speed of 10 ft/s and is rotating about the vertical axis at the constant rate $\omega_1 = 0.4$ rad/s. The boom is being raised at this instant at $\dot{\theta} = 0.2$ rad/s and $\ddot{\theta} = 0$. Determine the velocity and acceleration of end B on the crane at this instant.

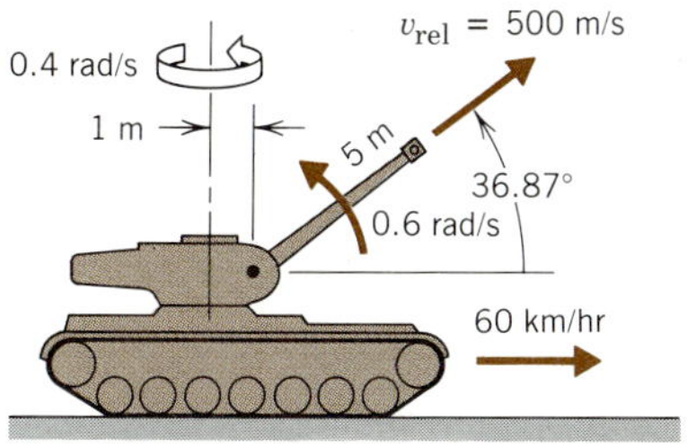

PROB. 15.83

***15.83** In the position shown, the turret on the tank is rotating about the vertical axis at 0.4 rad/s and the barrel is being raised at 0.6 rad/s; both rotation rates are constant. The tank has a constant forward velocity of 60 km/hr. If a cannon shell is fired at this instant with a muzzle speed of 500 m/s relative to the barrel, determine the velocity and acceleration of the shell immediately before it leaves the barrel.

***15.84** The disk of the gyroscope rotates about its own axis at 600 rev/min, and the gimbal support is rotating about the vertical axis at the constant rate of 10 rad/s. In the position where $\theta = 90°$ it is given that $\dot{\theta} = 5$ rad/s and $\ddot{\theta} = 0$. Determine the angular velocity and angular acceleration of the disk (a) using two moving reference frames as in Illustrative Problem 5, (b) using the formulas for Eulerian angles derived in Illustrative Problem 5.

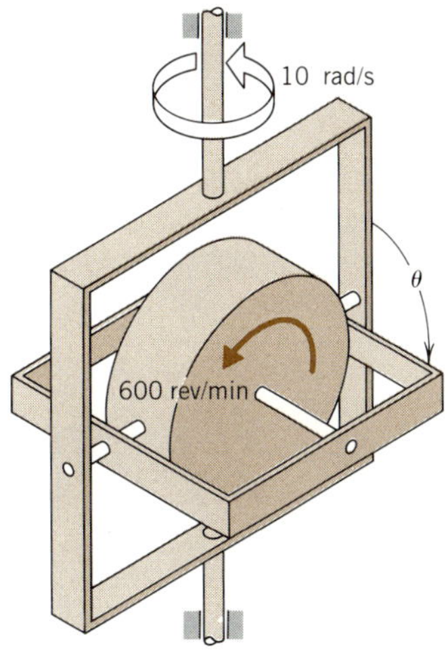

PROB. 15.84

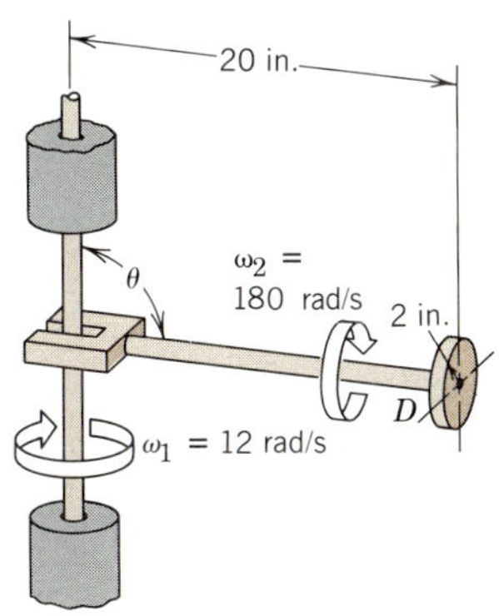

PROB. 15.86

***15.85** Given that $\dot{\theta} = -5$ rad/s and $\ddot{\theta} = 10$ rad/s^2 when $\theta = 60°$, solve Problem 15.84.

***15.86** The disk is spinning about its shaft at $\omega_2 = 180$ rad/s as the shaft rotates about the vertical axis at $\omega_1 = 12$ rad/s. In the position shown, $\theta = 90°$, $\dot{\theta} = 2$ rad/s, $\ddot{\theta} = 0$. Determine the velocity and acceleration of point *D*. *Hint:* Refer to Illustrative Problem 5.

CHAPTER 16

KINETICS OF A RIGID BODY IN THREE-DIMENSIONAL MOTION

OBJECTIVES

1. To employ the parallel axis theorems to transform moments and products of inertia between parallel coordinate systems.

2. To evaluate the mass, the location of the center of mass, and the moments and products of inertia of a composite body.

3. To evaluate the angular momentum of a body that has a specified angular motion.

4. To evaluate the rate of change of the angular momentum of a body relative to an allowable point for summing moments.

5. To identify the portion of the rate of change of the angular momentum that represents the gyroscopic moment, and to explain how such an effect arises.

6. To formulate and solve the equations of motion in order to evaluate the reactions and/or any unknown motion properties.

7. To evaluate the kinetic energy of a body in constrained motion and to use the work–energy principle to relate the motion of the body at two positions.

8. To use the linear and angular impulse-momentum principles to relate the motion of a body at two instants. Situations involving impulsive forces and conservation of angular momentum are of specific importance here.

Our objective in this chapter is to relate the three-dimensional motion of a rigid body to the forces that cause this motion. Similar to the kinetics studies of particles and rigid bodies in planar motion, this study of the laws of kinetics will consist of investigations of the equations of motion, the work–energy principle, and the linear and angular impulse-momentum principles. The new feature here is that we will be dealing with the most general case of motion of a rigid body.

Our study of three-dimensional rigid body kinematics in the previous chapter disclosed that it is necessary to determine both the motion of a point in the body and the angular motion of the body, similar to the case of a rigid body in planar motion. However, in contrast to planar motion, when a rigid body executes an arbitrary motion in three dimensions, its angular velocity can have a variable direction as well as a variable magnitude. We will see that this variable characteristic produces some fundamental phenomena that are unique to three-dimensional motion.

A. FUNDAMENTAL PRINCIPLES

Let us pause to recall the principles of statics. It was shown that there are six scalar equations of equilibrium when a rigid body is subjected to a three-dimensional force system. These equations represent the components of the vector equations obtained by requiring that the resultant of the force system satisfy $\sum \bar{F} = \bar{0}$, and that the resultant moment of the force system about an arbitrary point A satisfy $\sum \bar{M}_A = \bar{0}$.

Correspondingly, in this chapter we will find that there are six scalar equations of motion governing a rigid body undergoing an arbitrary three-dimensional motion. For this study we will employ the principles in Chapter 12 governing a system of particles. Forming the resultant of the forces external to the system leads to three scalar equations that are the components of

$$\boxed{\sum \bar{F} = m\bar{a}_G} \tag{1}$$

The mass m of the rigid body is considered to be the collection of an infinite number of infinitesimal masses dm filling the region $\mathscr{V}$ occupied by the body. Therefore

$$m = \int_{\mathscr{V}} dm$$

Equation (1) serves to describe the motion of one point in the body, the center of mass G. Therefore, the major task confronting us is the determination of the equations of motion governing the rotation of the body. The basis of this derivation is the angular impulse-momentum principle for a system of particles.

$$\sum \bar{M}_A = \frac{d}{dt} \bar{H}_A \equiv \dot{\bar{H}}_A \tag{2}$$

where

$$\bar{H}_A = \sum_{i=1}^{N} (\bar{r}_{i/A} \times m_i \bar{v}_{i/A}) \tag{3}$$

Remember that in order for eq. (2) to be valid, point A must be either a fixed point, the center of mass of the system, or a point accelerating toward the center of mass. See Chapter 12, Section I.2.

B. ANGULAR MOMENTUM OF A RIGID BODY

Equation (2) illustrates the importance of the vector $\bar{H}_A$: the angular momentum of a rigid body relative to point A. Our objective in this section is to develop a general formula for the angular momentum of a rigid body undergoing three-dimensional motion.

Figure 1 shows an infinitesimal particle dm within an arbitrary rigid body. In addition to the requirement that point A be one of the three special points mentioned in Section A, we further restrict the choice of this point to be a point that is fixed to the body. Then, because the body is rigid, we know that $\bar{r}_{i/A}$ is a vector with constant magnitude that rotates at the angular velocity $\bar{\omega}$ of the body. From our study of rigid body kinematics in Chapter 15, we have

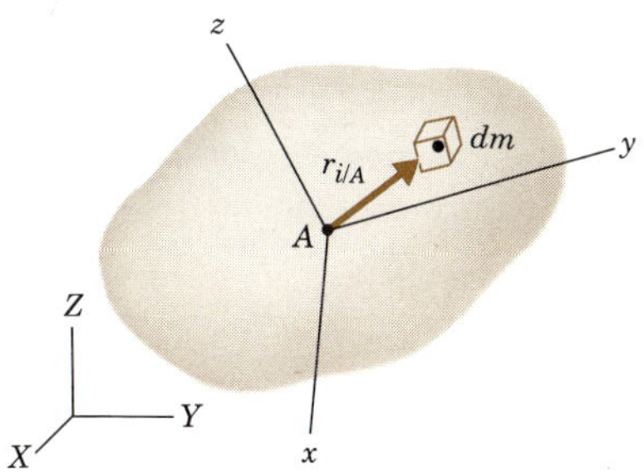

FIGURE 1

$$\bar{v}_{i/A} = \bar{v}_i - \bar{v}_A = \bar{\omega} \times \bar{r}_{i/A}$$

The contribution of each differential element dm to the value of $\bar{H}_A$ must be evaluated by an integral, rather than by a conventional summation process. Equation (3) therefore becomes

$$\bar{H}_A = \int_{\mathscr{V}} [\bar{r}_{i/A} \times (\bar{\omega} \times \bar{r}_{i/A})]\, dm \tag{4}$$

In order to proceed further, we must choose a coordinate system for describing the vector components and the boundary of the body. We could try to use the fixed reference system XYZ. Such a choice creates the awkward situation where the boundary of the body, in terms of the X, Y, Z coordinates, will change as a function of time. We avoid this difficulty by attaching a moving reference frame xyz to the body, so that all points in the body are fixed with respect to the xyz frame.

Let the origin of the xyz frame coincide with point A. Such a coordinate system is shown inFigure 1. (The question of a suitable choice for the orientation of the xyz frame will be discussed later.) The position vector of any mass element dm at point i is

$$\bar{r}_{i/A} = x\bar{i} + y\bar{j} + z\bar{k}$$

To be consistent, we express the angular velocity of the body in terms of its x, y, and z components. Thus

$$\bar{\omega} = \omega_x \bar{i} + \omega_y \bar{j} + \omega_z \bar{k}$$

Now, substituting $\bar{r}_{i/A}$ and $\bar{\omega}$ into eq. (4) and calculating the cross products leads to

$$\bar{H}_A = \int_{\mathscr{V}} \{[\omega_x(y^2+z^2) - \omega_y xy - \omega_z xz]\bar{i} + [\omega_y(z^2+x^2) - \omega_z yz - \omega_x xy]\bar{j} + [\omega_z(x^2+y^2) - \omega_x xz - \omega_z yz]\bar{k}\}\, dm$$

Recall that the angular velocity is a property of the motion of the body. The components of $\bar{\omega}$ do not depend on a particular location within the body and may therefore be removed from the integrand. As a result, the angular momentum equation may be written more compactly as

$$\bar{H}_A = (I_{xx}\omega_x - I_{xy}\omega_y - I_{xz}\omega_z)\bar{i} + (I_{yy}\omega_y - I_{xy}\omega_x - I_{yz}\omega_z)\bar{j} + (I_{zz}\omega_z - I_{xz}\omega_x - I_{yz}\omega_y)\bar{k} \tag{5}$$

The quantities I_{xx}, I_{yy}, and I_{zz} are the *moments of inertia* about the coordinate axes denoted by the repeated subscript. Their definitions are

$$I_{xx} = \int_{\mathscr{V}} (y^2+z^2)\,dm \qquad I_{yy} = \int_{\mathscr{V}} (x^2+z^2)\,dm \qquad I_{zz} = \int_{\mathscr{V}} (x^2+y^2)\,dm \tag{6}$$

The quantity I_{zz} was the only value that arose in the study of a rigid body in planar motion. There, rotation was only about the z axis. In contrast, the three moments of inertia I_{xx}, I_{yy}, and I_{zz} are important for three-dimensional motion because there can be rotation about all three axes.

The quantities I_{xy}, I_{yz}, and I_{xz} are the *products of inertia*. As we will see, these parameters are associated with a lack of symmetry of the mass distribution with respect to the coordinate axes. The definition of these parameters is

$$I_{xy} = \int_{\mathscr{V}} xy\,dm \qquad I_{xz} = \int_{\mathscr{V}} xz\,dm \qquad I_{yz} = \int_{\mathscr{V}} yz\,dm \tag{7}$$

Equations (6) and (7) highlight one of the benefits that result from attaching the xyz coordinate system to the body. Specifically, the entire boundary remains fixed with respect to the xyz axes. This means that the integrals over the region $\mathscr{V}$ have limits that do not change with time. Thus the moments and products of inertia are constant properties of the body.

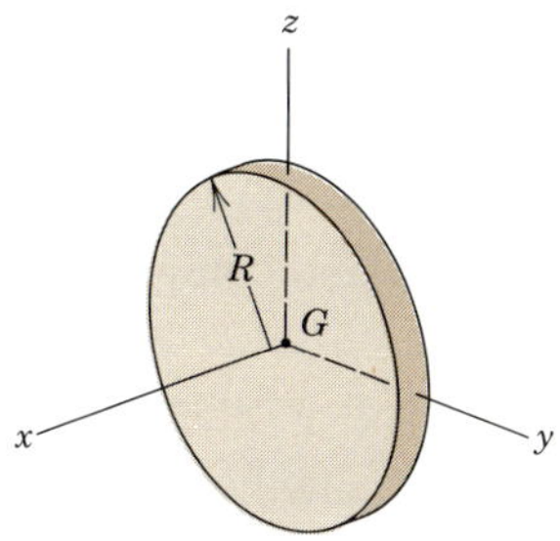

$I_{xx} = \frac{1}{2} mR^2$
$I_{yy} = I_{zz} = \frac{1}{4} mR^2$
$I_{xy} = I_{yz} = I_{xz} = 0$

FIGURE 2

We will find that a proper selection for the orientation of the xyz frame with respect to the body can greatly facilitate the calculation of the inertia properties. For instance, there is always an orientation of the xyz frame for which $I_{xy} = I_{yz} = I_{xz} = 0$. In such cases we shall refer to xyz as *principal axes*. Also, by properly orienting xyz, we will frequently be able to make use of tables that list the inertia properties of basic geometric shapes.

To demonstrate how the angular momentum of a body is determined, we shall now consider two examples of the rotation of a thin disk. For the inertia properties, recall that principal axes are those for which all products of inertia are zero. It is for such axes that inertia properties of standard geometric shapes are most often tabulated, as are most of the entries in Appendix B. Figure 2 shows such information for a thin disk. In the examples that follow, note how we select the axes to make use of this information.

EXAMPLE 1

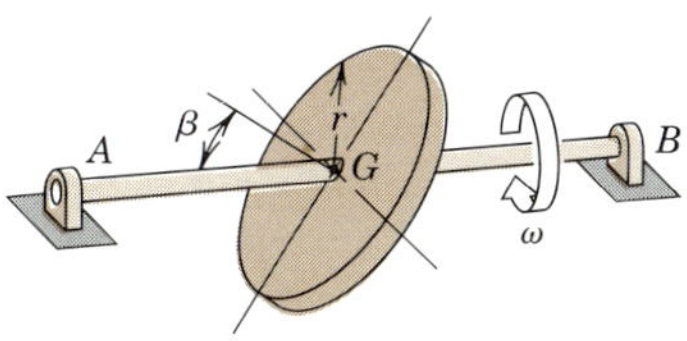

A thin disk of radius r is mounted obliquely on a rigid shaft, with the result that the center of mass G is on the axis of the shaft, but the axis of symmetry of the disk is not parallel to the shaft. If the shaft is rotating at the constant rate ω, what is the angular momentum of the disk about its center of mass?

Solution

In order to determine the angular momentum of the disk about its center of mass, we attach xyz to the disk with point G selected as the origin. Aligning the x axis with the axis of symmetry of the disk, as shown in the sketch, allows for the direct use of the given inertia properties. The fixed reference frame XYZ we select is one whose X axis is aligned with the fixed axis of rotation, that is, the shaft. Note that because xyz rotates with the disk, the X axis is always situated in the xy plane, with β being the angle between the x and X axes.

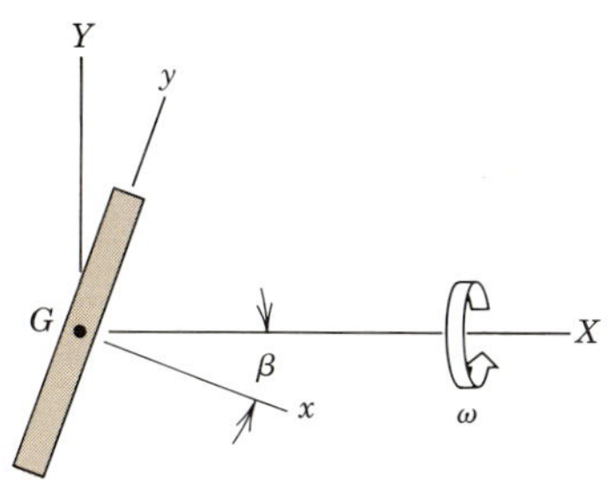

The inertia properties are known from Figure 2. Thus, we need only determine the components of angular velocity prior to evaluating the angular momentum. In terms of the chosen coordinate axes, we have

$$\bar{\omega} = \omega\bar{I}: \quad \text{always}$$
$$= \omega(\cos\beta\,\bar{i} + \sin\beta\,\bar{j})$$

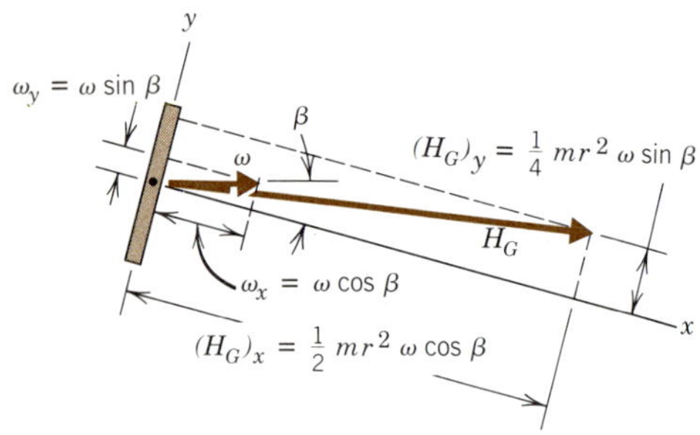

The second expression is also always valid, because β is a constant angle between $\bar{i}$ and $\bar{I}$. Thus the components of $\bar{\omega}$ are constant.

$$\omega_x = \omega \cos \beta, \qquad \omega_y = \omega \sin \beta, \qquad \omega_z = 0$$

Upon substituting the moments of inertia and components of $\bar{\omega}$ into eq. (5) we get

$$\bar{H}_G = mr^2\omega(\tfrac{1}{2} \cos \beta \,\bar{i} + \tfrac{1}{4} \sin \beta \,\bar{j})$$

The resulting angular velocity and angular momentum vectors are shown in the sketch. As can be seen, the angular momentum vector is not parallel to the angular velocity vector. The components of $\bar{H}_G$ with respect to xyz are constant because ω_x, ω_y, and ω_z are constant values. In general, a vector has constant components relative to a moving reference frame only when it is rotating in unison with that reference frame. Thus $\bar{H}_G$ for the disk rotates about the fixed shaft at angular velocity $\bar{\omega}$.

EXAMPLE 2

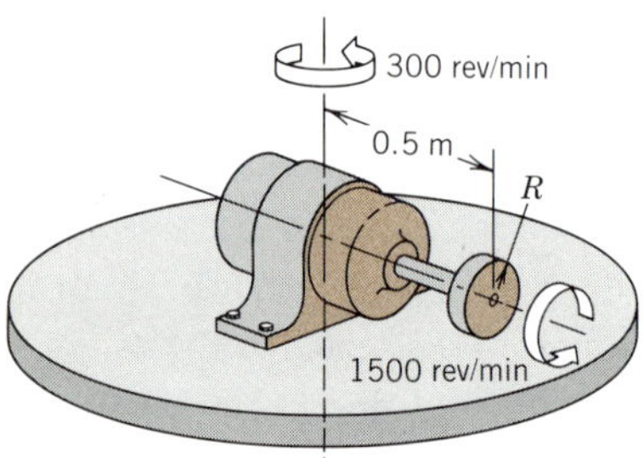

An electric motor is mounted horizontally on a turntable that rotates about the vertical axis at 300 rev/min as shown. The flywheel of the motor is concentrically mounted on the shaft of the motor, whose rate of rotation is 1500 rev/min. The center of mass of the flywheel is 0.5 m from the axis of the turntable. Approximate the flywheel as a thin disk of radius R to determine its angular momentum relative to its center of mass.

Solution

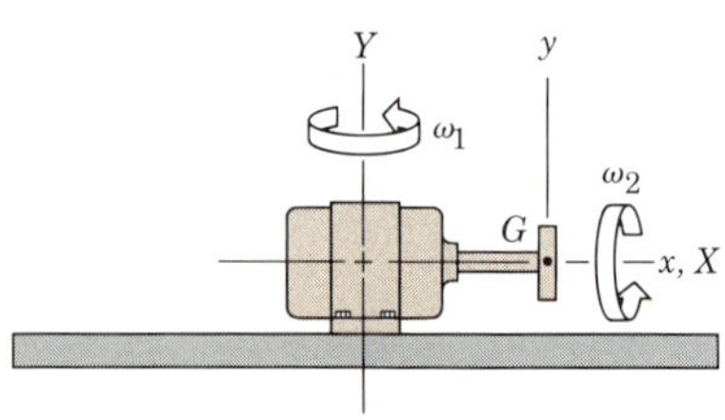

The angular momentum depends only on the angular motion of the body and the location of the origin. As always, we attach the xyz frame to the body with its origin coincident with the point of reference for the angular momentum, in this case the center of mass. We orient the xyz frame relative to the body for greatest convenience in describing the moments and products of inertia. Hence we let the x axis be the axis of the disk, so that the given inertia properties can be used directly. This still leaves the choice of the y and z axis undefined. We choose their instantaneous orientation for convenience in describing the rotation. Thus, the y axis at this instant is aligned with the fixed vertical axis of rotation. In contrast, the fixed Y axis always coincides with the vertical axis.

The inertia properties for a thin disk are given in Figure 2. In accord with the kinematical method developed in the previous chapter, the general expression for the angular velocity of the disk is

$$\bar{\omega} = \omega_1 \bar{J} + \omega_2 \bar{i}: \quad \text{always}$$

Because $\bar{J} = \bar{j}$ at the instant illustrated in the sketch, the instantaneous expression for $\bar{\omega}$ is

$$\bar{\omega} = \omega_1 \bar{j} + \omega_2 \bar{i}$$

Converting the given rotation rates to the proper units for angular rotation (radians) gives

$$\omega_x = \omega_2 = 1500\left(\frac{2\pi}{60}\right) = 50\pi \text{ rad/s}$$

$$\omega_y = \omega_1 = 300\left(\frac{2\pi}{60}\right) = 10\pi \text{ rad/s}$$

Then, from eq. (5) we find that

$$\bar{H}_G = \tfrac{1}{2}mR^2(50\pi)\bar{i} + \tfrac{1}{4}mR^2(10\pi)\bar{j}$$
$$= 2.5\pi mR^2(10\bar{i} + \bar{j})$$

The sketch of the resulting vectors $\bar{\omega}$ and $\bar{H}_G$ shows that the angular velocity and the angular momentum are not parallel, which is usual for three-dimensional systems.

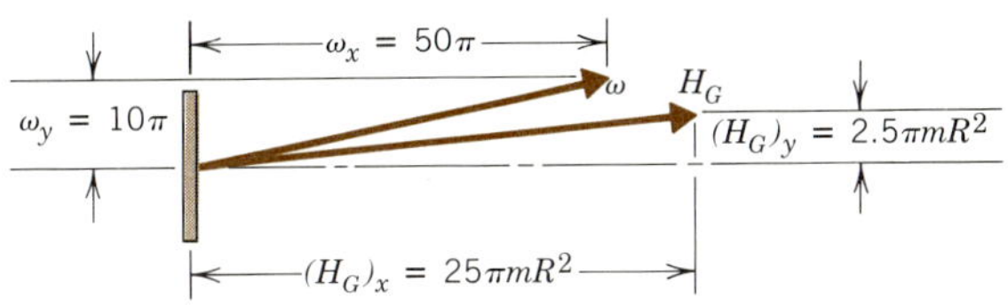

HOMEWORK PROBLEMS

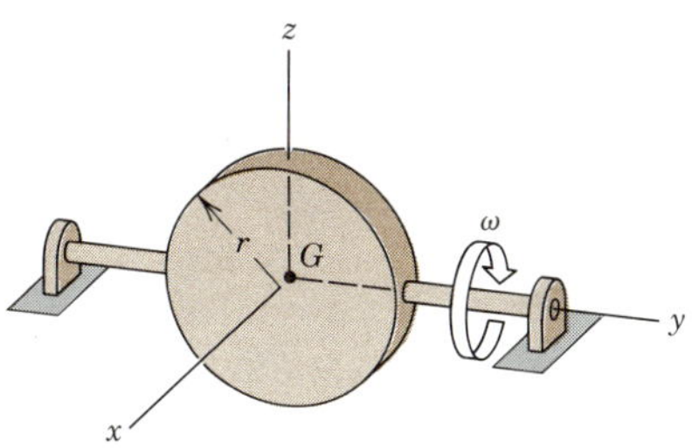

PROB. 16.1

16.1 A thin disk of mass m and radius r is mounted on a shaft whose axis intersects the center of the disk. The shaft is driven at the angular speed ω. Determine the angular momentum of the disk about its mass center. Sketch the resulting $\bar{H}_G$ and $\bar{\omega}$ vectors.

16.2 During a maneuver an earth satellite is spinning at the angular velocity $\bar{\omega} = 0.2\bar{i} + 0.8\bar{j} + 0.1\bar{k}$ rad/s. The xyz axes are principal axes for which the moments of inertia are $I_{xx} = I_{yy} = 8000$ lb-s^2-ft and $I_{zz} = 12{,}000$ lb-s^2-ft. Determine the angular momentum $\bar{H}_G$ of the space capsule about its mass center G. Sketch the resulting $\bar{H}_G$ and $\bar{\omega}$ vectors.

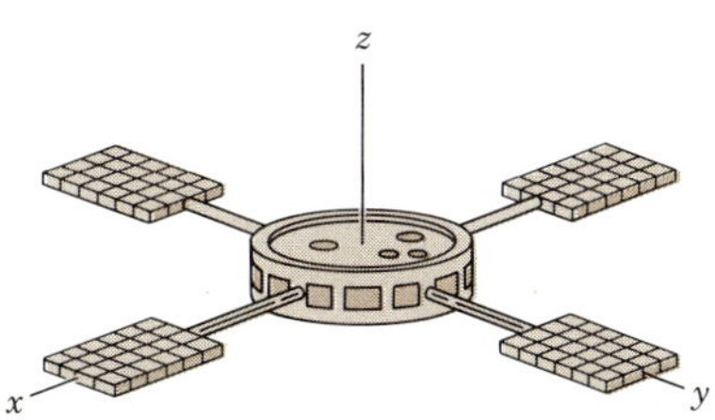

PROB. 16.2

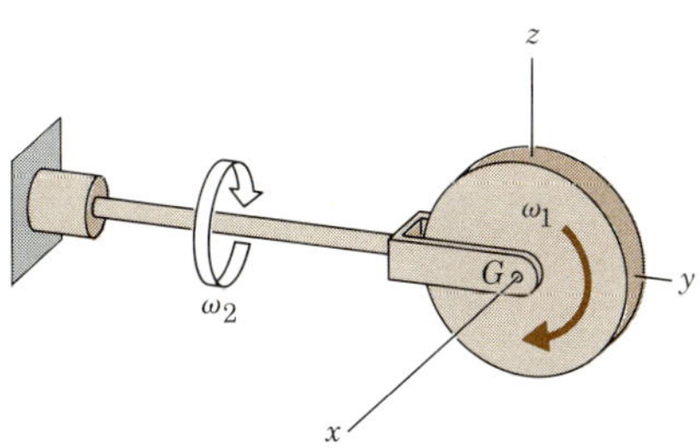

PROB. 16.3

16.3 The thin disk of mass 8 kg and radius $r = 500$ mm rotates at constant angular speed $\omega_1 = 25$ rad/s while the forked rod is driven at constant angular speed $\omega_2 = 10$ rev/min. Determine the angular momentum of the disk about its mass center. Sketch the resulting $\bar{H}_G$ and $\bar{\omega}$ vectors for the disk.

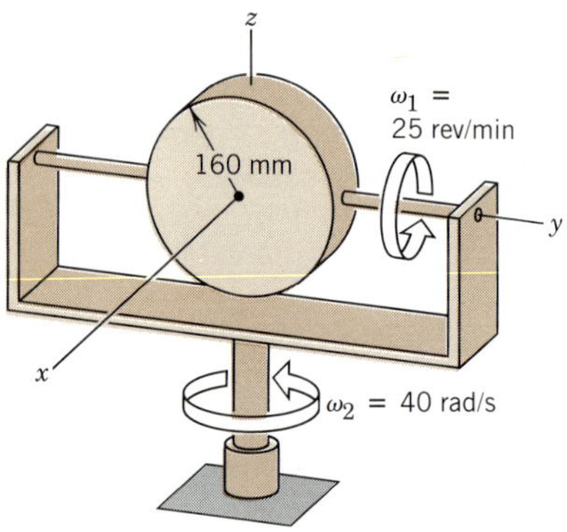

PROB. 16.4

16.4 A 6 kg thin disk of 160 mm radius is welded on a shaft mounted in a rotating frame. The disk rotates at a constant angular speed $\omega_1 = 25$ rev/min about the axis of the shaft, as the frame rotates at a constant angular speed $\omega_2 = 40$ rad/s. Determine the angular momentum of the disk about its mass center. Sketch the resulting $\bar{H}_G$ and $\bar{\omega}$ vectors.

16.5 The disk, which is welded to shaft AB, pivots downward in the vertical plane at angular speed $\dot{\theta} = 2$ rad/s, while the vertical shaft rotates at 50 rad/s. The disk weighs 8 oz. Determine the angular momentum $\bar{H}_G$ of the disk about its center of mass when $\theta = 90°$.

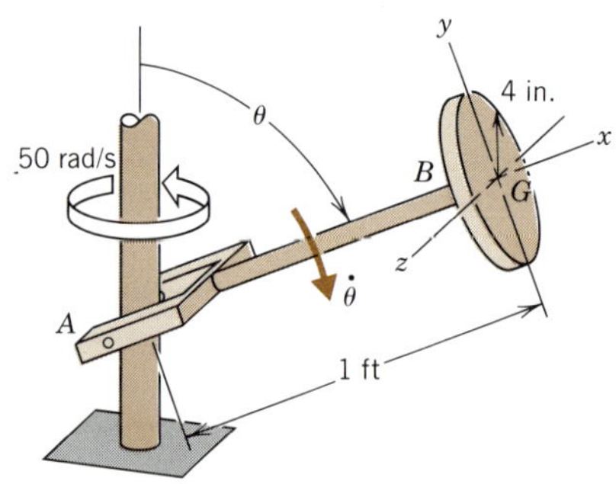

PROB. 16.5

16.6 Solve Problem 16.5 if the given value of $\dot{\theta}$ occurs at $\theta = 53.13°$.

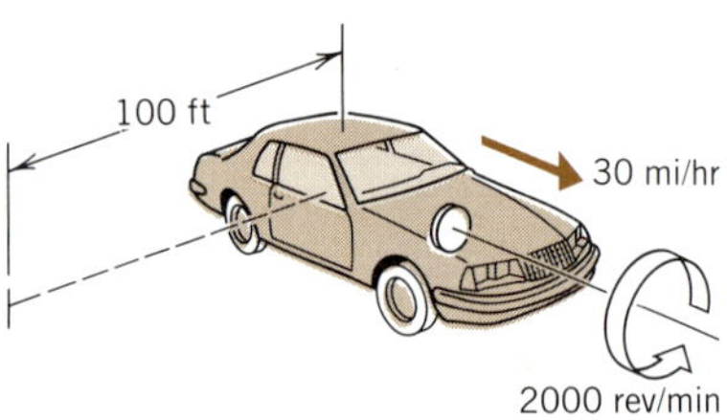

PROB. 16.7

16.7 The automobile is following a 100 ft circle at a speed of 30 mi/hr, while its engine rotates the flywheel at 2000 rev/min. The flywheel is a 40 lb disk whose diameter is 10 in. Determine the angular momentum of the flywheel about its center of mass.

16.8 Consider a circular disk of mass m and radius r rolling on the ground in a situation where the plane of the disk is oriented vertically. The center of the disk has speed v_O and there is no slipping. Determine the angular momentum of the disk about its mass center G when this point is following (a) a straight path, and (b) a circular path of radius R in the xy plane. Why are $\bar{H}_G$ and $\bar{\omega}$ parallel in the first case, but not in the second?

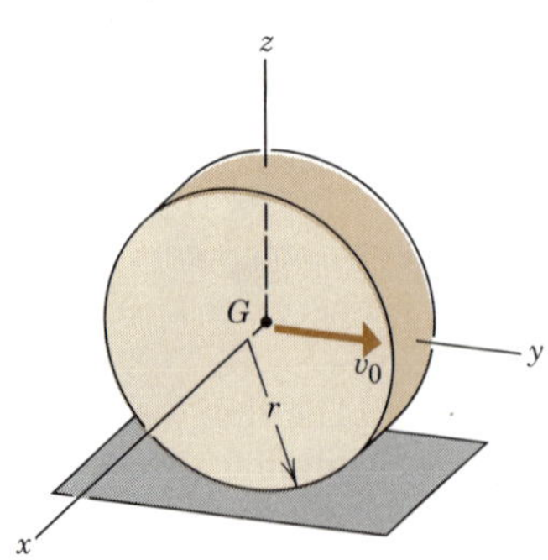

PROB. 16.8

***16.9** A disk of mass m and radius r is mounted on the rigid bar AB. The bar is welded to the vertical shaft, which is rotating at the angular speed ω_1 about the vertical axis. The disk is spinning with respect to bar AB at the angular speed $\omega_2 = 10\omega_1$ in the sense shown. Determine (a) $\bar{H}_G$ for the disk, and (b) the angle β between $\bar{H}_G$ and the vertical direction.

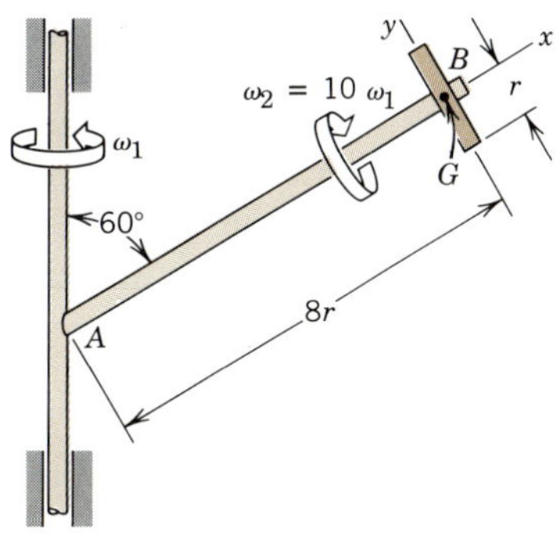

PROB. 16.9

C. MOMENTS AND PRODUCTS OF INERTIA

1. Physical Significance

From the discussion and examples in the previous section, it is obvious that knowledge of the inertia properties is crucial to the determination of the angular momentum vector. A question that arises is: What is the physical significance of the moments and products of inertia?

Let us first consider one of the moments of inertia, for instance, I_{zz}. Looking down the z axis at an arbitrary body, such as that in Figure 3, shows that $R_z = (x^2 + y^2)^{1/2}$ is the perpendicular distance from the z axis to the mass element dm whose coordinates are (x, y, z). Thus eq. (6) for I_{zz} may be written as

$$I_{zz} = \int_{\mathscr{V}} (x^2 + y^2)\, dm \equiv \int_{\mathscr{V}} R_z^2\, dm$$

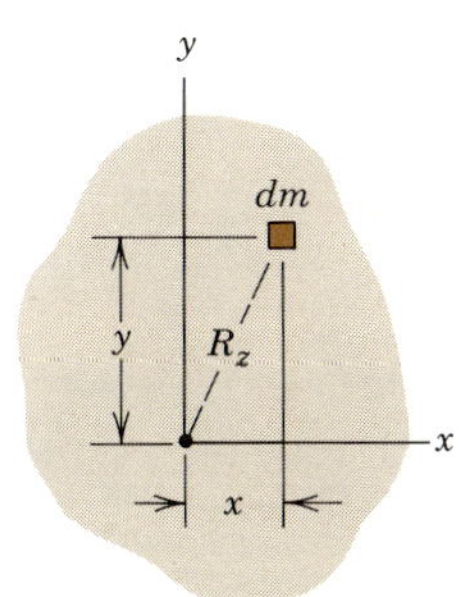

FIGURE 3

This is the same definition as the one employed in Chapter 14 to treat planar motion.

A new feature of the kinetics of rigid bodies in three-dimensional motion is that the moments of inertia about all three coordinate axes are important to the motion. However, the meaning of each term is unchanged. It describes how the mass is distributed about the corresponding axis. For instance, if two bodies have the same mass, the one whose mass is distributed further from the x axis has the greater value for I_{xx}.

One common way of describing a moment of inertia about an axis, such as the x axis, is to give the *radius of gyration* k_x. This quantity is the radius of a fictional circular ring having the same mass and moment of inertia about its centerline as the actual body it simulates. For such a ring, the mass is concentrated at the perimeter, so that

$$I_{xx} = mk_x^2 \tag{8}$$

where the x axis is normal to the plane of the ring, passing through the center. Similar definitions apply for the radii of gyration k_y and k_z.

Now let us turn our attention to the products of inertia by considering I_{xy} for the body of Figure 3. The definition in eq. (7) is

$$I_{xy} = \int_{\mathscr{V}} xy\,dm$$

which means that I_{xy} is the sum of the products of the x and y coordinates of a differential element and the mass of that element. Clearly, if the mass of a body is predominantly in one octant, then I_{xy} will have the sign associated with the product of the x and y values in that octant. For example, if a body is entirely situated in an octant where $x < 0$, $y > 0$, then I_{xy} will be negative. Generalizing this observation leads to the conclusion that the products of inertia describe how the mass of the body is distributed within the various octants of the xyz coordinate system.

An important case arises when we deal with bodies having a plane of symmetry. In Figure 4 we have drawn the cross section of a body that is symmetric about the xz plane. For this cross section in the yz plane at some value of x, we show a typical element dm_r in the right half of the body and its mirror image mass element dm_l in the left half of the body. The property of symmetry means that there is an element dm_l corresponding to every dm_r.

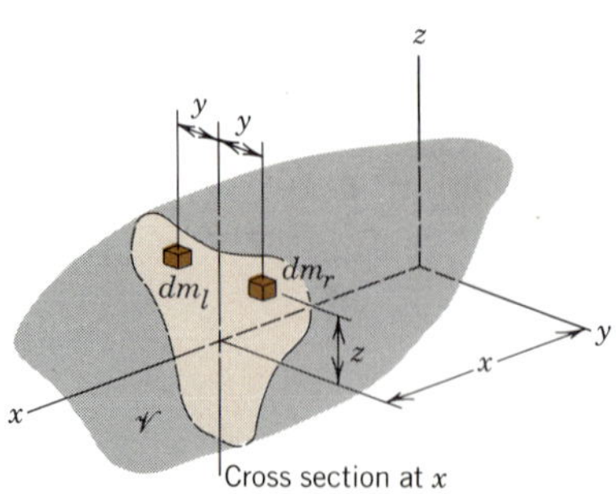

FIGURE 4

Thus, as is shown in Figure 4, if the coordinates of dm_r are (x, y, z), then the coordinates of dm_l must be $(x, -y, z)$. Hence the product $xy\, dm_l$ for the left element is the negative of the value $xy\, dm_r$ for the right element, and the two terms cancel each other in the integral over the region $\mathscr{V}$. As a result we find that $I_{xy} = 0$. Similar reasoning shows that $I_{yz} = 0$ also. From this we obtain the following important fact:

> If one of the coordinate axes is normal to the plane of symmetry of a body, with the origin of the coordinate system coincident with the plane of symmetry, the two products of inertia having this axis in the subscript are zero.

An important corollary of this statement describes the case of a body having two perpendicular planes of symmetry, such as a body of revolution.

> A coordinate system whose axes form two perpendicular planes of symmetry constitutes a principal set of axes, for which all products of inertia are zero.

The majority of systems encountered in practice have a basic geometric shape, such as a disk or a bar. To expedite computations, it is useful to tabulate the inertia properties for a variety of common shapes. A tabulation of this type may be found in Appendix B. *Any product of inertia that is not specified there can be shown to be zero due to symmetry.*

HOMEWORK PROBLEMS

16.10 For the coordinate system attached to the homogeneous body shown, use symmetry to identify which products of inertia are zero.

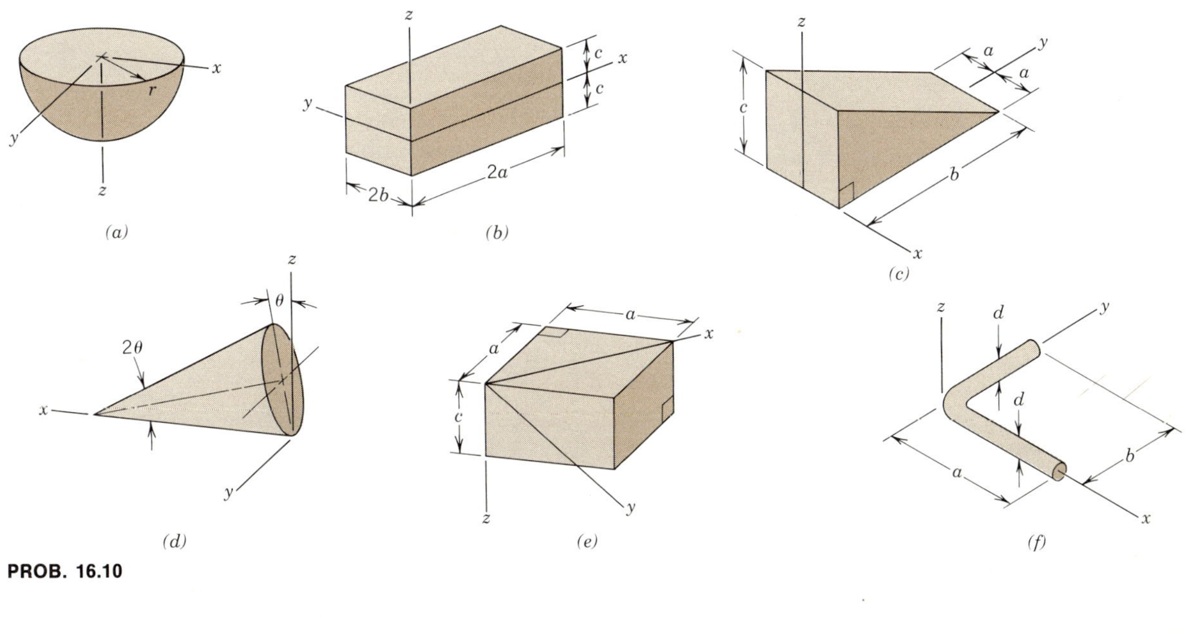

PROB. 16.10

16.11 Examine the symmetry of the homogeneous body shown to select a principal set of xyz axes whose origin is the center of mass G.

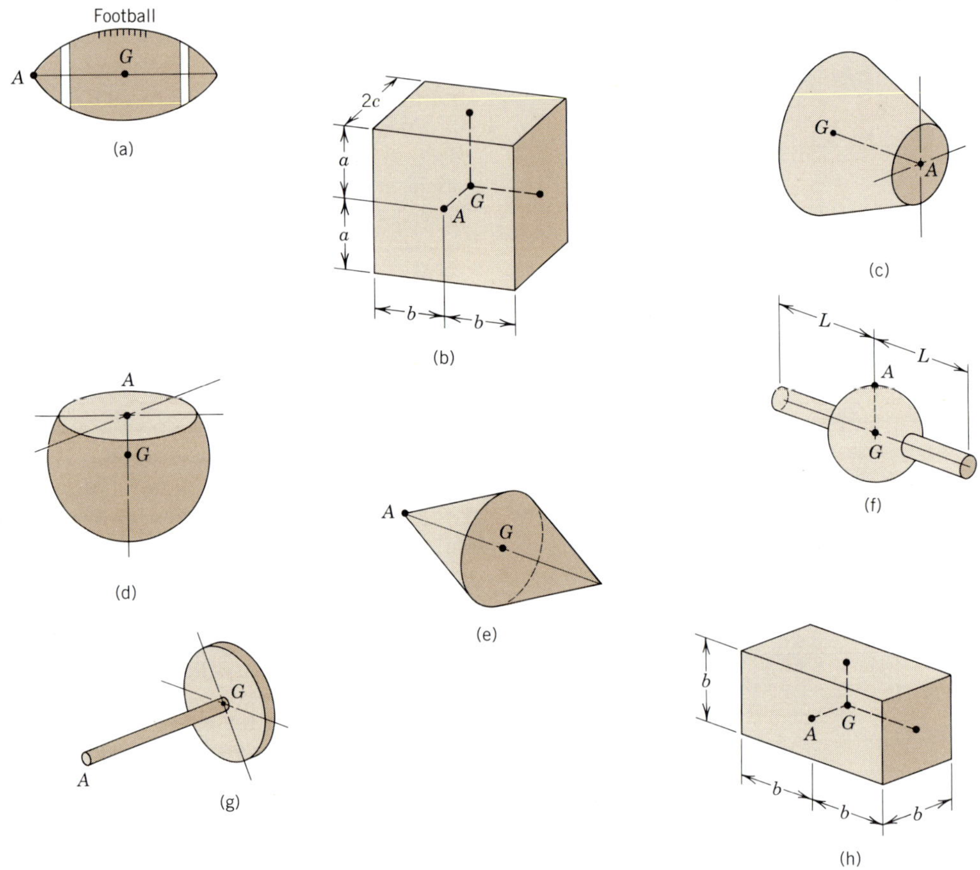

PROBS. 16.11 AND 16.12

16.12 Examine the symmetry of the body shown to select a principal set of xyz axes whose origin is point A.

16.13 The disk is welded obliquely to the vertical shaft. Sketch the principal axes having an origin at the center of mass G (a) for the disk, (b) for the shaft. (c) Are the principal axes for the system of the shaft and the disk obvious from symmetry? Explain your answer.

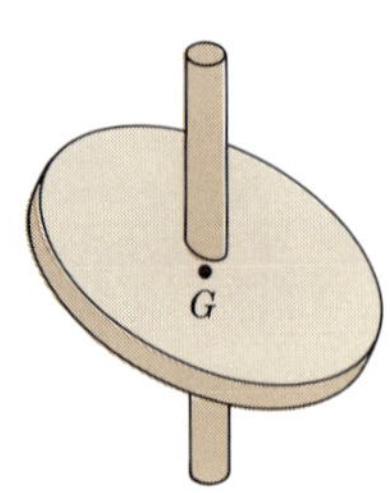

PROB. 16.13

2. Parallel Axis Theorems

Tabulations, such as Appendix B, are often not sufficient to determine the inertia properties of a body of interest. One reason for this limitation is that the moments and products of inertia given there are for a coordinate system whose origin is at the body's center of mass. There are numerous situations where a coordinate system having a different origin would be more suitable.

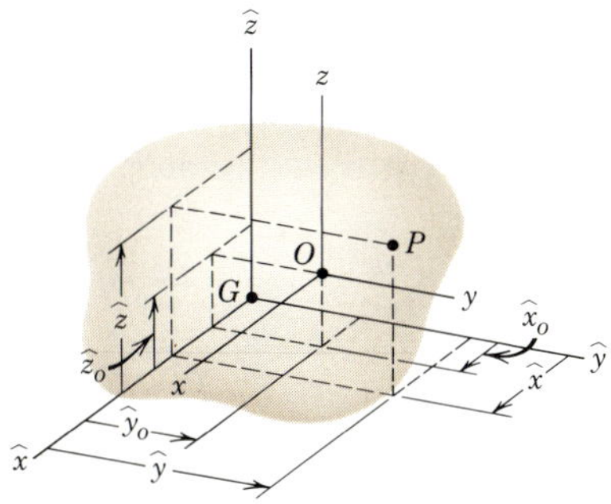

FIGURE 5

Figure 5 illustrates an arbitrary body. We consider the moment and products of inertia to be known quantities with respect to the coordinate system $\hat{x}\hat{y}\hat{z}$ whose origin is the center of mass G. The task is to determine the inertia properties of this body with respect to the coordinate system xyz whose axes are parallel to the $\hat{x}$, $\hat{y}$, and $\hat{z}$ axes, respectively. We denote the $(\hat{x}, \hat{y}, \hat{z})$ coordinates of the origin O of the xyz coordinate system by the values $(\hat{x}_O, \hat{y}_O, \hat{z}_O)$.

As can be seen from Figure 5, the (x, y, z) coordinates of any point are related to the $(\hat{x}, \hat{y}, \hat{z})$ coordinates by

$$x = \hat{x} - \hat{x}_O, \qquad y = \hat{y} - \hat{y}_O, \qquad z = \hat{z} - \hat{z}_O \tag{9}$$

Let us substitute these expressions into the definition of the moment of inertia I_{xx} given in eqs. (6).

$$I_{xx} = \int_{\mathscr{V}} (y^2 + z^2)\,dm = \int_{\mathscr{V}} [(\hat{y} - y_O)^2 + (\hat{z} - \hat{z}_O)^2]\,dm$$
$$= \int_{\mathscr{V}} (\hat{y}^2 + \hat{z}^2)\,dm - 2\hat{y}_O \int_{\mathscr{V}} \hat{y}\,dm - 2\hat{z}_O \int_{\mathscr{V}} \hat{z}\,dm + (\hat{y}_O^2 + \hat{z}_O^2)\int_{\mathscr{V}} dm$$

The first integral is, by definition, the moment of inertia $I_{\hat{x}\hat{x}}$. The second and third integrals are the first moments of mass. These integrals are zero, because the origin of the xyz frame is the center of mass, for which the vector form of the first moment of mass is

$$\int_{\mathscr{V}} \bar{r}_{i/G}\,dm = \int_{\mathscr{V}} (\hat{x}\bar{i} + \hat{y}\bar{j} + \hat{z}\bar{k})\,dm = m\bar{r}_{G/G} \equiv \bar{0}$$

In addition, the last integral is obviously m, the total mass of the body. Similar steps for the other moments of inertia yield the final form of the *parallel axis theorems for moments of inertia:*

$$\boxed{\begin{aligned} I_{xx} &= I_{\hat{x}\hat{x}} + m(\hat{y}_O^2 + \hat{z}_O^2) \\ I_{yy} &= I_{\hat{y}\hat{y}} + m(\hat{x}_O^2 + \hat{z}_O^2) \\ I_{zz} &= I_{\hat{z}\hat{z}} + m(\hat{x}_O^2 + \hat{y}_O^2) \end{aligned}} \tag{10}$$

This result takes on a simple meaning once we note that Figure 5 and the Pythagorean theorem indicate that the distances between the parallel pairs of axes are

$$d_x = (y_O^2 + z_O^2)^{1/2}$$
$$d_y = (x_O^2 + z_O^2)^{1/2}$$
$$d_z = (x_O^2 + y_O^2)^{1/2}$$

Thus eqs. (10) are the extension to three dimensions of the parallel axis theorem for planar bodies. Specifically, they state that

> The moment of inertia about any axis equals the moment of inertia about a parallel axis that *intersects the center of mass* plus the product of the mass and the square of the distance between the axes.

The determination of the products of inertia with respect to the xyz coordinate system proceeds in a similar manner. When we substitute the coordinate transformation given by eqs. (9) into eq. (7) for I_{xy}, we find that

$$I_{xy} = \int_{\mathscr{V}} xy\,dm = \int_{\mathscr{V}} (\hat{x} - \hat{x}_O)(\hat{y} - \hat{y}_O)\,dm$$
$$= \int_{\mathscr{V}} \hat{x}\hat{y}\,dm - \hat{x}_O \int_{\mathscr{V}} \hat{y}\,dm - \hat{y}_O \int_{\mathscr{V}} \hat{x}\,dm + \hat{x}_O\hat{y}_O \int_{\mathscr{V}} dm$$

The first integral is the definition of $I_{\hat{x}\hat{y}}$, and the last integral is the mass m. Also, the second and third integrals vanish because the center of mass is the origin of $\hat{x}\hat{y}\hat{z}$. Following similar steps for I_{xz} and I_{yz} yields the *parallel axis theorems for products of inertia:*

$$\begin{aligned} I_{xy} &= I_{\hat{x}\hat{y}} + m\hat{x}_O\hat{y}_O \\ I_{xz} &= I_{\hat{x}\hat{z}} + m\hat{x}_O\hat{z}_O \\ I_{yz} &= I_{\hat{y}\hat{z}} + m\hat{y}_O\hat{z}_O \end{aligned} \tag{11}$$

Proper utilization of both eqs. (10) and (11) requires that we remember that

> The origin of the $\hat{x}\hat{y}\hat{z}$ reference frame must be the center of mass of the body, and $(\hat{x}_O, \hat{y}_O, \hat{z}_O)$ represent the coordinates of the origin O of xyz with respect to $\hat{x}\hat{y}\hat{z}$.

This statement is of particular importance for assigning the coordinates $(\hat{x}_O, \hat{y}_O, \hat{z}_O)$ their proper algebraic signs when evaluating products of inertia.

3. Composite Bodies

Before we consider an example that illustrates the application of the parallel axis theorems, let us consider one more facet of the determination of the inertia properties, specifically, the matter of composite bodies. By a composite body we mean an arbitrary body whose form is composed of several basic shapes. An example of a composite body is shown in Figure 6, where the cylinder 1 and the rectangular parallelepiped 2 are the components of the composite body.

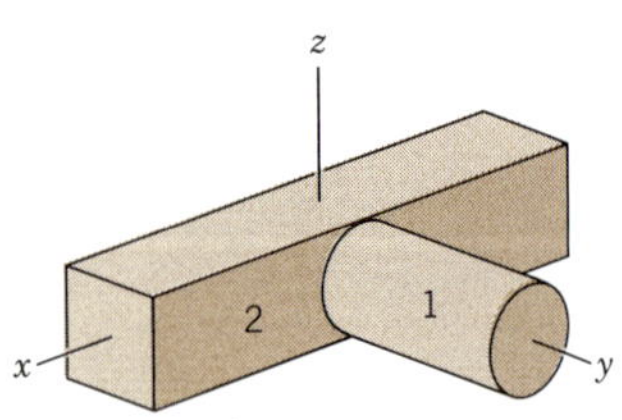

FIGURE 6

The basic mass properties of any body are its total mass, the location of the center of mass, and the moments and products of inertia with respect to a given coordinate system xyz (whose origin is not necessarily the center of mass). The method for locating the center of mass of an arbitrary composite body is a straightforward extension of the method in Chapter 14 for a planar composite body. Although some of you may have studied this topic as part of statics, we shall review it here for completeness.

The total mass m of the body is obviously the sum of the masses of the components, so that

$$m = \sum_{i=1}^{N} m_i \tag{12}$$

To describe the location of the center of mass G of the composite body, we select any convenient $x'y'z'$ coordinate system; often the origin of $x'y'z'$ coincides with the center of mass of one of the basic shapes. The location of point G is obtained from the first moment of mass. Let (x'_G, y'_G, z'_G) and $(x'_{G_i}, y'_{G_i}, z'_{G_i})$ denote the coordinates relative to the $x'y'z'$ system of point G and of the center of mass G_i of component body i, respectively. Then, to obtain the coordinates of point G, we have

$$mx'_G = \sum_{i=1}^{N} m_i x'_{G_i} \qquad my'_G = \sum_{i=1}^{N} m_i y'_{G_i} \qquad mz'_G = \sum_{i=1}^{N} m_i z'_{G_i} \tag{13}$$

The next step in the description of the mass characteristics of the composite body is the determination of the moments and products of inertia with respect to the coordinate system xyz whose origin is point A; generally, point A will be the point that was selected for the moment sum in eq. (2). To obtain the inertia properties, eqs. (6) and (7) may be decomposed into integrals over the regions $\mathscr{V}_i$ of the basic components. It follows that, when *evaluated for a given set of axes*, the inertia properties are additive, in which case

$$\begin{aligned} I_{xx} &= \sum_{i=1}^{N} (I_{xx})_i \qquad I_{yy} = \sum_{i=1}^{N} (I_{yy})_i \qquad I_{zz} = \sum_{i=1}^{N} (I_{zz})_i \\ I_{xy} &= \sum_{i=1}^{N} (I_{xy})_i \qquad I_{xz} = \sum_{i=1}^{N} (I_{xz})_i \qquad I_{yz} = \sum_{i=1}^{N} (I_{yz})_i \end{aligned} \tag{14}$$

When eqs. (14) are employed in conjunction with the inertia properties of basic shapes, it is important to remember that it will probably be necessary to use the parallel axis theorems for each shape in order to transform the basic properties to the desired xyz coordinate system.

All the aforementioned features of the parallel axis theorems and the methods for handling composite bodies are incorporated in the following example.

EXAMPLE 3

Locate the center of mass of the steel body shown and determine the inertia properties with respect to the given xyz coordinate axes.

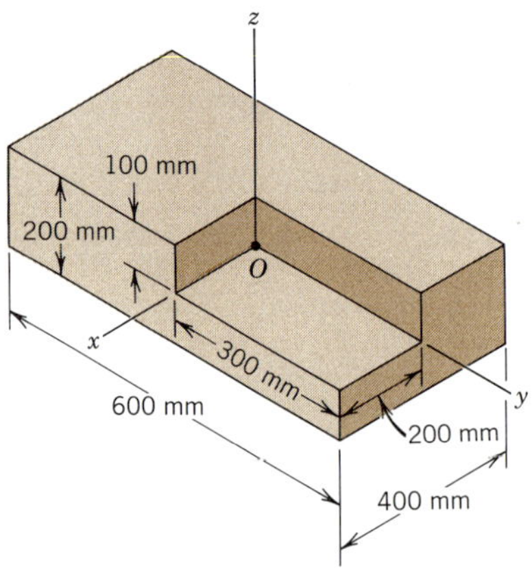

Solution

The composite body can most easily be depicted as a steel block, call it body 1, of dimensions 600 mm × 200 mm × 400 mm, from which a smaller block, call it body 2, of dimensions 300 mm × 100 mm × 200 mm, has been cut away. Hence, for this system we shall subtract the properties of body 2 with respect to the coordinate axes through O from those of body 1. In general, to determine the inertia properties of composite bodies we shall need coordinate axes $\hat{x}_i$, $\hat{y}_i$, $\hat{z}_i$ for each component shape i. The origin of $\hat{x}_i$, $\hat{y}_i$, $\hat{z}_i$ is the center of mass G_i of that shape, and each $\hat{x}_i$, $\hat{y}_i$, $\hat{z}_i$ system should be parallel to the xyz coordinate system of interest. In this problem the center of mass G_1 is concident with point O, as illustrated in the sketch.

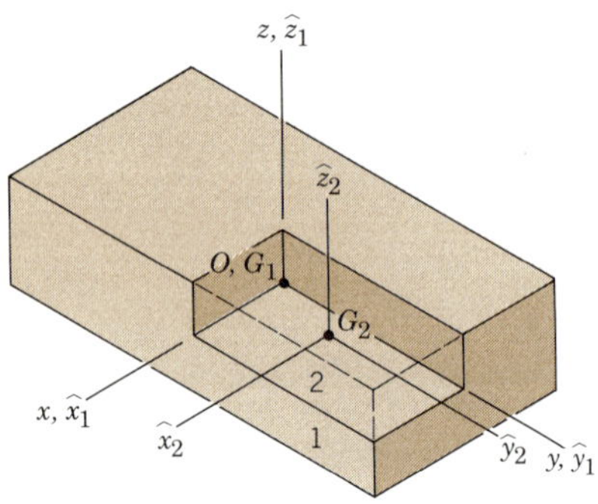

Our approach for composite bodies is to first describe each component body with the aid of Appendix B. The properties we compute for each are the mass $m_i \equiv \rho \mathscr{V}_i$, the coordinates of point G_i with respect to xyz, and the centroidal moments of inertia. Note that the coordinate system $\hat{x}_i\hat{y}_i\hat{z}_i$ for each body forms a set of principal axes because the $\hat{x}\hat{y}$, $\hat{x}\hat{z}$, and $\hat{y}\hat{z}$ planes are perpendicular planes of symmetry. The specific gravity of steel (from Appendix D) is 7.84, so the density of steel is $\rho = 7.84(1000)\text{ kg/m}^3$. Thus

Body 1:

$$m_1 = \rho \mathscr{V}_1 = 7.84(10^3)(0.4)(0.6)(0.2) = 376.3 \text{ kg}$$

$$(x_{G_1}, y_{G_1}, z_{G_1}) = (0, 0, 0)$$

$$(I_{\hat{x}_1\hat{x}_1})_1 = \tfrac{1}{12}m_1[(0.6)^2 + (0.2)^2] = 12.543 \text{ kg-m}^2$$

$$(I_{\hat{y}_1\hat{y}_1})_1 = \tfrac{1}{12}m_1[(0.4)^2 + (0.2)^2] = 6.272 \text{ kg-m}^2$$

$$(I_{\hat{z}_1\hat{z}_1})_1 = \tfrac{1}{12}m_1[(0.4)^2 + (0.6)^2] = 16.306 \text{ kg-m}^2$$

$$(I_{\hat{x}_1\hat{y}_1})_1 = (I_{\hat{x}_1\hat{z}_1})_1 = (I_{\hat{y}_1\hat{z}_1})_1 = 0$$

Body 2:

$$m_2 = \rho \mathcal{V}_2 = 7.84(10^3)(0.2)(0.3)(0.1) = 47.04$$

$$(x_{G_2}, y_{G_2}, z_{G_2}) = (+0.10, +0.15, +0.05) \text{ m}$$

$$(I_{\hat{x}_2\hat{x}_2})_2 = \tfrac{1}{12}m_2[(0.30)^2 + (0.10)^2] = 0.392 \text{ kg-m}^2$$

$$(I_{\hat{y}_2\hat{y}_2})_2 = \tfrac{1}{12}m_2[(0.20)^2 + (0.10)^2] = 0.196 \text{ kg-m}^2$$

$$(I_{\hat{z}_2\hat{z}_2})_2 = \tfrac{1}{12}m_2[(0.20)^2 + (0.30)^2] = 0.510 \text{ kg-m}^2$$

$$(I_{\hat{x}_2\hat{y}_2})_2 = (I_{\hat{x}_2\hat{y}_2})_2 = (I_{\hat{y}_2\hat{z}_2})_2 = 0$$

The next step is to evaluate the mass from eq. (12), and to then form the first moments of mass with respect to the xyz axes using eq. (13). For this, we *subtract* the contribution of body 2 because it is *removed* from body 1. Thus

$$m = m_1 - m_2 = 329.3 \text{ kg}$$

$$329.3x_G = m_1 x_{G_1} - m_2 x_{G_2} = 0 - 47.04(0.10)$$

$$329.3y_G = m_1 y_{G_1} - m_2 y_{G_2} = 0 - 47.04(0.15)$$

$$329.3z_G = m_1 z_{G_1} - m_2 z_{G_2} = 0 - 47.04(0.05)$$

$$(x_G, y_G, z_G) = (-0.01428, -0.02143, -0.00714) \text{ m}$$

Before we can combine the moments and products of inertia for the individual bodies, we must transform the tabulated properties for each body about its mass center to the value for the xyz system. For this we use the parallel axis theorems. Note that the xyz and $\hat{x}_1\hat{y}_1\hat{z}_1$ coordinate systems coincide. Thus the inertia properties of body 1 that we computed earlier are also those for the xyz coordinate system. For body 2 the coordinates of the origin O of xyz *with respect* to $\hat{x}_2\hat{y}_2\hat{z}_2$ are $(-0.10, -0.15, -0.05)$ m, so eqs. (6) and (7) give

$$(I_{\hat{x}\hat{x}})_2 = (I_{\hat{x}_2\hat{x}_2})_2 + m_2(\hat{y}_O^2 + \hat{z}_O^2) = 0.392 + 47.04[(-0.15)^2 + (-0.05)^2]$$
$$= 1.568 \text{ kg-m}^2$$

$$(I_{yy})_2 = (I_{\hat{y}_2\hat{y}_2})_2 + m_2[(-0.10)^2 + (-0.05)^2] = 0.784 \text{ kg-m}^2$$

$$(I_{zz})_2 = (I_{\hat{z}_2\hat{z}_2})_2 + m_2[(-0.10)^2 + (-0.15)^2] = 2.039 \text{ kg-m}^2$$

$$(I_{xy})_2 = (I_{\hat{x}_2\hat{y}_2})_2 + m_2\hat{x}_O\hat{y}_O = 0 + 47.04(-0.10)(-0.15) = 0.706 \text{ kg-m}^2$$

$$(I_{xz})_2 = (I_{\hat{x}_2\hat{z}_2})_2 + m_2(-0.10)(-0.05) = 0.235 \text{ kg-m}^2$$

$$(I_{yz})_2 = (I_{\hat{y}_2\hat{z}_2})_2 + m_2(-0.15)(-0.05) = 0.353 \text{ kg-m}^2$$

Once again, we subtract the properties of body 2, because it is removed from the larger body 1. Thus

$$I_{xx} = (I_{xx})_1 - (I_{xx})_2 = 12.543 - 1.568 = 10.975 \text{ kg-m}^2$$

$$I_{yy} = (I_{yy})_1 - (I_{yy})_2 = 5.488 \text{ kg-m}^2$$

$$I_{zz} = (I_{zz})_1 - (I_{zz})_2 = 14.267 \text{ kg-m}^2$$

$$I_{xy} = (I_{xy})_1 - (I_{xy})_2 = 0 - 0.706 = -0.706 \text{ kg-m}^2$$

$$I_{xz} = (I_{xz})_1 - (I_{xz})_2 = -0.235 \text{ kg-m}^2$$

$$I_{yz} = (I_{yz})_1 - (I_{yz})_2 = -0.353 \text{ kg-m}^2$$

HOMEWORK PROBLEMS

16.14–16.16 The homogeneous body shown has mass m. Determine the moments and products of inertia with respect to the indicated xyz coordinate system using the inertia properties given in Appendix B.

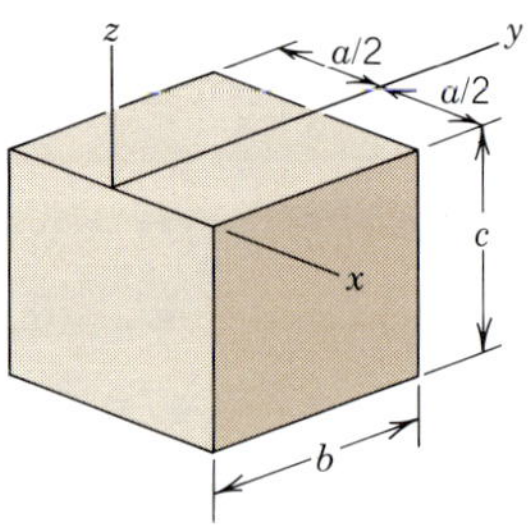

PROB. 16.14

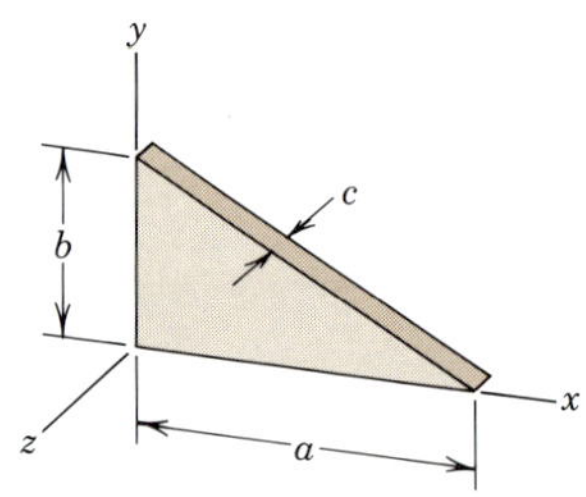

PROB. 16.15

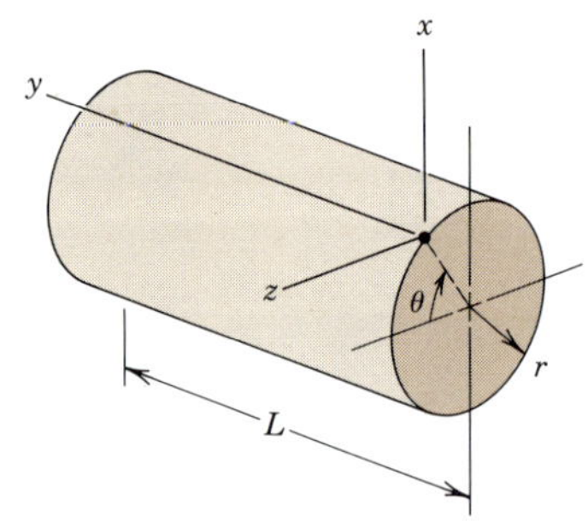

PROB. 16.16

16.17–16.20 For the aluminum body shown, determine the mass and the coordinates of the center of mass with respect to the given xyz coordinate system. Then determine I_{xx} and I_{xy}.

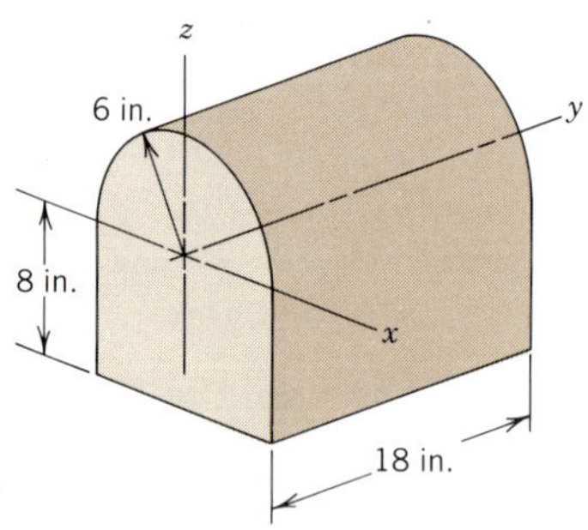

PROB. 16.17

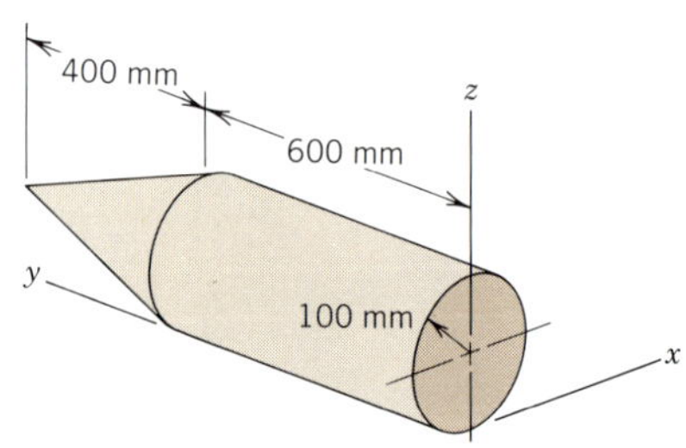

PROB. 16.18

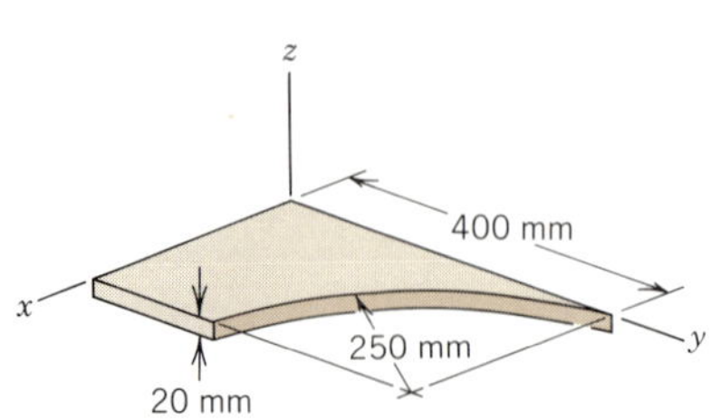

PROB. 16.19

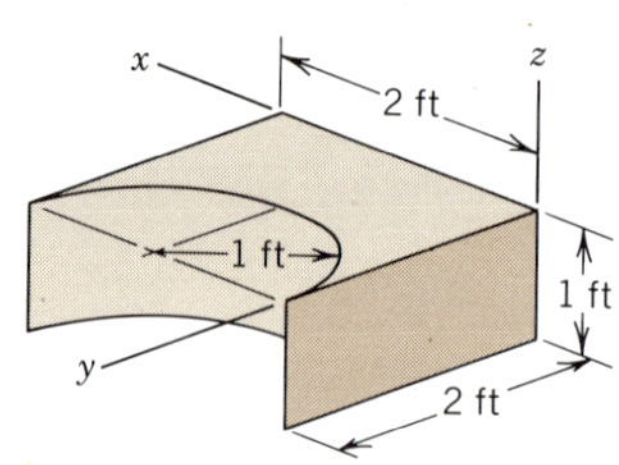

PROB. 16.20

16.21 Solve Problem 16.18 if the cylindrical section is composed of aluminum and the cone is composed of steel.

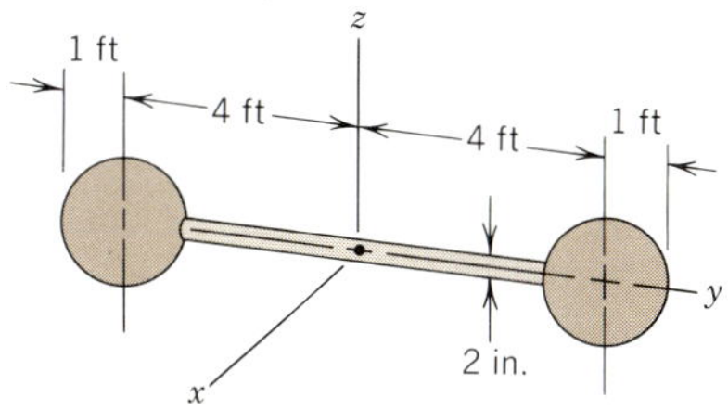

PROB. 16.22

16.22 Steel spheres are welded to each end of the circular steel bar. Determine the value of I_{zz} (a) considering the spheres and the bar to be three-dimensional bodies, (b) neglecting the mass of the bar and considering the spheres to be particles of the same mass as the actual spheres. What conclusions can you draw from these results?

16.23 Use the inertia properties for a hemisphere given in Appendix B to derive formulas for the moments of inertia of the hollow hemisphere with respect to the coordinate system shown, expressing the results in terms of the mass m, radius r, and thickness h. Then, by taking the limit as $h \to 0$ with constant m and r, derive the corresponding properties for a thin hemispherical shell.

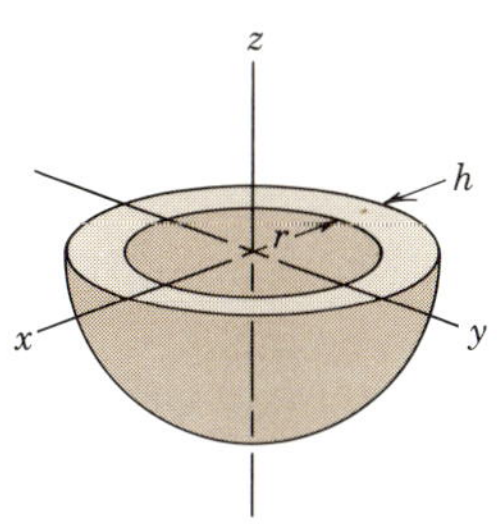

PROB. 16.23

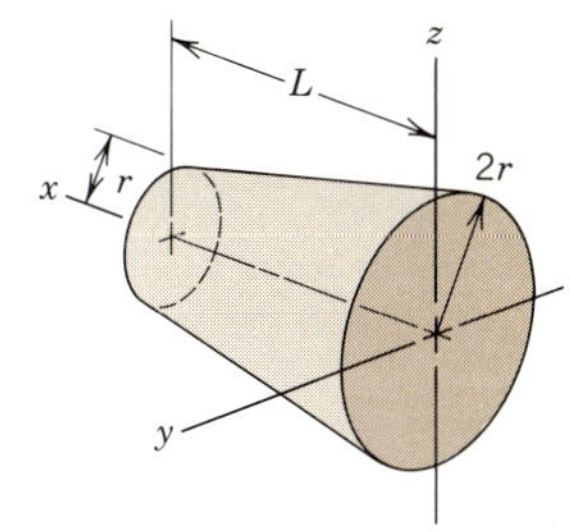

PROB. 16.24

16.24 Determine the moments and products of inertia of the truncated cone with respect to the given xyz coordinate system.

16.25 Determine the location of the mass center of the shelf and bracket shown and also determine the moments and products of inertia with respect to the xyz coordinate system shown. The two pieces are made of wood having a density of 750 kg/m^3.

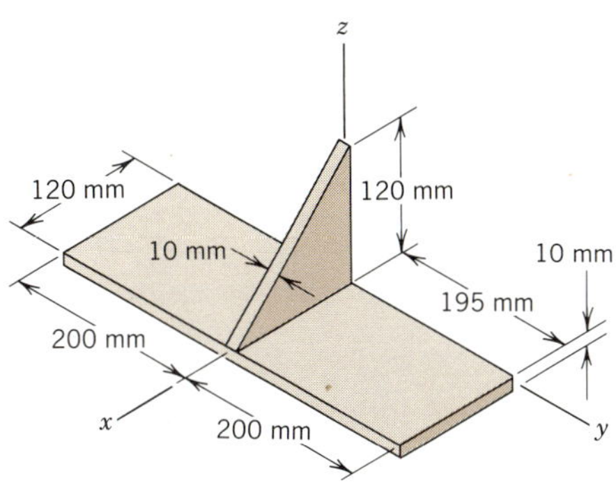

PROB. 16.25

16.26 Solve Problem 16.25 for the case where the rectangular shelf is composed of wood, and the triangular bracket is composed of steel.

16.27 For the body in Problem 16.18, locate the center of mass G, and then determine the principal moments of inertia for a coordinate system whose origin is point G.

***16.28** For the body in Problem 16.20, locate the center of mass G, and then determine the principal moments of inertia for a coordinate system whose origin is point G.

---

*4. Inertia Properties by Integration

Most of the bodies encountered in practice may be treated by employing tables of inertia properties, such as those contained in Appendix B, and the concept of a composite body. However, occasionally the shape of a body of interest might not resemble any of the entries in the appendix.

In that case we may determine the inertia properties by returning to their definitions as integrals over the region $\mathcal{V}$ occupied by the body.

The integration techniques employed in Chapter 14 were based on knowing that the body was symmetric with respect to the xy plane. The body we consider here has an arbitrary shape. The mass element that can describe such a body has infinitesimal dimensions in all directions. This leads to a triple integral over the range of the three position coordinates describing the location of points in the body. The two types of position coordinates we have employed are rectangular Cartesian and cylindrical coordinates. The volume element $d\mathcal{V}$ in each system is formed from differentials in each of the position coordinates.

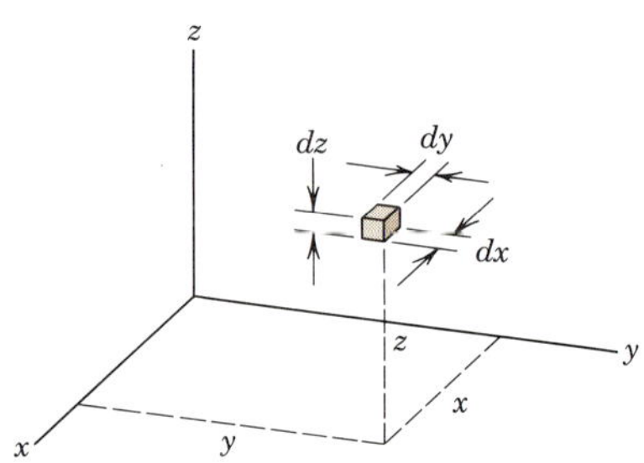

FIGURE 7

The element $d\mathcal{V}$ in Figure 7, which describes the situation for rectangular Cartesian coordinates, is a box whose edges are dx, dy, and dz in the direction of the corresponding axes. The mass dm of the element is the volume $d\mathcal{V}$ multiplied by the density of the point surrounded by the element (ρ need not be constant throughout the body), so

$$dm = \rho\, dx\, dy\, dz \tag{15}$$

Figure 8 depicts the situation for cylindrical coordinates. The contribution of higher-order differentials is negligible, so the volume element $d\mathcal{V}$ is now a body whose edges are dR in the radial direction, $R\,d\phi$ in the transverse direction, and dz in the axial direction. Thus the mass $dm = \rho\, d\mathcal{V}$ is

$$dm = \rho(R\,d\phi)(dR)(dz) \tag{16}$$

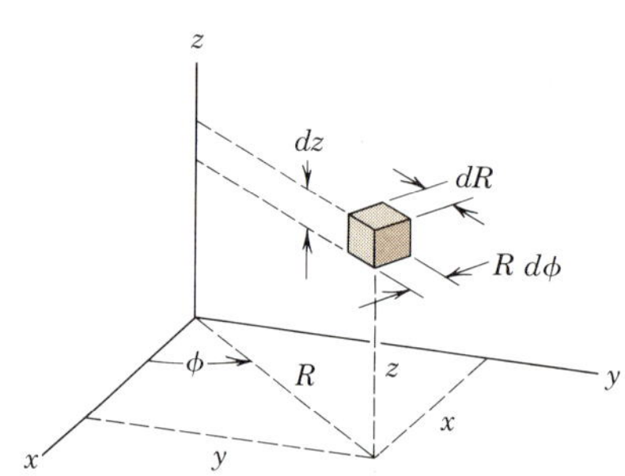

FIGURE 8

Another consideration is that the variables x and y occur in the integrands in eqs. (6) for the moments of inertia and in eqs. (7) for the products of inertia, as well as in the first moments of mass. These distances must be expressed in terms of the cylindrical coordinates. The coordinate transformations obtained from Figure 8 are

$$x = R\cos\phi, \qquad y = R\sin\phi \tag{17}$$

The decision to use either rectangular Cartesian or cylindrical coordinates should be based on identifying which coordinate system most readily defines the boundaries of the body. The triple integrals in either set of coordinates are formed by successively integrating over each coordinate (in any consistent order), with the ensuing coordinates kept constant. Covering the full range of the coordinate being integrated might involve variable integration limits; such limits are functions of the coordinates that have not been integrated. These features are displayed in the following two examples.

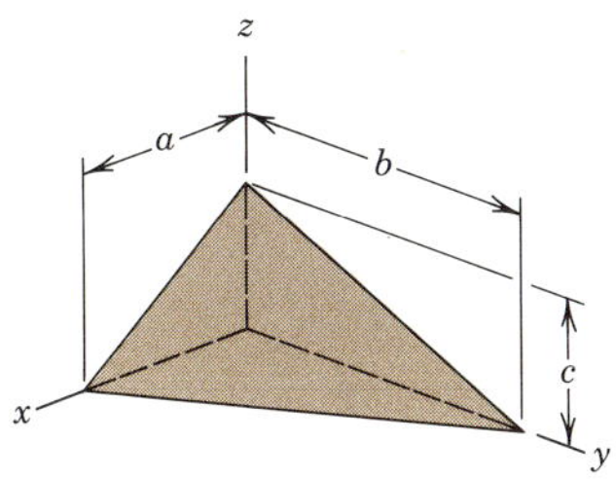

EXAMPLE 4

Three sides of the orthogonal tetrahedron coincide with the coordinate planes. The density is ρ. Determine the moment of inertia I_{xx} and the product of inertia I_{xy} of this body.

Solution

All surfaces of a tetrahedron are planar. We choose Cartesian coordinates for the formulation because the equation of any plane is linear in x, y, and z according to

$$z = k_1 x + k_2 y + k_3$$

The coordinate planes correspond to constant values of either x, y, or z. The coefficients k_i for the slanted surface may be found by noting that the points $(a, 0, 0)$, $(0, b, 0)$, and $(0, 0, c)$ are on the surface. Thus

$$0 = k_1 a + k_3, \qquad 0 = k_2 b + k_3, \qquad c = k_3$$

Solving for k_1, k_2, and k_3 yields

$$z = c\left(1 - \frac{x}{a} - \frac{y}{b}\right)$$

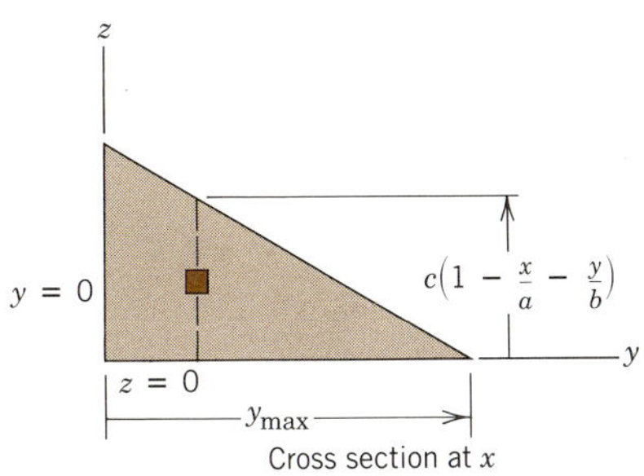

Cross section at x

Because the height z is known, let us draw a sketch of a cross section of the tetrahedron parallel to the yz plane. The value of x for this plane is arbitrary. The integration sequence depicted in the sketch is to first form a vertical strip by integrating over all z ranging from the lower surface $z = 0$, up to the slanted surface. Then all vertical strips lying on the cross section are accounted for by integrating from the left surface $y = 0$ to the point on the right where $y = y_{max}$ and $z = 0$. Thus at the point where $z = 0$

$$0 = c\left(1 - \frac{x}{a} - \frac{y_{max}}{b}\right)$$

from which we obtain the integration limit

$$y_{max} = b\left(1 - \frac{x}{a}\right)$$

Finally, all cross sections are included by integrating from the end $x = 0$ to the tip, where $x = a$.

The mass element is given by eq. (15). When the expression for dm is substituted into eq. (6) for I_{xx}, the differentials should be written in the order of integration. Thus

$$I_{xx} = \int_0^a \int_0^{b(1-x/a)} \int_0^{c(1-x/a-y/b)} (y^2 + z^2)(\rho\, dz\, dy\, dx)$$
$$= \rho \int_0^a \int_0^{b(1-x/a)} \left[y^2 z + \tfrac{1}{3} z^3\right]\Big|_0^{c(1-x/a-y/b)} dy\, dx$$
$$= \rho \int_0^a \int_0^{b(1-x/a)} \left[c\left(1 - \frac{x}{a}\right)y^2 - \frac{c}{b}y^3 + \frac{1}{3}c^3\left(1 - \frac{x}{a} - \frac{y}{b}\right)^3\right] dx\, dy$$
$$= \rho \int_0^a \left[\frac{c}{3}\left(1 - \frac{x}{a}\right)y^3 - \frac{c}{4b}y^4 - \frac{bc^3}{12}\left(1 - \frac{x}{a} - \frac{y}{b}\right)^4\right]\Bigg|_0^{b(1-x/a)} dx$$
$$= \rho \int_0^a \frac{1}{12}(b^3 c + bc^3)\left(1 - \frac{x}{a}\right)^4 dx$$
$$= \tfrac{1}{60}\rho abc(b^2 + c^2)$$

Equations (7) give the definition of I_{xy}. Aside from changing the integrand, the integration follows the same steps as for I_{xy}.

$$I_{xy} = \int_0^a \int_0^{b(1-x/a)} \int_0^{c(1-x/a-y/b)} xy(\rho\, dz\, dy\, dx)$$
$$= \rho \int_0^a \int_0^{b(1-x/a)} x\left[c\left(1 - \frac{x}{a}\right)y - \frac{c}{b}y^2\right] dy\, dx$$
$$= \rho \int_0^a x\left[\frac{1}{6}b^2 c\left(1 - \frac{x}{a}\right)^3\right] dx = \frac{1}{120}\rho a^2 b^2 c$$

EXAMPLE 5

Determine the moment of inertia I_{yy} and the product of inertia I_{yz} for the semi-cone of uniform density ρ.

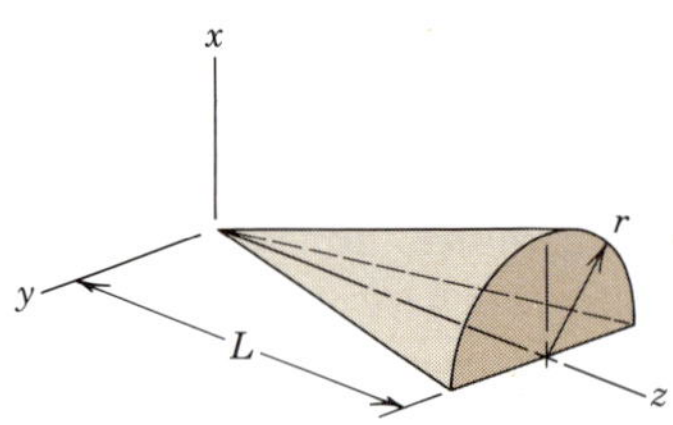

Solution

Any cross section of the cone parallel to the xy plane, such as the one in the sketch, forms a semicircle. Cylindrical coordinates, with z defined as the axial direction, readily define such a shape. The value of z is constant on the aforementioned cross section.

The transverse angle ϕ has been defined in the sketch relative to the x axis (as usual), so that

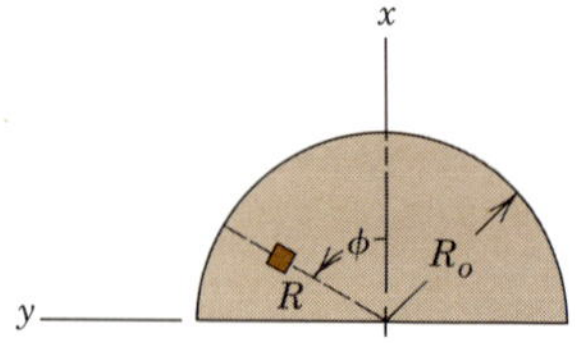

Cross section at z

$$x = R\cos\phi, \qquad y = R\sin\phi, \qquad dm = \rho(dR)(R\,d\phi)(dz)$$

All elements dm on the cross section are situated within the ranges $0 \le R \le R_0$, and $-\pi/2 \le \phi \le \pi/2$. The radius R_0 for the cross section is proportional

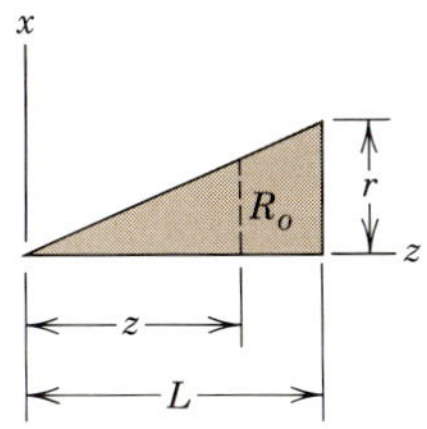

to the distance z, as can be seen in the sketch. Using the similar triangles appearing in this sketch yields

$$R_0 = \frac{r}{L} z$$

Note that the limits in the integration over the cross section are constant, so either R or ϕ may be chosen for the first integration. The scheme appearing in the sketch corresponds to integrating first over R. The last integration is over the range $0 \le z \le L$, which accounts for all of the cross sections.

When eqs. (6) and (7) are formulated, it is necessary to replace x and/or y by their polar coordinate representations. Thus

$$\begin{aligned}
I_{yy} &= \int_0^L \int_{-\pi/2}^{\pi/2} \int_0^{rz/L} (x^2 + z^2)(\rho R)(dR\, d\phi\, dz) \\
&= \rho \int_0^L \int_{-\pi/2}^{\pi/2} \int_0^{rz/L} (R^2 \cos^2 \phi + z^2) R\, dR\, d\phi\, dz \\
&= \rho \int_0^L \int_{-\pi/2}^{\pi/2} \left[\frac{1}{4}\left(\frac{rz}{L}\right)^4 \cos^2 \phi + \frac{1}{2}\left(\frac{rz}{L}\right)^2 z^2 \right] d\phi\, dz \\
&= \rho \int_0^L \left[\frac{1}{4}\frac{r^4}{L^4}\frac{\pi}{2} + \frac{1}{2}\frac{r^2}{L^2}\pi \right] z^4\, dz \\
&= \pi\rho r^2 L(\tfrac{1}{40}r^2 + \tfrac{1}{10}L^2)
\end{aligned}$$

$$\begin{aligned}
I_{xz} &= \int_0^L \int_{-\pi/2}^{\pi/2} \int_0^{rz/L} xz(\rho R)(dR\, d\phi\, dx) \\
&= \rho \int_0^L \int_{-\pi/2}^{\pi/2} \int_0^{rz/L} (R \cos \phi) z R\, dR\, d\phi\, dz \\
&= \rho \int_0^L \int_{-\pi/2}^{\pi/2} \frac{1}{3}\left(\frac{rz}{L}\right)^3 z \cos \phi\, d\phi\, dz \\
&= \rho \int_0^L \frac{1}{3}\frac{r^3}{L^3} z^4 [\sin \phi]\Big|_{-\pi/2}^{\pi/2} dz \\
&= \tfrac{2}{15}\rho r^3 L^2
\end{aligned}$$

HOMEWORK PROBLEMS

16.29 Determine by integration the moment of inertia of the uniform right circular cone of mass m about the y axis shown.

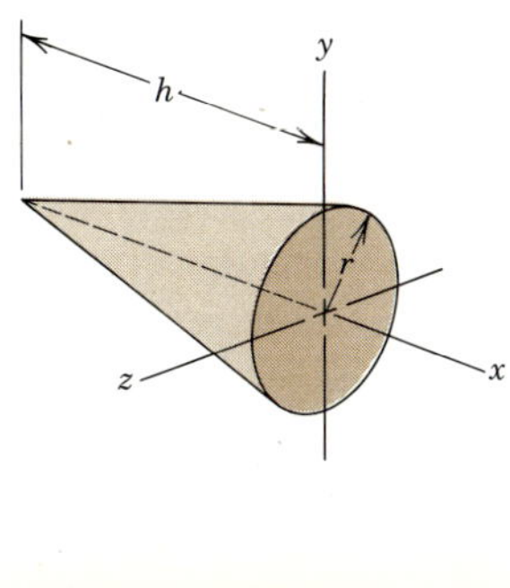

PROB. 16.29

16.30 Determine by integration the moments of inertia of the triangular prism of mass m about the axes shown.

16.31 Determine by integration the products of inertia of the triangular prism of mass m about the axes shown.

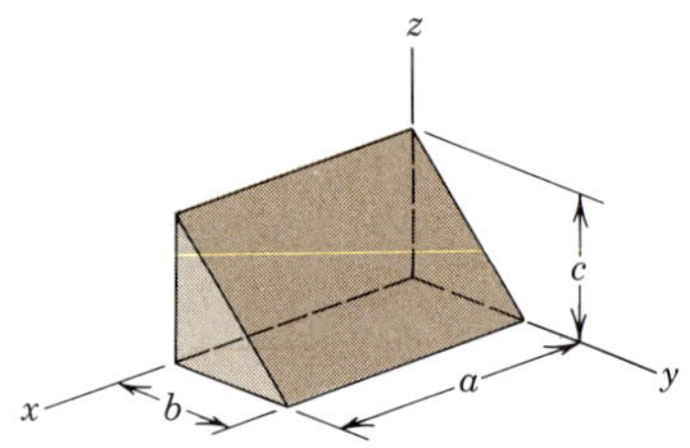

PROBS. 16.30 AND 16.31

16.32 The body shown is one-half of a parabolic cylinder. Its curved surface is defined by $z = c_1 y^2 + c_2$, where c_1 and c_2 are coefficients to be determined. The body is homogeneous with density ρ. Determine the corresponding values of I_{xx} and I_{yz}.

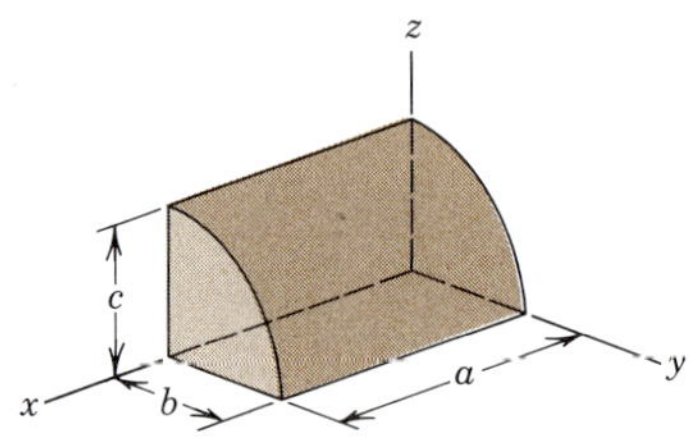

PROB. 16.32

16.33 For the body in Problem 16.32, determine I_{zz} and I_{yz}.

16.34 The upper surface of the body shown is a paraboloid of revolution defined by $z = (h/r^2)(x^2 + y^2)$. The density is ρ. Determine the values of I_{xx} and I_{zz} for this body.

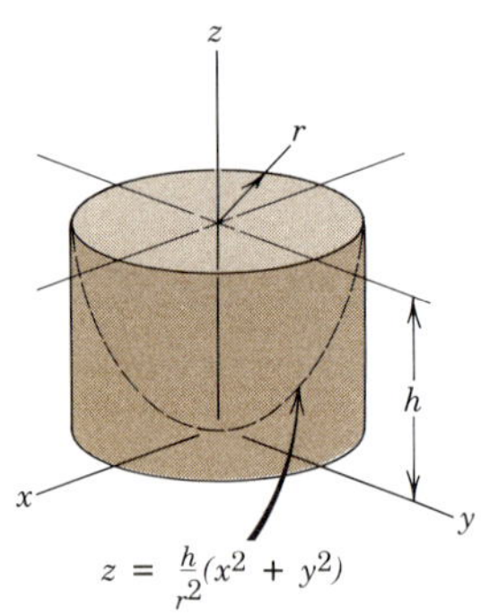

PROB. 16.34

16.35 The body shown is one-quarter of a paraboloid of revolution, for which $z = c(x^2 + y^2)$ along the curved surface. The density is ρ. Determine by integration the moments of inertia about the axes shown.

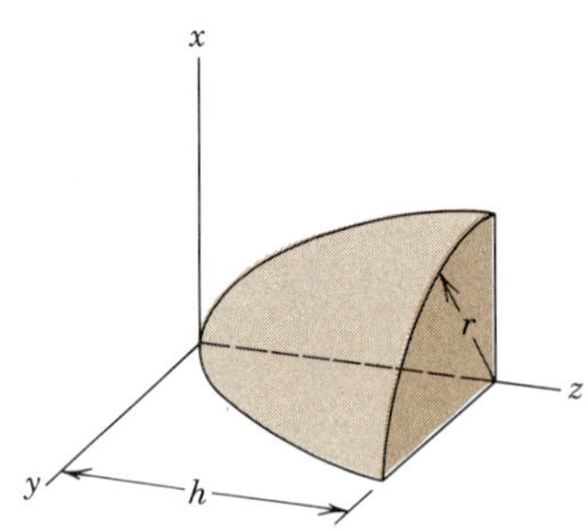

PROB. 16.35

16.36 Determine the products of inertia of the paraboloid of revolution in Problem 16.35 for the axes given there.

***16.37** The orthogonal pyramid has three surfaces that coincide with the coordinate planes and two slanted surfaces. The density is ρ. Determine the moment of inertia I_{xx} and the product of inertia I_{yz} of this body.

***16.38** The density of the pyramid in Problem 16.37 is given by $\rho = \rho_0 x/a$. Determine the mass and the moment of inertia I_{yy} of this body.

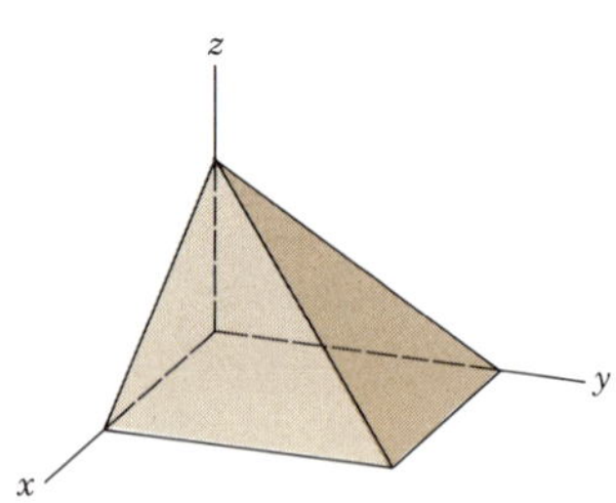

PROB. 16.37

*5. Rotation Transformation of Inertia Properties

The parallel axis theorems detail the transformation of the inertia properties for two coordinate systems having different origins but identical orientations. In some situations it may be necessary to transform the inertia properties between two coordinate systems having the same origin but different orientations.

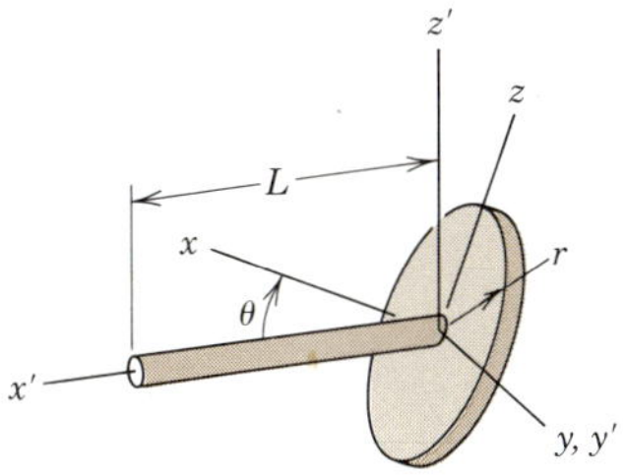

FIGURE 9

This is the case when we evaluate the properties of the system in Figure 9, where a thin disk of mass m_1 is mounted obliquely on a slender shaft of mass m_2. The properties of the disk and of the shaft with respect to their centroidal coordinate systems xyz and $x'y'z'$, respectively, are given in Appendix B. However, the properties of the individual bodies can be added to form the composite only when they are calculated with respect to the same coordinate system. Accomplishing this for the system in Figure 9 involves a rotational transformation between the xyz and $x'y'z'$ coordinate systems.

In the general situation we know the moments and products of inertia of a body with respect to one of the coordinate systems, say $x'y'z'$, and we wish to ascertain the equivalent inertia properties for the xyz coordinate system. Note that xyz is selected to have the same origin as $x'y'x'$, but an arbitrary orientation. The orientation of xyz with respect to $x'y'z'$ may be described by a set of direction cosines. Let $l_{pq'}$ ($\equiv l_{q'p}$) be the cosine of the angle between axis p of the xyz system and axis q' of the $x'y'z'$ system. For example, Figure 10 shows an arbitrary orientation of x with respect to the $x'y'z'$ frame, for which we have $l_{xx'} = \cos\alpha$, $l_{xy'} = \cos\beta$, and $l_{xz'} = \cos\gamma$.

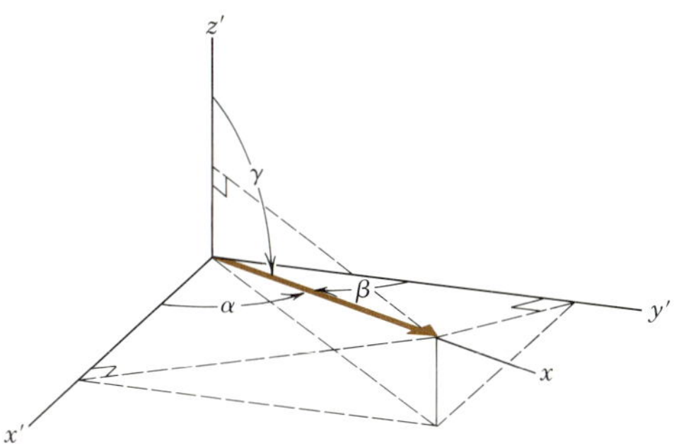

FIGURE 10

As can be seen in Figure 10, the unit vector $\bar{i}$ can be written in terms of its components in the $x'y'z'$ frame as

$$\begin{aligned}\bar{i} &= (\cos\alpha)\bar{i}' + (\cos\beta)\bar{j}' + (\cos\gamma)\bar{k}' \\ &\equiv l_{xx'}\bar{i}' + l_{xy'}\bar{j}' + l_{xz'}\bar{k}' \end{aligned} \tag{18a}$$

Relations for the unit vectors $\bar{j}$ or $\bar{k}$ may be obtained in a similar manner by considering Figure 10 to represent either the y or z axis. Thus

$$\bar{j} = l_{yx'}\bar{i}' + l_{yy'}\bar{j}' + l_{yz'}\bar{k}' \tag{18b}$$

$$\bar{k} = l_{zx'}\bar{i}' + l_{zy'}\bar{j}' + l_{zz'}\bar{k}' \tag{18c}$$

One use of these expressions for the unit vectors is to obtain relationships governing the direction cosines. Because the x, y, and z axes are mutually orthogonal,

$$\bar{i}\cdot\bar{i} = \bar{j}\cdot\bar{j} = \bar{k}\cdot\bar{k} = 1$$

$$\bar{i}\cdot\bar{j} = \bar{j}\cdot\bar{k} = \bar{i}\cdot\bar{k} = 0$$

Then in view of eqs. (18) and the fact that $x'y'z'$ also form a set of orthogonal axes, the foregoing relations require that

$$\begin{aligned}&(l_{px'})^2 + (l_{py'})^2 + (l_{pz'})^2 = 1 \qquad p = x, y, \text{ or } z \\ &I_{px'}l_{qx'} + l_{py'}l_{qy'} + l_{pz'}l_{qz'} = 0 \qquad p, q = x, y, \text{ or } z, \qquad p \neq q\end{aligned} \tag{19}$$

Equations (19) represent six relationships that the nine direction cosines must satisfy, so that only three of the direction cosines, such as $l_{xx'}$, $l_{yy'}$, and $l_{zz'}$, are independent quantities.

Let us now consider the matter of the computation of the moment of inertia I_{xx}. By definition,

$$I_{xx} = \int_{\mathscr{V}} (y^2 + z^2)\, dm$$

But $(y^2 + z^2)^{1/2} = R_x$ is the distance from the differential element at point i to the x axis. In other words, it is the component of $\bar{r}_{i/O}$ perpendicular to the x axis, as illustrated in Figure 11. From vector algebra we know that

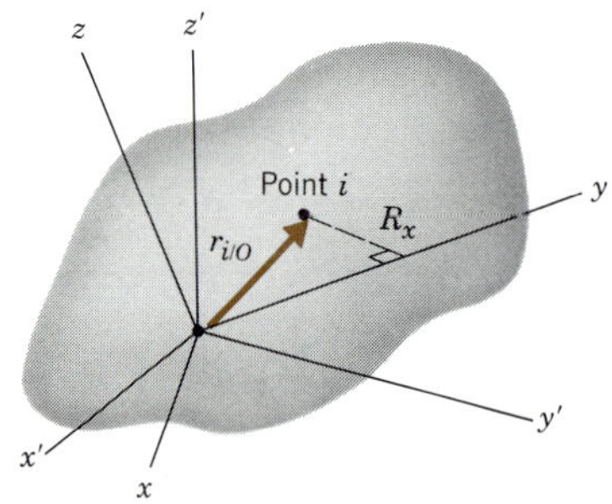

FIGURE 11

the component of a vector perpendicular to an axis is the magnitude of the cross product of the vector and the unit vector for that axis. Thus

$$R_x = |\bar{r}_{i/O} \times \bar{i}| = [(\bar{r}_{i/O} \times \bar{i}) \cdot (\bar{r}_{i/O} \times \bar{i})]^{1/2}$$

All components will be referred to the $x'y'z'$ coordinate system because we are considering the inertia properties to be known with respect to this reference frame. Thus describing $\bar{r}_{i/O}$ in terms of the (x', y', z') coordinates gives

$$\bar{r}_{i/O} = x'\bar{i}' + y'\bar{j}' + z'\bar{k}' \tag{20}$$

It follows from eq. (18a) that

$$\bar{r}_{i/O} \times \bar{i} = (l_{xz'}y' - l_{xy'}z')\bar{i}' + (l_{xx'}z' - l_{xz'}x')\bar{j}' + (l_{xy'}x' - l_{xx'}y')\bar{k}'$$

From this we form

$$\begin{aligned} R_x^2 &\equiv (\bar{r}_{i/O} \times \bar{i}) \cdot (\bar{r}_{i/O} \times \bar{i}) \\ &= (l_{xz'}y' - l_{xy'}z')^2 + (l_{xx'}z' - l_{xz'}x')^2 + (l_{xy'}x' - l_{xx'}y')^2 \end{aligned}$$

Collecting terms in this equation for R_x^2, yields

$$\begin{aligned} I_{xx} = \int_{\mathscr{V}} R_x^2\, dm = \int_{\mathscr{V}} \{&l_{xx'}^2[(y')^2 + (z')^2] + l_{xy'}^2[(x')^2 + (z')^2] \\ &+ l_{xz'}^2[(x')^2 + (y')] - 2l_{xx'}l_{xy'}x'y' - 2l_{xy}l_{xz'}y'z' - 2l_{xx'}l_{xz'}x'z'\}\, dm \end{aligned}$$

The direction cosines are constants for the integration over the region $\mathscr{V}$, so they may be factored out. The remaining integrals are then either moments or products of inertia with respect to $x'y'z'$.

Comparable steps in evaluating I_{yy} and I_{zz} would result in the permutation of the symbols in the foregoing expressions for I_{xx}. With p used to represent the rotated axis of interest, the general formula for the moments of inertia with respect to the xyz system is then

$$\begin{aligned} I_{pp} = {} & l_{px'}^2 I_{x'x'} + l_{py'}^2 I_{y'y'} + l_{pz'}^2 I_{z'z'} \\ & - 2 l_{px'} l_{py'} I_{x'y'} - 2 l_{py'} l_{pz'} I_{y'z'} \\ & - 2 l_{px'} l_{pz'} I_{x'z'} \qquad p = x, y, \text{ or } z \end{aligned} \tag{21}$$

Relating the products of inertia with respect to xyz to the $x'y'z'$ inertia properties also requires a transformation between the two sets of coordinates. For this we use the identities

$$x \equiv \bar{r}_{i/O} \cdot \bar{i}, \qquad y \equiv \bar{r}_{i/O} \cdot \bar{j}, \qquad z \equiv \bar{r}_{i/O} \cdot \bar{k}$$

When eq. (20) for $\bar{r}_{i/O}$ and eqs. (18) for the unit vectors are substituted into these identities, the result is

$$\begin{aligned} x &= l_{xx'} x' + l_{xy'} y' + l_{xz'} z' \\ y &= l_{yx'} x' + l_{yy'} y' + l_{yz'} z' \\ z &= l_{zx'} x' + l_{zy'} y' + l_{zz'} z' \end{aligned} \tag{22}$$

The combination xy occurs in the definition of I_{xy}. Equations (22) and the relations for the direction cosines in eqs. (19) give

$$\begin{aligned} xy = {} & l_{xx'} l_{yx'} (x')^2 + l_{xy'} l_{yy'} (y')^2 + l_{xz'} l_{yz'} (z')^2 + (l_{xx'} l_{yy'} + l_{xy'} l_{yx'}) x'y' \\ & + (l_{xx'} l_{yz'} + l_{xz'} l_{yx'}) x'z' + (l_{xy'} l_{yz'} + l_{xz'} l_{yy'}) y'z' \\ = {} & -(l_{xy'} l_{yy'} + l_{xz'} l_{yz'})(x')^2 - (l_{xx'} l_{yx'} + l_{xz'} l_{yz'})(y')^2 - (l_{xx'} l_{yx'} + l_{xy'} l_{yy'})(z')^2 \\ & + (l_{xx'} l_{yy'} + l_{xy'} l_{yx'}) x'y' + (l_{xy'} l_{yz'} + l_{xz'} l_{yy'}) y'z' + (l_{xz'} l_{yx'} + l_{xx'} l_{yz'}) x'z' \end{aligned}$$

These terms are regrouped when the product of inertia is formed, with the result that

$$\begin{aligned} I_{xy} &= \int_{\mathcal{V}} xy \, dm \\ &= \int_{\mathcal{V}} \{ -l_{xy'} l_{yy'} [(x')^2 + (z')^2] - l_{xz'} l_{yz'} [(x')^2 + (y')^2] \\ &\quad - l_{xx'} l_{yx'} [(y')^2 + (z')^2] + (l_{xx'} l_{yy'} + l_{xy'} l_{yx'}) x'y' \\ &\quad + (l_{xy'} l_{yz'} + l_{xz'} l_{yy'}) y'z' + (l_{xz'} l_{yx'} + l_{xx'} l_{yz'}) x'z' \} \, dm \end{aligned}$$

As was true for the moments of inertia, the integrals remaining after the direction cosines are factored out are the inertia properties with respect to $x'y'z'$.

Comparable steps for the other products of inertia lead to permutations of the symbols in the foregoing equation. Letting p and q represent x, y, or z, the general formula for the product of inertia with respect to the xyz system is

$$I_{pq} = (l_{px'}l_{qy'} + l_{py'}l_{qx'})I_{x'y'} + (l_{py'}l_{qz'} + l_{pz'}l_{qy'})I_{y'z'} + (l_{px'}l_{qz'} + l_{pz'}l_{qx'})I_{x'z'} - l_{px'}l_{qx'}I_{x'x'} - l_{py'}l_{qy'}I_{y'y'} - l_{pz'}l_{qz'}I_{z'z'} \qquad p, q = x, y, \text{ or } z, \text{ but } p \neq q \tag{23}$$

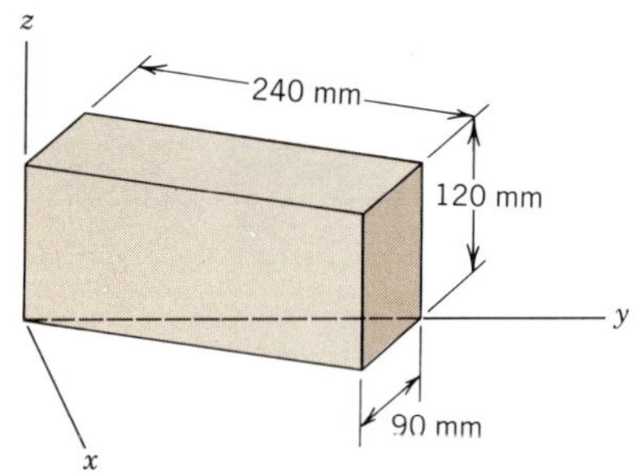

EXAMPLE 6

Determine the moment of inertia I_{yy} and the product of inertia I_{xy} of the 6 kg homogeneous box.

Solution

The inertia properties of the box with respect to the centroidal coordinate system $\hat{x}\hat{y}\hat{z}$ shown in the sketch are given in Appendix B. However, the rotation transformations require that the two coordinate systems have the same origin. The $x'y'z'$ system has been defined in the sketch to be parallel to $\hat{x}\hat{y}\hat{z}$, so that the parallel axis theorems may be employed.

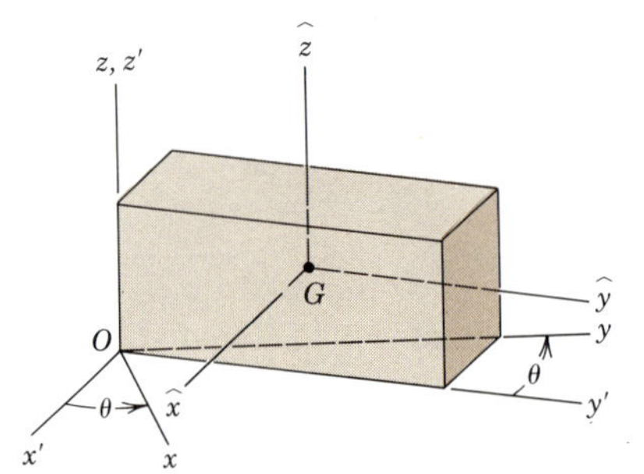

The $\hat{x}\hat{y}\hat{z}$ system forms principal axes (by symmetry) for which Appendix B gives

$$I_{\hat{x}\hat{x}} = \tfrac{1}{12}(6)[(0.24)^2 + (0.12)^2] = 0.036 \text{ kg-m}^2$$
$$I_{\hat{y}\hat{y}} = \tfrac{1}{12}(6)[(0.09)^2 + (0.12)^2] = 0.01125 \text{ kg-m}^2$$
$$I_{\hat{z}\hat{z}} = \tfrac{1}{12}(6)[(0.09)^2 + (0.24)^2] = 0.03285 \text{ kg-m}^2$$
$$I_{\hat{x}\hat{y}} = I_{\hat{x}\hat{z}} = I_{\hat{y}\hat{z}} = 0$$

The coordinates of the origin O with respect to $\hat{x}\hat{y}\hat{z}$ are $(0.045, -0.12, -0.06)$ meters, so the parallel axis theorems yield

$$I_{x'x'} = I_{\hat{x}\hat{x}} + m[(-0.12)^2 + (-0.06)^2] = 0.1440 \text{ kg-m}^2$$
$$I_{y'y'} = I_{\hat{y}\hat{y}} + m[(0.045)^2 + (-0.06)^2] = 0.0450 \text{ kg-m}^2$$
$$I_{z'z'} = I_{\hat{z}\hat{z}} + m[(0.045)^2 + (-0.12)^2] = 0.1314 \text{ kg-m}^2$$
$$I_{x'y'} = I_{\hat{x}\hat{y}} + m(0.045)(-0.12) = -0.0324 \text{ kg-m}^2$$
$$I_{x'z'} = I_{\hat{x}\hat{z}} + m(0.045)(-0.06) = -0.0162 \text{ kg-m}^2$$
$$I_{y'z'} = I_{\hat{y}z} + m(-0.12)(-0.06) = 0.0432 \text{ kg-m}^2$$

The rotation transformation may now be evaluated. The angle θ in the sketch describes the orientation of the x and y axes. It is found from the fact that the x axis is the diagonal of the lower face of the box. Thus

$$\theta = \tan^{-1}\left(\frac{0.09}{0.24}\right) = 20.56^\circ$$

Because the z and z' axes coincide, the direction cosines are

$$l_{xx'} = \cos\theta = 0.9363, \qquad l_{xy'} = \cos(90^\circ - \theta) = 0.3512, \qquad l_{xz'} = \cos 90^\circ = 0$$

$$l_{yx'} = \cos(90^\circ + \theta) = -0.3512, \qquad l_{yy'} = \cos\theta = 0.9363, \qquad l_{yz'} = \cos 90^\circ = 0$$

$$l_{zx'} = l_{zy'} = \cos 90^\circ = 0, \qquad l_{zz'} = \cos 0^\circ = 1$$

We use eq. (21) to compute I_{yy} by setting $p = y$ there. This yields

$$\begin{aligned} I_{yy} &= (-0.3512)^2(0.144) + (0.9363)^2(0.045) + 0 \\ &\quad - 2(-0.3512)(0.9363)(-0.0324) - 0 - 0 \\ &= 0.03590 \text{ kg-m}^2 \end{aligned}$$

Equation (23) yields I_{xy} by setting $p = x$ and $q = y$. The result is

$$\begin{aligned} I_{xy} &= [(0.9363)^2 - (0.3512)^2](-0.0324) + 0 + 0 \\ &\quad - (0.9363)(-0.3512)(0.1440) - (0.3512)(0.9363)(0.0450) - 0 \\ &= 0.00815 \text{ kg-m}^2 \end{aligned}$$

Additional Remarks

In the preceding example the inertia properties given in the tabulation were for principal axes because of the symmetry of the body. Also the problem statement requested only two inertia values, and the z and z' axes were identical. As a result, the calculations for the solution were not unduly complicated. It is not difficult to see that a rotation transformation for a more general problem can be quite involved. As a computational aid, the transformations defined by eqs. (21) and (23) may be written in either matrix or tensor form. Matrices and tensors are not presented in this textbook. However, either technique would greatly assist us in determining the principal axes of *any* body, not just symmetric ones. Principal axes have obvious usefulness, because of the simplifications that arise in the angular momentum eq. (5). The student who is interested in these topics is referred to higher-level textbooks in dynamics.

Here, we shall find that elementary problems in kinetics are more easily solved by using the most convenient set of axes, regardless of whether or not they are principal ones. Such an approach avoids the extra step of determining principal axes. In engineering practice, the decision to use principal axes should be carefully made, especially when formulating a problem for solution by electronic computer. The computational savings associated with the simplifications in the expression for the angular momentum resulting from principal axes must be weighed against the computations necessary to determine such axes.

HOMEWORK PROBLEMS

16.39 Determine I_{xx} and I_{xy} for the circular disk of mass m.

16.40 Determine the moments and products of inertia of the 200 lb rectangular plate with respect to the centroidal xyz coordinate system shown. The z axis is perpendicular to the plane of the plate.

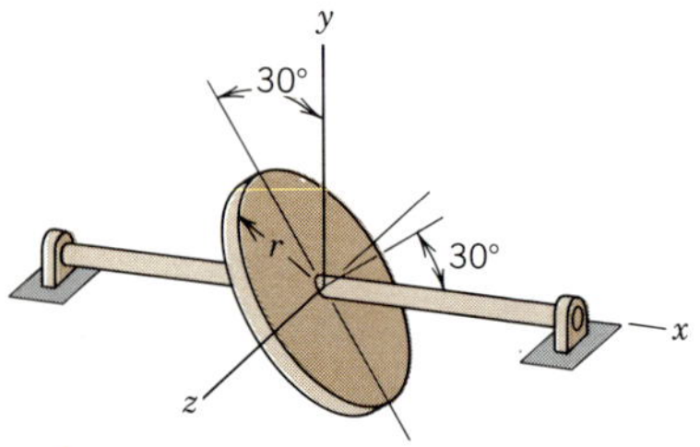

PROB. 16.39

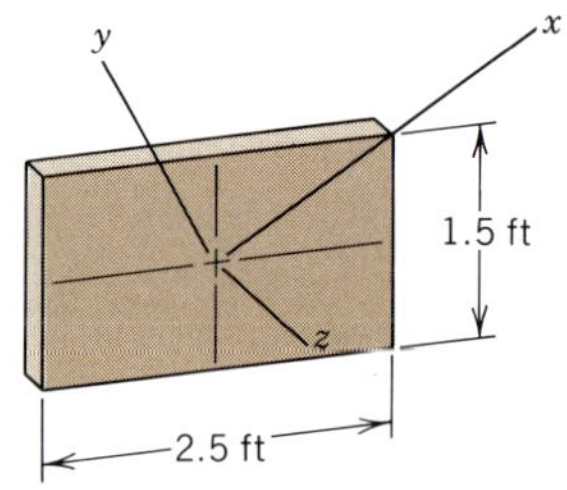

PROB. 16.40

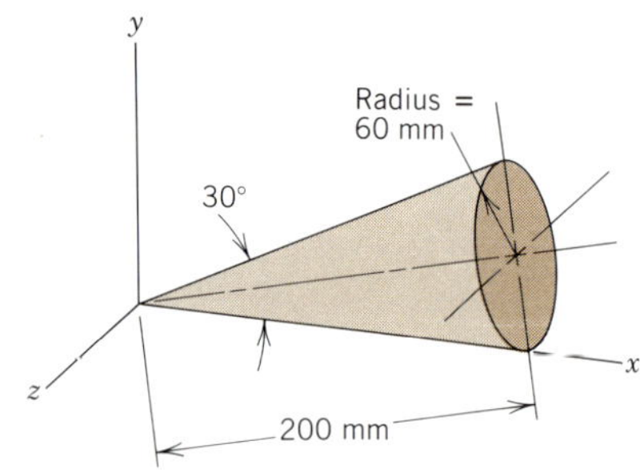

PROB. 16.41

16.41 Determine I_{xx} and I_{xy} for the 8 kg right circular cone.

16.42 Determine I_{xx} and I_{xy} for the cross bar assembly. Each bar weighs 1 lb/ft.

16.43 Determine $I_{y'y'}$ and $I_{x'y'}$ for the cross bar assembly. Each bar weighs 1 lb/ft.

16.44 The disk in Figure 9 has mass m_1, and the mass of the slender shaft is m_2. Derive an expression for I_{xx} of the composite in terms of m_1, m_2, r, L, and θ.

***16.45** The rectangular plate shown has a mass of 30 kg. Determine I_{xx} for the centroidal xyz coordinate system shown. The x axis is in the plane of the plate.

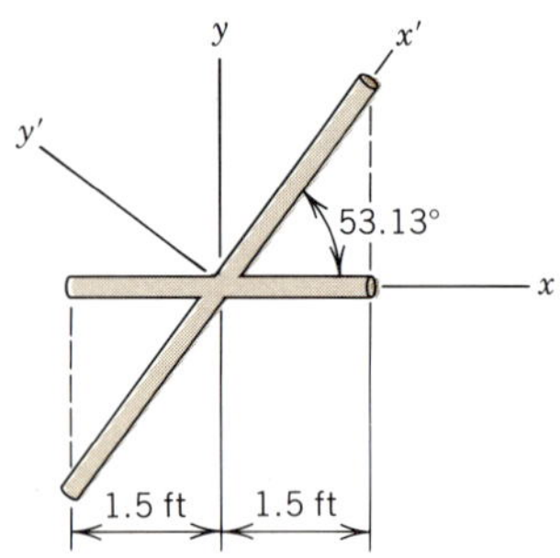

PROBS. 16.42 AND 16.43

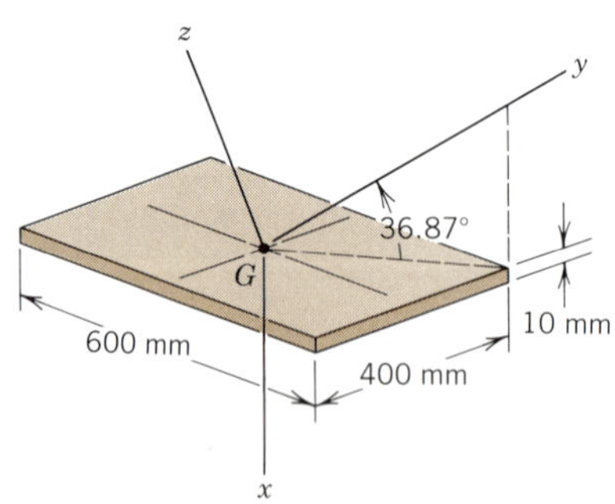

PROB. 16.45

***16.46** Determine I_{yz} for the rectangular plate and xyz coordinate system in Problem 16.45.

D. EQUATIONS OF MOTION

Now that we have studied how the inertia properties of a body are obtained, we shall continue with the development of the relationship between angular momentum and the moment exerted by the force system. The basic relations for this study are eqs. (5) and (2), which we have repeated below.

$$\begin{aligned}\bar{H}_A = (I_{xx}\omega_x - I_{xy}\omega_y - I_{xz}\omega_z)\bar{i} \\ + (I_{yy}\omega_y - I_{xy}\omega_x - I_{yz}\omega_z)\bar{j} \\ + (I_{zz}\omega_z - I_{xz}\omega_x - I_{yz}\omega_y)\bar{k}\end{aligned} \tag{5}$$

$$\sum \bar{M}_A = \dot{\bar{H}}_A \tag{2}$$

Because of their importance, let us recapitulate the criteria that were imposed on the xyz (moving) reference frame in developing these relations.

1. Select point A to be the center of mass of the body, or a fixed point in space that is also fixed relative to the body. (As we did for planar motion, we shall ignore the possibility of selecting a point in the body that is accelerating directly toward or away from the center of mass. Recall that eq. (2) is valid for such points.)
2. Select a moving reference frame xyz that is attached to the body with origin at point A. This choice for xyz ensures that the moments and products of inertia with respect to the reference frame are constants.
3. Orient the reference frame relative to the body in a manner that facilitates the determination of the moments and products of inertia. This orientation is usually suggested by the coordinate axes appearing in Appendix B.

In the case where a body undergoes a translational motion, the only allowable point for summing moments is the center of mass G. Then, because $\bar{\omega} \equiv \bar{0}$ for translation, the moment eq. (2) reduces to $\sum \bar{M}_G = \bar{0}$. On the other hand, when a body has rotational motion, eq. (5) describes $\bar{H}_A$ in terms of its $(\bar{i}, \bar{j}, \bar{k})$ components. Both the magnitudes of these components and the unit vectors may change with time. Therefore, the evaluation of $\dot{\bar{H}}_A$ requires that we employ the concept of differentiation with respect to a moving reference frame, as derived in Chapter 15. Recall that, for an arbitrary vector $\bar{B}$, the general formula for this procedure is

$$\frac{d\bar{B}}{dt} = \frac{\delta\bar{B}}{\delta t} + \bar{\omega} \times \bar{B} \tag{24}$$

Operating on $\bar{H}_A$ in this manner changes eq. (2) to

$$\sum \bar{M}_A = \dot{\bar{H}}_A = \frac{\delta\bar{H}_A}{\delta t} + \bar{\omega} \times \bar{H}_A \tag{25}$$

By definition, the components of $\delta\bar{H}_A/\delta t$ are the derivatives of the components of $\bar{H}_A$. In view of the fact that the moments and products of inertia are constants, differentiating the components of $\bar{H}_A$ in eq. (5) gives

$$\begin{aligned}\frac{\delta \bar{H}_A}{\delta t} &= (I_{xx}\dot{\omega}_x - I_{xy}\dot{\omega}_y - I_{xz}\dot{\omega}_z)\bar{i} \\ &+ (I_{yy}\dot{\omega}_y - I_{xy}\dot{\omega}_x - I_{yz}\dot{\omega}_z)\bar{j} \\ &+ (I_{zz}\dot{\omega}_z - I_{xz}\dot{\omega}_x - I_{yz}\dot{\omega}_y)\bar{k}\end{aligned} \tag{26}$$

The interpretation of this expression involves an important subtlety. In general, it is a difficult matter to differentiate the components of $\bar{\omega}$ because such an evaluation would require expressions for ω_x, ω_y, and ω_z that are valid for all time instants. However, in the approach we have developed, the xyz reference frame is attached to the body and therefore has the same angular velocity as the body. As a result, eq. (24) gives

$$\frac{d\bar{\omega}}{dt} \equiv \bar{\alpha} = \frac{\delta\bar{\omega}}{\delta t} + \bar{\omega} \times \bar{\omega} \equiv \frac{\delta\bar{\omega}}{\delta t}$$

or, in component form,

$$\alpha_x = \dot{\omega}_x \qquad \alpha_y = \dot{\omega}_y \qquad \alpha_z = \dot{\omega}_z$$

In other words, because the xyz frame is attached to the body, the components of the angular acceleration of the body are identical to the derivatives of the angular velocity components. This means that the procedures in Chapter 15 for determining $\bar{\omega}$ and $\bar{\alpha}$ provide the kinematical relations needed to formulate the moment equation. As a result, eq. (26) may be written in a more useful form as

$$\boxed{\begin{aligned}\frac{\delta \bar{H}_A}{\delta t} &= (I_{xx}\alpha_x - I_{xy}\alpha_y - I_{xz}\alpha_z)\bar{i} \\ &+ (I_{yy}\alpha_y - I_{xy}\alpha_x - I_{yz}\alpha_z)\bar{j} \\ &+ (I_{zz}\alpha_z - I_{xz}\alpha_x - I_{yz}\alpha_y)\bar{k}\end{aligned}} \tag{27}$$

Equations (5), (25), and (27) form the basic relationships between the resultant moment applied to a rigid body and the angular motion of the body. Clearly, the moment equation has three components, so that in conjunction with the components of eq. (1) governing the effect of the resultant force,

$$\boxed{\sum \bar{F} = m\bar{a}_G} \tag{1}$$

we have the six scalar equations of motion alluded to earlier.

In order to gain insight into the significance of the moment equation, let us consider what conditions must exist for eq. (25) to reduce to the familiar moment equation for planar motion. For an arbitrary body in planar motion parallel to the fixed XY plane, we select the z axis to be perpendicular to the plane. The rotation is always about the z axis, so

$$\bar{\omega} = \omega\bar{k}, \qquad \bar{\alpha} = \dot{\omega}\bar{k}$$

Equations (5) and (27) then give

$$\bar{H}_A = -I_{xz}\omega\bar{i} - I_{yz}\omega\bar{j} + I_{zz}\omega\bar{k}$$

$$\frac{\delta\bar{H}_A}{\delta t} = -I_{xz}\dot{\omega}\bar{i} - I_{yz}\dot{\omega}\bar{j} + I_{zz}\dot{\omega}\bar{k}$$

Substituting the foregoing expressions into the moment equation (25) yields the special relation for the case of planar motion,

$$\begin{aligned}\sum\bar{M}_A &= \dot{\omega}(-I_{xz}\bar{i} - I_{yz}\bar{j} + I_{zz}\bar{k}) + \omega\bar{k}\times\omega(-I_{xz}\bar{i} - I_{yz}\bar{j} + I_{zz}\bar{k})\\ &= -(I_{xz}\dot{\omega} - I_{yz}\omega^2)\bar{i} - (I_{yz}\dot{\omega} + I_{xz}\omega^2)\bar{j} + I_{zz}\dot{\omega}\bar{k}\end{aligned} \tag{28}$$

Notice that the $\bar{i}$ and $\bar{j}$ components in this equation cause it to be different from the moment equation $\sum\bar{M}_A = I_{zz}\dot{\omega}\bar{k}$ employed in Chapter 14. The $\bar{i}$ and $\bar{j}$ components in eq. (28) describe the resultant moments of the force system about the x and y axes necessary to constrain an arbitrary body to planar motion. This is the situation that arises when a body that is rotating about a fixed axis is *dynamically unbalanced* (as will be illustrated in Example 7 to follow). Only in situations where $I_{xz} = I_{yz} = 0$, that is, where the z axis is a principal axis, will these constraint moments be unnecessary to maintain a body in planar motion. This condition was assured in Chapter 14, because there we considered only bodies that are symmetric about the xy plane.

One of the major differences between planar motion of symmetric bodies and other cases of rigid body motion is apparent from the foregoing discussion. In the former case, the force system will exert a resultant moment only when the rate of rotation changes, $\dot{\omega} \neq 0$. In contrast, a resultant moment is required to maintain the angular motion in the general case, even when the rates of rotation are constant. This is evidenced in eq. (28) by the terms that depend on ω^2. In general, the portion of $\dot{\bar{H}}_A$ that depends only on the rates of rotation is called the *gyroscopic moment.*

Let us now return to the study of the equations of motion for a body in three-dimensional motion. In those cases where the chosen xyz reference system does not form principal axes, we will formulate the problem by evaluating $\bar{H}_A$ and $\delta\bar{H}_A/\delta t$ for the axes chosen, and then substitute these expressions into eq. (25) for $\sum\bar{M}_A$.

When the xyz system forms principal axes, the expressions for $\bar{H}_A$ and $\delta\bar{H}_A/\delta t$ simplify considerably. The form of eq. (25) in this case is called *Euler's equation of motion*, named after Leonhard Euler (1707–1783). Thus, *when the xyz coordinate system is a set of principal axes of the body*, the vector equation for the rotational motion of the body is

$$\boxed{\begin{aligned}\sum\bar{M}_A = &[I_{xx}\alpha_x - (I_{yy} - I_{zz})\omega_y\omega_z]\bar{i}\\ &+ [I_{yy}\alpha_y - (I_{zz} - I_{xx})\omega_x\omega_z]\bar{j}\\ &+ [I_{zz}\alpha_z - (I_{xx} - I_{yy})\omega_x\omega_y]\bar{k}\end{aligned}} \tag{29}$$

It should be evident that for problems where the xyz system forms a principal set of axes, eq. (29) involves fewer computations than eqs. (5), (25), and (27). However, the separate computation of $\bar{H}_A$ and $\delta\bar{H}_A/\delta t$ has the advantage of enabling us to gain insight about the underlying physics of the system being studied. Also, this approach is the one that must be employed if the xyz system is a nonprincipal set of axes.

EXAMPLE 7

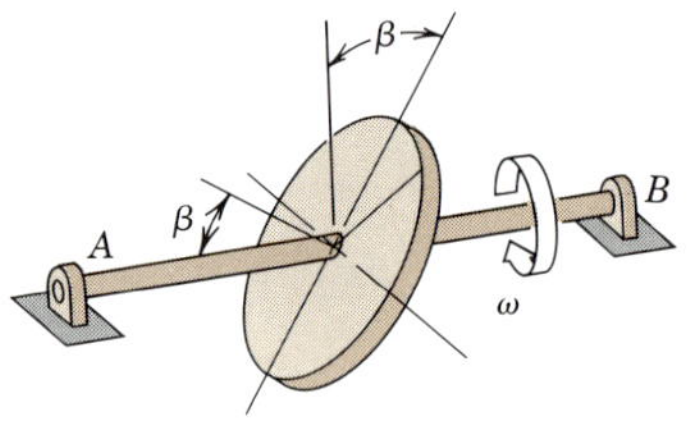

The obliquely mounted disk of Example 1 is illustrated again here. The rigid shaft, which is welded to the disk, is rotating at the constant angular speed ω. The bearings are frictionless and can apply only transverse shear forces to restrain the transverse motion of the shaft. Additionally, bearing A can exert a thrust force to restrain axial motion of the shaft. Determine (a) the gyroscopic moment exerted on the disk, and (b) the reaction forces exerted by the bearings due to the motion of the system.

Solution

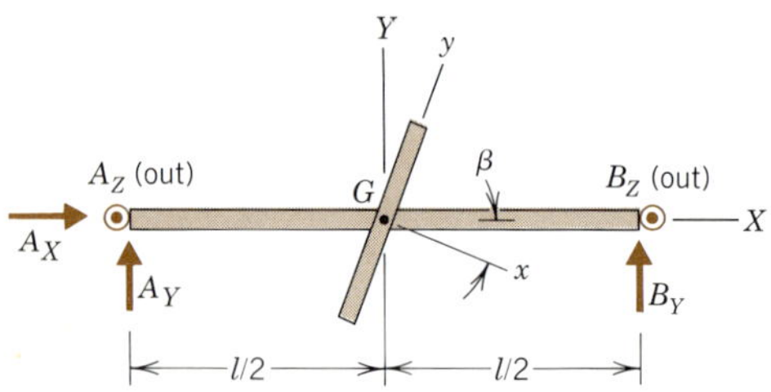

We begin with a free body diagram of the system that also shows the orientation of xyz relative to the system. The reactions at bearings A and B are represented in the free body diagram in terms of components parallel and perpendicular to the shaft. We have not included the gravity force because it is a static effect; if we sought the total reactive forces we would include it.

The xyz reference frame is the same as that used in Example 1, so that we may use the results for the vectors $\bar{\omega}$ and $\bar{H}_G$ derived there.

$$\bar{\omega} = \omega \bar{I} = \omega(\cos\beta\,\bar{i} + \sin\beta\,\bar{j}): \quad \text{always}$$

$$\bar{H}_G = \tfrac{1}{4} m r^2 \omega (2\cos\beta\,\bar{i} + \sin\beta\,\bar{j})$$

Both the rate of rotation ω and the unit vector $\bar{I}$ are constants, so we find $\bar{\alpha} = \dot{\bar{\omega}} = \bar{0}$. Then eq. (27) gives $\delta\bar{H}_G/\delta t = \bar{0}$, from which eq. (25) reduces to

$$\dot{\bar{H}}_G \equiv \frac{\delta \bar{H}_G}{\delta t} + \bar{\omega} \times \bar{H}_G = -\tfrac{1}{4} m r^2 \omega^2 \sin\beta\cos\beta\,\bar{k}$$

$$\equiv -\tfrac{1}{8} m r^2 \omega^2 \sin 2\beta\,\bar{k}$$

The value of $\dot{\bar{H}}_G$ given above represents the couple that must be exerted in order to rotate the disk at the constant rate ω. In other words, it is the gyroscopic moment requested in part (a).

Solving part (b) requires consideration of the external force system. Because the center of mass G is on the fixed axis of rotation, it has no acceleration, and

$$\sum \bar{F} = m\bar{a}_G = \bar{0}$$

This equation can be resolved into components with respect to any coordinate system. As usual, we will employ the moving xyz system. Summing like components of the forces in the free body diagram yields

$$\sum F_x = A_X \cos\beta - (A_Y + B_Y)\sin\beta = 0$$

$$\sum F_y = A_X \sin\beta + (A_Y + B_Y)\cos\beta = 0$$

$$\sum F_z = A_Z + B_Z = 0$$

The moment equation must be formulated in terms of xyz components, because those are the components for $\dot{\bar{H}}_G$. Cross products are useful for this step. Note that the reaction $A_X\bar{I}$ exerts no moment about point G.

$$\sum \bar{M}_G = \bar{r}_{A/G} \times \bar{F}_A + \bar{r}_{B/G} \times \bar{F}_B$$

$$= \frac{l}{2}(-\cos\beta\,\bar{i} - \sin\beta\,\bar{j}) \times (-A_Y \sin\beta\,\bar{i} + A_Y \cos\beta\,\bar{j} + A_Z\bar{k})$$

$$+ \frac{l}{2}(\cos\beta\,\bar{i} + \sin\beta\,\bar{j}) \times (-B_Y \sin\beta\,\bar{i} + B_Y \cos\beta\,\bar{j} + B_Z\bar{k})$$

$$= \frac{l}{2}\sin\beta(-A_Z + B_Z)\bar{i} + \frac{l}{2}\cos\beta(A_Z - B_Z)\bar{j}$$

$$+ \frac{l}{2}[(-\cos^2\beta - \sin^2\beta)(A_Y - B_Y)]\bar{k}$$

$$= \dot{\bar{H}}_G = -\tfrac{1}{8}mr^2\omega^2 \sin 2\beta\,\bar{k}$$

Matching like components yields

$$\bar{i}\text{ component:}\quad \frac{l}{2}\sin\beta(-A_Z + B_Z) = 0$$

$$\bar{j}\text{ component:}\quad \frac{l}{2}\cos\beta(A_Z - B_Z) = 0$$

$$\bar{k}\text{ component:}\quad -\frac{l}{2}(A_Y - B_Y) = -\tfrac{1}{8}mr^2\omega^2 \sin 2\beta$$

Simultaneous solution of the force and moment equations yields

$$A_X = A_Z = B_Z = 0$$

$$A_Y = -B_Y = \tfrac{1}{8}m\frac{r^2}{l}\omega^2 \sin 2\beta$$

The force components A_Y and B_Y are a consequence of the fact that $\bar{H}_G$ does not have a fixed direction. Example 1 showed that $\bar{H}_G$ rotates (precesses) about the shaft at angular velocity $\bar{\omega}$ in the manner illustrated in the sketch.

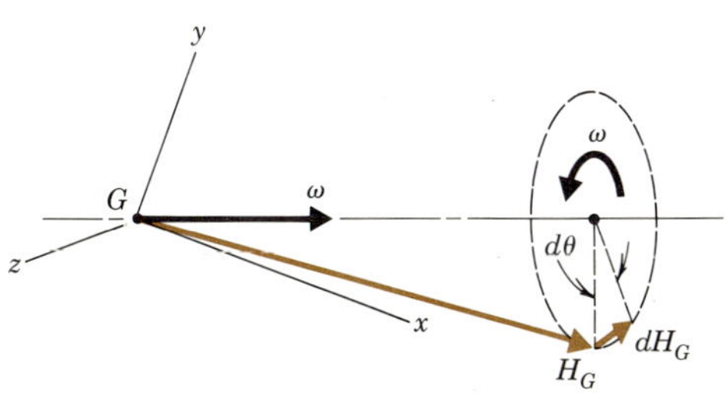

As the system rotates an infinitesimal amount $d\theta$ from its instantaneous position, the tip of $\bar{H}_G$ moves in the negative z direction as shown in the sketch. Dividing $d\bar{H}_G$ by dt shows that $\dot{\bar{H}}_G$ is in the negative z direction, consistent with the result obtained in part (a). Among the forces appearing in the free body diagram, only the reactions A_Y and B_Y can exert a moment about the z axis to produce the gyroscopic moment $\dot{\bar{H}}_G$.

We selected the orientation of xyz such that the yz plane was the plane of the disk at an arbitrary instant. For this problem the gyroscopic moment is always in the negative $\bar{k}$ direction, so that the reaction forces A_Y and B_Y must always lie in the xy plane. This leads to the conclusion that the reaction forces rotate with the motion of the disk.

It might surprise you that the system is balanced in the static sense (the center of mass is on the axis of rotation), and yet reaction forces are generated by the rotation of the system. This system is *dynamically unbalanced.* If we had solved this problem for an arbitrary body, rather than a disk, we could have shown that a system is dynamically balanced only when it is statically balanced *and* the fixed axis of rotation is a principal axis [see eq. (28)].

EXAMPLE 8

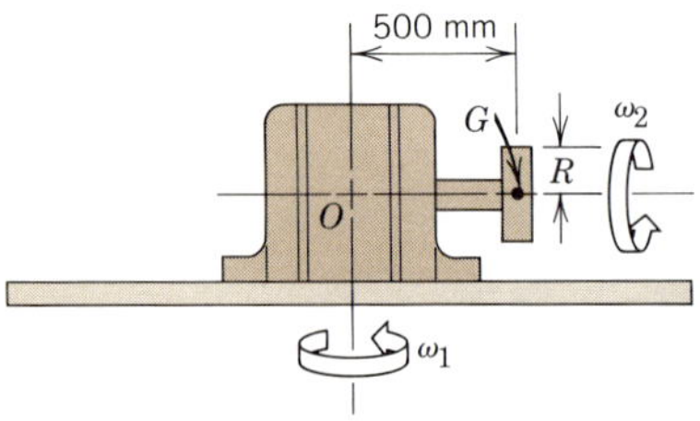

Let us now reconsider the system of Example 2 in which a disk is welded onto the shaft of a motor fixed to a turntable, as shown in the sketch. Initially, the turntable is rotating at the rate $\omega_1 = 10\pi$ rad/s and the motor is rotating at $\omega_2 = 50\pi$ rad/s. Determine the force-couple system exerted by the shaft on the disk at the instant when the turntable begins to decelerate at 5π rad/s^2. The mass of the disk is 2 kg, and $R = 200$ mm.

Solution

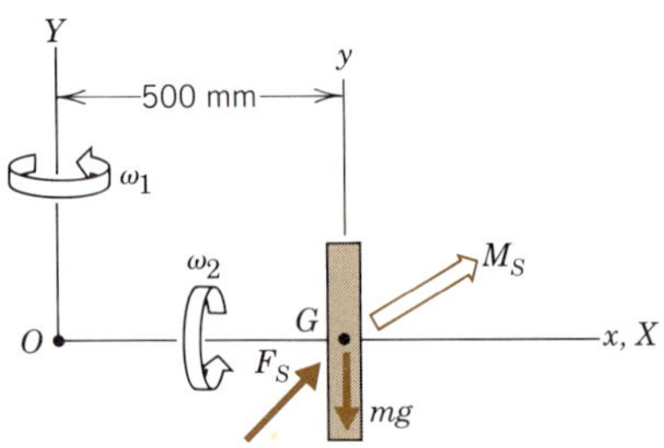

The shaft is fixed to the disk, so it restricts the movement of the point of connection between the bodies in all directions. The shaft also restricts the rotation of the disk about three coordinate axes. Therefore, the shaft exerts an unknown three-dimensional force $\bar{F}_S$ on the disk, and also an unknown three-dimensional couple $\bar{M}_S$. This is illustrated in the free body diagram. In this sketch we also show the xyz reference frame. The instantaneous orientation of xyz has been chosen for the convenience it affords in obtaining the inertia properties and in describing the angular velocity.

Expressions for $\bar{a}_G$ and $\dot{\bar{H}}_G$ will be needed. In order to understand the solution more fully, and also for computational efficiency, we shall retain all results in algebraic form until the completion of the problem. From the given information we know that $\omega_1 = 10\pi$ rad/s, $\omega_2 = 50\pi$ rad/s, $\dot{\omega}_1 = -5\pi$ rad/s^2 ($\dot{\omega}_1$ is negative because ω_1 is decreasing), and $\dot{\omega}_2 = 0$.

The value $\bar{a}_G$ is readily determined by noting that point G and fixed point O are points on the centerline of the shaft. This line is rotating about the vertical axis at the rate ω_1, so

$$\begin{aligned}\bar{a}_G &= \bar{a}_O + \dot{\omega}_1\bar{j} \times \bar{r}_{G/O} + (\omega_1\bar{j}) \times [(\omega_1\bar{j}) \times \bar{r}_{G/O}] \\ &= -0.5\dot{\omega}_1\bar{k} - 0.5\omega_1^2\bar{i}\end{aligned}$$

The angular velocity of the disk was obtained in Example 2. The angular acceleration must also be determined. The general expressions for these quantities are

$$\bar{\omega} = \omega_1\bar{J} + \omega_2\bar{i}: \quad \text{always}$$

$$\bar{\alpha} = \dot{\omega}_1\bar{J} + \dot{\omega}_2\bar{i} + \omega_2\dot{\bar{i}} = \dot{\omega}_1\bar{J} + \bar{0} + \omega_2(\bar{\omega} \times \bar{i})$$

The instantaneous expressions are obtained by noting that at the instant illustrated in the free body diagram, $\bar{J} = \bar{j}$, so that

$$\bar{\omega} = \omega_1 \bar{j} + \omega_2 \bar{i}: \quad \text{instantaneously}$$

$$\bar{\alpha} = \dot{\omega}_1 \bar{j} + \omega_2(\omega_1 \bar{j} + \omega_2 \bar{i}) \times \bar{i} = \dot{\omega}_1 \bar{j} - \omega_1 \omega_2 \bar{k}$$

The inertia properties of the disk were obtained in Example 2 directly from the tables in Appendix B. By symmetry, xyz are principal axes, and

$$I_{xx} = \tfrac{1}{2} mR^2, \qquad I_{yy} = I_{zz} = \tfrac{1}{4} mR^2$$

We next form the equations of motion. From the free body diagram and the results for $\bar{a}_G$, we have

$$\sum F_x = (F_S)_x = m(a_G)_x = -0.5m\omega_1^2$$

$$\sum F_y = (F_S)_y - mg = m(a_G)_y = 0$$

$$\sum F_z = (F_S)_z = m(a_G)_z = -0.5m\dot{\omega}_1$$

Thus

$$\bar{F}_S = -0.5m\omega_1^2 \bar{i} + mg\bar{j} - 0.5m\dot{\omega}_1 \bar{k}$$

Because xyz are principal axes, we may use this example to demonstrate the application of Euler's equation (29). By summing moments about point G and considering the disk to be sufficiently thin to allow us to consider the force $\bar{F}_S$ to be applied to point G, we find that

$$\begin{aligned} \sum M_{Gx} &= (M_S)_x = I_{xx}\alpha_x - (I_{yy} - I_{zz})\omega_y \omega_z \\ &= \tfrac{1}{2} mR^2(0) - (0)(\omega_1)(0) = 0 \end{aligned}$$

$$\begin{aligned} \sum M_{Gy} &= (M_S)_y = I_{yy}\alpha_y - (I_{zz} - I_{xx})\omega_z \omega_x \\ &= \tfrac{1}{4} mR^2 \dot{\omega}_1 - (\tfrac{1}{4} - \tfrac{1}{2}) mR^2(0)(\omega_2) = \tfrac{1}{4} mR^2 \dot{\omega}_1 \end{aligned}$$

$$\begin{aligned} \sum M_{Gz} &= (M_S)_z = I_{zz}\alpha_z - (I_{xx} - I_{yy})\omega_x \omega_y \\ &= \tfrac{1}{4} mR^2(-\omega_1 \omega_2) - (\tfrac{1}{2} - \tfrac{1}{4}) mR^2(\omega_2)(\omega_1) = -\tfrac{1}{2} mR^2 \omega_1 \omega_2 \end{aligned}$$

Thus

$$\bar{M}_S = mR^2(\tfrac{1}{4}\dot{\omega}_1 \bar{j} - \tfrac{1}{2}\omega_1 \omega_2 \bar{k})$$

Substituting the given values of ω_1, ω_2, $\dot{\omega}_1$, m, and R yields

$$\bar{F}_S = -987\bar{i} + 19.6\bar{j} + 15.7\bar{k} \text{ N}$$

$$\bar{M}_S = -0.3\bar{j} - 197.4\bar{k} \text{ N-m}$$

These answers, including the large component of $\bar{M}_s$ in the z direction have simple explanations. The shaft force $\bar{F}_s$ causes the center of mass to follow a circular path. Thus, for this problem $(F_s)_x$ imposes a normal acceleration, $(F_s)_y$ counterbalances gravity to produce planar motion, and $(F_s)_z$ imposes a tangential acceleration. The shaft couple $\bar{M}_s$ imparts to the disk the proper angular motion. The value of $(M_s)_y$ is independent of the speed of the motor ω_2. It represents the couple that must be applied to the disk if the turntable is to slow down. The large value of $(M_s)_z$ is the gyroscopic moment that must be applied to the disk if its angular motion is to be maintained. This is best illustrated by considering the result for $\bar{H}_G$ from Example 2, which is sketched again on the next page.

Because of the rotation ω_1 about the Y axis, the tip of $\bar{H}_G$ must move into the plane of the sketch, thus requiring that the force system apply a gyroscopic moment in the negative z direction. Once again, if we had relied on planar kinetics for intuition, we would have completely ignored the gyroscopic moment, thus missing a major effect in the system.

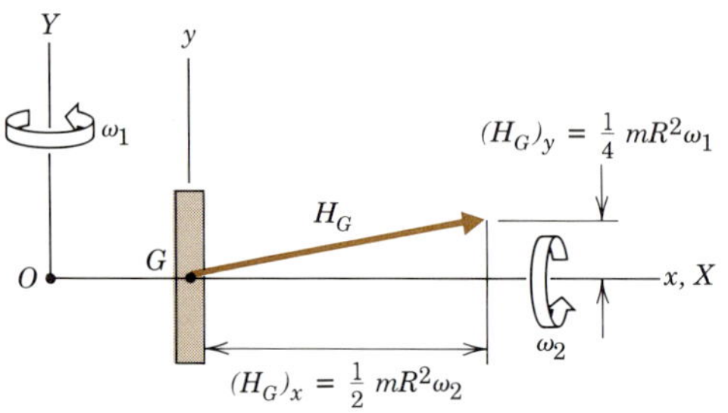

EXAMPLE 9

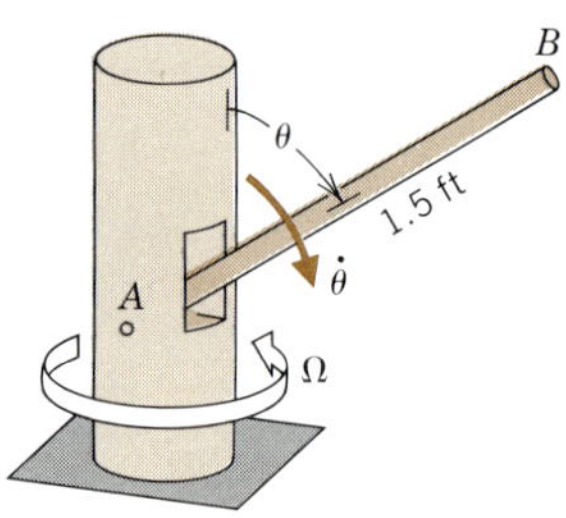

The 5 lb bar AB pivots in the vertical plane about horizontal pin A through the vertical shaft. The system is rotating about the vertical axis at the constant rate $\Omega = 2$ rad/s. When $\theta = 53.13°$, $\dot{\theta} = 10$ rad/s. Determine the corresponding value of $\ddot{\theta}$ and the couple reaction exerted on the bar by pin A in order to achieve this motion.

Solution

The starting point for any kinetics problem is a free body diagram. Pin A exerts an arbitrary force $\bar{F}_A$ to hold the end stationary. It also restricts the rotation of the bar about axes that are perpendicular to the axis of the pin. However, the pin does not prevent the angle θ from changing. Thus, $M_{Az} = 0$, so the couple $\bar{M}_A$ applied by the pin has no component perpendicular to the plane of the free body diagram.

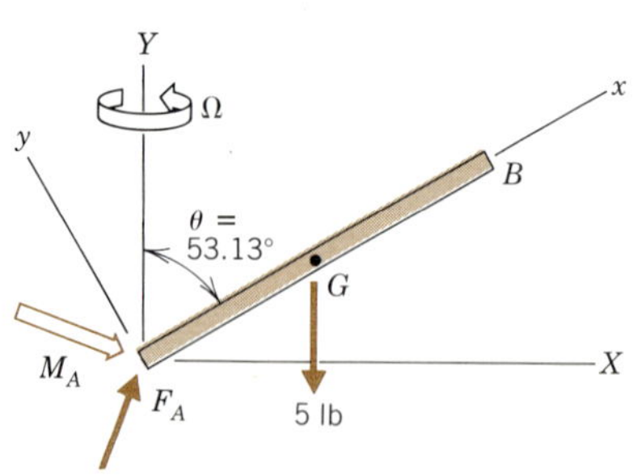

It will be noted that the origin of xyz has been placed at point A. This is an allowable point for summing moments because it is a point on bar AB that is fixed. The reason for selecting point A is that it enables us to eliminate the reaction $\bar{F}_A$ when formulating the moment equation. The orientation of

xyz matches that used to describe a slender bar in Appendix B. By symmetry xyz are principal axes and $I_{yy} \equiv I_{zz}$. Appendix B and the parallel axis theorem give

$$I_{xx} = 0, \qquad I_{yy} = I_{zz} = \frac{1}{12} ml^2 + m\left(\frac{l}{2}\right)^2 = \frac{1}{3} ml^2 = 0.11657 \text{ lb-s}^2\text{-ft}$$

Formulating $\dot{\bar{H}}_A$ requires that $\bar{\omega}$ and $\bar{\alpha}$ be evaluated. The rotation Ω is about the fixed Y axis, whereas the rotation $\dot{\theta}$ is about an axis perpendicular to the xy plane, that is, about the z axis. It is given that $\dot{\Omega} \equiv 0$, but $\ddot{\theta} \neq 0$. Thus

$$\bar{\omega} = \Omega \bar{J} + \dot{\theta}(-\bar{k}): \quad \text{always}$$

$$\bar{\alpha} = -\ddot{\theta}\bar{k} - \dot{\theta}\dot{\bar{k}} = -\ddot{\theta}\bar{k} - \dot{\theta}(\bar{\omega} \times \bar{k})$$

The instantaneous expressions are found by expressing $\bar{J}$ in terms of its xyz components when $\theta = 53.13^\circ$. Substituting the given values of Ω and $\dot{\theta}$ yields

$$\bar{J} = \cos 53.13^\circ \bar{i} + \sin 53.13^\circ \bar{j} = 0.6\bar{i} + 0.8\bar{j}$$

$$\bar{\omega} = 2(0.6\bar{i} + 0.8\bar{j}) - 10\bar{k} = 1.2\bar{i} + 1.6\bar{j} - 10\bar{k} \text{ rad/s}$$

$$\bar{\alpha} = -\ddot{\theta}\bar{k} - 10(1.2\bar{i} + 1.6\bar{j} - 10\bar{k}) \times \bar{k} = -16\bar{i} + 12\bar{j} - \ddot{\theta}\bar{k} \text{ rad/s}^2$$

We shall use this example to demonstrate the application of the basic equations (5), (25), and (27). Because xyz are principal axes, this gives

$$\begin{aligned}\bar{H}_A &= I_{xx}\omega_x\bar{i} + I_{yy}\omega_y\bar{j} + I_{zz}\omega_z\bar{k} \\ &= 0.11657(1.6\bar{j} - 10\bar{k}) = 0.18651\bar{j} - 1.1657\bar{k} \text{ lb-s-ft}\end{aligned}$$

$$\begin{aligned}\frac{\delta \bar{H}_A}{\delta t} &= I_{xx}\alpha_x\bar{i} + I_{yy}\alpha_y\bar{j} + I_{zz}\alpha_z\bar{k} \\ &= 0\bar{i} + 0.11657(12)\bar{j} + 0.11657(-\ddot{\theta})\bar{k}\end{aligned}$$

$$\begin{aligned}\dot{\bar{H}}_A &= \frac{\delta \bar{H}_A}{\delta t} + \bar{\omega} \times \bar{H}_A \\ &= (1.3988\bar{j} - 0.11657\ddot{\theta}\bar{k}) + (1.2\bar{i} + 1.6\bar{j} - 10\bar{k}) \times (0.18651\bar{j} - 1.1657\bar{k}) \\ &= 2.798\bar{j} + (-0.11657\ddot{\theta} + 0.2238)\bar{k} \text{ lb-ft}\end{aligned}$$

The free body diagram indicates that only gravity exerts a force about the z axis. Recalling that $M_{Az} \equiv 0$ and equating the moment sum about each axis to the corresponding component of $\dot{\bar{H}}_A$ then yields

$$\sum M_{Ax} = M_{Ax} = 0$$

$$\sum M_{Ay} = M_{Ay} = 2.798$$

$$\sum M_{Az} = -(5 \sin 53.13^\circ)(0.75) = -0.11657\ddot{\theta} + 0.2238$$

The solution to these equations is

$$\bar{M}_A = 2.80\bar{j} \text{ lb-ft}, \qquad \ddot{\theta} = 27.7 \text{ rad/s}^2$$

HOMEWORK PROBLEMS

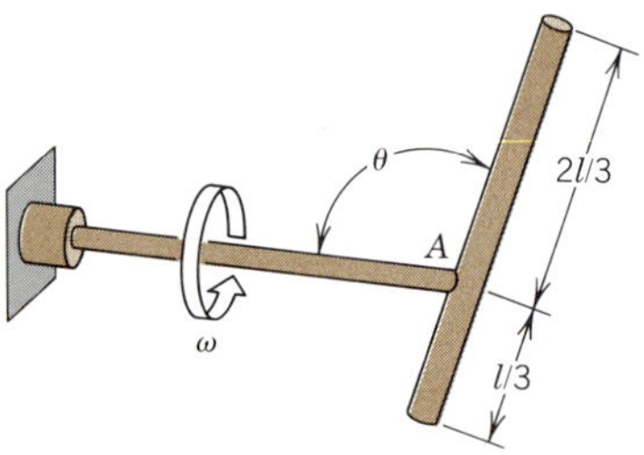

PROB. 16.48

16.47 Determine $\dot{\bar{H}}_G$ for the system in Problem 16.1 when $\dot{\omega} \neq 0$.

16.48 The slender bar of mass m and length l is welded to the horizontal shaft, which is rotating at a constant rate ω. For the slender bar determine $\bar{H}_G$ and $\dot{\bar{H}}_G$ in terms of the constant angle θ. Explain why $\dot{\bar{H}}_G = \bar{0}$ for $\theta = 0$ and $\theta = 90°$.

16.49 For the system in Problem 16.48 determine the total force-couple reaction at point A that the horizontal shaft exerts on the bar when the bar coincides with the vertical plane.

16.50 The 20 lb thin rectangular plate shown is welded diagonally onto the horizontal shaft AB. This shaft is rotating at the constant rate $\omega = 2400$ rev/min. Determine the gyroscopic moment $\dot{\bar{H}}_G$.

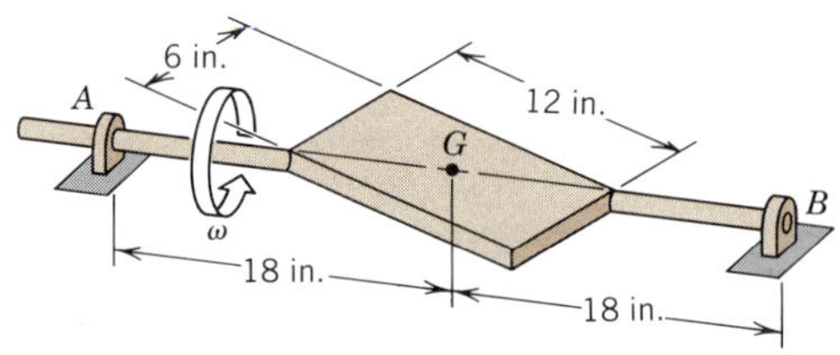

PROBS. 16.50 AND 16.51

16.51 The 20 lb thin rectangular plate shown is welded diagonally onto the horizontal shaft AB. Determine the reactions exerted by bearings A and B when the rectangular plate is rotating at the constant rate $\omega = 3000$ rev/min.

16.52 Two identical 40 kg disks are welded concentrically to a horizontal shaft that rotates at the angular rate ω. Identical small masses are attached to the perimeter of the disks. Is the system unbalanced, statically balanced only, or dynamically balanced when masses are added (a) nowhere, (b) at point A_1, (c) at points A_1 and A_2, (d) at points A_1 and C_2, (e) at points A_1, B_1, C_2, and D_2, (f) at points A_1, C_1, B_2, and D_2?

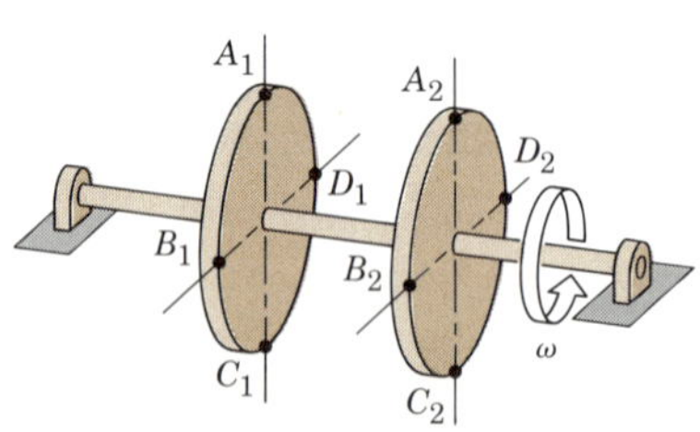

PROB. 16.52

16.53 Determine $\dot{\bar{H}}_G$ for the system in Problem 16.5.

16.54 Determine $\dot{\bar{H}}_G$ for the system in Problem 16.3.

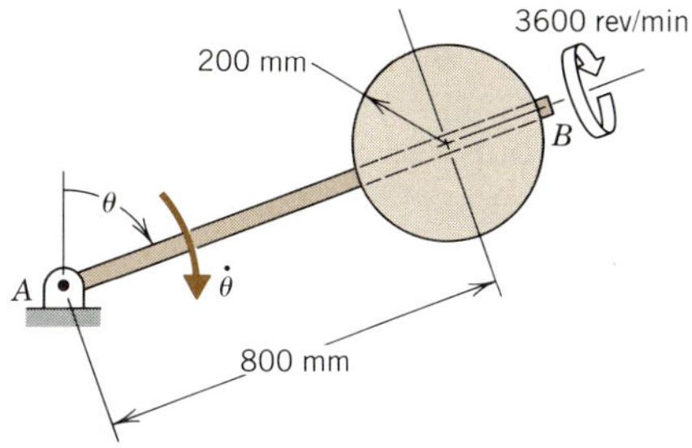

PROB. 16.55

16.55 The 200 mm radius sphere spins relative to bar AB at 3600 rev/min as the bar pivots about pin A. The mass of the sphere is 6 kg and the mass of the bar is negligible. When $\theta = 45°$, $\dot{\theta} = 5$ rad/s and $\ddot{\theta} = 10$ rad/s^2. Determine the corresponding values of $\bar{H}_G$ and $\dot{\bar{H}}_G$ for this system.

16.56 Determine $\dot{\bar{H}}_A$ for the system in Problem 16.55. Then use that result to evaluate the couple that must be applied to pivot A to obtain this motion.

16.57 The 2 lb disk is spinning at 15 rad/s about the axis of the massless shaft AB, which coincides with a diameter of the disk. The system is rotating about the vertical axis at 6 rad/s. Both rates of rotation are constant, and the shaft is horizontal ($\theta = 90°$). Determine the force-couple reaction exerted on the shaft at joint A when the disk is situated in the vertical plane, as shown.

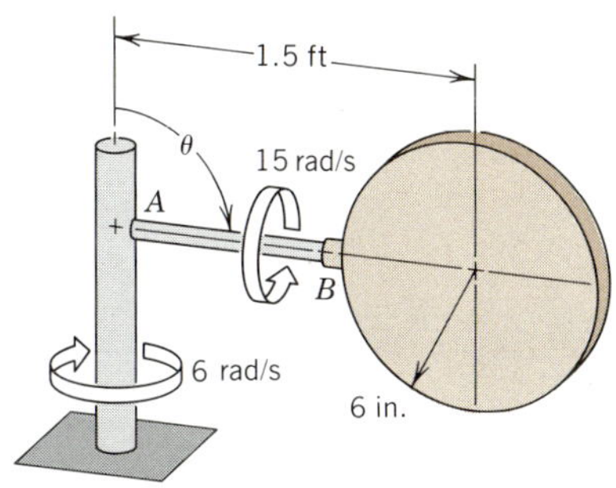

PROB. 16.57

***16.58** Solve Problem 16.57 for the case where shaft AB is inclined at $\theta = 60°$.

16.59 The two identical spheres of mass m and radius r rotate independently at rates ω_1 and ω_2 about shaft AB, as shown. This shaft is supported by a pin joint at its midpoint C and rotates about the vertical axis at the rate Ω. All rates of rotation are constant, and the mass of shaft AB is negligible. Determine (a) $\bar{H}_C$ and $\dot{\bar{H}}_C$ for each sphere, (b) the couple that must be applied at the pin joint to sustain this motion. Why is this couple zero when $\omega_1 = \omega_2$?

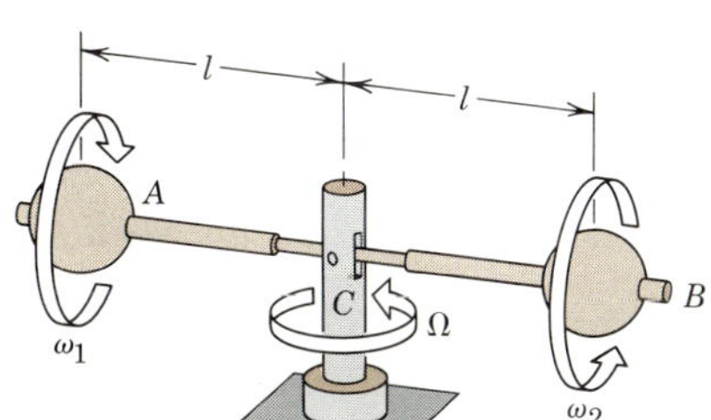

PROB. 16.59

E. PROBLEM SOLVING

The primary emphasis thus far has been evaluating the rate of change of the angular momentum. However, this is only one facet of the formulation of the equations of three-dimensional motion. The following steps, which may already be apparent from the Examples, are intended to give consistency to this formulation.

1. Draw a complete free body diagram of the system. In order to be sure that all reactions are properly described, remember that if the motion of a point is restricted in a certain direction, then there is a reaction force in that direction. Similarly, if the rotation of a body about a certain axis is restricted, then there is a reaction couple about that axis.
2. Write down, in equation form, all given information that does not appear in the free body diagram. List all quantities desired in the solution.
3. Choose a point A for summing moments. This point should be either the center of mass or a fixed point in space that is also fixed relative to the body.
4. Attach xyz axes to the body with origin at point A and orient xyz relative to the body so as to minimize any computations for inertia properties. If possible, orient xyz so as to make it possible to use the inertia properties in Appendix B and the parallel axis theorems. Select XYZ such that one of its axes coincides with the fixed axis of rotation. (There might not be such an axis in some problems, in which case XYZ is not needed.) Show both coordinate systems in the free body diagram in step 1.
5. Determine the inertia properties with respect to xyz.
6. Determine the angular velocity $\bar{\omega}$, the angular acceleration $\bar{\alpha}$, and the acceleration of the center of mass $\bar{a}_G$, all in terms of components with respect to xyz. In doing this, be sure to satisfy any kinematical constraints on the motion of the system. (It sometimes proves useful to treat the rotation rates and their derivatives algebraically when following this step.)
7. Compute the components of $\sum \bar{M}_A$, and equate them to the corresponding components of $\dot{\bar{H}}_A$. Substitute all known quantities into these equations. In general, use eqs. (5), (25), and (27) to perform this step. However, Euler's eq. (29) may be used instead if xyz form principal axes.
8. Equate each component of $\sum \bar{F}$ to the corresponding component of $m\bar{a}_G$.
9. Solve the equations.

Note that steps 7 and 8 include all the kinematics and kinematical constraints that were determined in step 6. Hence the number of equations and the number of unknowns at this point should balance. If not, the two most common conditions that may have been overlooked are (1) a kinematical constraint condition, or (2) a friction force relationship.

ILLUSTRATIVE PROBLEM 1

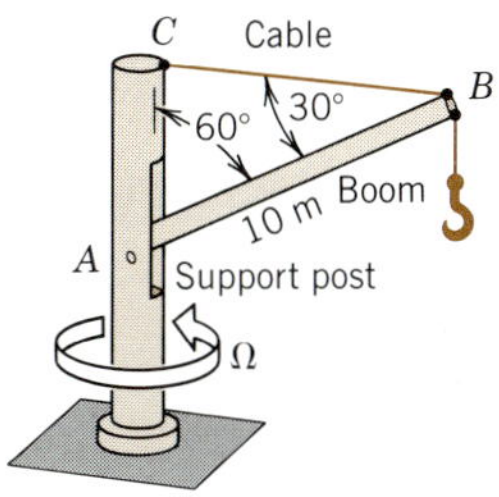

The crane shown in the sketch is initially at rest when the vertical support post is given a constant angular acceleration $\dot{\Omega} = 0.5$ rad/s^2. The boom, which has a mass of 600 kg, remains at 60° from the vertical during the motion. Determine the tension in the cable and the reactions at pivot A when $t = 4$ s.

Solution

Step 1 The cable tension appears as an external force in a free body diagram of the boom. To construct this diagram we note that the pin connection at point A constrains all motions of the boom except for rotations about the axis of the pin. It follows that the force-couple system representing the reaction at this joint consists of a three-dimensional force $\bar{F}_A$ and a moment $\bar{M}_A$ that has no component perpendicular to the plane formed by the post and the boom ($M_{Az} = 0$).

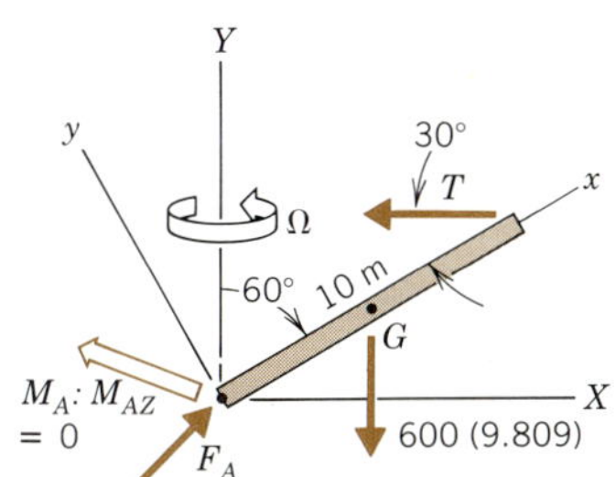

Step 2 The given information tells us that $\Omega = 0$ initially and $\dot{\Omega} = 0.5$ rad/s^2. Thus $\Omega = \dot{\Omega}t = 2$ rad/s when $t = 4$ s. We wish to find the tension T, as well as $\bar{F}_A$ and $\bar{M}_A$, when $t = 4$ s.

Step 3 We select the fixed point A for the origin of the moving reference system, thereby eliminating the reaction force $\bar{F}_A$ from the moment equation.

Step 4 Considering the boom to be a slender rod and selecting the x axis to coincide with the axis of the boom allows us to obtain the inertia properties directly from Appendix B. For convenience in describing the angular motion, we select the fixed Y axis to coincide with the vertical post and let the y axis lie the xY plane, as as shown in the free body diagram.

Step 5 The inertia properties of a slender rod in Appendix B and the parallel axis theorems yield

$$I_{xx} = 0$$

$$I_{yy} = I_{zz} = \frac{1}{12}mL^2 + m\left(\frac{L}{2}\right)^2 = \left(\frac{1}{12} + \frac{1}{4}\right)(600)10^2$$

$$= 2(10^4) \text{ kg-m}^2$$

$$I_{xy} = I_{xz} = I_{yz} = 0$$

Step 6 The boom is rotating about the fixed vertical axis, so that the general expressions for $\bar{\omega}$ and $\bar{\alpha}$ are

$$\bar{\omega} = \Omega \bar{J}, \qquad \bar{\alpha} = \dot{\Omega} \bar{J}: \quad \text{always}$$

Representing $\bar{J}$ in terms of its component with respect to xyz, and substituting the values for Ω and $\dot{\Omega}$ when $t = 4$ s gives

$$\bar{\omega} = 2(\cos 60^\circ \bar{i} + \sin 60^\circ \bar{j}) = 1.0\bar{i} + 1.7321\bar{j} \text{ rad/s}$$

$$\bar{\alpha} = 0.5(\cos 60^\circ \bar{i} + \sin 60^\circ \bar{j}) = 0.25\bar{i} + 0.4330\bar{j} \text{ rad/s}^2$$

Because point A is fixed, the kinematical relation for the acceleration of the center of mass of the boom is

$$\begin{aligned}\bar{a}_G &= \bar{a}_A + \bar{\alpha} \times \bar{r}_{G/A} + \bar{\omega} \times (\bar{\omega} \times \bar{r}_{G/A}) \\ &= (0.25\bar{i} + 0.4330\bar{j}) \times 5\bar{i} + (1.0\bar{i} + 1.7321\bar{j}) \times [(1.0\bar{i} + 1.7321\bar{j}) \times 5\bar{i}] \\ &= -15.0\bar{i} + 8.660\bar{j} - 2.165\bar{k} \text{ m/s}^2\end{aligned}$$

Step 7 We may use Euler's equation because xyz is a principal set of axes. Substituting the inertia values found in step 5 and the angular velocity and angular acceleration components found in step 6 into eq. (29) yields

$$\begin{aligned}\sum \bar{M}_A &= (0)\bar{i} + [2(10^4)(0.4330)]\bar{j} + [2(10^4)(1.0)(1.7321)]\bar{k} \\ &= 0.8660(10^4)\bar{j} + 3.464(10^4)\bar{k} \text{ N-m}\end{aligned}$$

The moment of the force system is now computed from the free body diagram. The reaction couple $\bar{M}_A$ has no component in the z direction, so this yields

$$\begin{aligned}\sum \bar{M}_A &= M_{Ax}\bar{i} + M_{Ay}\bar{j} + [(T \sin 30^\circ)(10) - 600(9.807)(\sin 60^\circ)5]\bar{k} \\ &= M_{Ax}\bar{i} + M_{Ay}\bar{j} + [5T - 2.548(10^4)]\bar{k} \text{ N-m}\end{aligned}$$

Matching the corresponding components of these expressions for $\sum \bar{M}_A$ gives

$$\sum M_{Ax} = M_{Ax} = 0$$

$$\sum M_{Ay} = M_{Ay} = 0.866(10^4) \text{ N-m}$$

$$\sum M_{Az} = 5T - 2.548(10^4) = 3.464(10^4) \text{ N-m}$$

Step 8 Equating the components of the resultant force found from the free body diagram to the components of $m\bar{a}_G$ found from step 6 yields

$$\sum F_x = F_{Ax} - 600(9.807) \cos 60^\circ - T \cos 30^\circ = 600(-15.0) \text{ N}$$

$$\sum F_y = F_{Ay} - 600(9.807) \sin 60^\circ + T \sin 30^\circ = 600(8.660) \text{ N}$$

$$\sum F_z = F_{Az} = 600(-2.165) \text{ N}$$

Step 9 The moment equations in step 7 yield the tension T and the couple $\bar{M}_A$.

$$T = 12.02 \text{ kN}, \qquad \bar{M}_A = 8.66\bar{j} \text{ kN-m}$$

Substituting the result for T into the force equations in step 8 then gives $\bar{F}_A$.

$$\bar{F}_A = 4.36\bar{i} + 4.28\bar{j} - 1.30\bar{k} \text{ kN}$$

ILLUSTRATIVE PROBLEM 2

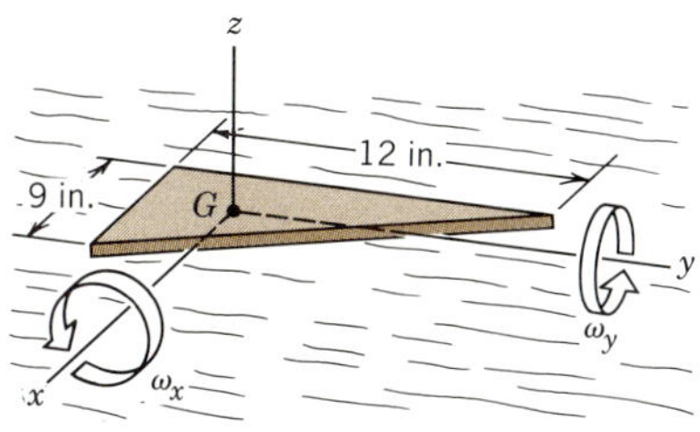

The 5 lb right triangular plate is floating on water. At the instant when the z axis is vertical, the plate is pitching about its x axis and rolling about its y axis at the maximum rates $\omega_x = 1.5$ rad/s and $\omega_y = 2.5$ rad/s, respectively. The center of mass G is accelerating upward at this instant at $0.3g$. Determine the force-couple system acting at point G that represents the resultant of the forces exerted by the water on the plate at this instant.

Solution

Step 1 Denote the specified force-couple system as $\bar{F}_G$ and $\bar{M}_G$, respectively. The other force appearing in the free body diagram is the weight. This force is parallel to the z axis at this instant because z is vertical.

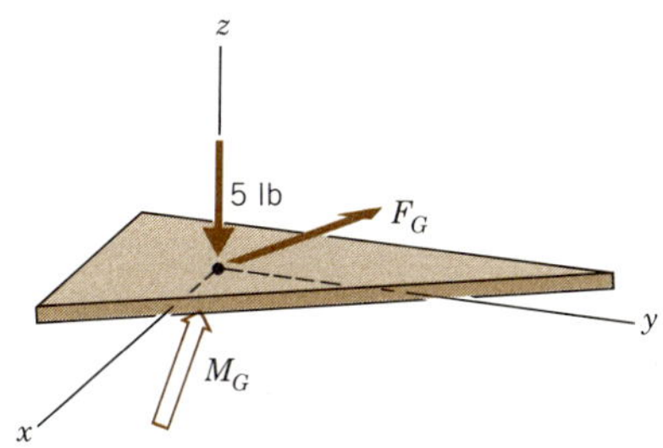

Step 2 It is given that $\bar{a}_G = 0.3(386.0)\bar{k}$ in./s², and that $\omega_x = 1.5$ rad/s and $\omega_y = 2.5$ rad/s. Also, $\dot{\omega}_x = \dot{\omega}_y = 0$ because the rates of rotation are specified to be the maximum values. (The yaw rate ω_z is not given, so it is considered to be zero.) We wish to evaluate $\bar{F}_G$ and $\bar{M}_G$.

Steps 3 and 4 The only allowable point in this system for summing moments is point G. The xyz coordinate system given in the problem statement matches the one used in Appendix B to describe a triangular plate, so it is the one we choose for the solution. There is no need for the fixed XYZ system, because the rotation rates are both about axes that are attached to the body.

Step 5 The moments of inertia obtained from Appendix B (set the dimension $a = 0$ there) are

$$I_{xx} = \tfrac{1}{18}\left(\tfrac{5}{386}\right)(12)^2 = 0.10363 \text{ lb-s}^2\text{-in.}$$

$$I_{yy} = \tfrac{1}{18}\left(\tfrac{5}{386}\right)(9)^2 = 0.05829 \text{ lb-s}^2\text{-in.}$$

$$I_{zz} = \tfrac{1}{18}\left(\tfrac{5}{386}\right)[(12)^2 + (9)^2] = 0.16192 \text{ lb-s}^2\text{-in.}$$

$$I_{xy} = \tfrac{1}{36}\left(\tfrac{5}{386}\right)(9)(-12) = -0.03886 \text{ lb-s}^2\text{-in.}$$

$$I_{yz} = I_{xz} = 0$$

where the last two products of inertia are zero because xy is a plane of symmetry.

Step 6 The given angular speeds ω_1 and ω_2, describe the rotation of the body about axes that are fixed to the body. Then, in view of the sense of the rotations shown in the given diagram, it is generally true that

$$\bar{\omega} = \omega_x \bar{i} + \omega_y \bar{j}: \quad \text{always}$$

$$\bar{\alpha} = \dot{\omega}_x \bar{i} + \dot{\omega}_y \bar{j} + \omega_x(\bar{\omega} \times \bar{i}) + \omega_y(\bar{\omega} \times \bar{j})$$

The instantaneous values are found by substituting the given rotation rates. Thus

$$\bar{\omega} = 1.5\bar{i} + 2.5\bar{j} \text{ rad/s}: \quad \text{instantaneously}$$

$$\bar{\alpha} = \bar{0} + \bar{0} + 1.5(1.5\bar{i} + 2.5\bar{j}) \times \bar{i} + 2.5(1.5\bar{i} + 2.5\bar{j}) \times \bar{j} = \bar{0}$$

There is no need to evaluate $\bar{a}_G$ because it was given.

Steps 7–9 The formulation of $\dot{\bar{H}}_G$ must employ eqs. (5), (25), and (27) because xyz are not principal axes. Thus

$$\begin{aligned}\bar{H}_G &= [0.10363(1.5) - (-0.03886)(2.5) - 0]\bar{i} \\ &\quad + [0.05829(2.5) - (-0.03886)(1.5) - 0]\bar{j} + 0\bar{k} \\ &= 0.2526\bar{i} + 0.2040\bar{j} \text{ lb-s-in.}\end{aligned}$$

$$\frac{\delta \bar{H}_G}{\delta t} = \bar{0}$$

$$\begin{aligned}\dot{\bar{H}}_G &= \frac{\delta \bar{H}_G}{\delta t} + \bar{\omega} \times \bar{H}_G = (1.5\bar{i} + 2.5\bar{j}) \times (0.2526\bar{i} + 0.2040\bar{j}) \\ &= -0.3255\bar{k} \text{ lb-in.}\end{aligned}$$

Only the couple $\bar{M}_G$ exerts a moment about point G. Thus

$$\sum \bar{M}_G = \bar{M}_G = \dot{\bar{H}}_G = -0.3255\bar{k} \text{ lb-in.}$$

In order to determine $\bar{F}_G$ the force sums are obtained from the free body diagram, and then equated to the corresponding components of $m\bar{a}_G$.

$$\sum F_x = (F_G)_x = m(a_G)_x = 0$$

$$\sum F_y = (F_G)_y = m(a_G)_y = 0$$

$$\sum F_z = (F_G)_z - 5 = m(a_G)_z = (\tfrac{5}{386})(0.3)(386)$$

Thus only $(F_G)_z \neq$ to 0.

$$\bar{F}_G = 6.5\bar{k} \text{ lb}$$

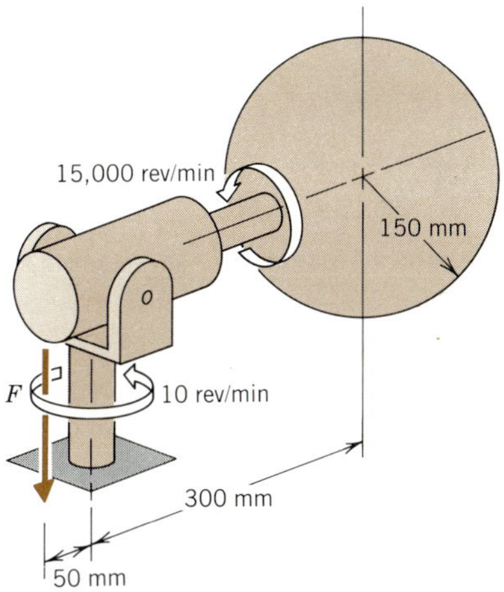

ILLUSTRATIVE PROBLEM 3

The motor is mounted on the yoke, which rotates about the vertical axis at 10 rev/min. The 500 gram sphere is being spun by the motor at 15,000 rev/min. Determine the dynamic portion of the vertical force $\bar{F}$ required to keep the axis of the motor horizontal during this motion.

Solution

Step 1 The desired force $\bar{F}$ is applied to the motor, so we must depict this body in a free body diagram. However, the sphere has angular momentum, so we need to consider this body also. We may combine the motor and the sphere in a single free body diagram for reasons that will be discussed shortly. Note that gravity has been omitted because it is a static force. The force components A_y, B_y, A_z, and B_z represent the horizontal and vertical reactions exerted by the pins of the yoke.

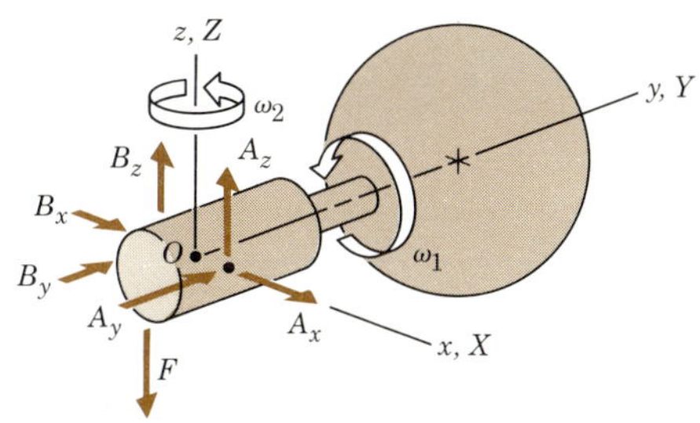

Step 2 Given that $m = 0.5$ kg for the sphere, and that $\omega_1 = 1.5(10^4)(2\pi/60)$ rad/s, $\omega_2 = 10(2\pi/60)$ rad/s, determine F.

Steps 3 and 4 Attach the xyz axes to the sphere. Fixed point O, where the axes of the horizontal and vertical shafts intersect, is an allowable point for the moment sum because its position relative to the sphere does not change. Note that this point is collinear with the line between the pins of the yoke. Using point O for the moment sum and aligning the x axis with the pins will allow us to eliminate the unknown reactions from the x component of the moment sum. The Z axis of the fixed coordinate system is aligned with the vertical axis of rotation.

Step 5 Symmetry indicates that xyz form principal axes. Then Appendix B for a sphere and the parallel axis theorems gives

$$I_{xx} = I_{zz} = \tfrac{2}{5}mR^2 + m(0.3)^2 = 4.95(10^{-2})\ \text{kg-m}^2$$

$$I_{yy} = \tfrac{2}{5}mR^2 = 4.5(10^{-3})\ \text{kg-m}^2$$

$$I_{xy} = I_{xz} = I_{yz} = 0$$

Step 6 The general expressions for the angular velocity and angular acceleration of the sphere are

$$\bar{\omega} = \omega_1(-\bar{j}) + \omega_2\bar{K}: \quad \text{always}$$

$$\bar{\alpha} = -\omega_1\dot{\bar{j}} = -\omega_1(\bar{\omega} \times \bar{j})$$

At the instant depicted in the sketch $\bar{K} = \bar{k}$. Substituting for ω_1 and ω_2 therefore results in

$$\bar{\omega} = -1570.8\bar{j} + 1.0472\bar{k} \text{ rad/s:} \quad \text{instantaneously}$$

$$\bar{\alpha} = -1570.8(-1570.8\bar{j} + 1.0472\bar{k}) \times \bar{j} = 1644.9\bar{i} \text{ rad/s}^2$$

There is no need to evaluate the acceleration of the center of mass because we will consider only the moment equation.

Steps 7–9 Both the sphere and the motor have angular momentum. However, $\bar{H}_O$ for the motor is constant because the motor is rotating at a constant rate about the Z axis, and it is symmetric about the xy plane. Thus the moment of the forces acting on the combined system must equal $\dot{\bar{H}}_O$ for the sphere. The force $\bar{F}$ exerts a moment about the x axis only; the corresponding component of $\dot{\bar{H}}_O$ may be found from Euler's eq. (29) because xyz are principal axes. Thus

$$\begin{aligned}\sum M_{Ax} = 0.05F &= I_{xx}\alpha_x - (I_{yy} - I_{zz})\omega_y\omega_z \\ &= 4.95(10^{-2})1644.9 - [4.5(10^{-3}) - 4.95(10^{-2})](-1570.8)(1.0472) \\ &= 7.402 \text{ N-m} \\ F &= 148 \text{ N}\end{aligned}$$

ILLUSTRATIVE PROBLEM 4

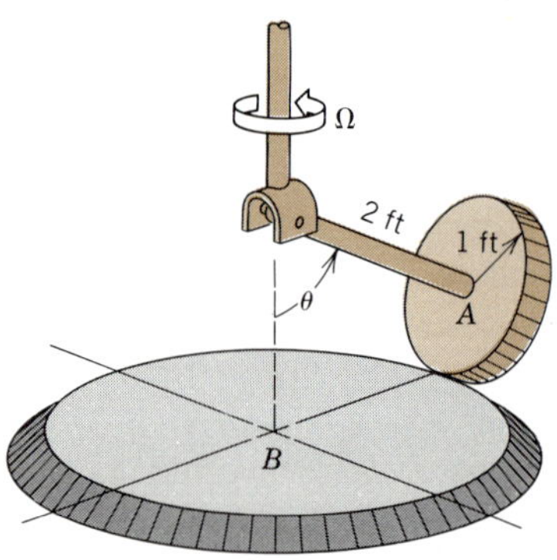

Gear A, which meshes with the fixed horizontal gear B, freely rotates about shaft OA. This shaft is attached to the vertical shaft by clevis joint O. The angular speed Ω of the vertical shaft is constant. Obtain an expression for the normal force exerted between the two gears in terms of Ω and the angle of inclination θ of the shaft. (In this particular design of the gears, the desired force is perpendicular to shaft OA.) The weight of gear A is 15 lb, and the weight of the shaft is negligible.

Solution

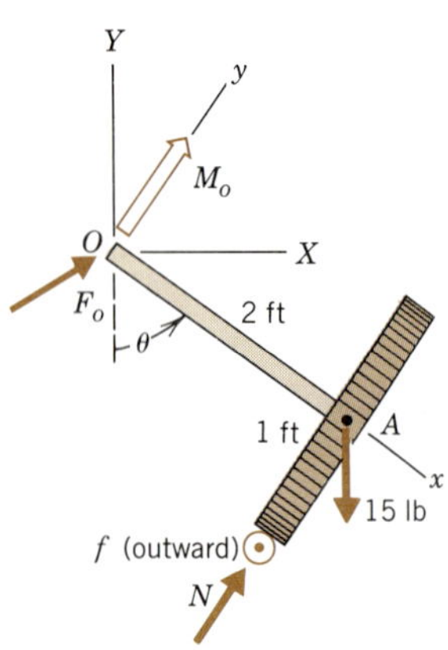

Step 1 For the free body diagram we isolate gear A and shaft OA in order to utilize the special features of a clevis joint. A side view of the system is most suitable. Acting at the clevis joint is an arbitrary (three-dimensional) force $\bar{F}_O$ that prevents point O from moving. The axis of the pin at this joint is perpendicular to the plane of the free body diagram, so there is no couple reaction about the z axis. Also, because there is no friction between gear A and shaft OA, there is no tendency for this shaft to twist about the x axis. Hence, the couple reaction $\bar{M}_O$ is oriented perpendicular to shaft OA in the vertical plane, as shown.

Step 2 Given that Ω is constant, the problem is to find the value of N.

Step 3 Point O is a fixed point in space that is also fixed relative to gear A. We select this point for summing moments, rather than the center of mass A, in order to eliminate the reaction $\bar{F}_O$ from the moment equation.

Step 4 We attach an xyz coordinate frame to gear A in order to describe the angular momentum of that body. We consider the gear to be a thin disk, and select the x axis to coincide with shaft OA, so that xyz will agree with the corresponding entry in Appendix B. The Y axis of the fixed coordinate system is oriented to coincide with the vertical axis of rotation.

Step 5 The disk is symmetric with respect to both the xy and xz planes, so xyz forms a set of principal axes. Thus,

$$I_{xy} = I_{xz} = I_{yz} = 0$$

From Appendix B and the parallel axis theorem, we find that the inertia properties relative to the xyz axes are

$$I_{xx} = \frac{1}{2} mR^2 = \frac{1}{2}\left(\frac{15}{32.17}\right)(1)^2 = 0.2331 \text{ lb-s}^2\text{-ft}$$

$$I_{yy} = I_{zz} = \frac{1}{4} mR^2 + m(2)^2 = \left(\frac{15}{32.17}\right)(1)^2\left(\frac{1}{4} + 4\right) = 1.9817 \text{ lb-s}^2\text{-ft}$$

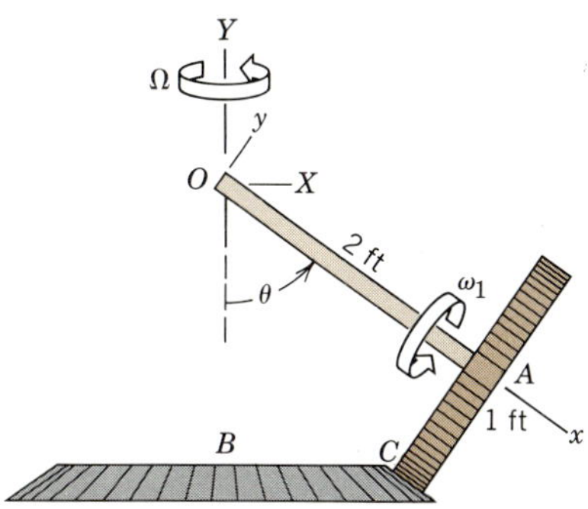

Step 6 The study of the kinematics of the system is begun by drawing a sketch that shows the angular speed ω_1 of gear A about shaft OA. The general expressions for the angular velocity and angular acceleration are

$$\bar{\omega} = \Omega\bar{J} + \omega_1\bar{i}: \quad \text{always}$$

$$\bar{\alpha} = \dot{\Omega}\bar{J} + \dot{\omega}_1\bar{i} + \omega_1(\bar{\omega} \times \bar{i})$$

At the instant of the sketch we have $\dot{\Omega} = 0$ and

$$\bar{J} = -\cos\theta\,\bar{i} + \sin\theta\,\bar{j}$$

Thus, the instantaneous values of $\bar{\omega}$ and $\bar{\alpha}$ are

$$\bar{\omega} = (-\Omega\cos\theta + \omega_1)\bar{i} + \Omega\sin\theta\,\bar{j}$$

$$\begin{aligned}\bar{\alpha} &= \dot{\omega}_1\bar{i} + \omega_1[(-\Omega\cos\theta + \omega_1)\bar{i} + \Omega\sin\theta\,\bar{j}] \times \bar{i} \\ &= \dot{\omega}_1\bar{i} - \omega_1\Omega\sin\theta\,\bar{k}\end{aligned}$$

An expression for ω_1 in terms of Ω and θ may be found by noting that $\bar{v}_C = \bar{0}$ for rolling without slipping of gear A over the fixed gear B. Pivot point O has a fixed position relative to the disk, so the velocity equation relating points O and C gives

$$\begin{aligned}\bar{v}_C = \bar{0} &= \bar{v}_O + \bar{\omega} \times \bar{r}_{C/O} \\ &= \bar{0} + [(-\Omega\cos\theta + \omega_1)\bar{i} + \Omega\sin\theta\,\bar{j}] \times (2\bar{i} - 1\bar{j}) \\ &= [-(-\Omega\cos\theta + \omega_1) - (\Omega\sin\theta)2]\bar{k}\end{aligned}$$

$$\bar{k} \text{ component:} \qquad 0 = \Omega\cos\theta - \omega_1 - 2\Omega\sin\theta$$

$$\omega_1 = \Omega(\cos\theta - 2\sin\theta)$$

This result shows that ω_1 is proportional to Ω, which means that if $\dot{\Omega} = 0$, then $\dot{\omega}_1 = 0$. Substituting these relations for ω_1 and $\dot{\omega}_1$ into the earlier expressions for $\bar{\omega}$ and $\bar{\alpha}$ then yields

$$\bar{\omega} = \Omega\sin\theta(-2\bar{i} + \bar{j})$$

$$\bar{\alpha} = -\Omega^2(\sin\theta\cos\theta - 2\sin^2\theta)\bar{k}$$

The determination of $\bar{a}_G$ is omitted because the equations of motion obtained from summing forces define only the force $\bar{F}_O$. We shall therefore also omit step 8.

Step 7 We may now compute $\dot{\bar{H}}_O$ by using eqs. (5), (25), and (27). The inertia properties in step 5 and the expressions for $\bar{\omega}$ and $\bar{\alpha}$ in step 6 yield

$$\begin{aligned}\bar{H}_O &= I_{xx}\omega_x\bar{i} + I_{yy}\omega_y\bar{j} \\ &= (0.2331)(-2\Omega\sin\theta)\bar{i} + 1.9817(\Omega\sin\theta)\bar{j} \\ &= \Omega\sin\theta(-0.4662\bar{i} + 1.9817\bar{j})\text{ lb-s-ft}\end{aligned}$$

$$\frac{\delta\bar{H}_O}{\delta t} = I_{zz}\alpha_z\bar{k} = -1.9817\Omega^2(\sin\theta\cos\theta - 2\sin^2\theta)\bar{k}$$

$$\begin{aligned}\dot{\bar{H}}_O &= \frac{\delta\bar{H}_O}{\delta t} + \bar{\omega}\times\bar{H}_O \\ &= -1.9817\Omega^2(\sin\theta\cos\theta \quad 2\sin^2\theta)\bar{k} \\ &\quad + \Omega\sin\theta(-2\bar{i} + \bar{j})\times(\Omega\sin\theta)(-0.4662\bar{i} + 1.9817\bar{j}) \\ &= (0.4662\Omega^2\sin^2\theta - 1.9817\Omega^2\sin\theta\cos\theta)\bar{k}\end{aligned}$$

We now use the free body diagram to evaluate the sum of the moments about the origin O.

$$\begin{aligned}\sum\bar{M}_O &= M_O\bar{j} + 2\bar{i}\times 15(\cos\theta\bar{i} - \sin\theta\bar{j}) + (2\bar{i} - \bar{j})\times(N\bar{j} + f\bar{k}) \\ &= -f\bar{i} + (M_O - 2f)\bar{j} + (-30\sin\theta + 2N)\bar{k}\end{aligned}$$

Step 8 Omit

Step 9 The value of N appears only in the z component of the moment sum. The result of matching this component of $\sum\bar{M}_O$ to the corresponding component of $\dot{\bar{H}}_O$ is

$$\bar{k}\text{ component:}\quad -30\sin\theta + 2N = (0.4662\Omega^2\sin^2\theta - 1.9817\Omega^2\sin\theta\cos\theta)$$

$$N = 15\sin\theta + \Omega^2(0.233\sin\theta - 0.991\cos\theta)\sin\theta$$

The first term on the right side of this result represents the static gravitational effect. The second term represents two opposing inertial effects. If gear A did not mesh with the fixed gear, then the angular speed ω_1 would be zero and the centripetal acceleration of point A would have the effect of decreasing the value of N. However, due to the rotation ω_1, a gyroscopic moment is created that has the effect of forcing the gear downward. If $\theta = \tan^{-1}(0.991/0.233) = 76.8°$, then the two dynamic effects counterbalance each other and N reduces to its static value $N = 15\sin\theta$.

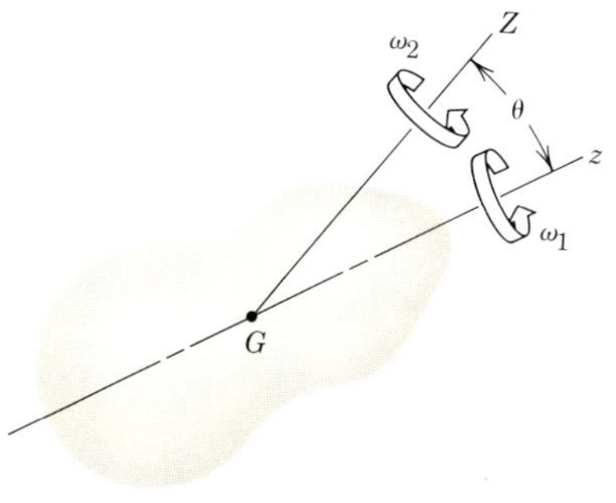

ILLUSTRATIVE PROBLEM 5

The body shown is in free flight with negligible air resistance. Its properties are arbitrary, with the exception that it is rotationally symmetric about the z axis. From the fact that the gravity force acts at the center of mass, it follows that $\sum \bar{M}_G = \dot{\bar{H}}_G = \bar{0}$, so the angular momentum $\bar{H}_G$ is constant. Letting Z be a fixed axis of arbitrary orientation, assume that the motion of the body consists of a constant spinning rotation ω_1 about the z axis, as the z axis itself rotates about the fixed Z axis at the constant rate ω_2 and the constant angle θ. Derive an expression for ω_2 in terms of ω_1 and θ such that the equation $\dot{\bar{H}}_G = \bar{0}$ is satisfied. Show that the corresponding $\bar{H}_G$ is a constant vector parallel to the Z axis. Also, derive an expression for the angle β between the angular velocity $\bar{\omega}$ and the z axis in terms of θ.

Solution

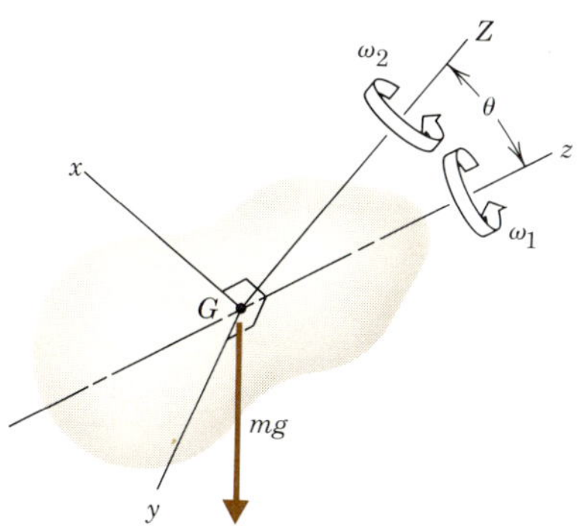

Step 1–4 From the given information we know that the only force is gravity. The free body diagram drawn here also shows the given information about the rotations. The body is in general motion, so the origin of the xyz axes is located at the center of mass. For convenience, the x axis is chosen such that the Z axis lies in the xz plane.

Step 5 For rotational symmetry, $I_{zz} = I_1$, $I_{xx} = I_{yy} = I_2$, where I_1, I_2, and the mass m are arbitrary constants. Also, xyz are principal axes.

Step 6 In the assumed motion, ω_1, ω_2, and θ are constant, so the general expressions for $\bar{\omega}$ and $\bar{\alpha}$ are

$$\bar{\omega} = \omega_2 \bar{K} + \omega_1 \bar{k}: \quad \text{always}$$

$$\bar{\alpha} = \omega_1(\bar{\omega} \times \bar{k})$$

At the instant depicted in the sketch in step 1,

$$\bar{K} = \sin\theta\,\bar{i} + \cos\theta\,\bar{k}$$

so the instantaneous expressions are

$$\bar{\omega} = \omega_2 \sin\theta\,\bar{i} + (\omega_1 + \omega_2\cos\theta)\bar{k}: \quad \text{instantaneously}$$

$$\begin{aligned}\bar{\alpha} &= \omega_1[\omega_2 \sin\theta\,\bar{i} + (\omega_1 + \omega_2\cos\theta)\bar{k}] \times \bar{k} \\ &= -\omega_1\omega_2 \sin\theta\,\bar{j}\end{aligned}$$

We omit considering $\bar{a}_G$, for we are not interested in it.

Step 7 Because xyz are principal axes, we may use Euler's equation (29). Substituting the inertia properties from step 5 and the values of $\bar{\omega}$ and $\bar{\alpha}$ found in step 6, we find that

$$\begin{aligned}\sum \bar{M}_A \equiv \bar{0} &= [I_2(-\omega_1\omega_2 \sin\theta) - (I_1 - I_2)(\omega_2 \sin\theta)(\omega_1 + \omega_2\cos\theta)]\bar{j} \\ &= \omega_2 \sin\theta[-I_1(\omega_1 + \omega_2\cos\theta) + I_2\omega_2\cos\theta]\bar{j}\end{aligned}$$

Step 8 The moment equation in step 7 is satisfied when either $\omega_2 = 0$, $\sin\theta = 0$, or

$$\omega_1 + \omega_2\cos\theta = \frac{I_2}{I_1}\omega_2\cos\theta$$

The first two possibilities mean that the only rotation is about the z axis. This is the axis of symmetry, so that $\bar{\omega}$ and $\bar{H}_G$ are then always in the z direction, making $\bar{H}_G$ a constant.

The more interesting possibility is the third one. Solving for ω_2 in that relation gives

$$\omega_2 = \frac{I_1\omega_1}{(I_2 - I_1)\cos\theta}$$

After this expression is substituted into $\bar{\omega}$, the angular momentum may be evaluated from eq. (5). Thus

$$\bar{\omega} = \frac{I_1\omega_1}{(I_2 - I_1)\cos\theta}\sin\theta\,\bar{i} + \left[\omega_1 + \frac{I_1\omega_1}{(I_2 - I_1)}\right]\bar{k}$$

$$= \frac{\omega_1}{(I_2 - I_1)}(I_1\tan\theta\,\bar{i} + I_2\bar{k})$$

$$\bar{H}_G = I_{xx}\omega_x\bar{i} + I_{zz}\omega_z\bar{k}$$

$$= I_2\frac{I_1\omega_1\tan\theta}{(I_2 - I_1)}\bar{i} + I_1\frac{I_2\omega_1}{(I_2 - I_1)}\bar{k}$$

$$= \frac{I_1 I_2}{(I_2 - I_1)}\omega_1(\tan\theta\,\bar{i} + \bar{k})$$

The magnitude of $\bar{H}_G$ is constant because its components are constant. We may check whether $\bar{H}_G$ is parallel to the Z axis by showing that $\bar{H}_G \times \bar{K} = \bar{0}$. Thus

$$\bar{H}_G \times \bar{K} = \frac{I_1 I_2}{(I_2 - I_1)}\omega_1(\tan\theta\,\bar{i} + \bar{k}) \times (\sin\theta\,\bar{i} + \cos\theta\,\bar{k})$$

$$= \frac{I_1 I_2}{(I_2 - I_1)}\omega_1\left(-\frac{\sin\theta}{\cos\theta}\cos\theta + \sin\theta\right)\bar{j} \equiv \bar{0}$$

Finally, the angle β between $\bar{\omega}$ and the z axis may be determined by drawing a sketch of the components of $\bar{\omega}$. This sketch shows that

$$\tan\beta = \frac{\omega_2\sin\theta}{(I_2/I_1)\,\omega_2\cos\theta} = \frac{I_1}{I_2}\tan\theta$$

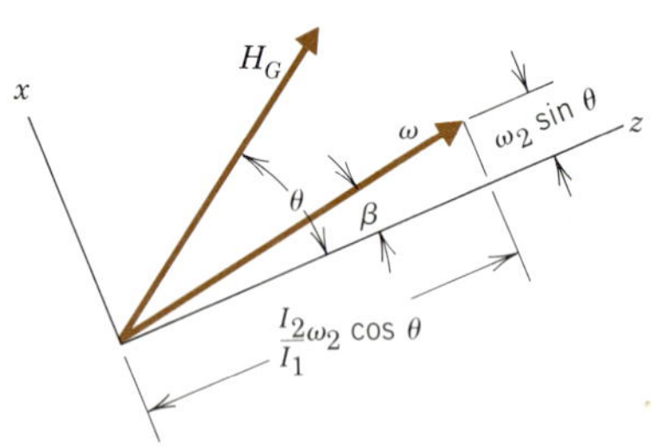

From these results we see that the axis of symmetry, the fixed angular momentum $\bar{H}_G$, and the angular velocity $\bar{\omega}$ are coplanar, with constant angles between them. This type of motion is called *steady free precession*; ω_2 is the rate of precession. There are two kinds of free precession: regular ($\beta < \theta$ and $\omega_2/\omega_1 > 0$) and retrograde ($\beta > \theta$ and $\omega_2/\omega_1 < 0$), depending on whether I_1/I_2 is less than one or not. (What happens when $I_2 = I_1$?) The regular case is exhibited by a poorly thrown football, whereas the retrograde case is typified by a wobbling disk. Various types of precessional motion are treated in advanced texts on dynamics.

HOMEWORK PROBLEMS

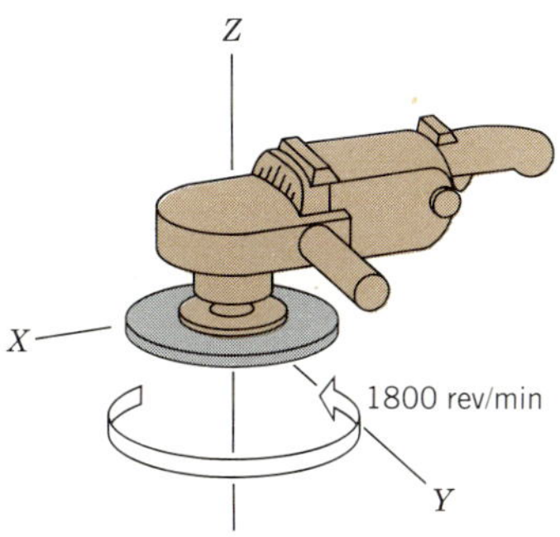

PROB. 16.60

16.60 The pad of a portable rotary sander rotates at 1800 rev/min. The pad may be approximated as a 1 lb thin disk of 8 in. diameter. When the plane of the pad is in the horizontal position shown, it is desired to rotate the sander (a) about the fixed X axis at 1 rad/s, (b) about the fixed Y axis at 1 rad/s. For each case of rotation, determine the couple that the person holding the sander must apply to produce the motion.

16.61 The 0.5 kg bicycle wheel is spinning about its axle at 600 rev/min, while the axle is rotating about the fixed X axis at the constant rate of 2 rad/s. Determine the reaction components A_x, A_z, B_x, and B_z at this instant. The wheel may be considered to be a thin ring.

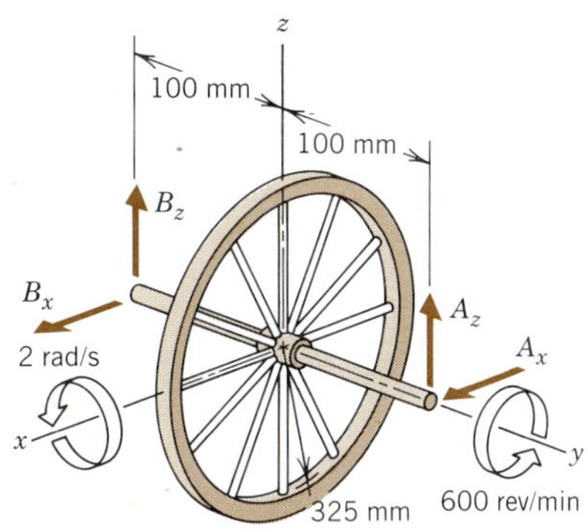

PROB. 16.61

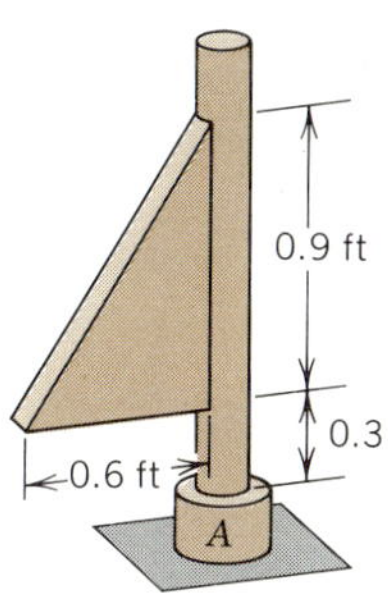

PROB. 16.62

16.62 The right triangular plate, which weighs 15 lb, spins about the vertical axis at 3600 rev/min. Determine the force-couple system exerted by bearing A.

16.63 and 16.64 The horizontal shaft that supports the 40 kg rigid bar is rotating at 150 rad/s. A torque $\bar{M}$ is applied to the shaft, with the result that the rate of rotation instantaneously begins to decrease at 50 rad/s^2. Neglecting the mass of the shaft, determine the torque $\bar{M}$ and the dynamic reactions exerted by the bearings at the instant after $\bar{M}$ is applied.

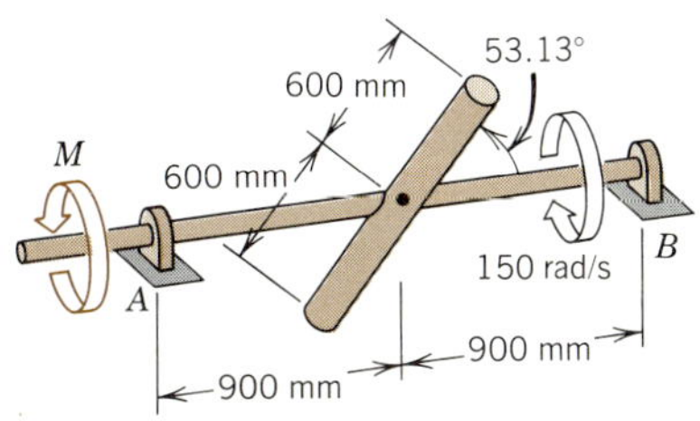

PROB. 16.63

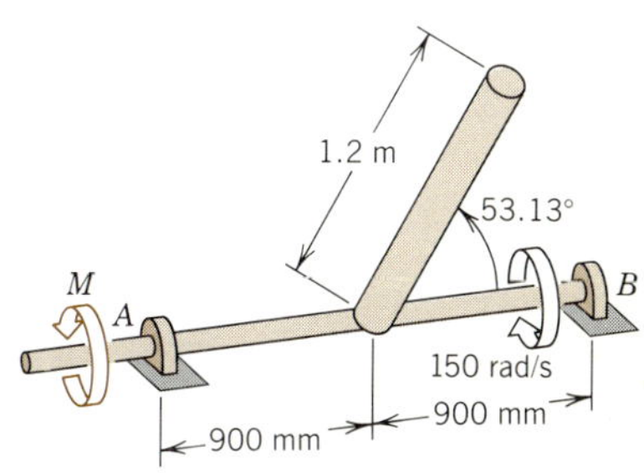

PROB. 16.64

16.65 and *16.66 The 2 lb bar spins around the vertical axis at the constant rate of 600 rad/s, such that the angle of inclination θ of the bar is a constant non-zero value. Determine θ.

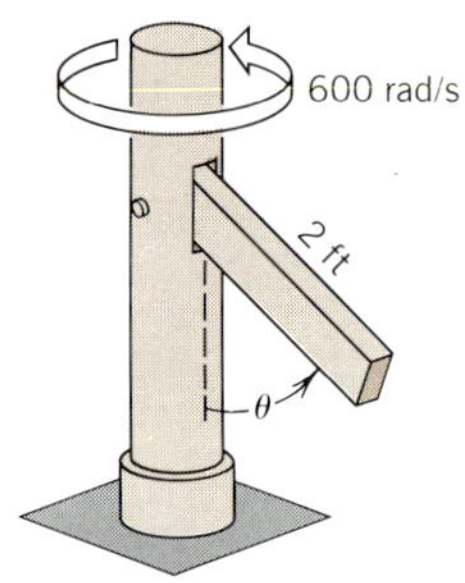

PROB. 16.65

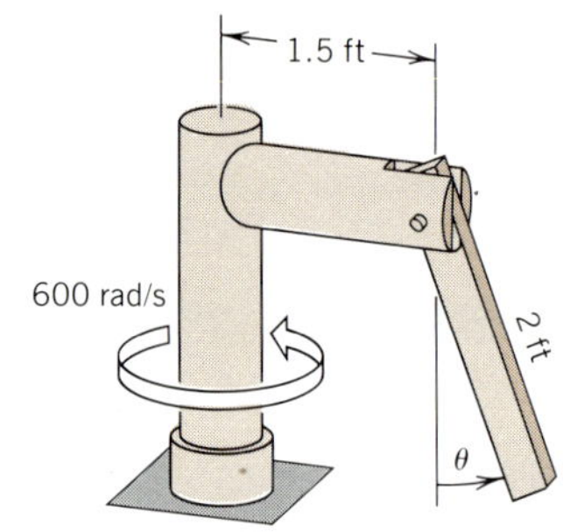

PROB. 16.66

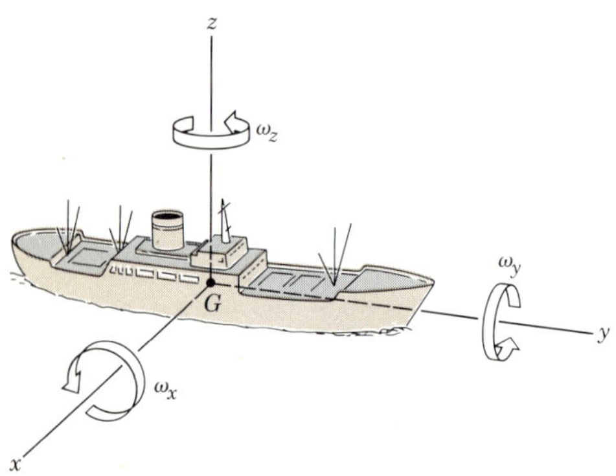

PROBS. 16.67 AND 16.68

16.67 Due to the action of ocean waves, a ship is pitching and rolling as its center of mass moves with a constant velocity. At a certain instant these rotation rates about the axes of the body-fixed xyz coordinate system are maximum at $\omega_x = 0.04$ rad/s and $\omega_y = 0.12$ rad/s. This coordinate system forms a set of principal axes with origin at the center of mass G. The ship weighs 50,000 tons, and the radii of gyration are $k_x = 200$ ft, $k_y = 20$ ft, and $k_z = 210$ ft. Determine the force-couple system at point G that is equivalent to the set of forces causing the motion.

16.68 The angles of rotation of the ship about the pitch and roll axes are given by $\theta_x = 0.05\sin(2t)$ and $\theta_y = 0.10\cos(5t)$, respectively, where the angles are measured in radians and t is in seconds. The center of mass has a constant velocity. Determine, as a function of time, the force-couple system at point G that is equivalent to the set of forces causing this motion. The given coordinate system forms a set of principal axes with origin at the center of mass G. The ship weighs 50,000 tons, and the radii of gyration are $k_x = 200$ ft, $k_y = 20$ ft, and $k_z = 210$ ft.

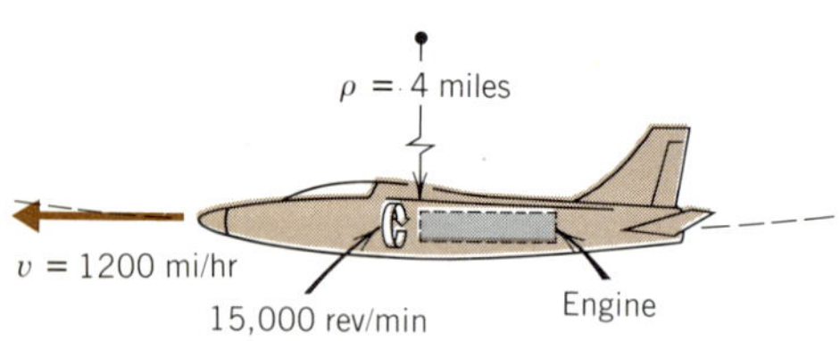

PROB. 16.69

16.69 A jet airplane at the bottom of a 4 mile radius vertical circle is traveling at a constant speed of 1200 mi/hr. The 200 lb rotor of its single fanjet engine has a radius of gyration of 8 in. and rotates at 15,000 rev/min counterclockwise as viewed from the front of the airplane. The rotor is mounted by bearings in front and back. (a) Determine $\dot{\bar{H}}_G$ for the rotor. (b) How are the aerodynamic control surfaces adjusted to provide this gyroscopic moment? (c) If the aerodynamic control surfaces were not adjusted, in which direction would the nose of the aircraft tend to shift?

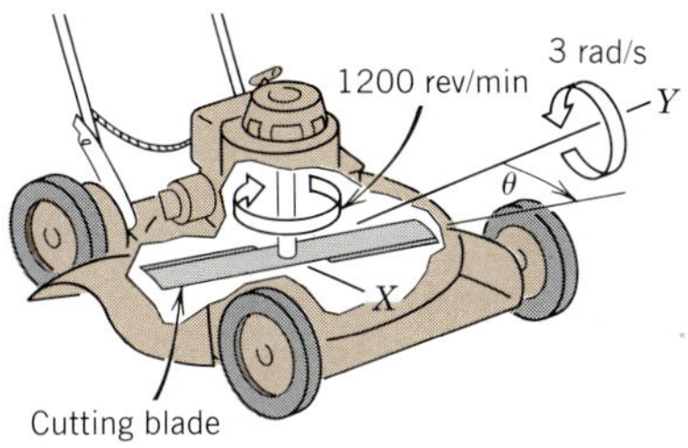

PROB. 16.70

16.70 The cutting blade of a rotary lawn mower is rotating at 1200 rev/min, clockwise as viewed from above. The blade is 600 mm long, has a mass of 1 kg, and is dynamically balanced as it rotates about its center. The lawn mower is in the horizontal position when it is rotated backward about its rear wheels at a constant rate of 3 rad/s. Determine the gyroscopic moment that develops in terms of the angle θ. How is this moment applied to the system of the mower and the blade?

16.71 The blades of an oscillating electric fan rotate with the motor shaft at the constant angular speed ω_1, as the motor housing rotates about the vertical axis at angular speed ω_2. The angle θ between the motor shaft and the vertical is constant, but it may be adjusted. (a) Obtain an expression for $\bar{H}_G$ of the fan blades in terms of ω_1, ω_2, θ, the mass m of the blade assembly, and the principal radii of gyration. (b) By examining the $\bar{H}_G$ and $\bar{\omega}$ vectors, explain why only the radius of gyration of the fan blades about the motor shaft need be known in the case when $\theta = 90°$.

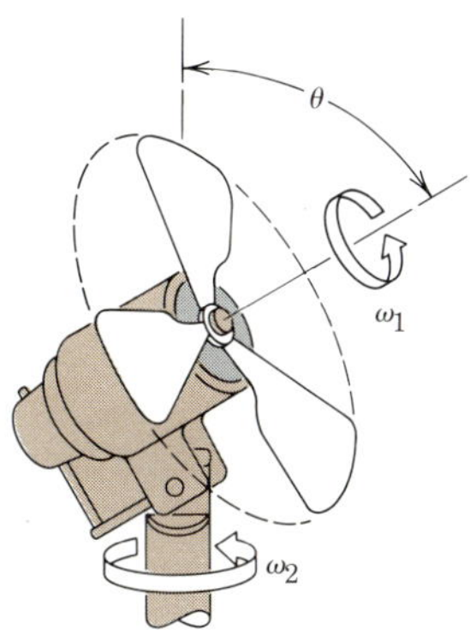

PROB. 16.71

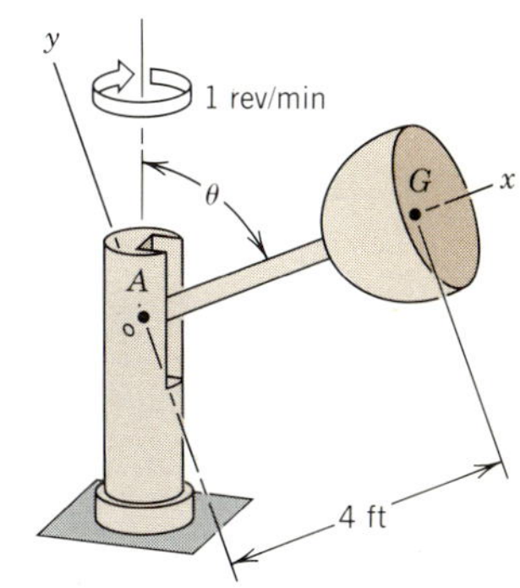

PROB. 16.72

16.72 A radar antenna rotates about the vertical axis at the constant rate of 1 rev/min as the angle of elevation at which the antenna is aimed varies according to $\theta = (\pi/3)(1 + \cos \pi t)$, where θ is in radians and t is in seconds. This antenna weighs 500 lb, and its radii of gyration with respect to the principal axes shown are $k_x = 2$ ft, $k_y = k_z = 3$ ft. Determine the force-couple reaction at support A when $t = 0.25$ s.

16.73 A 1200 kg light airplane is at the top of a vertical circle of 500 m radius, flying at a constant speed of 180 km/hr. At this instant its wings are horizontal and it is performing an aerobatic stunt by rolling about the longitudinal x axis at the constant rate of 0.2 rad/s, as shown. The radii of gyration about the principal xyz coordinate system shown are $k_x = 2$ m, $k_y = 3$ m, $k_z = 3.5$ m. Determine the force-couple system acting at the center of mass G that is equivalent to the aerodynamic forces on the aircraft. The inertia of the propellers is negligible.

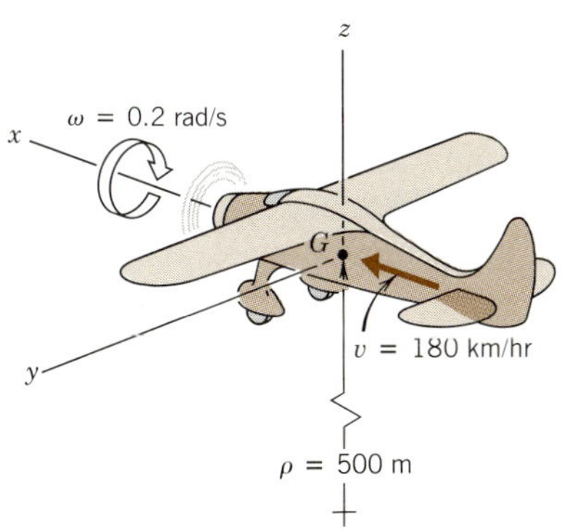

PROB. 16.73

16.74 In another maneuver, the aircraft in Problem 16.73 is rolling about its x axis at $\omega_x = 0.1$ rad/s, pitching about its y axis at $\omega_y = -0.2$ rad/s, and yawing about its z axis at $\omega_z = 0.05$ rad/s, where all rates are constant. Determine the aerodynamic moment about the center of mass for this maneuver. The inertia of the propeller is negligible.

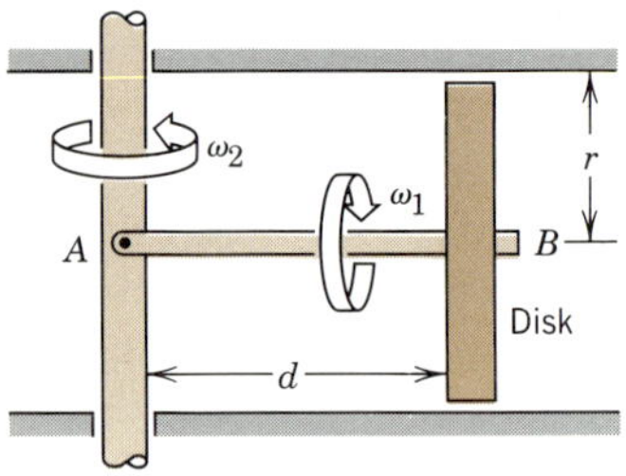

PROB. 16.75

16.75 A thin disk of radius r and mass m rotates freely at rate ω_1 about the lightweight rod AB. The rod is pinned to a vertical shaft that is rotating at the constant rate ω_2. The motion occurs in the region between two fixed horizontal surfaces, as shown in the cross-sectional view. The distance between these two surfaces is only slightly larger than the diameter of the disk, so rod AB may be considered always to be horizontal. Assuming that the disk is rolling without slipping over the lower surface, determine the value of the normal force exerted on the disk by the lower horizontal surface. Is there any value of ω_2 for which this force will be zero?

16.76 The disk in Problem 16.75 is rotating without slipping over the upper surface, thereby reversing the direction of the rotation ω_1 of the disk relative to rod AB. Determine the normal force exerted on the disk by the upper horizontal surface. Is there a minimum value of ω_2 for this rotation?

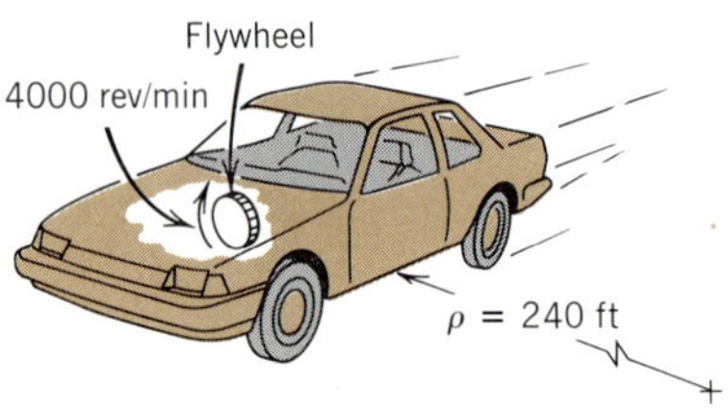

PROB. 16.77

16.77 An automobile is executing a left turn on a road of radius of curvature $\rho = 240$ ft while traveling at 75 mi/hr. The 20 lb flywheel is being driven by the engine at 4000 rev/min clockwise as viewed from the front of the vehicle. The radius of gyration of the flywheel about its axis of symmetry is known to be 1 ft, but the other radius of gyration is unknown. The wheelbase of the vehicle is 10 ft. Determine (a) the gyroscopic moment $\dot{\bar{H}}_G$ for the flywheel, and (b) the amount that the normal force exerted by the ground on each of the front tires increases or decreases to provide this gyroscopic moment. (c) Explain why only one radius of gyration need be known to solve this problem.

16.78 The tires of the automobile in Problem 16.77 weigh 50 lb each and have a diameter of 2.5 ft and radii of gyration of 1.5 ft and 1.0 ft about centroidal axes that are parallel and perpendicular to an axle. (a) Determine $\dot{\bar{H}}_G$ for each tire. (b) Determine the change in the normal reaction on each wheel necessary to produce the gyroscopic moment in part (a). The wheels on each axle are 6 ft apart.

16.79 A force $\bar{F}$ perpendicular to the 4 kg bar AB makes the angle of inclination θ increase at the constant rate $\dot{\theta} = 10$ rad/s. The horizontal shaft is rotating at 120 rev/min. Determine $\bar{F}$ for the instant when $\theta = 60°$. The bar is situated in the vertical plane at that instant.

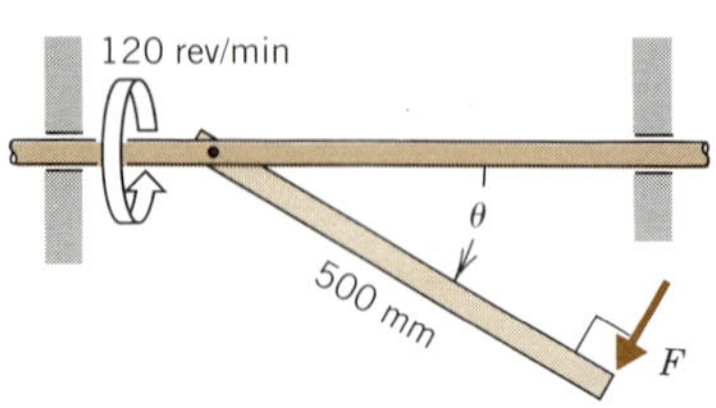

PROB. 16.79

16.80 The gyroscopic turn indicator consists of a 1 kg flywheel whose radii of gyration are $k_x = 50$ mm, $k_y = k_z = 40$ mm. The center of mass G of the flywheel coincides with the intersection of axes AB and CD. The flywheel spins about axis AB with respect to the supporting bracket at a constant rate $\omega = 10{,}000$ rev/min. This bracket is pivoted about the horizontal axis CD, so a couple $\bar{M}$ about axis CD is needed to hold the angle β constant. The whole device is rotating about the vertical axis at the constant rate $\Omega = 0.8$ rad/s, with $\beta = 0$. Determine the magnitude and sense of the couple $\bar{M}$.

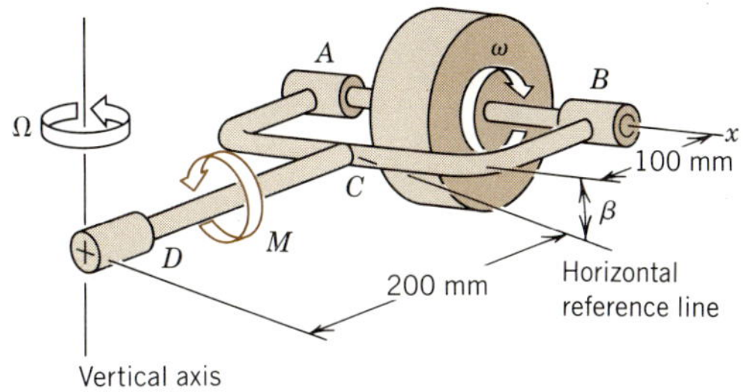

PROB. 16.80

***16.81** Solve Problem 16.80 for the case where $\beta = 30°$.

16.82 The 4 kg homogeneous cylinder rotates about shaft AB at $\omega_1 = 4200$ rev/min while the vertical shaft rotates at the constant rate ω_2. It is observed that the angle θ remains constant at $\theta = 90°$. Determine the magnitude and sense of ω_2. All bearings are smooth, and the mass of shaft AB is negligible.

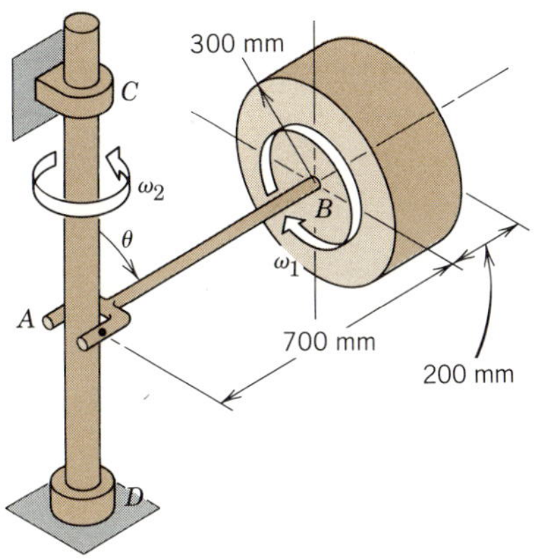

PROB. 16.82

16.83 A toy gyroscope consists of a spinning disk of mass m and radius r that spins about the lightweight arm AB at the constant angular speed ω_1. This arm is supported at point B and rotates about the vertical axis at angular speed ω_2. (a) If θ is constant, derive an equation for θ in terms of ω_1, ω_2, l, r, and m. (This type of motion is a steady precession about the vertical axis.) (b) Determine the relationship between ω_1 and ω_2 for $\theta = 90°$.

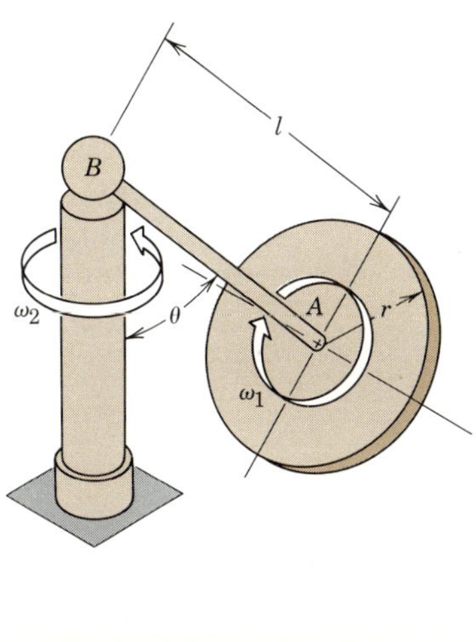

PROB. 16.83

16.84 For the toy gyroscope in Problem 16.83 $l = 16$ in., $r = 2$ in., $\omega_2 = 3$ rad/s, and $\theta = 75°$. What is the value of ω_1?

***16.85** A 10 N-m couple is applied to the vertical shaft, causing the system to accelerate from rest with an angular acceleration $\dot{\omega}$. The rectangular plate has a mass of 6 kg, and the mass of the shaft is negligible. Both bearings are smooth, and bearing B can sustain a thrust loading. Determine (a) the value of $\dot{\omega}$, and (b) the reactions at both bearings 5 s after the motion begins.

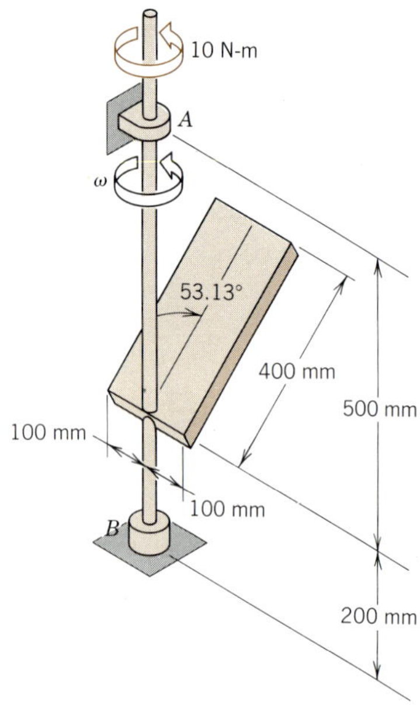

PROB. 16.85

16.86 A football is thrown poorly. When released, the axis of symmetry is 10° above the horizontal and the angular velocity has a magnitude of 250 rev/min at an angle of 15° above the axis of symmetry, as shown. The radius of gyration of the football about its axis of symmetry is half that about an axis perpendicular to its axis of symmetry passing through the center of mass. Determine (a) the angle between the axis of precession and the horizontal direction, (b) the rate of precession, and (c) the rate of spin. Is this regular or retrograde precession? *Hint:* Refer to Illustrative Problem 5.

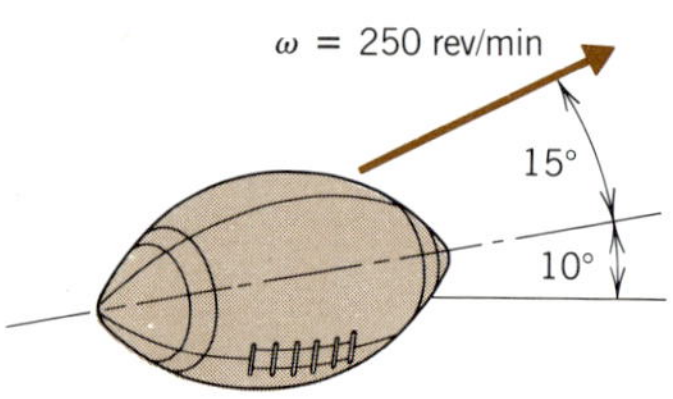

PROB. 16.86

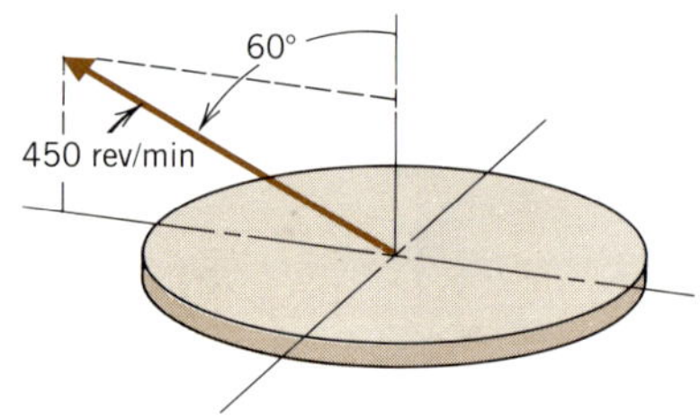

PROB. 16.87

16.87 A coin is thrown into the air. At a certain instant the coin is horizontal and the angular velocity is 450 rev/min in the direction shown. Determine (a) the angle between the precession axis and the centerline of the coin, (b) the rate of precession, and (c) the rate of spin. Is this regular or retrograde precession? *Hint:* Refer to Illustrative Problem 5.

16.88 The ore grinder consists of a 1000 lb disk mounted on arm AB, as shown. This arm is pinned to the vertical shaft, which is driven at a constant angular speed ω_1. The disk rotates freely on arm AB as it rolls without slipping over the interior surface of the vertical cylindrical container. Determine the minimum value for ω_1 required for contact to be maintained between the disk and the interior wall. Neglect the mass of arm AB.

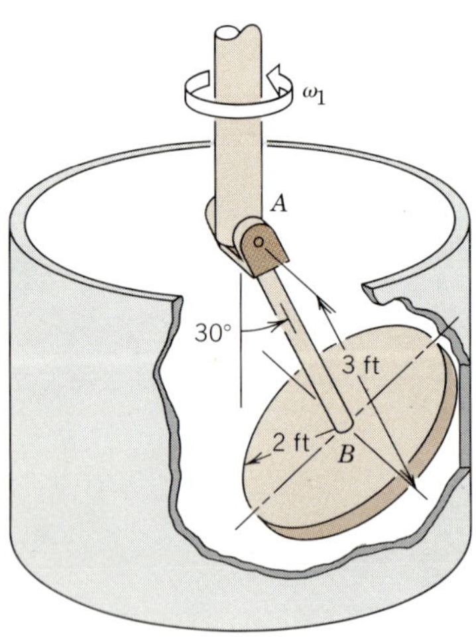

PROB. 16.88

***16.89** Solve Problem 16.88 if the disk is replaced by a 1000 lb sphere having a radius of 2 ft. The sphere is centered at point B, and the other dimensions given in the diagram are unchanged.

16.90 The homogeneous cone is rolling without slipping on a horizontal surface such that it makes 1 rev about the vertical axis every 2 seconds. The cone weighs 2 lb. Determine the force-couple system acting at the apex A that represents the reaction of the ground on the cone.

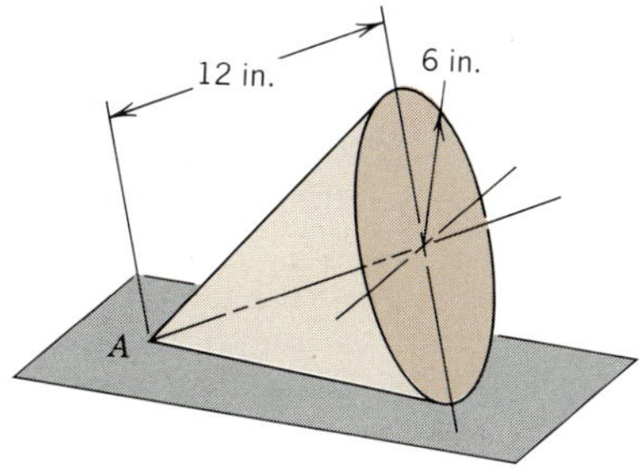

PROB. 16.90

***16.91** A coin of mass m and radius r rolls without slipping on a horizontal surface. The normal to the plane of the coin is inclined at a constant angle θ and the center of the coin follows a path of radius L. Determine the relationship between the constant speed v, the angle θ, and the radius r.

***16.92** The 4 kg disk is obliquely mounted on a shaft AB, which spins at the constant rate $\omega_2 = 50$ rad/s. The yoke supporting this shaft rotates about the vertical axis at the constant rate $\omega_1 = 10$ rad/s. Determine the force-couple system representing the reaction at bearing C. The mass of the shaft and of the yoke is negligible.

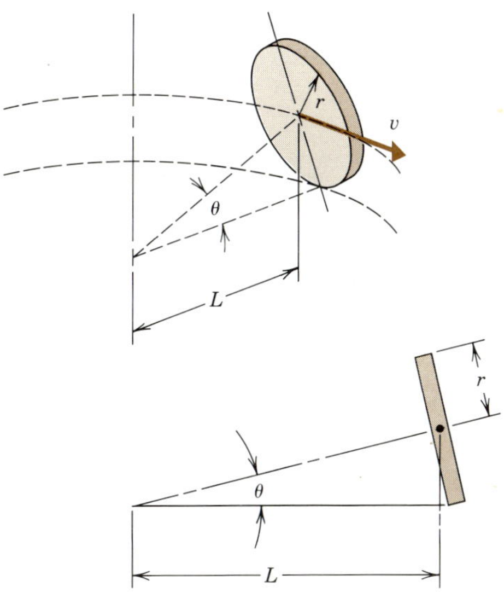

PROB. 16.91

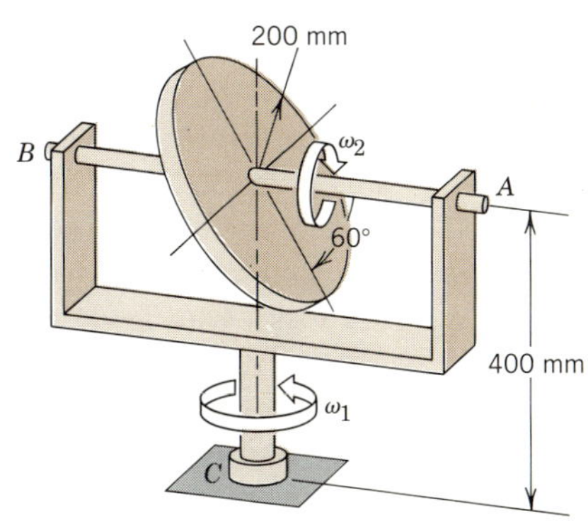

PROB. 16.92

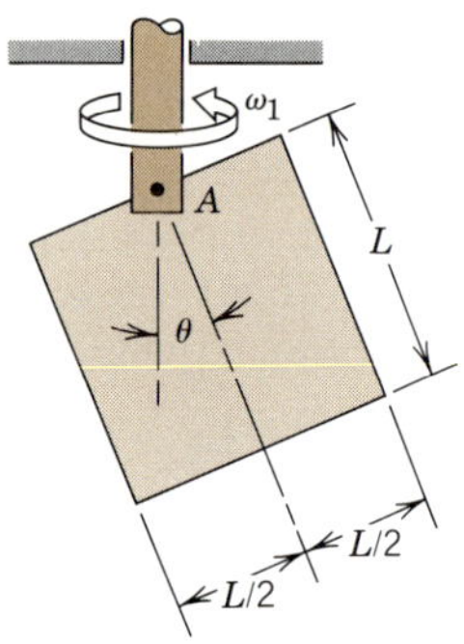

PROB. 16.93

***16.93** The thin rectangular plate is pinned at point A to the vertical shaft, which is rotating at the constant rate ω_1. (a) Considering the angle θ as a variable, derive a single differential equation of motion for this parameter. (b) Using the result of part (a), determine the relationship between ω_1 and the constant non-zero value that angle θ can be. (c) Letting θ be a very small angle, so that $\sin\theta \approx \theta$ and $\cos\theta \approx 1$, examine the result of part (a) to determine the maximum value of ω_1 for which θ will return to zero if slightly disturbed from the zero value. *Hint:* The equation in part (c) will resemble that for a pendulum.

***16.94** Two identical disks weighing 4 lb each are mounted perpendicularly on a lightweight shaft. Small eccentric weights, 2 oz each, are fastened antisymmetrically to the disks at locations A and B. The system rolls without slipping along the horizontal surface at a constant speed $v = 10$ ft/s. Determine the normal and friction forces exerted on each disk by the ground when the plane formed by the shaft and points A and B is as shown.

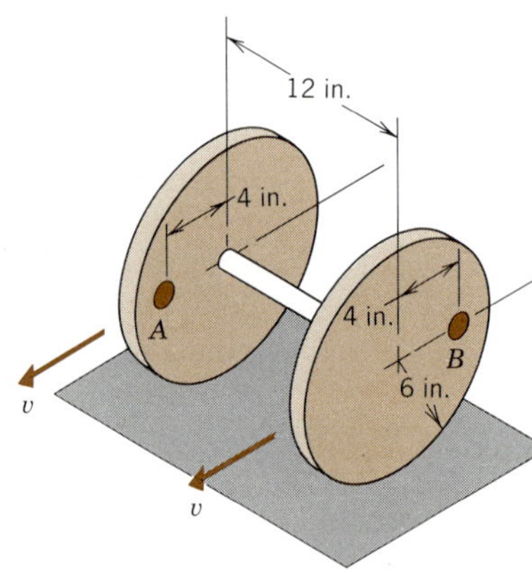

PROB. 16.94

F. ENERGY AND MOMENTUM PRINCIPLES

1. Kinetic Energy of a Rigid Body in Three-Dimensional Motion

The work–energy principle for the motion of particles was derived in Chapter 12, starting from Newton's second law. The applicability of this principal was extended in Chapter 14 to the planar motion of rigid bodies. That development involved rederiving the kinetic energy function for a particle, $T = \frac{1}{2}mv^2$, in order to account for the rotational characteristics of the body; the result was $T = \frac{1}{2}mv_G^2 + \frac{1}{2}I_{zz}\omega^2$. In this section we will extend the field of application of the work–energy principle to the general three-dimensional motion of a rigid body by deriving the kinetic energy function for a three-axis rotation. Nevertheless, the basic form of the work–energy equation remains unchanged, being

$$T_2 + V_2 = T_1 + V_1 + U_{1\to 2}^{(\mathrm{nc})} \tag{30}$$

In order to derive the new expression for the kinetic energy, we have depicted a general body in Figure 12, where xyz is a body-fixed coordinate

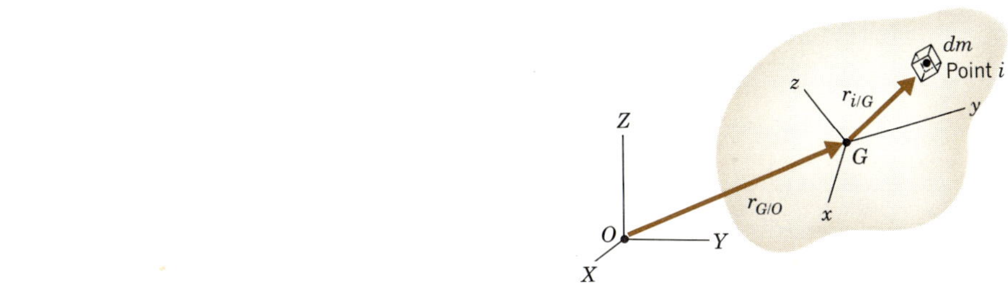

FIGURE 12

system, point i is the location of an infinitesimal element of mass, and point G is the center of mass of the rigid body. Summing the kinetic energy of each element of mass dm contained in the region $\mathscr{V}$ requires an integration, so

$$T = \frac{1}{2} \int_{\mathscr{V}} v_i^2 \, dm$$

The velocity of any point i is related to the velocity $\bar{v}_G$ of the center of mass and the angular velocity $\bar{\omega}$ of the body, according to

$$\bar{v}_i = \bar{v}_G + \bar{\omega} \times \bar{r}_{i/G}$$

The kinetic energy expression then becomes

$$\begin{aligned} T &= \tfrac{1}{2} \int_{\mathscr{V}} \bar{v}_i \cdot \bar{v}_i \, dm = \tfrac{1}{2} \int_{\mathscr{V}} (\bar{v}_G + \bar{\omega} \times \bar{r}_{i/G}) \cdot (\bar{v}_G + \bar{\omega} \times \bar{r}_{i/G}) \, dm \\ &= \tfrac{1}{2}(\bar{v}_G \cdot \bar{v}_G) \int_{\mathscr{V}} dm + \bar{v}_G \cdot \left(\bar{\omega} \times \int_{\mathscr{V}} \bar{r}_{i/G} \, dm\right) \\ &\quad + \tfrac{1}{2} \int_{\mathscr{V}} (\bar{\omega} \times \bar{r}_{i/G}) \cdot (\bar{\omega} \times \bar{r}_{i/G}) \, dm \end{aligned}$$

In the foregoing we were able to remove $\bar{v}_G$ and $\bar{\omega}$ from the first and second integrals because these quantities are properties of the overall motion of the body, and thus depend only on time. The first integral is readily recognizable as the mass of the body, and the second integral is zero because it is the first moment of mass about the center of mass. Thus, the kinetic energy expression reduces to

$$T = \tfrac{1}{2} m v_G^2 + \tfrac{1}{2} \int_{\mathscr{V}} (\bar{\omega} \times \bar{r}_{i/G}) \cdot (\bar{\omega} \times \bar{r}_{i/G}) \, dm \tag{31}$$

To proceed further, we express $\bar{\omega}$ and $\bar{r}_{i/G}$ in component form with respect to the body-fixed axes.

$$\begin{aligned} \bar{\omega} &= \omega_x \bar{i} + \omega_y \bar{j} + \omega_z \bar{k} \\ \bar{r}_{i/G} &= x\bar{i} + y\bar{j} + z\bar{k} \\ \bar{\omega} \times \bar{r}_{i/G} &= (\omega_y z - \omega_z y)\bar{i} + (\omega_z x - \omega_x z)\bar{j} + (\omega_x y - \omega_y x)\bar{k} \\ (\bar{\omega} \times \bar{r}_{i/G}) \cdot (\bar{\omega} \times \bar{r}_{i/G}) &= (y^2 + z^2)\omega_x^2 + (x^2 + z^2)\omega_y^2 + (x^2 + y^2)\omega_z^2 \\ &\quad - 2yz\omega_y\omega_z - 2xz\omega_x\omega_z - 2xy\omega_x\omega_y \end{aligned}$$

We now substitute the last expression into the kinetic energy eq. (31). Then the fact that the components of $\bar{\omega}$ are functions of time only enables us to factor these components out of the integrals. This leads to

$$T = \tfrac{1}{2}mv_G^2 + \tfrac{1}{2}\omega_x^2 \int_{\mathscr{V}} (y^2 + z^2)\,dm + \tfrac{1}{2}\omega_y^2 \int_{\mathscr{V}} (z^2 + x^2)\,dm$$
$$+ \tfrac{1}{2}\omega_z^2 \int_{\mathscr{V}} (x^2 + y^2)\,dm - \omega_x\omega_y \int_{\mathscr{V}} xy\,dm$$
$$- \omega_y\omega_z \int_{\mathscr{V}} yz\,dm - \omega_x\omega_z \int_{\mathscr{V}} xz\,dm$$

We now recognize the integrals as the inertia properties with respect to the xyz axes given in eqs. (6) and (7), so the kinetic energy expression reduces to

$$T = \tfrac{1}{2}mv_G^2 + \tfrac{1}{2}I_{xx}\omega_x^2 + \tfrac{1}{2}I_{yy}\omega_y^2 + \tfrac{1}{2}I_{zz}\omega_z^2 - I_{xy}\omega_x\omega_y - I_{yz}\omega_y\omega_z - I_{xz}\omega_x\omega_z \tag{32a}$$

The (alternative) vector form of this equation is

$$T = \tfrac{1}{2}m\bar{v}_G \cdot \bar{v}_G + \tfrac{1}{2}\bar{\omega} \cdot \bar{H}_G \tag{32b}$$

The vector expression for the kinetic energy of a rigid body emphasizes that the reference system employed for the computation of the inertia properties has its origin at the mass center. Clearly, if the axes are principal axes, the last three terms in eq. (32a) vanish.

From eq. (32b) it is evident that the total kinetic energy is the sum of the translational energy, $mv_G^2/2$, associated with the motion of the center of mass and the rotational energy, $\bar{\omega} \cdot \bar{H}_G/2$, associated with the rotation about the center of mass. In the case where the body is in planar motion parallel to the xy plane, the rotational kinetic energy reduces to $I_{zz}\omega^2/2$.

At this juncture let us retrace the steps in deriving eqs. (32). We began by selecting a body-fixed reference frame whose origin was the center of mass G. As a result, one of the terms in the initial expression for the kinetic energy was zero. Had we located the origin of the reference system at a general point, the derived expression for kinetic energy would be less convenient for computations. However, a special case arises when some point O, having a fixed position with respect to the body, has no velocity. In this case we may locate the origin of the body-fixed reference frame at point O, with the result that

$$\begin{aligned} T &= \tfrac{1}{2}I_{xx}\omega_x^2 + \tfrac{1}{2}I_{yy}\omega_y^2 + \tfrac{1}{2}I_{zz}\omega_z^2 \\ &\quad - I_{xy}\omega_x\omega_y - I_{xz}\omega_x\omega_z - I_{yz}\omega_y\omega_z \\ &= \tfrac{1}{2}\bar{\omega} \cdot \bar{H}_O \end{aligned} \tag{33}$$

The vector form of eq. (33) emphasizes that in this case we are considering the kinetic energy function to be the rotational kinetic energy about point O, which is at rest, and equally important, that the origin of the xyz reference frame is located at point O.

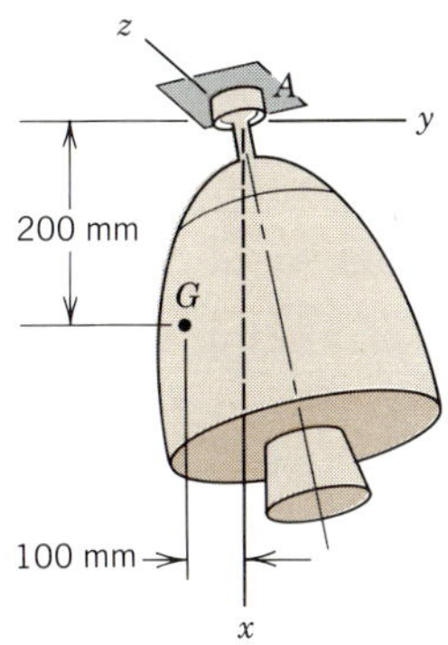

EXAMPLE 10

A 100 kg model of a satellite is being tested by mounting it in a ball-and-socket joint at point A. At the instant shown, point A is at rest and the angular velocity is $\bar{\omega} = 2\bar{i} - 3\bar{j} + 4\bar{k}$ rad/s, where $\bar{i}, \bar{j}, \bar{k}$ are the unit vectors of the set of principal axes fixed to the capsule with origin at point A. It is known that with respect to these axes, $I_{xx} = 30$ kg-m^2, $I_{yy} = 50$ kg-m^2, and $I_{zz} = 60$ kg-m^2. (a) Determine the kinetic energy of the satellite. (b) In a second experiment, point A is towed at a velocity $\bar{v}_A = -5\bar{i} + 2\bar{j} - \bar{k}$ m/s. Determine the kinetic energy of the satellite

Solution

Part a

In this case the ball joint at point A is fixed, so eq. (33) is applicable. We use that formulation because it does not require that we calculate the value of $\bar{v}_G$. The vector $\bar{\omega}$ was given as

$$\bar{\omega} = 2\bar{i} - 3\bar{j} + 4\bar{k} \text{ rad/s}$$

Equation (5) for the angular momentum gives

$$\begin{aligned}\bar{H}_A &= I_{xx}\omega_x\bar{i} + I_{yy}\omega_y\bar{j} + I_{zz}\omega_z\bar{k} \\ &= 30(2)\bar{i} + 50(-3)\bar{j} + 60(4)\bar{k} \\ &= 60\bar{i} - 150\bar{j} + 240\bar{k} \text{ kg-m}^2/\text{s}\end{aligned}$$

It follows from eq. (33) that

$$T = \tfrac{1}{2}\bar{\omega}\cdot\bar{H}_A = \tfrac{1}{2}(120 + 450 + 960) = 765 \text{ J}$$

Part b

In this case the model is in general motion, so we must employ either of eqs. (32). In order to determine $\bar{v}_G$ we use

$$\bar{v}_G = \bar{v}_A + \bar{\omega} \times \bar{r}_{G/A}$$

From the given diagram we see that

$$\bar{r}_{G/A} = 0.20\bar{i} - 0.10\bar{j} \text{ m}$$

so

$$\begin{aligned}\bar{v}_G &= (-5\bar{i} + 2\bar{j} - \bar{k}) + (2\bar{i} - 3\bar{j} + 4\bar{k}) \times (0.20\bar{i} - 0.10\bar{j}) \\ &= -4.60\bar{i} + 2.80\bar{j} - 0.60\bar{k} \text{ m/s}\end{aligned}$$

Let $\hat{x}\hat{y}\hat{z}$ denote the coordinate system whose origin is at the center of mass G, and whose axes are parallel to those of the given xyz system. The coordinates of point A with respect to $\hat{x}\hat{y}\hat{z}$ are $(-0.20, 0.10, 0)$m. Solving the parallel axis theorems for the centroidal properties yield

$$I_{\hat{x}\hat{x}} = I_{xx} - m(\hat{y}_A^2 + \hat{z}_A^2) = 30 - 100[(0.10)^2 + 0] = 29.0 \text{ kg-m}^2$$

$$I_{\hat{y}\hat{y}} = I_{yy} - m(\hat{x}_A^2 + \hat{z}_A^2) = 50 - 100[(-0.20)^2 + 0] = 46.0 \text{ kg-m}^2$$

$$I_{\hat{z}\hat{z}} = I_{zz} - m(\hat{x}_A^2 + \hat{y}_A^2)$$
$$= 60 - 100[(-0.20)^2 + (0.10)^2] = 55.0 \text{ kg-m}^2$$
$$I_{\hat{x}\hat{y}} = I_{xy} - m\hat{x}_A\hat{y}_A = 0 - 100(-0.20)(0.10) = 2.0 \text{ kg-m}^2$$
$$I_{\hat{y}\hat{z}} = I_{\hat{x}\hat{z}} = 0$$

With the inertia properties determined, we can evaluate the angular momentum $\bar{H}_G$ using eq. (5).

$$\bar{H}_G = (I_{xx}\omega_x - I_{xy}\omega_y)\bar{i} + (I_{yy}\omega_y - I_{xy}\omega_x)\bar{j} + I_{zz}\omega_z\bar{k}$$
$$= [29(2) - 2(-3)]\bar{i} + [46(-3) - 2(2)]\bar{j} + 55(4)\bar{k}$$
$$= 64\bar{i} - 142\bar{j} + 220\bar{k}$$

Thus

$$T = \tfrac{1}{2}m\bar{v}_G \cdot \bar{v}_G + \tfrac{1}{2}\bar{\omega} \cdot \bar{H}_G$$
$$= \tfrac{1}{2}(100)[(-4.60)^2 + (2.80)^2 + (-0.60)^2]$$
$$+ \tfrac{1}{2}[(2\bar{i} - 3\bar{j} + 4\bar{k}) \cdot (64\bar{i} - 142\bar{j} + 220\bar{k})]$$
$$= 2185 \text{ J}$$

The rotational portion of the kinetic energy is $\bar{\omega} \cdot \bar{H}_G/2 = 717$ J. This is less than that found in part (a), even though the angular velocity is the same, because the axes in part (a) had an origin that is not the center of mass.

EXAMPLE 11

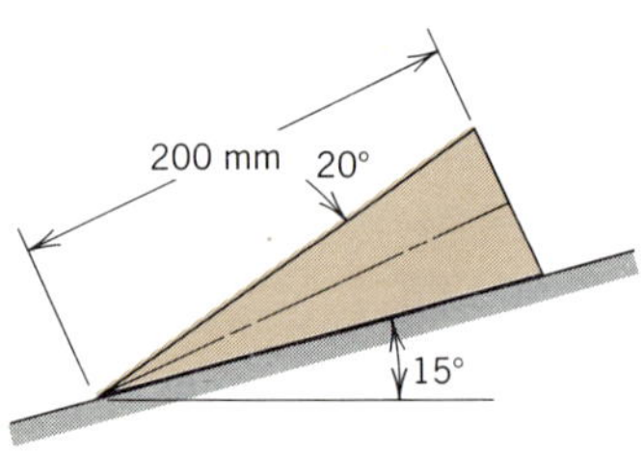

The 2 kg homogeneous cone is at rest on the 15° incline in the position shown. It is released and rolls without slipping down the incline. Determine the rate of rotation of the axis of the cone when it passes through the lowest position on the incline.

Solution

The solution will be formulated in terms of the work–energy principle because we wish to determine the motion of the cone in the final position for a given set of initial conditions. In order to evaluate the work, we draw a free body diagram of the cone in a general position.

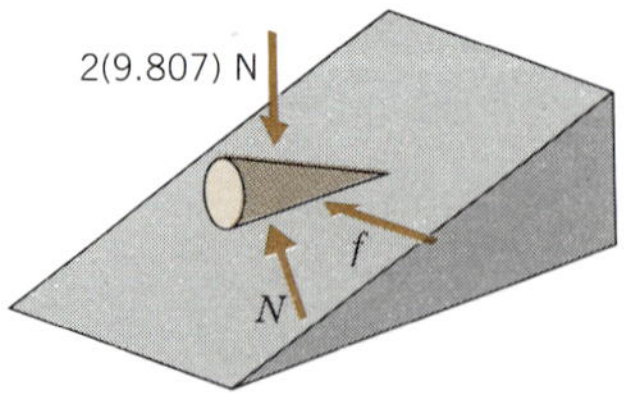

The friction force $\bar{f}$, parallel to the incline, is the constraint force that prevents slipping. As was shown in planar motion, such a friction force does no work. Also, the contact force $\bar{N}$ is normal to the plane. Therefore,

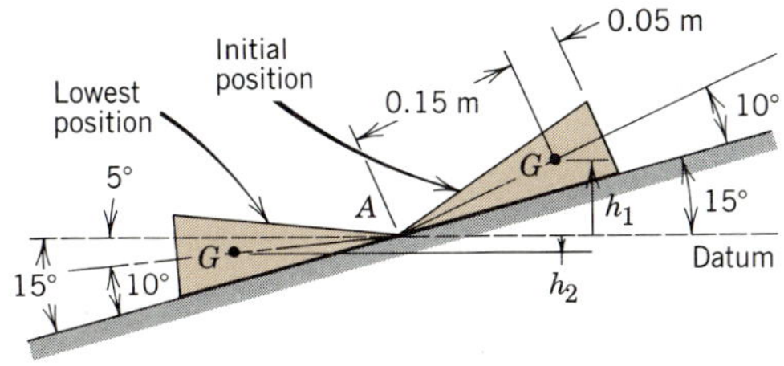

it also does no work. The only force remaining is gravity, which is conservative. It follows that $U^{(nc)}_{1\to 2} = 0$. Let us draw a sketch showing the initial and final positions of the cone. This sketch will assist in computing the potential energy of gravity. In the sketch the apex of the cone is considered to be fixed. Such a motion results when the cone does not slip relative to the incline.

The elevation of the apex A is convenient for a datum. The corresponding initial and final elevations of the center of mass are

$$h_1 = 0.150 \sin 25^\circ = 0.06340 \text{ m}$$

$$h_2 = -0.150 \sin 5^\circ = -0.01307 \text{ m}$$

Thus, the potential energy of gravity is

$$V_1 = mgh_1 = 2(9.807)(0.06340) = 1.2434 \text{ J}$$

$$V_2 = mgh_2 = 2(9.807)(-0.01307) = -0.2564 \text{ J}$$

The kinetic energy in the initial position is zero because the cone was at rest when released. The kinetic energy in the final position may be computed according to eq. (33) because the apex A is a point in the body that is fixed. We therefore define the xyz coordinate system to have point A as its origin, with its axes oriented parallel to those used in Appendix B to describe a cone. The tabulated inertia properties and the parallel axis theorems then give

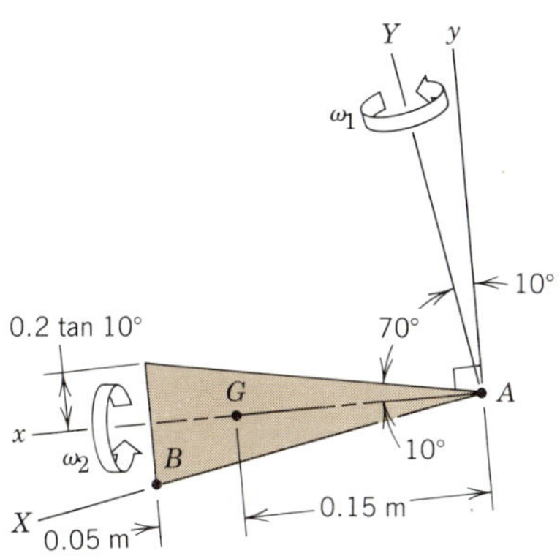

$$I_{xx} = \tfrac{3}{10}mr^2 = \tfrac{3}{10}(2)(0.2 \tan 10^\circ)^2 = 7.462(10^{-4}) \text{ kg-m}^2$$

$$\begin{aligned} I_{yy} = I_{zz} &= \tfrac{3}{80}m(4r^2 + h^2) + md_y^2 \\ &= \tfrac{3}{80}(2)[4(0.2 \tan 10^\circ)^2 + (0.2)^2] + (2)(0.15)^2 \\ &= 4.837(10^{-2}) \text{ kg-m}^2 \end{aligned}$$

$$I_{xy} = I_{xz} = I_{yz} = 0$$

To describe the rotation of the cone, we select a fixed reference system XYZ whose Y axis is normal to the incline, as shown in the last sketch. Then the rotation of the cone consists of a superposition of rotations about the Y axis at angular speed ω_1 and about the x axis at angular speed ω_2. Note that ω_1 is the value we seek. From this we find that the general expression for the angular velocity is

$$\bar{\omega} = \omega_1 \bar{J} - \omega_2 \bar{i}: \quad \text{always}$$

Using the diagram to obtain the instantaneous expression, we have

$$\begin{aligned} \bar{\omega} &= \omega_1(\sin 10^\circ \bar{i} + \cos 10^\circ \bar{j}) - \omega_2 \bar{i}: \quad \text{instantaneously} \\ &= (0.17365\omega_1 - \omega_2)\bar{i} + 0.9848\omega_1 \bar{j} \end{aligned}$$

The rates ω_1 and ω_2 are related by the condition of no slipping. This requires that any point in contact with the incline must have zero velocity. Selecting points A and B to satisfy this condition yields

$$\bar{v}_B = \bar{v}_A + \bar{\omega} \times \bar{r}_{B/A}$$

$$\bar{0} = \bar{0} + [(0.17365\omega_1 - \omega_2)\bar{i} + 0.9848\omega_1\bar{j}] \times (0.20\bar{i} - 0.20 \tan 10^\circ \bar{j})$$
$$= [(0.17365\omega_1 - \omega_2)(-0.03527) - 0.9848\omega_1(0.20)]\bar{k}$$

$$\omega_2 = 0.17365\omega_1 + \frac{0.9848(0.20)}{0.03527}\omega_1 = 5.759\omega_1$$

It then follows that the angular velocity is

$$\bar{\omega} = (0.17365\omega_1 - 5.579\omega_1)\bar{i} + 0.9848\omega_1\bar{j}$$
$$= \omega_1(-5.585\bar{i} + 0.9848\bar{j})$$

Now T_2 may be determined by substituting the inertia properties and the angular velocity components into eq. (33). This gives

$$T_2 = \tfrac{1}{2}(I_{xx}\omega_x^2 + I_{yy}\omega_y^2 + I_{zz}\omega_z^2)$$
$$= \tfrac{1}{2}(7.462)(10^{-4})(-5.585\omega_1)^2 + \tfrac{1}{2}(4.837)(10^{-2})(0.9848\omega_1)^2$$
$$= 3.509(10^{-2})\omega_1^2 \text{ J}$$

Finally, the energy principle for a conservative system gives

$$T_2 + V_2 = T_1 + V_1$$
$$3.509(10^{-2})\omega_1^2 - 0.2564 = 0 + 1.2434$$
$$\omega_1 = 6.54 \text{ rad/s}$$

HOMEWORK PROBLEMS

16.95 The homogeneous ellipsoid shown has a mass of 100 kg. Its center of mass is fixed and its instantaneous angular velocity is $\bar{\omega} = 5i + 4\bar{j} + 8\bar{k}$ rad/s. Determine the kinetic energy of the ellipsoid.

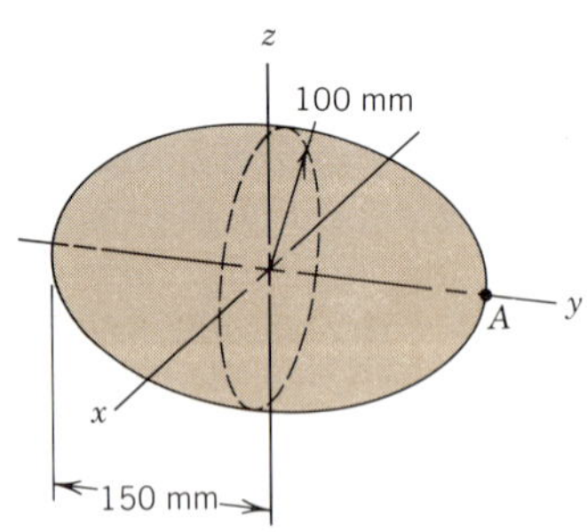

PROBS. 16.95 AND 16.96

16.96 The ellipsoid shown has a mass of 100 kg. Point A at the end of the major axis is a fixed point. The instantaneous angular velocity of the ellipsoid is $\bar{\omega} = 5\bar{i} + 4\bar{j} + 8\bar{k}$ rad/s. Determine the kinetic energy of the ellipsoid (a) using eq. (32), (b) using eq. (33).

16.97 The 5 kg gear A is welded to lightweight arm AB. The gear may be considered to be a uniform disk. Determine the maximum angular velocity of arm AB if the system is released from rest at $\theta = 0°$.

16.98 Solve Problem 16.97 for the case where arm AB has a mass of 2 kg.

16.99 The 3 lb bar screws into the single threaded vertical shaft whose pitch is 0.25 in. The bar, which fits loosely onto the shaft, is given an initial angular speed of 20 rad/s to make it descend the shaft. After 32 rev the bar comes to rest. Determine the average frictional couple opposing the motion of the bar.

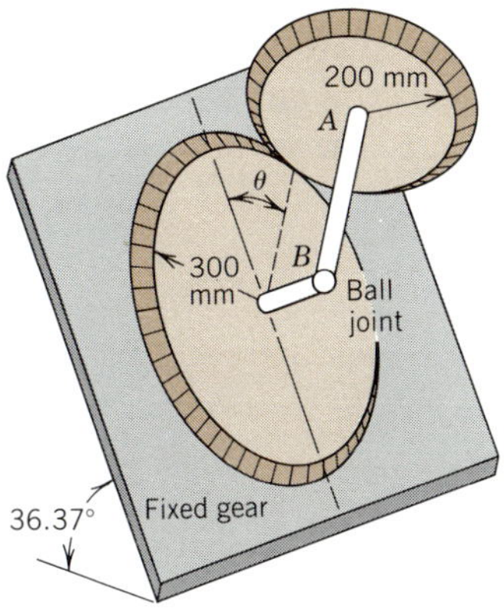

PROB. 16.97

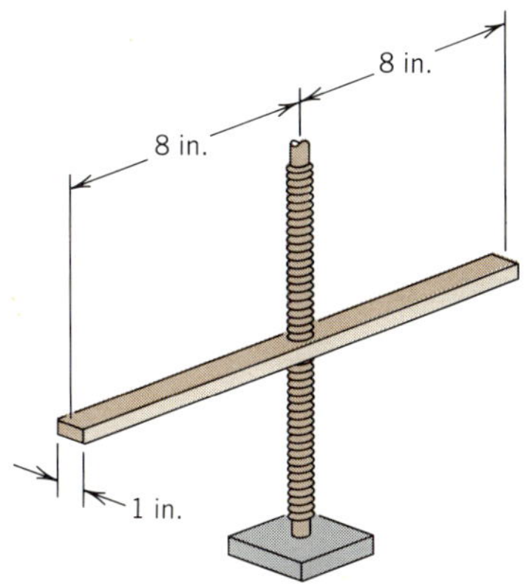

PROB. 16.99

16.100 The vertical shaft is rotating at 5 rad/s, thereby causing gear B to rotate over fixed gear C. Determine the kinetic energy by using eqs. (33). Treat gear B as a thin disk of mass 1.5 kg.

16.101 Determine the kinetic energy of the disk in Problem 16.100 by using either of eqs. (32).

16.102 The disk of mass m and radius r is rolling without slipping over the horizontal surface. The plane of the disk forms a constant angle θ with the vertical plane. The center C has a speed v as it follows a circular path of radius ρ. Derive an expression for the kinetic energy of the disk in terms of m, r, ρ, v, and θ.

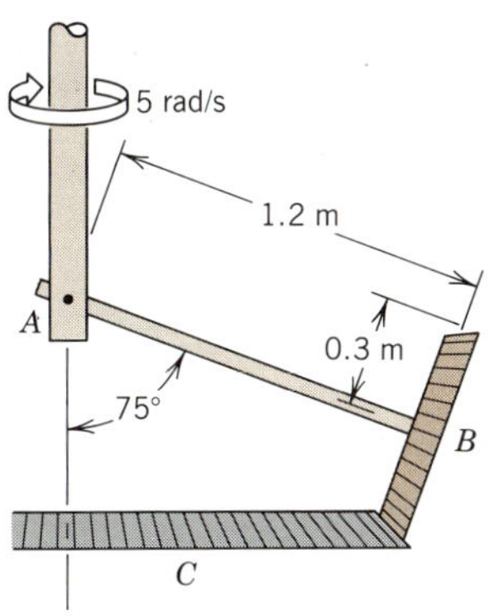

PROB. 16.100

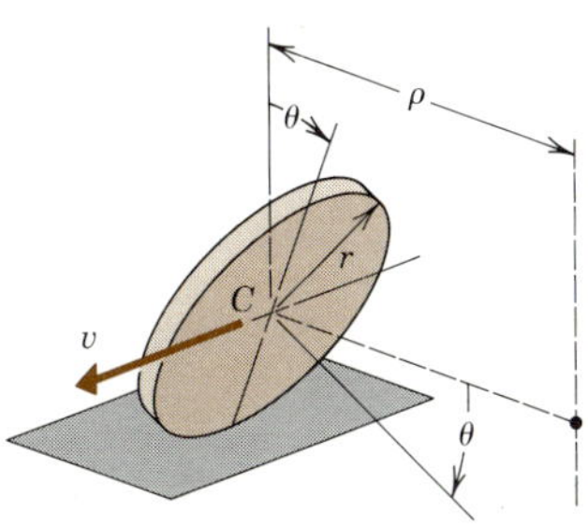

PROB. 16.102

16.103 and 16.104 A couple $\bar{M}$ is applied to the vertical shaft in order to hold the angular speed of this shaft constant at 8 rad/s. The 1.5 kg bar AB is released when the cord is cut, with the result that $\dot{\theta} = -2$ rad/s when $\theta = 30°$. Determine the work done by the couple $\bar{M}$ in this motion.

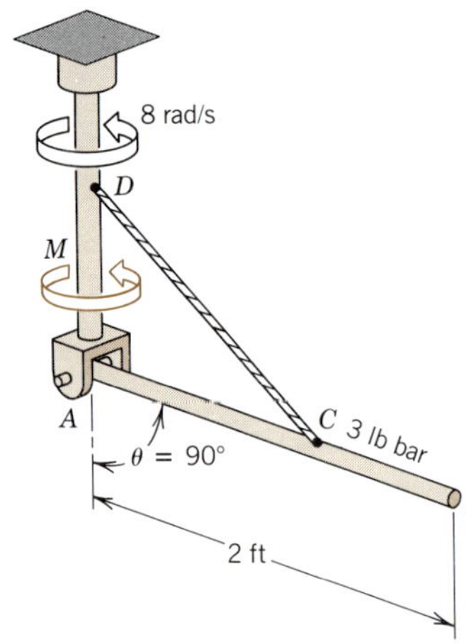

PROB. 16.103

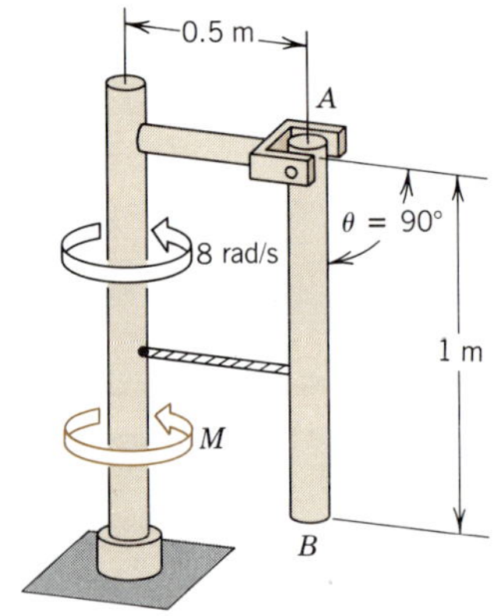

PROB. 16.104

16.105 Solve Problem 16.104 for the case where $\dot{\theta} = -2$ rad/s at $\theta = -30°$.

16.106 The 2 lb sphere rotates about arm AB at the constant rate $\omega_1 = 300$ rad/s. A 40 ft-lb torque is applied to the vertical shaft, causing it to rotate at the ω_2. The masses of the shafts are negligible, and $\omega_2 = 0$ initially. (a) Determine the value of ω_2 after the system has completed 5 rev about the vertical axis. (b) Compare the result of part (a) to that obtained when the angular speed ω_1 is zero.

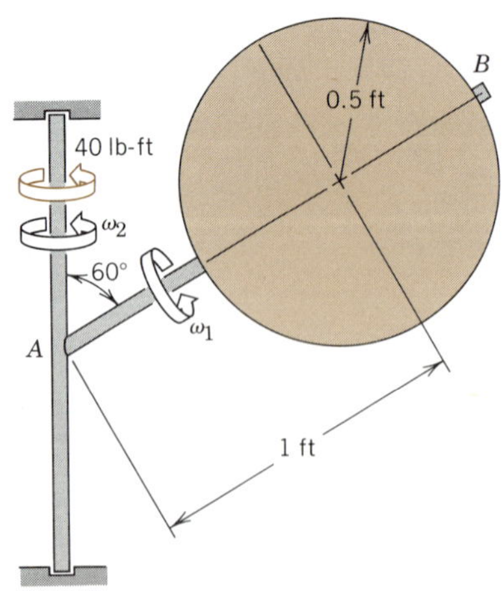

PROB. 16.106

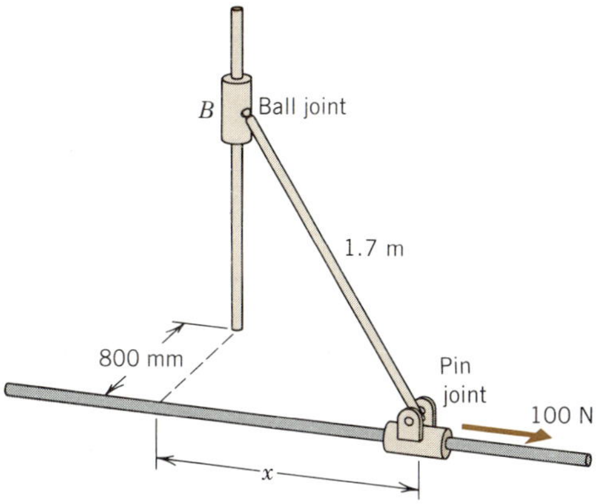

PROB. 16.107

***16.107** The 15 kg bar AB is attached at both ends to lightweight collars that slide over the smooth horizontal and vertical guides. A constant 100 N force, parallel to the horizontal guide, is applied to collar A as shown. If the system is released from rest at $x = 0$, determine the speed of collar A when $x = 1.2$ m.

***16.108** The uniform steel block is moving at 16 ft/s along the smooth angle irons when it encounters the 10° bend in one of the angle irons. Determine the angular velocity of the block after it has moved through a horizontal distance of 4 in. beyond the bend. Assume that edge AB, which rides in the straight angle iron, remains horizontal. *Hint:* If edge AB remains horizontal, then the general motion of the block consists of a rotation about this edge and a translation parallel to the edge. Also, the component of the velocity of corner C lying in the vertical plane must be parallel to the lower leg of the bent angle iron.

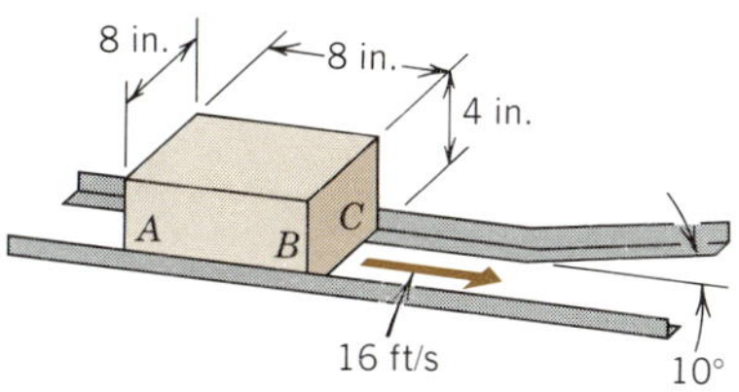

PROB. 16.108

2. Impulse-Momentum Principles

In symbolic form the linear and angular impulse-momentum principles are the same as presented in Chapter 12 for a system of independently moving particles and in Chapter 14 for a rigid body in planar motion. Specifically, the linear impulse-momentum principle is

$$m(\bar{v}_G)_2 = m(\bar{v}_G)_1 + \int_{t_1}^{t_2} \sum \bar{F}\, dt \tag{34}$$

and the angular impulse-momentum principle is

$$(\bar{H}_A)_2 = (\bar{H}_A)_1 + \int_{t_1}^{t_2} \sum \bar{M}_A\, dt \tag{35}$$

Both of the foregoing relations express the fact that the increase in the respective momentum quantities equals the corresponding impulse. In eq. (35) we restrict point A to be one of the allowable points for summing moments (the center of mass or a point in the body that is fixed). This restriction enables us to use eq. (5) to describe the angular momentum of the body.

The impulse-momentum equations given above are useful when the force sum in a specified direction or the moment about a specified axis is known. For example, if a body spins freely about a *fixed* axis, then the angular momentum about that axis is constant.

Equations (34) and (35) are of particular importance in the case of impulsive forces, that is, very large forces that act over a short interval of time. As was the case for forces of this type in particle and planar motion problems, the major effect of any impulsive force is described by its impulse and not by its actual variation with time.

EXAMPLE 12

The 3 in. cube, which weighs 2 lb, is welded to bar AB. This system rotates freely about the vertical axis at angular speed Ω as the bar descends. The initial position was $\theta_1 = 30°$, corresponding to $\dot{\theta}_1 = 0$ and $\Omega = 20$ rad/s. For the position $\theta_2 = 90°$ derive expressions for the corresponding values of Ω_2 and $\dot{\theta}_2$. The masses of bar AB and vertical shaft CD are negligible.

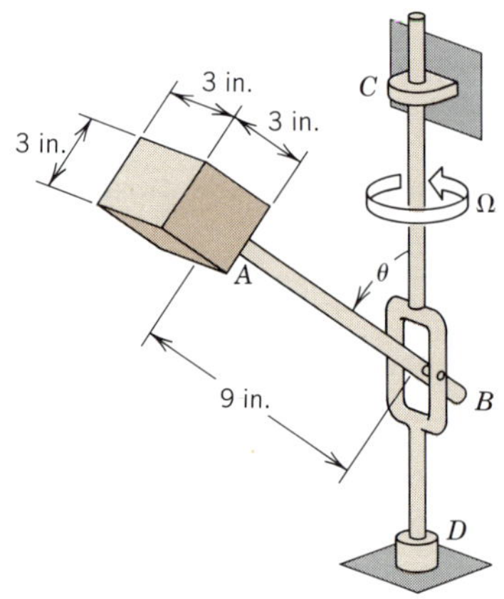

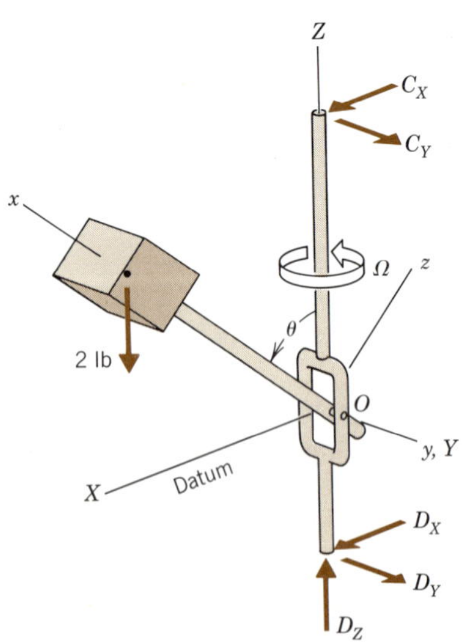

Solution

We draw a free body diagram of the entire system in order to investigate the rotation about the vertical axis. Aside from the reactions at bearings C and D, the only other force appearing in this diagram is the weight. The xyz axes, whose origin is the fixed point O, are oriented parallel to the coordinate system in Appendix B for a rectangular parallelepiped, and the fixed Z axis coincides with the vertical axis. The corresponding inertia properties are

$$I_{xx} = \frac{1}{12}\left(\frac{2 \text{ lb}}{386 \text{ in./s}^2}\right)[(3 \text{ in.})^2 + (3 \text{ in.})^2] = 7.772(10^{-3}) \text{ lb-s}^2\text{-in.}$$

$$I_{yy} = I_{zz} = \frac{1}{12}\left(\frac{2}{386}\right)[(3)^2 + (3)^2] + \left(\frac{2}{386}\right)\left(9 + \frac{3}{2}\right)^2 = 0.5790 \text{ lb-s}^2\text{-in.}$$

$$I_{xy} = I_{xz} = I_{yz} = 0$$

A subtle observation for this problem is that the force system exerts no moment about the (fixed) Z axis. This requires that $\bar{H}_O$ have a constant component in the Z direction. Also, the only force doing work is gravity. Thus, energy is conserved. Applying both principles will yield the relations required to determine $\dot{\theta}_2$ and Ω_2.

Because we will need to evaluate $\bar{\omega}$ and $\bar{H}_O$ for two positions, let us develop general expressions for these quantities. The total rotation of the cube is a combination of a rotation about the vertical axis at Ω and a rotation about the (moving) y axis at θ. Expressing $\bar{K}$ in terms of its components with respect to xyz yields

$$\begin{aligned}\bar{\omega} &= \Omega\bar{K} + \dot{\theta}\bar{j}: \quad \text{always}\\ &= \Omega(\cos\theta\,\bar{i} + \sin\theta\,\bar{k}) + \dot{\theta}\bar{j} = \Omega\cos\theta\,\bar{i} + \dot{\theta}\bar{j} + \Omega\sin\theta\,\bar{k}\end{aligned}$$

$$\begin{aligned}\bar{H}_O &= I_{xx}\omega_x\bar{i} + I_{yy}\omega_y\bar{j} + I_{zz}\omega_z\bar{k}\\ &= 7.772(10^{-3})\Omega\cos\theta\,\bar{i} + 0.5790\dot{\theta}\bar{j} + 0.5790\Omega\sin\theta\,\bar{k}\end{aligned}$$

A general expression for the vertical component of $\bar{H}_O$ may be found by using a dot product. Thus

$$\begin{aligned}H_{OZ} &= \bar{H}_O\cdot\bar{K} = \bar{H}_O\cdot(\cos\theta\,\bar{i} + \sin\theta\,\bar{k})\\ &= 7.772(10^{-3})\Omega\cos^2\theta + 0.5790\Omega\sin^2\theta\end{aligned}$$

The corresponding kinetic energy expression is found from eq. (33).

$$T = \tfrac{1}{2}\bar{\omega}\cdot\bar{H}_O = 3.886(10^{-3})\Omega^2\cos^2\theta + 0.2895\dot{\theta}^2 + 0.2895\Omega^2\sin^2\theta$$

Applying conservation of H_{OZ} between position 1 ($\theta = 30°, \Omega = 20$ rad/s, $\dot{\theta} = 0$) and position 2 ($\theta_2 = 90°$) yields

$$\begin{aligned}\text{at } \theta = 90°: \quad H_{OZ} &= 7.772(10^{-3})\Omega_2\cos^2(90°) + 0.5790\Omega_2\sin^2(90°)\\ &= 0.5790\Omega_2\end{aligned}$$

$$\begin{aligned}\text{at } \theta = 30°: \quad H_{OZ} &= 7.772(10^{-3})(20)\cos^2(30°) + 0.5790(20)\sin^2(30°)\\ &= 3.012 \text{ lb-s-in.}\end{aligned}$$

$$\begin{aligned}0.5790\Omega_2 &= 3.012 \text{ lb-s-in.}\\ \Omega_2 &= 5.201 \text{ rad/s}\end{aligned}$$

In order to formulate the work–energy equation, we define the elevation of the pin at origin O to be the datum, so the potential energy of gravity is

$$V_1 = 2 \text{ lb}(10.5\cos 30° \text{ in.}) = 18.187 \text{ in.-lb}, \qquad V_2 = 0$$

Then, because $U_{1\to 2}^{(nc)} = 0$

$$T_2 + V_2 = T_1 + V_1$$

$$\begin{aligned}&3.886(10^{-3})\Omega_2^2\cos^2(90°) + 0.2895\dot{\theta}_2^2 + 0.2895\Omega_2^2\sin^2(90°) + 0\\ &\quad = 3.886(10^{-3})(20)^2\cos^2(30°) + 0 + 0.2895(20)^2\sin^2(30°) + 18.187\end{aligned}$$

Substituting the earlier result $\Omega_2 = 5.201$ rad/s then yields

$$\dot{\theta}_2 = 11.8 \text{ rad/s}$$

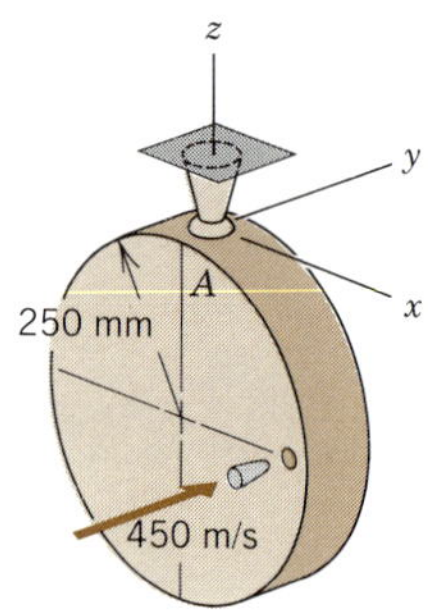

EXAMPLE 13

The 50 kg disk is supported by a ball-and-socket joint at A that permits rotation about all axes. The disk is struck at point B by the 200 gram bullet traveling at 450 m/s perpendicular to the plane of the disk. The bullet is fully embedded in the disk 50 ms after the instant of first contact. Determine (a) the angular velocity of the disk immediately after the impact, and (b) the average reaction force at joint A during the impact.

Solution

Clearly, this is an impulse-momentum problem. We begin with free body diagrams of the disk and of the bullet, in which we neglect the nonimpulsive weight force in comparison to the impact force $\bar{F}_B$ exerted between the bullet and the disk.

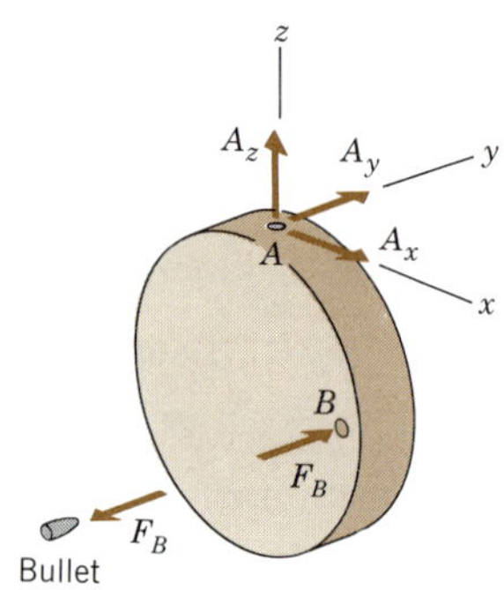

In order to eliminate the impulsive reaction force at point A, we formulate the solution in terms of the angular momentum of the disk about point A. The inertia properties are found from Appendix B and the parallel axis theorems to be

$$I_{xx} = \tfrac{1}{4}mr^2 + mr^2 = \tfrac{5}{4}(50)(0.25)^2 = 3.906 \text{ kg-m}^2$$
$$I_{yy} = \tfrac{1}{2}mr^2 + mr^2 = \tfrac{3}{2}(50)(0.25)^2 = 4.688 \text{ kg-m}^2$$
$$I_{zz} = \tfrac{1}{4}mr^2 = \tfrac{1}{4}(50)(0.25)^2 = 0.781 \text{ kg-m}^2$$

The disk was initially at rest, so $(\bar{H}_A)_1 = \bar{0}$. The angular velocity after the impact is not known, so

$$\bar{\omega}_2 = \omega_x \bar{i} + \omega_y \bar{j} + \omega_z \bar{k}$$
$$(\bar{H}_A)_2 = I_{xx}\omega_x \bar{i} + I_{yy}\omega_y \bar{j} + I_{zz}\omega_z \bar{k}$$
$$= 3.906\omega_x \bar{i} + 4.688\omega_y \bar{j} + 0.781\omega_z \bar{k}$$

In the free body diagram we observe that the force $\bar{F}_B$ exerted by the bullet is the reaction to the force $-\bar{F}_B$, which slows the bullet from its given initial velocity of $450\bar{j}$ m/s. The final velocity of the bullet is the same as that of point B where it is embedded. Thus its final velocity is

$$(\bar{v}_B)_2 = \bar{v}_A + \bar{\omega}_2 \times \bar{r}_{B/A} = \bar{0} + (\omega_x\bar{i} + \omega_y\bar{j} + \omega_z\bar{k}) \times (0.25\bar{i} - 0.25\bar{k})$$
$$= -0.25\omega_y\bar{i} + (0.25\omega_z + 0.25\omega_x)\bar{j} - 0.25\omega_y\bar{k}$$

The impulse momentum equation for the bullet then yields

$$0.2(\bar{v}_B)_2 = 0.2(\bar{v}_B)_1 + (-\bar{F}_B)\,\Delta t$$
$$\bar{F}_B\,\Delta t = +0.05\omega_y\bar{i} + [90 - 0.05(\omega_z + \omega_x)]\bar{j} + 0.05\omega_y\bar{k}$$

Substituting this expression for the impulse of $\bar{F}_B$ into the angular impulse-momentum equation for the disk results in

$$(\bar{H}_A)_2 = (\bar{H}_A)_1 + \int_0^t \bar{r}_{B/A} \times \bar{F}\,dt = \bar{0} + 0.25(\bar{i} - \bar{k}) \times \bar{F}_B\Delta t$$

$$3.906\omega_x\bar{i} + 4.688\omega_y\bar{j} + 0.781\omega_z\bar{k} = 0.25[90 - 0.05(\omega_z + \omega_x)]\bar{i} - 0.025\omega_y\bar{j}$$
$$+ 0.25[90 - 0.05(\omega_z + \omega_x)]\bar{k} \text{ N-m}$$

$$\bar{i}\text{ component:}\quad 3.906\omega_x = 22.5 - 0.0125(\omega_z + \omega_x)$$
$$\bar{j}\text{ component:}\quad 4.688\omega_y = -0.0250\omega_y$$
$$\bar{k}\text{ component:}\quad 0.781\omega_z = 22.5 - 0.0125(\omega_z + \omega_x)$$

The $\bar{j}$ component gives $\omega_y = 0$. Solving the $\bar{i}$ and $\bar{k}$ equations simultaneously yields

$$\omega_x = 5.65, \qquad \omega_z = 28.27$$

Thus

$$\bar{\omega}_2 = 5.65\bar{i} + 28.27\bar{k}\text{ rad/s}$$

Now that the motion of the system after the impact is known, the reactions may be evaluated. The linear impulse-momentum principle for the disk is suitable for this task. The initial velocity of the center of mass G was $(\bar{v}_G)_1 = \bar{0}$. The final velocity of point G is related to the result for ω_2, according to

$$(\bar{v}_G)_2 = \bar{v}_A + \bar{\omega}_2 \times \bar{r}_{G/A} = (5.65\bar{i} + 28.27\bar{k}) \times (-0.25\bar{k}) = 1.413\bar{j}\text{ ft/s}$$

The impulse of the force exerted by the bullet was described earlier in terms of ω_x, ω_y, and ω_z. Substituting the results for these values yields

$$\bar{F}_B\Delta t = \bar{0} + [90 - 0.05(5.65 + 28.27)]\bar{j} + \bar{0} = 88.30\bar{j}\text{ N-s}$$

Then the momentum principle yields

$$m(\bar{v}_G)_2 = m(\bar{v}_G)_1 + \bar{F}_B\,\Delta t + (A_x\bar{i} + A_y\bar{j} + A_z\bar{k})\,\Delta t$$
$$50(1.413\bar{j}) = \bar{0} + 88.30\bar{j} + (A_x\bar{i} + A_y\bar{j} + A_z\bar{k})(0.05)$$

$$\bar{i}\text{ component:}\quad A_x = 0$$
$$\bar{j}\text{ component:}\quad 50(1.413) = 88.30 + A_y(0.05)$$
$$\bar{k}\text{ component:}\quad A_z = 0$$

$$\bar{F}_A = A_y\bar{j} = -353\bar{j}\text{ N}$$

HOMEWORK PROBLEMS

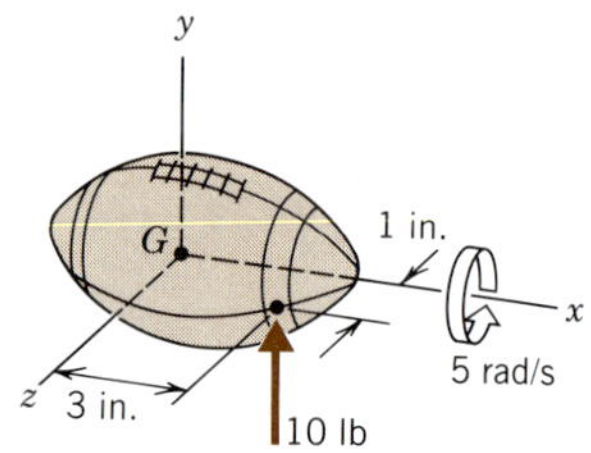

PROB. 16.109

16.109 A 2 lb football is in free flight, traveling horizontally at 60 ft/s as it spins about its x axis at 5 rad/s. A would-be receiver deflects the ball, applying a 10 lb force as shown for an interval of 0.10 s. Determine the linear and angular velocity of the football after it is deflected. The radii of gyration are $k_x = 2$ in., $k_y = k_z = 3$ in.

16.110 The 10 kg disk is suspended at point A by a cable. It is at rest when it is struck at point B by a hammer. The force $\bar{F}$ exerted by the hammer is perpendicular to the disk, with an average magnitude of 400 N during a 50 ms impact interval. Determine the velocity of the center of mass G and the angular velocity of the disk immediately after the impact.

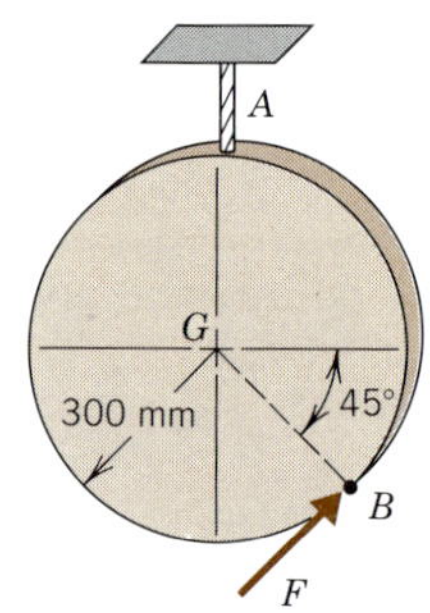

PROB. 16.110

16.111 Solve Problem 16.110 if the disk is supported by a ball-and-socket joint at point A instead of a cable. Also, determine the average reaction at point A during the impact.

16.112 In another case of rotation, it is observed that the largest value of θ for the system in Example 12 is $\theta_2 = 150°$, at which position the system is observed to be rotating about the vertical axis at 40 rad/s. For an initial position $\theta_1 = 60°$, determine the corresponding initial rotation rates $\dot{\theta}_1$ and Ω_1.

16.113 The 10 kg bar is supported by a yoke at the end of the vertical shaft, which rotates freely. In the initial position $\theta_1 = 36.87°$, $\Omega_1 = 1800$ rev/min. Determine the initial value of $\dot{\theta}_1$ required to make $\theta_2 = 120°$ the highest position attained by the bar. Also determine the corresponding angular speed Ω_2.

16.114 The 10 kg bar is supported by a yoke at the end of the vertical shaft, which rotates freely. The rotation rates are $\Omega = 600$ rev/min and $\dot{\theta} = 0$ in the initial position of $\theta = 150°$. Determine the position at which the rotation rate Ω is a minimum. What are the values of Ω and $\dot{\theta}$ at that position?

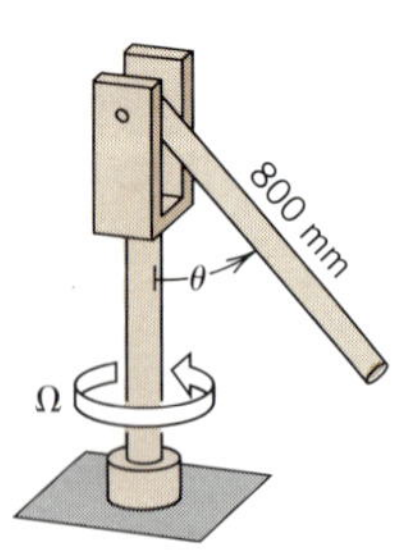

PROBS. 16.113 AND 16.114

16.115 and 16.116 The 50 lb semicircular plate is suspended by a ball-and-socket joint at point A. Initially it was in the vertical plane with the orientation shown, at which instant its angular velocity was $5(-\bar{i})$ rad/s. A 5000 lb force is exerted perpendicular to the plane of the plate at corner B, acting for a 8 ms interval. Determine (a) the angular velocity of the plate immediately after the force is removed, (b) the velocity of the center of mass immediately after the force is removed, and (c) the average reaction force at point A during the interval when the force acts.

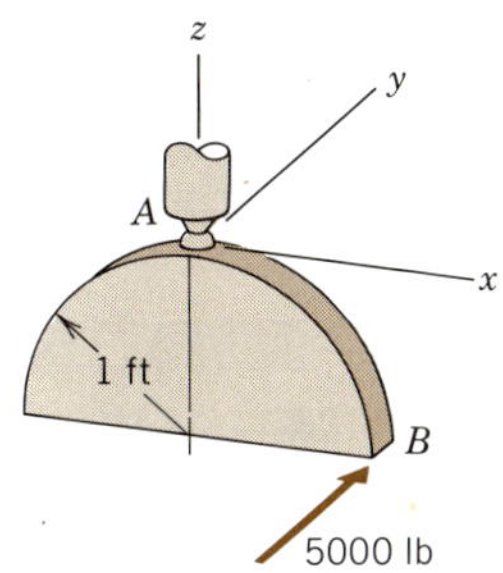

PROB. 16.115

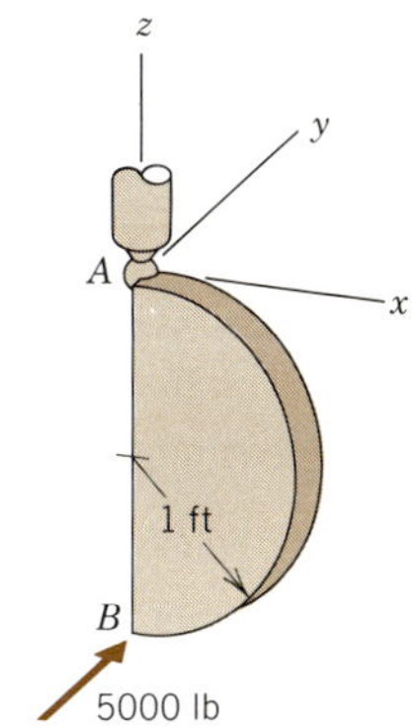

PROB. 16.116

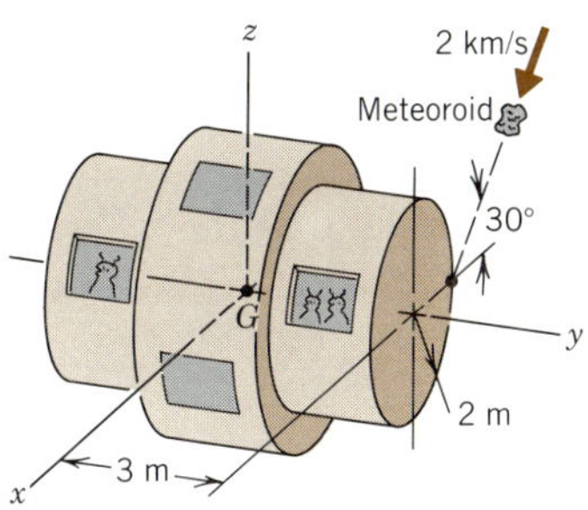

PROB. 16.117

16.117 The 12,000 kg earth satellite was initially in pure translation when it was hit by a 3 kg meteoroid traveling at a speed of 2 km/s relative to the space vehicle in the direction shown. The principal radii of gyration of the satellite are $k_x = k_z = 2$ m, $k_y = 1$ m. Determine the angular velocity of the satellite and the change in its velocity immediately after the impact. The meteor becomes embedded in the satellite at the point of first contact.

16.118 The angular velocity of the 10 lb disk is initially $\bar{\omega}_1 = -4\bar{k}$ rad/s when it is hit by a 3 oz bullet at point B, as shown. The bullet becomes embedded in the disk, with the result that the disk has no angular velocity component about the z axis after the impact. Determine (a) the initial speed v of the bullet to cause this change, and (b) the angular velocity of the disk after the impact.

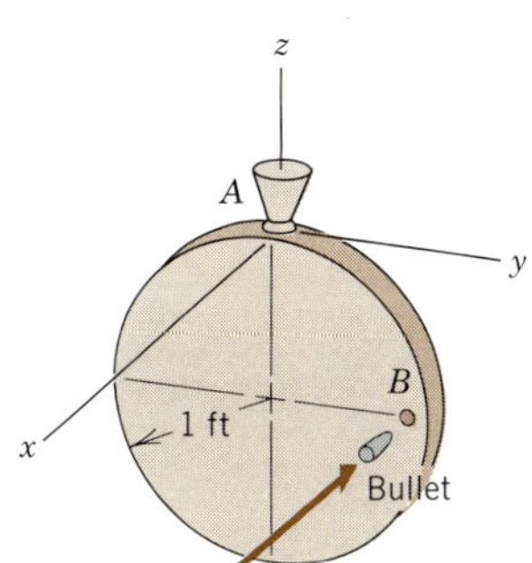

PROB. 16.118

***16.119 and *16.120** Gear A, whose mass is 2 kg, spins relative to shaft AC as it rolls over the stationary gear B. Shaft AC is pinned to the vertical shaft. Both shafts have negligible mass, and gear A may be considered to be a disk. A 50 N-m torque is applied to the vertical shaft when the system is at rest. Determine the angular speed Ω of the vertical shaft 10 s later.

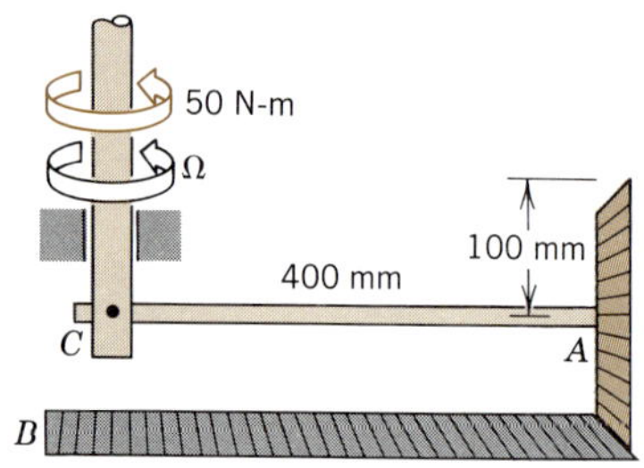

PROB. 16.119

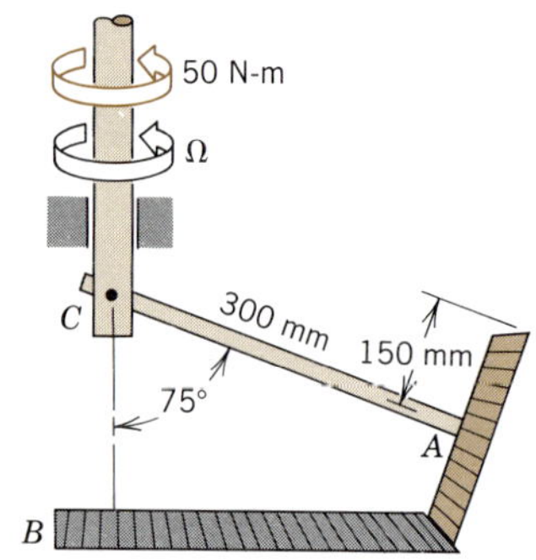

PROB. 16.120

***16.121** The semicircular plate, which weighs 10 lbs, is pinned to the vertical shaft, which rotates freely at angular speed Ω. Initially $\theta_1 = 0$ and $\dot{\theta}_1 = 0$, at which instant $\Omega_1 = 120$ rad/s. Determine Ω and $\dot{\theta}$ when $\theta = 15°$.

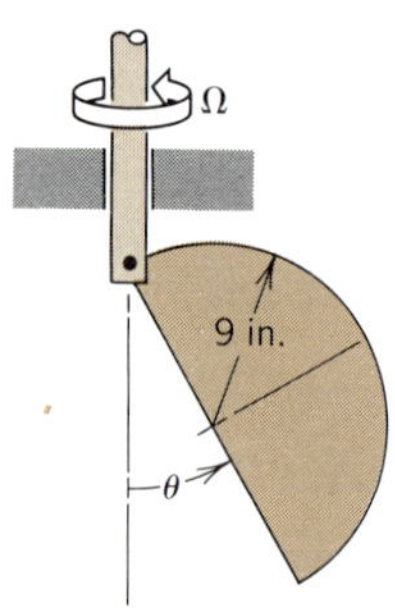

PROB. 16.121

16.122 The 2 kg semicircular plate is translating downward at 5 m/s when corner A strikes the ledge. Assuming that the collision is perfectly plastic, determine the velocity of the center of mass G and the angular velocity of the plate immediately after the collision.

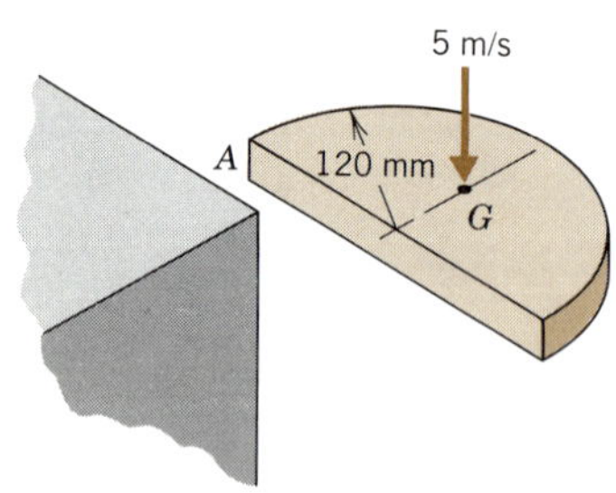

PROB. 16.122

***16.123** Solve Problem 16.122 if the coefficient of restitution for the collision is 0.50.

CHAPTER 17

VIBRATIONS

OBJECTIVES

1. To understand the significance of the static equilibrium position when the equations of motion of a one degree of freedom vibrating system are derived.

2. To evaluate the natural frequency, period, and free response of a one degree of freedom undamped system.

3. To evaluate the amplitude and phase angle of a response that consists of several harmonic terms having the same frequency.

4. To identify whether a system is underdamped, overdamped, or critically damped, and to use this knowledge to evaluate the free response of a damped vibrating system.

5. To use the damped period and logarithmic decrement of an underdamped system to evaluate system parameters, such as the dashpot constant and spring stiffness.

6. To evaluate the response of a system to harmonic excitation for given initial conditions, and to identify the portions of that response that are the transient response and the steady-state response.

7. To use the frequency–amplitude–phase angle relations for the steady-state response to harmonic excitation to evaluate the response for given system parameters or, conversely, to evaluate the system parameters when the properties of the response are given.

8. To determine the equation of motion and natural frequency of a rigid body system having one degree of freedom for small displacements from the static equilibrium position.

The general concern in dynamics is with the motion of a system and the forces that cause that motion. This chapter is devoted to a specialized dynamics investigation entitled *vibrations*, which is the study of the oscillatory motion that may occur in systems having a stable position of static equilibrium. This branch of dynamics is so important that no industrial firm would think of marketing its products, be they household appliances, electronic equipment, airplanes, or automobiles, before thoroughly testing them for vibratory effects.

A. SYSTEM MODEL

Consider any of the systems mentioned above, say an automobile. The objective in modeling a system is to create the simplest mathematical description that retains the essential features of the system. The automobile consists of a chassis suspended by springs, energy-dissipative shock absorbers, and pneumatic tires. A primary function of these components is to insultate the chassis from irregularities in the roadway.

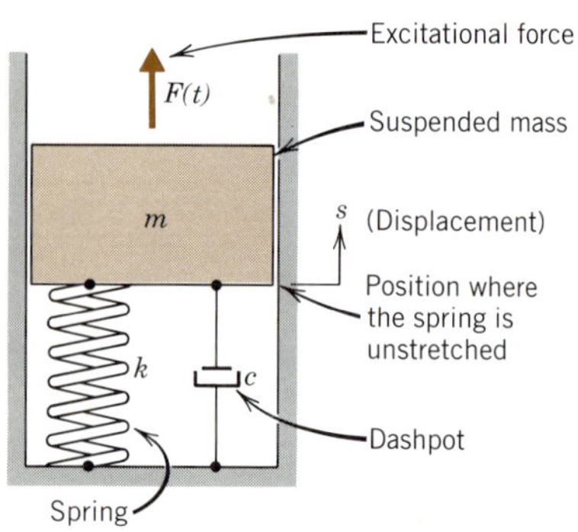

FIGURE 1

As a first attempt to investigate the vertical motion of an automobile, let us model the automobile as a mass suspended by a spring and a friction device, as shown in Figure 1. In this sketch $\bar{F}(t)$ is an excitational force that is causing the motion. The origin of this force could be an unbalanced rotation of the wheels or a rough highway.

The friction device is idealized as a linear viscous *dashpot* that produces a damping force opposing the motion. The pictorial representation of the dashpot in Figure 1 originates from a common design of these devices. As the piston (suspended mass) contained within the cylinder moves, it displaces a viscous fluid. The reason for using the term *linear* to describe the dashpot is that the resistance force it develops is considered to depend linearly on the velocity of the piston. This force is in the opposite sense from the velocity of the piston, so

$$\bar{f}_{\text{dashpot}} = -c\bar{v}_{\text{piston}} \tag{1}$$

where c is the *dashpot constant*, whose dimensions are FT/L. The force developed by a linear dashpot has been shown by experiments to be a good model for the dissipation occurring in the vibrations of many systems.

A prominent feature of the model appearing in Figure 1 is that only one scalar quantity, the vertical displacement s of the mass, need be known to locate the position of the body. Hence, the system of Figure 1 can be used to treat the motion of the center of mass of the automobile only in the vertical direction. A system whose movement is described by a single geometric quantity is said to have *one degree of freedom.* In our studies here, we shall consider only one degree of freedom systems.

B. GENERAL EQUATION OF MOTION

Let us begin to describe the movement of the block in Figure 1 by measuring the displacement s from the position where the spring has its original unstretched length. Then the elongation of the spring, Δ, is equal to s. The piston of the dashpot is moving upward through the cylinder at $v_{\text{piston}} = \dot{s}$. Thus the free body diagram for the system is as shown in Figure 2. Note that the sense of the spring force is depicted such that the force for positive s tends to return the mass to the position $s = 0$. As noted earlier, the dashpot force for positive $\dot{s}$ opposes the sense of the velocity with which the piston is moving through the cylinder.

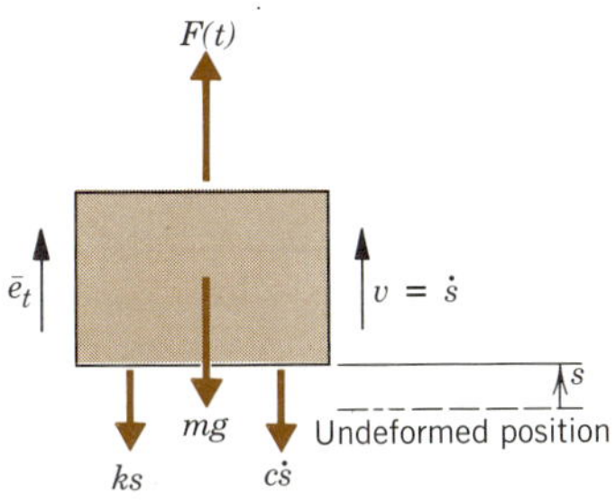

FIGURE 2

The upward acceleration of the body is $\dot{v} = \ddot{s}$, so the equation of motion for the block is

$$\sum F_t = F - mg - ks - c\dot{s} = m\dot{v} \equiv m\ddot{s}$$

$$m\ddot{s} + c\dot{s} + ks = F - mg \tag{2}$$

The foregoing is a second-order (the order of the highest derivative), linear, ordinary differential equation having constant coefficients. The right-hand side describes the static effect of gravity and the time dependent effect of the excitation $F(t)$. It is possible to isolate only the dynamic effect of $F(t)$ by redefining the geometrical quantity describing the movement. For this, note that when the body is in static equilibrium, $\dot{s} = \ddot{s} \equiv 0$, so eq. (2) yields

$$s_{\text{static}} = -\frac{mg}{k} \tag{3}$$

Instead of measuring movement from the position where the spring is undeformed, let us measure the movement of the block relative to the static equilibrium position; y will denote this movement. By definition, $y = 0$ when $s = s_{\text{static}}$. Then it follows that

$$y = s - s_{\text{static}} = s + \frac{mg}{k} \tag{4}$$

This relation is depicted in Figure 3.

The equation of motion governing the displacement y is obtained by solving eq. (4) for s, and then substituting that expression into eq. (2). Because mg/k is a constant, we find

$$m\ddot{y} + c\dot{y} + k\left(y - \frac{mg}{k}\right) = F - mg$$

$$\boxed{m\ddot{y} + c\dot{y} + ky = F} \tag{5}$$

Suppose that the original system in Figure 1 had been oriented in the horizontal plane. In that case, there would have been no component of the weight force in the direction of motion, and eqs. (2) and (5) would have identical forms. In effect, using the displacement y allows us to ignore the gravity force when we construct a free body diagram for the system in Figure 1, as we do in Figure 3.

This result may be generalized.

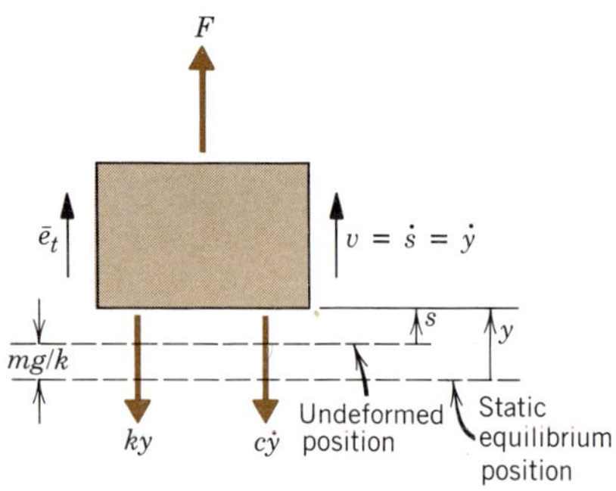

FIGURE 3

> When the displacement of a system is measured from its static equilibrium position, the weight of the system is canceled by the static force in the spring(s) associated with the static displacement. The equation of motion governing such a displacement may be obtained without considering the static loading.

The equation of motion (5) is sometimes referred to as the equation of a forced, damped, linear oscillator. The term *oscillator* refers to the tendency of the mass to move back and forth because of the interplay between the tendency of the spring force to return the mass to the static equilibrium position, and the tendency of the body to continue past this position due to its inertia. We will discuss various aspects of this type of motion in more detail in the following sections.

The solution of the differential equation (5) cannot generally be achieved by the method of separating variables, although we could use this method for the case where only c, k, or $F(t)$ is non-zero. In the following we shall employ another method, which is valid for all linear differential equations having constant coefficients.

C. FREE RESPONSE

An important type of motion arises when a system is *free* of any external excitation. The motion in this case is a result of the potential and/or kinetic energy that is associated with the system's initial conditions. Situations where the effects of energy dissipation are negligible are treated by omitting the dashpot from the representative system in Figure 1. The oscillator is then said to be *undamped*.

1. Free Undamped Motion

The case of free response is obtained by setting $F(t) = 0$, and the undamped case is obtained by setting $c = 0$. Thus, from the equation of motion (5), free undamped motion is represented as

$$m\ddot{y} + ky = 0 \tag{6}$$

The theory of linear differential equations states that the solution to the homogeneous eq. (6) must have the form

$$y = Ae^{\lambda t} \tag{7}$$

Substituting eq. (7) into eq. (6) gives

$$(m\lambda^2 + k)Ae^{\lambda t} = 0$$

The solution must be valid for all instants t, so we cancel the factor $e^{\lambda t}$. Also, $A = 0$ is a trivial solution. Thus, because k and m are positive numbers, the condition that eq. (7) be a nontrivial solution is satisfied when

$$\lambda^2 + \frac{k}{m} = 0 \quad \Rightarrow \quad \lambda = \pm\left(-\frac{k}{m}\right)^{1/2} = \pm i\left(\frac{k}{m}\right)^{1/2} \tag{8}$$

The symbol $i \equiv \sqrt{-1}$, so the values of λ are imaginary. In order to understand the significance of this result, recall the identity relating the rectangular coordinate and polar forms of a complex number (Euler's formula)

$$e^{ip} = \cos p + i \sin p$$

Equation (8) therefore suggests that the response $y(t)$ consists of sinusoidal terms with $p = (k/m)^{1/2}t$. Because the differential equation is second order, we must construct the general solution from two independent solutions. A suitable form is

$$\boxed{y = A\cos(\omega_n t) + B\sin(\omega_n t)} \tag{9}$$

where A and B are constants. The quantity ω_n is called the *circular natural frequency*

$$\boxed{\text{circular natural frequency} = \omega_n = \left(\frac{k}{m}\right)^{1/2}} \tag{10}$$

As we will see, eq. (10) conveys information about the rate at which the system oscillates naturally, that is, without any dissipation or external excitation. Its units are radians per second, which is the reason for using the word *circular* to describe it.

In general, the coefficients A and B are obtained by applying to eq. (9) the initial values of y and $\dot{y}$ at the initial time t_0. However, it is easier to understand the response $y(t)$ if we write it in an alternative form. For this, we note that $\omega_n t$ has units of radians. Considering $\omega_n t$ to be an angle suggests that the two terms in eq. (9) are like components of a vector.

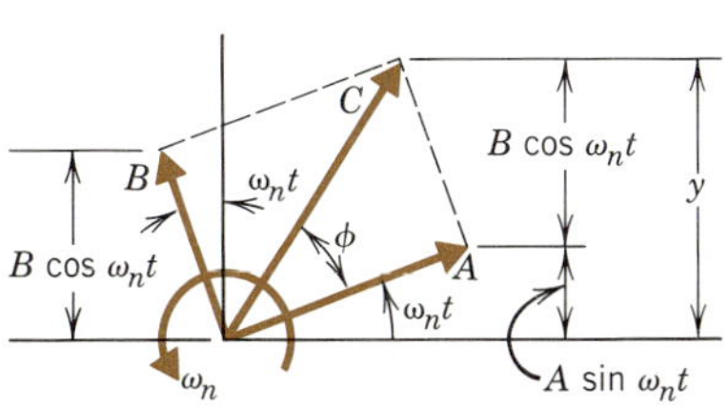

FIGURE 4

The angle $\omega_n t$ locating each vector component is defined such that all components rotate counterclockwise at angular speed ω_n. One possible representation, using vertical components, is shown in Figure 4 (other representations are also possible). This diagram shows that the two vector components associated with the terms in eq. (9) are equivalent to a vector $\bar{C}$ at an angle ϕ ahead of the $\sin(\omega_n t)$ term. The value of C is the *amplitude* of y(t), and the value of ϕ is called the *phase angle* because it describes the angle of the response relative to a *reference response*.

For the situation in Figure 4 the reference is a pure sine response, and ϕ is a *phase lead* because $y(t)$ is ahead of the reference. (Negative ϕ would be a *phase lag*.) The response $y(t)$ may be written in terms of these parameters as

$$y = C \sin(\omega_n t + \phi) \tag{11}$$

We see from Figure 4 that the amplitude and phase angle are related to the coefficients A and B by

$$C = (A^2 + B^2)^{1/2}, \qquad \phi = \tan^{-1}\left(\frac{B}{A}\right) \tag{12}$$

Note that the quadrant of ϕ must be selected carefully.

It is often more convenient to derive eqs. (12) without recourse to diagrams. This may be achieved by using the familiar trigonometric identities

$$\begin{aligned} \sin(a + b) &= \sin a \cos b + \cos a \sin b \\ \cos(a + b) &= \cos a \cos b - \sin a \sin b \end{aligned} \tag{13}$$

Applying the first of these identities to eq. (11) and then equating that result to eq. (9) yields

$$\begin{aligned} y &= C \cos\phi \sin(\omega_n t) + C \sin\phi \cos(\omega_n t) \\ &= A \sin(\omega_n t) + B \cos(\omega_n t) \end{aligned}$$

Matching like functions of time then yields

$$C \cos\phi = A, \qquad C \sin\phi = B \tag{14}$$

These relations are equivalent to eqs. (12); they emphasize the need to assign the quadrant for ϕ based on the signs of A and B.

A typical plot of y versus t is shown in Figure 5, where the amplitude C is the height of the lobes of the sinusoidal curve and ϕ/ω_n is the amount of time by which $y(t)$ occurs before a pure sine response. As exhibited in Figure 5, the response is cyclical, repeating itself whenever $\omega_n t$ is increased by 2π radians. The *period* τ is defined as the time for the response to pass through one cycle, so

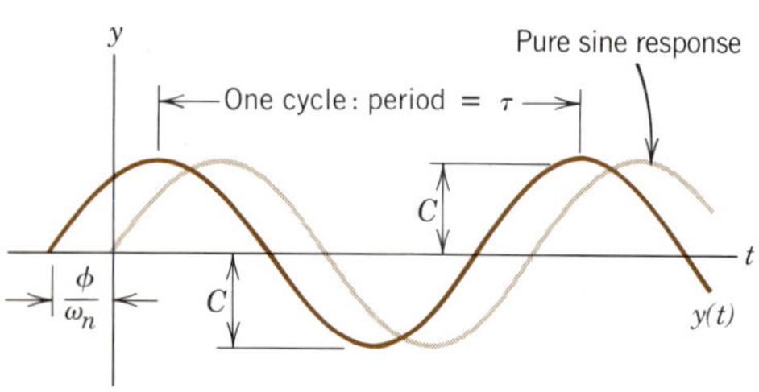

FIGURE 5

$$\boxed{\tau = 2\pi/\omega_n} \tag{15}$$

Another way of visualizing τ is to think of it as the time required for the vectors in Figure 4 to execute one revolution.

The period τ in most engineering systems is a small fraction of a second. In such situations it is easier to measure the reciprocal of τ, that is, the number of cycles that occur in one second. The unit of cycles per second is called a *hertz*, abbreviated Hz. The number of cycles per second executed by a system in its free response is its cyclical natural frequency,

$$\boxed{\text{cyclical natural frequency (hertz)} = \frac{1}{\tau} = \frac{\omega_n}{2\pi}} \tag{16}$$

It should be noted that although the natural frequency of a system is usually stated in units of hertz, the quantity ω_n to be used in mathematical operations must be in units of radians per second.

As a closure to this discussion, it is useful to introduce additional standard terminology. We already defined an oscillation as a fluctuation in a response about some mean value. A synonomous term that is widely used in mechanics is a *vibration*. Any response $x(t)$, such as that in Figure 6, that represents a fluctuation about some mean value is a vibration (or oscillation).

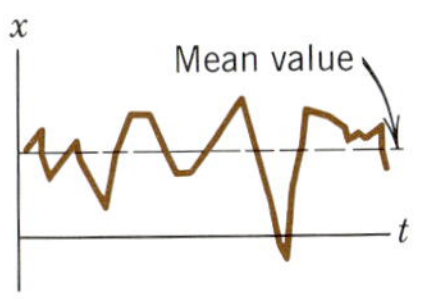

FIGURE 6

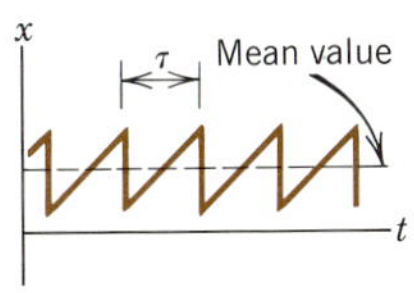

FIGURE 7

When the vibration has the form of a repeated pattern with a constant time interval τ, the oscillation is termed *periodic*. A typical periodic vibration appears in Figure 7.

A periodic vibration that has the shape of a sinusoidal curve is said to be *harmonic*. Thus, Figure 5 represents a harmonic vibration about a zero mean value. (The term *harmonic* to describe a sinusoidal variation in time stems from music, where pure tones are sinusoidal.)

EXAMPLE 1

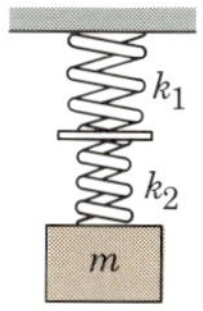

The mass m, which is suspended vertically, is given an initial displacement x_0 downward and released. Determine an equivalent spring constant for the system and the natural frequency ω_n of vertical vibration.

Solution

We begin with a free body diagram of the mass when it is in static equilibrium. In this state, the tensile force in both springs is

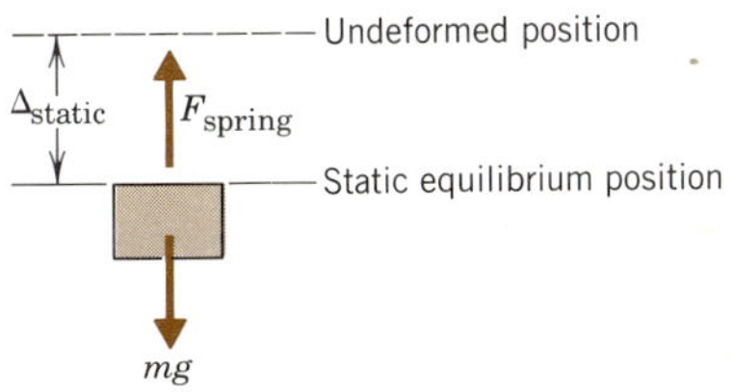

$$F_{\text{spring}} = mg$$

An equivalent spring is defined by

$$k_{\text{equiv}} \Delta_{\text{static}} \equiv F_{\text{spring}} = mg$$

so we need to relate Δ_{static} to the given parameters. For this we note that the elongation Δ_{static} is the sum of the elongations in the individual springs. Because F_{spring} is the same for both springs, we find

$$\Delta_{static} = \Delta_1 + \Delta_2 = \frac{F_{spring}}{k_1} + \frac{F_{spring}}{k_2} = mg\left(\frac{1}{k_1} + \frac{1}{k_2}\right)$$

Substituting this expression into the definition of k_{equiv} yields

$$k_{equiv}\left[mg\left(\frac{1}{k_1} + \frac{1}{k_2}\right)\right] \equiv mg$$

$$\frac{1}{k_{equiv}} = \frac{1}{k_1} + \frac{1}{k_2} \quad \text{or} \quad k_{equiv} \equiv \frac{k_1 k_2}{k_1 + k_2}$$

Once the equivalent stiffness is known, the natural frequency follows immediately.

$$\omega_n = \left(\frac{k_{equiv}}{m}\right)^{1/2} = \left[\frac{k_1 k_2}{m(k_1 + k_2)}\right]^{1/2}$$

EXAMPLE 2

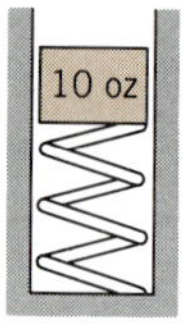

The 10 oz piston moves within the smooth cylinder under the restraint of the spring whose stiffness is 60 lb/in. The piston is released from a point 3 in. below the static equilibrium position, with an upward velocity of 20 ft/s. Determine the displacement $y(t)$ relative to the static equilibrium position. What are the amplitude and the phase angle relative to a pure sine response?

Solution

We define the sense of increasing displacement y to be upward, so a free body diagram of the piston for positive y shows a downward spring force. Note

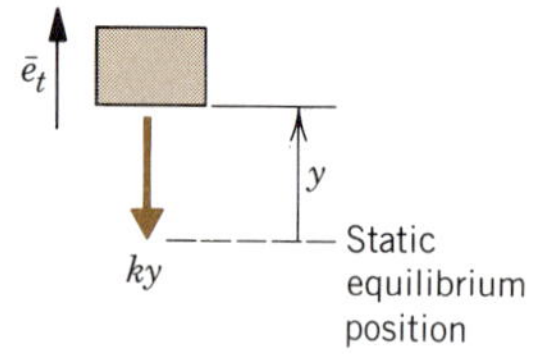

that all static forces are ignored in this diagram, because y is relative to the static equilibrium position. Newton's second law for the piston is

$$\sum F_t = -ky = m\ddot{y} \quad \Rightarrow \quad \ddot{y} + \frac{k}{m}y = 0$$

To evaluate the natural frequency we must convert all parameters to a common set of units. Choosing inches for length yields

$$m = \frac{(10/16)}{386} = 1.6192(10^{-3})\ \text{lb-s}^2/\text{in.}$$

The equation of motion here has the same form as the differential equation (6), so the natural frequency is

$$\omega_n = \left(\frac{k}{m}\right)^{1/2} = 192.50\ \text{rad/s}$$

Because the amplitude and phase angle are explicitly requested, we write the free vibration solution in the form of eq. (11)

$$y = C \sin(\omega_n t + \phi)$$

The values of C and ϕ must be chosen such that $y(t)$ satisfies the given initial conditions. With the instant of release denoted as $t = 0$, the initial conditions are

$$y|_{t=0} = -3\ \text{in.}, \qquad \dot{y}|_{t=0} = 20(12) = 240\ \text{in./s}$$

Note that the initial value of y is negative because the piston is released from below the static equilibrium position and positive displacement is upward. Matching the general response to these initial conditions then yields

$$y|_{t=0} = C \sin\phi = -3, \qquad \dot{y}|_{t=0} = \omega_n C \cos\phi = 240$$

Thus

$$C = \left[(-3)^2 + \left(\frac{240}{\omega_n}\right)^2\right]^{1/2} = 3.249\ \text{in.}$$

$$\phi = \tan^{-1}\left(\frac{-3}{240/\omega_n}\right) = -1.1769\ \text{rad}, \quad 1.96477\ \text{rad}$$

The two values of ϕ arise because the tangent of an angle is negative when the angle is in neither the second or fourth quadrant. However, the initial conditions indicated that $\sin\phi < 0$, $\cos\phi > 0$, which is satisfied only in the fourth quadrant. Thus

$$\phi = -1.1769\ \text{rad} = -67.43^\circ$$

where ϕ is a phase lag. (We could increase ϕ by 2π rad or 360° with equal correctness, which would result in a phase lead of larger magnitude.)
The response corresponding to these parameters is

$$y = 3.25 \sin(192.50t - 1.1769)\ \text{in.}$$

Again, note that the argument of the sine term must be in units of radians.

EXAMPLE 3

A response is given by $y = 10 \sin(50t) + 20 \cos(50t + \pi/4)$ mm. Determine the amplitude of y and the phase angle relative to a pure *cosine* response (a) using the concept of rotating vectors (b) using the trigonometric identities in eqs. (13).

Solution

Method a

The frequency of each term forming y is 50 rad/s, so the vector associated with each term will rotate at 50 rad/s counterclockwise. To construct the diagram we consider the sine term to be the component opposite a vertex angle of $50t$ rad. (Vertical components are used in the sketch.) The cosine term must also be a vertical component; we consider it to be the side adjacent to a vertex angle of $50t + \pi/4$ rad. This angle must be measured from the upward vertical axis, in order that both vertical components be additive. The sketch also depicts the angle for a pure cosine respone. The resultant C is the amplitude of y, and the angle ϕ in the sketch is the phase angle relative to a cosine response.

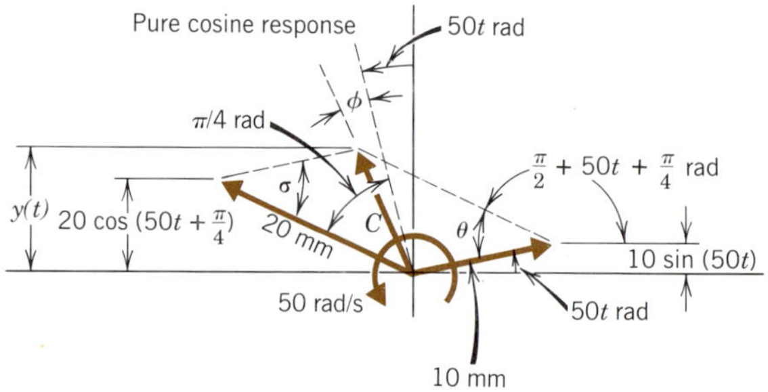

These parameters may be evaluated by trigonometry. The angle σ between the individual vectors is

$$\sigma = \pi - \left[\left(\frac{\pi}{2} + 50t + \frac{\pi}{4}\right) - 50t\right] = \frac{\pi}{4} \text{ rad}$$

The law of cosines then gives the amplitude

$$C^2 = (20)^2 + (10)^2 - 2(20)(10)\cos\frac{\pi}{4}$$

$$C = 14.736 \text{ mm}$$

The phase angle is found from the law of sines

$$\frac{\sin\left(\frac{\pi}{4} - \phi\right)}{10} = \frac{\sin\left(\frac{\pi}{4}\right)}{C} \quad \Rightarrow \quad \sin\left(\frac{\pi}{4} - \phi\right) = 0.4798$$

$$\phi = \frac{\pi}{4} - \sin^{-1}(0.4798) = 0.2849 \text{ rad} = 16.325^\circ$$

Method b

To obtain the amplitude and phase angle mathematically, we write the desired form of the response using a cosine function in order that the phase angle be with respect to a pure cosine response.

$$y = C\cos(50t + \phi)$$

We equate this expression to the given response and apply the identities to all terms that contain the sum of two angles. Thus

$$\begin{aligned} y &= C[\cos\phi\cos(50t) - \sin\phi\sin(50t)] \\ &= 10\sin(50t) + 20\left[\cos\left(\frac{\pi}{4}\right)\cos(50t) - \sin\left(\frac{\pi}{4}\right)\sin(50t)\right] \\ &= -4.142\sin(50t) + 14.142\cos(50t) \end{aligned}$$

Matching like functions of t on each side of the equation yields

$$C\cos\phi = 14.142, \qquad C\sin\phi = 4.142$$

$$C = (14.142)^2 + (4.142)^2]^{1/2} = 14.736 \text{ mm}$$

$$\phi = \tan^{-1}\left(\frac{4.142}{14.142}\right) = 0.2849 \text{ rad} = 16.325°$$

These results, of course, are the same as found in Method a. In situations where the rotating vectors are difficult to depict, the second method usually proves to be more direct.

EXAMPLE 4

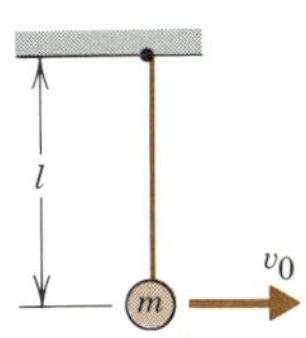

A pendulum consists of a small sphere having mass m that is suspended by a cable of length l. When $t = 0$ the pendulum passes the position $\theta = \theta_0$ with a speed v_0. Determine the period and amplitude of the swinging motion.

Solution

To determine the equation of motion (which is invariably the first step in a vibrations analysis) we draw a free body diagram, using polar coordinates. Remember that $\bar{e}_\theta$ is oriented in the direction of increasing θ. In this system, the component of the gravity force tangent to the path is $-mg\sin\theta$, which is not a constant. Hence gravity does not have a static effect that may be ignored in formulating the equation of motion.

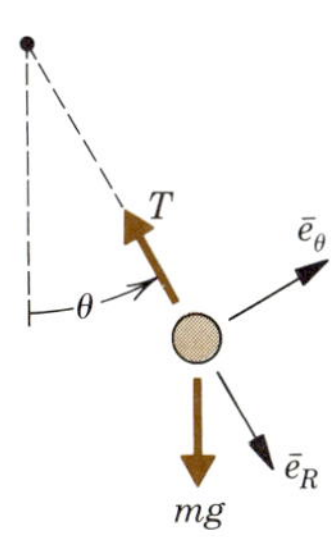

We are not interested in the value of the cable tension T, so we sum forces only in the transverse direction. The radial distance R has the constant value of l, so

$$\sum F_\theta = -mg \sin \theta = ma_R = m(R\ddot{\theta} + 2\dot{R}\dot{\theta}) = m\,l\,\ddot{\theta}$$

Simplifying the differential equation of motion gives

$$\ddot{\theta} + \frac{g}{l} \sin \theta = 0$$

This differential equation is nonlinear because multiplying θ by a factor does not multiply the sine term by the same factor. Nonlinear differential equations of motion arise frequently in engineering systems. Initial investigations usually employ the concept of linearization, in which the response is assumed to be sufficiently small so that we may neglect nonlinearities.

Limiting the angle θ to be very small means that $\sin \theta \simeq \theta$, so the equation of motion becomes

$$\ddot{\theta} + \frac{g}{l} \theta = 0$$

The natural frequency is

$$\omega_n = (g/l)^{1/2}$$

and the period is then

$$\tau = \frac{2\pi}{\omega_n} = 2\pi \left(\frac{l}{g}\right)^{1/2}$$

We could determine the amplitude by deriving a solution for $\theta(t)$ that satisfies the initial conditions, in the manner of Example 2. A simpler approach here is to use the principle of conservation of energy. In the position where θ has attained its maximum value, $\theta = \theta_{max}$, the velocity is zero. Selecting the pivot as the datum for measuring the gravitational potential energy then yields

$$T_1 = \tfrac{1}{2}mv_0^2, \qquad V_1 = mg(-l \cos \theta_0)$$

$$T_2 = 0, \qquad V_2 = mg(-l \cos \theta_{max})$$

$$T_2 + V_2 = T_1 + V_1$$

$$-mgl \cos \theta_{max} = \tfrac{1}{2}mv_0^2 - mgl \cos \theta_0$$

$$\theta_{max} = \cos^{-1}\left[\cos \theta_0 - \frac{v_0^2}{2gl}\right]^{1/2}$$

Finally, recall that we have assumed that θ is very small. Given values for the system's parameters, we could check this assumption by computing θ_{max} (in radians), and then compare the angle to $\sin \theta_{max}$.

HOMEWORK PROBLEMS

17.1 and 17.2 Determine the equation of motion governing the displacement of the mass from (a) the position where the spring is not deformed, and (b) the static equilibrium position.

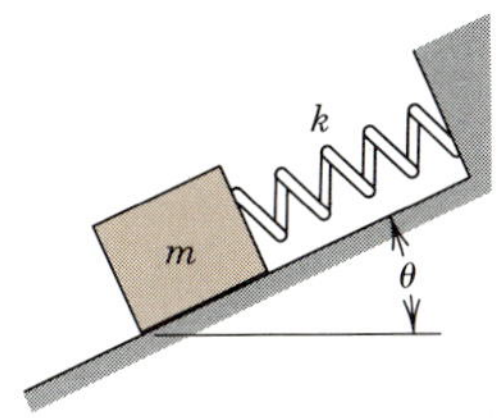

PROB. 17.1

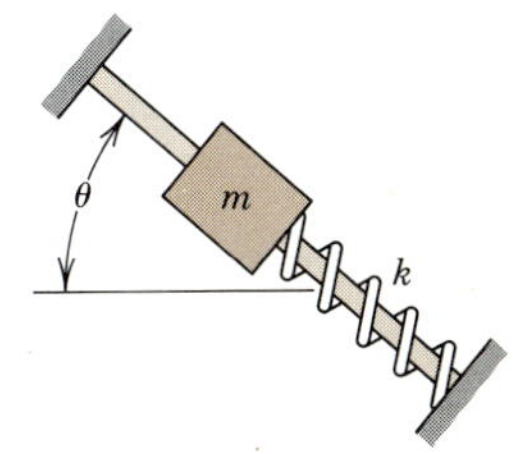

PROB. 17.2

17.3 and 17.4 Determine the ratio of k_2/k_1 required for the two systems shown to have the same natural frequency.

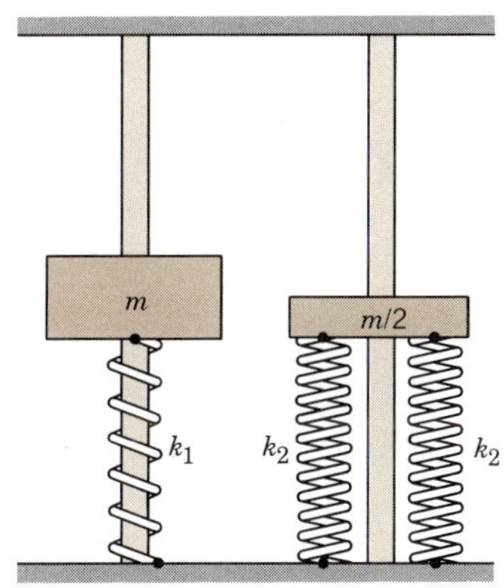

PROB. 17.3

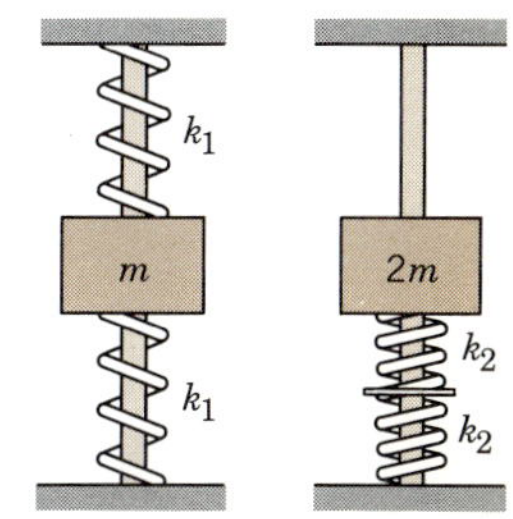

PROB. 17.4

17.5 Three different ways of supporting a piston of mass m that rides in a smooth vertical cylinder are shown. Each system has two identical springs of stiffness k. Determine the natural frequency of each system.

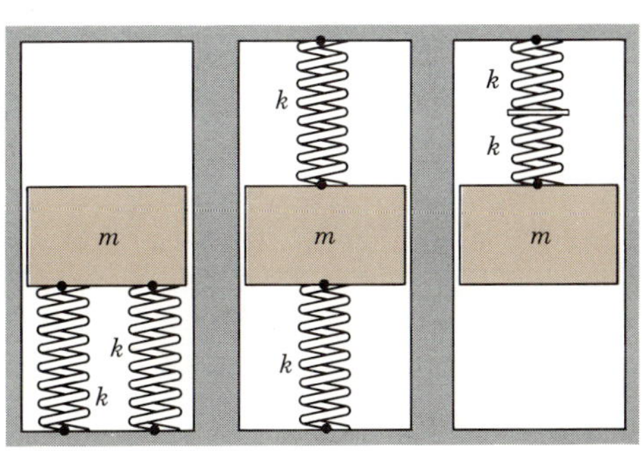

PROB. 17.5

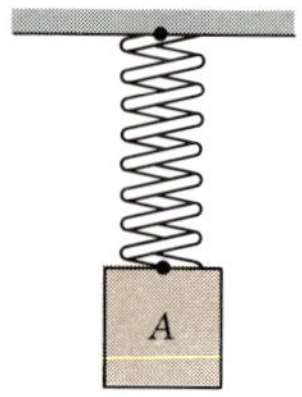

PROB. 17.6

17.6 The mass of block A and the stiffness of the spring are unknown. It is known that the system vibrates freely with a period of 12.5 ms. When a 4 kg block is fastened to block A, the period rises to 20 ms. Determine the mass of block A and the stiffness of the spring.

17.7 The mass m and the stiffness k_1 are unknown, and $k_2 = 400$ lb/ft. The period of free vibration of the system is 20 ms, and removing spring 2 lowers this period to 25 ms. Determine m and k_1.

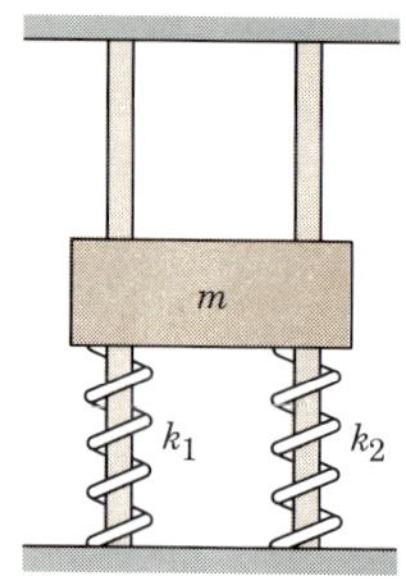

PROB. 17.7

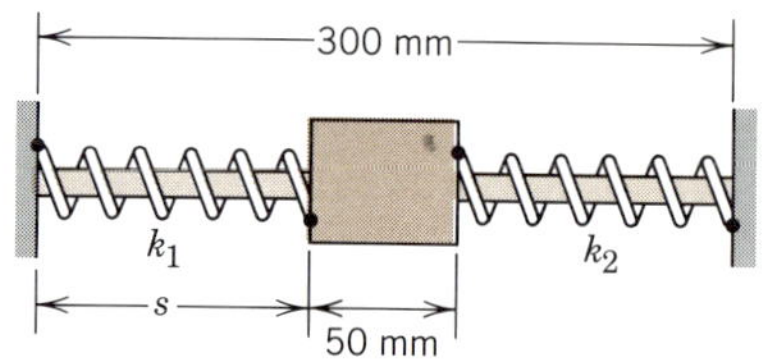

PROB. 17.8

17.8 Two pretensioned springs are attached to the 0.5 kg block. The unstretched length of each spring is 60 mm, and their stiffnesses are $k_1 = 2$ kN/m, $k_2 = 1$ kN/m. Determine (a) the value of the distance s corresponding to static equilibrium, and (b) the cyclical natural frequency of the system.

17.9 Solve Problem 17.8 for the case where the unstretched lengths are 80 mm for spring 1 and 120 mm for spring 2.

17.10 The 8 oz collar slides on the smooth horizontal bar. The spring has a stiffness of 2 lb/in. and the system is in the static equilibrium position when $s = 10$ in. (a) Determine the natural frequency in units of radians per second and hertz. (b) If the collar is released from rest at $s = 6$ in., determine where it will again come to rest and the elapsed time between these two positions.

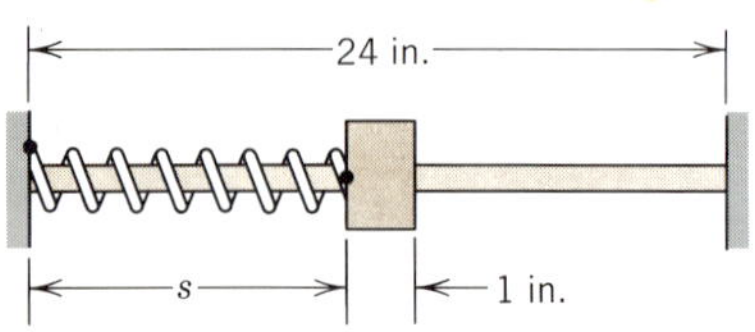

PROB. 17.10

17.11 The collar in Problem 17.10 has a velocity of 80 in./s to the left when it is at $s = 12$ in. Determine the elapsed time from this instant until the collar returns to the static equilibrium position.

17.12 The 2 kg block slides on the smooth horizontal surface under the restraint of the spring whose stiffness is 400 N/m and whose unstretched length is 1 m. The block is given an initial velocity of 5 m/s to the right at $s = 200$ mm. Determine (a) the response $x(t)$, where x is the displacement measured from the static equilibrium position, (b) the amplitude, period, and phase angle relative to a sine response for the response in part (a). (c) Would the answers to parts (a) and (b) change if the system was inclined upward to the left at 30° above the horizontal?

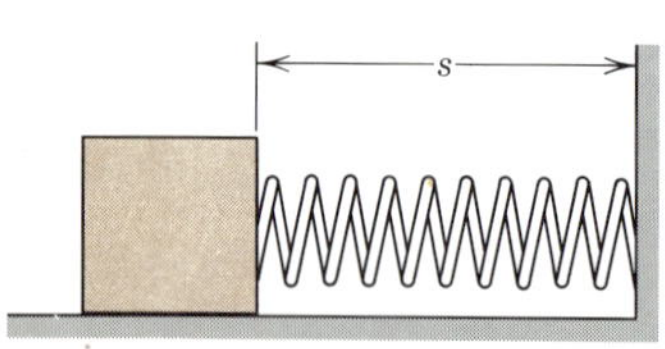

PROB. 17.12

17.13 The 1 lb collar C slides over the smooth rod AB, which lies in the vertical plane. The collar is released from rest at $\theta = 5°$. (a) Determine the amplitude and cyclical natural frequency of the subsequent motion. (b) Determine the time after release when the collar will reach $\theta = -2.5°$.

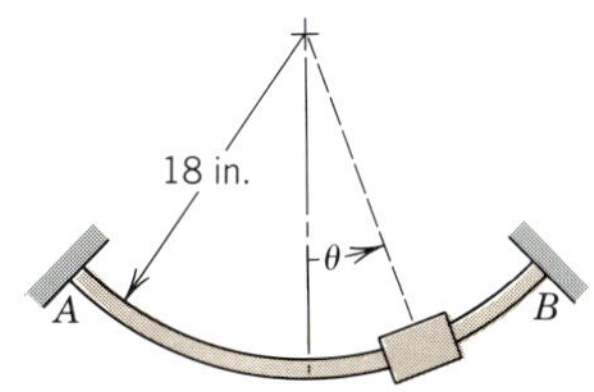

PROB. 17.13

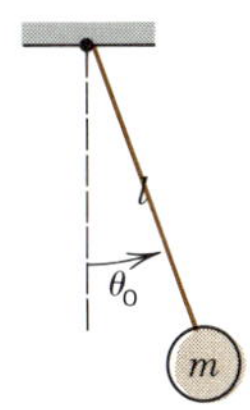

PROB. 17.14

17.14 The pendulum is released from rest at $\theta = \theta_0$. Determine the position and earliest time for which (a) the acceleration component tangent to the path is maximum, and (b) the acceleration component normal to the path is maximum.

17.15 A block of mass m is fastened to a lightweight (negligible mass) beam at its midpoint. Unloaded, the beam is straight. With the block attached, the static downward displacement of the midpoint of the beam is δ_{st}, as shown. (a) Derive an expression for the period of free vibration of the beam–mass combination. (b) If the system is held in the position where the beam is straight and then released from rest, what will be the amplitude of the subsequent motion? *Hint:* The beam acts, in effect, like a spring.

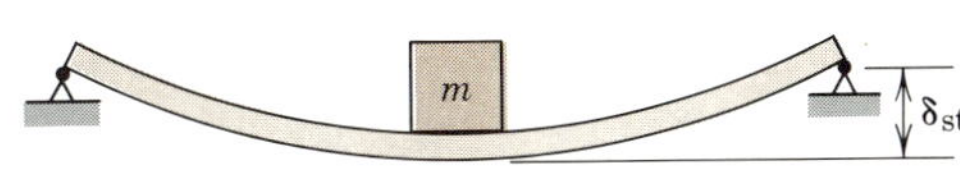

PROB. 17.15

17.16–17.19 Determine the amplitude and phase angle relative to a pure sine response for the responses listed below.

17.16 $y = 5\sin(10t) - 5\cos(10t)$ mm

17.17 $y = 10\cos(50t) - 15\sin(50t)$ in.

17.18 $y = 0.05\sin(200t - \pi/6) + 0.10\cos(200t)$ ft

***17.19** $y = 8\sin(10t - \pi/4) - 12\cos(10t + \pi/3)$ ft

17.20–17.22 For the responses listed below, determine (a) the amplitude and phase angle relative to a pure cosine response, (b) the earliest value of t for which $x = 0$, and (c) the earliest value of t for which $\dot{x} = 0$.

17.20 $x = 0.15\sin(40t) + 0.10\cos(40t)$ in.

17.21 $x = 2.5\cos(100t + 1.0) + 4.0\sin(100t)$ ft

17.22 $x = 18\sin(\Omega t + \pi/3) + 21\sin(\Omega t - 0.4\pi)$ mm

***17.23** The 200 gram sphere is pushed into the fixed curve tube at end A with a speed of 300 mm/s. Linearize the equation of motion in order to determine the time when the sphere will emerge from end B. Friction is negligible.

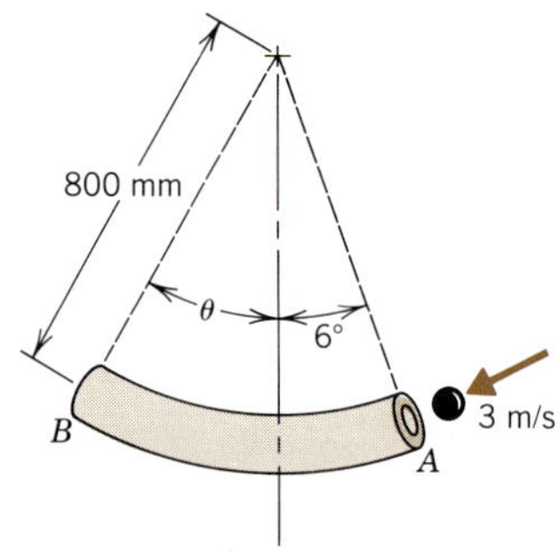

PROB. 17.23

***17.24** The speed governor shown consists of collar C of mass m that may slide along the smooth arm AB. The governor is actuated when the collar contacts end B. The rate of rotation about the vertical axis is Ω. The stiffness of the spring is k, and the unstretched length of the spring is $L/2$. Determine the equation of motion for the radial distance R and also determine the natural frequency of the system. Explain why $\Omega = (k/m)^{1/2}$ is critical for this system.

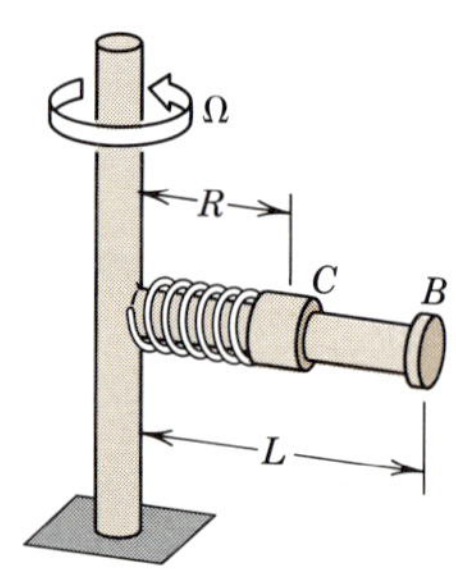

PROB. 17.24

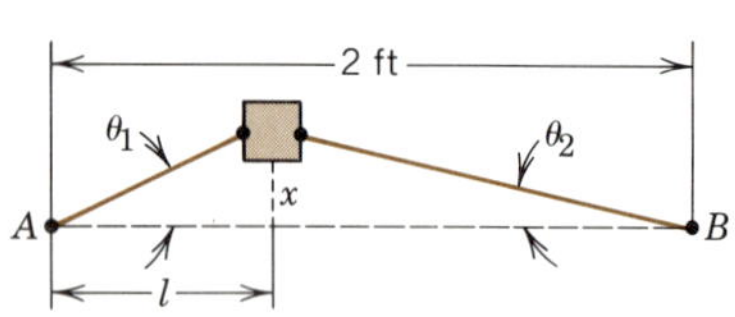

PROB. 17.25

***17.25** A 2 lb block slides on a smooth horizontal table under the restraint of the two cables. In the static equilibrium position the tensile force within the cables is 40 lb. Because of the (slight) elasticity of the cables, the block can execute small movements x in the direction perpendicular to line AB. (a) Determine the natural frequency of the system in terms of the length l of one of the cables. (b) Determine the value of l that minimizes the natural frequency and also determine this minimum frequency. *Hint:* Formulate the equation of motion in terms of the displacement x, considering the tension to be constant and the angles θ_1 and θ_2 to be very small.

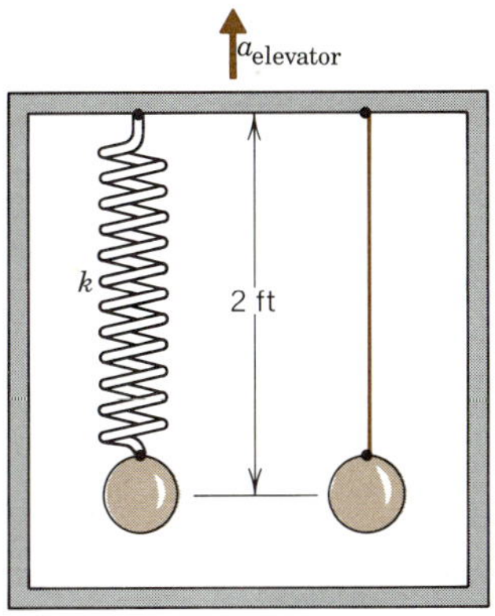

PROB. 17.26

***17.26** Two 1 lb spheres are suspended from the roof of an elevator, as shown. When the elevator is at rest, the period of extensional free vibration of the spring-supported sphere is the same as that for the rotation of the pendulum. (a) Determine the stiffness k of the spring. (b) Determine the period of each system when the elevator accelerates upward at $\dot{v} = 0.6g$.

17.27 The 800 gram block rests on the spring, but is not attached to it. The block is held at rest 20 mm below the static equilibrium position and then released, this distance being the largest initial displacement for which the block never loses contact with the spring. Determine (a) the stiffness of the spring, and (b) the displacement of the block from the static equilibrium position as a function of time. *Hint:* What is the total spring force when the block loses contact?

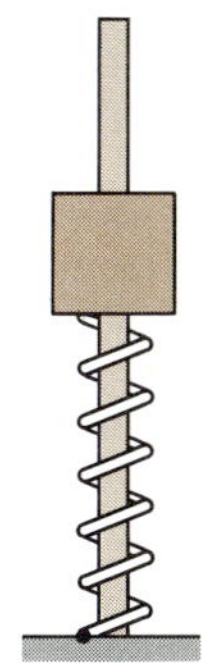

PROB. 17.27

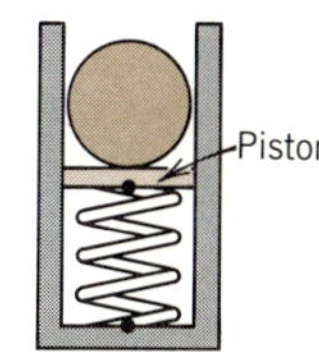

PROB. 17.28

***17.28** A 6 oz sphere is to be catapulted vertically into the air by the piston, which rides within the smooth cylinder. The piston weighs 2 oz, and the stiffness of the spring is 80 lb/ft. The piston is held 3 in. below the static equilibrium position and then released. Determine the distance the piston is above the static equilibrium position when the ball loses contact with the piston, and also the elapsed time from the instant of release. *Hint:* Draw a free body diagram of the sphere to determine the acceleration for which the normal force exerted between the piston and the sphere is zero.

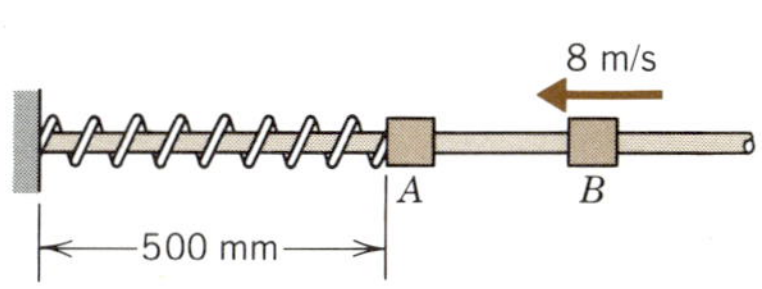

PROB. 17.29

17.29 Collars A and B, whose masses are 2 kg and 1 kg, respectively, may slide on the smooth horizontal bar. Collar A is initially at rest in the position shown. The spring that is attached to block A has a stiffness of 3.6 kN/m and an unstretched length of 500 mm. Collar B strikes collar A with a speed of 8 m/s. This collision is perfectly elastic. Determine the amplitude, frequency, and phase angle relative to a pure sine response for the motion of block A after the collision.

17.30 Solve Problem 17.29 if the coefficient of restitution for the collision is 0.50.

*17.31 The collar of mass m slides on the smooth horizontal bar. The spring, whose stiffness is k, has a pretension force T when the block is in the static equilibrium position. Derive the differential equation of motion that is valid for large values of x and then linearize this result to obtain the equation for small x. Does the value of k affect the latter equation? Explain your answer.

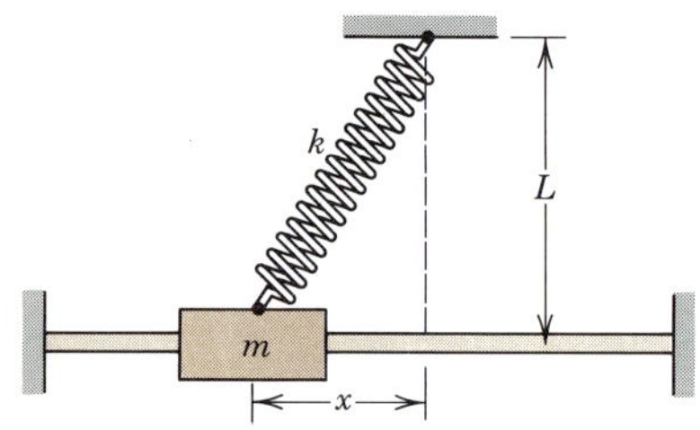

PROB. 17.31

---

2. Free Damped Motion

The model of an undamped oscillator that we studied in the previous section is an idealization, because it does not account for the energy dissipation that occurs in real systems. Here we shall investigate the effect of damping on the free vibration of a one degree of freedom system. As the response is derived, note that the natural frequency $\omega_n = \sqrt{k/m}$ of the undamped system proves to be an important parameter for the damped system.

The equation of motion for the standard system in Figure 1 when the excitation force is not present ($F = 0$) is found from eq. (5) to be

$$m\ddot{y} + c\dot{y} + ky = 0 \tag{17}$$

The harmonic solution for undamped free vibration does not satisfy this equation because of the presence of the first derivative resulting from damping. To solve eq. (17) we use the basic form for the solution of a linear differential equation.

$$y = Ae^{\lambda t} \tag{18}$$

Substituting this expression into eq. (17), dividing by m, and replacing k/m by ω_n^2 yields

$$\left(\lambda^2 + \frac{c}{m}\lambda + \omega_n^2\right)Ae^{\lambda t} = 0$$

As we did in the undamped case, we equate the coefficient of $Ae^{\lambda t}$ to zero. This yields a quadratic equation for λ, whose solution is

$$\lambda = -\frac{c}{2m} \pm \left[\left(\frac{c}{2m}\right)^2 - \omega_n^2\right]^{1/2} \tag{19}$$

Clearly the sign of the quantity inside the square root will affect the result. For convenience, we define a new parameter,

$$\beta \equiv \frac{c}{2m\omega_n} \equiv \frac{c}{2(km)^{1/2}} \tag{20}$$

so the two values for λ appearing above become

$$\lambda = -\omega_n\beta \pm \omega_n(\beta^2 - 1)^{1/2} \tag{21}$$

There are now three possibilities to consider, depending on whether β is less than, greater than, or equal to one.

Case a: $\beta < 1$. This is the situation where the coefficient of damping is small. For this case the square root in eq. (21) is imaginary, and the two values of λ are complex conjugates of each other; that is,

$$\lambda = -\omega_n\beta \pm i\omega_n(1 - \beta^2)^{1/2}$$

The two roots for λ correspond to two solutions in the form of eq. (18). As was true for the undamped response, conjugate imaginary exponents lead to sine and cosine terms. The real part of λ corresponds to an exponential factor that is common to both terms. The general solution is therefore

$$y = e^{-\omega_n\beta t}[A\cos(\omega_d t) + \beta\sin(\omega_d t)] \tag{22}$$

where the parameter ω_d, which is called the *damped natural frequency*, is given by

$$\omega_d = \omega_n(1 - \beta^2)^{1/2} \tag{23}$$

Like the natural frequency of the undamped system, ω_n, the parameter ω_d represents an oscillation rate. However, the damped response is *not* harmonic because of the exponential factor. An alternative form of the response may be written by combining the effects of the sine and cosine terms into an amplitude and phase angle. For example,

$$y = Ce^{-\beta\omega_n t}\sin(\omega_d t + \phi) \tag{24}$$

The coefficients A and B in eq. (22), or C and ϕ in eq. (24), are determined by satisfying the initial conditions. The combination of C and the exponential factor, in effect, represents an amplitude that decays with time. Note that the exponential factor decreases with time when $\beta > 0$, that is, when $c > 0$. This is the dissipation effect associated with the opposition of the dashpot force in eq. (1) to the motion.

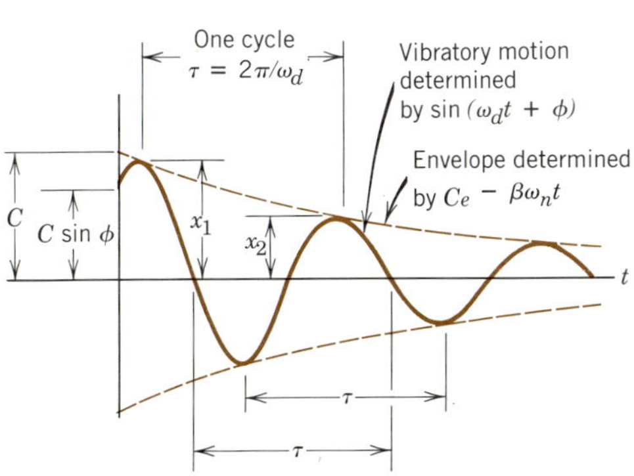

FIGURE 8

A typical $y(t)$ response is plotted in Figure 8, which graphically displays the nonperiodic nature of the response. The response is plotted there by plotting an envelope formed from the amplitude C and the exponential decay term. The frequency of oscillation within this envelope is the damped natural frequency ω_d. The corresponding period is τ_d, the *damped natural period*,

$$\tau_d = \frac{2\pi}{\omega_d} = \frac{2\pi}{\omega_n(1 - \beta^2)^{1/2}} \tag{25}$$

The fact that the response equals zero in intervals of length τ_d may be directly verified by noting that increasing t by τ_d increases the argument of the sine term by 2π. Thus if $\sin(\omega_d t + \phi) = 0$, then $\sin[\omega_d(t + \tau_d) + \phi] = 0$ also.

Proving that $\tau_d/2$ is also the time interval between a maximum value and the next minimum value (or conversely between a minimum and the next maximum) is the subject of Homework Problem 17.40. This feature of the response leads to a parameter that may be readily measured in an experiment, and then used to evaluate the coefficient β. Suppose that the maximum (or minimum) value y_1 is measured at time t_1. Then the next maximum (or minimum) response y_2 occurs when $t_2 = t_1 + \tau_d$. Because $\omega_d t_2 = \omega_d t_1 + 2\pi$, we have

$$\sin(\omega_d t_1 + \phi) = \sin(\omega_d t_2 + \phi)$$

Writing the two responses and taking their ratio gives

$$\frac{y_1}{y_2} = \frac{Ce^{-\beta\omega_n t_1}\sin(\omega_d t_1 + \phi)}{Ce^{-\beta\omega_n t_2}\sin(\omega_d t_2 + \phi)} = e^{\beta\omega_n(t_2 - t_1)} = e^{\beta\omega_n \tau_d}$$

Thus, in view of eqs. (23) and (25), we find

$$\delta \equiv \ln\left(\frac{y_1}{y_2}\right) = \frac{2\pi\beta}{(1 - \beta^2)^{1/2}} \tag{26}$$

where the quantity δ is called the *logarithmic decrement* of the system.

Case b: $\beta > 1$. The coefficient of damping in this situation is large. For this case the square root in eq. (21) is real, and hence the two values of λ are real. Corresponding to these two values of λ are two solutions, the sum of which is the general solution.

$$y = e^{-\omega_n \beta t}(Ae^{\omega_n\sqrt{\beta^2 - 1}t} + Be^{-\omega_n\sqrt{\beta^2 - 1}t}) \tag{27}$$

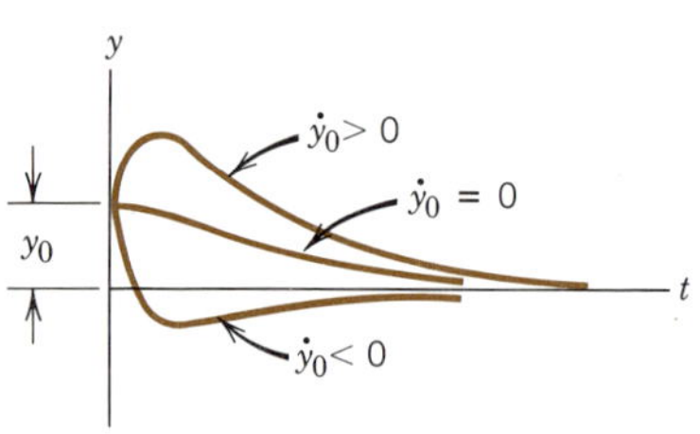

FIGURE 9

Noting that $(\beta^2 - 1)^{1/2}$ is less than β, the solution given by eq. (27) will decay with increasing time. Also, because the solution is a combination of exponential terms, the response is not oscillatory. This is exhibited in Figure 9, where the three curves show the effect of different types of initial conditions on the response.

Case c: $\beta = 1$. This represents a transition between the lightly damped situation ($\beta < 1$) and the heavily damped case ($\beta > 1$). Equation (21) now yields only one value for λ, which is

$$\lambda = -\omega_n \beta \equiv -\omega_n$$

However, the theory of linear differential equations tells us that there must be two distinct solutions to a second-order equation. The second solution is obtained by multiplying the basic exponential solution by a factor t. Hence

$$y = e^{-\omega_n t}(A + Bt) \tag{28}$$

Recalling that

$$\lim_{t \to \infty} (te^{-\omega_n t}) = 0$$

it follows that the response in this case also decays to zero. In fact, Figure 10 shows that the response in this case strongly resembles in all respects the response of Figure 9 for a more heavily damped system.

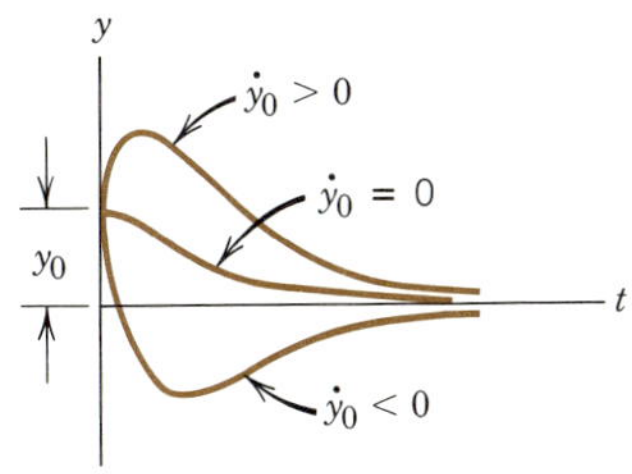

FIGURE 10

An interesting aspect of the response when $\beta = 1$ is that it represents the most rapid rate of decay to zero. There is a simple explanation of this. When $\beta < 1$, the dashpot force does not provide sufficient resistance to overcome the tendency of the spring–mass system to oscillate. In contrast, when $\beta > 1$, the dashpot force is so large that it competes strongly with the tendency of the spring to return the mass to the static equilibrium position.

This phenomenon, as well as the fact that $\beta = 1$ represents the transition from an oscillatory response when $\beta < 1$ to an exponentially decaying response when $\beta > 1$, makes the condition $\beta = 1$ a very important case. A system for which $\beta = 1$ is said to be *critically damped.* A system for which $\beta < 1$ is termed *underdamped* (or *lightly damped*), and one for which $\beta > 1$ is termed *overdamped* (or *heavily damped*). The critical damping coefficient is found from eq. (20) by setting $\beta = 1$, which gives

$$\boxed{c_{\text{cr}} \equiv 2\sqrt{km}} \tag{29}$$

Substituting this expression back into eq. (20) yields

$$\boxed{\beta = \frac{c}{c_{\text{cr}}}} \tag{30}$$

In other words, the value of β for a system is the ratio of the actual coefficient of damping of the system to the critical coefficient; it is called the *ratio of critical damping.*

A common example is the case of the shock absorbers of an automobile. When new, they do not permit the vehicle to bounce continuously when it encounters an irregularity in the roadway. The automobile is then overdamped. As the shock absorbers wear from age and use, the automobile passes from the overdamped situation, through critical damping, and into the case of underdamping. Disturbances in the roadway then result in the automobile bouncing freely. It is then time to install new shock absorbers.

EXAMPLE 5

A descending 1000 lb elevator car comes into contact with the combination spring–shock absorber shown. The subsequent vibration of the elevator car is observed to be an underdamped response, with peak upward displacements occurring in intervals of 1.2 s. Each successive peak value is 60% of the previous one. Determine the spring stiffness k and damping constant c.

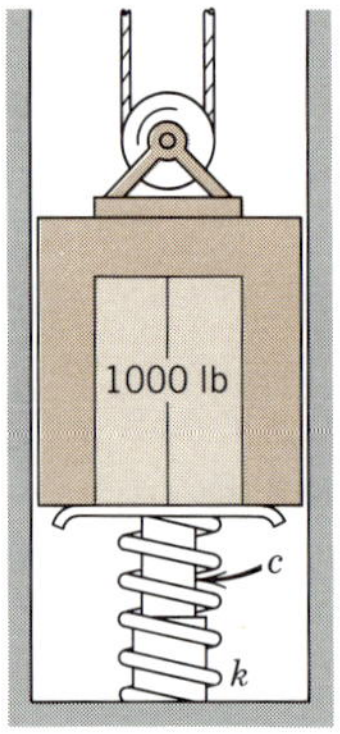

Solution

It is given that the damped period is $\tau_d = 1.2$ s, and that the logarithmic decrement is

$$\delta = \ln\left(\frac{x_1}{x_2}\right) = \ln\left(\frac{x_1}{0.60x_1}\right) = 0.5108$$

Equations (25) and (26) for an underdamped system indicate that

$$\tau_d = \frac{2\pi}{\omega_n(1-\beta^2)^{1/2}}, \qquad \delta = \frac{2\pi\beta}{(1-\beta^2)^{1/2}}$$

Solving the second relation for β yields

$$(1-\beta^2)\delta^2 = 4\pi^2\beta^2 \quad \Rightarrow \quad \beta = \frac{\delta}{(4\pi^2+\delta^2)^{1/2}} = 0.08103$$

Then

$$\omega_n = \frac{2\pi}{\tau_d(1-\beta^2)^{1/2}} \equiv \frac{\delta}{\beta\tau_d} = 5.253 \text{ rad/s}$$

The mass of the elevator is 1000/32.17 lb-s^2/ft, so the definitions of ω_n and β yield

$$k = \omega_n^2 m = 857.8 \text{ lb/ft}$$

$$c = 2m\omega_n\beta = 26.46 \text{ lb-s/ft}$$

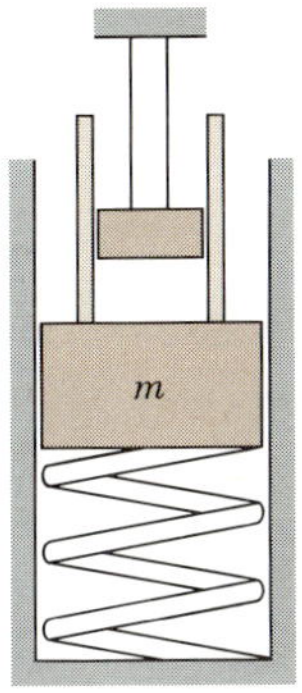

EXAMPLE 6

The block is at the static equilibrium position when it is given a downward velocity v. Assuming that the system is overdamped, derive an expression for the elapsed time before the block first comes to rest. What happens to this expression as the condition of critical damping is approached?

Solution

The equation of motion for the downward displacement y relative to the static equilibrium position is found from the free body diagram to be

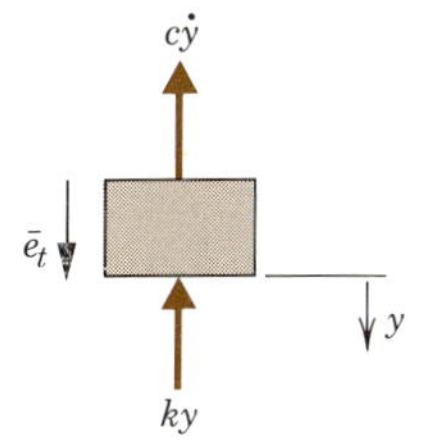

$$\sum F_t = -ky - c\dot{y} = m\ddot{y}$$

$$\ddot{y} + 2\beta\omega_n\dot{y} + \omega_n^2 y = 0, \qquad \omega_n = \left(\frac{k}{m}\right)^{1/2}, \qquad 2\beta\omega_n = \frac{c}{m}$$

The response of an overdamped system is known from eq. (27) to be of the form

$$y = Ae^{\lambda_1 t} + Be^{\lambda_2 t}$$

where

$$\lambda_1 = -\beta\omega_n + \omega_n(\beta^2 - 1)^{1/2}, \qquad \lambda_2 = -\beta\omega_n - \omega_n(\beta^2 - 1)^{1/2}$$

(The foregoing form, which uses λ_1 and λ_2 as parameters, is convenient for computations.) The coefficients A and B are found from the given initial conditions that $y = 0$ and $\dot{y} = v$ when $t = 0$. Thus

$$y|_{t=0} = A + B = 0$$

$$\dot{y}|_{t=0} = \lambda_1 A + \lambda_2 B = v$$

This yields

$$B = -A, \qquad A = \frac{v}{\lambda_1 - \lambda_2}$$

so the response is

$$y = \frac{v}{(\lambda_1 - \lambda_2)}(e^{\lambda_1 t} - e^{\lambda_2 t})$$

At an instant $t = T$ when $\dot{y} = 0$ it must be that

$$\dot{y} = \frac{v}{(\lambda_1 - \lambda_2)}(\lambda_1 e^{\lambda_1 T} - \lambda_2 e^{\lambda_2 T}) = 0$$

$$\lambda_1 e^{\lambda_1 T} = \lambda_2 e^{\lambda_2 T} \quad \Rightarrow \quad e^{(\lambda_1 - \lambda_2)T} = \frac{\lambda_2}{\lambda_1}$$

Solving for T and substituting for λ_1 and λ_2 yields

$$T = \frac{\ln(\lambda_2/\lambda_1)}{\lambda_1 - \lambda_2} = \frac{1}{2\omega_n(\beta^2 - 1)^{1/2}} \ln\left[\frac{\beta + (\beta^2 - 1)^{1/2}}{\beta - (\beta^2 - 1)^{1/2}}\right]$$

As the limiting value for critical damping $\beta = 1$ is approached, the argument of the logarithm approaches one. Thus, both the numerator and the denominator in the expression for T approach zero. The limit could be evaluated by L'Hôpital's rule. A simpler approach for the limit in this case is to use a series expansion based on the fundamental series

$$\ln\left(\frac{1+u}{1-u}\right) = 2(u + \tfrac{1}{3}u^3 + \cdots); \qquad u \ll 1$$

Let $u = (\beta^2 - 1)^{1/2}/\beta$ in the expression for T. Then

$$T = \frac{1}{2\omega_n(\beta^2 - 1)^{1/2}} \ln\left[\frac{1 + (\beta^2 - 1)^{1/2}/\beta}{1 - (\beta^2 - 1)^{1/2}/\beta}\right]$$

$$= \frac{1}{2\omega_n(\beta^2 - 1)^{1/2}}(2)\left\{\frac{(\beta^2 - 1)^{1/2}}{\beta} + \frac{1}{3}\frac{(\beta^2 - 1)^{3/2}}{\beta^3} + \cdots\right\}$$

$$= \frac{1}{\beta\omega_n}\left\{1 + \frac{1}{3}\frac{\beta^2 - 1}{\beta^2} + \cdots\right\}$$

$$\lim_{\beta \to 1} T = \frac{1}{\omega_n}$$

An interesting aspect of this result is that the value of T for $\beta \to 1$ is larger than the value for any $\beta > 1$. This is logical because decreasing β decreases the dashpot force $c\dot{y}$. Thus, the damping resistance developed to bring the block to rest decreases as $\beta \to 1$.

HOMEWORK PROBLEMS

17.32–17.34 Determine the equation of motion for the system shown. Find an expression for the critical damping coefficient.

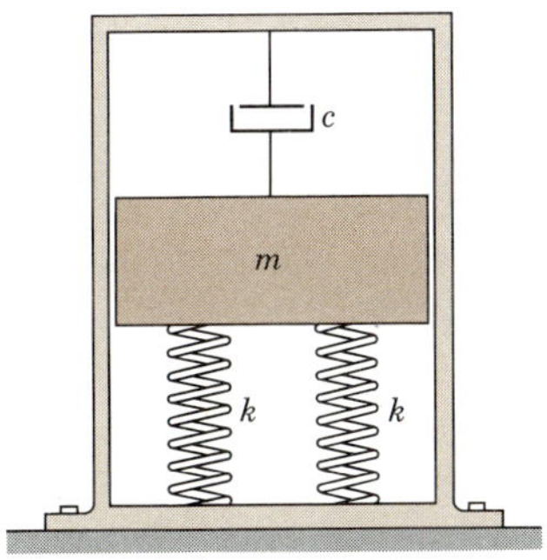

PROB. 17.32

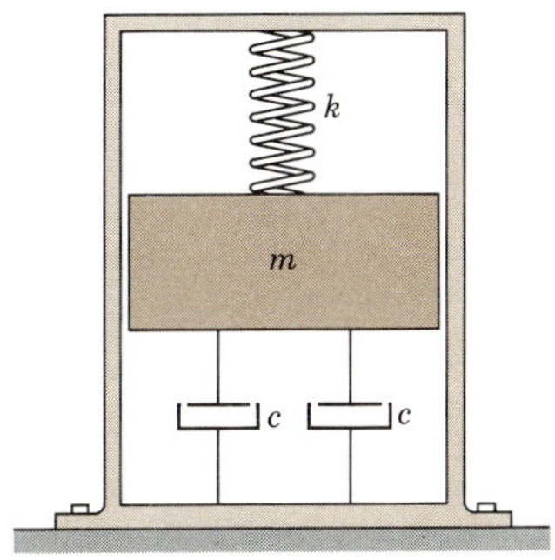

PROB. 17.33

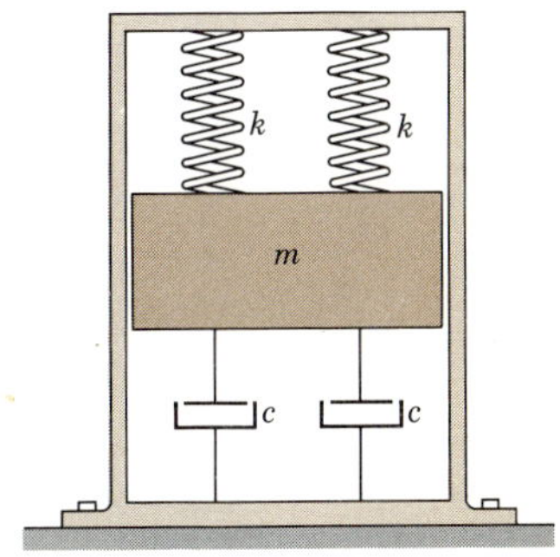

PROB. 17.34

17.35 An underdamped system whose ratio of critical damping is $\beta = 0.60$ is given an initial velocity v_0 when it is at the static equilibrium. position. The undamped natural frequency is 100 Hz. Determine the earliest time after release at which the system passes the static equilibrium position. Compare this time to the corresponding value for an undamped system.

17.36 Solve Problem 17.35 for the case where $\beta = 0.99$.

***17.37** A damped system for which $\omega_n = 50$ rad/s is released from rest with a positive initial displacement x_0. Determine the time required for the displacement to be $0.05x_0$ if the critical damping ratio is (a) $\beta = 2.010$, (b) $\beta = 1.005$.

17.38 Solve Problem 17.37 for the case of a critically damped system, $\beta = 1$.

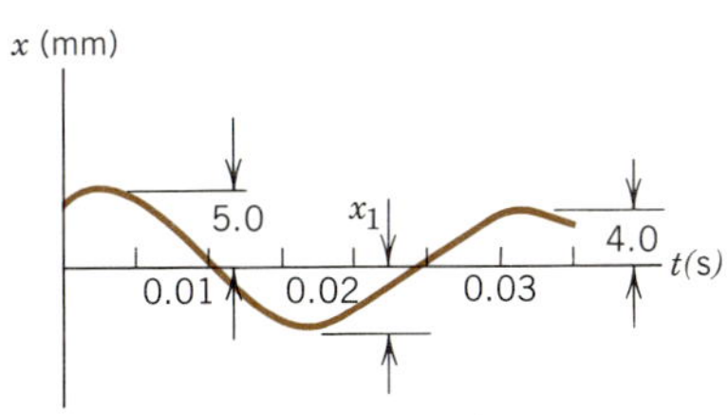

PROB. 17.39

17.39 The result of electronic measurement of the displacement of a system is shown. Determine the damped natural frequency, the ratio of critical damping, and the displacement x_1 indicated.

17.40 Prove that if an extreme value (i.e., maximum or minimum value) of the response of a one degree of freedom system occurs at an instant $t = t_1$, then the next extreme value occurs when $t = t_1 + \tau_d/2$, where τ_d is the damped natural frequency.

17.41 An alternative method for measuring the logarithmic decrement is to compare the maximum responses over many cycles of free vibration. Let x_1 and x_{n+1} denote the maximum displacements occurring at the beginning and end of n cycles of vibration. Show that the logarithmic decrement may be obtained from $\delta = (1/n)\ln(x_1/x_{n+1})$.

17.42 The free fall acceleration g on the surface of a planet can be determined from the vibrational properties of a one degree of freedom oscillator. It is observed that when the oscillator is aligned vertically, the static displacement is 3.2 mm. The underdamped response is observed to have successive maximum responses in intervals of 180 ms, with each peak 25% of the previous maximum. Determine the corresponding value of g.

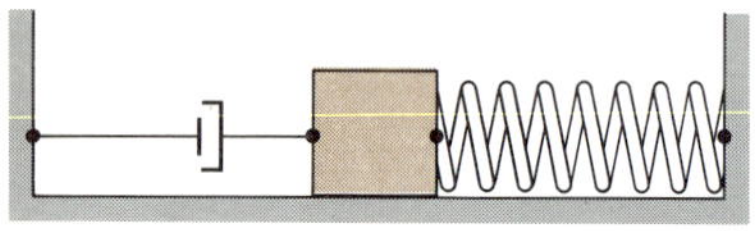

PROB. 17.43

17.43 The 12 oz block is restrained by a spring whose stiffness is 80 lb/ft and a dashpot whose constant is 2 lb-s/ft. Initially, the spring is elongated by 0.2 ft and the block is given a velocity of 8 ft/s to the right. Determine (a) the response as a function of time, (b) the time at which the elongation of the spring first attains a zero value.

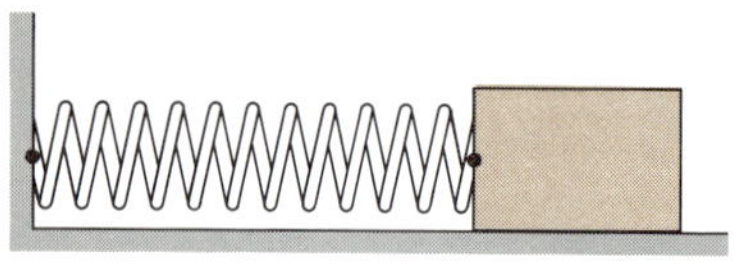

PROB. 17.44

17.44 The block slides on the horizontal surface, which is lubricated by oil, so the friction force is proportional to the velocity of the block. The system is underdamped. When the block is released from rest the spring is compressed a distance x_0. What percentage of the mechanical energy in the system at the beginning of a cycle is dissipated during that cycle? Is this percentage constant?

***17.45 and 17.46** Several dashpots acting on the mass in a one degree of freedom oscillator may be combined into an equivalent single dashpot, in a manner similar to that used to compute an equivalent spring stiffness. Determine the equivalent dashpot constant for the system shown.

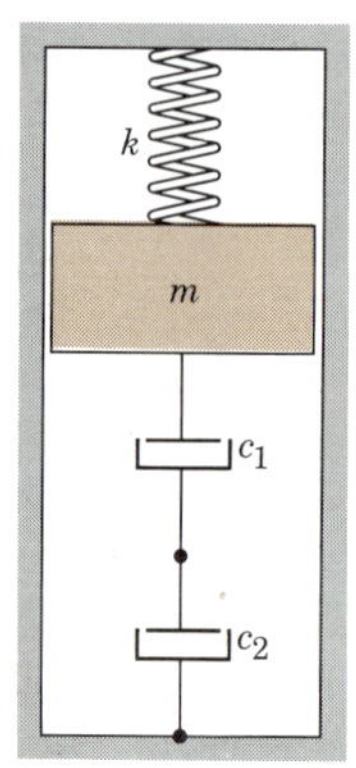

PROB. 17.45

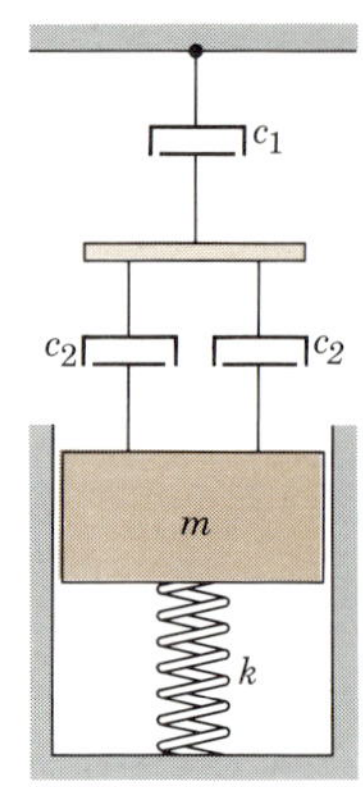

PROB. 17.46

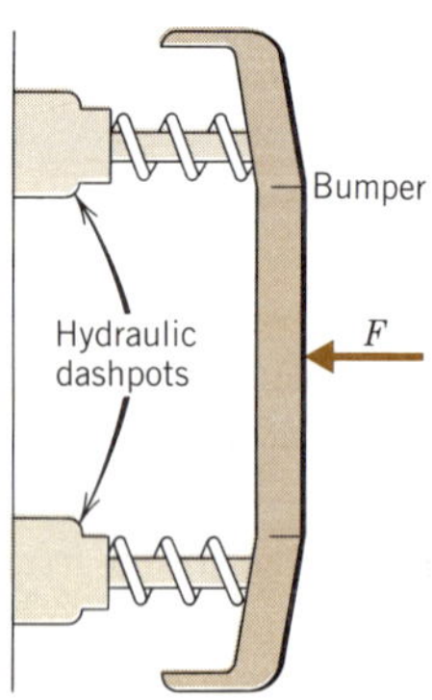

PROB. 17.47

17.47 In the shock-absorbing mechanism for the automobile bumper shown, two springs act concentrically with two hydraulic dashpots. The stiffness of each spring is 200 lb/ft, and the bumper weighs 80 lb. When subjected to the impulsive force $\bar{F}$, it is desired that the bumper compress in the shortest possible time, without subsequently oscillating when it rebounds. Determine the value of the damping constant μ required for each dashpot to accomplish this.

***17.48** In Problem 17.47 the damping constant for each dashpot is $c =$ 100 lb-s/ft. A collision force $\bar{F}$, which exerts an impulse of 400 lb-s over a very short time interval, is applied to the bumper. Determine (a) the displacement of the bumper as a function of time t, where t

is measured from the instant when $\bar{F}$ is applied, (b) the maximum distance the springs are compressed and the corresponding time t, and (c) the maximum compression of the springs if damping were not present.

***17.49** Solve Problem 17.48 for $c = 20$ lb-s/ft.

17.50 The 800 gram platform is supported by the spring-loaded shock absorber. The stiffness of the spring is 600 N/m and the dashpot constant is 20 N-s/m. The platform is at rest in the static equilibrium position when a 4 kg package that is falling at 5 m/s impacts against it. The collision is perfectly plastic, so the platform and the package move together in the subsequent motion. Determine the response after impact.

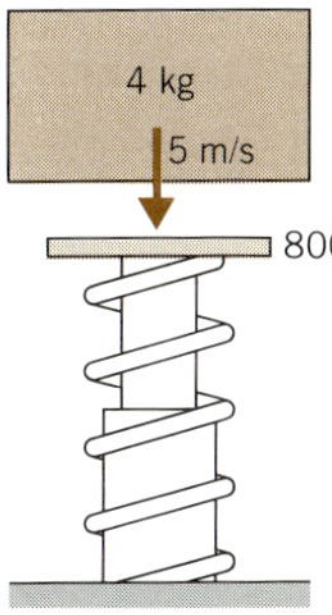

PROB. 17.50

D. FORCED VIBRATION—HARMONIC EXCITATION

Thus far we have concentrated on the free response of a system. We now wish to consider the effect of dynamic excitation of the system. The type of force we shall consider is a harmonic force, that is, one that varies sinusoidally with time. The reasons for studying such a forcing function are many. For example, if the response of a system to a harmonic excitation is known, it is possible to determine the response to any periodic excitation by using a Fourier series.

The frequency of the excitation, in units of radians per second, will be denoted as Ω, and the amplitude of the excitation as F. We shall represent the time dependence by a sine function. Thus

$$F(t) = F\sin(\Omega t)$$

Any other harmonic dependence may be obtained from this by replacing Ωt by Ωt plus a suitable phase angle. For example, a cosine term is

$$F(t) = F\sin(\Omega t + \pi/2) \equiv F\cos(\Omega t)$$

The general equation of motion (5) for this case of an excitation force is

$$m\ddot{y} + c\dot{y} + ky = F\sin(\Omega t) \tag{31}$$

There are two types of solutions for eq. (31). One, called the *particular solution* and denoted as y_p, corresponds to the nonhomogeneous (time-dependent) term in the differential equation. To this solution we add the complementary solution y_c, which is the solution corresponding to $F(t) = 0$. For eq. (31) the *complementary solution* has the form of the free response of an undamped, underdamped, overdamped, or critically damped system, depending on the values of m, c, and k. Thus, we shall focus attention here on the determination of the particular solution.

The most direct way to obtain a particular solution is to guess its form. Because y and its first and second derivatives must combine to give a sine term, we guess

$$\boxed{y_p = D\sin(\Omega t - \psi)} \tag{32}$$

The amplitude D and phase angle ψ are determined by making y_p satisfy eq. (31). The negative sign precedes the phase angle in order to emphasize that the phase angle is a lag, as will be proven shortly.

Upon substituting eq. (32) into eq. (31), we find that

$$(-m\Omega^2 + k)D\sin(\Omega t - \psi) + c\Omega D\cos(\Omega t - \psi) = F\sin(\Omega t)$$

The easiest way to proceed at this point is to manipulate the term on the right side using the trigonometric identity for the sine of the difference of two angles, as follows.

$$\begin{aligned}(-m\Omega^2 + k)D\sin(\Omega t - \psi) &+ c\Omega D\cos(\Omega t - \psi)\\ &= F\sin[\Omega t - \psi) + \psi]\\ &= F\sin(\Omega t - \psi)\cos\psi + F\cos(\Omega t - \psi)\sin\psi\end{aligned}$$

Equating the coefficients of the sine and cosine terms gives

$$(-m\Omega^2 + k)D = F\cos\psi$$

$$c\Omega D = F\sin\psi$$

Solving these equations for D and ψ in terms of F, Ω, m, c, and k yields

$$D = \frac{F}{\sqrt{(k - \Omega^2 m)^2 + c^2\Omega^2}} \qquad \tan\psi = \frac{c\Omega}{k - m\Omega^2} \tag{33}$$

This is the dimensional form of the *frequency–amplitude–phase angle relations*. It is useful at this juncture to check these relations for consistency of their dimensions. For this, we note that in terms of the symbols M, L, T, and F for mass, length, time, and force, respectively, that

$$k \sim F/L, \qquad c \sim F \cdot T/L, \qquad \Omega \sim 1/T$$

$$\Omega^2 m \sim M/T^2 = (F \cdot T^2/L)/T^2 = F/L$$

Making the appropriate substitutions in eqs. (33) we find that D has the dimensions of length and $\tan \psi$ is nondimensional.

Enlarging on the foregoing, it is possible to write eqs. (33) in nondimensional form. By doing so, the nondimensional frequency–amplitude–phase angle relations may be plotted in a manner that is descriptive of *all* one degree of freedom systems. We obtain this result by dividing the numerator and denominator of both of eqs. (33) by k. Then the definitions of the (circular) natural frequency, $\omega_n = (k/m)^{1/2}$, and the ratio of critical damping, $\beta = c/2(mk)^{1/2}$, are used to eliminate k/m and c. This yields

$$D = \frac{(F/k)}{\sqrt{(1 - \Omega^2/\omega_n^2)^2 + 4\beta^2\Omega^2/\omega_n^2}}$$
$$\psi = \tan^{-1}\left(\frac{2\beta\Omega/\omega_n}{1 - \Omega^2/\omega_n^2}\right) \tag{34}$$

There are a number of interesting results displayed in eqs. (34). For instance, notice that F/k is the static displacement that the mass would have if it were loaded by a static force of magnitude F. Thus, the fraction multiplying F/k in the equation for D is a *dynamic magnification factor* for the amplitude D. Furthermore, for a given set of system parameters, the amplitude and phase angle depend only on the ratio of the frequency of excitation to the natural frequency.

To study eqs. (34) further, let us sketch D and ψ as functions of Ω/ω_n, as is shown in Figures 11 and 12. Such plots are called *amplitude–frequency* (Figure 11) and *phase angle–frequency* (Figure 12) diagrams. They are very useful for designing systems to meet dynamic criteria, such as limitations on the amplitude of vibration.

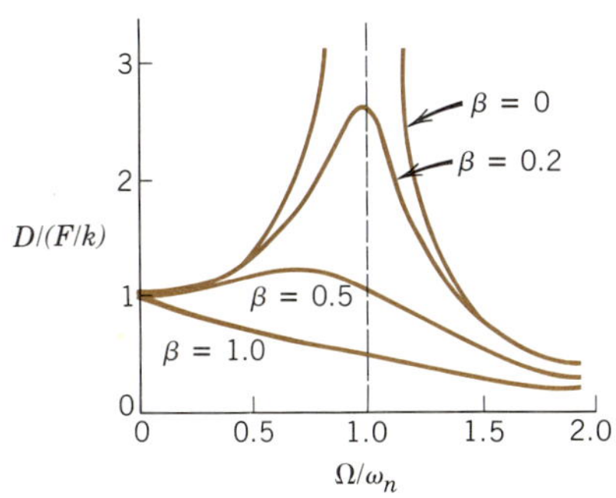

FIGURE 11

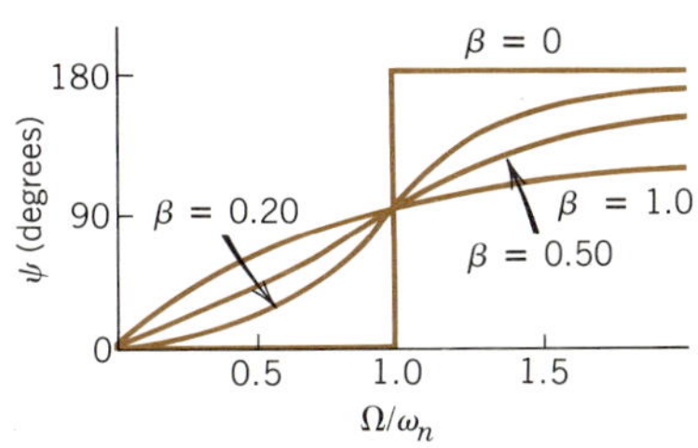

FIGURE 12

Focusing on Figure 11 for an undamped system ($\beta = 0$), notice that the value of D becomes infinite as the frequency of excitation approaches the natural frequency ($\Omega/\omega_n \to 1$). This situation, which is called *resonance*, is obviously quite dangerous. However, it is important to note that the harmonic particular solution, eq. (32), is *not valid* for resonance. The proper solution is described in Example 8, which follows.

When a system has a small amount of damping ($\beta \ll 1$), Figure 12 shows that for low frequencies of excitation the system behaves like a static one, because $D \simeq F/k$ in that case. The natural frequency is the frequency at which a system would prefer to vibrate freely. Excitations close to the natural frequency are in accord with this tendency, which explains the peak amplitude near $\Omega = \omega_n$. In contrast, very high frequencies ($\Omega \gg \omega_n$) lead to very high accelerations $\ddot{y}_p$. The inertial resistance of the moving mass in such a situation substantially lessens the amplitude of vibration.

A useful observation from eq. (34) is that $D/(F/k) = \frac{1}{2}\beta$ when $\Omega = \omega_n$. This relation is often employed to measure the ratio of critical damping in a system. Increasing the amount of damping lessens the maximum value of the amplitude D, and the frequency Ω for this maximum also decreases. Ultimately, when $\beta > \sqrt{2}/2$, the maximum amplitude occurs at $\Omega = 0$, and the amplitude for vibration at any Ω is less than the static displacement.

The phase angle for the undamped system is also of interest. When $\beta = 0$ and $\Omega < \omega_n$, Figure 12 indicates that $\psi = 0$. This means that the displacement y_p is positive whenever the force $F(t) = F \sin(\Omega t)$ is positive. We then say that the response and the excitation are *in phase*. As the resonance condition is passed, so that $\Omega > \omega_n$, ψ suddenly changes to 180°, so the displacement y_p is negative whenever $F(t) = F \sin(\Omega t)$ is positive. The response and the excitation are then 180° *out of phase*. Responses that are in phase and 180° out of phase from the excitation are depicted in Figures 13a and 13b.

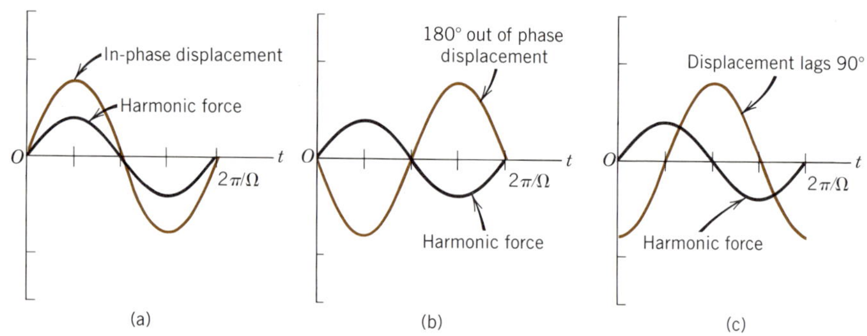

FIGURE 13

As shown in Figure 12, the value of ψ for a damped system ($\beta > 0$) is in the range $0 \le \psi < 180°$. This confirms that ψ represents a phase lag relative to the excitation force, as was anticipated in eq. (32). An interesting phenomenon pertaining to this phase lag arises when $\Omega = \omega_n$; the *response lags* 90° *behind the excitation* for any value of β. Figure 13c, which depicts such a response, shows that the displacement is maximum when the force is zero, and vice versa. A readily verified feature of the 90° phase lag is that the velocity is in phase with the force. This phenomenon is often used to experimentally identify the natural frequency of a system.

Let us now return to the task of determining the response of a system that is governed by eq. (31). For this discussion we shall consider the case of an underdamped system. Thus, the complementary solution is given by eq. (23), which in combination with the particular solution of eq. (32), gives

$$\begin{aligned} y &= y_c + y_p \\ &= e^{-\beta\omega_n t} C \sin(\omega_d t + \phi) + D \sin(\Omega t - \psi) \end{aligned} \tag{35}$$

A computational point to bear in mind is that the constants C and ϕ are found by using the full expression (35) to satisfy the initial conditions, not just the portion of eq. (35) that is the complementary solution.

Examining eq. (35), we see that as t becomes large, the first term, the complementary solution, goes to zero. Because of this effect, dynamicists call y_c the *transient* solution and y_p the *steady-state* solution. It is a steady-state vibration in the sense that after the transient decays the system will maintain a steady harmonic oscillation at the amplitude D as long as the harmonic force is applied. The energy put in through the forcing function then balances the energy lost through damping. Note too that the steady-state vibration takes place at the frequency of the forcing function. An exception occurs in the case of an undamped system, for then the complementary solution in eq. (35) does not decay. In reality, this is of little concern, because all systems have some damping, which causes the transient solution eventually to decay to zero. Thus, we see that it is correct to approximate a system as having no damping only if we are observing it over a comparatively small time interval.

EXAMPLE 7

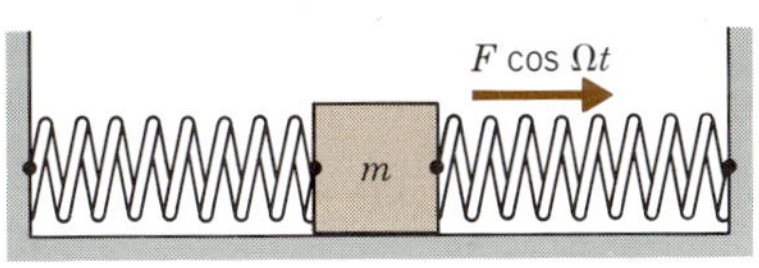

When $t = 0$, the system shown is at rest in its static equilibrium position, where the tension forces in each of the two springs both have the value T. At this instant a harmonic force $F\cos(\Omega t)$ is applied to the block. Neglecting friction, determine the subsequent motion of the block.

Solution

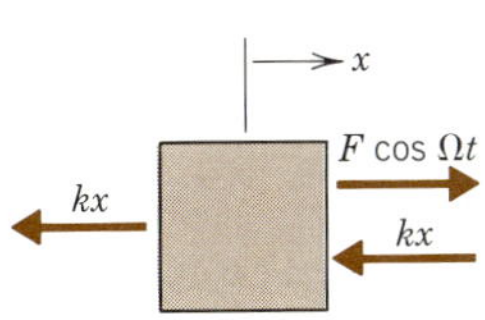

The static part of each spring force is the tension T. If the motion is described in terms of the displacement x from the static equilibrium position, these static forces may be ignored. The equation of motion corresponding to the free body diagram is

$$m\ddot{x} + 2kx = F\cos(\Omega t)$$

The solution of the equation of motion is obtained by adding the complementary and particular solutions for an undamped oscillator. The natural frequency of this system is

$$\omega_n = \sqrt{\frac{2k}{m}}$$

The form of the complementary solution given in eq. (9) is useful for satisfying initial conditions. Thus

$$x_c = A\cos(\omega_n t) + B\sin(\omega_n t)$$

The particular solution for harmonic excitation, given by eqs. (32) and (33), was derived for a sine force; the force in this problem is a cosine force. However, because this is simply a 90° difference in phase, we may directly employ the derived solution by shifting the response by 90°. This shift changes the sine function to a cosine function. The particular solution is therefore

$$x_{\mathrm{p}} = D\cos(\Omega t - \psi)$$

The amplitude D in eq. (32) is positive, so for $\beta = 0$ we have

$$D = \frac{F/2k}{|1 - \Omega^2/\omega_n^2|}$$

$$\psi = 0 \quad \text{if} \quad \Omega < \omega_n$$

$$\psi = 180^\circ \quad \text{if} \quad \Omega > \omega_n$$

It is useful in the case of no damping to remove the phase angle from the problem by setting $\psi \equiv 0$ and letting D be positive or negative. Recall that a phase lag of 180° is equivalent to a negative amplitude; see Figure 13b. It then follows that

$$x_{\mathrm{p}} = D\cos(\Omega t) \qquad \text{where} \qquad D = \frac{F/2k}{1 - \Omega^2/\omega_n^2}$$

It is a simple matter to verify, by substitution, that the foregoing is the solution.

Finally, we satisfy the initial conditions. From the given information we know that $x = 0$ and $\dot{x} = 0$ when $t = 0$. This gives

$$\begin{aligned} x(0) = 0 &= (x_{\mathrm{c}} + x_{\mathrm{p}})|_{t=0} \\ &= A\cos(\omega_n t) + B\sin(\omega_n t) + D\cos(\Omega t)|_{t=0} = A + D \\ \dot{x}(0) = 0 &= (\dot{x}_{\mathrm{c}} + \dot{x}_{\mathrm{p}})|_{t=0} \\ &= \omega_n[-A\sin(\omega_n t) + B\cos(\omega_n t)] - \Omega D\sin(\Omega t)|_{t=0} = \omega_n B \end{aligned}$$

Thus, $A = -D$ and $B = 0$, which means that there is a transient solution, even though the initial displacement and velocity are zero. This emphasizes the fact that the initial conditions must be satisfied after first combining the complementary and particular solutions. The response may now be written as

$$x = D[\cos(\Omega t) - \cos(\omega_n t)]$$

$$x = \left(\frac{F/2k}{1 - \Omega^2/\omega^2}\right)[\cos(\Omega t) - \cos(\omega_n t)]$$

EXAMPLE 8

For the system in Example 7, consider the case where the value of Ω is close to the value of ω_n. In order to express this proximity, let $\omega_n - \Omega = 2\Delta$, where Δ is a small quantity. Sketch $x(t)$ versus t for this case. What is the response in the limiting case of $\omega_n \equiv \Omega$?

Solution

The full response is known from Example 7 to be

$$x = \left(\frac{F/2k}{1 - \Omega^2/\omega_n^2}\right)[\cos(\Omega t) - \cos(\omega_n t)]$$

$$= \left(\frac{\omega_n^2 F/2k}{\omega_n^2 - \Omega^2}\right)[\cos(\Omega t) - \cos(\omega_n t)]$$

$$\equiv \frac{F}{m(\omega_n^2 - \Omega^2)}[\cos(\Omega t) - \cos(\omega_n t)]$$

In order to investigate the limiting cases recall the trigonometric identity

$$\cos\alpha - \cos\beta = 2\sin\left(\frac{\beta - \alpha}{2}\right)\sin\left(\frac{\beta + \alpha}{2}\right)$$

Using this identity, the displacement becomes

$$x = \frac{2F}{m(\omega_n^2 - \Omega^2)}\sin\left[\left(\frac{\omega_n + \Omega}{2}\right)t\right]\sin\left[\left(\frac{\omega_n - \Omega}{2}\right)t\right]$$

The definition of the parameter Δ permits us to eliminate Ω from the foregoing, because

$$\omega_n - \Omega = 2\Delta, \qquad \omega_n + \Omega = 2(\omega_n - \Delta)$$

$$\omega_n^2 - \Omega^2 = (\omega_n + \Omega)(\omega_n - \Omega) = 4\Delta(\omega_n - \Delta)$$

Thus x may be rewritten as

$$x = \frac{F}{2m\Delta(\omega_n - \Delta)}\sin(\Delta t)\sin[(\omega_n - \Delta)t]$$

Because Δ is small, the period of $\sin(\Delta t)$ is much larger than the period of $\sin[(\omega_n - \Delta)t]$. This suggests that we should consider the response to have a frequency of $\omega_n - \Delta$ and an amplitude that varies with the period of the $\sin(\Delta t)$ term. The previous expression for $\omega_n + \Omega$ leads to the conclusion that

$$x(t) = D(t)\sin(\Omega_{av} t)$$

where $\Omega_{av} = (\omega_n + \Omega)/2$ is the average of the excitation and free vibration frequencies. The term $D(t)$ represents a time-dependent amplitude

$$D(t) = \frac{F}{2m\Delta(\omega_n - \Delta)}\sin(\Delta t)$$

A typical response of this type is shown in the sketch. In music, it may be heard when two instruments play tones that are slightly different from each other. The combination is a single tone of the average frequency that alternately grows louder and then softer. This is often referred to as a *beating* response.

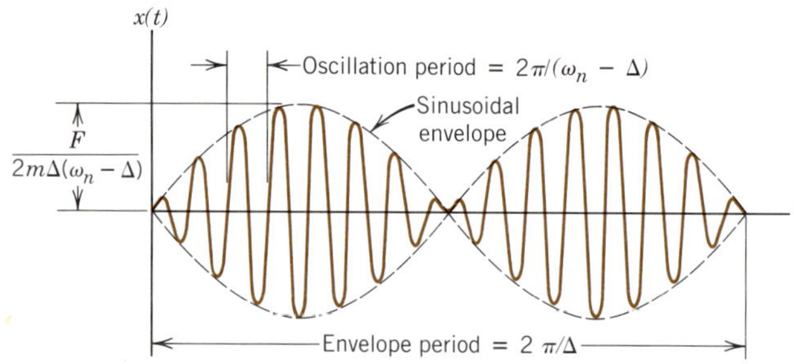

The limiting form of $x(t)$ when $\Delta \to 0$, that is, when $\Omega = \omega_n$, is found by L'Hôpital's rule. Because $\sin(\Delta t)/\Delta \to t$ as $\Delta \to 0$, we find that

$$x = \frac{F}{2m\omega_n} t \sin(\omega_n t)$$

The case where $\Omega = \omega_n$ in an undamped system is resonance. The above result shows that the resonant response consists of an oscillation at the natural frequency whose amplitude grows linearly with time. Only when $t \to \infty$ does the amplitude become infinite. This feature is displayed in our sketch of a resonant excitation.

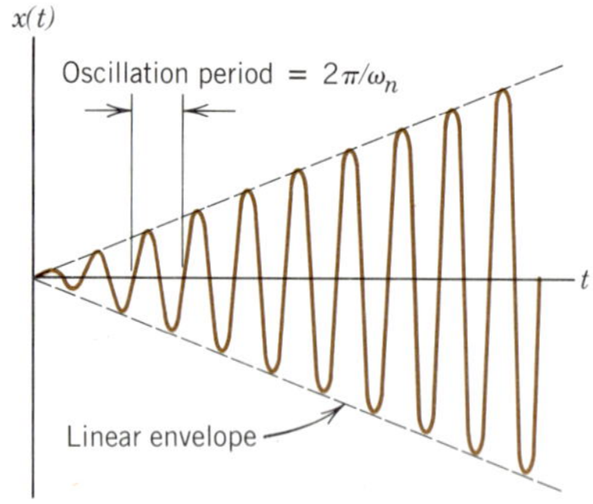

EXAMPLE 9

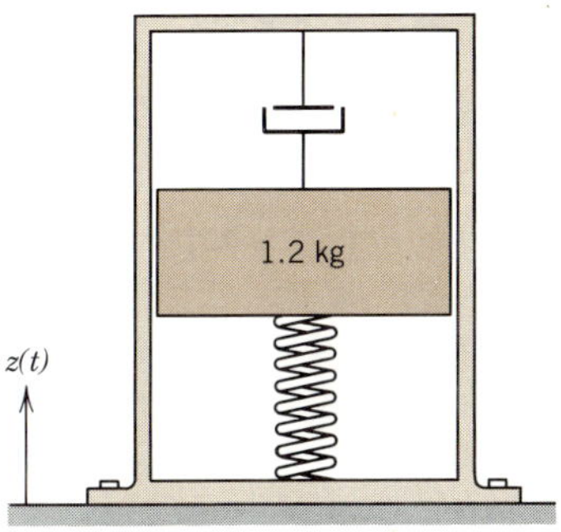

The sketch shows a model of a seismometer. When the ground is at rest, the undamped natural frequency of the system is 50 Hz and the ratio of critical damping is 0.40. The block has a mass of 1.2 kg. The vertical displacement of the ground is $z = 50 \sin(90\pi t)$ mm. Determine the amplitude and phase angle relative to a sine response for the displacement of the block from its static equilibrium position.

Solution

Let y be the vertical displacement of the block, with $y > 0$ meaning upward motion. Because z is positive upward also, the elongation of the spring is $\Delta = y - z$. Similarly, the rate at which the dashpot is increasing in length is $\dot{y} - \dot{z}$. The free body diagram accounts for these forces; it omits the static

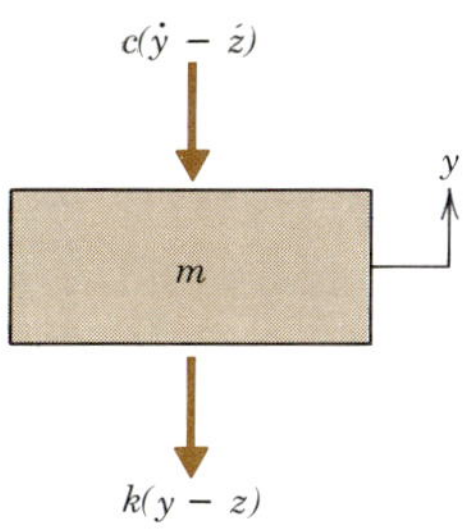

forces. The corresponding equation of motion is

$$\sum F_t = -k(y - z) - c(\dot{y} - \dot{z}) = m\ddot{y}$$
$$m\ddot{y} + c\dot{y} + ky = c\dot{z} + kz$$

Note that the portions of the dashpot and spring forces that are generated by the ground motion z are equivalent to the excitation force $F(t)$ in the general equation of motion (5). Because the values of m, ω_n, and β are given, we may eliminate the parameters c and k by using the definitions in eqs. (10) and (20). Converting from the cyclical to circular natural frequency then yields

$$\begin{aligned} F(t) &= c\dot{z} + kz \equiv 2m\omega_n\beta\dot{z} + m\omega_n^2 z \\ &= 2(1.2)(100\pi)(0.4)[(90\pi)0.05\cos(90\pi t)] + 1.2(100\pi)^2[0.05\sin(90\pi t)] \\ &= 4264\cos(90\pi t) + 5922\sin(90\pi t)\ \text{N} \end{aligned}$$

Because the response will be represented in terms of an amplitude and phase angle relative to a sine response, it is useful to represent the excitation in the same manner. Thus, we write

$$\begin{aligned} F(t) &= F\sin(90\pi t + \theta) \\ &\equiv F\sin\theta\cos(90\pi t) + F\cos\theta\sin(90\pi t) \end{aligned}$$

Matching this to the excitation force computed above yields

$$F \sin\theta = 4264, \qquad F\cos\theta = 5922$$
$$F = 7297 \text{ N}, \qquad \theta = 0.6240 \text{ rad}$$

Thus the equation of motion is

$$m\ddot{y} + c\dot{y} + ky = 7297 \sin(90\pi t + 0.6240)$$

Because the excitation has a phase angle, we introduce the same phase angle into the particular solution. This phase angle (which is like shifting the origin of time) is in addition to the phase lag ψ associated with the steady-state response. The form of the particular solution is therefore

$$y_p = D\sin(90\pi t + \theta - \psi)$$

The values of D and ψ may be determined from eqs. (34), for which we note that

$$\frac{F}{k} \equiv \frac{F}{m\omega_n^2} = \frac{7297}{1.2(100\pi)^2} = 0.06161 \text{ m}$$

The first of eqs. (34) with $\Omega/\omega_n = 90\pi/100\pi = 0.9$ yields

$$D = \frac{0.0616}{\{[1-(0.9)^2]^2 + 4(0.4)^2(0.9)^2\}^{1/2}} = 0.08272 \text{ m}$$

The corresponding value of ψ, obtained from the second of eqs. (34), is

$$\psi = \tan^{-1}\left[\frac{2(0.4)(0.9)}{1-(0.9)^2}\right] = 1.3128 \text{ rad}$$

In summary, the steady-state response is

$$y_p = 0.08272 \sin(90\pi t + 0.6240 - 1.3128) \text{ m}$$

$$\text{amplitude} = 82.7 \text{ mm}$$

$$\text{phase lag} = 1.3128 - 0.6240 = 0.6888 \text{ rad} = 39.5^\circ$$

HOMEWORK PROBLEMS

17.51 The 4 lb block is restrained by a spring whose stiffness is 50 lb/ft. The external force varies harmonically according to $F = 12\sin(16t)$, where F is in units of pounds and t is in seconds. At $t = 0$ the block is at rest at the static equilibrium position. Determine the response of the block and identify the transient and steady-state portions.

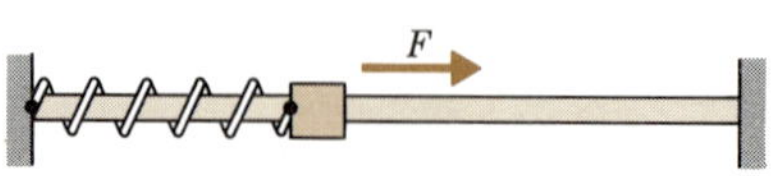

PROB. 17.51

17.52 Solve Problem 17.51 for the case where, at $t = 0$, the block is passing the static equilibrium position with a velocity of 2 ft/s to the right.

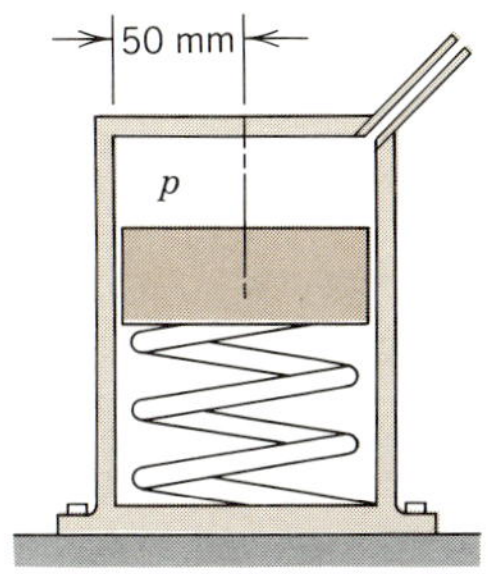

PROB. 17.53

17.53 The 600 gram piston, whose cross section is circular, is supported by the 15 N/m spring. The pressure fluctuation relative to atmospheric in the cylindrical chamber above the piston varies according to $p = 2\sin(4t)$ where p is in units of pascals and t is seconds. The block is initially at rest in the static equilibrium position. Determine the response of the piston. Identify the transient and steady-state portions of this response.

17.54 Solve Problem 17.53 for the case where the pressure in units of pascals depends on the time t in units of seconds according to $p = 1.5\cos(4.5t + \pi/3)$.

17.55–17.57 A harmonic driving force acts on a mass as shown. Determine the initial conditions for which the transient solution does not appear. That is, the complete solution describing the motion in this case contains the steady-state part only.

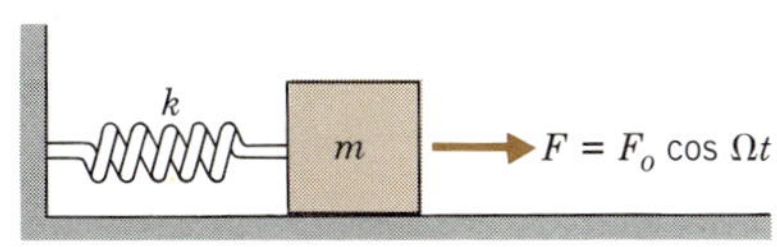

PROB. 17.55

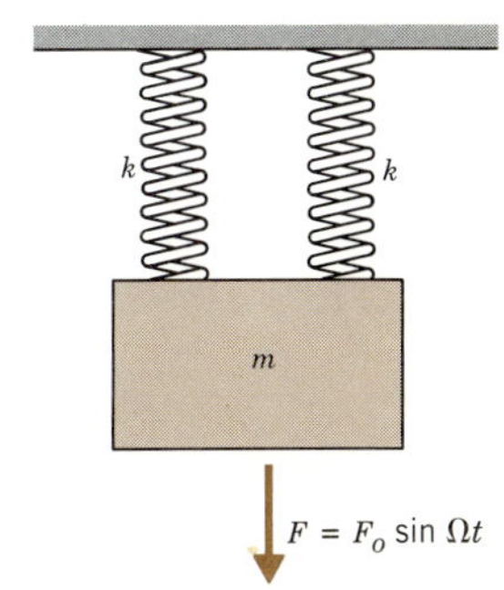

PROB. 17.56

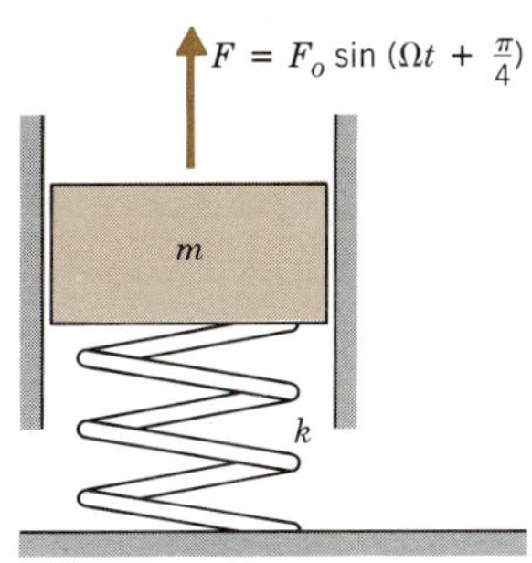

PROB. 17.57

17.58 The natural frequency of a one degree of freedom system having a mass of 50 kg is 200 Hz. When the frequency of a harmonic excitation is very close to zero, the amplitude of the steady-state response is 8 mm, and applying this force at 200 Hz produces a displacement of 32 mm. Determine the amplitude and phase lag of the response relative to the excitation when the frequency is 205 Hz.

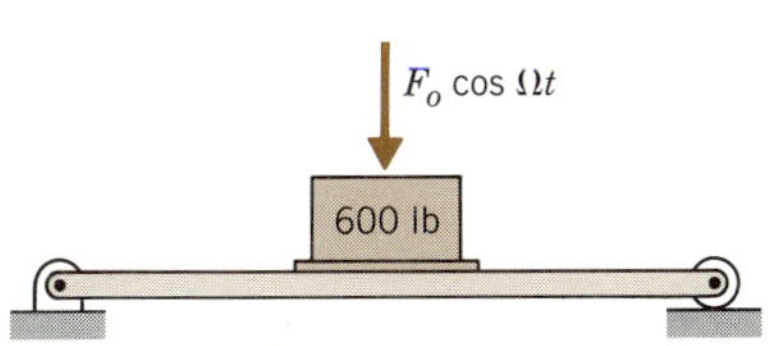

PROB. 17.59

17.59 The lightweight beam supports a 600 lb piece of equipment. The static displacement of the beam is 1.5 in. A harmonic force $F_0\cos(\Omega t)$ applied at $\Omega \simeq 0$ results in a steady-state displacement whose amplitude is 0.8 in. Damping is negligible. Determine (a) the value of F_0, and (b) the amplitude and phase lag relative to the excitation for the response at 90% of the natural frequency.

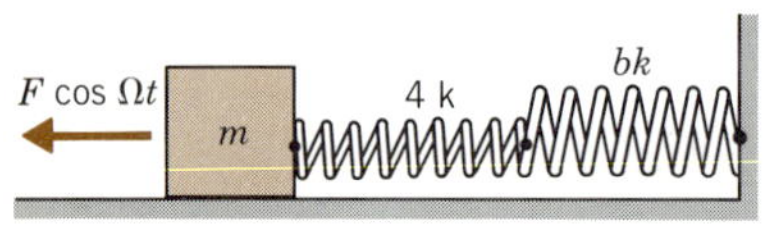

PROB. 17.61

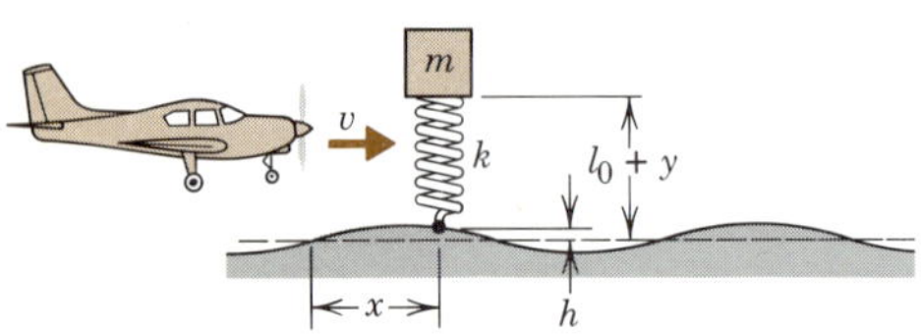

PROB. 17.65

17.60 Solve Problem 17.59 for the case where a dashpot is inserted under the equipment. This dashpot has the property that the amplitude of the steady-state response at the natural frequency is 6 in.

17.61 In the system shown, k is a basic stiffness unit for the series-connected springs. The system resonates when $\Omega = (k/m)^{1/2}$, and the amplitude of the steady-state response is mg/k when $\Omega = 2(k/m)^{1/2}$. Determine (a) the value of the factor b describing the stiffness of the second spring, and (b) the amplitude F of the force.

17.62 The excitation force in units of newtons acting on a one degree of freedom system is $F(t) = 50\cos(40\pi t) + 30\sin(40\pi t)$. The mass of the system is 3 kg, the natural frequency is 25 Hz, and the critical damping ratio is 0.20. Determine the amplitude and phase angle relative to a cosine response for the steady-state response to this excitation.

17.63 The excitation force in units of pounds acting on a one degree of freedom system is $F(t) = 200\sin(80\pi t) + 100\cos(80\pi t)$. The moving part of the system weighs 30 lb, the natural frequency of the system is 50 Hz, and the critical damping ratio is 0.10. Determine the amplitude of the steady-state response and the corresponding phase angle relative to a pure sine response.

17.64 After a harmonic excitation whose frequency is 16 Hz has been acting on a system for a long time, it is observed that the velocity of the moving mass is zero whenever the excitation is zero. It is also observed that the response lags 105° behind the excitation when the frequency is 18 Hz. Determine the natural frequency and ratio of critical damping for this system.

17.65 The runway of an airport can be approximated as a sinusoidal surface whose vertical elevation h is defined by $h = h_0 \sin(\pi x/L)$. An airplane of mass m lands smoothly on the runway with a speed v. Use the model shown to determine the critical landing speed at which the airplane would resonate. *Hint:* Let $x = vt$ in the expression for the elevation h.

17.66 The automobile gives the 500 lb trailer a constant horizontal velocity component v as the wheels of the trailer follow a wavy road. The z elevation of the road surface is related to the horizontal distance x traveled by the trailer according to $z = 0.4\sin(\pi x/2)$, where z and x are measured in feet. The trailer does not have shock absorbers, and each of the two springs supporting the trailer has a stiffness of 300 lb/in. Neglecting the mass and the radius of the wheels, determine (a) the differential equation of motion for the vertical displacement of the trailer, (b) the speed v at which the trailer would resonate, and (c) the amplitude of the vibration of the trailer when $v = 30$ ft/s. See the hint given in Problem 17.65.

PROB. 17.66

17.67 Solve Problem 17.66 if, in addition to the two springs, two shock absorbers whose constants are 5 lb-s/ft each are also connected between the axle and the body of the trailer.

PROB. 17.68

17.68 In order to isolate an electronic instrument from the vibratory motion of the floor, it is mounted on a platform that is supported by four springs (only two are shown). The mass of the instrument and the platform is 15 kg and the stiffness of each spring is 2.10 kN/m. The upward displacement of the floor is $z = 5\sin(\Omega t)$, where z is in millimeters. Determine the maximum acceleration $\ddot{y}$ of the instrument when the floor is vibrating at 3.6 Hz and at 4.0 Hz.

17.69 Solve Problem 17.68 if each of the springs is concentric with a dashpot whose constant is 20 N-s/m.

17.70 Solve Problem 17.68 if each of the springs is concentric with a dashpot whose constant is 200 N-s/m.

17.71 Instrumentation within the accelerometer shown measures the displacement x of the block relative to the surrounding case as an indication of the vertical displacement z of the floor. In a certain situation the block is initially at rest in the static equilibrium position when the base is given an upward acceleration $\ddot{z}$. (a) Determine the equation of motion governing x when there is no damping. (b) Use the equation in part (a) to determine the response $x(t)$ when a ground acceleration $\ddot{z} = a\sin(\Omega t)$ is applied to the system, where $\Omega = 0.95\omega_n$.

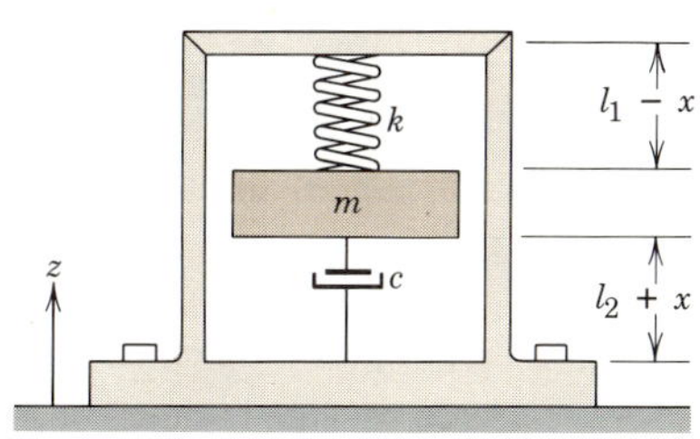

PROB. 17.71

***17.72** Solve Problem 17.71 for the case where the damping is 50% of critical.

***17.73** Solve Problem 17.71 for the case where the damping is 10% larger than critical.

***17.74** When a one degree of freedom system is subjected to a harmonic force having an amplitude of 50 N and a frequency of 100 Hz, the steady-state response is zero 2.5 ms after the force is zero. The amplitude of the response at this frequency is 50 mm, but applying a 50 N force statically produces a displacement of 4 mm. Determine the mass m, stiffness k, and dashpot constant c for this system.

***17.75** An excitation in pounds of $60\sin(\Omega t) + 80\cos(\Omega t)$ produces a steady-state response whose amplitude is 0.15 ft when $\Omega \simeq 0$. The natural frequency of the system is 20 Hz, and $\Omega = 18$ Hz produces a steady-state displacement that is in phase with a sine response. Determine the amplitude of the displacement when $\Omega = 18$ Hz.

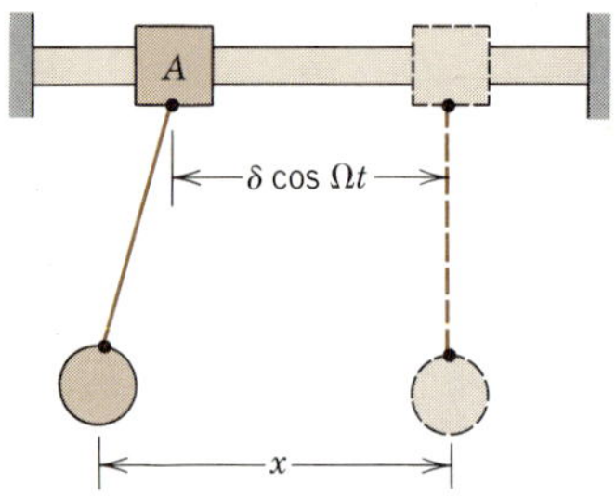

PROB. 17.76

*17.76 A simple pendulum formed from a particle of mass m at the end of a cable of length l is attached to collar A. The collar is given a harmonic displacement of $\delta \cos(\Omega t)$. For what range of values of Ω will the steady-state amplitude of the horizontal displacement z of the mass be (a) at least three times larger than δ? (b) less than δ?

*17.77 A system whose natural frequency is ω_n, and whose critical damping ratio is β is subjected to a harmonic excitation at frequency Ω. (a) Determine the value of Ω at which the amplitude of the steady-state displacement is a maximum. (b) What are the amplitude and phase angle at the frequency found in part (a)? (c) For what range of values of β will an increase in the value of Ω, starting from $\Omega = 0$, always result in a decrease in the steady state amplitude?

*17.78 A system whose natural frequency is ω_n and whose ratio of critical damping is β is subjected to a harmonic force $F_0 \cos(\Omega t)$. Determine (a) the instantaneous power developed by the force as a function of time in the steady-state motion of the system, (b) the total work done by the force in one cycle, (c) the average power developed by the force over a complete cycle. Compare the results of parts (a) and (c).

E. VIBRATION OF A RIGID BODY

The vibrating systems we have treated so far were successfully modeled as particles. However, as we saw in the basic studies of kinetics, in order to retain the essential features of a physical system it is frequently necessary to model the system as a rigid body. The techniques for investigating the vibration of a rigid body about its static equilibrium position are essentially the same as for a particle. A geometrical parameter is selected to represent the movement of the body, and the differential equation of motion for this parameter is derived according to the principles of dynamics. The free and forced vibration responses may then be determined by solving the resulting equation of motion.

There is a complication that arises when we account for the force of a spring that is attached to a specific point in a rotating body. In some respects this complication resembles that encountered in the study of the pendulum in Example 4. We saw there that, for large angles of rotation of the pendulum, the equation of motion is nonlinear, and thus difficult to solve. By restricting the angle of rotation to be a small value, we were able to linearize the equation. Similarly, it will be necessary for us to restrict our attention to situations in which a rigid body only undergoes small rotations and displacements, in order to have a linear equation of motion.

To see how such an approximation is made, consider the system in Figure 14a, where a slender bar may rotate in the vertical plane about the pin A. For small angles of rotation θ of the bar from the vertical position shown in Figure 14a, we may consider point C to move only horizontally, so, in essence, the springs maintain their horizontal orientations. Furthermore, for small θ, $\tan\theta \simeq \sin\theta \simeq \theta$. Hence, according to Figure 14b, we may approximate the deformation Δ of the springs as

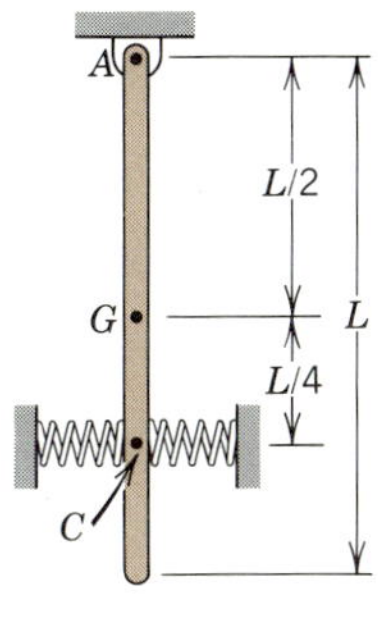

FIGURE 14a

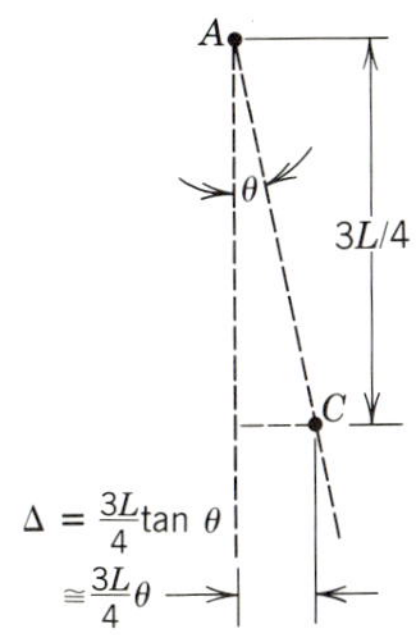

FIGURE 14b

$$\Delta \simeq \frac{3L}{4}\theta$$

Similarly the lever arm of the gravity force is approximated as $(L/2)\theta$.

For this approximate analysis, the free body diagram of the bar is as shown in Figure 15. Notice that the free body diagram was drawn for a general position of the body, but all linearizing approximations regarding the smallness of the motion are employed.

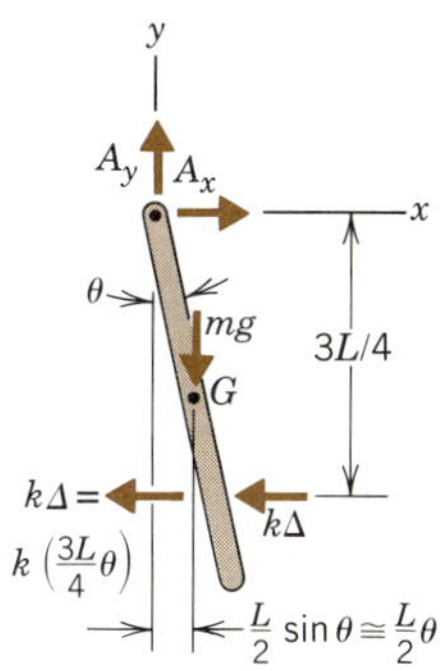

FIGURE 15

The equation of motion for this system may be obtained by summing moments about the z axis passing through the fixed point A, thereby eliminating the unknown reaction forces $\bar{A}_x$ and $\bar{A}_y$. Using the approximation for the length of the lever arm of the gravity force appearing in the free body diagrams yields

$$\sum M_{Az} = -mg\left(\frac{L}{2}\theta\right) - 2k\left(\frac{3L}{4}\theta\right)\left(\frac{3L}{4}\right) = I_{zz}\alpha = \tfrac{1}{3}mL^2(\ddot{\theta})$$

$$\tfrac{1}{3}mL^2\ddot{\theta} + (\tfrac{9}{8}kL^2 + \tfrac{1}{2}mgL)\theta = 0$$

A comparison of this result to the general equation of motion (5) shows that the swinging bar acts like a one degree of freedom, undamped oscillator in free motion. The position variable in this case is an angle of rotation, rather than a linear displacement. Thus the coefficient in the acceleration term is a moment of inertia, rather than a mass, and the coefficient in the position term is a torsional, rather than an extensional, spring stiffness.

It is often cumbersome to determine the differential equation of motion from the basic force and moment equations. Such a situation occurs for bodies in general motion, as well as for systems in which several bodies are interconnected. Both types of systems involve reaction forces that cannot be eliminated easily from the basic equations.

An alternative for treating such systems is to employ energy concepts. Here we will only consider conservative systems. Because external excitations and damping forces are nonconservative, this restriction means that we shall address only free undamped systems. The other effects, as well as systems featuring several degrees of freedom, may be addressed in an energy formulation known as *Lagrange's equations*. Such a formulation arises in the subject of *analytical mechanics*, which is treated in more advanced texts.

The basic concept in our formulation is that for a conservative system, the sum of the kinetic and potential energies at any instant is the constant mechanical energy E.

$$T + V = E \tag{36}$$

Differentiating eq. (36) with respect to time leads to the *power form* of the conservation of energy principle:

$$\boxed{\dot{T} + \dot{V} = 0} \tag{37}$$

The equation of motion for the system may be obtained directly from eq. (37). Doing so requires substitution of expressions for the kinetic and potential energies in terms of the geometrical parameter describing the movement of the system.

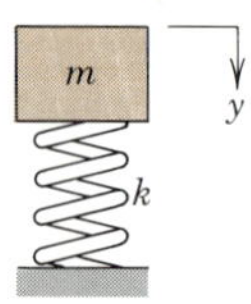

FIGURE 16

We shall develop the energy formulation for a conservative system by a series of examples. First let us consider the basic one degree of freedom oscillator shown in Figure 16. Let y be the downward displacement from the static equilibrium position, and let $y = 0$ be the datum for gravitational potential energy. Then a positive value of y means that the block is below the datum. For static equilibrium the spring is initially compressed a distance mg/k, so the total compression of the spring during any motion is $y + mg/k$. The energy expressions are

$$T = \tfrac{1}{2}mv^2 = \tfrac{1}{2}m\dot{y}^2$$

$$V = V_{\text{spring}} + V_{\text{gravity}} = \frac{1}{2}k\left(y + \frac{mg}{k}\right)^2 + mg(-y)$$

$$= \frac{1}{2}ky^2 + \frac{1}{2}k\left(2y\frac{mg}{k}\right) + \frac{1}{2}k\left(\frac{mg}{k}\right)^2 - mgy$$

$$= \tfrac{1}{2}ky^2 + \tfrac{1}{2}k(mg)^2$$

Substituting T and V into eq. (37) then yields

$$\frac{d}{dt}\left(\frac{1}{2}m\dot{y}^2\right) + \frac{d}{dt}\left[\frac{1}{2}ky^2 + \frac{1}{2k}(mg)^2\right] = 0$$

The results of carrying out the derivatives is

$$m\dot{y}\ddot{y} + ky\dot{y} = 0$$

Because $\dot{y}$ cannot be identically zero in any motion, that factor may be canceled, which then leaves

$$m\ddot{y} + ky = 0$$

This, of course, is the differential equation we obtained earlier using $\sum \bar{F} = m\bar{a}$.

An important observation regarding the foregoing is that considering the static gravity force mg and the corresponding spring deformation mg/k only added a constant to the potential energy V. This constant disappears in the evaluation of $\dot{V}$. Thus we could have ignored static effects when we first formed V. This observation is generally true for other systems, provided that the movement is measured with respect to the static equilibrium position, and that only small movements are to be considered.

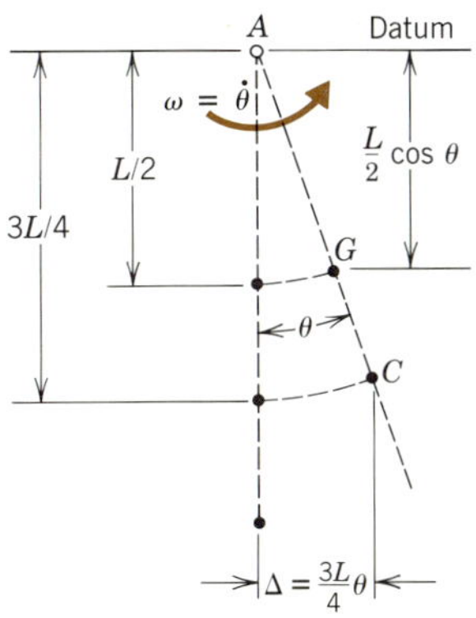

FIGURE 17

A more representative application of eq. (37) may be obtained by studying the system in Figure 14a. We have constructed, in Figure 17, a sketch showing all of the geometrical and kinematical properties required for relating the energies to the position angle θ. Note that the assumption that θ is small was used to describe the spring deformation Δ. However, it was not used for the elevation of the center of mass G. This is a subtle matter. Suppose we were to set $\cos\theta = 1$ because θ is small. Then the elevation of point G would seem to be constant, and the potential energy of gravity would also seem to be constant. Constant energy terms do not affect eq. (37). Hence, we (incorrectly) would have eliminated the effect of gravity. To avoid this question, it is best to introduce linearization for the gravity force *after* the derivatives in eq. (37) have been evaluated.

The bar is in pure rotation about end A, and the moment of inertia about that point is $I_A = mL^2/3$, so the kinetic energy is

$$T = \tfrac{1}{2}I_A\omega^2 = \tfrac{1}{6}mL^2\dot{\theta}^2$$

Adding the potential energy of the springs and gravity yields

$$V = 2\left(\frac{1}{2}k\Delta^2\right) + mgy_G$$
$$= k\left(\frac{3}{4}L\theta\right)^2 + mg\left(-\frac{L}{2}\cos\theta\right)$$

Substituting T and V in eq. (37) then yields

$$\frac{d}{dt}\left(\frac{1}{6}mL^2\dot{\theta}^2\right) + \frac{d}{dt}\left[\frac{9}{16}kL^2\theta^2 - \frac{1}{2}mgL\cos\theta\right]$$
$$\equiv \frac{1}{3}mL^2\dot{\theta}\ddot{\theta} + \frac{9}{8}kL^2\theta\dot{\theta} - \frac{1}{2}mgL(-\sin\theta)\dot{\theta} = 0$$

We cancel the non-zero factor $\dot{\theta}$. Also we may *now* linearize the effect of the gravity force by replacing $\sin\theta$ by θ in the equation of motion. Thus the equation of motion reduces to

$$\tfrac{1}{3}mL^2\ddot{\theta} + (\tfrac{9}{8}kL^2 + \tfrac{1}{2}mgL)\theta = 0$$

This is the same equation as the earlier one obtained from a moment equation.

EXAMPLE 10

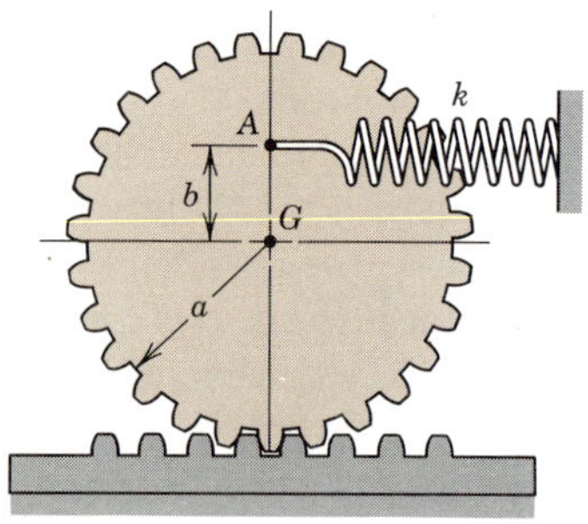

The gear shown in the diagram has a mass M and a radius of gyration r. It rolls over a fixed rack and is restrained by a spring. The gear is in static equilibrium in the position shown. Determine the natural frequency of the system by using the basic force and moment equations of motion.

Solution

For small motions of the gear, we can consider the spring to remain horizontal. Let us denote the displacement of the center of mass G by the variable s, with positive s representing a movement to the right. Such a movement will result in a compressive deformation of the spring by the quantity Δ, Hence in the free body diagram shown, the spring force is $k\Delta$ to the left.

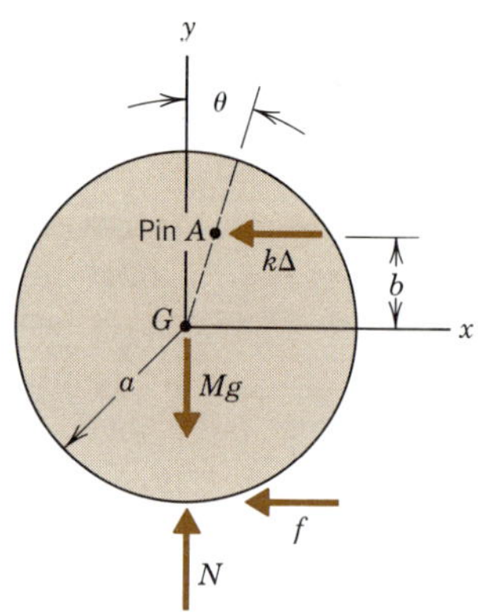

The gear is in general motion, so we sum moments about the center of mass. The basic equations of motion are therefore

$$\sum F_x = -k\Delta - f = M(a_G)_x = M\ddot{s}$$

$$\sum F_y = N - Mg = M(a_G)_y = 0$$

$$\sum M_{Gz} = (k\Delta)b - fa = I_{zz}\alpha = Mr^2\alpha$$

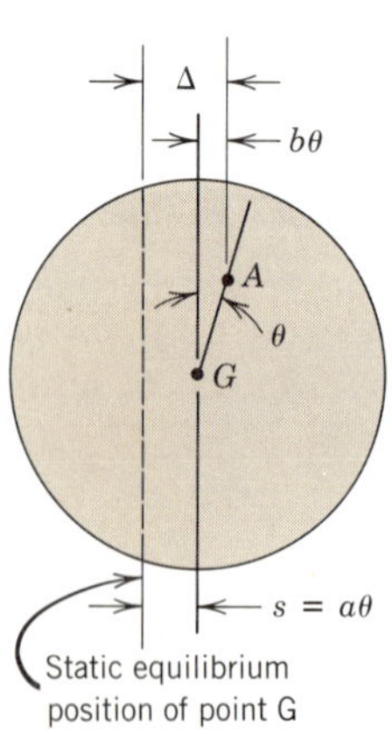

In these three equations there are five unknowns: Δ, f, s, N, and α. To find additional relations we use the kinematical constraint that the gear rolls without slipping, which enables us to relate the rotation and the deformation of the springs to the displacement s. We do this by sketching the displaced position of the gear. Note that in this sketch we employed the restriction of small displacement, allowing us to approximate the horizontal movement of point A with respect to point G as $b\theta$.

For rolling without slipping, the movement horizontally equals the arc length along the gear, so

$$\theta = \frac{s}{a}, \qquad \Delta = s + b\theta = s\left(1 + \frac{b}{a}\right)$$

A positive value of θ represents a clockwise rotation. However, positive angular acceleration α is counterclockwise for the xyz axes we are using. Therefore substituting these expressions into the basic equations of motion yields

$$-ks\left(1 + \frac{b}{a}\right) - f = M\ddot{s}$$

$$N = Mg$$

$$ks\left(1 + \frac{b}{a}\right)b - fa = -Mr^2 \frac{\ddot{s}}{a}$$

We wish to obtain a single differential equation governing the position s of the center of mass. Hence, solving for f from the first equation, and substituting that expression into the third equation yields

$$ks\left(1 + \frac{b}{a}\right)b + ks\left(1 + \frac{b}{a}\right)a + Ma\ddot{s} = -Mr^2 \frac{\ddot{s}}{a}$$

Dividing the equation by a and then collecting like terms results in

$$M\left(1 + \frac{r^2}{a^2}\right)\ddot{s} + k\left(1 + \frac{b}{a}\right)^2 s = 0$$

This equation has the same form as that for undamped free vibration of a general one degree of freedom system. The natural frequency is found by forming the ratio of the stiffness and inertia terms.

$$\omega_n = \left[\frac{k(1 + b/a)^2}{M(1 + r^2/a^2)}\right]^{1/2} = \frac{a + b}{(a^2 + r^2)^{1/2}} \left(\frac{k}{M}\right)^{1/2}$$

EXAMPLE 11

Determine the natural frequency of the system in Example 10 using the power form of the principle of conservation of energy.

Solution

The free body diagram of the disk and the kinematical analysis were detailed in Example 10. The normal and friction forces do no work because there is rolling without slipping, and the center of mass G remains at a constant elevation. Thus, this is a conservative system whose potential energy is stored in the spring. The potential energy is therefore

$$V = \tfrac{1}{2}k\Delta^2$$

The moment of inertia of the disk about its center of mass is Mr^2, and the disk is in general motion, so its kinetic energy is

$$T = \tfrac{1}{2}Mv_G^2 + \tfrac{1}{2}I_G\omega^2$$

The key step in an energy formulation for a one degree of freedom system is to relate all quantities to a single variable describing the movement.

The analysis in Example 10 used the horizontal displacement s of the center of mass. It was found there that

$$v_G = \dot{s}, \qquad \omega = \frac{\dot{s}}{a}, \qquad \Delta = s\left(1 + \frac{b}{a}\right)$$

The energy expressions therefore become

$$V = \frac{1}{2} ks^2\left(1 + \frac{b}{a}\right)^2$$

$$T = \frac{1}{2} M\dot{s}^2 + \frac{1}{2}(Mr^2)\frac{\dot{s}^2}{a^2} = \frac{1}{2} M\left(1 + \frac{r^2}{a^2}\right)\dot{s}^2$$

Substituting these expressions into eq. (37) then yields

$$\frac{d}{dt}\left[\frac{1}{2} M\left(1+\frac{r^2}{a^2}\right)\dot{s}^2\right] + \frac{d}{dt}\left[\frac{1}{2} ks^2\left(1+\frac{b}{a}\right)^2\right] \equiv M\left(1+\frac{r^2}{a^2}\right)\dot{s}\,\ddot{s} + ks\dot{s}\left(1+\frac{b}{a}\right)^2 = 0$$

Cancelling the non-zero factor $\dot{s}$ leaves the following differential equation of motion.

$$M\left(1 + \frac{r^2}{a^2}\right)\ddot{s} + k\left(1 + \frac{b}{a}\right)^2 s = 0$$

This is the same equation as the one derived in Example 10, so the natural frequency is as given there.

$$\omega_n = \frac{a + b}{(a^2 + r^2)^{1/2}}\left(\frac{k}{M}\right)^{1/2}$$

It is instructive to compare the two methods by which this problem was solved. The kinematical analysis was essentially the same for both, but the sign associated with each movement was of less concern in the energy formulation. Most important, the friction force, which was of no interest, never entered into the energy formulation. Recall that one of the primary assets of the energy formulation is that reactions generally are eliminated from the analysis.

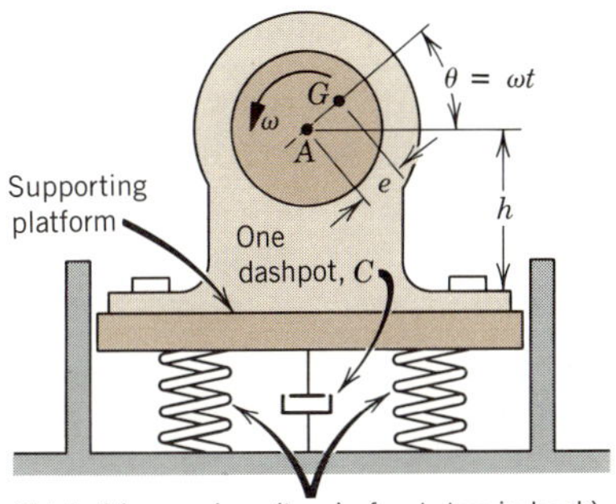

Total of four springs (two in front, two in back), Stiffness k for each

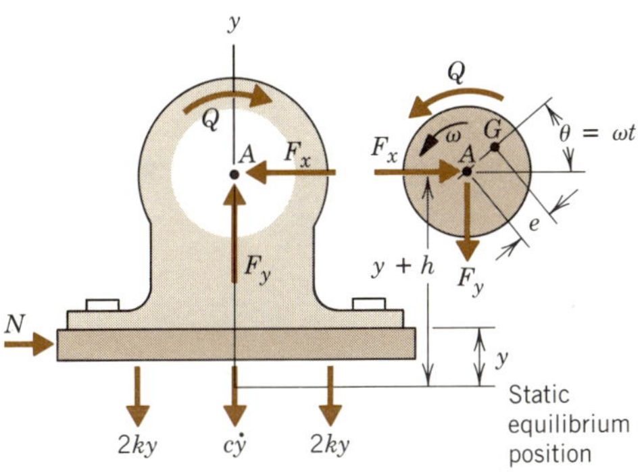

EXAMPLE 12

A common source of excitation is an imbalance when the center of mass of a rotating body does not coincide (i.e., is eccentric) with the axis of rotation. In the system shown the combined mass of the motor and the platform is M and the mass of the flywheel is m. The flywheel rotates at a constant angular speed ω, such that the angle of elevation from the motor shaft A to the center of mass G of the flywheel is $\theta = \omega t$. The distance from A to G is the eccentricity e. Derive the equation of motion for this system.

Solution

We cannot use the power form of the conservation of energy principle for two reasons. Obviously the dashpot exerts a nonconservative force. A much more subtle reason stems from the fact that the rotation rate is constant. Such a motion is achieved by the motor exerting a torque on the flywheel. Because the torque is not known in this problem, the power that must be supplied to the motor to turn the flywheel is also unknown. Hence we shall formulate the problem using the basic equations of motion for rigid bodies.

The system is composed of two moving bodies, one being the body of the motor and the platform, the other being the flywheel. Therefore, we draw a free body diagram for each system. The interaction exerted between them is the reaction $\bar{F}$ at the shaft, and the torque $\bar{Q}$ producing the rotation. Notice that we are measuring the vertical displacement y from the static equilibrium position, so we ignore the static effect of gravity. The equation for vertical motion of the motor and its framework is

$$\sum F_y = -4ky - c\dot{y} + F_y = Ma_y = M\ddot{y}$$

To determine the reaction component F_y, we sum forces in the vertical direction for the flywheel. This gives

$$\sum F_y = -F_y = m(a_G)_y = m\ddot{y}_G$$

where the negative sign for F_y arises because of Newton's third law. From the distances depicted in the free body diagram, we find that the vertical distance to point G from the static equilibrium position is

$$y_G = y + h + e \sin\theta = y + h + e\sin(\omega t)$$

Because the distance h is a constant, the acceleration is

$$\ddot{y}_G = \ddot{y} - e\omega^2 \sin(\omega t)$$

Hence the equation of motion for the flywheel becomes

$$\sum F_y = -F_y = m[\ddot{y} - e\omega^2 \sin(\omega t)]$$

The reaction F_y may be eliminated by adding this equation to the equation describing $\sum F_y$ for the flywheel. This yields

$$-4ky - c\dot{y} = M\ddot{y} + m[\ddot{y} - e\omega^2 \sin(\omega t)]$$

A simple rearrangement of terms yields the differential equation of motion for the system

$$(M + m)\ddot{y} + c\dot{y} + 4ky = em\omega^2 \sin(\omega t)$$

We see from this equation that the eccentric flywheel affects the system in two ways. Its mass m adds to the inertia of the system, and the fact that it is eccentric creates a harmonic excitation that rises as the square of the rotation rate. In essence, this excitation is the reaction corresponding to the force that must be applied to the flywheel in order to impart the centripetal acceleration $e\omega^2$ to the center of mass G of the flywheel.

Note that $M + m$, $4k$, and $em\omega^2$ are equivalent to the mass, stiffness, and force amplitude, respectively, for a general one degree of freedom system undergoing harmonic excitation. Consequently, we see that no new techniques are needed to solve for the response of a system in which there is a rotating imbalance.

HOMEWORK PROBLEMS

17.79–17.84 The (small) angle θ measures the rotation in the vertical plane of the system shown from its static equilibrium position. Use the force and moment equations of motion to derive the linear differential equation governing θ. Wherever applicable, assume that there is no slipping.

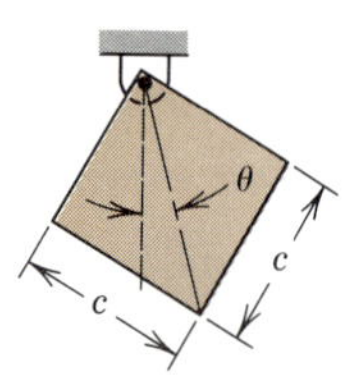

PROBS. 17.79 AND 17.85

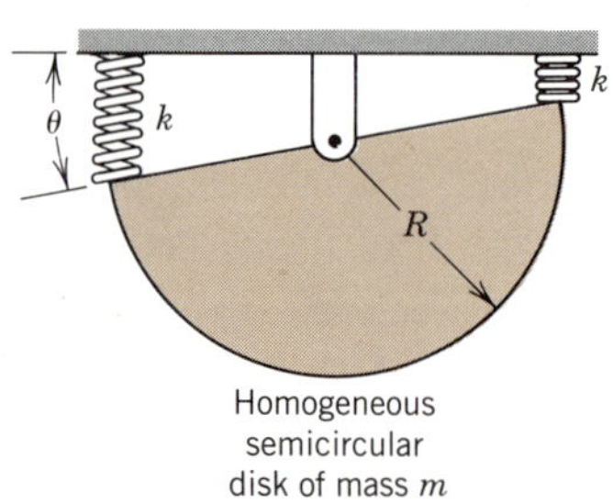

PROBS. 17.80 AND 17.86

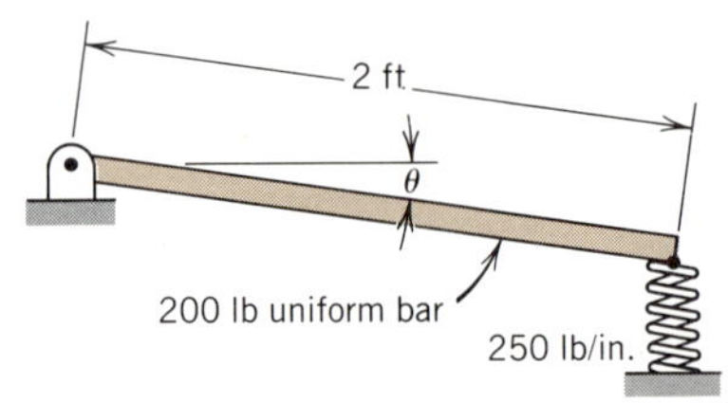

PROBS. 17.81 AND 17.87

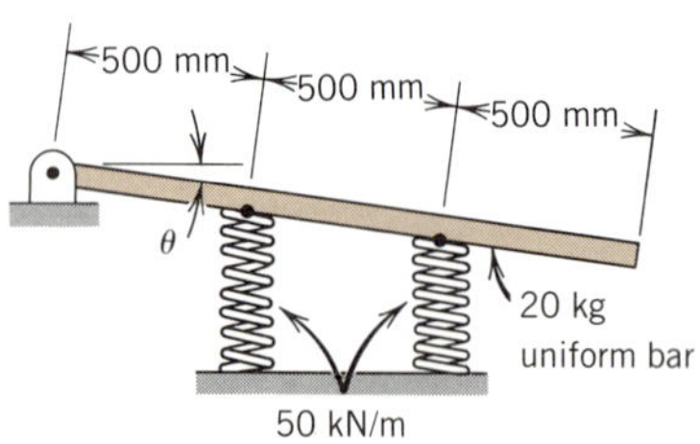

PROBS. 17.82 AND 17.88

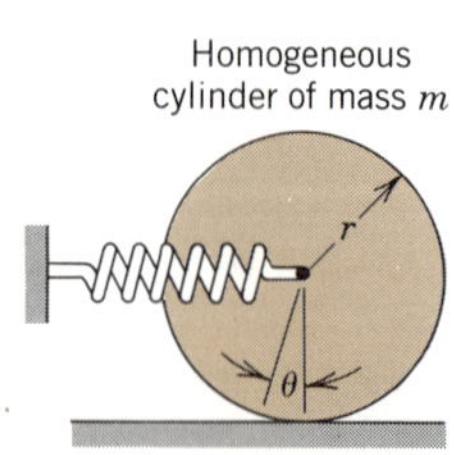

PROBS. 17.83 AND 17.89

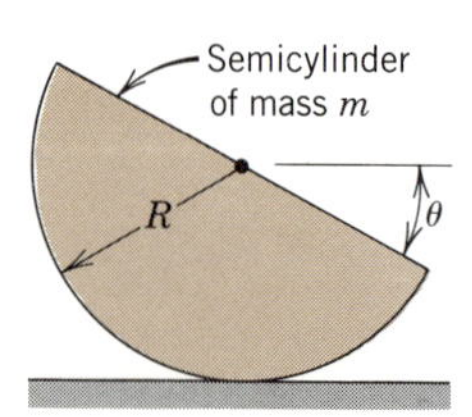

PROBS. 17.84 AND 17.90

17.85–17.90 The (small) angle θ measures the rotation in the vertical plane of the system shown from its static equilibrium position. Use the power form of the principle of conservation of energy to derive the linear differential equation governing θ. Wherever applicable, assume that there is no slipping.

17.91 In the system of Example 12, the eccentricity is $e = 1$ in., the flywheel weighs 1.2 lb, and the combined weight of the motor and the platform is 5 lb. The stiffness of each spring is 10 lb/ft. When the angular speed of the motor equals the natural frequency, the amplitude of the steady-state vertical displacement is 0.18 ft. Determine the dashpot constant c.

17.92 A flywheel is suspended from its center by a metal cable. When the flywheel is rotated by an angle θ about the axis of the cable, the cable exerts a restoring torque $k\theta$ opposite the sense of the rotation, due to its elasticity. The free vibration of the system is observed to have a period τ. Derive an expression for the mass moment of inertia of the flywheel about the axis of the cable.

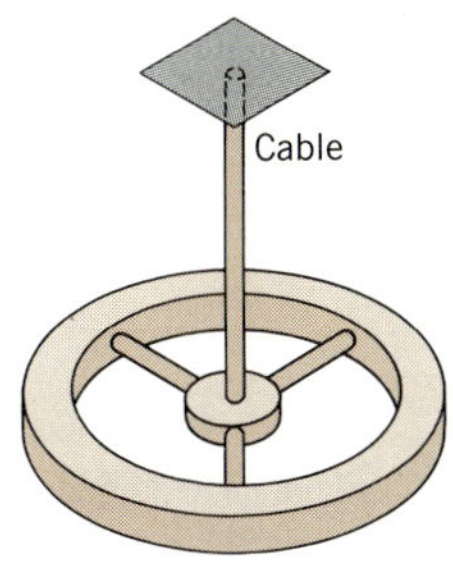

PROB. 17.92

17.93 A 50 kg electric motor is mounted on a lightweight beam. The flywheel of the motor has a mass of 1 kg, and its center of mass is 5 mm from the axis of rotation. Knowing that the static deflection of the beam when the motor and flywheel are at rest is 4 mm, determine (a) the angular velocity of the flywheel that will cause the system to resonate, and (b) the amplitude of the steady-state displacement when the angular speed of the flywheel is eight-tenths of the resonant speed found in part (a).

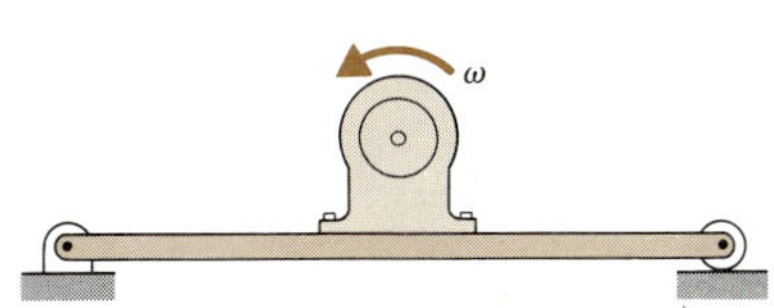

PROB. 17.93

17.94 It is desired to operate the electric motor in Problem 17.93 at the angular velocity that would cause the system to resonate. To do this, a dashpot is attached to the beam from the ground, directly under the motor. If the steady-state displacement of the motor in this condition is to be 10 mm, determine the required constant for the dashpot.

17.95–17.102 The system is shown in its static equilibrium position. Determine its natural frequency.

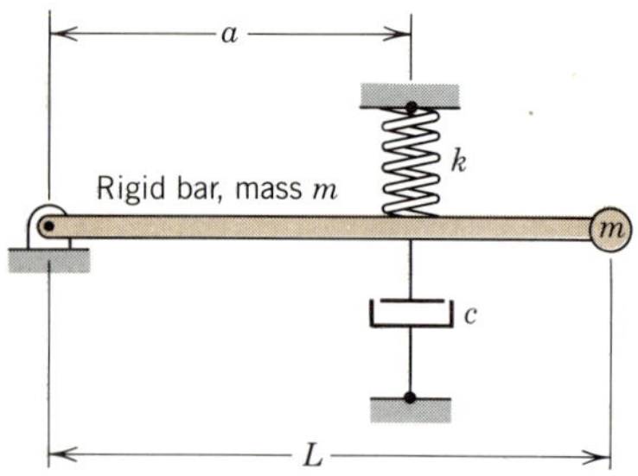

PROB. 17.95

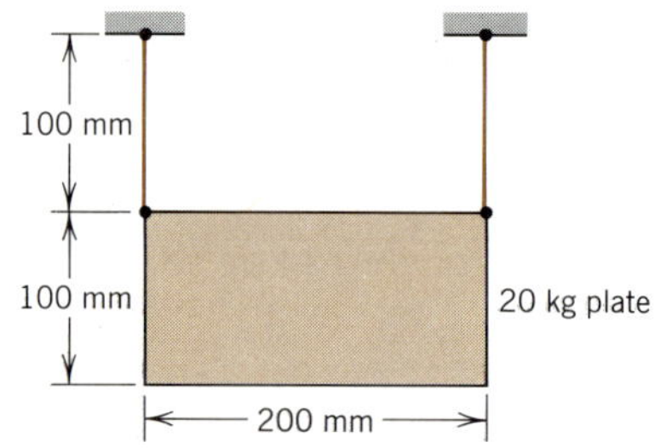

PROB. 17.96

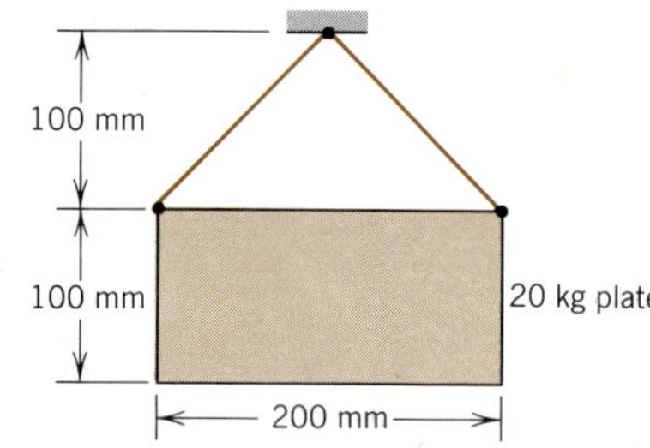

PROB. 17.97

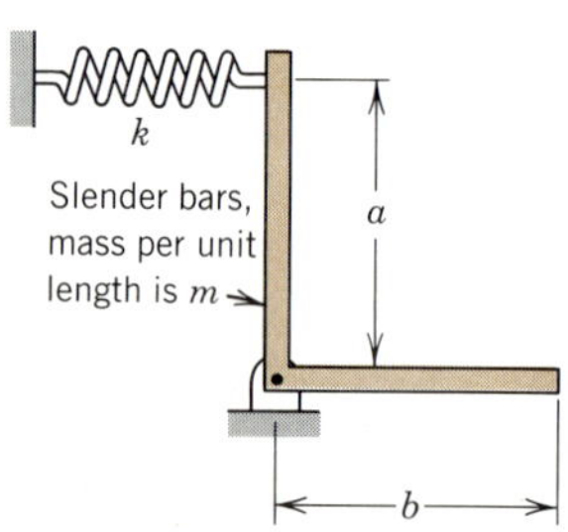

PROB. 17.98

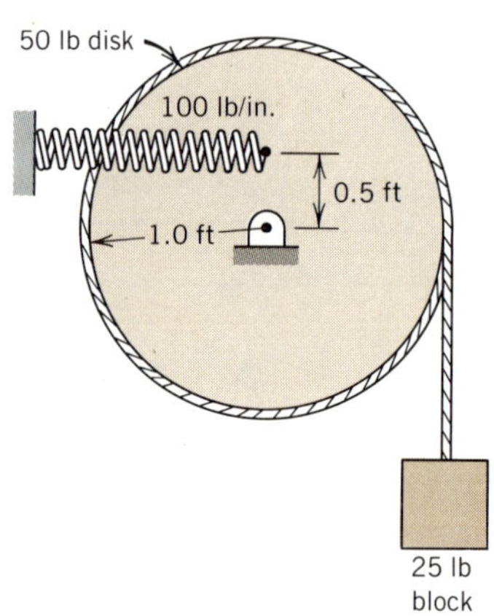

PROB. 17.99

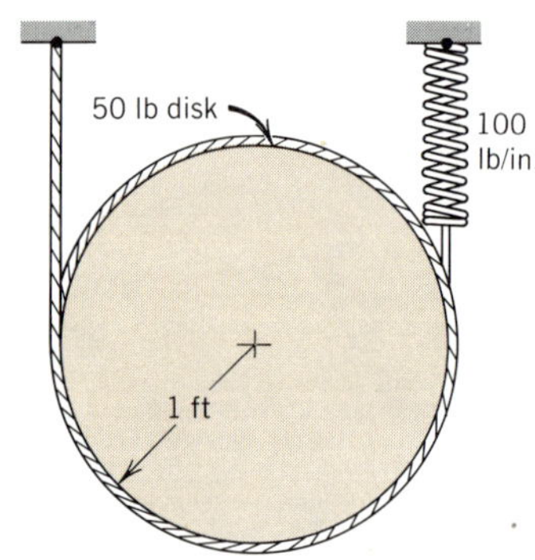

PROB. 17.100

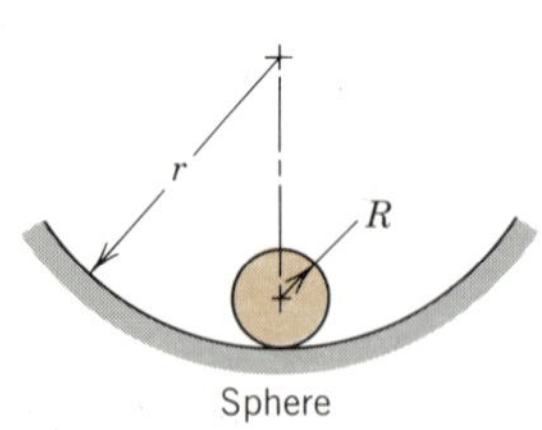

PROB. 17.101

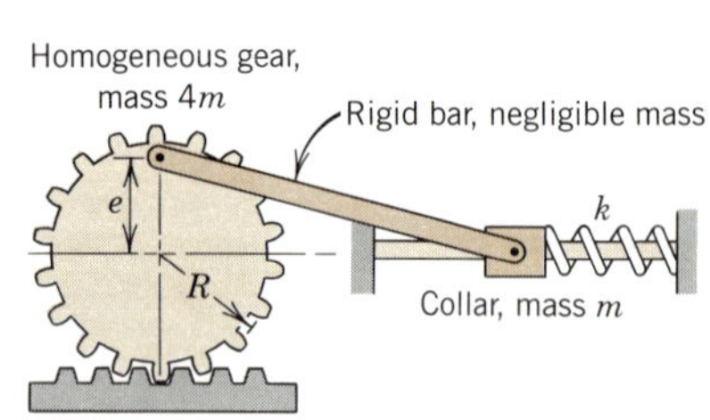

PROB. 17.102

***17.103** A motor mounted on a platform is rotating at 3000 rev/min. In the steady-state response, the motor is at its static equilibrium position when the center of mass of the flywheel is at its highest position. The corresponding amplitude of the displacement of the motor is 0.5 in. The motor weighs 40 lb, the flywheel weighs 10 lb, and the center of mass of the flywheel is eccentric by 0.3 in. Determine (a) the natural frequency and critical damping ratio, and (b) the amplitude of the response when the motor turns at 3600 rev/min.

17.104 The combined mass of a motor and its supporting platform is 50 kg, and the mass of the flywheel is 4 kg. An angular speed of 50 Hz for the flywheel produces a steady-state vertical displacement in which the motor reaches its highest elevation 5 ms after the center of mass of the flywheel reaches its highest elevation. The amplitude of the response in this case is 300 mm. It is also observed that when the angular speed is increased far beyond 50 Hz, the steady-state displacement approaches an amplitude of 5 mm. Determine (a) the eccentricity of the flywheel, (b) the natural frequency, and (c) the ratio of critical damping.

***17.105 and 17.106** Determine the natural frequency of each of the systems shown. Are there any values of the spring stiffness k for which either system will not oscillate freely about the vertical position? Discuss the meaning of these values of k with regard to the static stability of the vertical position.

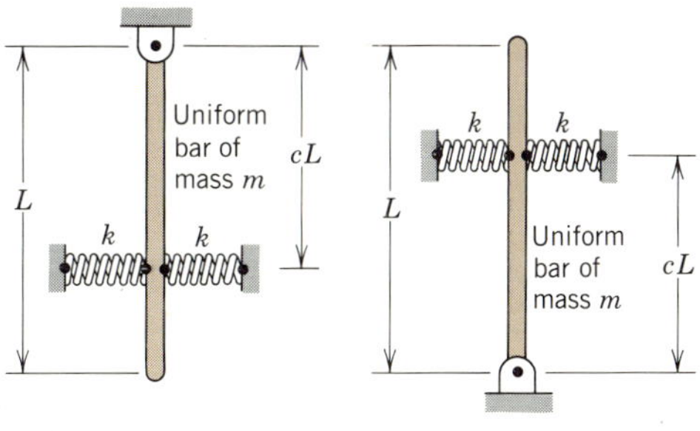

PROB. 17.105

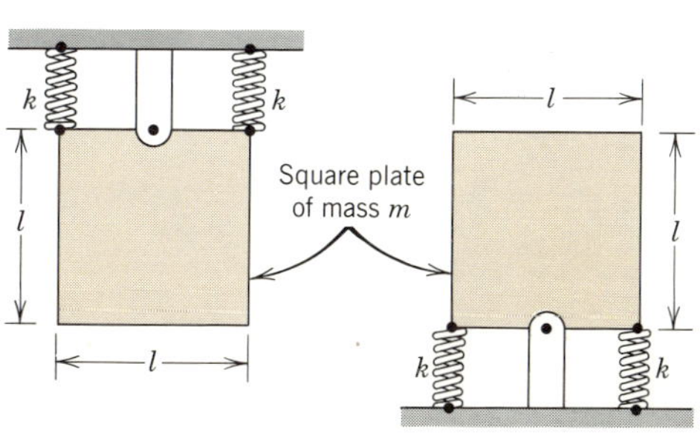

PROB. 17.106

APPENDIX A

SI UNITS

TABLE 1
CONVERSION FACTORS FROM U.S. CUSTOMARY UNITS

PHYSICAL QUANTITY	U.S. CUSTOMARY UNIT	= SI EQUIVALENT
		BASIC UNITS
Length	1 foot (ft)	= $3.048(10^{-1})$ meter (m)*
	1 inch (in.)	= $2.54(10^{-2})$ meter (m)*
	1 mile (U.S. statute)	= $1.6093(10^{3})$ meter (m)
Mass	1 slug ($lb\text{-}s^2/ft$)	= 1.4594(10) kilogram (kg)
	1 pound mass (lbm)	= $4.5359(10^{-1})$ kilogram (kg)
		DERIVED UNITS
Acceleration	1 foot/second^2 (ft/s^2)	= $3.048(10^{-1})$ meter/second^2 (m/s^2)*
	1 inch/second^2 ($in./s^2$)	= $2.54(10^{-2})$ meter/second^2 (m/s^2)*
Area	1 foot^2 (ft^2)	= $9.2903(10^{-2})$ meter^2 (m^2)
	1 inch^2 ($in.^2$)	= $6.4516(10^{-2})$ meter^2 (m^2)*
Density	1 slug/foot^3 ($lb\text{-}s^2/ft^4$)	= $5.1537(10^{2})$ kilogram/meter^3 (kg/m^3)
	1 pound mass/foot^3 (lbm/ft^3)	= 1.6018(10) kilogram/meter^3 (kg/m^3)
Energy and Work		(1 joule ≡ 1 meter-newton)
	1 foot-pound (ft-lb)	= 1.3558 joules (J)
	1 kilowatt-hour (kW-hr)	= $3.60(10^{6})$ joules (J)*
	1 British thermal unit (Btu)	= $1.0551(10^{3})$ joules (J)
Force		(1 newton ≡ 1 kilogram-meter/second^2)
	1 pound (lb)	= 4.4482 newtons (N)
	1 kip (1000 lb)	= $4.4482(10^{3})$ newtons (N)
Power		(1 watt ≡ 1 joule/second)
	1 foot-pound/second (ft-lb/s)	= 1.3558 watt (W)
	1 horsepower (hp)	= $7.4570(10^{2})$ watt (W)
Pressure and Stress		(1 pascal ≡ 1 newton/meter^2)
	1 pound/foot^2 (lb/ft^2)	= 4.7880(10) pascal (Pa)
	1 pound /inch^2 ($lb/in.^2$)	= $6.8948(10^{3})$ pascal (Pa)
	1 atmosphere (standard, 14.7 $lb/in.^2$)	= $1.0133(10^{5})$ pascal (Pa)
Speed	1 foot/second (ft/s)	= $3.048(10^{-1})$ meter/second (m/s)*
	1 mile/hr	= $4.4704(10^{-1})$ meter/second (m/s)
	1 mile/hr	= 1.6093 kilometer/hour (km/hr)
Volume	1 foot^3 (ft^3)	= $2.8317(10^{-2})$ meter^3 (m^3)
	1 inch^3 ($in.^3$)	= $1.6387(10^{-5})$ meter^3 (m^3)
	1 gallon (U.S. liquid)	= $3.7854(10^{-3})$ meter^3 (m^3)

* Denotes an exact factor.

TABLE 2
CONVERSION FACTORS FROM "OLD" METRIC UNITS

PHYSICAL QUANTITY	"OLD" METRIC UNIT	= SI EQUIVALENT
Energy	1 erg	$= 1.00(10^{-7})$ joule (J)*
Force	1 dyne	$= 1.00(10^{-5})$ newton (N)*
Length	1 angstrom	$= 1.00(10^{-10})$ meter (m)*
	1 micron	$= 1.00(10^{-6})$ meter (m)*
Pressure	1 bar	$= 1.00(10^{5})$ pascal (Pa)*
Volume	1 liter	$= 1.00(10^{-3})$ meter^3 (m^3)*

* Denotes an exact factor.

TABLE 3
SI PREFIXES

PREFIX	SYMBOL	FACTOR BY WHICH UNIT IS MULTIPLIED
tera*	T	10^{12}
giga*	G	10^{9}
mega*	M	10^{6}
kilo*	k	10^{3}
hecto	h	10^{2}
deka	da	10
deci	d	10^{-1}
centi	c	10^{-2}
milli*	m	10^{-3}
micro*	μ	10^{-6}
nano*	n	10^{-9}
pico*	p	10^{-12}
femto*	f	10^{-15}
atto*	a	10^{-18}

* Denotes preferred prefixes.

APPENDIX B

VOLUMES, CENTROIDS, AND MASS MOMENTS OF INERTIA

SHAPE	VOLUME	MOMENTS OF INERTIA
Uniform slender bar	$\mathscr{V} = L\mathscr{A}$	$I_{xx} = I_{zz} = \frac{1}{12}mL^2$ $I_{yy} = 0$
Rectangular parallelipiped	$\mathscr{V} = abc$	$I_{xx} = \frac{1}{12}m(b^2 + c^2)$ $I_{yy} = \frac{1}{12}m(a^2 + c^2)$ $I_{zz} = \frac{1}{12}m(a^2 + b^2)$

Uniform slender bar:

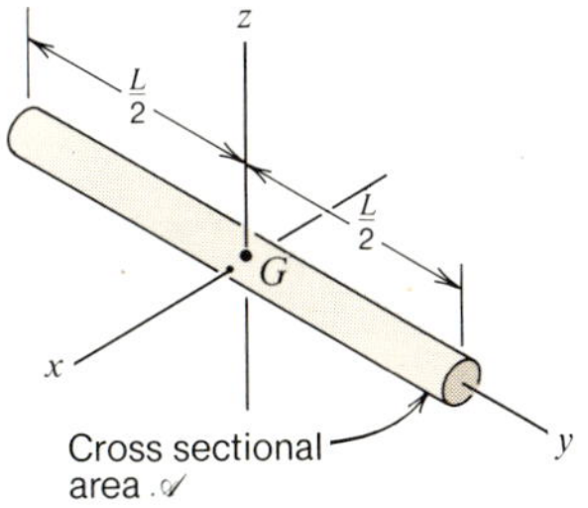

Rectangular parallelipiped:

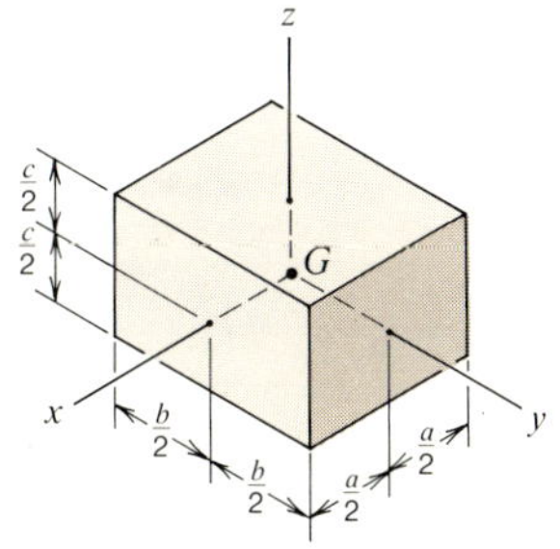

Cylinder

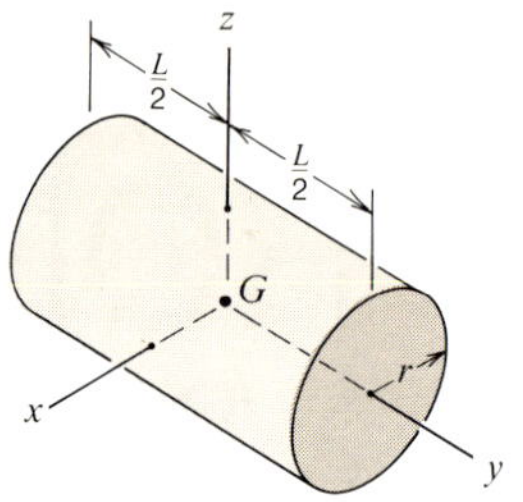

$\mathcal{V} = \pi r^2 L$

$$I_{xx} = I_{zz} = \tfrac{1}{12}m(3r^2 + L^2)$$
$$I_{yy} = \tfrac{1}{2}mr^2$$

Semicylinder

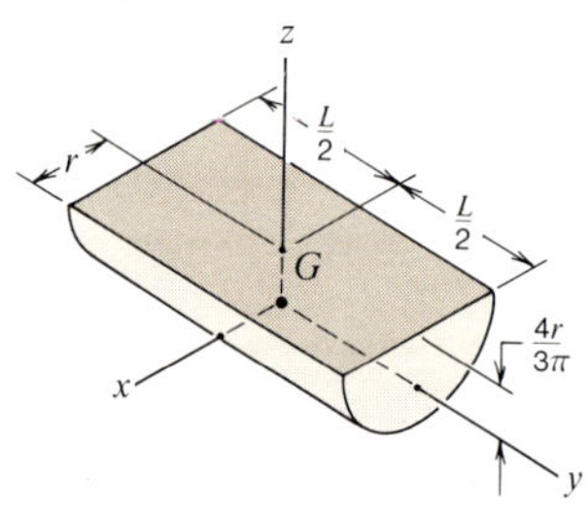

$\mathcal{V} = \tfrac{1}{2}\pi r^2 L$

$$I_{xx} = \left(\frac{9\pi^2 - 64}{36\pi^2}\right)mr^2 + \tfrac{1}{12}mL^2$$
$$I_{yy} = \left(\frac{9\pi^2 - 32}{18\pi^2}\right)mr^2$$
$$I_{zz} = \tfrac{1}{12}m(3r^2 + L^2)$$

Semicylindrical Shell

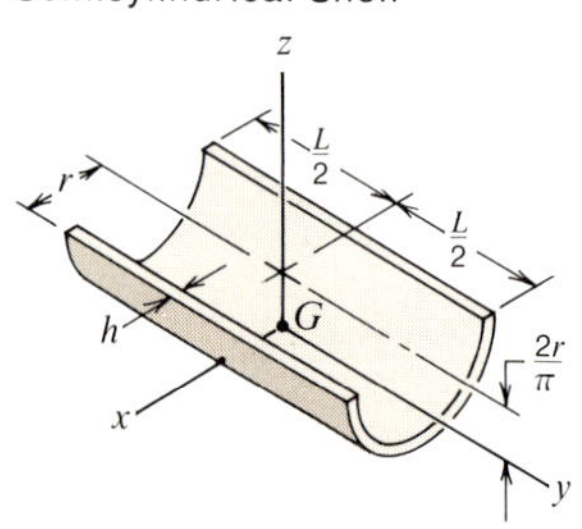

$\mathcal{V} = \pi r L h$

$$I_{xx} = \left(\frac{1}{2} - \frac{4}{\pi^2}\right)mr^2 + \tfrac{1}{12}mL^2$$
$$I_{yy} = \left(1 - \frac{4}{\pi^2}\right)mr^2$$
$$I_{zz} = \tfrac{1}{2}mr^2 + \tfrac{1}{12}mL^2$$

Sphere

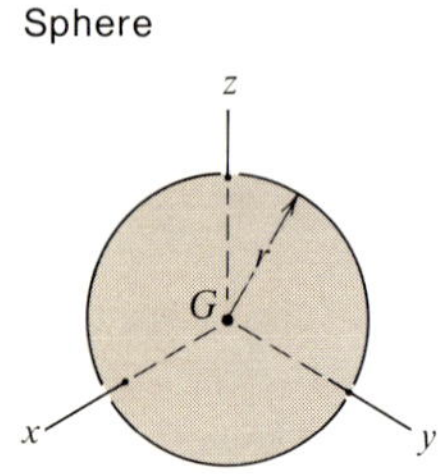

$\mathcal{V} = \tfrac{4}{3}\pi r^3$

$$I_{xx} = I_{yy} = I_{zz} = \tfrac{2}{5}mr^2$$

Hemisphere

$\mathcal{V} = \frac{2}{3}\pi r^3$

$I_{xx} = I_{zz} = \frac{83}{320} m r^2$

$I_{yy} = \frac{2}{5} m r^2$

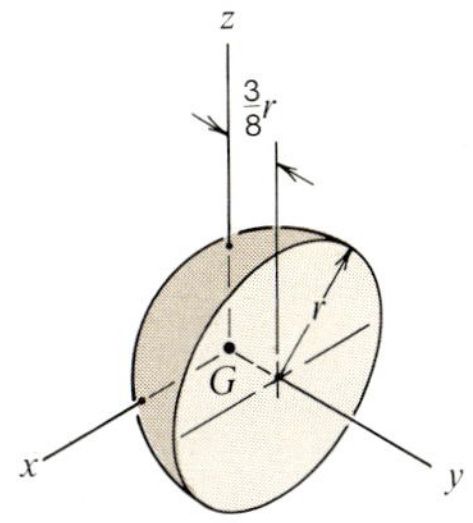

Hemispherical shell

$\mathcal{V} = 2\pi r^2 h$

$I_{xx} = I_{zz} = \frac{5}{12} m r^2$

$I_{yy} = \frac{2}{3} m r^2$

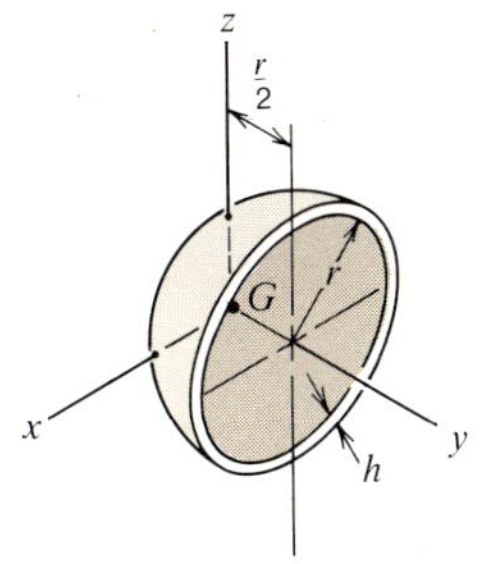

Semiellipsoid

$\mathcal{V} = \frac{2}{3}\pi abc$

$$I_{xx} = \frac{m}{320}(64b^2 + 19c^2)$$

$$I_{yy} = \frac{m}{320}(64a^2 + 19c^2)$$

$$I_{zz} = \frac{1}{5} m (a^2 + b^2)$$

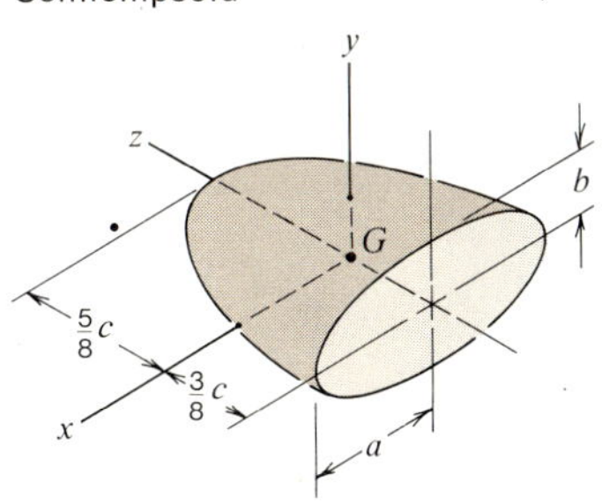

Cone

$\mathcal{V} = \frac{1}{3}\pi r^2 L$

$I_{xx} = I_{zz} = \frac{3}{80} m (4r^2 + L^2)$

$I_{yy} = \frac{3}{10} m r^2$

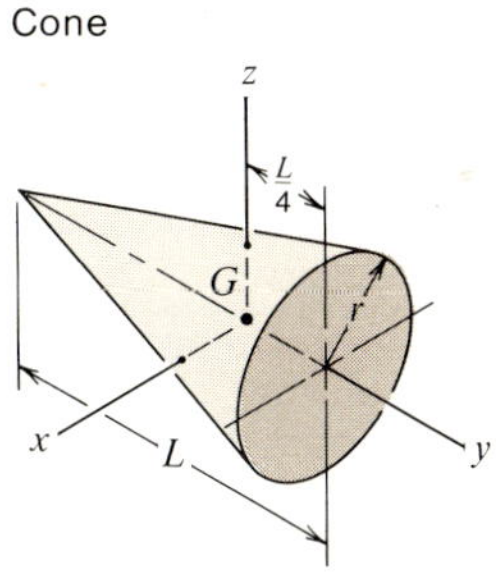

Semicone

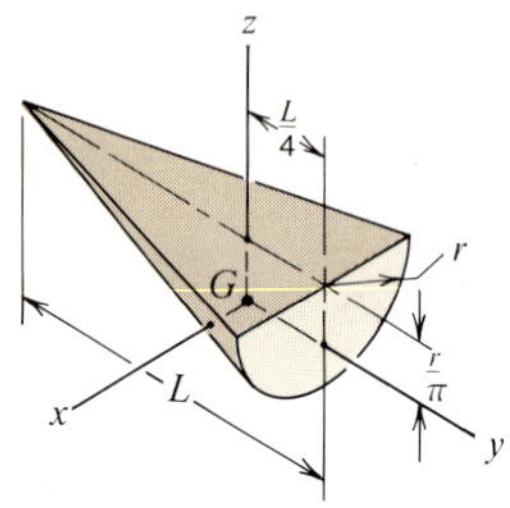

$$\mathcal{V} = \tfrac{1}{6}\pi r^2 L$$

$$I_{xx} = \left(\frac{3}{20} - \frac{1}{\pi^2}\right) m r^2 + \tfrac{3}{80} m L^2$$

$$I_{yy} = \left(\frac{3}{10} - \frac{1}{\pi^2}\right) m r^2$$

$$I_{zz} = \tfrac{3}{80} m(4r^2 + L^2)$$

$$I_{yz} = -\frac{1}{20\pi} m r L$$

Semiconical shell

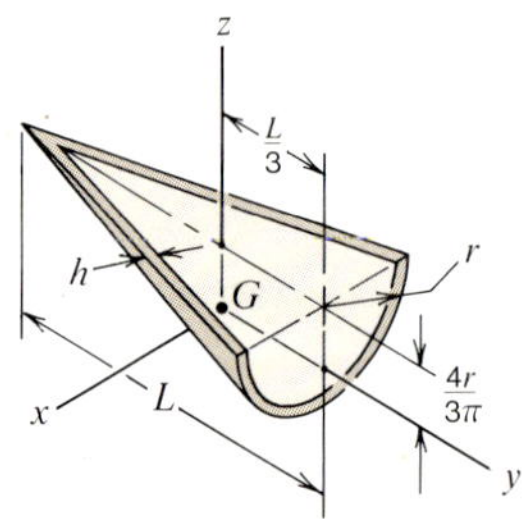

$$\mathcal{V} = \frac{\pi}{2} h r (r^2 + L^2)^{1/2}$$

$$I_{xx} = \left(\frac{1}{4} - \frac{4}{9\pi^2}\right) m r^2 + \tfrac{1}{18} m L^2$$

$$I_{yy} = \left(\frac{1}{2} - \frac{4}{9\pi^2}\right) m r^2$$

$$I_{zz} = \tfrac{1}{36} m(9r^2 + 2L^2)$$

$$I_{yz} = -\frac{1}{9\pi} m r L$$

Orthogonal tetrahedron

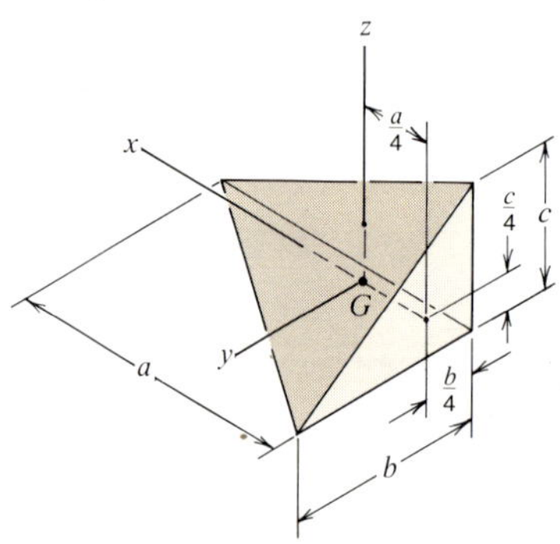

$$\mathcal{V} = \tfrac{1}{6} abc$$

$$I_{xx} = \tfrac{3}{80} m(b^2 + c^2)$$

$$I_{yy} = \tfrac{3}{80} m(a^2 + c^2)$$

$$I_{zz} = \tfrac{3}{80} m(a^2 + b^2)$$

$$I_{xy} = -\tfrac{1}{80} mab$$

$$I_{xz} = -\tfrac{1}{80} mac$$

$$I_{yz} = -\tfrac{1}{80} mbc$$

Right triangular prism

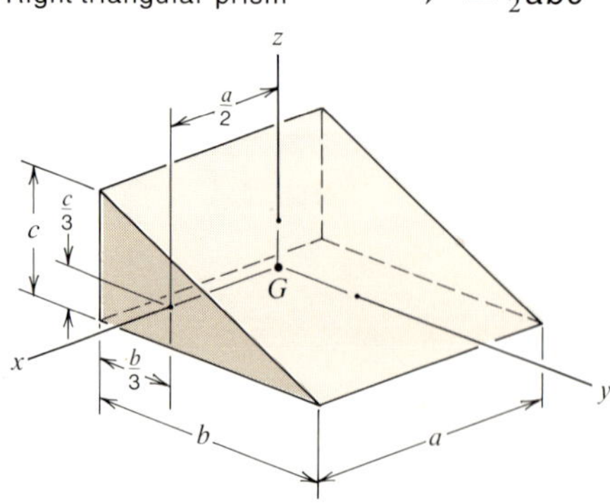

$$\mathcal{V} = \tfrac{1}{2} abc$$

$$I_{xx} = \tfrac{1}{36} m(3b^2 + 2a^2)$$

$$I_{yy} = \tfrac{1}{18} m(b^2 + c^2)$$

$$I_{zz} = \tfrac{1}{36} m(2a^2 + 3b^2)$$

$$I_{xz} = -\tfrac{1}{36} mac$$

Rectangular plate $\mathscr{V} = abh$

$I_{xx} = \frac{1}{12}mb^2$

$I_{yy} = \frac{1}{12}ma^2$

$I_{zz} = \frac{1}{12}m(a^2 + b^2)$

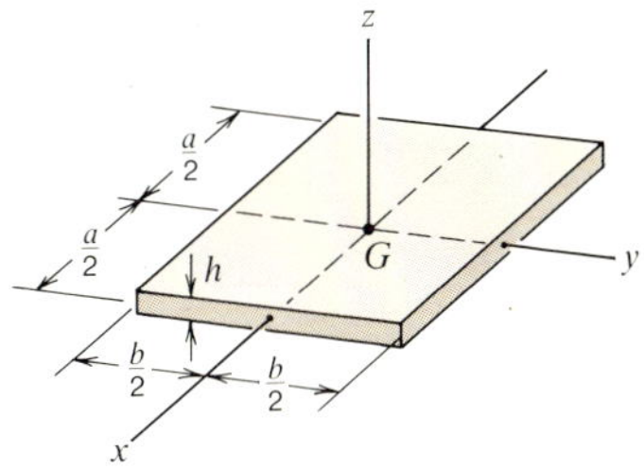

Triangular plate $\mathscr{V} = \frac{1}{2}bch$

$I_{xx} = \frac{1}{18}m(a^2 + b^2 - ab)$

$I_{yy} = \frac{1}{18}mc^2$

$I_{zz} = \frac{1}{18}m(a^2 + b^2 + c^2 - ab)$

$I_{xy} = \frac{1}{36}mc(2a - b)$

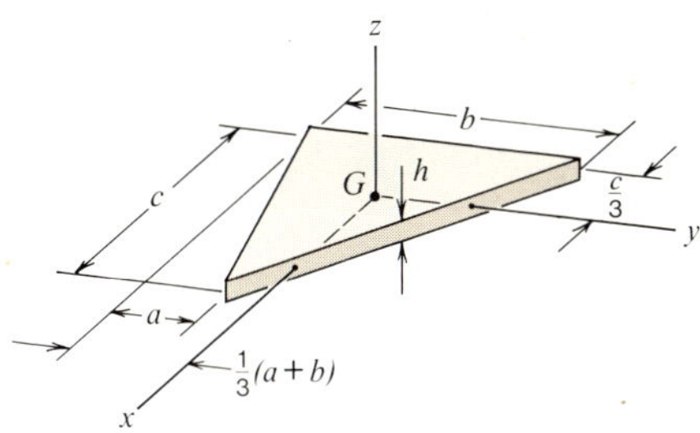

Thin circular ring $\mathscr{V} = 2\pi r\mathscr{A}$

$I_{xx} = I_{yy} = \frac{1}{2}mr^2$

$I_{zz} = mr^2$

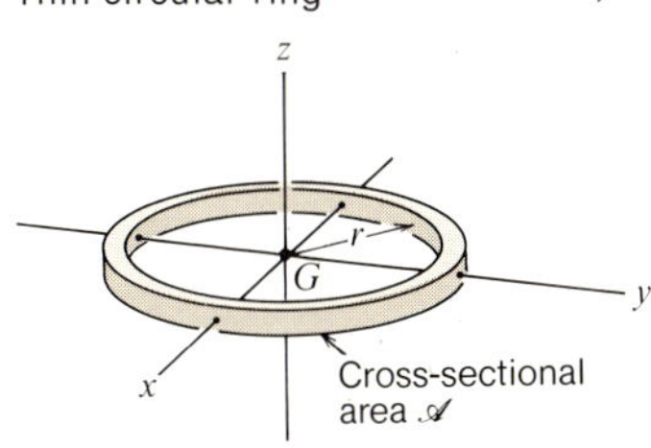

Circular plate $\mathscr{V} = \pi r^2 h$

$I_{xx} = I_{yy} = \frac{1}{4}mr^2$

$I_{zz} = \frac{1}{2}mr^2$

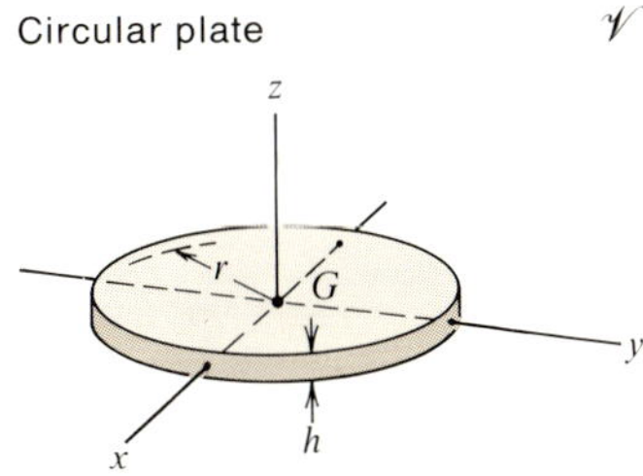

Quarter circular plate

$\mathscr{V} = \frac{\pi}{4} r^2 h$

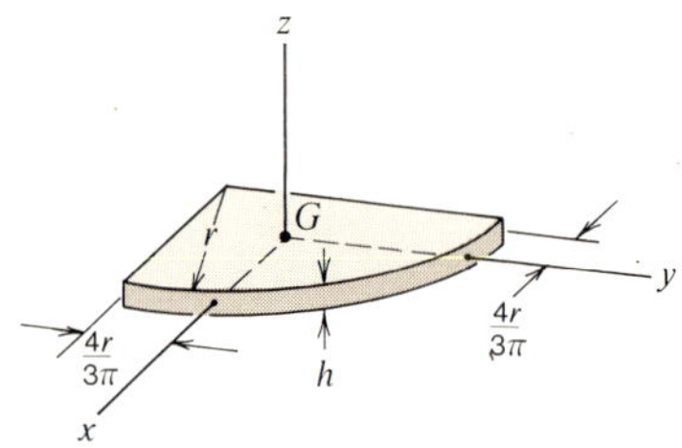

$$I_{xx} = I_{yy} = \left(\frac{9\pi^2 - 64}{36\pi^2}\right) mr^2$$

$$I_{zz} = \left(\frac{9\pi^2 - 64}{18\pi^2}\right) mr^2$$

$$I_{xy} = \left(\frac{9\pi - 32}{18\pi}\right) mr^2$$

Elliptical plate

$\mathscr{V} = \pi abh$

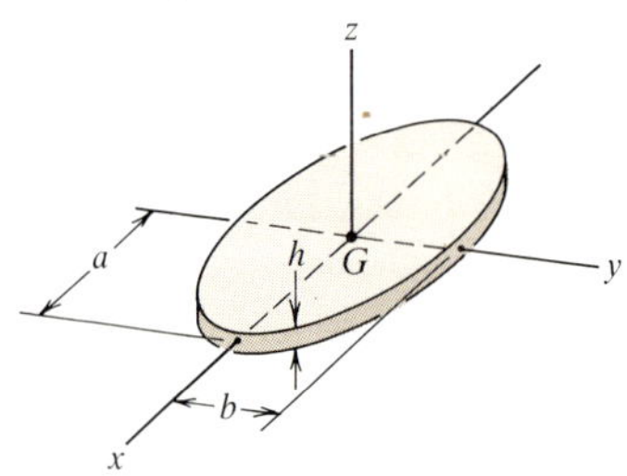

$$I_{xx} = \tfrac{1}{4} mb^2$$

$$I_{yy} = \tfrac{1}{4} ma^2$$

$$I_{zz} = \tfrac{1}{4} m(a^2 + b^2)$$

APPENDIX C

AREAS, CENTROIDS, AND AREA MOMENTS OF INERTIA

SHAPE	AREA	AREA MOMENT OF INERTIA FOR CENTROIDAL AXES ($J_{\hat{z}} = I_{\hat{x}} + I_{\hat{y}}$)
Rectangle	$\mathscr{A} = ab$	$I_{\hat{x}} = \frac{1}{12}ab^3$ $I_{\hat{y}} = \frac{1}{12}a^3b$

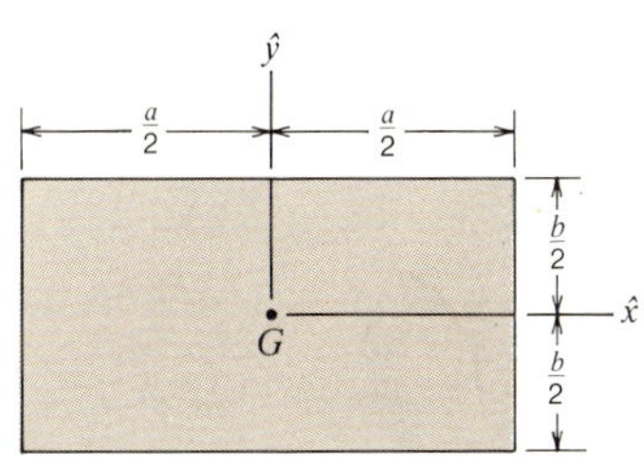

SHAPE	AREA	AREA MOMENT OF INERTIA FOR CENTROIDAL AXES
Right triangle	$\mathscr{A} = \frac{1}{2}ab$	$I_{\hat{x}} = \frac{1}{36}ab^3$ $I_{\hat{y}} = \frac{1}{36}a^3b$ $I_{\hat{x}\hat{y}} = -\frac{1}{72}a^2b^2$

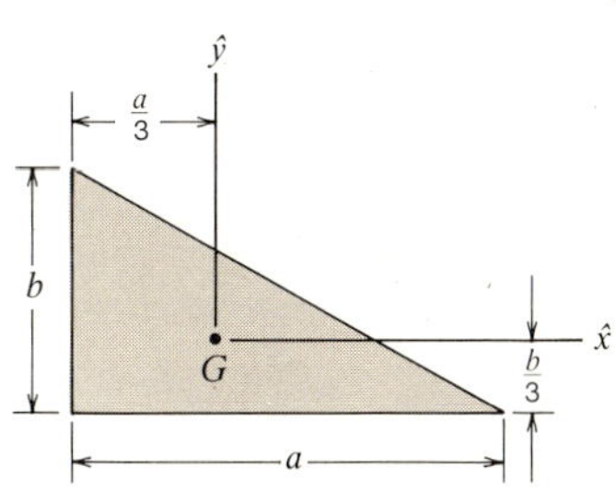

Scalene triangle

$\mathscr{A} = \frac{1}{2}ab$

$I_{\hat{x}} = \frac{1}{36}ab^3$

$I_{\hat{y}} = \frac{1}{36}ab(a^2 + c^2 - ac)$

$I_{\hat{x}\hat{y}} = \frac{1}{72}ab^2(2c - a)$

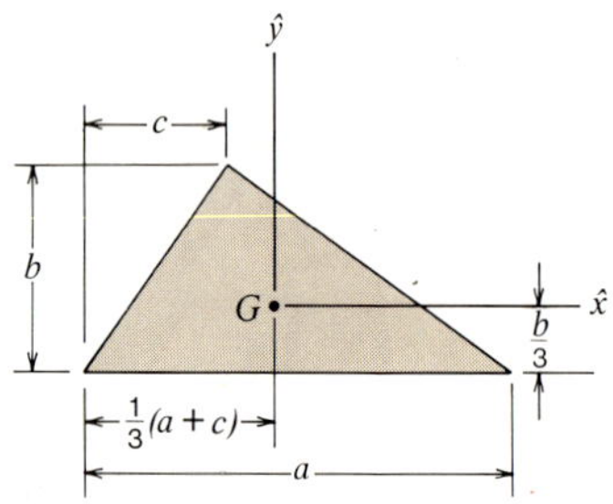

Circle

$\mathscr{A} = \pi r^2$

$I_{\hat{x}} = I_{\hat{y}} = \frac{1}{4}\pi r^4$

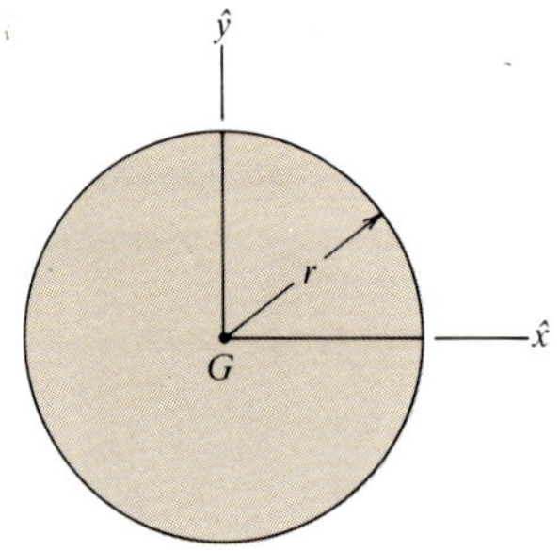

Quarter circle

$\mathscr{A} = \frac{1}{4}\pi r^2$

$$I_{\hat{x}} = I_{\hat{y}} = \left(\frac{9\pi^2 - 64}{144\pi}\right)r^4$$

$$I_{\hat{x}\hat{y}} = \left(\frac{9\pi - 32}{72\pi}\right)r^4$$

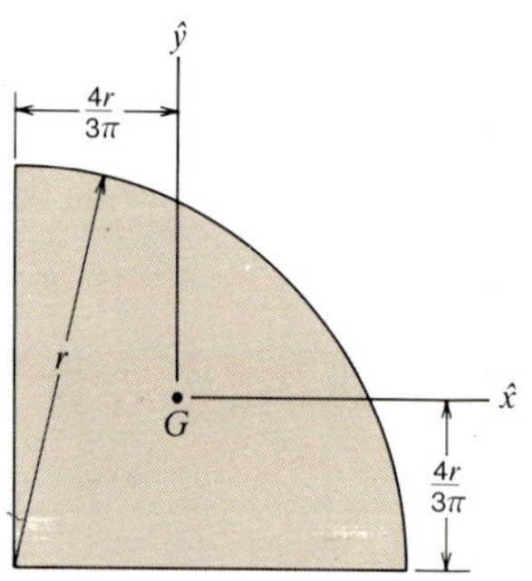

Circular sector

$\mathscr{A} = \frac{1}{2}\alpha r^2$

$I_{\hat{x}} = \frac{1}{8}(\alpha - \sin\alpha)r^4$

$$I_{\hat{y}} = \left[\frac{\alpha + \sin\alpha}{8} - \frac{4}{9\alpha}(1 - \cos\alpha)\right]r^4$$

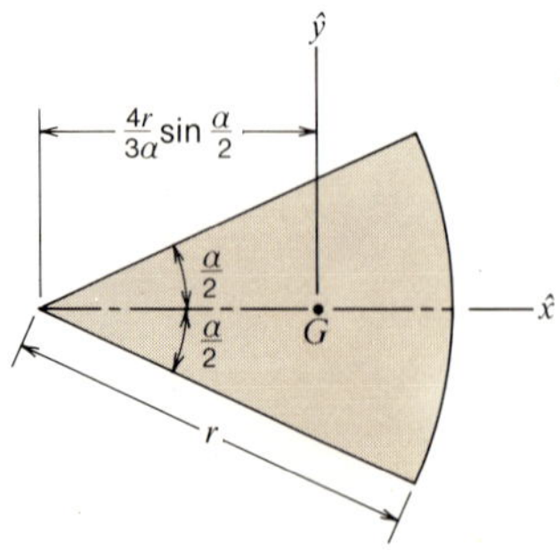

Circular arc $\mathscr{A} = \alpha r h$ $I_{\hat{x}} = \frac{1}{2}(\alpha - \sin\alpha) r^3 h$

$$I_{\hat{y}} = \left[\frac{\alpha + \sin\alpha}{2} - \frac{2}{\alpha}(1 - \cos\alpha)\right] r^3 h$$

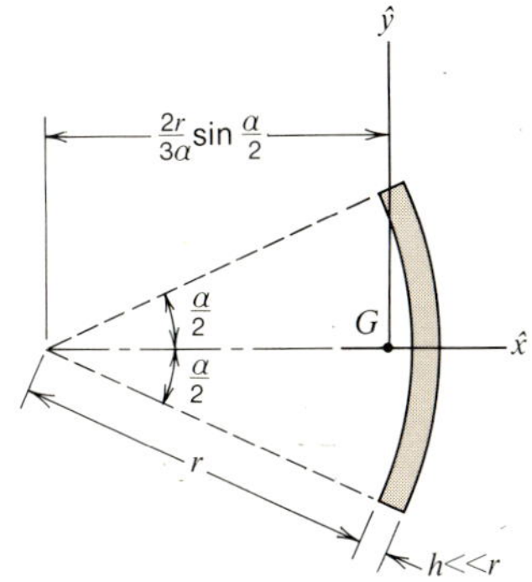

Ellipse $\mathscr{A} = \pi a b$ $I_{\hat{x}} = \frac{\pi}{4} a b^3$

$$I_{\hat{y}} = \frac{\pi}{4} a^3 b$$

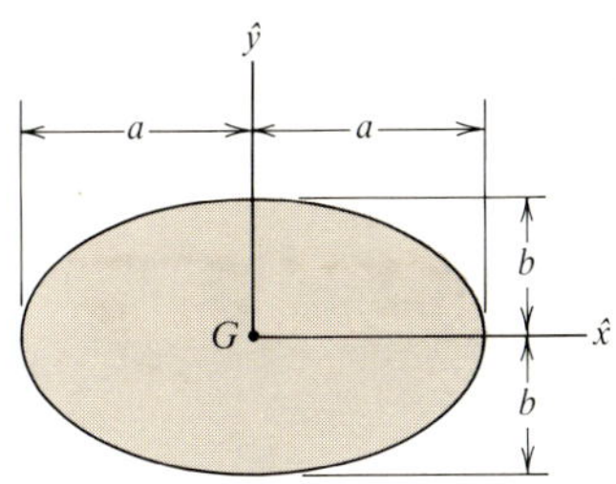

Quarter ellipse $\mathscr{A} = \frac{1}{4}\pi a b$ $I_{\hat{x}} = \left(\frac{9\pi^2 - 64}{144\pi}\right) a b^3$

$$I_{\hat{y}} = \left(\frac{9\pi^2 - 64}{144\pi}\right) a^3 b$$

$$I_{\hat{x}\hat{y}} = \left(\frac{9\pi - 32}{72\pi}\right) a^2 b^2$$

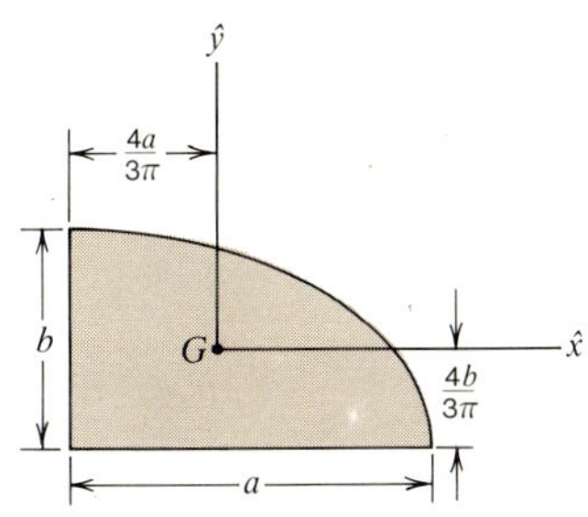

Parabolic section $\mathscr{A} = \frac{2}{3} a b$ $I_{\hat{x}} = \frac{8}{175} a b^3$

$$I_{\hat{y}} = \frac{19}{480} a^3 b$$

$$I_{\hat{x}\hat{y}} = -\frac{1}{60} a^2 b^2$$

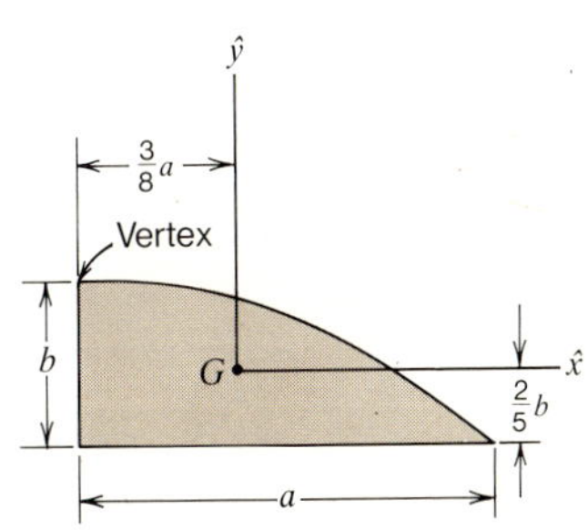

Parabolic spandrel $\mathscr{A} = \frac{1}{3}ab$

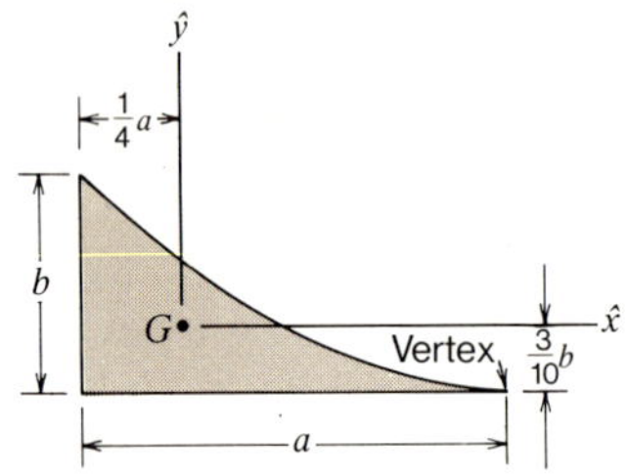

$I_{\hat{x}} = \frac{19}{1050}ab^3$

$I_{\hat{y}} = \frac{1}{80}a^3b$

$I_{\hat{x}\hat{y}} = -\frac{1}{120}a^2b^2$

APPENDIX D

DENSITIES

MATERIAL	SPECIFIC GRAVITY $(\rho/\rho_{water})^a$
Aluminum	2.69
Concrete (av.)	2.40
Copper	8.91
Earth (av. wet)	1.76
(av. dry)	1.28
Glass	2.60
Iron (cast)	7.21
Lead	11.38
Mercury	13.57
Oil (av.)	0.90
Steel	7.84
Water (fresh, liquid)	1.00
(ice)	0.90
Wood (soft pine)	0.48
(hard oak)	0.80

[a] $\rho_{water} = 1.00(10^3)$ kg/m^3.
$\gamma_{water} = \rho_{water} g = 62.4$ lb/ft^3.

ANSWERS TO EVEN-NUMBERED PROBLEMS

STATICS

1.2 $\bar{R} = 66.1$ lb at $33.41°$ left of vertical

1.4 $\bar{R} = 49.5$ lb at $28.52°$ left of vertical

1.6 $\bar{B} + \bar{C} = 21.50$ at $2.63°$ N of E,
$\bar{B} - \bar{C} = 16.61$ at $15.54°$ W of N

1.8 $\bar{A} + \bar{B} + \bar{C} = 23.3$ at $22.71°$ N of E
$\bar{A} - \bar{B} + \bar{C} = 9.17$ at 29.43 E of S

1.10 $\bar{A} + \bar{B} + \bar{C} + \bar{D} = 31.3$ at $32.52°$ E of N

1.12 $\bar{E} = 24.4$ at $10.64°$ E of S

1.14 (a) $1\ m^3 = 35.31\ ft^3$
(b) 1 km/hr = 0.9113 ft/s
(c) $1\ m/s^2 = 3.281\ ft/s^2$
(d) $1\ Pa = 0.02089\ lb/ft^2$

1.16 (a) $c_1 \sim 1$ (dimensionless)
(b) $c_1 \sim M/T^2$
(c) $c_1 \sim T$, $c_2 \sim T^4/L$
(d) $c_1 \sim$ degree/rad

1.18 (a) 1 td = 91.44 m
(b) 1 dash = 9 s
(c) 1 golf ball = 0.4504 N

1.20 $k = 26.3$ kN/m

1.22 $v \sim L^2/T$

1.24 $t_0 = 4$ s, $v_0 = 72.1$ ft/s

2.2 $F = 233$ lb, $\theta = 59.04°$

2.4 $\theta = 23.58°$, $F_H = 458$ lb (left)

2.6 $\alpha = 63.43°$, $P_H = 179$ N (left), $P_v = 358$ N (up)
or $\alpha = 116.57°$, $P_H = 179$ N (right), $P_v = 358$ N (up)

2.8 $F_{||} = 433$ lb (left and up), $F_{\perp} = 250$ lb (down and left)
$F_H = 465$ lb (left), $F_v = 183$ lb (down)

2.10 $\bar{F}_A = -\bar{F}_B = 80\bar{i} + 60\bar{j}$ lb ($\bar{i}$ right, $\bar{j}$ up)

2.12 $x = 0.6447$ m

2.14 $F_x = 223$ N, $F_y = -59.8$ N

2.16 $|\bar{Q}| = 20.6$ lb, $\alpha = 60.98°$, $\beta = 136.69°$, $T = 119.02°$

2.18 $\bar{F} = F(0.224\bar{i} + 0.837\bar{j} + 0.50\bar{k})$
$\alpha = 77.05°$, $\beta = 33.23°$, $\gamma = 60°$

2.20 (a) $\bar{F} = -242\bar{i} - 665\bar{j} + 707\bar{k}$ N
(b) $\alpha = 103.99°$, $\beta = 131.64°$, $\gamma = 45°$

2.22 $\beta = 133.25°$, $\gamma = 69.97°$

2.24 $\bar{F}_A = -\bar{F}_C = 33.3\bar{i} + 22.2\bar{j}$ lb
$\bar{F}_B = -\bar{F}_D = -33.3\bar{i} + 22.2\bar{j}$ lb

2.26 $\bar{F}_{AB} = -1.789\bar{i} + 0.894\bar{j}$ kN
$\bar{F}_{BC} = -2.4\bar{i} - 1.8\bar{j}$ kN ($\bar{i}$ right, $\bar{j}$ up)

2.28 $F_H = 183.5$ N (right), $F_v = 79.5$ N (down)

2.30 $\bar{F}_{AB} = -39.7\bar{i} - 139.1\bar{j} + 39.7\bar{k}$ lb
$\bar{F}_{CD} = -66.7\bar{i} - 116.7\bar{j} + 66.7\bar{k}$ lb

2.32 $F_{AB} = 244$ lb

2.34 $\bar{F}_C = 27.1\bar{i} - 24.4\bar{j} + 8.8\bar{k}$ N

2.36 $\bar{R} = 208$ lb at $51.67°$ down and right

2.38 $\bar{R} = 9.53$ tons at $10.46°$ up and right

2.40 $\alpha = 51.54°$, $\bar{R} = 5.44\bar{i}$ tons ($\bar{i}$ right)

2.42 (a) $F_3 = 93.8$ N, (b) $F_3 = -604$ N

2.44 $|\bar{F}| = 0.728$ kN, -5.41 kN

2.46 $\bar{R} = -1.85\bar{i} + 2.26\bar{j} - 3.02\bar{k}$ kN ($\bar{i} = \bar{e}_{C/B}$, $\bar{j}$ right)

2.48 $\bar{R} = -49.3\bar{j} - 186.9\bar{k}$ ($\bar{j} = \bar{e}_{D/C}$, $\bar{k}$ up)

2.50 $F_{AB} = 18.75$ lb, $F_{AC} = 15.31$ lb, $\bar{R} = 18.75$ lb (up)

2.52 $F_{AC} = 1.719$ kN, $F_{AD} = 1.613$ kN, $R = 0.957$ kN

2.54 $F_{AD} = 4.74$ kN, $F_{BD} = 9.13$ kN, $F_{CD} = 7.03$ kN

2.56 $\bar{F}_1 \times \bar{F}_2 = -41138\ \text{N}^2$

2.58 $R_n = -2.82$ kN

2.62 $\phi = 90°$

2.64 $\bar{F} \times \bar{e}_{B/D} = 384$ lb

2.66 $\bar{R}_{CD} = 215.9\bar{i} - 107.1\bar{j} + 78.6\bar{k}$ lb

2.68 $\bar{F}_n = 1250\bar{i} + 2165\bar{j} + 2500\bar{k}$ lb

2.70 $N_{left} = N_{right} = 1226$ N

2.72 $T = W/3$

2.74 $m_B = 10$ kg

2.76 $F = 97.6$ N, $\alpha = 61.12°$

2.78 $\alpha = 60°$, $\beta = 0°$, $F = 113.2$ N

2.80 $m_B = 6.93$ kg

2.82 $(N_A)_{wall} = 654$ N, $N_{AB} = 817$ N
$(N_B)_{floor} = 1471$ N, $(N_B)_{wall} = 654$ N

2.84 $F = 56.6$ N

2.86 $W_B = 72.7$ lb

2.88 $x = 15$ in.

2.90 $m_C = 0.0551$ kg

2.92 $W = 9.47$ lb

2.94 $F_{AD} = F_{BD} = F_{CD} = 378$ N

2.96 $F_{AB} = 0.746$ kN, $F_{AC} = 0.491$ kN, $F_{AD} = 0.368$ kN

2.98 $N_{left} = 8$ N, $N_{rear} = 6$ N, $N_{floor} = 27.1$ N

2.100 $F = 566$ N, $T = 836$ N

3.2 $(\bar{A} \times \bar{B}) \times \bar{C} = 22\bar{i} + 7\bar{j} - 2\bar{k}$
$\bar{A} \times (\bar{B} \times \bar{C}) = \bar{i} - 2\bar{j} - \bar{k}$

3.4 (a) $A_x = 7.5$, $B_z = 0.80$
(b) $A_x = 7.5$, $B_z = 0.80$

3.6 $\phi = 60.02°$, $\bar{e}_n = 0.433\bar{i} + 0.541\bar{j} + 0.721\bar{k}$

3.8 $z = 8.77$, $\bar{e}_n = 0.092\bar{i} - 0.925\bar{j} + 0.369\bar{k}$
or $z = -2.23$, $\bar{e}_n = 0.648\bar{i} - 0.480\bar{j} + 0.592\bar{k}$

3.10 $x_B = 1.367$ m, $\bar{e}_n = 0.344\bar{i} - 0.015\bar{j} + 0.939\bar{k}$

3.12 $\bar{M}_E = 4620$ lb-in. ccw

3.14 $\bar{M}_A = 27.5$ N-m ccw

3.16 (a) $\bar{M}_A = 16.07$ lb-ft cw
(b) $\bar{M}_A = 21.65$ lb-ft cw

3.18 $\bar{M}_A = 807$ lb-in. ccw

3.20 $\bar{M}_C = 4371$ lb-in. cw

3.22 $\theta = 53.13°$, $\bar{M}_C = 3.75$ kN-m cw
or $\theta = -126.87°$, $\bar{M}_C = 3.75$ kN-m ccw

3.24 $\bar{M}_A = -0.333\bar{k}$ kN-m
$\bar{M}_B = 0.166\bar{i} - 0.111\bar{j} - 0.333\bar{k}$ kN-m ($\bar{j}$ right, $\bar{k} = \bar{e}_{A/B}$)

3.26 $\bar{M}_B = -25.5\bar{i} - 17.0\bar{j} - 17.0\bar{k}$ kip-in. ($\bar{i} = \bar{e}_{D/C}$, $\bar{k} = \bar{e}_{B/C}$)

3.28 Cable AE: $\bar{M}_D = -5333\bar{j} + 10667\bar{k}$ lb-in.
Cable BE: $\bar{M}_D = 8944\bar{i} - 8944\bar{j} + 17888\bar{k}$ lb-in.

3.30 $\bar{M}_C = -0.36\bar{j} - 0.27\bar{k}$ kN-m
$\bar{M}_A = -0.18\bar{i} - 0.36\bar{j} - 0.27\bar{k}$ kN-m

3.32 $\bar{M}_A = -65.9\bar{i} + 152.1\bar{k}$ kN-m

3.34 $\bar{M}_{CD} = 8.96\bar{e}_{C/D}$ N-m

3.36 $\bar{M}_{CD} = 240(-\bar{i} + \bar{k})$ lb-in.

3.38 $\bar{M}_{AD} = -1.44\bar{j} + 1.08\bar{k}$ kN-m ($\bar{i} = \bar{e}_{A/B}$, $\bar{k}$ up)

3.40 $\bar{M}_{AB} = -19.2\bar{j} - 9.6\bar{k}$ kip-in. ($\bar{i} = \bar{e}_{D/C}$, $\bar{k} = \bar{e}_{B/C}$)

3.42 $\bar{M}_A = -8\bar{i} - 13.86\bar{j} + 10.39\bar{k}$ N − m ($\bar{j}$ parallel to right spoke, $\bar{k} = \bar{e}_{B/A}$) $r_A = 0.477$ m, $r_{AB} = 0.260$ m

3.44 $\bar{N}_A = 1.2$ kN down, $\bar{N}_B = 1.2$ kN up

3.46 $\bar{F}_A = 28.0$ N up, $\bar{F}_B = 28.0$ N down

3.48 $\bar{M} = 208$ N-m cw

3.50 $|\bar{M}| = 15.81$ kN-m, $\alpha = 71.57°$, $\beta = 161.57°$, $\gamma = 90°$

3.52 $\bar{M}_A = -0.20\bar{i} - 0.396\bar{k}$ N-m ($\bar{j}$ right, $\bar{k}$ up)

3.54 $\bar{F} = -1.20\bar{i} - 4.20\bar{j} + 3.92\bar{k}$ kips
$\bar{M} = -19.61\bar{i} - 6.36\bar{j} - 12.36\bar{k}$ kip-ft

3.56 $\bar{F} = 120\bar{i} - 160\bar{j}$ N, $\bar{M}_C = -134\bar{k}$ N-m ($\bar{i}$ right, $\bar{j}$ up)

3.58 $\bar{F} = 2\bar{i} - 3.46\bar{j}$ kips, $\bar{M} = -3.60\bar{k}$ kip-ft ($\bar{i}$ right, $\bar{j}$ up)

3.60 (a) and (b) $F = 300\bar{i} + 300\bar{j}$ N
(a) $\bar{M}_A = -78\bar{k}$ N-m (b) $\bar{M}_B = -108\bar{k}$ N-m
($\bar{i}$ right, $\bar{j}$ up)

3.62 $\bar{F} = 34.3\bar{i} - 5.18\bar{j}$ lb, $\bar{M} = 16.55\bar{k}$ lb-ft ($\bar{i}$ right, $\bar{j}$ up)

3.64 $F_{AC} = 172.9$ lb, $\bar{R} = -72\bar{i} - 522\bar{j}$ kN ($\bar{i}$ right, $\bar{j}$ up)

3.66 $\bar{F}_A = 400\bar{i}$ kN, $\bar{F}_B = -160\bar{i} - 320\bar{j}$ kN ($\bar{i}$ right, $\bar{j}$ up)

3.68 $\bar{F} = -250\bar{k}$ kN, $\bar{M} = -112.5\bar{i} - 250\bar{j}$ kN-m ($\bar{j}$ right, $\bar{k}$ up)

3.70 (a) and (b) $\bar{F} = 10\bar{i} + 20\bar{j} - 40\bar{k}$ kips
(a) $\bar{M}_A = -180\bar{j} - 120\bar{k}$ kip-in.
(b) $\bar{M}_B = -2400\bar{i} - 180\bar{j} - 720\bar{k}$ kip-in.
($\bar{j}$ right, $\bar{k}$ up)

3.72 $\bar{F} = -0.86\bar{i} - 6.15\bar{j} + 3.08\bar{k}$ kN
$\bar{M}_A = 5.53\bar{i} - 1.54\bar{j} - 1.54\bar{k}$ kN-m
($\bar{j} = \bar{e}_{B/A}$, $\bar{k} = \bar{e}_{D/A}$)

3.74 (a) and (b) $\bar{F} = -0.6\bar{i} + 2.5\bar{j} + 0.9\bar{k}$ kN
(a) $\bar{M}_A = -0.78\bar{k}$ kN-m, (b) $\bar{M}_B = -1.50\bar{i} - 0.36\bar{j} - 0.78\bar{k}$ kN-m ($\bar{j}$ right, $\bar{k}$ up)

3.76 $\bar{R} = 3$ kN down at 2.67 m from left end

3.78 $\bar{R} = 400$ N down at 0.45 m from left end

3.80 $\bar{R} = -10\bar{i} + 10\bar{j}$ N at 0.16m above A ($\bar{i}$ right, $\bar{j}$ up)

3.82 $\bar{R} = 0.18\bar{i} - 1.64\bar{j}$ at 3.33 m above A

3.84 $F = 49.4$ kN, $\alpha = 72.34°$

3.86 (a) $\bar{R} = 21.7$ lb at 64.03° below $\bar{e}_{B/A}$
(b) 9.41 in. right of A
(c) 13.30 in. right of D

3.88 $\alpha = -21.09°$, $\bar{R} = -9.33\bar{i} - 7.60\bar{j}$ kN ($\bar{i}$ right, $\bar{j}$ up)

3.90 $\bar{R} = 57$ kN down at 2.32 m forward and 2 m right of D

3.92 $\bar{R} = 12$ kips down at 0.722 ft back and 1.250 ft right of center

3.94 $\bar{F} = 100$ N down along front edge at 0.8 m right of 20 N force

3.116 $\bar{F}_A = -1.5\bar{i} + 3\bar{j}$ kN, $\bar{M}_A = 2.5\bar{k}$ kN-m ($\bar{i}$ left, $\bar{j}$ up)

3.118 $\bar{F}_A = -0.10\bar{i} - 11.95\bar{j}$ kN ($\bar{i}$ right, $\bar{j}$ up)

3.120 $m_B/m_A = (3 + \tan\theta)/(3 - \tan\theta)$

3.122 $T_1 = 6.83$ kN, $T_2 = 11.32$ kN, $\alpha = 83.66°$

3.124 $\bar{F}_A = 0.5\bar{j}$ N, $\bar{F}_B = -0.5\bar{j}$ N ($\bar{j}$ up)

3.126 $f = 36.7$ N up and left tangent to wheel, $N = 32.7$ N toward center

3.128 $T = 23.2$ kips, $\bar{F}_A = 22.4\bar{i} - 3\bar{j}$ kips ($\bar{i}$ right, $\bar{j}$ up)

3.130 $m = 29.2$ kg

3.132 $F_{tangent} = 86.6$ lb, $F_{normal} = 50$ lb

3.134 $F = 33.0$ N

3.136 (a) $T = 41.7$ lb, (b) $T = 166.7$ lb

3.138 $\bar{F}_A = 26.11$ lb (right)
$\bar{F}_B = 16.78°$ at 50° above left horizontal

3.140 $\bar{F}_A = 866\bar{i} + 600\bar{j}$ N ($\bar{i}$ right, $\bar{j}$ up)
$\bar{F}_B = 1000$ N at 30° below left horizontal

3.142 $\bar{F}_{upper} = -60\bar{i} - 58.9\bar{j}$ kN ($\bar{i}$ right, $\bar{j}$ up)
$\bar{F}_{lower} = 120$ kN at 60° above right horizontal

3.144 $\bar{F}_A = 120$ N left, $\bar{F}_B = 53.3$ N up, $T = 66.7$ N

3.146 $mgd \sin\theta - \text{k}R^2\theta = 0$

3.148 $N_A = 10$ lb on lower surface
$N_B = 13.33$ lb on upper surface
$N_C = 13.33$ lb on lower surface

3.150 $m_A/m_B = \cos(\alpha - \beta)/\cos(\alpha + \beta)$

3.152 $\cos\theta = (\sin^3\theta)L/(2R)$

3.154 $m = (2kL/g)\tan\theta$

3.156 $d = 0.5$ m

3.158 $\alpha = 30°$

3.160 $\bar{F}_A = -60\bar{i} - 80\bar{j} + 40\bar{k}$ kips
$\bar{M}_A = 307\bar{i} - 50\bar{j} + 360\bar{k}$ kip-ft
($\bar{j} = \bar{e}_{B/A}$, $\bar{k} = \bar{e}_{C/B}$)

3.162 $F_{BD} = F_{BE} = 147.1$ N

3.164 $\bar{F}_A = 206\bar{k}$ N, $\bar{F}_B = 55.3\bar{k}$ N, $\bar{F}_C = 130.8\bar{k}$ N ($\bar{k}$ up)

3.166 $T_{CE} = 79$ lb, $T_{CF} = 1300$ lb, $T_{DH} = 2167$ lb
$\bar{F}_A = 636\bar{i} + 3274\bar{j}$ lb
($\bar{i} = \bar{e}_{B/A}$, $\bar{j} = \bar{e}_{D/A}$)

3.168 $F = 77.38$ N, $\bar{F}_A = -52.3\bar{j} + 9.8\bar{k}$ N
$\bar{F}_B = 65.4\bar{i} + 65.4\bar{j} + 49.0\bar{k}$ N

3.170 $\bar{F}_A = -6\bar{i} + 8\bar{j} + 3\bar{k}$ kN
($\bar{i} = \bar{e}_{A/C}$, $k = \bar{e}_{A/B}$)

3.172 $F_t = 2$ lb, $\bar{F}_A = -2\bar{i} - 1.2\bar{j} + 3.4\bar{k}$ lb
($\bar{i}$ outward from wall, $\bar{k}$ up)

3.174 $\bar{F}_A = -37.5\bar{i} - 7.56\bar{j}$ N
$\bar{F}_C = 62.7\bar{i} - 10.1\bar{k}$ N
($\bar{i} = \bar{e}_{C/D}$, $\bar{j} = \bar{e}_{C/B}$)

3.176 $F = 12.26$ N

3.178 $T_{left} = 5.15$ kN, $T_{right} = 2.24$ kN
$\bar{F}_A = -1.75\bar{i} + 2.10\bar{j}$ kN, $\bar{F}_B = 2.10\bar{j}$ kN
($\bar{i}$ right, $\bar{k} = \bar{e}_{A/B}$)

3.180 improper constraint

4.2 $F_{AB} = 0.289$F (T), $F_{AC} = 0.577$F (C)
$F_{BC} = 1.155$F (T)

4.4 $F_{AB} = 41.6$ kN (T), $F_{BC} = 25.6$ kN (T)
$F_{BD} = 52.5$ kN (C), $F_{CD} = 16.0$ kN (C)

4.6 $F_{1-2} = 45$ kips (T), $F_{1-8} = 30$ kips (T)
$F_{2-3} = 22.5$ kips (T), $F_{2-7} = 30$ kips (T)
$F_{2-8} = 37.5$ kips (C), $F_{3-4} = 7.5$ kips (T)
$F_{3-6} = 20$ kips (T), $F_{3-7} = 25$ kips (C)
$F_{4-5} = 10$ kips (T), $F_{5-6} = 0$
$F_{6-7} = 7.5$ kips (C), $F_{7-8} = 22.5$ kips (C)

4.8 $F_{AG} = 37.7$ kN (T), $F_{BC} = 32$ kN (C)
$F_{BG} = 0$, $F_{CD} = 32$ kN (C)
$F_{CG} = 8$ kN (T), $F_{DE} = 12.8$ kN (C)
$F_{DF} = 8$ kN (C), $F_{DG} = 22.6$ kN (T)
$F_{EF} = 15.1$ kN (T), $F_{FG} = 12.8$ kN (T)

4.10 $F_{1-2} = 33.8$ kips (T), $F_{1-9} = 40.6$ kips (C)
$F_{2-3} = 33.8$ kips (T), $F_{2-9} = 0$
$F_{3-4} = 23.8$ kips (T), $F_{3-8} = 22.4$ kips (T)
$F_{3-9} = 0$, $F_{4-5} = 26.3$ kips (T)
$F_{4-7} = 5$ kips (C), $F_{4-8} = 5.6$ kips (T)
$F_{5-6} = 33.8$ kips (T), $F_{5-7} = 12.5$ kips (T)
$F_{6-7} = 40.6$ kips (C), $F_{7-8} = 31.6$ kips (C)
$F_{8-9} = 40.6$ kips (C)

4.12 $F_{AB} = 144$ kN (C), $F_{AH} = 187.4$ kN (T)
$F_{BC} = 144$ kN (C), $F_{BH} = 40$ kN (C)
$F_{CD} = 144$ kN (C), $F_{CF} = 31.2$ kN (T)
$F_{CG} = 120$ kN (C), $F_{CH} = 31.2$ kN (T)
$F_{DE} = 144$ kN (C), $F_{DF} = 0$
$F_{EF} = 187.4$ kN (T), $F_{FG} = 156.2$ kN (T)
$F_{GH} = 156.2$ kN (T)

4.14 $F_{1-2} = 16$ kN (T), $F_{1-8} = 7.21$ kN (T)
$F_{2-3} = 12$ kN (T), $F_{2-7} = 5.66$ kN (T)
$F_{2-8} = 8$ kN (C), $F_{3-4} = 8$ kN (T)
$F_{3-6} = 4.47$ kN (T), $F_{3-7} = 6$ kN (C)
$F_{4-5} = 8$ kN (T), $F_{4-6} = 4$ kN (C)
$F_{5-6} = 8.95$ kN (C). $F_{6-7} = 13.42$ kN (C)
$F_{7-8} = 17.89$ kN (C), $F_{8-9} = 22.4$ kN (C)

4.16 $F_{AB} = 66$ kN (T), $F_{AC} = 39.0$ kN (C)
$F_{AD} = 72.1$ kN (C), $F_{BC} = 39.0$ kN (C)
$F_{BD} = 72.1$ kN (C), $F_{CD} = 78.0$ kN (T)
$F_{DE} = 101.8$ kN (T)

4.18 $F_{AB} = 1.537$ kips (C), $F_{AC} = 1.044$ kips (C)
$F_{AD} = 1.537$ kips (C), $F_{BC} = 0.167$ kips (T)
$F_{BD} = 0.225$ kips (T), $F_{CD} = 0.167$ kips (T)

4.20 $F_{1-2} = 207$ lb (C), $F_{1-3} = 207$ lb (C)
$F_{1-4} = 75$ lb (T), $F_{2-3} = 75$ lb (T)
$F_{2-4} = 207$ lb (C), $F_{2-5} = 400$ lb (T)
$F_{3-4} = 207$ lb (C)

4.22 $F_{AB} = 45.0$ kN (T), $F_{AC} = 23.7$ kN (C)
$F_{AD} = 23.7$ kN (C), $F_{BC} = 7.5$ kN (T)
$F_{BD} = 7.5$ kN (T), $F_{BE} = 90$ kN (T)
$F_{BF} = 23.7$ kN (C), $F_{BG} = 23.7$ kN (C)
$F_{CD} = 0$, $F_{CF} = 22.5$ kN (C)
$F_{DF} = 0$, $F_{DG} = 22.5$ kN (C)

4.24 $F_{BC} = 0.333$ kN (T), $F_{BF} = 1.803$ kN (T)

4.26 $F_{CD} = 12$ kN (C), $F_{CE} = 11.31$ kN (C)

4.28 $F_{DE} = 0$ $F_{BF} = 23.1$ kN (T)

4.30 $F_{2-8} = 9.17$ kN (T), $F_{8-9} = 8.44$ kN (C)

4.32 $F_{3-4} = 23.8$ kips (T), $F_{3-8} = 22.4$ kips (T)
$F_{8-9} = 40.6$ kips (C)

4.34 $F_{5-11} = 354$ N (T), $F_{11-12} = 825$ N (T)

4.36 $F_{BH} = 4.33$ kips (C), $F_{BJ} = 0$

4.38 $F_{2-3} = 10.25$ kN (C), $F_{3-11} = 6.15$ kN (T)

4.40 $F_{5-8} = 15$ kN (C), $F_{7-10} = 5$ kN (C)

4.42 $F_{5-6} = 1.45$ kips (T), $F_{6-15} = 8.35$ kips (C)

4.44 $d = 5\text{m}{:}F_{CD} = 67.1$ kN (C), $F_{FG} = 35.0$ kN (T)
$d = 2.5$ m:improper constraint

4.46 $F_{5-6} = 7.50$ kN (C), $F_{6-8} = 6.00$ kN (T)

4.48 $F_{BF} = 0$, $F_{CG} = 21.3$ kN (T)

4.52 $\bar{F}_A = -333\bar{i} - 111\bar{j}$ lb, $\bar{F}_C = 333\bar{i} + 511\bar{j}$ lb
($\bar{i}$ right, $\bar{j}$ up)

4.54 $\bar{M}_{CD} = 500$ lb-ft ccw

4.56 On AC: $\bar{F}_A = 100\bar{i} + 200\bar{j}$ N, $\bar{F}_C = -100\bar{i} + 100\bar{j}$ N
($\bar{i}$ right, $\bar{j}$ up)

4.58 $B = 6.67P$, $|\bar{F}_A| = 7.49P$

4.60 $\bar{F}_A = 2$ kN up, $\bar{M}_A = 0.8$ kN-m ccw
$\bar{F}_B = 4$ kN up, $\bar{M}_B = 1.2$ kN-m cw

4.62 $\bar{F}_A = 520\bar{i} - 200\bar{j}$ lb, $\bar{F}_C = -520\bar{i} + 400\bar{j}$ lb
($\bar{i}$ right, $\bar{j}$ up)

4.64 $\bar{M} = 300$ lb-in. cw

4.66 $\bar{M}_C = 3450$ N-m cw

4.68 $T = 199.9$ N

4.70 On BD: $\bar{F}_B = 710\bar{i} - 753\bar{j}$ lb, $\bar{F}_D = -710\bar{i} + 313\bar{j}$ lb
($\bar{i}$ right, $\bar{j}$ up)

4.72 $\bar{F}_A = -1086\bar{i} - 857\bar{j}$ lb
$\bar{F}_B = 1917$ lb at 26.57° above i
$\bar{F}_C = -1029\bar{i}$ lb ($\bar{i}$ right, $\bar{j}$ up)

4.74 $\bar{F}_E = 2280\bar{i} + 455\bar{j}$ lb
$\bar{F}_F = -2280\bar{i} - 1785\bar{j}$ lb ($\bar{i}$ right, $\bar{j}$ up)

4.76 On AC: $\bar{F}_A = 362\bar{i} - 43\bar{j}$ N
$\bar{F}_B = 721$ N at 15° above $-\bar{i}$
$\bar{F}_C = 335\bar{i} - 143\bar{j}$ N
On CE: $\bar{F}_C = -335\bar{i} + 143\bar{j}$ N
$\bar{F}_D = 696\bar{i} - 387\bar{j}$ N
$\bar{F}_E = -362\bar{i} + 243\bar{j}$ N
($\bar{i}$ right, $\bar{j}$ up)

4.78 $T_{AE} = 23.3$ lb

4.80 $\bar{F}_C = 2$ kN up, $\bar{M}_C = 2.0$ kN-m ccw

4.82 $\bar{M}_2 = 2M_1(r_1 + r_2)/r_1$ cw
$\bar{M}_3 = M_1(r_1 + 2r_2)/r_1$ ccw

4.84 $\bar{F}_C = 7.17$ kN at 15° left of $\bar{j}$
$\bar{F}_D = 2.66\bar{i} - 2.46\bar{j}$ kN
($\bar{i}$ right, $\bar{j}$ up)

4.86 $\bar{F}_A = -1.667\bar{i} - 2.5\bar{j}$ kN
$\bar{F}_G = 1.667\bar{i} + 3.5\bar{j}$ kN
($\bar{i}$ right, $\bar{j}$ up)

4.88 $\bar{F} = 14.29$ N at 75° left of vertical
$\bar{F}_A = -50\bar{i} + 39.8\bar{j}$ N

4.90 $F_{spring} = 11.39$ kN (C)
$\bar{F}_A = -3.12\bar{i} - 1.25\bar{j}$ kN
$\bar{F}_C = -6.07\bar{i} + 5.75\bar{j}$ kN
($\bar{i}$ right, $\bar{j}$ up)

4.92 $F = 2 \sin\theta - \tan\theta$ kN

4.94 $F = 1.5P \tan\theta$

5.2 $\theta = 21.80°$

5.4 (a) $\bar{f} = 3.61$ N up the incline
(b) $\bar{f} = 9.95$ N down the incline

5.6 $\theta = 65.27°$

5.8 $m = 22.4$ kg

5.10 $m = 225$ kg

5.12 $\mu_B = 0.884$, block B tending to move down

5.14 $\bar{f}_A = 109.8$ N up, $\bar{f}_C = 32.1$ N right
$\mu_A = 3.42$, $\mu_B = 0.113$

5.16 All suitcases tend to slide together

5.18 $\bar{f} = 141.1$ N down, $\mu = 0.353$

5.20 $\bar{f}_A = 5.77$ N right, $\mu = 0.577$

5.22 $\bar{M} = 287$ lb-ft ccw

5.24 $\mu = 0.296$

5.26 (a) $\bar{f} = 384$ N down and right
$\bar{N} = 165$ N down and left
(b) $\mu = 2.33$

5.28 (a) $P = 265$ lb, (b) $f = 221$ lb up incline
(c) $\mu = 0.75$

5.30 $d = 1.543$ m

5.32 Is in static equilibrium

5.34 $\bar{P} = 153.9$ N up

5.36 $\bar{f} = 20$ lb to right, (b) $\mu = 0.693$

5.38 $\bar{M}_A = 38$ N-m cw

5.40 $P = 550$ N

5.42 Equilibrium if $28.40° < \theta < 44.05°$

5.44 $\mu_{left} = 0.333$, $\mu_{right} = 0.382$

5.46 $P = 58.8$ N

5.48 $h = 3.57$ ft: not possible

5.50 $P = 235$ N

5.52 $P = 475$ N

5.54 $P = 34.1$ N

5.56 $P = 22.4$ lb to left

5.58 (a) $d = 2$ m, (b) $P = 49.04$ N, (c) $P = 42.0$ N

5.60 $\beta > \tan^{-1}(\mu)$ for sliding
$\beta > \tan^{-1}(b/h)$ for tipping

5.62 $P = 1127$ N

5.64 $P = 165.8$ N

5.66 $P = 1947$ N

5.68 $P = 477$ N to right, self-locking

5.70 $P = 244$ newtons

5.72 (a) $\bar{F}_B = 6.31$ kN to right
(b) $\bar{F}_C = 2$ kN to left
$\bar{F}_A = 4.31$ kN to left

5.74 $P = 780$ N

5.76 self-locking

5.78 $\mu_A = 0.0124$, $\mu_B = 0.0875$

5.80 $\bar{P} = 219$ lb down

5.82 (a) $M = 1.842$ N-m, (b) $M = 1.047$ N-m

5.84 (a) $F = 5.58$ kips, (b) $F = 6.56$ kips

5.86 $M = 238$ lb-in.

5.88 $F_D = 273$ N

5.90 $M = 8.58$ N-m

5.92 (a) $\bar{M}_{AB} = 33.8\bar{i}$ lb-ft, (b) $\bar{M}_{AB} = -23.7\bar{i}$ lb-ft
($\bar{i}$ right)

5.94 3 turns

5.96 $m_B = 10.21$ kg

5.100 $\mu_1 = 0.533$

5.102 (a) $\bar{M}_A = 219$ lb-ft cw, (b) $\bar{M}_A = 164$ lb-ft cw

5.104 $M_A = 75$ N-m, $M_B = 300$ N-m, slipping at B first

5.106 $M_A = 24.1$ N-m, $M_B = 48.3$ N-m

6.2 $m = 8.83$ lb-s^2/ft, 0.769 ft above base

6.4 $m = 0.465$ lb-s^2/ft, 3.36 in. forward from back face

6.6 $m = 559$ lb-s^2/ft, (1.15 forward, 2.13 right, 1.83 up) ft from lower left rear corner

6.8 $m = 58.5$ kg, (0.310 right, 0.067 up) m from lower left edge

6.10 $m = 3.75$ lb-s^2/ft, (0.681 forward, 1.167 right, 0.380 up) ft from lower left rear corner

6.12 $m = 1.219(10^6)$ kg, (3.14 forward, 4.91 right, 3.93 up) m from lower left rear corner

6.14 $m = 87.6$ kg, (0.0118 left, 0.0141 up) m from center

6.16 $m = 3.39(10^4)$ kg, 2.35 m forward from right rear face, 0.56 m up from bottom

6.18 $m = 1.372(10^6)$ kg, (3.42 forward, 4.74 right, 3.79 up) m from lower left rear corner

6.20 $m = 92.5$ kg, (2.06 forward, 1.50 right, 1.065 up) ft from lower left rear corner

6.22 (1.288 left, 1.288 up) m from center of arc

6.24 $\mathscr{A} = 28.5$ in.^{2}, 2.97 in. down from top

6.26 $m = 15.05$ kg, (0.107 right, 0.283 up) m from lower left corner

6.28 $m = 4.06$ kg, (0.346 right, 0.177 up) m from lower left corner

6.30 $m = 28.5$ gram, (0.044 right, 0.254 up) m from lower left corner

6.32 $m = 1.494$ kg, (1.20 right, 0.53 up) m from lower left corner

6.34 $m = 74.2$ gram, (0.068 back, 0.025 right, 0.060 down) m from center of arc

6.36 $m = 271$ kg, 0.111 m up from bottom

6.38 $m = 0.110$ lb-s^2/ft, (2.61 right, 2.0 back, 0.35 up) from lower left front corner

6.40 $x_G = 3a/8$, $y_G = 2b/5$

6.42 $x_G = 1.614a$, $y_G = 0.708a$

6.44 $x_G = 4a/3\pi$, $y_G = 4b/3\pi$

6.46 $x_G = (2r/3\alpha)\sin\alpha$, $y_G = (2r/3\alpha)(1 - \cos\alpha)$

6.48 $x_G = 0$, $y_G = (2r/\alpha)\sin(\alpha/2)$

6.50 $x_G = 0.574a$, $y_G = 0.410a$

6.52 $x_G = y_G = 0$, $z_G = 2h/3$

6.54 $m = (3\pi/4)\rho_0 a^2 h$, $x_G = y_G = 0$, $z_G = 11h/15$

6.56 $x_G = z_G = 0$, $y_G = -a/2$

6.58 $x_G = a/4$, $y_G = b/4$, $z_G = c/4$

6.60
$$x_G = \frac{a}{2\pi(b/a)^2(3 - b/a)}\left\{3\left[\frac{\pi}{2} - \sin^{-1}\left(1 - \frac{b}{a}\right)\right] - \left(\frac{2b}{a} - \frac{b^2}{a^2}\right)^{1/2}\left(3 + \frac{b}{a} - 6\frac{b^2}{a^2} + 2\frac{b^3}{a^3}\right)\right\}$$
$$y_G = 0$$
$$z_G = \frac{b}{4}\left(\frac{8 - 3b/a}{3 - b/a}\right)$$

6.62 $x_G = -a/4$, $y_G = 0$, $z_G = \frac{5}{16}h$

6.64 $m = 0.505\rho_0 a^2/k$, $x_G = y_G = 0.330a$, $z_G = 0.323/k$

6.66 $x_G = y_G = z_G = a/2$

6.68 21.6 kN at 6 m from left support

6.70 576 N at 4 m from left support

6.72 1.9 kN at 4.86 m from left

6.74 9 kN at 6.75 m from right support

6.76 16 kN down at 2.67 m from left support

6.78 56.3 lb down at 0.52 ft left of left end

6.80 $\bar{R}_A = \bar{R}_B = 1.08$ kN up

6.82 $\bar{R}_A = 336$ N up, $\bar{R}_B = 240$ N up

6.84 $\bar{R}_A = 1.9$ kN up, $\bar{M}_A = 9.23$ kN-m ccw

6.86 $\bar{R}_B = 9$ kN up, $\bar{M}_B = 60.8$ kN-m cw

6.88 $\bar{R}_A = 10.67$ kN up, $\bar{R}_B = 5.33$ kN up

6.90 $\bar{R}_A = 59.3$ lb up, $\bar{R}_B = 2.9$ lb down

6.92 $R = q_0L/4$ at $L/5$ from right end

6.94 $R = 2q_0L/3$ at $3L/5$ from left end

6.96 $q_2/q_1 = 1$

6.98 $q_2/q_1 = 0.50$

6.100 $R = 25{,}900$ lb at 1.5 ft right, 1 ft down from top of pole

6.102 $R = 16{,}965$ lb at 2.55 ft right, 3 ft down from top of pole

6.104 $R = 5530$ lb at 2.13 ft right, 1.29 ft down from top of pole

6.106 $R = 9048$ lb, $T = 4800$ lb

6.108 $R = 134.8$ kN at 16.8 m from left edge

6.110 $R = 35.1$ kN at 3 m forward, 11.25 m right of left rear corner

6.112 $R = P_0ab/4$ at $x = 2a/3$, $y = 2b/3$

6.114 $R = 53.2$ kips at 4.62 ft along the wall from the bottom

6.116 $R = 25.3$ kips at 3.6 ft from the bottom

6.118 $\bar{R}_A = 15.16$ kips right, $\bar{R}_B = 10.11$ kips right

6.120 $T = 130$ lb

6.122 $F_{sp} = 25.2$ kN

6.124 $F_{sp} = 67.0$ kN

6.126 $T = 398$ lb

6.128 $R = 22.9$ kips at 3.30 ft from bottom

6.130 $\bar{R} = 35.3\bar{i} - 13.9\bar{j}$ kN ($\bar{i}$ right, $\bar{j}$ up)

6.132 $\bar{R} = 85.7$ kips at 38.15° below right horizontal

6.134 288 kN at 23.23° above right horizontal

6.136 $\bar{R}_A = 2995\bar{i} - 3850\bar{j}$ lb, $\bar{M}_A = -11{,}981\bar{k}$ lb-ft ($\bar{i}$ right, $\bar{j}$ up)

6.138 $T = 509$ lb

6.140 $F = 12.4$ kN

6.142 $F = 29.0$ kN

6.144 $\bar{R}_A = 1123$ lb left, $\bar{R}_B = 1764$ lb up

6.146 $\bar{R}_A = 2621$ lb left, $\bar{R}_B = 3262$ lb up

6.148 (a) $R = 196$ lb at 0.25 ft below the center of the circle
(b) $R = 1960$ lb at 0.025 ft below the center of the circle

6.150 (a) $R = 562$ lb at 0.833 ft above the bottom edge
(b) $R = 2246$ lb at 0.958 ft above the bottom edge

6.152 $P = 0.777$ kN

6.154 $P = 0.507$ kN

6.156 $T_A = T_B = 127$ lb, $T_C = 363$ lb

7.2 axial = 0, shear = 80 N, b.m. = 128 N-m

7.4 axial = −2 kN, shear = 1.732 kN, b.m. = 4.33 kN-m

7.6 axial = 6 kN, shear = 11 kN, b.m. = 11 kN-m

7.8 axial = −4.24 kN, shear = 1.061 kN, b.m. = 1.356 kN-m

7.10 axial = 0, shear = 7.32 N, b.m. = 14.64 N-m

7.12 axial = −5.50 kN, shear = 1.2 kN, b.m. = 1.6 kN-m

7.14 axial = −624 N, shear = 40 N, b.m. = 26.7 N-m

7.16 axial = −443 N, shear = 256 N, b.m. = 128 N-m

7.18 axial = −75 N, shear = 304 N, torsion = 20 N-m, b.m. = 226 N-m

7.20 axial = 6 kN, shear = 8.94 kN, torsion = 40 kN-m, b.m. = 37.7 kN-m

7.22 $V = Pa/L$ left of $\bar{P}$, $V = -Pb/L$ right of $\bar{P}$
$|M|_{max} = Pab/L$ at $\bar{P}$

7.24 $|V|_{max} = 23.3$ lb right of 30 lb load
$|M|_{max} = 70$ lb-ft at 30 lb load

7.26 $|V|_{max} = w_0L/2$ at right end
$|M|_{max} = w_0L^2/6$ at right end

7.28 $|V|_{max} = 264$ N at 0.7 m from left end
$|M|_{max} = 62.5$ N-m at 0.7 m from left end

7.30 $|V|_{max} = 750$ N at right end
$|M|_{max} = 980$ N at 2.45 m from left end

7.32 $|V|_{max} = 8$ kN right of 7 kN load
$|M|_{max} = 18$ kN-m at right end

7.34 $|V|_{max} = \mathrm{wL}$ at right end
$|M|_{max} = \mathrm{wL}^2/2$ at right end

7.36 $|V|_{max} = 23.3$ lb right of 30 lb load
$|\mathrm{M}|_{max} = 70$ lb-ft at 30 lb load

7.38 $|V|_{max} = 2867$ lb at 8 ft from left end
$|M|_{max} = 3735$ lb-ft at 8 ft from left end

7.40 $|V|_{max} = 264$ N at 0.7 m from left end
$|M|_{max} = 62.5$ N-m at 0.7 m from left end

7.42 $|V|_{max} = 1575$ lb at right end
$|M|_{max} = 5097$ lb-ft at 8.11 ft from left end

7.44 $|V|_{max} = 750$ N at right end
$|M|_{max} = 980$ N-m at 2.45 m from left end

7.46 gap in middle = 1 ft, right end = 4.5 ft,
$|M|_{max} = 22.5$ kip-ft

7.48 $h_c = 2.36$ ft, $T_{CD} = 8.52$ kN, $\theta_{CD} = 30.58°$

7.50 $h_c = 5.13$ m, $T_{AB} = 697$ N, $\theta_{AB} = 45°$

7.52 $T_{DE} = 9.05$ kN, $h_B = 1.754$ m, $h_D = 1.938$ m, $\ell = 11.88$ m

7.54 $d_B = 2.42$ ft, $d_D = 0.53$ ft, $T_{CD} = 91.6$ lb, $\theta_{DE} = 75.16°$

7.56 (a) $T_{max} = 389$ kN, $T_{min} = 333$ kN, (b) $x = 69.3$ m from right end

7.58 $h = 75$ ft, $T_{max} = 1.061(10^5)$ lb

7.60 $\bar{R} = 0.450(10^6)$ lb down

7.62 $T_{max} = 88.1$ kN, $T_{min} = 85.7$ kN, $\ell = 80.19$ m

7.64 $T_{max} = 6.31(10^4)$ lb, $\ell = 205.5$ ft

7.66 $h = 1.125$ m, $\ell = 50.76$ m

7.68 $h = 0.1605$ m, $\ell = 0.9188$ m

7.70 $T_{max} = 95.6$ lb, $T_{min} = 45.6$ lb, $\ell = 336.1$ ft

7.72 $\ell = 10.551$ m, $T_{min} = 10.79$ N, $T_{max} = 59.8$ N

7.74 $\ell = 80$ m (insignificant sag)

7.76 $h = 0.1226\ L$

7.78 $h = 0.173$ m, 0.328 m

8.2 $I_x = 1.827(10^{-3})\ \text{m}^4$

8.4 $I_x = 3520\ \text{in.}^4$

8.6 $I_y = 8.23(10^{-3})\ \text{m}^4$, $I_{xy} = 0$

8.8 $I_y = 7920\ \text{in.}^4$, $I_{xy} = 4676\ \text{in.}^4$

8.10 $J_z = \pi(r_2^4 - r_1^4)/8$, $k_z = [(r_2^2 + r_1^2)/2]^{\frac{1}{2}}$

8.12 $J_z = 3.20\ \text{m}^4$, $k_z = 1.352$ m

8.14 $I_x = I_y = \pi(r_2^4 - r_1^4)/16$, $I_{xy} = (r_2^4 - r_1^4)/8$

8.16 $I_x = 0.542\ \text{m}^4$, $I_y = 2.66\ \text{m}^4$, $I_{xy} = 0.969\ \text{m}^4$

8.18 $k_x = 0.1418$ m, $k_y = 0.1225$ m, $k_z = 0.1874$ m

8.20 $I_x = 6.67(10^{-5})\ \text{m}^4$, $J_z = 1.333(10^{-4})\ \text{m}^4$, $I_{xy} = 0$

8.22 $I_x = 1.440(10^{-2})\ \text{m}^4$, $J_z = 9.12(10^{-2})\ \text{m}^4$, $I_{xy} = 2.88(10^{-2})\ \text{m}^4$

8.24 $I_x = 0.994\ \text{m}^4$, $J_z = 2.59\ \text{m}^4$, $I_{xy} = 1.055\ \text{m}^4$

8.26 $I_x = 42.5\ \text{m}^4$, $J_z = 160.7\ \text{m}^4$, $I_{xy} = 54.8\ \text{m}^4$

8.28 $I_x = 7.87(10^{-3})\ \text{m}^4$, $J_z = 8.13(10^{-3})\ \text{m}^4$, $I_{xy} = 1.067(10^{-3})\ \text{m}^4$

8.30 $a = 2.47$ in., $c = 2.41$ in.

8.32 $\mathscr{A} = 0.25\ \text{m}^2$, $I_x = 2.25(10^{-2})\ \text{m}^4$, $I_{xy} = 0$, $I_{x'y'} = -5.0(10^{-3})\ \text{m}^4$

8.34 $I_x = 1.244(10^5)\ \text{in.}^4$, $k_x = 13.15$ in.

8.36 $I_x = 3.09\ \text{m}^4$, $k_x = 0.963$ m

8.38 $J_z = 3.49(10^5)\ \text{in.}^4$, $I_{xy} = 1.140(10^5)\ \text{in.}^4$

8.40 $J_z = 8.59\ \text{m}^4$, $I_{xy} = 3.36\ \text{m}^4$

8.42 $y_G = 1.565$ in., $I_x = 6.24(10^4)\ \text{in.}^4$

8.44 $y_G = 0.0816$ m, $I_x = 3.10(10^{-4})\ \text{m}^4$

8.46 centroid at 3.25 in. right, 4 in. up from lower left corner, $I_x = 63.6\ \text{in.}^4$, $I_y = 11.52\ \text{in}^4$, $J_z = 75.1\ \text{in.}^4$, $I_{xy} = 19.69\ \text{in.}^4$

8.48 centroid at 0.782 in. right, 1.282 in. up from lower left corner, $I_x = 3.96\ \text{in.}^4$, $I_y = 1.920\ \text{in.}^4$, $I_{xy} = -1.616\ \text{in.}^4$, $J_z = 5.88\ \text{in.}^4$

8.50 $I_{x'} = 11316\ \text{in.}^4$ $I_{y'} = 2900\ \text{in.}^4$, $I_{x'y'} = 727\ \text{in.}^4$

8.52 $I_{x'} = 0.1323\ \text{m}^4$, $I_{y'} = 0.1209\ \text{m}^4$, $I_{x'y'} = -0.0599\ \text{m}^4$

8.54 (a) $I_{x'} = 4724\ \text{in.}^4$, $I_{y'} = 9492\ \text{in.}^4$, $I_{x'y'} = -3543\ \text{in.}^4$
(b) $I_{x'} = 3827\ \text{in.}^4$, $I_{y'} = 10388\ \text{in.}^4$, $I_{x'y'} = -2734\ \text{in.}^4$

8.56 (a) $I_{x'} = 0.0642\ \text{m}^4$, $I_{y'} = 0.1711\ \text{m}^4$, $I_{x'y'} = 0.0308\ \text{m}^4$
(b) $I_{x'} = 0.0750\ \text{m}^4$, $I_{y'} = 0.1630\ \text{m}^4$, $I_{x'y'} = 0.0445\ \text{m}^4$

8.58 $I_{x'} = 0.267\ \text{m}^4$, $I_{y'} = 3.067\ \text{m}^4$, $I_{x'y'} = 0.200\ \text{m}^4$

8.60 $\tilde{\theta} = 25.10°$ ccw, $I_{\tilde{x}} = 11378\ \text{in.}^4$, $I_{\tilde{y}} = 2838\ \text{in.}^4$

8.62 $\tilde{\theta} = 21.90°$ cw, $I_{\tilde{x}} = 0.0560\ \text{m}^4$, $I_{\tilde{y}} = 0.1793\ \text{m}^4$

8.64 $\tilde{\theta} = 16.50°$ ccw, $I_{\tilde{x}} = 0.532\ \text{ft}^4$, $I_{\tilde{y}} = 22.56\ \text{ft}^4$

8.66 $\theta = 45°$, $I_{x'} = I_{y'} = 2.59(10^{-3})\ \text{m}^4$, $I_{x'y'} = 0.726(10^{-3})\ \text{m}^4$

8.68 $\theta = 22.5$ ccw, $I_{x'} = I_{y'} = 1.667\ \text{m}^4$, $I_{x'y'} = 1.414\ \text{m}^4$

9.2 $\delta U = (mgr_2 \sin\theta - kr_1^2\,\theta)\,\delta\theta$

9.4 $\delta U = -(1 + \cos^2\theta)(F\ell/2)\,\delta\theta$

9.6 $\delta U = -(F_2/2)[(1 + h^2/2\ell^2)/(1 - h^2/4\ell^2)^{\frac{1}{2}}]\delta h$

9.8 $\delta U = -[a/(4b^2 - a^2)^{\frac{1}{2}}](F_2 + 2F_3)\,\delta a$

9.10 $\delta U = -0.08\text{k}L^2\,\delta\theta$

9.12 $\theta = \sin^{-1}(k_1 r_1^2/mgr_2)$

9.14 (a) $M\cos(\theta/2) = (\text{m}/2)\cos\theta$, (b) $\theta = 65.07°$, (c) $M = \text{m}/2$

9.16 $F = (mg/2)\cot\theta$

9.18 $\theta = \sin^{-1}[2a/\ell)^{1/3}]$

9.20 $M = k\theta(r_1^2 r_3/r_2)$

9.22 $M_c = 3450$ N-m cw

9.24 $F_2/F_1 = 0.25\cot\theta$

9.26 $T = 0.5F[a/(L^2 - a^2)^{\frac{1}{2}}]$

9.28 $F = 706$ N

9.30 $k = P/[2\ell(2\cos\theta - 1)]$

9.32 $k = 200\cos\theta/[\sin\theta - \sqrt{2}\sin(\theta/2)]$ lb/ft

9.34 $P = 9.81\ \cos\theta/\cos(\theta/2)$ kN (C)

9.36 $\bar{B} = 983$ lb up

9.38 $\bar{M}_B = 1.60$ kN-m cw

9.40 $\bar{C} = 5.11$ kN up

9.42 $\sin(\theta/2) = 1/(4 - 2\ mg/kL)$

9.44 $\sin^3\theta = (2r/L)\cos\theta$

9.46 $k = (P/\ell)\cot\theta/(\cos\theta - 0.5)$

9.48 $k = 200\cos\theta/[\sin\theta - \sqrt{2}\sin(\theta/2)]$ lb/ft

9.50 $P = 2k\ell\,(\cot\theta)(\sqrt{2} - 2\sin\theta)$

9.52 $\theta = 90°$: unstable for all k
$\theta = \sin^{-1}(1 - 2mg/k\ell)$: stable if $k > mg/\ell$

9.54 (a) unstable, (b) stable, (c) neutral

9.56 $\theta = 0$: unstable, $\theta = 180°$: stable

9.58 $k = 6.46\ mga/\ell^2$: stable

9.60 $k = 1096$ N/m: stable

9.62 stable if $k > mga/2b^2$

ANSWERS TO EVEN-NUMBERED PROBLEMS DYNAMICS

10.2 (a) 1 km/hr = 0.911 ft/s
(b) 1 liter = 0.0353 ft^3
(c) 1 kg/m^2 = 0.00637 $lb\text{-}s^2/ft^3$

10.4 $m = 4378$ kg

10.6 (a) $a = 12\ m/s^2$
(b) $a = 43.2$ km/hr/s
(c) $a = 39.4\ ft/s^2$

10.8 (a) $c_1 \sim 1$
(b) $c_1 \sim M/T^2$
(c) $c_1 \sim 1/L$, $c_2 \sim T$, $c_3 \sim L/T^2$
(d) $c_1 \sim$ degrees/radians
(e) $c_1 \sim 1/T^2L$, $c_2 \sim 1/T^3$
(f) $c_1 \sim T/L^{1/2}$
(g) $c_1 \sim L/T^{1/2}$

10.10 1 basket = 0.399 kg

10.12 $\bar{r}_{B/A} = -24\bar{\imath} + 32\bar{\jmath}$ m

10.14 $\bar{r}_{B/A} = 3\bar{\imath} + 4\bar{\jmath} - 2.89\bar{k}$ in.

10.16 $\bar{e}_{B/A} = 0.520\bar{\imath} + 0.693\bar{\jmath} - 0.50\bar{k}$
$\alpha = 58.7°$, $\beta = 46.2°$, $\gamma = 120°$

10.18 $\bar{B} \cdot \bar{C} = 6$, $\bar{A} \cdot (\bar{B} + \bar{C}) = -14$
$(\bar{A} - 2\bar{B}) \cdot \bar{C} = -24$

10.20 $(\bar{A} \times \bar{B}) \times \bar{C} = -54\bar{\imath} + 24\bar{\jmath} + 12\bar{k}$

10.20 (*Continued*)
$\bar{A} \times (\bar{B} \times \bar{C}) = -52\bar{\imath} + 26\bar{\jmath} + 14\bar{k}$
$\bar{A} \cdot (\bar{B} \times \bar{C}) = -12$
10.22 $\bar{M}_C = 66.7\bar{\imath} + 106.7\bar{\jmath} + 196.6\bar{k}$ N-m
10.24 $\bar{R} = -892\bar{\imath} - 653\bar{\jmath} + 392\bar{k}$ lb
$\bar{M}_D = (-0.631\bar{\imath} - 0.642\bar{\jmath} - 1.785\bar{k})(10^5)$ lb-in.
($\bar{\jmath}$ right, $\bar{k}$ up)
10.26 (a) $\bar{R} = (-0.554\bar{\imath} + 0.185\bar{\jmath} - 1.139\bar{k})(10^4)$ lb
$\bar{M}_A = (-2.55\bar{\imath} + 6.0\bar{\jmath} + 2.22\bar{k})(10^4)$ lb-ft
(b) $\bar{M}_{AE} = -2.55(10^4)\bar{\imath}$ lb-ft
10.28 $\bar{F}_n = 362\bar{\imath} + 362\bar{\jmath} + 482\bar{k}$ N
$\bar{F}_t = 138\bar{\imath} + 504\bar{\jmath} - 482\bar{k}$ N
10.30 $\bar{r} = 6.31\bar{I} + 4.92\bar{J}$ m
10.32 $r_X = 7.75$ ft, $r_y = 6.25$ ft
10.34 $\bar{v} = -7.9\bar{\imath} - 18.0\bar{\jmath}$ m/s
10.36 $\bar{v}_{2-3} = 3152\bar{\imath} - 1522\bar{\jmath}$ m/s
$\bar{a} = 9912\bar{\imath} - 1512\bar{\jmath}$ m/s^2
10.38 $\bar{a} = -1.1\bar{\imath} - 0.7\bar{\jmath}$ m/s^2
11.2 $t = 2v_0/g$
11.4 $a = -1076$ ft/s^2, $t = 0.1364$ s
11.6 (a) $t = 6$ s, (b) $s = 180$ ft (right)
(c) $d = 225$ ft
11.8 $\bar{v}_0 = 31.25$ m/s east
$\bar{v}_f = 18.75$ m/s west
11.10 $a = 0.3$ m/s^2, $v_f = 29$ m/s
11.12 $t_{min} = 252$ s
11.14 $\bar{r} = 388\bar{\imath} + 98.9\bar{\jmath}$ m
$\bar{v} = -7.6\bar{\imath} + 29.6\bar{\jmath}$ m/s
11.16 $\bar{v} = 39.7\bar{\imath} - 29.8\bar{\jmath}$ ft/s
($\bar{\imath}$ right, $\bar{\jmath}$ up)
11.18 $\bar{r}_{C/A} = \bar{0}$, $\bar{v} = 400\bar{\imath}$ mm/s
11.20 $\dot{v} = 7.76$ m/s^2
11.22 $\theta_f = 45.8°$ N of E, $v = 40$ m/s
$\bar{a}_f = -2\bar{e}_t + 1.067\bar{e}_n$ m/s^2
11.24 $s = 400$ ft, $v = 60$ ft/s,
$\bar{a} = 4.5\bar{e}_n$ ft/s^2
11.26 $\rho = 65.0$ m
11.28 $v_0 = (8g\ell)^{1/2}$
11.30 $s = 0.125$ ft, $\bar{v} = 1.0\bar{e}_t$ ft/s
$\bar{a} = 0.667\bar{e}_n$ ft/s^2
11.32 (a) $\bar{a} = 0.01g\,\bar{e}_n$
(b) $\bar{a} = -0.1g\,\bar{e}_t$
(c) $\bar{a} = -0.05g\,\bar{e}_t + 0.0075g\,\bar{e}_n$
11.34 (a) $\dot{v} = 1257$ m/s^2,
$\bar{a} = 1257\bar{e}_t + 987\bar{e}_n$ m/s^2
11.36 $s = 1.25$ km
11.38 $\bar{a} = -4.90\bar{e}_t + 0.87\bar{e}_n$ m/s^2
11.40 $\bar{v} = 0.73\bar{\imath} - 2.91\bar{\jmath}$ ft/s
$\bar{a} = -0.249\bar{\imath} - 0.062\bar{\jmath}$ ft/s^2
11.42 $\bar{v} = -24.5\bar{\imath} - 12.3\bar{\jmath}$ m/s
$\bar{a} = -4.81\bar{\imath} + 3.60\bar{\jmath}$ m/s^2
11.44 $\bar{r}_{P/O} = -91.9\bar{\imath} - 91.9\bar{\jmath} + 97.5\bar{k}$ mm
$\bar{v} = 31.7\bar{\imath} - 38.5\bar{\jmath} + 3.7\bar{k}$ m/s
$\bar{a} = 14862\bar{\imath} + 12185\bar{\jmath} - 134\bar{k}$ m/s^2
11.46 $\bar{r} = 4\bar{\imath} + 3\bar{\jmath} - 1.2\bar{k}$ ft
$\bar{v} = 5.94\bar{\imath} - 7.92\bar{\jmath} + 1.39\bar{k}$ ft/s
$\bar{a} = -20.2\bar{\imath} - 5.74\bar{\jmath} + 17.78\bar{k}$ ft/s^2
11.48 $\bar{v} = 16.64$ m/s at 32.7° above x axis
$\bar{a} = 6.33$ m/s^2 at 71.6° above x axis
11.50 $xy = 16$, $\bar{v} = 1.36\bar{\imath} - 2.94\bar{\jmath}$ ft/s
$\bar{a} = 0.680\bar{\imath} + 1.472\bar{\jmath}$ ft/s^2
11.52 $\bar{v} = 0.544\bar{\imath} - 0.237\bar{\jmath}$ m/s
$\bar{a} = -4.93\bar{\imath} - 2.77\bar{\jmath}$ m/s^2
11.54 $\bar{a} = -\rho^2\bar{r}$
11.56 $\bar{v} = 6.00\bar{\imath} + 5.40\bar{\jmath} + 6.36\bar{k}$ m/s
$\bar{a} = 180\bar{\imath} + 162\bar{\jmath} + 159.1\bar{k}$ m/s^2
11.58 $\bar{v} = -0.747\bar{e}_R + 0.169\bar{e}_\phi$ m/s
$\bar{a} = -0.495\bar{e}_R - 0.747\bar{e}_\phi$ m/s^2
11.60 $\bar{v} = 5.34\bar{e}_R + 4.19\bar{e}_\phi$ ft/s
$\bar{a} = -67.0\bar{e}_R + 170.7\bar{e}_\phi$ ft/s^2
11.62 $\bar{v} = 20t\bar{e}_R + 150t^4\bar{e}_\phi + 8t^3\bar{k}$ m/s
$\bar{a} = (20 - 2250t^6)\bar{e}_R + 900t^3\bar{e}_\phi + 24t^2\bar{k}$ m/s^2
$R = 3.42\phi^{2/3}$, $z = 0.234\phi^{4/3}$
11.64 $\dot{\phi} = 0.148$ rad/s
$\bar{v}_B = 0.5\bar{e}_R + 6.67\bar{e}_\phi$ m/s
11.66 $\dot{R} = 9.40$ ft/s, $\ddot{R} = 23.1$ ft/s^2
$\dot{\phi} = 2.28$ rad/s, $\ddot{\phi} = -37.1$ rad/s^2
11.68 $\bar{a}_B = 3.9\bar{e}_R - 0.84\bar{e}_\phi - 3.84\bar{k}$ m/s^2
11.70 $v = 150$ ft/s
11.72 $v = 20$ m/s, $\dot{v} = 0$, $\rho = 40.8$ m
11.74 $\bar{v} = 64\bar{e}_t$ in./s
11.76 $\bar{v} = 64\bar{e}_t$ in./s
11.78 $a_t = 2.16(10^7)$ m/s^2
$a_n = 1.08(10^{10})$ m/s^2
$\rho = 3$ m
11.80 $\ddot{R} = 1.042$ ft/s^2
$\ddot{\phi} = -1.179$ rad/s^2
11.82 $v_A = 5$ m/s and $\dot{v}_A = 3$ m/s^2
or $v_A = 20$ m/s and $\dot{v}_A = 0$
11.84 $a = 60$ ft/s^2, $t = 1$ s
11.86 $|\bar{a}| = 1.589$ m/s^2
11.88 $\dot{v} = 104$ m/s^2
11.90 $xy = 4$, $\bar{v} = 6\bar{\imath} - 0.296\bar{\jmath}$ m/s
$\bar{a} = 2\bar{\imath} + 0.296\bar{\jmath}$ m/s^2
11.92 $\bar{v} = 20\bar{e}_R + 0.16\bar{e}_\phi$ m/s
$a = 125.7$ m/s^2
11.94 $\bar{v} = 50\bar{\imath} - 28.9\bar{\jmath}$ mm/s
$\bar{a} = -25\bar{\imath} + 14.4\bar{\jmath}$ mm/s^2
11.96 $\dot{R} = 10.21$ m/s, $\dot{\theta} = -0.0144$ rad/s
11.98 $v = 6.28$ m/s,
$\bar{a} = -29.4\bar{e}_t + 32.9\bar{e}_n$ m/s^2
11.100 (a) $a = -k_1x - k_2x^3$
(*b*) $x = 0$, $v = v_0$, $a = 0$
11.102 $\bar{v} = 50.8\bar{\imath} - 36.9\bar{\jmath} - 19.4\bar{k}$ mm/s
$\bar{a} = -11.6\bar{\imath} - 16.0\bar{\jmath} - 9.4\bar{k}$ mm/s^2

11.104 $\dot{\phi} = -0.5$ rad/s, $\ddot{\phi} = 0$
11.106 $\dot{v} = -1.528$ m/s^2, $\rho = 23670$ km
11.108 $\ddot{R} = 9.88$ m/s^2, $\dot{\phi} = 0$
11.110 (a) $\bar{a} = \bar{0}$
(b) $\bar{a} = 20\bar{e}_t + 0.00444(s^2 - 85.6s + 1780)\bar{e}_n$ in./s^2
(c) $\bar{a} = 20\bar{e}_t + 0.444(s - 35.6)\bar{e}_n$ in./s^2
11.112 $\bar{v} = -4.89\bar{\imath} - 11.53\bar{\jmath}$ m/s
$\bar{a} = -208\bar{\imath} + 170\bar{\jmath}$ m/s^2
($\bar{\imath}$ right, $\bar{\jmath}$ up)
11.114 $\bar{v} = 15.7\bar{\imath} - 2.3\bar{\jmath}$ ft/s
$\bar{a} = -29.0\bar{\imath} - 123.7\bar{\jmath}$ ft/s^2
11.116 $v = 18.0$ ft/s, $a_t = -183.8$ ft/s^2
$\rho = 1.50$ ft
11.118 $\bar{v} = 4.17\bar{e}_t$ m/s, $\bar{a} = 23.2\bar{e}_n$ m/s^2
11.120 $\bar{v} = 50\bar{\imath} - 10\bar{\jmath}$ mm/s,
$\bar{a} = -20\bar{\imath}$ mm/s^2
($\bar{\imath}$ right, $\bar{\jmath}$ up)
11.122 $\bar{a} = -(A^2/R^3)\bar{e}_R$
11.124 $v = 21.4$ m/s, $\dot{v} = 0.2$ m/s^2
$|\bar{a}| = 0.258$ m/s^2
11.126 $\bar{v}_{air} = 38.6$ km/hr, 11.51° W of N
11.128 $v_{boat} = 3.75$ ft/s, $v_{water} = 6.25$ ft/s
11.130 $\alpha = 55.4°$
11.132 $\bar{v}_{B/A} = -60\bar{\imath} + 52.0\bar{\jmath}$ km/hr
$\bar{a}_{B/A} = -4\bar{\imath} - 3.46\bar{\jmath}$ m/s^2
($\bar{\imath}$ right, $\bar{\jmath}$ up)
11.134 $\bar{v}_{A/B} = 25\bar{\imath}$ mi/hr
$\bar{a}_{A/B} = -8\bar{\imath} + 21.5\bar{\jmath}$ ft/s^2
($\bar{\imath}$ right, $\bar{\jmath}$ up)
11.136 $\bar{v}_{A/B} = -41.5\bar{\imath} + 15.5\bar{\jmath}$ m/s
$\bar{a}_{A/B} = -1.56\bar{\imath} + 5.08\bar{\jmath}$ m/s^2
($\bar{\imath}$ right, $\bar{\jmath}$ up)
11.138 $\bar{r}_{B/A} = -1250\bar{\imath} + 915\bar{\jmath}$ ft
$\bar{v}_{B/A} = 12.12\bar{\imath} - 7.0\bar{\jmath}$ ft/s
$\bar{a}_{B/A} = -0.908\bar{\imath} + 0.427\bar{\jmath}$ ft/s^2
($\bar{\imath}$ right, $\bar{\jmath}$ up)
11.140 $\bar{v}_{A/B} = 150.5\bar{\imath} - 40.3\bar{\jmath}$ ft/s
$\bar{a}_{A/B} = 5.42\bar{\imath} - 3.62\bar{\jmath}$ ft/s^2
($\bar{\imath}$ right, $\bar{\jmath}$ up)
11.142 $\bar{v} = 40\bar{\imath} + 115.5\bar{\jmath}$ mm/s
($\bar{\imath}$ right, $\bar{\jmath}$ up)
11.144 $\bar{v}_C = 18.66$ ft/s down
11.146 $\bar{v} = 16.46\bar{\imath} - 43.54\bar{\jmath}$ m/s
($\bar{\imath}$ right, $\bar{\jmath}$ up)
11.148 $\bar{v}_B = 30$ ft/s down
$\bar{a}_B = 15$ ft/s^2 up
11.150 $\bar{a}_B = 0.375g$ right
$\bar{a}_{B/A} = 0.125g$ left
11.152 $\bar{v} = 0.25$ m/s up
11.154 $\bar{v}_C = 2$ ft/s down
$\Delta s_C = 10$ ft down
11.156 (a) $\bar{v}_A = \frac{1}{2}v$ up
(b) $\bar{v}_B = \frac{1}{2}v$ up
11.158 $\bar{v}_B = 2.70$ m/s up
12.2 $\bar{v}_5 = 144.4\bar{e}_t$ ft/s, $\bar{v}_{15} = 300\bar{e}_t$ ft/s
12.4 $v_{max} = 625$ m/s, $t = 12.5$ s
12.6 $t = 2.02$ s
12.8 $\bar{v}_{10} = 39.2\bar{e}_t$ m/s, $\bar{v}_{20} = 29.4\bar{e}_t$ m/s
12.10 $\bar{v}_{10} = 12.7\bar{e}_t$ m/s, $\bar{v}_{20} = 32.7\bar{e}_t$ m/s
12.12 $v = 38.1$ m/s
12.14 $v = 2.91$ m/s
12.16 $s = 100.3$ ft
12.18 $v = 63.7$ ft/s
12.20 $v = 3.47$ m/s
12.22 $v = 17.94$ ft/s
12.24 $\Delta s = 5699$ ft
12.26 $\Delta s = .09153$ m
12.28 $v = 78.7$ m/s $= 79.5\%$ of v_{term}
12.30 $v = 14.79$ m/s $= 64.4\%$ of v_{term}
12.32 $s_{max} = 30.9$ m
12.34 (a) $\Delta s = 5.22$ mi, (b) $\Delta t = 25$ min
12.36 $\Delta s = 0.05(1 - \cos 5t)$ m
12.38 $s = 20t - t^3/15$ m
12.40 $s = 150 - \dfrac{600}{\pi}\cos\left(\dfrac{\pi t}{20}\right)$ ft
12.42 $t = 0.447$ s, $\dot{v} = -2.42$ m/s^2
12.44 $\bar{N} = 2.25\bar{e}_n$ lb,
$\bar{a} = 19.30\bar{e}_t + 90\bar{e}_n$ ft/s^2
12.46 (a) $v = 126.8$ ft/s
(b) $\dot{v} = -11.67$ ft/s^2
12.48 $\dot{v} = 5.88$ ft/s^2, $\rho = 51.0$ ft
12.50 $T = mg/\cos\theta$
$v = (g\ell \sin^2\theta/\cos\theta)^{1/2}$
12.52 $N_A = 3258$ lb, $T_A = 0$
$N_B = 2508$ lb, $T_B = 750$ lb
12.54 $v = 114.5$ km/hr
12.56 $f = 2680$ N out
12.58 $h_{max} = 6.37$ m, $t = 0.528$ s
12.60 $F_{AB} = 3.20$ N 30° below rt. horiz.
$F_D = 8.36$ N up
12.62 $F_{AB} = 472$ lb 71.6° above rt. horiz.
$F_D = 448$ lb left
12.64 $y = -\left(\dfrac{g}{2v_0^2\cos^2\beta}\right)x^2 + (\tan\beta)x$
$\bar{v} = v_0\cos\beta\,\bar{\imath} + \left(v_0\sin\beta - \dfrac{gx}{v_0\cos\beta}\right)\bar{\jmath}$
12.66 $\bar{N} = 1956$ N 53.13° below rt. horiz.
$\bar{F} = 1173$ N left
12.68 $\omega = 205$ rev/min
12.70 $d = 1.961$ m
12.72 $\bar{F} = 6.61\bar{e}_R$ N
12.74 $T = 2.44$ lb
$\bar{N} = 2.83$ lb 36.87° below left horiz.
12.76 $\omega = 6.73$ rad/s
12.78 $F = 841$ lb
12.80 $\Delta s = 92.3$ mm, $t = 92.3$ ms

12.82 $F_{traction} = 0.9m_1g + 0.3m_2g$
$F_{1-2} = 0.9m_1g$

12.84 $\bar{F}_{traction} = 16.33$ kN pushing train right
$F_{lt} = 4.33$ kN tension
$F_{rt} = 8.67$ kN tension

12.86 $F = 400$ N, $T = 120$ N

12.88 $\bar{v}_A = 65$ ft/s rt., $\bar{v}_B = 65$ ft/s dn. & rt.

12.90 $\bar{v}_A = 40.6$ ft/s rt.,
$\bar{v}_B = 81.1$ ft/s dn. & rt.

12.92 $\Delta s = 6.48$ ft

12.94 $T_{upper} = 13.87$ N
$T_{lower} = 2.69$ N

12.96 $v = \left[gh\left(\frac{r^2}{h^2} - 1\right)\right]^{1/2}$

12.98 $\bar{a} = -57.1\bar{e}_t + 200\bar{e}_n$ m/s^2
$\bar{N} = 54\bar{e}_n + 2.9\bar{e}_b$ N

12.100 $\mu_s = 0.672$

12.102 (a) $\dot{v}_A = -1.258$ m/s^2
(b) $\dot{v}_B = -10.51$ m/s^2

12.104 $|\bar{f}| = 548$ lb

12.106 $\theta = 37.85°$

12.108 $d_{min} = -0.008$ ft

12.110 $\bar{v}_B = 1827\bar{i} - 254\bar{j}$ ft/s

12.112 $\bar{a} = 14.52\bar{e}_t + 5\bar{e}_n$ ft/s^2

12.114 $v_h = 3.75$ m/s

12.116 $h = 20.1$ ft

12.118 $\omega = 58.0$ rad/s

12.120 $N_{lower} = 368$ lb
$N_{upper} = 638$ lb

12.122 $\ddot{R} = 27.1$ m/s^2

12.124 $\omega = 4.29$ rad/s, B will move out

12.126 $\bar{F}_{AB} = \bar{0}$, $\bar{F}_{groove} = 217$ N outward

12.128 (a) $\ddot{R} = 1288$ ft/s^2
(b) $\bar{N}_{hor} = 0.631$ lb up

12.130 $\bar{N} = 12.0\bar{e}_R$ lb

12.132 $\bar{F}_{AB} = 11.84$ N right

12.134 $U_{1\to2} = 1.8$ ft-lb

12.136 $U_{1\to2} = 43.7$ J

12.138 $U_{1\to2} = 58.8$ J

12.140 $U_{1\to2} = 10.54$ in.-lb

12.142 $U_{1\to2} = 7.48$ J

12.144 $U_{1\to2} = 4.0$ J

12.146 $F = 75.7$ lb

12.148 $U_{1\to2} = -7.11$ ft-lb

12.150 $U_{1\to2} = -13.99$ J

12.152 $U_{1\to2} = 5.61(10^9)$ ft-lb

12.154 $F = 34.3$ lb

12.156 (a) $U_{1\to2} = -1.961(10^6)$ J
(b) $U_{1\to2} = -4.903(10^5)$ J

12.158 $U_{A\to B} = 55.0$ J

12.160 (a) $V = \frac{1}{2}k_1\Delta^2 + \frac{1}{4}k_2\Delta^4$
(b) $\Delta = (-k_1/k_2)^{1/2}$ for $F = 0$
$\Delta = (-2k_1/k_2)^{1/2}$ for $V = 0$

12.162 $v = 36.0$ m/s

12.164 $h = 2.5\,R$

12.166 $v_0 = (4g\ell)^{1/2}$

12.168 (a) $v = 2.05$ m/s
(b) $\theta = 73.37°$

12.170 $\Delta = 1.773$ ft, $F_{sp} = 42.6$ lb

12.172 $\bar{v}_A = 40.1\bar{e}_t$ ft/s, $\bar{N}_A = 203\bar{e}_n$ lb
$\bar{v}_B = 35.9\bar{e}_t$ ft/s, $\bar{N}_B = -62.5\bar{e}_n$ lb

12.174 $v = 12.58$ m/s

12.176 $v = 17.37$ km/s

12.178 $v = 12.47$ ft/s

12.180 $v = 515$ in./s, $\bar{N} = 7.65\bar{e}_n$ lb

12.182 $F_{max} = 1.148(10^5)$ lb

12.184 $x = 6.52$ m

12.186 (a) $v = 40.6$ ft/s
(b) Does not reach D

12.188 $F = 10.78$ lb

12.190 $m_A = 3.64$ kg

12.192 $F = 1085$ lb

12.194 $\omega = 10.68$ rad/s

12.196 (a) & (b) $\bar{v} = 11.34$ ft/s down

12.198 Power $= 4.75(10^5)$ watt

12.200 (a) $v_{max} = 122.2$ ft/s
(b) $\dot{v} = 0.0658$ ft/s^2
(c) Power $= 3927$ HP to friction

12.202 (a) $v = 3.40$ m/s
(b) $\dot{v} = 5.88$ m/s^2

12.204 (a) $\dot{v} = 5$ m/s^2
(b) Power $= 10$ kw

12.206 $\dot{v}_A = -113.4$ ft/s^2

12.208 $t = 10.19$ s

12.210 $F_A = F_B = 777$ lb

12.212 (a) $F_A = 66.7$ N,
(b) $\bar{v}_B = 2$ m/s right
(c) $\Delta x_B = 9$ m

12.214 $F_{traction} = 3883$ lb forward
$F_{coupling} = 2935$ lb tension

12.216 $\Delta t = 15.3$ s
$F_{left} = 6.54$ kN (C),
$F_{right} = 6.54$ kN (T)

12.218 $t = 2.5$ s, $\bar{v} = 5\bar{i} - 17.5\bar{k}$ m/s

12.220 $\bar{F} = 62.7\bar{i} + 2.14\bar{j}$ lb ($\bar{i}$ rt, $\bar{j}$ up)

12.222 $\bar{v}_A = 31.5$ ft/s left
$\bar{F} = 490\bar{i} - 291\bar{j}$ lb on A
($\bar{i}$ left, $\bar{j}$ up)

12.224 $v = 1000$ m/s

12.226 $\bar{v}_A = -4.98\bar{i} - 7.25\bar{j} + 0.80\bar{k}$ m/s
$\bar{F} = 20\bar{i} + 1415\bar{j} + 797\bar{k}$ N

12.228 $v = 2.67$ ft/s, $F_{AB} = 9948$ lb (C)
$F_{BC} = 4974$ lb (C)

12.230 $\bar{v}_A = 0.333$ m/s left,
$\bar{v}_B = 0.667$ m/s right
$\Delta KE = -8000$ J

12.232 $\bar{v}_A = -4.67\bar{i}$ ft/s
$\bar{v}_B = -0.67\bar{i} + 17.32\bar{j}$ ft/s
($\bar{i}$ right, $\bar{j}$ up)

12.234 $\bar{v}_A = 0.67\bar{i}$ ft/s,
$\bar{v}_B = 8.67\bar{i} + 17.32\bar{j}$ ft/s
($\bar{i}$ right, $\bar{j}$ up)

12.236 (a) $\bar{v}_A = \bar{v}_B = 0$, $\bar{v}_C = v\bar{i}$, 2 collisions
(b) $\bar{v}_A = \bar{v}_B = \bar{v}_C = \frac{1}{3}v\bar{i}$, 2 collisions
($\bar{i}$ left, $\bar{j}$ up)

12.238 $d = 48$ mm, $\bar{v}_B = 4.56\bar{i}$ m/s
($\bar{i}$ intersects both centers)

12.240 $\bar{v}_A = 15.76\bar{i} + 7.88\bar{j}$ m/s
$\bar{v}_B = 4.26\bar{i} + 2.78\bar{j}$ m/s
($\bar{i}$ left, $\bar{j}$ up)

12.242 $\theta = \tan^{-1}(\epsilon^{1/2})$, $v_2 = \epsilon^{1/2}v_1$

12.246 $h_2 = \epsilon^2 h_1$, $d = (1 + \epsilon)\, v_1(2h_1/g)^{1/2}$

12.248 $\bar{v}_A = 9.15\bar{i} + 14.14\bar{j}$ ft/s
$\bar{v}_B = 1.46(-\bar{i} + \bar{j})$ ft/s
($\bar{i}$ 45° above left horiz., $\bar{j}$ 45° above rt. horiz.)

12.250 $v_R = 18.42$ ft/s, $v_\phi = 20$ ft/s

12.252 $h = 12{,}142$ km, $v = 3.39$ km/s

12.254 $v_\phi = 6$ ft/s, $v_R = 14.26$ ft/s

12.256 $R = 1.224$ m

12.258 (a) $v = 2.38$ m/s, (b) $v = 3.43$ m/s
(c) $U_{1\to 2} = 8.06$ J

12.260 $v_A = 4.39$ m/s, $v_B = 0.488$ m/s
$F = -3.90$ N

12.262 $\omega = 2.67$ rad/s, $v_B = 1.57$ m/s

12.264 $\omega = 377$ rad/s, $F = 11.66$ kN

12.266 $\bar{\omega} = 2.73$ rad/s ccw

12.268 $\ddot{\theta} + \dfrac{g}{2\ell}\sin\theta = 0$

12.270 $R\ddot{\theta} + 2\dot{R}\dot{\theta} = \dfrac{M}{mR} - g\cos\theta$
$\ddot{R} - R\dot{\theta}^2 = -g\sin\theta$

12.272 $1/R = 9.2066(10^{-9})\cos\phi + 3.288(10^{-8})$ ft^{-1}
$v_A = 1.548(10^4)$ ft/s,
$v_P = 2.75(10^4)$ ft/s

12.274 $1/R = 7.322(10^{-8})\sin\phi + 1.333(10^{-7})$ m^{-1}
$e = 0.5491$

12.276 $\psi = 36.56°$

12.278 (a) 16.95, (b) $h_{min} = 2.49(10^4)$ km,
(c) $\phi = 31.7°$

12.280 $\phi = 51.46°$

12.282 (a) $h_0 = 6.69(10^{11})$ ft²/s
(b) $E = -1.893(10^8)$ ft²/s²
(c) $e = 0.380$
(d) $v = 13{,}035$ ft/s
(e) $h = 3.04(10^7)$ ft

12.284 $\bar{v} = 15.37\bar{e}_\phi$ km/s

12.286 (a) $\bar{v} = 3.82\bar{e}_\phi$ km/s
(b) $\bar{v} = \pm 2.55\bar{e}_R + 6.37\bar{e}_\phi$ km/s

12.288 (a) $\bar{v} = 4.59\bar{e}_R + 10.86\bar{e}_\phi$ km/s
(b) $\bar{v} = 7.46\bar{e}_R + 4.89\bar{e}_\phi$ km/s

12.290 $\bar{v} = 3.67\bar{e}_R + 3.65\bar{e}_\phi$ km/s

12.292 $\tau = 53.9$ hrs

12.294 $\tau = 3.45$ hrs

12.296 $\Delta t = 178.77$ days

12.298 $\phi = 101.08°$

13.2 (a) $\bar{a} = 18.36\bar{i}$ in./s²
(b) $\bar{a} = -4.9\bar{i} - 355\bar{j}$ in./s²
(c) $\bar{a} = -874\bar{i} - 308\bar{j}$ in./s²
($\bar{i}$ right, $\bar{j}$ up)

13.4 $\bar{v} = -90.5\bar{i} + 34.8\bar{j}$ m/s
($\bar{i}$ right, $\bar{j}$ up)

13.6 $\bar{v} = 2.46\bar{i} + 3.20\bar{j}$ m/s
$\bar{a} = 60.2\bar{i} - 103.7\bar{j}$ m/s²
($\bar{i}$ right, $\bar{j}$ up)

13.8 $\bar{v} = -27.3\bar{i} + 3.1\bar{j}$ ft/s
($\bar{i}$ right, $\bar{j}$ up)

13.10 $\bar{\omega}_{BC} = 4.19$ rad/s cw
$\bar{v} = -2.85\ell\bar{i} - 5.28\ell\bar{j}$
($\bar{i}$ right, $\bar{j}$ up)

13.12 $\bar{v} = 18.75$ m/s right

13.14 (a) $\bar{v} = 4$ ft/s 45° below rt. horiz.
(b) $\bar{a} = 30.6$ ft/s² 45° above left horiz.

13.16 (a) $\bar{a}_B = 3026(\bar{i} - \bar{j})$ in./s²
(b) $\bar{a}_C = 1463\bar{i} - 1513\bar{j}$ in./s²
($\bar{i}$ right, $\bar{j}$ up)

13.18 $\bar{v} = 7.5$ ft/s up

13.20 $\bar{v} \doteq \bar{0}$, $\bar{a} = -133\bar{i} - 231\bar{j}$ mm/s²
($\bar{i}$ right, $\bar{j}$ up)

13.22 $\bar{a} = 85.3$ ft/s² left

13.24 $\bar{a}_A = -125.1\bar{i} - 93.8\bar{j}$ m/s²
$\bar{a}_C = 71.9\bar{i} - 199.8\bar{j}$ m/s²
($\bar{i} = \bar{e}_{A/B}$, $\bar{j} = \bar{e}_{C/B}$)

13.26 $\bar{v} = 27.7$ in./s at 25.6° cw from $\bar{e}_{C/B}$

13.28 $\bar{\omega} = 49.1$ rad/s cw
$\bar{v} = 16.27$ m/s at 75° above left horiz.

13.30 $\bar{v} = -25\bar{i} + 8.66\bar{j}$ ft/s
($\bar{i}$ right, $\bar{j}$ up)

13.32 $\bar{\omega} = 0.1183\bar{k}$ rad/s
$\bar{v} = 94.6\bar{i} - 54.6\bar{j}$ mm/s
($\bar{i}$ right, $\bar{j}$ up)

13.34 (a) i.c. is 1.0 in. below pin O
(b) $\bar{\omega} = 6.0$ rad/s ccw
(c) $\bar{v}_0 = 0.5$ ft/s left

13.36 $\bar{v} = 4$ m/s right

13.38 $\bar{v}_A = \bar{0}$, $\bar{v}_B = 0.9\bar{i} - 0.9\bar{j}$ m/s
$\bar{v}_C = -1.536\bar{i} - 0.636\bar{j}$ m/s
$\bar{a}_A = 18\bar{j}$ m/s², $\bar{a}_B = 18\bar{i}$ m/s²
$\bar{a}_C = 12.73(\bar{i} - \bar{j})$ m/s²
($\bar{i}$ right, $\bar{j}$ up)

13.40 $\bar{\omega} = 20$ rad/s ccw

13.42 $\bar{v}_A = 16\bar{i}$ ft/s
$\bar{v}_B = 4\bar{i} - 12\bar{j}$ ft/s
($\bar{i}$ right, $\bar{j}$ up)

13.44 $\bar{v}_O = 1.732$ ft/s right
$\bar{a}_O = 2.0$ ft/s^2 right

13.46 $\bar{v}_O = 25.1\bar{\imath}$ m/s, $\bar{v}_C = 50.3\bar{\imath}$ m/s,
$\bar{v}_D = 75.4\bar{\imath}$ m/s
$\bar{a}_O = \bar{0}$, $\bar{a}_C = -632\bar{j}$ m/s^2,
$\bar{a}_D = -1263\bar{j}$ m/s^2
($\bar{\imath}$ right, $\bar{j}$ up)

13.48 $\bar{v}_O = \omega r\bar{\imath}$, $\bar{a}_O = \dot{\omega} r\bar{\imath} + \dfrac{\omega^2 r^2}{(R - r)}\bar{j}$
($\bar{\imath} = (\bar{e}_t)_O$, $\bar{j} = (\bar{e}_n)_O$)

13.50 (a) $\bar{\alpha} = -5\bar{k}$ rad/s^2
(b) $\bar{a}_A = 7.5\bar{j}$ m/s^2,
$\bar{a}_B = -2.5\bar{\imath} - 15\bar{j}$ m/s^2
$\bar{a}_D = 20\bar{\imath} - 17.5\bar{j}$ m/s^2
($\bar{\imath} = (\bar{e}_t)_C$, $\bar{j} = (-\bar{e}_n)_C$)

13.52 (a) $\bar{\omega} = -100\bar{k}$ rad/s,
$\bar{\alpha} = 20.8\bar{k}$ rad/s^2
(b) $\bar{v}_C = 4\bar{\imath}$ m/s, $\bar{a}_C = 2.5\bar{\imath}$ m/s^2
$\bar{v}_D = 4\bar{\imath} - 12\bar{j}$ m/s,
$\bar{a}_D = 1198\bar{\imath} + 2.5\bar{j}$ m/s^2
($\bar{\imath}$ right, $\bar{j}$ up)

13.54 $\bar{\omega}_{BC} = 32.1$ rad/s cw,
$\bar{\omega}_{CD} = 33.7$ rad/s ccw

13.56 $\bar{\omega}_{BC} = 1.406$ rad/s ccw
$\bar{v}_C = 0.338$ m/s up

13.58 $\bar{v}_A = 6.74$ m/s right

13.60 $\bar{\omega}_{BC} = \bar{0}$, $\bar{\omega}_{CD} = 5$ rad/s ccw
$\bar{\alpha}_{BC} = 33.3$ rad/s^2 ccw, $\bar{\alpha}_{CD} = \bar{0}$

13.62 $\bar{\omega}_{BC} = \omega$ cw, $\bar{\omega}_{CD} = 2\omega$ ccw
$\bar{\alpha}_{BC} = 6\omega^2$ ccw, $\bar{\alpha}_{CD} = 14\omega^2$ cw

13.64 $\alpha_{AB} = 16.12$ rad/s^2 ccw

13.66 $\bar{a}_A = 1607$ m/s^2 right

13.68 $\bar{\omega}_{AC} = 4$ rad/s cw,
$\bar{\alpha}_{AC} = 27.7$ rad/s^2 cw

13.70 $\bar{v}_B = \dfrac{v_A}{4}$ up, $\bar{\omega}_{left} = \dfrac{v_A}{4R}$ ccw
$\bar{\omega}_{mid} = \dfrac{v_A}{2R}$ ccw, $\bar{\omega}_{rt} = \dfrac{v_A}{R}$ cw

13.72 $\omega = 2$ rad/s, $\dot{\omega} = 6.34$ rad/s^2

13.74 $\bar{v}_C = -0.5\omega r\bar{\imath} + 0.289\omega r\bar{j}$
$\bar{v}_E = 0.75\omega r\bar{\imath} + 0.1444\omega r\bar{j}$
($\bar{\imath}$ right, $\bar{j}$ up)

13.76 $\bar{\omega}_{spider} = 83.8$ rad/s ccw

13.78 $v = 33.9$ ft/s

13.80 $\bar{\alpha}_{BC} = 190$ rad/s^2 cw
$\bar{\alpha}_{CD} = 329$ rad/s^2 cw

13.82 $\bar{\omega}_{BC} = 40.8$ rad/s cw
$\bar{\omega}_{CD} = 31.4$ rad/s cw

13.84 $\bar{a}_G = -36\bar{\imath} - 15.3\bar{j}$ in./s^2
($\bar{\imath}$ right, $\bar{j}$ up)

13.86 $\bar{v} = 5225$ in./s right
$\bar{a} = 4.39(10^6)$ in./s^2 right

13.88 $\bar{v} = 1.0$ m/s right,
$\bar{a} = 7.02$ m/s^2 right

13.90 $\bar{v} = 5$ m/s right, $\bar{a} = 100$ m/s^2 right

14.2 $I_{zz} = \dfrac{1}{12}\rho abc(3a^2 + b^2)$

14.4 $I_{zz} = \rho abh\left(\dfrac{7}{48}a^2 + \dfrac{1}{12}b^2\right)$

14.6 $x_G = 0$, $y_G = \dfrac{2r}{3\theta}\sin\theta$
$I_{zz} = \dfrac{1}{2}mr^2$

14.8 $x_G = 0$, $y_G = \dfrac{2r}{3}\dfrac{\sin^3\theta}{\theta - \sin\theta\cos\theta}$
$I_{zz} = \dfrac{1}{12}mr^2$
$\times\left(\dfrac{12\theta - 4\sin 2\theta - \sin 4\theta}{2\theta - \sin 2\theta}\right)$

14.10 $m = \sigma\pi r$, $x_G = 0$, $y_G = 2r/\pi$
$I_{zz} = \sigma\pi r^3$

14.12 $z_G = \dfrac{2}{3}\ell$, $I_{zz} = \dfrac{1}{3}mr^2$

14.14 $I_{z'z'} = \dfrac{4}{3}mr^2$

14.16 $I_{\xi\xi} = \dfrac{1}{18}m(a^2 + b^2)$

14.18 $I_{\xi\xi} = \dfrac{1}{12}\sigma\ell^3$

14.20 $I_{z_2z_2} = 13.70$ kg-m^2

14.22 $I_{zz} = 0.001731$ kg-m^2

14.24 $I_{zz} = 0.0241$ kg-m^2

14.26 $I_{zz} = 3.41$ lb-s^2-ft

14.28 $m = 4.22$ lb-s^2/ft
G is 0.0444 ft left, 0.0222 ft below center
$I_{\xi\xi} = 3.16$ lb-s^2-ft

14.30 $m = 0.01565$ lb-s^2/in.
G is 19.74 in. left, 8.99 in. above lower rt. corner
$I_{\xi\xi} = 3.11$ lb-s^2-in.

14.32 $\dot{v} = -g$

14.34 $\bar{a} = 4.31$ ft/s^2 60° above rt. horiz.
$N_{upper} = 200$ lb, $N_{lower} = 1800$ lb

14.36 $\Delta s = 14.36$ ft

14.38 $\bar{a} = 3.63$ m/s^2 left

14.40 $\bar{a} = 13.93\bar{\imath} - 24.1\bar{j}$ ft/s^2
$F_1 = F_2 = 100$ lb
($\bar{\imath}$ left, $\bar{j}$ up)

14.42 $\bar{a} = 16.09\bar{\imath} + 18\bar{j}$ ft/s^2
$F_1 = F_2 = 285$ lb
($\bar{\imath}$ 30° below left horiz., $\bar{k}$ in)

14.44 (a) $h = 1.002$ m
(b) $\bar{a} = 9.56$ m/s^2 right

14.46 $\bar{a} = 25.7$ ft/s^2 53.13° below right horiz.
$F_{lt} = 1.11$ lb, $F_{rt} = 28.9$ lb

14.48 $M = 0.1507$ lb-ft
14.50 (a) $\bar{\alpha} = 16$ rad/s² cw
(b) $\bar{\alpha} = 8.81$ rad/s² cw
14.52 $\bar{\alpha} = 20.9$ rad/s² cw
$\bar{A} = 5.28$ lb up
14.54 $q = \dfrac{L}{3}$, $\bar{\alpha} = \dfrac{6P}{mL}$ ccw
14.56 $I_{zz} = m_a R^2$
14.58 $\bar{F} = 4003$ N up
14.60 $\bar{\alpha} = 2.96$ rad/s² cw
$\bar{A} = 175$ lbs left, 950 lbs down
14.62 $\bar{\alpha} = 7.71$ rad/s² ccw
$\bar{A} = 26.4$ N rt., 70.5 N up
14.64 (a) $M = 2.45m$ N-m
(b) $t = 6.41$ s
14.66 $F_{left} = 2201$ lb, $F_{rt} = 2001$ lb
$\bar{\alpha} = 2.29$ rad/s² cw
14.68 $F_A = 960$ lb, $F_B = 1848$ lb
14.70 (a) $\Delta F = 3491$ kN
(b) $\bar{a} = 65.1$ m/s² up
14.72 (a) $\bar{a} = 10.72$ ft/s² right
(b) $\bar{f} = 16.67$ lb left
(c) $\mu = 0.1667$
14.74 $\bar{a} = -10.17\bar{\imath} - 3.81\bar{\jmath}$ ft/s²
$F_{lt} = 2.31$ lb, $F_{rt} = 2.63$ lb
($\bar{\imath}$ right, $\bar{\jmath}$ up)
14.76 $\bar{a}_G = 5.23(\bar{e}_t)_G + 20(\bar{e}_n)_G$ m/s²
$\bar{N} = 64.7\bar{e}_n$ N, $\bar{f} = -6.54\bar{e}_t$ N
14.78 $\bar{a}_C = 1.967$ m/s² right,
$\bar{a}_{rack} = 1.312$ m/s² left
14.80 $\bar{\alpha} = 201$ rad/s² ccw,
$\bar{a}_O = 13.40$ ft/s² right
14.82 (a) $\mu_s = 0.3216$
(b) $\bar{\alpha} = 0.289g/R$ cw
14.84 $\bar{a}_G = 1.657$ m/s² lt,
$\bar{\alpha} = 8.29$ rad/s² ccw
14.86 $\bar{M}_C = 28.9$ lb-in. ccw, $F_A = 23.4$ lb,
$F_B = 19.4$ lb
14.88 $\bar{a}_G = 6.92$ m/s² down, $F_A = 144.2$ N
14.90 (a) $v_G = 17.86$ ft/s
(b) $\Delta s = 23.8$ ft
14.92 $\bar{F} = 154.2$ left, $\bar{N}_A = 95.8$ N down,
$\bar{N}_B = 151.4$ N up
14.94 $\bar{\alpha} = 10.44$ rad/s² cw, $\bar{N}_A = 30.5$ N
up, $\bar{N}_B = 16.5$ N down
14.96 $T = 189.5$ J
14.98 $T = 456$ ft-lb
14.100 $T = 1350$ ft-lb
14.102 $T = 0.966\omega^2$
14.104 $\omega = 3.95$ rad/s
14.106 $v = 9.38$ ft/s
14.108 $F = 11.96$ lb
14.110 $\bar{M}_A = 4.58$ lb-ft cw
14.112 $\bar{M} = 4.85$ N-m ccw
14.114 $v_A = 30.1$ ft/s
14.116 $v = 3.34$ ft/s
14.118 $v = 5.77$ m/s
14.120 $\omega_{AB} = 8.53$ rad/s
14.122 $\omega = 0.818(g/\ell)^{1/2}$
14.124 $F = 261$ N
14.126 $v_B = 0.889$ ft/s
14.128 $t = 246$ s
14.130 (a) $F_{cable} = 5.93$ lb
(b) $\bar{F}_A = 9.41$ lb right
14.132 (a) $\bar{v} = 0.557$ m/s left
(b) $N_A = 127$ N, $N_B = 363$ N
14.134 $\bar{v} = 0.672$ m/s down
14.136 $h = 0.45$ m, $\bar{\omega} = 40$ rad/s cw
14.138 $\bar{M}_A = 0.1495$ lb-in. cw
14.140 $F = 10$ kN, $\bar{F}_A = \bar{0}$
14.142 (a) $\omega = 11.31$ rad/s,
(b) $U_{1\to 2} = 173.0$ ft-lb
14.144 $h = 42$ mm, $\bar{v} = 10$ m/s rt
14.146 $t = 15.62$ s
14.148 (a) $\bar{v} = 8.07$ m/s 30° above rt. horiz.
(b) $F_{lt} = 504$ lb, $F_{middle} = 224$ lb,
$F_{rt} = 0$
14.150 $\bar{\omega} = 7.5$ rad/s ccw, $F = 58.6$ kN
14.152 (a) $\bar{\omega} = 4$ rad/s cw
(b) $\bar{F}_C = 290$ lb 45° above lt. horiz.
(c) $\bar{F}_A = 103$ lb left, 205 lb dn.
14.154 $\bar{v}_A = \bar{0}$, $\bar{v}_C = 10.18$ ft/s right
14.156 $v = 1.906$ m/s
14.158 $\bar{v}_G = 7.71$ m/s left,
$\bar{\omega} = 16.71$ rad/s cw
14.160 $\bar{v}_G = 8$ m/s right, $\omega = 100$ rad/s cw
14.162 $\bar{v}_G = 8.54$ m/s down,
$\omega = 14.99$ rad/s ccw
15.2 $\bar{v}_A = 6\bar{\imath} - 2\bar{\jmath}$ in./s ($\bar{\imath}$ right, $\bar{\jmath}$ up)
15.4 (a) $\bar{v} = 4\bar{\imath}$ m/s,
(b) $\bar{v} = -4\bar{\jmath} - 10\bar{k}$ m/s
($\bar{\imath}$ right, $\bar{\jmath}$ up)
15.6 $\bar{v} = 0.369\bar{\imath} + 0.169\bar{k}$ m/s
($\bar{\imath}$ 60° above rt. horiz., $\bar{\jmath}$ 30° above lt. horiz.)
15.8 $\bar{v} = 10\bar{\jmath} - 75.4\bar{k}$ ft/s ($\bar{\imath}$ right, $\bar{\jmath}$ up)
15.10 $\bar{v} = -6.98\bar{\jmath} - 0.60\bar{k}$ m/s
($\bar{\imath}$ 72.4° rt. of vert., $\bar{\jmath}$ 17.6° lt. of vert.)
15.12 $\bar{\omega} = -5\bar{\imath} - 94.3\bar{\jmath}$ rad/s
($\bar{\imath}$ left along pin, $\bar{\jmath}$ right along shaft)
15.14 $\bar{a} = -1100\bar{\imath} + 300\bar{\jmath}$ in./s
($\bar{\imath}$ right, $\bar{\jmath}$ up)
15.16 $\bar{a} = -789.6\bar{\imath} + 197.4\bar{\jmath}$ m/s²
($\bar{\imath}$ right, $\bar{\jmath}$ up)
15.18 $\bar{a} = 216$ ft/s² left
15.20 $\bar{a} = -9.75\bar{\jmath} - 40\bar{k}$ ft/s²
($\bar{\imath}$ down, $\bar{k}$ left along CC)
15.22 $\bar{a} = 140$ m/s² right
15.24 $\bar{a} = -5.27\bar{\imath} - 14.51\bar{\jmath} - 7.38\bar{k}$ m/s²

15.26 $\bar{a} = -16.35\bar{\imath} + 21.96\bar{\jmath} - 0.88\bar{k}$ m/s^2 ($\bar{\imath}$ 72.4° rt. of vert., $\bar{\jmath}$ 17.6° lt. of vert.)

15.28 $\bar{N} = 14710$ N vert., 7.6 N west

15.30 $\bar{a}_{rel} = 5.16(10^{-3})$ ft/s^2 south

15.32 $\bar{v}_{rel} = -34.7\bar{\imath} + 27.8\bar{\jmath}$ m/s

15.34 $a_{rel} = 0.282$ ft/s^2 west, 32.17 ft/s^2 vert. down

15.36 $\bar{v} = 209\bar{\imath} + 363\bar{\jmath} + 363\bar{k}$ ft/s
$\bar{a} = (-0.570\bar{\imath} - 0.987\bar{\jmath} + 1.316\bar{k})(10^5)$ ft/s^2
($\bar{\jmath} = \bar{e}_{C/A}$, $\bar{k} = \bar{e}_{C/B}$)

15.38 $\bar{a} = -106\bar{\imath} + 40\bar{\jmath} - 60\bar{k}$ m/s^2

15.40 (a) $\bar{v}_{rel} = -40\bar{\imath} + 30\bar{\jmath}$ ft/s
(b) $\bar{a}_{rel} = -60\bar{\imath} - 160\bar{\jmath} - 32.17\bar{k}$ ft/s^2
($\bar{\imath} = \bar{e}_{B/C}$, $\bar{k}$ up vert.)

15.42 (a) $\bar{v}_{B/A} = -54.6\bar{\imath} + 7.2\bar{\jmath}$ m/s
$\bar{a}_{B/A} = -0.399\bar{\imath} + 0.224\bar{\jmath}$ m/s^2
(b) $(\bar{v}_{B/A})_{rel} = -56.6\bar{\imath} + 15.2\bar{\jmath}$ m/s
$(\bar{a}_{B/A})_{rel} = -1.78\bar{\imath} - 6.64\bar{\jmath}$ m/s^2
($\bar{\imath} = (\bar{e}_t)_A$, $\bar{\jmath} = -(\bar{e}_n)_A$)

15.44 (a) $\bar{a} = -7.5(10^{-4})\bar{\imath} + 4.3(10^{-4})\bar{\jmath} - 64.95\bar{k}$ m/s^2
(b) $N_y = 1.70$ N, $N_z = -12.99$ N

15.46 (a) $\bar{v}_B = -8\bar{\imath} - \bar{k}$ m/s,
$\bar{a}_B = -650\bar{\jmath}$ m/s^2
(b) $\bar{v}_C = 5.67\bar{\imath} + 5.67\bar{\jmath} + 0.71\bar{k}$ m/s,
$\bar{a}_C = -453\bar{\imath} + 460\bar{\jmath} - 113\bar{k}$ m/s^2
($\bar{\imath}$ right, $\bar{\jmath}$ up)

15.48 $\bar{v}_C = 2\bar{\imath} - 10\bar{\jmath} - 20.9\bar{k}$ m/s
$\bar{a}_C = 100\bar{\imath} - 2215\bar{\jmath}$ m/s^2
($\bar{\imath} = \bar{e}_{B/A}$, $\bar{\jmath} = \bar{e}_{C/B}$)

15.50 $\bar{\alpha} = 138.2$ rad/s^2 cw

15.52 (a) $\bar{\omega} = -50\bar{\imath} + 10j$ rad/s,
$\bar{\alpha} = 500\bar{k}$ rad/s^2
(b) $\bar{v}_C = -5.0\bar{k}$ m/s,
$\bar{a}_C = -75\bar{\imath} - 125\bar{\jmath}$ m/s^2
($\bar{\imath} = \bar{e}_{B/A}$, $\bar{\jmath} = \bar{e}_{C/B}$)

15.54 $\bar{v}_{A/G} = -4.5\bar{k}$ ft/s
$\bar{a}_{A/G} = -1.35\bar{\imath} - 2.25\bar{\jmath} + 6.0\bar{k}$ ft/s^2

15.56 $\bar{v} = 1.2\bar{\imath} + 5.3\bar{\jmath} + 0.6\bar{k}$ m/s
$\bar{a} = 0.198\bar{\imath} - 0.344\bar{\jmath} - 0.104\bar{k}$ m/s^2

15.58 $\bar{v} = 1.238$ m/s down

15.60 (a) $\omega_2 = \dfrac{L}{R}\,\omega_1$, $\dot{\omega}_2 = \dfrac{L}{R}\,\dot{\omega}_1$
(b) $\bar{\omega} = \omega_1\left(\dfrac{L}{R}\,\bar{\imath} + \bar{\jmath}\right)$
$\bar{\alpha} = \dot{\omega}_1\left(\dfrac{L}{R}\,\bar{\imath} + \bar{\jmath}\right) - \dfrac{L}{R}\,\omega_1^2\bar{k}$
($\bar{\imath}$ right, $\bar{\jmath}$ up)

15.62 $\bar{v} = 600\bar{\imath} + 150\bar{\jmath} - 290\bar{k}$ ft/s
$\bar{a} = -97\bar{\imath} + 203\bar{\jmath} - 402\bar{k}$ ft/s^2

15.64 $\bar{v} = \dot{r}\bar{e}_r + r\dot{\theta}\bar{e}_\theta + r\dot{\phi}\sin\theta\,\bar{e}_\phi$
$\bar{a} = (\ddot{r} - r\dot{\theta}^2 - r\dot{\phi}^2\sin^2\theta)\bar{e}_r + (r\ddot{\phi}\sin\theta + 2\dot{r}\dot{\phi}\sin\theta + 2r\dot{\phi}\dot{\theta}\cos\theta)\bar{e}_\phi + (r\ddot{\theta} + 2\dot{r}\dot{\theta} - r\dot{\phi}^2\sin\theta\cos\theta)\bar{e}_\theta$

15.66 (a) $\omega_C = 167.6$ rad/s
(b) $\bar{\omega} = -335\bar{\imath} + 168\bar{\jmath}$ rad/s
$\bar{\alpha} = 5.61(10^4)\bar{k}$ rad/s^2
($\bar{\imath}$ right, $\bar{\jmath}$ up)

15.68 $\bar{v}_B = 398.5\bar{\imath} + 35.6\bar{\jmath}$ m/s
$\bar{a}_B = 32.6\bar{\imath} - 49.7\bar{\jmath}$ m/s^2
($\bar{\imath} = (\bar{e}_t)_A$, $\bar{\jmath} = -(\bar{e}_n)_A$)

15.70 (a) $\bar{\omega} = -1047\bar{\imath} - 0.235\bar{\jmath}$ rad/s,
$\bar{\alpha} = -246\bar{k}$ rad/s^2
(b) $\bar{v} = 1.64\bar{\imath} - 0.60\bar{\jmath} - 2095\bar{k}$ in./s
$\bar{a} = 984\bar{\imath} - 2.194(10^6)\bar{\jmath} + 0.44\bar{k}$ in./s^2
($\bar{\imath} = \bar{e}_{B/A}$, $\bar{\jmath} = \bar{e}_{D/C}$)

15.72 $\bar{v} = 5\bar{\imath} + 5\bar{\jmath}$ m/s
($\bar{\imath} = \bar{e}_{A/C}$, $\bar{\jmath} = \bar{e}_{D/C}$)

15.74 $\bar{v}_{rel} = 888$ in./s 45° above rt. horiz.
$\bar{\omega}_A = 70.6$ rad/s cw

15.76 (a) $\omega_1 = 9.43$ rad/s
(b) $\bar{\omega} = -8.38\bar{\imath} + 2.96\bar{\jmath}$ rad/s,
$\bar{\alpha} = 27.9\bar{k}$ rad/s^2
($\bar{\imath}$ along centerline, $\bar{\jmath}$ up, parallel to base)

15.78 $\bar{v} = 0.8\bar{\imath} + 1.8\bar{k}$ m/s
($\bar{\imath}$ right, $\bar{\jmath}$ up)

15.80 $\bar{v} = 80$ ft/s backward

15.82 $\bar{v} = -19.3\bar{\imath} + 23.0\bar{\jmath} - 48.0\bar{k}$ ft/s
$\bar{a} = -23.8\bar{\imath} - 3.9\bar{\jmath} + 15.4\bar{k}$ ft/s^2
($\bar{\imath}$ right, $\bar{\jmath}$ up)

15.84 $\bar{\omega} = -5\bar{\imath} + 62.8\bar{\jmath} + 10\bar{k}$ rad/s
$\bar{\alpha} = -628\bar{\imath} - 50\bar{\jmath} - 314\bar{k}$ rad/s^2
($\bar{\jmath}$ right along centerline of disk, $\bar{k}$ up)

15.86 $\bar{v} = 240\bar{\imath} - 24\bar{\jmath} - 400\bar{k}$ in./s
$\bar{a} = -6.51(10^4)\bar{\imath} - 4.40(10^3)\bar{\jmath}$ in./s^2
($\bar{\jmath}$ right along shaft, $\bar{k}$ up)

16.2 $\bar{H}_G = 1600\bar{\imath} + 6400\bar{\jmath} + 1200\bar{k}$ lb-s-ft

16.4 $\bar{H}_G = 0.1005\bar{\imath} + 1.536\bar{\jmath}$ kg-m^2/s

16.6 $\bar{H}_G = 0.0259\bar{\imath} + 0.0173\bar{\jmath} - 0.00086\bar{k}$ lb-s-ft

16.8 (a) $\bar{H}_G = -\dfrac{1}{2}mrv_0\bar{\imath}$,
(b) $\bar{H}_G = -\dfrac{1}{2}mrv_0\bar{\imath} + \dfrac{1}{4}m\dfrac{r^2}{R}v_0\bar{k}$

16.10 (a) $I_{xy} = I_{xz} = I_{yz} = 0$,
(b) $I_{xz} = I_{yz} = 0$
(c) $I_{xy} = I_{xz} = 0$, (d) $I_{xy} = I_{yz} = 0$
(e) $I_{xy} = I_{yz} = 0$, (f) $I_{xz} = I_{yz} = 0$

16.14 $I_{xx} = \frac{1}{3} m(b^2 + c^2)$,

$I_{yy} = \frac{1}{12} m(a^2 + 4c^2)$

$I_{zz} = \frac{1}{12} m(a^2 + 4b^2)$, $I_{xy} = I_{xz} = 0$

$I_{yz} = -\frac{1}{4} mbc$

16.16 $I_{xx} = \frac{1}{12} m[4L^2 + (3 + 12\cos^2\theta)r^2]$

$I_{yy} = \frac{3}{2} mr^2$, $I_{zz} = \frac{1}{12} m[4L^2 + (3 + 12\sin^2\theta)r^2]$

$I_{xy} = -\frac{1}{2} mrL\sin\theta$,

$I_{xz} = \frac{1}{2} mr^2 \sin 2\theta$

$I_{yz} = -\frac{1}{2} mrL\cos\theta$

16.18 $m = 61.97$ kg, G is 0.373 m from base along centerline

$I_{xx} = 12.44$ kg-m^2, $I_{xy} = 0$

16.20 $m = 12.68$ lb-s^2/ft, $x_G = 1$ ft,

$y_G = 0.628$ ft, $z_G = -0.5$ ft

$I_{xx} = 11.13$ lb-s^2-ft, $I_{xy} = 7.96$ lb-s^2-ft

16.22 (a) $I_{zz} = 2095$ lb-s^2-ft

(b) $I_{zz} = 2038$ lb-s^2-ft

16.24 $I_{xx} = 3.10\pi\rho r^4 L$

$I_{yy} = I_{zz} = \pi\rho r^2 L\left(\frac{31}{20} r^2 + \frac{8}{15} L^2\right)$

$I_{xy} = I_{xz} = I_{yz} = 0$

16.26 $x_G = 47.8$ mm, $y_G = 0$,

$z_G = 22.5$ mm

$I_{xx} = 0.00617$ kg-m^2,

$I_{yy} = 0.00445$ kg-m^2

$I_{zz} = 0.00788$ kg-m^2

$I_{xy} = I_{yz} = 0$, $I_{xz} = 0.00057$ kg-m^2

16.28 G is on both midplanes 0.628 ft from back surface

$I_{\hat{x}\hat{x}} = 2.97$ lb-s^2-ft, $I_{\hat{y}\hat{y}} = 5.96$ lb-s^2-ft

$I_{\hat{z}\hat{z}} = 6.82$ lb-s^2-ft

($\bar{j}$ left, $\bar{k}$ up)

16.30 $I_{xx} = \frac{1}{6} m(b^2 + c^2)$,

$I_{yy} = \frac{1}{6} m(2a^2 + c^2)$

$I_{zz} = \frac{1}{6} m(2a^2 + b^2)$

16.32 $I_{xx} = \frac{2}{105}\rho abc(7b^2 + 8c^2)$

$I_{yz} = \frac{1}{12}\rho ab^2c^2$

16.34 $I_{xx} = \frac{1}{12}\pi\rho hr^2(2r^2 + h^2)$,

$I_{zz} = \frac{1}{3}\pi\rho hr^4$

16.36 $I_{xy} = \frac{1}{24}\rho hr^4$, $I_{xz} = I_{yz} = \frac{2}{21}\rho h^2 r^3$

16.38 $m = \frac{1}{8}\rho_0 abc$,

$I_{yy} = \frac{1}{240}\rho_0 abc(10a^2 + c^2)$

16.40 $I_{xx} = 1.714$ lb-s^2-ft, $I_{yy} = 2.69$ lb-s^2-ft

$I_{zz} = 4.404$ lb-s^2-ft, $I_{xy} = I_{yz} = 0$

$I_{xy} = -0.914$ lb-s^2-ft

16.42 $I_{xx} = 0.0207$ lb-s^2-ft

$I_{xy} = 0.0155$ lb-s^2-ft

16.44 $I_{xx} = \frac{1}{2} m_1 R^2 + \frac{1}{3} m_2 L^2 \sin^2\theta$

16.46 $I_{yz} = -0.358$ kg-m^2

16.48 $\bar{H}_G = \frac{1}{12} m\ell^2\omega \sin\theta\,\bar{i}$

$\dot{\bar{H}}_G = \frac{1}{24} m\ell^2\omega^2 \sin 2\theta\,\bar{k}$

($\bar{j}$ upward along bar, $\bar{k}$ out of page)

16.50 $\bar{H}_G = 11781\bar{k}$ lb-in. ($\bar{k}$ up $\perp$ to plate)

16.52 (a) balanced,

(b) & (c) unbalanced statically

(d) & (e) unbalanced dynamically

(f) balanced

16.54 $\dot{\bar{H}}_G = -26.2\bar{k}$ N-m

16.56 $\dot{\bar{H}}_G = 181.0\bar{j} - 39.4\bar{k}$ N-m

$\bar{M}_A = 181.0\bar{j} - 6.1\bar{k}$ N-m

($\bar{i} = \bar{e}_{B/A}$, $\bar{k}$ out of page)

16.58 $\bar{F}_A = -1.52\bar{i} + 3.17\bar{j}$ lb

$\bar{M}_A = 5.38\bar{k}$ lb-ft

($\bar{i} = \bar{e}_{B/A}$, $\bar{j}$ up)

16.60 (a) $\bar{M}_G = -0.323\bar{j}$ lb-ft

(b) $\bar{M} = 0.323\bar{i}$ lb-ft

16.62 $\bar{F}_A = -1.325(10^4)\bar{i} + 15\bar{j}$ lb

$\bar{M}_A = 6.97(10^3)\bar{k}$ lb-ft

($\bar{i}$ left along base, $\bar{j}$ up)

16.64 $M = 614$ N-m

$\bar{F}_A = -101\bar{J} - 224\bar{K}$ N

$\bar{F}_B = -331\bar{J} - 736\bar{K}$ N

($\bar{I} = \bar{e}_{B/A}$, $\bar{J}$ up)

16.66 $\theta = 89.998°$

16.68 $\bar{F} = \bar{0}$

$\bar{M} = [-24.9(\sin 2t)\bar{i} - 3.1(\cos 5t)\bar{j} + 6.2(\sin 5t)(\cos 2t)\bar{k}](10^9)$ lb-ft

16.70 $\dot{\bar{H}}_G = 22.6\cos\theta\,\bar{i} + 0.27\sin\theta\cos\theta\,\bar{k}$ N-m

($\bar{j}$ right along blade, $\bar{k}$ up)

16.72 $\bar{F}_A = -444.6\bar{i} + 33.8\bar{j} + 6.5\bar{k}$ lb

$\bar{M}_A = -14.8\bar{i} - 11.4\bar{j} + 930.7\bar{k}$ lb-ft

16.74 $\overline{M} = -39.0\bar{\imath} - 49.5\bar{\jmath} - 120.0\bar{k}$ N-m

16.76 $N = m\left(\frac{1}{2} r\omega_2^2 - g\right)$, $\omega_{min} = \left(\frac{2g}{r}\right)^{1/2}$

16.78 (a) $\dot{\overline{H}}_G = 141.0$ lb-ft parallel to $\bar{v}$
(b) $\Delta N = 47.1$ lb

16.80 $\overline{M} = 2.09\bar{e}_{D/C}$ N-m

16.82 $\omega_2 = 0.396$ rad/s in sense shown

16.84 $\omega_1 = 373$ rad/s

16.86 (a) precession axis 57.0° above horiz.
(b) $\omega_{precess} = 9.27$ rad/s,
(c) $\omega_{spin} = 18.97$ rad/s

16.88 $(\omega_1)_{min} = 3.08$ rad/s

16.90 $\overline{F}_A = 0.526\bar{\imath} + 1.973\bar{\jmath}$ lb
$\overline{M}_A = 18.86\bar{k}$ lb-in.
($\bar{\imath} = \bar{e}_{G/A}$, $\bar{\jmath}$ upward)

16.92 $\overline{F}_C = 39.2\overline{K}$ N, $\overline{M}_C = 71.6\overline{I}$ N-m
($\overline{J} = \bar{e}_{A/B}$, $\overline{K}$ up)

16.94 $N_A = N_B = 4.125$ lb
$\overline{F}_B = 0.518$ lb in dir. of $\bar{v}$,
$\overline{F}_A = -\overline{F}_B$

16.96 $T = 132.25$ J

16.98 $\omega = 7.30$ rad/s

16.100 $T = 38.2$ J

16.102 $T = \frac{1}{2} mv^2 \left[\frac{3}{2} + \frac{1}{4}\left(\frac{r\cos\theta}{\rho}\right)^2\right]$

16.104 $U_{1\to 2} = 37.5$ J

16.106 (a) $\omega_2 = 201$ rad/s
(b) $\omega_2 = 218$ rad/s

16.108 $\bar{\omega} = 4.21$ rad/s in dir. of $\bar{v}$

16.110 (a) $\bar{v}_G = -2\bar{k}$ m/s,
(b) $\bar{\omega} = 18.86(\bar{\imath} + \bar{\jmath})$ rad/s
($\bar{\imath}$ right, $\bar{\jmath}$ up)

16.112 $\dot{\theta}_1 = 13.17$ rad/s, $\Omega_1 = 13.81$ rad/s

16.114 $\theta = 90°$, $\Omega = 15.71$ rad/s,
$\dot{\theta} = 27.8$ rad/s

16.116 (a) $\bar{\omega} = 92.2\bar{\imath} - 165.0\bar{k}$ rad/s
(b) $\bar{v}_G = 22.2\bar{\jmath}$ ft/s
(c) $\overline{F}_A = 279\bar{\jmath} + 50\bar{k}$ lb

16.118 (a) $v = 52.5$ ft/s,
(b) $\bar{\omega}_2 = 0.80\bar{\jmath}$ rad/s

16.120 $\Omega = 1905.4$ rad/s

16.122 $\bar{\omega} = -33.5\bar{\imath} - 22.0\bar{k}$ rad/s,
$\bar{v}_G = -4.34\bar{\jmath}$ m/s
($\bar{\imath}$ right, $\bar{\jmath}$ up)

17.2 (a) $m\ddot{x} + kx = mg\sin\theta$
(b) $m\ddot{y} + ky = 0$

17.4 $k_2/k_1 = 8$

17.6 $m = 2.56$ kg, $k = 648$ kN/m

17.8 (a) $s_0 = 103.3$ mm,
(b) $\omega_n = 12.33$ Hz

17.10 (a) $\omega_n = 39.3$ rad/s $= 6.25$ Hz
(b) $t = 0.080$ s, $s = 14$ in.

17.12 (a) $x = 0.875 \sin(14.142t - 1.987)$ m
(b) $C = 0.875$ m, $\tau = 0.444$ s,
$\phi = 1.987$ rad (lag)
(c) no change

17.14 (a) $\theta = -\theta_0$, $t = \pi(\ell/g)^{1/2}$
(b) $\theta = 0$, $t = \frac{1}{2}\pi(\ell/g)^{1/2}$

17.16 $C = 7.07$ mm, $\phi = \pi/4$ rad (lag)

17.18 $C = 0.0866$ ft, $\phi = 1.047$ rad (lead)

17.20 (a) $C = 0.1803$ ft,
$\phi = 0.983$ rad (lag)
(b) $t = 0.0638$ s, (c) $t = 0.0246$ s

17.22 (a) $C = 16.10$ mm,
$\phi = 1.847$ rad (lag)
(b) $t = 0.276/\Omega$, (c) $t = 1.847/\Omega$

17.24 $m\ddot{R} + (k - m\Omega^2)R = kL/2$,
$\omega_n = \left(\frac{k}{m} - \Omega^2\right)^{1/2}$

17.26 (a) $k = 0.5$ lb/ft, (b) $\tau_{lt} = 1.567$ s,
$\tau_{rt} = 1.239$ s

17.28 $x = 0.075$ in., $t = 0.0222$ s

17.30 $|x| = 94.3$ mm, $\omega_n = 42.4$ rad/s,
$\phi = 0$

17.32 $m\ddot{x} + c\dot{x} + 2kx = 0$, $c_{cr} = 2(2km)^{1/2}$

17.34 $m\ddot{x} + 2c\dot{x} + 2kx = 0$, $c_{cr} = (2km)^{1/2}$

17.36 $t = 0.0354$ s (damped), 0.005 s (undamped)

17.38 $t = 0.0949$ s

17.42 $g = 4.09$ m/s²

17.44 $\Delta E = 100\left[1 - \exp\left(-\frac{4\pi\beta}{(1 - \beta^2)^{1/2}}\right)\right]\%$

17.46 $C_{eq} = \frac{2C_1C_2}{C_1 + 2C_2}$

17.48 (a) $x = 2.108[\exp(-2.05t) - \exp(-78.4t)]$ ft
(b) $x = 1.861$ ft, $t = 0.0477$ s
(c) $x = 12.683$ ft

17.50 $y = \exp(-2.08t)[0.255 \sin(10.99t) - 0.654 \cos(10.99t)]$ m

17.52 $x = 0.661 \sin(16t) - 0.527 \sin(20.05t)$ ft

17.54 $x = 0.527 \cos\left(4.5t + \frac{\pi}{3}\right) - 0.262 \cos(5t) + 0.410 \sin(5t)$ m

17.56 $x(0) = 0$, $\dot{x}(0) = \frac{F_0\Omega}{2k - m\Omega^2}$

17.58 $C = 0.0306$ m, $\phi = 101.2°$

17.60 (a) $F_0 = 320$ lb,
(b) $|x| = 3.56$ in., $\phi = 32.3°$

17.62 $|x| = 1.635$ mm, $\phi = 1.267$ rad (lag)

17.64 $\omega_n = 32\pi$ rad/s, $\beta = 0.441$

17.66 (a) $15.54\ddot{y} + 7200y = 2880 \sin(\pi vt/2)$

(b) $v = 13.70$ ft/s,
(c) $|y| = 0.1054$ in.

17.68 $a_{max} = 29.6\text{ m/s}^2$ for 3.6 Hz
$a_{max} = 24.7\text{ m/s}^2$ for 4.0 Hz

17.70 $a_{max} = 2.82\text{ m/s}^2$ for 3.6 Hz
$a_{max} = 3.42\text{ m/s}^2$ for 4.0 Hz

17.72 (a) $m\ddot{x} + c\dot{x} + kx = -m\ddot{z}$
(b) $x/a = 0.945 \sin(0.95\omega_n t - 1.469) + \exp(-0.5\omega_n t) \times [0.437 \sin(0.866\omega_n t) + 0.940 \cos(0.866\omega_n t)]$

17.74 $m = 0.0317$ kg, $k = 12.5$ kN/m, $c = 1.592$ N-s/m

17.76 (a) $\left(\frac{2}{3}\frac{g}{\ell}\right)^{1/2} < \Omega < \left(\frac{4}{3}\frac{g}{\ell}\right)^{1/2}$
(b) $\Omega > \left(2\frac{g}{\ell}\right)^{1/2}$

17.78 (a) $Power = \frac{1}{2}F_0\Omega|x|[\sin\phi - \sin(2\Omega t - \phi)]$
(b) $U_{1\to2} = \pi F_0|x| \sin\phi$
(c) $(Power)_{aver} = \frac{1}{2}F_0\Omega|x| \sin\phi$

17.80 $\frac{1}{2} mR^2\ddot{\theta} + \left(2kR^2 + \frac{4}{3\pi} mgR\right)\theta = 0$

17.82 $15\ddot{\theta} + 62500\,\theta = 0$

17.84 $\left(\frac{3}{2} - \frac{8}{3\pi}\right)R\ddot{\theta} + \frac{4}{3\pi} g\theta = 0$

17.86 See prob. 17.80

17.88 See prob. 17.82

17.90 See prob. 17.84

17.92 $I_{zz} = \frac{\tau^2}{4\pi^2} k$

17.94 $c = 1263$ N-s/m

17.96 $\omega_n = 9.90$ rad/s

17.98 $\omega_n = \left[\frac{3}{2}\left(2\frac{k}{m} - g\right)\frac{a^2}{(a^3 + b^3)}\right]^{1/2}$

17.100 $\omega = 45.4$ rad/s

17.102 $\omega_n = \left[\frac{(R + e)^2}{(7R^2 + 2Re + e^2)}\frac{k}{m}\right]^{1/2}$

17.104 (a) $e = 5$ mm, (b) $\omega_n = 100\pi$ rad/s,
(c) $\beta = 0.00833$

17.106 left: $\omega_n = \left[\frac{6}{5}\left(\frac{k}{m} + \frac{g}{\ell}\right)\right]^{1/2}$,
always oscillates
right: $\omega_n = \left[\frac{6}{5}\left(\frac{k}{m} - \frac{g}{\ell}\right)\right]^{1/2}$,
unstable if $k < \frac{mg}{\ell}$

INDEX

CONVERSION FACTORS

ALPHABETICAL LIST OF UNITS
U.S. CUSTOMARY UNITS TO SI UNITS

To convert from	to	Multiply by
atmosphere	pascal (Pa)	1.013 25 (10^5)
bar	pascal (Pa)	1.000 000 (10^5)*
British thermal unit	joule (J)	1.055 056 (10^3)
Btu/hour	watt (W)	2.930 711 (10^{-1})
centimeter of mercury (0°C)	pascal (Pa)	1.333 22 (10^3)
dyne	newton (N)	1.000 000 (10^{-5})*
dyne-centimeter	newton-meter (N-m)	1.000 000 (10^{-7})*
dyne/centimeter^2	pascal (Pa)	1.000 000 (10^{-1})*
erg	joule (J)	1.000 000 (10^{-7})*
erg/second	watt (W)	1.000 000 (10^{-7})*
foot	meter (m)	3.048 000 (10^{-1})*
foot^2	meter^2 (m^2)	9.290 304 (10^{-2})*
foot^3	meter^3 (m^3)	2.831 685 (10^{-2})
foot/second	meter/second (m/s)	3.048 000 (10^{-1})*
foot-pound	joule (J)	1.355 818
foot-pound/second	watt (W)	1.355 818
foot/second^2	meter/second^2 (m/s^2)	3.048 000 (10^{-1})*
gallon (U.S. liquid)	meter^3 (m^3)	3.785 412 (10^{-3})
gram/centimeter^3	kilogram/meter^3 (kg/m^3)	1.000 000 (10^3)*
horsepower (550 ft-lb/s)	watt (W)	7.456 999 (10^2)
inch	meter (m)	2.540 000 (10^{-2})*
inch^2	meter^2 (m^2)	6.451 600 (10^{-4})*

*Denotes an exact quantity.